IFMBE Proceedings

Volume 41

Series Editor

Ratko Magjarevic

Deputy Editors

Fatimah Binti Ibrahim
Igor Lacković
Piotr Ładyżyński
Emilio Sacristan Rock

For further volumes:
http://www.springer.com/series/7403

The International Federation for Medical and Biological Engineering, IFMBE, is a federation of national and transnational organizations representing internationally the interests of medical and biological engineering and sciences. The IFMBE is a non-profit organization fostering the creation, dissemination and application of medical and biological engineering knowledge and the management of technology for improved health and quality of life. Its activities include participation in the formulation of public policy and the dissemination of information through publications and forums. Within the field of medical, clinical, and biological engineering, IFMBE's aims are to encourage research and the application of knowledge, and to disseminate information and promote collaboration. The objectives of the IFMBE are scientific, technological, literary, and educational.

The IFMBE is a WHO accredited NGO covering the full range of biomedical and clinical engineering, healthcare, healthcare technology and management. It is representing through its 60 member societies some 120.000 professionals involved in the various issues of improved health and health care delivery.

IFMBE Officers
President: Ratko Magjarevic, Vice-President: James Goh
Past-President: Herbert Voigt
Treasurer: Marc Nyssen, Secretary-General: Shankhar M. Krishnan
http://www.ifmbe.org

Laura M. Roa Romero
Editor

XIII Mediterranean Conference on Medical and Biological Engineering and Computing 2013

MEDICON 2013, 25–28 September 2013, Seville, Spain

Volume 1

Editor
Laura M. Roa Romero
Biomedical Engineering Group
Engineering School of the University of Seville
Sevilla
Spain

ISSN 1680-0737　　ISSN 1433-9277　(electronic)
ISBN 978-3-319-00845-5　ISBN 978-3-319-00846-2　(eBook)
Printed in 2 Volumes
DOI 10.1007/978-3-319-00846-2
Springer Cham Heidelberg New York Dordrecht London

Library of Congress Control Number: 2013949465

© Springer International Publishing Switzerland 2014

This work is subject to copyright. All rights are reserved by the Publisher, whether the whole or part of the material is concerned, specifically the rights of translation, reprinting, reuse of illustrations, recitation, broadcasting, reproduction on microfilms or in any other physical way, and transmission or information storage and retrieval, electronic adaptation, computer software, or by similar or dissimilar methodology now known or hereafter developed. Exempted from this legal reservation are brief excerpts in connection with reviews or scholarly analysis or material supplied specifically for the purpose of being entered and executed on a computer system, for exclusive use by the purchaser of the work. Duplication of this publication or parts thereof is permitted only under the provisions of the Copyright Law of the Publisher's location, in its current version, and permission for use must always be obtained from Springer. Permissions for use may be obtained through RightsLink at the Copyright Clearance Center. Violations are liable to prosecution under the respective Copyright Law.
The use of general descriptive names, registered names, trademarks, service marks, etc. in this publication does not imply, even in the absence of a specific statement, that such names are exempt from the relevant protective laws and regulations and therefore free for general use.
While the advice and information in this book are believed to be true and accurate at the date of publication, neither the authors nor the editors nor the publisher can accept any legal responsibility for any errors or omissions that may be made. The publisher makes no warranty, express or implied, with respect to the material contained herein.

The IFMBE Proceedings is an Official Publication of the International Federation for Medical and Biological Engineering (IFMBE)

Printed on acid-free paper

Springer is part of Springer Science+Business Media (www.springer.com)

Welcome Message

It is a great pleasure and honor to welcome all participants in the XIII Mediterranean Conference on Medical and Biological Engineering and Computing (MEDICON 2013), which will be held in Seville, the capital of the Spanish region of Andalusia, from September 25 to 28, 2013.

MEDICON is a regional conference with a long tradition and high scientific level, which is organized every three years in a Mediterranean country under the umbrella of the International Federation of Medical and Biological Engineering (IFMBE).

The general theme of MEDICON 2013 is "Research and Development of Technology for Sustainable Healthcare". This decade is being characterized by the appearance and use of emergent technologies under development. This situation has produced a tremendous impact on Medicine and Biology from which it is expected an unparalleled evolution in these disciplines towards novel concept and practices. The consequence will be a significant improvement in health care and well-fare, i.e. the shift from a reactive medicine to a preventive medicine. This shift implies that the citizen will play an important role in the healthcare delivery process, what requires a comprehensive and personalized assistance. In this context, society will meet emerging media, incorporated to all objects, capable of providing a seamless, adaptive, anticipatory, unobtrusive and pervasive assistance. The challenge will be to remove current barriers related to the lack of knowledge required to produce new opportunities for all the society, while new paradigms are created for this inclusive society to be socially and economically sustainable, and respectful with the environment. In this way, the conference program will be focused on the convergence of biomedical engineering topics ranging from formalized theory through experimental science and technological development to practical clinical applications.

In spite of all the problems that many Mediterranean countries are currently facing due to the financial and political crisis, we are proud of the great diffusion of this Conference, with participants coming from all the continents. This participation is a sign that the Biomedical Engineering Community can contribute to overcome the challenges that are affecting society. I would like to share this success with all the members of the committees, institutions, participants, and individual persons and that have made possible the celebration of this event.

Seville is a millenary town with a rich artistic heritage as a consequence of the fusion of the three cultures. The conjunction of the Spanish weather, history and natural resources have shaped it as a modern city with an outstanding projection at the levels of technological research and industry, and biomedical research. All the improvements made on the city during the last decades, have allowed Seville to become a cosmopolitan and easy-to-get destination.

I am sure that you will enjoy a wonderful experience of visiting Seville during your stay.

Laura M. Roa   Conferencechair, MEDICON 2013

Preface

We are pleased to introduce the Proceedings of the 13th Mediterranean Conference on Medical and Biological Engineering and Computing (MEDICON 2013), that are published as the 39th volume of the IFMBE Proceedings, published by Springer under the sponsorship of the International Federation for Medical and Biological Engineering (IFMBE).

Over 500 papers have been submitted for review by the International Program Committee/International Scientific Committee, and the program is structured under ten themes covering the latest advances on Medical and Biological Engineering and Computing. These Proceedings are structured as follows: the first section includes the conference committees, themes and tracks, and parallel activities comprising special sessions, workshops and round tables; next, keynote speakers with their plenary lectures and invited papers; finally, the papers selected for presentation at the conference. The indexing of the Proceedings are eased by two registers for authors and themes and tracks.

We are grateful to Grupo Pacifico Technical Secretary and to Springer for their help and support for the edition of these Proceedings.

We wish that this volume of IFMBE Proceedings will serve as a valuable source of information for professionals, scholars, researchers and students as well as a reference for the state-of-the-art in this period of rapid development and changes in our fields of Medical and Biological Engineering and Computing.

Laura M. Roa Romero
Igor Lackovic
Kang-Ping Lin
Javier Reina-Tosina
Publication Committee

Conference Committee

Conference Chair

Laura M. Roa — Universidad de Sevilla (Spain)

Local Organizing Committee Chair

Javier Reina-Tosina — Universidad de Sevilla (Spain)

Publication Committee

Laura M. Roa — Universidad de Sevilla (Spain)
Igor Lackovic — University of Zagreb (Croatia)
Kang-Ping Lin — Chung Yuan Christian University (Taiwan)
Javier Reina-Tosina — Universidad de Sevilla (Spain)

Young Investigator Competition Committee

Marcello Bracale — Università degli Studi di Napoli Federico II (Italy)
Fumihiko Kajiya — Kawasaki University of Medical Welfare (Japan)
Nicolas Pallikarakis — University of Patras (Greece)
Joe Barbenel — University of Strathclyde (UK)
Enrique J. Gómez — Universidad Politécnica de Madrid (Spain)

Local Organizing Committee

José A. Milán — Hospital Universitario Virgen Macarena (Spain)
Isabel Román — Universidad de Sevilla (Spain)
Jorge Calvillo-Arbizu — Universidad de Sevilla (Spain)
David Naranjo-Hernández — Universidad de Sevilla, CIBER BBN (Spain)
Amparo Callejón — Universidad de Sevilla (Spain)
Miguel Ángel García — Universidad de Sevilla (Spain)
Alejandro Talaminos — Universidad de Sevilla (Spain)
Gerardo Barbarov — Universidad de Sevilla (Spain)
Alfonso Lara — Hospital Universitario Virgen Macarena (Spain)

Regional Conference Organizing Committee

Associazione Italiana di Ingegnaria Medica e Biologica

Leandro Pecchia — University of Nottingham (UK)
Paolo Melillo — Università degli Studi di Napoli Federico II (Italy)

Croatian Medical and Biological Engineering Society

Igor Lackovic	University of Zagreb (Croatia)
Vedran Bilas	University of Zagreb (Croatia)

Cyprus Association of Medical Physics and Biomedical Engineering

Prodromos A. Kaplanis	University of Cyprus (Cyprus)
Constantinos Pattichis	University of Cyprus (Cyprus)

European Alliance for Medical and Biological Engineering and Science

Birgit Glasmacher	Leibniz Universitaet Hannover (Germany)

Greek Society for Biomedical Engineering

Nicolas Pallikarakis	University of Patras (Greece)
Panagiotis Bamidis	Aristotle University of Thessaloniki (Greece)

IEEE-Engineering in Medicine and Biology Society

Donna Hudson	UCSF Fresno (USA)

Israel Society for Medical and Biological Engineering

Haim Azhari	Technion-IIT (Israel)
Idit Avrahami	Ariel University (Israel)

Italian Association of Clinical Engineers

Paola Freda	Paola Freda, AReSSPiemonte - Agenzia Regionale per i Servizi Sanitari (Itlay)
Paolo Lago	Fondazione IRCCS - Policlinico San Matteo Pavia (Italy)

Société Française de Génie Biologique et Médical

Catherine Marque	Université de Technologie de Compiègne (France)
Véronique Migonney	Université Paris 13 (France)

Sociedade Portuguesa de Egenharia Biomedica

Mario F. Secca	Universidade Nova de Lisboa (Portugal)
Pedro M.C. Vieira	Universidade Nova de Lisboa (Portugal)

Slovene Society for Medical and Biological Engineering

Damijan Miklavcic	University of Ljubljana (Slovenia)
Tomaz Vrtovec	University of Ljubljana (Slovenia)

Spanish Society of Biomedical Engineering

Enrique J. Gómez	Universidad Politécnica de Madrid (Spain)
Raimon Jané	Universitat Politècnica de Catalunya (Spain)

International Program Committee

Zulfiqur Ali	Teesside University (England)
Metin Akay	University of Houston (USA)

Conference Committee

Ahmad Taher Azar	Misr University for Science and Technology (Egypt)
Haim Azhari	Technion-IIT Haifa (Israel)
Panagiotis Bamidis	Aristotle University of Thessaloniki (Greece)
José López Barneo	Universidad de Sevilla (Spain)
José Becerra	Universidad de Málaga (Spain)
Anna M. Bianchi	Politecnico di Milano (Italy)
Paolo Bonato	Harvard Medical School (USA)
Lodewijk Bos	ICMCC (Holland)
Marcello Bracale	Università degli Studi di Napoli Federico II (Italy)
Saide Calil	State University of Campinas (Brazil)
Pere Caminal	Universidad Politécnica de Cataluña (Spain)
Alicia Casals	Universidad Politécnica de Cataluña (Spain)
Bernardo Celda	Universidad de Valencia (Spain)
Sergio Cerutti	Politecnico di Milano (Italy)
Walter H. Chang	Chung Yuan Christian University (Taiwan)
Febo Cincotti	Sapienza Università di Roma (Italy)
Jean Louis Coatrieux	Université de Rennes 1 (France)
Yadin David	Engineering Consultants LLC (USA)
José M. Delgado-García	Universidad Pablo de Olavide (Spain)
Manuel Desco	Universidad Carlos III (Spain)
Manuel Doblaré	Universidad de Zaragoza (Spain)
José M. Ferrero-Jr	Universidad Politécnica de Valencia (Spain)
Mario Forjaz Secca	Universidade Nova de Lisboa (Portugal)
Dimitris Fotiadis	University of Ioannina (Greece)
Alejandro Frangi	Universitat Pompeu Fabra (Spain)
Enrique Gómez	Universidad Politécnica de Madrid (Spain)
Tomás Gómez-Cía	Hospital Universitario Virgen del Rocío (Spain)
Juan Guerrero	Universidad de Valencia (Spain)
Carlos H. Salvador	Instituto de Salud Carlos III (Spain)
Elena Hernando	Universidad Politécnica de Madrid (Spain)
Roberto Hornero	Universidad de Valladolid (Spain)
Gerhard Holzapfel	Technische Universität Graz (Austria)
Donna Hudson	UCSF Fresno (USA)
Helmut Hutten	Technische Universität Graz (Austria)
Antonio Fernando C. Infantosi	Federal University of Rio de Janeiro (Brazil)
Robert Istepanian	Kingston University (England)
Christopher James	University of Warwick (England)
Raimon Jané	Universidad Politécnica de Cataluña (Spain)
Akos Jobbagy	Budapest University of Technology and Economics (Hungary)
Fumihiko Kajiya	Kawasaki University of Medical Welfare (Japan)
Roger Kamm	Massachusetts Institute of Technology (USA)
Gyeon-Man Kim	CEIT (Spain)
Pablo Laguna	Universidad de Zaragoza (Spain)
Thomas Lango	SINTEF Health Research Trondheim (Norway)
Alberto Leiva	Hospital Sant Pau (Spain)
Nigel Lovell	University of New South Wales (Australia)
Ratko Magjarevich	University of Zagreb (Croatia)
Nicos Maglaveras	Aristotle University of Thessaloniki (Greece)
Luca Mainardi	Politecnico di Milano (Italy)

Quality Based Adaptation of Signal Analysis Software in Pregnancy Home Care System 559
J. Wrobel, K. Horoba, J. Jezewski, T. Kupka, M. Jezewski, T. Przybyla

A Recovery of FHR Signal in the Embedded Space 563
T. Przybyła, T. Pander, J. Wróbel, R. Czabański, D. Roj, A. Matonia

Online Drawings for Dementia Diagnose: In-Air and Pressure Information Analysis 567
Marcos Faundez-Zanuy, Enric Sesa-Nogueras, Josep Roure-Alcobé, Josep Garré-Olmo, Karmele Lopez-de-Ipiña, Jordi Solé-Casals

Spontaneous Speech and Emotional Response Modeling Based on One-Class Classifier Oriented to Alzheimer Disease Diagnosis 571
K. Lopez-de-Ipiña, J.B. Alonso, N. Barroso, J. Solé-Casals, M. Ecay-Torres, P. Martinez-Lage, F. Zelarain, H. Egiraun, C.M. Travieso

Complexity of Epileptiform Activity in a Neuronal Network and Pharmacological Intervention 575
D. Abásolo, L. González, Y. Chen

Evaluation of Different Handwriting Teaching Methods by Kinematic Analysis 579
A. Accardo, M. Genna, I. Perrone, P. Ceschia, C. Mandarino

Analysis of MEG Activity across the Life Span Using Statistical Complexity 583
J. Poza, C. Gómez, M. García, A. Bachiller, A. Fernández, R. Hornero

Impact of Device Settings and Spontaneous Breathing during IPV in CF Patients 587
E. Fornasa, A. Accardo, M. Ajcevic, R. Sartori, F. Poli

Estimation of Respiratory Mechanics Parameters during HFPV 591
M. Ajcevic, A. Accardo, E. Fornasa, U. Lucangelo

Signal Source Estimation Inside Brain Using Switching Voltage Divider 595
Yusuke Sakaue, Shima Okada, Masaaki Makikawa

A Phase-Space Based Algorithm for Detecting Different Types of Artefacts 599
A. Brignol, T. Al-ani

Automatic Classification of Respiratory Sounds Phenotypes in COPD Exacerbations 603
Daniel S. Morillo, M.A. Fernández Granero, A. León, L.F. Crespo

Response Detection in Narrow-Band EEG Using Signal-Driven Non-Periodic Stimulation 607
M. Cagy, A.F.C. Infantosi

EMG-Based Analysis of Treadmill and Ground Walking in Distal Leg Muscles 611
F. Di Nardo, S. Fioretti

Temporal Variation of Local Fluorescence Sources in the Photodynamic Process 615
I. Salas-García, F. Fanjul-Vélez, N. Ortega-Quijano, J.L. Arce-Diego

Statistical Analysis of EMG Signal Acquired from Tibialis Anterior during Gait 619
F. Di Nardo, A. Mengarelli, G. Ghetti, S. Fioretti

Auto-Mutual Information Function for Predicting Pain Responses in EEG Signals during Sedation 623
U. Melia, M. Vallverdú, M. Jospin, E.W. Jensen, J.F. Valencia, F. Clariá, P.L. Gambus, P. Caminal

Prony's Method for the Analysis of mfVEP Signals 627
A.J. Fernández-Rodríguez, L. de Santiago, R. Blanco, C. Amo, R. Barea, J.M. Miguel-Jiménez,
J.M. Rodríguez-Ascariz, E.M. Sánchez-Morla, M. Ortiz, L. Boquete

EEG Denoising Based on Empirical Mode Decomposition and Mutual Information 631
A. Mert, A. Akan

Decomposition Analysis of Digital Volume Pulse Signal Using Multi-Model Fitting 635
Sheng-Cheng Huang, Hao-Yu Jan, Geng-Hong Lin, Wen-Chen Lin, Kang-Ping Lin

A Neural Minimum Input Model to Reconstruct the Electrical Cortical Activity 639
S. Conforto, I. Bernabucci, N. Accornero, M. Bertollo, C. Robazza, S. Comani, M. Schmid, T. D'Alessio

Improved Splines Fitting of Intervertebral Motion by Local Smoothing Variation 643
P. Bifulco, M. Cesarelli, G. D'Addio, M. Romano

Effects of Wavelets Analysis on Power Spectral Distributions in Laser Doppler Flowmetry Time Series 647
G. D'Addio, M. Cesarelli, P. Bifulco, L. Iuppariello, G. Faiella, D. Lapi, A. Colantuoni

Outliers Detection and Processing in CTG Monitoring 651
M. Romano, G. Faiella, P. Bifulco, G. D'Addio, F. Clemente, M. Cesarelli

Changes in Heart Rate Variability Associated with Moderate Alcohol Consumption 655
A. Fratini, P. Bifulco, F. Clemente, M. Sansone, M. Cesarelli

EEG Rhythm Analysis Using Stochastic Relevance 658
L. Duque-Muñoz, C.A. Aguirre-Echeverry, G. Castellanos-Domínguez

Optimize ncRNA Targeting: A Signal Analysis Based Approach 662
N. Maggi, P. Arrigo, C. Ruggiero

Mathematical Modelling of Melanoma Collective Cell Migration 666
J.V. Gallinaro, C.M.G. Marques, F.M. Azevedo, D.O.H. Suzuki

Synthetic Atrial Electrogram Generator 670
M.W. Rivolta, L.T. Mainardi, R. Sassi, V.D.A. Corino

EMG Topographic Image Enhancement Using Multi Scale Filtering 674
K. Ullah, B. Afsharipour, R. Merletti

Model Based Estimates of Gain between Systolic Blood Pressure and Heart-Rate Obtained from Only Inspiratory or
Expiratory Periods 678
D.S. Fonseca, A. Beda, A.M.F.L. Miranda de Sá, D.M. Simpson

Dielectric Properties of Dentin between 100 Hz and 1 MHz Compared to Electrically Similar Body Tissues 682
T. Marjanović, I. Lacković

Hypoglycaemia-Related EEG Changes Assessed by Approximate Entropy 686
C. Fabris, A.S. Sejling, G. Sparacino, A. Goljahani, J. Duun-Henriksen, L.S. Remvig, C. Cobelli, C.B. Juhl

Non-linear Indices of Heart Rate Variability in Heart Failure Patients during Sleep 690
R. Cabiddu, S. Mariani, J. Henriques, S. Cerutti, A.M. Bianchi

Title	Page
Detrended Fluctuation Analysis of EEG in Depression *M. Bachmann, A. Suhhova, J. Lass, K. Aadamsoo, Ü. Võhma, H. Hinrikus*	694
Can Distance Measures Based on Lempel-Ziv Complexity Help in the Detection of Alzheimer's Disease from Electroencephalograms? *S. Simons, D. Abásolo*	698
Predictive Value on Neurological Outcome of Early EEG Signal Analysis in Brain Injured Patients *A. Accardo, M. Cusenza, L. Prisco, F. Monti, A. Draisci, W. Calligaris*	702
Lempel-Ziv Complexity Analysis of Local Field Potentials in Different Vigilance States with Different Coarse-Graining Techniques *D. Abásolo, R. Morgado da Silva, S. Simons, G. Tononi, C. Cirelli, V.V. Vyazovskiy*	706
Heart Rate Variability in Pregnant Women before Programmed Cesarean Intervention *Juan Bolea, Raquel Bailón, Eva Rovira, Jose María Remartínez, Pablo Laguna, Augusto Navarro*	710
Fractal Changes in the Long-Range Correlation and Loss of Signal Complexity in Infant's Heart Rate Variability with Clinical Sepsis *E. Godoy, J. López, L. Bermúdez, A. Ferrer, N. García, C. García Vicent, E.F. Lurbe, J. Saiz*	714
Estimation of Coupling and Directionality between Signals Applied to Physiological Uterine EMG Model and Real EHG Signals *A. Diab, M. Hassan, J. Laforêt, B. Karlsson, C. Marque*	718
Dynalets: A New Tool for Biological Signal Processing *J. Demongeot, O. Hansen, A. Hamie*	722
Cyclostationarity-Based Estimation of the Foetus Subspace Dimension from ECG Recordings *M. Haritopoulos, J. Roussel, C. Capdessus, A.K. Nandi*	726
Feature Extraction Based on Discriminative Alternating Regression *C.O. Sakar, O. Kursun, F. Gurgen*	730
Influence of Signal Preprocessing on ICA-Based EEG Decomposition *Z. Zakeri, S. Assecondi, A.P. Bagshaw, T.N. Arvanitis*	734
Muscle Synergies Underlying Voluntary Anteroposterior Sway Movements *S. Piazza, D. Torricelli, I.M. Alguacil Diego, R. Cano De La Cuerda, F. Molina Rueda, F.M. Rivas Montero, F. Barroso, J.L. Pons*	738
Recognition of Brain Structures from MER-Signals Using Dynamic MFCC Analysis and a HMC Classifier *Mauricio Holguin, German A. Holguin, Hernán Darío Vargas Cardona, Genaro Daza, Enrique Guijarro, Alvaro Orozco*	742
Feed-Forward Neural Network Architectures Based on Extreme Learning Machine for Parkinson's Disease Diagnosis *P.J. García-Laencina, G. Rodríguez-Bermúdez, J. Roca-Dorda*	746
Breast Tissue Microarray Classification Based on Texture and Frequential Features *M.M. Fernández-Carrobles, G. Bueno, O. Déniz, M. García-Rojo*	750
Fuzzy System for Retrospective Evaluation of the Fetal State *R. Czabanski, J. Wrobel, K. Horoba, J. Jezewski, A. Matonia*	754

Advanced Processing of sEMG Signals for User Independent Gesture Recognition A. Doswald, F. Carrino, F. Ringeval	758
Adaptive Classification Framework for Multiclass Motor Imagery-Based BCI L.F. Nicolas-Alonso, R. Corralejo, D. Álvarez, R. Hornero	762
Computer Program for Automatic Identification of Artifacts in Impedance Cardiography Signals Recorded during Ambulatory Hemodynamic Monitoring P. Piskulak, G. Cybulski, W. Niewiadomski, T. Pałko	766
Comparison between Artificial Neural Networks and Discriminant Functions for Automatic Detection of Epileptiform Discharges C.F. Boos, G.R. Scolaro, F.M. Azevedo	770
Classification of Early Autism Based on HPLC Data T. Kristensen	774
Feature Selection Techniques in Uterine Electrohysterography Signal D. Alamedine, M. Khalil, C. Marque	779
Studying Functional Brain Networks to Understand Mathematical Thinking: A Graph-Theoretical Approach Georgios Bamparopoulos, Manousos A. Klados, Nikolaos Papathanasiou, Ioannis Antoniou, Sifis Micheloyannis, Panagiotis D. Bamidis	783
A Short Review on Emotional Recognition Based on Biosignal Pattern Analysis Manousos A. Klados, Charalampos Styliadis, Panagiotis D. Bamidis	787
Matching Pursuit with Asymmetric Functions for Signal Decomposition and Parameterization K.J. Blinowska, W.W. Jedrzejczak, K. Kwaskiewicz	791
Analysis of Intracranial Pressure Signals Using the Spectral Turbulence M. García, J. Poza, D. Santamarta, D. Abásolo, R. Hornero	795
Graph-Theoretical Analysis in Schizophrenia Performing an Auditory Oddball Task A. Bachiller, J. Poza, C. Gómez, V. Molina, V. Suazo, A. Díez, R. Hornero	799
Spectral Parameters from Pressure Bed Sensor Respiratory Signal to Discriminate Sleep Epochs with Respiratory Events Giulia Tacchino, Guillermina Guerrero, Juha M. Kortelainen, Anna M. Bianchi	803
Wavelet Energy and Wavelet Entropy as a New Analysis Approach in Spontaneous Fluctuations of Pupil Size Study – Preliminary Research W. Nowak, E. Szul-Pietrzak, A. Hachol	807
Rhythm Extraction Using Spectral–Splitting for Epileptic Seizure Detection L.M. Sepúlveda, J.D. Martínez, L. Duque, C.D. Acosta, G. Castellanos	811
EEG Metrics Evaluation in Simultaneous EEG-fMRI Olfactory Experiment Eva Manzanedo, Ana Beatriz Solana, Elena Molina, Ricardo Bruña, Susana Borromeo, Juan Antonio Hernández-Tamames, Francisco del Pozo	815
An Emboli Detection System Based on Dual Tree Complex Wavelet Transform G. Serbes, B.E. Sakar, N. Aydin, H.O. Gulcur	819

AALUMO: A User Model Ontology for Ambient Assisted Living Services Supported in Next-Generation
Networks 1217
   P.A. Moreno, M.E. Hernando, E.J. Gómez

Internet of Things for Wellbeing – Pilot Case of a Smart Health Cardio Belt 1221
   E. Kovatcheva, R. Nikolov, M. Madjarova, A. Chikalanov

Advances in Modelling of Epithelial to Mesenchymal Transition 1225
   R. Summers, T. Abdulla, J.-M. Schleich

Glucose-Level Interpolation for Determining Glucose Distribution Delay 1229
   Tomas Koutny

A Supervised SOM Approach to Stratify Cardiovascular Risk in Dialysis Patients 1233
   J. Ion Titapiccolo, M. Ferrario, S. Cerutti, C. Barbieri, F. Mari, E. Gatti, M.G. Signorini

Logistic Regression Models for Predicting Resistance to HIV Protease Inhibitor Nelfinavir 1237
   L.M. Raposo, M.B. Arruda, R.M. Brindeiro, F.F. Nobre

Application of Special Parametric Methods to Model Survival Data 1241
   J. Holčík, K. Opršalová

Protein Function Prediction Based on Protein-Protein Interactions: A Comparative Study 1245
   K.S. Ahmed, S.M. El-Metwally

Inhibitory Regulation by microRNAs and Circular RNAs 1250
   J. Demongeot, H. Hazgui, J. Escoffier, C. Arnoult

Public Electronic Health Record Platform Compliant with the ISO EN13606 Standard as Support to Research
Groups 1254
   *R. Sánchez de Madariaga, J. Cáceres Tello, A. Muñoz Carrero, O. Moreno Gil, I. Velázquez Aza, A. Castro Serrano,
   R. Somolinos Cristóbal*

Development of a Visual Editor for the Definition of HL7 CDA Archetypes 1258
   David Moner, José Alberto Maldonado, Diego Boscá, Alejandro Mañas, Montserrat Robles

homeRuleML Version 2.1: A Revised and Extended Version of the homeRuleML Concept 1262
   H.A. McDonald, C.D. Nugent, J. Hallberg, D.D. Finlay, G. Moore

Detailed Clinical Models Governance System in a Regional EHR Project 1266
   D. Bosca, L. Marco, D. Moner, J.A. Maldonado, L. Insa, M. Robles

An Extensible Free Software Platform for Managing Image-Based Clinical Trials 1270
   M.A. Laguna, N. Malpica, J.A. Hernández-Tamames

Reuse of Clinical Information: Integrating Primary Care and Clinical Research through a Bidirectional Standard
Interface 1274
   Paolo Fraccaro, Valeria Pupella, Roberta Gazzarata, Mauro Giacomini

Archetype-Based Solution to Tele-Monitor Patients with Chronic Diseases 1278
   *Juan Mario Rodríguez, Carlos Cavero Barca, Paolo Emilio Puddu, John Gialelis, Petros Chondros,
   Dimitris Karadimas, Kevin Keene, Jan-Marc Verlinden*

Connecting HL7 with Software Analysis: A Model-Based Approach 1282
  A. Martínez-García, M.J. Escalona, C.L. Parra-Calderón

Integrating the EN/ISO 13606 Standards into an EN/ISO 12967 Based Architecture 1286
  J. Calvillo, I. Román, L.M. Roa

Design of a Semantic Service for Management of Multi-domain Health Alarms 1290
  J. Calvillo, I. Román, L.M. Roa

SILAM: Integrating Laboratory Information System within the Liguria Region Electronic Health Record 1294
  Alessandro Tagliati, Valeria Pupella, Roberta Gazzarata, Mauro Giacomini

A Model for Measuring Open Access Adoption and Usage Behaviour of Health Sciences Faculty Members 1298
  E.T. Lwoga, F. Questier

EHR Anonymising System Based on the ISO/EN 13606 Norm 1302
  *R. Somolinos, A. Muñoz, M.E. Hernando, M. Pascual, J. Cáceres, R. Sánchez-de-Madariaga, J.A. Fragua,
  M. Carmona, A.L. Castro, O. Moreno, C.H. Salvador*

The Status of Information Systems in the Hospitals of the Greek National Health System 1306
  George Aggelinos, Sokratis Katsikas

Operating Room Efficiency Improving through Data Management 1310
  P. Perger, M. Buccioli, V. Agnoletti, E. Padovani, G. Gambale

A Semantically Enriched Architecture for an Italian Laboratory Terminology System 1314
  Silvia Canepa, Sabrina Roggerone, Valeria Pupella, Roberta Gazzarata, Mauro Giacomini

A National Electronic Health Record System for Cyprus 1318
  K.C. Neokleous, E.C. Schiza, E. Salameh, K. Palazis, C.N. Schizas

Information Driven Care Pathways and Procedures 1322
  R.J. Dickinson, R.I. Kitney

Decision Making in Screening Diagnostics E-Medicine 1326
  G. Balodis, I. Markovica, Z. Markovics, D. Matisone

Using Social Network Apps as Social Sensors for Health Monitoring 1330
  I. Pagkalos, L. Petrou

Advanced Medical Expert Support Tool (A-MEST): EHR-Based Integration of Multiple Risk Assessment Solutions for
Congestive Heart Failure Patients 1334
  *Carlos Cavero Barca, Juan Mario Rodríguez, Paolo Emilio Puddu, Mitja Luštrek, Božidara Cvetković,
  Maurizio Bordone, Eduardo Soudah, Aitor Moreno, Pedro de la Peña, Alberto Rugnone, Francesco Foresti,
  Elena Tamburini*

Modeling and Implementing a Signal Persistence Manager for Shared Biosignal Storage and Processing 1338
  S. Pirola, E. Opri, A.M. Bianchi, S. Marceglia

Heart Rate Variability for Automatic Assessment of Congestive Heart Failure Severity 1342
  P. Melillo, E. Pacifici, A. Orrico, E. Iadanza, L. Pecchia

Support System for the Evaluation of the Posterior Capsule Opacificaction Degree 1346
  D. Ruiz, L. González, A. Soriano

A Custom Decision-Support Information System for Structural and Technological Analysis in Healthcare............ 1350
   A. Luschi, L. Marzi, R. Miniati, E. Iadanza

Performance Assessment of a Clinical Decision Support System for Analysis of Heart Failure..................... 1354
   G. Guidi, P. Melillo, M.C. Pettenati, M. Milli, E. Iadanza

Self-reporting for Bipolar Patients through Smartphone 1358
   P. Berg Andersen, A. Babic

Comorbidities Modeling for Supporting Integrated Care in Chronic Cardiorenal Disease 1362
   E. Kaldoudi, N. Dovrolis

Overall Survival Prediction for Women Breast Cancer Using Ensemble Methods and Incomplete Clinical Data 1366
   Pedro Henriques Abreu, Hugo Amaro, Daniel Castro Silva, Penousal Machado, Miguel Henriques Abreu,
   Noémia Afonso, António Dourado

Automatic Blood Glucose Classification for Gestational Diabetes with Feature Selection: Decision Trees vs. Neural
Networks 1370
   E. Caballero-Ruiz, G. García-Sáez, M. Rigla, M. Balsells, B. Pons, M. Morillo, E.J. Gómez, M.E. Hernando

On the Global Optimization of the Beam Angle Optimization Problem in Intensity-Modulated Radiation Therapy 1374
   H. Rocha, J.M. Dias, B.C. Ferreira, M.C. Lopes

Effective Supervised Knowledge Extraction for an mHealth System for Fall Detection 1378
   G. Sannino, I. De Falco, G. De Pietro

Chemoprophylaxis Application for Meningococcal Disease for Android Devices 1382
   M. Parejo-Bellido, E. Dorronzoro-Zubiete, M. Zurbarán, F.J. Sánchez-Laguna, L. Fernández-Luque,
   A.A. Muñoz-Macho

Automation of Evaluation Protocols of Stand-to-Sit Activity in Expert System 1386
   M.J. Cunha, G. Cardozo, F. de Azevedo

Digital Diary for Persons with Psychological Disorders Using Interaction Design 1390
   H. Sørheim, A. Babic

A Novel Data-Mining Platform to Monitor the Outcomes of Erlontinib (Tarceva) Using Social Media................ 1394
   A. Akay, A. Dragomir, B.E. Erlandsson

Advanced Networked Modular Personal Dosimetry System 1398
   R. Chil, L.M. Fraile, J. Vaquero, E. Picado, A. Rodriguez-Moreno, M.C. Rodriguez-Sanchez, S. Borromeo,
   M. Desco, J.M. Udías, J.J. Vaquero

Methods for Personalized Diagnostics 1402
   D.L. Hudson, M.E. Cohen

The Prediction of Blood Pressure Changes by the Habit of Walking....................................... 1406
   Toshiyo Tammura, Soichi Maeno, Yutaka Kimira, Yuichi Kimura, Takumu Hattori, Kotaro Minato

Parallel Workflows to Personalize Clinical Guidelines Recommendations: Application to Gestational Diabetes
Mellitus 1409
   G. García-Sáez, M. Rigla, E. Shalom, M. Peleg, E. Caballero, E.J. Gómez, M.E. Hernando

Case Based Reasoning in a Web Based Decision Support System for Thoracic Surgery 1413
 A. Babic, B. Peterzen, U. Lönn, H.C. Ahn

Data Mining in Cancer Registries: A Case for Design Studies 1417
 G. Kanza, A. Babic

Software Prototype for Triage and Instructional for Homecare Patients 1421
 L. Boom, N. Escobar, C. Ruales, L. López

Analysis and Impact of Breast and Colorectal Cancer Groups on Social Networks 1425
 I. De la Torre, B. Martínez, M. López-Coronado

Health Apps for the Most Prevalent Conditions 1430
 B. Martínez-Pérez, I. de la Torre-Díez, M. López-Coronado

A Mobile Remote Monitoring Service for Measuring Fetal Heart Rate 1435
 G. Lanzola, I. Secci, S. Scarpellini, A. Fanelli, G. Magenes, M.G. Signorini

Investigating Methods for Increasing the Adoption of Social Media amongst Carers for the Elderly 1439
 Kyle Boyd, Chris Nugent, Mark Donnelly, Raymond Bond, Roy Sterritt, Lorraine Gibson

AMELIE: Authoring Multimedia-Enhanced Learning Interactive Environment for e-Health Contents 1443
 P. Sánchez-González, I. Oropesa, P. Moreno-Sánchez, J.M. Martínez-Moreno, J. García-Novoa, E.J. Gómez

mHealth: Cognitive Telerehabilitation of Patients with Acquired Brain Damage 1447
 C. Suárez-Mejías, M. Parejo, M.J. Zarco, A. Naranjo, C. Echevarría, J. Barros, R. Díez, G. Escobar, M. Elena,
 C.L. Parra

Mobile Telemedicine Screening Complex 1451
 J. Lauznis, Z. Markovics, I. Markovica

Radial-Basis-Function Based Prediction of COPD Exacerbations.. 1455
 M.A. Fernández Granero, Daniel S. Morillo, A. León, M.A. López Gordo, L.F. Crespo

Research Benefits of Using Interoperability Standards in Remote Command and Control to Implement Personal Health
Devices 1459
 H.G. Barrón-González, Student Member, IEEE, M. Martínez-Espronceda, Member, IEEE, S. Led, L. Serrano,
 Senior Member, IEEE

Low Cost, Modular and Scalable Remote Monitoring Healthcare Platform Using 8-Bit Microcontrollers 1464
 F.B. Cosentino, J. Marino-Neto, F.M. Azevedo

Adaptive Healthcare Pathway for Diabetes Disease Management 1468
 Giuseppe Fico, Alessio Fioravanti, Maria Teresa Arredondo, Chiara Diazzi, Giovanni Arcuri, Claudio Conti,
 Giampiero Pirini

Real Time Health Remote Monitoring Systems in Rural Areas 1473
 M.G. Sánchez, R. Nocelo López, J.A. Gay-Fernández

Detecting Accelerometer Placement to Improve Activity Classification 1477
 Ian Cleland, Chris D. Nugent, Dewar D. Finlay, Roger Armitage

Incorporating the Rehabilitation of Parkinson's Disease in the Play for Health Platform Using a Body
Area Network 1481
    F. Tous, P. Ferriol, M.A. Alcalde, M. Melià, B. Milosevic, M. Hardegger, D. Roggen

Activity Classification Using 3-Axis Accelerometer Wearing on Wrist for the Elderly 1485
    D.I. Shin, S.K. Joo, J.H. Song, S.J. Huh

Intelligent Chair Sensor – Classification and Correction of Sitting Posture 1489
    L. Martins, R. Lucena, J. Belo, R. Almeida, C. Quaresma, A.P. Jesus, P. Vieira

Non-invasive System for Mechanical Arterial Pulse Wave Measurements 1493
    Mikko Peltokangas, Jarmo Verho, Timo Salpavaara, Antti Vehkaoja

Automatic Identification of Sensor Localization on the Upper Extremity 1497
    S. Lambrecht, J.L. Pons

Part X: Medical Devices and Sensors

A New Device to Assess Gait Velocity at Home 1503
    R. Jaber, A. Chkeir, D.J. Hewson, J. Duchêne

Magnetic Induction-Based Sensor for Vital Sign Detection 1507
    H. Mahdavi, J. Rosell-Ferrer

Unconstrained Night-Time Heart Rate Monitoring with Capacitive Electrodes 1511
    A. Vehkaoja, A. Salo, M. Peltokangas, J. Verho, T. Salpavaara, J. Lekkala

Wrist-Worn Accelerometer to Detect Postural Transitions and Walking Patterns 1515
    V. Ahanathapillai, J.D. Amor, M. Tadeusiak, C.J. James

Comparative Measurement of the Head Orientation Using Camera System and Gyroscope System 1519
    P. Kutilek, O. Cakrt, J. Hejda, R. Cerny

A Noninvasive Method of Measuring Force-Frequency Relations to Evaluate Cardiac Contractile State of Patients
during Exercise for Cardiac Rehabilitation 1523
    M. Tanaka, M. Sugawara, Y. Ogasawara, I. Suminoe, T. Izumi, K. Niki, F. Kajiya

A Phase Lock Loop (PLL) System for Frequency Variation Tracking during General Anesthesia 1527
    C.A. Teixeira, B. Direito, A. Dourado, M.P. Santos, M.C. Loureiro

Evaluation of Arterial Properties through Acceleration Photoplethysmogram 1531
    R. Gonzalez, A. Manzo, E. Cardenas, J. Herrera, F. Martinez, J. Gomis, J. Saiz

Comparing over Ground Turning and Walking on Rotating Treadmill 1535
    A. Olenšek, J. Pavčič, Z. Matjačić

Displacement Measurement of a Medical Instrument Inside the Human Body 1539
    D.A. Fotiadis, A. Astaras, A. Kalfas, K. Papathanasiou, P. Bamidis

"Algorithmically Smart" Continuous Glucose Sensor Concept for Diabetes Monitoring 1543
    G. Sparacino, A. Facchinetti, C. Zecchin, C. Cobelli

Piezoresistive Goniometer Network for Sensing Gloves 1547
    G. Dalle Mura, F. Lorussi, A. Tognetti, G. Anania, N. Carbonaro, M. Pacelli, R. Paradiso, D. De Rossi

Preliminary Study of Pressure Distribution in Diabetic Subjects, in Early Stages 1551
   M.L. Zequera, L. Garavito, W. Sandham, Á Rodríguez, J.A. Alvarado, C.A. Wilches

System for Precise Measurement of Head and Shoulders Position 1555
   J. Hejda, P. Kutílek, J. Hozman, R. Černý

An Online Program for the Diagnosis and Rehabilitation of Patients with Cochlear Implants 1559
   T. Lopez-Soto, A. Castillo-Armero

Analysis of Anomalies in Bioimpedance Models for the Estimation of Body Composition 1563
   D. Naranjo, L.M. Roa, L.J. Reina, M.A. Estudillo, N. Aresté, A. Lara, J.A. Milán

Risk Analysis and Measurement Uncertainty in the Manufacturing Process of Medical Devices 1567
   P.E. Baru, N.M. Roman, C. Rusu

Development of an Equipment to Detect and Quantify Muscular Spasticity: SpastiMed – A New Solution 1571
   V. Fernandes, I. Clemente, C. Quaresma, P. Vieira

Comparison between Two Exercise Systems for Rodents 1575
   T. Ödman, N. Ödman, E. Rabotchi, S. Åkervall, M. Lindén

Tripolar Flexible Concentric Ring Electrode Printed with Inkjet Technology for ECG Recording 1579
   Y. Ye-Lin, E. Senent, G. Prats-Boluda, E. Garcia-Breijo, J.V. Lidon, J. Garcia-Casado

Measurement System for Pupil Size Variability Study 1583
   W. Nowak, A. Żarowska, E. Szul-Pietrzak, A. Hachoł

Patellar Reflex Measurement System with Tapping Force Controlled 1587
   Y. Salazar-Muñoz, L.A. Ruano-Calderón, J.A. Leyva, O.S. Martínez, O. García-Cano, B. García-Caballero

A Bioelectric Model of pH in Wound Healing 1591
   M. Amparo Callejón, Laura M. Roa, Javier Reina-Tosina

Part XI: Molecular, Cellular and Tissue Engineering and Biomaterials

Competitive Adsorption of Albumin, Fibronectin and Collagen Type I on Different Biomaterial Surfaces:
A QCM-D Study 1597
   H. Felgueiras, V. Migonney, S. Sommerfeld, N.S. Murthy, J. Kohn

Biomimetic Poly(NaSS) Grafted on Ti6Al4V: Effect of Pre-adsorbed Selected Proteins on the MC3T3-E1 Osteoblastic
Development 1601
   H. Felgueiras, V. Migonney

Bioactive Intraocular Lens – A Strategy to Control Secondary Cataract 1605
   Yi-Shiang Huang, Virginie Bertrand, Dimitriya Bozukova, Christophe Pagnoulle, Edwin De Pauw,
   Marie-Claire De Pauw-Gillet, Marie-Christine Durrieu

Ultrathin Films by LbL Self-assembly for Biomimetic Coatings of Implants 1609
   M. Giulianelli, L. Pastorino, R. Ferretti, C. Ruggiero

Nanohelical Shape and Periodicity Dictate Stem Cell Fate in a Synthetic Matrix 1613
   R.K. Das, O.F. Zouani, R. Oda, M.-C. Durrieu

The Role of Adult Stem Cells on Microvascular Tube Stabilization 1617
  O.F. Zouani, Y. Lei, M.-C. Durrieu

Processes of Gamma Radiolysis of Soluble Collagen in the Aspect of the Scaffolds Preparation 1622
  K. Pietrucha

Building Up and Characterization of Multi-component Collagen-Based Scaffolds..................................... 1626
  K. Pietrucha

Fabrication of Gelatin/Bioactive Glass Hybrid Scaffolds for Bone Tissue-Engineering 1630
  S. Borrego-González, J. Becerra, A. Díaz-Cuenca

Collagen-Targeted BMPs for Bone Healing 1634
  P.M. Arrabal, R. Visser, L. Santos-Ruiz, J. Becerra, M. Cifuentes

Layer by Layer: Designing Scaffolds for Cardiovscular Tissues 1638
  B. Glasmacher, A. Repanas A., O. Gryshkov, F. AL Halabi, T. Rittinghaus, R. Kortlepel, S. Wienecke,
  M. Müller, H. Zernetsch

Part XII: Neural and Rehabilitation Engineering

Voice Activity Detection from Electrocorticographic Signals 1643
  V.G. Kanas, I. Mporas, H.L. Benz, N. Huang, N.V. Thakor, K. Sgarbas, A. Bezerianos, N.E. Crone

Assessment of an Assistive P300–Based Brain Computer Interface by Users with Severe Disabilities 1647
  R. Corralejo, L.F. Nicolás-Alonso, D. Álvarez, R. Hornero

Single-Trial Detection of the Event-Related Desynchronization to Locate with Temporal Precision the Onset of
Voluntary Movements in Stroke Patients 1651
  J. Ibáñez, M.D. del Castillo, J.I. Serrano, F. Molina Rueda, E. Monge Pereira, F.M. Rivas Montero,
  J.C. Miangolarra Page, J.L. Pons

Hybrid Brain Computer Interface Based on Gaming Technology: An Approach with Emotiv EEG
and Microsoft Kinect 1655
  Nuno André da Silva, Ricardo Maximiano, Hugo Alexandre Ferreira

Rehabilitation Using a Brain Computer Interface Based on Movement Related Cortical Potentials -A Review 1659
  K. Dremstrup, I.K. Niazi, M. Jochumsen, N. Jiang, N. Mrachacz-Kersting, D. Farina

Numerical Simulation of Deep Transcranial Magnetic Stimulation by Multiple Circular Coils 1663
  M. Lu, S. Ueno

Volitional Intention and Proprioceptive Feedback in Healthy and Stroke Subjects 1667
  M. Gandolla, S. Ferrante, F. Molteni, E. Guanziroli, T. Frattini, A. Martegani, G. Ferrigno, A. Pedrocchi, N.S. Ward

Nonlinear Relationship between Perception of Deep Pain and Medial Prefrontal Cortex Response Is Related to
Sympathovagal Balance 1671
  R. Sclocco, M.L. Loggia, R.G. Garcia, R. Edwards, J. Kim, S. Cerutti, A.M. Bianchi, V. Napadow, R. Barbieri

Dynamics of Learning in the Open Loop VOR 1675
  P. Colagiorgio, G. Bertolini, C. Bockisch, D. Straumann, S. Ramat

Numerical Modeling of Optical Radiation Propagation in a Realistic Model of Adult Human Head 1679
    N. Ortega-Quijano, F. Fanjul-Vélez, I. Salas-García, J.L. Arce-Diego

Individual Noise Responses and Task Performance and Their Mutual Relationships 1683
    T. Niioka, S. Ohnuki

Is the Software Package Embedding Dynamic Causal Modeling Robust? 1686
    P. Tayaranian Hosseini, S. Wang, S.L. Bell, J. Brinton, D.M. Simpson

The Focuses of Pathology of Electrical Activity of Brain in Assessment of Origin of Syncope 1690
    K. Peczalski, T. Palko, D. Wojciechowski, W. Jernajczyk, N. Golnik, Z. Dunajski

Neuroanatomic-Based Detection Algorithm for Automatic Labeling of Brain Structures in Brain Injury 1694
    M. Luna, F. Gayá, A. García-Molina, L.M. González, C. Cáceres, M. Bernabeu, T. Roig, A. Pascual-Leone,
    J.M. Tormos, E.J. Gómez

Brain Activity Characterization Induced by Alcoholic Addiction. Spectral and Causality Analysis of Brain Areas
Related to Control and Reinforcement of Impulsivity 1698
    J. Guerrero, A. Rosado, M. Bataller, J.V. Francés, T. Iakymchuk, A. Luque-García, V. Teruel-Martí, J. Martínez-Ricós

A Novel RMS Method for Presenting the Difference between Evoked Responses in MEG/EEG – Theory
and Simulation 1702
    I. Nemoto, M. Kawakatsu

The Influence of Neuronal Density on Network Activity: A Methodological Study 1706
    E. Biffi, G. Regalia, A. Menegon, G. Ferrigno, A. Pedrocchi

Effective Connectivity Patterns Associated with P300 Unmask Differences in the Level of Attention/Cognition between
Normal and Disabled Subjects 1710
    S.I. Dimitriadis, Yu Sun, N.A. Laskaris, N. Thakor, A. Bezerianos

A General Purpose Approach to BCI Feature Computation Based on a Genetic Algorithm: Preliminary Results 1714
    S. Ramat, N. Caramia

Modular Control of Crouch Gait in Spastic Cerebral Palsy 1718
    D. Torricelli, M. Pajaro, S. Lerma, E. Marquez, I. Martinez, F. Barroso, J.L. Pons

Time-Frequency Analysis of Error Related Activity in Anterior Cingulate Cortex 1722
    G. Mijatovic, T. Loncar Turukalo, E. Procyk, D. Bajic

Depth-Sensitive Algorithm to Localize Sources Using Minimum Norm Estimations 1726
    B. Pinto, A.C. Sousa, C. Quintão

Measurement of Gait Movements of a Hemiplegic Subject with Wireless Inertial Sensor System before and after
Robotic-Assisted Gait Training in a Day 1730
    Takashi Watanabe, Jun Shibasaki

Glenohumeral Kinetics in Gait with Crutches during Reciprocal and Swing-Through Gait: A Case Study 1734
    E. Perez-Rizo, R. Casado-Lopez, V. Lozano-Berrio, M. Solis-Mozos, M. Nieto-Diaz, S. Martin-Majarres, J.L. Pons,
    A. Gil-Agudo

A Feasibility Study to Elicit Tactile Sensations by Electrical Stimulation 1738
    S.H. Hwang, J. Ara, T. Song, G. Khang

Dystonia: Altered Sensorimotor Control and Vibro-tactile EMG-Based Biofeedback Effects 1742
  C. Casellato, S. Maggioni, F. Lunardini, M. Bertucco, A. Pedrocchi, T.D. Sanger

Closed-Loop Modulation of a Notch-Filter Stimulation Strategy for Tremor Management with a Neuroprosthesis 1747
  J.A. Gallego, E. Rocon, J.M. Belda-Lois, J.L. Pons

Objective Metrics for Functional Evaluation of Upper Limb during the ADL of Drinking: Application in SCI 1751
  A. de los Reyes-Guzmán, I. Dimbwadyo-Terrer, S. Pérez-Nombela, F. Trincado, D. Torricelli, A. Gil-Agudo

Efficacy of TtB-Based Visual Biofeedback in Upright Stance Trials 1755
  C. D'Anna, D. Bibbo, M. Goffredo, M. Schmid, S. Conforto

A Data-Globe and Immersive Virtual Reality Environment for Upper Limb Rehabilitation after Spinal Cord Injury 1759
  Ana de los Reyes-Guzman, Iris Dimbwadyo-Terrer, Fernando Trincado-Alonso, Miguel A. Aznar, Cesar Alcubilla,
  Soraya Pérez-Nombela, Antonio del Ama-Espinosa, Begoña Polonio-López, Ángel Gil-Agudo

Wearable Navigation Aids for Visually Impaired People Based on Vibrotactile Skin Stimuli 1763
  M. Reyes Adame, K. Möller, E. Seemann

Brain Injury MRI Simulator Based on Theoretical Models of Neuroanatomic Damage 1767
  L.M. González, M. Luna, A. García-Molina, C. Cáceres, J.M. Tormos, E.J. Gómez

Haptic Feedback Affects Movement Regularity of Upper Extremity Movements in Elderly Adults 1771
  M. Schmid, I. Bernabucci, S. Comani, S. Conforto, B. D'Elia, B. Fida, T. D'Alessio

Dysfunctional Profile for Patients in Physical Neurorehabilitation of Upper Limb 1775
  M.A. Villán-Villán, R. Pérez-Rodríguez, C. Gómez, E. Opisso, J.M. Tormos, J. Medina, E.J. Gómez Aguilera

Video-Based Tasks for Emotional Processing Rehabilitation in Schizophrenia 1779
  R. Caballero-Hernández, A. Vila-Forcén, S. Fernandez-Gonzalo, J.M. Martínez-Moreno, M. Turon,
  R. Sánchez-Carrión, E.J. Gómez

Preliminary Experiment with a Neglect Test 1783
  C. Lassfolk, M. Linnavuo, S. Talvitie, M. Hietanen, R. Sepponen

A Subject-Driven Arm Exoskeleton to Support Daily Life Activities 1787
  E. Ambrosini, S. Ferrante, M. Rossini, F. Molteni, G. Ferrigno, A. Pedrocchi

Evaluation of a Novel Modular Upper Limb Neuroprosthesis for Daily Life Support 1791
  S. Ferrante, E. Ambrosini, M. Rossini, F. Molteni, M. Bulgheroni, G. Ferrigno, A. Pedrocchi

Evaluating Spatial Characteristics of Upper-Limb Movements from EMG Signals 1795
  O. Urra, A. Casals, R. Jané

An Adaptive Rreal-Time Algorithm to Detect Gait Events Using Inertial Sensors 1799
  N. Chia Bejarano, E. Ambrosini, A. Pedrocchi, G. Ferrigno, M. Monticone, S. Ferrante

A Graphical Tool for Designing Interactive Video Cognitive Rehabilitation Therapies 1803
  J.M. Martínez-Moreno, P. Sánchez-González, A. García, S. González, C. Cáceres, R. Sánchez-Carrión, T. Roig,
  J.M. Tormos, E.J. Gómez

Part XIII: Young Investigator Competition

Motion-Related VEPs Elicited by Dynamic Virtual Stimulation 1809
    P.J.G. Da Silva, B.P. Rosa, M. Cagy, A.F.C. Infantosi

A Distributed Middleware for the Assistance on the Prevention of Peritonitis in CKD 1813
    M.A. Estudillo-Valderrama, A. Talaminos-Barroso, L.M. Roa, D. Naranjo-Hernández, L.J. Reina-Tosina

Bioactive Nanoimprint Lithography: A Study of Human Mesenchymal Stem Cell Behavior and Fate 1817
    Z.A. Cheng, O.F. Zouani, K. Glinel, A.M. Jonas, M.-C. Durrieu

Automated Normalized Cut Segmentation of Aortic Root in CT Angiography 1821
    Mustafa Elattar, Esther Wiegerinck, Nils Planken, Ed vanbavel, Hans van Assen, Jan Baan Jr., Henk Marquering

Classification Methods from Heart Rate Variability to Assist in SAHS Diagnosis 1825
    J. Gómez-Pilar, G.C. Gutiérrez-Tobal, D. Álvarez, F. del Campo, R. Hornero

AdaBoost Classification to Detect Sleep Apnea from Airflow Recordings 1829
    G.C. Gutiérrez-Tobal, D. Álvarez, J. Gómez-Pilar, F. del Campo, R. Hornero

Effect of Electric Field and Temperature in E.Coli Viability 1833
    A.M. Oliva, A. Homs, E. Torrents, A. Juarez, J. Samitier

3D Shape Landmark Correspondence by Minimum Description Length and Local Linear Regularization 1837
    M. Valenti, C. Chen, E. De Momi, G. Ferrigno, G. Zheng

β-Band Peak in Local Field Potentials as a Marker of Clinical Improvement in Parkinson's Disease after Deep Brain
Stimulation 1841
    P.D. Frangou, K.P. Michmizos, P. Stathis, D. Sakas, K.S. Nikita

Ergonomics during the Use of LESS Instruments in Basic Tasks: 2 Articulated vs. 1 Straight and 1 Articulated
Graspers 1845
    M. Lucas-Hernández, F.J. Pérez-Duarte, A.M. Matos-Azevedo, J.B. Pagador, F.M. Sánchez-Margallo

Network-Based Modular Markers of Aging Across Different Tissues... 1849
    Aristidis G. Vrahatis, Konstantina Dimitrakopoulou, Georgios N. Dimitrakopoulos, Kyriakos N. Sgarbas,
    Athanasios K. Tsakalidis, Anastasios Bezerianos

Experimental Characterization of Active Antennas for Body Sensor Networks 1853
    D. Naranjo, L.M. Roa, L.J. Reina, G. Barbarov, A. Callejón

Part XIV: Special Sessions

Models of Arrhythmogenesis in Myocardial Infarction 1859
    Natalia A. Trayanova

Multiscale-Multiphysics Models of Ventricular Electromechanics – Computational Modeling, Parametrization and
Experimental Validation 1864
    G. Plank, A.J. Prassl, R. Arnold, Y. Rezk, T.E. Fastl, E. Hofer, C.M. Augustin

Effect of Purkinje-Myocyte Junctions on Transmural Action Potential Duration Profiles 1868
    R. Walton, O. Bernus, E.J. Vigmond

Morphometry and Characterization of Electrograms in the Cavotricuspid Isthmus in Rabbit Hearts during Autonomic Sinus Rhythm 1871
    E. Hofer, D. Sanchez-Quintana, R. Arnold

Cavotricuspid Isthmus: Anatomy and Electrophysiology Features: Its Evaluation before Radiofrequency Ablation 1875
    D. Sánchez-Quintana, J.A. Cabrera

Preoperative Prognosis of Atrial Fibrillation Concomitant Surgery Outcome after the Blanking Period 1879
    A. Hernández, R. Alcaraz, F. Hornero, J.J. Rieta

A Decision Support System for Operating Theatre in Hospitals: Improving Safety, Efficiency and Clinical Continuity 1883
    R. Miniati, G. Cecconi, F. Frosini, F. Dori, G. Biffi Gentili, F. Petrucci, S. Franchi, R. Gusinu

LICENSE: Web Application for Monitoring and Controlling Hospitals' Status with Respect to Legislative Standards 1887
    E. Iadanza, L. Ottaviani, G. Guidi, A. Luschi, F. Terzaghi

Relation among Breathing Pattern, Sleep Posture and BMI during Sleep Detected by Body Motion Wave 1891
    Junýa Wada, Tadashi Yajima, Takenori Imamatsu, Hiroaki Okawai

Sophisticated Rate Control of Respiration and Pulse during Sleep Studied by Body Motion Wave 1895
    H. Okawai, T. Yajima, T. Imamatsu, J. Wada

A Tool for Patient Data Recovering Aimed to Machine Learning Supervised Training 1899
    G. Guidi, M.C. Pettenati, M. Milli, E. Iadanza

The Potential of Machine-to-Machine Communications for Developing the Next Generation of Mobile Device Monitoring Systems 1903
    R.S.H. Istepanian, B. Woodward, D.J. Mulvaney, S. Datta, P. Harvey, A.L. Vyas, O. Farooq

A Study on Perception of Managing Diabetes Mellitus through Social Networking in the Kingdom of Saudi Arabia 1907
    T.M. Alanzi, R.S.H. Istepanian, N. Philip, A. Sungoor

Validity of Smartphone Accelerometers for Assessing Energy Expenditure during Fast Running 1911
    C. Easton, N. Philip, A. Aleksandravicius, J. Pawlak, D.J. Muggeridge, P.A. Domene, R.S.H. Istepanian

Medical Quality of Service Analysis of Ultrasound Video Streaming over LTE Networks 1915
    Ali Alinejad, R.S.H. Istepanian, N. Philip

Smart Social Robotics for 4G-Health Applications 1919
    R.S.H. Istepanian, A. Good, N. Philip

Regional Cooperation in the Development of Biomedical Engineering for Developing Countries 1923
    K.I. Nkuma-Udah, G.I. Ndubuka, E.E.C. Agoha

Author Index 1927

Keyword Index 1939

Part I
Keynote Speakers

Design, Development, Training and Use of Medical Devices; with Practical Examples from Cardiovascular Medicine and Surgery

Alan Murray

Newcastle University, Newcastle upon Tyne, UK

Abstract—The important role of medical engineers in the design and development of medical devices is without question. The clinical need for such devices has grown year by year, and will continue to grow. The design and development of useful, effective and safe devices depends on a good understanding of clinical use and a careful follow up of devices in clinical use to ensure they do work as planned and are safe.

Keywords—Medical device, design, development, training, use.

I. INTRODUCTION

In all disciplines of medicine and surgery, the use of medical devices is essential. Without them the success of modern medicine would be much poorer. Users want these devices to work as they expect, but this does not always happen. Serious problem are rare, but difficulties are all too common. This paper reviews some of the issues associated with medical devices. It takes a specific interest from a medical engineering standpoint.

The main factors discussed here are:

- Design
- Development
- Training
- Use

II. BACKGROUND TO MEDICAL DEVICES

The term "medical device" covers an enormous range and number of devices used in hospitals and clinics, from a simple wooden spatula, to mechanical stethoscopes, to electronic monitors, to complex servo-controlled anaesthetic machines. These are given just as examples. Increasingly medical devices are being used in patients' homes, and also by members of the public seeking to promote their own health.

The definition of a medical device is precise and used internationally. In Europe a device is defined as any instrument, apparatus, appliance, material or other article, whether used alone or in combination, including the software necessary for the proper application, intended by the manufacturer to be used for human beings for the purpose of:

diagnosis, prevention, monitoring, treatment or alleviation of disease,

diagnosis, monitoring, treatment, alleviation of or compensation for an injury or handicap,

investigation, replacement or modification of the anatomy or of a physiological process,

control of conception,

and which does not achieve its principal intended action in or on the human body by pharmacological, immunological or metabolic means, but which may be assisted by such means.

So it can be seen that the definition is very wide. Medical device manufacturers must follow all required legislation. Much of this legislation is to ensure that the devices work as specified, are described accurately and without exaggerated claims, and are safe.

When a device does not do what it is expected to do it can present a safety hazard, whether it just stops working or whether it does something unexpected. In both these cases the fault can lie in the device itself, or with the user who might not understand fully how the device works or has received little training or simply assumes the device should do something it was never designed to do.

Since safety is such a major concern, many international bodies are working to improve clinical safety, including safety with medical devices. The World Health Organization has issued many useful documents and statements. Their view of medical device safety can be summarized as "Medical devices should be designed and manufactured in such a way that, when used under the conditions and for the purposes intended …. they will not compromise the clinical condition or the safety of patients, or the safety and health of users …. provided that any risks ….constitute acceptable risks when weighed against the benefits to the patient" [1].

Throughout the world organizations have been set up to record, monitor and act on healthcare safety. In the UK a document "Building a Memory" gives a definition of a patient safety incident as "Any unintended or unexpected incident that could have or did lead to harm for one or more patients" [2]. It is vital to realise that incidents where no harm came to patients or staff must be reviewed, as the potential for harm was there, and a similar incident could easily lead to serious harm. The document cited above was titled "Building a memory" because such events should not be forgotten. They should be used to prevent as far as possible any repetition of similar incidents.

An example of a device that continues to have problems throughout the world is diathermy or electrocautery which has received much redevelopment, but is still relatively complex to use, with users failing to use it correctly resulting in patient burns [3].

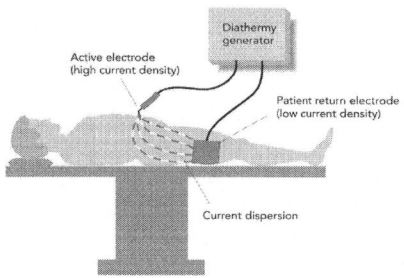

Fig. 1 Diathermy / electrocautery with potential for burns.

So what makes a good medical device? Firstly it must make an important contribution to clinical care. Then in addition it must be easy to use, always do what is expected, and be intuitive to use.

So before examining problems it is worth reflection on the positive benefits of medical devices in healthcare. As an example, in the UK document "Engaging Clinicians" we are told that every day more than 1 million patients are treated safely and successfully in the British healthcare system, but that evidence also tells us that in complex health care systems things will and do go wrong, no matter how dedicated and professional the staff [4]. However, it is worth reflecting on a report from The Health Foundation [5]. They reviewed surgical operations, where medical devices are always in use and where there is the potential for harm. The report highlights issues over equipment and medical devices, and found that problems were more common than had been expected. In nearly one on five operations, the equipment was either faulty, missing, used incorrectly, or the staff in the operating theatre did not know how to use it.

III. DEVICE PROBLEMS

Taking the UK as an example, with a population of over 60 M, over 1M incidents are reported by hospitals and medical clinics each [6]. These take into account drug and other problems as well as device incidents. Overall, devices feature in between 3 and 4% of reported incidents. Of device incidents, approximately 300 deaths each year occur associated with a medical device, but not necessarily caused by the device [7].

IV. DESIGN, DEVELOPMENT, TRAINING AND USE

This paper concentrates specifically on what we can learn from problems and difficulties faced by medical users, and reviews problems that have actually happened in clinical use.

A. Design

The design of the device is perhaps the most important aspect, as at this stage the designer can make sure the device should be useful and easy to use. The designer is the one who theoretically could be the expert user, but who may never use it in clinical practice, and may not be unaware of external factors influencing its use. The device may be so complex in its operation that it needs significant training to remember how to use it successfully. Many designers try to include almost every function they think might be useful, when some may never actually be used.

I remember a cardiac electrophysiology stimulator that had so many switches and controls that none of the cardiologists felt comfortable using it. There were even occasions during clinical procedures when a bioengineer who had studied the different modes and so had a mental picture of the device's function had to be called for assistance. Needless to say, that device is no longer available.

Many clinicians and nurses working in intensive care may remember early computerised monitors that had so many layers of display staff became completely lost in how to operate them. These devices quickly thereafter installed a button to take users back to the first screen; a simple, but important advance. Thankfully since then designers have thought much harder about the interface between the user and the device.

Poor user concentration results in inappropriate device settings, but often with better device design such problems could have been avoided. User misunderstanding is also often related to device design, and could have been avoided by better consideration of how the device would be used in often hectic clinical conditions. An example is a problem that still occurs is over the setting of filters on electrocardiographs that can distort the ECG, producing ST segment distortion that can be mistaken for ischaemia. It needs to be clear when these devices are set into a monitoring mode and not in their diagnostic setting, and more importantly any print out needs to be very clearly labelled to indicate when the ECG recording was not made in the usual diagnostic mode with its appropriate filters. Such a user fault, which is really a design fault, can result in inappropriate therapy.

The vast majority of electronic medical devices contain a microprocessor, and in many cases several microprocessors. These require computer programs that are difficult to test, and it is probably impossible to be absolutely sure that the device will not find its way into an unplanned mode. Rigorous computer design methods must be followed. Since this has much to do with development, more will be said about this subsequently. However, at the design stage it is important to specify clearly each function and its interrelationships with other functions.

B. Development

Development is when the design is transferred to a useable device, initially as a prototype. This is where there are opportunities to be creative, but where the creativity should not deviate from the initial design without a clear change to the design. Otherwise there is potential confusion on how the devise should operate.

The internal layout of components can prevent problems that could be created by external effects. External interference has been known to result in the mode of pacemakers or implantable defibrillators being accidentally changed. To this day you still see warning notices telling patients with pacemakers not to stand near specific areas, such as at shop entrances and exits where anti theft devices are in use, and at airport security areas. Clearly there is still room for improved developments.

Sometimes situations arise that were never in the design, and devices simply fail. This has happened with defibrillators. Manufacturers are usually able to rectify these problems with new software, but it does emphasise the need to have careful and well planned development.

Computer program development is an exceptionally difficult area and needs great care with well agreed procedures for development and testing. With the inevitable increase in the use of microprocessors and the increasing complexity of medical equipment this will become even more important than it is currently. Many researchers have no training in safe computer code development and need to accept that any code they generate in testing the device concept will need to be reengineered, and so researchers need to be particularly careful in documenting fully every step in their procedure. This is not easy for developers that want to get quickly to the end point.

There have been many examples of devices in recent years that have had to be withdrawn, or more usually reprogrammed, after some users experience problems. Such problems are often rare, but when they occur they can be serious, such as when a device suddenly stops functioning during a critical clinical procedure. This has happened with external defibrillators when they refused to deliver a shock for no apparent reason.

C. Training

No matter how simple a device, training is important. It is very easy for the user to assume that it is easy to understand, when this may not be the case. It must be remembered that training should be genuine training and not just an exercise in logging that some training has been given, no matter how effective that training. Designers and developers have an important role here, not in the actual training, but in ensuring that their device is so simple and straightforward to use, that its use is intuitive and training will almost not be necessary.

All devices will have a user manual, and designers and developers will have a major input into this. It is usually referred to as the "Instructions for use". If this is difficult to write the device may be more complex than it needs to be, so writing the first draft of these Instructions before and during and not after the device development is important, as complexities uncovered can still influence device design.

D. Use

Devices are built to be used, and so to get to this point is the goal. However it is not the end. The use of the device needs to be monitored, primarily so that safety problems can be tackled. This allows devices with significant problems to be withdrawn, and enables corrective action with others to be taken. So monitoring problems is vital, and manufacturers are required a follow a Vigilance procedure in which they record, report to relevant authorities, and introduce corrective action.

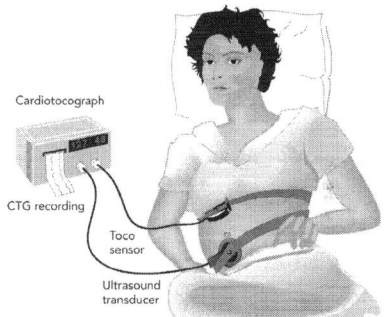

Fig. 2 Electrocardiotocograph.

User expectations are often strong. If a device gives a result, it will be assumed that it is correct. There are many examples where this may be wrong. One example that sadly keeps recurring is the use of electrocardiotocograph devices for fetal monitoring. It is known that these devices can lock onto the maternal heartbeat and present double that frequency as the fetal heart rate. Because the rate can be similar to that expected, a problem may not be noted, even when the fetus is in danger. Research and development will undoubtedly improve these devices, but in the meantime users need to realise that devices are never perfect. This is a difficult situation, but one in which design, development and training are just as important as use. However, whenever a medical device is in clinical use, monitoring what happen with the device in clinical use is vital.

Most developed counties in the world have some reporting system for reporting device problems. These systems have varying effectiveness. In the UK where most hospitals are within the National Health Service, hospital staff are required to report device safety issues to the Medicines and Healthcare products Regulatory Agency (MHRA). The MHRA acts as the Competent Authority for the EU medical device regulations. Because of the formal reporting system it provides effective feedback, letting hospitals act to improve safety, no matter the cause. Reporting tends to work most effectively when incidents are serious, especially with those resulting in injury or death. It is not always easy to know whether the device caused the injury, but the association is important to note, because if similar reports are received from other hospitals this may allow the cause to be found. The MHRA rightly publish data on all incidents, whether there is the potential for a device to be implicated or not. Examples in this review can be found in their publications [8].

Device failures that occur outside a hospital or clinic are harder to handle. If they occur many years later and are seen as an isolated failure there can be a tendency not to report. If every failure could be reported a better predictor could be developed. Clinicians who see failures may see only one failure in their life, but if these could be recorded, action could be taken earlier from combined results.

A notable example was over a heart valve failure when a weld began to fail resulting in the valve failure and death. Professor Bertil Jacobson when he acted as scientific advisor in medical engineering to the Swedish Board of Health and Welfare saw such events recur at a national level and realised that something had to be done to lessen such events by learning from them. This was the instigation for the book "Medical Devices: Use and Safety" which collected and published 140 case histories of medical device incidents from all over the world, covering every type of device. Sadly many involved serious injury or death. One common thread was that device incidents usually had many interrelated causes. They were what are known as system problems [9].

Being aware of device problems is important for all medical engineers involved with the design and development of medical devices. Such knowledge helps not just with the specific device they have produced, but brings awareness about problems encountered in daily medical care which should influence design and development of any device.

ACKNOWLEDGMENT

I acknowledge Professor Bertil Jacobson of the Karolinska Institute in Stockholm who sadly died in 2009. He was a great and enthusiastic advocate for patient safety, and a meticulous co-author of "Medical Devices: Use and Safety". The two illustrations in this paper are taken from this book. Any of its many illustrations in the book can be freely reproduced for teaching with acknowledgement.

REFERENCES

1. World Health Organization (2003) Medical device regulations: global overview and guiding principles
2. National Patient Safety Agency (2005) Building a memory: preventing harm, reducing risks and improving patient safety. The first report of the National Reporting and Learning System and the Patient Safety Observatory. UK.
3. Khambete N, Kelkar-Khambete A, Desurkar V, Murray A (2010) Safety of medical equipment: a review of hospital safety testing in the Indian City of Pune. In: Appropriate Healthcare Technologies for Developing Countries. Institution of Engineering and Technology, London, 1-4
4. National Patient Safety Agency (2005) Engaging clinicians; a resource pack to help you promote patient safety and the reporting of incidents amongst healthcare staff in training. UK.
5. The Health Foundation. Evidence in brief: how safe are clinical systems? Primary research into the reliability of systems within seven NHS organisations and ideas for improvement (2010) The Health Foundation, UK.
6. http://www.nrls.npsa.nhs.uk/
7. Medicines and Healthcare products Regulatory Agency. Report on devices adverse incidents 2010 (2011) UK Department of Health
8. http://www.mhra.gov.uk/Publications/Safetyguidance/DeviceBulletins/
9. Jacobson B, Murray A (2007). Medical devices: use and safety. Elsevier; Edinburgh (reprinted as South Asia edition 2011)

Author: Prof Alan Murray
Institute: Newcastle University
City: Newcastle upon Tyne
Country: UK
Email: alan.murray@ncl.ac.uk

Part II
Invited Presentations

David Dewhurst – Biomedical Engineer and IFMBE Pioneer

R.L.G. Kirsner[1] and J.S. McKenzie[2]

[1] Department of Electronic Engineering, La Trobe University Bundoora, VIC 3083, Australia
[2] Department of Physiology, The University of Melbourne, Parkville VIC 3010, Australia

Abstract—David Dewhurst was an idiosyncratic, warm-hearted and cheerful man who played a major part in establishing bio-medical engineering in Australia. He had a long association with the IFMBE, was President from 1968-1971 and chaired the 9th ICMBE held in Melbourne. Initially a Classicist destined for the church, he studied physiology and electronics after his return from World War 2 and ran an innovative medical instrumentation group in the Department of Physiology at The University of Melbourne. With the purchase of a minicomputer in 1965, the first in Melbourne, David and his group developed hardware and software for biological signal analysis and carried out a number of studies of human arm reflexes. His strong interest in electrical safety and electrocution led him to play a prominent role in the development of electrical safety standards, with Australia leading the world in this. He was proud to be a member of a government panel evaluating medical technology and advising the Australian Government on the introduction of new medical techniques such as MRI. David also co-supervised the development and construction of the first cochlear ear implant. A gifted teacher and excellent communicator, he wrote widely on technical and non-technical topics including his "On the Real Axis" series produced for the IFMBE.

David John Dewhurst was born in 1919 in country Victoria, the son of an Anglican minister. As a boy, he was interested in electronics and amateur radio, and from his little workshop under the stairs, he was the fixer of the family wireless set. But his aim was to enter the church, like his father and grandfather, and to this end he studied Classics at The University of Melbourne, graduating with an Honours Arts degree.

DJD in 1940

Then came the Second World War, and from June 1940 David served with the Australian Imperial Forces, first in the Middle East. He was in the Corps of Signals, often attached to anti-aircraft artillery, and he picked up some useful skills. For example, when Australian troops occupied Beirut in 1941 after battling Vichy forces, his unit had to learn quickly to use the mechanical predictors of French 75 mm AA cannons, al-though the manuals had been destroyed by the previous owners. David used such events in later life to illustrate points of principle about good biomedical engineering, such as the ability to adapt to a lack of "suitable tools" - an army term for standard repair-kits, which often went missing.

He was evacuated in early 1942 with hepatitis A, and back in Australia, he was redirected to Air Support Control Signals, and was attached to an anti-aircraft brigade. In March 1945 he was seconded to an army trade school in the former Marconi School of Wireless in Sydney, where he trained as an Instructor.

This wartime experience brought about a change of career. In 1946, after demobilisation, he began a science degree, majoring in physiology and electronics.

In the same year, a young medical graduate, Dr Ian McCallum, was employed in the Department of Physiology. He remained only three years, but he set up its first electrophysiology unit, complete with workshop and an electrically-shielded room for electrophysiological recording. His unit produced various pieces of electronics mainly for research purposes, as well as spending a lot of time recording clinical ECGs and EEGs using a fairly primitive 6 channel EEG machine.

As a student, David spent a long vacation working in McCallum's unit. His surviving reports are investigations of pieces of equipment bought by the department that had never worked properly; in several cases he emphasized that "this design can never work and the department should require the manufacturers either to replace it or to give the department's money back". In at least one of these cases the attached document suggests that his advice was followed.

David graduated at the end of 1948 and, following McCallum's resignation early in 1949, he took over the electrophysiology unit while working on his MSc and demonstrating in the Physiology Department. His 1952 MSc thesis was in two parts, one entitled "A Critique of Electric Modellism in Tissue", the other "The Design of an Oscilloscope for Clinical Research". He was starting to apply his electronic knowledge to biological measurement.

In 1952 he was appointed to a recently vacated lectureship in Physiology and, along with his teaching load, he quickly transformed the small electrophysiology laboratory into a renowned centre of medical instrumentation. This was located mainly in two aged preparation rooms behind the Medical School lecture theatres. Together they were known as "the Shielded Room", a name that became legendary.

Rebecca Radar Display

The post-war years were lean years for the university and he liked to quote Rutherford, "We had no money so we had to think". What was available and cheap was army surplus equipment, and one of David's early large-scale successes was the modification of 30 Rebecca [REcognition of BEacons] Radar displays – the airborne end of a system of direction-finding equipment designed to assist the air-drop delivery of supplies to army or resistance units in occupied territory. He transformed these into was 30 physiological monitors for undergraduate classes. And to drive them, he built preamplifiers using ex-army valves. Many of the valves either didn't work at all or were unreliable, so he bought several hundred to ensure that he had enough good ones to keep 30 preamplifiers working. As a result of this work, every undergraduate student-pair recorded sciatic nerve action potentials, as well as muscle responses, with their own set-up.

David also designed and built apparatus for his (and his colleagues') research using cheap army disposals components together with sound design. A 1957 University publication reports on his design for a cardiac defibrillator and his electronic equipment for recording physiological variables during surgery.

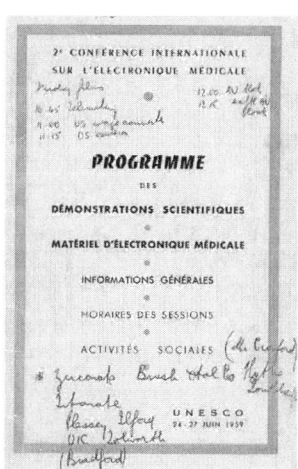
2nd ICME program

Late in 1958, he completed his PhD thesis on "The Significance of Electrical Parameters in Tissue" and he was awarded the degree in 1959. The year was a seminal one for him, marking a significant development in his career. He and his family spent a sabbatical year in Cambridge, where he worked in Sir Bryan Matthews' laboratory in the Department of Physiology, and the year had two lasting outcomes.

First, Matthews and his colleagues were studying the relationship between the firing rate of single motor units of a muscle and the force applied to the muscle. David became very involved in this work and after his return to Australia he designed the famous "muscle puller" which, in ever-improved versions, applied an increment or decrement of force to a subject's elbow.

Secondly David attended the 2nd conference of the then fledgling International Society for Medical Electronics, in Paris. This led to a lasting involvement with the society, later to become the IFMBE.

He also joined the UK Biological Engineering Society, where he made a number of long-lasting friendships with luminaries in the area such as Jack Perkins, Heinz Wolff, Dennis Hill and Keith Copeland.

Back in Australia in 1960, David was made Senior Lecturer in Biophysics and was promoted to Reader in Biophysics in 1964.

As already mentioned, David built the "muscle puller" that applied an increment or decrement of force to a subject's elbow and measured the resulting movement and EMG in one or more of the relevant muscles. David and his group used it extensively to analyse rapid skilled movements by a human subject.

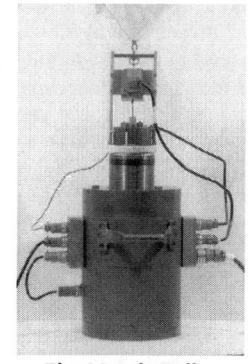
The Muscle Puller

Then in 1965, David organized the purchase of one of the first minicomputers in Australia, a PDP8. With it, his group could to acquire and process electrical and mechanical physiological data in real time. Together they developed considerable in-house software and hardware expertise, which made possible on-line signal processing and coherent averaging. For the muscle puller experiments, this meant that they could produce histograms of motor unit activity and study consistency of responses and the effects of nerve blocks.

DJD & the PDP8, 1967

David was an expert PDP8 programmer well ahead of the field in Australia, and he passed his skills onto many others. Ever the teacher, he produced a programming manual that went through a number of editions.

In 1961 David started an extension course on medical electronics for biological researchers. It covered both theory and practical construction, and by the end of the course, the students had a thorough grounding in basic electronics and transducers, and had built a small oscilloscope and a Geiger counter. These students included cardiologists, neurologists and surgeons from Melbourne's teaching hospitals, a number of them senior clinicians and medical researchers, and they had a major impact on the perception and understanding of the role that biomedical engineering can play in clinical practice and research, and on the practice of high technology medicine.

The course's high standing led to Pergamon publishing the course notes as a book in 1966; then, as the course evolved to match the rapid advances in electronics into the nineteen seventies, a revised and extended version was published in 1976. It was in demand for many years.

As mentioned, David became interested in international biomedical engineering activities during his time in Cambridge in 1959.

He attended some of the first international meetings on medical and biological engineering in the early 60s and became very involved in activities of the IFMBE. In those early years, it was a small group who knew each other well and became good friends.

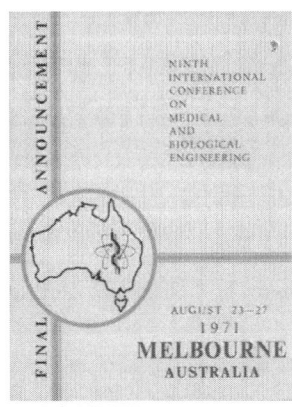
9th ICMBE program

In 1965, David was elected to the Administrative Council of the Federation, and was President from 1969 to 1972. During his 3-year presidency, Melbourne hosted the 9th International Conference on Medical & Biological Engineering in 1971, with David firmly at the helm and, as always, the genial host. The venue was The University of Melbourne's brand new medical school.

The Conference was a great success. Delegates came from the UK, Japan, Russia, the US, Czechoslovakia... For a number of us in David's group, it was an exciting introduction to international biomedical activities and lasting friendships.

In 1979, David was made an Honorary Life Member of the IFMBE, but his international involvement spread beyond the Federation; he was one of the Foundation Fellows of the Biological Engineering Society in the UK and he was also awarded the IEEE's Centennial Medal.

He didn't ignore the local scene. His enthusiasm for all aspects of medical electronics and biomedical engineering was contagious and led to the formation of the Society for Medical & Biological Engineering (Vic) in 1959, and in 1967, the Australian Federation for Medical & Biological Engineering. With the AFMBE established, Australian biomedical engineering was then able to affiliate formally with the IFMBE.

David's interest in improving the organisation and status of professional biomedical engineering in Australia was a major factor in the formation, in 1967, of the Institution for Biomedical Engineering (Australia), which was subsumed into Engineers Australia's College of Biomedical Engineers in 1994.

Always the classicist, David had an encyclopaedic and eclectic memory and frequently illustrated his conversations and teaching with quotes from Homer, Chaucer, Shakespeare, Lewis Carroll, Gilbert and Sullivan etc, etc. Biomedical engineers across the globe read his idiosyncratic and always delightful column in the IFMBE newsletter, "On the Real Axis", which he wrote from 1977 to 1988. The

IFMBE Book Launch, Kyoto, 1991

articles defy easy categorisation – observations on the practice of biomedical engineering, the human relationships involved, the ethics, with odd interpolations about wombats, bludgers, Omar Khayyam, steam engines, Horace, the plague.... The column had a wide following, and in 1991 the International Federation published a selection of the essays as a book.

He was delighted that some of the articles were translated and published in a Swedish biomedical journal.

Australia had electrical safety standards for electromedical equipment in the early 1970s, well in advance of international standards, and this was very largely due to David's efforts. He took an active role in Australian Standards for many years, as member and chairman of various standards committees. He remained, throughout his professional life, intensely concerned about the electrical safety of medical instruments, and had cautionary tales to tell of accidental electrocution due to poor design or maintenance, as of the patient who sat down for a rest on an improperly insulated grid supposedly protecting the power source of an early X-ray machine. The initial impetus was his understanding of the physiology of cardiac fibrillation and electrocution. Together with a medical colleague, he investigated the parameters of fatal current flow affecting the heart, and contributed decisively to the establishment of safety standards on a physiological basis.

David began to identify more and more with Biomedical Engineering as a profession, especially after moving from

First cochlear ear implant, 1975

the Physiology Department to the Electrical Engineering Department in 1975. It was there that he collaborated in the design of the first cochlear ear implant, which was designed in the Electrical Engineering Department as a PhD project under his supervision. The original implant was really state-of-the art, utilised CMOS technology and lasted almost 5 years.

This was an important and significant project and I believe that neither David nor the Electrical Engineering Department received anywhere near adequate recognition of their major achievement and their contribution to the success of the bionic ear project.

David was very involved in the early work of the National Health Technology Advisory Panel and was proud of the part he played in technology evaluation and the introduction of high technology medical techniques such as Magnetic Resonance Imaging.

And finally there was FRED. David conceived FRED, or Friendly Electronic Device, as a way of utilising microprocessor technology to make an interactive teaching device for disabled people who cannot use a computer keyboard.

FRED involved a pursuit task where the player used a lever to move an image of a toy bear on a screen to keep it inside a square that moved periodically. The speed of the game and the type of lever could be changed to suit the abilities of the player – each move could take an hour if necessary. One disabled person said that it was the first time in his life that he had managed to complete something without someone else coming in and finishing it for him.

FRED's genesis was inextricably bound up with David and Marjorie's experience with their disabled son Peter, and the acute understanding that this fostered of the needs of people with a disability. Although FRED's hardware changed radically in the 15 years after the first prototype units were tested, the basic design philosophy remained unchanged – the device must always reinforce success and never failure. FRED gave a degree of independence to many severely disabled people, and convinced them that they could indeed achieve something. Unfortunately, David

DJD & FRED ca. 1982

never succeeded in finding the commercial support needed to make the project viable. It was also being designed at a time of enormous advances in computer hardware and software, and it was overtaken by technology.

David Dewhurst's work in this field was widely recognized. When he was made a Member of the Order of Australia in 1990, the citation read "For services to biomedical engineering for people with disabilities".

David retired from the university in 1985, and moved with Marjorie to his much loved holiday house in the seaside town of Portarlington. His workshop went with him, and he continued for years to work on FRED and other biomedical projects. He died on 4th March 1996 aged 77.

Three characteristics of David's stand out across his lifetime:

First, he was an excellent communicator who could talk to anyone. It was, I suspect, a natural ability possibly reinforced by his wartime experience. He knew everyone. He established a group in the Physiology Department in which ideas flourished.

Second, he was a gifted teacher who had the capacity to get to the essence of a complex subject. An example of this, of course, was the way that he taught medical electronics to medical researchers and clinicians.

Third, he was an excellent organiser. He could see where a new organization was needed and how to go about putting it in place, as he did with the IBME.

His legacy is the change of perception that he brought about of the role of biomedical engineers in research and in hospitals and the way he brought Australia into the international biomedical engineering community.

He had friends in many disciplines and was widely known and respected throughout the international biomedical engineering community. From a personal perspective, David kick-started my interest in international biomedical engineering and the IFMBE, and he was an excellent sounding board for new ideas and constitutional problems.

He was someone with whom I could discuss problems, and a good friend. It was a great privilege to have known him.

Did Jan Swammerdam Do the First Electric Stimulation over 100 Years before Luigi Galvani?

J.A. Malmivuo[1], J. Honkonen[2], and K.E. Wendel[3]

[1] Department of Electronics, Aalto University, Espoo, Finland
[2] Idman Airfield Lighting, Vantaa, Finland
[3] Department of Electronics and Communication, Tampere University of Technology, Tampere, Finland

Abstract—It is generally believed that the first experiment on bimetallic electric stimulation of living body was made by Luigi Galvani with frog leg in 1786. Galvani however succeeded to produce electric stimulation of the frog leg already in 1781 with electricity produced with electric machine. It has been suggested by Rowbottom and Susskind that the first bimetallic stimulation of the frog muscle was done already in 1664 by Jan Swammerdam, who apparently did not understand the mechanism of this stimulation.

We made an analysis of Swammerdam's experiment and repeated it with frog leg. We found that it is theoretically thinking very probable that Swammerdam's experiment produced bimetallic electric stimulation of the muscle. We verified this with successful experiment. We believe that our work verifies that the first documented bimetallic electric stimulation of muscle was made by Jan Swammerdam in 1664.

Keywords—Bimetallic stimulation, frog muscle.

I. INTRODUCTION

The most famous experiments in neuromuscular stimulation were performed by Luigi Galvani, professor of anatomy at the University of Bologna [1]. He succeeded to produce electric stimulation of the frog leg in 1781 with electricity produced with electric machine. In September 1786, Galvani was trying to obtain contractions from atmospheric electricity during calm weather. He suspended frog preparations from an iron railing in his garden by brass hooks inserted through the spinal cord. Galvani happened to press the hook against the railing when the leg was also in contact with it. Observing frequent contractions, he repeated the experiment in a closed room. He placed the frog leg on an iron plate and pressed the brass hook against the plate, and muscular contractions occurred. [2]

Continuing these experiments systematically, Galvani found that when the nerve and the muscle of a frog were simultaneously touched with a bimetallic arch of copper and zinc, a contraction of the muscle was produced. This experiment is often cited as the classic study to demonstrate the existence of bioelectricity [3 p. 39]. However, it is possible that Jan Swammerdam had already conducted similar experiments in 1664 [3, 4].

II. SWAMMERDAM'S EXPERIMENT

The first carefully documented scientific experiments in neuromuscular physiology were conducted by Jan Swammerdam (Dutch, 1637-80). At that time it was believed that contraction of a muscle was caused by the flow of "animal spirits" or "nervous fluid" along the nerve to the muscle. In 1664, Swammerdam conducted experiments to study the muscle volume changes during contraction, Fig. 1.

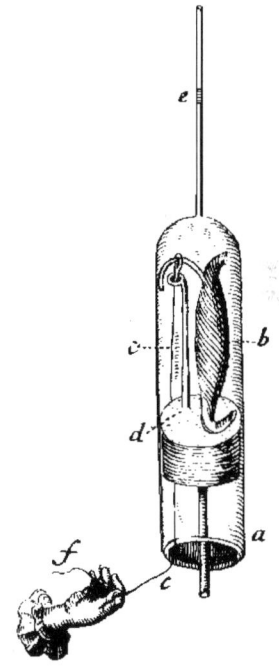

Fig. 1 Swammeram's experiment

Swammerdam placed a frog muscle (b) into a glass vessel (a). When contraction of the muscle was initiated by stimulation of its motor nerve, a water droplet (e) in a narrow tube, projecting from the vessel, did not move, indicating that the muscle did not expand. Thus, the contraction could not be a consequence of inflow of nervous fluid.

In many similar experiments, Swammerdam stimulated the motor nerve by pinching it. In fact, in this experiment stimulation was achieved by pulling the nerve with a wire (c) made of silver (*filium argenteum*) against a loop (d) made of brass (*filium aeneum*). According to the principles of electrochemistry, the dissimilar metals in this experiment, which are embedded in the electrolyte provided by the tissue, are the origin of an electromotive force (emf) and an associated electric current. The latter flows through the metals and the tissue, and is responsible for the stimulation (activation) of the nerve in this tissue preparation.

It is believed that this was the first documented experiment of motor nerve stimulation resulting from an emf generated at a bimetallic junction [4]. Swammerdam apparently did not understand that neuromuscular excitation is an electric phenomenon. Some authors interpret the aforementioned stimulation to have resulted actually from the mechanical stretching of the nerve. The results of this experiment were published posthumously in 1738 [5].

III. THEORETICAL BACKGROUND

A. Metals Used in This Work

Swammerdam used brass and silver to stimulate frog's muscle. Brass is an alloy of copper and zinc. In the late 17th century calamine brass was still used where zinc-percentage was about 15 % to 30 %. New smelting methods were developed during the 18th century and nowadays the zinc-percentage for normal common brass is about 37 % [6, 7]. Dezincification occurs in brass when it is in electrolyte with another metal. Zinc is selectively leached from brass alloy and copper remains. This changes the mechanical properties of brass [8]. Theoretical half-cell potential for brass is from -400 millivolts to -260 millivolts referenced to Saturated Calomel Electrode, SCE. This range is for all brasses. There is no reference of half-cell potential for calamine brass, but it is pure zinc-copper alloy so it must be something between -400 mV and -260 mV [9].

Silver is precious metal which has the highest electrical and thermal conductivity of any metal. Processes of making silver and the purity of silver have not drastically changed after 17th century. Theoretical half-cell potential for silver is from -150 mV to -100 mV referenced to SCE [9]. Voltage difference between brass and silver is then from 110 mV to 300 mV. Temperature affects on reaction speed. In higher temperatures the reaction speed rises and therefore the voltage between the metals rises [10].

Theoretical voltage differences for different metal pairs are shown in Table 1. These metals are iron, copper, silver and brass. Cathodic metal is on the left and anodic metal is on the right. These values are calculated from the maximum and minimum half-cell potentials for each metal [9, 11, 12].

Table 1 Calculated theoretical voltage between different metal pairs in electrolyte [9, 11, 12]

Metal Pair	Voltage [mV]
Copper – Brass	-110...100
Silver – Copper	150...270
Silver – Brass	110...300
Copper – Iron	780...960
Brass – Iron	750...1000
Silver – Iron	840...1240

Voltage differences were calculated with different metal pairs for comparing them to the later measurements. Later on the stimulation is tested with different metal pairs to achieve different voltages for stimulation. Two of these metal pairs have lower voltage difference than brass and silver and three of them have higher voltage difference.

B. Frog's Muscle

Jan Swammerdam probably used common frog, *Rana temporaria* in his experiment. In our experiment oriental fire-bellied toad, *Bombina orientalis* is used instead. They are from different families and they differ from each other at size [13]. It is assumed, that between these frog species threshold voltages of muscles and electrical properties do not differ from each other radically.

Rheobase is the minimal strength of an electrical stimulus that is able to cause contraction of a muscle. The average rheobase for frog's sartorius muscle occurs at -51.9 mV when the average resting potential is -92.9 mV [14]. The voltage needed to raise the membrane potential from resting potential to the rheobase is called threshold voltage. The calculated threshold voltage is then 41 mV. These measurements were made with frog's sartorius muscle.

Voltage clamp method was used in the experiment by Adrian et al [14]. In voltage clamp method measuring electrode is put inside the nerve [1]. In this work the measuring of the threshold voltage is however achieved by stimulating the surface of the nerve. This is done with two silver electrodes and altering the voltage between them. This measurement can be done because the bimetallic stimulations are all made on the surface of the nerve. In Swammerdam's experiment brass and silver wires were also touching only the surface of the nerve.

The electrical properties of frog skin and blood are similar to human skin and blood [15]. To approximate the voltages between different metal pairs in frog's nerve before the actual measurements with the frog are conducted, we can measure voltages between different metal pairs in physiological saline solution. This is because the physiological saline solution has the same electrical properties than bodily fluids.

IV. MEASUREMENTS WITH METALS

Voltage measurements between different metals in physiological saline solution were made to compare them to the theoretical values. Fifty milliliters of physiological saline solution was poured in to a glass beaker. Then the silver wire and the brass wire were placed in to the solution. Voltage meter was connected to the wires with alligator clips, silver to + and brass to -. Five voltage readings between brass and silver were taken. Between each measurement wires were cleaned, dried and put back in the solution. Physiological saline solution was in room temperature in each of the measurements.

Measurements with other metal combinations were also made. This is because we can try the stimulation of frog's muscle with different voltages from different metal pairs. These metals were iron, copper, brass and silver. Voltages between each of the metal pairs were measured. This was repeated five times with each of the metal pairs. Between each measurement wires were cleaned, dried and put back in the solution.

Results from the measurements between different metal pairs are shown in Table 2. These are average values from five different measurements with each of the metal pairs. All values are in the range of theoretical values shown previously in Table 1.

Table 2 Voltage between different metal pairs in physiological saline solution.

Metal Pair	Voltage [mV]
Copper – Brass	52
Silver – Copper	176
Silver – Brass	242
Copper – Iron	853
Brass – Iron	799
Silver – Iron	1050

Voltage between brass and silver seems to be high enough to activate frog's muscle, because the theoretical threshold voltage for the frog's muscle is 41 mV [14]. Calculated voltage difference from measured half-cell values was 223 mV and the measured voltage in physiological saline solution at room temperature was 242 mV. Both of these voltages are over five times the theoretical threshold voltage necessary to stimulate the frog's sartorius muscle.

V. MEASUREMENTS WITH FROG'S NERVE AND MUSCLE

A. Preparation

Preparation of the frog was done just before the measurements. First the frog's right leg was prepared and experiments were made with it. Then the left leg was prepared and experiments were repeated.

There is no specific information on which muscle Jan Swammerdam used in his experiment. There is only a mention that he used thigh muscle, but not exactly which one of the thigh muscles [16]. The theoretical threshold voltage used in this work was obtained with frog's sartorius muscle, which is one of the thigh muscles [14]. In this experiment gastrocnemius muscle was used because the sciatic nerve connected to the muscle was more easily isolated from the leg than the nerve connected to the sartorius muscle. The threshold voltage should be the same for these nerves and muscles.

B. Testing of Bimetallic Stimulation

Bimetallic stimulation with frog's muscle was tested with different metal pairs. Nerve was put over the metal wire and it was stimulated with the other metal wire. Both wires were constantly in contact with the nerve and stimulation was achieved by connecting the metal wires together.

All metal pairs were able to stimulate the frog's muscle except one, brass and copper. The measured voltage between brass and copper in physiological saline solution was 52 mV and the theoretical voltage was from -110 mV to 100 mV. This was the lowest voltage between the different metal pairs used in this work so it can be assumed that the stimulation measurements were correct. The most important metal pair in this work, brass and silver worked fine. Also the voltage between silver and copper was enough to stimulate frog's muscle. The voltage between silver and copper is lower than the voltage between silver and brass. It seems that the threshold voltage for frog's muscle is then something between 52 mV and 176 mV. These voltages are measured in physiological saline solution, so the real voltages in nerve between different metal pairs have to be measured before any further conclusions about the threshold voltage can be made.

C. Voltage Measurements with Different Metal Pairs in Nerve

Voltages between different metal pairs in frog's nerve were measured with a voltage meter which was connected to different metal wires. Metal wires were then put on the nerve without touching each other. Voltage was then measured between the metals three times with each of the metal pairs. Other metal wire was taken off from the nerve between the measurements. These measurements were repeated with frog's another nerve similarly.

Table 3 Success of stimulation and average voltages between different metal pairs in nerve.

Metal Pair	Stimulation	Voltage [mV]
Copper – Brass	No	48
Silver – Copper	Yes	106
Silver – Brass	Yes	147
Copper – Iron	Yes	723
Brass – Iron	Yes	745
Silver – Iron	Yes	842

The results from the measurements for both frog's muscles are shown in Table 3. Average voltage values are calculated from the measured ones. These voltages are lower than the measured voltages in physiological saline solution. The electrical properties of frog's nerve and physiological saline solution are not quite the same. These lower voltages may be also due to the impurities on the surfaces of the metals. Metal-chloride ions on metal wires can effect on voltage. Nevertheless these voltages between different metal pairs excluding copper and brass are still enough to stimulate the frog's muscle. After these measurements with different metal pairs in frog's nerve it seems that the threshold voltage for frog's muscle is somewhere between 48 mV and 106 mV.

D. Threshold voltage for frog's nerve and muscle

Real threshold voltage for frog's nerve and muscle was measured by connecting the nerve to DC-voltage supply. Silver wires were used as electrodes. One electrode was constantly connected to the nerve and the nerve was stimulated with the other electrode by touching the nerve with it.

The measured threshold voltage for the frog's muscle was 50 mV ± 1 mV. The voltage between silver and brass in nerve was 147 mV. That is almost three times the measured threshold voltage.

VI. RESULTS AND CONCLUSIONS

Voltages between different metal pairs and stimulation occurrence were shown in Table 3. The threshold voltage for frog's muscle was about 50 mV. The voltage between brass and silver in nerve was 147 mV. That is almost three times the voltage needed to stimulate the frog's muscle.

The stimulation also worked with silver and copper which have smaller voltage difference than silver and brass. In minimum the theoretical voltage between brass and silver is 110 mV. That is still over twice as much as the measured threshold voltage. The bimetallic stimulation was easy to achieve comparing it to the mechanical stimulation. After these results we can say that it is highly probable that Jan Swammerdam did the first reported bimetallic stimulation over one hundred years before Luigi Galvani.

ACKNOWLEDGMENT

Financial support from the Ragnar Granit Foundation is greatly acknowledged.

REFERENCES

[1] Malmivuo J, Plonsey R (1995) Bioelectromagnetism - principles and applications of bioelectric and biomagnetic fields. New York. Oxford University Press. http://www.bem.fi/book
[2] Galvani L (1791) De viribus electricitatis in motu musculari. Commentarius. De Bononiesi Scientarium et Ertium Instituto atque Academia Commentarii 7: 363-418
[3] Rowbottom M, Susskind C (1984) Electricity and Medicine. History of Their Interaction, 303 pp. San Francisco Press, San Francisco
[4] Brazier M (1959) The historical development of neurophysiology. In Handbook of Physiology. Section I: Neurophysiology, Vol. I, eds. Field I, Magoun H, Hall V, pp. 1-58, American Physiological Society, Washington
[5] Swammerdam J (1738) Biblia Naturae, Vol. 2, ed. H. Boerhaave, Leyden
[6] Brady G, Clauser H (1986) Materials handbook, 12th edition, McGraw-Hill, New York, 112-114
[7] Tylecote R (1992) A History of Metallurgy, 2nd edition, The Institute of Materials, London, 95--121
[8] Callister W Jr (2006) Materials Science and Engineering: An Introduction, 6th edition, Wiley-VCH, 584-585
[9] Galvanic Series: Corrosion potentials in Flowing Seawater. Referred 25.04.2013, http://www.corrosionsource.com/FreeContent/1/Galvanic+Series
[10] Zanello P (2003) Inorganic Electrochemistry. Theory, Practice and Application, The Royal Society of Chemistry, Cambridge, 141—143
[11] Lide D (1975) CRC handbook of chemistry and physics, 56th edition, Boca Raton, D-141
[12] Milazzo G (1978) Tables of standard electrode potentials, Wiley, Chichester, 141--143
[13] Halliday T, Adler K (1986) World of Animals: The encyclopedia of Reptiles and Amphibians, Equinox Ltd, Oxford, 192-215
[14] Hodgkin A, Adrian R, Chandler W (1969) The kinetics of mechanical activation in frog muscle, Journal of Physiology, 204(1), 207-230
[15] Schwartz J-L, Mealing G (1985) Dielectric properties of frog tissues in vivo and in vitro, Phys. Med. Biol., 30, 117-124
[16] Cobb M (2002) Exorcizing the animal spirits: Jan Swammerdam on nerve function. Nature Reviews: Neuroscience, 3:395—400

Corresponding Author

Author: Jaakko Malmivuo
Institute: Department of Electronics, Aalto University
Street: Meritullinkatu 16 A 5
City: 00170 Helsinki
Country: Finland
Email: *jaakko.malmivuo@aalto.fi*

Part III
Biomechanics, Robotics and Minimal Invasive Surgery

Shear Stress Rapidly Alters the Physical Properties of Vascular Endothelial Cell Membranes by Decreasing Their Lipid Order and Increasing Their Fluidity

K. Yamamoto[1], Akira Kamiya[1], and Joji Ando[2]

[1] Laboratory of System Physiology/Department of Biomedical Engineering,
Graduate School of Medicine, The University of Tokyo, Tokyo, Japan
[2] Laboratory of Biomedical Engineering, School of Medicine, Dokkyo Medical University, Tochigi, Japan

Abstract—Vascular endothelial cells (ECs) sense shear stress generated by flowing blood and alter their functions that play important roles in vascular homeostasis and pathophysiology. A unique feature of shear-stress-sensing is the rapid activation of many different types of membrane–bound molecules, including receptors, ion channels, and adhesion proteins, but the mechanisms remain unknown. One hypothesis is that shear stress might alter the physical properties of EC membranes, which lead to the activation of various membrane-bound molecules. To determine how shear stress influences the cell membrane, cultured human pulmonary artery ECs were exposed to shear stress in a flow-loading apparatus and examined for changes in membrane lipid order and fluidity by means of Laurdan two-photon imaging and FRAP measurements. Upon shear stress stimulation, the lipid order of EC membranes rapidly decreased in an intensity-dependent manner, and caveolar membrane domains changed form the liquid-ordered state to the liquid-disordered state. Notably, a similar decrease in lipid order occurred when artificial membranes of giant unilamellar vesicles composed of cholesterol and phospholipids were exposed to shear stress, suggesting that this is a physical phenomenon. Membrane fluidity increased over the entire EC membranes in response to shear stress. Addition of cholesterol to ECs abolished the effects of shear stress on the membrane lipid order and fluidity, and it markedly suppressed ATP release which is a well-known EC response to shear stress and is involved in shear stress Ca^{2+} signaling. These findings indicate that EC membranes, especially caveolar membrane domains, rapidly respond to shear stress by changing their physical properties, including their lipid phase order and fluidity, and that these changes may be linked to shear-stress-sensing and response mechanisms.

Keywords—Endothelial cell, shear stress, caveolae, plasma membrane, lipid order, membrane fluidity.

I. Introduction

The vascular endothelial cells (ECs) that line the inner surface of blood vessels are constantly exposed to shear stress, a mechanical force generated by flowing blood. ECs respond to shear stress by changing their morphology, functions, and gene expression, and these EC responses play important roles in maintaining the homeostasis of the circulatory system, including in controlling blood pressure, blood flow-dependent vasodilation, and vascular remodeling [1]. Impairment of EC responses to shear stress leads to a variety of vascular diseases, however, how ECs sense shear stress as a signal and transmit it into the cell interior remains largely unknown. Previous work has revealed that shear-stress-sensing is mediated by rapid activation of various membrane molecules and microdomains, and these findings suggested the presence of mechanisms by which shear stress is able to activate a variety of membrane molecules and microdomains almost simultaneously. Since changes in the physical properties of the plasma membrane have been shown to affect the activity of membrane molecules, it has been hypothesized that the EC membrane itself plays a crucial role in the mechanisms responsible for shear-stress-sensing. Although EC membrane fluidity has been demonstrated to increase under flow conditions [2, 3], the influences that shear stress exert on the physical properties of the plasma membrane are not fully understood.

To investigate how shear stress influences the lipid order of EC membranes, in the present study we exposed cultured ECs to controlled levels of shear stress in a flow-loading device and examined their plasma membranes for changes in Laurdan fluorescence in a real-time manner. We also examined giant unilamellar vesicles by Laurdan imaging to determine whether shear stress affects the lipid order of artificial lipid-bilayer membranes. In addition, we studied the effects of shear stress on membrane fluidity by means of FRAP measurements.

II. Mateials and Methods

Cell culture: Human pulmonary artery ECs (HPAECs) were obtained from Clonetics and grown on a 1% gelatin-coated tissue culture flask in M199 supplemented with 15% FBS, 2 mM L-glutamine, 50 μg/ml heparin, and 30 μg/ml EC growth factor. The cells used in the experiments conducted in this study were in the 7th and 10th passage.

Flow-loading experiments: A parallel-plate-type apparatus (FCS2, Bioptechs) was used to apply laminar shear stress to the cells, as previously described [4]. The intensity of the shear stress (τ, dynes/cm^2) acting on the EC layer was calculated by using the formula: $\tau = 6\mu Q/a^2 b$, where μ is the viscosity of the perfusate, Q is the flow volume, and a and b are the cross-sectional dimensions of the flow path.

Preparation of giant unilamellar vesicles: GUVs were prepared essentially as described [5]. Dipalmitoylphosphatidylcholine (DPPC), dioleoylphosphatidylcholine (DOPC), and cholesterol were mixed in a molar ratio of 2:2:1 in a chloroform/methanol mixture (2:1, v/v) to prepare a lipid solution that had a total lipid content of 1 mg. Laurdan was added to the lipid solution in a Laurdan to total lipid molar ratio of 1:500. The mixtures were dried with a rotary evaporator to form a thin lipid film. A 5 mL volume of prewarmed water was gently added, and the flask was incubated at 65°C overnight to produce GUVs. The GUVs were applied to a coverglass coated with 0.1% poly (L-lysine) solution to attach them to the coverglass. The GUVs were exposed to shear stress in the FCS2 chamber at 23°C and examined by Laurdan imaging.

Laurdan two-photon microscoy: HPAECs growing on coverglasses were incubated for 30 minutes at 37°C in culture medium containing 10 μM Laurdan dye under a gas mixture of 95% air-5% CO_2. After rinsing the cell monolayer with HBSS, the coverglass was placed in a flow-loading apparatus and then mounted on the stage of a confocal laser-scanning microscope (TCS SP2 AOBS) equipped with a two-photon laser (MaiTai BB). Laurdan fluorescence was excited with 770 nm wavelength light, and the emitted light was collected in the 400-460 nm range and 470-530 nm range for the two channels, respectively. General polarization (GP) was calculated by using the formula as follows:

$$GP = (I_{400-460} - G \times I_{470-530}) / (I_{400-460} + G \times I_{470-530}),$$

where I is light intensity, and G is a correction factor calculated through two variables: a known GP value for Laurdan in DMSO at 22°C and the GP value of the Laurdan stock solution (500 μM) determined in each experiment. GP images were analyzed and pseudocolored with ImageJ 1.44n software. GP is an indicator of the hydration depth of the bilayer interior, and it has been used to indirectly estimate changes in membrane lipid order.

FRAP measurements: HPAECs were loaded with 2.5 mM 1, 1'-dihexadecyl-3, 3,3',3'-tetramethylindocarbocyanine perchlorate ($DiIC_{18}$) for 30 minutes at 37°C in a 95% air-5% CO_2 gas mixture. After washing the coverglass with adherent cells in HBSS, it was mounted in a flow-loading apparatus and placed on the stage of a confocal laser-scanning microscope equipped with an Ar/ArKr laser and HeNe laser. A defined region was bleached at full laser power using the 488 nm line of the laser, and a recovery of fluorescence was monitored by scanning the bleached region at low laser power. Half-life time was calculated by using the formula: $t_{1/2} = -\ln(0.5)/k$ where k is a parameter obtained from the curve-fitting, and the diffusion coefficient (D) is calculated by using the formula: $D = \omega^2/4 \cdot t_{1/2}$, where ω is the effective radius of the focused laser beam.

Immunohistochemistry: Cells were fixed with 4% paraformaldehyde and maintained in 1% normal bovine serum albumin to block nonspecific protein binding sites. The cells were then incubated with rabbit anti-caveolin-1 polyclonal antibody, and after washing they were incubated with Alexa Fluor 594 goat anti-rabbit IgG. Stained cells were photographed through a confocal fluorescence microscope.

Quantification of extracellular ATP: The ATP released by the HPAECs was measured by means of a luciferin-luciferase assay. The cells were exposed to a stepwise increase in shear stress at 37°C, and after collecting the perfusate every 20 seconds, 100-μL aliquots were applied to the ATP assay system (Toyo Ink). The ATP assay mixture (luciferase, D-luciferin, and bovine serum albumin) was injected into each sample, and its relative light intensity was recorded for 10 seconds in a Lumat LB 9501 luminometer (Berthold) at room temperature. A calibration curve for ATP concentrations was obtained for each experiment by using the same batch of luciferin-luciferase reagents, and the total number of moles of ATP released per second per 10^6 cells was calculated.

III. RESULTS AND DISCUSSION

A. Liquid-Ordered State and Liquid-Disordered State Coexist in an EC Membrane and High GP Regions Coincide with Caveolar Membrane Domains

Images of Laurdan-labeled human pulmonary artery ECs (HPAECs) were collected at different focal depths, and generalized polarization (GP) images were obtained. All GP images are shown in pseudo color. The GP images clearly showed a heterogeneous distribution of GP values across the cell surface (Fig. 1). It is noteworthy that high-GP regions tended to be located at the cell periphery.

After the Laurdan imaging, ECs were immunostained with an antibody against caveolin-1, a marker protein for cholesterol-rich membrane microdomains called caveolae. Caveolin-1 was unevenly distributed over the cell surface

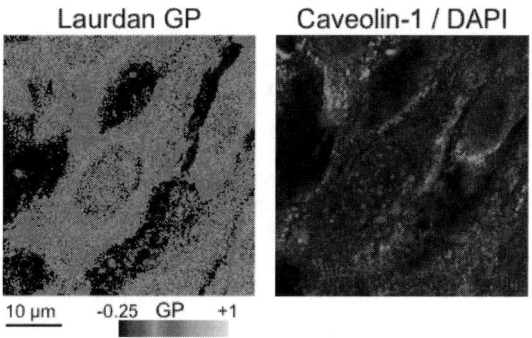

Fig. 1 Laurdan GP images of living HPAECs and caveolin-1 distribution. After Laurdan imaging, cells were immunostained with an antibody to caveolin-1, a marker protein for caveolae, and the cell nuclei were stained with DAPI. The high-GP regions coincided with caveolin-1-rich regions.

and was concentrated at specific parts of the cell periphery. Comparison between the Laurdan GP images and caveolin-1 distribution revealed that the high-GP regions of the plasma membranes almost perfectly coincided with their caveolar membrane domains.

B. Shear Stress Rapidly Decreases the Lipid Order of Plasma Membrane

Laurdan-labeled HPAECs were exposed to shear stress in a flow-loading apparatus and examined for changes in membrane lipid order. As soon as shear stress was applied, GP values decreased over the entire plasma membrane, and the decrease was especially marked in the high-GP regions (Fig. 2A). The temporal changes in lipid order were quantified by placing ROIs in the high-GP regions and the low-GP regions. The GP values in both regions began to decrease immediately after the application of shear stress, continued to decrease with time, and then increased after the shear stress ceased (Fig. 2B, C). These findings indicate that EC membranes rapidly decrease their lipid order in response to shear stress, and that shear stress shifts caveolar membrane domains from the liquid-ordered state to the liquid-disordered state.

C. The Lipid Order of Arttificial Membranes Decreases in Response to Shear Stress

Laurdan-labeled GUVs that had attached to a poly(L-lysine)-coated coverglass were exposed to shear stress and examined for changes in the lipid order of the artificial lipid bilayer membranes. The heterogeneity seems to be attributable to phase separation between the liquid-ordered phase and liquid-disordered phase, and the high-GP regions appear to represent lipid-ordered membrane regions, not caveolar domains. Upon shear stress stimulation, membrane lipid order decreased over the entire membrane, and the extent of the high-GP regions diminished (Fig. 3A). The temporal changes in lipid order were quantified by placing ROIs in the high-GP regions and the low-GP regions. The GP values in both regions began to decrease immediately after the application of shear stress and continued to decrease with time, and they remained at the decreased levels even after the cessation of shear stress (Fig. 3B). The data obtained from many GUVs confirmed that shear stress decreases lipid order in both high-GP and low-GP regions (Fig. 3C). These findings suggest that the lipid order response to shear stress occurs in artificial membranes as well as in the plasma membranes of living cells.

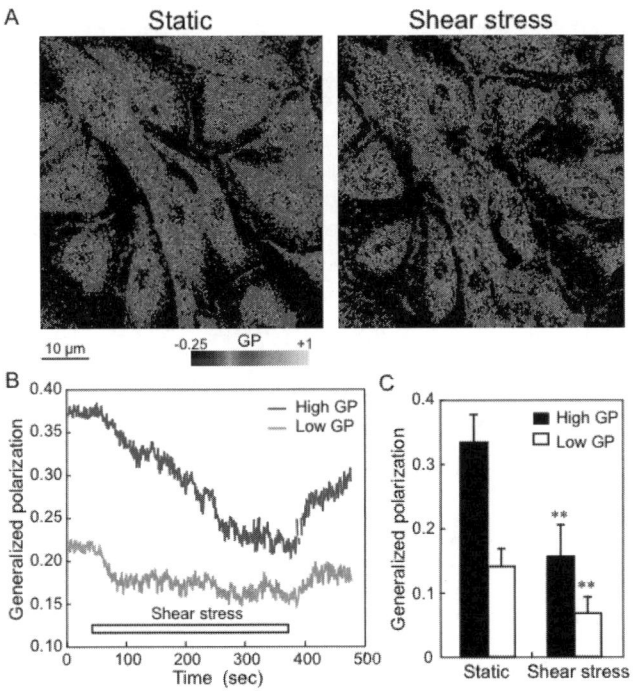

Fig. 2 Effects of shear stress on the lipid order of EC membranes. A, GP images before and 5 minutes after application of shear stress (15 dynes/cm^2). B, temporal GP value changes in the high-GP region and the low-GP region. C, quantitative analysis of the shear-induced changes in GP values. Values are means ± s.d. of 25 cells. **$P<0.01$ vs. static control.

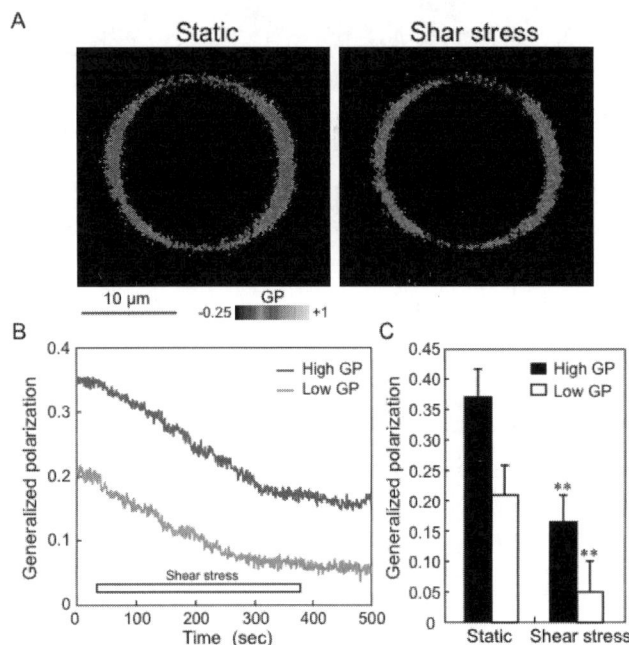

Fig. 3 Effects of shear stress on artificial lipid-bilayer membranes. A, GP images before and 5 minutes after application of shear stress (15 dynes/cm^2). B, temporal GP value changes. The changes were quantified by placing regions of interest on a high-GP region and a low-GP region in a cross-sectional GP image. C, quantitative analysis of the shear-induced changes in GP values. Values are means ± s.d. of the data obtained in 12 GUVs. **$P<0.01$ vs. static control.

D. Shear Stress Increases Membrane Fluidity in ECs

Membrane fluidity is related to the mobility of membrane proteins within the lipid bilayer and has been used to assess the physical properties of plasma membranes. To investigate the effects of shear stress on membrane fluidity, FRAP measurements were made when ECs were exposed to shear stress. FRAP measurements were made after Laurdan imaging. ROIs were placed in the DiI fluorescent images by referring to the Laurdan image, and ROI-1 was set in a high-GP region and ROI-2 in a low-GP region. Calculation of diffusion coefficients based on the FRAP curves showed that shear stress significantly increased the membrane fluidity in both regions of the EC membranes (Fig. 4B). Addition of cholesterol to ECs decreased the diffusion coefficients in both regions. These findings indicated that EC membrane fluidity is sensitive to shear stress as well as to cholesterol.

E. Membrane Lipid Order Affects EC Responses to Shear Stress

To determine whether changes in the membrane lipid order influence EC responses to shear stress, HPAECs that had been treated or not treated with cholesterol were exposed to shear stress and examined for ATP release, a well-known EC response to shear stress that is involved in Ca^{2+} signaling of shear stress [6]. The amount of ATP released by ECs was measured by means of a luciferin-luciferase assay. ECs released ATP dose-dependently in response to shear stress, and the responses were markedly suppressed by treating cells with cholesterol (Fig. 4C), suggesting that the degree of membrane lipid order affects EC responses to shear stress.

IV. CONCLUSIONS

The results of the present study demonstrated that shear stress alters the physical properties of EC membranes by rapidly decreasing their lipid order. The lipid order response is a physical phenomenon that does not involve participation by any membrane proteins, the cytoskeleton, or biological activities of living cells. Changes in the physical properties of EC membranes, such as in their membrane lipid order and fluidity, especially at caveolar membrane domains, may be linked to their shear stress-sensing and response mechanisms.

ACKNOWLEDGMENT

This work was partly supported by Grants-in-Aid for Scientific Research from the Ministry of Education, Culture, Sports, Science and Technology to J. A. and K. Y., Grants-in-Aid from the Japan Science and Technology Agency, and Canon Foundation to K. Y..

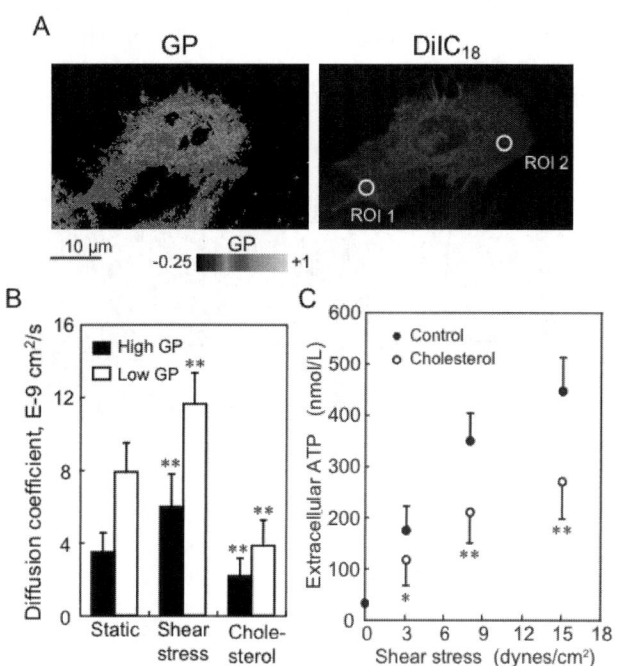

Fig. 4 Effects of shear stress on EC membrane fluidity. A, a typical cell sequentially examined by Laurdan imaging and FRAP measurements. Regions of interest were placed on the high-GP regions (ROI-1) and on the low-GP regions (ROI-2). B, quantitative analysis of changes in membrane fluidity. FRAP measurements were performed in cells under static conditions (Static), cells exposed to shear stress, and cells treated with cholesterol (100 μM, 4 hours). The diffusion coefficients were based on the FRAP curves. Values are means ± s.d. of the data obtained in 20 cells. C, effects of cholesterol on shear-stress-induced ATP release. Values are means ± s.d. of the values obtained in 15 cells. *$P<0.05$, **$P<0.01$ vs. control.

REFERENCES

1. Ando J, Yamamoto K. (2009) Vascular mechanobiology: endothelial cell responses to fluid shear stress. Circ J 73:1983-92
2. Haidekker MA, L'Heureux N, Frangos JA (2000) Fluid shear stress increases membrane fluidity in endothelial cells: a study with DCVJ fluorescence. Am J Physiol Heart Circ Physiol 278:H1401-6
3. Butler PJ, Norwich G, Weinbaum S et al (2001) Shear stress induces a time- and position-dependent increase in endothelial cell membrane fluidity. Am J Physiol Cell Physiol 280:C962-9
4. Yamamoto K, Furuya K, Nakamura M et al (2011) Visualization of flow-induced ATP release and triggering of Ca^{2+} waves at caveolae in vascular endothelial cells. J Cell Sci 124:3477-83
5. Akashi K, Miyata H, Itoh H et al (1996) Preparation of giant liposomes in physiological conditions and their characterization under an optical microscope. Biophys J 71:3242-50
6. Yamamoto K, Sokabe T, Ohura N, et al (2003) Endogenously released ATP mediates shear stress-induced Ca^{2+} influx into pulmonary artery endothelial cells. Am J Physiol Heart Circ Physiol 285:H793-803

An In Vitro Analysis of the Influence of the Arterial Stiffness on the Aortic Flow Using Three-Dimensional Particle Tracking Velocimetry

U. Gülan[1], B. Lüthi[1], M. Holzner[1], A. Liberzon[2], A. Tsinober[2], and W. Kinzelbach[1]

[1] Institute of Environmental Engineering, ETH Zurich, Zurich, Switzerland
[2] School of Mechanical Engineering, Tel Aviv University, Tel Aviv, Israel

Abstract—A three-dimensional pulsatile aortic flow in a human ascending aorta is investigated in-vitro in this paper. A non-intrusive measurement technique, 3D Particle Tracking Velocimetry (3D-PTV), has been applied to the anatomically accurate silicon replicas. A compliant and a stiff aortic model were analyzed to better understand the influence of the arterial stiffness. The realistic models are transparent which allows optical access to the investigation domain. Our results showed that increasing the arterial stiffness considerably increases the systolic velocity and hence mean kinetic energy. Quite strikingly, the turbulent kinetic energy is about one order of magnitude higher in the stiffer model during the deceleration phase which manifests that a blood element is exposed to higher shear stresses in the stiffer model. Moreover, we found that the compliant model introduces pressure oscillations during the diastolic phase which are associated with the Windkessel effect.

Keywords— 3D-PTV, ascending aorta, image processing, arterial stiffness, Lagrangian flow field.

I. INTRODUCTION

The aorta, which is the largest artery in the cardiovascular system, is the origin of the systemic circulation. It is a flexible artery with a high distensibility. The elastic walls allow the aorta to respond to the flow and pressure variations. During the systolic phase, the aorta stores almost half of the stroke volume ejected by the left ventricle by distending the walls. The stored blood is forwarded to the peripherial circulation contracting the arterial walls during the diastolic phase which is linked to the so-called "Windkessel effect" [1].

Stiffness is defined as the resistance to deformation [2]. Calcification and reduction in elastic fibers cause an increase in stiffness of the artery. Compliance is the inverse of stiffness and refers to the ratio of the change in volume to the pressure difference during the systolic and diastolic phases [1]. The compliance is a function of the wall properties, i.e. thickness, diameter and elastic modulus of the artery. Compliance of the aortic wall is thus believed to eminently influence the temporal and spatial evolution of the flow. Ageing and hypertension are claimed as the primary factors responsible for the increase in arterial stiffness [3]. From clinical point of view, measuring stiffness is a crucial marker as a predictor of cardiovascular morbidity [2]. Aortic stiffness may provoke abnormalities in blood flow in the cardiovascular system. It is found that aortic stiffness and atherosclerosis are highly correlated [4]. Accordingly, further analysis on the influence of the aortic stiffness on the aortic flow and the evolution of turbulent kinetic energy is needed to provide an informative and complimentary data for clinical purposes.

Phase Contrast Magnetic Resonance Imaging (PC-MRI) has been applied in-vivo and in-vitro to assess the flow field and turbulent velocity fluctuations. However, long scan times, low temporal resolution and low signal-to-noise ratio limits the capability of PC-MRI. Recently, 3D Particle Tracking Velocimetry (3D-PTV), an image based measurement technique, has been applied to extract the Eulerian and Lagrangian flow fields in an ascending aorta [5]. With the ability of 3D-PTV assessing flow velocities, velocity derivatives and Lagrangian trajectories simultaneously, it has become one of the standards for validation of in-vitro measurements [6].

The aim of this study is to shed a light on the influence of the aortic stiffness on the flow and fluctuating fields which can be translated into clinical application. For this purpose, 3D-PTV measurements were performed in anatomically realistic, rigid and compliant aortic replicas mimicking realistic flow conditions.

II. METHOD

3D-PTV is a non-intrusive measurement technique based on imaging of flow tracers. The experimental setup is explained in detail in the study of Gülan et al. [5]. Concisely, the experimental setup comprises an aortic replica, a Ventricular Assist Device (VAD, MEDOS, Germany) to mimic the function of the heart, a pneumatical pump system to drive the VAD and the optical part comprising a high speed camera (Photron SA5, Japan), an image splitter, mirrors and an Ar-Ion laser as light source as shown in Figure 1. The geometries of rigid and compliant aortic phantoms were obtained from high-resolution MRI scan of a healthy patient. The replicas were manufactured by Elastrat (Geneva, Switzerland). The high-speed camera was synchronized with the pump system to trigger the recordings in the beginning of every pulse. A novel dynamic calibration tool for image-based measurement techniques, the so-called "dumbbell calibration" was applied to calibrate the camera positions [5].

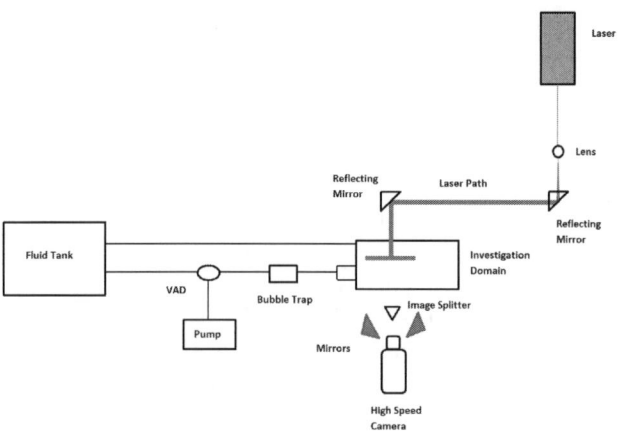

Fig. 1 Schematic view of the experiment setup [5]

3D-PTV allows to track particles in mediums with different refractive indexes. A mixture of glycerin, water and NaCl was used as working fluid to minimize the optical distortions by matching the refractive index of silicon. The parameters used in the measurements are presented in Table 1. In total, 35 heartbeats were recorded where a single heartbeat corresponds to a pulse length of 1.15s and data was phase averaged and mapped onto an Eulerian grid [5].

Table 1 Working fluid properties and flow parameters

Parameter	Unit	Rigid Model	Compliant Model
Stroke volume	ml	54	54
Density	g/cm^3	1.2	1.2
Viscosity	cm^2/s	4.85×10^{-6}	4.85×10^{-6}
Peak Reynolds number		3584	2323

III. RESULTS

Aortic flow is a three-dimensional, complex and transitional flow. Generally, we found that pulsatility, the curved geometry of the aorta and its non-stationary boundaries induced a 3D flow with the presence of strong secondary flows. Figure 2 qualitatively describes the influence of arterial stiffness on the flow field. The comparison of Lagrangian trajectories along the compliant (left) and rigid (right) ascending aortas for a time interval of 0.1s during the deceleration phase is shown in Figure 2. As it is seen from the figure, the particle trajectories are mainly oriented along the axial direction and pronounced helical flow features are present at the inner wall in the compliant model. On the other hand, flow is more disorganized and coherent helical pattern vanishes in the rigid model which manifests that stiffness has considerable influence on the flow field.

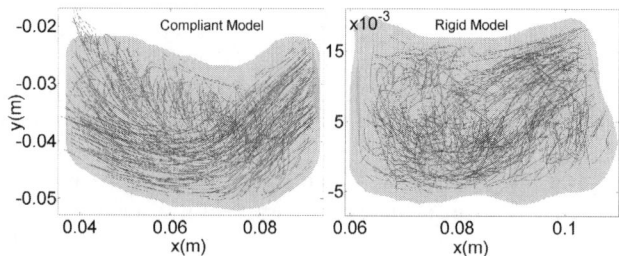

Fig. 2 Lagrangian trajectories along the compliant model (left) and rigid model (right) during the deceleration phase (Blue-coded trajectories refer to antegrade flow and the red-coded trajectories refer to retrograde flow)

Figure 3 shows the temporal behavior of the volumetric flux for the rigid and the compliant model. Both curves show a steep increase, peak and steep decrease during systole, where the peak value is higher for the rigid model. During diastole, there is a qualitative difference where the compliant model shows oscillations. These oscillations are due to the compliance of the walls, i.e. the aorta contracts and ejects fluid. This is analyzed in more detail in the following.

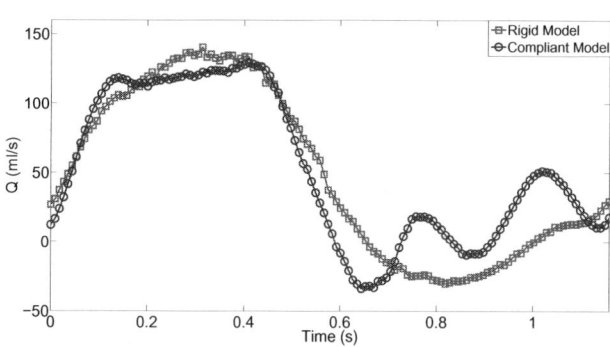

Fig. 3 Temporal evolution of phase averaged volumetric flux for both rigid and compliant models

To better understand the factors responsible for the flux oscillations, it is interesting to decompose the budget of the change of the volumetric flux. The variation of volumetric flux over time can be written as;

$$\underbrace{\frac{\partial Q}{\partial t}}_{\substack{\text{volumetric}\\ \text{flux change}}} = \frac{\partial AU}{\partial t} = U \underbrace{\frac{\partial A}{\partial t}}_{\substack{\text{areal}\\ \text{change}}} + A \underbrace{\frac{\partial U}{\partial t}}_{\substack{\text{local}\\ \text{acceler-}\\ \text{ation}}} \quad (1)$$

where Q(x) is the volumetric flux, U(x,t) is the streamwise velocity averaged over the cross section, A(x,t) is the cross sectional area. The temporal change of cross sectional area is zero for the rigid model and hence the volumetric flux change is directly proportional to the local acceleration. As shown in Figure 4, the local acceleration is mainly responsible for the change in volumetric flux for the compliant model and the influence of the cross sectional change is negligible.

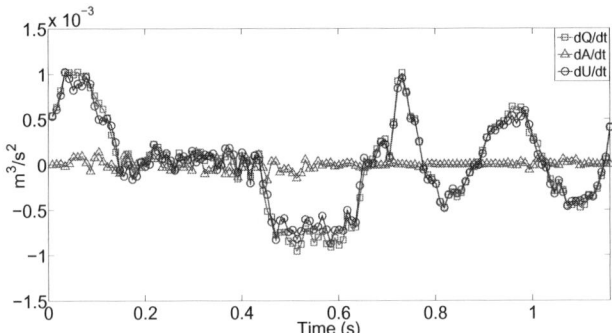

Fig. 4 Rate of the change of volumetric flux and its decomposition at the mid cross section of the compliant model

As we observed that local acceleration has dominant influence on the volumetric flux, it is logical to investigate the budget of local acceleration to highlight the factors responsible for the variation of the local acceleration. Accordingly, we analyze the phase averaged one dimensional Navier-Stokes equation averaged over the cross section.

$$\underbrace{\frac{\partial \tilde{U}}{\partial t}}_{\substack{\text{local ac-}\\ \text{celeration}}} = \underbrace{-\alpha \tilde{U} \frac{\partial \tilde{U}}{\partial x}}_{\substack{\text{convective}\\ \text{term}}} - \underbrace{\frac{1}{\rho} \frac{\partial \tilde{P}}{\partial x}}_{\substack{\text{pressure}\\ \text{gradient}}} - \underbrace{\tilde{\tau}_0/R_h \rho}_{\substack{\text{wall fric-}\\ \text{tion}}}$$
$$+ \underbrace{g\sin\theta}_{\substack{\text{gravitational}\\ \text{accelera-}\\ \text{tion}}} - \underbrace{\frac{1}{A} \int \langle (\tilde{u} \cdot \nabla) \tilde{u} \rangle_\phi dA}_{\text{Reynolds stresses}} \quad (2)$$

where $\tilde{U}$ is the streamwise velocity component averaged over the cross section, $\tilde{P}$ is the pressure averaged over the cross section, g is the gravitational acceleration, $\tilde{\tau}_0$ is the wall shear stress, R_h is the hydrodynamic radius, $\tilde{u}$ is the turbulent fluctuating velocity, α is convective coefficient and the angle brackets denote the average over different pulses. We found that convective acceleration and pressure gradient terms have the strongest influence on the local acceleration. In Figure 5, we present the temporal evolution of local acceleration and convective acceleration for both rigid and compliant models.

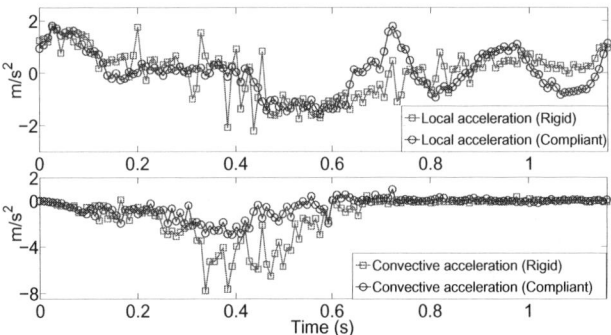

Fig. 5 Comparison of local and convective acceleration for the rigid and compliant models

As it is seen from the figure, local acceleration becomes negative during the deceleration phase by means of the existence of the retrograde flow for both models. During the diastole, the change in local acceleration is small in the rigid model which is consistent with Figure 3. Furthermore, we found that convective acceleration is higher during the systolic phase in the rigid model which is associated with the high mean velocity for the rigid model.

To provide a complimentary result to Figure 5, we exhibit the temporal evolution of pressure gradient for both models in Figure 6.

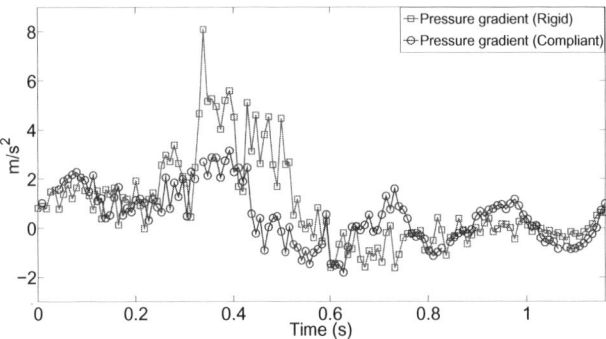

Fig. 6 Temporal evolution of pressure gradient for both models

Since there is no considerable change in convective acceleration during the diastolic phase (Figure 5, bottom), pressure gradient is responsible for the oscillations during the diastole in the compliant model which is linked to the Windkessel effect. Moreover, rigid model shows higher pressure gradient during the systolic phase.

Cardiovascular flow is typically laminar. However, blood flow becomes turbulent in larger arteries such as aorta [5]. As turbulent flow induces higher shear stress on a blood element compared to the laminar flow, it is important to investigate the evolution of turbulent kinetic energy to estimate the transport inefficiency. We exhibit the temporal evolution of mean kinetic energy (MKE) and turbulent kinetic energy (TKE) averaged over the entire ascending aorta in Figure 7. It is seen that MKE in the rigid model is higher than that in the compliant model during peak systole which points that stiffness increases systolic velocity. Furthermore, TKE of the rigid model is about one order of magnitude higher than that of the compliant model during the deceleration phase when TKE reaches a maximum. It manifests that stiffer aorta provokes pronounced production of TKE.

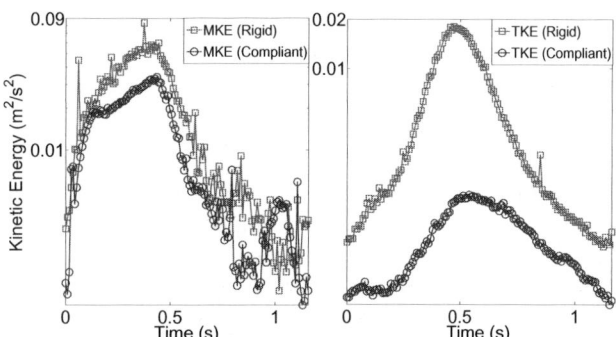

Fig. 7 Time evolution of mean turbulent kinetic energy (MKE) (left) and turbulent kinetic energy (TKE) (right) for both rigid and compliant models

IV. CONCLUSION

We presented an in-vitro investigation of the pulsatile aortic blood flow in an ascending aorta to highlight the influence of the arterial stiffness on the phase averaged and fluctuating velocity fields. An image-based measurement technique, 3D-PTV, has been applied to the anatomically realistic aortic replicas with matching index of refraction to avoid the optical distortions. Clinically realistic flow conditions have been mimicked. It is concluded that aortic distensibility plays a crucial role on the flow pattern. We found that increasing arterial stiffness decreases the influence of the Windkessel effect which is associated with the capability of blood storage of the peripheral flow during the systolic phase. We also observed that stiffer aorta leads to an increase in systolic velocity and hence an increase in convective acceleration and mean kinetic energy, a decrease in diastolic pressure gradient, an increase in systolic pressure gradient and less pressure oscillations during the diastolic phase. Analyzing the temporal evolution of the turbulent kinetic energy for two different models showed that the arterial stiffness noticeably increases turbulent kinetic energy during the entire heartbeat. The maximum turbulent kinetic energy level is reached during the deceleration phase for both models. A clear distinction between the rigid and the compliant models arises during the deceleration phase in terms of fluctuating velocities. The stiffer aorta introduces high fluctuating velocity and hence high production of turbulent kinetic energy. It is known that arterial stiffness is related to the atherosclerosis [2, 3, 4]. In the pathological flow conditions, red blood cells and platelets become more vulnerable as a result of high shear stress acting on a blood element. In future, this work will be extended towards the analysis of small scale stress analysis on a blood element to better understand the response of the cardiovascular system to the arterial stiffness.

ACKNOWLEDGEMENTS

This work was supported by ETH Research Grant ETH-24 08-2. Support from COST Action MP0806 is kindly acknowledged.

REFERENCES

1. Belz G G. Elastic properties and Windkessel function of the human aorta Cardiovascular Drugs and Therapy. 1995;9:73-83.
2. Cavalcante J L, Lima J A C, Redheuil A, Al-Mallah M H. Aortic stiffness: current understanding and future directions Journal of the American College of Cardiology. 2011;57:1511-1522.
3. London G M, Guerin A P. Influence of arterial pulse and reflected waves on blood pressure and cardiac function Am Heart J. 1999;138:220-224.
4. Popele N M Van, Mattace-Raso F U, Vliegenthart R, et al. Aortic stiffness is associated with atherosclerosis of the coronary arteries in older adults: the Rotterdam Study Journal of Hypertension. 2006;24:2371-2376.
5. Gülan U, Lüthi B, Holzner M, Liberzon A, Tsinober A, Kinzelbach W. Experimental study of aortic flow in the ascending aorta via Particle Tracking Velocimetry Experiments in Fluids. 2012;53:1469-1485.
6. Knobloch V, Binter C, Gulan U, Boesiger P, Kozerke S. Assessment of 3D velocity vector fields and turbulent kinetic energy in a realistic aortic phantom using multi-point variable-density velocity encoding J Cardiovasc Magn Reson. 2012;14:W50.

Author: Utku Gülan
Institute: IfU, ETH Zurich
Street: Wolfgang Pauli Str-15, HIL G 34.2
City: Zurich
Country: Switzerland
Email: guelan@ifu.baug.ethz.ch

A New Method for Coronary Artery Mechanical Properties

Omer Shalev[1], Dar Weiss[1], Yigal Kassif[3], Rami Haj-Ali[2], and Shmuel Einav[1,4]

[1] Biomedical Engineering
[2] Mechanical Engineering
[3] Sheba Medical Center, Tel-Aviv University, Israel
[4] Stony |Brook University, USA

Abstract—The wall properties of coronary arteries are important to the success of vascular interventions. They are also important in the attempts to construct numerical models of the coronary tree as a tool for the inteventionalists to plan the procedures. In the current study we measured and calculated the inelastic mechanical properties of coronary vessels and laid the ground for a working numerical model.

I. INTRODUCTIONS

The coronary arteries extend from the aorta to the heart walls supplying oxygenated rich and nutrient filled blood to the heart muscle including the atria, ventricles, and septum of the heart. The lack of reliable mechanical data on coronary arteries and, more specifically, on their wall properties hampers the application of numerical models and simulations to vascular problems, and precludes physicians from knowing in advance the response of coronary arteries to the different interventions. Preliminary attempts have been made to explore their mechanical properties almost exclusively on animals. These attempts have shown a great difference between their characteristics and of other arteries.

II. METHODS

In order to further investigate their mechanical properties we devised a novel system that mimics arterial pulsatile flow, while, separately measure the different axial properties. Eight porcine coronary arteries (Fig. 1) were harvested and utilized for tensile testing (Instron 5582 with 100N load cell). Each artery was mounted on a specially designed mounting device; both its ends were attached to adapters, connecting them to a closed flow loop. The diameter of each arterial segment was monitored and measured under a high-resolution CCD camera (Labvision Imager E-lite 2M). Digital Image Correlation (DIC) was used for strain measurement (Fig. 2). DIC measures surface deformations by the usage of digital images. The method used was "Lucas–Kanade". It finds the measure of match between fixed-size feature windows in the previous and current frame by using the least square of the intensity differences over the windows. [Kanade, 1985].

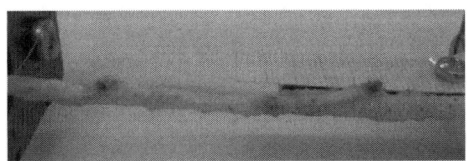

Fig. 1: Porcine Coronary.

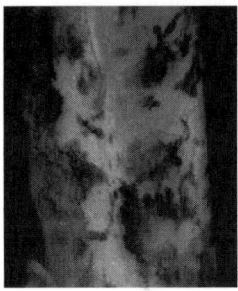

Fig. 2: Porcine Coronary artery image before analysis.

III. RESULTS

The average initial external diameter of the coronary arteries was 2.8 ± 0.14 mm and the average initial wall thickness 0.169 ± 0.04 mm. We have obtained a linear relationship between the axial strain in the coronaries and the applied pressure (Fig. 3). The material properties of the examined coronary arteries were both non-linear and highly anisotropic.

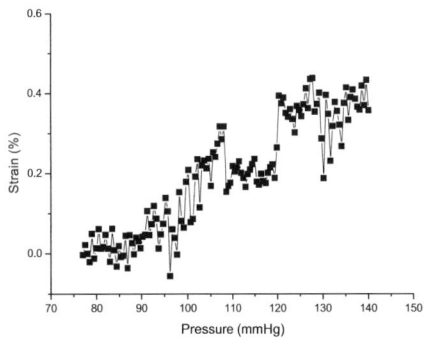

Fig. 3: Strain vs. dimensionless pressure.

IV. DISCUSSION

The porcine coronary artery is of the muscular type, which means that the media consists mainly of smooth muscle cells. The mechanical behavior of the coronary wall is therefore expected to be elastic. The actual mechanical characteristics of these vessels are determined by its components and the structural interrelationship between the components. We expect our results to be important in numerical simulations of the coronary system.

REFERENCES

Kanade et al, Artificial Intelligence, 2, 1985.

Numerical Analysis of a Novel External Support Device for Vein Bypass Grafts

T. Meirson[1], E. Orion[2], G. Bolotin[3], M. Brand[4], and I. Avrahami[4]

[1] Medical Engineering, Afeka Academic College of Engineering, Tel Aviv, Israel
[2] Vascular Graft Solutions Ltd, Israel
[3] Cardiac Surgery, Rambam Health Care Campus, Haifa, Israel
[4] Mechanical Engineering and Mechatronics, Ariel University, Ariel, Israel

Abstract—This paper presents a numerical analysis of a novel external support device for coronary artery bypass graft (CABG). The analysis includes the fluid and the structural dynamics of vein grafts with and without an external support device (ESD). Six different vein grafts geometries were modeled under time-dependent physiological conditions. The hemodynamics and the stresses developed at the graft wall were analyzed on patient based geometries 12 months post-transplantation.

Results: veins under constriction of ESD were associated with higher wall shear stresses, lower shear gradients, uniform flow and lower structural effective stresses.

In conclusion, these findings suggest that constricted graft with ESD improves hemodynamic and structural factors which are correlated to graft failure following CABG procedure.

Keywords—ESD, CABG, saphenous vein, CFD, FEM.

I. INTRODUCTION

Since its first successful use in the 1960's, the saphenous vein graft (SVG) has been one of the most commonly used conduits in coronary artery bypass graft (CABG) surgery. Some of the main advantages of using the saphenous vein include its ease of accessibility, its ease of handling, and the number of grafts, typically three, that can be constructed from a single vein (see illustration in Figure 1).

Despite these advantages and the widespread use of saphenous veins in CABG surgery, its poor long term patency rates are still a big limitation of CABG. Five years post CABG, 35-40% of the vein grafts are occluded while many of the patent grafts suffer from severe disease [1]. The main pathological mechanism behind failing vein grafts is intimal hyperplasia (IH), mainly triggered by wall structures and anatomic dimensions that are ill-suited to accommodate arterial hemodynamics. Two specific factors identified to contribute to vascular remodeling and IH. The first one is increase in pressure exerted on the vein. Venous pressure is usually 3-7 mmHg. After graft transplantation, the arterial pressure exerted on the graft is increased to 80-120 mmHg. This dramatic increase in pressure is resulted in significant circumferential stresses developed in the graft which lead to smooth muscle cells proliferation and Intimal hyperplasia.

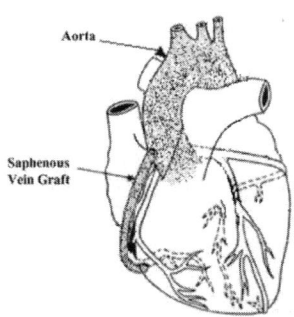

Fig. 1 An illustration of GSV as a CABG

The second factor involves the wall shear stresses. Since the graft's lumen diameter is larger than the artery, and due to the graft's dilatation, the graft hemodynamic is characterized with slow, sluggish and disturbed flow with lower wall shear stress. It is well proven that hemodynamic factors are involved in the formation and development of IH. These hemodynamic factors include low and oscillating wall shear stress (WSS) [2], large spatial wall shear stress gradient (WSSG) [3], and long residence time of blood cells [4]. Several studies have indicated the correlation between these hemodynamic factors and localized sites of intimal thickening (IT) in numerical studies, the improvement of these hemodynamic factors has been utilized as an indicator to achieve superior geometries and higher patency rates for bypass grafts [4].

In search of mechanisms behind focal intimal proliferation, irregular luminal dimensions were shown to play a key role. By causing eddy blood flow, which is associated with areas of low fluid shear stress and increased shear gradients, luminal irregularities are main culprits of focal intimal hyperplasia. To eliminate the fluid dynamic triggers of both diffuse and focal intimal hyperplasia, the concept of a constrictive external support mesh for vein grafts was suggested as early as in 1963 [5]. By constricting the vein graft diameter, these meshes were expected to mitigate both diffuse intimal hyperplasia, through higher-flow velocity and thus higher shear forces, and focal intimal hyperplasia, through the elimination of luminal irregularities. Examples of external support device s are shown in Figure 2 and Figure 3.

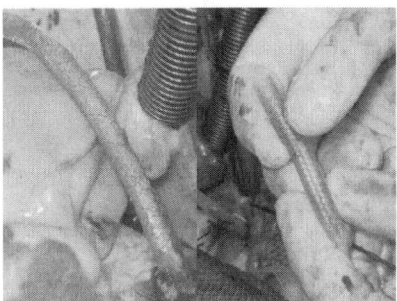

Fig. 2 The FLUENT External support device (ESD) (Vascular Graft Solutions, Tel Aviv, Israel)

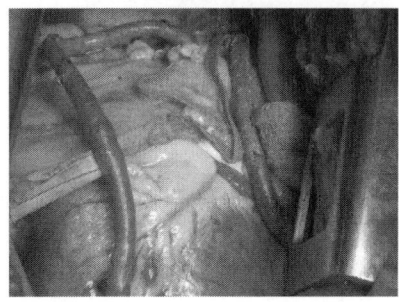

Fig. 3 Image of an unsupported (left) and supported grafts (right)

External support for venous graft has been applied in order to improve hemodynamics and increase patency rates. The goal of this study was to assess numerically the effect of the external support for the venous graft. In order to do that, patient's based models of venous graft and external support device were built and run under physiological conditions. The simulations incorporated both the fluid and solid domain and the stress and flow regimes were analyzed and compared.

II. METHODS

The analysis included 8 models of the fluid and structural domains of graft geometries of the saphenous vein bypass graft, as listed in Table 1. The models were created based on 6 angiographies of saphenous vein grafts bypassed to the right coronary artery (RCA) 12 months post implantation shown in Figure 4. In the figure, models A,B and C are unsupported and models D, E and F are supported grafts with ESD. These angiographies were taken from the Venous External Support Trial (VEST). The models included six models of the fluid domains (3 with ESD and 3 without ESD), and two models of the structural domain (one with ESD and the other without ESD).

The models were built using image analysis procedures, assuming planar and axisymmetric grafts. An example of the numerical model is shown in Figure 5.

Table 1 List of models studied

Case name	Domain	ESD
A_f	Fluid	Yes
A_s	Solid	Yes
B	Fluid	Yes
C	Fluid	Yes
D_f	Fluid	No
D_s	Solid	No
E	Fluid	No
F	Fluid	No

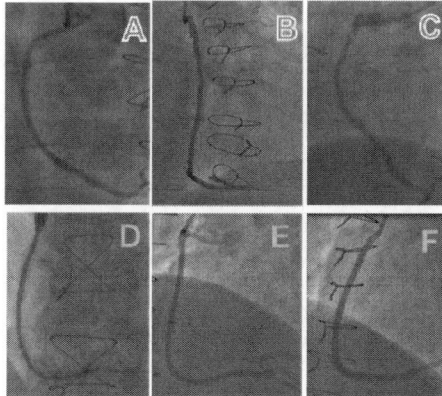

Fig. 4 Angiographic scans of unsupported (A,B,C) and supported grafts (D,E,F) 12 months post-transplantation

The blood was assumed Newtonian and homogenous with viscosity of 3.5 cP and density of $1.05 gr/cm^3$. The flow was assumed laminar. The graft's wall was assumed homogenous, linearly elastic and isotropic, density of $1.6 gr/cm^3$, and Poisson's ratio of 0.499. The 12 months post-plantation graft wall was modeled with constant wall thickness of 0.065cm and elastic modulus of 4.2MPa.

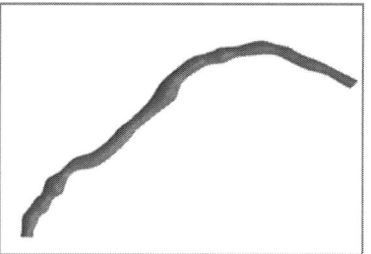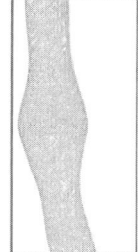

Fig. 5 An example of a graft model and a magnified view at the model mesh

Physiological flow conditions were imposed for all the models, with average coronary flow of 100 mL/min, heart rate of 75 BPM and aortic pressure of 80/120 mmHg. The boundary conditions for the fluid domains included time-dependent inlet coronary flow velocity. The boundary conditions for the structural domain included time-dependent aortic pressure. The fluid and solid domains were calculated separately, neglecting the effect of wall shear stresses on stresses developed in the wall, and the effect of wall motion on the flow.

Mesh resolution tests revealed that meshes with ~140,000 elements had less than 7% error in maximal stresses both for fluid and solid domains. All simulations solved two cycles, using time steps of 0.1sec.

III. RESULTS

Results include the time-dependent velocity, pressure and shear stress distribution in the flow fields, and the effective stress and displacements distribution in the solid domains. Examples of the results for the unsupported cases are shown in Figure 6.

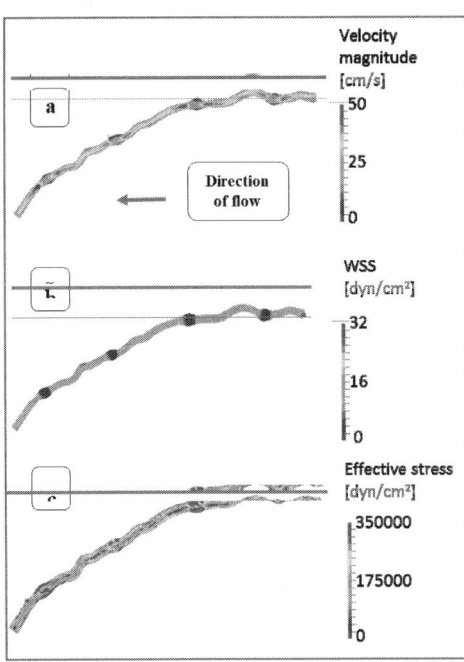

Fig. 6 Example of simulations results: (a) Velocity magnitudes (b) Wall shear stress (c) Effective stress at the wall

Examples of magnified views of velocity vectors for the unsupported case C are shown in Figure 7. The results depict disturbed flow in the unsupported cases and vortexes around the dilated areas. Furthermore low velocities are found near the walls and are more eminent at the dilatations. No vortexes were found in the supported grafts.

For comparison between supported and unsupported cases, Figure 8 shows effected stresses and WSS during peak flow for the six graft's models. The upper row results of A, B and C depicts the unsupported grafts, while the lower row results of D, E and F, depicts the supported grafts. In the unsupported cases, significant stress fluctuations are prominent at dilatations, while in the supported cases the stress distributions are relatively uniform.

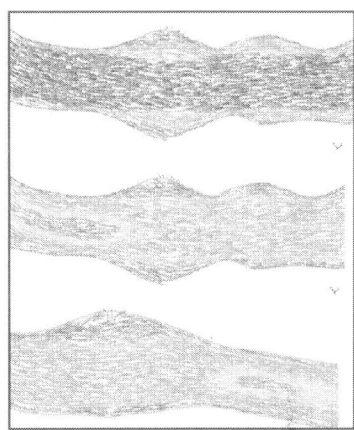

Fig. 7 Examples of velocity vectors in local dilatations of unsupported grafts – showing disturbed flow regimes

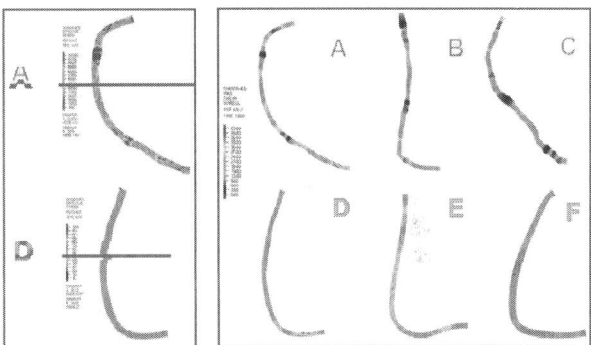

Fig. 8 Comparison of unsupported (top) and supported (bottom) grafts at peak diastole. Left: effective stress at the graft wall, Right – WSS at the fluid domain

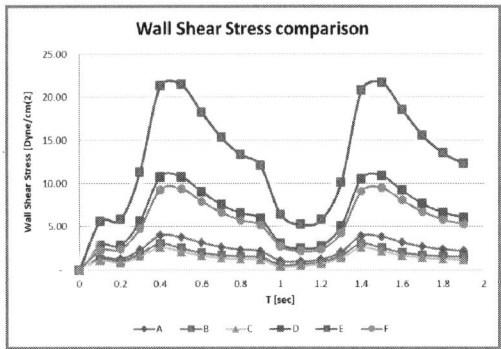

Fig. 9 Time-dependent WSS at the region with the lowest stresses of each model

Figure 9 shows a comparison plot of time-dependent values of WSS at regions with the lowest stresses of each model. One may notice that the unsupported models A, B and C have lower values than the supported grafts D, E and F. The model with the lowest stresses is C, and the highest is E. The minimal stresses of model E is ~10 times larger than model C.

IV. DISCUSSION

The results delineate the effect of local luminal irregularities on the graft's hemodynamics and wall stresses. Low velocity magnitudes, reverse flow and velocity profiles with low values are found at the areas with irregular diameter size such as a sinus and sharp curved areas.

Distinct differences in values of WSS and effective stresses are found between supported and unsupported grafts. The unsupported grafts are characterized by low WSS with fluctuations at the dilatation areas, and high stresses at the narrow areas. In comparison, the supported grafts are characterized with higher WSS along the graft and more uniform stress distribution. Disturbed and reverse flow with vortex formation was observed in the unsupported models, indicating increased probability of vortex formation in dilated areas. No vortexes or disturbed flow was seen on any of the supported grafts.

The effective stress results depict elevated stresses at the dilatations at the unsupported graft, while significantly lower and more uniform stresses are found for the supported case.

These findings suggest that focal intimal hyperplasia would target the areas with luminal irregularities, especially the dilatations where low WSS are obtained, high WSS gradient and disturbed flow. In addition the elevated WSS stress found in unsupported grafts may lead to further vein remodeling and intimal hyperplasia.

The elevated stresses calculated for the unsupported models at the dilatations, imply that these grafts are prone vein remodeling and increased risk of intimal hyperplasia compared to supported grafts.

These findings suggest that constricted graft with ESD improves hemodynamic and mechanical factors that are involved in intimal hyperplasia. The clinical significance of preventing intimal hyperplasia is expressed by maintaining high patency rates and preventing graft failure.

V. CONCLUSIONS

In order to improve patency rates and decrease graft failure following CABG surgery procedure, the leading causes of intimal hyperplasia should be treated. This investigation helps to understand and establish the advantages of the external mesh support and to evaluate the effects of the external support on the leading causes of graft failure.

REFERENCES

1. Sabik, J.F. and E.H. Blackstone, *Coronary Artery Bypass Graft Patency and Competitive Flow.* Journal of the American College of Cardiology, 2008. **51**(2): p. 126-128.
2. Ethier, C.R., *Computational modeling of mass transfer and links to atherosclerosis.* Annals of Biomedical Engineering, 2002. **30**(4): p. 461-471.
3. Lei, M., C. Kleinstreuer, and G. Truskey, *A focal stress gradient-dependent mass transfer mechanism for atherogenesis in branching arteries.* Medical Engineering & Physics, 1996. **18**(4): p. 326-332.
4. Kleinstreuer, C., M. Lei, and J. Archie Jr, *Flow input waveform effects on the temporal and spatial wall shear stress gradients in a femoral graft-artery connector.* Journal of biomechanical engineering, 1996. **118**(4): p. 506.
5. Human, P., et al., *Dimensional analysis of human saphenous vein grafts: Implications for external mesh support.* The Journal of Thoracic and Cardiovascular Surgery, 2009. **137**(5): p. 1101-1108.

Hemodynamical Aspects of Endovascular Repair for Aortic Arch Aneurisms

A. Nardi[1], M. Brand[1], M. Halak[2], M. Ratan[3], D. Silverberg[2], and I. Avrahami[1]

[1] Ariel Biomechanics Center, Ariel University, Ariel, Israel
[2] Vascular Surgery, the Chaim Sheba Medical Center, Tel Hashomer, Israel
[3] Medical Engineering, Afeka Academic College of Engineering, Tel Aviv, Israel

Abstract—The presented study is focused on the hemodynamics aspects of thoracic aortic aneurysm and approaches for restoring hemodynamics in the aortic arch. The study includes numerical investigation of the aortic arch hemodynamics of a healthy aorta, aorta with aneurysm, and of two endovascular repairing procedures. The first endovascular repair approach is the total aortic arch hybrid debranching. The second implantation uses chimney graft technique.

The analysis includes the fluid dynamics in the aorta and branching arteries under time-dependent physiological conditions. The results show the effect of aneurysm on blood flow in the descending aorta and in aortic arch side branches. In the aneurysmatic case, the aneurysm provokes a highly disturbed flow and large recirculation regions, especially during diastole. Out of the two endovascular techniques, the hybrid procedure was found preferred from hemodynamics point of view, with less disturbed and recirculating regions.

Although the chimney procedure requires less manufacturing times and cost, it is associated with higher risks rate, and therefore, it is recommended only for emergency cases. This study may shade light on the hemodynamic factors for these complications, and provide insights on ways to improve the procedure.

Keywords—Aortic Arch Aneurism, endovascular repair, chimney vs. hybrid techniques, CFD.

I. INTRODUCTION

Aortic arch aneurysm is a rare condition but carries a high risk of rupture. The actuarial 5-year survival of non-treated patients is only 13% with many patients dying from aortic rupture. Aneurysmal disease that involves the entire aortic arch is especially prone to extensive involvement because it is due to diffuse aortic dissection or medial degenerative disease in most cases.

Aortic arch aneurysms are considered complex aortic pathologies require coverage of one or more aortic arch vessels, and are usually repaired with open surgery.

The introduction of endovascular stent graft technology has attended in a new era in therapy for diseases of the aortic arch and descending thoracic aorta. The use of endovascular stent grafting offers a less invasive alternative to open surgical repair. This 'radical and revolutionary' stent graft technology has been accompanied by simultaneous advances in noninvasive imaging technology and materials technology.

Endovascular operative technique offers two distinct advantages over conventional open surgery for repair of aortic arch aneurysm. First, the endovascular graft is inserted through remote, easily accessible arteries outside the body cavity, whereas the open operation often requires a combination of median sternotomy and lateral thoracotomy for insertion. Second, the endovascular graft can be deployed without interrupting blood flow, whereas open repair usually requires aortic cross-clamping and hypothermic circulatory arrest.

The challenge in endovascular repair of the aortic arch, as in surgical repair, is to maintain blood flow to the brain and side branches in the sealing zone of the stent-graft.

Several approaches were introduced to overcome this challenge. The main two approaches considered are the graft procedures using fenestrations or chimney technique (e.g. Chimney of Innominate Artery, see Figure 1a), or the total hybrid debranching procedures (see Figure 1b).

In the chimney graft technique[1], a covered stent is deployed parallel to the main aortic stent-graft, protruding somewhat proximally, like a chimney, to preserve flow to a vital side branch (e.g. the Innominate artery (IA), or the left Subclavian artery (LSA)). This technique requires bypass connections between the side branches, for example, bypass between the IA and the LSA and between the LSA and the left common carotid artery (LCCA), as shown in Figure 1a.

The chimney graft technique allows the use of standard off-the-shelf stent-grafts to instantly treat lesions with inadequate fixation zones, providing an alternative to fenestrated stent-grafts in urgent cases, in aneurysms with challenging neck morphology, and for reconstituting an aortic side branch unintentionally compromised during endovascular repair.

In the hybrid total aortic arch debranching (TAAD), a bifurcated Dacron graft is connected to the ascending aorta using a proximal end-to-side anastomosis. The deployment of the endograft is done after transposition of epiaortic vessels as shown in Figure 1b.

Both approaches were proven to be technically feasible with the high short-term technical success rate and relatively favorable rates of perioperative outcomes for aortic arch pathologies. Long-term outcomes remain undefined[2, 3].

The TAAD technique is considered to have better performance; however it uses custom-made devices associated with long manufacturing times and increased costs. [4]. The chimney technique has the advantage of applying available off-the-shelf devices, being technically less demanding and can reduce or eliminate the need for surgical bypass. However, in high-risk patients it is associated with a relevant morbidity, mortality, and reintervention rate. Therefore, it is recommended only for patients not suitable for conventional aortic arch repair or emergency cases at present [5, 6].

In this study we explore the hemodynamic behavior of the aortic arch for healthy, aneurysmatic and treated cases with the two treatments approaches.

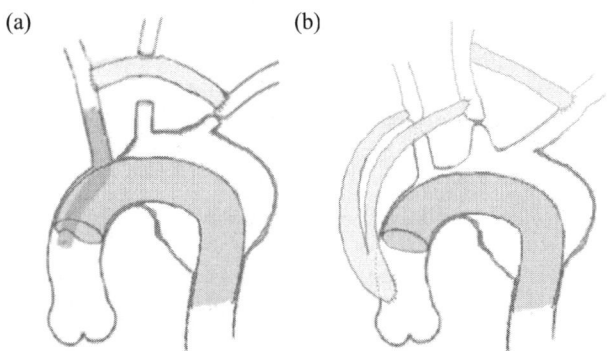

Fig. 1 An illustration of Chimney of Innominate Artery (CIA) – left, and Total Aortic Arch Debranching (TAAD) – right

II. METHODS

The analysis included four models of the time-dependent fluid domains of healthy and aneurysmatic arch, and of the CIA and TAAD endovascular techniques.

The simulations used computational fluid dynamics (CFD) methods to delineate the time-dependent flow dynamics in the aortic arch and the abdominal aorta with and without aneurysm. The geometric models of the four cases are shown in Figure 2.

The flow and pressure fields in the lumen were calculated by numerically solving the equations governing momentum and continuity in the fluid domain:

$$\nabla \cdot \mathbf{V} = 0$$
$$\rho \frac{D\mathbf{V}}{Dt} = -\nabla p + \mu \nabla^2 \mathbf{V} + \rho \mathbf{g}$$
(1)

where p is static pressure, t is time, $\mathbf{V}$ is velocity vector, ρ and μ are density and the dynamic viscosity of blood, and $\mathbf{g}$ is vector of gravity. Blood was assumed homogenous, incompressible (with density $\rho = 1$ gr/mL), and Newtonian (with viscosity $\mu = 3.5$ cP). The flow was assumed laminar. Gravity of $\mathbf{g}=981$ cm/s^2 was employed.

The boundary conditions were set according to typical physiological conditions with pressure of 120/80 mmHg, heart rate of 75 BPM, and average cardiac output of CO=5L/min (shown in Figure 3).

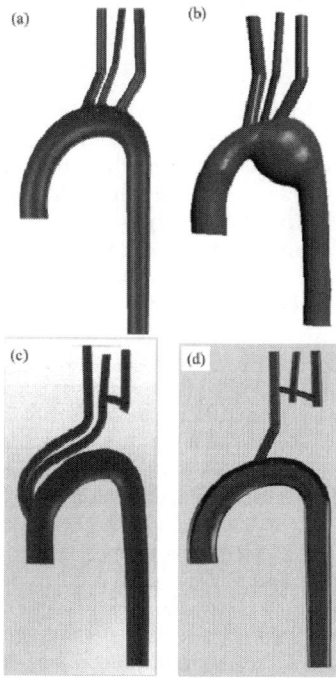

Fig. 2 The geometrical models: (a) healthy case, (b) with aneurism, (c) CIA and (d) TAAD procedures

Time-dependent flow rates were imposed at the aortic root inlet, and vessels outlets, as described in Figure 4.

The commercial software ADINA (ADINA R&D Inc., MA) was used to solve the set of fluid and structure equations using the finite-element scheme. The numerical meshes for the fluid domains of the pre- and post-grafted aneurysm models consisted of about 800,000 tetrahedral elements each. For each case, four cardiac cycles were calculated (0 sec<t<3.2 sec) with total 800 time steps per cycle. The results of the third cycle were fully periodic.

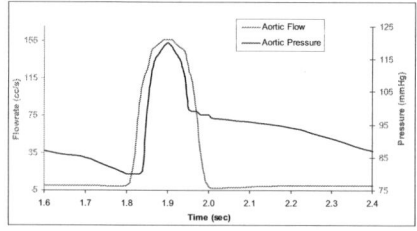

Fig. 3 Imposed inlet aortic flow and outlet aortic pressure as a function of time

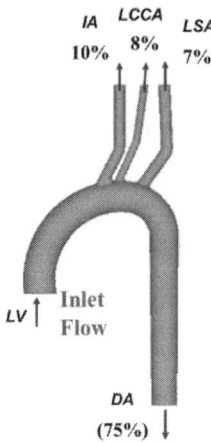

Fig. 4 The imposed boundary conditions: Flow distribution between side branches. (*IA*- Innominate artery; *LSA* - left Subclavian artery; *LCCA* - left common carotid artery; *DA* - descending aorta)

III. RESULTS AND DISCUSSION

The resulted flow field as calculated for the fluid domains for healthy and aneurysmatic aorta during the third calculated cycle (1.6 sec<t<2.4 sec) are shown in Figure 5. Figure 5a show pressure distributions at peak systole (t=1.9 sec). Figure 5b show velocity vector plots during diastole (t=2.2 sec). Figure 5c show Lagrangian tracking pathlines of 300 particles released in the ascending aorta or aortic root during end diastole (t=2.4 sec).

The results clearly depict the effect of aneurysm on blood flow. The aneurysm provokes a highly disturbed flow and large recirculation regions (lower figures) in comparison to the cases without aneurysms (upper figures). The disturbed flow observed in the aneurysm sacs, especially during diastole, imply poor particle washout from the aneurysm sac and regions of stagnation in the center of the circulation zones – where activated platelets can accumulate and aggregate to form thromboembolism. Moreover, the non-oriented flows in the aneurysms require flow particles to change their direction and velocity on their way downstream. In the healthy case without aneurysm, the flow field is significantly smoother and particles are well directed towards the side arteries in a uniform flow profile, even at the last stage of diastole. This increase in pressure drop in the aneurysmatic aorta may further intensify the risk for wall rupture [7].

Figure 6 and Figure 7 show the result for the calculated flow field in the CIA and TAAD endovascular repair techniques, respectively. The flow in the CIA case is clearly more disturbed, and the number of local impair hemodynamic sites is larger than in the TAAD case. Unlike in the

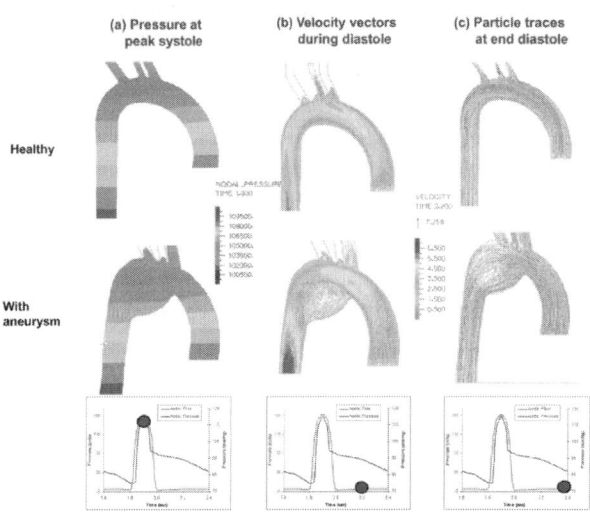

Fig. 5 Results for flow field as calculated for the healthy (top) and aneurysmatic (bottom) aortic arch: (a) pressure distribution in the models during peak systole (t=1.9 sec) (in dyn/cm2), (b) velocity vectors plot (in cm/s) during diastole (t=2.2 sec), and (c) traces of particles released in the ascending aorta during end diastole (t=2.4 sec)

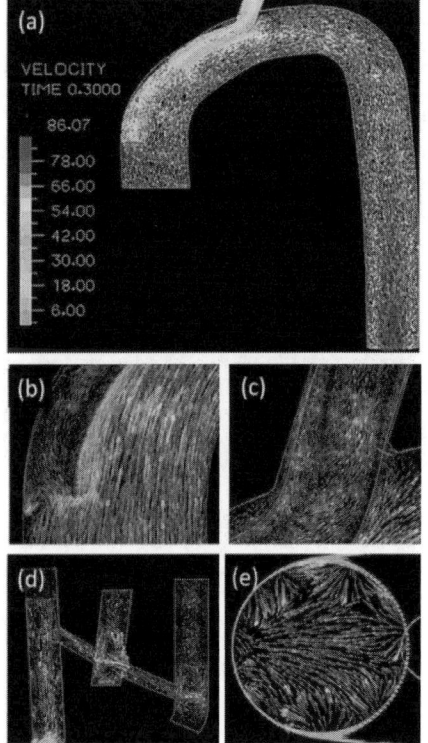

Fig. 6 Velocity vectors in the CIA model: (a) flow in the total model, (b) chimney inlet, (c) chimney curve, (d) chimney outlet, (e) transverse cut of the aortic arch

bifurcated Dacron graft of the TAAD case, the flow in the chimney covered stent is poor and undirected. In addition, the extra bypass connection required for the CIA procedure is another source for poor hemodynamics. Although the TAAD case has a major vortex formation near the proximal anastomosis, this vortex presents only during diastole and it is unstable and temporary.

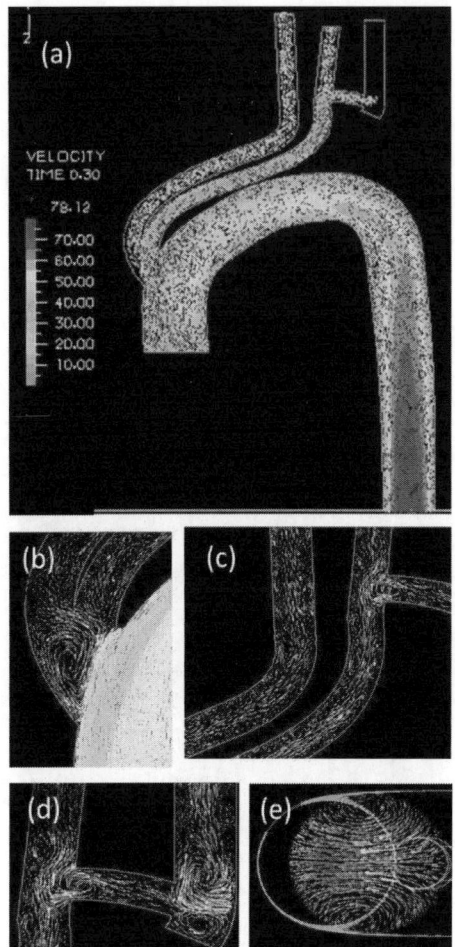

Fig. 7 Velocity vectors in the TAAD model: (a) flow in the total model, (b) debranching inlet, (c) LCCA to LSA bypass, (d) connection to the BT & LCCA arteries, (e) transverse cut of the aortic arch.

IV. CONCLUSIONS

This study delineates the poor hemodynamics of the aortic arch aneurysm, and presents the hemodynamic advantages of the hybrid TAAD technique.

Although the chimney procedure requires less manufacturing times and cost, it is associated with higher risks rate, and therefore, it is recommended only for emergency cases. This study may shade light on the hemodynamic factors for these complications, and provide insights on ways to improve the procedure.

REFERENCES

1. Ohrlander, T., et al., *The chimney graft: a technique for preserving or rescuing aortic branch vessels in stent-graft sealing zones.* Journal Information, 2008. **15**(4).
2. Yang, J., et al., *Endovascular chimney technique of aortic arch pathologies: a systematic review.* Annals of Vascular Surgery, 2012. **26**(7): p. 1014-1021.
3. Cires, G., et al., *Endovascular debranching of the aortic arch during thoracic endograft repair.* Journal of Vascular Surgery, 2011. **53**(6): p. 1485-1491.
4. Yoshida, R., et al., *Total endovascular debranching of the aortic arch.* European Journal Of Vascular And Endovascular Surgery, 2011. **42**(5): p. 627-630.
5. Geisbüsch, P., et al., *Complications after aortic arch hybrid repair.* Journal of Vascular Surgery, 2011. **53**(4): p. 935-941.
6. Sugiura, K., et al., *The applicability of chimney grafts in the aortic arch.* Journal of Cardiovascular Surgery, 2009. **50**(4): p. 475-481.
7. Brand, M., et al., *Clinical, Hemodynamical and Mechanical Aspects of Aortic Aneurysms and Endovascular Repair.* Cardiology Research and Clinical Developments 2013: Nova Publisher

Optical Tracking System Integration into IORT Treatment Planning System

E. Marinetto[1], V. García-Vázquez[1,3], J.A. Santos-Miranda[4,5], F. Calvo[4,5], M. Valdivieso[6], C. Illana[6], M. Desco[1,2,3], and J. Pascau[1,2,3]

[1] Instituto de Investigación Sanitaria Gregorio Marañón, Unidad de Medicina y Cirugía Experimental, Madrid, Spain
[2] Dept. Bioingeniería e Ingeniería Aeroespacial, Universidad Carlos III de Madrid, Madrid, Spain
[3] Centro de Investigación Biomédica en Red de Salud Mental (CIBERSAM), Madrid, Spain
[4] Dept. de Oncología, Hospital General Universitario Gregorio Marañón, Madrid, Spain
[5] Universidad Complutense de Madrid, Madrid, Spain
[6] GMV, Madrid, Spain

Abstract—**Intra-Operative Radiation Therapy (IORT) is a technique that combines surgery and adjuvant radiation directly applied to a post-resected tumor bed by means of a specific applicator docked to the linear accelerator. An IORT Treatment Planning System (TPS) allows radiation oncologists to plan the treatment using computed tomography (CT) images of the patients. The integration of an optical tracking system into the IORT scenario would allow guiding the radiation applicator according to the planned treatment. In this paper the complete integration of an optical tracking system into the TPS and the experimental set-up in the real IORT scenario is described, showing the feasibility of applicator guidance in IORT procedures.**

Keywords—**IORT, Optical Tracking, surgery planning.**

I. INTRODUCTION

Intra-Operative Radiation Therapy (IORT) is a technique that combines surgery and adjuvant radiation to reduce local cancer recurrence. A single-fraction dose of radiation is directly delivered to the residual tumor or tumor bed at the end of a surgical resection by means of specific applicators connected to a conventional or mobile linear accelerator [1]. Radiation oncologists plan the treatment before surgery. During planning, oncologists choose parameters such as applicator diameter, position, bevel-angle or radiation energy based on their clinical experience. However, they often modify the original plan due to anatomical changes of the patient during surgery.

A novel IORT treatment planning system (*radiance®*, GMV, Madrid, Spain) allows radiation oncologists to design treatment approach using computed tomography (CT) images of the patients. This Treatment Planning System (TPS) simulates the insertion of a selected applicator in the area of interest and estimates the dose distribution in the tumor bed and organs at risk [2]. In a previous work we have shown that applicator position and orientation can be recorded using optical tracking [3]. This pose and motion data can be used to display the virtual applicator over the CT image in real time. Oncologists may exploit this feature to guide the applicator during surgery according to the planned treatment. Furthermore, they can re-estimate the treatment parameters, evaluate possible procedural complications and assess final delivered dose depending on the current intra-operative findings.

The workflow described in [3] is a proof of concept that demonstrated the applicator tracking feasibility. This tracking must be fully integrated into the TPS, including the registration steps, in order to perform IORT guidance in real time. The main purpose of this work is to describe the complete integration of an optical tracking system into the TPS and the experimental set-up in the real IORT scenario (operating theater in Hospital Gregorio Marañón, Madrid, Spain).

II. MATERIAL

A. Optical Tracking

Optical tracking systems provide real-time pose data of different *tools* inside a tracked 3D volume. Most commercial systems use a small number of cameras to acquire the position data [4, 5] assuming direct line-of-sight from tools to cameras. However, maintaining this line-of-sight during optical tracking is hard to achieve in the IORT scenario. To avoid this problem some systems use higher number of cameras.

In this work we have used an Optitrack (NaturalPoint, Inc.) tracking system. Optitrack comprises a group of infrared (IR) cameras surrounding the tracked volume (Fig. 1 left). Cameras have a LED ring that emits IR light towards the scene. Optical markers (Fig. 1 right), attached to the instruments, reflect this light. Cameras filter the IR light and provide the real-time position of each optical marker. Thus, the system can track several tools inside the volume using different spatial arrangement of these markers, also called 'rigid-bodies'.

Other advantage of the selected tracking system is its flexibility, since the user can choose the number and distribution of the cameras around the area of interest. However, a system calibration is needed in order to establish the position of every camera. Calibration is carried out by a special rigid-body called OptiWand, that consists of three aligned optical markers. The user has to wave the OptiWand inside the tracking volume while Optitrack software acquires the OptiWand pose data. After acquisition, the system calculates extrinsic (location and orientation) and intrinsic (lens distortion, focal length, skew, etc.) parameters of each camera and saves them in a calibration file [6].

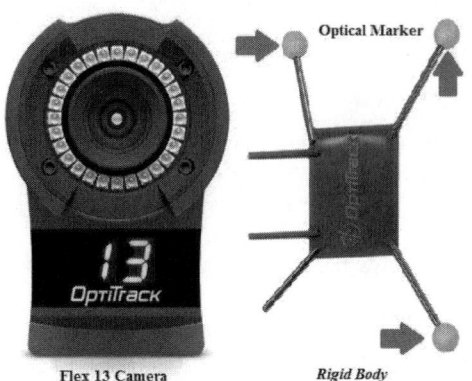

Fig. 1 Left: Flex-13 Optitrack camera. Right: 3-Markers *rigid-body*. Red arrows point to IR-reflecting markers

B. Tracking Tools

Two Specific tools were tracked in our setting: a pointer to select reference landmarks on the patient and the IORT applicator.

The pointer used (Fig. 2) was made of polyoxymethylene (POM), and included six optical markers. The tip position is obtained from the location of these markers. This tool was used to select landmarks on the patient which, when aligned with the corresponding points identified on the CT image, enables the calculation of the patient-to-image rigid transformation matrix.

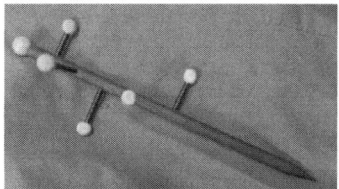

Fig. 2 Pointer Tool with six IR-reflecting markers

The applicator is the tool that radiation oncologists employ to deliver the radiation during IORT. A rigid-body composed of four optical markers was designed and attached to the applicator in order to track its position during the procedure (Fig. 3). One important feature of this rigid-body is that it can be detached from the applicator. In this way, once the radiation oncologist has selected the final applicator position inside the patient, he can detach the rigid-body from the applicator, which is afterwards docked to the linear accelerator for radiation delivery. Tracking the applicator, clinicians can check different procedure plans and calculate the delivered dose before deciding the optimal treatment setting.

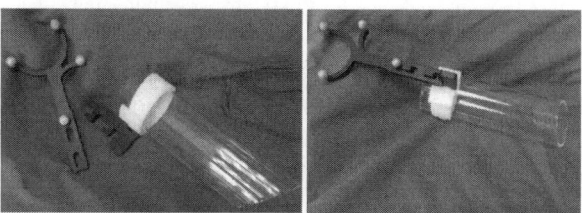

Fig. 3 Rigid-body for IORT applicator

C. Phantom

With the purpose of simulating a real IORT case, we used a plastic skeleton phantom (Medical Simulator S. L., Madrid, Spain). Ten ball-shaped metal markers were arranged along skeleton chest (diameter 1.5 mm, Suremark Inc., UK), to be used as registration landmarks. The phantom was immobilized on an X-ray transparent stretcher and a CT image was acquired using a Toshiba Aquilion LB scanner (512x512x801 pixels, 1.2 x 1.2 x 1 mm). Since metal markers are radiopaque, they can be clearly identified on the CT image for the registration step (patient-to-image registration).

D. Operating Room Systems

The operating theater for IORT procedures at Hospital Gregorio Marañón has been modified to include all the necessary elements for IORT pre-planning and tracking. Two screens (Fig. 4-C) were placed allowing radiation oncologists to check applicator position from TPS in real time. Two recording cameras (Fig. 4-E) and an operating lamp with video camera (Fig. 4-D) were also installed to record the IORT procedure.

We designed an L-shaped structure (Fig. 4-A) for the optical tracking system set up (Fig. 4-D). Two L-shaped structures were used, and a total of eight cameras were installed –four cameras along each L-structure (Fig. 4-B)-. Structures were installed along the ceiling in order to surround the patient volume for tracking.

A computer, connected to the screens, was placed outside the operating room. We installed the TPS and the tracking software on the computer, which had also access to Hospital Information System and Picture Archiving and Communication System. With this solution, oncologists can check patient information, display the pre-planning and track the applicator position in the TPS from the OR during a real IORT procedure.

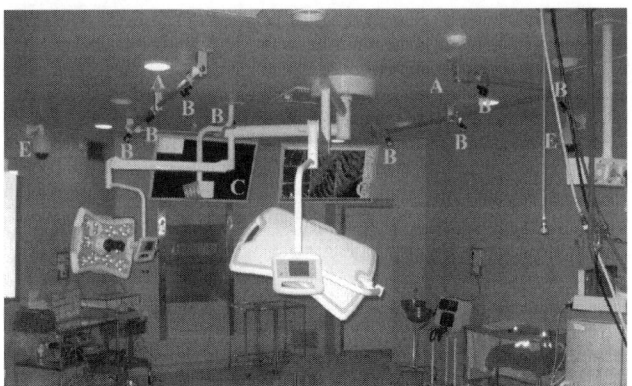

Fig. 4 A) L-structures for optical tracking system installation. B) IR cameras. C) Information screens. D) Surgery video recording lamp. E) Video recording cameras

III. METHODS

A. Tracking Interface

Optitrack tracking system provides an API (Application Programming Interface) to control the cameras, load calibration parameters and extract pose data. It uses quaternions nomenclature to describe rigid-bodies orientation. Quaternions are four-dimensional vectors commonly written as:

$$q_x \cdot i + q_y \cdot j + q_z \cdot k + q_w \qquad (1)$$

$$i^2 = j^2 = k^2 = ijk = -1 \qquad (2)$$

$$ij = k \;,\; jk = i \;,\; ki = j \qquad (3)$$

The parameters $[q_x, q_y, q_z]$ correspond to the rotation axis and q_w to the rotation angle from an initial orientation of the rigid-body. On the other hand, TPS uses transformation matrix nomenclature. Thus, a conversion from API rotation data to TPS is needed. Eq. 4 describes the required conversion.

$$T = \begin{bmatrix} 1 - 2q_y^2 - 2q_z^2 & 2q_xq_y + 2q_wq_z & 2q_xq_z - 2q_wq_y & t_x \\ 2q_xq_y - 2q_wq_z & 1 - 2q_x^2 - 2q_z^2 & 2q_yq_z + 2q_wq_x & t_y \\ 2q_xq_z + 2q_wq_y & 2q_yq_z - 2q_wq_x & 1 - 2q_x^2 - 2q_y^2 & t_z \\ 0 & 0 & 0 & 1 \end{bmatrix} \qquad (4)$$

Note that the last column of matrix 'T' corresponds to translation parameters, which does not need conversion because the API provides them directly.

We developed a wrapping API library comprising two classes. Class *OptControl* implements the necessary functions to:

- Start and stop the tracking system for recording pose data.
- Load calibration files.
- Check periodically the state of the system.
- Calculate the transformation matrix between both coordinate systems using a classical point-based registration algorithm.

Class *OptTool* is an abstract class that refers to all the tracked tools. Pointer and applicator rigid-bodies are instances of this class. To make the obtained tracking data more robust, *OptTool* calculates an average of 2000 pose samples and performs the conversion from API to TPS nomenclature.

The developed library was included into TPS software and a tracking interface was implemented, allowing the user to load a calibration file, extract point locations and track the applicator position.

B. Experimental Set-Up

For the purpose of checking the full-functioning integration, we used the skeleton phantom described above.

The calibration process of the tracking camera system was carried out and the parameters were loaded on the tracking interface. The acquired CT image was loaded in the TPS and metal landmarks were identified, recording their positions with the tracking interface. Thereafter, the phantom was moved to the operating room and metal markers were located using the pointer tool. This process was also guided by the implemented tracking interface. Once metal markers position was acquired, tracking interface calculated the patient-to-image rigid transformation matrix. From this point, treatment planning system was ready to obtain the real-time applicator pose data, placing the virtual applicator on the corresponding position over the CT image.

A radiation oncologist placed the applicator along different positions over the phantom in order to evaluate the full tracking integration into the TPS.

IV. RESULTS

After registration, both systems worked together showing the real applicator position over the phantom CT image as

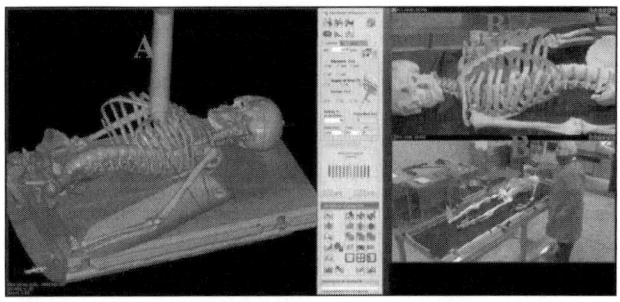

Fig. 5 Applicator Tracking. A) Virtual applicator over CT image in *TPS*. B) Applicator real position over phantom

we can see in the Fig. 5. While the oncologist was modifying the applicator position, the virtual applicator followed the motion for all tested positions.

V. DISCUSSION AND CONCLUSIONS

In this work we present the modifications carried out on the IORT operating theater with specific equipment (screens, video cameras for recording and optical tracking system) that allows radiation oncologists to check the preplanned IORT procedures and to adapt them to surgical findings. The proposed tracking scheme makes dose distribution recalculation possible during applicator placement. Moreover, it helps to fulfill, check and adapt the pre-planning on-site. Recording video cameras provide a way to gather information about the case i.e. surgery plan or applicator location.

One limitation is the need of a technician in the operating room for system setup and calibration. Thus, improvements in technician interface have to be considered.

The complete operating room design –comprising the optical tracking system, screens and video recording cameras– also becomes a powerful learning tool that, together with the IORT TPS (*radiance*), will help new oncologists in learning this beneficial procedure for cancer treatment.

Finally, we can conclude that the optical tracking system integration into the TPS is feasible for the applicator guidance in IORT procedures.

ACKNOWLEDGMENT

This work was partially supported by the Spanish Ministry of Economy and Competitiveness (TEC2010-21619-C04, IPT-300000-2010-003, IPT-2012-0401-300000) and ERD Funds.

REFERENCES

1. Calvo, F.A., R.M. Meirino, and R. Orecchia, *Intraoperative radiation therapy: First part: Rationale and techniques.* Critical Reviews in Oncology/Hematology, 2006. **59**(2): p. 106-115.
2. Pascau, J., et al., An Innovative Tool for Intraoperative Electron Beam Radiotherapy Simulation and Planning: Description and Initial Evaluation by Radiation Oncologists. Int J Radiat Oncol, 2012. **83**(2): p. e287-e295.
3. García-Vázquez, V., et al., Towards a real scenario in intraoperative electron radiation therapy, in Int J CARS. 2012. p. S64-S65.
4. Polaris, N., Northern Digital Polaris tracking system. 2004.
5. Optotrak, N., Optotrak Certus Spatial Measurement. 2007.
6. Heikkila, J. and O. Silven. A four-step camera calibration procedure with implicit image correction. in Computer Vision and Pattern Recognition, 1997. Proceedings., 1997 IEEE Computer Society Conference on. 1997: IEEE.

Author: Eugenio Marinetto Carrillo
Institute: Instituto de Investigación Sanitaria Gregorio Marañón
Street: Doctor Esquerdo 46, 28007
City: Madrid
Country: Spain
Email: emarinetto@hggm.es

Comparative Study of Two Laparoscopic Instrument Tracker Designs for Motion Analysis and Image-Guided Surgery: A Technical Evaluation

J.A. Sánchez-Margallo[1], F.M. Sánchez-Margallo[2], I. Oropesa[3,4], M. Lucas[1], J. Moreno[5], and E.J. Gómez[3,4]

[1] Bioengineering and Health Technologies Unit, Jesús Usón Minimally Invasive Surgery Centre, Cáceres, Spain
[2] Laparoscopy Unit, Jesús Usón Minimally Invasive Surgery Centre, Cáceres, Spain
[3] Bioengineering and Telemedicine Centre, ETSI Telecomunication, Universidad Politécnica de Madrid, Madrid, Spain
[4] Networking Research Centre on Bioengineering, Biomaterials and Nanomedicine (CIBER-BBN), Zaragoza, Spain
[5] Laboratory of Robotics and Artificial Vision, Universidad de Extremadura, Cáceres, Spain

Abstract—Laparoscopic instrument tracking systems are a key element in image-guided interventions, which requires high accuracy to be used in a real surgical scenario. In addition, these systems are a suitable option for objective assessment of laparoscopic technical skills based on instrument motion analysis. This study presents a new approach that improves the accuracy of a previously presented system, which applies an optical pose tracking system to laparoscopic practice. A design enhancement of the artificial markers placed on the laparoscopic instrument as well as an improvement of the calibration process are presented as a means to achieve more accurate results. A technical evaluation has been performed in order to compare the accuracy between the previous design and the new approach. Results show a remarkable improvement in the fluctuation error throughout the measurement platform. Moreover, the accumulated distance error and the inclination error have been improved. The tilt range covered by the system is the same for both approaches, from 90º to 7.5º. The relative position error is better for the new approach mainly at close distances to the camera system.

Keywords—Laparoscopic tool tracking, Optical pose tracker, Motion analysis, Motion analysis, Image-guided intervention.

I. INTRODUCTION

Image-guided interventions (IGI) have been greatly expanded by the advances in medical imaging and computing power over the past 20 years, driven by the surgical aim of progressively provide less invasive and harmful treatments. Tracking systems are an essential component of IGI systems for determining the spatial relationship between the surgical instruments, the anatomy, and the preoperative information. This is a useful tool for surgeons when the surgical instrument is outside the field of view, obscured by artifacts or occlusions, or when the instrument cannot be detected by the imaging system [1]. However, as of today, further efforts are needed in order to provide affordable and accurate tracking systems for laparoscopic instruments that can be used in a real interventional site.

On the other hand, minimally invasive surgery is a high demanding surgical approach concerning technical requirements for the surgeon, which must be trained in order to perform a safe surgical intervention. Traditional surgical education in minimally invasive surgery is commonly based on subjective criteria to quantify and evaluate surgical abilities. However, researchers, surgeons and associations are increasingly demanding the development of more objective training and assessment tools that can accredit surgeons as technically competent [2]. As has been reported in the literature, laparoscopic instrument motion analysis can be a suitable solution for developing automatic objective assessment tools for assessment of surgical technical skills.

In a previous work, we proposed a tracking system in order to address these two concerns [3]. It applies a third generation optical pose tracker (MicronTracker® Hx60; Claron Technology Inc., Toronto, CAN) to laparoscopic practice for both motion analysis of laparoscopic instruments for surgical assessment and image-guided applications.

This first version enabled tracking with real laparoscopic instruments while allowing users to grip and use the instruments in a natural way. Reported accuracy results showed stable but low positional accuracy to track the instrument tip. These results could be enough for objective assessment of skills based on instrument motion; however, accuracy is a crucial issue in order to use tracking systems in IGI.

The main objective of this study is to improve the accuracy of this laparoscopic instrument tracking system based on a third generation optical pose tracker. We hypothesize that this accuracy improvement could be achieved with a design enhancement of the artificial markers placed on the laparoscopic instrument and an improvement of the calibration process. To this end, a comparative study of positional accuracy between the previous design and the new approach is presented.

II. MATERIALS AND METHODS

A. System Description

Departing from the first version of the tracking system, a new support for three artificial markers (one at the front and two in both sides) was designed in order to track each laparoscopic instrument. This support was placed on the handle of each instrument to avoid not disturbing the natural use of the laparoscopic instruments. The material chosen for the support (ABS polymer) is lightweight (12 g) and tough.

In order to increase the system accuracy of the previous design [3], equation (1) was taken into account to estimate the error in computing the position of the instrument tip (provided by the manufacturer). This equation shows that the only two possible ways for error reduction are increasing the distance between markers placed on the support (l) or decreasing the distance between the support and the instrument tip (d). The latter option entails placing the markers on the instrument shaft or inserting a foreign body inside the patient or simulator. Consequently, it was decided to increase the separation between markers (Fig. 1a).

$$e_{tip} \approx e_m + 1.5 \cdot e_m \cdot d/l \qquad (1)$$

$$e_m = e_{calibration} + e_{jitter} = 0.35mm + 0.14mm = 0.49mm$$

One of the challenges in computing the laparoscopic instrument tip is that it is difficult to determine the exact position of its centre point. The tip of each laparoscopic instrument is different and with irregular shape. To solve this problem an additional support for laparoscopic instruments was designed (Fig. 1b), which covers the distal part and provides us a fixed point of reference (P') to be used during the calibration process.

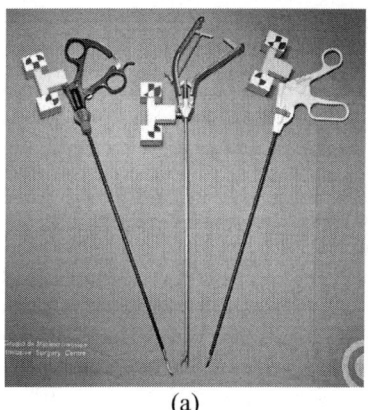
(a)

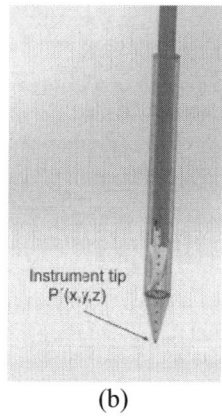
(b)

Fig. 1 (a) Set of laparoscopic instruments with the new marker supports. (b) Support for computing the instrument tip with regard to the central axis of the instrument shaft

For the calibration process the aforementioned support (Fig. 1b) was used, as well as a calibration plate under the tip in order to compute the transformation matrices from the markers to the same point of reference (P').

B. Technical Evaluation

An adaptation of the methodology defined by Hummel et al. [4] has been used to verify the positional accuracy and quantify the effects of noise of both tracking designs. A laparoscopic dissector (Richard Wolf GmbH, Knittlingen, Germany) was used to perform all technical evaluation tests. The camera system was placed at 600 mm from the working area and its height was established at 340 mm, which is approximately the same as the height of the markers on the instrument.

To technically validate the new system design and compare the accuracy results with the previous approach, two measurement platforms were developed. One platform for positional accuracy assessment was built with modified bricks and building plates of LEGO® (LEGO 6176 DUPLO Basic Bricks), which provides precise measures. The size of each basic piece used is 31.75 mm (Fig. 2a). To test inclination accuracy, a similar platform to the one proposed for the first version of the tracker was used [3]. The platform features 13 positions for the instrument at intervals of 7.5 degrees (Fig. 2b). Its configuration has been adapted to be used with the additional support for the instrument tip (Fig. 1b), and therefore provide a stable position of the instrument tip throughout the evaluation test.

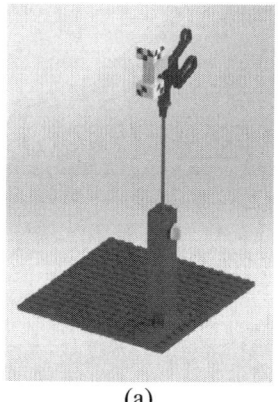
(a)

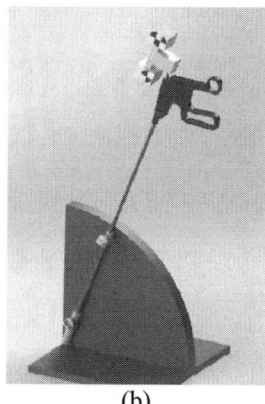
(b)

Fig. 2 Evaluation platforms. (a) Platform for the positional assessment. (b) Platform for the inclination assessment

For the evaluation test of positional accuracy the instrument tip was placed in 64 positions on the measurement platform distributed as a grid. For the inclination test the instrument was placed at each degree of inclination and with the tip fixed at the same origin. For both evaluation tests the position of the instrument tip was recorded during 10 seconds at each position.

The fluctuation error at each position was computed by the root mean square error (RMSE). The accumulated distance error was computed by means of obtaining the distances from the first position of the measurement platform $P(i,1)_{i=1,2,...,8}$ to all other column positions $P(i,j)_{j=1,2,...,8}$, and comparing them with the real distances. Relative position errors were computed by comparing the Euclidean distances reported by the tracking system to the known physical distances on the measurement platforms. Possible distances on the platform were computed with regard to multiples of a Displacement Unit (DU) of 31.75 mm.

III. RESULTS

Figure 3 shows the graphs for the fluctuation error for both designs. The maximum fluctuation error for the new design (0.536 mm) is considerably lower than the previous approach (2.991 mm). In addition, for the new design this error is more stable throughout the working area than the previous one.

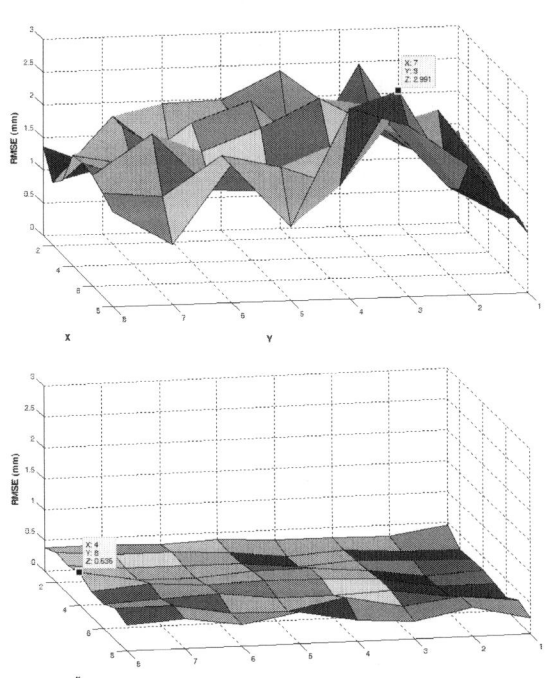

Fig. 3 Fluctuation error. (Top) Previous design. (Bottom) New design

As we expected, the accumulated distance error increases with the distance from the camera system for both designs (Fig. 4). However, the new design has a more linear distribution and its maximum error is lower (8.205 mm for the previous design versus 5.448 mm for the new approach). Both designs present their maximum errors at the last row of the measurement platform.

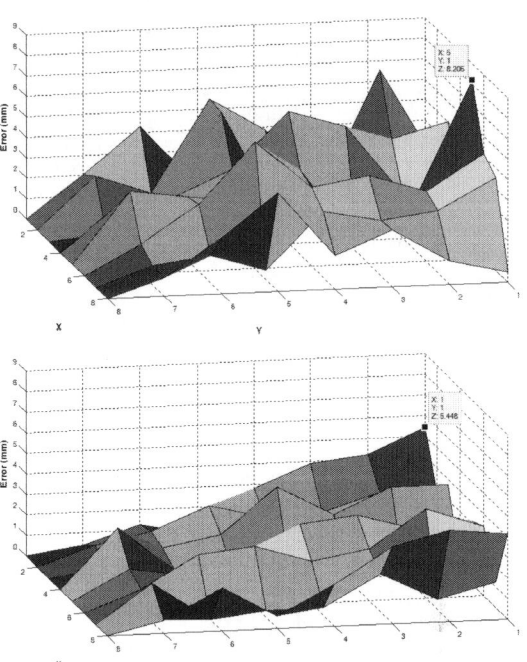

Fig. 4 Accumulated distance error. (Top) Previous design. (Bottom) New design

Table 1 Relative error (in mm). DU: Displacement Unit (31.75 mm). SD: Standard Deviation

Distance	Previous design		New design	
	Mean	Max.	Mean	Max.
1*DU	2.149 ± 1.713	7.782	0.933 ± 0.725	1.778
2*DU	2.377 ± 1.767	8.702	1.165 ± 0.857	2.198
3*DU	2.372 ± 1.879	6.976	1.540 ± 1.034	2.873
4*DU	2.500 ± 1.648	6.470	1.953 ± 1.059	3.362
5*DU	2.467 ± 1.802	9.668	2.238 ± 1.083	3.797
6*DU	3.140 ± 1.671	7.142	2.437 ± 1.315	4.851
7*DU	3.373 ± 2.057	8.205	2.898 ± 1.329	4.935

Table 1 shows the relative error rates. In general, this error is lower for the new design for all analyzed distances. This reduction is more noticeable at short distances (from 1*DU to 4*DU). Moreover, maximum errors at each evaluated distance are decreased.

The tilt range covered by both systems goes from 90 to 7.5 degrees. For the new design there is a reduction of both fluctuation error and accumulated distance error at each position. For both designs the maximum fluctuation error and accumulated distance error take place at the most horizontal position of the instrument (7.5°).

Table 2 Inclination error (in mm). RMSE: Root mean square error. ADE: Accumulated distance error

Degree	Previous design		New design	
	RMSE	ADE	RMSE	ADE
90.0	0.568	-	0.266	-
82.5	0.589	1.807	0.339	0.124
75.0	0.688	1.136	0.269	0.245
67.5	0.756	1.382	0.293	0.513
60.0	0.640	1.328	0.354	0.935
52.5	0.592	1.014	0.352	0.731
45.0	0.692	0.651	0.416	0.923
37.5	0.615	1.988	0.437	0.979
30.0	0.431	1.907	0.399	1.067
22.5	0.532	0.581	0.491	1.122
15.0	0.737	1.294	0.477	1.292
7.5	0.880	2.775	0.583	1.337

IV. DISCUSSION

This study presents the improved version of an original approach to laparoscopic instrument tracking based on a third dimensional optical pose tracker. This kind of tracking systems is fully passive and use light in the visible spectrum to identify targets. In general, they are more affordable than commercial IR-based systems, have no wires hanging from the instrument, and suffer no interference from metallic objects as in the case of electromagnetic tracking systems.

The new approach improves the accuracy results of a previous design [3], and therefore enables it to be used for motion analysis of laparoscopic instruments in training/ assessment and IGI applications. This system design does not interfere with the natural use of the instrument during surgery and nor increase its weight. It tries to address the need of tracking systems more robust and transparent for the user. Nevertheless, the use of video-based tracking systems has some challenges that need to be tackled. As reported by Maier-Hein et al. [5] these systems have some difficulties concerning poor lighting conditions and fast movements. Possible solutions to these limitations will be sought combined with computer vision techniques and pattern recognition techniques [6]. Moreover, to avoid line-of-sight obstructions, ergonomic studies are being carried out to determine optimal placement of the camera system in OR settings.

Despite these shortcomings, presented evaluation tests give proof of its accuracy and robustness, improving even the results from the first prototype presented [3]. Current works are focusing both on the analysis of the dynamic positional error of the system, as well as a complete validation study of the tracking system in a real scenario for MIS skills assessment.

V. CONCLUSIONS

The presented work offers a new tracking solution of laparoscopic instruments for objective evaluation of surgical technical skills and IGI. The system does not disturb the natural use of the surgical instruments. Both the design of the artificial markers on the instrument and the calibration process has been optimized as a means to improve the positional accuracy of the system with regard to a previous approach. Results have shown that this new design provides a general reduction both of positional and inclination error. The next step in order to comprehensively assess this system for clinical applications will be to analyze the dynamic positional error and its validation in a real scenario for training and assessment of MIS skills.

ACKNOWLEDGMENT

This study was supported in part by Gobierno de Extremadura, Consejería de Empleo, Empresa e Innovación, and the European Social Fund.

REFERENCES

1. K Cleary and TM Peters (2010) Image-guided interventions: technology review and clinical applications. Annu Rev Biomed Eng 12: 119-42.
2. van Hove PD, Tuijthof GJM, Verdaasdonk EG et al. (2010) Objective assessment of technical surgical skills. Br J Surg 97(7): 972-987.
3. Sánchez-Margallo JA, Sánchez-Margallo FM, Pagador JB et al. (2013) Technical Evaluation of a Third Generation Optical Pose Tracker for Motion Analysis and Image-Guided Surgery. Lect Notes Comput Sci 7761:75-82.
4. Hummel JB, Bax MR, Figl ML et al. (2005) Design and application of an assessment protocol for electromagnetic. Med Phys 32(7): 2371–9.
5. Maier-Hein L, Franz A, Meinzer HP et al. (2008) Comparative assessment of optical tracking systems for soft tissue navigation with fiducial needles. Proc. SPIE. vol. 6918, San Jose, CA, USA, 2008, pp 69181Z-69181Z-9.
6. Sánchez-Margallo JA, Sánchez-Margallo FM, Pagador JB et al. (2011) Video-based assistance system for training in minimally invasive surgery. Minim Invasive Ther Allied Technol 20(4): 197-205.

Author: Juan Alberto Sánchez Margallo
Institute: Jesús Usón Minimally Invasive Surgery Centre
Street: Ctra N-521, km 41,8. 10071
City: Cáceres
Country: Spain
Email: jasanchez@ccmijesususon.com

A Decision Support System Applied to Lipodystrophy Based on Virtual Reality and Rapid Prototyping

G. Gómez Ciriza[1], C. Suárez Mejías[1], C. Parra Calderón[1], T. Gómez Cía[2], and R. López García[2]

[1] Technological Innovation Group, Virgen del Rocío University Hospital, Seville, Spain
[2] Plastic and Reconstructive Surgery Department, Virgen del Rocío University Hospital, Seville, Spain

Abstract—In this paper, we present a medicine decision support system based on medical image processing, virtual reality and rapid prototyping techniques, providing surgeons a computer-aided diagnosis tool that allows evaluation of volumetric changes in lipodystrophy. It also provides a decision support system for surgical planning and intraoperative guidelines to perform the intervention.

This system began to be developed in our hospital in 2005 under a research, development and innovation project called VirSSPA. The system is composed by a surface scanner, VirSSPA software and physical biomodels generated by rapid prototyping in our installations. VirSSPA allows surgeons to generate and modify a virtual model of the patient and control any change in the facial contours.

As a result we have improved the quality of care using innovative techniques such as those described above.

Keywords—virtual reality, surgery, software, planning, lipofilling, rapid prototyping.

I. INTRODUCTION

Congenital lipodystrophy is a very rare autosomal recessive skin condition, characterized by an extreme scarcity of fat in the subcutaneous tissues [1]. Only 250 cases of the condition have been reported, and it is estimated that it occurs in 1 in 10 million people worldwide [2]. HIV patients are also affected by this localized fat loss and they are treated in our hospital with the same methodology, the progressive improvement of antiretroviral drugs is causing a decrease of the population of patients affected by this problem [3]. The diagnosis of lipodystrophy is complex, especially in its early stages. Until now the methods used for diagnosis have been subjective such as the self-assessment of patient or doctor observation and objective such as photographs, anthropometric measurements, bioelectrical impedance Analysis (BIA). Other techniques are based on the medical image like Dual Energy X-Ray Absorptiometry (DEXA), the Computed tomography (CT), magnetic resonance imaging (MRI). Some of these techniques are invasive for patient because they are based in ionizing radiation.

The facial contours are very complex and highly variable between individuals. Monitoring minimum volumetric changes over long periods of time in this population is particularly difficult. For this reason, an economical and noninvasive technique to measure facial changes is necessary. In this paper, we present a decision support system based in a software application that allows surgeons to detect changes in facial contour a to plan de surgical intervention. The software began to be developed by clinicians and engineers in our hospital in 2005 under a research, development and innovation project called VirSSPA.

II. MATERIALS AND METHODS

VirSSPA works with 3D images of patient's facial contour obtained by CTA or surface scanning, this second technology is noninvasive and has not high cost. The aim is to register the facial topology and compare it with previous images of the patient or with the corrected facial contour. In VirSSPA, once surgeons have detected and measured the changes, can simulate and plan the surgery. The correction volume calculated allows the surgeons to plan a personalized lipofilling procedure to correct irregularities and grooves in facial contours using autologous fat. It is one of the most common methods used for facial lipoatrophy correction. Some indications for lipofilling are sunken cheeks, the disappearance of fat from the cheekbones, deep grooves running from the nose to the corners of the mouth, and in some instances of lines between the lower eyelids and the cheek. For lipofilling the necessary fat is obtained by a limited liposuction through one or several 3-5 mm incisions. The fat is processed and injected where planned. The fat is evenly distributed into the area by injecting minimum amounts in the tissues so that the injected fat is well surrounded by healthy tissue. This ensures that the transplanted fat remains in contact with the surrounding tissues which must supply it with oxygen and nutrients. Possible risks of lipofilling are asymmetry, irregularities, overcorrection and infection. All this risks are what we are trying to reduce employing the presented system. VirSSPA has been piloted in different specialties and was introduced in clinical practice in the surgical planning sessions, after its efficiency and effectiveness had been proven in more than 700 planned surgical cases [4-7]. The 3D models obtained with VirSSPA also can be exported to an industrial manufacturing format to be worked out in rapid prototyping machines. As result, we can obtain a 3D personalized pieces made with rapid prototyping in a plastic polymer. This pieces, once sterilized, can be used as intraoperative reference for different surgeries.

VirSSPA was designed by surgeons of different surgical specialties and engineers of the Technological Innovation Group both from Virgen del Rocío University Hospital with the collaboration of School of Engineering of Seville and other enterprises. VirSSPA was developed in C++ and it generates the 3D model of patients using radiological images in DICOM format or surface images by scanner. So that surgeons can plan the surgical intervention on virtual reality using VirSSPA, first the image is pre-processed, then it is segmented and finally the 3D model of selected tissue is generated in VirSSPA. VirSSPA allows surgeons to work with radiological images from different capture machines due to preprocessing [8]. The preprocessing is composed of the following steps: first, the radiological image is standardized, then the image is filtered to avoid any noise and finally an enhancement of contrast is made to increase the difference between tissues. This pre-process is transparent to users. After preprocessing, segmentation is applied to the image [9,10]. The segmentation is the process that allows you to select a region of interest inside a tissue. In VirSSPA, mainly there are three segmentation methods. The first uses a thresholding method. For this, in VirSSPA surgeons use a thresholding bar. It represents the different grey level that a radiological image has. The surgeon selects in the bar the range of grey levels of interest tissue and it is seen in green. The second method is based on region growing by seeds. With this method, Surgeons put seeds on the radiological images and with inclusion criteria, the tissue is selected. The third method is an innovative semiautomatic algorithm. This algorithm is a self-assessed adaptive region growing segmentation algorithm. In this algorithm seeds are provided manually but the typical tolerance parameter, that determines if a new pixel is included in the segmented region, is managed internally using a measure of the varying contrast of the growing region, Specifically, in the iteration i of the region growing, a region Ri has been segmented according to a tolerance parameter τi so that neighboring pixels to region are included if:

$$|f-f_i'| \leq \tau_i \quad (1)$$

where f is the grey-level of a neighboring pixel and f_i' is the average grey-level in region Ri. Then in iteration i, the following contrast parameter is assessed:

$$C(R_i) = |f_{il}' - f_{io}'| / |f_{il}' + f_{io}'| \quad (2)$$

where f_{il}' is the inner border of Ri and f_{io}' is the outer border. The region i is the optimum segmented region, Ropt, if its contrast parameter is a significant local maximum:

$$C(R_{i-1}) < C(R_i) < C(R_{i+1}) \quad (3)$$

If does not match this requirement the tolerance parameter is increased and a new region is segmented. After the tissue segmentation, the 3D model is generated using an algorithm based on Marching Cubes [11] and a graphic engine designed by OpenGL.

One of the inputs that have been used more in recent months has been the surface scanner. This is a very interesting technology for some fields of medicine, such as plastic and reconstructive surgery, where a large number of interventions are performed to improve body contour defects.

The scanner incorporates a software to manipulate the image captured with different tools. Some of the most used are image alignment, erasing (to remove noise or unwanted surface), Smoothing, etc. These tools are very useful in combination with VirSSPA to manipulate images and reach satisfactory results.

The hospital has set up a workspace in which a set of rapid prototyping machines have been installed. These machines are being used to support surgeons in some of the cases that have been planned in VirSSPA. Different applications have been identified for which these models can be useful, for example, intraoperative reference, bench surgery, educational, etc. [12]. In this case, the technology employed is the fusion deposition modeling (FDM), whereby the 3D part is built layer by layer from the base to the upper end by a head that deposits a melted polymer filament that solidifies on the previous layer. The parts obtained by this technique can be sterilized in the hospital's facilities using ethylene oxide, then bagged and sent to the operating room where they will be used by the surgeon as reference for the intervention.

III. RESULTS

The case model presented is a patient in our hospital with severe facial lipoatrophy, for which a lipofilling surgery with autologous fat extracted and centrifuged in the operating room were performed. For planning this surgery it was used a surface scanner of the older sister as an ideal facial contour. Using the above tools (VirSSPA software and scanner) a 3D mesh was adapted to the contour of the patient and an ideal filling volume was designed using as inner face the skin of the patient and as outer face the skin of the sister. This volume provides the exact area to be intervened and the volume needed to make a good correction [13]. To improve the accuracy of the surgery two 3D pieces were manufactured in plastic resin with rapid prototyping technology.

On the figures 1 and 2 it can be observed the masks constructed as ideal contour for the patient and the fat volume necessary to make the correction designed. These 3D pieces were manufactured in Hospital facilities enabled to support clinicians and surgeons with the technological solutions presented.

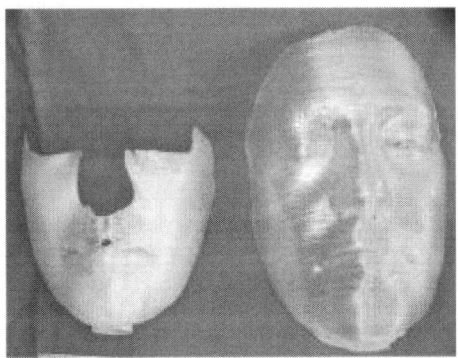

Fig. 1 Fat volume and external contour masks

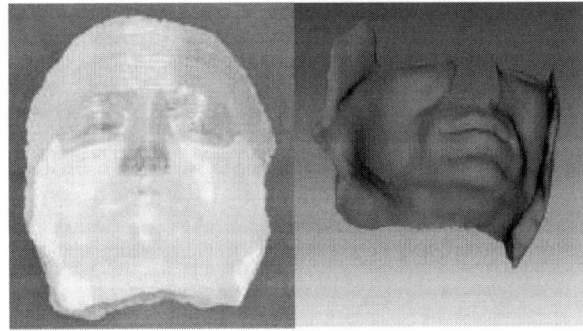

Fig. 2 Masks superposition and virtual 3D view of the fat filling mask

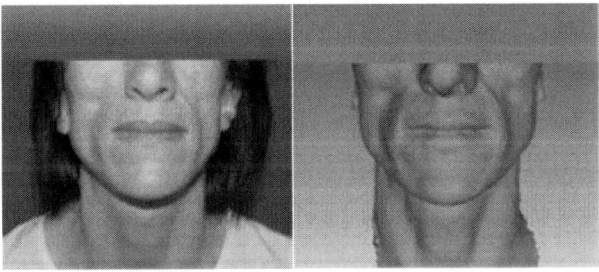

Fig. 3 Preoperative photographic and scanner images

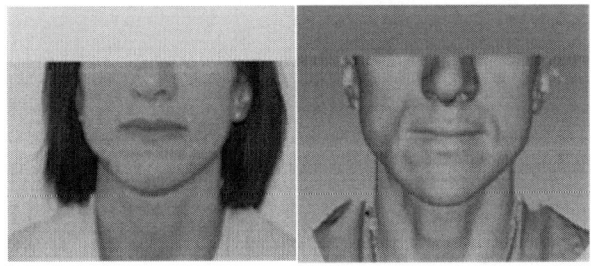

Fig. 4 Postoperative photographic and scanner images

IV. DISCUSSION

For the system validation, it was done a prospective study composed by ten HIV lipodystrophy patients in which a lipofilling procedure was done. These patients had 3D images before and after the surgical intervention. The 3D images were obtained with surface scanner and CTA. The volume changes measured between both 3D images in each patient were calculated and compared with the real volume that surgeon injected in each patient. The postoperative clinical control was performed at 2, 6, 12 and 18 months. The injected volume was obtained from the clinical history. The pre-and postoperative images obtained from the surface scanner and CTA were manipulated to calculate the volume from the virtual image. To do this, the preoperative and postoperative images were aligned taking into consideration some fixed anatomical landmarks that were identified (Hall headset, chin, nose, forehead). Analysis of data was performed using the Statistical Package for the Social Sciences MAC version 21.0. Quantitative variables were represented as mean and standard deviation, or median and interquartile range as follow a normal distribution or not, respectively. To validate the correlation between the different measurements, the Intraclass Correlation Coefficient (ICC) is calculated (Table 1). Its values range from 0 and 1, so that an ICC near 1 indicates a strong correlation or matching between both measurements systems and would therefore potentially interchangeable. We interpret ICC values > 0.7 are indicative of a strong correlation between measurements analyzed, and therefore system would be considered valid.

Table 1 Results obtained in the study.

	ICC	95 % Confidence Interval		F test with true zero value	
		Lower Limit	Upper Limit	Value	gL1
Individual measurement	0,997	0,968	1,0	581,994	4
Average measurement	0,998	0,983	1,0	581,994	4

With these results, we can affirm that VirSSPA and 3D physical models obtained with rapid prototyping can be used for surgical planning of the lipofilling procedure.

The surface scanner that we use is based on white light emitted and captured with the machine when the reading is being made so that there is no ionizing radiation to the patient.

Being a safe technology, fast and low cost, the number of patients who have been benefited from the incorporation of this technology has grown steadily.

As a result of this work a new research, development and innovation project has begun. The objective of this new project is to design and develop a software that allows surgeons to predict automatically the result of lipofilling surgery and calculates the volume to inject to correct the patient´s defect. In this project Innovation Technologies Group, Plastic and Reconstructive Surgery Department from our hospital and the Signal Processes Department of the School of Engineering from Seville University are involved.

V. CONCLUSIONS

In this paper we present a noninvasive system with no additional cost for monitoring volumetric changes in patients with facial lipodystrophy. This system is based in VirSSPA software. VirSSPA works with medical images obtained with a surface scanner and virtual reality to provide surgeons a computer-aided diagnosis and surgical planning tool. The surgical planning is personalized for each patient, decreasing the time spent and potential risks during the surgery..

The quality of care is improved with this system, because surgeons had a decision support tool. Making decisions using virtual models allows to optimize the use of available information as well as promotes professional exchange of knowledge and facilitates sharing experiences in a collaborative space.

ACKNOWLEDGEMENTS

VirSSPA was developed at Virgen del Rocío, University Hospital in collaboration with Signal Process Department (Engineering School) from Seville. VirSSPA was financed by the Andalusian Department of Health, Spain. This project also is made under the framework of RETICS, Red ITEMAS funded by the Instituto de Salud Carlos III. It is identified with the code PI09/90518 and CIBER BBN. AYRA is the commercial version of VirSSPA.

REFERENCES

[1] W.D. James, T.G. Berger et al, "Andrews' Diseases of the Skin: clinical Dermatology", *Saunders* Elsevier, 2006, pp. 495.
[2] A. Garg, "Acquired and inherited lipodystrophies", *The New England Journal of Medicine,* 350 (12), 2004, pp.1220-1234.
[3] Fiorenza, C. G., Chou, S. H., & Mantzoros, C. S. (2010). Lipodystrophy: pathophysiology and advances in treatment. Nature Reviews Endocrinology, 7(3), 137-150.
[4] P. Gacto Sánchez et al, "A three-dimensional virtual reality model for limb re-construction in burned patients", *BURNS: The Journal of the International Society of Burn Injuries*, 2008, pp.
[5] T. Gómez-Cía et al, "The virtual reality tool VirSSPA in planning DIEP microsurgical breast reconstruction", *International Journal Radiology of Computer Assisted Radiology and Surgery*, Editor Springer Berlin, vol 4, nº 4, 2009.pp.
[6] P. Gacto-Sánchez et al, "Use of a three-dimensional virtual reality model for preoperative imaging in DIEP flap breast reconstruction" *Journal of Surgical Research,* vol. 162, nº1, 2010, pp.140-147.
[7] P. Gacto-Sánchez et al, "Computerised tomography angiography with VirSSPA 3D software for perforator navigation improves perioperative outcomes in DIEP flap breast reconstruction", *Journal of Plastic and Reconstructive Surgery*, vol. 125, 2010, pp. 24-31.
[8] Gonzalez and Woods "Digital Image Processing", 3rd Ed. Prentice Hall, 2008.
[9] C. Suárez-Mejías et al, "VirSSPA - A virtual reality tool for surgical planning workflow", *International Journal Radiology of Computer Assisted Radiology and Surgery*, Springer vol. 4, 2009, pp. 133-139.
[10] I. García Fenoll et al. "Algoritmo de Segmentación 3d Basado en Crecimiento de Regiones Por Tolerancia Adaptativa y Optimización de Contraste". *Actas del XXIII Symposium Nacional de la Unión Científica de Radio*. Ursi'09, 2009, pp. 51-51.
[11] E. W. Lorencsen et al, "Marching cubes: A HD 3D surface construction algorithm", *Computer graphics*, V.21, nº 4, 2004, pp. 163-169.
[12] Torres K, Staskiewicz G, Sniezynski M, Drop A, Maciejewski R: Application of rapid prototyping techniques for modelling of anatomical structures in medical training and education. Folia Morphol (Warsz) 2011, 70:1-4.
[13] Sanghera B, Amis A, McGurk M: Preliminary study of potential for rapid prototype and surface scanned radiotherapy facemask production technique. J Med Eng Technol 2002, 26:16-21.

Author: C. Suárez Mejías
Institute: Tech. Innovation Group, Virgen del Rocío University Hospital
Street: Manuel Siurot s/n
City: Seville
Country: SPAIN
Email: cristina.suarez.exts@juntadeandalucia.es

Comparison between Optical and MRI Trajectories in Stereotactic Neurosurgery

K. Wårdell[1], N. Haj-Hosseini[1], and S. Hemm[1,2]

[1] Department of Biomedical Engineering, Linköping University, Sweden
[2] Institute for Medical and Analytical Technologies, University of Applied Sciences and Art Northwestern Switzerland, Basel

Abstract—Deep brain stimulation (DBS) is an effective treatment for movement disorders e.g. Parkinson's disease. Thin electrodes are implanted into the deep brain structures by means of stereotactic technique and electrical stimulations are delivered to the brain tissue. Accuracy and safety during the implantation is important for optimal stimulation effect and minimization of bleedings. In addition to microelectrode recording and impedance measurements, intraoperative optical measurements using laser Doppler perfusion monitoring (LDPM) have previously been suggested as guidance tool during stereotactic DBS implantations.

In this study we compare optical trajectories, recorded with LDPM ranging from cortex towards the subthalamic nucleus (STN), to the corresponding magnetic resonance imaging (MRI) trajectories. Inversed gray scales from the T2-weighted MRI were used for comparison with the total light intensity (TLI) representing tissue grayness. Both curves followed a general tendency with a deep dip in the vicinity to the left ventricle. MRI trajectories might help in predicting the optical trajectory but further studies including more data and fine tuning of the comparative methodology are required.

Keywords—Magnetic resonance imaging, Laser Doppler perfusion monitoring, Deep brain stimulation, Stereotaxy.

I. INTRODUCTION

Brain stimulation is a common therapy for relief of movement disorders such as Parkinson's disease, dystonia and essential tremor [1, 2].Stereotactic neurosurgical procedures are used to implant DBS electrodes [3]. For optimal clinical outcome with minimal side effects the implantation should be done with high precision, accuracy and safety. In addition to preoperative stereotactic target identification in MR and CT images, intra-operative measurements are often used in order to find and confirm the target region. Microelectrode recordings (MER) are commonly used as intraoperative measurement technique. MER, however, increases the risk of bleedings [4] and methods with high ability to guide with reduced bleeding risk are therefore preferable. In recent years optical measurements have been introduced as an alternative intraoperative guidance technique [5-10].

Both laser Doppler perfusion monitoring (LDPM) [9, 10] and diffuse reflectance spectroscopy (DRS) [5-8] have proven to be useful in localization of in-vivo grey-white tissue boundaries by means of processing of the reflected light. During the surgical procedure, a thin probe with smoothly rounded tip and optical fibers along the shaft is carefully inserted, manually or with a hand driven mechanical device, along the precalculated trajectory. At the same time as the optical recordings are done in order to identify the tissue type, a tract is created for the DBS electrode to be implanted. Investigations with this method have so far been done along more than 100 trajectories used for stereotactic DBS implantation [11].

In a recent study typical optical trajectories towards two common DBS targets, the subthalamic nucleus (STN) and the venteromediate nucleus of the thalamus (Vim), have been defined by use of LDPM [10]. As MRI is the most common imaging tool in order to identify the target and to plan the trajectory in the preplanning process, it would be of interest to investigate if the preoperative MRI trajectory could be used as a predictor for the optical trajectory.

The aim of this study was therefore to set up a methodology for comparing optical trajectories with the preplanned MRI-intensity along the implantation path. For the investigation a typical DBS target was used, the STN.

II. MATERIAL AND METHODS

Optical data recorded with LDPM and processed to optical trajectories were used for comparison with MR images from a stereotactic DBS implantation procedure.

A. Laser Doppler Perfusion Monitoring

The system for intracerebral recordings of TLI comprises a laser Doppler perfusion monitor (PF 5010, Perimed AB, Sweden), a specially designed probe with four optical fibers connected with a 4 m long optical cable to a personal computer with software for data acquisition, analysis and presentation. The outer dimensions of the probe (l = 190 mm, ϕ_{shaft}=2.2 mm, ϕ_{tip}=1.5 mm) was designed to fit with

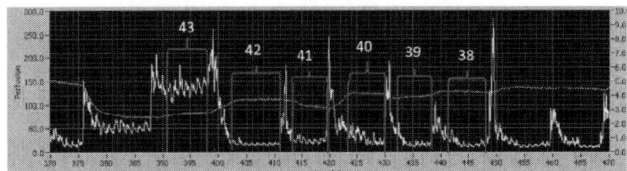

Fig. 1 Part of the TLI curve (blue) and corresponding perfusion values (yellow). The perfusion peaks were used as references for the analysis. Figures corresponds to the number of millimeters from the target.

the Leksell® Stereotactic System (LSS, Elekta Instrument AB, Sweden). Prior to surgery, the probe and the cable were sterilized according to the STERRAD® protocol.

B. Preoperative Planning, Imaging and Measurements

Stereotactic CT imaging (slice thickness 1 mm, GE Lightspeed Ultra, GE Healthcare, UK) was performed after placement of the LSS on the patients head. Direct anatomical targeting of the STN was done on T2-weighted MR images (slice thickness 2 mm, 1.5 Tesla, Philips Intera. The Netherlands). The trajectories to the targets were calculated by image fusion of the preoperative MRI and CT scans by using Leksell Surgiplan® (Elekta Instrument AB, Sweden). During surgery (Ethically approved, M182-04, T54-09) the optical probe connected to the LDPM system was inserted in 1 mm steps by means of a mechanical device designed to fit the LSS [10]. The TLI reflecting the tissue grayness was recorded at each mm along the bilateral trajectories from cortex towards the target in the STN.

C. Calculation of Optical Trajectory

At each measurement site the average TLI over a 5-10s recording was calculated (Fig. 1). Movement artefacts in the simultaneously recorded LDPM perfusion signal were used as reference for the analysis and made it possible to identify the respective analysis regions of the TLI signal.

The corresponding anatomy was identified with Surgiplan on pre- and postoperative CT-images co-registered to the Schaltenbrandt-Wahren atlas. The TLI values were normalized to the average of eight consecutive white matter measurements along the internal capsule (IC). The respective values were plotted as a curve and related to the anatomy [10]. In the curves the target point i.e. STN was set to 0 mm.

D. Calculation of MRI Trajectory

In order to obtain the intensities of the trajectories in the MR images, these were stacked to render a 3D image model. The Leksell® coordinates and angles used during surgery were transformed to image coordinates using the fiducials generated by the in the MR-fused CT images. MR trajectories were then obtained by extracting the intensities of the voxels with the coordinates of a line in the 3D space that extended from the target towards the entry point in the cortex (Fig. 2). The values one mm ahead of the corresponding TLI site were averaged to one measurement value and normalized by the corresponding white matter tract in the IC. The lengths of the trajectories were set equal to the distance of the measured optical trajectories.

E. Comparison between Optical and MRI Trajectories

As the coordinates of the extracted MR voxel intensity points have a lower resolution than the optical measurement points, the coordinates and the corresponding intensities were interpolated using 3D cubic function and then down sampled to provide a point at each mm distance from the target. The MR trajectories (white matter < gray matter intensity) were down-scaled and inversed for an easier visual comparison with the optical TLI values (white matter > gray matter intensity). For comparison, both the MRI and optical trajectories were plotted in the same diagram. Data analysis was done in MatLab (MathWorks™).

III. RESULTS

The calculated optical and MRI trajectories are presented in Fig. 3. On both sides MRI intensities follow the general tendancy of the optical measurements even if there seems to be a slight shift. TLI curves on both sides and the MRI intensity curve on the right side start with low values in the cortex. They increase when entering white matter. As seen in the MRI (Fig. 2), the left trajectory touches the ventricle. This is seen as a deep dip in the curves about 25-27 mm from the target. On the right, where the trajectory passes bidgre between the Caudate nucleus and the Putamen a less prononced dip can be seen. While on the optical trajectory the target area shows in general lower values compared to white matter, this difference can not be seen on MRI trajectories: the gray value level in the target area corrsponds to or is even higher than the white matter level.

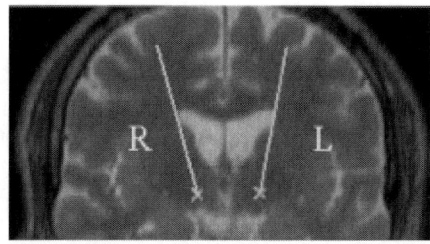

Fig. 2 T2-weighted MR image slice of the trajectories plane. The lines show the calculated MRI-trajectory and the markers show the target points.

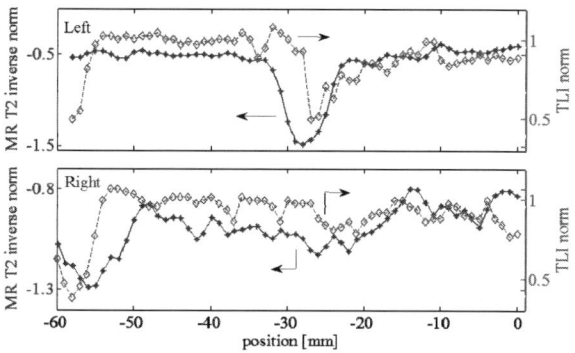

Fig. 3 Comparison of optical and MR voxel intensities. Each marker corresponds to the value at each mm along the trajectory, with STN at 0 mm.

IV. CONCLUSIONS

A methodology for prediction of optical trajectories by means of MRI prior to DBS implantation is suggested. It could be shown that the general tendency of both data curves corresponds but that values differ in the target area compared to white matter. A possible reason can be the resolution. While the optical measurements have a resolution around 1 mm, the MR was recorded with 2 mm thickness and therefore interpolated before presented. Other potential sources of error that may contribute to the deviation may be in- homogeneities in the magnetic field along the trajectory, voxels containing a mixture of two or more tissue types or brain shift especially for the second trajectory (here the right one). The image fusion between non-stereotactic MRI and stereotactic CT can be another source of error due to the necessary interpolations. In order to use MR images as predictor for optical measurements the methodology needs further fine tuning and evaluation on more data. In this study the artefact arising in the perfusion signal during the probe insertion was only used as reference for the data analysis. Simultaneous measurements of microvascular perfusion and TLI have previously been presented [10]. As the purpose of this study was to compare the TLI with the MRI-trajectory no additional comparison with the perfusion signal was done. In the future is could be interesting to study the relation between all three trajectory features.

ACKNOWLEDEGMENTS

The authors would like to thank Ali Aghajani for appreciated work with the MR trajectory software code and the Staff at the Neurosurgical Department, Linköping University Hospital for help with the data acquisition.

REFERENCES

1. A. L. Benabid, S. Chabardes, J. Mitrofanis, and P. Pollak, "Deep brain stimulation of the subthalamic nucleus for the treatment of Parkinson's disease," *Lancet Neurol*, vol. 8, pp. 67-81, Jan 2009.
2. M. D. Johnson, H. H. Lim, T. I. Netoff, A. T. Connolly, N. Johnson, A. Roy*, et al.*, "Neuromodulation for brain disorders: challenges and opportunities," *IEEE Trans Biomed Eng*, vol. 60, pp. 610-24, Mar 2013.
3. S. Hemm and K. Wårdell, "Stereotactic implantation of deep brain stimulation electrodes: a review of technical systems, methods and emerging tools," *Med Biol Eng Comput*, vol. 48, pp. 611-24, Jul 2010.
4. L. Zrinzo, T. Foltynie, P. Limousin, and M. I. Hariz, "Reducing hemorrhagic complications in functional neurosurgery: a large case series and systematic literature review," *J Neurosurg*, vol. 116, pp. 84-94, Jan 2012.
5. C. A. Giller, H. L. Liu, P. Gurnani, S. Victor, U. Yasdani, and D. C. German, "Validation of a near-infrared probe for detection of thin intracranial white matter structures," *J Neurosurg*, vol. 98, pp. 1299-1306, 2003.
6. C. A. Giller, H. Liu, D. C. German, D. Kashyap, and R. B. Dewey, "A stereotactic near-infrared probe for localization during functional neurosurgical procedures: further experience," *J Neurosurg*, vol. 110, pp. 263-73, Feb 2009.
7. J. Antonsson, O. Eriksson, P. Blomstedt, A. T. Bergenheim, I. H. M, J. Richter, P. Zsigmond and K. Wårdell, "Diffuse reflectance spectroscopy measurements for tissue-type discrimination during deep brain stimulation," *J Neural Eng*, vol. 5, pp. 185-90, Jun 2008.
8. J. D. Johansson, P. Blomstedt, N. Haj-Hosseini, A. T. Bergenheim, O. Eriksson, and K. Wårdell, "Combined Diffuse Light Reflectance and Electrical Impedance Measurements as a Navigation Aid in Deep Brain Surgery," *Stereotact Funct Neurosurg*, vol. 87, pp. 105-113, Feb 18 2009.
9. K. Wårdell, P. Blomstedt, J. Richter, J. Antonsson, O. Eriksson, P. Zsigmond, A. T. Bergenheim and M.I. Hariz. "Intracerebral microvascular measurements during deep brain stimulation implantation using laser Doppler perfusion monitoring," *Stereotact Funct Neurosurg*, vol. 85, pp. 279-86, 2007.
10. K. Wårdell, P. Zsigmond, J. Richter, and S. Hemm, "Relationship between laser Doppler signals and anatomy during deep brain stimulation electrode implantation toward the ventral intermediate nucleus and subthalamic nucleus," *Neurosurgery*, vol. 72, pp. ons127-40, Jun 2013.
11. K. Wårdell, "Experience from optical measurements during 100 DBS implantations," presented at the XXth Congress of the European Society for Stereotactic and Functional Neurosurgery, Cascais/Lisbon, Portugal, 2012.

Author: Karin Wårdell
Institute: Department of Biomedical Engineering, Linköping University
City: Linköping
Country: Sweden
Email: karin.wardell@liu.se

EVA: Endoscopic Video Analysis of the Surgical Scene for the Assessment of MIS Psychomotor Skills

I. Oropesa[1,2], P. Sánchez-González[1,2], J.A. Sánchez-Margallo[3], J. García-Novoa[1,2], F.M. Sánchez-Margallo[3], E.J. Gómez[1,2]

[1] Bioengineering and Telemedicine Centre (GBT), ETSI Telecomunicación, Universidad Politécnica de Madrid UPM), Madrid, Spain
[2] Networking Research Center on Bioengineering, Biomaterials and Nanomedicine (CIBER-BBN), Zaragoza, Spain
[3] Jesús Usón Minimally Invasive Surgery Centre (JUMISC), Cáceres, Spain

Abstract—The present work covers the first validation efforts of the EVA Tracking System for the assessment of minimally invasive surgery (MIS) psychomotor skills. Instrument movements were recorded for 42 surgeons (4 expert, 22 residents, 16 novice medical students) and analyzed for a box trainer peg transfer task. Construct validation was established for 7/9 motion analysis parameters (MAPs). Concurrent validation was determined for 8/9 MAPs against the TrEndo Tracking System. Finally, automatic determination of surgical proficiency based on the MAPs was sought by 3 different approaches to supervised classification (LDA, SVM, ANFIS), with accuracy results of 61.9%, 83.3% and 80.9% respectively. Results not only reflect on the validation of EVA for skills' assessment, but also on the relevance of motion analysis of instruments in the determination of surgical competence.

Keywords—MIS, competence, assessment, EVA, TrEndo.

I. INTRODUCTION

Skills' acquisition in minimally invasive surgery (MIS) is gradually adapting from the mentor-apprentice-based Halstedian paradigm towards structured, objective training and assessment programs, where direct involvement of residents in real surgeries is delayed until becoming proficient in the required skills. Several motivators can be identified behind this: media and public awareness towards medical errors, the need to reduce costs in hospitals, or the overloaded schedules of surgeons [1].

In this context, the first stages of basic psychomotor training take place in controlled laboratory settings by means of box trainers and virtual reality simulators [2]. The incorporation of tracking technologies allows them to capture data on the laparoscopic instruments' movements when performing an exercise. This data, when handled properly, may yield a series of motion analysis parameters (MAPs) providing useful, objective information on performance [3].

Tracking technologies for box trainers and VR simulators have traditionally relied on sensor-based systems, based on optical, electromagnetic or mechanic technologies [4]. However, their use may modify the ergonomics and constrain movements of the instruments, altering the users' experience and performance. Moreover, transfer of these technologies for training and assessment of skills in the OR is compromised as they are often bulky, are not easily sterilized, may require a clear line of sight (optical devices) or be affected by ferromagnetic materials (electromagnetic devices), and thus present errors in position tracking [5].

The present work covers the first validation efforts of the EVA Tracking System for the assessment of MIS psychomotor skills. In its current implementation, the tracking algorithm allows offline determination of the 3D position of the laparoscopic instruments with respect to the endoscope, based solely on their physical and geometrical characteristics. The paper will briefly describe the working principles of the algorithm, as well as some of the findings of this technology when applied to a real training scenario.

II. MATERIALS AND METHODS

A. Working Principles

Tracking of the instruments in EVA works under the following assumptions:

- Performing color-based segmentation and applying an edge filter, instruments may be isolated from the video frame. Ideally, a clean image showing the instrument edges yields two neighboring peaks in Hough space. Detecting these maxima provides information regarding borders' position on the screen [6].
- In case of there being two instruments, it is typical in a box trainer setting that they will have opposing entrance angles. In this case, a division of Hough space for positive and negative values of theta can isolate detection of the left and right instruments respectively.
- Detection of the instrument's tip in the screen can be performed by color-based gradient analysis along its bisecting line (obtained from borders' information) [7].
- 3D reconstruction of the tip and orientation information is obtained by means of the geometrical properties of the instrument [8].

B. Data Acquisition and Preparation

A series of experiments were conducted to validate the use of EVA for the assessment of MIS psychomotor skills.

Forty-two participants (4 expert surgeons, 22 residents, 16 novice medical students) performed a box trainer task at the skillslab of the Leiden University Medical Centre (LUMC, Leiden, The Netherlands), where the goal was to place a number of chickpeas into different target holes using one laparoscopic grasper. Video recordings of performance were taken from the laparoscopic feed for offline analysis with EVA. Simultaneously, instrument movements were recorded with an optical sensor-based tracking system developed at the Delft University of Technology (TUDELFT, Delft, The Netherlands), the TrEndo [9]. Nine MAPs were recorded using both tracking systems (Table 1).

Table 1 MAPs definition.

MAP	Definition
Time	Total time to perform a task (s).
Idle time	% of time where the instrument is considered to be still (speed<5mm/s).
Path length	Total path covered by the instrument (m).
Depth	Total path covered in the instrument's axis direction (m).
Average speed	Rate of change of the instrument's position (mm/s).
Average acceleration	Rate of change of the instrument's velocity (mm/s^2).
Motion smoothness	Jerky movements caused by abrupt changes in acceleration (m/s^3).
Economy of area	Ratio between maximum area covered by the instrument on the task surface and path length (-).
Economy of volume	Ratio between maximum volume covered by the instrument on the setting and path length (-).

Since the purpose of the experiment was to validate the task for evaluation purposes, no prior trials were allowed. A brief explanation of the task's objectives was given to participants to let them infer, based on their own experience and skills, the best strategy to perform it.

C. Construct and Concurrent Validation

Construct validation was performed along the three experience groups (Kruskal-Wallis analysis) and in pairs (Mann-Whitney test) to measure statistical significant differences between MAPs. Differences were considered significant at $p<0.05$.

Concurrent validation was performed between EVA and TrEndo. Pearson's correlation (ρ) was employed to measure the degree of concurrence. Values between 0.4-0.7 were considered as medium correlating values, whilst values >0.7 showed strong correlation between MAPs [10].

D. Supervised Classification of Performance

Three supervised classification techniques were applied to discern whether surgical experience may be derived from performance in box trainer tasks. Participants were arranged according to the number of surgeries performed: more than 10 (Experienced, Ex: 4 experts + 14 residents) and less than 10 (Non-experienced, NEx: 16 students + 8 residents). Based on the 9 MAPs, classifiers yielded a binary output according to performance: {S: skilled, NS: not skilled}.

MAPs were normalized and principal component analysis (PCA) applied to reduce the number of input dimensions. Three different classifiers were tested: linear discriminant analysis (LDA), support vector machines (SVM) and artificial neuro fuzzy inference systems (ANFIS).

Leave one out cross validation was performed, in which data from all subjects was employed to train the classifiers except for one, who was used for validation. The process is repeated until each subject has been used both for training and testing. The following parameters were sought:

- *Accuracy:* % of subjects correctly classified according to the input categories.
- *Sensitivity (S):* % of Ex classified as S.
- *Specificity (E):* % of NEx classified as NS.
- *RMSE:* Main error measurement between expected (Ex, NEx) and predicted (S, NS) values per classifier.

Additionally, receiver operator curves (ROC) were obtained, plotting specificity vs. sensitivity based on the posterior probability of each classifier.

Significant differences between the three classifiers were sought by means of Cochran's Q test ($p<0.05$) to determine the degree with which each classifier is coherent with the rest when evaluating a participant. For a more intuitive representation of this idea, classifier plots were made for each technique, in order to show the expected and predicted values, as well as their posterior probability.

Finally, significant differences between classification results obtained by using TrEndo and EVA for data acquisition were measured by means of McNemar's test ($p<0.05$).

III. RESULTS

A. Construct and Concurrent Validation

Figure 1 graphically reflects results for construct and concurrent validation. Construct validation was obtained along the three groups for time, path length, depth, average speed, average acceleration and economy of area/volume.

Paired comparisons showed significant differences for all valid MAPs between students and residents/experts except for average speed, where differences occurred only between novice medical students and residents.

For concurrent validation, all MAPs except motion smoothness obtained values of $\rho>0.7$, reflecting a strong correlation between values obtained with TrEndo and EVA.

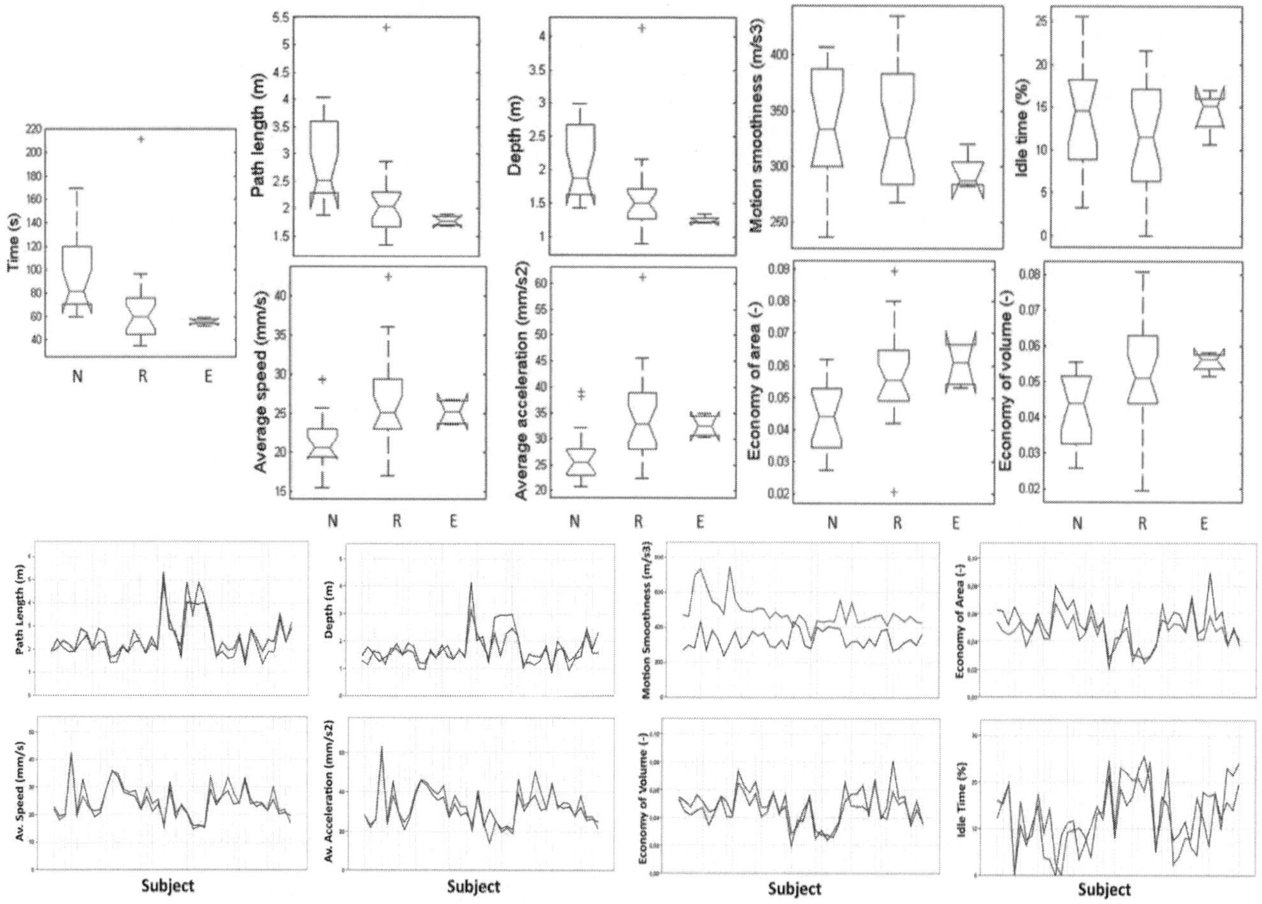

Fig. 1 Top: Construct validation: Results are expressed as notched box diagrams, in which every box distinguishes lower, median and upper quartile value. Significance is shown where the notched sections of the boxes do not overlap each other. N: Novice medical students; R: Resident; E: Experts. Bottom: Concurrent validation scores per subject, in order of participation (x-axis). Black: Scores obtained by EVA. Red: Scores obtained by TrEndo.

B. Classification Results

SVM and ANFIS showed the highest accuracy rates (Fig.2; Table 2) at 83.3% and 80.9% respectively, and a better identification of NEx subjects (91.7% and 87.5% respectively). On the other hand, the three classifiers performed similarly on classification of Ex surgeons (72.2%). Differences according to Cochran's test were considered significant (p = 0.02). Visual inspection of classifiers' plots (Fig. 3) and accuracy results point at the higher number of misclassifications for LDA as possible cause. McNemar test for each classifier revealed no significant differences between using the EVA and TrEndo data.

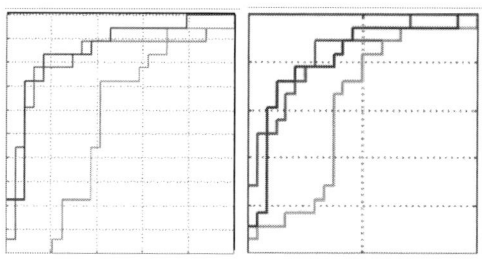

Fig. 2 ROC curves. Left: EVA. Right: TrEndo. Green: LDA; Blue: SVM; Red: ANFIS. X-axis: (1-E), Y-axis: S.

Table 2 EVA classification results for the proposed classifiers (boldface). In parenthesis, homologous results obtained by TrEndo are given.

Classifier	Accuracy	RMSE	S	E
LDA	**61.9** (61.9)	**0.62** (0.62)	**72.2** (66.7)	**54.2** (58.3)
SVM	**83.3** (76.2)	**0.41** (0.48)	**72.2** (72.2)	**91.7** (79.2)
ANFIS	**80.9** (76.2)	**0.44** (0.48)	**72.2** (66.7)	**87.5** (83.3)

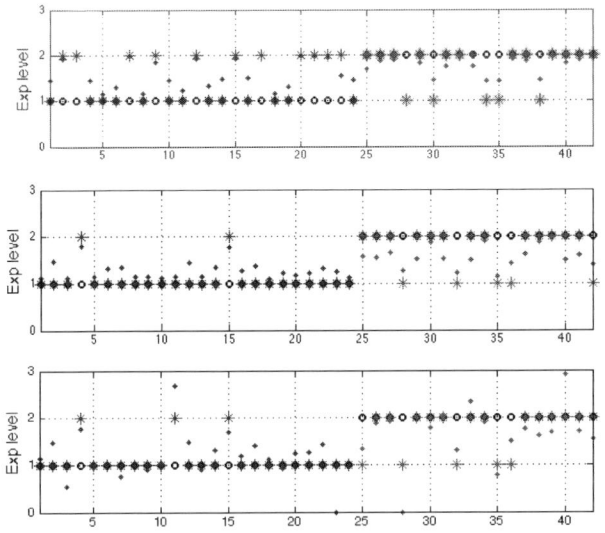

Fig. 3 Classifier plots: Top to bottom: LDA, SVM, ANFIS. Default classes (NEx=1; Ex=2) are shown as circles (o). Output levels (NS=1; S=2) are defined. Probabilistic outputs are shown as dots (•). A threshold for classification is set at 1.5, results given as asterisks (*). Blue: NEx; Red: Ex

IV. DISCUSSION

This study covers the first validation efforts of the EVA Tracking System applied to MIS psychomotor skills' assessment. Results not only give proof of its usefulness for this purpose, but serve as confirmation on the importance of motion analysis when determining surgical competence.

Construct validation was established for 7 out of 9 possible MAPs. To a greater or lesser extent most of them have been featured in the literature and validated for different tasks and abilities, and thus this study helps corroborate their relevance for assessment purposes [2]. More importantly, concurrent validation was established with data registered by the gold standard system employed, the TrEndo. In this sense, only motion smoothness did not present significant differences between systems. A possible reason for this may reside in the post-processing stage of EVA, which effectively applies low pass filtering of the signal that dampens the influence of movements' jerkiness.

The relevance of motion analysis manifests also in the correlation between experience and expertise detected by the supervised classifiers. While it is true that SVM and ANFIS performed better than LDA for this specific task, the former fact is more important at this point of our research rather than finding an optimal classifier (if indeed there is one). However, certain trends in the data (e.g.: subjects misclassified for 2 or 3 classifiers) suggest that other factors besides prior experience may be conditioning performance of the task, whether subject- (musical aptitudes, stress, etc.) or setting-related (box trainer, endoscope position, etc.).

V. CONCLUSIONS

The EVA Tracking System has been proven valid for assessment of MIS psychomotor skills. The current MATLAB implementation is being migrated to C++/OpenCV to increase processing speed and allow for real time tracking, and research continues in order to increase robustness of the segmentation stages. Our final goal is to achieve a system combining real time tracking and intelligent data analysis to provide immediate, relevant feedback on performance in a stress-free, patient-safe environment.

ACKNOWLEDGMENTS

Authors kindly thank the Minimally Invasive Surgery and Interventional Techniques group (TUDELFT) for their support with TrEndo, as well as the LUMC staff for their support during the validation experiments. Authors participate under partial funding of LLP-Leonardo da Vinci project MISTELA (528125-LLP-1-2012-1-UK).

REFERENCES

1. Tsuda S, Scott D, Doyle J et al. (2009) Surgical Skills Training and Simulation. Curr Prob Surg 46: 271-370.
2. Oropesa I, Sanchez-Gonzalez P, Lamata P et al. (2011) Methods and Tools for Objective Assessment of Psychomotor Skills in Laparoscopic Surgery. J Surg Res 171: e81-e95.
3. van Hove PD, Tuijthof GJM, Verdaasdonk EGG et al. (2010) Objective assessment of technical surgical skills. Brit J Surg 97: 972-987.
4. Chmarra MK, Grimbergen CA, Dankelman J (2007) Systems for tracking minimally invasive surgical instruments. Minim Invasive Ther Allied Technol 16: 328-340
5. Peters TM (2006) Image-guidance for surgical procedures. Phys Med Biol 51(14): R505-40
6. Oropesa I, Sánchez-Margallo JA, Sánchez-González P et al. (2012) Comparative study of two video-based laparoscopic instrument tracking algorithms for image guided surgical applications. Proc. II Joint Workshop New Technologies for Computer/Robot Assisted Surgery.
7. Allen BF, Kasper F, Nataneli G et al. (2011) Visual tracking of laparoscopic instruments in standard training environments. Stud Health Technol Inform 163: 11-17.
8. Cano AM, Lamata P, Gayá et al. (2006) New methods for video-based tracking of laparoscopic tools. In: Harders M, Székely G (eds) ISBMS 2006, LNCS, vol 4072. Springer, Heidelberg, 142–149.

9. Chmarra MK, Bakker NH, Grimbergen CA et al. (2006) TrEndo, a device for tracking minimally invasive surgical instruments in training setups. Sens Actuat: A-Physical 126: 328-334.
10. Aggarwal R, Grantcharov T, Moorthy K et al. (2007) An evaluation of the feasibility, validity, and reliability of laparoscopic skills assessment in the operating room. Ann Surg 245: 992-999.

Corresponding author:
Author: Ignacio Oropesa
Institute: Biomedical Engineering and Telemedicine Centre, ETSI Telecomunicación, Universidad Politécnica de Madrid
Street: Avda. Complutense, 30
City: Madrid
Country: Spain
Email: ioropesa@gbt.tfo.upm.es

Quantitative Evaluation of a Real-Time Non-rigid Registration of a Parametric Model of the Aorta for a VR-Based Catheterization Guidance System

P. Fontanilla-Arranz[1], B. Rodriguez-Vila[1], H. Fontenelle[2], J. Tarjuelo-Gutiérrez[1], O.J. Elle[2,3], and E.J. Gómez[1]

[1] Bioengineering and Telemedicine Centre, ETSI de Telecomunicación, Universidad Politécnica de Madrid, Madrid, España
{pfontanilla,brvila,egomez}@gbt.tfo.upm.es
[2] The Intervention Center, Oslo University Hospital, Oslo, Norway
hugues.fontenelle@rr-research.no
[3] Department of Informatics, University of Oslo, Oslo, Norway
oelle@ous-hf.no

Abstract—**This work presents a real time non-rigid registration method for a parametric model of the aorta based on computer animation techniques. The endovascular system is modeled as a polygonal mesh and a skeletal structure, both calculated automatically from segmented pre-operative images. The skeleton, registered using linear regression methods, supports an adaptive triangular mesh, both adapting in real time to the longitudinal deformations and changes affecting the aorta during the procedure. These changes are offered to the surgeon as part of a virtual-reality catheterization guidance system. The initial evaluation shows that the algorithm and implementation are able to handle up to 5 situation updates per second while maintaining a mean registration error of 1,012 mm and a maximum of 4.346 mm, thus effectively supporting real-time navigation.**

Keywords—**Catheterization, navigation, virtual-reality, non-rigid registration, real-time.**

I. INTRODUCTION

A catheter is medical devices that can be inserted in the body to treat diseases or perform a surgical procedure. Catheterization refers to the use of or insertion of a catheter into a body cavity, duct, or vessel for cardiovascular, urological, gastrointestinal, neurovascular, or ophthalmic applications. In this paper we focus on aortic catheterization, for example for the deployment of a stent, balloon or valve in the descending aorta, although the method is readily applicable for other procedures.

The goal of a catheterization guidance system is to provide the surgeon with a updated navigable 3D vision with the tools and anatomic structures present in an aortic catheterization: the catheter, the aorta and its branches.

A key component of these systems is the registration algorithm [1], which calculates the geometrical relation between the pre- and intraoperative images. The quality of the registration is fundamental for the correct positioning of the catheter as well as precise navigation.

There are two basic approaches to the registration problem: rigid registration algorithms, used for affine transformation in static structures, and non-rigid algorithms, used in deformable structures such as organic tissue [2].

Registration methods used in projects such as ARIS*ER [3] are of the rigid kind, and therefore limited to non-dynamic structures, such as the brain or the bones.

However, the aorta suffers two classes of motions and deformations during a catheterization, due to the heart beats, respiratory movements and the catheter insertion itself. These are longitudinal (modifying the medial line of the artery) and transversal distortions, mainly due to an asymmetric pulsatile distension of the artery.

All these changes must be reflected in the aorta model by means of a complete registration, which provides a reliable, up-to-date image of the current vessel configuration. In this work a non-rigid, real-time approach to the modeling of the longitudinal deformations is presented.

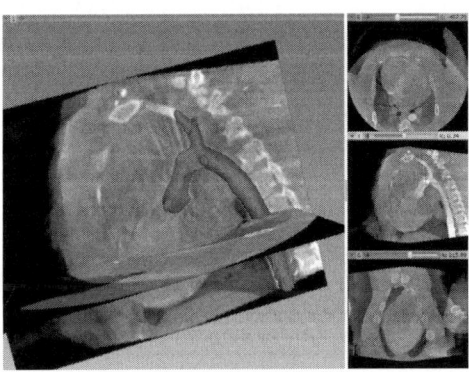

Fig. 1 Axial, sagittal and coronal views of the DynaCT of the pig, and 3D model of the segmented lumen.

II. MATERIALS

This research work is integrated into the SCATh platform, which is a virtual reality (VR) based platform aimed to minimize the ionizing radiation during the catheter navigation in aortic catheterization.

The images used for the evaluation of the method were obtained in the angio-room at the Intervention Center, Oslo University Hospital, Oslo, Norway during different stages of a vascular intervention using a porcine model. Three different

DynaCT (cone-beam computed tomography) with contrast, acquired with breath-hold in mid-phase, were selected. The lumen of the aorta was segmented in each DynaCT using the open-source Slicer 4 software [5] (see Figure 1).

III. METHODOLOGY

From the patient pre-operative images to the deformable model, the proposed methodology follows the steps depicted in Figure 2:

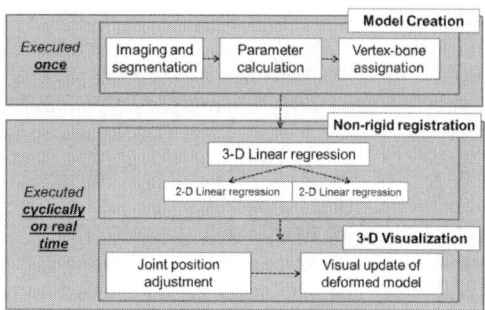

Fig. 2 Workflow, from MRI to deformed model

A. Model Creation

The aorta model is based on the computer animation technique named skeletal animation [6], normally used to represent vertebrate characters. The model is composed by two parts: the *skin*, a polygonal surface which represents the exterior of the model; and the *skeleton*, a set of straight lines (bones) which approximate the centerline of the volume. Figure 3 presents an example of the skeleton (blue line) and the mesh (pink surface) of one of the models used.

The model is loaded in a graphical engine [7], which will be in charge of rendering it and executing the deformations and changes that take place during the intervention. This animation is only valid to represent longitudinal changes in the aorta, which are the ones modeled.

The following characteristics of skeletal animation have direct influence in the registration algorithm:

- The tessellated surface (skin) is attached to the skeleton in a process called skinning. Each vertex of the polygon set is assigned to one or more bones. This assignation is then used by the graphical engine to know which vertices should be affected when the skeleton is modified.
- The skeleton is a hierarchical chain of straight segments, called bones, calculated from the centerline of the segmented images. The point where two bones are linked is called *joint*.

The full process of the model creation is explained in detail in [8].

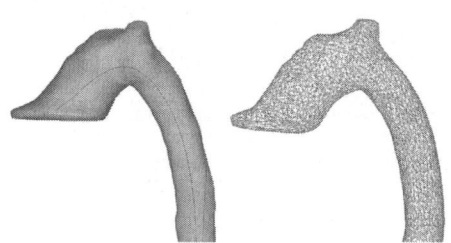

Fig. 3 Left: full model with skeleton in blue. Right: Mesh of the model.

B. Non-rigid Registration

The objective of the non-rigid registration process is to adjust the model configuration to the intraoperative information received during the procedure.

The intraoperative aorta centerlines are sent to the registration algorithm as series of 3-D points, samples of the interpolated curve. The full skeleton is updated by iteratively finding the position for each bone that best fits the provided discrete curve.

The registration process is composed of two phases: (1) determining the samples of the intraoperative curve that must be fitted by each bone and (2) finding the best position for that bone.

1. Bone-sample assignation

The link joining each bone of the model's skeleton to his *parent* (the joint) is fixed, and should the parent bone move, the child bone is forced to move along to maintain the union. In addition, the length of the bones must not be changed during the model manipulation, or the skin will not be properly updated afterwards, causing visual glitches. Therefore, as one end of the bone is fixed by its father and the length is constant, the only task needed to adjust the whole bone position is to find the optimal location for the other end.

Each bone will be able to fit a certain amount of samples, limited again by its length and its link to the parent joint. During the model creation, the bones are created as long as possible, but always under a user-set maximum length parameter. Therefore, in this step the bone is placed so that the maximum number of samples fit under it. Once this set of points is determined, the optimal location for the bone can be calculated in the next step.

2. 3-D Linear Regression

The bone-samples set is, under certain restrictions, akin to a set of discrete points and a (limited) straight line. From a mathematical standpoint, the problem of fitting the bone to the samples can be approached as a 3-D linear regression.

The presented modified algorithm is performed as two 2-D linear regressions in cascade. On broad terms, it involves the following: first obtaining the 2-D linear fitting of 2 dimensions of the data set (X and Y), and then using the projection of the solution line and the remaining coordinate (Z) to perform another 2-D linear fitting. This will provide the director cosines of the straight line that is oriented as the fitted bone should be. Measuring from the proximal end (a point found as the distal end of the previous bone, by the same procedure) a distance equal to the current bone length, the final position for the distal end of the current bone is set.

As a result, the bone is fitted, providing in turn one fixed point for the next bone to be repositioned. This process is performed sequentially for each bone, eventually updating the whole skeleton. Once the new position for every bone has been found, the graphical engine must be informed so it can apply the appropriate deformation to the skin: the positions of the joints are set to the new ones, and the engine is ordered to update the model. This calculates new positions for all the vertices, providing the updated aspect of the model which will follow the registered skeleton.

The Hausdorff distance, a standard metric for evaluating distances between meshes [9], has been used to evaluate the non-rigid registration performance.

IV. RESULTS AND DISCUSSION

For this assessment, an offline registration of the porcine aorta models was implemented. The 3D models shown in Figure 4 represent different moments of the same intervention, and therefore different states of deformation. Each of these models was registered against the other two, providing six registrations in total. For each model, its centerline was extracted, and then used as the target centerline for the other two models. In this way, it is tested that a model of any state of the patient aorta is deformable to represent other states that the vessel will undergo during the intervention.

Fig. 4 Models of 3 different stages of the porcine aorta during the intervention.

One of the registrations is shown in Figure 5. The "patient-specific" preoperative 3D model of the aorta is shown using two orthogonal views: from the right side (left column) and from the anterior side (right column).The upper row shows the 3D model in its original shape, while the lower row presents the deformed model using the new registered skeleton. The old centerline (pink spheres) and the new centerline (yellow spheres) are shown to facilitate the visualization of the changes introduced during the procedure.

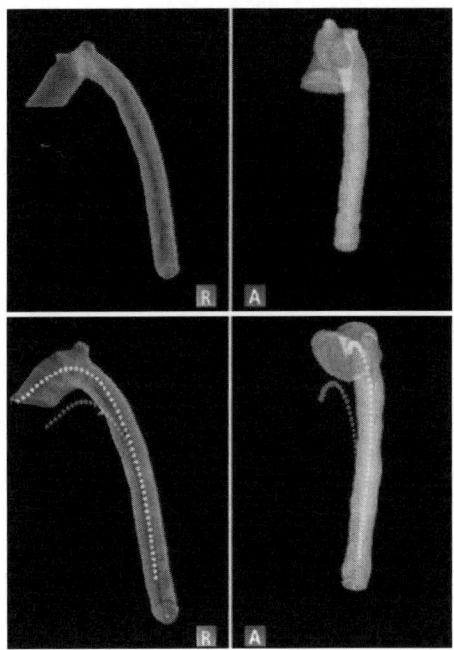

Fig. 5 Top, preoperative model before registration. Bottom, preoperative (blue) model after being fitted to the centerline of the red model. Previous centerline shown in pink, registered centerline shown in yellow

For these models, composed of up to 50 bones (joints and links), and using centerlines divided in 300 samples, the registration takes less than 100 ms on a desktop computer, enough to be considered as real-time. However, currently, the new centerline is extracted from a DynaCT, an intraoperative image which takes 8 seconds. As a consequence, this centerline extraction remains the main bottleneck of the process for real-time purposes.

The results from the quantitative evaluation are shown in Table 1. The average mean error for the 6 cases is slightly over 1 mm, while the maximum error is around 5 mm in the worst case.

Table 1 Hausdorff distance values for the 6 registrations, in millimeters.

	Minimum	Maximum	Mean	Std dev
A to B	0,0021	5,3671	0,7698	1,0135
A to C	0,0031	3,3108	1,0155	0,6179
B to A	0,0037	2,7507	0,5138	0,4282
B to C	0,0004	5,1530	1,2096	0,7787
C to A	0,0062	4,6630	1,3581	0,9235
C to B	0,0048	4,8333	1,2072	0,8612
Average	0,0034	4,3463	1,0123	0,7705

Figure 5 represents these results, overlaying the error values over the 3D volumes. It seems to the authors that the registration values are below 1 mm except in certain regions: the lower part of the descending aorta and ascending aorta, and around the brachyocephalic artery. These errors may be due to the original segmentation differences, and excluding these regions the registration is excellent.

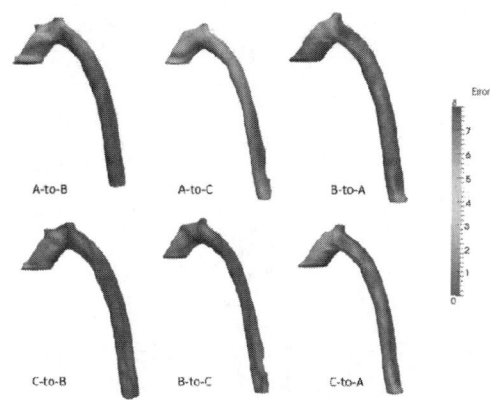

Fig. 6 Graphical representation of the Hausdorff distance over the 3D geometrical model

V. CONCLUSIONS

This work presents a non-rigid registration method suitable for real-time navigation. The quantitative evaluation of the method shows that the modeling of the longitudinal deformations is enough for a correct matching of simple anatomies like the aorta, obtaining a great improvement compared to the use of a rigid registration method.

This registration approach is integrated into the SCATh platform, a catheterization navigation platform based on virtual reality. The updated deformed state of the models allows the surgeon to know the exact position of the tracked catheter inside the vessel, potentially increasing the effectiveness and safety of the catheterization procedures.

Future works involve the centerline extraction using intraoperative information, such as intra-vascular imaging and real-time sensing. In addition, more complex models can be created, modeling the outward arteries with additional branches in the skeleton. We propose that this will reduce the registration error for bifurcation areas such as the top of the arch.

ACKNOWLEDGMENT

The authors gratefully acknowledge the financial support of the European Commission under the 7th Framework Programme in the frame of project SCATh, FP7-ICT-2009-4-248782. They also would like to thank A. Dore, S.Lin and V. Meiser and F.Gaya for their help and advice.

REFERENCES

1. Intracardiac echo-guided image integration: optimizing strategies for registration. T.S. Fhamy, H. Mlcochova, O.M. Wazni, et al. 2007, J Cardiovasc Electrophysiology, págs. 18:276-82.
2. J.B.A.Maintz, M. ,. (1998). A survey of medical image registration vol 2, Nº 1 . Medical Image Analysis, 1-37.
3. ARIS*ER consortium, project main page. Retrieved February 2013 from http://www.ariser.info/
4. Sette, M., E.V. Poorten, J. Vander Sloten, A. Dore, B. Rodriguez-Vila, H. Fontenelle, D.M. Pierce, G. Leo, M. Vatteroni and V. Meiser, Smart Catheterization Project, The Virtual Physiological Human Conference(VPH 2012), London, England, September 18{20, 2012.
5. Fedorov A., Beichel R., Kalpathy-Cramer J., Finet J., Fillion-Robin J-C., Pujol S., Bauer C., Jennings D., Fennessy F., Sonka M., Buatti J., Aylward S.R., Miller J.V., Pieper S., Kikinis R. 3D Slicer as an Image Computing Platform for the Quantitative Imaging Network. Magn Reson Imaging. 2012 Nov;30(9):1323-41. PMID: 22770690.
6. Real-time 3D Character Animation with Visual C++. Lever, N. (2002). Focal Press.
7. Irrlicht documentation. (s.f.). Retrieved 12th November 2011, from http://irrlicht.sourceforge.net/docu/
8. Automatic generation of real-time deformable parametric model of the aorta for a VR-based catheterism guidance system. P. Fontanilla Arranz, B. Rodríguez Vila, P. Sánchez-González,E.J.Gómez Aguilera, A. Dore, M. Sette, J. Vander Sloten. Int J CARS Proceedings (2012) 7 (Suppl 1): S 161-162
9. Mesh: Measuring errors between surfaces using the Hausdorff distance. Aspert, Nicolas, Diego Santa-Cruz, and Touradj Ebrahimi. Multimedia and Expo, 2002. ICME'02. Proceedings. 2002 IEEE International Conference on. Vol. 1. IEEE, 2002.

Transtibial Amputee Gait: Kinematics and Temporal-Spatial Analysis

A.E.K. Ferreira[1,2], E.B. Neves[1], A.G. Melanda[2], A.C. Pauleto[2], D.D. Iucksch[2], L.A.M. Knaut[2], R.M. da Silva[2], and R.F.M. da Cunha[2]

[1] Federal Technological University of Paraná, Graduate Program in Biomedical Engineering, Curitiba, Brazil
[2] Ana Carolina Moura Xavier Hospital Rehabilitation Center (CHR), Curitiba, Brazil

Abstract—Transtibial amputees gait patterns are widely studied. Usually, kinematic and temporal-spatial parameters data are used to investigate their gait pattern. The Gait Profile Score (GPS) and the Movement analysis Profile (MAP) are new tolls that summarize kinematics data in one single number. The aim of this study was to use GPS, Movement analysis Profile (MAP) and temporal-spatial parameters to quantify gait deviations of a homogeneous group of transtibial amputees, using the same prosthetic components and that were rehabilitated in a specific center. Besides, it was observed the correlation between GPS scores and temporal-spatial parameters. Five unilateral traumatic transtibial amputees participated on this study. All the participants used KBM (Kondylen Bettung Münster) prosthetic fitting and solid ankle cushion heel (SACH) foot. Kinematic and temporal-spatial data were assessed through 3D gait analysis. All analyzed variables presented deviations compared with normal expected values. Prosthetic limb GPS score was larger than intact limb GPS score as well as step length with the prosthetic leg was longer than with the intact one. Time of single support with the intact limb was longer than that with the prosthetic limb. The largest gait variable scores (GVS) were in the hip flexion/extension for the prosthetic limb, knee flexion/extension for the intact limb, and hip rotation for both. The strongest correlation occurred between overall GPS and prosthetic step length, overall GPS and time of single support with the prosthetic limb, prosthetic limb GPS and prosthetic step length, and between prosthetic limb GPS and time of single support with the prosthetic limb. The GPS, MAP and temporal-spatial parameters were useful in quantifying gait deviation on transtibial amputees. GPS scores were increased and temporal-spatial parameters values were lower than that found in health subjects.

Keywords—Gait Profile Score, transtibial amputees, kinematics, temporal-spatial parameters, gait deviation.

I. INTRODUCTION

Many studies used 3D gait analysis (3DGA) to investigate transtibial amputees' ambulation [1-3]. They described some kinematic and temporal-spatial deviations often found on this population, such as decreased self-selected walking velocity longest step length shorter single support duration with the prosthetic limb, increased hip and knee flexion during swing phase, low range of motion of the prosthetic ankle [1, 2, 4]. Some of them compared types of prosthetic feet and sockets [3, 5]; others did not considered prosthetic components on their analysis [1, 6]. All of them tried to find some deviation pattern for this group. Findings present consistent results that make it possible to understand the strategies developed by these patients during walking [7]. Methodological inconsistence of the studies and diversity of gait parameters used to describe their ambulation make it difficult to distinguish gait patterns usually adopted for transtibial amputees [4, 7].

New tolls that can be used to analyze transtibial amputees gait are the Movement Analysis Profile (MAP) and the Gait Profile Score (GPS)[6, 8]. GPS is a gait summary measure that is obtained by the calculation of the root mean square (RMS) difference between subject's kinematic data and data from a person with no gait pathology [9]. GPS summarizes gait kinematic data to a single number, measured in degrees. It helps clinicians to understand quickly the magnitude of kinematic problems. First the RMS of nine kinematic variables (pelvic tilt, pelvic obliquity, pelvic rotation, hip flexion/extension, hip abduction/adduction, hip rotation, knee flexion/extension, ankle dorsi/plantar flexion and foot progression) are calculated for left and right sides. Each value is called Gait Variable Score (GVS) for one specific kinematic variable. The GVS for these nine kinematic variables for the right and left legs are combined to form a bar chart, called MAP [10]. It can be used to highlight where patients have specific gait problems. Left and right GPS scores are the RMS average of the nine GVS for the right and left sides. An overall GPS score is obtained by the RMS average of all GVS. An increased value of GVS and GPS, compared to people without gait pathology, indicates more gait deviation [9, 10].

One group of researchers studied the use of the MAP and GPS with lower limb amputees [6, 8]. On the first study they tested the ability of these tolls to detect asymmetries and differentiate between two levels of amputation [8]. On the other one they assessed the suitability of gait summary measures for use with this population [6]. Both studies concluded that MAP and GPS can be applied to quantify and identify gait deviations among lower limb amputees [6, 8]. The authors included in their sample transtibial and transfemoral amputees with vascular and trauma etiology of amputation. Besides, they did not controlled prosthetic components and rehabilitation process.

The aim of this study was to use GPS, MAP and temporal-spatial parameters to quantify gait deviations of a homogeneous group of transtibial amputees, which used the same prosthetic components and which were rehabilitated in a specific center.

II. MATERIALS AND METHODS

A. Subjects

Five unilateral traumatic transtibial amputees (four men and one woman) participated on this study. They were recruited from a group of patients that were rehabilitated at Ana Carolina Moura Xavier Hospital Rehabilitation Center (CHR), Curitiba, Brazil, to ensure that all of them received the same treatment in both preprosthetic and post-prosthetic stages. Then, they provided their informed consent. All the participants used KBM (Kondylen Bettung Münster) prosthetic fitting and solid ankle cushion heel (SACH) foot. Exclusion criteria were amputation in upper limbs or another lower limb, less than three months consistent prosthesis use, use of ambulation aids, other cause of amputation and muscular, neurological or/and circulatory diseases that affect gait pattern.

B. Procedure

3DGA was captured in the Gait Laboratory at CHR. Before capturing gait data, patients' anthropometric data, range of motion and muscular strength were measured. Individuals wore a pair of shorts and a vest and reflexive markers (20-mm diameter) were placed on anatomical and prosthetic corresponding landmarkers according to the Helen Hayes marker set. Kinematic data was captured by 6 cameras Hawks (Motion Analysis Corporation, Santa Rosa, CA) at 60 Hz. First, subjects were placed in the center of the walkway and a static trial was collected. Then, medium markers from the knees and the ankles were removed. Patients walked across a 10-m walk-way at their self-selected speed. At least 10 trials for each subject were collected.

C. Data Analysis

Trials collected for each subject were edited with the software Cortex 1.1.4.368 (Motion Analysis Corporation, Santa Rosa, CA) and the one which represented better patient's gait pattern was chosen to the analysis. Ortrotrak 6.5.1 software (Motion Analysis Corporation, Santa Rosa, CA) was used to calculate kinematics data and the six most representative trials were used to calculate GPS (Gait Profile Score), which were calculated according to the authors [9]. All correlations were performed with SPSS version 20.0 and the level of significance set at $p < 0.05$.

III. RESULTS

The sample consisted of one woman and four men, with mean age of 46.2 ($\pm$ 6.94) years. All of them lost the limb because of a trauma, four due to motor vehicle accidents and one due to an accident during sporting practice. The average time since amputation was 8.6 ($\pm$ 8.2) years.

The Table 1 and Table 2 show the results of GPS, GVS and the temporal-spatial parameters of prosthetic and intact limbs.

Table 1 Average and Standard Deviation (SD) for GPS scores and Temporal-spatial parameters

Amputees characteristics	Average $\pm$ SD
Overall GPS	9,59 $\pm$ 1,38
Prosthetic limb GPS	9,32 $\pm$ 2,05
Intact limb GPS	9,02 $\pm$ 1,00
Velocity (cm/s)	89,66 $\pm$ 16,16
Cadence (steps/min)	91,55 $\pm$ 8,25
Step Width (cm)	16,43 $\pm$ 3,03
Prosthetic limb step length (cm)	61,04 $\pm$ 9,19
Intact limb step length (cm)	55,8 $\pm$ 6,7
Prosthetic limb Single Support (% cycle)	31,24 $\pm$ 2,13
Intact limb Single Support (% cycle)	37,8 $\pm$ 1,55

Table 2 Average and Standard Deviation of GVS scores for the nine kinematic variables

	Prosthetic limb	Intact limb
Pelvic Tilt	6.72 $\pm$ 4.01	6.72 $\pm$ 4.01
Hip flexion/extension	10.26 $\pm$ 4.80	9.57 $\pm$ 4.36
Knee flexion/extension	8.24 $\pm$ 3.64	10.54 $\pm$ 2.44
Ankle dors/plant flexion	9.31 $\pm$ 1.12	8.61 $\pm$ 1.91
Pelvic obliquity	4.47 $\pm$ 2.66	4.47 $\pm$ 2.66
Hip abduction/adduction	6.5 $\pm$ 3.15	5.97 $\pm$ 2.99
Pelvic rotation	8.97 $\pm$ 2.21	8.97 $\pm$ 2.21
Hip rotation	14.54 $\pm$ 3.88	10.26 $\pm$ 6.43
Foot progression	8.46 $\pm$ 2.92	9.45 $\pm$ 2.86

Table 3 shows how many normal GVS scores each subject got, its velocity and step length for the prosthetic and intact limbs. Individual D got 7 normal GVS scores and had better values for velocity (114.90 cm/s) and step length (74.10 and 67.27 cm/s) than the others.

Table 3 Number of normal GVS scores, velocity and step length for the five subjects

Subject	Numbers of normal GVS	Velocity (cm/s)	Step Length (cm)	
			Prosthetic limb	Intact limb
A	2	79.20	65.50	52.60
B	1	72.60	49.74	49.91
C	2	89.21	58.08	55.01
D	7	114.90	74.10	67.27
E	0	92.40	57.79	54.22

Correlation coefficients (Spearman's rho) between GPS scores and temporal-spatial parameters and their statistical significance (*p*-value) are summarized in Table 4. Overall GPS was moderately correlated ($r = -0.50$) with velocity and single support for the intact limb. Moreover, GPS correlations with step length and single support for the prosthetic limb were strong ($r = -0.70$). GPS score with the prosthetic limb was strongly correlated ($r = -0.70$) with step length and single support for the prosthetic side and moderately correlated ($r = 0.50$) with single support for the intact leg. Correlations didn't show significance for the sample size of this present study.

Table 4 Spearman's correlation (rho) and statistical significance (*p*-value) between GPS scores and temporal-spatial parameters

		Spearman's rho	*p-value*
Overall GPS	Velocity	-0.50	0.196
	Step length Pro	-0.70	0.094
	Step Length Int	-0.30	0.312
	Single support Pro	-0.70	0.094
	Single support Int	0.50	0.196
Prosthetic limb GPS	Step length Pro	-0.70	0.094
	Step Length Int	-0.30	0.312
	Single support Pro	-0.70	0.094
	Single support Int	0.50	0.196
Intact limb GPS	Step length Pro	-0.30	0.312
	Step Length Int	0.00	0.500
	Single support Pro	-0.30	0.312
	Single support Int	-0.20	0.374

IV. DISCUSSION

Kinematic and temporal-spatial parameters data are usually used to describe and quantify transtibial amputees' deviations [4]. These were some of the first variables used to describe the ambulation of this population [5, 7]. These data are used to determine gait deviations, to analyze the effectiveness of rehabilitation programs, to detect prosthetic alignment problems and to help to define the end of the rehabilitation process.

Participants presented temporal-spatial parameters deviations compatible with previous studies [7]. They presented walking speed average of 89.66 (16.16) cm/s, while able-bodied individuals walk at 124.63 cm/s. Besides, their mean step lengths (61.04 and 55.8 cm) were short compared with the 65.20 cm expected for individuals with no gait pathology. So, markedly, subjects' functional ability was poorer than that of healthy individuals.

Similar to previous findings, subjects gait were asymmetrical [11, 12]. Step length with amputated limb was greater (61.04 cm) than that with the intact limb (55.8 cm) and single support with intact leg (37.8% of the gait cycle) was longer than with the prosthetic leg (31.24 %). In amputees, the difference of stance duration contributes to asymmetry as well as deficiencies associated with prostheses fitting and components [4]. Due to these factors, transtibial amputees fell less confident to load over prosthetic limb and, also, tend to increase the base of support. In this study individuals presented a step width of 16.43 cm (± 3.03), while in able-bodied subjects the base of support measures about 12 cm.

Two studies used GPS to quantify gait deviations in transtibial amputees [6, 8]. However, their sample consisted of traumatic and vascular transtibial amputees, which present poorer gait prognosis compared with traumatic amputees, and have a variety of other pathologies that can influence negatively the power of the study's findings. In addition, they did not control prosthetic components. Different types of feet and prosthetic fitting influence differently some gait parameters [3-5].

On the present study the sample size consisted of only traumatic transtibial amputees, using KBM socket and SACH foot. Besides that, all participants were rehabilitated in the same center, following the same preprosthetic and post-prosthetic protocol. All of these inclusion criteria had the objective of minimize mistakes in the gait analyzes, due to variation in the sample.

Overall, prosthetic and intact limbs GPS scores (9.59°, 9.32° and 9.02°, respectively) were higher than the expected for individuals with no gait pathology (5.6°, 5.3° and 5.3°) [9]. Previous studies obtained similar results [6, 8]. On both studies, prosthetic limb GPS was higher than that with the intact one. The first study found a GPS score of 12.3° for the amputated leg and 11.4° for the opposite leg [8]. On the one, GPS scores were 7.1° for prosthetic limb and 6.3° for the intact limb [6].

This asymmetry reflects the influence of the natural ankle function absence on the gait pattern in transtibial amputees. That fact became worst with the use of SACH foot, which present less range of motion compared with others feet. In addition, this kind of foot prolongs the time which the amputated leg maintain a heel-only contact, which increases asymmetry and results in a period of instability with this limb [4].

On the study of Kark et al [8], pelvic tilt, hip flexion/extension and knee flexion/extension were the kinematic variables that obtained the highest scores, what represent that they presented larger deviations from normal. Our study obtained similar results, differentiating from that just in the hip rotation variable, which had the major GVS score (14.54° on the prosthetic limb). Problems with prosthetic alignment and the difficulty to position knee and ankle landmarkers can cause this deviation. An increased internal or external rotation of the prosthetic foot can influence the final hip rotation angle. Furthermore, the positions of knee and ankle landmarkers are used to calculate the rotation center of these joints, which are used to calculate hip's joint center.

On the present study, subject D had more GVS with normal values (7) and, at the same time, presented the best functional ability, represented by a velocity of 114.9 cm/s and values of 74.10 and 67.27 cm for step length. However, the opposite pattern was not observed. This indicates that these findings could suggest a strong relationship between the GPS and amputees' functional ability but, they should be more explored at the next phase of the research.

Overall GPS was moderately correlated with self-selected velocity (r = -0.50), but this correlation did not show significance. Another study found a better correlation between these two variables (r = -0.70) [6], which could be attributed to a larger sample size of it. The strongest correlation occurred between overall GPS and prosthetic step length, overall GPS and time of single support with the prosthetic limb, prosthetic limb GPS and prosthetic step length, and between prosthetic limb GPS and time of single support with the prosthetic limb (r = -0.70).

The poor statistical significance of this relationship may be due to sample size of this study. The increase of sample size can increase the significance of these correlations.

V. CONCLUSION

The GPS, MAP and temporal-spatial parameters were useful in quantifying gait deviation on transtibial amputees. GPS scores were increased and temporal-spatial parameters values were lower than that found in able-bodied subjects. The poor correlation observed between GPS and temporal-spatial parameters may be due to small sample.

ACKNOWLEDGMENT

We would like to thank CNPq, CAPES and Araucaria Foundation for financial support to our participation in this event.

REFERENCES

[1] N. Vanicek, S. Strike, L. McNaughton, and R. Polman, "Gait patterns in transtibial amputee fallers vs. non-fallers: Biomechanical differences during level walking," *Gait & Posture*, vol. 29, pp. 415-420, 2009.

[2] K. Parker, E. Hanada, and J. Adderson, "Gait variability and regularity of people with transtibial amputations," *Gait & Posture*, vol. 37, pp. 269-273, 2013.

[3] G. N. S. Marinakis, "Interlimb symmetry of traumatic unilateral transtibial amputees wearing two different prosthetic feet in the early rehabilitation stage," *Journal of Rehabilitation Research & Development*, vol. 41, pp. 581-589, 2004.

[4] Y. Sagawa, K. Turcot, S. Armand, A. Thevenon, N. Vuillerme, and E. Watelain, "Biomechanics and physiological parameters during gait in lower-limb amputees: a systematic review," *Gait & Posture*, vol. 33, p. 511, 2011.

[5] H. van der Linde, C. J. Hofstad, A. C. H. Geurts, K. Postema, J. H. B. Geertzen, and J. van Limbeek, "A systematic literature review of the effect of different prosthetic components on human functioning with a lower-limb prosthesis," *Journal of Rehabilitation Research and Development*, vol. 41, pp. 555-570, Jul-Aug 2004.

[6] L. Kark, D. Vickers, A. McIntosh, and A. Simmons, "Use of gait summary measures with lower limb amputees," *Gait & Posture*, vol. 35, pp. 238-243, 2012.

[7] A. S. O. D. C. Soares, E. Y. Yamaguti, L. Mochizuki, A. C. Amadio, and J. C. Serrão, "Biomechanical parameters of gait among transtibial amputees: a review," *São Paulo medical journal = Revista paulista de medicina*, vol. 127, p. 302, 2009.

[8] L. Kark, D. Vickers, A. Simmons, and A. McIntosh, "Using the movement analysis profile with lower limb amputees," *Gait & Posture*, vol. 30, Supplement 2, pp. S42-S43, 11// 2009.

[9] R. Baker, J. L. McGinley, M. H. Schwartz, S. Beynon, A. Rozumalski, H. K. Graham, and O. Tirosh, "The Gait Profile Score and Movement Analysis Profile," *Gait & Posture*, vol. 30, pp. 265-269, 2009.

[10] S. Beynon, J. L. McGinley, F. Dobson, and R. Baker, "Correlations of the Gait Profile Score and the Movement Analysis Profile relative to clinical judgments," *Gait & Posture*, vol. 32, pp. 129-132, 2010.

[11] E. Isakov, O. Keren, and N. Benjuya, "Trans-tibial amputee gait: time-distance parameters and EMG activity," *Prosthet Orthot Int*, vol. 24(3), pp. 216-220, 2000.

[12] S. A. G. Po-Fu Su, Robert D. Lipschutz, Todd A. Kuiken, "Gait characteristics of persons with bilateral transtibial amputations," *Journal of Rehabilitation Research and Development*, vol. 44, pp. 491-502, 2007.

Corresponding author:

Author: E.B. Neves
Institute: Federal Technological University of Paraná
Street: 3165, Sete de Setembro
City: Curitiba,
Country: Brazil
Email: borbaneves@hotmail.com

Development and User Assessment of a Body-Machine Interface for a Hybrid-Controlled 6-Degree of Freedom Robotic Arm (MERCURY)

N. Moustakas[1,2], A. Athanasiou[1,3], P. Kartsidis[1], P.D. Bamidis[1], and A. Astaras[1,2]

[1] Lab of Medical Informatics, Medical School, Aristotle University of Thessaloniki (AUTH), Thessaloniki, Greece
[2] Dept. of Automation, Alexander Technological Educational Institute of Thessaloniki (ATEITH), Thessaloniki, Greece
[3] Dept. of Neurosurgery, Papageorgiou General Hospital, Thessaloniki, Greece

Abstract—This paper presents the development, pilot testing and user assessment results for a body-machine interface (BMI) designed to control a 6-degree of freedom robotic arm, developed by our research team. The BMI was designed to be wearable, immersive and intuitive, constituting the first part of a hybrid real-time user interface. A total of 34 volunteers participated in this study, performing two sets of three tasks in which they controlled the robotic arm, a) within direct line of sight and b) through a video link. All participants completed questionnaires to evaluate their technological background, familiarization with informatics, electronics, robotics and video teleconferencing. At this point of development the system does not capture brainwaves or electric neural input, it simply captures the motion of the operator's arm. The complete MERCURY prototype system is still under development and additionally comprises a wearable, wireless brain-computer interface (BCI) headset. The BCI headset is currently being integrated into the system and has not yet been pilot tested. The complete hybrid-interface system is primarily intended for research into human-computer interfaces, neurophysiological experiments, as well as industrial applications requiring immersive remote control of robotic machinery.

Keywords—**Body-machine interface, brain-computer interface, robotic, user assessment.**

I. Introduction

Brain machine interfaces (BMIs) are systems that allow users to control external devices using movement of their body, their goal being to map a user's residual motor skills into efficient patterns of control [1]. Brain Computer Interfaces (BCIs) are interactive systems that permit users to control external devices using their thought. BCI research and development is a relatively recent occurrence, however commercial applications of BCIs are already on the market [2]. Rehabilitation and motor restoration for patients with severe neurological impairment are commonly mentioned in the literature as possible BCI applications [3].

The use of brainwaves to control robotic devices has already produced promising clinical results [4]. Various research teams have demonstrated that invasive BCIs can be used to restore a certain degree of motor functions and even provide high accuracy control of robotic prosthetic arms [5]. However, in order for such BCI-controlled robotic applications to achieve ethical and end-user acceptance for everyday tasks, the use of non-invasive, unobtrusive and relatively low-cost systems is required.

We have designed, partly implemented and tested MERCURY, a hybrid BCI-BMI robotic system, to investigate the capabilities and limitations for combining these technologies, primarily for medical and biomedical engineering applications [6]. All components used for MERCURY have been designed, implemented and tested by our research and development team.

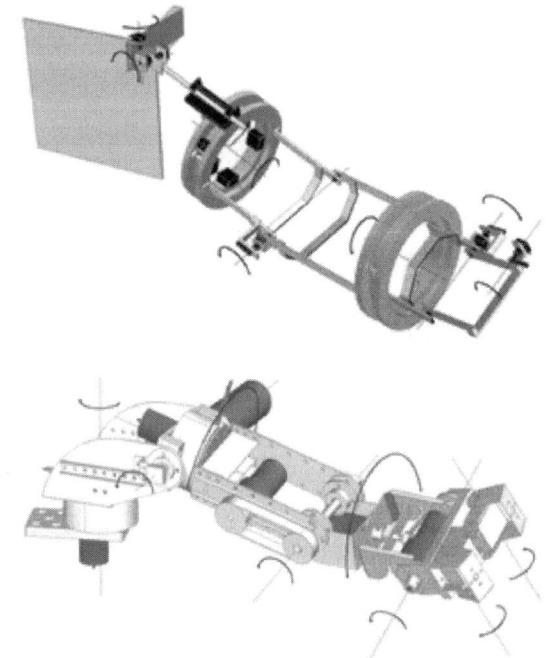

Fig. 1 Computer aided design diagram of the 6 degree of freedom MERCURY BMI (top) and robotic arm (bottom)

Design requirements included rapid and smooth robotic movement that approximates the natural movement of a human operator's arm; an intuitive and immersive interface, portability, scalability, and relatively low cost (less than 3000€). As a primary application, we targeted investigative neurophysiological experimentation in which an operator remotely controls a 6-degree of freedom (DOF) robotic arm using their arm movement, their brainwaves or both. Moreover, a pilot study was designed and is presented in this paper in order to provide early indications regarding user acceptance of the BMI modality.

II. MATERIALS AND METHODS

A total of 34 people participated in the MERCURY pilot study, of whom 25 (73.5%) were men and 9 (26.5%) were women. The mean participant age was 26.71 ($\sigma = \pm\ 8.77$). Each session lasted approximately half an hour. None of the participants were part of the design team or had previous experience in tele-robotic operation. All experimental sessions were video recorded with the participant's approval. Participant performance in the various experimental tasks was evaluated using time to completion, the number of failed attempts and a subjective mental effort self-assessment rating provided by the participants.

A. Initial Briefing, Assessment and Familiarization with BMI Robotic Control

Participants received an initial verbal briefing followed by a video demonstration of the operator and robot arm in action. They were subsequently presented with the hardware prototype and were asked to complete the *Godspeed* questionnaire [7] to gauge their attitude towards the robotic setup. The *Godspeed* questionnaire is a general assessment tool covering human attitude towards robotics, several aspects of which were not relevant to MERCURY.

During the next stage, the BMI harness was custom-fitted to the participant's arm using interchangeable spacer rubber foam. All subjects wore the BMI harness on their right arm, regardless of which hand they considered to be dominant or more dexterous in everyday activities. Pilot testing engineers ensured that each subject retained complete freedom of movement while wearing the harness.

The robotic arm was placed on a table at the front lower right area of the participant's field of view, to facilitate the impression of immersive control. The participant was given 3 minutes of free experimentation in order to discover and familiarize themselves with the robotic arm's response to the movements of their right arm.

B. Task Assignment and Assessment

Upon conclusion of the familiarization period, participants were asked to use the BMI to control the robotic arm and perform 3 tasks: knock an object off the table (task 1), pick and place the object on an elevated platform (task 2) and pick and place the object through a ring on an elevated platform (task 3). Dropping the object was noted in the experimental notebook and an experimenter restored it to the upright position. Dropping the object 3 times was considered a task failure and the participant was encouraged to move on to the next task.

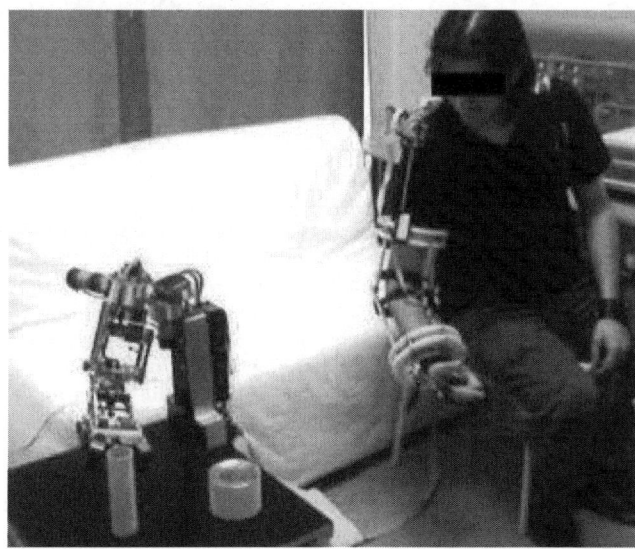

Fig. 2 Depiction of the MERCURY experimental setup used for qualitative assessment of the BMI. The robotic arm is pictured on the left side, controlled by a pilot operator wearing the harness on the right

Having completed the initial 3 tasks, a divider screen was placed between the operator and the robotic arm. The participant was consequently asked to perform the same 3 tasks using a live video feed from a personal computer webcam, which they viewed through a 12-inch computer screen. The video link was fast with no perceptible intermittent frame rates, however it still adversely affected depth perception for most participants and rendered the experimental tasks more difficult. This effect could have been partly overcome with improved lighting, a higher resolution or 3D camera and a larger viewing screen, but in this pilot study a basic hardware setup was deliberately chosen to render experimental tasks as challenging as possible.

Upon completion of each task, the subject was asked to rate the required effort on a standardized *Subjective Mental Effort Questionnaire* (SMEQ) [8], which was translated in the participants' native language (Greek) for the purpose of this study. The SMEQ is a cognitive workload scale ranging from 0 (no effort at all) to 150 points (exceptional amount of effort) and has been used as a tool in various laboratory and field studies [9].

C. Debriefing and Qualitative Assessment

Debriefing involved collecting some personal information about the participant and having them complete some further questionnaires. The first questionnaire was the Locally

Experienced Discomfort (LED) scale [10]. The LED scale facilitates communication of physical discomfort experienced by the participant, prompting them to assign a numeric value to several locations on a chart outline of the upper body. This value ranges from extreme comfort (0) to extreme discomfort (10). The LED scale was translated into the participants' native language (Greek) for the purposes of this study.

In addition, all participants completed the Godspeed questionnaire for a second time, this time taking into account their recent experience in using the BMI to control the robotic arm. Comparisons between the two sets of answers by the same participant are considered indicative of the difference between their perceived and empirically verified beliefs about the MERCURY robotic system. This aspect was deemed beyond the scope of the current paper and therefore the relevant analysis is not presented here.

Finally, each participant completed a questionnaire developed specifically for this pilot study. It involves a total of 36 multiple-choice statements which can be answered on a scale ranging from 1 (strongly disagree) to 5 (strongly agree), including a "do not know" option. Positive and negative statements were interlaced in order to minimize hasty careless answers. The statements gauged the participants' previous experience in robotics, tele-control devices and the extent of technology use in everyday activities (mobile phones, computers, game consoles, the internet, etc). Of particular importance for this paper were statements used to evaluate the participants' evaluation of the MERCURY BMI interface from the point of view of intuitive and immersive control.

D. Statistical Analysis

Statistical analysis was based on the number of attempts per task, time to completion and correlations between subsets of participants and their performance at the experimental tasks. Apart from calculating average and mean values, a Pearson chi squared method was used to test for statistical significance among possible correlations.

In order to test for meaningful statistical correlations, subsets were selected among participants based on the following criteria: sex, professional background in electronics or related fields, self-declared degree of familiarization with technological tools and automations in daily activities.

III. RESULTS AND DISCUSSION

When operating the robotic arm at direct line of sight, the participants had almost 100% success rate on all 3 experimental tasks (table 1). Results differed when participants were asked to complete the same 3 tasks through a low quality video link comprising a PC web camera and 12-inch LCD screen (table 2).

Statistical analysis revealed that the subsets of participants who were male, who had training in engineering and/or related sciences, and those with a high level of technological familiarization in their daily activities were more successful at the most demanding final task assignments, yet equally successful at the same assignments when working at direct line of sight.

Table 1 Task completion when participants performed the assigned experimental tasks looking directly at the robotic arm

Direct line of sight control		Task 1 completed		Task 2 completed		Task 3 completed	
		Yes	No	Yes	No	Yes	No
Gender	M	25	0	25	0	24	1
	F	9	0	8	1	9	0
Professional background in electronics	Yes	18	0	18	0	17	1
	No	16	0	15	1	16	0
Technological familiarization	Low	15	0	14	1	14	1
	High	19	0	19	0	19	0

Table 2 Task completions when participants performed the assigned experimental tasks via video link

Remote control via video link		Task 1 completed		Task 2 completed		Task 3 completed	
		Yes	No	Yes	No	Yes	No
Gender	M	25	0	17	8	18	7
	F	9	0	1	8	3	6
Professional background in electronics	Yes	18	0	14	4	14	4
	No	16	0	4	12	7	9
Technology familiarization	Low	15	0	4	11	5	10
	High	19	0	14	5	16	3

Further statistical analysis showed that -on average- participants agreed that the MERCURY interface was intuitive to use and produced an immersive remote control experience (table 3). We consider this a significant result, given that the experimental setup comprised a proof of concept prototype, the initial familiarization period was deliberately maintained brief (3 minutes) and the participants were assigned to operate the robotic arm through a low quality video link.

Table 3 Participant evaluation for statements "I found control of the robotic arm to be intuitive" (S1) and "After some time, the robotic arm felt like an extension of my own body" (S2)

	Answers		Participant %	
	S1	S2	S1	S2
Strongly Disagree	0	0	0	0
Disagree	3	6	8.8	17.6
Neutral	4	14	11.8	41.2
Agree	21	13	61.8	38.2
Strongly Agree	6	1	17.6	2.9
Total	34	34	100.0	100.0

As expected, those participants who claimed to be the most familiar with information and electronics technology in their daily lives were more likely to agree with the statement that the MERCURY BMI offers an immersive experience. With regard to the same statement, differences were far less pronounced between the two sexes and when comparing groups with and without professional training in electronics-related disciplines.

As for the LED scale the majority of the participants reported discomfort on their wrist, shoulder and the cubit, without any statistical significant differences in respect to the aforementioned subsets. In regard to the SMEQ score there were significant differences among the six tasks. The first task in both settings was considered effortless, while the second and third tasks were demanding. There is no statistically significant difference among the three aforementioned subsets with respect to the SMEQ score.

Table 4 Mean and SD values of SMEQ score

SMEQ	Direct line of sight control			Remote control via video		
	Task 1	Task 2	Task 3	Task 1	Task 2	Task 3
Mean	16	65	68	11	84	81
SD	19	27	28	11	29	30

IV. CONCLUSIONS

It is possible to design a wearable, immersive BMI and a robotic arm copying a human operator's movement on a tight budget (material costs <3000€). The robotic arm setup presented in this paper is the first development stage of a BCI-BMI hybrid system and experimental results indicate that the BMI is comparable in size, dexterity, speed and smoothness of movement to the human operator's arm.

Qualitative testing has indicated that the design goals for intuitive and immersive remote control have been achieved.

While conclusions from the results of this pilot study were encouraging, they are based on a low number of participants and a relatively narrow age distribution. Further testing is required in order to better determine the competitive advantages of the MERCURY BMI and extract meaningful design improvements for future prototypes.

Further work will focus on additional user assessment testing, as well as on integrating the existing robotic setup into a BCI-BMI hybrid system for comparative experiments. Apart from serving as a scientific experimental platform, improved versions of MERCURY are expected to find application in medical prosthetics, telematic surgery and real-time robotic control in dangerous and safety-critical environments.

REFERENCES

1. Casadio M, Pressman A, Acosta S et al. (2011) Body machine interface: remapping motor skills after spinal cord injury. 2011 IEEE Int Conf Rehab Rob (ICORR), Switzerland, 2011 DOI 10.1109/ICORR.2011.5975383
2. Nicolas-Alonso LF, Gomez-Gil J. (2012) Brain computer interfaces, a review. Sensors 12:1211–1279.
3. Athanasiou A, Bamidis PD. (2010) A review on brain computer interfaces: contemporary achievements and future goals towards movement restoration. Aristotle University Medical Journal 37:35–44
4. Galán F, Nuttin M. (2008) A brain-actuated wheelchair: asynchronous and non-invasive brain–computer interfaces for continuous control of robots. Clin Neurophysiol, 119:2159–2169
5. Yanagisawa T, Hirata M, Saitoh Y et al. (2011) Real-time control of a prosthetic hand using human electrocorticography signals. J Neurosurg 114:1715 –1722
6. Astaras A, Moustakas N, Athanasiou A et al. (2013) Towards brain computer interface control of a 6-degree of freedom robotic arm using dry EEG electrodes. Adv Hum Comput Interact 2013:641074 DOI 10.1155/2013/641074
7. Bartneck C, Kulić D, Croft E et al. (2009) Measurement instruments for the anthropomorphism, animacy, likeability, perceived intelligence, and perceived safety of robots. Int J Soc Rob 1:71–81
8. Zijlstra FRH, van Doorn L. (1985) The construction of a scale to measure perceived effort. Department of Philosophy and Social Sciences, Delft University of Technology, Delft
9. van der Schatte Olivier RH, van't Hullenaar C, Ruurda JP et al (2009) Ergonomics, user comfort, and performance in standard and robot-assisted laparoscopic surgery. Surg Endosc 23:1365–1371
10. Corlett E, Bishop R. (1976) A technique for measuring postural discomfort. Ergonomics 19:175–182

Author: Nikolaos Moustakas
Institute: Medical School, Aristotle University of Thessaloniki
Street: Ag. Dimitriou, 54124
City: Thessaloniki
Country: Greece
Email: moustakas.nikolaos@gmail.com

EMG and Kinematics Assessment of Postural Responses during Balance Perturbation on a 3D Robotic Platform: Preliminary Results in Children with Hemiplegia

C. De Marchis[1], F. Patané[2,3], M. Petrarca[3], S. Carniel[3], M. Schmid[1], S. Conforto[1], E. Castelli[3], P. Cappa[2,3], and T. D'Alessio[1]

[1] Biolab[3], Department of Engineering, University Roma TRE, Rome, Italy
[2] Department of Mechanical and Aerospace Engineering, "Sapienza" University of Rome, Italy
[3] Pediatric Neuro-Rehabilitation Division, Children's Hospital "Bambino Gesù" IRCCS, (Rome), Italy

Abstract—Dynamic posturography has been proposed as a valuable tool for the assessment of balance impairments. This paper compares the postural responses of healthy and hemiplegic children while keeping balance on a 3D robotic perturbed platform, addressed as Rotobit[3D]. Dynamic postural responses are assessed by using surface EMG, recorded from four muscles of both legs, and lower limb kinematics. These preliminary results show that the used protocol is able to highlight significant differences in the postural responses, in terms of postural asymmetries between the less affected and more affected side for both electrophysiological and kinematic recordings.

Keywords—Dynamic Posturography, Robotic Platform, Postural Response, Hemiplegia, EMG.

I. INTRODUCTION

The control of standing is a complex task for the nervous system and involves the control of many muscles all over the body segments [1]. Balance impairments are present in a wide range of pathologies, including neurological diseases and musculoskeletal disorders. Quantifying balance impairments is relevant for both diagnostic and therapeutic purposes, in order to develop optimal treatment strategies in the perspective of improving the knowledge of the mechanisms underlying diseases [2]. Quantitative posturographic analysis has been proposed with the aim of overcoming the drawbacks of traditional clinical tests. These tests are generally based on subjective scoring scales, are affected by variability in the test conditions and cannot reveal details of the underlying pathologic status. Some efforts have been provided to standardize posturographic protocols [3][4] in order to allow the comparison and the follow-up of patients.

Dynamic posturography allows to actively manipulate the posture or balance, and to evaluate the subject's response using quantitative indexes. Experimentally induced balance perturbations can be achieved through the use of robotic platforms, which are traditionally based on translational perturbations in multiple directions [5] or, more recently, on impedance-controlled rotational perturbations as a paradigm for diagnosis and rehabilitation [6][7][8]. Dynamic postural control in hemiparetic subjects has revealed asymmetries and an increased risk of fall towards the affected side [9], so that the use of dynamic posturography could be a valuable tool to assess the degree of impairment [10]. While the postural control of adult population has been assessed in many studies [2], the development of postural strategies in pediatric population has shown significant variations with age [11], often amplified when observed during concurrent tasks [12].

The combined analysis of kinematics and electrophysiological measurements could provide a complete description of the postural responses, could highlight differences in the sensorimotor control between pathologic and normal population and, at the same time, can monitor the ongoing development of the postural control strategies in pediatric age.

This preliminary study explores the feasibility of using surface EMG and lower limb kinematics, recorded while participants maintain the upright posture in dynamic posturography delivered by Rotobit[3D] [6]. The aim of the present study is the assessment of differences in the postural responses in a population of children with different forms of hemiplegia with respect to control subjects, according to the indications provided by the Evidence Based Medicine. EMG-related indexes are used to characterize the responses of the two studied population samples.

II. MATERIALS AND METHODS

A. Participants and Procedure

Two normally developed children (ND) (12 and 14 years) and 2 children with hemiplegia (CH) (8 and 9 years) participated.

The exercise consists of maintaining an upright posture while a perturbation is applied by means of Rotobit[3D]. Two different kinds of perturbation have been applied. The feet are positioned in parallel with a width equal to the shoulder projection on the terrain. The axis of ankle joint is aligned with platform axis of rotation in order to minimize any push effects on the feet. A contact between the hands of the participant and of the trainer is allowed in order to avoid falls and to guarantee the trial completion for all participants, especially for children with balance disturbances.

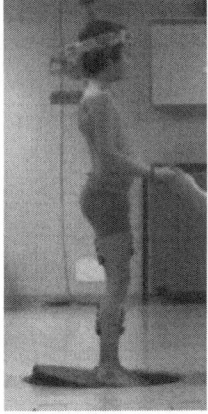 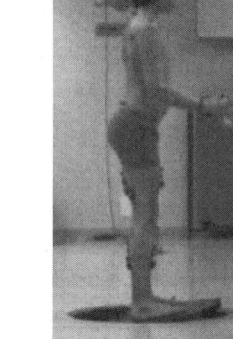

- forward tilt *- backward tilt*

The perturbation began with a slow dorsi/flexion movement of 12° in order to reach initial position. After 3 s of pause the platform moved 24° forward then, after a new pause of 2 s, moved 24° backward and, after a pause of 3 s, the platform came back to horizontal position. A rest period is permitted before the successive trial. Every trial is repeated four times for each perturbation condition (forwards and backwards), and velocity (mean velocities equal to 30°/s, 48°/s and 80°/s).

Rotobit^{3D}

Children stood at the center of the platform. Each perturbation was applied under computer control with three different durations: 800 ms, 500 ms, and 300 ms.

B. Data Acquisition

Surface EMG was recorded from the following 4 muscular groups: hamstrings (HS), quadriceps (QD), triceps surae (TS) and tibialis (TA). EMG was acquired through the wireless Vawe system (Cometa, IT) and sampled at 1 kHz. Kinematics was captured by VICON system (MX 8-camera-workstation, Nexus 1.7 software, 200 Hz, PlugIn-Gait marker set). All the trials were video-recorded in the sagittal and frontal planes.

C. Data Processing and Statistical Analysis

Each EMG signal was low-pass filtered at 400 Hz to reduce noise and processed to remove low frequency movement artifacts as in [13]. Then it was full-wave rectified and low-pass filtered with a cut-off frequency of 3 Hz (3rd order Butterworth digital filter) in order to extract the linear envelope. Each envelope profile was amplitude normalized to the maximum value of the envelope within each trial.

For each trial and for each muscle, the EMG envelope within the period of the perturbation was time interpolated on 100 data points, representative of equally spaced integer percentages of the perturbation period (Figure 2). Together with the EMG envelopes, also the knee flexion and hip flexion angles from the left and right legs were time interpolated within the perturbation interval. Ankle plantar-flexion and dorsi-flexion angles were omitted since the kinematic profile showed negligible variations respect to the platform motion, and the temporal evolution might not be indicative of the motor control strategies. The Pearson's correlation coefficient r between the muscle responses and between the joint angles (DoFs) of the two lower limbs was taken into account as a measure of symmetry, since it can highlight differences in both shape and timing of the postural responses between legs.

The Pearson's correlation coefficient r between legs for each considered parameter underwent a 1-way ANOVA to check whether this parameter is globally affected by the pathology, before undergoing a N-way ANOVA having trials (forward vs. backward), durations (300 ms, 500 ms and 800 ms) and conditions (control vs. hemiplegic) as factors, in order to assess which muscle/trial/duration combination is able to better highlight possible differences between healthy and pathologic status. We hypothesized that the asymmetric contribution of a certain muscle could be more visible in one of the trials or under one particular duration of the perturbation.

III. RESULTS

We performed a 1-way ANOVA with conditions (ND vs. CH) as factors, including all the other parameters (such as duration, trial and muscles), A global statistically significant difference between the two samples was found (Table I). As it can be seen, on the whole the response is more symmetric for the ND (high r level, indicating high similarity in the muscle activation profiles and in the joint angles for both shape and timing), than for the CH ($r < 0.4$ and $r < 0.5$ for muscles and DoFs, respectively).

Table 1 Inter-limb Pearson correlation coefficient (mean ± standard deviation) of the muscular responses and joint angles. Underlined values indicate significant difference between the two populations (p < 0.05).

	CH	ND
Muscles	r = 0.32 ± 0.03	r = 0.69 ± 0.02
DoFs	r = 0.47 ± 0.05	r = 0.81 ± 0.03

A 2-way ANOVA with trials (forward vs. backward) and conditions (ND vs. CH) within each muscle (HS, QD, TS, TA) was performed.

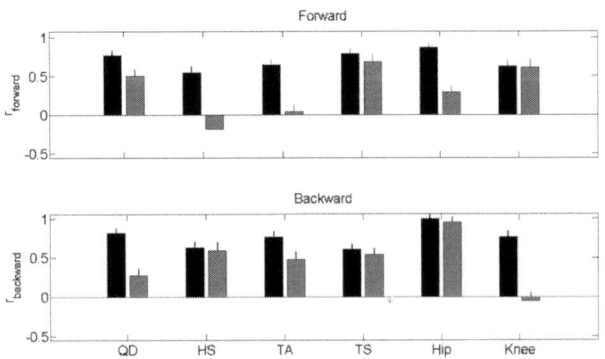

Fig. 1 Effect of perturbation direction on the postural response symmetry of muscles (QD, HS, TA and TS) and joint angles (hip and knee); r is reported as mean (solid bar) + standard deviation (thick line). Black and grey refer to control and hemiplegic subjects, respectively.

Forward perturbations induce asymmetries in the muscular responses of the hamstrings and tibialis in CH (low r values), while backward perturbations highlight differences in the quadriceps muscle responses (Figure 1). Triceps surae does not show a significant difference between conditions across the different trials. For what concerns the coordination between the joint angles of the two lower limbs, forward perturbation highlights asymmetries in the hip joint flexion in hemiplegic subjects, while the forward perturbation induces an asymmetric response in the knee flexion angle. The statistical analysis of the response symmetry can be extended by adding information related to the duration of the perturbation (300 ms, 500 ms, 800 ms). From a preliminary analysis on those muscles showing asymmetrical response in hemiplegic subjects (i.e. QD, HS and TA), the 500 ms perturbation maximizes differences in response asymmetries (Figure 3).

IV. DISCUSSION AND CONCLUSIONS

The use of dynamic posturography with rotational paradigms, combining electrophysiological and kinematic symmetry index, highlights differences in the motor control strategies of a population of CH when compared to ND. The used protocol is able to significantly discriminate the postural responses between the two studied samples.

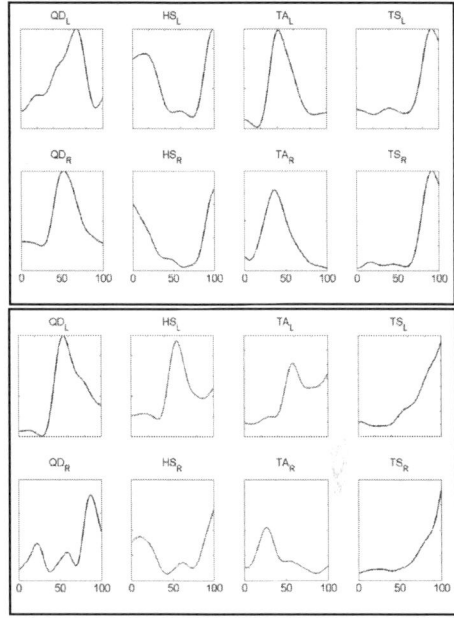

Fig. 2 Example of muscular response for a 800 ms tilt forward perturbation. Differences in muscular responses between a healthy (upper plot) and a hemiplegic subject (lower plot) are shown. Red traces in the lower plot highlight a visible asymmetry in muscular response between right and left leg. y-axis reports muscle activation. x-axis is referred to the equally spaced integer percentages of the perturbation period.

While ND did not show asymmetry, some differences emerged in CH. Forward perturbations around the frontal plane induced significant between-legs asymmetries in the hamstrings and in the tibialis anterior muscle. This action was reflected in an asymmetric behaviour of the hip flexion angles. Backward perturbations highlighted an asymmetric response of the knee-extensor muscles, which was accompanied by an asymmetric response of the knee flexion angles. No significant differences between ND and CH were found in the response of the Triceps Surae muscle group. We hypothesize that the perturbation could be compensated at the ankle level, and that the participants could not be able to follow and to compensate the perturbation in real time, while the mechanical redundancy of body configuration allowed using alternative solutions. In particular, what we observed is usually a knee strategy for forward perturbations and a hip strategy for backward ones.

Although an asymmetric response was generally present in the muscular adjustment by CH, this asymmetry was more evident in the 500 ms and 800 ms perturbation, when compared to the 300 ms, which, on the contrary, highlights asymmetries in the joint angles. Thus, 300 ms may be considered as a challenging perturbation. With respect to this perturbation, the main findings are the following: (i) this perturbation represents the more difficult task as evidenced by the kinematic measures; (ii) it is solved with a general co-contraction, i.e. by increasing the stiffness. In this specific case EMG may not be suitable as an index to assess the chosen motor strategy. From a rehabilitation perspective, the difficulty of the more challenging condition (300 ms) may hinder its use in physical therapy practice.

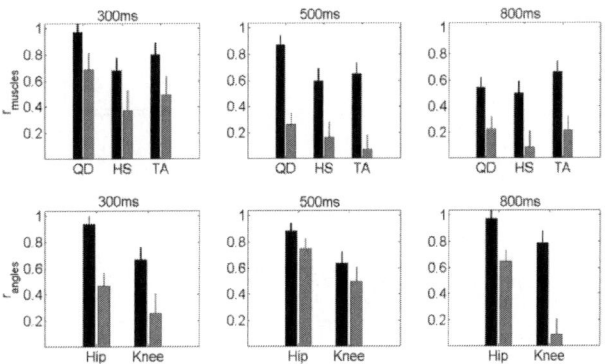

Fig. 3 effect of perturbation duration on the postural response symmetry of muscles (upper plots) and joint angles (lower plots); r is reported as mean (solid bar) + standard deviation (thick line). Black and grey refer to control and hemiplegic subjects, respectively.

Differences in EMG and whole body kinematics can reveal whether there is a deficit in the CNS or the response reflects a mechanical constraint due postural changes [14].

In order to further characterize the pathologies and to assess the development of postural motor control in children, future studies should aim at analyzing further aspects of motor control, i. e. a possible effect of motor adaptation to the perturbation after multiple repetitions of the exercise in both control and pathologic children, or the effects of learning after long training periods. The previous objectives would benefit from the analysis of additional indexes related to the motor response, such as the latency in the response revealed by the EMG onset after perturbation, the whole body coordination in a reduced dimensionality to reveal coordination mechanisms among kinematic DoFs, or even the analysis of the underlying muscle synergies characterizing postural responses [15], together with their possible changes with adaptation and training [16].

Acknowledgments

This work was supported by the Department of Innovation, Health and Social Politics, Italian Ministry of Welfare, under Grant 2007 "Pilot study on a novel typology of medical devices: Apparatus for dynamic and closed-loop posturography".

References

1. Balasubramaniam, R., & Wing, A. M. (2002). The dynamics of standing balance. Trends in cognitive sciences, 6(12), 531-536.
2. Visser, J. E., Carpenter, M. G., van der Kooij, H., Bloem, B. R. (2008). The clinical utility of posturography. Clinical Neurophysiology, 119(11), 2424-2436.
3. Schmid, M., Conforto, S., Camomilla, V., Cappozzo, A., D'Alessio, T. (2002). The sensitivity of posturographic parameters to acquisition settings. Medical Engineering and Physics, 24(9):623-31.
4. Giacomozzi, C., Macellari, V., Benvenuti, F., Chiari, L., Conforto, S., Della Croce, U., Fioretti, S., Morasso, P., Tonini, A. (2002). Towards the standardisation of clinical protocols in posturography, Gait & Posture, 16 (Suppl. 1), p. S22.
5. Brown, L. A., Jensen, J. L., Korff, T., Woollacott, M. H. (2001). The translating platform paradigm: perturbation displacement waveform alters the postural response. Gait & posture, 14(3), 256-263.
6. Cappa, P., Jackson, J. L., Patane, F. (2010). Moment Measurement Accuracy of a Parallel Spherical Robot for Dynamic Posturography. Biomedical Engineering, IEEE Transactions on, 57(5), 1198-1208.
7. Patanè, F., Cappa, P. (2011). A 3-DOF parallel robot with spherical motion for the rehabilitation and evaluation of balance performance. Neural Systems and Rehabilitation Engineering, IEEE Transactions on, 19(2), 157-166.
8. Commissaris, D. A. C. M., Nieuwenhuijzen, P. H. J. A., Overeem, S., De Vos, A., Duysens, J. E. J., Bloem, B. R. (2002). Dynamic posturography using a new movable multidirectional platform driven by gravity. Journal of neuroscience methods, 113(1), 73-84.
9. Ikai, T., Kamikubo, T., Takehara, I., Nishi, M., Miyano, S. (2003). Dynamic postural control in patients with hemiparesis. American journal of physical medicine & rehabilitation, 82(6), 463-469.
10. Allum, J. H. J., Bloem, B. R., Carpenter, M. G., & Honegger, F. (2001). Differential diagnosis of proprioceptive and vestibular deficits using dynamic support-surface posturography. Gait & posture, 14(3), 217-226.
11. Schmid, M., Conforto, S., Lopez, L., Renzi, P., D'Alessio, T. (2005). The development of postural strategies in children: a factorial design study. Journal of NeuroEngineering and Rehabilitation, 2(1), 29.
12. Schmid, M., Conforto S., Lopez L., D'Alessio T., (2007). Cognitive load affects postural control in children. Experimental Brain Research, 179:375–385.
13. Conforto, S., D'Alessio, T., Pignatelli, S. (1999). Optimal rejection of movement artefacts from myoelectric signals by means of a wavelet filtering procedure, Journal of Electromyography and Kinesiology, 9(1):47-57.
14. Burtner, P. A., Qualls, C., Woollacott, M. H. (1998). Muscle activation characteristics of stance balance control in children with spastic cerebral palsy. Gait & posture, 8(3), 163-174.
15. Torres-Oviedo, G., & Ting, L. H. (2007). Muscle synergies characterizing human postural responses. Journal of neurophysiology, 98(4), 2144-2156.
16. De Marchis, C., Schmid, M., Bibbo, D., Castronovo, A. M., D'Alessio, T., Conforto, S. (2013) Feedback of mechanical effectiveness induces adaptations in motor modules during cycling. Frontiers in Computational Neuroscience, 7, 35.

Author: Cristiano De Marchis
Institute: Biolab3, Department of Engineering, Roma TRE University
Street: Via Vito Volterra 62
City: 00146, Rome
Country: Italy
Email: cristiano.demarchis@uniroma3.it

Simulation-Based Planification Tool for an Assistance-as-Needed Upper Limb Neurorehabilitation Robotic Orthosis

R. Pérez-Rodríguez[1,2], C. Rodríguez[3], F. Molina[1,2], C. Gómez[4], E. Opisso[4], J.M. Tormos[4], J. Medina[4], and E.J. Gómez[1,2]

[1] Bioengineering and Telemedicine Centre, ETSI Telecomunicación, Universidad Politécnica de Madrid, Madrid, Spain
[2] Centro de Investigación Biomédica en Red en Bioingeniería, Biomateriales y Nanomedicina, Madrid, Spain
[3] Cartif Foundation, Valladolid, Spain
[4] Functional neurorehabilitation services, Institut de Neurorehabilitació Guttmann, Badalona, Spain

Abstract—This research work proposes a simulation-based planification tool for the selection of the most suitable configuration parameters of an assistance-as-needed controlled upper limb neurorehabilitation orthosis. The selection of these parameters is performed by applying planification algorithms that analyze the results of a set of simulation batteries. Obtained results demonstrate that the proposed system is able to find those assistance-as-needed configuration parameters that match a selected clinical criterion for a specific patient.

Keywords—Planification tool, simulation, assistance-as-needed, neurorehabilitation, upper limb, activities of the daily living.

I. INTRODUCTION

One of the main objectives of neurorehabilitation is to provide patients with the capacity to perform specific Activities of the Daily Living (ADL) required for an independent life, taking into account that continual practice of fundamentally inappropriate compensatory strategies may be a critical factor limiting recovery after brain damage [1,2]. Although traditional physical therapy can enhance functional recovery after stroke, robotic devices may offer more intensive practice opportunities without increasing time spent on supervision by the treating therapist [3]. This, along with the assertion that traditional therapies are expensive and likely dosage dependant, have caused a remarkable increase in research aimed at creating, controlling and using robotic devices [4,5]. An ideal behavior of these systems would consist in emulating real therapists by providing anticipated force feedback to the patients in order to encourage and modulate neural plasticity.

To provide patients with ADL-based functional rehabilitation under the assistance-as-needed paradigm [6] (which means to assist the subject only as much as is needed to accomplish the task) and without the presence of a therapist but under his/her supervision, is one of the main challenges of the current neurorehabilitation technologies. These assistance-as-needed strategies face one crucial challenge: the adequate definition of the desired limb trajectories regarding space and time that the robot must generate to assist the user during the exercise [7].

Several approaches to the assistance-as-needed paradigm can be found in the scientific literature. Some robotic systems provide an assistance that is proportional to the deviation of the patient given a predefined trajectory. Well known examples of this control strategy are MIT-MANUS [8], MIME [9], GENTLE/S [10], ARMin [11], L-EXOS [12], ReoGo [13] or NeReBot [14]. These assistive robotic therapy controllers focus on the following idea: when the subject moves along a desired trajectory (and an artificially created virtual tunnel), the robot should not intervene, and if the participant deviates from the desired trajectory, the robot must create a restoring force [15]. An anticipatory assistance-as-needed control algorithm that tries to solve the limitations of the aforementioned devices have been recently proposed [16,17].

Dynamic control systems, that are able to adapt to the current needs of the patient based on online performance measurements, can be also found in the scientific literature. The basis of these control strategies is to adapt their configuration parameters tuning the system to the subject changing needs. Riener et al. [18] developed such system for gait rehabilitation by recognizing the patient intention and adapting the level of assistance to the subject's contribution. Regarding the upper limb, only inter-session parameter adaptation methods, that allow the selection of the working parameters once a previous performance measurement is available [19,20], can be found. Finally, some assistance strategies introduce a forgetting factor to keep a challenging assistance level for the patient in order to avoid slacking [9,21,22].

Nowadays, there are not many robotic devices specifically oriented for practicing ADLs, being the ADLER orthosis [23] the most relevant amongst them. This device assists the patient along programmed ADL trajectories providing customized forces through three active.

A full and more exhaustive review of rehabilitation robotics control strategies can be found in [15].

The main goal of this research work is the utilization of a multijoint upper limb robotic orthosis simulator that, under the assistance-as-needed neurorehabilitation paradigm, allows the determination of the most suitable working parameters for a given patient and clinical criterion. A real use case would correspond to that one where before the patient starts a rehabilitation session using a system consisting on a robotic orthosis (along with a VR device to provide the patient with both force and audiovisual feedback), the simulator would be used to help the therapist with the selection of the most appropriated working parameters in order to attain an established clinical criterion.

II. METHODOLOGY

A. Assistance-as-Needed Control Algorithm

The intelligent assistance-as-needed control layer proposed in the authors' previous work [16,17] is based on the computational model of motor control proposed by Shadmehr et al. [24], where the key component are the cerebellum forward models [25,26]: they predict the state of the limb allowing one to act on this estimate of state rather than relying solely on a delayed sensory feedback. In this fashion, the anticipatory assistance-as-needed control algorithm consists in 3 subsystems:

1. A biomechanical motion prediction subsystem that estimates the subject trajectory depending on a previously created dysfunctional profile; this subsystem is composed of 3 modules: a minimum-jerk [27] trajectory generator, a Multilayer Perceptron (MLP)-based inverse kinematics solver module [28] and a dysfunctional adaptation module
2. A decision subsystem that, considering the motion prediction and based on a fuzzy Bayesian decision system [29], determines the alternative that maximizes utility (provide force-feedback or not). The decision is made departing from the measurement of the so called adaptability coefficient over the prediction K_p [17]
3. A command generation subsystem that modulates the orthosis rigidity (κ_i) depending on the assistance decisions [17]. Value κ_i gets increased by an assistance factor (f_a) every time an assistance command takes place and gets decreased by a forgetting factor (f_f) when it is not necessary to physically assist the patient [17]. Here, it is important to remark that the control algorithm analyzes every DoF independently

B. Assistance-as-Needed Simulator

The robotic simulator features a simplified 3 DoF impedance controlled model [30] with the DH configuration [31] presented in Table 1, where L_i values correspond to the segment lengths. This simplified model allows making modifications and seeing the robot behavior in real time. Besides, the orthosis also features a haptic control layer which is responsible for transforming the otherwise rigid, non-backdrivable mechanism, into a compliant robot by means of a joint space impedance controller. The haptic control layer is also capable of rendering several primitives like virtual springs, dampers, walls, and force fields, within the device's workspace boundaries [17].

Table 1 DH parameters of the 3 DoF anthropomorphic (previously rotated $\pi/2$ to be compliant with the world coordinate system)

Link	a_i	α_i	d_i	θ_i	Offset
1	L_1	$-\pi/2$	0	θ_1	27
2	L_2	0	0	θ_2	25
3	L_3	0	0	θ_3	24,5

C. External Torques Simulator

To carry out a full simulation it is also necessary to emulate the external torques that the patient provides to the robotic orthosis. To calculate these moments of force, a dysfunctional report associated to each patient is created [17]. This report contains information of previous ADL executions, and is used to estimate the external torques applied by the patient in such a way that they are proportional to the angular difference between the current biomechanical configuration and its equivalent in the recorded motion [17].

D. Planification Tool

Given the assistance-as-needed simulator, the following parameters can be configured to adapt its behavior to the specific clinical needs: (1) ADL duration, (2) permissivity associated to the calculation of the adaptability coefficient, (3) minimum assistance per DoF, (4) assistance utility per DoF (given by the utility matrix), (5) assistance factor and (6) forgetting factor.

The planification tool will decide the most efficient robot configuration for a given patient and clinical criterion. With this aim, a set of simulations batteries has been designed: the speed values have been set to both normal speed and the speed that the subjects took to complete the task without any assistance; permissivity values of 0 and 10 have been used; the minimum rigidity has been set either to 0 or to a 10% of the maximum that the orthosis could provide; and the forgetting and assistance factors have been set to have two possible values, 0.005 or 0.01.

For each simulation, Pearson correlation coefficient (C), Root Mean Squared Error (RMSE) and assistance percentage (considering that a value of 100% would correspond to the case of providing the maximum rigidity value during all the time instants) are calculated between the patient motion (obtained throughout the simulation process) and a reference model [32] to be used as the metrics to determine the working configuration.

Up to this point of the research, the planification tool is able to determine the most suitable assistance-as-needed configuration for two clinical targets determined by the Institut Guttmann clinical experts: (1) maximize similarity, in which the configuration of the simulation that obtains the highest correlation coefficient is selected, and (2) minimize assistance, where the configuration of the simulation that minimizes the assistance provided to de patient is selected.

III. EXPERIMENTAL WORK

Two different ADLs designed by therapists from the Institut Guttmann Neurorehabilitation Hospital have been used to validate the planification tool: 'serving water from a jar' and 'picking up a bottle'. To build the motion models, data from 73 healthy subjects, 34 men and 39 women with a mean age of 37.97±12.44 years old were captured for the 'serving water from a jar' ADL and from 40 healthy subjects, 17 men and 23 women with a mean age of 30.45±5.25 years old for the 'picking up a bottle' ADL. Both ADLs have been partitioned into several identifiable states taking into account both the ADL landmarks and its associated angular features obtaining a state-chart diagram for each task. In this way, 'serving from a jar' ADL has been fragmented into 6 time intervals and 'picking a bottle from a shelf' into 3. For each time interval, the most suitable assistance-as-needed configuration is determined (for a given patient and clinical criterion).

Five patients, whose data are shown in Table 2, participated in this study. Dysfunctional reports of these non-healthy subjects have been created.

Table 2 Patient data (Et.=Etiology; Fc.=Focality; FM=Fugl-Meyer; M=Male; F=Female; lA=Arm length; lF=Forearm length; lT=Total length; II=Isquemic ictus; HI=Hemorrhagic ictus; E=Encephalopathy; L=Left; R=Right)

ID	Gender	Age	Height	l_A	l_F	l_T	Et.	Fc.	FM
P1	M	69	182	32,5	27	62	II	L	55
P2	F	46	167	27,5	25	59,5	E	R	62
P3	F	64	156	29	24,5	58,5	HI	L	58
P4	M	63	163	31	24	65	HI	R	42
P5	F	47	149	31	19	55,5	II	L	55

IV. RESULTS AND DISCUSSION

Table 3 shows the obtained results. It can be evidenced how the assistance-as-needed control algorithm, and, consequently the robotic orthosis, modify their behavior depending on both the patient and the therapist needs. When the selected clinical criterion is to maximize similarity the highest correlation is obtained at the cost of raising the assistance percentage; when the system operates under the minimize assistance criterion, the assistance percentage is the lowest, but also the similarity gets impoverished. From the simulation results it can be appreciated that only in three cases (out of 30) the planification tool does not attain the given clinical criterion. This is due to simplicity of the algorithms applied to select the working parameters that do not take into account the cross influence between the DoFs. However, these algorithms are sufficient to show the great potential of the robotic simulator as a planification tool.

Table 3 Simulation results (MS=Maximize similarity; mA=Minimize assistance; C=Correlation coefficient; RMSE=Root Mean Squared Error)

		Serving water from a jar									Picking up a bottle								
		C			RMSE			% assistance			C			RMSE			% assistance		
		fexS	abdS	fexE	fexS	abdS	fexE	fexS	abdS	fexE	fexS	abdS	fexE	fexS	abdS	fexE	fexS	abdS	fexE
MS	P1	0,98	0,97	0,92	3,34	1,95	4,85	53,31	37,03	38,77	0,99	0,99	0,96	0,33	0,04	1,31	21,47	27,51	31,90
	P2	0,98	0,97	0,97	2,58	3,04	4,69	73,94	63,80	94,31	0,99	0,91	0,92	0,93	1,81	5,03	64,62	10	17,19
	P3	0,94	0,99	0,96	1,06	0,81	5,44	18,01	43,06	6	0,99	0,99	0,86	1,02	0,45	0,37	64,37	60,80	11,34
	P4	0,99	0,99	0,99	0,38	0,92	0,83	38,94	53,26	81,30	1	0,90	0,95	0,51	4,69	2,47	88,45	51,24	38,03
	P5	0,96	0,97	0,92	2,09	0,60	0,34	31,35	28,97	41,11	0,99	0,88	0,53	5,44	12,42	12,39	86,07	6,76	16,44
mA	P1	0,94	0,92	0,89	7,06	2,96	6,44	6,64	30,96	18,24	0,98	0,96	0,84	1,89	0,39	4,05	3,61	14,31	5,35
	P2	0,93	0,96	0,97	5,32	3,79	5,01	42,30	47,99	88,75	0,96	0,77	0,89	4,36	3,18	7,39	0,90	0	0,00
	P3	0,94	0,91	0,89	0,86	0,82	6,30	3,17	49,20	19,08	0,99	0,96	0,86	1,74	2,14	2,90	55,40	3,65	3,40
	P4	0,97	0,98	0,95	0,32	0,05	2,39	16,84	44,96	60,53	1	0,82	0,92	0,26	4,50	3,51	76,78	0,00	27,54
	P5	0,92	0,93	0,57	3,30	0,04	3,63	13,61	36,04	7,52	0,95	0,98	0,77	4,52	3,07	4,89	48,77	73,64	58,76

V. CONCLUSIONS

In this research, the authors present a simulation-based planification tool for the selection of the most suitable configuration parameters of an assistance-as-needed controlled upper limb neurorehabilitation orthosis. Obtained results demonstrate that, by performing simulations of previously created configuration batteries, the planification tool is able to find those assistance-as-needed configuration parameters that match the selected clinical criterion given a specific patient (and his/her associated dysfunctional profile).

Future work will mainly address the definition of a wider range of clinical criteria (such as enhance patient range of motion, etc.) as well as the design and development of the corresponding parameter selection algorithms and the consideration of the cross influence between the DoFs.

Acknowledgments

This research work was partially funded by CDTI (project: REHABILITA; CIN/1559/2009), Spanish Government.

References

1. Carr J, Shepherd R: A motor learning model for stroke rehabilitation. Physiotherapy 1989, 75:372-380.
2. Davies P: Right in the middle: selective trunk activity in the treatment of adult hemiplegia. Berlin: Springer Verlag 1990.
3. Dobkin BH: Strategies for stroke rehabilitation. The Lancet Neurology 2004, 3(9):528-536.
4. Wolbrecht E, Chan V, Reinkensmeyer D, Bobrow J: Optimizing Compliant, Model-Based Robotic Assistance to Promote Neurorehabilitation. IEEE Transactions on Neural Systems and Rehabilitation Engineering 2008, 16(3):286-297.
5. Conesa L, Costa U, Morales E, Edwards DJ, Cortes M, Leon D, Bernabeu M, Medina J: An observational report of intensive robotic and manual gait training in sub-acute stroke. Journal of NeuroEngineering and Rehabilitation 2012, 9:13.
6. Emken J, Bobrow J, Reinkensmeyer D: Robotic movement training as an optimization problem: designing a controller that assists only as needed. In Rehabilitation Robotics, 2005. ICORR 2005. 9th International Conference on 2005:307-312.
7. Belda-Lois JM, Mena-del Horno S, Bermejo-Bosch I, Moreno JC, Pons JL, Farina D, Iosa M, Molinari M, Tamburella F, Ramos A, et al.: Rehabilitation of gait after stroke: a review towards a top-down approach. Journal of neuroengineering and rehabilitation 2011, 8:66.
8. Krebs H, Palazzolo J, Dipietro L, Ferraro M, Krol J, Rannekleiv K, Volpe B, Hogan N: Rehabilitation robotics: Performance-based progressive robot-assisted therapy. Autonomous Robots 2003, 15:7-20.
9. Lum PS, Burgar CG, Shor PC, Majmundar M, Van der Loos M: Robot-assisted movement training compared with conventional therapy techniques for the rehabilitation of upper-limb motor function after stroke. Archives of Physical Medicine and Rehabilitation 2002, 83(7):952-959.
10. Loureiro R, Harwin W: Reach & grasp therapy: design and control of a 9-DOF robotic neurorehabilitation system. In Rehabilitation Robotics, 2007. ICORR 2007. IEEE 10th International Conference on 2007:757-763.
11. Nef T, Mihelj M, Riener R: ARMin: a robot for patient-cooperative arm therapy. Medical & Biological Engineering & Computing 2007, 45(9):887-900.
12. Montagner A, Frisoli A, Borelli L, Procopio C, Bergamasco M, Carboncini M, Rossi B: A pilot clinical studyon robotic assisted rehabilitation in VR with an arm exoskeleton device. In Virtual Rehabilitation, 2007 2007:57-64.
13. Bovolenta F, Sale P, Dall'Armi V, Clerici P, Franceschini M: Robot-aided therapy for upper limbs in patients with stroke-related lesions. Brief report of a clinical experience. Journal of NeuroEngineering and Rehabilitation 2011, 8:18.
14. Rosati G, Gallina P, Masiero S: Design, Implementation and Clinical Tests of a Wire-Based Robot for Neurorehabilitation. IEEE Transactions on Neural Systems and Rehabilitation Engineering 2007, 15(4):560-569.
15. Marchal-Crespo L, Reinkensmeyer D: Review of control strategies for robotic movement training after neurologic injury. Journal of NeuroEngineering and Rehabilitation 2009, 6:20-35.
16. Pérez-Rodríguez R, Costa U, Rodríguez C, Cáceres C, Tormos JM, Medina J, Gómez EJ: Assistance-as-needed robotic control algorithm for physical neurorehabilitation. In International Conference on Recent Advances in Neurorehabilitation, Valencia (Spain) 2013.
17. Pérez-Rodríguez R: Metodologías de modelado, monitorización y asistencia robótica en neurorrehabilitación funcional de extremidad superior. ETSI Telecomunicación, Universidad Politécnica de Madrid 2013.
18. Riener R, Lunenburger L, Jezernik S, Anderschitz M, Colombo G, Dietz V: Patient-Cooperative Strategies for Robot-Aided Treadmill Training: First Experimental Results. IEEE Transactions on Neural Systems and Rehabilitation Engineering 2005, 13(3):380-394.
19. Krebs H, Palazzolo J, Dipietro L, Ferraro M, Krol J, Rannekleiv K, Volpe B, Hogan N: Rehabilitation robotics: Performance-based progressive robot-assisted therapy. Autonomous Robots 2003, 15:7-20.
20. Kahn L, Rymer W, Reinkensmeyer D: Adaptive assistance for guided force training in chronic stroke. In Engineering in Medicine and Biology Society, 2004. IEMBS'04. 26th Annual International Conference of the IEEE, Volume 1 2004:2722-2725.
21. Wolbrecht E, Chan V, Le V, Cramer S, Reinkensmeyer D, Bobrow J: Real-time computer modeling of weakness following stroke optimizes robotic assistance for movement therapy. In Neural Engineering, 2007. CNE'07. 3rd International IEEE/EMBS Conference on 2007:152-158.
22. Mihelj M, Nef T, Riener R: A novel paradigm for patient-cooperative control of upper-limb rehabilitation robots. Advanced Robotics 2007, 21(8):843-867.
23. Johnson MJ, Wisneski K, Anderson J, Nathan DE, Strachota E: Task-oriented and Purposeful Robot Assisted Therapy. International Journal of Advanced Robotics Systems, Vienna, Austria 2007:978-973.
24. Shadmehr R: Computational Approaches to Motor Control. Encyclopedia of Neuroscience 2009, 3: 9-17.
25. Wolpert D: Internal models in the cerebellum. Trends in cognitive sciences 1998, 2(9):338-347.
26. Miall R, Wolpert DM: Forward Models for Physiological Motor Control. Neural Networks 1996, 9(8):1265-1279.
27. Flash T, Hogan N: The coordination of arm movements: an experimentally confirmed mathematical model. The journal of Neuroscience 1985, 5(7):1688-1703.
28. Pérez-Rodríguez R, Marcano-Cedeño A, Costa r, Solana J, Cáceres C, Opisso E, Tormos JM, Medina J, Gómez EJ: Inverse Kinematics of a 6 DoF Human Upper Limb using ANFIS and ANN for anticipatory actuation in ADL-based physical Neurorehabilitation. Expert Systems with Applications 2012, 39(10): 9612–9622.
29. Ross T: Fuzzy Logic with Engineering Applications. U.K.: John Wiley & Sons, 2nd edition 2004.
30. Hogan N: Impedance Control: An Approach to Manipulation. Journal of Dynamic Systems, Measurement, and Control 1985, 107(March):1-23.
31. Denavit J, Hartenberg R: A kinematic notation for lower-pair mechanisms based on matrices. Transactions of ASME 1955, 22(77):215-221.
32. Costa U, Opisso E, P_erez R, Tormos JM, Medina J: 3D motion analisys of activities of daily living: implication in neurorehabilitation. In International Gait and Clinical Movement Analysis Conference, Miami (USA) 2010.

Author: Rodrigo Pérez-Rodríguez
Institute: Bioengineering and Telemedicine Centre, ETSI Telecomunicación, Universidad Politécnica de Madrid
Street: Avda. Complutense 30, 28040
City: Madrid
Country: Spain
Email: rperez@gbt.tfo.upm.es

Total Joint Replacement: Biomaterials for Application in the Temporomandibular Joint

N. Fakih-Gomez[1], L.M. Gonzalez-Perez[1], and B. Gonzalez Perez-Somarriba[2]

[1] Virgen del Rocio University Hospital/ Department of Maxillofacial Surgery, Seville, Spain
[2] School of Engineering, University of Seville, Spain

Abstract—End-stage temporo-mandibular joint (TMJ) disease presents a challenging biomechanical problem. The primary function of joint replacement surgery is to relieve pain and restore function, which includes transmitting physiological loads and the provision of both a physiological range of movement and an articulation with minimum friction and wear. It has been demonstrated that the use of appropriate biomaterials and design parameters can decrease material wear and increase the longevity of TMJ replacement (TMJR) devices. Therefore, as with any implanted functioning biomechanical device, surgical revisions may be necessary to remove or replace the articulating components due to material wear or failure. The purpose of this study is to evaluate the criteria for the successful use of TMJR devices and thereby establish a rationale for the use of these devices in the long-term management of end-stage TMJ disorders.

Keywords—Cobalt-chromium alloys for surgical implants, ultra-high molecular weight polyethylene for surgical implants, titanium alloys for surgical implants, total joint replacement, temporomandibular joint.

I. Introduction

The history of alloplastic joint reconstruction has been characterized by multiple failures based on inappropriate design, lack of attention to biomechanical principles, and ignorance of what already has been documented in the biomaterials literature [1]. TMJR is a biomechanical rather than a biological solution to end-stage TMJ disease. TMJR have been used clinically for over 15 years in its present form, and remains today as one of the most successful applications of prosthetic TMJ surgery. The number of TMJR procedures is increasing at a significant rate. The success of these procedures, the increased longevity of the population, the demand for increased quality of life and more active lifestyles, the earlier onset and diagnosis of degenerative diseases, and the correction of congenital anomalies mean that TMJR is now undertaken in a broad age-range of patients. This has placed increased demands on both, the design and performance of the prostheses [2].

The physical environment into which the joint replacement is implanted is extremely challenging. Not only does it have particular chemical, biochemical, biological and biomechanical characteristics, but also the fact that the tissue surrounding the prosthetic components remains living means that the joint replacement interface and environment can undergo continual change along time. These changes are not only related to the natural ageing of the patient, but also can occur in response to the function and properties of the prosthetic device itself. This results in a complex interactive biological and biomechanical environment involving the living tissue and prosthetic joint in the body which can determine the lifetime of the replacement joint. Over the years it has proven the difficulty to predict preclinically many of these interactions, and it has only been as a result of clinical experience and research that particular clinical failure and success scenarios have emerged. This has resulted in more rigorous and demanding requirements for joint replacement designs and materials (Table 1) [3].

Table 1 Biomaterials for application in TMJ

Metals	Polymers
316L stainless steel	Polyurethane (PE)
Co-Cr alloys	Ultra-high molecular weight polyethylene (UHMWPE)
Titanium	
Titanium-6 Aluminum-4 Vanadium Alloy	Polymethylmethacrylate (PMMA)

Despite the bone resorption and adverse tissue reactions initially reported in the early 1990s with Proplast-Teflon and poly-tetra-fluoro-ethylene TMJ prostheses, it became clear that wear debris was the major cause of osteolysis and loosening in joint replacement. Studies of retrieved tissues showed an abundance of micron and submicron-sized wear particles, which were also found in laboratory wear studies. These particles were shown to stimulate macrophages to release osteolytic cytokines, which lead to osteolysis and bone resorption [4]. More recent studies have shown that the submicron wear particles (0.1–1 μm) are more reactive than larger particles (>1 μm) and that the release of osteolytic cytokines is dependent on the volumetric concentration of the submicron wear particles. This has led to the study of the volume distribution of the wear particles in different size ranges to analyze their osteolytic potential [5].

Many of these devices are supported by pre-clinical simulation tests which indicate improved performance compared to traditional technologies. The ultimate test is the long-term clinical follow-up. Until this is established there will always remain a degree of uncertainty surrounding any new joint replacement technology.

The objective of this article is to evaluate the criteria for the successful use of TMJR devices and thereby establishing a rationale for the use of TMJR devices in the long-term management of end-stage TMJ disorders.

II. MATERIAL AND METHODS

21 patients (16 females, 5 males) involving 24 joints (18 unilateral, 3 bilateral) were operated on, and 24 total joint prostheses (Biomet Microfixation TMJ Replacement System, Jacksonville, FL, USA) were fitted. The mandibular component was manufactured from Cobalt-28 Chromium-6 Molybdenum Alloy conforming to ASTM F1537 standard with a roughened titanium alloy plasma spray porous coating on the host bone side of the ramus plate for increased bony integration. The fossa prosthesis was made of Ultra High Molecular Weight Polyethylene (UHMWPE) conforming to ASTM F648 standard. The system's screws were made of Titanium-6 Aluminum-4 Vanadium Alloy conforming to ASTM F136 standard. The mean age at surgery was 55years ±10.5 (Mean ± SD). All patients had: 1) history of persistent and significant pain and functional impairment; 2) clinically and radiographically documented end-stage TMJ disease. All patients were included in a 5-year prospective follow-up study. Change in pain intensity (preoperative vs. current) was measured by two methods: pain experience (1 to 10) and pain intensity (visual analog scale, 1 to 10). Jaw opening, chewing ability and joint noise were evaluated (visual analog scales, 1 to 10). Surgical morbidity and implant survival were documented. The study received prior approval from the Committee for Clinical Ethics, and informed consent was obtained from all patients before inclusion in the study.

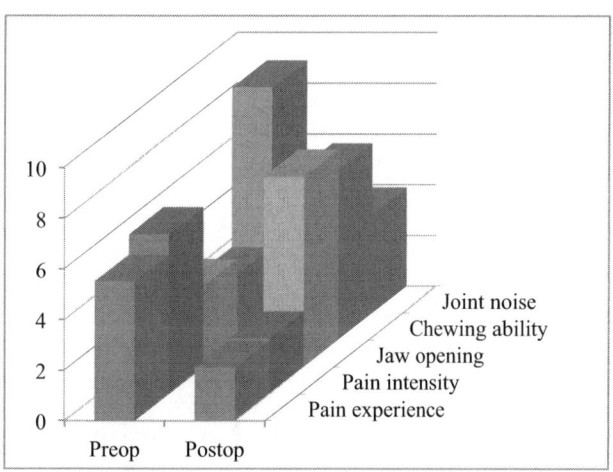

Fig. 1 Clinical changes after TMJR

III. RESULTS

After treatment pain experience, pain intensity and joint noise were reduced 6.1±1.5, 6.5±2.3 and 5.7±1.3, respectively. Jaw opening and chewing ability were improved by 5.8±1.3 and 6.2±1.5, respectively (Fig. 1). The statistical analysis was carried out with SPSS 15.0 software. None of the implants were removed during the study period. Patients' satisfaction with the clinical outcome was 9 on a scale of 1 to 10.

IV. DISCUSSION

TMJR is a complex interactive biomechanical and biological system, and, while extensive preclinical development and evaluation is now undertaken, long-term clinical results remain the ultimate test for new designs, materials and bearings. It should be noted that none of our 21 patients have clinical histories well beyond 5 years and the proposed improved wear performance and reduction in osteolytic potential has still to be demonstrated in clinical practice [6]. It should be recognized that although bearing surface performance has traditionally been compared by measuring volumetric wear, the osteolytic potential of the bearing is dependent on the nature of the wear debris. Size, morphology, chemistry, its distribution function and volumetric concentration all have to be considered in the analysis of the performance of different bearing materials.

Improvements in design can have considerable impact on function [7]. Design and material wear characteristics related to longevity must be considered in relation to the following factors:

A. TMJR and Established Criteria for Successful Alloplastic TMJ Devices

There are two categories of TMJR devices approved for implantation: off-the-shelf (*stock*) devices which the surgeon has to 'make fit' at implantation, and patient-fitted (*custom*) devices which are 'made to fit' in each specific case. Stock TMJR systems with multiple 'make fit' choices, designed and manufactured from either thin cast Co–Cr fossa or all UHMWPE fossa components, utilizing cast Cr–Co ramus/condyle components, can pose multiple design and material issues.

From 2010 onwards, nearly all TMJ prostheses implanted in our country have polyethylene glenoid fossa cups (Fig. 2). This prompted considerable research into factors that caused acceleration of the wear of polyethylene as well as the development of alternative bearing surfaces and new technologies to reduce wear and osteolysis. During the first decade of large joint (hip, knee) replacement, the majority of polyethylene components were sterilized using gamma irradiation in the presence of air. During the early 1990s it emerged that the irradiation, which causes chain scission and free radicals, renders the material unstable and subject to oxidative degradation causing a reduction in its mechanical properties and an increase in the wear rate. As well as a higher wear rate, the oxidized materials also produce small particles with greater osteolytic potential. The majority of condylar heads was constructed from polished metal alloys which were shown to become scratched and damaged, resulting in accelerated wear. The widespread recognition of the role of polyethylene wear debris-induced osteolysis in the long-term failure of TMJ prostheses has led to a new generation of designs and bearing materials for TMJR [7].

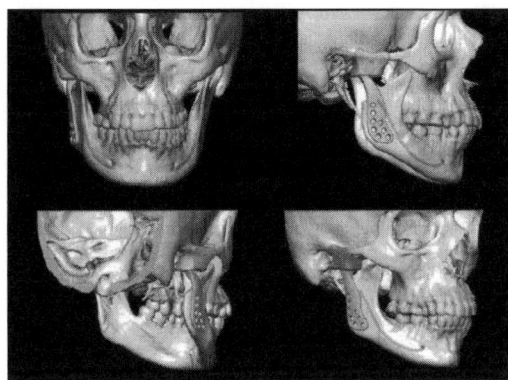

Fig. 2: Custom-made TMJR designed with CAD-CAM technology from a computed tomography scan may be indicated as in this case of end-stage TMJ disease.

B. *The Components of any TMJR Device Must Be Stable in Situ at Implantation*

The prosthetic materials must be anatomical in shape, be securely fixed to the surrounding bone, and remain securely fixed throughout the patient's lifespan. Preferably, the prosthetic components should be implanted with minimum bone resection. The TMJR has functional movements that are unconstrained. Stresses and strains directly or eccentrically vectored against an incomplete or inadequate component to host-bone interface during TMJR create wear. Unstable, thin, cast Co–Cr fossa cyclically loaded by the metal condylar head can lead to micromotion, galling, fretting corrosion, component screw loosening and/or thin cast metal fossa component fatigue and fracture. Cold flow is the property which allows UHMWPE under loading to develop alteration of shape rather than particulation. In TMJR, this property dictates that the stable component of a TMJR articulation (i.e. the glenoid fossa) is held in position and stabilized by a stronger material (metal) [8].

Custom TMJR fossa components are designed and manufactured to material specifications. Further, the metallic component of a custom fossa offers solid structure through which the zygomatic arch fixation screws pass. Stock TMJR devices with an UHMWPE flange screw fixation design have the potential to develop material cold flow around the screw holes or fracture should micromotion occur if the surgeon cannot or does not make the fossa component fit properly. Cold flow of the resultant screw fixation hole can lead to loosening of the stock fossa fixation screws and increased micromotion under repetitive loading, resulting in device failure [1].

C. *The Materials Must Be Biocompatible and Able to Withstand the Forces of Mandibular Function*

The biomaterials from which the implant is made must be biocompatible, and any wear particles produced must also be compatible with the body and not cause adverse biological reactions. The joint replacements have to be compatible with a range of different patient anatomies and geometries and typically a range of different sizes is necessary. Similarly, the bone quality of patients is quite variable and the methods of fixation have to be able to accommodate different bone interface conditions.

Employing the most advantageous physical characteristics of biocompatible materials is an essential consideration in the design and manufacture of any TMJR device. Cr-Co-Mb, with its relatively high carbon content, contributes to its strength, polishability, and biocompatibility. Its excellent wear characteristics when articulated against an UHMWPE presently make it the standard for the non-moveable articulating surface of most orthopaedic total joint replacement devices [4].

Cobalt-based alloys were initially used as an orthopedic biomaterial because they were more corrosion-resistant than stainless steel. Cast Cr–Co, often employed in the manufacture of stock TMJR devices, is biomechanically inferior to any wrought alloy. Metallurgical flaws such as inclusions and porosity found in cast Cr–Co components have been associated with the fatigue failure of metal-on-metal prostheses. These flaws may also lead to the failure of Cr–Co TMJR components, resulting in noxious metallic debris (metalosis) found in adjacent tissues. UHMWPE is a linear unbranched polyethylene chain with a molecular weight of more than one million. Testing over one decade of use in TMJR has led to the conclusion that UHMWPE is considered to have excellent wear and fatigue resistance for a polymeric material. To date, no cases of UHMWPE particulation-related osteolysis have been reported in the TMJR literature [4].

D. TMJR Devices Must Be Designed to Withstand Loads Delivered over the Full Range of Function of the TMJ to Be Replaced

Stock fossa components are designed without a posterior stop to prevent the TMJR device condyle from displacing posteriorly. If the stock condyle is not perfectly aligned in the centre of the stock fossa, medio-laterally and/or antero-posteriorly, the condyle can be displaced posteriorly, and can impinge on the tympanic plate and/or the auditory canal. This can result in pain and mandibular dysfunction and facial deformity. There is also the potential for infection in relation with a pressure-related perforation associated with the auditory canal. This is of special concern when using a stock TMJR in combination with another surgical procedure. The custom TMJR fossa has a posterior stop, alleviating this concern [6]. Since the components of a custom TMJR interface so well with the host bone and the screw fixation is stable from implantation, mandibular function can begin immediately after implantation (Fig. 3).

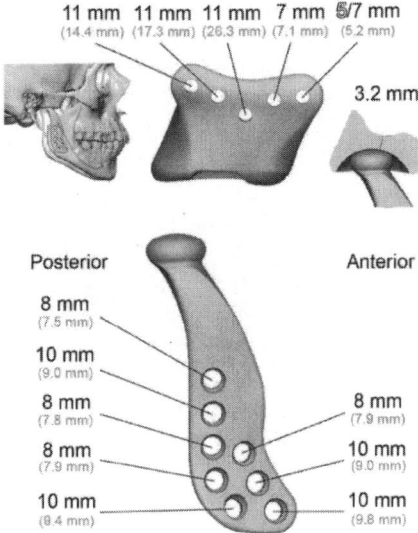

Fig. 3 3D photo processing techniques may be used to fabricate patient-specific alloplastic devices for improved compliance and efficacy

V. Conclusions

TMJR has been one of the major successes in TMJ surgery over the last decade. The surgical placement of a TMJ prosthesis significantly reduces pain and dysfunction secondary to advanced disease. Clinical success, long-term results, and increased expectation and lifetimes of patients have driven the need for improved materials, bearing surfaces, and new designs. It has been demonstrated that the use of appropriate biomaterials and design parameters can decrease material wear and increase the longevity of TMJR devices. Tried and tested stabilized polyethylene bearings articulating against polished metal condylar components have a very high probability of providing more than 10 years successful clinical use. Different designs, materials and bearings are available for clinical use in large joints (hip, knee). However, when used in the TMJ, the potential long-term uncertainties outweigh the benefits, and the new technological solutions require rigorous and effective clinical follow-up. The result of this case series supports further investigation of this form of surgical treatment in a rigorously controlled prospective analysis.

References

1. Driemel O, Braun S, Müller-Richter UD et al. (2009) Historical development of alloplastic temporomandibular joint replacement after 1945 and state of the art. Int J Oral Maxillofac Surg 38: 909-920
2. Quinn PD (2000) Total Temporomandibular Joint Reconstruction. Oral Maxillofac Surg Clin North Am 12: 93-104
3. Kashi A, Saha S, Christensen RW (2006) Temporomandibular joint disorders: artificial joint replacements and future research needs. J Long Term Eff Med Implants 16: 459-474
4. Giannakopoulos HE, Sinn DP, Quinn PD (2012) Biomet Microfixation Temporomandibular Joint Replacement System: A 3-Year Follow-Up Study of Patients Treated During 1995 to 2005. J Oral Maxillofac Surg 70: 787-794
5. Machon V, Hirjak D, Beno M et al. (2012) Total alloplastic temporomandibular joint replacement: the Czech-Slovak initial experience. Int J Oral Maxillofac Surg 41: 514–517
6. Voiner J, Yu J, Deitrich P et al. (2011) Analysis of mandibular motion following unilateral and bilateral alloplastic TMJ reconstruction. Int J Oral Maxillofac Surg 40: 569–571
7. Westermark A, Leiggener C, Aagaard E et al. (2011) Histological findings in soft tissues around temporomandibular joint prostheses after up to eight years of function. Int J Oral Maxillofac Surg 40: 18–25
8. Linsen SS, Reich RH, Teschke M (2012) Mandibular Kinematics in Patients with Alloplastic Total Temporomandibular Joint Replacement—A Prospective Study. J Oral Maxillofac Surg 70: 2057-2064

Author´s address:
N. Fakih-Gomez MD and L.M. Gonzalez-Perez MD, DDS
Department of Maxillofacial Surgery. Virgen del Rocio University Hospital
Avenida Manuel Siurot s/n. 41012 Sevilla. Spain.
Email: lumigon@telefonica.net

Kinematic Indexes' Reproducibility of Horizontal Reaching Movements

G. D'Addio[1], L. Iuppariello[2], M. Romano[2], F. Lullo[1], N. Pappone[1], and M. Cesarelli[2]

[1] S. Maugeri Foundation, Rehabilitation Institute of Telese Telese Terme, Italy
[2] Dept. of Biomedical, Electronics and TLC Engineering University of Naples, "Federico II", Naples, Italy

Abstract—Upper limb reaching movements are the most used motor task both in diagnosis and rehabilitation treatments of several movement disorders of the arm and the shoulder of various central and peripheral etiology. One of the most appealing features of these new technologies consists in the possibility to record and measure motions and mechanical indexes, allowing quantitative kinematics evaluations, while traditional clinical scales permit only qualitative and potentially disagreeing evaluations, since carried out by different therapists. Few papers have actually addressed this issue and aim of the study is the assessment the reproducibility of kinematics indexes of upper arm horizontal reaching movements in normal subjects during RMT. We studied 10 normal subjects (35±8 year old, males). Each subject underwent to 2 sessions, repeating each reaching protocol at four different target amplitudes of 15°, 20°, 25° and 30° and at three different target velocity, respectively of 20°/s, 30°/s and 40°/s, with at least intervals of 15' of resting time between each trial. Reaching movements have been performed by a shoulder rehabilitation, the Multi-Joint-System (MJS) of the Tecnobody, equipped by a four freedom ranges mechanical arm, was used. Although all parameters generally showed a poor reproducibility, it can be noted that both kurtosys and symmetry indexes showed the lower bias values, ranging between 8 to 9%, while both symmetry and smoothness indexes showed doubled bias values. The kurtosys is also the parameter with the lower standard deviation of the bias and so it can be considered as the more reproducible kinematic index between the studied parameters. Bland-Altman plots showed that the scatter of all kinematic indexes measures around the bias line is not constant but getting larger as the average gets higher, like for example for the kurtosys, or getting larger as the average gets lower, like for example for the skewness.

Keywords—upper limb, kinematics, reproducibility, robot – mediated – therapy, neurological rehabilitation.

I. INTRODUCTION

Upper limb reaching movements are surely the most used motor task both in diagnosis and rehabilitation treatments of several movement disorders of the arm and the shoulder of various central and peripheral etiology. This kind of movements are at the base of the use of most of the new rehabilitative technologies, as robot mediated therapies, virtual reality and motion capture systems in exergaming contest. But one of the most appealing features of these new technologies consists in the possibility to record and measure motions and mechanical indexes, allowing quantitative kinematics evaluations, while traditional clinical scales permit only qualitative and potentially disagreeing evaluations, since carried out by different therapists [1].

Kinematic analysis can yield quantitative data about movement patterns that can help clinicians to better address the rehabilitation protocols providing information not captured using clinical measures, and usable as reliable outcome measures in clinical upper limb rehabilitative settings.

Despite fundamental to this scope is the knowledge of the uncertainty about kinematics indexes measurements, very few paper [2] have actually addressed the problem of reproducibility of these indexes; perhaps because kinematics variables are task specific so that reliability should be interpreted in the context of task requirements, depending on type, amplitude and velocity of the motor task and because has not yet reached a common consensus on kinematics indexes and algorithms to use.

Aim of the paper is the study of the reproducibility of kinematics indexes during upper arm horizontal reaching movements at different amplitude and velocity target in normal subjects during robot assisted therapy.

II. MATERIALS AND METHODS

A. Study Population and Motor Task

We studied 10 normal subjects (35±8 year old, males). Each subject underwent to 2 sessions, repeating each reaching protocol at four different target amplitudes of 15°, 20°, 25° and 30° and at three different target velocity, respectively of 20°/s, 30°/s and 40°/s, with at least intervals of 15' of resting time between each trial.

Reaching movements have been performed by a shoulder rehabilitation, the Multi-Joint-System (MJS) of the Tecnobody, equipped by a four freedom ranges mechanical arm, was used. (Fig.1). During the exercise, subjects were asked to seat on the ergonomic robot chair with the trunk erected, neck straight fixing the central green starting point on the front monitor. The arm under test holding the robot grip by the hand in a position parallel to the floor at 90° with the trunk, the arm not under test on side handle close to the seat. The task required each subject to move from the center position to the right/left target, returning to the center, with a sequence of two movements measured and performed by their dominant arm and whose further details have already been described [3-5].

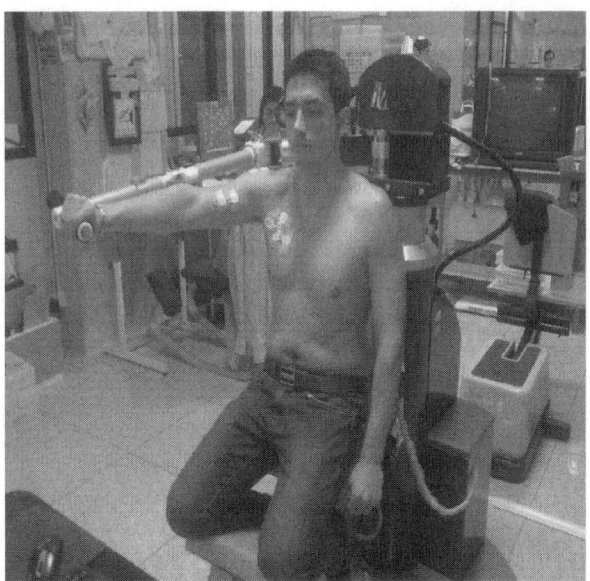

Fig. 1 MJS Robot for upper arm, rehabilitation. The picture shows a subject sit on the ergonomic chair of the robot making his exercise

We assumed the handle position right on the target if within 1 cm radius from the target circle for more than 100 ms, as indicated by an audiovisual biofeedback.

B. Movement Detection Algorithm

Spatial coordinates of the handle position were recorded with a 1/10° degree resolution and a 20 Hz sampling rate.

Movements were automatically detected by a threshold level on the velocity profile by means of a moving average derivative filter with trade-off features between a low-pass filtering and theoretical derivative high-pass transfer function. Movement's onset/end times were calculated as the first velocity zero crossing respectively before/after the movement detection interval. Since also small handle jerks, causing high noise on the velocity profile, can lead to false detections, movements associated to velocity values lower than 30% of the peak velocity or narrower than 0.5 s were excluded.

C. Quantitative Kinematic Analysis

Quantitative kinematic analysis of the reaching movements has been described both by local morphological indexes and by global mathematical indexes.
The main series of the local morphological indexes consists on amplitude and duration of the position signal, mean and peak velocity and simmetry coefficient, calculated as the ratio between the time interval from the peak of the velocity to the end of the movement (deceleration time) and from the onset to the peak of the velocity (acceleration time).

In addition to this main series, movements have been described by their smoothness. This parameter is based on the minimum jerk theory stating that any movement will have maximum smoothness when the magnitude of the J parameter given by Eq.(1) is minimized over the duration of the movement.

$$J = \int_0^d \left| \frac{d^3 x}{dt^3} \right| dt \quad (1)$$

Jerk is the rate of the change of acceleration with respect to time, (third time derivative of the position) and some studies conclude that humans by nature tend to minimize the jerk parameter over the duration of the reaching movement of the arm [6].

All these local indexes represent a synthetic morphological description of the movement depending on the onset/end movement detection times. A different way to describe each movement is to consider global mathematical indexes based on the consideration of the bell-shaped gaussian-like morphology of the velocity profile of the movement [7]. On this basis, the signal can be described statistically like a probability density function respectively by means of k-order moments or k-orders central moments of Eq.(2) and (3).

$$M_k = E[X^k] \quad (2)$$

$$M_k = E[(X - M_1)^k] \quad (3)$$

Of particular interest appear the third and fourth order central moment, respectively named skewness and kurtosis. The skewness coefficient describes the simmetry of the shape, with a zero value in case of simmetry, a positive or a negative value respectively in case of a right or left asymmetry. The kurtosis coefficient describes the flatness of the shape, with a zero value (normokurtosis) in case of a gaussian bell-shape flatness, a positive (leptokurtosis) or a negative (platikurtosis) value respectively in case of a shape more pointed or flatter than a Gaussian bell-shape. This approach has the advantage to not depend on the level of noise or on the movements detection threshold since their values depend only globally from the signal shape allowing to obtain indexes more standardized an comparable between different study populations.

D. Statistical Kinematic Analysis

Reproducibility of all kinematics indexes has been studied by Bland-Altaman technique, plotting the percentual difference between the two repeated measurements against their mean value, which can be assumed as the best estimate that we have of the true value.

Removing the variation between subjects, Bland-Altman plots describe the bias and the standard deviation respectively as the average and the standard deviation of the difference between the two measures.

The last one is used to calculate the limits of agreement, computed as the mean bias plus or minus 1.96 times its standard deviation, accordingly to definition of repeatability coefficient. Any feature measure should lie within the limits of agreement approximately the 95% of the time.

III. RESULTS

We studied a total of two repeated sets of 240 horizontal reaching movements, with means and standard deviation at four amplitude's values and at three different target velocity as shown in Table 1

Table 1 Horizontal reaching movements

	Ampl°	velocity 20°/s	velocity 30°/s	velocity 40°/s
Symmetry	15°	3.6 ± 3.7	1.2 ± 0.8	1.5 ± 0.7
	20°	1.1 ± 0.6	1.1 ± 0.4	1.1 ± 0.4
	25°	1.6 ± 0.9	1.6 ± 1.0	1.4 ± 0.8
	30°	1.7 ± 0.8	1.4 ± 0.8	1.5 ± 0.7
Skewness	15°	-0.1 ± 0.5	0.1 ± 0.4	-0.04 ± 0.56
	20°	0.2 ± 0.3	0.2 ± 0.4	0.1 ± 0.4
	25°	0.2 ± 0.4	0.2 ± 0.4	0.1 ± 0.4
	30°	0.1 ± 0.4	0.04 ± 0.43	-0.03 ± 0.40
Kurtosys	15°	2.4 ± 0.5	1.8 ± 0.3	2.0 ± 0.5
	20°	1.9 ± 0.4	1.8 ± 0.4	1.8 ± 0.4
	25°	2.0 ± 0.6	1.8 ± 0.3	1.7 ± 0.4
	30°	2.1 ± 0.5	0.6 ± 0.4	1.7 ± 0.3
Smoothness	15°	0.3 ± 0.2	0.2 ± 0.2	0.3 ± 0.3
	20°	0.4 ± 0.4	0.5 ± 0.4	0.8 ± 0.7
	25°	0.4 ± 0.3	0.4 ± 0.2	0.8 ± 0.6
	30°	0.4 ± 0.2	0.6 ± 0.4	0.8 ± 0.6

Table 2 Bias and standard deviation of bias for all reaching movements

	Simmetry	Skewness	Kurtosys	Smoothness
Bias	16.1	-9.23	-8.31	-17.5
SD of Bias	63.8	97.45	25.5	74

Fig 2 to 5 show Bland-Altman plots of studied kinematics indexes for reaching movements performed at all the three tested target velocity and at all the amplitude's values.

Bias and standard deviation of bias for all reaching movements are reported in Table 2.

IV. CONCLUSIONS

Results allow to point out the following findings.

Although all parameters generally showed a poor reproducibility, it can be noted that both kurtosys and symmetry indexes showed the lower bias values, ranging between 8 to 9%, while both symmetry and smoothness indexes showed doubled bias values. The kurtosys is also the parameter with the lower standard deviation of the bias and so it can be considered as the more reproducible kinematic index between the studied parameters.

Bland-Altman plots showed that the scatter of all kinematic indexes measures around the bias line is not constant but getting larger as the average gets higher, like for example for the kurtosys, or getting larger as the average gets lower, like for example for the skewness. This indicates that both the movements very large or very fast that those very short or very slow show the poorest reproducibility. This suggest to limits protocols for the evaluation of the kinematic performance of the upper arm rather to horizontal reaching movements of medium velocity, ranging around 30°/s, and medium amplitude reaching movements, ranging between 20° to 25°. Kinematic analysis is a useful tool for investigating upper limb movement through objective description of movement quality for a specific task, nevertheless only reliable kinematic variables should be used as outcome measures in clinical trials. To overcome the high variability of these indexes and their general poor reproducibility, future developments must be addressed in the direction of specifically oriented protocols characterized by a

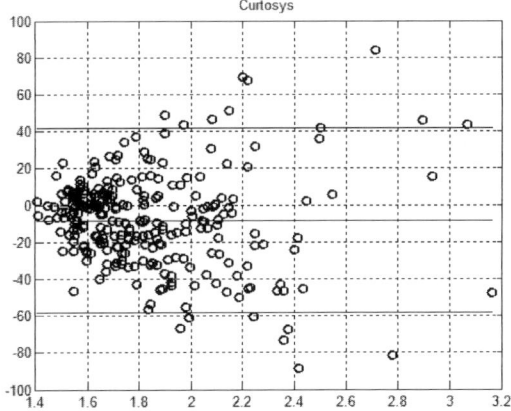

Fig. 2 Bland-Altman plot of Kurtosys for upper arm horizontal reaching movements at different amplitude and velocity target. Percentual values of the differences on Y axis. Mean values between measures on X axis. Bias and standard deviation of bias as superimposed continuous lines.

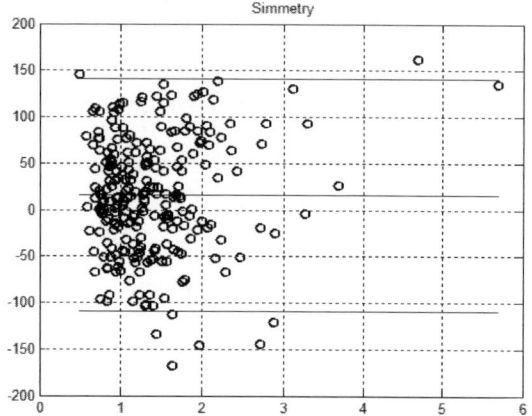

Fig. 3 Bland-Altman plot of Simmetry for upper arm horizontal reaching movements at different amplitude and velocity target

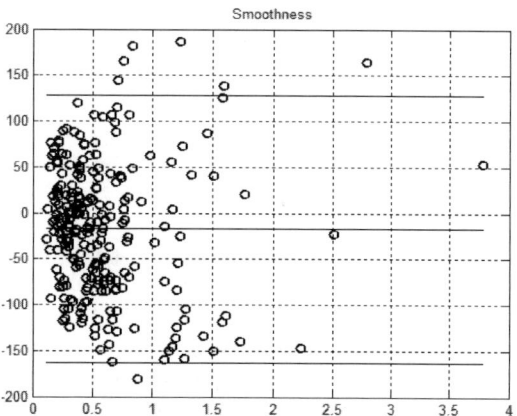

Fig. 4 Bland-Altman plot of Skewness for upper arm horizontal reaching movements at different amplitude and velocity target. Percentual values of the differences on Y axis. Mean values between measures on X axis. Bias and standard deviation of bias as superimposed continuous lines.

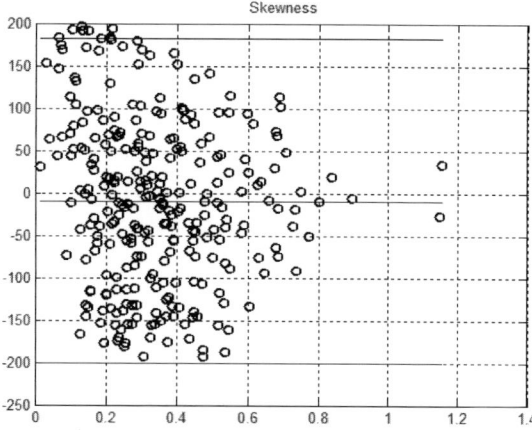

Fig. 5 Bland-Altman plot of Smoothness for upper arm horizontal reaching movements at different amplitude and velocity target.

set of homogeneous repeated reaching movements. This could allow to obtain a better reliable movement profile trough averaging and normalizations algorithms, like for example is achieved for gait analysis studies, considering the evaluation of kurtosis and skewness indexes rather applied to the average of a sequence of medium amplitude and velocity horizontal reaching movements.

REFERENCES

1. Krebs HI, Volpe BT, Aisen ML et al (2000). Increasing productivity and quality of care: robot-aided neuro-rehabilitation. J Rehabil Res Dev;37:639-52.
2. D'Addio G., Cesarelli M., Lullo F., Clemente F., De Nunzio A., Pappone N. (2012). Reproducibility of Kinematics Indexes of Upper Arm Reaching Movement in Robot Assisted Therapy; IEEE Proceedings of the International Symposium on Medical Measurements and Applications. ISBN 978-1-4673-0880-9:pp 184-187.
3. Cesarelli M, D'Addio G, Romano M, De Nunzio, A.M, Pappone N. (2011).Kinematics patterns of upper arm reaching movement in robot-mediated therapy. Proceedings of IEEE International Workshop on Medical Measurements and Applications Proceedings, ISBN: 978-1-4244-9336-4, pp   465 – 468.
4. D'Addio G., Cesarelli M., Romano M., Faiella G., Lullo F.,Pappone N.. Kinematic and EMG Patterns Evaluation of Upper Arm Reaching Movements in Robot Assisted Therapy. IEEE Proceedings of The Fourth IEEE RAS/EMBS International Conference on Biomedical Robotics and Biomechatronics 2012,ISBN: 978-1-4577-1199-2, pp 1383- 1387.
5. D'Addio G., Cesarelli M., Romano M., De Nunzio A., Lullo F. and Pappone N.. EMG Patterns in Robot Assisted Reaching Movements of Upper Arm. 5th European Conference of the International Federation for Medical and Biological Engineering IFMBE Proceedings Volume 37, 2012, pp 749-752
6. T. Flash, N. Hogan. Coordination of arm movements: an experimentally confirmed mathematical model. J. Neuros; 5:1688-1703, 1985.
7. Morasso, Spatial control of arm movements. Exp Brain Res (1981) 42, pp: 223-227.

Author: Mario CESARELLI
Institute: D.I.E.T.I., University   "Federico II"
Street: via Claudio, 21
City: Naples
Country: Italy
Email: cesarell @unina.it

Handling Disturbances on Planned Trajectories in Robotic Rehabilitation Therapies

V. Rajasekaran[1], J. Aranda[1,2], and A. Casals[1,2]

[1] Institute for Bioengineering of Catalonia, Barcelona, Spain
[2] Technical University of Catalonia, Barcelona-Tech, Barcelona, Spain

Abstract—Robotic rehabilitation therapies are an emerging tool in the field of Neurorehabilitation in order to achieve an effective therapeutic development in the patient. In this paper, the role of disturbances caused by muscle synergies or unpredictable effects of artificial stimulation in muscles during rehabilitation therapies is analyzed. In terms of gait assistance it is also important to maintain synchronized movements to ensure a dynamically stable gait. Although, disturbances affecting joints are corrected by a force control approach, we define two methods to ensure stability and synchronization of joint movements in the trajectory to be followed. The performance of the presented methods is evaluated in comparison with a preplanned trajectory to be followed by the patients.

Keywords—Exoskeleton, force control, gait assistance, neurorobot, trajectory planning.

I. INTRODUCTION

Rehabilitation therapy using robotics has been recently a topic of interest for many rehabilitation therapists and also for researchers working in the field of medical robotics. Rehabilitation using robots not only involves the mode of training patients but also involves gait assistance for spinal cord injury (SCI) and stroke individuals. Studies on the effectiveness of robotic neurorehabilitation have proven to be beneficial in impairment measures but are not so effective in terms of functional outcomes [1]. Robots are also used to study the process of motor learning in healthy subjects and this helps in introducing force perturbations that induce large trajectory errors to which the subjects must then adapt. This approach has allowed scientists to test several hypotheses about the computational mechanisms of motor control and motor learning [2]. Co-operative control architecture has been developed for the *Lokomat* which enforces the active force contribution of the patient. This control strategy is driven by patient and helps in accomplishing the free walking movements in order to maximize the progress of the patient [3]. Recently there is an increase in the interest of developing active orthoses capable of adding and controlling power at the joints. Many active exoskeletons have been developed for active gait restoration with considerable variation in actuator and sensing technologies, and control strategies. There are still some limitations to overcome in providing effective gait compensation [4]. A therapeutic robot *Physiotherabot* for the lower limbs of patient has been developed to react based on the feedback data from the patient. The robot adapts to patients performance by rearranging its joint positions [5]. A new promising approach in the field of robotic rehabilitation is the combined effect of FES technology and exoskeletons. This approach is intended to overcome the drawbacks of individual approaches by combining them in a single system [6].

In this paper, we will be analyzing the methods to handle the unpredictable forces or disturbances appearing along the execution of a learned trajectory pattern. Disturbances can be caused by the patient's muscle synergies in terms of spasticity or plasticity or due to the unpredictable effects of artificial electrical stimulation of muscles (FES). Muscle stimulation techniques are usually considered to assist the patient in achieving the goal but at some point these stimulations can cause too large forces which affect the pattern to be followed. Prolonged stimulations in the muscles also lead to muscle fatigue, which also affects the output pattern to be followed. In terms of gait assistance it is important to follow the pattern with time constraints and synchronized joint movements. An occurrence of these disturbing forces in any joint affects the robot from following the planned trajectory. An effect in one joint obviously affects the trajectory of the other joints due to the dynamics and the co-relation among them. Planning and re-planning a trajectory based on the effect of disturbance, plays an important role in all the rehabilitation studies and especially in gait assistance with time constraint. Time constraints are necessary in order to handle the equilibrium of the patient and this will respectively help in gait assistance.

II. METHODS

Gait assistance for individuals with spinal cord injury (SCI) and stroke need some assistive tools to aid them maintaining the equilibrium. Some of the tools like FES induced assistance, may also cause disturbances like muscle synergies and residual forces while following a pre-planned trajectory. An occurrence of these types of disturbances becomes a common scenario in most neurorehabilitation therapies. These unpredicted forces or disturbances can be impulsive signals caused by FES induced motion in the therapy and sinusoidal or continuous due to muscle synergies. In this work, we test two methods to handle the effect of these disturbances in a pre-planned trajectory and to study the way to avoid their influence in patient's therapies. The main aim of the work **presented** is to avoid the influence of disturbances in therapies by providing the required assistive and resistive forces to the affected joints and also to minimize the effect of disturbances in the other joints. When one of the joint trajectories is affected by a perturbation, it is important to analyze the way to prevent the disturbance affecting other joints. Another important factor is to maintain the synchronized movement in all the joints while following the planned trajectory. In gait assistance, it is critical to maintain the synchronization of the joints and also the time constraints in order to maintain the equilibrium. The following are the two methods by which we try to achieve our goal of handling the effects of unpredictable forces while following a preplanned trajectory: A classical Cartesian space based control and a Variable input speed method.

A. Cartesian Position Control

The use of Jacobian matrix is a classical approach to evaluate the error in joint space by transforming them to the Cartesian space. As a first approach, the errors in joint positions are given by monitoring the actual positions against reference positions continuously. These errors are then transformed to the Cartesian space by using the direct kinematic transformations. The transformed error from the joint space into Cartesian coordinates provides the required joint forces to avoid disturbances. In the presence of disturbance, the Jacobian matrix is used to compute the joint torques by considering errors in Cartesian space. Computed torque is added to the joint torques in order to avoid the unpredictable effects in the other joints and also to follow the preplanned trajectory. The input joint torques to be applied are calculated by the following equation.

$$\tau = J^T(\theta) * (K_1 E_x + K_2 \int E_x) \quad (1)$$

Where, τ is the Joint torque, J^T the Transposed Jacobian matrix of the joint coordinates, E_x the Error in Cartesian coordinates and K_1, K_2 are Constants.

Our initial approach was to generate forces in the joint space by considering them proportional to the error in the Cartesian space. However, an integral function of the error signal is included in order to avoid the steady state error caused in the output pattern.

B. Variable Input Rate

The occurrence of disturbances in a single joint changes the planned trajectory and affects the other joints due to the dynamics of the exoskeleton. The forces in the joints can be assistive or resistive based on the direction of the force and trajectory to be followed.

Synchronization of joints can be obtained by varying their speed of the joints while trying to handle the disturbances. In this approach, when a disturbance is detected, a constant stiffness is applied to the disturbed joint while varying the speed of other joints. A disturbance is considered when the difference between actual and reference trajectory goes beyond a given threshold. The joint speeds are controlled by varying the input rate of position orders to the joint controller. As shown in Fig. 1, we have two possibilities based on the effect of the disturbance on the trajectory, depending on its direction: in the same direction or against movement. The direction of the disturbance can be defined by evaluating the interaction forces.

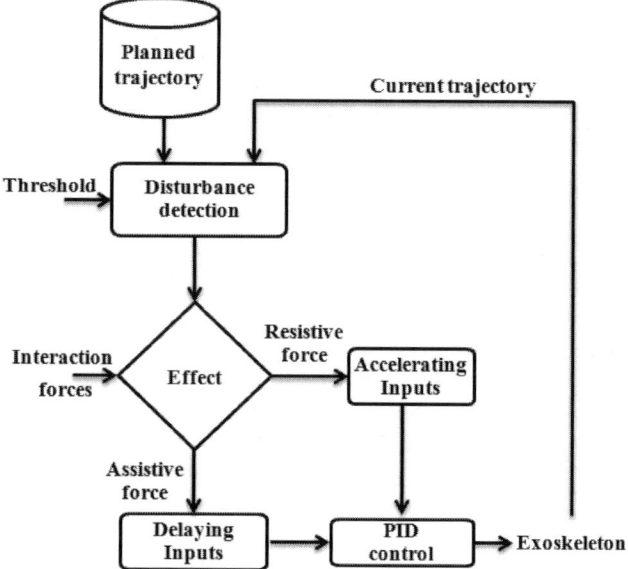

Fig. 1 Schematic representation of the Variable input rate approach

Two possible cases appear in this strategy: Assistive and Resistive interaction forces.

1. Assistive interaction forces:

This force is significant when the disturbance is against the direction of the joint movement, reducing its execution speed. In order to synchronize the joint movements, the input rate of the undisturbed joints is decreased, while maintaining the coordination between the joints.

2. Resistive interaction forces:

This force is characterized by the disturbance being on the same direction of the joint movement, increasing its execution speed. The joint movements are synchronized by increasing or accelerating the input rate to the undisturbed joints, while maintaining the synchronization. The joint errors are monitored and the torque acting on the joint is directly proportional to the error in position. This is given by the following equation:

$$\tau = K.(E_\theta) \quad (2)$$

Where, τ is the applied torque in the joint, K is the stiffness constant and E_θ the error in position.

The disturbed joint can affect the movement of the other joints due to the dynamics between them. So by increasing the speed of the input to the other joints we can see the differences in these joint movements. Increasing the speed in turn results in the assistance of the robot to the patient and this will result in the guiding of the patient to follow the recorded trajectory.

III. EXPERIMENTAL SETUP

The exoskeleton which has been used in this demonstration was built as part of the HYPER project. Fig. 2 shows the exoskeleton built with 3 degrees of freedom in each leg. The exoskeleton is a 6 joint planar robot developed in order to assist the patients with neurological disorders like individuals with SCI and stroke.

Fig. 2 6-dof Exoskeleton developed as part of the HYPER project

Each joint of the exoskeleton has 1 dof revolute joint which can be driven by position and torque control respectively based on the therapy requirements. The mechanical parameters of the exoskeleton can be adjusted depending on the anthropomorphic parameters of each individual. Fig. 3 shows the Simulink model of the exoskeleton developed based on the real mechanical considerations of the exoskeleton. This Simulink model has been used in this analysis to verify the suitable method for correcting the disturbances efficiently. The model consists of 3 joints each driven by a 1 dof revolute actuator which in turn is actuated by an implicit force control mechanism. The force orders are evaluated based on the proportional error in joints. The joints of the model are driven by PID control, based on the force applied which in turn is based on the error in its position.

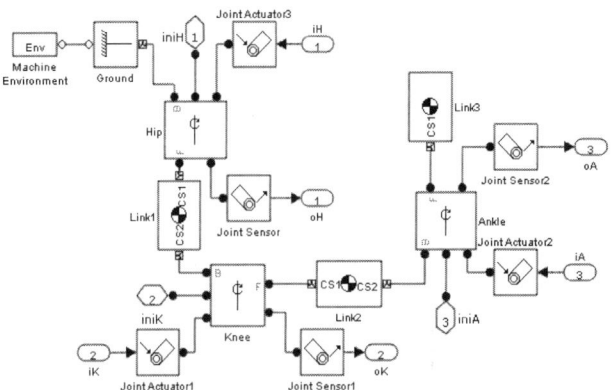

Fig. 3 Simulink model of the Exoskeleton

In order to evaluate the methods presented, a sinusoidal or continuous signal was introduced as a disturbance to one of the joint. The sinusoidal signal represents the muscle synergies within the limit of -10N to +10N. A change in behavior of the system was observed as a result of the dynamical evolution of the disturbances to other joints. When a disturbance is observed in one of the joints, the joint position orders are fed into one of the methods for handling the disturbances. The disturbance signal is applied at the same frequency in order to evaluate the efficiency of the methods explained above.

IV. RESULTS

The results are compared with the planned trajectory which is to be followed by the patient. The performances of the two methods are evaluated with respect to the position errors between the planned and current trajectories. As explained earlier, the synchronized joint movements and time constraints play an important role in gait assistance. From Fig. 4, we can see that there is a significant difference in the planned and current trajectories of hip joint. This difference in behavior is due to the application of disturbances in the hip joint of the exoskeleton model. The disturbance is applied in the hip joint for the evaluation of both methods. With the Cartesian control approach, there is a steady state error present in the trajectory and this is also verified by the differences in position. The variable input rate approach follows the planned trajectory more efficiently and this is evident by the minimum position error in all the joints. In the disturbed hip joint, the maximum position error is in the range -2 to 2 in case of Variable input rate and -10 to 5 in terms of Cartesian position control. The maximum position error is observed at the same time instant in both methods (Fig. 5). With the variable rate control, there is a time delay of 2-5ms in the undisturbed joints and this helps in maintaining the synchronized movements along the trajectory.

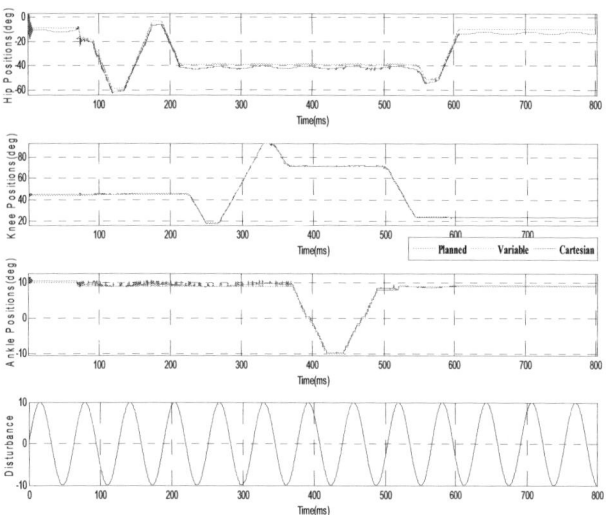

Fig. 4 Comparison of planned trajectory, variable input rate and Jacobian for Hip, Knee and Ankle joint

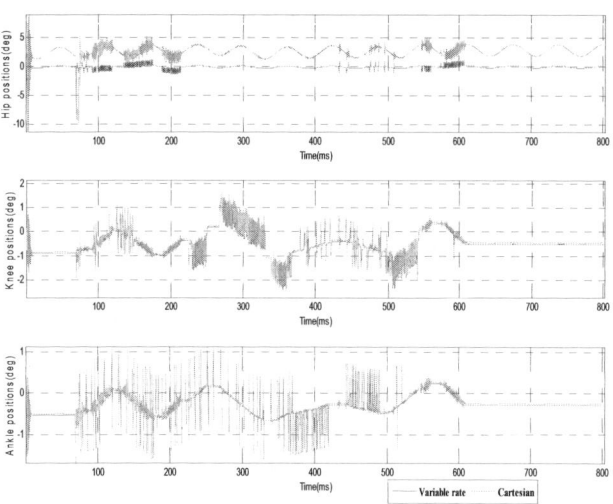

Fig. 5 Error in positions in Hip, Knee and Ankle joints

V. CONCLUSIONS

The unpredictable forces affecting the trajectories in therapies can be limited by applying a high stiffness on the disturbed joint. But there is a need to inhibit these residual forces from affecting the dynamic stability in gait assistance and also the synchronization of the joint movements in following a trajectory. As expected the variable input rate approach was able to follow the planned trajectory with minimum position errors and maximum efficiency. The experiment confirmed that the time delay in the undisturbed joint trajectory is minimum and thus facilitating the synchronized movement in the joints.

ACKNOWLEDGMENT

This work was supported by grants from the 7th Framework programme grant CSD2009-00067 CONSOLIDER INGENIO 2010 HYPER (Hybrid Neuroprosthetic and Neurorobotic devices for functional compensation and rehabilitation of motor disorders) project.

REFERENCES

1. Huang VS and Krakauer J W (2009) Robotic neurorehabilitation: a computational motor learning perspective. Journal of Neuroengineering and Rehabilitation.
2. Shadmehr R and Wise S (2007) The computational neurobiology of reaching and pointing- a foundation for motor learning. Network:Computation in Neural Systems, Informa UK Ltd.
3. Bernhardt M, Frey M,Colombo G and Riener R(2005) Hybrid Force-Position control yields cooperative behaviour of the rehabilitation robot LOKOMAT.9th International Conference on Rehabilitation Robotics,Chicago,IL,USA.
4. Waldner A, Werner C and Hesse S(2008) Robot assisted therapy in neurorehabilitation. Europa Medico Physica.
5. Erhan A and Adli MA(2011) The design and control of a therapeutic exercise robot for lower limb rehabilitation:Physiotherabot. Mechatronics pp 509-522, 2011.
6. Del-Ama AJ, Koutsou AD, Moreno J C, de-los-Reyes A, Angel GA and Pons JL(2012) Review of hybrid exoskeletons to restore gait following spinal cord injury.VA Healthcare vol. 49,pp 497-514, 2012.

Author: V.Rajasekaran

Institute: Institute for Bioengineering of Catalonia
City: Barcelona
Country: Spain
Email: vrajasekaran@ibecbarcelona.eu

Computational Modelling of the Shape Deviations of the Sphere Surfaces of Ceramic Heads of Hip Joint Replacement

V. Fuis[1], Koukal[2], Z. Florian[2], and P. Janicek[1]

[1] Centre of Mechatronics – Institute of Thermomechanics AS CR and Institute of Solid Mechanics, Mechatronics and Biomechanics, Faculty of Mechanical Engineering, Brno University of Technology, Brno, Czech Republic
[2] Institute of Solid Mechanics, Mechatronics and Biomechanics, Faculty of Mechanical Engineering, Brno University of Technology, Brno, Czech Republic

Abstract—The problem of the damage of the bioimplants is very important today. In the endoprosthesis surgery there is a large percentage of implant defects, which cause the failure of the whole prosthesis. One kind of the total hip replacement functionality loss is acetabular cup pull-off from pelvis bone. This paper is aimed at manufacture perturbations (shape deviations) analysis as one of the possible reasons to this kind of failure. Dimension and of geometry manufacturing perturbations (roundness) were analyzed in detail. It was found, that these perturbations affect considered values of contact pressure and frictional moment. Contact pressure and frictional moment are quantities affecting replacement success and durability.

Keywords—Total hip replacement, shape deviations, roundness, contact pressure, frictional moment.

I. INTRODUCTION

The paper is aimed at problems of total hip replacements (THR). It is engaged with manufacture perturbations (fabrication tolerances) impacts of femoral head and acetabular cup upon to failure of functionality of THR [2], [3], [7]. It is necessary to prevent the loss of THR functionality, as it's accompanied with painfulness and then the re-operation and convalescence is necessary [10].

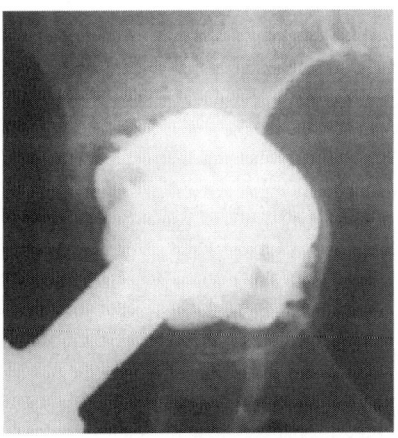

Fig. 1 Pull-off acetabular cup from pelvis bone [1]

THR function loss can be caused by several reasons: inappropriate material [11], geometry, combination of materials, surface finish treatment, shape deviations [6], [8], [9], [12] etc. Clinic study of 600 THR's with ceramic-ceramic contact between femoral head and acetabular cups showed there was 20% failure in period 1977–89 [4]. One sort of THR functionality loss is pull-off of the acetabular cup from pelvis bone (Fig. 1).

Fig. 2 Acetabular component made by Walter-Motorlet Co [1]

Medicals at the Faculty Hospital Brno are interested in impacts of manufacture perturbations as a clinical aspect [1]. During the five year period, they had many failed THR due to pull-off there. There are two types of the cups where the number of the failure is greater than 25%. One of this type (Walter Motorlet Cup – Fig. 2) was modeled in FEM system ANSYS to find the causes of the cup's failures.

Before modeling used and unused acetabular cups and femoral heads were analysed and measured [5]. It was found, that inner spheric surface of cups shows non-uniform abrasion and the inner cup's diameter is smaller than the head's diameter [5]. It causes the interference fit between the head, the head is pressed in to the cup and due to the friction the head with the stem are bonded with the cup as shown in Fig. 3.

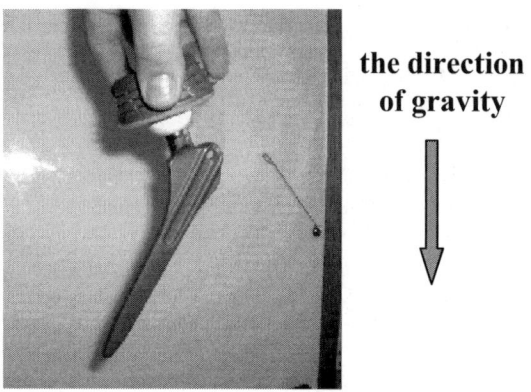

Fig. 3 Head with the stem are bonded to the Walter-Motorlet cup [5]

II. COMPUTATIONAL MODELLING INPUTS

The parametric geometry model of the system was created due to the possibility to study influences of perturbations of the shape deviations. Model consists of the following parts (Fig. 4): 1. pelvis bone with cortical and cancellous part, 2. acetabular shell (titanium), 3. acetabular cup (UHMW polyethylene) and 4. femoral head (ceramic).

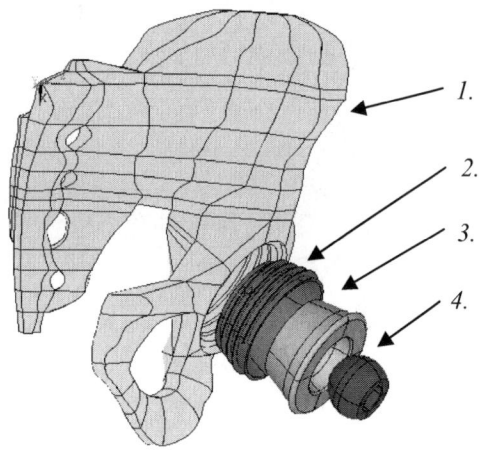

Fig. 4 Components of the analysed system

Variable parameters were:

- inner cup's diameter – from 31.7 mm to 32.4 mm (from interference fit (overlap) 0.3 mm to clearance fit (gap) 0.4 mm - nominal diameter is 32 mm),
- femoral head roundness – from 0 mm to 0.2 mm,
- position of the roundness deviations – three values of the angle β (0°, 45°, 90°) of the roundness perturbations (hill – R_2 and valley - R_1) changed against the unchanged position of the loading force F (Fig. 5).

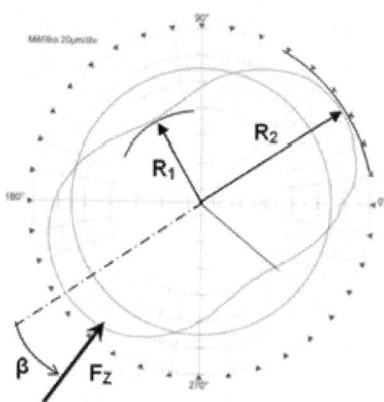

Fig. 5 Measured deviations of the roundness of the inner cup's sphere surface and the position of the loading against the shape deviations (angle β)

Linear isotropic material models were used for all components which is characterized by the modulus of elasticity E and Poisson's ratio μ (ceramic head E = 3.9e5 MPa, μ = 0.23, polyethylene cup E = 1e3 MPa, μ = 0.4, titanium shell E = 1e5 MPa, μ = 0.3, cancellous bone E = 2e3 MPa, μ = 0.25 and cortical bone E = 1.4e4 MPa, μ = 0.3).

The first contact pair was modelled by creating contact elements between femoral head and acetabular cup, with coefficient of friction f = 0.1. The second contact pair was created between acetabular cup and shell.

There are two loading steps. The first loading is realized by the force F = 2500 N which is applied in the centre of femoral ball head. The frictional moment is calculated in the second loading step - the femoral head is 30° rotated and the reaction forces represent the wanted frictional moment.

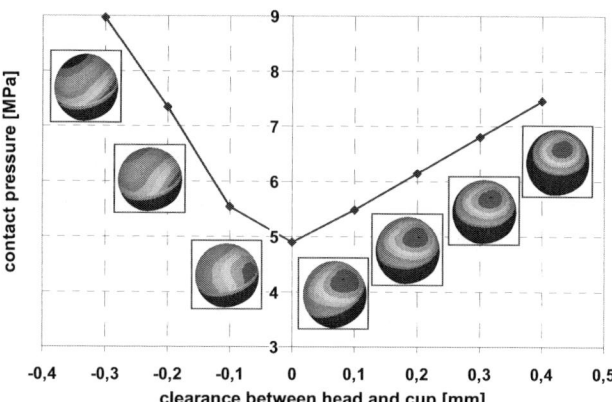

Fig. 6 The maximum value of the contact pressure for different head – cup's diameters (overlap and the gap) with the contact pressure isolines

III. RESULTS AND DISCUSSION

Stress and deformation of the head and the cup are significantly influenced by the situation in the contact (overlap – interference fit, the same diameters and the gap – clearance fit). Maximum value of the contact pressure between the head and the cup is shown in Fig. 6 and the frictional moment is shown in Fig. 7. The contact region is different for the overlap and for the gap between the head and the cup – distribution of the contact pressure is shown in the small figures in Figs. 6 and 7.

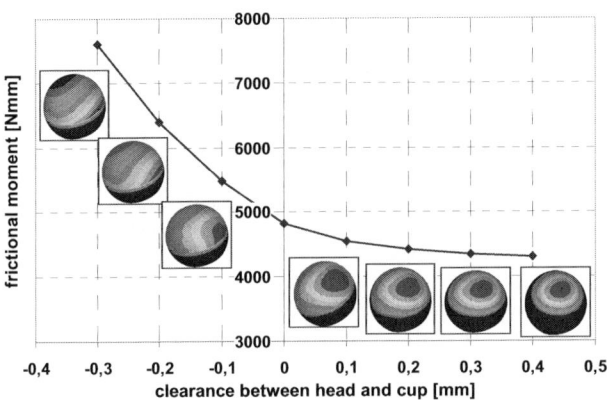

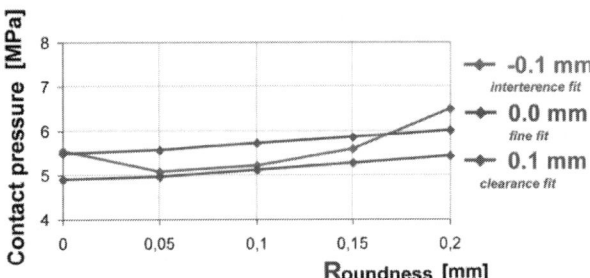

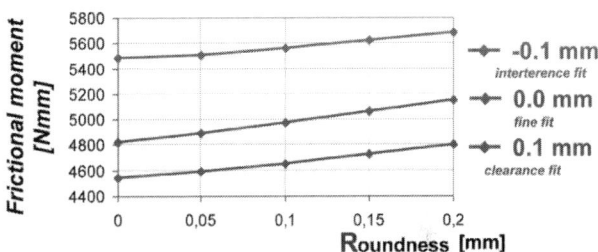

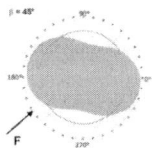

Fig. 9 Contact pressure and frictional moment for different values of the roundness of the femoral head - angle $\beta = 45°$

Fig. 7 The frictional moment for different head – cup's diameters (overlap and the gap) with the contact pressure isolines

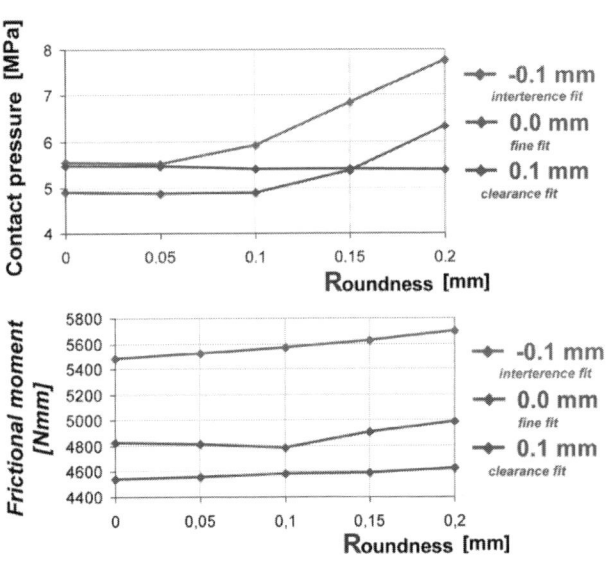

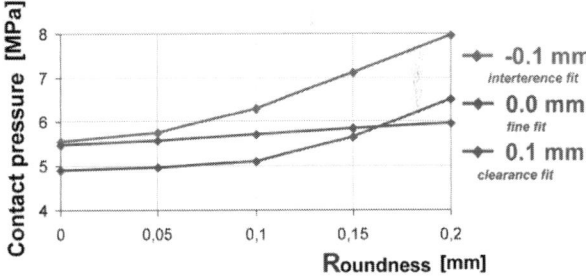

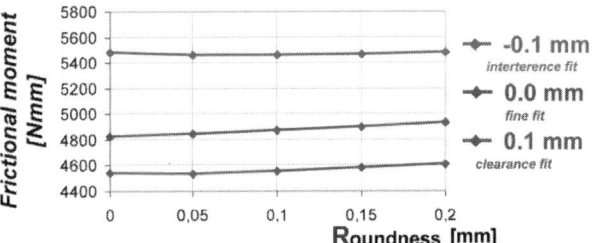

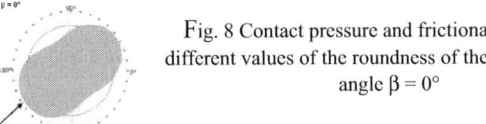

Fig. 8 Contact pressure and frictional moment for different values of the roundness of the femoral head - angle $\beta = 0°$

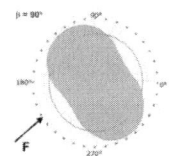

Fig. 10 Contact pressure and frictional moment for different values of the roundness of the femoral head - angle $\beta = 90°$

To study influence of geometric fabrication tolerances the roundness the femoral head was varied, ranging between 0.05 mm to 0.2 mm (difference between R_1 and R_2 in Fig. 5). Position β of the roundness perturbation (hill) was changed against the loading force position (Fig. 5). The assumed angle β values were 0°, 45° and 90°. Clearance between femoral head and cup was changed, with values - 0.1 mm (interference fit (overlap)), 0.0 mm (fine fit) and 0.1 mm (clearance fit (gap)). Values of contact pressure and frictional moment for mentioned values of the angle β are shown in the Figs. 8 - 10.

It was found that with clearance increase from 0 mm (fine fit) value of contact pressure increase linearly in the investigated interval (Fig. 6). Value of contact pressure increased about 10% from 5 MPa with 0.1 mm clearance increase from 0 mm. The value of frictional moment slightly decrease (Fig. 7) with clearance increase. This is caused by decrease of the contact surface. With clearance decrease from 0 mm (fine fit), the value of contact pressure rise with approximately double speed than with increase of clearance from 0 mm (Fig. 6) and the value of frictional moment steeply increase (Fig. 7) about 10% with 0.1 mm decrease. This is due to increase of the circumferential tension, caused by negative clearance (interference).

After considering the effect of varying the entities diameter, the influence of roundness perturbations was assessed. It was found, that with roundness perturbation increase the contact pressure and the frictional moment rise, regardless the clearance value or roundness perturbation position against the loading force (Figs. 8 - 10). Bellow the value of roundness perturbation 0.05 mm the values of contact pressure and frictional moment change slightly. Above the value of roundness perturbation 0.05 mm results change steeply.

With the change of roundness perturbation to 0.2 mm, the value of contact pressure increase 45% in the case of interference fit, 30% in the case of fine fit and 10% in the case of clearance fit. Value of frictional moment increases 5% maximum in all types of fit.

Position of roundness perturbation affects the contact pressure value, especially in case of interference or fine fit. Highest values of the contact pressure were found at extreme positions (0°, 90°) of the perturbation (hill).

IV. CONCLUSION

It was found that manufacture perturbations influence the values of contact pressure and frictional moment between femoral head and acetabular cup. Both contact pressure and frictional moment values affect the success and durability of the replacement, therefore it's recommended to minimize their vales. From obtained results it can be considered that the most durable type of fit is fine, or slightly towards clearance. The reasons are minimum value of contact pressure and not big influence of clearance value to the frictional moment in the clearance type of fit. The value of roundness should not be higher than 0.05 mm.

ACKNOWLEDGMENT

The research has been supported by the project of the Czech Science Foundation GA CR nr. 13-34632S.

REFERENCES

1. Krbcc M, Fuis V, Florian Z (2001) Pull-off of the uncemented cups of total hip replacement from pelvis bone, National Congress ČSOT, Prague, Czech Rep., 2001, p. 108
2. Charnley J, Kamangar A, Longfield M D (1969) The Optimum Size of Prosthetic Heads in Relation to the Wear of Plastic Socket in Total Replacement of the Hip, Med & Biol Eng. vol. 7, iss. 1, pp 31 – 39
3. Fuis V, Janicek P (2002) Stress and reliability analyses of damaged ceramic femoral heads, Conference on Damage and Fracture Mechanics, Maui Hawaii, Structures and Materials vol. 12, pp 475-485
4. Gualtieri G, Calderoni P, Ferruzzi A et al (2001) Twenty Years Follow-up in Ceramic-Ceramic THR at Rizzoli Orthopaedic Institute, 6th International BIOLOX Symposium, pp.13-17
5. Fuis V, Koukal M, Florian Z (2011) Shape Deviations of the Contact Areas of the Total Hip Replacement, 9th International Conference on Mechatronics Location: Warsaw, Poland, pp 203-212
6. Fuis V, Malek M, Janicek P (2011) Probability of destruction of Ceramics using Weibull's Theory, 17th International Conference on Engineering Mechanics, Svratka, Czech Rep., pp 155-158
7. Teoh S H, Chan W H, Thampuran R (2001) An Elasto-plastic Finite Element Model for Polyethylene Wear in Total Hip Arthroplasty, J Biomech 2002, 35: 323-330
8. Fuis V., Janicek P, Houfek L (2008) Stress and Reliability Analyses of the Hip Joint Endoprosthesis Ceramic Head with Macro and Micro Shape Deviations, Proc. vol. 23, Issue 1-3, International Conference on Biomedical Engineering, Singapore, pp 1580-1583
9. Fuis V (2004) Stress and reliability analyses of ceramic femoral heads with 3D manufacturing inaccuracies, 11th World Congress in Mechanism and Machine Science, Tianjin, China, pp 2197-2201
10. Yew A, Jagatia M, Ensaff H, Jin Z M (2003) Analysis of Contact Mechanics in McKee-Farrar Metal-on-Metal Hip Implants, Journal of Engineering in Medicine, 217, 333-340
11. Fuis V., Navrat T., Hlavon P. et al. (2006) Reliability of the Ceramic Head of the Total Hip Joint Endoprosthesis Using Weibull's Weakest-link Theory, IFMBE Proc. vol. 14, World Congress on Medical Physics and Biomedical Engineering, Seoul, South Korea, pp 2941-2944
12. Fuis V (2009) Tensile Stress Analysis of the Ceramic Head with Micro and Macro Shape Deviations of the Contact Areas, RECENT ADVANCES IN MECHATRONICS: 2008-2009, Proc. International Conference on Mechatronics, Brno, Czech Rep., pp 425-430

Author: Vladimir Fuis
Institute: Institute of Thermomechanics AS CR – branch Brno
Street: Technicka 2
City: Brno
Country: Czech Republic
Email: fuis@fme.vutbr.cz

An Innovative Multisensor Controlled Prosthetic Hand

N. Carbonaro[1], G. Anania[1], M. Bacchereti[3], G. Donati[3], L. Ferretti[3], G. Pellicci[3], G. Parrini[3], N. Vitetta[3], D. De Rossi[1,2], and A. Tognetti[1,2]

[1] Research Center E.Piaggio, University of Pisa, Via Diotisalvi 2, Pisa, Italy
[2] Information Engineering Department, University of Pisa, Via Caruso 2, Pisa, Italy
[3] Fabrica Machinale s.r.l., Via Giuntini 13, Navacchio, Pisa, Italy

Abstract—This article reports the design, realisation and preliminary testing of an innovative multisensor controlled prosthetic hand. The mechanical design is strongly oriented to satisfy the requirements of delivery a penta-digital prosthetic hand with reduced hand size, low weight, independent finger movement and bio-mimetic shape. Moreover, an ad hoc sensing architecture has been designed including traditional EMG system together with in socket force measurement and inertial sensing. This sensing architecture is intended to reduce the socket complexity in terms of adaptation to different patient forearms, without loosing the possibility of performing multiple hand grips. An innovative concept introducing inertial sensing in order to select appropriate robotic hand configurations is also described. A first prototype of the prosthetic hand has been realised and preliminary tests are reported.

Keywords— hand prosthesis, force myography, inertial sensing, biomimetic.

I. INTRODUCTION

The analysis of commercial prosthetic hand available on the market points out the absence of products with reduced hand size, low weight, penta-digital mechanical solution with independent finger movements and bio-mimetic shape. Manufacturers such as Touch Bionics and RSL steeper produce penta-digital hand but only of a big sizes (I-limb [1] and Be-Bionic [2]). Otto Bock [3] produces little hands (even for children) but not with independent movements on the fingers. Another important aspect is the high cost of upper prosthetic devices such as electronic hands or electrical finger nowadays used in partial hand amputation (for example amputation of finger I, II and/or III). To obtain cost reduction is necessary have components for production with scale's economies. Commercially available prosthetic hands are controlled through electromyographic (EMG) signals and patients can set only a few of different hand configurations with the contraction of agonist and/or antagonist limb's muscle. The signal is gathered using electrodes conveniently positioned in the prosthetic socket using residual muscular activities of patient's forearm. This technique has the advantage to allow a "natural" control for the patient but with the limitation of providing only one practical degree of freedom (DOF) of prosthesis actuation, basically opening and closing the robotic hand in a fixed and predefined configuration. Different studies have been done to evaluate more independent channels of control signals from EMG with advanced signal processing techniques [4, 5, 6, 7]. Moreover, several alternative signal sources have been considered in literature: the mechanical force exerted by a tunnel muscle or tendon [8], the acoustic signal generated from muscle activity [9], the morphological changes of residual limb tissues [10]. Moreover, even electroencephalogram (EEG) and neuronal signal have been considered [11, 12]. Many of these scientific works have been limited to laboratory experiments, but are quite far from commercialisation.

The aim of this study is to design and develop a market oriented prosthetic hand (that hereinafter will be called *Tiny-Hand*) having reduced dimension, weight and cost, good biomimetic reproduction of human hand, finger design for use also in a partial amputation and with sufficient grip force. Concerning the control strategy, the Tiny-Hand main requirements were the possibility to perform different kind of hand grips/positions to improve the manipulation experience, without increasing too much the system complexity and the overall costs. To this aim an ad hoc sensing architecture has beed designed, developed and tested. In particular, traditional EMG system has been considered together with in socket force sensing and inertial sensing. In socket pressure sensing has been included to monitor the so called Force Myography (FMG). FMG uses a set of force sensors to measure the pressure on the forearm caused by the muscular activity. FMG signal has been used in few pilot works to extract the hand neuromuscular volition [13, 14]. In the present work in socket force distribution and conventional EMG have been used at the same time to encode subject/patient movement intention. The advantage of the FMG is the reduced sensor cost and complexity with respect to EMG and the possibility to spread the sensor in redundant configuration in order to reduce the personalization effort of the prosthesis. The inertial sensor on the forearm has been included to enable the patient to select different hand grips in relation to the forearm orientation in space. Moreover, the possibility to select particular hand configuration in correspondence to peculiar dynamic movements of the forearm has been introduced. The Tiny-Hand prototype has been realised in the frame of the project MANOROB funded by Regione Toscana (Italy) and it has been patented (Italian patent application n. PI2013A000004).

II. MECHANICAL DESIGN

Tiny-Hand (see Fig. 1) is a penta-digital prosthetic device with passive thumb's abduction-adduction. It has 11 DOFs actuated by five motors, hand's dimension (perimeter around the knuckles) is a x-small size for women (about 6 inches). The hand's weight is about 220 grams and the wrist's design has an elliptical geometry to obtain a more natural shape instead of the circular one usually proposed on similar types of prosthesis. Tiny-Hand is a prosthetic device designed for women and adolescents patients, with a great bio-mimetic reproduction and with important advantages relative to cost production. To reach the described requirements a particular attention was applied on the finger design. The DC motor was located directly on the finger to allow the use of the prosthesis also for partial amputation. Moreover, regarding finger kinematic morphology the use of cylindrical joint appeared the best solution to apply. In fact the use of complex kinematic solutions, as for example multi-links system, are difficult to obtain in so a such small devices, providing also great complication during product assembly and maintenance. The number of DOFs for a prosthetic finger could be 1 or 2: in the first solution the flexion/extension of a mono-phalanx can allow limited types of grip and is related to a more simple device with lower cost; on the other hand the use of a system with two phalanges (proximal one and a second that replicate medial and distal ones) allow to reach a better bio-mimetic aspect and a better compliance with different grips. The solution selected use 2 DOFs and the distal phalanx is under actuated through the proximal one. Another important aspect that has been considered during finger design is the stiffness of the system. In case of impacts it is crucial to avoid repercussion on the patient's stump through the support of the same prosthetic device. Moreover, the stiffness of fingers in extension movement has to be quite high to grip different objects in power and precision grips. Finally, a peculiar finger design, consisting in an internal irreversibility of the mechanical transmission, is needed to reduce energy consumption during grips and to allow an high autonomy of the prosthesis. The result of all these considerations is the design of the finger reported in Fig. 2.

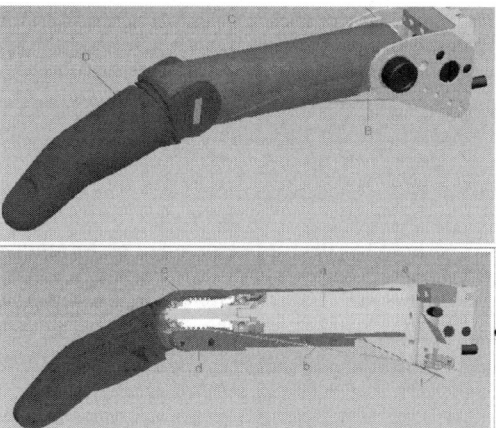

Fig. 2 Tiny-Hand finger components and its sagittal section

III. SENSING SYSTEM AND CONTROL STRATEGY

The Tiny-Hand sensing system is composed of a standard EMG device (produced by Otto Bock), a set of force sensitive resistors for FMG extraction (FSR, Flexiforce A201 produced by Tekscan), an inertial sensor (ADXL330 produced by Analog Devices) and the dedicated acquisition electronics directly connected to the hand controller through a serial link. The acquired signals are elaborated and fused in real time to extract the patient intention in terms of opening/closing the hand and grip selection. The block diagram of the developed control strategy is reported in Fig. 3. The elaboration of FMG signal allows the detection of hand activation, causing the transition from flat neutral position to the selected grip. FMG is acquired trough 8 FSR sensors distributed uniformly around the prosthesis socket placed in the carpal flexor area (Fig. 4 shows the realised preliminary demonstrator). The in-socket force due to the muscular activity of the carpal flexors is continuously tracked by the FSR sensors. The event of hand activation is obtained by an adaptive peak detector algorithm working without predefined thresholds in order to fit with the different physical characteristic of the patients. The FSR lightness, reduced thickness and dimensions allows to

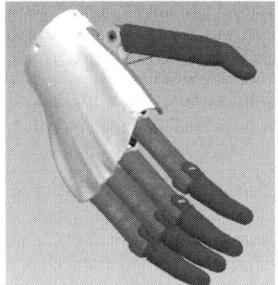 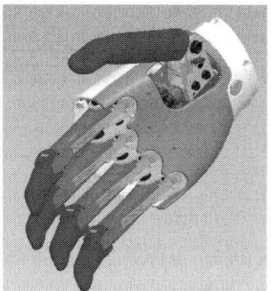

Fig. 1 Tiny-Hand CAD model

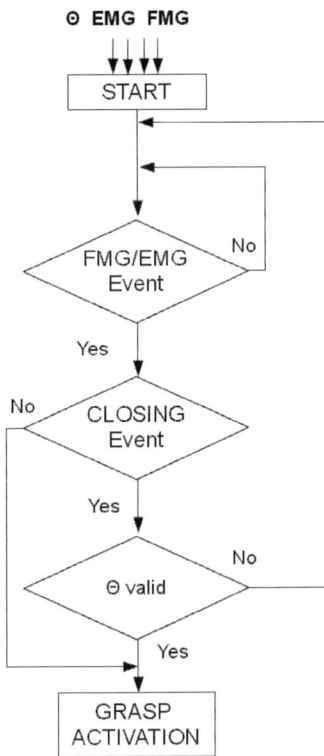

Fig. 3 Block diagram of the developed algorithm. *Theta* is the forearm orientation gather from the inertial sensor

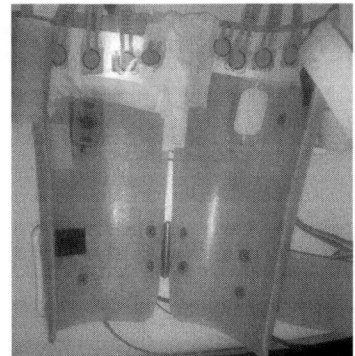

Fig. 4 Sensor arrangement in the prosthesis socket

realise a redundant distribution of force sensors in the area of interest. In this way the precise positioning of in-socket FSR is not needed, thus reducing the socket complexity in terms of adaptation to different bodily structures of the patients (in contrast with EMG sensor positioning that has to be patient specific). The EMG sensor is used to discriminate the intention of deactivating the hand grip (i.e. transition from the performed grip to the neural flat hand position). The sensor is placed on the carpal extensors and the acquired signal is elaborated by means of standard EMG treatment algorithms. This sensing architecture, using FMG on the flexor and EMG on the extensor, has been experimentally identified by using, in the system design phase, a more complex prototype based on two EMG sensors (placed on both flexor and extensor muscles) and the FSR configuration previously described. These experiments have pointed out a close correlation between the FSR relived in the flexor area and the correspondent EMG sensor and a low correlation of the FMG and EMG taken on the extensor. Fig. 5 shows the signals acquired in the identification phase on a patient performing continuos cycles of alternate activations of the carpal flexors and extensors.

Another innovation of the proposed control strategy is the direct correlation between the arm spatial orientation, detected by the inertial sensor, and the grip selection. Practically, if the patient activate the hand through the carpal flexor movement detected by the FSR sensors, the performed robotic hand grip changes according to the forearm orientation. In the first trials, five different functional grasps were associated to different forearm orientations. Moreover, an addiction hand configuration can be activated if the inertial device recognised a peculiar dynamic movement pattern of the forearm (in this first example a forearm shake was selected). In the first Tyny-hand prototype this modality was linked to a pointing with the forefinger (possibly useful for PC keyboard usage). This configuration can be deactivated by the recognition of the same activation movement pattern. Before the final integration, the sensing system was linked to a tridimensional hand model replicating the robotic hand functionalities (shown in the monitor present in Fig. 6).

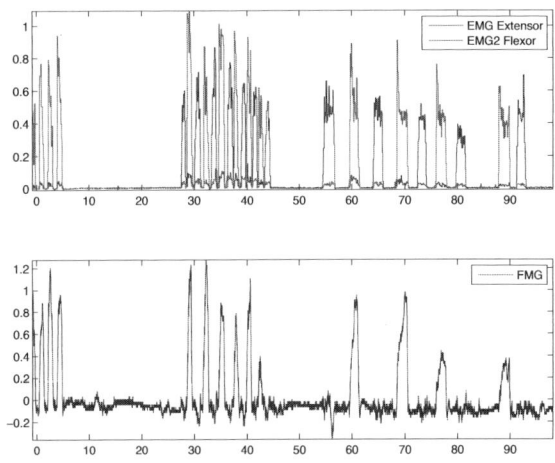

Fig. 5 EMG and FMG signals acquired during the design phase

IV. PRELIMINARY RESULTS

The first prototype of the Tiny-Hand has been realised consisting of both the mechatronic hand and the described sensing system integrated togheter (Fig. 6). The prototype has been preliminary tested on ten non patient subjects, with the prosthetic hand fixed on a bench as shown in Fig. 5. The subjects were asked to arbitrary choose a certain number of hand grips and to perform the necessary movements to select them. Before the test a brief training phase was performed where the subjects were free to learn the prosthesis usage. The Tyny-Hand performances were evaluated in terms of number of recognised grip intentions. The test, although preliminary, has demonstrated encouraging results in terms of potentiality and robustness to different subject physical structures (above 90% of the hand grips were effectively recognised).

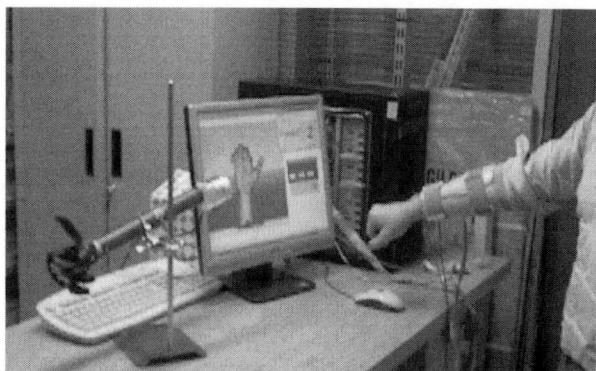

Fig. 6 Tiny-hand prototype during the preliminary testing phase

V. CONCLUSION

The paper shows the prototype realisation and the preliminary testing of an innovative prosthetic device. This penta-digital mechatronic hand is characterised by reduced dimension, weight and cost and good bio-mimetic reproduction. It can be adapted to small hands, such as the ones of children and women, and even to partial amputation. From the control strategy point of view, an innovative multi-sensing system has been developed. The main characteristic of the realised sensing system is the reduced complexity and the possibility to perform multiple hand grips. The realised prototype has been tested on normal subjects with encouraging results. Future activities will be an extensive validation phase and an accurate analysis of the signal processing strategy.

ACKNOWLEDGEMENTS

This research has been supported by the project Mano Robotica (MANOROB, N. 73523) funded by the Regione Toscana (Italy) through the BANDO UNICO R&S (2008), Linea 1.5-1.6 (Codice CUP Cipe: D57I10000730007).

REFERENCES

1. www.touchbionics.com
2. bebionic.com
3. www.ottobock.com
4. Chan F.H.Y., Yang Yong-Sheng, Lam F.K., Zhang Yuan-Ting, Parker P.A.. Fuzzy EMG classification for prosthesis control *Rehabilitation Engineering, IEEE Transactions on.* 2000;8:305-311.
5. Hudgins B., Parker P., Scott R.N.. A new strategy for multifunction myoelectric control *Biomedical Engineering, IEEE Transactions on.* 1993;40:82-94.
6. Micera Silvestro, Sabatini Angelo M, Dario Paolo, Rossi Bruno. A hybrid approach to EMG pattern analysis for classification of arm movements using statistical and fuzzy techniques *Medical engineering & physics.* 1999;21:303–311.
7. Englehart K., Hudgins B.. A robust, real-time control scheme for multifunction myoelectric control *Biomedical Engineering, IEEE Transactions on.* 2003;50:848-854.
8. Weir RF, Heckathorne CW, Childress DS. Cineplasty as a control input for externally powered prosthetic components. *Journal of rehabilitation research and development.* 2001;38:357.
9. Barry DT, Leonard Jr JA, Gitter AJ, Ball RD, others . Acoustic myography as a control signal for an externally powered prosthesis. *Archives of physical medicine and rehabilitation.* 1986;67:267.
10. Craelius William. The bionic man: restoring mobility *Science.* 2002;295:1018–1021.
11. Heasman JM, Scott TRD, Kirkup L, Flynn RY, Vare VA, Gschwind CR. Control of a hand grasp neuroprosthesis using an electroencephalogram-triggered switch: demonstration of improvements in performance using wavepacket analysis *Medical and Biological Engineering and Computing.* 2002;40:588-593.
12. Nicolelis Miguel AL. Actions from thoughts *Nature.* 2001;409:403–407.
13. Kenney LPJ, Lisitsa I, Bowker P, Heath GH, Howard D. Dimensional change in muscle as a control signal for powered upper limb prostheses: a pilot study *Medical engineering & physics.* 1999;21:589–597.
14. Wininger Michael, Kim N, Craelius William. Pressure signature of the forearm as a predictor of grip force *Journal of Rehabilitation Research and Development.* 2008;45:883–892.

Author: Alessandro Tognetti
Institute: Research Center E.Piaggio, University of Pisa
Street: Via Diotisalvi 2
City: Pisa
Country: Italy
Email: a.tognetti@centropiaggio.unipi.it

Robot-Assisted Surgical Platform for Controlled Bone Drilling: Experiments on Temperature Monitoring for Assessment of Thermal Bone Necrosis

M.A. Landeira Freire[1], J.C. Ramos González[2], and E. Sánchez Tapia[1]

[1] CEIT - Applied Mechanics Department, San Sebastian, Spain
[2] TECNUN (University of Navarra) - Thermal and Fluids Engineering Division, San Sebastian, Spain

Abstract—One of the most serious problems encountered during bone-drilling is the generation of heat, as the increase in temperature of the drilling site may result in thermal injury due to bone necrosis phenomena. So developing a method for predicting, controlling and preventing bone ischemic states during surgical interventions would be a great improvement, regarding adequate prostheses fixation and patient recovery. Implementing a predictive equivalent thermal circuit for necrosis assessment involves performing a number of experimental measurements for model parameter identification.

Thereby, several drilling experiments on bovine bone have been conducted, using a Robot-Assisted Surgical platform developed at CEIT, where the temperature evolution of the drilled area has been monitored immediately post-drilling, using Infrared thermography. The maximum temperature values reached for different sets of machining parameters have been established and natural convective cooling patterns have been characterized by obtaining an average value of the thermal time constant of the system.

Keywords—**Robot Surgery, Drilling Parameters, Necrosis.**

I. INTRODUCTION

Among the different machining processes that are regularly applied during surgical interventions, drilling is probably the most frequent process since in various clinical specialties it serves as a preliminary step in the insertion of pins and screws during the reduction, repair or fixation of bone fractures and installation of prosthetic devices [1].

However, several key problems are often reported in the literature [2, 3, 4]: difficulties in maintaining freehand control of the drill, even using a drill guide; attaining geometric accuracy in hole size, location and orientation in order to ensure a suitably rigid mechanical fixation or the *slippage* problem, where the drill bit tends to "walk" or slip on the bone surface as it begins to cut the work piece.

Some of the above issues can be addressed by introducing technological developments that include robotic systems for assistance during surgical procedures; these are known as Robot-Assisted Surgical (RAS) platforms [5, 6]. They have the potential to improve the safety and effectiveness of conventional surgical procedures by ensuring levels of precision, dexterity and safety that are unattainable by other means. Customized drilling tools are designed for these platforms in order to perform complex drilling operations safely, yielding real-time control architectures.

However the heat generation during bone machining, produced by the plastic shear deformation of the bone chips and the friction between the drilling tool and the bone, is still an open issue [1, 3]. In general, many surgical procedures depend on the degree of stability of the screws that fix the implanted device to the bone; consequently, the loss of bone at the drilling site may lead to implant failure. Heat generation during drilling causes a substantial increase in the temperature of the drilled area, and this may result in thermal injury due to ischemic bone necrosis phenomena, which cause an undesirable increase in bone resorption around the implanted screws. The threshold for considering thermal necrosis has been defined as temperature values beyond 47 °C lasting for 1 minute [2, 7].

There are different ways in which to maintain osteonecrosis levels to a minimum. It is extremely important to define sets of machining parameters that are suitable for surgical procedures and to relate the effect of the machining parameters to the temperature increase at the drilling site, the most common effective parameters being cutting speed (N_s), feed-rate (f_r), applied drill force, tool type and tool tip geometry [7]. Also, the beneficial effect of external saline irrigation, regarding the immediate reduction of temperature at the drilling site, is well documented in literature [8].

Introducing a real-time predictive control model for assessment of bone necrosis, such as an equivalent thermal impedance circuit, could be a valuable tool during surgical interventions to avoid possible prostheses fixation problems. In order to build the mathematical model of the predictive thermal circuit, a system parameter estimation process must be performed from experimental data measurements.

Therefore, the main goal of this work is to perform several *in vitro* drilling experiments on bovine bone, using a Robot-Assisted Surgical (RAS) platform together with a customized drilling tool, both developed at CEIT, where bone temperature is monitored immediately post-drilling by means of Infrared (IR) thermography equipment. Maximum temperatures attained during drilling, expected at the bottom of the drilled hole [9], are calculated from acquired data. In addition, natural convective cooling patterns of the system are obtained from the temperature evolution and the time code of the IR recording; characterizing the thermal system relaxation through its time constant is essential as this provides the needed information for later system parameter identification.

The remaining sections are arranged as following: section II details the methodology and experimental procedure; results from the experiments are given in section III and discussed in section IV. Finally, section V presents the conclusions of the study and future work.

II. MATERIALS AND METHODS

A. Description of the Surgical Platform

The RAS platform developed at CEIT was conceived to provide assistance during transpedicular fixation surgical procedures [6]. The robot used in this platform is a PA10-7C (Mitsubishi Heavy Industries, Japan), an industrial robotic arm with 7 degrees of freedom that offers a high level of dexterity thanks to its redundant joint (Figure 1a).

The software platform chosen for executing the robot control loop is the Real Time Application Interface (RTAI) for the Linux OS. RTAI is an extension of the Linux kernel that enables real-time capabilities in the OS, allowing a dual-kernel structure: the standard Linux kernel and real-time kernel. The data acquisition board used is the Sensoray-626.

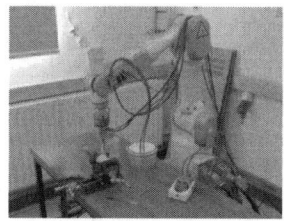 

(a) RAS platform (b) Surgical Tool

Fig. 1 RAS platform and drilling tool prototype developed at CEIT

A customized drilling tool (Figure 1b) was designed for performing the drilling procedure. The implemented hardware architecture is presented in Figure 2.

A conventional, stainless steel twist drill-bit that is 3.2 [mm] in diameter and 70 [mm] long and has a point angle of 90°, was used in the drilling experiments.

B. Bone Sample Harvesting and Preparation

Bone specimens from adult bovine cattle were collected from a local slaughterhouse and preserved in a refrigeration chamber. Samples were obtained from a diaphyseal distal radius, which was chosen as a suitable long bone for drilling. The thickness of the cortical bone of the test work pieces ranged from 7 to 9 mm.

Prior to conducting the bone drilling experiments, a surgical scalpel was used to clean the bone samples of tissue remnants and marrow; later the samples were submerged in a sodium hypochlorite bath (30 min), a hydrogen peroxide bath (2 hour) and preserved in saline buffer at room temperature.

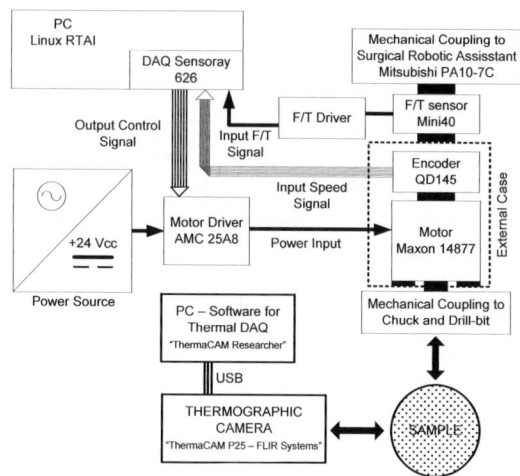

Fig. 2 Hardware architecture of the drilling tool and experimental setup

C. Measuring Equipment

As has been previously stated, temperature distribution measurements were performed using IR thermographic equipment. IR cameras detect radiation in the infrared range of the electromagnetic spectrum and produce images of that radiation, called *thermograms* or *temperature maps*.

Measurements were taken with the ThermaCAM P25 (FLIR systems, Sweden) thermographic camera, with a focal plane array of non-refrigerated micro-bolometers of 320x240 sensors (see Figure 3a). The camera was placed orthogonal to the bone drilling surface, at a distance of 33 [cm].

The emissivity calibration of the bone (ε_{bone}) was performed by using a piece of black tape stuck to the bone surface (knowing *a priori* that ε_{tape}= *0.93*) and a *datalog* with two thermocouples in order to measure the bone-tape surface temperature and the environment temperature. It was determined that ε_{bone}= *0.41*.

 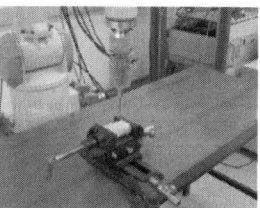

(a) ThermaCAM P25 (b) Experimental Setup

Fig. 3 IR camera and experimental setup

D. Testing Procedure

The set of parameters used in this work to carry out the bone drilling experiments (see Table 1) were determined by taking into account the experience of other research groups whose experiments have been described in the literature [2, 7].

Sample A was tested after being removed from the saline buffer solution for three hours, and sample B was tested after immediate removal from the saline buffer solution.

Bone work pieces were held using a clamp, which allowed for adequate placement of the bone probe with respect to the robotic arm (see Figure 3b) for orthogonal drilling.

For each experiment, the temperature measurement was performed immediately after the withdrawal of the drilling tool. This methodology will be discussed in section IV.

Table 1 Experimental parameters for bone drilling

Test	Sample	Feed Rate [mm/s]	Cutting Speed [rpm]
1	A	0.5	3000
2	A	0.2	1500
3	A	0.2	3000
4	A	0.5	1500
5	B	0.2	3000
6	B	0.2	1500
7	B	0.5	3000
8	B	0.5	1500

III. EXPERIMENTAL RESULTS

Temperature distribution maps of the drilled bone samples were obtained using ThermaCAM Researcher® software (FLIR systems). Figure 4 shows various thermograms from some of the tests.

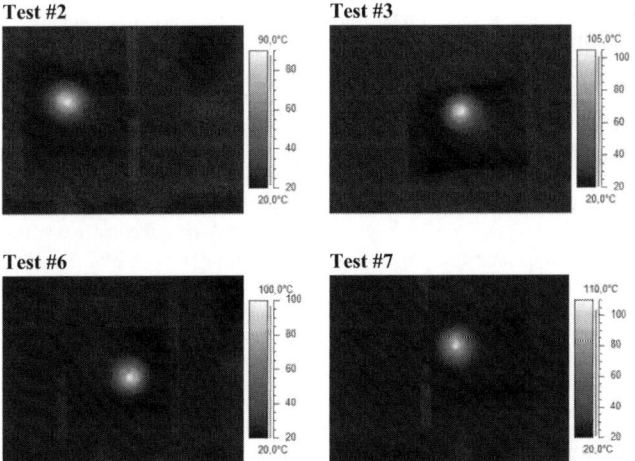

Fig. 4 Temperature maps obtained from Tests 2, 3, 6 and 7

Using the available image processing tools in this software, the immediate post-drilling 'maximum temperature value' was calculated by averaging the values of a 1 mm diameter circular Region of Interest (ROI) which contained the peak temperature found in each image. Results for all the experiments are presented in Figures 5 and 6, where the influence of the drilling parameters and the *in vitro* conditions of the samples is shown.

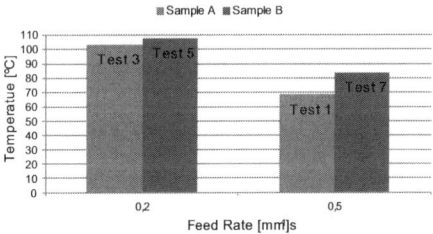

Fig. 5 Average maximum temperature in drilling area, $N_s = 3000$ [rpm]

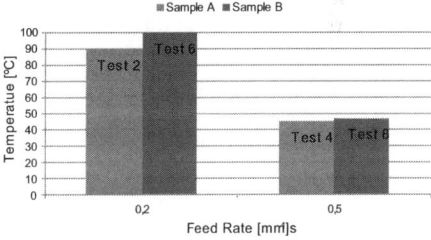

Fig. 6 Average maximum temperature in drilling area, $N_s = 1500$ [rpm]

The characteristic natural convection curves were obtained from each experimental test data, being known the time code of the acquired images, by means of regression analysis. The purpose of carrying out this analysis was to obtain the relaxation time constant of the thermal system in order to characterize the *in vitro* cooling process.

$$T(t) = T_{env} + \left(T_0 - T_{ref}\right) \cdot e^{\frac{-t}{\tau_b}} \quad (1)$$

Experimental cooling curves follow the exponential fitting pattern presented on Equation 1, where $T(t)$ is the temperature evolution, T_{env} is the environment temperature, T_{ref} is the bone reference temperature, T_0 is the initial temperature of the heated system and τ_b corresponds to the relaxation time constant, measured in seconds. According to thermocouple measurements, the environmental temperature during the experiments was $T_{env}=21.5$ [°C]; also, the bone reference temperature has been considered the same as T_{env}.

Once all time constants were obtained for the different experiments, a statistical analysis was performed to obtain the mean and standard deviation values for τ_b. It was determined that $\tau_b=45.02\pm15.81$ [s], yielding an average correlation coefficient of $R^2=0.994$.

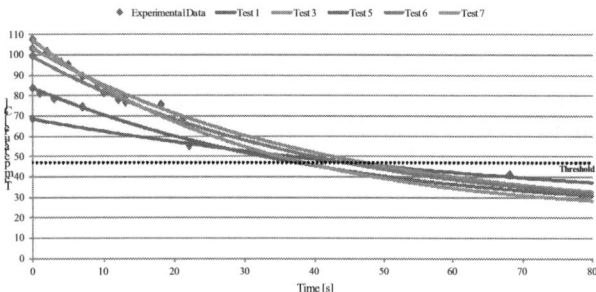

Fig. 7 Exponential fitting for Tests 1, 3, 5, 6 and 7

IV. DISCUSSION

The influence of the feed rate is clearly a determining factor in temperature increase: for a higher feed rate, there was a large reduction in temperature increase (average reduction of 40%), though caution is recommended in case it is not possible for the tool to remove bone material at such a rate. The difference in the spindle cutting speed between experiments also proved that temperature rise is lower for lower cutting speeds (average reduction of 25%). It was noted that for wet sample (B), the temperature rise was slightly higher than for dry sample (A).

Slippage problems were observed during the experiments, though they were minimized because of the use of the RAS platform; also, increasing the feed rate helped minimize the slippage effect. Geometric accuracy and tool guidance were greatly improved over the preliminary manual tests that were performed prior to this experimental work and thus were not considered of interest.

The use of thermocouples was discarded because it is an invasive method which requires an extremely accurate definition *a priori* of the thermocouple and the drilled hole relative locations, thus limiting the number and type of experiments to be performed on a sample. Additionally, data obtained from a thermocouple would consist of indirect measurements of the considered ROI, at the base of the hole.

However, since thermal radiation is a process by which electromagnetic radiation is emitted by a heated surface, the IR camera is able to monitor only surface temperatures; so it is not possible either to directly measure the temperature of the bottom of the hole during the drilling process. Nevertheless, assuming that bone is a system with high thermal inertia, it was postulated *a priori* that natural convective cooling over time should be slow, as shown in Figure 7. Based on this, we decided that it was suitable to perform the drilling operation and then place the IR camera immediately after finishing in order acquiring the images of the maximum initial temperature and the cooling process. In order to assess bone necrosis, it seemed reasonable that for such high temperatures such small inexactitudes (up to 6-8 [°C] over an initial temperature of 100 [°C]) were acceptable.

More experimental testing needs to be performed for increasing the statistical significance of τ_b. Different in vitro conditions should also be tested as in actual operating theaters conditions are more adverse (higher reference temperature, worse convection properties...), thus increasing the value of τ_b and introducing the need for irrigation, which was not considered here given the actual *in vitro* testing conditions.

V. CONCLUSSIONS AND FUTURE WORK

Several bone drilling experiments were successfully performed using our RAS platform. Thermograms were acquired using IR equipment and the influence of different drilling parameters has been shown. The time constant of the thermal system has been obtained through statistical analysis, yielding a highly significant correlation coefficient.

This work corresponds with the experimental stage of a broader project, where a predictive model for assessment of thermal bone necrosis will be derived, provided that more tests will be conducted in order to gain statistical significance.

REFERENCES

1. Wiggins K L, Malkin S (1976) Drilling of Bone. J Biomechanics 9(9):553-559.
2. Augustin G, Davila S et al. (2008) Thermal osteonecrosis and bone drilling parameters revisited. Arch Orthop Trauma Surg 128(1): 71-77.
3. Udiljak, T., D. Ciglar et al. (2007). Investigation into Bone Drilling and Thermal Bone Necrosis. Advances in Production Engineering and Management 2(3): 103-112.
4. Karmani S, Lam F (2004) The design and function of surgical drills and K-wires. Current Orthopaedics 18(6):484-490.
5. Ortmaeir T, Weiss H et al. (2006) Experiments on robot-assisted navigated drilling and milling of bones for pedicle screw placement. Int J Med Robotics Comput Assist Surg 2(4):350:363.
6. Melo J, Bertelsen A, Borro D, Sanchez E (2011) Implementation of a cooperative human-robot system for transpedicular fixation surgery, Proceedings CASEIB 2011, Cáceres, Spain, 2011.
7. Karaca F, Aksakal B, Kom M (2011) Influence of orthopaedic drilling parameters on temperature and histopathology of bone tibia: An in vitro study. Med Eng Phys 33(10):1221-1227.
8. Benington I C, Biagioni P A, Briggs J, Sheridan S, Lamey P-J, (2001) Thermal changes observed at implant sites during internal and external irrigation. Clin Oral Impl Res 13(3): 293-297
9. Lee J, Rabin Y, Ozdoganlar O B (2011) A new thermal model for bone drilling with applications to orthopaedic surgery. Med Eng Phys 33(10):1234-1244

Corresponding author:

Author: Martín Alfonso Landeira Freire
Institute: CEIT – TECNUN/UNAV
Address: Paseo Manuel Lardizábal, 15 - C.P. 20018, San Sebastián
Country: Spain
Email: mlandeira@ceit.es

A Kinematic Analysis of the Hand Function

J. Martin-Martin[1] and A.I. Cuesta-Vargas[1,2]

[1] Department of Physiotherapy, University of Malaga, Spain
[2] School of Clinical Science, Faculty of Health Science, Queensland University Technology, Australia

Abstract—Background: the evaluation of the hand function is an essential element within the clinical practice. The usual assessments are focus on the ability to perform activities of daily life. The inclusion of instruments to measure kinematic variables provides a new approach to the assessment. Inertial sensors adapted to the hand could be used as a complementary instrument to the traditional assessment. Material: clinimetric assessment (Upper Limb Functional Index, Quick Dash), antrophometric variables (eight and weight), dynamometry (palm preasure) was taken. Functional analysis was made with Acceleglove system for the right hand and computer system. The glove has six acceleration sensor, one on each finger and another one on the reverse palm. Method: analytic, transversal approach. Ten healthy subject made six task on evaluation table (tripod pinch, lateral pinch and tip pinch, extension grip, spherical grip and power grip). Each task was made and measure three times, the second one was analyze for the results section. A Matlab script was created for the analysis of each movement and detection phase based on module vector. Results: The module acceleration vector offers useful information of the hand function. The data analysis obtained during the performance of functional gestures allows to identify five different phases within the movement, three static phase and tow dynamic, each module vector was allied to one task. Conclusion: module vector variables could be used for the analysis of the different task made by the hand. Inertial sensor could be use as a complement for the traditional assessment system.

Keywords—kinematics, hand, assessment, inertial sensor.

I. INTRODUCTION

The deterioration of the function of the hand directly affects the development of everyday life [1]. Between the different systems used for the evaluation of the hand, the questionnaires and functional tests are the most used [2].

Self reported questionnaires that are made for the patients offer a subjective vision of the state of health by identifying the own capability to perform various tasks [3]. Similarly the rating scales that are filled in by the therapists depending on the capabilities of the subjects, provide a limited range of responses, referred to the difficulty to perform certain actions [4]. A third element that it not used in the evaluation of the hand is the real performance of the development in the task, this could be the main outcome variable, despite of the time.

The deficiencies identified between the different systems of valuation in the hand, requires a specific job in evaluations to fill those gaps and to obtain a joint vision of all affected items. In this aspect, the use of new technologies such as complementary elements to the conventional evaluations can be of great utility.

One of the technological developments that greater application can have use as a complement to the valuation is the analysis of the kinematics of the movement through the use of inertial sensors [5].

The analysis of the kinematic variables provides information relative to the speed, trajectory, accelerations, and angles, among others. These tools have been used as instruments of measures in different pathologies such as Parkinson's disease [6], cerebral palsy [7], or stroke [8]. In addition have been integrated in virtual environments for functional recovery [9].

The use of the sensors on the hand, has been facilitated by the production of gloves equipped with such technology that offer the possibility of registering the different variables [10-12]. The gloves have already been used in different studies [13,14].

Therefore the purpose of this study is to assess the use of inertial sensors accelerometer type in the hand as a complementary tool to the existing functional assessments. Thus to obtain new variables that provide greater information about the function of the hand.

II. METHOD

A. Design

Quantitative, non-experimental, analytic, transversal approach, aimed at detecting functionality variables of the functional task.

B. Subjects

Ten healthy subjects from the University of Malaga, took part in this study. The inclusion criteria were: age range between 18 and 36 years, no previous health issues, no impairment to upper right limb mobility, no affection skin, right-handed, informed of the study and written consent obtained. The exclusion criteria were: left-handedness, disability locomotive and any other which did not meet with the inclusion criteria.

C. Material

The instruments use for the data collection were classified in four groups: a) anthropometric variables: height and weight using the procedure described [15] b) dynamometry; c) clinical variables: the Upper Limb Functional Index (ULFI) [16] and the QuickDASH [17,18]; d) monitorized variable with accelerometer using the Acceleglove device (AnthroTronix,Inc) [10].

The dynamometer used was the Jamar Hydraulic Hand Dynamometer manufactured by Sammons Preston Rolyan [19] activated with palm pressure. The force of palm pressure was measured in kilograms/cm^2.

The AcceleGlove is a nylon/lycra glove equipped with six inertial accelerometer sensors, one on the back of each finger on the middle phalanx and a sixth sensor on the back of the palm. The software used for recording and capturing data was the Acceleglove Visualizer supplied by the manufacturer. The sampling rate of the device was 120 Hz. Each accelerometer (thumb, index, middle, ring, pinky and palm) has three axis positions (X,Y,Z) with a precision range of ±1,5g. The axis correlation of the glove is illustrated in Figure 1.

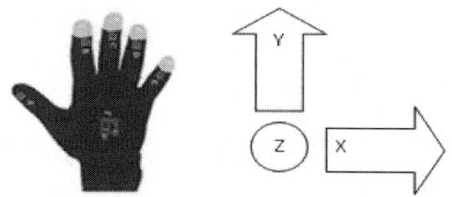

Fig. 1 Acceleglove output signal convention (top view, right hand).

On the basis of acceleration and time data, two new indirect variables were calculated: time; obtained in seconds based on the Unix measurement recorded by the device 1 January 1970 [20]. Module Vector Acceleration; expressed as the following formula: $\sqrt{X^2+Y^2+Z^2}\sqrt{X^2+Y^2+Z^2}$. The operation was performed on the "x", "y" and "z" axes of each of the accelerometers corresponding to each sensor.

D. Method

The participants performed the functional gesture of the hand (tip pinch, tripod pinch, lateral pinch, force grip, extension grip and spherical grip). Each gesture was repeated three consecutive times and measurements taken. The subjects performed the test while seated on a 50 cm high chair, with a straight back and the arm held close to the body with the elbow bent at 90°. The assessment table was placed opposite the subject on a flat surface 75 cm high. A reference mark was placed on the assessment table on which the middle finger of the right hand was situated prior to commencing the test. Each of the subjects remained in the aforementioned position for a period of 4 seconds, after which a warning sound signalled them to move their hand to the area indicated to carry out the gesture. Four seconds after the first signal, a second warning sound signalled them to return to the initial position. The procedure was performed in a series of three repetitions.

E. Acquisition and Processing of Data

Based on the aforementioned protocol the different gesture of the hand were parametrized and allied with a main module vector. The thumb vector was linked with tip and lateral pinch, index vector with tripod pinch and extension grip; palm vector with the force and spherical grip.

While performing each gesture five sub-phases were identified due to significant variations produced by the acceleration vectors. A Matlab[21] script was been created based on the recognition of numeric patterns, identifying in three different sections the most repeated values on a range data. The stability range was determined around the value that was repeated more times in each static section on one module vector. This range was defined by the production of ten consecutive records with approximate values around the most repeated over a range of ±2 units based on a smooth original signal.

These variations have a direct correlation with the various sub-phases of the gesture (T1-T5), corresponding to the movements and positions adopted by the hand (static or dynamic). In the static phase the main module vector of each movement remained a constant acceleration, in the dynamic phase the module vector did not remain.

The sequencing of the phases was: T1 or repose (static), the hand remains static awaiting the sound signal; T2 or calibration (dynamic), the hand moves to the area indicated to perform the task gesture, T3 or success (static), the hand performs the task in the indicated area; T4 or return (dynamic), the hand moves to the initial position, T5 or repose (static), the hand remains static in the reference mark.

Figure 2 represents the temporal spectrum of a subject while performing the tip pinch gesture based on the resulting module thumb vector of the ACC values throughout the sequence, based on data produced by Sigmaplot [22]. This values have been compressed in order to obtain a more uniform curve.

III. RESULTS

Table 1 provides the analysis of the sample based on: anthropometric variable (age, height and weight), clinical variables (ULFI and QuickDash) and dynamometric values of the ten healthy subjects.

A Kinematic Analysis of the Hand Function

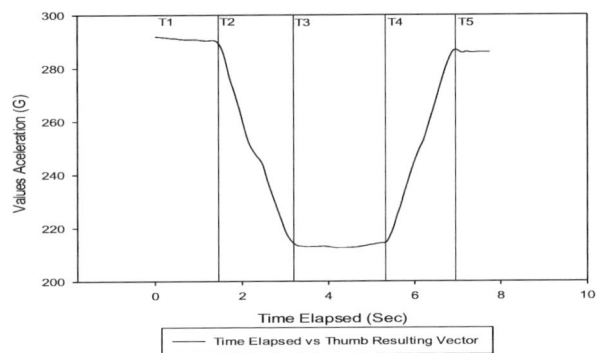

Fig. 2 Tip Pinch by phase based on Thumb module Vector

Table 1 Descriptive of the sample

	N	Mean ± SD
Age	10	26,80 ± 3.67
Height (m)	10	1,68 ± .107
Weight (kg)	10	65,80 ± 16.00
Dinamo Max Ext	10	38,60 ± 13.06
Dinamo Max Flex	10	36,40 ± 11.68
ULFI	10	,85 ± 1.79
QuickDASH	10	4,31 ± 10,30
QuickDash Work	10	2,50 ± 6.06
QuickDash Sport	10	9,37 ± 15.38

The descriptive analysis based on the variation of the module vector values on the static phases of the movement detected, values offered by each module vectors in the terminal pinch, table 2.

Table 2 Terminal pinch in static phase

	T1	T3	T5
Thumb	8,65±5,11	10,39±4,512	11,94±6.46
Index	7,78±6,25	16,19±9,05	11,02±7.37
Middle	9,14±14,20	18,20±23,86	10,53±5.32
Ring	10,03±15,49	15,23±13,43	11,04±9.88
Pinky	8,83±10,72	11,93±7,82	10,43±6.44
Palm	8,62±6,99	6,93±2,48	18,98±35.05

Variation of the vector values in terminal pinch (mean± SD).

Table 3 shows the descriptive results of the variation of the acceleration (mean ± SD) in relation with the task done and the director vector of the same, refereed to static phase of the movement. Variation values offered by module vector were static when it´s lower than 15 unit (g).

Figure three shows the range values of the module vector in the terminal pinch, calculated on basis to the maximum values, minimum and means of all values based on the results obtained by the participants. These values are obtained by extraction from each one of them in each unit of time measurement, for the tip pinch. In the graph the dotted line reflects the maximum possible values, the line continues the average value obtained and the striped line the minimum values.

Table 3 Static phase of the movement, task and director vector

	T1	T3	T5
Tip Pinch (Thumb)	8.65±5.11	10.39±4.51	11.94±6.46
Lateral (Thumb)	10.57±9.87	11.56±3.53	11.21±8.24
Tripod (Index)	6.32±5.33	7.35±1.31	6.94±1.40
Extension (Index)	5.33±1.37	8.72±1.52	10.11±4.65
Grip (Palm)	6.81±4.49	8.66±4.23	12.53±5.68
Spherical (Palm)	6.77±3.70	7.02±1.41	9.80±5.48

The variation of the acceleration has been obtained between the difference of maximum and minimum values of each periods. SD (65%). Unit (g).

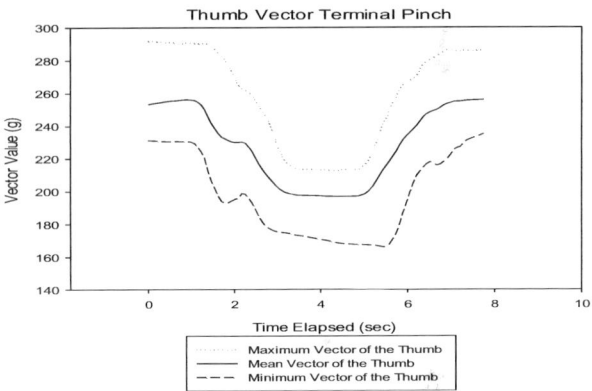

Fig. 3 range values of Thumb Vector in terminal punch

IV. DISCUSSION

The designed protocol is valid for describing the different functional task of the hand, based on descriptive variables obtained using inertial sensors.

The use of peripheral devices (Acceleglove) for the parameterization in real-time of the task movement, enabled it to be fragmented into various phases. The fragmentation of the results obtained from the resulting vector was a key element in describing and predicting the movement.

Phases T2 and T4 (Figure 2) did not follow a standard pattern in their graphic and numerical representations due to the different combinations possible at the approximation phase (T2) and return phase (T4). In other words, depending on the subject who performed the approximation movement, this took place either by first flexing the elbow, placing the shoulder on a flat surface, or flexing the wrist, amongst others.

There are various studies in which analyzes the function of the hand from different perspectives. Lui X et al.[23] analyzed the different areas of existing contact and the direction vector produced in the hand relation to their involvement in the different hand tasks.

The motion capture systems have been used to reproduce different movement of the human body, based in ergonomics applications [24] or for obtained new numerical variables of the movement based of reflective marker system [25]. Periods of reach, grip and return equivalent to T2, T3 and T4 in this study, have been valued by electromyography and kinematic variables by other authors at arm and shoulder complex [26], getting an important new variables for the analysis of them such as tangential velocity of the wrist relative to the time of completion of the task.

The inertial sensors also have been used in the analysis of the effects produced by the pathologies that affect the upper limb of the human body as it is Parkinson's disease [27] or stroke survive [28].

The module vectors obtained on the basis of the results of the inertial sensors are valid for the fragmentation of the movement in temporal phases, based on the variation of the data.

REFERENCES

[1] R. D. Rondinelli, W. Dunn, K. M. Hassanein, C. A. Keesling, S. C. Meredith, T. L. Schulz, and N. J. Lawrence, (1997)"A simulation of hand impairments: effects on upper extremity function and implications toward medical impairment rating and disability determination," *Arch Phys Med Rehabil*, vol. 78, no. 12, pp. 1358–1363.

[2] A. E. Bialocerkowski, K. A. Grimmer, and G. I. Bain, (2000)"A systematic review of the content and quality of wrist outcome instruments," *Int J Qual Health Care*, vol. 12, no. 2, pp. 149–157.

[3] C. Metcalf, J. Adams, J. Burridge, V. Yule, and P. Chappell, (2007)"A review of clinical upper limb assessments within the framework of the WHO ICF," *Musculoskeletal Care*, vol. 5, no. 3, pp. 160–173.

[4] K. Schoneveld, H. Wittink, and T. Takken, (2009)"Clinimetric evaluation of measurement tools used in hand therapy to assess activity and participation," *J Hand Ther*, vol. 22, no. 3, pp. 221–235; quiz 236.

[5] A. Cuesta-Vargas, A. Galán-Mercant, and J. Williams,(2010) "The use of inertial sensors system for human motion analysis," *Physical Therapy Reviews*, vol. 15, no. 6, pp. 462–473.

[6] A. Machowska-Majchrzak, K. Pierzchała, S. Pietraszek, B. Łabuz-Roszak, and W. Bartman,(2012) "The usefulness of accelerometric registration with assessment of tremor parameters and their symmetry in differential diagnosis of parkinsonian, essential and cerebellar tremor," *Neurol. Neurochir. Pol.*, vol. 46, no. 2, pp. 145–156.

[7] A. C. de Campos, N. A. C. F. Rocha, and G. J. P. Savelsbergh, (2009) "Reaching and grasping movements in infants at risk: a review," *Res Dev Disabil*, vol. 30, no. 5, pp. 819–826.

[8] A. Timmermans, H. Seelen, R. Geers, P. R. Saini, S. Winter, J. Te Vrugt, and H. Kingma, (2010)"Sensor-Based Arm Skill Training in Chronic Stroke Patients: Results on Treatment Outcome, Patient Motivation and System Usability," *IEEE Trans Neural Syst Rehabil Eng*, Apr.

[9] B. Peñasco-Martín, A. de los Reyes-Guzmán, Á. Gil-Agudo, A. Bernal-Sahún, B. Pérez-Aguilar, and A. I. de la Peña-González, (2012) "Application of virtual reality in the motor aspects of neurorehabilitation," *Rev Neurol*, vol. 51, no. 8, pp. 481–488.

[10] "AcceleGlove." [Online]. Available: http://www.acceleglove.com/.

[11] "MediTouch - HandTutor™." [Online]. Available: http://www.meditouch.co.il/index.aspx?id=2433.

[12] "Overview | cyberglovesystems.com." [Online]. Available: http://www.cyberglovesystems.com/products/cyberglove-III/overview.

[13] J. L. Hernandez-Rebollar, N. Kyriakopoulos, and R. W. Lindeman, "The AcceleGlove: a whole-hand input device for virtual reality," in *ACM SIGGRAPH 2002 conference abstracts and applications*, New York, NY, USA, 2002, pp. 259–259.

[14] E. Carmeli, S. Peleg, G. Bartur, E. Elbo, and J.-J. Vatine, (2011) "HandTutor™ enhanced hand rehabilitation after stroke--a pilot study," *Physiother Res Int*, vol. 16, no. 4, pp. 191–200.

[15] T. O. M. M.-J. (Arthur S. and L. Carter), (2006) *International Standards for Anthropometric Assessment*. International Society for the Advancement of Kinanthropometry.

[16] C. P. Gabel, L. A. Michener, B. Burkett, and A. Neller, (2006) "The Upper Limb Functional Index: development and determination of reliability, validity, and responsiveness," *J Hand Ther*, vol. 19, no. 3, pp. 328–348; quiz 349.

[17] M. T. Hervás, M. J. Navarro Collado, S. Peiró, J. L. Rodrigo Pérez, P. López Matéu, and I. Martínez Tello, (2006)"Spanish version of the DASH questionnaire. Cross-cultural adaptation, reliability, validity and responsiveness" *Med Clin (Barc)*, vol. 127, no. 12, pp. 441–447.

[18] "The QuickDASH | DASH." [Online]. Available: http://www.dash.iwh.on.ca/quickdash.

[19] "Patterson Medical - Evaluation." [Online]. Available: http://www.pattersonmedical.com/app.aspx?cmd=getProduct&key=IF_921002866.

[20] B. LLC, *Time Measurement Systems: Iso 860, (2010) Metric Time, Unix Time, 12-Hour Clock, 24-Hour Clock, Decimal Time, Thai Six-Hour Clock*. General Books LLC.

[21] "MathWorks España - MATLAB - El lenguaje de cálculo técnico." [Online]. Available: http://www.mathworks.es/products/matlab/.

[22] "Sigmaplot Exact Graphs and Data Analysis." [Online]. Available: http://www.sigmaplot.com/.

[23] X. Liu and Q. Zhan, (2012) "Description of the human hand grasp using graph theory," *Med Eng Phys*.

[24] N. Vignais and F. Marin, (2011) "Modelling the musculoskeletal system of the hand and forearm for ergonomic applications," *Computer Methods in Biomechanics and Biomedical Engineering*, vol. 14, no. sup1, pp. 75–76.

[25] "Motion Capture Systems from Vicon." [Online]. Available: http://www.vicon.com/.

[26] J. Jacquier-Bret, N. Rezzoug, J. M. Vallier, H. Tournebise, and P. Gorce, (2009) "Reach to grasp kinematics and EMG analysis of C6 quadriplegic subjects," *Conf Proc IEEE Eng Med Biol Soc*, vol. 2009, pp. 5934–5937.

[27] J. D. Hoffman and J. McNames, (2011) "Objective measure of upper extremity motor impairment in Parkinson's disease with inertial sensors," *Conf Proc IEEE Eng Med Biol Soc*, vol. 2011, pp. 4378–4381.

[28] C. Calautti, P. S. Jones, N. Persaud, J.-Y. Guincestre, M. Naccarato, E. A. Warburton, and J.-C. Baron, (2006) "Quantification of index tapping regularity after stroke with tri-axial accelerometry," *Brain Res. Bull.*, vol. 70, no. 1, pp. 1–7.

Author: Jaime Martin-Martin
Institute: University of Malaga
Street: Paseo Martiricos, s/n
City: Malaga
Country: Spain
Email: jaimemartinmartin@gmail.com

Single Incision Laparoscopic Surgery Using a Miniature Robotic System

I. Rivas-Blanco, M. Cuevas-Rodriguez, E. Bauzano, J. Gomez-deGabriel, and V.F. Muñoz

System Engineering and Automation, University of Malaga, Spain

Abstract—This paper presents a robotic system aimed at solving the main drawbacks of Single Incision Laparoscopic Surgery. The system is composed of a miniature camera robot, a lighting robot to provide efficient illumination to the scene, and a robotic grasper. These devices are introduced into the abdominal cavity through the single port, and are attached to the abdominal wall by magnetic interaction. Two external robotic arms, at which end effector the magnetic holders are attached, are used to guide the internal devices along the abdominal wall. Camera and lighting robots are handled by voice commands, whereas the robotic grasper is teleoperated with a haptic device. An in-vitro experiment to compare the advantages of using this system versus a traditional procedure is developed.

Keywords—Laparoscopic surgery, miniature robots, SILS.

I. INTRODUCTION

Single Incision Laparoscopic Surgery (SILS) is a recently developed laparoscopic technique aimed at reducing the invasiveness of the traditional procedures. Rather than four to five incisions, SILS is developed through a single incision at the entry port, whereby all surgical instruments and the laparoscope are introduced. To this end, new access devices have been developed, such as the Triport™, the SILS™ Port, the Uni-X™ Single Port System, and the Airseal ™ [1]. These devices consist of a single plastic 2 to 3 cm disk holding the working ports, connected to a plastic ring by a clear plastic sheath [2]. Benefits of this new generation method include less incisional pain, better patient's recovery thanks to lower postoperative narcotic requirements and complications, improved cosmetics, and higher patient satisfaction [3]. Despite its many advantages, this method presents some challenges that cannot be overlooked. Close proximity of the laparoscope and instruments, as they are sharing the same port, entails a loss of triangulation between the camera and the working ports, and limits the range of motion [4].

Many researches are addressing the solution to these problems via the development of miniature robots, which are introduced into the abdominal cavity through the entry port. The positioning and fixation of these robots to the abdominal wall is done by suturing [5], by needle locking [6]-[8], and by magnetic interaction. The main advantage of using magnetic anchoring versus suturing or needle locking is the possibility of easily displacing the robot along the abdominal wall by displacing the external magnetic source.

Researches of the University of Texas have successfully performed laparoscopic nephrectomy and appendectomy in two human patients using a magnetically anchored camera system [9]. Lehman et al. [10] proposed a magnetic miniature robot with two arms fitted with cautery and forceps end effectors. The efficacy of this robot was demonstrated in a porcine model in three nonsurvivable procedures (abdominal exploration, bowel manipulation, and cholecystectomy). In Simi et al. [11], a wired miniature vision platform based on a magnetically activated robot has been tested in a female pig. The fixation of the robot to the abdominal wall is done with a set of external permanent magnets, moved by hand. A similar conception is used in [12], where the internal unit consists of an actuated triangular-shaped magnetic frame which allows the anchoring of several robotic units at the same time. In Terry et al. [13], an internal device fixed to the single-port access through a rigid ring and cantilever bar, or by an external magnetic handle, has been tested in a porcine model in a cholecystectomy procedure.

This paper presents a robotic system for single incision laparoscopic procedures composed of a miniature camera robot, a lighting robot, and a robotic grasper, all of them with magnetic anchoring to the abdominal wall. Unlike previous work, external magnets are attached to the end effector of two robotic arms, one to guide the camera and lighting robots, and the other one to handle the magnetic grasper. This system has been tested in an in-vitro experiment aimed at comparing the advantages of using the robotic system versus a traditional procedure.

II. MATERIALS AND METHODS

A. Robotic System

Figure 1 shows the global control architecture of the robotic system proposed in this work, which is composed of a camera robot, a lighting robot, and a robotic grasper. The internal devices are provided with a set of magnets, so once inside the abdomen, they are attached to the abdominal wall through magnetic interaction with three magnetic holders. These holders are handled by two external robotic arms, one to handle both camera and lighting robots, and the other one to handle the robotic grasper. The camera robot is controlled by voice commands, which enables the surgeon to guide the camera along the abdominal wall, as well as controlling pan and tilt motion. The lighting robot provides an appropriate illumination of the operating area. This device can also be controlled by voice commands, or handled by hand motion of its magnetic holder. Finally, the robotic grasper is teleoperated with a haptic device which controls the movements of the robotic arm in charge of handling the grasper.

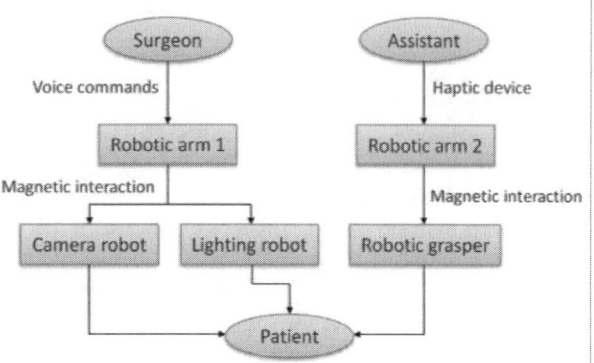

Fig. 1 Global control architecture

Figure 2 shows the distribution of the system in the abdominal cavity. All the internal devices are wireless and are fully inserted into the abdominal cavity through the entry port. One of the advantages of the system is that the camera robot can be moved away from the entry port, avoiding the problem of the loss of triangulation due to the proximity of the camera and the surgical instruments. Furthermore, as the camera can be moved along the abdominal wall, more camera angles than with traditional laparoscopes can be obtained, providing the surgeon with additional viewpoints of the abdominal cavity. On the other hand, as all robots are wireless, the number of instruments sharing the single port is reduced, avoiding the problem of the limitation in the range of motion of the instruments handled by the surgeon. Moreover, in procedures which require the use of three different surgical instruments at the same time, as a fundoplication, the robotic grasper avoids the need of an additional incision. Next, the different components of the system are described.

The camera robot's magnetic holder is composed of two parts: the magnetic holder itself and a wrist attachment component (Figure 3). Coupling between the wrist attachment and the magnetic holder allows the possibility of releasing the magnetic holder once the camera robot has been placed in a desired location. This way, both camera and lighting robots can be handled by the same manipulator. Moreover, this system allows both automatic guidance with a manipulator or dragging the robot by hand motion of the magnetic holder. Miniature camera robot has two DOF to orientate the camera towards different areas of the abdomen: pan and tilt. Pan motion is obtained by rotating the nth joint of the manipulator, as the internal magnets of the camera robot follow the movement of the magnetic holder. The tilt DOF is actuated by a linear motor in the wrist attachment component, which displaces magnet A in Figure 3. This piece is stuck to the internal magnet B of the magnetic holder. Displacement of magnet B causes a displacement of a cylindrical magnet (C) located inside the camera robot, which is connected to the camera and therefore provokes a change in its inclination. This way, the actuation of the motor generates a tilt motion of the camera, which allows a variation of ± 42° in the camera angle.

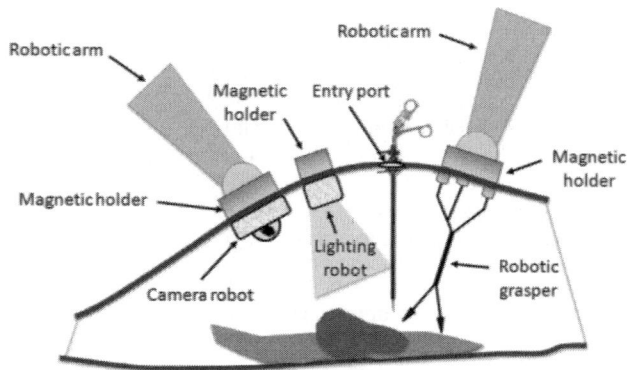

Fig. 2 Robotic system overview

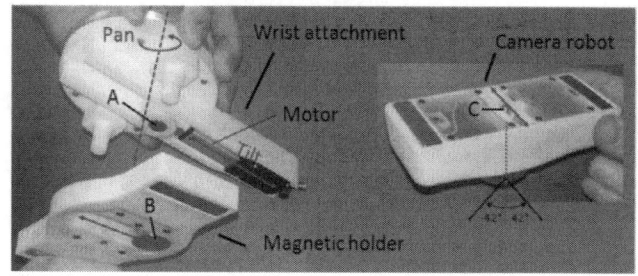

Fig. 3 Camera robot pan and tilt motion

Figure 4 (a) shows the first prototype of the miniature camera robot, of dimensions 12 cm length by 4 cm width. A wireless camera is used to acquire the images of the abdominal cavity, which are transmitted to an external receiver. Additionally to tilt and pan motion, software zoom can be used to focus the image on a smaller area. To illuminate the scene, a set of six white LEDs are distributed around the camera, each emitting 6.8 candela with a 20 mA current. Seven 9 volts batteries (250 mAh) are used to supply the power required by the camera and the LEDs system, avoiding the need of wires to supply the system. The autonomy of the system is around 4 hours.

Although the camera robot is provided with a set of LEDs, they don't provide as much illumination to the scene as a traditional laparoscope. Including a larger number of LEDs would require more power to feed the system, which

Single Incision Laparoscopic Surgery Using a Miniature Robotic System

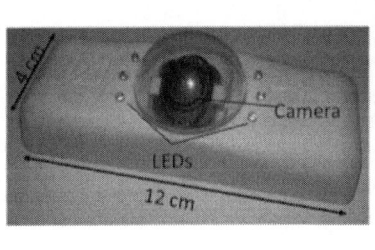
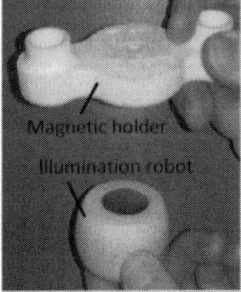

(a) Camera robot (b) Lighting robot

Fig. 4 First camera and lighting robots prototypes

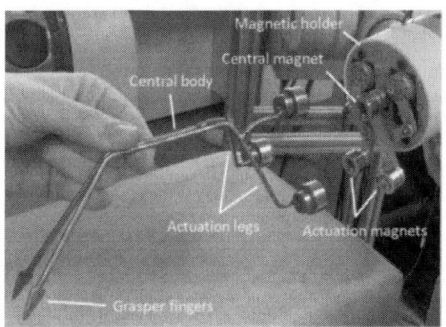

Fig. 5 First prototype of the robotic grasper

would enlarge the size of the robot. Therefore, a lighting robot is included in the system in order to properly illuminate the operating area. Moreover, having an independent source of light makes it possible to change the direction of the light depending on the surgeon's preferences. Figure 4 (b) shows the first lighting robot prototype, of dimensions 35 mm (diameter) x 20 mm (height). As mentioned before, the robot is guided along the abdominal wall with an external robotic arm, which is coupled to the magnetic holder. Additionally, the magnetic holder can be dragged by hand motion. Future work will include autonomous positioning of the lighting robot in order to release the surgeon of this task. LEDs of the lighting robot are the same type of the camera robot ones. Two 3 volts button cell lithium batteries (225 mAh) are used to feed the system, providing the robot with almost two hours of autonomy, which is enough for a typical minimally invasive procedure, as a cholecystectomy.

The third component of the robotic system is the robotic grasper, which first prototype 8 cm length is shown in Figure 5. The magnetic holder is composed of three permanent magnets: a central one, which remains fixed, and two actuation magnets, which motion is actuated by two motors contained in the magnetic holder. Approaching and distancing of the actuation magnets provoke the movement of the actuation legs of the robotic grasper, as two permanent magnets are attached at their tips. As each actuation leg is connected with a grasper finger, approaching and distancing of them provoke the closing and opening of the grasper, respectively. As the motors which actuate the grasper are placed at the magnetic holder, there is no need of wires in the internal device. The magnetic holder is attached at the end effector of an external robotic arm, so the grasper can be teleoperated with a haptic device that controls the movements of the robotic arm. As the abdominal wall is a flexible surface, orientation of the grasper can be done by pushing the magnetic holder against the abdominal wall. Thus, a change in the orientation of the holder provokes a change in the orientation of the grasper.

B. Experiments

The system described above has been tested in an in-vitro environment, where a gall bladder with a tumor inside has been simulated (Figure 6). The experiment consists on retracting the gall bladder to have access to the tumor, cutting the tumor, and finally retracting the tumor from the patient. The aim of the experiment is to compare the advantages of using the robotic system versus a traditional laparoscopic procedure. In the traditional environment, the surgeon cuts and retracts the tumor using two surgical instruments, while a human assistant holds the laparoscopic camera and retracts the gall bladder. In such situation, an extra incision is required in order to introduce the surgical tool used by the assistant, as the camera and the instruments handled by the surgeon are introduced through the SILS access device. On the other hand, in the robotic environment (shown in Figure 6) the traditional laparoscope and the grasper to retract the gall bladder are replaced by the camera and lighting robots, and the robotic grasper, respectively. In this second situation, the assistant handles the robotic grasper with a haptic device (placed out of the scene depicted in Figure 6) while the surgeon guides the camera and lighting robots by means of voice commands.

III. RESULTS

The surgical procedure has been developed by five non-experience surgeons. Table 1 shows the task time in both the traditional and the robotic environments, starting with the retraction of the gall bladder, and finishing with the retraction of the tumor. As it can be seen, the execution time is very similar in both cases. In the robotic environment, the surgeon employs less time in cutting and retracting the tumor, as the range of motion of the surgical tools he/she handles is not limited by the laparoscopic camera. However, the assistant employs more time on retracting the gall bladder than with a traditional grasper, as he/she has to get used to handle the haptic device. On the other hand, all surgeons stated to be more dexterous in the handling of the surgical tools without the laparoscopic camera entering through the same port. Moreover, the use of the robotic grasper avoids the need of an extra incision for the grasper handled by the assistant.

Table 1 Traditional environment versus robotic environment

Surgeon	Traditional environment	Robotic environment
Surgeon 1	2 minutes 36 seconds	2 minutes 55 seconds
Surgeon 2	3 minutes 18 seconds	2 minutes 59 seconds
Surgeon 3	2 minutes 47 seconds	2 minutes 41 seconds
Surgeon 4	3 minutes 07 seconds	2 minutes 43 seconds
Surgeon 5	2 minutes 55 seconds	3 minutes 08 seconds

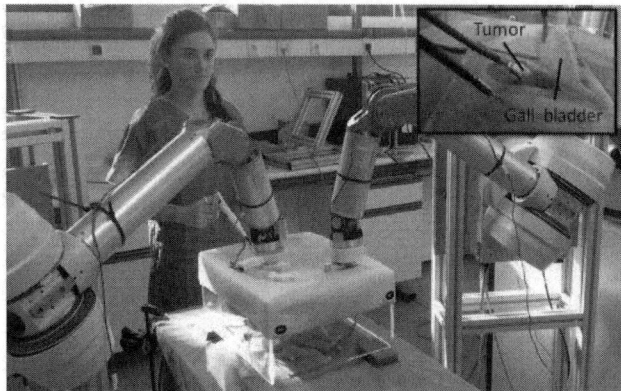

Fig. 6 Surgical procedure with the robotic system

IV. DISCUSSION

The robotic system proposed in this work represents a solution to the main drawbacks of SILS procedures. On the one hand, the problem of the loss of triangulation due to the proximity of the camera and the surgical tools is avoided thanks to a camera robot that can be moved away from the entry port. As there is no need of introducing a laparoscopic camera, the range of motion of the instruments handled by the surgeon is not limited by its presence, so surgeons can operate in a more comfortable way. Moreover, the use of the robotic grasper avoids the need of an extra incision in the procedures in which more than two surgical instruments are required at the same time. The system has been tested in an in-vitro environment to compare the advantages of using the robotic devices versus a traditional laparoscopic procedure.

REFERENCES

1. J.R. Romanelli, T.B. Roshek III, D.C. Lynn, and D.B. Earle, "Single-port laparoscopic cholecystectomy: initial experience," Surgical Endoscopy, Vol. 24, pp. 1374-1379, 2010.
2. S. Agrawal, A. Shaw, and Y. Soon, "Single-port laparoscopic totally extraperitoneal inguinal hernia repair with the TriPort system: initial experience," Surgical Endoscopy, Vol. 24, pp. 952-956, 2010.
3. S.R. Philipp, B.W. Miedema, and K. Thaler, "Single-incision laparoscopic cholecystectomy using conventional instruments: early experience in comparison with the gold standard," Journal of the American College of Surgeons, Vol. 209, No. 5, pp. 632-637, 2009.
4. P. Bucher, F. Pugin, and P. Morel, "Single port access lapaoscopic right hemiclectomy," International Journal Colorectal Dis., vol. 23, pp. 1013-1016, 2010.
5. T. Hu, P.K. Allen, N.J. Hogle, and D.L. Fowler, "Insertable surgical imaging device with pan, tilt, zoom and lighting", 2008 IEEE International Conference on Robotics and Automation, pp. 2948-2953, May 2008.
6. T. Kawahara, T. Takaki, I. Ishii, and M. Okajima, "Development of a broad-view camera system for minimally invasive surgery", 2010 IEEE/RSJ Internatinal Conference on Intelligent Robots and Systems, pp. 2810-2815, October 2010.
7. C.A. Castro, S. Smith, A. Alqassis, T. Ketterl, Y. Sun, S. Ross, A. Rosemurgy, P.P. Savage, and R.D. Gitlin, "MARVEL: a wireless miniature anchored robotic videoscope for expedited laparoscopy", 2010 IEEE Internatinal Conference on Robotics and Automation RiverCentre, pp. 2926-2931, May 2012.
8. Y. Sun, A. Anderson, C. Castro, B. Lin, R. Gitlin, S. Ross, and A. Rosemurgy, "Virtually transparent epidermal imagery for laparoendoscopic single-site surgery", 33rd Annual International Conference of the IEEE EMBS, pp. 2107-2110, 2011.
9. J. A. Cadeddu, R. Fernandez, M. Desai, R. Bergs, C. Tracy, S. J. Tang, P. Rao, M. Desai, and D. Scott, "Novel magnetically guided intraabdominal camera to facilitate laparoendoscopic single-site surgery: Initial human experience," Surg. Endosc., vol. 23, pp. 1894–1899, 2009.
10. A.C. Lehman, J. Dumpert, N.A. Wood, L. Redden, A.Q. Visty, S. Farritor, B. Varnell, and D. Oleynikov, "Natural orifice cholecystectomy using a miniature robot," Surgical Endoscopy Journal, Vol. 23, pp. 260-266, 2009.
11. M. Simi, M. Silvestri, C. Cavallotti, M. Vatteroni, P. Valdastri, A. Menciassi, and P. Dario, "Magnetically Activated Stereoscopic Vision System for Laparoscopic Single-Site Surgery," in IEEE/ASME Transactions on Mechatronics, Vol. 18, No.3, pp. 1140-1151, June 2013
12. S. Tognarelli, M. Salerno, G. Tortora, C. Quaglia, P. Daria, and A. Menciassi, "An endoluminal robotic platform for minimally invasive surgery", The fourth IEEE RAS/EMBS International Conference on Biomedical Robotics and Biomechatronics, pp. 7-12, June 2012.
13. B.S. Terry, Z.C. Mills, J.A. Schoen, and M.E. Rentschler, "Single-Port-Access Surgery with a Novel Magnet Camera System", IEEE Transactions on Biomedical Engineering, Vol. 59, No. 4, April 2012e

Characterization of Anastomosis Techniques for Robot Assisted Surgery

Jordi Campos[1,2], Enric Laporte[2], Gabriel Gili[2], Carlos Peñas[2], Alicia Casals[1,3], and Josep Amat[1]

[1] Universitat Politècnica de Catalunya (UPC), Barcelona, Spain
[2] Corporació Sanitària Parc Taulí (CSPT), Sabadell, Spain
[3] Institut de Bioenginyeria de Catalunya (IBEC), Barcelona, Spain

Abstract—A connection between two vessels or other tubular structures is known as anastomosis, one of the most common procedures in its field but, at the same time, one of the most complex suture-based techniques. This procedure requires not only a lot of skill and dexterity but also a lot of attention and plenty of concentration from the surgeon. This makes many of the actions to be performed irregularly, exposing the patient to human error resulting from the monotony. On the other hand, the field of robotics has earned itself a place in medicine, especially as assistants during a surgical intervention. Even so, medical robotics is quite young and still has not done much in the field of vessel anastomosis. Therefore, this paper presents a preliminary study of the most common suturing techniques, taking into account their typology and performance, within all the possible anastomosis procedures known. Subsequently, a detailed study of workflow and actions during an anastomosis is made, obtaining a diagram for each of the suturing techniques studied. This allows analyzing all procedures and to create a tool to find those actions and repeated tasks and/or common in all of them, indicating which of these are potential candidates for an automation study. This preliminary work focuses on finding where robotics can help to avoid rutinary tasks, which can be learned in a mechanical level and therefore, relatively easy to be automated using a robotic system or to assist the surgeon in certain tasks that need a lot of skill and attention.

Keywords—Suture, Robot Assisted Surgery, Robot Knotting.

I. INTRODUCTION

Suturing and knot tying are the most universal wound closure methods, a necessary skill that surgeons must possess. They are among the most difficult tasks in surgery and consume a significant percentage of operating time in an intervention.

Human dexterity and physiology have certain inherent characteristics, such as the hand and fingers size in relation to the operating field, and some limitations like precision, tremor or reach. In order to minimize or suppress some of these human limitations, surgical robotics can play a major role. Despite its importance in surgery and its technical difficulties, there is little research on suturing in the field of robotic surgery.

The main objective of this study is to classify the different techniques of anastomosis, looking for those actions susceptible of being assisted by a robotic system, aiming to improve both the surgeon working conditions and patient safety and postoperative recovery. The study consists of three phases. The first analyses the existing suturing techniques for all anastomosis common interventions. Subsequently, a study for each kind of intervention relying on a specific technique is conducted, generating a diagram of the decisions, actions and interactions performed during the course of an anastomosis. And finally, these diagrams are compared, looking for common, frequent or complex actions, tasks, variables or environments to evaluate the robotizing needs.

II. BACKGROUND IN SUTURING AND ROBOTIC SURGERY

A. State of the Art in Suturing Techniques

The history of medical sutures dates back nearly five thousand years, but the greatest evolutionary leap forward was made between the 19th and the 20th centuries[4], and has evolved exponentially in this 21st century. From suture needles made of bone, ivory or iron with threads from plants in ancient Egypt, progress has led to monofilament silks and biodegradable and absorbable synthetic materials today, passing through animal tissues such as tendons, ligaments or guts used as thread. Starting at the basic physiological aspects of human body vessels[1], the research and study of the literature related to suture techniques in surgery has considered both, the field of general surgery[10] and also vascular surgery in more detail[15]. In the latter, two parallel techniques evolve differently: traditional sutures, with research mainly in the materials for both needle and filament[6], and sutureless anastomosis in which the challenge to be faced lies both in materials for the connectors and in the technique itself[3]. A base study has been done to establish a solid knowledge basis about the various existing suturing techniques and types of procedures currently in practice.

B. Robot knotting in Surgery

Several approaches and attempts to study and develop suture manipulation in Robot Assisted Surgery (RAS) have been performed, although none is yet suitable for real clinical practice. Kang et al.[8] and Yuel et al.[13] delved into the basic understanding of the mathematical basis behind the knot itself. Although using a different approach, many of the concepts discussed will become a guide in some development processes of this research. Tian et al.[14] proposed the concept of suture manipulation within the knotting process, not only focusing on the movements involved to perform it, but also dealing with aspects concerning manipulation of the whole suturing thread, treating the suture itself in a wider sense.

Although the above examples come from a purely engineering approach, it is also necessary to consider those concerns and needs arising from the daily activity of surgeons. As regards the relationship between new technologies applied to surgery and surgeons, the most common demands among surgeons towards robotic systems are:

- To perform repetitive tasks, that may be rutinary or require too much attention, either autonomously or in assisted way.
- Perform tasks that, despite having the necessary knowledge, the surgeon cannot execute them comfortable and efficiently due to constraints related either to his/her perception, dexterity or accessibility.

This study aims to bring engineering closer to the real surgical needs, by creating a tool that analyses common surgical tasks in anastomosis in order to identify those medical actions that may benefit from robot assistance.

III. IDENTIFICATION OF KNOTS AND SUTURE TYPOLOGIES

A preliminary study to classify each type of knot or suture considers: approaching vessels or direct connection (approximation), stitching vessels with a fold to improve blood flow (eversion) or with knots emplaced within the vessel (inverted)[5][16] (Figure 1).

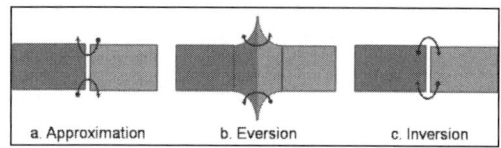

Fig. 1 Knot types

Apart from the three groups of knots discussed, three criteria for the technique selection have been also considered. The first considers simple sutures: performing stitch by stitch individually. The second considers continuous sutures: the thread is passed in its entirety and closed with a single junction knot. And last but not least, if the suture is recurrent: a stitch or thread pass relies somehow on the previous one (Figure2).

All of them are analyzed in relation to the four possible combinations of anastomosis procedures and their variations: cut or incision closure (patch), end-to-side anastomosis, distal (or side-to-side) anastomosis and end-to-end anastomosis[2] (Figure 3).

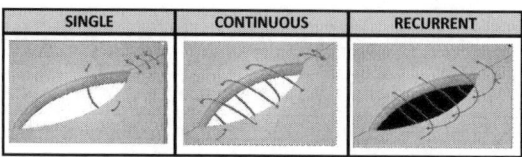

Fig. 2 Knotting techniques

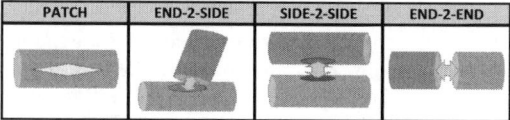

Fig. 3 Types of Anastomosis Procedures

After classifying sutures and knots and after studying the applicable techniques, some of those techniques were immediately discarded, since the needle does not completely cross the tissue, but the thread is beard within the tissue thickness, feature not applicable to the anastomosis due to the vessel typology. The resulting classification is shown in Table1. To perform this classification, all different kinds of knots were taken and related to all variants of the four basic techniques for anastomosis, considering whether they have pre-suture, direct suture or angled suture stages, as indicated on the second row. If a knot is used for a technique, it is marked on in the corresponding check box. Therefore, a preliminary sifting is obtained of the most used knots depending on the procedure. Likewise, sutureless anastomosis techniques[3],[9],[12] were also analyzed and classified, considering both, the materials used and whether the technique itself requires special equipment or devices to perform the four same procedures mentioned above. From a purely medical standpoint, obviously each type of anastomosis may prefer one technique over another in order to minimize the inherent risks of such an intervention[7].

Table 1 Classification of Anastomotic Closure

TYPE	TECH.	PATCH		END-TO-SIDE		SIDE-TO-SIDE	END-TO-END		
		Direct Closing	Patch Graft	Pre-Suture	Direct Closing	Pre-Suture	Pre-Suture	Frontal Direct	Diagonal Direct
APROXIMATION	Close Single	X	X		X			X	X
	Wide Simple	X	X		X			X	X
	X-Cross	X	X		X			X	X
	Simple Cont.	X	X	X	X	X	X	X	X
	Reverdin Cont.	X	X	X	X	X	X	X	X
EVERSION	Rec. H. U-Shape	X	X		X			X	X
	Rec. V. U-Shape	X	X		X			X	X
	Cont. H. Matress	X	X	X	X	X	X	X	X
INVERTED	Schmieden	X	X	X	X	X	X	X	X
	Wide Recurrent	X	X		X			X	X
	Conelli	X	X	X	X	X	X	X	X

*Cont. = Continuous, *V. = Vertical, *H. = Horizontal, *Rec = Recurrent

Current robotic systems used nowadays increase the surgeon's skills, but at the expense of volume, set-up and reaction time, maintenance and, above all, their price. Therefore, a link between the medical characteristics associated with anastomosis interventions and the mechanical capabilities as a supporting robotic system can be set. However, although robot interaction with deformable tissues is still an open research field and an important factor in suturing, this matter is not yet considered in this preliminary study.

Although robot systems are very versatile and progress every day, dazzling with their cutting-edge technological capabilities, there are still many aspects that make them quite inefficient. As studied in Marecik et al.[11], major differences have been encountered in various suturing tasks performed both with a robotic system and by traditional manual methods. Advantages, disadvantages and problems associated with each of the techniques in each used procedure were identified, but never further analyzed.

For this reason, this study is focused on finding common ground between all the studied techniques or procedures, in order to identify the subtasks a suture action is composed of. By splitting its complexity into smaller parts, some subtasks may be identified as potential candidates to be more easily executed with robot assistance.

IV. WORKFLOW FOR PROCESS CHARACTERIZATION

According to their typologies and taking as basis the classification of suture techniques, the detailed analysis of all the actions involved in anastomosis processes is related to each of the studied techniques. Then, a flowchart for each intervention is generated, taking into account the tissue adaptation phases, before and after anastomosis, and every step or decision taken. As an example, the flowchart of an end-to-side anastomosis procedure is presented, using a double-needle thread and a continuous suture (Figure 4). In this example, all subroutines are presented as blocks, and within each of them, the tasks and actions with their relevant instructions. With this distribution, one can see better both the interconnection between tasks and the criticality of the order in which they are performed.

The following step corresponds to the detailed characterization of subtasks, as shown in the diagram. To this end, all their inputs were taken into consideration, whether physical variables related to the patient or to the surgeon, related to surgical instruments and devices or related to the surgical operating environment. This analysis does not only focus on the tasks, seen from the human point of view, but also considers the context wherein they are performed: the baseline scenario, the modifications applied and the resulting stage. Thus, this methodology considers all the variables associated to each task, whether technical, medical or structural.

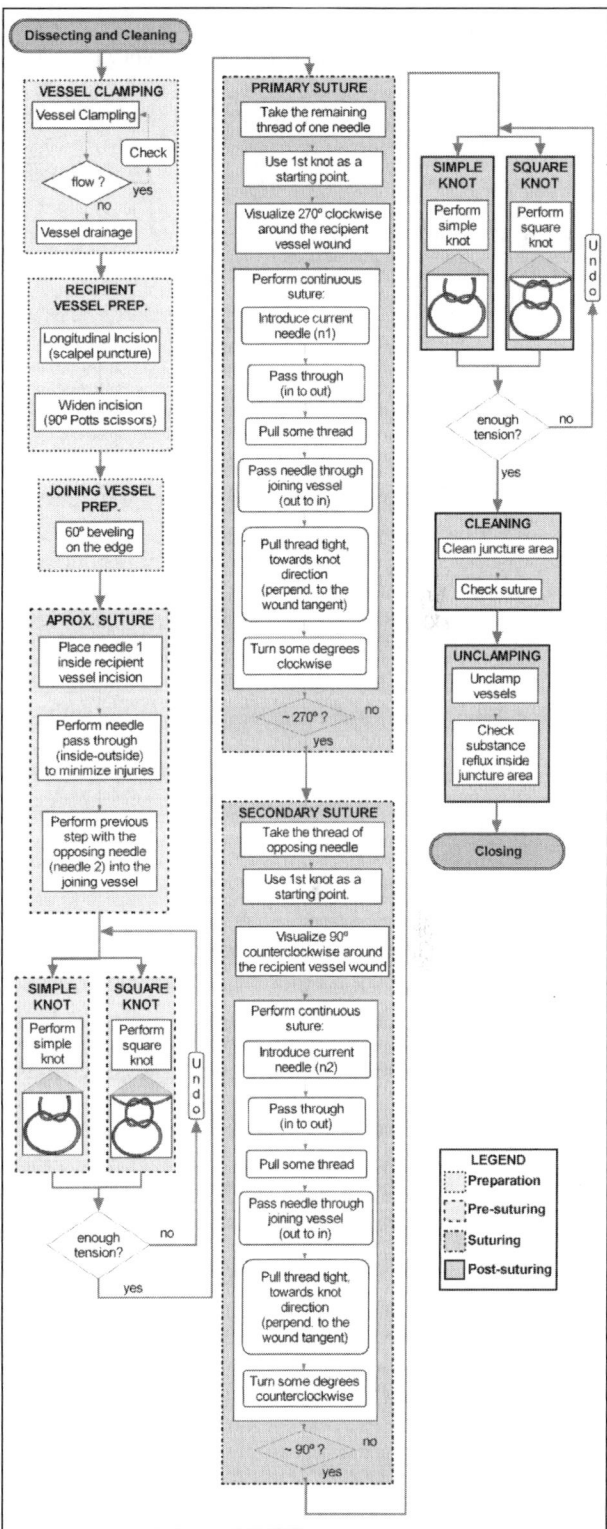

Fig. 4 Workflow of an End-to-Side Double Needled Continuous Pre-Sutured Anastomosis

This analysis took also into account the inner evolution of all variables and the resulting outputs of every task. Thus, by having the input and output layout, the analyzed task can be modeled. This type of characterization eases finding resembling tasks, procedures or working environments in each of the studied techniques, being able to establish some general common blocks. These blocks will be analyzed in more detail in order to see if they are good candidates for automation.

V. CONCLUSIONS

After analyzing and comparing several suturing and knotting techniques, both with and without thread, and fulfill their own workflow, some similarities among them were found. As examples, most of the procedures share some common actions during the operating field conditioning and tissue preparation process, regardless of which suture technique is used. Within the suturing phase, actions during the needle manipulation are very similar, varying only in some controllable variables such as the insertion angle or the number and distance between insertions (knots). This common ground creates the opportunity to think of suturing as a whole configurable unit, with small variations depending on which technique is used.

Focusing on the thread-based techniques, the insertion of the needle at a particular angle is a critical action that has to ensure the structural integrity of the perforated vessel. The distance from the insertion point to the edge is repeatedly found in almost all the analyzed procedures. Furthermore, the surgeon encounters several restrictions while performing the suture, almost all related to needle insertion or connector placement into the vessels. As an example, the working space to perform a sutured anastomosis decreases as the percentage of connected vessel increases, reaching situations where the visible range of the operating space tends to be null causing the surgeon to perform the suture almost blindly and instinctively. Combining the above mentioned drawbacks (needle precision and of the difficulty of implementation), a specific need is found where robotics can intervene. The complexity of these tasks makes autonomous operation an impossible objective. However, a possible help to overcome these common drawbacks is to be studied from this preliminary analysis. Some potential assistance to surgeons could rely on predefining trajectories with virtual fixtures, calculated according to the movements to be made depending on the type of suture technique and the procedure of choice. Those constrained or guided trajectories would absorb any deviation greater than a predefined margin, consistently guiding the surgeon's motion through the perfect execution, even if working at the microscale or under adverse visibility conditions. Other aids could be a change of scale or a pattern based supervision. Therefore, at the technological level, the presented arrangement of the task in suturing facilitates finding the drawbacks and main problems directly related to human limits or environmental conditions. This perspective will ease considerably the programming and implementation of an action, task or sequence into a RAS system.

From this preliminary study, the next step is to quantify the degree of automation of such actions and assign parameters and variables to them. Once established, the design, programming and implementation of virtual specific aids for motion control will begin.

ACKNOWLEDGMENT

This work has been done in the frame of projects DPI2011-2966001, 02 and 03, in the frame of the Spanish Research Program, MINECO and with FEDER funds, EC.

REFERENCES

1. Guthrie, C.C. (1908). "Some Physiologic Aspects of Blood Vessel Surgery". JAMA, 51:1658
2. Seidenberg, B., Hurwitt, E.S., and Carton, C.A. (1958). "The Technique of Anastomosing Small Arteries". Surg., Gynec. And Obst., 106:743
3. Seidenberg, B., Hurwitt, E.S. (1962). "Non-Suture Techniques for Vascular Anastomosis" Progress in Cardiovascular Surgery, 125:130
4. Mackenzie,D. (1971). "The History of Sutures". The Scottish Society of the History of Medicine, 23th Annual Meeting, pp: 158-168.
5. Díaz-Bertrana Sánchez, C. (1990). Practical Suture Handbook. Universitat Autònoma de Barcelona
6. Moy R.L., Waldman B., Hein D.W. (1992). "A review of sutures and suturing techniques". JDerma. Surg. and Onco. 18(9):785-95.
7. Chen Y.X., Chen L.E., Seaber A.V., Urbaniak J.R. (2001). "Comparison of continuous and interrupted suture techniques in microvascular anastomosis". The Journal of Hand Surgery, 26(3):530-9
8. H. Kang, and J.T. Wen,. (2002). "Robotic Knot Tying for Minimally Invasive Surgeries". IRoS Conference, pp. 1421-1426.
9. Zeebregts C.J., Heijmen R.H., van den Dungen J.J., van Schilfgaarde R. (2003). "Non-suture methods of vascular anastomosis". The British Journal of Surgery, 90(3):261-71
10. Fischer, J.E., Bland, K.I. (2007). Mastery of Surgery 5th Edition. Ed. Lippincott Williams & Wilkins
11. Marecik, S.J., Chaudhry,V. (2007). "A comparison of robotic, laparoscopic, and hand-sewn intestinal sutured anastomosis". The American Journal of Surgery, 193:349–355
12. Tozzi, P. (2007). Sutureless Anastomosis: Secrets for Success. Ed. Springer
13. Yuel, L., Caot, Y., Wang, S., Wang,H. (2007). "Twisting Knot Tying Method of Suture: A Novel Method for Robotic Knot Tying". IEEE - International Conference on Complex Medical Engineering, 87-91
14. Tian, X., Kazanzides,P., Taylor,R., Kumar,R. (2009). "Suture knot manipulation with a robot" Surgical Robotics European Summer School, Montpelier
15. Zierler, R.E. (2011). Vascular Surgery Lecture Notes: Clinical Medicine for Engineers. Mayo Clinic
16. Wheeless III, C.R., Serafin, D. (2012). "Technique for Microanastomosis" Wheeless Textbook of Orthopedics, Duke University.

Considering Civil Liability as a Safety Criteria for Cognitive Surgical Robots

E. Berges[1] and A. Casals[1,2]

[1] Robotics group, Institute for Bioengineering of Catalonia, Barcelona, Spain
[2] Universitat Politecnica de Catalunya, BarcelonaTech, Spain

Abstract—One of the challenges of the robotics community is to develop robots that behave more and more autonomously. Therefore, it is necessary to establish new design criteria, as well as more complex methodologies supporting the analysis of associated risks. The procedure described in this paper includes civil liability as an additional criterion to validate the safety of a surgical robot. In order to understand the concept, a methodology is presented through the description of a simple case. This work aims to establish the basis for a further implementation.

Keywords—Design methodology, Product development, Product liability, Safety, Robotic surgery, Cognitive robotics.

I. INTRODUCTION

Operating rooms are dynamic scenarios without trivial routines, what makes that surgeries can never be replicated. The evolution of robotic surgery aims to facilitate or even enhance the abilities of surgeons and, therefore, let the devices perform complex tasks, whether operated or supervised by humans at present, but expected to become increasingly more implicated in surgical procedures. Apart from standard features as accuracy, tremor correction or augmented reality, robots will help surgeons to take decisions by providing them relevant information, or even by anticipating to certain actions predicted from the perception of human intention. Robots endowed with cognitive capacities will increase their capability to take own decisions. Although extensive literature about robotic systems design has been published (some relevant references are described in section II), there are no clear strategies to prevent risks when robots become highly autonomous. This paper presents a novel methodology, which includes the legal assessment as an additional criterion to locate the origin of potential anomalies.

II. SAFETY AND SYSTEM CHARACTERIZATION

In this section, several criteria to prioritize the characterization of a robotic system are shown, having all of them in common the need to avoid injuries to people sharing the same workspace, as it is the case of surgical robots.

A. Measuring and Limiting Human-Robot Contact Forces

Surgical instruments are continuously in touch with patients, so algorithms of collision avoidance are not always useful. It is necessary to teach the robot the difference between contact and impact. The robot has to understand if the movement that is doing or going to do will harm a human in case of collision or sustained contact. The system classifies the type of contact and evaluates the risks involved according to relevant parameters, as severity or occurrence probability. Many authors have developed theories about how to measure the consequences of a robot colliding with humans and objectify them by means of representative indicators. In [1, 2] the force levels and dynamics of a robot hitting a human without causing pain in order to establish affordable limits is studied. In fact, researches belonging to this group and cited by [3] have been oriented to design mechanisms and control systems to avoid human-robot collisions or minimize them to a level of non-hazardous injuries.

B. Design Based on Experimental Data Acquisition

Another criterion to look after human safety in a surgical procedure consists of analysing surgeon's dynamics while operating, so that the robot learns the limits of positions, movements and the corresponding forces. Human movements are parametrized and measured as much as possible, so that the robot can be modelled afterwards to accurately replicate them. In [4] a methodology focused on surgeon needs is described. It starts with experimental data acquisition of kinematics and dynamics of surgical tools performing tasks in different modalities. The measurements obtained consist of positions, forces and torques of several elements, considering whether tools and tissues are in contact or not. The results are used to define the characteristics of the robotic arm to be designed, both for the mechanical parts and for the software architecture.

C. Design Based on Human-Robot Interaction (HRI)

Concepts like autonomy, cognition and context awareness have become usual terms in ontologies, where robots are active and self-governing devices interacting with humans. Up to date, the studies related to design robotic systems based on HRI consider that the robot and the humans understand a given situation by following the Perception-Cognition-Action; hence both can share a common interaction, called the human-robot interface. The main objective of the system design is to achieve the largest common area of communication. HRI evaluation is very common when the surgical system consists of robots co-manipulating tools with the surgeon. An example of this new paradigm is described in [5], where two approaches of HRI are compared. For a robot to understand the environment, it is necessary a previous task of defining potential scenarios and the resulting consequences of combined interactions. In this line, in [6] a case of tools co-manipulation in a surgical scenario is proposed. This breakdown allows the designer to list the set of conditions to be satisfied by the system in each case.

D. Risk Analysis

A risk analysis has to be done imperatively for any system interacting with humans. One of the most widely used in any industrial process is the Failure Mode and Effect Analysis (FMEA, or FMECA in case of evaluating also the criticality) [7]. Designers search all potential failures of any system component and evaluate the consequences in a bottom-up [8] analysis, anticipating a reaction to undesired situations and considering the system redesign. Another similar method is called the Fault Tree Analysis (FTA), which starts from the main process and decomposes it into smaller ones to detect its potential failures. The complexity of an autonomous surgical robot leads to a huge number of situations, the associated behaviors of which cannot be directly foreseen. Therefore, the methodology described in this paper follows a FTA top-down analysis. Both methods use standard indicators to measure the risks associated to each potential failure and specific software to handle the variables are usually required. In order to measure objectively the concept of risk or hazard the IEC61508 (originally IEC1508) categorizes it according to its severity and probability of occurrence [9], though in engineering the concept of detectability is also considered. The need of testing and validating the software that controls complex systems has also to be mentioned. For this purpose, there are specific regulatory requirements to demonstrate the effectiveness and safety of the related devices. In [10], an iterative technique focused on software development to avoid potential risks and satisfy the IEC1508 is described.

III. CIVIL LIABILITY AS ADDITIONAL SAFETY ASSESSMENT

In this section, some legal aspects regarding medical devices are briefly exposed in order to introduce the methodology presented in this paper.

A. Legal Introduction

A surgical robotic system is legally considered a medical device. Therefore, there already exist standards which regulate a proper design and commercialization requirements. In Europe, the regulatory framework is based on the Council Directive 93/42/EEC on medical devices, which prioritizes the human health and safety. Unfortunately, these regulations are not only too general when describing safety constraints, but also the technology has evolved significantly since the date they were written. Even if a product fulfils all the regulatory requirements, the manufacturer is still responsible in case that the product is defective. Although each country applies internal civil laws, apparently they have only slight differences among EU partners and have in common that laws involving medical devices are split into two clearly assigned groups, depending on whether the defect comes from the product or from the medical service. The following methodology aims to help minimizing ambiguity when boundaries are not clear due to autonomous behaviors of surgical robots.

B. Methodology Description

According to the surgical needs and expected functionality of a robotic system, a first approach of robot design is done, taking into account conventional criteria (available technology, optimized workflows, economical budget,...). There are no constraints to choose any initial architecture and workflow of a system. The complete set of elements participating in a surgical procedure, hereinafter named *components* have to be defined and noted as C_i ($i = 1$ to total number of components), likewise all functional interactions among them, hereinafter named *relationships* and noted $R_{j,k}$ (meaning any interaction between component C_j and C_k). The robot is going to interact with the environment, so conventional surgical tools and human actors are also considered components, and the way that clinical staff makes use of robots can be considered relationships. The following steps to check and validate the safety of the system are: breakdown of robot's surgical tasks, failure analysis and study of the legal framework in each case. Depending on which step a failure is detected, civil liability assignment may vary; hence the designer has additional inputs to decide whether the system needs to be redesigned or if it is a matter of information workflow to be improved. This analysis can be represented by a three axis graphic Fig. 2, which provides an easy detection of failures and correspondences with potential causes. The axes represent the three categories related through this methodology: system and workflows, risk analysis and legal analysis, as described below.

Considering Civil Liability as a Safety Criteria for Cognitive Surgical Robots

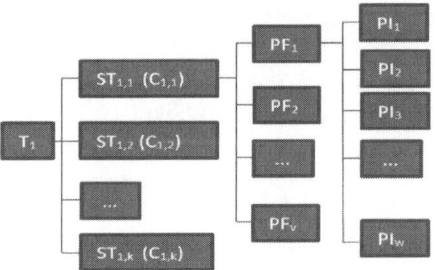

Fig. 1 FTA scheme, including task, subtasks, potential failures and injuries.

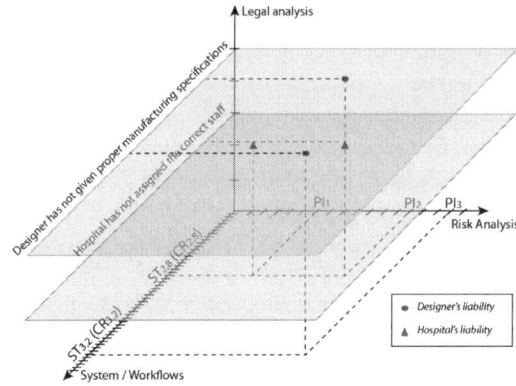

Fig. 2 Graphical representation of the three axis relationship methodology.

The first axis contains information related to the robotic system and the surgical process. The use of robot along the surgical procedure is characterized by the definition of the task it is designed for. Moreover, mechanical, electrical and software specifications are reflected through the above mentioned *components*, as well as the internal and external interactions among all relevant elements, by means of their *relationships*. The tasks T_n ($n = 1$ to total number of main surgical tasks) to be accomplished by the robot during the surgical procedure are listed; then divided into subtasks $ST_{r,s}$ ($s = 1$ to total number of subtasks included in task T_r) and so, down to the desired level of detail. Each minimal subtask, for example $ST_{x,y}$ will be characterized by its components and relationships (from all the C_i and $R_{j,k}$ defined at the original system setup) required to perform the corresponding action. Each particular group of components and relationships are noted by the same subscript as the subtask to which they belong ($CR_{x,y}$). The second axis is the result of the risk analysis applied to the initial system design. For each and every subtask defined above, there might be a certain number of potential failures PF_l ($l = 1$ to total number of potential failures) which might cause hazardous effects leading to a detriment of human safety PI_m ($m = 1$ to total number of potential injuries). In present medical devices, this analysis is done according to human knowledge and previous experiences of similar systems, hence only foreseeable situations will be detected. In future developments where robots are gradually endowed with cognition, powerful software will be necessary to match a larger number of variables (multiple combinations of working components, tasks assignments, environmental dynamics, human intentions, etc.) and determine potential risks through algorithms based on artificial intelligence. In Fig. 1, the geometric progression of finding potential injuries that may occur during a robotic surgery is shown.

Up to this point, the methodology is able to determine which actions of the robot may cause damages. However, they are not all necessarily a consequence of a bad design or system failure, therefore a new axis is added that corresponds to a legal analysis. This new axis complements the previous information by adding a new set of potential scenarios, assigning responsibilities to all actors involved along the chain, from the system design until the surgeon being the final user. Consequently, the combinations substantially grow, reinforcing the idea of needing sophisticated software to properly manage them. Based on information workflow, a general classification of responsibility levels can be the following:

1. Designer has (not) properly verified the robotic system.
2. Designer has (not) properly given the system specifications to the Manufacturer.
3. Manufacturer has (not) properly trained or given instructions to Hospital's staff.
4. Hospital has (not) properly assigned the staff to carry out the surgical intervention.
5. Surgeon has (not) properly informed the Patient about advantages and risks of undergoing surgery by a cognitive robotic system.

At present, any damage caused by a defective product (responsibilities 1 to 3) is a manufacturer's (or designer's) fault. Then, most efforts from designers should be focused on achieving a system that is entirely safe, if the requirements from 3 to 5 are satisfied. The novelty of introducing this third axis helps designers to evaluate what are the risks, the avoidance of which depends on elements that they can control, since all variables are finally related.

An example of a simple case is shown in Fig. 2, in which three potential injuries have been detected. Two of them belong to $ST_{2,8}$ and the third to $ST_{3,2}$. Further analysis should reveal different possibilities about the origin of these potential injuries. Taking only into consideration the technological and health axis, the redesign of the robotic system and the improvement of subtasks $ST_{2,8}$ and $ST_{3,2}$ seems to be necessary, but the addition of the third axis shows that PI_1 would have happened due to a bad decision of the hospital. PI_3 could mean that the robotic system does not reach the desired performance, due to the lack of precise specifications given by the designer to the manufacturer. Finally, in PI_2, more than one agent can be responsible for the hazard. From this information, the designer will rethink the components and relationships included in $CR_{2,8}$ and, probably, remark the requirement of appropriate staff in the user guide instructions for $ST_{2,8}$. The frequency analysis in each axis should help the software evaluate which are the most recommendable corrections. If many PI are involved in the same subtask, a larger probability of system redesign will be indicated. Many ST involved in a single PI requires further evaluations (grouping of subtasks and redefinition of tasks at a higher level, correlation with other axes could indicate the need of proposing some specific actions). A singular case could be that a single PI occurs in all ST, meaning there is an inherent system failure. The frequency in the third axis provides valuable information. The location of points density in a vertical indicates a higher or lower probability of responsibility, from designers to users.

IV. CONCLUSIONS

The methodology described is a proposal to be gradually developed, increasing its level of complexity as cognitive capacities are being endowed to robotic surgical systems. Present robotic systems still allow the analysis to be carried out by humans, but, as stated in previous sections, autonomous robots working in dynamic environments lead to a number of uncountable potential situations. Not only there may be a huge number of tasks and subtasks to be defined, but an increasing complexity appears from the fact that tasks might not always be sequential, but simultaneous. The same concept may take place with the relationships among components. Probably, two components may behave differently one respect to the other depending on third parties acting or not, or depending on the robot's assigned task, etc. Moreover, the suggested information workflow to determine liabilities is not trivial, either. Apart from being very difficult to demonstrate each step (under the legal point of view), simultaneity also may happen. The number of operations and considerations are too high to be manually handled, so this methodology needs to be implemented using cognitive-based algorithms. On the other hand, the development of powerful software tool which can relate so many variables and learn from previous experiences will help foresee those unpredictable situations and help define new laws and regulations, which are at present too far from evaluating ambiguous outcomes that artificial intelligence may bring.

ACKNOWLEDGEMENT

This work has been developed in the frame of the Coordinated Action European Robotic Surgery (Eurosurge), under the 7th work programme FP7 ICT-2011-2-1, project number 288233.

REFERENCES

1. Yamada Y., Hirasawa Y., Huang S., Umetani Y., Suita K.. Human-robot contact in the safeguarding space Mechatronics, IEEE/ASME Transactions on. 1997;2:230-236.
2. Ikuta Koji, Ishii Hideki, Nokata Makoto. Safety Evaluation Method of Design and Control for Human-Care Robots I. J. Robotic Res.. 2003;22:281-298.
3. Haddadin Sami, Haddadin Simon, Khoury Augusto, et al. On making robots understand safety: Embedding injury knowledge into control. I. J. Robotic Res.. 2012;31:1578-1602.
4. Lum Mitchell J. H., Trimble Denny, Rosen Jacob, et al. Multidisciplinary approach for developing a new minimally invasive surgical robot system in In Proceedings of the 2006 BioRob Conference 2006.
5. Kim M., Oh K., Choi J., Jung J., Kim Y.. User-Centered HRI: HRI Research Methodology for Designers;1010 of Intelligent Systems, Control and Automation: Science and Engineering:13-33. Springer Netherlands 2011.
6. Giralt X Amat J. Human robot interaction analysis during robotic co-manipulation of medical instruments Conf. of Society for Medical Innovation and Technology, Barcelona. 2012.
7. Korb W Birkfellner W Boesecke R Figl M Fuerst M Kettenbach J Vogler A Hassfeld S Kornreif G.. Risk analysis and safety assessment in surgical robotics: A case study on a biopsy robot Minimally Invasive Therapy and Allied Technologies. 2005;14:23-31.
8. Kazanzides P., Fichtinger G., Hager G.D., Okamura A.M., Whitcomb L.L., Taylor R.H.. Surgical and Interventional Robotics - Core Concepts, Technology, and Design [Tutorial] Robotics Automation Magazine, IEEE. 2008;15:122-130.
9. Brazendale John. IEC 1508: Functional Safety: Safety-Related Systems in Software Engineering Standards Symposium, 1995. (ISESS95) Experience and Practice, Proceedings., Second IEEE International:8-17 1995.
10. Varley P.. Techniques for development of safety-related software for surgical robots Information Technology in Biomedicine, IEEE Transactions on. 1999;3:261-267.

Author: EDUARD BERGES MARTIN
Institute: INSTITUTE FOR BIOENGINEERING OF CATALONIA (IBEC)
Street: BALDIRI REIXAC, 10
City: BARCELONA 08028
Country: SPAIN
Email: eberges@ibecbarcelona.eu

Kinematic Quantification of Gait Asymmetry in Patients with Elastic Ankle Wrap Based on Cyclograms

P. Kutilek[1], J. Hejda[1], and Z. Svoboda[2]

[1] Faculty of Biomedical Engineering, Czech Technical University in Prague, Czech Republic
[2] Faculty of Physical Culture, Palacky University of Olomouc, Olomouc, Czech Republic

Abstract—This article focuses on gait asymmetry evaluated using synchronized bilateral cyclograms. The bilateral hip-hip and knee-knee cyclograms has never been used before to study the gait asymmetry in patients with figure-eight elastic ankle wrap. The cyclograms were created for patients with foot drop, to quantify gait symmetry before and immediately after application of the ankle wrap. The application of method of the identification of the asymmetry based on the inclination angle of synchronized cyclograms was tested. The kinematic symmetry index was used as a comparative method. The methods based on the inclination angle and symmetry index gave different results. The reason of different results is that the symmetry index depends on discrete variables and is unable to reflect the asymmetry as it evolves over a complete gait cycle. The inclination angle of the hip-hip and knee-knee cyclogram depends on the evolutions of the hip and knee joint angles over time. Only in the case of symmetry index of hip joint angles is significant difference between the data measured before and after application of ankle wrap.

Keywords—**human walking, figure-eight elastic ankle wrap, cyclogram, asymmetry, foot drop.**

I. INTRODUCTION

Several methods can be used in physiotherapy for identifying defects in bipedal walking before and after application of a figure-eight elastic ankle wrap. The widely-used method for studying gait behavior is gait phase analysis [1]. For a study of gait we focus on method based on cyclograms (i.e. angle-angle diagrams or cyclokinograms). The creation of cyclograms is based on gait angles that are objective and well suited for statistical study [2]-[4].

At present, algebraic indices and statistical parameters are used [5], [6]. Algebraic indices include also the symmetry index (SI). The indices usually depend on discrete variables and are unable to reflect the asymmetry as it evolves over a complete gait cycle. Statistical methods such as paired t-test can be used to evaluate the gait symmetry [5], [7], but their computation is more complex and interpretation is less transparent. In 2003, Goswami introduced a technique based on bilateral synchronized cyclograms [5], [8]. The orientation of the synchronized bilateral cyclogram is geometric parameters used to evaluate the symmetry [5], [9], [10]. However, synchronized bilateral cyclogram has never been used before to study the gait asymmetry before and after application of the figure-eight elastic ankle wrap. The ankle wrap is used in the treatment and rehabilitation, for example, of patients with foot drop, [11], [12]. In this paper, we present application of cyclograms for quantification of gait asymmetry before and after application of the figure-eight elastic ankle wrap. The bilateral hip-hip and knee-knee cyclograms has never been used before to study the gait asymmetry in patients with figure-eight elastic ankle wrap.

II. MATERIALS AND METHODS

To create and study synchronized bilateral cyclograms, model of the human body created in MatLab Simulink and SimMechanic software was used, [3], [13]. The movement of the model of a body is controlled by data measured by MoCap system, which identifies the position and orientation of body segments. As a MoCap system, we used medical camera system - Lukotronic AS 200 (Lutz Mechatronic Technology e.U.), and eight IR markers placed on the following anatomical points (left and right side) on the person being measured: malleolus lateralis, epicondylus lateralis, trochanter major and spina iliaca anterior superior, [3], [9]. The markers are placed in accordance with the recommendation of the manufacturer of the MoCap system. We also used the MatLab Simulink to identify the joint angles [3], [13], [14]. Using these techniques we can record and study the movement in a 2D sagittal plane.

After the measurement and identification of joint angles, [13], [14], the method based on synchronization of two angle plots to obtain bilateral hip-hip and knee-knee cyclograms was used, [5]. The heel-touchdown and forefoot/foot flat-touchdown was used to synchronize the two plots. For a symmetric gait a properly synchronized twin trajectories from corresponding joints should be identical, and synchronized bilateral cyclogram should lie on "ideal" symmetry line at an angle of 45° [5]. The orientation of cyclogram was used to evaluate the symmetry, [5]. The two-dimensional cyclogram represents a set of states, and thus the Linear regression is used to determine the value of the inclination (i.e. cyclogram orientation). Simple linear regression fits a straight line through the set of points i.e. states, [14], [15]. The polynomial equation of the regression line is $\beta_R = a_{0k} + a_{1k} \cdot \beta_L$ in the case of knee-knee cyclogram, and

$\alpha_R = a_{0k} + a_{1k} \cdot \alpha_L$ in the case of hip-hip cyclogram. The a_{0i} and a_{1i} are parameters identified by the least squares estimator, [14], [15]. Therefore, the slopes of the regression lines, i.e. inclination angles (θ) of cyclograms are obtained from $tan\theta_k = a_{1k}$ and $tan\theta_h = a_{1h}$.

The symmetry index was used as a comparative method to quantify the symmetry of bipedal walking, [16], [17]. The SI is defined as the ratio of the hip (h) or knee (k) joint angles on the left side (L) to the right side (R), or vice versa, i.e. $SI_h = ROM_{hL}/ROM_{hR}$ and $SI_k = ROM_{kL}/ROM_{kR}$. The ROM is the "range of motion" of the joint of left leg or right leg in the sagittal plane. After the measurements, the SIs and the inclination angles of cyclogram are calculated.

The principal axis of the cyclogram is inclined at an angle θ, the range of values is $0° \leq \theta \leq 90°$. The range of values of the SI is $(0, +\infty)$. The value of the θ and SI can be smaller or larger than the ideal value (i.e. $\theta = 45°$ and SI=1). Thus, it is difficult to interpret the relationship between the θ and SI. In order to determine the relationship between the θ and SI, mathematical technique assuming a value of the SI close to 1 was used. The difference between identified θ of the cyclogram and ideal inclination angle is $\Delta\theta = |45° - \theta|$. Obviously, the greater the $\Delta\theta$, the greater the asymmetry. It is suggested that if $\theta \leq 45°$ then new value is $\theta = 45° + \Delta\theta$, and thus $45° \leq \theta \leq 90°$. The difference between identified SI and ideal SI is $\Delta SI = |1 - SI|$. Obviously, the greater the ΔSI, the greater the asymmetry in ROMs. It is suggested that if $SI \geq 1$ then new value is $SI = 1 - \Delta SI$, and thus the range of values of the SI is $(-\infty, 1)$.

The statistical analysis of SI and θ of the patients without orthosis and with ankle wrap was also performed using the MatLab. First, we used the measured data to illustrate the relationship between the SI and θ. We calculated the regression equation of the regression line, coefficient of determination and Pearson product-moment correlation. Second, we calculated the median, first quartile (Q1) and third quartile (Q3). We used the descriptive statistic to illustrate the relationship between the SI and θ of the patients without and with ankle wrap. The Wilcoxon signed rank test (significance level was set at p<0.05) was used to assess the significance of the differences between values obtained before and immediately after the application of the ankle wrap.

III. MEASUREMENT

The set of data was measured on nine adult patients (age of 46 ± 15 years) with foot drop recruited from patients of the Rehabilitation Center Kladruby (Kladruby u Vlasimi, the Czech Republic). The walking speed (on a treadmill) was 2.3 km/h.

First, patients were asked to walk without the ankle wrap, and the joint angles (L/R hip angles and L/R knee angles in

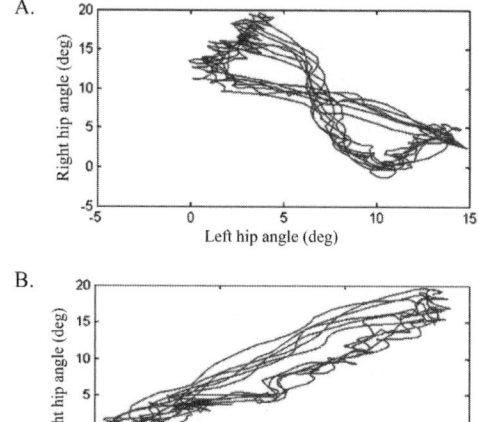

Fig. 1 Example of an unsynchronized bilateral hip-hip cyclograms (A.) and the synchronized bilateral cyclogram (B.).

the sagittal plane) were measured on all subjects. After that, the physician applied the appropriate figure-eight elastic ankle wrap and patients were asked to walk immediately after the application of the ankle wrap. We used ProCare DS Ankle Wrap (made by DonJoy, Inc.). The wrap is composed of low profile elastic contact closure figure-eight straps which provide medial/lateral and arch support without adding bulk and fits right or left foot.

The MoCap system was used to record the data about the movements of body segments. Nine subjects were measured two times: without and with ankle wrap. After obtaining the measured data, joint angles are identified, and unsynchronized and synchronized bilateral cyclograms are created in Matlab software, [3], [5], [8], Fig.1. The deviations of the angles are also caused by a tremor, and thus it is suitable to create a cyclogram by using unfiltered data. We can visualize the data in synchronized bilateral cyclograms with the principal axis and symmetry line, see example Fig.2.

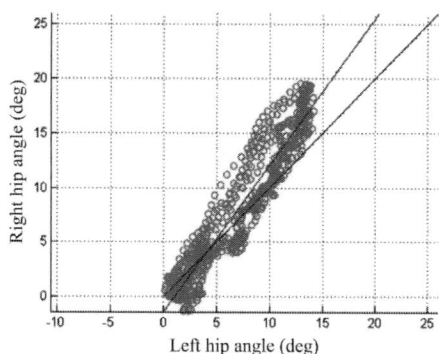

Fig. 2 Example of a synchronized bilateral cyclogram with the principal axis (black) and symmetry line (blue).

Circular marks on the trajectory, Fig.2, correspond to the moments of measurement (sampling frequency 60 Hz).

IV. RESULTS

The described methods are used for identifying the θ and the SI, Table 1 and Table 2. The Fig. 3 shows the relationship between the SI and θ of hip joint angles, where the regression equation of the regression line is $\theta=74.9-27.6 \cdot SI$ and squared value (i.e. coefficient of determination) is 0.41. The Pearson product-moment correlation coefficient is 0.64. The Fig. 4 shows the relationship between the SI and θ of knee joint angles, where the regression equation is $\theta=59.4-11.9 \cdot SI$ and squared value is 0.18. The Pearson product-moment correlation coefficient is 0.42.

In the case of hip joints of patients, correlation coefficient (0.64) is statistically significantly different from zero and indicates a moderately strong correlation between the SI and θ. In the case of knee joints of patients, correlation coefficient (0.42) is not very significant and indicates a weak correlation. The results also indicate that both methods based on θ and SI show different results.

Also, the Wilcoxon test was used. In the case of hip angle of patients with the ankle wrap, the results did not demonstrate significant difference in θ (p=0.91) and show some

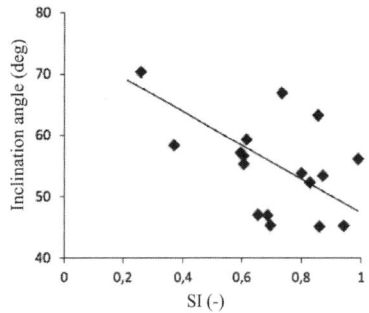

Fig. 3 The relationship between the inclination angle and the symmetry index of the hip joint angles.

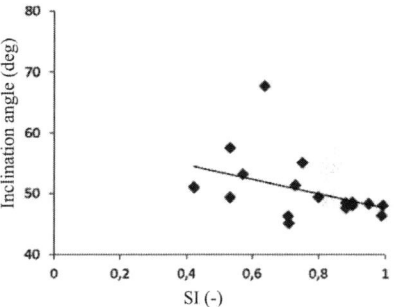

Fig. 4 The relationship between the inclination angle and the symmetry index of the knee joint angles.

Table 1 Inclination angle and SI of the hip-hip cyclogram before and after the application of ankle wrap.

patient	θ (deg)		SI (-)	
	without wrap	ankle wrap	without wrap	ankle wrap
1.	56.6	53.4	0.61	0.87
2.	47.0	46.9	0.66	0.69
3.	57.2	59.4	0.60	0.62
4.	58.4	55.3	0.37	0.61
5.	52.3	63.3	0.83	0.86
6.	45.2	56.1	0.94	0.99
7.	70.4	80.2	0.26	0.21
8.	66.9	53.8	0.73	0.80
9.	45.4	45.1	0.70	0.86

Table 2 Inclination angle and SI of the knee-knee cyclogram before and after the application of ankle wrap.

patient	θ (deg)		SI (-)	
	without wrap	ankle wrap	without wrap	ankle wrap
1.	48.4	46.2	0.88	0.71
2.	47.6	48.5	0.88	0.90
3.	67.6	45.1	0.64	0.71
4.	51.3	53.1	0.73	0.57
5.	57.5	50.9	0.53	0.42
6.	49.4	48.3	0.80	0.95
7.	49.3	51.1	0.53	0.42
8.	48.0	46.3	0.99	0.99
9.	48.0	55.0	0.90	0.75

difference in SI (p=0.03). In the case of knee angle of patients with the figure-eight elastic ankle wrap, the results did not demonstrate significant difference in SI (p=0.20) and θ (p=0.65). We also calculated the median, Q1 and Q3 to illustrate the relationship between the SI of the patients without and with ankle wrap, and between the θ of the patients without and with ankle wrap, see Table 3.

Except for the SI of hip joint angles, all other calculated p-values were greater than the significance level (p<0.05). Therefore, we do not reject the null hypothesis, and there is no significant difference between the data measured before and the data measured after the application of the ankle wrap. Only in the case of SI of hip angles, we can say that there is significant difference between the data measured before and immediately after application of ankle wrap, see Table 3. Thus, the SI indicates significant improvement of

Table 3 The θ and SI of the patients without and with orthosis.

	Hip joint angles				Knee joint angles			
	without wrap		ankle wrap		without wrap		ankle wrap	
	θ (deg)	SI (-)	θ (deg)	SI (-)	θ (deg)	SI (-)	θ (deg)	SI (-)
Median	56.6	0.66	55.3	0.80	49.3	0.80	48.5	0.71
Q1	47.1	0.60	53.4	0.62	48.0	0.64	46.3	0.57
Q3	58.4	0.73	59.4	0.86	51.3	0.88	51.1	0.90

the gait symmetry. In other cases, some data measured after the application of ankle wrap indicate the slight increase or decrease of the median of the θ or SI. In the case of the θ of hip-hip and knee-knee cyclograms, the median of the θ of patients with ankle wrap is lower than median of the θ of the patients without wrap. It could generally show some improvement in gait symmetry, but it is negligible (according to the Wilcoxon signed rank test).

V. DISCUSSION

Only the method based on SI identified significant improvement of the gait symmetry after the application of the ankle wrap, in the case of the hip joint angles. The main reason of this effect is that the SI depends on maximum range of angles and is unable to reflect the asymmetry as it evolves over a complete gait cycle. The inclination angle of the hip-hip and knee-knee cyclogram depends on the complete gait cycle, i.e. evolutions of the angles over time. In the cases of the inclination angle of bilateral cyclograms, however, significant difference between the data measured before and immediately after application of ankle wrap is not identified. Thus, the synchronized bilateral cyclograms do not show improvement in the symmetry of evolution of the joint angles over time. Hence we can say that the method based on inclination angle is preferable because the physicians may notice changes in ROM but they could have a problem with evaluation of improvement during movement.

The main reason of the slight improvement in gait symmetry is the short application time of the ankle wrap. The patients were measured immediately after the application of the orthoses, and subjects were not accustomed to using them. Also, the physician applied the ankle wrap without using a MoCap system, and only the maximum joint angles used to evaluate the walking performance.

VI. CONCLUSION

According to the method described here, synchronized bilateral hip-hip and knee-knee cyclograms can be used to study of the gait asymmetry, but did not identify significant improvement of the gait symmetry after the application of the figure-eight elastic ankle wrap. A slight improvement is measured, but it is negligible. The inclination angles of the synchronized bilateral hip-hip and knee-knee cyclograms have never been used or compared before in the study of patients with figure-eight elastic ankle wrap. In the future study, we plan to measure more patients with other types of orthoses and evaluate the gait symmetry during a longer period of therapy.

ACKNOWLEDGMENT

This research has been supported by project SGS 13/091/OHK4/1T/17 CTU Prague.

REFERENCES

1. Gage R, Hicks R (1989) Gait Analysis in Prosthetics. Clin Prost Orthot 9:17-21
2. Grieve DW (1969) The assessment of gait. Physiotherapy 55:452–60.
3. Kutilek P, Viteckova S (2012) Prediction of Lower Extremity Movement by Cyclograms. Acta Polytech 52:51-60
4. Hajny O, Farkasova B (2010) A Study of Gait and Posture with the Use of Cyclograms. Acta Polytech 50:48-51
5. Goswami A (2003) Kinematic quantification of gait symmetry based on bilateral cyclograms. Proc. XIXth Congress of the International Society of Biomechanics, Dunedin, New Zealand, 2003, pp 1-6
6. Herzog W, Nigg BM, Read LJ et al (1989) Asymmetries in ground reaction force patterns in normal human gait. Med Sci Sport Exer 21:110-114 DOI 10.1249/00005768-198902000-00020
7. Pierotti SE, Brand RA, Gabel RH et al (1991) Are leg electromyogram profiles symmetrical? J Orthop Res 9:720-729 DOI 10.1002/jor.1100090512
8. Goswami A (1998) New Gait Parameterization Technique by Means of Cyclogram Moments: Application to Human Slope Walking. Gait Posture 8:15-36 DOI 10.1016/S0966-6362(98)00014-9
9. Kutilek P, Farkasova B (2011) Prediction of Lower Extremities Movement by Angle-Angle Diagrams and Neural Networks. Acta Bioeng Biomech 13:57-65
10. Goldstein H (1980) Classical Mechanics. Addison-Wesley, Boston
11. Whittle M (2002) Gait analysis: an introduction. Elsevier, Amsterdam
12. Wong M, Wong D, Wong A. (2010) A Review of Ankle Foot Orthotic Interventions for Patients with Stroke. Internet Journal of Rehabilitation 1:1
13. Kutilek P, Hajny O (2010) Study of Human Walking by SimMechanics. Proc. 18th Technical Computing, Bratislava, Slovakia, 2010, pp 1-10
14. Kutilek P, Hozman J. (2011) Prediction of lower extremities movement using characteristics of angle-angle diagrams and artificial intelligence. Proc. of 3rd Int. Conf. on E-Health and Bioengineering, Iasi, Romania, 2011, pp 1-4
15. Seber G, Lee A (1977) Linear Regression Analysis. John Wiley & Sons, New York
16. Zifchock R, Davis I, Higginson J et al (2008) The symmetry angle: A novel, robust method of quantifying asymmetry. Gait Posture 27:622-627 DOI 10.1016/j.gaitpost.2007.08.006
17. Horvath M, Tihanyi T, Tihanyi J. (2001) Kinematic and kinetic analyses of gait patterns in hemiplegic patients. Phys Educ Sport 1:25-35

Author: Patrik Kutilek
Institute: CTU FBME in Prague
Street: Sq. Sitna 3105
City: Kladno
Country: Czech Republic
Email: kutilek@fbmi.cvut.cz

Fat Percentage Equation for Children with Cerebral Palsy: A Novel Approach

E.B. Neves[1,3], E. Krueger[1,2], B.R. Rosário[1], M.C.N. Oliveira[1], S. Pol[1], and W.L. Ripka[2]

[1] Campos de Andrade University Center, Vitória Research Center, Curitiba, Brazil
[2] Federal Technological University of Paraná, Graduate Program in Electrical Engineering and Computer Science, Curitiba, Brazil
[3] Federal Technological University of Paraná, Graduate Program in Biomedical Engineering, Curitiba, Brazil

Abstract—Nutritional changes are commonly related to children with cerebral palsy (CP), since these are problems as: dysphagia, vomiting and gastrointestinal reflux. For nutritional monitoring, body composition has become an important tool and has been present in this population assessments routine. In this sense, the aim of this study was to develop an equation for body composition estimate for children with cerebral palsy, aged between 5 and 6 years, using skinfolds, from the results of body composition obtained by Dual Energy X-rays Absormetry (DEXA). The study included 10 male children with cerebral palsy, aged between five and six-years-old, who participated in the physical therapy intensive program in Vitória research center, Curitiba, Brazil. Participants were assessed by: the Gross Motor Function Classification System; DEXA; and skinfold thickness. The skinfolds that showed better correlation with the total fat percentage and segmental fat percentage were: biceps and abdominal. The results of the regression analysis obtained equations to estimate the fat percentage by skinfold thickness with R^2: 1.000 and 0.971, to seven and two skinfold thickness, respectively. In this sense, it is recommended to use the equation that utilizes the biceps and abdominal skinfolds thickness to estimate fat percentage of children with cerebral palsy, as this presents good estimation indicators.

Keywords—Cerebral Palsy, Children, Anthropometry, DEXA, skinfold thickness.

I. INTRODUCTION

The nutritional assessment status is a fundamental tool for analysis of dysfunctions in children's health [1]. Nutritional changes are commonly related to children with cerebral palsy (CP), since these are problems as: dysphagia, vomiting and gastrointestinal reflux [2-4], these complications can lead to malnutrition or obesity.

For nutritional monitoring, body composition has become an important tool and has been present in this population assessments routine [3, 5, 6]. Among the reference methods one can cite: deuterium dilution, hydrostatic weighing and Dual Energy X-rays Absormetry (DEXA)[7, 8]. However, these methods have limitations regarding its application due to the high cost and complex procedures for data collection, often unfeasible for individuals with CP [3].

In order to avoid these limitations, the literature evaluates skinfolds as a good tool to estimate the fat percentage [8]. The method is often used in combination with the Slaughter equation [8, 9].

The Slaughter equation was initially developed to estimate fat percentage of healthy children aged 7-18 years old [10]. Studies tested its adaptation for children with CP and found correlations ranging from 0.406 to 0.690 for fat mass estimative [3, 8, 9]. These studies pointed to the need to developing specific equations for estimating body fat in children with CP, and also for children with CP in the age groups up to seven years old, since the specificity of a skinfold equation helps in reducing the error estimative and increases the reliability of the information obtained.

In this sense, the aim of this study was to develop an equation for body composition estimate for children with cerebral palsy, aged between 5 and 6 years old, using skinfolds, from the results of body composition obtained by DEXA.

II. MATERIALS AND METHODS

The study included 10 male children with cerebral palsy, aged between five and six-years-old, who participated in the physical therapy intensive program in Vitória research center, Curitiba, Brazil. The study was approved by the institutional ethical committee. Informed consent was obtained from each participant's parent or legal guardian.

Participants were assessed for functional motor impairment using the Gross Motor Function Classification System (GMFCS), which levels range from I to V, with individuals at level V having the greatest motor impairment. The sample was composed by children classified into levels II to IV (group with disability) [8].

For anthropometric measurements were used the following materials: graduated in centimeters and decimeters of centimeters, scientific skinfold (Cescorf) calibrated, digital scales (Wiso W801) with a capacity of 0-180 kg and graduation of 100g; stadiometer (WCS Woody Compact). It was collected measures of points: triceps (TR), biceps brachial (BI), suprailiac (SI), subscapular (SB), abdominal (AB), medial calf (MC), and thigh (TH). The thickness of skinfold was measured three times for each point and then calculated the average. Slaughter equation [10] was used to calculate the fat percentage (%Fat) by skinfold measures of children (Equation 1):

$$\%Fat = 1.21(TR + SB) - 0.008\,(TR + SB)^2 - 1.7 \quad (1)$$

Where: TR (Triceps thickness), SB (Subscapular thickness).

The assessment by DEXA was performed with the device Lunar Prodigy Advance (GE Healthcare - General Electric Company), with Encore 2011 GE Healthcare software that use NHANES III table, for body composition assessment of the CETAC laboratory, Curitiba, Brazil. The equipment calibration was performed daily using a "Phantom". Descriptive statistics, the Spearman *rho* correlations test, the Wilcoxon Signed-Rank test and linear regression analysis was used for the data analysis. All calculations were performed with SPSS version 20.0 and the level of significance set at p <0.05.

III. RESULTS

Table 1 shows the values for age, Gross Motor Function Classification System, total body mass, height and fat percentage, obtained by DEXA and by skinfolds in combination with the Slaughter equation.

Table 1 Descriptive analysis of the variables: age, Gross Motor Function Classification System (GMFCS), body mass, height, fat percentage obtained by DEXA (% Fat DEXA) and fat percentage obtained skinfolds in combination with the Slaughter equation (% Fat Slaughter).

	N	Minimum	Maximum	Average	Standard Deviation
Age (yrs)	10	5.00	6.60	5.53	0.57
GMFCS (I-V)	10	II	V	-	-
Body Mass (kg)	10	12.00	18.00	15.00	2.12
Height (m)	10	1.00	1.16	1.09	0.06
% Fat DEXA*	10	9.70	29.30	15.01	6.00
% Fat Slaughter*	10	6.93	22.73	11.09	4.73

* significant difference between %Fat DEXA and %Fat Slaughter by Wilcoxon Signed-Rank test (p<0.05).

The Spearman *rho* correlations between % Fat DEXA and % Fat Slaughter was 0.826, p = 0.003. Table 2 shows the correlations between the skinfold thickness of anatomical points and segmental fat percentage obtained by DEXA.

Table 2 Spearman *rho* correlations between the values of skinfold thickness (for each anatomical site) and segmental fat percentage, obtained by DEXA.

	TR	BI	SI	SB	AB	MC	TH
%Fat Arm	0.804*	0.882*	-	-	-	-	-
% Fat Legs	-	-	-	-	-	0.246	0.863*
% Fat Trunk	-	-	0.747*	0.706*	0.447*	-	-
% Fat Total	0.831*	0.853*	0.610	0.647*	0.423	0.213	0.835*

Where: TR (triceps), BI (Biceps brachial), SI (Suprailiac), SB (Subscapular), AB (Abdominal), MC (medial calf), TH (thigh). * p < 0.05.

Table 3 presents the results of the regression analysis to obtain equations to estimate the fat percentage by skinfold thickness, considering DEXA as the gold standard.

Table 3 Result of the regression analysis to obtain equations to estimate the fat percentage by skinfold thickness, considering DEXA as the gold standard.

	R^2	p	Equation
7_ST	1.000	0.000	%Fat_7ST = -80.791 + 0.513BM + 62.373HE - 1.589TR + 9.302BI + 4.630SI + 1.831TH - 0.576AB -1.625MC - 5.268SB
3_ST	0.890	0.047	%Fat_3ST = 7.735 + 1.275BM -25.698HE + 2.876BI - 0.347TH + 2.265SI
2_ST	0.883	0.015	%Fat_2STsi = 16.813 + 1.229BM − 31.891HE + 1.867BI + 1.866SI
2_ST	0.971	0.003	%Fat_2STab = 42.274 − 86.213HE + 3.352BM + 5.836BI − 1.532AB

Where: ST (Skinfold thickness), TR (Triceps), BI (Biceps brachial), SI (Suprailiac), TH (thigh), AB (Abdominal), MC (medial calf), SB (Subscapular), HE (Height), BM (Body Mass).

IV. DISCUSSION

The skinfolds thickness that showed better correlation with the total fat percentage and segmental fat percentage were: biceps and thigh (Table 2). However, during the linear regression analysis, the best statistical results were obtained with the skinfolds: biceps and abdominal, according to Table 3. This fact may be related with the body distribution of fat mass in male children and the source (anatomical segment) of the terms of equation, because, the equations %Fat_3ST, %Fat_2STsi and %Fat_2Stab regards information of trunk and members.

All equations developed in this study presented higher correlation with DEXA than the ones obtained by Slaughter equation with DEXA. It reinforces the importance of using specific equations for each population.

In the present study, the best results of fat percentage were estimated by equations %Fat_7ST and %Fat_2STab presented in Table 3 with R^2: 1.000 and 0.971, respectively. In a similar study, Liu et al. [9] studied the estimative of fat percentage by bioelectrical impedance and skinfolds thickness, in children with CP, also considering DEXA as the gold standard. The age average of participants was 10 years old (4 – 17 years old) and the correlations coefficient between fat percentage obtained by dual-energy X-ray absorptiometry and that obtained by skinfold thickness were: 0.891 and 0.758, to two and four skinfold, respectively [9].

For Liu et al. [9], there was no advantage in using equations with four skinfold measures over equations with two skinfold. Thus, this study reinforces those researchers' considerations because it also found very similar statistical indicators between equations that use two and seven skinfolds (%Fat_2Stab, R^2=0.971 and %Fat_7ST, R^2=1.000).

Kuperminc et al. [11] assessed and compared the DEXA and anthropometric data of body composition in children with CP. The researchers evaluated children with GMFCS scale between III-V (our study II-V, Table 1). According to the authors, only anthropometric measures (Body Mass Index - BMI) were poor predictors of fat percentage for children with CP. The researchers found low correlation (R^2=0.27) between fat percentage (DEXA) and BMI. Due to these results, they recommended the use of others methods of measuring (skinfold) to the development of equations in order to improve the power of estimative [8]. In this study, the power of estimates was high, with R^2 ranging from 0.883 to 1.000.

The results found in this study are consistent with others studies that assessed children with CP by: anthropometric measures, doubly labeled water and DEXA, and found correlations between fat percentage obtained by skinfold and the reference method (doubly labeled water and DEXA) ranging from 0.758 to 0.891 [9, 12]. It is worth emphasizing that the ability of skinfold assessor is very important for the accuracy of the estimatives of equations produced from these data.

This study is limited by the small numbers evaluated, but it adds to the little information available regarding the anthropometry of these children. These data support the need of new studies to define an equation with large application and based on a large sample of children with CP.

V. CONCLUSION

This study developed four equations for predicting body fat percentage for children with cerebral palsy. All equations developed in this study presented higher correlation with DEXA than the ones obtained by Slaughter equation with DEXA. For clinical application, it is recommended to use the equation that includes the skinfolds of biceps and abdominal.

ACKNOWLEDGMENT

We would like to thank CNPq, CAPES and Araucaria Foundation for financial support to our participation in this event.

REFERENCES

[1] T. Ogunlesi, M. Ogundeyi, O. Ogunfowora, and A. Olowu, "Socio-clinical issues in cerebral palsy in Sagamu, Nigeria," *South African Journal of Child Health*, vol. 2, 2008.

[2] H. Y. Tomoum, N. B. Badawy, N. E. Hassan, and K. M. Alian, "Anthropometry and body composition analysis in children with cerebral palsy," *Clinical nutrition*, vol. 29, pp. 477-481, 2010.

[3] R. Rieken, E. A. Calis, D. Tibboel, H. M. Evenhuis, and C. Penning, "Validation of skinfold measurements and bioelectrical impedance analysis in children with severe cerebral palsy: a review," *Clinical nutrition*, vol. 29, pp. 217-221, 2010.

[4] E. A. Calis, R. Veugelers, J. J. Sheppard, D. Tibboel, H. M. Evenhuis, and C. Penning, "Dysphagia in children with severe generalized cerebral palsy and intellectual disability," *Developmental Medicine & Child Neurology*, vol. 50, pp. 625-630, 2008.

[5] E. B. Neves, E. M. Scheeren, C. R. Chiarello, A. C. M. S. Costin, and L. P. G. Mascarenhas, "O PediaSuit™ na reabilitação da diplegia espástica: um estudo de caso," *Lecturas, Educación Física y Deportes, Revista Digital*, vol. 15, pp. 1-9, 2012.

[6] E. B. Neves, "Trends in Neuropediatric Physical Therapy," *Frontiers in Public Health*, vol. 1, pp. 1-2, 2013.

[7] D. S. Freedman, C. L. Ogden, H. M. Blanck, L. G. Borrud, and W. H. Dietz, "The Abilities of Body Mass Index and Skinfold Thicknesses to Identify Children with Low or Elevated Levels of Dual-Energy X-Ray Absorptiometry– Determined Body Fatness," *The Journal of pediatrics*, 2013.

[8] M. J. Gurka, M. N. Kuperminc, M. G. Busby, J. A. Bennis, R. I. Grossberg, C. M. Houlihan, R. D. Stevenson, and R. C. Henderson, "Assessment and correction of skinfold thickness equations in estimating body fat in children with cerebral palsy," *Developmental Medicine & Child Neurology*, vol. 52, pp. e35-e41, 2010.

[9] L.-F. Liu, R. Roberts, L. Moyer-Mileur, and L. Samson-Fang, "Determination of body composition in children with cerebral palsy: bioelectrical impedance analysis and anthropometry vs dual-energy x-ray absorptiometry," *Journal of the American Dietetic Association*, vol. 105, pp. 794-797, 2005.

[10] M. H. Slaughter, T. Lohman, R. A. Boileau, C. Horswill, R. Stillman, M. Van Loan, and D. Bemben, "Skinfold equations for estimation of body fatness in children and youth," *Human biology*, vol. 60, pp. 709-723, 1988.

[11] M. N. Kuperminc, M. J. Gurka, J. A. Bennis, M. G. Busby, R. I. Grossberg, R. C. Henderson, and R. D. Stevenson, "Anthropometric measures: poor predictors of body fat in children with moderate to severe cerebral palsy," *Developmental Medicine & Child Neurology,* vol. 52, pp. 824-830, 2010.

[12] R. Van Den Berg-Emons, M. Van Baak, and K. R. Westerterp, "Are skinfold measurements suitable to compare body fat between children with spastic cerebral palsy and healthy controls?," *Developmental medicine and child neurology,* vol. 40, p. 335, 1998.

Corresponding author:

Author: E.B. Neves
Institute: Vitória Research Center
Street: 1, João Scuissiato
City: Curitiba,
Country: Brazil
Email: borbaneves@hotmail.com

Functional and Structural of the Erector Spinae Muscle during Isometric Lumbar Extension

M. González-Sánchez[1] and A.I. Cuesta-Vargas[1,2]

[1] Department of Physiotherapy and Psychiatry, University of Malaga, Spain
[2] School of Clinical Science, Faculty of Health Science, Queensland University Technology, Australia

Abstract—**Study Design: cross-sectional study. Objectives: to compare erector spinae (ES) muscle fatigue between chronic non-specific lower back pain (CNLBP) sufferers and healthy subjects from a biomechanical perspective during fatiguing isometric lumbar extensions. Background: paraspinal muscle maximal contraction and fatigue are used as a functional predictor for disabilities. The simplest method to determine muscle fatigue is by evaluating the evolution during specific contractions, such as isometric contractions. There are no studies that evaluate the evolution of the ES muscle during fatiguing isometric lumbar extensions and analyse functional and architectural variables. Methods: In a pre-calibrated system, participants performed a maximal isometric extension of the lumbar spine for 5 and 30 seconds. Functional variables (torque and muscle activation) and architecture (pennation angle and muscle thickness) were measured using a load cell, surface electromyography and ultrasound, respectively. The results were normalised and a reliability study of the ultrasound measurement was made. Results: The ultrasound measurements were highly reliable, with Cronbach's alpha values ranging from 0.951 0.981. All measured variables shown significant differences before and after fatiguing isometric lumbar extension. Conclusion: During a lumbar isometric extension test, architecture and functional variables of the ES muscle could be analised using ultrasound, surface EMG and load cell. In adition, during an endurance test, ES muscle suffers an acute effect on architectural and functional variables.**

Keywords—**Ultrasound, EMG, Torque, Fatigue.**

I. INTRODUCTION

The erector spinae (ES) muscle is mainly responsible of lumbar extension, also contributing to the lateral tilt of the trunk and maintenance of posture [1]. These functions should serve as motivation for further study of this important muscle in the lower back.

The lower back is the area that has been proven most prone to fatigue in efforts requiring lumbar extension (more so than the upper back, hamstrings and buttocks) [2].

A comprehensive study of the muscular strength and endurance specific to the lumbar region could be very relevant for understand the behavior of the main extension lumbar muscle [3].

Neuromuscular fatigue is considered a very complex phenomenon that is due to several causes: central fatigue, fatigue of the neuromuscular junction and muscle fatigue [4]. Muscle fatigue has been defined as the state that temporarily limits the ability to produce work to a certain intensity caused by the work itself [5,6] or as any exercise that causes a reduction in the maximum capacity to generate force or power. One of the easiest ways to determine muscle fatigue is to assess the evolution of the work over a specific time through isometric contractions, ergonomically supported loads, etc. [3]. These two variables, maximum strength and muscular endurance have been used as predictors of disability in musculoskeletal pathologies [7,8].

In biomechanics, ultrasound has been used to assess morphological changes in muscle thickness [8-12], muscle fibres [13-15], pennation angle [16,17] and even the cross-sectional area [18,19]. Studies have shown that the reliability of ultrasound in the paraspinal muscles varies between moderate and excellent at presenting intraclass correlation index values ranging between 0.72 and 0.98 [20].

Local monitoring of muscle is possible using surface electromyography (sEMG) [2,21]. These analytical instruments have expanded greatly in recent years because of the advantages they possess, including being noninvasive, they can be applied in situ, they allow monitoring of changes in muscle activation (surface electromyography) when the muscle performs defined work, they are instruments with high reliability, they can focus on a particular muscle, and they are relatively inexpensive compared with other analysis tools [21].

No studies have been found that describe the response of the ES muscles after MVC and a fatigue test during isometric lumbar extension using sEMG and ultrasound simultaneously as measuring instruments to extract architectural and functional variables.

The objective of this study was to describe and to analyze architectural variables (muscle thickness and pennation angle) and functional aspects (muscle activation and torque) of ES muscle when performing a muscular endurance test and maximal voluntary contraction during isometric lumbar extension.

II. MATERIALS AND METHODS

A. Design

Cross-sectional study, which analyses the response of the ES muscle during maximal isometric contraction of 5 and 30 seconds duration on a pre-calibrated system.

Participants: Participants: The criteria for inclusion were that subjects were adults aged between 18 and 65 years. The criteria for exclusion were pain in the nerve root/radicular pain, BMI ≥ 35, infection, neoplasms, metastases, osteoporosis, arthritis, scoliosis or any asymmetry evident in the spinal column, pregnancy, cognitive impairment from any cause and inability to complete the proposed exercise.

All subjects signed an informed consent, stating that they had received a careful explanation of the purpose and design of the study.

The study was approved by the Tribunal of Review of Human Subjects at the University of Malaga. Data were handled in accordance with the ethical standards of the Helsinki Declaration of 1975, as revised in 2000 [22].

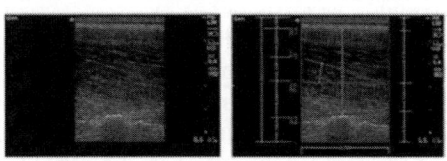

Fig. 1 Example of architecture variables measured from ultrasound image.

B. Procedure

a) Electromyographic record

Two bipolar surface electromyography sensors (Datalog Biometrics, England) were placed bilaterally. Each sensor has two electrodes separated by one centimetre. The skin was sterilised with alcohol and a neutral adhesive specifically designed to adhere the sensor was used.

EMG was collected from the ES muscle group, placing the sensor 2 cm lateral from the line of the spinous processes at L3-L4 level.

EMG was sampled at 1000 Hz (Datalog Biometrics, England) and analysed using a customised Datalink 3.0 programme. Before processing the signal, a low pass filter was used (1 kHz cutoff frequency) to reduce high frequency noise.

b) Ultrasonography record

Architectural variable (thickness and pennation angle) measurements were made from the ES muscle ultrasound pictures.

The ultrasound machine used (SonoSite M-Turbo, SonoSite Inc., Bothell, WA, USA) has a linear translator and a frequency range of 6 13 MHz for ultrasound imaging. The head used was 5 cm wide. The image had a depth of 6.5 cm. The measurements were taken bilaterally, placing the head at the level of L_3-L_4, at 3 cm from the line of the spinous processes. Each ultrasound image was analysed to obtain the two muscle architecture variables considered in this study: pennation angle and thickness. An example is shown in Figure 1.

c) Torque record.

To record the torque of each participant, a computerised load cell (Real Power, Globus Italia) was positioned between two chains. One chain was placed on the arm of the machine that guided the gesture on which the subject would push during the execution of movement, while the other was fixed to the wall in front of the machine where the test was being run.

The angles between the chain and the thrust direction and between the chain and the floor were zero at the time of the isometric contraction.

C. Experimental protocol

Each participant was placed in the precalibrated machine with invincible resistance (figure 2). A chain, in the middle of which was placed a load cell (computerised Real Power, Globus, Italy), was used to measure the torque created by the participant during the test execution. The participant then performed three MVCs for 5 s, with two minutes rest between each execution. Each subject could freely practice the movement before the MVCs and equivalent verbal encouragement was given to all participants. During the execution of the three maximum isometric contractions, electromyographic, ultrasound and torque (using the load cell) recordings were taken. The maximum value recorded after three repetitions was considered the value of the MVC and served as a reference for normalisation of the values recorded later in the fatigue test.

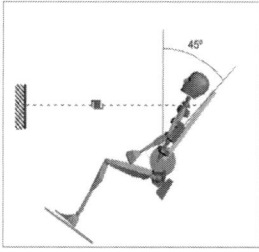

Fig. 2 Scheme of the subject's position and ejecution of the gesture analised.

For the fatigue test, the same protocol was repeated increasing the time up to 30 seconds the maximum isometric extension and up to 240 seconds the rest time between repetitions. During this test, the differences in the variables considered in this study (muscle activation, force moment, muscle thickness and pennation angle) were calculated from measurements taken at the beginning and end of the test. Before being analysed, the data were treated as and then normalised based on the MVC.

The subject performed the execution of maximal isometric extensions, at 45° from the vertical with the lumbar spine in a neutral position. The thighs and hips were attached to the machine by straps. A position execution scheme of the maximum isometric extension is shown in Figure 1.

d) Data analysis

The analysis consisted of a descriptive analysis of each variable and a calculation of the mean difference, between pre and post values recorded during endurance test, by Student's t test for independent data.

For each architecture variable, the three measures taken during MVC isometric contraction were used to calculated the internal consistency (Cronbach's alpha) of the measure, together with the 95% confidence interval.

For statistical analysis, the Statistical Package for Social Sciences program (SPSS; version 17.0 for Windows; Illinois, USA) was used.

III. RESULTS

The participants were 46 healthy subjects (21 men and 25 women). The mean age was 30.39 years (± 7.79), average height 170.52 m (± 16.93), average weight 73.59kg (± 21.20) and body mass index (BMI) 23.68 (±3.15).

Table 1 shows the main values of the functional and architectural variables during the MVC for 5 seconds.

Table 1 Results variables of MVC for 5 sec

	MEAN (SD)
Torque (N·m)	62.26 (±19.47)
Angle Right (°)	13.67 (±2.29)
Angle Left (°)	12.65 (±2.19)
Thickness Right (mm)	30.6 (±6.1)
Thickness Left (mm)	31.2 (±5.9)
N	46

Table 2 shows the difference between variables at the start and end the test. These results were normalised to the MVC, so the difference could be represented as a percentage change in the variable. It could be observed that functional and architectural variables underwent significant changes during the execution of the test.

Table 2 Differences between the start and end of the fatigue test

	Healthy Subjects			Sig. (bilateral)
	Mean	CI (95%)		
		Sup	Inf	
Torque (N·m)	**0.20**	0.20	0.21	**0.000**
Pennation Angle (°)	**0.47**	0.44	0.51	**0.000**
Muscle Thickness (cm)	**0.18**	0.16	0.20	**0.000**
Muscle Activation (mV)	**0.21**	0.18	0.25	**0.000**

The results after the reliability test performed on thickness and pennation angle (architectural variables) showed a very high stability measure, with Cronbach's alpha values for right angle 0.951 (95%CI: 0.920 - 0.971), for left angle 0.981 (95%CI: 0.969 - 0.989), for right thickness 0.969 (95%CI: 0.963 - 0.981) and for left thickness 0.971 (95%CI: 0.966 - 0.978).

IV. DISCUSION

To our knowledge, this is the first study that analised the behavior of the ES muscle when performing a maximal voluntary contraction and an endurance test of lumbar extension, using sEMG, ultrasound and load cell simultaneously.

We also did not find studies measuring the angle of muscle pennation. Some studies have measured the thickness of this muscle using the same method as that employed here (placing the head longitudinally to the muscle) [23,24]. Despite placing the head in the same way and taking measurements at the same point of the lumbar spine (L_3), the average values obtained in the two studies differed from those observed by us.

The ES average thickness values on the right and left, 30.6mm (±6.1) and 31.2mm (± 5.9), respectively, were different from the corresponding values observed in the study by Watanabe et al. [23], where the value was 33.9mm (± 8.4), and Masuda et al.[24], where the mean was 39.4 (±4.2mm). One explanation for the difference could be found at the position of the subjects when the ultrasound image was recorded. Participants in the studies of Watanabe et al [23] and Masuda et al. [24] were in full lumbar extension, whereas in our trial, they stopped 45° from the vertical. The authors of the other two studies obtained the same trend of, starting from a position of maximum flexion, muscle thickness increasing progressively as it approached its maximum extension position, where the highest muscle thickness was recorded.

No studies were found that uses ultrasound as a method to analyse the changes undergone by the ES muscle during a fatigue test. However, several studies have demonstrated by ultrasound the acute effects of muscular fatigue in the biceps brachii [25] and the vastus lateralis due to fatigue [26,27]. They show what appears to be muscle fatigue, with the muscle architecture parameters considered in this study (pennation angle and thickness) showing a significant decrease after the test.

V. Conclusion

During a lumbar isometric extension test, architecture and functional variables of the ES muscle could be analised using ultrasound, surface EMG and load cell. In adition, during an endurance test, ES muscle suffers an acute effect on architectural and functional variables.

References

1. Bogduk, N., Macintosh, J.E., Pearcy, M.J. (1992) A universal model of the lumbar back muscles in the upright position. Spine 17, 897–913.
2. Larivière C, da Silva RA, Arsenault AB, et al. (2010) Specificity of a back muscle exercise machine in healthy and low back pain subjects. Med Sci Sports Exerc Mar;42(3):592-9.
3. Larivière C, Bilodeau M, Forget R, et al. (2010) Poor back muscle endurance is related to pain catastrophizing in patients with chronic low back pain. Spine (Phila Pa 1976) 35(22):E1178-1186.
4. Merletti R, Rainoldi A, Farina D. (2004) Myoelectric manifestations of muscle fatigue. In: Merletti R, Parker P, eds. Electromyography–Physiology, Engineering, and Noninvasive Applications, New Jersey, USA: John Wiley & Sons, Inc 233–258.
5. Heimer S. Fatigue. (1987) In: Medved R, ed., Sports Medicine, second ed 147–151.
6. Vollestad NK. (1997) Measurement of human muscle fatigue. J Neurosci Meth 74(2):219–227.
7. Granacher U, Gruber M, Gollhofer A. Force production capacity and functional reflex activity in young and elderly men. Aging Clin. Exp. Res. 2010; 22:374–382.
8. Biering-Sorensen F. (1984) Physical measurements as risk indicators for low-back trouble over a one-year period. Spine 9(2):106–19.
9. Luo X, Lynn George M, Kakouras I, et al. (2003) Reliability, validity and responsiveness of the short form 12–item survey (SF–12) in patients with back pain. Spine (Phila Pa 1976) 1:1739–45.
10. Morse CI, Thom JM, Birch KM, et al. (2005) Changes in triceps surae muscle architecture with sarcopenia. Acta Physiol Scand 183(3):291-298.
11. Narici M, Maganaris C. (2006) Muscle architecture and adaptations to functional requirements. In Bottinelli R, Reggiani C, eds. Skeletal Muscle Plasticity in Health and Disease. Amsterdam, The Netherlands: Springer 265–288.
12. Visser M, Goodpaster BH, Kritchevsky SB, et al. (2005) Muscle mass, muscle strength, and muscle fat infiltration as predictors of incident mobility limitations in well-functioning older persons. J Gerontol A Biol Sci Med Sci 60(3):324-333.
13. Okita M, Nakano J, Kataoka H, et al. (2009) Effects of therapeutic ultrasound on joint mobility and collagen fibril arrangement in the endomysium of immobilized rat soleus muscle. Ultrasound Med Biol 35, 237–44.
14. Maganaris CN. (2001) Force–length characteristics of in vivo human skeletal muscle. Acta Physiol Scand, 172, 279–85.
15. Ichinose Y, Kawakami Y, Ito M, et al. (1997) Estimation of active force-length characteristics of human vastus lateralis muscle. Acta Anat 159, 78 – 83.
16. Kawakam Y, Abe T, Fukunaga T. (1993) Muscle–fiber pennation angles are greater in hypertrophied than in normal muscles. J Appl Physiol 74, 2740–4.
17. Mahlfeld K, Franke J, Awiszus F. (2004) Postcontraction changes of muscle architecture in human quadriceps muscle. Muscle Nerve, 29, 597–600.
18. Kanehisa H, Ikegawa S, Tsunoda N et al. (1999) Strength and cross-sectional areas of reciprocal muscle groups in the upper arm and thigh during adolescence. Int J Sports Med 16, 54–60.
19. Narici MV, Binzoni T, Hiltbrand E, et al. (1996) In vivo human gastrocnemius architecture with changing joint angle at rest and during graded isometric contraction. J Physiol-London, 496, 287–97.
20. Stokes M, Hides J, Elliott J, et al. (2007) Rehabilitative ultrasound imaging of the posterior paraspinal muscles. JOSPT 37(10), 581-595.
21. Cifrek M, Medved V, Tonković S, et al. (2009) Surface EMG based muscle fatigue evaluation in biomechanics. Clin Biomech (Bristol, Avon). 24(4):327-40.
22. Declaracion De Helsinki De La Asociacion Medica Mundial - principios éticos para las investigaciones médicas en seres humanos [homepage on the Internet]. [cited 11/5/2009]. http://www.wma.net/es/30publications/10policies/b3/index.html. Accessed 21 May, 2012.
23. Watanabe K, Miyamoto K, Masuda T, et al. (2004) Use of ultrasonography to evaluate thickness of the erector spinae muscle in maximum flexion and extension of the lumbar spine. Spine (Phila Pa 1976). 29(13):1472-1477.
24. Masuda T, Miyamoto K, Oguri K et al. (2005) Relationship between the thickness and hemodynamics of the erector spinae muscles in various lumbar curvatures. Clin Biomech (Bristol, Avon) 20(3): 247-53.
25. Shi J, Zheng YP, Huang QH, et al. (2008) Continuous monitoring of sonomyography, electromyography and torque generated by normal upper arm muscles during isometric contraction: sonomyography assessment for arm muscles. IEEE Trans Biomed Eng Mar;55(3): 1191-1198.
26. Brancaccio P, Limongelli FM, D'Aponte A, et al. (2008) Changes in skeletal muscle architecture following a cycloergometer test to exhaustion in athletes. J Sci Med Sport 11(6):538-541.
27. Csapo R, Alegre LM, Baron R. (2011) Time kinetics of acute changes in muscle architecture in response to resistance exercise. J Sci Med Sport 14(3):270-274.

Author: Manuel González Sánchez
Institute: University of Malaga
Street: Paseo Martiricos, s/n
City: Malaga
Country: Spain
Email: mgsa23@uma.es

On Evaluation of Shoulder Roundness by Use of Single Camera Photogrammetry

A. Katashev[1], E. Shishlova[2], and V. Vendina[1]

[1] Riga Technical University/Institute of Biomedical Engineering and Nanotechnology, Riga, Latvia
[2] Riga Technical University/Department of Sport, Riga, Latvia

Abstract—Slouch back is the posture abnormality that becomes especially widespread among young people due to excessive use of computers. Shoulder index is convenient measure to evaluate rounded shoulder. It is calculated as the ratio of shoulder width measured between acromial processes from the front side of patient to the shoulder arch, measured from the back of the patient. The paper presents an attempt to evaluate shoulder index and shoulder roundness by means of photogrammetric measurements instead of measurements "in nature", with the further goal to implement developed technique in health monitoring system. For this one used set of anatomical landmarks, easily defined from the photograph. Unfortunately, proposed measurements did not allow evaluate neither shoulder arches nor shoulder index itself quantitatively. The front shoulder angle, calculates using quasiacromion – anathomical landmark defined from the side view photograph measurements – may be used to test shoulder roundness with sensitivity 0.81 and specifity 0.69. In females, two qualitative parameters may be used to test shoulder roundness: armpit protraction ($S = 0.86$, $Sp = 0.80$) and armpit tilt angle ($S = 1.00$, $Sp = 0.90$). Besides, this test does not work in males.

Keywords—**Rounded shoulder, Posture, Shoulder index, Photogrammetry.**

I. INTRODUCTION

Slouch back is the posture abnormality that becomes especially widespread among young people due to excessive use of computers [1-3]. Alongside, posture deformities are often associated with various pain syndromes [4, 5]. In literature the problem of the slough back is associated with kyphosis and round shoulder, the latter is accessed mainly by two measures: distance between acromion and supporting surface in supine position [6] and angle between plumb line and direction C7 – acromion in lateral view. Alternative measure is sex – independent shoulder index, proposed by V. Velitchenko [7]. This index is calculated as ratio of shoulder width measured between acromial processes from the front side of patient to the shoulder arch, measured from the back of the patient (see below).

Besides to measuring tape and caliper, tools and instrumentation for posture evaluation and measurements of the anthropometrics parameters include expensive 3D scanners, stereophotography, marker-based systems etc. Although there are commercially available products (for instance, TEMPLO ™ by Contemplas [8]), new systems are under development, too [9]. The tendency is to integrate anthropometrical measurements into broader screening system, often having expert functions. The present work is the part of such project [10]. It is targeted on evaluation of anatomical / posture parameters on the base of 2D images, obtained with single USB camera.

Single camera photogrammetry demonstrates reasonable accuracy in evaluation of posture via linear or angular measurements [11]. Besides, such measures, as waist, hips etc circumferences generally require 3D object model and a question of accuracy of circumferential measures calculation from 2D images arises. It was demonstrated, that circumferential measures may be calculated using two person's images, say lateral and frontal, but the accuracy depends on the shape, used to replace real body cross section [12, 13]. Whichever approach is used, the relationship between "natural" and "photogrammetric" measures has to be established.

The goal of the present research is to find a way to measure the shoulder index from set of patient 2D images.

II. METHODS

43 young volunteers attended amateur swimming section and aged 8 – 25 years old was enrolled to the study. Nearly all the volunteers had more or less expressed posture problems and were not professional swimmers. The data on composition of the research group is summarized at the Table 1.

Table 1 Composition of the studied sample group

Age, years	Females	Males
8 - 16	9	11
18 - 25	8	15
Total	17	26

For each subject, the distance between shoulders acromion processes was measured using the flexible measurement tape. Two measurements, with the tape went over the subject's chest and over the subject's back (Fig.1) was made, the first provided front shoulder arch while the second – back shoulder arch. Shoulder index was calculated as

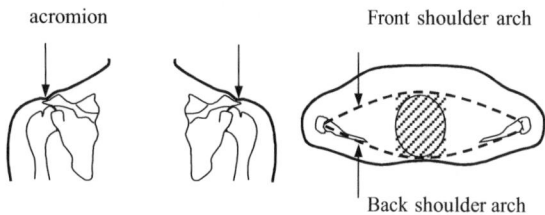

Fig. 1 Measurements made to calculate the shoulder index

$$SI = \frac{Front\ shoulder\ arch}{Back\ Shoulder\ arch} \times 100\% \qquad (1)$$

For normal posture, the values of index are within 96 – 100% following [7]. Besides, other sources propose other limits. In the present paper, the subjects were classified as having round shoulder if the index was less than 90% [14].

Subjects were photographed in upright position, standing on the 50 cm wide mat used for further image calibration. Front and left side view images were obtained by means of wide angle lens USB camera (TheImagingSource® DMK 31BU03, lens with focal length 2.3 mm), placed 2 m from the center of the mat at the height of 1.5 m. Camera lens distortion was corrected, following [15]. Subjects were instructed to keep their ordinary posture without excessive strain.

Exact position of acromion is poorly determined from the photo, unless it is not marked previously by manual examination. Therefore landmarks that may be easily identified at the image have been selected [Fig.2]. At the front view, it was right and left axilla folds F1 and F2. The length of the line F1F2 in a first approach may be associated with the shoulder width. At the left side view, the visible spinous process of C7 was marked as a point where back silhouette line changes curvature. Point A is the visible process of the apex of the chest bone, marked at the point where the front silhouette line changes curvature. Point B marks back silhouette line at the vertical level of point A, so the line AB is horizontal. Points C and D mark anterior and posterior ends of the axilla fold; points E and F indicate width of the arm just above the elbow joint.

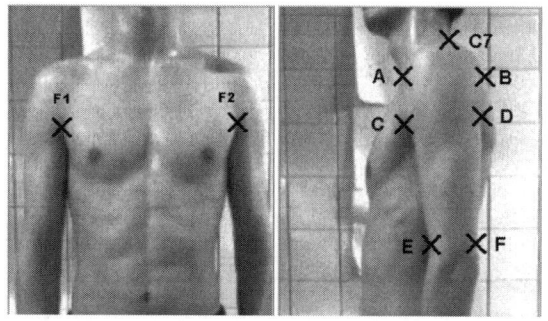

Fig. 2 Anatomic landmarks used in the study (description in the text)

Selected landmarks were used to compose number of parameters, either numerical or qualitative. Numerical parameters was further compared with one measured from the subjects, qualitative parameters was associated with rounded shoulder condition.

For calculations, the real position of acromion was replaced with *"quasiacromion"* point G, constructed as crossing point of the line segment AB and the middle line of the arm. The latter, in turn, was defined as line drawn through the middles of segments CD and EF (Fig.3a). Auxiliary line segment F1' F2', having the same length as F1F2, was drawn through G perpendicularly to the sagittal plane (Fig 3b).

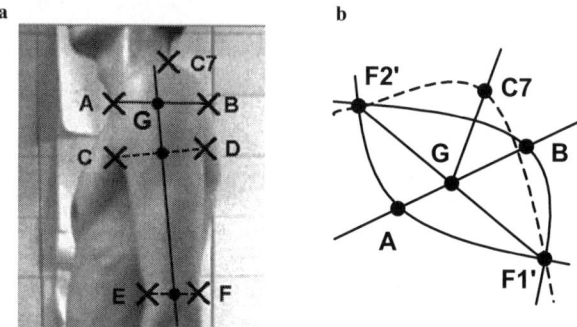

Fig. 3 Determination of quasiacromion point G (a) and scheme (b) used for estimation of shoulder arcs (description in the text)

Front shoulder arch length was evaluated as a length of the parabolic line, drawn through points F1', A and F2'.

Two estimations of the *back shoulder arch length* was used: one by the length of the parabolic line, drawn from F1' to F2' through point B and another through C7 (Fig.3b).

Shoulder protraction was estimated by the *forward shoulder angle* (FSA) between plumb line and segment directed from C7 to quasiacromion. This technique is similar to one described elsewhere (see, for instance, [16]). The question, in what extent the measure, obtained with such a way will be close to or differs from one, calculated using real acromion, is out of scope of the present paper.

Qualitative parameter, that characterize relative position of the points A and C is related to *protraction of armpit*. For protracted armpit, point C is distal to the point A.

Another qualitative parameter is related with the tilt of the line CD between anterior and posterior axilla folds. The *armpit tilt angle* is positive if posterior fold (Fig.3, point D) situated caudally as compared with anterior fold (point C).

III. RESULTS AND DISCUSSION

Fig.4 demonstrates relationships between parameters (front and rear shoulder arches and shoulder index), calculated using photograph and measured by tape. Estimation of the rear shoulder arch at the Fig.4b was made by line drawn through point B (Fig. 3b). One have to note, that estimation using line that goes through point C7 gives practically same picture. Strong correlation was found both for front and rear arches measurements: correlation coefficient between calculated and measured results was 0.84 ($P < 10^{-10}$) for the front shoulder arc and 0.80 ($P < 10^{-10}$) and 0.82 ($P < 10^{-10}$) for rear shoulder arch, calculated by the first (line F1'BF2') and second (line F1'C7F2') way, respectively. Despite of the strong correlation, data, calculated from the photograph poorly coincides with ones measured by tape: the difference may reach 7 cm. Such a high uncertainty makes proposed method for the shoulder arch evaluation by photograph impractical.

The situation with evaluation of the shoulder index is even worse. The correlation coefficients between calculated and measured indexes are 0.28 ($P = 0.07$) for F1'BF2' and 0.29 ($P = 0.06$) for F1'C7F2' estimation. Such a low correlation, although barely significant, leads to conclusion, that proposed method may not be used for evaluation of the shoulder index.

Moderate negative correlation ($r = -0.36$, $P = 0.02$) exists between FSA, calculated from the photographs and measured shoulder index. Such correlation does not allow evaluation of the shoulder index directly from the FSA data. Besides, some test – type estimation may be established: the value of shoulder angle greater then pre-defined threshold means existence of the round shoulder deformity. Assuming evaluation by the shoulder index indicates true subject condition, ROC curve of such test was constructed and analyzed. Threshold value of FSA equal to 24° provides best test parameters: sensitivity 0.81 and specifity 0.69.

Table 2 and 3 demonstrates contingency tables for armpit protraction and armpit tilt angle. Data was analyzed separately for men and women; association between proposed parameters and shoulder index was evaluated using two-sided Fisher's exact probability test.

Association of the proposed qualitative parameters appears to be sex-dependent. Armpit protraction is strongly associated with the rounded shoulder in female group, but poor association exists in the male group. For females, armpit protraction test sensitivity is 0.86 and specifity 0.80. Despite of low Fisher's test P-value for the whole group ($P = 0.06$), use of this test for both males and females seems unpractical.

Table 2 Contingency table for armpit protraction test

	No protraction	Protracted	Total	P-value
Females				
SI < 90% (Rounded)	1	6	7	
SI ≥ 90% (Normal)	8	2	10	0.015
Total	9	8	17	
Males				
SI < 90% (Rounded)	6	8	14	
SI ≥ 90% (Normal)	9	3	12	0.13
Total	15	11	26	
Females & Males				
SI < 90% (Rounded)	7	14	21	
SI ≥ 90% (Normal)	17	5	22	0.006
Total	24	19	43	

SI – shoulder index

Similar results were obtained for the armpit tilt angle test: in females, it is strongly associated with the round shoulder, besides there is no such association in men group. In females, armpit tilt test had sensitivity 1 and specifity 0.90.

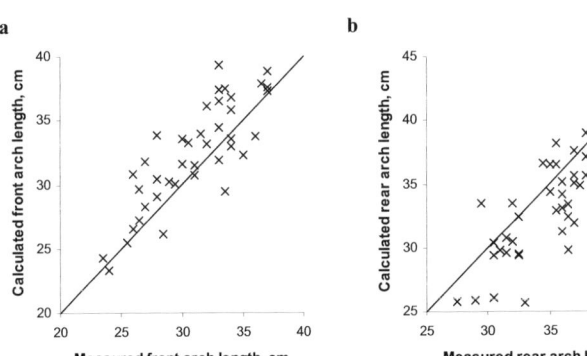
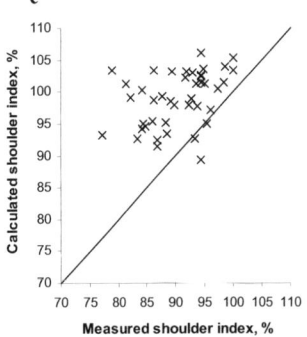

Fig. 4 Correlation between parameters calculated from image and measured "in nature": front shoulder arc (a), rear shoulder arc (b) and shoulder index (c) Rear arch length was evaluated by drawing parabolic line from F1' to F2' through point B. Estimation using line drawn through C7 looks similar.

Table 3 Contingency table for armpit tilt angle test

	ATA < 0	ATA ≥ 0	Total	P-value
Females				
SI < 90% (Rounded)	7	-	7	
SI ≥ 90% (Normal)	1	9	10	0.0004
Total	8	9	17	
Males				
SI < 90% (Rounded)	3	11	14	
SI ≥ 90% (Normal)	4	8	12	0.66
Total	7	19	26	
Females & Males				
SI < 90% (Rounded)	10	11	21	
SI ≥ 90% (Normal)	5	17	22	0.12
Total	15	28	43	

ATA – armpit tilt angle

Further division of subject subgroups by ages (see Table 1) showed, that established associations persist in both age groups.

IV. CONCLUSIONS

Proposed "quasiacromion" method does not allow accurate photogrammetric evaluation of neither shoulder index nor front or rear shoulder arch. Despite of the strong correlation between real and measured shoulder arch lengths, the inaccuracy of the shoulder arch measurement may reach 7 cm. Correlation between real and measured shoulder index is too poor ($r = 0.29$; $P = 0.06$)

Moderate correlation exists between shoulder index and forward shoulder angle, measured photogrammetrycally using "quasiacromion" landmark. With the upper threshold of 24° for normal shoulder, forward shoulder angle test indicates rounded shoulder condition with sensitivity 0.81 and specifity 0.69.

Both armpit protraction and armpit tilt parameters are good indicators of the rounded shoulder in females. For this group, armpit tilt test ($S = 1.00$, $Sp = 0.90$) had better characteristics, then armpit protraction test ($S = 0.86$, $Sp = 0.80$) Besides, both tests are not applicable for men population.

ACKNOWLEDGMENT

This research has been financed in the framework of the European Regional Development Fund Project "Mobile Telemedicine Screening Complex", agreement Nr.2011/ 0007/ 2DP/ 2.1.1.1.0/ 10/ APIA/ VIAA/ 008.

REFERENCES

1. Duļevska I, Umbraško S, Boka S, Gavričenkova L, Cēderstrēma Z, Žagare R, Kažoka Dz, Sirmulis M (2005) Investigation of the Physical Development of the Children and Youth - Historical Review. Eesti Arst, Tartu, Lisa 6: 55-57
2. Umbrashko S (2005) Types of Schoolshildren's Postures and Feet Parameters at the turn of the Century. Thesis, Riga (in Latvian)
3. Breen R, Pyper S, Rusk Y, Dockrell S (2007) An investigation of children's posture and discomfort during computer use. Ergonomics, 50 (10): 1582–1592
4. Griegel-Morris P, Larson K, Mueller-Klaus K, Oatis C A, (1992). Incidence of common postural abnormalities in the cervical, shoulder, and thoracic regions and their association with pain in two age groups of healthy subjects. Phys Ther 72(6): 425–431
5. Dolphens M, Cagnie B, Coorevits P, Vanderstraeten G, Cardon G, D'Hooge R, Danneels L, (2012). Sagittal standing posture and its association with spinal pain: A school-based epidemiological study of 1196 flemish adolescents before age at peak height velocity. Spine 37(19): 1657–1666
6. Sahrman S (2002) Diagnosis and Treatment of Movement Impairment Syndromes. Mosby, St. Louis
7. V Velitchenko (2000) Fizkul'tura dlya oslablennykh detej (Physical training for weakened children). Terra-Sport, Moscow (in Russian)
8. TEMPLO: Motion Analysis at http://www.contemplas.com
9. Noll M, Tarragô Candotti C, Vieira A, Fagundes Loss J (2012) Back Pain and Body Posture Evaluation Instrument (BackPEI): development, content validation and reproducibility. Int J Public Health, pp. 1–8 (in press).
10. Markovitch Z, Lauznis J, Balodis G, Katashev A, Markovitcha I (2013) Development of new mobile telemedicine screening complex. IFMBE Proc. 38: 31–34
11. Ferreira, E. A, Duarte M, Maldonado E. P, Bersanetti A. A, Marques A. P (2011) Quantitative assessment of postural alignment in young adults based on photographs of anterior, posterior, and lateral views," J Manip Physiol Ther 34 (6):371–380
12. Chi-Yuen Hung P, Witana C.P, Goonetilleke R.S (2004) Anthropometric measurements from photographic images. In: Work with Computing Systems, Khalid H.M., Helander M.G., Yeo A.W. (Eds.), Kuala Lumpur: Damai Sciences, pp.764 - 769
13. Celinskis D, Katashev A, Zemite V, (2013) On calculation of the human circumferential measures using wide angle lens photogrammetry system, Proc of BIOMDLORE 2013, Palanga, Lithuania, 2013 (in press)
14. Health navigator at http://health-navigator.ru (in Russian)
15. Celinskis D, Katashev A (2013) On criteria for wide-angle lens distortion correction for photogrammetric applications. IFMBE Proc 38: 153–158
16. Thigpen C, Padua D, Michener L, Guskiewicz K, Giuliani C, Keener J.D, Stergiou N (2010) Head and shoulder posture affect scapular mechanics and muscle activity in overhead tasks. J Electromyography Kinesiology 20(4): 701–9

Address of the corresponding author:

Author: Alexei Katashev
Institute: BIN institute, Riga Technical University
Street: Kalku street 1
City: Riga
Country: Latvia
Email: katashev@latnet.lv

Effects of Superimposed Electrical Stimulation Training on Vertical Jump Performance: A Comparison Study between Men and Women

D.C. Costa, M.N. Souza, and A.V. Pino

Biomedical Engineering Program/COPPE, Federal University of Rio de Janeiro, Rio de Janeiro, Brazil

Abstract—Vertical jump height is a relevant feature in the assessment of muscle power; besides, jumps are essential in sports like volleyball, basketball and gymnastics. Training protocols combining Neuromuscular Electrical Stimulation (NMES) to voluntary contractions are valuable tools to improve jump height, although differences are expected according to gender. This paper aimed to evaluate the effects of a 4-week training protocol superimposing NMES to squats, into the vertical jump height of healthy, non-athlete men and women. Six women and eight men, age between 18 and 35, volunteered the study, being divided into groups according to gender, and subdivided into a group that would receive NMES and the control group. All the volunteers performed 5 series of 10 squats along the four weeks; a series of 10 Counter-Movement Jump and other of 10 Squat Jump were performed before and after the training protocol, in order to evaluate jump improvement. The results indicate that the superimposed training protocol improves men's jump height, although no results have been found for women.

Keywords—Electrical Stimulation, voluntary contractions, superimposed technique, vertical jump height.

I. INTRODUCTION

The vertical jump height is an important tool for the analysis and quantification of lower limbs motor skills and power. Not only does it help pointing new talents in sports and [1, 2], but it also helps evaluating lower limbs power in sedentary people [3, 4]. Also, the development of jump height can contribute to better performance in sports like volleyball, basketball and gymnastics, among others [1, 2, 5]. Jump height is closely related to take-off power [6], which involves muscle strength and contraction velocity.

Making use of NMES associated to voluntary contractions may lead to better physiological and neural adaptations, leading to gains in muscle mass, force, velocity, power, activation and resistance [7–9], caused by differences in muscle recruitment that contribute to motor optimization [7, 8]. This association of techniques shows a better postural control and improvement in complex dynamic movements, as the vertical jump, for both athletes and sedentary people.

The superimposed technique, widely investigated by Paillard [10–13], consists in applying NMES at the same time that a voluntary contraction is performed. The NMES superimposed to voluntary contractions shows positive effects in strength gains, including the lower limbs [8, 12, 14], being able to recruit more muscle fibers during eccentric contractions and to maintain recruitment, for concentric and eccentric contractions, for a longer time than conventional exercises could do [10], [15].

Women tend to have lower sensitive and higher motor thresholds than men [16, 17] probably related to the different fat tissue thickness between genders. Thus, the choice of optimal current doses is paramount to the effectiveness of motor NMES treatments [17].

Changes in responses to NMES superimposed to voluntary contractions are expected due to the differences in electrical stimulation thresholds between men and women; however, the differences in the effectiveness of the same technique applied to both men and women, adjusting only the current values, are not clear yet.

The main goal of the present study was to evaluate the effects of a superimposed NMES training protocol in the improvement of the vertical jump height for both men and women.

II. MATHERIALS AND METHODS

This study consisted on applying NMES simultaneously to quadriceps femoris muscle concentric and eccentric contractions, during squat movements. Sessions occurred three times a week, on alternate days, during 4 weeks.

A. Volunteers

Six women and eight men volunteered the study. Those subjects were non-athletes, age between 18 and 35, with no diseases, and good tolerance to electrical stimulations. Those volunteers were divided into four groups: two groups consisting of 4 men and another of 3 women, respectively called Male NMES group and Female NMES group, received the superimposed technique. Other two groups, composed by 4 men and 3 women, just performed the voluntary contractions and were named Male control group and Female control group. The volunteers were present at least 75% of the sessions and did not absent twice at the same week. The study has been approved by the local institutional committee (CAAE: 00908912.1.0000.5257).

B. Intervention Protocol

All the groups performed the same voluntary exercise, which consisted on 5 series of 10 squats (Fig. 1), knees flexed 90° (angular velocity about 90°/sec), in a total of 20 seconds per series and a rest period of 40 seconds; upper limbs were maintained parallel to the ground, in order to help body balance. Before each session, all volunteers performed lower limbs stretching.

Fig. 1 Squats with upper limbs parallel to the ground.

For the NMES groups, biphasic symmetric square pulses generated by Neurodyn II (Ibramed, Brazil) were applied bilaterally, over the quadriceps femoris muscles, through eight 5×5 cm^2 skin electrodes (model CF5050, Valutrode®, Axelgaard, USA). The negative electrodes were placed over the motor points of vastus lateralis and vastus medialis muscles and positive electrodes were set next to the knees (Fig. 2). The muscles motor points were localized every session, with a negative electrode placed next to the knees while a "pen electrode", delivering a positive stimulus of 1 Hz, 300 μs and 20 mA, scanned the motor point area until find the maximal involuntary muscle contraction.

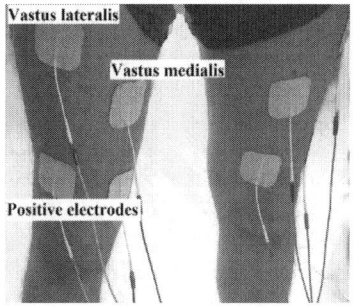

Fig. 2 Self-adhesive electrodes placed over vastus lateralis and vastus medialis motor points.

The parameters selected for NMES training protocol were pulse width 400 μs and frequency of 75 Hz. The intensity was set previous to each session, according to the maximum tolerated by each volunteer, with no voluntary contractions superimposed at all. Along the session, the current was raised up according to changes in comfort threshold, allowing the depolarization of new and deeper muscle fibers, following the basic principles of a progressive training program [18, 19].

C. Evaluation Protocol

The vertical jump height was recorded three days before the beginning of the intervention protocol (Week-0) and three days after its end (Week-4), through a 200 Hz cinemetry system (BTS Smart-D, BTS *Bioengineering,* Italy). To capture the movements, reflexive markers were placed over anatomical points, according to Fig. 3. The volunteers performed 10 Counter-Movement Jumps (CMJ) and 10 Squat Jumps (SJ), with a 10 seconds interval between repetitions and a rest interval of 2 minutes between CMJ and SJ.

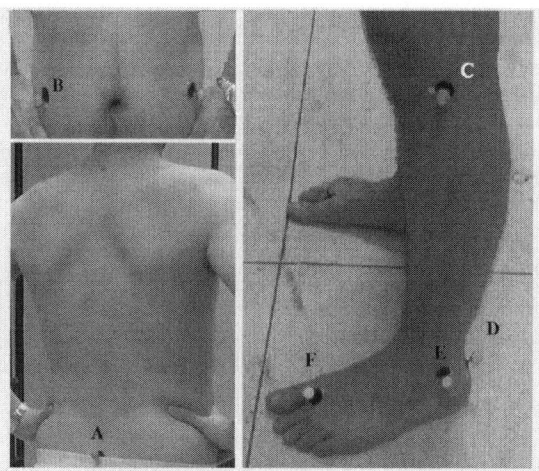

Fig. 3 Reflexive markers placed over sacrum (A) e bilaterally over anterior superior ilia spine (B), lateral condyle (C), calcaneus (D), lateral malleolus (E) and first metatarsophalangeal (F).

D. Data Processing and Analysis

Cinemetry data were processed previously in SMART *Analyser* software (BTS *Bioengineering,* Italy) and later in MATLAB® R2006a (MathWorks, USA), in order to get absolute coordinates of jump height. For each volunteer, those values were divided by the mean of jump height obtained in the first week, in order to normalize the data. After that, the Wilcoxon Signed Rank Test was performed in R, 64 bits, version 2.15.0 (The R Foundation for Statistical Computing), to compare the results for each group tested.

III. RESULTS

The plots in Fig. 4 depict the normalized values of jump height for the Male group, either NMES group or Control Group. A p-value < 0.001 was accepted as significant and signalized (*) in the plots. The normalized values of CMJ and SJ for both Female NMES group and control group, Week-0 and Week-4, are shown in Fig. 5.

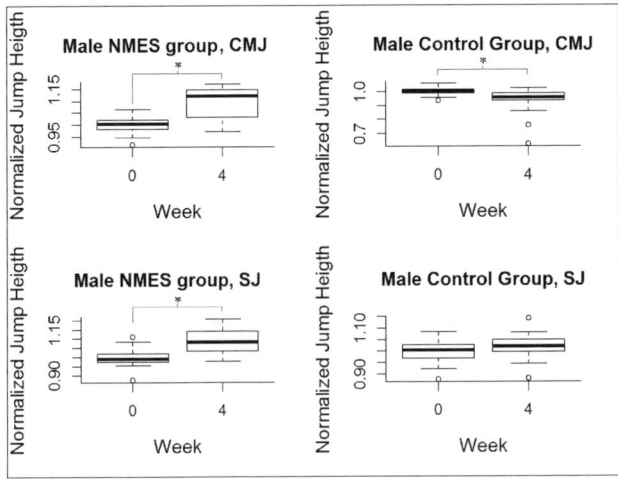

Fig. 4 Vertical jump height for the male volunteers. *p-value < 0.001.

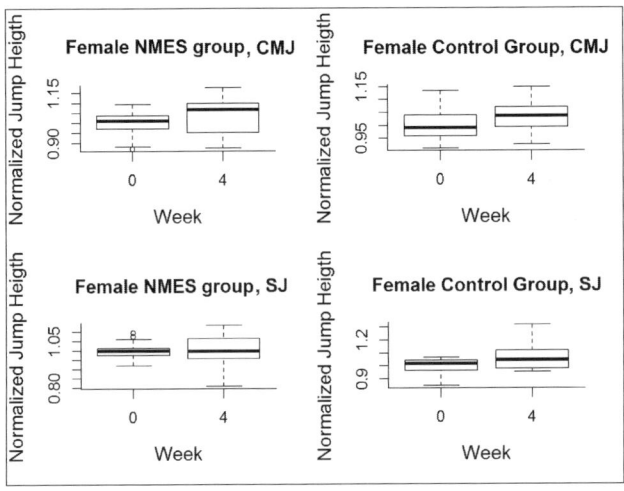

Fig. 5 Vertical jump height for the female volunteers.

Male NMES group got significant increase in jump heights, both for CMJ and SJ, with p-values < 0.001. The Male control group showed a significant decrease in jump height for Week-4 (p-value < 0.001), while no significant differences could be seen for SJ before and after the intervention protocol (p-value ≥ 0.19).

Neither Female NMES group nor Female control group got significant increases in jump height; the former group achieved p-values higher than 0.08 for CMJ and 0.48 for SJ, while the latter got p-values ≥ 0.004 for SJ and p-value ≥ 0.01 for CMJ.

Current values increased about 20 mA from the initial to the final session for all the volunteers who received NMES.

IV. DISCUSSION

The present study evaluated the effects over the jump heights of men and women after a training protocol involving NMES superimposed to voluntary muscle contractions.

The relevant increase in jump height of both CMJ and SJ for the Male NMES group, with p-values <0.001, may be related to neuromuscular adaptations induced by NMES [9], as no muscle hypertrophy is expected neither in a small period of NMES training [20], nor in a conventional training without load. Those neuromuscular adaptations concerns increase in muscle activation and changes in muscle morphology [18, 20, 21], which leads to increases in muscle force and contraction velocity when it comes to switch myosin heavy chain from type-1 to type-2A [22]. It was not expected that an increase in jump height could be seen in 4-week training in the control group, because muscle hypertrophy, strongly related to muscle force, depends on load training [23], and no changes in muscle morphology were expected [23].

A reduction in CMJ was observed for Male Control Group, probably due to a loss of force not related to the training protocol; indeed, the loss of fast-fatiguing fibers due to a sedentary life style may have impaired elastic energy storage to the CMJ, since fast-fatiguing fibers can store more elastic energy than slow-type fibers [6].

On the other hand, no differences were found for both Female groups (p ≥ 0.001). One of the explanations for the different results between men and women could be the different sensitive thresholds for electrical stimulation between genders [16, 17], that could lead to differences in muscle activation below the pain threshold. If by increasing NMES intensity delivered to muscles, there is an increase of the recruitment of deeper fibers [18], leading a greater muscle contraction, one can expect that for men, the greater activation allowed due to their better tolerance to NMES [16, 17] is the responsible for the differences seen in this study, in terms of vertical displacement gains.

One way of getting better results to NMES training, in case of pain threshold, is investigating the effects of a longer threshold adjustment period in the training results, increasing the tolerance to electrical stimulation, with no exercise superimposed, since the perception of the current intensity decreases and becomes more comfortable after a few minutes of NMES [24]. This procedure could help in obtaining a better muscle contraction response, because the higher the activation, the stronger is the muscle recruitment. Other studies may be carried to investigate the effects of either increasing the weeks of the training protocol here described or the period of signal accommodation in women's jump height.

V. CONCLUSIONS

The four-week training superimposing NMES to squat movements increases jump height for men, being useful as a training protocol when results are needed in a short period of time. Differences between genders show the need of investigating the relationship between current tolerated by women and the efficacy of the superimposed technique, when it comes to a short period of training.

ACKNOWLEDGMENT

The authors thank the Brazilian agencies CNPQ and CAPES for the financial support to this research.

REFERENCES

1. Deley G, Cometti C, Fatnassi A, et al. (2011) Effects of combined electromyostimulation and gymnastics training in prepubertal girls. Journal of Strength and Conditioning Research 25:520–526.
2. Maffiuletti NA, Gometti C, Amiridis IG, et al. (2000) The Effects of Electromyostimulation Training and Basketball Practice on Muscle Strength and Jumping Ability. International Journal of Sports Medicine 21:437,443. doi: 10.1055/s-2000-3837
3. Bosco C, Luhtanen P, Komi P (1983) A simple method for measurement of mechanical power in jumping. Eur J Appl Physiol 50:273–282.
4. Markovic G, Dizdar D, Jukic I, Cardinale M (2004) Reliability and factorial validity of squat and countermovement jump tests. Journal of Strength and Conditioning Research 18:551–555.
5. Herrero JA, Izquierdo M, Maffiuletti N (2006) Electrostimulation and plyometric training effects on jumping and sprint time. International Journal of Sports Medicine 27:533–539.
6. Hamill J, Knutzen KM (2012) Bases biomecanicas do movimento humano. Manole, São Paulo (SP)
7. Dehail P, Duclos C, Barat M (2008) Electrical stimulation and muscle strengthening. Annales de Readaptation et de Medecine Physique 51:441–451.
8. Paillard T (2008) Combined application of neuromuscular electrical stimulation and voluntary muscular contractions. Sports Medicine 38:161–177.
9. Maffiuletti NA, Zory R, Miotti D, et al. (2006) Neuromuscular Adaptations to Electrostimulation Resistance Training. American Journal of Physical Medicine & Rehabilitation 85:167–175. doi: 10.1097/01.phm.0000197570.03343.18
10. Paillard T, Lafont C, Dupui P (2005) Effects of electrical stimulation onto posturokinetic activities in healthy elderly subjects. Science and Sports 20:95–98.
11. Paillard T, Noé F, Edeline O (2005) Effets neuromusculaires de l'électrostimulation transcutanée surimposée et combinée à l'activité volontaire : une revue. Annales de Réadaptation et de Médecine Physique 48:126–137. doi: 10.1016/j.annrmp.2004.10.001
12. Paillard T, Noé F, Passelergue P, Dupui P (2005) Electrical stimulation superimposed onto voluntary muscular contraction. Sports medicine 35:951–966.
13. Paillard T, Margnes E, Maitre J, et al. (2010) Electrical stimulation superimposed onto voluntary muscular contraction reduces deterioration of both postural control and quadriceps femoris muscle strength. Neuroscience 165:1471–1475.
14. Gondin J, Cozzone PJ, Bendahan D (2011) Is high-frequency neuromuscular electrical stimulation a suitable tool for muscle performance improvement in both healthy humans and athletes? European Journal of Applied Physiology 111:2473–2487. doi: 10.1007/s00421-011-2101-2
15. Paillard T, Lafont C, Pérès C, et al. (2005) L'électrostimulation surimposée à la contraction musculaire volontaire présente-t-elle un intérêt physiologique chez les sujets âgés? Annales de Réadaptation et de Médecine Physique 48:20–28. doi: 10.1016/j.annrmp.2004.08.005
16. Maffiuletti NA, Herrero AJ, Jubeau M, et al. (2008) Differences in electrical stimulation thresholds between men and women. Annals of Neurology 63:507–512.
17. Maffiuletti NA, Morelli A, Martin A, et al. (2011) Effect of gender and obesity on electrical current thresholds. Muscle and Nerve 44:202–207.
18. Maffiuletti NA (2010) Physiological and methodological considerations for the use of neuromuscular electrical stimulation. European Journal of Applied Physiology 110:223–234. doi: 10.1007/s00421-010-1502-y
19. Maffiuletti NA, Bramanti J, Jubeau M, et al. (2009) Feasibility and efficacy of progressive electrostimulation strength training for competitive tennis players. J Strength Cond Res 23:677–682. doi: 10.1519/JSC.0b013e318196b784
20. Gondin J, Guette M, Ballay Y, Martin A (2005) Electromyostimulation training effects on neural drive and muscle architecture. Med Sci Sports Exerc 37:1291–1299.
21. Pérez M, Lucia A, Rivero J-L, et al. (2002) Effects of transcutaneous short-term electrical stimulation on M. vastus lateralis characteristics of healthy young men. Pflügers Arch - Eur J Physiol 443:866–874. doi: 10.1007/s00424-001-0769-6
22. Minetto MA, Botter A, Bottinelli O, et al. (2013) Variability in Muscle Adaptation to Electrical Stimulation.
23. M.d MNL, M.D BMK, Stanton BA (2006) BERNE & LEVY. Fisiología + Student consult, 4a ed. ©2006 Últ. Reimpr. 2006. Elsevier España
24. Kitchen S (2003) Eletroterapia: Prática Baseada em Evidências, 11.ª ed. Manole, São Paulo

Author: Denise da Conceição da Costa
Institute: Universidade Federal do Rio de Janeiro (COPPE/UFRJ)
Street: Centro de Tecnologia, Bloco H, Sala 330, Ilha do Fundão
City: Rio de Janeiro
Country: Brazil
Email: denisecst@gmail.com

Attentional Focus and Functional Connectivity in Cycling: An EEG Case Study

S. Comani[1], S. Di Fronso[1], E. Filho[1], A.M. Castronovo[2], M. Schmid[1,2], L. Bortoli[1],
S. Conforto[1,2], C. Robazza[1], and M. Bertollo[1]

[1] BIND - Behavioral Imaging and Neural Dynamics Center, University "G. d'Annunzio" of Chieti-Pescara (Italy)
[2] Engineering Department, University Roma TRE, Rome (Italy)

Abstract—This study aimed to test the efficacy of associative and dissociative attention-based strategies derived from the MAP model to improve performance in endurance activity, and to verify whether specific cortical functional networks are associated with the different types of performance foreseen in the MAP model. The findings from one cyclist support the hypothesis that dissociative strategies induce electrophysiological conditions facilitating flow performance states, which are mainly characterized by extensive functional connectivity across all brain areas in the alpha band. Associative strategies do not seem to conform to this framework, although focusing on the core components of action minimized the awareness of unpleasant afferent feedback, thus delaying detrimental increments of perception of effort, which manifest as a predominant frontal-motor coupling in the alpha band and fronto-occipital coupling in the beta band.

Keywords—MAP model, EEG coherence, performance, cycling.

I. INTRODUCTION

A psychobiological account of endurance fatigue is a relatively novel approach in sport and exercise science. The conventional model of endurance fatigue, which focuses on the muscular and cardiovascular systems, has been replaced by a psychobiological account of fatigue [1]. According to this new framework, the perception of effort during dynamic exercise is not determined by afferent feedback from small-diameter muscles, the heart and the lungs, but rather by a task disengagement that occurs (a) when the effort required by a given task/test is equal to the maximum effort that the subject is willing to exert, or (b) when the subject believes to have exerted a true maximal effort, and thus the continuation of the exercise is perceived as impossible [2].

According to this theory, exhaustion is not related to the organic inability to continue a performance, but rather to the individual decision of "giving up". Empirical findings suggest that the neuromuscular system is actually able to continue the task [1], and that exhaustion is mainly a safety mechanism like thirst, pain and hunger. Therefore, the underlying psychological and physiological mechanisms linked to the central perception of fatigue are viewed as survival-seeking behaviors. It is possible to implement associative and dissociative attention-based strategies to modify these behaviors and help the athletes to alter their perception of effort and improve their performance [3,4].

Recently, Bortoli and colleagues [5] have developed a Multi-Action Plan (MAP) interventional model within which specific strategies (e.g., action-focused coping) have been implemented to help the athletes to maintain optimal performance also during high-fatigue or stressful situations. Within this framework, combinations of distinct performance levels (i.e., optimal/suboptimal) and attentional demands (i.e., automatic/controlled) can be classified into four categories: optimal-automatic (type 1), optimal-controlled (type 2), suboptimal-controlled (type 3), and suboptimal-automatic (type 4). The authors have also found that specific behavioral and psychophysiological patterns underlie the four types of performance in shooters and dart-throwers [6].

During the last five years, studies have suggested that brain cortical activity is influenced by exercise preference, mode, and intensity during cycling [7,8], or have documented an increased communication between the mid/anterior insular area and the motor cortex in fatigue-induced states during cycling exercise [9].

On the basis of these frameworks and investigations, we conducted a single-subject study to test the efficacy of the psychological strategies derived from the MAP model in an endurance activity, and to verify whether specific cortical functional networks are associated with the different types of performance foreseen in the MAP model.

II. METHOD

Participant. A 20-year-old road-cycling athlete participated in the study. After being briefed on the general purpose of the intervention, the athlete agreed to participate in the study and signed a written informed consent. The study was conducted in accordance with the local ethical guidelines, and conformed to the declaration of Helsinki.

Procedure. Five visits to the laboratory were planned, with inter-visit intervals of 48 to 72 hours. During the first visit, the participant received standard instructions about the use of the Rating of Perceived Exertion (RPE) using the CR 10 scale [10], and performed an incremental test to determine his anaerobic threshold (AT) and individual optimal pedaling rate (IOPR) in revolutions/minute (rpm). During the second visit, the precision of the estimated AT and IOPR was verified, and a reference EEG was acquired during a time-to-exhaustion test. The time-to-exhaustion interval is defined as the maximum interval for which the subject can maintain an exercise intensity equal to AT +10%, and/or after which he reaches volitional exhaustion. During the last three visits, the participant performed the time-to-exhaustion test while the EEG was recorded. During each visit, one of three different strategies was randomly assigned. The strategies (see Figure 1) required that the athlete kept his focus of attention on: (1) a metronome that reproduced his IOPR (dissociative strategy), to attain a type 1 performance state typified by movement automaticity and fluidity and optimal-pleasant emotions (type 1 manipulation); (2) his IOPR (associative strategy on the core component of action), to achieve a type 2 performance condition characterized by focused attention on the relevant aspects of the action and optimal-unpleasant emotions (type 2 manipulation); (3) muscle fatigue feelings (associative strategy on internal feelings), to induce a type 3 dysfunctional performance state (type 3 manipulation).

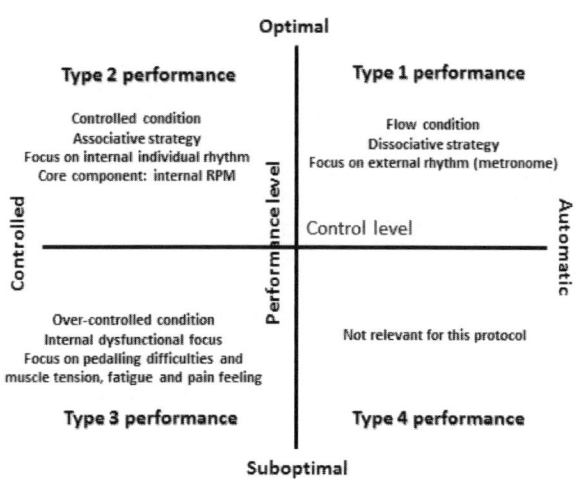

Fig. 1 Performance categorization according to the MAP model

Incremental test. After a warm-up (4 minutes cycling at 75 watt), Oxygen uptake (VO2) and Carbon Dioxide production (VCO2) were measured using an incremental protocol on a Monark Cycle Ergometer (939 E). AT was measured using the V-Slope method [11]. Pedal rate was maintained at 70 rpm, and the workload power output, initially set at 75 W, was step-wise increased by 25 W every 2 minutes until exhaustion. Heart rate, VO2 and VCO2 were continuously monitored using the Schiller CS 200 system.

Time-to-exhaustion test at individual constant load. During the entire duration of the time-to-exhaustion test, the EEG was continuously recorded. The participant's AT was 250 W, and his IOPR was 101 rpm. After a resting period (no movement) of 2 minutes and a warm-up period of 4 minutes on the cycle ergometer at 125 W, the athlete performed a constant load exhaustive test at 250 W with pedal rate at 101 rpm. After exhaustion (i.e. Borg level 12), there was a recovery period of 4 minutes at 80 W, and a further resting period of 2 minutes. RPE scores were collected 5 s before the end of each 1-minute sub-period. A manipulation check questionnaire was administered at the end of acquisition to verify adherence to the experimental assignments.

EEG recording and processing. The EEG was recorded using a 32-channel EEG system by ANT (Advanced Neuro Technology, Enshede, Netherlands) that uses a waveguard cap with 32 active electrodes for movement compensation positioned over the scalp according to the 10-20 system. The montage used a common electrical reference, with ground electrode located between Fpz and Fz. The electrode impedance was kept below 10 kΩ, and sampling frequency was set at 512 Hz.

The EEG data were band pass filtered between 0.1 Hz and 50 Hz. Epochs showing instrumental, ocular, and muscular artifacts were detected using a threshold of ±200 μV and further by visual inspection, and ultimately eliminated using the ASA software [12].

Coherence analysis was performed to detect functional connectivity patterns in the EEG data in relation to the different manipulation types (see *Procedure*). The complex coherence in a given frequency band between two signals x_1 and x_2 (namely the signals from two electrodes) was calculated as the cross spectrum between the signals, normalized by the square root of the product of the power spectrum of the two signals. Given that coherence is a normalized measure of the correlation between two signals in the frequency domain, its values can vary from 0 to 1. Only coherence values > 0.8 were retained. Coherence maps were calculated for the alpha (8-12 Hz) and beta (13-30 Hz) bands by averaging on EEG signal blocks of 125 ms duration, epoched on 10 s intervals subsequent to each RPE evaluation throughout the EEG acquisition, for a total of about 80 blocks.

III. RESULTS

The analysis of the RPE scores collected during the time-to-exhaustion test (Figure 2) revealed that the best performance was achieved when a type 1 manipulation was used. This means that the athlete had the lowest perception of effort during the exercise (lowest RPE scores), and therefore performed the time-to-exhaustion test at individual constant load for the longest time interval. Type 2 manipulation produced higher RPE scores during the first half of the test, and a shorter time-to-exhaustion time interval. Type 3 manipulation produced the highest RPE scores and the shortest time-to-exhaustion time interval.

The analysis of the RPE scores collected during the recovery phase (Figure 2) only show that the slowest recovery period occurred after type 2 manipulation.

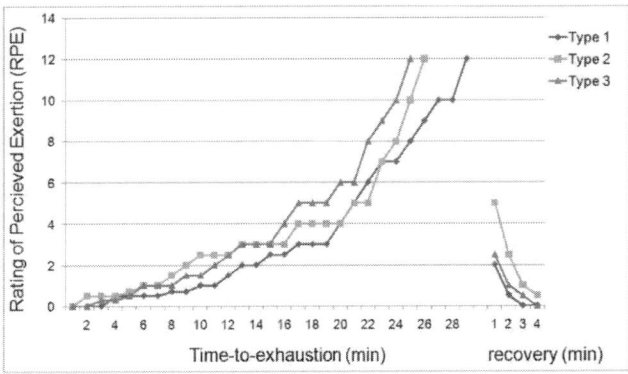

Fig. 2: RPE values during time-to-exhaustion test and recovery phase for the different types of manipulations.

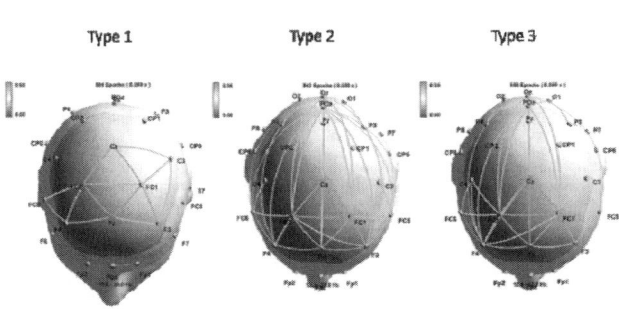

Fig. 3 Topographic maps of the Fisher-Z transformed coherence amplitude values in the alpha (upper panel) and beta (lower panel) bands

The coherence analysis of the EEG signals recorded during the time-to-exhaustion tests performed using the three manipulations (Figure 3) showed a strong and diffused cortico-cortical functional connectivity in the alpha band for type 1 manipulation, while type 2 and type 3 manipulations were characterized by a strongly localized coupling limited to the prefrontal motor network.

On the other hand, the coherence analysis in the beta band highlighted a more intense and diffused cortico-cortical functional connectivity for type 2 and, to a lesser extent, for type 3 manipulations, while type 1 manipulation featured a localized coupling limited to the frontal and central areas. Interestingly, functional connectivity was bilateral in all manipulations and for both frequency bands.

IV. DISCUSSION AND CONCLUSIONS

The rating of perceived exertion (RPE) is essential to measure the individual awareness of the central motor commands to the locomotor and respiratory muscles [1]. The results of the RPE monitoring in this cyclist support the notion that the best performance is achieved when a dissociative strategy is adopted (i.e. when the focus of attention is kept external, as in type 1 manipulation). Conversely, an associative strategy focusing on individual internal rhythms (type 2 manipulation) or centered on muscle activity and fatigue feelings (type 3 manipulation) did not facilitate movement fluidity or led to suboptimal performance.

Additionally, cortical functional connectivity has been assessed through EEG coherence, which has been extensively used to study the neural mechanisms underlying perceptive, cognitive, and motor processes [13]. In general, high coherence values suggest an effective information exchange between brain areas. Cortical coherence may thus help to address the issue of functional communication in the cortex during exercise, leading to more accurately describe the neural substrates of optimal attentional focus, hence of flow performance states and optimal performance [13].

It has been demonstrated that fronto-occipital and interfrontal coherence in the alpha band may discern the brain functional states related to different arousal levels [14]. The alpha coherence map obtained for the type 1 manipulation suggests that a dissociative strategy engages an extensive functional connectivity across all brain areas. This result is consistent with the observation that lower arousal states are accompanied by a higher alpha power broadly in the cortex. Furthermore, it has been suggested that significant frontal EEG coherence indicates a greater functional coupling of the frontal circuits involved in goal-directed behavior and in the neuronal implementation of self-evaluation [15]. To this extent Berchicci and colleagues [16] recently posited that the perception of fatigue may lead to a local reorganization of the prefrontal-motor network. Likewise, the bilateral functional coupling of the frontal areas in the alpha band found for the associative strategies might be explained as a neural expression of the requirement to keep the attentional focus on the core components of action or on the muscle fatigue feelings.

The fact that all our coherence maps were bilateral is in contrast with the notion that expert athletes (e.g., skilled marksmen) engage in less cortico-cortical communication in the alpha and beta bands, with a strong contro-lateral activation in the temporal and motor cortex, which implies a reduced involvement of cognition [17]. However, it is worth noting that our results do not refer to a unilateral and more mental sport such as shooting or archery, but rather to a bilateral sport (cycling) that requires sustained movement and effort despite the automatic nature of this performance.

This specific feature of cycling is likely to be at the origin of the coherence maps obtained for the beta band. Indeed, a persistent signal power reduction in the beta band is observed for sustained movements, consistently with the beta coherence map observed for type 1 manipulation, which shows a functional connectivity limited to the frontal-motor network. On the other hand, an increased signal power in the beta band is associated with resistance to movement or its voluntary suppression, as well as with negative feelings of emotional "intensity", as in the case of an athlete "trying too hard" in an attempt to cope with fatigue or to concentrate on the internal core components of action [18]. This condition is mirrored in the beta coherence maps obtained for type 2 and type 3 manipulations, which highlight the ongoing cognitive processing underlying the associative strategy and the motivational and affective traits of withdrawing actions [19].

Overall, our results indicate that specific functional connectivity patterns may be associated with different types of intervention strategies. Our findings support the hypothesis that dissociative strategies tend to prompt optimal performance states (flow) and specific electrophysiological patterns that lead to superior performance. Associative strategies do not seem to conform to this framework. However, focusing on the core components of action minimizes the awareness of unpleasant afferent feedback, manifested as a predominant frontal-motor coupling in the alpha band and fronto-occipital coupling in the beta band, thus delaying the detrimental increments of the perception of effort.

REFERENCES

1. Marcora, SM (2009). Perception of effort during exercise is independent of afferent feedback from skeletal muscles, heart and lungs. J appl physiol 106: 2060-2062.
2. Marcora, SM (2008). Do we really need a central governor to explain brain regulation of exercise performance? Eur j appl physiol 104: 939-931.
3. Birrer D, Morgan G (2010). Psychological skills training as a way to enhance an athlete's performance in high-intensity sports Scand J Sports Sci & Med DOI: 10.1111/j.1600-0838.2010.01188.x
4. Castronovo A.M., Conforto S, Schmid M, Bibbo D, D'Alessio T (2013). How to Assess Performance in Cycling: the Multivariate Nature of Influencing Factors and Related Indicators. Front. Physiol. 4:116. doi: 10.3389/fphys.2013.00116.
5. Bortoli L, Bertollo M, Hanin Y, Robazza C (2012). Striving for excellence: A multi-action plan intervention model for shooters. Psychol Sport Exerc 13:693-701.
6. Bertollo M, Bortoli L, Gramaccioni G, Hanin Y, Comani S, Robazza C (2013). Behavioural and psychophysiological correlates of athletic performance: A test of the multi-action plan model. Appl Psychophys Biof, 38:91-99 doi: 10.1007/s10484-013-9211-z
7. Brümmer V, Schneider S, Abel T, Vogt T, Strüder HK (2011). Brain cortical activity is influenced by exercise mode and intensity. Med Sci Sports Exerc, 43 1863-72. doi: 10.1249/MSS.0b013e3182172a6f.
8. Schneider S, Brummer V, Abel T, Askew CD, Struder HK (2009). Changes in brain cortical activity measured by EEG are related to individual exercise preferences. Physiol Behav.;98: 447–52.
9. Hilty L, Langer N, Pascual-Marqui R, Boutellier U, Lutz K (2011). Fatigue-induced increase in intracortical communication between mid/anterior insular and motor cortex during cycling exercise. Eur J Neurosci, 34(12):2035-42. doi: 10.1111/j.1460-9568.2011.07909.x.
10. Borg, G, Borg, E (2001). A new generation of scaling methods: Level-anchored ratio scaling. Psychologica. 28, 15-45.
11. Wasserman K, Stringer WW, Casaburi R, Koike A, Cooper CB (1994). Determination of the anaerobic threshold by gas exchange: biochemical considerations, methodology and physiological effects. Z Kardiol. 83 Suppl 3:1-12.
12. Zanow F, Knosche TR (2004). ASA—Advanced source analysis of continuous and event-related EEG/MEG signals. Brain Topog 16:287–290.
13. Del Percio C, Iacoboni M, Lizio R, Marzano N, Infarinato F, Vecchio F, Bertollo M, Robazza C, Comani S, Limatola C, Babiloni C (2011). Functional coupling of parietal alpha rhythms is enhanced in athletes before visuomotor performance: a coherence electroencephalographic study. Neuroscience. 175:198-211.
14. Cantero JL, Atienza M, Salas RM, Gómez CM (1999). Alpha EEG coherence in different brain states: an electrophysiological index of the arousal level in human subjects. Neurosci Lett, Aug 27;271(3):167-70.
15. Travis F, Tecce J, Arenander A, Wallace RK (2002). Patterns of EEG coherence, power, and contingent negative variation characterize the integration of transcendental and waking states. Biol Psychol, Nov;61(3):293-319.
16. Berchicci M, Menotti F, Macaluso A, Di Russo F (2013). The neurophysiology of central and peripheral fatigue during sub-maximal lower limb isometric contractions. Front. Hum. Neurosci. 7:135. doi: 10.3389/fnhum.2013.00135
17. Deeny SP, Haufler AJ, Saffer M, Hatfield BD. (2009). Electroencephalographic coherence during visuomotor performance: a comparison of cortico-cortical communication in experts and novices. J Mot Behav. 4:106-16. doi:10.3200/JMBR.41.2.106-116.
18. Wilson WS, Thompson M, Thompson L, Thompson J, Fallahpour K (2011). Introduction to EEG biofeedback neurofeedback. In B. Strack, M. Linden, & V. S. Wilson (Eds.), Biofeedback & neurofeedback applications in sport psychology (pp. 175-198). Wheat Ridge, CO: American Association Psychophysiology & Biofeedback.
19. Davidson RJ (1993). Cerebral asymmetry and emotion: Conceptual and methodological conundrums. Cognition & Emotion, 7, 115-138.

Author: Silvia Comani
Institute: BIND – Behavioral Imaging and Neural Dynamics Center
Street: Via dei Vestini, 33
City: 66100 Chieti
Country: Italy Email: comani@unich.it

ERD/ERS Patterns of Shooting Performance within the Multi-Action Plan Model

S. Comani[1], L. Bortoli[1], S. Di Fronso[1], E. Fiho[1], C. De Marchis[1,2], M. Schmid[1,2],
S. Conforto[1,2], C. Robazza[1], and M. Bertollo[1]

[1] BIND - Behavioral Imaging and Neural Dynamics Center, University "G. d'Annunzio" of Chieti-Pescara (Italy)
[2] Engineering Department, University Roma TRE, Rome (Italy)

Abstract—The multi-action plan (MAP) model reflects the notion that different psychophysiological states underlie distinct performance-related experiences. Previous empirical evidence suggested that attentional focus, affective states, and psycho-physiological patterns differ among optimal-automatic (type 1), optimal-controlled (type 2), suboptimal-controlled (type 3), and suboptimal-automatic (type 4) performance experiences.

The purpose of this study was to test the cortical patterns correlated to the performance categories conceptualized within the MAP model.

Three elite pistol shooters (age range 16-30 years), members of the Italian Shooting Team and with extensive international experience, participated in the study. Participants performed 120 air-pistol shots at 10 meters from an official target. After each shot, they reported perceived control and accuracy levels on a 0-11 scale. Objective performance scores were also gathered. Electroencephalographic (EEG) activity was recorded with a 32 channel system (ANT). High alpha band ERD/ERS analysis during the three seconds preceding each shot and at shot's release was performed. Findings revealed differences in cortical activity related to performance categories. In particular, type 1 and type 4 performance were characterized by a clear relative decrease in signal power (ERS) at shot's release involving the central areas and the contralateral parietal and occipital areas, but differed for the cortical activity patterns before the shot. No ERS pattern was observed at shot's release in type 2 performance, while a interesting relative increase of signal power (ERD) occurred in the frontal and occipital areas just before the shot, similarly to what occurred in type 1 performance.

Our preliminary results suggest that lower cortical activation at shot's release is associated with an automatic performance, partially supporting the "neural efficiency hypothesis". Additionally, the analysis of the cortical activations related to the performance-related experiences defined in the MAP model supports the hypothesis that distinct neural activation patterns are associated with the control and performance levels.

Keywords—ERD/ERS, MAP model, EEG, performance, shooting.

I. INTRODUCTION

The analysis of psycho-bio-social mechanisms underlying optimal performance experiences has received great attention in the domain of sport science [1]. Current avenues of research involve the use of multi-methods that target diverse structural components (e.g., emotional processes, cognitive functioning, motor behavior) underlying human performance [2]. To this extent, Bortoli and colleagues [3] have recently developed a multi-action plan (MAP) intervention model to help elite level shooters in improving, stabilizing and optimizing their performance during practice and competition. According to the authors, behavioral patterns underlying distinct performance levels (i.e., optimal-suboptimal) and attentional demands (i.e., automatic-controlled) may be classified into four categories: optimal-automatic (type 1), optimal-controlled (type 2), suboptimal-controlled (type 3), and suboptimal-automatic (type 4) (Figure 1).

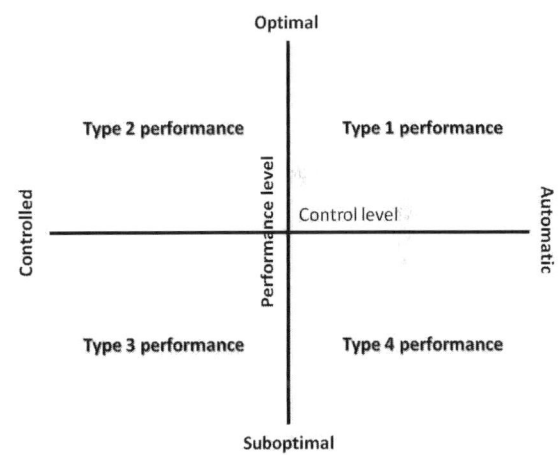

Fig. 1 Performance categorization according to MAP model

In a subsequent study, Bertollo and colleagues [4] assessed specific affective, behavioral, psychophysiological (e.g., skin conductance responses, heart rate), and postural trends on shooting and dart-throwing performance. Findings provided further empirical support to the 2 × 2 (optimal/suboptimal × automatic/controlled) conceptualization.

Neurophysiological mechanisms in general, and cortical activity in particular, are at the core of an integrated view of human performance [5]. Electroencephalographic (EEG) and event-related potential (ERP) measurements have been essential in shaping our understanding of skilled performance in sports [6]. For instance, "economy of effort" mechanisms have been widely studied with a neurophysiological approach, especially in precision sports and self-paced tasks (e.g., target shooting, golf putting, dart-throwing, and archery) [5, 7, 8, 9]. Altogether, skilled performance in sports has been associated with decreased cortical activation (i.e., "the economy of effort principle" or "the neural efficiency hypothesis") [10].

In order to better analyze cortical functioning and the "neural efficiency hypothesis", Event Related Desynchronization/Synchronization (ERD/ERS) analysis has been used. ERD and ERS are relative measures of local cortical activity at a certain time and location. It has been found that an event-related reduction of signal power in the alpha and beta bands localized in the sensorimotor areas denotes cortical information processing related to voluntary movement. On the other hand, a visual input results not only in a power reduction of occipital alpha rhythms but also in an enhancement of central alpha rhythms, while the opposite is found during self-paced, voluntary hand movement [11]. Research using ERD/ERS analysis can therefore unravel new information about the neurodynamics of cortical networks. This analysis, recently applied to analyze shooting performance [9], has revealed that differences in the alpha band occur between expert and novices, and between best and worst performance.

In particular, EEG studies on attentional control and emotional content have focused on either comparing athletes and non-athletes, or two skill levels within sports (i.e., "the expert-novice paradigm"). While nomothetic investigations have greatly advanced our understanding of the "brain states" and physiological correlates of optimal experiences in sports, idiographic studies of experts are particularly important to advance our understanding of the mechanisms underlying expertise in a given domain (i.e., "the expert performance approach"). Noteworthy are studies on experts who are recognized through objective performance markers rather than subjective assessment methods, such as peers voting [12]. The individual zones of optimal functioning [1] framework and, more recently, the idiographic affective probabilistic zones methodology are further well-known concepts shaped through idiosyncratic analysis and single-case designs [13,14].

The purpose of the present study was to assess the ERD/ERS patterns underpinning the performance categories conceptualized within the MAP model. We aimed to test the following hypotheses: (1) optimal-automatic performance experiences (type 1) are characterized by an instinctive-like minimal conscious control, high level of energy matching task demands, and cortical arousal synchronized with the event (shot); (2) optimal-controlled experiences (type 2) are characterized by consciously focused control, with a compensatory high level of energy, and a cortical arousal higher than in type 1; (3) suboptimal-automatic experiences (type 3) are characterized by high level of conscious control, task irrelevant focus, energy misuse, with cortical activity desynchronized with the event; and (4) suboptimal-controlled experiences (type 4) are characterized by minimal conscious control, low level of energy, with a cortical activity synchronized with the event.

II. METHOD

Participants: Three elite pistol shooters (age range 16-30 years), members of the Italian Shooting Team and with extensive international experience, agreed to participate in the study and signed a written informed consent. The study was conducted in accordance with the local ethical guidelines, and conformed to the declaration of Helsinki.

Procedure: Participants were asked to identify the core components of their "chain of action" that they deemed fundamental for optimal performance, and then asked to choose one element (i.e., an idiosyncratic core component) deemed fundamental in order to optimally perform. Afterward, they performed a total of 120 air-pistol shots, being "free to relax" between consecutive shots, and to shoot when they felt "ready to go" (average inter-shot interval of about 1 minute). The distance between the shooter and the target was 10 m, and the diameter of the target was 6 cm, in accordance with the international shooting competition rules (www.issf-sports.org/theissf/rules/english_rulebook.ashx).
An electronic scoring target was used to (automatically) record the shooting scores. Noteworthy, online shot information was initially concealed from the athletes because we were interested in assessing their perceived accuracy. Hence, after each shot, the athletes were asked to report their perceived shooting score (ranging from 0 to 10.9). They also reported (a) the control level of the idiosyncratic core component of action, and (b) the accuracy level related to the execution of the core component (from 0 to 11 on a Borg scale). After this evaluation, the actual shooting score (i.e. the objective performance) of each shot was made available to the shooter. Beside self-evaluation, the electroencephalogram was recorded using a 32 channels EEG ASAlab system and the waveguard cap by ANT (Advanced Neuro Technology, Enshede, Netherlands).

EEG Recordings: EEG data were continuously recorded (sampling frequency: 1024 Hz) from the 32 scalp electrodes (active electrodes for movement compensation) positioned over the whole scalp according to the 10-20 system. EEG signals were recorded with common reference; the ground electrode was positioned between Fpz and Fz; electrode impedance was kept below 5 kΩ.

A device based on acoustic technology (cardio-microphone and Powerlab 16/30, ADInstruments, Australia) was used to identify the instant of shot release. Acoustic signals were acquired with a sampling frequency of 1 kHz.

Preliminary data analysis: EEG data were band-pass filtered between 0.01 to 40 Hz, segmented into single epochs of 10 s duration, with each epoch starting at -6 s and ending at +4 s with respect to t=0 (i.e. the instant when the shot was released). Data epochs showing instrumental, ocular and muscular artifacts were identified, both via automatic detection (maintaining only the signal with amplitude between -150 μV and 150 μV) and by visual inspection, and excluded from further analysis [15]. Accordingly, only the epochs free from artifacts were considered in the analysis. The shooting results and the control levels exerted by each shooter on his/her core components of action were used to categorize the EEG epochs according to the four types of performance foreseen in the MAP model.

ERD/ERS analysis: To quantify the event-related changes in the high alpha power, the individual alpha ERD/ERS maps were calculated following the procedure proposed by Zanow and colleagues [15], where ERD and ERS are defined, respectively, as the percent increase and decrease of signal power as compared to the baseline. This definition is opposite to that proposed by Pfurtscheller and da Silva [11]. The Hilbert transform was performed before ERD/ERS analysis. Then, for a given frequency band, ERD/ERS maps were calculated for a given interval of interest as the percent variation of the signal power with respect to the power of the baseline signal in each EEG channel.

We calculated the ERD/ERS maps for the high alpha band, defined as the frequency band from the individual alpha frequency (IAF) to the IAF+2Hz. Given that the IAF of our athletes were very similar (9.9, 10 and 9.9 Hz), we considered the frequency band 10-12 Hz for all of them.

Three intervals of interest, each of 1 s duration, were considered during the 3 s preceding each shot (i.e. from -3 to 0 s, for t_{shot}=0), whereas the baseline signal was epoched from -5 to -4 s before the shot. Periods before -5 s were not suitable because of body movements, small adjustments of head/trunk, and respiration.

For each participant and for each group of EEG epochs categorized according to the types of performance foreseen in the MAP model, the baseline signals were averaged to reduce background noise before ERD/ERS calculation. Similarly, averaged ERD/ERS maps for each interval of interest were obtained, for each participant and for each group of EEG epochs, from the single ERD/ERS maps. Finally, for each type of performance the individual ERD/ERS map were averaged across subjects to account for the cortical patterns underlying the four MAP model types.

III. RESULTS

Findings revealed differences in cortical activity with respect to performance types (Figure 2). In particular, type 1 and type 4 performances (optimal-automatic and suboptimal-automatic) seem to be characterized by a similar relative decrease in signal power (ERS) at shot's release, involving the central areas and the contralateral parietal and occipital areas. Conversely, they differ for the ERD pattern before the shot, involving prefrontal, frontal and occipital areas in type 1 performance, and prefrontal areas in type 4 performance. In type 2 performance (optimal-controlled), no ERS pattern is observed at shot's release, while a clear relative increase of signal power (ERD) occurs in the central areas 3 s before the shot and in the frontal and occipital areas at -2 and -1 s. Type 3 performance (suboptimal-controlled) seems to be characterized by a relative decrease in signal power (ERS) in the contralateral frontal and parietal areas at shot's release, with a high relative increase of signal power (ERD) in the central areas just 1 s before the shot.

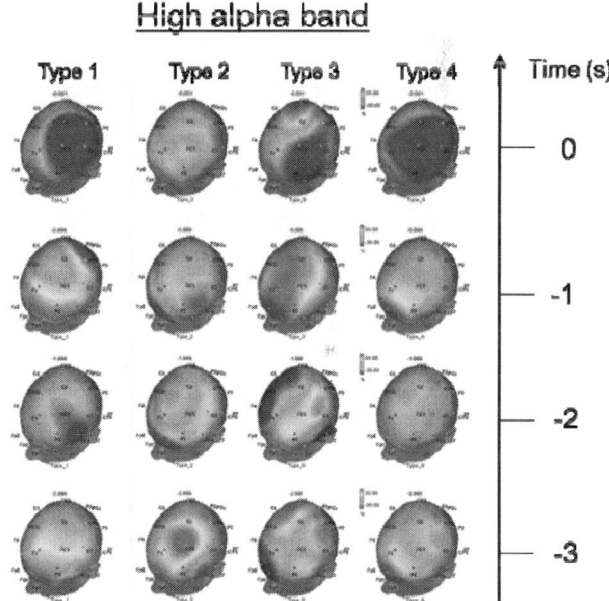

Fig. 2 Average ERD/ERS maps in the high alpha band (10-12 Hz) at shot release (t=0) and during the three seconds preceding it, categorized for the four types of performance. Color scale: maximum ERD and ERS are coded in Red and Blue, respectively

IV. DISCUSSION AND CONCLUSIONS

Overall, our preliminary findings for type 1 and type 4 performances support the notion that lower cortical activation at shot's release is associated with an automatic performance, corresponding to a state of quiescence of the organism, task relevant focus of attention, and movement automaticity and fluidity. This notion is partially in line with the "neural efficiency hypothesis", which links optimal performance to optimal psychophysiological conditions, as found only in type 1 performance. Indeed, our results indicate that optimal outcomes, derived from a type 1 and not from a type 4 performance state, are rather determined by the patterns of cortical activation during the seconds preceding the shot, for which differences between the two performance states were found. In particular, one second before the shot type 1 performance state featured an increased signal power that involved not only the prefrontal areas, as in type 4 performance, but also the frontal and occipital areas, which are related to focused attention.

The hypothesis that the activation of the frontal and occipital areas during the seconds preceding the shot might be involved in optimal performance, is supported by the results obtained for type 2 performance (optimal-controlled), which indicate that the athletes can attain good performance levels also without total movement automaticity and fluidity, when they focus their attention on their core components of action. In this case, the athletes use consciously focused control that was mirrored, in our results, in an over-activation of the central cortical areas at -3 s, and in the activation of the frontal and occipital areas during the 2 s preceding the shot, as in type 1 performance. No such pattern was found in type 3 performance (suboptimal-controlled).

In conclusion, our findings echo the notion that functional performance states are plausible within a broad range of behavioral and physiological antecedents. However, the analysis of the cortical activations related to the psychophysiological states underlying distinct performance-related experiences, as defined in the MAP model, has brought new insights on the neural correlates of the action control and performance levels separately.

ACKNOWLEDGMENT

We thank the Olympics athletes of the Italian Shooting Team for their participation in the study.

REFERENCES

1. Hanin Y. (2007) Emotions in Sport: Current issues and perspectives. In G. Tenenbaum & R.C. Eklund. Handbook of Sport Psychology 3rd ed. (pp. 31-58). Hoboken, NJ: John Wiley & Sons.
2. Tenenbaum G, Hatfield BD, Eklund RC, Land WM, Calmeiro L, Razon S, Schack T. (2009) A conceptual framework for studying emotions-cognitions-performance linkage under conditions that vary in perceived pressure. Prog Brain Res 174:159-78. doi: 10.1016/S0079-6123(09)01314-4.
3. Bortoli L, Bertollo M, Hanin Y, Robazza C (2012) Striving for excellence: A multi-action plan intervention model for shooters. Psychol Sport Exerc 13:693-701.
4. Bertollo M, Bortoli L, Gramaccioni G, Hanin Y, Comani S, Robazza C (2013) Behavioural and psychophysiological correlates of athletic performance: A test of the multi-action plan model. Appl Psychophys Biof. [Epub ahead of print] DOI 10.1007/s10484-013-9211-z
5. Hatfield BD, Kerick SE (2007) The psychology of superior sport performance: A cognitive and affective neuroscience perspective. In R. C. Eklund & G. Tenenbaum (Eds.), Handbook of sport psychology 3rd Ed (84-109). John Wiley & Sons Inc.
6. Thompson T, Steffert T, Ros T, Leach J, Gruzelier J. (2008) EEG applications for sport and performance. Methods, 45(4):279-288.
7. Goodman S, Haufler A, Shim JK, Hatfield B. (2009) Regular and random components in aiming-point trajectory during rifle aiming and shooting J Mot Behav, 41: 367-382.
8. Del Percio C, Iacoboni M, Lizio R, et al. (2011) Functional coupling of parietal alpha rhythms is enhanced in athletes before visuomotor performance: a coherence electroencephalographic study. Neuroscience, 175: 198-211.
9. Del Percio C., Babiloni C., Bertollo, et al. (2009) Visuo-attentional and sensorimotor alpha rhythms are related to visuo-motor performance in athletes. Hum Brain Mapp, 30: 3527–3540. doi:10.1002/hbm.20776.
10. Hatfield BD, Hillman CH (2001): The psychophysiology of sport: A mechanistic understanding of the psychology of superior performance. In: Singer RN, Hausenblas HA, Janelle CM, editors. Handbook of Sport Psychology. New York (NY): Wiley: 362–388.
11. Pfurtscheller G, Lopes da Silva FH (1999) Event-related EEG/MEG synchronization and desynchronization: Basic principles. Clin Neurophysiol 110: 1842–1857.
12. Ericsson KA. (2006) Protocol analysis and expert thought: Concurrent verbalizations of thinking during experts' performance on representative task. In KA Ericsson, N Charness, P Feltovich, and RR Hoffman, (Eds.). Cambridge handbook of expertise and expert performance (223-242). Cambridge, UK: Cambridge University Press.
13. Johnson MB , Edmonds WA, Moraes LC, Medeiros Filho E, Tenenbaum G. (2007). Linking affect and performance of an international level archer incorporating an idiosyncratic probabilistic method. Psychol Sport Exerc, 8, 317–335. DOI: 10.1016/j.psychsport.2006.05.004
14. Bertollo M, Robazza C, Falasca WN. et al. (2012). Temporal pattern of pre-shooting psychophysiological states in elite athletes: A probabilistic approach. Psychol Sport Exerc, 13, 91-98. doi:10.1016/j.psychsport.2011.09.005
15. Zanow F, Knosche TR (2004). ASA—Advanced source analysis of continuous and event-related EEG/MEG signals. Brain Topog 16:287–290.

Author: Silvia Comani
Institute: BIND – Behavioral Imaging and Neural Dynamics Center
Street: Via dei Vestini, 33
City: 66100 Chieti
Country: Italy
Email: comani@unich.it

Upper Limb Joint Torque Distribution Resulting from the Flat Tennis Serve Impact Force

W.D. Masinghe[1], T. Nanayakkara[2], and G. Collier[1]

[1] Faculty of Science Engineering and Computing, Kingston University London, UK
[2] Department of Informatics, King's College London, UK

Abstract—The serve is a vital part of a tennis player's game. Serve has to be fast and powerful to enhance the chance of winning the initial point. The extreme intensity and severity of the ball racket collision involved at the service often lead to various upper limb injuries. Detailed biomechanical models of the human upper arm are needed to study how the impulse force propagates to individual joint or muscle level. This study investigates the torque distribution resulting from the impact force of the flat service stroke based on a detailed 9-Degrees of Freedom (DoF) kinematic model. Five trials were sampled from a professional level tennis player during the flat service scenario. A Qualisis motion capturing system with six cameras was used to capture 3 Dimensional (3D) position data of the subject, at a sampling rate of 240 Hz. The torque distribution was quantitatively analyzed using a Jacobian matrix. The results of the analysis reflected that the highest torque was apportioned to sterno-clavicular joint followed by elbow, sterno-clavicular joint and wrist joint respectively. A statistical based sensitivity analysis has been carried out on upper limb joints. The proximal radioulnar Joint was found to be the most sensitive joint with respect to torque distribution.

Keywords—Tennis serve, Upper limb, Joint torque, Biomechanics.

I. INTRODUCTION

No other stroke is important as the service stroke in the game of tennis. It is the only stroke in which players have full control over the outcome. To gain the advantage, the players tend to serve at higher speeds which may often exceed physical endurance limits. From a biomechanical point of view, inefficient motion patterns of various joints can either be detrimental to the speed and the spin of the ball or may even increase the risk of injury [1]. Unfortunately, most people do not appreciate the intensity and severity of these collisions on the body until there is an injury. The aetiology of musculoskeletal injuries generally has a kinetic mechanical cause [2].

Epidemiological studies have reported a range of injuries in tennis. Elite players commonly suffer from shoulder and medial elbow injuries. Reference [3] reported 56% of competitive players suffered rotator cuff and impingement injuries to the shoulder. Overuse injuries are particularly likely to involve the upper extremity given the repetitive stroke technique required [4], including rotator cuff tendinopathy, tennis elbow and wrist problems [5]. Many stake holders involved in the game of Tennis know that the impact of the ball with the strings is an impulsive event, usually lasting between 4 and 8 one one-thousandths of a second. One to two milliseconds (ms) later the shock wave of the collision reaches the player's hand [6].

A better understanding of the physical demands placed upon the body during the impact of the tennis serve will enable the players and coaches to optimize the service stroke and also health professionals to design more optimal injury prevention, treatment and rehabilitation programs.

A number of studies have carried out on tennis serve biomechanics. Most biomechanical studies of the tennis serve have been limited to kinematic studies [7], [8]. However after the year 2000 several studies have been carried out in relation with dynamics of the tennis arm. Authors in [9] have carried out a study on changes in angular momentum during the tennis serve. Joint loading during the tennis serve has been studied in [10]. However, the study in [10] was limited only to shoulder and elbow loading. A more comprehensive analysis on shoulder joint loading during tennis flat serves and kick serve has been carried out in [11].

To-date no in-depth study has been done to analyze the joint torque distribution resulting from the impact force. A comprehensive joint torque distribution analysis for the tennis flat serve is presented in this paper based on a kinematic model with 9 revolute joints.

II. METHOD

A. 9-DoF Biomechanical Model

The human upper limb is a quite complex structure, composed of three chained modules, the shoulder girdle, the elbow and the wrist. Each module can be modeled as consisting of revolute joints since their translations are negligible compared to rotations. In biomechanical literature, the human arm represented with three rigid segments, connected by frictionless joints with a total of 7-DoF, is the generally accepted model [12], [13].

A mechanism with 9-DoF [14] has been used, where 2 revolute joints are located at the sterno-clavicular Joint (P0), 3 located at the gleno-humeral Joint (P1), one at the elbow (P2), one at the proxial radioulnar (P2) and 2 at the wrist

(P3). For simplicity, a single position (P2) was taken to represent the elbow joint and proxial radioulnar Joint. The 9-DoF model enables detailed analysis of the motion of the upper human limb including the synchronous motion of sub-joints of the shoulder [14]. See Fig 1.

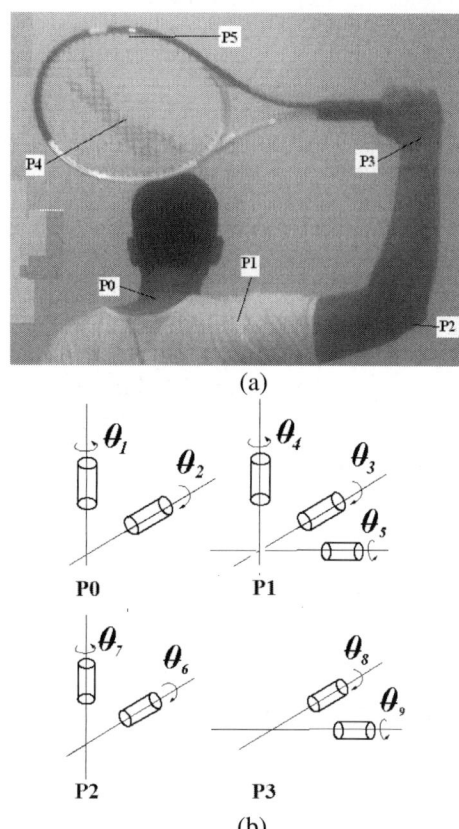

Fig. 1 (a) Key body reference joint locations
(b) The revolute joints and the direction of rotations

The order of Euler angle rotations in the model is as follows: protraction and retraction of the scapula (P0), elevation and depression of the scapula (P0), transverse flexion and transverse abduction of the shoulder (P1), shoulder flexion and extension (P1), shoulder external and internal rotation (P1), elbow flexion and extension (P2), elbow pronation and supination (P2), wrist abduction and adduction (P3), wrist flexion and extension (P3).

Standard DH parameters were used to obtain transformation matrices. A novel inverse kinematic algorithm proposed by author [14] was used to determine the joint angles.

B. Data Collection

3-D positions were captured by a Qualisis Motion Capture system, by placing passive reflective markers at six key positions (P0, P1, P2, P3, P4 and P5). A UK county level right handed tennis player with a height of 200.5 cm and weight of 96.4kg was volunteered to provide samples for the study. This study was approved by the Kingston University research ethics committee. After adequate warming up, the subject was asked to carry out a sequence of Tennis flat serves within the constraints of the laboratory environment. Five sample strokes were taken in to the analysis.

C. Impact Force Calculation

The area under the force-time graph represents the impact of a tennis ball. Hence, provided the change in velocity (Δv) and the mass of the ball (m) are known, the average impact forces ($F_{average}$) resulting during the contact time period (Δt) can be calculated as follows;

$$\int_{t_0}^{t_0+\Delta t} F dt = Impact = m\Delta v \quad (1)$$

$$F_{average} = \frac{m\Delta V}{\Delta t} \quad (2)$$

The average impact force of the ball is assumed to be fully exerted by the racket during the racket ball collision setting. An equal and opposite force was taken as the racket impact force. The spring force of the strings was disregarded for simplicity.

D. Jacobian and Torque Distribution

Considering the workspace of the upper limb, the work done in Cartesian space is equal to the work done in Joint space. Hence;

$$F^T . \delta x = \tau^T . \delta \theta \quad (3)$$

If Jacobian is defined as; $J = \frac{\partial x}{\partial \theta}$

$$F^T J \delta \theta = \tau^T \delta \theta \quad (4)$$

F – Impact Force vector; τ – Joint torque vector;
δx – linear displacement vector;
$\delta \theta$ – joint angle displacement vector

$$F^T J = \tau^T \quad (5)$$
$$\tau = J^T F \quad (6)$$

For the study 3X1 impact force vector matrix, 3X9 Jacobian vector matrix and 9X1 joint torque vector matrix were defined. Custom defined programs were coded using Matlab to obtain the joint angles, the Jacobian matrix and the Joint torque distribution resulting from the Force.

III. RESULTS

A. Pre and Post Velocities

The velocity profile for pre and post impact was obtained by tracking the ball movement. The obtained pre impact and post impact ball mean velocity values for the samples are given in table 1.

Table 1 Pre and Post ball mean velocities

Direction	Impact ball Velocity(ms^{-1})	
	Pre impact	Post Impact
Antero-posterior(Y)	0.31	44.84
Lateral (X)	-0.96	-9.12
Vertical (Z)	-1.53	-10.60

The velocity results were compared with the results presented in [15], and found our resultant mean velocity value was 8% lower. However, the velocity values in [15] involved elite tennis players including the world rank one at that time.

B. Contact Time

The contact time was calculated using a high speed Qualisis Camera system that sampled frames every 0.5 ms.

The study presented in [16] estimated ball-and racquet contact durations in tennis forehand and backhand strokes using strain gauges. He reported that the average ball-and-racquet contact duration ranged from 0.006 to 0.012 seconds. The mean value obtained was 0.008s and it was within the range suggested by [16].

C. Force

Mass of the ball was taken as 0.0565 kg [17]. Using eq. 2 the force vector (F_o) with respect to the original framework were calculated. See Table 2.

Table 2 Impact force components

Direction	Directional Force (N)
Antero-posterior(Y)	+354.33
Lateral(X)	-64.97
Vertical(Z)	-72.14

The force vector (F_n) with respect to the end effector framework for each sample were calculated as follows;

$$F_0 = T_o^n F_n \quad (7)$$
$$F_n = (T_o^n)^{-1} F_o \quad (8)$$

Where, T_o^n: Transformation matrix(4x4) from origin to the end effector.

The resultant mean value of the impact is 363.4 N. In [18] the peak impact force had been measured using a strain gauge on a fixed tennis racquet and expressed the peak impact force as a function of ball velocity before the impact. Results presented in [18] suggest a peak impact force of 377 N, which is notably similar to the calculated resultant impact force in our study.

D. Torque Distribution

A novel inverse kinematic approach suggested in [14] has been used to calculate the instantaneous joint angles. Using the joint angles and Jacobian the joint torques were calculated. See Table 3.

Table 3 Joint torque distribution

Joint	Movement	Torque(Nm)	
		Mean	SD
Sterno-Clavicular (P0)	Protraction/Retraction	-245.67	17.01
	Elevation/Depression	184.47	15.75
Gleno-Humeral (P1)	Flexion/Extension	-0.29	0.05
	Abduction/Adduction	-110.50	15.67
	Int/ext rotation	-84.85	12.75
Elbow(P2)	Flexion /Extension	-110.08	24.00
Proximal Radioulnar(P2)	Pronation/Supination	-109.59	33.54
Wrist(P3)	Ulna/Radial deviation	-65.11	19.04
	Flexion/Extension	-42.99	11.69

IV. ANALYSIS AND DISCUSSION

The resultant torques at each joint position were calculated. The highest mean resultant torque of 307Nm was observed at sterno-clavicular joint. At the gleno-humeral, the mean resultant torque recorded was 139 Nm. The torques related to abduction and internal rotation articulations have contributed largely for this. At the elbow, a torque of 110 Nm was recorded with a maximum value of 147 Nm. A torque with mean magnitude of 109Nm is apportioned to radio-ulnar joint.

A sensitivity analysis has been carried out by keeping the mean force vector fixed and varying the joint angles one at a time. For each angle, fifty random sample values were obtained based on the mean and standard deviation values of the respective angle θ_x.

$$\theta_x = \theta_{x,mean} + r\theta_{x,SD} \quad (10)$$

Where; r- A randomly generated number with SD of 0.1

The results were plotted in graphs. As per the obtained distribution graphs the proximal radioulnar joint was found

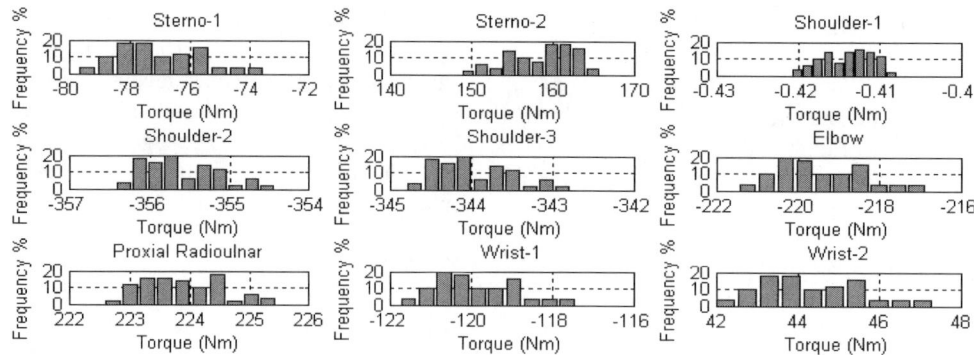

Fig. 2 Joint torque distribution variation for different proximal radioulnar joint angle values

as the highest sensitive joint during the tennis serve. The joint torque distribution variation for proximal radiounlar is shown in Fig. 2.

Identifying the torques necessary to prevent shoulder injury was found in a past study done by reference [19]. Reference [19] proposed that anything over 50Nm gives a high risk for upper arm injuries. In our study, we found all of joint torque resultant values comprehensively exceed this limit. But, the scope of our joint torque distribution analysis was limited to the joint torques resulting from the static impact force. The effects of the related dynamic forces will have to be taken into account to calculate the actual resultant joint torques.

Also, to improve the accuracy and standard deviations of collected data, more samples have to be studied.

V. V. CONCLUSION

The joint torque distribution resulting from the impact force was analysed based on a detailed 9-DoF biomechanical model. The 9-DoF model provides better prosthesis and more realistic androids of upper limb movements, compared to standard 7-DoF models. The impact shock wave dramatically increases joint torques on portions of the players' hand. The clinical and performance impacts of the findings of this research should be further investigated.

VI. REFERENCES

1. Kibler W, Van der Meer D (2001) Mastering the Kinetic Chain. In: World-Class Tennis Technique, Ed: P. Roetert and J. Groppel Champaign, USA: Human Kinetics:99-113
2. Whiting W, Zernicke (1998) Biomechanics of Musculoskeletal Injury, 1st edition: Human Kinetics. Champaign, USA
3. Hill J (1983), "Epidemiologic perspective on shoulder injuries," J. Clin Sport Med, vol. 2,no. 2: 241-246
4. Raikova R (1992)," A general approach for modeling and mathematical investigation of the human upper limb," J. Biomechanics, vol. 25:857-867
5. Peat M (1986), "Functional Anatomy of the Shoulder Complex," J. American Physical Therapy Association, vol. 66:1855-1865
6. Knudson D (2001)," What happens at impact and why it can hurt," M. TennisPro magazine, Sept. 2001
7. Elliott B, Marshall R, (1995)," Contributions of upper limb segment rotations during the power serve in tennis." J. Applied Biomechanics, vol. 11:433- 442
8. Sprigings R et al. (1994), "A three-dimensional kinematic method for determining the effectiveness of arm segment rotations in producing racquet-head speed," J. Biomechanics, vol. 27: 245-254
9. Bahamonde R E (2000), 'Changes in angular momentum during the tennis serve',J. Sports Sciences, vol. 18,no. 8:579-592
10. Elliott B et al. (2003), "Technique effects on upper limb loading in the tennis serve," J. Science and Medicine in Sport , vol. 6:76-87
11. Reid M et al. (2007) "Shoulder joint loading in the high performance flat and kick tennis serves," J. Sports Med, vol. 41:884–889
12. Desmurget M, Prablanc C (1997), "Postural control of three-dimensional prehension movements," J. Neurophysiology, vol. 77, no. 1:452-464
13. Lemay MA (1996)," A dynamic model for simulating movements of the elbow, forearm, and wrist.," J. Biomechanics, vol. 29:1319-1330
14. Masinghe W, Nanayakkara T, Collier G , Ordys A, "A Novel Approach to Determine the Inverse Kinematics of a Human Upper Limb Model with 9 Degrees of Freedom,", In Proc. of the IEEE IEMBC Conference, 2012. Lankawi, Malaysia: 525–530
15. Chow J et al. (2011), "Comparing the pre- and post-impact ball and racquet kinematics of elite tennis players' first and second serves: a preliminary study," J. Sports Sciences, vol.21,no.7:529-37
16. Bernhang AM et al. (1974) Tennis elbow: a biomechanical approach. J. Sports Med, vol. 2: 235-258
17. BBC Sports academy: Tennis, [Online]. Available at <http://news.bbc.co.uk/sportacademy/hi/sa/tennis/default.html>
18. Hatze H (1976),"Forces and duration of impact and grip tightness during the tennis stroke,"J. Med Sci Sports, vol. 8: 88-95
19. Dillman et al. (1995),"What do we know about body mechanics involved in tennis skills?" In H. Krahl, H Pieper, B. Kibler & P. Renstrom(Eds), Tennis: Sports Medicine and Science, Society for Tennis Medicine and Science. Auglage:6-11

Corresponding author: Waruna D Masinghe
Institute: Kingston University
City: London
Country: United Kingdom
Email: W.Masinghe@kingston.ac.uk

The Fatigue Vector: A New Bi-dimensional Parameter for Muscular Fatigue Analysis

S. Conforto[1,2], A.M. Castronovo[1], C. De Marchis[1,2], M. Schmid[1,2], M. Bertollo[2], C. Robazza[2], S. Comani[2], and T. D'Alessio[1]

[1] Department of Engineering, University of Roma TRE, via Vito Volterra 62, 00146 Rome, Italy
[2] BIND - Behavioral Imaging and Neural Dynamics Center, University "G. D'Annunzio" of Chieti-Pescara, Italy

Abstract—The aim of this study is to introduce a new parameter for fatigue investigations, which relies on a bi-dimensional analysis of sEMG signals in temporal and spectral domains. The new parameter, the *Fatigue Vector*, is defined in a space domain whose coordinates are the amplitude and the mean spectral frequency of the sEMG signal. The performance of the *Fatigue Vector* has been compared to those of classical parameters. The analysis has been carried on signals recorded from Rectus Femoris, Vastus Lateralis and Vastus Medialis during knee extension repetitions performed until exhaustion. The task was repeated twice with different biomechanical loads in order to test the muscular activity variations with respect to the different force demands.

The performance of the *Fatigue Vector* in assessing the occurrence of muscular fatigue are promising, and the obtained preliminary results open an interesting scenario for the application of this parameter to several fields.

Keywords—Neuromuscular fatigue, sEMG, biomechanics, isometric contractions, quadriceps.

I. Introduction

Localized muscle fatigue is generally referred to as a task-induced phenomenon, which consists of losing the ability to maintain or generate a force during sustained sub-maximal or maximal contractions [1]. Muscular fatigue is produced by changes at the neuromuscular junction, due to the decreased release of Ca^{2+} ions, which causes: 1) inhibition of the development, 2) reduction of the amplitude of the mechanical twitch [2], and 3) decrease of the conduction velocity (CV) in the muscle fibers. The altered Ca^{2++} ions release is due to changes in extracellular pH [3], that depends on the same central mechanisms driving the impaired neuromuscular propagation and the reduced discharge frequency of the spinal motor neurons [4].

These physiological changes are reflected in the surface ElectroMyoGraphic (sEMG) signal, where some modifications of the amplitude and frequency characteristics have been noticed and have been defined as electrical signs of muscular fatigue. It has been observed that the sEMG amplitude increases to maintain the required level of force during isometric contractions [5,6], so justifying a linear relationship between exerted force and signal amplitude [1,5]. However, also a non-linear trend has been observed during fatigue, due to the high dependence of the phenomenon on the task performed [7,8]. The spectrum of the sEMG signal shifts to lower frequencies during both sub-maximal and maximal contractions [9], due to the CV decrease.

Even if the physiological meaning of the electrical signs is not completely defined yet [8, 10], sEMG feature changes can be quantified by indices such as: Average Rectified Value or Root Mean Square (RMS), which have been widely used as signal amplitude estimators, and Mean or Median Frequency, or even higher order spectral moments [11,12].

During the last years, the reliability of these indices has been extensively discussed, and several methods have been developed to improve their consistency [13,14]. However, the electrical indices are not fully accepted as robust probes for detecting and monitoring muscular fatigue because of a missing standardization in the processing (due to the non-stationary nature of sEMG signal during fatigue) and controversial outcomes from experimental studies, especially in the case of dynamic (i.e. non-isometric) contractions [15].

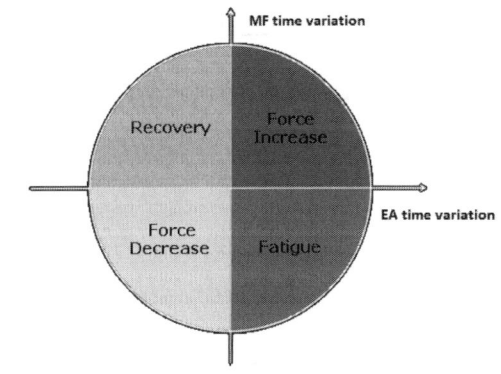

Fig. 1 Bi-dimensional space for muscular status representation proposed by Luttman and colleagues [16]. EA stands for muscular Electrical Activity, while MF stands for Mean spectral Frequency

A different use of the electrical indices has been proposed with the Joint Analysis of EMG Spectrum and Amplitude (JASA) by Luttman and colleagues [16], who considered the information provided simultaneously by a pair of estimators (that is, one in the amplitude and the other in the frequency domain). This approach has the purpose to simultaneously take into account the electrical modifications that are induced by the time-varying conditions of muscular fatigue and force request during motor tasks. By adopting a bi-dimensional representation, Luttman and colleagues [16] suggest a four-class coding of the muscular status (force increase, force decrease, fatigue and recovery from fatigue) that is well represented in the diagram reported in Figure 1. On the basis of the JASA approach, it would be crucial not only to map the muscular status space, but also to define an indicator including both domains. The aim of this work is to introduce a new parameter, the Fatigue Vector, which relies on the bi-dimensional analysis of the time and spectral domains of sEMG signals. The behavior of this new parameter and its dependence on biomechanical load is showed.

II. Materials and Methods

Participants - Four healthy right-handed volunteers (age: 32.5±8.38 yrs, height: 175±6.3cm, weight: 72±9.1 kg) participated in the study after being briefed on the overall procedure, including possible risks and discomfort. All volunteers signed a written informed consent and none reported any history of knee pathology or surgery. The study was conducted in accordance with the local ethical guidelines, and conformed to the Declaration of Helsinki.

Experimental Protocol – The experimental protocol was performed on a leg extension machine (Leg Extension ROM, Technogym). The subject was seated on the chair in an upright position in order to align the axis of rotation of the machine moving arm with the lateral femoral condyle. The participants visited the laboratory three times. During the first visit they familiarized with the instrumentation, were informed about the protocol, and performed 5 s isometric leg extensions of the dominant leg with incremental load until the maximum tolerable load (ML) was reached. The ML value was then used during the next two visits, during which the subjects performed a task consisting in repetitions of the following protocol until exhaustion: 1 isometric leg extension of 7 s duration, 5 s of rest, 10 dynamic leg extension movements, 5 s of rest. The task was preceded by two maximal contractions to determine the Maximum Voluntary Contraction (MVC) value, used to normalize the isometric sEMG data.

The task was performed with two different loads during the second and the third visits, 20% and 70% of ML, respectively. These two biomechanical conditions will be referred to henceforth as Low Intensity Exercise (LIE at 20% ML) and High Intensity Exercise (HIE at 70% ML). The task was repeated until the task failure point was reached, defined as the instant when subjects stopped the exercise voluntarily because of exhaustion or muscular pain.

Electromyographic data – sEMG signals were acquired from three muscles of the quadriceps of the right leg (Rectus Femoris (RF), Vastus Lateralis (VL) and Vastus Medialis (VM)) by means of a wireless EMG system provided with 8 bipolar channels (FREEMG 300, BTS Bioengineering SpA, Garbagnate Milanese, Italy). Sampling frequency was 1 kHz and signals were digitized via a 14 bit AD converter. Before applying the Ag/AgCl electrodes, skin was shaved and cleaned in order to control skin impedance. All electrodes were placed according to SENIAM recommendations [17]. The onset and offset of the muscular activations were determined with a detector optimized with respect to the Signal-to-Noise Ratio (SNR) [18]. In this work, sEMG data were analyzed off-line and only the epochs related to the isometric bursts were considered. sEMG data were band-pass filtered with a 3rd order Butterworth digital filter in the range [20-450] Hz. Data were normalized with respect to MVC values. For each isometric burst, two values were extracted: the muscular Electrical Activity (**EA**) and the Mean spectral Frequency (**MF**). **EA** was calculated as the mean value of the linear envelope (extracted by the RMS operator), and **MF** as the normalized first order moment of the power, which was estimated using an autoregressive model of order p =10 [19]. Then, for each burst we obtained a couple of parameters {EA_i,MF_i} that is representative of one point in the {EA,MF} space. Each point was then expressed in terms of percentage variation with respect to the values {EA_1,MF_1} calculated for the first isometric burst, as in [16] (Equation 1):

$$\begin{cases} \Delta EA_i\% = \dfrac{EA_i - EA_1}{EA_1} * 100 \\ \Delta MF_i\% = \dfrac{MF_i - MF_1}{MF_1} * 100 \end{cases} \quad (1)$$

In this study, the points representative of the first and the last bursts of the task performed in both the LIE and HIE conditions were compared.

The Fatigue Vector – In the {EA,MF} space, we considered the distance between the origin of the axes and each point obtained as explained before. Then, we transformed the points in the polar domain, obtaining what we called the Fatigue Vector, which is characterized by a magnitude ρ and a phase ϑ, calculated according to Equation 2:

$$\rho_i = \sqrt{\Delta EA\%_i^2 + \Delta F\%_i^2}$$
$$\vartheta_i = \dfrac{2 * a\tan\left(\dfrac{\Delta F\%_i^2}{\sqrt{\Delta F\%_i + \Delta EA\%_i} + \Delta EA\%_i}\right)}{\pi} * 180 \quad (2)$$

In particular, ρ increases when EA_i and MF_i vary with respect to the initial values, so representing a modification of the muscular conditions during the task. On the other hand, ϑ contains the sign of the variations, so allowing to define the type of ongoing muscle modification. For example, two vectors having the same magnitude but different phases will express two different muscular modifications (i.e. force increase vs fatigue, fatigue vs force decrease, etc).

Statistical Analysis – All the extracted features, **EA**, **MF**, **ΔEA%**, **ΔMF%**, ρ, ϑ, underwent descriptive analysis. In particular, we compared **EA** and **MF** for the initial and final phases of the isometric exercise (IN vs. END) in both biomechanical conditions (LIE vs. HIE). The **ΔEA%**, **ΔMF%** and the new features ρ and ϑ were compared for the load conditions. The comparisons were performed by means of a Student's t test, with significance level set at $p < 0.05$.

III. RESULTS

Temporal and spectral indicators, EA and MF – The amplitude parameter **EA** shows no significant difference between IN and END phases in RF. Significant differences are observed for VL and VM during LIE ($p<0.05$). The frequency parameter **MF** shows significant differences between IN and END phases in all muscles during HIE ($p<0.05$), while differences are observed only in VM during the LIE condition. The results are summarized in Table 1.

Table 1 EA and MF values (mean±standard deviation) across subjects for the different phases (IN vs. END) and load conditions (LIE vs. HIE). Significance is reported as †n.c, * p<0.05, ** p <0.001

Muscle	EA			MF			Load
	IN	END		IN	END		
RF	0.11±0.02	0.30±0.2	†	102.5±10.7	93.62±1.94	†	LIE
	0.42±0.28	0.44±0.22	†	106.4±8.45	86.5±7.6	*	HIE
VL	0.16±0.06	0.3±0.07	*	102.6±26.9	96.7±22.6	†	LIE
	0.46±0.32	0.42±0.32	†	104.8±26.5	85.6±20.3	*	HIE
VM	0.12±0.05	0.27±0.08	*	96.07±13.5	88.8±9.08	*	LIE
	0.37±0.29	0.34±0.29	†	95.5±15.03	82.9±9.06	*	HIE

Table 2 ΔEA% and ΔF% mean±standard deviation values across subjects for load conditions (LIE vs. HIE). Significance:†n.c, * p<0.05, ** p <0.001

Muscle	ΔEA%		ΔMF%		Load
RF	217.07±279.67	†	-7.55±12.35	*	LIE
	25.88±83.92		-18.65±5.67		HIE
VL	125.46±152.22	†	-5.12±5.25	*	LIE
	-4.09±18.35		-17.24±10.02		HIE
VM	188.68±244.2	†	-7.18±4.19	*	LIE
	-8.29±19.60		-12.69±4.44		HIE

Normalized percent values of temporal and spectral indicators, ΔEA% and ΔMF% – The temporal indicator **ΔEA%** didn't show significant difference between loads for any muscle, while **ΔMF%** is significantly different between the two biomechanical conditions for all muscles ($p<0.05$).

The Fatigue Vector – The magnitude ρ of the Fatigue Vector shows no significant differences between the LIE and the HIE conditions ($p>0.05$). On the other hand, the phase ϑ is significantly different between the two conditions for all muscles ($p<0.05$).

Table 3 ρ and ϑ mean±standard deviation values across subjects for load conditions (LIE vs. HIE). Significance: †n.c, * p<0.05, ** p <0.001

Muscle	ρ			ϑ	
	LIE	HIE		LIE	HIE
RF	218.6±278.5	61.13±58.7	†	-10.5±13.45	-85.43±67.15
VL	126.37±151.42	23.11±11.98	†	-11.06±12.90	-108.05±40.7
VM	151.38±227.5	21.12±10.75	†	-6.56±7.78	-105.2±53.6

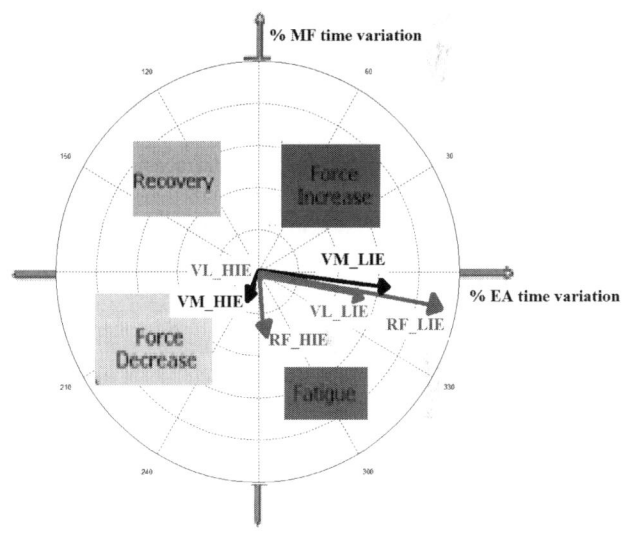

Fig. 2 Representation on the bi-dimensional domain of the Fatigue Vector for all the three muscles and the two conditions

IV. DISCUSSION AND CONCLUSIONS

In this study we analyzed the sEMG signals with single domain indicators and with a new parameter suitable for fatigue analysis, the Fatigue Vector, which accounts for both temporal and spectral features of the sEMG signals.

During the LIE condition, sEMG amplitude tends to increase in the VL and VM muscles ($p<0.05$), while no differences are observed during the HIE condition. This outcome can be explained by the number of isometric contractions performed during HIE, which is lower than in LIE because of an early exhaustion, and in terms of recruitment mechanisms that, for HIE, are constant in order to satisfy the load request, as outlined by the initial values of **EA**. The mean frequency of the sEMG signal tends to significantly decrease during HIE for all muscles ($p<0.05$) as a result of the sensitivity of the parameter to fatigue occurrence. During the LIE condition, only VM showed a significant difference between the beginning and the end of the exercise. The single-parameter analysis does not clearly detect the occurrence of fatigue, but simply outlines that **EA** seems less sensitive to force requirements but more sensitive to task duration than **MF**, which, instead, had an opposite behavior ($p<0.05$), probably due to the high force demand that induces a fast accumulation of acid ions in the muscle fibers, which slows down the muscular conduction velocity and modifies the spectral content of the sEMG signal. Therefore, the use of single parameters does not seem to provide robust information about the occurrence and evolution of neuromuscular fatigue in the studied experimental conditions. When the normalized parameters are used, no temporal reference is available, so that only a comparison between the biomechanical conditions (LIE vs. HIE) can be performed. The fatigue vector provides a better explanation of the behavior of sEMG characteristics. ρ does not change with respect to the biomechanical loads, because the simultaneous variations of the Δ**EA%**, Δ**MF%** are not so large. ϑ, which represents the orientation of the fatigue vector, is significantly different across conditions ($p<0.05$), so outlining a link between the load and the variations of the muscular status during the task (Figure 2).

Further studies on a larger population are needed to validate these preliminary conclusions. Future developments will include the analysis of dynamic fatiguing contractions [20], which will open a wide scenario dealing with the relationships between exercise performance and motor control strategies [21, 22, 23].

I. REFERENCES

[1] Bigland-Ritchie B, Johansson R, Lippold OCJ et al. (1983). Changes in motor neuron firing rates during sustained maximal voluntary contractions. *J Physiol*, **340**(1): 335-46.

[2] Place N, Bruton JD, Westerblad H (2008). Mechanisms of fatigue induced by isometric contractions in exercising humans and in isolated mouse single muscle fibers. *Proc Aust Physiol Soc*, **39**: 115-22

[3] Gandevia, S. C. (1995). *Fatigue: neural and muscular mechanisms*. Vol. **384**. Plenum Publishing Corporation.

[4] Gandevia SC (2001). Spinal and Supraspinal Factors in Human Muscle Fatigue. *Physiol Rev*, **81**(4):1725-89.

[5] Moritani T, Muro M (1987). Motor unit activity and surface electromyogram power spectrum during increasing force of contraction. *Eur J Appl Physiol Occup Physiol*, **56**(3):260-5.

[6] Bigland-Ritchie B (1981). EMG/Force relations and fatigue of human voluntary contractions. *Exerc Sport Sci Rev*, **9**(1):75-118.

[7] Fuglevand AJ, Zackowsky Km, Huey Ka et al. (1993). Impairment of neuromuscular propagation during human fatiguing contractions at submaximal forces. *J Physiol*, **460**(1):549-72.

[8] Dideriksen JL, Farina D, Enoka R (2010). Influence of fatigue on the simulated relation between the amplitude of the surface electromyogram and muscle force. *Phil Trans R Soc A*, **368**(1920):2765-81.

[9] Arendt-Nielsen L, Mills KR (1988). Muscle fiber conduction velocity, mean power frequency, mean EMG voltage and force during submaxiaml fatiguing contractions of human quadriceps. *Eur J Appl Physiol Occup Physiol*, **58**(1-2):20-5.

[10] Farina D (2008). Counterpoint: Spectral properties of the surface EMG do not provide information about motor unit recruitment and muscle fiber type. *J Appl Physiol*, **105**(5):1673-4.

[11] Vøllestad NK (1997). Measurements of human muscle fatigue. *J Neurosci Methods*, **74**(2): 219-27.

[12] Moritani T, Nagata A, Muro M (1982). Electromyographic manifestations of muscular fatigue. *Med Sci Sport Exerc*, **14**(3):198-202.

[13] Potvin JR, Bent LR (1997). A validation of techniques using surface EMG signals from dynamic contractions to quantify muscle fatigue during repetitive taks. *J Electromyogr Kinesiol*, **7**(2):131-9.

[14] Conforto S, D'Alessio T (1999). Real time monitoring of muscular fatigue from dynamic surface myoelectric signals using a complex covariance approach. *Med Eng Phys* **21**(4):225-34.

[15] Arendt-Nielsen L, Sinkjaer T (1991) Quantification of human dynamic muscle fatigue by electromyography and kinematic profiles, J. Electromyogr Kinesiol 1:1-8.

[16] Luttmann A, Jäger M, Laurig W, (2000) Electromyographical indication of muscular fatigue in occupational field studies. Int. J. Industr. Ergonomics 25:645-660.

[17] Hermens HJ, Freiks B, Disselhorst-Lug C et al. (2000). Development of recommendations for sEMG sensors and sensor placement procedures. *J Electromyogr Kinesiol*, **10**(5):361-74.

[18] Severini G, Conforto S, Schmid M, D'Alessio T (2012). Novel formulation of a double threshold algorithm for the estimation of muscle activation intervals designed for variable SNR environments. *J Electromyogr Kinesiol*, **22**(6):878-85

[19] Farina D, Merletti R (2000). Comparison of algorithms for estimation of EMG variables during voluntary isometric *contractions*. *J Electromyogr Kinesiol*, **10**(5):337-49

[20] Castronovo AM, De Marchis C, Bibbo D, Conforto S, Schmid M, D'Alessio T (2012). Neuromuscular adaptations during submaximal prolonged cycling. In Proceedings EMBS, 2012 Annual International Conference of the IEEE: 3612-3615.

[21] Castronovo AM, Conforto S, Schmid M, Bibbo D, D'Alessio T (2013) How to Assess Performance in Cycling: the Multivariate Nature of Influencing Factors and Related Indicators. Front. Physiol. DOI: 10.3389/fphys.2013.00116.

[22] De Marchis C, Castronovo AM, Bibbo D, Schmid M, Conforto S (2012) Muscle synergies are consistent when pedaling under different biomechanical demands. In Proceedings EMBS, 2012 Annual International Conference of the IEEE: 3308-3311.

[23] De Marchis C, Schmid M, Bibbo D, Castronovo AM, D'Alessio T, Conforto S (2013) Feedback of mechanical effectiveness induces adaptations in motor modules during cycling. Frontiers in computational neuroscience, 7.

Corresponding author
Author: Silvia Conforto
Institute: Department of Engineering, University of Roma TRE
Street: via Vito Volterra 62
City: 00146, Rome
Country: Italy
Email: silvia.conforto@uniroma3.it

Kinetic Analysis of Manual Wheelchair Propulsion in Athletes and Users with Spinal Cord Injury

M. Solís-Mozos[1], A. del Ama-Espinosa[1], B. Crespo-Ruiz[1], E. Pérez-Rizo[1], J.F. Jimenez-Díaz[2], and A. Gil-Agudo[1]

[1] Biomechanics and Technical Aids Department, National Hospital for Spinal Cord Injury, Toledo, Spain
[2] Laboratory of Performance and Sports Rehabilitation, Faculty of Sport Science, University of Castilla La Mancha Toledo, Spain

Abstract—During the last years, manual wheelchair propulsion in daily life and sports is increasingly being studied. We analyzed shoulder joint kinetics while propelling a wheelchair placed on a treadmill and compare shoulder joint net forces and moments of wheelchair propulsion on the treadmill in athlete group and non-athlete group. Ten subjects with thoracic spinal cord injury participated in this studied. A kinematic analysis system consisting of four camcorders (Kinescan-IBV) and a kinetic device that registered the contact force of the hand on the pushrim (SmartWheel) were used in to different protocols using a pulley system. The propulsion is different in the same conditions in both groups. Significant increases in shoulder joint loading were observed in wheelchair athlete group.

Keywords—Biomechanics, treadmill, wheelchair propulsion, pulley system, spinal cord injury.

I. INTRODUCTION

People with spinal cord injury (SCI) often rely on their ability to propel a manual wheelchair for independent mobility [1]. Upper extremity in persons with SCI is important for activities of daily living. Shoulder pain is a well-known problem in manual wheelchair users [2]. The incidence of shoulder pain in people with SCI who use wheelchairs ranges from 30% to 73% [3].

Wheelchair propulsion biomechanical analysis yields pertinent information to identify the factors that predispose to such injuries. When conducting biomechanical analysis of wheelchair propulsion, the aim is to obtain a laboratory setup that reproduces as closely as possible the conditions of wheelchair propulsion that users encounter in real life.

Most of the studies have been done by wheelchair exercise testing, using ergometer [4-7], treadmill [8-13] or comparing different conditions. Regarding experiments conducted on a treadmill, several strategies can be chosen to set the workload: changing speed [8,9,12], changing the slope [10,11], or changing the resistance with the use of a pulley system [13] to study the effects on the propulsion variables.

The wheelchair propulsion on a treadmill is the situation considers more similar to real-life propulsion on a smooth surface [12,14].

Previous investigators have studied shoulder pain in athlete or non-athlete wheelchair users, and the prevalence is high in both [2].

We undertook this study to analyze temporospatial variables and kinetics shoulder variables while propelling a wheelchair on treadmill, comparing shoulder joint net forces and moments using a pulley system on a treadmill between a wheelchair's athlete group and a group of wheelchair's non-athlete in the same conditions.

Further, it is hypothesized the kinetics shoulder variables is different in wheelchair's athlete and non-athletes wheelchair's when propelling in the same conditions.

II. MATERIALS AND METHODS

A. Subjects

Ten wheelchair users with SCI participated in the study. The subjects were divided in two groups depending on the number of hours of sports practice a week. Eligible participants met the following criteria:

- Complete SCI (T1-L5), ASIA A or B.
- At least 18 months since the spinal cord injury.
- Not having recent upper extremities injuries with medical contraindications.
- Experience in wheelchair propulsion.
- Over 18 years and under 45 years.
- In one group are required to practice sport over 6 hours a week.

The local ethics committee approved the study and all subjects gave their written informed consent. The guidelines of the declaration of Helsinki were followed in every case. The demographic characteristics of these patients are showed in Table 1.

Table 1 Subject demographics in two groups (mean ± s.d.)

	Wheelchair athlete	Wheelchair non-athlete
n	5	5
Sex	5 Male	5 Male
Age (years)	35.4 (2.71)	33.2 (7.66)
Height (m)	71.3 (9.24)	66.74 (11.23)
Weight (kg)	1.80 (0.02)	1.75 (0.04)
Injury level	T1-T12	T1-T12

B. Kinematics

Three dimensional markers position were collected at 50 Hz sampling frequency with four camcorders (Kinescan-IBV, Instituto Biomecánico de Valencia, Valencia, Spain). Spatial marker coordinates were smoothed out using a mobile means procedure.

The right upper limb was used for kinematic analysis. Reflective markers were located following ISB recommendations to define local reference systems on the hand, forearm and arm [15]. Twenty-two reflective markers were positioned on the following bony landmarks: 3 on the trunk (C7 spinous process, right and left acromio-clavicular joint), 4 on the arm, 2 on the forearm, 3 on the hand, 2 clusters of three markers each on the upper arm and on the lower forearm and 4 on the wheelchair. The axes of this reference system have been described in detail elsewhere [3].

C. Kinetics

The wheels of the wheelchair replaced by two SmartWheel (Three Rivers Holdings, LLC, Mesa, AZ, USA). The Smart-Wheel measures three-dimensional forces and torques applied to the pushrim at 240 Hz. A synchronization pulse to the Kinescan-IBV was used to trigger the start of kinetic and kinematic collection. Marker spatial coordinates were interpolated by cubic spline to synchronize with the kinetic data [3].

Five consecutive cycles were selected from the 20s data recording and each cycle was analyzed separately to obtain the mean value of the 5 cycles. The right wheel was used for the kinetic analysis. All subjects were right-handed.

D. Biomechanical Model

A model of inverse dynamics was created to identify the net forces and moments exerted on the shoulder joint complex obtained from anthropometric data, pushrim kinetics and upper limb kinematics. In this model, body segments were considered as rigid bodies with revolution geometrics of uniform constant density. The analysis was focused on the shoulder.

E. Experiment Design and Data Collection

A standard wheelchair, Action 3 Invacare, was used for all subjects. The experiment was performed on a treadmill (Bonte Zwolle BV, BO Systems, Netherlands) of suitable dimensions for wheelchair used. Power output and speed was determined in the form of a drag test, in which the drag force of the wheelchair's user system was measured [14].

We use two different standardized protocols. One of this, the constant protocol, of 20 minutes duration, with an individualized speed and load maintained throughout the whole test, to support 20w. The other test, a slope protocol, which the speed is maintained and the load was increased 5w each two minutes, starting the first two minutes at 20w. The order of the protocol was randomized and the time between both was 48 hours.

The subjects then were instructed to propel the wheelchair on the treadmill at speed determined for the drag test during a five-minute adaptation period. After this period, a 1-s static video capture was taken to calibrate the marker model.

F. Statistical Analysis

For each variable, descriptive analyses (mean ± s.d.) were computed for each subject. Analyses were made with SPSS 17.0 (SPSS Inc., Chicago, IL, USA). Data were analyzed with non-parametric test. To evaluate the difference between groups in the wheelchair propulsion techniques in different protocols a Wilconxon Signed Rank Test was performed. Level of significance was set at $p<0.05$.

III. RESULTS

Temporospatial variables results are different (Table 2). Changes in the propulsion values were found between groups. Considering the constant protocol, contact angle was increased ($p<0.05$), as well as the propulsion angle ($p<0.05$) in the wheelchair athlete and decreased the release angle ($p<0.05$) and the cadence (Table 2).

The kinetic shoulder joint variables in both groups are listed in Table 3 and Table 4. During the push phase maximal (lateral) Fz peak values were higher in the wheelchair athlete ($p<0.05$) in the constant protocol (Table 3). However, no significant differences were found between groups in peak shoulder moments (Table 4).

Table 2 Temporospatial and kinetic variables in Constant Protocol vs Slope Protocol (mean ± s.d)

	Cadence (cycles/min)	Contact angle (degrees)	Release angle (degrees)	Propulsion angle (degrees)	Peak total force (ftot)	Peak tangential force Ft (N)	Peak propulsion moment Mp (Nm)
Constant Protocol							
Athletes	0.62 (0.07)	117.66 (6.87)*	38.10 (3.99)*	79.55 (8.01)*	64.87 (5.82)	58.80 (9.24)	15.11 (2.37)
Non-athletes	0.73 (0.18)	106.51 (6.29)	51.27 (7.31)	55.24 (8.39)	69.27 (6.41)	58.70 (8.03)	15.08 (2.06)
Slope Protocol							
Athletes	0.67 (0.05)	108.46 (6.01)	27.38 (5.47)	81.08 (7)	110.89 (20.04)	89.96 (11.33)	23.12 (2.91)
Non-athletes	0.84 (0.19)	103.80 (8.64)	30.33 (7.13)	73.47 (14.59)	114.49 (23.44)	95.89 (16.51)	24.64 (4.24)

* Significantly different between two group ($p<0.05$).

Table 3 Peak shoulder forces and moments acting on shoulder joint in Constant Protocol during the Push phase (mean ± s.d.)

	Maximum value	Minimum value		Maximum value	Minimum value
Fx (N) (+ anterior, - posterior)			Mx (Nm) (+ adduction, - abduction)		
Athletes	50.16 (10.70)	-35.38 (7.08)		2.10 (2.12)	-5.98 (2.67)
Non-athletes	38.27 (13.58)	-43.21 (12.82)		0.38 (2.20)	-6.36 (5.49)
Fy (N) (+ superior, - inferior)			My (Nm) (+ rot int, - rot exter)		
Athletes	-8.14 (5.76)	-56.08 (11.81)		3.04 (1.42)	-3.40 (0.85)
Non-athletes	0.49 (14)	-48.59 (15.32)		4.07 (3.24)	-2.01 (1.23)
Fz (N) (+ lateral, - medial)			Mz (Nm) (+ flexion, - extension)		
Athletes	14.60 (6.01)*	-11.78 (3.69)		13.50 (1.01)	-9.12 (1.52)
Non-athletes	7.28 (2.31)*	-20.63 (19.16)		12.91 (2.79)	-9.27 (3.40)

* Significantly different between two group ($p<0.05$)

Table 4 Peak shoulder forces and moments acting on shoulder joint in Slope Protocol the Push phase (mean ± s.d.)

	Maximum value	Minimum value		Maximum value	Minimum value
Fx (N) (+ anterior, - posterior)			Mx (Nm) (+ adduction, - abduction)		
Athletes	60.56 (14.72)	-69.32 (6.97)		8.30 (7.69)	-9.95 (6.41)
Non-athletes	53.77 (10.22)	-81.35 (22.75)		2.88 (3.61)	-8.30 (3.83)
Fy (N) (+ superior, - inferior)			My (Nm) (+ rot int, - rot exter)		
Athletes	22.42 (21.38)	-69.65 (25.84)		3.94 (1.78)	-7.25 (2.44)
Non-athletes	21.70 (13.44)	-66.42 (26.02)		5.57 (2.53)	-5.13 (1.79)
Fz (N) (+ lateral, - medial)			Mz (Nm) (+ flexion, - extension)		
Athletes	22.89 (10.52)	-14.93 (6.03)		25.22 (3.70)	-15.03 (8.39)
Non-athletes	11.27 (7.51)	-19.70 (10.20)		25.54 (7.37)	-15.55 (8.59)

IV. DISCUSSION

Our hypotheses were confirmed. The propulsion in the same conditions is different in temporospatial variables and in Fz peak during the push phase in the constant protocol.

To our knowledge, this is the first study in wheelchair manual propulsion used a treadmill pulley system.

In previous studies with wheelchair users, when increased the speed the cadence and propulsion angle was increased [6,12]. Groups that used a roller ergometer and groups that used a dynamometer to measure shoulder load produced in propulsion for wheelchair athletes found differences in cadence, time of contact, time of propulsion angle and the forces applied [6,16]. Our findings were similar to studies that used other equipment, an ergometer or a dynamometer.

Comparing our groups, we found that using a constant protocol, the group of athletes used a propulsion in with hands were longer with the pushrim than the non-athlete group. We believe that this because the athletes have a greater training and a greater use of the wheelchair and, therefore, its propulsion in conditions becomes much more comfortable. They use fewer contacts in the rim that the group of non-athletes.

However, when we used the slope protocol, we did not find differences in propulsion patterns between the two groups. We believed that this is probably because the propulsion condition were difficult to be adapted by the users.

V. CONCLUSIONS

Differences were found between wheelchair's athlete and wheelchair's non-athlete in a treadmill with a pulley system.

In the constant protocol, the contact angle and the propulsion angle was increased in wheelchair athletes and the release angle and the cadence decreased versus non-athlete. Fz peak (lateral) values during the push phase were higher in the wheelchair athletes.

ACKNOWLEDGMENT

This work was part of a project financed by VI National Plan for Scientific Research, Development and Technological Innovation 2008-2011 (MICINN) DEP 2011-29222-C02-02, which does not have any commercial interested in the results of this investigation.

REFERENCES

1. Collinger J, Boninger M, Koontz A, et al. (2008). Shoulder biomechanics during the push phase of wheelchair propulsion: a multisite study of persons with paraplegia. Arch Phys Med Rehabil 89: 667-676.
2. Fullerton H, Bockardt J, Alfano A. (2003). Shoulder pain: a comparison of wheelchair athletes and nonathletic wheelchair users. Med Sci Sports Exerc 35: 1598-1961.
3. Gil Agudo A, del Ama A, Pérez E, et al. (2010). Shoulder joint kinetics during wheelchair propulsion on a treadmill at two different speeds in spinal cord injury patients. Spinal Cord 48: 290-296.
4. de Groot S, de Bruin M, Noomen SP, et al. (2008) Mechanical efficiency and propulsion technique after 7 weeks of low-intensity wheelchair training. Clin Biomech 23: 434-441.
5. De Groot S, Veeger D, Peter A, et al. Wheelchair propulsion technique and mechanical efficiency after 3 week of practice. Med Sci Sports Exerc 34: 756-766.
6. Veeger HEJ, Rozendaal LA, van der Helm FCT. (2002) Load on the shoulder in low intensity wheelchair propulsion. Clin Biomech 17: 211-218.
7. De Groot S Veeger HEJ, Hollander AP, et al. (2004) Effect of wheelchair stroke pattern on mechanical efficiency. Am J Phys Med Rehabil 83: 640-649.
8. Ambrosio F, Boninger M, Souza A, et al. (2005) Biomechanics and strength of manual wheelchair users. J Spinal Cord Med 28: 407-414.
9. Rankin J, Richter W, Neptune R. (2011) Individual muscle contribution to push and recovery subtask during wheelchair propulsion. Journal of Biomechanics 44: 1246-1252.
10. Van der Woude L, Botden E, Vriend I, et al. (1997) Mechanical advantage in wheelchair level propulsion: effect on physical strain and efficiency. Jof Rehabil Research and Development 34: 286-294.
11. Richter W, Rodriguez R, Kevin ME, et al. (2007) Consequences of a cross slope on wheelchair handrim biomechanics. Arch Phys Med Rehabil 88: 76-80.
12. Richter W, Rodriguez R, Woods KR, et al. (2007). Stroke pattern and handrim biomechanics for level and uphill wheelchair propulsion at self-selected speeds. Arch Phys Med Rehabil 88: 81-87.
13. Van Drongelen S, Arnet U, Veeger D, et al. (2013). Effect of workload setting on propulsion technique in handrim wheelchair propulsion. Med Eng & Phys 35: 283-288.
14. Van der Woude L, Veeger H, Dallmeijer AJ et al. (2001) Biomechanics and physiology in active manual wheelchair propulsion. Med Eng Phy 23: 713-733.
15. Wu G, van der Helm F, Veeger HEJ, Makhsous M, et al. (2005). ISB recommendation on definitions of joint coordinate systems of various joints for the reporting of human joint motion. Part II: shoulder, elbow, wrist and hand. J Biomechanics 38: 981-992.
16. Boninger M, Cooper R, Robertson R, et al. (1997) Wrist biomechanics during two speeds of wheelchair propulsion: an analysis using a local coordinate system. Arch Phys Med Rehabil 78: 364-372.

Corresponding author:

Author: Marta Solís Mozos
Institute: National Hospital for Spinal Cord Injury-Biomechanics
Street: Finca la Peraleda s/n
City: Toledo
Country: Spain
Email: msolism@sescam.jccm.es

Multimodal MRI Evaluation of Physiological Changes on Leg Muscles due to Fatigue after Intense Exercise

Mario Forjaz Secca[1,2], Sergio S. Alves[1], Ana Rita Pereira[3], José Nuno Alves[3], Filipa Joao[4], Antonio P. Veloso[4], Michael Noseworthy[5,6], Nuno Jalles Tavares[2], and Cristina Meneses[2]

[1] Physics Department, Cefitec, Monte de Caparica, Portugal
[2] Ressonancia Magnética - Caselas, Lisboa, Lisboa, Portugal
[3] Physics Department, Monte de Caparica, Portugal
[4] Laboratório de Biomecânica, Cruz Quebrada, Lisboa, Portugal
[5] Imaging Research Centre, St. Joseph's Healthcare, Hamilton, Ontario, Canada
[6] School of Biomedical Engineering, McMaster University, Hamilton, Ontario, Canada

Abstract—DTI, BOLD and T2 can give us non-invasive information on the functioning of muscles in real time. We used a multimodal approach, that included DTI, BOLD and T2 measurements, to evaluate the physiological changes in several leg muscles, and were able to assess that the soleus and gastrocnemius were the most involved muscles when our volunteers were submitted to intense one legged jump physical exercise conducive to fatigue. DTI, BOLD and T2 can give us non-invasive information on the functioning of muscles in real time. We used a multimodal approach, that included DTI, BOLD and T2 measurements, to evaluate the physiological changes in several leg muscles, and were able to assess that the soleus and gastrocnemius were the most involved muscles when our volunteers were submitted to intense one legged jump physical exercise conducive to fatigue.

Keywords—MRI, Biomechanics, Function, Sports performance.

I. INTRODUCTION

In MRI, several imaging techniques like DTI[1,2], BOLD[3] and T2[4] can give us non-invasive information on the functioning of muscles in real time. We used a multimodal approach, that included the acquisition of DTI, BOLD and T2 measurements, to evaluate the physiological changes in several leg muscles, and were able to assess that the soleus and gastrocnemius were the most involved muscles when our volunteers were submitted to intense one legged jump physical exercise conducive to fatigue.

II. METHODOLOGY

Seven young, healthy volunteers who regularly practiced physical exercise, were studied on a 1,5T GE Signa HDxt. BOLD, DTI and T2 images were acquired at rest, prior to exercise, and immediately following exercise outside of the scanner. The exercise consisted on one-legged jumps, in a standing position, until complete exhaustion was reached, which required around 2 minutes.

The axial sequences acquired, covering the most voluminous part of the leg, were the following:

- Anatomical Proton Density (PD) (54 slices; Slice Thickness = 3.9 mm; TE = 7.6 ms; TR = 4140 ms), EP DTI (54 slices; Slice Thickness = 3.9 mm; b = 400 s/mm^2; TE = 72.2 ms; TR = 10000 ms; NEX = 4; 16 gradient directions),
- BOLD Gradient-Echo EPI (54 slices, 60 temporal points, TE/TR = 35/3622 ms, 64x64 matrix, and 3.9 mm thick, 0 mm spacing),
- T2-weighted FSE (54 slices; TR = 2000 ms; TE = 105.8 ms; Slice Thickness = 3.9 mm).

The immobilization of the scanning leg was carried out using a purpose built support, which avoided muscles pressing against the table, and prevented deformation of the muscle cross section.

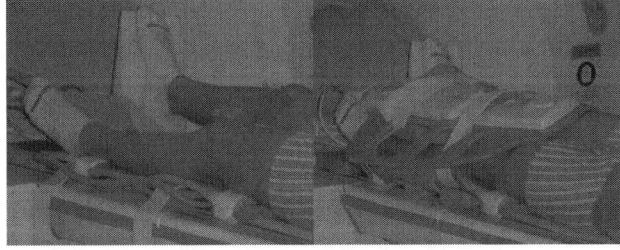

Fig. 1 Left: Leg of the subject in the supine position resting on the custom made structure, to avoid compression of the leg muscles against the MRI table. Right: MRI pelvic coil in position on the leg.

Because of the short time taken for the muscle to recover after fatigue and to make sure that both the post-exercise DTI images and BOLD images were acquired right after the exercise, we acquired the images for the same volunteer in two different sessions: one with the DTI data acquired right after the exercise and the other with the BOLD followed by T2 data acquired right after the exercise.

The ROI analysis of the Tibialis Anterior (TA), the Gastrocnemius Medial (GM) and Lateral (GL) and the Soleus (Sol) muscles was performed using TrackVis (Martinos Center,USA), for the DTI data, and Osirix, for the BOLD and T2 data, taking care to avoid inclusion of blood vessels.

III. RESULTS

On average, the FA values were observed to increase after exercise in the Gastrocnemius Medial (4,47%) and Gastrocnemius Lateral (1,80%) and decrease in the Tibialis Anterior (-1,67%) and Soleus (-5,48%). The ADC measured increased in the Tibialis Anterior (9,90%), but this effect was more intense on the Gastrocnemius Medial (23,10%), Gastrocnemius Lateral (27,62%) and Sol (25,27%).

During rest the BOLD signal intensity remained constant for all muscles studied. In all cases, the values at rest were normalized to 0%. After exercise, all values increased, with maximal change seen in lateral and Gastrocnemius Medial, and minimal (although positive) change in Tibialis Anterior.

Regarding T2, there was a percent gain of muscle T2 values after exercise for all muscles analyzed. Tibialis Anterior had an increase in T2 of 0,01%, Gastrocnemius Medial 4,56%, Soleus 3,86% and Gastrocnemius Lateral 2,93%.

The percentage gains for the different muscles and the different techniques are shown in Figure 2.

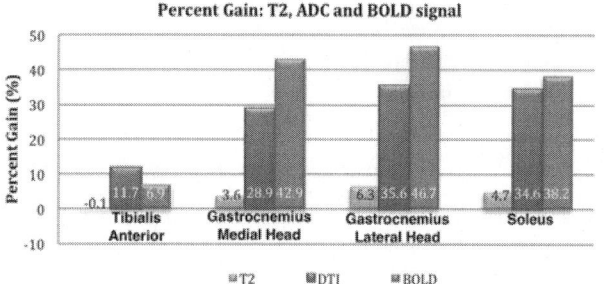

Fig. 2 Percentage gains for the different muscles and the different techniques

IV. CONCLUSIONS

As we can see there is a good agreement for the different techniques for the type of exercise performed: the soleus and gastrocnemius showed the greatest change in ADC, BOLD signal and T2. The ADC and eigenvalues increase shows us that restriction of water diffusion decreased on all directions. The BOLD signal change is consistent with increased hyperemia, typical with this type of exhaustive exercise. The highest percent gain of T2, observed for Gastrocnemius Medial is also consistent with the fatigue exercise. The different MRI parameters give us different physiological information, which will be used to further study the physiology of exercise.

ACKNOWLEDGMENT

We acknowledge funding from FCT through a grant PTDC/DES/103178/2008.

REFERENCES

1. Heemskerk,A.M et al. Current Medical Imaging Reviews Vol3: 152-160.
2. Galban,C.J. et al. Eur J Appl Physiol 93:253-262.
3. Noseworthy, M.D., Bulte, D.P., Alfonsi, J., 2003, Semin Musculoskel R, 307-315
4. Noriyuk, T.T2 mapping of muscle activity using Ultrafast Imaging. (2011) Magn Reson Med Sci, Vol. 10, 2:85-91

Address of the corresponding author:

Author: Mário Forjaz Secca
Institute: Departamento de Física, Faculdade de Ciências e Tecnologia, Universidade Nova de Lisboa,
Street: Quinta da Torre, 2829-516 Caparica
City: Caparica
Country: Portugal
Email: mfs@fct.unl.pt

Part IV
Biomedical Imaging and Processing

Design and Implementation of a Bipolar Current Source for MREIT Applications

H.H. Eroglu[1,2], B.M. Eyüboglu[2], and C. Göksu[2]

[1] TSK Rehabilitation and Care Center, Ankara, Turkey
[2] Department of Electrical and Electronics Engineering, Middle East Technical University, Ankara, Turkey

Abstract—In this study, an adjustable bipolar current source topology for magnetic resonance electrical impedance tomography (MREIT) applications is designed and implemented. The implemented current source is composed of a microcontroller and digital to analog converter (DAC) based signal generator, sourcing and sinking current mirrors and power electronics converters. By means of the signal generator topology, amplitude and pulse-width of the output current is adjusted and the generated waveform is applied to the load by utilizing current mirrors. The high voltage requirement of the current supply is provided by the power electronic converters. The overall design is easy to built-up and implemented with low-cost components. Although the implemented current supply is designed for MREIT applications, it could also be used in various biomedical applications where an adjustable bipolar current supply is necessary.

Keywords—Bipolar current supply, microcontroller, digital to analog converter, power electronics converter, current mirror.

I. INTRODUCTION

Magnetic resonance electrical impedance tomography (MREIT) is an imaging modality that provides cross sectional conductivity distributions of the objects being investigated [1]. By means of investigating the conductivity images of biological tissues, valuable diagnostic information could be obtained [2]. For example, studying the conductivity contrast seems to be an effective tool for detecting pathological conditions in biological tissues [3]. MREIT is based on injecting electrical current during a magnetic resonance imaging (MRI) experiment. The injected current induces a magnetic flux density in the direction of the main field of the MRI system, and the induced magnetic flux density results in a phase change in the signal acquired by the MRI system. Therefore, by utilizing the MRI phase images and known applied current, conductivity distributions of the objects could be obtained which is also known as the inverse problem of MREIT [1].

In order to conduct MREIT experiments, a current source compatible with various MRI systems is necessary and it should provide current in pulse form which should also be in harmony with the applied MRI pulse sequence. Furthermore, the amplitude and timing of the current source should be adjusted with high resolution, since the artifacts in the applied current waveforms will degrade the quality of MR phase images [4]. In this study, an adjustable bipolar current source topology which could be used in MREIT applications is designed and implemented. The implemented current source is composed of a microcontroller and digital to analog converter (DAC) based signal generator topology, sourcing and sinking current mirrors, and, power electronics converters. By means of the signal generator structure, amplitude and pulse-width of the output current is adjusted and the generated signal waveform is applied to the load by utilizing the current mirrors. The high voltage requirement of the current source is provided by the power electronic converters. The overall design is easy to built-up and implemented with low-cost components. Although the implemented current source is designed for MREIT applications, it could also be used in various biomedical applications where an adjustable bipolar current source is necessary.

After explaining the main aspects of the study, the remaining parts of the paper could be expressed as follows. In the next section, named as operating principles and hardware topology, the implementation details are presented. Consequently, the experimental performance of the current source with a 10 kilo-ohms load is evaluated in the results and discussion section.

II. OPERATING PRINCIPLES AND HARDWARE TOPOLOGY

In MREIT experiments, the use of spin echo pulse sequence with bipolar current is very common [1]. The basic MREIT pulse sequence is shown in Figure 1. By means of the pulse sequence shown in Figure 1, an injected current related phase (Φ_c) is introduced in MR phase images as shown in (1). By using (1), the magnetic flux density due to the applied current (B_{jz}) could be calculated [1]. By utilizing B_{jz} information and Ampere's law, current density (J) expression could be obtained as shown in (2).

$$\phi_c = \gamma \times B_{jz}(x,y) \times T_c \qquad (1)$$

$$J = \mu_0^{-1}(\nabla \times B_j) \qquad (2)$$

Exact current density information could be obtained by using the applied current related flux density (B_j) in all

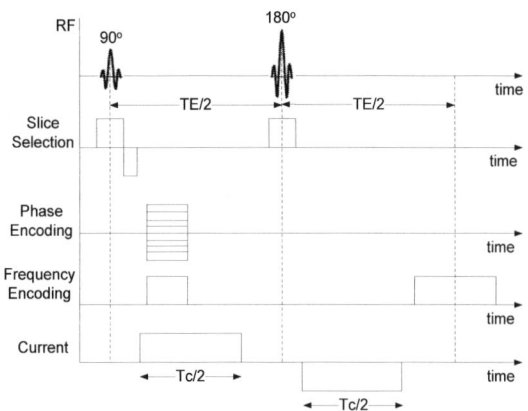

Fig. 1 MREIT pulse sequence.

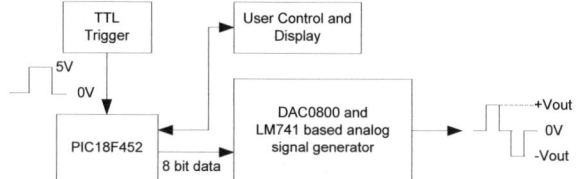

Fig. 3 Block diagram of the signal generator.

orientations. However, in practice only the z-oriented flux density (B_{jz}) could be used. There are different methods for obtaining conductivity distributions from the measured B_{jz} data [2]. Most of the methods use "J" or "B_{jz}" data in addition with boundary conditions, and solve a numerical electromagnetic problem iteratively in order to determine the unknown conductivity distribution. In this study, it is focused on how to construct a bipolar current source topology that can be used in MREIT applications. As mentioned in the introductory part, the proposed current source topology is composed of a signal generator, sourcing and sinking current mirrors, and, power electronics converters. The block diagram of the proposed current source is shown in Figure 2.

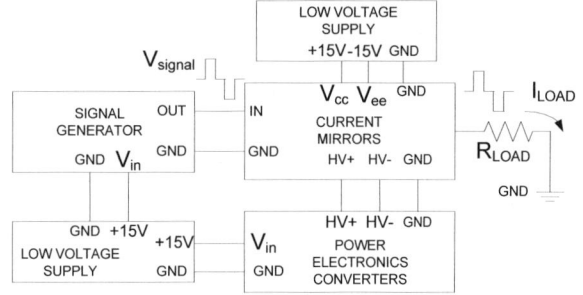

Fig. 2 Block diagram of the bipolar current source topology.

As shown in Figure 2, the first part of the current source topology is the signal generator which is composed of a microcontroller, DAC and an op-amp. The block diagram of the signal generator is shown in Figure 3. As shown in Figure 3, a PIC18F452 microcontroller is used for generating the eight bit binary sequence for the DAC circuit.

The microcontroller is also used for displaying the system status and user controls. The user could adjust the amplitude and pulse-width of the output current via this interface. The current waveform shown in Figure 1 is programmed inside the microcontroller. The amplitude of the waveform can be adjusted in 1mA step-size and the pulse-width can be adjusted in 1ms step size via the push-buttons and the selected parameters are displayed in an alphanumeric liquid crystal display (LCD). The generated current waveform is triggered by the rising edge of a TTL signal which represents the firing of 90° radio frequency (RF) pulse. The eight bit digital data is converted to an analog signal in ±12V range via a DAC circuit. For the DAC circuit, DAC0800 and LM741 based symmetric offset binary operation technique [5] is utilized. The output of the signal generator is fed to the input of the current mirror topology which is shown in Figure 4. As shown in Figure 4, the topology is composed of sourcing and sinking current mirrors triggered by op-amp based buffer circuits [6]. When the input signal (V_{signal}) is positive, npn type BJT Q_{b1} is triggered and conducts current. The amplitude of the current is shown in (3).

$$I = \frac{V_{signal}}{R_{adj1}} \quad (3)$$

The adjusted current in the buffer circuit is copied by the sourcing current mirror which is composed of pnp type BJTs "$Q1, Q2, Q3$". The current is applied to the load between the high voltage (HV+) and ground (GND). The amplitude of the output current is limited by the HV+ voltage level as shown in (4). Similar argument is applicable for the sinking current. For this case, the input signal level is negative and pnp type BJT, "Q_{b2}" is triggered and provides the negative current reference. The negative current is copied by the sinking current mirror composed of npn type BJTs "$Q4, Q5, Q6$" and applied to the load between GND and negative high voltage level (HV-).

$$I_{load} = \begin{cases} I \dots\dots\dots\dots\dots\dots I \times R_{load} < V_{HV+} \\ V_{HV+}/R_{load} \dots I \times R_{load} > V_{HV+} \end{cases}, \quad I = \frac{V_{signal}}{R_{adj1}} \quad (4)$$

As seen in (4), the current source topology requires positive and negative high voltage supplies (HV+ and HV-). For example, in order to source 20 mA to a 5kΩ load, at least a voltage level of 100V is required according to (4). This

Design and Implementation of a Bipolar Current Source for MREIT Applications

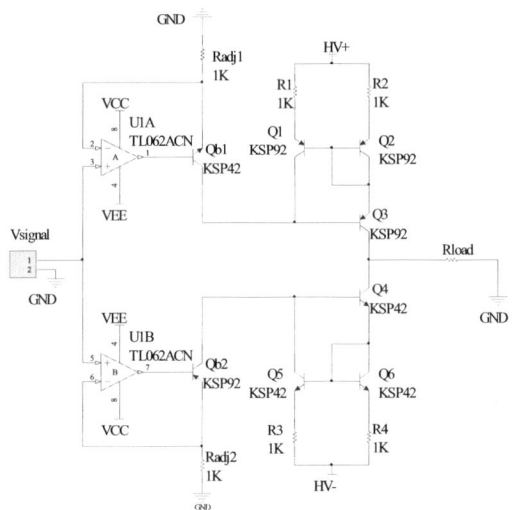

Fig. 4 Schematic diagram of the current mirror topology.

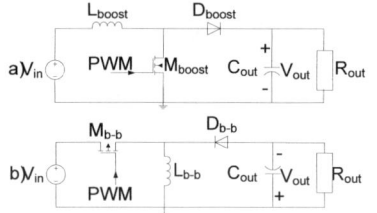

Fig. 5 Power electronics converters, (a) boost converter, (b) buck-boost converter

$$S = \begin{cases} PWM(D = 0.9),...if\ Vout < V_{ref} \\ 0,........................if\ Vout \geq V_{ref} \end{cases} \quad (7)$$

PI controller. Afterwards, the output of the PI controller (*d*) is fed into a limiter and duty cycle (*D*) value is obtained. Consequently, *D* is fed into a PWM block and the resultant switching signal (*S*) is applied to the buck-boost converter.

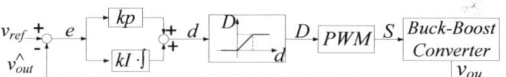

Fig. 6 Block diagram of the linear control for buck-boost converter

The implemented signal generator, current mirrors, and power electronics converters are integrated in order to construct the bipolar current source topology which is shown in Figure 7. In the next section, experimental results of the current source are presented.

III. RESULTS AND DISCUSSION

The output of the current source is investigated with a 10 kΩ load. A TTL signal is used as a trigger source for the current pulses. Simulation results for different current amplitude (I_{ref}) and pulse-width references (T_c) are shown in Figure 8. When the waveforms shown in Figure 8 are investigated, it is seen that programmed I_{ref} values are exactly seen at the output since the amplitude of the measured output voltage is 20V and 40V for the I_{ref} values of 2mA and 4mA, respectively. The ripple levels at the measured output voltage are below 1% of the reference voltages. The programmed T_c values are also clearly observed in the output waveforms.

When the experimental results are considered, it is seen that the current waveform designed for the spin echo pulse sequence could be generated with the implemented current source topology easily. The user could adjust the amplitude and pulse-width of the waveform and the adjustments are seen at the output current with high resolution (1 mA, 1ms

requirement is provided by the power electronic converters. In this study, boost converter is used for HV+ requirement, whereas, buck-boost converter is used for HV- requirement which are shown in Figure 5. By switching the power semiconductors properly (M_{boost}, M_{b-b}), boost converter provides output voltage that is higher than the input voltage. On the other hand, buck-boost converter could provide output voltage that could be higher or lower than the input voltage. In addition, the buck-boost converter's output voltage has negative polarity. For a standard pulse width modulation (PWM) switching signal, the relationship between the input voltage and output voltages of the boost, and buck-boost converters are shown in (5-6), respectively [7]. In (5-6), "D" represents the duty cycle of the PWM signal and it is a value between zero and one.

$$V_{out\_boost} = (1-D)^{-1} \times V_{in} \quad (5)$$

$$V_{out\_buck\_boost} = D \times (1-D)^{-1} \times V_{in} \quad (6)$$

For the control of boost converter hysteresis control is implemented in a PIC18F452 microcontroller. With this control, the output voltage of the converter (V_{out}) is compared with the reference voltage (V_{ref}) which is set to 120V, and the switching signal (*S*) is varied according to (7) [8].

For the control of the buck-boost converter, a proportional integral (I) type linear control is implemented in a PIC18F452 microcontroller. The block diagram of the implemented linear control is shown in Figure 6. As shown in Figure 6, at first, V_{out} is measured and compared with V_{ref} which is set to -120V. Then, the error signal (*e*) is fed into a

Fig. 7 Implemented bipolar current source topology.

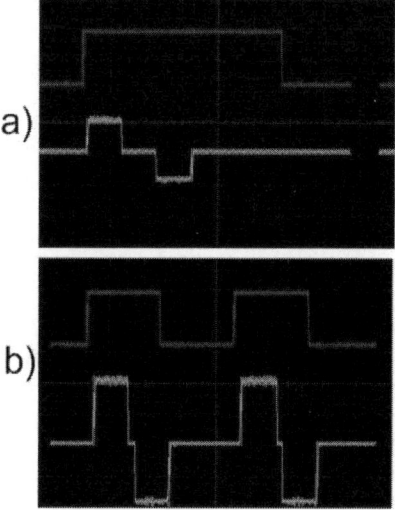

Fig. 8: Oscillograms for the output current, a) I_{ref}=2mA, T_c=2s, (20V/div,1s/div), b) I_{ref}=4mA, T_c=1s, (20V/div,500ms/div).

step size) and low noise. The magnitude of the output current ripple is lower than 5 % of the applied current reference. From the user interface, the current amplitude can be adjusted with 1mA step size. However, with the proposed DAC based signal generator topology, the amplitude of the output current could be adjusted with a step size as small as 0.1 mA. By utilizing the proposed current source topology, the undesirable effects due to the output current waveform degradations could be minimized in MREIT experiments. The proposed current source topology provides single channel operation. However, in most of the MREIT applications, multi-channel current supplies are required in order to obtain current distributions in different orientations especially in anisotropic reconstruction applications. Therefore, it is needed to increase the number of channels of the current source. This could be achieved easily, by using the current mirror topology for each channel and using ground referenced switches for the negative electrode of each channel.

IV. CONCLUSION

In this study, an adjustable bipolar current source for MREIT applications is designed and implemented. By using the proposed design, the amplitude and pulse-width of the output current can be controlled with high resolution (1 mA, 1 ms step size) and low noise. It is observed that the ripple level in output current is below 5 % of the reference current. As it is mentioned, the proposed hardware topology is designed for MREIT applications. However, it could be used in various biomedical applications where bipolar current sources are required.

ACKNOWLEDGMENT

This study is partially funded by METU-BAP 07-02-2012-101 and TUBITAK 113E301.

REFERENCES

1. Eyüboğlu B.M , (2006) Electrical Impedance Imaging: Injected Current Electrical Impedance Imaging,. In WILEY-Encyclopedia of Biomedical Engineering (Metin Akay, ed.), Vol.2, pp.1195-1205
2. Bodenstein M, Davi M, Markstaller K, (2009) Principles of electrical impedance tomography and its clinical application, Crit Care Med., Vol:37, No:2, pp:713-24
3. Hahn G, Just A et al. (2006) Imaging pathologic pulmonary air and fluid accumulation by functional and absolute EIT, Physiological Measurement Vol:27 No:5 pp:187-198
4. Kim Y.T, Yoo P.J. et al., (2010) Development of a low noise MREIT current source. J. Phys.: Conf. Ser. 224 012150 DOI:10.1088/1742-6596/224/1/012150
5. Texas Instruments, (2006) DAC0800/DAC0802 8 bit digital to analog converters datasheet, SNAS538B
6. Wu H.C., Young S.T. et al. (2000) A versatile multichannel direct-synthesized electrical stimulator for FES applications. Instrumentation and Measurement, IEEE Transactions on Vol:51 No:1 pp:2-9
7. Mohan N., Undeland T., Robbins W.P. (2003) Power electronics, converters applications, and design. Third edition, USA.
8. Eroğlu H.H. and Eyüboğlu B.M., (2012) Design and implementation of a monopolar constant current stimulator. Proc of BİYOMUT 2012 pp 283-286

Author: Hasan Hüseyin EROĞLU
Institute: TSK Rehabilitation and Care Center, Ankara / Turkey
Email: hheroglu@gata.edu.tr

Automatic Detection of Heart Center in Late Gadolinium Enhanced MRI

K. Engan[1], V. Naranjo[2], T. Eftestøl[1], L. Woie[3], A. Schuchter[1], and S. Ørn[3]

[1] Dept. of Electrical Eng. and Comp. Science, University of Stavanger, Stavanger, Norway
kjersti.engan@uis.no
[2] Labhuman, I3BH, Universitat Politècnica de València, Valencia, Spain
[3] Stavanger University Hospital, Stavanger, Norway

Abstract— **This paper presents and compares two methods for automatic detection of the center of the heart in Late Gadolinium Enhanced Cardiac Magnetic Resonance Imaging (LGE-CMR). We consider this as the first step in a process to automatically segment the myocardial muscle in the MRI slices. Most methods for myocardial segmentation in current use are semiautomatic. A region of interest, or a point inside the heart, has to be manually fed to the algorithm. In this work all the consecutive slices of the left ventricle, short axis view, is taken into account when finding the most probable heart center, and no manually cropping of a region of interest is done. One of the methods uses a combination of morphological pre processing and the circular Hough transform. The other method uses morphological preprocessing and grey level distance function. Both methods use information from all consecutive slices to improve the results compared to a slice-by-slice approach. We define the Heart Center Index (HCI) as a performance measure, and report results on a set of 54 patients with 6-12 slices per patient.**

Keywords—**LGE-CMR, Hough transform, morphology, gray level distance, automatic segmentation.**

I. INTRODUCTION

Automatic segmentation of the myocardial muscle in Late Gadolinium Enhanced Cardiac Magnetic Resonance (LGE-CMR) images of patients that have had myocardial infarction is a difficult problem. The contrast agent (gadolinium) is used to make the scarred areas visible in the MRI images, and they will appear brighter than the healthy myocardial muscle. In many hospitals today the segmentation of the myocardial muscle, as well as the segmentation of the scarred areas, are performed manually or semiautomatically by expert cardiologists. This can be time-consuming work, and the results will have a degree of inter- and intra-observer variability. There have been attempts to solve the problem of automatic segmentation of the myocardial muscle reported in the literature, but to the authors knowledge a fully automatic method for segmenting LGE-CMR images does not exist today. Bruijne and Nielsen [1] presented a semiautomatic method that requires manually placed landmarks on the epi- and endocardial contour. Spreeuwers and Breeuwer [2] presented a semiautomatic method using coupled active contours, and Folkesson et. al. [3] presented a method for segmentation of the myocardial muscle in LGE-CMR using statistical classification and geodesic active regions. They claim their method is fully automatic but at the same time they seem to use a cropped version of the image quite close to the heart muscle and with the heart centered. Heiberg et.al. [4] provide a freely available software for cardiovascular image analysis, including a segmentation part. Some initial input is necessary, providing a region of interest around the heart.

This work is addressing the problem of automatically finding the (approximated) center of the left ventricle in the short axis slices using the stack of available short axis slices to utilize the 3D information available. This is fully automatic with the entire, non-cropped, version of the LGE-CMR images as input. We see this as an important first step in an algorithm for automatic segmentation of the myocardial muscle. The authors are not aware of any published work addressing the same problem for comparison, but the paper will compare two different algorithms based on different ideas.

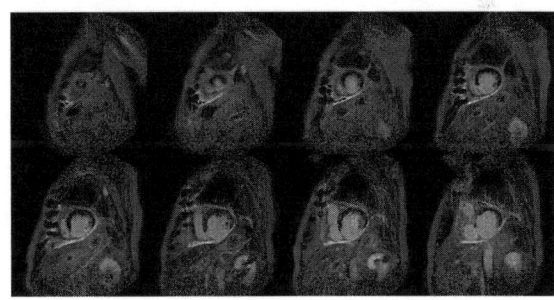

Fig. 1 Consequtive MRI slices of an example patient with myocardial scar.

The Department of Cardiology at Stavanger University Hospital provided LG enhanced CMR images of 54 patients, all with previous myocardial infarction. An example of all the images from one patient is seen in Figure 1. CMR was performed with a 1.5 T Philips Intera R 8.3. A gadolinium based contrast agent was administered intravenously at a dosage of 0.25 mmol/kg. LGE images were acquired 10-15 minutes later, using an inversion recovery prepared T1 weighted gradient echo sequence with a pixel size of $0.82 \times 0.82 \text{mm}^2$, covering the whole ventricle with short-axis slices of 10 mm thickness, without inter-slice gaps. Healthy myocardium appears very dark in CMR images, and concentrating on the left ventricle short axis slices, the myocardium has the shape of a ring on each slice. However, the edges of the heart in LGE-CMR images where the patient has a scar in the myocardium are sometimes very weak or non-existing since the scarred areas will take intensity levels close to the blood pool or the surrounding areas. In some slices the scar can cover almost all the myocardium, whereas other slices can have a much smaller scarred part, and some slices have no visible scar. Thus, in both our proposed methods we look both at the slice by slice information and also information from the collection of the short axis slices to improve the accuracy of the algorithms.

II. Proposed Methods

A. Mathematical Definitions

Let $f^i(x)$, $i = 1 \ldots N_{slice}$ represent the MRI slices of a patient. $x = [x_{row}\ x_{column}]^T$ is the pixel position, and i represents the slice number. We define the morphological operations [5] of dilation and erosion as $[\delta_B f](x)$ and $[\varepsilon_B f](x)$ respectively, where $B(x)$ is a structuring element, and opening and closing as $\gamma_B(f) = \delta_B(\varepsilon_B(f))$ and $\phi_B(f) = \varepsilon_B(\delta_B(f))$ respectively. Morphological center, $(\beta(f))$, is a morphological filter used in our preprocessing algorithms defined as:

$$\beta_{B_{mc}}(f)(x) = \min(\max(f(x), f_1(x)), f_2(x)) \quad (1)$$

where $f_1(x) = \gamma_{B_{mc}}(\phi_{B_{mc}}(\gamma_{B_{mc}}(f)))$ and $f_2(x) = \phi_{B_{mc}}(\gamma_{B_{mc}}(\phi_{B_{mc}}(f)))$. The morphological gradient of an image $f(x)$ is defined as: $\rho(x) = \delta_{B_\rho}(f) - \varepsilon_{B_\rho}(f)$, where B_ρ is the elementary structure element. The algebraic opening called area opening is defined as: $\gamma_\lambda = \max_i\{\gamma_{B_i}|B_i \text{ is connected and card}(B_i) = \lambda\}$, and it will, in short, remove bright objects smaller than λ pixels. In geodesic transformations two images are required, the reference $f(x)$ and the marker $g(x)$. The morphological reconstruction or reconstruction by dilation is the successive geodesic dilation of the marker regarding the reference up to idempotence:

$$\gamma^{rec}(g,f) = \delta_g^{(i)}(f) \text{ so that } \delta_g^{(i+1)}(f) = \delta_g^{(i)}(f), \quad (2)$$

where $\delta_g^{(i)}(f) = \delta_g^{(1)} \delta_g^{(i-1)}(f)$ and $\delta_g^{(1)} = \min(\delta_B(f), g)$. Using the morphological reconstruction, we can define the close-hole operator which fills all holes in a gray-scale image $f(x)$ that do not touch the image border, $f_{border}(x)$, defined as: $f_{border}(x) = f(x), \text{for } x \in \{x_{border}\}$ and 0 everywhere else.

$$\phi^{ch}(f) = [\gamma^{rec}(f^c(x), f_{border}(x))], \quad (3)$$

where $f^c(x)$ denotes the inverse of the image $f(x)$. The reconstruction can be used to suppress all image objects connected to the image borders, $f_{border}(x)$, defining the clear-border operation: $\phi^{cb}(f) = f - \gamma^{rec}(f(x), f_{border}(x))$.

B. Method 1 (M1): Circular Hough transform

The shape of the left ventricle is approximately as a ring in slices with no visible scar, and a ring with occlusion in slices with visible scar, thus the inner wall (endocardium) and outer wall (epicardium) might possibly be detected by the circular Hough transform (CHT). The Hough transform was originally suggested for binary (edge) images, but can also be used for grey level images [6],[7].

Algorithm 1. Finding heart center based on CHT, M1.

Data: CMR images $f^i(x) \in \mathscr{R}^{N \times M}, i = 1, \ldots N_{slice}$
Result: Coordinates of heart center,
$C_{M1}^i \in \mathscr{R}^2, i = 1, \ldots N_{slice}$

initialization: $B_{mc}, \lambda, B_c, B_o$;
for $i \leftarrow 1$ **to** N_{slice} **do**
$\quad f_{prep}^i(x) \leftarrow \gamma_\lambda((\beta_{B_{mc}}(f)))$;
$\quad \rho^i(x) \leftarrow \varepsilon_{B_\rho}(f_{prep}^i) - \delta_{B_\rho}(f_{prep}^i)$;
$\quad \rho^i(x) \leftarrow \phi^{cb}(\rho^i(x))$;
$\quad I_{CHTm}^i \leftarrow CHT(\rho^i(x), [r_{min}, r_{max}])$;
end
$I_{CHTm} \leftarrow \sum_{i=1}^{N_{slice}} I_{CHTm}^i$;
$I_{CHTsc} \leftarrow normalize(\phi^{cb}(\phi_{B_o}(\gamma_{B_c}(I_{CHTm}))))$;
for $i \leftarrow 1$ **to** N_{slice} **do**
$\quad I_{CHTf}^i \leftarrow I_{CHTsc} \times f^i(x)$;
end
$I_{Center} \leftarrow \sum_{i=1}^{N_{slice}} I_{CHTf}^i$;
$I_{Center} \leftarrow I_{Center} \times P_{prior}$;
for $i \leftarrow 1$ **to** N_{slice} **do**
$\quad I_{Center}^i \leftarrow I_{Center} \times f^i(x)$;
$\quad C_{M1}^i \leftarrow \max(I_{Center}^i)$;
end

The outline of the algorithm is as follows: Firstly all slices of a patient are jointly normalized. A morphological preprocessing is conducted by applying the morphological center function to reduce noise on each image slice, $f^i(x)$, followed by the area opening operator. Thereafter the morphological gradient is found, and the clear border operation is performed on the gradient images, $\rho^i(x)$. The CHT is performed for the known area of radius-of-interest based on a priori knowledge of the image resolution and typical heart size. $I_{CHTm}^i = CHT(\rho^i(x), [r_{min}, r_{max}])$. We collect the output in an image I^i_{CHTm} with the same size as the input image $\rho^i(x)$ and with the metric output from CHT placed at the corresponding center position, indicating how strong the possibility of that position being a center of a circle with a radius within $[r_{min}, r_{max}]$. There are other circular shaped parts that will be visible in some of the slices as well. Typically this will vary from slice to slice. The position of the heart center will not vary much between consecutive slices. Thus averaging over the output from the Hough transform over all the slices will produce an output, I_{CHTm} with a stronger probability of revealing the true (averaged) center point. To connect close possible center-points and remove areas connected to the border I_{CHTm} is exposed to a morphological closing, followed by opening and by the clear border operation and a normalization giving I_{CHTsc}. Some slices in some patients will have other circular shaped parts, often dark. To exclude these the Hademar product (denoted by $\times$) of I_{CHTsc} and a scaled version of the normalized slices, $f^i(x)$ is found and averaged, favoring circular shapes with bright centers. The output is subsequently Hademar multiplied by P_{prior}, a 2D Gaussian function modeling the prior probability of the heart being somewhat in the center of the slices. In a very last step, to make it possible to find different heart centers for different slices since this might vary a little, a new Hademar multiplication with the individual slices is done, resulting in I^i_{Center}, where the coordinates of the maximum value is interpreted as the heart center coordinates of that slice for method 1: C^i_{M1}. The algorithm outline is seen Algorithm 1.

C. Method 2 (M2): Gray-Level Centroid

The core idea of the second method is to take advantage of the fact that the blood pool at the inside of the endocardium is a quite large and bright area in the CMR images, as well as being relatively centered. This can be exploited by using the gray-level centroid (GL-Centroid, GLC). The GL-centroid of a gray-level image can be calculated based on the generalized distance function (GDF)[11]. The idea of this centroid is to take into account not only the pixel position but also its gray level. This has the advantage of avoiding the process of thresholding or segmentation needed to calculate the centroid of a binary region. The GDF to a set $g(x)$ in an image $f(x)$ is denoted $d_f(g)$. The GDL-centroid of $f(x)$ is the position where $d_f(g)$ is maximum, being $g(x) = f_{border}(x)$, where $f_{border}(x)$ is the image border.

Method 2, depicted in Algorithm 2, is a three-step algorithm based on the gray-level distance function. To assure the inside of the endocardium is the largest bright area in each slice, the CMR slices ($f_i(x)$) are firstly subjected to a

---

Algorithm 2. Finding heart center based on GLC, M2.

Data: CMR images $f^i(x) \in \mathcal{R}^{N \times M}, i = 1, \ldots N_{slice}$
Result: Coordinates of heart center,
$\quad C^i_{M2} \in \mathcal{R}^2, i = 1, \ldots N_{slice}$
for $i \leftarrow 1$ **to** N_{slice} **do**
$\quad CH \leftarrow \phi^{ch}((f^i(x))^c)$ Close-holes of negative ;
$\quad RCH \leftarrow CH - (f^i(x))^c$;
$\quad GDF_i \leftarrow d_{RCH}(f_{border}(x))$ Gray-level distance ;
$\quad C_i(l) \leftarrow \arg_x[(GDF_i) \geq 0.8 * \max(GDF_i)]$ Centroid ;
end
$GDF \leftarrow \sum_{i=1}^{N_{slice}} GDF_i$;
$C_o \leftarrow \text{argmax}_x(GDF)$ Global centroid ;
for $i \leftarrow 1$ **to** N_{slice} **do**
$\quad l_0 \leftarrow \text{argmin}_l(\text{distance}(C_o, C_i(l)))$;
$\quad C^i_{M2} \leftarrow C_i(l_0)$;
end

---

preprocessing step based on the close-holes operator, obtaining the close-hole residue, RCH. Thereafter the GDF of RCH to the image borders is calculated per slice, giving GDF_i. The position of the maximum of GDF_i is defined as the GLC, but to increase robustness all the positions where this distance is higher than 80% of its maximum is collected in a position vector $C_i(l)$ per slice. In the final step the global centroid C_0 is calculated as the position of the maximum average GDF over all the slices. Finally, the position of each slice centroid is estimated as the closest point to C_0 among the possible found centroids $C_i(l)$, resulting in C^i_{M2}.

D. Evaluation Method

To evaluate the performance of the proposed algorithms for finding the heart center a ground truth marking is needed. The endo- and epicardium were manually marked by two expert cardiologists from Stavanger University Hospital. The true center, C^i_e, is defined as the centroid of the epicardium marking, and the approximated radius, R^i_{epi} as the radius of the circle that would have the same area as contained within the epicardium, both defined for each slice. We define the Heart Center Index (HCI) for Method x as:

$$HCI^i_{Mx} = \frac{\sqrt{(C^i_{Mx}(1) - C^i_e(1))^2 + (C^i_{Mx}(2) - C^i_e(2))^2}}{R^i_{epi}} \quad (4)$$

$$HCI_{Mx} = \frac{1}{N_{slice}} \sum_{i=1}^{N_{slice}} HCI^i_{Mx} \quad (5)$$

A perfect match with the ground truth center will give HCI=0. As long as the proposed center lies within the epicardium, $HCI \in [0,1]$, but if the proposed center lies outside the heart, $HCI > 1$.

III. Experiments and Results

All experiments were conducted in Matlab. We have available 54 patients, all with myocardial scars. The Circular Hough transform was done using phase-coding with the available Matlab function *imfindcircles*. The parameters of Method 1 was empirically chosen to be: $B_{mc} = square(10)$, $\lambda = 200$, $B_c = disk(5)$, $B_o = disk(2)$. The HCI results for all the patients for both methods are plotted in Figure 2. A zoomed in region of all the slices of an example patient is depicted in Figure 3. Here the true center (centroid from the doctors markings of the epicardium) is seen as a green circle, results from M1 is plotted as red stars, and from M2 as yellow diamonds. The total mean and standard deviation of the calculated HCI for the two methods are as follows: $\overline{HCI}_{M1} = 0.275$, $\sigma_{HCI_{M1}} = 0.113$ and $\overline{HCI}_{M2} = 0.247$, $\sigma_{HCI_{M2}} = 0.265$.

IV. Discussion and Conclusion

The myocardial scar appears bright whereas the myocardial muscle appears dark. Thus the tasks of segmenting the myocardial muscle and of finding the approximated heart center are much harder in patients with large myocardial scars. However, utilizing the information from the consecutive slices, as well as prior knowledge concerning approximated heart size both methods are performing very well in finding the approximated heart center. Method 1 is somewhat less computationally demanding, but more conceptually complex and with more parameters (structuring elements) that needs to be set. Method 2 is more elegant with no parameters, and also produces slightly better average results. However Method 2 fails on one of the patients indicating that it is somewhat less robust than Method 1. We have now ongoing work on segmentation of the myocardial muscle using the reliable heart center from these algorithms as an important step.

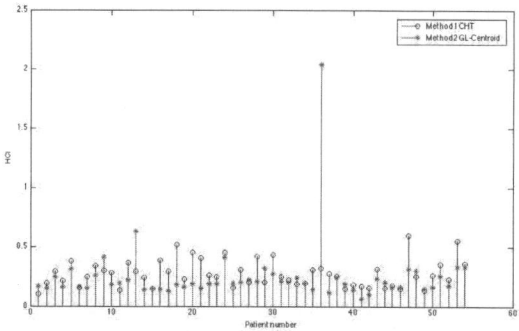

Fig. 2 Results plotted as HCI versus patient number. Blue circles: Method 1, CHT based. Red stars: Method 2, GL-Centroid based.

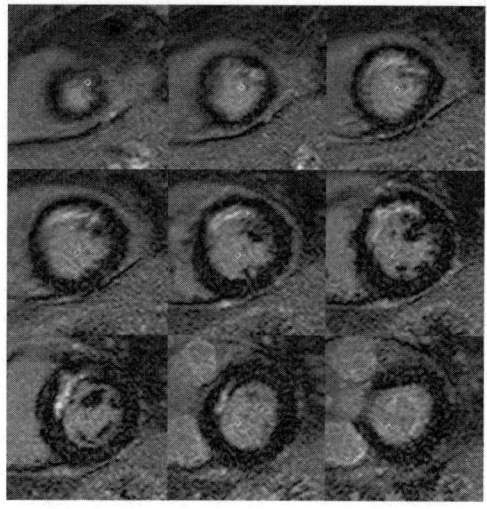

Fig. 3 All slices from one patient, zoomed in. Green circle: ground truth, red star: M1 (CHT based), yellow diamond: M2 (GLC based).

References

1. Bruijne M., Nielsen M.. Shape particle filtering for image segmentation in *Proc. of MICCAI*;3216:168–175 2004.
2. Spreeuwers L., Breeuwer M.. Detection of left ventricular and epi- and endocardial borders using coupled active contours in *Proc. Comput. Assisted Radiol. Surg.*:11471152 2003.
3. Folkesson J., E.Samset , R.Y.Kwong , C.F.Westin . Unifying Statisti-cal Classification and Geodesic Active Regions for Segmentation of Cardiac MRI *IEEE Transactions on Information Technology in Bi-omedicine*. 2008;12.
4. Heiberg E., Sjgren J., Ugander M., Carlsson M., Engblom H., Arheden H.. Design and Validation of Segment a Freely Available Software for Cardiovascular Image Analysis in *BMC Medical Imaging*;10:1 2010.
5. Serra Jean. *Image Analysis and Mathematical Morphology*;1. London: Ac. Press, 1989.
6. R.O.Duda , Hart P.E.. Use of the Hough transformation to detect lines and curves in pictures *Commun. Assoc. Comput. Match.* 1972;15: 11–15.
7. Yuen H.K, Princen J., Illingworth J., Kittler J.. Comparative study of Hough transform methods for circle finding *Image and Vision Computing*. 1990;8:71–77.

The Application of Highly Accelerated MR Acquisition Techniques to Imaging the Peripheral Vasculature[*]

Stephen J. Riederer[**]

The MR Research Laboratory, Departments of Radiology and Biomedical Engineering, Mayo Clinic,
Rochester MN 55905 USA
riederer@mayo.edu

Abstract—Peripheral arterial disease (PAD) is a major health problem worldwide. Due to its high prevalence as well as the potential for effective treatment it is critical to diagnose and characterize PAD. Methods based on magnetic resonance imaging (MRI) have long been under study for assessing the vascular system but have been limited because of long scan times and limited spatial resolution. However, just in the last several years a number of techniques have allowed major improvements in the time vs. spatial resolution tradeoff. In this work it is shown that the combination of acquisition techniques, coil technology, and acceleration methods now allows imaging the peripheral vasculature with 1 mm isotropic resolution with acquisitions that can capture the arterial phase of transit of contrast-enhanced blood. Special purpose receiver coil arrays are also used which retain high SNR in spite of the high acceleration. In vivo results are presented to illustrate the general applicability of the method.

I. INTRODUCTION

THE imaging of the vascular lumen, or "angiography" has long been of interest to clinicians. In the last several decades a number of methods have been developed for doing this including digital x-ray imaging [1], ultrasound [2], and computed tomography [3] modalities. Along with these, researchers have worked on MRI-based techniques for vascular imaging or "angiography," and these techniques have included those based on phase differences of the measured signal [4] as well as those attributable to the magnitude of the signal magnetization or so-called "time-of-flight" [5] effects. In the mid-1990s other investigators suggested the use of an MR contrast material for imaging the vasculature, generally subsequent to intravenous administration [6]. This general approach is referred to as "contrast-enhanced MR angiography" or "CE-MRA." The basis of the method is inject a contrast agent into the blood, and the paramagnetic properties of the agent cause a marked reduction in the T1 relaxation time of the blood, allowing significant signal enhancement. CE-MRA, particularly when done with a three-dimensional (3D) acquisition, has many favorable properties, including applicability to all vascular regions of the body, a 3D image format allowing arbitrary direction of projection and selection of sub-volumes, lack of ionizing radiation, small dosages of the contrast material, and negligible risk due to the intravenous vs. intraarterial administration. However, this is offset by the principal intrinsic disadvantage of MRI, namely an extended acquisition time. In initial work the acquisition time was typically the order of a minute for routine imaging of the renal or carotid arteries. Since the introduction of CE-MRA there has been steady improvement in the speed as well as other aspects of the technique. In addition to the desire to provide reduced acquisition time, other early issues included the need to synchronize the data acquisition to the arterial phase of the contrast bolus, and the desirability for minimal venous enhancement. These issues were addressed with a number of improvements, including the development of short repetition time (TR) gradient echo pulse sequences, non-real-time [7] and real-time [8, 9] means for accurate synchronization, and development of special purpose ordering of the data acquisition in which the low-spatial-frequency image components are acquired earliest in the acquisition, allowing extended acquisition times well into the venous phase [10] but without objectionable venous signal. CE-MRA is now widely used in imaging multiple vascular territories in conjunction with single-arterial-phase methods [11].

Due to the broad range of arrival times patient-to-patient in the peripheral vasculature after contrast injection, as well as the potential for unusual filling patterns in cases of vessel occlusion or other pathology, one may often want to image the time-varying nature of the contrast-enhanced blood as it advances through the vasculature. This can allow, for example, determination of unusual filling patterns, identification of the specific feeder vessels of vascular malformations, and choice of a candidate vessel as a potential graft site. However, because time spent in acquiring data for a new image might be alternatively spent in improving the spatial resolution of a single image, forming a new image necessarily prevents use of the time to

---

[*] This work was supported in part by NIH Grants EB000212, HL070620, and RR018898.
[**] Corresponding author.

sample high frequency Fourier coefficients of k-space to generate an image with high spatial resolution. This illustrates the fundamental tradeoff between spatial and temporal resolution as studied, for example, with the TRICKS technique [12].

The purpose of this work is to describe recent developments in time-resolved contrast-enhanced MR angiography (CE-MRA) which now permit very high quality imaging of the peripheral vasculature. The engineering enablers of the method are 2D parallel acquisition techniques which allow high (R>8) acceleration factors as facilitated by high-coil-count multi-element receiver coil arrays specifically designed for 2D parallel acquisition. In the next sections these elements are discussed individually, and their integration is illustrated in results from in vivo studies.

II. METHODS

A. MR Data Acquisition

Images for this work were generated using 3DFT data acquisition. With this method readout is doing every repetition interval along the k_X direction, and phase encoding is done along both directions of the k_Y-k_Z plane. The final 3D image is formed by 3D Fourier transformation of the (k_X, k_Y, k_Z) or "k-space" data into (x, y, z) space. Sampling within the k_Y-k_Z plane is done on a rectilinear grid, and the process is referred to as "Cartesian" sampling. The specific data technique used for this work used the method of Cartesian Acquisition with Projection Reconstruction-like sampling (CAPR) as discussed in detail in [13]. The technical properties of this have been studied in detail using computer-controlled phantoms [14]. Like many methods for performing time-resolved CE-MRA, the technique of view sharing is employed in which not all k-space data are replaced from one image in the time series to the next; i.e. some k-space data are "shared." This in turn provides many options for how the acquired data should be sorted for image formation. However, as shown in Ref. [14], it is critical to have consistent k-space sampling from frame to frame, compact sampling of the k_Y-k_Z-space center within each frame, and sampling of the k-space center should be performed near the end of the frame time so as to minimize artifactual signal appearing in advance of the leading edge of the contrast bolus, referred to as "anticipation artifact." Also central to the ability for short acquisition times are the use of both 2D SENSE acceleration [15], with acceleration factors of R = 8 or higher, as well as 2D homodyne acquisition, which exploits the conjugate symmetry properties in Fourier space of the magnetization [16], allowing another factor of about 1.8 in speed. The combination of 2D SENSE and homodyne provides acceleration factors of R_{NET} = 14.4 or more. The importance of this is that 3D image sets with higher spatial resolution and requiring a shorter acquisition time are now possible vs. performance available a decade ago.

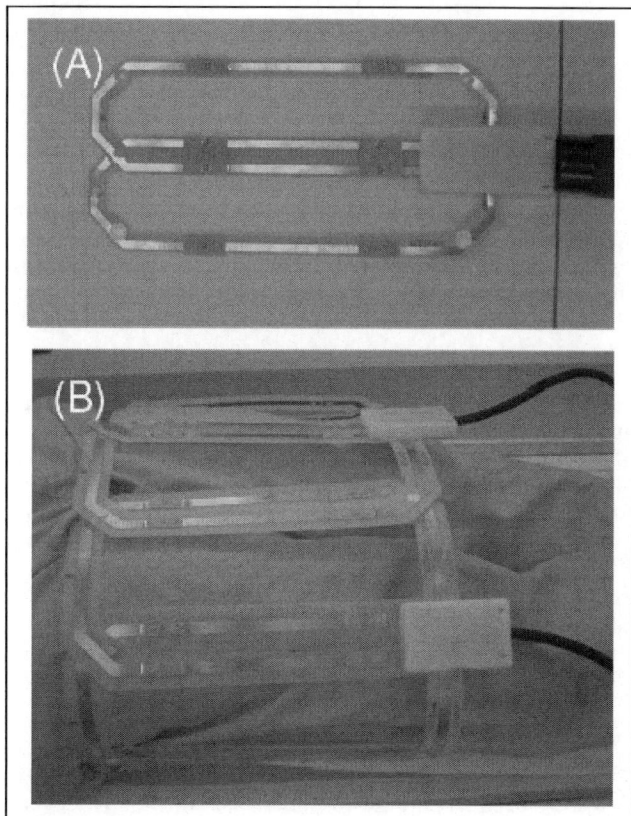

Fig. 1 (A) Photograph of a two-element module for forming a multi-element receiver array. (B) Placement of eight-element array around a patient for imaging the vasculature of the thighs.

B. Receiver Coil Arrays for CE-MRA

The development of receiver coil arrays for the calves for 2D SENSE-accelerated CE-MRA has been discussed previously in Ref. [15]. One major teaching of that work was that because the two directions of the parallel acquisition for the coronal-format CE-MRA acquisition lie within the transverse plane, the placement of coil elements circumferentially around the calves provides both good SNR and low g-factors. Similar coils for other regions of anatomy have also been developed, such as the brain, feet, hands, and abdomen. The design of each coil array is similar. First, a basic element size is specified, with the length chosen to allow imaging along the extent of the superior/inferior(S/I) field-of-view (FOV). The width is selected to provide moderate falloff of sensitivity in the transverse direction into the patient anatomy. Next, two

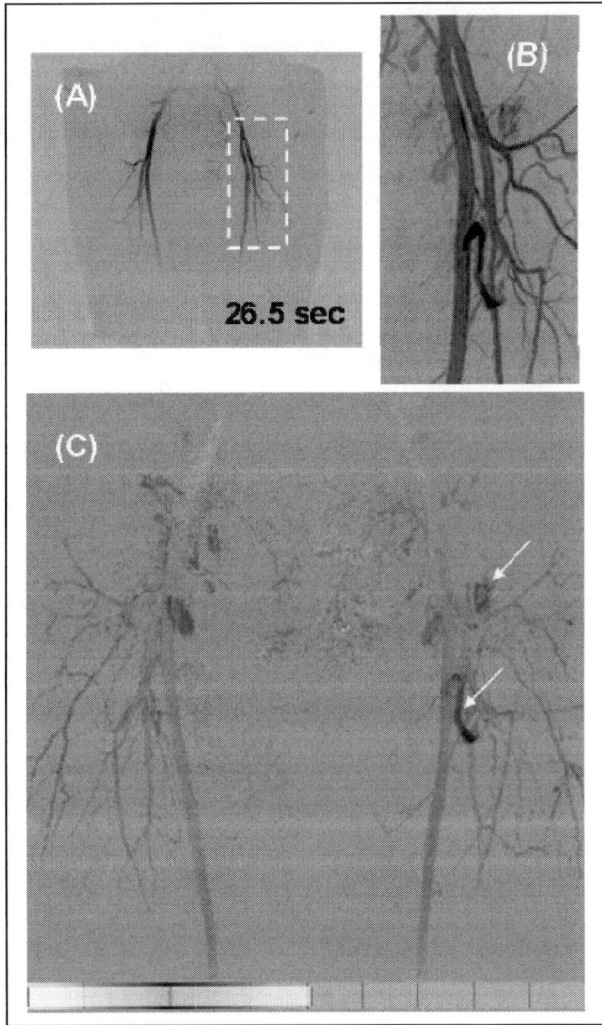

Fig. 2 Results from a patient having a multi-nidus vascular malformation of the thigh. See text for detailed description.

placement of a circumferential array composed of four two-element modules, eight elements total, placed around a patient for imaging the thighs. This approach works effectively for 2D parallel acquisition.

C. In Vivo Experiments

The CAPR acquisition sequence was used in conjunction with 2D SENSE parallel acquisition and 2D homodyne reconstruction in imaging various vascular regions of volunteers as well as in patients for whom CE-MRA was clinically indicated. The study was done under a protocol approved by the Institutional Review Board of our institution. Written consent was obtained from all volunteers. The CAPR acquisition was performed using a fast spoiled gradient echo pulse sequence with the typical parameters: repetition time (TR) 5.85 msec; echo time (TE) 2.7 msec; flip angle 30°; bandwidth ±62.5 kHz; and sampling of a full 400-point echo along the frequency encoding direction.

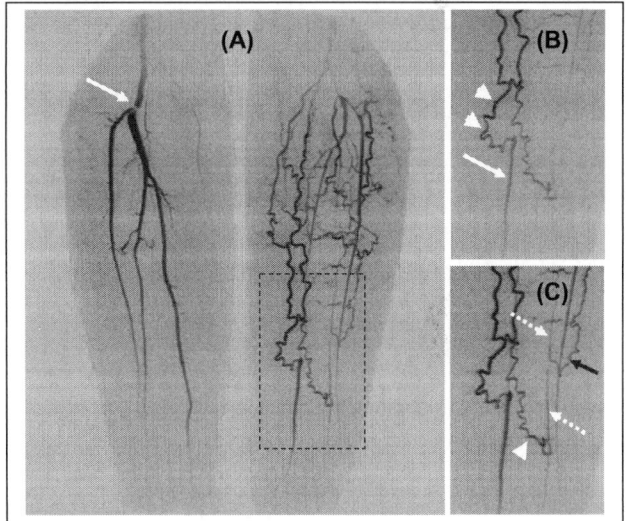

Fig. 3 Results from a patient study of the vasculature of the calves. See text for detailed description.

such elements are overlapped to minimize mutual inductance and firmly attached to form a two-element module. Third, a sufficient number of modules are then attached together into an extended linear array for placement around the region under study, with the leading and trailing modules of the array attached to each other. This arrangement results in a so-called circumferential array. The modular design allows for patients of different sizes. Additional two-element modules are simply inserted into the linear array to allow the circumference to accommodate the patient.

Figure 1 illustrates this for the specific case of imaging the thighs. Fig. 1A shows a two-element module, with the long axis designed to be placed along the long axis of the patient, in this case about 28 cm long. Fig. 1B shows

Prior to the contrast-enhanced run a calibration scan was performed using a similar pulse sequence but with twofold reduction of spatial resolution along both the Y and Z directions vs. that used for the actual CE-MRA run. The information collected in this scan is necessary for performing the SENSE algebraic unfolding. For the contrast-enhanced run for each subject no more than 20 mL of Multihance (gadobenate dimeglumine, Bracco Diagnostics, Princeton NJ) was injected into an arm vein at a rate of 3 mL/sec followed by a 20 mL saline flush also administered at 3 mL/sec. Both were applied using a power injector (Spectris, Medrad, Indianola PA). Prior to the actual injection the CAPR sequence was initiated and at least one contrast-free 3D image was acquired of the region under study. For each subject the CAPR sequence was repeated for several minutes, until 36 image sets were acquired at the chosen frame time interval. Reconstructions were performed on a custom system interfaced to the MRI scanner [16], and typically all reconstructions including Maximum Intensity Projection (MIP) images along the slice select direction were completed within two minutes after the end of the CE-MRA run. The specific multicoil array and the SENSE acceleration were selected according to the targeted study.

III. RESULTS

The CAPR method is now routinely used to image patients suspected of having vascular disease. Sample results from two in vivo studies are shown in the figures. Fig. 2 is from a patient having a vascular malformation in the thighs. Fig. 2A shows the MIP from an image taken from the time series and acquired 26.5 sec post-injection depicting the vasculature. The region of concern is highlighted in the white box. An enlargement of the image made 37.1 sec post-inj shows the region in more detail. Finally, Fig. 2C shows a time-of-arrival (TOA) map of the same region. In this image the arrival time of the contrast material post injection was determined for each pixel in the image as observed in the 3D time series. That time was then converted into a color, and that pixel was then displayed in that color. The color scale used was red-to-blue, with red representing early arrival and blue late arrival. Figure 2B clearly portrays the malformation with the late-arriving blood in the blue color.

Figure 3 shows results from a different patient for whom it was necessary to determine the flow patterns in the calves. Fig. 3A shows a MIP from one of the time frames. A lesion is seen on the patient's right side (arrow) at the distal end of the popliteal artery. Enlargements of a frame made one 5 sec frame earlier (B) and same frame (C) as (A) show how the native left posterior tibial artery (B, arrow) is fed by a collateral vessel (B, arrowheads). Next frame (C) shows how the left peroneal artery (dashed arrows) is fed by collateral vessels (arrowhead and black arrow). This detailed information is allowed by the combination of the high spatial resolution (1 mm isotropic) as well as the good temporal resolution (5 sec frame time, 17 sec temporal footprint).

IV. DISCUSSION

We have described the principal elements of a method for generating high spatiotemporal resolution 3D contrast-enhanced MR angiograms of the peripheral vasculature. The technical enablers of the method are view sharing, 2D acceleration methods using both 2D SENSE and 2D homodyne techniques, customized modular multi-element coil arrays allowing circumferential placement around the patient, all of which allow high speed image acquisition with adequate signal-to-noise ratio (SNR) to permit routine diagnostic interpretation. We have demonstrated the applicability of this method in sample patient studies of the thighs and calves.

Whenever acceleration of this magnitude is performed a major concern is the loss of SNR due to: (i) use of less data in the reconstruction, and (ii) any noise amplification intrinsic to the unfolding process. These appear to be well addressed by the family of receiver coils which can be sized and placed around the targeted vasculature on a patient-specific basis.

Future directions of this work include application to the vasculature of the abdomen, in which case due to the complication of respiratory motion breathhold acquisition is desired. Another ongoing area of study is to perform imaging for more than just one region of the peripheral vasculature; i.e. from abdomen all the way to the feet. This can be accomplished by using high speed reconstruction to allow the operator to view the transit of the contrast material in real time and trigger advance of the patient table. This will necessarily require additional considerations for receiver coil number and placement, as well as acquisition speed beyond that reported here.

In summary, recent developments in CE-MRA offer unprecedented spatiotemporal resolution in imaging the peripheral vasculature, and with these improvements CE-MRA provides considerable advantages over other imaging modalities.

REFERENCES

[1] R. A. Kruger, C. A. Mistretta, T. L. Houk, S. J. Riederer, C. G. Shaw, M. M. Goodsitt, A. B. Crummy, W. Zweibel, J. C. Lancaster, G. C. Rowe, and D. Flemming, "Computerized fluoroscopy in real-time for noninvasive visualization of the cardiovascular system," *Radiology*, vol. 130, pp. 49-57, 1979.

[2] N. Jacobs, E. Grand, and D. Schellinger, "Duplex carotid sonography: criteria for stenosis, accuracy, and pitfalls," *Radiology*, vol. 154, pp. 385-391, 1985.

[3] G. D. Rubin, M. D. Dake, S. A. Napel, C. H. McDonnell, and R. B. Jeffrey, "Three-dimensional spiral CT angiography of the abdomen: initial clinical experience," *Radiology*, vol. 186, pp. 147-152, 1993.

[4] C. L. Dumoulin and H. R. Hart, "Magnetic resonance angiography," *Radiology*, vol. 161, pp. 717-720, 1986.

[5] G. Laub and W. Kaiser, "MR angiography with gradient motion refocusing," *J Comput Assist Tomogr*, vol. 12, pp. 377-382, 1988.

[6] M. R. Prince, E. K. Yucel, J. A. Kaufman, D. C. Harrison, and S. C. Geller, "Dynamic gadolinium-enhanced 3D abdominal MR arteriography," *J Magn Reson Img*, vol. 3, pp. 877-881, 1993.

[7] J. P. Earls, N. M. Rofsky, D. R. DeCorato, G. A. Krinsky, and J. C. Weinreb, "Hepatic arterial-phase dynamic gadolinium-enhanced MR imaging: optimization with a test examination and a power injector," *Radiology*, vol. 202, pp. 268-273, 1997.

[8] T. K. F. Foo, M. Saranathan, M. R. Prince, and T. L. Chenevert, "Automated detection of bolus arrival and initiation of data acquisition in fast, three-dimensional, gadoliuium-enhanced MR angiography," *Radiology*, vol. 203, pp. 275-280, 1997.

[9] A. H. Wilman, S. J. Riederer, B. F. King, J. P. Debbins, P. J. Rossman, and R. L. Ehman, "Fluoroscopically-triggered contrast-enhanced three-dimensional MR angiography with elliptical centric view order: application to the renal arteries," *Radiology*, vol. 205, pp. 137-146, 1997.

[10] A. H. Wilman and S. J. Riederer, "Improved centric phase encoding orders for three dimensional magnetization prepared MR angiography," *Magn Reson Med*, vol. 36, pp. 384-392, 1996.

[11] M. R. Prince, T. M. Grist, and J. F. Debatin, *3D Contrast MR Angiography*. Berlin: Springer, 2003.

[12] F. R. Korosec, R. Frayne, T. M. Grist, and C. A. Mistretta, "Time-resolved contrast-enhanced 3D MR angiography," *Magn Reson Med*, vol. 36, pp. 345-351, 1996.

[13] C. R. Haider, H. H. Hu, N. G. Campeau, J. Huston III, and S. J. Riederer, "3D high temporal and spatial resolution contrast-enhanced MR angiography of the whole brain," *Magn Reson Med*, vol. 60, pp. 749-760, 2008.

[14] P. M. Mostardi, C. R. Haider, P. J. Rossman, E. A. Borisch, and S. J. Riederer, "Controlled experimental study depicting moving objects in view-shared time-resolved MRA," *Magn Reson Med*, vol. 62, pp. 85-95, 2009.

[15] M. Weiger, K. P. Pruessmann, and P. Boesiger, "2D SENSE for faster 3D MRI," *Magma*, vol. 14, pp. 10-19, 2002.

[16] D. C. Noll, D. G. Nishimura, and A. Macovski, "Homodyne detection in magnetic resonance imaging," *IEEE Trans Med Img*, vol. 10(2), pp. 154-163, 1991.

A New Label Fusion Method Using Graph Cuts: Application to Hippocampus Segmentation

C. Platero, M.C. Tobar, J. Sanguino, and O. Velasco

Applied Bioengineering Group, Technical University of Madrid, Spain

Abstract—The aim of this paper is to develop a probabilistic modeling framework for the segmentation of structures of interest from a collection of atlases. Given a subset of registered atlases into the target image for a particular Region of Interest (ROI), a statistical model of appearance and shape is computed for fusing the labels. Segmentations are obtained by minimizing an energy function associated with the proposed model, using a graph-cut technique. We test different label fusion methods on publicly available MR images of human brains.

Keywords— Label fusion, atlas-based segmentation, hippocampus segmentation.

I. Introduction

Automatic segmentation of subcortical structures in human brain MR images plays a crucial role in clinical practice. Specifically, hippocampus segmentation is an important tool for the study of neurodegenerative diseases. Nowadays, brain MR images have poor quality due to their inherently low spatial resolution, insufficient tissue contrast and ambiguous tissue intensity distributions. To overcome these difficulties, many approaches have been proposed. Atlas-based segmentation has become a standard technique to identify structures from brain MR images. An atlas, in the context of this paper, is an image in one modality with its respective labeling (usually generated by manual segmentation). Segmentations with a single atlas are intrinsically biased towards the shape and the appearance of a subject. Several studies have shown that approaches which incorporate properties of a group of atlases outperform the use of a single atlas [1, 2]. There are two different atlas-based segmentation strategies using multiple atlases: a) Probabilistic atlas and b) Multi-atlas segmentation. In a probabilistic atlas, the information from atlases is combined into a mathematical model in a common coordinate system. The advantage is that, once the probabilistic atlas has been generated, only a single registration is required for obtaining the segmentation. However, this method depends on the success of a single registration. An alternative strategy is to register each atlas to the target image separately. The main benefit of the multi-atlas segmentation approach is that the effect of the errors associated with any single atlas propagation can be reduced in the process of combination. Similar to the probabilistic atlas, the transferred atlases are used to build a model for segmenting the target image. This process is often called label fusion. The main drawback of the multi-atlas segmentation is the computational complexity. However, not all atlases have to be registered into the target image [2]. Aljabar et al [3] showed that an atlas selection framework is required for ranking the atlases and fixing a number of atlases to be fused which depends on the application. These studies also indicate that the similarity between the target image and the atlas is a crucial factor for improving registration and segmentation accuracies. Furthermore, brain images show different structures of interest to be segmented. Therefore, a region-wise approach is more appropriate [4]. This can be achieved by dividing the image into multiples anatomically meaningful regions. Once defined the ROIs, a ranking of atlases is calculated. The transferred labels, which belong to the selected atlases, are fused into the ROI of the target image. The fusion of the propagated segmentations can be achieved in different ways: STAPLE [5], majority voting rule or minimization of an energy function with intensity and prior terms [6]. Recent works have shown that statistical models from the registered atlases can improve the segmentation quality [7, 8]. The aim of this paper is to develop a probabilistic modeling framework for the segmentation of structures of interest from a collection of atlases. The paper is organized as follows. In Section 2, the label fusion method is presented. Experiments for the hippocampus segmentation are described in Section 3. Conclusions are presented in Section 4.

II. Label Fusion Method

We present a label fusion method based on minimizing an energy function by using graph-cut technique. This energy function incorporates terms of appearance and shape, which are estimated from the training atlases. Other authors have previously used this framework [6, 9]. Our label fusion method has the following differences from previous proposals: a) An appearance generative model based on multiple features extracted from each pixel and its neighborhood, b) A label prior probability is estimated by using a weighted voting method [7] and c) A spatial regularizer that minimizes the surface of separation between two different labels [10].

Consider a set of N training atlases for each ROI $\{A_i\}_{i=1,\ldots,N} = \{I_i, S_i\}_{i=1,\ldots,N}$ and a target image I, where $I_i : \Omega_i \subset \mathbb{R}^n \to \mathbb{R}$, $n = 3$ and $S_i : \Omega_i \subset \mathbb{R}^n \to \{0,1\}$ are the label maps. We assign to $S(x) = 1$ the foreground pixels and $S(x) = 0$ to the background pixels. We denote $\Phi_i : \Omega \to \Omega_i$ to be the spatial mapping from the target image coordinates to the coordinates of i-th training subject. For simplicity, we assume that $\{\Phi_i\}_{i=1,\ldots,N}$ have been pre-computed using a pairwise registration procedure. This assumption allow us to shorthand $\mathbb{A} = \{\tilde{S}_i = S_i \circ \Phi_i, \tilde{I}_i = I_i \circ \Phi_i\}_{i=1,\ldots,N}$ as the training set into the common coordinates. The segmentation of an image I, based on image intensities and prior knowledge, is computed by the minimization of an energy function.

$$S = \arg\min_S E^{\mathbb{A}}(S), \qquad E^{\mathbb{A}}(S) = E_B^{\mathbb{A}}(S) + E_F(S), \quad (1)$$

where the term $E_B^{\mathbb{A}}(S)$ is derived from $\mathbb{A}$ using the framework of Bayesian estimation theory and $E_F(S)$ is associated with an image-based Finsler metric.

A. Probabilistic Model

To find the MAP estimation is equivalent to minimize the following energy function where the Bayes theorem is applied

$$E_B^{\mathbb{A}}(S) = -\log(p(S|I;\mathbb{A})) = -\log\left(\frac{p(I|S;\mathbb{A})p(S;\mathbb{A},I)}{p(I;\mathbb{A})}\right).$$

We assume that the observed intensities of I are independent random variables. The image likelihood $p(I|S;\mathbb{A})$ can then be written as a product of the likelihoods of the individual pixels: $p(I|S;\mathbb{A}) = \prod_{x \in \Omega} p(I(x)|S(x);\mathbb{A})$. Usually, the intensity distribution is modeled by a mixture of Gaussians [11]. Alternatively, we use a multivariate Gaussian distribution for each pixel and for each label [4]: $p(I(x)|l;\mathbb{A}) = \frac{1}{(2\pi)^{f/2}|\Sigma_l(x)|^{1/2}} \exp(-\frac{1}{2}(I(x) - \mu_l(x))^T \Sigma_l^{-1}(x)(I(x) - \mu_l(x)))$, where $l \in \{0,1\}$, μ is the mean, Σ is the covariance matrix and f is the dimension of the feature space. The effect of sample size has to be considered on feature selection. The means and covariance matrices are estimated by using a variable number of samples $\#Q_l(x)$, where $Q_l(x) = \{i|\tilde{S}_i(x) = l\}$. The number of observations requires from each of the two class to ensure that the classification error is bounded relative to a infinite number of samples depend on f ($f \le \frac{1}{5}\min(\#Q_0(x), \#Q_1(x))$ for $f \le 8$ [12]). To get that the gaussian parameters are the least biased, it is used a neighborhood system around the pixel for obtaining more samples. The gaussian parameters are computed from $\mathbb{A}$:

$$\mu_l(x) = \frac{\sum_{y \in \mathcal{N}(x)} \sum_{i \in Q_l(y)} \tilde{I}_i(y)}{\sum_{y \in \mathcal{N}(x)} \#Q_l(y)} \quad (2)$$

and

$$\Sigma_l(x) = \frac{\sum_{y \in \mathcal{N}(x)} \sum_{i \in Q_l(y)} (\tilde{I}_i(y) - \mu_l(x))(\tilde{I}_i(y) - \mu_l(x))^T}{\sum_{y \in \mathcal{N}(x)} \#Q_l(y) - 1}. \quad (3)$$

Further, f is variable in each pixel. For each pixel is analyzed the correlation matrix. It only selects uncorrelated features in runtime.

The label prior probability $p(S;\mathbb{A},I)$ models the joint probability of all pixels in a particular label configuration. Instead, we assume that the prior probability that pixel x has label l only depends on its position and the similarity between I and $\tilde{I}_i$: $p(S;\mathbb{A},I) = \prod_{x \in \Omega} p(S(x);\mathbb{A},I)$. For each pixel x and each label $l \in \{0,1\}$, we define

$$p(S(x) = l;\mathbb{A},I) = \frac{\sum_{i \in Q_l(x)} m(I(x), \tilde{I}_i(x))^q}{\sum_{l=0}^{1} \sum_{i \in Q_l(x)} m(I(x), \tilde{I}_i(x))^q} \quad (4)$$

where $m(I(x), \tilde{I}_i(x))$ is a global or local similarity measure between the target image and the registered atlas image and q is an associated gain exponent [7].

B. Spatial Regularization

Following the work of Boykov and Kolmogorov [10], the smoothness term E_F of the energy function is defined from a Finsler metric. These authors decomposed the energy into E_R and E_f with weights $\lambda_1, \lambda_2 \in \mathbb{R}, \lambda_1 \ge 0$, that is,

$$E_F(S) = \lambda_1 E_R(S) + \lambda_2 E_f(S).$$

The first part minimizes the segmentation surface by a Riemannian metric and the second one takes into account the orientation of the segmentation surface in the metric. We consider that the isotropic Riemannian metric from the image is defined by $D(x) = g(\|\nabla I(x)\|)\mathbb{I}$, where $\mathbb{I}$ is an identity matrix, $g(x) = (\exp(-x/\gamma))^{1/3}$ and γ is estimated by the average of $\|\nabla I(x)\|$. The energies are defined by

$$E_R(S) = \sum_x \sum_{y, \{x,y\} \in \mathcal{N}} \omega_x^R(y)(1-S(x))S(y),$$

$$E_f(S) = \sum_x \sum_{y, \{x,y\} \in \mathcal{N}} \omega_x^f(y)(S(x)(1-S(y)) - S(y)(1-S(x))),$$

where $\mathcal{N}$ is a neighborhood system, $\omega_x^R(y) = \frac{g(\|\nabla I(x)\|)}{\|x-y\|}$ and $\omega_x^f(y)$ is the component of the vector $\nabla I(x)$ along the vector defined by x and y.

C. Optimization

For the min-cut/max-flow algorithms, the energy to be minimized is represented by a weighted graph, $\mathcal{G} = \langle \mathcal{V}, \mathcal{E} \rangle$, with two special nodes, namely the source s and the sink t. The rest of nodes represents the pixels of the image. The set of edges is denoted by $\mathscr{E} = \mathscr{E}_{\mathcal{N}} \cup \mathscr{E}_{\mathcal{T}}$, where $\mathscr{E}_{\mathcal{N}}$ denotes the set of pixel-to-pixel edges in the defined neighborhood system (n-links) and $\mathscr{E}_{\mathcal{T}}$ denotes the set of pixel-to-terminal s or t edges (t-links). We assign a nonnegative cost $c(x,y)$ to each edge $(x,y) \in \mathscr{E}$. In the considered energy, the cost of a t-link is defined by the prior probabilities and the coefficients ω_x^f, while the cost of a n-link is given by the coefficients ω_x^R. It easily follows that for all x and y with $(x,y) \in \mathcal{N}$

$$c(s,x) = -\log(p(S(x)=1|I;\mathbb{A})) + \lambda_2 \sum_{y,\{x,y\}\in \mathcal{N}} \omega_x^f(y),$$

$$c(x,t) = -\log(p(S(x)=0|I;\mathbb{A})) + \lambda_2 \sum_{y,\{x,y\}\in \mathcal{N}} \omega_y^f(x),$$

$$c(x,y) = \lambda_1 (\omega_x^R(y) + \omega_y^R(x)). \quad (5)$$

III. EXPERIMENTS WITH BRAIN MR DATA

To evaluate the performance of the different label fusion methods, we employ an available database of T1-weighted MR images of epileptic and nonepileptic subjects [13]. Images were acquired by using two MR imaging systems with field strengths (1.5 T and 3.0T) and thus have different resolutions (0.78 x 2 x 0.78 mm^3 and 0.39 x 2 x 0.39 mm^3). All atlases are skull-stripped using BET [14]. An atlas (HFH_021) is selected as a reference to which all atlases are then co-registered with an affine transformation using FLIRT [15]. After spatial normalization, a region is defined for each structure studied (left and right hippocampus) as the minimum bounding box containing the structure for all training atlases expanded by three pixels along each dimension. The size of these ROIs is in the range of 105 x 38 x 115 pixels. The target image is also normalized and parceled by using BET, FLIRT and predefined ROIs. For each ROI, the atlases are ranked based on their similarities with the target image. Then, the selected atlases are co-registered non-rigidly to the ROI of the query image using a B-spline registration with an isotropic grid spacing of 3.0 mm. All nonrigid registration are computed using *Elastix* [16]. The negative Mutual Information (MI) is used as the cost function. Finally, the transferred labels are fused and an inverse affine transformation is applied to return the segmentation into the native space.

A. Setting Parameters

The proposed label fusion method has several input parameters: a) Similarity measures, b) Scalar features for the appearance term, c) The Lagrange multipliers λ_1 and λ_2 of the energy function and d) The optimal number of atlases to be fused. To set these parameters, twenty five leave-one-out segmentations on the training atlases are performed to determinate the tunable parameters. These parameters are varied in certain ranges and their effects are measured by the overlap between the resulting segmentation and the ground truth. Dice coefficient is chosen as a measure of the segmentation overlaps. The parameters are adjusted to give the highest values of Dice coefficient.

MI is used as the similarity measure. For the weighted voting rule, a semi-local strategy is used to calculate the similarity for each registered atlas. A mask image is built by joining all transferred label images. This mask image is used to define the domain for measuring the similarity between the image target and the registered atlases. This strategy is specially suitable when contrast between neighboring structures is low [7], as in the hippocampus.

For each T1-weighted MR image, the following features are calculated: intensity, gradients, laplacians, curvatures and local entropies in different scales. Some of these features are not invariants in gray level so an intensity normalization is applied to the registered atlas images by histogram matching. Spatial derivatives are implemented by Gaussian-derivative filters. In our experiments, the optimal scale is $\sigma = 2$ for the Guassian masks. Bhattacharyya distances and Dice coefficients are used to identify those features which are more important in discrimination among labels. These scalar features are the intensity, the gradient module and the local entropy. To estimate the statistical parameters of the appearance term and since the number of samples for any label is low, a 26-neighborhood system in the sagittal plane is tuned and applied to the equations (2) and (3). In runtime, a matrix of correlation coefficients is calculated for building the quadratic classifiers. The features, whose correlation coefficients are below 0.6 in absolute value, are considered independents and are used in the classifier. The dimension of the feature space is variable for each pixel and can be 3, 2 or 1.

The Lagrange multipliers λ_1 and λ_2 of the energy function are tuned by Dice evaluation. We have observed that the Riemannian metric is more influential than the surface orientation term in the optimization process. Considering 3D grid-graphs with 6 neighborhood system in (5), the edge weights are calculated with $\lambda_1 = 4$ and $\lambda_2 = 1$. The computational burden is reduced by calculating the edge weights of the n-link only for pixels whose labels have uncertainty.

In the atlas selection framework, once the atlases are ranked by MI in the whole ROI, for all segmentation methods we employ a leave-one-out validation strategy, where an optimal number of atlases is tuned. According to the ROI and the segmentation method, the number of the fused atlases varies from 6 to 15.

Table 1 Correlation for manual and automatic volumes and mean and standard deviation values of the five quality measures: correlation (r), Dice coefficient (m_1), relative absolute volume difference (m_2), average symmetric surface distance (m_3), root mean square symmetric surface distance (m_4), and maximum symmetric surface distance (m_5).

Type		r	m_1	m_2	m_3 [mm]	m_4 [mm]	m_5 [mm]
STAPLE	LH	0.48	0.722 ± 0.121	0.22 ± 0.19	0.79 ± 0.37	1.03 ± 0.44	5.26 ± 1.73
	RH	0.52	0.737 ± 0.063	0.24 ± 0.15	0.74 ± 0.12	0.97 ± 0.22	5.17 ± 1.53
Majority Voting	LH	0.36	0.722 ± 0.127	−0.15 ± 0.14	0.77 ± 0.37	0.99 ± 0.46	4.62 ± 1.68
	RH	0.51	0.744 ± 0.073	0.11 ± 0.09	0.70 ± 0.14	0.91 ± 0.21	4.41 ± 1.21
Weighted Voting	LH	0.57	0.732 ± 0.069	−0.16 ± 0.12	0.72 ± 0.17	0.96 ± 0.28	4.74 ± 1.70
	RH	0.63	0.757 ± 0.045	−0.12 ± 0.09	0.66 ± 0.06	0.84 ± 0.10	4.06 ± 0.85
Our approach 1F	LH	0.66	0.753 ± 0.073	0.22 ± 0.15	0.74 ± 0.13	0.98 ± 0.20	6.38 ± 1.66
	RH	0.77	0.773 ± 0.055	0.20 ± 0.11	0.70 ± 0.10	0.91 ± 0.21	5.41 ± 1.56
Our approach 3F	LH	0.70	0.754 ± 0.061	0.13 ± 0.11	0.74 ± 0.11	0.99 ± 0.21	6.18 ± 1.59
	RH	0.79	0.778 ± 0.050	0.11 ± 0.09	0.69 ± 0.10	0.92 ± 0.22	5.06 ± 1.92

B. Results

We compare five label fusion methods: STAPLE, Majority Voting (MV), Weighted Voting (WM) and two methods that we derive from our proposal. In the appearance term, we consider either a singular feature using the intensity (1F) or the proposal with multi-features (3F). The same parameters are applied to WM with respect to our proposal in label prior (semi-local, MI and $q = 4$). The performances of these approaches are evaluated by comparing six measures for the cases of left (LH) and right (RH) hippocampus segmentations. Table 1 gives the correlation for manual and automatic volumes and the mean and standard deviation values of the five quality measures for each method and each ROI. The minus sign in m_2 indicates a result of under-segmentation. Our scores on MV are slightly worse than those given in [13]. This could be because it has not been applied any manual correction in BET. A paired t-test is applied in Dice distributions between MV and the other methods with the following results: STAPLE (LH $p = 0.888$, RH $p = 0.105$), WV (LH $p = 0.515$, RH $p = 0.124$), 1F (LH $p = 0.056$, RH $p = 0.035$) and 3F (LH $p = 0.028$, RH $p = 0.008$).

IV. CONCLUSIONS

We introduce a new label fusion method. It combines an appearance generative model based on multiple features with a label prior using a weighted voting method and a spatial regularizer that minimizes the surface of separation between two different labels. The proposed combination provides high accuracy in segmentation, it shows significant improvements in relation to the conventional framework and also the best correlation between manual and automatic volumes. The proposed method is generic and could be incorporated to other applications.

REFERENCES

1. Rohlfing T., Brandt R., Menzel R., Russakoff D., Maurer C.. Quo vadis, atlas-based segmentation? *Handbook of Biomedical Image Analysis.* 2005:435–486.
2. Heckemann R.A., Hajnal J.V., Aljabar P., Rueckert D., Hammers A.. Automatic anatomical brain MRI segmentation combining label propagation and decision fusion *NeuroImage.* 2006;33:115–126.
3. Aljabar P., Heckemann RA, Hammers A., Hajnal JV, Rueckert D.. Multi-atlas based segmentation of brain images: Atlas selection and its effect on accuracy *NeuroImage.* 2009;46:726–738.
4. Han X., Fischl B.. Atlas renormalization for improved brain MR image segmentation across scanner platforms *Medical Imaging, IEEE Transactions on.* 2007;26:479–486.
5. Warfield S.K., Zou K.H., Wells W.M.. Simultaneous truth and performance level estimation (STAPLE): an algorithm for the validation of image segmentation *Medical Imaging, IEEE Transactions on.* 2004;23:903–921.
6. Lijn F., Heijer T., Breteler M., Niessen W.J.. Hippocampus segmentation in MR images using atlas registration, voxel classification, and graph cuts *NeuroImage.* 2008;43:708–720.
7. Artaechevarria X., Muñoz-Barrutia A., Solorzano C.. Combination strategies in multi-atlas image segmentation: Application to brain MR data *Medical Imaging, IEEE Transactions on.* 2009;28:1266–1277.
8. Sabuncu M.R., Yeo B.T.T., Van Leemput K., Fischl B., Golland P.. A generative model for image segmentation based on label fusion *Medical Imaging, IEEE Transactions on.* 2010;29:1714–1729.
9. Wolz Robin, Heckemann Rolf A, Aljabar Paul, et al. Measurement of hippocampal atrophy using 4D graph-cut segmentation: application to ADNI *NeuroImage.* 2010;52:109.
10. Kolmogorov V., Boykov Y.. What metrics can be approximated by geocuts, or global optimization of length/area and flux in *Computer Vision, 2005. ICCV 2005. Tenth IEEE International Conference on*;1:564–571 2005.
11. Pohl K.M., Fisher J., Grimson W.E.L., Kikinis R., Wells W.M.. A Bayesian model for joint segmentation and registration *NeuroImage.* 2006;31:228–239.
12. Raudys S.J., Jain A.K.. Small sample size effects in statistical pattern recognition: Recommendations for practitioners *IEEE Transactions on pattern analysis and machine intelligence.* 1991;13:252–264.
13. Jafari-Khouzani Kourosh, Elisevich Kost V, Patel Suresh, Soltanian-Zadeh Hamid. Dataset of magnetic resonance images of nonepileptic subjects and temporal lobe epilepsy patients for validation of hippocampal segmentation techniques *Neuroinformatics.* 2011;9:335–346.
14. Smith Stephen M. Fast robust automated brain extraction *Human brain mapping.* 2002;17:143–155.
15. Jenkinson Mark, Bannister Peter, Brady Michael, Smith Stephen, others . Improved optimization for the robust and accurate linear registration and motion correction of brain images *Neuroimage.* 2002;17:825–841.
16. Klein S., Staring M., Murphy K., Viergever M.A., Pluim J.P.W.. elastix: a toolbox for intensity-based medical image registration *Medical imaging, IEEE transactions on.* 2010;29.

Computer Aided Decision Support Tool for Rectal Cancer TNM Staging Using MRI

A. Torrado-Carvajal[1], T. Martin Fernandez-Gallardo[2], and N. Malpica[1]

[1] Department of Electronics, Universidad Rey Juan Carlos, Mostoles, España
[2] Diagnostic Imaging, Hospital Universitario de Fuenlabrada, Fuenlabrada, España

Abstract—Rectal cancer (RC) is associated with a poor prognosis because of the risk both for metastases and for local recurrence after total mesorectal excision (TME) surgery. Preoperative assessment (cTNM) for neoadjuvant therapy (radiation therapy and/or chemotherapy) is very important to minimize recurrence rates. Therefore, the challenge for preoperative imaging in RC is to determine the different risk of recurrence in different patients.

TME has changed several strategies for treatment in RC patients in developed countries. Additionally the introduction of national training programs, such as the "Audited teaching program for the treatment of rectal cancer in Spain" (Viking Project), have decreased local recurrence and mortality rates and while increased survival.

These programs aim at assessing TME including an evaluation of the use of MRI for preoperative assessment. This situation explains the increasing interest in computer aided diagnosis (CAD) tools for this pathology. However, MRI processing in RC is notoriously difficult due to the uncertainty of signal intensity changes along the mesorectal fascia and the different criteria used to predict nodal involvement.

This paper presents a complete computer aided decision support tool for rectal cancer TNM staging following the "Viking Project" recommendations. We have applied image processing to extract and quantify the extension of the primary tumor (T staging) as well as to characterize and classify the lymph nodes to predict nodal involvement (N staging).

Our tool includes tools for: (1) segmentation of the main structures of the mesorectum (lumen, primary tumor and mesorectal fascia) for T staging, (2) segmentation, feature extraction, and classification of the local lymph nodes for N staging, (3) 3D rendering of the segmented structures for surgery planning, and (4) automatic report generation.

Accuracy of the results has been assessed with an expert radiologist. Remaining image processing challenges are indicated and some directions for future research are given.

Keywords—Computer aided diagnosis, feature extraction, image segmentation, image classification, rectal cancer.

I. INTRODUCTION

Rectal cancer (RC) is considered a public health problem due to its high incidence and mortality, especially in developed countries [1]. However, it is often curable if it is detected early.

RC stages are set according to the system Tumor-Node-Metastasis (TNM), agreeing with the International Union for Cancer Control (UICC) criteria [2]. Preoperative assessment of tumor dissemination and local lymph node involvement before surgery (cTNM) are of the utmost importance. Depending on the staging, a more aggressive treatment or another will be applied.

Treatment depends on these stages, the state of general health of the patient, age, medical history and his tolerance for specific medications and treatments, but may include surgery, chemotherapy, and/or radiotherapy. Incomplete removal of the lateral spread of the tumor is accepted as the cause of most of these recurrences [3-4]. Surgical techniques for the treatment of RC were standardized with the introduction of total mesorectal excision (TME) [5], which has proven to be the technique with a lower rate of local recurrence.

One of the major noninvasive tools to assess and characterize accurately this type of cancer is the study of magnetic resonance imaging (MRI) studies [6]. The MRI and Rectal Cancer European Equivalence (MERCURY) study showed MRI is reproducible and allows patients to be selected on this basis for preoperative treatment. These findings have led to the introduction of national training programs to establish some requirements in the treatment of RC [7-9]. The "Audited teaching program for the treatment of rectal cancer in Spain" (Viking Project), has proved to decrease local recurrence and mortality rates while increased survival [10].

Additionally, recent years have seen the emergence of a number of computer aided diagnosis (CAD) applications in other medical frameworks [11-13]. However, there is no CAD application to assist TNM reporting at the moment.

For all the reasons described above, we have developed a computer aided decision support tool for rectal cancer TNM staging using MRI. Our tool includes

- Basic study information
- T staging: segmentation of the main structures of the mesorectum (lumen, tumor and mesorectal fascia)
- N staging: segmentation and classification of nodes
- 3D rendering of the structures for surgery planning
- Automatic "rtf" report generation

II. MATERIALS AND METHODS

A. Patient Database

Our patient database consists of 125 MRI studies of RC patients from the Hospital Universitario de Fuenlabrada

(HUF). The HUF participates with another 57 centers in Spain in the "Audited Teaching Program for the Treatment of Rectal Cancer in Spain" (Viking Project), promoted by the Spanish Association of Surgeons (ACS).

B. MRI Acquisition

MRI images at HUF are acquired following the Viking Project protocol. Imaging is performed on a 1.5T scanner (Signa R HDX; GE Medical Systems, Milwaukee, WI) with the parameters shown in table 1, with the following steps:

- A first T2-weighted Fast Spin Echo (FSE) sequence in the three directions is acquired: sagittal, axial and coronal to the body (table 1: FSE T2 sequences).
- Once the tumor is located, high resolution T2 sequences perpendicular to the major axis of the tumor are acquired (table 1: high resolution sequence).
- In the case of low rectal tumors, a volume parallel to the axis of the anal canal is also acquired.

In this work we use only the high resolution T2 sequences, in particular, the perpendicular to the major axis of the primary tumor.

C. Segmentation

The nature of the pelvic cavity is a challenge to apply automatic segmentation algorithms [14]. The constitution of the pelvic organs makes a big difference between different images, making difficult to model the problem. Additionally, the histograms of the images show that the different tissues present biased distributions (Fig. 1).

a) T Staging:

The main prognostic factor is to assure that the primary tumor is controlled. Segmentation and measurement of the lumen, the primary tumor and the mesorectal fascia (MF) establishes its extension in the pelvic cavity.

The lumen –the inside space of the intestine– presents a low signal intensity in the MRI images. Neighbor voxels of colorectal tissue differ enough to apply region growing segmentation from a seed point (Fig. 2 left).

The tumor is established around the rectum, swelling its walls. We have applied the expectation-maximization (EM) algorithm to the Gaussian smoothed volume to segment the rectum and the tumor (Fig. 2.center).

The MF is the most difficult structure to segment. In several points of the MF there is almost no difference between its limit and the mesorectum. Furthermore, changes in the intensity are variable along the MF. In this case, we have performed manual segmentation (Fig. 2. right).

b) N Staging:

The presence and number of involved lymph nodes has an influence on the prognosis of the patient [15-16]. Most of the published studies about MRI of RC have used the nodes size as a criterion to identify nodal involvement, even though there is little consistency to discriminate between benign and malignant nodes [17-20]. An important number of subsequent studies emphasize the irregularity of the edges of the lymph nodes as well as the homogeneity of their signal in identifying suspicious metastatic nodes [21-22]. These cases illustrate the problems for a radiologist to accurately stage the nodal status.

Previous work in our group has proved the viability of segmentation, feature extraction and machine learning for computer aided classification [23]. Other groups have also demonstrated the use of a computer algorithm to quantitatively analyze morphological features for this task [24].

Table 1 Sequences for GE Signa 1.X Version 9.1 1.5-T

	FSE T2	High Resolution Sequences	
Plane	Sagittal, coronal, axial	Coronal	Axial
Matrix	512x256	256x256	
FOV	24 cm	16 cm	
Slice (gap)	5mm (1 mm)	3 mm (0.1 mm)	
TR	6000	4000	
TE	130	85	
Echo Train Length	12	12	
Bandwidth	20.83	20.83	
Frequency Direction	A/P	H/F	A/P
Phase Encoding	H/F	R/L	
Saturation Bands	H, F, A	H, F, A	

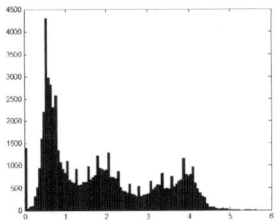
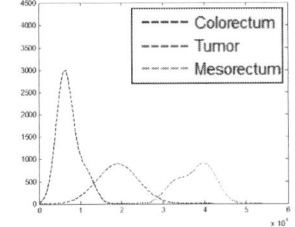

Fig. 1: Histogram of the pelvic cavity (left) and estimated probability density functions (PDFs) for three different tissue classes (right).

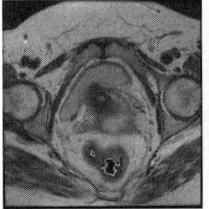
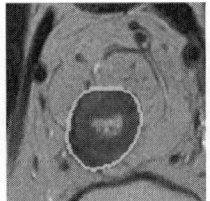
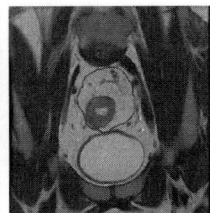

Fig. 2 Segmentation of the lumen (left), the rectum and the primary tumor (center), and the mesorectal fascia (left).

III. RESULTS

We have developed the first computer-aided decision support tool for rectal cancer TNM staging using MRI studies (Fig. 3). This tool has been developed in Matlab and includes functionality to apply the previous algorithms.

Our tool has six main modules, following the Viking project criteria:

1. MRI viewer: it shows the MRI images loaded from DICOM or NIfTI data. Images are normalized to adjust contrast when loaded.
2. Exam information: according to the Viking Project criteria, every report has to include the patient ID, the MRI date, and the technical satisfaction of the exam.
3. T staging: this tool allows segmentation of the different structures of the mesorectum. Different buttons are related to the structures explained in section II.C.a.
4. N staging: resulting information from semi-automatic node segmentation, feature extraction and classification is included in this tool. It also includes the position of the node respect to the MF and its minimum distance.
5. 3D viewer: the segmentation information is rendered and showed in this tool. It allows anatomical reconstructions that can be used to plan the surgery, allowing 3D navigation.
6. Reporting: data from RC studies has to be reported including the required information in the Viking Project. Our application can generate this report in ".rtf" format, and launch automatically the installed office suite word processor.

IV. DISCUSSION

Medical image processing for rectal cancer TNM staging using MRI studies is a difficult task due to intra-patient and inter-patient factors.

On one hand, anatomical differences between patients introduce high variability. On the other hand, it involves the detection and evaluation of diverse difficult structures in the image; especially thin and small structures such as the MF or lymph nodes.

We have shown that it is possible to apply automated methods to approximate the problem, so medical image processing can quantitatively report rectal cancer TNM staging studies.

As a decision support tool, the radiologist guides the workflow. The qualitative assessment of the application and its results by the experts has been very good.

The design of our application and the analysis process are based on the national standards introduced by the Viking project. Thus, the application can be used in the clinical practice of other hospitals.

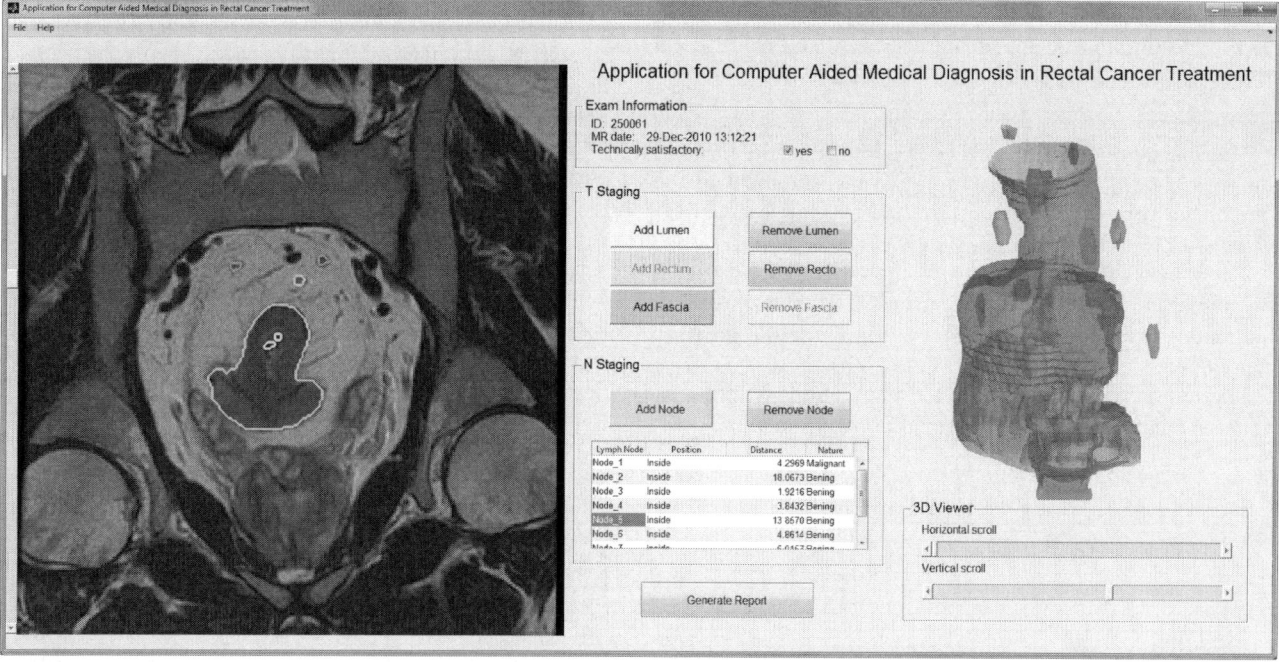

Fig. 3 Computer aided decision support tool for rectal cancer TNM staging using MRI studies. This tool includes MRI – DICOM and NIfTI – viewer (left), exam information according to the Viking Project (center up), T staging segmentation (center middle up), N staging segmentation and classification (center middle down), report generation (center down), and 3D viewer (right).

V. CONCLUSIONS

RC is an important and growing medical problem. MRI is considered as the major noninvasive tool to detect and characterize this type of cancer. While clinicians gain valuable information about the RC stage through the analysis of MRI images, automated image analysis can be helpful for them as a tool for decision support, especially in quantifying the need of neoadjuvant prior to surgery.

We have developed the first computer-aided application for rectal cancer TNM staging and reporting

Our goals have included semi-automatic segmentation of the main structures of the mesorectum, and lymph nodes and their automatic classification. Currently we are working to improve the segmentation of the main structures for the study of RC. Additionally, we believe that the results of the node classification can be improved.

The development of this application has supposed a proof of concept to check the feasibility of the application of automatic methods of medical image processing to the RC problem, and the establishment of a foundation for future development.

Future research may improve performance so as to develop a complete Computer Aided Application system for fully automated rectal identification and classification on MRI exams.

REFERENCES

1. Eurostat web page (European Commission), Database of Public Health at http://ec.europa.eu/eurostat (Consulted: April 2013).
2. Union for International Cancer Control (2009) TNM Classification of Malignant Tumors, 7th edition, Wiley-Blackwell.
3. Adam IJ, Mohamdee MO, Martin IG et al. (1994). Role of circumferential margin involvement in the local recurrence of rectal cancer. The Lancet 344(8924):707-711.
4. Quirke P, Durdey P, Dixon MF, and Williams NS (1986). Local recurrence of rectal adenocarcinoma due to inadequate surgical resection. Histopathological study of lateral tumour spread and surgical excision. The Lancet 2(8514):996-999.
5. Heald RJ and Ryall RD (1986). Recurrence and survival after total mesorectal excision for rectal cancer. The Lancet 1(8496):1479-1482.
6. Beets-Tan RG, and Beets GL (2004). Rectal Cancer: Review with Emphasis on MR Imaging1. Radiology, 232(2):335-346.
7. Burton S, Brown G, Daniels IR, Norman AR, et al. (2006). MRI directed multidisciplinary team preoperative treatment strategy: the way to eliminate positive circumferential margins?. British. J. Cancer, 94(3):351-357.
8. Glimelius B, and Oliveira J (2009). Rectal cancer: ESMO clinical recommendations for diagnosis, treatment and follow-up. Ann. Oncol., 20(Suppl 4):54-56.
9. The Association of Coloproctology of Great Britain and Ireland (2007). Guidelines for the management of colorectal cancer, 3rd edition. London, UK: Royal College of Surgeons of Ireland.
10. Codina-Cazador A, Espín E, Biondo S et al. (2007). Audited teaching program for the treatment of rectal cancer in Spain: results of the first year. Cir. Esp., 82(4):209.
11. Charbonnier JP, Smit EJ, Viergever MA et al. (2013). Computer-aided diagnosis of acute ischemic stroke based on cerebral hypoperfusion using 4D CT angiography. In SPIE Medical Imaging (pp. 867012-867012). International Society for Optics and Photonics.
12. Hambrock T, Vos PC, Hulsbergen–van de Kaa C A, et al. (2013). Prostate Cancer: Computer-aided Diagnosis with Multiparametric 3-T MR Imaging-Effect on Observer Performance. Radiology, 266(2):521-530.
13. Morillo DS, Jiménez AL, and Moreno SA (2013). Computer-aided diagnosis of pneumonia in patients with chronic obstructive pulmonary disease. J. Am. Med. Inform. Assoc., DOI:10.1136/ amiajnl-2012-001171.
14. Joshi N, Bond S, and Brady M (2010). The segmentation of colorectal MRI images. Med. Image Anal., 14(4):494-509.
15. Wolmark N, Fisher B, and Wieand, HS (1986). The prognostic value of the modifications of the Dukes' C class of colorectal cancer. An analysis of the NSABP clinical trials. Ann. Surg., 203(2):115.
16. Tang R, Wang JY, Chen JS, et al. (1995). Survival impact of lymph node metastasis in TNM stage III carcinoma of the colon and rectum. J. Am. Coll. Surgeons, 180(6):705.
17. Kusunoki M, Yanagi H, Kamikonya N, et al. (1994). Preoperative detection of local extension of carcinoma of the rectum using magnetic resonance imaging. J. Am. Coll. Surgeons, 179(6):653.
18. Indinnimeo M, Grasso RF, Cicchini C, et al. (1996). Endorectal magnetic resonance imaging in the preoperative staging of rectal tumors. Int. Surg., 81(4):419.
19. Zerhouni EA, Rutter C, Hamilton SR, et al. (1996). CT and MR imaging in the staging of colorectal carcinoma: report of the Radiology Diagnostic Oncology Group II. Radiology, 200(2):443-451.
20. Vogl TJ, Pegios W, Mack MG, et al. (1997). Accuracy of staging rectal tumors with contrast-enhanced transrectal MR imaging. Am. J. Roentgenol., 168(6):1427-1434.
21. Schnall MD, Furth EE, Rosato EF, and Kressel HY (1994). Rectal tumor stage: correlation of endorectal MR imaging and pathologic findings. Radiology, 190(3):709-714.
22. Brown G, Richards CJ, Bourne MW, et al. (2003). Morphologic Predictors of Lymph Node Status in Rectal Cancer with Use of High-Spatial-Resolution MR Imaging with Histopathologic Comparison1. Radiology, 227(2):371-377.
23. Torrado-Carvajal A, Malpica N, and Martin Fernandez-Gallardo T, (2011). Aportaciones del análisis de imagen en el estudio del carcinoma de recto mediante imágenes de resonancia magnética. Libro de Actas del XXIX Congreso Anual de la Sociedad Española de Ingeniería Biomédica – CASEIB2011 (pp. 325-328).
24. Tse DML, Joshi N, Anderson EM, Brady M, and Gleeson FV (2012). A computer-aided algorithm to quantitatively predict lymph node status on MRI in rectal cancer. Brit. J. Radiol., 85(1017):1272-1278.

Author: Angel Torrado Carvajal
Institute: Department of Electronics – Universidad Rey Juan Carlos
Street: C/ Tulipan s/n
City: Mostoles
Country: Spain
Email: angel.torrado@urjc.es

Volume of Tissue Activated in Patients with Parkinson's Disease from West-Center Region of Colombia Treated with Deep Brain Stimulation

Hernán Darío Vargas Cardona[1], Mauricio A. Álvarez[1], Genaro Daza[2], Enrique Guijarro[3], and Álvaro Orozco[1]

[1] Department of Electrical Engineering, Universidad Tecnológica de Pereira, Pereira, Colombia
[2] Instituto de Epilepsia y Parkinson del Eje Cafetero, Pereira, Colombia
[3] I3BH, Universitat Politécnica de Valencia, Valencia, Spain

Abstract—**Parkinson's Disease (PD) is a group of neurological disorders caused by a deficiency of dopamine in a brain structure called Substantia Nigra Reticulata (SNR). The neurosurgical procedure known as Deep Brain Stimulation (DBS) of Subthalamic Nucleus (STN) has proved to be an effective treatment for this disease. DBS modulates neural activity with electric fields. However, the mechanisms regulating the therapeutic effects of DBS are not clear, in fact there is not a full knowledge about the voltage distribution generated in the brain by the stimulating electrodes, what is usually named as Volume of Tissue Activated (VTA). VTA is useful to find the optimal parameters of stimulation, that allow the neurosurgeons to get the best clinical outcomes and minimal side effects. In this work, we applied a methodology to estimate the VTA, based on analysis of diffusion tensor images (DTI) of PD patients from the west-center region of Colombia. DTI are used to characterize the conductivity of brain structures relevant in DBS. The results are consistent to similar studies in other populations.**

Keywords—**Volume of Tissue Activated (VTA), Parkinson's Disease (PD), Deep Brain Stimulation (DBS) , Subthalamic Nucleus(STN), Diffusion Tensor Images (DTI).**

I. INTRODUCTION

Parkinson's Disease (PD) is a progressive degenerative condition of the central nervous system, and it manifest as muscle rigidity, hypokinesia, bradykinesia and tremor [1]. The causes of the disease are not fully known. But, it is known that there is a significant deterioration of the cells of a brain structure known as the Substantia Nigra Reticulata (SNR). This problem, leads to a decrease in dopamine levels, producing an effect on other structures such as the Subthalamic Nucleus (STN) and the Globus Pallidus (GPi). STN and GPi play a major role on movement control of human beings [2]. Since, the drug side effects become a problem for patients, surgical treatment is an alternative for the treatment of PD. Deep brain stimulation (DBS) of Subthalamic Nucleus (STN) is the most common surgical procedure for the treatment of Parkinson's disease [1].

Although the purpose of DBS is to modulate neural activity with electric fields, there is not a full knowledge about the voltage distribution generated in the brain by the stimulating electrodes. In addition, there are many questions related to the consequences that DBS generates in the nervous system [3]. Also, it is very difficult to predict with accuracy, which areas are directly affected by stimulation, and it can not easily be determined the Volume of Tissue Activated (VTA) in the brain. VTA is the amount of brain tissue which presents excitation or electrical response to stimulation of the electrodes [4]. Currently, the therapeutic benefits and side effects of DBS have become an important field of research. Previous clinical studies have established that therapeutic results of DBS depend strongly on three factors: 1). selection of appropriate patients [5], 2). precise surgical placement of electrodes in the brain target for an established movement disorder [6] and 3). specific evaluation of stimulation parameters for each patient [7]. Other studies about VTA estimation examined direct relationships between stimulation configurable parameters (voltage, frequency, pulse width, contact impedance) and therapeutic response to DBS [8, 9].

In this work, we applied a simulation methodology to estimate the VTA, using diffusion tensor images (DTI) obtained from several patients with Parkinson's Disease located in the west-central region of Colombia. VTA estimation depends on the stimulation parameters such as amplitude, frequency, pulse width, impedance and brain conductivity.

II. MATERIALS

A. MRI-DTI Database

Diffusion tensor measurements of five subjects with PD were taken at Cedicaf diagnostic imaging center located in Pereira, Colombia. They use a 1.5 Tesla resonator. 96 slices were obtained with a 512×512 resolution, giving $1 \times 1 \times 1$ mm^3 voxels. These images followed the acquisition protocols designed by the *Instituto de Epilepsia y Parkinson del Eje Cafetero*, located also at Pereira. Patients are informed of the research purposes and they sign an informed consent.

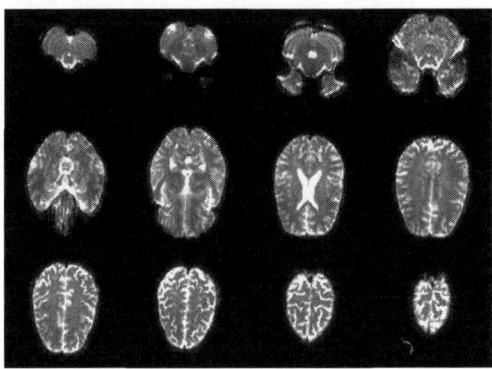

Fig. 1 Sequence of 12 unprocessed images with diffusion gradients. Gradient coordinates are (x,y,z)=0.707, 0.000, 0.007. The images correspond to axial axis.

B. Software

1. **Matlab R2009a:** This software was used for MRI-DTI analysis and calculation of the diffusion Tensor **D**. *Matlab* is a language for high-level technical computing for algorithm development, data visualization, data analysis, and numerical computation.
2. **3D Slicer:** It is a free software with open architecture for medical image analysis and scientific visualization. It also allows medical image registration, processing of MRI-DTI and volumetric reconstruction. *3D slicer* and full documentation can be downloaded free at http://www.slicer.org/.
3. **Comsol Multiphysics 4.3:** VTA Estimation through finite elements is done with *Comsol Multiphysics 4.3*. This software simulates any physical system.
4. **Visual Studio 2008:** 3D visualization model is developed in C++ using Visual Studio 2008 and VTK library (Visualization Toolkit), which allows the development of graphics interfaces.

III. METHODS

Project development was based on cooperation of the *Research Group on Automatica (Universidad Tecnológica de Pereira)* and the *Instituto de Epilepsia y Parkinson del eje cafetero*. MRI-DTI acquisition is performed by a group of medical specialists conformed of neurologists, neurosurgeon and neurophysiologists.

A. Calculation of Diffusion Tensor D

A MRI-DTI typically contains one image for each gradient direction and a reference image for each slice. In this case, we select K=6 gradient directions and 96 slices that include relevant structures: Subthalamic Nucleus (STN), Thalamus (Thal) and Substantia Nigra Reticulata (SNR). The total dataset contains $(6+1) \times 96 = 672$ images. The result is a diffusion tensor matrix of 5 dimensions: $\mathbf{Q}[i, j, row, column, slice]$, where i=1,2,3 is the ith row element, j=1,2,3 is the jth column element, both are indexed in the tensor. Each voxel has a tensor $\mathbf{D} \in \mathbb{R}^{3X3}$.

B. Diffusion-Conductivity Transformation

Diffusion values obtained in DTI analysis are converted to conductivity using the scalar transformation[10] given by:

$$\sigma = \frac{k(S \bullet sec)}{mm^3} \mathbf{D} \qquad (1)$$

where **D** is the individual tensor of each voxel, $k = 0.844 \pm 0.0545$, $S : Siemens$, $sec : seconds$, $mm^3 : cubic\ milimeters$.

C. VTA Estimation

Using the stimulation parameters (amplitude, frequency, pulse width, impedance) and tissue conductivity, we estimate the VTA and its spacial distribution. The result is a three-dimensional matrix that contains electrical activation values and each dimension is a spatial axis (x, y, z). The brain electrical activation is described mathematically through the Poisson equation [11]:

$$\nabla \bullet \sigma \nabla V_e = -I \qquad (2)$$

where V_e: voltage, I: stimulation current and σ: brain tissue conductivity. Equation 2 is solved by approximation using finite elements models [11].

IV. EXPERIMENTAL RESULTS

Once we obtain the conductivity of the brain structures, we use *3D slicer* to locate regions of interest (ROIs) on MRI-DTI. This process is made with the NAC-Brain Atlas, available at http://www.spl.harvard.edu/publications/item/view/2037. In PD the ROIs are Thalamus, Subthalamic Nucleus and Substantia Nigra Reticulata. Diffusion results are shown in Table 1.

We use Equation 1 and table 1 for obtaining the conductivity values in table 2. Finally, we solve the equation 2 via approximation by finite elements models using Comsol Multiphysics 4.3 software. We estimate anisotropic models of Thalamus and Subthalamic Nucleus.

Table 1 Diffusion Values for Regions of Interest. Ant: Anterior, Lat: Lateral, Post: Posterior, Med: Medial.

	$D_{xx}(mm^2/s) \cdot 10^{-3}$	$D_{yy}(mm^2/s) \cdot 10^{-3}$	$D_{zz}(mm^2/s) \cdot 10^{-3}$	Average Diffusion $\langle D \rangle (mm^2/s) \cdot 10^{-3}$
Thal (Ant)	(0.49 ± 0.01)	(0.61 ± 0.07)	(0.35 ± 0.06)	(0.48 ± 0.06)
Thal (Lat)	(0.44 ± 0.05)	(0.37 ± 0.05)	(0.57 ± 0.06)	(0.46 ± 0.05)
Thal (Post)	(1.63 ± 0.19)	(1.82 ± 0.12)	(1.48 ± 0.14)	(1.65 ± 0.15)
Thal (Med)	(0.44 ± 0.05)	(0.33 ± 0.08)	(0.37 ± 0.08)	(0.38 ± 0.07)
STN	(0.33 ± 0.06)	(0.57 ± 0.08)	(0.25 ± 0.09)	(0.38 ± 0.08)
SNR	(0.35 ± 0.09)	(0.44 ± 0.10)	(0.59 ± 0.08)	(0.46 ± 0.08)

Table 2 Conductivity Values for Regions of Interest. Ant: Anterior, Lat: Lateral, Post: Posterior, Med: Medial.

	$\sigma_{xx}(S/m)$	$\sigma_{yy}(S/m)$	$\sigma_{zz}(S/m)$	Average Conductivity $\langle \sigma \rangle (S/m)$
Thal (Ant)	(0.39 ± 0.11)	(0.49 ± 0.12)	(0.28 ± 0.11)	(0.38 ± 0.11)
Thal (Lat)	(0.35 ± 0.10)	(0.29 ± 0.11)	(0.45 ± 0.11)	(0.37 ± 0.10)
Thal (Post)	(1.38 ± 0.24)	(1.54 ± 0.17)	(1.25 ± 0.19)	(1.39 ± 0.20)
Thal (Med)	(0.35 ± 0.11)	(0.26 ± 0.13)	(0.29 ± 0.13)	(0.30 ± 0.12)
STN	(0.26 ± 0.11)	(0.45 ± 0.13)	(0.20 ± 0.14)	(0.28 ± 0.13)
SNR	(0.28 ± 0.14)	(0.35 ± 0.15)	(0.47 ± 0.13)	(0.37 ± 0.13)

A. Anisotropic Models of Thalamus and Subthalamic Nucleus

We simulated DBS with current stimulation from 20 uA to 200uA and a monopolar electrode. We estimate the VTA as a function of DBS parameters and conductivity values from table 2. The simulation results for thalamic and subthalamic stimulation are observed in figures 2,3,4.

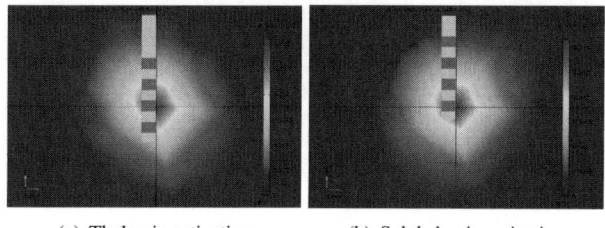

(a) Thalamic activation   (b) Subthalamic activation

Fig. 2 a). Electrical activation for Thalamic stimulation in XY plane. The maximum voltage value is 3.09 mv; b). Electrical activation for Subthalamic Nucleus stimulation in XY plane. The maximum voltage value is 3.15 mv. Both simulations are done for a 40 μA current stimulation.

B. 3D Visualization Model

We developed a three-dimensional graphical model for reconstruction of relevant brain structures in surgery for Parkinson's Disease. This model allows the inclusion of stimulation electrode and its motion in space. We developed the interface in C++ language using the VTK (Visualization Toolkit) library. In figure 5 we ilustrate a full simulation of a DBS in

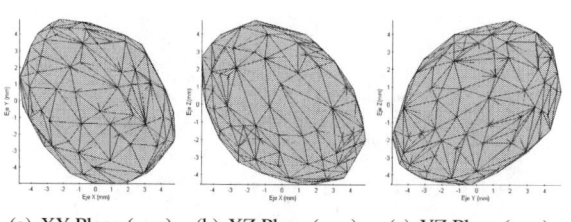

(a) XY Plane (mm)   (b) XZ Plane (mm)   (c) YZ Plane (mm)

Fig. 3 VTA in two-dimensional planes for anisotropic model of Thalamus.

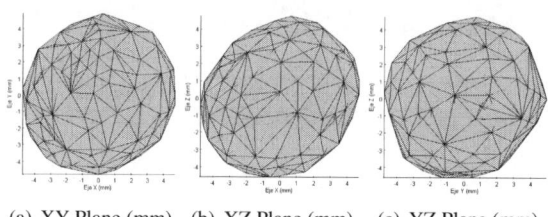

(a) XY Plane (mm)   (b) XZ Plane (mm)   (c) YZ Plane (mm)

Fig. 4 VTA in two-dimensional planes for anisotropic model of Subthalamic Nucleus.

STN (gray color). In this escene appears the electrode (gray color cilinder), Thalamus (yellow color) and the VTA generated by a 40 μA current stimulation (brown color).

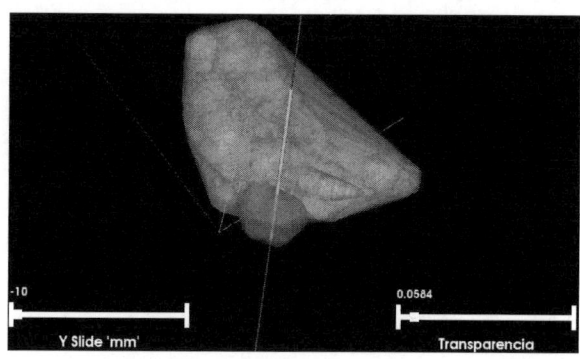

Fig. 5 Full simulation of VTA generated by a 40 μA current stimulation.

V. DISCUSSION AND CONCLUSIONS

The most important issues are raised in the following points,

1. We acquired a database with diffusion tensor images from 5 patients with Parkinson's disease in the west-center region in Colombia. DTI is very useful for characterizing the electrical conductivity of brain structures. Also, these images have a great potential in research

about diagnosis and treatment of neurodegenerative diseases.
2. We applied a simulation methodology for estimating the Volume of Tissue Activated (VTA), generated by deep brain stimulation in patients with Parkinson's disease. The methodology is based on analysis of diffusion tensor images for electrical characterization of Subthalamic Nucleus, Thalamus and Substantia Nigra Reticulata. Preliminary results are consistent with previous similar works in terms of quantity and geometry of the VTA [9, 11].
3. Inclusion of diffusion tensors in the model has improved the estimation of brain tissue conductivity, as they allow to calculate a conductivity value for each voxel. The escalar conversion developed by [10], is an efective approximation from diffusion to conductivity. Conductivity values (see table 2) are consistent with reported measurements for cerebrospinal fluid (1.54 S/m) and basal ganglia (0.3 S/m) obtained in isotropic models [12]. In comparison with isotropic models of VTA, the anisotropic model is more accurate because STN and Thal are composed of anisotropic tissue (white matter)[9].
4. VTA estimation is described mathematically by Poisson equation (Eq. 2). We solve this differential equation by finite element approximation. The anisotropic models of Thal and STN show slightly irregular geometries (see figs. 2,3,4). But, volumetric expansion of both models is very similar. The volume radio is between 4 mm and 5 mm (40 uA current stimulation). The shape and amount of VTA did not change with modification in frequency of pulses.
5. We developed a graphic model for 3D reconstruction of brain structures, the electrode intraoperative and VTA. The interface allows interactive visualization from several angles of electrode displacement in basal ganglia and the VTA generated by deep brain stimulation (see fig. 5).
6. In this work, we applied a realist methodology for observe the VTA incidence in the human brain. VTA analysis is useful for correlate stimulation parameters with therapeutic outcomes and side effects produced by DBS in patients with PD. Although, these results are preliminary, this model will be improved with medical image registration, segmentation techniques and clinical validation.

As future work, we propose:

- Patient-specific analysis for correlating VTA with therapeutic outcomes and side effects generated by DBS in patients with PD. Also, we want to find the optimal parameters of DBS in each patient. This procedure requires a long time between the surgery stage and patient's rehabilitation. Therefore, this research will be continued with financial support of Colciencias and medical support of the *Instituto de Epilepsia y Parkinson del eje cafetero*.

ACKNOWLEDGEMENTS

This research is developed under the projects: *"Desarrollo de un sistema automático de mapeo cerebral y monitoreo intraoperatorio cortical y profundo: aplicación neurocirugía"* and *"Desarrollo de un sistema efectivo y apropiado de estimación de volumen de tejido activo para el mejoramiento de los resultados terapéuticos en pacientes con enfermedad de Parkinson intervenidos quirúrgicamente"*, both financed by Colciencias with codes 111045426008 and 111056934461 respectively. GDS author was supported by *"Patrimonio autónomo fondo nacional de financiamiento para la ciencia, la tecnología y la innovación, Francisco José de Caldas"*

REFERENCES

1. Obeso JA, Olanow CW, Rodriguez-Oroz MC, Krack P, Kumar R, Lang AE. Deep-brain stimulation of the subthalamic nucleus or the pars interna of the globus pallidus in Parkinson's disease N Engl J Med. 2001;345(13):956-963.
2. Neurological Disorders NINDS (National Institute, Stroke) . Parkinson's Disease: Challenges, Progress, and Promise 2004:6-10.
3. Nowinski WL, Belov D, Pollak P, Benabid AL. Statistical analysis of 168 bilateral subthalamic nucleus implantations by means of the probabilistic functional atlas Neurosurgery. 2005;57:319-30.
4. Johnson MD, Miocinovic S, McIntyre CC, Vitek JL. Mechanisms and targets of deep brain stimulation in movement disorders Neurotherapeutics. 2008;5(2):294-308.
5. Walter BL, Vitek JL. Surgical treatment for Parkinson's disease Lancet Neurol. 2004;3(12):719-728.
6. Machado A, Rezai AR, Kopell BH, Gross RE, Sharan AD, Benabid AL. Deep brain stimulation for Parkinson's disease: surgical technique and perioperative management Mov Disord. 2006;21:247-258.
7. Volkmann J, Moro E, Pahwa R. Basic algorithms for the programming of deep brain stimulation in Parkinson's disease Mov Disord. 2006;21(S14):284-299.
8. McIntyre Cameron C, Butson Christopher R. Tissue and electrode capacitance reduce neural activation volumes during deep brain stimulation Clinical Neurophysiology. 2005;116:2490-2500.
9. Butson C, Maks C, Walter B, Vitek J, McIntyre CC. Patient-Specific Analysis of the Volume of Tissue Activated During Deep Brain Stimulation Neuroimage. 2007;34(2):661-670.
10. Tuch D.S, Wedeen V.J, Dale A.M, George J.S, Belliveau J.W. Conductivity tensor mapping of the human brain using diffusion tensor MRI Proc Natl Acad Sci USA. 2001;98:11697–11701.
11. Miocinovic S, Lempka S, Russo G, et al. Experimental and theoretical characterization of the voltaje distribution generated by deep brain stimulation Exp Neurol. 2009;216(1):166-176.
12. Roth B.J. Steady-state point-source stimulation of a nerve containing axons with an arbitrary distribution of diameters Med. Biol. Eng. Comput. 1992;30:103-108.

Computer-Aided Diagnosis of Abdominal Aortic Aneurysm after Endovascular Repair Using Active Learning Segmentation and Texture Analysis

G. García[1], J. Maiora[2], A. Tapia[1], M. Graña[3], and M. De Blas[4]

[1] Systems Engineering and Automatic Control Department –EUP, University of the Basque Country, San Sebastián, Spain
[2] Electronic Technology Department –EUP, University of the Basque Country, San Sebastián, Spain
[3] Computational Intelligence Group, University of the Basque Country, San Sebastián, Spain
[4] Radiology Department, Donostia Hospital, San Sebastián, Spain

Abstract—Endovascular repair is a minimal invasive alternative to open surgical therapy. From a long term perspective, complications such as prostheses displacement or leaks inside the aneurysm sac (endoleaks) could appear influencing the evolution of treatment. The objective of this work is to develop a Computer-Aided Diagnosis system (CAD) for an automated classification of EndoVascular Aneurysm Repair (EVAR) progression from Computed Tomography Angiography (CTA) images. The CAD system is based on the extraction of texture features from segmented thrombus aneurysm samples and a posterior classification. Image segmentation is produced by an interactive Active Learning procedure based on Random Forest (RF) classification.. An initial set of labeled points are required to start training an RF. The human operator is presented with the most uncertain unlabeled voxels to select some of them for inclusion in the training set, retraining the RF classifier. Three texture-analysis methods such as the gray level co-occurrence matrix (GLCM), the gray level run length matrix (GLRLM), and the gray level difference method (GLDM), were applied to each ROI to obtain texture features. Classification of the ROI is carried out by an ensemble of SVM classifiers (ECs). The final decision is based on the application of a voting scheme across the outputs of the individual SVMs. The performance of the classifier was evaluated using 10-fold cross validation and the average of accuracy (93.84% ± 0.29), sensitivity (94.72 ± 0.23), and specificity (91.48 ± 0.35) results.

Keywords—Aneurysm, EVAR, segmentation, texture features, ensembles of classifiers.

I. INTRODUCTION

The endovascular prostheses in Abdominal Aortic Aneurysm has proven to be an effective technique to reduce the pressure and rupture risk of aneurysm, offering shorter post-operation recovery than open surgical repair. The Endovascular Aneurysm Repair (EVAR) treatment, is a percutaneous image-guided endovascular procedure in which a stent graft is inserted into the aneurysm cavity. Once the stent is placed, the blood clots around the metallic mesh forcing the blood flux through the stent and reducing the pressure on the aneurysm walls consequently. Nevertheless, in a long term perspective different complications such as prostheses displacement or leaks inside the aneurysm sac (endoleaks) could appear provoking a pressure elevation and increasing the danger of rupture consequently. Due to this, periodic follow-up scans of the prosthesis behaviour are necessary. At present, contrast enhanced Computed Tomographic Angiography (CTA) is the most commonly used examination for imaging surveillance [1]. On the other hand, the post-operation analysis is quite crude as it involves manually measuring different physical parameters of the aneurysm cavity [2]. According to these measurements, the evolution of the aneurysm can be split up into two main categories: in favourable evolution a reduction of the diameter of the aneurysm sac can be observed. This means that the aneurysm has been correctly excluded from the circulation. In unfavourable evolution a growth of the aneurysm diameter is observed and endoleaks can sometimes be detected thanks to contrast in the CT images. According to our previous studies [3], texture thrombus in favourable shrinking aneurysms differs from unfavourable expanding or stable ones. A CAD system based on texture analysis can help clinicians to determine the evolution of mass thrombus inside the aneurysm sac, particularly in ambiguous aneurysm evolution cases. This is clinically important because a more complete assessment of EVAR progression would be useful in re-defining patients management [4].

In recent years, many efforts have been put into the developing of Computer Aided Diagnosis (CAD) systems based on image processing methods. The principal motivation for the research on this kind of systems has been to assist the clinicians on the analysis of medical images, [5]. In many occasions this analysis implies the detection or measurement of subtle differences, usually difficult to appreciate by visual inspection even for experienced radiologist. CAD systems have been successfully utilized in a wide range of medical applications [6-8]. A particular field inside the image processing methods for CAD systems is the so named texture-based analysis. This analysis studies, not only the variation of the pixel intensity values along the image but also the possible spatial arrangement of them, and the more or less periodic repetition of such arrangement (primitives). From this point of view texture analysis can help on the functional characterization of different kind of organs, tissues, etc, at the evolution of disease. The textures features obtained from the analysis can be fed as inputs for a deterministic or probabilistic classifier, which assign each sample with its specific class.

Prior to the texture analisys, a segmentation of the AAA thrombus is required. This is a challenging task due to the low contrast of signal intensity values between the AAA thrombus and its surrounding tissue. Furthermore, the AAA thrombus shows great shape variability, both intra and inter-subjects, so that little prior information is available to guide the segmentation. General reviews of blood vessel segmentation methods are given in [9].

AAA thrombus segmentation methods reported in the literature need a lot of human interaction or *a priori* information one way or the other [10-12]. The approach followed in this paper for AAA thrombus segmentation is to build a voxel classifier into AAA thrombus or background classes. An The Active Learning system returns to the user the unlabeled voxels whose classification outcome is most uncertain with the current classifier. After manual labeling by the user, voxels are included into the training set and the classifier is trained again [13]. After segmentation, thrombus samples were manually extracted by experiment radiologist.

Three statistical texture methods, namely the Gray Level Co-ocurrence Matrix (GLCM) [14], the Gray Level Run Length Matrix (GLRLM) [15] and the Gray Level Difference Method (GLDM) [16] were applied to the thrombus samples. The validity of these methods has been proved in many studies [17-19]. In the last years, multiple classifier architecture has been proposed to improve the performance of CAD systems [20][21]. In multiple classifier systems, also called ensembles of classifiers (EC), an initial prediction is made by several separate classifiers and then fused into one final assessment through a combining strategy [22].

Our purpose in this study is to develop a preliminary support system based on textures analysis which provides clinicians with complementary information for correctly classifying EVAR treatment evolution. This is clinically important because a more complete assessment of EVAR progression would be useful in re-defining management pathways for patients, particularly when new treatment options are available [4].

II. MATERIALS AND METHODS

A. Dataset

The CTA image scans used in this work were obtained by a group of 3 experienced radiologists from the Vascular Surgery Unit and Interventional Radiology Department of the Donostia Hospital. These CTA images belong to the scan studies of 90 patients with ages ranging from 65 to 93 years, conducted over a maximum of 5 years in fixed periods of 6-12 months. The total patient set was selected by each one of the radiologist, unaware of each others results. In order to generate ground-truth data sets only cases with complete agreement on classification among all radiologists were selected. Balanced samples sets for training and testing the classifier system were obtained. All the studies were taken from the abdominal area with a spatial resolution of 512×512 pixels and 12-bit gray-level at the WW400 y WL40 window and 5 mm thickness in DICOM format. For each patient a maximum of three ROIs (15x15 pixels) from aneurysm thrombus were extracted from different slices, resulting in a total of 270 ROIs. Half of them corresponded to the "favourable evolution" group and the rest to the "unfavourable" one.

B. Active Learning Thrombus Segmentation

The goal is to classify image voxels into two classes, the target region and the background. The feature vector associated to each pixel for its classification is computed using information from its neighboring voxels, applying linear and/or non-linear filtering. In this paper the features initially associated with CTA voxels are: its coordinates in the data domain grid, the voxel intensity, the mean, variance, maximum and minimum of the voxel neighborhood, for different values of the neighborhood radius $(1,2,4...2^n)$.

A first step towards the practical feasibility of the approach (meaning affordable computation times) is the selection of the most informative features, reducing data dimensionality. In this paper feature selection is done on the basis of the variable importance [23].

Active Learning [24] focuses on the interaction between the user and the classifier. Let $X = \{\mathbf{x}_i, y_i\}_{i=1}^{l}$ be a training set consisting of labeled samples, with $\mathbf{x}_i \in \mathbb{R}^d$ and $y_i \in \{1, ..., N\}$. Let be $U = \{\mathbf{x}_i\}_{i=l+1}^{l+u} \in \mathbb{R}^d$ the *pool of candidates*, with $u \gg l$, corresponding to the set of unlabeled voxels to be classified. In a given iteration t, the Active Learning algorithm selects from the pool U^t the q candidates that will, at the same time, maximize the gain in performance and reduce the uncertainty of the classification model when added to the current training set X^t.

C. Texture Analysis – Feature Extraction

Three conventional texture-analysis methods such as the gray level co-occurrence matrix (GLCM) [14], the gray level run length matrix (GLRLM) [15] and the gray level difference method (GLDM) [16], were applied to each ROI to obtain texture features. Regarding the GLCM, features have been calculated at distance 1 due to the reduced size of the aneurysm samples. Initially, the assumption of an isotropic texture distribution inside the aneurysm sac was considered. Consequently averaging over the four angular directions was computed. To reduce the influence of random noise on texture features, the number of gray levels was reduced to 16 prior to the accumulation of the matrix. From GLCM matrix a set of 13 features has been evaluated. Regarding the GLRLM, averaging over the four angular directions was computed and the number of gray levels has been kept in 16, equal than in the GLCM method in order to make both methods comparable. In our case, 11 features has been calculated. Concerning the GLDM, d distance was considered equal to 1 and 5 textural features were measured. The assumption of an isotropic texture distribution inside the aneurysm sac was also considered, consequently averaging over the four angular directions was computed.

D. Classification

The classification scheme (Fig 1) consists of an ensemble of three SVMs [25], each using as input one of the three feature vector types (GLCM, GRLM, and GLDM). A SVM with 13 inputs for GLCM method, another with 11 inputs for GLRLM method, and other one with 5 inputs for GLDM method were implemented. All the textural features were normalized by the sample mean and standard deviation of the data set before being fed to the SVM. The final decision of the system is generated by combining the diagnostic outputs of individual SVMs through a majority voting scheme [22].

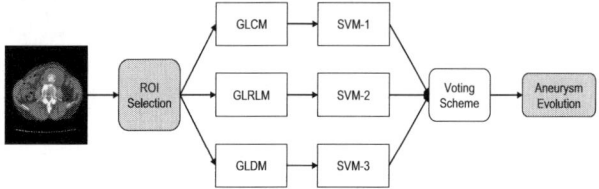

Fig. 1 Implemented CAD scheme.

III. RESULTS

In order to evaluate the potential of texture analysis based CAD systems to classify between the two types of aneurysm evolution the development and validation of the SVM structure has been based on the 10-fold cross validation method. To improve the evaluation of performance of the feature sets, the 10-fold cross validation was repeated 6 times, averaging the results. The average of total percent of correctly classified cases (accuracy) for all the trials was used as an estimate of the performance of each classifier. Table 1 shows the accuracy values (mean ± standard deviation) estimated by the 10-fold cross validation of the testing sets of individual classifiers and the ensemble of them.

Table 1 Mean and standard deviation values for classification Accuracy (in %), Sensitivity (%) and Specifity (%) testing sets of individual classifiers and the ensemble of them.

Classifier	Accuracy (%)	Sensitivity (%)	Specificity (%)
GLCM	91.37± 0.24	93.33± 0.40	86.96± 0.31
GLRLM	81.12 ± 0.53	83.01 ± 0.48	76.83 ± 0.51
GLDM	88.47 ± 0.31	90.41 ± 0.29	82.38 ± 0.43
Ensemble of Classifiers	93.84 ± 0.29	94.72 ± 0.23	91.48 ± 0.35

From the Table 1 it is shown that the ensemble of classifiers architecture yielded to the best accuracy performance (93.84% ± 0.29) followed by a 91.37% ± 0.24, 88.47 ± 0.31, and 81.12 ± 0.53, obtained from GLCM, GLRLM and GLDM texture methods respectively.

IV. CONCLUSIONS

In the present study we have analysed the capacity of CAD systems based on textures analysis for classifying evolutions experimented by aorta aneurysms treated by EVAR.

We apply an Active Learning approach for training RF classifiers for the segmentation of the thrombus in EVAR. A high classification accuracy is obtained in all slices and CTA volumes tested after four iterations adding five voxels at each iteration.

Texture analysis has been applied by three different methods, GLCM, GRLM, and GLDM. At the same time an ensemble of three SVMs classifiers have been proposed for the classification system. As a conclusion, it can be considered that the CAD system proposed could be used as a second opinion tool for assisting physicians in characterization of aneurysm evolution after EVAR treatment. However, taking into account the limited number of patients, further investigations will be needed to assess the potential of the system as a clinical tool.

REFERENCES

1. Thompson MM: Controlling the expansion of abdominal Aneurysm. Br. J. Surg 2003; 90:897-898.
2. S. William Stavropoulos and Sridhar R. Charagundla: Imaging Techniques for Detection and Management of Endoleaks after Endovascular Aortic Aneurysm Repair, Radiology, June 2007; 243: 641 - 655.
3. G. García, J. Maiora, A. Tapia, M. De Blas. Evaluation of texture for classification of abdominal aortic aneurysm after endovascular repair. Journal of Medical Imaging. 2012; 25(3); 369-376.

4. Bashir, Mustafa R., Ferral, Hector, Jacobs, Chad, McCarthy, Walter, Goldin, Marshall. Endoleaks After Endovascular Abdominal Aortic Aneurysm Repair: Management Strategies According to CT Findings Am. J. Roentgenol. 2009 192: W178-186.
5. Computer-Aided Diagnosis in Medical Imaging: Historical Review, Current Status and Future Potential. Kunio Doi. Comput Med Imaging Graph. 2007; 31(4-5): 198–211.
6. Morton MJ, Whaley DH, Brandt KR, Amrami KK. Screening mammograms: interpretation with computer-aided detection-prospective evaluation. Radiology 2006;239:375–383.
7. Boniha L, Kobayashi E, Castellano G, Coelho G, Tinois E, Cendes F, et al. Texture analysis of hippocampal sclerosis. Epilepsia 2003;44: 1546–50.
8. Arimura H, Li Q, Korogi Y, Hirai T, Abe H, Yamashita Y, Katsuragawa S, Ikeda R, Doi K. Automated computerized scheme for detection of unruptured intracranial aneurysms in threedimensional MRA. Acad Radiol 2004;11:1093–1104.
9. D. Lesage, E. D. Angelini, I. Bloch, and G. Funka-Lea, "A review of 3D vessel lumen segmentation techniques: Models, features and extraction schemes," *Med. Image Anal.*, vol. 13, no. 6, pp. 819 – 845, 2009.
10. M. de Bruijne, B. van Ginneken, M. A. Viergever, and W. J. Niessen, "Interactive segmentation of abdominal aortic aneurysms in CTA images.," *Med Image Anal*, vol. 8, no. 2, pp. 127–138, Jun. 2004.
11. S. Demirci, G. Lejeune, and N. Navab, "Hybrid Deformable Model for Aneurysm Segmentation.," in *ISBI'09*, 2009, pp. 33–36.
12. M. F. ans S. Esses, L. Joskowicz, and J. Sosna, "An iterative model-constrained graph-cut algorithm for Abdominal Aortic Aneurysm thrombus segmentation.," in *7th IEEE Int. Symposium on Biomedical Imaging (ISBI 2010)*, 2010.
13. J. Maiora and M. Graña, "Abdominal CTA Image Analisys Through Active Learning and Decision Random Forests: Aplication to AAA Segmentatio," in *IJCNN*, 2012.
14. Haralick, R., Shanmugam, K., Dinstein, I.: Textural features for image classification. IEEE Transactions on Systems, Man, and Cybernetics SMC-3 (1973), 610–621.
15. R. M. M. Galloway, "Texture analysis using gray level run lengths," Comput. Graphic. Image Processing, vol. 4, pp. 172–179, 1975.
16. J. S. Weszka, C. R. Dyer, and A. Rosenfeld, "A comparative study of texture measures for terrain classification," *IEEE Trans. Syst., Man, Cybern.*, vol. SMC-6, pp. 269–285, Apr. 1976.
17. Mir, A.H.; Hanmandlu, M.; Tandon, S.N.; , "Texture analysis of CT images," *Engineering in Medicine and Biology Magazine, IEEE*, vol.14, no.6, pp.781-786, Nov/Dec 1995.
18. Gibbs P, Turnbull LW. Textural analysis of contrast-enhanced MR images of the breast. Magn Reson Med 2003; 50:92–98.
19. Nikita A, Nikita KS, Mougiakakou SG, Valavanis IK. In: Evaluation of texture features in hepatic tissue characterization from non-enhanced CT images. Engineering in medicine and biology society, 2007. EMBS 2007. 29th annual international conference of the IEEE; 2007. p. 3741-4.
20. Sboner A, Eccher C, Blanzieri E, Bauer P, Cristofolini M, Zumiani G, et al. A multiple classifier system for early melanoma diagnosis. Artif Intell Med 2003;27(1):29—44.
21. Jerebko AK, Malley JD, Franaszek M, Summers RM. Multiple neural network classification scheme for detection of colonic polyps in CT colonography data sets. Acad Radiol 2003;10(2):154—60.
22. Roli F, Giacinto G, Vernazza G. Methods for designing multiple classifier systems. In: Roli F, Kittler J, editors. Proceedings of 2nd international workshop on multiple classifier systems in volume 2096 of lecture notes in computer science. London, UK: Springer Verlag; 2001. p. 78—87.
23. R. Genuer, J.-M. Poggi, and C. Tuleau-Malot, "Variable selection using random forests," *Pattern Recognit. Lett.*, vol. 31, no. 14, pp. 2225–2236, Oct. 2010.
24. D. Tuia, M. Volpi, L. Copa, M. Kanevski, and J. Munoz-Mari, "A Survey of Active Learning Algorithms for Supervised Remote Sensing Image Classification," *Sel. Top. Signal Process. Ieee J.*, vol. 5, no. 3, pp. 606 –617, Jun. 2011.
25. Cortes and Vapnik, 1995 C. Cortes and V. Vapnik, Support-vector networks, *Machine Learning* 20 (2) (1995), pp. 273–297.

Assessment of Magnetic Field in the Surroundings of Magnetic Resonance Systems: Risks for Professional Staff

V.M. Febles Santana[1], J.A. Hernández Armas[1], M.A. Martín Díaz[1], S. de Miguel Bilbao[2], J.C. Fernández de Aldecoa[1], and V. Ramos González[2]

[1] Department of Engineering, Canary University Hospital Consortium, La Laguna (Santa Cruz de Tenerife), Spain
[2] Telemedicine and e-Health Research Unit, Health Institute Carlos III, Madrid, Spain

Abstract—The objective of this study is the assessment of the exposure levels of the static magnetic field from three equipments of resonance magnetic imaging (RMI) installed in annexed areas of the Canary University Hospital Consortium (CUHC).

The measurements were performed in predefined points following the methodology established by this research group in previous works.

The result of the measurements allows concluding that for distances greater than 1 meter from the diagnostic equipment, the field levels are lower than 20 mT. This involves that the most restrictive thresholds that are established in the Spanish and European standards are never reached, so the emissions from the RMI equipments are safe for the practitioners using these diagnostic devices.

Keywords—resonance magnetic imaging, electromagnetic fields, exposure threshold, occupational risk prevention.

I. INTRODUCTION

Since the time the X-rays began to be used for diagnostic imaging, those in charge have had an ongoing concern for monitoring the health of workers exposed to new physical agents used in medicine. Therefore, legislators have had a significant concern to measure, assess and control the risks to users and professionals under the action of ionizing radiation, promoting protective measures and standards to regulate and limit the use of these medical devices.

Current European legislation about the exposure of professionals to electromagnetic fields in relation to the assessment of occupational risk prevention, aims to become more and more stringent. However, those far, a final agreement about the field levels in the proximity of MRI equipments has not been reached [1]. While the law does not change, for now, the duty of those in charge of these systems is to monitor the health of professionals and to ensure that exposure levels do not exceed the limits in force.

In the case of the electromagnetic fields (EMF) generated by the system of magnetic resonance used in the diagnostic by imaging, the legislator proposes warnings about the thresholds to ensure the prevention of overexposure of health professionals, although there is no evidence that low levels of magnetic fields may cause adverse effects.

In Spanish legislation, Royal Decree 1066/2001 establishes the conditions for protection of public radio, the radio emission restrictions, and measures of health protection against radio emissions [2]. This decree establishes the limit of 40 mT in the case of general public exposure to static magnetic fields (0 Hz).

Also in Spain, the General Health Law 14/1986 [3] attributed to the health administration the competence of sanitary control of products, components or forms of energy that can be harmful to human health. The Royal Decree 1450/2000 [4] establishes the power of assessment, prevention and health monitoring of non-ionizing radiation, both for workers and for the general public.

The European Directive 2004/40/EC establishes the minimum requirements for the protection of the health of workers exposed to EMF [5]. Indications are provided to define allowable working areas where risks related to the action of non-ionizing radiation can be harmful to normal activity.

While the Spanish legislation establishes a value of 40 mT as the threshold for exposure of the general population to static magnetic fields, in the European Directive, for the moment, the threshold of magnetic induction is 200 mT. If this value is exceeded, actions must be taken to protect the exposed workers from the occupational risks.

The main interest of this work is to carry out the measurements, registration and control of the static EMF levels in the proximity of MRI equipment. The evaluated equipments were three systems located in an annexed space of the Canary University Hospital Consortium (CUHC). The objectives of this work are the following:

- To compare the obtained values of magnetic induction of the static field (0 Hz), with values provided by the manufacturer that were ensured in the processes of production and installation.
- To analyze the levels of static magnetic field comparing them with the thresholds of the exposure of the general public established in the current Spanish legislation.
- To evaluate the compliance with European standards regarding the occupational risks of the professionals exposed to static magnetic fields.

II. METHOD AND MATERIALS

The three evaluated imaging diagnostic systems are managed by the public company Medical Institute from Tenerife (IMETISA). The three analyzed equipments are identified as RM-1, RM-2, RM-3. RM-1 and RM-2 and located in the room in the second basement of an annexed space of the CUHC, and RM-3 is located in the second basement of the building of the hospital. The model and brand of each of the equipment are the following: RM-1: EXCITE HD 3.0T, GE Healthcare; RM-2: SIGNA EXCITE MR/i 1.5T, GE Healthcare; and RM-3: SIGNA HORIZON 5x of 1.5 T, GE Medical Systems.

The considered unit of the static magnetic field is militesla (mT). The measurements were carried out with a 4048 F. W. Bell gaussmeter or magnetometer, equipped with a T4048-001flat probe.

The magnetic induction was measured in the spaces where equipments are located in the points belonging to a predefined grid of 5 x 5 m^2, precisely in the vertices of each square within the grid. The levels of the magnetic field were also measured in several rooms around the space dedicated to the MRI, such as: control stations, waiting rooms, patient preparation areas, etc…, in these spaces a predefined grid of 2.5 x 2.5 m^2 was used. All in accordance with the methodology established by this research group in previous works [6], [7].

In total 167 measurements were performed. Fig. 1 is the map of the location of RM-1 and RM-2, and Fig.2 is the map of the location of RM-3. All maps have been obtained with AUTOCAD software. The points of measurement are marked, and are identified with a letter and a number.

III. RESULTS

It has been observed that the obtained values in the rooms where the RMI equipments are not located are always lower than 0.2 mT. This implies that the levels of exposure to static magnetic fields are negligible and far below the Spanish and European standards, respectively, for workers and for the general public. In the case of the rooms where the RMI equipments are located, the obtained values are always lower than 20 mT, considering the following distances from the casing of the gantry: more than one metre for RM of 3 T, and 0.5 metres for the 1.5 T equipments.

The positions close to the geometrical centre of the magnet are practically occupied only by the patient during the development of the diagnostic test [8]. This would mean that the active and passive shielding, designed and installed by the manufacturer, operates according to expectations.

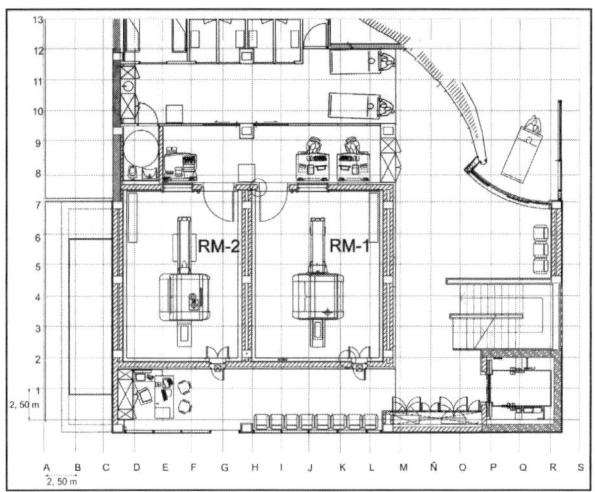

Fig. 1 Map of the spaces where the equipments RM-1 and RM-2 are located.

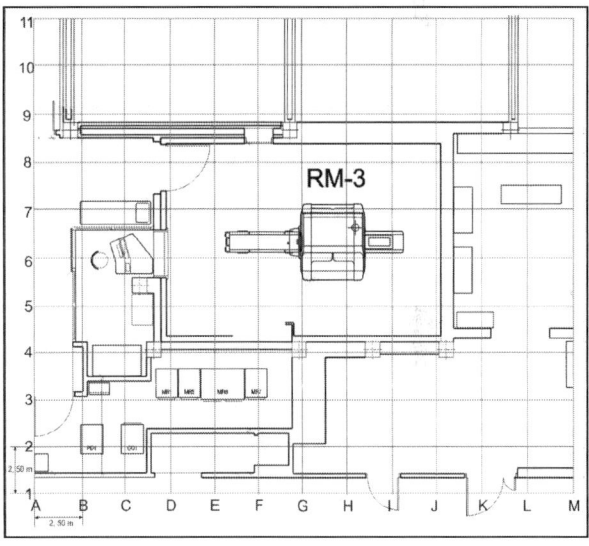

Fig. 2 Map of the spaces where the equipment RM-3 is located.

Furthermore, the obtained measurements of the magnetic induction match the iso-field lines reflected in the data sheet provided by the manufacturer.

Table 1 shows the obtained values inside the rooms where RM-1 and RM-2 are located. The gaussmeter exceeds its measurement range in points very close to the gantry of the RMI, indicating with the symbol ">" that the field value is greater than 20 mT. Evidently it was not possible to measure at points coincident with the gantry, these points are shown in the table with an X.

Table 1 Values of the static magnetic field in the room where the RM-1 and RM-2 equipments are located

Magnetic field values (mT)						
Point	2	3	4	5	6	7
D	0,99	1,49	2,15	2,15	1,22	0,23
E	8,91	>20	X	>20	7,11	0,83
F	13,1	>20	X	>20	6,28	1,17
G	1,53	1,29	2,42	4,36	1,15	0,48
H	0,58	0,65	0,86	0,78	0,34	0,4
I	16,2	>20	>20	>20	5,42	1,54
J	17,43	>20	X	>20	14,56	2,4
K	9,77	>20	X	>20	12,92	1,89
L	1,08	4,36	3,91	6,58	3,35	0,92

Table 2 shows the measured static field values in the surroundings rooms of RM-1 and RM-2, such as: the equipment operators rooms, work rooms of medical practitioners, toilets, changing rooms, waiting rooms or parking stretchers with patients waiting to enter the diagnostic rooms, etc.... The values of the magnetic induction are lower than 0.20 mT, being in most cases of the order of hundredths of mT (0.01 to 0.03 mT).

Table 2 Values of the static magnetic field measured outside of the room where RM-1 and RM-2 equipments are located.

Magnetic field values (mT)							
Point	1	3	5	7	9	11	13
A	0,01	0,01	0,01	0,01	0,01	0,01	0,01
C	0,01	0,01	0,01	0,01	0,01	0,01	0,01
E	0,14	-	-	-	0,04	0,03	0,03
G	0,15	-	-	-	0,09	0,03	0,02
I	0,08	-	-	-	0,08	0,03	0,01
K	0,20	-	-	-	0,13	0,01	0,01
M	0,03	0,03	0,10	0,01	0,01	0,03	0,03
O	0,03	0,04	0,03	0,03	0,01	0,04	0,03
Q	Lift	0,01	0,04	0,03	0,04	0,04	0,02
S	0,01	0,01	0,01	0,01	0,01	0,01	0,01

Tables 3 and 4 contain the measurements in the surrounding rooms and in the room where RM-3 is located. This location is in the second basement of the building of the hospital, far from RM-1 and RM-2.

The measured data was transferred to the software Surfer 8 to draw 2D colour maps according to the previously measured levels of static magnetic field in the different rooms. Fig. 3 is a colour map of the levels of static magnetic field generated by the equipments RM-1 and RM-2 and Fig. 4 is a colour map of the levels of static magnetic field generated by the equipments RM-3.

Table 3 Values of the static magnetic field in the room where the RM-3 equipment is located

Magnetic field values (mT)				
Point	5	6	7	8
D	0,24	0,33	0,27	0,10
E	0,92	2,20	1,86	0,29
F	1,19	16,07	5,6	0,44
G	10,57	X	X	0,77
H	2,1	X	X	0,88
I	2,50	18,0	16,99	0,63
J	0,54	3,82	5,90	0,28

Table 4 Values of the static magnetic field measured outside of the room where RM-3 equipment is located

Magnetic field values (mT)						
Point	1	3	5	7	9	11
A	0,03	0,08	0,08	0,07	0,04	0,04
C	0,05	0,02	0,12	0,10	0,05	0,04
E	0,04	0,02	-	-	0,09	0,03
G	0,04	0,05	-	-	0,06	0,03
I	0,03	0,04	-	-	0,05	0,03
K	0,03	0,02	0,07	0,05	0,04	0,03
M	0,03	0,04	0,05	0,03	0,07	0,05

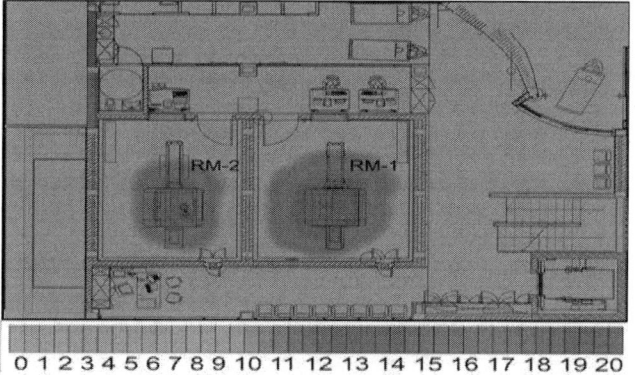

Fig. 3 Colour map of levels of static magnetic field in mT of the equipments RM-1 and RM-2.

Analyzing Fig. 3 and Fig. 4, it can be detected a rapid increase of magnetic induction level in the proximity of the gantry for each of the equipment. The levels decrease immediately in a very short distance from the equipments. The colour changes abruptly from red (values of 20 mT) to light green (values lower than 10 mT), so the number of values corresponding to dark green (values from 10 mT to 15 mT) is nearly nonexistent.

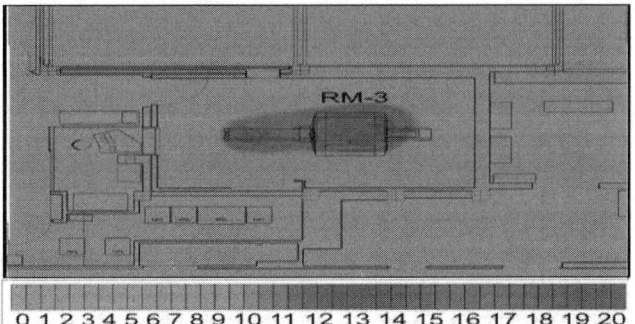

Fig. 4: Colour map of levels of static magnetic field in mT of the equipment RM-3.

Examining the results in the surroundings rooms, where the equipments are not located, the measured values are considerably low.

IV. CONCLUSION

- Magnetic induction levels of the static field (0 Hz) are less than 20 mT inside the rooms where the MRI equipments are located. But this limit is exceeded in the positions very close to the gantry where the permanent magnet of the own equipment itself is situated (less than 1 meter for 3 T equipment and less than 0.5 for the 1.5 T equipment).
- In the adjacent rooms, the field levels are always below 0.20 mT, and in general, the order of hundredths of mT (between 0.01 and 0.03 mT).
- The active and passive shielding installed by the manufacters is highly effective. The area around MRI equipments, where the levels of the static magnetic field are more significant, is perfectly defined.
- The iso-field lines of magnetic induction that have been obtained, roughly correspond to those provided by the manufacturer in their technical data sheets.
- The value established by the Spanish legislation as the limit for the general population exposure to electromagnetic fields of 0 Hz, is 40 mT. This value is reached in locations very close to the gantry of the MRI equipment, where only the patients are located during the diagnostic tests.
- For equipment tested in this work the static field levels are not above 200 mT, which is the limit set by European legislation to ensure the protection of exposed workers in terms of occupational risk protection.

V. ACKNOWLEDGMENT

Authors want to thank the valuable cooperation of the staff of GE Healthcare in the Canary Islands, and the staff of the public company IMETISA.

REFERENCES

1. Guibelalde E. Exposiciones ocupacionales a campos electromagnéticos en la proximidad de equipos de resonancia magnética para uso clínico. Estado actual de la Directiva Europea 2004/40/EC. *Revista de Física Médica,* vol 9 (1), 2008, pp 25-32.
2. Royal Decree 1066/2001, of 28 September, which approves the Regulation establishing the conditions for protection of public radio, radio emission restrictions and measures of health protection against radio emission
3. Law 14/1986, of 25 April, General Health
4. Royal Decree 1450/2000, of July 28, on the basic organic structure of the Ministry of Health
5. Directive 2004/40/EC of the European Parliament and of the Council of 29 April 2004 on the minimum health and safety requirements regarding the exposure of workers to the risks arising from physical agents (electromagnetic fields)
6. Febles VM, Placeres JM, Ascanio Velázquez, et.al. Convivencia de señales electromagnéticas en medios hospitalarios. *SEEIC-2007*, Córdoba, 2007, pp. 16.
7. Hernández JA, Carranza N, García J, et. al. Metodología para el establecimiento de Mapas de Niveles de Intensidad de Campos Electromagnéticos en Hospitales. *CASEIB 2009*, Cádiz, 2009 pp. 244
8. Carranza N, Febles V, Hernández JA, et. al, Patient safety and electromagnetic protection: a review. *Health Physics*, vol 100 (5), 2011, pp 530-541.

Author: Victor Febles Santana
Institute: Department of Engineering, CUHC
Street: s/n La Cuesta
City: La Laguna (Santa Cruz de Tenerife)
Country: Spain
Email: vfebsan@gobiernodecanarias.org

Morphological Characterization of the Human Calvarium in Relation to the Diploic and Cranial Thickness Utilizing X-Ray Computed Microtomography

E. Larsson[1,2,3], F. Brun[1,2], G. Tromba[2], P. Cataldi[4], K. Uvdal[3], and A. Accardo[1]

[1] Department of Architecture and Engineering, University of Trieste, Trieste, Italy
[2] Sincrotrone Trieste S.C.p.A., Basovizza, Trieste, Italy
[3] Department of Physics, Chemistry and Biology, Linköping University, Linköping, Sweden
[4] Department of Pathological Anatomy, Azienda per i Servizi Sanitari n. 5 – "Bassa Friulana", Italy

Abstract—When attempting to establish accurate models for the human diploe, micro-scale morphological differences in the four main areas of the calvaria could also be considered. In this study, X-ray computed microtomography (μ-CT) images were analyzed in order to quantitatively characterize the micro-architecture of the human calvarium diploe. A bone specimen from each area of the skull (temporal, frontal, parietal and occipital) was extracted from a set of 5 human donors and each specimen was characterized in terms of density, specific surface area, trabecular thickness, trabecular spacing. The obtained results revealed that subject-individual structural differences could be related with the diploic as well as the total cranial thickness of the human skull bones. Some tendencies of dependency could also be made with respect to the age of the subject. A consideration of these individual variations can improve traditional models that assume equal conditions throughout the skull.

Keywords—computed microtomography, image processing, image analysis, cranial bones, calvarium diploe.

I. INTRODUCTION

The characterization of human calvarial bone structures is of interest for areas such as anthropology, forensic medicine, craniofacial surgery [2], calvarial bone grafts development [5] and for refinement of EEG (electroencephalogram) source localizations of brain activity, where the measurements are dependent on the conductivity and anatomical properties of the skull [6,7].

Sabancioğullari et. al stated that the results in the literature related to the investigation of the total cranial thickness and its relation to age or gender are conflicting and pointed out some main reasons for this, e.g. a lack of sample group, different measuring points or diseases affecting cranial bones [2]. Lynnerup et. al. studied biopsies of skull bones by X-ray projection imaging, in order to estimate the diploic thickness instead [4] and compared it to their earlier physical measurements [3] of the total cranial thickness and found a strong correlation between the two [4]. Sabancioğullari et. al also measured the diploic thickness based on MRI data and found a statistically significant correlation for an increase of the diploic thickness with age [2].

They also reported that the diploic thickness was lower in parietal bone, than in the frontal and occipital bone for both male and female patient.

Todd et al. reported that up to the age of 60 years, slight increases of the cranial thickness can be detected, but that the cranial thickness alone is not enough to estimate the age of a patient [8].

Apart from the already debated correlation between age and the diploic and total cranial thickness, we believe that these length scale measures can be explained by investigating the micro-architecture of the trabecular bone, present in the inner part of the diploe.

In this work, biopsy bone specimens of cranial bones (temporal, parietal, frontal and occipital) from 5 human donors were scanned using computed microtomography (μ-CT). The obtained three-dimensional (3D) images were then analyzed with a set of quantitative descriptors, in order to detect structural differences in-between the skull bone regions for each subject. The focus of this work was also to relate these structural properties with the total cranial and diploic thickness.

II. MATERIALS AND METHODS

A. Materials Characterization

A bone specimen from each region of the skull – temporal (T), parietal (P), frontal (F), and occipital (O) was extracted from 5 human donors[1] in apparent health status (3 females and 2 males in the ages 50-81 years). The samples were stored in 10% formalin solution until the time of acquisition. Details for each considered subject are reported in Table 1.

B. Micro-CT Image Acquisition

The samples were scanned at the TOMOLAB μ-CT facility (www.elettra.trieste.it/Labs/TOMOLAB). The samples

---

[1] The current Italian mortuary police regulation does not require an approval from the ethical committee in case of bone sample for diagnostic examination. The study falls within this condition. Each sample, after examination, is returned.

Table 1 Information about each subject.

Subject	Gender	Age	Birth-Death
F70	Female	70	1941-2011
F59	Female	59	1953-2012
F50	Female	50	1960-2010
M81	Male	81	1930-2011
M58	Male	58	1953-2011

were rotated over an angle of 360° and 1800 tomographic projections were acquired. The operating parameters were adjusted with respect to the varying sample thickness, in order to permit sufficient penetration of the X-rays (H_V = 80-120 kVp, I = 69-88 μA, Al-filter: 0.75-1.25 mm). The source-to-object distance and the object-to-detector distance were chosen between 80-150 mm and 100-205 mm respectively, with respect to the sample width. The obtained spatial resolution of the images was between 10-13 μm. Fig. 1 reports a μ-CT slice section from the occipital region for all subjects.

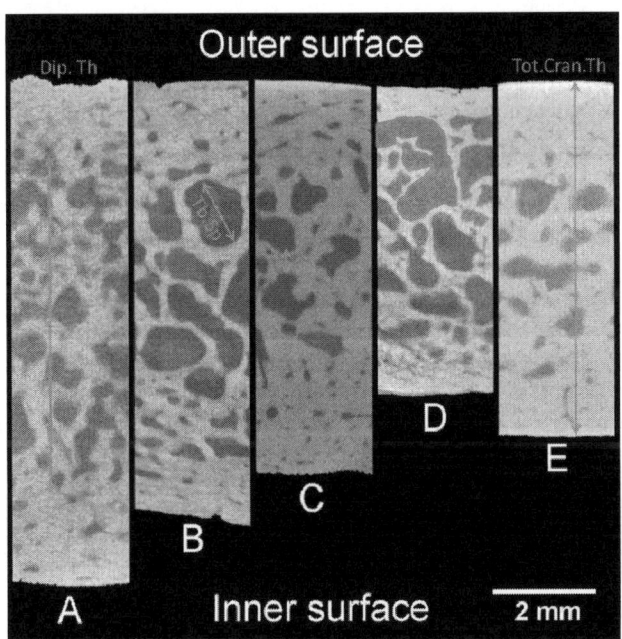

Fig. 1 A μ-CT slice section from the occipital region for subject A) F70, B) F59, C) F50, D) M81 and E) M58. The inner surface was originally facing the cranial cavity, containing the brain. The bone marrow (diploe) is seen in gray and bone in white. Please note the increment of the Tot.Cran.Th and Dip.Th with age for subject A)-C), as well as the difference of BV/TV and BS/TV of bone for subject A)-E).

C. Image Processing and Analysis

The image processing protocol and analysis steps used in this present study has already been established and described in detail for subject F50 [1] and is here briefly summarized. 4 representative VOIs (Volume of Interest) per sample of the size: 2.7 × 2.7 × 2.7 mm³ containing only trabecular bone (found inside the diploe) were chosen far away from sample borders. Each VOI was investigated with a set of quantitative parameters, e.g. bone volume to total volume (BV/TV), bone surface to total volume (BS/TV), trabecular thickness (Tb.Th), trabecular spacing (Tb.Sp) using the Pore3D software library [1].

For each sample the average total cranial thickness was estimated by measuring the minimum and maximum thickness (in total 18 measurements) on 9 μ-CT slices separated with 1.0 mm. The slices above were also used to calculate the Minimal Intensity Projection (minMIP) image, which was then used to estimate the average diploic thickness based on 10 measurements. Both length scale measures were estimated using ImageJ (rsbweb.nih.gov/ij).

III. RESULTS AND DISCUSSION

Fig. 2 shows the mean and standard deviation of the quantitative parameters and length measures for the considered subjects and skull bone regions T, P, F, O. For the BV/TV and Tb.Th parameter a very similar trend occurs for all given regions and subjects, thus hypothesizing that the BV/TV and Tb.Th parameters are proportional to each other. The highest recorded BV/TV and Tb.Th parameter can be seen for the youngest subject (F50), meanwhile the oldest subject (M81) and an intermediate subject in age (F59) demonstrate the lowest values, with overlapping error bars for region F and O for the BV/TV parameter. Moreover, the values in the parietal region of sample M81 are closer to the BV/TV and Tb.Th values reported by the younger subjects F70 and M58 for the same skull bone region, with some overlapping error bars.

Further more, the two female subjects F50 and F70 present their peak values for region P and F, meanwhile higher values are seen for the youngest subject in all bone regions, thus proposing and age dependency. However, this theory is not consistent, due to the recorded values of subject F59, which reports the lowest values among the female subjects, with some mentionable overlapping error bars for region O, with subject F70.

For the male subjects M81 and M58 the reported values for region T and P are very close, with some overlapping error bars, thus suggesting that for these bones no age dependency can be seen. However, for region F and O subject M58 reports clearly higher values, than for M81, thus suggesting an age dependency effect, for these two bone regions.

The BS/TV and Tb.Sp parameter shows rather similar trends and it can be concluded that these parameters are proportional to each other, as well as inversely proportional to the two former described parameters BV/TV and Tb.Th.

For the male subjects M81 and M58 similar values are found in region T and P, with overlapping errobars, meanwhile the highest values now can be seen for the oldest subject M81, in region F and O, due to the inverted relationship with the former two parameters.

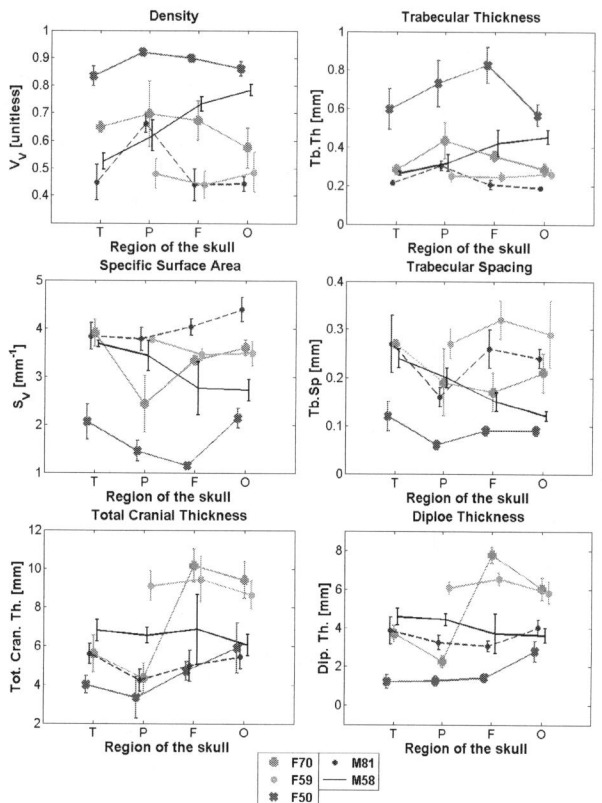

Fig. 2: Quantitative analysis graphs of skull bones for the parameters a) BV/TV, b) BS/TV, c) Tb.Th, d) Tb.Sp, e) Tot.Cran.Th and f) Dip.Th for the considered subjects. (The abbreviations are interpreted as T=temporal, F=frontal, P=parietal and O= occipital bone region).

The female subjects F50 and F70 follow the same type of trend for the BS/TV and Tb.Sp parameter, but with a mentionable wider errorbar in region P. However, subject F59 exhibits some of the highest recorded values of all the subjects for the Tb.Sp parameter.

The total cranial and diploic thickness are clearly proportional to each other (Fig.2), with an average ratio (for all bone regions) of 2.9 (F50), 1.7 (M58), 1.6 (F70) and 1.5 for both (M81) and (F59). These two length measures also showed slight tendencies of being proportional to the BS/TV and Tb.Sp parameter and thereby inversely proportional to the BV/TV and Tb.Th parameters. This seems reasonable for subject: F50, M58 and F59, thus suggesting that bones with a thinner cranial and diploic thickness as for subject (F50), weigh this up with generating more dense bones (BV/TV), in order to be able to bare the same amount of implied stress on the bones. On the contrary, thicker cranial bones (F59), reduces their BV/TV property, in order to reduce their weight. The same effect can be seen in subject F70 for region F and O, which present thicker cranial and diploic thicknesses and a lower BV/TV property, with all error bars overlapping with subject F59 in region O.

Further more, the sufficiently large increase of the total cranial thickness for subject F70 and F50 for region F and O can be further explained by the observed decrease of the their BV/TV value.

Subject F70 also presents a similar behavior as subject M81 in region T and P. However, in general subject M81 does not respect the previous rule, by showing both lower values of diploic/cranial thicknesses and a lower BV/TV-value, probably due to poor bone status, related to age.

Subject M58 on the other hand reports intermediate values, thus proposing a rather normal trade off between cranial/diploic thicknesses and the BV/TV property.

The mean values and standard deviations of the quantitative parameters and length scale measures (Fig. 2) were utilized to perform a One-way ANOVA, with Tukey 90% simultaneous confidence intervals, in order to reveal any significant differences in-between the given skull bone regions: T, P, F, O. Fig. 3 shows the reliability of the mean population parameter being captured within the observed interval.

In order to obtain an easy and understandable measure of the amount of structural differences of the skull bones for a given subject, the number of significant intervals were divided with the total amount of intervals, which revealed that most significant differences were found in subject M58 (24/36=66.7%), M81 (19/30=52.7%), F50 (18/36=50.0%) and F70 (17/36=47.2%). A lower amount of significant differences were detected in subject F59 (3/18=16.7%).

The sensitivity of each quantitative parameter and length scale measure, used to detect differences between the skull bone regions for all the considered subjects, revealed that the Dip.Th and Tot.Cran.Th detected the highest amount of differences, with 70.4% and 66.7% respectively, meanwhile the BS/TV detected 48.1%, BV/TV and TbTth 40.7% each and Tb.Sp 33.3%.

The sensitivity was also calculated based on each specific skull bone interval and revealed that the structural parameters (BV/TV, BS/TV, Tb.Th, and Tb.Sp) detected between 50-80% of structural differences for all the considered subjects for the bone region intervals F-T, P-T and O-P. A lower amount of structural differences (0-50%) were detected in the bone regions P-F, O-F and O-P. The length scale measures (Tot.Cran.Th and Dip.Th) were able to detect more than 50-80% of length differences for the considered subjects for all the bone region intervals F-T, P-T, O-T, P-F, O-F and O-P.

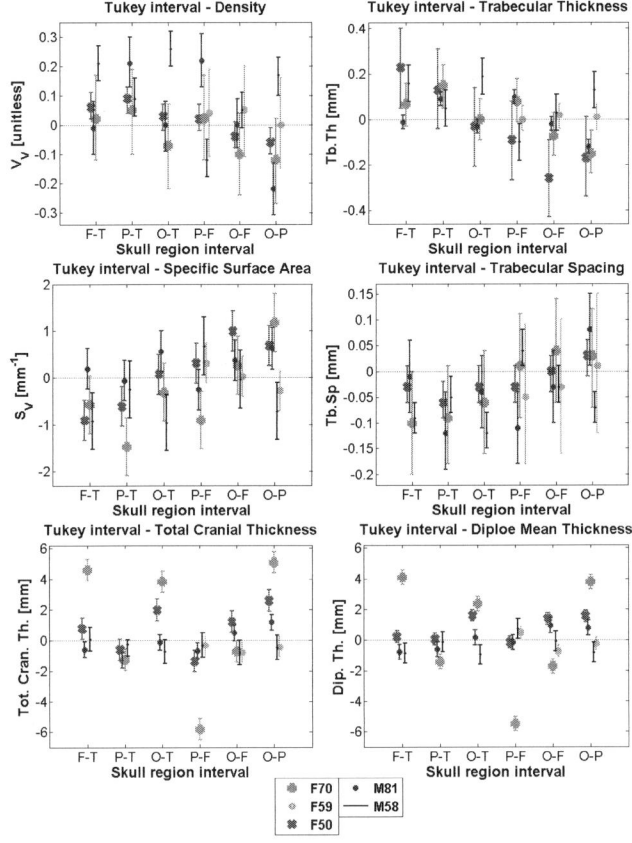

Fig. 3 One-way ANOVA with Tukey 90% simultaneous confidence intervals of each considered parameter. The individual confidence level was 97.50% for the parameters BV/TV, BS/TV, Tb.Th, Tb.Sp for all 5 subjects. For the Tot.Cran.Th parameter the individual confidence level was 97.74% for subject F70, F50, M81, M58 and 95.93% for F59. For the Dip.Th the individual confidence level was 97.70 for subject F70, F50, M81, M58 and 95.87 for F59. Please note that statistically significant intervals do not cross the "zero" (y-axis).

IV. CONCLUSIONS

In this work, a quantitative characterization of the microarchitecture of the human calvarium diploe was performed by means of X-ray computed microtomography. A general correlation between the total cranial thickness and the diploic architecture with age and sex is a long time debated issue with contradicting results. A tendency of age dependency could be detected in this work for certain bone regions and subjects. It could also be shown that some bone regions in average are different from each other, in terms of size and morphology. A rather logical relation between the measured quantitative bone morphological parameters with respect to the length scale measures of total cranial and diploic thicknesses could also be established. Future work based on the consideration of more subjects will better confirm and clarify these preliminary findings.

ACKNOWLEDGMENT

This work was supported by the Ministry of Foreign Affairs, Directorate General for Cultural Promotion and Cooperation in Italy and by the Ministry of Education and Research in Sweden, through the Executive Programme on Scientific and Technological Cooperation between the Italian Republic and the Kingdom of Sweden for the years 2010-2013.

REFERENCES

1. E. Larsson et al. Quantification of structural differences in the human calvarium diploe by means of x-ray computed microtomography image analysis: a case study, IFMBE Proceedings, 37, pp. 599-602, 2011.
2. V. Sabancioğullari et al, Diploe thickness and cranial dimensions in males and females in mid-Anatolian population: An MRI study, Forensic Science International, 219 (1-3), pp. 289.e1-289.e7, 2012.
3. N. Lynnerup et al. Cranial thickness in relation to age, sex and general body build in a Danish forensic sample, Forensic Science International, 117 (1-2), pp. 45-51, 2001.
4. N. Lynnerup et. al, Thickness of the human cranial diploe in relation to age, sex and general body build, Head Face Med. 20 (1) 13, 2005.
5. Hosseinnejad et. al, Modelling and tissue engineering of three layers of calvarial bone as a biomimetic scaffold, Journal of Biomimetics, Biomaterials, and Tissue Engineering, 15, pp. 37-53, 2012
6. R. Hoekema et al., Measurement of the conductivity of skull, temporarily removed during epilepsy surgery, Brain Topography, 16 (1), pp. 29-38, 2003.
7. G. Huiskamp, Interindividual variability of skull conductivity: an EEG-MEG analysis. International Journal of Bioelectromagnetism, 10(1):25-30, 2008.
8. T.W. Todd, Thickness of the male white cranium, Anat. Rec. 27 (5), 245–256, 1924.
9. F. Brun et al. Pore3D: A software library for quantitative analysis of porous media, Nuclear Instruments and Methods in Physics Research, Section A, 615(3):326-332, 2010.

Author: Emanuel Larsson
Institute: SYRMEP Beamline, Sincrotrone Trieste S.C.p.A.
Street: Strada Statale 14 - km 163,5 in AREA Science Park
City: 34149 - Basovizza, Trieste,
Country: Italy
Email: emanuel.larsson@elettra.trieste.it

A Programmable Current Source for MRCDI & MREIT Applications

C. Göksu, B.M. Eyüboğlu, and H.H. Eroğlu*

Department of Electrical and Electronics Engineering, Middle East Technical University, Ankara, Turkey

Abstract—In this study, a four-channel programmable current source, to be used in magnetic resonance current density imaging (MRCDI) and magnetic resonance electrical impedance tomography (MREIT) applications, is designed and implemented. The current source is composed of a microcontroller unit, a digital to analog convertor, a step-up DC-DC convertor, and a voltage to current (V-I) convertor with a current steering topology. TFT-LCD screen in the microcontroller unit provides a user-friendly graphical user interface (GUI), in which the amplitude, pulse-width, and application time of the output current can be adjusted for each channel. The step-up DC-DC convertor is used to satisfy the high voltage requirement of the current source. It operates with a feedback taken from V-I convertor in order to prevent excess-power consumption during the low level output current injection. The programmed output current is injected to the load by the V-I convertor with a current steering topology. The amplitude and temporal resolution of the output current are significantly high.

Keywords—Programmable current source, microcontroller, step-up DC-DC convertor, voltage to current convertor, current steering.

I. INTRODUCTION

Electrical properties of the biological tissues vary spatially and imaging of the electrical properties of tissues may provide significant diagnostic information. For instance, conductivity images can be used for tumor identification. Besides, current density distribution may provide useful information in research and development of electrical stimulation, electrosurgery, defibrillation, and cardiac pacing devices. MRCDI and MREIT are two imaging modalities based on the current density distribution and the conductivity variation in biological tissues [1-2].

Both in MRCDI and MREIT, external current is injected to the object of interest. When external current, which is synchronized with the MRI imaging sequence, is injected to the object, a magnetic flux density contribution parallel to the main magnetic field of the MRI scanner is induced. This additional magnetic flux density causes a phase accumulation in the obtained MRI signal. Therefore, the MRI phase images obtained with external current injection are correlated with the current density distribution inside the object [1-2].

Exact knowledge of the amplitude and pulse-width of the injected current are substantial for obtaining a current density distribution with high resolution. If resolution of the obtained current density distribution is low, quality of the MRCDI and MREIT output images may be degraded. In addition, as signal to noise ratio (SNR) of these output images are correlated with the induced phase, the SNR can be improved by increasing the amplitude and pulse-width of the injected current [3]. It is well known that electrical properties of biological tissues are frequency dependent [4]. Therefore, a variable frequency programmable current source is required for MRCDI and MREIT applications. So far MRCDI / MREIT current source designs and experimentation are demonstrated by Topal et al., TI Oh et al., and MR Cantaş [5-6-7]. Also, M. Goharian et al. designed a microcontroller controlled current driver for CDI [8].

The aim and main aspects of the study is given so far and the remaining parts of the paper will explain first the fundamentals of MRCDI and MREIT, then the operating principles and the hardware topology of the implemented current source. Finally experimental results with a 1kOhm resistor and the discussion of the experimental results are presented.

II. THEORY

In MRCDI and MREIT applications, it is very common to use a spin-echo pulse sequence with an externally applied current. The corresponding pulse sequence and the external current waveform are shown in Figure 1 [3].

Considering the MRI basics, a complex MR image obtained with a spin-echo pulse sequence with and without current injection can be expressed as,

$$M_c(x,y) = M(x,y) \exp(j\gamma Bt + j\emptyset_c) \quad (1)$$

$$M_{cj}(x,y) = M_j(x,y) \exp(j\gamma Bt + j\gamma B_j(x,y)T_c + j\emptyset_c) \quad (2)$$

where, "$M(x,y)$" is the continuous real transverse magnetization, "B" is the inhomogeneity component of the magnetic field without external current injection, "$M_j(x,y)$" is the continuous real transverse magnetization, "$B_j(x,y)$" is the inhomogeneity component of the magnetic field due to the external current with current injection, "$\emptyset_c$" is the constant phase term and "T_c" is the duration of the applied current [3].

As "$|M(x,y)|$" and "$|M_j(x,y)|$" are equal, the normalized phase introduced by the injected external current, "$\emptyset_{jn}$", can be expressed as shown in (3) [3]. Then using "$B_j(x,y)$", the current density can be expressed as shown in (4) by using Ampere's Law [3]

---

* TSK Rehabilitation and Care Center, Ankara, Turkey.

$$\emptyset_{jn}(x,y) = \gamma B_j(x,y) T_c \quad (3)$$

$$\vec{J} = (\nabla \times \vec{B})/\mu_0 \quad (4).$$

Based on the current density distribution, the conductivity distribution can be obtained by using different reconstruction algorithms.

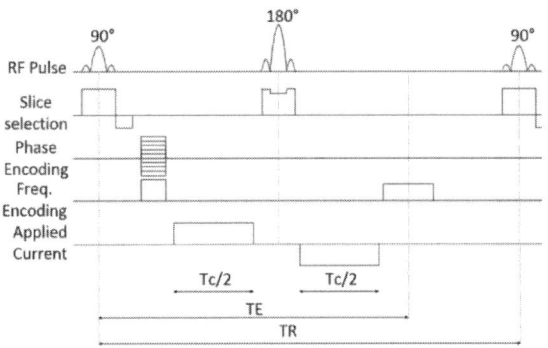

Fig. 1 Spin-echo pulse sequence with the external current waveform

III. OPERATING PRINCIPLES, HARDWARE AND SOFTWARE TOPOLOGY

The implemented current source is composed of a microcontroller unit, a digital to analog convertor, a step-up DC-DC convertor, and a voltage to current (V-I) convertor with a current steering topology. The block diagram of the implemented current source is shown in Figure 2.

The microcontroller board is an integrated development system composed of a TFT-LCD touchpad screen, a USB connector, and digital signal controller (DSC) functions controlled by a DSPIC33FJ family DSC [9]. A user-friendly GUI of the microcontroller unit is designed with the Visual TFT software. By using the GUI, the amplitude, pulse-width, and frequency of the output current can be adjusted in 0-220mA range with 1mA increments, in 0-999msec range with 1msec intervals and in 0-300Hz range with 1Hz increments, respectively. In addition, the GUI provides the selection of the output channels.

After determination of the output current parameters and selection of the channels, an 8-bit digital code is generated and sent to a typical digital to analog convertor (DAC) application circuit and converted to an analog signal, "V_{DAC}" in 0-3.3V range [10]. This analog signal determines the amplitude of the output current.

The schematic diagram of the V-I convertor with a current steering topology is shown in Figure 3. The op-amp is used to drive the power MOSFET and the output current is forced to be regulated to the current on the sensing resistor, "R_{CS}" [8-11]. Therefore, the magnitude of the output current can be expressed as shown in (5)

$$I_{out} = (V_{DAC})/R_{CS} \quad (5)$$

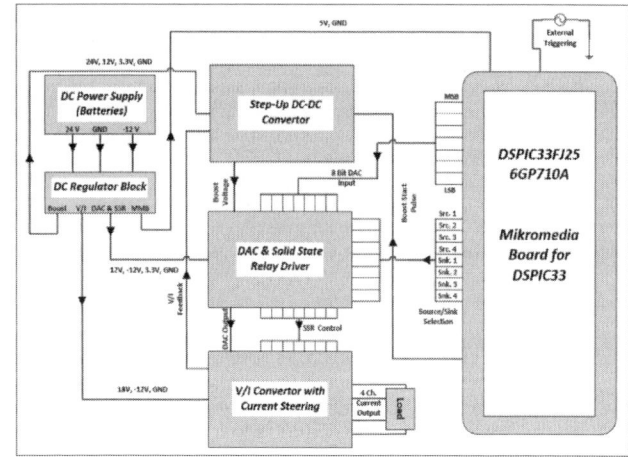

Fig. 2 The block diagram of the overall system

The current steering topology is used to select the sourcing and the sinking channels. By means of this topology, direction of the output current is determined by activation of the channels which is satisfied by solid state relays (SSR).

By this way, difficulty of driving high side switches is eliminated. SSR's are operated at the fastest state, because they are driven by buffer amplifiers where the driving current is set to the maximum forward current of SSR's, 25mA [12]. SSR's are managed by an 8-bit controlling signal produced by the evaluation board. The controlling signal is synchronized with a 3.3V TTL signal which simulates the 90° RF pulse.

The current source is designed to be capable of producing output current as high as 220mA to 1kOhm load. Physically, this requires high voltages, at least 220V. The high voltage requirement of the current source, "V_{BOOST}", is satisfied by a step-up DC-DC convertor. The schematic diagram of the step-up DC-DC convertor is shown in Figure 4.

A hysteresis control based output voltage regulation mechanism is implemented in a PIC24 family microcontroller in order to control the output voltage of the step-up DC-DC converter. The gain of the system can be expressed as shown in (6), where "D" is the duty cycle of the pulse width modulation (PWM) signal [13]

$$V_{OUT}/V_{IN} = 1/(1-D) \quad (6)$$

As it is mentioned, the current source is designed as programmable. Therefore, the current source can also produce low amplitude output currents. In these cases, generating high voltages may degrade the efficiency of this topology and increases the power consumption in the step-up DC-DC convertor significantly. In order to solve this problem, the step-up DC-DC convertor controller software is designed. Based on this software, at first, the required minimum voltage supply is determined and the power convertor regulates

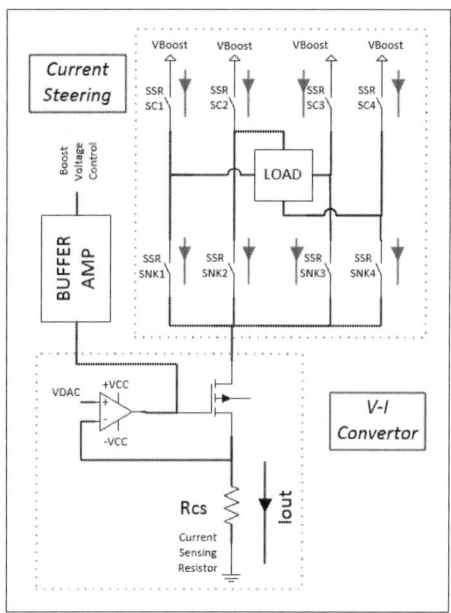

Fig. 3 The schematic diagram of the V-I convertor with current steering

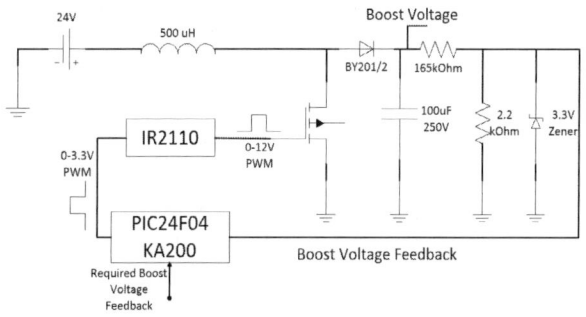

Fig. 4 The schematic diagram of the step-up DC-DC convertor

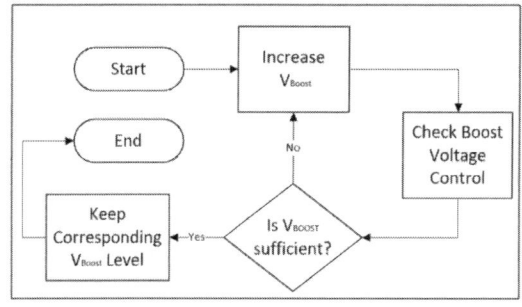

Fig. 5 The flow chart of step-up DC-DC convertor controller

"V_{BOOST}" to the corresponding value. The topology is based on the feedback taken from the V-I convertor. As long as the op-amp operates in positive saturation, "V_{BOOST}" is continuously increased. When it starts to operate in linear region "V_{BOOST}" is stabilized. The flow chart of the step-up DC-DC convertor controller software is shown in Figure 5.

IV. EXPERIMENTAL RESULTS AND DISCUSSION

The photograph of the implemented current source is shown in Figure 6. The current source is implemented as MR compatible in order to be used for MRCDI and MREIT applications. In addition, the designed GUI of the microcontroller unit is easy to use and provides various selection parameters to the user. The appearance of the designed GUI is shown in Figure 7.

The experimental results are obtained with a 1kOhm resistive load. The measured output current amplitudes and the corresponding error values are tabulated in Table 1. The maximum amplitude error of the measurement is 1mA. Also even in 1mA current injection pattern, the error is less than 10%. In addition, the amplitude step-size of the design is 0.859mA.

The measured output current pulse-widths and the corresponding error values are tabulated in Table 2. The maximum temporal error of the measurement is 6.6%. Also even in 5msec pulse-width pattern the error is 6.4%. In addition, the temporal step-size of the design is 1msec. Although the temporal step-size of the design is 1msec, given error values are experimental and correspond to SSR latency.

The oscillograms for the 50mA-DC and 50mA-50Hz output currents with 100msec and 50msec forward and inverse pulse-widths cases are shown in Figure 8, respectively. The ripple level of the output currents are less than 3.5% of the peak value of the set current. The experimental results demonstrate that the desired current pattern shown in Figure 1 is created efficiently by the implemented current source.

The implemented current source software and hardware topology can easily be improved for the purpose of increasing the number of output channels and changing the step-size of the step-up DC-DC convertor voltage control level. By this way the efficiency of the current source may be increased and it may be used in various multi-channel applications more efficiently.

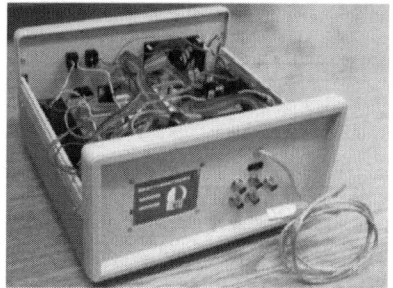

Fig. 6 The photograph of the implemented current source

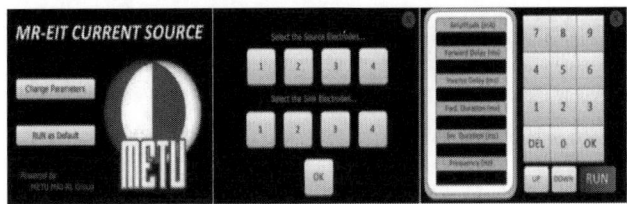

Fig. 7 The appearance of the designed GUI

Table 1 Measured output current amplitude

Input Amplitude	Meas. Amplitude	Amp. Error
mA	0.90mA	0.10mA
0mA	49.00mA	1.00mA
00mA	100.80mA	0.80mA
50mA	151.00mA	1mA

Table 2 Measured output current pulse-width

Input Pulse Width	Meas. Pulse Width	Pulse Width Error
5ms	4.68ms	0.32ms
10ms	9.34ms	0.66ms
50ms	47.20ms	2.80ms
100ms	96.80ms	3.20ms

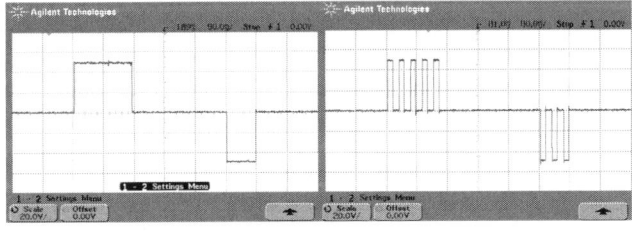

Fig. 8 The oscillograms for 50mA-DC and 50mA-50Hz output currents

V. CONCLUSION

In this study, a four-channel programmable current source for MRCDI and MREIT applications is designed and implemented. The maximum step-up DC-DC convertor voltage is 250V. Output current does not depend on the load and maximum current capability is 250V / Z_{LOAD}. The maximum frequency of output current is 300Hz without any distortion. The amplitude and temporal step-sizes are 0.859mA and 1msec, respectively. In addition, the maximum amplitude and temporal error in output current are less than 1mA and 6.6%. The current source topology is open to any hardware or software improvements. The overall success of the current source may be proven after its use in MRCDI and MREIT experiments in future studies. In addition, the implemented current source may be used in other bioelectrical imaging and stimulation applications.

ACKNOWLEDGEMENT

This study is partially funded by METU-BAP 07-02-2012-101 and TUBITAK 113E301.

REFERENCES

1. Eyüboğlu BM (2006) Magnetic resonance-electrical impedance tomography. WILEY-Encyclopedia of biomedical engineering (Metin Akay, ed.) 4:2154-2162
2. Eyüboğlu BM (2006) Magnetic resonance-current density imaging. WILEY-Encyclopedia of biomedical engineering (Metin Akay,ed.) 2:1195-1205
3. Eyüboğlu BM, Reddy R, Leigh JS (1998) Imaging electrical current density using nuclear magnetic resonance. Elektrik 6:201-214
4. Cole KS, Cole RH (1941) Dispersion and absorption in dielectrics: alternating current characteristics. J. Chem. Phys. 9:341-351
5. Oh TI, Cho Y, Hwang YK, Oh SH, Woo EJ, Lee SY (2006) Improved current source design to measure induced magnetic flux density distributions in MREIT. J. Biomed. Eng. Res. 27:30-37
6. Topal T, Değirmenci E, Boyacıoğlu R, Arpınar VE, Eyüboğlu BM (2010) Current source design for MREIT technique and its experimental application, BIYOMUT Proc., 15th National Biomedical Engineering Meeting, Antalya, Turkey, 2010, pp 1-4
7. Cantaş MR (2012) Modified 3D sensitivity matrix method and use of multichannel current source for magnetic resonance electrical impedance tomography (MREIT). Bilkent University, Turkey.
8. Goharian M, Chin K, Moran GR (2005) A novel microcontroller current driver design for current density imaging. ISMRM Proc. 13th Scientific Meeting and Exhibition, South Beach, Miami, Florida, USA, 2005, p 2352
9. Mikroelektronika (2011) MikroMedia for DSPIC33 user manual
10. Texas Instruments (2006) DAC0800/DAC0802 8 bit digital to analog converters datasheet, SNAS538B
11. Analog Devices (2010-2011) Versatile high precision programmable current sources using DACs, Op-Amps and MOSFET transistors, circuit note, CN-0151
12. International Rectifier (2003) Series PVX6012 datasheet, No. PD 10046-D
13. Eroğlu HH and Eyüboğlu BM (2012) Design and implementation of a monopolar constant current stimulator. BIYOMUT Proc., 17th National Biomedical Engineering Meeting, İstanbul, Turkey, 2012, pp 283-286

Author: Cihan GÖKSU
Institute: Middle East Technical University, Electrical and Electronics Engineering Department, Ankara / Turkey, 06800
Email: cigoksu@metu.edu.tr

How Does Compressed Sensing Affect Activation Maps in Rat fMRI?

C. Chavarrias[1], J.F.P.J. Abascal[1,2], P. Montesinos[1,2], and M. Desco[1,2]

[1] Instituto de Investigación Sanitaria Gregorio Marañón, Madrid, Spain
[2] Departamento de Bioingeniería e Ingeniería Aeroespacial, Universidad Carlos III de Madrid, Madrid, Spain

Abstract—The acceleration of dynamic MRI studies such as fMRI would be a killing application of the recently developed compressed sensing techniques. We present the application of a spatiotemporal total variation reconstruction algorithm to undersampled rat fMRI series. In addition to the t maps extraction, a quantitative analysis of the percentage signal change has been performed to assess the maximum acceleration feasible. Undersampling up to 50%(x2 acceleration) can be applied without significant percentage signal change loss, and the activation maps preserve their shape within that range. To our knowledge, this is the first work performing quantitative analysis in compressed sensing reconstructions of rat fMRI data.

Keywords—**fMRI, BOLD, compressed sensing, undersampling.**

I. INTRODUCTION

fMRI has become one of the most useful imaging techniques on brain research. Its use on rodents is being very helpful to understand physiological and biochemical processes underlying brain activity. Dedicated high field scanners provide the challenging resolution required for small animal imaging, but the use of strong fields is not only restricted to this context since it has also reached human research.

One of its drawbacks, however, is the need of statistical mapping in order to account for the poor SNR (signal-to-noise ratio), which subsequently implies the acquisition of long volume series. The limited availability of scanners and the patient discomfort during long sessions have pushed acceleration techniques research further in recent years.

Compressed sensing has proved to be an efficient solution for rapid MR imaging [1]. It allows the reconstrstruction of undersampled images under certain assumptions: radom undersampling for incoherent aliasing, a sparse transform domain and a non linear specific reconstructor.

Its performance even improves with the number of dimensions in which the signal is undersampled, which makes dynamic imaging in general (and cardiac imaging or fMRI in particular) the most promising application [1-2].

Although compressed sensing on structural images has been assessed, its indirect effects on statistical mapping have not been thoroughly studied yet [1-5]. The aim of this work was to quantitatively describe the evolution of both the area and the mean percentage signal change values inside the somatosensory contralateral cortex of a rat according to the amount of data undersampled from complete fMRI studies of rat brain.

II. MATERIALS AND METHODS

A. Acquisition

The original raw images were acquired from a male Wistar rat (300g aprox.) brain following the protocol described in [6] with a 7T Bruker Biospec 70/20 scanner. Final spatial resolution was 0.3 x 0.3 x 2 mm with a 0.1 mm gap between slices. Left forelimb was stimulated with sensorial electrical stimuli (1 mA, 9Hz, 0.3 ms duration). The fMRI series consisted of 115 SE-EPI (TR/TE = 3000/30 ms) volumes acquired using a block design paradigm. The resting block comprised 15 images (45 s), whereas the stimulation block comprised 5 images (15 s). The first and last blocks corresponded to resting periods.

B. Standard Deviation Estimation

The reference reconstruction of fully sampled data was given by Bruker inverse Fourier transform. The slice containing the target primary somatosensory cortex (S1) was extracted from the dataset and was subject to 3 different instances (J1, J2 and J3) of temporal jackknifing in order to obtain a rough estimate of the typical standard deviation for both the activation maximum percentage signal change and area. The jackknifing was applied preserving 3 samples out of the initial 5 volumes at each stimulation block and 12 samples out of the original 15 volumes at resting blocks, generating a total series length of 87 volumes instead of the original 115 (see Figure 1).

C. Compressed Sensing Simulation: Undersampling

One of the reduced datasets (J1) was chosen as a new completely sampled reference and was undersampled in k space with a different phase encoding pattern (following a variable polynomial distribution [1] with higher density in the k space center) for each temporal point, resulting in a pseudo-randomized undersampling in both phase and t directions. The series was undersampled in 10 different runs to preserve 5% (x20), 10% (x10), 20% (x5), 30% (x3.33), 40% (x2.5), 50% (x2), 60% (x1.67), 70% (x1.42), 80% (x1.25), and 90% (x1.11) of its data (namely [U5,U10...U90]). See Figure 1 for a methodological scheme.

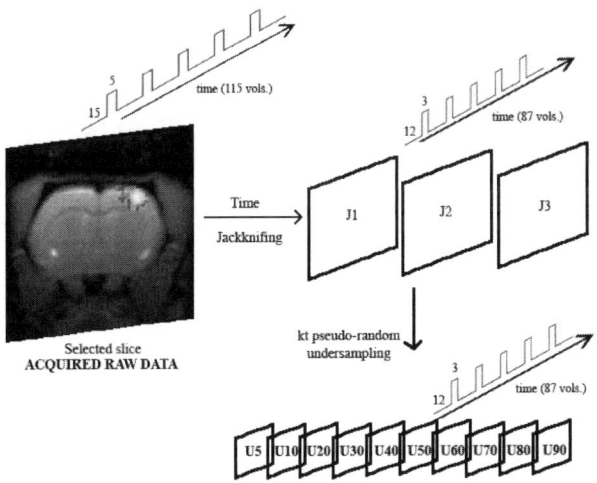

Fig. 1 Scheme of the data treatment followed in order to have an estimation of the fMRI signal change and area standard deviations, from the raw data to the undersampled series.

D. Compressed Sensing Reconstruction

The image reconstruction is performed through a constrained optimization algorithm which minimizes the spatiotemporal total variation and follows the expression:

$$\min_u \|\nabla u\|_1 + \|D_t u\|_1 \quad \text{such that } Fu = f, \quad (1)$$

where $\nabla = (D_x, D_y)$ is the gradient, and D_x, D_y and D_t are the spatial and time derivatives respectively, and $\|\cdot\|_1$ is the L1 norm, $\|\nabla u\|_1$ is the spatial total variation, and $\|D_t u\|_1$ is the time total variation.

Reformulation via Split Bregman method yields

$$\min_{d_x, d_y, d_t, u} \|(d_x, d_y)\|_1 + \|d_t\|_1 + \frac{\mu}{2}\|Fu - f^k\|_2^2 + \frac{\lambda}{2}\|d_x - D_x u - b_x^k\|_2^2 \quad (2)$$
$$+ \frac{\lambda}{2}\|d_y - D_y u - b_y^k\|_2^2 + \frac{\lambda}{2}\|d_t - D_t u - b_t^k\|_2^2$$

where each b_i represents the Bregman iteration that imposes a constraint, and the new variables, d_i, allow decoupling L2 and L1 functionals, leading to analytical solutions that can be solved efficiently [7-8].

E. fMRI Analysis

The final statistical t-maps were obtained through a standard analysis[9] for all the original raw central slice extracted from the acquisition, the 3 jackknifing slice series (J1, J2 and J3) resulting from section *B)* and the 10 undersampled slice series obtained from one of them (J1) at section *C)*. T test with uncorrected p<0.05 and cluster size=12 yielded the statistical maps, which were overlaid on the first slice of the series. The area, as well as the maximum (SC_{max}) and mean percentage signal change (calculated following [10]) inside the primary somatosensorial cortex (S1) were saved to disk. This was performed via a manually segmented binary mask.

All these steps were performed through an *in-house* interface tool which makes use of several functions from the SPM package (*Statistical Parametric Mapping*, The Wellcome Trust Centre for Neuroimaging), as well as from the CBMGmosaic (Northwestern Cognitive Brain Mapping Group) and SPMMouse toolboxes (Wolfson Brain Imaging Centre, University of Cambridge [11]).

III. RESULTS

A. Maps Resulting from Jackknifing

Figure 2 shows the fully sampled data map and the three maps (J1, J2, and J3) resulting from jackknifing.

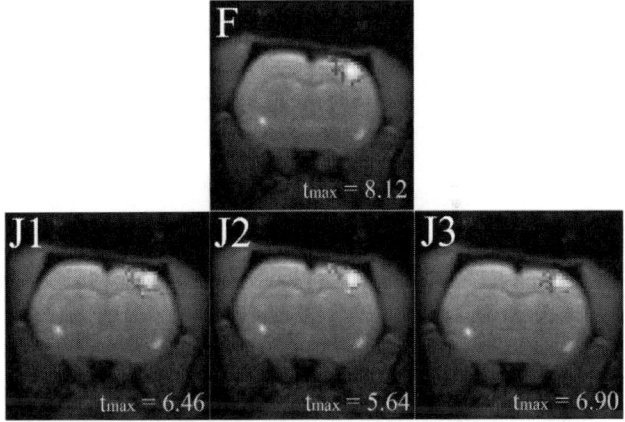

Fig. 2 Upper row: fully sampled dataset (F) activation map (series length=115). Lower row: maps (J1, J2 and J3) resulting from the three pseudorandom temporal jackifing (series length=87).

Slight area variations are observed between the three jackknifed maps, as well as the expected decrease (around 30%) in the maximum t-value in comparison to the fully sampled map caused by the loss of temporal information produced by the discard of 28 temporal samples in each series.

B. Maps Resulting from Simulated Compressed Sensing

Activation maps (positive in hot colorscale and negative in cool colorscale, uncorrected p<0.05 and cluster size threshold=12) corresponding to undersampled data series obtained from J1, the first jackknifing dataset are shown in

Figure3. Their maximum t value is shown in yellow color, and was always located in the contralateral S1 cortex until 30% data preservation, where the maximum lies in a different region. The S1 area shows BOLD contrast until 10%(x10) of the initial J1 data are preserved, when it suddenly disappears. However, for 30%(x3.33) of data and below, the activation in S1 loses its original shape, and other areas different from the expected S1 excel the threshold.

The undersampling was performed from the J1 series to be able to observe whether the variations on the statistical maps due to undersampling were inside the typical signal variation range or not, so that quantifications falling outside the range would indicate an abnormal behavior of the statistics, indicating the usage limit for the compressed sensing approach.

A. Mean Percentage Signal Changes in S1 Cortex

The mean percentage signal change was calculated for the region of interest, contralateral S1, in every dataset: the full dataset, J1, J2, J3 and the undersampled U90, U80, ..., U10 and U5 following [10].

The fully sampled map showed a mean of 0.809% signal change. Figure 4 shows the mean signal change for the undersampled series. Values at S1 in U40, U30, U20 and U10 differed significantly ($p<0.05$) from those from the J1, J2 and J3, taken as the reference. U5 showed even a higher difference ($p<0.01$).

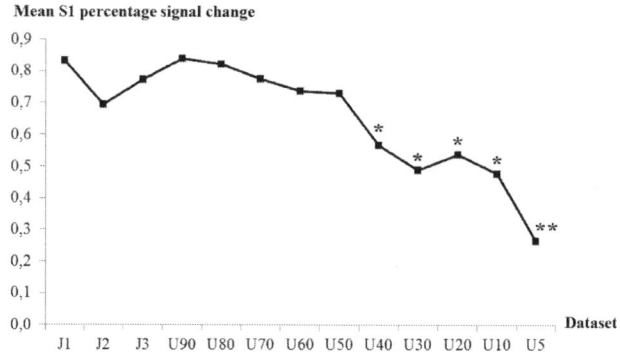

Fig. 4 Mean percentage signal change observed in the contralateral S1 cortex in the maps with respect to the undersampling applied. *$p<0.05$, **$p<0.01$.

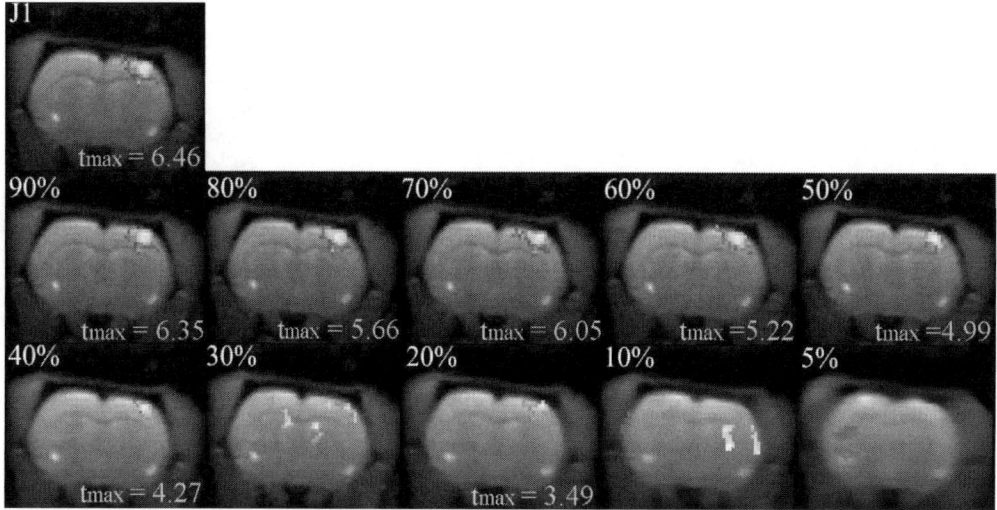

Fig. 3 Activation maps (positive in hot colorscale and negative in cool colorscale, uncorrected $p<0.05$ and cluster size threshold=12) corresponding to undersampled data series obtained from J1, the first jackknifing dataset. Their maximum t value is shown in yellow color, and was always located in the contralateral S1 cortex except for the 30, 10 and 5% data preservation, where the maximum lies in a different region. The S1 area shows BOLD contrast until 10% of the initial J1 data are preserved, when it suddenly disappears, but from 30% of data the shape of the activation differs substantially from the complete J1 map. For strong undersamplings other areas different from the expected S1 excel the threshold.

IV. DISCUSSION

There have already been several attempts to apply compressed sensing to fMRI [2-5, 12-13], but to our knowledge, this is the first time that a compressed sensing reconstruction has been applied to rodent fMRI series, proving that a x2 acceleration is feasible in this context. In previous works performed in human studies, the accelerations factors obtained ranged from 2 to 4x [3-4, 12], but a direct comparison is not possible since the acquisition geometries and reconstructions are very different from ours.

In addition, our reconstruction method combines spatial and temporal total variation [14], which takes advantage of the temporal redundancy, whereas most of the previous works on the field were based on the minimization of the wavelet L1 norm of each time frame independently.

Furthermore, the use of Split Bregman decomposition provides an efficient solution of constrained L1 optimization problems [7].

The method described here has proved to compensate for the loss of BOLD contrast until 40%(x2.5) of the original data was preserved. Previous works mainly show the resulting activation maps, lacking any quantitative analysis of the performance with different data undersamplings (except for [4]). We successfully assessed it quantitatively in our context, but the percentage signal change analysis could be completed with further area assessment via ROC curves for example[12]. However, visual inspection of the maps shows consistent results up to 40% (x2.5) data reduction.

REFERENCES

1. Lustig M, Donoho D, Pauly JM. (2007) Sparse MRI: The application of compressed sensing for rapid MR imaging. Magnetic Resonance in Medicine; 58(6): 1182-95.
2. Jung H, Sung K, Nayak KS, Kim EY, Ye JC. (2009) k-t FOCUSS: a general compressed sensing framework for high resolution dynamic MRI. Magn Reson Med; 61(1): 103-16.
3. Avdal J, Kristoffersen A, Håberg A, Goa PE. (2012) Functional MRI employing Compressed Sensing and separation of signal and noise in k-space, 20th Annual Meeting of ISMRM, Melbourne, Australia.
4. Holland D, Liu C, Song X, Mazerolle E, Stevens M, Sederman A, Gladden L, D'Arcy R, Bowen C, Beyea S (2013) Compressed sensing reconstruction improves sensitivity of variable density spiral fMRI. Magnetic Resonance in Medicine, DOI: 10.1002/mrm.24621.
5. Hugger T, Zahneisen B, LeVan P, Lee K, Lee H, Zaitsev M, Hennig J (2011) Fast undersampled functional magnetic resonance imaging using nonlinear regularized parallel image reconstruction. PLoS One **6**, DOI: 10.1371/journal.pone.0028822.
6. Weber R, Ramos-Cabrer P, Wiedermann D, van Camp N, Hoehn M. (2006) A fully noninvasive and robust experimental protocol for longitudinal fMRI studies in the rat. Neuroimage; 29(4): 1303-10.
7. Goldstein T, Osher S. (2009) The Split Bregman Method for L1-Regularized Problems. Siam Journal on Imaging Sciences; 2(2): 323-343.
8. Abascal J, Chamorro-Servent J, Aguirre J, Arridge S, Correia T, Ripoll J, Vaquero JJ, Desco M. (2011) Fluorescence diffuse optical tomography using the split Bregman method. Medical Physics; 38(11): 6275-6284.
9. Friston K, Ashburner J, Kiebel S, Nichols T, Penny W. Statistical Parametric Mapping: The Analysis of Functional Brain Images, ed. K Friston, et al. 2007: Academic Press, Elsevier.
10. Brett M, Anton JL, Valabregue R, B PJ. (2002) Region of interest analysis using an SPM toolbox, 8th International Conference on Functional Mapping of the Human Brain, Sendai, Japan.
11. Sawiak SJ, Wood NI, Williams GB, Morton AJ, Carpenter TA. (2009) Voxel-based morphometry in the R6/2 transgenic mouse reveals differences between genotypes not seen with manual 2D morphometry. Neurobiology of Disease; 33(1): 20-27.
12. Jung H, Ye J. (2009) Performance evaluation of accelerated functional MRI acquisition using compressed sensing, IEEE International Symposium on Biomedical Imaging: From Nano to Macro, 2009. ISBI '09. .
13. Jeromin O, Pattichis MS, Calhoun VD. (2012) Optimal compressed sensing reconstructions of fMRI using 2D deterministic and stochastic sampling geometries. Biomed Eng Online; 11: 25.
14. Montesinos P, Abascal JFPJ, Chamorro J, Chavarrias C, Benito M, Vaquero JJ, Desco M. (2011) High-resolution dynamic cardiac MRI on small animals using reconstruction based on Split Bregman methodology, Nuclear Science Symposium and Medical Imaging Conference (NSS/MIC), 2011 IEEE.

Author: Cristina Chavarrías
Institute: Instituto de Investigación Sanitaria Gregorio Marañón
Street: Dr. Esquerdo 46
City: Madrid
Country: Spain
Email: cchavarrias@hggm.es

Resting State Functional Connectivity Analysis of Multiple Sclerosis and Neuromyelitis Optica Using Graph Theory

E. Eqlimi[1,2], N. Riyahi Alam[1], M.A. Sahraian[2], A. Eshaghi[2], S. Riyahi Alam[2], H. Ghanaati[3], K. Firouznia[3], and E. Karami[4]

[1] Department of Biomedical Engineering and Medical Physics, Tehran University of Medical Sciences, Tehran, Iran
[2] Sina MS Research Center, Sina Hospital, Tehran University of Medical Sciences, Tehran, Iran
[3] Department of Radiology, Medical Imaging Center, Imam Khomeini Hospital, Tehran University of Medical Sciences, Tehran, Iran
[4] Departments of Engineering, Tehran Shomal Payame Noor University, Tehran, Iran

Abstract—The aim of this study is to investigate resting state functional connectivity in brain tissue using fMRI data in order to differentiating Neuromyelitis Optica (NMO) from Multiple Sclerosis (MS). In this method, normalized mutual information accompanied with graph theoretical analysis and community structure measures were used to differentiating. The fMRI time series were extracted for 264 nodes in which were selected based on neurological principals. Normalized mutual information (NMI) between each two time series was calculated and constructed NMI based connectivity matrix. The graph of pairwise mutual information is thresholded such that the top 3471 un-directed links are preserved. The graphs are analyzed graph theoretic measures and community structure procedure was applied in order to modularity detection. There are no significant differences between NMO and MS patients in terms of classical graph theoretic measures. However a significant difference was found in modularity statistics between NMO and MS patients.

Keywords—fMRI, Resting State Functional Connectivity, MS, NMO, Graph Theory.

I. INTRODUCTION

The human brain is always busy even during resting state. When brain is in resting state, there are a lot of anatomical areas show a vast amount of spontaneous neuronal activity and functionally linked to each other. There are a coherence behavior between some networks in brain during rest that are said "Resting State Networks". Interest in investigating intrinsic low-frequency fluctuations in resting-state brain activity is steadily growing. The functionality linked resting state reflects the underlying structural connectivity architecture of the human brain. The functional connectivity is neuronal activity between anatomically separate brain regions[1, 2].

Multiple Sclerosis is an inflammatory disease in which the fatty myelin sheathes around the axons of the brain and spinal cords are damaged, leading to demyelination and scarring as well as a broad spectrum of signs and symptoms. The publications about functional connectivity networks for multiple sclerosis are limited. Neuromyelitis Optica (NMO) an idiopathic CNS demyelinating disease is characterized by severe attacks of optic neuritis and myelitis. NMO pathology differs from that of classic multiple sclerosis (MS). Relapsing remitting MS patients have an unpredictable rhythm of attacking and improvement. Most diagnostic assessment is based on T2-weighted MRI and gadolinium-enhanced images[3]. However the structural MRI has some drawbacks such as the normal appearing brain tissues was not captured, the lesions of MS are not always specific, T2 hyperintensities are histologically unspecified since inflammation and demyelination as well as axonal damage and gliosis have similar signal characteristics and the correlation of lesion load and clinically significant impairment is poor[4]. Functional magnetic resonance imaging (fMRI) can capture the underlying and hidden lesions. Blood oxygen level dependent (BOLD) signal is an indirect measure for detecting underlying degenerations. Multiple Sclerosis lesions altered the organization of functional brain network[5].

The human brain is considered to be the most complex object in the universe. Graph theory-based approaches model the brain as a complex network represented graphically by a collection of nodes and edges. Graph theory is a natural framework for the mathematical representation of complex networks. Recently, graph theory has attracted considerable attention in brain network research because it provides a powerful way to quantitatively describe the topological organization of brain connectivity. Most graph theory – based analysis for brain functional connectivity is based on temporal correlation between ROI activities. However the correlation measure has just an insight for second order statistic[6] . A current obstacle to the graph-based study of functional brain organization is that it very difficult to define the individual nodes that make up a brain network. Most approaches are not meant to correspond to macroscopic "units" of brain organization, and thus there is no direct reason to believe that these approaches result in well-formed nodes[5, 7].

In this paper we used well- formed nodes for defining the vertices of functional connectivity graph. We obtained average BOLD signal in these well- formed regions and then measured pairwise normalized mutual information between the time series of nodes. Then, we performed an optimization method for modularity detection in graphs and used detected modules as discriminative features.

II. MATERIALS AND METHODS

A. Subjects

Twenty five clinically stable patients with relapsing-Remitting multiple sclerosis and twenty nine Neuromyelitis Optica patients were selected for this cross-sectional study. All patients were recruited at "Sina MS Research Center, Sina hospital, Tehran, Iran" from 2009 to 2012. Inclusion criteria for patients included MS diagnosis by 2005 revised McDonald criteria for the MS group and NMO diagnosis by 2006 revised Wingerchuk's criteria for the NMO group.

B. Data Acquisition

Structural and functional MRI data was acquired on a Siemens 3 T platform. The resting- state functional magnetic imaging data were acquired using gradient echo planar (TR=2/2 s, TE=30 ms, FA= 90°, matrix size=64×64, voxel size=3 ×3×3mm^3, 40 slices per volume, slice thickness=3 mm, slice sequence: interleaved, 200 volumes).The structural images were acquired using a high resolution three dimensional T1-weighted MPRAGE sequence (TR=2.53s, TE=3.44 ms, FA=7°, matrix size=256×256, voxel size=1×1×1 mm^3, 176 slices, slice thickness=1 mm).

C. Image Preprocessing

Longitude magnetization last to reach steady state, so the 5 first scans of each acquisition were discarded for equilibration the magnetic field and 195 volumes were kept. All the preprocessing steps were performed using spm8.They included slice timing correction with first slice reference for the temporal difference complementation in acquisition among different slices, within subject realignment to mean of volumes in order to correcting patient motion, between-subject spatial normalization with the corresponding T1-volume and warping into a standard stereo space at a resolution of 3 × 3 × 3 mm^3, using the Montreal Neurological Institute (MNI) EPI template in SPM8, smoothing by an isotropic Gaussian kernel with full width and half width 8 mm.

D. Extracting Time Series

The standard approaches for defining the nodes of graph in resting-state functional analysis is based on selecting a lot of random voxels, large extracted ROIs form anatomical atlases. These methods don't consider the corresponding of nodes with functional units in brain. We used the 264 putative were defined according with neurobiological principles (Table 1). These selected nodes are included default mode, dorsal attention, ventral attention, Fronto-parietal task-control networks and spanned cerebral cortex, sub cortical structures, and the cerebellum. The time series were derived from each selected ROI and region-averaged time series were obtained.

Table 1 Selected Functional Network for Defining Graph Nodes and the Number of Nodes in Each Network. Default Mode Network Has Maximum Contribution

Functional Networks	Number of Nodes
Uncertain	28
Sensory/Somatomotor Hand	30
Sensory/Somatomotor Mouth	5
Cingulo-Opercular Task Control	14
Auditory	13
Default Mode	58
Memory Retrieval	5
Ventral Attention	9
Visual	31
Fronto-Parietal Task Control	25
Salience	18
Subcortical	13
Cerebellar	4
Dorsal Attention	11

E. Normalized Mutual Information

After defining the nodes of graph, the weights of pair wise links in each graph between the nodes were determined in terms of normalized mutual information. Indeed, functional connectivity of each region with remained areas was measured based on mutual information. Since the interaction between brain regions is very complex, it's necessary to measure the interactions regardless of statistical order interaction. Formally, the mutual information of two random variables X and Y can be defined as:

$$I(x,y) = \sum_{y \varepsilon Y} \sum_{x \varepsilon x} p(x,y) \log \left(\frac{p(x,y)}{p(x)p(y)}\right) \quad (1)$$

In order to estimating joint probability density function ($p(x, y)$), we used the histogram approach.

After computing pairwise normalized mutual information between the regions, a 264 × 264 connectivity matrix was obtained for each subject. The connectivity matrix is considered as weights of links between nodes in the graph.

F. Functional Connectivity Matrices

A complex network can be considered as a binary and undirected graph. Functional connectivity is defined based on strength of interaction between brain areas and means the undirected and binary graphs can capture functional connectivity. After thresholding and binarization of mutual information based- functional connectivity and removing self-self connection (diagonal elements), the symmetric adjacency matrix was obtain for each subject. The graphs can constructed based on the adjacency matrix, where each ROI is correspond to a node and presence of a edge between two nodes depends on the value of corresponding element in adjacency matrix, where 1 means, there is a edge between nodes and 0 means no edge.

G. Graph Theoretical Analysis

After constructing the binary and undirected graph for each participant, the graph theoretical analysis was performed in two participating groups. In most graph theoretical analyses, several measures like clustering are calculated for each subject and are comprised with random graphs. We calculated features for each graph and comprised with Erdos-Renyi random graph with preserved density as the constructed graphs. Table 2 shows the definitions of graph theoretical measures that we used in this paper. A schematic of our method has been shown in Fig. 1.

Table 2 The definitions of graph theoretical measures

Threshold	Threshold (Th) is preserving a proportion (Th) of the strongest weights. All other weights and all weights on the main diagonal (self-self connections) are set to 0.
Characteristic path length	Characteristic path length (CP) is the average shortest path length between all pairs of nodes in the network.
Clustering coefficient	Clustering coefficient(C) is the fraction of the node neighbors that are also neighbors of each other.
Transitivity	Transitivity (T) is a classical version of the clustering coefficient with normalized collectively and consequently.
Giant Component	Giant Component (G) is Largest connected component in the graphs relative to the complete graph.
Assortativity	Assortativity (A) is Correlation coefficient between the degrees of all nodes on two opposite ends of a link.
Nodal betweenness centrality	Nodal betweenness centrality (N) is the Fraction of all shortest paths in the network that pass through a given node.

III. RESULTS

We calculated characteristic path length(CP) [8], average clustering coefficient(C) [8], transitivity (T)[9], giant component (G) [10],Assortativity (A) [11] and average nodal betweenness centrality (N) [12].

There were no significant differences between the normal and MS group demonstrating these features do not appear to be affected by the disease. We found, there are significant differences for the Normal and MS groups vs. random graphs. Table 3 shows the non-parametric test of difference in median between the groups.

We performed optimal community structure and modularity procedures [13]. There are significant differences between normal and MS group in terms of modularity measure. When 5% of the strongest weights were preserved, we obtained best classification performance between healthy control and MS groups and also between MS and NMO Group separately, but the discrimination between healthy control and NMO isn't acceptable in terms of modularity measure (Fig. 2).

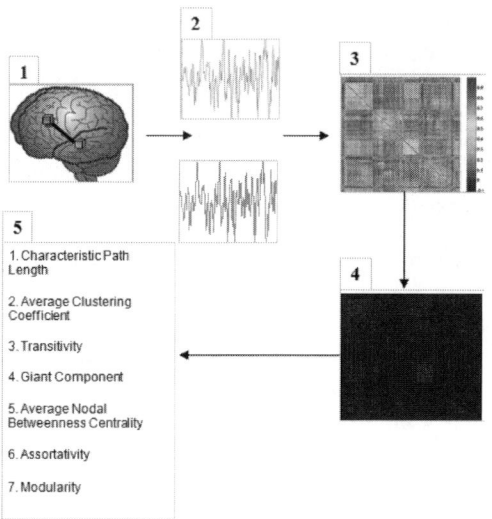

Fig. 1 The proposed framework.

Step1: we defined 264 nodes according with neurobiological principles.
Step2: The time series of each node was extracted.
Step3: Pairwise normalized mutual information (NMI) were calculated between the time courses of nodes, yielding 264*264 connectivity matrix for each subject.
Step4: The graphs of pairwise mutual information were thresholded such that the top 3471 un-directed links were preserved.
Step5: The graphs were analyzed with graph theoretic measures and community structure procedure was applied in order to modularity detection

Table 3 Non-parametric test of difference in median between the groups in terms of graph theoretical measures. H: healthy control, M: MS, N: NMO, R: Random

P-value	N	A	G	T	C	CP
H vs. M	0.279	0.470	0.231	0.399	0.434	0.098
M vs. N	0.333	0.059	0.391	0.285	0.201	0.107
N vs. H	0.326	0.053	0.310	0.364	0.277	0.479
(H,M,N) vs. R	<0.001	<0.001	<0.001	<0.001	<0.001	<0.001

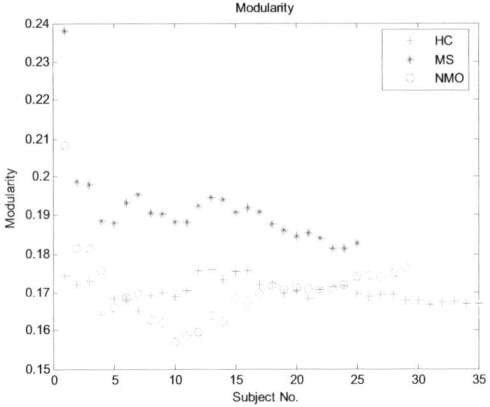

Fig. 2 Modularity values of each graph for three groups, HC (Healthy Control), Multiple Sclerosis (MS), Neuromyelitis Optica (NMO)

IV. CONCLUSIONS

Resting state functional connectivity analysis based on normalized mutual information graphs (NMI) using fMRI is capable of reliable classification of healthy controls and patients with MS on the basis of modularity measure. Increased modularity measure in patients with MS vs. healthy subjects reveals separated functional networks as compared to healthy controls. This fact is correct for discrimination between NMO and MS group, but NMI graphs based on modularity measure are not capable of acceptable discriminating. These lead to an efficient resting state functional connectivity discrimination between MS and healthy control group and also between MS and NMO.

We concluded the increment of modularity in NMI graphs of MS and NMO group is not equal and this augmentation is more than NMO. Indeed, the number of modulated networks in MS group is very more than both of healthy control and NMO group.

Selection of graph nodes, threshold in binary graph construction and weight criteria of graph links are effective parameters in classification.

REFERENCES

1. M. P. Van Den Heuvel, *et al.*, "Functionally linked resting - state networks reflect the underlying structural connectivity architecture of the human brain," *Human brain mapping*, vol. 30, pp. 3127-3141, 2009.
2. M. De Luca, *et al.*, "fMRI resting state networks define distinct modes of long-distance interactions in the human brain," *Neuroimage*, vol. 29, pp. 1359-1367, 2006.
3. C. H. Polman, *et al.*, "Diagnostic criteria for multiple sclerosis: 2010 revisions to the McDonald criteria," *Annals of neurology*, vol. 69, pp. 292-302, 2011.
4. J. Richiardi, *et al.*, "Classifying minimally-disabled multiple sclerosis patients from resting-state functional connectivity," *Neuroimage*, 2012.
5. J. D. Power, *et al.*, "Functional Network Organization of the Human Brain," *Neuron*, vol. 72, pp. 665-678, 2011.
6. M. Mørup, *et al.*, "Infinite relational modeling of functional connectivity in resting state fmri," *Neural Information Processing Systems 23*, 2010.
7. G. S. Wig, *et al.*, "Concepts and principles in the analysis of brain networks," *Annals of the New York Academy of Sciences*, vol. 1224, pp. 126-146, 2011.
8. D. Watts and S. Strogatz, "The small world problem," *Collective Dynamics of Small-World Networks*, vol. 393, pp. 440-442, 1998.
9. M. E. J. Newman, "The structure and function of complex networks," *SIAM review*, vol. 45, pp. 167-256, 2003.
10. M. E. J. Newman, *et al.*, "Random graphs with arbitrary degree distributions and their applications," *Physical Review E*, vol. 64, p. 026118, 2001.
11. M. E. Newman, "Assortative mixing in networks," *Physical review letters*, vol. 89, p. 208701, 2002.
12. L. C. Freeman, "A set of measures of centrality based on betweenness," *Sociometry*, pp. 35-41, 1977.
13. M. E. J. Newman, "Modularity and community structure in networks," *Proceedings of the National Academy of Sciences*, vol. 103, pp. 8577-8582, 2006.

Author: Ehsan Eqlimi
Institute: Department of Biomedical Engineering and Medical Physics, Tehran University of Medical Sciences
Street: 16 Azar
City: Tehran
Country: Iran
Email: Eghlimi@razi.tums.ac.ir

Quantitative Evaluation of Patient-Specific Conforming Hexahedral Meshes of Abdominal Aortic Aneurysms and Intraluminal Thrombus Generated from MRI

J. Tarjuelo-Gutierrez[1,2], B. Rodriguez-Vila[1,2], D.M. Pierce[3], T. Fastl[3], and E.J. Gomez[1,2]

[1] Bioengineering and Telemedicine Centre, ETSI de Telecomunicación, Universidad Politécnica de Madrid, Madrid, Spain
[2] Networking Research Center on Bioengineering, Biomaterials and Nanomedicine (CIBER-BBN), Zaragoza, Spain
[3] Institute of Biomechanics, Center of Biomedical Engineering, Graz University of Technology, Graz, Austria

Abstract—A novel method for generating patient-specific high quality conforming hexahedral meshes is presented. The meshes are directly obtained from the segmentation of patient magnetic resonance (MR) images of abdominal aortic aneurysms (AAA). The MRI permits distinguishing between structures of interest in soft tissue. Being so, the contours of the lumen, the aortic wall and the intraluminal thrombus (ILT) are available and thus the meshes represent the actual anatomy of the patient's aneurysm, including the layered morphologies of these structures. Most AAAs are located in the lower part of the aorta and the upper section of the iliac arteries, where the inherent tortuosity of the anatomy and the presence of the ILT makes the generation of high-quality elements at the bifurcation is a challenging task. In this work we propose a novel approach for building quadrilateral meshes for each surface of the sectioned geometry, and generating conforming hexahedral meshes by combining the quadrilateral meshes. Conforming hexahedral meshes are created for the wall and the ILT. The resulting elements are evaluated on four patients' datasets using the *Scaled Jacobian* metric. Hexahedral meshes of 25,000 elements with 94.8% of elements well-suited for FE analysis are generated.

Keywords—magnetic resonance imaging, abdominal aortic aneurysm, intraluminal thrombus, conforming hexahedral meshes, finite element analysis.

I. INTRODUCTION

An aortic aneurysm is a localized dilation of the aorta that can be found anywhere in the artery, the most common being the abdominal aortic aneurysms (AAA) [1]. Weakening of the aortic wall is one of the biggest risks associated with this particular disease, which can lead to rupture or dissection of the artery. It is possible that blood stagnates in the dilation, inducing intraluminal thrombus (ILT) formation [2]. Three different structures may be observed in the aorta zone where an aneurysm is present: AAA wall, ILT and lumen.

Evaluating the rupture risk is a critical task due to the high mortality associated to this pathology [3]. Solutions based on the finite element method (FEM) and fluid–structure interaction (FSI) modeling can be used to evaluate rupture risk [4]. FE simulations for AAAs can be patient-specific, and permit modeling complex stress states that include the effect of the ILT. They also improve local environment representation based on preoperative anatomical images.

Generating an appropriate FE mesh is a pre-requisite for applying several numerical techniques, including those based on FEM [5]. Hexahedral meshes are preferred to tetrahedral ones due to: first, a model with the same volume is comprised of significantly less hexahedra than one made of tetrahedra; and second, there is an aspect ratio distortion present in tetrahedral solutions and not in hexahedral solutions [6], what makes hexahedra more suitable for applying FEM. However, in the specific case of vascular structures the use of hexahedral meshes is specially challenging due to the complex 3D branching topology. Hence, a specific algorithm is proposed to work with the geometry of the particular case study obtained from the patients' preoperative imaging. In our context, an aneurysm is characterized by its location and shape. About 90% of abdominal aneurysms are located below the renal arteries [1]. Furthermore, around two out of three abdominal aneurysms are located not only in the aorta, but extend to one or both iliac arteries [7]. Thereby, the geometry under study can be roughly described by a tubular structure in the shape of an inverted Y, distorted by the aneurysm and ILT wherever they are present.

Most publications in literature which focus on this topic deal with planar bifurcations, i.e. centerlines evolving in a single plane [8-9] which do not represent the actual anatomy of the structures, but refined approximations. De Santis et al [10] propose a method for modeling structures evolving out of a single plane, where conforming meshes of the wall and the lumen are presented, but the outer contour of the wall cannot be segmented and is reconstructed from anatomical data.

In this study we present a novel and robust procedure for generating hexahedral conforming meshes of the aortic wall and the ILT with high-quality elements at the bifurcation, suitable for FEM analysis of stress states. The developed algorithm takes into account the evolution of the vessel's centerline out of a single plane and is constructed directly from segmented images, not needing reconstructed triangular surfaces.

II. MATERIAL AND METHODS

A technique for generating two conforming hexahedral meshes, one for the AAA wall and another one for the ILT, is presented. Starting from the manual segmentation of MR images, three quadrilateral meshes are constructed: one for the lumen, another for the aortic wall, and another one for the ILT, wherever it is present. These quadrilateral meshes are combined in order to construct the final hexahedral meshes. Generating regularly shaped hexahedral elements at irregular anatomic structures such as bifurcations is a challenging task. In this work we present a novel method to handle the meshing at bifurcation producing high-quality elements. The algorithm has been implemented using MATLAB R2009a (Mathworks Inc., Natick, MA, USA) and the resulting meshes have been evaluated by applying the *Scaled Jacobian* metric.

A. Image Segmentation

A novel MRI acquisition protocol which provides high contrast between the aortic wall and the ILT is used. This protocol makes it possible to distinguish and segment the different structures of interest. Hence, three 3D binary images of the wall, the ILT and the lumen are obtained for each patient using the ITK-SNAP software [11].

B. Quadrilateral Meshes Generation

The initial phase for modeling the bifurcation consists of dividing it into three different vessels, following the idea proposed by Lee [12]. Once the vessels have been separated, each one is divided into two sections that are modeled independently, so the quality and number of elements can be tuned at each section, depending on the anatomical characteristics of each section. By doing so, the computational cost of the process is minimized as the density of elements is increased only where necessary. The tuning is controlled by a set of parameters that divide the sections in the longitudinal and circumferential directions.

Three quadrilateral meshes are constructed for the external face of the aortic wall (Fig. 1a), internal face of the aortic wall (Fig. 1b) and the lumen (Fig. 1c), after determining the sections and setting the parameters. The process for generating each mesh is detailed next. The initial axial contours are divided according to the circumferential parameters, which establish points equally spaced over the contours. The union of the corresponding points from each axial contour results in a set of longitudinal lines. Then these lines are divided by applying the longitudinal parameters. In this case, the resulting points are not equally spaced along the longitudinal lines due to the dilation of the bifurcation. They are placed further apart the closer they are to the bifurcation. Again, the union of corresponding points results in new set of circumferential lines. The combination of longitudinal and circumferential lines provides the quadrilateral mesh. In order to build hexahedral meshes from the quadrilateral ones, a set of center points must be available. These points are established as midpoints of the extreme longitudinal lines.

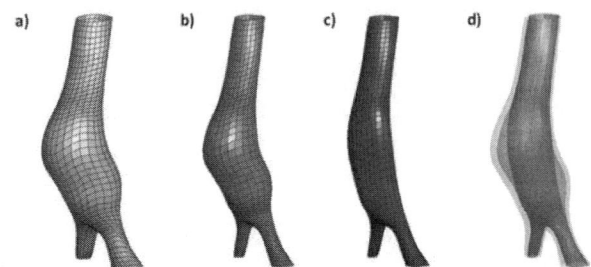

Fig. 1 Quadrilaterals meshes of a) external face of the wall b) internal face of the wall c) lumen. d) Hexahedral meshes of the wall and the ILT

C. Hexahedral Meshes Generation

The necessary elements for the construction of hexahedral meshes are the central points and the three quadrilateral meshes; one of them is taken as reference. The choice of the reference mesh is arbitrary: in this case we have chosen the wall's external face mesh. In order to avoid the formation of distorted hexahedra, the quadrilaterals of the remaining meshes must be refined. This improvement, performed using the ray casting technique, is based on the reference mesh. The hexahedral mesh of the wall is built by joining the quadrilaterals of the external and internal faces of the wall. The process is repeated likewise for the generation of the hexahedral mesh of the ILT, taking as reference in this case the just refined quadrilateral mesh. Any hexahedra with no volume (wherever the ILT does not exist, so the internal wall's face is equal to the lumen) are removed (Fig. 1d).

D. Evaluation of Mesh Quality

Within the context of the FE method, the quality of the finite elements obtained from the mesh generation greatly affects both the convergence of the simulations and the resulting approximations to the solutions of the governing partial differential equations. Additionally, accuracy in the representation of the true, patient-specific in vivo geometry also influences the applicability of the results [13]. Metrics for mesh quality must detect inverted elements (elements which generate meaningless results) and provide an estimate of the mesh's fitness for use in numerical simulations.

For the analysis for solid structures the *Scaled Jacobian* is a common quality metric [10]. *Scaled Jacobian* takes the range [-1,1] for a hexahedral element, -1 corresponding to the worst possible elements and +1 the best possible ones. The *Scaled Jacobian* applies only to regular hexahedral elements having eight (distinct) vertices at different spatial locations in 3D space. We use the open source program Paraview (Kitware, Inc., Clifton Park, New York, USA) to evaluate the *Scaled Jacobian*, among other quality metrics.

III. RESULTS

The performance of the technique described in the methodology was tested over a set of four MRI studies of patients presenting AAAs near the bifurcation of the iliac arteries. We are able to generate conforming hexahedral meshes, e.g. containing 25,000 elements in approximately 30 sec using a PC with 3 GHz Core Duo Pentium, 8 GB RAM and a 64 bit OS.

Fig. 2 illustrates in a qualitative way the resulting conforming hexahedral meshes of the arterial wall, while Fig. 3. illustrates their corresponding ILTs (not to scale). The meshes in Figs. 2 and 3 are shaded according to the *Scaled Jacobian* of the mesh elements.

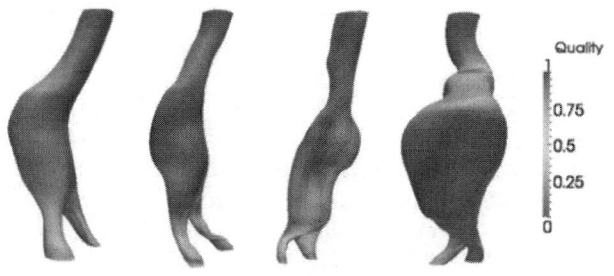

Fig. 2 Results for four representative Abdominal Aortic Aneurysms (AAAs) walls using the *Scaled Jacobian* as a measure of element quality.

Table 1 presents a quantitative description of the element's quality distribution. The defined ranges for this distribution represent from the unacceptable results (<0.0) until the excellent quality elements (0.8-1.0). Both the distribution of the *Scaled Jacobian* values for the separated layers of the AAA walls (intima, media and adventitita), and the separated layers of the ILTs (luminal, medial and abluminal) are illustrated for the four representative MRI studies. The representation is in terms of the arithmetic means and the standard deviations. Elements with *Scaled Jacobian* values in the range of 0.5 to 1.0 are well-suited to FE analysis [9].

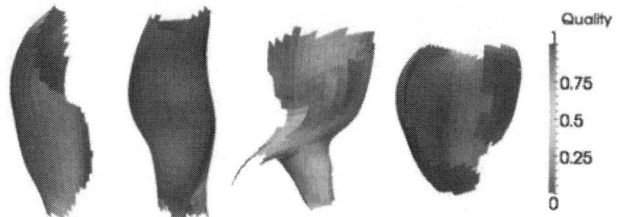

Fig. 3 Results for four representative Abdominal Aortic Aneurysms (AAAs) intraluminal thrombi (ILTs) using the *Scaled Jacobian* as a measure of element quality.

IV. DISCUSSION

Negative values signify invalid (inverted) elements ill-suited to FE analysis. For the ILTs, the element quality is generally at a very high level: there are no elements with negative *Scaled Jacobian* values and 94.8% of the total elements exhibit *Scaled Jacobian* values greater than 0.5. Thrombi of different shapes and located at different heights of the artery are represented. Importantly, note that the (layered) wall shapes are generally more complex than the

Table 1: Distribution frequencies (%) of scaled Jacobian values for individual layers in AAA wall and AAA thrombus, expressed as the arithmetic mean (AM) and the standard deviation (SD) of the patient set

Scaled Jacobian	<0		0.0-0.2		0.2-0.4		0.4-0.6		0.6-0.8		0.8-1.0	
	AM	SD	AM	SD	AM	SD	AM	SD	AM	SD	AM	SD
Adventitia	-	-	2.05	3.42	7.86	5.70	18.65	3.27	36.36	2.53	35.08	7.21
Media	-	-	1.52	3.24	6.97	6.05	19.18	3.66	37.17	1.96	35.16	7.81
Intima	-	-	1.29	2.80	6.34	6.22	19.74	4.06	37.81	2.84	34.82	7.95
Abluminal	-	-	0.83	1.85	3.49	4.33	16.97	11.41	35.42	12.21	43.29	25.81
Medial	-	-	1.05	2.35	2.01	3.13	13.16	11.07	34.44	15.19	49.35	28.75
Luminal	-	-	1.43	2.44	2.05	2.22	11.21	11.90	37.39	14.85	53.92	29.46

thrombi shapes. Nonetheless, the elements of the AAA wall are also of high quality: 87.8% of elements in the wall exhibit *Scaled Jacobian* values greater than 0.5; meanwhile only 0.22% fall below 0.2. In the examples shown, we use approximately 25,000 elements per mesh (suitable for FE analysis in less than one week).

The relatively low quality elements do not occur in the region of the bifurcation, but rather at regions where the surfaces of the original geometry lie relatively parallel to the original MR Imaging planes, due either to the presence of the ILT or to the tortuosity inherent to the arteries, as illustrated in Fig. 4.

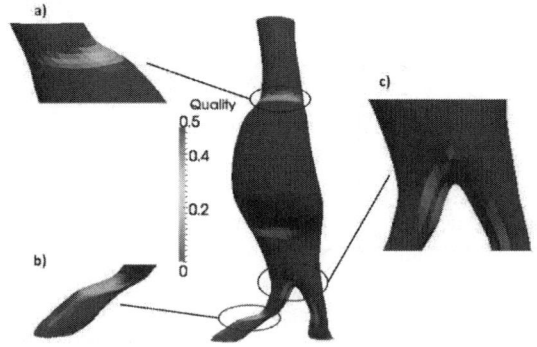

Fig. 4 Representative result for a patient-specific AAA: *Scaled Jacobian* shown with range 0.0 – 0.5. Relatively low quality elements do not occur in the region of the bifurcation (c), but rather at regions where the surfaces of the geometry lie relatively parallel to the original MRI planes(a-b).

V. CONCLUSIONS

In this work we propose a novel procedure for generating patient-specific conforming hexahedral meshes of AAAs and their corresponding thrombi with high-quality elements, particularly in the area of the bifurcation into the iliac arteries. These structures are divided into their layered morphologies. By doing so we provide novel input for simulations, as the layers have different histological and mechanical properties, representing the actual anatomy of the structures.

ACKNOWLEDGMENT

The authors gratefully acknowledge the financial support of the EC under the 7th Framework Programme in the frame of project SCATh, FP7-ICT-2009-4-248782.

REFERENCES

1. Fauci AS, Braunwald E, Kasper DL, Hauser SL, Longo DL, Jameson JL, Loscalzo J (2012) Harrison's Manual of Medicine. 18th ed. McGrawHil, New York, NY
2. Gloviczki P, Ricotta JJII (2007) Aneurysmal vascular disease. In: Townsend CM, Beauchamp RD, Evers BM, Mattox KL (eds) Sabiston Textbook of Surgery. 18th ed., Philadelphia, PA, Saunders Elsevier: chap 65
3. European cardiovascular disease statistics (2012). published by the European Heart Network
4. Venkatasubramaniam AK, Fagan MJ, Mehta T, Mylankal KJ, Ray B, Kuhan G, Chetter IC, McCollum PT (2004) A comparative study of aortic wall stress using finite element analysis for ruptured and non-ruptured abdominal aortic aneurysms. Eur J Vasc Endovasc Surg 28(2):168–176
5. Ho-Le K (1988) Finite element mesh generation methods: a review and classification. Comput Aided Design 20(1):27-38
6. Ramos A, Simões JA (2006) Tetrahedral versus hexahedral finite elements in numerical modeling of the proximal femur. Med Eng Phys. 28(9):916–924
7. Wassef M, Baxter BT, Chisholm RL, Dalman RL, Fillinger MF, Heinecke J (2001) Pathogenesis of abdominal aortic aneurysms: a multidisciplinary research program supported by the National Heart, Lung, and Blood Institute. J Vasc Surg 34(4):730-738
8. Antiga L, Ene-Iordache B, Caverni L, Cornalba GP, Remuzzi A (2002) Geometric reconstruction for computational mesh generation of arterial bifurcations from CT angiography. Comput Med Imaging Graph 26(4):227–235
9. Antiga L, Steinman D (2004) Robust and objective decomposition and mapping of bifurcating vessels. IEEE Trans Med Imag 23(6):704–713
10. De Santis G, De Beule M, Segers P, Verdonck P, Verhegghe B (2011) Patient-specific computational haemodynamics: generation of structured and conformal hexahedral meshes from triangulated surfaces of vascular bifurcations. Comput Methods Biomech Biomed Engin 14(9):797–802
11. Yushkevich PA, Piven J, Hazlett HC, Smith RG, Ho S, Gee JC, Gerig G (2006) User-guided 3D active contour segmentation of anatomical structures: Significantly improved efficiency and reliability Neuroimage 1(3):1116-28.
12. Lee SE, Piersol N, Loth F, Fischer P, Leaf G, Smith B, Yedevalli R, Yardimci A, Alperin N, Schwartz L (2000) Automated Mesh Generation of an Arterial Bifurcation Base upon in vivo MR Images. In: Engineering in Medicine and Biology Society. Proceedings of the 22nd Annual International Conference of the IEEE, Chicago, IL, Vol.1, pp 719-722
13. Knupp P (2007) Remarks on mesh quality. In: Proceedings of AIAA 45th Aerospace Sciences Meeting and Exhibit, Reno, NV

Author: Jaime Tarjuelo Gutiérrez
Institute: Bioengineering and Telemedicine Centre, ETSI de Telecomunicación, Universidad Politécnica de Madrid
Street: Avenida Complutense 30
City: Madrid
Country: Spain
Email: jtarjuelo@gbt.tfo.upm.es

Compressed Sensing for Cardiac MRI Cine Sequences: A Real Implementation on a Small-Animal Scanner

P. Montesinos[1,2], J.F.P.J. Abascal[1,2], C. Chavarrías[2], J.J. Vaquero[1,2], and M. Desco[1,2,3]

[1] Departamento de Bioingeniería e Ingeniería Aeroespacial, Universidad Carlos III de Madrid, Spain
[2] Instituto de Investigación Sanitaria Gregorio Marañón (IiSGM), Madrid, Spain
[3] Centro de Investigación Biomédica En Red de Salud Mental, CIBERSAM, Madrid, Spain

Abstract—Over the lasts years many works have addressed the potential of compressed sensing techniques to accelerate acquisition of cardiac MRI. However, most of these works claimed the achievement of acceleration factors solely based on simulated data or on fully sampled acquisition data retrospectively undersampled. In this work the practical feasibility of compressed sensing acquisitions for the acceleration of cardiac cine imaging in small animals is proved with real acquisitions. Our experiments using a combined spatiotemporal technique confirm that high acceleration factors of about 10 are feasible. Future work involving the optimization of undersampling patterns and cardiac artifact correction might further improve this acceleration factor.

Keywords—Compressed sensing, undersampling pattern, data acquisition, Split Bregman, cardiac cine MRI.

I. INTRODUCTION

Compressed sensing (CS) techniques have revolutionarily changed the field of accelerating magnetic resonance studies. These techniques have special impact on the field of dynamic imaging and particularly in cardiovascular imaging, where the tradeoff between acquisition time, spatiotemporal resolution and signal-to-noise ratio is particularly challenging.

Compressed sensing enables image reconstruction from pseudo-randomly undersampled data by means of a nonlinear reconstruction and an appropriate sparsifying transform [1].

In cardiac MRI many works have focused on prospective cardiac cine sequences, reporting different degrees of acceleration obtained with different reconstruction algorithms and sparsity transforms [2-7]. Very few works focus on small-animal cardiac cine sequences [8-10], which is our area of interest.

Among cardiac cine sequences, self-gated cine sequences avoid the use of electrodes to monitor ECG signal by acquiring navigator echoes. Navigator data serve to retrospectively classify the continuously acquired data into different time frames, according to its value at each temporal instant [11]. IntraGateFLASH, used in this work, is an example of these self-gated sequences.

With regard to the compressed sensing reconstruction approach, the gradient is the most common sparsifying transform, that leads to the minimization of the L1-norm of the image gradient, the so-called total variation (TV). In this work we propose to exploit the intrinsic temporal sparsity of cine images minimizing the total variation across both the spatial and temporal dimensions (ST-TV).

Among the wide variety of reconstruction methods, Split Bregman has proved to be computationally very efficient to solve L1-norm constrained optimization problems [12]. This reconstruction methodology was applied to small-animal self-gated cardiac cine sequences in a previous work [13], where acceleration factors up to 15 were found to be feasible. That study, as most of compressed sensing works, was based on acquiring fully-sampled data to which undersampling patterns were retrospectively applied.

Despite the great amount of compressed sensing works published in the past years, very few in the literature report compressed sensing accelerations based on actually acquired data [14-16] and they do not focus on the comparison between simulation results and real compressed sensing acquisition.

The aim of this work is to verify the feasibility of the previously reported acceleration factors by implementing compressed sensing acquisitions for self-gated cardiac cines.

II. THEORY

A. Reconstruction Algorithm

As mentioned before, in this work we use the spatiotemporal gradient as sparsifying transform that, according to the compressed sensing formulation, leads to the minimization of spatiotemporal TV subject to a data constraint:

$$\min_u \|\nabla u\|_2 + \|D_t u\|_1 \text{ such that } Fu = f \quad (1)$$

where $\nabla = (D_x, D_y)$ is the gradient, and D_x, D_y and D_t are the spatial and time derivatives respectively, $\|\cdot\|_1$ is the L1-norm, $\|\nabla u\|_2$ is the isotropic spatial total variation [12], and $\|D_t u\|_1$ is the temporal total variation.

The Split Bregman formulation can be used to solve (1), defining the following equivalent problem

$$\min_{d_x,d_y,d_t,u} \|(d_x,d_y)\|_2 + \|d_t\|_1 + \frac{\mu}{2}\|Fu-f^k\|_2^2$$
$$+\frac{\lambda}{2}\|d_x - D_x u - b_x^k\|_2^2 + \frac{\lambda}{2}\|d_y - D_y u - b_y^k\|_2^2 \quad (2)$$
$$+\frac{\lambda}{2}\|d_t - D_t u - b_t^k\|_2^2,$$

where each b_i represents the Bregman iteration that imposes a constraint, and the new variables, d_i, allow decoupling L2 and L1 functionals [12, 17].

B. Undersampling Patterns

The undersampling patterns were created based on a variable probability density function that gives higher probability of being preserved to lines according to its distance to the k-space center (Fig. 1).

The benefits of including randomization across time in the undersampling patterns have been well proven [13] and thus the undersampling patterns applied in both acquisition and retrospective simulation vary in both time and phase encoding direction.

III. METHODS

A. Implementation of the Compressed Sensing Acquisition

For this work we used a Bruker 70/20 USR MRI scanner with software ParaVision 5.0.

In this machine sequences are defined by means of programmable components called methods (sets of files that provide and assign adequate values for acquisition parameters).

We modified the pulse program (.ppg) of an IntraGateFLASH sequence in such a way that the list of phase encoding lines acquired was no longer linear but corresponded to a certain randomly selected undersampling pattern.

B. Acquired Dataset

The modified sequence used the following acquisition parameters:

TE = 2.43 ms, TR = 8 ms, number of frames = 8, matrix size = 192x192, FOV = 4.8x4.8 cm, and slice thickness = 1.2 mm. In the case of the fully sampled image, the number of phase encoding repetitions was 200, acquired in a total time of 5 min 7 s.

Images were acquired using a linear coil resonator for transmission and a dedicated four-element cardiac phased array coil for reception.

To validate the proposed method and to determine the highest acceleration factor achievable for this real application, we performed different data acquisitions for percentages of undersampling ranging from 30% (x3.3) to 5% (x20) of the complete data (acceleration factor shown within parenthesis).

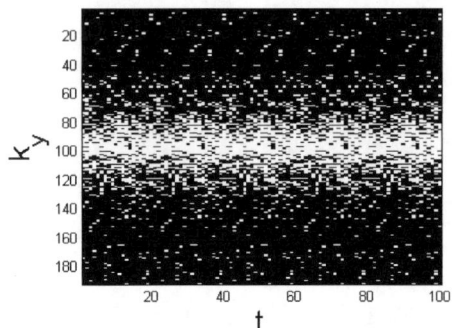

Fig. 1 Undersampling pattern generated from a variable probability density function (20% data). Black lines represent eliminated or non-acquired phased-encoding lines from the original complete data

The undersampling patterns previously optimized using a simulated study [13] had to be adapted due to implementation requirements. Because of programming restrictions, the number of different temporal undersampling patters had to be limited to 20 different patterns repeated over time, as it can be seen in Figure 1. The repetition rate of these patterns did not correlate with the same frequency of frames in such a way that, after data sorting and averaging, each frame had a different undersampling pattern. Figure 1 shows an example of undersampling pattern for 20% of data.

As in the case of the previous simulated work [13], all the elements of the array were separately reconstructed and then combined.

IV. RESULTS

The experiments confirmed the feasibility of the implementation of the CS technique. The forecasted reduction in acquisition times corresponding to each acceleration factor was successfully verified.

Figure 2 shows a fully sampled image and retrospective undersampled compressed sensing reconstructions for different percentages of undersampled data ranging from 20% (x5) to 10% (x10).

Figure 3 shows images directly acquired with compressed sensing for different percentages of undersampled data from 20% (x5) to 10% (x10); larger accelerations resulted in reconstructions with artifacts. We must point out that with real data the correspondence between frames of different images cannot be perfect because they belong to different acquisitions, and thus the matching between frames is only approximate.

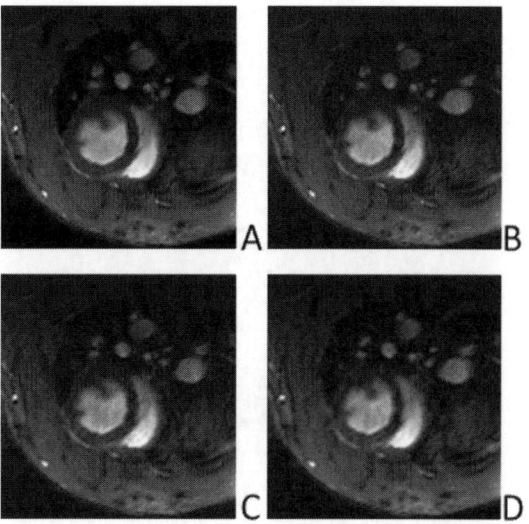

Fig. 2 Reconstructed images for different undersampling patterns and acceleration factors in retrospectively simulated study. A) Fully-sampled image. B) 20% (x5) of preserved data. C) 15% (x7) of preserved data. D) 10% (x10) of preserved data

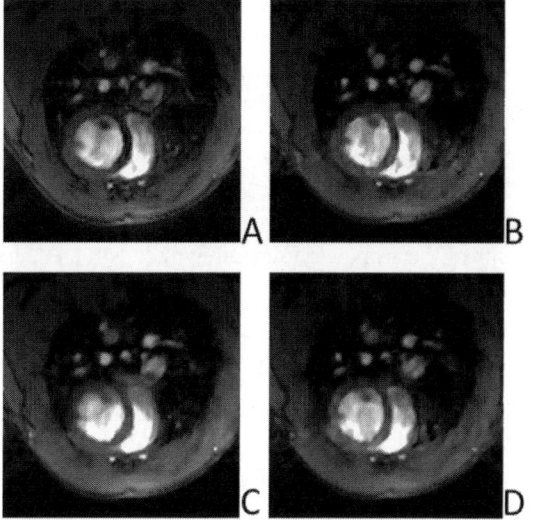

Fig. 3 Reconstructed images for different undersampling patterns and acceleration factors experimentally acquired. A) Fully-sampled image. B) 20% (x5) of preserved data. C) 15% (x7) of preserved data. D) 10% (x10) of preserved data

V. DISCUSION

The feasibility of implementing compressed sensing acquisition for cardiac cine MRI sequences was successfully proven in this work.

Acceleration factors up to 10 or 15 were feasible for retrospectively simulated undersampled data (Figure 2). In the case of actually acquired undersampled data (Figure 3) acceleration factors around 10 were achievable.

With both simulated and acquired undersampled data, larger acceleration factors increase the smoothness on the image and make more conspicuous the presence of some artifacts. However, for the accelerations claimed these artifacts do not alter the diagnostic value of the image, which is mostly determined by the inner and outer delimitation of the myocardium [18].

For the same amount of preserved data the quality of the simulated undersampled data is slightly better than the one achieved with actually acquired undersampled data. This difference might be explained by the fact that the undersampling pattern in the simulation was optimized *ad hoc* for the dataset -the best undersampling pattern was chosen among several simulations- while in the case of acquired compressed sensing only one undersampling pattern can be acquired and evaluated.

Another problem with the real acquisition is that the number of random patterns could not be higher than 20, as compared with the 200 patterns applied to simulated data.

In order to further optimize the acquisition parameters a more detailed study of undersampling patterns is warranted. Also, studies involving a larger sample of subjects with both health and unhealthy subjects are required to claim that a certain acceleration factor is feasible for in vivo routine studies.

VI. CONCLUSIONS

In this work we demonstrate that compressed sensing acquisitions are feasible for cardiac cine acquisitions. Acceleration factors around 10 are achievable, close to those obtained in previous simulations.

ACKNOWLEDGMENT

This work was partially funded by the Spanish Ministry of Economy and Competitiveness projects RECAVA RD07/0014/2009; RD12/0042/0057 and AMIT CEN-20101014, CDTI-CENIT program.

REFERENCES

1. Lustig, M., D. Donoho, and J.M. Pauly, Sparse MRI: The application of compressed sensing for rapid MR imaging. Magnetic Resonance in Medicine, 2007. 58(6): p. 1182-1195.
2. Montefusco, L.B., et al., A fast compressed sensing approach to 3D MR image reconstruction. IEEE Trans Med Imaging, 2011. 30(5): p. 1064-75.
3. Christodoulou, A.G., et al., High-Resolution Cardiac MRI Using Partially Separable Functions and Weighted Spatial Smoothness Regularization, in 2010 Annual International Conference of the Ieee Engineering in Medicine and Biology Society, IEEE: New York. p. 871-874.

4. Lustig, M., et al., k-t SPARSE : High frame rate dynamic MRI exploiting spatio-temporal sparsity. Proceedings of the 14th Annual Meeting of ISMRM, Seattle, Washington, USA, 2006: p. 2420.
5. Pedersen, H., et al., k-t PCA: temporally constrained k-t BLAST reconstruction using principal component analysis. Magn Reson Med, 2009. 62(3): p. 706-16.
6. Usman, M., et al., k-t Group sparse: a method for accelerating dynamic MRI. Magn Reson Med, 2011. 66(4): p. 1163-76.
7. Usman, M., et al., A computationally efficient OMP-based compressed sensing reconstruction for dynamic MRI. Physics in Medicine and Biology, 2011. 56(7): p. N99-N114.
8. Motaal, A.G., et al., Accelerated high-frame-rate mouse heart cine-MRI using compressed sensing reconstruction. NMR in Biomedicine, 2013. 26(4): p. 451-457.
9. Wech, T., et al., Accelerating Cine-MR Imaging in Mouse Hearts Using Compressed Sensing. Journal of Magnetic Resonance Imaging, 2011. 34(5): p. 1072-1079.
10. Wech, T., et al., Highly accelerated cardiac functional MRI in rodent hearts using compressed sensing and parallel imaging at 9.4T. Journal of Cardiovascular Magnetic Resonance, 2012. 14(Suppl 1): p. P65.
11. Bovens, S.M., et al., Evaluation of infarcted murine heart function: comparison of prospectively triggered with self-gated MRI. NMR in Biomedicine, 2011. 24(3): p. 307-315.
12. Goldstein, T. and S. Osher, The Split Bregman Method for L1-Regularized Problems. SIAM Journal on Imaging Sciences, 2009. 2(2): p. 323-343.
13. Montesinos, P., et al., Application of the compressed sensing technique to self-gated cardiac cine sequences in small animals. Subbmited to Magnetic Resonance in Medicine.
14. Han, S., et al., Temporal/spatial resolution improvement of in vivo DCE-MRI with compressed sensing-optimized FLASH. Magn Reson Imaging, 2012. 30(6): p. 741-52.
15. Vasanawala, S.S., et al. Practical parallel imaging compressed sensing MRI: Summary of two years of experience in accelerating body MRI of pediatric patients. in Biomedical Imaging: From Nano to Macro, 2011 IEEE International Symposium on. 2011.
16. Haldar, J.P., D. Hernando, and Z.P. Liang, Compressed-sensing MRI with random encoding. IEEE Trans Med Imaging, 2011. 30(4): p. 893-903.
17. Cai, J., S. Osher, and Z. Shen, Split Bregman Methods and Frame Based Image Restoration. Multiscale Modeling & Simulation, 2010. 8(2): p. 337-369.
18. Hankiewicz, J.H., et al., Principal strain changes precede ventricular wall thinning during transition to heart failure in a mouse model of dilated cardiomyopathy. Am J Physiol Heart Circ Physiol, 2008. 294(1): p. H330-6.

Author: Paula Montesinos
Institute: Universidad Carlos III de Madrid.
Street: Avda. De la Universidad, 30
City: Leganés (Madrid)
Country: Spain
Email: pmontesinos@mce.hggm.es

The Importance of a Valid Reference Region for Intensity Normalization of Perfusion MR Studies in Early Alzheimer's Disease

M. Lacalle-Aurioles[1,2,3], Y. Alemán-Gómez[2,3], J. Guzmán-De-Villoria[4], J. Olazarán[5], I. Cruz[5], J.M. Mateos-Pérez[2,3], M.E. Martino[3], and M. Desco[1,2,3]

[1] Departamento de Bioingeniería e Ingeniería Aeroespacial, Universidad Carlos III de Madrid, Madrid, Spain
[2] Centro de Investigación en Red de Salud Mental (CIBERSAM), Madrid, Spain
[3] Instituto de Investigación Sanitaria Gregorio Marañón, Madrid, Spain
[4] Servicio de Radiodiagnóstico, Hospital General Universitario Gregorio Marañon, Madrid, Spain
[5] Servicio de Neurología de la conducta, Hospital General Universitario Gregorio Marañon, Madrid, Spain

Abstract—Magnetic resonance imaging (MRI) of perfusion could represent a powerful tool in the characterization and tracking of Alzheimer's disease (AD). Brain perfusion presents a large physiological variability across subjects due to biological factors that make it difficult the detection of perfusion abnormalities in comparative analysis; this fact makes necessary a step of intensity normalization of perfusion images. The cerebellum is the most commonly used reference region in perfusion studies in AD patients using nuclear medicine techniques, since it has been reported to provide unbiased estimations. This knowledge has been directly extrapolated to perfusion studies with MRI, but no reports evaluate the consequences of using different normalization regions in MRI studies, and the cerebellum has not been yet confirmed as an optimal reference region. The purpose of this study, is to address the effect of using three reference regions, cerebellum, whole-brain white matter, and whole-brain cortical gray matter in the normalization of cerebral blood flow (CBF) parametric maps, based on a comparative analysis between patients with stable mild cognitive impairment (MCI), patients with AD, and healthy controls. Our results suggest that normalization by whole-brain cortical gray matter enables a more sensitive detection of perfusion abnormalities in AD. The cerebellum, therefore, is not the best reference region in MRI studies of early stages of AD.

Keywords—MRI, Perfusion-weighted imaging, Cerebral blood flow, Intensity normalization, Reference region, Alzheimer's disease.

I. INTRODUCTION

Perfusion MRI is a promising alternative to nuclear medicine in early detection of AD [1]. The large physiological variability between subjects due to uncontrolled biological and experimental factors [2] makes normalization necessary for comparative analysis. The most widely used intensity normalization method computes the ratio of a region of interest (ROI) value to the average perfusion value of all voxels within a reference region. While the cerebellum is the most commonly used region for normalization in nuclear medicine studies [3-5], there are not previous reports confirming the cerebellum as the optimal reference region in perfusion-weighted MRI measurements.

In this study, we address the effect of using three different regions for normalization (whole-brain cortical gray matter, whole-brain white matter, and the cerebellum in a comparative analysis of CBF parametric maps between stable MCI patients, AD patients, and healthy controls.

II. MATERIAL AND METHODS

A. Subject

Three groups of subjects were included in the study: the AD group (mean age=74.86) comprised 12 patients recruited with probable AD at stage 1 in the Clinical Dementia Rating Scale (CDR) and 16 subjects recruited at the stage of MCI (CDR=0.5) that converted to probable AD after 2 years of follow-up. The MCI group (mean age=70.87, CDR=0.5) comprised 15 patients who did not convert to AD in the same follow-up period. Control group (mean age=71.65, CDR=0) comprised 20 healthy subjects.

The participants were recruited prospectively in the behavioural neurology clinic of a teaching hospital. The local ethics committee approved the study and an informed consent was obtained from all subjects.

B. Image Acquisition

The study included a volumetric scan used for tissue segmentation (T1-weighted 3D gradient echo, FA=30°; TR=16 ms, TE=4.6 ms; matrix size=256x256; FOV= 256x256 mm; and 100 slices with slice thickness=1.5 mm).

Perfusion-weighted images (PWI) were obtained using an echo-planar imaging sequence (EPI factor=61, FA=40°; TR=1439 ms, TE=30 ms; matrix size=128x128; FOV=230x230 mm; section thickness=5 mm) after the injection of a bolus of gadolinium chelate, a dynamic susceptibility contrast. CBF maps were obtained according to the following equation:

$$CBF = \frac{CBV}{MTT}$$

Where

$$CBV = \frac{\int C_m(t)}{\int AIF(t)} ; MTT = \frac{\int C(t)}{C_{max}}$$

In the previous equation, there are two different concentrations: the concentration that can be measured directly from the image data ($C_m(t)$) and the concentration resulting from a deconvolution operation with the arterial input function (AIF). Both concentrations can be expressed mathematically as

$$C_m(t) = -\ln\frac{S(t)}{S_0} ; C(t) = C_m(t) \otimes^{-1} AIF(t)$$

Where $S(t)$ is the MRI image signal, S_0 is the signal in the first six frames, when the contrast bolus still is not present and $\otimes^{-1}$ represents the deconvolution operation.

The AIF was calculated automatically using the method detailed in [6]. In order to remove noise and other undesired second-order effects, the contrast curves where fitted to a gamma function according to the linearization method proposed in [7].

C. Image Analysis

The T1 scans were processed with the VBM8 toolbox for the SPM8 package to obtain the skull-stripped 'p0' image that consists of brain tissue classified into gray matter (GM), white matter (WM), and cerebrospinal fluid (CSF). The package FreeSurfer version 4.5.1. (Martinos Center for Biomedical Imaging) was used to obtain the volume per ROI. Giral-based ROIs were defined according to the Desikan-Killiany atlas [8]. Whole-brain WM and cerebellum were segmented according to [9]. CBF parametric maps were coregistered with the T1-weighted images using mutual information methods [10]. After coregistration, masks of every ROI were applied to CBF maps and an average CBF value per ROI was then computed.

D. Statistical Analysis

We used ANOVA models to 1) discard differences between groups in the reference regions studied, 2) to confirm that no group-bias existed in those regions and 3) to find differences between groups in brain lobes (frontal, parietal, temporal and occipital) before and after normalization process.

III. RESULTS

We did not find significant differences between patients groups and controls in any of the reference regions chosen for this study (figure 1).

Absolute CBF data showed no statistically significant differences between patients and controls groups in any brain lobe. Differences in perfusion values between groups were detected only after normalization of data. CBF data normalized by whole-brain cortical GM showed significantly higher mean values in the right medial temporal lobe and lower CBF mean values in parietal lobes in AD patients than in the control group. When normalizing by whole-brain WM and cerebellum differences between AD patients and controls appeared only in the right medial temporal lobe (Table 1), showing the AD group higher CBF mean values than controls. We found no significant differences between patients with stable MCI and controls.

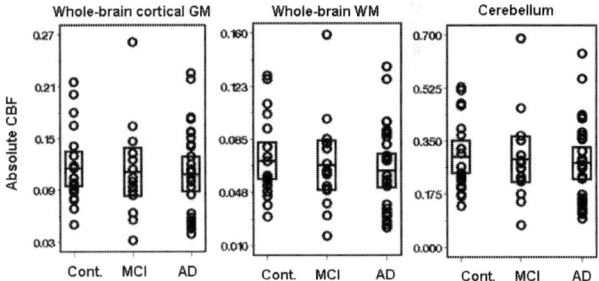

Fig. 1 Scatter plots of CBF values for controls, patients with mild cognitive impairment, and patients with Alzheimer's disease in both parietal lobes and right medial temporal lobe. Absolute CBF units are given in ml of blood/100 g of tissue/min. Cont, controls; MCI, mild cognitive impairment; AD, Alzheimer's disease. Bar shows two standard deviations below and above the mean (horizontal line).

Table 1: ANOVA p values (p) for mean CBF values per ROI, normalized by the three reference regions studied: CBFcgm, cerebral blood flow of whole-brain cortical gray matter; CBFcer, cerebral blood flow of cerebellum; CBFwm, cerebral blood flow of whole-brain white matter

	Right medial temporal lobe	Right parietal lobe	Left parietal lobe
	p	p	p
CBFcer	<0.05	-	-
CBFwm	<0.01	-	-
CBFcgm	<0.05	<0.01	<0.01

IV. DISCUSSION AND CONCLUSION

The first criterion for a valid reference region is to show stable values across subjects and no group differences [11]. Therefore, the absence of significant differences between groups in the three regions proposed in this paper makes them potentially good reference regions. However, only the normalization of the ROIs by whole-brain cortical GM showed the pattern of perfusion abnormalities characteristic of AD [12].

In summary, our results suggest that normalization of CBF parametric maps by whole-brain cortical GM enables a more sensitive detection of perfusion abnormalities in AD. Our results also suggest the absence of perfusion abnormalities in our sample of stable MCI.

ACKNOWLEDGEMENT

This work was partially funded by AMIT Project (Programa CÉNIT. Ministerio de Economía y Competitividad, Spain)

REFERENCES

1. Luckhaus, C., et al., *A novel MRI-biomarker candidate for Alzheimer's disease composed of regional brain volume and perfusion variables.* European Journal of Neurology, 2010. **17**(12): p. 1437-1444.
2. Diamant, M., et al., *Twenty-four-hour non-invasive monitoring of systemic haemodynamics and cerebral blood flow velocity in healthy humans.* Acta Physiol Scand, 2002. **175**(1): p. 1-9.
3. Talbot, P.R., et al., *Choice of reference region in the quantification of single-photon emission tomography in primary degenerative dementia.* Eur J Nucl Med, 1994. **21**(6): p. 503-8.
4. Karbe, H., et al., *Quantification of functional deficit in Alzheimer's disease using a computer-assisted mapping program for 99mTc-HMPAO SPECT.* Neuroradiology, 1994. **36**(1): p. 1-6.
5. Soonawala, D., et al., *Statistical parametric mapping of (99m)Tc-HMPAO-SPECT images for the diagnosis of Alzheimer's disease: normalizing to cerebellar tracer uptake.* Neuroimage, 2002. **17**(3): p. 1193-202.
6. Rempp, K.A., et al., *Quantification of regional cerebral blood flow and volume with dynamic susceptibility contrast-enhanced MR imaging.* Radiology, 1994. **193**(3): p. 637-41.
7. Li, X., et al., *Adaptive total linear least square method for quantification of mean transit time in brain perfusion MRI.* Magn Reson Imaging, 2003. **21**(5): p. 503-10.
8. Desikan, R.S., et al., *An automated labeling system for subdividing the human cerebral cortex on MRI scans into gyral based regions of interest.* Neuroimage, 2006. **31**(3): p. 968-80.
9. Fischl, B., et al., *Whole brain segmentation: automated labeling of neuroanatomical structures in the human brain.* Neuron, 2002. **33**(3): p. 341-55.
10. Collignon, A., et al., *Automated multi-modality image registration based on information theory*, in *Information Processing in Medical Imaging.* 1995, Kluwer Academic Publishers: Amsterdam.
11. Borghammer, P., et al., *Normalization in PET group comparison studies--the importance of a valid reference region.* Neuroimage, 2008. **40**(2): p. 529-40.
12. Alsop, D.C., et al., *Hippocampal hyperperfusion in Alzheimer's disease.* Neuroimage, 2008. **42**(4): p. 1267-74.

Author: María La Calle Aurioles
Institute: Instituto de Investigación Sanitaria Gregorio Marañón
Street: Dr. Esquerdo 46
City: Madrid
Country: Spain
Email: mlacalle@hggm.es

A Full Automatic Method for the Soft Tissues Sarcoma Treatment Response Based on Fuzzy Logic

E. Montin[1], A. Messina[2], and L.T. Mainardi[1]

[1] Politecnico di Milano / Dipartimento di Elettronica Informazione e Bioingegneria, Milan, Italy
[2] Istituto nazionale dei tumori di Milano / Dipartimento di radiodiagnostica per immagini unità 1, Milan, Italy

Abstract—Aim of this study was to develop a full automatic method for the soft tissue sarcoma (STS) identification and its evaluation during chemotherapy (CT) treatment, based on diffusion MRI.

This procedure includes two main phases, the first one is a registration step in order to compose a coregistered set of images concerning morphological and functional images pre and post CT, registered to the pre-treatment T1. The second phase is a fuzzy characterization of STS diffusive parameter along with a fuzzy inference step for the evaluation of treatment response on the characterized area.

The results of this procedure are two membership degree maps which measure the probability for each segmented pixel to be responding or not to CT. These two maps could assist radiologists during follow-up assessment, by automatically extracting the lesion volume, report lesion composition, and measure the uncertainty of the estimate, in order to manage the intrinsic and well-known heterogeneity of STS.

Many studies demonstrated the prognostic values of diffusion MRI and the apparent diffusion coefficient (ADC) in the assessment of cellularity changes during therapy and their correlation with lesion response.

In this scenario, the proposed framework, allows for a total unsupervised identification and probabilistic evaluation of STS treatment response based on diffusion MRI, mapping the local changes of ADC during therapy rather than describe the whole lesion by a statistical moment.

Keywords—MRI, DWI, Diffusion, Cancer treatment response, Image registration.

I. INTRODUCTION

Soft tissue sarcomas (STS) are a heterogeneous group of tumors that pose significant diagnostic and therapeutic challenges for clinical care and medical research [1]. The assessment of treatment response in STS following chemotherapy (CT) or radiation (RT) is one of the most important aspects of patient care, as therapeutic options and the timing of surgery may vary depending on the success of response [2]. The primary purpose of diagnostic imaging subsequent to CT or RT in STS is to assess patient response to therapy. Diffusion weighted magnetic resonance imaging (DWI) techniques, have been advocated for this purpose [11]. In fact as a non-invasive technology with no ionizing radiation, DWI offers unique contributions to the determination and monitoring of the therapeutic response in STS. DWI is an integral part of the multi-modality approach to the assessment of STS diagnosis and treatment.

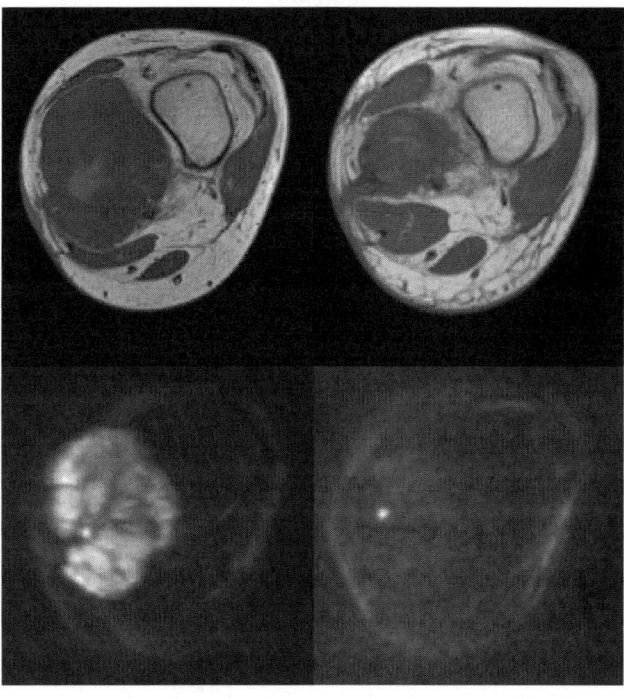

Fig. 1 STS morphological (top) and DWI b1000 s/mm² images (bottom) of pre (left) and post (right) treatment. After CT lesions decrease their volume. DWI shows the heterogeneity of the lesions which make difficult to quantitative evaluate the response.

STSs assessment response is made usually with RECIST [3] and CHOI [4] criterion which account only for volume changes in morphological images.

Many articles proved the prognostic value of the ADC but also its intrinsic variability and heterogeneity within the lesion [14], this poses several problem on the selection of the region of interest for its quantification as mean value.

In this scenario we developed a method for the characterization of ADC values in the lesion by three discrete types able to locally describe the tissue composition pre and post treatment, in order to infer a map of tissues changes which could visualize, analyses and report the therapy effect without any operator interaction.

II. MATERIALS AND METHODS

The suggested method can be subdivided in two main phases, an intra and inter registration step and a fuzzy segmentation through fuzzy logic, as reported in figure 2, and explained in the next sessions.

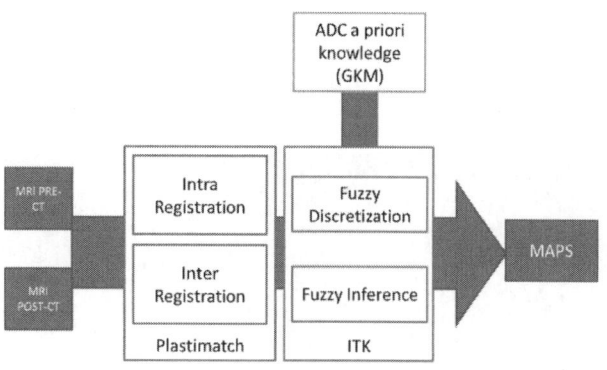

Fig. 2 Two step registration / segmentation framework.

A. Imaging Protocol

For each patient two subsequent exams have been considered, a pre CT and an early reassessment exam (usually after three months) as shown in Fig.1. Each exam included morphological and functional sequences: a T1 turbo spin echo (TSE) images pre and post contrast, (TR 530, TE 14, in plan resolution 0.7 x 0.7 mm, slice thickness 5 mm and gap 1 mm, FOV 184 x 230 mm) and a DWI based on echo planar imaging sequence (using 4 b-values (50, 400, 800, 1000 s/mm²) TR 6500, TE 78, in plan resolution 1.9 x1.9 mm, average 4, no gap, FOV 308 x 379).

B. Image Processing and Apparent Diffusion Coefficient Computation (ADC)

The main aim of the first processing step is to realign the functional images, the DWI, on the pre and post treatment exam to allow the evaluation of the ADC changes in tissues due to therapy. Thus two registrations are needed: an intra-exam and inter-exam ones. The first one realigns the DWI image on the morphological one within the same exam; the second registers the post treatment morphological image on the pre-treatment one. These steps allow superimposing pre and posting treatment DWI for the fuzzy logic analysis.

All the registrations were performed using plastimatch 1.5.11 [5], an open source software for image computation, whose main focus is high-performance volumetric registration of medical images.

Intra exam registration of functional images

For the DWI image registration on morphological T1 (fixed) the lower b-values image (floating) was used and then the estimated vector field was applied to the other b-values images. The lowest b-values image was used as floating image because it maintains the highest tissue information, allowing for the best registration [6].

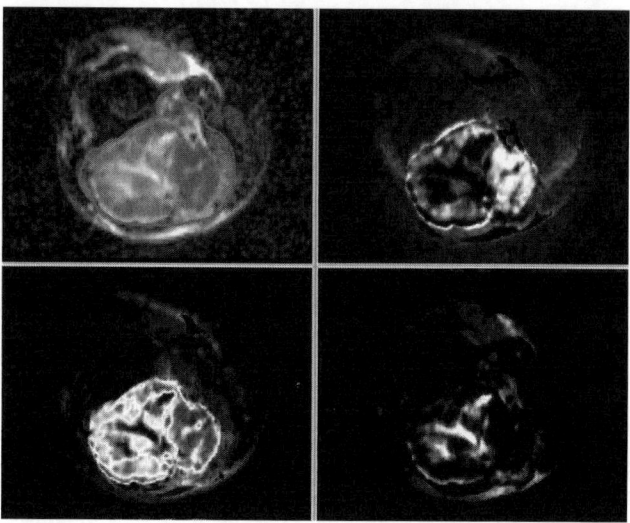

Fig. 3 Fuzzy segmentation of ADC values: top left the ADC map, right top, bottom left and bottom right the low, medium and high diffusion membership degrees label of the slice, on the first image is countered the lesions. Each membership degree map is rescaled from 0 to 100, these values represent the probability of each voxel to belong to the label.

Each registration step is composed by three subsequent registrations which estimate a translation, an affine and finally a non-rigid transformation (based on b-spline) using mutual information (MI) as similarity measure [6][7].

Inter exam registration for the follow up evaluation

To follow the evolution of each single voxel between pre and post treatment exams a further registration step is needed. The post treatment T1 image was therefore registered with the pre-treatment T1 exam.

The registration step was carried out by a combination of translation, affine and non-rigid registration using mean square error (MSE) as similarity measure. MSE was used because sequences were obtained using the same TSE sequence with constant TR and TE. The obtained transformation was then used to warp all the functional images and the derived map of the second exam onto the pre-treatment one.

Parametric map estimation

DWI images through the ADC can reveal the water movement freedom in extracellular space of the district, this implies that the typical abnormal growth of cells number inside the lesions drags to a restriction of the extracellular area which reflects on a slow loss of signal in DWI images and therefore a low ADC value. ADC was calculated using a least square fitting of the logarithm of the b-values exponential decrease [8].

C. Fuzzy Logic Segmentation and Classification

After the image registration step a set of functional and morphological co-registered images (pre and post-treatment) was obtained. Among them the ADC map was used to classify the lesion according to radiology standards [12], which identify area of restricted, medium and high diffusive tissue and suggest the typical values.

To provide an automatic identification of these three areas inside the lesion a fuzzy set classification was used. Three membership functions were defined using Gaussian Kernels centered in 600, 1200 and 2000 mm²/s, respectively; all the kernels had a standard deviation of 300 mm²/s, these values arose from a Gaussian kernel model estimation, initialized by literature values (800, 1500, 2500 mm²/s) [12], optimized by expectation maximization on a subset of the lesion defined by a threshold on the highest b-values images. Results of this procedure are reported in Fig 3, which shows an example of the fuzzy segmentation outcomes: the ADC map (top left) and the three membership degree maps for restricted, medium and high diffusive tissue, report the probability of each pixel to belong to the three tissue types. These maps range from 0 to 100, where 100 represents the maximum probability to belong to a particular discrete fuzzy set label.

To evaluate the therapy response, this fuzzy discretization method was applied on both pre- and post-treatment exams and then the six computed maps were combined considering a rise of the diffusive parameter as a positive response [13], as reported in table 1.

Table 1 Table of rules for the cancer response.

ADC	POST_LOW	POST_MEDIUM	POST HIGH
PRE_LOW	NR	R	R
PRE_MEDIUM	NR	NR	R
PRE_HIGH	NR	NR	R

NR non responding, R responding to therapy.

Mamdani's " min implication rules " [9] was applied to this six discretized membership degree maps and used for the calculation of the output probabilistic maps of the responding and non-responding part of the lesion.

For every voxels the response map was defined as:

$$Response = max\ (\ min\ (\ PRE\_LOW,\ POST\_MEDIUM\ ),$$
$$min\ (PRE\_LOW, POST\_HIGH),$$
$$min(PRE\_MEDIUM,\ POST\_HIGH\ ),$$
$$min\ (\ PRE\_HIGH, POST\_HIGH\ )\ ).$$

And similarly the non-responding:

$$Non\text{-}response = max\ (\ min(PRE\_LOW,\ POST\_LOW\ ),$$
$$min(PRE\_MEDIUM, POST\_LOW),$$
$$min(PRE\_MEDIUM,\ POST\_MEDIUM\ ),$$
$$min\ (\ PRE\_HIGH, POST\_LOW\ ),$$
$$min\ (PRE\_HIGH,\ POST\_MEDIUM)).$$

The results of this last step are two probabilistic maps which measure the membership degree of each pixel to responding or non-responding label, these probability like maps were rescaled from 0 to 100.

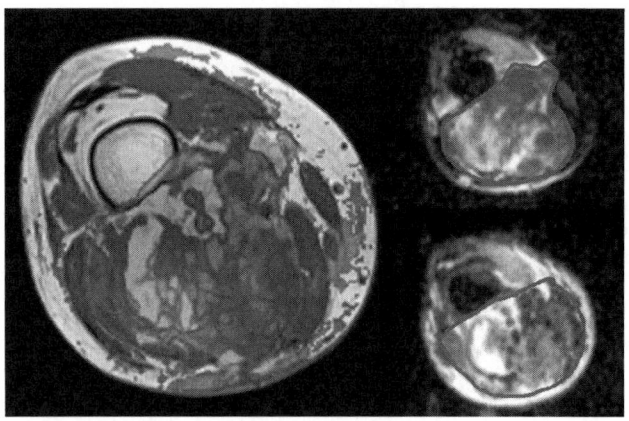

Fig. 4 On top right and bottom right ADC map on the pre and post treatment, in red the contour of the lesion. Left image is the pre-treatment T1 with on the red channel the positive response map and on blue one the non-responding pixels, green part of the lesion is the mixed part, where there's both a response and a non-response of the tissue, the membership degree maps assist the radiologist during the evaluation of patient therapy.

III. RESULTS

An example of the output of the fuzzy inference step is shown in Fig 4, the images display a mixed membership degree map combining the two membership functions to form a RGB map with response membership degree on red channel and non-responding membership degree on blue channel.

As we can see, although tumors extension is relatively the same between the pre and post treatment (see the red board line on Fig.4), the ADC values inside the lesion are different. Indeed during the therapy, the cellular density of the lesions changes and this reflects on the variation of the diffusion functional parameter values [13]. A rise of the ADC values usually implies a positive response of the neoplasia while a decrease is considered as a negative response[12]. Thus, local changes in ADC values in the two time points report the effect of the treatment on that area.

In this context, the application of this methodology is able to enhance the different pattern response of the therapy, also in this heterogeneous tumors, and reports many feature of the lesion including cellular composition, pre and post treatment probability of response and global lesion volume. The maps obtained by the automatic method were visually evaluated by clinicians and radiologists, after the anatomo-patologist exam, who established a good correlation with the real patient's state and lesions composition.

Furthermore this visualization allows the radiologists to evaluate the early cancer treatment response, suggesting which part of the tumor is responding. A measure of the reliability of this judgment is given by the membership degree.

IV. CONCLUSIONS

As expected, diffusion plays a key role in tumor response [10]; indeed post CT tissue variations modify the diffusivity on cellular ensemble implying changes on ADC values.

The tissue composition pre and post treatment maps allow, to get a sense of global changes in tissues due to therapy effect, this information integration, is modeled in the automatic fuzzy inference step. The final response map can assist the radiologist in the therapy evaluation, managing the intrinsic STS heterogeneity and the low resolution and signal to noise ratio of the DWI techniques, focusing the attention of the operator in fewer local points.

STS are very heterogeneous tumors, then a statistical moment on hand made ROI could not be considerate the best estimator of the lesion diffusion and pose problems on where to locate it. By thresholding the response maps to a reliable degree is possible to retrospectively evaluate the pre and post therapy ADC values subdivided in the three types and finally correlate these values with the biomarker exposed in the literatures, providing an accurate description of the phenomenon.

This method supplies a total unsupervised analysis of tumor treatment response since its early stages and has potential to become a useful method for diagnosis and follow up of STS, easily extendable to other tissues tumors.

REFERENCES

1. Wang X, Jacobs MA, Fayad L (2011). Therapeutic response inmusculoskeletal soft tissue sarcomas: evaluation by MRI; NMR Biomed. 2011 Jul; 24(6):750-63. Doi: 10.1002/nbm.1731.
2. Grimer R, Judson Ian, Peake D, and Seddon B. (2010) Guidelines for the Management of Soft Tissue Sarcomas. Sarcoma 15 pages, doi:10.1155/2010/506182
3. Stacchiotti S, Collini P, Messina A, Morosi C, Barisella M,M Bertulli R, Piovesan C, Dileo P, Torri V, Gronchi A, Casali PG.(2009) High-grade soft-tissue sarcomas: tumor response assessment - pilot study to assess the correlation between radiologic and pathologic response by using RECIST and Choi criteria. Radiology; 251(2):447-56. Doi: 10.1148/radiol.2512081403.
4. De Sanctis R, Bertuzzi A, Magnoni P, Giordano L, Gasco M, LutmanR. (2012) Superiority of choi vs recist criteria in evaluating outcome of advanced soft tissue sarcoma (STS) patients treated with sorafenib; 37 ESMO congress 2012.
5. James A Shackleford, NagarajanKandasamy, Gregory C Sharp (2010) On developing B-spline registration algorithms for multi-core processors, Physics in Medicine and Biology, Vol 55, No 21, pp 6329-6351.
6. Montin E, Potepan P, Mainardi LT, (2012) A registration framework for evaluation of T1 and T2 and DWI signal intensities in multiple myeloma, BIOSIGNALS, 563-566
7. Veronese F, Montin E, Potepan P, Mainardi LT, (2012) Quantitative characterization and identification of lymph nodes and nasopharyngeal carcinoma by co registered magnetic resonance images, doi:10.1109 /EMBC.2012.6347198
8. Burdette, Jonathan H.; Elster, Allen D.; Ricci, Peter E. (1998) Calculation of Apparent Diffusion Coefficients (ADCs) in Brain Using Two-Point and Six-Point Methods. Journal of Computer Assisted Tomography. 22(5):792-794.
9. E. H. Mamdani. (1977) Applications of Fuzzy Set Theory to Control Systems: A Survey," in Fuzzy Automata and Decision Processes, M.Gupta, G. N. Saridis and B. R. Gaines, eds., North-Holland, North New York, pp. 1-13
10. Oka K, Yakushiji T, Sato H et al (2009) The value of diffusion weighted imaging for monitoring the chemotherapeutic response of osteosarcoma: a comparison between average apparent diffusion coefficient and minimum apparent diffusion coefficient. Skeletal Radiol39:141-146, DOI 10.1007/s00256-009-0830-7
11. Dudeck, O.; Zeile, M.; Pink, D.; Pech, M.; Tunn, P.-U.; Reichardt, P.; Ludwig, W.-D. & Hamm. (2008) Diffusion-weighted magnetic resonance imaging allows monitoring of anticancer treatment effects in patients with soft-tissue sarcomas; *Journal of Magnetic Resonance Imaging, 27*, 1109-1113
12. Costa, Flavia Martins and Ferreira, Elisa Carvalho and Vianna, Evandro Miguelote. (2011) Diffusion-Weighted Magnetic Resonance Imaging for the Evaluation of Musculoskeletal Tumors Magnetic Resonance Imaging Clinics of *North America,19*, 159-180
13. Thoeny, Harriet C. and Ross, Brian D. (2010) ;Predicting and monitoring cancer treatment response with diffusion-weighted; MRI*Journal of Magnetic Resonance Imaging, 32*, 2-16.
14. Koh, D.-M. & Collins, D. J.(2007); Diffusion-Weighted MRI in the Body: Applications and Challenges in Oncology American Journal of Roentgenology,188, 1622-1635

Author:	Eros Montin
Institute:	Politecnico di Milano / Dipartimento di Elettronica Informazione e Bioingegneria, Milan, Italy
Street:	Via Golgi 39
City:	Milano
Country:	Italy
Email:	eros.montin@polimi.it

Reconstruction of DSC-MRI Data from Sparse Data Exploiting Temporal Redundancy and Contrast Localization

D. Boschetto[2], M. Castellaro[1], P. Di Prima[1], A. Bertoldo[1], and E. Grisan[1]

[1] Department of Information Engineering, University of Padova, Padova, Italy
[2] Pattern Recognition and Image Analysis Unit, IMT Institute for Advanced Studies Lucca, Lucca, Italy

Abstract—In order to asses brain perfusion, one of the available methods is the estimation of parameters such as cerebral blood flow (CBF), cerebral blood volume (CBV) and mean transit time (MTT) from Dynamic Susceptibility Contrast-MRI (DSC-MRI). This estimation requires both high temporal resolution to capture the rapid tracer kinetic, and high spatial resolution to detect small impairments and reliably discriminate boundaries. With this in mind, we propose a compressed sensing approach to decrease the acquisition time without sacrificing the reconstruction, especially in the region affected by tracer passage. To this end we propose the utilization of an available TV-L1-L2 minimization scheme with a novel additional term that introduce the information on the volume at baseline (no tracer). We show on simulated data the benefit of such a scheme, that is able to achieve an accurate reconstruction even at high acceleration (x16), with a RMSE of 2.8, 10 times lower than the error obtained with the original reconstruction.

Keywords—compressed sensing, MRI, dynamic susceptibility, DSC-MRI, contrast kinetics.

I. INTRODUCTION

A well evaluated technique to asses brain perfusion is the Dynamic Susceptibility Contrast-MRI (DSC-MRI). This method is based on a set of T2*-weighted images acquired before, during and after the injection of a contrast agent (gadolinium-DTPA). The reduction of the transverse tissue relaxation time (T2*) caused by the tracer can be detected as an attenuation in the MRI signal. DSC-MRI allows the analysis of several perfusion parameters, namely the cerebral blood flow (CBF), the cerebral blood volume (CBV) and the mean transit time (MTT) [1, 2, 3]. The most used technique to extract this parameters involves several step and the crucial step is a deconvolution operation. DSC-MRI requires in order 1) high temporal resolution to better describe the rapid tracer kinetic, 2) high spatial resolution to detect small impairments and reliably discriminate boundaries and 3) high signal-to-noise ratios to reduce the ill-conditioning problem of the deconvolution step. With such impacting constraints, DSC-MRI could potentially gain from lower data acquisition requirements. Clinically, faster sampling could allow accurate arterial input function estimation or application of tracer kinetic models that require higher temporal resolution than is currently available. Compressed Sensing (CS) is a novel technique used to decrease the acquisition time or increase reconstruction fidelity [4, 5]. CS allow to overcome the limit of Shannon theorem and reconstruct the signal from a small sample set of measurements under the sparsity hypothesis of the signal itself. To increase the impact of CS is common to use a wavelet transform that is known to sparsify the representation of the signal. In cardiac or angio-MRI another proposed technique requires to explore the sparsity in the temporal domain, decreasing the acquisition time or increasing the reconstruction quality [6]. The application of CS to DSC and its influence on quantitative parameter estimation is gaining attention, but it is just beginning to be explored [7].

Mainly, the use of CS within this technique is related only to decrease the acquisition time of each volume. No consideration of the sparsity of the contrast agent distribution, its impact on the acquisition, and the ability to reconstruct the contrast kinetics on the potentially small regions where involved by its passage have been investigated. On one side, the tracer passage will change the slice intensity profile from one temporal sample to the next one, on the other the mutual information between two temporal samples is extremely redundant, with the exception of the components introduced by the tracer. We try to exploit this redundancy to constrain the reconstruction, so to concentrate the influence of the information gathered from different temporal samples to be concentrated near the more perfused area, that will experience a larger attenuation in the MR signal.

II. METHODS

The main idea in this work is the efficient application of compressed sensing theory to the DSC-MRI. In order to do this, an existing tool for the reconstruction of partial Fourier data in MRI has been adapted and modified. This tool is called *RecPF* and it has been developed by Yang et al. [8]. This algorithm uses an alternating direction method to minimize a TV-L1-L2 objective function (1) to reconstruct the image.

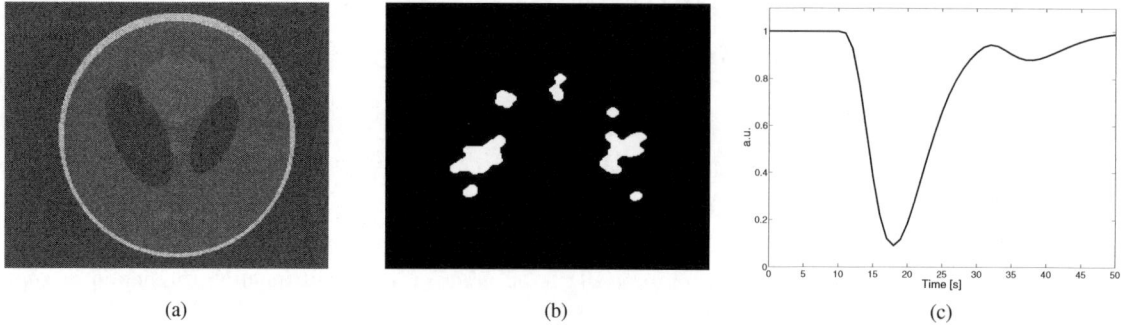

Fig. 1 Baseline image (a), mask indicating the regions with added tracer signal (b), and tracer kinetic model (c)

$$\hat{u} = \min_u \sum_i \|D_i u\|_2 + \sum_i |\psi_i^\top u| + \|\mathcal{F}_p u - f_p\|_2^2 \quad (1)$$

Given an $N \times N$ image represented as a row vector u in the space domain, and indicating with D_i the discrete differential operator, the first term $\sum_i \|D_i u\|_2$ represents the total variation (TV) of the image, that is a first order gradient containing structure edges information, and whose minimization enforces a piecewise smooth estimate of the image.

Indicating with ψ_i the discrete wavelet transform matrix [9], the second term $\sum_i |\psi_i^\top u|$ is the ℓ_1-norm of the wavelet representation of the image: this term is widely used in compressed sensing applications since it promotes the signal sparsity [5]. Finally, indicating with $\mathcal{F}_p$ the $p \times N^2$ compressed sensing matrix mapping the image into the $p < N^2$ acquired frequencies, the term $\|\mathcal{F}_p u - f_p\|_2^2$ being the ℓ_2-norm in the Fourier domain, expresses the fidelity of the reconstruction to all of the Fourier partial data acquired. By relaxing the ℓ_1 terms with a quadratic approximation, the function can be minimized with an alternate minimization iterative scheme. The solution of the minimization is explicit, and after a Fourier inverse transformation the final reconstructed image is obtained. The new method developed is aimed at enhancing the acceleration capabilities given by this algorithm with the application to the particular context of the DSC-MRI scan. Since this kind of acquisitions is highly redundant in terms of spatial information, the idea is to exploit this redundancy in the form of a prior information or baseline. Two regions of pixels can be distinguished: in the first one the signal can be approximated as stationary, because it is not affected by tracer kinetics; in the other one, less numerous in terms of voxels, the signal varies according to the attenuation related to the contrast agent passage. By estimating a spatial mask which approximately separates these two regions, it is possible to exploit the prior knowledge gathered from the baseline acquisition, concentrating the influence of the newly acquired data on the reconstruction onto the voxels that are affected by the gadolinium passage. By this means, it is possible to further reduce the data (undersampling) needed for the reconstruction, that means a reduction in the time needed to acquire a single slice in the MR sequence, without causing data loss compared to the standard method applied to a traditional non-dynamic MRI.

Given a baseline image acquired at time $t = 0$, and an image u^T to be reconstructed at time $t = T > 0$ acquiring $p \ll N^2$ samples of the K-space, we add to the Eq. 1 a fourth term penalizing the distance of the reconstructed image u_T from u_0. Since we need to constrain this penalty only for the pixels where there is negligible tracer passage, we estimate a mask M similarly to what has been proposed in [10]. To this end we set a threshold θ and estimate the mask M:

$$M = \|u^0 - u^T\|_2 > \theta \quad (2)$$

We then add a term to Eq. 1:

$$\sum_i \|\psi^\top u_i^0 - \psi^\top u_i^T\|_2, \quad i \notin M \quad (3)$$

In order increase the smoothness in the reconstructed image, we use a relaxed version of Eq.3, where a weighted average between u^0 and u^T is computed for each $i \notin C$. Finally, in order to exploit the same minimization strategy proposed in [8], we firs introduce the auxiliary variable:

$$\xi_i = \begin{cases} \psi^\top \left((1-\alpha)u_i^0 - \alpha u_i^T\right) & i \in M \\ \psi^\top u_i^0 & i \notin M \end{cases} \quad (4)$$

with $0 < \alpha < 1$, so to obtain the following minimization problem that can be solved with an alternate iterative scheme:

$$\hat{u} = \min_u \sum_i \|D_i u\|_2 + \sum_i |\psi_i^\top u| + \|\mathcal{F}_p u - f_p\|_2^2 + \sum_i |\xi_i|. \quad (5)$$

We call this method *RecPFbase*, having added the baseline information to the original objective function. The added term is an ℓ_1-norm that minimizes the difference between the reconstructed image and the baseline only in the estimated stationary region; the term is expressed in the wavelet domain, due to the necessity of conveying the localization information of the mask. The contribution of the least square expansion of this term can be simply added to the iterative scheme proposed in [8].

III. SIMULATED DATA

In order to evaluate the performance of the original method proposed in [8] and those proposed here, we simulated a set of a DSC-MRI sequence. To this purpose, we used the a 256×256 pixels Shepp-Logan phantom [11], commonly used in the simulations in the field of MRI. To the phantom we added white noise to decrease the signal to noise ratio (SNR). Moreover, to simulate the local transit of gadolinium, we defined some regions where the intensities change in time mimick the attenuation produced by gadolinium in certain areas of the brain. The simulation of the passage of the medium contrast was thus obtained adding to the predefined regions a monovariate gamma function $y_{gad}(t)$ with fixed parameters with a secondary gamma function $y_{rec}(t)$ representing the recirculation [12]:

$$y_{gad}(t) = (t-10)^4 e^{-\frac{t-10}{2}}, \quad (6)$$

$$y_{rec}(t) = 0.1(t-20)^4 e^{-\frac{t-20}{2}}, \quad (7)$$

$$y(t) = \begin{cases} 1 - y_{gad}(t) - y_{rec}(t) & t > 10 \\ 1 & t < 10. \end{cases} \quad (8)$$

By setting the variance of the noise so to obtain a level of signal to noise ratio of 15 dB, the baseline noisy image, the mask indicating the regions with tracer transit, and function modeling the tracer kinetics are represented in Fig. 1. As for the K-space sampling scheme, we follow what has been proposed by [10], taking random frequency samples with a linearly decreasing density from the zero frequency to the maximum available frequency.

IV. RESULTS

We tested the reconstruction performance of the original and of the proposed algorithm applying different undersampling levels, by taking 1/4, 1/8 and 1/16 (corresponding acceleration factors of 4, 8 and 16) of the complete K-space domain composed of 256^2 coefficients. The reconstruction error has been calculated as the average pixel-wise root mean square error, then averaged over all time. Being $\hat{u}^t$ the

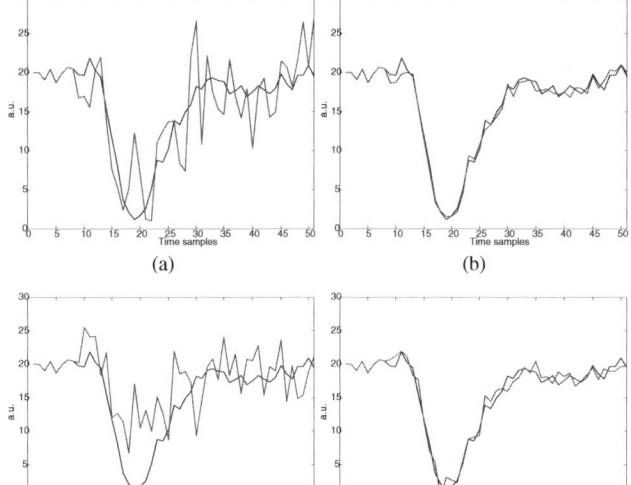

Fig. 2 Comparison of the temporal signal from the simulated data (blue line) and from the reconstructed sequence (red line) using original RecPF method (a)(c), and RecPFbase (b)(d) with acceleration 4 (first row) and 16 (second row)

Table 1 Mean reconstruction over all times and all pixels obtained with the different methods under different acceleration factors

	RMSE	
Method	RecPFbase	RecPF
Acceleration 4	2.5515	9.1435
Acceleration 8	2.7123	18.2003
Acceleration 16	2.7908	20.6354

estimated reconstruction, and u^t the original simulated image at time t, we have:

$$RMSE = \frac{1}{T} \sum_{t=1}^{T} \left(\sqrt{\frac{1}{N} \sum_i (\hat{u}_i^t - u_i^t)^2} \right). \quad (9)$$

The results are reported in Tab. 1, showing the improvement in the reconstruction accuracy with respect to the original RecPF algorithm by considering the baseline information. Moreover, the mean squared error does not deteriorate increasing the acceleration factor (reducing the number of k-space samples). Representative reconstructed images at the first undersampled image (time T_1) and at the maximum attenuation (time T_2) are reported in Fig. 3. Moreover, in order to evaluate the ability of the reconstruction algorithms to maintain the tracer kinetics information (usually

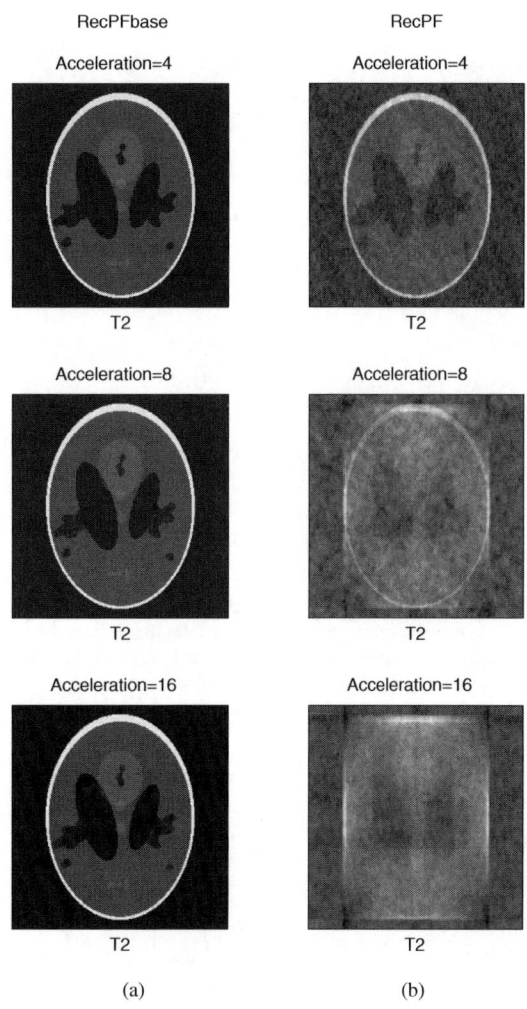

Fig. 3 Representative reconstruction of the simulated data at time T_2 (peak attenuation from contrast agent) with the proposed method RecPFbase (left column), and the original RecPF (right column), under different accelerations.

involving small regions of the image), we show the reconstructed temporal patterns in a selected region of interest (ROI) of dimension 3×3 pixels superimposed to the original simulated temporal signal in Fig. 2.

V. CONCLUSION

In the paper we propose a compressed-sensing scheme for accelerating the time samples acquisition in DSC-MRI, without sacrificing the reconstruction accuracy. Building on the $TV - L1 - L2$ objective function minimization, we added a fidelity term constraining the reconstructed image to be close to the baseline (no contrast) in the region where does not appear to arrive any contrast agent. By introducing the constraint in the wavelet domain, it allows good localization at the same time allowing the utilization of the same fast minimization method proposed for the original $TV - L1 - L2$ method. We show the performance of the reconstructions on simulated data, with respect to the algorithm not including the baseline information. The proposed algorithm achieves lower error, without increasing substantially at higher acceleration factors.

REFERENCES

1. Østergaard Leif, Weisskoff Robert M., Chesler David A., Gyldensted Carsten, Rosen Bruce R.. High resolution measurement of cerebral blood flow using intravascular tracer bolus passages. Part I: Mathematical approach and statistical analysis *Magnetic Resonance in Medicine.* 1996;36:715–725.
2. Østergaard L, Sorensen AG, Kwong KK, Weisskoff RM, Gyldensted C, Rosen BR.. High resolution measurement of cerebral blood flow using intravascular tracer bolus passages. Part II: Experimental comparison and preliminary results *Magnetic Resonance in Medicine.* 1996;36:726–736.
3. Calamante F, DL Thomas, Pell GS, Wiersma J, Turner R. Measuring cerebral blood flow using magnetic resonance imaging techniques. *J Cereb Blood Flow Metab.* 1999;19:701–735.
4. Donoho DL. Compressed Sensing *IEEE Transactions on Information Theory.* 2006;52:1289 - 1306.
5. Lustig Michael, Donoho David, Pauly John M.. Sparse MRI: The application of compressed sensing for rapid MR imaging *Magnetic Resonance in Medicine.* 2007;58:1182–1195.
6. Lustig M., Santos J.M., Donoho D.L., Pauly J.M.. k-t SPARSE: High frame rate dynamic MRI exploiting spatio-temporal sparsity in *Proceedings of the 14th Annual Meeting of ISMRM*:2420ISMRM 2006.
7. Smith D.S., Li Xia, Gambrell J.V., et al. Robustness of Quantitative Compressive Sensing MRI: The Effect of Random Undersampling Patterns on Derived Parameters for DCE- and DSC-MRI *Medical Imaging, IEEE Transactions on.* 2012;31:504-511.
8. Yang Junfeng, Zhang Yin, Yin Wotao. A Fast Alternating Direction Method for TV-L1-L2 Signal Reconstruction From Partial Fourier Data *Selected Topics in Signal Processing, IEEE Journal of.* 2010;4:288-297.
9. He L., Carin L.. Exploiting structure in wavelet-based bayesian compressive sensing *IEEE Transactions on Signal Processing.* 2009;57:3488 - 3497.
10. Sung K., Daniel B.L., Hargreaves B.A.. Location Constrained Approximate Message Passing (LCAMP) Algorithm for Compressed Sensing in *Proceedings of the 19th Annual Meeting of ISMRM*:72ISMRM 2011.
11. Shepp L.A., Logan B.F.. The Fourier reconstruction of a head section *IEEE Transactions on Nuclear Science.* 1974;21:21-43.
12. D. Peruzzo M. Calabrese E. Veronese F. Rinaldi V. Bernerdi A. Favaretto P. Gallo A. Bertoldo. Heterogeneity of cortical lesions in multiple sclerosis: an MRI perfusion study. *J Cereb Blood Flow Metab..* 2013;33:457-63.

The Influence of Slice Orientation in 3D CBCT Images on Measurements of Anatomical Structures

W. Jacquet[1], E. Nyssen[2], C. Politis[3], B. Vande Vannet[4], and P. Bottenberg[5]

[1] Vrije Universiteit Brussel, Vakgroepen SOPA-COPR-EDWE, Brussel, België
[2] Vrije Universiteit Brussel, Vakgroep Elektronica en Informatica - ETRO, Brussel, België
[3] KU Leuven, Stomatologie en Maxillo-faciale Heelkunde, Leuven, België
[4] Vrije Universiteit Brussel, Stomatologie, Orthodontie en Parandontologie - SOPA, Brussel, België
[5] Vrije Universiteit Brussel, Conservatieve Tandheelkunden en Prosthodontics - COPR, Brussel, België

Abstract—Longitudinal anatomical structures can be studied through slicing perpendicularly to their general orientation. The digital nature of 3D cone beam computed tomography (CBCT) images creates a need to define the general orientation of such structures. The aim is to investigate the influence of slice orientation on the measurements of thickness and overall width of a mandible. Most often the mandibular structure is explored by browsing through coronal views. A more sophisticated approach consists of the study of slices perpendicular to the general shape of the mandible obtained through tracing the mandibular arch in an occlusal view.

Slicing perpendicularly to the mandibular canal containing the inferior alveolar nerve (IAN) is presented as a third slicing strategy. The latter mimics the slicing approach used in classical anatomic studies of the mandible. When comparing all three strategies, significant differences in measurements of the buccal, lingual and inferior thickness of the walls can be detected. In 69% of cases, the difference in overall width at the level of the IAN exceeds one mm between measurements on a slice constructed perpendiculary to the IAN and a coronal slice.

Keywords—CBCT, Inferior Alveolar Nerve (IAN), measurements, mandible, anatomy.

I. INTRODUCTION

Planning orthodontic interventions and craniofacial surgery requires the accurate assessment of positions and measurements as well as the thickness of bony structures. At our laboratory earlier work was already dedicated in this context on the appropriate representation of the inferior alveolar nerve (IAN) and its surrounding structures for visual assessment. [1] Placing dental implants is limited by the presence of adequate bone volume permitting their anchorage. Therefore, an adequate measurement is of the utmost importance in the planning of osseointegrated implants. Cone beam computed tomographical (CBCT) images provide 3D images that allow for accurate linear measurements of the craniofacial complex. [2] [3] All linear measurements require the identification of point pairs in the 3D volume. When the measurement concerns the assessment of the distance between feature points – anatomical sites – it is possible to find and indicate these points, navigating through different pseudo 3D representations of the anatomical structures. [4] [5] [6] The measurement of height, width, thickness and position of longitudinal structures is less trivial and depends heavily on its orientation. For representational purposes the measurement is most often performed on slices defined within the 3D model. The orientations of the slices have to be chosen carefully. Slicing a longitudinal structure at an angle other than 90º will increase the measured thickness. Non-perpendicular slicing will therefore result in an overestimate of the available bone volume. Suomalainen et al., in a study exploring the accuracy of linear measurement using CBCT, placed the mandible in such a way during acquisition that the slices were expected to be perpendicular to the orientation of the mandibular section under study. [7]

When assessing structures in the mandible two slice orientations are considered the most natural ones. Coronal slices are readily available without need for interpolation and provide good views to qualitatively assess the mandible. However, it must be clear that coronal slices are only perpendicular to the mandible on rare occasions. Moreover, slice views depend on the positioning of the patient during image acquisition. A more appropriate approach is the indication of the center line of the mandible defining a mandibular arch using a horizontal slice as for the definition of a simulated panoramic image. Slices are produced perpendicularly to this center line on regular intervals along the center line. However, a level of dependency on the acquisition process still remains and the angle between the occlusal plane and the orientation of the mandible might influence measurements of the height of structures. The approach we propose is to follow the IAN tracing and to slice perpendicularly to the orientation of this tracing. As such, measurements become independent of the positioning of the patient during acquisition of the image and

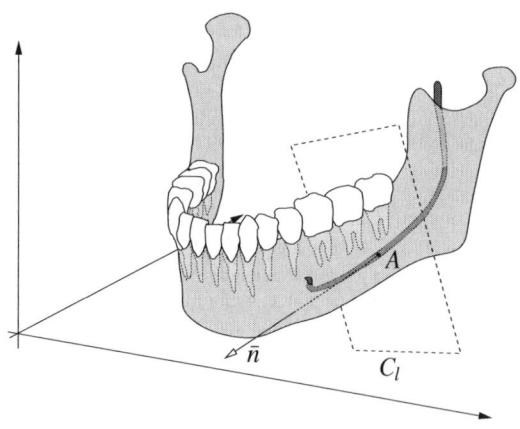

Fig. 1 A cutting plane C_l is constructed at 24mm from the mental foramen along the IAN perpendicular to the orientation of the IAN.

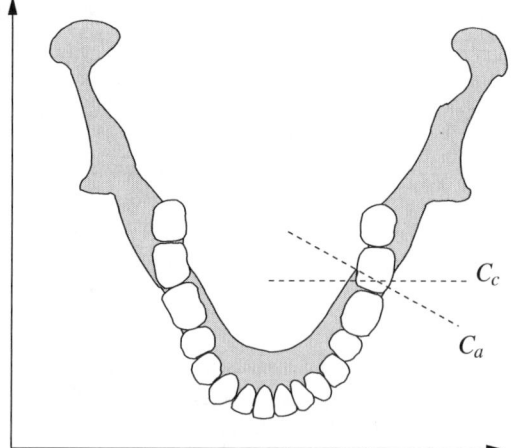

Fig. 2 Two additional cutting planes C_c and C_a are constructed through the IAN point, situated at a distance of 24mm from the mental foramen. One C_c is a coronal plane and the other a plane perpendicular to the mandibular arch and the occlusal plane C_a.

might provide measurements more in line with histological sections. [8] [9] In what follows, we compare several measurements of thickness and the overall width based on the different slicing strategies.

II. MATERIAL AND METHODS

A total of 32 pre-operative CBCT images of the mandible were obtained from patients files together with manual IAN canal tracings. The left and right side of a single mandible were treated as separate cases ($N = 64$). The nerve canals were traced based on eight to twelve points starting from the mental foramen up to the mandibular foramen. The orientation of the IAN was obtained through a least squares fit of a line through a re-sampled version of the tracing. The re-sampling was performed along the IAN canal tracing with steps of 2mm, between the points at a distance of 10mm and 40mm from the mental foramen. A cutting plane (Line) C_l is constructed to contain an IAN point, situated at a distance of 24mm from the mental foramen, and perpendicular to the orientation of the IAN – see Fig. 1. Two additional cutting planes C_c and C_a are considered: the coronal plane C_c (Coro) through this point and a plane perpendicular to the mandibular arch and the occlusal plane (Arch) C_a – see Fig. 2.

For all slices, six point were indicated allowing for the calculation of the buccal w_b and lingual w_l thickness of the bone – see Fig. 3. The center of the image corresponds to the IAN tracing. The thickness w_i of the bone at the inferior side of the mandible was measured as well – see Fig. 3.

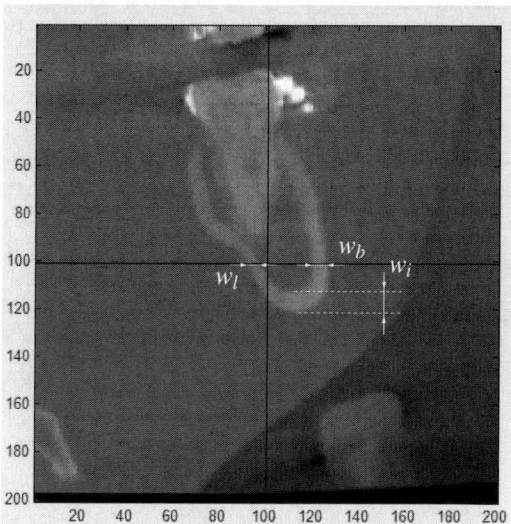

Fig. 3 Measurements in the cutting plane.

III. RESULTS

Descriptive statistics are presented in table 1. Average thickness ranges from 1.98mm up to 3.72mm. Lingual and Buccal measurements in the line (C_l) and arch (C_a) slices seem to be quite similar.

As predicted highly significant differences were found for lingual and buccal measurements between all measurement types except between the lingual and buccal measurements in the line slices C_l and arch slices C_a. The expected number of patients with a difference above 1mm ranges from five

Table 1			
Thickness			
		Mean (mm)	SD
Buccal	Coro	2,38	0,74
	Line	2,04	0,71
	Arch	2,04	0,76
Lingual	Coro	2,36	0,71
	Line	1,98	0,68
	Arch	1,98	0,60
Inferior	Coro	3,72	0,78
	Line	3,28	0,57
	Arch	3,66	0,60

Table 2				
Difference in thickness				
		Mean (mm)	SD	% > 1mm
Buccal	Coro - Line	0.34*	0.41	5%
	Coro - Arch	0.34*	0.49	9%
	Line - Arch	0.00	0.33	0%
Lingual	Coro - Line	0.38*	0.44	8%
	Coro - Arch	0.38*	0.44	8%
	Line - Arch	0.00	0.32	0%
Inferior	Coro - Line	0.44*	0.52	14%
	Coro - Arch	0.06	0.44	2%
	Line - Arch	−0.39*	0.39	0%
Overall width	Coro - Line	1.47*	0.93	69%
	Coro - Arch	−0.16**	0.40	2%
	Line - Arch	−0.56**	1.36	37%

* significant t-test $\alpha = 0.001$ Bonferonni corrected.
** significant t-test $\alpha = 0.05$ Bonferonni corrected.

percent up to eight percent. When considering the inferior structure, only the difference between measurements in the coronal plane C_c and the arch slice C_a did not differ significantly. Fourteen percent of cases are expected to have a difference exceeding 1mm between the measurement of the inferior bone thickness in the coronal plane C_c and the line slice C_l. When considering the overall width, all methods differ ($\alpha = 0.05$), the difference between coronal and line measurements being the most explicit. The predicted number of patients with a difference more than 1mm exceeds 69%.

IV. DISCUSSION AND CONCLUSIONS

As recognized in literature the slice orientation and patient orientation influences measurements when dealing with longitudinal structures. [7] Our measurements show that slice orientation when measuring thickness of bone structures of the mandible is extremely important for clinical practice. Differences can raise up to more than one mm for a significant number of patients depending on whether slices based on orientations of the anatomical structure are used. An additional advantage of using slices based on the anatomical structure itself is the elimination of the dependency on the positioning of the patient during image acquisition. As such an IAN tracing can be used to define the orientation of the mandible.

Conclusions

- Slice orientation is extremely important for measurement of bone thickness of the mandible.
- The mandibular canal containing the IAN is a good candidate for slice orientation allowing for acquisition independent measurement of bone thickness.

Future research has to include the study of the IAN tracing on the measurements as well as a multi rater reliability test-retest study.

REFERENCES

1. Jacquet,W. Nyssen, E. Sung, Y. De Munter, S. Sijbers, J. and Politis, S. . ANUTSA: an image unfolding technique for the visualization and quantification of inferior alveolar nerve tracings in 3D CBCT *J. Craniofac. Surg.*. 2013. Accepted for publication.
2. Ludlow J.B., Laster W.S., See M., Bailey L.J., Hershey H.G.. Accuracy of measurements of mandibular anatomy in cone beam computed tomography images *Oral Surg. Oral Med. Oral Pathol. Oral Radiol. Endod.*. 2007;103:534542.
3. Mischkowski, R.A. Pulsfort, R. Ritter, L. Neugebauer, J. Brochhagen, H.G. Keeve, E. and Zoller, J.E. . Geometric accuracy of a newly developed cone-beam device for maxillofacial imaging *Oral Surg. Oral Med. Oral Pathol. Oral Radiol. Endod.*. 2007;104:551559.
4. Hilgers M.L., ScarfeW.C., Scheetz J.P., Farman A.G.. Accuracy of linear temporomandibular joint measurements with cone beam computed tomography and digital cephalometric radiography *Am. J. Orthod. Dentofacial. Orthop.*. 2005;128:803-811.
5. Stratemann S.A., Huang J.C., Maki K., Miller A.J., Hatcher D.C.. Comparison of cone beam computed tomography imaging with physical measures *Dentomaxillofac. Radiol.*. 2008;37:80-93.
6. Hassan B. van der Stelt P., Sanderink G.. Accuracy of three-dimensional measurements obtained from cone beam computed tomography surfaceerendered images for cephalometric analysis: influence of patient scanning position *Eur. J. Orthodont.*. 2009;31:129134.

7. Suomalainen A., Vehmas T., Kortesniemi M., Robinson S., Peltola J.. Accuracy of linear measurements using dental cone beam and conventional multislice computed tomography *Dentomaxillofac. Radiol.*. 2008;37:10-17.
8. Kilic C., Kamburogglu K., Ozen T., et al. The position of the mandibular canal and histologic feature of the inferior alveolar nerve *Clin. Anat.*. 2010;23:3442.
9. Kqiku L.,Weiglein A.H., Pertl C., Biblekaj R., Stadtler P.. Histology and intramandibular course of the inferior alveolar nerve *Clin. Oral Investig.*. 2010.

Author: W. Jacquet
Institute: Vrije Universiteit Brussel
Street: Pleinlaan 2
City: Brussel 1050
Country: België
Email: wolfgang.jacquet@vub.ac.be

A Prior-Based Image Variation (PRIVA) Approach Applied to Motion-Based Compressed Sensing Cardiac Cine MRI

J.F.P.J. Abascal[1,2], P. Montesinos[1,2], E. Marinetto[1,2], J. Pascau[1,2], J.J. Vaquero[1,2], and M. Desco[1,2,3]

[1] Departamento de Bioingeniería e Ingeniería Aeroespacial, Universidad Carlos III de Madrid, Madrid, Spain
[2] Instituto de Investigación Sanitaria Gregorio Marañón (IiSGM), Madrid, Spain
[3] Centro de Investigación en Red de Salud Mental (CIBERSAM), Madrid, Spain

Abstract—**Motion-Based Reconstruction (MBR) and Prior-Based Reconstruction (PBR) are compressed sensing approaches for cardiac cine MRI that achieve high acceleration factors by exploiting temporal sparsity based on a prior image. It would be appealing to reconstruct only the image variation with respect to the prior image, as this is sparser than the image itself, but this leads to low signal-to-noise ratio and has been generally avoided.**

In this work we propose a novel PRior based Image VAriation (PRIVA) reconstruction method that overcomes problems previously encountered with image variation reconstruction by using an image splitting approach. We tested PRIVA in combination with a MBR method (PRIVA-MBR), where motion is estimated from a prior image and then PRIVA is used to reconstruct the image variation with respect to the prior image. The prior image was reconstructed with the SpatioTemporal Total Variation (ST-TV) method. We analyzed PRIVA-MBR in terms of solution error and maximum achievable acceleration factor that maintained acceptable image quality for a prospective cardiac cine study in small animal MRI.

The prior, given by ST-TV, presented temporal-blurring effects at high acceleration factors. PRIVA-MBR, which takes motion into account, corrected for these effects, achieving acceleration factors of x7. In conclusion, we have validated that image variation reconstruction is feasible using PRIVA, and we have shown that the PRIVA-MBR method leads to high acceleration factors for cardiac cine MRI.

Keywords—**Image reconstruction, compressed sensing, Split Bregman, motion estimation.**

I. INTRODUCTION

Compressed sensing accelerates cardiac cine MRI by randomly undersampling the acquisition process and by using a nonlinear reconstruction algorithm that imposes the reconstructed image to be sparse in a transformed domain [1-3]. Selection of the transformed domain is done based on the available a-priori information for a specific application. For dynamic applications a prior image is usually available, so it would be interesting to recover only the image variation with respect to the prior image, which is sparser than the image itself.

A straightforward possibility is the direct recovery of the sparse image variation by subtracting the contribution of the prior image to the data. However, this approach has been proven to lead to low SNR results [4]. In addition, adding the reconstructed image variation to the reference image can produce artifacts, and therefore direct recovery of image variation is usually by-passed and the entire image is recovered instead.

There are two main approaches for gated cardiac reconstruction that exploit temporal sparsity by using a prior image: Prior-Based Reconstruction (PBR) and Motion-Based Reconstruction (MBR). PBR, usually adopted in x-ray-CT, assumes that a high quality prior image is available and that the entire image is sparse when subtracted from the prior image [5,6]. This approach generally works well when there is a high quality image available, allowing the recovered image to maintain the prior texture. However, this method does not allow the reconstructed image to evolve far from the prior.

MBR, which has been previously used for cardiac cine MRI [4,7], is a two-step approach where motion is estimated in a first step from a prior image and images are reconstructed in a second step using the estimated motion. In MBR, the prior image is used for estimating motion but it is not used for the image reconstruction step. Thus, MBR is preferred to PBR when a high quality prior is not available.

We propose a PRrior-based Image Variation (PRIVA) reconstruction, a method that directly recovers the image variation using an image splitting approach that overcomes the problems previously encountered of low data SNR. PRIVA can be combined with PBR and MBR approaches. Here, we combine PRIVA with a MBR (PRIVA-MBR). In a first step motion is estimated based on a prior image given by the SpatioTemporal Total Variation (ST-TV) method [8,9], and in a second step PRIVA is used to reconstruct the image variation with respect to the prior image. Motion is modeled using a Free-Form Deformation (FFD) method based on cubic B-splines functions in order to preserve motion smoothness [10,11]. Solution error and acceleration factors for PRIVA-MBR are compared to those given by ST-TV, for a prospective cardiac cine MRI scan in small animal.

II. METHODS

A. PRIVA Method

Compressed sensing enables accurate image reconstructions from randomly undersampled data [12,13] by using convex optimization and imposing sparsity of the solution in some transformed domain, subject to a data fidelity constraint. Sparsity can be enforced by minimizing the L_1-norm of the solution in the transformed domain. Being F the undersampled Fourier operator, f the measured k-space, u the target image, and assuming that there is a prior image u_p, such that the image variation $v=u-u_p$ is sparse under the transform Ψ, PRIVA reconstructs the image variation as

$$\min_v \|\Psi v\|_1 \text{ such that } \|F(u_p + v) - f\|_2^2 \leq \sigma^2 \quad (1)$$

where σ is accounts for noise in the data.

B. PRIVA-MBR Method

In dynamic applications sparsity can be exploited in both spatial and temporal domain. A newer approach for dynamic applications is MBR, which estimates the motion T in a first step and reconstructs images u in a second step using the previously estimated motion. The motion estimation step is solved based on a previous reconstructed image, u_p, given here by the ST-TV method. For the image reconstruction step, we incorporate PRIVA to reconstruct only the image variation with respect to the prior.

The proposed method PRIVA-MBR is described in Table 1. T represents an operator that encodes the motion between frames, relating pixels across frames, and R(T) is a regularization factor that enforces smoothness. Φ is the symmelet wavelet transform that imposes sparsity to the image variation. Ψ is the spatial gradient, which leads to spatial TV, ensuring that the entire image is smooth, avoiding artifacts and stabilizing the solution [14]. u and v represent image concatenation for all frames, i.e. $u=[u_1^T,\ldots,u_m^T]^T$, where m is the number of frames.

Table 1 Pseudocode for PRIVA-MBR method.

$$u^0 = u_p$$
For $n = 0 : n_{it}$
1st step - motion estimation :
$$T_i^n = \min_{T_i} \|T_i(u_i^n) - u_{i-1}^n\|_2^2 + R(T_i)$$
2nd step - reconstruction :
$$v^n = \min_v \gamma \|\Phi v\|_1 + (1-\alpha)\|\Psi(u_p + v)\|_1 +$$
$$+ \alpha \|T^n(u_p + v)\|_1 \text{ st.} \|F(u_p + v) - f\|_2^2 \leq \sigma^2$$
$$u^n = u_p + v^n$$
$$u_p = u^n$$
end

Motion T was estimated using a Free-Form Deformation (FFD) model, which is based on cubic B-splines in order to preserve motion smoothness [10,11]. FFD-based registration method was implemented using the software available in MATLAB Central (Dirk-Jan Kroon; B-spline grid, image and point based registration; 2012, retrieved from <http://www.mathworks.com/matlabcentral/fileexchange/20057-b-spline-grid-image-and-point-based-registration>).

The image reconstruction step was solved using the Split Bregman formulation [15,16].

We remark that both the motion estimation step and the reconstruction step are iterative processes and the outer loop (n in Table 1) accounts for the number of times that motion is estimated.

C. Test Data Set and Undersampling

A rat cardiac cine FLASH sequence, synchronized with ECG and respiration, was acquired with a 7 T Bruker Biospec 70/20 scanner using a linear coil resonator for transmission and a dedicated four-element cardiac phased array coil for reception, with the following parameters: TR/TE=6.36/2.18 ms, Flip angle=15°, number of frames=16, matrix size=192x192, FOV=4x4 cm, slice thickness=1 mm, NA=4, and Tacq=5 min. Animals were treated according to the European Communities Council Directive (86/609/EEC) and with the approval of the institutional Animal Experimentation Ethics Committee.

Quasi-random spatially varying undersampling patterns were generated by adapting the methodology proposed by Lustig *et al.* [13], where randomization is preformed across both phase encoding direction and through the temporal dimension [9].

D. Analysis and Comparison of Methods

PRIVA-MBR was compared to the prior image given by ST-TV to analyze the improvement provided by taking into account motion across consecutive frames. Comparison was done in terms of the relative solution error norm within a region-of-interest (zoomed region in Fig. 1), adopting the full data reconstruction as gold-standard (Fig. 1). We tested undersampling factors of 10% (x10), 15% (x7), 20% (x5) and 30 % (x3) from the complete data (acceleration factor shown within parenthesis).

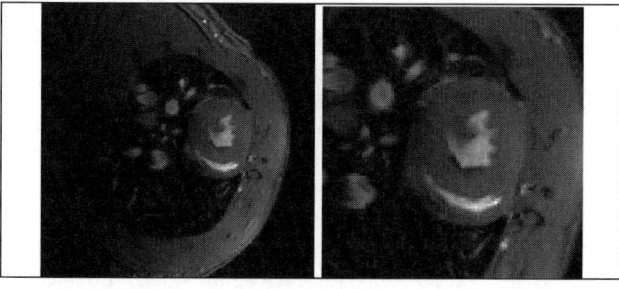

Fig. 1 Reconstruction and zoom of fully sampled data corresponding to a systole frame, used as target image.

III. RESULTS

As an example, Fig. 2 shows the prior image, given by ST-TV, and both the image variation and the entire image reconstructed by PRIVA-MBR, for acceleration x7.

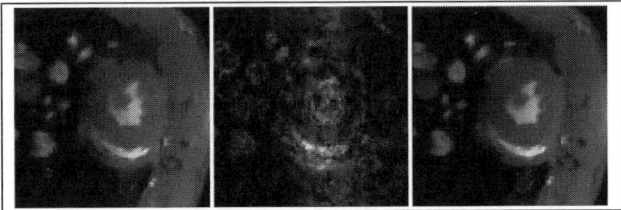

Fig. 2 Zoom of prior image (given by ST-TV) and both image variation and entire image reconstructed by PRIVA-MBR, for a frame at systole and acceleration x7. These three images correspond to u_p, v and u in Table 1.

PRIVA-MBR improves ST-TV in terms of solution error for all accelerations (Fig. 3).

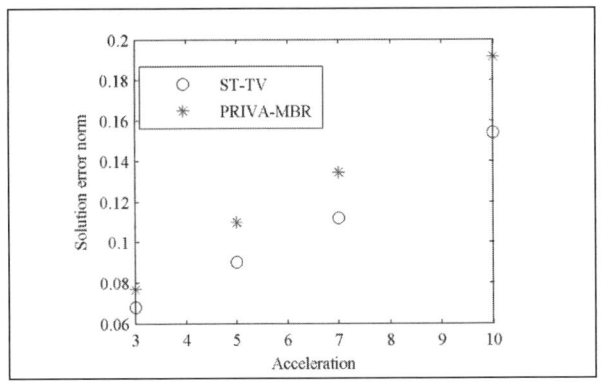

Fig. 3 Solution error norm for ST-TV and PRIVA-MBR for different acceleration factors, where each value is the mean error over all frames.

Reconstructed images with ST-TV and PRIVA-MBR are shown in Fig. 4. ST-TV leads to temporal-blurring artifacts at the inner and external parts of the myocardium and to a cartoonish-like pattern, which increases with the acceleration factor. PRIVA-MBR corrects temporal-blurring artifacts for all accelerations, but the cartoonish-like pattern is still slightly apparent for acceleration x10. Overall, an optimum solution is yielded by PRIVA-MBR for acceleration x7.

Reconstruction parameters of PRIVA-MBR (Table 1) have been empirically chosen. The most relevant are $\alpha=0.7$ and $\gamma=0.01$, which impose sparsity, and $n_{it}=1$ as further iterations did not provide significant improvements.

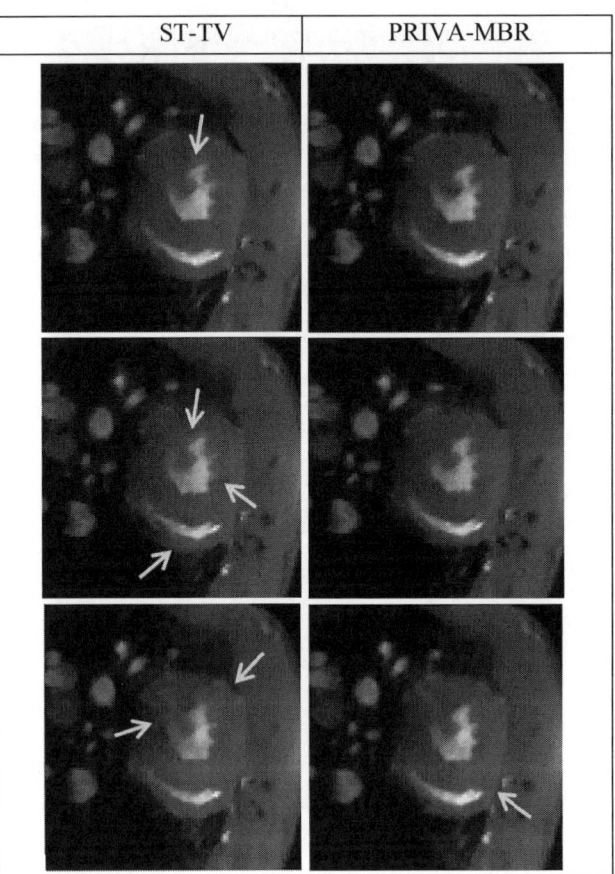

Fig. 4 Reconstructed images with ST-TV (first column) and PRIVA-MBR (second column) for accelerations of x5, x7 and x10 (rows from top to bottom).

IV. DISCUSSION AND CONCLUSIONS

We propose PRIVA, a novel image splitting approach for the recovery of the image variation with respect to a previous reconstruction. We validated PRIVA for solving the second step of a motion-based reconstruction method (PRIVA-MBR), testing it with small-animal cardiac cine MRI. PRIVA-MBR was superior to the commonly used ST-TV method, correcting temporal blurring, and leading to accelerating factors of x7.

Previous compressed sensing studies in small animal using prospective gating focused on mice, achieving accelerations of x3 [17] and of x9 in combination with parallel imaging [18], so they are not directly comparable to our study on rats.

There are some limitations in our work. Data has been retrospectively simulated from images and undersampling patterns have not been optimized, so slightly different accelerations may be found in practice. In addition, we have shown that image variation reconstruction is feasible using PRIVA and that it can be applied to MBR methods [4] and PBR methods [5,6,19]. However, this is a proof of concept and the benefit of applying PRIVA to PBR and MBR needs further quantification. Some of the potential benefits of using PRIVA are the reconstruction of only the image variation instead of the entire image, the incorporation of a prior image no necessarily of high quality, imposing sparsity on both the entire image and the image variation, and convergence in less iterations. Besides, this method is particularly relevant for alternating methods based on the Split Bregman formulation, for which there is not a straightforward way to initialize the method with a previous reconstruction.

Hence, we have shown that PRIVA method can be used to recover image variation and that in combination with motion-based reconstruction methods provides high accelerations for cardiac cine MRI.

ACKNOWLEDGMENT

This work was partially funded by the Spanish Ministry of Economy and Competitiveness RECAVA-RETIC network (RD07/0014/2009, RD12/0042/0057), and AMIT project (CEN-20101014) from the CDTI-CENIT program.

REFERENCES

1. Lustig M, Santos JM, Donoho D et al. (2006) k-t SPARSE : High frame rate dynamic MRI exploiting spatio-temporal sparsity. Proc ISMRM of the 14th Scientific Meeting, pp 2420.
2. Jung H, Sung K, Nayak KS, et al. (2009) k-t FOCUSS: a general compressed sensing framework for high resolution dynamic MRI. Magn Reson Med 61(1):103-116.
3. Montefusco LB, Lazzaro D, Papi S et al. (2011) A fast compressed sensing approach to 3D MR image reconstruction. IEEE Trans Med Imaging 30(5):1064-1075.
4. Bilen C, Wang Y, Selesnick I (2012) Compressed Sensing for Moving Imagery in Medical Imaging. Report.
5. Chen G H, Tang J and Leng S (2008) Prior image constrained compressed sensing (PICCS): A method to accurately reconstruct dynamic CT images from highly undersampled projection data sets. Med Phys Lett 35(2): 660-663.
6. Abascal J, Sisniega A, Chavarrías C et al. (2012) Investigation of different Compressed Sensing Approaches for Respiratory Gating in Small Animal CT. Proc of IEEE Nuclear Science Symposium and Medical Imaging Conference Record, pp 3344-3346.
7. Asif MS, Hamilton L, Brummer M et al. (2012) Motion-adaptive spatio-temporal regularization for accelerated dynamic MRI. Magn Reson Med (in press).
8. Goud S, Hu Y, Jacob M (2010) Real-time cardiac MRI using low-rank and sparsity penalties. Proc. of 7th IEEE International Symposium on Biomedical Imaging, New York, pp 988-991.
9. Montesinos P, Abascal JFP-J, Chamorro J et al. (2011) High-resolution dynamic cardiac MRI on small animals using reconstruction based on Split Bregman methodology. Proc of IEEE Nucl Sci, pp 3462-3464.
10. Rueckert D, Sonoda LI, Hayes C, et al. (1999) Nonrigid registration using free-form deformations: application to breast MR images. IEEE Trans Med Imag 18(8):712-721.
11. Hill DL, Batchelor PG, Holden M et al. (2001) Medical image registration. Phys Med Biol, 46(3):R1-45.
12. Candes EJ, Romberg J, Tao T (2006) Robust uncertainty principles: Exact signal reconstruction from highly incomplete frequency information. IEEE Trans Inf Theory 52(2):489-509.
13. Lustig M, Donoho D, Pauly JM (2007) Sparse MRI: The application of compressed sensing for rapid MR imaging. Magn Reson Med 58(6):1182-1195.
14. Starck J L, Elad M and Donoho D L (2005) Image decomposition via the combination of sparse representations and a variational approach. IEEE Trans Image Process 14(10): 1570-1582.
15. Goldstein T and Osher S (2009) The Split Bregman Method for L1 Regularized Problems. SIAM J. Imaging Sci (2): 323-343.
16. Abascal J, Chamorro-Servent J, Aguirre J et al. (2011) Fluorescence diffuse optical tomography using the split Bregman method. Med Phys 38(11): 6275-6284.
17. Wech, T., Lemke, A., Medway et al. (2011) Accelerating cine-MR imaging in mouse hearts using compressed sensing. J Magn Reson Imaging 34: 1072–1079.
18. Tobias Wech, Craig A Lygate, Stefan Neubauer, et al. (2012) Highly accelerated cardiac functional MRI in rodent hearts using compressed sensing and parallel imaging at 9.4T. J Cardiov Magn Reson 14(1): 65.
19. Lauzier PT, Tang J, Chen GH (2012) Prior image constrained compressed sensing: implementation and performance evaluation. Med Phys 39(1): 66-80.

Author: Juan Felipe Pérez-Juste Abascal
Institute: Departamento de Bioingeniería e Ingeniería Aeroespacial, Universidad Carlos III de Madrid
Street: Avenida de la Universidad 30
City: Leganés (Madrid)
Country: Spain
Email: jabascal@hggm.es

Symmetry Based Computer Aided Segmentation of Occluded Cerebral Arteries on CT Angiography

Emilie M.M. Santos[1,2,3,4], Henk A. Marquering[3,4], Olvert A. Berkhemer[3], Wim van der Zwam[5], Aad van der Lugt[1], Charles B. Majoie[3], and Wiro J. Niessen[1,2,6]

[1] Dept. of Radiology, Erasmus MC, Rotterdam, The Netherlands
[2] Dept. of Medical Informatics, Erasmus MC, Rotterdam, The Netherlands
[3] Dept. of Radiology, AMC, Amsterdam, The Netherlands
[4] Dept. of Biomedical Engineering and Physics, AMC, Amsterdam, The Netherlands
[5] Dept. of Radiology, Maastricht University Medical Centre, Maastricht, The Netherlands
[6] Faculty of Applied Sciences, Delft University of Technology, Delft, The Netherlands

Abstract—Thrombus volume and length are important predictors of success of recanalization in patients with acute ischemic stroke. The lack of contrast between low density thrombus and surrounding brain tissue in CT images makes manual delineation a difficult and time consuming task. We developed and evaluated an automated lumen and thrombus segmentation pipeline on CT angiography. The method generates a shape prior from the segmentation of the contralateral artery, mirror symmetry and local image intensity. The accuracy and interobserver variability were assessed using Bland Altman analysis and the calculation of Pearson´s correlation coefficient. The accuracy of the method was comparable to the interobserver variation in manual delineations. To our knowledge, this method is the first automated thrombus segmentation on base-line CT angiography and therefore opens a way to automatic thrombus characterization for treatment planning.

Keywords—Ischemic Stroke, Automated segmentation, CT Angiography, Computer aid, Symmetry.

I. INTRODUCTION

According to the World Health Organization, 15 million people suffer from a stroke worldwide each year. Approximately 70% of strokes are caused by the interruption of the blood supply to the brain, usually because a blood vessel is blocked by a thrombus [1]. Treatment consists of chemically dissolving or mechanically extracting the embolic particles such that reflow is achieved. Numerous studies proved that treatment efficiency can be predicted by thrombus characteristics [2], [3]. One of the major limitations of intravenous thrombolysis is related to resistance due to the composition and to the amount of thrombotic material to be dissolved. Therefore, assessing thrombus length and volume are clinically relevant. There are currently two approaches allowing quantification of the thrombus: measurement of hyperdenses thrombi on Non-Contrast CT (NCCT) [4] and the quantification of absence of contrast-enhanced lumen in CT angiography. It should be noted that low density thrombi cannot be assessed in NCCT images because of the lack of contrast between the thrombus and surrounding tissue. The reported prevalence of hyperdense thrombi ranges from 2% to 58% [5] depending on the thrombus location. This means that for many patients the thrombus dimensions cannot be assessed on NCCT. The assessment of the extent of a thrombus on CTA can be performed on thick-slab maximum-intensity-projection (MIP) (see Fig. 1). However, this method is known to be prone to overestimation since it assumes that contrast has reached the distal end of the thrombus, which is not the case in around 51% of the patients [6]. The Clot Burden Score (CBS) has been proposed as a semi-quantitative measurement related to the volume and location of the thrombus. This score is a rather crude measure since the actual length and volume of the thrombus are not included [7].

As such there is currently no reliable method available to measure the thrombus length and the volume for all patients. In this paper we developed and evaluate a method that enables thrombus segmentation on CTA. This method addresses the lack of contrast at the site of the thrombus by using anatomical information from the contralateral (CLL) side of the brain.

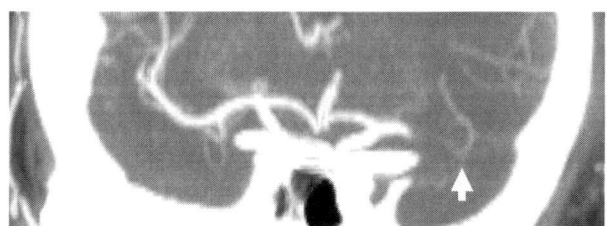

Fig. 1 Example of a MIP CTA, the arrow shows the contrast highlight of the distal part of the thrombus.

II. MATERIALS AND METHODS

A. Method

The proposed thrombus segmentation consists of three steps: (1) contralateral lumen segmentation; (2) assuming symmetry of the brain, the segmented vasculature is mapped to the affected side by mirroring in the midline and subsequently adjusted to local image information; (3) based on the contralateral segmentation, a luminal mask of the occluded segment is generated, which is used as an initialization for an intensity based thrombus segmentation. The entire image processing pipeline is implemented in MeVisLab®.

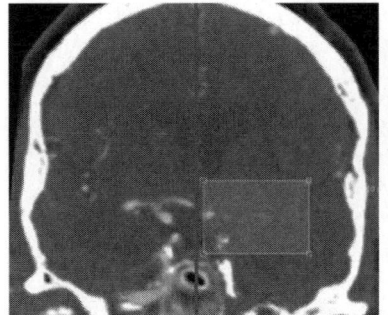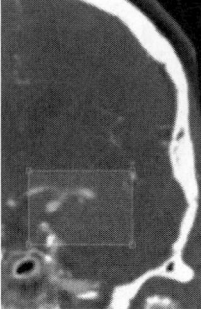

Fig. 2 *Left*: Example of the result of the rigid registration to find the midplane of the brain (red line), with a manual drawing of the area of interest (yellow area). *Right*: results of the mirroring the contralateral (CL) side of the skull.

The lumen contours of the contralateral vasculature are segmented using a two point optimal path calculation on a cost image (generated using a vesselness filter) [8] and a graph-cuts technique using a robust kernel regression [9], (Fig 3).

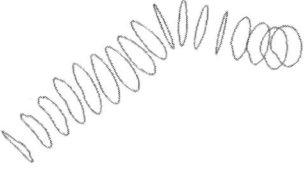

Fig. 3 Example of a segmentation of the ICA/ M1 lumen of the unaffected side.

By performing a 3D-Bspline registration of the mirroring of the contralateral image with the occluded side, we can obtain an initial estimation of the centerline through the Occluded Segment (OS). This centerline allows a segmentation of the contrast enhanced OS. Because the intensity of high density thrombi (~ 50 to 80 Hounsfield Units (HU)) is always much lower than contrast enhanced lumen (~250 to 300 HU), this segmentation never included the thrombus. At the positions along the centerline where the ratio of the contralateral radii is smaller than 0.90, thrombus is assumed (Fig. 4) and the OS contour is replaced by a circle with the radius of the CLL side if the position is located in the thrombus, (Fig 5).

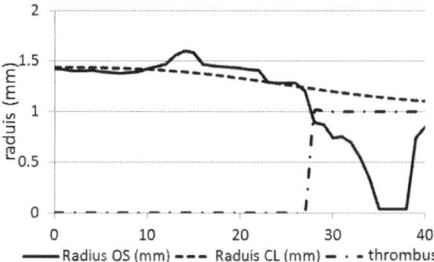

Fig. 4 Plot of the vessel radius (OS and CLL) as a function of the position along the centerline. The radius of the CLL along the segment is smoothed by a weighted Gaussian kernel regression [10].

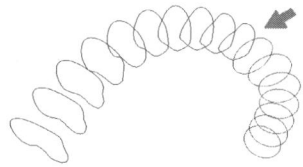

Fig. 5 Example of 3D contour of the OS lumen, where at the thrombus position the initial contour is replaced by a circle with the radius of the CLL side (starting at the arrow).

A seeded intensity based region growing (IRG) is performed, initialized automatically at the centerline (Fig. 6). The IRG method utilises two thresholds; T_{max} corresponding to the maximum intensity of the CLL contrast-enhanced lumen (CLLCL) (previously segmented) and T_{min} that is the average of the intensity of the voxel in the close neighbor of the CLLCL. Mathematical morphology is performed to smooth the contours of the thrombus.

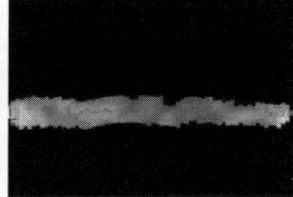

Fig. 6 Example of a Multi Planer Reformat view of clotted M1, before background masking (generated from the lumen contour) and IRG (*Left*) and after IRG (*Right*). The background mask prevents the IRG from leaking into the surrounding tissue.

B. Parameter Settings

Parameter settings were optimized for agreement with observer 1 on a training set of 4 patients. The used parameters are presented in Table 1.

Table 1 Parameters.

Step (Parameters)	Values (unit)
Vesselness filter (σ_{low}, σ_{up})	1.0, 3.5
Graph cut (σi_{nside}, $\sigma_{outside}$)	20 (HU), 20 (HU)
Radial Reg. (fraction$_{outside}$, P$_{outside}$, blur$\sigma_{distance}$)	0.75 , 0.5, 2 (mm)
Weighed Gaussian Reg. (Iteration, σ_{radius})	2, 0.5 (mm)
3D-BSpline registration (Grid spacing, grey level)	2 (mm), 32 (bit)
Mathematical morphology (opening, dilatation (3D kernel (2n+1)))	n=1(voxel), n=1(voxel)

C. Reference Standards and Statistical Analysis

To evaluate our method, we compared the thrombus segmentation with manual measurements.

The thrombus was manually delineated by 3 expert neuroradiologists, each with more than 10 years of experience, blinded from all clinical information. The observers had access to BaseLine (BL)-CTA and BL-NCCT images of the patient. To support the manual delineation a custom made tool was developed in MeVisLab®. A manual centerline was obtained by interpolating a polynomial curve through manually placed points in the thrombus on BL-NCCT scans. The thrombus contours were delineated on a stretched-vessel view through the occluded segment. To determine the length of the thrombus, a point proximal and a point distal to the thrombus were placed. The automated thrombus volume and the length measurements were compared with the interobserver variability of manual observers, using Bland & Altman analysis and the calculation of the Pearson correlation coefficient (PCC).

D. Patient Selection

For all patients with a clinical diagnosis of acute proximal ischemic stroke admitted between December 2010 and March 2012, all BL-CTA and BL-NCCT images data were retrospectively collected from our Dutch multi-center database. Exclusion criteria were low quality scans, slice thickness >2 mm, incomplete volume of interest and craniotomy. Image data of 62 patients were collected and anonymised before analysis. After exclusion, 31 patients with hyperdense thrombi on their BL-NCCT were included. 14 patients were delineated both by observer 1 and 2 (interobserver group), 2 and 15 patients were delineated by observer 1 and 3 respectively. Four patients from the interobserver group were randomly included in the training set for parameter optimization and thus were not used for the validation.

E. Accuracy and robustness

The average thrombi volume and length for the automated method and the observers were 208 ± 157 mm³ for and 20.8 ± 12.2 mm, 258 ± 224 mm³ and 19.8 ± 12.6 mm respectively. The PCCs between the automated method and observer 1 were 0.79 and 0.91 for the volume and length respectively. The mean measurement difference (bias) was 50 (the limit of agreement (LoA) -227 to 326) mm³ for the volume, and –1.0 (LoA, -11.8 to 9.8) mm for the length (see also Table 2 and Figure 7).

Table 2 Comparison of observers and the automated measurement.

	Average ± SD (min,max)		PCC (95% CI)	Average Difference (Absolute,SD)	unit
	Observers	Automated			
Volume	258 ± 224 (26,735)	208 ± 157 (27, 604)	0.79 (0.53, 1.05)	50 ± 183 (84, 113)	mm³
Length	19.8 ± 12.6 (2.2, 57.4)	20.8 ± 12.2 (2.7, 48.8)	0.91 (0.73, 1.08)	1.0 ± 5.8 (3.9, 4.0)	mm

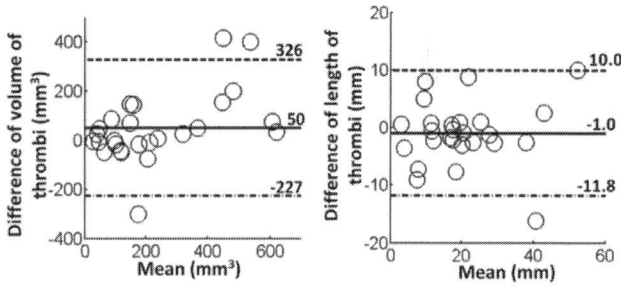

Fig. 7 Bland-Altman plot comparing volume and length between all observers and the automated measurements.

F. Interobserver Variability

The mean thrombus volume was 145 ± 110 mm³ and 158 ± 176 mm³ for observer 1 and 2 respectively. The average thrombi length was 17.2 ± 8.9 mm and 16.5 ± 11.4 mm for observer 1 and 2 respectively. The PCC between observers 1 and 2 was 0.80 (95%CI: 0.19-0.81) and 0.81 (95%CI: 0.27-1.00) for thrombus volume and length respectively. The mean difference between observers was 13 mm³ (LoA, – 215 to 242 mm³) for volume and -0.8 (LoA, -14.5 to 12.9) mm for length.

Table 3 Comparison of two manual references

	Average ± SD (min,max)		PCC (95% CI)	Average Difference (Absolute,SD)	unit
	Observer 1	Observer 2			
Volume	158 ± 176 (27,656)	145 ± 110 (34, 412)	0.80 (0.30, 0.86)	13 ± 117 (70, 91)	mm³
Length	16.4 ± 11.4 (3.0, 44.4)	17.2 ± 8.9 (6.9, 37.4)	0.81 (0.33, 0.89)	-0.78 ± 7.0 (5.5, 4.0)	mm

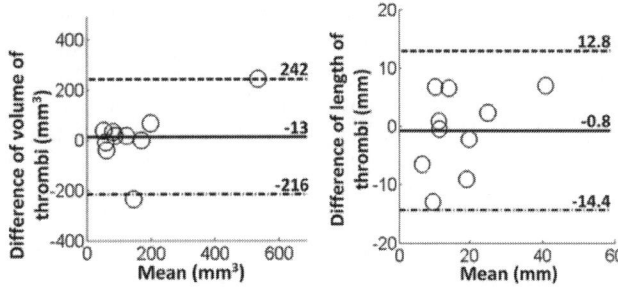

Fig. 8 Bland-Altman plot comparing volume and length between observer 1 and observer 2.

III. DISCUSSION

We have presented an automated method for the quantification of thrombus volume and length using an automated based on symmetry supervised thrombus segmentation in BL-CTA. The agreement of the automated method with manual delineations was comparable to the variation between the observers.

The automated segmentation required only a few minutes (5-7 minutes) and minimal user interactions while the manual delineation took in average 20 minutes per patient. It had successfully been applied to proximal cerebral arteries in all ischemic stroke patients in our study population. Our test dataset was limited to a small amount of patients (26) due to the fact that we had to select patients that presented with a high-density thrombus on BL-NCCT because the observers had to manually delineate the thrombi, which is too difficult for low density thrombi. Our method is able to segment low density thrombi. However, the validation could not be performed because the manual observers could not delineate them. Due to the proximity of the skull to the arteries the 3D Bspline registration (step 2.1) could place one of the centerline curve-defining control points in the surrounding skull. A simple correction consisted in manually suppressing control points that obviously deviates from the centerline and takes only few seconds. The method was based on the hypothesis that the cerebral proximal arteries had the same radius at the same position and presented an anatomy that allows CLL segmentation. Among our dataset, CLL segmentation was always possible.

IV. CONCLUSION

We have presented the first method to semi-automatically segment thrombi in intracranial arteries in acute ischemic stroke patients on BL-CTA. This automated method was based on symmetry supervised thrombus segmentation in BL-CTA and showed good accuracy. The automated measurement correlated well with manual measurements. This automated thrombus segmentation on CTA is therefore promising and opens a way to automatic thrombus characterization for treatment planning.

ACKNOWLEDGMENT

The authors wish to thank for the support from the Technology Foundation STW, The Netherlands under Grant 11632.

REFERENCES

1. Foulkes M, Wolf P et al. (1988) The Stroke Data Bank: design, methods, and baseline characteristics. Stroke, v. 19, no. 5, p. 547-554.
2. Marder V, Chute D et al. (2006) Analysis of thrombi retrieved from cerebral arteries of patients with acute ischemic stroke. Stroke, vol. 37, p. 2086–2093.
3. Molina C et al. (2005) Imaging the clot: does clot appearance predict the efficacy of thrombolysis? Stroke, v. 36, p. 2333–2334.
4. Riedel C, Jensen U et al. (2010) Assessment of Thrombus in Acute Middle Cerebral Artery Occlusion Using Thin-Slice Nonenhanced Computed Tomography Reconstructions. Stroke, p 1659-64.
5. Leys D, Pruvo J et al. (1992) Prevalence and significance of hyperdense middle cerebral artery in acute stroke. Stroke, p. 317–324
6. Christoforidis G, Mohammad Y et al. (2005) Angiographic assessment of pial collaterals as a prognostic indicator following intra-arterial thrombolysis for acute ischemic stroke. AJNR Am J Neuroradiol;26:1789–97
7. Puetz V, Dzialowski I et al. (2008) Intracranial thrombus extent predicts clinical outcome, final infarct size and hemorrhagic transformation in ischemic stroke: the Clot Burden Score Stroke, n. 3, p. 230–36.
8. Metz C, Schaap M et al. (2008) Two point Minimum Cost Path Approach for CTA coronary centerline Extraction. The MIDAS Journal, MICCAI Workshop: Grand Challenge Coronary Artery Tracking.
9. Schaap M, van Walsum T et al. (2011)Robust Shape Regression for Supervised Vessel Segmentation and its Application to Coronary Segmentation in CTA. Med. Im., IEEE Transactions, vol. 30, no. 11, pp. 1974 - 1986.
10. Shazad R, van Walsum T et al. (2012) Automatic detection, quantification and lumen segmentation of the coronary arteries using two-point centerline extraction scheme. In: Proc. of MICCAI Workshop 3D Cardiovascular Imaging a MICCAI Segmentation Challenge.

Author: Emilie M.M. Santos
Institute: Erasmus Medisch Centrum, BIGR, Rotterdam
Street: Dr. Molewaterplein 50/60
City: 3015 GE Rotterdam
Country: Netherlands
Email: e.santos@erasmusmc.nl

Automatic Segmentation of Gray Matter Multiple Sclerosis Lesions on DIR Images

E. Veronese[1], M. Calabrese[3], A. Favaretto[2], P. Gallo[2], A. Bertoldo[1], and E. Grisan[1]

[1] Department of Information Engineering, University of Padova, Padova, Italy
[2] Department of Neurosciences, University of Padova, Padova, Italy
[3] Department of Neurological, Neuropsychological, Morphological and Movement Sciences, University of Verona, Verona, Italy

Abstract—**Multiple Sclerosis (MS) is a chronic inflammatory-demyelinating disease that affects both white and gray matter (GM). GM lesions have been demonstrated to play a major role in the physical and cognitive disability and in the disease progression. The diagnosis and monitoring of the disease is mainly based on magnetic resonance imaging (MRI). Lesions identification needs visual detection performed by experienced graders, a process that is always time consuming, error prone and operator dependent. We present a technique to automatically estimate GM lesion load from double inversion recovery (DIR) MRI sequences. We tested the proposed algorithm on DIR sequences acquired from 50 MS patients. Regions corresponding to probable GM lesions were manually labeled to provide a reference. The resulting automatic lesion load estimate provides a correlation of 98.5% with manual lesion number, and of 99.3% with manual lesion volume.**

Keywords—**Multiple sclerosis, Double inversion recovery, Lesion segmentation.**

I. Introduction

Multiple sclerosis (MS) is an inflammatory demyelinating disease of the central nervous system with a usual onset on young adults, causing morphological and structural changes to the brain and leading to physical disability and cognitive impairment. Its diagnosis and patient follow-up can be helped by an evaluation of the lesions load, which can be studied thanks to MRI sequences. Although since the '60s there have been evidences of cortical demyelinated lesions (CLs) in autopsied brains in more than 90% of patients with chronic MS ([1]), traditionally lesions in the white matter (WM) have been regarded as the most important pathologic feature in MS. Only over the last few years, the role of CLs in the pathophysiology of MS has gained increasing attention ([2, 3, 4, 5]), despite the higher difficulty to detect them with conventional imaging techniques.

In order to accurately detect the anatomic border between the cortex and subcortical white matter the use of double inversion-recovery (DIR) MR sequence has been proposed. This sequence, proposed in ([6]) and subsequently used by ([7]), consists of two inversion pulses preceding a conventional spin-echo sequence; suitable choice of inversion times allows the suppression of either CSF and WM, so to image the cortex alone, proving useful in providing a better gray-white matter differentiation than other sequences [8, 9] Still, CLs identification is based on visual detection, a process that is both time consuming and strongly operator dependent. Besides, given that DIR is not a conventional MR sequence, it would take time for the human rater to gain the experience to easily and confidently evaluate the acquired images. As pointed out in Geurts et al. ([10]) where a consensus has been established for the identification of GM lesions on DIR sequence, readers should become acquainted with this sequence and with its artifacts, which are typically present in the insula and in the correspondence of cortical vessels. In this paper we present a new method to automatically segment gray matter multiple sclerosis lesions, based on the use of DIR sequence. It provides an extremely good capability of detecting actual CLs, together with repeatability short computational time.

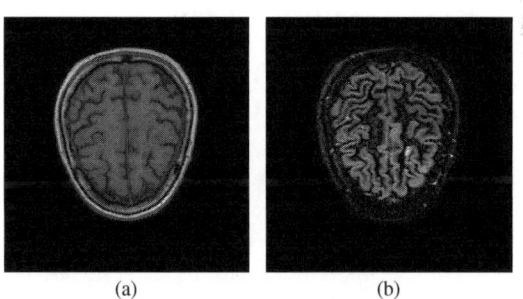

Fig. 1 Different acquisition modalities and appearance of MR data from the same patient and approximately the same brain position. In the leftmost panel a T1 MR slice (a), in the rightmost panel a DIR slice (c).

II. Materials

Fifty patients (37 women and 13 men overall mean age 34.0 ± 10.7 years, range 18-65 years) affected by Relapsing Remitting Multiple Sclerosis (RRMS) who consecutively presented to the Multiple Sclerosis Centre of Veneto Region from June to October 2009 were included in the study. The mean disease duration was 11 ± 7.5 years, range 2-31 years. The study was approved by the local Ethic Committee and written informed consent was obtained from all subjects.

Examinations were performed at the Euganea Medica Medical Center (Albignasego (Padova), Italy) with a 1.5T MR imager (Achieva, Philips Medical Systems, Best, The Netherlands) with 33 mT/m power gradient, using a 16 channel head coil. MRI examination protocol included a DIR sequence (TR = 15631 ms, TE = 25 ms, TI = 3400 ms, 50 contiguous axial slices, images with a field of viwe of 250x250, a matrix of 256x256, in-plane resolution of 1 mm^2, with a thickness = 3 mm). An example of a typical DIR slice along with its corresponding T1-weighted slice is shown in Fig. 1.

In our study, MS lesions were manually identified and outlined by two experienced neurologists (M.C. and A.F.) who were not aware of the results of our automatic method. They were asked to manually highlight MS cortical lesions in DIR images with possible visual inspection of the corresponding FLAIR and T1-weighted images. The voxels corresponding to a cortical lesion as defined in [10] were manually identified. When the two human raters provided different evaluation of a cortical region, they were asked to reach a consensus.The final results were used as gold standard for the evaluation of the performance of our algorithm.

The average number of annotated lesions was 8.48 per subject, with a range going from a minimum of 0 (no lesion) to a maximum of 43.

III. METHODS

The algorithm performs at first a pre-processing step, in order to provide a skull-stripped version of the brain. Then, exploiting the relatively high contrast of DIR images between gray and white matter, we obtain a segmentation of the GM. Then, candidate lesions are identified by finding locally hyperintense regions, which are finally classified as true lesions by a supervised classifier using texture features, similarly to what has been proposed to segment brain tissues in healthy subjects. It is worth noting that both the extraction of the contour of the brain and the segmentation of the GM do not make use of any anatomical *a priori* information, i.e., we do not exploit image registration on brain atlases. Moreover, we do not use commonly available brain tissue segmentation tools, such as FSL ([11]), SPM8 ([12]) or Freesurfer ([13]), because they have been designed for general use, so they might not be optimized for application to images from patients with MS disease ([14, 15, 16]): in particular the segmentation of MR images of MS patients executed with such tools causes the misclassification of the voxels belonging to MS lesions, either as WM or as GM or as partial volume effect ([14]).

The proposed algorithm can be used on the whole 3D data, or separately slice by slice. In our implementation, the preprocessing step (skull stripping and GM segmentation) is used in 3D modality in order to exploit the z-connectivity of the brain structures, whereas we chose to detect the lesions in 2D modality, separately on each slice. We used a 2D inspection since the slice thickness of DIR images increases the risk of linking together hyperintense regions belonging to different structures.

A. Data Preprocessing

The DIR sequence is specifically designed to enhance gray matter tissue while attenuating both white matter and cerebro-spinal fluid ([6]), so that the highest intensities in the data are given by gray matter, bones and lesions. By evaluating the Otsu's threshold th ([17]) on the voxels' intensities, considering only the voxels brighter than th identifies the skull and the gray matter as disconnected regions, plus a number of noisy voxels. After the application of a morphological closing obtained with a ball-shaped structuring element of radius $R = 1$ voxels, the largest connected region is identified as gray matter. The binary skull stripped image is then used to mask the original data, in order to take into account only voxels belonging to GM.

Furthermore, in order to enhance the contrast of the DIR sequence and to smooth the noise, we apply the algorithm proposed in [18] using as structuring element for the dilation and the erosion a disk of radius $R = 3$ voxels.

B. Candidate Lesion Segmentation

Given the enhanced slice F_k, we estimate the local mean $I_{\mu,k}(x,y)$ and standard deviation $I_{\sigma,k}(x,y)$ of the gray matter around each voxel, considering a neighborhood of NxM voxels. Then, the locally hyperintense voxels are identified as those fulfilling the inequality:

$$F_k(x,y) - I_{\mu,k}(x,y) \geq \theta_{glob} I_{\sigma,k}(x,y) \quad (1)$$

where θ_{glob} is a parameter with which it is possible to balance the sensitivity and the specificity of the detection phase. Then, and all regions smaller than 4 voxels are discarded in accordance with the recommendations in [10].

In order to detect only the regions that show a real lesion-like appearance, a further classification at the region-level is needed. First of all, the mean intensity of the gray matter in each slice is computed:

$$\mu_{gm,k} = E[F_k(x-y) - I_{\mu,k}(x,y)] \quad (2)$$

Then, for each candidate region $L_{cand,kj}$, the mean $\mu_{kj,int}$, the maximum value $\max_{kj,int}$ and standard deviation $\sigma_{kj,int}$ of its intensities, and the mean $\mu_{kj,ext}$ and standard deviation $\sigma_{kj,ext}$ on the intensities of the cortical voxels belonging to its contour are evaluated. The external surrounding is obtained with a dilation of $L_{cand,kj}$ with a structuring element of radius 1 voxel, paying attention not to include in the external neighborhood voxels that might belong to other candidate lesions.

In order to discard all candidates whose region-wide statistics mark them clearly as non-lesion, we will consider as candidate lesions those regions that are brighter with respect to the mean cortical gray level, and, as further condition, that are brighter than the gray matter in their immediate surrounding, and finally that show a sufficiently high maximum intensity:

$$\text{Cond. 1} \quad \frac{\mu_{kj,int}}{\mu_{gm,k}} > \theta_1 \quad (3)$$

$$\text{Cond. 2} \quad \frac{\mu_{kj,ext} - \mu_{kj,int}}{\sigma_{kj,ext}} > \theta_2 \quad (4)$$

$$\text{Cond. 3} \quad \max_{kj,int} > \theta_3 \quad (5)$$

where the θ_i are user-defined parameters.

C. Lesion Classification

Due to the natural variability of gray matter tissue, method described in Sec. Bprovides a number of candidate lesions that can be high. To improve the specificity of the identification, a support vector machine (SVM) classifier ([19]), with radial basis function is trained. Each candidate lesion is represented by means of a set of features, composed by statistics on the intensities in the region or on its surroundings (mean intensity, standard deviation, kurtosis, skewness, maximum value, minimum value), by morphological measures (area, perimeter, solidity), and by the normalized histograms of local binary patterns ([20]) of the candidate's voxels.

The performance of the SVM classification is assessed on the 46 volumes of the test set by means of a leave-one-out procedure, so that at each round the classifier is trained on the candidate lesions identified in 45 subjects, and tested on the candidates segmented in the remaining subject.

A representative result of the identified lesions is shown in Fig. 2.

D. Performance Metrics

To test the capability of our algorithm in finding correct CLs, and to assess the performance of the whole identification pipeline (candidates identification and lesions classification). We computed the number of true positive (TP), false positive (FP) and false negative (FN). From these, we evaluated the sensitivity index $TP/(TP+FN)$ and fraction of false positives ($FP/(TP+FN)$) on the 46 subjects of the test set using the leave-one-out procedure.

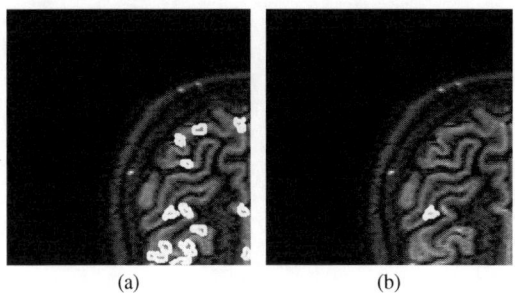

Fig. 2 Representative results of the candidate lesion identification (a) in a DIR image, and the final lesion identification (b).

IV. RESULTS

A. Parameters Setup

On four subjects out of the 50 available, the parameters for the identification of the candidate lesions were chosen as to guarantee a high sensitivity and a reduced number of false positives are $N = M = 51$, $\theta_{glob} = 1.8$, $\theta_1 = 0.15$, $\theta_2 = 0.4$ and $\theta_3 = 0.05$. The Support Vector Machine (SVM) lesion classifier was evaluated using a leave-one-out procedure, using one volume as test set and the remaining as training at each round.the Paper.

B. Experimental Results

On the remaining 46 volumes of the validation set, a total number of 397 lesions were manually detected by the raters, who were blind to the result of our algorithm. Of the 397 true, 389 were correctly recognized by the candidate detection stage, providing a percentage of sensitivity of 98.0%. When the candidate lesions were classified with the SVM, the mean sensitivity using the leave-one-out validation decreased to 94.2%. However, the number of false positives was reduced from 12 times the number of true lesions to thrice the number of true lesions. Interestingly, the number of false positives has a Pearson correlation coefficient of 0.92 ($p < 10^{-10}$) with the number of real lesions as can be seen in Fig. 3, suggesting a correlation between the number of true lesions and the appearance of gray matter in terms of intensity variability.

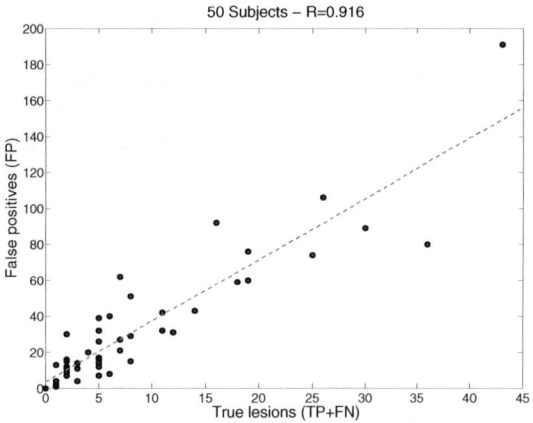

Fig. 3 Correlation between the true lesions present in a volume (true positives and false negatives) and those erroneously identified (false positives).

V. CONCLUSION

To the best of our knowledge the diagnosis and monitoring of the disease is mainly based on visual detection of MRI lesions ([21]), a process that is always time consuming, error prone and operator dependent. Moreover the identification of CLs requires an accurate training of the operator in order to avoid any lesion misclassification, even for the detection of WM lesions, that are usually much more prominent and easily recognizable than cortical lesions. Here we have developed an algorithm that provides an automatic segmentation of cortical MS lesions on double inversion recovery (DIR) MRI sequences. It provides a great sensitivity, thus speeding up the usual procedure of lesions segmentation, which is developed manually by experienced rater.

ACKNOWLEDGEMENTS

The authors wish to thank Euganea Medica Medical Center, Padova (Italy), Center for their continuous support in the DIR-MR acquisition and in the machine set-up

REFERENCES

1. Brownell B., Hughes J.T.. Distribution of plaques in the cerebrum in multiple sclerosis *J Neurol Neurosurg Psychiatry*. 1962:315-320.
2. Bo L., Geurts J.J., Valk P., Polman C., Barkhof F.. Lack of correlation between cortical demyelination and white matter pathologic changes in multiple sclerosis *Arch Neurol*. 2007;64:76-80.
3. Catalaa I., Fulton J.C., Zhang X., et al. MR imaging quantitation of gray matter involvement in multiple sclerosis and its correlation with disability measures and neurocognitive testing *Am J Neuroradiol*. 1999:16131618.
4. Calabrese M., Gallo P.. Magnetic resonance evidence of cortical onset of multiple sclerosis *Multiple Sclerosis*. 2009:933-941.
5. Calabrese M., Agosta F., Rinaldi F., et al. Cortical lesions and atrophy associated with cognitive impairment in relapsing-remitting multiple sclerosis *Arch Neurol*. 2009;66:1144-1150.
6. Redpath T.W., Smith F.W.. Technical note: use of a double inversion recovery pulse sequence to image selectively grey or white brain matter *The British Journal of Radiology*. 1994:1258-1263.
7. Bedell B.J., Narayana P.A.. Implementation and evaluation of a new pulse sequence for rapid acquisition of double inversion recovery images for simultaneous suppression of white matter and CSF *J magn Reson Imaging*. 1998:544-547.
8. Turetscher K., Wunderbaldinger P., Bankier A.A.. Double inversion recovery imaging of the brain; initial experience and comparison with fluid attenuated inversion recovery imaging *Magn Reson Imaging*. 1998:127-135.
9. Geurts J.J.G., Pouwels P.J.W., Uitdehaag B.M.J., Polman C.H., Barkhof F., Castelijns J.A.. Intracortical lesions in multiple sclerosis: improved detection with 3D double inversion recovery MR imaging *Radiology*. 2005:254-260.
10. Geurts J.J.G., Roosendaal S.D., Calabrese M., et al. Consensus recommendations for MS cortical lesion scoring using double inversion recovery MRI. Neurology. 2011;76:418-424.
11. Smith S.M., Jenkinson M., Woolrich M.W., et al. Advances in functional and structural MR image analysis and implementation as FSL *NeuroImage*. 2004:S208-S219.
12. Ashburner J., Friston K.J.. Unified segmentation *NeuroImage*. 2005;26:839-851.
13. Fischl B., Salat D.H., van der Kouwe A., et al. Sequence-independent segmentation of magnetic resonance images Neuroimage. 2004:S69-S84.
14. Nakamura K., Fisher E.. Segmentation of brain magnetic resonance images for measurement of gray matter atrophy in multiple sclerosis patients *NeuroImage*. 2009:769-776.
15. Chard D.T., Jackson J.S.,Miller D.H.,Wheeler-Kingshott C.A.. Reducing the impact of white matter lesions on automated measures of brain gray and white matter volumes. JMagn Reson Imaging. 2010:223-228.
16. Gelineau-Morel R., Tomassini V., Jenkinson M., Johansen-Berg H., Matthews P.M., Palace J.. The effect of hypointense white matter lesions on automated gray matter segmentation in multiple sclerosis. Hum Brain Mapp.. 2012;33:28022814.
17. Otsu N.. A threshold selection method from gray-level histograms *IEEE Transactions on Systems, Man, and Cybernetics*. 1979;9:62-66.
18. Souplet J.C., Lebrun C., Ayache N., Malandain G.. An automatic segmentation of T2-FLAIR multiple sclerosis lesions in *The MIDAS Journal - MS Lesion Segmentation (MICCAI 2008 Workshop)*(New York, NY, USA) 2008.
19. Cortes C., Vapnik V.. Support-vector networks Machine learning. 1995:273-297.
20. Ojala T, Pietikainen M., Maenpaa T.. Multiresolution gray-scale and rotation invariant texture classification with local binary patterns Pattern Analysis and Machine Intelligence, IEEE Transactions on. 2002;24:971987.
21. McDonald W.I., Compston A., Edan G., et al. Recommended diagnostic criteria for multiple sclerosis: guidelines from the International Panel on the diagnosis of multiple sclerosis *Annals of Neurology*. 2001;50:121-127.

Calibration of a C-arm X-Ray System for Its Use in Tomography

C. de Molina[2], J. Pascau[1,2,3], M. Desco[1,2,3], and M. Abella[2]

[1] Instituto de Investigación Sanitaria Gregorio Marañón, Madrid
[2] Dept. Bioingeniería e Ingeniería Aeroespacial, Universidad Carlos III de Madrid, España
[3] Centro de investigación en red en salud mental (CIBERSAM), Madrid

Abstract—**C-arm enables a great variety of movements, and its characteristic structure makes it possible its use in intraoperative cases, as the arc can be located around a patient lying in the bed. Such a device is conventionally intended to obtain planar images with no depth information. Our group is evaluating the use of the C-arm to provide 3D imaging guide to Intraoperative Radiation Therapy procedures. For the use of the C-arm in computerized tomography, mechanical calibration becomes an important issue, as many parameters influence the quality of the reconstruction: center of rotation, detector position, distance between detector and source, etc. In this work, we present preliminary results of the implementation of a calibration method needed to generate tomographic images with a C-arm system.**

Keywords—**C-arm, calibration, mechanical misalignments. X-ray, tomography.**

I. INTRODUCTION

The goal of Intraoperative Radiation Therapy (IORT) treatments is the irradiation of post-resected tumor beds or partially resected tumors avoiding, as possible, affecting critical organs [1]. The choice of the volume of irradiation and the total dose for the tumor are the basis for planning external radiotherapy, done on a CT image. Given the difficulties derived from the change of the anatomy during surgery, the IORT planning is not an obvious task. Our group has developed a radiation therapy planning tool [2, 3] to assess the surgical scenario beforehand and to optimize the decision-making process and treatment location. This tool shows 3D images based on a previous CT, simulates the tumor resection and the location of the applicator in order to predict the dose distribution to the target volume. The complete evaluation of this tool would need an imaging technique to assess the real situation after surgery, following the concept of Image Guided Radiation Therapy (IGRT), which includes the acquisition of 2D and 3D images before, during and after radiation [4, 5].

The mobile fluoroscopic system known as C-Arm, consists of two units, the X-ray generator and the detector (image intensifier or flat panel) mounted on an arc-shaped wheeled base, together with a workstation used to visualize, store, and manipulate the images. The C-arm enables a great variety of movements, and its characteristic structure makes it possible its use in intraoperative cases, as the arc can be located around a patient lying in the bed. It is used to obtain intraoperative planar images in different surgical procedures such as cardiology, orthopedics, and urology.

We propose to use a C-Arm as a tomograph to obtain 3D images of the patient during IORT. In that sense, the C-Arm will work analogously to a CT scanner with cone-beam geometry. The use of a C-arm for tomography presents several challenges: it may have mechanical strains and looseness in the detector (image intensifier) and the movements of source and detector may differ from a circular path. In order to obtain good quality images, it is necessary to evaluate the effects of these non-idealities and to perform an exhaustive calibration of the system, not needed when it is used for planar imaging [6].

There are different methods to perform the mechanical calibration of a cone-beam tomograph. A possible approach is based on the so called "camera model" calibration, that calculates a projection matrix. This option was rejected because of the reported method instability [6] and the fact that it does not provide the center of rotation, as it is focused on projective systems. Cho et al [6] presented a method specifically designed for C-arm systems that provides all the geometric parameters describing the geometry of a cone-beam system. These parameters are estimated at any angular position of the C-arm by using a single projection of a simple phantom.

In previous work, we presented the effects of the possible misalignments of the image detection panel of the C-Arm [7] and showed the necessity of calibrating this equipment in order to generate 3D images. In this work, we present the implementation of a calibration method that characterizes the geometry of the system. Results have been obtained with simulated data using the misalignment simulation previously tool developed by our group [7].

II. MATERIALS AND METHODS

We have implemented a calibration method based on the algorithm Cho et al. described in [6], adapted to the C-arm available at our laboratory: the C-Arm Model SIREMOBIL Compact L of Siemens (Figure 1, left). The method uses the projection of a calibration phantom with two circular patterns formed by eight ball bearings (with 2 mm of diameter) symmetrically located in a cylinder with 70 mm of diameter and high (Figure 1, right). This phantom has been sized to cover most of the detector field of view (11.5 cm of diameter) provided by the image intensifier.

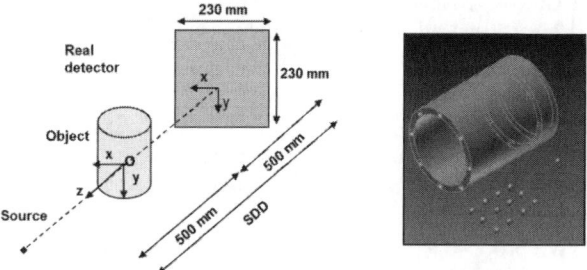

Fig. 1 Scheme of the geometry of the system (left) and design of the calibration phantom (right)

The calibration method estimates the ellipses formed by the balls on each projection and finds out the system parameters based on geometrical relationships between the ellipses. These parameters are the following (Figure 2): detector rotation (skew), inclination angles (pitch and roll), piercing point location (projection of the center of the calibration phantom), SDD (source to detector distance) and source and detector position.

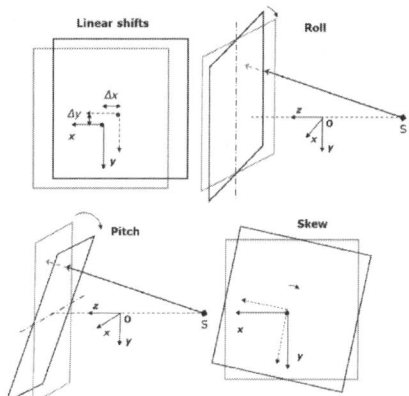

Fig. 2 Scheme of the mechanical misalignments in the detector panel

For each projection, the different steps of the algorithm, implemented using Matlab R2008a, calculates the following: 1) ball bearings segmentation, 2) center positions of ball bearings, 3) ellipses parameters, 4) piercing point/x-y offset, 5) skew angle and 6) converging point/inclination angles (θ and φ), 7) source position and 8) detector position.

First of all, the ball bearings are segmented by thresholding and the centers of mass of the balls are calculated. The ellipses formed by the centers of mass are described using the polynomial equation:

$$p_0 x_i^2 + y_i^2 - 2p_1 x_i - 2p_2 y_i + 2p_3 x_i y_i + p_4 = 0 \quad (1)$$

To calculate p_i parameters in equation (1), we solve the following system:

$$\begin{pmatrix} x_0^2 & -2x_0 & -2y_0 & 2x_0 y_0 & 1 \\ & & \vdots & & \\ x_7^2 & -2x_7 & -2y_7 & 2x_7 y_7 & 1 \end{pmatrix} \begin{pmatrix} p_0 \\ \vdots \\ p_7 \end{pmatrix} = \begin{pmatrix} -x_0^2 \\ \vdots \\ -y_7^2 \end{pmatrix} \quad (2)$$

where x_i, y_i are the coordinates of the centers of mass of the balls calculated previously. We need to describe the ellipses in the same form as in [6] with:

$$a(u - u_0)^2 + b(v - v_0)^2 + 2c(u - u_0)(v - v_0) = 1 \quad (3)$$

where (u_0, v_0) is the center of the ellipse and a, b, c are parameters that describe the shape. We obtain these parameters using the relations described in [8]:

$$u_0 = \frac{(p_1 - p_2 p_3)}{(p_0 - p_3^2)}, \quad v_0 = \frac{(p_0 p_2 - p_1 p_3)}{(p_0 - p_3^2)},$$
$$a = \frac{p_0}{(p_0 u_0^2 + v_0^2 + 2p_3 u_0 v_0 - p_4)}, \quad (4)$$

Left panel of Figure 3 shows the fitted ellipses on a projection of the calibration phantom with no misalignments.

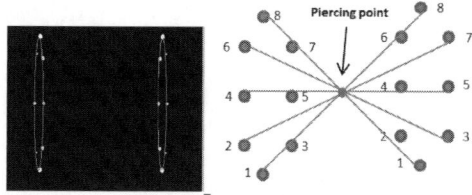

Fig. 3 Drawing of the fitted ellipses over the projections of the balls (left) and scheme of the piercing point in the projection by the intersection of the lines (right)

To calculate the possible horizontal and vertical displacements of the detector panel, known as x and y-offset, we estimate the piercing point, that is, the projection of the center of the calibration phantom on the detector plane. To this end, we obtain the intersection of all the lines that connect opposite balls from the two circular patterns (see Figure 3 right). Then, the x- and y-offsets are obtained from the difference between this point and the center of the image.

The skew of the detector panel, η, is obtained from the equations that relate the radius of the short axis of the ellipses with the distance from the point P_a (zero-short axis ellipse, produced by the projection of the central slice of the phantom) to the center of each ellipse (see Figure 5 of [6]).

$$\eta = [P_a^1 \cdot \tau_1 + P_2^a \cdot \tau_2] / P_1^2 \quad (5)$$

where τ_i are the rotation angles of both ellipses, obtained from the parameters calculated before with the expression:

$$\tau = (1/2) \operatorname{acot}(a - b/2c) \quad (6)$$

Next, the algorithm calculates both inclination angles of the detector. When the pitch angle θ is zero, both ellipses have the same shape and the lines that connects the extremes balls of the two circular patterns are parallel (see Figure 3, right). When θ is different from zero, one ellipse is extended and the other is shortened [7] as illustrated in Figure 4.

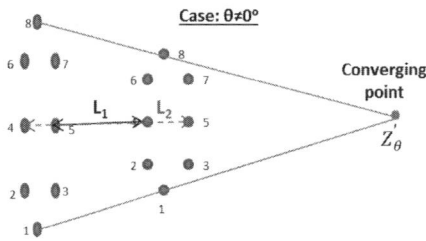

Fig. 4 Scheme of the projection when the pitch angle θ is not zero, being the converging point the intersection of the lines

Following geometrical considerations we can obtain the pitch angle with the following expression:

$$\theta = \operatorname{asin}\left[\frac{Z_S}{Z'_\theta}\right], \qquad Z_S = \left[\frac{2 \operatorname{rad} L_1 L_2}{[l(L_2 - L_1)]}\right] \qquad (7)$$

where Z'_θ is the converging point along the equivalent z axis with inclination (Figure 4), L_1, L_2 are the distances between the two ellipses (see Figure 4), Z_S is the SDD in Figure 1, and rad and l are is the radius and the distance between the two circular patterns in the real phantom.

We can calculate the roll angle, φ, from the ellipses parameters with the following equations:

$$\sin(\varphi) = -c_1 \beth_1 / 2(a_1) - c_2 \beth_2 / 2(a_2) \qquad (8)$$

where $\beth_k$ are intermediate parameters used in each ellipse parameter calculation:

$$\beth_k = \frac{Z'_S a_1 \sqrt{a_1}}{\sqrt{a_1 b_1 + a_1^2 b_1 (Z'_S)^2 - c_1^2}}, \qquad k = 1,2 \qquad (9)$$

$Z'_S = Z_S \cos(\varphi) \cos(\theta) + Y_S / \cos(\theta)$, $Y_S = \cos(\theta) P_0^a$

To obtain the roll angle, we solve the optimization problem that minimizes the difference between the two terms in equation (8) using the Nelder-Mead simplex method.

Then, the source position $P'_S = (X'_S, Y'_S, Z'_S)$ in the detector plane with all the misalignments is calculated using the rotation matrix R' described in equation (5) of [6] that converts the coordinates in the ideal detector, P_S, to those in the detector with misalignments after x and y-offset correction:

$$P'_S = R_I P_S \qquad (10)$$

Finally, we calculate the detector position $(0, Y_d, Z_d)$ in the ideal detector plane using these equations:

$$Z_d = \frac{Z_S(L_1 - l)}{L_1} + \operatorname{rad} \qquad (11)$$

$$Y_d = Y_S / Z_S \, Z_d \qquad (12)$$

In order to assess the method, we have simulated the mechanical misalignments of the detector in the projections (286x286 pixels) of the calibration phantom using the simulation tool previously developed by our group [7].

We evaluated the stability of the calculation of the parameters (x-y offset, skew, roll and pitch) along different projections (360-degree scan) simulating linear shifts and skew of the detector. We also studied the accuracy of the algorithm measuring the percentage of the difference between the estimated value and the simulated one for a set of different misalignment values.

III. RESULTS

Figure 5 shows the stability of the different geometrical parameters calculated with the algorithm (x- and y-offset of 3.6 mm, skew angle of 3°, inclination angles (roll and pitch of 0°) along different projection angles (360° scan with a step of 1°).

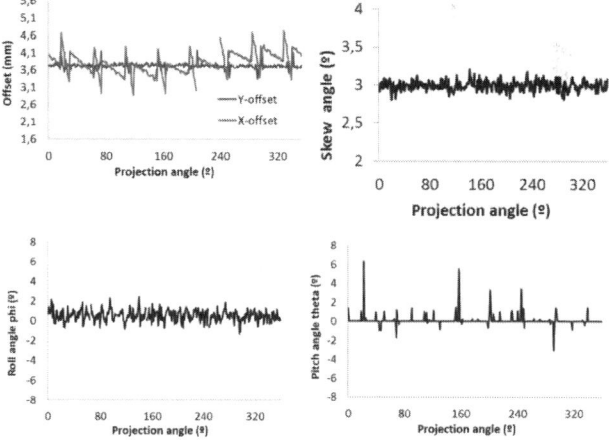

Fig. 5 Geometrical values estimated for different projection angles. The scales of these graphs are fitted to the tolerances values studied before in [7].

Figure 6 shows the error measured in one projection (0°) of the estimated parameters for different values.

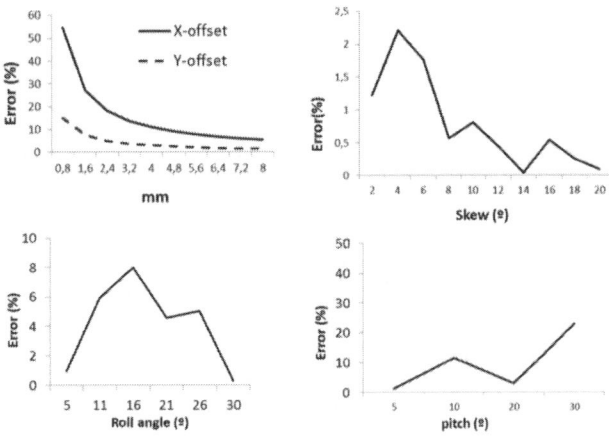

Fig. 6 Percentage of error of the estimation of the calibration parameters for different simulated values

IV. CONCLUSIONS

We have presented the implementation of a calibration method for a C-arm system to be used as a tomograph, based on the work by Cho et al. [6]. The method is based on a simple calibration phantom that can be placed inside the field of view without special positioning care. It provides the piercing point location, the x and y-offset, the detector rotation angle η (skew), the inclination angles (pitch and roll), the SDD, and the source and detector positions of the system to correct all the misalignments in the reconstruction process.

We evaluated the method using simulated data. The x-offset parameter estimation varies up to 1.1 mm for different projections and exceeds the tolerance studied in [7] in the 3% of the projections when it is 3.6 mm. For small values, the error is big but preliminary results with real data indicate that the common values are above 3.6 mm (error below 12%). In order to solve this problem, we will eliminate the projections with a high error. On the other hand, the estimation of the y-offset is very stable for all the projections and the error is always below 15% (in the worst case).

In the case of the skew, the stability of its estimation is also good enough (less than 1°) and error estimation is below the 3%. The variation of the roll angle estimation is under the tolerance (8°) for all projections [7] and the error does not exceed the 8%. Errors for pitch angles up to 20 degrees are below 10%, angles over this value are not expected in a C-arm equipment. For a few number of projections the estimation of the pitch angle presents a big error probably due to the wrong calculation of the mass centers when the projection of the balls overlap. As future work, we propose to improve the calculation of these mass centers. Also, we will evaluate the calibration method with real data extracted with the C-arm described before.

ACKNOWLEDGMENT

This work was partially funded by AMIT project (CEN-20101014) from the CDTI-CENIT program, RECAVA-RETIC Network (RD09/0077/00087), projects TEC2010-21619-C04-01 and TEC2011-28972-C02-01 from Spanish Ministerio de Ciencia e Innovación, ARTEMIS program (S2009/DPI-1802) from Spanish Comunidad de Madrid, and IPT-300000-2010-003. INNPACTO from Ministerio de Ciencia e Innovación.

REFERENCES

1. Calvo F., et al., Radioterapia intraoperatoria: desarrollo metodológico y experiencia clínica inicial. Oncología 1997;20(7):435-43. Oncología, 1997. 20(7): p. 435-43
2. Desco, M., et al., Simulated Surgery on Computed Tomography and Magnetic Resonance Images: An Aid for Intraoperative Radiotherapy. Comput Aided Surg, 1997. 2(6): p. 333-9.
3. Pascau, J., et al., An Innovative Tool for Intraoperative Electron Beam Radiotherapy Simulation and Planning: Description and Initial Evaluation by Radiation Oncologists. Journal of Radiation Oncology, Biology, Physics, 2012. 83(2): p. e287-e295.
4. Dawson, L.A. and D.A. Jaffray, Advances in image-guided radiation therapy. J Clin Oncol, 2007. 25(8): p. 938-46.
5. Xing, L., et al., Overview of image-guided radiation therapy. Med Dosim, 2006. 31(2): p. 91-112.
6. Cho, Y., et al., Accurate technique for complete geometric calibration of cone-beam computed tomography systems. Med. Phys., 2005. 32(4): p. 968-83.
7. M Paraíso, C de Molina, J Pascau, M Desco, M Abella. Evaluation of the effect of calibration accuracy in a C-arm for its use in tomography. *"Evaluation of the effect of calibration accuracy in a C-arm for its use in tomography"*. Libro de Actas XXX CASEIB 2012, s.p., 2012
8. Noo, F., R. Clackdoyle, C. Mennessier, T. A. White and T. J. Roney (2000). "An analytic method based on identification of ellipse parameters for scanner calibration in conebeam tomography." Phys. Med. Biol. 45: 3489-3508.

Author: Claudia de Molina Gómez
Institute: Dep. De Bioingeniería e Ingeniería Aeroespacial. UC3M
Street: Avenida de la Universidad 30
City: Leganés (Madrid)
Country: Spain
Email: cmolina@hggm.es

Quantitative Assessment of Prenatal Aortic Wall Thickness in Gestational Diabetes

Elisa Veronese[1], Silvia Visentin[2], Marius Linguraru[3], Erich Cosmi[2], and Enrico Grisan[1]

[1] Department of Information Engineering, University of Padova, Via Gradenigo 6/b, 35100 Padova, Italy
[2] Department of Woman and Child Health, University Hospital of Padova, Via Giustiniani 3, 35128 Padova, Italy
[3] Children's National Medical Center Sheikh Zayed Institute for Pediatric Surgical Innovation, 111 Michigan Avenue NW, Washington DC 20010, USA

Abstract— **Intrauterine environment, and especially a mismatch between the early and later-life environments, is thought to induce epigenetic and morphological changes that may manifest in later life as an increased vulnerability to non communicable diseases as diabetes. Prenatal events, such as intrauterine growth restriction, as well as an increased risk of developing diabetes and cardiovascular alterations, have been shown to be associated with an increased intima-media thickness (aIMT) of the abdominal aorta in the fetus. To date its measure, has been performed manually on ultrasound fetal images by skilled practitioners. We present an automatic algorithm that identifies abdominal aorta and estimates its diameter and thickness from routine third trimester ultrasonographic fetal data, providing a correlation between end-diastole aIMT automatic and manual measures of 0.96, with a mean error of 0.02 mm, and a relative error of 3%.**

Moreover, we show preliminary results on the application of the algorithm, proving a significant difference aIMT values between fetuses with normal weight and those with normal weight but gestational diabetic mother.

Keywords—**Segmentation, Mixture models, B-mode ultrasound, Fetal ultrasound, Intima-media thickness, Gestational diabetes, Diabetes.**

I. INTRODUCTION

According to the so-called Barker hypothesis [1] several adult diseases originate through adaptations of the fetus when it is undernourished. Low birth weight, caused either by preterm birth and/or intrauterine growth restriction (IUGR), was recently known to be associated with increased rates of cardiovascular disease and non-insulin dependent diabetes in adult life [2, 3]. It is well established that infants who had IUGR have a thicker aorta, suggesting that prenatal events (e.g. impaired fetal growth) might be associated with structural changes in the main vessels [4]. Hence, the measurement of abdominal aortic intima-media thickness (aIMT) in children becomes a sensitive marker of atherosclerosis risk [5]. In [4] aIMT of the abdominal aorta in newborn infants with low birth weight is compared with normal controls. In IUGR newborns aIMT was significantly greater than in the controls. Recently, [6, 7] studied for the first time aIMT in fetuses and then in the same infants after a mean follow up of 18 months. These studies showed that aIMT measurements in IUGR fetuses were inversely related to estimated fetal weight, showing that low birth weight and Doppler abnormalities may be correlated to an altered vascular structure causing possible endothelial damage. On this basis, we hypothesize that other nutrients and metabolic alterations (such as gestational diabetes) may lead to structural changes and developmental programming that can be assessed by quantitatively measuring phenotypic markers such as aIMT.

To date, aIMT measure has been performed manually by skilled practitioners, thus being susceptible to intra- and inter-operator variability. Given the far greater difficulty in analyzing fetal vessels and in obtaining good quality US images, computer aided image analysis and measurement has not been performed. In order to reduce the measurement variability and to produce results comparable for longitudinal and multicentric studies, the aIMT has to be measured in a consistent manner at the same point of the cardiac cycle. In [5] all measures were taken at end-diastole of the cardiac cycle, determined as the time of maximal expansion of the vessel during the entire cardiac cycle. In the present study we propose an automatic algorithm that identifies the abdominal aorta and estimates its diameter, derives the cardiac cycle, and finally provides the aIMT value. This technique is intended not only to make the procedure faster, but also to create reproducible results less susceptible to intra and inter operator variability. We finally apply the developed method in the study of intima-media thickness of fetuses with normal weight and those with normal weight but gestational diabetic mother.

II. MATERIALS

37 videos were obtained in the University Hospital of Padova (Padova, Italy), Department of Woman and Child Health from women undergoing routine US examinations during pregnancy. The study was approved by the local ethical committee and all women gave written informed consent. Aortic diameter and aIMT measurements were manually determined by a consensus of skilled practitioner (E.C. and S.V.) reviewing off-line the acquired videos blind

from the automatic results. The data were acquired from fetuses at a mean gestational age of 32 weeks (range 30 to 34 weeks) using US machine equipped with a 5.5 to 7 MHz linear array transducer (Antares, Siemens Medical Solutions, Mountain View, CA). 23 fetuses were with appropriate weight for gestational age (AGA group), 14 were with appropriate weight for their gestational age but with a gestational diabetic mother (Diab group). Intima media thickness and diameter were manually measured in a coronal or sagittal view of the fetus at the dorsal arterial wall of the most distal 15 mm of the abdominal aorta sampled below the renal arteries and above the iliac arteries. Gain settings were used to optimize image quality. The vessel was visualized in a longitudinal view of the fetus, making sure it was a maximal longitudinal section of the vessel (it contains the vessel diameter), and tilting the transducer to obtain an angle of insonation as close to 0° as possible and always less than 30°.

III. METHODS

A. Aorta Segmentation

The operator is asked to manually insert a reference line L_k in the kth frame f_k, to be placed inside the abdominal aorta, parallel to it, covering the portion of the vessel where all measurements are likely to be taken. From this line the system first estimates the vessel boundaries, i.e. lumen-intima borders along the aorta. Being s the linear coordinate along the line from every point $P(s)$ of coordinates $(x(s), y(s))$ along the manually inserted line, the intensity signal i_s corresponding to a segment centered in $P(s)$ and perpendicular to the line is extracted. Starting from $P(s)$ and moving first upward toward the superior wall and then downward toward the inferior wall, the first bright peaks $S_{sup}(s)$ and $S_{inf}(s)$ in the intensity are considered: these points correspond to hyperechoic muscular tissues composing the media layer surrounding the aorta: the border between blood and intima layer (i.e. the vessel boundary) is found as the point where the ultrasound intensity signal raises consistently from the lumen (low intensity) to the corresponding peak $S_{sup}(s)$ and $S_{inf}(s)$. The distance between positions of the two lumen-intima estimated borders provides an estimate $D(s)$ of the vessel diameter at the point $P(s)$. After all the points along the line L_i have been processed, the aorta diameter D_k of the analyzed frame f_k is obtained as the mean of all $D(s)$ All remaining frames are processed using the same starting line if they are deemed processable by the verification of two conditions, which aim at assessing the similarity of the grey level within the lumen and across the vessel boundaries: this will avoid the processing of frames in which the US image of the aorta disappears or is corrupted. Given the kth frame f_k, the intensity profile $i_{L,k}$ along the line $L_k = L_1$, and the intensity profile across its middle point $i_{M,k}$, and given the same intensity profile extracted from the subsequent frame f_{k+1}, we compute the difference of the corresponding profiles and evaluate the conditions:

$$\|i_{L,k} - i_{L,k+1}\| < h_1 \quad (1)$$
$$\|i_{M,k} - i_{M,k+1}\| < h_2 \quad (2)$$

with the $\|x\|$ indicating the euclidean norm, and with h_1 and h_2 being two thresholds, empirically chosen. If at least one of the conditions is not satisfied (which might mean that either the fetus or the operator made a sudden movement during the acquisition, thus causing subsequent frames to change abruptly), the user is asked to re-insert a reference line L_{k+1} inside the aorta or to discard the frame. Changing the values of h_1 and h_2 set the trade off between processing a higher number of frames and measuring aIMT in possibly unreliable positions. Set h_1 and h_2 to lower values will result in an increase interaction of the operator who will be asked more frequently to manually reinsert a reference line.

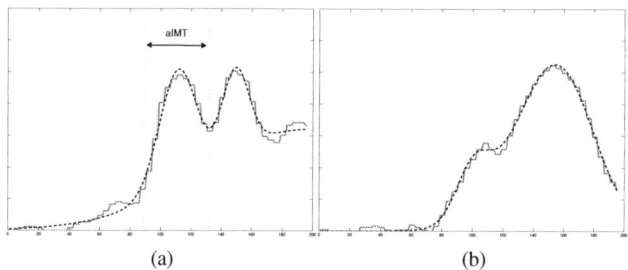

Fig. 1 Mixture of gaussian fit on the gray levels across along the aorta wall, with a successful estimate of the aIMT (left panel), and where the estimated tonaca media boundary were rejected (right panel)

Given the estimated diameters $D(s)$ along the line L_k in the frame f_k, the aIMT has to be estimated at each point $P(s)$. We use the gray level profile extracted from a line centered in $(x(s), y(s))$, perpendicular to L_k and with length equal to $(1.5 \cdot D_k/2$ pixels on both sides. According to the definition in [8], we consider as aIMT the distance between the leading edge of the blood-intima interface, and the leading edge of the media-adventitia interface on the far wall of the vessel. Since the edges we are interested in are rarely well definite, but rather appear as a local intensity minimum between two bright peaks related to the intima-media and adventitia layers, we choose to model the profile intensities comprising lumen, intima-media and adventitia as a mixture of three gaussians, along with the method proposed in [9]:

$$GM(g) = \sum_{j=1}^{3} z_i \cdot e^{-0.5 \cdot \left(\frac{g-\mu_j}{\sigma_j}\right)^2} \quad (3)$$

where μ_j is the mean of the jth gaussian, σ_j its standard deviation and z_j its mixing proportion. The mixture is usually composed of a large-variance gaussian $G_1(g)$, representing the low-frequency variation of the intensities across the aorta edge, and two small-variance components $G_2(g)$ and $G_3(g)$, modeling the vessel wall layers. By this means, we can identify the media-adventitia (MA) interface as the local minimum $\hat{g}_{MA}$ of $GM(g)$ in the interval $g \in [\mu_2, \mu_3]$. In order to obtain a reliable measure, we further impose that:

$$\min(GM(\mu_1) - GM(\hat{g}_{MA}), GM(\mu_2) - GM(\hat{g})_{MA}) \leq T_{rel} \quad (4)$$

where T_{rel} is a user defined threshold used to set the sensitivity to small variations in the intensity profile (see Fig. 1). We set $T_{rel} = 0.05$. We then refine the blood-intima (BI) interface position by evaluating the position $\hat{g}_{BI}$ for which $GM(\hat{g}_{BI}) = 0.5 * GM(\mu_1)$, in the interval $g \in [0, \mu_1]$ For the kth frame, and for each of the $P(s)$ points in the ROI, we thus obtain a value for the aorta diameter $ad_{k,P}$ and a value for the IMT:

$$aIMT_{k,p} = \hat{g}_{MA,k,p} - \hat{g}_{BI,k,p} \quad (5)$$

so that we can compute the sample mean of the aIMT $\mu_{aIMT}(k)$.

B. Cardiac Cycle Measurement Synchronization

The proposed system estimates in each processable frame f_k a mean value for the aortic diameter D_k, and a mean intima-media thicknesses $aIMT_k$. In order to obtain measures comparable with the literature and corresponding to the same end-diastolic phase of the cardiac cycle [5] a further step is needed. The diameter and aIMT estimated values vary with the cardiac cycle, as it can be seen in Fig. 2 The diameter variation in time is modeled as a sinusoidal function $aDiam(t) = A \cdot sin(ft + \phi)$, that is fit on the data $\mu_{aD}(k)$ through a nonlinear least squares procedure. The frames corresponding to the maxima of the estimated sinusoid represent the end-diastolic cardiac phase, whereas the minima are the end-systolic frames. Being interested in the mean end-diastolic values for the comparison with the manual values, and being the end-diastolic frames those corresponding to $k = (\pi/2 - \phi)/f + 2m\pi/f$, with $m \in \mathcal{N}$, we obtain:

$$aIMT_{dia} = \frac{1}{K} \sum_{k=0}^{K} (\mu_{aIMT}(k)) \quad (6)$$

$$aDiam_{dia} = \frac{1}{K} \sum_{k=0}^{K} (\mu_{aD}(k)) \quad (7)$$

A for the end systolic phase, we estimate the values $aIMT_{sys}$ and $aDiam_{sys}$ by considering $k = (\pi/2 - \phi)/f + (2m+1)\pi/f$, with $m \in \mathcal{N}$

RESULTS

The proposed automatic method for the analysis of the fetal abdominal aorta from US data was evaluated against the manually annotated frames. The correlation between the manual and automatic estimation of aIMT is 0.96, with a mean error of 0.02 mm corresponding to a relative error of 3%, showing an excellent accordance as can be visually appreciated in Fig. 3. Since the aIMT values depend on the dimension of the aorta (represented by its diameters), their value might be deceiving e.g. when including in the AGA population both single and twin pregnancies. To tackle the problem, along with the $aIMT_{dia}$ and $aDiam_{dia}$ values, we

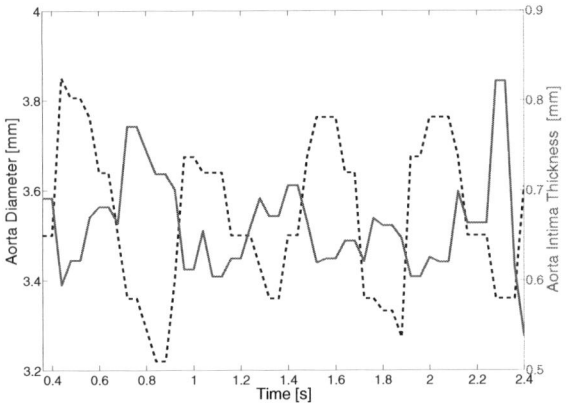

Fig. 2 Mean aorta diameter (dashed black line) and mean aIMT (solid gray line), estimated in 78 subsequent frames of a representative US video. For visual purposes the values of aIMT are scaled to be superimposed to aorta diameter values

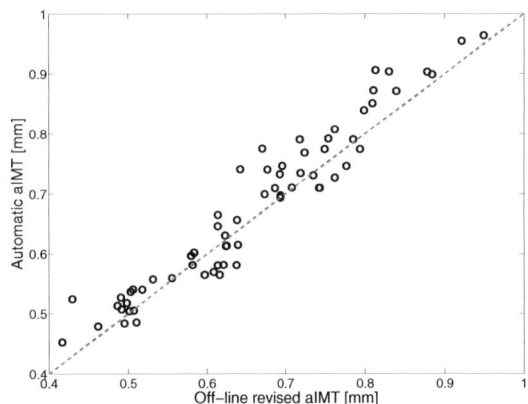

Fig. 3 Manual versus automatic estimation of aIMT on the 95 reviewed frames

computed also the intima-media relative thickness of the aorta wall with respect to the aorta diameter $aIMRT_{dia} = aIMT_{dia}/aDiam_{dia}$. We report the values of the computed parameters divided into the two fetal populations analysed at the end diastolic phase. in Tab. 1. When considering the whole set of data available (37 videos), no significant differences in the obtained estimated were found with respect to the diameter, confirming the assumption of the two populations being substantially normal weighting, whereas aIMT was significantly different between AGA and Diab fetuses ($p < 0.01$) both at diastolic and systolic phases as it can be seen in Fig. 4. When normalizing the aIMT by the diameter to obtain aIMRT, significant differences were found between AGA and Diab fetuses ($p < 0.01$) at the distolic phase only.

Table 1 Diameter and aIMT estimation performed in the two groups of fetuses during the the end-diastole frames. All measures are in millimeters

	aDiam [mm]		aIMT [mm]		aIMRT [mm]	
	Mean	Std	Mean	Std	Mean	Std
AGA	4.22	0.65	0.73	0.14	0.17	0.03
Diab	4.53	0.77	0.92	0.19	0.21	0.03

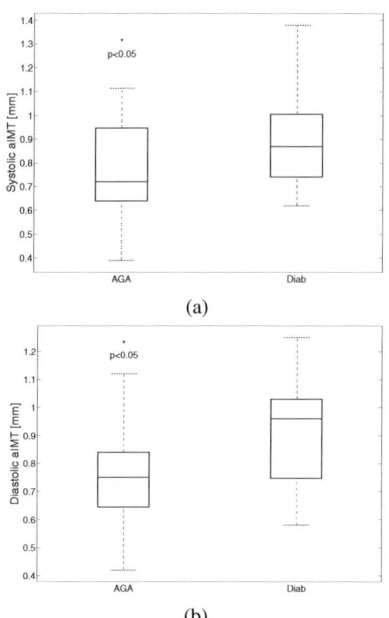

Fig. 4 Boxplot showing diastolic aIMT (left panel) and systolic aIMT (right panel) distributions. On each box the median as the central mark, the 25th and 75th percentiles as the edges of the box, and using the whiskers to show the extension of the most extreme data points not considered outliers. Significant differences between Diab, the AGA fetuses are reported.

IV. CONCLUSION

Aortic diameter and aIMT are usually manually measured by skilled practitioners in videos recorded during routine third trimester ultrasonographic fetal biometry. Here we propose a semiautomatic algorithm to identify the abdominal aorta and to estimate its diameter in US fetal images. Our technique provides a measure for the aortic diameter, derives the cardiac cycle, and finally provides the aIMT value. Manual and automated assessments resulted to be highly correlated, proving that the proposed method is adequate and reliable for the measure of important structures and suggesting that its application to derive morphology measurements would faster ultrasonographic fetal biometry.

We show the application of this method to a pilot study comparing aIMT values in fetuses with normal weight for their gestational age and fetuses with normal weigh but gestational diabetic mother, yielding a significant difference hinting for some underlying developmental programming adaptation to the different intrauterine environment.

REFERENCES

1. Barker D.J.P.. Maternal nutrition, Fetal nutrition, and Disease in Later Life *Nutritions.* 1997;13:807.
2. McMillen I.C., Robinson J.S.. Developmental origins of the metabolic syndrome: prediction, plasticity, and programming *Physiological Reviews.* 2005;85:571-633.
3. Langley-Evans S.C.. Nutritional programming of disease: unravelling the mechanism *Journal of Anatomy.* 2009;215:36-51.
4. Skilton M. R., Evans N., Griffiths K. A., Harmer J. A., Celermajer D. S.. Aortic wall thickness in newborns with intrauterine growth restriction *Lancet.* 2005;365:1484-6.
5. Litwin M., Niemirska A.. Intima-media thickness measurements in children with cardiovascular risk factors *Pediatric Nephrology.* 2009;24:707-719.
6. Cosmi E., Visentin S., Fanelli T., Mautone A. J., Zanardo V.. Aortic intima media thickness in fetuses and children with intrauterine growth restriction *Obstetrics and Gynecology.* 2009;114:1109-1114.
7. Visentin S., Grisan E., Zanardo V., et al. Developmental programming of cardiovascular risk in intrauterine growth restricted twin fetuses and aorta intima thickness *J. Ultrasound Med..* 2013;32:279-284.
8. Koklu E., Kurtoglu S., Akcakus M., Yikilmaz A., Coskun A., Gunes T.. Intima-media thickness of the abdominal aorta of neonates with different gestational ages *Journal of Clinical Ultrasound.* 2007;35:491-497.
9. Veronese E., Cosmi E., Visentin S., Poletti E., Grisan E.. Estimation of fetal aorta intima-media thickness from ultrasound examination in *MICCAI Workshop on Perinatal and Paediatric Imaging: PaPI 2012*(Nice) 2012.

Fast Anisotropic Speckle Filter for Ultrasound Medical Images

G. Ramos-Llordén[1], G. Vegas-Sánchez-Ferrero[1], S. Aja-Fernández[1],
M. Martín-Fernández[1], and C. Alberola-López[1]

[1] Laboratorio de Procesado de Imagen (LPI), Universidad de Valladolid, Valladolid, Spain

Abstract—Low contrast in ultrasound images caused by the granular pattern (speckle) makes difficult computational analysis and diagnosis. Thus, speckle filtering is a common step before computed automatic analysis. However, speckle depends on the inner echo-morphology of tissue and it should be removed without over-filtering relevant details for diagnostic purposes. Some methods were proposed to preserve important details by means of anisotropic diffusion schemes. However, they require to solve an evolutionary partial differential equation, which needs considerable computation time that makes this kind of filters impractical for real-time scenarios. Additionally, there is no rational criteria to select the optimal stop criteria. Some other detail preserving filters are based in the Non-Local means philosophy, however they involves an even higher computational cost.

In this work we propose a fast anisotropic speckle filter which makes use of speckle statistics to preserve the tissue echo-morphology while the speckle is properly removed from regions without clinical relevant features, such as blood in heart cavities. The implementation is based on an anisotropic and spatially variant Gaussian kernel whose covariance depends on structural information of tissues. The proposed implementation is computationally efficient, where no stop criterion is needed. Results confirmed the low computation cost compared to diffusion and Non-Local means based filters. Quantitative evaluation with synthetic data also confirmed the better performance of the filter compared to other state of the art methods.

Keywords—**Ultrasound, Speckle, Fast, Filter, Anisotropy.**

I. Introduction

Medical ultrasound (US) imaging has become a widely diagnostic technique for human organs such as heart, kidney, prostate, coronary arterial, etc. The noninvasive nature together with low cost and real time capability make US imaging a powerful computational assisted technique. As an example, the study of mitral insufficiency is undertaken using strain rate estimation in [1]. In a similar way, segmentation of tissues in the kidney are possible due to US images [2]. However, computational analysis is a hard task in US, due to low contrast between tissues. This is because of scattering, which appears in US acquisition due to the echo pulse interaction with structures smaller than the wavelength producing a granular pattern typically known as Speckle [3]. As a pre-processed step before computed automatic analysis, filtering is usually performed for preserving tissue morphology while noise is removed. However speckle statistics depend on tissue echo-morphology, so considering speckle as noise is not a correct assumption [4]. Many filters proposed do not consider this particularity and therefore tends to over-filter the image, reducing essential diagnostic tissue features. They consider speckle as a multiplicative noise which is an over-simplified and sometimes unrealistic assumption [5].

Some of the speckle filters are based on the diffusion equation described by evolutionary partial differential equations (PDE) [6, 7, 8], which are designed to enhance edges while performing filtering in homogeneous regions. Diffusion filters implementation are based on discretization schemes of the heat equation where the diffusion tensor is calculated from the anatomical structures of the image [9]. In [8], the structure information is obtained from a statistical characterization of tissues in US images which provide the most probable edges of tissues. The implementation of diffusion methods needs several iterations to obtain a satisfactory filtering. Each iteration requires to solve at least a linear system by semi or full implicit schemes, which results in a considerable computation time that make this kind of filters impractical for real-time scenarios. Furthermore, there is no rational criteria to select the optimal stop time for the iterative scheme, so an heuristic approach is usually applied.

Another widely used speckle filter for US images is the extension of the Non-Local Means filter to speckle statistics [10]. In this case, the mathematical framework is different from diffusion schemes, but the computational cost is even higher due to the exhaustive patch searching (even considering the optimization techniques proposed in [10]).

Due to the necessity of a computationally efficient method for removing the speckle while preserving structural details, in this work we propose a fast anisotropic speckle filter which can distinguish between tissues due to speckle statistics and preserves the tissue echo-morphology while filtering in regions without clinical relevant features, such as blood in heart cavities. The relevant structure information is estimated by means of statistical tissue characterization which leads to the estimation of the most probable boundaries of tissues [8]. The Implementation is based on a spatially variant Gaussian kernel whose covariance matrix depends on the structural information of tissues. Furthermore, the implementation is computationally efficient because it is based on a fast decomposition of non-stationary Gaussian kernels, where no stop criterion is needed. In summary, the main contributions of this work are: 1) To provide an efficient anisotropic filter which distinguishes between tissues. 2) The implementation of a fast decomposition of non-stationary Gaussian Kernels whose Covariances depend on the statistics of tissues.

The outline of this paper is as follows. The mathematical formulation of the proposed method and the efficient implementation are described in Section II. In Section III, the proposed filter is compared to other common state-of-the-art filters in synthetic and real scenarios. Finally, some conclusions and future lines are provided.

II. METHOD

We denote the noisy image as $f(\mathbf{r})$ where $\mathbf{r} = (x,y)^T$ and $\mathbf{D}(\mathbf{r})$ as the diffusion tensor obtained from the statistical characterization as proposed in [8], whose diagonalization is:

$$\mathbf{D}(\mathbf{r}) = E(\mathbf{r})\Sigma(\mathbf{r})E^T(\mathbf{r}) \qquad (1)$$

where $E(\mathbf{r})$ is the matrix whose columns are the vectors $\mathbf{u}(\mathbf{r})$, $\mathbf{v}(\mathbf{r})$ with orientation along the most probable contours of $f(\mathbf{r})$ and orthogonal to them respectively. $\Sigma(\mathbf{r})$ is a diagonal matrix with components $\sigma_u(\mathbf{r})$ and $\sigma_v(\mathbf{r})$.

The filtered image $g(\mathbf{r})$, is obtained performing the following spatially variant convolution:

$$g(\mathbf{r}) = \int\int \frac{1}{2\pi\sqrt{|\Sigma(\mathbf{r})|}} e^{-\frac{\mathbf{r}'^T \mathbf{D}^{-1}(\mathbf{r})\mathbf{r}'}{2}} f(\mathbf{r}-\mathbf{r}')d\mathbf{r}', \qquad (2)$$

where $\mathbf{r}' = (x',y')$.

Note that this convolution is a separable operator in the base defined by vectors $\mathbf{u}(\mathbf{r})$, $\mathbf{v}(\mathbf{r})$. So, for each $\mathbf{s} = E^t\mathbf{r}$, the filtered image is:

$$g(\mathbf{r}) = \frac{1}{2\pi\sqrt{|\Sigma(\mathbf{r})|}} e^{-\frac{v^2}{2\sigma_v(\mathbf{s})^2}} \otimes_v e^{-\frac{u^2}{2\sigma_u(\mathbf{r})^2}} \otimes_u f(\mathbf{s}(\mathbf{r})) \qquad (3)$$

where $\otimes_u$ is the convolution operator in the direction of the eigenvector $\mathbf{u}$.

Note that the calculation of Eq. (2) in the (x,y) discretized domain needs the computation of the discrete samples of a two dimensional Gaussian kernel with spatially variant parameters. Yet the implementation of the convolution in the (u,v) space is not practical because it involves many interpolations due to the need of evaluating $f(\mathbf{s}(\mathbf{r}))$ for each $\mathbf{r}$.

In order to obtain an efficient computation of Eq. (2), we adopt the non-orthogonal discrete decomposition proposed by Geusebroek et. al. in [11] where a two-dimensional Gaussian filtering along θ direction is decomposed into two one-dimensional Gaussian filters. Specifically, the decomposition is based on one Gaussian filter along x-direction with σ_x standard deviation followed by another one in the direction described by line $t = x\cos\phi + y\sin\phi$, where ϕ is the angle between t-line and x-axis and σ_ϕ is the standard deviation of the Gaussian along ϕ direction. Their mathematical relationships are as follows [11]:

$$\tan\phi = \frac{\sigma_v^2\cos^2\theta + \sigma_u^2\sin^2\theta}{(\sigma_u^2 - \sigma_v^2)\cos\theta\sin\theta} \qquad (4)$$

$$\sigma_\phi = \frac{1}{\sin\phi}\sqrt{\sigma_v^2\cos^2\theta + \sigma_u^2\sin^2\theta} \qquad (5)$$

$$\sigma_x = \frac{\sigma_u\sigma_v}{\sqrt{\sigma_v^2\cos^2\theta + \sigma_u^2\sin^2\theta}} \qquad (6)$$

We extend the decomposition for spatially variant Gaussian filter described in Eq. (3), where both σ_v and σ_u depend on $\mathbf{r}$. Thus, ϕ, σ_ϕ and σ_x also depend on $\mathbf{r}$.

The computation of a discrete convolution along x-direction may be prohibitively expensive since one need at least $4\sigma_x(\mathbf{r})$ coefficients for an acceptable error [12]. Then, we used the recursive implementation proposed by Deriche in [12] which just needs 16 products and 25 sums for a Gaussian approach of fourth order.

Finally, the spatially anisotropic Gaussian convolution is performed by the convolution along the t-line. Again, the recursive method can be applied in this direction, which involves the evaluation of the image $f(\mathbf{s}(\mathbf{r}))$ along the t-line. Note that the evaluation needs just one interpolation, whereas two interpolations are needed when the convolution is performed in the conventional way described by Eq. 2. Thus, the number of computations are dramatically reduced. To meet the requirements of an efficient implementation, linear and bilinear interpolation are the most suitable options. In this work, we adopted the bilinear one.

III. EXPERIMENTS

In order to validate the proposed filter we carry out real and synthetic experiments. In both cases, we applied the statistical characterization of tissues proposed in [8], which leads to a definition of a diffusion tensor $\mathbf{D}(\mathbf{r})$ according the most probable edges of tissues. This definition requires obtaining probability maps of the noisy image. The implementation of the anisotropic Gaussian convolution along x and t directions was performed by using Deriche's method [12]. A bilinear scheme was chosen for the interpolation along the t-line.

We compare the proposed filter with the Matlab implementation of the Optimized Bayesian Non-Local Means (OBNLM), which has been massively used for speckle filtering during the last years, and the Detail Preserving Anisotropic Diffusion Filter (DPAD) [7], which is an example of diffusion filter that preserves tissue details. The implementations were download from the corresponding authors web ([1],[2]) respectively.

Two different comparative measures were used for the synthetic experiment: the Mean Square Error (MSE) and the Structural Similarity Measure Index (SSIM) [13]. The former allows to compare filters performance considering a phantom as the unknown parameter to be estimated. The latter penalizes losing edge and structural details (the closer to 1, the better the filter). Additionally, the computational cost was also compared by means of the CPU time needed for the filtering process. Experiments were carried out with an Intel Core i7-3610QM CPU @ 2.30 Ghz processor.

The synthetic US image was obtained from the realistic kidney image from Field II [14]. The image was sampled following a fan arrangement and then contaminated by simulating the speckle as a random walk stochastic process. The reconstruction was performed as in [5], where the noisy image is interpolated to obtain the Cartesian arrangement. The kidney image intensity is encoded as the standard deviation of scatterers within the resolution cell. The diffusion tensor, $\mathbf{D}(\mathbf{r})$, is obtained by following the method proposed in [8], where the probability of belonging to each tissue class is derived from a Gamma distribution as it was proposed for interpolated fully formed speckle [5]. The number of tissue classes were fixed to four [15].

The parameter optimization of each filter was carried out by maximizing the SSIM measure in a simulated image of the explained phantom. Then, the validation was performed for 30 independent realizations of the same phantom. Numerical results are shown in Table 1.

Some examples are shown in Fig. 1 where the true image and the polar sampling are represented (Fig. 1.a-b) and the resulting noisy image (Fig. 1.c). The filtered image obtained with the OBNLM method (Fig. 1.d) shows a good filtering in blood regions since it is expected that those homogeneous regions are properly weighted by the Non-Local Means method, however the noise is not properly remove from the tissue regions. In the case of the DPAD filter (Fig. 1.e), though DPAD preserves fine details, it generates undesired spurious for both visual and automatic analysis.

These conclusions are supported by numerical results. The proposed filter obtains the lowest MSE and SSIM average.

---

[1] http://www.bic.mni.mcgill.ca/PersonalCoupepierrick/OBNLMFilter
[2] http://www.lpi.tel.uva.es/ santi/personal/download.htm

Table 1 Numerical results for Synthetic Experiment (mean value ± standard deviation). Optimized parameters for OBNLM: $M = 6$, $\alpha = 3$, $h = 0.8$; DPAD: $\Delta_t = 0.005$, $n_{iter} = 130$; Proposed method: $\mathbf{D}(\mathbf{r})$ defined as [8], Four tissue classes.

Filter	MSE×10^{-3}	SSIM	CPU-Time(s)
Noisy	12.3±0.2388	0.7199±0.0014	–
OBNLM	0.63±0.3216	0.8447±0.0057	16.6245±0.5312
DPAD	0.48±0.1642	0.8546±0.0026	19.1803±0.8239
Proposed	**0.38±0.1368**	**0.8828±0.0019**	**5.5510±0.3973**

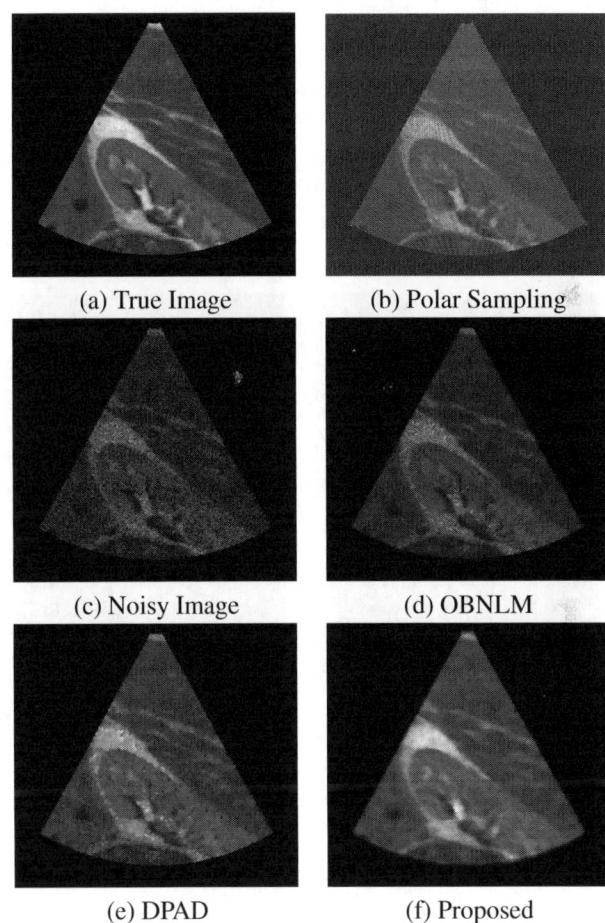

Fig. 1 Synthetic Kidney Experiment.

Besides, CPU time is considerable lower than the DPAD and OBNLM corresponding time. Additionally, the standard deviation for each measure is the smallest in the proposed filter implementation. This indicate the robustness of the method.

In the case of real experiments, a real image of the heart was considered (1024 × 768 and 8 bits). It was acquired from a clinical machine Philips Medical Systems iE33 with the software PMS5.1 Ultrasound iE33 4.0.1.357. The parameters used for each filter were those obtained for the kidney, where the SSIM measure was optimized. The proposed filter used a Gamma distribution for the characterization of tissues. The number of tissue classes was fixed to two (myocardial tissue and blood). The results are depicted in Fig. 2.

As we can observe, though OBNLM (Fig. 2.b) preserves the valve morphology, it does not perform enough filtering in the region near to the ultrasound beam origin (Top-center of the image) and creates artificial borders in the tissue due to the enhancement effect of the granular pattern of speckle (Bottom-center of the image). The DPAD method (Fig. 2.c) also shows artificial borders in myocardial tissue due to an over-filtering effect and speckle in blood regions is not removed. The proposed filter (Fig. 2.d) preserves contour definition along myocardium tissue, and performs a considerable filtering in cavities. Besides, the computational time needed was 17.7s in comparison to 32.5s and 74.9s for OBNLM and DPAD respectively. This result confirms that the proposed implementation obtains good results for a reasonably low computational cost.

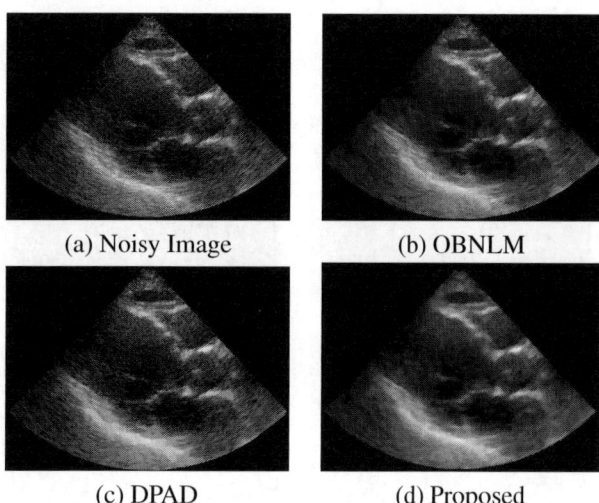

(a) Noisy Image (b) OBNLM
(c) DPAD (d) Proposed

Fig. 2 Ecocardiography Experiment.

IV. CONCLUSION

The proposed method achieves to filter homogeneous regions where no medical information is expected to be found, e.g. blood, and at the same time preserves tissue morphology as valve contour or myocardium structures in US images. Besides, the implementation can be done efficiently due to the decomposition of spatially variant convolution into non-orthogonal axes. Then, the proposed filter is suitable for real-time necessities which is a considerable advantage over diffusion or Non-Local Means based filters. Numerical results support the statements pointed out before. Our filter outperforms DPAD and OBNLM results for SSIM, MSE quality measures and also the computational cost.

REFERENCES

1. Curiale A.H, Vegas-Sanchez-Ferrero G., Perez-Sanz T., Aja-Fernandez S.. Strain rate tensor estimation from echocardiography for quantitative assessment of functional mitral regurgitation in *IEEE International Symposium on Biomedical Imaging: From Nano to Macro, 2013* 2013.
2. Xie J., Jiang Y., Tsui H.-T.. Segmentation of kidney from ultrasound images based on texture and shape priors *Medical Imaging, IEEE Trans. on.* 2005;24:45-57.
3. Burckhardt C.B.. Speckle in ultrasound B-mode scans *Sonics and Ultrasonics, IEEE Trans. on.* 1978;25:1-6.
4. Thijssen J. M., Oosterveld B. J.. Texture in tissue echograms: speckle or information *American Institute of Ultrasound in Medicine 9.* 1990:215–229.
5. Vegas-Sánchez-Ferrero G, Martín-Martínez D., Aja-Fernández S., Palencia C.. On the influence of interpolation on probabilistic models for ultrasonic images in *IEEE International Symposium on Biomedical Imaging: From Nano to Macro*:292 -295 2010.
6. Y. Yongjian, Acton S.T.. Speckle reducing anisotropic diffusion *Image Processing, IEEE Trans. on.* 2002;11:1260 - 1270.
7. Aja-Fernandez S., Alberola-Lopez C.. On the estimation of the coefficient of variation for anisotropic diffusion speckle filtering *Image Processing, IEEE Trans. on.* 2006;15:2694 -2701.
8. Vegas-Sanchez-Ferrero G., Aja-Fernandez S., Martin-Fernandez M., Frangi A. F., Palencia C.. Probabilistic-driven oriented speckle reducing anisotropic diffusion with application to cardiac ultrasonic images in *13th international conference on Medical image computing and computer-assisted intervention: Part I*(MICCAI'10):518–525 2010.
9. Weickert J.. *Anisotropic Diffusion in Image Processing.* B.G Teubner (Stuttgart) 1998.
10. Coupe P., Hellier P., Kervrann C., Barillot C.. Nonlocal Means-Based Speckle Filtering for Ultrasound Images *Image Processing, IEEE Trans. on.* 2009;18:2221-2229.
11. Geusebroek J.-M., Smeulders A.W.M., Weijer J.. Fast anisotropic Gauss filtering *Image Processing, IEEE Trans. on.* 2003;12:938-943.
12. Deriche R. Recursively implementing the gaussian and its derivatives tech. rep.INRIA, Unit de Recherche Sophia-Antipolis 1993.
13. Wang Z., Bovik A.C., Sheikh H.R., Simoncelli E.P.. Image quality assessment: from error visibility to structural similarity *Image Processing, IEEE Trans. on.* 2004;13:600-612.
14. Jensen J. A.. FIELD: A Program for Simulating Ultrasound Systems in *10th Nordicbaltic Conference On Biomedical Imaging, vol. 4, Supplement 1, Part 1:351–353*:351–353 1996.
15. Vegas-Sanchez-Ferrero G., Aja-Fernandez S., Palencia C., Martin-Fernandez M.. A Generalized Gamma Mixture Model for Ultrasonic Tissue Characterization *Computational and Mathematical Methods in Medicine.* 2012;2012:25.

Author: Gabriel Ramos Llorden
Institute: Laboratorio de Procesado de Imagen (LPI)
Street: E.T.S.I. Telecomunicacion, Campus Miguel Delibes s/n
City: Valladolid
Country: Spain
Email: gramllo@lpi.tel.uva.es

Parallel Implementation of a X-Ray Tomography Reconstruction Algorithm for High-Resolution Studies

J. Garcia[1], M. Abella[2,3], C. de Molina[2,3], E. Liria[1], F. Isaila[1], J. Carretero[1], and M. Desco[2,3,4]

[1] Computer Architecture and Communication Area, University Carlos III, Madrid, Spain
[2] Bioengineering and Aerospace Engineering Dept., University Carlos III, Madrid, Spain
[3] Instituto de Investigacion Sanitaria Gregorio Marañon (IiSGM), Madrid, Spain
[4] Centro de Investigacion en Red de Salud Mental (CIBERSAM), Madrid, Spain

Abstract— Most small-animal X-ray computed tomography (CT) scanners are based on cone-beam geometry with a flat-panel detector orbiting in a circular trajectory. Image reconstruction in these systems is usually performed by approximate methods based on the algorithm proposed by Feldkamp, Davis and Kress (FDK). Currently there is a strong need to speed-up the reconstruction of CT data in order to extend its clinical applications. The evolution of the semiconductor detector panels has resulted in an increase of detector elements density, which produces a higher amount of data to process. This work focuses on both standard and future high-resolution studies, in which multiple level of parallelism will be needed in the reconstruction process. In addition, this paper addresses the future challenges of processing high-resolution images in many-core and distributed architectures.

Keywords—Reconstruction, Image processing, Parallel architectures, MPI, CUDA.

I. INTRODUCTION

Many small animal X-ray computed tomography (CT) scanners are based on cone-beam geometry with a flat-panel detector orbiting in a circular trajectory. Image reconstruction in these systems is usually performed based on the algorithm proposed by Feldkamp et al. because of its straightforward implementation and computational efficiency [1]. With the evolution of the technology, the acquisition time has been reduced. On the one hand, the evolution of the detector panels has resulted in an increase of detector elements density, which produces a higher amount of data to process [2]. Together with this increase of data, there is a need of faster reconstructions to address the newest uses of CT: planning and monitoring in radiotherapy, image assisted surgery, and other clinical applications required the real time imaging [3]. On the other hand, some recent advances in algorithms have not been exploited yet at the full potential in high performance implementations, which represents a barrier for extending the use of this technology [3]. All this motivates the need to look for optimizations that can handle the increasing complexity and demand of the reconstruction task.

The main goal of this work is to enhance the Mangoose++ application in terms of performance, scalability, and high resource utilization in both heterogeneous multi-GPU and distributed systems. First, we target to improve the application performance in terms of speed-up by applying a combination of distributed and many-core architectures. Second, the solution has to scale on heterogeneous multi-GPU systems. The main contributions of this paper are the following. First, we propose a novel implementation of Mangoose++ for multiple GPU and distributed architectures which focuses on high-resolution studies. Second, for achieving a high computational resource utilization of heterogeneous resources, we employ a hierarchical approach based on multi-node and multi-GPU problem decomposition. Third, we address the current performance limitations of I/O subsystems by hiding I/O latency through overlapping computation and I/O, specially for big data sets.

II. RELATED WORK

There is a large number of works that focus on optimization of image processing algorithms based on GPGPUs (General-Purpose Graphics Processing Unit). Lee et al. focus in the importance of parameters such as cache hits, memory transfers CPU/GPU, and thread block size [4]. Other research work show that the backprojection time is approximately equal to the total execution time [5]. As far as we know, Zhue et al. were the first authors in take into account the overall execution time, considering other factors such as asynchronous writes, number of GPUs, and more [6]. The authors demonstrated the benefits of hiding the volume store. As we show in Section IV, this feature is especially important for large resolution studies (e.g. 4096^2 projection images). Scherl et al. show reconstruction time for 6 GPU families [7]. The authors discuss about the performance benefits of the latest models in terms of device memory bandwidth.

Additionally, they state that the current trade-off is based on memory bandwidth and computing resources. We demonstrate that the memory capacity is more important than the memory bandwidth itself.

Zhang et al. present a parallel execution flow for multiple GPUs [8]. The workflow aims to increase the utilization factor of GPUs by running multiple kernels concurrently. Our solution differs from this work by overlapping GPU computation and memory transfers between host and GPU. Other works take into account the GPU memory bandwidth [9]. Page-locked allocated memory (pinned memory) makes copies between CPU and GPU faster. In our evaluation section we demonstrate that the combination of independent optimizations, in this case pinned memory, can affect the overall behaviour, especially when other optimizations are enabled.

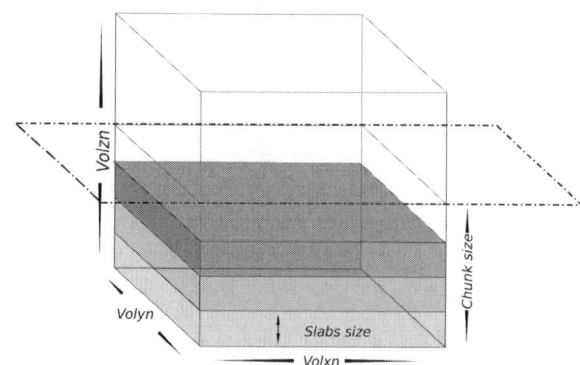

Fig. 1 Volumes layout. The volume is divided into smaller sub-volumes which are assigned to MPI processes and GPU devices.

III. DESIGN OVERVIEW

In this section we present an overview of the design and implementation of the Mangoose++ application for large high-resolution studies. The main goal is to improve the flexibility of tuning the performance and scalability of the application, while achieving a high degree of resource utilization.

The Mangoose++ architecture consists of seven phases: read I/O (R), scatter (S), filtering (F), texture upload (T), backprojection (B), gather (G), and write I/O (W). These phases can be partially overlapped (functional parallelism) and there is a lot of available concurrency in each phase (as shown in Figure 2). The main challenge is how to efficiently deploy this application on an architecture, while efficiently exploiting the high degree of available parallelism and the access to the memory hierarchy. In the next subsection we discuss the parallelization strategy employed.

In Figure 1 we represent the proposed data layout. The resulting volume is divided into sub-volumes, given the *chunk size* parameter. These sub-volumes are assigned, first to a MPI node (coarse-grain) and second to specific GPU device (fain-grain). This solution allows exploit all level of parallelism.

A. Implementation Details

The implementation leverages both application-level pipeline parallelism (i.e. the functional parallelism between phases of the application) and the hardware-level parallelism (i.e. CPU multi-core parallelism, multiple GPU parallelism, inner GPU multi-core parallelism). The read I/O phase brings 2D projections from a file into the main memory. The scattering phase (S) transfers data from main memory to the memory devices. Our implementation provides the option of using pinned memory in S phase. The filtering phase (F) iteratively processes a number of 2D projections. The computation is decomposed into groups of projections, which are processed in a loop, whose iterations are scheduled by OpenMP. Each iteration consists of four pipelined steps, which are implemented as CUDA kernels: 1) direct FFT based on CUFFT (the Fast Fourier Transform library provided by NVidia) 2) multiplication 3) inverse FFT based on CUFFT 4) weighting. All the devices execute the filtering kernel concurrently over all the projections set. As we demonstrate in the Section IV, this solution is faster than the memory transfer need by each device.

In the upload texture phase (T), the previous processed projections are uploaded into the device texture memory. This phase needs an additional memory transfer in order to store temporally the filtered projections into the host memory, mainly motivated by the memory limitation of the GPU devices. A reduced number of projections are copied in the texture memory (slabs), depending of the projection size and the total capacity of the device. In case of low resolution studies, this phase could be replaced by a direct memory copy inside the F phase.

The result produced by F phase in used as input to the backprojection (B) phase. The B phase input is passed through the GPU memory, i.e. it does not involve any transfer between host and devices. In the same fashion as the F phase, the B phase is decomposed into groups of projections (slabs), which are processed in a loop, taking in advantage of the device shared memory. Each iteration consists of two pipelined steps: the interpolation and the proper backprojection. The interpolation is based on a CUDA kernel, which leverages the GPU texturing hardware for improving the performance. The

2D projections are interpolated fast by fetching them through CUDA arrays bound to texture units. The proper backprojection is a CUDA kernel, which aggregates the interpolated 2D projection into a partial volume calculation.

The aggregation phase (A) transfers sub-volume calculations from memory of the devices to the host memory. This phase is optimized by write-combining page-locked memory, which frees up the CPU's L1 and L2 cache resources, making more cache available to the rest of the application [10]. Finally, the write phase (W) stores the resulting values to a file.

The reconstruction process iterate the phases G, T, B, and W until all the sub-volumes are generated and stored. The Figure 2 aims to show this iterative process. As the figure depicted, most of the phases can be executed concurrently, overlapping memory (G and T), CPU (B), and I/O (W).

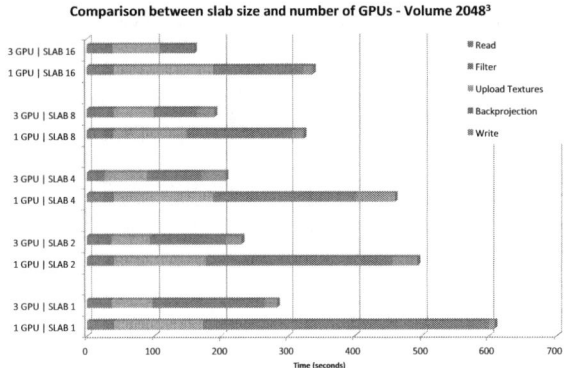

Fig. 3 Breakdown of application time for reconstructing 3D volumes with 2048^3 resolution from 2D projections of size 2048^2. We provide results from 1 up to 16 slabs for executions with one and three GPUs.

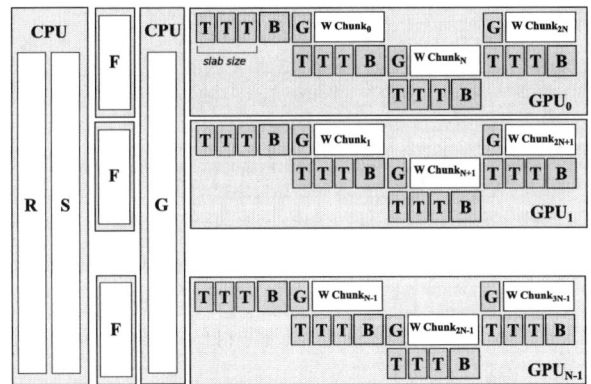

Fig. 2 Mangoose++ workflow. The main thread read the projection files (R). All the GPUs are in charge of filtering projections (F). After all slabs are uploaded into the GPU (T), the backprojection kernel is called (B).

IV. EXPERIMENTAL RESULTS

Data were acquired with the CT subsystem of an ARGUS PET/CT [11], based on cone-beam geometry and circular trajectory. In order to provide standard timings, we have used four standard resolution studies (pixel size of 0.05 mm), with 360 each one, of projection sizes 512^2, 1024^2, 2048^2, and 4096^2 pixels respectively. The distributed system is composed by two computers equipped with two Intel Xeon E5640 (2.67 GHz quad-core processors) and 64 GBytes of RAM, interconnected with a 10Gbit Ethernet network. The motherboard supports up to four PCIe 2.0 x16 buses. We have evaluated two GPU models (GTX 470 and GRX 680). The source code was compiled with Intel 10 compiler with all optimization flags activated. In order to provide realistic results, all caches are flushed before run the experiments.

For the MPI-based experiments the storage solution is the PVFS2 [12] parallel file system version 2.8, using four I/O servers. The files were striped over all I/O servers round-robin with a block of 4 MBytes. The MPI distribution was MPICH 3.0.1 [13]. We ran all tests with one process per compute node.

A. Multiple GPU

This section demonstrate the effect of GPUs shared memory for large standard resolution studies. In Figure 3 we plot the total execution time breakdown for a 2048^3 pixels reconstruction, varing the number of GPUs and the slab size. As we show, the slab size is a key factor to speed up the backprojection phase. The slab size is limited by the amount of available memory of the devices. In this experiment, the slab size is defined statically, resulting a limitation factor for non-homogeneous systems. In the future we plan to design adaptive methods that adjusts this parameter in terms of the total available GPU memory.

B. MPI-Based for High-Definition Studies

In this subsection we present a scalability evaluation of our solution. In Figure 4 we present result of our hybrid MPI and CUDA solution. We scale the volume size from 512^3 up to 4096^3 pixels. In all the experiments the chunk size is configured in terms of the maximum memory capacity of the GPU.

As we plot the the figure, our solution scale as we increase the problem size, especially for high resolution studies of 4096^3 pixels. The most limiting factor for large studies is the texture upload phase. This is because of the limited memory size of current GPU devices. However the texture memory utilization speeds up the backprojection phase due to the

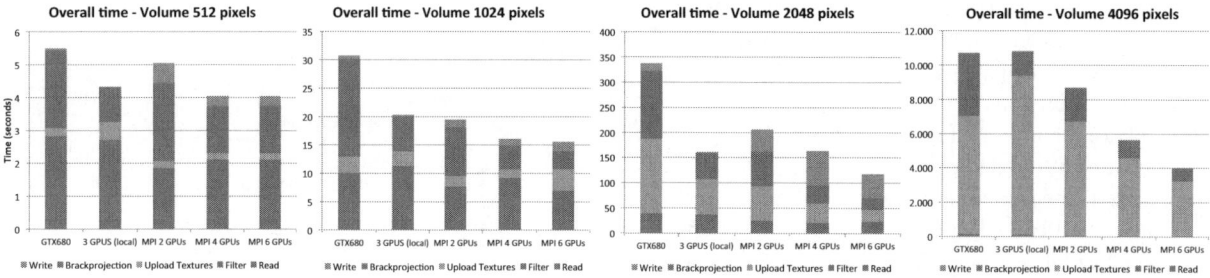

Fig. 4 Breakdown of application time for reconstructing 3D volumes from 512^3 to 4096^3 pixels. We provide the results for a single node with 1 and 3 GPUs and a MPI+CUDA based solution with 2, 4, and 6 aggregate GPUs respectively.

high cache hit ratio. Additionally, we demonstrate that our solution is able to overlap computation and the write stage in all the cases. This contribution is especially important in case of large high-resolution volumes, in which the W phase writes up to 256 GB of data in the bigger case.

V. CONCLUSIONS

In this work, we have presented an efficient modular implementation of a FDK-based algorithm for cone-beam CT. The main disadvantage of having different modules for filtering and backprojection is the inability to exploit synergies between the different stages, leading to a significant increase in data transfer between CPU and GPU. However, the modularity approach enables an efficient replacement algorithms implementation, facilitating the adaptability of the proposed solution to new architectures and incoming devices. We have demonstrated that the amount of memory in the GPU devices is a key factor for future high-resolution studies.

ACKNOWLEDGEMENTS

This work was partially funded by the Spanish Ministry of Science and Technology under the grant TIN2010-16497, the AMIT project (CEN-20101014) from the CDTI-CENIT program, RECAVA-RETIC Network (RD07/0014/2009), projects TEC2010-21619-C04-01, TEC2011-28972-C02-01, and PI11/00616 from the Spanish Ministerio de Ciencia e Innovacion, ARTEMIS program (S2009/DPI-1802), from the Comunidad de Madrid.

REFERENCES

1. Feldkamp L. A., Davis L. C., Kress J. W.. Practical cone-beam algorithm J. Opt. Soc. Am. A. 1984;1:612619.
2. Zhao Xing, Hu Jing-Jing, Zhang Peng. GPU-based 3D cone-beam CT image reconstruction for large data volume Journal of Biomedical Imaging. 2009;2009:8:18:8.
3. Xu Fang, Mueller Klaus. Real-time 3D computed tomographic reconstruction using commodity graphics hardware Physics in Medicine and Biology. 2007;52:3405.
4. Lee Daren, Dinov Ivo, Dong Bin, Gutman Boris, Yanovsky Igor, Toga Arthur W.. CUDA optimization strategies for compute- and memory-bound neuroimaging algorithms Computer Methods and Programs in Biomedicine. 2012;106:175 - 187.
5. Mukherjeet S., Moore N., Brock J., Leeser M.. CUDA and OpenCL implementations of 3D CT reconstruction for biomedical imaging in IEEE Conference on High Performance Extreme Computing (HPEC), 2012:1-6 2012.
6. Zhu Yining, Zhao Yunsong, Zhao Xing. A multi-thread scheduling method for 3D CT image reconstruction using multi-GPU Journal of X-Ray Science and Technology. 2012;20:187197.
7. Scherl Holger, Kowarschik Markus, Hofmann Hannes G., Keck Benjamin, Hornegger Joachim. Evaluation of state-of-the-art hardware architectures for fast cone-beam CT reconstruction Parallel Computing. 2012;38:111 - 124.
8. Zhang Hanming, Yan Bin, Lu Lizhong, Li Lei, Liu Yongjun. High performance parallel backprojection on multi-GPU in 9th International Conference on Fuzzy Systems and Knowledge Discovery (FSKD), 2012:2693-2696 2012.
9. Papenhausen E., Zheng Z., Mueller K.. GPU-accelerated backprojection revisited: Squeezing performance by careful tuning in Workshop on High Performance Image Reconstruction (HPIR):1922 2011.
10. NVIDIA Corporation . NVIDIA CUDA Compute Unified Device Architecture Programming Guide. NVIDIA Corporation 2007.
11. Vaquero J.J., Redondo S., Lage E., et al. Assessment of a New High-Performance Small-Animal X-Ray Tomograph IEEE Transactions on Nuclear Science. 2008;55:898 -905.
12. Ligon W.B., Ross R.B.. An Overview of the Parallel Virtual File System in Proceedings of the Extreme Linux Workshop 1999.
13. Message Passing Interface ForumMPI2: Extensions to the Message Passing Interface 1997.

Stacked Models for Efficient Annotation of Brain Tissues in MR Volumes

Fabio Aiolli[1], Michele Donini[1], Enea Poletti[2], and Enrico Grisan[2]

[1] Department of Mathematics, University of Padova, Padova, Italy
[2] Department of Information Engineering, University of Padova, Padova, Italy

Abstract—Magnetic resonance imaging (MRI) allows the acquisition of high-resolution images of the brain. The diagnosis of various brain illnesses is supported by the distinguished analysis of the different kind of brain tissues, which imply their segmentation and classification. Brain MRI is organized in volumes composed by millions of voxels (at least 65.536 per slice, for at least 50 slices), hence the problem of labeling of brain tissue classes in the composition of atlases and ground truth references, which are needed for the training and the validation of machine-learning methods employed for brain segmentation. We propose a stacking classification scheme that does not require any other anatomical information to identify the 3 classes, gray matter (GM), white matter (WM) and Cerebro-Spinal Fluid (CSF). We employed two different MR sequences: fluid attenuated inversion recovery (FLAIR) and double inversion recovery (DIR). The former highlights both gray matter (GM) and white matter (WM), the latter highlights GM alone. Features are extracted using a local multi-scale texture analysis, computed for each pixel of the DIR and FLAIR sequences. The 9 textures considered are average, standard deviation, kurtosis, entropy, contrast, correlation, energy, homogeneity, and skewness, evaluated on a neighborhood of 3x3, 5x5, and 7x7 pixels. A stacked classifier is proposed exploiting the a priori knowledge about DIR and FLAIR features. Results highlight a significative improvement in classification performance with respect to using all the features in a state-of-the-art single classifier.

Keywords—Stacked learning, Magnetic Resonance Imaging, Annotation, Textures, Brain.

I. INTRODUCTION

Magnetic resonance imaging (MRI) allows the acquisition of high-resolution images of the brain. The diagnosis of various brain illnesses, is supported by the distinguished analysis of the white matter (WM), gray matter (GM) and cerebro-spinal fluid (CSF). In this work we present a semi-supervised method to segment WM, GM, and CSF from MRI data that combines DIR and FLAIR scans, without exploiting any anatomical a priori information, and with the specific objective of preserving the lesions belonging to their correct tissue.

There exist widely available and commonly used brain tissue segmentation software, such as the segmentation tool in SPM [1] and FAST in FSL [2], which use both intensity and a priori anatomic information. However, having been designed for general use, they are not necessarily optimized for specific pulse sequences or for application to images from patients with a specific disease. For example, as observed in [3, 4], when used to segment MR images of MS patients, these tools occur in misclassification of MS lesions as GM due to overlapping intensities, which then requires time-consuming manual editing and introduces operator variability into the measurements.

Hence, manual delineation remains the *gold standard* procedure in studies where brain segmentation of MR data sets is required, especially when dealing with specific populations (e.g. [5, 6, 7, 8, 9, 10]). However, it is expert dependent, observer demanding and time consuming, and essentially not transferable. Automated techniques are necessary to overcome these obstacles, especially when large cohorts of data sets are involved [5]. Moreover, given the growing interest in translational studies in neuroscience, the need for building annotated *gold standard* segmentation and atlases on non human data ([11, 12]) has further stressed the demand for automatic techniques or fast annotation methods.

We propose a supervised classification method that exploits the texture information of the brain tissue provided by the two sequences FLAIR and DIR [13] (1). The former is characterized by the suppression of CFS and by the consequent enhancement of both GM and WM, which are however difficult to be distinguished one from the other; the latter has two inversion recoveries that allow suppressing the contribution of both CSF and WM, thus enhancing GM. We thus avoid using T1-w sequence, which, even if characterized by high spatial resolution, proved inadequate for tissue segmentation when brain lesions are present. In addition, the method does not need population-derived location-based priors, registration to template space, or explicit bias field modeling.

II. MATERIAL

Twenty-four slices (256×256), from $z = 20$ to $z = 44$, from both DIR and FLAIR sequences acquired on a patient affected by MS [14], have been taken into account. In each slide, the three classes GM, WM and CSF were manually labeled.

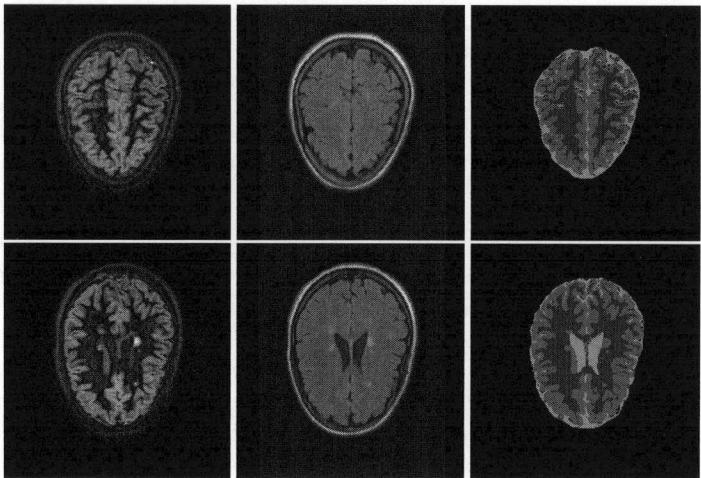

Fig. 1 (Leftmost column) two slices of the DIR sequence; (central column) the corresponding FLAIR slices; (rightmost column) manual ground truth segmentations provided by the experts

III. METHODS

A. Feature Extraction

The rationale of the feature extraction approach is to use the peculiar texture characteristic of a pixel neighborhood in order to obtain information about the pixel tissue class [13]. Image texture analysis has been subject of intense study and has been employed in a variety of applications; however, there is no general agreement upon definition of texture. For our specific application, we assume that a region in an image has a constant texture if a set of local statistics or other local properties of the picture function are almost constant.

Depending on the number of pixels defining the local feature, the statistical methods can be respectively classified as 1st-order, 2nd-order and higher-order statistics. 1st-order statistics measure the likelihood of observing a specific gray value at a random location in the image (hence directly computable from the image histogram). 2nd-order statistics measure the likelihood of observing a specific pair of gray values in a randomly placed dipole of pixels (computable from the gray level co-occurrence matrices (GLCM) [15]). Method proposed in [16] employed three 1st-order statistics: skewness-, median-, and median absolute deviation-based textures, which, on T1-w images, are approximately independent of bias field and of scanner gain. In order to increase the discriminability of the classes, and at the same time couple at best with the double source of information at disposal (i.e., the DIR and the FLAIR sequences), we opted to employ as features four 1st-order statistics and five 2nd-order statistics. The 1st-order statistics considered in this work are mean, standard deviation, skewness, and kurtosis, while the 2nd-order ones are contrast, correlation, homogeneity, entropy and energy.

We extract the local texture information at 3 different scales from blocks of $N \times N$ pixel, with $N = 3, 5,$ and 7. For each 2nd-order texture, 4 GLCMs are constructed, with $d = (d_x, d_y) \in \{(0,1),(1,1),(1,0),(1,-1)\}$. Then, to make the textures invariant to rotation, the obtained matrices are averaged over the 4 angles. Since the feature extraction is performed on both DIR and FLAIR images, the final feature vector associated to each pixel is composed by 56 values (2 original sequence pixel values, plus 9 textures $\times$ 3 scales $\times$ 2 sequences).

B. Classification

State-of-the-art algorithms typically cast the multi-class problem of classifying CSF, GM and WM with a *one against all* technique. The classifier system consists of three binary classifiers. The classifier h_c is trained with all the available labeled examples giving a positive label to examples of class c and a negative label to examples of other classes. All the features for each image (DIR and FLAIR) are used to build these three classifiers. The prediction is made by comparing the scores of the three classifiers and predicting the class whose corresponding classifier maximizes this score.

However, once considering the *a priori* knowledge we have about our specific problem, thinking in this way can be counter intuitive. In fact, using the approach described above, we are not exploiting the fact that different types of features (DIR and FLAIR) contain diverse information and each one is naturally tailored to a more specific task. FLAIR based features highlight both gray matter (GM) and white matter

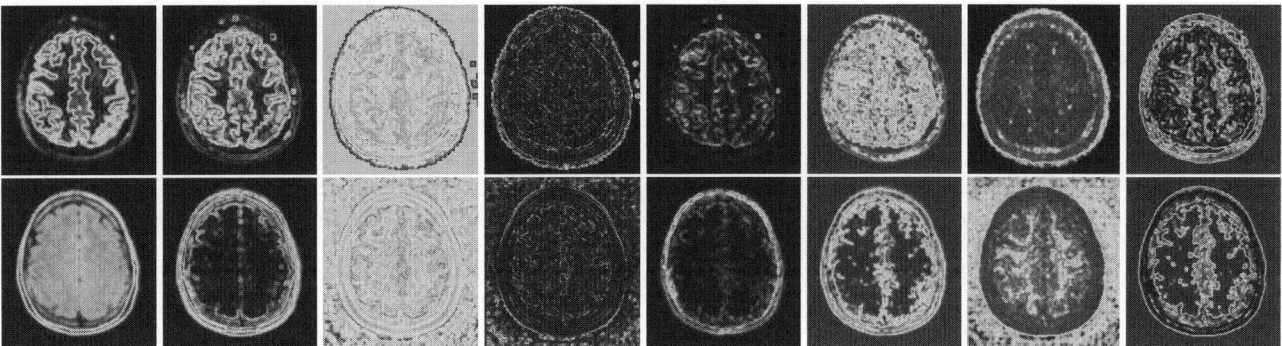

Fig. 2 *Upper row:* 8 textures extracted from the DIR slice. *Lower row:* 8 textures extracted from the FLAIR slice. *Columns, from left to right:* mean, standard deviation, skewness, kurtosis, contrast, homogeneity, entropy, energy. The scale employed is 5 (5×5 pixel block analysis). Original DIR and FLAIR slices are the one in Fig. 1; the slice is at $z = 33$.

(WM), then they can be more useful to discriminate CSF and not-CSF (GM or WM). On the other side, DIR based features highlight GM alone and thus can be more useful to discriminate between GM and WM when we already know that a particular voxel is not of class CSF.

Here, we consider a *two-step stacked* system that creates, in a first step, a binary classifier h_{CSF} using the FLAIR image features only to select the subset of voxel corresponding to CSF. In a second step, it creates a binary classifier h_{WM} that selects the WM from the not-CSF part of voxel. The remaining part of voxels are classified as GM.

C. Experimental Setting

We compared three classification settings using a dataset consisting of ten manually labeled slices. Specifically,

1. One-Against-All (OAA, [17]), where each binary classifier is trained with all features (DIR and FLAIR)
2. Stacked All Features (SAF, [18]), where stacking is performed as described above and all the features are used in both levels
3. Stacked Disjoint Features (SDF), where stacking is performed as described above. FLAIR features used in the first level only and DIR features used in the second level.

An *SVM-like* classifier [19] with *RBF* kernel ($\gamma = 0.01$) has been used for all different settings. On each experiment:

1. We randomly select few (5) labelled examples for each class (CSF, GM and WM).
2. We train all SVM binary classifiers using these labeled data as training data.
3. We classify all the unlabeled data.

We have repeated the steps above for 1000 times to increase the significance of our experiments and we calculated the average accuracy and standard deviation.

IV. RESULTS

Results for all the three settings are summarized in the following table.

Algorithm	Accuracy	StdDev
One against all	68.821%	0.05243
Two-step (all the features)	68.870%	0.04961
Two-step (*a priori* knowledge)	70.403%	0.05742

We can see that our two-step stacking algorithm has a significantly better accuracy than state-of-the-art methods. We have also demonstrated that a significant improvement can be obtained by using *a priori* knowledge on the task at hand. Moreover, the baseline (OAA) method requires the training of three binary classifiers and all the features, while our two-step algorithm needs only two binary classifiers, each one working with a halved number features. The proposed approach provided both better results and better computational performance.

In order to support further our proposal on how to deal with this kind of *a priori* knowledge, another experiment was performed reversing the order of the two classifiers in our two-step algorithm. As expected, we obtained a significant decrease in performance in this case.

V. DISCUSSION AND CONCLUSION

We have shown an effective way to inject a priori knowledge about the different nature of MR sequences in a stacking

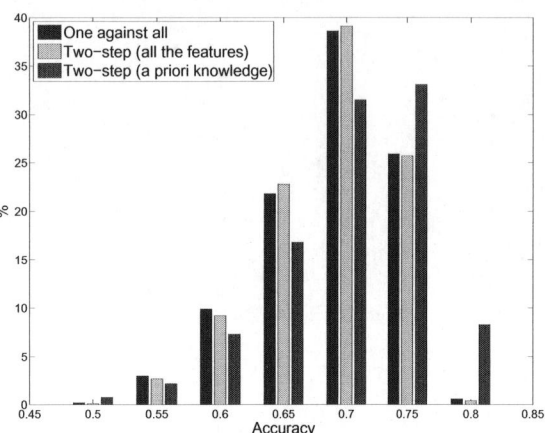

Fig. 3 Histogram of accuracy

model for brain segmentation. In the future, we plan to improve our two-step algorithm in two principal ways. Firstly, by exploiting the existing topology among voxels given by their physical closeness. For this task we can create a graph representation of the brain containing the topological information. Preliminary experiments have shown an improvement in the results even using horizontal topological information only. Secondly, we also plan to study *active learning* algorithms to guide the initial selection of manual labeling. For example, we could study an active learning algorithm that selects the best voxels for manual labeling. The histogram in Fig. 3 shows how the accuracy strongly depends on the initial choice of voxels. Interestingly, in our two-step algorithm, we observe a larger number of cases in which the accuracy is over 80% with respect to the other methods. So we hope to give active learning algorithms able to choose the best voxels eligible for the training set, in an unsupervised manner.

REFERENCES

1. Ashburner J., Friston K.J.. Unified segmentation *NeuroImage*. 2005;26:839-851. cited By (since 1996) 1330.
2. Smith S.M., Jenkinson M., Woolrich M.W., et al. Advances in functional and structural MR image analysis and implementation as FSL *NeuroImage*. 2004;23:S208-S219. cited By (since 1996) 1935.
3. Nakamura K., Fisher E.. Segmentation of brain magnetic resonance images for measurement of gray matter atrophy in multiple sclerosis patients *NeuroImage*. 2009;44:769-776. cited By (since 1996) 22.
4. Gelineau-Morel R., Tomassini V., Jenkinson M.. The effect of hypointense white matter lesions on automated gray matter segmentation in multiple sclerosis *Hum Brain Mapp*. 2011. cited By (since 1996) 4.
5. Gousias Ioannis S, Rueckert Daniel, Heckemann Rolf a, et al. Automatic segmentation of brain MRIs of 2-year-olds into 83 regions of interest. *NeuroImage*. 2008;40:672–84.
6. Shi Feng, Yap Pew-Thian, Fan Yong, Gilmore John H, Lin Weili, Shen Dinggang. Construction of multi-region-multi-reference atlases for neonatal brain MRI segmentation. *NeuroImage*. 2010;51:684–93.
7. Tang Yuchun, Hojatkashani Cornelius, Dinov Ivo D, et al. The construction of a Chinese MRI brain atlas: a morphometric comparison study between Chinese and Caucasian cohorts. *NeuroImage*. 2010;51:33–41.
8. Habas Piotr a, Kim Kio, Corbett-Detig James M, et al. A spatiotemporal atlas of MR intensity, tissue probability and shape of the fetal brain with application to segmentation. *NeuroImage*. 2010;53:460–70.
9. Gholipour Ali, Akhondi-Asl Alireza, Estroff Judy a, Warfield Simon K. Multi-atlas multi-shape segmentation of fetal brain MRI for volumetric and morphometric analysis of ventriculomegaly. *NeuroImage*. 2012;60:1819–31.
10. Serag Ahmed, Aljabar Paul, Ball Gareth, et al. Construction of a consistent high-definition spatio-temporal atlas of the developing brain using adaptive kernel regression. *NeuroImage*. 2012;59:2255–65.
11. McLaren Donald G, Kosmatka Kristopher J, Oakes Terrance R, et al. A population-average MRI-based atlas collection of the rhesus macaque. *NeuroImage*. 2009;45:52–9.
12. Calabrese Evan, Badea Alexandra, Watson Charles, Johnson G Allan. A quantitative magnetic resonance histology atlas of postnatal rat brain development with regional estimates of growth and variability. *NeuroImage*. 2013;71C:196–206.
13. Poletti E., Veronese E., Calabrese M., Bertoldo A., Grisan E.. Supervised classification of brain tissues through local multi-scale texture analysis by coupling DIR and FLAIR MR sequences *Proceedings of SPIE*. 2012;8314:83142T.
14. Veronese E., Poletti E., Calabrese M., Bertoldo A., Grisan E.. Unsupervised Segmentation of Brain Tissues using Multiphase Level Sets on Multiple MRI Sequences in *Intelligent Systems and Control/742: Computational Bioscience*ACTA Press 2011.
15. Haralick Robert M.. STATISTICAL AND STRUCTURAL APPROACHES TO TEXTURE. *Proc IEEE*. 1979;67:786-804. cited By (since 1996) 1640.
16. Vovk A., Cox R.W., Stare J., Suput D., Saad Z.S.. Segmentation priors from local image properties: Without using bias field correction, location-based templates, or registration *NeuroImage*. 2010. cited By (since 1996) 1.
17. Allwein Erin L., Schapire Robert E., Singer Yoram. Reducing Multiclass to Binary: A Unifying Approach for Margin Classifiers *Journal of Machine Learning Research*. 2000;1:113-141.
18. Cohen William W.. Stacked sequential learning in *International Joint Conference on Artificial Intelligence*:671-676 2005.
19. Aiolli F., Martino G. Da San, Sperduti. A.. A Kernel Method for the Optimization of the Margin Distribution (Prague, Czech Republic) 2008.

Author: Fabio Aiolli
Institute: Dept of Mathematics, University of Padova
Street: Via Trieste, 63
City: Padova
Country: Italy
Email: aiolli@math.unipd.it

Mumford-Shah Based Unsupervised Segmentation of Brain Tissue on MR Images

A. Cevik[1] and B.M. Eyuboglu[1,2]

[1] Biomedical Engineering Graduate Program, Graduate School of Natural and Applied Sciences
[2] Department of Electrical and Electronics Engineering, Middle East Technical University, 06800, Ankara, Turkey

Abstract—Automated segmentation of different tissues on medical images is a crucial concept for medical image analysis. In this study, unsupervised image segmentation problem is generalized as a Mumford-Shah energy minimization problem, and several solution proposals for the problem are investigated. Ambrosio-Tortorelli approximation method is implemented, and the performance of the algorithm on magnetic resonance (MR) images of brain is evaluated. First image used in the experiments is chosen among the ones which contain an edema formation due to a brain tumor, and the second one belongs to a healthy subject on which gray matter/white matter segmentation is aimed. Acquired results are presented in visual, tabular and numerical forms. Results and performance are discussed and quantitatively evaluated.

Keywords—Medical image processing, unsupervised image segmentation, brain tumor segmentation, gray matter/white matter segmentation.

I. INTRODUCTION

Medical image analysis is a critical task in clinical radiology, since results gained by the analysis lead physicians for diagnosis, treatment planning, and verification of administered treatment.

Medical images require sequential application of several image post-processing techniques -such as restoration, regularization, segmentation and registration- in order to be used for quantification and analysis of intended features. These features may correspond to the properties of specific parts of the segmented image -like normal tissues, edemas, tumors, or lesions- as well as they may correspond to any statistical property over the entire image domain or over parts of it.

There are many image segmentation algorithms in literature which can be directly used (or adopted into a form that can be used) in medical image domains. However, effective use of many of these methods requires remarkable amount of manual interaction. This situation creates several negations such as difficulty in use and diversity on acquired results. Use of classification based segmentation algorithms requires prior knowledge about class labels, in our case distinct regions of the image. On the other hand, purely clustering based methods suffer from high dependency to the image properties such as noise and/or texture characteristics. Therefore, use of segmentation algorithms falling directly into one of these classes generally involves input dependency, which leads to high standard deviation in results gathered with different parameter sets or images.

In scope of this study, Mumford-Shah based definition of segmentation problem [1] is introduced, a previously established solution to the emphasized problem, approximation of Ambrosio and Tortorelli [2] is implemented and applied on selected brain MR images for unsupervised segmentation of abnormal and different tissues, and the experimental results are analyzed.

Proposed method has previously been used for PET reconstruction, blood cell segmentation [4], denoising and segmentation of brain MR images for white matter-gray matter separation [5], and vascular segmentation and skeletonization [6].

II. BACKGROUND

A. Mumford-Shah Functional

Mumford and Shah described the segmentation problem as minimization of the energy functional:

$$E = \beta \iint_R (u-z)^2 + \alpha \iint_{R-B} |\nabla u|^2 + \iint_R l(B) \quad (1)$$

Equation (1) defines an energy functional which is formed by summation of three terms. Here, z and u are the functions which are representing the original image and the piecewise smooth (segmented) image, respectively. R denotes the complete image domain. Hence, $\iint_R (u-z)^2$ multiplied by a weighting factor of β (a positive real constant), constitutes a measure for dissimilarity of the segmented image to the original one. Therefore, the first term on the right hand side of (1) is usually called as "*data fidelity term*".

B stands for set of points which compose the boundaries of the segmented image. As a consequence, $R - B$ domain, over which $|\nabla u|^2$ is integrated, is the set of non-boundary points in the segmented image. $|\nabla u|^2$ is an inverse measure for smoothness inside the partitions separated with image boundaries. Ideally (for a purely cartoonized image), second term on the right hand side of (1) is equal to zero. Multiplied by a positive real weighting factor α, the term is ordinarily named as "*regularization term*".

Third building block of Mumford-Shah energy functional gives the total length of the boundaries over the segmented image. Boundaries are the regions which create highest gradient over the piecewise smooth image domain. Observing from linear diffusion perspective, value of the third term approaches to zero with time (in more realistic terms, with iterations), and finally vanishes. This banishes the image from its original state. Therefore, minimizing the third term contradicts with minimization of the data fidelity term. Hence, minimization of the Mumford-Shah energy functional as a whole, transfers the image into an equilibrium state between restoration and regularization processes.

B. Proposed Solutions

Several procedures are proposed for minimization of Mumford-Shah energy functional in references [7,8] and [9]. Referred studies involve utilization of simulated annealing, graph cut algorithms, level set (spline) methods, convex relaxation approaches, and finite-difference discretization for image segmentation based on minimization of Mumford-Shah functional.

Each of these algorithms work well in practice although they have various important drawbacks, such as converging to local minimums, not allowing open boundary formation, and excessiveness of the number of iterations to reach a convergence criterion.

C. Ambrosio-Tortorelli Approximation

In [2], Ambrosio and Tortorelli proposed an approximation for the Mumford-Shah energy functional (Equation (1)), which allows formation of open boundaries, in this sense, is more appropriate with the nature of the original energy, respectively. They proposed to replace the boundary-set term by means of defining a 2D function v and designed the phase field energy functional given below:

$$L_{v,\rho} = \iint_R \{\rho|\nabla v|^2 + (1-v)^2/4\rho\}d\boldsymbol{x}. \quad (2)$$

In (2), ρ denotes a small positive real number and $\boldsymbol{x}$ is the coordinate vector over 2D image domain.

Although function v cannot be expressed mathematically in an explicit manner, it can be defined and perceived as given by (3):

$$\lim_{\rho \to 0} \frac{1}{2}\iint\{\rho|\nabla v|^2 + v^2/\rho\} = length(B). \quad (3)$$

If the edge term $\iint_R length(B)$ in original Mumford-Shah energy functional is replaced with the phase field energy term $L_{v,\rho}$ given in (2), and the resulting equation is reorganized such that all terms of integration fall onto the same domain, we end up with the Ambrosio-Tortorelli approximation of Mumford-Shah energy functional, E_{AT} given in (4):

$$E_{AT} = \iint \begin{bmatrix} \beta(u-z)^2 + \alpha|\nabla u|^2(1-v)^2 + \\ (1/2)\{\rho|\nabla v|^2 + (v^2/\rho)\} \end{bmatrix} d\boldsymbol{x}. \quad (4)$$

Equation (4) has discrete numerical solutions for u and v both of which can be factorized in partial differential equation (PDE) form. The reader is referred to [13, 14] for PDE forms and to [10] for detailed derivations of implicit expressions for u and v.

III. EXPERIMENTAL RESULTS & DISCUSSION

Image segmentation experiments are done with two brain MRI images. The first image is a 1.5 Tesla T2-weighted FLAIR MRI image[1] (size=288x288 pixels, TR=9000 ms, TE=100 ms, pixel spacing=0.79861/0.79861 mm, flip angle=90°, and acquisition parameters=256/191) which belongs to a subject with edema related to brain tumor. A former study with brain MR images which involve MS (multiple sclerosis) lesion formation is previously performed and the results are published in [14]. The second image -which is chosen randomly from the MRI image database of LONI[2]- (T1-weighted MRI) belongs to a healthy male subject, and automated separation of gray matter and white matter of the brain is aimed.

A. Edema Related to Brain Tumor

Edema formation related to brain tumor is often associated with neurologic dysfunction and lowered quality of life. Therefore, it is crucial for the physician to be able to follow-up its progression. Cross-sectional area of the region with edema on a 2D image is one of the most discriminative features for the evaluation of progress. However, manual techniques necessitate more labor, as well as they're more prone to errors originating from subjective assessment.

Although the image in Figure 1 (a) contains an easily observable edema, it is not that straightforward to determine the exact position of boundary pixels, because of the smoothness of boundaries and complex shape of the region.

Aim of the experiments is to test the convergence of the algorithm in image domain with meaningful results. Algorithm is applied on selected region of interest (ROI) shown in Fig 1 (b) with the input parameters presented on first three rows of Table 1. Resulting image is given in Figure 1 (c), which is a piecewise-smooth version of the original image. Boundaries of brain and tumor can easily be examined from the boundary-set given in Figure 1 (e) and they are marked by green curve on Figure 1 (d). Figure 1 (f) shows the region inside the boundary surrounding the brain, but outside the tumor region.

---

[1] PHILIPS Medical Systems.PHILIPS DICOM Image Server.[Online].
[2] Laboratory of Neuro Imaging at UCLA.(2009) Segmentation Validation Engine.[Online].

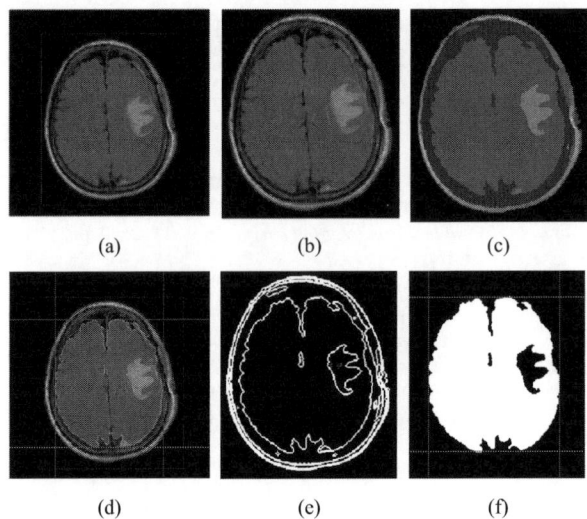

Fig. 1 (a) Original brain MR image with edema (b) Selected ROI (c) Segmented image (d) ROI, rectangular area surrounding brain tissue and region boundaries are labeled (e) Visual representation of v function (f) Binary representation of the selected area.

Table 1 Information regarding segmentation of image with edema

INPUTS	Regularization Factor	100	
	Data Fidelity Factor	10	
	Edge Complexity Term	0.05	
	Iteration Count	20000	
OUTPUTS	SSD (Rate of change)	$1.237e^{-003}$	$9.695e^{-006}$
	Mean Value (ROI)	0.195	0.195
	Standard Deviation (ROI)	0.164	0.156
	Entropy (ROI)	6.090	4.189
	Total MS Energy	$2.263e^{005}$	$6.231e^{004}$

Table 2 Information regarding distance and area computation

METADATA	Pixel X-Size	0.79861 mm
	Pixel Y-Size	0.79861 mm
	Slice Thickness	5 mm
COMPUTATIONS	Max Distance (X-Axis)	40 px
		31.9444 mm
	Max Distance (Y-Axis)	57 px
		45.5208 mm
	Cross-Sectional Area	1154 px
		735.9978 mm^2
	Volume On Slice	3679.9888 mm^3

Lower part of Table 1 shows statistical outputs of the procedure. Since Neumann Boundary Condition [11] is considered in the algorithm design stage, whole process appears to be mean preserving. Stopping condition is defined as Sum of Squared Differences (SSD) being smaller than e^{-5}. Decrease in standard deviation and entropy values are reasonable since the aim of the process is lowering the complexity by eliminating noise and texture components over the image. Total Mumford-Shah energy is also reduced significantly.

Two images given in Figure 2 are produced by using same set of input parameters, but just altering the selected region at the last step. This time, region representing the edema is selected and extracted. Computations of horizontal and vertical maximum distances along the abnormal region, cross-sectional area, and volume on slice are handled automatically using DICOM metadata parameters. Results are given on Table 2 in both pixel units and real metrics.

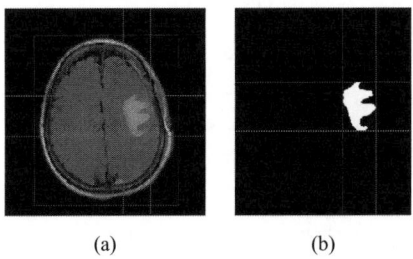

Fig. 2 (a) ROI and boundaries are labeled for edema region (b) Binary representation of the edema region.

B. Gray Matter/White Matter Segmentation

For the purpose of performance evaluation, an original brain MRI image (Figure 3 (a)) and its ground truth segmented version (Figure 3 (c)) are acquired from Segmentation Validation Engine of Laboratory of Neuroimaging at UCLA. The challenge here is to extract the white matter region from the image in selected ROI.

Algorithm is applied for the ROI shown by blue rectangle on Figure 3 (b). Resulting boundary set is labeled on the same figure. Figure 3 (d) shows the white matter of the right hemisphere of the brain in binary form.

In [12], several metrics for validation of segmentation is introduced. 6 of them, namely, Jaccard Similarity, Dice Coefficient, Sensitivity, Specificity, False Negative Rate (FNR), and False Positive Rate (FPR) are selected among those metrics. Values for aforementioned metrics are calculated using the number of true positives, false positives, true negatives, and false negatives by pixel by pixel comparison of ground truth image and the resulting image

Acquired results of performance evaluation step are presented on Table 3. As observed, total ratio of false decisions to all decisions is under 0.08. It is also seen that value of dice coefficient is above 0.9, and values of sensitivity and specificity are higher than acceptable level.

IV. CONCLUSIONS

One of the most crucial steps in the analysis of brain MRI images for tissue abnormalities is the image

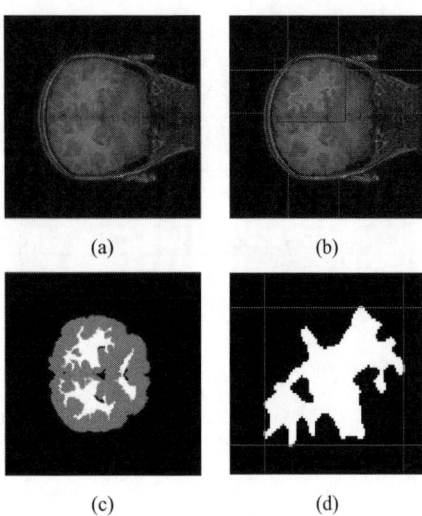

Fig. 3 (a) Original brain MR image of healthy subject (b) Boundary set and the ROI is marked (c) Ground truth segmented image (d) Binary result of white matter segmentation.

Table 3 Performance evaluation measures for gray matter/white matter separation

Jaccard	0.851810	Specificity	0.969430
Dice	0.919970	FNR	0.045817
Sensitivity	0.954182	FPR	0.030574

segmentation step. Efficient use of an unsupervised algorithm increases accuracy and robustness. It also enables non-specialized physicians to acquire standardized results independent from input variance. A variant of Mumford-Shah based segmentation is applied on sample brain MR images in order to question and observe these facts.

It is observed that Ambrosio-Tortorelli approximation to Mumford-Shah based segmentation is appropriate for segmentation of soft tissues, since it results with an accuracy of higher than 92% and sensitivity and specificity of higher than 95%. It is also important for the algorithm to allow open boundary formation, preserve the average gray-level, and provide robustness by constituting a balance between regularization and restoration on the basis of degree of complexity in the image domain.

Search for relations between data fidelity and regularization constants and optimality criteria in selection of those are considered as our following studies in the field.

ACKNOWLEDGMENT

This study is based on Alper Çevik's M.Sc. thesis and B. Murat Eyüboğlu is the thesis supervisor. This research project has been supported by the Graduate School of Natural and Applied Sciences, METU Scientific Research Fund "BAP 07-02-2012-101".

REFERENCES

1. Mumford, D. and Shah, J., "Optimal approximations by piecewise smooth functions and associated variational problems," *Communications on Pure and Applied Mathematics*, vol. 42, no. 5, pp. 577--685, 1989.
2. Tortorelli, V.M. andAmbrosio, L., "Approximation of functional depending on jumps by elliptic functional via t-convergence," *Communications on Pure and Applied Mathematics*, vol. 999-1036, p. 43, 1990.
3. Zhou, J.,Shu, H., Xia, T.,Luo, L., "PET Image Reconstruction Using Mumford-Shah Regularization Coupled with L^1Data Fitting," *Engineering in Medicine and Biology Society, 2005. IEEE-EMBS 2005. 27th Annual International Conference of the*, vol., no., pp.1905-1908, 17-18 Jan. 2006
4. Lin, P., Yan, X.,Zheng, C., Yang, Y., "Medical image segmentation based on Mumford-Shah model," *Communications, Circuits and Systems, 2004. ICCCAS 2004. 2004 International Conference on*, vol.2, no., pp. 942- 945 Vol.2, 27-29 June 2004
5. Du, X., Bui, T.D., "Image segmentation based on the Mumford-Shah model and its variations," *Biomedical Imaging: From Nano to Macro, 2008. ISBI 2008. 5th IEEE International Symposium on*, vol., no., pp.109-112, 14-17 May 2008
6. Lam, B.S.Y., Yan, H., "Blood Vessel Extraction Based on Mumford Shah Model and Skeletonization," *Machine Learning and Cybernetics, 2006 International Conference on*, vol., no., pp.4227-4232, 13-16 Aug. 2006
7. Cremers, D., Tischhauser, F., Weickert, J. and Schnorr, C., "Diffusion snakes: introducing statistical shape knowledge into the Mumford-Shah functional," *J. OF COMPUTER VISION*, vol. 50, 2002.
8. Vese, L.A. and Chan, T.F., "A Multiphase Level Set Framework for Image Segmentation Using the Mumford and Shah Model," *International Journal of Computer Vision*, vol. 50, pp. 271-293, 2002.
9. Chambolle, A., "Finite-differences discretizations of the Mumford-Shah functional," *M2AN*, vol. 33, no. 2, pp. 261-288, 1999.
10. Erdem, E., Sancar-Yilmaz, A. and Tari, S., "Mumford-Shah regularizer with spatial coherence," in *Proceedings of the 1st international conference on Scale space and variational methods in computer vision*, Berlin, Heidelberg, pp. 545-555, 2007.
11. Cheng, A. and Cheng D. T.,"Heritage and early history of the boundary element method", *Engineering Analysis with Boundary Elements*, 29, 268–302. 2005.
12. Shattuck, D. W., Prasad G., Mirza M., Narr K. L., and Toga A. W., "Online resource for validation of brain segmentation methods," *NeuroImage*, vol. 45, no. 2, pp. 431-439, 2009.
13. Cevik, A.,"A Medical Image Processing and Analysis Framework,"Master's thesis, Biomedical Engineering Graduate Program,Middle East Technical University, Ankara, Turkey, 2011.
14. Cevik, A., Eyuboglu B. M., "Doku Anomalisi İçeren Beyin MR İmgeleri Üzerinde Mumford-Shah Tabanlı Bölütleme," EMO Bilimsel Dergi, vol. 1, no. 2, pp. 103-107, 2011.

Author: Alper Çevik
Institute: Middle East Technical University
Street/City/Country: METU/Ankara/Turkey
Email: alper.cevik@metu.edu.tr

Validation of a Computer Aided Segmentation System for Retinography

M. Baroni[1], P. Fortunato[2], L. Pollazzi[2], and A. La Torre[2]

[1] Dept of Information Engineering, University of Florence, Italy
[2] Dept of Oto-Neuro-Ophthalmological Surgical Sciences, University of Florence, Italy

Abstract—In spite of the huge literature on angiography, some problems are still open to discussion, such as segmentation of entire vascular networks. In the present work a new computer approach is developed in two stages, with the aim of improving the analysis and comparison of retina vessel images in the follow up of patients. The first stage adopts multiscale filtering to detect objects of different sizes: a two scale Laplacian of Gaussian scheme is used with the related sigma values chosen according to the smallest and greatest vessel widths. An approximate segmentation is achieved simply by means of the Laplacian sign. The interpretation stage is application-specific and accomplishes classification and quantitative analysis. The skeleton of the binary structures is subdivided in vessel segments, their features (intensity, position, length and width) are fed into an artificial neural network (ANN), after back-propagation training. The segments classified as vessels are assembled into the retinal vascular tree by rule-based tracking, starting from optic disc (OD). Results are evaluated on STARE and DRIVE data bases. Accuracy is 95% and the false positive rate is decreased to about 1%, lower than literature values.

Keywords—angiography, retina, vessel segmentation, artificial neural network, tracking.

I. INTRODUCTION

Many pathologies may affect blood vessels in all the body districts, which can be investigated by a wide range of angiographic modalities. Accordingly, many computer methods [1] have been developed to support medical research and routine examinations and to improve visual inspection of angiograms. The first step of these methods is segmentation: vessels are detected from a structured and noisy background. Then, vessel features are measured, such as width, length, branching angles and tortuosity [2]. Unfortunately, computer segmentation is not always fully automatic nor it warrant optimal performances. Accuracy depends on image quality: acquisition and pathological artifacts can not be always distinguished by blood vessels, so the analysis results are biased by the false positive rate.

Our specific interest regards retinal vessels that can be observed directly and non-invasively with color fundus photos, i.e. their green channel that has the best contrast. A computer vision approach detects candidate vessels with multiscale filtering, splits them into segments, that are then classified by a trained ANN and tracked from optic nerve heads (OD), so as to reduce false positives. The proposed approach is validated with STARE [3] and DRIVE [4] retinal databases, as they also provide vessel segmentation by two observers.

II. METHODS

A. Image Pre-processing and Segmentation

Among previous vessel segmentation methods, here we refer to matched filters [3], classification of Gaussian [4] or Wavelet feautres [5] or to their adaptive thresholding [6].

In this work, retinal images have been enhanced by Laplacian of Gaussian (LoG) filtering, implemented in its fast, separable form, as mono-dimensional convolutions (indicated with *, instead of 2D convolution ⊗), according to Marr & Hildreth:

$$LoG(x,y) = \frac{\partial^2}{\partial x^2}[I(x,y) \otimes G(x,y)] + \frac{\partial^2}{\partial y^2}[I(x,y) \otimes G(x,y)] =$$
$$= G(x) * [I(x,y) * \frac{\partial^2}{\partial y^2} G(y)] + G(y) * [I(x,y) * \frac{\partial^2}{\partial x^2} G(x)]$$

where I(x,y) is the image to be processed and G(x) or G(y) are 1D Gaussian functions, defined as follows:

$$G(x) = \frac{1}{\sqrt{2\pi\sigma^2}} \exp(\frac{-x^2}{2\sigma^2})$$

Gaussian filters can be tuned by varying the sigma parameter and they respond to vessels as well as to spurious objects, so that the extracted features should be examined by a subsequent classifier. On the other hand, vessel width varies from tens of pixels to sub-pixel level. It is well known how such filters preserve the objects of size tuned to their scale, whereas other objects are smoothed and smeared, till to cancel the finest ones.

Therefore, multiscale filtering has been adopted to simulate the visual perception of human observers: edges are detected at a coarse scale and then are localized at the finest scale. The greater and smaller sigma values have been chosen according to the expected maximal and minimal vessel width, respectively. The outputs of LoG filters with these scale values are very similar to the response of a Gaussian matched filter, because they extract the whole vessel width. Using smaller sigma values, or applying these values to larger vessels, results in extracting the edges of vessels rather than whole vessels. Two dimensional Laplacian changes sign across boundaries and therefore segmentation is achieved by a unique threshold: pixels with negative Laplacian are marked as vessels (white), and the other pixels as background (black).

A novel scheme is used to combine the two scale outputs: a bit-wise AND is operated at pixel level, between the thresholded coarse and fine scale Laplacian images. In this way, the small vessels that are detected also at the coarse scale, preserve their original width, whereas the noisy structures detected only at the fine scale are canceled. As a possible drawback of this scheme, larger vessels may exhibit spurious holes, where their width is over the filter scale. To avoid this and other artifacts in the subsequent skeletonizing step, simple median filtering is used, as a cleaning procedure on the segmented images, according to the following sequence:

$$I_r = [\,(I_{\sigma1} \otimes m_5) \cap I_{\sigma2}\,] \otimes m_3$$

where I_r is the resulting segmented image, $I_{\sigma1}$ and $I_{\sigma2}$ are the binary fine and coarse scale Laplacian images, which are combined by means of a pixel-by-pixel AND operator; finally, m_5 and m_3 indicate median filters, with 5x5 and 3x3 windows, respectively.

Then, a classical thinning algorithm is used to extract the skeleton of the low level vascular tree. Thinning deletes border points without destroying connectivity and its result is a connected, topological preserving central axis of the vascular tree. Three types of significant points are detected in the skeleton: terminal points, bifurcation points and crossing points. Based on these points, skeletons are partitioned into single segments. Two neighboring bifurcation points are merged into one crossing points [2].

B. Vessel Classification and Tracking

The candidate vessels, first extracted by the LoG sign, must then be classified. From experts' descriptions of vascular trees, the most relevant features for classification can be derived. Specifically, each skeleton vessel segment is tracked from an outermost point to the other one, by using a chain code, and the following measurements are made: length, width, inner and outer intensity, position with respect to the optic disc. This analysis module gives a length value, L (along the skeleton segment), and mean and s.d. values of width, Wm and Wsd (estimated at every three skeleton points as the shortest path across the binary vessel).

Moreover, an average value of inner intensity, I, is computed through region growing on the green image within each single binary region. To try a distinction between true vessels and false positives, also an outer intensity index, O, is evaluated as follows. The average gray level, O1 and O2, and their s.d. are computed for both the regions besides each candidate binary region, with equal length and half width. If $\delta=|O1-O2|>$ s.d., the outer intensity index, O, is set to $-\delta$, indicating a possible false positive (e.g. due to OD border, between the bright disc and the dark retinal background); otherwise O is set to (O1+O2)/2. This value indicates a possible true vessel or an artifact between two pathological structures (e.g. exudates) which can be discriminated when the latter are much brighter than normal retinal background. Finally, by using the two outermost points of each segment, two distances from OD, d1 and d2, are computed for each candidate vessel segment. To this aim, the brightest and greatest blob has been detected by smoothing the green images and by matching a 2D Gaussian to yield approximate locations of OD.

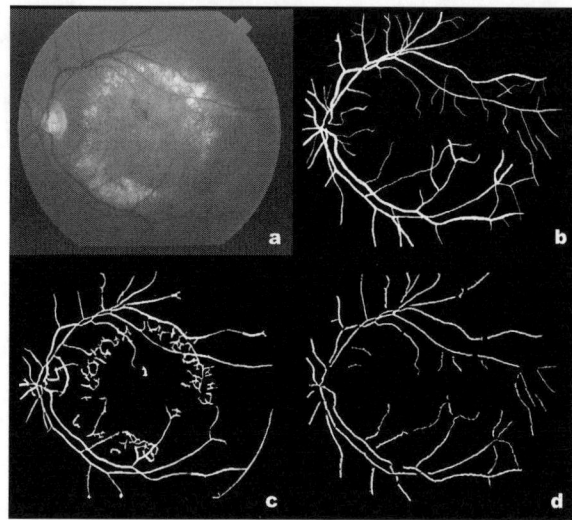

Fig. 1 a) Green channel of an abnormal fundus image of STARE database. b) manual segmentation. c) segmentation by the Hoover's method [3]. d) segmentation by the proposed method (LoG-ANN).

For each candidate vessel we have seven features that are fed into an artificial neural network (ANN) to make the classification. Among the various ANN classifiers, a three layers feed-forward architecture has been chosen and trained through the back-propagation algorithm. The input layer has seven neurons, according to the aforementioned feature vectors, and the output layer has one neuron. The number n of hidden layer neurons has been empirically determined, in order to maximize the mean square error during training and testing as well as to improve generalization. To the latter aim, the number of the vessel segments in the training set is taken greater than the number of parameters in the network.

The binary segments classified as vessels are then assembled into their vascular tree by a tracking algorithm: they are sorted based on the distance from the optic disc (OD) and tracked by chain code, starting from their terminal point nearer to OD. According to this distance sorting, the vessel segments are labeled as roots (parents) of partial vessel trees. For each parent, its branches (daughters) are then tracked from the current bifurcation point till to another bifurcation point or till to their terminal point. The daughters of every parent vessel, v_k, are numbered as $2v_k$ and $2v_{k+1}$ in order to track the tree in either direction.

Tracking continues iteratively until no other daughters are detected and no other vessel segment has remained unlabeled and identified as a new parent. The final step of tracking examines again the labeled vessels, trying to connect them if their first terminal point (the terminal point nearer to OD) is within a short distance (r=20) from the second terminal point (the farther one from OD) of another labeled segment. This tracking is similar to [2] but it does not need user interaction and the r distance allows to follow interrupted vessels or disconnected branches, which may turn out from image artifacts due to processing or pathology. Finally, the vessel trees are filled from the labeled skeletons. From these vessel data, further geometrical and topological analysis may be accomplished.

III. EXPERIMENTAL EVALUATION AND RESULTS

Ad hoc software was written using a C++ programming environment and a neural network toolkit, developed beforehand at our lab. To devise the optimal settings of the proposed method, several experiments were undertaken. The best performance of Gaussian smoothing for optic disk (OD) matching was obtained with a sigma value of 5, whereas the smaller and greater sigma values for vessel segmentation have been chosen equal to 1 and 3, respectively. A human expert manually labeled the candidate vessel/nonvessel segments; their features were computed, normalized in the range [0,1], and fed into the ANN. The training set includes one half (10) of the STARE images and one half (20) of the DRIVE images. The other halves of both databases are included in the test set. The maximal number of epochs was 3000 and the goal mean square error was 0.005. The best network configuration has a number n = 20 of hidden layer neurons, and it was simulated with the test set for performance evaluation.

The STARE and DRIVE public databases provided a most valuable reference for these activities. Specifically, the proposed method is evaluated by computing sensitivity, specificity and accuracy at pixel level with respect to manually segmented vessels available in both databases. The first observer segmentation is taken as ground truth. Any pixel which is classified as vessel in both the ground truth and segmented image is accumulated among the true positives; any pixel classified as vessel in the segmented image but not in ground truth is counted among the false positives.

Illustrative segmentation results are shown in fig.1.

Table 1 presents the average values of specificity, sensitivity and accuracy computed on the DRIVE test images by different methods.

Table 1 Performance (%) of vessel segmentation (DRIVE database)

method	specificity	sensitivity	accuracy
2nd observer	97.25	77.61	94.73
Staal et al.[4]	97.73	71.94	94.42
Soares et al.[5]	97.88	72.83	94.66
Zhang et al.[6]	97.24	71.20	93.82
LoG-ANN	98.61	65.16.00	94.93

Table 2 Performance (%) of vessel segmentation (STARE database)

method	specificity	sensitivity	accuracy
2nd observer	93.90	89.49	93.54
Hoover et al.[3]	95.67	67.51	92.67
Staal et al.[4]	98.10	69.70	95.16
Soares et al.[5]	97.48	71.65	94.80
Zhang et al.[6]	97.53	71.77	94.84
LoG-ANN	98.72	63.06	94.78

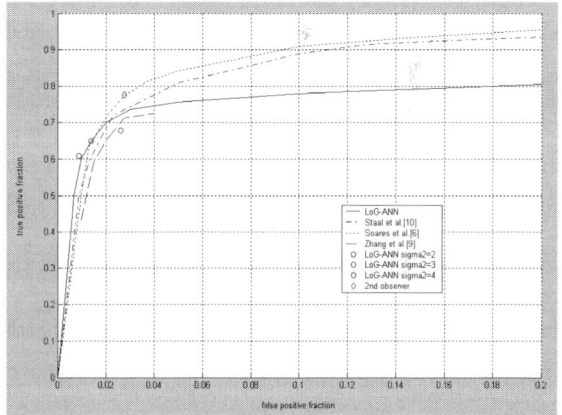

Fig. 2: The ROC curve for DRIVE database when False Positive Fraction is less than 0.2 for LoG-ANN, compared with three other published ROC curves and with manual segmentation. The circle markers refer to the best LoG-ANN results of multiscale filtering with the greater sigma value equal to 2, 3 or 4.

The same results of the appropriate methods applied to STARE database are presented in Table 2. Total accuracy, computed on both databases, was nearly 0.95 (0.9485 ± 0.003), which is comparable with the best accuracy values reported in literature. However, the false positive rate was very close to 1%, so that specificity is better than 98%, whereas the values reported in literature are not better than 96 or 97 %. On the other hand, sensitivity is less than 70% and cannot reach the higher values reported in some papers (over 72% for [5,6], about 77% and 90% for the 2nd observer in DRIVE and STARE images respectively).

For a more complete comparison to previous works, ROC curves have been produced for both DRIVE and STARE databases by varying the sigma parameter of LoG filtering and the thresholds of the ANN results. The aforementioned performance values have been achieved by using the 0.5 value as the decision ANN threshold. At a threshold of 1 the false detection rate is zero, but no vessels are detected, while at a threshold of 0 all the candidate vessels are detected. Thus, this threshold is a natural choice to produce a ROC curve. Figure 2 and 3 show the ROC curves obtained by this and other methods [4-6] for test DRIVE and STARE images. To make the display clearer, results are shown for false positive fractions less than 0.2. Results for different values of the coarse-scale sigma parameter are shown only for 0.5 decision threshold. Varying the fine-scale sigma has not produced good results.

The performance of LoG-ANN method is similar to Zhang et al.[6]; it seems that these enhancement filters cannot detect all the vessels as segmented by the first observer. However, both ROC curves exhibit a portion higher than other methods. Finally, it is worth noting the optimal capability in discriminating between vessels and other spurious structures, as visible in pathological eyes (figure 1).

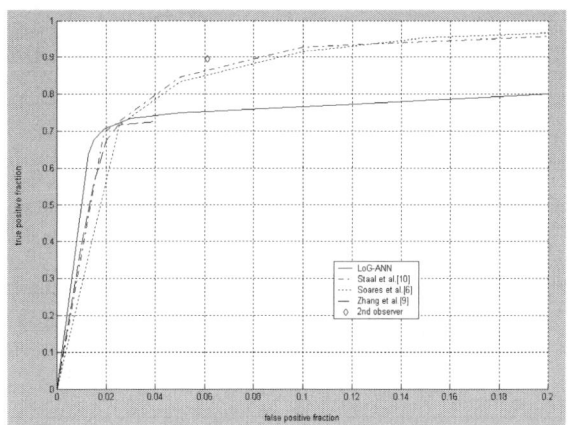

Fig. 3 The ROC curve for STARE database when FP < 0.2 for the proposed method (LoG-ANN) compared with three other published ROC curves and with the second observer's manual segmentation

IV. DISCUSSION

While the first stage of the proposed segmentation method detects candidate vessel segments, the second stage measures specific vessel features, recognizes them and re-assembles the vascular tree. ANN classifier was chosen as it is robust and able to recognize simple vessel features. High sensitivity is not reached: however, a low false positive rate in vessel segmentation is very important in many applications. For example, in retinopathy of prematuriy, where therapy is decided according to normal or abnormal width and tortuosity of retinal vessels, it is essential to make these measurements on the whole length of principal vessels, whereas their thinner branches may give minor information.

In conclusion, though developed for retinal images, this method can be applied to other tree-like vascular images as soon as a starting point is detected.

REFERENCES

1. Fraza M, Remagninoa P, Hoppea A, et al. (2012) Blood vessel segmentation methodologies in retinal images–A survey. Computer Meth.&Progr.Biomed. 108: 407–433.
2. Martinez-Perez ME, Hughes AD, et al. (2002) Retinal Vascular Tree Morphology: A Semi-Automatic Quantification. IEEE Trans Biomed Eng. 49: 912-917.
3. Hoover A, Kouznetsova V, Goldbaum M. (2000) Locating blood vessels in retinal images by piecewise threshold probing of a matched filter response. IEEE Trans Med Imaging. 19:203–210.
4. Staal J, Abramoff MD, Niemeijer M, Viergever MA, van Ginneken B. (2004) Ridge-based vessel segmentation in color images for the retina. IEEE Trans Med Imaging. 23:501–509.
5. Soares JVB, Leandro JG, et al.(2006) Retinal Vessel Segmentation Using the 2-D Gabor Wavelet and Supervised Classification. IEEE Trans Med Imaging. 23:1214-1222.
6. Zhang B, Zhang L, Karray F. (2010) Retinal Vessel extraction by matched filter with first-order derivative of Gaussian. Computers in Biology and Medicine. 40:438–445.

Author: Maurizio Baroni
Institute: Dept of Information Engineering
Street: via Santa Marta 3
City: Firenze, 50139
Country: Italy
Email: maurizio.baroni@unifi.it

Computer-Assisted System for Hypertensive Risk Determination through Fundus Image Processing

S. Morales[1], V. Naranjo[1], F. López-Mir[1], A. Navea[2], and M. Alcañiz[1,3]

[1] Instituto Interuniversitario de Investigación en Bioingeniería y Tecnología Orientada al Ser Humano, Universitat Politècnica de València, I3BH/LabHuman, Camino de Vera s/n, 46022 Valencia, Spain
[2] Fundación Oftalmológica del Mediterráneo, Valencia, Spain
[3] Ciber, Fisiopatología de Obesidad y Nutrición, CB06/03 Instituto de Salud Carlos III, Spain

Abstract—From a fundus image, the system proposed in this paper automatically detects retinal vessels and measures some geometrical properties on them such as diameter and bifurcation angles. Its goal is to establish objective relations between different vessels, thus being able to determine cardiovascular risk or other diseases, as well as to monitor their progression and response to different treatments. The system has been used in a double-blind study where its sensibility, specificity and reproducibility to discriminate between health fundus (without cardiovascular risk) and hypertensive patients has been evaluated in contrast to the expert ophthalmologist opinion obtained through visual inspection of the fundus image. An improvement of almost a 20% has been achieved comparing the system results with the clinical visual classification.

Keywords—**Retinal vessels, Retinal vascular tree, Vessel caliber, Bifurcation angles, Hypertension.**

I. INTRODUCTION

Retinal vasculature is able to indicate the status of other vessels of the human body. Classically, its study is included in the standard screening of any patients with cardiovascular risk and other diseases in which the vessels may be altered inasmuch as it is a non-invasive or minimally invasive procedure.

Nowadays, retinopathies associated with systemic diseases such as diabetes and hypertension are increasingly affecting population. A direct, regular and complete ophthalmologic examination seems to be the best approach for an assessment of the risk population [1]. However, population growth, ageing, physical inactivity and rising levels of obesity are contributing factors to increase this type of diseases, which cause the number of ophthalmologists for the assessment by direct examination of the population risk is a limiting factor [2].

Due to high resolution of digital fundus images, they can be automatically processed providing invaluable help to clinicians in diagnosis and disease prevention. Most attempts to automate the process of interpretation of retinal vascular imaging are focused on a specific disease, diabetic retinopathy, a disease of high incidence and a significant risk of blindness that occupies a very important part of the medical-surgical activity of the ophthalmologic resources. In some of these studies it has been possible to relate the evolution of the disease and the positive or negative response to treatment with retinal vessel caliber. However, there is not too much experience in the use of these methods to evaluate other types of vascular pathology. Vascular changes produced in systemic diseases usually induce particular modifications in the vessels, such as changes in the angle of intersection between arteries and veins, and changes in the vessel calibers. Based on these facts, a system capable of detecting the retinal vessels and measuring some geometrical properties from a fundus image has been developed. The goal of the proposed system is to establish objective relations between the different vessels, to determine cardiovascular risk or other diseases, as well as to monitor progress and response to different treatments. This tool has been applied in a clinical study in order to evaluate sensibility, specificity and reproducibility of the developed system to discriminate between a normal vascularization and cardiovascular pathology in contrast to the opinion of an expert ophthalmologist obtained through visual inspection of the fundus image.

The rest of the paper is organized as follows: in Section II the proposed method is described, including segmentation process, labeling of retinal vascular tree and geometric measures. Section III shows the system validation through a double-blind study. Finally, Section IV provides discussion and Section V conclusions and some future work lines.

II. METHOD

To characterize any retinal morphological changes, a segmentation process is always necessary as first step. Afterwards, detected vessels must be labeled as a means to be able to perform desired measures on them and to quantify these changes. The main stages involved in the presented system are illustrated in Fig. 1, both image acquisition and processing of this image, including the pre and post processing steps. Fig. 2 shows the aspect of the implemented tool.

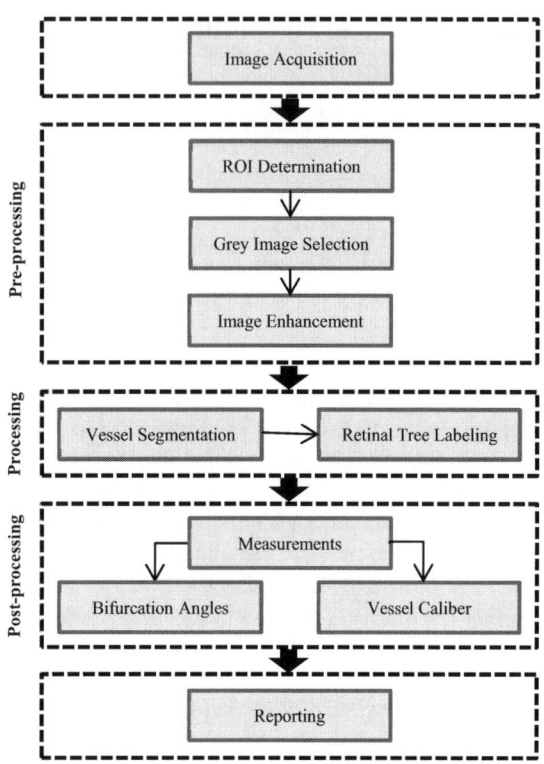

Fig. 1 Block diagram of the presented system

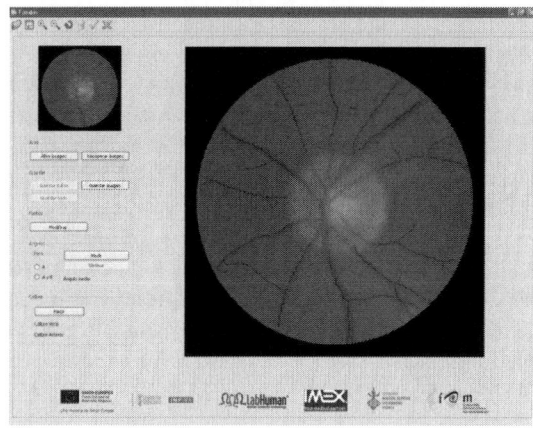

Fig. 2 Aspect of the implemented tool

A. Pre-processing

Based on a standard protocol, most of software, which compute measurements of retinal vessel calibers [3-5], are focused on a specific region of interest (ROI) of the fundus image. This area is concentric to the optic disc and it is related to with its diameter. So, for that reason, the proposed system detects in first place the optic disc automatically [6] in order to be able to determine the ROI where all measures will be performed (Figure 3b). It must be stressed that the mentioned state-of-the-art software [3-5] do not detect automatically retinal vessels and either measure subsequently geometrical properties on them but are trained graders who measure the diameters of all arterioles and venules coursing through the specified area.

Although fundus images are RGB images, the system is drawn on monochrome images for its processing. They are obtained from the green band because this band provides an improved visibility of retinal blood vessels. Afterwards, an image enhancement [7] is applied to improve even more, if it is possible, their visibility.

B. Processing

The segmentation method used by the system is based on mathematical morphology [8], curvature evaluation and k-means clustering [9] for the detection of vascular tree [10]. Once the vessels have been detected, they must be labeled.

Retinal vascular tree labeling is focused on obtaining the skeleton of vascular tree, detecting significant points (terminal, bifurcation and crossing points) and a tracking process [10], necessary steps to perform the desired measures later (Figure 3c).

C. Post-processing

Certain geometric measurements of blood vessels can help to establish whether they have undergone morphological changes over time and facilitate diagnose illness. The next parameters have been chosen due to the fact that they have particular interest for the early hypertension detection:

Bifurcation angle: Angle formed by the daughter branches for each bifurcation point. The branches are fitted for straight lines by least-squares into a circular window centered on these points (Figure 3d).

Vessel caliber: It has been estimated as two times the average of the geodesic distance [8] calculated from the skeleton points of the branch to the edge of the corresponding vessel (Figure 3e).

III. RESULTS

For system validation, a set of 67 fundus images was used. These images belong to a private database of the Fundación Oftalmológica del Mediterráneo (Spain), which contains color images of 2048 x 1536 pixels.

The implemented system was installed in this organization in order to be used by its clinicians in their daily practicing and be able to carry out a clinic validation of the developed software in base on a double-blind study. The study was performed on 67 patients between 33 and 73 years old. Among them, a group of "control" without previous known pathology and another group "study" of

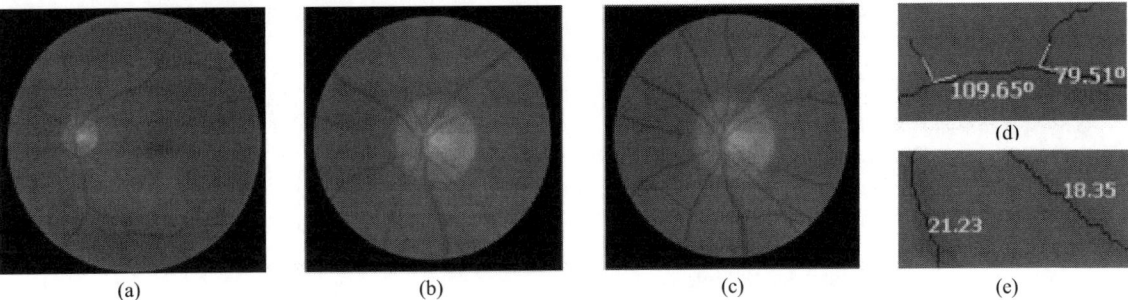

Fig. 3 Segmentation steps: (a) Original fundus image, (b) ROI determination, (c) Vessel segmentation, (d) Bifurcation Angles and (e) Vessel caliber

hypertensive patients treated or not of more than 5 years of evolution, were established. Their fundus images were evaluated twice. Once, by qualified ophthalmologists who determined the presence or absence of vascular alteration through visual inspection of the original image, and another time using the developed system. A patient was considered hypertensive if the clinician detected in their fundus at least one sign of pathological arteriovenous crossing. With the implemented system, bifurcation angles, inside of an existing region of the original image concentric to the optic disc, were measured along with the caliber of a vein and an artery manually selected and situated at the same distance from the optic disc. Subsequently, a statistical analysis on different variables, extracted from data provided by the system, was conducted to see if any of them were able to discriminate whether a patient belonged to the group of hypertensive or without cardiovascular disease. Statistical Package for the Social Sciences (SPSS, IBM SPSS Data Collection) version 17.0 was used to this purpose.

Two parallel statistical studies were performed, one analyzing the variables related to the caliber and other analyzing those related to the bifurcation angles. The reason for the distinction was that only fundus images with at least five measured bifurcation angles were considered valid for this specific study. In both, the comparison between groups "control" and "study" was conducted using one-way ANOVA, where the dependent variables were each of the parameters calculated from the data provided by the system (vein caliber, artery caliber, relative caliber of the vein and artery (A/V ratio), average of the branching angles, median angles, deviation angles, variance angles, minimum angle, maximum angle and difference between maximum and minimum angle) and the factor was the group formed by the "control" (no known disease) and "study" (hypertension) subgroups.

Previously to the analysis, the normality of the calculated parameters was checked using the Kolmogorov-Smirnov test. From the obtained values, it can be concluded that only the caliber of the artery ($F(1;66) = 4:471$; $p < 0:05$) and the ratio between the caliber of the vein and the artery ($F(1;60) = 4:161$; $p < 0:05$) show statistically significant differences between the "control" and "study" subgroups. Afterwards, using only these parameters, the optimal threshold to separate both classes was established, and then, the sensibility and specificity of the system to discriminate between healthy patients and with hypertensive pathology was calculated.

$$\text{Sensibility} = \frac{TP}{TP+FN} = \frac{22}{22+17} = 56.41\% \quad (1)$$

$$\text{Specificity} = \frac{TN}{TN+FP} = \frac{19}{19+9} = 67.86\% \quad (2)$$

where TP;FN;TN;FP are true positives, false negatives, true negatives and false positives, respectively.

IV. DISCUSSION

On the one hand, with the system validation, it has been demonstrated that the caliber of the arteries and the relative caliber of the veins and arteries show significant differences when patients are discriminated between healthy and hypertensive. Despite these results, it cannot be concluded that bifurcation angles are not significant for this purpose because only a 54% of the fundus of the dataset could be analyzed due to the fact that the rest of them contains less than 5 angles per image. So, the region, where the measures are taken, should be enlarged or the database increased to repeat the same analysis.

On the other hand, although the values of sensibility and specificity of the system are not too much high in the hypertension discrimination, it must be stressed that they improve the results achieved by clinicians by visual inspection of the fundus. In particular, the sensibility and the specificity are improved in almost a 20%.

With regard to the obtained measures, they are accurate and reliable but also dependent on a correct skeleton detection and significant point classification.

V. CONCLUSION

From a fundus image, the implemented system automatically detects blood vessels of a specific region of the image. Moreover, it allows to measure bifurcation angles found and to select branches to know their caliber. These data facilitate expert medical diagnosis and study of the progression of a disease. In particular, this system has been used by expert ophthalmologists to help them to discriminate between a normal vascularization and cardiovascular pathology. It has improved in almost a 20% their sensibility and specificity in hypertension detection achieved by direct visual inspection of the fundus.

About future work lines, the system will be applied to analyze the retinal microvascular architecture of children with low birthweight in order to establish the relationship between the measurements obtained in these children and diseases such as hypertension and cardiovascular problems in adult life, using it as a prognostic marker of cardiovascular risk. The base of this study (carried out by the Department of Pediatrics of General Hospital of Valencia (Spain)) is that, bifurcation angles, which are determined at birth, are predictors of future development of hypertension and cardiovascular disease. Moreover, it should be repeated the performed analysis of the angles enlarging the region of measures or increasing the database, as it has been mentioned in the previous Section. Regarding the values of sensibility and specificity of the system, they could be improved if other classifiers were used, such as SVM, among others, or when the database was increased. In addition to that, the study to correlate segmentation parameters and hypertensive pathologies will be widen.

ACKNOWLEDGMENT

This work has been funded by the project IMIDTA/2010/47 and partially by projects Consolider-C (SEJ2006 14301/PSIC),"CIBER of Physiopathology of Obesity and Nutrition, an initiative of ISCIII" and Excellence Research Program PROMETEO (Generalitat Valenciana. Conselleria de Educación, 2008-157). We would like to express our deep gratitude to Imex Clinic S.L., the Department of Paediatrics of General Hospital of Valencia and the Fundación Oftalmológica del Mediterráneo for its participation in the project.

REFERENCES

1. H. M. Herbert, K. Jordan, and D. W. Flanagan (2003), "Is screening with digital imaging using one retinal view adequate?," Eye (Lond), vol. 17, no. 4, pp. 497–500.
2. L. Varma, G. Prakash, and H. K. Tewari (2002), "Diabetic retinopathy: time for action. no complacency, please!," Bulletin of the World Health Organization, vol. 80, pp. 419.
3. T. Y Wong, M. D. Knudtson, R. Klein, B. E. K. Klein, S. M. Meuer, and L. D. Hubbard (2004). Computer assisted measurement of retinal vessel diameters in the beaver dam eye study: methodology, correlation between eyes, and effect of refractive errors. Ophthalmology, 111(6):1183–90.
4. M. D. Knudtson, K. E. Lee, L. D. Hubbard, T. Y. Y. Wong, R. Klein, and B. E. Klein (2003). Revised formulas for summarizing retinal vessel diameters. Current eye research, 27(3):143–149.
5. L. D. Hubbard, R. J. Brothers, W. N. King, L. X. Clegg, R. Klein, L. S. Cooper, A. R. Sharrett, M. D. Davis, and J. Cai (1999). Methods for evaluation of retinal microvascular abnormalities associated with hypertension/sclerosis in the atherosclerosis risk in communities study. Ophthalmology, 106(12):2269–2280.
6. S. Morales, V. Naranjo, J. Angulo, M. Alcañiz (2013). Automatic detection of optic disc based on PCA and mathematical morphology. IEEE Trans Med Imaging, 32(4):786-96.
7. T. Walter and J.C. Klein (2002), "A computational approach to diagnosis of diabetic retinopathy," in 6th Conference on Systemics, Cybernetics and Informatics (SCI).
8. P. Soille (2003). Morphological Image Analysis: Principles and Applications. Springer-Verlag New York, Inc., 2nd edition.
9. J. B. MacQueen (1967). Some methods for classification and analysis of multivariate observations. In Proc. Of the fifth Berkeley Symposium on Mathematical Statistics and Probability, volume 1, pages 281–297.
10. S. Morales, V. Naranjo, J. Angulo, J.J. Fuertes, M. Alcañiz (2012). Segmentation and analysis of retinal vascular tree from fundus image processing. International Conference on Bio-inspired Systems and Signal Processing (BIOSIGNALS), pp. 321 – 324.

Interventional 2D-3D Registration in the Presence of Occlusion

S. Demirci[1], F. Manstad-Hulaas[2], and N. Navab[1]

[1] Computer Aided Medical Procedures (CAMP), Technical University of Munich, Munich, Germany
[2] Radiology Department, St Olavs Hospital, Trondheim, Norway

Abstract— It is crucial that 2D-3D medical image registration algorithms meet special requirements in terms of accuracy and robustness in order to be applied during clinical interventions. Existing algorithms may be affected by image dissimilarities introduced by medical instruments visible only in the interventional 2D image. Based on our previous results, we present a fully automatic framework for disocclusion-based 2D-3D registration technique that detects the occlusion and performs a matching on the reconstructed 2D image. Compared to earlier approaches, the proposed algorithm is fully automatic and therefore highly appropriate for clinical application. Our technique is validated on synthetic and real interventional data and compared with conventional methods. Results prove that disocclusion-based registration yield higher accuracy and robustness and outperforms existing approaches in terms of speed.

Keywords—medical image registration, 2D-3D, disocclusion, stent.

I. INTRODUCTION

One of the key challenges of computer-assisted surgery is accurate and robust 2D-3D registration of medical images. In particular the fusion of preoperative Computed Tomography (CT) volumes and interventional X-ray images was found to be crucial for improving the guidance of medical instruments under realtime X-ray imaging [1, 2]. However, navigation under 2D angiography imaging is still common clinical practice.

In interventional images important image information might be occluded by medical instruments such as catheters, grafts, and probes. Such occlusions severely affect the robustness of existing 2D-3D registration algorithms [3, 4].

Although 2D-3D medical image registration based on intensity information has been the focus of many different research projects throughout the last two decades [2], the problem of dissimilarities and occlusion during interventional settings has been tried to being solved via robust similarity measures, such as Gradient Difference [3] and binary masks [5]. However, the following problems are associated to these. Robust similarity measures identify all outliers as equal and do not make any difference between the unknown image noise from the "known" occlusion. Binary masks, in contrast,

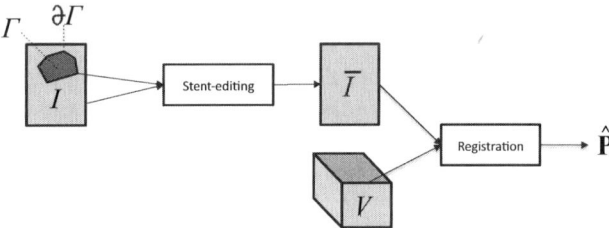

Fig. 1 Automatic Framework

use a model for the occluded region, but require the user to outline this region. In a more recent work [6, 4], we evaluated the benefit of a new disocclusion technique, *Stent-editing*, for improving 2D-3D registration. We obtained promising results, but the approach included heavy manual interaction, for which the method was not suitable for clinical application.

In this paper, we show the potential of disocclusion for interventional 2D-3D registration. We therefore integrate *Stent-editing* into a fully automatic medical image registration framework and compare the results against the conventional registration method using the popular Gradient Difference [3] measure. In section II , we introduce our registration framework and recapitulate the *Stent-editing* technique with adapted automatic initialization. Our experimental results are given in section III. Finally, the paper concludes with section IV.

II. METHODOLOGY

In this section, we detail our fully automatic framework for disocclusion-based 2D-3D registration for medical images. Furthermore, we recapitulate the idea of *Stent-editing* and introduce a novel strategy for automatic initialization.

A. Registration Framework

In this paper, we propose disocclusion-based registration to solve the problem of occlusions. As visualized in Fig. 1, our pipeline first initiates a reconstruction of the occluded image part and then performs a 2D-3D registration of the reconstructed 2D image and the 3D preoperative volume employing a gradient-based similarity measure.

Given a 3D volume V and a 2D image I, the problem of estimating the optimal projective transformation $\hat{P}$ of V such that its projection perfectly aligns with I in terms of a certain similarity measure $\mathfrak{S}$ is described by rigid-body 2D-3D registration:

$$\hat{\mathbf{P}} = \arg\min_{\mathbf{P}} \mathfrak{S}(\mathbf{P} \circ V, I) \quad (1)$$

where $\circ$ denotes the application of projection $\mathbf{P}$ to volume V, in particular the multiplication of $\mathbf{P}$ with every image vector of V. The projection transformation $\mathbf{P} = K[R|t]$ consists of the 6-DOF extrinsic parameters $[R|t]$ for rotation (r_x, r_y, r_z) and translation (t_x, t_y, t_z) of the 3D volume and the 4-DOF intrinsic imaging parameters K of the pinhole projection model [7].

In our practical implementation of the 2D-3D registration procedure, we realized the projection $\mathbf{P} \circ V$ through the concept of *Digitally Reconstructed Radiographs* (DRR) that are produced by casting virtual rays through a 3D CT volume. We use GPU (Graphics Processing Unit) accelerated raycasting [8] with a single render pass for DRR generation and employ a 1D transfer function [9] for the conversion from CT houndsfield units to X-Ray attenuation values.

For our intended medical application, we use the visible bony structure to align an interventional X-ray to the preoperative CTA. We therefore chose to employ Gradient Correlation [10] as similarity measure $\mathfrak{S}$.

By a number of tests, we found the *Downhill Simplex* method to give most suitable results for solving the optimization of Equ. (1).

B. Stent-editing

Let $I : \Omega \longrightarrow \mathbb{R}$ define an image on domain $\Omega \subset \mathbb{R}^2$. In the following, we denote by $\Gamma \subset \Omega$ the occluded image part and by $\partial \Gamma$ its boundary (as depicted in Fig. 1).

Stent-editing intends a reconstruction of the occluded region Γ via a certain guidance field $g : \Gamma \to \mathbb{R}^2$. The reconstructed image part $\bar{I} : \Gamma \to \mathbb{R}$ is then computed such that it best resembles g which can be expressed by the variational problem

$$\min_{\bar{I}} \int_{\Gamma} \|\nabla \bar{I} - g\|_2^2 \, dx, \text{ subject to } \bar{I}|_{\partial \Gamma} = I|_{\partial \Gamma}. \quad (2)$$

Hence, we are seeking for an image function $\bar{I}$ that is equal to I on the boundary of Γ and whose gradient is close, with respect to the L_2-norm, to the guidance field g. The reconstructed pixel values inside Γ can be then obtained by

$$I(x_\Gamma, y_\Gamma) = \bar{I}(x_\Gamma, y_\Gamma). \quad (3)$$

For the success of this method, we need to carefully choose an appropriate guidance field g. As the abdominal aorta is almost parallel to the spinal column, the thin wires of the stent graft are also more or less parallel to it. Thus all gradients of these wires are nearly orthogonal to the upper and lower borders of the vertebrae. In order to obtain an approximation of the stent graft's centerline direction, we can compute the principal direction v of Ω using i.e. *principal component analysis* (PCA). If we assume that v and its orthogonal complement $v^\perp$ are normalized, we can decompose ∇I as follows:

$$\nabla I(x) = \langle \nabla I(x), v \rangle \cdot v + \langle \nabla I(x), v^\perp \rangle \cdot v^\perp, \quad (4)$$

where $\langle \cdot, \cdot \rangle$ denotes the scalar product. The second term on the right hand side of (4) contains now all undesired gradients that belong to the wires of the stent graft. Hence, a suitable guidance field is given by

$$g(x) = \langle \nabla I(x), v \rangle \cdot v. \quad (5)$$

By this choice of the guidance field we keep as much structural information as possible while avoiding the creation of unreasonable intensity information at the same time.

C. Automatic Initialization

The previously recapitulated *Stent-editing* technique requires an outline of the occluded region as input. Here, we present a method that automatically extracts a region of interest containing the entire stent.

Fig. 2 displays our initialization scheme. The stent region S is extracted by first subtracting thick curvilinear structures (Frangi filter for scales $5-6$) from thin curvilinear structures (Frangi filter for scale 2) for only highlighting the stent wires. Subsequent employment of a median filter for noise removal and mean filter for dominant region extraction leads to the desired image region that contains the stent graft. Instead of the Frangi filter, any filter can be used that highlights curvilinear structures of selected sizes.

III. EXPERIMENTS

A set of experiments was carried out in order to investigate the accuracy, robustness, and speed of the proposed disocclusion-based 2D-3D registration.

Table 1 Image dimensions

Modality	Size (pixel)	Resolution (mm)
CTA	512 x 512 x 398	0.59 x 0.59 x 1.25
X-ray	308 x 982	0.15 x 0.15
DRR	308 x 982	1.00 x 1.00

Interventional 2D-3D Registration in the Presence of Occlusion

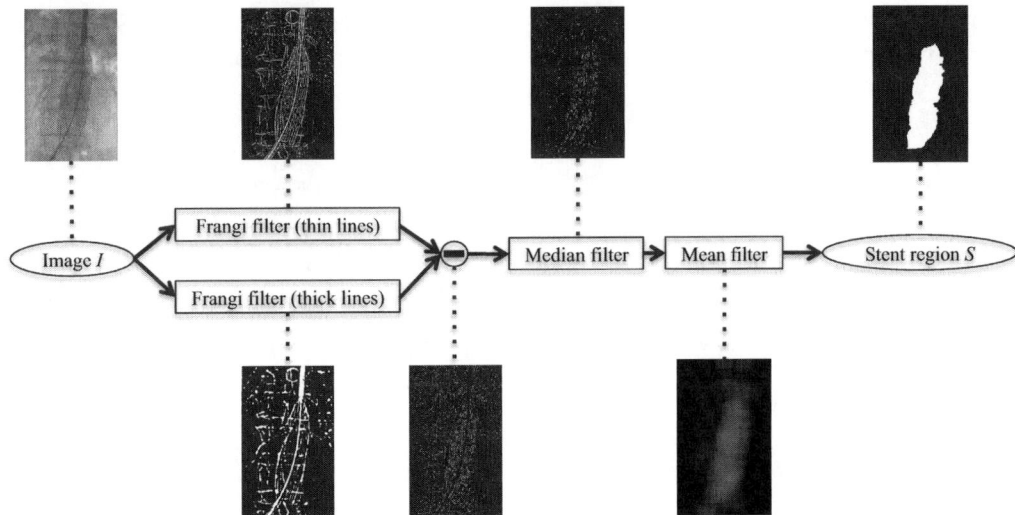

Fig. 2 Automatic scheme for outlining occluded region (here stent region S)

A. Experimental Setup

All registrations were carried out on a Windows XP Professional 2002 PC with Intel Core 2 CPU, 2.4 GHz, 2 GB RAM combined with 512 MB RAM on a GeForce 8800 GTS Graphics card. Except of the DRR generation, the algorithms were not optimized for speed.

For this experiment, a real abdominal Computed Tomography Angiography (CTA) scan acquired by a SIEMENS Somatom Definition scanner, was provided by our partner clinic. The synthetic experimental setup further consisted of a self-produced DRR from an arbitrary projection including a digitally inserted stent graft [11] such that the bone structure is partially hidden. In addition, an intraoperative fluoroscopy image of the same patient showing occlusion by medical instruments (Fig. 2) was provided by our partner clinic. The data had been acquired using a fully mounted Siemens AXIOM Artis dTA angiography suite with dynaCT software that gives information about the focus setting of the X-ray machine, from which we are able to calculate a ground truth camera calibration matrix K [4]. The respective image dimensions are given in Table 1.

For a thorough validation and presentation of the results, we employed a slightly adapted standardized evaluation methodology [12]. Thereby, given a ground-truth projection matrix $\mathbf{P}_{gt}$ and a set of 10 anatomical landmark points $Q = \{\mathbf{q}_i\}$ ($i = 1,...,10$) extracted within the respective CT volume (including the occluded region) by a clinical expert, two error measures for each experiment are calculated: *mean target registration error* (*mTRE*) and *mean projection distance* (*mPD*). We further computed 12 intervals containing each 12 different starting positions for R, t yielding a total number of 144 registrations per evaluation case and experiment. Next to the error measurements $mTRE, mPD$ for accuracy inspection, we also calculated *success rate* which indicates the percentage of registrations yielding a *mTRE* of less than 10mm (success threshold), and *capture range* which is 95% of the largest *mPD* of a successful registration. As success rate and capture range highly depend on the compactness of error distribution, their values are influenced by the selected success threshold. Due to this dependency, only the error measurements are significant accuracy indicators.

B. Results

We evaluated the robustness, accuracy, and speed of the proposed registration framework (see Fig. 1) in comparison to the conventional method [3] and our previous approach [4] including manual annotation of occluded region. Tables 2 and 3 show error measures $mTRE, mPD$ as well as success rate, capture range, and performance for each of the methods. It is interesting to see that disocclusion can improve the accuracy and robustness of 2D-3D registration only for

Table 2 Evaluation study in artificial scenario with real volume

	Disocclusion-based		Conventional [3]
	Manual [4]	Automatic	
mTRE	3.39 ± 6.44	1.89 ± 4.02	2.01 ± 2.01
success rate (%)	95.83	98.17	94.44
capture range (mm)	3.80	6.65	5.08
mPD	12.96 ± 28.76	8.93 ± 22.41	5.08 ± 10.12
success rate (%)	84.03	89.99	85.67
capture range (mm)	50.30	51.25	50.30
speed (sec/run)	11.99	1.99	3.38

Table 3 Evaluation study in real scenario

	Disocclusion-based		Conventional [3]
	Manual [4]	Automatic	
mTRE	9.31 ± 7.15	8.52 ± 3.21	9.32 ± 5.36
success rate (%)	93.06	95.54	56.25
capture range (mm)	1.90	0.95	0.95
mPD	51.66 ± 48.28	45.24 ± 23.33	59.85 ± 49.55
success rate (%)	34.03	52.89	15.97
capture range (mm)	23.75	10.02	19.95
speed (sec/run)	11.68	1.68	4.12

real interventional images. Here, *mTRE* as well as *mPD* values could be significantly reduced by both disocclusion-based methods including manual and automatic occlusion outlining (see Table 3). In Table 2, however, the conventional method [3] slightly outperforms at least our previous disocclusion-based method using manual outlining. The reason here may be that a digital projection of CT volumes introduces different noise than the analog projection by X-rays. The conventional method [3] using the robust similarity measure *Gradient Difference* can better handle this type of noise. However it has severe problems when applied in a clinical setting. This observation is further backed by the comparison of performance values for all experimental setups (see Tables 2 and 3. Whereas the conventional method [3] took 3-4 seconds in average for one registration, our novel automatic disocclusion framework was able to decrease this to 1.5-2 seconds. It is also worth noticing that the automatic outlining of the occluded region has a direct influence on the error measures of the subsequent registration. This can be explained by the fact that our automatic initialization procedure optimally separates the occluded image part from the non-occluded and thereby makes sure that only a very small amount of original image data is lost during textitStent-editing reconstruction.

IV. CONCLUSION

In this paper, we have introduced a fully automatic framework for disocclusion-based 2D-3D registration of medical images showing stents. First the occluded region is outlined by an automatic preprocessing method employing extraction of curvilinear structures and reconstructed using our previously introduced *Stent-editing* technique. A robust 2D-3D image registration algorithm is then computed on the processed interventional image and a preoperative 3D scan of the patient. We performed an evaluation study comparing the new automatic framework to our previously presented disocclusion-based registration method employing manual annotation, and to a conventional method using Gradient Difference measure [3]. Our experimental setup consisted of a patient Computed Tomography Angiography (CTA) scan as well as two different 2D datasets, one for synthetic and one for real scenario. We could successfully show that our automatic apporach is able to outperform the other test methods in terms of accuracy, robustness, and speed.

ACKNOWLEDGEMENTS

We would like to thank Asbjørn Ødegård (MD), Department of Radiology, St Olavs Hospital, Trondheim, Norway, for providing interventional image data.

REFERENCES

1. Groher M., Zikic D., Navab N.. Deformable 2D-3D Registration of Vascular Structures in a One View Scenario *IEEE Trans Med Imag.* 2009;28:847–860.
2. Markelj P., Tomaževič D., Likar B., Pernuš F.. A review of 3D/2D registration methods for image-guided interventions *Med Image Anal.* 2012;16:642-661.
3. Penney G. P., Weese J., Little J. A.., Desmedt P., Hill D. L. G., Hawkes D. J.. A comparison of similarity measures for use in 2-D-3-D medical image registration *IEEE Trans Med Imag.* 1998;17:586 – 595.
4. Demirci S., Baust M., Kutter O., Manstad-Hulaas F., Eckstein H.-H., Navab N.. Disocclusion-based 2D-3D Registration for Angiographic Interventions *Computers in Biology and Medicine.* 2013;43:312-322.
5. Kaneko S., Satoh Y., Igarashi S.. Using selective correlation coefficient for robust image registration *Pattern Recogn.* 2003;36:1165-1173.
6. Baust M., Demirci S., Navab N.. Stent Graft Removal for Improving 2D-3D Registration in *6th IEEE International Symposium on Biomedical Imaging: From Nano to Macro (ISBI)*:1203–1206 2009.
7. Hartley R., Zisserman A.. *Multiple View Geometry.* Cambridge University Presssecond ed. 2003.
8. Engel Klaus, Hadwiger Markus, Kniss Joe M., Rezk-Salama Christoph, Weiskopf Daniel. *Real-Time Volume Graphics.* AK Peters, Ltd. 2006.
9. Khamene A., Bloch P., Wein W., Svatos M., Sauer F.. Automatic Registration of Portal Images and Volumetric CT for Patient Positioning in Radiation Therapy *Med Image Anal.* 2006;10:96-112.
10. Brown L. M. G., Boult T. E.. Registration of planar film radiographs with computed tomography in *Workshop on Mathematical Methods in Biomedical Image Analysis*:42-51IEEE Computer Society 1996.
11. Demirci S., Bigdelou A., Wang L., et al. 3D Stent Recovery from One X-Ray Projection in *Medical Image Computing and Computer-Assisted Intervention (MICCAI), Part I* (Fichtinger Gabor, Martel Anne L., Peters Terry M.., eds.);6891 of *LNCS*:178-185Springer 2011.
12. Kraats E., Penney G., Tomaževič D., Walsum T., Niessen W.. Standardized Evaluation Methodology for 2-D-3-D Registration *IEEE Trans Med Imag.* 2005;24:1177-1189.

Author: Dr. Stefanie Demirci
Institute: Institut fuer Informatik I16, Technische Universitaet Muenchen
Street: Boltzmannstr. 3
City: Garching
Country: Germany
Email: stefanie.demirci@tum.de

Marker-Controlled Watershed for Volume Countouring in PET Images

V. Naranjo[1], F. López-Mir[1], C. Marín[1], S. Morales[1], J.J. Fuertes[1], and E. Villanueva[1]

[1] Instituto Interuniversitario de Investigacin en Bioingeniera y Tecnologa Orientada al Ser Humano, Universitat Politcnica de Valncia, I3BH/LabHuman, Camino de Vera s/n, 46022 Valencia, Spain

Abstract— **In this paper an automated method for volume contouring in PET images is presented. It is a slice-by-slice approach based on marker-controlled watershed segmentation applied to gradient image. A preprocessing step, based on geodesic transformations, is proposed to get a well defined boundary gradient of the region to be segmented (the tumour in our case), since the watershed results are improved. Moreover, a scheme of marker selection is proposed taking into account a priori segmentation knowledge from previous segmented slices. The method has been validated along with other 32 methods in a wide study [1] using phantom and real data. Regarding its degree of interactivity, our method obtained the highest accuracy results in patient's data ($A^* = 0.694$) and similar results ($A^* = 0.670$) in the case of phantom's data.**

Keywords— **Positron emission tomography, mathematical morphology, image segmentation, watershed transformation.**

I. INTRODUCTION

Positron Emission Tomography (PET) is a low invasive imaging modality which is used for the diagnosis and treatment of different diseases, such as the detection of tumour pathologies. First, a short-life radiotracer is injected into the patient's body [2]. The most commonly radiotracer used in clinical oncology for PET is 2-[18F] fluoro-2-deoxy-Dglocose (FDG) since it is a sign of tissue metabolic activity, but other isotopes can be used for the detection of other molecules [3]. This tracer is a glucose analog that is taken up by glucose-using cells and phosphorylated by hexokinase (whose mitochondrial form is greatly elevated in rapidly growing malignant tumours). Thus, the FDG uptake is increased in tumour cells compared to healthy tissues, emitting positron energy that a gamma camera detects. In this way, a 3D image can be reconstructed using that information. Therefore, FDG-PET is a medical technique with low resolution but more sensitive for discovering initial tumour cells than conventional computer tomography (CT) or magnetic resonance (MR), which are image modalities with high spatial resolution. The different features of both techniques make them complementary for diagnosis and treatment [4].

Intracranial neoplasm is a disease with 17.500 new cases per year in USA and uniform age distribution. The segmentation of these lesions is important for tumour characterization (diameter, volume, location, etc). With this information available, personal treatments and objective protocols can be carried out. PET studies cause some efforts to be focused on registration process with MR/CT, where tumors can be better segmented although some problem may appear: First, and more important is that some lesions are not visible in anatomical images. These lesions are small and located in specific zones in their first pathological stages. [5]. So, the tumour PET segmentation is required although the registration has been made. The second problem is the validation of the registration methods that ensure the correct correlation of two multimodal studies with different resolutions. This evaluation is carried out in a subjective manner [6]. The third problem is that sometimes the clinician does not have the two studies and the registration is not possible. These problems and others, turn the direct tumour segmentation in PET images into an important and necessary task. The methods for the segmentation of tumours in PET images can be classified in: thresholding methods, region growing algorithms, watershed-based approaches, gradient-based methods and hybrids and pipeline methods [1]. Thresholding methods are based on pixel intensity for classifying a tissue as health/unhealthy. Region growing methods take into account the pixel intensity and regional properties of the neighbours (as intensity, connection, etc.) for tumour classification; gradient-based algorithms are based on the difference of intensity between different structures, and hybrids and pipeline methods use a concatenation of different filters for tumour segmentation. Finally, watershed-based algorithms are based on gradient properties and pixel intensity. The main goal in PET segmentation is to delimit the tumour accurately, but other characteristics as the computational cost, the easy initialization and the automation are also important when selecting the final method for clinical purposes. In this paper, a method for PET tumour segmentation in brain with the features mentioned above is deeply described. The results of this method are also presented but a more precise description of the evaluation process and the comparison with other methods are presented in [1].

II. METHOD

A. Tools

In this section, some morphological operators used in our algorithm will be briefly described. More detailed explanation can be found in [7]. The two basic morphological operators are dilation, $\delta_B(f)(\mathbf{x})$, and erosion, $\varepsilon_B(f)(\mathbf{x})$, being $f(\mathbf{x})$ a grayscale image, $\mathbf{x}$ the pixel position and $B(\mathbf{x})$ the structuring element (shape probe) centered at pixel $\mathbf{x}$. From these basic operators, their combinations opening and closing can be defined as $\gamma_B(f) = \delta_B(\varepsilon_B(f))$ and $\varphi_B(f) = \varepsilon_B(\delta_B(f))$, respectively. The morphological gradient is defined as: $\rho(x) = \delta_{B_\rho}(f) - \varepsilon_{B_\rho}(f)$, where B_ρ is the elementary structure element and the area opening, which removes clear objects smaller than λ, can be defined as:

$$\gamma_\lambda = \max_i \{\gamma_{B_i} | B_i \text{ is connected and card}(B_i) = \lambda\}.$$

Other kind of morphological operators are called geodesic transformations where two images are required, the reference g and the marker f. In this work we use the reconstruction by erosion, which is the successive geodesic erosion of the marker regarding the reference up to idempotence:

$$\varphi^{rec}(g,f) = \varepsilon_g^{(i)}(f) \text{ so that } \varepsilon_g^{(i+1)}(f) = \varepsilon_g^{(i)}(f),$$

being $\varepsilon_g^{(i)}(f) = \varepsilon_g^{(1)} \varepsilon_g^{(i-1)}(f)$ and $\varepsilon_g^{(1)} = \max(\varepsilon_B(f), g)$.

The core of the segmentation approach presented in this paper is the marker-controlled watershed [7]. The success of a based-watershed segmentation is achieved if the image minima are inside the objects of interest and its maxima correspond to the object boundaries. For that reason, it is typically applied to gradient image. However, due mainly to the noise which causes the presence of a great deal of minima, a problem of over-segmentation usually appears in the result. One approach to solve this problem is to use the marker-controlled watershed, which consists of changing artificially the minima of the gradient. The new minima are defined in set of markers indicating the objects of interest (internal markers) and the background (external markers). So, the challenges of a watershed based algorithm are finding a set of proper markers and using an input image (gradient image) where the boundaries of the regions to be segmented are highly contrasted.

B. Algorithm

The proposed method is a slice-by-slice algorithm fully automated, as described in section B1. During the initialization phase presented in section B2, the user is required to select a slice where the tumour is visible as much as possible, and this slice is processed in a different way than the rest of the 3D data.

B1 General Algorithm

The block diagram depicted in Fig. 1 illustrates the slice-by-slice algorithm. The condition to segment a slice is that the energy of the region to be segmented decreases less than a 50% with regard to the previous slice. This energy is obtained as follows:

$$E = \sum_\mathbf{x} (slice_i(\mathbf{x}) \times mask_{i-1}(\mathbf{x}))^2,$$

where $mask_{i-1}$ is the previous slice ($slice_{i-1}$) segmentation result. The segmentation core of each slice is the marker-controlled watershed that needs two input parameters, the gradient and the set of markers.

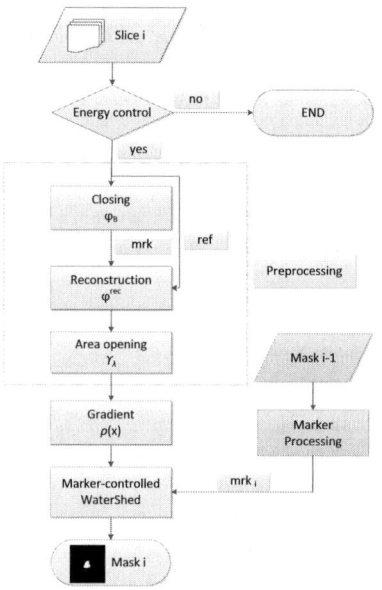

Fig. 1 Block diagram of the slice-by-slice algorithm.

Obtaining the gradient: The image must be preprocessed to obtain an image with well-defined borders (specifically in the region of interest). The preprocessing step (see Fig. 1) consists of closing by reconstruction the image with the aim of removing the dark areas inside the tumoural region. After that, an area open is applied to remove the clear regions with an area smaller than the estimation of the smallest projected tumour area (in the axial axis) from the image. At this point, the gradient of this preprocessed image is introduced as input in the marked-controlled watershed. Fig. 2 shows the results for different algorithm steps for two slices of the same patient (one row per slice). In the figure, the original slice (a), the preprocessed image gradient (b) and this gradient with the markers can be appreciated.

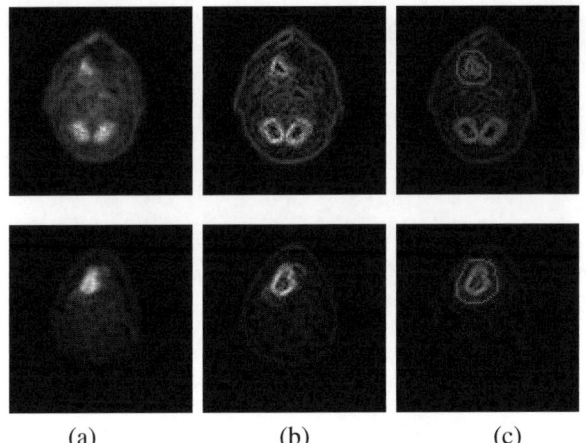

Fig. 2 Results of different general algorithm steps for two slices (one slice by row): (a) Original image, (b) Gradient of preprocessed image, (c) Gradient with markers superimposed.

Obtaining the set of markers: To obtain the set of markers, the segmentation result of the previous slice ($mask_{i-1}$) has to be processed. The external marker will be the perimeter of the dilated ($mask_{i-1}$) with a structuring element of size equal to the maximum tumour growing expected between slices (the selection of this parameter is not critical). To obtain the internal marker, the centroid of ($mask_{i-1}$) is computed.

B2 Initialization

The process of the first slice differs from the general process described above only in the selection of the markers. In the case of the first processed slice the mask from where the markers are obtained is the result of thresholding the preprocessed image. The selection of the threshold is automated using the Otsu's algoritihm [8]. After the selected initial slice is processed, the general algorithm is run upwards and downwards the initial slice until the stopping energy criterion is reached.

III. RESULTS

A. Data Description

The algorithm proposed in this paper has been assessed along with other 32 methods in the comparative study presented in [1]. For that, four sets of images have been used: two images of a tumour phantom and two clinical PET images of different head-and-neck cancer patients. In all of them, the metabolic tracer F-Fluorodeoxyglucose (FDG) and a hybrid PET/CT scanner (GE Discovery) were used, as in routine for cancer diagnosis and treatment planning. The tumour phantom contains glass compartments of irregular shapes, based on cancer of the oral cavity and lymph node metastasis, filled with low concentrations of FDG [1]. In the case of patient's images, a large tumour of the oral cavity and a small tumour of the larynx were selected from two different patients, along with a metastatic lymph node. In the case of real images, the CT images have also been provided, although our method do not use them. The groundtruth sets and the images of this challenge are available online [9] and widely described in [1].

B. Evaluation Parameters

As it was explained in [1], several metrics were used to evaluate the goodness of the different methods. AUC' is a modification of the method Area Under the Curve based on receiver operating characteristic (ROC) analysis. Another measure used is a variant of the Hausdorff distance (HD') [10] that uses the reference surface to calculate Hausdorff distance and takes the maximum for any point on the surface of the minimum distances from that point to any point on the reference surface. Both HD' and AUC' are normalized to range between 0 and 1 to facilitate the comparisons. Another parameter used is the standard metric of Dice similarity coefficient (DSC) [1]. AUC' and DSC evaluate the volumetric agreement and HD' the surface displacement. Moreover a synthetic composite accuracy metric has been defined as:

$$A^* = 0.5\,AUC' + 0.25\,DSC + 0.25\,HD'.$$

A^* weights in the same proportion probabilistic (AUC') and deterministic measurements (HD' and DSC). It penalizes deviation from the 'true' absolute volume.

C. Results

Fig. 3 shows some segmentation results for different slices of one patient. Here, we can observe how the method obtains

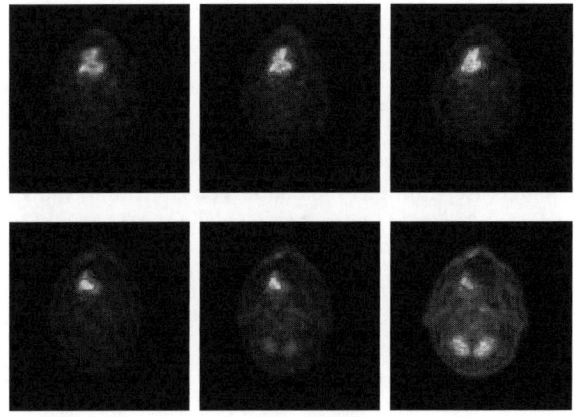

Fig. 3 Segmentation results.

Table 1 Method results (Average values and standard deviations): Ph (Phantoms) and Pt (Patients).

	$\overline{AUC'}(\sqrt{\sigma_{AUC'}})$	$\overline{HD'}(\sqrt{\sigma_{HD'}})$	$\overline{DSC}(\sqrt{\sigma_{DSC}})$	$\overline{A^*}(\sqrt{\sigma_{A^*}})$
Ph	0.670(0.202)	0.376(0.239)	0.575(0.119)	0.573(0.196)
Pt	0.872(0.02)	0.334(0.225)	0.687(0.108)	0.694(0.139)

Table 2 Method result comparison according to A^* (Ph (Phantoms) and Pt (Patients))

	ours	PL^f	$T1^a$
Ph	0.670(0.202)	0.776 (0.172)	0.4(0.902)
Pt	0.694(0.139)	0.623 (0.166)	0.692(0.140)

a good approximation of the tumour contour and it is able to distinguish even if other brain structures appear in the image with similar gray levels (see last image).

With regards to quantitative measurements, the proposed method has obtained the results presented in Table 1 using the metrics explained above. In this table the results for different metrics (average values and standard deviation) are provided, differentiating between phantom results (Ph) and patient results (Pt). As it can be observed, the results are significantly better in the case of real images.

IV. DISCUSSION AND CONCLUSION

The main contribution of the presented approach is the low degree of interactivity (only an easy initialization step is required), low computational cost and promising results. This method has been evaluated in an exhaustive study [1] being compared with other 32 methods depending on several parameters: the accuracy, the degree of interactivity with the user and the use of other kind of image information (CT images). Methods were classified in groups depending on the degree of interactivity (max,high,mid,low,none) and the use of CT (high, low,none). Our method was classified into the group: interactivity = low and CT use = none. Compared with methods in the same conditions, it is, with the same degree of interactivity and CT use, our method is the best in the case of patient's data, as it can be observed in Table 2. In this table, the results in terms of A^* for the best methods of the same group as ours are shown in the cases of phantoms (Ph) and patients (Pt). Method PL^f is a multi-segmentation method which uses different methods for segmentation refinement. $T1^a$ is an automatic thresholding algorithm [1]. In the global ranking obtained in this study, the results of our method for patient's data are very close to the results obtained by other methods but with higher interactivity degree. Regarding computational cost, the algorithm was tested on an Intel core i5 @ 2.80 GHz, with a RAM of 2 GHz and Windows 7 (32 bits) using Matlab R2012a. The average processing time of a tumour segmentation (20 slices of 256×256 pixels) was less than 5 seconds.

ACKNOWLEDGEMENTS

This work has been supported by Centro para el Desarrollo Tecnolgico Industrial (CDTI) under the project Oncotic (IDI-20101153).

REFERENCES

1. Shepherd T., Teras M., Beichel R. R., et al. Comparative Study With New Accuracy Metrics for Target Volume Contouring in PET Image Guided Radiation Therapy *Medical Imaging, IEEE Transactions on.* 2012;31:2006-2024.
2. Han Anqin, Xue Jie, Hu Man, Zheng Jinsong, Wang Xiaohui. Clinical value of 18F-FDG PET-CT in detecting primary tumor for patients with carcinoma of unknown primary *Cancer Epidemiology.* 2012;36: 470 - 475.
3. Ogawa Toshihide, Kanno I., Shishido F., et al. Clinical Value of Pet with 18F-Fluorodeoxyglucose and L-Methyl-11C-Methionine for Diagnosis of Recurrent Brain Tumor and Radiation Injury *Acta Radiologica.* 1991;32:197-202.
4. G Antoch, FM Vogt, LS Freudenberg, al . WHole-body dual-modality pet/ct and whole-body mri for tumor staging in oncology *JAMA.* 2003;290:3199-3206.
5. Fuertes Juan Jose, Naranjo Valery, Verdu-Monedero Rafael, Bernabeu Angela, Lopez-Mir Fernando, al . Incorporation of a Variational Registration Method into a Spectroscopy Tool in *Proceding of IWBBIO 2013, Granada, march 2013.*
6. Gong S.J., O'Keefe G.J., Scott A.M.. Comparison and Evaluation of PET/CT Image Registration in *Engineering in Medicine and Biology Society, 2005.*:1599-1603 2005.
7. Soille Pierre. *Morphological Image Analysis: Principles and Applications.* Secaucus, NJ, USA: Springer-Verlag New York, Inc.2 ed. 2003.
8. A Threshold Selection Method from Gray-Level Histograms *Systems, Man and Cybernetics, IEEE Transactions on.* 1979;9:62-66.
9. Shepherd T.. Contour evaluation, Tumour phantom for contour evaluation 2012. Available at: http://www.turkupetcentre.fi/files/tumourphantom/ 2012.
10. Huttenlocher D.P., Klanderman G.A., Rucklidge W.J.. Comparing images using the Hausdorff distance *Pattern Analysis and Machine Intelligence, IEEE Transactions on.* 1993;15:850-863.

Predictive and Populational Model for Alzheimer's Disease Using Structural Neuroimaging

D. López-Rodríguez and A. García-Linares

Brain Dynamics, Málaga, Spain

Abstract—This paper aims populational modeling of volumetric degeneration of the gray matter due to Alzheimer's disease, to establish the parameters of degeneration, and to contrast the state of an individual with respect to that model. In this way, you can get an early diagnosis of the disease.

We have used 2100 structural magnetic resonance images (sMRI), classified by sex (1097 M/1003 F), and corresponding to healthy people (C, Controls) and with Dementia (AD, Alzheimer's Disease), between 18 and 96 años (M-C: 59.44±24, F-C: 60.75±22.79/ M-AD: 75.35±7.07, F-C: 74.23±8.02), from public domain databases.

The SMRI processing methodology uses filtering, segmentation algorithms, and the calculation of parameters such as cortical thickness or volume. Furthermore, registration was performed on each subject with a standard template and a 116 anatomical structures atlas in which the above parameters are calculated.

It was possible to establish which structural changes the brain undergoes when affected by Alzheimer's disease, according to criteria of loss of volume or gray matter thickness, relative to healthy subjects: paracentral lobe, angular gyrus, calcarine sulcus ($p<0.001$), among others. some rules Have also been developed for classifying (error <9%) a given subject as normal or with Alzheimer with a given probability.

We generated a model of Alzheimer's disease, using statistical techniques and imaging processing. This study shows the brain areas that atrophy faster with the disease, and in what sequence they do.

Keywords—Mild Cognitive Impairment, Early Diagnosis, Neuroimage, Computational Intelligence, Validation.

I. INTRODUCTION

Alzheimer's disease (AD) is a neurodegenerative disease, progressive and irreversible, characterized by cognitive impairment, whose onset occurs in adulthood, especially in people over 65 years. In developed countries it is considered the most common cause of neurodegenerative dementia. Among the population over 65 years, it is the second leading cause of death and the leading cause of dependency. The AD has already become a public health problem that will worsen in the coming decades.

This disease may begin to develop up to ten years before clinical symptoms start. Current diagnostic criteria for the AD require cognitive deficits and dementia are relevant, ie when the diagnosis is made, there are already a large number of affected brain areas and a major neuropathological damage.

A high percentage of cases are diagnosed and treated when the AD is in mild and moderate stages of dementia. Currently there are effective treatments for both mild stage dementia and moderate, but the results indicate that once diagnosed it is too late to recover lost cognitive functions or control its progression [1,2,3].

In recent years there have been a significant number of articles aimed at the early detection of AD, before the onset of dementia and be too late to reverse the course of this disease. Thus, currently the major research lines are aimed at the identification of biomarkers in cerebrospinal fluid (CSF) and neuroimaging.

With regard to the search for chemical biomarkers present in CSF several studies have shown a correlation between AD and AB42 concentration, total tau protein, etc. All these markers involve an important advance in this area, but its invasive nature could be an issue of some importance [4].

But it is the Neuroimaging which, thanks to advances in technology, can provide reliable information on the structure and brain function.

The Magnetic Resonance Imaging (MRI) provides structural and functional information. A major research activity, ultimately being able to speak of "Cortical Thickness" or coactivation patterns, Difussion Tensor Imaging, etc.. Reasonable costs, its complete safety for the patient, scientific and technical improvements continued, and its wide dissemination in hospitals make it, from our point of view, a great candidate for use in the determination of the phases of early AD.

In addition, Information Technology and Telecommunications (ICT) provide a wide range of possibilities, including the possibility of tools and techniques that automatically provide objective data on the structural and functional state of the brain of the subject or subjects under study. From these data, one can obtain, also automatically, a pattern of degeneration in the case of AD and early AD.

In this context, the most appropriate tools for generating computer models (automatic) are those included within the so-called Computational Intelligence and Data Mining.

These techniques are designed to extract information and knowledge from raw data. The information generated from the data (results of a neuroimaging study, for example) can be easily interpreted by a human expert as a causal model formed by logical rules. These rules express the cause and effect relationships between the different variables of the system and the study's conclusion. For example, a study by MRI provides a number of parameters measured in the image, such as cortical thickness and volume of internal structures of the brain. The computer model can generate rules that associate different values of these parameters with the subject's mental state, indicating so those brain structures associated with that state of mind.

The objectives of this work are summarized in:
a) Propose a robust methodology imaging processing, so that the selected algorithms can commit the slightest error, and thus get ensure statistical validity of the results of neuroimaging studies.
b) Develop a population model of the structural changes that occur in the progression of Alzheimer's disease from its initial stages and mild cognitive dementia to the most advanced form of the disease.
c) Develop and implement an automated system for the early detection of early AD, by processing neuroimaging, and building automated tools and objective techniques based on Artificial Intelligence and Data Mining.
d) Validate the results of the proposed model, determining their sensitivity, specificity and predictive value.

II. MATERIALS AND METHODS

The MRI scans structural used in this study were obtained from databases with public access, such as ADNI [5](Alzheimer's Disease Neuroimaging Initiative) and OASIS [6] (Open Access Series of Imaging Studies), corresponding to normal subjects and AD in various stages.

The first one corresponds to an initiative of the LONI (Laboratory of Neuro Imaging) at the University of California at Los Angeles (UCLA). Its purpose is to collect and make accessible to all researchers who wish a number of brain imaging from both normal subjects and subjects with AD in different stages of evolution.

On the other hand, OASIS is a project to make freely available to the scientific community a series of cerebral magnetic resonance imaging. The purpose of collecting and distributing freely MRI data sets, hopes to facilitate future discoveries in basic neuroscience and clinical research. OASIS is a project carried out at the Howard Hughes Medical Institute (HHMI) at Harvard University in the Neuroinformatics Research Group (NRG) at the School of Medicine of the University of Washington and the Network of Biomedical Informatics Research (BIRN). Despite following heterogeneous protocols in the acquisition of brain images, all follow a strict quality control.

Of these subjects, we know a set of data for cataloging, depending on their sex, age, preference for right or left hand, apart from a description of the subject's mental state based on criteria such as GDS (Global Deterioration Scale), CDR (Clinical Dementia Rating) or MMSE (Mini Mental State Examination).

It has been selected a set of MRI studies coming from these databases, which served as training and test of knowledge extraction algorithms used for generating classifier system. The distribution of subjects by sex and mental condition (normal, mild dementia) can be seen in Table 1 below:

Table 1 Data from the study subjects, per database used and total

	N	NORMAL	AD
ADNI	1514	766	748
OASIS	586	421	165
TOTAL	2100	1187	913

Brain imaging postprocessing was performed of the subjects mentioned above, using advanced techniques of quantification and morphometry of cerebral gray matter.

To ensure the greatest possible accuracy of the above parameters, special attention has been paid to the use of segmentation algorithms with low error and accurate registration techniques.

First, the segmentation is the classification of all image voxels, based on significant labels as "gray matter", "white matter" or even more specific, like "hippocampus", "thalamus" ... Errors in this phase result in the thickness or volume of different brain areas can not be calculated exactly, as it would take into account voxels not really corresponding with the type of tissue indicated by the segmentation. The algorithms used to segment brain volumes contain an error rate of around 12%. The algorithm developed for this study reduces the error of it to 8%, beating most common algorithms in this field, such as FAST[8], from package FSL [9], SPM [10], or FreeSurfer [11], as explained in the review article [7].

Furthermore, registering a subject brain volume with standard atlas is essential to label different anatomical regions of the brain. Any errors at this stage would cause parameters to be assigned to incorrect anatomical areas. They are therefore fundamental in population studies, to be able to compare the data obtained from a given subject with a certain population segment.

Our new implementation of Thirion algorithm [12,13], which, unlike other methods, registers the entire volume (yielding much more information, and more precise than using only brain surface) with IBASPM atlas [14] in MNI space coordinates (MNI152 template specifically, of the Montreal Neurological Institute).

The use of the template MNI152 [15] allows labeling of 116 anatomical structures, both cortical and subcortical, 4 parameters calculated for each of them: average volume (cc) Average thickness (mm) and average fractal dimension, and local density (both measured dimensionless). Some of these parameters to specific brain regions are considered biomarkers for various diseases, according to studies published in journals [16].

Furthermore, it has been calculated, for each structure, the ratio between volume and total brain volume, dimensionless measure which indicates the percentage of the brain that occupies the structure. This makes a total of 580 parameters for each studied brain volume.

To create the system predictor of early AD, it has been used the implementation of decision trees using the method C4.5 [17] with MultiBoost technique. Using MultiBoosting is generated a series of decision trees for the same set of training, but random weights are assigned to each subject in the database in each of the various trees generated. The final classification is a combination by voting from different individual classifiers.

III. RESULTS

A. Degenerative Model with Respect to Age

The differences between healthy subjects and subjects with Alzheimer's disease, in terms of volumetric parameters in gray matter, follows a time course very characteristic. The highest volumetric differences occur in the early stages of the disease, whereas, with age, the differences tend to cancel.

Table 2 lists the major areas of the brain with structural changes, showing the significance (p-value) according to the age segment. You can see that the differences are more significant in the early stages, while in the following, differences cease to be.

B. Classifier and Predictive System

To evaluate this system, data obtained from the brain image processing of each subject is used as input along with study-specific data provided by the databases of images used. The output of the system represents the state of the subject ("normal" or "early AD").

The generated model may be expressed as simple logical rules. As an example, one of the rules generated by the system:

IF (AGE > 55) AND (TEMPORAL MIDDLE GYRUS THICKNESS <= 2.448985)
THEN CLASS = AD [PROBABILITY = 0.946]

In Table 3 below, we show the results of correct classification, sensitivity and specificity obtained by our algorithm.

Table 3 Validation Results.

Parameter	Value
Accuracy	91.48%
Sensitivity	90.80%
Specificity	92.30%

We see that the correct classification rate is over 90%, which makes the predictive power of the proposed model and algorithm is at least comparable to a mammogram [18].

IV. DISCUSSION

Given the need for a diagnostic method, we have developed a system capable of detecting early AD using MRI analysis with a high success rate, non-invasive, fast and objective. As sources of brain studies have used the ADNI and OASIS databases. The distribution of subjects by sex, age and condition is found in Table 1.

The first step was to generate a population model of the structural differences between healthy subjects and AD

Table 2 List of structures and significance (p) of the volumetric difference between healthy subjects and AD.

STRUCTURE	50	60	70	80
Paracentral Lobe (Right)	0.000	0.000	0.000	0.030
Paracentral Lobe (Left)	0.000	0.000	0.000	0.029
Postcentral Gyrus (Right)	0.000	0.000	0.000	0.003
Postcentral Gyrus (Left)	0.000	0.002	0.000	0.019
Precentral Gyrus (Right)	0.000	0.000	0.000	0.050
Angular Gyrus (Right)	0.000	0.000	0.001	0.016
Angular Gyrus (Left)	0.000	0.000	0.011	0.003
Calcarine Sulcus (Left)	0.000	0.000	0.043	0.134
Cuneus (Left)	0.000	0.000	0.043	0.057
Inferior Occipital Gyrus (Right)	0.000	0.005	0.008	0.007

patients. We have established which brain regions are those with the greatest differences (from the statistical point of view) in the evolution of AD compared to normal subjects, see Table 2. Also, we can see that the normal and AD in the elderly make these structural differences tend to disappear.

This degeneration model serves as the basis for the system of support to early diagnosis.

As a result of the validation on our database, we proved that the system's accuracy in classifying subjects as "healthy" or "AD" exceeds 90%, see Table 3. This indicates that it is a method with a high power predictive diagnosis, so that it could be considered as a screening method for early detection of AD even presymptomatic.

V. CONCLUSIONS

A good clinical history and application of a range of neuropsychological tests to assess the degree of impairment of various cognitive domains are of great importance for the diagnosis of AD. But this usually happens clinically when the AD is obvious (even in the slightest degree) and dementia has already appeared, and though there are effective treatments for the early stages of dementia, it is not possible to recover lost cognitive functions, or slow the progression of disease.

Hence, the importance of early detection of AD is undeniable, from the point of view both clinical and socio-economic. Methods for early and accurate diagnosis of early stage of AD would, on one hand, delay or at least mitigate the loss of cognitive functions, and second, more effective development of drugs which only possible target so far, has been to change the course of this disease.

As future work, we see the integration of the results of neuroimaging processing with neuropsychological test results to create a population model, and predictive of AD from more information, correlating structural findings with cognitive / behavioral.

REFERENCES

1. Franz, C.E., et al., When help becomes a hindrance: mental health referral systems as barriers to care for primary care physicians treating patients with Alzheimer's disease. Am, J. Geriatr Psychiatry, 2010. **18**(7): p. 576-85.
2. Molinuevo JL, Berthier ML, and R. L., Donepezil provides greater benefits in mild compared to moderate Alzheimer's disease: implications for early diagnosis and treatment. Arch Gerontol, Geriatr, 2009.
3. Rountree SD, et al., Persistent treatment with cholinesterase inhibitors and or memantine slows clinical progression of Alzheimer disease Alzheimers Res Ther, 2009: p. 1-7.
4. Shaw LM, V.H., Knapik-Czajka M, et al., *Cerebrospinal fluid biomarker signature in Alzheimer's disease neuroimaging initiative subjects.* Ann Neurol, 2009. **65**: p. 403-413.Smith J, Jones M Jr, Houghton L et al. (1999) Future of health insurance. N Engl J Med 965:325–329
5. Mueller SG, W.M., Thal LJ, Petersen RC, Jack CR, Jagust W, Trojanowski JQ, Toga AW, Becket L Ways toward an early diagnosis in Alzheimer's disease: The Alzheimer's Disease Neuroimaging Initiative (ADNI), Alzheimer's Dementia. 2005. **1**, 55-66.
6. Marcus, D.S., et al., Open Access Series of Imaging Studies (OASIS): cross-sectional MRI data in young, middle aged, nondemented, and demented older adults. J Cogn Neurosci, 2007. **19**(9): p. 1498-507.
7. Frederick Klauschen, A.G., Vincent Barra, Andreas Meyer-Lindenberg, and Arvid Lundervold, Evaluation of automated brain mr image segmentation and volumetry methods. Human Brain Mapping, 2009. 30: p. 1310-1327.
8. Y. Zhang, M.B., and S. Smith, Segmentation of brain mr images through a hidden markov random field model and the expectation maximization algorithm. IEEE Trans. on Medical Imaging, 2001.
9. S.M. Smith, M.J., M.W. Woolrich, C.F. Beckmann, T.E.J. Behrens, H. Johansen-Berg, P.R. Bannister, M. De Luca, I. Drobnjak, D.E. Flitney, R. Niazy, J. Saunders, J. Vickers, Y. Zhang, N. De Stefano, J.M. Brady, and P.M. Matthews, Advances in functional and structural mr image analysis and implementation as fsl. NeuroImage, 2004. 23: p. 208-219.
10. Ashburner, J. and K.J. Friston, Unified segmentation. NeuroImage, 2005. 26: p. 839-851.
11. Fischl, B., M.I. Sereno, and A.M. Dale, *Cortical surface-based analysis ii: Inflation, flattening, and a surface-based coordinate system.* Neuroimage, 1999. **9**: p. 195-207.
12. Yeo, B.T.T., et al., Spherical demons: Fast surface registration. MICCAI (1). 2008. 745-753.
13. Thirion., J.-P., Non-rigid matching using demons, in CVPR1996. p. 245-251.
14. Available from: http://www.thomaskoenig.ch/Lester/ibaspm.htm.
15. Grabner, G., et al., Symmetric atlasing and model based segmentation: an application to the hippocampus in older adults, 2006. p. 58-66.
16. Prins, N.D. and J.C.v. Swieten, *Alzheimer Disease: MRI and CSF biomarkers in AD - accuracy and temporal change.* Nature Reviews Neurology, 2010. **6**: p. 650-651.South J, Blass B (2001) The future of modern genomics. Blackwell, London
17. Quinlan, J.R., *C4.5: Programs for Machine Learning.* 1993: Morgan Kaufmann, Publishers.
18. Pisano, E.D., et al., *Diagnostic accuracy of digital versus film mammography: Exploratory analysis of selected population subgroups in DMIST.* Radiology, 2008. **246**(2): p. 376-383.

Author: Domingo López-Rodríguez
Institute: Brain Dynamics
Street: Severo Ochoa, 34
City: Málaga
Country: Spain
Email: domingo.lopez@brain-dynamics.es

Accurate Cortical Bone Detection in Peripheral Quantitative Computed Tomography Images

T. Cervinka[1,2], J. Hyttinen[1,2], and H. Sievänen[3,4]

[1] BioMediTech, Tampere, Finland
[2] Department of Electronics and Communications Engineering, Tampere University of Technology, Tampere, Finland
[3] Bone Research Group, UKK Institute, Tampere, Finland
[4] Pirkanmaa Hospital District, Science Center, Tampere, Finland

Abstract—An accurate assessment of the whole bone strength is an essential goal in clinical bone research. This is, however, possible only with accurate and precise description of actual bone geometry. In this paper, we introduce a refined automated approach of OBS method for accurate segmentation of cortical bone cross-sectional area. The approach employs morphological operations and utilization of two fixed thresholds. For comparison, the standard OBS and method based on level set evolution (DRLSE) were evaluated. The performance of used methods was tested on in vivo peripheral quantitative computed tomography (pQCT) images of distal tibia. As to the detection of cortical bone geometry in pQCT images, the new refined OBS method performed reasonably well and was indicating somewhat more consistent results in comparison to standard OBS and DRLSE based method.

Keywords—Bone strength, cortical bone, pQCT, segmentation.

I. INTRODUCTION

Osteoporosis and associated fragility fractures are a common health problem in aging population that forms a considerable socioeconomic burden on health care systems. Osteoporotic fragility fractures are attributable both to interaction between external, mostly fall-induced loading and bone fragility. The bone fragility, in turn, is largely determined by the bone structure and its particulars [1] and the structural analysis of bones may indeed improve the identification of individual susceptible to fragility fractures [2].

The current image processing and analysis algorithms used in clinical bone research are mostly based on median filtering of image data and application of density thresholds [3, 4]. This practice comes not only from technical simplicity but also from the need for reproducible results. This is, however, not always guaranteed by currently used threshold-based analyses [5, 6], mainly due to the partial volume effect, relatively low signal to noise ratio, and presence of movement artifacts. Obviously, both accurate and precise description of relevant bone structural changes depends on the image quality provided by the used imaging modality. Subsequent image processing cannot overcome inherent limitations of image data e.g. due to limited resolution. Many sophisticated algorithms are also not considered for clinical bone research because of the need for large operator involvement (increasing the risk of decreased reproducibility of results) or high sensitivity to noise. However, any improvement in accurate and precise description of the bone structure is an essential goal for appropriate assessment of the whole bone strength [7].

In this study, we introduce a refined automated segmentation method of a recent OBS segmentation method [8] that provided an accurate detection of cortical bone in images yielded by the peripheral quantitative computed tomography (pQCT). Whereas pQCT systems lack sufficient spatial resolution to capture specific structural traits in comparison with present high-resolution pQCT (HR-pQCT) systems, they are still more available and offer reasonable assessment of the bone cross-sectional geometry and separation it into trabecular and cortical compartments [9].

The performance of the novel approach will be compared to results of manual segmentation (considered as gold standard in vivo analysis in this study), OBS method and recently introduced variational level set formulation for image segmentation [10].

II. MATERIALS AND METHODS

A. The pQCT in vivo Data

In this study, the mean from 2 repeated pQCT scans (XCT 3000, Stratec Medizintechnil GmbH, Pforzheim, Germany) of the distal tibia from 25 volunteers was used in the analysis to reduce random variation so that the differences between different approaches could be better determined. The repeated pQCT scans were obtained from a precision study that was carried out as a part of quality assurance procedure of our bone densitometry unit. Informed consent was obtained from the subjects and the in-house review board approved the study protocol. Both scans were performed on the same day with repositioning according to our standard procedure [9]. The pixel size of the pQCT image was 0.5x0.5 mm, the slice thickness was 2.5 mm.

B. Manual Segmentation

The manual segmentation of the cortical cross-sectional area (CoA) in raw unprocessed pQCT images was performed by 3 experts in segmenting radiological images. The mean from these 3 blinded independent analyses was considered the gold standard of CoA. The manual segmentation was implemented by in-house software developed for segmentation and visualization of radiological.

C. OBS Method

The OBS method for pQCT bone image segmentation was recently introduced by Cervinka et al. 2012 [8]. In short, the method comprises a simple delineation procedure of the outer boundary of cortical bone (based on fast tracing of the peak values in the first derivation of image data) and a subsequent radial shrinking procedure of detected cortical pixels along the outer boundary until the next local maximum of the first derivation, considered as the inner cortical boundary is found.

D. Variational Level Set Based Segmentation

In addition to OBS method, a sophisticated segmentation method based on distance regularized level set evolution (DRLSE) developed by Li et al [10] was performed. The DRLSE algorithm, in contrast to conventional level set formulations, allows elimination of re-initialization; use of large time steps to significantly speed up curve evolution, while ensuring numerical accuracy; and a computationally efficient simple implementation.

E. Enhanced OBS Method

The current OBS algorithm was refined by incorporating morphological inner cortical bone boundary correction that deals with possible "bulgy" contour caused by blurred or lost edges of cortical bone in image data, and then by utilization of two fixed density thresholds (set to 180 and 661mg/cm^3) to correct and enhance cortical bone detection. The morphological correction is triggered by exceeding assumed maximal cortical bone thickness (6 pixels) in detected cortical bone in pQCT image of distal tibia. Algorithm detects places where "leaking" of detected cortical shell, caused by blurred edges (mainly a result of insufficient resolution level, partial volume effect or movement artifacts), appeared and based on a priori knowledge, thin (in sub-pixel measure) cortical bone can be assumed. From so detected pixels, taken as seed points, the morphological erosion operation is repeatedly performed till the cortical thickness in detected place reaches 1 pixel thin layer.

Table 1 Descriptive data of CoA, Correlation, Dice coefficient (mean, SD) as obtained from different segmentation methods

Manual segmentation	CoA		
	150.8 (19.2)		
Observed traits	OBS method	Enhanced OBS	DRLSE
CoA	153.0 (14.7)	156.5 (15.1)	172.6 (15.3)
Correlation	0.88	0.87	0.76
Dice coef.	0.83 (0.04)	0.85 (0.04)	0.81 (0.04)
Range of Dice coef.	0.71 - 0.9	0.77 - 0.93	0.71 - 0.88

Besides morphological refinement, two fixed threshold levels are further utilized for detection of falsely detected cortical bone. The lower threshold (180mg/cm^3 – density level determining that one quarter of the pixel contains bone tissue) provides the second refinement of detected cortical bone in places where insufficient resolution level and partial volume effect artificially increase thickness of detected cortical shell (although the correct thickness is in sub-pixel measure), however, results in decrease of bone mineral density range in such a place. Therefore, pixels within detected cortical shell that have bone mineral density below the lower threshold are detected and the cortical shell thickness in these places is then set to 1px thin layer. Application of the higher threshold (661mg/cm^3) allows further enhancement (or correction of previously applied refinement steps) of accuracy in detection of cortical bone shell by incorporating of all pixel that belong to cortical bone compartment with high probability (their bone mineral density values are above the higher threshold) but are not detected by the original OBS method [8]. The threshold value was selected based on previous findings of Kontulainen et al. 2007 [6]: the results of simple threshold based cortical bone analysis, with utilization of above mentioned value, agreed well with bone histomorphomotery results.

F. Statistical Analysis

As descriptive statistics, the means and standard deviation (SD) are given. Besides the segmented CoA, the Pearson correlation coefficients and Dice similarity coefficients were determined as relevant measures of CoA segmentation accuracy. The Dice similarity coefficient was calculated as follows:

$$Dice\ coef. = \frac{2p}{2p+d+q} \quad (1)$$

where p is the number of pixels where the cortical bone was detected in both, results from manual segmentation and the method of interest, d and q indicate the number of pixels that are detected only either in results from manual

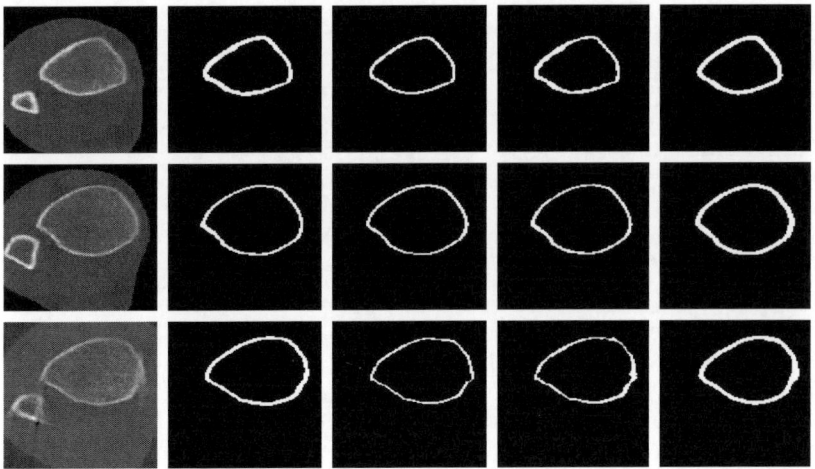

Fig. 1 Examples of segmentation performance of particular approaches on different data sets, from left to right: Original pQCT images of distal tibia, results of manual segmentation, results of original OBS method, results of enhanced OBS method, and results of DRLSE algorithm

segmentation or the method of interest, respectively. In addition, the Bland-Altman plots illustrating the differences between the manual segmentation and the method of interest are shown for CoA results.

III. RESULTS

Differences in the distal tibia CoA, Pearson correlation coefficient and Dice similarity coefficient from various segmentation methods are shown in Table 1.

In general, all tested methods show high correlation of CoA segmentation results in comparison to the manual segmentation, however, the sophisticated DRLSE based segmentation results in nearly 15% overestimation of the CoA in comparison to 2% and 4% for the original OBS method and the enhanced OBS method, respectively. The difference between CoA and correlation results, in comparison to manual segmentation, obtained from original OBS method and enhanced OBS method are marginal. Nevertheless, Dice similarity coefficient shows an improvement in agreement of CoA's segmented by enhanced OBS method in comparison to manual segmentation. This improvement in segmentation performance can be clearly seen on various examples in Fig. 1. However, as shown by the example in the third line of Fig. 1., the enhanced OBS method can fail to correctly segment CoA when severe artifacts are presented in image data.

Further, the Bland-Altman plots illustrating the mean agreement of the CoA assessment between different segmentation methods and the manual segmentation are shown in Fig. 2.

IV. DISUSSION

In this work, a novel modification of automatic segmentation method OBS [8] designed for accurate assessment of the cortical bone geometry from pQCT images was introduced. The method comprises the original OBS algorithm, utilization of morphologically based correction of inner cortical bone border over segmentation due to blurred edges of cortical shell and utilization of two fixed threshold values that further refine final cortical bone detection. The aim of present study was to investigate whether the enhanced OBS method can improve relatively high performance of the originally developed OBS method in the *in vivo* assessment of tibial cortical bone geometry [8]. In addition, an advanced level set based approach (DRLSE [10]) was used as an example of more sophisticated image processing algorithms. The enhanced OBS method showed similar performance in assessment of standard geometrical trait (CoA) analysis in comparison to the original OBS method [8], however, the Dice measure of similarity suggested an improvement that could lead to more consistent cortical bone detection.

Recently, the cortical bone geometry has been pinpointed as one of most important determinants of the whole bone strength [11, 12]. Evidently, incorrect determination of the cortical geometry can compromise the estimation of the whole bone strength in clinical research. Therefore, the accurate assessment of relevant cortical bone geometry traits is necessary for the reliable estimation of the whole bone strength.

The present study showed that there seems to be no large improvement between CoA analysis performed by original

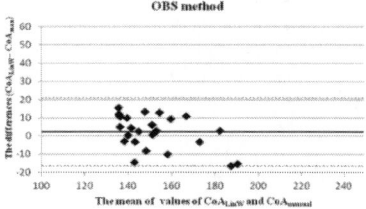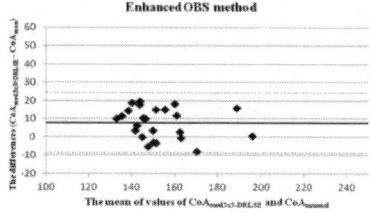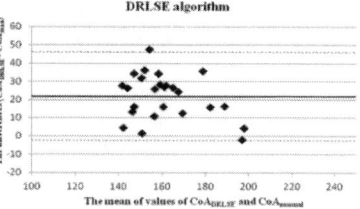

Fig. 2 Bland-Altman plots for each CoA analysis approach

and enhanced OBS method (Table 1 and Fig. 2), although they both outperform the DRLSE approach. However, the Dice similarity coefficients (Table 1) and results presented in Fig. 1. suggest a reasonable improvement in correct cortical bone detection performed by enhanced OBS method, although, the correct detection may fail with presence of artifacts that radically change the quality of image data.

There are some issues that need further discussion. The utilization of manual segmentation as a gold standard may be questioned. Obviously, the performance of human raters largely depends on previous experience with analyzing of image data and can substantially vary between raters as can be seen in Fig. 2. Therefore, in this study we used the mean of 3 independent manual segmentations of cortical bone as the gold standard to reduce possible variance in determination of CoA. However, to reveal the correct accuracy of presented enhanced OBS method and reveal whether the method can outperform even results of the manual segmentation, further analyses based on known, realistic model data need to be performed.

V. CONCLUSIONS

This study showed that the enhanced OBS algorithm performed reasonably well as the Dice's measure of similarity of segmented cortical bones was somewhat greater indicating more consistent, even anatomically sector wise, cortical bone geometry assessment of pQCT images. In broader perspective, more reliable cortical analysis may result in improved estimation of the whole bone strength.

ACKNOWLEDGMENT

We would like to thank to Markus Hannula and Nathaniel Narra for their help with manual assessing of the cortical bone compartment. Further, we acknowledge the Finnish Cultural Foundation and Research School of Tampere University of Technology for financial support of the doctoral studies for TC. This study was also supported by the Competitive Research Funding of the Tampere University Hospital, Pirkanmaa Hosptital District (Grant 9K121).

REFERENCES

1. Jarvinen TLN, Sievanen H, Jokihaara J et al. (2005) Revival of Bone Strength: The Bottom Line. J Bone Min Res 20:717-720
2. Mayhew P, Kaptoge S, Loveridge N et al (2004). Discrimination between cases of hip fracture and controls is improved by hip structural analysis compared to areal bone mineral density: An ex vivo study of the femoral neck. Bone 34:352-361
3. Hangartner TN (2007) Thresholding technique for accurate analysis of density and geometry in QCT, pQCT and microCT images. J Musculoskelet Neuronal Interact 7:9-16
4. Veitch SW, Findlay SC, Ingle BM et al. (2004) Accuracy and precision of peripheral quantitative computed tomography measurements at the tibial metaphysis. J Clin Densitom 7:209-217
5. Ward KA, Adams JE, Hangartner TN (2005) Recommendations for thresholds for cortical bone geometry and density measurement by peripheral quantitative computed tomography. Calcif Tissue Int 77:275-280
6. Kontulainen S, Liu D, Manske S et al. (2007) Analyzing cortical bone cross-sectional geometry by peripheral QCT: comparison with bone histomorphometry. J Clin Densitom 10:86-92
7. Sievanen H, Kannus P, Jarvinen TL (2007) Bone quality: an empty term. PLoS Med 4:e27
8. Cervinka T, Hyttinen J, Sievanen H (2012) Threshold-Free Automatic Detection of Cortical Bone Geometry by Peripheral Quantitative Computed Tomography. J Clin Denstom 15:413-421
9. Sievänen H, Koskue V, Rauhio A et al. (1998) Peripheral quantitative computed tomography in human long bones: evaluation of in vitro and in vivo precision. J Bone Miner Res 13:871-882
10. Li C, Xu C, Gui C et al. (2010) Distance regularized level set evolution and its application to image segmentation. IEEE Trans Image Process 19:3243-3254.
11. Melton LJ 3rd, Riggs BL, Keaveny TM et al. (2007) Structural determinants of vertebral fracture risk. J Bone Miner Res 22:1885-1892
12. Holzer G, von Skrbensky G, Holzer LA et al. (2009) Hip fractures and the contribution of cortical versus trabecular bone to femoral neck strength. J Bone Miner Res 24:468-474.

Author:	Tomas Cervinka
Institute:	BioMediTech, Dep. of Elect. and Com. Engineering
Street:	Biokatu 6, 4th floor
City:	Tampere
Country:	Finland
Email:	tomas.cervinka@tut.fi

Spatial Aliasing and EMG Amplitude in Time and Space: Simulated Action Potential Maps

B. Afsharipour, K. Ullah, and R. Merletti

Laboratory for Engineering of the Neuromuscular System, Politecnico di Torino, Torino, Italy

Abstract—**Average Rectified Value (ARV) and Root Mean Square (RMS) are the two amplitude indicators that are commonly used in the field of EMG. These two indicators are compared a) analytically for a one dimensional single sinusoid, sum of sinusoids, two dimensional sinusoids, and b) numerically by simulating a high density detection system for sampling the distribution of propagating surface action potentials generated by a muscle motor unit (MU). Results show that the RMS does not depend on the sampling rate while ARV does, even when the sampling theorem is satisfied. This is important since, in high density surface EMG detection systems (HDsEMG), the inter electrode distance (IED) samples the surface potentials often below the Nyquist frequency, generating aliasing in space. The largest IED that avoids spatial aliasing for the simulated MU is 10mm. IEDs below this value are recommended for experimental measurements.**

Keywords—**EMG Amplitude, ARV, RMS, Spatial aliasing, Optimal IED.**

I. Introduction

Surface EMG (sEMG) signals are the algebraic summation of the motor unit action potentials (MUAP), occurring within the detection volume of the electrodes. Since MUAPs are not synchronized, constructive and destructive superimpositions occur [1]. Theoretically this problem can be avoided by recording from each motor unit separately. Although this is practically unfeasible, it suggests the use of multiple, spatially distributed EMG channels, collecting different views of separate sources that could then be separated.

A two dimensional electrode grid including NxM electrodes (equally spaced in "x" and "y" direction), amplifiers and recording tools, make the EMG imaging technique applicable. High-density sEMG (HDsEMG) grids collect NxM monopolar EMG signals over a relatively small surface. Each electrode may be conceived as a pixel p with coordinates (x , y). sEMG distribution versus time can be considered as a movie whose frames represent the instantaneous amplitude of the NxM channels.

The amplitude indicators (root mean square (RMS) and average rectified values (ARV) for continuous and sampled signals) in time and space, are discussed in this study. The main research question is: Is one indicator preferable to the other? Spatial aliasing comes from sampling the distribution of surface EMG in space and is related to the inter electrode distance (IED). This issue and the issue of image truncation have not yet been addressed in the literature. In this work the study of ARV and RMS is carried out from the simplest cases (one dimensional (1D) single sinusoid) toward more complicated signals (1D sum of sinusoids), 2D sinusoids, and simulated action potential (AP) signals in space (1D and 2D). Spatial aliasing and inter electrode distance are also discussed for a simulated single fiber and a motor unit.

II. Materials and Method

EMG amplitude estimation is the task of best estimating the standard deviation of a zero mean colored random process in additive noise. This estimation problem has been studied in time, for several decades using a full wave rectifier followed by a low-pass filter [2, 3]. Since any signal can be described as the summation of sine and cosine functions in Fourier series, this study started from the case of a single sinusoid in space or time and evolved toward simulated motor unit action potentials sampled over the skin by a two dimensional detection grid. ARV and RMS of 1D and 2D sinusoids are analytically discussed in sections A, B, C, and D. In section E, the issue of spatial aliasing for a simulated muscle fiber action potential and the simulated motor unit is addressed.

Aliasing refers to the distortion or artifact, when the signal reconstructed from its samples, is different from the original continuous signal. In this section spatial aliasing and its effect in simulated surface action potential is discussed to provide estimates of the spatial Nyquist frequency. HDsEMG electrodes are small (1 mm^2 to 25 mm^2). For the sake of simplicity, they will be considered in this section as point-like. Two cases were considered:

Case a) To study the effect of sampling in space, a previously developed model [4] was used to simulate the monopolar potential distribution generated by the propagating action potential of a single muscle fiber parallel to the skin placed at different depth in the muscle. Skin and fat layers thickness were considered 1 mm and 3 mm respectively.

The detection system was defined as 128x128 electrodes (grid) with 1mm inter electrode distance (IED) as the reference. The center of detection grid was placed over the neuromuscular junction. The fiber length was considered 100mm (upper semi fiber length = 55mm, lower semi fiber length = 45mm). The muscle anisotropy ratio is 5. Monopolar skin potentials were simulated.

Case b) A motor unit with circular territory (radius = 15mm) including 150 fibers (uniform distribution) was simulated. The detecting area over the skin surface was defined 128x128mm^2. The lower and upper semi fiber lengths were considered 60mm and 65mm respectively. The mean fiber depth in the muscle was 15.5mm and the innervation zone (IZ) spread was 10mm. The upper and lower spreads of tendon regions were 10mm and 8mm respectively. All other model parameters are as in case a).

III. RESULTS AND DISCUSSION

A. Average Rectified Value of a Single Sinusoid (ARV)

The ARV of a continuous sinusoid is phase independent and given by:

$$ARV = \frac{1}{T}\int_{-\frac{\varphi T}{2\pi}}^{T-\frac{\varphi T}{2\pi}} |A \sin\left(\frac{2\pi t}{T} + \varphi\right)| dt = \frac{2A}{\pi} \quad (1)$$

where A is amplitude, T is the period, $f_0 = 1/T$ and φ is the phase. In the discrete case, the sampling interval is $\Delta t = T/N$, $F_{samp.} = 1/\Delta t = N/T$, where T [s] is one period, N is the number of samples per period, and n is the sample index $\in$ [0 to N-1]. The ARV is given by:

$$ARV = \frac{2}{N}\sum_{n=0}^{\left|\frac{N}{2}-1\right|} A \sin(2\pi n/N) = \frac{2A}{N}\left(\frac{\cos(\pi/N)}{\sin(\pi/N)}\right) \quad (2)$$

with $\alpha = \frac{2\pi}{N}$ = the sampling interval in radiant:

$$lim_{N\to\infty} ARV = \lim_{\alpha\to 0}\left(\frac{2A}{N}\left(\frac{\cos(\alpha/2)}{\sin(\alpha/2)}\right)\right) = \frac{2A}{\lim_{\alpha\to 0}\left(N\frac{\sin(\alpha/2)}{\cos(\alpha/2)}\right)} = \frac{4A}{N\alpha} = \frac{2A}{\pi};$$

We consider the general case in which N is a rational number and study the RMS and ARV when the signal is sampled slightly above or below the Nyquist frequency. In order to do that, we need to consider an integer number of periods K, such that KN is an integer. Therefore, KN is the total number of samples in K periods and consequently, the time interval between samples will be $\Delta t = KT/KN$, $F_{samp} = 1/\Delta t$ and $n \in [0, KN-1]$ is the sample index. Considering $\alpha = 2\pi/N$, ARV is:

$$ARV = \frac{1}{KN}(\sum_{n=0}^{KN-1}|A \sin(2n\alpha)|) =$$

$$\frac{1}{KN}\left(\sum_{n=0}^{\left|\frac{KN}{2}-1\right|} A\sin(2n\alpha) + \sum_{\left|n=\frac{KN}{2}\right|}^{KN-1}(-A\sin(2n\alpha))\right)$$

Applying the series summation $\sum_{n=0}^{N}\sin(nx) = \frac{\sin(Nx/2)\sin((N+1)x/2)}{\sin(x/2)}$, the ARV can be expressed as:

$$ARV = 2A \sin\left(\left|\frac{N}{2} - 1\right|\left(\frac{\alpha}{2}\right)\right)/(N\sin(\alpha/2)) \quad (3)$$

In conclusion, the ARV depends on the total number of samples that are taken in K periods and approaches its expected value $2A/\pi$ as $N \to \infty$ (estimation error <1% is obtained for N>18).

B. Root Mean Square of a Single Sinusoid

RMS^2 of a sinusoid in the continuous case is:

$$RMS^2 = \frac{1}{T}\int_{-\frac{\varphi T}{2\pi}}^{T-\frac{\varphi T}{2\pi}} A^2 \sin^2\left(2\pi f_0 t + \varphi\right) dt = \frac{A}{2} \quad (4)$$

By discretizing (4) and applying trigonometric properties, RMS^2 can be rewritten as:

$$RMS^2 = \frac{A^2}{2} - \frac{A^2}{2KN}\sum_{n=0}^{KN-1} Real\left[e^{j2\left(\frac{2\pi n}{N}+\varphi\right)}\right] \quad (5)$$

where the general case is considered, when N is a rational number and KN is an integer representing the number of samples in K periods. It can be shown analytically from eq. (5) that at the frequency of signal (f_0) and the Nyquist frequency $(2f_0)$ the RMS is phase dependent and is: $RMS = A\sin(\varphi)$, $0 \leq \varphi < 2\pi$ while at any other sampling frequency the $RMS^2 = A^2/2$.

These calculations reveal that the RMS of a single sinusoid is correct even at sampling frequencies below the Nyquist frequency, except at the frequency of sinusoid and its Nyquist frequency, and does not depend on the number of samples while the ARV is a function of N and behaves as eq.3.

C. Root Mean Square of Sum of Sinusoids

A finite series of Fourier harmonics can be defined mathematically as follows:

$$x(t) = \sum_{i=1}^{M} A_i \sin(2\pi f_i t + \varphi_i) \quad (6)$$

where M is the integer number of individual sinusoids, $f_i = if_0$ is the frequency (i[th] harmonic) and A_i [a.u.] is the amplitude of the ith sinusoid. φ_i is the phase [rad] of the ith sinusoid and t [s] is the time. In the continuous case it can be shown that:

$$RMS^2 = \int_0^T \left(\sum_{i=1}^M A_i \sin(2\pi f_i t + \varphi_i)\right)^2 dt = \sum_{i=1}^M RMS_i^2 \quad (7)$$

In the discrete case, The RMS^2 of (7) is calculated as follows:

$$RMS^2 = \frac{1}{N}\sum_{i=1}^M \sum_{n=0}^{N-1} \left(A_i e^{j\left(\frac{2\pi f_i n}{N} + \varphi_i\right)}\right)^2 +$$
$$\frac{1}{N}\sum_{n=0}^{N-1} \sum_{m=1}^M \sum_{l=1}^M A_m A_l e^{j\left(\frac{2\pi f_m n}{N} + \varphi_m\right)} e^{j\left(\frac{2\pi f_l n}{N} + \varphi_l\right)}, l \neq m$$

that can be summarized in two conditions mentioned below where $Ny.freq.$ is the Nyquist frequency:

$F_{samp.} > Ny.freq. \rightarrow RMS^2 = \sum_{i=1}^M RMS_i^2$

$F_{samp.} \leq Ny.freq.$:

if $\forall m, \forall l, m \neq l, F_{samp.} = f_m + f_l$: $RMS^2 \neq \sum_{i=1}^M RMS_i^2$

if $\forall m, \forall l, m \neq l, F_{samp.} \neq f_m + f_l$: $RMS^2 = \sum_{i=1}^M RMS_i^2$

D. RMS and ARV of a 2D Sinusoid

Consider an image $f(x,y)$ obtained by multiplication of two sinusoids along x and y direction such that:

$$f(x,y) = A_x \sin(2\pi f_x x + \varphi_x) * A_y \sin(2\pi f_y y + \varphi_y) \quad (8)$$

where f_x and f_y are the frequencies [cycles/m], A_x, A_y are the amplitudes [a.u.], and φ_x, φ_y are the phases [rad] of the sinusoids along x and y directions respectively. When the sampling theorem is satisfied, the RMS^2 of the image is computed as (9):

$$RMS^2 = \frac{1}{AB}\left(\int_0^A \int_0^B f^2(x,y) dx dy\right) = RMS_x^2 * RMS_y^2 \quad (9)$$

where A and B are specifying the region (that is being sampled along x and y respectively) containing an integer number of periods. RMS and ARV versus sampling frequencies of an image defined as $f(x,y) = 2\sin(2\pi 5x + \pi/3) * 3\sin(2\pi 3y + \pi/4)$ are shown in Fig.1. There are no changes in spatial RMS values for any sampling frequency greater than the Nyquist rate, while for the ARV this is not true. The ARV estimation error is <1% for N>18 samples/period, a condition usually not met in EMG spatial sampling.

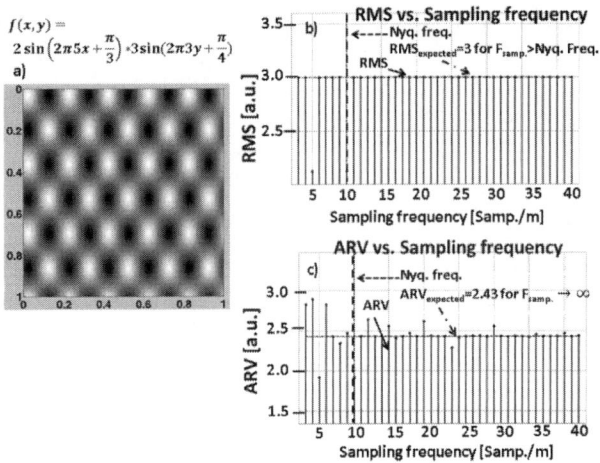

Fig. 1 a) Example of sinusoid image defined as $f(x,y) = 2\sin(2\pi 5x + \pi/3) * 3\sin(2\pi 3y + \pi/4)$; b) RMS and c) ARV of the image versus sampling frequency (same for x and y). The Nyquist frequency is 10 samples/m for y and 6 samples/m for x.

E. Spatial Aliasing in Potentials Detected over the Skin

Results of case a) described in section II and presented in [5] show that when a single fiber surface potential distribution is sampled with 200 and 100 samples/m (i.e., IED = 5mm and 10mm) aliasing is negligible for IED=5mm but not for IED=10mm. The highest image spatial frequency of interest is near 100 cycles/m. Therefore, the sampling frequency must be higher than 200 cycles/m, that is, IED < 5mm. For a potential travelling at CV = 4 m/s, 100 cycles/m corresponds to a time domain frequency of 400 Hz, which is also the first zero of a differential spatial filter with 10 mm IED (first Lindstrom dip). Estimates of the RMS or the ARV of the image, computation of the image center of gravity, as well as any interpolation or segmentation process, are all affected by the spatial sampling frequency and truncation effects [5]. Results of case b) described in the Materials and Method are shown in panels a), b), and c) of Fig.2 and presenting a motor unit surface potential distribution detected with 1mm, 5mm, and 10mm IED systems over a surface of 128x128 mm² respectively. Panels d) and e) of Fig.2 show one cycle of the amplitude spectrum in space for sampling frequency = 200 and 100 samples/m. Panel f) in Fig2. depicts the ARV and RMS of the sampled surface potentials versus the IED. Panel e) in Fig. 2 shows the effect of aliasing for a simulated motor unit while for a superficial single fiber action potential (case a) aliasing started from smaller IED = 5mm [5]. Case a) might not happen in real measurements where more than a fiber exists in a muscle. In a real condition, the sEMG is the summation of the action potentials from recruited number of motor units that provides a smoother distribution over the skin surface. The HDsEMG electrodes were considered point-like, while in real measurement they have physical dimensions. Physical dimensions of an electrode imply a smoothing (low-pass) effect that is a two-dimensional spatial filter transfer function, which is a 2D sinc function for rectangular and a Bessel function for a circular electrode [2]. These functions could be considered as antialiasing filters in space with cutoff frequencies $f_c \approx$ 319, 80, and 64 cycles/m (corresponding to 1276, 320, and 256Hz in time domain considering conduction velocity = 4m/s) introduced by a 1x1, 4x4, and 5x5mm^2 square electrodes respectively. The pattern behavior seen in panel f) of Fig.2 for both ARV and RMS is due to the truncation of the EMG image that provides dc variable components to the sampled image. Finally, it should be noted that propagating action potentials are considered. Images showing the end of fiber effect must also be considered and will be addressed in future works.

IV. CONCLUSION

Results show that the RMS is independent from sampling frequency f_s for $f_s > f_{Nyq.}$, while ARV is dependent on f_s even for $f_s > f_{Nyq}$. These results suggest that RMS should be preferred as an amplitude indicator either in time or space. Spatial aliasing appears when the IED increases in a HDsEMG detection system. For a simulated motor unit including 150 fibers uniformly distributed in a circular territory (radius=15mm), the largest IED to avoid spatial aliasing is about 10mm and this value is recommended as maximum IED for HDsEMG recordings.

ACKNOWLEDGEMENT

The authors would like to appreciate Dr. Loredana Lo Conte for her guidelines. This study was supported by Lagrange fellowship and by EPA (Ergonomics Prevention and Environment) s.r.l.

REFERENCES

1. Staudenmann D, Idsart K, Daffertshofer A et al. (2006) Improving EMG-based muscle force estimation by using a high-density EMG grid and principal component analysis. Biomed. Eng. IEEE Trans. 53(4):712-719
2. Merletti R, Prker P A (2004) Electromyography Physiology, Engineering, and Noninvasive Applications. IEEE PRESS
3. Inman V T, Ralston H J, Saunders J et al. (1952) Relation of human electromyogram to muscular tension. Electroencephalography and Clinical Neurophysiology 4(2):187-194
4. Farina D, Merletti R (2001) A novel approach for precise simulation of the EMG signal detected by surface electrodes. Biomed. Eng. IEEE Trans. 48(6):637-646
5. Merletti R, Afsharipour B, Piervirgili G (2012) High Density Surface EMG Technology, ICNR Proc. Vol2, Converging Clinical and Engineering Research on Neurorehabilitation, Toledo, Spain, 2012, pp 1207-1210

Corresponding author: Roberto Merletti

Institute: Laboratory for Engineering of the Neuromuscular System, Politecnico di Torino
Street: Via Cavalli 22/H-10138
City: Torino
Country: Italy
Email: roberto.merletti@polito.it, babak.afsharipour@polito.it

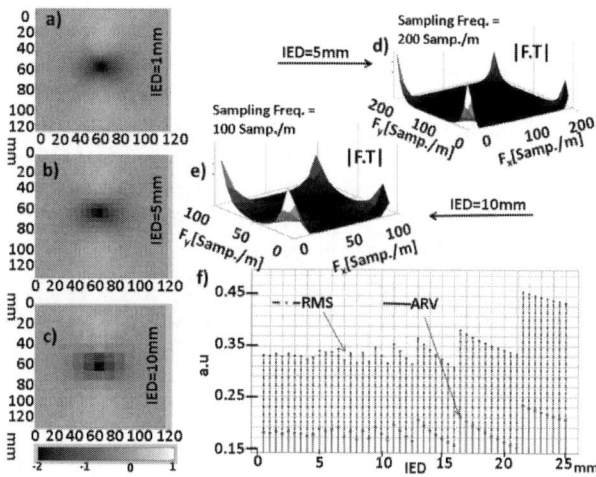

Fig. 2 a), b) and c). Distribution of the simulated monopolar surface potential in the propagation phase produced by a single Motor Unit including 150 fibers uniformly distributed in the circular territory (mean depth of the fibers in the muscle = 15.5mm, territory radius=15 mm), 1 mm skin and 3mm subcutaneous tissue thickness. The action potential is in the propagating phase. The 128 x 128 mm skin surface is sampled at 1000 samples/m (reference case, IED=1mm), 200 samples/m (IED= 5 mm), and 100 samples/m (IED=10 mm). Panels d) and e) magnitude of the 2-D Fourier transform for the last two cases, displayed up to the sampling frequency represented as gray levels; Black = 0. Panel f) shows the ARV and RMS of sampled distribution of simulated motor unit surface potential versus inter electrode distance.

Lacunarity-Based Inherent Texture Correlation Approach for Wireless Capsule Endoscopy Image Analysis

V.S. Charisis, L.J. Hadjileontiadis, and G.D. Sergiadis

Aristotle University of Thessaloniki/Department of Electrical and Computer Engineering, Thessaloniki, Greece

Abstract—Wireless capsule endoscopy (WCE) is a novel technology that offers non-invasive visual inspection of the digestive tract (DT) and especially small bowel. However, the revision of the large amount of images produced is highly time-consuming and prone to human error. This weakness was the rationale to propose a novel strategy for automatic detection of WCE images related to ulcer, one of the most common findings of DT. This paper introduces a new texture extraction method based on the inherent texture correlation measure and lacunarity index. WCE data are pre-processed with Bidimensional Ensemble Empirical Mode Decomposition to reveal their congenital structure primitives, and also to adaptively reconstruct new refined images. Classification results demonstrated promising classification accuracy (89.1%) exhibiting high potential towards further research in this field.

Keywords—feature extraction, multiscale analysis, ulcer.

I. INTRODUCTION

The novel Wireless Capsule Endoscopy (WCE) [1] has marked a revolution in the field of gastroenterology, beginning an era of comfortable visualization of the entire gastrointestinal (GI) track, even the middle part of small intestine [2]. The innovative endoscopic capsule travels through the GI tract capturing and wirelessly transmitting 55000 frames. At the end of the examination these images, which are temporarily stored in a wearable recorder, are downloaded to a computer and reviewed by the physician. The burdensome task of image inspection costs 1 to 2 hours of intense labor for an experienced physician who needs to stay focused for such a long time. Moreover, abnormalities may appear in only one or two frames and be missed easily due to oversight. Researchers, inspired by these weaknesses, attempt to alleviate gastroenterologists' workload and the examination costs by developing computer-assisting systems.

Various approaches, as recorded in the literature, are geared to this notion. Automatic detection of abnormalities is the most prevalently adopted concept. The techniques proposed for tumor and polyp detection entail texture spectrum along and neural networks [3], [4], co-occurrence matrix-based geometric and texture features [5], SUSAN detectors along with Gabor filters [6], wavelet-based local binary pattern analysis (LBP) [7] and rotation invariant uniform LBP [4]. Ulcer detection efforts are based on color and texture characteristics elicited from several color models. In particular, curvelet-based LBP [8], color moment invariants in conjunction with LBP and Log-Gabor filters along with statistical measures[9], texture spectrum and unit number transforms [10], color channel histograms [11] and color wavelet covariance [12] have contributed to this direction. Moreover, ulcer recognition is achieved by processing WCE images in RGB, HSV and CIE Lab color spaces with bidimensional ensemble empirical mode decomposition (BEEMD) [13] and by extracting efficient color-texture features through a dual intrinsic higher-order correlation (IHC) - lacunarity (LAC) scheme [14].

In this study the aim is to augment the work presented in [14] and propose a novel approach for automatic detection of ulcer, one of the most common GI diseases. Thus, a new LAC-based Inherent Texture Correlation (L-InTeCo) scheme for WCE image analysis is introduced. WCE images are decomposed by BEEMD to their innate components, called intrinsic mode functions (IMFs) in order to disclose the important structures of the images. Then, L-InTeCo scheme elicits efficient color texture characteristics by analyzing: a) the second and higher-order correlation between the intrinsic texture primitives of the WCE data, and b) the pixel intensity distribution within adaptively refined [15] WCE data. At last, texture analysis output is classified into healthy or ulcerous with the aid of Support Vector Machine.

II. MATERIALS AND METHODS

A. Materials

The proposed scheme was realized in MATLAB 7.9. The WCE images used in this study for the development and assessment of the proposed approach were drawn from ten patients with ulcerous diseases, such as NSAID ulcerations and Crohn's disease. The dataset consists of 170 ulcer and 170 normal images. Two experts reviewed the endoscopic video and manually isolated regions of interest (ROI). Moreover, the ulcer ROIs were equally divided into two classes: easy (distinct ulcers of high severity) and hard (hard to detect, low severity ulcers). It must be highlighted that the normal ROIs include both simple and confusing healthy tissue (folds, villus etc) in order to hamper the detection process.

B. Bidimensional Ensemble Empirical Mode Decomposition

In 1998, Huang et al. introduced Empirical Mode Decomposition (EMD), an innovative procedure for adaptive, data-driven, time - frequency decomposition and analysis of nonlinear nonstationary signals without the requisite of a priori basis functions. EMD is designed to divulge, through a sifting technique, the oscillatory components that are inherent in every signal, namely Intrinsic Mode Functions (IMFs). A disadvantage of EMD that results in inaccurate IMFs is mode mixing phenomenon that occurs when a signal has unevenly distributed extrema. This issue is addressed by ensemble EMD (EEMD), a noise-assisted version of EMD. EEMD applies EMD on multiple distorted (by Gaussian white noise) copies of the original signal and the IMFs are calculated by averaging the IMFs of the copies. Bidimensional EEMD (BEEMD) [16] is a two-dimensional (2D) approach of EEMD for the analysis of 2D signals. BEEMD, as an alternative space-frequency analysis technique for efficient multi-scale image processing and pattern extraction [17], decomposes an image to its inherent spatial frequency components (2D IMFs) that reveal its textural structure. BEEMD is implemented by transplanting the idea of EEMD algorithm to 2D data.

C. Differential Lacunarity Analysis

Lacunarity is a fractal property that was introduced to discriminate surfaces that share the same fractal dimension. More specifically, lacunarity evaluates the distribution of gap sizes along datasets. Heterogeneous, high-lacunarity sets contain gaps of divergent lengths while sets with even gaps are homogeneous and present low lacunarity. Nevertheless, a heterogeneous set at small scales can be considered homogeneous when observed at larger scales and vice versa. From this perspective, lacunarity can be regarded as a scale dependent measure of heterogeneity of texture. The most simple, yet efficient, algorithm to calculate lacunarity is the "gliding box algorithm" (GBA) [18], suitable for binary data. However, most real life image analysis applications need to elicit texture data from images that cannot be thresholded to binary. To this end, Dong [19] introduced differential lacunarity (LAC) for grayscale image analysis. The calculation of LAC is based on a differential box counting algorithm that utilizes a gliding window w (w x w pixels) similar to the one used in GBA and a gliding box r (r x r pixels, r<w), essential for the calculation of "window mass".

D. Correlation Measurement

The most familiar measure of dependence between two elements z_i and z_j of a vector $Z=[z_1,...,z_k]$ is Pearson product-moment correlation coefficient. The second-order correlation measure between z_i, z_j is defined as:

$$r_{ij} = \text{cov}[\phi(z_i), z_j]/\sqrt{\text{cov}[\phi(z_i)]\text{cov}[z_j]}, i \neq j, \quad (1)$$

where $\Phi(z_i) = z_i$, cov is the covariance and $0<|r_{ij}|<1$. In case $\Phi(z_i)$ denotes a nonlinear function, i.e. $\Phi(z_i) = z_i^2 + z_i^3$, (1) defines a higher-order correlation measure (hr_{ij}) between z_i, z_j. Let D_i and HD_i denote the second and higher-order overall correlation measurements, of z_i relative to all the other components z_j of Z [20]. D_i and HD_i are defined as:

$$D_i = \sum_j (r_{ij})^2, j=1:k, \quad HD_i = \sum_j (hr_{ij})^2, j=1:k \quad (2)$$

Vectors $\mathbf{D}=[D_1,...,D_k]$ and $\mathbf{HD}=[HD_1,...,HD_k]$ express the inter-relationships between the components of Z and have been widely used for blind source separation problems [21]. In this work, $\mathbf{D}$ and $\mathbf{HD}$ measures are employed in order to investigate how the texture dependencies within WCE data alter according to the type of tissue depicted.

E. The Proposed Approach

The aim of this work is the color-texture-based detection of ulcer tissue from WCE data. The color-texture concept was motivated by physicians' clinical practice, where the color and texture properties of WCE data are utilized for reaching a diagnosis. A flow chart of the proposed approach is shown in Fig. 1. At first, WCE images are analysed to their color components. It has been shown [13] that Red-Green-Blue (RGB) is one of the most efficient spaces for WCE-based ulcer detection. In the second step, BEEMD is employed in order to decompose WCE images into their intrinsic structural components (IMFs) and achieve a dual objective, refinement and pattern-extraction facilitation. WCE images due to the ambient conditions and the hardware used are likely to exhibit misleading content that needs to be eliminated. A refined image is reconstructed through an adaptive process [15] where only the most informative IMFs are selected based on gradient criteria of the IMF intensity distribution curves. The next step involves texture features extraction through LAC-based Inherent Texture Correlation and LAC analysis (L-InTeCo). InTeCo applies

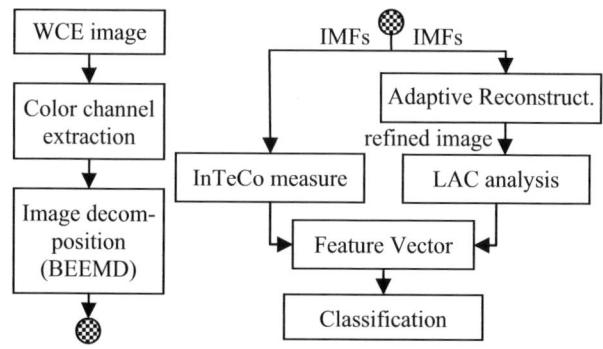

Fig. 1 Flow chart of the proposed L-InTeCo detection scheme

LAC analysis on the individual IMFs in order to extract and quantify the inherent textural components of the images. LAC is calculated for constant box size (r=4) and various window sizes (w=5-30) resulting in a LAC-w curve. Then, second and higher-order correlation (D_i, HD_i) between the LAC curves of IMFs and residue are computed in order to investigate the relationships between the image structural components. The overall correlation vectors D and HD are calculated by (2). Apart from InTeCo, LAC analysis is applied on the adaptively refined image for multi-scale texture features extraction, based on the pixel intensity distribution [13]. The concept of reducing the feature space dimension without omitting crucial information introduces the essence of modeling LAC-w curve by the hyperbola function [14]. The combination of the three hyperbola parameters (*a*, *b*, *c*) (global behavior) with the first five LAC values (local behavior) delivers optimum results with significant FV dimension reduction (69%). The texture features and correlation measurements obtained from the previous step are combined in a feature vector (FV). Various FV scenarios are examined in order to draw more conclusive results about the efficiency of the ulcer information extracted. More specifically, FV embodies either the texture correlation measurements, or a combination of the correlation measurements and LAC values. At last, the tissue classification procedure into healthy and ulcerous takes place by Support Vector Machine (SVM) with radial basis function.

III. RESULTS

The performance of the proposed L-InTeCo scheme is evaluated through the experimental results derived from the application of the introduced approach to our dataset. To this end, results from the various texture feature scenarios applied not only on both ulcer cases (easy, hard), but also on their concatenation (total), are presented in this section.

Table 1 presents the classification performance of the proposed approach when the FV contains the inherent texture correlation data (D, HD, both (D+HD)) extracted from green channel (G) exclusively. A related study has shown that the majority of ulcer information lies on G plane [13], rendering it more efficient for analysis than the other channels since it exhibits improved overall performance [14]. The highest classification rates achieved for easy, hard and total ulcer cases are 89.8%, 78.7% and 85.3%, respectively. As far as the FV scenario is concerned, higher-order correlation data are more informative than second-order correlation data by achieving 3.5 to 6.5 percentage points (pp) higher accuracy for all ulcer cases. The combination of D and HD is proven to be even more powerful (1.1 pp, 3.2 pp, 0.9 pp higher accuracy than HD for easy, hard and total cases respectively) delivering the best performance. Table 2

Table 1 Classification results (%) for InTeCo feature vector

SVM	Classification Scenario								
	Easy			Hard			Total		
	D	HD	D+HD	D	HD	D+HD	D	HD	D+HD
Acc.	85.2	88.7	89.8	69.1	75.5	78.7	77.9	84.4	85.3
Sens.	82.5	89.9	90.9	64.5	72.9	73.1	82.5	86.8	88.7
Spec.	88.2	87.3	89.1	74.0	78.4	83.8	73.2	81.9	81.8

Table 2 Classification results (%) for L-InTeCo feature vector

SVM	Classification Scenario								
	Easy			Hard			Total		
	D+LAC	HD+LAC	D+HD+LAC	D+LAC	HD+LAC	D+HD+LAC	D+LAC	HD+LAC	D+HD+LAC
Acc.	88.4	90.8	93.1	73.5	81.8	83.9	86.0	87.3	89.1
Sens.	89.2	93.6	94.5	68.9	78.3	79.6	85.9	88.4	89.7
Spec.	87.5	88.7	91.9	78.7	86.0	88.7	86.1	86.2	88.4

Table 3 Comparison of L-InTeCo with AR and IHC-LAC approaches

SVM	Classification Scenario								
	Easy			Hard			Total		
	L-InTeCo	IHC-LAC	AR	L-InTeCo	IHC-LAC	AR	L-InTeCo	IHC-LAC	AR
Acc.	93.1	94.5	92.4	83.9	78.1	81.1	89.1	87.2	86.0
Sens.	94.5	95.0	94.1	79.6	80.4	80.9	89.7	88.5	86.5
Spec.	91.9	93.8	90.6	88.7	75.7	81.2	88.4	85.8	85.4

presents the classification performance rates when engaging the enhanced FV that contains inherent texture correlation and LAC information. It is clear that the usage of LAC-based texture features from the adaptively refined images significantly improves the overall performance of the proposed scheme by 2.1 pp to 9.1 pp, depending on the ulcer case and FV scenario. Once again, the highest classification accuracy (93.1% - easy, 83.9% - hard, 89.1% - total) was achieved by the combination of second and higher-order correlation. In order to further testify the performance of the proposed approach, L-InTeCo is compared to a simple adaptive refinement scheme (AR) and IHC-LAC strategy [14]. AR scheme is similar to the one currently proposed but without InTeCo step. IHC-LAC utilizes the higher-order correlation between IMFs and the LAC curves of denoised WCE images. The denoising procedure is accomplished by removing two high-frequency IMFs. The best classification results for each method are presented in Table 3.

IV. DISCUSSION

By observing the results in Table 1 for the three ulcer cases, it becomes clear that automatic detection of severe ulcerations is far more successful than the detection of mild ulcerations. This is blatantly depicted by the contrasting sensitivity (73.1%) and specificity (83.8%) rates of the hard case that highlight the unsurprising misinterpretation of a large part of ulcer images as normal. The partly eroded mucosa and poorly shaped boundaries in conjunction with the presence of confusing normal tissue are responsible for the misclassification. The results of FV scenarios suggest that the majority of ulcer information resides in the higher-order dependencies between the intrinsic texture components. The combination of D and HD data is even more efficient, yet the improvement is limited considering the doubling in size of the FV. The hard case is an exception where even the less competent second-order correlation information plays a major role to the final results.

The superiority of the enhanced FV of L-InTeCo scheme is depicted by the classification rates of Table 2. These results are more important for the easy and total cases, considering the high sensitivity rates (94.5%-easy, 89.7%-total) that imply a good balance between ulcer and healthy tissue discrimination. Regarding the total case, it is remarkable that the additional LAC features extracted from the refined images do not ameliorate greatly the performance of the proposed scheme (accuracy rises from 85.3% to 89.1%). On the contrary, there is a major contribution of 5.2 pp to the hard case while the least gain lies upon easy case. Hence, it can be assumed that adaptive reconstruction of WCE images, by discarding misleading IMFs, facilitates vital feature extraction for the discrimination of low severity ulcers.

The comparison of L-InTeCo to AR and IHC-LAC approaches is interesting. L-InTeCo achieves the highest classification rates for the total case although the improvement is not compelling. On the contrary, L-InTeCo is far more efficient for the hard case (2.8 pp and 5.8 pp higher accuracy than AR and IHC-LAC), whereas AR presents the highest sensitivity. The most competent approach for the easy case is IHC-LAC implying that dependencies between raw IMFs are more capable than dependencies between the LAC curves of IMFs when ulcer regions are clearly depicted. Nevertheless, IHC-LAC superiority is marginal.

V. CONCLUSIONS

This work presents an innovative L-InTeCo scheme for ulcer detection based on adaptively refined WCE images. Novel texture correlation data, combined with lacunarity based texture features, enable an advanced and rather promising overall performance (89.1% acc.). However, the experimental dataset should be expanded and more GI diseases should be considered in order to evaluate the approach from a more general perspective and pave the way for its use in a prospective automatic WCE diagnosis system.

REFERENCES

1. Iddan G et al. (2000) Wireless capsule endoscopy. Nature 405(6785):405-417
2. Pennazio M (2005) Diagnosis of small-bowel diseases in the era of capsule endoscopy. Expert Rev. Med. Dev. 2(5):587-598
3. Kodogiannis V et al. (2007) A neuro-fuzzy-based system for detecting abnormal patterns in wireless-capsule endoscopic images. Neurocomputing 70:704-717
4. Li B, Meng M Q H (2012) Comparison of several texture features for tumor detection in CE images. J. Med. Syst. 36:2463-2469
5. Karkanis S A et al. (2003) Computer-aided tumor detection in endoscopic video using color wavelet features. IEEE Trans. Inf. Tech. Biomed. 7:141-152
6. Karargyris A, Bourbakis N (2011) Detection of small bowel polyps and ulcers in wireless capsule endoscopy videos. IEEE Trans. Biomed. Eng. 58(10):2777-2786
7. Li B, Meng M Q H, Lau J Y W (2012) Computer-aided small bowel tumor detection for capsule endoscopy. Artif. Intel. Med. 52:11-16
8. Li B, Meng M Q H (2009) Texture analysis for ulcer detection in capsule endoscopy images. Image Vision Comput. 27:1336-1342
9. Chen D et al. (2011) A novel strategy to label abnormalities for wireless capsule endoscopy frames sequence. Proc. IEEE Int. Conf. Inf. Autom., Shenzen, China, pp 379-383
10. Kodogiannis V et al. (2007) The usage of soft-computing methodologies in interpreting capsule endoscopy. Eng. Appl. Artif. Intel. 20:539-553
11. Li B, Meng M Q H (2007) Analysis of the gastrointestinal status from wireless capsule endoscopy images using local color feature. Proc. Int. Conf. Inf. Acq., Jeju City, Korea, pp 553-557
12. Liu X et al. (2012) A new approach to detecting ulcer and bleeding in Wireless capsule endoscopy images. Proc. IEEE Int. Conf. Biomed. Health Inform., Hong Kong, China, pp 737-740
13. Charisis V S et al. (2012) Capsule endoscopy image analysis using texture information from various colour models. Comput. Methods Programs Biomed. 107(1):61-74
14. Charisis V S et al. (2012) Intrinsic higher-order correlation and lacunarity analysis for WCE-based ulcer classification. Proc. IEEE Int. Symp. Comput. Based Med. Syst., Rome, Italy
15. Charisis V S et al. (2012) Enhanced ulcer recognition from capsule endoscopic images using texture analysis in New advances in the basic and clinical gastroenterology, Brzowski T (ed.). InTech, Croatia
16. Wu Z, Huang N E, Chen X (2009) The multidimensional ensemble empirical mode decomposition method. Adv Adapt Data Analys 1:339-372
17. Nunes J C et al. (2003) Image analysis by bidimensional empirical mode decomposition. Image Vision Comput. 21:1019-1026
18. Allain C, Coitre M (1991) Characterizing the lacunarity of random and deterministic fractal sets. Phys. Rev. A 44(6):3552-3558
19. Dong P (2000) Test of a new lacunarity estimation method for image texture analysis. Int. J. Remote Sens. 21(17):3369-3373
20. Sun T U et al. (2008) An efficient noise reduction algorithm using empirical mode decomposition and correlation measurement. Proc. Int. Symp. Intel. Signal Process. Com. Syst. Bangkok, Thailand
21. Lou S T, Zhang X D (2003) Fuzzy-based learning rate determination for blind source separation. IEEE Trans. Fuzzy Syst. 11(3):375-383

Author: Leontios J. Hadjileontiadis
Institute: Aristotle University of Thessaloniki
Street: University Campus, GR 54124
City: Thessaloniki
Country: Greece
Email: leontios@auth.gr

Tracer Kinetic Modeling with R for Batch Processing of Dynamic PET Studies

J.M. Mateos-Pérez[1,2], M. Desco[2,3], and J.J. Vaquero[3]

[1] Centro de Investigación en Red de Salud Mental (CIBERSAM), Madrid, Spain
[2] Instituto de Investigación Sanitaria Gregorio Marañón, Madrid, Spain
[3] Departamento de Bioingeniería e Ingeniería Aeroespacial, Universidad Carlos III de Madrid, Madrid, Spain

Abstract—Dynamic PET imaging can provide information regarding metabolism, blood flow or other physiological properties of biological tissues beyond what static imaging may offer. A common technique for analyzing these images is compartmental modeling, whose mathematical resolution generally involves solving a set of differential equations using a non-linear least squares algorithm. This paper presents a guide that shows how the R programming language can be used to perform this analysis. To test the suitability of the proposed implementations, a comparison with the PMOD software package has been carried out. We show that the R programming language provides easy to use libraries that can be combined to implement solutions for different compartmental models whose results match those obtained using a commercial tool.

Keywords—PET, dynamic imaging, R, kinetic analysis, compartmental modeling

I. Introduction

Positron emission tomography is an imaging technique that measures the amount of radioactive tracer within a given tissue or organ. While a single PET image provides useful functional information, the dynamic acquisition of several images after the tracer injection enables the study other parameters such as the myocardial blood flow [1, 2], cerebral metabolic rate of glucose [3, 4] or drug-receptor interaction [5], to cite just a few examples.

Dynamic acquisitions are analyzed by analyzing the temporal information through time-activity curves (TACs) for each tissue of interest, usually obtained via a segmentation process that involves regions of interest (ROIs) drawn directly over the image by an experienced operator. The input function, that is, the amount of tracer present in arterial blood, must also be obtained. While the most reliable method is the blood sampling using an arterial line [3], it is also common to use image-derived input functions in order to avoid invasive procedures [6]; though this approach depends on the presence of large blood pools in the acquired volume [7]. Arterial measurements may be necessary nonetheless in order to obtain absolute values of some parameters.

Once the TACs have been obtained, the system can be modeled in several ways. For a voxel-wise analysis, simple linear methods can be used; for instance, the cerebral metabolic rate for glucose can be computed reliably using graphical methods such as the Patlak plot [8]. These methods are fast and therefore allow for computing parametric maps of the brain. In any case, the most general methods involve some kind of non-linear fit of the measured data to a compartmental model; this fitting process is used to obtain the rate constants of tracer exchange between the different compartments.

The R programming language (http://www.r-project.org/) is gaining increasing notoriety as a data analysis and statistical tool. Its main strengths are to be free software and the numerous packages available that extend the base system with additional features. This programming language can be used to implement the compartmental models equations in a way that allows the user to analyze several studies in bulk, regardless of the imaging analysis tool used for the segmentation process.

In this paper we show the implementation of two compartmental models in R and validate them against the popular PMOD software package (PMOD Technologies, Zurich, Switzerland). The developed functions are provided free of charge (see section III.A) and interested researchers are encouraged to test them and modify them to suit their purposes.

II. Material and Methods

A. Hardware and Software Specifications

All the code in this paper has been developed in the R programming language, v. 2.15.2, using the RStudio environment (http://www.rstudio.com), v. 0.97.312, on an Intel (Intel Corp., Santa Clara, California, United States) Xeon CPU with 8 GB or RAM running Windows 7 Professional (Microsoft Corp., Redmond, Washington, United States) as operating system.

B. TAC Extraction

It is beyond the scope of this paper to describe the segmentation process necessary to obtain the relevant TACs for each tissue and the input function. The only requisite for the functions developed is that the applied segmentation yields an ASCII file reporting the activity for each region and, optionally, also the frame times. Anyway, these times can be added afterwards as some systems do not include this information in the segmentation result files.

C. One-Tissue Compartment Model for ^{82}Rb Cardiac Studies

This section presents a step-by-step implementation of the R functions that are necessary to solve the compartmental model described in [1], of which only the relevant equations will be presented here. ^{82}Rb kinetic behavior can be described by a one-compartment model. The differential equation that defines this model has the following solution:

$$C_m(t) = K_1 e^{-k_2 t} \otimes C_a(t) \quad (1)$$

In the above equation, $C_m(t)$ is the tissue activity (the myocardium, in this case), $C_a(t)$ is the input function, K_1 and k_2 are the rate constants and $\otimes$ is the convolution operator. Considering that the PET signal consists of a mixture of the tissue signal plus a fraction of blood, the equation that needs to be solved is

$$C_{PET}(t) = v_B \cdot C_a(t) + (1 - v_B) \cdot C_m(t) \quad (2)$$

$C_{PET}(t)$ is the TAC obtained from the relevant ROI and v_B is the fraction of blood in the tissue. This tracer does not show a linear relationship between the K_1 and the myocardial blood flow (MBF) parameter. Instead, a Renkin-Crone equation must be used to approximate the extraction:

$$K_1 = (1 - a e^{-\frac{b}{MBF}}) \cdot MBF \quad (3)$$

The values for the constants a and b have been obtained experimentally and their values are 0.77 [unitless] and 0.63 [ml/min/g], respectively [1].

In order to solve Eq. (2), three pieces of information are needed: tissue activity curve, input function and sampling times.

The following function shows the implementation of the model in 14 lines of code plus one small auxiliary function that implements Eq. (3). Code comments have been removed due to space constraints.

```
1 lortie.model <- function(ca, FLV,
                           flow, k2, tend) {
2   tsample <- tend
3   len <- length(tsample)
4   time.vector <- 0:tsample[len]
5   ca.inter <- approx(tsample, ca,
                       xout = time.vector,
                       rule = 2,
                       method = "linear")$y
6   K1 <- flow2k1(flow)
7   sol <- convolve(K1*exp(-k2 * time.vector),
                    rev(ca.inter), type = "o")
8   return(FLV * ca.inter[tsample + 1] +
          (1 - FLV) * sol[tsample + 1])
9 }

10 flow2k1 <- function(flow) {
11   a <- 0.77
12   b <- 0.63
13   return((1 - a*exp(-b/flow))*flow)
14 }
```

Several remarks should be made about the code above. Since in most PET acquisition the first frames are shorter than the last ones, the input function must be interpolated for the convolution operation to work properly; in this case, a linear interpolation was used. Frames that fall outside the original sampling times are also linearly extrapolated. The activity acquired during a single frame is considered a punctual measurement located at the end of that frame acquisition time. It has to be taken into account that the conventional units used are ml/min/g for K_1 and MBF, and 1/min for k_2. As the *convolve* function in R actually returns the correlation, the input function is reversed.

Eq. (2) is then solved using a Levenberg-Marquardt algorithm (chosen because it is the default fitting option in PMOD, the software taken as our gold standard). For practical purposes, instead of fitting K_1 and k_2, MBF and k_2 are fitted, and then K_1 is obtained using Eq. (3).

The actual fitting process is done using the previous functions and the *minpack.lm* R package (http://cran.r-project.org/web/packages/minpack.lm/index.html).

The following piece of code shows the actual implementation of the fitting process.

```
1   lortie.fit <- function(ca, cm, time,
                           time_end, plot = FALSE,
                           plot.t = NULL) {
2   library(minpack.lm)
3   fit <- nlsLM(cm ~ lortie.model(ca, FLV, flow,
                                   k2, time_end),
                 start = list(FLV = 0.1,
                              flow = 0.5,
                              k2 = 0.1),
                 weights = time_end - time,
                 lower = c(FLV = 0, flow = 0,
                           k2 = 0),
                 upper = c(FLV = 1, flow = 8,
                           k2 = 8))
4   kparms <- coef(fit)
5   kparms <- c(flow2k1(kparms[2]),
                kparms[c(2, 3, 1)])
6   names(kparms) <- c("K1", "flow", "k2", "FLV")
7   if (plot == TRUE) {
8       plot(c(time, time),
             c(cm, predict(fit)),
             type = "n",
             xlab = "Time [s]",
             ylab = "Activity [kBq/cc]",
```

```
                main = plot.t)
9       grid()
10      lines(time, predict(fit, time), type = "o",
            pch = 19)
12      points(time, cm)
13      legend("bottomright",
               c("Measured data", "Model"),
               lty = c(NA, 1), pch = c(1, 19),
               inset = 0.02)
14   }
15   return(kparms)
16 }
```

The function above uses the myocardium TAC to fit the previously implemented model. This fitting function uses sensible initial values (close to the ones expected, or at least within the same order of magnitude) and sets lower and upper limits to help the algorithm to converge quickly. The weights for the fitting process are proportional to frame length. Two optional keywords can be used to plot the result and visually check the goodness of the fit.

Up to this point, the produced R code can analyze individual studies. Finally, the following function can be used to batch-process several studies placed in the same directory.

```
1   batch.process <- function(path) {
2       kinetics <- lapply(dir(path),
                    function(x) {
                        k <- read.data(x)
                        lortie.fit(k$lv,
                                   k$myo,
                                   k$start,
                                   k$end
                                   plot = TRUE,
                                   plot.t = x)
                    })
3       return(do.call(rbind, kinetics))
4   }
```

This function not only returns the fitted parameters for all the studies in the directory, but also plots the fit for each study. The *read.data* function simply reads each study from the ASCII files that contains the segmentation results and is not shown here.

D. *Two-Tissue Compartment Model*

The two-tissue compartment model implements the equations that can be found in [9]. In this case, the differential equations that describe the system are:

$$\frac{dC_f(t)}{dt} = K_1 \cdot C_a(t) - (k_2 + k_3) \cdot C_f(t) + k_4 \cdot C_m(t)$$
$$\frac{dC_m(t)}{dt} = k_3 \cdot C_f(t) - k_4 \cdot C_m(t) \tag{4}$$

E. *Data Analysis*

Results obtained for both one-tissue and two-tissue models were tested against the PMOD package. For the one-tissue compartment model, we analyzed 8 cardiac studies using ^{82}Rb in both rest and stress conditions. The segmentations employed in this comparison were automatically performed by PMOD. For the two-tissue compartment model, we studied 10 rat tumors using ^{68}Ga-DOTA-TOC with a two-tissue compartment model. The results were compared by computing Pearson correlation scores and Bland-Altman plots. P-values lower than 0.05 are considered statistically significant.

III. RESULTS

A. *Code Downloads*

All the code presented in this paper can be obtained in https://github.com/HGGM-LIM/tracerkinetic as part of *traceRkinetic*, a still-in-development R library for kinetic analysis.

B. *Validation*

The kinetic parameters for all the studies modeled with one compartment where obtained in 0.34 s (n = 16); the two-compartment model was solved in 10.05 s (n = 10). Plots as the one shown in Figure 1 are generated for every study for visual checking of the fitting result.

The relationship between the rate parameters obtained with both tools is shown in Figure 2. There is an excellent agreement in the results, either taken altogether, as shown in the figure, or individually per rate parameter ($R^2 > 0.99$, $p < 0.001$). The Bland-Altman plots (not shown here) do not display any evident bias in the results.

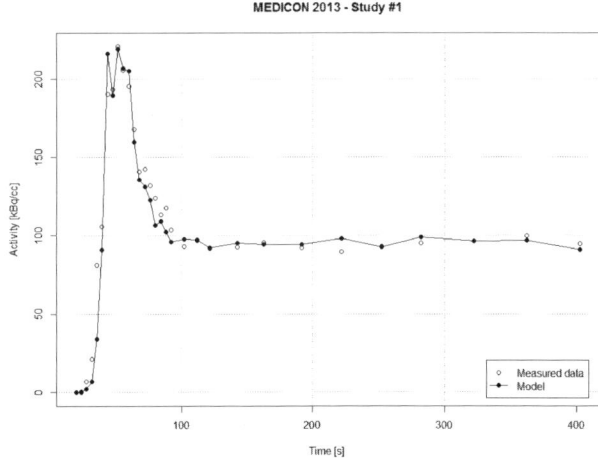

Fig. 1 Plot showing the fitting result for the one-compartmental model for one ^{82}Rb study.

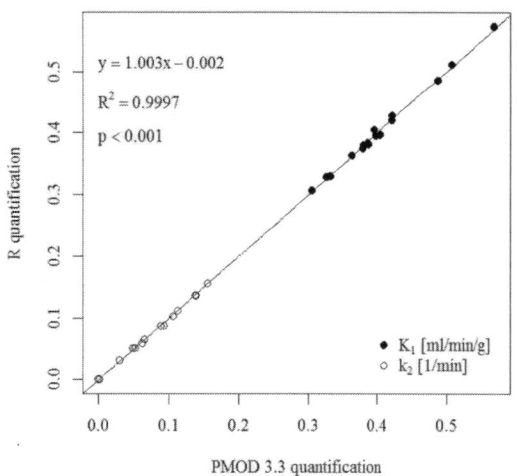

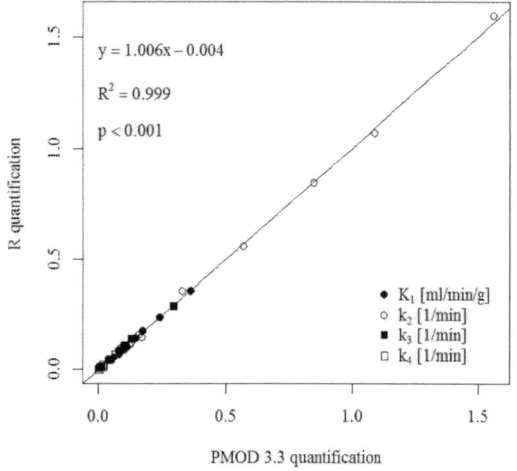

Fig. 2 Comparison between the rate parameters obtained using PMOD 3.3 and the presented R implementation for the one and two-compartment models. The straight line is the identity line.

IV. DISCUSSION AND CONCLUSION

In this paper we present an open R implementation of two compartmental models widely used in kinetic analysis of PET studies. This implementation can be used to solve studies that have been previously segmented using available imaging tools. The implementation has been tested on a limited number of studies and tracers to prove that the results are comparable to those obtained with a commercial, widely-used tool.

The developed code has been published in an open repository as part of a kinetic modeling library currently under development.

Due to the small sample tested, it cannot be reliably said that they can be used as a drop-in replacement. In any case, the potential of R to help researchers to process a great number of dynamic studies in little time has been shown.

ACKNOWLEDGEMENT

This work was supported by projects S2009/DPI-1802 (ARTEMIS), Comunidad de Madrid; Tecnologías de imagen molecular avanzadas (AMIT), CEN-20101014. Programa CENIT. CDTI. Ministerio de Ciencia e Innovación; and RECAVA (Red Temática de Enfermedades Cardiovasculares), RD07/0014/2009. Subprograma RETICS. Ministerio de Ciencia e Innovación.

REFERENCES

1. Lortie M, Beanlands RSB, Yoshinaga K, et al. (2007) Quantification of myocardial blood flow with 82Rb dynamic PET imaging. European journal of nuclear medicine and molecular imaging 34:1765–74. doi: 10.1007/s00259-007-0478-2
2. Wu H-M, Hoh CK, Buxton DB, et al. (1995) Quantification of Myocardial Blood Flow Using Dynamic Nitrogen-13-Ammonia PET Studies and Factor Analysis of Dynamic Structures. Journal of Nuclear Medicine 36:2087–2093.
3. Huisman MC, Van Golen LW, Hoetjes NJ, et al. (2012) Cerebral blood flow and glucose metabolism in healthy volunteers measured using a high-resolution PET scanner. EJNMMI Research 2:63. doi: 10.1186/2191-219X-2-63
4. Alf MF, Wyss MT, Buck A, et al. (2012) Quantification of Brain Glucose Metabolism by 18F-FDG PET with Real-Time Arterial and Image-Derived Input Function in Mice. Journal of nuclear medicine: official publication, Society of Nuclear Medicine. doi: 10.2967/jnumed.112.107474
5. Hostetler ED, Sanabria-Bohórquez S, Eng W, et al. (2012) Evaluation of [(18)F]MK-0911, a Positron Emission Tomography (PET) Tracer for Opioid Receptor-Like 1 (ORL1), in Rhesus Monkey and Human. NeuroImage. doi: 10.1016/j.neuroimage.2012.11.053
6. Fung EK, Carson RE (2013) Cerebral blood flow with [(15)O]water PET studies using an image-derived input function and MR-defined carotid centerlines. Physics in medicine and biology 58:1903–23. doi: 10.1088/0031-9155/58/6/1903
7. Zanotti-Fregonara P, Chen K, Liow J-S, et al. (2011) Image-derived input function for brain PET studies: many challenges and few opportunities. Journal of cerebral blood flow and metabolism: official journal of the International Society of Cerebral Blood Flow and Metabolism 31:1986–98. doi: 10.1038/jcbfm.2011.107
8. Patlak C, Blasberg R, Fenstermacher J (1983) Graphical evaluation of blood-to-brain transfer constants from multiple-time uptake data. J Cereb Blood Flow Metab
9. Zaidi H (2005) Quantitative Analysis in Nuclear Medicine Imaging. Springer, New York

Tensor Radial Lengths for Mammographic Image Enhancement

S.E. Chatzistergos, I.I. Andreadis, and K.S. Nikita

School of Electrical and Computer Engineering, National Technical University of Athens, Athens, Greece

Abstract—Proper enhancement of mammographic images in order to reveal diagnostically critical information is a very important task in everyday clinical practice. As the volume of screening mammograms increases so does the importance of algorithms that can reveal tumors or other kind of lesions. In this paper a mammogram enhancement method, inspired from the concepts of tensor image representation and tensor scale, is presented. In particular a number of tensor radial lengths is defined at each image location and their mean value is then subtracted from the original image. The proposed method was tested on a dataset of 192 images containing mass lesions taken from the Digital Database for Screening Mammography providing quite promising results. Furthermore the enhancement performance of the proposed method was compared with the enhancement performance of Contrast Limited Adaptive Histogram Equalization, Histogram Equalization and Unsharp Masking methods and clearly outperformed them.

Keywords—image enhancement, mammogram, mass lesion, tensor image.

I. INTRODUCTION

Mammography is the main imaging technique for the detection and diagnosis of breast cancer, however about 10% of all cancerous lesions is missed by radiologists due to false interpretation of mammographic images [1]. Reading mammograms is known to be a very demanding job for radiologists, since judgments depend on training, experience, and subjective criteria [2]. Furthermore even well-trained experts may present an inter-observer variation rate of 65–75% [3]. To this end, a variety of Computer Aided Diagnosis (CAD) systems were developed aiming at providing a more robust and consistent interpretation of mammographic images. Thus, enhancement of mammographic images has become a very important task since it can reveal diagnostically critical information that otherwise would have gone unnoticed and also serve as an important preprocessing step to many CAD systems increasing their effectiveness.

Several algorithms for mammogram enhancement have been developed recently. According to [4] these methods can be categorized in frequency-domain methods and spatial-domain methods.

Frequency-domain methods are based on a multiscale sub-band representation of the image [5] or fuzzy logic theory [6]. The sub-band representation is performed using various types of classical wavelets, curvelets, and even contourlets [7]. Next the transform coefficients are modified using various techniques like soft thresholding [8] or nonlinear filtering [9]. The modified coefficients are then used to reconstruct the enhanced image.

Fuzzy set theory has been used to enhance image contrast since it is suitable to deal with the uncertainty associated with the definition of image edges, boundaries and contrast [4]. Furthermore, it is quite common fuzzy logic to be used along with the sub-band representation [10], while the combination of fuzzy logic with Tsallis entropy has also been proposed [11].

However, frequency-domain methods present limitations, since they may enhance image globally but often fail to enhance adequately local regions [4].

Spatial-domain methods are based on non linear filtering [12], adaptive neighborhood [13] or Unsharp Masking (UM) [14]. Non-linear filtering has the ability to preserve edge's information and discard (blur) all other regions. A good example of non-linear filtering is anisotropic filtering. Adaptive neighborhood methods on the other hand use the information of small image segments (neighborhoods) to adapt the way they treat the original image. Finally in the UM technique, a high-pass filtered, scaled version of an image is added to the image itself thus enhancing edge and detail information.

In the present work, an image enhancement method for mammographic images is presented based on the notion of tensor representation of an image and more specifically the notion of tensor scale. The rest of the paper is organized as follows. In section 2 a brief description of tensor scale is provided. The proposed modification of tensor scale along with the image enhancement process are given in sections 3 and 4, followed by experimental results presented in section 5.

II. TENSOR SCALE

A tensor is generally N-th order array while vectors and matrices can be seen as first and second order tensors respectively. Tensor representation of an image or generally an object has become popular nowadays especially for dimensionality reduction and classification. Tensor provides a better representation for image space by avoiding information loss in vectorization [16]. Two common tensor image representations are tensor scale and structure tensor.

In order to represent an image using the structure tensor representation, a 2x2 array is defined for every pixel of the original image. This 2x2 array is formed from the partial derivatives of a certain image point convolved with a smoothing curve, usually Gaussian. Structure tensor is commonly combined with anisotropic filtering for image enhancement purposes [18].

Tensor scale is actually a local morphometric parameter yielding a unified representation of structure size, orientation, and anisotropy [19]. *Tensor scale* at any image point *p* is the parametric representation of the largest ellipse centered at *p* that is contained in the same homogeneous region under a predefined criterion [19]. To do that a fixed number of radially outward sample lines originating from point p are used. The optimum edge point is located at each sample line. On every pair of radially opposite sample lines the distance of the edge points from the point p is compared and the smaller one is used for both lines and therefore edge points are repositioned. Tensor scale for point p is calculated by finding the best fitting eclipse on the repositioned edge points.

III. Tensor Radial Lengths

In order to perform the mammographic image enhancement task a modification of tensor scale is used. What differentiate the proposed method from classical tensor scale is the fact that a different "homogeneous" region is defined for every image pixel and also the fact that we don't proceed to the calculation of the best fitting ellipse but rather stop at radials calculation.

To be more specific, at each image location (x0, y0) a gray value threshold is defined. This threshold is defined after a certain value (K_p) is subtracted from the image gray value at that location. Different ways to define K_p where evaluated, with the best results being achieved when K_p was defined as percentage of the global maximum image value (1). Parameter α is a user defined parameter with values in the range (0,1), while I represents the image gray level values.

$$K_p = \alpha * \max(I) \quad (1)$$

The thresholded image produced for a certain image point (x0,y0) provides the required from tensor scale method, "homogeneous" region on which the tensor scale radials for the specific point (x0,y0) point are defined. This way each point of the original image is represented by a fixed number of radial lengths. In other words a tensor representation of the original image is created.

The whole process of radial lengths calculation can also be seen as a process where a radial originating from point (x0,y0) and at specific angle, relative to image axes, begins to propagate or else increase in size. This propagation continues until an "obstacle" or "barrier" is found. This happens when the radial meets an image point with gray value smaller than the gray value of point (x0, y0) minus K_p.

Since each image location is now represented by a number of radial lengths, a number of characteristics based on these lengths, can be calculated at each image location. The easiest thing to do is to calculate the mean value of radial lengths at each image location (2).

$$m_{RL}^{(x_0,y_0)} = \frac{\sum_{i=1}^{N} r_i^{(x_0,y_0)}}{N} \quad (2)$$

where N is the total number of radial lengths r_i calculated around point (x0,y0).

This procedure gave a new image m_{RL} which provides a better illustration of mammogram's underlying structure (see Fig. 1). The basic idea is that images can be thought of as consisting of two basic regions according to the underlying breast tissue: one region consisting mainly of fatty tissue which generally appears dark and one region consisting mainly of dense tissue that appears lighter than the fatty tissue regions. Image points that belong to the fatty tissue regions would generally have lower gray values, thus their radial lengths would be quite large since they wouldn't find any "obstacle" as they propagate.

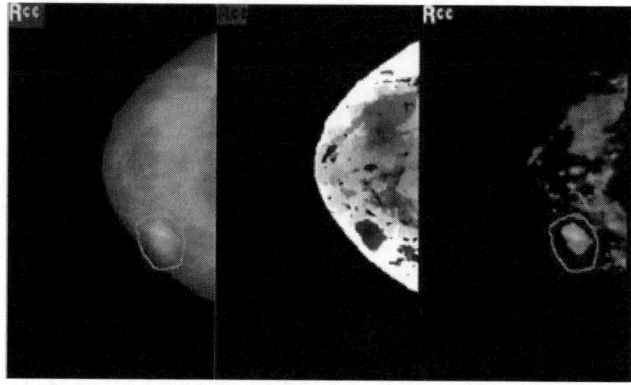

Fig. 1 Mammogram enhancement. Original image (left), mean of tensor radial lengths (m_{RL}) (middle), enhanced image (right) (a=0.95).

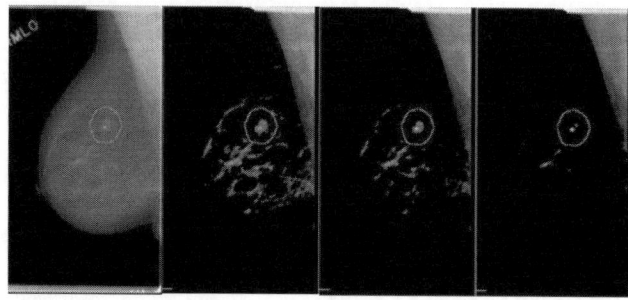

Fig. 2 Effect of parameter α on final enhanced image. From left to right: original image, α=0.97, α=0.95, α=0.90.

On the other hand image points belonging to the dense tissue regions would generally have smaller values since it would be more likely for them to find "obstacles" as they propagate.

IV. MAMMOGRAPHIC IMAGE ENHANCEMENT

In order to perform the actual image enhancement process we begin by calculating the mean value of radial lengths at each image location (x_0, y_0) and thus create a new image referred to as $m_{RL}(2)$.

This image is scaled in such a way that image's max value equals to 1 while min value equals to 0 (m_{RL}^S). m_{RL}^S is then subtracted from the similarly scaled original image I^S, according to the following equation.

$$I_{enh} = I^S - m_{RL}^S \qquad (3)$$

Through this subtraction, gray level values at points that belong to fatty tissue regions are suppressed since radial lengths at these regions are expected to be large, while gray level values at dense tissue regions are preserved since radial lengths at these regions are expected to be small.

Both m_{RL}^S and I^S would take values in the [0,1] range and thus I_{enh} would be in the range [-1,1]. Since all diagnostically critical information is located in regions with large gray values, I_{enh} values that are lower than zero are discarded. Thus the produced image would take values in the range [0,1].

Fig. 2 shows the effect of the parameter α on the final enhanced image. As expected, large values of α result in overestimation of the dense breast region while small values of α result in under-estimation of the dense breast region.

V. EVALUATION

In order to evaluate the performance of the proposed method in enhancing mammographic images and revealing diagnostically critical information we used a number of images from the Digital Database for Screening Mammography (DDSM) [20]. Images in DDSM are divided in categories according to the kind of lesion they contain. For our purposes we selected a number of 192 images containing mass lesions of varying subtlety.

We have compared the performance of the proposed mammogram enhancement algorithm with Contrast Limited Adaptive Histogram Equalization (CLAHE), Histogram Equalization (HE) and Unsharp Masking (UM). All these are well known image enhancement method, used excessively in medical image enhancement tasks [14, 21, 22].

To measure the enhancement performance of the proposed algorithm we used the following performance measures: D measure [23], region contrast (RC) [24] and measure of enhancement by entropy (EMEE) [25]. In order to use the D measure, two image regions had to be defined, one containing the actual mass lesion (Target) and one containing the breast region around the mass (Background). In order to do that, the mass annotations that accompany DDSM images were used. These annotations provide a rough outline of the mass region. Using an active contour method [26] we were able to define the exact mass region (Target), while image region belonging to the image annotation but not to the mass region was used as the Background. However, the Active Contour method failed to define the exact mass region in all images and therefore 72 of them had to removed and ignored during the performance evaluation step. Methods RC and EMEE don't require the actual mass region to be defined and therefore a rectangular region containing the annotated region from all 192 images was used.

Comparison of the enhancement performance of the proposed method with the methods used as gold standard is given in table 1.

Table 1 Enhancement comparison results for various values of α

	Parameter α	CLAHE	HE	UM
D	0.98	64,17	68,33	84,17
	0.97	**65,83**	**70,83**	**85,83**
	0.96	65,83	71,67	85
	0.94	64,17	69,17	85
RC	0.98	**99,48**	**91,67**	45,31
	0.97	**97,92**	**89,06**	37,5
	0.96	94,79	81,25	28,65
	0.94	82,81	72,4	10,94
EMEE	0.98	**80,21**	**73,96**	**84,38**
	0.97	**76,56**	**69,79**	**80,21**
	0.96	71,88	69,27	75
	0.94	64,58	64,58	71,88

The provided results is the percent of cases were the proposed algorithm outperformed comparison enhancement method (columns) using the various performance metrics (rows).

VI. CONCLUSIONS

A mammogram enhancement method based on the concept of tensor scale has been presented. The method uses the mean value of tensor radial lengths and the basic idea is that regions with large gray values (dense breast regions) would have smaller radial lengths than regions with lower gray values (fatty breast regions). Comparison of the proposed method with a number of popular enhancement methods using various enhancement performance metrics, has shown that the proposed method provides superior results, except for UM according to RC metric, where the proposed method gave better enhancement results in 45% of all cases.

REFERENCES

1. Sundaram, M., et al., *Histogram Modified Local Contrast Enhancement for mammogram images.* Applied Soft Computing, 2011. **11**(8): p. 5809-5816.
2. Cheng, H.D., et al., Approaches for automated detection and classification of masses in mammograms. Pattern Recognition, 2006. **39**(4): p. 646-668.
3. Skaane, P., K. Engedal, and A. Skjennald, Interobserver variation in the interpretation of breast imaging. Comparison of mammography, ultrasonography, and both combined in the interpretation of palpable noncalcified breast masses. Acta Radiol, 1997. **38**(4 Pt 1): p. 497-502.
4. Panetta, K., et al., *Nonlinear Unsharp Masking for Mammogram Enhancement.* Information Technology in Biomedicine, IEEE Transactions on, 2011. **15**(6): p. 918-928.
5. Mencattini, A., et al., *Mammographic Images Enhancement and Denoising for Breast Cancer Detection Using Dyadic Wavelet Processing.* Instrumentation and Measurement, IEEE Transactions on, 2008. **57**(7): p. 1422-1430.
6. Bhattacharya, M. and A. Das. Fuzzy Logic Based Segmentation of Microcalcification in Breast Using Digital Mammograms Considering Multiresolution. in Machine Vision and Image Processing Conference, 2007. IMVIP 2007. International. 2007.
7. Malar, E., et al. A comparative study on mammographic image denoising technique using wavelet, curvelet and contourlet transforms. in Machine Vision and Image Processing (MVIP), 2012 International Conference on. 2012.
8. Sakellaropoulos, P., L. Costaridou, and G. Panayiotakis. An adaptive wavelet-based method for mammographic image enhancement. in Digital Signal Processing, 2002. DSP 2002. 2002 14th International Conference on. 2002.
9. Chun-Ming, C. and A. Laine, *Coherence of multiscale features for enhancement of digital mammograms.* Information Technology in Biomedicine, IEEE Transactions on, 1999. **3**(1): p. 32-46.
10. Chen, C.H. and G.G. Lee. A multiresolution wavelet analysis of digital mammograms. in Pattern Recognition, 1996., Proceedings of the 13th International Conference on. 1996.
11. Mohanalin, J., P.K. Kalra, and N. Kumar. Tsallis Entropy Based Contrast Enhancement of Microcalcifications. in Signal Acquisition and Processing, 2009. ICSAP 2009. International Conference on. 2009.
12. George, J. and S.P. Indu. Fast Adaptive Anisotropic Filtering for Medical Image Enhancement. in Signal Processing and Information Technology, 2008. ISSPIT 2008. IEEE International Symposium on. 2008.
13. Yajie, S., et al. Effect of Adaptive-Neighborhood Contrast Enhancement on the Extraction of the Breast Skin-Line in Mammograms. in Engineering in Medicine and Biology Society, 2005. IEEE-EMBS 2005. 27th Annual International Conference of the. 2005.
14. Tian, X. The application of adaptive unsharp mask algorithm in medical image enhancement. in Cross Strait Quad-Regional Radio Science and Wireless Technology Conference (CSQRWC), 2011. 2011.
15. Wang, C., et al., Image representation using Laplacian regularized nonnegative tensor factorization. Pattern Recognition, 2011. **44**(10–11): p. 2516-2526.
16. George, J. and S.P. Indu. Color image enhancement and denoising using an optimized filternet based Local Structure Tensor analysis. in Signal Processing, 2008. ICSP 2008. 9th International Conference on. 2008.
17. Saha, P.K., Tensor scale: A local morphometric parameter with applications to computer vision and image processing. Computer Vision and Image Understanding, 2005. **99**(3): p. 384-413.
18. Michael Heath, K.B., Daniel Kopans, Richard Moore and W. Philip Kegelmeyer, *The Digital Database for Screening Mammography.* Proceedings of the Fifth International Workshop on Digital Mammography, 2001: p. 212-218.
19. Kurt, B., V.V. Nabiyev, and K. Turhan. Medical images enhancement by using anisotropic filter and CLAHE. in Innovations in Intelligent Systems and Applications (INISTA), 2012 International Symposium on. 2012.
20. Siddharth, R. Gupta, and V. Bhateja. An improved Unsharp Masking algorithm for enhancement of mammographic masses. in Engineering and Systems (SCES), 2012 Students Conference on. 2012.
21. Singh, S. and K. Bovis, *An evaluation of contrast enhancement techniques for mammographic breast masses.* Information Technology in Biomedicine, IEEE Transactions on, 2005. **9**(1): p. 109-119.
22. Jinshan, T., L. Xiaoming, and S. Qingling, *A Direct Image Contrast Enhancement Algorithm in the Wavelet Domain for Screening Mammograms.* Selected Topics in Signal Processing, IEEE Journal of, 2009. **3**(1): p. 74-80.
23. Agaian, S.S., K. Panetta, and A.M. Grigoryan, *Transform-based image enhancement algorithms with performance measure.* Image Processing, IEEE Transactions on, 2001. **10**(3): p. 367-382.
24. Chunming, L., et al., *Distance Regularized Level Set Evolution and Its Application to Image Segmentation.* Image Processing, IEEE Transactions on, 2010. **19**(12): p. 3243-3254.

Author: Sevastianos Chatzistergos
Institute: National Technical University of Athens
Street: 9, Iroon Polytechniou
City: Athens
Country: Greece
Email: schatzist@biosim.ntua.gr

Magnetic Resonance Texture Analysis: Optimal Feature Selection in Classifying Child Brain Tumors

Suchada Tantisatirapong[1], Nigel P. Davies[2,1,3], Daniel Rodriguez[4], Laurence Abernethy[5], Dorothee P. Auer[4,6], C.A. Clark[7,8], Richard Grundy[4,6], Tim Jaspan[6], Darren Hargrave[8], Lesley MacPherson[2], Martin O. Leach[9], Geoffrey S. Payne[9], Barry L. Pizer[5], Andrew C. Peet[1,3], and Theodoros N. Arvanitis[1,3]

[1] University of Birmingham, Birmingham, United Kingdom
[2] University Hospitals Birmingham NHS Foundation Trust, Birmingham, United Kingdom
[3] Birmingham Children's Hospital NHS Foundation Trust, Birmingham, United Kingdom
[4] University of Nottingham, Nottingham, United Kingdom
[5] Alder Hey Children's NHS Foundation Trust, Liverpool, United Kingdom
[6] University Hospital Nottingham, Nottingham, United Kingdom
[7] University College London, London, United Kingdom
[8] Great Ormond Street Hospital, London, United Kingdom
[9] The Institute of Cancer Research and Royal Marsden Hospital, Sutton, United Kingdom

Abstract—Textural feature based classification has shown that magnetic resonance images can characterize histological brain tumor types. Feature selection is an important process to acquire a robust textural feature subset and enhance classification rate. This work investigates two different feature selection techniques; principal component analysis (PCA), and the combination of max-relevance and min-redundancy (mRMR) and feedforward selection. We validated these techniques based on a multi-center dataset of pediatric brain tumor types; medulloblastoma, pilocytic astrocytoma and ependymoma, and investigated the accuracy of tumor classification, based on textural features of diffusion and conventional MR images.

Keywords—PCA, mRMR, feedforward selection, pediatric brain tumors, diffusion and conventional MR images.

I. INTRODUCTION

Conventional Magnetic Resonance Imaging (MRI) is an essential part of clinical diagnosis and offers superior structural imaging of anatomical elements compared with diffusion MRI. However, conventional MRI offers limited functional information of brain tissue. Diffusion MRI, a functional imaging technique, allows examination of tissue microstructure and provides complementary structural visualization. Diffusion weighted imaging (DWI) and diffusion tensor imaging (DTI), has been utilized as a biomarker in various neuro-pathological diseases [1]. A sensitization of DTI to subtle disturbances in white-matter tracts has revealed the neural tracts and tumor infiltration and invasion on healthy tissue [2]. Diffusion MR images would provide promising information for discriminating brain tumor types.

Texture analysis has been widely used to extract information from the MR images [3, 4]. Dimensionality reduction is commonly required to avoid over-fitting and to obtain the optimal feature subset. Choosing an appropriate feature selection approach is important because an inappropriate use of feature selection can distort classifier performance when larger datasets are used [5]. Feature selection is broadly divided into three main methods which are filter, wrapper and embedded approaches. Filter methods do not require learning algorithms; for example PCA, Fisher score and mutual information based feature selection. Wrapper and embedded methods use a learning algorithm to score feature subsets according to their discriminative rate as the selection criterion [6]. Wrapper methods can provide more accurate solutions than filter methods [7], but in general are more computationally expensive because the induction algorithm must be evaluated over each feature set. The embedded methods perform feature selection during the process of training. The most common embedded methods are such as least absolute shrinkage and selection operator (Lasso) and ℓ1-norm regularized Support Vector Machines [8]. In this study, we examine the unsupervised PCA and the combination of maximum relevance and minimum redundancy (mRMR) [7] and feedforward selection.

Support vector machines (SVMs), originally proposed by Vapnik et al 1992 [9], have shown superb performance at binary classification [10] and have been widely used in medical image analysis. SVMs tend to be more robust with smaller standard error compared to artificial neural network (ANN) training [11] and linear discriminant analysis (LDA), which is prone to non-singular value decomposition error. We applied the SVM as a validating classifier for the comparison of the two feature selection techniques.

II. MATERIAL AND METHODS

A. Material

Three types of pre-treatment brain tumors have been considered: medulloblastoma (MB), pilocytic astrocytoma (PA) and ependymoma (EP). These cases were obtained from the

Table 1 Number of subjects for each image type

Image type	MB	PA	EP	Resolution (x, y, z)
ADC (DWI)	16	21	11	(0.8, 0.8, 4) – (1.8, 1.8, 7.5)
MD (DTI)	19	17	8	(0.94, 0.94, 2.5) – (2, 2, 6.5)
FA (DTI)	19	17	8	(0.94, 0.94, 2.5) – (2, 2, 6.5)
T2w	25	34	15	(0.4, 0.4, 4) – (0.9, 0.9, 7.5)
FLAIR	20	19	9	(0.45, 0.45, 3) – (1, 1, 6.5)
T1w pre-contrast	21	24	12	(0.4, 0.4, 4.4) – (0.9, 0.9, 7.25)
T1w post-contrast	21	28	15	(0.4, 0.4, 4.4) – (0.9, 0.9, 7.25)

Children's Cancer and Leukemia Group (CCLG) database. Apparent Diffusion Coefficient (ADC) maps reconstructed from DWI, Mean Diffusivity (MD) and Fractional Anisotropy (FA) maps derived from DTI and conventional MR images were acquired from four centers using a 1.5T GE, 1.5T Siemens, 1.5T and 3T Phillips scanners, following a common protocol defined by the CCLG Functional Imaging Group [12]. The number of MB, PA and EP cases used in this study is shown in Table 1.

B. Data-preprocessing

The processing pipeline is conducted according to the work presented by Tantisatirapong et al 2013 [13] including eddy current correction, skull stripping, Diffusion MR image reconstruction, registration, intensity normalization, segmentation and texture analysis. Due to protocol variation at the different centers and multiple slices acquired for the analysis, an image type would be included according to the slice spacing. If the registered images have the slice spacing greater than the T2w image (the reference image for registration and segmentation) and the difference is more than 20% by calculating from 100 x $(s_1-s_2)/(s_1+s_2)$, where s_1 is a T2w's slice spacing and s_2 is a current image's slice spacing, the images are excluded. The number of gray levels applied in the texture analysis affects the discrimination outcome. Inappropriate quantization provides unsatisfactory results and too many gray levels yields longer processing times. We set the number of gray levels for co-occurrence and run-length matrices to be 9 for all MR image types.

C. Feature Selection

1. Principal Component Analysis (PCA)

PCA is an unsupervised feature selection method which transforms observed data to an orthogonal subspace such that the transformed variables are linearly uncorrelated (called principal components). The transformation arranges the principal components (PCs) in descending order in a way that the first PC has the largest variance; the second PC has the second largest variance and so on. In this study, we used the number of PCs that provides the maximum balanced accuracy (an average accuracy obtained from MB, PA and EP).

2. Maximum Relevance and Minimum Redundancy (mRMR)

The mRMR method applies mutual information theory in order to arrange features according to the criterion of maximum relevance (max-relevance) and minimum redundancy (min-redundancy). Max-relevance criteria search for features which satisfy the maximum relevant between individual features and class. The max relevant features could be highly redundant, because the features are highly dependent on each other. As a result, the power of discrimination would not change much. The highly redundant features are filtered by applying the min-redundancy condition to obtain the first m^{th} features. To obtain the highest accuracy, feedforward selection technique is used to search the optimal combination within the top-ranked 20th features. If any feature subset provides the same accuracy, the first combination feature subset is selected. For scoring feature in feedforwad selection, the accuracy is calculated by using balanced accuracy values.

D. Classification

Support vector machines (SVMs) compute the maximum marginal separation line between two classes. By using a kernel function, it maps the original features into higher dimensional space where it computes a hyperplane that maximizes the distance from the hyperplane to the training data in each class. Having found such a hyperplane, the SVM can then predict the classification of unlabeled test data by mapping it into the feature space and checking on which side of the separating plane the test data lies. We used the LIBSVM tools [14] with the function of C-SVC, and a linear kernel. For multi-classification, the 'one-against-one' approach implemented within LIBSVM was applied. Due to the small dataset, the validation of the classification was evaluated by using a leave-one-out validation technique.

III. RESULTS AND DISCUSSION

The number of principal components used is an important parameter in PCA analysis, with diffusion images requiring less components to achieve maximum classification rates than conventional images (see Fig. 1). This shows that the textural features derived from diffusion MR images contain the stronger feature subset in the top-ranked of PCs. In this study, features selected by using the PCA do not

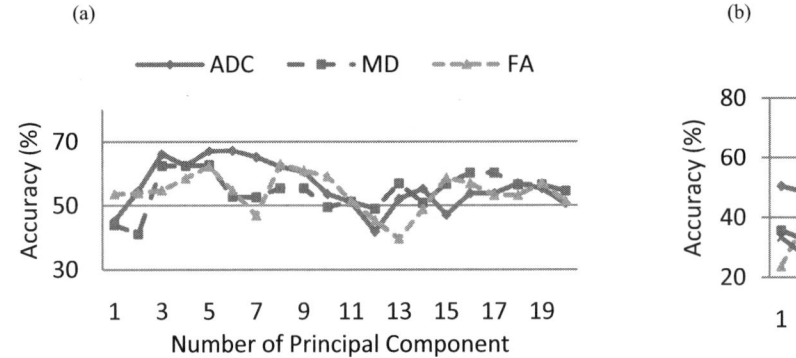
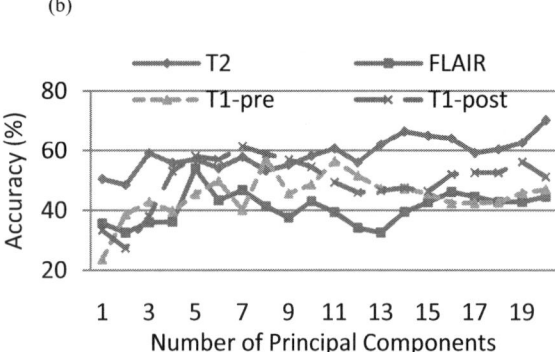

Fig. 1 Plot of number of principal component against classification accuracy for a) diffusion MR images b) conventional MR images

provide obviously better result than using the original feature set (see Fig. 2). The fact is that the PCA transforms all the observation data to obtain the maximum variance and still retains all variables to generate the PCs. Some outliers or noise, contributed by less separable textural features in the PCs, can affect the classification outcome.

Combining mRMR and feedforward selection scheme apparently enhances the discrimination performance for all of MR image types (see Fig. 2). The feature subset selected from this technique is similar to the finding presented by Krier et al 2006 [15]. The group of most individually relevant features does not always provide the superior prediction result. Adding some less relevant features to the features having high discriminant index, contributes more to the accuracy of the classifier than using the top-ranked relevant features only (see Table 2).

The prediction rate for each tumor type (MB, PA and EP), as well as the overall accuracy, are higher in ADC and MD image based analysis (see Fig. 2). Textural features based on ADC and MD provides the most promising result compared to those based on other MR modalities. The result indicated a good separation between MB and PA. However improvement in the discrimination of EP from the other two groups still needs to be achieved.

Table 2 Feature ID selected from mRMR and feedforward selection for each MR image type

Image type	ID of the top-ranked 20th features selected by using mRMR	ID of the feature subset of the top 20th features selected by using feedforward selection
ADC (DWI)	[65, 62, 1, 97, 143, 121, 52, 85, 113, 49, 7, 9, 11, 123, 120, 94, 70, 42, 8, 57]	1-H:mean, 11-R:GLN $0°$, 8-A:kurtosis, 9-R:SRE $0°$, 65-C:svarh $0°$, 57-C:cshad $0°$, 42-R:SRE $135°$, 85-C:svarh $45°$
MD (DTI)	[1, 11, 80, 64, 56, 65, 5, 84, 2, 100, 104, 38, 12, 124, 129, 102, 6, 85, 24, 53]	1-H:mean, 56-C:cprom $0°$, 12-R:RLN $0°$
FA (DTI)	[142, 35, 81, 9, 134, 57, 70, 144, 28, 100, 44, 7, 21, 46, 91, 90, 19, 117, 139, 42]	134-Wavelet, 35-R:RP $90°$, 28-R:SRHGE $45°$, 142-Wavelet
T2w	[97, 86, 53, 139, 101, 77, 129, 10, 9, 117, 63, 126, 6, 22, 73, 8, 57, 131, 69, 83]	97-C:cshad $90°$, 129-C:infl h $135°$, 117-C:cshad $135°$, 57-C:cshad $0°$, 139-Wavelet, 8-A:kurtosis, 22-R:GLN $45°$, 10-R:LRE $0°$, 6-A:variance, 77-C:cshad $45°$, 131-C:indnc $135°$, 101-C:homom $90°$, 73-C:autoc $45°$, 69-C:infl h $0°$, 53-C:autoc $0°$
FLAIR	[88, 44, 97, 94, 33, 27, 61, 129, 28, 3, 16, 7, 108, 6, 24, 17, 142, 38, 128, 74]	15-R:HGRE $0°$, 97-C:cshad $90°$, 44-R:GLN $135°$, 3-H:skewness, 74-C:corrp $45°$, 88-C:denth $45°$, 95-C:contr $135°$
T1w pre-contrast	[3, 60, 129, 2, 12, 145, 97, 102, 8, 56, 55, 17, 16, 57, 19, 135, 76, 1, 80, 23]	3-H:skewness, 102-C:maxpr $90°$, 97-C:cshad $90°$, 145-Wavelet, 76-C:cprom $45°$, 129-C:infl h $135°$, 60-C:entropy $0°$, 57-C:cshad $0°$
T1w post-contrast	[57, 11, 96, 49, 143, 2, 47, 69, 22, 77, 30, 27, 100, 82, 56, 117, 144, 12, 17, 41]	56-C:cprom $0°$, 22-R:GLN $45°$, 47-R:LGRE $135°$, 11-R:GLN $0°$, 96-C:cprom $90°$, 144-Wavelet, 30-R:LRHGE $45°$, 77-C:cshad $45°$, 49-R:SRLGE $135°$, 57-C:cshad $0°$, 41-R:LRHGE $90°$, 117-C:cshad $135°$

Abbreviation: H-histogram, A-absolute gradient, R-gray-level run length matrix, C-gray-level cooccurrence matrix, GLN-gray-level nonuniformity, SRE-short run emphasis, SRLGE-short run low gray-level emphasis, SRHGE-short run high gray-level emphasis, LRHGE-long run high gray-level emphasis, SRLGE-short run low gray-level emphasis, RLN-run length nonuniformity, LRE-long run empasis, HGRE-High Gray-Level Run Emphasis, RP-run percentage, svarh-sum variance, cshad-cluster shade, cprom-cluster prominence, infl h-information measure of correlation, homom-homogeneity, indnc-inverse difference normalized, autoc-autocorrelation, denth-difference entropy, corrp-correlation, maxpr-maximum probability, contr-contrast 13],[16,17].

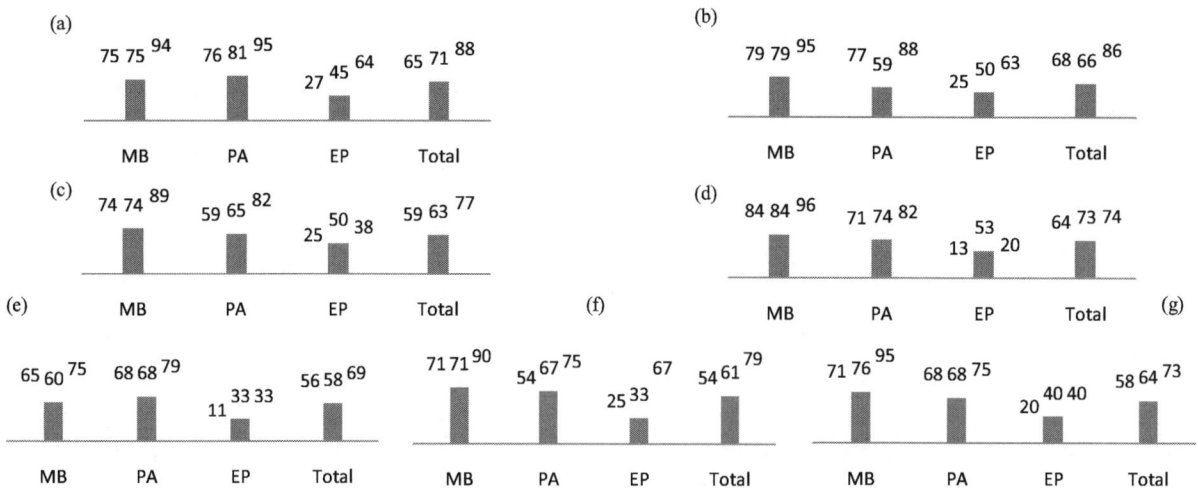

Fig. 2 The comparison of feature selection methods against the original feature based on (a) ADC, (b) MD, (c) FA, (d) T2, (e) FLAIR (f) T1w-pre contrast (g) T1w-post contrast. The legend: Original ■ PCA ⋟ mRMR+Feedforward

IV. CONCLUSIONS

For all image types, combining mRMR and feedforward selection provides a textural feature subset that can better discriminate the three tumor types than using the original features or the transformed features extracted by using the PCA. With this feature selection approach, classification based on ADC and MD image analysis outperforms other types of MR images. Future work involves the investigation of fusing textural features, based on multi-modal MR images, in order to boost up the prediction performance of classifiers.

ACKNOWLEDGMENT

ST and TNA would like to acknowledge the financial support from the National Science and Technology Development Agency (NSTDA), Thailand. NPD is supported by a CSO/NIHR Healthcare Scientist Research Fellowship. DH, MOL, GSP acknowledge the support received from the CRUK and EPSRC Cancer Imaging Centre in association with the MRC and Department of Health (England) grant C1060/A10334, also NHS funding to the NIHR Biomedical Research Centre and NIHR Clinical Research Facility. MOL is an NIHR Senior Investigator. All authors acknowledge the support received from CRUK/EPSRC/MRC/NIHR Cancer Imaging Programme grant C7809/A10342.

REFERENCES

1. Oksuzler Y F, Cakmakci H, Kurul S et al. (2005) Diagnostic Value of Diffusion-Weighted Magnetic Resonance Imaging in Pediatric Cerebral Diseases. Pediatr Neurol 32:325-333
2. Price S, Jena, R, Burnet N et al. (2007) Predicting patterns of glioma recurrence using diffusion tensor imaging. Eur Radiol 17:1675-1684
3. Castellano G, Bonilha L, Li M et al. (2004) Texture analysis of medical images. Clin Radiol 59:1061-1069
4. Kassner A., and Thornhill, R. E. Texture analysis: a review of neurologic MR imaging applications. AJNR AM J Neuroradiol 31: 809-816
5. Smialowski P, Frishman D, Kramer S (2010) Pitfalls of supervised feature selection. Bioinformatics 26:440-443
6. Guyon I, Elisseeff A (2003) An introduction to variable and feature selection. J Mach Learn Res 3:1157-1182
7. Peng H, Fulmi L, Ding C (2005) Feature selection based on mutual information criteria of max-dependency, max-relevance, and min-redundancy. IEEE Trans. Pattern Anal. Mach. Intell. 27:1226-1238
8. Lijun Z, Chun C, Jiajun B et al. (2012) A Unified Feature and Instance Selection Framework Using Optimum Experimental Design. IEEE Trans. Image Process. 21:2379-2388
9. Boser B, Guyon L, Vapnik V (1992) A training algorithm for optimal margin classifiers. In the 5th Annual Workshop on Comp. Learning Theory, pp. 144-152, ACM, Pittsburgh, PA, USA, pp. 144-152
10. Li T, Zhu S, Ogihara M (2006) Using discriminant analysis for multi-class classification: an experimental investigation. Knowl. Inf. Syst. 10:453-472
11. Byvatov E, Fechner U, Sadowski J et al. (2003) Comparison of support vector machine and artificial neural network systems for drug/nondrug classification. J. Chem. Inf. Comput. Sci. 43:1882-1889
12. http://www.cclgfig.bham.ac.uk/protocol.aspx
13. Tantisatirapong S, Davies N P, Abernethy L et al. (2013) Automated processing pipeline for texture analysis of childhood brain tumours based on multimodal magnetic resonance imaging. In the IASTED BioMed 2013, ACTA, Innsbruck, Austria, pp. 376-383,
14. Chang C-C, Lin C-J (2011) LIBSVM: A library for support vector machines. ACM Trans. Intell. Syst. Technol. 2:1-27
15. Krier C, François D, Wertz V et al. (2006) Feature Scoring by Mutual Information for Classification of Mass Spectra. In FLINS 2006, 7th Int. FLINS Conf. on Appl. AI, pp. 557-564
16. Xiaoou T (1998) Texture information in run-length matrices. IEEE Trans. Image Process. 7, 1602-1609
17. Haralick R M, Shanmugam K, Dinstein I H (1973) Textural Features for Image Classification. IEEE Trans. Syst. Man Cybern. Syst. 3, 610-621

Image Segmentation for Treatment Planning of Electroporation-Based Treatments of Tumors

M. Marcan[1], D. Pavliha[1], R. Magjarevic[2], and D. Miklavcic[1]

[1] Faculty of Electrical Engineering, University of Ljubljana, Ljubljana, Slovenia
[2] Faculty of Electrical Engineering and Computing, University of Zagreb, Zagreb, Croatia

Abstract—Electroporation is a term describing an increase of cell membrane permeability due to exposure of a cell to a sufficiently high electric field. The described effect is used in treatment of tumors, whether in combination with chemotherapeutic drugs (electrochemotherapy) or as a non-thermal ablation method (non-thermal irreversible electroporation). In order for the treatment to be successful it is necessary to achieve complete coverage of the tumor volume by a sufficiently high electric field. In cases of applying electroporation-based treatments on deep-seated solid tumors the complete tumor coverage is predicted through treatment planning. In the core of treatment planning are numerical calculations based on real patient geometry acquired through segmentation of medical images.

Our aim is to create an integrated treatment planning procedure to be used for electroporation-based treatments. The procedure includes image segmentation methods for images of various modalities and imaging parameters. Therefore, we aim towards creating image segmentation methods that would be as robust as possible with least user interaction. Our first task was performing image segmentation in order to facilitate treatment planning for electrochemotherapy of colorectal metastases in the liver. The aim was to segment hepatic vessels with a diameter size > 3 mm which should not be damaged during insertion of electrodes. We implemented two methods: the first based on region growing and the second combining multi-scale filtering with automatic thresholding. We also performed the validation of implemented methods on phantoms regarding the sizes of objects which are of interest in the treatment planning procedure. The mean errors of the first and second method in the worst case were 62,67% and 0,14%, respectively. These results indicate that the second method might be accurate enough to be used in the treatment planning. Still, additional validation based on radiologists' segmentation needs to be performed.

Keywords—electroporation, electrochemotherapy, treatment planning, medical image segmentation, image segmentation accuracy.

I. INTRODUCTION

Electroporation occurs when a living cell is exposed to a sufficiently high electric field, which causes changes in the cell membrane structure and the increase of membrane permeability [1]. The increased membrane permeability enhances introduction of molecules into the cell which naturally cross the membrane with difficulties or not at all. Reversible electroporation, where the membrane eventually returns to initial state, is used in electrochemotherapy (ECT) of tumors to enhance toxicity of chemotherapeutic drugs [2]. Irreversible electroporation, which results in permanent changes in the membrane structures, is also being used in treating tumors as a non-thermal ablative technique. This technique is called non-thermal irreversible electroporation (N-TIRE) [3].

In order to successfully apply electroporation-based treatments, it is necessary to ensure complete coverage of the tumor by sufficiently high electric field [4]. This requirement can be met by performing patient-specific treatment planning. The process of treatment planning includes generation of patient geometry through segmentation of medical images and numerical calculations of the electric field distribution based on that geometry [5]. To simplify the treatment planning and electroporation-based treatments from the clinician's point of view, we have created a web-based software for treatment planning (www.ECTplan.com). The main aim of the software is to minimize user interaction in the treatment planning process and make the treatment planning robust with respect to the source of medical images. As a first step in achieving this aim, we have developed algorithms for segmentation of liver, liver tumors and hepatic vessels to support ECT of colorectal metastases in the liver [6]. The liver segmentation algorithms have already been evaluated on 15 clinical cases and the results of this evaluation are considered for publication. For validation of vessel segmentation algorithms we performed a study on phantoms taking into account different diameters of vessels which are critical in ECT.

II. MATERIALS AND METHODS

A. Vessel Segmentation

We developed and evaluated two methods for segmentation of hepatic vessels. The first method is based solely on image intensity and isolates the vessels by region growing which is complemented by thorough pre-processing. The second method relies on knowledge about properties of second-order structures in the image which are incorporated into filters and applied at multiple scales.

Segmentation Based on Region Growing

Before image segmentation, appropriate pre-processing of the images is necessary. First, the images are transformed using a sigmoid function according to the Volume-of-Interest (VOI) parameters Window-Center (WW) and Window-Width (WW) that are stored in the DICOM header of every slice. The sigmoid transformation is required for pre-processing to be successful, since without the sigmoid transformation the intensity distribution of the voxels is inappropriate for segmentation. Then, image data is filtered to eliminate the possible noise using an average and a Gaussian blur filter ($\sigma=3$), both with window sizes of 3 by 3 pixels. Finally, the intensities are thresholded using a fixed threshold value of 20.000 which was obtained empirically.

After the pre-processing stage, image data are prepared for segmentation. The segmentation is performed as a simple two-dimensional region-growing algorithm on all the pre-processed images. Hence, the feature condition of the region growing algorithm is the intensity of the pre-processed tissue [7]. Segmentation is initiated by the user who manually places a starting seed on an image and, then, the desired segment (i.e. vessel), is extracted from the image.

Segmentation Based on Multi-scale Vessel Filtering

In the preprocessing phase of the second segmentation method, we used multi-scale vessel enhancement filter described in [8]. The values of factor of scaling were chosen so that they correspond to the range from 4-8 mm. The vessel enhancement was performed on original images which were masked by the liver mask obtained from the liver segmentation process.

The images filtered by vesselness filter are thresholded using an algorithm based on minimizing intra-class variance [9]. The results of thresholding are, then, pruned through morphological opening on a 2D level. The purpose of this step is to eliminate all objects which roughly have a diameter smaller than 4 mm, since we are only interested in obtaining vessels larger than that.

In order to reconnect possible disconnected segments, iterative 3D region growing is applied on the results of thresholded segmentation. The final segmentation step consists of another morphological opening, this time in 3D with the task of pruning remaining disconnected segments.

B. Validation on Phantoms

For the validation of the segmentation procedure, we have created simple phantoms which are models of hepatic vessels of various sizes in various positions. We have filled plastic cups with a 0.5% agarose. We inserted glass tubes filled with physiological solution into the gel in order to resemble the hepatic vessels. The tubes used were of three different diameter sizes: 4 mm, 6 mm and 8 mm. We used two tubes of each size in two different positions, which resulted with six phantoms in total. The first position was perpendicular to the cup bottom as seen in Figure 1.a, while the second position was angular, as seen in Figure 1.b.

Fig. 1 Phantom positions. A. Perpendicular to the bottom of the cup. B. Forming an angle with respect to the bottom of the cup

All of the six phantoms were imaged simultaneously inside a 1.5 T Siemens Avanto MRI device. The imaging was performed at two different resolutions: 1.56 mm by 1.56 mm and 1.04 mm by 1.04 mm with slice thickness of 2 mm in both cases.

C. Error Metrics

For *reference vessel area* we created a theoretical model which observes different ways in which a perfect circle can be positioned on a grid of certain size, depending on the circle size and the grid size. The need for such theoretical model is caused by *partial volume effect*. The illustration of our theoretical model for the case of a 4 mm vessel on a 1.56 mm by 1.56 mm grid is presented in Figure 2.

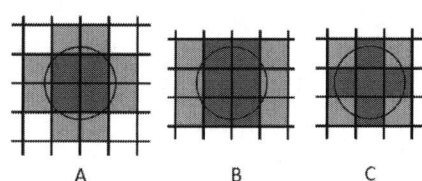

Fig. 2 Theoretical model of reference vessel area. A. Vessel center in the pixel point. B. Vessel center in the middle of the pixel edge. C. Vessel center in the middle of the pixel

Based on theoretical models, we defined two referential area values for each of the three possible positions of the vessel center. The first value, the *optimal vessel area* is the number of all pixels which contain at least half of the vessel tissue (pixels colored darker in Figure 2). The second reference value, the *maximum vessel area* is the number of all pixels that contain any amount of the vessel tissue (pixels colored darker + pixels colored lighter in Figure 2).

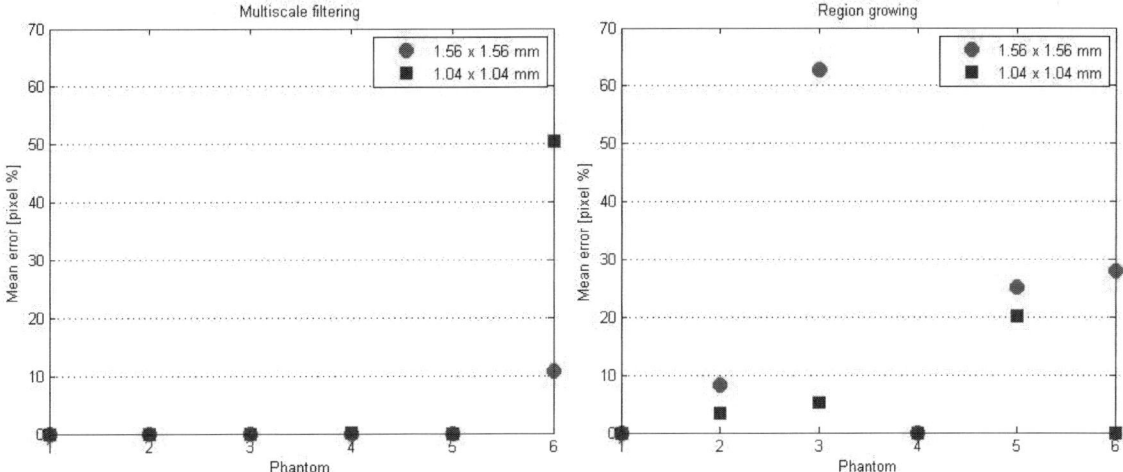

Fig. 3 Mean relative area error for different phantoms and image resolutions. A. Segmentation method based on region growing. B. Segmentation method based on multiscale filtering

We express the error as the number of pixel contained in segmented vessel area which do not fall in the range between *optimal vessel area* and *maximum vessel area*, and normalize it with the expected number of pixel calculated from the measured dimensions of the phantom.

III. RESULTS

Relative area error defined in previous section is systematically represented in Figure 3. Mean values of relative area error per individual phantom are shown, dependent on tube diameter and position, grouped by image resolution. Figure 3.a shows results for the segmentation method based on region growing. Figure 3.b shows results for the segmentation method based on multiscale filtering. Information about size and position of each of the six phantoms is given in Table 1, where marks for position are P for perpendicular and A for angular with respect to the bottom of the cup. In both Figures 3.a and 3.b, the results are grouped per different image resolution: 1.56 mm x 1.56 mm and 1.04 mm x 1.04 mm.

Table 1 Sizes and positions of phantoms. P - Perpendicular, A - angular

Phantom	1	2	3	4	5	6
Diameter [mm]	8	6	4	8	6	4
Position	P	P	P	A	A	A

IV. DISCUSSION

In order to support minimizing user interaction in the process of treatment planning through a dedicated treatment planning software, we need to focus on minimizing user interaction in image segmentation as well. At the same time, the segmentation algorithms need to be robust regardless of the image source and modality to enable wide use of the treatment planning software for electroporation-based treatments in various clinical centers. The demands for least user interaction and robustness should not, however, impact the accuracy of segmentation.

The main issue of assessment of segmentation accuracy is defining a gold standard against which to compare the segmentation [10]. The most often used approach is comparison against segmentation performed by one or more expert radiologists. Radiologist segmentation, however, cannot be considered completely correct as it is prone to inter- and intra-observer variability [10]. The second approach for segmentation validation relies on phantoms with precisely measured dimensions. Although the drawback of this approach is its inability to capture full complexity of the human body it is still useful in observing segmentation algorithm performance in detail.

The other difficulty in assessing the segmentation accuracy is that there is neither standardized data to use for as gold standard nor any defined standard error metrics. It is therefore very difficult to single out a segmentation method as 'the best', even for a specific object such as hepatic vessels. Contrary to most of the studies on segmentation accuracy [11], we decided not to strive to achieve perfect accuracy; we aim at a method accurate enough to support treatment planning of electroporation-based treatments. In the case of ECT of colorectal metastases we do not need to segment all of the hepatic vessels, but only the vessels with diameter size > 3 mm which were identified as critical by surgeons. Having this requirement in mind, we chose to design phantoms which would model vessels no smaller than 4 mm in diameter, as those were the smallest commercially available tubes to model vessels.

The study performed on the designed phantoms showed that the vessel segmentation method based on multiscale filtering with its maximum mean relative error of 0,14% could be considered accurate enough to be used to segment vessels larger than 4 mm in diameter, which covers all of the large hepatic vessels that are critical during ECT in the liver. The method based on region growing provided higher mean error in segmenting phantoms of 6 mm and 4 mm in diameter size for the larger image resolution, which for the final score resulted in maximum mean relative error of 62,67%. The reason for better performance of the method in images of higher resolution is the partial volume effect. This is also supported by the observation made from the results, that the relative error increases as the phantom diameter (and also the number of pixel with which we normalize the error) decreases. The only phantom in the study which was standing out with the resulting error was the last phantom, a 4 mm tube inserted into the agar under an angle. After a closer inspection of the original images it was noted that the reason for this abnormality was caused by a rupture in the agar gel, which indicates this phantom should be excluded while determining the feasibility of the algorithm based on error rate.

It is however still necessary to perform additional study on clinical data with radiologist segmentation as a gold standard to assess the performance of the algorithms on real-case data. Such study would also take into account the constraint of segmenting only vessels that are larger than 3 mm in diameter. To fully confirm the validity of our segmentation for use for treatment planning of electroporation-based treatments, a sensitivity study using numerical modeling should also be performed. Such sensitivity study would be based on comparing electrical field distribution obtained from numerical calculations for gold-standard segmentation and for segmentation with errors. The results of the study would provide information about the level of error that is tolerable and does not affect the resulting distribution of the electric field.

ACKNOWLEDGMENT

This work was supported by the Slovenian Research Agency (ARRS). Research was conducted in the scope of the Electroporation in Biology and Medicine (EBAM) European Associated Laboratory (LEA). This paper is resulting in part from COST TD1104 Action networking efforts (www.electroporation.net). The author would like to thank Igor Fučkan for his assistance in MRI imaging of the phantoms.

REFERENCES

1. Kotnik T, Pucihar G, Miklavcic D (2010) Induced transmembrane voltage and its correlation with electroporation-mediated molecular transport. J Membr Biol 236:3–13
2. Sersa G, Miklavcic D, Cemazar M et al. (2008) Electrochemotherapy in treatment of tumours. EJSO 34:232–240
3. Davalos R, Mir L M, Rubinsky B (2005) Tissue Ablation with Irreversible Electroporation. Ann Biomed Eng 33:223–231
4. Miklavcic D, Corovic S, Pucihar G et al. (2006) Importance of tumour coverage by sufficiently high local electric field for effective electrochemotherapy. EJC Suppl 4:45–51
5. Pavliha D, Kos B, Zupanic A et al. (2012) Patient-specific treatment planning of electrochemotherapy: Procedure design and possible pitfalls. Bioelectrochemistry 87:265–273
6. Edhemovic I, Gadzijev E M, Brecelj E et al. (2011) Electrochemotherapy: a new technological approach in treatment of metastases in the liver. TCRT 10:475–485
7. Mancas M, Gosselin B (2005) Segmentation using a region-growing thresholding, Proceedings of SPIE. vol. 5672, Image Processing: Algorithms and Systems IV, San Jose, CA, USA, 2005, pp 388–398
8. Frangi A F, Niessen W J, Vincken K L et al. (1998) Multiscale vessel enhancement filtering, Proceedings of Medical Image Computing and Computer-Assisted Intervention, Cambridge, MA, USA, 1998, pp 130–137
9. Otsu N (1979) A Threshold Selection Method from Gray-Level Histograms. IEEE Trans Syst Man Cybern B Cybern 9:62–66
10. Chalana V, Kim Y (1997) A methodology for evaluation of boundary detection algorithms on medical images. IEEE Trans Med Imaging 16:642–652
11. Heimann T, Van Ginneken B, Styner M A et al. (2009) Comparison and evaluation of methods for liver segmentation from CT datasets. IEEE Trans Med Imaging 28:1251–1265

Corresponding author:

Author: Marija Marcan
Institute: Faculty of Electrical Engineering, University of Ljubljana
Street: Trzaska cesta 25
City: Ljubljana
Country: Slovenia
Email: marija.marcan@fe.uni-lj.si

Detection of Retinal Vessel Bifurcations by Means of Multiple Orientation Estimation Based on Regularized Morphological Openings

Álvar-Ginés Legaz-Aparicio[1], Rafael Verdú-Monedero[1], Juan Morales-Sánchez[1], Jorge Larrey-Ruiz[1], and Jesús Angulo[2]

[1] Universidad Politécnica de Cartagena, Dept. de Tecnologías de la Información y las Comunicaciones, Cartagena, Spain
[2] MINES Paristech, CMM-Centre de Morphologie Mathématique, Mathématiques et Systèmes, Fontainebleau Cedex, France

Abstract—This paper describes an approach to detect vessel bifurcations and crossovers in retinal images. This approach is based on an novel method to estimate multiple orientations at each pixel of a gray image. The main orientations are provided by directional openings whose outputs are regularized in order to extend the orientation information to the whole image. The detection of vessel bifurcations is based on the coexistence of two or more than two different main orientations in the same pixel. Results on retinal images show the robustness and accuracy of the proposed method to detect vessel bifurcations.

Keywords—multiple orientation estimation, retinal fundus, vessel bifurcations.

I. INTRODUCTION

The analysis of retinal vascular structures has a great use in medical analysis. Specifically, the identification and study of bifurcations and crossovers has great significance in cardiovascular diseases as well as in their early detection [1]. There are two approaches to detect automatically vessel bifurcations: methods based on geometrical features and methods based on models [2]. The method proposed in this paper belongs to the first group, it depends on the segmentation of the retinal image.

The segmentation of the vessel tree can be performed by matched filters [3], by region growing and scale-space analysis [4], by mathematical morphology [5], etc. Regarding the detection of vessel bifurcations and crossovers on segmented retinal images it can be done, e.g., by using a set of trainable keypoint detectors and a bank of Gabor filters [6], or with matched filters [7]. This paper addresses the detection of vessel bifurcations by analyzing a vectorial orientation field provided by the regularization of directional morphological openings.

The morphological orientation field [8] is given by a directional signature for each pixel using a series of directional openings with a line segment. Then, the orientation of a pixel is defined as the one associated to the directional opening which produces the maximal value of signature at the pixel. Nevertheless, the original approach from [8] does not deal with the multiple orientation case: locally, pixels in natural images can be associated to more than one orientation,

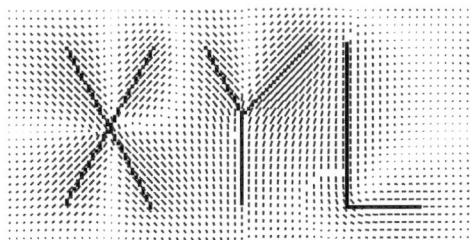

Fig. 1 Orientation vector field obtained by ASGVF [12] using $\eta=1$. The size of the image is 128×64 pixels.

e.g., crossing lines, corners and junctions (also known as X-, L- and Y-junctions, respectively). To determine not only the main direction but all the significant orientations, the directional signature is analyzed in the present work using multiple peak detection on the curve interpolated by b-splines.

This paper is organized as follows: Section II-A addresses the method to estimate multiple orientations at each pixel and II-B describes the approach followed to detect the bifurcations and crossovers in retinal images. In Section III we apply the method in a retinal image, and finally, Section IV closes the paper with the conclusions and future work.

II. PROPOSED METHOD

In previous works of the authors, the orientation field was based on the average squared gradient (ASG) (see e.g. [10]). This orientation field is valid in images where only one main orientation exists at each pixel. When a pixel is close to edges with different orientations, the ASG provides an orientation which is a linear combination of these near orientations (as can be seen in Fig.1). The method proposed here to estimate multiple orientations is depicted in Fig. 2. This approach differs from [11], where the multiple main orientations are estimated by analyzing a block of the image, whereas in this paper the multiple orientations are estimated at each pixel.

A. Multiple Orientation Estimation

Let $f(\mathbf{x}) : E \to \mathbb{R}$ be a gray-level image, where the support space is $E \subset \mathbb{Z}^2$ and the pixel coordinates are $\mathbf{x} = (x,y)$. Let us define $g(\mathbf{x})$ as the absolute value of the gradient of $f(\mathbf{x})$, i.e., $g(\mathbf{x}) = \|\nabla f(\mathbf{x})\|$.

The directional opening of $g(\mathbf{x})$ by a linear (symmetric) structuring element (SE) of length l and direction θ is defined as the directional erosion of g by $L^{\theta,l}$ followed by the directional dilation with the same SE [10]:

$$\gamma_{L^{\theta,l}}(g)(\mathbf{x}) = \delta_{L^{\theta,l}}[\varepsilon_{L^{\theta,l}}(g)](\mathbf{x}). \quad (1)$$

We remind that the subgraph of the opened image by $L^{\theta,l}$ is equivalent to the union of the translations of the linear SE when it fits the subgraph of the original image [13], or in other terms, bright linear image structures that cannot contain $L^{\theta,l}$ are removed by the opening.

The proposed orientation model is based on a decomposition of the gradient information by families of linear openings. Let $\{\gamma_{L^{\theta_i,l}}\}_{i \in I}$ be a family of linear openings of length l according to a particular discretization of the orientation space $\{\theta_i\}_{i \in I}$. In mathematical morphology theory, this parameterized family of openings is called a granulometry [13]. The granulometry is applied to the gradient image $g(\mathbf{x})$, i.e., for each orientation an opened image of the intensity gradient information is obtained. In addition, the accumulation by supremum of the openings (which has also the properties of an opening [13], i.e., idempotency and anti-extensivity) produces also an approximation of $g(\mathbf{x})$ at linear scale l, denoted $a_l^{\Theta}(g)(\mathbf{x})$:

$$a_l^{\Theta}(g)(\mathbf{x}) = \bigvee_{i \in I} \gamma_{L^{\theta_i,l}}(g)(\mathbf{x}). \quad (2)$$

The accumulation image $a_l^{\Theta}(g)(\mathbf{x})$ extracts the linear structures of $g(\mathbf{x})$ whose length is longer than l. Hence, the residue between $g(\mathbf{x})$ and $a_l^{\Theta}(g)(\mathbf{x})$,

$$r_l^{\Theta}(g)(\mathbf{x}) = g(\mathbf{x}) - a_l^{\Theta}(g)(\mathbf{x}), \quad (3)$$

yields the image structures of length shorter than l, and $r_l^{\Theta}(g)(\mathbf{x})$ might be used in a multiscale approach as the source image in the following stage with a smaller l (see Fig. 3).

When dealing with discrete images, it is important to remember the effects of discretization. As described in [12], the angular resolution of a structuring element of length l is $\Delta\theta = \frac{90}{l-1}$ degrees, therefore the angles that can be used are $\theta_i = i\Delta\theta$, $i \in I = [0, 2(l-1)-1]$. A long structuring element allows for a larger angular resolution (i.e., more directions) but the structures to be detected have to be bigger. On the other hand, a smaller structuring element offers fewer directions but it allows get into small details in the image.

In the next step of the proposed method a regularization is performed at each one of the directional openings (depicted as H_η in Fig. 2). The regularization diffuses the orientation information and avoid angle mismatches due to noise. Each

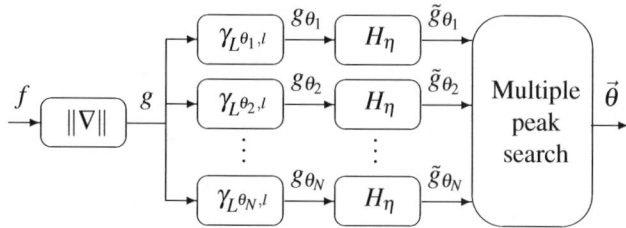

Fig. 2 Flowchart of the multiple orientation estimation method.

one of the regularized openings, $\tilde{g}_{\theta_i}$, minimizes the following energy functional

$$\mathcal{E}(\tilde{g}_{\theta_i}) = \mathcal{D}(\tilde{g}_{\theta_i}) + \alpha \mathcal{S}(\tilde{g}_{\theta_i}), \quad (4)$$

where $\mathcal{D}$ represents a distance measure given by the square difference between the original and the regularized directional opening, weighted by the squared value of the first one,

$$\mathcal{D}(\tilde{g}_{\theta_i}) = \frac{1}{2} \int_E \|g_{\theta_i}\|^2 \|\tilde{g}_{\theta_i} - g_{\theta_i}\|^2 dx dy, \quad (5)$$

where E is the image support. The energy term $\mathcal{S}$ is the regularization term, it determines the smoothness of the regularized openings and represents the energy of the second order derivatives of the signal:

$$\mathcal{S}(\mathbf{v}) = \frac{1}{2} \int_E \|\Delta \tilde{g}_{\theta_i}\|^2 dx dy, \quad (6)$$

where $\Delta = \partial^2/\partial x^2 + \partial^2/\partial y^2$ is the Laplacian operator. The positive parameter α governs the trade off between the fit to g_{θ_i} and the smoothness of the solution $\tilde{g}_{\theta_i}$ (see e.g. [14]). According to the calculus of variations, the regularized opening $\tilde{g}_{\theta_i}$ minimizing Eq. (4) is necessarily a solution of the Euler–Lagrange equation:

$$(\tilde{g}_{\theta_i} - g_{\theta_i})|g_{\theta_i}|^2 + \alpha \Delta^2 \tilde{g}_{\theta_i} = 0. \quad (7)$$

These equations can be solved by adding an artificial time and computing the steady-state solution. The solution is obtained by the following iterative implementation in the frequency domain

$$\tilde{G}^n_{\theta_i} = H_\eta(\tilde{G}^{n-1}_{\theta_i} + F^{n-1}) \quad (8)$$

where $\tilde{G}^n_{\theta_i}$ and $\tilde{G}^{n-1}_{\theta_i}$ are the 2D discrete Fourier transforms (DFT's) of their respective signals in the spatial domain, superscript n denotes the iteration index, F^{n-1} is the 2D DFT of $f^{n-1} = (g_{\theta_i} - \tilde{g}_{\theta_i})|g_{\theta_i}|^2$, H_η is the sampling of the low-pass filter

$$H_\eta(\omega_1, \omega_2) = \frac{\eta}{\eta + 4(2 - \cos\omega_1 - \cos\omega_2)^2}. \quad (9)$$

Once the directional openings have been regularized, the directional signature at pixel **x** is defined as

$$s_{\mathbf{x};l}(i) = \tilde{g}_{\theta_i}(\mathbf{x}). \tag{10}$$

$s_{\mathbf{x};l}(i)$ is interpolated using cubic b-splines and its detected maxima correspond to the multiple orientations existing at pixel **x**. Collecting all the orientations estimated at all the pixels in the image provides the multidimensional vector field $\vec{\theta}(\mathbf{x})$ (see Fig. 6).

B. Detection of Bifurcations and Crossovers

Previous method can be applied on a gray image or on a binary image. However, most of the methods which detect bifurcations use a segmented, and therefore binary, image. In this paper we use a segmented image which contains the vessel tree to estimate the multiple orientation field, this orientation vector field is used then to detect the existing bifurcations and crossovers.

The detection procedure takes into account only the two main orientations at each pixel. The method considers that a bifurcation or crossover exists if two conditions happens. The first conditions is that the magnitude of the regularized opening has to be greater than a threshold, $|\tilde{g}_{\theta_i}| > th$. The second condition to be fulfilled is that two main orientations must differ more than $2\Delta\theta$ to avoid false positives due to high curvature of the vessels.

III. RESULTS

We have used a segmented retinal image from the DRIVE database [9] to evaluate the performance of the proposed method. Given that the size of this image is 565×584 pixels, the directional openings use an oriented linear structuring element of 7 pixels. This length comes from the tradeoff between the curvature of the vessels and the angular resolution of the structuring element (this length provides $\Delta\theta = 15°$ and produces a filter bank with 12 branches). The regularization of the directional openings is performed using $\eta=1$ and 10 iterations of the procedure given in eq. (8). Finally, the threshold th used to consider that a pixel belongs to a bifurcation or crossover is the 25% of the maximum intensity value.

As can be seen in Fig. 7, the proposed method detects all the bifurcations and crossovers, however, vessel with high curvature are also detected. In Fig. 8 a close-up of the retinal image with the multiple orientation vector field is shown.

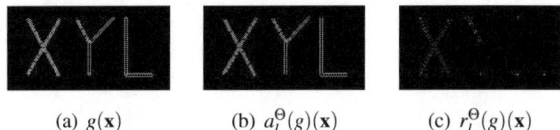

(a) $g(\mathbf{x})$ (b) $a_l^\Theta(g)(\mathbf{x})$ (c) $r_l^\Theta(g)(\mathbf{x})$

Fig. 3 a) Absolute value of the gradient of $f(\mathbf{x})$, b) accumulation by supremum of directional openings, c) residue of the orientation apertures.

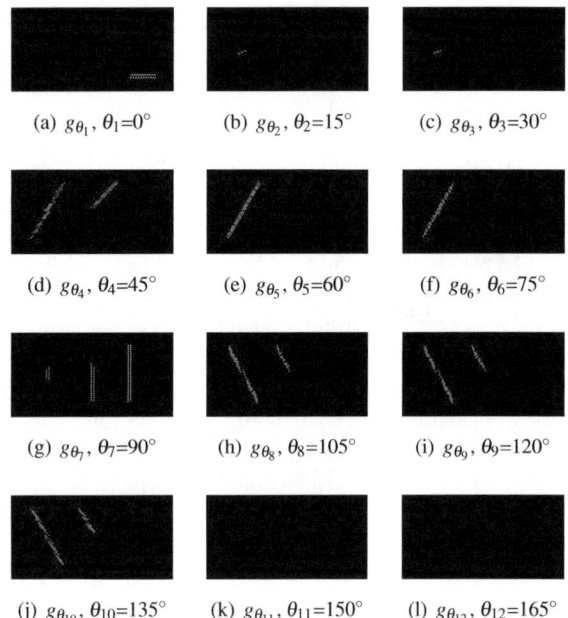

(a) g_{θ_1}, $\theta_1=0°$ (b) g_{θ_2}, $\theta_2=15°$ (c) g_{θ_3}, $\theta_3=30°$

(d) g_{θ_4}, $\theta_4=45°$ (e) g_{θ_5}, $\theta_5=60°$ (f) g_{θ_6}, $\theta_6=75°$

(g) g_{θ_7}, $\theta_7=90°$ (h) g_{θ_8}, $\theta_8=105°$ (i) g_{θ_9}, $\theta_9=120°$

(j) $g_{\theta_{10}}$, $\theta_{10}=135°$ (k) $g_{\theta_{11}}$, $\theta_{11}=150°$ (l) $g_{\theta_{12}}$, $\theta_{12}=165°$

Fig. 4 Directional openings of $g(\mathbf{x})$ with $l=7$.

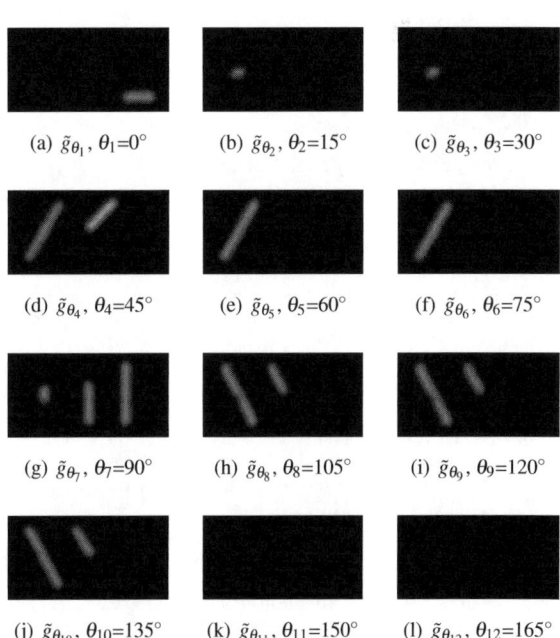

(a) $\tilde{g}_{\theta_1}$, $\theta_1=0°$ (b) $\tilde{g}_{\theta_2}$, $\theta_2=15°$ (c) $\tilde{g}_{\theta_3}$, $\theta_3=30°$

(d) $\tilde{g}_{\theta_4}$, $\theta_4=45°$ (e) $\tilde{g}_{\theta_5}$, $\theta_5=60°$ (f) $\tilde{g}_{\theta_6}$, $\theta_6=75°$

(g) $\tilde{g}_{\theta_7}$, $\theta_7=90°$ (h) $\tilde{g}_{\theta_8}$, $\theta_8=105°$ (i) $\tilde{g}_{\theta_9}$, $\theta_9=120°$

(j) $\tilde{g}_{\theta_{10}}$, $\theta_{10}=135°$ (k) $\tilde{g}_{\theta_{11}}$, $\theta_{11}=150°$ (l) $\tilde{g}_{\theta_{12}}$, $\theta_{12}=165°$

Fig. 5 Regularized directional openings using $\eta=1$.

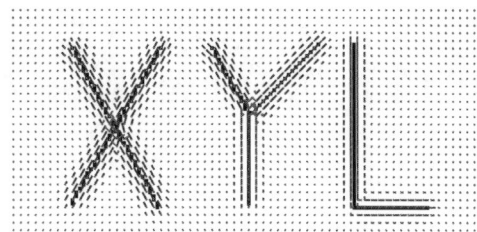

Fig. 6 Multiple orientation vector field $\vec{\theta}$ obtained by the proposed method.

IV. CONCLUSIONS

This paper has presented a method for the detection of bifurcations and crossovers based on a novel estimation of multiple orientations. The robustness of the morphological operators to provide the orientation vector field produces a correct detection of all bifurcations and crossovers, but also high curvature vessels are wrongly detected. Note that due to the regularization of the directional openings, resulting in an orientation discrimination, the proposed method is able to detect bifurcations with gaps which do not preserve completely the connectivity of the bifurcations.

As future work, we will evaluate the performance of the proposed method on all retinal images from DRIVE [9] and STARE [15] databases. Comparisons with other state-of-art methods as, e.g., [6], also will be done. Another task for further research and development is the analysis and post-processing of the results in order to detect vessels with high curvature and then try to reduce the false positives.

ACKNOWLEDGEMENTS

This work is supported by the Spanish Ministerio de Ciencia e Innovación, under grant TEC2009-12675.

REFERENCES

1. Witt N., et al. Abnormalities of retinal microvascular structure and risk of mortality from ischemic heart disease and stroke *Hypertension.* 2006;47:975-981.
2. Abramoff M.D., Garvin M.K., Sonka M.. Retinal Imaging and Image Analysis *Biomedical Engineering, IEEE Reviews in.* 2010;3:169-208.
3. Chaudhuri S., Chatterjee S., Katz N., Nelson M., Goldbaum M.. Detection of blood vessels in retinal images using two-dimensional matched filters *IEEE Trans. Medical Imaging.* 1989;8:263-269.
4. Martínez-Pérez M. et al. Scale-space analysis for the characterisation of retinal blood vessels *MICCAI.* 1999:90-97.
5. Zana F., Klein J.-C.. Segmentation of vessel-like patterns using mathematical morphology and curvature evaluation *IEEE Trans. on Image Processing.* 2001;10:1010-1019.
6. Azzopardi G., Petkov N. Automatic detection of vascular bifurcations in segmented retinal images using trainable COSFIRE filters in *Pattern Recognition Letters*;34:922-933 2013.

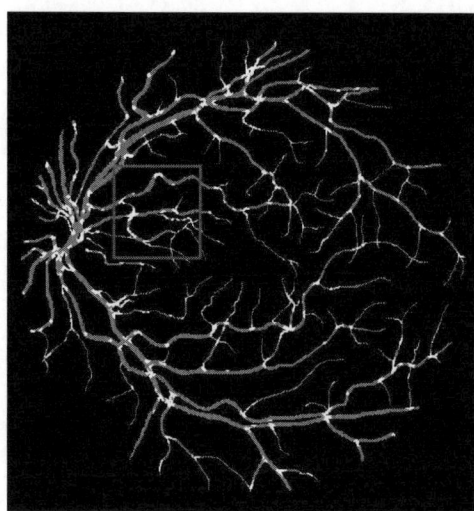

Fig. 7 Segmented retinal image *01_manual1* from DRIVE database [9] with detected bifurcations and crossovers.

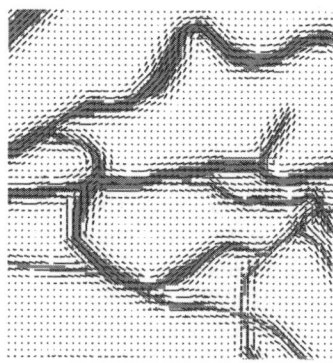

Fig. 8 Close up of the red square depicted on Fig. 7 with the multiple orientation vector field.

7. Ardizzone E., Pirrone R., Gambino O., Radosta Salvatore. Blood vessels and feature points detection on retinal images *Engineering in Medicine and Biology Society.* 2008:2246-2249.
8. Soille P., Talbot H.. Directional Morphological Filtering *IEEE Trans. on Pattern Analysis and Machine Intelligence.* 2001;23:1313-1329.
9. Staal J. et al. Ridge-based vessel segmentation in color images of the retina *IEEE Trans Medical Imaging.* 2004;23:501-509.
10. Verdú-Monedero R., Angulo J., Serra J. Anisotropic Morphological Filters With Spatially-Variant Structuring Elements Based on Image-Dependent Gradient Fields *IEEE Trans. Image Processing.* 2011;20:200-212.
11. Angulo J., Verdú-Monedero R., Morales-Sánchez J. Multiscale local multiple orientation estimation using Mathematical Morphology and B-spline interpolation *ISPA.* 2011:575 - 578.
12. Verdú-Monedero R., Angulo J. Spatially-Variant Directional Mathematical Morphology Operators Based on a Diffused Average Squared Gradient Field *LNCS: ACIVS.* 2008;5259:542-553.
13. Soille Pierre. *Morphological Image Analysis.* Springer-Verlag 1999.
14. Engl H., Hanke Martin, Neubauer A.. *Regularization of Inverse Problems*;375 of *Mathematics and Its Applications.* Springer 2000.
15. Hoover A., Kouznetsova V., Goldbaum M.. Locating blood vessels in retinal images by piecewise threshold probing of a matched filter response *IEEE Trans on Medical Imaging.* 2000;19:203-210.

Segmentation of Basal Nuclei and Anatomical Brain Structures Using Support Vector Machines

A. Bosnjak[1], R. Villegas[1], G. Montilla[1], and I. Jara[2]

[1] Centro de Procesamiento de Imágenes, Universidad de Carabobo, Valencia, Venezuela
[2] Hospital Metropolitano del Norte, Valencia, Venezuela

Abstract—Segmentation of structures inside of the brain are essential for planning computer assisted surgery. Structures such as basal nuclei are difficult to detect in MR images because they have fuzzy edges, and exhibit few changes on the gray level intensities relative to the anatomical structures surroundings. The traditional techniques of image processing thus cannot be used to segment the basal nuclei. We propose a new processing pipeline conformed by five modules: 1) Image acquisition on DICOM format using MRI inside clinics or hospitals. 2) An image pre-processing for improvement on its contrast in a specific window. 3) An anisotropic filter applied to these images. 4) An automatic selection of input vectors for the Support Vector Machine applied. This selection expands radially, where the operator give only a center point of the structures to be detected. This software searches in a radial direction for the supporting vectors that belong to the basal nucleus, and for those corresponding to the background, based on the histogram. 5) Once the Support Vector Machine is trained in the previous module, the generated model serves to classify the image. Finally, the structures are properly detected and segmented, and are then validated by a neuro-anatomist by comparing the segmentation made by our software with a manual segmentation. In conclusion, we propose a new processing pipeline that includes a classifier using Support Vector Machines (SVM) for segmentation that can be used for image guided surgery.

Keywords—Image Segmentation, Support Vector Machine, (SVM), Basal nuclei, Image Guided Surgery.

I. INTRODUCTION

Image segmentation has a key role in clinical diagnosis. An ideal method of segmentation has some basic properties, such as minimal operator interaction, fast calculation, accuracy and noise-robust results through the image variability for different patients. Many studies have been done that have focused on segmentation of X-ray imaging, MR images, CT scan, and ultrasound. There has not been however, one method that is ideal according to the above properties. Some traditional methods such as regional growth, form-based and statistical methods, work well, only when the segmentation structures have a defined gray level [1] [2].

Haegelen et al. [3], compared three recently segmentation methods, such as: Automatic Nonlinear Image Matching and Anatomical Labeling (ANIMAL) Symmetric Image Normalization - (SyN), and one patch-label fusion technique with the manual segmentation made by an specialist. The automatic labeling of deep brain structures is compared to the manual segmentation. Haegelen [3] used the Dice – kappa metrics and center of gravity to evaluate the quality of the segmentation. These metrics are useful for comparing the quantity of common pixels among the segmented structures made by the automatic and manual methods.

Basal nuclei (figure 1), also known as basal ganglia, is a set of anatomical structures located deep in the cerebral cortex, between the anterior and the medial zones of the brain. Their main components are the striatum, globus pallidus, subthalamic nucleus and substantia nigra. The striatum is composed by the caudate nucleus and putamen.

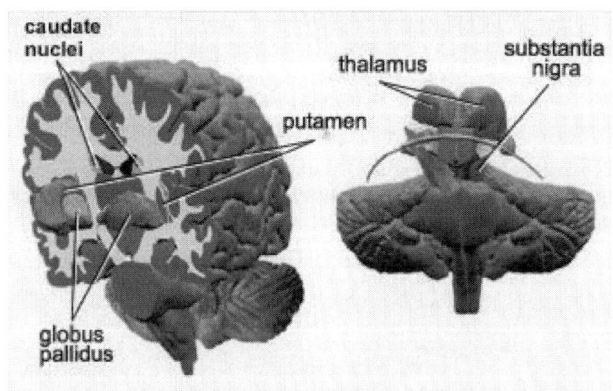

Fig. 1 Localization of caudate nuclei, globus pallidus, and putamen. (source: The Brain From Top to Bottom at http://thebrain.mcgill.ca/flash/i/i_06/i_06_cr/i_06_cr_mou/i_06_cr_mou.html)

Functionality of these structures is associated to the generation of stimulatory and inhibitory signals, which are essential to control and coordinate motor activities, as well as for regulatory processes of movements, behaviors and emotions. Basal nuclei are key elements in treatment of pathologies of the nervous system, such as Parkinson's or Huntington's disease, whose symptoms include the difficult performance of motor functions, producing from writhing movements, abnormal postures, to essential tremor and Tourette syndrome.

Minimally invasive surgeries oriented to treat this kind of neurodegenerative disease, such as pallidotomy or deep brain stimulation (DBS), are based on performing therapeutic lesions or implanting electrodes for electrical stimulation in specific zones of the basal nuclei. During the planning phase of these procedures, neurosurgeons use magnetic resonance (MR) images of the patient in order to locate the nuclei and select the structures that will be the target of the surgical approach.

Basal nuclei are not clearly observed on MR images because they have fuzzy borders and show fewer changes in the gray level intensities relative to neighboring anatomical structures. Therefore, traditional image processing techniques fail to detect them and generally require auxiliary assistance in order to locate and segment their edges.

II. METHODS: SINGLE BINARY CLASSIFIER (SVC)

The SVM was constructed by Vapnik [4] as the learning machines which minimize the classification error, finding the hyperplane for maximum margin separating the two classes in the featured space. The binary classification is presented as follow.

Given a set of points in the input space $\{\mathbf{x}_i\} \subset \Re^n \; i=1,\cdots,l$ and a function $\Psi: \mathbf{x}_i \to y_i \quad y_i \in \{-1,1\}$ which assigns to the points one of two possible values, Vapnik [4] proposed to project the problem to another space (feature space) using a transformation $\Phi: \Re^n \to \Re^m$. In the featured space the classes are lineally separable by a hyperplane of maximum margin. This proposal is presented in figure 2, and the optimization problem is defined by the following equations [5].

$$\min_{w,b,\xi} \frac{1}{2}\|w\|^2 + C\sum_{i=1}^{l}\xi_i \quad (1)$$

$$y_i\left(\mathbf{w}^T\Phi(\mathbf{x}_i)+b\right) \geq 1-\xi_i \quad (2)$$

$$\xi_i \geq 0 \quad i=1,...,l \quad (3)$$

Figure 2 is considered a planar function (Distance Function) in the featured space. This function is extended in the featured space and it is null over the hyperplane of maximum separation. It can assign arbitrarily a value of +1 to distance function over the nearest points to the optimum hyperplane, which we called border vectors. Vectors can also be allowed a distance function $1-\xi_i$, which we called outliers. The other vectors behind the two planes of distance 1 are called the interior points. The variable w (gradient of the distance function) adjusts the smoothness of the function. A minimum value of w gives the maximum smoothness, and a maximum separation between two classes, since the real distance between the two planes of distance

function 1 and -1 is $2/\|w\|$. Equation (2) expresses all the points which are projected behind the planes of distance 1 except the border vectors and the outliers. Equation (1) presents a multi-objective minimization problem that involves the magnitude of w (coefficient of smoothness or gradient) and the sum of the errors.

Equations (4) - (5) provide the dual problem from the Lagrangian. Equation (4) shows the term $K(\vec{x_i},\vec{x_j})$ that represents the scalar product in the featured space. Equation (6) represents the distance function in the featured space, and it can also be drawn as the input space. This is the decision function of the classifier. The zero level surface of this function will be used in solving the 3D modeling problem.

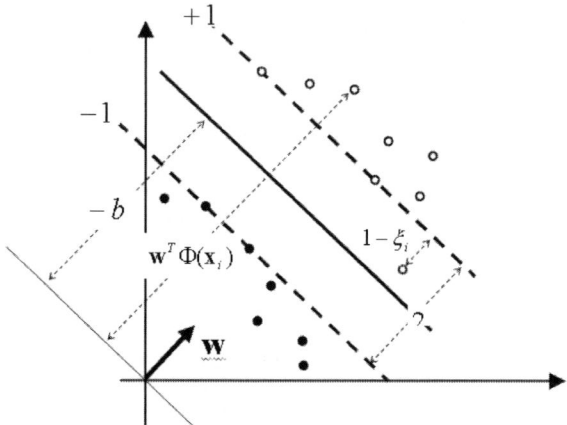

Fig. 2 Points and hyperplane in the feature space

$$\min_{\alpha} \frac{1}{2}\sum_{j=1}^{l}\alpha_i\alpha_j\gamma_i\gamma_j K(x_i,x_j) - \sum_{i=1}^{l}\alpha_i \quad (4)$$

$$\sum_{i=1}^{l}\alpha_i\gamma_i = 0 \quad (5)$$

$$D(x) = \sum_{i=1}^{l}\alpha_i\gamma_i K(x_i,x) + b \quad (6)$$

III. GENERAL SCHEME OF THIS PROJECT

This work is divided into several modules from image acquisition to its classification as shown in the diagram of Figure 3. All studies were conducted with MRI. The general scheme of this project follows these methods: 1) We developed a DICOM image reader and DICOMDIR directory files reader because they allow organized access to images and information sets [6]. This software was developed using the DCMTK 2012 library [7], which has several years of evolution and continuous use in medical applications. It can also be attached to any other software under development that requires the handling of DICOM images and

DICOMDIR directory files. 2) We made a contrast enhancement of each image based on the histogram, because the structures of interest were blurred. This pre-processing step should be performed because these images do not use fully the gray level at 12 bits range. 3) We made a pre-processing filter corresponding to the anisotropic filter inside of the 5 x 5 pixels window. 4) It picks up the candidate vectors radially as explained below. 5) A procedure that trains the Support Vector Machine, and 6) Finally, the classification performed is based on an image test file ('test') that uses all pixels of the image and determines whether it belong to a class 1 = "belongs to basal nuclei", or to class 2 = "corresponds to the background of the image".

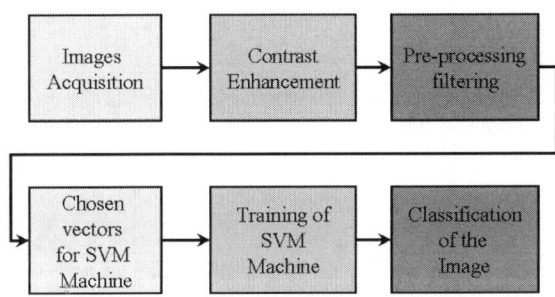

Fig. 3 Developed modules for segmentation of Basal Nuclei and other structures

A. Initialization of Segmentation

The user selects a seed within the structure of interest, in this case inside of the center of the basal nuclei. The software calculates the mean and variance radially from the center outwardly corresponding from the selected point to the edges of the structure. The candidate vector that belongs to the edge is calculated when there is a significant variation on the profile of the histogram. A dipole is placed on the edge, which corresponds to the two training vectors. The interior vector is placed just inside the edge of the structure to be segmented. The exterior vector is placed outside the structure. Figure 4 shows the selected vectors inside and outside of the structure of interest.

B. Extraction of Textured Features

The model chosen in this paper for the segmentation of basal nuclei tissues is based on two indicators which measure the homogeneity and contrast of texture: the average in a window of 5 × 5 pixels, and the variance calculated at position (x, y) following the arrangements of the windows used for the Nagao filter. The average is obtained by averaging the gray scale level in a 25 pixels centered window, computed according to equation (7). The directional variance is computed at each of the selected masks in a 5 x 5 pixels window using equation (8) for each of eight possible directions where the central pixels are these red points and those blue points shown in Figure 5.

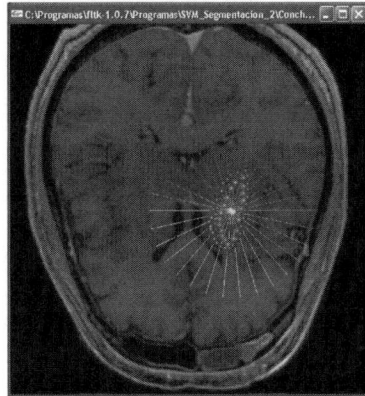

Fig. 4 Selection of the training vectors in a radial search from the center of the structure of interest

$$\mu = \frac{1}{M.N}\sum_{i=0}^{M-1}\sum_{j=0}^{N-1} I(i,j) \qquad (7)$$

$$\sigma_d^2 = \frac{1}{\#\in d}\sum_{i,j\in d}[I(i,j) - \mu_d]^2 \qquad (8)$$

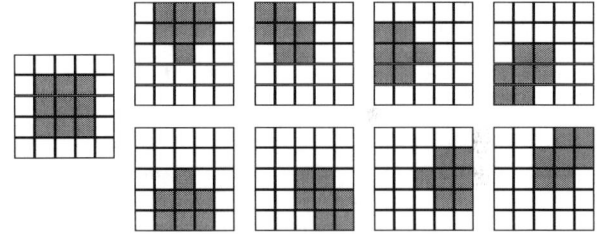

Fig. 5 Nine masks used to implement the Nagao filter

Nagao [8] proposed those sub-regions that take into account the orientation of the contour. Nagao [8] described this filter as a rotating bar around the pixel (x, y) that detects the orientation of a mask where the variance of the gray level is minimal. For this implementation, Nagao picked a neighborhood of 5 x 5 pixels, which is subdivided into nine sub-regions as shown in figure 5. We calculate then the mean and the variance over the area represented by a nine masks of figure 5. Each of the variances calculated on the Nagao's window are used as texture components for the Support Vector Machine as shows in figure 6.

C. Training of the v-SVM Machine

According to the model previously defined, each training set is represented by a vector which contains: the position (*x, y*), the mean μ in a 5 *x* 5 pixels window, and eight textural components computed as explained above. The label of the class is also included in the vector of features. This label determines whether it belongs to basal nuclei or not. Figure 6.

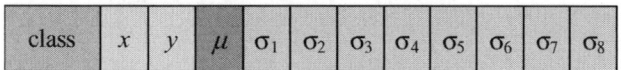

Fig. 6 Structure of features vector for the training set

The training of the ν-SVM machine is done by adjusting the parameters nu (ν) and gamma (γ) of the SVM, using a Gaussian kernel. When training is complete, feature vectors associated with the pixels of the brain's image are used to query the SVM to determine whether the new pixel belongs or not to the basal nuclei.

IV. EXPERIMENTAL RESULTS

To evaluate the methodology presented, several experiments were performed in order to detect the *globus pallidus* and *putamen*. The images chosen were taken from brain imaging studies of male and female patients aged between 42 and 75 years old. Sagittal, axial and coronal MR images at a resolution of 256 x 256 and 512 x 512 pixels were used.

For training, a point is chosen within the structure that we wish to detect. This software handles the selection of the interior class of the structure to be segmented, and searches the outer class based on the difference of the radial gradient relative to the central point selected. We assembled the training vectors using the positions (x, y), the gray level averaged over a 5x5 pixels window placed on the position x, y, and eight (8) directional variances calculated in windows proposed by the Nagao filter.

The training set was inputted to a ν-SVC machine with RBF kernel, provided by the LibSVM library from National Taiwan University [9]. Parameters ν and γ were adjusted in a way that the number of support vectors and the classification error is minimized.

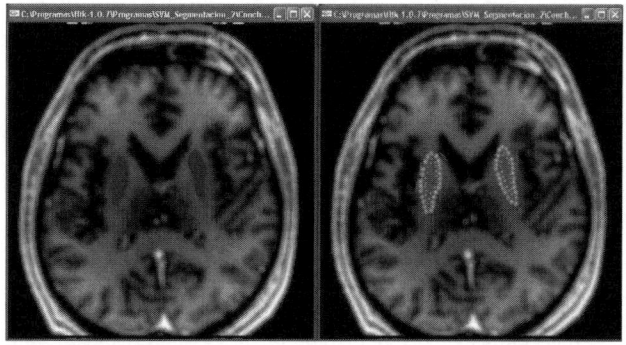

Fig. 7 Experimental results showing the structures manually segmented (left), and segmented structures (right) obtained with C = 300 and γ= 55.0

V. CONCLUSIONS

We developed a semi-automatic method of segmentation with a minimal interaction. The user selects the position inside the area corresponding to the basal nuclei and this software calculates the textured parameters, and trains the SVM machine. The preliminary results of the segmentation of the basal nuclei performed using our method is shown on Figure 7. We can see the detection of the basal nuclei and the comparison between the segmentation performed manually by a specialist.

With this technique, there is a need to adjust the model's parameters, which is done by trial and error. After adjusting the parameters for one type of acquisition, these parameters can be used in the study of other images of different patients that have been made with the same equipment. We are researching for methods of error evaluation in segmenting images with noise, applying the above processing pipeline.

REFERENCES

1. Zhang J, Ma K-K, Meng Er, Chong V. (2010) "Tumor Segmentation from Magnetic Resonance Imaging by Learning via one-Class Support Vector Machine". Report of INRIA-00548532, Dec 20, 2010.
2. LI Q-Y, Zhang S.X, Tan L.W, Qiu M.G. (2009) "Reconstruction and Application of Digital Brain Model on Chinese Visible Human (CVH)". O. Dössel and W.C. Schlegel (Eds): World Congress 2009, IFMBE Proceedings 25/IV, pp. 1119-1122.
3. Haegelen C., Coupé P., Fonov V., Guizard N., Morando X., Collins DL. (2013) "Automated Segmentation of Basal Ganglia and deep brain structures in MRI of Parkinson's disease". International Journal of Computer Assisted Radiology and Surgery. 2013 Jan; 8 (1) pp. 99-110 DOI 10.1007/s11548-012-0675-8
4. Vapnik V (1998) "Statistical Learning Theory". John Wiley & Sons, Inc, New Jersey, USA.
5. Chen P-H, Lin C-J, Schölkopf B. (2005). "A tutorial on -support vector machines". Appl. Stochastic Models Bus Ind, 21:111-136.
6. Villegas R, Montilla G, Villegas H. "A Software Tool for Reading DICOM Directory Files". User Centered Design for Medical Visualization. DOI: 10.4018/978-1-59904-777-5.ch018
7. DCMTK 2012, "Digital imaging and communications in medicine tool kit". DICOM toolkit software documentation. Oldenburger Forschungs und Entwicklungsinstitut für Informatik-Werkzeuge und Systeme (OFFIS). Retrived on July 2, 2013, from http://dicom.offis.de/dcmtk.php.en
8. Nagao M., Matsuyama T. (1979) "Edge Preserving Smoothing". Computer Graphics and Image Processing. Vol. 9, pp. 394-407.
9. Chang C-C, Lin C-J (2010) LIBSVM: "A library for support vector machines". accessed at http://www.csie.ntu.edu.tw/~cjlin/libsvm

Author:	Antonio BOSNJAK
Institute:	Centro de Procesamiento de Imágenes.
Street:	Facultad de Ingeniería, Universidad de Carabobo.
City:	Valencia.
Country:	Venezuela
Email:	antoniobosnjak@yahoo.fr

Color Analysis in Retinography: Glaucoma Image Detection

Roberto Román Morán[1], Rafael Barea Navarro[1], Luciano Boquete Vázquez[1],
Elena López Guillén[1], Jaime Campos Pavón[2], Lucía de Pablo Gómez de Liaño[2],
David Escot Bocanegra[3], Luis de Santiago[1], and Miguel Ortiz[1]

[1] Electronic Department, Alcala University, Spain
bertoroman@gmail.com, {barea,boquete}@depeca.uah.es
[2] Oftalmology Service, Hospital 12 Octubre, Spain
jaimecampospavon@yahoo.es, kinga84@hotmail.com
[3] Detectability laboratory and electronic war – INTA - Torrejón de Ardoz, Madrid
escotbd@inta.es

Abstract—Glaucoma is a leading cause of permanent blindness. Glaucoma has an initial phase difficult to diagnose, so different measures are used to evaluate it. One of them is a retinography exam which is used to determinate the ratio between the cup and the disc areas, and vein shifting in disc perimeter. A system that allows different measures and calculate them automatically, could be useful to medical community. In this paper it shows a study about papilla and cup segmentation, their ratio and vein shifting ratio and their classification using neural networks. Our system is able to classify the glaucoma automatically with a sensitivity and specifity of 96,77% and 68,2% respectively.

I. MOTIVATION

Glaucoma is a progressive optic neuropathy producing structural changes in the optic nerve head reflected in the visual field [1]. Its asymptomatic phase is a challenge for medical team, because even though there is no cure for glaucoma yet, early detection and prevention is the only way to avoid total loss of vision. It could be detected usig different tests: Oftalmoscopy, Tonometry and Perimetry [2]. One of them is ratio between the papilla and the cup.

Apply these tests to hundreds of patients by qualified teams, would require a lot of time and great dedication to review the retinographies obtained with the special devices. A computer added diagnosis system in retina, presents advantages preventing loss of vision, saving on time and saving economic resources. With this motivation in mind, we present in this paper different techniques which have been developeded to help in Glaucoma detection. Several reports in detection have been developed. Ulieru et al. [3] presented a neuro-fuzzy method obtaining 75,8% sensitivity in Glaucoma classification. Parfitt et al. [4] used a neural network apply to topographics retina images with 84,4% of sensitivity. Chan et al. [5] made a statistic study of different classifiers (SVM, MLP, L-QDA, MOG) discussing advantages and problems of every one in detail. Nayak & Rajendra et al. [6] made a threshold study based in standar desviation of cup and disc and locate veins distribution, obteining 100% sensitivity in their study. Hatanaka & Noudo [7] made a color study, segmenting cup with intensity profiles, obteining 80% sensitivity and 85% specifity.

In this study, presents a helping system to diagnosis Glaucoma analizing two features: ratio between the cup/disc and ratio in vein shifting in nasal-temporal zone. It supports developing neural networks to automate the analysis.

II. METODOLOGY

This study is based on a previous work in optic disc localization [8] where 99% of sensitivy was obtained. This is demonstrated in Fig.1.

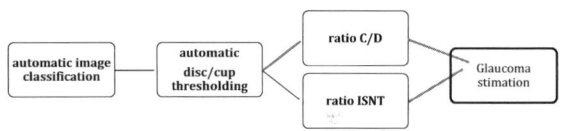

Fig. 1 Estimation image glaucoma strategy

Fundus images were taken in oftalmology service of "12 de Octubre" Hospital in Madrid [9]. We have used 53 images JPEG 768x576 taken with a TRC.50IX retinal camera: 22 normal images and 31 glaucom images. All of them with disc and cup marked from oftalmology team in hospital (Fig. 2). This marked will be our Gold Standard in the study.

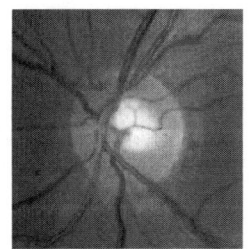 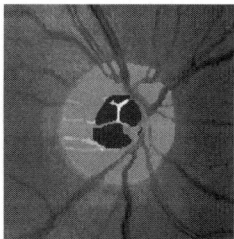

Fig. 2 Image with papilla and cup marked

Once *optic disc* has been located, a 200x200 pixels window is achieved with the centre in the middle of the papilla. This window is classified (first manually then with a neural network) as **normal** if the optic disc can be differentiated from fundus or as **saturated** if optic disc can not be differentiated from fundus. The next phase, a thresholding is made (first manually then with a neural network) to segmentate the papilla and the cup. For normal images, R channel will be used for the *papilla* and G channel por the *cup*. For saturated image, Y channel will be used for the *papilla* and G cannel for the *cup*. With this values of *papilla and cup,* ratio will be calculated. Then a estimation of veins shifting in nasal/temporal areas in *disc* will be developed. Using this information will estimate the presence of glaucoma in retinography.

III. IMAGE ANALYSIS

A. Image Classification

Non uniform ilumination is especially a problem in retinography because of a combination of factors (geometrics features of eye, electronic limits of adquisition devices, etc..) [10], [11].

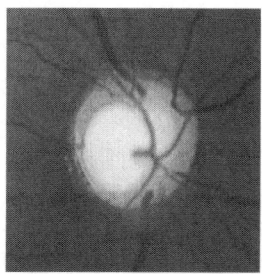 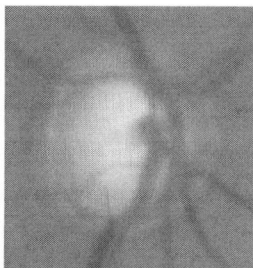

Fig. 3 Normal and saturated images.

Because of this problems, some images have good definition and others are saturated and papilla/cup get wrong with fundus retinography. In Fig. 3 can observe examples of this images.

Analyzing manually average RGB levels in each window, can determinate threshold to decide if an image is normal or saturated. In our case, this threshold is determinated when average R channel have values over 0.78 and standard deviation (0.1,0.2). Once an image is identified as normal o saturated, will be processed in different form. This values has been used to train an artificial neural nertwork [12] [13] to separate automatically both kind of images.

B. Color Threshold: Ratio Cup-Papilla

In this part, medical manual marked will be used and thus compare it with color threshold for the papilla and the cup. This process is made in different color spaces: RGB, CMYK, YCbCr, HSV.

The process is a illumination threshold between (0,1) pixel by pixel in each color space with a morfology process to erase veins. Algorithm performance was evaluated by measuring the overlapping degree between the true OD regions in "gold standard" images and the approximated regions obtained with the described approach. The proposal by Lalonde et al. [14] was used with this purpose: an overlapping score S is defined to measure the common area between a true OD region T and a detected region D as:

$$S = \frac{\text{Area}(T \cap D)}{\text{Area}(T \cup D)}$$

Using this measure, its proved that maximums of S are found for normal images in R channel (papilla) and G channel (cup). For saturated images in Y (papilla) and G (cup) such as is shown in Fig. 4. This values (mean of R, G, B and standard deviation of R, G) has been used to train an artificial neural nertwork to calculate automatically what is the threshold for disc and cup in each image and each channel. For cup, in non-Glaucoma images, S-parameter is worse (values under 70%). One reason for that behavior is because our normal images cup is not clear in Green channel. Smax for R channel are reached over 0,7 and Smax for Y channel are reached over 0,5.

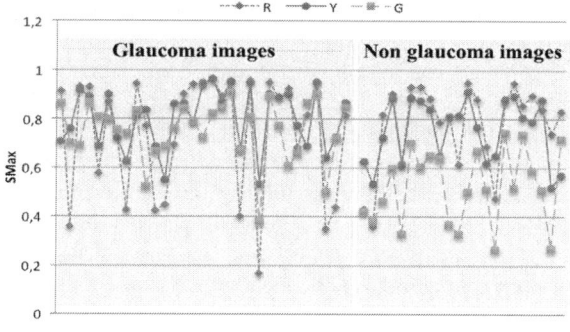

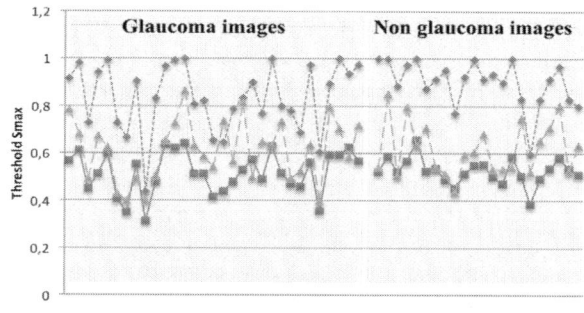

Fig. 4 Up Smáx (R-Y-G). Down threshold for SMáx.

Saturated images Smax is, generally, better in Y than R as you can see in Fig. 5. For that reason, Y channel is used in that kind of images.

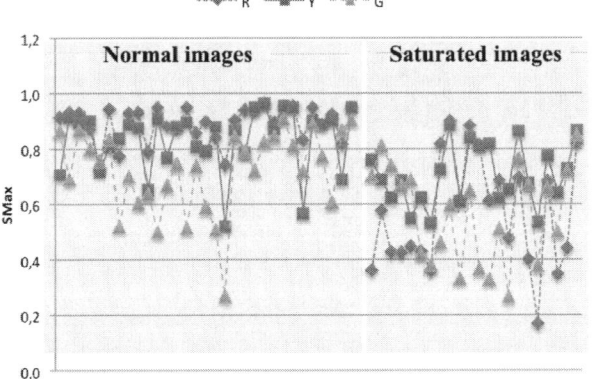

Fig. 5 Smax. Normal and saturated images

Once the papilla and the cup areas are calculated, ratio area and ratio heights of every area has been made and compare with Gold Standard.

C. Veins Shifting

A neighborhood filter has been used [15] to segmentate only veins in the image.

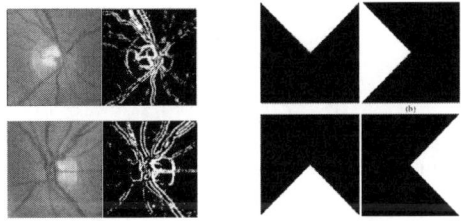

Fig. 6 Veins extraction and ISNT mask

Using the masks shows in Fig 6, veins distribution in nasal-temporal-superior-inferior region (ISNT rule) will be achieved. The target is identify those optic disc whose veins are shifting towars nasal/temporal region. Normally veins are concentrated in superior and inferior regions of optic disc [16]. A shifting in the head of optic nerve produces rising of the veins area in nasal and temporal regions. Making ratio between sum of vein areas in superior/inferior regions and sum of vein areas in nasal/temporal regions, this is smaller in glaucoma images than normal images. In Fig 7. Ratio media for glaucoma image is 1,39 and for normal image is 1,46.

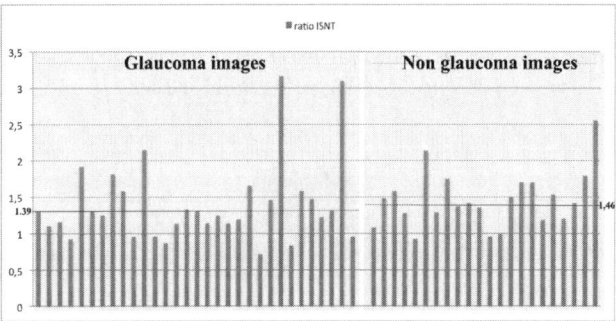

Fig. 7 ISNT of veins

IV. RESULTS

Once images have been classified in normal or saturated, papilla and cup have ben segmented. This result is compared with Golden Standard marked for ophtalmologist using S-parameter. Next, ratio between cup and papilla is calculated in two different methods. The first of them is a ratio between cup height and papilla height. The second method is a ratio between cup area and papilla area. To give a linear measure of last one, square of this ratio area has been made.

In Fig.8 - 9 you can see segmented results of normal and saturated image showed in Fig. 5. Papilla result above and cup result below. Second column shows in red ophtalmologist marked. Third column shows results of our algorithm. Last one shows S – parameter results.

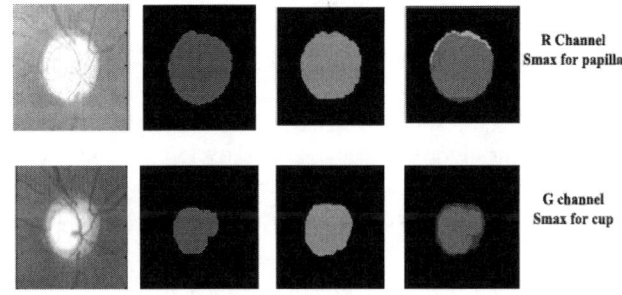

Fig. 8 Papilla and cup segmentation for normal image

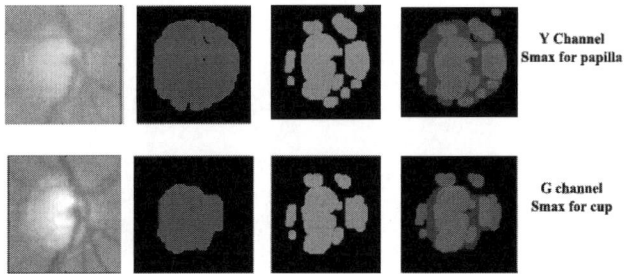

Fig. 9 Papilla and cup segmentation for saturated image

Fig.10 shows an example of papilla and cup segmented. Area and height of papilla/cup is shown to calculate differents ratios and compare with Gold Standar

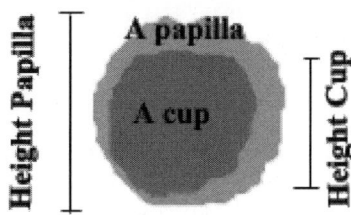

Fig. 10 Area and heights of Papilla and cup

As far as S is concerned, area and heights ratio have been made comparing with gold standard ratio of medical team, obtaining results in Fig.11

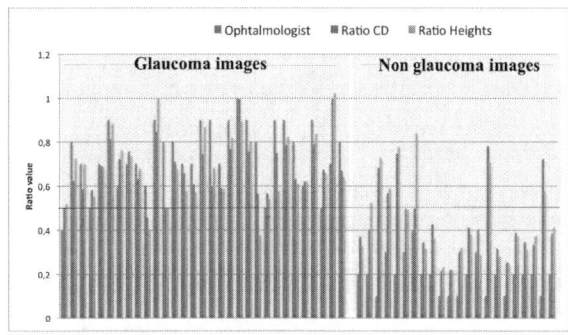

Fig. 11 Manual ratio, area ratio (square) and heights ratio.

Our ratio threshold to detec glaucom images is 0.5. In this case our algorithm has a sensitivity to detect Glaucom images of **96,77%**. On the other hand, to detect non-glaucoma images it has a specifity of **68,2%** as you can see in Table 1.

Table 1 Sensibility and specifity

	samples	Detected	Percentage
Normal	22	15	68,2%
Glaucoma	31	30	96,77%

Finally, results of NNA to classify images have been 100% detecting saturated image and 88% detecting normal image. Results of NNA to automate cup threshold in 14 sample images are shown in Fig. 12. Results of NNA to automate papilla threshold are similars. However our number of images to trained them is limited and their results are very variables and unstables.

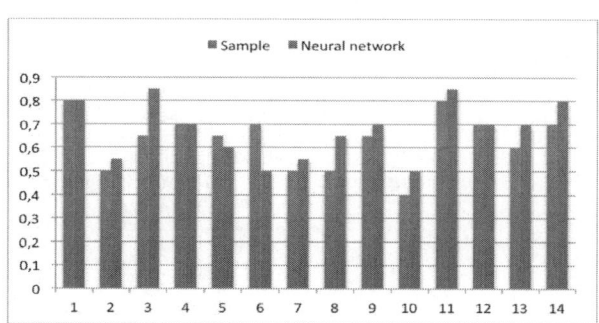

Fig. 12 ANN cup threshold results

V. DISCUSSION

Low specifity in the method could be attributable to normal images in our 53 group. There are not enough shines in Green channel for cup area and our system does not segmentate it very well. Added to this, our database is varied as regards to color. This feature makes harder a development of general and uniform method for all of them. ANN are useful to automate analysis but it needs a large quantities of images to improve its results and be trained better than now.

Values obtained in ISNT ratio validate that ratio in glaucoma is minor than normal image but the threshold between groups is small. It's not a good key risk indicator of Glaucoma with our group of images. A database with different images, better contrast and without colour variety, could improve our results.

VI. CONCLUSIONS

A helping system of glaucoma diagnosis analyzing different color spaces an vein shifting have been presented. Values of 96,77% of sensibilitiy and 68,2% of sensitivity have been obtained. With values of image classification and manual values of threshold obtained for papilla and cup, 3 ANN have been trained to automate the analysis.

Values obtained in ISNT ratio validate that ratio in Glaucoma (1,39) is minor than normal image (1,46) but the threshold is small. It's not a good key risk indicator of Glaucoma in this system. Results of NNAs are promising but it needs a large quantities of images to improve its results.

ACKNOWLEDGMENT

This research has been partially supported by the Ministerio de Ciencia e Innovacion (Spain), project: "Advanced analysis of Multifocal ERG and Visual Evoked Potentials applied to the diagnosis of optic neuropathies", with reference: TEC2011-26066.

REFERENCES

[1] Hitchings, R. A., and Spaeth, G. L., The optic disc in glaucoma, ii: correlation of appearance of the optic disc with the visual field. Br. J. Ophthalmol. 61:107–113, 1977. doi:10.1136/bjo.61.2.107.

[2] www.clevelandsightcenter.org.

[3] Ulieru, M et al, Application of soft computing methods to the diagnosis and prediction of glaucoma. Proc. IEEE Int. Conf. Syst. Man Cybern.5:3641–3645, 2000.

[4] Parfitt et al. The detection of glaucoma using an artificial neural network. Proceedings of 17th Annual Conference IEEE Engineering in Medicine and Biology, 1, 847–848, 1995.

[5] Chan, K., Lee et al. Comparison of machine learning and traditional classifiers in glaucoma diagnosis. IEEE Trans. Biomed. Eng. 49:9963–974, 2002. doi:10.1109/TBME.2002.802012.

[6] Jagadish Nayak & Rajendra Acharya "Automated Diagnosis of Glaucoma Using Digital Fundus Images" J Med Syst (2009) 33:337–346 DOI 10.1007/s10916-008-9195-z Received: 11 June 2008 / Accepted: 22 July 2008 / Published online: 9 August 2008

[7] Yuji Hatanaka et al. Vertical cup-to-disc ratio measurement for diagnosis of glaucoma on fundus images Medical Imaging 2010: Computer-Aided Diagnosis, edited by Nico Karssemeijer, Ronald M. Summers Proc. of SPIE Vol. 7624, 76243

[8] Román R, Barea R, Boquete L, Campos J, De Pablo L, Escot D. "Localización automática del disco óptico en retinografías" Libro de Actas XXX CASEIB 2012. ISBN – 10: 84-616-2147-6

[9] Servicio de oftalmología Hospital 12 de octubre. http://www.madrid.org/cs/Satellite?cid=1142638647463&language=es&pagename=Hospital12Octubre%2FPage%2FH12O_contenidoFinal

[10] Toby Berk et al. (August 1982). "A human factors study of color notation systems for computer graphics". Communications of the ACM 25(8): 547–550.

[11] Sinthanayothin C et al. Automated localization of the optic disc, fovea, and retinal blood vessels from digital colour fundus images.Br J Ophthalmol 1999;83: 902-910.

[12] Haykin, S., Neural networks a comprehensive foundation. 2ndedn. Pearson Education, 1999.

[13] Pajares G, De la cruz García. "Ejercicios resueltos de visión por computador. Editorial Rama. 2007

[14] M. Lalonde et al. "Fast and robust optic disk detection using pyramidal decomposition and Hausdorff-based template matching," IEEE Trans. Med. Imag., vol. 20, no. 11, pp. 1193–1200, Nov. 2001.

[15] Omer Demirkaya, Musa Asyali, Prasana Shaoo. "Image Processing with Matlab – Applications in Medecin and Biology"

[16] Greaney, M. J., Hoffman, D. C., Garway-Heath, D. F., et al.Comparison of optic nerve imaging methods to distinguish normal eyes from those with glaucoma. Invest. Ophthalmol. Vis. Sci. 43:140–145, 2002

An Image Analysis System for the Assessment of Retinal Microcirculation in Hypertension and Its Clinical Evaluation

X. Zabulis[1], A. Triantafyllou[2], P. Karamaounas[1], C. Zamboulis[2], and S. Douma[2]

[1] Institute of Computer Science, Foundation for Research and Technology - Hellas
[2] 2nd Propaedeutic Dept. of Int. Medicine, Hippokration General Hospital,
Aristotle University of Thessaloniki, Thessaloniki, Greece

Abstract— A system for the assessment of hypertension through the measurement of retinal vessels in fundoscopy images, is presented. The proposed approach employs multiple image analysis methods, in an integrated system that is used in clinical practice. Automating the measurement process enables the conduct of a clinical study that, for the first time, shows the correlation between macrovascular and microvascular alterations, based on numerous measurements acquired by this system. Experience and perspectives gained from clinical usage and evaluation are reported.

I. INTRODUCTION

Cardiovascular disease is a leading cause of mortality and morbidity, with cardiovascular events accounting for about one-third of the recorded deaths worldwide [4]. Hypertension (high blood pressure) is one of the most important, highly prevalent and reversible cardiovascular risk factors and the World Health Organization has rated hypertension as one of the most important causes of premature death worldwide. Until recently, cardiovascular disease, which clinically manifests mainly as myocardial infarction and stroke, was considered to involve primarily large vessels. Recent research has highlighted the crucial involvement of the microcirculation in many cardiovascular conditions [15, 14]. The retina is an accessible "window" through which the microcirculation can be depicted directly and repeatedly through time [7]. As new assessment methods and software automation become widely available, abnormalities of microcirculatory function can be detected non-invasively and cost-efficiently through fundoscopy.

During the last years, automated vessel detection software provided the opportunity to investigate the clinical and prognostic significance of retinal vessel alterations, as well as the pathophysiological mechanisms linking them to hypertension and cardiovascular diseases. Thorough assessment of these alterations over time, during the course of the disease and after the administration of appropriate treatment is considered still inadequate. The lack of relevant software (able to match different images of the same patient, obtained at different time points) might explain why there are only a few

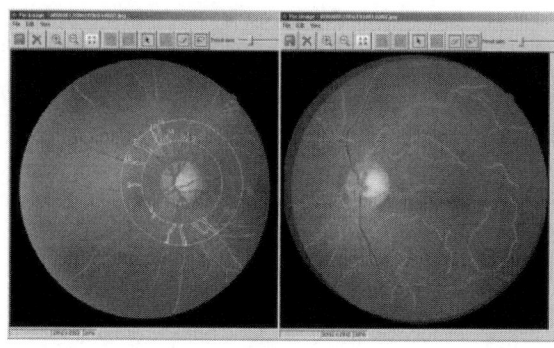

Fig. 1 User interface components for vessel measurement (left) and comparison of two registered diagnostic images, of the same patient (right).

studies, most of them with major methodological limitations, investigating whether the eligible anti-hypertensive treatment can reverse retinal alterations and reduce patient cardiovascular risk [17, 22].

The key measurement, in this work, is the width of vessels as it indicates a remodeling of vessel walls: as blood pressure rises vessel walls undergo a narrowing of the internal lumen and, in turn, vessel image width. Despite ongoing research in retinal image analysis (see [2] for a review), there is a disparity between current state-of-the-art and the software applications available to the healthcare professional. Currently, the clinical practitioner lacks tools that find vessels and measure them, in a highly usable and automatic manner. Furthermore, progress marked in vessel segmentation and width measurement, has not yet been matched by progress in monitoring patient status along time, so as to associate and compare measurements from different examinations, i.e. screening and follow-up.

In this work, we propose an image analysis system that extends state-of-the art in the automatic assessment of hypertensive signs in fundoscopy images. Fig. 1, shows two central panels of the proposed system. The left presents vessel detection and measurement around the detected optical disk for computation of a diagnostic metric called ArterioVenus Ratio (AVR). The right shows an image registration process that enables the comparison of measurements from different examinations, at the same anatomical locations. Besides encapsulating state-of-the-art methods for image segmentation and measurement, the proposed system employs image analysis methods in the user interface, to increase usability in tasks such as the selection of vessels to measure and the comparison of their dimensions.

The remainder of this paper is organized as follows. In Sec. II related work is reviewed. In Sec. III the proposed system is presented and in, Sec. IV, the conducted clinical study is described. Clinical perspectives, system evaluation, and directions for future work are provided in Sec. V.

II. RELATED WORK

Simple fundoscopy modalities are accompanied by software that performs basic observation and file management functions, such as the "2D/3D Non-mydriatic Retinal Camera / Analysis System" (*Kowa Optimed*), the "DRS non-mydriatic fundus camera" (*CenterVue*), the "VISUCAM-PRO" (*Carl Zeiss Meditec*), and the "IMAGEnet" (*Topcon*).

More advanced systems perform "generic" image analysis in the sense that they detect anatomical features, such as blood vessels, the optic disc, and the fovea [6, 19]. Methods of such systems can be the basis for a wide-range of diagnostic studies, but their integration with additional context-based techniques is still required in clinical practice.

Despite the lack of systems for the detection of hypertensive signs in retinal images, progress has been made in the domain of diabetes. Pertinent applications are not directly relevant, but include functionalities useful for the assessment of retinal microcirculation in hypertensive patients. The "Diabetic Retinopathy Risk Analyzer" (*VisionQuest Biomedical*), provides automatic AVR computation and tools for artery-vein classification. "Retinal Analysis Tool" (*ADCIS*) includes semi-automatic tools for vessel segmentation but focuses on diabetic retinopathy and age-related macular degeneration. "SIRIUS" [12] is an online-system for lesion detection and AVR computation. "DR-CAD" [1] is a screening system for diabetes that uses machine learning to detect vessels, optic disk, and lesions.

Retinal image analysis systems that provide registration functionalities focus on the spatial domain. I.e. "DualAlign i2k Retina" (*Topcon Medical Systems*) and "AutoMontage" (*MediVision Medical Imaging*) stitch retinal images together to create a "montage": a composite image showing a larger region of the retina. However, there is lack of work in temporal image registration. The need, in this case, is to register two images of the same anatomical location, acquired over widely separated time instances (i.e. 6 months). The goal is to compare changes, due to the manifestation of a disease or the reversal of symptoms due to the administered treatment.

Our work in [10] was evaluated on benchmark datasets to provide state-of-the-art accuracy in vessel segmentation. In *this work*, we capitalize on this accuracy in the context of a clinical study, to capture variations in vessel width and AVR values that are linked to hypertension. Besides incorporating state-of-the-art in segmentation algorithms, this work employs image analysis in a highly usable user interface, to increase the automation of examination. Furthermore, it provides the capability of temporal image registration to avail comparisons of patient status in initial and follow-up examinations.

III. THE PROPOSED SYSTEM

The proposed system avails image analysis methods for vessel representation and measurement and provides diagnostic tools based on these methods. A graphical user interface (GUI) integrates them in a way useful to clinical practice, including visualization and inspection of vessel representations, as well as, the ability to edit them, in order for the healthcare professional to compensate for potential errors of image analysis or ambiguous vessel imaging.

A. Vessel Representation

Specific vessel representation is provided by the binary segmentation image, obtained by the method in [10]. To achieve system modularity, a CORBA-based Interface Definition Language (IDL) is used to invoke the segmentation method. This way, the segmentation method can be updated as state-of-the-art evolves. Also, alternative methods can be invoked for images that are acquired by different modalities (i.e. fluoroangiography).

At a more abstract level, lies the skeletonization of vessels, by which vessels are represented through their medial axes. The skeletonized segmentation image captures the geometry of vasculature, such as the vessel "stem", end-points, junction-points, and vessel overlaps. This skeleton is treated as a topological graph which exhibits mainly tree structure, but includes some circles as some vessels may overlap. Bifurcation points are discovered, through the degree of nodes being 3 and vessel overlaps are characterized by a degree of 4. An IDL-based invocation of the method enables its future upgrade. Measurement of vessel widths is performed as in [10], once for each skeleton point, associating the corresponding measurement to it.

Retinal anatomy is considered by detection of the optical disk. Vessel measurements are assessed with respect to their distance to the optical center, a property of diagnostic significance (see Sec. Q. We have enhanced the Mean-Shift

[3] based method in [10], by requiring that the optical disk contains the brightest image region, thus avoiding local minima during optimization. Also, initializing optimization in the vicinity of this region reduces convergence time.

In this work, several pixel-based operations employed by segmentation and skeletonization (i.e. convolution) were re-implemented in the CUDA programming language to parallelize and accelerate their execution. Indicatively, an image of 3000×3000 pixels is processed in $\approx 0.4\,sec$, from tens of seconds required in [10].

Once representations are available they are stored in system files. An XML-indexed file structure is maintained to organize the files of all patients, as each patient may be subject to several examinations, and a GUI component assists the user to find, gather, and sort patient examinations.

B. Visual Inspection and Editing

Inspection and editing of vessel representations is availed through a GUI. This interface includes conventional tools, by which the user can observe image details. The magnification (zoom) function is accompanied with a navigation panel (see Fig. 3) which shows the image thumbnail and enables fast access to different image locations, by clicking on pertinent locations on the thumbnail. Superimposition of representations on the image facilitates their inspection. In addition, brightness/contrast adjustment controls facilitate the inspection of faintly imaged vessels. Most importantly, the interface includes tools enabling interactive measurements and, also, avails the ability to compensate for segmentation errors as image segmentation.

To ease vessel measurement, a variety of selection tools assist the user to select the vessels of interest. The pertinent tools access the obtained representations and, in particular, the graph that represents the skeleton of the vasculature. All selection operations are performed using the mouse device. The simplest case is to select a vessel point, which is considered as the closest skeleton point (or graph node) indicated by the user. To select a vessel segment, two skeleton points are indicated by the user. All possible paths between these two points are recursively found and the shortest one is considered as the result.

The AVR metric averages widths of multiple vessels, at a range of distances $[d_1, d_2]$ from optical center. To achieve the selection, skeleton points at distance d_1 are found. Then, graph paths are recursively grown at the direction away from the disk center, with the recursion ending when distance d_2 is reached. Duplicate paths that may be selected, due to branching points, are thereafter discarded. Finally, paths are sorted in a clockwise order around the optical disk to provide intuitive enumeration in the GUI.

Measurements can be exported in multiple ways. An exhaustive output yields vessel widths per skeleton point, organized in the selected entities. Furthermore, cumulative or per vessel mean average, variance and median of widths can be exported, organized in a spreadsheet format.

The GUI also enables the editing of the extracted representations. The purpose of this component is twofold: (1) provide a way to correct segmentation or skeletonization errors and, thereby, improve measurement accuracy and (2) acquire ground truth results, to be able to compare the accuracy of different segmentation and skeletonization algorithms. The segmentation result can be edited by adding and deleting pixels of the segmentation image, much like in drawing applications, using size-configurable, pencil and eraser tools, as well as, region growing tools. The GUI facilitates editing by enabling zoom in details, storing intermediate results, multiple-level undo, etc. While the segmentation is edited, the skeleton representation and width measurements are automatically recomputed, for the new segmentation. Fig. 2, illustrates the process, for two cases of segmentation errors. In the top row, segmentation underestimates the spatial extent of a faintly imaged vessel (leftmost two images, show initial segmentation and skeleton). By editing the segmentation with a region growing tool the error is recovered (rightmost two images). In the bottom row, two overlapping vessels yield an ambiguous segmentation result, spuriously merging the two vessels into one (left: original image segment, middle: segmentation result). User intervention disambiguates the result into separate vessels (right) providing a more accurate representation and, in turn, AVR estimation. Other representations are simpler to edit, such as the optical disk which can be moved and scaled.

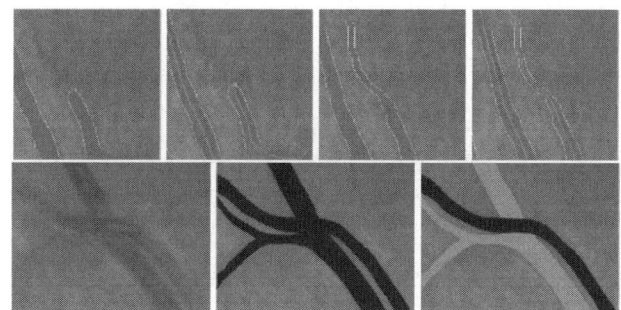

Fig. 2 Editing segmentation results automatically corrects skeletonization (see text).

C. Diagnostic Tools

Calculation of the ratio of Central Retinal Arterial Equivalent (CRAE) over Central Retinal Venous Equivalent (CRVE) is a diagnostic metric for the assessment of hypertensive retinopathy [13]. Its calculation is available through the proposed system. Its value is determined by the mean widths of arteries and veins within a region of interest which, in turn, requires the classification of the accounted vessels into veins or arteries. This classification is provided by the user through a user interface component that facilitates the process.

Registration of retinal images enables the comparison of vessel width between the initial and follow-up examinations at the level of individual vessels, because it provides the spatial correspondence of the pertinent representations and measurements. Fig. 1 (right) shows a GUI component where images are displayed registered and transparently overlapped for a more extensive comparison. Fig. 3 illustrates the indication of corresponding anatomical points in two images acquired at separate examinations. The user can navigate easily in the magnification through the navigation panels, which appear on the bottom right of each UI panel. (in this figure we have, additionally, superimposed minifications of the entire images above these navigation panels, for better reference).

The proposed system achieved retinal image registration similarly to [9], which registers retinal images as 3D curved surfaces. However, to cope with tissue change, the whole retinal texture is compared and only coarse scale features are considered by the cost function of the matching process. Registration provides a pixelwise mapping between the two images and, thus, in the interface the user indicates a point of measurement in one image and the system provides the corresponding measurement at the other.

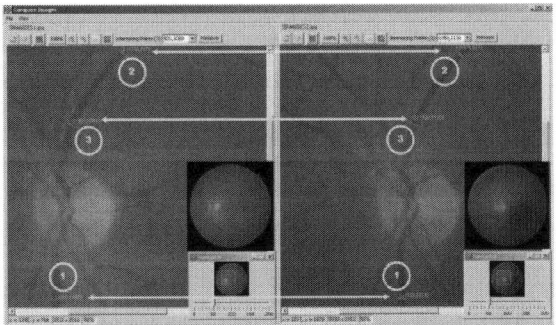

Fig. 3 Retinal image registration (see text).

IV. CLINICAL STUDY

Data derived from studies conducted in pre-hypertensive patients and in healthy volunteers with a family history of hypertension show that the above changes in microcirculation might precede the development of hypertension and be partially responsible for the establishment of hypertension as well [8]. At the same time, an abundance of data suggests that aortic stiffness is a subclinical index of atherosclerosis [23], associated, independently of other known risk factors, with increased incidence of hypertension [5] and cardiovascular and total morbidity and mortality [20]. Although there is evidence that aortic stiffness is related with microangiopathy in organs such as the brain and kidneys [11], as well as that hypertensive retinopathy is associated with other markers of hypertensive target-organ damage, such as microalbuminuria, renal impairment and left ventricular hypertrophy [16], data on the relationship between arterial stiffness and retinal vascular calibers, remains scarce. The aim of this study was to use the above mentioned software to evaluate for the first time the relationship between aortic stiffness (estimated by augmentation index) and retinal arteriolar narrowing (estimated by retinal vessel diameter), in a naive (with no other comorbidities), except for high blood pressure, adult population and in healthy subjects.

A. Methods

Consecutive hypertensive patients attending the Hypertension Unit of the 2nd Propaedeutic Department of Internal Medicine, Aristotle University, Thessaloniki, Greece were included in the study. The normotensive group was recruited from the Internal Medicine Outpatient Clinic from adults admitted for regular check-up. Procedures in the study protocol were performed according to institutional guidelines and the principles of the Helsinki declaration and were approved by the Ethics Committee of our University. All participants gave written informed consent.

According to the research protocol all participants underwent clinical examination, recording of demographic and anthropometric characteristics and collection of fasting blood samples to determine values of serum lipids. Blood pressure (BP) was measured with oscillometric technique (Omron device 705IT), using standard methodology, after 15 minutes of rest in a seated position. Hypertension was defined as BP $\geq 140/90$ mmHg. Microcirculation assessment was based on retinal vessels diameter analysis, from retinal photographs obtained by a non mydriatic digital fundus camera (NIDEK AFC-230/210) and analyzed with the software thoroughly described above. Macrocirculation, and specifically arterial stiffness, was estimated using the augmentation index (AI) corrected for the heart rate (AI75) and obtained from the radial artery pressure waveform via applanation tonometry (Sphygmocor device). Augmentation index was defined as the difference between the second and the first systolic peak (P2-P1) expressed as a percentage of the pulse pressure (PP).

Statistical analysis was performed with the Statistical Package for Social Sciences (SPSS) 19. Student t test or Mann Whitney test was used to estimate differences between

Table 1 Baseline characteristics of the study population

	Total (n=105)	Hypertensives (n=70)	Normotensives (n=35)	p value
Age, years	45.5 ± 10.8	45.5 ± 11.1	45.5 ± 10.4	0.792
Sex (Male%)	67.6	77.1	46.9	0.003
Body Mass Index, Kg/m^2	27.1 ± 4.1	27.4 ± 4.1	26.4 ± 4.1	0.186
Height	174.5 ± 9.7	175.2 ± 9.7	172.81 ± 9.61	0.254
Smoking (yes %)	38.1	41.4	31.3	0.326
Systolic blood pressure, mmHg	140.3 ± 18.6	149.2 ± 14.1	120.1 ± 9.2	< 0.001
Diastolic blood pressure, mmHg	89.4 ± 11.8	94.74 ± 8.8	77.13 ± 8.2	< 0.001
Augmentation Index 75%	22.05 ± 12.4	23.62 ± 11.1	19.9 ± 14.8	0.002
Central retinal arteriolar equivalent (μm)	89.4 ± 10.9	87.2 ± 10.5	94.3 ± 10.5	0.032
Central retinal venular equivalent (μm)	119.3 ± 15.5	120.2 ± 16.1	117.1 ± 13.9	0.352
ArterioVenus Ratio	0.757 ± 0.109	0.733 ± 0.106	0.811 ± 0.099	< 0.001
Low Density Lipoproteins	133.4 ± 40.3	139.04 ± 40.1	120.81 ± 38.5	0.036

mean values. Comparison of frequencies was performed by Pearson chi square test. Multivariate linear regression analysis was used to explore the relationship between retinal vascular calibers and AI, while controlling for other covariates. A probability value of p≤0.05 was considered statistically significant.

B. Results

In total, 105 individuals, aged 18 to 68 years old, participated in the study, 70 of them being hypertensives and 35 healthy controls. Baseline demographic and clinical characteristics of the study population are depicted in Table 1. As expected, hypertensives had statistically significant higher systolic and diastolic BP, higher LDL levels, as well as signs of hypertensive retinopathy, such as narrower retinal arteries and a lower AVR. Linear regression analyses (adjusted R square=0.466 , R square=0.525, p¡0.001) revealed that augmentation index was significantly associated with age (Unstandardized Coefficient (UC): 0.42, p¡0.001), sex (UC:5.35, p=0.041), smoking (UC:-3.78, p=0.048), height (UC:-0.11, p=0.017), diastolic BP (UC: 0.291, p=0.015) and all retinal parameters, CRAE (UC:-1.16, p=0.026), CRVE (UC:1.16, p=0.026) and AVR (UC:-152.2, p=0.015), independently of other factors (weight (UC:-0.35, p=0.080), systolic BP (UC:0.003, p=0.964) and LDL (UC:0.004, p=0.872)).

V. CONCLUSIONS

To the best of our knowledge, this is the first study that uses automated retinal vessel evaluation software to assess the relationship between quantitative retinal vascular signs and augmentation index (AI) in an adult, otherwise healthy, hypertensive and normotensive population. The results of the study showed that hypertensive patients exhibit increased arterial stiffness, estimated by AI, and have higher rates of mild hypertensive retinopathy, compared to normotensives. Retinopathy affected mainly the arteries, as CRAE and AVR but not CRVE differed significantly between the two groups. These results are in agreement with data derived from large epidemiological and cross sectional studies [16, 21, 18]. However, the most important finding of this study is the statistically significant linear correlation, independently of age, BMI, smoking, LDL and blood pressure levels between AI and the diameter of the retinal vascular calibers, in a naive population with no other known diseases, except for high blood pressure. Future studies should focus on the clinical significance of this association in the field of cardiovascular disease.

In the course of conducting this study, several conclusions were drawn regarding the usability of the proposed software in the context of performing clinical studies. Some of them have already been taken into account from the formative evaluation of early prototypes of this work [10], while others are planned for future work. In this context, the most useful features of the software were found to be the following.

Improved processing speed enabled conservation of user time and enabled the analysis of images during the patient examination. As images could be processed on the fly, healthcare professionals were not required to plan ahead for this analysis, facilitating the inclusion of the examination in the clinical workflow. In this examination, automation of vessel selection tasks was found to be important again, due to user time conservation. The selection of vessels around the optical disk, also enables the automatic application of measurement protocols. Finally, organized output of vessel measurements in a spreadsheet increased the automation by which statistics over multiple patients were computed.

Regarding the GUI, contrast adjustments, were found useful so that the user can inspect vessels easier and evaluate the extracted representations. Zoom and, in particular, the navigation panel accelerate inspection and editing of segmentation errors, while keyboard shortcuts and toolbar icons contributed to the ergonomy of interaction.

A future extension targets the increment of segmentation accuracy, as this would directly reduce user interaction and relax the requirement for user expertise. Despite that the employed algorithm is competitive on benchmark datasets [10] we are pursuing further research. Future work also includes the automatic classification of vessels into veins and arteries, as it still requires user expertise.

The conclusion is that without the employed software it would not have been possible to accurately conduct the reported study, as measuring the vessels manually would not only have been immensely time-consuming but would, also, have been subjective and prone to human error. The ability to register images acquired at different time instances enables a next, follow-up study which will avail information of the reversal of symptoms after patient treatment.

ACKNOWLEDGMENTS

This work was supported by the FORTH-ICS internal RTD Programme "Ambient Intelligence and Smart Environments".

REFERENCES

[1] M. Abramoff et al. Evaluation of a system for automatic detection of diabetic retinopathy from color fundus photographs in a large population of patients with diabetes. *Diabetes Care*, 31(2):193–198, 2008.

[2] M. Abramoff et al. Retinal imaging and image analysis. *Biomedical Engineering, IEEE Reviews in*, 3:169–208, 2010.

[3] Y. Cheng. Mean shift, mode seeking, and clustering. *PAMI*, 17(8):790–799, 1995.

[4] B. Dahlof. Cardiovascular disease risk factors: Epidemiology and risk assessment. *American Journal of Cardiology*, 105(1):3–9, 2010.

[5] J. Dernellis and M. Panaretou. Aortic stiffness is an independent predictor of progression to hypertension in nonhypertensive subjects. *Hypertension*, 45(3):426–431, 2005.

[6] N. El-Bendary et al. ARIAS: Automated retinal image analysis system. In *Soft Computing Models in Industrial and Environmental Applications*, pages 67–76, 2011.

[7] A. Grosso et al. Hypertensive retinopathy revisited: some answers, more questions. *British Journal of Ophthalmology*, 89(12):1646–1654, 2005.

[8] M. Ikram et al. Retinal vessel diameters and risk of hypertension: the Rotterdam study. *Hypertension*, 47(2):189–194, 2006.

[9] Y. Lin and G. Medioni. Retinal image registration from 2D to 3D. In *CVPR*, 2008.

[10] G. Manikis, V. Sakkalis, X. Zabulis, P. Karamaounas, A. Triantafyllou, S. Douma, C. Zamboulis, and K. Marias. An image analysis framework for the early assessment of hypertensive retinopathy signs. In *IEEE International Conference on e-Health and Bioengineering*, 2011.

[11] M. O' Rourke and M. Safar. Relationship between aortic stiffening and microvascular disease in brain and kidney: cause and logic of therapy. *Hypertension*, 46(1):200–204, 2005.

[12] M. Ortega et al. SIRIUS: A web-based system for retinal image analysis. *I. J. Medical Informatics*, 79(10):722–732, 2010.

[13] J. Parr and G. Spears. Mathematic relationships between the width of a retinal artery and the widths of its branches. *American Journal of Ophthalmology*, 77(4):478–483, 1974.

[14] T. Sairenchi et al. Mild retinopathy is a risk factor for cardiovascular mortality in Japanese with and without hypertension: the Ibaraki prefectural health study. *Circulation*, 124(23):2502–11, 2011.

[15] H. Struijker-Boudier et al. Evaluation of the microcirculation in hypertension and cardiovascular disease. *European Heart Journal*, 28(23):2834–2840, 2007.

[16] C. Sun et al. Retinal vascular caliber: systemic, environmental, and genetic associations. *Survey of Ophthalmology*, 54(1):74–95, 2009.

[17] S. Thom et al. Differential effects of antihypertensive treatment on the retinal microcirculation: An anglo-scandinavian cardiac outcomes trial substudy. *Hypertension*, 54(2):405–408, 2009.

[18] A. Triantafyllou, M. Doumas, P. Anyfanti, E. Gkaliagkousi, X. Zabulis, K. Petidis, E. Gavriilaki, P. Karamaounas, V. Gkolias, A. Pyrpasopoulou, A. Haidich, C. Zamboulis, and S. Douma. Divergent retinal vascular abnormalities in normotensive persons and patients with never-treated, masked, white coat hypertension. *American Journal of Hypertension*, 2013.

[19] C. Tsai et al. Automated retinal image analysis over the internet. *IEEE Transactions on Information Technology in Biomedicine*, 12(4):480–487, 07 2008.

[20] C. Vlachopoulos et al. Prediction of cardiovascular events and all-cause mortality with arterial stiffness: a systematic review and meta-analysis. *Journal of the American College of Cardiology*, 55(13):1318–1327, 2010.

[21] T. Wong and R. McIntosh. Systemic associations of retinal microvascular signs: a review of recent population-based studies. *Ophthalmic Physiological Optics*, 25(3):195–204, 2005.

[22] T. Wong and P. Mitchell. Hypertensive retinopathy. *New England Journal of Medicine*, 351(22):2310–2317, 2004.

[23] S. Zieman et al. Mechanisms, pathophysiology, and therapy of arterial stiffness. *Arteriosclerosis, Thrombosis, and Vascular Biology*, 25(5):932–943, 2005.

Automatic Selection of CT Perfusion Datasets Unsuitable for CTP Analysis due to Head Movement

F. Fahmi[1], H.A. Marquering[1,2], G.J. Streekstra[1], L.F.M. Beenen[2], N.Y. Janssen[1], A. Riordan[3], H. De Jong[3], C.B.L. Majoie[2], and E. van Bavel[1]

[1] Department of Biomedical Engineering and Physics, Academic Medical Centre, Amsterdam, The Netherlands
[2] Department of Radiology, Academic Medical Centre, Amsterdam, The Netherlands
[3] Department of Radiology, University Medical Center Utrecht, Utrecht, The Netherlands

Abstract—CT Brain Perfusion imaging (CTP) is a diagnostic tool for initial evaluation of acute ischemic stroke patients. Head movement of the patients during acquisition limits its applicability. CTP data with excessive head movement must be excluded or corrected for accurate **CTP** analysis. Instead of manual selection by visual inspection, this study provides an automatic method to select unsuitable CTP data subject to excessive head movement. We propose a 3D image-registration based movement measurement that provides 6 rigid transformation parameters: 3 rotation angles and 3 translations. This method is based on the registration of CTP datasets with a non contrast CT image with a larger volume of interest as reference, which is always available as part of a standard protocol for stroke patients. All parameters from the 3D registration are compared to a set of threshold value to objectively decide whether the CTP dataset suitable for accurate CTP analysis or not. Thresholds for unacceptable head movement were derived using controlled movement experiments with CTP phantom data. Validation was done by comparing the automatic selection of unsuitable data with radiologists' manual selection using binary classification analysis. The accuracy of the method was 77% with a high sensitivity (95%) and fair specificity (56%). Since all these processes are carried automatically, it assures that clinical decision are not based upon faulty CTP analysis of data with head movement and it saves time in acute time-critically situations for acute ischemic stroke patients.

Keywords—**CT Brain Perfusion, Computer Aided System, Ischemic Stroke.**

I. INTRODUCTION

CT Brain Perfusion imaging (CTP) is evolving towards a promising diagnostic tool for initial evaluation of acute ischemic stroke patients. In CTP images, areas with brain perfusion defects can be detected after the onset of clinical symptoms and can facilitate the differentiation between the irreversibly damaged infarct core and the potentially reversibly damaged infarct penumbra, which is important in choosing the most suitable therapy [1,2].

The patient's head movement during scanning, however, limits the applicability of CTP. It is known that head movement is a common problem during CTP acquisition of stroke patients[3,4]. Since CTP analysis assumes that a specific location in the images is associated with a single anatomical position, it is likely that head movement deteriorates CTP analysis results, as shown in Figure 1.

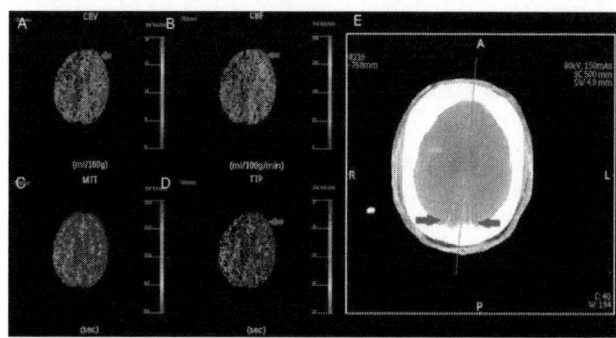

Fig. 1 Example of a perfusion map disturbed by head movement: the red arrow points out artifacts in Cerebral Blood Flow Map(**a**), Cerebral Blood Volume Map (**b**), Mean Transit Time Map (**c**), Time To Peak Map (**d**), and Summary Map (**e**) where double falx cerebri notified by the two red arrows.

In clinical practice, a radiologist must first visually check the CTP datasets to conclude whether it contains head movement and decide whether the data set is suitable for perfusion analysis. If the data set is approved, a registration commonly available in software analysis package is applied to correct for small motions. Alternatively, certain time frames of the CTP acquisition that could disturb the perfusion analysis are removed. The visual inspection is performed for 25 time frames on average and it is conducted in 2D maximum intensity projections. Next to being time consuming, the selection of out-of plane head movement acquisitions with this method is observer dependent. Furthermore, it relies on the subjective interpretation whether CTP data sets are considered suitable for analysis.

The goal of this study is to develop and validate an automatic selection method for unsuitable CTP data subject to excessive head movement. To this end, we perform a phantom study to estimate threshold values of the head motion parameters beyond which unacceptable deviations of perfusion results occur. These threshold parameters are combined with an image registration based motion detection method to come to an automated selection of CTP acquisitions with excessive head motion.

II. MATERIAL AND METHODS

A. CTP Data

We collected CTP data from patients who were suspected of acute ischemic stroke and underwent CTP. Non-contrast CT was performed on all these patients. All CT image acquisitions were performed on a sliding gantry 64 slice Siemens scanner (Somatom Sensation 64, Siemens Medical Systems, Erlangen Germany) in the emergency room. The non contrast CT (ncCT) scan of the brain was acquired at admission using 120 kV and 300-375 mAs, slice thickness was 5mm. In case severe head movement occurred during ncCT scanning and this was noticed by a technician, this CT was repeated. For the CTP acquisition, 40 ml iopromide (Ultravist 320; Bayer HealthCare Pharmaceuticals, Pine Brook, New Jersey) was infused at 4 ml/s using an 18 gauge canula in the right antecubital vein followed by 40 ml NaCl 0.9% bolus. Acquisition and reconstruction parameters were: 80 kV tube voltage, 150 mAs, collimation 24x1.2 mm, FOV 300 mm, reconstructed slice width of 4.8 mm. During acquisition, a standard foam headrest was used to provide patient with comfortable position and to minimize the head movement.

B. Quantification of Head Movement

The range and direction of head movement were quantified by 3D registration of every time frame in the CTP data set with the ncCT admission image data of the same patient. An ncCT scan is always present for patients suspected of stroke therefore no additional scan was required. The ncCT is scanned in a much shorter time period than the CTP acquisition. Therefore no, or only minimal, head movement is expected during the ncCT acquisition. Furthermore, it covers a larger volume than the CTP time frames, thus it is suitable for registration.

The rigid registration was applied resulting in 6 motion parameters: three angles of rotations: pitch, roll, and yaw; and three spatial translations in the x-, y- and z-direction (Figure 2). The registration was performed using Elastix[5], with the normalized correlation coefficient as a similarity measure and gradient descent algorithm to solve the optimization problem.

The extent of movement was determined for each time frame, providing 25x6 temporal motion parameter values, one for each time frame, using the first volume as a reference.

The range of each motion parameter was defined as the largest value subtracted by the smallest value for the whole CTP acquisition.

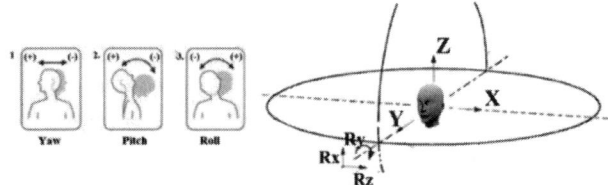

Fig. 2 Head movement parameters

C. Phantom Study

CTP phantom data were obtained from a digital hybrid phantom, which was based on a combination of CT images of an anthropomorphic head phantom with clinically acquired MRI brain images to quantify the different tissues. These data were combined with processed high resolution 7T MRI images to include healthy and diseased brain parenchyma, as well as the cerebral vascular system. Time attenuation curves emulating contrast bolus passage based on perfusion as observed in clinical studies were added. This resulted in a dynamic 3D, noise-free, non contrast-enhanced CT volume of 104 thin slices (0.8mm), 25 time-frames with a 2 seconds interval[6]. The volume of infarct core and penumbra were created subsequently by applying mask to the digital hybrid phantom data.

To simulate head movement, the CTP phantom data were rotated and translated using *Transformix*[5]. Translations and rotations were performed along and around each coordinate axis. The simulated rotation angles were -10 to +10 degrees around the z-axis (yaw), and -5 to +5 degree for the x-axis (pitch) and y- axis (roll), with steps of one degree. The translations were set from -10 to +10 mm for all three axes. The head movement was simulated in the time frames 8 to 20.

The original and transformed CTP datasets were processed by a trained operator (FF) using Philips Extended Brilliance Workspace version 3.5 Brain CT Perfusion Package (Philips Healthcare, Cleveland, OH) to obtain the perfusion maps. We calculated the volume similarity and the spatial agreement between summary maps generated from both rotated and original phantom data, using the Dice Similarity Coefficient (DSC), showing agreement not only for size but also location of infarct core and penumbra in between summary maps. Rotation angles and translations where the DSC was 0.7 were used as thresholds below which a CTP data set was marked as unsuitable.

D. Validation

Two emergency radiologists (CM and LFB, both with more than 10 years experiences) qualitatively graded the severity of the patient's head movement in selected CTP datasets in one of 2 categories: Group 1 (Accepted) consisted of CTP data with head movement considered suitable to be corrected by the registration available in the CTP analysis software and group 2 (Rejected) consisted of CTP datasets with severe movement, which is expected to affect the CTP analysis in a major part of the brain and should be excluded. This manual selection of CTP data was used as ground truth for validation of automatic selection method.

For the automatic selection method, the decision to reject or accept a CTP dataset was based on the range of its movement parameters as the result of the registration process. If one or more parameter from the 6 movement parameters exceeded the threshold value, it was rejected. Otherwise the CTP dataset was accepted.

E. Statistical Analysis

We use a binary / binomial classification test in comparing the automatic selection result and manual selection result to evaluate the performance of this method. True positive rate, false positive rate, true negative rate and false negative rate were measured to estimate the diagnostic accuracy of accepting CTP data by the automatic method compared to manual selection by radiologists. Furthermore, we calculate diagnostic accuracy, sensitivity and specificity of our automatic method.

III. RESULT

We selected 35 CTP datasets as suspected with head movement to be included in the analysis. From these data, radiologist rejected 18 CTP dataset (group 2) for its severe movement, observed during visual inspection.

A. Motion Parameters

The mean value of the rotation angles of all CTP datasets was $4.2^0 \pm 4.7^0$, $5.6^0 \pm 10.8^0$ and $9.5^0 \pm 12.3^0$ for roll, pitch and yaw respectively. The mean value of the translation was 2.8 ± 3.0 mm, 1.6 ± 1.3 mm and 17 ± 16 mm for translation in x, y and z direction. The largest rotation recorded was in-plane (yaw) movement. Figure 3 shows an example presenting the dynamic behavior of motion parameters profiles for each time frame during CTP acquisition.

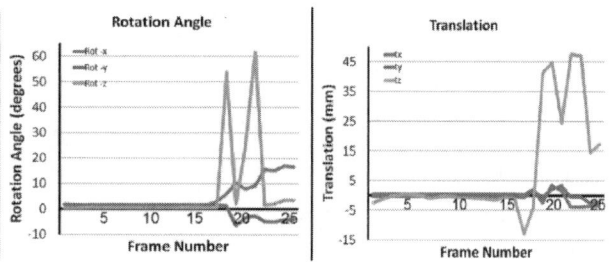

Fig. 3 Examples of movement parameters for 25 time frames as a result of the head movement quantification

B. Phantom Study Result

Figure 4 shows that the registration available in CTP software package can only correct the motion within narrow ranges of simulated rotation angles: -7^0 to 7 for yaw, -1^0 to 1 for pitch, and -2^0 to 2^0 for roll. Beyond these ranges the DSC is below 0.7 (Fig. 4 A-C)

Translation in x and y is perfectly corrected, meanwhile for translation in z- direction, the available registration technique seems not working properly even for small translations of 2 mm (Fig. 4D). These values were also used as threshold to automatically accept or reject CTP data based on its movement profile.

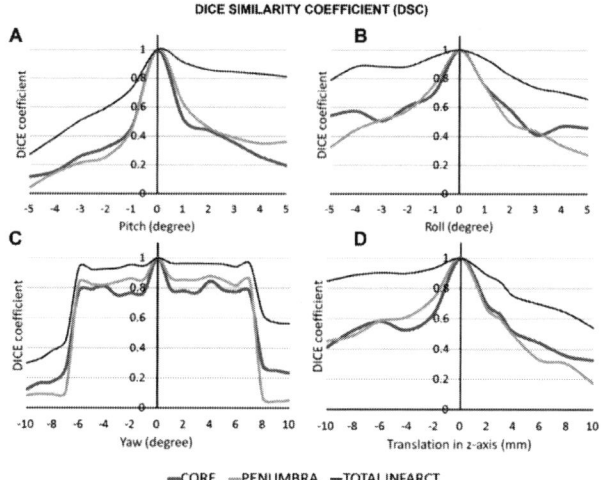

Fig. 4 DSC for simulation phantom study *(adapted from Fahmi et al; Submitted for Publication,2013)*

C. Validation Result

From our 35 datasets, the automatic method detected 25 datasets with excessive head movement to be rejected (Table 1):

Table 1 Validation Result

Automatic \ Manual	Rejected	Accepted	Total
Rejected	18	7	25 PPV 72%
Accepted	1	9	10 NPV 90%
Total	19 SEN 95%	16 SPC 56%	35

PPV = Positive Predictive Value, NPV = Negative Predictive Value, SEN=Sensitivity, SPC = Specificity

The diagnostic accuracy of our method was 77.2% (95%CI: 63.3%-91.1%), with a sensitivity of 94.7% (95%CI: 87.3%-102%) and a specificity of 56.3% (95%CI: 39.9%-72.7%).

IV. DISCUSSIONS

We presented a method for automatic selection of unsuitable CTP data sets based on detection of motion between time frames of a CTP data set and results of simulations of CT perfusion imaging. Performance of this automatic selection method was evaluated by comparison with manual qualitative classification of head movement severity by radiologist. To our knowledge this is the first study that quantitatively analyses head movement during CTP acquisition using available image data only.

Based on simulations with the digital hybrid phantom, it was shown that head movement can be dealt with for only a small range of movement by the standard CTP analysis software. We have shown that head movement strongly alters size and position of infarct core and penumbra in the CTP analysis. The range of movement parameters for which an accurate CTP analysis result can be expected were (-1^0, 1^0) for pitch, (-2^0, 2^0) for roll and (-7^0, 7^0) for yaw and a translation in z-axis less than 2mm. These values are defined as threshold values to accept or reject CTP data, and might be different for other software packages. Individual phantom study must be conducted for other software packages when the automatic selection method will be applied.

The sensitivity of the proposed method is high, while the specificity is lower. This characteristic is more preferable for such a computer aided system where the radiologists are expected to check the suspected CTP data anyhow. It was shown that the automatic detection of suspicious data sets resulted in a potential rejection of more data than the selection of the radiologists. The threshold values used in this study could be optimized for every CTP methods and scanners. However, this was beyond the scope of this study.

The presented method has the potential to be integrated in clinical practice such that any movement during CTP acquisition will be automatically detected allowing radiologists to remove suspicious image data from the CTP analysis.

V. CONCLUSIONS

We presented a method that automatically select CTP datasets that are subject to excessive head movement. The accuracy of the method was 77% with a high sensitivity (95%) and fair specificity (56%). It supports the accuracy of CTP analysis and assures that clinical decision is not based upon faulty image data due to excessive head movement. Since the selection of suspicious data is performed fully automatic, this automatic selection method could save time in acute time-critically situations for acute ischemic stroke patients.

ACKNOWLEDGMENT

This work has been supported by LP3M University of Sumatera Utara and RS Pendidikan USU; through Directorate General of Higher Education (DIKTI) Ministry of National Education Indonesia.

REFERENCES

1. Wintermark M, Flanders AE, Velthuis B, Meuli R, van Leeuwen M, Goldsher D, et al. Perfusion-ct assessment of infarct core and penumbra: Receiver operating characteristic curve analysis in 130 patients suspected of acute hemispheric stroke. *Stroke*. 2006;37:979-985
2. Murphy BD, Fox AJ, Lee DH, Sahlas DJ, Black SE, Hogan MJ, et al. Identification of penumbra and infarct in acute ischemic stroke using computed tomography perfusion-derived blood flow and blood volume measurements. *Stroke*. 2006;37:1771-1777
3. Sesay M, Jeannin A, Moonen CT, Dousset V, Maurette P. Pharmacological control of head motion during cerebral blood flow imaging with ct or mri. *J Neuroradiol*. 2009;36:170-173
4. Wang XC, Gao PY, Lin Y, Ma L, Guan r, Xue J, et al. Clinical value of computed tomography perfusion source images in acute stroke. *Neurological research*. 2009;31:1079-1083
5. Klein S, Staring M, Murphy K, Viergever MA, Pluim JP. Elastix: A toolbox for intensity-based medical image registration. *IEEE Trans Med Imaging*. 2010;29:196-205
6. Riordan AJ, Prokop M, Viergever MA, Dankbaar JW, Smit EJ, de Jong HW. Validation of ct brain perfusion methods using a realistic dynamic head phantom. *Med Phys*. 2011;38:3212-3221

The address of the corresponding author:
    Author: F. Fahmi
    Institute: Department of Biomedical Engineering, Academic Medical Centre, Amsterdam The Netherlands
    Street: Meibergdreef 9 1105AZ
    City: Amsterdam
    Country: The Netherlands
    Email: f.fahmi@amc.uva.nl

An In-Vitro Model of Cardiac Fibrillation with Different Degrees of Complexity

A.M. Climent[1], M.S. Guillem[2], P. Lee[3], C. Bollensdorff[4], F. Atienza[1], M.E. Fernández-Santos[1], R. Sanz-Ruiz[1], P.L. Sánchez[1], and F. Fernández-Avilés[1]

[1] Department of Cardiology, Hospital General Universitario Gregorio Marañón, Madrid, Spain
[2] Instituto ITACA, Universidad Politécnica de Valencia, Valencia, Spain
[3] Essel Research, Toronto, Canada
[4] Qatar Cardiovascular Research Center, Qatar Foundation, Doha, Qatar

Abstract—The difficulty in obtaining research models of chronic atrial fibrillation (AF) is one of the main constraints to make progress in the elucidation of the mechanisms of perpetuation of this arrhythmia. The aim of this study is to present a new model of in vitro AF with different degrees of complexity similar to those observed during AF with varying degrees of structural remodeling. In the present study, optical calcium mapping is used to characterize the degree of disorganization of the reentrant activity in HL-1 confluent cell cultures. Specifically, the number of simultaneous rotors is linked with the time and degree of inhomogeneity of cell cultures. HL-1 cell cultures showed electrophysiological characteristics dependent on their degree of confluency resembling processes with varying degrees of complexity as it occurs during AF with varying degrees of structural remodeling. This model could be useful for studying the effect of remodeling on the fibrillatory process and to evaluate new antiarrhythmic drugs.

Keywords—Atrial Fibrillation, Optical Mapping, HL-1 cells, Electrophysiology.

I. INTRODUCTION

Animal and in-vitro experiments have allowed the postulation that fibrillatory processes can be maintained by functional reentrant activity. Those models have been useful to develop effective treatments in patents suffering paroxysmal atrial fibrillation (AF). However, the difficulty in obtaining research models of chronic AF is one of the main constraints to progress in elucidating the mechanisms of perpetuation of this arrhythmia and the development of effective therapies.

A main limitations to develop in-vitro models of AF is the lack of adult human cardiac cell lines. Nevertheless, HL-1 cell line was derived from the mouse atrial cardiomyocytes tumor linage AT-1 [1] and presents the phenotype of an adult atrial cardiomyocyte. Many different aspects of HL-1 cells have been studied based on genetic, immunohistochemical, pharmacological and electrophysiological techniques [2]. Most of those studies were focus on single cell and cardiac biology, however, HL-1 cells have demonstrated to generate viable confluent population of HL-1 cells that present cell interconnection and global electrophysiological behavior that allows wave conductions and reentries [3]. In addition, this cell line present important advantage over primary cultures like its ability to indefinitely proliferate and to display remodeling [4].

The aim of this study is to present a new in vitro model of AF based in HL-1 cells with different degrees of complexity similar to those observed during AF with varying degrees of structural remodeling.

In order to achieve this objective, optical mapping calcium imaging is used to characterize the degree of disorganization of the reentrant activity in HL-1 confluent cell cultures. The number of simultaneous rotors is linked with the time and degree of inhomogeneity of cell cultures. Here, we demonstrated that HL-1 cells exhibit different degrees of tissue and electrophysiological complexity depending on the time of culture. In addition, different response of cell cultures to Verapamil infusion, an L-type calcium channel blocker that already has demonstrated the reduction of fibrillatory disorganization in HL-1 cells [5], has been evaluated.

II. METHODS

HL-1 cell culture. HL-1 cells, from Dr. William Claycomb (Lousiana State University Medical Center, New Orelans, LA), are a continuously proliferating cardiomyocyte cell line derived from mouse atrial tumors with adult cardiac phenotypes that express $Na+$, $K+$ and $Ca2+$ channels [1]. Cells were maintained, grown and proliferated according to the protocol supplied by Dr. Claycomb's laboratory. Briefly, HL-1 cells were cultured on gelatin and fibronectin-coated tissue-culture flasks in Claycomb medium (Sigma- Aldrich, St. Louis, MO) supplemented with 10% fetal bovine serum (FBS, Sigma- Aldrich, St. Louis, MO), 100 U/mL, 100 mg/mL penicillin/streptomycin (Invitrogen, Carlsbad, CA), 2 mM L-glutamine (Invitrogen, Carlsbad, CA), and 0.1 mM norepinephrine (SigmaeAldrich, St. Louis, MO). HL-1 cells were plated in 60-mm tissue culture dishes at a density of approximately 200.000 cells per dish and cultured at 37ºC and 5% $CO2$. Fresh medium was supplied every 2-3 days. Cell cultures were divided into two groups: (A) cultured during 6.1±1.3 days (N=10) and (B) cultured during 11.7±0.5 days (N=8).

Calcium Dye Loading: For transient calcium imagining, HL-1 cell cultures were stained by immersion in Claycomb culture medium with a solution of Rhod-2 AM (Ca2+ sensitive probe, TEFLabs, Inc, Austin, TX. USA) dissolved in DMSO at 1mM (3.3 µl per ml of culture medium) and Probenecid (TEFLabs, Inc, Austin, TX, USA) at 420µM for 30 minutes under incubation conditions. After dye incubation, culture medium was changed to fresh modified Krebs solution at 36.5ºC (containing, in mM: NaCl, 120; NaHCO3, 25; CaCl2, 1.8; KCl, 5.4; MgCl2, 1; glucose, 5.5; H2O4PNa•H2O, 1.2; bubbled with oxygen; pH 7.4). All chemicals were obtained from Sigma-Aldrich (Dorset, UK) or Fisher Scientific Inc. (New Jersey, USA).

Optical mapping: Experiments were performed in a dark room to avoid photobleaching and dye washout. In order to excite rhod-2 AM, cell cultures were illuminated with a filtered green LED light source: LED: CBT-90-G (peak power output 58 W; peak wavelength 524 nm; Luminus Devices), plano-convex lens (LA1951; focal length= 25.4 mm; Thorlabs) and a green excitation filter (D540/25X; Chromo Technology. Two LEDs with similar characteristics were used to achieve a homogeneous illumination. Fluorescence was recorded with an electron- multiplied charge-coupled device (EMCCD; Evolve-128: 128x128, 24x24µm-square pixels, 16 bit; Photometrics, Tucson, AZ, USA), with and emission filter (ET585/50-800/200M; Chroma Technology) suitable for rhod-2AM placed in front of the lens. Camera was manually focused to cover the entire dish with the maximum number of pixels.

An eight-processor microcontroller (Propeller-Chip; Parallax, Rocklin, CA, USA) was used to control/coordinate all major components. Control software for time-critical tasks was written in the microcontroller's assembly language to ensure a time resolution of 50 ns. Communication with a desktop computer occurred via USB interface (UM245R; Future Technology Devices, Glasgow, United Kingdom) [6]. Both LED lights microcontroller and CCD camera were controlled from custom made MATLAB software using Micro-Manager Open Source Software [7].

Immunohistochemistry and microscopy. In a subset of cell cultures, cell nuclei were labeled with 4′,6-diamidino-2-phenylindole, DAPI (D8417, 1 mg/ml stock solution in ultrapure water, Sigma-Aldrich) according to the stain fabricant protocol. Cells were imaged using a Leica DMI3000 inverted microscope equipped with UV Leica EL6000 Lamp and excitation (BP 450-490) and emission (LP 515) filter combination. Images were acquired with a Leica DFC310FX camera. In addition to fluorescence images, bright field microscopy of the same region was acquired.

Image pairs (i.e. bright field image and DAPI images) were used to estimate the correlation between the number of nuclei measured in DAPI images and the homogeneity measurement proposed by Haralik et al based on the gray level co-occurrence matrix [8] of bright field images.

Experimental protocol. For all cell cultures of both groups, two seconds optical mapping recordings were acquired at basal conditions, after the administration of Verapamil 4µM dissolved in Krebs solution and after washing the drug with fresh Krebs solution. In addition to calcium imaging, 5 bright field microscopy images randomly distributed around the dish were acquired to calculate the homogeneity of the cell culture.

Calcium Image processing. Custom software written in MATLAB (MathWorks) was used to perform optical mapping image processing. Specifically, the disorganization of a fibrillatory process of each dish was estimated as the mean number of simultaneous functional reentrant activities. Those reentries were automatically identified as singularity points that remain stable by using phase map analysis. Phase maps of each movie were obtained by calculating the instantaneous phase signal of each individual pixel of the movie according to the equation:

$$\text{Phase signal} = \angle(\text{HT}(\text{EG})) \quad (1)$$

where $\angle( )$ is the phase operator and $\text{HT}( )$ is the Hilbert Transform. The phase signal has a range between 0 and 2π and represents the relative delay of each signal in one period. A singularity point was defined as the point in a phase map which is surrounded by phases from 0 to 2π. The phase value was obtained in 3 circles around each evaluated point. To define one point as a singularity point, the phase in at least two of these three circles should achieve: (a) phase transitions between two consecutive pixels should not seceed 0.6π; and (b) the phase must be monotonically increasing or decreasing. Once singularity points were identified at each frame, they were connected in time and space (with a minimum distance required for connection of 5 pixels and 5 samples) into rotors. Unstable rotors with duration of less than 100ms discarded. Finally the mean number of simultaneous functional reentries was calculated as the summed duration of all stable rotors divided by the total duration of the movie.

Statistical analysis. Data are presented as mean±standard deviation. Cross correlation was used to estimate the relation between the number of nuclei in the DAPI images and the homogeneity of bright field images. The Student's t-test was performed in order to compare (1) the mean number of simultaneous rotors, (2) the effects of Verapamil infusion and (3) the homogeneity of bright field images between both cell culture groups.

III. RESULTS

In Figure 1, bright field and immunohistochemistry microscopy images of two illustrative examples of HL-1 cell cultures with different degrees of complexity are depicted. Homogeneity in bright filed images and the number of nuclei in DAPI images presented a correlation of 0.98 ($p<0.01$). This result allowed us to estimate the degree of complexity of each individual cell culture with bright field images without the need for immunohistochemistry staining.

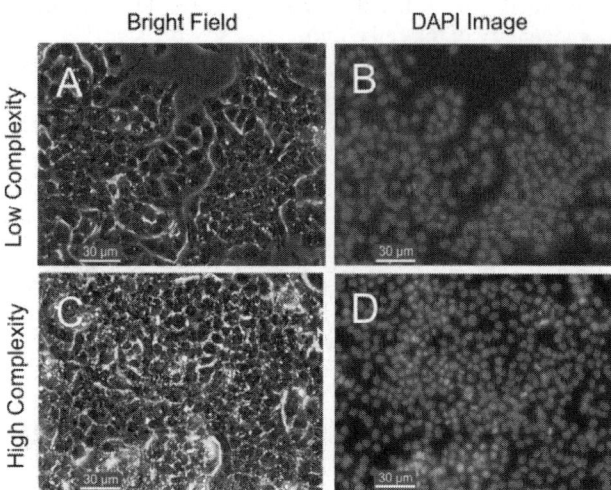

Fig. 1 Comparison between bright field images (panels A and C) and DAPI images (panels B and D) in two representative examples of low (panels A and B) and high cell culture complexity (panels C and D).

In Figure 2, normalized calcium transient and phase map images of two representative cell cultures from group A and B are shown. As it can be observed in the example with a low complexity cell culture (panels A and B), a main rotor located in the upper-left region of the dish generated relatively regular wavefronts that covered almost the entire dish. However, notice that at the periphery of the dish small wavebreaks produced a significant number of singularity points. In a dish imaged after more culture days (Fig. 2 C and D), several small wavefronts and singularity points can be observed without a clear predominat rotor. In this case the number of simultaneous rotors was much higher.

In Figure 3 the amount of simultaneous rotors before and after Verapamil administration for all cell cultures of both groups are summarized. As it can be observed, dishes imaged after less culture days (group A) had fewer simultaneous rotors than those of group B (i.e. 16.7±5.1 vs. 26.8±3.6, $p <0.01$). The amount of rotors at baseline showed an inverse correlation equal to with the degree of homogeneity in the bright field microscopy images ($\rho=0.7$, $p <0.01$).

Verapamil infusion produced a significant reduction in the amount of reentries in both groups, however that reduction was more efficient for group A versus group B (i.e. 63±31% vs. 35±25%, $p <0.05$). After 5 minutes of Verapamil infusion, reentries ended on 5 dishes of group A versus none of the dished in group B ($p <0.001$). All dishes recover reentrant patterns after drug washing.

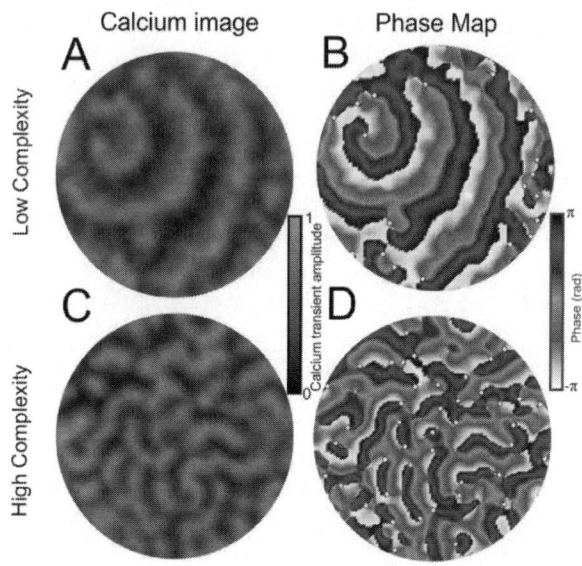

Fig. 2 Comparison between calcium transient maps (panels A and C) and phase maps (panels B and D) in two representative examples of low (panels A and B) and high cell culture complexity (panels C and D). Notice that in phase maps panels singularity points are mark with white dots.

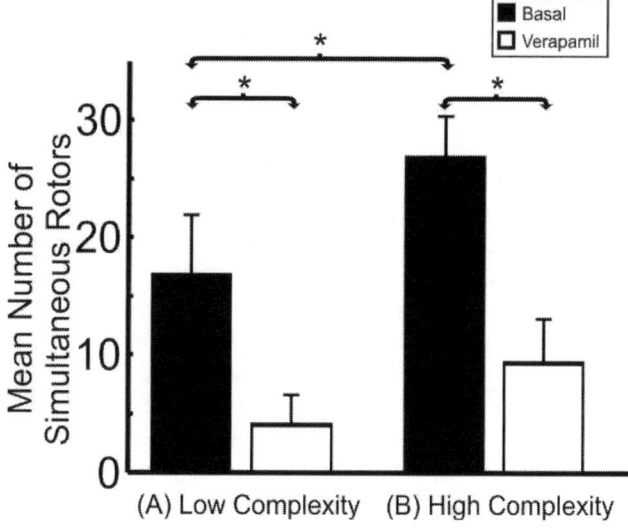

Fig. 3 Mean number of simultaneous rotors for group A (Low complexity) and group B (High complexity) during basal conditions and after Verapamil administration. (*, $p<0.001$).

IV. DISCUSSION

The main finding of the present study is that HL-1 cell cultures can present different electrophysiological characteristics and drug responses according to the level of complexity and confluency. Our results demonstrated that the number of days in culture was correlated with the level of complexity of the cell culture and with the amount of rotors. In addition, the degree of effectiveness of an antiarrhythmic drug likes Verapamil depending on the days of confluency. Specifically, our results indicated that when a fibrillatory process is relatively simple, it can be pharmacologically finished, however, when the amount of rotors and wavebreaks is sufficiently high, probably due to the complexity of the cardiac tissue and its structural remodelling, the same drug dose has a lower effectiveness. These results are equivalent to the different clinical outcome between patients suffering paroxysmal or persistent AF.

Since the development of HL-1 cell line by Claycomb et al. [1], this remains the only available cell line that can proliferate indefinitely and present phenotype of adult cardiomyocytes. Thanks to this, HL-1 cells had been extensively used to evaluate the cardiac biological behavior, protein expression and electrophysiological mechanisms [2]. However, it has not been until recently that the ability of HL-1 cells to generate viable confluent cultures that maintain wavefronts and reentries has been shown [3,9]. Nevertheless, this is the first study in which complexity of reentries and wavefronts has been linked with the HL-1 cell cultures inhomogeneities. Specific mechanisms that generate does inhomogeneities remain unclear, however, those modifications have been demonstrated to display characteristics of molecular remodeling as found in atrial tissue from patients with chronic AF [3,4].

Limitations. Cell culture inhomogeneities have been measured from bright field images by using image processing techniques. Although it may present an important limitation over the immunohistochemistry evaluation, this new technique for the automatic estimation of HL-1 cell cultures inhomogeneity has been demonstrated to have a high correlation with the number of nuclei detected in DAPI images. The presented methodology allows the characterization of cardiac myocyte cultures without the need of expensive antibodies and time consuming protocols.

The present study has been developed with murine atrial cells. Right now it is the only existent myocardial cell line. However, recent development of human cardiac myocytes with mature electrophysiological behaviour from induced pluripotencial stem cells [10] may allow the use of in-vitro models of human cells in the near future.

V. CONCLUSION

HL-1 cell cultures showed electrophysiological characteristics dependent on their degree of confluency resembling varying complexity processes that occur during AF with varying structural remodeling. This model could be useful for studying the effect of remodeling on the fibrillatory process and to evaluate new antiarrhythmic drugs.

ACKNOWLEDGMENT

This work was supported by MINECO under the project SABIO (PLE2009-0152) and Universitat Politècnica de València through its research initiative program.

REFERENCES

1. Claycomb WC, Lanson NA Jr, Stallworth BS et al. (1998) HL-1 cells: a cardiac muscle cell line that contracts and retains phenotypic characteristics of the adult cardiomyocyte. Proc Natl Acad Sci 17;95(6):2979-84.
2. White SM, Constantin PE, Claycomb WC. Cardiac physiology at the cellular level: use of cultured HL-1 cardiomyocytes for studies of cardiac muscle cell structure and function. Am J Physiol Heart Circ Physiol. 286(3):H823-9.
3. Hong JH, Choi JH, Kim TY et al. (2008) Spiral reentry waves in confluent layer of HL-1 cardiomyocyte cell lines. Biochem Biophys Res Commun. 26;377(4):1269-73.
4. Tsai CT, Chiang FT, Chen WP et al. (2011) Angiotensin II induces complex fractionated electrogram in a cultured atrial myocyte monolayer mediated by calcium and sodium-calcium exchanger. Cell Calcium 49(1):1-11.
5. Lee P, Yan P, Ewart P et al. (2012) Simultaneous measurement and modulation of multiple physiological parameters in the isolated heart using optical techniques. Pflugers Arch. 464(4):403-14.
6. Stuurman N, Edelstein A, Amodaj N et al (2010) Computer Control of Microscopes Using µManager. Curr Protoc Mol Biol. 14.20.1-14.20.17
7. Haralick RM, Shanmugam K, Dinstein I (1973) Textural Features for Image Classification. IEEE Trans, on Systems, Man., and Cybernetics 3:610-621
8. Tsai CT, Chiang FT, Tseng CD et al. (2011) Mechanical stretch of atrial myocyte monolayer decreases sarcoplasmic reticulum calcium adenosine triphosphatase expression and increases susceptibility to repolarization alternans. J Am Coll Cardiol. 8;58(20):2106-15.
9. Brundel BJ, Kampinga HH, Henning RH (2004) Calpain inhibition prevents pacing-induced cellular remodeling in a HL-1 myocyte model for atrial fibrillation. Cardiovasc Res. 1;62(3):521-8.
10. Lee P, Klos M, Bollensdorff C et al. (2012) Simultaneous voltage and calcium mapping of genetically purified human induced pluripotent stem cell-derived cardiac myocyte monolayers. Circ Res. 8;110(12):1556-63.

Author: Andreu M. Climent
Institute: Hospital General Universitario Gregrorio Marañon
Street: c\ O'Donnell 48
City: Madrid
Country: Spain
Email: acliment@cardiovascularcelltherapy.com

Semi-automatic Segmentation of Sacrum in Computer Tomography Studies for Intraoperative Radiation Therapy

E. Marinetto[1], I. Balsa-Lozano[1], J. Lansdown[3], J.A. Santos-Miranda[4,5], M. Valdivieso[6], M. Desco[2,1], and J. Pascau[2,1]

[1] Unidad de Medicina y Cirugía Experimental, Instituto de Investigación Sanitaria Gregorio Marañón, CIBERSAM, Madrid, Spain
[2] Departamento de Bioingeniería e Ingeniería Aeroespacial, Universidad Carlos III de Madrid, Madrid, Spain
[3] John Hopkins University, Baltimore, Maryland, USA
[4] Departamento de Oncología, Hospital General Universitario Gregorio Marañón, Madrid, Spain
[5] Universidad Complutense de Madrid, Madrid, Spain
[6] GMV, Madrid, Spain

Abstract—Intra-Operative Radiation Therapy (IORT) is a technique that combines surgery and adjuvant radiation. It is applied directly to a post-resected tumor bed by means of a specific applicator docked to the linear accelerator. IORT's primary purpose is to reduce local cancer recurrence. Separating healthy tissues from the tumor bed is essential in order to maximize treatment efficacy and to avoid damage to organs at risk. Current image segmentation during planning procedures is done manually, using computed tomography studies, by an expert radiation oncologist. This procedure can be time consuming. The authors present an automated procedure that would expedite the process of segmentation, specifically tested in rectal cancer cases where the sacrum is the area at risk. A segmentation algorithm was developed based on the creation of an averaged image (template) that was manually segmented. The algorithm geometrically transforms template segmentation by registering it to the input images. Preliminary results demonstrate the feasibility of the proposed approach.

Keywords—**Automatic Segmentation, IORT, Sacrum.**

I. INTRODUCTION

Intra-Operative Radiation Therapy (IORT) is a procedure aimed at reducing local cancer recurrence. A single-fraction dose of radiation is delivered directly to the residual tumor or tumor bed at the end of a surgical resection by means of specific applicators connected to a conventional or mobile linear accelerator [1]. Radiation oncologists plan the treatment before surgery. During planning, they manually segment and identify the organs at risk and estimate the tumor bed that is to be irradiated.

A novel IORT treatment planning system (*radiance*®, GMV, Madrid, Spain) allows radiation oncologists to design the treatment strategy using computed tomography (CT) studies of the patients. This Treatment Planning System (TPS) simulates the insertion of a selected applicator in the area of interest and estimates the dose distribution in the tumor bed and organs at risk [2].

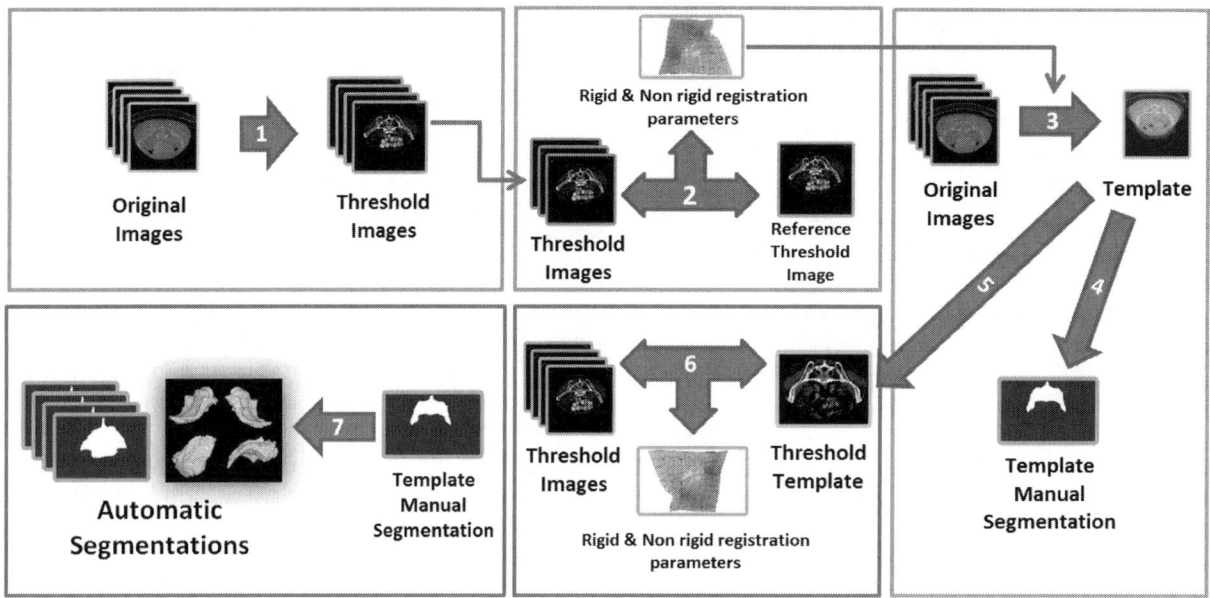

Fig. 1 Automatic segmentation workflow. Green: Template Creation. Orange: Template segmentation. Red: Automatic Segmentation

The segmentation of each organ at risk is an important step, since it will affect the estimated dose and consequently the treatment plan, and is also time-consuming. In rectal cancer –one of the most common IORT indications – radiation oncologists must segment and identify the sacrum as an organ at risk, as the tumor bed lies between the sacrum and the applicator bevel.

The purpose of this study is to describe a new semi-automatic algorithm for sacrum segmentation to assist radiation oncologists during IORT planning. The proposed method has been evaluated in four cases, using expert manual segmentation as the gold standard.

II. MATERIAL & METHODS

The segmentation algorithm is based on constructing an average image (template) from a set of patient CT studies. This template will be used as a model image for automatic segmentation. The template is carefully manually segmented (Figure 1, orange-framed steps) and this mask is transferred to patient images using image registration to obtain the final masks automatically (Figure 1, red-framed).

Section A describes the template creation using bone thresholding and image registration, and section B presents the automatic segmentation algorithm. Figure 1 illustrates the whole process.

A. Template Creation

For the creation of a template image we used four CT studies acquired using a Philips Brilliance 16 CT scanner (512x512x801 pixels, 0.84x0.84x2 mm), from four male IORT rectum cases.

Template creation can be divided into three steps. First, a bone thresholding is done on the images in order to focus the second registration step on the areas of interest. Next, images are spatially normalized using image registration and finally, original images are geometrically transformed (using previous registration parameters) and averaged to create the template image.

a. Bone thresholding

This study is focused on the segmentation of sacrum, part of the dorsal spine. The sacrum is denser than soft tissue. Based on this idea, the template creation procedure generates a bone tissue mask (Figure 1, 1) by thresholding with a value of 1200 (Hounsfield Units, HU), which was optimized empirically. Pixels with intensities below the threshold value were set to zero.

During the registration, the algorithm omits the zero-valued pixels in the cost-function calculation, thus focusing on denser areas along the images.

b. Registration

A study was randomly selected from the previously thresholded set. Then, all studies were registered to this reference one, and registration parameters were stored (Figure1, 2). These parameters were then applied to original studies. In this way, original studies became spatially-normalized and the template was created by averaging them (Figure 1, 3).

All studies registrations included an initial rigid registration followed by a non-rigid registration (Fast Fluid Deformations, FFD) step, and were performed with the *niftyreg* software package (version 1.3.9) available from [*sourceforge.net/projects/niftyreg/*]. Rigid and non-rigid registration schemes used are described in [3, 4]. The rigid registration process was performed with *reg_aladin* command (built into the *niftyreg* package) with default parameters. For non-rigid registrations, *reg_f3d* command was run, specifying sum of squared differences as cost-function, which is appropriate for intra modality registrations.

The algorithm was run in an 8-cores AMD Opteron 6238 x2.6 GHz with 12GB RAM and Debian 6.04 OS for all cases.

B. Segmentation Algorithm

Once the template is available, an expert segments the sacrum on the template image following the procedure described in section II-C (Figure 1, step 4). The step-by-step automatic segmentation algorithm is described below:

1. Template is thresholded employing the same threshold value of 1200 (HU) (Figure 1, 5).
2. Template is registered to each thresholded original study (Figure 1, 6). Registration parameters are stored.
3. Registration parameters are applied to the manual segmentation of the template image resulting in the desired automatic segmentations (Figure 1, 7).

C. Evaluation

For the evaluation of the algorithm we employed the same 4 studies described in II-A section for the template creation. Manual sacrum segmentations of the images were extracted by an expert using ITK-SNAP [5]. Sacrum bone was manually segmented slice by slice along the transverse plane. From cranial to caudal, limits were defined by the proximal region of the body of L5 vertebra and the distal region of the coccyx bone. On the other hand, lateral limits were established by transverse processes of L5 in the cranial region followed by the body of L5, sacroiliac joints, and the lateral edges of the sacrum and coccyx along the caudal region. From anterior to posterior, the limits were set from the surface of the L5 body, sacrum and coccyx to the sacral crest and the dorsal surface of the coccyx.

Once manual segmentation was accomplished along the transverse plane, it was corrected along the sagittal plane to obtain the final segmented mask.

Jaccard similarity index (Eq. 1) was used to compare automatic to manual segmentations [6].

$$J(A,B) = \frac{|A \cap B|}{|A \cup B|} \quad (1)$$

To assess if the automatic segmentations are correct for estimating the IORT dose in the sacrum, a radiation oncologist planned the IORT treatment for each case. Images were loaded into the treatment planning system (Fig. 2) and a radiation oncologist planned the treatment with automatic and manual segmentations of the sacrum. All treatment parameters were the same for both plans, only sacrum masks were different. Dose-volume histograms were extracted [7].

Figure 3 shows a manual and automatic segmentation overlapped on the evaluation set of 4 studies, for visual inspection.

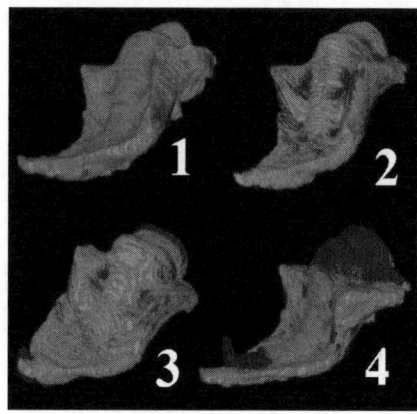

Fig. 3 Manual (blue) and Automatic (green) segmentations overlapping each validation study.

Figure 4 represents the dose-volume histograms extracted for each case.

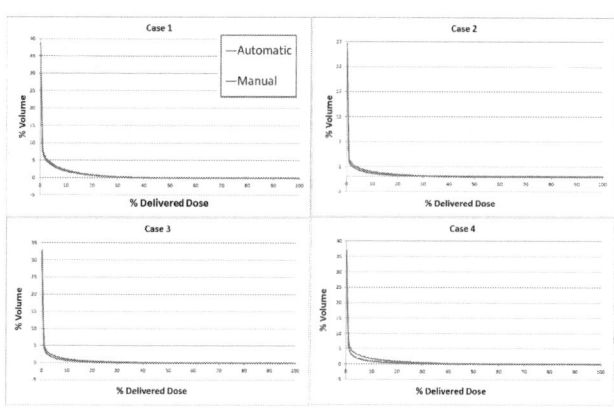

Fig. 4 Simulated dose Volume Histograms for each case. Red: manually segmented sacrum. Blue: Automatically segmented sacrum

Figure 5 shows the absolute difference between manual and automatic segmentation dose-volume histograms for each case. They were calculated by subtraction of data from figure 4.

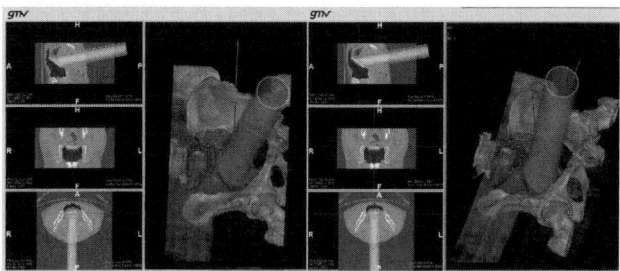

Fig 2 IORT simulation over the masked CT study. Left: manual sacrum segmentation, right: automatic sacrum segmentation.

III. RESULTS

The Jaccard index for each case is showed in table 1.

Table 1 Jaccard Similarity Index

Jaccard Index			
Case 1	Case 2	Case 3	Case 4
0,8376	0,7757	0,8244	0,7811
Jaccard Average			
0,8047			

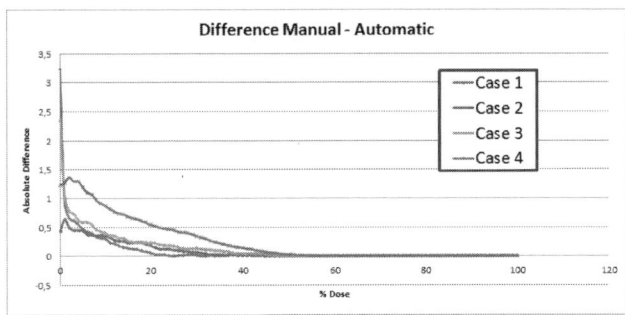

Fig. 5 Absolute difference of the % Volume - % Dose curves for each case.

IV. DISCUSSION AND CONCLUSIONS

This work presents an algorithm for sacrum segmentation of CT images, based on image registration. This algorithm shows clinically adequate Jaccard similarity indexes (table 1) for the four cases. The visual comparison of manually and automatic segmented masks (figure 3) suggests that the resulting masks are adequate throughout the sacrum structure. The quantitative dose study of the treatment planning system confirms that the manual and automatic segmentation similarities are acceptable for use in IORT clinical cases.

The main limitation of our work is that the same images used to create the template image were used as evaluation dataset. Thus, the performance of the proposed approach with new images that have not contributed to the template creation should be further studied.

One drawback is the need for a manual segmentation of the template mask, which must be created by experienced personnel. However, this step has to be performed only once. Segmenting a new patient would not require any further user interaction.

The accuracy of the algorithm has been assessed by comparing with manual segmentations. Thus, our preliminary results suggest that the presented method will help IORT planning providing a clinically valid segmentation of the sacrum.

V. ACKNOWLEDGMENT

This work was partially supported by the Spanish Ministry of Economy and Competitiveness (TEC2010-21619-C04, IPT-300000-2010-003, IPT-2012-0401-300000) and ERD Funds.

VI. REFERENCES

1. Calvo, F.A., R.M. Meirino, and R. Orecchia, *Intraoperative radiation therapy: First part: Rationale and techniques.* Critical Reviews in Oncology/Hematology, 2006. **59**(2): p. 106-115.
2. Pascau, J., et al., *An Innovative Tool for Intraoperative Electron Beam Radiotherapy Simulation and Planning: Description and Initial Evaluation by Radiation Oncologists.* Int J Radiat Oncol, 2012. **83**(2): p. e287-e295.
3. Yushkevich, P.A., et al., *User-guided 3D active contour segmentation of anatomical structures: significantly improved efficiency and reliability.* Neuroimage, 2006. **31**(3): p. 1116-1128.
4. Modat, M., et al., *Fast free-form deformation using graphics processing units.* Computer Methods and Programs in Biomedicine, 2010. **98**(3): p. 278-284.
5. Rueckert, D., et al., *Nonrigid registration using free-form deformations: application to breast MR images.* Medical Imaging, IEEE Transactions on, 1999. **18**(8): p. 712-721.
6. Real, R. and J.M. Vargas, *The probabilistic basis of Jaccard's index of similarity.* Systematic Biology, 1996. **45**(3): p. 380-385.
7. Boersma, L.J., et al., *Estimation of the incidence of late bladder and rectum complications after high-dose (70-78 GY) conformal radiotherapy for prostate cancer, using dose-volume histograms.* International journal of radiation oncology, biology, physics, 1998. **41**(1): p. 83.

Author: Eugenio Marinetto
Institute: Instituto de Investigación Sanitaria Gregorio Marañón
Street: Doctor Esquerdo 46, 28007
City: Madrid
Country: Spain
Email: emarinetto@hggm.es

Anatomical Discovery: Finding Organs in the Neighborhood of the Liver

C. Oyarzun Laura[1], K. Drechsler[1], and S. Wesarg[1]

[1] Fraunhofer IGD, Department Cognitive Computing & Medical Imaging, Fraunhoferstrasse 5, 64283, Darmstadt, Germany

Abstract— **Image segmentation and registration algorithms are fundamental to assist medical doctors for better treatment of the patients. To this end accuracy in the results given by those algorithms is crucial. The surroundings of the organ to be segmented or registered can provide additional information that at the end improves the result. In this paper a novel algorithm to detect the organs that surround the liver is introduced. Even though our work is focused on the liver, the algorithm could be extended to other parts of the body. The algorithm has been tested in 24 clinical CT datasets. In addition to this, an example application is introduced for which the detection is a useful tool.**

Keywords—**detection, registration, initialization, resection.**

I. INTRODUCTION

Thanks to the advances in medical imaging patients can be treated in a more effective way. Therefore applications are used that need accurate segmentation and registration algorithms to work properly. To have prior knowledge on the surroundings of the structure to be segmented or registered can improve the results of those algorithms.

A series of authors have proposed algorithms for the detection of organs in the body. Viola et al. [1] proposed an efficient algorithm for face detection using a cascade of classifiers. Based on their work Zhan et al. [2] generated a learning-based localizer for the detection of organs in whole body CT scans. Both Pauly et al. [3] and Criminisi et al. [4] developed algorithms based on regression techniques. The former detects multiple organs in MR Dixon sequences and the latter in CT volumes. In both cases the algorithm is trained to obtain an accurate predictor for organ localization. All those algorithms provide information on the location of several organs. However, they do not give information on which organ is adjacent to another one in a concrete point. This is the knowledge that could increase the accuracy of certain segmentation and registration algorithms. This was first introduced by Ling et al. [5] who determined a series of so called patches representing the organs adjacent to the liver at its boundary, and he used this to improve the liver segmentation results. The disadvantage of this approach is that it depends on some landmarks at the liver surface that should exist in every liver. The landmarks of previously annotated data will determine the organ adjacent to the liver in a concrete position. This makes the algorithm useless in cases of pathological liver (e.g. liver that underwent resection surgeries, or liver that developed hypertrophy) since in those situations part of the landmarks would be missing in some of the datasets. The number of liver resections per year is increasing [6] making this situations not unusual.

To solve this problem we use samples of the organs that surround the liver and knowledge on their location to carry out our detection algorithm. The liver is represented as a mesh and the output of the detection is the organ that is adjacent to the liver at each point of the mesh. No corresponding landmarks are necessary between different livers which solves the problem of the probability-based detection by Ling et al. [5]. We tested the algorithm in 24 clinical CTs and introduce an example application. Thus, our contribution that goes beyond the state of the art is an organ detection algorithm that:

- is able to work on pathological liver,
- does not need correspondences common to every liver,
- does not require a model atlas.

In addition to this the output of the detection algorithm will be used for a landmark-based initialization algorithm that it is able to align livers that have overcome big resections.

II. FEATURE DETECTION

Generation of Reference Histograms. We assume that the liver is surrounded by six organs, namely heart (yellow), lung (red), ribcage (green), spine (white), kidney (pink) and intestine (blue) (Figure 1). The first step of the algorithm is to take 3D image samples of those organs. These images are manually cropped from a CT image of the abdomen. Their size is the largest possible volume so that they do not contain information of any adjacent organ. Then the histograms corresponding to those images are automatically calculated. Therefore the intensity of every voxel v in the sample image is analyzed. The histograms are divided in a certain number of bins. Each bin contains the number of voxels seen in the image which intensity is between $bin_{i_{min}}$ and $bin_{i_{max}}$. These values will differ from bin to bin and represent groups of

Anatomical Discovery: Finding Organs in the Neighborhood of the Liver

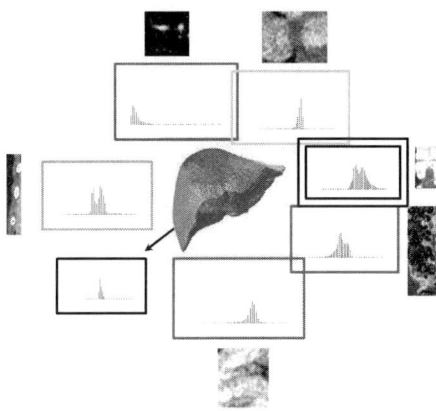

Fig. 1 Liver surface surrounded by the lung (red), heart (yellow), ribcage (green), spine (white), kidney (pink) and intestine (blue) image samples and corresponding histograms. Black: histogram from one of the surface points.

Table 1 Angles that are used to decide if the histogram under study should be compared to a specific organ or not.

Organ	Angle		
	x	y	z
Heart	$> 10, < 135$	< 90	$\geq 25, < 135$
Lung	-	≤ 90	< 160
Kidney	≤ 90	$> 100, \leq 180$	≤ 135
Ribcage	> 90	$< 135, > 45$	-
Spine	< 45	≤ 180	≤ 90
Intestine	-	$> 90, < 270$	-

intensities, e.g. the first bin could contain the voxels with intensity values from $bin_{1_{min}} = 0$ to $bin_{1_{max}} = 10$. The second bin bin_2 would then start at intensity values $bin_{2_{min}} = 10$. The value of the histogram of a certain organ H_{o_n} and in a concrete bin_i is given by:

$$H_{o_n}(bin_i) = \begin{cases} F_{t-1} + 1 & \text{if } bin_{i_{min}} \leq I(v) \leq bin_{i_{max}} \\ F_{t-1} & \text{rest} \end{cases}, \quad (1)$$

where F_{t-1} is the number of voxels contained in the bin in the previous step $t-1$, given that when $t=0$ $F_{t-1} = 0$ and $I(v)$ returns the intensity of the current voxel. Finally all the histograms are normalized. The set of reference histograms H_o that contains every H_{o_n} is used every time the algorithm is applied to a new dataset. Only one dataset has been used to extract the reference histograms, this showed to give good results in the evaluation on 24 datasets. However, if necessary more datasets could be used so that the histograms represent more accurately the intensities of the organs.

Detection. Before the detection, the liver is segmented [7]. Then a surface mesh of the liver is created. The algorithm iteratively visits every point on the surface (Q_i). In each iteration a normal vector ($\vec{n}_i$) to the surface is generated and a series of points (P_j) are sampled that lie on $\vec{n}_i$ and outside the liver:

$$P_j = j \frac{\vec{n}_i}{\|\vec{n}_i\|} d, \quad (2)$$

where $\frac{\vec{n}_i}{\|\vec{n}_i\|}$ is the unit normal vector to the surface and d is the distance between sampled points. P is the set that contains all P_j of the current iteration. $P = \{P_j : j \in [0,k]\}$ being k the number of points to be sampled.

A new histogram is generated (H_p) using the intensity values of P. The goal now is to find the organ that has the most similar intensity profile compared to H_p. Due to the fact that the position of the organs with respect to the liver has not a big interpatient difference, it is not necessary to compare H_p with all the histograms in H_o. It will suffice to compare it with those organs that are located within a certain angle from the world coordinate axis (Table 1). For example, if $\vec{n}_i$ is pointing in the direction from liver to head, the algorithm will compare H_p with the histograms of heart and lung. The range of angles has been determined after evaluating the detection results in 24 clinical CTs of the liver and adjusting them to get the best possible results. Note that the evaluation with resected or hypertrophic livers might change the range of those angles to make them more tolerant to organ movements if necessary. However the detection algorithm will still work after proper adjustment of the ranges.

The similarity (S) between two histograms is calculated as the sum of the differences between the organ histogram (H_{o_n}) and H_p (Equation 3). The difference of both histograms without any region restriction would result in a similarity with an absolute value different from zero, even though the current organ is the correct one. The reason for this is that the number of voxels visited to generate H_p is low and therefore there will be empty bins as opposite as in H_{o_n}. To solve this problem we restrict the calculations to those bins bin_i for which the frequency in H_p is not zero, thus to the region: $A = \{\forall bin_i : H_p(bin_i) \neq 0 \wedge i \in N \wedge 0 \leq i \leq c\}$, being c the number of bins of the histogram and N the set of natural numbers. Mathematically:

$$S = w \sum_{bin_i \in A} (H_{o_n}(bin_i) - H_p(bin_i)). \quad (3)$$

There are some organs that lead to comparable S (e.g. spine and kidney). Their location related to the liver and their histograms are similar which can lead to wrong assignments. To solve this problem a weighting term w is added,

$$w = \begin{cases} 0.1 & \text{if } r_1 \leq R \leq r_2 \\ 1 & \text{rest} \end{cases}. \quad (4)$$

Every organ has a range (r_1, r_2) of bins in which the most likely intensity values are concentrated. w will favor a histogram H_{o_n} if the maximum frequency bin (R) of H_p is in its range. The concrete value of w could be set to another one as far as the desired effect of favoring the assignment of one organ against another one is achieved. After all points Q_i are visited they will have the label value corresponding to the organ they belong to. To remove outliers we detect the biggest connected component per organ. The voxels not belonging to this component are stored for further reassignment. The neighborhood of those stored voxels is investigated and they are assigned to the most frequently found organ.

III. EXAMPLE APPLICATION: INITIALIZATION

Registration algorithms require an initial alignment (*initialization*) of the images to work properly. This task is harder in cases of left (LH) or right (RH) hepatectomy and extended left (ELH) or right (ERH) hepatectomies as in these cases big parts of the liver are missing in one of the datasets. We assume that for every kind of resection [8] both the preoperative data and the postoperative data contain either the area with the connection between heart and intestine (RH and ERH) or the area with the connection between ribcage and intestine (LH and ELH). If none of the mentioned areas remain the algorithm could still work by using the areas connecting other organs.

Resection of Right Lobe of the Liver. In this case the feature used for the alignment will be the connection between heart and intestine. Figure 2 will help to clarify the process:

- Thresholding: The voxels labeled as organs not visible in the resected liver are removed from the preoperative liver (red and green pixels in the first step of Figure 2).
- Region of interest (ROI) detection: The ROI is reduced automatically setting the lower limit of the region in x direction to that of the heart area (red circle in Figure 2). Then the bounding box of the ROI is calculated (yellow).
- Initialization: The upper limits of the intestine bounding boxes are matched (violet circle in Figure 2).
- The 2D slice in the middle of the dataset (sagittal view) is chosen and the center of its bounding box is calculated. Both centers are matched (third step in Figure 2).

Resection of Left Lobe of the Liver. When the left lobe of the liver is resected the initialization algorithm uses the area in which intestine and ribcage join. Now the ROI will be calculated using the upper limit of the ribcage. For the initialization step, the lower limits of both intestine bounding

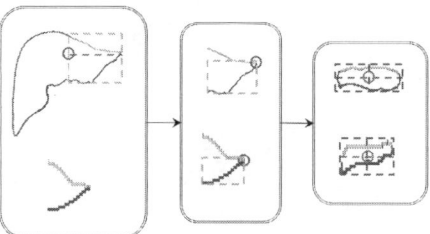

Fig. 2 Initialization process after resection of the right lobe of the liver.

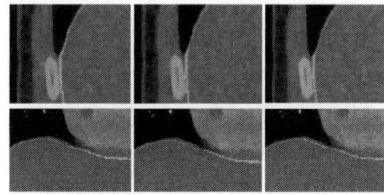

Fig. 3 Effect of decreasing P_j: 20, 10 and 5. Area where the liver meets the lung and ribcage/heart (top row/bottom row).

boxes are matched and after that the centers of the middle slice in sagittal view are merged.

IV. RESULTS

Clinical CT data of the abdomen has been used for evaluation. For each one of the 12 patients, 2 CT were acquired in different moments.

Evaluation of the Organ Detection. There are two values that can influence the accuracy of the results: the number of bins of the histograms and the number of sample points, k. As k decreases (Figure 3) the ribcage (green) will be more accurately detected. The opposite happens with the heart (yellow). A solution for this is to use 20 sample points and apply a refinement to the results in the areas known as being problematic like the one near the ribcage.

The influence of both values in the results has been quantitatively evaluated by careful visual inspection of the datasets. Taking into account that the accuracy of the results in the 24 datasets was similar (Figure 4) we randomly selected 8 of them and for each one we chose 3 slices: one with a bigger number of mismatches ($< 15\%$), one with a small number of mismatches ($< 5\%$) and an intermediate one. Note that those containing $\approx 15\%$ of mismatches belong to slides in which the spine was mismatched and only in those patients for which the spine is not adjacent to the liver. This could be solved by using a threshold in the similarity calculation to determine if the spine is adjacent or not. The mean error E has been calculated with the equation: $E = \frac{\sum_{i=0}^{s} e(\%)}{s*100}$, where $e = m/w$, m is the number of wrong assignments, w is the

Table 2 Effect of the variation in the number of P_j (5, 10 and 20) and in the number of bins per histogram.

50 bin			100 bin		
5 pt.	10 pt.	20 pt.	5 pt.	10 pt.	20 pt.
16.23	8.92	6.93	17.72	9.63	7.2

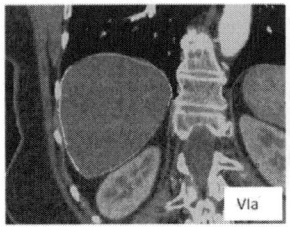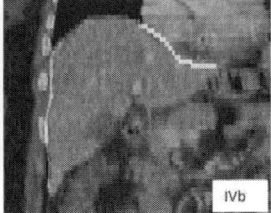

Fig. 4 Results of organ detection in two patients: lung (red), heart (yellow), kidney (pink), intestine (blue), ribcage (green) and spine (white).

Table 3 Mean, maximum and minimum distances after initialization (mm).

	Mean	Max	Min
Right hepatectomy	7.34	21.74	0.34
Lelft hepatectomy	8	34.89	0.35

total number of assignments in that slice and s is the number of the slices used for calculation. The process is repeated for every dataset, changing the number of bins (50 and 100) and k (5, 10 and 20). As k increases the number of mismatches decreases (Table 2).

Evaluation of the Initialization. The initialization has been evaluated using the same 12 pairs of datasets. We first segmented the liver in every dataset and carried out the organ detection. Due to the lack of real data of resected livers, the data for the initialization has been simulated. Since our goal is to show the robustness of the initialization algorithm against big resections we cropped one of the segmented and labeled livers from every pair of datasets to simulate ELH and ERH. Then the resected liver has been initialized to the corresponding preoperative data which contained additional real breathing deformations with respect to the resected one and the distances between both surfaces are calculated. Figure 5 shows the qualitative evaluation results of the initialization. The top row shows the preoperative image aligned to the resected liver surface (pink). The middle and bottom rows show the results in three of the datasets in the case of ERH and ELH respectively. The reddisher the color the bigger the distance. One can see that in ERH the distances are bigger, this is due to the fact that that area uses to contain bigger deformations than the right lobe. This is also quantitatively reflected in Table 3.

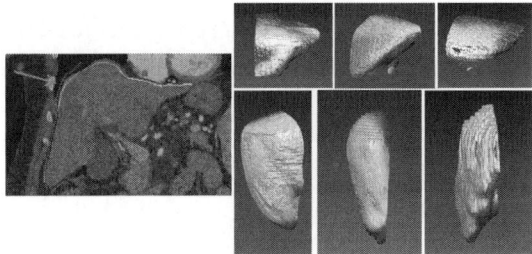

Fig. 5 Results of the initialization. Left: Resected data (pink) overlapped with preoperative data. Right top and bottom: Distances between surfaces. The reddisher the color the bigger the distance.

V. CONCLUSION

We presented a novel algorithm to automatically detect the organs adjacent to the liver. We used a surface mesh of the liver and a weighted comparison of histograms. The knowledge of the surroundings of an organ can improve the results of registration or segmentation algorithms. As opposite to state of the art algorithms our approach does not need any correspondences with reference datasets which makes it appropriate for pathological organs. In addition to this, we do not require previously annotated data. The detection algorithm has been evaluated in 24 clinical CTs and an example application has been introduced in which preoperative and postoperative data are initialized. The automatic segmentation of clinical resected data will be addressed in the near future.

REFERENCES

1. Viola P., Jones M. J.. Robust real-time face detection in *IJCV* 2004.
2. Zhan Y., Zhou X. S., Peng Z., Krishnan A.. Active scheduling of organ detection and segmentation in whole-body medical images in *MICCAI* 2008.
3. Pauly O., Glocker B., Criminisi A., et al. Fast multiple organ detection and localization in whole-body MR dixon sequences in *MICCAI* 2011.
4. Criminisi A., Shotton J., Robertson D., Konukoglu E.. Regression forests for efficient anatomy detection and localization in CT studies in *MICCAI Workshop MVC* 2011.
5. Ling H., Zhou S. K., Zheng Y., Georgescu B., Suehling M., Comaniciu D.. Hierarchical, learning-based automatic liver segmentation in *IEEE CVPR* 2008.
6. Bentrem D. J., DeMatteo R. P., Blumgart L. H.. Surgical therapy for metastatic disease to the liver in *Annual Review of Medicine*;56 2005.
7. Erdt M., Steger S., Kirschner M., Wesarg S.. Fast automatic liver segmentation combining learned shape priors with observed shape deviation in *IEEE CBMS* 2010.
8. Belghiti J., Clavien P. A., Gadzijev E., et al. The Brisbane 2000 terminology of liver anatomy and resections in *HPB*;2 2000.

A Comparative Study of Different Methods for Pigmented Lesion Classification Based on Color and Texture Features

J.A. Pérez-Carrasco, B. Acha, and C. Serrano

Dpto. Teoría de la Señal, ETSI, Universidad de Sevilla, España
jperez2@us.es, {bacha,cserrano}us.es

Abstract—This paper presents a method that classifies color dermoscopic images into their different dermatologic patterns. CSGV (composite subband gradient vector) is used to represent each pattern. CSGV is obtained from the gradient vectors generated from the different sub-images in a wavelet decomposition. Classification results are compared with those obtained applying Gabor filters and Markov Random Field (MRF). Classification is performed with a fuzzy-ARTMAP. Performance is analysed in *L*a*b** and *RGB* color spaces. *L*a*b** provides the best results (81,25%).There are two main contributions in this work. The first one is to combine features from already existing methods with color information, in the *RGB* or *L*a*b** representations. The second contribution is that a large study of different features applied to the specific problem of dermoscopic pattern classification has been implemented.

Keywords—Dermatology, Gabor, gradient, neural network, Wavelet.

I. INTRODUCTION

In the last years skin lesions and especially malignant melanoma are becoming serious problems all around the world. Still today, early diagnosis and prompt surgical excision of the primary cancer are the best and almost unique solutions. When a lesion has been considered melanocytic, in order to classify it into benign or malign, four different approaches are the most commonly used: the ABCD rule of dermoscopy, the 7-point checklist, the Menzies method, and Pattern Analysis [1]. In the last years new tools have helped to early diagnosis of melanoma. Among them, the most important are Dermoscopy [2] and Computer Assisted Diagnosis (CAD) systems [3][4] [5]. Some comparisons between different classification techniques can be found in [6].

In the present paper, our purpose is to perform a classification of the different global patterns that a lesion can present. Following the CASH (Color, Architectural order, Symmetry of pattern and Homogeneity) criteria [7], skin lesions can be classified as reticular or network pattern, globular, cobblestone, homogeneous, starburst, parallel and multicomponent [4]. An illustration of the patterns listed above is presented in Fig. 1.

In the present work, dermoscopic images are classified into their global patterns via wavelets [5][9] and color information in the two color spaces RGB and L*a*b*. A fuzzy ARTMAP neural network is employed as classifier [10].

II. IMPLEMENTATION

The wavelet-based algorithm proposed in this paper consists of two stages. The first one consists in the extraction of a gradient-based feature vector from the input image.

The second stage consists in classifying the image into one of the five different patterns analyzed in this work: reticular, globular, cobblestone, homogeneous and parallel. Aimed to implement this classification, the feature vectors are supplied to a fuzzy ARTMAP neural network [10].

For comparison purposes, we have also implemented the feature extraction stage using Gabor filters and a model-based technique using Markov Random Fields [4].

A. Creation of a Gradient-Based Feature Vector

The first step in our algorithm is to decompose an image using wavelets [12]. In our prototype system, we choose *DAUB4* as the wavelet basis because it has the best average performance [13]. As a consequence, each plane of the original image in the *L*a*b** or in the *RGB* color space can be decomposed into four sub-images, *LL, LH, HL* and *HH*. After obtaining the sub-images, we construct a gradient vector [14] for each sub-image. A gradient image, generated by a gradient operator, reflects the magnitude and direction of maximal change at each pixel of an input image. A number of gradient operators such as the popular Sobel operator can be used for generating gradient images. Assume that there are 360 directions, by summing up the magnitude value in the same direction at different pixels, a histogram of gradient directions with 360 bins can be compiled (called gradient vector). To reduce the length of a gradient vector and the sensitivity due to small changes in image orientation, every successive *k* directions can be grouped together to form one bin. Therefore, the total number of bins and the length of a gradient vector will be *360/k* (in our experiments, a *k* value of 90 provided the best results). Let *S1, S2, S3* and *S4* be the gradient vectors associated with sub-images *LL, LH, HL*, and *HH*, respectively. We construct the

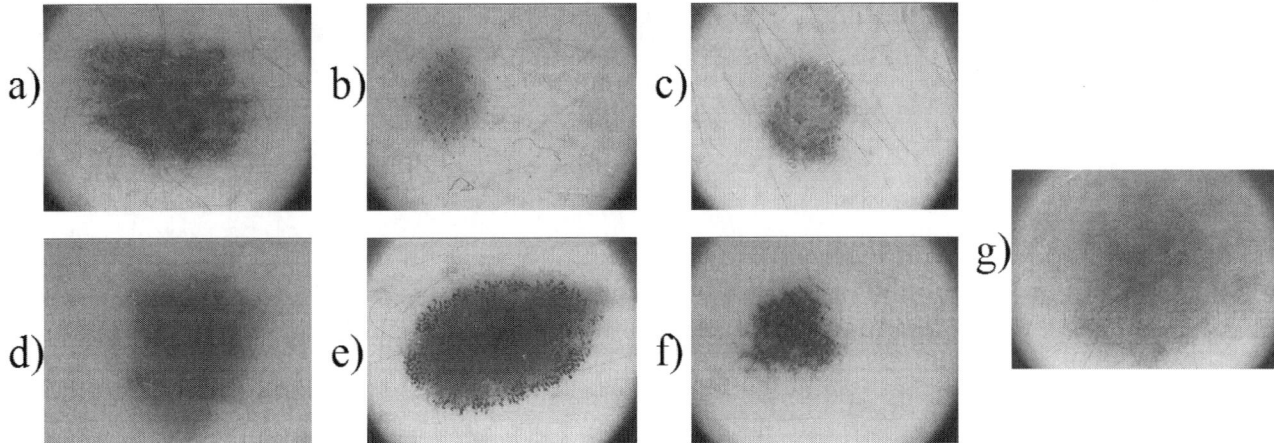

Fig. 1 D Skin lesions presenting each one of the 7 possible patterns: (a) Reticular pattern, (b) Globular, (c) Cobblestone, (d) Homogeneous, (e) Starburst, (f) Parallel, (g) Multicomponent (reticular, globular and homogeneous).

CSGV vector, called the composite subband gradient vector [14] by annexing the four gradient vectors in the following way:

$$CSGV = [S1 \quad S2 \quad S3 \quad S4] \quad (1)$$

We have computed different feature vectors using the *RGB* color space in four different scenarios. In the first one the gray-level image is decomposed using wavelets and the corresponding *CSGV* vector is extracted(*fwavL* with 16 features). In the second scenario, we use only the mean values of the three color planes (*f_color* with 3 features). The third one includes the 16 features of first scenario and incorporates the mean values of the image in the three different color planes (*fwavL_color* with 19 features). Finally, in the fourth scenario, a wavelet decomposition is applied to each one of the three color planes (*fwav_RGB* with 48 features). Equivalently, $L^*a^*b^*$ color space have also been considered. In this sense, *fwav_LAB* would be the correspondent version to *fwav_RGB* when the $L^*a^*b^*$ color space is considered. In Table 1 all the different vector vectors that have been analysed are described.

B. Creation of a Gabor-Based Feature Vector

Texture has also been analysed applying Manjunath's Gabor wavelet features for pattern analysis [15]. Each image is filtered with a set of 24 Gabor filters (4 scales and 6 orientations), and from each filtered image W_{mn}, we compute the mean, μ_{mn}, and standard deviation, σ_{mn}, of the pixel values distribution, where m represents the scale and n the orientation. Again, in order to incorporate color information, results obtained for two different color spaces ($L^*a^*b^*$ and RGB) and in three different scenarios have been compared. The first scenario is the classification from a feature vector extracted from the filtered gray-level image: mean and standard deviation for each band (fgabL with 48 features in total). The second scenario includes the 48 features of the first scenario and incorporates the mean values of the image in the three color planes (fgabL_color with 51 features in total). Finally, in the third scenario, the three color planes are filtered with Gabor filters (fgab_RGB or fgab_LAB with 144 features).

C. Creation of a MRF-Based Feature Vector

In [4], C. Serrano and B. Acha constructed an algorithm based on Markov Random Fields to classify skin lesions. In their model, an image can be described by a finite symmetric conditional model (*FSCM*) as follows

$$g_s = \mu_s + \sum_{t \in \eta_g} \beta_{s,t}[(g_{s+t} - \mu_{s+t}) + (g_{s-t} - \mu_{s-t})] + e_s \quad (2)$$

Table 1 Dice, jaccard, Sensitivity and Positive Predictive value coefficients for the three cases under analysis

	VECTOR CREATION	RATE L*a*b* (%)	RATE L*a*b* (%)
fwavL	[CSGVL]	81,25%	81,25%
f_color	[µL µa µb] [µR µG µB]	56,36%	56,36%
fwavL_color	[CSGVL µL µa µb] [CSGVL µR µG µB]	73,75%	73,75%
fwav_LAB / fwav_RGB	[CSGVL CSGVa CSGVb] [CSGVR CSGVG CSGVB]	71,25%	71,25%
fgabL	$[\mu_{mn}\ \sigma_{mn}]_{m=0,...3;\ n=0,...5}$	63,64%	63,64%
fgabL_color	$[\mu_{mn}\ \sigma_{mn}\ \mu_L\ \mu_a\ \mu_b]_{m=0,...3;\ n=0,...5}$ // $[\mu_{mn}\ \sigma_{mn}\ \mu_R\ \mu_G\ \mu_B]_{m=0,...3;\ n=0,...5}$	59,09%	59,09%
fgab_LAB / fgab_RGB	$[\mu_{mnL}\ \sigma_{mnL}\ \mu_{mna}\ \sigma_{mna}\ \mu_{mnb}\ \sigma_{mnb}]_{m=0,...3;\ n=0,...5}$ // $[\mu_{mnR}\ \sigma_{mnR}\ \mu_{mnG}\ \sigma_{mnG}\ \mu_{mnB}\ \sigma_{mnB}]_{m=0,...3;\ n=0,...5}$	75,45%	75,45%
fMRF	$(\mu_L, \mu_a, \mu_b, \sigma_L^2, \sigma_a^2, \sigma_b^2, \beta_{L,t}, \beta_{a,t}, \beta_{b,t})_g$	59,09%	59,09%

where $\eta_g = \{(0,1),(1,0),(1,1),(-1,1)\}$ is the set of shift vectors corresponding to the second order neighborhood system, μ_s is the mean of the pixels in a window centered in site s, $\{\beta_{s,t} : t \in \eta_g\}$ is the set of correlation coefficients associated with the set of translations from the central site s, and $\{e_s\}$ is a stationary Gaussian noise sequence with variance σ_s^2. Then, each image is characterized by a unique parameter vector in the L*a*b* color space. Therefore, the parameter vector is formed by 18 components

$$f_{MRF} = (\mu_L, \mu_a, \mu_b, \sigma_L^2, \sigma_a^2, \sigma_b^2, \beta_{L,t}, \beta_{a,t}, \beta_{b,t} : t \in \eta_g) \quad (3)$$

D. Feature Classification Using a Fuzzy ARTMAP Neural Network

Once the feature vector has been created, it is utilized as input to a *fuzzy ARTMAP* neural network [10]. *Fuzzy ARTMAP* [10] is a class of neural network architectures that perform incremental supervised learning of recognition categories and multidimensional maps in response to input vectors presented in arbitrary order.

III. RESULTS

The proposed feature vectors have been tested on a database containing 76 samples of 76 images corresponding to the following patterns: reticular, globular, cobblestone, homogeneous and parallel. We have performed 10-fold cross-validation [16]: at each iteration 90% of the total set of images has been used to train (68 images) and 10% to validate (8 images).

The different feature vectors obtained from the wavelet decomposition, Gabor filtering and Markov Random Fields [4] in the *L*a*b** and *RGB* color spaces have been evaluated. To this end, a classification via a *fuzzy ARTMAP* neural network have been perform with these feature vectors as input. Table 1 shows the results together with the features contained in each feature vector. Wavelet decomposition of the *L** color plane provided the best results (81.25%). From this experiment it could be infer that gray-level spatial distribution is the determinant feature for pattern classification.

A K-Nearest-Neighbour (KNN) classifier and a single-layer backpropagation neural network were also employed to classify into the different global patterns in order to analyse the good performance of the *fuzzy ARTMAP* neural network. In most of the cases recognition rates were lower than those obtained using *fuzzy ARTMAP* as classifier (and less than 60% in many cases).

IV. CONCLUSIONS

In this paper, a classification algorithm which applied filter-based modeling is presented. Different techniques that already exist have been extended to color images using RGB and L^*, a^* and b^* representations. In order to create a feature vector, L^*, a^* and b^* planes have been decomposed with *DAUB4* wavelets. First-order statistics derived from the different subbands constituted the feature vector, which is classified using a *fuzzy ARTMAT* neural network. A large number of comparisons with other filter-based and model-based feature vectors were developed and wavelet features provided the best results individually. In addition, results indicate that $L^*a^*b^*$ color space is better than *RGB* for classification purposes.

There are two main contributions in this work. The first one is to combine features obtained from already existing methods with color information, in the RGB or L*a*b* representations. The second contribution is that a large study of different features applied to the specific problem of dermoscopic pattern classification has been implemented.

In future implementations a larger database will be used, which will allow us to better evaluate our implementation. Furthermore, a combination of the features obtained with the different techniques described above will be analyzed in order to improve the classification rate.

ACKNOWLEDGMENT

This research has been supported by TEC2010-21619-C04-02.

REFERENCES

1. Braun R. P. et al (2005), Dermoscopy of pigmented skin lesions, J. Am. Acad. Dermatol. 52:109-121.
2. Westerhoff K., McCarthy W., Menzies S. W. (2000), Increase in the sensitivity for melanoma diagnosis by primary care physicians using skin surface microscopy, Br. J. Dermatology, 143:1016-1020.
3. Binder M. et al (1998) Epiluminescence microscopy-based classification of pigmented skin lesions using computerized image analysis and an artificial neural network, Melanoma Res, 8: 261-266.
4. Serrano C., Acha B. (2009) Pattern analysis of dermoscopic images based on Markov random fields, Pattern Recognition 42(6): 1052-1057, 2009.
5. Surowka G., Grzesiak-Kopec K., (2007) Different Learning Paradigms for the Classification of Melanoid Skin Lesions Using Wavelets, Conf Proc IEEE Eng Med Biol Soc, 2007, pp. 3136-3139.
6. Lisboa P. J., Taktak A. F. (2006) The use of artificial neural networks in decision support in cancer: A systematic review, Neural Networks, 19(4): 408–415.
7. Braun R. P. et al (2005) Dermoscopy of pigmented skin lesions, J. Am. Acad. Dermatol., 52:109-121.
8. Speis A., Healey G., (1996) Feature extraction for texture discrimination via random field models with random spatial interaction, IEEE Trans. Image Process., 5(4):635–645
9. Randen T., HusZy J. H., (1999) Filtering for texture classification: a comparative study, IEEE Trans. Pattern Anal. Mach. Intell. 21(4):291-310.
10. Carpenter A., (1992), Fuzzy ARTMAP: A Neural Network Architecture for Incremental Supervised Learning of Analog Multidimensional Maps, IEEE Transactions on Neural Networks, 3(5):698–713.
11. Fukunaga, K., (1990) Introduction to Statistical Pattern Recognition, 2nd edition, Morgan Kaufmann (Academic Press), San Diego, CA, 1990.
12. Chang T., Kuo C.-C.J., (1993) Texture analysis and classification with treestructured wavelet transform, IEEE Trans. Image Process., 2(4): 429–441.
13. Manian V., Vasquez R. (1998) Scaled and rotated texture classification using a class of basis functions, Pattern Recognition, 31: 1937–1948.
14. Huang P. W., Dai S. K., (2003) Image retrieval by texture similarity, Pattern Recognition, 36(3): 665-679.
15. Manjunath B. S., Ma W. Y.,(1996) Texture features for browsing and retrieval of image data, IEEE Trans. Pattern Anal. Mach. Intell., 18(8): 837–842.
16. Duda, R.O. et al., Pattern Classification, Wiley, New York, 2001.

Corresponding author:

Author: Jose-Antonio Pérez-Carrasco
Institute: Dpto. Teoría de la Señal, ETSI, Universidad de Sevilla
Street: Camino de los Descubrimientos, s/n., 41092
City: Seville
Country: Spain
Email: jperez2@us.es

Automatic Burn Depth Estimation from Psychophysical Experiment Data

Begoña Acha[1], Tomás Gómez-Cía[2,3], Irene Fondón[1], and Carmen Serrano[1]

[1] Dpto. Teoría de la Señal y Comunicaciones, University of Seville, Seville, Spain
[2] Virgen del Rocío Hospital, Seville, Spain
[3] Networking Research Center on Bioengineering, Biomaterials and Nanomedicine (CIBER-BBN)

Abstract—In this paper a psychophysical experiment and a Multidimensional Scaling (MDS) analysis are undergone to determine the physical characteristics that physicians employ to diagnose a burn depth. Subsequently, these characteristics are translated into mathematical features, correlated with these physical characteristics analysis. Finally, they are introduced to a Support Vector Machine (SVM) classifier. Results validate the ability of the mathematical features extracted from the psychophysical experiment to classify burns into their depths.

Keywords—burn, color, multidimensional scaling, CAD.

I. Introduction

For a favorable evolution of a burn injury, it is essential to initiate the correct first treatment [1]. As the cost of maintaining a burn treatment unit is high, it would be desirable to have an objective or automatic system to give a first assessment at primary health-care centers, where there is a lack of specialists [2, 3]. The determination of burn depth and whether a wound will heal spontaneously within 21 days is around 50% for inexperienced surgeons as reported by Hlava [4]. The clinical estimates of burn depth arise to the range from 64 to 76% for experienced surgeons [5]. This fact justifies the utility of a computer-aided diagnosis (CAD) tool in the clinical practice.

Severity of a burn patient depends upon three main factors: extension and location, age and burn depth. These three factors control the life expectancy, first treatment and the eventual requirement to referral to a burn center [6-8].

Burns are often classified according to their depth into four groups [9]:

1. First-degree burns or epidermal burns are usually red, dry, and painful.
2. Superficial dermal burns are moist, red and weeping; blanches with pressure. They are painful to air and temperature.
3. Deep dermal burns are wet or waxy dry; variable color (patchy to cheesy white to red) and do not blanch with pressure.
4. Full-thickness burns are generally leathery in consistency, dry, insensate, and waxy. These wounds will not spontaneously heal.

First degree burns are not considered in order to assess the percentage of burnt body surface area and thus the subsequent treatment. They correspond to sun burns that do not normally reach the Burn Unit and heal spontaneously. So, in clinical practice there are three main degrees of burns (superficial dermal, deep dermal and full-thickness). Therefore, first degree burns were out of the scope of this paper and if present in a photograph they are joined to healthy skin.

On the other hand, there is a two-groups classification essential for a plastic surgeon: those burns that require excision and those that heal spontaneously. Epidermal and superficial dermal burns are expected to heal on its own within 21 days and deep dermal and full thickness burns require surgery.

Previous attempts to automatically assess burn depth from digital photographs have been published with encouraging results (82.26% success rate) [10].

II. Material and Methods

A. Materials

The psychophysical experiment was performed with 20 burn images. They were acquired with a Canon EOS 300D in the first-aid room at the Burn Unit from Virgen del Rocío Hospital in Seville (Spain). Camera characterization was performed to maintain colors in the photographs faithful to real colors [10]. The image pixels were converted into sRGB in this characterization procedure. Then, "sRGB Color Space Profile" was chosen as the ICC screen profile for the display. In this sense, the colors displayed were expected to be the same as the ones captured by the camera.

74 new images were acquired in the same conditions to be employed as test images. Physicians (plastic surgeons affiliated to the Burn Unit) determined their burn diagnosis by directly observing the burn at the admission in all cases, that is, following the standard protocol established in the Unit. Physicians confirmed their diagnosis after one week or ten days of evolution.

The image acquisition protocol was developed by an interdisciplinary group formed by burn specialized physicians and technicians [10].

B. Scheme of the Procedure

The objective of this study was to identify a set of features useful to determine the burn depth according to experts. Subsequently these features were employed to develop a tool to classify burns into their depths. The procedure followed to develop a tool to classify burns consists of three main steps:

1) Psychophysical experiment. Images were presented in pairs to 8 plastic surgeons affiliated to the Burn Unit of Virgen del Rocío Hospital, in Seville (Spain). The experiment followed the rules dictated in Rec. ITU-R BT.500-10 [11]. As a result a similarity matrix, where each element represents the similarity between two images, was obtained.

2) Multidimensional Scaling. Images are represented in a multidimensional space where proximities between images are correlated with similarities obtained in step 1). The dimensionality for this space was determined with the stress formula [12].

3) Interpretation of the coordinates and clasification. In this phase matematical features correlated to the coordinates have been identified and these features are introduced into an SVM to classify.

C. Materials

This section describes the subjective experiment we have performed in order to obtain new characteristics to describe a burn [13,14].

The experiment was conducted with the help of eight plastic surgeons affiliated to the Burn Unit of Virgen del Rocío Hospital, Seville (Spain). Each of them underwent the test separately, but each session had the same structure based on the ITU-R BT.500-10 Recommendation [11].

During the experiment 190 different pairs of images, from the 20 images combined in pair, were presented to each doctors and they have to punctuate the similarity between them. We can see an example of these pairs in Fig.2.

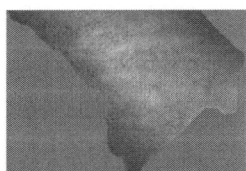 

Fig. 1 An example of the pairs of images that physicians have to compare

A medical expert, who did not undergo the experiment, selected the 20 images from the database. The reason is that we need representative images in order to include all the possible burn appearances.

At the end of the experiment a similarity matrix, Δ, was obtained. Let be I the number of images, $I=20$ in this experiment. Then the similarity matrix will have sions 20×20. Its elements, δ_{ij}, are the proximity or similarity values. δ_{ij} with $i,j=1,2,..,I$ and $i \neq j$ are the score assigned in the experiment, that is, the level of similarity between image i and image j according to the expert.

D. Multidimensional Scaling

Multidimensional Scaling (MDS) is a multivariant analysis method [12] to analyze similarity matrixes for a graphical representation of the data that would lead to a better interpretation of them. If the input is a *similarity matrix* Δ, with dimension $I \times I$, the output is a spatial representation consisting in a geometrical configuration of the I points, each point corresponding to one object, a burn image in this case. The objective of MDS is to find the best configuration of points $x_1,...,x_I$ in a space of dimension P which will represent the objects. If we define d_{ij} as the Euclidean distance between point i and point j in the MDS space, a matrix D whose elements are d_{ij}, for $i,j=1,...,20$ can be constructed. The objective is to find the configuration X_I that attained a matrix D highly correlated with matrix Δ.

E. Interpretation of the Coordinates and Classification

This aim was achieved in three substeps:

1) A brainstorming was designed to infer the physical meaning for each coordinate.

The 20 images were shown in a 3D projection: coordinate 1 versus coordinate 2 on one side, and coordinate 1 versus coordinate 3 on the other side. 6 experts in Digital Image Processing take part in the brainstorming. They were asked to identify physical features in the images that motivated their location in each coordinate.

As a result of the brainstorming the three coordinates were identified as:

Dimension 1: This dimension was labeled as *amount of pink in the image* and *vividness* or *moisture* in the skin.

Dimension 2: This dimension was considered as *texture of the color*. That is, images with low values in coordinate 2 have very homogeneous colors while images with high values in this coordinate present inhomogeneous colors.

Dimension 3: This dimension was interpreted as related to *colorfulness*.

2) Translation of these features into numerical parameters.

Dimension 1: This dimension is directly related to color, more specifically with hue and saturation. Therefore, correlation between this coordinate and different measures of these two color attributes were tested. More specifically, chroma and hue of CIE $L^*a^*b^*$ and CIE $L^*u^*v^*$ color spaces and hue and saturation of HSV color space were tested.

In addition, *moisture* can be related to the presence of glitters in the photograph. Thus, outliers in the L^* component or jointly in the R, G and B components were potential features to measure this physical characteristic. More specifically, if $\mathbf{I}_{RGB}(m,n) = [I_R(m,n), I_G(m,n), I_B(m,n)]$ is the image with color represented in the RGB color space then:

$$o_{L^*} = \frac{1}{|Q|} \sum_{(m,n) \in Q} (I_{L^*}(m,n) > 90)$$

$$o_{RGB} = \frac{1}{|Q|} \sum_{(m,n) \in Q} (I_R(m,n) > 240) \,\&\, (I_G(m,n) > 240) \,\&\, (I_B(m,n) > 240)$$

where o_{L^*} and o_{RGB} represent the number of outliers in the L^* component and in the RGB components respectively, normalized by the area of the burn.

Dimension 2: The texture referred in this dimension does not correspond with spatial repetition or special arrangement of its parts but rather with variability of the colors. Thus first-order statistical parameters in a^* and b^* coordinates may suffice. Specifically, as features potentially correlated with this coordinate we have taken skewness of a^* and b^* (sk_{a^*}, sk_{b^*}), kurtosis of a^* and b^* (k_{a^*} and k_{b^*}), and the angular variance of hue (v_{h^*}), calculated according to [31]:

$$v_{h^*} = 1 - \frac{1}{|Q^*|} \sum_{(m,n) \in Q^*} 1 - \cos\left(\arctan\left(\frac{I_{b^*}(m,n)}{I_{a^*}(m,n)}\right) - h^*\right)$$

where Q^* is the set of pixel that, not being considered outliers, belong to the burn and h^* is the angular hue mean within the burn except for pixels that are outliers.

Dimension 3: Chroma is a perceptual attribute traditionally related to colorfulness in Color Science. $C^*_{a^*b^*}$ is a numerical parameter that tries to measure this colorfulness. In addition, we could observed that colorful sensation was also related with the histogram of L^*: a long right tail in L^* distribution produces a *vividness* sensation. This has been measured according to quartile skewness proposed in [32], where left and right tail measures are defined as measures of skewness that are applied to the half of the probability mass lying to the left, respectively the right, side of the median of F, denoted as $F^{-1}(0.5)$, being F a continuous univariate distribution.

$$SK_{0.25} = \frac{\left(F^{-1}(1-0.25) - F^{-1}(0.5)\right) - \left(F^{-1}(0.5) - F^{-1}(0.25)\right)}{F^{-1}(1-0.25) - F^{-1}(0.25)}$$

Likewise peakedness of color components is also a sign of vividness. Thus, kurtosis in a^* and b^* have been tested as potential features related to dimension 3.

All these numerical features, that had been intuitively related to the three coordinates, have been analyzed to test if in fact they are correlated to them. To this aim, correlation coefficient and p-value have been calculated. Each p-value is the probability of getting a correlation as large as the observed value by random chance, when the true correlation is zero. These two coefficients are shown in Tables 1-3. According to their p-value, the eight features more correlated with the three coordinates in the MDS analysis were selected. s_{HSV} was not selected, although it has a low p-value, because it exhibited a very high cross-correlation with $C^*_{a^*b^*}$.

Table 1 Correlation of some features with dimension 1

Feature	Correlation coefficient	p-value
$C^*_{a^*b^*}$	-0.6275	0.0031
$s_{u^*v^*}$	-0.1753	0.4597
s_{HSV}	-0.6145	0.0039
h	0.3477	0.1331
o_{L^*}	-0.3346	0.1493
o_{RGB}	-0.3694	0.1090

Table 2 Correlation of some features with dimension 2

Feature	Correlation coefficient	p-value
v_{h^*}	0.3428	0.1389
sk_{a^*}	0.4179	0.0667
sk_{b^*}	0.3123	0.1801
k_{a^*}	0.2030	0.3908
k_{b^*}	-0.2142	0.3645

Table 3 Correlation of some features with dimension 3

Feature	Correlation coefficient	p-value
C^*_{a*b*}	0.5507	0.0119
$SK_{0.25}$	-0.4386	0.0531
k_{a*}	0.6558	0.0017
k_{b*}	0.0869	0.7156

3) Classification.

The eight features, selected because they were correlated with the three coordinates in the MDS analysis (Table 4), were employed as inputs to a Support Vector Machine (SVM).

Table 4 The eight features identified in the psychophysical experiment

C^*_{a*b*}	o_{RGB}	h	v_{h^*}	sk_{a*}	sk_{b*}	$SK_{0.25}$	k_{a*}

III. RESULTS

The SVM was trained with the 20 images used in the experiment. Afterwards, the remaining 74 images were employed to analyze the classification performance. The SVM has been applied with a Radial Basis Function (RBF) kernel. Sequential Forward Selection (SFS) and Sequential Backward Selection (SBS) were performed for feature selection. In order to apply SBS and SFS, different subsets of features are the inputs to the SVM and the classification error for each subset of features is calculated when the SVM classifies 74 different images. The subset which provides less classification error is selected. After applying both methods SVM chose three features out of the initial eight: C^*_{a*b*}, v_{h^*} and sk_{b*}. The results are summarized in Table 5.

Table 5 Classification results with a SVM

	Success rate
Classification into three burn depths	75,7%
Classification into two burn depths	82,4%
Sensitivity (S)	0.80
Specificity (E)	0.86
PPV	0.86
NPV	0.79

IV. CONCLUSIONS

As first conclusion, the eight mathematical features identified as correlated with plastic surgeons perception successfully classify burns into their depths. As second conclusion, SVM with the eight identified features successfully classifies burns.

With the psychophysical experiment we have discovered physical features that physicians used in order to diagnose a burn, and we have successfully translated them into numerical values that are able to classify burns into their depths.

ACKNOWLEDGMENT

The authors would like to thank Manuel Sosa and to the Burn Unit of the Virgen del Rocío Hospital from Seville (Spain). This work has been partially supported by project TEC2010-21619-C04-02 (CYCIT, Spain).

REFERENCES

1. B.S. Atiyeh, S.W. Gunn, and S. N. Hayek, "State of the Art in Burn Treatment," *World Journal of Surgery*, vol. 29, pp. 131-148, 2005.
2. L. Roa, T. Gómez-Cía, B. Acha, and C. Serrano, "Digital imaging in remote diagnosis of burns," *Burns*, vol. 25, no. 7, pp. 617-624, 1999.
3. C. Serrano, B. Acha, and J.I. Acha, "Segmentation of burn images based on color and texture information," *SPIE Int. Symp. on Med. Imaging*, San Diego (CA, USA), vol. 5032, pp. 1543-1550, 2003.
4. P. Hlava, J. Moserova, and R. Konigova, "Validity of clinical assessment of the depth of a thermal injury", *Acta Chir Plast*, vol. 25, pp. 202-208, 1983.
5. D. M. Heimbach, M. A. Afromowitz, L. H. Engrav, J. A. Marvin, B. Perry, "Burn depth estimation- man or machine," *Journal of Trauma and Acute Care Surgery*, vol. 24, pp. 373-378, 1984.
6. A. Franco, *Manual de Tratamiento de las Quemaduras*, Liade, Madrid (Spain), 1985.
7. http://www.burnsurgery.org/Betaweb/Modules/initial/bsinitialsec11.htm, (16/5/2012).
8. F. Harbrand, C. Schrank, G. Henckel-Donnersmarck, et al., "Integration of pre-existing conditions on the mortality of burn patients and the precision of predictive admission-scoring systems," *Burns*, vol. 18, pp. 368-372, 1997.
9. J. A. Clarke, *A Colour atlas of burn injuries*, Chapman & Hall Medical, London, 1992.
10. B. Acha, C. Serrano, J. I. Acha, and L. M. Roa, "Segmentation and classification of burn images by color and texture information," *Journal of Biomedical Optics*, vol. 10, no. 3, pp. 1-11, 2005.
11. Rec. ITU-R BT.500-10, "Methodology for the subjective evaluation of the quality of television images".
12. J. B. Kruskal, and M. Wish, *Multidimensional Scaling*. Madison, WI: Bell Telephone Laboratories, Inc. , 1978.
13. I. Fondón, B. Acha, C. Serrano, and M. Sosa, "New Characteristics for the Classification of Burns: Experimental Study," *Lecture Notes in Computer Science*, vol. 4142, pp. 502-512, 2006.
14. B.Acha, C.Serrano, I.Fondón, T.Gómez-Cía, "Burn depth analysis using Multidimensional Scaling applied to psychophysical experiment dat", *IEEE Trans. on Image Processing*, 2013, in press.

Author: Begoña Acha
Institute: Dpto. Teoría de la Señal y Comunicaciones. Univ. of Seville
Street: Camino de los Descubrimientos, s/n. 41092
City: Seville
Country: Spain
Email: bacha@us.es

Segmentation of Retroperitoneal Tumors Using Fast Continuous Max-Flow Algorithm

J.A. Pérez-Carrasco[1], C. Suárez-Mejías[2], C. Serrano[1], J.L. López-Guerra[2], and B. Acha[1]

[1] Dpto. Teoría de la Señal, ETSI, Universidad de Sevilla, España
jperez2@us.es, {cserrano,bacha}us.es
[2] Grupo de Innovación Tecnológica, Hospital Universitario Virgen del Rocío de Sevilla, Sevilla, España
{cristina.suarez.exts,josel.lopez.guerra.sspa}@juntadeandalucia.es

Abstract—In this paper a new algorithm for the segmentation of retroperitoneal tumors is presented. In this paper an accumulated gradient distance is proposed as a new regional term in the continuous max-flow algorithm. In this preliminary study, three CT images have been segmented and compared with manual segmentations. *DICE, Jaccard, Sensitivity* and *Postive Predictive Value (PPV)* indexes have been computed and their values show promising results.

Keywords—continuous maxflow, convex optimization, retroperitoneal tumors, segmentation.

I. INTRODUCTION

Retroperitoneal masses or lymph nodes are a diverse group of benign and malignant tumors that arise within the retroperitoneal space but outside the major organs in this space. Among the retroperitoneal masses, 70%–80% are malignant in nature, and these account for 0.1%–0.2% of all malignancies in the body [1]. Although Computed Tomography (CT) and Magnetic Resonance (RM) imaging can demonstrate important characteristics of these tumors, diagnosis is often challenging for radiologists, oncologists and surgeons. The retroperitoneal space, as shown in Figure 1, is hidden toward the back of the abdomen where organs are quite mobile. Thus, the retroperitoneal tumors do not have an established pattern and can grow quite large and differently, moving organs out of their path, before being discovered.

Diagnostic challenges include precise localization of the lesion, determination of the extent of invasion, and characterization of the specific pathological type. Lymph nodes or masses are routinely assessed by clinicians for monitoring disease progress and effectiveness of the cancer treatment. The use of images has improved the quality and accuracy of the process. In particular, detection and segmentation is important for cancer staging and surgical and treatment planning. When the cancer treatment is successful, the tumor decreases in size. Since finding the lymph nodes or masses is time consuming and highly dependent on the observer's experience, a system for detection and measurement is desirable.

Different abdominal tumors have been automatically segmented in the literature. In [2] the authors developed an algorithm for automatic segmentation of solid lymph nodes in axillaries and pelvic regions from CT images based on learning method. Authors pointed out that the algorithm has more mistakes in the pelvic area due their location inside the abdomen. Many false positives arise from the intestines and vessel bifurcations. Luo used active contour for segmentation of liver tumor [3] and Subramanya et al [4] employed a combination using watershed and active contour for brain tumors segmentation. However, active curve evolution can drive to local optima of the minimization energy function and suffers from high sensitivity to initialization.

Graph cut techniques consist in discrete minimization of the objective energy. They have the advantage that they guarantee the global optima in nearly real time. Thus this technique has been applied successfully in recent works [5][6]. In this sense, Zhang et al [7] use an prior elliptical shape in a modified graph cut for cervical lymph nodes on sonograms. They made a qualitative and quantitative evaluation and demonstrated that the incorporation of prior elliptical shape improved segmentation results. Suzuki and Feulner et al [8][9] use graph-cut for lymph nodes in the mediastinum. They proposed a system that reaches a detection rate of 60.9% at 6.1 false alarms per volume and they state that this is the best result compared with the state of the art in mediastinal lymph node detection. In [10][11] authors also applied graph-cut to the problems of segmentation of pulmonary and brain tumors.

In the literature, segmented tumors are mainly located in brain, lungs and neck, but, to the best of our knowledge, there is not any reference in the literature that focuses on the segmentation of retroperitoneal tumors. Retroperitoneal tumor segmentation has the added complexity that tumor intensity overlaps intensity of surrounding organs and no prior knowledge about its shape can be applied. In this paper, we face this challenge and get good results after a clinical validation. The methodology applied is based on continuous convex relaxation [12][13] which share the advantages of both active curves and graph cuts [13]. A recent study [13] showed that, in 3D grids, convex relaxation approaches outperform graph cuts in regard to speed and accuracy.

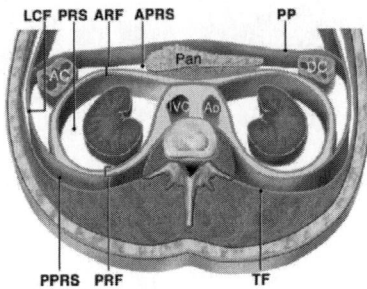

Fig. 1 Drawing of the anatomy of the retroperitoneal space, where APRS stands for Anterior Pararenal Space, PP Parietal Peritoneum, ARF Anterior Renal Fascia, Pan is Pancreas, AC Ascending Colon, DC Descending Colon, PPRS Posterior Pararenal Space, PRF Posterior Renal Fascia, TF Transversalis Fascia, PRS Perirenal, Ao aorta, IVC Inferior Vena Cava and LCF Lateroconal Fascia.

II. Methodology

The methodology introduced in this approach basically consists of four stages: a pre-processing stage, the computation of the accumulated gradient distance volume, a segmentation stage using a fast continuous max-flow algorithm and a final post-processing stage to improve slightly the segmen-tation result.

1.- Preprocessing Stage. In this preprocessing step, contrast between the tumor and the background is increased. Hounsfield values in an abdominal CT have a wide range, which needs to be compressed to the rank [0,1] to make feasible the implementation of semiautomatic algorithms. Therefore, an exponential law has been used to enhance the contrast. Thus, values close to the mean of the Hounsfield values inside the tumor will have an approximate lineal mapping whereas those values far from the mean will be closely saturated to 0 or 1. The enhanced output is obtained as

$$output = \frac{1}{1+e^{-pac/level}} \quad (1)$$

where *pac* is the original volume and *level* is the average value inside the tumor. This value is computed using a reduced number N (usually 3) of manually segmented slices. To facilitate the interaction with the surgeon and to speed-up the algorithm execution, this manual segmentation is only required to be approximated.

2.- Computation of the *accumulated gradient distance volume*. In this step, firstly the gradient of the volume GV is computed in the three directions x, y and z. Once the gradient has been obtained we proceed to compute the accumulated gradient distance volume (***GDV***) or accumulated gradient distance from the manually segmented tumor to the rest of the volume measured according to the gradients that have to be surpassed to reach the tumor. To do this, the generalized distance function algorithm (***GDF***) [14] is used. This algorithm is focused on modifying the classic two-pass sequential distance function [15] so that: (1) edge cost is taken into account; (2) raster and anti-raster scans of the image are iterated until stability. Let $N^+(p)$ (resp., $N^-(p)$) be the 26-connected neighbourhood of voxel p scanned before p (resp., after p) in a raster scan, GV the gradient volume computed in the previous step and $C_f(p,q) = GV(p) + GV(q)$ the associated cost to two neighboring voxels p and q, the algorithm of GDF proceeds as follows:

1. Initialize result volume d_f as : $d_f(p) = 0$ if p belongs to the tumor in the manually segmented slices and $d_f(p) = +\infty$ otherwise.

2. Iterate until stability, for each pixel p:
   Scan image in raster order:

$$d_f \leftarrow \min\{d_f(p), \min\{d_f(q) + C_f(p,q), q \in N^+(p)\}\} \quad (1)$$

   Scan image in anti-raster order:

$$d_f \leftarrow \min\{d_f(p), \min\{d_f(q) + C_f(p,q), q \in N^-(p)\}\} \quad (3)$$

The accumulated gradient distance volume is quite useful as it provides low values in the tumor area and high values in the outside. This happens because, in spite of only a few slices being manually segmented, the accumulated gradient within the tumor in the rest of the slices is low due to the absence of high gradient voxels inside the tumor.

3.- Max-flow based Segmentation Stage. In this stage, the accumulated gradient distance image will be utilized to carry out the segmentation. For this purpose the algorithm implemented by J. Yuan [16] has been employed. This algorithm solves the 3D multi-region image segmentation problem (Potts Model), based on the fast continuous max-flow method (CMF) proposed by J. Yuan et al [12]. Given the continuous image domain Ω (3D volume in our case), we assume that there are two terminals, the source s and the sink t. We assume that for each image position $x \in \Omega$, there are three concerning flows: the source flow $p_s(x) \in \Re$ directed from the source s to x, the sink flow $p_t(x) \in \Re$ directed from x to the sink t and the spatial flow field $p(x) \in \Re^2$.

The three flow fields are constrained by capacities:

$$p_s(x) \leq C_s(x), p_t(x) \leq C_t(x), |p(x)| \leq C(x); \forall x \in \Omega \quad (4)$$

In addition, for $\forall x \in \Omega$, all flows are conserved, i.e.

$$p_t - p_s + div\, p = 0, \forall x \in \Omega \qquad (5)$$

Therefore, the corresponding max-flow problem is formulated by maximizing the total flow from the source:

$$\max_{p_s, p_t, p} \int_\Omega p_s dx \qquad (6)$$

subject to flow constraints (4) and (5).

Yuan et al [12] proved that such a continuous max-flow formulation (12) is equivalent to the continuous s-t min-cut problem [17][18] as follows:

$$\min_{u(x) \in [0,1]} \int_\Omega (1-u) C_s dx + \int_\Omega u C_t dx + \int_\Omega C(x) |\nabla u| dx \qquad (7)$$

Actually, (7) just gives the dual model to (6) and the labeling function $u(x)$ is the multiplier to the flow conservation condition (5). Furthermore, an efficient and reliable max-flow based algorithm can be built up through (6). In [16] regional terms introduced to the segmentation algorithm in equation (7) are based on gray-level information. On the other hand, in the algorithm proposed here, the regional term provided as input to the continuous max-flow algorithm is the accumulated gradient distance volume GDV computed in the previous stage. In this way, we are not analyzing the gray level information but the contour information. Conversely, as we are not considering a gradient but an accumulated gradient, noise effect are avoided. Thus, in the implementation proposed the terms C_s and C_t are computed using the information provided by the GDV as follows:

$$\begin{aligned} C_s &= GDV \\ C_t &= GDV - 0.16 \end{aligned} \qquad (8)$$

The constant 0.16 in Eq. (8) is the average mean value of GDV outside the tumor for the different cases that have been analyzed. In addition, the algorithm penalizes an excessive area of the surface limiting the segmented volume. This penalization term is the most right term in Eq. (7). This penalization is modeled by a penalization function depending on the gradient along the surface, so that if the gradient is high along the surface, a large area of the surface is not penalized. The penalization function is computed as follows:

$$C(x) = \frac{b}{1 + a \cdot |\nabla f(x)|} \quad a, b > 0 \qquad (9)$$

where f is the volume, and parameters a and b control the importance of the gradient in the penalization function.

4.- Postprocessing. Once the max-flow segmentation has been performed, morphological operations of dilation and erosion are carried out in order to make the final segmentation more accurate.

Fig. 2 shows images obtained at different steps of the algorithm. It can be observed that the segmentation provided by the algorithm (*Fig. f*)) and the one provided manually by the surgeon (*Fig. g*)) match quite well, showing the effectiveness of the algorithm here described.

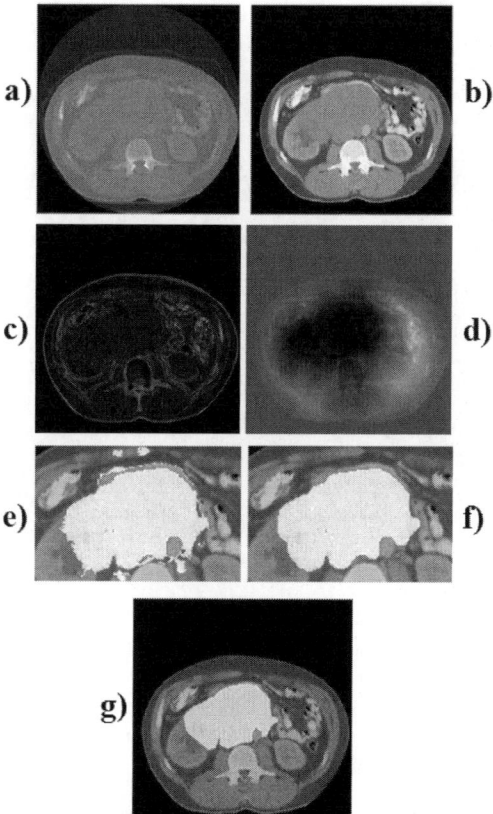

Fig. 2 a) a slice of the patient before the enhancement operation, b) same slice after the contrast enhancement operation, c) gradient image corresponding to one slice in one of the volumes, d) gradient distance image corresponding to the same slice, e) a detail of segmented result provided by the max-flow algorithm, f) a detail of the same segmented slice after the implementation of the morphological operation, g) manual segmentation provided by the surgeon in the same slice.

III. EXPERIMENTAL RESULTS

In our experiments, three cases corresponding to three different patients have been used. Manual segmentations of all the slices were provided in order to measure the performance of our algorithm. In the three cases, the results obtained were satisfactory to the surgeon and some objective measures to analyze the performance were computed. These measures are the *Jaccard* [19], *Dice* [20], *Sensitivity (S)* and

Positive Predictive Value (PPV) parameters. Both *Jaccard* and *DICE* coefficients measure the set agreement in terms of false positive, false negative, true negative and true positive counts. Table I shows these coefficients for the three different cases analyzed. Note that values higher than 0.8 were obtained.

IV. CONCLUSIONS

In this paper a semiautomatic algorithm for retroperitoneal tumor segmentation has been developed. The segmentation of such tumors is a complicated task because they present intensities overlapping with those of surrounding tissues. To the best of our knowledge, segmentation of retroperitoneal masses has not been addressed in the literature. In this paper, a new segmentation algorithm, based on continuous max-flow optimization, has been proposed to solve this problem. This algorithm has been assessed using three CT images manually segmented, and several coefficients have been computed to measure its performance. Results obtained are over 0.8, which indicates neither the FP rate nor the FN rate are high. These results have to be analyzed under the assumption that exact segmentation of the tumors is a difficult task even for an expert in tumors because of its diffuse borders. Thus different surgeons should segment the tumors for a better assessment of the algorithm. This is a future task. Likewise, further validation, with a higher number of manual segmentations for more cases will be performed in future publications. Finally, regional energies that take into account both edge and grayscale information will be minimized with the max flow algorithm in order to outperform the current algorithm.

ACKNOWLEDGMENT

This research has been supported by TEC2010-21619-C04-02.

REFERENCES

1. Rajiah P, Sinha R, et al (2011) Imaging of Uncommon Retroperitoneal Masses. RadioGraphics 31:049-976 10.1148/rg.314095132.
2. Barbu A., Suehling M., Xun x. and et al (2012) Automatic Detection and Segmentation of lymph Nodes from CT Data. IEEE Transactions on medical imaging 31: 241-250
3. Luo Z (2011) Segmentation of liver tumor with local C-V level set, IEEE2011, 20011, pp. 7660-7663.
4. Subramanya B, Sanjeev R, Kunte (2010) A mixed model based on Watershed and active contour algorithms for brain tumor segmentation, International Conference on Advances in Recent Technologies in Communication and Computing, IEEE, pp. 398-400.
5. Boykov Y, Funka G (2009) Graph cuts and efficient N-D image segmentation. Int. J. Comput. Vision, 70: 109-131.
6. Boykov Y, Kolmogorov V (2004) An experimental comparison of min-cut/max-flow algorithms for energy minimization in vision. IEEE Trans. Pattern Anal. Mach. Intell., 26: 1124-1137.
7. Zhang J, Wang Y, Shi X (2009) An improved graph cut segmentation method for cervical lymph nodes on sonograms and its relationship with node's shape assessment, Computerized Medical Imaging and Graphics 33:602-607.
8. Suzuki K, (2011) Segmentation based features for lymph node detection from 3D Chest CT, LNC, Machine Learning in Medical Imaging - Second International Workshop, MLMI 2011, Held in Conjunction with MICCAI 2011,7009:91-99
9. Feulner J, Zhou S, Hammon M et al (2012) Lymph node detection and segmentation in chest CT data using discriminative learning and a spatial prior, Medical Image Analysis, DOI 10.1016/j.media.2012.11.001
10. Song Q, Chen M, Bai J et al (2011), Surface region context in optimal multi-objet Graph Based Segmentation: Robust Delineation of Pulmonary Tumors, IPMI, 2011 pp. 61-72.
11. Chen V, Ruan S (2010) Graph cut segmentation technique for MRI brain tumor extraction, Image Processing Theory, Tool and Applications.
12. Yuan J, Bae E et al (2010) A study on continuous max-flow and min-cut approaches, CVPR, San Francisco, CA, 2010, pp 2217-2224.
13. Punithakumar K, Yuan J, et al (2012), A Convex Max-Flow Approach to Distribution-Based Figure-Ground Separation, SIAM J. Imaging Sciences 5(4):1333-1354
14. Vincent, L. Vincent (1998) Minimal path algorithms for the robust detection of linear features in gray images, ISSM Proc. Fourth international symposium on Mathematical morphology and its applications to image and signal processing, Amsterdam, The Netherlands, 1998, pp. 331–338.
15. Rosenfeld A, Pfaltz J L (1968), Distance functions on digital pictures. Pattern Recognition, 1(1):33–61.
16. Continuous Max Flow at https://sites.google.com/site/wwwjingyuan/continuous-max-flow.
17. Nikolova, M., Esedoglu, S., Chan, T.F (2006), Algorithms for finding global minimizers of image segmentation and denoising models. SIAM J. App. Math. 66:1632–1648.
18. Bresson, X., Esedoglu, S et al (2007), Fast global minimization of the active contour/snake model. Journal of Mathematical Imaging and Vision, 28:151–167.
19. Jaccard, P. (1901), Étude comparative de la distribution florale dans une portion des Alpes et des Jura. Bulletin de la Société Vaudoise des Sciences Naturelles, 37: 547–579
20. DICE coefficient at http://sve.loni.ucla.edu/instructions/metrics/dice/

Corresponding author:

Author: Carmen Serrano
Institute: Dpto. Teoría de la Señal, ETSI, Universidad de Sevilla
Street: Camino de los Descubrimientos, s/n. Isla de la Cartuja, 41092
City: Seville
Country: Spain
Email: cserrano@us.es

Evaluation of the Visibility of the Color and Monochrome Road Map by the Searching Distance and Time of a Locus of Eye Movement

Takakazu Kobayashi[1], Tatsuya Oizumi[2], Kazushige Kimura[2], and Katsumi Sugiyama[1]

[1] Shibaura Institute of Technology, Department of Electronic Engineering
[2] Graduate School of Engineering and Science, Shibaura Institute of Technology and Science

Abstract—The purpose of this study is to investigate visibility of the color and monochrome map by total searching distance and time of a locus of eye movement exploring the visual target. We found that it is easily to find out the visual target on the color map compared to the monochrome map, In single target total searching distance and search time of color map are 4.5 times and 5.7 times shorter than those of monochrome map respectively. In two targets total searching distance and search time of color map are 3.4 times and 2.4 times shorter than those of monochrome map respectively. In two targets we can find out faster than single target both color and monochrome map. These results suggest that the visual target of color map is easily to find out and recognized compared to monochrome map due to hue and saturation.

Keywords—Evaluation of visibility, color and monochrome map, locus, eye movement, searching distance and time.

I. INTRODUCTION

Despite the great amount of research on eye movements of viewing different types of visual scene such as character, picture, pattern, real scene [1-5], a little is known about the mechanism of eye movement finding out the visual target in complicate picture.

Recently, an opportunity to see a map tends to increase by the spread of mobile information terminals. Generally speaking about the map, anyone needs to see easily and to search quickly the visual target on the map. Jeff B. Pelz et al. [5] studied scan path for narrow area and path finding sequence in map reading task based on the trace of eye movement. About an intelligible map Ninomiya [6] studied an abbreviation map by generation algorithm for mobile information terminals.

When we search a road map, we have experience that the map printed by color is clear and easily to find out the visual target, however, the map copied to monochrome is hard to find out it. It is not clear why the visual target of the color map easily find out and the color map is more impressive than monochrome map. For example, the same thing can be seen in the complicate pictures such as cerebral dissection. When we search the visual target on the map, we have question which is easier to find out on the color or monochrome map? how we search the target on the map?, Is there any difference of way to search between the color and monochrome map? Our motivation to the research is to know and understand about these questions. In the present study, to evaluate visibility of the color and monochrome map quantitatively, we measured total searching distance and search time by pursuing the locus of eye movement of search the visual target and assessed visibility of color and monochrome road map.

II. METHODS

A. Eye Movement

There are three kinds of assessing eye movements, ductions, versions or vergences. Smooth pursuit eye movement occurs closely follow a moving object and saccadic eye movement occurs quick movement up to 600 deg/s and jumps from one point to another. Here we focused later eye movement to search the visual target in static road map.

B. Pupil Measurement

In this study we recorded eye movement by an infrared sensitive CCD camera module with LED (Sharp, MK-0323E), 270 k pixels, resolution 542 ×492 and 30 fps. In this purpose, we do not need to use high speed camera. From the difference in reflectance of the iris and pupil, binarization of the gray scale component was performed by setting appropriate threshold. Video image of the pupil was taken by a capture board (Sky Digital, SKY-CXHDMI) and data were transferred to personal computer.

C. Image Processing

The center of gravity of the pupil was determined by calculating center of whole pixels at each frame of image. The locus of center as eye movement has been converted to a number of coordinates of each frame from the recorded video by above procedure. The locus of eye movement is drawn on the map. Total searching distance was added up whole line from start to end, and the search time was counted by time indicated computer screen. Figure 1 shows the flow chart of image processing. Figure 5 shows an example of road map superimposed the locus of eye movement.

D. Experimental Procedure

In this study we measured the eye movement by using eye image measurement system own made (Figure 2). The subjects were selected from 6 healthy male and female students (age: 22-24). The road map is placed via half mirror in front of 300 mm away from the subjects who are fixed face by head fixation apparatus, the CCD camera having infrared sensitive was put at 200 mm in front of the subjects in the direction of perpendicular to the mirror. We showed alternatively color and monochrome road maps in

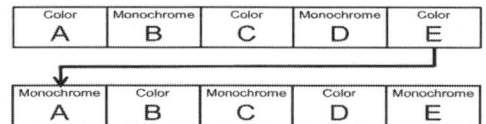

Fig. 3 The order of measurement to show the visual target on the map

maps, each 5 color and 5 monochrome map written in Japanese respectively. Used visual target here was convenient store written by symbol. In the order of measurement the same type of map is set to non continuous as shown in figure 3. Before start of the experiments, the subjects practiced to explore the target by using another map.

III. RESULTS

To make sure whether measurement system works well or not, subjects looked line of quadrilateral on white paper of same size of map, starting from top left to one round along clockwise direction. In figure 4 the locus of eye movement was seen as almost rectangular shape from upper left along clockwise direction. We could confirm that the locus of eye movement has linearity, so measurement system works properly.

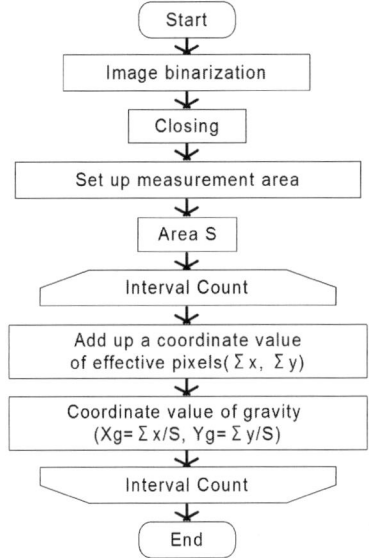

Fig. 1 Flow chart of image processing in the eye movement. The locus (Coordinates) of the eye movement is measured by the procedure of the flow chart shown above.

Fig. 4 Confirmation in performance of measurement system by looking rectangular line on the white paper. Subject started to look from upper left to one round along the direction of clockwise.

A. Single Visual Target

The locus of eye movement in the color map and monochrome map in the case of one target, are shown in figure 5A and B. In these examples, when the subjects explore the target, the locus of eye movement has begun from near the center of a map. In color map subject is search surrounding area on the map. On the other hand in monochrome map subject is search upper central area of the map.

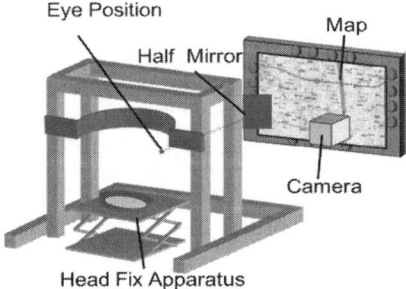

Fig. 2 Experimental setup for measuring eye movement. A jaw and the brow are placed and fixed to a head fixed stand. A right eye looks at the road map through a half mirror. An eye movement is taken with a CCD camera in the direction of perpendicular to the mirror.

Average total searching distance and average search time in the case of single target are shown in table 1 and figure 6A and 6B. Bar in the figures shows standard deviation. Average total searching distance of the locus of eye movement in the monochrome map becomes about 4.5 times longer than that in the color map. On the other hand average search time on the monochrome map has taken about 5.7 times longer than that in the color map.

disorder to avoid previous history, because if we show same monochrome map after color map, the subjects memorize previous color map. Therefore subjects were measured with five different maps. We have carried out in a total of 10

B. Two Visual Targets

The locus of eye movement in the color map and the monochrome map in case of two targets, are shown in figures 7A and B. In these examples, when the subject explores for the targets, in each case the locus of eye movement is more complicated, it takes time to find out targets and starting point is random. The average total searching distance and search time of the locus of eye movement in the case of two targets are shown in table 1 and figures 8A and B.

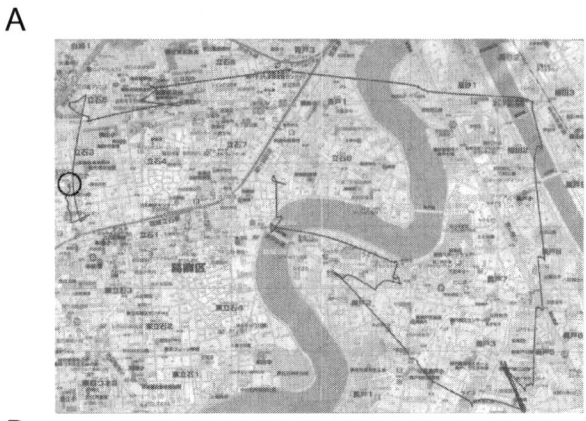

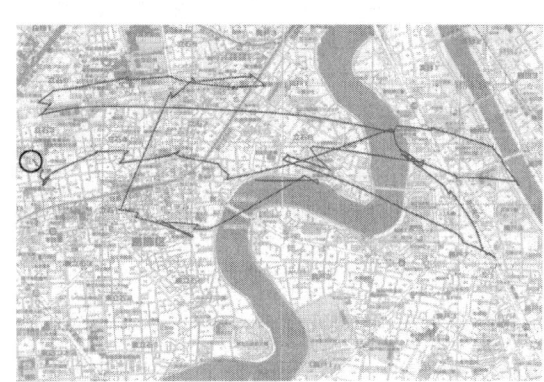

Fig. 5 Example of locus of eye movement for search single the visual target of the map. A: color map. B: monochrome map. The visual target is convenience store later encircled.

In figure 7A, subject searches the visual targets along the main road. Average total searching distance of the locus of eye movement of a monochrome map becomes about 3.4 times longer than that of a color map. Average search time of monochrome map becomes 2.4 times longer than that of color map. These values are small compare to those in color map. In monochrome map the value of search time of map with two targets is 0.65 times smaller than that with single target. In the case of two the visual targets, it is found that search time is unexpectedly short.

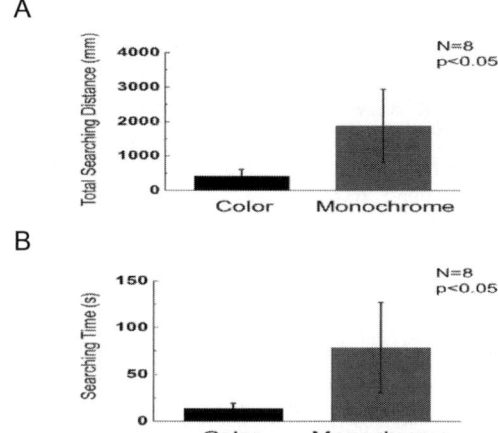

Fig. 6 Total searching distance (A) and search time (B) of the locus of the eye movement in single target on the map.

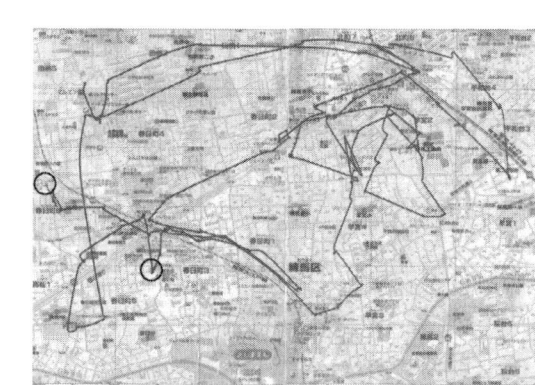

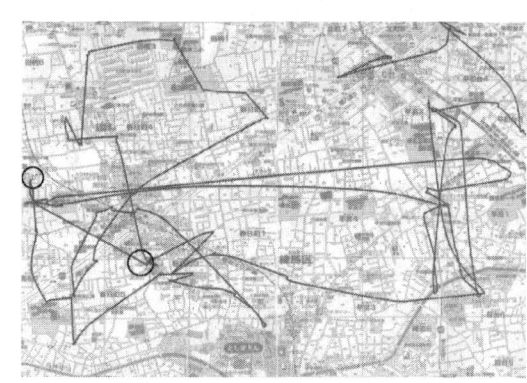

Fig. 7 Example of locus of eye movement for search two the visual targets of the map. A: color map. B. monochrome map. The visual target is convenience store later encircled.

IV. DISCUSSION

The purpose of this study was to evaluate visibility of the color and monochrome map quantitatively. We could

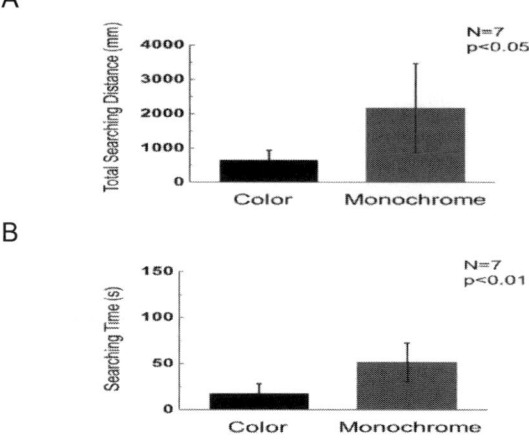

Fig. 8 Total searching distance (A) and serching time (B) of the locus of eye movement in two targets on the map.

Table 1 Total searching distance and search time

		Total search Distance (mm)	Search time (s)
Single target	Color	417±190	13.9±5.7
	Monochrome	1879±1069	78.6±48.4
Two targets	Color	637±302	17.1±10.8
	Monchrome	2161±1297	51.1±21.1

Numbers of experimental example are 8 in single the visual target and 7 in two the visual targets, and each value shows mean and standard deviation.

accomplish this purpose to measure total searching distance and total search time by pursuing the locus of eye movement with accuracy of 0.15mm/pixel. We found that both total searching distance and total search time in the color map are remarkably shorter than those in the monochrome map. This means that we can easily find out the targets in color map comparing to monochrome map about 3-4 times quick. This fact is also true for the case of pictures like complicated brain map.

In the case of monochrome map, it is hard to find out target on the map, subjects search every place where expected targets do not exist. Therefore it is considered that there is difference between monochrome and color map about total searching distance and search time.

In the case of single target, total searching distance and search time were accidentally found shorter and faster in monochrome map than in color map, it does not occur such case if we increase number of the targets. It is also found that in two targets total searching distance and search time longer than those in single target, but it does not take 2 times. Because finding possibility increases in two targets compare to single target.

In recent brain physiology, it is found how color visual information is processed in the cortex. From the V1 blobs, color information is sent to cells in the second visual area, V2. The cells in V2 that are most strongly color tuned are clustered in the "thin stripes" that, like the blobs in V1, stain for the enzyme cytochrome oxidase. Neurons in V2 then synapse onto cells in the extended V4. This area includes not only V4, but two other areas in the posterior inferior temporal cortex, anterior to area V3, the dorsal posterior inferior temporal cortex, and posterior TEO. Area V4 is exclusively dedicated to color, but this has since been shown not to be the case. Color processing in the extended V4 occurs in millimeter-sized color modules called globs. However why the visual target in color map be able to see clearly and to find out easily compare to monochrome map is still unknown.

V. CONCLUSION

In this study we investigated visibility of the color and monochrome road map by total searching distance and search time of the visual target in the locus of eye movement. We found that it is easily to find out the visual target on the color map compared to the monochrome map, In single target total searching distance and search time of monochrome map are 4.5 times and 5.7 times longer than those of color map respectively. In two targets total searching distance and search time of monochrome map are 3.4 times and 2.4 times longer than those of color map respectively.

REFERENCES

[1] A. L. Yarbzls (1967) Eye Movements and Vision. Plenum Publishing Corporation.
[2] K. Rayner (1998) Eye Movements in Reading and Information Processing: 20 Years of Research. Psychological Bulletin Vol. 124, No. 3, 372-422,
[3] M. F. Land (2006) Eye movements and the control of actions in everyday life. Progress in Retinal and Eye Research 25, 296–324,
[4] B. W. Tatler, M. M. Hayhoe, M. F. Land, and D. H. Ballard (2011) Eye guidance in natural vision: Reinterprreting salience. Journal of Vision 11(5), 5. 1-23
[5] J. B. Pelz*, R. L. Canosa, and D. Kucharczyk (2000) Portable Eye tracking: A Study of Natural Eye Movements Human vision and electronic imaging V.SPIE proceedings
[6] N.Ninomiya, N.Togawa, M.Yanagisawa, and T.Ohtsuki (2008) A deformed map generation algorithm for small displays based on cognitive science and its stochastic evaluation. Inst. of Elec. Inf. and Commun. Eng . J91-A 869-882

Author: Takakazu Kobayashi
Institute: Shibaura Institute of technology
Street: 3-7-5 Toyosu, Koto-ku
City: Tokyo
Country: Japan

3D Segmentation of MRI of the Liver Using Support Vector Machine

J.L. Moyano-Cuevas[1], A. Plaza[2], I. Dopido[2], J.B. Pagador[1],
J.A. Sanchez-Margallo[1], L.F. Sánchez[1], and F.M. Sánchez-Margallo[3]

[1] BTS Unit, Jesus Usón Minimally Invasive Surgery Centre, BTS, Cáceres, Spain
[2] Hipercomp, University Extremadura, Cáceres, Spain
[3] Scientific Director, Jesús Usón Minimally Invasive Surgery Centre, Cáceres, Spain

Abstract—In this paper, we propose a semi-supervised method for segmentation of the liver in three-dimensional (3D) magnetic resonance images (MRI), based on a Support Vector Machine (SVM) classifier. For segmentation, two classes have been considered: 'Liver' and 'Background'. Firstly, an anisotropic diffusion filter is applied to eliminate noise in the image and generate a multi-band image. Then a method based on an edge detector is used to select the training set. This method minimizes the user intervention during the training process of the SVM. Finally, the 3D volume of the image is segmented by the SVM classifier. The experiments on real MRI have shown that the proposed method allows segmenting the liver with high accuracy.

Keywords—MRI, Segmentation, 3D, Liver, Support Vector Machine.

I. INTRODUCTION

Magnetic resonance imaging (MRI) is crucial for diagnosis and treatment of liver diseases. Liver imaging has multiple applications such as measurements of liver volume [1]; diagnosis and quantification of tumours and other diseases [2]; surgical planning prior to hepatic resection or surgical navigation systems [3]. All these applications need to previously solve the problem of the segmentation of the liver. This is a complicated task because of the noise of MRI, the variations of grey level inside the liver, the morphological alterations and similarity of grey levels with other neighbouring structures. All these reasons lead to the point that many research groups are currently focused on the segmentation of the liver. Therefore, this paper proposes a method that allows segmenting the liver in MRI with minimal user intervention.

Many algorithms have been described in the literature for segmenting the liver in medical images. These algorithms can be classified into deformable models, clustering, shape models and classifiers. Deformable models in combination with other algorithms provide high accuracy in the segmentation of the liver in CT images [4,5], however it is difficult to apply them to magnetic resonance imaging (MRI) due to the variation of the grey levels. Therefore, the number of studies using these methods in MRI is low in this type of images. Other algorithms used for liver segmentation are the statistical shape models [6]. These methods provide a high accuracy on CT images but have the same problems as deformable models when applied to MRI. Besides, the use of complex shapes might increase the execution time of the method.

Other methods used for liver segmentation are clustering methods where the fuzzy c-means method (FCM) can be highlighted. This method has been applied to classify different types of anatomical. For the liver segmetation Farrahet et al. applied a clustering algorithm with hight accuracy [7], however this work requires a high user intervention.

Graph cut, probabilistic maps, region growing and K-nearest-neighbour, among many others, are reviewed by Heimann et al. [8] who gathered the results of 22 different segmentation methods applied to CT images. Even though, the application of these methods to MRI is less frequent.

Finally, other algorithms also used for medical imaging segmentation are the classifiers. Supervised classifiers group the set of methods that require the user intervention during the training period of the classifier. These supervised classification algorithms are sensitive to the initial conditions so in order to guarantee the correctness of the results, the training set must contain enough samples and the samples should be representative of the classes to be segmented. Despite these limitations, if the training set is suitable, these methods provide accurate results for different classification problems. The SVM method has been successfully applied in different disciplines. In MRI this classifier has been applied favourably in the brain, such as the work by Zhan et al. [9] and Ruan [10] who propose a method based on the SVM for tumours segmentation in the brain with high accuracy. For 3D volume segmentation of the liver, the SVM classifier has provided good results in CT images [11]. However, this method has not been used in MRI for segmentation the 3D volume of the liver.

In this context and due to the problems of segmenting the liver in MRI on one side and the advantages of SVM (high accuracy and less training requirements) against other classifiers, we propose a method based on SVM algorithm for 3D volume segmentation of the liver in MRI.

II. METHOD

The proposed method presents the segmentation of the liver volume slice by slice. Figure 1 shows the four steps that comprise the method. The first step corresponds to the pre-processing of the image by applying an anisotropic diffusion filtering for noise reduction and generation of the multi-spectral image. The second step includes a method to semi-automatically choose the training set based on an edge detector. After defining the training set and the features, the SVM training is performed. Afterwards, the classification of the input image is carried out. This methodology implies that the SVM has to be trained for each of the images to be classified.

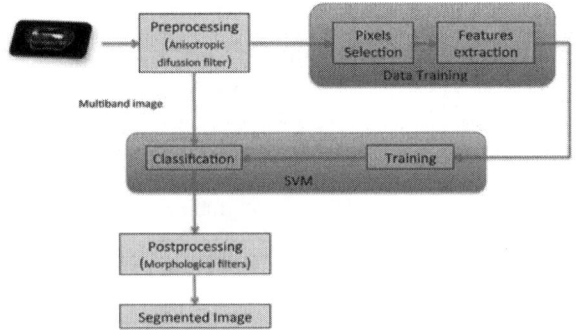

Fig 1 Overview of the proposed liver segmentation method

A. Pre-processing

An anisotropic diffusion filter was applied to the original image to remove noise and create a multiband image for classification. Therefore, the methodology proposed in this paper generates a multiband image formed by two images: the original image and an image obtained after applying an anisotropic diffusion filter described by Perona-Malik [12]. This filter is widely used in MRI enhancement because it removes image noise and smooths small image artefacts without removing significant parts of the image content. The anisotropic parameters used are a integration constant equal to 1/7 and gradient threshold of 25 and iteration number equal to 25. These values provided the best results for the MR images included in the study.

In summary, this step results in a noiseless image used to define the training set features and the creation of a multiband image for input in the classifier.

B. Pre-processing

Training the SVM requires 1) defining the features that better represent the different classes to be segmented, and 2) selecting the set of pixels of the image for which the value of these features will be extracted and that will be used for the SVM training.

1) Training features

In the proposed method, four features are used to train the SVM. These features characterize each pixel that will comprise the training set:
- Position of the pixel (x, y coordinates in the 2D image).
- Grey level of the pixel from the original image;
- Grey level of the pixel from the filtered image;

These features have been selected after carrying out several preliminary tests with different images. The position of the pixel in the image is the most representative feature and the one that better improve the results. However, choosing the position of the pixel as feature has also some drawbacks when segmenting the volume of the liver, due to the fact that the location of the liver within the image changes throughout the MRI sequence. To solve this problem the SVM is trained for each slice.

2) Selection of the training set

In this step, the proposed method semi-automatically selects a set of pixels of the image to train the SVM. Once these pixels are selected, the features previously described are extracted for each one of the pixels, resulting in a complete training set. Being a supervised classifier, the expert must manually select representative points of the regions to be segmented and tag them with the appropriate class label for each one of the MR images where the liver can be seen. The number of pixels to be manually selected for each of the classes to train is usually high to achieve proper results. This process is repeated for all the images used for training, which makes it a time-consuming and cumbersome task.

In order to minimize the user intervention during the SVM training process, a methodology based on an edge detection method has been used. In first place, a regular Canny edge detector was applied to the filtered image to obtain a binary image with all edges of image. It is in this image where the expert user has to identify all regions corresponding to the liver. For this purpose, a pixel must be selected anywhere within the area of the liver, considering as many regions as the liver is divided into. Once the user has identified all portions of the liver, the points of the training set are automatically selected and tagged with one of the two defined labels: 'Liver' if they are within any of the regions identified by the expert or 'Background' in any other case. 4% of the pixels of the images are included in the training set. This threshold has been empirically obtained. Therefore and with this method, the user only has to select a single point in each one of the regions corresponding to the liver the image, which is usually only one.

One problem that can occur applying this method is that not all edges of the liver are properly detected due to the similarities of the liver with other adjacent anatomical structures. This might cause that the selected points for training are mislabelled as 'Liver' class when the pixel actually corresponds to 'Background'. To minimize this problem, a dilation filter has been applied to the binary image so leakage around the edges of the liver is decreased. The radius of this filter is 1 pixel.

As a result of this step, a data training set is obtained and expressed as a matrix with so many columns as selected pixels and so many rows as defined features.

C. Support Vector Machine

The proposed methodology for 3D volume segmentation of the liver is based on the Support Vector Machine (SVM).

This supervised classifier was described by Cortes and Vapnik [13]. Given a set of n training pairs $\{(x_i, y_i)\}$, where $x_i \in R^n$ are the points of the image and $y_i \in \{-1,1\}$ are the corresponding labels and $i=1,\ldots,n$; the objective of the SVM is to find a hyperplane that maximizes the distance between both classes. Therefore, the problem of SVM is to optimize the following expresion

$$\frac{1}{2}\mathbf{w}^T\mathbf{w} + c\sum_{i=1}^{n}\xi_i$$

where ε is a non-negative variable introduced to denote error, c is a constant and $\mathbf{w}$ can be expressed as a linear combination of support vectors as:

$$\mathbf{w} = \sum_{i=1}^{n} y_i \alpha_i \varphi(\mathbf{x}_i)$$

Different kernel functions have been described in the literature. In this paper, a Radial Basis Function kernel function was used and it is defined as:

$$K(\mathbf{x}, \mathbf{x}_i) = e^{(-\gamma \|\mathbf{x} - \mathbf{x}_i\|^2)}$$

D. Post-processing

Finally two morphological filters are applied to refine the results provided by the classifier. In first place, a dilate filter is applied to reduce islands inside the liver. On the resulting image, an erode filter is applied to eliminate the pixels introduced by the dilate filter at the same time that islands outside the liver are reduced. The radius of the dilate and erode filters are equal to 1 and 2 pixels, respectively.

III. RESULTS

In our study, the liver of four pigs were scanned with a magnetic resonance of 1.5 Teslas (Intera Philips, Netherlands) using a T2-weithed sequence with a 2 mm thickness slice and gap equal to 0 mm. As results of these studies, four MRI volumes of the liver were obtained. The SVM training and classification is implemented based on the LIBSVM toolkit [14].

To quantitatively validate the method, test images with known ground truth have been used. An expert radiologist manually segmented the liver in all images of the four datasets included in the study. All manually segmented images were compared with the results provided by the supervised method proposed in this paper.

To validate the accuracy of the method, the following metrics were used: True positive rate (TPR), Dice similarity coefficient (DSC), Volumetric overlap error (VOE), Relative absolute volume difference (RVD) and Average symmetric surface distance (ASSD). This metrics were described in [19].

Table 1 shows the quantitative results of the proposed method for the four analysed volumes. Mean TPR is 94.30%, with similar values for all analysed studies, implying a low under-segmentation. However, values obtained for VOE and positive values of RVD denote that the method provides a higher over-segmentation, with an average value of 19.13% and 10.94%, respectively. The DICE coefficient is very close to 90% in all analysed studies but lower than the TPR value, due to the over-segmentation showed by the method. With respect to differences in surface, an average ASSD of 5.26 mm is obtained. Therefore there are no major differences between the contour traced by the expert and the obtained by method.

Table 1 Quantitative metrics for assessment of the method for each of the four analysed volumes.

Item	TPR%	DSC%	VOE%	RVD%	ASSD%
Volume A	95,91	88,50	20,63	16,76	4,16
Volume B	94,12	89,90	18,34	9,38	5,94
Volume C	92,99	89,64	18,78	7,48	5,72
Volume D	94,20	89,64	18,77	10,15	5,25

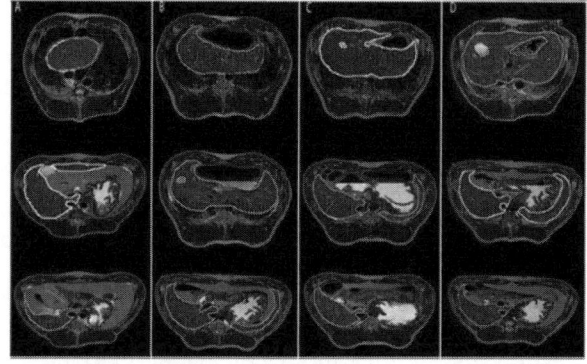

Fig. 2 Example of the results provided by the proposed method for the four analysed volumes analysed (A-D) in three different slices. The white contour represents the manual segmentation and the red contour represents the semi-automatic segmentation

Figure 2 shows the results of the segmentation of the liver for eight slices corresponding to two of the four analysed sequences. In these figures, the contour resulting after the segmentation has been overlaid to the original image in order to visually check the differences between the liver and the segmented region. The similarity of the liver with adjacent anatomical structures can be observed in these figures.

IV. CONCLUSION

Segmentation of magnetic resonance images (MRI) of the liver is complicated due to variations of grey levels, the high component of noise and the similarity of the liver to other adjacent organs and anatomical structures.

This paper presents a method based on the Support Vector Machine (SVM) classifier for segmentation of 3D volume of the liver from MRI. The proposed method integrates the use of 1) an anisotropic diffusion filter to eliminate noise and generate a multiband image; 2) a Canny edge detector to facilitate the selection of the points for training; and finally 3) a SVM classifier to segment the image. This method has been quantitatively validated with actual liver MRI images.

The results of this validation show that the proposed method identifies the liver 3D volume on MRI, reducing user intervention during the training stage of the classifier. The main drawback of the proposed method is the over-segmentation. The similarity between the anatomical structures contiguous to liver and the liver itself is the major cause of this problem. This feature makes the proposed edge detector to fail to detect some parts of the contour of the liver, therefore labelling as 'Liver' pixels belonging to the background. The dilate filter applied to the image minimizes this problem. However, and thanks to the other features include in the training, the SVM provides accurate segmentation results.

This work sets the groundwork for liver segmentation based on SVM. Due to this fact, the execution time for this first approach has been elevated as a consequence of the training phase of the SVM. Future work will be focused on optimize the running time using technics allows paralleling the training stage. In addition, new training features will be selected in order to increase the precision of the algorithm.

ACKNOWLEDGMENT

This work has been founded by Riteca project, Red de Investigación rtansfronteriza Extremadura, Centro y Alentejo, is co-founded by "Fondo Europeo de Desarrollo Regional" (FEDER) and "Programa Operativo de Cooperación Transfronteriza España-Portugal" (POCTEP) 2007-2013.

REFERENCES

1. Sánchez-Margallo FM, Moyano-Cuevas JL, Latorre R et al. (2011) Anatomical changes due to pneumoperitoneum analyzed by MRI: an experimental study in pigs. Surg Radiol Anat 33(5): 389-396.
2. Goryawala M, Guillen MR, Cabrerizo M, et al. (2012) A 3-D liver segmentation method with parallel computing for selective internal radiation therapy. IEEE Trans Inf Technol Biomed 16(1):62-69
3. Maeda T, Hong J, Konishi K et al. (2009) Tumor ablation therapy of liver cancers with an open magnetic resonance imaging-based navigation system. Surg Endosc 23(5):1048-1053.
4. Alomari RS, Kompalli S, Chaudhary V (2008) Segmentation of the liver from abdominal CT using markov random field model and GVF snakes. Proccedings International Conference on Complex, Intelligent and Software Intensive Systems, Barcelona, Sapin, 2008, pp 293–298
5. Li M, Lei Z (2008) Liver contour extraction using snake and initial boundary auto-generation. Proc International Conference on Computer Science and Software Engineering, Shangai, pp 2669-2672.
6. Heimann T, Meinzer HP (2009) Statistical shape models for 3D medical image segmentation: A review. Med Image Anal 13(4): 543-564.
7. Farraher SW, Jara H, Chang K, et al (2005) Radiology Liver and Spleen Volumetry with Quantitative MR Imaging and Dual-Space Clustering Segmentation. Radiology 237:322-328.
8. Heimann T, van Ginneken B, Styner MA et al (2009) Comparison and evaluation of methods for liver segmentation from CT datasets. IEEE Trans Med Imaging 28(8):1251-1265
9. Zhang N, Ruan S, Lebonvallet S, et al. (2011) Kernel feature selection to fuse multi-spectral MRI images for brain tumor segmentation. Comput Vis Image Underst 115(2): 256-269.
10. Ruan S, Lebonvallet S, Merabet A, Constans JM (2007) Tumor segmetation from a multispectral MRI Images by using SVM. Proc International Symposium on Biomedical Imaging: From Nano to Macro, Arlington, USA, pp. 1236-1239
11. Luo S, Hu Q, He X, et al. (2009) Automatic liver parenchyma segmentation from abdominal CT images using support vector machines. In: Proc International Conference on Complex Medical Engineering, Tempe, USA, pp. 1-5.
12. Perona P, Malik J (1990) Scale-Space and Edge Detection Using Anisotropic Diffusion. IEEE Trans Pattern Anal Mach Intell, 12(7):629-639
13. Cortes C, Vapnik V (1995) Support vector networks. Mach Learn 20:1-25.
14. Chang CC, Lin CJ (2011) LIBSVM: a library for support vector machines. ACM Transactions on Intelligent Systems and Technology 2(27):1-27

Author: Jose Luis Moyano
Institute: Centro de Cirugía de Mínima Invasión
Street: N-521 Km 41,8
City: Cáceres
Country: Sapin
Email: jmoyano@ccmijesususon.com

Interpolation Based Deformation Model for Minimally Invasive Beating Heart Surgery

A.I. Aviles[1] and A. Casals[1,2]

[1] Universitat Politècnica de Catalunya– BarcelonaTech, Jordi Girona 1-3 K2M Building, 08034 Barcelona, Spain
[2] Institute for Bioengineering of Catalonia, Robotics Lab, Baldiri Reixac 4-6, 08028 Barcelona, Spain

Abstract—Heart motion compensation is a key issue in medical robotics due to the benefits that minimally invasive beating heart surgery offers over traditional cardiac surgery. Although different proposals have been presented, nowadays, there is not yet a suitable solution working in real clinical environments due to the lack of robustness of existing methods. The process of heart motion estimation required to produce the compensation actions can be tackled as a process of three iterative steps. The first based on generating a deformation model from the processing of a video sequence of the beating heart. The selection of a deformation model is crucial in the sense that it has to offer both valuable information and good computational performance. These characteristics are required when the reaction time has a significant repercussion over the system behavior, as in this case. This paper, presents a computational analysis of deformation model based on interpolation methods. In particular, wavelet and thin-plate splines are evaluated. The significance of this study relies on the fact that it is a reference starting point of reference for creating both a common framework and a robust solution. In addition, the obtained results will contribute to increase the robustness from the initial stage of the solution.

Keywords—Deformation model, wavelets, computer performance, radial basis functions, interpolation methods.

I. INTRODUCTION

In last years, minimally invasive beating heart surgery has been a focus of attention in robotics due to the advantages that it offers over traditional cardiac surgery [1,2]: shorter rehabilitation, quicker recovery, reduction of both risk for neurological injury and blood transfusions or even improve cosmetics etc. Consequently, this surgery is a key factor for improving quality of human life because according to the World Health Organization (WHO) [3], cardiovascular diseases are the first cause of death throughout the world.

Regardless of the benefits of beating heart surgery, two disturbances affect this kind of procedure: heart motion and breathing. Thus, stabilization is a challenge because the surgeon has to deal with a constant moving target. Even though, small devices have been proposed as a manual solution for stabilizing the area of interest over the heart, a residual motion is still significant [4]. Therefore, a research goal is compensating the motion of the heart with the objective of offering the surgeon the sensation of being working in a static area.

Compensation implies the need of knowing the heart movement law or tracking its movement in real-time. Thus, the solution of this problem relies on computer vision. The first research in this line was proposed by Nakamura [5] in which the concept of heartbeat synchronization was proposed. But the major drawback in this approach was the use of artificial markers which are impractical due to the complexity of fixing them over the tissue. Later on, Ortmaier [6] was the first in proposing a solution based on natural markers; their research was a point of reference for realistic solutions. Although, different approaches have been proposed to solve the problem for example those based on [7,8,9]: texture, shape from shading or radial basis functions; the major difficulty is related to the lack of robustness. Thus, there is not a real solution capable of working in real conditions. In this regards, this research is oriented to overcome the limitations of all techniques presented up to now.

This paper is organized as follows. In Section II the description of the global problem related to heart motion compensation is presented. In Section III, the paper goes deep in the study of the deformation model presenting a mathematical basis of two different approaches, while Section IV compares their performance. Finally, in Section V a discussion of this research leads to some conclusions.

II. HEART MOTION COMPENSATION

The procedure to follow for heart motion compensation oriented to minimally invasive surgery can be seen as a minimization problem where three parts are involved in an iterative way: i) deformation model, ii) generation of the objective function and iii) the optimization technique (see Fig. 1). Mathematically, the problem involves two images, the former a fixed or reference image $F: \Omega_F \subset \mathbb{R}^d$, and the latter acquired image (i.e. from laparoscope) $A: \Omega_A \subset \mathbb{R}^d$, and the goal is to find the optimal solution for:

$$arg\ min_T\ D\big(F, T(A)\big) + P(T) \quad (1)$$

Where D denotes how aligned are the two images while P is related to the regularization term that makes reference to

Hadamard's concept. This problem becomes more complex due to various facts: small workspace, variable lighting, glossy area and liquids, deformable tissue and hardware limitation (i.e. camera integrated in the endoscope). Also, the necessity of obtaining a solution in real-time which is critical in this kind of procedures becomes an additional challenge.

Therefore, as a three-part decomposition problem, this paper focusses on the first step related to deformation model. The selection of the adequate technique to generate a deformation model has a significant repercussion over the global performance. It allows us to describe the nature of the changes on the tissue. For this reason, deformation theory has a significant role in different areas; mainly, because different elements in real world can be physically interpreted through their deformations. The first proposal for using deformation theory in the context of computer vision was presented in [10]. From then, the advances achieved in this area have been successfully applied in medicine.

According to the classification proposed in [11], and its extension presented in [12], deformation models are classified into three main categories: i) based on physical models, ii) based on interpolation and approximation theory, and iii) based on knowledge. In spite of the fact that the first category has been widely applied in the medical area, it presents disadvantages such as inaccurate deformation and complex computation [13]. On the other hand, despite models in the third category are able to deal with complex deformations their computation is still a problem. For these reasons, this analysis is based on the second category due to its ability to offer valuable information during the image transformation, requiring at the same time a reasonable computational time.

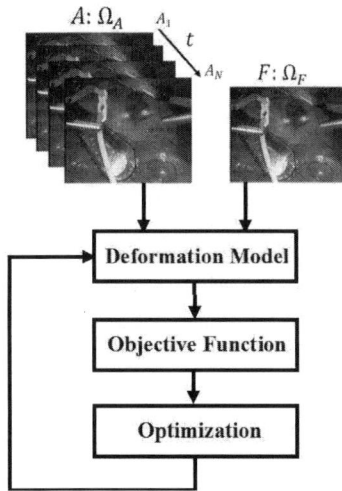

Fig. 1 Compensation of heart motion as a three steps loop iterative process

III. DEFORMATION MODEL

From the different subcategories of deformation models based on interpolation, two of them are analyzed: thin-plate spline (TPS - radial basis functions) and wavelets (basis functions from signal processing). They were selected because of the valuable information obtained during the transformation with the use of an acceptable number of parameters. In this section, the mathematical basis of the two categories is presented. Referring to (1), the transformation T applied to every pixel in 2-dimensional images is represented by $T(x,y) = (x,y) + u(x,y)$ (2) where $u(x,y)$ denotes the deformation field.

A. Deformation Model Based on Thin-Plate Splines

After Duchon presented TPS in [16], this approach has been extended to different domains. In particular, TPS has been deeply studied in the context of medical applications. Especially, the work presented by Bookstein in [17] was pioneered in this area. In it, TPS was proposed for non-rigid registration. Since then, TPS has become a common selection in problems where deformations need to be represented.

The idea behind this approach is that given a set of points (or control points) in both reference and acquired images a deformation function can be determined [18]. In other words, TPS allows mapping pixel information from A to F through minimization of the bending energy. Mathematically, it can be seen as finding the displacement field $u(x,y)$ that minimizes function C [19], represented by:

$$C = \iint_{R^2} (u_{xx}^2 + 2u_{xy}^2 + u_{yy}^2)dxdy \qquad (3)$$

A factor to bear in mind is the number of control points to be defined. This number has a significant repercussion over the time-consuming but a good representation depends on their density. Thus, it is important to find a balance that allows obtaining a good representation with a small number of points.

B. Deformation Model Based on Wavelets

The wavelet transform (WT) is a tool that cuts up data, functions or operators into different frequency components, and then studies each component with a resolution matched to its scale [14]. WT has had a fast expansion due to the advantages that can offer over traditional analysis (i.e. Fourier) and its properties in the context of computer vision. Compared with WT, Fourier analysis has two main disadvantages: it works well only with stationary signals and it is localized only in frequency domain. Even if Short-Time Fourier Transform (STFT) was presented as a solution to previous, it is difficult to define the size of the window that provides good resolution in both time and frequency. Therefore, WT has been presented as solution to overcome all drawbacks presented in both FT and STFT. That is, WT allows a time-frequency analysis with a better representation because of Multiresolution Analysis (MRA).

Discrete Wavelet Transform (DWT) is used because it allows avoiding redundant representation. In this regards, the representation of displacement field $u(x,y)$ to obtain the transformation in (2) is given by the addition of approximation (the first term in (4)) and details (second term in (4)). In general, it is represented by:

$$u(x,y) = \sum_k c_{jn}(k)\varphi_{jn,k}(x,y) + \sum_j \sum_k d(j,k)\psi_{j,k}(x,y) \quad (4)$$

Where space location and scale are represented by j and k. $\varphi_k(x,y)$ refers to scaling function and $\psi_{j,k}(x,y)$ denotes wavelet function while both c, d are the wavelet coefficients to be calculated. In general, the coefficients can be obtained by inner products but according to [15] the most efficient way to implement them is using filter banks which is an advantage for the implementation part. In Fig. 2 the process to calculate the coefficients at next level of resolution using filter banks can be seen. In there, L and H denote the low and high filters respectively.

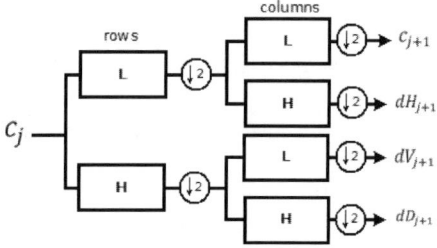

Fig. 2 2D-DWT implementation based on filter banks

IV. EXPERIMENTAL RESULTS

In this section, an evaluation of two types of deformation models is carried out: wavelet and thin-plate splines. In particular, within WT four families are evaluated: daubechies-5, coiflet-5, quadratic b-splines and biorthogonal-2.6. They were selected because their different properties allow us to conduct a complete analysis (i.e. semi-orthogonal, orthogonal, optimal time-frequency localization etc.). On the other hand, all experiments were assessed under the same conditions, using a PC with Intel Core i7-2630QM CPU 2.00GHz, 8.00GB RAM under windows 7 operating system. The deformation models have been evaluated on the data set of Hamlyn Centre Laparoscopic at Imperial College of London [20]. This sequence consists of 900 frames of 355*285 pixels size, corresponding to the cardiac surface affected by respiration and cardiac motion.

Fig. 3 shows a frame of the sequence with the obtained displacement fields. Fig. 3a corresponds to the TPS approach using eight control points, and Fig. 3b using WT with three levels of resolution (j). These values were found to provide a good quality of information with respect to the size of the frame. Finally, with the main objective of selecting the best deformation model that offers valuable information with a good time performance in Fig. 4 an evaluation of the amount of time spent in processing the two approaches is presented. Clearly, the global performance allows identifying the potential candidate models to use in motion compensation.

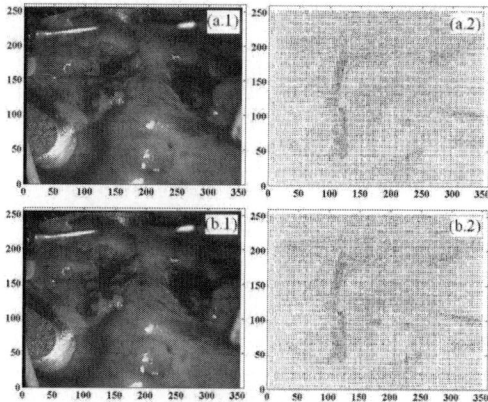

Fig. 3 The displacement field obtained in a frame a.1-b.1 using TPS can be seen in a.2 while in b.2 using WT.

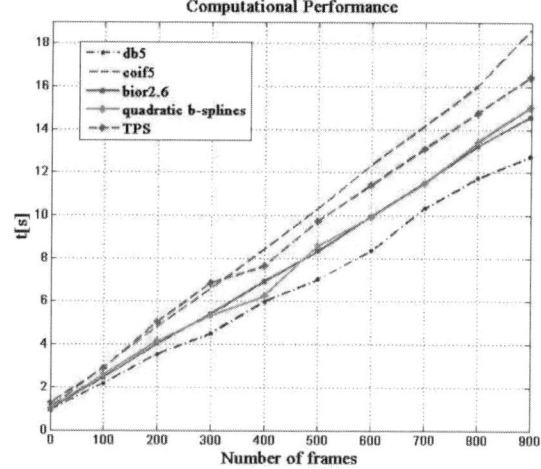

Fig. 4 Computational performance obtained from four wavelet families and TPS approaches

V. CONCLUSIONS

Heart motion compensation is considered a still open and challenging problem in medical robotics, in this sense, with the main goal of generating a realistic solution we analyze one of the most significant parts of our methodology, the deformation model. That is, we presented a study of deformation model based on interpolation methods. The importance of these results are focused in the fact that they are a point of reference to increase the robustness. In particular, two approaches have been evaluated: TPS and WT. According with the results, WT demonstrated a better computational performance in comparison with TPS. In addition, WT can be easily oriented to parallel computation to decrease even more the time-consuming, and gives the possibility of using only few coefficients for obtaining a good approximation. In contrast, TPS tends to use a dense set of control points for obtaining a good result. This is a problem in the sense that this kind of application is oriented to work on real-time with a large amount of information. Therefore, even if db5 has shown the best computational result, we have selected b-spline wavelet as our deformation model because it offers additional properties such as optimal time-frequency localization, extreme regularity etc. In conclusion, WT approach is a perfect match to this application due to it offers both valuable information and good computational time.

ACKNOWLEDGMENT

This work has been done in the frame of project DPI2011-29660-Co4-01 and DPI2011-29660-C04-03, under the Spanish Research Program, MINECO and with FEDER funds, EC. This research is supported by a FPU national scholarship from the Spanish Ministry of Education with reference FPU12/01943.

REFERENCES

1. Livesay J.J. (2003) The benefits of off-pump coronary bypass: A reality or an illusion? Texas Heart Institute Journal.
2. Karamanoukian H.L.(2002) Decreased incidence of postoperative stroke following on-pump coronary artery bypass. Journal of the American College of Cardiology.
3. World Health Organization at http://www.who.int/en/
4. Lemma A., Mangini A., Redaelli A. and Acocella F. (2005) Do cardiac stabilizers really stabilize? experimental quantitative analysis of mechanical stabilization Interactive CardioVascular and Thoracic Surgery.
5. Nakamura Y, Kishi K. and Kawakami H. (2001) Heartbeat Synchronization for Robotic Cardiac Surgery. International Conference on Robotics and Automation, Seoul, Korea, pp. 2014-2019
6. Ortmaier T., Groger M., Boehm D.H., Falk V. and Hirzinger G. (2005) Motion estimation in beating heart surgery. IEEE Trans. Biomed. Eng. 1729–1740
7. Noce A., Triboulet J. and Poignet P. (2007) Efficient tracking of the heart using textures. Proceedings of the 29th Annual International Conference of the IEEE EMBS
8. Lo B., Chung A.J., Stoyanov D., Mylonas G. and Yang G.Z.(2008) Real-time intra-operative 3D tissue deformation recovery. In: Proceedings of IEEE International Symposium on Biomedical Imaging, 1387–1390.
9. Richa R., Poignet P. and Liu C. (2009) Three-dimensional motion tracking for beating heart surgery using a thin-plate spline deformable model. The International Journal of Robotics Research
10. Terzopoulos D. and Fleischer K. (1988). Deformable models. The Visual Computer, 306–331
11. Holden M., (2008) A review of geometric transformations for nonrigid body registration. Medical Imaging, IEEE Transactions on 27, 111-128
12. Sotiras A., Davatazikosy C. and Paragios N. (2012) Deformable Medical Image Registration: A Survey. Research Report Num. 7919, INRIA, France
13. Tseng D.C. and Lin J.Y. (2000) A Hybrid physical deformation modeling for laparoscopic surgery simulation, In proceedings of the 22 annual EMBS international conference, Chicago IL.
14. Daubechics I. (1992) Ten Lectures on Wavelets. SIAM, Philadelphia, PA.
15. Mallat S.G. (1989) A Theory for Multiresolution Signal Decomposition: The Wavelet Representation. IEEE Transactions on Pattern Analysis and Machine Intelligence Vol. II. No. 7
16. Duchon J. (1976) Splines minimizing rotation invariant semi-norms in Sobolev spaces. In: Constructive Theory of Functions of Several Variables, pp. 85-100.
17. Bookstein F.L. (1989) Principal Warps: Thin-Plate Splines and the Decomposition of Deformations. IEEE Transaction on Pattern Analysis and Machine Intelligence.
18. Bookstein F. L. and Green W. D. K. (1993) A feature space for edges in image with landmarks, Journal of Mathematical Imaging and Vision, 3:231-261, 1993
19. Johnson H.J. and Christensen G.E. (2001) Landmark and intensity-based, consistent thin-plate spline image registration, Springer-Verlang Berlin Heidelberg, pp. 329-343
20. Stoyanov D., Mylonas G., Deligianni F., Darzi A. and Yang G.Z. (2005) Soft-tissue Motion Tracking and Structure Estimation for Robotic Assisted MIS Procedures. Medical Image Computing and Computer Assisted Interventions, vol. 2, pp. 139-146

The address of the corresponding author:

Author: ANGELICA IVONE AVILES RIVERO
Institute: UNIVERSITAT POLITÈCNICA DE CATALUNYA
Street: C/ JORDI GIRONA 1-3, K2M BUILDING
City: BARCELONA
Country: SPAIN
Email: angelica.ivone.aviles@upc.edu

A Framework for Automatic Detection of Lumen-Endothelium Border in Intracoronary OCT Image Sequences

G. Cheimariotis[1], V. Koutkias[1], I. Chouvarda[1], K. Toutouzas[2], Y.S. Chatzizisis[3], A. Giannopoulos[3], M. Riga[2], A. Antoniadis[3], C. Doulaverakis[4], I. Tsampoulatidis[4], I. Kompatsiaris[4], C. Stefanadis[2], G. Giannoglou[3], and N. Maglaveras[1]

[1] Lab of Medical Informatics, Medical School, Aristotle University, Thessaloniki, Greece
[2] 1st Department of Cardiology, Athens Medical School, Hippokration Hospital, Athens, Greece
[3] Cardiovascular Engineering and Atherosclerosis Laboratory, 1st Cardiology Department, AHEPA University Hospital, Medical School, Aristotle University, Thessaloniki, Greece
[4] Information Technologies Institute, Center for Research and Technology-Hellas, Thessaloniki, Greece

Abstract—Intracoronary optical coherence tomography (OCT) is increasingly being used for real-time visualization of coronary arteries aiming to help in the identification of high-risk atherosclerotic plaques associated with geometrical and morphological features of the arterial wall. This paper presents a framework towards the automatic detection of the inner wall of the coronary artery (lumen-endothelium border) in intracoronary OCT image sequences by employing a multi-step image processing method. The major focus of this work was to address difficult cases that are frequently met in intracoronary OCT, e.g. images with small/big branches, multiple branches, blood presence, calcifications, artifacts, etc. We present each step employed and the results obtained both in qualitative and quantitative terms. The proposed segmentation algorithm has been proven very efficient in the majority of the examined cases.

Keywords—Intracoronary Optical Coherence Tomography (OCT), Lumen-Endothelium Border, Image Processing, Automatic Segmentation, Evaluation.

I. INTRODUCTION

Intracoronary optical coherence tomography (OCT) is an evolving diagnostic tool that allows for the visualization of the coronary arterial wall in real time and may significantly help in identifying new markers of atherosclerosis risk associated with the geometry and the morphology of the arterial wall [1]. In order to identify locations in the artery with stenosis or regions at high risk for adverse events, the internal contour of the artery wall has to be assessed. The latter gives the area of a cross section of the lumen, but it is also used for three-dimensional (3D) reconstruction of the arterial lumen geometry [2, 3]. Manual segmentation of the contour is laborious, time consuming and dependent on the subjective evaluation of an expert. On the other hand, automatic segmentation of intracoronary OCT images is very challenging, as the images have a wide variety of difficulties (e.g., artifacts, blood presence, calcifications, etc.) that have to be handled [4].

In this paper, we present our work towards the automatic detection of the inner wall contour of the coronary vessel (lumen-endothelium border) through a multi-step segmentation procedure by considering also images with the difficulties mentioned before, which are often excluded from such studies [5], and have been recognized as a major research challenge in other works as well [6, 7]. The core idea of the approach is that the task of identifying the internal contour of the wall is easier to accomplish by distinguishing in the image the relatively homogeneous and bright structure of the artery wall, with accuracy as far as the internal contour is concerned. Thus, the proposed method is based on the selection (via an iterative algorithm) and the application of a threshold on the image that makes it binary, in order to distinguish the bright areas of interest. Morphological functions retain only those areas which could be part of the wall. The inner contour in a radial scan (the so-called A-line) is located at the point where the scan (starting from the catheter) meets a bright shape (i.e., a transition from black to white). In case of multiple transitions, the transition that corresponds to the luminal-endothelium border is determined through the application of heuristics.

Once the initial assessment of the internal contour is performed, side branches of the vessel and calcifications are identified with statistical control of the slope of the curve which represents the internal contour, and then appropriate corrections are applied. A heuristic method is used for the same purpose, in order to handle the special case of large branches. The assessment of the algorithm in different image sequences has been performed to illustrate the effectiveness of the proposed method.

The paper is structured as follows. In section II we refer to the image sequences that have been analyzed, and then present the proposed segmentation algorithm and the |metrics employed for its evaluation. In section III we present both qualitative (sample segmented images) and quantitative (based on the defined metrics) results of our work. In section IV we discuss the limitations of our work, future work directions, as well as the conclusions of the current study.

II. METHODS

A. Data of the Study

In the scope of this work, we elaborated on a corpus of two image sets, one for development purposes, i.e. to define the heuristics, and the other for testing. The development set is completely independent from the test set. In particular, the development set comprised of 4 subsets:

- 21 images, resolution 1024x1024,
- 21 images, resolution 512x512,
- 100 images, resolution 512x512, and
- 165 images, resolution 512x512,

while the test set consisted of 109 images with 256x256 resolution.

In all cases, image acquisition was performed through a Frequency Domain OCT modality (FD-OCT C7-XRT OCT Intravascular Imaging System, Westford, MA). The pullback speed was 20 mm/sec, while the axial resolution was 15 μm and the maximum frame rate was 100 frames/sec. For our analysis, pullback segments were employed that were considered of clinical interest. In addition, the elaborated sequences included artifacts due to various reasons, e.g. recording errors, as well as other parts like presence of blood, calcium, etc., which all introduce high challenges for a correct and spatially meaningful segmentation. Figure 1 depicts an example source image.

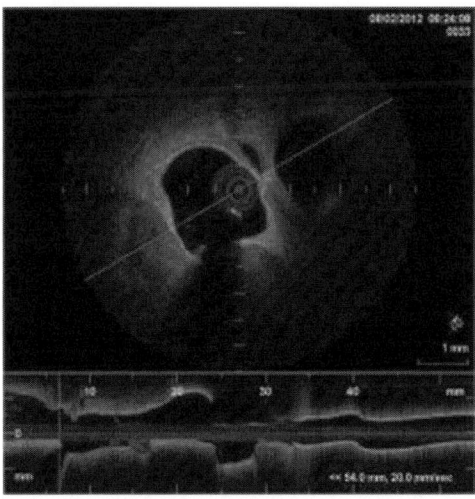

Fig. 1 Example input image.

B. Segmentation Framework

The elaborated segmentation framework includes the following consecutive steps (S1-S4):

S1. *Preprocessing*: This involves various substeps like grayscale image transformation, median filtering, smoothing, removal of calibration markers, image transformation into polar coordinates (Figure 2) [8], as well as removal of the catheter artifact.

S2. *Main processing and extraction of the candidate contour points*: The image is first transformed in binary format by applying a local filtering scheme. Then, morphological filters are applied (opening/closing), resulting in the extraction of the candidate contour points (Figure 2(a)). Next, further controls are introduced to take into account specific image characteristics that affect the segmentation outcome.

S3. *Corrections of Contour Gaps and Handling Branches*: Side branches of the vessel and calcifications are identified with statistical control of internal contour gradient taking into account its expected continuity, and appropriate corrections are applied (Figure 2(b)-(c)). A heuristic method is used for the same purpose in the special case of large branches.

S4. *Final outcome*: There are various cases in which the resulting contour does not have the smoothness that the vessel is expected to have. In this step contour smoothing is performed by applying a series of low-pass Haar filters [9].

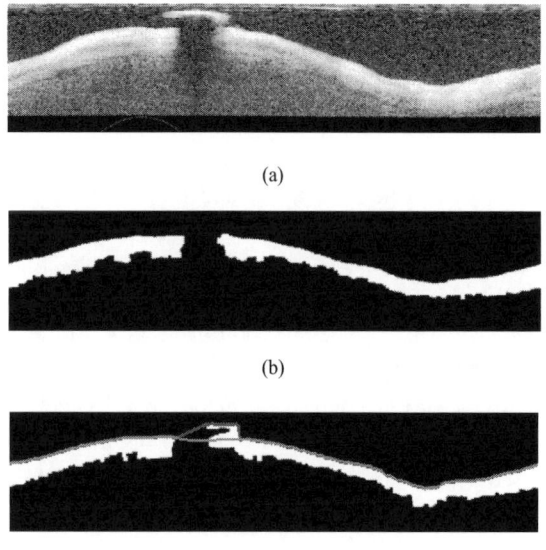

Fig. 2 (a) Polar coordinates representation of an OCT image.
(b) Outcome of the morphological filtering and noise reduction.
(c) Two candidates for contour correction; the adopted solution is depicted in green as it is producing a smoother outcome.

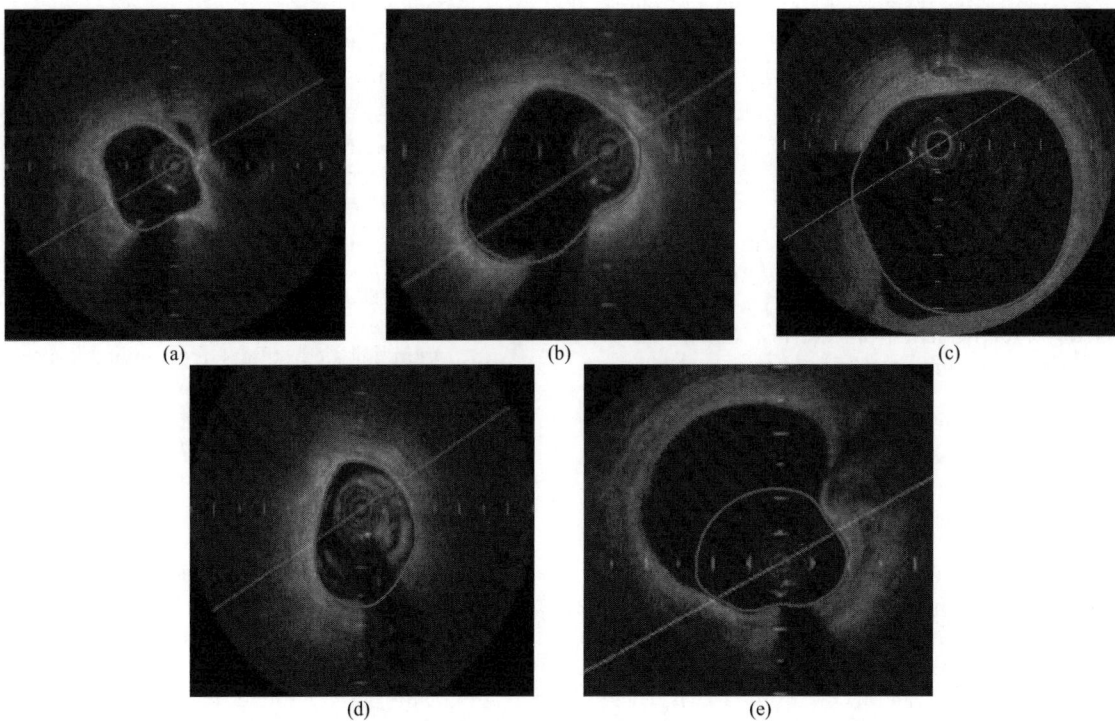

Fig. 3 Indicative segmentation results: (a) result for the image depicted in Fig. 2, (b) detection in an image with one branch, (c) image with two branches, (d) image with significant noise, and (e) unsuccessful segmentation due to a big branch.

C. Evaluation Metrics

The evaluation of the algorithm was based on the comparison with the gold standard, i.e. the contours manually defined by clinical experts. Besides qualitative assessment (i.e. comparative view of the two contours), a quantitative assessment was performed by calculating and comparing characteristic shape metrics applied on a frame by frame basis. These metrics are:

- *Area* (mm^2) defined by the detected closed contour,
- *Perimeter* (mm) of the detected contour, and
- *Centroid* (pixels), i.e. the coordinates (x,y) of the center of mass defined by the detected contour.

These metrics are indicative for quantifying such an assessment and they were considered adequate at this stage of our study.

III. RESULTS

A. Qualitative Assessment

The image sequences described in section II.A have been segmented using the proposed algorithm. The obtained results illustrate satisfactory outcomes in most of the cases, as presented in Figure 3. The Figure depicts various typical cases, both successful ones, i.e. an image with one small branch, a case with two branches, an image with significant noise, and unsuccessful, i.e. an image with a big branch.

Table 1 Statistical differences aiming to assess the error between the manually defined and the automatically obtained contours (KS-test: Kolmogorov–Smirnov test, NS: non-significant difference, Mean relative error = mean(difference)/ground_truth).

	Area (mm^2)	Perimeter (mm)	Centroid x (pixel)	Centroid y (pixel)
Mean	0.1754	-0.4384	-1.9712	-1.4679
Standard deviation	0.6069	0.5807	1.8228	2.7500
Skewness	3.8031	2.5361	0.3230	-3.4421
Mean relative error	0.0173	-0.0489	-0.0152	-0.0178
25-50-75 quartiles	-0.0083 / 0.1533 / 0.3246	-0.5971 / -0.4475 / -0.2786	-2.6472 / -2.1200 / -1.4167	-1.7624 / -0.8213 / -0.3765
KS-test	NS p=0.7284	p<0.001	NS p=0.1279	NS p=0.3088

B. Quantitative Assessment

Table 1 depicts basic statistics that were calculated by comparing the manually defined (gold standard) with the automatically obtained contours for the test set, according to the parameters presented in section II.C. As it is shown, the results are in good accordance between the reference and the proposed method. In particular, small differences are shown in the area and the centroid (less than 2 pixels), while a significant difference is shown in the perimeter (still less than 5% on average with mean relative error 0.0489). These features illustrate the potential of the proposed method, although further and more in depth evaluation has to be conducted.

IV. DISCUSSION

In this paper we presented a framework for detecting the inner wall contour of the coronary artery (lumen-endothelium border) in intracoronary OCT image sequences. Our major focus was on addressing difficult cases that are rather frequent in images obtained from the OCT modality, i.e. images with small/big branches, multiple branches, blood presence, calcifications, artifacts etc. In cases where such difficulties are not present, the proposed algorithm has proven very efficient.

In addition, in various cases where some special features are present in the images, the algorithm behaves satisfactorily. Its major shortcoming originates from the heuristics that have been employed in some cases, which may result in contradictory outcomes. Thus, it is a major future work direction to elaborate on the generalization of these heuristics to the extent possible. This will enable us to develop a segmentation framework that is robust to the morphological variations among images. Employing machine learning techniques and expert knowledge to determine the values of the algorithm's parameters, as well as employing contour information in the segmentation between consecutive frames could contribute in the above aim.

The systematic evaluation of the segmentation algorithm, through the analysis of a wider corpus of images and the comparison of the outcomes with those obtained via manual segmentation performed by experts [10], is another future work direction that will quantify the potential of the proposed approach and its adoption in clinical practice as a support tool. In this respect, our next steps include the development of a graphical user interface and integration of the segmentation outcomes with a CAD tool that will enable 3D visualization of the arteries.

REFERENCES

1. Brezinski ME (2011) Current capabilities and challenges for Optical Coherence Tomography as a high-impact cardiovascular imaging modality. Circulation 123(25):2913–2915
2. Bezerra HG, Costa MA, Guagliumi G, Rollins AM, Simon DI (2009) Intracoronary Optical Coherence Tomography: A comprehensive review: Clinical and Research Applications. J Am Coll Cardiol Intv 2(11):1035–1046
3. Jang IK, Bouma BE, Kang DH et al. (2002) Visualization of coronary atherosclerotic plaques in patients using optical coherence tomography: comparison with intravascular ultrasound. J Am Coll Cardiol 39(4):604–609
4. Prati F, Regar E, Mintz GS et al. (2010) Expert review document on methodology, terminology, and clinical applications of optical coherence tomography: physical principles, methodology of image acquisition, and clinical application for assessment of coronary arteries and atherosclerosis. Eur Heart J 31(4):401–415
5. Tanimoto S, Rodriguez-Granillo G, Barlis P et al. (2008) A novel approach for quantitative analysis of intracoronary optical coherence tomography: high inter-observer agreement with computer-assisted contour detection. Catheter Cardiovasc Interv 72(2):228–235
6. Ughi GJ, Adriaenssens T, Onsea K, Kayaert P, Dubois C, Sinnaeve P, Coosemans M, Desmet W, D'hooge J (2011) Automatic segmentation of in-vivo intra-coronary optical coherence tomography images to assess stent strut apposition and coverage. Int J Cardiovasc Imaging 28(2):229–241
7. Tsantis S, Kagadis GC, Katsanos K, Karnabatidis D, Bourantas G, Nikiforidis GC (2012) Automatic vessel lumen segmentation and stent strut detection in intravascular optical coherence tomography. Med Phys 39(1):503–513
8. Nixon MS, Aguado AS (2008) Feature Extraction and Image Processing, 2nd Edition, Academic Press
9. Papadogiorgaki M, Mezaris V, Chatzizisis YS, Giannoglou DG, Kompatsiaris I (2008) Image analysis techniques for automated IVUS contour detection. Ultrasound Med Biol 34(9):1482–1498
10. Giannoglou GD, Chatzizisis YS, Koutkias V, Kompatsiaris I, Papadogiorgaki M, Mezaris V, Parissi E, Diamantopoulos P, Strintzis MG, Maglaveras N, Parcharidis GE, Louridas GE (2007) A novel active contour model for fully automated segmentation of Intravascular Ultrasound images: In-vivo validation in human coronary arteries. Comput Biol Med 37(09): 1292–1302

Author: Nicos Maglaveras
Institute: Aristotle University, Lab of Medical Informatics
City: Thessaloniki
Country: Greece
Email: nicmag@med.auth.gr

Learning Optimal Matched Filters for Retinal Vessel Segmentation with ADA-Boost

E. Poletti and E. Grisan

Department of Information Engineering, University of Padova, Padova, Italy

Abstract—Retinopathy of Prematurity (ROP) is an eye disease that affects premature infants. Its signs are tortuosity and dilation of retinal vessels, which are subjectively evaluated by clinicians for the diagnosis and the follow-up of the disease. The availability of algorithms for vascular segmentation would allow vessel geometrical characterization, and hence the quantitative and objective clinical evaluation of the these signs. Unfortunately, algorithms designed for adults' fundus images do not work well in infants' fundus images, due to their very low quality. At variance with available methods, we propose a data-driven approach, in which the system learns an array of optimal discriminative convolution kernels, to be employed in a ADA-boost supervised classification. The array is employed as a rotating bank of matched filters, whose response is used by the boosted linear classifier to provide a classification of each image pixel into the two classes of interest (vessel/background). In order to test the generality of the approach, we assessed the performance of the proposed method both on adults' fundus images using the DRIVE dataset, and also on infants' images by cross-validation on a dataset of 20 images acquired with a RetCam fundus camera. Average accuracy and Matthews' correlation coefficient are respectively 0.94 and 0.69 for DRIVE and 0.98 and 0.66 for the Retcam dataset with respect to the manual ground truth references.

Keywords—Retinopathy of Prematurity, Vessel Segmentation, Retinal Fundus, ADA Boost, Supervised Classification, Filter learning.

I. INTRODUCTION

Retinopathy of Prematurity (ROP) is a serious eye disease that affects premature infants [1]. Shortage in the availability of ophthalmologists able to provide ROP diagnostic examinations is a substantial barrier to ensuring worldwide appropriate ROP care. One solution would be to perform a wide screening with retinal photography by automatizing the image assessment. Wide-field retinal cameras (130° of field of view, as the RetCam by Clarity Medical Systems, CA, USA) are commercially available devices whose clinical value have been evaluated in several studies [2]: they allow inspecting the most peripheral area of the eye, where vessels grow during the last weeks of gestation. However, their effectiveness is limited by the unavailability of effective automated analysis tools and of quantitative clinical grading procedures.

The main drawbacks of RetCam images with respect to images provided by the standard fundus cameras used for adult patients, are: *i)* presence of interlacing artifacts, as images are actually single frames extracted from video; *ii)* narrow blood vessels, due to the wide-field of view coupled with the 640x480 pixel resolution; *iii)* non uniform illumination in the captured wide field of view; *iv)* high visibility of choroidal vessels, related to the lack of pigmentation of the infant choroid.

All these aspects make the automatic analysis of RetCam images quite challenging: the best-performer algorithms designed for adults' fundus images do not work well in infants' fundus images. A custom techniques is therefore necessary to successfully trace the vasculature. We had previously tackled this problem with a supervised classification method that used features extracted from both texture analysis and *vesselness* measures [3]. Although it shows good performance, its usability is limited by the high computational power it needs (more than 10 minutes for feature extraction and ~5 minutes for prediction per image).

A way to reduce the analysis computational time is to use an unsupervised method, e.g., a multiple-oriented matched filter approach like in [4], where the Gaussian kernel of the filter is meant to represent the average profile of a vessel. Unfortunately this kind of approaches showed performance largely inferior to supervised approaches [5].

In this work we present a data-driven approach in which the features to be used with an ADA-boost classifier are learned as optimal matched filters from the training set. A direct benefit of such an approach is that it do not need any feature extraction step. We demonstrated its effectiveness by testing it on both adults' and infants' image fundus datasets.

II. MATERIAL

Twenty RetCam images, with 130° field of view and 640×480 pixels size, provided by Clarity Medical Systems, CA, USA, were acquired in premature infants with different severity of ROP. A manual segmentation of the vessel network was provided by an author (EP) and used to evaluate the performance of the algorithm proposed for vessel extraction.

In order to provide also a comparison between the effectiveness of our learning scheme (learnt optimal matched used with ADA-boost) and the standard Gaussian matched filters [4], we tested our algorithms also on the public datasets DRIVE [5].

III. METHODS

Following the approach proposed in [6] we aim at estimating a set of discriminative convolution kernel (or linear weak learner) allowing a linear classification of each image pixel into the two classes of interest through ADA-boost.

In subsection III.A the design of the optimal linear weak learner (LWL) is described. Their boosting and the building of the strong classifier are described in subsection III.B. Training and testing setup is described in subsection III.C.

A. Optimal Linear Weak Learner (LWL)

We consider a binary classification problem, for which we have a set of N training data $(x_1,y_1), \ldots, (x_N,y_N)$ where x_i is an image described as a $\mathbb{R}^W$ dimensional vector and y_i is a label that can take values in $\{-1,1\}$.

Being $p(x,y)$ the unknown joint probability distribution density of (x,y), and being $H:\mathbb{R}^W \to \{-1,1\}$ a classifier function, the optimal classifier H is the one that minimizes the expected risk:

$$Err(H) = \sum_{y=\{-1,1\}} \int p(x,y)\chi(H(x) \neq y)dx \quad (1)$$

with $\chi(\cdot)$ the indicator function. The unknown probability density $p(x,y)$ can be written as a Parzen window estimate, using a Gaussian kernel $g_\Sigma(\cdot)$ with zero mean and covariance Σ, resulting in the smoothed expected risk:

$$Err(H) = \frac{1}{N}\sum_{i=1}^{N} \int g_\Sigma(x-x_i)y_i H(x)dx \quad (2)$$

A classifier H can be designed by additively combining several so-called *weak classifiers*, each of them doing at least better than the chance. The idea is that through a proper linear combination of these weak classifiers it is possible to boost their performance [7].

We then define M weak classifer $h_m:\mathbb{R}^W \to \{-1,1\}$ and the resulting strong classifier as

$$H_M = \sum_{m=1}^{M} \alpha_m g_\Sigma * h_m(x)) \quad (3)$$

with the parameters $\alpha_1, \ldots, \alpha_m \in \mathbb{R}$.

The weak classifiers h_m can be automatically and efficiently estimated, as shown in [6], if they can be computed as a scalar product:

$$h_m = sign(\gamma_0 + \gamma_1 \cdot x) \quad (4)$$

$h_m(x) = sign(\gamma_0 + \gamma_1 \cdot x)$ with γ_0 and $\gamma_1 \in \mathbb{R}\gamma_1 \in \mathbb{R}^W$. In this case the empirical error to be minimized becomes:

$$Err_{pw}(h_m) = \frac{1}{2} - \frac{1}{2N}\sum_{i=1}^{N} y_i erf\left(\frac{\gamma_0 + \gamma_1 \cdot x}{\sqrt{2\gamma_1^T \Sigma \gamma_1}}\right) \quad (5)$$

and the parameters $\hat{\gamma}_0$ and $\hat{\gamma}_1$ yielding the minimum $Err_{pw}(H)$ can be obtained *via* gradient descent optimization.

B. Boosting LWL

Given N training samples and an initial value $D_1(i)=1/N$, for each weak learner h_m, with $m\in\{1, \ldots, M\}$, we consider:

$$\varepsilon_m = P_{i \sim D_m}[h_m(x_i) \neq y_i] \quad (6)$$

$$\alpha_m = \frac{1}{2}\log\frac{1-\varepsilon_m}{\varepsilon_m} \quad (7)$$

$$D_{m+1}(i) = D_m(i)\frac{e^{-\alpha_m y_i h_m(x_i)}}{z_m} \quad (8)$$

where z_m is a normalization factor, $D_m(i)$ is the relative weight of each training sample x_i and ε_m is the error weighted by D_m of the weak classifier h_m.

Fig. 1 From top-left to bottom-right, the first 49 linear weak learners/matched filters learned from the DRIVE training set, showed at their horizontal orientation.

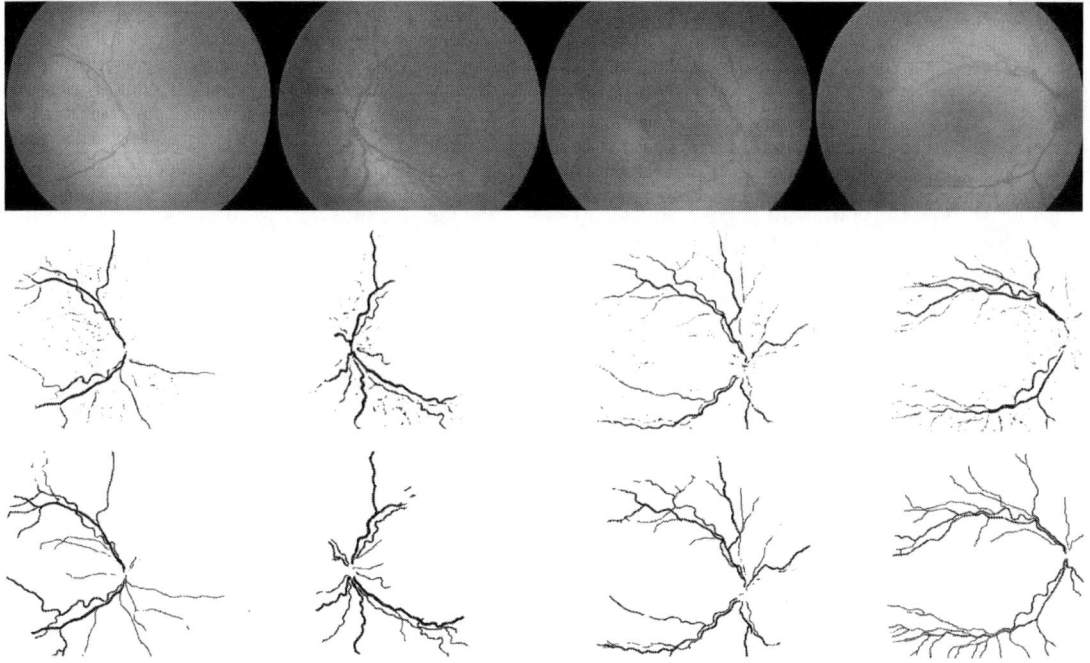

Fig. 1 In the first line four original Retcam image are shown as example.
In the second line and in the third line are respectively the results of the automatic segmentation and the ground truth references.

The boosting procedure is then:

1. Start with weights $D_1(i)$
2. Repeat for all $m \in \{1, ..., M\}$
   a) Estimate the optimal weak learner parameter $\hat{\gamma}_0$ and $\hat{\gamma}_1$ minimizing $Err_{pw}(h_m)$ (Eq 2)
   b) Compute ε_m
   c) Compute α_m
   d) Update the weights $D_{m+1}(i)$
3. Evaluate the strong classifier $H_M(x)$ (Eq. 3)

C. Training and Testing Setups

We estimated $M=50$ *weak learners* (Fig. 1) by randomly chosing 3000 pixels belonging to the class 1 (vessels) and 3000 relative to the class -1 (background) from each image of the training set. For each selected pixel, we defined the x_i (see sec. III.A) as the square window neighborhood of 15x15 pixels centered on the selected pixel. In order to reduce the intra-class variability of x_i, each 15x15 window was preventively rotated so as to have its main *direction* horizontally oriented. The *direction* of a window is identified by computing the eigen-decomposition of its Hessian (as in [8]), and by choosing the orientation of the eigenvector associated to the main eigenvalue.

As for the testing is concerned, we employed the estimated weak learners as they were an array of rotating matched filter templates [4]. We convolved each pixel of the testing image for R different rotated versions of the M weak learners (at the orientations $r\pi/R$ radians, with $r \in \{0, ..., R-1\}$), obtaining then R different $H_{M,r}(x)$. The estimated class for each pixel i is then simply:

$$H_M(x_i) = \sum_{r=0}^{R-1} H_{M,r}(x_i) \qquad (9)$$

IV. RESULTS

In order to measure the effectiveness of the obtained optimal matched filters with respect to the standard Gaussian-kernel ones [4], we trained/tested our method on the DRIVE dataset [5]. Training accuracy was 0.98, while testing accuracy reached the value of 0.94, against the 0.88 obtained in [4] with matched filters, and the best results of 0.96 obtained in [9].

As regard the Retcam dataset, we performed a 2-fold cross validation, each fold composed by 10 images. We computed the average accuracy, sensitivity, specificity, positive predicted value, negative predicted value, and Matthews' correlation coefficient, with their respective standard deviations (Table 1).

Table 1 Performance indexes computed for the Retcam images.

	Accuracy	Sensitivity	Specificity	Positive predicted value	Negative predicted value	Matthews' correlation coefficient
Formulation	(TP+TN)/(P+N)	TP/P	TN/N	TP/$\dot{P}$	TN/$\dot{N}$	(TP·TN)+(FP·FN)/$\sqrt{P \cdot N \cdot \dot{P} \cdot \dot{N}}$
Supervised classification based on textures analysis and vesselness measures [3].	0.97 (0.00)	**0.60** (0.04)	0.99 (0.00)	0.77 (0.10)	0.98 (0.01)	**0.66** (0.04)
Proposed method training set: Retcam testing set: Retcam	**0.98** (0.00)	0.51 (0.06)	0.99 (0.00)	0.76 (0.04)	0.98 (0.01)	0.62 (0.04)
Proposed method training set: DRIVE testing set: Retcam	**0.98** (0.00)	0.57 (0.07)	0.99 (0.00)	**0.78** (0.04)	**0.99** (0.00)	**0.66** (0.05)

P, N, $\dot{P}$ and $\dot{N}$ stand respectively for positives, negatives, estimated positives, and estimated negatives, while TP, TN, FP and FN stand respectively for true positive, true negative, false positive, and false negative.

In order to show the robustness and the generalization capability of the proposed approach, we provided in Table 1 also its performance in testing on the Retcam dataset (whole 20 images) when trained on the DRIVE's 20 training images. The time required for the classification of one image (from both DRIVE and Retcam) is ~15 seconds.

V. DISCUSSION AND CONCLUSION

The improvement in accuracy of 0.06 (0.94 *vs* 0.88) obtained with our method in DRIVE shows its superior capability in learning discriminating patterns of vessels compared to the standard Gaussian formulation of the matched filters.

The performance in the Retcam dataset is very robust despite the high variability of the 20 images (e.g., see Fig. 2), which show different contrast, illumination, and choroidal transparency. If compared to the ones obtained by [3] on the same dataset (Table 1, first row), our method shows a better accuracy and specificity, at the expense of the sensitivity. Moreover, the computational time required by the proposed method is about 60 times lower than the one proposed in [3] (~15 seconds *vs* ~15 minutes).

It is worth mentioning that the performance on Retcam dataset is better when the learning process is accomplished on the DRIVE dataset. This fact may sound odd, as the two datasets are composed by images acquired in two totally different environments. Nevertheless it can be explained by these two observation: first, Retcam images have poorer quality and higher variability than DRIVE images; second, infants' images have a high visibility of choroidal vessels, which are considered background although having an aspect similar to retinal vessels. These two factors together make the learning of a model of the vessel pattern more fruitful in the DRIVE dataset, where the vessel class have less intra-class variance. In our opinion, the cross-dataset discriminative power showed by the proposed method highlights its capability in providing an array of filters that are a good generalization of the vessel model.

REFERENCES

1. Committee for the Classification of Retinopathy of Prematurity, The International Classification of Retinopathy of Prematurity revisited. Arch Ophthalmol, 2005.
2. B. Lorenz et al., "Wide-field digital imaging based telemedicine for screening for acute retinopathy of prematurity (ROP). Six-year results of a multicenter field study", Graefes Arch Clin Ophthalmol. (2009); 247(9):1251-1262.
3. Poletti E, Ruggeri A. Segmentation of vessels through supervised classification in wide-field retina images of infants with Retinopathy of Prematurity, 25th International Symposium on CBMS, IEEE, 2012
4. S. Chaudhuri eta al., "Detection of Blood Vessels in Retinal Images Using Two-Dimensional Matched Filters". IEEE Transactions on Medical Imaging, 1989.
5. J.J. Staal, M.D. Abramoff, M. Niemeijer, M.A. Viergever, B. van Ginneken, "Ridge based vessel segmentation in color images of the retina", IEEE Transactions on Medical Imaging, 2004.
6. A. Vedaldi, P. Favaro, and E. Grisan Boosting Invariance and Efficiency in Supervised Learning, , in Proceedings of the International Conference on Computer Vision (ICCV), 2007
7. J. H. Friedman, T. Hastie, and R. Tibshirani. Special invited paper. additive logistic regression: A statistical view of boosting. The Annals of Statistics, 28(2):337–374, 2000.
8. A. Frangi et al., "Multiscale vessel enhancement filtering". In Proc. 1st MICCAI, 1998.
9. W. S. Oliveira, I. R. Tsang, G. D. C. Cavalcanti, "Retinal Vessel Segmentation Using Average of Synthetic Exact Filters and Hessian Matrix", IEEE International Conference on Image Processing, 2012.

Author: Enrico Grisan
Institute: Dept of Information Engineering, University of Padova
Street: via G.Gradenigo 6/B
City: Padova
Country: Italy
Email: enrico.grisan@unipd.it

Using Optical Flow in Motion Analysis for Evaluation of Active Music Therapy

M. Suzuki[1], S. Kataoka[2], and E. Shimokawa[3]

[1] School of Information Environment, Tokyo Denki University, JSMBE, Chiba, Japan
[2] HALCA Laboratory, JSMBE, Tokyo, Japan
[3] Rehabilitation Section, Saitama Medical Center for Children with Disabled, Saitama, Japan

Abstract—In this study, we focused on the disabled children to introduce their voluntary motion. Motion analysis of their active music therapy is important for evaluation and planning, but the quantitative method is not well established yet. Since the therapists are too busy to spend time for analysis, more automatic tool is needed than the conventional motion analysis software. Therefore we investigated the use of optical flow in analyzing the client's motion from the recorded video of music therapy. For this purpose, Lucas-Kanade method was used to calculate flow vectors. This algorithm needs feature points to be detected before flow calculation. Such features are limited to the ROI, that is, a part of the child's body. Therefore mask image is introduced to specify the area for feature detection. OpenCV was used to develop experimental software and 12 video images of 5 subjects were processed. From the experimental results, the use of mask image to configure ROI was found to be effective. Though the correlation between results of optical flow and that of the conventional motion analysis was low in some cases, the timing of the detected motions was agreeable enough. These results may be effective for quantitative evaluation and intuitive visualization of active music therapy.

Keywords—Motion Analysis, Optical Flow, Music Therapy.

I. INTRODUCTION

Nowadays music therapy is widely performed, for people with various needs. As a kind of health care, this music therapy is required to be evident based. Since music therapy puts emphasis on psychological interaction between the client and the therapist, its evaluation tends to be qualitative and providing quantitative evidence is not established yet. For this purpose, generally some rating scales are used [1]. Other measurement methods are also investigated, for example, multi-channel audio signals [2], video recorded facial action [3] and motion analysis [4].

In this study, our target is the active music therapy for the children with some difficulties to introduce their voluntary motion, among the many kinds of music therapy. During the session, the client child is being promoted to play some musical instruments to the therapist's talking, singing and piano playing. The major concern in this therapy is the client's motion, so that motion analysis of the recorded video is chosen as the evaluation method of the preformed session. For quantizing how much the voluntary motion was introduced, we can draw the graph of the amplitude of the motion to grasp the trend of the whole session at a glance. It helps the therapists evaluate and plan the music therapy program.

The problem about applying motion analysis is that the situation during the therapy is not well arranged for the automatic image processing. The conventional motion analysis software uses "mean shift" algorithm [5] that needs a clear target to track easily. But the client sometimes dislikes attaching markers for tracking to his/her body. Even if the markers can be attached, occlusion may often happen because of the diversity of the client's motion. In that case, the tracking his/her body must be done by manual operation. This is quite time consuming since the recorded time period is rather long. The therapists are too busy to do this, and want to know the result as soon as possible.

Therefore, by focusing on detecting the timing of the client's motion rather than its amplitude, we considered that more automatic processing could be realized. For this purpose, optical flow was investigated in our study.

II. USING OPTICAL FLOW

Basically, optical flow algorithm detects many moving points in the image sequence, and calculate flow vector at these points automatically. Since the number of detected points is pretty larger than that of the target in "mean shift" algorithm, optical flow algorithm can be robust against occlusion. Even if some points are occluded, rest of points are still enough to calculate flow vector. Therefore we decided to use the optical flow algorithm.

This optical flow algorithm is classified into two types. One calculates "dense" flow over the whole image area, and another handles "sparse" flow at specified points. Since our concern is limited area of the video image, that is, ROI (Region Of Interest), latter "sparse" type algorithm is considered to be appropriate for tracking a part of the client's body. Therefore we introduce Lucas-Kanade method [6] which is the most typical algorithm of this type.

Prior to calculating flow vectors, Lucas-Kanade method requires that feature points for tracking are detected. Good feature points should have edge information both horizontally and vertically. Shi's algorithm [7] is based on the gradient of the image and appropriate for detecting such feature points.

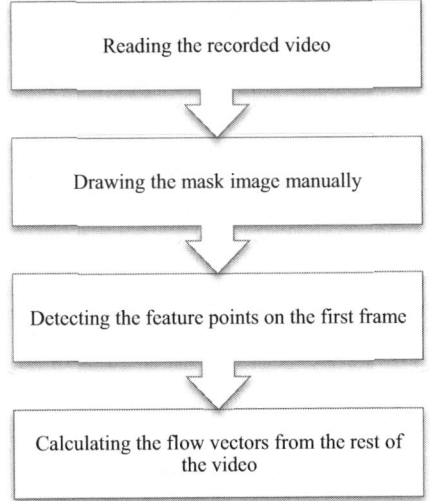

Fig. 1 Diagram of Processing of Optical Flow

Because the ROI in the recorded video image, for example, the client's upper limb, is just a part of the field of view, we have to limit the area for detecting the feature points. Generally ROI is configured by geometric form like a rectangle or circle, but the ROI in this study is a part of the client's body as described before. So the form of ROI is too free to be fitted with such a simple shape. For this reason, we introduced the "mask" image for defining the area to detect the feature points. At this moment, mask image is drawn by manually because the automatic method to determine the ROI area is not established yet. Fig. 1 shows the schematic diagram of these processing.

III. DEVELOPED SOFTWARE

To apply optical flow to evaluation of active music therapy, the experimental software was developed using OpenCV [8] and Microsoft Visual C++ 2008. It has the following functions for processing the recorded video image of the therapy. Fig. 2 shows the example images of these functions.

a) Reading video file and defining time section to be processed

Usually therapists are interested in some part of the recorded long video, so start and end frame of the time section can be specified.

b) Drawing mask image by mouse operation

Mask image is created by manual operation in this software. The user can draw or erase a filled circle by clicking the mouse button on the displayed source image (Fig. 2a). Its radius is variable from 5 pixel to 30 pixel. Example mask image is shown in Fig. 2b).

a) Source image

b) Mask image (blue area)

c) Detected feature points (green dots)

d) A scene while calculating flow vectors showing tracked points (cyan circles) and mean point (red dot)

Fig. 2 Example images of the process

c) Selecting the conversion method of grayscale or saturation for making single channel image

Detecting feature points and calculating optical flow must be done on single channel image, though the source video is 3 channels (RGB) color image. Though grayscale image is used as the single channel image generally, we considered that also saturation channel in HSV color space could be effective in some cases. Therefore source image is converted to either grayscale image or saturation image before detecting feature points. This can be switched by the user.

d) Detecting feature points, calculating optical flow and saving the result in a text file

These functions were implemented simply using the modules of OpenCV. Fig. 2c) shows an example image of detecting feature points and Fig. 2d) shows one of the scenes while calculating optical flow.

All feature points are averaged in each frame. In saving the results, the x-y coordinates of the mean point are stored as the measured results of the specified time section. This mean represents the approximate position of the ROI.

IV. METHODS AND RESULTS

To investigate the feasibility of the developed software, recorded video images of 5 child subjects were used. 2 of them were Rett syndrome, another 2 were cerebral palsy and one was developmental disease.

A motion analysis software (Move-Tr/2D ver. 7; Library Co., Ltd., Japan) using mean shift algorithm was used to compare with the developed software. The subject's wrist was taken as ROI. The relative distance from the start position was measured in both methods, and the correlation factor was calculated.

12 video images were processed in this study. In two cases, saturation was selected to detect feature points, because of better results than grayscale. Resulted correlation factors of 12 videos were 0.455 to 0.978, as summarized in Table 1.

Fig. 3 is the one example that shows good correlation between the Lucas-Kanade algorithm and the mean shift method. The worst case is shown in Fig. 4. Ideally, the

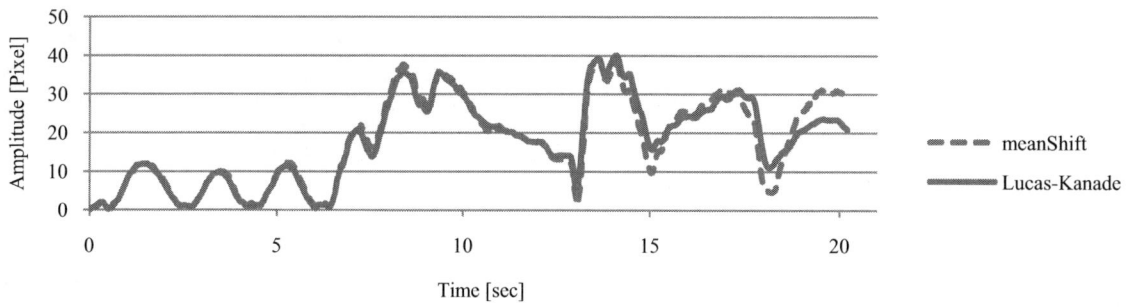

Fig. 3 Result of the subject A case 2 (r = 0.971)

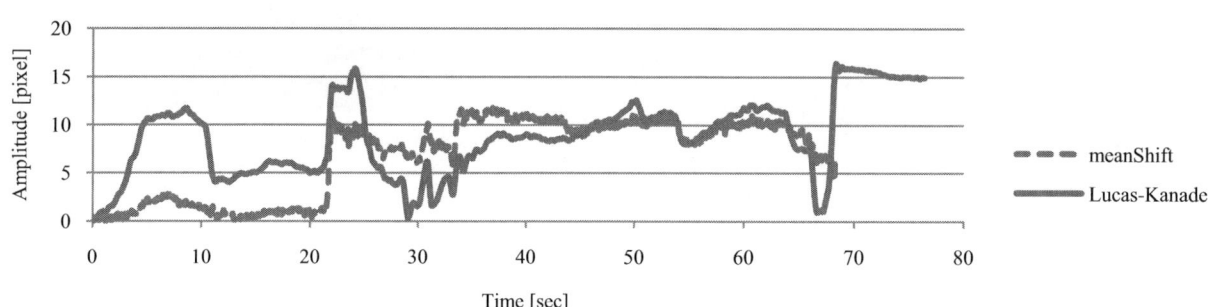

Fig. 4 Result of the subject B case 2 (r = 0.455)

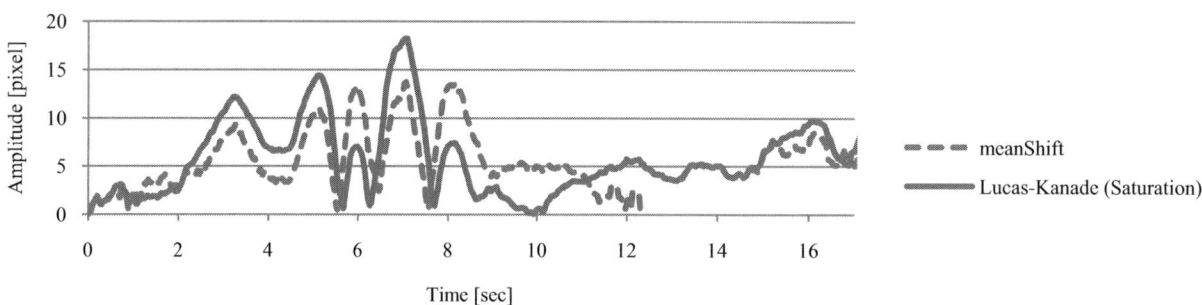

Fig. 5 Result of the subject C case 1 (r = 0.658, Occlusion occurred from 12.3 to 14.7 sec)

feature points must be detected uniformly over the specified ROI area. In the case that an error occurred, it is considered that the detection of feature points was deviated. Though, the timing of the most peak was same in both methods even if the correlation factor was low.

Fig. 5 is the example case of occlusion. The Lucas-Kanade method could measure the approximate result continuously, while the conventional motion analysis failed to trace the wrist of this subject.

Table 1 Correlation of results between Lucas-Kanade and Mean shift

Subject	Case#	Duration [Sec]	Grayscale / Saturation	Detected feature points	Correlation factor
A (Rett syndrome)	1	67.5	G	5	0.875
A	2	20.2	G	10	0.971
B (Rett syndrome)	1	76.4	G	8	0.534
B	2	76.4	G	5	0.455
C (cerebral palsy)	1	20.2	Saturation	4	0.658
C	2	20.2	Saturation	14	0.963
C	3	16.7	G	9	0.520
C	4	16.7	G	7	0.975
C	5	36.8	G	9	0.965
D (cerebral palsy)	1	58.7	G	16	0.781
D	2	36.7	G	12	0.964
E (developmental disease)	1	12.9	G	9	0.978

V. DISCUSSION

The experimental results suggest that the optical flow algorithm with mask image can analyze the timing of the motion rather than the amplitude. This is still beneficial in case that an approximate but intuitive visualization is needed. Therefore our developed software may be used for screening of long time recorded videos.

As a spin-off, such a result was clarified to be beneficial for the client's family, because the quantitative material can make them feel relieved and reduce their stress.

The mask image is important to get good results. If a mask image is large, more feature points will be detected, but unrelated points to the ROI also will be included in detected results. This may cause larger error for tracking results. On the other hand, a small mask image can detect only the points associated with ROI, but detected points may be insufficient for tracking. In this case, occlusion may happen even in optical flow and some of the tracking results may be missing. Therefore, it is preferable to propose a candidate of the appropriate mask image. Further investigation on relation between mask shape and measured result is needed.

VI. CONCLUSIONS

In this study, we investigated the use of optical flow in analyzing the client's motion from the recorded video of active music therapy. By introducing mask image to detection of feature points, the ROI of the video image was successfully specified as a free shape. Lucas-Kanade method was found to be applicable to detect the timing of the client's motion.

Since our final goal is providing music therapists a useful tool to evaluate the therapy, we are planning to release it as open source software and to collect opinions about usability from more users in the future.

ACKNOWLEDGMENT

The authors thank the subjects and their family for admission of using the recorded video during the music therapy.

REFERENCES

1. Pacchetti C, Mancini F, Aglieri R et al (2000) Active Music Therapy in Parkinson's Disease: An Integrative Method for Motor and Emotional Rehabilitation. Psychosom Med 62:386–393
2. Streeter E (2010) Computer Aided Music Therapy Evaluation: Investigating and Testing The Music Therapy Logbook Prototype 1 System. Ph.D. thesis, Univ. of York
3. Ragneskog H, Asplund K, Kiglgren M, Norberg A (2001) Individualized music played for agitated patients with dementia: Analysis of video-recorded sessions. Intl J Nurs Prac 7:146-155
4. Go T, Mitani A (2009) A qualitative motion analysis study of voluntary hand movement induced by music in patients with Rett syndrome. Neuropsychiatr Dis Treat 5:499-503
5. Comaniciu D, Meer P (2002) Mean shift: A robust approach toward feature space analysis. IEEE Trans. Patt. Anal. Machine Intell 24:603-619
6. Lucas BD, Kanade T (1981) An iterative image registration technique with an application to stereo vision. Proc. Of the 1981 DARPA Imaging Understanding Workshop 121-130
7. Shi J, Tomasi C (1994) Good features to track. Proc CVPR'94 593-600
8. OpenCV at http://opencv.org/

Author: Makoto Suzuki
Institute: School of Information Environment, Tokyo Denki University
Street: 2-1200 Muzai-gakuendai
City: Inzai-shi, Chiba
Country: Japan
Email: msuzuki@mail.dendai.ac.jp

Application of Gaussian Mixture Models with Expectation Maximization in Bacterial Colonies Image Segmentation for Automated Counting and Identification

I. Silva Maretić and I. Lacković

University of Zagreb, Faculty of Electrical Engineering and Computing, Zagreb, Croatia

Abstract—This paper presents an approach to identification of homogenous bacterial colonies grown on agar in Petri dishes. The aim of the work was to recognize regions in digital images of Petri dishes were homogenous bacterial colonies were developed as well as to estimate their size. Isolation of bacterial cultures' region from the dish was achieved by image segmentation based on histogram analysis. The histogram was parameterized using Gaussian Mixture Model with Expectation Maximization. This algorithm gave a good estimation of the actual gray level distribution and was able to separate merging distribution of two different objects. However, it performed poorly with the presence of outliers. The algorithm performance was also dependent on initial model and the number of Gaussians chosen. Overall, for images taken under controlled conditions the application of Gaussian Mixture model with Expectation Maximization proved to be successful and efficient approach to image segmentation of bacterial colonies. Final step was separation of circular-like colonies from non-circular ones. This was achieved using appropriate shape identification techniques. Validation of the proposed segmentation process was demonstrated using images from the microbiology laboratory, some artificially generated images and images downloaded from the Internet.

Keywords—Image segmentation, Object recognition, Colony counting, Gaussian Mixture Model, Expectation-maximization algorithm

I. INTRODUCTION

Determination of the number of colonies (colony forming units, CFU) that grow over a certain period on agar in Petri dishes is a standard method in microbiological analysis. Current manual procedures to analyze vast amounts of microbiological data are time consuming, prone to large variability and are of great cost. Reducing the amount of human intervention in the data analysis is crucial in order to cope with the increasing volume of data and to achieve more objective and quantitatively accurate measurements as well as to obtain repeatable results.

The process of preparing, storing and identification of bacterial samples grown on agar in Petri dishes involves placing the sample, collected from various sources, onto a sterilized Petri dish that contains a growth medium (agar plus nutrients) and streaking the sample across the surface of the agar. The dish is then covered with a lid to avoid contamination from outside environment and placed inside an incubator where it remains stored for some period of time under controlled temperature and other conditions. At some point during the incubation, the number of bacteria will be such that visible distinct individual colonies will be formed in that area [1].

Often laboratory staff needs to open the incubator and check stored dishes for presence of homogeneous bacterial colonies, which can then be sent for analysis (if the purpose of the culture is to identify bacteria type contained in the sample) or identify possible sample contamination (as to determine if the culture has only single type of bacteria - called pure culture). During the process of preparing the bacterial sample and checking for signs of homogeneous colonies, even with proper care taken to ensure minimum exposure, there is a possibility of sample contamination from external organisms which is undesirable (and sometimes can render the whole culture useless).

Automatizing this process not only removes the task of manually checking every individual sample, but also can reduce the probability of sample contamination by placing the entire process in an isolated sterile environment. A computerized incubator could prepare the samples, store and automatically check for results without having to remove or expose samples from within its chamber.

For this to be possible it is necessary to develop an algorithm to identify regions of homogeneous bacterial colonies, mark their location and estimate their size. Once this information is known we can further inspect each colony as to extract desired information (color, shape, texture, etc.) or sample the colonies' location and prepare it to be sent for further analysis.

Although various semi-automated and automated counting processes based on different algorithms and methods have been developed and also tested and a few systems are commercially available, not one of these has managed to establish itself as the standard method [2-6]. A central problem in many studies, and often regarded as the cornerstone of image analysis, is image segmentation.

In this work we present image segmentation approach based on histogram analysis where image histogram is modeled as Gaussian mixture and the parameters of the Gaussian mixture model are estimated with Expectation Maximization (EM) algorithm.

II. IMAGE PROCESSING AND ANALYSIS

The process of extracting homogenous bacterial colonies from digital images of Petri dishes can be divided in to four steps:
- Pre-processing;
- Segmentation;
- Labeling and rating clusters;
- Characteristics extraction.

A. Pre-processing

a) Image Conversion from RGB to Grayscale

Since color information is not relevant for the segmentation algorithm we applied the default Matlab mapping of the R (red), G (green) and B (blue) components of the color image to obtain a grayscale image.

b) Filtering:

Before the segmentation process it is desirable to apply a filter to the grayscale image as to reduce the amount of noise present, if any. Since the size of homogenous colonies in the image is considerably smaller then the size of the entire image (average size 700×800 pixels) we restricted the kernel size of a filter to 3×3 as to avoid losing information on these areas. A choice between three types of filters was implemented: simple symmetric moving average filter, Gaussian low-pass filter and median filter.

c) Image Histogram:

As the last pre-processing step, image histogram was extracted in order to observe the distribution of gray level intensities present in the image.

B. Segmentation

After the image has been pre-processed and its histogram extracted the next step is to divide the image into regions according to a pre-defied criteria.

In this work segmentation was achieved using gray level thresholding based on histogram analysis. Thresholding is the process of dividing an image into regions where each region is formed by a group of pixels whose gray level intensity falls within region's bounds.

For the task of identifying regions of homogeneous colonies present in the image the assumptions that the gray level distribution of pixels corresponding to the dish can be approximated to a Gaussian distribution with mean μ_P and variance σ_P was made. This assumption expects that gray level intensities of pixels belonging to the surface of the dish are confined to the dish area. This allows grouping pixels belonging to the dish regardless of gray level intensities of the background or bacterial culture.

a) Gaussian Mixture Model with Expectation Maximization

A Gaussian Mixture Model (GMM) is a parametric probability density function represented as a weighted sum of Gaussian component densities. GMM parameters are estimated from training data using the iterative Expectation-Maximization (EM) algorithm or Maximum A Posteriori (MAP) estimation from a well-trained prior model [7-10]. Expectation Maximization (EM) algorithm is the most popular technique used to determine the parameters of a mixture with an *a priori* given number of components.

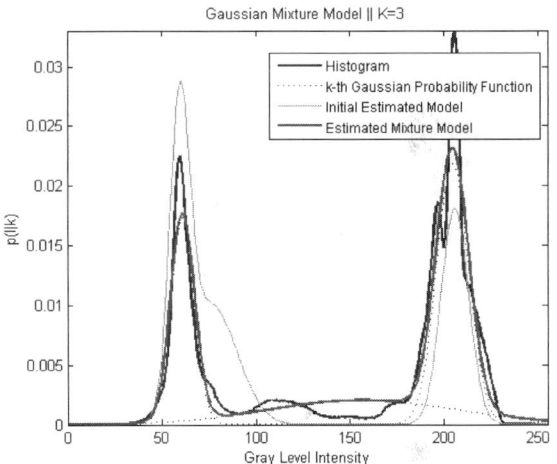

Fig. 1 Estimated Gaussian Mixture Model for a given distribution with 3 Gaussian probability functions.

The method consists of two steps which are repeated interchangeably and, at each step *i*, calculates the parameters of the Gaussian functions such that the likelihood of the model approaches a local maximum.

The first step is called the E-step. This is part of an expectation step where the model of distribution of data X is assumed to be known and, using this model, calculates the probability of the given model to have generated the actual distribution X.

$$q^{(i)}(n|k) = \frac{p_k^{(i)} \cdot f(x_n; \mu_k^{(i)}, \sigma_k^{(i)})}{\sum_K p_k^{(i)} \cdot f(x_n; \mu_k^{(i)}, \sigma_k^{(i)})} \quad (1)$$

The next step is a maximization step, called M-step. In this step the values of parameters of the assumed model, i.e. $\theta=\{p_k,\mu_k,\sigma_k\}$, are updated in a way that increases the likelihood of the model to have generated the actual data distribution.

$$\mu_k^{(i+1)} = \frac{\sum_N q^{(i)}(n|k) \cdot x_n}{\sum_N q^{(i)}(n|k)} \quad (2)$$

$$\sigma_k^{(i+1)} = \sqrt{\frac{\sum_N q^{(i)}(n|k) \cdot (x_n - \mu_k^{(i+1)})^2}{\sum_N q^{(i)}(n|k)}} \quad (3)$$

$$p_k^{(i+1)} = \frac{1}{N} \sum_N q_k^{(i)}(n|k) \quad (4)$$

Repeating E-step and M-step the likelihood of the model (defined by θ) approaches a local maximum such that the joint distribution of Gaussians, given by p_k, μ_k and σ_k, has most likely generated the actual distribution of data X. For practical application, however, is sufficient to stop the algorithm once a small step between the previous value of $\theta^{(i)}$ and $\theta^{(i+1)}$ is reached.

In order to extract information within the dish area, first the full mask of the dish needs to be reconstructed. This was achieved by image closing and image fill. After obtaining the reconstructed shape of the dish a logical AND operation was performed between the reconstructed shape and the inverse of the partial binary mask constructed from segmenting the dishes' region (Fig. 3).

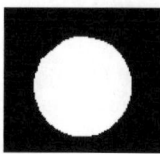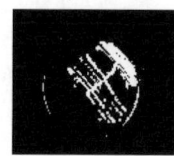

Fig. 3 Steps to obtain the content of the Petridish. I. Partial binary mask; II. Reconstructed mask; III. Dish content mask

C. Labeling and Cluster Rating

After a binary mask corresponding to pixels belonging to bacterial culture within the area of a dish is obtained, cluster of pixels having an approximated circular shape needs to be identified. Those clusters have the highest probability to represent a homogeneous bacterial colony present in the sample.

To achieve this first it is necessary to group pixels which are connected to each other by a neighborhood relation. This process is called image labeling.

After identifying each individual group of pixels (or clusters) an estimate of its shape, based on characteristics extracted from its geometrical form, can then be calculated. This process is called rating.

Labeling consists on determining which pixels are connected to each other by a defined neighborhood relation (4-neighborhood or 8-neighborhood). Pixels that are connected to each other form a cluster; each cluster is then assigned a label; each label represents an object in the image.

The process of labeling is done by a two pass algorithm called connected-component labeling. This is a simple top-down algorithm which consists of checking each pixel's neighborhood and assigning the current pixel to an already existing group or creating a new group.

The rating system used was designed to classify shapes according to its similarity to a circle. The rates are normalized to the interval [0 1], where objects with rating equal to 1 represents a perfect circle. As the rating decreases the object starts to lose its circular shape until it reaches the end of the scale 0 when it becomes a line.

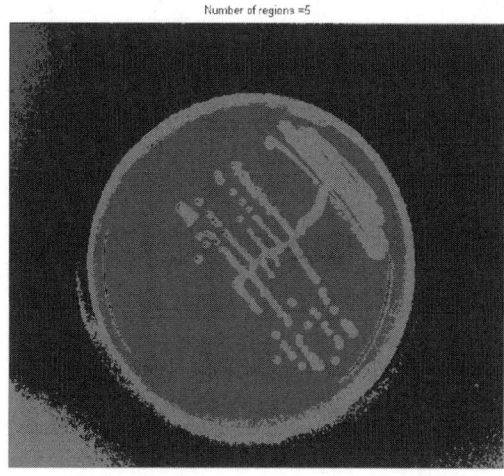

Fig. 2 Example of segmentation using GMM with EM algorithm. Each color coded region represents a cluster of pixels belonging to a different Gaussian distribution.

b) Binary Mask Creation for Dish Content Extraction

As bacterial cultures can vary in color and, consequently, in gray level intensities, extraction of its region by direct application of histogram thresholding is out of question. To add a bit of generality to the algorithm no restriction on gray level intensities' distribution of pixels belonging to the image's background was imposed. As mentioned in section B, pixel's gray level intensities belonging to the dish was assumed to have a Gaussian distribution.

Based on this assumption a binary mask was constructed which has value "1" wherever the location of the pixel in the binary mask corresponds to the location of a pixel in the grayscale image whose value is within range of dishes' group boundaries, and "0" everywhere else.

D. Characteristics Extraction

After each group of pixel representing an object in the image has been rated, those with highest ratings can be selected as they are more likely to represent circular shaped objects. The work of identifying the growth region of homogeneous bacterial colonies present on a Petri dish is completed. The location of each colony is now known, as well as the set of pixels which forms the image of the colony.

III. RESULTS

Fig. 4 shows an example of the obtained results using the segmentation algorithm proposed in this work. Table 1 presents the radius, the size and the rating for each identified homogenous colony depicted in Fig 4.

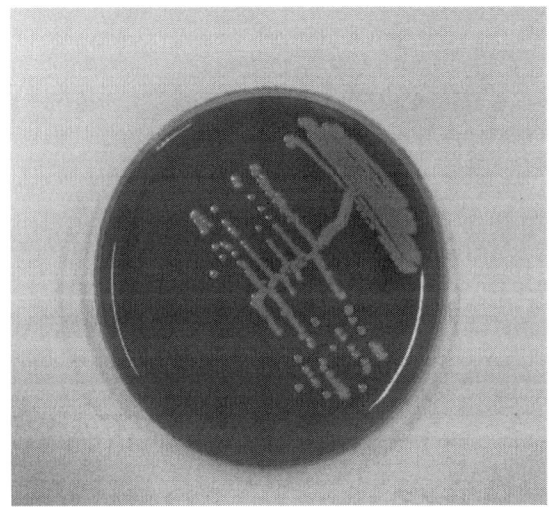

Fig. 4 Result obtained from segmenting a Petri dish containing homogenous bacterial colonies (depicted with blue circles)

Table 1 Identified homogenous circular regions in Fig. 4. Each row is composed of estimated radius (in pixels), estimated size of a colony (in pixels) and estimated rating ([0-1], where 1 represents a perfect circle).

No.	Radius [px]	Size [px]	Rating	No.	Radius [px]	Size [px]	Rating
1	8	196	0.93	9	8	198	0.96
2	8	189	0.89	10	8	203	0.85
3	7	146	0.91	11	7	167	0.93
4	7	147	0.96	12	7	146	0.95
5	6	122	0.93	13	8	213	0.89
6	6	125	0.92	14	8	190	0.94
7	6	98	0.91	15	4	44	0.92
8	8	183	0.93				

IV. CONCLUSIONS

With the approach presented in this work the growth regions of homogeneous bacterial colonies were able to be identified in images of Petri dishes containing bacterial cultures. The segmentation process was carried out using GMM with EM. The segmentation method demonstrated to be invariant to the distribution of gray level intensities belonging to the image background and bacterial culture, only assuming a Gaussian distribution of intensities corresponding to the dish surface. The process of reconstructing the shape of the dish allowed isolation of the bacterial sample without the need to make assumptions about the background's and bacterial culture's gray level composition, as well as the shape of the dish, making it a reliable method for content extraction within the dish area. Even with the respective limitations, the method gave satisfactory results without making too many assumptions about the scene, showing that it is possible to implement computer vision to recognize homogeneous bacterial colonies.

ACKNOWLEDGMENT

The authors thank dr. Ivanka Lerotić, the Head of the Department of microbiology at the Institute of Public Health in Zagreb County for providing Petri dishes containing bacterial colonies. This work was in part financed by the Ministry of Science, Education and Sport, grant no. 036-0362979-1554.

REFERENCES

1. Engelkirk PG, Duben-Engelkirk (2010) Burton's Microbiology for the Health Sciences 9th ed. Lippincott Williams &Wilkins, New York
2. Mukherjee DP, Pal A, Sarma SE et al. (1995) Bacterial colony counting using distance transform. Int J Bio-Med Comput. 38:131–40
3. Barber PR, Vojnovic B, Kelly J et al. (2001) Automated counting of mammalian cell colonies. Phys Med Biol 46:63–76
4. Marotz J, Lübbert C, Eisenbeiß W (2001) Effective object recognition for automated counting of colonies in Petri dishes (automated colony counting). Comput Meth Prog Biomed 66:183-198
5. Bernar R, Kanduser M, Pernus F (2001) Model-based automated detection of mammalian cell colonies. Phys Med Biol 46:3061-3072
6. Brugger SD, Baumberger C, Jost M et al. (2012) Automated counting of bacterial colony forming units on agar plates. PLoS ONE 7(3): e33695. doi:10.1371/journal.pone.0033695
7. McLachlan GJ, T. Krishnan T (2008) The EM Algorithm and Extensions (Wiley Series in Probability and Statistics) 2nd ed, Wiley-Interscience
8. McLachlan G, Peel D (2000) Finite Mixture Models (Wiley Series in Probability and Statistics), Wiley-Interscience
9. Verbeek J, Vlassis N Krose B (2003) Efficient Greedy Learning of Gaussian Mixture Models, Neural Comput 15(2):469-485 doi:10.1162/089976603762553004
10. Shi JQ, Choi T (2011), Gaussian Process Regression Analysis for Functional Data, Chapman and Hall/CRC

Author: Igor Lacković
Institute: University of Zagreb, Faculty of Electrical Engineering and Computing
Street: Unska 3
City: Zagreb
Country: Croatia
Email: igor.lackovic@fer.hr

Need of Multimodal SPECT/MRI for Tracking of ^{111}In-Labeled Human Mesenchymal Stem Cells in Neuroblastoma Tumor-Bearing Mice

L. Cussó[1,2,3], I. Mirones[4], S. Peña-Zalbidea[2], L. Lopez-Sánchez[1,2], V. García-Vázquez[3,2], J. García-Castro[4], and M. Desco[2,1]

[1] Departamento de Bioingeniería e Ingeniería Aeroespacial, Universidad Carlos III de Madrid, Spain
[2] Instituto de Investigación Sanitaria Gregorio Marañón, Madrid, Spain
[3] Centro de Investigación Biomédica en Red de Salud Mental (CIBERSAM), Madrid, Spain
[4] Unidad de Biotecnología Celular (Área de Genética Humana, IIER), Instituto de Salud Carlos III, Madrid, Spain

Abstract—Tumor-homing is a complex, multistep process used by many cells, like mesenchymal stem cells (MSCs), to travel from a distant location to a tumor. The purpose of this study is to investigate the applicability of ^{111}In-oxine for tracking human MSCs in vivo with single photon emission computed tomography (SPECT) and magnetic resonance imaging (MRI). ^{111}In labeled hMSCs (10^6 cells) were infused intraperitoneally in neuroblastoma tumor-bearing mice, and SPECT/MRI images were performed 24 and 48 hour afterwards. High initial hMSC localization occurred in the abdomen cavity and 48 hour later hMSC-injected animals showed tumor uptake. MRI information was essential to properly define Regions of Interest on the images. Tracking ^{111}In labeled hMSC combining SPECT and MRI is feasible and may be transferable to human studies.

Keywords—Mesenchymal stem cells, 111-Indium, SPECT/MRI, homing, multimodality.

I. INTRODUCTION

Human bone marrow-derived mesenchymal stem cells (hMSCs) have the potential to home to tumors including gliomas, carcinomas and many other metastatic tumors from a large variety of administration routes. The ability of injected cells to actively home tumors supports the use of hMSCs as therapeutic vehicles. The use of the corresponding imaging modality provides a novel, noninvasive method for serially track and quantify the fate of administered hMSCs in vivo, however not all these approaches are applicable to humans [1, 2].

The present work takes places in the framework of a research line to evaluate a new experimental cancer treatment called CELYVIR [3]. It consists of autologous hMSC infected with a conditional replication oncolytic adenovirus (ICOVIR5). In this framework, it is of the most importance to verify that stem cells actually migrate to the neuroblastoma tumor using tools that enable translation to human studies.

The purpose of this study was to detect homing of hMSCs in tumor-bearing mice combining single photon emission computed tomography (SPECT) with molecular resonance imaging (MRI).

II. MATERIALS AND METHODS

A. ^{111}In-Oxine Labeled hMSCs

hMSCs were isolated from bone marrow aspirate of patients who gave informed consent. For labeling, 6×10^6 cells were incubated with 10 Bq/cell ^{111}In-oxine following Gildehaus et al. protocol. [4].

B. Animal Model

All animal studies were approved by the Institutional Animal Care and Use Committee. Five adult male BalbC-SCID immunodeficient mice with a subcutaneous xenograft of a neuroblastoma cell line (NB 1691) were used in this study. Four mice received 1-1.3 million ^{111}In-labeled hMSC in 0.5 mL PBS intraperitoneally (i.p.). One control animal received 3.3 MBq of free ^{111}In-oxine i.p.

C. SPECT/MRI Imaging and Registration

Multimodality imaging was performed 24 and 48 hours after the administration with a small animal SPECT scanner (uSPECT, MILabs, The Netherlands) and a preclinical MRI system (7T Bruker Biospin, Bruker, Germany). In order to register the SPECT and MRI images, each animal was placed on a homemade multimodal bed surrounded by three capillaries filled with a mixture of ^{99m}Tc and CuSO$_4$, visible in both modalities. Using different SPECT radiotracer for the fiducials allowed us to reconstruct separately the capillaries and animal images, using the respective radioisotope photopeaks (140 keV for 99mTc and 171 + 245 keV for 111In). Other SPECT acquisition parameters were 0.4 mm isotropic voxel size and to 2 hours of acquisition time.

MRI was acquired using a TurboRARE T2 sequence with TR/TE = 3425/15.9 ms, and field of view (FOV) 6 x 4.13 cm. The spatial transformation to align SPECT and MRI images was obtained by matching the corresponding fiducials of both modalities with a method analogous to that described. [5].

D. Image Analysis

Regions of interest (ROIs) for tumor and reference (abdomen) were selected on the MRI at each time point. These ROIs were applied to the SPECT images in order to obtain the mean activity, corrected by the decay factor.

III. RESULTS

Figure 1 shows axial views of the fusion of SPECT and registered MRI images of an hMSCs animal (A) versus a control animal (B), 48 hours after the administration.

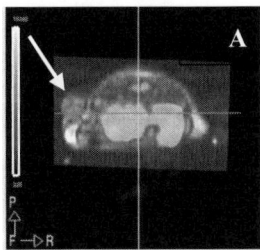
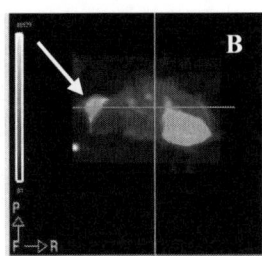

Fig. 1 Abdominal cavity shows activity in both animals, what hinders the visualization of hMSCs. A) hMSCs animal shows uptake inside the tumor area (white arrow). B) Control animal tumor does not show ^{111}In uptake

Figure 2 shows the mean activity in tumor and reference (abdomen) ROIs, 24 and 48 hours after the injection. hMSCs–injected animals showed an uptake increase in the tumor area at 48 hours, while control just showed some activity at 24 hours (perhaps due to spill over from the abdominal cavity), decreasing at 48 hours.

IV. DISCUSSION AND CONCLUSIONS

In our study, the use of intraperitoneal administration reduces animal mortality but the radioactivity from the testicles and abdominal cavity, close to the tumor area, produced inconvenient spillover together with a very difficult identification of the tumor ROI, similarly to what Kraitchman et al. reported regarding lung retention and cardiac uptake in myocardial-infarcted animals [6]. For this reason, the availability of an MRI anatomical template proved to be critical in order to properly identify the different regions, from which the dynamic behavior of the ROI activity allowed us to ascertain that actual uptake was taking place in the tumor.

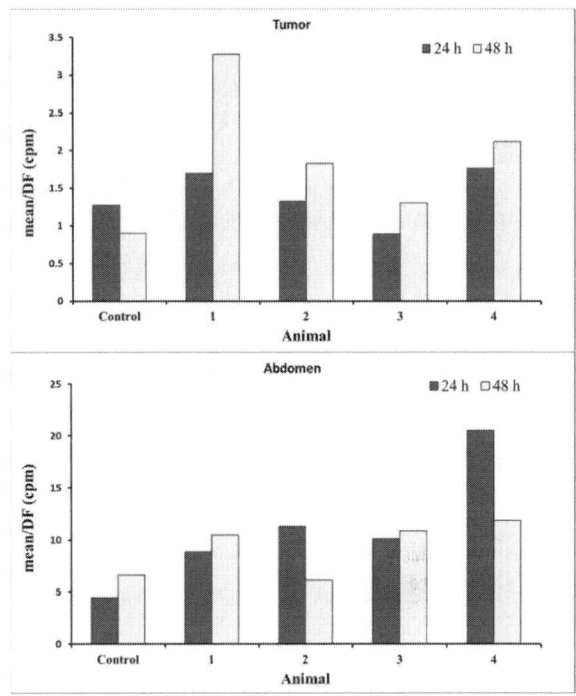

Fig. 2 Mean decay-corrected (dc) activity. Top: Tumor uptake increases 48 hours after the ^{111}In-labeled hMSCs injection. Bottom: reference region show similar uptake at both time points. Cells in animals 2 and 4 could migrate from the abdomen to different organs

In conclusion, we have been able to detect homing of hMSCs towards a neuroblastoma tumor in 24/48 hours. using a multimodality SPECT+MRI approach which proved to be essential for obtaining meaningful results. The procedure followed is basically translational to human patients.

ACKNOWLEDGMENT

This work was funded in part by grants from MINECO (PLE2009-0115), Red Tematica de Investigacion Cooperativa en Cancer (RTICC/ISCIII; RD12/0036/0027) and the Madrid Regional Government (S-BIO-0204-2006– MesenCAM and P2010/BMD-2420-CellCAM) and the Ministerio de Ciencia e Innovación (CEN-20101014) in Spain

REFERENCES

1. Rodriguez-Porcel M, Wu JC, and Gambhir SS (2009) Molecular imaging of stem cells, T.S.C.R. Community, Editor. Stembook.
2. Reagan MR and Kaplan DL (2011) Concise review: Mesenchymal stem cell tumor-homing: detection methods in disease model systems. Stem Cells. 29(6): p. 920-7.

3. Garcia-Castro J, et al. (2010) Treatment of metastatic neuroblastoma with systemic oncolytic virotherapy delivered by autologous mesenchymal stem cells: an exploratory study. Cancer Gene Ther. 17(7): p. 476-83.
4. Gildehaus FJ, et al. (2011) Impact of Indium-111 Oxine Labelling on Viability of Human Mesenchymal Stem Cells In Vitro, and 3D Cell-Tracking Using SPECT/CT In Vivo. Molecular Imaging and Biology. 13(6): p. 1204-1214.
5. Pascau J, et al. (2012) A method for small-animal PET/CT alignment calibration. Phys Med Biol. 57(12): p. N199-207.
6. Kraitchman DL, et al. (2005) Dynamic imaging of allogeneic mesenchymal stem cells trafficking to myocardial infarction. Circulation. 112(10): p. 1451-61.

Author: Lorena Cussó
Institute: Universidad Carlos III de Madrid
Street: Avda. Universidad, 30
City: Madrid
Country: Spain
Email: lcusso@hggm.es

First Steps towards a USE System for Non-invasive Thyroid Exploration

J. Rodriguez[1], O. Moreno-Perez[2], and J. Sabater-Navarro[1]

[1] nBio Research Group, Miguel Hernandez University, Elche, Spain
[2] Hospital General de Alicante, Alicante, Spain

Abstract—About 50% of the general population presents nodules in the thyroid gland. Although 90% of such nodules are benign, the cytological examination of material obtained by fine-needle aspiration (FNA) shows a 30% of not conclusive results. Apart from FNA, clinical data and echographic patterns are very limited in the prediction of malignancy in indeterminate nodules. Recently, ultrasound elastography (USE) has been proposed to study the elasticity of nodules by measuring its deformation under the application of a gentle pressure with the ultrasound (US) probe. Several recent studies show the effectiveness of USE for predicting malignancy of unselected thyroid nodules. These studies found a very high sensitivity, specificity, positive predictive value and negative predictive value for the detection of thyroid cancer. USE is a recent technique that is implemented in a few commercial US machines and thus very expensive. The system presented in this paper is a very affordable alternative to these USE commercial options, giving USE capabilities to a standard US machine. This system has been developed to be installed on every commercial US machine, and to be easy and little time consuming to operate. The system includes a complete database where clinical data for the patients and different tests and echographic data for the nodules can be stored for further studies.

Keywords—Thyroid, cancer, nodule, ultrasonography, elastography.

I. INTRODUCTION

Nodules in the thyroid gland are a common finding in the general population. 5% of subjects present nodules that are palpable. This percentage increases to 50% if thyroid ultrasound is used [1-3]. Fortunately, over 90% of such nodules are benign. Hashimoto's thyroiditis, which is the most common cause of hypothyroidism, is associated with an increased risk of thyroid nodules. Iodine deficiency is also known to cause thyroid nodules.

The best single test for differentiating malignant from benign nodules is the cytological examination of material obtained by fine-needle aspiration (FNA). This test has a high sensitivity and specificity [4]. The problem with FNA cytology is that about 15% of specimens are non-diagnostic [5], and about 20% are indeterminate [6]. Therefore, about 30% of samples tested with FNA are not conclusive [2]. Cystic or hemorrhagic lesions that do not provide a sufficient number of cells for diagnosis may present a non-diagnostic cytology. Moreover, a considerable proportion of solid nodules fail to provide sufficient material for cytological analysis [4-5]. Additionally, in up to 20% of all FNA specimens, cytology is classified as indeterminate even though the collected material is adequate [7].

Aside from FNA, there are multiple indicators of malignancy like clinical data and echographic patterns. Clinical data is very limited in the prediction of malignancy in indeterminate lesions. Sex, presence of nodules larger tan 3 cm, TPO, single nodule or multinodular gland, none of these indicators made a significant statistical difference. In the same way, echographic pattern was poorly predictive of malignancy with the exception of spot microcalcifications [8].

Thyroid cancers have a harder consistency than benign thyroid nodules. US elastography (USE) has been proposed to study the stiffness of nodules by measuring the degree of elasticity under the application of an external force, thus differentiating malignant from benign lesions. Several studies have proven USE as an effective tool for predicting malignancy of unselected thyroid nodules. USE was also effective in the presurgical selection of thyroid nodules with indeterminate or not diagnostic cytology. Some studies found for the detection of thyroid cancer a sensitivity of 97%, a specificity of 100%, a positive predictive value of 100%, and a negative predictive value of 98% [8-9]. As USE is a very recent technique, the US machines that can perform it are few and very expensive.

The system introduced here is an add-on for any commercial US machine, providing them USE capabilities and a complete data base for patient clinical data and individual nodule data. iThyroid makes possible for any hospital or clinic to have this powerful tool for a small outlay of money.

The system presented here is named iThyroid and is part of a family of US based medical oriented apps being developed at University Miguel Hernández [10]. The final goal of those apps is to integrate them on a complete surgical system [11] that could be used in the global process of surgical operation. iThyroid is part of the diagnosis and pre-planning tools.

II. MATERIALS AND METHODS

iThyroid system was designed under the premises that it should work in every hospital or clinic and that it should be

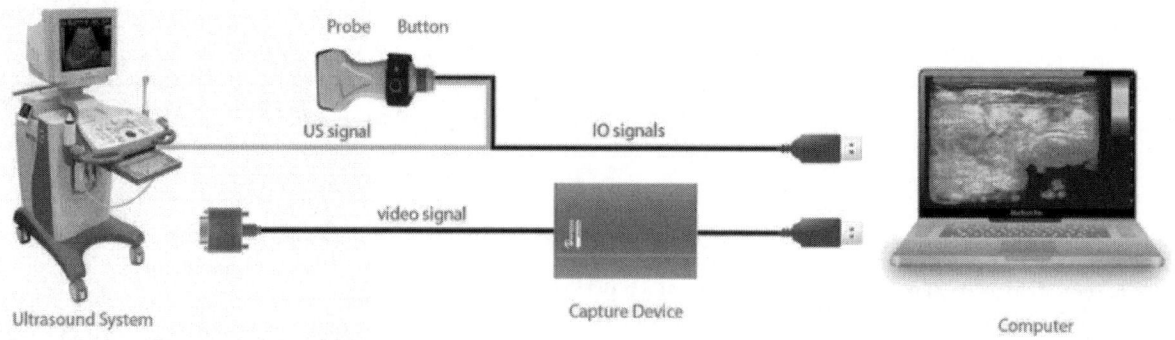

Fig. 1 iThyroid system diagram.

robust and inexpensive. For this reason, the hardware was carefully selected to meet these requirements and thus work with almost any standard US machine in the market.

The basic idea behind the system is to capture two US images of the nodule applying different forces to the probe, and then apply an algorithm to calculate the elasticity of the nodule under pressure.

The hardware of the system consisted of:

- Chison 8800 Digital Ultrasound System with a plane probe. (Any other US machine can be used instead).
- Apple Mac Pro with MacOSX10.7. (Any Mac computer can be used instead)
- Video capture device Epiphan DVI2USB.
- PhidgetInterfaceKit 2/2/2.

The hardware is distributed as follows: a US machine with a planar probe, a Mac computer and a container box with all the other hardware needed (video capture device, VGA splitter, interface kit, USB hub, cables, leds and buttons). These elements are connected as shown in Figure 1.

To send the capture signal to the computer, a button and a led have to be installed in the US probe. This is achieved with a special Velcro strap that has this elements installed as shown in Figure 2.

The button and the signaling LED were implemented using a Phidget Interface Kit 2/2/2. This interface kit has two analog inputs (not used), two digital inputs one of which is used for the button, and two digital outputs one of which is used for the signaling LED [12].

To capture the video from the US machine, an Epiphan **DVI2USB** device was chosen. It is the only external, dual-mode (VGA & DVI/HDMI) digital video capture device that can capture and broadcast **diagnostic-quality images and videos** from a VGA, DVI, HDMI video source such as

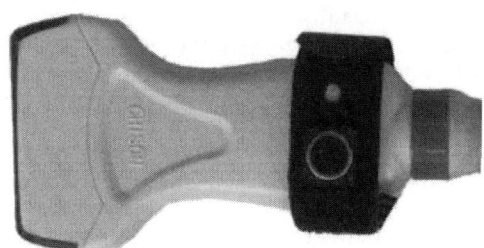

Fig. 2 Probe with the Velcro strap attached.

an US machine, and transfer to a USB port on a computer. Virtually all US machines have VGA, DVI or HDMI outputs, so DVI2USB fits them all. In case that the US Machine does not have a specific video output, the video signal can be obtained from the monitor VGA or HDMI plug using a VGA splitter. Epiphan's active VGA Splitter GLI was chosen for this purpose. It splits a VGA source into one Ground Loop Isolated VGA port, and a second fault-tolerant pass through VGA output port. Unlike traditional Y-splitter cables and active splitters, this splitter connects to the existing VGA cable, intercepts the VGA signals, amplifies the signals, and then generates a second ground loop isolated VGA output port [13].

A. iThyroid Software

iThyroid application was programmed using XCode 4 [14] in a OSX 10.7 environment. It makes use of OpenCV (Open Source Computer Vision) libraries [15] for elastography calculations and image algorithms. The database is programmed with Core Data Technologies by Apple due to its reliability (Figure 3). iThyroid application is designed to work with Epiphan USB video capturers to get the images of the US machine and with Phidgets sensors to get I/O signals.

The algorithm used to calculate the USE image was the Gunnar-Farneback optical flow algorithm, that belongs to the OpenCV's motion analysis and object tracking family [16]. The Gunnar-Farneback algorithm was applied to every pixel in the US images.

iThyroid application was developed to be easy and intuitive to operate. The processes that the doctors have to perform were thought to consume the minimum time possible. Thus, the number of mouse clicks to perform actions was reduced to the minimum.

iThyroid application consists of four modules: Patient module, Data Base module, Elastography module and Configuration module. The Patient module is were patients are added to and selected from the database as well as their clinical data.

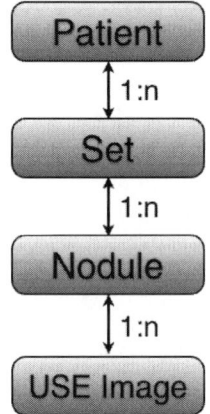

Fig. 3 Database diagram.

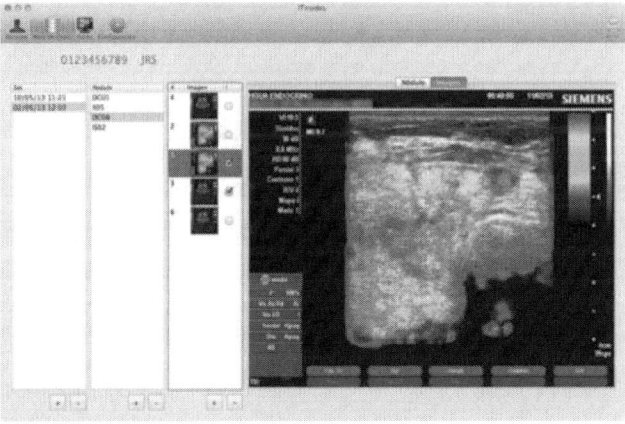

Fig. 4 Database window.

In the Database module, sets (different explorations) are added to the patients, as well as the nodules found with their characteristics and USE images. In this module is possible to revise the images obtained by the Elastography module. These images can be marked as important or delete them if they are not necessary. A screenshot of the Database module is shown in Figure 4.

The Elastography module is where the US images are captured to compute the USE image. The USE image can be displayed in three ways: monochrome coded, vectors or color coded (Figure 5). The color coded presentation is the most popular.

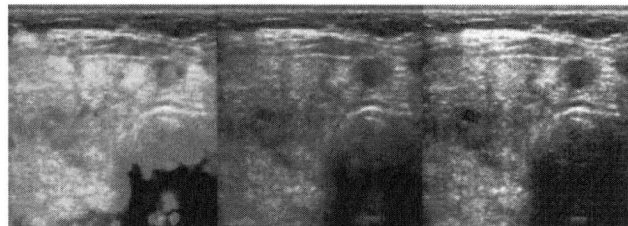

Fig. 5 Left: color coded USE, center: monochrome coded USE, right: vector coded USE.

The procedure for the USE image generation is performed as follows:

The US probe is positioned over the nodule under study and then the button attached to the probe is pressed. This will capture one US image that will be shown in the window marked with '1'.

The US probe is gently pressed against the nodule and the button attached to the probe is pressed. This will capture another US image that will be shown in the window marked with '2'. Immediately the USE image will be calculated using the US images previously captured and the result will be shown in the window marked with 'Elasto'. This USE image will be automatically stored in the database for that nodule.

To reset the Windows and start with another nodule or repeat the procedure with the same, the button attached to the US probe has to be pressed one last time.

An Elasto module window is shown in Figure 6.

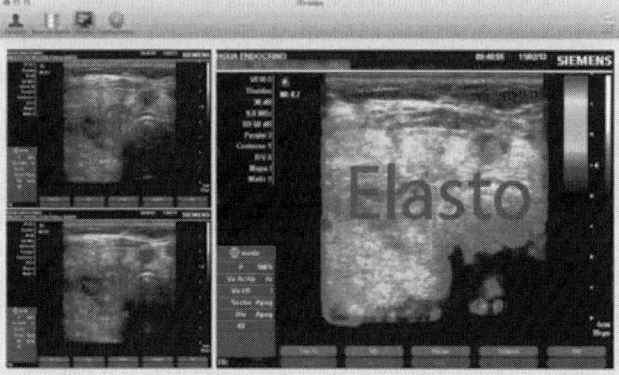

Fig. 6 Elasto window.

In the Configuration module, the different parameters for the USE algorithm can be configured. These settings have to be done only the first time that the iThyroid system is mounted with a new US machine.

Finally, it is also possible to print selected USE images as well as export clinical and nodule data for individual or all the patients in the data base for further statistical studies.

III. RESULTS AND CONCLUSIONS

At the present time, iThyroid system is being tested at the Hospital General de Alicante, having a great acceptance by doctors. The goal is to generate a 1000 patients database to evaluate the specificity and sensibility of the elastographic algorithm used.

At the end of the trial period some questionnaires will be given to the doctors to know if the system is friendly to use and if it presents an advantage compared to the traditional methods.

ACKNOWLEDGMENT

Authors acknowledge the support of the Spanish Ministerio de Economia y Competitividad through the action DPI2010-21126-C03-02.

REFERENCES

1. Rago et al. (2010) Real-Time Elastosonography: Useful Tool for Refining the Presurgical Diagnosis in Thyroid Nodules with Indeterminate or Nondiagnostic Cytology. J Clin Endocrinol Metab 95(12):5274–5280
2. Cooper DS, Doherty GM et al. (2009) American Thyroid Association (ATA) Guidelines Taskforce on Thyroid Nodules and Differentiated Thyroid Cancer. Revised American Thyroid Association management guidelines for patients with thyroid nodules and differentiated thy- roid cancer. Thyroid 19:1167–1214
3. Gharib H, Papini E, Paschke R 2008 Thyroid nodules: a review of current guidelines, practices, and prospects. Eur J Endocrinol 159: 493–505
4. Castro MR, Gharib H 2003 Thyroid fine-needle aspiration biopsy: progress, practice, and pitfalls. Endocr Pract 9:128–136
5. Redman R, Zalaznick H, Mazzaferri EL, Massoll NA 2006 The impact of assessing specimen adequacy and number of needle passes for fine-needle aspiration biopsy of thyroid nodules. Thyroid 16: 55–60
6. Rago et al. 2007 Combined clinical, thyroid ultrasound and cytological features help to predict thyroid malignancy in follicular and Hurthle cell thyroid lesions: results from a series of 505 consecutive patients. Clin Endocrinol (Oxf) 66:13–20
7. Lewis CM, Chang KP et al. 2009 Thyroid fine-needle aspiration biopsy: variability in reporting. Thy- roid 19:717–723
8. RagoT,SantiniF et al. (2007) Elastosonography: new developments in ultrasound for predicting malignancy in thyroid nodules. J Clin Endocrinol Metab 92:2917–2922
9. Tranquart F, Bleuzen A, Pierre-Renoult P, Chabrolle C, Sam Giao M, Lecomte P 2008 Elastosonography of thyroid lesions. J Radiol 89: 35–39
10. J. Rodriguez, J.-M. Sabater, J.-L. Ruiz, A. Soto, N. Garcia, iProstate, a mathematical predictive model-based, 3D-rendering tool to visualize the location and extent of prostate cancer, International Journal of Medical Robotics and Computer Assisted Surgery 7 (1) (2011) 71–84.
11. J.M. Sabater, N. Garcia, C. Perez, Kinematics of a robotic 3UPS1S spherical wrist designed for laparoscopic applications, International Journal of Medical Robotics and Computer Assisted Surgery 6 (3) (2010) 291–300.
12. Phidgets. Products for USB sensing and control. Retrieved from: Phidgets: http://www.phidgets.com/, 2012, May.
13. Epiphan Systems. Video capture, recording and broadcasting. Retrieved from Epiphan Systems: http://www.epiphan.com/, 2012, June.
14. Xcode. Developer: Xcode 4. Retrieved from Apple developer: http://developer.apple.com/xcode/, 2011, June.
15. OpenCV. Open Source Computer Vision. Retrieved from OpenCV: http://opencv.org, 2012, June.
16. Farneback, G. (2001) Very high accuracy velocity estimation using orientation tensors, parametric motion, and simultaneous segmentation of the motion field International Conference on Computer Vision,. ICCV 2001. Proceedings. Eighth IEEE 171 - 177 vol.1

Registration of Small-Animal SPECT/MRI Studies for Tracking Human Mesenchymal Stem Cells

V. García-Vázquez[1,2], L. Cussó[1,2,3], J. Chamorro-Servent[2,3], I. Mirones[4], J. García-Castro[4], L. López-Sánchez[2,3], S. Peña-Zalbidea[2], P. Montesinos[2,3], C. Chavarrías[2], J. Pascau[2,3], and M. Desco[2,3]

[1] Centro de Investigación Biomédica en Red de Salud Mental (CIBERSAM), Madrid, Spain
[2] Instituto de Investigación Sanitaria Gregorio Marañón, Madrid, Spain
[3] Departamento de Bioingeniería e Ingeniería Aeroespacial, Universidad Carlos III de Madrid, Spain
[4] Unidad de Biotecnología Celular (Área de Genética Humana, IIER), Instituto de Salud Carlos III, Madrid, Spain

Abstract—Single-photon emission computed tomography (SPECT) offers the possibility to study the biodistribution of ^{111}In-labeled human mesenchymal stem cells (hMSCs). This functional information can be complemented by magnetic resonance imaging (MRI) that provides the anatomical context to the SPECT image. However, acquiring such studies with separate SPECT/MRI systems requires a multimodality registration in order to fuse both images. In this work, a special bed was designed based on three non-coplanar glass capillary tubes filled with a mixture of ^{99m}Tc and $CuSO_4$, visible in both modalities. Moreover, it was necessary to design a semi-automated rigid registration procedure based on matching these rods by minimizing the distance between corresponding lines. The maximum target registration error achieved was 0.9 mm, in the range of SPECT resolution.

Keywords—Registration, SPECT/MRI, mouse, multimodality.

I. INTRODUCTION

The present work is part of a research line that required the tracking of intraperitoneally injected human mesenchymal stem cells (hMSCs) in mice, which migrate away from the peritoneal cavity, homing in tumor beds. Labeling hMSCs with ^{111}In-oxine provides a noninvasive method to longitudinally track and quantify the fate of administered hMSCs in vivo [1]. Single-photon emission computed tomography (SPECT) allows following and measuring the biodistribution of ^{111}In-labeled hMSCs. However, the use of intraperitoneal (i.p.) administration might artifactually increase the radioactivity in the tumor located on mouse hind limbs, due to spillover effect from abdominal cavity and testicles. Moreover, boundaries of tumor or other organs of interest are challenging to delineate from SPECT images, what may bias the actual uptake in these regions. Magnetic resonance imaging (MRI) provides an anatomical context to SPECT images that greatly facilitates the identification of regions of interest (ROIs).

The use of separate modality scans require a careful transfer of the animal followed by image registration. The alignment accuracy depends on the resolution of both modalities and on possible image artifacts. Small-animal SPECT studies have a spatial resolution of about 1 mm^3 with a pinhole rat collimator that might be different from the resolution of a small-animal MRI 2-dimensional (2D) multislice acquisition. Moreover, a high magnetic field might increase geometric deformation artifacts due to field inhomogeneities.

Prospective alignment based on external markers makes the procedure independent of image contents. Procedures to correct centering errors have been previously proposed for adjacent and simultaneous systems, by using phantoms with a special configuration of tubes filled with a contrast agent for MRI and a radioisotope for nuclear imaging [2,3]. Similarly, [4] proposed a simple and affordable phantom with three non-coplanar glass capillary tubes and an automatic registration method based on minimizing the distance between corresponding lines [5]. This methodology showed reliability and robustness for PET/CT alignment but it is not suitable for our setting because the capillaries were placed centered in the acquisition field of view (FOV) and do not allow performing SPECT/MRI images of both animal and capillaries.

Another solution has been proposed for separate SPECT/MRI devices [6], but it is not adequate for our studies as the mouse was placed on a flexible plastic curved sheet whose curvature might position the animal limbs too close to the abdominal cavity, thus increasing the spillover in the tumor. The registration was based on several landmarks from the rigid loops described by a curved catheter fixed on that sheet. These external markers had to be manually identified and did not cover the whole mouse body opposite to the bed, thus increasing the registration error in this region.

The purpose of this work was to develop a mouse bed using capillary rubes as external markers for SPECT/MRI registration, and to assess the accuracy of the alignment using a line-based registration method analogous to that described in [4].

II. MATERIALS AND METHODS

A. Mouse Bed

The designed multimodal bed was a planar surface made of methacrylate with a 2 cm wide bridge on it (Figure 1). Three non-coplanar glass capillary tubes (micropipettes BLAUBRAND intraMARK Cat no 708744 with 100 μl/1.68 mm external diameter) were arranged at different heights and orientations on the bed (Figure 1). This configuration surrounded the abdominal area and the posterior hind limbs, regions of interest of our studies. The capillaries were filled with a mixture of ^{99m}Tc and $CuSO_4$, visible in both modalities.

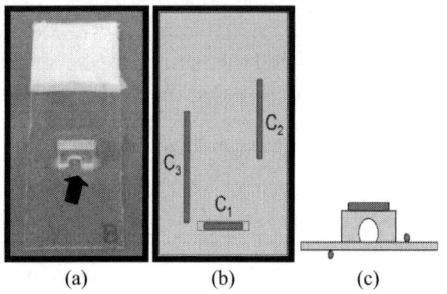

Fig. 1 Mouse bed. (a) Planar surface with the 2cm-bridge (black arrow) through which the mouse tail was passed. (b) Top view sketch of the bed showing the three capillary tubes in red (C1, 15 mm in length; C2, 25 mm in length; and C3, 35 mm in length) (c) Front view

B. SPECT/MRI Registration

The capillary tubes (rods) used as fiducials for the SPECT/MRI registration were segmented in both modality images using a region growing algorithm, as they were easily distinguishable from the background. Then, the points of each rod were adjusted to a line with a method based on principal component analysis (PCA).

The three lines in the SPECT image defined the line set $A = \{A_n\}$ with $n = 1,2,3$ where $A_n = (a_n, \hat{b}_n, l_n)$ corresponded to a line segment represented by its centroid a_n, its director vector $\hat{b}_n$ and its length l_n. Similarly, MRI lines were represented by the line set $X = \{X_n\}$ with $X_n = (x_n, \hat{y}_n, l_n)$. The best match between the two line sets was obtained by minimizing the distance $M(A, T, X)$ between the MRI and the transformed SPECT line over all possible rigid transformations $T(t, R)$ [5]:

$$M(A,T,X) = \sum_{n=1}^{3}[l_n\|a_n - t - Rx_n\|^2 + l_n^3(1 - \hat{b}_n^T R\hat{y}_n)/6] \quad (1)$$

where l_n was set equal to 100 for $n = 1,2,3$; t was the translation vector; and R the rotation matrix. The latter was estimated as $R = UV^T$ where U and V were the unitary matrixes of the singular value decomposition (SVD) of the following covariance matrix:

$$S = \sum_{n=1}^{3}[l_n(a_n - \tilde{a})(x_n - \tilde{x})^T + l_n^3 \hat{b}_n R\hat{y}_n^T/12] \quad (2)$$

with $\tilde{a} = \frac{1}{W}\sum_{n=1}^{3} l_n a_n$, $\tilde{x} = \frac{1}{W}\sum_{n=1}^{3} l_n x_n$ and $W = \sum_{n=1}^{3} l_n$. The translation vector was obtained from $t = \tilde{a} - R\tilde{x}$.

C. Validation

The design was validated with two experiments, one to assess the accuracy of the SPECT/MRI alignment and another one to verify the usability with ^{111}In-labeled hMSCs. Animal procedures were performed in compliance with the European Communities Council Directive of 22 September 2010 (2010/63/EU) and were approved by the Institutional Animal Care and Use Committee of the Hospital General Universitario Gregorio Marañón.

D. First Experiment

A glass capillary tube filled with the same mixture as in the external markers was fixed to the superior left hind limb of a mouse, emulating a tumor in that location. The mouse was anesthetized and attached to the multimodal bed by its fore and hind limbs using surgical tapes.

Multimodality imaging was performed using a small-animal SPECT scanner (uSPECT, MILabs, The Netherlands) and a preclinical MRI system (Biospin 7T, Bruker, Germany). In both modalities, the FOV covered the capillary tubes, the abdominal area and the hind limbs. The SPECT acquisition parameters were a 0.4 mm isotropic voxel size and 6 min acquisition time. After completion of the SPECT imaging, the bed was carefully transferred from the SPECT scanner to the MRI system. A TurboRARE T2 sequence (coronal) was acquired with a quadrature volume coil and the following parameters: TR/TE = 3545/48 ms, matrix 465 x 256, FOV 5 x 3.5 cm^2, slice thickness 1.1 mm, scan time 4 minutes.

The SPECT/MRI registration was evaluated by means of two different measurements at the capillary tube (target) that emulated the tumor in the mouse. This reference was segmented with the same algorithm as the fiducial rods, but was neglected in the registration process. Segmentation and registration steps were repeated three times. The first registration error measurement was the root mean square (rms) error between the positions of the points of the SPECT rod transformed to the *MRI space* and their corresponding MRI points. This correspondence was obtained by finding the closest points (minimum Euclidean distance criteria) as in the matching step of the iterative closest point (ICP) algorithm [7]. The second registration error measurement was based on adjusting the target to a line and applying the distance $M(A, T, X)$, the cost function used in the SPECT/MRI registration, to these lines ($n = 1$

and $l_n = 1$). This error takes into account the difference between the centroids of both point sets and their director vectors.

The spatial distribution of the registration error was obtained by calculating the target registration error (TRE) for a rigid point-based alignment [8]:

$$\langle TRE^2(\boldsymbol{r}) \rangle \approx \frac{\langle FRE^2 \rangle}{N-2}\left(1 + \frac{1}{3}\sum_{k=1}^{3}\frac{d_k^2}{f_k^2}\right) \quad (3)$$

where FRE is the fiducial registration error (the rms distance between the N MRI points and the transformed SPECT points used in the registration step), d_k is the distance of the point represented by coordinates $\boldsymbol{r}$ from principal axis k, and f_k the rms distance of the fiducials from principal axis k. Constants FRE and $N-2$ were removed from (3).

E. Second Experiment

A BalbC-SCID immunodeficient mouse was implanted with a subcutaneous xenograft of a neuroblastoma cell line (NB 1691) and received 1 million ^{111}In-labeled hMSCs in 0.5 mL Phosphate-Buffered Saline (PBS) i.p. one hour before the SPECT/MRI session. hMSCs were labeled following the protocol described in [1]. Capillary tubes set-up, mouse immobilization, anesthesia and SPECT/MRI imaging was similar to those in the first experiment, except for SPECT acquisition time (1 hour). In this case, the capillary tubes and the animal image were separately reconstructed using their respective radioisotope photopeaks (140 keV for ^{99m}Tc and 171 + 245 keV for ^{111}In). In this way, there was no spillover effect in the tumor from the capillary tubes.

III. RESULTS

Figure 2 shows the points from the MRI and the transformed SPECT capillary tubes (both registration and target rods) after the SPECT/MRI alignment. Table 1 shows the target rms error and the distance $M(A,T,X)$ for the three repetitions. Figure 3 illustrates the spatial distribution of the TRE. The regions of interest for our research (the tumor represented by the target rod and the abdominal cavity) are inside the volume where the TRE is lower. Figure 4 shows the fusion of the SPECT and MRI images of the tumor-implanted mouse. ^{111}In-labeled hMSCs can be detected in the abdominal cavity, as well as in the tumor. Note how the activity is higher in the abdomen than in the tumor, thus justifying the interest of the MRI study.

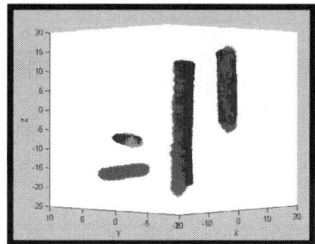

Fig. 2 SPECT/MRI registration: MRI capillary tubes (in blue) and transformed SPECT capillary tubes (external markers in red and target in green). Axis X is the left-right axis; axis Y, the anterior-posterior axis; and axis Z, along the mouse bed. All axes are in mm

Table 1 Target registration errors

Repetition	rms error	$M(A,T,X)$
1	0.3 mm	0.7 mm
2	0.4 mm	0.9 mm
3	0.4 mm	0.8 mm

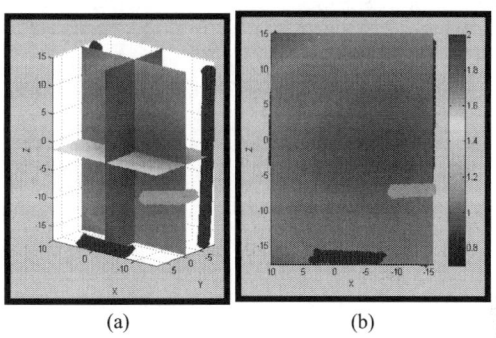

Fig. 3 Spatial distribution of the registration error (mm). Colors of rods and axes as in Figure 2

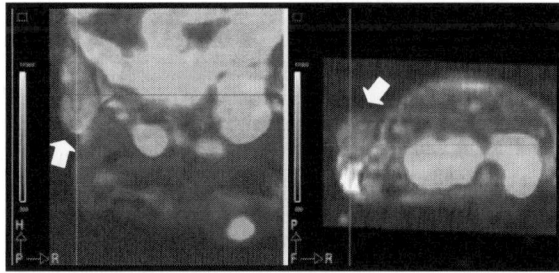

Fig. 4 SPECT/MRI fusion of the cell tracking experiment images. Arrows point to the tumor. Left to right: coronal and axial views

IV. Discussion

This work presents a semi-automated rigid method based on matching lines aimed to register small-animal SPECT/MRI studies. The approach resulted in target registration errors close to the SPECT spatial resolution (< 0.9 mm).

SPECT provides functional volumetric information about the distribution of ^{111}In-labeled hMSCs but it lacks anatomical references related to the tissue uptake of radiotracer. The spillover effect from the adjacent regions to the tumor makes difficult to delineate the tumor boundaries from the SPECT image. A registered MRI of the same animal allows better delineation of the ROIs and that segmentation can be fused with the SPECT image in order to better quantify ROIs uptake.

In terms of accuracy, our registration results are in accordance with previous literature. In [6] it was reported a mean TRE of 0.43 mm at the intersection of two crossing tubes (phantom) using a 3D MRI sequence (isotropic voxel of 0.5 mm, MRI system of 1.5 T). Since we have aligned in vivo animal studies, uncontrolled movements and respiratory motion may influence our results. Regarding performance on in vivo mice studies, in [2] a registration error of 0.93 mm was reported choosing anatomic references with a 3D MRI sequence (isotropic voxel of 0.5 mm, MRI system of 0.1 T), in accordance with our results.

Connecting the SPECT bed to MRI dock (or vice versa) [6] may be an alternative approach to our proposal, but our solution is simple, affordable and allows acquiring simultaneous SPECT/MRI images from different animals.

V. Conclusions

Tracking ^{111}In-labeled hMSCs in mice using separate SPECT/MRI systems requires a multimodality registration in order to fuse both images. A special mouse bed was designed together with a semi-automated rigid registration procedure, based on matching three non-coplanar glass capillary tubes. The procedure showed registration errors in the range of SPECT resolution, and demonstrated its usability for hMSCs tracking studies in mice.

Acknowledgment

This research was supported by Ministerio de Ciencia e Innovación (CEN-20101014, PI10/02986, TEC2011-28972-C02-01).

References

1. Gildehaus F J, Haasters F, Drosse I et al. (2011) Impact of Indium-111 oxine labelling on viability of human mesenchymal stem cells in vitro, and 3D cell-tracking using SPECT/CT in vivo. Mol Imaging Biol 13(6):1204-1214 DOI 10.1007/s11307-010-0439-1
2. Goetz C, Breton E, Choquet P et al. (2008) SPECT low-field MRI system for small-animal imaging. J Nucl Med 49(1):88-93 DOI 10.2967/jnumed.107.044313
3. Ng TS, Procissi D, Wu Y et al. (2010) A robust coregistration method for in vivo studies using a first generation simultaneous PET/MR scanner. Med Phys 37(5):1995-2003 DOI 10.1118/1.3369447
4. Pascau J, Vaquero JJ, Chamorro-Servent J, Rodríguez-Ruano A, Desco M (2012) A method for small-animal PET/CT alignment calibration. Phys Med Biol 57(12):N199-207 DOI 10.1088/0031-9155/57/12/N199
5. Kamgar-Parsi B, Kamgar-Parsi B (2004) Algorithms for matching 3D line sets. IEEE Trans Pattern Anal Mach Intell 26(5):582-93 DOI 10.1109/TPAMI.2004.1273930
6. J-P Dillensegera, B Guillaudb, C Goetza et al. (2013) Coregistration of datasets from a micro-SPECT/CT and a preclinical 1.5 T MRI. Nuclear Instruments and Methods in Physics Research A 702: 144-147 DOI: 10.1016/j.nima.2012.08.023
7. P J Besl, N D McKay (1992) A method for registration of 3-D shapes. IEEE Trans Pattern Anal Mach Intell 14(2): 239-256 DOI 10.1109/34.121791
8. J M Fitzpatrick, J B West, C R Jr. Maurer (1998) Predicting error in rigid-body point-based registration. IEEE Trans Med Imaging 17(5):694-702 DOI 10.1109/42.736021

Author: Verónica García-Vázquez
Institute: Centro de Investigación Biomédica en Red de Salud Mental
Street: Dr. Esquerdo 46, 28007
City: Madrid
Country Spain
Email: vgarcia@hggm.es

Segregation of Emotional Function in Subcortical Structures: MEG Evidence

C. Styliadis[1], A.A. Ioannides[2], P.D. Bamidis[1], and C. Papadelis[3]

[1] Laboratory of Medical Informatics, School of Medicine, Aristotle University of Thessaloniki, GREECE
[2] Laboratory for Human Brain Dynamics, AAI Scientific Cultural Services Ltd., Nicosia, CYPRUS
[3] Department of Neurology, Boston Children's Hospital, Harvard Medical School, Waltham, USA

Abstract—Arousal and valence are the primary dimensions of human emotion. However, the degree to which these dimensions correlate to complex subcortical structures (i.e. amygdala, cerebellum) that are anatomically homogeneous is still elusive. Magnetoencephalography (MEG) recordings were performed on 12 healthy individuals exposed to affective stimuli from the International Affective Picture System (IAPS) collection using a 2 (Valence levels) x 2 (Arousal levels) design. Source power was estimated using a beamformer within 1-30 and 30-100 Hz bands. Activations referring to the subcortical sub-regions were defined through probabilistic cytoarchitectonic maps (PCMs). Within the 1-30 Hz band, right laterobasal (LB) amygdala activity mediates negative valence (elicited by unpleasant stimuli) while left centromedial (CM) activity correlates to the interaction of valence by arousal (arousing pleasant stimuli). Within the 30-100 Hz band, cerebellar VIIIa lobule of the Vermis and left hemispheric VIIa Crus II lobule activity mediate high arousal while left hemispheric V lobule correlates to the interaction of valence by arousal (arousing pleasant stimuli). Our results support that distinct sub-regional subcortical activity responds specifically to valence and arousal dimensions as well as combinations of the two, pronouncing the sophisticated nature of emotion. Given the anatomical interconnections between amygdala and cerebellum, future studies may focus on the interplay of their specific sub-regions.

Keywords—magnetoencephalography (MEG), arousal, valence, interaction, subcortical.

I. INTRODUCTION

Emotions are often regarded as a representation of arousal and valence, namely the two independent emotional dimensions [1]. Research on the two dimensions' neural substrates has gradually shifted from initial propositions of a separate representation for arousal and valence, to a more complex representation which involves correlation of many cortical and subcortical structures to arousal, valence and their interaction [2], [3].

A single brain region may participate in a plethora of emotions while a single emotion may provoke activity in multiple regions [4]. Complex subcortical structures (i.e. amygdala, cerebellum) that are anatomically homogeneous have a consistent and multivariable participation in emotional processing.

For instance, the amygdala is anatomically distinguished in 13 nuclei of even more sub-nuclei [5] that form three major sub-regions, namely the laterobasal (LB), the centromedial (CM) and the superficial (SF) amygdala [6]. Novel functional research on the amygdala sub-regions suggests that they mediate behavioral, emotional, and physiological responses [7].

Similarly, the cerebellum is distinguished in two hemispheres joined across the vermis and its three lobes (anterior, posterior, flocculonodular), which consist of ten clearly defined lobules designated I to X [8]. Cerebellum's posterior lobe and particularly its hemispheric lobules VI and VII serve non-motor cognitive processing while a midline cerebellar structure, (i.e. the vermis) which is considered the limbic cerebellum, serves affective processes [9], [10] and [11].

Treating these structures in emotion as single homogenous rather than complex ones produces ambivalent functional findings due to the distinct functional specificity of the sub-regions of amygdala and cerebellum and their differential reactivity to arousal and valence. For instance, both the amygdala and the cerebellum activity have been correlated to arousal and valence encoding [3], [12], [13], [14], leaving answered questions of which exact sub-region plays these roles.

Our results show that within the 1-30 Hz band, right laterobasal (LB) amygdala activity mediates negative valence (elicited by unpleasant stimuli) while left centromedial (CM) activity correlates to the interaction of valence by arousal (arousing pleasant stimuli). Within the 30-100 Hz band, cerebellar VIIIa lobule of the Vermis and left hemispheric VIIa Crus II lobule activity mediate high arousal while left hemispheric V lobule correlates to the interaction of valence by arousal (arousing pleasant stimuli). Thus, addressing the subcortical structures sub-regional functional specificity aids in disentangling their exact nature in emotional processing. Their functional role in emotional processing will receive more attention in future neuroimaging studies that will make use of probabilistic cytoarchitectonic maps (PCMs). Our aim is to characterize the response of these subcortical structures to affective stimuli by assessing the differentiated functional role of their sub-regions in the emotional processing.

II. MATERIALS AND METHODS

Information on the materials and methods are thoroughly discussed in [15]. Here we provide selected information on the achievement of our results.

A. Participants

Twelve healthy individuals (7 males, mean age 30.8 ± 5.3, range 23 to 40 years, 5 females mean age 27.8±5.3, range 21 to 35 years) participated in the study. All participants had normal or corrected-to-normal visual acuity.

B. Affective Stimuli

Participants passively viewed pictures from the International Affective Picture System (IAPS) collection [16] on a homogenous black background. Four stimuli sets that manipulated the level of arousal within pleasant and unpleasant pictures [17] were used; pleasant and high arousing (PHA), pleasant and low arousing (PLA), unpleasant and high arousing (UHA) and unpleasant and low arousing (ULA).

C. Experimental Paradigm

Participants were given a simple instruction; fixate on the cross, stay still and do not blink during the recordings. A random design of two runs each of 80 trials (20 stimuli per category) was used. Each run initiated with the projection of a fixation cross centered on screen, at 40 x 40 pixels resolution for a pseudo-randomized interval of 1.5 ± 0.2 s. Trials were projected centered on screen at 400 x 400 pixels resolution for 1 s along with the fixation cross. Each run lasted 220 s resulting in a total duration of 440 s.

D. Coregistration

Coregistration results were manually checked and regarded as accurate if the mean distance between the surface of the head and face derived from the 3D camera, the 3D digitizer and the anatomical scan was less than 2 mm. Realistic geometry head models were reconstructed from T1-weighted MRIs for each participant.

E. Data Acquisition

The participant's head position was registered with localization coils (nasion and pre-auricular) at the beginning and end of each measurement. DC (direct current) offset was removed. The recorded MEG signal was visually inspected for possible artifacts and bad channels inferring noise (100mV, >5pT between minimum and maximum) were removed. Markers in the MEG data were synchronized to the onset of each visual stimulus by luminance detection.

F. SAM Analysis

Synthetic Aperture Magnetometry (SAM), based on the beamformer approach [18], was used to analyze task-related activation differences in 1-30 and 30-100 Hz. We employed the dual state; the active state was defined as the 1 s time window following the stimulus onset, and the control state (baseline) as the 1 s time window preceding the stimulus onset. The dual-state SAM output for PHA, PLA, UHA and ULA conditions was the contrast between the active state and the control state. SAM images from the two runs were averaged to generate a single SAM image per subject per condition [19], [20]. A multiple local-sphere model served as head model. The resultant volumetric maps were overlaid on the individual subject's structural MRI.

G. Group Level Analysis of MEG Source Activity

The resulting SAM images were normalized into the Montreal Neurological Institute (MNI) space. A factorial design of repeated ANOVA measures having stimulus valence (pleasant/unpleasant) and arousal (high/low) as the within-participants factors assessed the mean activation across participants for valence, arousal and their interaction. Statistical maps for the effects of valence and arousal had 32 degrees of freedom with a confidence level set at $p<0.001$ (uncorrected), minimum was thirty contiguous voxels.

H. Probabilistic Maps

The sub-regions of amygdala and cerebellum were identified using probabilistic cytoarchitectonic maps [6], [21]. The PCMs are based on histological analysis of ten human post-mortem brains. Their use benefits in identifying regions that are not conspicuous in structural images. The PCMs are freely available through the anatomy toolbox [22].

III. RESULTS

Figure 1 shows the group analysis MEG results within 1-30 Hz, ($p<0.001$, uncorrected) for the effect of unpleasant pictures which showed a pattern of neural responses in the right laterobasal amygdala (LB) (Fig.1a) and the interaction effect (arousing pleasant stimuli), which revealed activation in the left centromedial amygdala (CM) (Fig.1b).

Figure 2 shows the group analysis MEG results within 30-100 Hz, (p<0.001, uncorrected), overlaid on the cerebellum atlas, for high arousal (Fig. 2a) and its interaction by pleasant valence (Fig. 2b). Our results indicate significant cerebellar activity attributed to high arousal (Vermis of lobule VIIIa and left hemispheric VIIa Crus II and to arousing pleasant stimuli (left hemispheric V). Table 1 presents the anatomical location of the maxima of statistical significant differences for all contrasts in the MNI coordinate system, the cytoarchitecture of the amygdala and the cerebellum as well as their cluster size and corresponding statistical value.

Table 1 Local statistical maxima in activated complex subcortical structures in emotional processing

Structure	H	PCM	x	y	z	CS	T
Amygdala	R	LB	30	-4	-24	38	3.92
Amygdala	L	CM	-24	-12	-4	42	3.31
Cerebellum	-	Vermis (VIIIa)	0	-70	-40	105	6.55
Cerebellum	L	VIIa Crus II (Hem)	-16	-86	-35	260	6.39
Cerebellum	L	V (Hem)	-12	-50	-14	62	5.20

Note: Results are superimposed on standardized MNI coordinates; H, hemisphere; PCM, probabilistic cytoarchitectonic maps; x, left/right; y, anterior/posterior; z, superior/inferior; CS, cluster size in number of activated voxels; T, t-values for each peak; L, left; R, right; significant at p < 0.001 (uncorrected).

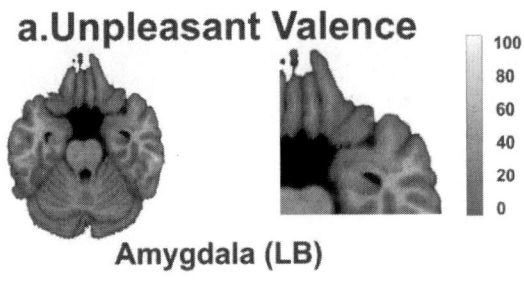

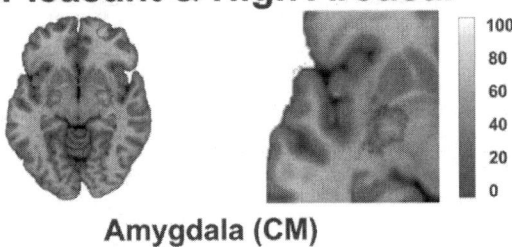
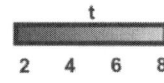

Fig. 1 Group parametric maps within 1-30 Hz (a) Unpleasant valence activates the right laterobasal (LB) amygdala. (b) High arousing and pleasant stimuli activates the left centromedial (CM) amygdala.

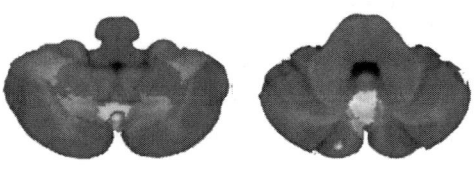
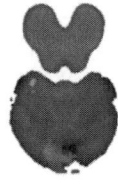
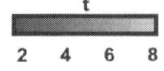

Fig. 2 Group parametric maps within 30-100 Hz (a) High arousal activates the vermis (VIIIa) and the left hemispheric VIIa Crus II lobule. (b) High arousing and pleasant stimuli activates the left hemispheric V lobule.

IV. DISCUSSION

Complex subcortical structures responses to valence and arousal, has on many occasions provided ambiguous results. We treated amygdala and cerebellum as complex structures and performed our results interpretation under the anatomical specificity of their sub-regions. This approach may aid in sorting out the precise nature of sub-regional subcortical response to the affective stimuli characteristics. Indeed, our results support that amygdala's as well as cerebellum's anatomical sub-regions serve different functions in emotional processing. Right (LB) and left (CM) amygdala activity was found to mediate valence (effect of unpleasant pictures) and its interaction by arousal (arousing pleasant pictures) respectively. Cerebellum mediates high arousal encoding at VIIIa lobule of the vermis lobule and left hemispheric VIIa Crus II lobule and also sub-serves high arousal's interaction to pleasant valence at the left hemispheric lobule V. These findings indicate distinct functional roles for the subcortical structures of these anatomical sub-regions. Even though there is vast knowledge on these structures when considered as homogenous ones, the precise contribution of their sub-regions is still elusive. Their function is generally acknowledged in emotional processing and yet it is more specific than previously thought. Our results contribute by attributing significant processes of emotion to subcortical sub-regions.

V. CONCLUSIONS

Here we shared MEG evidence on amygdala's and cerebellum's distinct functional roles on emotion. Given the anatomical interconnections between amygdala and cerebellum, future studies may focus on the interplay of their specific sub-regions.

ACKNOWLEDGMENT

Data acquisition of the modalities described as well as a part of the analysis was conducted at the Laboratory for Human Brain Dynamics (1998-2009), Brain Science Institute (BSI), RIKEN, Japan. The main part of the analysis was conducted at the Laboratory for Human Brain Dynamics, in Nicosia, Cyprus. Authoring of the study was completed at the Laboratory of Medical Informatics, Aristotle University of Thessaloniki, in Thessaloniki, Greece. This study was funded by the Operational Program "Education and Lifelong Learning" of the Greek Ministry of Education and Religious Affairs, Culture and Sports (ref. number 2012ΣΕ24580284). Extensive neurophysiologic discussion on the amygdala findings can be found in [23]. Our group's discussion on the cerebellum is currently under review.

REFERENCES

1. Russell JA (1980) A circumplex model of affect. Journal of Personality and Social Psychology 39:1161-1178
2. Nielen MMA, Heslenfeld DJ, Heinen K, et al. (2009) Distinct brain systems underlie the processing of valence and arousal of affective pictures. Brain and Cognition 71: 387–396
3. Colibazzi T, Posner J, Wang Z, et al. (2010) Neural systems subserving valence and arousal during the experience of induced emotions. Emotion 10:377-389
4. Poldrack RA (2010) Mapping mental function to brain structure: how can cognitive neuroimaging succeed? Perspect. Psychol. Sci. 5: 753–761
5. Sah P, Faber ES, Lopez DA et al. (2003) The amygdaloid complex: anatomy and physiology. Physiol Rev. 83: 803–834
6. Amunts K, Kedo O, Kindler M et al. (2005) Cytoarchitectonic mapping of the human amygdala, hippocampal region and entorhinal cortex: inter- subject variability and probability maps. Anat.Embryol. (Berl) 210: 343–352
7. Ball T, Rahm B, Eickhoff SB et al. (2007) Response properties of human amygdala subregions: evidence based on functional MRI combined with probabilistic anatomical maps PLoS ONE 2, e307
8. Schmahmann JD, Doyon J, Toga AW et al. (2000) MRI Atlas of the Human Cerebellum (San Diego, CA: Academic Press)
9. Heath RG (1977). Modulation of emotion with a brain pacemaker: Treatment for intractable psychiatric illness J Nerv Ment Dis:165:300–17
10. Schmahmann JD (1991) An emerging concept: The cerebellar contribution to higher function Arch. Neurol. 48: 1178–1187
11. Schmahmann JD (2004) Disorders of the cerebellum Ataxia, dysmetria of thought, and the cerebellar cognitive affective syndrome J. Neuropsychiatry Clin. Neurosci. 16: 367–378
12. Winston JS, Gottfried J, Kilner JM, et al. (2005) Integrated neural representations of odor intensity and affective valence in human amygdala Journal of Neuroscience 25, 8903–7
13. Lewis PA, Critchley HD, Rotshtein P, et al. (2007). Neural correlates of processing valence and arousal in affective words. Cerebral Cortex. 17, 742–748.
14. Posner J, Russell JA, Gerber A, et al. (2009) The neurophysiological bases of emotion: An fMRI stuy of the affective circumplex using emotion-denoting words Human Brain Mapping 30:883–895
15. Styliadis C, Papadelis C, Bamidis PD (2010) Neuroimaging of emotional activation: Issues on Experimental Methodology, Analysis and Statistics , XII Mediterranean Conference on Medical and Bibliological Engineering and Computing - Medicon, 27-30 May, Porto Carras, Marmaras, Greece.
16. Lang PJ, Bradley MM, Cuthbert BN (2005) International affective picture system (IAPS): Instruction manual and affective ratings. Technical Report A-6, The Center for Research in Psychophysiology, University of Florida
17. Cuthbert BN, Bradley MM, Lang PJ (1996) Probing picture perception: activation and emotion. Psychophysiology 33: 103–111,
18. Robinson SE, Vrba J (1999) Functional neuroimaging by synthetic aperture magnetometry (SAM). In: Yoshimoto T, Kotani M, Kuriki S, Karibe H, Nakasato N, editors. Recent advances in biomagnetism. Sendai: Tohoku University Press 302–305
19. Singh KD, Barnes G, Hillebrand A, et al. (2002) Task-related changes in cortical synchronization are spatially coincident with the hemodynamic response Neuroimage 16:103–114.
20. Singh KD, Barnes G, Hillebrand A (2003). Group imaging of task-related changes in cortical synchronisation using non-parametric permutation testing Neuroimage 19:1589–1601
21. Diedrichsen J, Balsters JH, Flavell J, et al. (2009). A probabilistic atlas of the human cerebellum. Neuroimage.
22. Eickhoff S, Stephan,KE, Mohlberg H, et al. (2005) A new SPM toolbox for combining probabilistic cytoarchitectonic maps and functional imaging data. NeuroImage. 25: 1325-1335.05
23. Styliadis C, Ioannides AA, Bamidis PD, & Papadelis C (2013) Amygdala responses to valence and its interaction by arousal revealed by MEG. International Journal of Psychophysiology. Epub ahead of print]

Author: Styliadis Charalampos
Institute: Laboratory of Medical Informatics, School of Medicine, Aristotle University of Thessaloniki
Street: P.O. Box 323, 54124
City: Thessaloniki
Country: Greece
Email: styliadis@hotmail.com

AFM Multimode Imaging and Nanoindetation Method for Assessing Collagen Nanoscale Thin Films Heterogeneity

A. Stylianou, S.V. Kontomaris, D. Yova, and G. Balogiannis

Biomedical Optics & Applied Biophysics Lab, School of Electrical and Computer Engineering,
National Technical University of Athens, Athens, Greece

Abstract—Atomic Force Microscopy (AFM) operates in a variety of modes and techniques. AFM can offer a wide range of information, from topography to mechanical properties of surfaces and interfaces, including those of biomaterials and scaffolds. In this paper it was sought to gain insights of structural and mechanical heterogeneity of collagen fibers in thin films by combining AFM multimode imaging, including phase imaging, with quantitative measurements through nanoindetation. Due to its filamentous shape and its associative properties collagen type I is a very promising molecule for the development of nanostructures, scaffolds in tissue engineering and nanobiomaterials Since collagen based materials nanotopography and mechanical properties can influence cell-biomaterial/scaffold interactions it is of crucial importance to characterize its surface and heterogeneity in the nanoscale. Among the different collagen-based biomaterials, collagen thin films are of great interest since they possess unique properties and can be used for forming novel biomaterials or for covering non-biological surfaces in order to offer them biocompatibility. The results demonstrated that the overlap and gap region on collagen fibers (D-periodicity) yield a significant phase contrast, due to different mechanical properties. In addition, phase contrast was also demonstrated in kinks (areas where collagen fibers changes abruptly direction) which provides evidence that collagen fiber shell and core possess different properties. The quantitative measurements with nanoindentation method confirmed the heterogeneity of collagen fibers D-periodicity since overlapping zones were characterized by a higher Young modulus (0.7 GPa) than the gap zones (0.46 GPa). The correlation between the heterogeneous structure and the mechanical properties of collagen fibrils in thin films will enable the design and development of biomaterials and tissue scaffolds with improved properties, as well as it will enable the investigation of cell response on different nanoscale features.

Keywords—AFM, Phase Imaging, Nanoindentation, Collagen, D-band.

I. INTRODUCTION

Atomic Force Microscopies (AFMs) record interactions between a probe and a surface and permit quantitative, high-resolution, non-destructive imaging of surfaces, including biological ones [1]. In addition they operate in many different modes, offering a wide range of information about biomaterials properties [2,3], from topography to mechanical properties [4,5]. Among the different modes, AFM can operate in tapping mode, which is a 'gentle' imaging mode particularly useful for imaging soft and fragile samples, like biopolymer thin films. In this mode the lateral forces in the sample are circumvented and the frictional forces are minimized [1,4]. In tapping mode, topography and phase images are simultaneously acquire so as to obtain different properties of the sample [6]. On the other hand, phase imaging goes beyond topographical features and is appropriate for detecting variations in composition, friction, viscoelasticity, producing contrast on heterogeneous samples [7].

In addition, AFM is increasingly being used to measure the mechanical properties of the sample in nanoscale using nanoindentation method. Nanoindentation is a technique used in testing mechanical properties of materials among which and biological [8-10]. The quantitative parameter for the determination of material elastic properties is stiffness which describes the relation between an applied, nondestructive load and the resultant deformation. Stiffness can be used for the determination of the mechanical heterogeneity of materials in different nanoregions [10].

Type I collagen is the most abundant protein in mammals and it provides structural support for the tissues as a result of its highly organized fibrils [11]. Due to its filamentous shape and its associative properties it is a very promising molecule for the development of nanostructures, scaffolds in tissue engineering and nanobiomaterials [12]. The surface properties of biomaterials play an important role in biomedicine as the majority of biological reactions occur on surfaces/interfaces and their quality could limit biomaterials applications. Furthermore, biomaterials nano-surface characteristics affects cell adhesion and proliferation [13,14]. For instance, collagen surfaces with align fibers or porous structure have better performance and can guide cells growth [15].

The purpose of this paper was to investigate the structural and mechanical heterogeneity of collagen fibers in thin films by combining AFM multimode imaging with quantitative measurements through nanoindetation. Special attention was given in order to characterize collagen nanoscale features, like the characteristic D-band repletion, of approximately 67 nm, which is consisted by gap and overlapping regions [11]. The investigation was performed in collagen thin films since these films are of great interest in the fields of biomaterials and tissue engineering since they possess unique nanoscale characteristics [4,16]. They are useful to address a variety of biological issues, including cell morphology and the influence of surface properties on intracellular signaling and can be used to cover non-biological surfaces offering them biocompatibility. In addition, thin films can be imaged with high resolution with AFM [17-19].

II. MATERIALS AND METHODS

A. Collagen Thin Film Formation

Type I collagen from bovine Achilles tendon (Fluka 27662) was dissolved in acetic acid (0.5 M) in a final concentration of 8 mg/ml and stored in 4 °C for 24h. The solution was then homogenized at 24000 rpm (IKA T18 Basic) and stored in 4 °C as the stock solution. Part of the collagen solution (50 µl) was flushed on fresh cleaved mica discs (71856-01, Electron Microscopy Science) and spin coated (WS-400B-6NPP/LITE Laurell Technologies) for 40 sec at 6000 rpm in order to form random oriented thin films.

B. Atomic Force Microscopy

AFM experiments (imaging and indentation tests) were carried out using a commercial microscope (CP II, Veeco) in contact and intermittent (also named tapping) mode in air, at room temperature. For tapping mode typical AFM probes (MPP-11123-10, Veeco) were used and topography (or height) and phase images were acquired simultaneously. In Phase Imaging the detector signal is the phase φ, of the cantilever oscillations relative to the phase of drive signal.

The AFM nanoindentation experiments were performed with pyramidal tips (MLCT; Veeco) and load – indentation curves were acquired. The formula between elastic modulus and stiffness can be written in the following form:

$$E = \frac{\sqrt{\pi}}{2}\left(1 - v^2\right)\frac{S}{\sqrt{A}} \quad (1)$$

where, E is the Young modulus, ν the Poisson's ratio, S the contact stiffness, and A an area function related to the effective cross-sectional or projecting area of the indenter. For all calculations, the Poisson ratios were assumed to be 0.2. This value is consistent with the literature [20]. This equation provides a very general relation that can be applied to any axisymmetric indenter [10,20]. Moreover, it can be used not only for elastic but also for elastic – plastic contact as well [10,21]. For the calculation of the elastic modulus values an algorithm was developed using Matlab (The MathWorks, Natick, MA).

The AFM image processing was made by using the image analysis software that accompanied the AFM system DI SPMLab NT ver.60.2, IP-Image Processing and Data Analysis ver.2.1.15 (Veeco) and the freeware scanning probe microscopy software WSxM 5.0 dev.2.1 [22].

III. RESULTS AND DISCUSSION

A. AFM Multimode Imaging of Collagen Thin Films

In this section AFM imaging in different modes was applied in order to characterize thin collagen films surface heterogeneity. Figure 1 demonstrates a quite large area (18x16 µm) of a collagen thin film in 2 different image types, topography and phase, respectively. It can be seen that the films consist of random oriented collagen fibers/fibrils with diameters of some hundred nm. In Figure 1a the image shows the topography and shape of the surface features of the thin film area. Roughness measurements showed that the surface had quite large Root-Mean-Squared Roughness (Rrms), ~31 nm, due to the fibrous structure of collagen fibers. Surface roughness is of great interest as it strongly affects the interaction between biomaterials and the cells due to an increase of the surface which is available for cell adhesion and growth [23]. Figure 1b displays the phase image of the same area. The image presents high phase contrast which indicates that the films are characterized by different properties. In addition, as Figure 2 highlights, collagen thin films were consisting of collagen fibers with the characteristic D-band of the 67 nm, which is a unique characteristic of the natural occurring fibers. The 67 nm band is the periodicity observed in natural collagen fibrils and indicates that in our films the collagen molecules in the fibrils are ordered in a native-like manner [11,24].

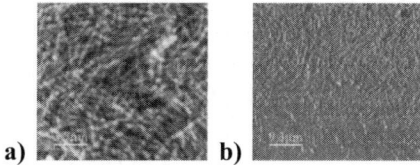

Fig. 1 Multimode AFM images of collagen thin film. a) Topography and b) Phase image of the same area of the collagen film.

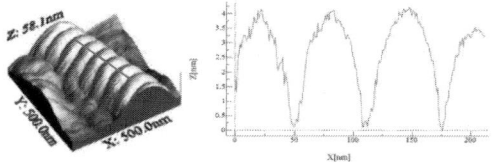

Fig. 2 The D-band of collagen fibrils. 3D Topography AFM image of a collagen fibril with D-band (Left) and the height profile showing the periodicity of ~67 nm (Right)

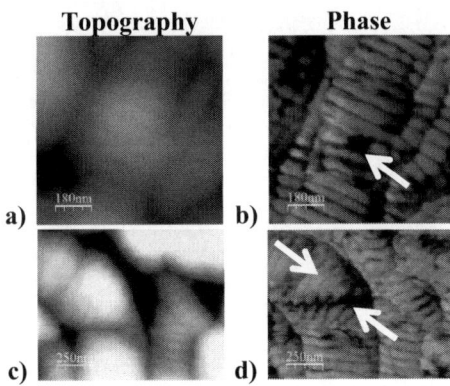

Fig. 3 AFM Topography (a,c) and the relevant Phase images (b,d) of collagen fibers. The images demonstrates the D-periodicity (a,b) and a kick (c,d).

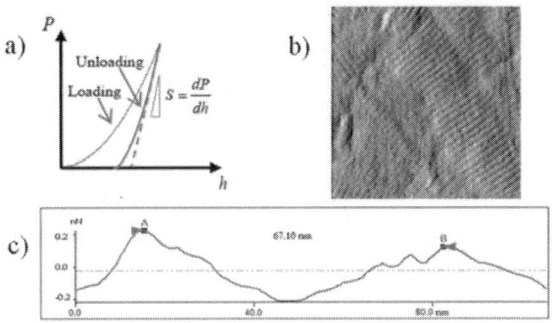

Fig. 4 a) Schematic illustration of load (P) –indentation (h) curves, b) Image of collagen fibril. The D band periodicity is obvious. The size of the image is 2μm×2μm, c) Height profile diagram of collagen fibril. The D band periodicity is approximately 67nm.

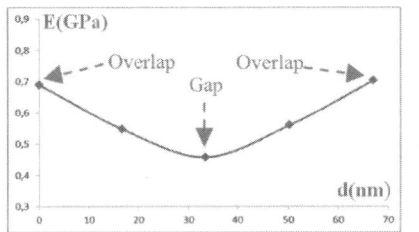

Fig. 5 Young modulus values ranged between 0.46GPa (gap regions) to 0.7GPa (overlapping regions).

Furthermore, high resolution images of smaller areas were acquired to investigate individual fibers structure heterogeneity (Fig.3). Figures 3a,b present the D-periodicity of a collagen fiber. In the phase image (Fig.3b) it can be clearly seen that there is a significant contrast between gap and overlap regions, which demonstrates that the two regions have probably different mechanical properties. Heterogeneity was also reported to induce greater phase contrast in the overlap regions than the more elastic gap zones in collagen fibers from dentin [25]. In addition, collagen fiber characteristics, like the destruction of the D-periodicity along the fiber (arrow in Fig. 3b), can be observed in in phase images. On the other hand, in topography image (Fig.3a) these characteristics were not highlighted.

Finally, collagen fibers with kinks (areas where collagen fibers changes abruptly direction) were also imaged (Fig. 3c,d). The arrows in the phase image show a straight and kink part of the same collagen fiber (Fig. 3d). The different contrast between the two parts demonstrates that fiber possesses different properties in these parts. These kinks have been reported to provide evidence that collagen fibers are characterized by a stiffer shell than the core, and collagen fibers are compared to tubes [26,27]. Consequently, the kinks are a result of strong mechanical deformation in tube-like structures. The different phase contrast that was demonstrated in this paper comes to support the tube-like structure of collagen fibers.

B. AFM Nanoindetation of Collagen Thin Films

It is known that the heterogeneous structure of collagen fibrils has an effect on their mechanical properties. This heterogeneity has been determined by obtaining 5 load – indentation curves. A Load - indentation curve, is used for determining the material contact stiffness (S), which can be obtained from the initial unloading part of the curve as it shown in Figure 4a. However, stiffness is not a suitable parameter for the characterization of the material elasticity because its value depends on the deformation geometry. Hence, the material elasticity must be characterized by its Young modulus value which can be determined using Oliver – Pharr analysis [4,9]. Five curves were obtained within a characteristic D period of 67nm in order to calculate the Young modulus for the gap and the overlapping regions (Figure 4b,c). Each load – indentation curve, was the average of 10 curves. The first curve was obtained on the left overlapping region (point A in Figure 4c) and corresponds to Young modulus value of 0.7 GPa. Moreover 4 curves were obtained and each one was at a distance of ~17 nm from the previous, so that the first and the last curve were within a D-period. As it is shown in Figure 5 the Young modulus value decreases from the overlapping region (0.7GPa) to (0.46GPa) at the gap region.

The results, which are in correlation with those obtained by S. Strasser [26], demonstrated that the D – band periodicity in the axial direction of the collagen fibril is responsible for the different mechanical properties on different nanoregions of the same fibril.

IV. CONCLUSIONS

In this paper it was shown that the combination of AFM multimode imaging and nanoindentation method can offer supplementary information collagen nanoscale heterogeneity in collagen thin films are characterized by heterogeneity in the nanoscale. The gap and overlap regions of the D-band along collagen fibers were found to have both different phase contrast and elastic modulus. Moreover, the kinks which are located in collagen fibers were found to have different phase contrast than that of the straight parts. Type I collagen is an anisotropic material not only due to its structural organization but also for its mechanical properties. The correlation between the heterogeneous structure and the mechanical properties of collagen fibrils in thin films will enable the design and development of biomaterials and scaffolds with improved properties. Finally, the clarification of collagen heterogeneity in nanoscale in thin films will enable investigation and correlation of cell response on different nanoscale features.

V. ACKNOWLEDGMENT

This research has been co-financed by the European Union (European Social Fund-ESF) and Greek national funds through the Operational Program "Education and Lifelong Learning" of the National Strategic Reference Framework (NSRF)-Research Funding Program: Heracleitus II. Investigating in knowledge society through the European Social Fund.

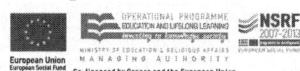

REFERENCES

1. Gadegaard N (2006) Atomic force microscopy in biology: Technology and techniques. Biotech Histochem 81:87-97
2. Stylianou A.,Yova , D, Politopoulos K (2012) Atomic Force Microscopy Quantitative and Qualitative Nanoscale Characterization of Collagen Thin Films. Proc of 5th ETNDT, Ioannina, Greece, 2012, pp. 415-420
3. Stylianou A, Politopoulos, K, Yova D (2011) Atomic Force Microscopy Imaging of the Nanoscale Assembly of Type I Collagen on Controlled Polystyrene Particles Surface. IFMBE Proc. vol.37, 5th European IFMBE Conference, Budapest, Hungary, 2011, pp 1058-1061
4. Stylianou, A.,Yova, D.(in press) Surface Nanoscale Imaging of Collagen Thin Films by Atomic Force Microscopy. Mat Sci Eng C doi.org/10.1016/j.msec.2013.03.029
5. Kontomaris SV, Stylianou A, Yova D, Politopoulos K (2012) Mechanical Properties of Collagen Fibrils on Thin Films by Atomic Force Microscopy Nanoindentation. Proc. of the 12th International Conference on Bioinformatics & Bioengineering (BIBE) art no. 6399742, Larnaca, Cyprus, 2012, pp 608-613
6. Morris VJ, Kirby AR, Gunning AP (2008) Atomic Force Microscopy for Biologists , fourth ed., Imperial College Press, London
7. Habelitz S, Balooch M, Marshall SJ et al. (2002) In situ atomic force microscopy of partially demineralized human dentin collagen fibrils. J Struct Biol 138:227-236
8. Stolz M, Raiteri R, Daniels AU et al. (2004) Dynamic elastic modulus of porcine articular cartilage determined at two different levels of tissue organization by indentation–type atomic force microscopy. Biophys J 86:3269–3283
9. Stolz M, Gottardi R, Raiteri R et al. (2009) Early detection of aging cartilage and osteoarthritis in mice and patient samples using atomic force microscopy. Nat Nanotechnol 4:186–192
10. Oliver WC, Pharr GM (2004) Measurement of hardness and elastic modulus by instrumented indentation: Advances in understanding, and refinements to methodology. J Mater Res 19:3-20
11. Fratzl P (2008) Collagen: Structure and Mechanics. Springer, New York
12. Hasirce V, Vrana E, Zorlutuna P et al. (2006) Nanobiomaterials: a review of the existing science and technology, and new approaches. J Biomater Sci Polymer Edn 17:1241-1268
13. Phong HQ, Wang S, Wang M (2010) Cell behaviors on micro-patterned porous thin films. Mater Sci Eng B Solid State Adv Technol 169:94-100
14. Tay CY, Irvine SA, Boey FYC et al. (2011) Micro-/nano-engineered cellular responses for soft tissue engineering and biomedical applications. Small 7:1361-1378
15. Brouwer KM, van Rensch P, Harbers VE et al. (2011) Evaluation of methods for the construction of collagenous scaffolds with a radial pore structure for tissue engineering. J Tissue Eng Regenerative Med 5:501-504
16. Lu J.T, Lee C.J et al. (2007) Thin collagen film scaffolds for retinal epithelial cell culture. Biomaterilas 28:1486-1494
17. Stylianou A,Yova D, Politopoulos K (2012) Atomic Force Microscopy Surface Nanocharacterization of UV-Irradiated Collagen Thin Films. Proc. of the 12th International Conference on Bioinformatics & Bioengineering (BIBE) art no. 6399742, Larnaca, Cyprus, 2012, pp 602-607
18. Stylianou A, Politopoulos K, Kyriazi M, Yova D (2011) Combined information from AFM imaging and SHG signal analysis of collagen thin films. Biomed Signal Proces 6 (3):307-313
19. Stylianou A, Kontomaris SB, Kyriazi M, Yova D.(2010) Surface characterization of collagen films by atomic force microscopy. IFMBE Proc. vol. 29, 12th Mediterranean Conference on Med. & Biol. Eng. & Com., MEDICON 2010; Chalkidiki, Greece, 2010, pp 612-615
20. Chung KH, Bhadriraju K, Spurlin TA et al. (2010) Nanomechanical Properties of Thin Films of Type I Collagen Fibrils. Langmuir 26:3629–3636
21. Fischer-Cripps AC (2009) The IBIS Handbook of Nanoindentation. Forestville NSW 2087 Australia, Fischer-Cripps Laboratories Pty Ltd
22. Horcas I., Fernández R., Gómez-Rodríguez JM et al. (2007) WSXM: A software for scanning probe microscopy and a tool for nanotechnology. Rev Sci Instrum 78(1), art. no. 013705
23. Covani, U. Giacomelli L, Krajewski A et al. (2007) Biomaterials for orthopedics: A roughness analysis by atomic force microscopy. J. Biomed Mate. Res 82A: 723-730
24. Shoulders MD, Raines RT (2009) Collagen structure and stability. Annu Rev Biochem 78:929-958
25. Bertassoni LE, Marshall GW, Swain MV (2012) Mechanical heterogeineity of dentin at different length scales as determined by AFM phase contrast. Micron 43:1364-1371
26. Strasser S, Zink A, Janko M et al. (2007) Structural investigations on native collagen type I fibrils using AFM. Biochem Biophys Res Commun 354:27-32
27. Gutsmann T, Fantner GE, Venturoni M et al. (2003) Evidence that collagen fibrils in tendons are inhomogeneously structured in a tube-like manner. Biophys J 84:2593-2598

Author: Stylianou Andreas
Institute: National Technical University of Athens, Athens
Street: Iroon Polytexneiou 9, Zografou Campus, 15780
City: Athens
Country: Greece
Email: styliand@mail.ntua.gr

Segmentation of Corneal Endothelial Cells Contour through Classification of Individual Component Signatures

E. Poletti and A. Ruggeri

Department of Information Engineering, University of Padova, Padova, Italy

Abstract—Corneal images acquired by in-vivo specular microscopy provide clinical information on the cornea endothelium health state. At present, the analysis is based on manual or semi-automatic methods and the segmentation of a large number of endothelial cells is required for a meaningful estimation of the clinical parameters (cell density, pleomorphism, polymegethism). For the practical application in clinical settings, a computerized method capable to fully automatize the segmentation procedure would be needed.

We propose here a supervised classification scheme for the segmentation of endothelium cells. In order to detect the cell contour polygon, i.e., its three components as vertexes, sides and body, a multi-scale 2-dimensional matched filter approach is employed. Three kernels have been specifically designed to the detect the three cell components' *signatures*, which are then used as features to train a Support Vector Machine classifier and to provide the final segmentation of the cells.

Performance of the proposed method is assessed by computing on a set of 20 images the differences in the three clinical parameters estimated from automatic segmentation and from manual segmentation. The results confirm that the automated system is capable to provide reliable estimates of these important clinical parameters.

Keywords—Image Segmentation, Image Classification, Matched Filters, Corneal Endothelium, Microscopy.

I. INTRODUCTION

The analysis of the main morphometric parameters of corneal endothelium provides clinical information capable to describe the cornea health state. Namely, *endothelial cell density* (ECD), *polymegethism* (differences in cell size expressed as fractional standard deviation of cell areas), and *pleomorphism* or *hexagonality coefficient* (fraction of hexagonal cells over the total number of cells) are commonly used as parameters to quantitatively characterize the endothelial cells' condition.

In order to make this analysis practical in clinical settings, a computerized method that fully automates the segmentation procedure would be needed. The fundamental problem with automated endothelial analysis is to correctly identify the cells' contour, a necessary prerequisite for the estimation of the clinical parameters [1]. Several computer programs have been proposed to accomplish this task [2], even if to the best of our knowledge they are only semi-automated, or work in a non-clinical context, e.g., with stained cells [3]. In the former case, the cell border detection provided by the computer needs to be revised by the user to correct inaccuracies by manual adjustment.

It has been reported that at least 75 cells per image should be evaluated for a reliable estimation of clinical parameters [4]. Although manual correction improves the accuracy of this estimation, it is tedious and time-consuming and therefore usually impractical in a clinical setting. This often leads the user to reduce the number of outlined cells to just a few tens, greatly affecting the accuracy of estimated parameters and thus the reliability of the clinical outcome.

We propose here a reliable, fully automated algorithm for the segmentation of the endothelial cell contours.

II. MATERIAL

Thirty images of corneal endothelium were acquired with a specular endothelial microscope (*SP-3000P, Topcon Co., Japan*) from both healthy and pathological subjects. The images cover an area of 0.25 x 0.5 mm^2 and were saved as 8-bits, 240 x 480 pixels grayscale images (Fig. 1).

In order to assess the accuracy of the morphometric parameters estimated by the computerized procedure, ground truth reference values were obtained by estimating the parameters on manually segmented images. Since in the peripheral regions of the image the poor quality often prevents a reliable detection of cell contours, a region of interest (ROI) was manually defined. In each image, the ROI covers an area of about 0.1 mm^2, which includes an average of 220 cells, enough to allow a reliable estimation of the morphometric parameters. For each image, all visible cell contours inside the ROI were manually traced with care by using a public-domain image manipulation program (GIMP v. 2.8, http://www.gimp.org), so as to outline the polygonal shape of each cell (see Fig 3-b).

Ten images from our data set were randomly chosen and used to develop the method, e.g., train classifier, whereas the remaining 20 images were used only for validation. All the original and the manually segmented images are publicly available for download at *http://bioimlab.dei.unipd.it*.

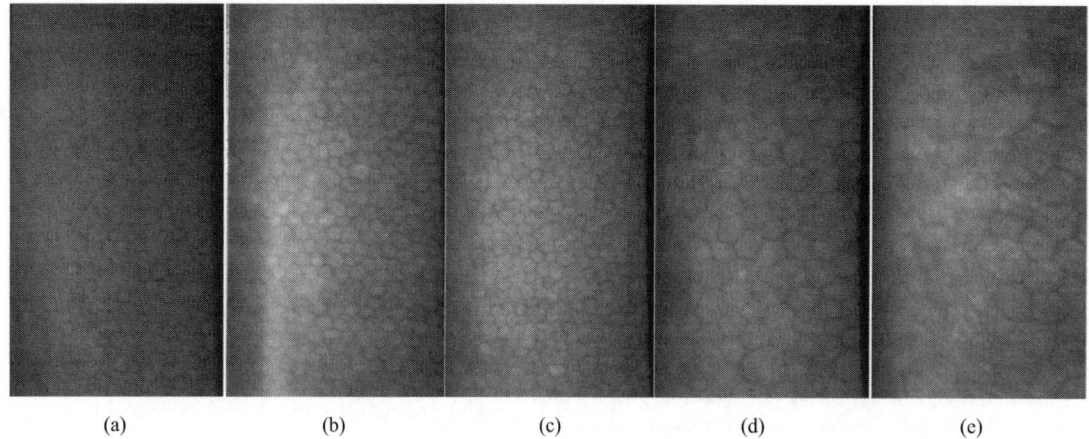

Fig. 1 Corneal endothelium images: (a) normal subject – good quality image, (b) normal subject – image with illumination drift artifact, (c) subject with high *polymegethism*, (d) subject with low ECD, and (e) subject with very low ECD.

III. METHODS

Cells appear in the image as relatively regular polygons with different sizes, orientations and numbers of sides. The rationale of the proposed approach is to extract, from each pixel neighborhood, three key features, which we call *signatures*, corresponding to the three different structures that define a cell: the vertex, the side, and the inner body of the cell itself. These three *signatures* are then used as features to train a classifier, used to label the pixel as cell vertex, side, or body.

The three *signatures* are extracted by means of a multi-scale 2-dimensional matched filter approach, in which each pixel in the image is convolved with three customized kernels at different scale and orientations. Each of the kernels has been specifically designed to provide a high response when its position, scale and orientation match those of the corresponding structure (vertex, side, or body).

A. The Cell Vertex Detector

The vertex of a cell polygon is characterized by tree dark line segments that have a pixel in common, describing a Y-shaped pattern. Albeit the angles between each pair of segments are different for each vertex, we can reasonably assume they belong to the interval $[\pi/2, 2\pi]$.

In order to extract the cell vertex signature, we designed a matched filter with a 2-dimensional kernel defined by the sum of four 2-dimensional isometric Gaussian functions (Fig. 2). Let assume that the center of the filter at scale s is at coordinates (x_0, y_0). The first Gaussian function, $G_0(s)$, has an amplitude of 2, is set at the center of the filter, and its standard deviation is $\sigma=2s$:

$$G_0(s) = 2\mathrm{N}\big((x_0, y_0), ((2s)^2, (2s)^2)\big) \quad (1)$$

The other three Gaussian functions, G_1, G_2, and G_3, have amplitude of -1, are positioned at a distance $s/2$ from the

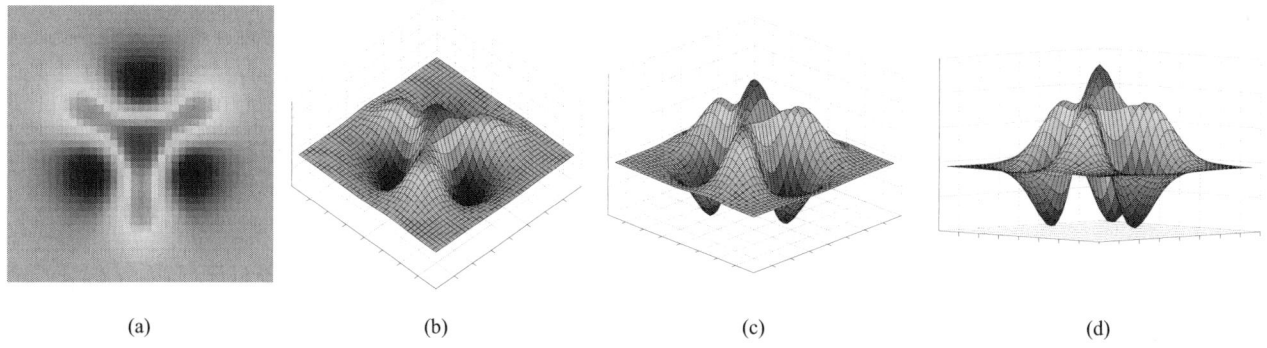

Fig. 2 Kernel of the matched filter for the identification of cell vertexes:
(a) intensity image; (b) isometric view with -45° of azimuth and vertical elevation of 60°, (c) 30°, and (d) 5°.

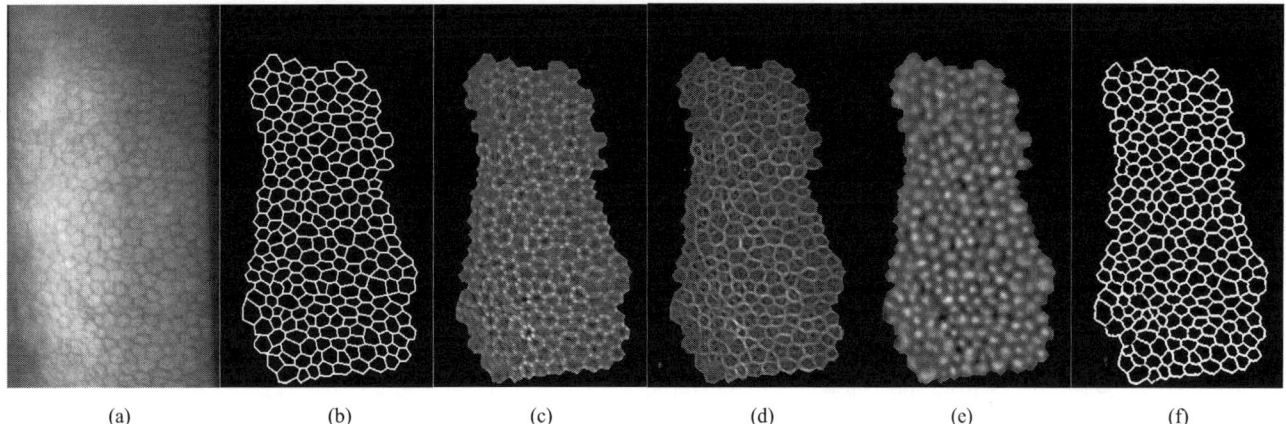

Fig. 1 (a) original Image; (b) ground truth manual segmentation inside the ROI; max response between all rotation of the convolution between the original image and (c) the vertex detector at s=4, (d) cell side detector at s=11, and (e) cell body detector at σ=5. (f) automatic classification outcome after the removal of spurious branches.

center (x_0, y_0) at the three different angular positions $\vartheta \in \{\pi/2, 5\pi/4, 7\pi/4\}$, and have a standard deviation $\sigma = s$:

$$G_n(s) = -N\left(\left(x_0 + \frac{s}{2}\cos\vartheta_n, y_0 + \frac{s}{2}\sin\vartheta_n\right), (s^2, s^2)\right) \quad (2)$$

with $n \in \{1, 2, 3\}$. The vertex detector at scale s is then defined as $VD(s) = \sum_{n=\{0,1,2,3\}} G_n(s)$.

Convolutions of the image with $VD(s)$ at different scales ($s = 3, 4, 5, 6, 7, 18$ pixels) and different rotations (18 for each scale) have been used to detect most of the possible occurrences of cell vertexes (Fig. 3-c). Eventually, for each pixel in the image the maximum response among all scales and rotations is recorded and taken as the cell vertex signature for the classifier.

B. The Cell Side Detector

In order to improve the vessel segmentation in retinal images, several papers introduced measures of "vesselness", which describe the likelihood of an elongated structure to be a vessel [5]. We propose here a similar measure to indicate the likelihood of an elongated structure to be the side of a cell. The method we adopted is based on the *line operator*, which computes the convolution of the image with a filter that represents a line passing through the center of the filter, at multiple orientations. In our implementation, 36 oriented lines with length of 5, 7, 9, 11 pixels have been employed. The cell side signature is the maximum response among the convolution with the three line operators (see Fig. 3-d).

C. The Cell Body Detector

The body of a cell appears as a bright area enclosed within dark lines. In order to extract its signature, we designed a 2-dimensional Laplacian of Gaussian matched filter. As in the previous cases, convolution has been carried out at different scales of the filter (σ = 5, 6, 7 pixels, with σ = standard deviation of the original Gaussian prior to the Laplacian operator). The maximum response among the scales was recorded and used as cell body signature.

D. Cell Contour Classification and Segmentation

We employed as classifier a Support Vector Machine (SVM) from the MATLAB® libSVM library [6]. Training was carried out with a Radial Basis Function kernel. In order to find the optimal parameters of the SVM model (γ and penalty parameter c) we performed a grid search on an exponential grid, with both γ and c ranging from 2^{-6} to 2^6. The parameter pair chosen is the one providing the best accuracy in a 2-fold cross-validation, using only a randomly chosen 0.5% of pixels from the 10 images used for training.

The binary classified images (white for vertexes and sides, black for body, see Fig. 3-f) were skeletonized to extract the cell contours as 1-pixel thick lines. To cope with possible spurious branches present in the skeleton, a post processing procedure based on morphological operation was also carried out.

IV. RESULTS

From both sets of segmented images (automated and manual), area and number of sides of all detected cells were straightforwardly derived. From these quantities, the estimation of ECD, pleomorphism, and polymegethism was carried out. ECD was computed as the sum of individual cell areas divided by the total number of cells; pleomorphism was computed by counting for each cell the number of neighboring cells (cells along the border of the ROI were excluded from this computation) and taking the percentage of cells with hexagonal shape; polymegethism was computed as the fractional standard deviation of all cell areas.

$$\text{pleomorphism} = \frac{\text{\# of exagonal cells}}{\text{\# of cells}} \cdot 100 \quad (3)$$

$$\text{polymegethism} = \frac{\text{sd(cells' area)}}{\text{mean(cells' area)}} \cdot 100 \quad (4)$$

Table 1 reports the statistics (mean, standard deviation, minimum and maximum) of the differences in morphometric parameters as estimated from the automatic segmentation and from the manual one.

The average run time for a complete image analysis with the current Matlab® prototype is 7 s.

V. DISCUSSION AND CONCLUSION

We presented here a system for the estimation of cornea endothelium morphometric parameters that requires no user intervention. The estimates of the parameters provided by the proposed algorithm are in very good agreement with ground truth, obtained with a careful manual analysis.

As regards ECD, its estimation can actually be performed with acceptable accuracy by many computerized systems, as the presence of some errors in cell detection is of limited impact on the final ECD estimation. On the contrary, as already noted by several authors, e.g. [7], the quantitative estimation of pleomorphism and polymegethism is significantly affected by errors in contour detection even in few cells, making the reliable estimation of these parameters quite difficult. A possible solution is the manual correction of cell borders, but this involves a significant amount of work, e.g., on about 50 to 75 per cent of cell borders [7].

For the reasons explained above, in large experimental studies with hundreds of images, the experimenter usually restricts him/herself to measure only ECD. The system we propose can be extremely valuable in these studies, as it allows to reliably estimate also the other two morphometric parameters. Moreover, it can process images containing hundreds (and not tens) of cells and thus provide a much higher accuracy for the estimated parameters. This could be further increased by performing the analysis on several images per subject, each positioned at slightly different locations in central cornea, and then averaging the results.

The present version of the proposed system, although being still a research prototype with no computational optimization, has a run time of 5~10 seconds per image, allowing a multi-image analysis to be performed in less than one minute per subject.

Table 1 Summary results for ECD, pleomorphism, and polymegethism estimations. Differences between automatic and manual assessment of the parameters are expressed as difference (diff), its percent (diff %), absolute difference (abs diff) and its percent (abs diff %). ECD values are in cells/mm^2, pleomorphism and polymegethism are per cent values.

ECD	diff	diff %	abs diff	abs diff %
mean	28.81	1.14 %	96.35	3.11 %
sd	109.15	3.68 %	58.83	2.29 %
min	-151.00	-4.53 %	18.00	0.57 %
max	245.00	10.37 %	245.00	10.37 %
Pleomor-phism	diff	diff %	abs diff	abs diff %
mean	-2,57	-4,41 %	4,21	7,32 %
sd	4,31	7,42 %	2,72	4,57 %
min	-11,80	-16,21 %	0,00	0,00 %
max	6,90	11,48 %	11,80	16,21 %
Poly-megethism	diff	diff %	abs diff	abs diff %
mean	-5,17	-13,40 %	5,72	14,78 %
sd	3,60	8,58 %	2,64	5,90 %
min	-12,20	-24,30 %	1,10	2,20 %
max	3,80	9,00 %	12,20	24,30 %

REFERENCES

1. Imre L, Nagymihaly A. Reliability and reproducibility of corneal endothelial image analysis by in vivo confocal microscopy, Graefes Arch Clin Exp Ophthalmol, 2001
2. Patel DV, McGhee CN. Quantitative analysis of in vivo confocal microscopy images: A review, Surv Ophthalmol, 2013
3. Ruggeri A et al. A system for the automatic estimation of morphometric parameters of corneal endothelium in alizarine red-stained images, Br J Ophthalmol, 2010
4. Doughty MJ, Muller A, Zaman ML. Assessment of the reliability of human corneal endothelial cell-density estimates using a noncontact specular microscope, Cornea, 2000
5. Poletti E, Ruggeri A. Segmentation of vessels through supervised classification in wide-field retina images of infants with Retinopathy of Prematurity, 25th International Symposium on CBMS, IEEE, 2012
6. Chang CC, Lin CJ. LIBSVM: a library for support vector machines, ACM TIST, 2011.
7. Doughty MJ, Aakre BM. Further analysis of assessments of the coefficient of variation of corneal endothelial cell areas from specular microscopic images, Clin Exp Optom, 2008

Author: Alfredo Ruggeri
Institute: Dept. of Information Engineering, University of Padova
Street: Via Gradenigo, 6/B
City: Padova
Country: Italy
Email: alfredo.ruggeri@unipd.it

Split Bregman-Singular Value Analysis Approach to Solve the Compressed Sensing Problem of Fluorescence Diffuse Optical Tomography

J. Chamorro-Servent[1,2], J.F.P.J. Abascal[1,2], J. Ripoll[1,2], J.J. Vaquero[1,2], and M. Desco[1,2,3]

[1] Departamento de Bioingeniería e Ingeniería Aeroespacial, Universidad Carlos III de Madrid, Spain
[2] Instituto de Investigación Sanitaria Gregorio Marañón, Madrid, Spain
[3] Centro de Investigación Sanitaria en Red de Salud Mental (CIBERSAM), Madrid, Spain

Abstract—Compressed Sensing (CS) techniques are becoming increasingly popular to speed up data acquisition in many modalities. However, most of CS theory is devoted to undetermined problems and there are few contributions that apply it to ill-posed problems. In this work we present a novel approach to CS for fluorescence diffuse optical tomography (fDOT), named the Split Bregman-Singular Value Analysis (SB-SVA) iterative method. This approach is based on the combination of Split Bregman (SB) algorithm to solve CS problems with a theorem about the effect of ill-conditioning on L_1 regularization. Our method restricts the solution reached at each SB iteration to a determined space where the singular values of forward matrix and the sparsity structure of each iteration solution combine in a beneficial manner. Taking Battle-Lemarie basis for wavelet transform, where fDOT is sparse, we tested the method with fDOT simulated and experimental data, and found improvement with respect to the results of standard SB algorithm.

Keywords—compressed sensing, Split Bregman, singular value analysis, fluorescence, tomography.

I. INTRODUCTION

The goal of CS theory is to recover an image encoding the minimum possible information. In the last years, CS has been applied to different areas such as medical imaging, geophysical imaging, signal theory, matching learning, astrophysics, etc.

However, most of CS theory is mainly devoted to undetermined problems and there are few contributions that apply it to ill-posed problems.

If we consider a vector image $x \in \mathbb{R}^n$, a measurement vector $b \in \mathbb{R}^m$ and a matrix $A \in \mathbb{R}^{m \times n}$ that relates x and b in a linear way, $Ax = b$, being $m \leq n$, the CS theory asserts that image can be recovered when the following conditions are fulfilled:

- The image $x \in \mathbb{R}^n$ is itself sparse or can be sparsified using an orthogonal transformation T, such that $z = Tx$ and z is sparse. AT is named the sensing matrix.

- The recovery of the unknown image is performed using a nonlinear reconstruction scheme that enhances sparsity, for instance involving the use of L_0, L_1 or TV regularization for the resulting sparse image x.

To verify the uniqueness of the solution it is common to study the spark of system matrix (smallest number of linear dependent columns) [1], the mutual coherence (measure of the worst similarity between the matrix columns) [2] or the Restricted Isometry Property (RIP) condition (preservation of Euclidian distances between k-sparse vectors) [3-4].

If the selected submatrices of the sensing matrix are highly ill-posed, there are singular values (s.v.) close (or equal) to zero and then RIP condition is violated. Something similar happens with mutual coherence, which is linked to the condition number of the submatrices of the sensing matrix. Since RIP and mutual coherence are sufficient but not necessary conditions, their violation does not mean that we cannot obtain a solution.

When dealing with undetermined systems, many studies make use of RIP and mutual coherence to create appropriated sensing matrices [4-5]. However, this cannot be extrapolated when considering an ill-posed problem without first well-conditioning it.

In a recent paper [6], the authors formulated a theorem stating that the efficiency of a L_1 regularization problem depends on how the sparsity of the true solution and the s.v. of the forward matrix relate. This is particularly important for ill-posed problems, where near-zero s.v. are involved.

Some authors have applied CS techniques to ill-posed fDOT or DOT problems [7-9]. Initial works used a transformation to sparsify the image. Süzen et al. [7] considered the discrete Fourier transform but, as they pointed out, a study of optimal sparse expansion of the investigated signal was not considered in their simulations. Ducros et al. [8] investigated the compression ability of different wavelets for the acquired fluorescence images and concluded that Battle-Lemarie functions achieve good compression of fluorescent images with the least degradation, as compared to other bases (Haar, Daubechies, Battle-Lemarie, Coiflet, Symlets).

In [9], authors stated that image is itself sparse, and they focused their research on reducing the coherence of the fDOT forward matrix based on the fact that sparse signals can be recovered exactly from an undetermined system when the underlying forward matrix is incoherent [5]. Thus, indirectly, they were well-conditioning the sensing matrix (forward matrix in this case).

However, the preconditioning used in [9] depends on the forward matrix; if the sensing matrix is different from the forward matrix (a transformed basis is involved), a new preconditioning have to be defined.

The goal of this work is to present a novel approach, that solves the CS problem while automatically provides well-conditioning of the sensing matrix, independently of whether a transformed basis is involved or not.

II. METHOD

A. The Novel Approach, SB-SVA

Based on the combination of SB algorithm to solve CS problems [10] with the theorem about the effect of ill-conditioning on L_1 regularization presented in [6], we propose a novel approach, named SB-SVA, to solve the problem

$$\min_f \frac{\mu}{2}\|Wf - d\|_2^2 + \frac{\lambda}{2}\|Tf\|_1, \text{ s.t. } f \geq 0 \quad (1)$$

where W is the fDOT forward matrix, f is a vector representing the concentration of fluorophore at each voxel, d is a vector containing the acquired measurements and T is a transform to a space where f has a sparse representation. In our case T was chosen as Battle-Lemarie spline wavelet transform based on [8].

The above cited theorem [6] states that in the highly ill-conditioned case and in presence of significant noise, as it is the case of fDOT problems, no small singular value of W can be tolerated in the set $Tf_{true} \neq 0$ (also known as the support of f_{true}), being f_{true} the true solution.

Thus, for each iteration of SB, we restrict the solution, f^{it}, to the subspace where the "sparsity of Tf^{it} and s.v. of W combine in a beneficial manner", not allowing small s.v. of W in the set $Tf^{it} \neq 0$ or support of f^{it}.

Therefore, being Ω the image space and Ω_{KerNz} the subset of the null-space of W corresponding to the support of f^{it}, we restrict the transformation T to the space $\Omega \setminus \Omega_{KerNz}$ as follows:

$$\begin{cases} T_j = T_j & \text{if } T_j f_j^{it} \in \Omega \setminus \Omega_{KerNz} \\ T_j = 0 & \text{if } T_j f_j^{it} \in \Omega_{KerNz} \end{cases} \quad (2)$$

where the subscript $j = 1,...,n$ indicates voxel indices.

For the T wavelet transformation we scale samples of wavelet transforms following a geometric sequence of ratio two. We have to take this into account when restricting the transformation T to the space $\Omega \setminus \Omega_{KerNz}$.

B. Phantom Simulated and Experimental Data

We tested our algorithm using simulated and experimental fDOT data.

a) Experimental Data

A 10-mm thick slab-shaped phantom was built using a resin base with added titanium dioxide and India ink to provide a reduced scattering coefficient of $\mu_s' = 0.8\text{mm}^{-1}$ and an absorption coefficient of $\mu_a = 0.01\text{mm}^{-1}$, as described in [12]. A 5-mm diameter cylinder hole was drilled and filled with a fluid that matched the optical properties of the resin [13], mixed with Alexa fluor 700 1μM (Invitrogen, Carlsbad, California, USA). The fDOT fluorescence and excitation data were acquired with a non-contact parallel plate fDOT scanner [14] using 9x9 source positions and 9x9 detector positions over a 12x12 mm^2 surface.

b) Simulated Data

An equivalent phantom was simulated. For the simulation of the excitation and fluorescent photon density and the construction of the forward matrix we used the TOAST toolbox [15], adapted for fDOT. Sources were modelled as isotropic point sources (located just below the surface). Measurements were modelled by a Gaussian kernel centered at the detector location. The number of sources, number of detectors, and the surface covered by them matched those used with the experimental setup. Phantom was simulated by using a fine finite element mesh (145000 nodes). The average intensity for the forward matrix was reconstructed on a coarser finite element mesh (55000 nodes) and mapped into a uniform mesh of 20x20x10 voxels.

The simulation was perturbed with different levels of additive Gaussian noise: 1%, 3%, and 5%.

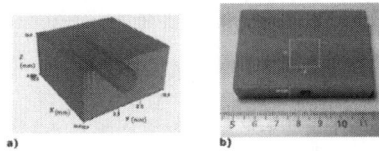

Fig. 1 a) Finite element model corresponding to the simulated phantom.
b) Image of the experimental phantom.

C. Comparison between SB and SB-SVA

The benchmark employed was SB. In order to compare SB-SVA with SB, we reconstructed simulated and experimental data with both methods and we obtained the y-profiles of central z-slice of different reconstructions.

In simulated data profiles were normalized by the average of highest voxel values in the corresponding reconstructions within a region of interest around fluorescent target.

III. RESULTS

Figures 2 and 3 show z-slices of simulated and experimental data reconstructed with SB and SB-SVA using a Battle-Lemarie wavelet basis transform. Improvement of SB-SVA against the benchmark employed, SB, is noticeable for both simulated and experimental phantom data.

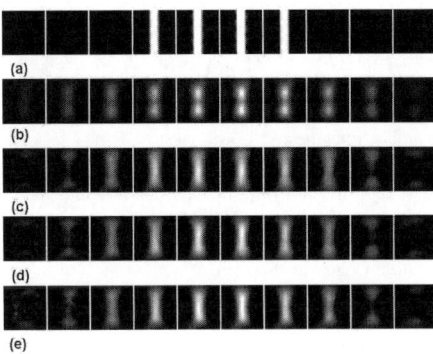

Fig. 2 Simulated data: 1mm-z-slices (y-x planes) of (a) Target, (b) Reconstruction of noise-free data by SB, (c) Reconstruction of noise-free data by SB-SVA, (d) Reconstruction of 3% noise data by SB-SVA, (e) Reconstruction of 5% noise data by SB-SVA.

Furthermore, for simulated data, SB used around 81.25% of total voxels (3250 of 4000), while SB-SVA used only 63.35% of total voxels (2534 of 4000), given its restriction to $\Omega \setminus \Omega_{KerNz}$ space.

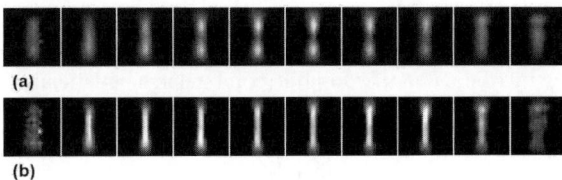

Fig. 3 1mm-z-slices (y-x planes) of experimental data reconstructions using: (a) by SB, (b) by SB-SVA.

Profiles of both simulated and experimental phantom data showed significant improvement provided by SB-SVA versus SB.

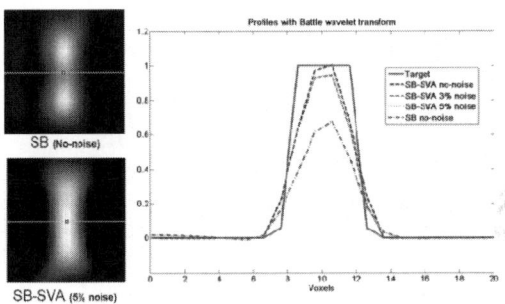

Fig. 4 y-profile of target (red) against y-profiles of SB reconstructions (green) and SB-SVA reconstructions of simulated data for different level of noise (blue).

For simulated data, no significant differences appear between different noise values.

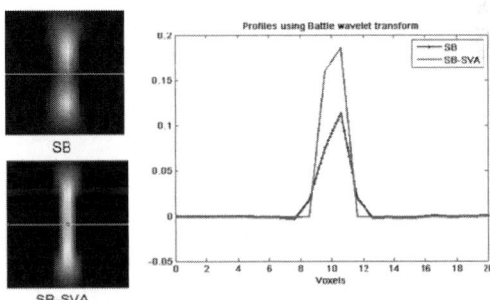

Fig. 5 y-profile of reconstruction by SB (blue) versus y-profile of SB-SVA (red) reconstruction of experimental data.

IV. DISCUSSION AND CONCLUSION

In this work, we propose a novel CS reconstruction method for fDOT, named SB-SVA, based on SB algorithm and a theorem about the effect of ill-conditioning on L_1

regularization. At each iteration of SB, we restrict the solution to a space where the s.v. of the forward matrix and the sparsity of the iterative solution combine in a beneficial way.

SB-SVA improved reconstruction images in terms of image quality and imaging profiles against SB (figures 1-4), even using fewer voxels than SB.

SB-SVA is, indirectly, constructing appropriated sensing matrices while finding the solution, due to our restriction to $\Omega \setminus \Omega_{KerNz}$ space of the solution in each iteration. Note that this restriction is, somehow, eliminating some of the columns of sensing matrix WT corresponding to $Ker(W)$. Thus, indirectly:

- It is reducing the similarity between the sensing matrix columns, that is, reducing the mutual coherence.
- It is well-conditioning the Gram matrix ($G = (WT)^T WT$) while reducing its condition number related with RIP.

Note that given how the restriction is built, if we change the transformation, the singular value decomposition of forward matrix does not require recalculation. Furthermore, the restriction is performed automatically.

Besides, SB-SVA is a simple and efficient algorithm since it is based on SB that has been shown its effectiveness for solving L_1-based regularization problems making it possible to split the minimization of L_1 and L_2 functionals.

Although SB-SVA provides significant improvements in terms of image quality, z-location is no optimal neither with SB-SVA nor with the benchmark employed, SB. This poor localization is due to the low resolution of fDOT in the axis perpendicular to the plates (z).

To conclude, when dealing with ill-conditioned fDOT problem, SB-SVA improves SB reconstructions in terms of image quality while it provides simultaneous well-conditioning of the sensing matrix.

ACKNOWLEDGMENT

This work was supported by Spanish Ministry of science and Innovation (FPI program, TEC2008-06715), European Regional Development Funds (FEDER), and CDTI under the CENIT program (AMIT project, cen-20101014).

REFERENCES

1. Donoho, D.L. and Tanner, J. (2006). Counting faces of randomly-projected polytopes when then projection radically lowers dimension. Tech. Rep. 2006-11, Stanford University.
2. Donoho, D.L. and Tanner, J. (2006). Counting faces of randomly-projected polytopes when then projection radically lowers dimension. Tech. Rep. 2006-11, Stanford University.
3. Mallat, S.G. and Zhang, Z. (1993). Matching pursuits with time-frequency dictionaries. Signal Processing, IEEE Transactions on. 41(12): p. 3397-3415.
4. Candes, E.J. and Romberg, J.K. (2005). Signal recovery from random projections. Electronic Imaging 2005. International Society for Optics and Photonics.
5. Theodoridis, S., Kopsinis, Y., and Slavakis, K. (2012). Sparsity-Aware Learning and Compressed Sensing: An Overview. arXiv preprint arXiv:1211.5231.
6. Elad, M (2007). Optimized projections for compressed sensing. IEEE Transactions on Signal Processing. 55(12): p.5695-5702.
7. van den Doel, K., Ascher, U., and Haber, E. (2012). The lost honour of l2-based regularization. Submitted.
8. Süzen, M., Giannoula, A. and Durduran, T. (2010). Compressed sensing in diffuse optical tomography. Optics Express. 18(23): p. 23676-23690.
9. Ducros, N., et al. (2013). Fluorescence molecular tomography of an animal model using structured light rotating view acquisition. Journal of biomedical optics. 18(2): p. 020503-020503.
10. Jin, A., et al. (2012). Preconditioning of the fluorescence diffuse optical tomography sensing matrix based on compressive sensing. Optics letters. 37(20): p. 4326-4328.
11. Goldstein, T. and Osher, S. (2009). The Split Bregman Method for L1 Regularized Problems. SIAM Journal on Imaging Sciences. 2(2): p. 323-343.
12. Buckheit, J., et al., WaveLab850 http://www-stat. stanford. edu/~ wavelab vol, Version.
13. Boas, D.A. (1996). Diffusion Photon Probes of Structural and Dynamical Properties of Turbid Media: Theory and Biomedical Applications. University of Pennsylvania: Philadelphia.
14. Cubeddu, R., et al. (1997). A solid tissue phantom for photon migration studies. Phys Med Biol, 42(10): p. 1971-9.
15. Aguirre, J. (2012). Estudios sobre la tomografía óptica difusiva de fluorescencia. PhD thesis UC3M.
16. Schweiger, M., et al. (1995). The finite element method for the propagation of light in scattering media: boundary and source conditions. Medical Physics. **22**(11): p. 1779-92.

Author: Judit Chamorro-Servent.
Institute: Dep. de Bioingeniería e Ingeniería Aeroespacial. UC3M.
Street: Avenida de la Universidad 30.
City: Leganés (Madrid).
Country: Spain.
Email: jchamorro@mce.hggm.es

Development of an Optical Coherence Tomograph for Small Animal Retinal Imaging

Susana F. Silva[1,2], José P. Domingues[1,2], José Agnelo[1,2], António Miguel Morgado[1,2], and Rui Bernardes[1,3]

[1] IBILI – Institute for Biomedical Imaging and Life Sciences, Faculty of Medicine, University of Coimbra, Portugal
[2] Department of Physics, Faculty of Sciences and Technology, University of Coimbra, Portugal
[3] Centre of New Technologies for Medicine, AIBILI, Coimbra, Portugal

Abstract—Optical Coherence Tomography (OCT) has been a helpful tool for ophthalmology diagnosis since its first appearance during the early 1990s. It is based on the optical interference phenomenon and has found major applications in the live tissue imaging. Particularly important is its capacity of producing high-resolution cross-sectional images of non-homogeneous tissues such as the ocular retina. The development of a high speed OCT, based on laser swept-source, for retinal imaging in small animals is our main goal. Preliminary results allow us to conclude that the system meet the requirements to produce OCT images of living tissues.

Keywords—OCT, Swept Source, High Speed Data Acquisition, Fast Fourier Transform, Retinal Imaging.

I. INTRODUCTION

Optical Coherence Tomography (OCT) is an optical technique based on the light interference phenomenon [1]. It is widely used in the field of ophthalmology as a diagnostic tool by allowing cross-sectional real-time imaging of ocular structures such as the retina [2, 3].

Currently, two main types of OCT systems can be considered: Time-Domain OCT and Fourier-Domain OCT [1, 4]. Time-Domain was the first to be developed and its working principle is quite similar to a Michelson interferometer [5]. A low-coherence light source directs the light into a beamsplitter which divides it in two paths: reference and sample. The light in the reference path will be backreflected by a reference mirror. The sample will also backreflect light coming from the beamsplitter. In order to occur interference, both sample and reference mirror must be placed at similar distances from the beamsplitter. This way, the superposition of the two backreflections is assured. A photodetector will collect the resultant interference signal. By moving the position of the reference mirror, a depth profile (A-scan) of the sample can be acquired [1].

In Fourier-Domain OCT, the interference profile is acquired with dependence on the frequency and not on the time. The reference mirror is fixed since the reference path is measured by the spectral response of the interferometer. This information is encoded on an interferogram and the reflectivity profile of the sample can be retrieved through Fourier analysis. Signal-to-noise ratio as well as optical layout simplicity is improved in FD-OCT [6]. Two methods for FD-OCT are available: Spectral-Domain OCT (SD-OCT) and Swept-Source OCT (SS-OCT). SD-OCT uses a broadband light to illuminate the interferometer and a spectrometer to separate the spectral components at the output before the CCD camera. As for SS-OCT, the spectrometer and the camera are replaced by an InGaAs dual balanced photodiode detector and its source, with a tunable narrowband, is swept over a broad range of optical frequencies (swept-source). This type of OCT presents advantages: reduced fringe washout, improved sensitivity with imaging depth, longer image range and higher detection efficiencies [5, 7]. Also SS-OCT has the advantage of a simpler optical design compared to SD-OCT [8].

In the field of biomedical research, small animals are very often used to develop, validate and test new techniques and therapies. The main purpose of this project is to develop a high speed swept source OCT for retinal imaging in small animals. Since imaging can give researchers the means to understand physiology, pathology and phenotypes of intact living systems similar to human beings, this instrument will be a valuable tool for research on retinal physiology. It will also be used as a development platform for new OCT instrumentation and methods [9].

The modular system under development will have the main components separated in several blocks: swept laser source, fiber optic paths and couplers, reference and sample arms with respective collimators and objectives, balanced light detector and fast data acquisition board.

II. THEORY AND SIMULATION

A simulation of the OCT processing equations was performed in Matlab®.

Considering an open-air OCT system with a beamsplitter to separate a reference and sample paths, the following equation describes the interference signal:

$$I(\omega, \Delta z) = T_r T_s S(\omega)|H(\omega)|^2 + T_r T_s S(\omega) + 2T_r T_s \Re\{S(\omega)H(\omega)e^{-i\varphi(\Delta z)}\} \qquad (1)$$

Where T_r and T_s are the beam-splitter transmissivity for reference and sample arms, respectively, $S(\omega)$ is the source spectrum, $H(\omega)$ the sample response and $\varphi(\Delta z)$ the phase accumulated in translating the reference mirror by a geometric distance Δz.

For the special case of SS-OCT, where there is no moving parts ($\Delta z=0$), and assuming an ideal beam-splitter with $T_r=T_s=0.5$, equation (1) can be rewritten as:

$$I(\omega) = \frac{1}{4}S(\omega)\{H(\omega) + 1\}^2 \qquad (2)$$

If Fourier transform is applied to equation (2), the signal dependence on time can be achieved:

$$I(t) = FT\{I(\omega)\} \qquad (3)$$

Actually, $I(\omega)$ is a set of discrete data points related with the intensity detected during laser sweep. This N-points array undergoes Fast-Fourier analysis and the resultant axial resolution can be approximated to [1]:

$$\delta z = \frac{1}{2n}\frac{\lambda_0^2}{\Delta\lambda} \qquad (4)$$

Where n is the average refraction index of the sample and λ_0 the source central wavelength.

Different sources and samples were simulated using those equations. Figures 1 and 2 present the results for a multilayer sample with refraction indexes from 1 to 1.5, central wavelength of 800 nm and bandwidth of 100 nm.

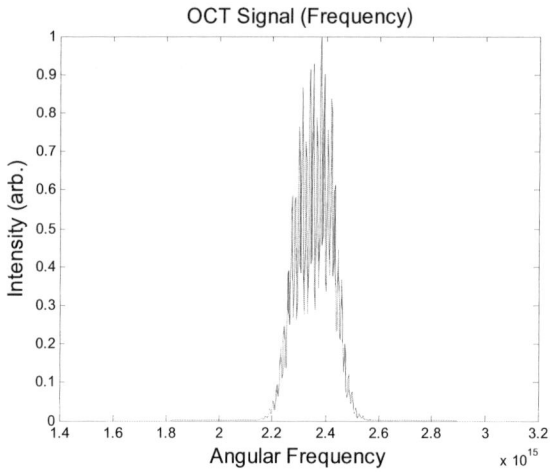

Fig. 1 FD-OCT signal for multilayer sample.

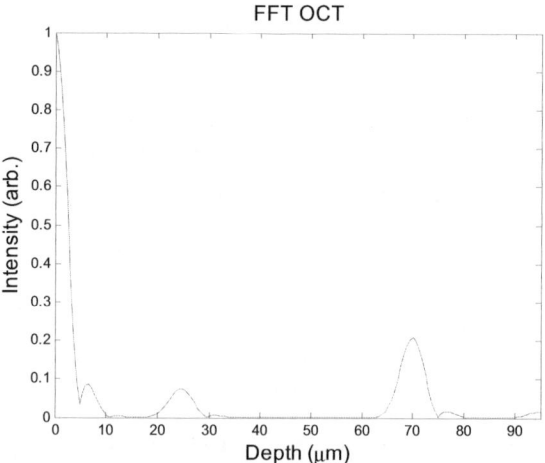

Fig. 2 Fourier transform of the signal shown in Fig.1. This corresponds to an axial scan of the multilayer sample (A-scan).

wavelength of 1060 nm, bandwidth 110 nm and sweep frequency of 100 kHz; a fast multi I/O 400 MSPS acquisition board X5-400M (Innovative Integration, Simi Valley, California, USA) with two A/D and two D/A channels; a InGaAs balanced amplified photodetector PDB145C (Thorlabs GmbH, Munich, Germany). Simplest layout is depicted in Figure 3 and was used to better understand and handle variables that affect the quality of interference signals.

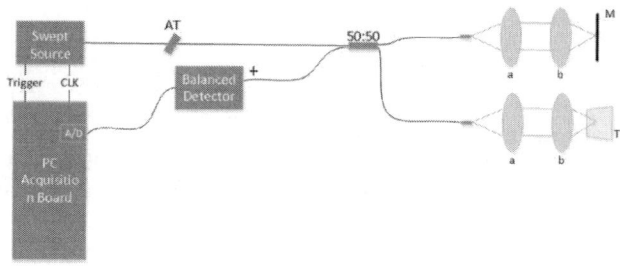

Fig. 3 – Schematics of the OCT unbalanced setup: a- collimators (F220APC-1064*), b-objective (LSM03-BB*), AT- variable attenuator (V0A1064-APC-SM*), M – gold coated reference mirror (M01*) and T-sample
* Thorlabs GmbH, Munich, Germany

The final configuration of the OCT system is represented in Figure 4. It is based on a dual balanced detection scheme and uses additional optical components. This type of detection presents advantages over the unbalanced scheme such as the rejection of common-mode intensity noise and the consequent boost in the dynamic range [10]. Additional optical components (circulator and additional coupler) guarantee that interference occurs only once before the balanced detector.

III. IMPLEMENTATION

The SS-OCT system implemented includes a commercially available swept source laser AXP50125-3 1060nm (Axsun Technologies, Billerica, MA, USA) with a central

Development of an Optical Coherence Tomograph for Small Animal Retinal Imaging

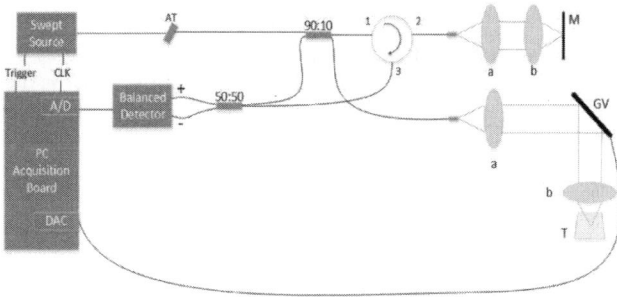

Fig. 4 Schematics of the OCT balanced setup: a –collimators (F220APC-1064*), b- objective (LSM03-BB*) AT- variable attenuator (V0A1064-APC-SM*), c –optical circulator (CIR1064-APC*), M-gold coated reference mirror (M01*), GV- galvanometers, T-sample.
*Thorlabs GmbH, Munich, Germany

A fully customized software program based on object-oriented programming language with Microsoft Visual C++ environment for 64-bit Microsoft Windows operating system has been developed. Innovative Integration libraries made available to customers are being used for data acquisition and hardware control, whereas data processing and OCT imaging are executed and produced on the graphics processing unit (GPU) through NVIDIAS's Compute Unified Device Architecture (CUDA) technology.

To perform B or C-scans over a sample, the laser beam must be directed along one and two axes, respectively. For that purpose, a Dual-Axis Scanning Galvo System GVS002 (Thorlabs GmbH, Munich, Germany) is used and controlled by one of the D/A channels available, allowing a maximum scan angle of 20º. With current hardware implementation, about 2048 points in sagittal plane (A-scan) are digitized in 10 µs and a maximum of 200 points in 2 ms are allowed in the transverse plane (B-scan). Consequently, 200 points in the coronal plane (C-scan) are obtained in 400 ms. From this, it can be concluded that the time needed for the final image construction is dependent on required resolution.

Apart from a customized software interface, other parameters are necessary to achieve synchronism between the acquisition system and the laser emission. The laser source has an embedded Mach-Zehnder interferometer (MZI) to provide a so-called optical clock signal. This signal presents maxima and minima equally spaced in the optical frequency domain (k-space). The difference between two maxima is defined by the free spectral range of MZI. Linearized fringe signals with equal k-spacing can be achieved by clocking the high speed A/D channel of the acquisition board with the clock provided by the source. The Fourier transform analysis can be directly applied on the acquired data. Moreover, the laser source also provides a trigger signal which is connected to the SYNC port of the acquisition board. This signal is responsible for starting the I/O module.

IV. RESULTS

Since the final configuration described in Figure 4 is not yet fully implemented, the results described on the next paragraphs refer to the simple unbalanced scheme and are considered preliminary. The Fourier post-processing analysis was performed with a Matlab® script. The first evidences that interference occurs were demonstrated using a simple mirror as a sample. By moving the position of this mirror in the axial direction, different patterns of fringes could be acquired and graphically represented. Figure 5 represents two fringe patterns acquired with an axial difference of 100 µm.

The total axial scan of the mirror covered a distance of 500 µm. After applying the fast Fourier transform it could be demonstrated that the position of the peak (correspondent to the sample mirror) varies linearly with the varying distance of the sample. The graphic is represented on Figure 6.

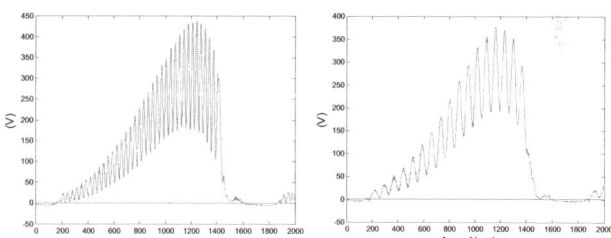

Fig. 5 Interference Patterns spaced by 100 µm.

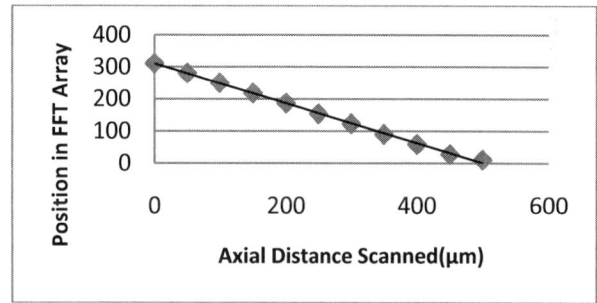

Fig. 6 Relation between the position in the FFT Array and the Axial Relation Scanned (R^2=0.9985).

A resolution target with equally spaced black lines (10 lpmm – Edmund Optics, East Gloucester Pike, Barrington, USA) was also tested. A lateral scan (B-scan) was performed covering a distance of 300 µm. From the result depicted in Figure 7, it can be concluded that the system was able to reproduce the pattern present on the target.

A simple lamella suspended in the air was also used as a test sample and the two interfaces present, air with lamella on the top and on the bottom, were detected.

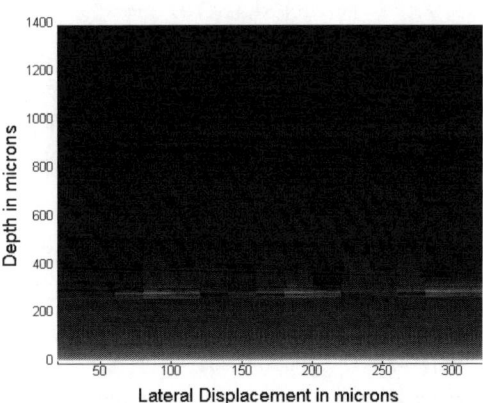

Fig. 7 OCT Image of 10 lpmm resolution target.

Finally we used a stepped target (Figure 8). This target is made of small steps and is used for calibration of commercial OCT systems. A simple B-scan was performed on a short path on the target and the result is depicted in Figure 9. There we can see a difference in depth that corresponds to one step of the target.

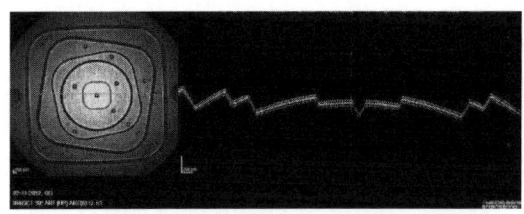

Fig. 8 Image of the stepped target obtained by Spectralis® OCT Heidelberg Engineering, Germany.

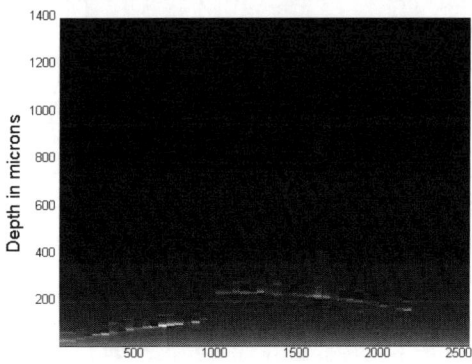

Fig. 9 OCT Image of a 2.5 mm length B-scan of the stepped target obtained by the system under development.

V. CONCLUSIONS

From the achieved preliminary results, it can be concluded that the concept is consistent with the final requirements of the system regarding acquisition speed, axial and lateral resolution and sensitivity. The different tests were performed in several different days and conditions, with their results being similar, showing that the system has adequate reproducibility.

ACKNOWLEDGMENT

This work was supported by FEDER, through the Programa Operacional Factores de Competitividade – COMPETE and by National funds through FCT – Fundação para a Ciência e Tecnologia in the frame of the project PTDC-SAL-ENB-122128-210 and also by Pest-C/SAUL/UI3282/2011.

REFERENCES

1. Tomlins P H and Wang R K (2005) Theory, developments and applications of optical coherence tomography. Journal of Physics: Appl. Phys. 38, pp. 2519-2535 DOI 10.1088/0022-3727/38/15/002
2. Targowski, P Wojtkowski M et al (2004) Complex spectral OCT in human eye imaging in vivo. Optics Communications 229 pp. 79-84.
3. Marschall S, Sander B et al (2011) Optical coherence tomography – current technology and applications in clinical and biomedical research. Anal Bioanal Chem 400 pp. 2699-2720 DOI 10.1007/s00216-011-5008-1
4. Huang D, Swanson C et al (1991) Optical Coherence Tomography. Science, 254(5035) pp. 1178-1181
5. Popescu D P, Choo-Smith, LP (2011) Optical Coherence Tomography: fundamental principles, instrumental designs and biomedical applications. Biophys Rev 3:155-169 DOI 10.1007/s12551-011-0054-7
6. F. de Boer, J., Cense, B. et al (2003) Improved signal-to-noise ratio in spectral-domain compared with time-domain optical coherence tomography. Optics Letters Vol. 28 No. 21, pp. 2067-2069
7. Potsaid B, Baumann B et al (2010) Ultrahigh speed 1050nm swept source/Fourier domain OCT retinal and anterior segment imaging at 100,000 to 400,000 axial scans per second. Optics Express Vol. 18 No.19, pp. 20029-20048
8. Nguyen, V. Duc, Weiss, N. et al (2012) Integrated-optics-based swept-source optical coherence tomography. Optics Letters Vol. 37 No. 23, pp.4820- 4822
9. Petersen-Jones, SM (1998) Animal models of human retinal dystrophies. Eye 12:566-570.
10. Liao, D et al (2004) Limits to Performance Improvement Provided by Balanced Interferometers and Balanced Detection in OCT/OCM Instruments. Proc. SPIE 5316:467-472

Author: Susana Figueiredo e Silva
Institute: IBILI – Institute for Biomedical Imaging and Life Sciences, Faculty of Medicine, University of Coimbra
Street: Azinhaga de Santa Comba, Celas
City: Coimbra
Country: Portugal
Email: susana.fig.silva@gmail.com

Inhomogeneous Modification of Cardiac Electrophysiological Properties due to Flecainide Administration

A.M. Climent[1], M.S. Guillem[2], P. Lee[3], C. Bollensdorff[4], F. Atienza[1], M.E. Fernández-Santos[1], R. Sanz-Ruiz[1], P.L. Sánchez[1], and F. Fernández-Avilés[1]

[1] Department of Cardiology, Hospital General Universitario Gregorio Marañón, Madrid, Spain
[2] Instituto ITACA, Universidad Politécnica de Valencia, Valencia, Spain
[3] Essel Research, Toronto, Canada
[4] Qatar Cardiovascular Research Center, Qatar Foundation, Doha, Qatar

Abstract—Brugada syndrome is characterized by a genetic mutation that leads to a reduction in sodium inward current in phase 0 of the cardiac action potential. However, electrophysiological abnormalities that underlie the phenotypical presentation of this disease remain unclear. The objective of this study was to characterize the spatial electrophysiological changes produced by a reduction in sodium current. Epicardial transmembrane voltage optical maps of 10 Wistar rat hearts isolated and perfused in a Langendorff system were analyzed. For each heart, mean transmembrane action potential duration (APD_{50}) and action potential up-stroke rise time (RT) were measured in the right ventricle (RV) right ventricular outflow tract (RVOT) and left ventricle (LV). Measurements were obtained at baseline and after injection of Flecainide 10uM, a potent sodium channel blocker. Flecainide injection produced a significant increase in APD duration and in the RT at each region with respect to basal conditions. However, the increase in the duration of the APD was higher in the RV compared to the LV (17.34 ± 6.67ms vs. 7.09 ± 6.39ms, p <.01). This increase inverted the APD gradient between RV and LV. By contrast, the increase in RT was significantly higher in the LV than in the RV (3.37 ± 1.84 vs. 2.80 ± 2.11ms, p <0.05). According to our results, Flecainide leads to inhomogeneous modification of electrophysiological properties of the heart and may therefore produce heterogeneities likely to initiate and/or maintain a fibrillatory process.

Keywords—Brugada Sindrome, Optical Mapping, Isolated Langendorff, Flecainide.

I. INTRODUCTION

Brugada syndrome (BrS) is a heritable pathology that may cause sudden cardiac death in young adults with apparently structurally normal hearts [1]. BrS is a channelopathy that has been linkedto a genetic mutation in SCN5A gen that leads to a reduction in sodium inward current in phase 0 of the action potential. Clinical manifestations of BrS is characterized by a coved-type ST in right precordial ECG leads that has been linked to electrophysiological heterogeneities in the right ventricular outflow tract (RVOT). However the electrophysiological changes that characterize the phenotype are transient and although several hypotheses have been proposed to explain those transitory ECG manifestations, specific mechanisms remain unclear [2].

The objective of this study was to characterize the electrophysiological changes produced by reducing the sodium current at different locations of the heart and try to clarify to potential generation of electrophysiological inhomogeneous modifications that could account for an increased arrhythmogenesis.

II. METHODS

Experimental model: Ten Wistar rat (367± 87 gr) hearts were isolated after median thoracotomy. The protocol was reviewed and approved by the Animal Care and Use Committees of Hospital Gregorio Marañon, Madrid, Spain. Specifically, 1,000 units of heparin were perfused and then rats were anesthetized with 250mg/kg sodium pentobarbital. Hearts were excised and immersed in 4°C cardioplegic solution containing (in mM): 140 NaCl; 5.4 KCl; 1 $MgCl_2$; 5 HEPES; 11 Glucose; 1.8 $CaCl_2$ with a pH of 7.4. Hearts were perfused at a constant rate of 5.5±0.5 mL/min with Tyrode's solution at 36.5°C (containing, in mM: NaCl, 120; $NaHCO_3$, 25; $CaCl_2$, 1.8; KCl, 5.4; $MgCl_2$, 1; glucose, 5.5; $H_2O_4PNa·H_2O$, 1.2; bubbled with carbogen; pH 7.4). All chemicals were obtained from Sigma-Aldrich (Dorset, UK), unless otherwise stated.

Optical mapping: For transmembrane voltage imaging, hearts were loaded with dye injected into the aortic cannula for coronary perfusion without recirculation. Specifically a 20-uL bolus of di-4-ANEPPS (Biotium, Inc. Hayward, CA, USA) of 4.16 mM (in DMSO) was applied over 5 minutes. Experiments were performed in a dark room to avoid photobleaching and dye washout. In order to excite di-4-ANEPPS, hearts were illuminated with a filtered green LED light source: LED: CBT-90-G (peak power output 58 W; peak wavelength 524 nm; Luminus Devices), plano-convex lens (LA1951; focal length= 25.4 mm; Thorlabs) and a green excitation filter (D540/25X; Chromo Technology).

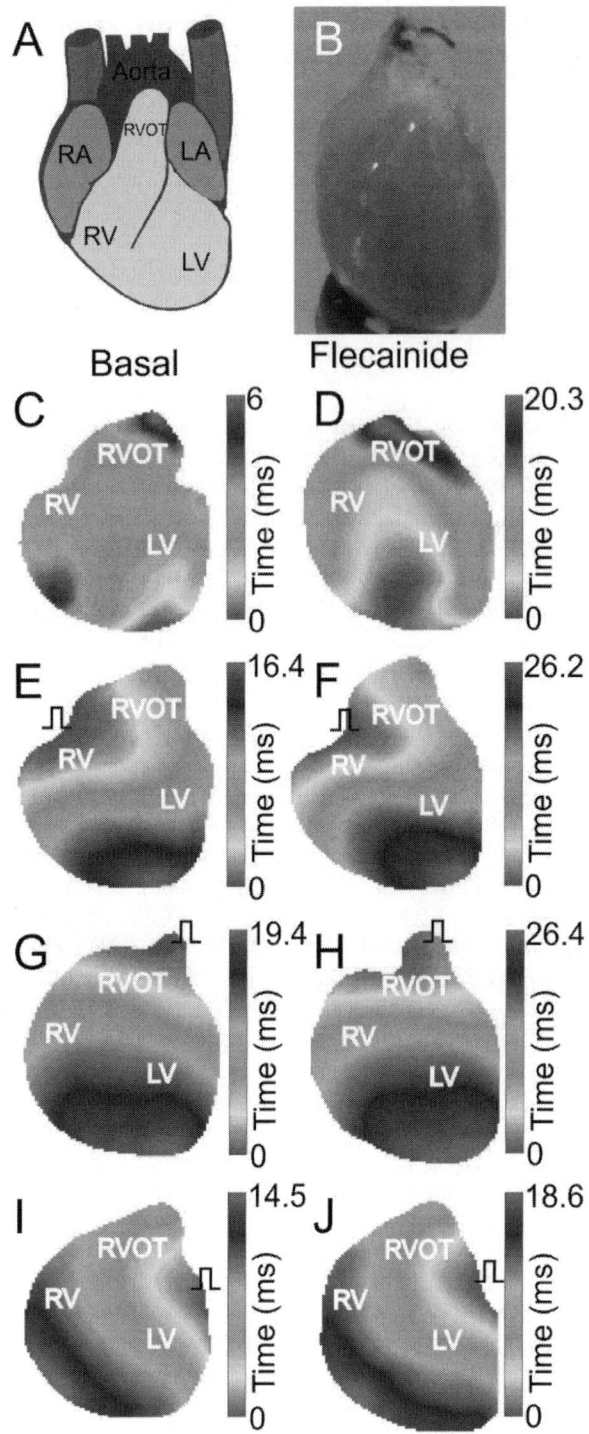

Fig. 1 Isochronal maps with and without infusion of Flecainide.
(A) Schematic representation of the chambers mapped, (B) image of one rat heart. Panels C-J: isochronal maps during sinus rhythm and stimulation from the right ventricle, RV, Right Ventricle Outflow Tract, RVOT, and Left Ventricle, LV before (C, E, G, I) and after Flecainide injection (D, F, H, J).

Two LEDs with the same characteristics were used in order to achieve a homogeneous illumination. Fluorescence was recorded with an electron- multiplied charge-coupled device (EMCCD; Evolve-128: 128x128, 24x24μm-square pixels, 16 bit; Photometrics, Tucson, AZ, USA), with an emission filter (HQ580LP; Chroma Technology) suitable for di-4-ANEPPS, placed in front of the lens. Camera was manually focused to cover the entire heart with the maximum number of pixels. Frame rate of the camera was fixed to 530 frames per second.

An eight-processor microcontroller (Propeller-Chip; Parallax, Rocklin, CA, USA) was used to control/coordinate all major components. Control software for time-critical tasks was written in the microcontroller's assembly language to ensure a time resolution of 50 ns. Communication with a desktop computer occurred via USB interface (UM245R; Future Technology Devices, Glasgow, United Kingdom) [3]. Both the LED lights and the CCD camera were controlled by custom made MATLAB software which made use of Micro-Manager Open Source Software [4].

Experimental protocol. The RVOT was located at the center of the field of view of the camera (Fig. 1). Posterior views of RV and LV were also visible. In order to reduce motion artifact, hearts were rested with a pair electrodes used to monitor the cardiac electrical activity. Excitation-contraction coupling was blocked for imaging purposes with blebbistatin (abcam, Cambridge, UK), using a concentration of 10 μM.

Four seconds optical mapping recordings were acquired during sinus rhythm and during pacing (250ms cycle length, biphasic 2ms pacing pulse) from the right ventricle (RV), left ventricle (LV) and RVOT. All measurements were performed in basal conditions and after partial sodium channel blockade by injection of Flecainide 10μM.

Image processing. Custom software written in MATLAB (MathWorks) was used to perform optical mapping image processing. Specifically, for each 4 second movie a template matching technique was used to obtain an averaged transmembrane potential wave of each individual pixel. For each pixel, optical action potential duration was estimated as time between the maximum upslope and the 50% of the repolarization time (APD_{50}). In addition, action potential upstroke rise time (RT) was measured from the time interval between 10% and 90% of action potential upstroke. In order to estimate the mean value of calculated parameters in each of the three analyzed regions (RV, RVOT and LV), values of APD_{50} and RT for all pixels around the stimulation area were averaged (i.e. 10% of the total mapped region).

Statistical analysis. Data are presented as mean±standard deviation. The paired Student's t-test was performed to compare APD_{50} and RT before and after the injection of Flecainide and to compare their duration between cardiac regions.

III. RESULTS

In Figure 1 representative examples of the effects of Flecainide in the activation sequence during sinus rhythm and during regular pacing from the RV, RVOT and LV are depicted. As expected, during sinus rhythm the apex was the first region to be depolarized whereas the RVOT was last both during basal conditions and after Flecainide injection. As it can be observed, the reduction in sodium current conductivity produced a significant prolongation on the conduction velocity and thus in the time required for a complete depolarization of the epicardium. This prolongation did not alter the propagation pattern neither during sinus rhythm nor during pacing protocols. Nevertheless, conduction velocity reduction was more evident during sinus rhythm than during regular stimulation. This observation can be attributed to an epicardial propagation close to the stimulation electrode during pacing.

In Figure 2, APD_{50} and RT with and without Flecainide administration for all the experiments are summarized together with representative optical action potentials for the three epicardial regions studied. Notice that the displayed action potential morphology is characteristic of rat hearts which do not present plateau and consequently have shorter APDs than larger mammalians.

As expected during a reduction of sodium current conductivity and subsequent reduction of upstroke velocity during myocyte depolarization, a significant increment of RT was measured during Flecainide administration in all three analyzed regions (Panel 2A). However, notice that the prolongation of the RT was not homogenous: RT prolongation was larger in the LV than in the RV (i.e. 3.37±1.84ms vs. 2.80±2.11ms, p<0.05).

As it can be observed in panel 2B, the reduction of sodium channel conductivity also produced a significant APD prolongation in all cardiac regions. As for the RT, notice that APD_{50} prolongations were not homogeneous throughout the epicardial tissue: prolongation of the APD_{50} was larger in the RV than in the LV (i.e. 17.34±6.67ms vs. 7.09±6.39ms, p<0.01). This inhomogeneous APD prolongation produced an APD gradient (LV vs. RV) reversal: in basal conditions the LV presented longer APDs than the RV but after Flecainide administration RV APDs were larger than LV APDs.

According to our results, APD_{50} and RT of optical action potentials from the RVOT and from the anterior region of the RV did not presented significant differences neither during basal conditions nor after Flecainide administration.

Another important remark that should be taken into account is that the APD prolongation was larger than the RT increment in all three regions (i.e. 11.4±11.1ms vs. 4.32±4.31ms, p<0.001). It may imply that the action potential prolongation was not only linked to a reduction in the upstroke velocity, but also to a probably second effect triggered by the sodium channel reduction.

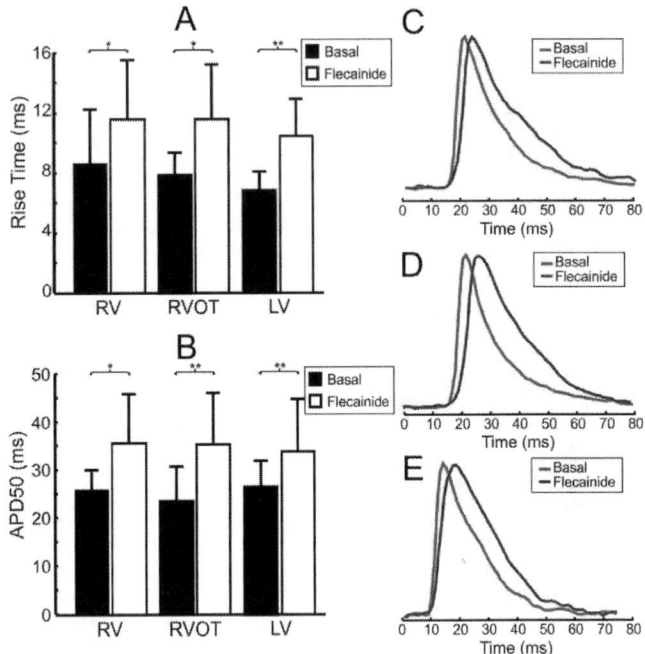

Fig. 2. Rise time (Panel A) and Action Potential Duration (panel B) before and after Flecainide administration in the three analyzed regions of the heart (i.e. Right Ventricle, RV, Right Ventricle Outflow Tract, RVOT, and Left Ventricle, LV). Representative action potential before and after Flecainide administration during RV stimulation (Panel C), RVOT stimulation (Panel D) and LV stimulation (Panel E) (*, p<0.05; **, p<0.01)

IV. DISCUSSION

The main finding of the present study is that a reduction in the conductivity of sodium current by Flecainide administration produced an inhomogeneous modification of the electrophysiological properties of epicedial tissue. Specifically, our results suggest that a larger increment of RT in the LV respect to the RV may introduce conduction velocity heterogeneities. These modifications together with a gradient reversal in APD between LV and RV after Flecainide administration may increase the susceptibility to suffer ventricular arrhythmias.

Presented results are in agreement with recent results showing a nonuniformly distribution of Na+ channels expression levels in the right and left ventricle in guinea pig [5-6] and sheep [7]. In fact, a higher expression of sodium channel proteins in the LV respect the RV may underline a more relevant role of Na^+ channels in the depolarization phase of LV and consequently a larger increase of RT due to Flecainide administration. However, this reduction in the RT of the LV that maybe linked to a reduction in the conduction velocity cannot explain the well described relevant role of the RV and more specifically the RVOT in the electrophysiological characterization of BrS [8]. Nevertheless, the significantly larger increase of the APD in the RV could be in agreement with the repolarization hypothesis that attempts to explain the electrocardiographic manifestation of BrS [2]. Unfortunately, rat ventricular cardiac action potentials do not present a prominent plateau which can be observed in human cardiomyocites, which limits the extrapolation of present results to humans. However present results suggest that a reduction in sodium current may trigger action potential morphology modifications not only linked with the expression of Na+ channels. In fact, as demonstrated by Veeraraghavan et al. [6] heterogeneities in the expression of other ion channel proteins such as Kir2.1 related to I_{K1} may significantly modify conduction velocity gradients between left and right ventricle. In addition to potassium channel gradients, a reduction of sodium channel conductivity has been linked to modifications in the calcium homeostasis [9] which heterogeneity is well described as an important arrhythmia trigger.

Limitations. Electrophysiological parameters showed in the present study were measured from optical mapping recordings. This technique presents important advantages over electrical mapping such as high density mapping, transmembrane potential measurement, etc. However, optical mapping measurements present some limitations that must be taken into account. Notice that optical action potentials represent spatially integrated responses of cells confined within the three-dimensional field of view of each pixel. Therefore, rise times measured in optical action potentials are a function of both the conduction velocity and field of view of each pixel [10] which may explain the slower RT of optical mapping action potentials as compared to single cell recordings.

Electromechanical uncouplers may modify electrophysiological characteristics and thus alter our measurements. However, in the present study we used Blebbistatin which has demonstrated to have minimal effects on APD and rise time as compared with other electromechanical uncouplers like 2,3-butanedione monoxime (BDM) [11].

V. CONCLUSION

Flecainide leads to inhomogeneous modification of electrophysiological properties of the epicardial tissue and may therefore produce heterogeneities likely to initiate and/or maintain fibrillatory process.

ACKNOWLEDGMENT

This work was supported by MINECO under the project SABIO (PLE2009-0152) and Universitat Politècnica de València through its research initiative program.

REFERENCES

1. Antzelevitch C., Brugada P., Brugada J. et al. (2005) Brugada syndrome: from cell to bedside. Curr Probl Cardiol. 30:9-54.
2. Wilde AAM., Postema PG, Di Diego JM, et al. (2010) The pathophysiological mechanism underlying Brugada syndrome: depolarization versus repolarization. J Mol Cell Cardiol. 49:543-553
3. Lee P, Yan P, Ewart P et al. (2012) Simultaneous measurement and modulation of multiple physiological parameters in the isolated heart using optical techniques. Pflugers Arch. 464(4):403-14.
4. Stuurman N, Edelstein A, Amodaj N et al (2010) Computer Control of Microscopes Using µManager. Curr Protoc Mol Biol. 14.20.1-14.20.17
5. Osadchii OE, Soltysinska E, Olesen SP. (2011). Na+ channel distribution and electrophysiological heterogeneities in guinea pig ventricular wall. Am J Physiol Heart Circ Physiol. 300(3):H989-1002.
6. Veeraraghavan R, Poelzing S. (2008). Mechanisms underlying increased right ventricular conduction sensitivity to flecainide challenge. Cardiovasc Res. 1;77(4):749-56.
7. Fahmi AI, Patel M, Stevens EB, et al. (2001). The sodium channel beta-subunit SCN3b modulates the kinetics of SCN5a and is expressed heterogeneously in sheep heart. J Physiol. 15;537(3):693-700.
8. Martin CA, Grace AA, Huang CL. (2011) Spatial and temporal heterogeneities are localized to the right ventricular outflow tract in a heterozygotic Scn5a mouse model. Am J Physiol Heart Circ Physiol. 300(2):H605-16.
9. Yan G.X., Antzetlevitch C. (1999) Cellular basis for the Brugada syndrome and other mechanisms of arrhythmogenesis associated with ST- segment elevation. Circulation;100:1660-1666.
10. Efimov I.R., Cheng Y. (2002) Optical mapping of cardiac stimulation: fluorescent imaging with photodiode array. Quantitative Cardiac Electrophysiology by Candido Cabo and David S. Rosenbaum. Taylor & Francis, London 555-591.
11. Lou Q, Li W, Efimov IR.(2012) The role of dynamic instability and wavelength in arrhythmia maintenance as revealed by panoramic imaging with blebbistatin vs. 2,3-butanedione monoxime. Am J Physiol Heart Circ Physiol.;302(1):H262-9.

Author: Andreu M. Climent
Institute: Hospital General Universitario Gregrorio Marañon
Street: O'Donnell 48
City: Madrid
Country: Spain
Email: acliment@cardiovascularcelltherapy.com

Predictive Analysis of Photoacoustic Tomography Images in Dermatological Liposarcoma

F. Fanjul-Vélez, D. Martín-Ruiz, N. Ortega-Quijano, I. Salas-García, and J.L. Arce-Diego

Applied Optical Techniques Group, TEISA Department, University of Cantabria, Santander, Spain

Abstract—Photoacoustic Tomography is a novel and promising imaging technique for tissue diagnosis. It employs an optical radiation for tissue irradiation, and detects the generated acoustic waves. By means of this approach it is capable of obtaining high resolution images at depths beyond the limits of purely optical techniques. The study of the potential efficiency of these techniques for diagnosis requires a predictive model for photoacoustic images. In this work we present a predictive complex model for Photoacoustic Tomography images. The model implements optical propagation by means of a Monte Carlo approach, initial pressure distribution and image reconstruction by the k-space method. We apply the method to dermatological liposarcoma at several states, and evaluate the validity of the images for diagnosis.

Keywords—Photoacoustic Tomography, k-space method, Monte Carlo method, optical diagnosis, sarcoma.

I. INTRODUCTION

Photoacoustic Tomography is called to revolutionize tissue diagnosis in the next decades [1], when a moderate penetration depth and high resolutions are required. This technique is capable of improving the penetration depth of optical techniques [2], while maintaining a high resolution in the images. The combination of an optical source with the internal generation of acoustic waves that are externally detected makes this technique non-invasive.

An adequate analysis of Photoacoustic Tomography requires a deep knowledge of the processes involved. A predictive analysis tool is highly desired, as in this way the efficiency for particular applications could be estimated. The limitations of the techniques could also be seen.

In this work we develop a predictive model for photoacoustic images and apply it to dermatological liposarcoma. Several stages of liposarcoma are modeled and the images are compared for judging their diagnostic validity.

The work is organized as follows. Section II describes the basics of Photoacoustic Tomography. Section III deals with the predictive analysis, including optical propagation, initial pressure distribution and image reconstruction by the k-space method. Section IV shows the application to liposarcoma and the results obtained. Finally Section V contains the conclusions of the work.

II. PHOTOACOUSTIC TOMOGRAPHY

The term photoacoustic tomography (PAT) refers to imaging based on the photoacoustic effect. In PAT the object is irradiated by a short pulsed laser beam under certain confinement conditions. Some of the light is absorbed by the object and partially converted into heat. The heat is then converted into a pressure rise via thermoelastic expansion. The pressure rise is propagated as an ultrasonic wave (Photoacoustic Wave), detected by transducers and used to form an image [1].

High-resolution optical imaging beyond the soft depth limit (1 mm), sometimes referred to as superdepth optical imaging, remained a void until PAT was developed. None of the commercially available optical ballistic imaging modalities (including confocal microscopy, two-photon microscopy and optical coherence tomography) can penetrate into scattering biological tissue beyond the soft depth limit. Furthermore PAT has a high endogenous contrast given by the local absorption.

The motivation driving the development of PAT is to overcome the poor spatial resolution of diffuse optical tomography and the soft depth limit of existing high resolution optical imaging.

III. PREDICTIVE ANALYSIS OF PHOTOACOUSTIC TOMOGRAPHY

A. Optical Propagation in Biological Tissue

Many of the problems with practical interest often imply a great variety of optical sources, tissues and complex geometries. Analytic solutions for realistic scenarios are usually complicated. These cases are solved using numerical techniques. When using the RT (radiation transport) model the most widely used approximation is the Monte Carlo method (MC) which refers to a broad class of methods that apply random numbers in the process of solving the problem [2].

The algorithms for the implementation of the basic elements in a MC simulation are quite simple. Therefore the technique has a great flexibility and is widely applicable in practical RT problems. The simulation is based on random paths the photons follow as they travel through tissue. The step size and angular deviation per dispersion event are obtained by statistically sampling the probability distributions. As the number of photons increases, the distribution of all paths produces an increasingly accurate approximation of the RT problem. The actual number of photons required for a realistic result depends on the details of the simulation.

In our case we will use the Monte Carlo Simulation Package (Modeling of Photon Transport in Multi-layered Tissues and Conv) [3],[4]. The first step is to define the optical properties [2] and thickness of each layer. After that, all we have to do is define the laser beam (type, radius and total energy of the pulse) and obtain the convolved data.

B. Initial Pressure Distribution

Two conditions must be met for the optimal generation of optoacoustic signals: thermal and stress confinement [5].

The timescale for heat dissipation of absorbed EM energy can be estimated by:

$$\tau_{th} \sim L_p/4D_T \quad (1)$$

Where L_p is the characteristic dimension of the heated volume and D_T the thermic diffusivity of the sample. The length of thermal diffusion during the pulse's period (τ_p) after his absorption can be estimated by

$$\delta_T = 2T\sqrt{D_T \tau_p} \quad (2)$$

The laser pulsewidth τ_p must be shorter than τ_{th} to generate photoacoustic signals efficiently. This condition is usually referred to as thermal confinement, making heat conduction negligible during the laser pulse.

The time that characterizes the stress wave propagation can be estimated by:

$$\tau_s = L_p/v_m \quad (3)$$

where v_m is the speed of sound in the medium.

The laser pulsewidth τ_p must be shorter than τ_s so a high thermoelastic pressure is achieved easily and stress propagation is negligible during the pulse.

We can express the fractional volume expansion on laser excitation as:

$$\frac{\Delta V}{V} = -\frac{1}{\gamma}\Delta P + \beta \Delta T \quad (4)$$

where $\beta(K^{-1})$ is the thermal coefficient of volumetrical expansion, $\gamma(Pa^{-1})$ is the isothermal compressibility coefficient; $\Delta P\ (Pa)$ and $\Delta T(K)$ represent the increase in pressure and temperature respectively.

The isothermal compressibility γ coefficient can be expressed as:

$$\gamma = \frac{C_p}{\rho v_m^2 C_V} \quad (5)$$

Here $\rho\left(\frac{kg}{m^3}\right)$ is density, while C_p and $C_V\left(\frac{J}{Kg}\right)$ denote the specific heat capacities at constant pressure and temperature, respectively.

We can rewrite Eq (4) to obtain the pressure increase:

$$\Delta P = -\frac{1}{\gamma}\frac{\Delta V}{V} + \frac{1}{\gamma}\beta \Delta T \quad (6)$$

If the laser excitation is in both thermal confinement and stress confinement, the fractional volume expansion will be negligible, $\Delta V/V \approx 0$, and we will have:

$$\Delta P \approx \frac{\beta \Delta T}{\gamma} \quad (7)$$

which can be rewritten as

$$\Delta P \approx \frac{\beta}{\gamma \rho C_V}E_T \quad (8)$$

knowing that in thermal confinement

$$\Delta T = \frac{E_T}{\rho C_V} \quad (9)$$

We define the Grüneisen parameter (dimensionless), which indicates the fraction of incident optical energy converted into acoustic energy by the photoacoustic effect, as:

$$\Gamma \equiv \frac{\beta}{\gamma \rho C_V} = \frac{\beta v_m^2}{C_P} \quad (10)$$

Substituting it into Eq.(8), we have:

$$\Delta P = \Gamma E_T = \Gamma \mu_{abs} F \quad (11)$$

where F(J/cm^2) is the fluence (energy density) of the laser source.

The initial pressure distribution can be also expressed as a function of distance z to the surface of the sample:

$$p_0(z) = \Gamma \mu_{abs} F_0 e^{-\mu_{eff} z} \quad (12)$$

Having obtained the 2-D absorption and fluence spatial maps previously, we only need to create a spatial matrix with the Grüneisen parameters ([6], [7]) of the tissues to produce the initial pressure distribution.

C. Photoacoustic Image Reconstruction

We will use k-wave [8], [9], an acoustic toolbox for Matlab©, for our photoacoustic image reconstruction. K-wave uses the k-space method to solve the system of coupled acoustic equations, in order to reduce memory and the number of time steps required for accurate simulations.

This approach combines the spectral calculation of spatial derivatives with a temporal propagator expressed in the spatial frequency domain or k-space.

In a standard finite difference scheme, spatial gradients are computed locally based on the function values at neighbouring grid points. In the simplest case, the gradient of the field can be estimated using linear interpolation. A better estimate of the gradient can be obtained by fitting a higher-order polynomial to a greater number of grid points and calculating the derivative of the polynomial. The more points used, the higher the degree of polynomial required, and the more accurate the estimate of the derivative. The Fourier collocation spectral method takes this idea further and fits a Fourier series to all the data. It is therefore sometimes referred to as a global, rather than local, method. There are two significant advantages of using Fourier series. First, the amplitudes of the Fourier components can be calculated efficiently using the Fast Fourier Transform (FFT). Second, the basis functions are sinusoidal, so only two grid points (or nodes) per wavelength are theoretically required, rather than the six to ten required in other methods.

For our simulations we will use a modified script (corresponding to 2D Iterative Image Improvement Using Time Reversal Example) able to implement an image reconstruction with our initial pressure distribution.

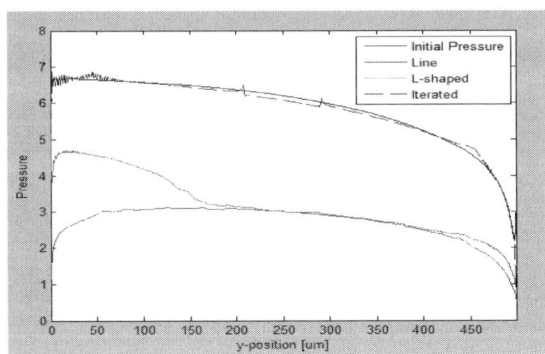

Fig. 1 Comparison between the initial pressure distribution and the reconstruction methods used.

In the first reconstruction, an image is formed from data recorded on a line array using time reversal. In the second, an image is formed from data recorded on an L-shaped sensor array (also using time reversal). Finally, this image is improved iteratively. Examples of the initial pressure distribution estimations by the reconstruction methods exposed can be appreciated in Figure 1.

This improvement is achieved by adding an additional set of (imaginary) sensors so that the sensor array encloses the region of interest. The missing data for these 'sensors' is then estimated using the forward model with the best image obtained so far. The original data and the estimated data are then used together to form an image, which - at least within the region satisfying the visibility (audibility) condition - will be an improvement on the previous image. The estimate of the missing data and subsequent image reconstruction may then be iterated for even greater image improvement.

IV. APPLICATION TO LIPOSARCOMAS

A sarcoma is a type of cancer that develops from certain tissues, like bone or muscle. There are 2 main types of sarcoma: bone sarcomas and soft tissue sarcomas. Soft tissue sarcomas can develop from soft tissues like fat, muscle, nerves, fibrous tissues, blood vessels, or deep skin tissues. They can be found in any part of the body. Most of them develop in the arms or legs [10]. is the second most common soft tissue sarcoma following malignant fibrous histiocytoma, accounting for approximately 16% to 18% of all soft tissue tumors and having an estimated annual incidence of 2.5 per million. Several subtypes are described, ranging from lesions nearly entirely composed of mature adipose tissue to tumors with very sparse fatty elements. Malignant fatty tumors are classified into four main groups: well-differentiated liposarcoma/atypical liposarcoma; dedifferentiated liposarcoma; myxoid liposarcoma; and pleomorphic liposarcoma. We will concentrate in intramuscular liposarcomas in striated muscle.

We defined several multi-layered models recreating different stages of a sarcoma and obtain 2-D absorption and fluence spatial maps. Afterwards we calculated the initial pressure distribution and tried to reconstruct the photoacoustic image with several reconstruction methods.

The simulations with a sample of healthy tissue will be comprised of striated muscle and a very thin (0.1 mm) layer of fatty tissue. After that we will begin to increase both in number and thickness the fatty tumoral layers to illustrate the development of the type of liposarcomas and liposomas previously mentioned.

Figure 2 shows the true image and four reconstructions for an early state sarcoma. As it can be clearly appreciated, both the L-shaped and Line methods do not adequately reflect the structure of the sarcoma. This fact could lead to false negative diagnosis. However, the iterated method shows clearly the striated structure in depth, indicative of sarcoma. This demonstrates the validity of Photoacoustic Tomography in this application.

Figure 3 shows a fully developed sarcoma, with a deeper and more complex layered structure. Again the L-shaped and Line reconstructions cannot reproduce the original pressure distribution, but so does the iterated method. If the sarcoma goes even deeper, the reconstruction procedure starts to fail, even with the iterated method.

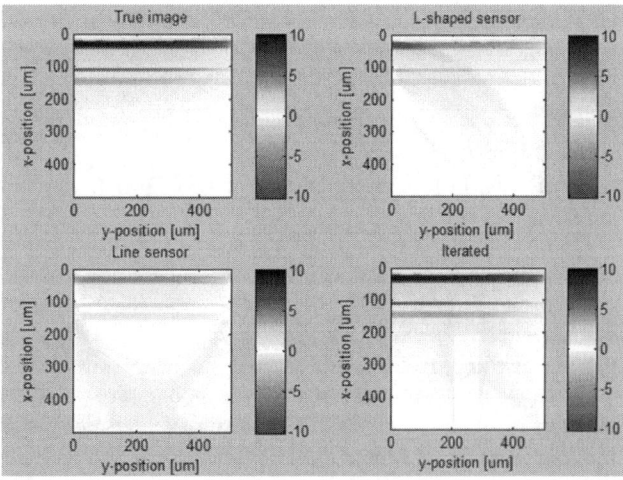

Fig. 2 Sarcoma in an early stage: original distribution (upper left) and the reconstructions performed, L-shaped (upper right), Line (lower left) and Iterated (lower right).

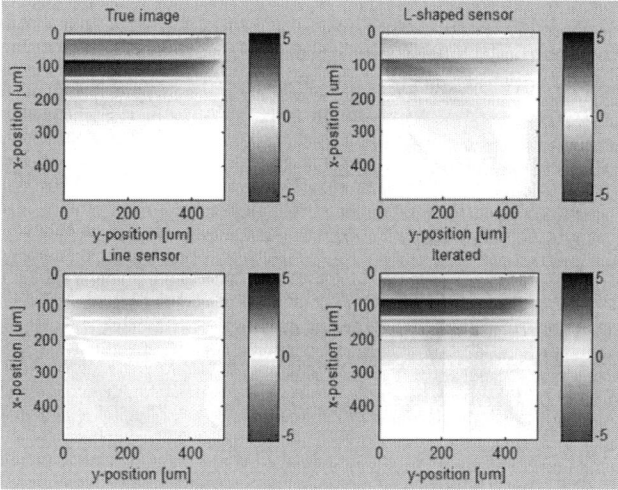

Fig. 3 Sarcoma fully developed and the reconstructions perfomed: original distribution (upper left), L-shaped (upper right), Line (lower left) and Iterated (lower right).

V. CONCLUSIONS

In this work a predictive analysis of Photoacoustic Tomography has been presented. This analysis allows the evaluation of its validity for particular applications. The predictive model is based on a Monte Carlo approach for optical propagation, the initial pressure calculation, and the k-space method for the image reconstruction.

The analysis was applied to intramuscular liposarcomas in different stages. The results show that the iterated method is the more appropriate for detecting malignant tissue, particularly deep into tissue. The structure shown by the images can be employed for non-invasive liposarcoma diagnosis.

ACKNOWLEDGMENTS

This work has been partially supported by the project MAT2012-38664-C02-01 of the Spanish Ministery of Economy and Competitiveness, and by San Cándido Foundation.

REFERENCES

1. M. H. Xu and L. V. Wang, "Photoacoustic imaging in biomedicine," Rev. Sci. Instrum. **77**, 041101 (2006).
2. Biomedical Photonics Handbook Edited by Tuan Vo-Dinh CRC Press 2003 Print ISBN: 978-0- 8493-1116-1 eBook ISBN: 978-0-203-00899-7
3. L.-H. Wang, S. L. Jacques, and L.-Q. Zheng, "MCML - Monte Carlo modeling of photon transport in multi-layered tissues," Computer Methods and Programs in Biomedicine 47, 131-146 (1995).
4. L.-H. Wang, S. L. Jacques, and L.-Q. Zheng, "CONV - Convolution for responses to a finite diameter photon beam incident on multilayered tissues," Computer Methods and Programs in Biomedicine 54, 141-150 (1997).
5. L. V. Wang and H. Wu, *Biomedical Optics: Principles and Imaging* (Wiley, Hoboken, NJ, 2007).
6. Pulsed-microwave-induced thermoacoustic tomography: Filtered backprojection in a circular measurement configuration Minghua Xu and Lihong V. Wang *Optical Imaging Laboratory, Biomedical Engineering Program, Texas A&M University,3120 TAMU, College Station, Texas 77843-3120* DOI: 10.1118/1.1493778
7. The challenges for quantitative photoacoustic imaging B. T. Cox, J. G. Laufer and P. C. Beard.Department of Medical Physics and Bioengineering, University College London, Gower Street, London WC1E 6BT, UK
8. B. E. Treeby and B. T. Cox, "k-Wave: MATLAB toolbox for the simulation and reconstruction of photoacoustic wave-fields," *J. Biomed. Opt.,* vol. 15, no. 2, p. 021314, 2010.
9. B. E. Treeby, J. Jaros, A. P. Rendell, and B. T. Cox, "Modeling nonlinear ultrasound propagation in heterogeneous media with power law absorption using a k-space pseudospectral method," *J. Acoust. Soc. Am.,* vol. 131, no. 6, pp. 4324-4336, 2012.
10. Imaging of Soft Tissue Tumors, 2nd ed Mark J. Kransdorf and Mark D. Murphey Philadelphia, Pa: Lippincott Williams & Wilkins, 2006. ISBN 0-7817-4771-6. Lock I, Jerov M, Scovith S (2003) Future of modeling and simulation, IFMBE Proc. vol. 4, World Congress on Med. Phys. & Biomed. Eng., Sydney, Australia, 2003, pp 789–792.

Author: Dr. Félix Fanjul-Vélez
Institute: University of Cantabria
Street: ETSII y de Telecomunicación, Av. de los Castros s/n, 39005
City: Santander
Country: Spain
Email: fanjulf@unican.es

Monte Carlo Simulation of Radiation through the Human Retina Using Geant4

G. Lopes, D. Tendeiro, J.P. Santos, and P. Vieira

Centro de Física Atómica, CFA, Departamento de Física, Faculdade de Ciências e Tecnologia, FCT,
Universidade Nova de Lisboa, 2829-516 Caparica, Portugal

Abstract—Age-related macular degeneration (AMD) is a chronic and irreversible eye disease. It causes damage to the retina by accumulation of extracellular materials, which leads to loss of visual acuity. One of the most common signs of AMD is the appearance of drusen, yellowish irregular nodules. Because of its irreversible character, a timely diagnosis is important for early treatment. The development of new ophthalmic diagnostic methods implies appropriate knowledge concerning the mode of propagation of light in the eye. In this work, it was developed the necessary simulation tools to study the mechanisms of the interaction of optical light with the retina. To accomplish this goal it was used the Monte Carlo (MC) simulation technique.

Keywords—**Human retina, Optical light, Monte Carlo.**

I. INTRODUCTION

In the Western Hemisphere, age-related macular degeneration (AMD) is a major cause of blindness in the population over 50 years [1]. The incidence and prevalence of AMD have been increasing, either by general aging of the population, either by the increasing of situations involved in its onset. One of the most common signs of AMD is the appearance of drusen in the retina, yellowish irregular nodules when the natural light it focuses. In an earlier stage of development, these structures are very difficult to detect by conventional techniques. The diagnosis of this disease is most often made using qualitative methods as the method of Amsler and angiography. Thus, it is important to build a diagnostic method to identify these structures more quickly and accurately.

The development of new ophthalmic diagnostic methods implies appropriate knowledge concerning the mode of propagation of light in the eye. In this work, it was developed the necessary simulation tools to study the mechanisms of the interaction of optical light with the retina. To accomplish this goal it was used the Monte Carlo (MC) simulation technique [2].

The MC is a powerful tool for simulating the way a photon interacts with biological tissues. In this work, it was devoted special attention on how the light interacts within the retina and how this interaction can provide us information about any anomalous structure existing there. Accordingly, emphasis is given to the dispersion pattern of the retina.

Several physics based MC codes have been developed and applied for radiation research, such as the EGS4 [3], the FLUKA [4], and the Geant4 [5] used in this work. The Geant4 is a collection of C++ class libraries, originally developed for high-energy physics detector simulation, that has found extensive use in the analysis of radiation and particle effects on low-medium-energy physics [6].

II. METHOD

A. Model

The simulation of the travel of the light through the human retina was performed using Geant4 (version 4.9.4) and its low energy package. This package, widely used in many different fields, such as nuclear and high-energy physics, medical physics and astrophysics, includes the photoelectric effect, Compton scattering, Rayleigh scattering, gamma conversion, Bremsstrahlung, ionization and fluorescence of excitation of atom [7].

In Geant4, it is necessary to build a particular virtual simulation scene by defining the geometry and composition of the media, the particles and the physical processes. The kernel, considering the material properties and the selected physical processes, tracks all interactions of the primary and secondary particles throughout the virtual structures.

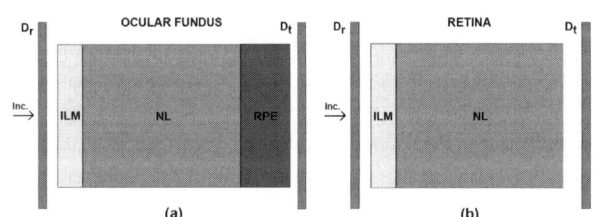

Fig. 1 Schematic systems used for the study of reflectance, transmittance and absorbance of ocular fundus (a) and retina (b). Dr and Dt are the photon detectors for the reflected and transmitted, respectively. ILM, NL and RPE stand for internal limiting membrane, neural layer and retinal pigment epithelium, respectively.

The virtual simulation scene of this work consists of a parallelepipedic world volume, filled with air, which has inside of it other volumes of different sizes and shapes. In Fig. 1 is represented schematically the internal limiting membrane (ILM), neural layer (NL) and retinal pigment epithelium (RPE) in the systems used for the study of reflectance, transmittance and absorbance of ocular fundus (a) and retina (b), as well as the used detectors.

B. Scattering Description

As mentioned previously, the aim of this work was to study the interaction of low energy, or optical, photons inside the human retina. It was chosen the energy of 2 eV (or 620 nm) because it describes the light used for observing the fundus, since it registers lower absorption in RPE layer [8].

In Geant4, the catalogue of optical processes includes reflection and refraction at medium boundaries, bulk absorption and Rayleigh scattering. Rayleigh scattering refers, after Lord Rayleigh (John Strutt) who was the first scientist to study quantitatively the phenomenon of the single scattering by particles with a diameter smaller than the wavelength of the incident light.

After the scattering process, photons travelling in an incident direction are scattered in a new direction that could be described by the scattering coefficient, μ_s. This coefficient is in general dependent of the incident direction of the scattered photons, and therefore the scattered intensity distribution depends on the incident direction. This situation is described by the scattering phase function p, which is defined as the normalized differential scattering coefficient.

The mean cosine of the scattering angle θ (angle between the incident and scattered directions) over the p distribution defines the anisotropy parameter g. It ranges from 1 to –1, for forward and backward scattering, respectively. In isotropic scattering, it is equal to zero. The modelization of this situation is frequently done by the reduced scattering coefficient μ'_s, also called transport scattering coefficient, which is given by $\mu'_s = (1-g)\mu_s$ [9].

In tissue samples that have a thickness greater than 10 μm, the interaction of scattered waves between neighboring particles cannot be ignored and multiple scattering of light becomes significant. This phenomenon may be described within the radiative transfer theory framework, by ignoring the wavelike behavior of light and describing the transport of photons through the absorption and scattered processes, being characterized by the absorption, μ_a, and reduced scattering, μ'_s, coefficients, respectively. Under these circumstances, a stochastic approach, like the Monte Carlo implemented in Geant4, becomes appropriate to model light propagation [10].

In this work, the scattered photon direction was obtained following the above-mentioned stochastic approach that uses the reduced scattered coefficient that describes the medium anisotropy together with the function $(1+\cos^2(\theta))$ implemented in Geant4.

C. Optical Parameters

The various tissues of the analyzed systems have different optical characteristics that were simulated in Geant4 by using optical parameters. The reflection and refraction of photons at tissue boundaries are characterized by the index of refraction, which relates to reduction of the speed of light in a medium.

The optical parameters for each material used in this work were taken from Ref. [11] and are listed in Table 1. These values were measured by Hammer et al. [12] for a wavelength of 633 nm.

Table 1 Optical parameters used in the simulations. The index of refraction of the medium external to the retina, vitreous, is 1.336.

Parameters	ILM	NL	PRE
Index of refraction (n)	1.355	1.355	1.400
Absorption coefficient (μ_a)	0.158	0.158	88.067
Reduced scattering coefficient (μ_s)	0.775	0.775	18.580
Thickness (μm)	2.0	200.0	12.0

D. Simulation Details

The incident beam was generated outside retina, with a width at half maximum of approximately 50 μm and the format of a circular beam of Gaussian profile (typical laser profile).

The calculations were performed on a cluster with Linux environment, composed of one AMD Opteron 275 at 2.2 GHz as the master node and nine Intel Core2 Quad Q6600 at 2.4 GHz as computing nodes. Each simulation ran in a single node and, for example, the time to simulate 10 million photons was about 2 hours.

III. RESULTS

With the aim of simulating the scanning laser beam to identify the presence of a spherical drusen in the retina, it was considered three situations, or runs, corresponding to three photon beam displacements from the drusen in the same axis, namely at the distances 100 μm (Run 0), 50 μm (Run 1) and 0 (Run 2).

Furthermore, in order to study how the variation of the scattering and absorption coefficients of drusen affects the spatial distribution of light reflected in the retina, drusen were implemented with five absorption and scattering coefficients, which are listed in Table 2.

In Figures 1, 2 and 3 are presented the scattering patterns and the number of reflected photons of the retina with the five drusen for the three runs. The analysis of these figures reveals clearly the presence of Drusen 1, 2 and 5 in Runs 1

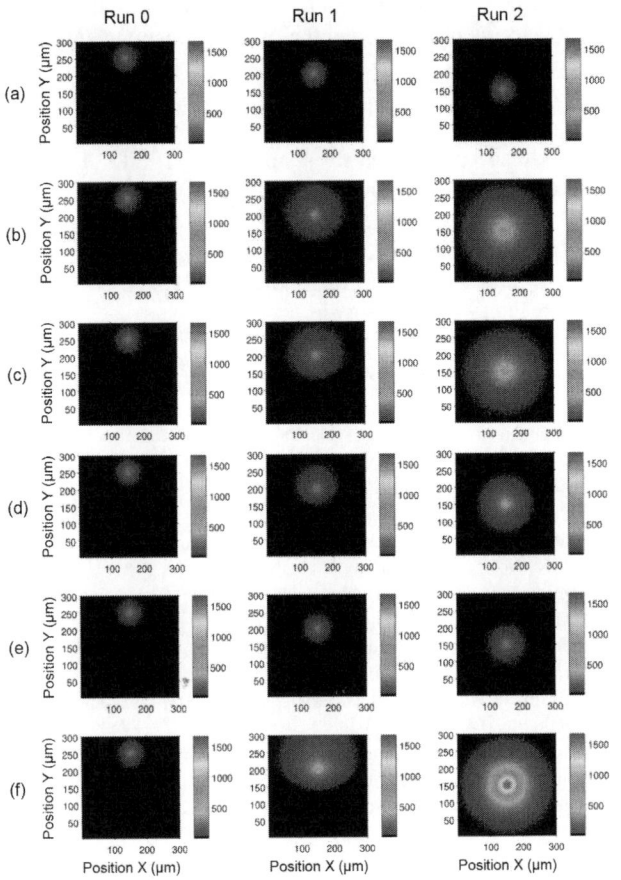

Fig. 2 Dispersion patterns of drusen with five different optical properties. (a) without drusen; (b) drusen 1; (c) drusen 2; (d) drusen 3; (e) drusen 4; (f) drusen 5.

and 2. Furthermore, we realize that there is no noticeable difference between the patterns of Drusen 1 and 2, which means that the absorption coefficient variation of drusen does not have great influence in the dispersion pattern.

Table 2 Drusen optical parameters.

	μ_a (cm^{-1})	μ'_s (cm^{-1})
Drusen 1	2.06	242,0
Drusen 2	20.6	242.0
Drusen 3	206.0	242.0
Drusen 4	20.6	24.2
Drusen 5	20.6	2420.0

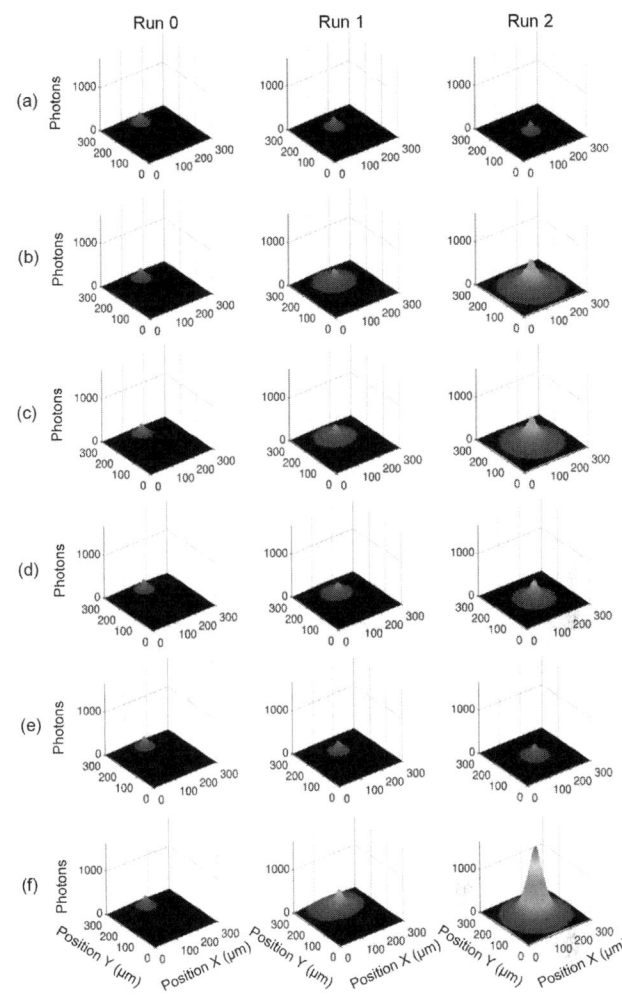

Fig. 3 Number of reflected photons by drusen with five different optical properties. (a) without drusen; (b) drusen 1; (c) drusen 2; (d) drusen 3; (e) drusen 4; (f) drusen 5.

The Drusen 4, meanwhile, shows patterns with greater resemblance to those presented by the retina without drusen. This demonstrates the strong dependence of the dispersion pattern with the reduced scattering coefficient, i.e., the lower capacity of scattering light.

Finally, the Drusen 5 results exhibit a clear evidence of the presence of disease in the retina. Observing the profile of the Run 1 in Figure 3 is evident a greater dispersion of light to the left indicating the presence of a drusen in the retina.

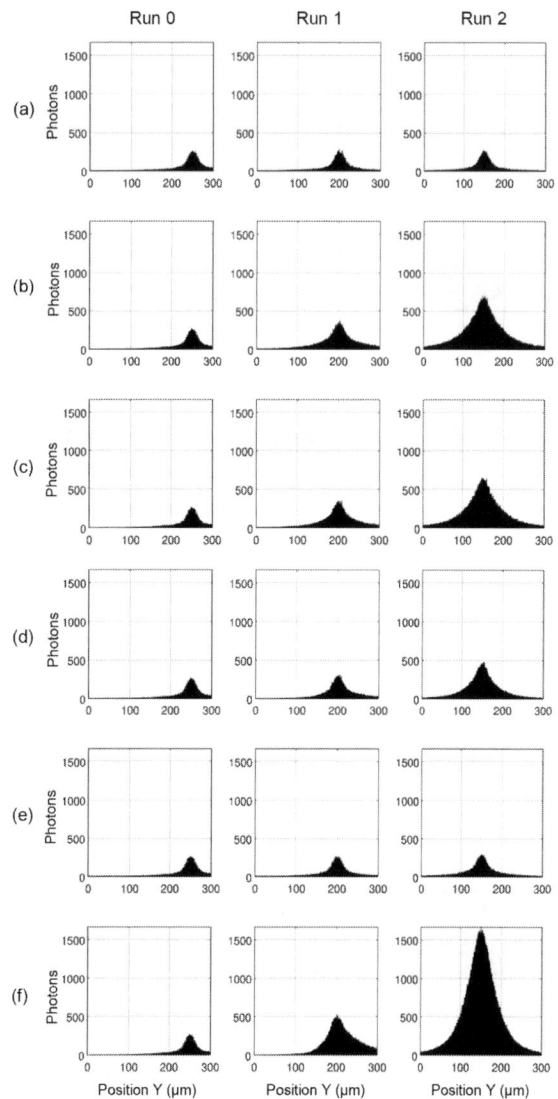

Fig. 4 Number of reflected photons by the drusen with five different absorption coefficients and scattering. (a) without drusen; (b) drusen 1; (c) drusen 2; (d) drusen 3; (e) drusen 4; (f) drusen 5.

IV. CONCLUSIONS

The aim of this work was to simulate a photon beam interacting with a retina with different drusen configurations using the Geant4 platform, and obtain the dispersion patterns of the reflected beam in order to evaluate the possibility of getting information about drusen's presence.

The results denote that the presence of drusen in the retina influences the dispersion pattern of the reflected beam, which gives important and necessary indications to the development of new ophthalmic diagnostic methods.

ACKNOWLEDGMENT

This research was partly supported by FCT - Fundação para a Ciência e a Tecnologia (Portugal), through Projects No. PEstOE/FIS/UI0303/2011 and No. PTDC/FIS/117606/2010, financed by the European Community Fund FEDER through the COMPETE – Competitiveness Factors Operational.

REFERENCES

1. Kinnunen K, Petrovski G, Moe MC, Berta A, Kaarniranta K (2012) Molecular mechanisms of retinal pigment epithelium damage and development of age-related macular degeneration. Acta Ophthalmologica 90(4):299-309
2. Preece SJ, Claridge E (2002) Monte Carlo modelling of the spectral reflectance of the human eye. Phys. Med. Biol. 47, 2863
3. Nelson WR, Hirayama H, Rogers DWO (1985) The Egs4 Code System, Stanford University
4. Ferrari A, Sala PR, Fassò A, Ranft J (2005) FLUKA: a Multi-Particle Transport Code (Program Version 2005) (CERN, Geneva)
5. Agostinelli S, Allison J, Amako K, et al. (2003) Nucl. Instrum. and Meth. Phys. A 506:250-303
6. Santina G, Nieminen P, Evansa H, et al. (2003) New Geant4 based simulation tools for space radiation shielding and effects analysis. Nuclear Physics B - Proceedings Supplements 125:69-74
7. Geant4.Web.Cern.Ch (2012)
8. Spaide RF, Curcio CA (2010) Drusen characterization with multimodal imaging. Retina 30(9):1441-1454
9. Rovati L, Cattini S, Viola F, Staurenghi G (2008) Measuring dynamics of scattering centers in the ocular fundus. International Journal on Smart Sensing and Intelligent Systems 1:799-811
10. Firbank M, Arridge SR, Schweiger M, Delpy DT (1996) An investigation of light transport through scattering bodies with non-scattering regions. Phys. Med. Biol. 41:767-783
11. Boas DA, Pitris C, Ramanujam N (2011) Handbook of Biomedical Optics (CRC Press)
12. Hammer M, Roggan A, Schweitzer D, G. Muller G (1995) Optical properties of ocular fundus tissues--an in vitro study using the double-integrating-sphere technique and inverse Monte Carlo simulation. Phys. Med. Biol. 40(6):963-78

Author: Pedro Vieira
Institute: Faculdade de Ciências e Tecnologia, Universidade Nova de Lisboa
Street: Quinta da Torre P-2829-516
City: Caparica
Country: Portugal
Email: pmv@fct.unl.pt

Semiautomatic Evaluation of Crypt Architecture and Vessel Morphology in Confocal Microendoscopy: Application to Ulcerative Colitis

E. Veronese[1], E. Poletti[1], A. Buda[2], G. Hatem[2], S. Facchin[2], G.C. Sturniolo[2], and E. Grisan[1]

[1] Department of Information Engineering, University of Padova, Padova, Italy
[2] Department of Surgery, Oncology and Gastroenterology, University of Padova, Padova, Italy

Abstract—**Neoangiogenesis plays a central role in both the initiation and perpetuation of the inflammatory response during chronic intestinal inflammation. However, limited data is available on the microvascular and crypt architecture during remission phases. In this study we have evaluated the intestinal mucosa of UC patients in clinical and endoscopic remission by probe-based confocal endomicroscopy (p-CLE) and quantified microvessel tortuosity and crypt architecture by a semiautomated analysis. The system firstly automatically identifies the crypts visible in the images, and estimates the graph connecting their centers in order to estimate the crypt architecture. Hyperfluorescence in the peri-crypt region is evaluated, and leakge is automatically detected. Length and tortuosity of the manually inserted vessels assess the vascular state. 6 patients with active UC and 2 with inactive UC were evaluated acquiring 19 set of p-CLE images with matched biopsies, showing significant differences in crypt architecture, vessel morphology and hyperfluorescence patterns.**

Keywords—Ulcerative colitis, automatic segmentation, confocal laser endomicroscopy, crypts, vessel, tortuosity.

I. Introduction

Ulcerative colitis (UC) is a chronic disease characterized by intense in?ammation of the mucosa limited to the colon, with bloody diarrhea, urgency to defecate, and general malaise [1]. It involves the rectum in about 95% of cases and may extend proximally in a symmetrical, circumferential, and uninterrupted pattern to involve parts or all of the large intestine [2]. Patients with UC have a higher risk of developing colorectal cancer. It is considered that this risk is a result of persistent inflammation of the colon [3]. Recent guidelines mention colonoscopy as a tool for screening this high-risk population.

The inflammatory changes in UC are represented by crypt architecture alterations with normal cells inside [4]. The chronic inflammatory infiltrate produces thickening of lamina propria and the distance between crypts becomes larger. All these aspects are accompanied by increased vasculature of the mucosa (neoangiogenesis), which plays a central role in both the initiation and perpetuation of the inflammatory response during chronic intestinal inflammation. Microscopic tissue evaluation is considered the gold standard for the final diagnosis of most diseases in gastroenterology.

Confocal laser endomicroscopy (CLE) has been recently proposed as a new technique that allows in vivo histologic assessment of mucosa during ongoing endoscopy, retrieving simultaneous endoscopic and endomicroscopic information and images in real-time [3]. It allows imaging of the mucosal layer, including epithelial cells and the lamina propria. The targets of endomicroscopic examination can be the cells, vascular structures, and/or tissue patterns [4]. Given its high resolution, CLE can be used to study in vivo both the vasculature of the mucosa and the architecture of crypts. Furthermore, CLE can be used in association with targeted contrast-agents, for instance fluorescein sodium (5 ml of a 10% solution, intravenous administration) [5] to enhance structures of interest as colonic pit patterns, surface epithelial cells, connective tissue matrix of the lamina propria, blood vessels and red blood cells.

The evaluation of vessel morphology and functionality has seen and increasing interest in the last few years, in order to assess and grade different inflammatory and neoplastic changes in the gastro-intestinal mucosa [6, 7, 8, 9, 10], as well as the study of crypts shape and architecture [11, 12, 13]. In particular the parameters that have been proposed are related to the regular arrangement and size of crypts, to the spaces between crypts and to the dilation of crypt openings.

II. Materials

8 patients with ulcerative colitis, 6 in the active phase and 2 in the non-active phase were studied. Their intestinal mucosa has been evaluated by probe-based confocal endomicroscopy (p-CLE, Mauna Kea Technologies, SA). 20 sites were imaged and the corresponding biopsies were taken for pathological assessment and UC remitting patients were clustered into either active or inactive disease according to histology. The Cellvizio system provided the mosaic of the acquired images (see representative images Fig. 1 that were ten exported for the quantitative analysis.

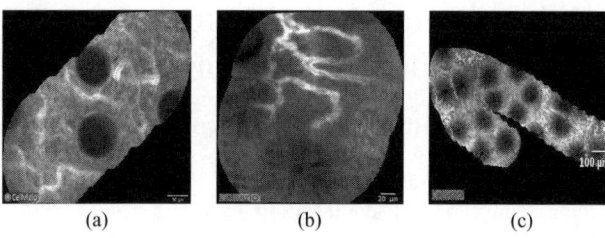

Fig. 1 Representative images of the colonic mucosa acquired with a pCLE system and montaged.

III. METHODS

Images were then analyzed in order to evaluate different parameters: crypt diameter, intercrypt distance, microvessel tortuosity and pericrypt fluorescence. In particular, a GUI has been designed using Matlab (2012a, The Mathworks, Inc., Natick, Massachusetts, US), to perform a semiautomated analysis which required the interactive identification of vessels.

A. Automatic Crypt Identification

Since crypts usually appears as dark ovoidal shapes within the images, a rough identification of their position may be achieved by the application of a thresholding operation on the gray-level intensities. After having evaluated the mean image intensity μ_i and its standard deviation σ_i, we identify the regions of image I whose intensity is below the value $\theta = \mu_i - \sigma_i$. The resulting binary image is subject to a morphological opening with a round structuring element od radius N, where N should be comparable to the typical radius of the imaged crypts. Then, the remaining connected components are considered to be candidate crypts and are separately processed to refine their borders. Given the kth connected component C_k, its centroid $P_k(x,y)$ is evaluated, and the algorithm considered a square window W centered in P_k. All pixels included in the window are classified by means of a k-means clustering based on the gray level intensity of the pixels. In particular, given that all pixels in an image can be roughly divided into three classes (i.e. very bright pixels in the correspondence of fluorescein, very dark pixels in the correspondence of crypts, and grayish pixels in the correspondence of cell bodies), we set the number of cluster to three. Using the obtained values of the centroid C_1, C_2, C_3 of each of the three clusters, a threshold θ is computed as $\theta = 0.75 * C_1 1 + 0.25 * C_2$, where C_1 and C_2 are the value of the centroid of the first (i.e. the darkest) and second (i.e. the mean) cluster, respectively. Only those pixels inside W and darker than θ are labeled as belonging to the crypt.

A morphological closing with a disk-shaped structuring element or radius r=25 is then applied to regularize the crypt's contours and to remove noisy pixels.

Finally, the peri-crypt region, is estimated taking the crown surrounding each crypt boundary and distant 0.40 μm from it, as can be seen in Fig. 2

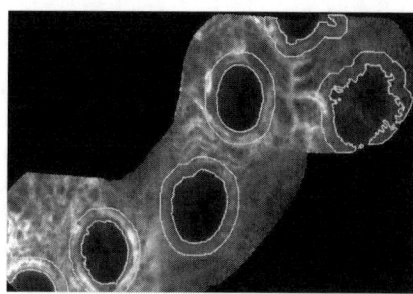

Fig. 2 Automatically identified crypt borders and peri-crypt limits.

B. Crypt Architecture Evaluation

When all crypts have been inserted, a set of parameters are estimated. In particular, for each crypt the algorithm computes the area, the centroid's coordinates, and the value of eccentricity. Besides, given the set of all the crypts, the algorithm computes their distance (both between their centroids and between their borders) using the Delaunay triangulation, and the regularity of the distribution is evaluated considering the mean distance along the arches of the triangulated graph connecting the crypt centers (Fig. 3). The crypt architecture is then characterized by four indexes. The first is the mean distance among crypt centers $\mu_{cry,cent}$, the second the mean intercrypt distance $\mu_{cry,inter}$ calculated as the mean distance along the triangulated arches between crypt borders. Finally the mean crypt area $\mu_{cry,area}$ and mean crypt eccentricity $\mu_{cry,ecc}$ are evaluated.

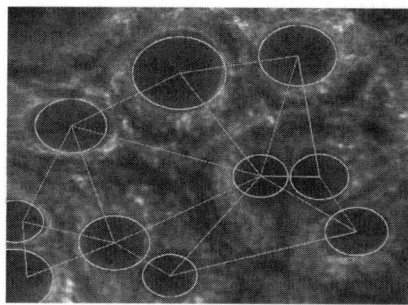

Fig. 3 Identified crypts (yellow) in a confocal microendoscopic image with with the identified graph connecting their centers (orange arches).

C. Semiautomatic Vessel Identification and Evaluation

The tracking of the mucosa vasculature is performed through an interactive tool, allowing the user to define and analyze any number of vessel in the image. For each vessel, the user is requested to locate the ending points (the beginning and the end) of a new vessel to be considered, by clicking on two different pixel in the image. The system will draw a straight line connecting these two points. At this stage the user can add other points of the vessel that will be inserted between the two ending points: the previously draw line becomes a smoothed cubic-spline that connects all the inserted points. If the user needs to change the position of any inserted point, he/she can click it and drag it into a new position. A right-click on any point will remove it from the spline. The display of the spline will be updated real-time every time a new point is inserted/moved/deleted by the user. When the whole spline curve lies inside the vessel, the procedure is done, and a new vessel can be inserted. For each vessel it is calculated its length and the tortuosity index computed according to the algorithm described in [14], yielding results as those represented in Fig. 4. The calculated indexes for each values are then summarized for image-wide assessment evaluating the mean tortuosity μ_{tort} and the mean length μ_{len}

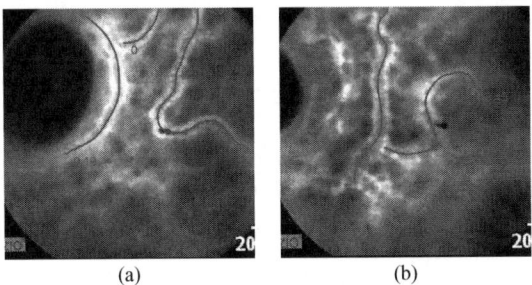

Fig. 4 Manually identified vessels with the corresponding tortuosity value.

D. Fluorescein Leakage Evaluation

Fluorescein leakage is an important marker of vessel blood barrier impairment so that its identification an quantification provide an important marker on the mucosal tissue and on the vessel state. To this end, the regions around vessels and crypts as described in Sec. A are masked so not to be considered in the leakage detection. Then, the remaining pixels of the image are analyzed to identify the areas with intensity greater than a threshold θ_{leak} set to the 90th percentile of the image intensity distribution. The parameter FA_{leak} related to the identified fluoresceine leakage is the percentage of the selected area with respect to the image area.

Moreover, since the per-crypt fluorescence seems to be a sensitive marker of the crypt condition, we evaluate three additional parameters. The first $\mu_{peri,all}$ is the mean intensity in the peri-crypt regions, the second $\mu_{peri,m}$ is the mean intensity of the regions surrounding the crypts with intensity higher than the image mean, and finally the third $\mu_{peri,\theta}$ is is the mean intensity of the regions surrounding the crypts with intensity higher than θ_{leak}.

IV. RESULTS

The proposed computer-aided system for the evaluation of colonic mucosa is able to detect the crypts with a sensitivity of 87% and without false positive (specificity 100%). For the great majority, the undetected crypts are represented by those at the margin of the field of view, often only partially images. Representative results of analysed images are shown in Fig. 5.

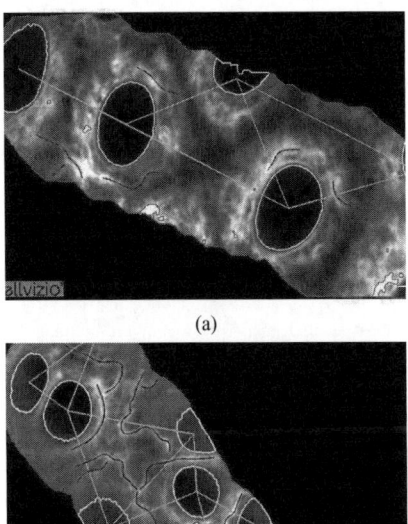

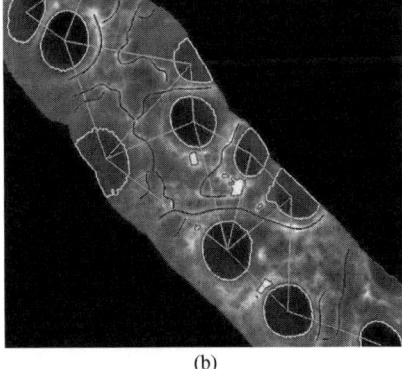

Fig. 5 Results of the computer assisted analysis. Identified crypts are outlined in yellow, with the orange graph connecting their centers. Red-bordered regions are detected leakage areas, whereas the blue lines are the manually inserted vessels.

The quantitative results of the evaluated parameters across the images of the two population analysed (active and non active ulcerative colitis) are reported separately in Tab. 1 for the crypt shape and architecture, in Tab. 2 for vessel morphology, and in Tab. 3 for leakage-related parameters either calculated in the peri-crypt regions or in the mucosa excluding vessels and crypt regions.

Table 1 Mean crypt architecture parameters on the image of active versus non-active ulcerative colitis. Statistical significance is reported.

	Mean Crypt Parameters			
	$\mu_{cry,cent}$ μm	$\mu_{cry,inter}$ μm	$\mu_{cry,area}$ μm^2	$\mu_{cry,ecc}$
Active	177	102	5578	0.56
Non-active	155	68	6496	0.61
p-value	0.05	¡0.01	0.07	ns

Table 2 Mean vessel morphology parameters on the image of active versus non-active ulcerative colitis. Statistical significance is reported.

	Vessel Parameters	
	μ_{tort}	μ_{len}
Active	1.55	103
Non-active	0.77	136
p-value	0.04	ns

Table 3 Mean fluoresceine-related parameters on the image of active versus non-active ulcerative colitis. Statistical significance is reported.

	Mean Leakage Parameters			
	FA_{leak}	$\mu_{peri,all}$	$\mu_{peri,m}$	$\mu_{peri,\theta}$
Active	0.53%	0.41	0.17	0.05
Non-active	0.28%	0.38	0.12	0.03
p-value	0.06	ns	0.05	ns 0.03

V. CONCLUSION

We presented a computer aided system for the analysis of probe-based confocal microendoscopy images, for the characterization of mucosal changes in ulcerative colitis. In particular, the proposes system is able to reliably and automatically identify the colonic crypts and the leakage regions. It evaluates shapes and distribution parameter of the crypts, as well as tortuosity and length of the manually inserted vessels. Additional fluorescein-related parameters are evaluated. The computer assisted analysis of p-CLE images seems able to detect intramucosal changes of microvessels and crypt architecture in the presence of subtle neangiogenesis and inflammation in endoscopic normal appearing mucosa.

REFERENCES

1. Carter MJ, Lobo AJ, Travis SP. Guidelines for the management of inflammatory bowel disease in adults. *Gut.* 1999;53:V1-V16.
2. Kornbluth A, Sachar DB. Ulcerative Colitis Practice Guidelines in Adults: American College of Gastroenterology, Practice Parameters Committee *Am J Gastroenterol.* 2010;105:501-523.
3. Gheonea DI, Saftoiu A, Ciurea T, Popescu C, Georgescu CV, Malos A. Confocal laser endomicroscopy of the colon *J Gastrointestin Liver Dis..* 2010;19:207-11.
4. Kiesslich R, Galle PR, MF Neurath. *Atlas of Endomicroscopy.* Heidelberg: Springer Medizin Verlag 2007.
5. Hoffman A, Goetz M, Vieth M, Galle PR, Neurath MF, Kiesslich R. Confocal laser endomicroscopy: technical status and current indications *Endoscopy.* 2006;38:1275-1283.
6. Becker V., Vieth M., Bajbouj M., Schmid R. M., Meining A.. Confocal laser scanning fluorescence microscopy for in vivo determination of microvessel density in Barrett's esophagus *Endoscopy.* 2008;40:888891.
7. Lin K.Y., Maricevich M., Bardeesy N., Weissleder R., Mahmood U.. In vivo quantitative microvasculature phenotype imaging of healthy and malignant tissues using a fiber-optic confocal laser microprobe *Translational Oncology.* 2008;1:84-94.
8. Loeser C.S., Robert M.E., Mennone A., Nathanson M.H., Jamidar P.. Confocal Endomicroscopic Examination of Malignant Biliary Strictures and Histologic Correlation With Lymphatics *J Clin Gastroenterol.* 2011;45:246-252.
9. Goetz M.. Confocal Laser Endomicroscopy: Applications in Clinical and Translational Science A Comprehensive Review *ISRN Pathology.* 2012;2012:13 pages.
10. Neumann H, Vieth M, Atreya R, et al. Assessment of Crohn's disease activity by confocal laser endomicroscopy *Inflamm Bowel Dis..* 2012;18:2261-9.
11. Watanabe Osamu, Ando Takafumi, Maeda Osamu, et al. Confocal endomicroscopy in patients with ulcerative colitis *Journal of Gastroenterology and Hepatology.* 2008;23:S286–S290.
12. Goetz M., Kiesslich R.. Confocal Endomicroscopy: In Vivo Diagnosis of Neoplastic Lesions of the Gastrointestinal Tract *Anticancer Research.* 2008;28:353-360.
13. Li CQ, Xie XJ, Yu T, et al. Classification of Inflammation Activity in Ulcerative Colitis by Confocal Laser Endomicroscopy *Am J Gastroenterol.* 2010;105:1391-6.
14. Grisan E., Foracchia M., Ruggeri A.. A Novel Method for the Automatic Grading of Retinal Vessel Tortuosity *Medical Imaging, IEEE Transactions on.* 2008;27:310-319.

Author: Enrico Grisan
Institute: Department of Information Engineering, University of Padova
Street: Via Gradenigo 6/b
City: Padova
Country: Italy
Email: enrico.grisan@dei.unipd.it

3D Reconstruction of Skin Surface Using an Improved Shape-from-Shading Technique

G. Balogiannis, D. Yova, and K. Politopoulos

Biomedical Optics & Applied Biophysics Lab-School of Electrical
and Computer Engineering-National Technical University of Athens, Athens, Greece

Abstract—Due to the heterogeneity and complexity of the skin, photo-shading techniques are suitable for 3D reconstruction in cases where small scale (<1mm) features are present. That is because high scattering surfaces, like the skin, diffuse the projected light patterns used in range scanners, and thus reduce their accuracy. Shape-from-Shading (SfS) algorithms are free from such drawbacks, as they use shading information and exploit the full resolution of the imaging system. In this paper an improved SfS method, called Shape-from-Isophotes, is introduced. This method aims to overcome most of the disadvantages of classical SfS methods, like numerical instability. The SfI improvements include the addition of a constraint deriving from the properties of isophote curves, and the decomposition of the original shading image into two frequency bands. The first feature attempts to increase the numerical stability of the equation system, and the second to reconstruct both small scale and large scale surface variations with the same accuracy. These new features combined with the Lambertian model for skin reflectance aim to create an accurate reproduction of the human skin surface, both in three-dimensional shape and reflective behavior.

Keywords—3D reconstruction, Skin optics, Shape-from-Shading, Isophotes.

I. INTRODUCTION

This paper aims to present an improved optical technique for skin surface reconstruction. The human skin has a complex 3D shape with many details and its three dimensional representation may provide significant information about its condition. It can be considered as a high scattering surface with a specular component. These unique optical features play a significant role in determining which imaging method can be used to obtain optimum results. When it comes to biological tissue, not all reconstruction methods have the same performance [1][2].

More specifically, in active optical methods, the high scattering that occur on biological tissues reduces the projected structure (laser line or pattern) sharpness, and thus degrades the method overall accuracy. Additionally, in the case of laser scanning, the long duration of the data acquisition proves not to be very practical for *in vivo* experiments [3]. In passive stereo vision methods, the detection of correspondence points is difficult in the case of the homogeneous texture of the skin [4].

The techniques to recover shape from shading images of the surface are called Shape-from-X techniques - where X can be shading, stereo, motion, texture, e.a. It is proven that SfX methods are more suitable in the case of high scattering media, like the human skin, which strongly diffuse any projected patterns and reduce their sharpness. The SfX methods do not have such restrictions, as they exploit the full resolution provided by the optical system and the CCD sensor. In addition, when it comes to living subjects SfS outmatch other range scanning methods, as the image acquisition process is significantly faster, which is important when even the slightest movement is unavoidable.

Shape-from-Shading (SfS) deals with the recovery of shape from a gradual variation of shading in a single image. Amongst SfX methods, Shape-from-Shading is more simple and easy to implement, as it requires only one shading image to perform the reconstruction. A simple model of image formation is the Lambertian model, in which the gray level at a pixel in the image depends on the light source direction and the surface normal.

The surface shape is described in terms of the surface normal, thus a linear equation with three unknowns has to be solved. Given that the surface shape is described in terms of the surface gradient, a non-linear equation with two unknowns is derived. Thus, the Lambertian constraint by itself cannot provide a solution to the problem and more constraints are required to have a unique solution [5][6][7]. Shape-from-Shading methods offer additional constraints in order to solve the system of equations. However, these assumptions are often mathematically unstable and are gradient descent based, so they cannot apply on continuous smooth surfaces [8][9].

In this paper an improved SfS method, i.e. Shape-from-Isophotes (SfI), is proposed. This method intends to overcome most of the problems that trouble traditional SfS methods. An additional constraint, originating from the properties of the brightness gradient on continuous surfaces, is used. The frequency dynamic range is decomposed in two bands, in order to reconstruct the full range of the object details. This allows as to adjust the algorithm parameters and reconstruct the two frequency bands separately, with the same accuracy.

II. MATERIALS AND METHODS

The binocular machine vision system developed in Biomedical Optics and Applied Biophysics Lab was used for the experiments 10. It includes one 1/1.8" color CCD PointGrey Scorpion SCOR-20SO camera. The image size was 1600×1200 pixels, with pixel size of 4.4×4.4 μm.

A. Theoretical background

Reconstruction methods aim to retrieve the 3D information from one or more 2D shading images. The recovered shape can be expressed as a two variable depth function $z(x,y)$. Conventionally, the z-axis is oriented along the optical axis of the imaging system.

In the developed method the equations of orthographic projection were used, which match with the properties of the telecentric lens:

$$x = mX, y = mY, m = -f/z_o \qquad (1)$$

where f is the lens focal length and z_o is the mean distance between the object and the sensor. m symbolizes the lens' magnification [11][12].

The choice of the reflectance model depends on the properties of the examined surface, and is crucial for the success of the reconstruction method. There is no universal reflectance theory to cover all known surfaces, thus each surface is approximated by a different reflectance equation [13][14].

For a scene consisting of a Lambertian surface illuminated by a single distant (parallel ray) light source, the observed image intensity I can be simply written as:

$$I = K_D \cos(\theta_i) = K_D (\mathbf{S} \cdot \hat{\mathbf{n}}) \qquad (2)$$

where K_D is the composite albedo, θ_i the angle of incidence i.e. the angle between the direction of the incident light and the surface normal. The angle θ_i can be expressed as the product of a two unit column vectors, $\mathbf{S}$, describing the incident direction, and $\hat{\mathbf{n}}$, the surface normal. The normal $\hat{\mathbf{n}}$ has components:

$$\hat{\mathbf{n}}(x,y,z) = \frac{1}{\|\hat{\mathbf{n}}\|}\left(\frac{\partial z(x,y)}{\partial x} \cdot \hat{i} + \frac{\partial z(x,y)}{\partial y} \cdot \hat{j} + \hat{k}\right) \qquad (3)$$

B. Method Description

The skin has a very complex superficial structure, which includes large scale (low frequency) and small scale (high frequency) features. As a result of the skin's complexity, it is difficult to find a generic method that can be applied to the image and produce satisfactory results for all frequency scales. The solution to this problem is to decompose the surface into low frequency and high frequency bands applying the analysis to each frequency band separately. The reconstruction method composes of two parts: a) the part that deals with the low frequencies of the object, which will be called *low-frequency module*, and b) the part that deals with the higher frequencies, which will be called *high-frequency module*.

The low-frequency module produces a rough estimate of the surface, which will provide initial values for the high-frequency reconstruction module. The reconstruction of the high frequency features will be done locally, fact that allows us to use the local surface conditions in order to minimize the residual error.

A standard technique used to reconstruct 3D surfaces in SfS is to estimate the normal vectors $\hat{\mathbf{n}}(x,y)$ for every point (x,y) of the shading image and then integrate the normal vectors along a specific path. In order to calculate the normal, we need one constraint for each unknown introduced by each vector. A standard constraint is introduced by the illumination equation (2). In order to derive additional constraints from the 2D image, some beneficial geometric properties of the isophote curves were exploited.

The image dynamic range is divided into a number of N_{is} brightness levels. Each brightness level corresponds to one isophote curve. Obviously, N_{is} cannot be greater than the total number of the brightness levels provided by the image quantization (8-bit or 256 levels for grayscale bitmaps) [15][16].

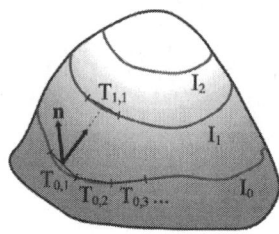

Fig. 1 Picture showing illuminated surface with isophote curves I_i and tangent vectors $\mathbf{T}_{i,k}$ illustrated.

In order to start the 3D reconstruction process, a set of initial values is needed. These initial values must be points of the shading image where the z-coordinates are known. In order to find such points, an isophote curve that is close to a brightness local maximum is chosen. The z-coordinates of the curve points can be approximated due the assumption that in a small range around this point, the surface gradient may be dependable to the brightness gradient [16]. The choice of the initial values is crucial for the overall accuracy and therefore they must be chosen carefully.

The reconstruction starts from the low-pass copy of the shading image (lower frequencies) – which represents the image copy with the fewer details – and the 3D reconstructed result provides an initial rough estimation of the surface three-dimensional shape. This rough estimation will provide the initial values for the higher level of reconstruction in the high-frequency module. The full range reconstructed result will be the sum of the results of the two frequency modules:

$$z_{tot} = z_{low} + z_{high} \qquad (4)$$

The low-pass copy of the surface results after the convolution of the original image with a Gaussian-like weighting function $g(m,n)$ [17]:

$$G(i,j) = g * L_p \Rightarrow G(i,j) = \sum_m \sum_n g(m,n) L_p(i+m, j+n)$$

We assume a distant point light source, with known direction $\mathbf{S} = [S_x, S_y, S_z]$. Assuming that the Lambertian reflectance model is valid, the *Lambertian constraint* is introduced:

$$\mathbf{S} \cdot \hat{\mathbf{n}} = I_i \Rightarrow S_x \cdot n_x + S_y \cdot n_y + S_z \cdot n_z = I_i \qquad (5)$$

where I_i is the isophote brightness value. In order to further reduce the number of unknown parameters, the fact that all vectors $\hat{\mathbf{n}}$ that are perpendicular to a surface are de facto vertical to any real curve that is tangent to the same surface, was taken into consideration. This introduces the verticality constraint:

$$\mathbf{T} \cdot \hat{\mathbf{n}} = 0 \Rightarrow T_x \cdot n_x + T_y \cdot n_y + T_z \cdot n_z = 0 \qquad (6)$$

The third equation needed for the solution of the system derives from the unity of the $\hat{\mathbf{n}}$ vector:

$$n_x^2 + n_y^2 + n_z^2 = 1 \qquad (7)$$

Solving equations (5),(6),(7) the normal vectors $\hat{\mathbf{n}}$ along the isophote curve I_i can be computed. By integrating the normals along the specific path of the isophote the absolute depth values z for all the points of the curve are calculated. Proceeding to the next isophote until the full surface is reconstructed.

III. EXPERIMENTAL RESULTS AND DISCUSSION

The algorithm was implemented on real skin images of human and mice. Images of skin with small scale wrinkles and epidermal ridges, like the ones shown in Fig. 2 & 3, oppose many challenges to 3D reconstruction, because the reflectance conditions differ from one sample to the other and they involve a very wide spectrum of surface frequencies.

The proposed method has the advantage to obtain stereo images *in vivo* using an improved Shape-from-Shading (i.e. Shape-from-Isophotes) algorithm. The fact that only one shading image is required for the reconstruction, makes the algorithm quick, robust and suitable for living samples. This system can obtain in vivo images without being affected by minor movements caused by breathing. The imaging apparatus used for the reconstruction is not complicated, as it uses only one camera and a light source. Since there is no stereo camera system, no camera calibration is needed.

The reconstruction of the human skin has shown the true potential of the technique. The epidermal ridges have an average depth of 100 μm [18], and therefore the depth resolution of the model is at least of the same magnitude.

The improved Shape-from-Isophotes method manages to avoid the numerical instabilities of the other Shape-from-Shading techniques and the stochastic nature of the traditional Shape-from-Isophotes methods, by introducing an additional equation deriving from the properties of the isophote curves. This is an improvement of the statistic nature of the traditional Shape-from-Isophotes techniques, which

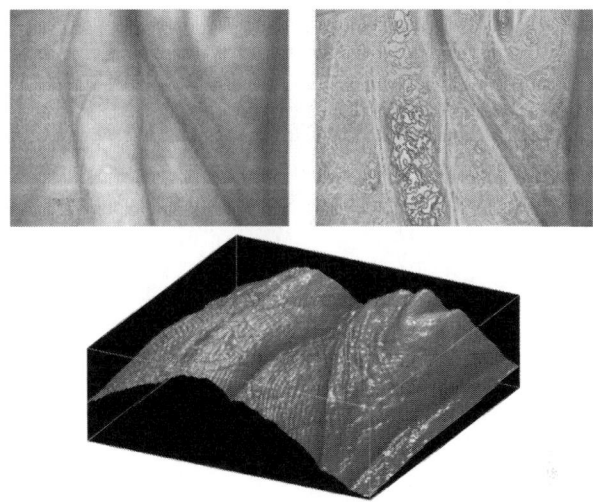

Fig. 2 Top-left: Shading image of human skin surface. Top-right: pseudo chromatic isophote map of the surface. Bottom: Reconstructed result in RGB color.

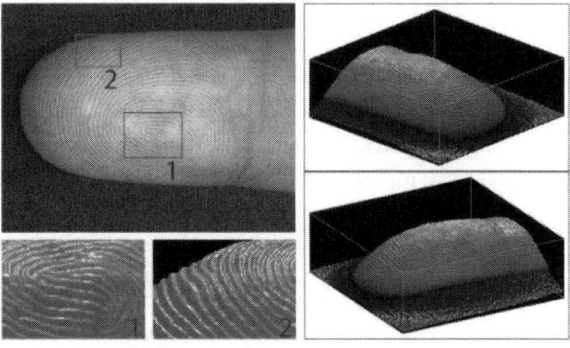

Fig. 3 3D reconstruction of human fingertip. The low-frequency module reconstructs the large scale features (finger's curvature) while the high-frequency module reconstructs the small scale details (epidermal ridges).

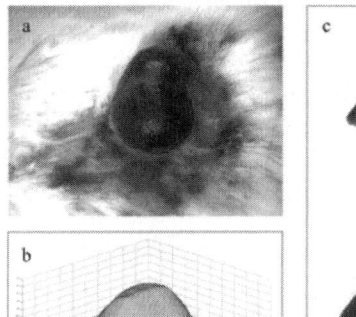
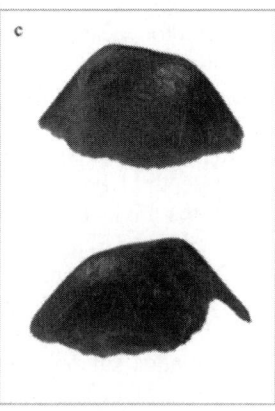

Fig. 4 3D reconstruction of tumor inserted under the skin of a mouse. (a) Color image showing the body of the mouse with the tumor apparent (red outline). (b): Reconstruction of tumor. (c): Views of the reconstructed surface with real color data.

use the isophote curves only for normal integration and not for normal computing. This equation makes the normal vector system of equations well defined and directly solvable. No minimization constraints are needed and the solution is unique.

Additionally, the described method manages to have satisfactory results with surfaces with both small and large scale variation surfaces. The decomposition of the surface into two frequency bands allows us to treat the two modules separately and minimize the error in the small scale details of the surface.

IV. CONCLUSION

A new Shape-from-Shading method for recovering surface normal was developed. It includes two innovative features: the definition of the equation system and the two frequency band decomposition. Thus, unlike minimization methods, it doesn't suffer from the common numerical instabilities of the SfS techniques. Also because the propagation direction is decoupled from the gradient direction it is less error prone than characteristic strip methods in areas where the gradient is zero. The two band decomposition allows us to reconstruct a wide range of frequencies with the same accuracy. The quantitative error analysis showed an improved performance with average error of less than 7%. The errors are attributed to noise and to the fact that real images do not fully satisfy the simplifying assumptions (i.e. interreflections, not truly distant light sources, etc.). Future work includes experiments using more complex surfaces and reflectance maps.

REFERENCES

1. A. N. Bashkatov, E. A. Genina, V. V. Tuchin (2011), "Optical properties of skin subcutaneous, and muscle tissues: a review", J Innov Opt Health Sci., Vol. 4, No. 1, pp. 9-38.
2. Ling Li, Carmen So-ling Ng (2009), "Rendering human skin using a multi-layer reflection model", Int J Math, Issue 1, Volume 3, pp. 44-53.
3. J. C. Rodríguez-Quiñonez, O. Sergiyenko, V. Tyrsa, L. C. Básaca-Preciado, M. Rivas-Lopez, D. Hernández-Balbuena, M. Peña-Cabrera, "3D Body & Medical Scanners' Technologies: Methodology and Spatial Discriminations", Optoelectronic Devices and Properties, 2011, Prof. Oleg Sergiyenko (Ed.), ISBN: 978-953-307-204-3, InTech, DOI: 10.5772/16233, pp. 307-323.
4. V. Niola, C. Rossi, S. Savino, S. Strano (2011), "A method for the calibration of a 3-D laser scanner", Robot Cim-Int Manuf 27 pp. 479-484.
5. Kong, Fan-Hui, "A New Method of Inspection Based on Shape-from-Shading", IEE P-Vis Image Sign, Volume: 2, pp. 291-294.
6. Jean-Denis Durou, Maurizio Falcone, Manuela Sagona (2008), "Numerical methods for Shape-from-Shading: A new survey with benchmarks", Comput Vis Image Und, 109, pp. 22–43.
7. Ruo Zhang, Ping-Sing Tsai, James Edwin Cryer, Mubarak Shah (1999), "Shape from Shading: A Survey", IEEE T Pattern Anal.
8. B.K.P. Horn, "Robot Vision", MIT Press, Cambridge, Massachusetts, 1986.
9. M. Visentini - Scarzanella, D. Stoyanov, G.-Z. Yang, "Metric Depth Recovery from Monocular Images Using Shape-from-Shading and Specularities", Proceedings - International Conference on Image Processing, ICIP, art. no. 6466786, Orlando, FL, 2012, pp. 25-28.
10. D Gorpas., K.Politopoulos, D. Yova (2007), "A Binocular Machine Vision System for Three-Dimensional Surface Measurement of Small Objects", Comput Med Imag Grap, 31: pp. 625-637.
11. O. Faugeras, "Three-Dimensional Computer Vision", MIT Press, 1993.
12. O. Lee, G. Lee, J. Oh, M. Kim, C. Oh (2010), "An optimized in vivo multiple–baseline stereo imaging system for skin wrinkles", Opt Commun 283 (23), pp. 4840-4845.
13. M. Oren and S.K. Nayar (1994), "Generalization of Lambert's Reflectance Model", Siggraph, Jul, pp. 239-246.
14. K.J. Dana, B. Van-Ginneken, S. K. Nayar, J. J. Koenderink (1999), "Reflectance and Texture of Real-World Surfaces", Acm T Graphic (TOG), Vol.18, No.1, Jan, pp. 1-34.
15. V. Dragnea, E. Angelopoulou (2005), "Direct shape from isophotes", Proceedings of BenCOS2005, Volume XXXVI-3/W36, pp. 1-6.
16. J. Lichtenauer, E. Hendriks, M. Reinders (2005), "Isophote Properties as Features for Object Detection", Proceedings of IEEE International Conference on Computer Vision and Pattern Recognition, pp. 649-654.
17. P. J. Burt, E. H. Adelson (1983), "The Laplacian pyramid as a compact image code", IEEE Trans Commun, Com-31, pp. 532-540.
18. M. Kucken, A. C. Newell (2005), "Fingerprint formation", J Theor Biol 235 pp. 71–83.

Author: Giorgos Balogiannis
Institute: National Technical University of Athens
Street: 9 Iroon Polytexneiou, Zografou Campus, 15780
City: Athens
Country: Greece
Email: gmpalog@central.ntua.gr

QuantiDOPA: A Quantification Software for Dopaminergic Neurotransmission SPECT

A. Niñerola[1], B. Marti[1], O. Esteban[2], X. Planes[3], A.F. Frangi[3], M.J. Ledesma-Carbayo[2], A. Santos[2], A. Cot[1], F. Lomeña[4], J. Pavia[1], and D. Ros[1]

[1] GIBUB, Universitat de Barcelona - CIBER-BBN, Barcelona, Spain
[2] BIT, Universidad Politécnica de Madrid - CIBER-BBN, Madrid, Spain
[3] CISTIB, Universitat Pompeu Fabra - CIBER-BBN, Barcelona, Spain
[4] Hospital Clínic de Barcelona- CIBER-SAM, Barcelona, Spain

Abstract—Quantification of neurotransmission Single-Photon Emission Computed Tomography (SPECT) studies of the dopaminergic system can be used to track, stage and facilitate early diagnosis of the disease. The aim of this study was to implement QuantiDOPA, a semi-automatic quantification software of application in clinical routine to reconstruct and quantify neurotransmission SPECT studies using radioligands which bind the dopamine transporter (DAT). To this end, a workflow oriented framework for the biomedical imaging (GIMIAS) was employed. QuantiDOPA allows the user to perform a semi-automatic quantification of striatal uptake by following three stages: reconstruction, normalization and quantification. QuantiDOPA is a useful tool for semi-automatic quantification in DAT SPECT imaging and it has revealed simple and flexible.

Keywords—Quantification, dopaminergic system, Parkinson's disease, SPECT, GIMIAS.

I. Introduction

Parkinson's disease is a neurological disorder associated with the loss of dopaminergic neurons. A number of ^{99m}Tc-agents and ^{123}I-agents have been developed for Single-Photon Emission Computed Tomography (SPECT) imaging to study presynaptic dopamine transporter (DAT) binding. Quantification of DAT SPECT imaging can be used to discriminate parkinsonian syndromes from other movement disorders. Quantification also facilitates an early diagnosis of the disease [1], to follow up its progression [2] and assess the effects of treatment strategies [3].

In clinical routine there is a need for quantification methods of DAT SPECT studies. This work deals with the implementation of QuantiDOPA, a semi-automatic quantification software for dopaminergic neurotransmission SPECT studies.

II. Material and Methods

This medical application was developed using GIMIAS [4], a workflow oriented framework for the biomedical imaging and modeling of prototypes in the context of the Virtual Physiological Human. It is an open source framework distributed under a BSD license, developed in C++ and is based on robust open source libraries such as VTK, ITK, MITK, DCMTK, NETGEN and TETGEN, among others.

GIMIAS relies on these libraries for visualization, user interaction, data access, image and mesh processing, and is designed to easily integrate proprietary algorithms and new libraries that broaden the functions of the framework. GIMIAS is currently collaborating to develop and adopt standards that could favour interoperability with similar software in the context of the VPHNoE [5]. It also collaborates with CTK [6] (Common Toolkit), an international effort to develop interoperable software. The goal of GIMIAS is to reduce the time needed to develop a clinical prototype.

Each application implemented on GIMIAS, named plug-in, represents an specific process.

QuantiDOPA is a GIMIAS plug-in whose goal is to obtain from the gammacamera projections the striatum quantified value. To this end three modules have been developed: reconstruction, normalization and quantification.

A. Reconstruction

This is an optional module that allows us to standardize the reconstruction protocol of projections from different gammacameras. Preprocessing of projections with different smoothing filters can be performed before reconstruction. Currently, the reconstruction method is based on the filtered backprojection (FBP) algorithm with different smoothing filters.

B. Normalization

The reconstructed image is normalized to a SPECT template, using two separate steps. In the first step, a 9 parameters 3D affine transformation is used. The second step provides a fine adjustment of the striatum separately for each hemisphere. This is achieved by calculating the local correlation coefficient [7] between the reconstructed striatum and the SPECT template striatum. This procedure is done for each hemisphere.

The software allows working with a database of different SPECT templates.

C. Quantification

In order to calculate the mean activity values on the normalized SPECT image, QuantiDOPA allows us to select different regions of interest (ROIs). These ROIs were defined following two different approaches which are commonly used in nuclear medicine departments:

- Standardised 3D ROIs map, consisting on an automatic volumetric definition of the ROIs based on the morphology of the striatum according to the Automated Anatomical Labelling (AAL) map [8].
- 3D cylindrical ROIs [9] placed on the striatum area by the user. Eight ROIs are considered, one in the caudate and three in the putamen of each hemisphere. In order to facilitate cylinders placement, a 2D image consisting of the sum of eight central slices of the reconstructed image is provided.

Mean activity values are obtained in ROIs located in the striatum and in a background region to obtain the Specific Uptake Ratio (SUR), a parameter that quantifies the specific radioligand uptake in the striatal volume. SUR is defined as:

$$SUR = \frac{S-B}{B} \quad (1)$$

where S and B are the concentration activity in the striatum and in a background reference region of non-specific uptake, respectively.

After calculation, the global SUR value in the striatum is displayed on the screen. The individual SUR values in each ROI allows us to obtain the specific uptake in each of the four striatal regions (caudate and putamen for both hemispheres). All this information is stored in a data file.

Figure 1 schematically shows the workflow of the entire image processing included in QuantiDOPA.

III. RESULTS

QuantiDOPA allows the user to perform a semi-automatic quantification of striatal uptake by following the three stages mentioned above:

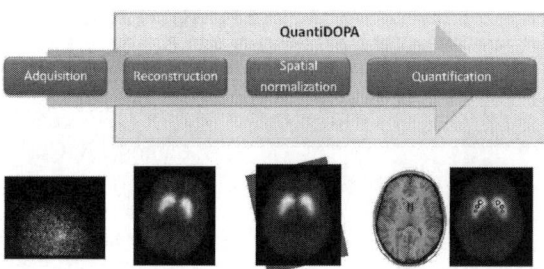

Fig. 1 Image processing workflow

A. Reconstruction

In this module the user only needs to check the parameters related to the acquisition and reconstruction, in order to run the FBP reconstruction.

Figure 2 shows and example of a reconstructed image using a 2D Butterwoth filter.

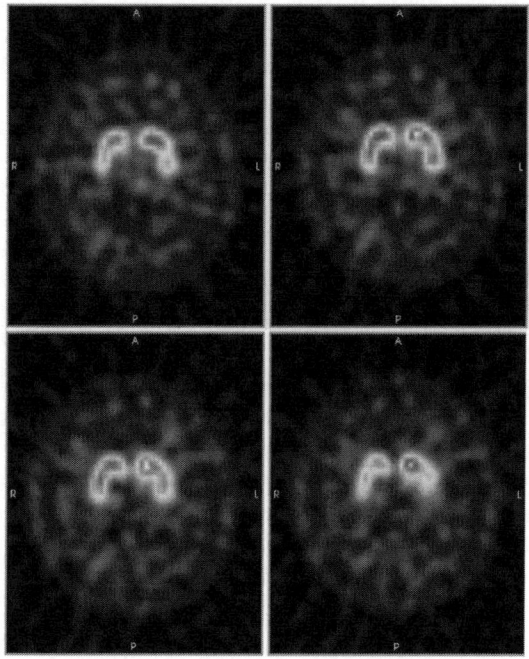

Fig. 2 Reconstructed SPECT image

B. Normalization

The image obtained in the previous module is automatically set as an input data for the normalization module. Before starting the normalization process, the user selects which template wants to normalize to.

Figure 3 shows axial, sagittal, coronal and 3D views of the SPECT study displayed in figure 2, after the two normalization steps.

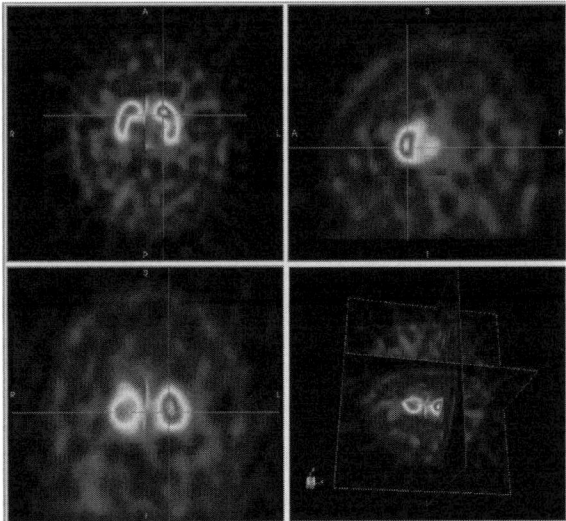

Fig. 3 Axial (top left), sagittal (top right), coronal (bottom left) and 3D view (bottom right) of the normalized SPECT image

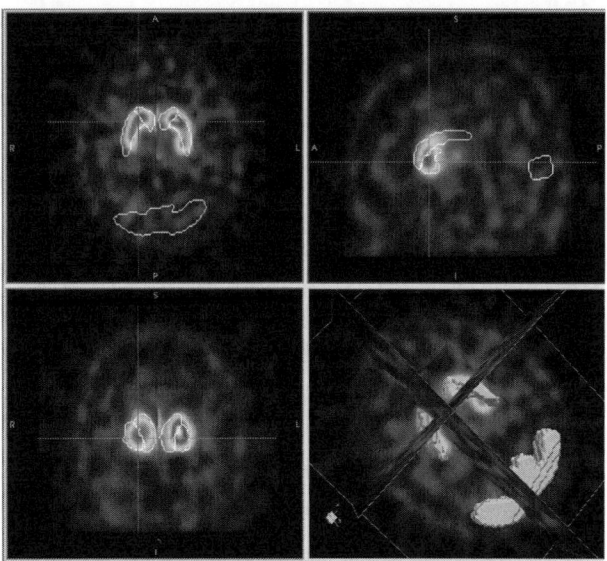

Fig. 4 Axial (top left), sagittal (top right), coronal (bottom left) and 3D view (bottom right) of the normalized SPECT image with AALs ROIs

C. Quantification

The normalized SPECT image obtained in the previous stage, is set as the input data of the quantification module.

ROIs based on AALs are automatically placed on the normalization study as shown in figure 4. In the case of cylinders, the user is allowed to slightly modify the positions and to change the radius and height. Figure 5 shows a spatially normalized SPECT study with cylindrical ROIs on the striatum and the background reference region. The sum of central slices of the reconstructed image with cylindrical ROIs is also displayed.

QuantiDOPA is being used in a clinical trial to obtain SUR values from healthy volunteers in six Spanish hospitals.

IV. CONCLUSIONS

QuantiDOPA is a useful tool for semi-automatic quantification in DAT SPECT imaging and it has revealed simple and flexible.

ACKNOWLEDGEMENTS

This work was supported in part by Spain's Ministry of Science & Innovation (SAF2009-08076), CDTI-CENIT (AMIT project), Fondo de Investigaciones Sanitarias (PI12-00390) and CIBER-BBN (MIND-t project).

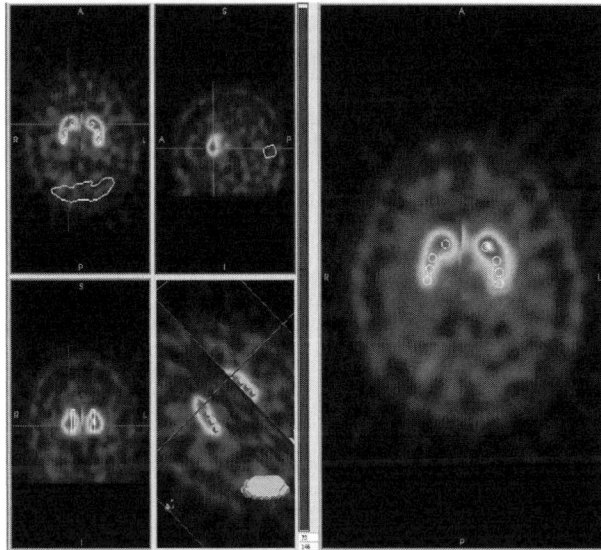

Fig. 5 Axial (top left), sagittal (top right), coronal (bottom left) and 3D view (bottom right) of the normalized SPECT image with cylindrical ROIs

REFERENCES

1. Booij J, Tissingh G, Winogrodzka A, Royen EA. Imaging of the dopaminergic neurotransmission system using single-photon emission tomography and positron emission tomography in patients with parkinsonism *Eur J Nucl Med.* 1999;26:171-182.
2. Seibyl JP, Marek K, Sheff K, et al. Test/retest reproducibility of iodine-123-betaCIT SPECT brain measurement of dopamine transporters in Parkinson's patients *J Nucl Med.* 1997;38:1453-1459.
3. Stoof JC, Winogrodzka A, Muiswinkel FL, et al. Leads for the development of neuroprotective treatment in Parkinson's disease and brain imaging *Eur J Pharmacol.* 1999;375:75-86.

4. Larrabide I, Omedas P, Martelli Y, et al. GIMIAS: An Open Source Framework for Efficient Development of Research Tools and Clinical Prototypes *Lecture Notes in Computer Science.* 2009;5528:417426.
5. Virtual Physiological Human Network of Excellence at http://www.vph-noe.eu
6. Common Toolkit at http://www.commontk.org
7. Ros D, Espinosa M, Setoain JF, Falcn C, Lomea FJ, Pava J. Evaluation of algorithms for the registration of 99Tcm-HMPAO brain SPET studies *Nucl Med Communications.* 1999;20:227-236.
8. Tzourio-Mazoyer N, Landeau B, Papathanassiou D, et al. Automated Anatomical Labeling of Activations in SPM Using a Macroscopic Anatomical Parcellation of the MNI MRI Single-Subject Brain *NeuroImage.* 2002;15:273-289.
9. Walker Z, Costa DC, Walker RWHH, et al. Differentiation of dementia with Lewy bodies from Alzheimers disease using a dopaminergic presynaptic ligand *J Neurol Neurosurg Psychiatry.* 2002;73:134-140.

Author: Aida Niñerola Baizán
Institute: GIBUB, Universitat de Barcelona - CIBER-BBN
Street: Casanova 143 - Planta 5 (Ala Nord)
City: Barcelona
Country: Spain
Email: aninerola@ub.edu

Spatial Normalization in Voxel-Wise Analysis of FDG-PET Brain Images

M.E. Martino[1], V. García-Vázquez[1,2], M. Lacalle-Aurioles[3], J. Olazarán[4],
J. Guzmán de Villoria[5], I. Cruz[4], J.L. Carreras[6], and M. Desco[1,3]

[1] Instituto de Investigación Sanitaria Gregorio Marañón, Madrid, Spain
[2] Centro de Investigación Biomédica en Red de Salud Mental (CIBERSAM), Madrid, Spain
[3] Departamento de Bioingeniería e Ingeniería Aeroespacial, Universidad Carlos III de Madrid, Spain
[4] Servicio de Neurología, Hospital General Universitario Gregorio Marañón, Madrid, Spain
[5] Servicio de Radiodiagnóstico, Hospital General Universitario Gregorio Marañón, Madrid, Spain
[6] Servicio de Medicina Nuclear Hospital Clínico San Carlos Universidad Complutense, Madrid, Spain

Abstract—Spatial normalization is a preliminary step in any PET image analysis based on statistical parametric mapping (SPM). This step consists in applying a spatial deformation to match each PET scan to an anatomical reference template The purpose of this study was to evaluate the effect of using different methods of spatial normalization on the results of SPM analysis of 18F-FDG PET images by comparing controls and patients diagnosed with mild cognitive impairment (MCI) that converted to probable Alzheimer's disease (AD) after two years of follow-up. We performed an SPM analysis between the two groups using three spatial normalization methods: 1) MRI-DARTEL 2) MRI-SPM8 and 3) FDG-SPM8. MRI-DARTEL and MRI-SPM8 combine structural and functional images, while FDG-SPM8 is based only on functional images. The results obtained with the three methods were consistent in terms of the pattern of hypometabolism detected in the patient group. However, MRI-DARTEL was the method more consistent with the patterns previously reported in the literature. These results suggest that MRI-DARTEL is the most accurate and powerful method for spatial normalization in SPM analysis of 18F-FDG PET images. Normalization based solely on functional imaging shows less sensitivity to detect significant differences.

Keywords—**Spatial normalization, FDG-PET, DARTEL, Voxel-wise analysis, AD.**

I. INTRODUCTION

Statistical parametric mapping (SPM) (http://www.fil.ion.ucl.ac.uk/spm, Wellcome Department of Imaging Neuroscience, Institute of Neurology, UCL, London, UK), is one of the most widely used software packages for voxel-wise analysis of functional images such as PET images [1]. In this analysis the spatial normalization is a very important step that consists in applying a spatial deformation to match each PET scan to an anatomical reference template. Although these transformations can be obtained from 18F-FDG PET images, more reliability can be expected when MRI images of every subject are available. In this case the normalization parameters can be determined from the structural image and then applied to the co-registered 18F-FDG PET image. Among the normalization methods for MRI, DARTEL, based on diffeomorphic registration algorithms, is theoretically one of the most powerful and robust [2].

We performed a study to investigate the effect of using three different spatial normalization procedures on the outcome of SPM analysis of 18F-FDG PET images by comparing control subjects and patients diagnosed with mild cognitive impairment (MCI) at the moment of obtaining of 18F-FDG PET and MRI images, who converted to probable Alzheimer's disease (AD) after two years of follow-up. The quality of the results was interpreted by comparison with the well-known pattern of cerebral hypometabolism characteristic of AD patients described in the literature [3-4].

II. MATERIALS AND METHODS

A. Patients and Controls

We studied a total of 20 subjects (9 patients and 11 controls) who were recruited at the Behavioral Neurology Unit of Hospital General Universitario Gregorio Marañón (Madrid, Spain). The control group included relatives and caregivers. All participants signed an informed consent, and the local ethics committee approved the study.

Both patients and controls underwent neurological assessments at the Behavioral Neurology Unit. Mental status was evaluated using Folstein's 'mini-mental state evaluation' test (MMSE) [5]. Neurological testing was completed using a battery of tests to characterize cognitive deficits and their severity.

The 9 patients (mean age, 69.56; SD ± 6.37 years; 6 women) were all diagnosed with MCI after a formal neuropsychological examination and had a clinical dementia rating (CDR) score of 0.5 [6]. However, they did not meet the criteria for dementia of DSM-IV-TR (the Diagnostic and Statistical Manual of Mental Disorders, Fourth Edition, Text Revision) at the moment of obtaining 18F-FDG PET and MRI images. These patients converted to probable AD at two years follow-up.

The 11 controls (mean age, 70.27; SD± 7.56 years; 6 women) had no cognitive disorders or clinically relevant conditions or other circumstances that could interfere with cognitive performance. To be included in the study, they had to score above 24 in the adapted Spanish version of the MMSE.

B. Imaging Protocol

1. 18F-FDG PET images were obtained using a PET/CT GEMINI system (Philips Medical Systems) 30 minutes after the injection of 370 MBq of 18F-FDG with the subject at rest and eyes open in a dark room. Matrix size was 128 x 128 x 90 voxel with a voxel size 2 x 2 x 2 mm.
2. MRI data were acquired on a Philips Intera 1.5 T MR scanner (Philips Medical Systems, Best, The Netherlands). The imaging protocol include a volumetric T1-weighted 3D gradient echo sequence (flip angle = 30°; TR = 16 ms, TE = 4.6 ms; matrix size = 256 x 256; voxel size = 1 x 1 x 1.5 mm; FOV = 256 mm).

C. Image Processing

Prior to the statistical analysis, images were processed with SPM8 software package in four steps: 1) co-registration of 18F-FDG PET images of each subject to their corresponding T1-weighted MRI using mutual information algorithms available in SPM8 and an affine transformation; 2) spatial normalization using the three different approaches under comparison (see details below); 3) smoothing: 8 x 8 x 8 mm FWHM filter; 4) global normalization: proportional scaling, as this method is considered the most suitable for intersubject PET imaging [7].

1. Spatial Normalization

We compared three methods of spatial normalization: two methods that make use of structural MRI data to estimate the normalization parameters (MRI-SPM8 and MRI-DARTEL) and the standard method for PET (FDG-SPM8), based only on information from the functional images. All the processing steps were implemented using routines from the SPM8 software.

FDG-SPM8 has long been the standard procedure for analyzing PET scans using voxel-wise statistical analysis [8]. This normalization obtains the spatial transformation required for each PET image to match a PET template in MNI space. In this study, we used a specific template for FDG (Fig. 1). The construction of the template is described in [9].

The *MRI-SPM8* procedure incorporates anatomical information from each subject's MRI in order to achieve better normalization than using only functional (PET) data. This method computes the spatial non-linear transformation required to normalize the T1 images of each subject to the MNI-152 template using the "Normalize" function of SPM8. This template is included in SPM8. The transformation parameters obtained are then saved and applied to each co-registered PET image (Fig. 1).

MRI-DARTEL incorporates anatomical information from the MRI and applies the DARTEL registration tools from SPM8 to create a study-specific MRI template from MRI images of both patients and controls. The template is then normalized to the MNI space. This procedure of spatial normalization consists of four steps: 1) tissue segmentation of the T1 images for each subject using the new segmentation tool of SPM8; 2) construction of a study-specific template of each tissue type in MNI space; 3) estimation of the non-linear transformation required to warp grey and white matter images from step 1 to match the study-specific template; 4) the transformation parameters obtained are then applied to each co-registered PET (Fig. 1).

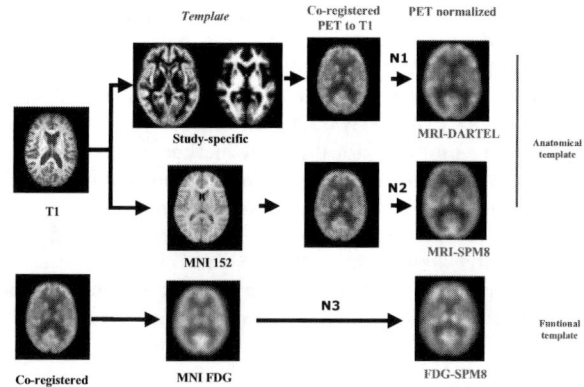

Fig. 1 Spatial normalization methods used in the study. MRI-DARTEL: Normalization of MRI based on a study-specific template generated using the DARTEL algorithm; MRI-SMP8: Normalization of MRI based on the MNI-152 template and a standard SPM8 algorithm; FDG-SPM8: Normalization of PET based on the FDG-PET template and a standard SPM8 algorithm; N1-N3: In all cases geometrical transformations were later applied to co-registered PET images.

D. Statistical Analysis

Following the four processing steps described, we generated three different data sets and we conducted three separate SPM analyses, one for each normalization method, based on a general linear model (GLM). The results of Student's t-tests at each voxel were used to generate statistical parametric maps of group differences showing regions where patients had reduced glucose metabolism as compared with the control group. We heuristically selected the threshold parameters p-value < 0.03 without multiple comparisons correction and a minimum extent of 150 voxels, (K = 150).

III. RESULTS

E. FDG-SPM8

The MCI group showed a statistically significant reduction in glucose metabolism in the right frontal, bilateral parietal cortex, and right precuneus (Table 1 and Fig. 2).

Table 1 Summary of results for the three different spatial normalization methods. MCI to AD, patients converted to probable AD after two years of follow-up. V: number of voxels; t: t statistic.

MCI to AD < CONTROLS	FDG - SPM8		MRI - SPM8		MRI - DARTEL	
Brain region	V	t	V	t	V	t
Right frontal	1,188	3.64	765	3.23	1,902	3.35
Left frontal	-	-	-	-	365	3.15
Right parahippocampal gyrus	-	-	231	2.97	225	2.55
Left parahippocampal gyrus	-	-	-	-	207	2.24
Right parietal	348	3.4	-	-	3,277	4.8
Left parietal	1,166	4.3	473	2.41	1,322	4.06
Right precuneus	220	2.70	-	-	1,750	2.60
Left precuneus	-	-	120	2.60	259	2.7
Right cingulate	-	-	-	-	2,425	3.72
Left cingulate	-	-	116	2.56	1,437	3.35
Left occipital	-	-	290	3.42	253	3.32

F. MRI-SPM8

MRI-SPM8 yielded the same regions as FDG-SPM except of right parietal and left precunues but with additional significant clusters. Compared with controls, statistically significant reduced glucose metabolism was observed in the patient group in the right parahippocampal gyrus, left precuneus, cingulate and occipital cortex (Table 1 and Fig. 2).

G. MRI-DARTEL

MRI-DARTEL detected statistically significant hypometabolism in the same regions as FDG-SPM8 and MRI-SPM8, as well as in the left frontal and parahippocampal gyrus, and right precuneus and cingulate (Table 1, Fig. 2).

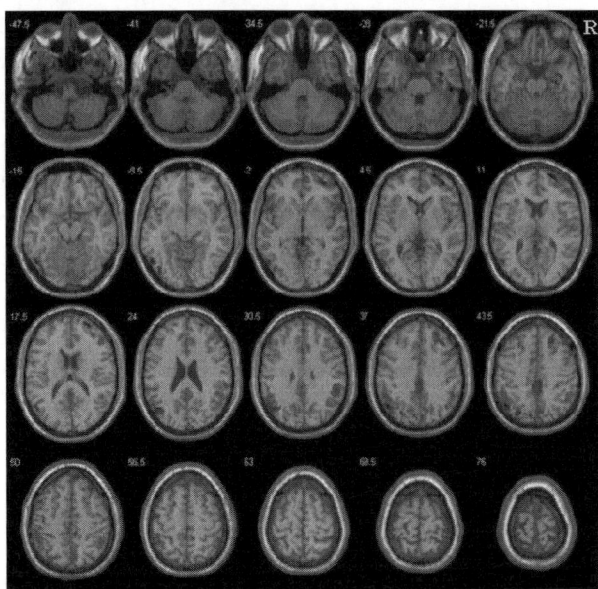

Fig. 2 Results obtained with three different spatial normalization methods: MRI-DARTEL, MRI-SPM8, and FDG-SPM8. Color blobs show the regions where patients had significant glucose hypometabolism (p < 0.03, uncorrected, extent threshold of 150 voxels). Significant clusters superimposed on the template T1 image showing a slices from z = -47.5 mm to z = 76 mm using a 6.5-mm increment. R, right.

IV. DISCUSSION

Patients who evolve to AD disease are supposed to show an incipient pattern of AD, which could be early detected if the imaging tools used are sensitive enough. Our results described above show that SPM analysis based on MRI-DARTEL normalization detects a larger number of regions with hypometabolism in the patient group than in the control, which corresponds with the pattern of hypometabolism describe in literature. However, when MRI-SPM8 or FDG-SPM8 is used, that pattern is incomplete (Table 1 and Fig. 2). The greater anatomical detail and better spatial resolution of MRI compared with FDG PET functional images may account for the more accurate spatial normalization obtained when using templates based on the structural images [10]. Using anatomical information for spatial normalization (MRI-DARTEL and MRI-SPM8 in the present study) ensures independence between the registration process and the subsequent statistical analysis. This independence is not possible when functional images are used for both spatial normalization and statistical analysis [11]. Moreover, in cases of pathological abnormalities of brain anatomy, intersubject registration based on the functional images is less accurate, resulting in increased noise, dispersion of voxel values and less power in the statistical analysis [9] [12].

V. CONCLUSIONS

The choice of spatial normalization method may considerably affect the outcome of SPM analyses of 18F-FDG PET studies. When only functional information was used for spatial normalization, we detected fewer regions with glucose hypometabolism, suggesting lack of statistical power. Therefore, results from FDG PET studies obtained without anatomical MRI images should be interpreted with caution, as they may underestimate pathological findings in patients. Our results suggest that the DARTEL algorithm is the most statistically powerful method of the three tested

ACKNOWLEDGMENT

This study was supported by Spanish Ministerio de Ciencia e Innovación, AMIT Programa CENIT. M.E. Martino was supported by the Contrato Río Hortega (Ministerio de Economía y Competitividad, Instituto de Salud Carlos III). None of the authors have conflicts of interest to declare.

REFERENCES

1. Frackowiak RSJ, Friston KJ, Frith CD, Dolan RJ, Mazziotta JC (1997) Human Brain Function. Academic Press, San Diego.
2. Ashburner J (2007) A fast diffeomorphic image registration algorithm. NeuroImage 38(1):95-113
3. Herholz K (2003) PET studies in dementia. Ann Nucl Med 17(2) : 79-89.
4. Mosconi L, Mistur R, Switalski R et al (2009) FDG-PET changes in brain glucose metabolism from normal cognition to pathologically verified Alzheimer's disease. Eur J Nucl Med Mol Imaging 36(5): 811-22. DOI 10.1007/s00259-008-1039-z
5. Folstein MF, Folstein SE, McHugh PR (1975) "Mini-mental state": A practical method for grading the cognitive state of patients for the clinician. J Psychiatr Res 12(3):189-98. DOI 10.1016/0022-3956(75)90026-6
6. Hughes CP, Berg L, Danziger WL et al (1982) A new clinical scale for the staging of dementia. Br J Psychiatry 140 : 566-72.
7. Acton PD, Friston KJ (1998) Statistical parametric mapping in functional neuroimaging: beyond PET and fMRI activation studies. Eur J Nucl Med Mol Imaging 25(7): 663-7.
8. Ashburner J, Chun-Chuan C, Guillaume F, Henson R, Kiebel S, Kilner J, Litvak, Vladimir Moran, Rosalyn Penny, Will Klaas, Stephan Chloe, Hutton Volkmar, Glauche Jeremie, Mattout Christophe, Phillips (2009) SPM8 Manual. 384 p.
9. Gispert JD, Pascau J, Reig S et al (2003) Influence of the normalization template on the outcome of statistical parametric mapping of PET scans. NeuroImage 19(3): 601-12. DOI 10.1016/s1053-8119(03)00072-7
10. Ashburner J, Friston KJ (1999) Nonlinear spatial normalization using basis functions. Hum Brain Mapp 7(4): 254-66.
11. Bookstein FL (2001) "Voxel-Based Morphometry" should not be used with Imperfectly Registered Images. NeuroImage 14(6): 1454-62. DOI 10.1006/nimg.2001.0770
12. Reig S, Penedo M, Gispert JD et al (2007) Impact of ventricular enlargement on the measurement of metabolic activity in spatially normalized PET. NeuroImage 35(2): 748-58. DOI 10.1016/j.neuroimage.2006.12.

Author: María Elena Martino
Institute: Instituto de Investigación Sanitaria Gregorio Marañón
Street: Doctor Esquerdo 46, 28007
City: Madrid
Country: Spain
Email: emartino@hggm.es

EndoTOFPET-US a High Resolution Endoscopic PET-US Scanner Used for Pancreatic and Prostatic Clinical Exams

C. Zorraquino

On behalf of EndoTOFPET-US Collaboration

Abstract—The EndoTOFPET-US collaboration aims at the development of a multimodal imaging technique for endoscopic pancreas and prostate exams that combines the benefits of high resolution metabolic information from PET and anatomical information from ultrasound (US). The project is supported by an FP7 grant from the European Union, financing the development and exploitation of a tool for studies of biomarkers specific to prostate and pancreatic cancers.

The development of two novel detectors is required, a PET head extension for a commercial US endoscope placed close to the region of interest (ROI) and a PET plate over the patient's abdomen in coincidence with the PET head. Technological challenges include: 1 mm image spatial resolution (SR), an unprecedented 200ps Coincidence Time Resolution (CTR) for enhanced background rejection, online tracking of both detectors and image reconstruction of images with partial volume information from an asymmetric geometry.

EndoTOFPET-US aims to push current limitations on full-body multimodal PET detectors by means of cutting-edge technology in a novel (ROI specific) configuration. Such a system requires innovative solutions in different technologies now under development. Promising results obtained with prototypes indicate that the design goals can be achieved.

Keywords—PET, multimodal imaging.

I. INTRODUCTION

Prostate cancer is the most common cancer among males and pancreatic cancer is most often detected on an advanced state of development. Therefore, they have a high mortality rate and thus a big effort is being placed on the design and exploitation of new specific biomarkers for them.

EndoTOFPET-US collaboration [1] was conceived to cope with these needs by means of a new multimodal PET-US imaging technology. A novel scanner that breaks with the traditional whole body scheme by means of a ROI specific configuration thanks to the incorporation of a detector head in a commercial endoscopy probe.

The complete detector architecture is defined by a PET head extension for a commercial US endoscope placed close to the ROI and a PET plate placed over the patient's abdomen in coincidence with the PET head [2].

The PET head requires heavy miniaturization; it houses scintillating crystals, photo-detectors and readout electronics in a small volume of 23 mm in diameter. Crystals are coupled to a fully digital silicon photomultiplier (SiPM) offering single SPAD readout.

The external plate uses 4096 crystals coupled to commercial SiPM arrays which are read by a fast and low-noise Application-Specific Integrated Circuit (ASIC) developed by the collaboration. The ASIC is able to measure Time-Of-Flight (TOF) of the first photoelectron. FPGAs concentrate event data sent by ASICs and transmit this information to an external Data Acquisition System (DAQ) where gamma events data from probe and plate will meet.

When compared with state of the art PET systems, where latest detectors provide a SR of 3-5mm and a CTR of ~500ps, EndoTOFPET-US will be able to detect smaller lesions thanks to its 1mm SR and to give precise information of lesion's morphology thanks to its improved image sharpness given by its 200ps CTR. On the other hand, such a ROI specific configuration in a PET scanner has a main drawback: the intrinsic reduction of sensitivity due to the diminution of the Field of View when compared to full-body or full-rotation PET systems. This issue constitutes a challenge for the image reconstruction calling for innovative algorithms.

Additionally, in a short term this detector will provide physicians with a tool able to test latest biomarkers for pancreatic/prostatic cancer.

II. CRYSTALS AND OPTICS

A. Crystal Matrices

Prototypes of probe and plate scintillating crystal matrices have been realized and they are currently being characterized. They can be observed in Figure 1, which show the probe matrix on the left and the plate matrix on the right.

Two different probe geometries are being considered for the two different clinical exams. The standard probe crystal matrix is composed of 162 individual pixels of $0.71\times0.71\times10$ mm^3 LYSO:Ce fibers. The prostatic version accepts two of these matrices.

On the other hand, the external plate houses a big surface of crystals in an arrangement of 256 matrices of 4x4 LYSO:Ce crystal pixels. The external plate does not only contain more crystals but also bigger crystals, 3.5x3.5x15 mm^3 each channel. The pixels have been designed in order to optimize their scintillating properties such as light yield, rise time and decay time.

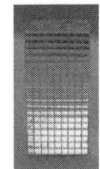

Fig. 1 Probe and plate crystal matrices

A test setup has been prepared for crystals and SiPMs characterization where a probe crystal matrix is coupled to SiPMs read by CERN proprietary electronics in one side and a plate matrix is placed on the front side on the same light-extraction configuration; measurements confirm the possibility of reaching a CTR of 238 ps [3], see Figure 2.

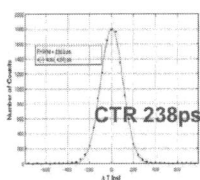

Fig. 2 CTR measurement on crystals characterization

B. Focusing Optics

The probe needs a customized design of a digital SiPM (dSiPM). Mechanical constrains lead to an important dead area on the active surface (more details on this concern are exposed on section III.B). Therefore, an intermediate stage Micro Optical Element (MOE) is being designed to concentrate the light on the active areas of the dSiPM boosting the Photo-Detection Efficiency (PDE). Figure 3 shows the first prototype where focusing prisms can be observed.

Fig. 3 MOE prototype

Experimental measurements demonstrate that light concentration is not enough to improve light collection. Last studies on light angular distribution of the crystal conclude that light dispersion is too broad for the current MOE.

New simulations are been carried out using the most recent information on crystals light angular distribution. New designs follow a new approach where light collimation and concentration are combined.

Furthermore, cutting-edge technics on crystal surface are being studied and prototyped: holographic techniques to achieve light collimation and nanostructures to boost light extraction.

III. PHOTO-DETECTION

A. External Plate SiPM

The external plate makes use of 256 commercial SiPM 4x4 matrices, see Figure 4 left picture. First feasibility tests on time resolution, Dark Count Rate (DCR) and cross talk demonstrate a good performance of the device under the system conditions. Figure 4 right picture, shows experimental results on CTR tests where a CTR close to 240 ps is achieved and hence it demonstrates that this SiPM is suitable for the EndoTOFPET-US Project.

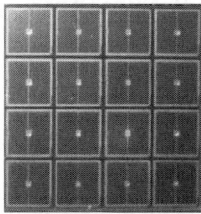
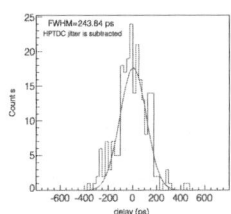

Fig. 4 SiPM matrix and CTR measure obtained on its characterization

B. Probe dSiPM

The collaboration faces a true challenge on the design of a fully digital dSiPM small enough to be integrated in the probe. First prototypes reach over 50% fill factor leading to a poor PDE that will be boosted by the MOE. Figure 5 shows a photomicrograph of the chip published in [4] with a detail of the pixels and the interconnection used to bring timing-sensitive signals to the on-chip time-to-digital converters at the bottom of the chip.

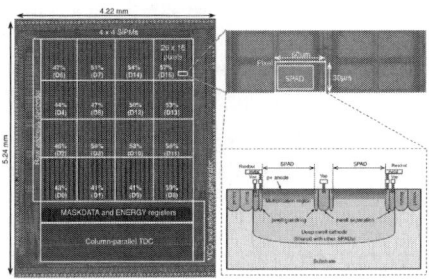

Fig. 5 dSiPM chip microphotograph and fill factor detail

A pixel masking procedure has been designed and integrated on the device in order to inhibit the pixels with higher DCR (screamers) at the expense of PDE reduction. Figure 6 exposes an analysis made over the 16 SiPM modules on chip where the tradeoff between PDE and DCR is shown as a function masking percentage. Results demonstrate that when a 10% masking is applied, a PDE of 10% can be reached along with a DCR of 10 MHz. Nevertheless, these tests where performed under realistic temperature conditions (20ºC) taking into account medical procedure limitations. However, the possibility of efficient thermal isolation on the probe is being studied in order to be able to cool down the device for an important performance improvement.

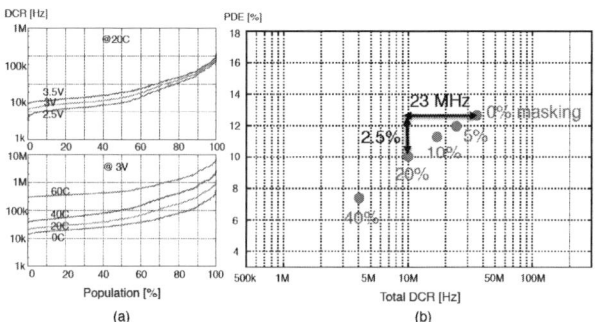

Fig. 6 (a) Cumulative DCR plot with various excess bias voltages and temperature for the 'D15' SiPM. (b) The relation between DCR and PDE for various SiPMs at 3 V excess bias and 20 ∘C.

IV. DATA ACQISITION SYSTEM

The DAQ is distributed among two different on-detector front-end (FE) electronics (endoscopic probe and abdominal plate) and the off-detector back-end (BE) electronics on the reconstruction workstation [4]. DAQ purpose does not just consist on data readout but it also takes care of configuration and monitoring of the different devices on the system.

A. Frontend Electronics

a) Endoscopic probe:

Probe electronics requires heavy miniaturization, and thus a small FPGA has been chosen to perform the dSiPM readout: a Lattice Semiconductor ice40 ultra low power FPGA. This FPGA processes and filters dSiPM data (L1-Probe Trigger Stage) on the FE providing a data stream towards the BE electronics with a rate up to 200kHz.

b) External plate:

Abdominal plate SiPM arrays are read by very fast 64-channels ASICs able to trigger on the first photoelectron (L1-Plate Trigger Stage), at a rate of 100kHz per channel. Two ASIC designs have been developed by the collaboration: STiCv2 [5] and TOFPET [6]. The former one has been prototyped in a preliminary 16-channel version. For the latter one a final 64-channel version is currently under test in two independent test setups that allow for chip characterization of both a board-bonded version and a 128-channel system-in-a-package (SiP) BGA version of the chip. The socket-based setup allows for systematic production qualification. Both ASIC versions (STiCv2 and TOFPET) are currently being characterized with functional tests using arrays of MPPCs (commercial SiPMs) and the crystal matrices selected for the external plate.

Xilinx Spartan-6 LX FPGAs on several FE boards (FEBs) concentrate event data from the ASICs on the FEB and transmit them to the BE electronics resulting in an overall data rate up to 40MHz. Nevertheless, DAQ system has been designed to be flexible and expansible. Therefore, the FPGA make use of Multi-Gigabit Transceivers (MGT) on the connection with the DAQ BE enabling event information transmission at data rates much higher than needed. First experimental test demonstrate that 1.6 Gbps can be used maintaining a 0 Bit Error Rate (BER) during 24 hours continuous acquisitions.

Another result of design flexibility are the two different topologies that the FEB firmware accepts. Being able to configure the DAQ architecture in such a way that FEBs are connected in a master-slave chain configuration (linearly increasing the 40MHz data rate on each chain connection), or in a configuration where each FEB is connected independently to the DAQ BE.

B. Backend Eelectronics

DAQ BE electronics is implemented in a board integrated in the image reconstruction workstation which is connected to the FEBs through HDMI fast point to point links. The interface PC - DAQ board is based on a PCIe motherboard connection that provides error free data transmission at 500 MBps in average as experimental tests confirm.

A Xilinx Virtex-4 FPGA on the BE board combines data from probe and plate. It looks for coarse temporal coincidences (L2 Trigger Stage) and procures a reduced data rate for fast processing on the reconstruction software where fine temporal coincidences (L3 Trigger Stage) are found.

V. TRACKING

In a system where two detector heads that move independently are considered, a precise tracking system is fundamental for accurate positioning and orientation of imaging probe tip with respect to the external plate. A good tracking will provide precise line-of-response for image reconstruction.

Three different tracking systems are being considered, namely: optical, electromagnetic (EM) and robotic. Each of them has its own limitations and benefits and consequently a combination of them is being considered. For example, optical tracking needs clear line-of-sight and EM tracking is highly sensitive to interferences; hence, a combination of different techniques will highly improve accuracy in a dynamic scenario where conditions change.

First experimental measurements, Table 1, show a good tracking accuracy and they demonstrate the benefit of using optimization techniques developed by the collaboration.

Table 1 Opto-EM Tracking accuracy

Measurement type	Translational mean error (mm)	Rotational mean error (degrees)
Static	1.44	1.66
Dynamic	6.7	2.76
Optimized	4.12	1.46

VI. Image reconstruction

One of the most ambitious topics in this detector is the reconstruction of images with partial volume information from an asymmetric geometry. Moreover, anatomical constrains limit the mobility of the detector inner head and hence a full rotation is not a feasible solution as it is for other non-ring scanners.

Therefore, any improvement in the overall detector sensitivity will be decisive for the final image quality in terms of Signal-to-Noise Ratio (SNR). However, simulation results demonstrate that some parameters have a stronger influence than others. An asymmetric geometry implies asymmetric needs. For example, the external plate detector geometry has almost no influence on image resolution and Depth-Of-Interaction (DOI) information is not necessary. On the other hand, any improvement in the internal probe sensitivity will drastically ameliorate image resolution.

Image reconstruction does not only need to be accurate but also as fast as possible. Therefore, parallelized GPU techniques are used in order to speed up image reconstruction, i.e. being able to obtain higher resolution images in shorter times.

Preliminary Geant4 simulations prove that the reconstruction SW is able to provide 1 mm SR under optimal conditions (full rotation around source) [8], see Figure 7.

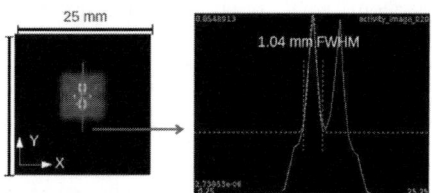

Fig. 7 Simulation results on Image spatial resolution

VII. Conclusions

EndoTOFPET-US aims to push current limitations on full-body multimodal PET detectors by means of cutting-edge technology in a novel (ROI specific) configuration. Such a system requires innovative solutions in different technologies now under development. Promising results obtained with prototypes indicate that the design goals can be achieved.

Acknowledgment

This work has been partially supported by a Marie Curie Early Initial Training Network Fellowship of the European Community's Seventh Framework Program under contract number (PITN-GA-2011-289355-PicoSEC-MCNet)

References

1. EndoTOFPET-US Proposal: "Novel multimodal endoscopic probes for simultaneous PET/ultrasound imaging for imageguided interventions", FP7/2007-2013 under Grant Agreement No. 256984.
2. T. Meyer et al., "Endo-TOFPET-US: A Multimodal Ultrasonic Probe Featuring Time of Flight PET in Diagnostic and Therapeutic Endoscopy", Nuclear Instruments and Methods in Physics Research.
3. E. Auffray et al., "Design and Performance of Detector Modules for the Endoscopic PET Probe for the FP7 Project EndoTOFPET-US", conference record IEEE NSS/MIC October2012.
4. S. Mandai and E. Charbon, "Multi-Channel Digital SiPMs: Concept, Analysis and Implementation", IEEE Nuclear Science Symposium, Anaheim, Oct. 2012.
5. R Bugalho et al., "EndoTOFPET-US Data Acquisition System", iWoRID 2012, JINST.
6. W. Shen et al., "STiC - a Mixed Mode Chip for SiPM ToF Applications", NSS/MIC 2012 IEEE.
7. M. D. Rolo et al., "A 64-Channel ASIC for TOFPET Applications", NSS/MIC 2012 IEEE.
8. B Frisch et al., "Combining Endoscopic Ultrasound with Time-Of-Flight PET: the EndoTOFPET-US Project", Nuclear Instruments and Methods A, in press.

Author:	Carlos Zorraquino Gastón
Institute:	LIP
Street:	Av. Elias Garcia, 14-1
City:	Lisbon
Country:	Portugal
Email:	carlos.zorraquino.gaston@cern.ch

Light Emission Efficiency of Gd₃Al₂Ga₃O₁₂:Ce (GAGG:Ce) Single Crystal under X-Ray Radiographic Conditions

I.E. Seferis[1], C.M. Michail[2], S.L. David[2], A. Bakas[3], N.I. Kalivas[2], G.P. Fountos[2], G.S. Panayiotakis[1], K. Kourkoutas[4], I.S. Kandarakis[2], and I.G. Valais[2]

[1] Department of Medical Physics, Faculty of Medicine, University of Patras, 265 00 Patras, Greece
[2] Technological Educational Institute (TEI) of Athens, Department of Medical Instruments Technology, Ag. Spyridonos, 12210, Athens, Greece
[3] Technological Educational Institute (TEI) of Athens, Department of Medical Radiological Technology, Faculty of Health and Caring Professions, Ag. Spyridonos, 12210, Athens, Greece
[4] Technological Educational Institution of Athens Department of Physics, Chemistry and Materials Technology, 12210, Egaleo, Greece

Abstract—Inorganic scintillators, in either powder or crystal form, are used in indirect digital radiography as X-ray to light converters coupled to electronic optical sensors (photodiodes, CCDs, CMOS). A recent developed single crystal scintillator with high density, fast scintillation response and high light yield is the Ce doped Gadolinium Aluminum Gallium Garnet (Gd₃Al₂Ga₃O₁₂:Ce or GAGG:Ce). The purpose of the present study was to evaluate GAGG:Ce single crystal scintillator under general radiographic imaging conditions. The two GAGG:Ce crystals used in this study were supplied by Furukawa Co Ltd with dimensions 10×10×10 mm³ and 10×10×20 mm³. Parameters such as the Absolute Luminescence Efficiency-AE (light energy flux over exposure rate) and the light spectral compatibility to electronic optical sensors (Effective Efficiency) were investigated under X-ray excitation in the radiographic energy range from 50 to 125 kVp. Results showed that the absolute efficiency of GAGG:Ce scintillator crystals increased with X-ray tube voltage. The emission spectrum of scintillator examined in the present study is well matched with the spectral sensitivities of the optical photon detectors often employed in radiation detectors. Taking these results into account GAGG:Ce crystal scintillator could potentially be considered for further research for use in X-ray medical imaging.

Keywords—GAGG:Ce, X-ray imaging, Absolute Efficiency, Maching Factor.

I. Introduction

In most X-ray and γ-ray applications inorganic scintillators are employed as radiation in light converters [1,2]. These applications demands scintillators with high light yield, good energy resolution, high effective atomic number, fast scintillation response and chemical stability [3]. Cerium-doped (Ce^{3+}) scintillators have a much faster response than most currently employed materials [4]. Cerium doped Gadolinium Oxyorthosilicate (Gd_2SiO_5 or GSO) and Lutetium Oxyorthosilicate, (Lu_2SiO_5:Ce or LSO:Ce) are fast emitting scintillators employed mainly in PET detectors [1]. However, LSO is of high cost and GSO shows relatively lower light yield (8000 ph/MeV).

Oxide materials based on Garnet structure are promising candidates as scintillators, because of well mastered technology developed for laser hosts and other applications, optical transparency and easy doping by rare-earth elements such as Lutetium or Yttrium [5]. Ce doped Gadolinium Aluminum Gallium Garnet ($Gd_3Al_2Ga_3O_{12}$:Ce or GAGG:Ce) is the most recent developed single crystal scintillator with high density, fast scintillation response and high light yield [3,5,6].

In this study we investigate the light emission characteristics of GAGG:Ce single-crystal scintillator under X-ray excitation. Parameters such as the Absolute Luminescence Efficiency-AE (light energy flux over exposure rate) and the Spectral Matching Factor - a_s (light spectral compatibility to electronic optical sensors) were investigated.

II. Materials and Methods

A. Calculations

The Absolute Efficiency was determined by measuring the light energy flux (Ψ_λ) emitted by the irradiated screen and dividing by the incident exposure rate ($\dot{X}$) measured at the screen surface [1,7].

$$AE = \Psi_\lambda / \dot{X} \qquad (1)$$

AE is expressed in efficiency units (EU= $\mu W \cdot m^{-2} / mR \cdot s$)

The spectral compatibility between the emitted scintillator light and the spectral sensitivities of Silicon (Si) photodiodes, CCD and CMOS [8] optical sensors was calculated by the Spectral Matching Factor (a_S) according to [9]:

$$\alpha_s = \int S_P(\lambda) S_D(\lambda) d\lambda / \int S_P(\lambda) d\lambda \qquad (2)$$

where $S_P(\lambda)$ is the spectrum of the light emitted by the scintillator and $S_D(\lambda)$ is the spectral sensitivity of the optical photon detector.

The Effective Efficiency (*neff*) has been defined by [10]:

$$n_{eff} = AE \times \alpha_s \qquad (3)$$

Where α_s is the spectral matching factor defining the emitted light spectrum compatibility with the spectral sensitivity of photodetectors, obtained from equation (2).

B. Experiments

The two GAGG:Ce crystals used in this study were supplied by Furukawa Co Ltd with dimensions 10×10×10 mm and 10×10×20 mm. The crystals were irradiated using a BMI General Medical Merate tube with rotating Tungsten anode and inherent filtration equivalent to 2 mm Al, with energies ranging from 50 to 125 kVp. This unit was used to simulate the conditions for general purpose radiography. Appropriate beam filtering of 21 mm Al was applied to simulate X-ray beam hardening by the human body.

For the determination of AE and *neff* X-ray exposures and light flux measurements were performed. The exposure rate was measured at the crystal's position using a RTI Piranha dosimeter (RTI electronics, Sweden). Light energy flux measurements were performed using a light integration sphere (Oriel 70451) coupled to a photomultiplier (EMI 9798B), equipped with an extended sensitivity S-20 photocathode and connected to a Cary 401 vibrating reed electrometer.

The Spectral Matching Factor was determined using equation (2). The emitted light spectrum $S_P(\lambda)$ of GAGG:Ce crystal was measured using the Ocean Optics optical grating spectrometer (Ocean Optics Inc., S2000). Spectral sensitivity $S_d(\lambda)$ data for various optical photon detectors were obtained from corresponding manufacturer's (Hamamatsu, EMI, etc.) datasheets. Seventeen optical photon detectors currently used in a large variety of imaging detectors (digital radiography, fluoroscopy, computed tomography, nuclear medicine, etc.) and their SMF with GAGG:Ce crystal spectrum was examined (Table 1).

III. RESULTS AND DISCUSSION

Fig. 1 shows the measured light emission spectrum of the GAGG:Ce crystal under X-ray excitation along with the spectral sensitivity of the extended S20 photocathode, used for the AE measurement. The peak value of the light spectrum was found at 530 nm.

Table 1 shows the variation of GAGG:Ce crystal spectral maching factor with several photo-detectors. The spectrum was found to be well matched with the spectral sensitivity curves of most optical detectors considered in this study.

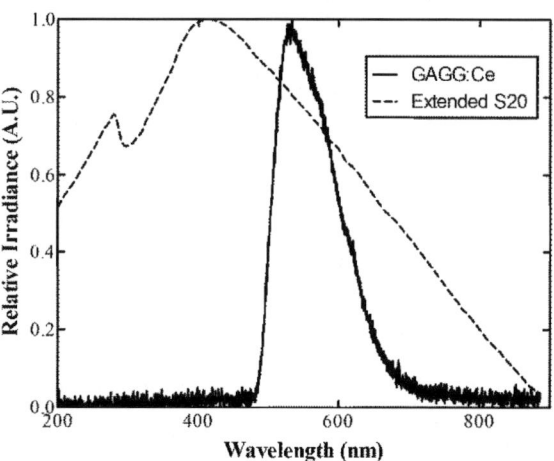

Fig. 1. Emission spectrum of GAGG:Ce single crystal.

Table 1 Spectral matching factors.

Optical sensor	SMF
CCD broadband AR coating	0.949
CCD IR AR coating	0.817
Hybrid CMOS NIR AR coating	0.874
Hybrid CMOS blue	0.985
Monolithic 0.5	0.941
a-Si passivated	0.803
a-Si non-passivated	0.954
CCD with ITO gates with micro	0.870
CCD with ITO gates	0.792
CCD with polygates	0.665
CCD no poly-gates LoD	0.860
CCD with traditional poly gates	0.889
CMOS Pgate	0.813
CMOS RadEye HR	0.925
GaAs	0.975
GaAsP	0.871
E-S20	0.715

Fig. 2 shows the variation of AE of the GAGG:Ce crystals with X-ray tube voltage in the range from 50 to 125 kVp. AE was found to increase as the X-ray tube voltage and the crystal thickness were increased. The absolute efficiency values of 20 mm thick crystal are higher than those of the corresponding 10 mm thick. Thicker crystals can detect more high energy X-ray photons, without having the problems of self absorbance of powder scintillators. From 70 to 80 kVp is observed a 50% increase in efficiency units. This is due to Gd absorption edge (50.2 keV).

Figure 3 shows the effective luminescence efficiency of the 10×10×20 cm^3 GAGG:Ce crystal scintillator with various digital optical detectors. Effective efficiency was found with high values for some electronic optical sensors such as CCDs and CMOS, due to the excellent spectral compatibility with the green wavelength range.

IV. CONCLUSIONS

In conclusion, results of the present study showed that the Absolute Efficiency of GAGG:Ce scintillator crystals increased with X-ray tube voltage under general X-ray conditions. The emission spectrum of scintillator examined in our study is well matched with the spectral sensitivities of the optical photon detectors often employed in radiation detectors. Taking into account the efficiency and the emission Spectral Compatibility of GAGG:Ce scintillator could be potentially considered for use in digital X-ray medical imaging.

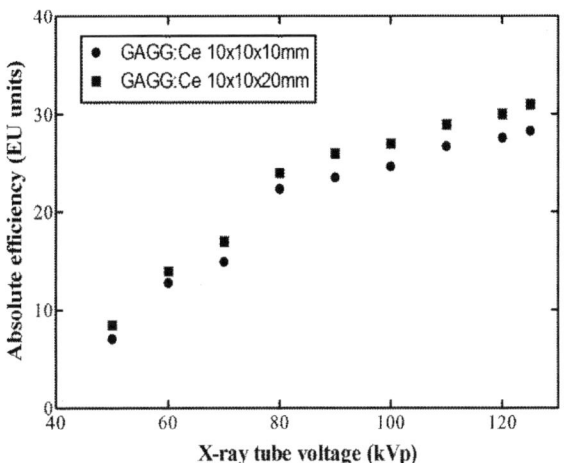

Fig. 2. AE variation with X-ray tube energy of GAGG:Ce crystals.

ACKNOWLEDGMENT

This research has been co-funded by the European Union (European Social Fund) and Greek national resources under the framework of the "Archimedes III: Funding of Research Groups in TEI of Athens" project of the "Education & Lifelong Learning" Operational Program.

REFERENCES

1. Valais I G et al. (2006) Evaluation of the light emission efficiency of LYSO:Ce scintillator under X-ray excitation for possible applications in medical imaging Nucl. Instr. and Meth. Phys. Res. A 569:201-204.
2. van Eijk C W E (2002) Inorganic scintillators in medical imaging Phys. Med. Biol. 47:R85-R106.
3. Iwanowska J et al. (2013) Performance of cerium-doped Gd$_3$Al$_2$Ga$_3$O$_{12}$ (GAGG:Ce) scintillator Nucl. Instr. and Meth. Phys. Res. A 712:34-40.
4. Michail C et al. (2009) Imaging performance and light emission efficiency of Lu$_2$SiO$_5$:Ce (LSO:Ce) powder scintillator under X-ray mammographic conditions in gamma-ray spectrometry Appl Phys B 95:131-139.
5. Kamada K et al. (2012) 2 inch diameter single crystal growth and scintillation properties of Ce:Gd$_3$Al$_2$Ga$_3$O$_{12}$ Journal of Crystal Growth 352: 88-90.
6. Kobayashi M et al. (2012) significantly different pulse shapes for γ and a-rays in Gd$_3$Al$_2$Ga$_3$O$_{12}$:Ce^{3+} scintillating crystals Nucl. Instr. and Meth. Phys. Res. A 694: 91-94.
7. Michail C et al. (2009) A comparative investigation of Lu$_2$SiO$_5$:Ce and Gd$_2$O$_2$S:Eu powder scintillators for use in x-ray mammography detectors, Meas. Sci. Technol. 20: 104008.

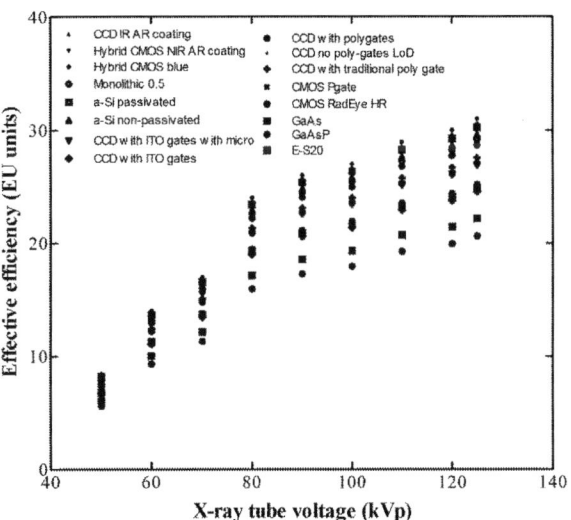

Fig. 3. n_{eff} variation with X-ray tube energy of 10×10×20 cm^3 GAGG:Ce crystals.

8. Michail C M et al. (2011) Experimental and Theoretical Evaluation of a High Resolution CMOS Based Detector Under X-Ray Imaging Conditions IEEE Trans. Nucl. Sci. 58(1):314-322.
9. Kandarakis I et al. (2001) X-ray luminescence of ZnSCdS:Au,Cu phosphor using X-ray beams for medical applications Nucl. Instr. and Meth. Phys. Res. B 179(2): 215-224.
10. Kandarakis I et al. (1997) Evaluating x-ray detectors for radiographic applications: A comparison of ZnSCdS:Ag with Gd_2O_2S:Tb and Y_2O_2S:Tb screens, Phys. Med. Biol. 42: 1351-1373.

Author: Ioannis Valais
Institute: Technological Educational Institute of Athens
Street: Ag. Spyridonos
City: Athens
Country: Greece
Email: valais@teiath.gr

Calcification Detection Optimization in Dual Energy Mammography: Influence of the X-Ray Spectra

V. Koukou[1], N. Martini[1], G. Fountos[2], P. Sotiropoulou[1], C. Michail[2], I. Valais[2], I. Kandarakis[2], and G. Nikiforidis[1]

[1] Medical School University of Patras/ Department of Medical Physics, Patras, Greece
[2] Technological Educational Institution of Athens/Department of Medical Instruments Technology, Athens, Greece

Abstract—Breast cancer may manifest as microcalcifications (μCs) in X-ray mammography. However, the detection and visualization of μCs are often obscured by the overlapping tissue structures. Dual-Energy imaging technique offers an alternative approach for imaging and visualizing μCs. With this technique, a high- and a low-energy image is acquired and their differences are used to "cancel" out the background tissue structures. In this report, various combinations of maximum tube voltages, 23-30kVp for low energy and 50-60kVp for high energy, and filter thicknesses were examined in order to obtain quasi-monochromatic spectra. The filters applied to Tungsten (W) spectra were selected according to their K-edge (K-edge filtering). The impact of these imaging parameters on the calcification signal-to-noise ratio (SNR) for a fixed entrance exposure was studied. The optimization of the study was accomplished by the maximization of the calcification SNR ($SNR_{tc} \geq 3$) and the minimization of the coefficient of variation of the incident photons, for the minimum μC size that can be detected. The best results were obtained from filtered spectra at 30kVp and 60kVp for low and high energy respectively. These spectra were filtered with 300 μm Cd and 800 μm Sm with mean energy values 23.9 and 42.2 keV for low and high energy respectively.

Keywords—dual-energy mammography, SNR, K-edge filtering, microcalcifications.

I. INTRODUCTION

Screening and diagnosis in X-ray mammography rely on the detection and visualization of microcalcifications (μCs) and/or soft tissue masses. The early detection of breast cancer has been shown to decrease breast cancer mortality [1]. Microcalcifications are composed mainly of calcium with attenuation properties greater than that of soft tissue. The detection and visualization of μCs are relatively easy over a uniform tissue background, but limited by the "clutter" due to overlapping tissue background (glandular tissue, vessels, and soft tissue masses in the breast). Depending upon the degree of clutter and the contrast of μCs, it may be difficult to detect a microcalcification, even though there may be sufficient contrast-to-noise ratio (CNR). In Dual-Energy subtraction imaging, high- and low-energy images are separately acquired and "subtracted" from each other in a weighted fashion to cancel out the cluttered tissue structure so as to decrease the obscurity from overlapping tissue structures [1]. In this simulation study, the influence of various imaging parameters (polyenergetic X-ray spectrum, filter materials, filter thickness, μCs size, and breast thickness) on the calcification SNR (SNR_{tc}) was studied. The calculations are based on the acquisition of quasi-monochromatic spectra by applying filters, based on their K-edge (K-edge filtering). The analysis presented here permits the choice to be optimized such that some critical parameter is minimized (or maximized). For a *fixed entrance exposure*, optimization was based on: 1) *Maximizing* the calcification SNR (SNR_{tc}). The SNR_{tc} threshold was set to be 3 ($SNR_{tc} \geq 3$). 2) *Minimization* of the coefficient of variation of incident photons (CV_{Iinc}). The CV_{Iinc} should not exceed the maximum threshold of 0.3% ($CV_{Iinc} < 0.3\%$), which corresponds to 10^5 incident photons.

II. MATERIALS AND METHODS

1. Dual Energy Microcalcification Imaging Technique

An analytical model was developed in order to determine the calcification SNR (SNR_{tc}) assuming only Poisson distribution [1]. Assume that along the X-ray path, a compressed breast is composed of adipose tissue of thickness t_a, glandular tissue of thickness t_g, and a cubic microcalcification of thickness t_c. The total breast tissue thickness, T (cm) is given by [1]:

$$T = t_a + t_g + t_c \quad (1)$$

Assuming that polyenergetic X-rays are used, the mean measured signals in the low- and high-energy images, S_l and S_h, can be expressed as:

$$S = \int dE \cdot R_j \cdot d^2 \cdot \Phi_j(E) \cdot e^{-\frac{\mu}{\rho_a}(E)\rho_a t_a - \frac{\mu}{\rho_g}(E)\rho_g t_g - \frac{\mu}{\rho_c}(E)\rho_c t_c} A(E) \cdot Q(E) \quad (2)$$

where, R_l and R_h are the unattenuated low- and high-energy X-ray exposures in mR at the detector plane, d is the pixel size in centimeters (cm), $\Phi_l(E)$ and $\Phi_h(E)$ are the unattenuated low- and high-energy photon fluence per unit exposure per unit energy (photons/cm² keV mR) at the detector input, A(E) is the quantum detection efficiency of the detector as a function of photon energy E(keV), and Q(E) is the detector response function and represents the signal generated by each detected X-ray photon as a function of photon energy E(keV). The SNR of the subtraction signal, tc, can then be expressed as follows:

$$SNR_{tc} = \frac{t_c}{\sqrt{\sigma_{tc}^2}} \quad (3)$$

where σ_{tc} represents noise level in the calcification subtraction image.

2. Mammographic Spectra

The dual energy method was adapted to polyenergetic spectra. Unfiltered spectra were obtained from Boone et al. for Tungsten (W) anode [2]. Filtering of X-ray spectra was implemented by determining the optimum combination and thickness of these filters. Also an inherent filtration of approximately 0.5 mm Beryllium (Be) window was used in order to obtain the unmodified spectra. The normalization of the filtered spectra at standard exposure conditions was implemented in terms of X-ray air KERMA at the entrance of the patient [3]. In the low energy beam (23-30 kVp) five different filter materials (Rh, Ag, Mo, Pd and Cd) were applied. In the high energy beam (50-60 kVp) four lanthanide filters (Ce, Nd, Sm, Eu) were applied on the basis of their K-edge. The filter thicknesses varied from 100 to 1000 μm. Computer simulations provided values for the following quality parameters: (i) mean energy (ME), (ii) Full Width at Half Maximum (FWHM), (iii) total counts (TC), (iv) Root Mean Square Error (RMSE) [4] and Coefficient of Variation of incident photons (CV_{Iinc}).

3. Calculation of Parameters of the Detector System

In this study, terbium-doped gadolinium oxysulfide (Gd_2O_2S:Tb) was considered to be the scintillator material coupled to CMOS photodiode array (RadEye HR). The density of Gd_2O_2S:Tb is 7.34 g/cm³. The surface density, W, (density multiplied by the thickness of the material) of Gd_2O_2S:Tb used in this study is 0.03394 g/cm² (approximately 34 mg/cm²). The sensor pitch is 22.5 μm [3]. The Quantum Detection Efficiency (QDE), $A(E)$, and the matching factor (α_s), $Q(E)$, were calculated according to Michail et al [5].

4. X-ray Attenuation Coefficients

The elemental compositions of adipose (0.93gr/cm³) and glandular (1.04gr/cm³) breast tissue, the microcalcifications-$CaCO_3$ (2.93gr/cm³), and the Gd_2O_2S:Tb scintillator (7.34gr/cm³) were used to calculate the mass-attenuation coefficients (μ/ρ) using data published by NISTIR (National Institute of Standards and Technology Interagency Report) [6]. The mass attenuation coefficient values (μ/ρ) were then multiplied by the density (ρ) of the appropriate material to obtain the linear attenuation coefficients (μ). Note that the mass attenuation coefficients were then replaced by *effective mass attenuation coefficients*.

5. X-ray Exposure Considerations

The total exposure (after filtration of the X-ray beams) of dual-energy image acquisition was kept at 684 mR, corresponding to 6 mSv [7]. For simplicity, the calcification SNR was computed by normalizing the total exposure to be 684 mR. The results can be easily extrapolated to other exposure values. The optimal distribution of the exposure between the low- and high-energy images was studied by computing the calcification SNR (SNR_{tc}) in the subtraction image signals as a function of the "low-energy exposure ratio," defined as the ratio of the low-energy exposure to the total exposure.

6. Calculation of SNR_{tc}

The simulation process in order to calculate the calcification SNRs (SNR_{tc}) is the following: (i) Considering a compressed breast thickness of 5 cm, 50% adipose and 50% glandular tissue composition, and low-/ high-energy input tungsten spectra at 30/60kVp, respectively. The SNR in the dual-energy μC image signal, SNR_{tc}, was then computed as a function of the low-energy exposure ratio to determine the minimum μC size that would yield an acceptable SNR ≈ 3 or higher ($SNR_{tc} \geq 3$). The microcalcification sizes examined were 100 to 300 μm (in 50 μm increments). (ii) Using low-/ high-energy input tungsten spectra at 30/60kVp, respectively, a 100 μm μC size, and 50% adipose and 50% glandular tissue composition, SNR_{tc} was computed as a function of the low-energy exposure ratio for various compressed breast thickness ranging from 4 to 7 cm (in 1 cm increments).

III. RESULTS

In order to obtain a spectrum with a high proportion of X-rays in the energy bands using a K-edge filter, a filter material with an appropriate K-edge must be selected. It is evident that the thicker the filter, the narrower the spectrum, and the greater the mean energy for the same kV. Simulation results concerning the evaluation of K-edge filtered W anode spectra are shown below. Indicative examples of filter performance in mammographic spectra are given in Table 1.

Table 1 Mammographic filter performance.

kVp	Filter	Thickness (μm)	Mean Energy (keV)	FWHM (keV)
30	Pd	200	22.45	2
30	Cd	200	23.76	3
50	Nd	800	39.16	6
60	Eu	900	42.91	8

Fig. 1 shows indicative filtered X-ray spectra studied from a tungsten anode tube, generated at two different kVp potentials. It can be clearly seen the effect of filtration with 200 μm Pd at 30kVp spectrum and 900 μm Nd at 60kVp spectrum. The effect of the K-edge filtering is obvious and these filters remove significant proportions of X-rays at both higher and lower energies of the spectra.

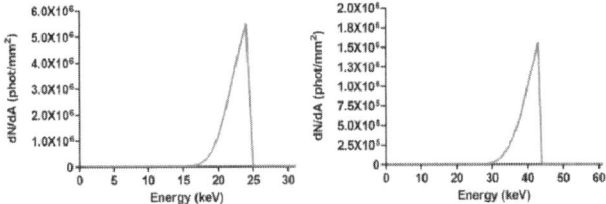

Fig 1. W anode spectra at 30 kVp modified with 200 μm Pd and 60kVp with 900 μm Nd, respectively.

Fig. 2 & 3 represent CV_{Iinc} as a function of surface density (density multiplied by the thickness of the material) for the low- and high- energy filters respectively. Cd minimizes CV_{Iinc} over the surface density range from 0.1 to 0.3g/cm^2. Sm minimizes CV_{Iinc} over the surface density range from 0.1 to 0.6g/cm^2. The SNR_{tc} was calculated as a function of the low-energy exposure ratio, for various μC sizes (100–300 μm), low-/ high-energy filtered spectra at 30/60kVp with Cd-300 μm and Sm-800 μm respectively, and assuming a 50% adipose and 50% glandular tissue composition and a breast thickness of 5 cm. It has long been recognized that, for the detection of a μC, the object SNR must exceed some minimum value. In this study we adopt a threshold value of 3 for SNR_{min}. In Fig. 4 a SNR of 3 is achieved for a μC size of 100 μm. Thus, a 100 μm μC size was used in all subsequent computations, as it may be considered as the minimum detectable μC size in dual-energy imaging. It is evident that as the μC size increases, the SNR_{tc} values increase. The highest SNR value is 10.49 for 300 μm μC size. In Fig. 5 SNR_{tc} is plotted as a function of the low-energy exposure ratio for a compressed breast thickness between 4 and 7 cm at 30/60 kVp filtered low/high spectra with Cd-300 μm and Sm-800 μm, respectively and a μC size of 100 μm. SNR_{tc} decreased with the increase of breast thickness. As the breast thickness increases, the number of X-ray photons reaching the detector decreases, which lead to SNR_{tc} reduction.

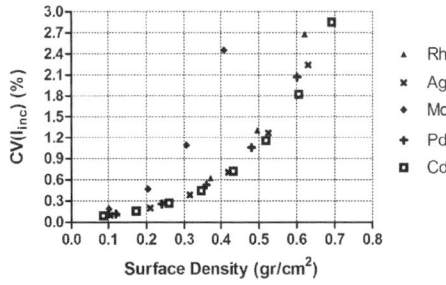

Fig 2. CV_{Iinc} as a function of surface density for all low energy filters at 30kVp.

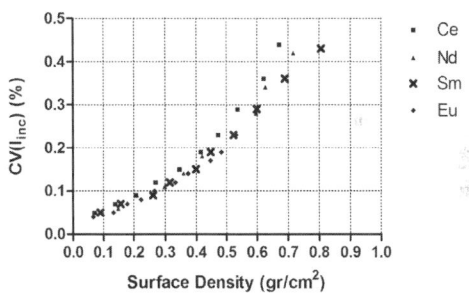

Fig 3. CV_{Iinc} as a function of surface density for all high energy filters at 60kVp.

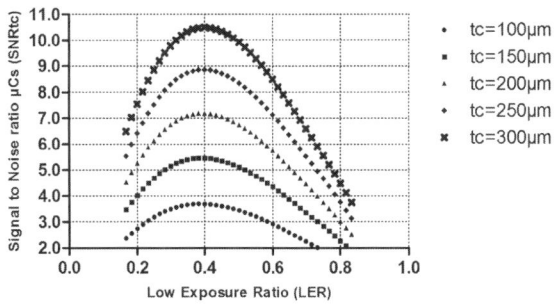

Fig 4. Plots of the SNR_{tc} as a function of the LER for various μC sizes.

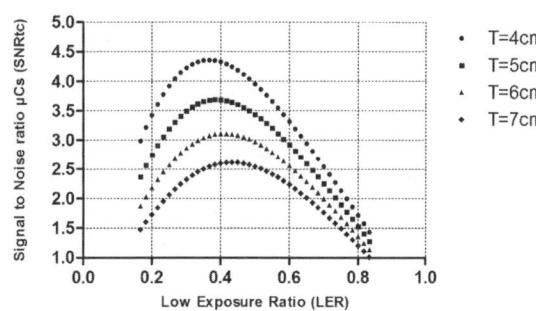

Fig 5. Plots of the SNR$_{tc}$ as a function of the LER for various breast thicknesses.

IV. DISCUSSION

The influence of various imaging parameters (polyenergetic X-ray spectrum, filter materials, filter thickness, μCs size, and breast thickness) on the calcification SNR was studied. For a 5 cm thick breast composed of 50% adipose and 50% glandular tissue, a μC size of 100 μm yielded a SNR$_{tc}$ of 3.69 with a 30/60kVp low- /high-energy spectra filtered with Cd and Sm for low and high energy respectively. Simulations, also, studied the effect of the breast thickness on SNR$_{tc}$. LER may be a function of the μC size and the breast thickness. In the case of a 7cm breast a 100 μm μC cannot be detected due to the increased X-ray attenuation. The optimal range varied with the compressed breast thickness from 0.35-0.48 for a 6cm breast, 0.23-0.58 for a 5cm breast, and 0.18-0.63 for a 4 cm breast. Based on the overlap of these ranges, we have concluded that for LER between 0.35 and 0.48, the estimated SNR$_{tc}$ values fulfill the criterion of SNR$_{tc}$ ≥ 3.

ACKNOWLEDGMENT

This research has been co-funded by the European Union (European Social Fund) and Greek national resources under the framework of the "Archimedes III: Funding of Research Groups in TEI of Athens" project of the "Education & Lifelong Learning" Operational Programme.

REFERENCES

1. Lemacks M, Kappadath S et al (2002) A dual-energy technique for microcalcification imaging in digital mammography-A signal-to-noise analysis. Med. Phys. 29:1739-1751
2. Boone J, Fewell T and Jennings R (1997) Molybdenum, rhodium, and tungsten anode spectral models using interpolating polynomials with application to mammography. Med. Phys. 24:1863-73
3. Michail C, Spyropouloy et al (2011) Experimental and Theoretical Evaluation of a High Resolution CMOS Based Detector under X-Ray Imaging Conditions. IEEE 58:314-322
4. Fountos G, Michail C et al (2012) A novel easy-to-use phantom for the determination of MTF in SPECT scanners. Med. Phys. 39:1561-1570.
5. Michail C. Fountos G et al (2010) Light emission efficiency and imaging performance of Gd$_2$O$_2$S:Eu powder scintillator under x-ray radiography conditions. Med. Phys. 37:3694-3703
6. Hubbel J and Seltzer M. (1995) Tables of X-ray mass attenuation coefficients and mass energy absorption coefficients 1 keV to 20MeV for elements Z=1 to 92 and 48 additional substances of dosimetric interest. NISTIR 5632
7. European Guidelines on Quality Criteria for Diagnostic Radiographic images, European Commission, EUR 16260 En, June 1996

Author: George Fountos
Institute: Technological Educational Institution of Athens
Street: Ag. Spyridonos
City: Athens
Country: Greece
Email: gfoun@teiath.gr

X-Ray Spectra for Bone Quality Assessment Using Energy Dispersive Counting and Imaging Detectors with Dual Energy Method

P. Sotiropoulou[1], G. Fountos[2], N. Martini[1], V. Koukou[1], C. Michail[2], I. Valais[2], I. Kandarakis[2], and G. Nikiforidis[1]

[1] Medical School University of Patras/ Department of Medical Physics, Patras, Greece
[2] Technological Educational Institution of Athens/Department of Medical Instruments Technology, Athens, Greece

Abstract—The aim of the present study was the optimization of dual energy x-ray spectra through the estimation of the Coefficient of Variation (CV). The simulation of monoenergetic x-ray beams provides the optimum dual energy pairs minimizing the CV. Single and double exposure methods were used in order to obtain polyenergetic spectra. K-edge filtering was applied to provide quasi-monochromatic beams with equal mean energy to the corresponding monochromatic ones. Furthermore the optimization for the obtained incident spectra was based on a limiting number of counts equal to 10^5 ($CV_{inc} \leq 0.3\%$) with minimum peak spectral width. The optimum results for single exposure method obtained from 80kVp with added beam filtration of 850μm Ce achieving a CV value of 0.21%. Whereas for the double exposure technique a CV value of 0.68% was achieved using 50 and 120kVp with added filtration of 400μm Ce and 1000μm Yb respectively. These CV values indicate that this method can be used as a diagnostic tool of great importance in bone quality assessment.

Keywords—**K-edge techniques, dual energy x-ray, Ca/P mass ratio, bone assessment.**

I. INTRODUCTION

Bone mineral density (BMD) is the gold standard for osteoporosis prognosis which primarily measures the quantity of bone in the skeleton. However, BMD overlooks subtle aspects of bone properties [1]. The improved knowledge of the different parameters of the bone quality in the determination of skeletal health, will improve the understanding of the pathogenesis and treatment of bone diseases. It is, therefore, important to determine what parameters contribute to decreased bone quality in diseased tissue. The in vivo determination of the Calcium-to-Phosphorous (Ca/P) mass ratio is an important intrinsic parameter for bone quality.

In this study, we present a method for optimum dual x-ray energy spectra selection determining the Ca/P mass ratio in the finger. The determination of the Ca/P ratios can be acquired with a wide selection of x-ray beam pairs generating from single and double exposure. Filtering material and thicknesses of the filter was used for x-ray spectrum modification in order to achieve the two different x-ray beams. The filter's selection was based on its K-edge filtering properties. The single exposure method can be used to photon counting, whereas the double exposure (kVp switching) can be used in both photon counting and energy integrating detectors (flat panel, CMOS [2]). The analysis presented here permits the selection of spectra to be optimized when a critical parameter is minimized. In this study, the reproducibility error of Ca/P ratio ($CV_{Ca/P}$) is selected to be minimized.

Goodness of spectra in terms of peak width narrowing with sufficient number of photons is investigated. For this study mean energy, Full Width at Half Maximum (FWHM), mean energy (ME), Root Mean Square Error (RMSE) of the spectral width, and Coefficient of Variation (CV) of incident beam intensity (CV_{inc}) measurement in each spectra peak is determined.

II. MATERIALS AND METHODS

A. Simulation Input Functions and Parameters

The theory upon which the in vivo determination of Ca/P ratio in the finger was based, is similar to that described by Fountos et al. [3,4]. The equations used, as well as the method in order to obtain the Ca/P mass ratio and the coefficient of variation (CV) due to counting statistics, have been described in detail [3]. The Ca/P mass ratio is obtained assuming that there is a three-component system: Ca, PO₄ and water. The reproducibility error, CV, of the Ca/P mass ratio, was calculated using propagation of error analysis, assuming Poisson distribution for incident beam photon detection. The CV is provided by:

$$CV_{\frac{Ca}{P}}^2 = \left(\frac{1}{I_{W_{(E_l)}}} + \frac{1}{I_{b_{(E_l)}}}\right)\left(\frac{(\Delta M_{E_h PO_4})^2}{(Y_l \Delta M_{E_h PO_4} - Y_h \Delta M_{E_l PO_4})^2} + \frac{(\Delta M_{E_h Ca})^2}{(-Y_h \Delta M_{E_l Ca} + Y_l \Delta M_{E_h Ca})^2}\right) * 100^2$$

$$+\left(\frac{1}{I_{W_{(E_h)}}}+\frac{1}{I_{b_{(E_h)}}}\right)\left(\frac{\left(\Delta M_{E_l PO_4}\right)^2}{\left(Y_l \Delta M_{E_h PO_4}-Y_h \Delta M_{E_l PO_4}\right)^2}+\frac{\left(\Delta M_{E_h Ca}\right)^2}{\left(-Y_h \Delta M_{E_l Ca}+Y_l \Delta M_{E_h Ca}\right)^2}\right)*100^2 \quad (1)$$

where $Y_i = \ln(I_{wi}/I_{bi})$ and $\Delta M_{Ca}(E_i) = M_{Ei,Ca}d_{Ca} - M_{Ei,w}d_w$, $\Delta M_{Ei,PO_4} = M_{Ei,PO_4}d_{PO_4} - M_{Ei,w}d_w$ with $i=\ell,h$ indicating low and high energy. M_{Ca}, M_{PO_4}, M_w, d_{Ca}, d_{PO_4}, d_w are the energy dependent mass attenuation coefficients and densities of Ca, PO_4, and water. I_{wi} and I_{bi} is the attenuated intensity of the beam passing through constant thickness material, composed by soft tissue and Ca, PO_4, water respectively.

The mass attenuation coefficients for each pair of energies are calculated using the data from Hubbell [5]. Combination of energies examined is from 15 to 120 keV in steps of 1 keV. In this simulation the optimum monoenergetic energy pair, which minimizes the CV, is provided. It is not necessary to make an assumption about the quantum flux for a monochromatic source since the CV is inversely proportional on the quantum flux. The aim is to substitute the optimal monochromatic dual energy radiation with a broader modified X-ray spectrum at the same mean energies. Then, the counting detection is generated by both the optimum energy pair and suboptimal spectral contributions. A CV of less than 2% could be obtained within the energy range of 23-40 keV, in combination with energies ranging 60-99 keV. It is possible to achieve the above energies with a variety of filters, such as Cd, Ho, Sm, Yb and Ce, with K edges ranging from 26 to 80 keV. Furthermore, the substitution of the mass attenuation coefficients with the effective mass attenuation coefficients is essential in order to calculate the CV. The energy-dependent effective mass attenuation coefficients can be calculated by:

$$\mu/\rho_{eff} = \frac{\sum_{Emin}^{Emax} I_{filt}(E)\mu/\rho(E)}{\sum_{Emin}^{Emax} I_{filt}(E)} \quad (2)$$

where E_{min} and E_{max} are the minimum and maximum energy of the incident spectrum and $\mu/\rho(E)$ is the energy-dependent mass attenuation coefficient of the filter.

B. Generation of Quasi-monochromatic Beams Using k-Edge Filtering

Different methods have been developed to approximate monochromatic beams i.e. to provide quasi-monochromatic beams.

The more common approach is to shape the emitted spectrum into two relatively narrow energy bands from a single exposure using K-edge filtering. This method is based on filter materials with a K-edge close to the cut-off energy. The filter can absorb sufficiently x-ray photons, whose energy is equivalent to, or slightly higher than the K-edge, while x-ray absorption level dramatically changes above this energy. Only a single exposure is required because both energies are present simultaneously in the radiation beam. This method requires photon counting energy dispersive detectors that allow energy peak discrimination and counting.

Imaging detectors can be used, instead, allowing faster examinations and region-of-interest (ROI) selection. In this case, double exposure technique is required, where two sequential measurements at different kVps, typically with different beam filters, mechanically moved in and out of the beam between exposures. The added filtration selectively removes lower-energy photons from the high-kV beam.

For filter selection, filtration material (Z), filter thickness, and the kVp are examined. A constraint of $CV_{inc} \leq 0.3\%$ in x-ray output due to the filtration, which corresponds in a limiting number of counts equal to 10^5, is adapted. The unfiltered spectra are obtained from Boone et al. [6] for Tungsten (W) anode with various tube voltages in the range of 40 to 120 kV. For x-ray incident beam 10^8 total counts in air are considered in each case and additional beam filtration with thicknesses ranging from $100\,\mu m$ to $1200\,\mu m$, providing spectra of different mean energies and peak widths. Detectors used are cadmium zinc telluride (CZT) photon detector as a counting energy dispersive detector and CMOS based imaging detector as energy integrating detector.

III. RESULTS

The modified spectrum that accomplice the requirements for total counts and spectral width (in terms of FWHM and RMSE) that minimizes the CV is determined. In Table 1 the results for single exposure method are presented in terms of filter material (Ft), thickness (Th), mean photon energy for low (LE) and high (HE) energy peak, FWHM, RMSE, CV_{inc} and $CV_{Ca/P}$. The x-ray spectrum at 80 kVp filtered with 850μm Cerium (Ce) indicated the best performance characteristics for the determination of $CV_{Ca/P}$. Thus, 37 and 67 keV mean photon energies was obtained achieving a CV of 0.21%. FWHM is in range of 2 for LE while for HE is in range of 14. This indicates that HE incident spectrum is broader than LE spectrum and has less number of photons ($CV_{inc}=0.030$ and $CV_{inc}=0.020$ respectively).

Table 1 Single exposure results for Ce, Ho and Cd with CVCa/P<2%

Ft		Th (μm)	kVp	ME keV	FWHM	Rel. RMSE	CV_{Iinc} (%)	$CV_{Ca/P}$ (%)
Ce	LE	850	80	37	2	0.245	0.021	0.21
	HE			67	14	0.182	0.030	
	LE	900	100	37	2	0.155	0.020	0.30
	HE			79	27	0.163	0.015	
	LE	900	120	38	2	0.152	0.018	0.31
	HE			90	35	1.977	0.008	
Ho	LE	1000	120	51	4	0.246	0.016	0.77
	HE			99	25	1.038	0.021	
Cd	LE	800	70	25	1	0.031	0.207	1.74
	HE			60	11	0.392	0.049	

Fig. 1 shows the modified W anode spectra at 70, 80 and 120 kVp that provides $CV_{Ca/P}$< 2% and the corresponding mean energies falling into the monoenergetic optimum energy range. The different material of the filters used, combined with the different x-ray tube peak potential, has an impact on the spectra on both mean energy values and total counts.

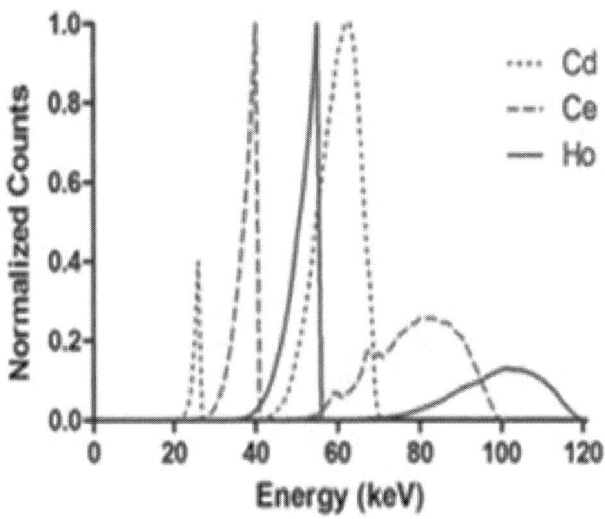

Fig. 1 Single exposure spectra at 80kVp with 850 μm Ce, 120kVp with 900 μm Ho and 70Vp with 800 μm Cd.

In Table 2 the results of double exposure method are presented. Filter material, thickness and x-ray tube peak potential affect the mean photon energy (LE and HE), FWHM, CV_{inc} and $CV_{Ca/P}$. The minimum $CV_{Ca/P}$ is obtained from combination of 50kVp Ce filtered spectrum (LE) and 120kVp Yb filtered spectrum (HE), with 400 and 1000 μm, respectively. Fig. 2 shows the corresponding low and high W anode spectra. Thus, 35 and 63 keV mean photon energies was obtained achieving a CV of 0.68%. FWHM is in range of 8 for LE while for HE is in range of 4. This indicates that LE incident spectrum is broader than HE spectrum and has less number of photons (CV_{inc}=0.011 and CV_{inc}=0.007 respectively).

Table 2 Double exposure results for Ce/Yb, Cd/Yb, Cd/Gd and Ce/Gd with CVCa/P<2%

	Ft	Th (μm)	kVp	ME	FWHM	CVI_{inc} (%)	$CV_{Ca/P}$ (%)
LE	Ce	400	50	35	8	0.011	0.68*
HE	Yb	1000	120	63	4	0.007	
		1100				0.008	1.05**
LE	Cd	200	40	24	3	0.014	0.82*
HE	Yb	1000	120	63	4	0.007	
		1100				0.008	1.04**
LE	Cd	200	40	24	3	0.014	1.37
HE	Gd	600	90	49	6	0.009	
LE	Ce	400	50	35	8	0.011	1.42
HE	Gd	600	90	49	6	0.009	

* CV value which results from the LE filter thickness combined with the HE smaller value of filter thickness.
** CV value which results from the LE filter thickness combined with the HE greater value of filter thickness.

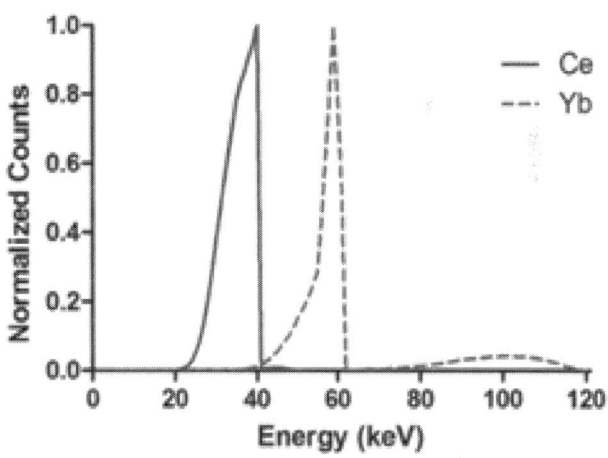

Fig. 2 Double exposure spectra at 50kVp with 400 μm Ce and 120kVp with 1000 μm Yb provides the minimum 0.68% CV value.

IV. DISCUSSION

The simulation of monoenergetic beams provides a physical insight and the method's theoretical peak performance indicating the x-ray spectrum properties.

Thicker filters lead to narrower spectra that approach the ideal monochromatic spectra. However, thicker filters require a much higher tube power which increase tube loading to maintain sufficient number of photons. Ignoring commercial tube power limitations, the precision improvement with increasing filtration continues increasing.

As long as the tube voltage is kept below 65 kV, the influence of the Ka emission line of the W anode at 59.3 keV is not significant and the resulting spectra consist of narrow energy peaks. Higher tube voltages result in rather broader spectra with a pronounced superimposed peak at 59.3 keV and, at even higher voltages, a second peak at 67.2 keV which is the Kb emission line of W. This can be partially resolved by thicker filters, while decreasing total counts.

In single exposure, when using K-edge filtering, by increasing thickness the average spectral separation increases. In double exposure, the spectral separation can be improved by selecting filters with appropriate K-edges and increasing their thicknesses. Accurate and precise assessment of bone properties other than BMD may be of importance in both research and clinical practice. To improve the reliability of fracture risk assessment, it is necessary to improve the diagnostic possibilities of modern imaging and non-imaging systems.

V. CONCLUSIONS

In this study, optimization of dual energy x-ray spectra through the estimation of the Coefficient of Variation was performed in order to improve the precision of Ca/P ratio.

Diagnosis of osteoporosis with x-ray depends on the determination of critical parameters related with bone quantity or quality. The dual energy method proposed in this work provides an alternative and useful approach to the diagnosis of osteoporosis. An estimated CV value of 0.2% can be achieved in Ca/P ratio determination indicating that this method can be used as a diagnostic tool in clinical practice.

A future goal is the experimental evaluation of the method and the implementation in other parts of body such as radius, pelvic and spine which are of great importance in bone quality assessment.

ACKNOWLEDGMENT

This research has been co-funded by the European Union (European Social Fund) and Greek national resources under the framework of the "Archimedes III: Funding of Research Groups in TEI of Athens" project of the "Education & Lifelong Learning" Operational Programme.

REFERENCES

1. Ruppel et al., (2008) "The effect of the microscopic and nanoscale structure on bone fragility", Osteoporos Int 19:1251–1265
2. Michail C, Spyropouloy et al (2011) Experimental and Theoretical Evaluation of a High Resolution CMOS Based Detector under X-Ray Imaging Conditions. IEEE 58:314-322
3. Fountos et al., (1997) "The skeletal calcium/phosphorus ratio: A new in vivo method of determination", Med. Phys. **24**, 1303-1310
4. Fountos et al., (1999) In vivo measurement of radius calcium/phosphorus ratio by x-ray absorptiometry. Appl. Radiat. Isot. 51, 273–278.
5. Hubbel J and Seltzer M. (1995) Tables of X-ray mass attenuation coefficients and mass energy absorption coefficients 1 keV to 20MeV for elements Z=1 to 92 and 48 additional substances of dosimetric interest. NISTIR 5632
6. Boone et al., (1997) "An accurate method for computer-generating tungsten anode X-ray spectra from 30 to 140 kV", Med. Phys. 24, 1661-1670

Author: George Fountos
Institute: Technological Educational Institution of Athens
Street: Ag. Spyridonos
City: Athens
Country: Greece
Email: gfoun@teiath.gr

CDMAM Phantom Optimized for Digital Mammography Quality Control by Automatic Image Readout

M.J. Floor-Westerdijk, W.N.J.M. Colier, and R.J.M. van der Burght

Artinis Medical Systems BV, Zetten, The Netherlands

Abstract—The CDMAM 3.4 phantom is the standard used phantom for quality control in digital mammography. Since digital mammography is improving and automatic image readout is becoming more popular, the CDMAM 4.0 phantom has been developed. The phantom contains more diameters and optimized gold disc thicknesses for each diameter. Six prototype CDMAM 4.0 phantoms and six CDMAM 3.4 phantoms were imaged and analyzed by the automatic readout software CDCOM. Psychometric curves and CD Curves were determined. The number of gold discs having relevant detection probabilities increased with 240% for the CDMAM 4.0. Both phantoms show similar CD Curves, but the CDMAM 4.0 phantoms have less variations at the outer diameters therefore giving more accurate results.

Keywords—CDMAM phantom, digital mammography, quality control.

I. INTRODUCTION

The contrast detail (CD) phantom for mammography "CDMAM" is an effective [1], successful quality control phantom, commonly known and used worldwide . The phantom was originally developed by the Radboud University, Nijmegen, The Netherlands [2] more than 20 years ago. The initial phantom (CDMAM 3.2) was primarily developed as an image quality control phantom for screen-film mammography. About 15 years ago the first digital (CR and DR) mammography systems were introduced. For optimal usage of the CDMAM phantom for both screen-film and digital mammography a change of the specifications was necessary, which resulted in the CDMAM 3.4 [3]. Compared to version 3.2 version 3.4 contains smaller diameters and an extended thickness range. The CDMAM 3.4 phantom has been brought commercially onto the market about 10 years ago by Artinis Medical Systems BV, Zetten, The Netherlands. This phantom is used for image quality evaluation in the "European Guidelines for quality assurance in breast cancer screening and diagnosis" [4] and obligatory for quality control in many countries.

The phantom was developed for image assessment with observer readings. However, the use of automatic readout and scoring of digital CDMAM images using the software tool CDCOM [5] as standardized method for image quality control has become more and more popular. The CDCOM software automatically identifies the gold discs on digital images of the CDMAM phantom and is available from the EUREF website [6]. The CDCOM program attempts to correctly locate the position of the gold discs by first identifying the grid position and subsequently the gold discs positions. The probability of gold disc detection by CDCOM is between 25% and 100% by the four-choice-method [7]. The gold disc thicknesses having 62.5% detection probability are the threshold thicknesses and are determined by applying psychometric curve fitting by a least squares procedure over the probabilities [7,8]. Gold discs having a detection probability in the relevant area, above 25% and below 100%, are mainly characterizing the psychometric curve. If a larger number of gold discs can be used for psychometric curve fitting, this will improve the accuracy of the fit and consequently the determination of the threshold thicknesses.

The increased use of automatic scoring of digital CDMAM images as standardized method for image quality control and the increased performance of modern digital mammography systems required the development of a new version of the CDMAM phantom; the CDMAM 4.0 (Fig. 1).

The CDMAM 4.0

As automatic scoring makes time consuming manual readout obviate, the CDMAM 4.0 could be developed with an increased number of gold discs in the relevant areas for psychometric curve fitting. Not only where there added more diameters but also the thicknesses where optimized for each individual diameter. The number of gold discs increased from 410 to 672. Consequently the contrast detail curve of the mammography system under evaluation can be determined with a higher accuracy.

Phantom Design

The phantom design is kept as close as possible to the CDMAM 3.4 for ease of use of the phantom and the translation ratio from automatic to human observer scores [5,10,11,12]. Like the CDMAM 3.4 the aluminum base consists of Al 1050 (with 99.5% purity) and a thickness of 0.5 mm. The aluminum plate is placed in a body, which consists of a 5 mm thick Plexiglas (PMMA) plate. In the body a cavity with a depth of 2 mm has been milled, to accommodate the aluminum base with the gold-disks. The grid is not rotated for a more efficient use of the phantom area and to minimize effect of the heel effect on the CD Curve [13]. The phantom is divided into four blocks for

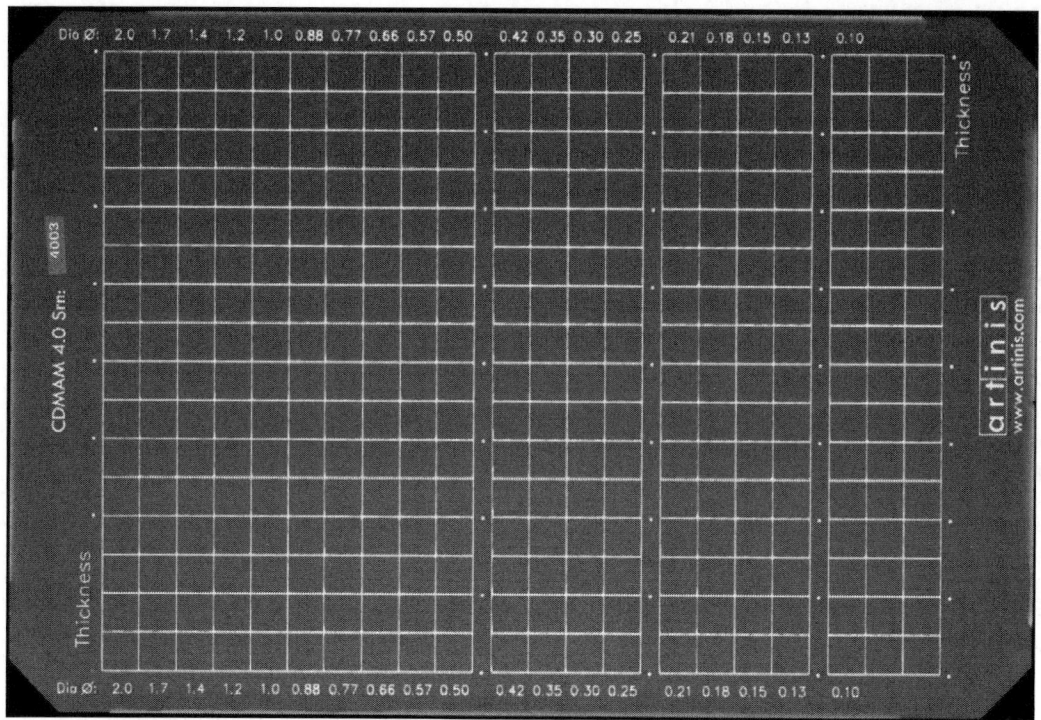

Fig. 1 The newly developed CDMAM phantom with increased number of gold discs and thicknesses in order to increase the accuracy of automatic read-out software and comply with modern mammography systems

Table 1 Nominal Thicknesses of CDMAM 4.0 in μm.

	Diameter (mm)																			
	2	*1.7*	*1.4*	*1.2*	*1*	*0.88*	*0.77*	*0.66*	*0.57*	*0.5*	*0.42*	*0.35*	*0.3*	*0.25*	*0.21*	*0.18*	*0.15*	*0.13*	*0.1*	*0.09*	*0.08*
16	0.103	0.105	0.106	0.106	0.129	0.147	0.168	0.191	0.208	0.240	0.264	0.341	0.446	0.549	0.675	0.849	1.195	1.396	2.434	2.633	2.800
15	0.094	0.096	0.098	0.097	0.109	0.128	0.148	0.170	0.192	0.210	0.239	0.308	0.401	0.493	0.601	0.763	1.105	1.310	2.228	2.411	2.601
14	0.087	0.089	0.090	0.090	0.099	0.109	0.130	0.149	0.171	0.193	0.208	0.269	0.351	0.448	0.554	0.681	0.964	1.205	2.019	2.179	2.380
13	0.078	0.078	0.079	0.079	0.091	0.100	0.110	0.130	0.150	0.171	0.191	0.242	0.312	0.401	0.493	0.606	0.865	1.111	1.854	2.000	2.180
12	0.069	0.069	0.071	0.071	0.080	0.091	0.100	0.110	0.130	0.150	0.169	0.209	0.272	0.350	0.442	0.554	0.777	0.970	1.680	1.830	1.958
11	0.056	0.056	0.057	0.057	0.071	0.080	0.092	0.100	0.110	0.130	0.148	0.192	0.243	0.310	0.396	0.490	0.690	0.860	1.560	1.655	1.787
10	0.048	0.048	0.049	0.049	0.057	0.071	0.080	0.091	0.100	0.109	0.129	0.170	0.210	0.270	0.344	0.440	0.610	0.770	1.440	1.531	1.623
9	0.042	0.042	0.043	0.043	0.049	0.057	0.071	0.080	0.091	0.100	0.108	0.147	0.190	0.240	0.304	0.390	0.550	0.678	1.330	1.415	1.508
8	0.037	0.037	0.038	0.038	0.043	0.049	0.057	0.070	0.080	0.090	0.098	0.127	0.167	0.206	0.264	0.338	0.485	0.598	1.210	1.304	1.386
7	0.032	0.032	0.033	0.033	0.038	0.043	0.049	0.057	0.070	0.079	0.089	0.107	0.144	0.186	0.234	0.297	0.429	0.538	1.100	1.184	1.273
6	0.027	0.027	0.028	0.028	0.033	0.038	0.043	0.049	0.056	0.070	0.077	0.096	0.125	0.163	0.200	0.256	0.379	0.467	0.947	1.081	1.165
5	0.023	0.024	0.024	0.024	0.028	0.033	0.038	0.042	0.048	0.056	0.069	0.088	0.104	0.142	0.181	0.227	0.327	0.416	0.833	0.928	1.047
4	0.020	0.021	0.021	0.021	0.024	0.028	0.032	0.037	0.042	0.048	0.055	0.076	0.093	0.121	0.158	0.194	0.285	0.366	0.731	0.810	0.890
3	0.017	0.018	0.018	0.018	0.021	0.024	0.027	0.032	0.037	0.041	0.047	0.067	0.085	0.101	0.136	0.173	0.245	0.313	0.640	0.716	0.782
2	0.015	0.015	0.015	0.015	0.018	0.021	0.024	0.027	0.032	0.036	0.040	0.053	0.073	0.090	0.116	0.150	0.216	0.273	0.557	0.617	0.684
1	0.012	0.012	0.012	0.012	0.015	0.018	0.021	0.023	0.027	0.031	0.035	0.045	0.064	0.081	0.096	0.129	0.165	0.231	0.493	0.535	0.590

production purposes and inspection and to create space needed for reference discs. These reference discs are placed on the sides and between the blocks and used by CDCOM readout software to accurately localize the position of the gold discs.

Diameter Range

Within the dimensions of the CDMAM 3.4 phantom the number of different diameters has been optimized. In total 21 different diameters are used with a logarithmic linear increase from 0.08 mm to 2.0 mm. The diameter of 0.06 mm which is present in the CDMAM 3.4 phantom has been skipped as this diameter has no practical added value. The diameter tolerance is ± 0.005 um (Mean ± 2SD).

Thickness Range

The phantom contains 16 optimized thicknesses per diameter. For the 2.0 mm diameter the gold disc thickness varies between 0.012 μm and 0.105 μm. For the 0.08 mm diameter the gold discs thickness varies between 0.590 μm and 2.800 μm (Table 1). The thicknesses are optimized to be used in conjunction with the European Guidelines containing the threshold gold discs thicknesses and for enlarging the number of discs which can be used for the fit of the psychometric curve. The thickness tolerance is ± 3% (mean ± 2 SD).

Objective

The objective of this study is to show the added value of the CDMAM 4.0 compared to the current available CDMAM 3.4 for digital mammography quality control using automatic readout and that the CDMAM 4.0 will give similar but more accurate results.

II. METHOD AND MATERIALS

We manufactured 6 prototype CDMAM 4.0 phantoms. Each of these prototypes were imaged 16 times at a Lorad Selena, Hologic Inc., USA (32 kVp, 76 mAs, W/Rh). The phantoms were placed between 4 plates of 10 mm PMMA with a 1% thickness accuracy and slightly moved between making the images. Also 6 CDMAM 3.4 phantoms were imaged 16 times under the same conditions.

Images of the CDMAM 3.4 phantom were readout automatically and scored by CDCOM v1.6. Images of the CDMAM 4.0 phantom were readout automatically and scored by an adapted version of CDCOM v1.6 because of the different grid orientations, the reference discs and gold disc positions. This CDCOM locates the reference discs and uses them for an accurate localization of the contrast detail gold discs. For both versions of the phantom the detection probabilities were read into Matlab (Mathworks Inc., USA), the psychometric curves were calculated by applying a least squares procedure [7,8]. At the 62.5% detection probability the threshold thicknesses were determined and used to calculate the CD Curves by a third order polynomial. For CDMAM 3.4 only diameters between 0.1 mm and 1.0 mm were included [5] and for CDMAM 4.0 all diameters were included.

III. RESULTS

We found that the CDMAM 3.4 had 108 of the 410 gold discs in the relevant area for the psychometric curve fitting. The CDMAM 4.0 had 500 of the 672 gold discs in the relevant area, which is an increase of 240%. The number of gold discs in the relevant area is shown in table 2 per diameter.

Table 2. The number of gold discs in the relevant area for the psychometric curve versus the total number of gold discs of the CDMAM 3.4 and the CDMAM 4.0 per diameter for which the threshold contrast is defined in the European Guidelines.

Diameter (mm)	2.0	1.0	0.5	0.25	0.1
CDMAM 3.4	2/9	4/13	7/16	9/15	9/11
CDMAM 4.0	10/16	12/16	12/16	12/16	10/16

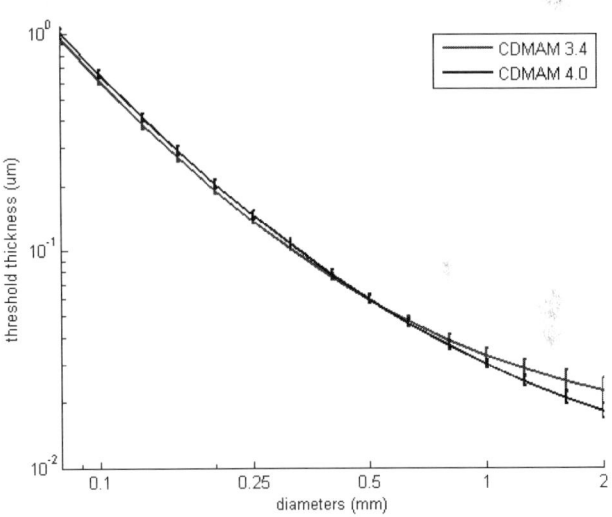

Fig. 2 Mean CD Curve of the CDMAM 3.4 phantom (blue) and the CDMAM 4.0 phantom (black) with standard deviations over the 6 phantoms

The CDMAM 3.4 and 4.0 show similar CD Curves (Fig. 2, Table 3). Differences between the CD Curves are within ±1 Standard deviation (Std) over both sets of 6 phantoms. However the standard deviation are smaller for most diameters of CDMAM 4.0, especially at the outer diameters

IV. DISCUSSION

The CDMAM 4.0 phantom gives similar CD Curves as the CDMAM 3.4 phantoms but with a smaller variation over the outer diameters. The main cause is the

Table 3. Mean threshold thicknesses (μm) of CDMAM 4.0 and CDMAM 3.4 phantoms, the standard deviation (μm) over the 6 phantoms and the fraction of the standard deviation versus the mean threshold thickness (%). The threshold thicknesses for the CDMAM 4.0 phantom were determined by interpolation over the diameters of the CDMAM 3.4 using the third order polynomial for comparison with the CDMAM 3.4 phantom.

	diameters	2	1.6	1.25	1	0.8	0.63	0.5	0.4	0.31	0.25	0.2	0.16	0.13	0.1	0.08	0.06
Threshold thickness CDMAM 3.4	mean	0.021	0.024	0.028	0.033	0.040	0.049	0.061	0.077	0.103	0.135	0.185	0.262	0.373	0.612	0.973	1.866
	std	0.004	0.004	0.004	0.003	0.003	0.003	0.002	0.002	0.003	0.005	0.010	0.018	0.035	0.080	0.171	0.456
	% std/mean	21%	17%	13%	10%	8%	5%	4%	3%	3%	4%	5%	7%	9%	13%	18%	24%
Threshold thickness CDMAM 4.0	mean	0.018	0.021	0.025	0.030	0.037	0.046	0.060	0.078	0.108	0.146	0.203	0.290	0.413	0.665	1.025	1.856
	Std	0.002	0.001	0.001	0.002	0.002	0.002	0.003	0.003	0.004	0.006	0.010	0.016	0.022	0.033	0.051	0.150
	% std/mean	9%	7%	6%	6%	5%	5%	5%	4%	4%	4%	5%	5%	5%	5%	5%	8%

extrapolation of the CD Curve of the CDMAM 3.4 phantoms over the outer diameters. Applying a Gaussian filter as suggested by [14] reduces these variations for both phantoms, but score filtering is not standardized.

The standard PMMA plates used for thickness compensation with the CDMAM 3.4 phantom have a thickness accuracy of ± 10% as described in the European Guidelines. For CDMAM 4.0 it is strongly recommend to use PMMA plates with an accuracy of ± 1% reducing the entrance dose levels differences and consequently increasing the accuracy of the threshold thickness determination.

By the large increase of the number of gold discs in the relevant area of the psychometric curves it is expected that the number of required images needed for significant reliable results might be reduced in the future.

The current thickness thresholds were at the achievable limits of the European Guidelines, but acquired with a high dose compared to the clinical doses used for the simulated breast thickness and at a high performing mammography system. As it is expected that mammography systems will improve in the future getting better contrast detail detection quality, the CDMAM 4.0 will stay a useful phantom because of results of this study and the optimized gold disc thicknesses.

V. CONCLUSIONS

The CDMAM 4.0 is an accurate phantom for contrast detail image quality control in digital mammography using automatic readout. By the increase of the number of available and relevant gold discs, diameters and optimized thicknesses the phantom gives more accurate thresholds than the CDMAM 3.4.

REFERENCES

1. Warren L, Mackenzie A, Cooke J et al. (2012) Mammography calcification cluster detection and threshold gold thickness measurement. Proc. of SPIE 8313, Medical Imaging 2012, Physics of Medical Imaging, 83130J1-83130J10
2. Bijkerk K, Thijssen M (1993) The CDMAM-phantom: A contrast-detail phantom specially for mammography. Radiol 185: 395
3. Thijssen M, Veldkamp W, Van Engen R et al. (2000) Comparison of the detectability of small details in a film-screen and a digital mammography system by the imaging of a new CDMAM-phantom. Digital mammography IDDM, 666-672
4. Van Engen R, Young K, Bosmans H et al. (2006) The European protocol for the quality control of the physical and technical aspects of mammography screening, part B: Digital Mammography, European Guidelines for Quality Assurance in Breast Cancer Screening and Diagnosis, 4th edition. Luxembourg, European Commission.
5. Visser R, Karssemeijer N (2008) CDCOM Manual: software for automatic readout of CDMAM 3.4 images
6. EUREF, www.EUREF.org
7. Karssemeijer N, Thijssen M (1996) Determination of contrast-detail curves of mammography systems by automatic image analysis. Proc. of the 3th international workshop on digital mammography, Digital Mammography '96, 155-160
8. Veldkamp W, Thijssen M, Karssemeijer N 2003. The value of scatter removal by a grid in full field digital mammography. Medical Physics 30: 1712-1718
9. Young K, Alsager A, Oduko et al. (2008) Evaluation of software for reading images of the CDMAM test object to assess digital mammography systems. Proc. SPIE 6913, Physics of Medical Imaging Imaging, 69131C1-C11
10. Young K, Cook J, Oduko H, Bosmans H (2006) Comparison of software and human observers in reading images of the CDMAM test object to assess digital mammography systems. Proc. SPIE, Medical Imaging, 1-13
11. Fletcher-Heath L, Van Metter R (2005) Quantifying the performance of human and software CDMAM phantom image observers for the qualification of digital mammography systems. Proc. of SPIE 5745, 486-498
12. Figl M, Hoffmann R, Kaar M et al (2011) Factors for conversion between human and automatic read-outs of CDMAM images. Medical Physics 38, 5090–5094
13. Thijssen M (1993) The assessment and control of image quality in diagnostic radiology. Thesis. University of Nijmegen (Dutch)
14. Young K, Johnson B, Bosmans H, Van Engen R (2005) Development of minimum standards for image quality and dose in digital mammography. Proc. 7th IWDM 2004, 149-154

Author: dr. ir. W.N.J.M. Colier
Institute: Artinis Medical Systems bv
Street: St. Walburg 4
City: Zetten
Country: The Netherlands
Email: willy@artinis.com

Experimental Evaluation of a High Resolution CMOS Digital Imaging Detector Coupled to Structured CsI Scintillators for Medical Imaging Applications

C.M. Michail[1], I.G. Valais[1], I.E. Seferis[2], F. Stromatia[3], E. Kounadi[4],
G.P. Fountos[1], and I.S. Kandarakis[1]

[1] Technological Educational Institute (TEI) of Athens/Department of Medical Instruments Technology,
Ag. Spyridonos, 12210, Athens, Greece
[2] Department of Medical Physics, Faculty of Medicine, University of Patras, 265 00 Patras, Greece
[3] Radiology and Nuclear Medicine Department, "IASO" General Hospital, Mesogion 264, 15562, Holargos, Greece
[4] Health Ministry, SEYYP, Pireos 205 str, 11853, Athens, Greece

Abstract—The purpose of the present study was to investigate the fundamental imaging performance of CsI scintillating screens specially treated for a CMOS digital imaging sensor, in terms of the Modulation Transfer Function (MTF), Normalized Noise Power Spectrum (NNPS) and the Detective Quantum Efficiency (DQE) in the general radiography energy range. The CMOS sensor was coupled to two columnar CsI:Tl scintillator screens obtained from the same manufacturer with thicknesses of 140 and 170 μm, respectively, which were placed in direct contact with the photodiode array. A CMOS photodiode array was used as an optical photon detector. The MTF was measured using the slanted-edge method while NNPS was determined by 2D Fourier transforming of uniformly exposed images. Both parameters were assessed by irradiation under the RQA-5 beam quality (IEC 62220-1). The DQE was assessed from the measured MTF, NNPS and the direct entrance surface air-Kerma (ESAK) obtained from X-ray spectra measurement with a portable CdTe detector. The detector response function was linear for the exposure range under investigation. The MTF of the present system was found higher than previously published MTF data for a 48 μm CMOS sensor, in the low and medium frequency ranges. DQE was found comparable, while the NNPS appeared to be higher in the 0–10 cycles/mm frequency range. The imaging performance of CsI:Tl scintillating screens in combination to the high resolution CMOS sensor, that was investigated in the present study, could be used in digital imaging systems in order to reduce exposure to patients.

Keywords—CMOS, Digital imaging, Image quality, structured scintillators.

I. INTRODUCTION

In recent years, CMOS technology replaced CCDs in almost all the imaging field applications, as well as in X-ray imaging [1]. CMOS based detectors provide high resolution, high frame rate imaging and, in association with scintillating screens, have been increasingly investigated for medical imaging applications [2]. CsI:Tl are widely used in many medical imaging applications since they exhibit high gamma ray detection efficiency and light yield (66000 photons/MeV) [3-5]. Its emission spectrum having the maximum at about 545 nm, well situated within the spectral sensitivities of optical detectors frequently employed in radiation detectors [6-8]. Calorimeters made of CsI scintillators, which achieve the best energy resolution for photons and electrons, as well as high energy physics experiments and non-destructive evaluation are based on the fact that the radiation hardness of CsI are retained at acceptable levels after the absorption of radiation doses up to a few hundred rads [9-12]. The most important feature of CsI:Tl scintillators is their ability to grow in columnar structure. This reduces the lateral light spread and promotes the directional light passing due to the total internal reflection on column boundaries [13]. CsI:Tl thickness can be increased while maintaining high spatial resolution, X-ray stopping power and conversion efficiency, providing also better detective quantum efficiency (DQE) than powder phosphors over all spatial frequencies [13]. The task of image formation in X-ray radiation at sufficient resolution and low radiation dose is important for medical purposes.

In the present study a CMOS sensor [14] was coupled to two different CsI:Tl phosphor screens, prepared by RMD [15] on demand by our laboratory, and the detector was evaluated under general radiography conditions. The investigation of the imaging performance of this detector was achieved by experimental assessment of the Signal Transfer Property (STP), the Modulation Transfer Function (MTF), the Normalized Noise Power Spectrum (NNPS) and the DQE. The experimental method was based on the guidelines published by the International Electrotechnical Commission (IEC). In addition the results of the present work have been compared with the results of previously studied CMOS sensors coupled to CsI:Tl and Gd_2O_2S:Tb phosphor screens [14, 17].

II. MATERIALS AND METHODS

A. Phosphor Screens

The CsI:Tl scintillator screens were provided by RMD [15] with thicknesses of 140 and 170 μm. The phosphor was used in the form of thin layers (test screens) to simulate the intensifying screen employed in X-ray radiography.

B. CMOS Sensor

For the detector under investigation, the CsI:Tl phosphor screens were manually coupled to an optical readout device including a CMOS Remote RadEye HR photodiode pixel array [18]. The CMOS photodiode array consists of 1200x1600 pixels with 22.5 μm pixel spacing. The CsI:Tl screens were directly overlaid onto the active area of the CMOS photodiode array and held by using a thin Polyurethane foam layer for compression between the screen and a 1-mm-thick Graphite cover. The RQA 5 (70 kVp) beam quality was used according to the IEC standards [16]. IEC standard X-ray spectrum was achieved by adding 21 mm Al filtration. The half value layer was calculated and found 7.1 mm. A Siemens AXIOM Aristos MX/VX X-ray tube with a rotating Tungsten anode (nominal 2.5 mm Aluminum (Al) inherent filtration at 80 kVp) was used [19]. The added filtration was placed as close as possible to the source.

C. X-Ray Spectra Measurement

A portable Amptek XR-100T X-ray spectrometer [20], based on a CdTe crystal detector was used for direct diagnostic X-Ray spectra measurements [14,21]. X-ray spectra were corrected by the inverse square law at the CMOS detector plane. In order to minimize pile-up distortions a dedicated collimation system (1 mm thick collimators with 400 μm diameter) was used. In addition the measured X-ray spectra were corrected for the efficiency of the CdTe detector. The air Kerma value K_α at the surface of the detector was calculated according to [14, 21]:

$$K_\alpha = 0.00869 \times \sum_{E_{min}}^{E_0} (1.83 \times 10^{-6} \times \Phi_0(E) \times E \times (\mu_{en}(E)/\rho)_{air}) \quad (1)$$

Where Φ_0 is the measured X-ray spectrum value (photons/mm^2) at energy E. $(\mu_{en}(E)/\rho)_{air}$ is the X-ray mass energy absorption coefficient of air at energy E [22].

D. Image Quality

D.1. Signal Transfer Property (STP)

The Signal Transfer Property (STP), or detector response, states the relationship between mean pixel value (MPV) and ESAK at the detector surface. This relationship was obtained by plotting pixel values (PV) versus ESAK at the detector, as described in the IEC method [16]. The system's response curve was fitted using a linear equation of the form:

$$MPV = \alpha + b \times K_\alpha \quad (2)$$

Where α and b are fit parameters. From the slope of the system's response curve, the value of the gain factor (G) was obtained. This value is relating MPV to the incident exposure at the detector (in digital units per μGy) [14]. The magnitude of the pixel offset at zero air Kerma was also estimated [14].

D.2. MTF

The MTF was measured using the slanted-edge technique, following the procedures described in IEC standard [23,24]. A PTW Freiburg tungsten edge test device was used to obtain the slanted edge images. Images of the edge, placed at a slight angle, were obtained. The edge spread function (ESF) was calculated by the extraction of an 1×1 cm^2 ROI with the edge roughly at the center. The angle of the edge was then determined using a simple linear least squares fit and the 2D image data were re-projected around the angled edge [23] to form an ESF with a bin spacing of 0.1 pixels. The ESF was smoothed with a median filter of five bins to reduce high frequency noise. Then the ESF was differentiated to obtain the line spread function (LSF) [14]. Finally, the normalized LSF was Fourier transformed to give the pre-sampling MTF.

D.3. NNPS

The NPS was calculated according to [16]. For each ROI, the PV were converted into air kerma units with equation 2. The area of analysis was subsequently divided into sub-images of 1024×1024 pixels. Half overlapping ROIs with a size of 128×128 pixels were then taken from the sub-images [16]. A total of 128 ROIs were taken from each flood image. For all the ROIs taken from each image 2D FFT of each ROI was calculated and added to the NPS ensemble. NNPS was obtained by dividing NPS by the square of the corresponding K_α and afterwards the ensemble average was obtained.

D.4. DQE

The DQE of the system was calculated by equation (3):

$$DQE(u) = MTF^2(u) / K_\alpha \times q \times NNPS(u) \quad (3)$$

Where q is the number of photons per unit Kerma (μGy) per mm^2, determined by dividing the number of photons per mm^2 (measured with the portable X-ray spectrometer) with the corresponding air Kerma value (μGy) [21].

III. RESULTS AND DISCUSSION

The system reproducibility was verified by measuring X-ray spectra several times. Entrance surface air Kerma was calculated by relation (1) using the Amptek XR-100T X-ray spectrometer measurements and was found 34.1 μGy.

Figure 1 shows the STP of the CMOS sensor combined with the 140 and 170 μm CsI:Tl screens. The detector was found to have a linear response, covering the whole exposure range (R^2 greater than 0.997 and 0.982 for the 140 and 170 μm CsI:Tl screens). Using flat-field images the gain factors were determined by linear regression to be G=10.45 and 13.93 digital units per μGy for the 140 and 170 μm CsI:Tl screens.

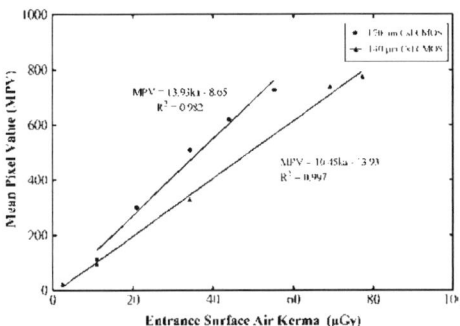

Fig. 1 STP curves for the RQA 5 beam quality.

Figure 2 shows measured in this study and previously published MTF curves of a CMOS sensor with 48 μm pixel pitch coupled to two CsI:Tl screen with thickness of 150 μm (the one optimized for high resolution-HR) [13], is also shown for comparison purposes.

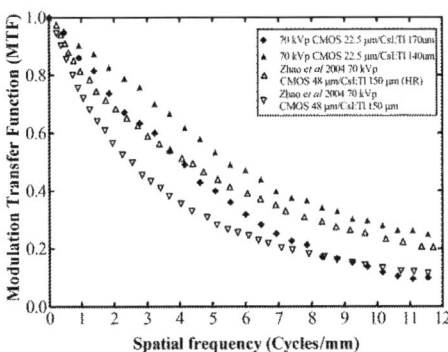

Fig. 2 Comparison of the MTFs of the CMOS/CsI:Tl sensor under investigation and a previously published CMOS/CsI:Tl sensor under the RQA 5 beam quality.

These curves have been also obtained under the RQA 5 beam quality. As it can be depicted from Fig. 2, the MTF of the 48 μm CMOS sensor [13] coupled to the 150-HR μm CsI:Tl screen is lower in the whole spatial frequency range than the MTF of the 22.5 μm CMOS sensor (investigated in the present study) coupled to the 140 μm CsI:Tl screen. This is due to the combined effects of screen thickness/sensor pixel size.

Figures 3 shows extracted 1D NNPS, for the u direction. NNPS curves are falling with increasing spatial frequency. This is attributed to the fact that in the CMOS system, a 2D correction is applied reducing the structured noise in each image. The rather slow variation of the NNPS may be attributed to the fact that MTF remains high up and in the higher frequency range. For comparison purposes, NNPS of a CMOS sensor measured in a previous study is also shown in Fig. 3, obtained at an exposure level of 31.05 μGy [13]. The noise levels of the CMOS sensor under investigation is lower than that of the previously published study [13], even though to the fact that the exposure levels used to obtain these curves are almost the same.

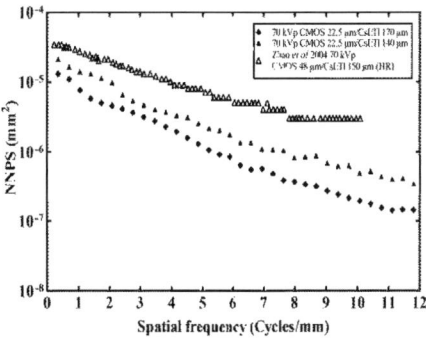

Fig. 3 Comparison of the NNPS curves between the CMOS/CsI:Tl sensor under investigation and a previously published CMOS/CsI:Tl sensor under RQA 5 beam quality.

Figure 4 show DQE curves of the sensor under investigation and for comparison purposes, DQE of the 48 μm pixel size CMOS sensor [17] (RQA 5) at an exposure level

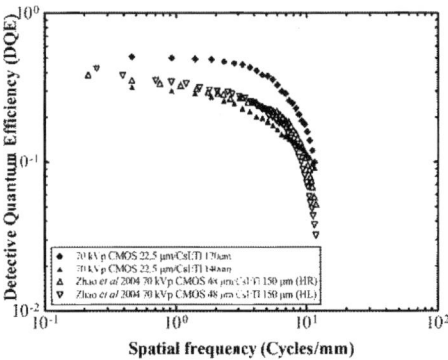

Fig. 4 Comparison of the DQE curves between the CMOS/CsI:Tl sensor under investigation and a previously published CMOS/CsI:Tl sensor under RQA 5 beam quality.

of 31.05 μGy. The DQE values of the previously published study [17] were lower in the whole spatial frequency range than those of the CMOS/170 μm CsI of this study and comparable with the corresponding CMOS/140 μm. This is mainly due to the screen thickness and secondary to the pixel size of the CMOS sensor.

IV. CONCLUSIONS

In the present study the imaging performance of a high resolution CMOS based imaging sensor combined with specially treated CsI screens was investigated in terms of MTF, NNPS and DQE in the general radiography energy range. Results showed that the CsI/CMOS sensor combination has better image quality parameters compared to previously published data Gd_2O_2S:Tb and CsI/CMOS sensor combinations. This phosphor/digital sensor combination could be used in digital imaging systems in order to reduce exposure to patients.

ACKNOWLEDGMENT

This research has been co-funded by the European Union (European Social Fund) and Greek national resources under the framework of the "Archimedes III: Funding of Research Groups in TEI of Athens" project of the "Education & Lifelong Learning" Operational Programme.

Authors wish to thank 'RMD A Dynasil Company' for providing the specially treated CsI screens.

REFERENCES

1. Rocha J G et al. (2010) Pixel Readout Circuit for X-Ray Imagers, IEEE Sensors 10(11):1740-1745.
2. Endrizzi M et al. (2013) CMOS APS detector characterization for quantitative X-ray imaging Nucl. Instr. and Meth. Phys. Res. A 703:26-32.
3. Valais I G et al. (2007) Luminescence properties of $(Lu,Y)_2SiO_5$:Ce and Gd_2SiO_5:Ce single crystal scintillators under x-ray excitation, for use in medical imaging systems IEEE Trans. Nucl. Sci. 54(1):11-18.
4. Bertolini M et al. (2012) A comparison of digital radiography systems in terms of effective detective quantum efficiency Med. Phys. 39(5):2617-2627.
5. Michail C M et al. (2009) A comparative investigation of Lu_2SiO_5:Ce and Gd_2O_2S:Eu powder scintillators for use in x-ray mammography detectors Meas. Sci. Technol. 20:104008.
6. Valais I et al. (2007) A systematic study of the performance of the CsI:Tl single-crystal scintillator under X-ray excitation Nucl. Instr. and Meth. Phys. Res. A. 571(1-2):343-345.
7. Nagarkar V V et al. (2004) High speed digital radiography using structured CsI screens Nucl. Instr. Meth. Phys. Res. B. 213:476.
8. Liu X and Shaw C C (2004) a-Si:H CsI Tl flat-panel versus computed radiography for chest imaging applications: image quality metrics measurement Med. Phys. 31(1):98-110.
9. Kocak F et al. (2011) Signal fluctuations in crystal-APD systems Nucl. Instr. and Meth. Phys. Res. A 648:128-130.
10. Woo T, Kim T (2012) Light collection enhancement of the digital x-ray detector using Gd_2O_2S:Tb and CsI:Tl phosphors in the aspect of nano-scale light Radiation Physics and Chemistry 81(1):12-15.
11. Tsuji K et al. (2010) X-ray Spectrometry Anal. Chem. 82:4950-4987.
12. Beylin D M et al. (2005) Study of the radiation hardness of CsI(Tl) scintillation crystals Nucl. Instr. and Meth. Phys. Res. A 541:501-515.
13. Zhao W et al. (2004) X-ray imaging performance of structured cesium iodide scintillators Med. Phys. 31(9):2594-2605.
14. Michail C M et al. (2011) Experimental and Theoretical Evaluation of a High Resolution CMOS Based Detector Under X-Ray Imaging Conditions IEEE Trans. Nucl. Sci. 58(1):314-322.
15. RMD A dynasil company, Watertown, MA [Online]. Available: http://rmdinc.com/about-us/dynasil-corporate-profile/
16. *Medical Electrical Equipment-Characteristics of Digital X-Ray Imaging Devices*, "Determination of the Detective Quantum Efficiency-Mammography Detectors" IEC, International Electrotechnical Commission, Geneva, Switzerland, 2005, IEC 62220-1-2.
17. Cho M K et al. (2008) Measurements of X-ray Imaging, Performance of Granular Phosphors With Direct-Coupled CMOS Sensors IEEE Trans. Nucl. Sci. 55(3):1338-1343.
18. Remote RadEye Systems Rad-icon Imaging Corporation a division of Dalsa, Sunnyvale, CA [Online]. Available: http://www.rad-icon.com/products-remote.php
19. Siemens Healthcare, [Online]. Available: http://www.medical.siemens.com/siemens/en_US/rg_marcom_FBAs/files/news/Sports_Special/AXIOM_Aristos_family_broshure.pdf
20. X-ray & Gamma Ray Detector, XR-100T-CdTe, Amptek, Bedford, MA [Online]. Available: http://www.amptek.com/xr100cdt.html
21. Michail C M et al. (2011) Evaluation of the Red Emitting Gd_2O_2S:Eu Powder Scintillators for use in Indirect X-Ray Digital Mammography Detectors IEEE Trans. Nucl. Sci. 50:2503-2511.
22. Greening J R (1985) Fundamentals of Radiation Dosimetry London, U.K.: Institute of Physics, pp. 56-64.
23. Marshall N W (2006a) A comparison between objective and subjective image quality measurements for a full field digital mammography system Phys. Med. Biol. 51:2441-2463.
24. Michail C M et al. (2010) Light emission efficiency and imaging performance of Gd_2O_2S:Eu powder scintillator under x-ray radiography conditions Med. Phys.37:3694-3703.

Author: Ioannis Valais
Institute: Technological Educational Institute (TEI) of Athens
Street: Ag. Spyridonos
City: Athens
Country: Greece
Email: valais@teiath.gr

Estimation of Contrast Agent Motion from Cerebral Angiograms for Assessing Hemodynamic Alterations Following Aneurysm Treatment by Flow Diversion

T. Benz[1], M. Kowarschik[1,2], J. Endres[3], P. Maday[1], T. Redel[2], and N. Navab[1]

[1] Computer Aided Medical Procedures (CAMP), Technische Universität München, Munich, Germany
[2] Angiography & Interventional X-Ray Systems, Siemens AG, Healthcare Sector, Forchheim, Germany
[3] Pattern Recognition Lab, Department of Computer Science, Friedrich-Alexander-Universität Erlangen-Nürnberg, Erlangen, Germany

Abstract—Quantitative assessment of hemodynamic changes induced by flow diversion could aid clinical decision making in the treatment of cerebral aneurysms by flow diversion. We propose to recover dense contrast motion fields from digital subtraction angiography (DSA) series by employing optical flow estimation based on conservation of mass and second-order regularization. In a retrospective study, metrics derived from the obtained motion fields are assessed. Promising initial results suggest further investigations in a clinical setting.

Keywords— cerebral aneurysm, flow diversion, digital subtraction angiography, optical flow.

I. INTRODUCTION

Cerebral aneurysms are abnormal dilatations of blood vessels in the brain that harbour the risk of rupture. Aneurysmal rupture causes subarachnoid hemorrhage, a major cause of morbidity and mortality throughout the world [1]. Several treatment options for the management of cerebral aneurysms are available today, including endovascular and open surgery techniques [2]. Flow diversion is an emerging therapeutic strategy that aims at restoring the natural healthy vessel anatomy by placing a very tightly woven stent (flow diverter) into the parent vessel across the orifice of the aneurysm [3]. Once in place, the flow diverter (FD) redirects the blood flow away from the aneurysm and back into the normal vessel, thereby inducing flow stagnation and promoting aneurysm thrombosis. In addition, the FD provides a scaffold for neointimal tissue overgrowth that eventually seals off the aneurysm from the main circulation and restores the vessel wall integrity.

During FD treatment the hemodynamic situation in the diseased vessel segment has to be appraised in order to decide whether the induced flow disruption is sufficient for aneurysm occlusion or if additional therapeutic measures have to be conducted. Current practice is to evaluate the efficacy of FD deployments by visual assessment of angiograms acquired during treatment. Such subjective analysis is hard to reproduce, not quantifiable and its predictive value cannot be validated easily. The development of image-based quantitative metrics that capture hemodynamic modifications induced by flow diversion therapy is thus desirable for advancing evidence-based medicine and implementing standard clinical guidelines.

II. RELATED WORKS

Several angiography-based approaches towards assessing hemodynamics in cerebral aneurysms have been proposed. In [4], a grading scheme for evaluating flow conditions in aneurysms treated with FDs is introduced. Such grading scheme improves upon a completely subjective analysis but is still only semi-quantitative. Other approaches rely on the analysis of parameters derived from curves representing the average gray value over time within a region of interest (ROI) of a DSA series [5]. Most curve parameter-based methods are objective and quantifiable but omit the analysis of direct hemodynamic parameters. Related to our approach, in [6] an optical flow (OF) [7] estimator similar to [8] is employed to recover contrast agent motion inside the aneurysm and on the parent artery. For normalization purposes, a 3DRA volume is used to obtain a volumetric flow estimate inside a ROI on the parent artery.

In this paper, we propose to estimate dense contrast motion fields from DSA series by employing an OF estimator that relies on the conservation of mass constraint and incorporates a regularization term that is suggested to be advantageous for recovering fluid flows in transmittance imagery [9]. To distil information from the blood velocity estimates, we propose several parameters that aim at capturing flow alterations induced by FD deployment.

III. THEORY AND METHODS

The dynamic behaviour of contrast agent in a DSA series can be assumed to be closely coupled to the flow of blood in the observed vasculature. To quantify blood flow based on DSA series, it seems thus appropriate to estimate contrast agent motion in the respective vessel segment. Obviously, spatial resolution constraints prohibit the tracking of individual particles. Given a modest injection rate, however, the contrast medium diluted in the bloodstream exhibits local brightness patterns whose apparent motions can be estimated by means of specific OF algorithms. This perceived motion can be regarded as being dependent on the actual 3D flow of blood. Therefore, estimating 2D velocity fields by means of OF algorithms can be used to gain insight into blood flow conditions in the diseased vessel segment.

OF estimators aim at recovering the perceived 2D motion of image features in a scene by computing a displacement field $\vec{u} = (u_1, u_2)^T$ of corresponding pixels from at least two subsequent images $I(\vec{x}, t)$ and $I(\vec{x}, t+1)$ of the scene. The classic 2D OF constraint assumes constant brightness of two corresponding pixels, i.e.

$$I(\vec{x}, t) = I(\vec{x} + \vec{u}, t+1). \tag{1}$$

By linearization through a first-order *Taylor series* expansion of the right-hand side of Eq. 1, we arrive at

$$I(\vec{x}, t) = I(\vec{x}, t) + \nabla I(\vec{x}, t) \cdot \vec{u} + \frac{\partial I(\vec{x}, t)}{\partial t}, \tag{2}$$

or in shorthand notation

$$I_t + \nabla I \cdot \vec{u} = 0, \tag{3}$$

where $\nabla I = (I_x, I_y)$ and $I_\bullet = \partial I / \partial \bullet$. In case of image sequences comprising well-defined objects and salient features the brightness constancy assumption is reasonable. In the context of recovering OF from transmittance imagery of fluid flow (e.g., DSA series), brightness constancy is easily violated. In this specific case, it was suggested to replace the OF data term (Eq. 3) by a more physically justified constraint based on conservation of mass [10, 8]. Indeed, the amount attenuation of a fluid is generally related to the density of a physical quantity dispersed in the fluid. In case of DSA series, the density of contrast agent particles represent this quantity. The contrast agent is advected by the blood flow, a transport process that respects the conservation of mass constraint. Thus, the continuity equation can be set up for the density ρ of contrast agent particles:

$$\frac{\partial \rho}{\partial t} + \nabla \cdot (\rho \vec{V}) = 0, \tag{4}$$

where $\nabla \cdot$ is the divergence operator and $\vec{V}$ is the 3D velocity field. The continuity equation in this form simply states that the temporal change in density is equal to the flux of density through the boundary of an infinitesimal volume. For image formation processes similar to X-ray imaging, it can be shown [10] that the continuity equation is still satisfied in 2D, when replacing the density ρ with the image intensity I:

$$I_t + \nabla \cdot (I\vec{u}) = 0, \tag{5}$$

with $\vec{u}$ being the density-weighted average along an imaging ray and assuming no normal flow out of the specimen [11]. In contrast to Eq. 3, Eq. 5 represents a physically justified data term that is suitable for recovering fluid flow from angiographic imagery.

For each pixel, Eq. 5 needs to be solved for two unknowns (u_1 and u_2), rendering it ill-posed. In order to obtain a well-posed problem, additional constraints need to be considered. Typically *spatial coherence* is assumed, meaning neighbouring pixels share the same velocity. This implies a smoothly varying motion field estimate. Regularization in variational optimization schemes is often conducted by searching for solutions that minimize the continuous energy functional

$$\int_\Omega \|\nabla u_1\|^2 + \|\nabla u_2\|^2 dx dy \tag{6}$$

on the image domain Ω. It can be shown [12], that the minimization of (6) is equivalent to the minimization of

$$\int_\Omega \operatorname{div}(\vec{u})^2 + \operatorname{curl}(\vec{u})^2 dx dy, \tag{7}$$

a first-order div-curl regularizer, where $\operatorname{div}(\vec{u}) = \nabla \cdot \vec{u} = \partial u_1/\partial x + \partial u_2/\partial y$ and $\operatorname{curl}(\vec{u}) = \omega_3 = \partial u_2/\partial x - \partial u_1/\partial y$, i.e. the z-component of $\vec{\omega} = \nabla \times (u_1, u_2, 0) = (0, 0, \omega_3)$. Therefore, OF estimators that minimize (6) (e.g., [8]) penalize non-laminar flow, where $\operatorname{div}(\vec{u}) \neq 0$ and $\operatorname{curl}(\vec{u}) \neq 0$. In the case of fluid flow, where regions with highly non-laminar flows can be as prevalent as regions with laminar flows, this is likely disadvantageous. In this work, we therefore follow [9], proposing the use of a second-order div-curl regularizer [13]

$$\int_\Omega \|\nabla \operatorname{div}(\vec{u})\|^2 + \|\nabla \operatorname{curl}(\vec{u})\|^2 dx dy \tag{8}$$

in addition to using the data term given in Eq. 5. In order to preserve natural flow discontinuities, a soft penalty function is employed that allows for local deviations from the constraints imposed upon the flow field. To ease numerical computation, auxiliary variables are introduced, and to address large displacements between consecutive frames, optimization is embedded within a multiresolution scheme [9].

In the following, $\vec{u} = \vec{u}(\vec{x}, t)$ is the vector field representing the estimated OF between a DSA frame at time t and a consecutive frame at time $t+1$ ($t = 1, \ldots, N-1$) at a pixel location $\vec{x}$. Given a ROI delineated in each frame of a DSA series and a velocity estimate $\vec{u}$ corresponding to the DSA series, the average velocity magnitude (speed) within the ROI over a certain time window was computed as

$$\mathbf{V}_{\text{avg}}(\vec{u}) = \frac{1}{|T||ROI|} \sum_{t \in T} \sum_{\vec{x} \in ROI} \|\vec{u}\|, \tag{9}$$

where *ROI* was a set of pixels that were covered by the selected ROI and $T \subseteq \{1,\ldots,N-1\}$. A temporally varying ROI is of interest if the anatomy that is covered by the ROI changes significantly in between frames. In the same manner, metrics for the median, $\mathbf{V}_{\text{med}}$, and maximum speed, $\mathbf{V}_{\text{max}}$, were derived.

Two more quantities, the average angular velocity magnitude (**AV**) and the average divergence (**D**), could be derived as per the above setting. Vorticity is a measure of the *local rotation* (or spin) of a fluid and is the curl of the corresponding velocity field (here $\vec{u}$). Thus, the average angular velocity magnitude was defined as

$$\mathbf{AV}(\vec{u}) = \frac{1}{|T||ROI|} \sum_{t \in T} \sum_{\vec{x} \in ROI} |\text{curl}(\vec{u})|, \qquad (10)$$

The average divergence, **D**, was defined analogously.

Based on the five parameters $\mathbf{V}_{\text{avg}}$, $\mathbf{V}_{\text{med}}$, $\mathbf{V}_{\text{max}}$, **AV**, and **D**, ratios of the same quantities obtained from DSA series acquired before and after FD treatment were derived. In case of $\mathbf{V}_{\text{avg}}$ this post-pre ratio, $\mathbf{VR}_{\text{avg}}$, was

$$\mathbf{VR}_{\text{avg}} = \frac{\mathbf{V}_{\text{avg}}^{\text{post}}}{\mathbf{V}_{\text{avg}}^{\text{pre}}}, \qquad (11)$$

where $\mathbf{V}_{\text{avg}}^{\text{pre}}$ and $\mathbf{V}_{\text{avg}}^{\text{post}}$ are the average velocity magnitude in a ROI derived from a DSA series acquired before (*pre*) and after (*post*) treatment (cf. Eq. 9), respectively. The two ROIs are assumed to be covering the same anatomy before and after treatment. Furthermore, the time frames chosen for the computation of the pre- and post-treatment parameter need to correspond to each other. For instance, the whole series or a certain number of heart cycles can be considered in both cases. For the remaining five parameters, post-pre ratios $\mathbf{VR}_{\text{med}}$, $\mathbf{VR}_{\text{max}}$, **AVR**, and **DR**, were defined in an equal manner.

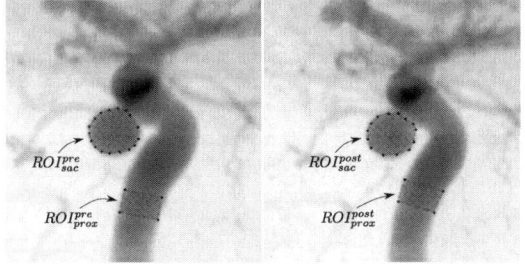

Fig. 1 Exemplary selection of ROIs in maximum opacification plots of DSA series acquired before (*pre, left*) and after (*post, right*) FD deployment.

By delineating two different ROIs in the same DSA series, one covering the inside of the aneurysm and one covering parts of the parent artery, proximal to the aneurysm (cf. Fig. 1), *inflow normalized* post-pre ratios could be computed. In case of $\mathbf{VR}_{\text{avg}}$ this equated to

$$\mathbf{VR}_{\text{avg}}^{\text{norm}} = \mathbf{VR}_{\text{avg}}^{\text{sac}} \cdot \left(\mathbf{VR}_{\text{avg}}^{\text{prox}}\right)^{-1}. \qquad (12)$$

In the same manner, normalized ratios were defined for $\mathbf{VR}_{\text{med}}$ and $\mathbf{VR}_{\text{max}}$, yielding $\mathbf{VR}_{\text{med}}^{\text{norm}}$ and $\mathbf{VR}_{\text{max}}^{\text{norm}}$, respectively.

Using the techniques described above, a first retrospective feasibility study was conducted. We obtained and evaluated OF estimates of 17 pre-post pairs of DSA series of cerebral aneurysms (cases A1 to A17; 34 DSA series in total). ROIs were delineated manually and it was taken care of that the same anatomic regions were covered by the ROIs in the DSA series acquired before and after treatment, respectively. Furthermore, the ROI on the parent artery was chosen to cover a vessel segment that exhibited little flow perpendicular to the imaging plane. Three pairs of series (A1, A2, A17) were acquired with a frame rate of 30 fps, the rest (A3-A16) with 15 fps. Case A17 corresponded to a synthetic data set [14]. All series were spatially cropped to a rectangular region around the aneurysm in which OF estimation was conducted. The pairs A13 and A14 were extracted from the same biplane DSA runs.

IV. RESULTS AND DISCUSSION

OF estimation was successfully conducted in all cases. In the parent artery, the estimated flow fields were mostly smooth and dominated by laminar flow components. Inside the aneurysm, the estimated flow fields were often highly non-laminar and converging and diverging flow areas as well as rotational flows could be observed, with sharp boundaries in between. Regions in DSA frames where the spatial gradient information was low due to a homogeneous distribution of contrast medium, turned out problematic. Incorporating explicit regularization for flow in proximity to vessel walls could prove advantageous in such situations. In general, however, the estimated motion seemed to be consistent with the apparent motion of contrast agent patterns.

Evaluation of the proposed metrics indicated, on average, a clear reduction of the average estimated blood speed inside the aneurysm, after FD deployment: $\overline{\mathbf{VR}_{\text{avg}}} = 80\% \pm 16\%$ (cf. Fig. 2). Inflow normalization markedly affected metrics related to estimated velocity magnitudes (e.g., $\overline{\mathbf{VR}_{\text{max}}} = 111\% \pm 32\%$, $\overline{\mathbf{VR}_{\text{max}}^{\text{norm}}} = 0.98\% \pm 0.25\%$). Speed metrics computed from DSA series corresponding to the same biplane DSA acquisition were in very good agreement, indicating a certain robustness of the proposed parameters towards viewpoint variations. In conclusion, the speed-based metrics yielded reassuring results with, considering the diversity of the data set and the retrospective nature of the study,

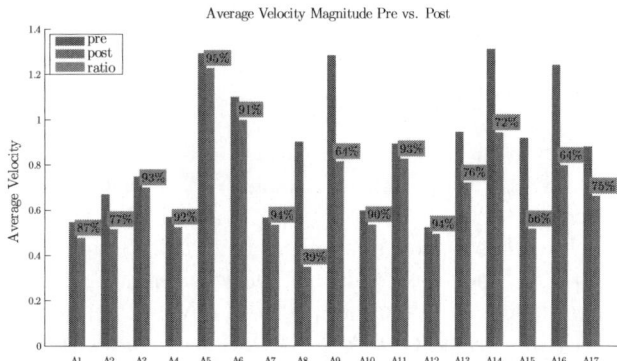

Fig. 2 Average estimated velocity magnitude, V_{avg}, in the aneurysm before (pre, *red*) and after (post, *blue*) FD treatment, and the corresponding post-pre ratio VR_{avg} (ratio, *green*).

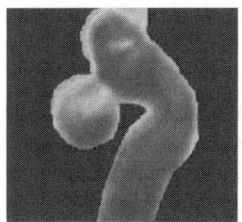
(a) Case A16 pre.

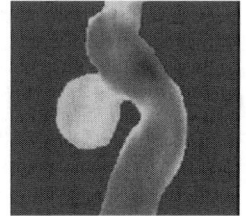
(b) Case A16 post.

Fig. 3 Average estimated velocity magnitude plots of case A16. Both images were normalized with respect to a ROI on the parent artery, proximal to the aneurysm. VR_{avg}^{norm} was 74% in this case.

reasonably low variance. For a comparative visualization of, for instance, the $\overline{VR_{avg}^{norm}}$ parameter, it seems suitable to, for each pixel, color code the average estimated velocity magnitude over the whole DSA series (cf. Fig. 3). Evaluation of the rotation and divergence metrics was not as clear-cut. On average, the results with respect to the **AVR** parameter were within reasonable bounds but strongly affected by outliers ($\overline{AVR} = 106\% \pm 121\%$, median 59%). We reckon that a more precise spatial confinement of the considered ROI may pose an improvement for the $\overline{AVR}$ quantity. The distribution of results obtained by **DR** was even broader, with only the median being inconspicuous ($\overline{DR} = 384\% \pm 765\%$, median 101%).

V. CONCLUSION

We proposed to employ second-order div-curl regularization for the recovery of contrast motion in DSA series and presented metrics for assessing hemodynamic changes induced by flow diverter treatment of cerebral aneurysms. A first feasibility study warrants future research efforts to establish the validity of the proposed quantities in a clinical context.

ACKNOWLEDGEMENTS

We would like to give special thanks to Dr. Saruhan Çekirge, Hacettepe University Hospital, Ankara, Turkey, and Dr. Demetrius Lopes, Rush University Medical Center, Chicago, Illinois, U.S.A.

REFERENCES

1. Suarez JI, Tarr RW, Selman WR. Aneurysmal subarachnoid hemorrhage *New England Journal of Medicine.* 2006;354:387–396.
2. Wiebers DO. Unruptured intracranial aneurysms: natural history, clinical outcome, and risks of surgical and endovascular treatment *The Lancet.* 2003;362:103–110.
3. Pierot L. Flow diverter stents in the treatment of intracranial aneurysms: Where are we? *Journal of Neuroradiology.* 2011;38:40–46.
4. Grunwald IQ, Kamran M, Corkill RA, et al. Simple measurement of aneurysm residual after treatment: the SMART scale for evaluation of intracranial aneurysms treated with flow diverters *Acta Neurochirurgica.* 2012;154:21–26.
5. Struffert T, Ott S, Kowarschik M, et al. Measurement of quantifiable parameters by time-density curves in the elastase-induced aneurysm model: first results in the comparison of a flow diverter and a conventional aneurysm stent *European Radiology.* 2013;23:521–527.
6. Pereira VM, Bonnefous O, Ouared R, et al. A DSA-Based Method Using Contrast-Motion Estimation for the Assessment of the Intra-Aneurysmal Flow Changes Induced by Flow-Diverter Stents *American Journal of Neuroradiology.* 2013 (Epub ahead of print).
7. Horn BKP, Schunck BG. Determining optical flow *Artificial Intelligence.* 1981;17:185–203.
8. Wildes RP, Amabile MJ, Lanzillotto AM, Leu TS. Recovering estimates of fluid flow from image sequence data *Computer Vision and Image Understanding.* 2000;80:246–266.
9. Corpetti T, Mémin É, Pérez P. Dense estimation of fluid flows *Pattern Analysis and Machine Intelligence.* 2002;24:365–380.
10. Fitzpatrick JM. The existence of geometrical density-image transformations corresponding to object motion *Computer Vision, Graphics, and Image Processing.* 1988;44:155–174.
11. Liu T, Shen L. Fluid flow and optical flow *Journal of Fluid Mechanics.* 2008;614:1.
12. Corpetti T, Mémin É, Pérez P. Dense fluid flow estimation *IRISA, Technical Report No. 1352.* 2000.
13. Suter D. Motion estimation and vector splines in *IEEE Computer Vision and Pattern Recognition*;6:939–942 1994.
14. Endres J, Redel T, Kowarschik M, Hutter J, Hornegger J, Doerfler A. Virtual angiography using CFD simulations based on patient-specific parameter optimization in *IEEE International Symposium on Biomedical Imaging*;9:1200–1203 2012.

Author:	Tobias Benz, M.Sc.
Institute:	Technische Universität München, Department I-16
Street:	Boltzmannstr. 3
City:	85748 Garching (Munich)
Country:	Germany
Email:	benzt@in.tum.de

Graphical User Interface for Breast Tomosynthesis Reconstructions: An Application Using Anisotropic Diffusion Filtering

A. Malliori[1], K. Bliznakova[2], A. Daskalaki[1], and N. Pallikarakis[1]

[1] Biomedical Technology Unit, Dept of Medical Physics, University of Patras, Greece
[2] Department of Electronics and Microelectronics, Technical University of Varna, Bulgaria

Abstract—Breast tomosynthesis is a three-dimensional x-ray imaging technique of the breast that on the basis of a set of angular projections taken over a limited arc around the breast can provide tomographic images at variable heights. Nowadays, breast tomosynthesis is Food and Drug Administration (FDA) approved for use in breast cancer screening as there are few clinical systems available and under development for use in patient examinations. In parallel with the development of breast tomosynthesis systems, optimization work is actively performed, related to image acquisition geometries and reconstruction algorithms in general. The Biomedical Technology Unit at the University of Patras, Greece, has a 25 years experience in developing and optimizing image acquisition geometries and reconstruction algorithms with focus on the specific field of breast tomosynthesis. As a result, reconstruction algorithms combined with different filters that can be used prior or after the reconstruction of the image in order to remove the noise have been developed. To facilitate further research in this field, we developed a platform with a Graphical User Interface dedicated to Breast Tomosynthesis. In this study, the platform has been exploited to optimize the parameters of a nonlinear anisotropic diffusion filter used in a breast tomosynthesis application.

Keywords—reconstruction algorithms, breast tomosynthesis, graphical user interface, anisotropic diffusion filtering.

I. INTRODUCTION

Digital Tomosynthesis is a tomographic technique that permits acquisition of a limited number of projections over a limited arc with dose levels not exceeding those of conventional mammography. The two-dimensional projections can provide three-dimensional information of the imaged object using generally simple reconstruction algorithms [1].

The Biomedical Technology Unit at the University of Patras, is well established in the field of tomosynthesis and has been recognized as one of the pioneers of this technique with broad research activities [2-4]. The last decade, we focused our research on the development and optimization of acquisition geometries and reconstruction algorithms for breast tomosynthesis [5,6]. In case of breast, tomosynthesis has proved the potential to extract additional diagnostic information in comparison to conventional mammographic images [7]. A limited number of x-ray breast images is acquired over a limited angular range. These images are then used with tomosynthesis reconstruction algorithms to produce three-dimensional structural information of the breast in which the influence of overlapping tissues is greatly reduced and the accuracy of cancer detection is increased, particularly for masses in dense breasts [8]. As a result of our investigation on tomosynthesis, several algorithms used to reconstruct images acquired at different geometries have been developed. Our latest study was related to the optimization of tomosynthesis scanning geometry and algorithms for volume reconstruction in case of breast towards improved visualization of breast features: masses and microcalcifications [9]. To facilitate further research we combine the available algorithms under a Graphical User Interface (GUI), dedicated for research in breast tomosynthesis. This paper reports on the design, implementation and use of a GUI dedicated for research in breast tomosynthesis.

II. MATHERIALS AND METHODS

A. Graphical Interface Breast Tomosynthesis Platform

A block-diagram of the Breast Tomosynthesis (BT) platform that has been developed in Matlab (The MathWorks, Inc.) is shown in Fig.1.

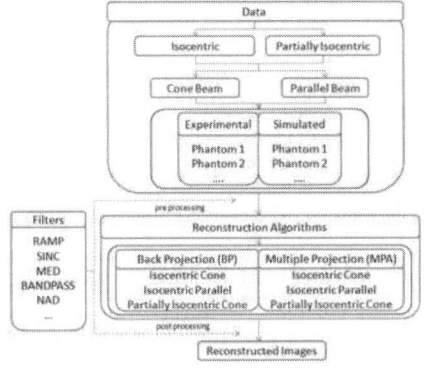

Fig.1 Block-diagram of the BT platform

In short the GUI allows reconstruction of breast images that are simulated or experimentally acquired with both isocentric and partial isocentric rotation (stable detector) and parallel/cone beam. The *Reconstruction Algorithm* field

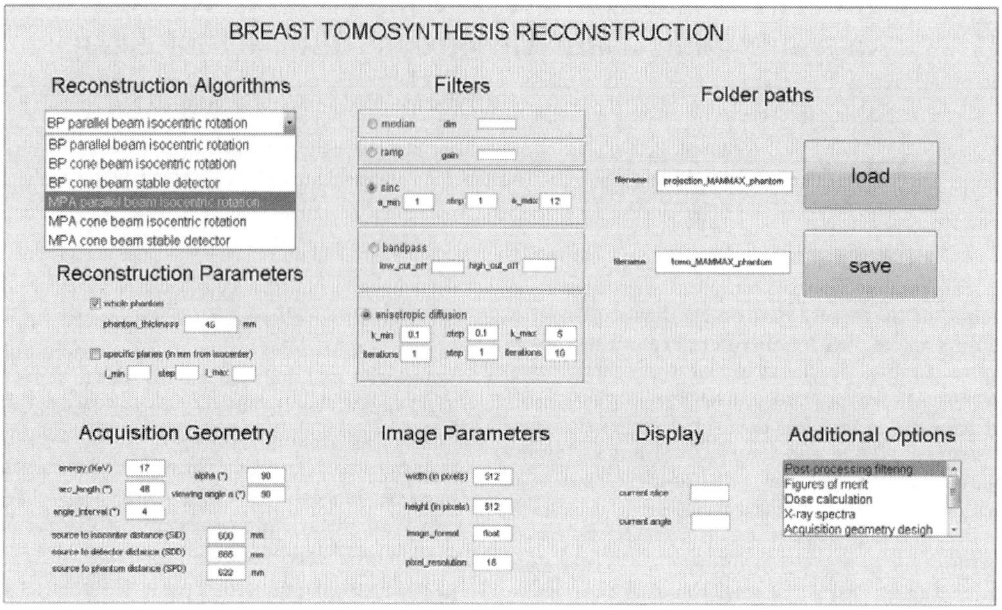

Fig. 2 Graphical User Interface of the BT platform The platform may easily be extended to incorporate more reconstruction algorithms and filters and consists of four parts: Data, Filters, Reconstruction Algorithms and Reconstructed Images. The BT interface (Fig. 2) enables the easy and fast import of acquisition and reconstruction parameters, the selection of the reconstruction algorithm and filtering method.

contains the Back Projection and the Multiple Projection Algorithm (MPA) [2] for isocentric rotation (parallel and cone beam) and for partially isocentric rotation (cone beam). These reconstruction algorithms may be combined with any of the five filters – median, ramp, sinc, bandpass, nonlinear anisotropic diffusion, available for pre-processing of projection images or post-processing of reconstructed tomograms. The first four filters were previously reported in studies that concerned the breast tomosynthesis using simulated and experimental data acquired at synchrotron facilities [9, 10]. The nonlinear anisotropic diffusion (NAD) filter has been developed in the framework of this study and described in next section.

In the field *Additional Options*, post-processing filtering, image evaluation using different figures of merit, image visualization and processing, dose calculation etc. can be accomplished.

B. Use of GUI for Breast Tomosynthesis Research

The developed GUI has been used in the current study with the aim to optimize a NAD filter combined with MPA for visualizing low contrast breast features, such as masses. For this purpose the NAD filter has been developed based on the diffusion function (Eq. 1) as proposed by Perona and Malik [11] and applied on breast projection images.

$$c(x,y) = \exp\left[-0.5\left(\frac{|\Delta u|}{\kappa}\right)^2\right] \quad (1)$$

Nonlinear anisotropic diffusion filtering is based on nonlinear evolution partial differential equations u, and seeks to improve images by removing noise while preserving details and even enhancing edges.

Several combinations of the filter (contrast) parameter κ for different number of iterations were tested in order to reduce the noise in the image and at the same time improve the visualization of the feature of interest. The feature was a 6 mm low-contrast mass inside a 4.5 cm thick breast phantom with highly heterogeneous background. 13 projection images were acquired for an arc length of 48°, at ELETTRA Synchrotron facilities, using a monochromatic beam of 17 keV. The reconstruction algorithm used on the filtered projections was MPA for parallel beam isocentric rotation.

The tomosynthesis images obtained with this filter were compared to tomosynthesis images obtained with a ramp and a 3 by 3 median filter. Further, the image quality was evaluated in terms of Contrast to Noise Ratio (CNR), Contrast (C) as well as profiles taken along the studied mass.

The CNR and C were calculated as:

$$CNR = \frac{|N_{feature} - N_{bkg}|}{\sigma_{bkg}} \quad \text{and} \quad C = \frac{|N_{feature} - N_{bkg}|}{|N_{feature} + N_{bkg}|} \quad (2)$$

In these equations, $N_{feature}$ is the mean pixel value of a circular region of interest (ROI) of diameter 5mm inside the feature, while N_{bkg} is the mean pixel value calculated for a large ROI in the background area, and σ_{bkg} is the standard deviation of this background ROI.

III. RESULTS

The developed GUI was used to test and optimize the NAD filter prior to MPA reconstruction. Figure 3 shows the CNR and C graphs of the 6mm mass for different κ values (0.01-0.7) and number of iterations (5-40).

For the feature under evaluation, there was a progressive improvement observed both visually and quantitatively for κ values in the range of 0.01-0.1 and then a much slower improvement up to κ=0.5. This value was selected for κ, as it resulted in optimal mass visibility and maximum CNR and C measurements (Fig. 3). Beyond this κ value, there was no further improvement in the figures of merit or the feature visualization. A subsequent increase in CNR and C was also observed with increasing the number of iterations. A selection of 30 iterations resulted in optimal mass visibility and a good trade between CNR and C values. Further increase of iterations (iter>30), caused blurring of the image which was visually observed and quantitatively demonstrated by a reduction of the image C [Fig. 3 (b)].

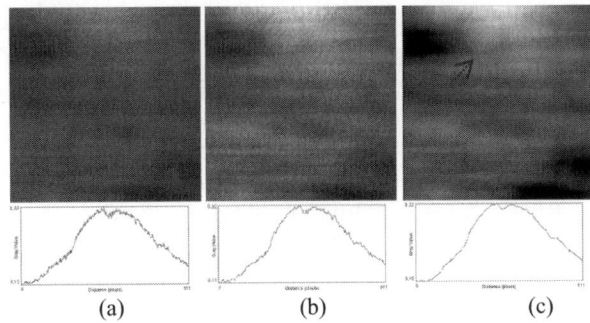

Fig.4 Tomosynthetic images reconstructed with (a) simple (no filter applied) MPA, (b) filtered with a median filter (FMPA_M) and (c) with NAD (FMPA_NAD) and the corresponding profiles of the 6mm mass.

while for the case of FMPA_NAD_S, where a sinc filter (Eq. 3) was used after NAD, the optimal α parameter reported in our previous work, i.e. α=12 was considered.

$$F(\omega_x) = \left|\frac{2}{\alpha}\sin\left(\frac{\alpha\omega_x}{2}\right)\right|\left(\frac{\sin\left(\frac{\alpha\omega_x}{2}\right)}{\frac{\alpha\omega_x}{2}}\right)^2 \quad (3)$$

Highest CNR was achieved using F_MPA_NAD and FMPA_NAD_S, while the later resulted into an improvement of the contrast value by 3 times compared to simple MPA (Table 1). As expected, the CNR of the low contrast feature is lowest for the case of using ramp prior to MPA. Ramp is a high-pass filter that can enhance the edges and increase contrast but also allows the high frequency noise in the image to remain while median is a smoothing filter that can improve the visualization of low contrast features such as breast masses. Sinc filter can be adjusted to behave as high- or low-pass and when used along with NAD can remove the noise while preserving the features' boundaries.

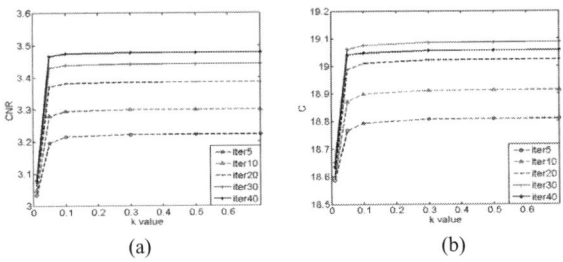

Fig. 3 CNR and C graphs of the 6mm mass for different combination of the parameters of the NAD filter.

Figures 4(a-c) show tomosynthesis slices of the breast mass, reconstructed with MPA applied for unfiltered [Fig. 4(a)] and filtered projection images and the corresponding mass profiles below each image. Figure 4(b) depicts a tomogram obtained with filtered projection images using a median filter (FMPA_M) while Fig. 4(c) using nonlinear anisotropic diffusion filtering (FMPA_NAD) for the selected values of κ=0.5 and 30 iterations.

The subjective comparison shows that the FMPA_NAD demonstrates better low contrast appearance and detection compared to the other two. An improved mass profile was also achieved [Fig. 4(c)] indicating noise reduction and a good combination of image smoothing and sharp feature boundaries.

A comparison between CNR and C values, calculated for the 6mm low-contrast feature reconstructed with the use of different filters is summarized in Table 1. For FMPA_NAD the optimized values of κ=0.5 and 30 iterations were used

Table 1 CNR and C values for the 6mm low contrast feature using different filtering methods in combination with MPA.

Filtering method	CNR	C
- (simple MPA)	2.64	18.15
Ramp (FMPA_R)	0.10	41.15
Median (FMPA_M)	3.09	18.60
Anisotropic Diffusion (FMPA_NAD)	3.44	19.08
Anisotropic Diffusion and sinc (FMPA_NAD_S)	3.25	63.13

Figure 5 shows a set of tomograms of the mass reconstructed with MPA and filtered with ramp and sinc using α=6 and α=12 (first row) and combinations of using NAD filter prior to ramp and sinc (FMPA_NAD_R, FMPA_NAD_S) with 5 iterations (second row) and 30 iterations (third row), respectively. NAD carried out with more processing iterations (third row) results in smoothing of the image and suppression of the superimposed background. The combination with sinc filter (FMPA_NAD_S) improves the visualization of the mass under evaluation and enables better detection of a second mass with much lower x-ray contrast that appears on the right of the mass under evaluation (Fig. 5(c), third row).

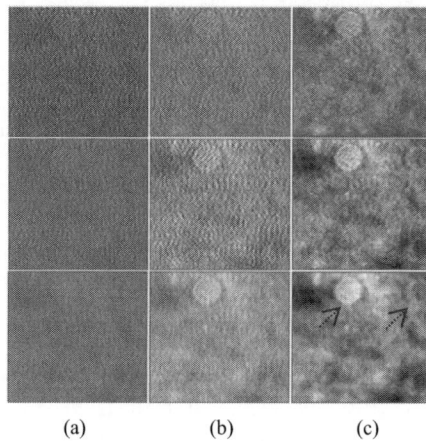

Fig. 5 Tomosynthetic images of the 6 mm mass filtered with: (a) ramp filter (first row), ramp filter after NAD with $\kappa=0.5$ and 5 iterations (second row), ramp filter after NAD with $\kappa=0.5$ and 30 iterations (third row), (b) sinc filter with $\alpha=6$ (first row), sinc filter with $\alpha=6$ after NAD with $\kappa=0.5$ and 5 iterations (second row), sinc filter with $\alpha=6$ after NAD with $\kappa=0.5$ and 30 iterations (third row), (c) sinc filter with $\alpha=12$ (first row), sinc filter with $\alpha=12$ after NAD with $\kappa=0.5$ and 5 iterations (second row), sinc filter with $\alpha=12$ after NAD with $\kappa=0.5$ and 30 iterations (third row).

This study was facilitated by the use of the GUI. By exploring all possible values for the parameters describing the NAD filter, we were able for a short period to find the optimal values of the filter and to apply alone or in combination with other filters prior to the reconstruction resulting, namely in improved visualization of large low contrast breast features.

IV. CONCLUSIONS

Nonlinear anisotropic diffusion filtering can improve image quality by removing noise while preserving details and even enhancing edges of the lesions which is critical for BT since its major advantage lies in the improvement of the detection and characterization of such low-contrast masses, particularly when they are spiculated.

This application reveals the usefulness and further developing and exploring of this GUI that allows access and exploitation of tomosynthesis algorithms for reconstruction of breast projection images obtained at different imaging conditions. Currently, the platform is used to evaluate reconstruction techniques and filtering methods applied on simulated and experimental cone beam data acquired with polychromatic spectra.

ACKNOWLEDGMENT

This research has been co-financed by the European Union (European Social Fund-ESF) and Greek national funds through the Operational Program "Education and Lifelong Learning" of the National Strategic Reference Framework (NSRF)-Research Funding Program: Heracleitus II.

REFERENCES

1. Niklason L T, Christian B T, Niklason L E et al. (1997) Digital tomosynthesis in breast imaging. Radiology 205:399–406
2. Kolitsi Z, Panayiotakis G, Anastassopoulos V et al. (1992) A multiple projection method for digital tomosynthesis. Med. Phys. 19(4):1045-1050
3. Kolitsi Z, Panayiotakis G, and Pallikarakis N (1993) A method for selective removal of out-of-plane structures in digital tomosynthesis. Med. Phys. 20:47-50
4. Badea C, Kolitsi Z, Pallikarakis N (2001) Image quality in extended arc filtered digital tomosynthesis Acta Radiol 42:244-248
5. Bliznakova K, Kolitsi Z, Speller R D et al. (2010) Evaluation of digital breast tomosynthesis reconstruction algorithms using synchrotron radiation in standard geometry. Med. Phys. 37:1893-1903
6. Malliori A, Bliznakova K, Pallikarakis N (2010) A computer-based platform for Digital breast Tomosynthesis simulation studies, Proceedings of the IEEE/EMBS Region 8 International Conference on Information Technology Applications in Biomedicine, ITAB, Corfu, Greece, 2010
7. Dobbins J T, Godfrey D J (2003) Digital x-ray tomosynthesis: current state of the art and clinical potential. Phys Med Biol. 48:R65-106
8. Gong X, Glick S J, Liu B et al. (2006) A computer simulation study comparing lesion detection accuracy with digital mammography, breast tomosynthesis, and cone-beam CT breast imaging. Med. Phys. 33:1041–1052
9. Malliori A, Bliznakova K, Dermitzakis A et al. (2013) Evaluation of the effect of acquisition parameters on image quality in digital breast tomosynthesis: Simulation studies, IFMBE Proc. vol. 39, World Congress on Med. Phys. & Biomed. Eng., Beijing, China, 2012, pp. 2211-2214
10. Malliori A, Bliznakova K, Speller R D et al. (2012) Image quality evaluation of breast tomosynthesis with synchrotron radiation. Med. Phys. 39 (9):5621-5634
11. Perona P, Malik J (1987) Scale Space and Edge Detection Using Anisotropic Diffusion, Proc. IEEE Comp. Soc. Workshop on Computer Vision, Miami Beach, 1987, IEEE Computer Society Press, Washington, pp. 16 – 22

Author: Anthi Malliori
Institute: University of Patras
Street: Rio
City: Patra
Country: Greece
Email: anmall@upatras.gr

Image Analysis of Breast Reconstruction with Silicone Gel Implant

A. Daskalaki[1], K. Bliznakova[2], and N. Pallikarakis[1]

[1] University of Patras/Biomedical Technology Unit, Dept of Medical Physics, Patras, Greece
[2] University of Varna/Department of Electronics and Microelectronics, Varna, Bulgaria

Abstract—The successful use of silicone gel implants and their acceptance from the medical community and the increased number of surgeries for breast reconstruction, necessitate a close examination of their effects in mammographic image quality. The insertion of silicone gel implants, because of their higher atomic numbers and attenuation coefficient compared with breast tissue, obscure a large area of the image acquired. Such an effect can reduce the ability of early detection of breast cancer. The objective of this work is to examine the effects which come along with a silicone gel insertion in the detection of cancer lesions using mammography. The study was performed with three software phantoms with breast equivalent tissue and silicone gel material, using Monte Carlo X-ray imaging simulations. From the images acquired the Contrast to Noise Ratio (CNR) of the main mammographic findings has been analyzed. Furthermore, the effect of increasing the incident beam energy, as well as the implant thickness on the visibility of breast lesions and the image quality of the complete mammogram have been studied. Results showed a major reduction of visibility in the area of the implant and adjacent to it as well as a reduction of CNR values of about 50%.

Keywords—silicone gel implant, digital tomosynthesis, mammography, Monte Carlo simulation.

I. INTRODUCTION

Women with silicone breast implants undergo X-ray imaging to detect early breast cancer just as women without breast implants. Several characteristics of implants and the techniques of their placement affect imaging evaluations. The presence of silicone gel-filled implants may interfere with standard mammography since silicone is radiopaque, and the physical presence of the implant compresses fat and glandular tissues, creating more homogeneous dense tissue that frequently lacks the contrast needed to detect subtle early features associated with breast cancer. Cohen et al. [1] concluded that silicone gel implants obscure portions of the breast. In one of the first actual estimates, Wolfe [2] reported breast tissue non visibility in the presence of silicone gel implants at about 25%. Silverstein et al. [3] calculated the area of mammographically visualized breast tissue before and after augmentation mammoplasty. Anterior breast tissue was seen better with displacement mammography while posterior breast tissue was seen better with compression mammography. Moreover in the case of implant insertion more tissue may be imaged in a smaller space that causes more superposition of structures, resulting in poor image quality.

Typically, prostheses used for breast reconstruction and augmentation contain a material with higher atomic number (Z) than breast tissue and this may affect the image quality and the dose distribution within the breast. Silicone filled implants have a low x-ray transmission in the region of the implant, which is reported to interfere with the detection of small masses in the breast. Artur et al. [4] refer to the problem resulting from using silicone gel to fill mammary prostheses and additionally reports that the detection of tumors adjacent to silicone gel implants requires special mammography procedures.

The objective of this work is to investigate the effects which come along with a silicone gel insertion. In order to perform our research we modeled three breast tissue equivalent phantoms with silicone gel prosthesis. Mammography images were simulated with Monte Carlo code.

II. MATTERIAL AND METHODS

A. Phantom Composition

Three mathematical breast tissue equivalent phantoms were created. The first two phantoms, shown in Fig. 1, are rectangular parallelepipeds with dimensions 50 (width) x 100 (length) x 45 mm (thickness), filled with a mixture of 50% adipose tissue and 50% glandular tissue. Artifacts which simulate the main findings in mammography such as microcalcifications and breast masses were inserted into the base material. These were two sets of 6 $CaCO_3$ spheres of radius 0.138 mm and 0.4 mm and a set of three water spheres of radius 1mm, 1.5mm and 2mm respectively. The implant was modeled in the form of a semi-ellipsoid with dimensions 20x46x20 mm filled with silicone gel ($CH_3[Si(CH_3)O]_4Si(CH_3)_3$). The phantom shown on the left side in Fig. 1 does not contain implant while the one on the right is modeled with silicone gel. Both phantoms are voxel based and have been created with the XRayImagingSimulator [5] with a voxel resolution of 0.2mm.

The third model is a step-wedge phantom, shown in Fig. 2, composed of 18 adjacent cuboids, modeled from a silicone gel, with a thickness in the range from 2mm to 36mm.

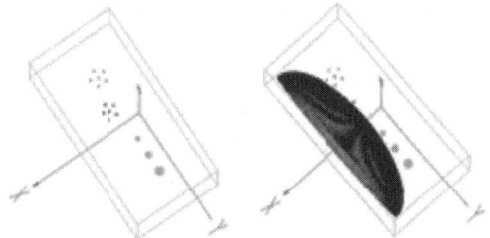

Fig. 1 The first phantom on the left with the breast equivalent tissue, two groups of microcalcifications and one group of three masses, on the right the second phantom with the silicone gel insertion.

The step wedge is placed in a 40 mm thick block, filled with a mixture of 50% adipose and 50% glandular tissue. A small CaCO$_3$ sphere of 0.2mm radius has been placed at a distance of 2mm under each cuboid. The phantom has been converted to voxel-based with a voxel resolution of 0.2mm.

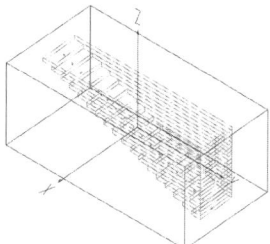

Fig. 2 The third phantom with breast equivalent tissue, different cuboid thicknesses of silicone gel and 18 microcalcifications placed under each one.

B. Image Acquisition

Mammographic projection images of the three phantoms were simulated with the XRayImagingSimulator. Projection images of size 770×770 pixels and resolution 49 pixel/mm^2 were simulated with source to isocenter distance (SID) 600mm, source to detector distance (SDD) 650mm. In addition, a breast tomosynthesis acquisition was also simulated. For this purpose, 25 projections acquired at 2^0 increments in an acquisition arc of 50^0, were generated. Tomosynthesis images were obtained by reconstructing preprocessed with a ramp filter projection images by Multiple Projection Algorithm [6].

Monochromatic incident beams in the area of 20 to 30 keV were simulated with the Monte Carlo irradiation code from the XRayImagingSimulator. The incident photon flux was 5x10^5 photons per pixel. The detector was assumed to have a 100% efficiency, i.e. to absorb all incoming photons.

C. Image Quality Evaluation Measurements

The visibility of mammographic features in all images was assessed visually and in terms of Contrast to Noise Ratio (CNR). The CNR is defined as:

$$CNR = \frac{|\bar{I}_f - \bar{I}_b|}{\sigma_b}$$

where $\bar{I}_f$ is the average value of the feature, $\bar{I}_b$ is the average value of the background area and σ_b is the standard deviation of this background area.

For the group of three masses, the average value $\bar{I}_f$ was calculated over a rectangle region of interest (ROI) inside the features with areas 272, 143 and 42 pixels. For the two groups of microcalcifications, the ROI was 9 and 1 pixel, respectively, while the average background value $\bar{I}_b$ was defined in a rectangle area of 2625 pixels located nearby the implant.

For the simulations with the third phantom, the average value $\bar{I}_f$ was calculated over a rectangle ROI of 6 pixels inside each microcalcification that was detectable, while the average background value $\bar{I}_b$ was defined in a rectangle area of 176 pixels in the corresponding silicone block.

III. RESULTS

A. Effect of Silicone Implant on Lesion Detection

This simulation study compares images obtained in mammography and tomosynthesis mode. The simulated planar mammography images of the first two phantoms without and with the silicone implant are depicted in Fig. 3 and Fig. 4.

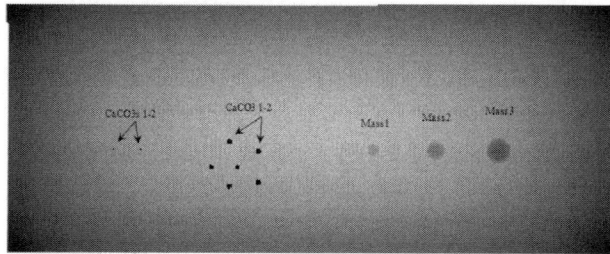

Fig. 3 2D projection image of the breast equivalent phantom

Fig. 4 2D projection image of the phantom with the silicone implant

Tomosynthesis image with reconstructed set of microcalcifications is shown in Fig. 5. Quantitatively, CNR values for various features measured under different imaging are summarized in Table 1.

Fig. 5: Tomosynthesis image of the phantom with the implant that corresponds to a plane, where microcalcifications are in focus

Table 1 CNR values for the main features of the acquired images.

CNR	No Implant	Implant 2D	Implant DTS
Silicone gel	-	92.95	1.42
Mass1	16.13	5.73	7.33
Mass2	10.25	4.34	4.50
Mass3	6.05	3.54	3.42
CaCO3-1	110.30	57.99	112.00
CaCO3-2	110.40	59.53	122.75
CaCO3s-1	75.41	37.85	170.25
CaCO3s-2	77.94	38.58	169.67

B. Effect of Incident Beam Energy on the Visibility of Mammographic Features in a Breast Phantom with Implant

Calculated CNR values for each feature on planar images obtained in the range 20 keV to 30 keV are listed in Table 2. Moreover, in order to evaluate better the influence of the implant on the visibility of the main mammographic findings, a line profile taken across each image vertically has been analyzed. The comparison of the profiles taken from images that were acquired with 20 keV with and without implant shows a reduction of 16% in intensity values in the area of the breast equivalent tissue. The reduction in this region and adjacent to the implant is a result of the presence of the silicone gel.

C. Effect of Silicone Implant Thickness at Different Beam Energies

This simulation study aimed to find the detection limits for microcalcifications under silicone gel of different thickness at different energies. In this simulation study, the third phantom has been used. Similarly to the previous two studies, we calculated the CNR of the small calcifications that should appear on mammography images. Fig. 6 illustrates the simulated projection image of the step-wedge phantom at 30keV. A combination of visual assessment and CNR shows that the last detected microcalcification is characterized with a CNR value of 4 and is placed under a silicone gel slab with a thickness of 10 mm for 20keV, 14mm for 24keV, 16mm for 26keV, 18mm for 28keV and 20mm for 30keV.

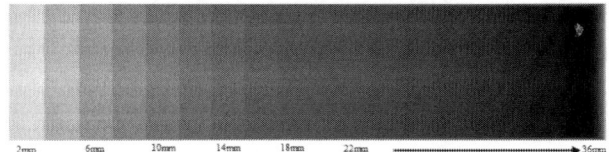

Fig. 6: 2D projection image of the step-wedge phantom at 30keV

In details, comparison of CNR values of a microcalcification with size of 400 μm with the silicone thickness and incident beam energy is shown in Fig. 7. As expected, increasing the energy results in improved feature detection even for thicker implants.

IV. DISCUSSION

The three simulation studies showed that the silicone gel filled implants introduced in breast, influence the quality of the images in terms of lower intensities, contrast and visibility of mammographic features. The first simulation study showed that microcalcifications which have been placed under the implant were not detected on the planar mammographic images. The area of the breast equivalent which could be imaged in the case of 2D image was 228424 pixels while an area of 2576 pixels that corresponds to the implant is totally obscured. The CNR values for all the findings are reduced more than half in case the implant is introduced in the phantom. Improved microcalcification detection is observed when tomosynthesis is used. The later, results also in higher CNR values (Table 1).

Table 2: CNR values for the main findings at different energies

Energies-CNR	20.0	24.0	26.0	28.0	30.0
Silicone gel	92.9	137.4	204.1	323.5	213.4
Mass1	5.7	6.8	10.6	19.0	14.5
Mass2	4.3	4.3	5.9	9.4	6.6
Mass3	3.5	3.3	3.7	4.6	2.2
CaCO3-1	58.0	63.4	81.0	112.0	65.2
CaCO3-2	59.5	64.3	81.6	110.7	63.3
CaCO3s-1	37.9	38.5	49.7	69.3	42.1
CaCO3s-2	38.6	40.1	51.9	75.3	47.5

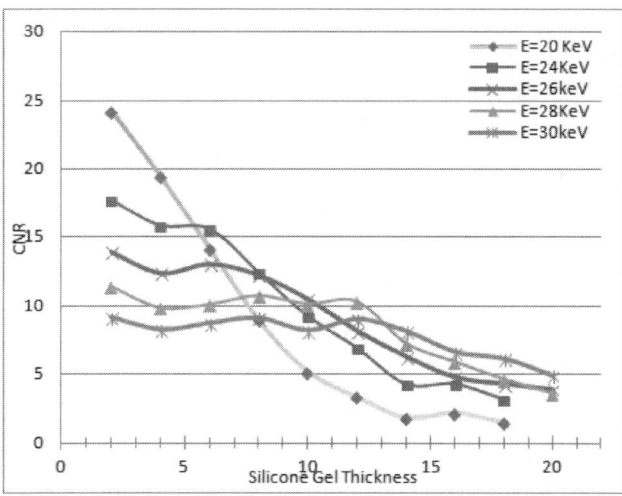

Fig. 7: CNR of mammographic features located below different thickness of implant

Tomosynthesis is a three dimensional imaging technique that recently entered in the clinical practice of imaging the breast. Specifically, limited number of x-ray images is acquired from each breast over a limited angular range with dose levels not exceeding those of conventional mammograms. These images are then used with tomosynthesis reconstruction algorithms to produce three-dimensional structural information of the breast in which the influence of overlapping tissues is greatly reduced and the accuracy of cancer detection is increased. Therefore, the tomosynthesis image shown in Fig. 5 shows larger breast equivalent area. Specifically, the tomosynthesis image with microcalcifications in focus demonstrated an increase of 826 pixels in the visualized breast area compared to the 2D projection image. Such increase may be critical, since it could reveal important information as in Fig. 5 where one microcalcification could be more clearly visualized in comparison to the 2D image, where the same microcalcification was hidden in the silicone texture and therefore impossible to be detected. In addition, tomosynthesis imaging technique resulted in increased CNR values especially for the microcalcifications owing higher CNR values than the corresponding ones for the case of phantom without the silicone implant.

The second simulation study showed that increasing the energy of the incident beam in the range 20keV to 30 keV resulted in improvement of the detection of microcalcification located below the silicone gel. In particular, the microcalcification is more clearly depicted at energies higher than 26keV. These observations are well confirmed by the improved CNR values (Table 2) as the incident beam energy increases. The optimal incident energy at which the CNRs for all studied features, demonstrated highest value was 28 keV, afterwards a sudden reduction of the CNR measurements was observed.

Finally, the last simulation study showed that increasing the incident energy resulted in improved feature detectability due to the more penetrating beam. We can observe that for each energy, the curve that represents the contrast of the feature at different background areas change forms. Each curve has different shape revealing important information about the image quality for different implant thicknesses. At the same time such information has to be considered in the terms of the dose that the patient is going to undergo.

V. CONCLUSIONS

This study analyzed the effect of silicone gel insertion in terms of image quality on the detectability of typical mammographic findings such as breast masses and microcalcifications. A set of Monte Carlo simulations were carried out with three breast tissue equivalent phantoms. Analysis of the images showed a reduction in CNR values up to 50 %, in case of using incident beam energy of 20 keV. Such a reduction in the contrast of breast lesions could be critical for breast cancer screening.

ACKNOWLEDGMENT

This research has been co-financed by the European Union (European Social Fund – ESF) and Greek national funds through the Operational Program "Education and Lifelong Learning" of the National Strategic Reference Framework (NSRF) - Research Funding Program: Aristeia. Investing in knowledge society through the European Social Fund.

REFERENCES

1. Cohen BE, Biggs TM, Cronin ED, Collins J and DR. Assessment and longevity of the silicone gel breast implant. Plast. Reconstr. Surg. 1997 May, 99, (6): 1597-1601.
2. Wolfe JN. On mammography in the presence of breast implants [letter]. Plast. Reconstr. Surg. 1978 Aug, 62, (2): 286.
3. Silverstein MK, Handel N, Gamagami P, Waisman E, and Gierson ED. Mammographic measurements before and after augmentation mammoplasty. Plast. Reconstr. Surg. 1990a Dec, 86, (6): 1126-1130.
4. Arthur AB II, Richard AG and Robert AE (1991). Radiolucent prosthetic gel. Plast. Reconstr. Surg .87, 885-892.
5. Bliznakova K et al (2010) Experimental validation of a radiographic simulation code using breast phantom for X-ray imaging. Comput. Biol. Med. 40(2) pp. 208-214 1903
6. Malliori A, Bliznakova K, Speller RD, Horrocks JA, Rigon L, Tromba G, Pallikarakis N. (2012) Image quality evaluation of breast tomosynthesis with synchrotron radiation Medical Physics, 39 (9), pp. 5621-5634.

Author: Anastasia Daskalaki
Institute: University of Patras
Street: Rio
City: Patra
Country: Greece
Email: dasnatasa@gmail.com

Quantitative Analysis of Marker Segmentation for C-Arm Pose Based Navigation

T. Steger and S. Wesarg

Fraunhofer Institute for Computer Graphics Research IGD, Darmstadt, Germany

Abstract—Intraoperative C-arm fluoroscopy is used for better instrument guidance during bronchoscopy. Unfortunately, C-arm images do not provide depth information. But, offering 3D instrument localization would enable faster and more accurate guidance of the bronchoscope. Using the C-arm pose, this can be achieved by combining intraoperative fluoroscopy with a preoperative CT. Thus, the 3D position of the bronchoscope tip inside the bronchial tree can be located and visualized.

We developed a marker plate for C-arm pose estimation, which is placed on the patient table. The markers are made of steel and appear in two different shapes: spheres and sticks. Detecting the markers is essential for the C-arm pose estimation method. In this work, we present and evaluate two detection methods for detecting the projected markers on the fluoroscopy images. Tests on cadaver images showed very good results regarding robustness and precision: For circles and lines, 80% and 85%, respectively, of all visible markers were detected, whereas only 1% and 3%, respectively, of all detected markers were missegmented.

Keywords— **segmentation, circle detection, line detection, C-arm pose estimation, fluoroscopy, bronchoscopy.**

I. INTRODUCTION

Lung lesions can be recognized on chest X-rays and CT scans. However, only a biopsy of the dubious tissue can confirm lung cancer. This is usually done by bronchoscopy. A flexible tube, equipped with a light source and a camera at the tip, is inserted through the patient's mouth or nose and guided through the bronchial tree to the target site. The bronchoscope's channel is then used to place a biopsy instrument at the target and extract the tissue. As the bronchocopic camera can only provide an endoluminal view of the airways, the intervention is supported by C-arm fluoroscopy for imaging the instruments inside the thorax. Unfortunately, the bronchial tree is not clearly visible on these images and 3D information is entirely missing. Consequently, providing a clear view of the airways on the fluoroscopy and localizing the instrument inside a 3D reconstruction of the bronchial tree, would result in a more accurate and faster intraoperative guidance of the bronchoscope.

A common method for tracking the 3D position of the surgical instruments during the intervention is electromagnetic (EM) tracking, e.g. [1]. But, EM sensors are sensitive to

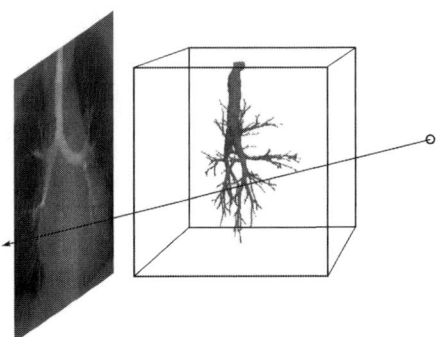

Fig. 1 Ray from X-ray source via airways (CT) to imaged point (X-ray)

metallic objects, which are, on the other hand, always present in operation rooms. Furthermore, used instruments have to be replaced expensively after each use. We presented a method using a specifically designed marker plate, which is easily fixated on the operation table [2]. Using this plate, the C-arm position and orientation (*pose*) during image acquisition can be estimated. Given a preoperatively acquired CT scan of the patient and an intraoperative fluoroscopy of the instrument inside the airways, these pose parameters can be used to generate a virtual ray from the X-ray source through the CT to the instrument's tip on the fluoroscopy (see Fig. 1). In that way, the current position of the instrument is narrowed down to the intersections between the ray and the bronchial tree in the CT volume.

Pose estimation strongly depends on accurate and robust segmentation of the calibration plate markers. Marker detection is especially challenging with overlaying anatomy or operational instruments. Also, different marker sizes and shapes after projection have to be considered. In [3], we briefly introduced marker detection methods suitable for the named constraints and the different marker types present on the marker plate. We evaluated them in combination with our pose estimation algorithm. In this paper, we improve these two marker detection methods, and, for the first time, analyze their results independent of C-arm pose estimation.

II. MATERIALS AND METHODS

First, we briefly describe the design of our calibration plate and the main features of our C-arm pose estimation method (details can be found in [3]). This is important to understand the encountered challenges of and the developed solutions to our marker detection problem. The second section explains both challenges in detail.

A. Pose Estimation Pattern

During fiducial-based C-arm pose estimation finding the mapping between the 2D points on the fluoroscopy and their corresponding 3D points on the calibration plate is vital. Also, it is neccessary that a minimum number of markers is both visible and detectable on the fluoroscopy to enable pose estimation. We decided to develop a marker plate, which is planar, made of acryl and has a size of 100 $cm \times 51$ cm. Thus, it can be easily fixated between patient and table. Even though similar constructs for different applications already exist, e.g. [4], only our method is suitable for bronchoscopy as it covers a thorax-sized field of view. Because of this broad view, using various marker shapes was out of question as far too many types would be needed. Forming these partly complex types from steel and especially robustly detecting them would not be feasible. They would also change their shape after projection from differing angles. Therefore, we decided to use simple uniform shapes, which are easily moldable and even in the presence of overlaying anatomy and after projection automatically detectable on fluoroscopy: spheres and sticks. So, inside the acrylic plate steel spheres with a diameter of 3 mm as well as sticks with a thickness of 1.6 mm and $6 - 12$ mm length are inserted in a well-defined pattern (see Fig. 2).

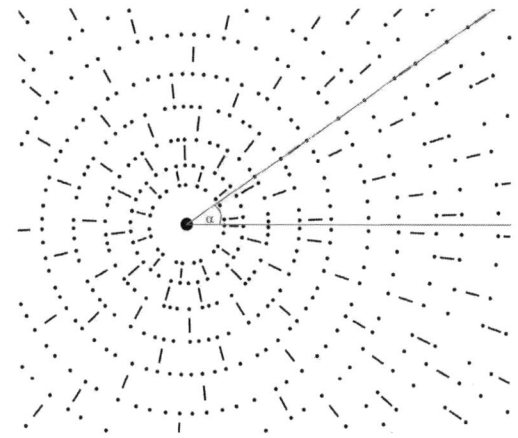

Fig. 2 Detail of the marker pattern used for C-arm pose estimation: Horizontal line (*blue*), line defined by several sticks (*red*) and angle defined by this line (α)

To find the correct mapping of the spheres from world (3D) to image (2D) coordinate system, our method uses the projective invariant *cross ratio* on collinear points and concurrent lines. Cross-ratio defines a relation between point distances or angles. As its value remains the same after central projection of the points or angles, it is particularly suitable for our purpose. Eq. 1 gives its definition for four points (*left*) and four angles (*right*) (see also [5]).

$$(ABCD) := \frac{(ABC)}{(ABD)} := \frac{\frac{\overline{AC}}{\overline{BC}}}{\frac{\overline{AD}}{\overline{BD}}} \qquad (abcd) := \frac{\frac{\sin(\sphericalangle ac)}{\sin(\sphericalangle bc)}}{\frac{\sin(\sphericalangle ad)}{\sin(\sphericalangle bd)}} \qquad (1)$$

The markers are placed in a well-defined pattern, which assures that neighboring points and lines have a unique cross-ratio value. While the center of a sphere defines a point, a stick defines the line, which passes lengthwise through it. The angle between this line and the horizontal gives the angle used for cross-ratio calculation (see Fig. 2). Clearly, successful 3D-2D mapping requires accurate detection of these markers.

B. Marker Segmentation

There are several challenges in the clinical scenario, which our marker detection method has to overcome. The spheres and sticks on the marker plate are imaged as circles and longish quadrangles on the fluoroscopy. Depending on the table height, they will appear in different sizes. This means, that it is not sufficient to search for circles and rectangles of predefined shape and size. Furthermore, the density of the markers in the image varies for different parts of the plate because of its radial pattern. Thus, it is not possible to notice unsegmented markers based on the overall count of detected markers or the space between them. Still, the biggest challenge is superposed patient anatomy or surgical instruments on the images. These structures not only reduce the contrast between markers and background, but may also cover the markers completely.

We developed two different methods for segmenting the markers on the fluoroscopy. The sticks are projected as longish rectangles, but for calculating the cross-ratio value it is sufficient to detect the lines lengthwise through these rectangles. Furthermore, for each line, there are always several sticks and it is enough to detect only one stick per line. The spheres are projected as circles. Here, only the center of these structures is interesting. The following describes both methods for finding points and angles defined by lines.

B1 Point Detection

1. *ROI definition:* The contrast considerably differs in different parts of the image. Therefore, the image (1024 $px \times 1024$ px) is divided into several rectangular

partly overlapping ROIs (300 $px \times 300\ px$). The following steps are executed separately for each of these ROIs.

2. *Gamma correction:* A gamma lookup table is used to transform the image and enhance the contrast ($\gamma = 2.5$). Thus, the dark areas caused by overlayed anatomy are brightened up and the markers are easier to detect.

3. *Canny-Edge detection:* First, a median filter is applied. Then, edges are segmented with a Canny-Edge-detector [6]. The resulting contours are closed to create connected polygonal lines.

4. *Arc segmentation:* The polygonal lines are simplified using polygon approximation by the RamerDouglasPeucker algorithm [7]. The resulting segments are classified into either lines or arcs (see Fig. 3(a)).

5. *Circle approximation:* All arc contours are used to approximate circles (see. Fig. 3(a)). For this, the geometrical distances between contour and circle points are minimized. To abate outliers, the contour points are weighted. In the next step, center and radius of the approximated circles is determined.

6. *Median of radiuses:* The median of the radiuses of all detected circles is calculated and then used to discard outliers, i.e. circles, which are strikingly smaller or bigger than this median.

B2 Line Detection

1. *Region segmentation:* To obtain a high-pass filtered image, the original image is subtracted from a smoothed version of it. By applying threshold filtering and determining connected components, several regions are found.

2. *Filling:* If the region area is within a predefined interval, potential hollow areas inside will be filled, providing these hollow areas also lie within a certain threshold. Afterwards a dilatation with a rectangular structuring element follows to merge neighboring regions.

3. *Skeletonization:* The next step is skeletonization of the emphasized longish, quadrangular regions, resulting in a set of line segments.

4. *Line approximation:* These line segments are separated into line pieces according to their curvature (see. Fig. 3(b)). For this, a connecting line between start and end point is created and depending on the distance to all line points, the segment will be split or not.

5. *Angle and length:* Now, the angles and lengths of the line pieces are calculated. Only line pieces exceeding a certain minimum length are chosen from the set.

6. *Intersecting circles:* Using the calculated angle, each line piece can be overlayed by a thin rectangle, which is twice as long as the line piece. As every stick on the marker plate is framed by two spheres, a line is selected, if the rectangle intersects circles on both sides (see. Fig. 3(b)).

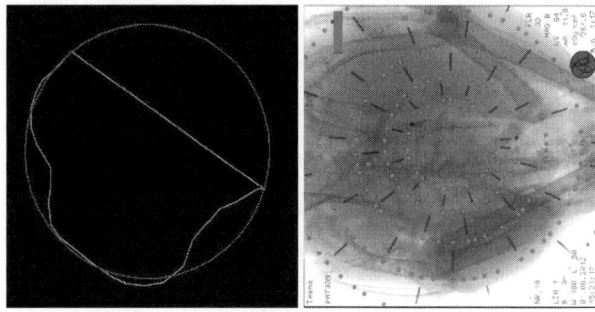

(a) Point detection: *Left:* Segmented arc (*blue*) and circle approximation (*violet*) *Right:* Detected points (*green*)

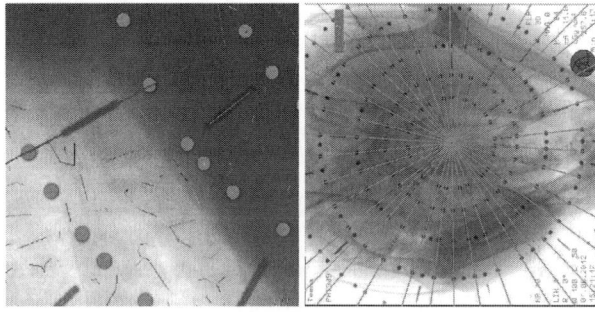

(b) Line detection: *Left:* Intersecting circles *Right:* Detected lines (*green*)

Fig. 3 Exemplary intermediate steps and results of both marker detection methods on turkey cadaver fluoroscopy image.

7. *Angle correction:* As the beforehand calculated angles might be inaccurate, these angles are corrected using the framing circles (see. Fig. 3(b)). Also, angles, which appear twice, because they are defined by multiple detected sticks, are discarded in this step.

8. *Intersecting center point:* The center point of the pattern is determined by using a clustering algorithm on the intersection points of all detected lines. Only the lines intersecting this center point are selected.

III. EXPERIMENTS AND RESULTS

Our marker segmentation method was evaluated on clinically acquired fluoroscopy images. The test data was created on a Ziehm C-arm system with a flat panel detector. Thus, the images were not radially distorted. The marker plate was placed on a patient table and a turkey cadaver with bones and bowels was placed on the marker plate. The cadaver had a size of about 30 $cm \times 16\ cm \times 25\ cm$. 34 fluoroscopy images from different positions and angles were acquired. As our aim was to test the method even in the presence of interfering anatomy, we ensured that on each test image the turkey

Table 1 Mean and standard deviation of point and line detection tests

	Spheres	Sticks	Angles
$Count_{visible}$	160.6($\pm$35.6)	40.2($\pm$7.8)	23.4($\pm$4.5)
$Count_{detected}$	140.1($\pm$28.4)	33.4($\pm$7.9)	18.7($\pm$4.3)
$Count_{missegmented}$	3.7($\pm$5.1)	0.2($\pm$3.8)	0.2($\pm$0.6)
$Count_{unsegmented}$	24.2($\pm$12.9)	13.9($\pm$4.7)	4.8($\pm$2.5)
$Rate_{false\_pos}$	3%($\pm$5%)	0.7%($\pm$10%)	1%($\pm$3%)
$Rate_{false\_neg}$	15%($\pm$6%)	35%($\pm$9%)	20%($\pm$10%)

anatomy was covering the imaging area. Pose estimation was executed on each image and the resulting transformation was used to reproject all source markers. Thus, we have a ground truth and are able to reliably count the visible markers on the fluoroscopy.

On all images, the marker segmentation method for both markers was applied. For each image, we counted how many markers were detected overall, how many were actually visible on the image, how many were falsely detected (missegmentations) and how many were not detected even though present on the image (unsegmented markers). Using these numbers, we also calculated the rate between missegmented and detected markers (*false positive rate, FPR*) and between unsegmented and visible markers (*false negative rate, FNR*) for each image. Then, we averaged all these values for the whole test data set. The same data was calculated for angles, which are defined by the stick markers.

Table 1 shows the results of the point and line detection tests. For a $1024 \times 1024\ px$ image, point detection took $151\ ms$ and line detection $57\ ms$ on average on a Intel Core i7 2.93 GHz machine.

IV. DISCUSSION

Point segmentation achieves an extremely low *FPR* (3%) and thus, proves its outstanding precision. Our subsequent pose estimation method can easily cope with such few missegmentations. They can be unmasked, if they do not lie on detected lines. As we always calculate cross-ratio on four neighbouring collinear points, unsegmented markers are only problematic, if lying in between such a group of four. Therefore, after mapping the points to the line they lie on, another marker search run is executed on this line. Thus, we ensure that every marker in a cross-ratio group is found. This means, that a *FNR* of 15% can safely be overcome. On the whole, our point segmentation method provides fully satisfactory results regarding sensitivity and presicion.

The method for line detection produces a *FPR* of about 1%, which is negligibly low and, thus, fully satisfying. The relatively high *FNR* is partly due to the fact, that sticks at the borders of the image cannot be detected by design. These sticks do not have framing points on both sides and thus, will be sorted out by our method. Still, this step is important to recognize false lines. Leaving these sticks out and considering only the calculated angles, which only need one stick per line, we achieve a *FNR* of about 20%. We argue that this is sufficiently robust and precise for our subsequent pose estimation method, which, strictly speaking, only needs $4-5$ correct angles per image.

V. CONCLUSIONS

We developed and evaluated two different detection methods for spherical and longish steel markers on a fluoroscopy image. The spheres and sticks are arranged on an acrylic plate specifically designed for C-arm pose estimation. Evaluation showed that point detection (spheres) was very sensitive ($FNR = 15\%$) and remarkably precise ($FPR = 3\%$). Line detection (sticks) was extremely precise ($FPR = 1\%$). It delivered good results regarding robustness ($FNR = 20\%$), also. Therefore, these algorithms are fully satisfying for subsequent C-arm pose estimation. Based on this pose, the current 3D position of the bronchoscope inside the airways can be reliably calculated. Thus, instrument guidance to the target lesion is be considerably facilitated for the physician.

REFERENCES

1. Schwarz Y. et al.. Real-time electromagnetic navigation bronchoscopy to peripheral lung lesions using overlaid CT images: the first human study *Chest.* 2006;129:988–994.
2. Steger T. et al.. Navigated bronchoscopy using intraoperative fluoroscopy and preoperative CT *Proc. IEEE ISBI.* 2012;9:1220 -1223.
3. Steger T. et al.. Marker detection evaluation by phantom and cadaver experiments for C-arm pose estimation pattern *Proc. SPIE Med Imag.* 2013;8671:86711V-86711V-9.
4. Moult E. et al.. Automatic C-arm Pose Estimation via 2D/3D Hybrid Registration of a Radiographic Fiducial *Proc. SPIE Med Imag.* 2011;7964:79642S.
5. Milne J.J.. *An elementary treatise on cross-ratio geometry with historical notes.* Cambridge University Press 1911.
6. Canny J. A computational approach to edge detection *IEEE TPAMI.* 1986;8:679–698.
7. Ramer U.. An iterative procedure for the polygonal approximation of plane curves *CGIP.* 1972;1:244 - 256.

Iterative Dual-Energy Material Decomposition for Slow kVp Switching: A Compressed Sensing Approach

A. Sisniega[1,2], J. Abascal[1,2], M. Abella[1,2], J. Chamorro[1,2], M. Desco[1,2,3], and J.J. Vaquero[1,2]

[1] Departamento de Bioingeniería e Ingeniería Aeroespacial, Universidad Carlos III de Madrid, Madrid, Spain
[2] Instituto de Investigación Sanitaria Gregorio Marañón, Madrid, Spain
[3] Centro de Investigación Biomédica en Red de Salud Mental (CIBERSAM), Madrid, Spain

Abstract—**Dual energy x-ray computed tomography (DECT) is a valuable research tool for both clinical and preclinical studies and a field of extensive research. Current implementations of DECT rely on the acquisition of two datasets either by using two source-detector pairs or by rapidly switching the kVp of the x-ray source between consecutive projections. Both alternatives require specific hardware, only available in a small number of systems.**

DECT data can also be acquired using standard hardware and minimizing acquisition time and dose by slowly modulating the kVp of the source to obtain the whole dataset in a single rotation. However, in this case, highly undersampled, non-registered data are obtained for the high and low-energy sinograms, depending on the slew rate used for the kVp modulation.

We propose a novel iterative method for raw data DECT material decomposition for slow modulation kVp by using a compressed sensing approach. Information provided by the slow switching kVp data at intermediate kVp values is used to generate a prior term containing the edges of the different regions in the sample. The problem is efficiently solved by the use of the Split-Bregman method.

Preliminary experiments on simulated data show promising results for the decomposition of slow modulated kVp DECT data into two material basis sinograms, obtaining acceptable error in the decomposed dataset from highly undersampled data.

Keywords—**Compressed-sensing, Dual-Energy, CBCT.**

I. INTRODUCTION

Dual energy x-ray computed tomography (DECT) allows the decomposition of CT data into basis materials –e.g bone and soft-tissue, or tissue and iodinated contrast– by acquiring the same sample at two different x-ray source voltages. The representation of the CT data into basis materials instead of attenuation coefficients makes it possible to quantitatively differentiate tissues within the body or to reduce artifacts arising from the polychromatic nature of the x-ray spectrum by building monochromatic images from the material decomposition.

Current approximations for DECT data acquisition either make use of two source-detector pairs [1] or switch the kVp [2] of the x-ray source between two values for alternative projections. Both approaches require dedicated hardware, only available in few systems. Another approach for obtaining DECT data consists in performing two acquisitions at two different kVps, one immediately after the other. In this case there is an increase of acquisition time and possible mismatch between the two datasets due to movement of the patient between acquisitions. It would be desirable to obtain DECT data using conventional hardware and slowly modulating the kVp of the source.

A previous approach for DECT data decomposition using slow kVp modulation has been proposed in [3]. In this approach, the authors proposed to independently reconstruct each of the two volumes (i.e. for the two extreme kVp values) using a prior image compressed sensing reconstruction method (PICCS) with the reconstruction of the mixed kVp dataset as prior term. The material decomposition is obtained by means of a weighted sum of the two reconstructed volumes, approach usually known as image domain decomposition.

While image domain decomposition has been largely used in the literature [4], the material images obtained contain non-idealities, such as beam hardening, that can be suppressed by directly generating the material decomposition from the acquired raw data [5].

We present here an iterative approach for the direct reconstruction of the decomposition into two materials from slow modulation kVp DECT data, using a compressed sensing term and prior knowledge obtained from the complete dataset containing intermediate kVp values.

II. MATERIALS AND METHODS

A. Data Acquisition

For the slow kVp modulation DECT image acquisition we follow the approach by Szczykutowicz et al. [3]. The kVp value of the x-ray source is continuously varied between two values while the system rotates around the subject, following a given modulation function which in this stage of the work has been set to a triangular wave. Therefore, the outcome of the acquisition process is a sinogram containing two groups of angular projections useful for the

DECT data decomposition, i.e. those at extreme kVp values, mixed with intermediate kVp projections. The acquisition process is illustrated in Fig. 1.

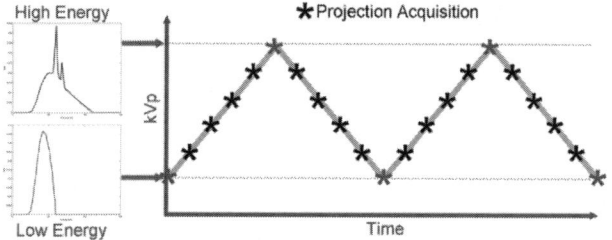

Fig. 1. Scheme of the data acquisition protocol for slow kVp switching DECT data.

B. Measurement and Object Model

Following the definitions of Noh et al. [6] and Huh et al. [5], we assume $m = 1,...,M_0$ sets of $i = 1,...,N_d$ polychromatic x-ray transmission line integral measurements, yielding

$$f_{mi} = -\log \int p_m(\varepsilon) \exp(-\sum_L s_{li} \beta_l(\varepsilon)) d\varepsilon \quad (1)$$

where $p_m(\varepsilon)$ is the normalized source spectrum for the mth acquisition (i.e. kVp) and energy bin ε.

The integral of the linear attenuation coefficient is represented using a basis material decomposition on L_0 materials, given by:

$$\int_{x_i} \mu(\vec{x},\varepsilon) dx = \sum_{l=1}^{L_0} \beta_l(\varepsilon) \int_{x_i} \rho_l(\vec{x}) dx = \sum_{l=1}^{L_0} \beta_l(\varepsilon) s_{li} \quad (2)$$

where $\beta_l(\varepsilon)$ is the mass attenuation coefficient and $\rho_l(x)$ is the unknown density map of the lth material type. In the present work we set $M_0 = L_0 = 2$.

The goal of the presented reconstruction method is the estimation of the density spatial distribution for each of the materials, in a voxelized volume, given the relating function, $f_{mi}(s_i)$, and that the measurements f are highly undersampled. For this reason, we redefine the line integrals s_{li} as the result of a matrix operation which accounts for the projection of the voxelized values on a detection bin i, yielding,

$$s_{li} = [A\rho_l]_i \quad (3)$$

where A is the projection matrix.

C. Algorithm for Material Decomposition

Let F be the undersampled sinogram data, and let T be a transformed sparse domain for the density distribution. Thus, we can formulate the restoration algorithm as

$$\min_\rho \|T\rho\|_1, \text{ such that } f(A\rho) = F,$$

where T is selected as a gradient operator weighted by the prior edge image w, yielding the weighted total variation (TV) as proposed in [7]. The selection of TV for T filters out noise while keeping edges and avoids the diffusion in the direction perpendicular to edges of the prior data.

The prior information is obtained from a regular FDK reconstruction of the complete acquired sinogram, including intermediate kVp values. The weights (w) multiplying the TV values are obtained using the following expression:

$$w = \exp\left((-\nabla \mu / \sigma)^2\right) \quad (4)$$

where μ are the attenuation values provided by the FDK reconstruction and σ is an empirical parameter governing the amount of smoothing allowed in the data.

This problem can be efficiently solved using the Split Bregman formulation [8, 9] by including new variables, dx, and dy that allow for the splitting,

$$\min_{d_x,d_y,\rho} \|(d_x,d_y)\|_1 + \frac{\mu}{2}\|f(A\rho) - F^k\|_2^2 + \frac{\lambda}{2}\|d_x - wD_x\rho - b_x^k\|_2^2 + \frac{\lambda}{2}\|d_y - wD_y\rho - b_y^k\|_2^2 \quad (5)$$

where the variables b_i represent the Bregman iterations that impose the constraints by adding the error back into the constraints. Solution of ρ only involves 2-norm functionals and is solved using a standard L-BFGS algorithm. Solution of dx, dy, is found using shrinkage formulas as described in [8, 9].

D. Algorithm Evaluation

The performance of the algorithm was evaluated using simulated data obtained from a polychromatic implementation of a Siddon projector, described in [10].

For the data simulation we used a synthetic phantom consisting of a uniform region of soft tissue with two circular bone inserts, as shown in Fig. 2. The physical properties of the materials for the simulation were obtained from the Penelope database [11].

We simulated a set of slow kVp switching DECT data, for extreme x-ray source voltage values of 60 and 110 kVp with 2 mm Al and 0.2 mm Cu added filtration. The set of spectra was simulated using the Spektr implementation [12] of the TASMIP model [13].

The voltage was modulated by a triangular wave as described in section II.A. The triangular wave had a period of 10 projections. We simulated a total of 360 projections, thus providing 18 projections for each extreme kVp value, i.e. 5 % of the data used in regular DECT for the same total number of projections.

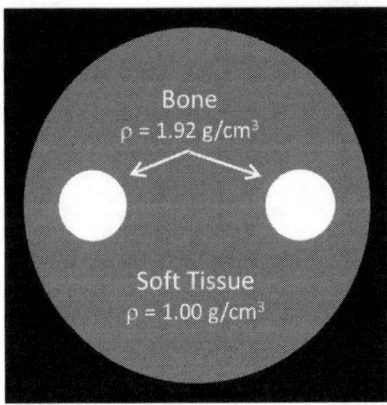

Fig. 2. Synthetic phantom used for the evaluation of the algorithm. The grey area was formed by soft-tissue while the two white inserts were made of bone.

The starting point for the algorithm was selected as the FDK reconstruction of the conventional DECT material separation, obtained from the interpolation of the extreme kVp sinograms.

III. RESULTS

Fig. 3 shows the results of the DECT material separation after 10 and 15 iterations of the algorithm. RMSE errors in the density maps after 10 iterations are of 0.16 and 0.09 for the soft-tissue and bone density maps, respectively. The errors are of 0.15 and 0.09 after 15 iterations.

IV. DISCUSSION AND CONCLUSIONS

We presented a novel method to obtain the material decomposition of DECT data from slow kVp switching acquisitions. The results on simulated data show that a reasonable image quality can be obtained from highly undersampled data (5 %), opening the possibility of using DECT on conventional CT and micro-CT systems, avoiding the need of dedicated hardware or performing two acquisitions.

In addition to the possibility of using the presented method in conventional systems, the requirement of performing one single acquisition and the possibility of including statistical models for the noise in the data (see, e.g., the approach by Stayman et al. [14] for undersampled CT data reconstruction) offer a potential for dose reduction in DECT.

While the presented method showed promising results, the performance in real scenarios has still to be evaluated; the presence of data degrading effects, such as noise or scatter, could challenge the performance of the algorithm. Also, the appropriateness of the sparsifying transform (T) has to be assessed, since its selection has been shown to significantly impact the results obtained by similar methods in CT reconstruction [15].

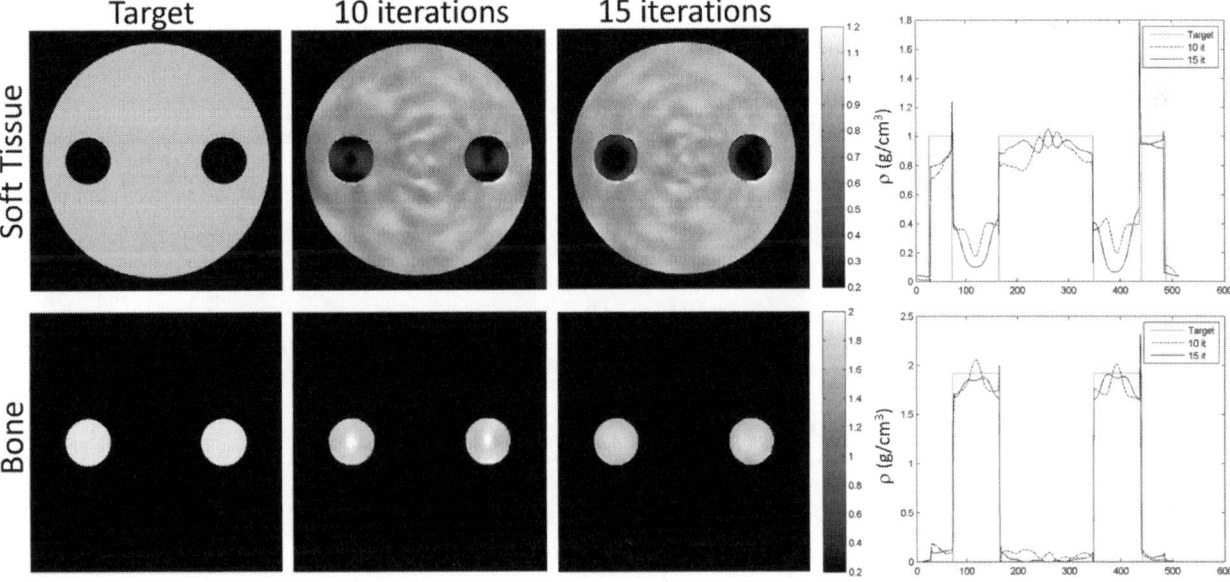

Fig. 3. Results of the material decomposition method after 10 and 15 iterations of the algorithm, compared to the target decomposition (i.e. the segmentation of the synthetic phantom in Fig.2 into bone and soft-tissue) in left column. Profile data across the central row of the image allow the quantitative evaluation of the algorithm performance.

Acknowledgment

This work was partially funded by AMIT project (CEN-20101014) from the CDTI-CENIT program, RECAVA-RETIC Network, projects TEC2010-21619-C04-01, TEC2011-28972-C02-01 and PI11/00616 from Spanish Ministerio de Ciencia e Innovación, ARTEMIS program (S2009/DPI-1802) from Spanish Comunidad de Madrid and EC-ERDF program, and PreDiCT-TB project (115337-1) from the 3rd IMI call - EC and EFPIA.

A. Sisniega is funded by an FPU grant from the Spanish Ministerio de Educación.

References

1. C. T. Badea, S. Johnston, B. Johnson, M. Lin, L. W. Hedlund, and G. A. Johnson, "A dual micro-CT system for small animal imaging," *Proc. SPIE*, **6913**, 691342, 2008
2. W.A Kalender., E. Klotz, and L. Kostaridou., "An algorithm for noise suppression in dual energy CT material density images," *IEEE Transactions on Medical Imaging*, **7**, 218-224, 1988
3. T. P. Szczykutowicz and G.-H. Chen, "Dual energy CT using slow kVp switching acquisition and prior image constrained compressed sensing.," *Physics in medicine and biology*, **55**, 6411-6429, 2010
4. G. J. Gang, W. Zbijewski, J. W. Stayman, and J. H. Siewerdsen, "Cascaded systems analysis of noise and detectability in dual-energy cone-beam CT," *Med Phys*, 39, 5145, 2012
5. W. Huh, J.A. Fessler, A.M. Alessio, P.E. Kinahan, "Fast kVp-switching dual energy CT for PET attenuation correction," *IEEE 2009 Nuclear Science Symposium Conference Record NSS/MIC*, 2510-2515, 2009
6. J. Noh and J. Fessler, "Statistical sinogram restoration in dual-energy CT for PET attenuation correction," *IEEE Transactions on Medical Imaging*, **28**, 1688-1702, 2009
7. Z. Tian, X. Jia, K. Yuan, T. Pan, and S.B. Jiang, "Low dose CT reconstruction via edge-preserving total variation regularization," *Phy. Med Biol*,**56**, 5949, 2011
8. T. Goldstein and S. Osher, "The Split Bregman Method for L1-Regularized Problems," *SIAM Journal on Imaging Sciences*, **2**, 323, 2009
9. J.F.P.-J. Abascal, J. Chamorro-Servent, J. Aguirre, S. Arridge, T. Correia, J. Ripoll, J.J. Vaquero, and M. Desco ., "Fluorescence diffuse optical tomography using the split Bregman method," *Med Phys*, **38**, 6275, 2011
10. A. Sisniega, W. Zbijewski, A. Badal, I.S. Kyprianou, J.W. Stayman, J.J. Vaquero, and J.H. Siewerdsen, "Monte Carlo study of the effects of system geometry and antiscatter grids on cone-beam CT scatter distributions," *Med Phys*, **40**, 051915, 2013
11. F. Salvat, J.M. Fernández-Varea, and J. Sempau, "PENELOPE - Code System for Monte Carlo Simulation of Electron and Photon Transport," presented at the Nuclear Energy Agency OECD, Issy-les-Moulineaux, 2006
12. J.H. Siewerdsen, A.M. Waese, D.J. Moseley, S. Richard, and D.A. Jaffray, "Spektr: a computational tool for x-ray spectral analysis and imaging system optimization," *Med Phys*, **31**, 3057-67, 2004
13. J.M. Boone and J.A. Seibert, "Accurate method for computer-generating tungsten anode x-ray spectra from 30 to 140 kV," *Med Phys*, **24**, 1661-1670, 1997
14. J.W. Stayman, W. Zbijewski, Y. Otake, S. Schafer, J. Lee, J.L. Prince, and J.H. Siewerdsen, "Penalized-Likelihood Reconstruction for Sparse Data Acquisitions with Unregistered Prior Images and Compressed Sensing Penalties," *SPIE Medical Imaging*, Orlando, **7961**, 79611L-1-6, 2011
15. J.F. Abascal, A. Sisniega, C. Chavarrías, J.J. Vaquero, M. Desco, and M. Abella. "Investigation of Different Compressed Sensing Approaches for Respiratory Gating in Small Animal CT". Abstract Book IEEE NSS/MIC 2012, 431-432, 2012

Author: Alejandro Sisniega
Institute: Departamento de Bioingeniería e Ingeniería Aeroespacial. Universidad Carlos III de Madrid
Street: Av. Universidad 30
City: Leganés, Madrid
Country: Spain
Email: alejandro.sisniega@uc3m.es

Brain Gray – White Matter Discrimination in Dual Energy CT Imaging: A Simulation Feasibility Study

A. Dermitzakis[1], G. Gatzounis[2], and K. Bliznakova[3]

[1] Biomedical Technology Unit, Dept of Medical Physics, University of Patras, Rio, Greece
[2] Department of Neurosurgery, University of Patras, Patras, Greece
[3] Department of Electronics and Microelectronics, Ttechnical University of Varna, Bulgaria

Abstract—The aim of this study is to investigate the feasibility of using dual-energy (DE) imaging for discrimination of gray-white matter with brain CT. Initially, an optimization study was accomplished that aimed to find the optimal low and high beam energy spectra for the DE application. Simulations were carried out with 181 noise free brain dual-energy CT (DECT) images, simulated with an in-house developed software simulator for several monochromatic and polychromatic spectra within a full gantry acquisition arc of 360^0. Ten monochromatic beams with energy from 20 keV to 110 keV and four polychromatic beam spectra 80, 100, 100, 140 kVp were simulated. The software brain phantom was the Digital Brain Phantom II, that is voxel-based model of the brain with no tumor insertion. To obtain DE projections, a non-linear algorithm for combining low and high energy images was exploited. Further, the optimal parameters of this algorithm were determined. The optimization study showed that the maximum contrast in brain DECT is achieved for the low/high energy combination of 100/110 keV in case of monochromatic beams and 80/100 kVp for polychromatic beams. For these cases, the contrast improvement was of the order of 12 and 7 times fold, respectively, compared to brain CT images acquired at high energies. The visual observation demonstrated superiority of the brain DECT compared to single energy brain CT images in terms of improvement in brain tissue differentiations. Evaluation of line profiles taken through different regions of interest on single and brain DECT images showed an excellent gray and white matter tissue discrimination.

Keywords—Brain Imaging, X-Ray Imaging Simulation, Dual Energy.

I. INTRODUCTION

A Computed Tomography (CT) brain scan is performed to inspect the structures of the brain and evaluate for the presence of any pathology such as bone abnormalities, mass or tumor, hemorrhage and fluid collection, trauma etc. However, brain CT demonstrates limitations in the possibility to depict the difference between gray and white matter. The visualization of these tissues is important for many pathologies such us Alzheimer's disease [1], dementia [2], aging disorders [3], low grade gliomas etc. in which MRI has proved more sensitive in detecting white matter lesion than CT [4]. Also the extend of white matter damage after traumatic brain injury is underestimated in conventional CT [5]. Although Dual Energy CT (DECT) was first investigated more than 3 decades ago from Alvarez and Macovski [6,7], the first commercially available DECT scanner was manufactured in 2006 but up to now the use of dual energy (DE) brain CT is restricted in intracerebral hemorrhage differentiation [8], bone removal and angiography [9]. A DECT approach may result in better gray to white tissue discrimination and facilitate inspection of brain disorders, specifically low contrast tumors hidden in the brain parenchyma.

The aim of this study is to investigate the feasibility of using DE imaging for brain CT. For this purpose, an optimization study has been initially carried out that aimed to find the optimal low and high energy incident spectra for DE application. Investigations were carried out with simulated DE brain CT noise free images generated with several monochromatic and polychromatic spectra from a software voxel-based brain phantom with no tumor insertion. For DE application, a non-linear subtraction algorithm for combining low and high energy images was exploited. CT brain slices of the obtained dual energy images were calculated using filtered multiple projection algorithm. Further, the optimal parameters of this algorithm have been determined. Brain DECT images were evaluated visually and quantitatively in comparison to single energy brain CT images. Both evaluations showed superiority of the DE brain CT compared to single energy CT images, as the former demonstrated an excellent grey and white matter tissue discrimination.

II. MATERIALS AND METHODS

A. Phantom

The phantom used in this study is the Digital Brain Phantom II [10]. This phantom has been created using 27 T1-weighted, 12 PD-weighted, 12 T2-weighted MRI scans, 1 CT scan and 1 MR Angiography from a single subject. The phantom is provided in the form of three-dimensional matrices of 10 brain tissue volumes, with each voxel having a value between 0 and 255. This value reflects weather the tissues in the respective voxel are evident or not. For this study volumes of gray, white, skull and cerebrospinal fluid were taken into consideration. In order to use the phantom in simulations of imaging techniques based on x rays, x-ray attenuation coefficients of the selected tissues were introduced. These attenuation coefficients were calculated according to Hubbell and Seltzer [11] for each incident energy. The resulted volumes were combined into a single 3D brain phantom, using weighted summing method.

B. Dual Energy Subtraction Algorithm

The subtraction algorithm used is based on the theory of Alvarez [6] and Macovski [7] and Lehmann et al. [12] The process followed is the decomposition of the attenuation coefficients (μ) of each pixel in the projection images of both low and high energy spectra, into a pair R_G and R_W of the basis materials Gray and White matter following the equation:

$$R_G = [\mu(high)\mu_W(low) - \mu(low)\mu_W(high)]/a \quad (1)$$

$$R_W = [\mu(low)\mu_G(high) - \mu(high)\mu_G(low)]/a \quad (2)$$

$$a = \mu_W(low)\mu_G(high) - \mu_W(high)\mu_G(low) \quad (3)$$

Subtracted image is created based on tissue cancelation performed with the following equation:

$$DE = R_G \sin(\varphi) + R_W \cos(\varphi) \quad (4)$$

The parameter φ is the Tissue Cancelation Angle (TCA) the variation of which changes the contrast/cancelation of the two basis tissues.

C. Simulated Images

Projection images were generated with the in-house developed XrayImagingSimulator [13] developed at the Biomedical Technology Unit (BITU). Photon scattering and detector response were not simulated, resulting in noise free projections. The simulated brain CT acquisition geometry included rotation of the x-ray source and detector in a full arc of 360^0 around the brain phantom. The simulated imaging protocol per incident photon beam used 181 synthetic images acquired at an increment of 2^0. Image resolution was limited to the brain phantom resolution. Therefore, simulated projections were generated with a size of 256 x 256 pixels and a pixel resolution of 1mm^2. These images represent the two-dimensional distributions of the line integrals of the attenuation coefficient along the paths of the x-rays, registered at the detector surface. To obtain photon based images, the number of photons that reach the surface of the detector was calculated taking into account the incident photon fluence. The later depends on the dose to the brain and therefore from the incident air kerma. According to Cohnen et. al. [14], the surface dose to the brain for a typical brain CT examination varies from 66.0 to 83.2 mGy. In this study, we set the dose to 83 mGy.

The following incident photon beams were simulated: ten monochromatic beams with energy from 20 keV to 110 keV and four polychromatic beams 80, 100, 100, 140 kVp. Figure 1 depicts two of the used incident spectra: at 80 and at 100 kVp, respectively. All of these four spectra are actually available at a commercially available DECT scanner.

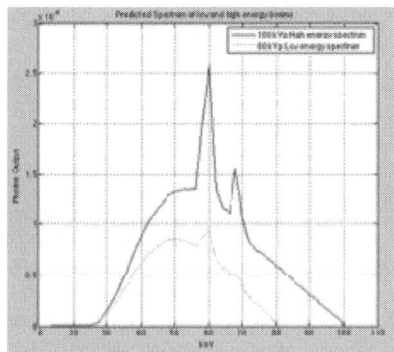

Fig. 1 Polyenergetic spectra of 80 and 100 kVp beams

The energies of the monochromatic beams vary from 20 to 100 keV for low-energy and from 20 to 110 keV for the high-energy beams resulting in 81 DE combinations. Polychromatic images are obtained through a weighted fusion of monochromatic images generated for energies sampled from the corresponding energy spectrum.

For each energy combination, the incident air kerma and therefore the incident photon fluence were calculated. Thereafter, the synthetic images are transformed to images that represent the number of photons absorbed in the detector.

Further, images of each low and high energy pair were calculated. In case of monochromatic beams, a total of 729 DE combinations were produced. These combinations consist of 81 different low and high energy pairs combinations for each one of which nine weightings between low and high energy have been performed.

In the particular case of polychromatic spectra, the basis materials attenuation that is used in the subtraction algorithm described in 2.2, weighted attenuation coefficient was calculated for each energy at all energy spectra. Tomograms were reconstructed using the filtered Multiple Projection Algorithm [15] applied on the obtained 180 DE images.

D. Evaluation

In order to quantitatively evaluate the brain DECT images, a Figure Of Merit (FOM) was defined, based on data extracted from several line profiles taken at different regions of interests (ROI). This FOM is referred as Line Contrast (LC). Specifically, for each brain DECT image, three different line profiles were drawn as shown in figure

2. In each line profile, 5 ROIs were set in the regions where there is a transition from white to gray matter. The line profiles are shifted and normalized to max value in order to be objective in all energy combinations, and the maximum difference within the regions of interest is calculated for each ROI. Then the LC is calculated by averaging the differences along the ROIs.

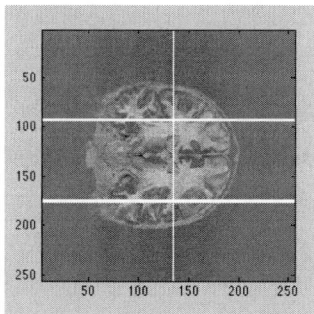

Fig. 2 Three different line profiles in a brain slice

Line Contrast calculations were performed for all 729 low/high monochromatic and 6 polychromatic energy combinations, as each one consists of 360 brain DECT images obtained for all 360 TCAs. Initially, the optimal TCA per energy combination is found and thereafter a comparison between the 729 LC results in the optimal low/high energy combination that would give a maximum contrast differentiation between the tissues that belong to the brain parenchyma.

III. RESULTS

A. Dual Energy Brain CT Image Optimization

Table 1 summarizes the first ten optimal low/high energy pair combinations of monochromatic beams and their corresponding optimal TCA. Greatest LC values and therefore superior image quality between gray and white matter is achieved for the energy pair 100 keV (low) / 110 keV (high), followed by the pairs 60/100 and 60/70 keV. Although the first two energy pairs may be beneficial in terms of reduced dose imparted to the brain, due to the high energy of the x-rays, it may not be technically achievable with the currently available CT scanners. The third energy pair 60 keV (low)/ 70 keV (high) is therefore considered as the most feasible combination and will be used in the rest of the paper for further comparisons.

These results were obtained for dose proportions set to 41.5 mGy to both 'low' and 'high' energy images. The initial experimentation with the other proportions per energy pair showed that there is no significant difference in the image quality and therefore this proportion between low and high energy images was adopted.

Table 1 Combinations of low and high energies monochromatic beams that result in the top 10 LC factor

Rank	Low Energy keV	High Energy keV	Line Contrast	Angle
1	100	110	0.4785	94
2	60	100	0.4584	167
3	60	70	0.4469	101
4	40	60	0.4283	155
5	50	60	0.4219	80
6	60	90	0.4122	83
7	80	100	0.3948	254
8	60	110	0.3591	197
9	50	90	0.3590	101
10	70	90	0.3342	21

Similarly, table 2 presents the results for DE brain CT images obtained from polychromatic beam pairs as the results are sorted by the highest LC value. Optimal combination is achieved for the combination 80 kVp (52 keV mean) and 100 kVp (59.5 keV mean), shown in figure 1.

Table 2 Combinations of low and high energies polychromatic beams sorted according LC factor

Rank	Low Energy kVp	High Energy kVp	Line Contrast	Angle
1	80	100	0.2784	311
2	80	140	0.2384	158
3	80	120	0.2266	180
4	120	140	0.2223	312
5	100	120	0.1877	164
6	100	140	0.1774	152

As it is seen for both monochromatic and polychromatic beams, the angle at which the maximum gray white contrast is achieved varies with the energy of the low and the high energy image. In addition, as excepted, the contrast between gray and white matter is higher on brain DECT images obtained with monochromatic beams compared to images obtained with polychromatic spectra by a factor of 2.

B. Visual Evaluation

The plots of the line profiles taken across the dual and single monochromatic energy images are shown together with the corresponding images for the optimal energies in figure 3. The quality of the image increases considerably in the brain DECT images where the white matter can be clearly distinguished in the brain parenchyma. In terms of Line Contrast the brain DECT image demonstrated improvement of factor of 12.6 compared with both low and high energy images.

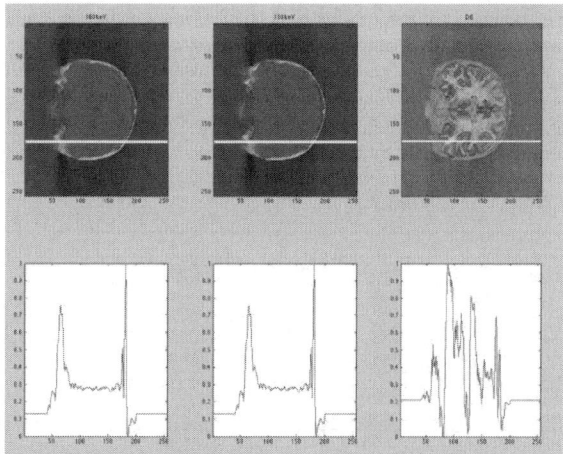

Fig. 3 Line profiles of single and dual monochromatic energy images for 100 and 110 keV beams

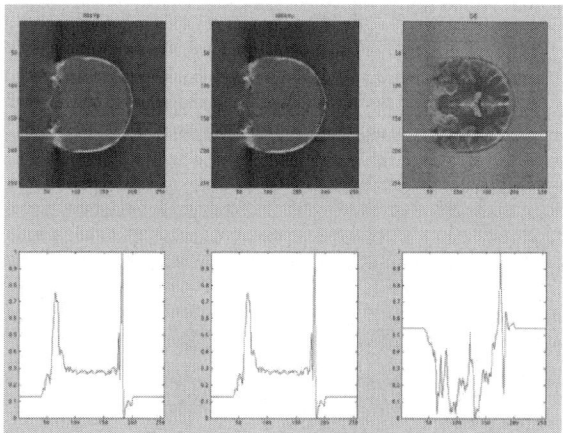

Fig. 4 Line profiles of single and dual polychromatic energy images for 80 and 100 kVp beams

Figure 4 depicts the comparison of line profiles taken across the brain CT images obtained from the optimal DE pair (80 kVp and 100 kVp) and single polychromatic energy images. Similarly to the monochromatic case, in the case of polychromatic beams the contrast of white and gray matter in the brain parenchyma is greatly increased in the brain DECT image. The LC of brain DECT image is improved with a factor of 7 compared with both low and high energy images. The improvement in this case is reduced almost half of that for monochromatic beam.

IV. CONCLUSION - FUTURE WORK

The simulation study demonstrated that the use of dual energy may be of great advantage for the visualization of the very small x-ray attenuation differences of soft tissues within the brain parenchyma when using brain CT modality. Currently a study that takes into account different noise sources such as photon scatter and complete detector response is under development. Future work will also concern the evaluation of the ability of this technique to detect brain tumors such as gliomas in their early stage as well experimental validation study with real data.

REFERENCES

1. Erkinjuntti T et al. (1987) Do white matter changes on MRI and CT differentiate vascular dementia from Alzheimer's disease? J Neurol Neurosur Ps 50:37–42
2. Johnson KA et al. (1987) Comparison of magnetic resonance and roentgen ray computed tomography in dementia. Arch Neuro 44:1075–1080
3. George AE et al. (1986) Leukoencephalopathy in normal and pathologic aging: 2. MRI of brain lucencies. AJNR 7:567–570
4. Van Swieten JC et al. (1990) Grading white matter lesions on CT and MRI: a simple scale. J Neurol Neurosur Ps 53(12):1080–1083
5. Kinnunen KM et al. (2011) White matter damage and cognitive impairment after traumatic brain injury. Brain 134(2):449–463
6. Alvarez RE et al. (1976) Energy-selective reconstruction in X-ray computerized tomography. Phys Med Biol 21:733-744
7. Macovski A et al. (1976) Energy dependent reconstruction in X-ray computerized tomography. Comput Biol Med 6:325–336
8. Gupta R et al. (2010) Evaluation of Dual-Energy CT for differentiating intracerebral hemorrhage from iodinated contrast material staining. Radiology 257:205-211
9. Vogl TJ et al. (2010) Different imaging techniques in the head and neck: Assets and drawbacks. World J Radiol 2(6):224-229
10. Aubert-Broche B et al (2006) A new improved version of the realistic digital brain phantom. NeuroImage 32(1):138–145
11. Hubbel JH and Seltzer SM (1995) Tables of x-ray mass attenuation coefficients and mass energy-absorption coefficients 1keV to 20 MeV for elements Z = 1 to 92 and 48 additional substances of dosimetric interest. NISTIR 5632, Gaithersburg
12. Lehmann la, Alvarez RE, Macovski A et al (1981) Generalized image combinations in dual KVP digital radiography. Med Phys 8(5):659–667
13. Bliznakova K, Kolitsi Z, Pallikarakis N (2006), Dual-Energy Mammography: Simulation Studies. Physics in Medicine an Biology, 51(18): 4497-4515
14. Cohnen M, Fischer H, Hamacher J et al (2000) CT of the head by use of reduced current and kilovoltage: Relationship between image quality and dose reduction. Am J Neuroradiol 21:1654-1660
15. Kolitsi Z, Panayiotakis G, Anastassopoulos V et al. (1992) A multiple projection method for digital tomosynthesis. Med Phys 19(4): 1045-1050

Author: Dermitzakis Aris
Institute: Biomedical Technology Unit, Dept of Medical Physics
Street: Patras University Campus
City: Rio, Patras
Country: Greece
Email: arisderm@gmail.com

Part V
Biomedical Signal Processing

Removal of Low Frequency Interferences from Electrocardio Signal Based on Transform of TP Segment Samples

O.V. Melnik and A.A. Mikheev

Department of Biomedical and Micro Electronics, Ryazan State Radioengineering University, Ryazan, Russia

Abstract—The questions of elimination additive low-frequency noise affecting to electrocardio signal such baseline drift and powerline interference are discussed. This paper describes novel methods for separation interference form electrocardio signal based on filtering the sequence of ECS samples taken at the TP segment in each cycle of heart rate. It is shown that in this case all of spectral components of electrocardio signal are unchanged and when converting the sample TP segments in the group of samples is possible to extend the frequency band of baseline drift which is eliminated.

I. INTRODUCTION

For correct evaluation of amplitude and time parameters of electrocardio signal (ECS) is required to have a "clean" signal that is exempt from superimposed interference when the desired information remains undistorted.

Basic methods for removing baseline drift, powerline interference and other types of disturbance widely used in practice are summarized in the book of R. M. Rangayyan "Biomedical Signal Analysis", IEEE / Wiley, 2002 [1].

Then removing the interference must to keep, as a rule, the amplitude-time parameters and the shape of the processed electrocardio signals. However, fluctuation noise and powerline interference filtering use lowpass filter (LPF) removes part of the spectral components of the ECS, which results in distortion of the rapidly changing elements (QRS-complex). When filtering baseline drift of electrocardio signal using high-pass filters (HPF) [2] forms of low-frequency components are distorted, in particular ST-segment, which adversely affects the subsequent identification of diagnostic features.

In addition to filtering one of the common ways to eliminate baseline drift is the interpolation of samples taken at the PQ-segment. These methods allow saving the form of ECS elements but have a fundamental limitation on the highest frequency in the spectrum of eliminated baseline drift. This frequency is theoretically can not be more than half of heart rate (HR), which in this case is the sampling frequency of the baseline drift signal. In real cases the limiting frequency of baseline drift can not exceed one-tenth of the heart rate. Also, PQ-segment is not always present in electrocardio signals clearly expressed and lying on the isoline, which makes the selection of samples for interpolation difficult.

All of these reasons lead to finding new ways to remove the additive low-frequency noise providing:

1) the preservation of all the spectral components of ECS while eliminating baseline drift and 50 or 60 Hz sinusoidal interference;

2) maximum increase of limit of removing baseline drift frequency (in the limit to the heart rate).

II. MATERIALS AND METHODS

One of the effective way to extract the signal of baseline drift is to filter the ECS sample sequence taken at TP segment in each cycle of heart rate. TP segment corresponds to the electrical diastole of the heart and should be laid on the zero line. Any TP segment deviation from the isoline indicates the presence of interference at electrocardio signal. Such interference may be referred to baseline drift, power frequency noice and fluctuation noise.

Sometimes after the T-wave U-wave presents. However, due to its very low amplitude and the rare occurrence U-wave practically no effect on the processing of ECS, which aims to eliminate the additive noise.

Authors were among the first in Russia who systematically started using TP segment samples in ECS processing procedures [3-5] in developing methods for removing noise.

This trend is the use of TP segment of ECS samples was further developed in the works connected with the elimination of baseline drift [6, 7] and aimed to determination the maximum bit of analog-to-digital conversion on the analysis of fluctuation noise samples [8].

To determine the feasibility of the tasks outlined above relating to the conservation of informative ECS spectral components and expansion of bandwidth of baseline drift elimination, regression analysis methods were used, methods of transformation of the spectra and spectral analysis.

III. RESULTS

TP segment duration significantly depends on the changes in heart rate and decreases with its increasing. In this regard, regression analysis according to the duration of TP segment of the heart rate was carried out with the objective of determining the maximum heart rate, which still allows to take samples from TP segment. More than 900 ECS records was analyzed.

The results lead to the conclusion that it is possible to use the TP-segment samples at a heart rate of 120 beats / min. This condition is satisfied the following cases:

- When the survey of patients conducted in a clinic or hospital, that is, in relatively calm conditions, even if the patients expressed a moderate tachycardia;
- During the screening;
- Under the control of the human condition in the course of employment with moderate physical activity that does not lead to an excessive increase in heart rate (over 120 beats per minute).

Use of Single TP Segment Samples

Fig. 1 shows the process of selection of TP-segment samples of electrocardio signal. ECS enters the input key unit 1 and unit 2 TP segment detection at each cycle of heart rate, which generates control pulses for key device. The period T of the pulse repetition frequency determined by heart rate, pulse duration τ is chosen based on the conditions task. If it is necessary to select at TP segment samples of powerline interference τ is divisible by one or more periods of noise. In the absence of power interference at the output of a key unit 1 in each cycle of heart rate samples 3 of baseline drift appear. If, in addition to the baseline drift power-frequency interference effect to ECS, at the output of 1 key device will be mix of baseline drift and power influence 4. At Fig. 1 τ is the duration equal two period of industrial frequency interference.

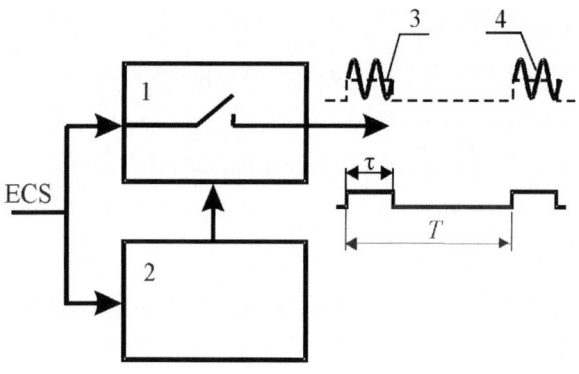

1 – key unit; 2 – TP segment detection unit; 3 – samples of baseline drift, taken at TP segment; 4 – samples of mix of baseline drift and powerline interference, taken at TP segment

Fig. 1 TP segment samples detection

At Fig. 2 and Fig.3 shown parts of the spectrum of TP segments sequence exposed at Fig. 1. Heart rate is assumed to 60 beats / min which corresponds to T=1. Fig. 2 illustrated a low-frequency part of the spectrum, Fig. 3 - band of the spectrum near the frequency of industrial noise (in Russia industrial frequency is 50 Hz).

If the sequence of samples of mix signal of baseline drift and powerline interference pass through a low-pass filter, the output filter will select all the spectral components of the isoline drift (shown at Fig. 2, Area 1), falling within its bandwidth (gain-frequency characteristic of LPF shown at Fig. 2 as dotted line). After suitable amplification extracted signal of baseline drift is subtracted from the pre-detainees to the deceleration time of filter original ECS, thus freeing it from the influence of baseline drift.

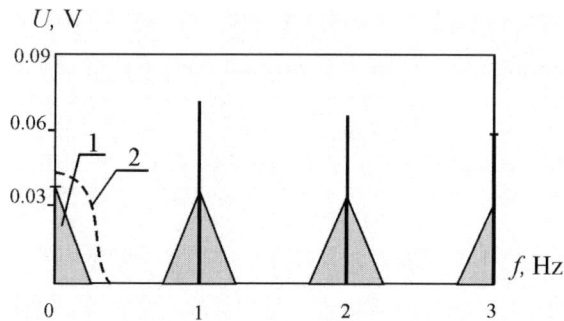

1 – zero spectral zone, contained components of baseline drift; 2 – gain-frequency characteristic of LPF

Fig. 2 Detection of baseline drift signal use LPF

Likewise, when applying a mix of baseline drift and powerline interference to the input of bandpass filter with center frequency equal to the frequency of industrial noise at the filter output will be selected component of the spectrum (indicated by 1 at Fig. 3) and its frequency will be equal to the frequency of powerline interference. The rest of the spectral components do not fall within the passband of the filter 2 and will be suppressed. After suitable amplification extracted signal is subtracted from the ECS, pre-detainees to the slowing-down time of filter, freeing ECS from the effects of powerline interference.

Once again, we note that in both cases, the filter is not distorted spectral components of the ECS as filtered samples are a mixture of baseline drift and powerline noise are taken at TP segment corresponding to the absence of electrical activity of the heart.

Use of Group of Samples Taken at TP Segment

At Fig. 2 shown that the maximum frequency in the spectrum of baseline drift can not exceed half of the heart

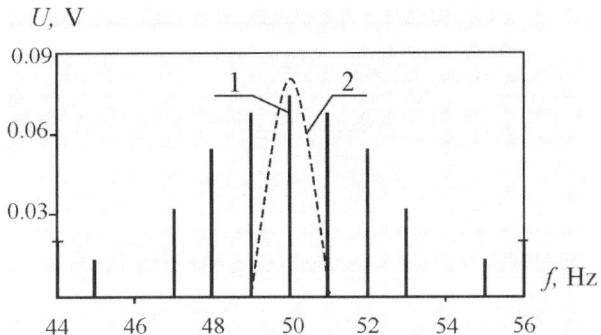

1 – spectral component of 50 Hz sinusoidal interference;
2 – gain-frequency characteristic of bandpass filter

Fig. 3 Detection of powerline interference

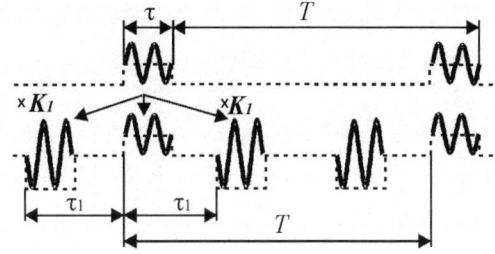

Fig. 4. Transformation of TP segment sample to the group of samples

rate which determinate the sampling frequency of interference signal. At the same time, the closer the frequency of baseline drift to half the sampling frequency, the stiffer requirements for band sharpness of the gain-frequency of low-pass filter characteristics. Also very stringent requirements are placed to the steepness of the decline of the gain-frequency response of the bandpass filter (Fig. 3) releasing the powerline interference signal.

It is possible to provide a significant reduction of requirements to the filters and extended of upper band of frequencies of baseline drift allocated to the heart rate frequency due to the transformation of each TP segment sample in the group of samples realized according to certain rules [9].

In general, to original sample added a pair of samples: one sample of each pair being disposed on the left of the original at τ_i distance, and the second - from the right at the same distance τ_i. Additional samples are multiplied by scaling factors K_i. When you add a pair of samples in the spectrum of a sequence of groups of samples, one spectral zone with any number k can be suppressed. When adding two pairs of additional samples may be suppressed any two spectral zones, etc.

In the efforts to extend the frequency range of allocated baseline drift is of interest to the suppression of spectral bands in the spectrum of the TP segment sample sequence, starting with the first spectral zone.

The process of conversion of single TP-segment sample to the group with a pair of additional samples with a suppressed spectral zone is illustrated at Fig. 4.

The scaling factor K_1 is determined by solving the equation

$$1 + 2 \cdot K_1 \cdot \cos\left(2 \cdot \pi \cdot k \cdot \frac{\tau_1}{T}\right) = 0$$

where $k = 1$ - number of suppressed spectral band.

If TP segment samples comprise a mix of baseline drift and powerline interference, then along with a first spectral band are suppressed even two spectral zones, one of which is located on the left of the industrial frequency, and the second - on the right. This case is shown at Fig. 5. By the crosses marked the suppressed spectral components.

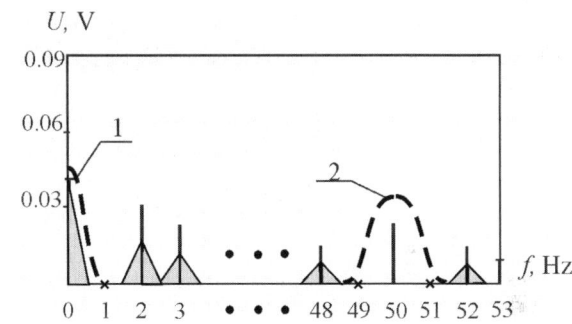

1 – gain-frequency characteristic of LPF; 2 – gain-frequency characteristic of bandpass filter

Fig. 5 The spectrum of a sequence of groups of samples with first spectral band suppressed

Thus, the suppression of the first spectral band in the spectrum of TP segment sample sequence provides both the extension of the frequency band of allocated baseline drift, reducing the requirements for the steepness of the decline of gain-frequency characteristic of the filters, as well as reliable release of the powerline interference in the event of instability its main harmonic frequency.

Proposed approach to allocate disturbance when using a digital filtering allow to use only low pass filter. This is achieved by the decimation of lowpass filter impulse characteristic. Retained only meaningful M weights, where $M = F_d / F_n$, F_d - sample rate, F_n - the frequency of power line, and coefficients located between the remaining are replaced by zeros. At the same filter than the bandwidth in the lower frequency passband appear at frequencies that are multiples of the power line. It provides simultaneous selection a signal of baseline drift and noise at power line frequency and number of its harmonics.

As in the case of single TP segments samples, when filtering of groups of samples taken at TP segment are not affected to the spectral components of the electrocardio signal, and only the additive noise signals allocated.

IV. CONCLUSIONS

Thus, we propose a new approach to elimination of low-frequency noise such baseline drift and powerline interference. It is based on detection of TP segment of electrocardio signal corresponding to electrical diastole of the heart and at the absence of interference with zero potential. The non-zero TP segment samples are samples of interference. The methods for allocation the interference based on filtering of sequences of a single TP segment sample and groups of samples are proposed.

The present methods provide:

- The preservation of all of the spectral components of electrocardio signal while eliminating baseline drift and powerline interference;
- Increasing the maximum frequency of the remove baseline drift signal to heart rate.

REFERENCES

1. Rangaraj M. Rangayyan (2002) Biomedical Signal Analysis. IEEE/Wiley, 516 p.
2. Kravchenko V.F., Popov A.Y. (1996) Sampling and digital filtering of electrocardiogram // Foreign electronics 1:38-44
3. RF Patent №2195164. A method for detection the beginning of the cardiac cycle and device for its implementation /A.A. Mikheev// Discovery. Inventions. 2002. №36
4. RF Patent №2251968. A method for baseline drift detection and device for its implementation /A. A. Mikheev, G. I. Nechaev// Discovery. Inventions. 2005. №14
5. Melnik O.V., Mikheev A.A., Nechaev G.I. (2005) Baseline drift detection. Biomedical technology and electronics. 1-2:26-30
6. Blinov P.A. (2009) Detection of Cardiac Complexes Using Cyclic Convolution. Biomedical Engineering. Vol. 43. № 6. pp. 274-275
7. RF Patent №2417050. Device for removal of isoline drift /S.P. Panko, A.V. Michurov// Discovery. Inventions. 2011. №12
8. Kazantzev A.P. (2008) Information analysis and transformation of electrocardiosignals for mobile telemedicine. Biomedical electronics 7: 22-28
9. V.V. Karasev, A.A. Mikheev, G.I. Nechaev (1996) Measuring systems for rotating components and mechanisms. Energoatomizdat, Moscow, 176 p.

Author: O.V. Melnik
Institute: Ryazan State Radioengineering University
Street: Gagarina, 59/1
City: Ryazan
Country: Russia
Email: omela111@yandex.ru

Author: A.A. Mikheev
Institute: Ryazan State Radioengineering University
Street: Gagarina, 59/1
City: Ryazan
Country: Russia
Email: maa0312@yandex.ru

Noise Reduction Using 2D Anisotropic Diffusion Filter in Inverse Electrocardiography

A. Mazloumi Gavgani and Y. Serinagaoglu Dogrusoz

Middle East Technical University / Electrical and Electronics Engineering
Department, Ankara, Turkey

Abstract—Noise in biomedical measurements has a negative effect on the accuracy of the solutions obtained. Therefore various filtering methods have been proposed in literature to reduce the effects of noise. In our previous studies we have demonstrated the application of one dimensional Anisotropic Diffusion Filter (ADF) to suppress the effects of measurement and geometric noise in body surface potential measurements (BSPM) with the goal of improving the corresponding solutions of the inverse problem of electrocardiography (ECG). In this study, we used two dimensional AD filter to cancel the noise on the body surface potentials and we compared the results to our previous observations. We used unfiltered and filtered BSPMs to estimate the epicardial potential distributions.

Keywords—Inverse problem of electrocardiography, Anistropic Diffusion Filters.

I. INTRODUCTION

Inverse problem of electrocardiography (ECG) can be described as the estimation of cardiac sources using potential measurements obtained on the body surface potential measurements (BSPM). However, due to attenuation within the body, inverse problem of ECG is highly ill-posed [1] which implies that a small amount of noise would make the problem unstable and the solution will not be reliable. This ill-posedness increases with the number of parameters in the desired solution. In order to overcome the ill-posed nature of the problem, regularization and statistical inversion techniques have been widely used in the literature. Some of these methods include Bayesian Maximum A Posteriori estimation [2], Laplacian weighted minimum norm [3], and recent studies have employed spatio-temporal approaches such as the multiple constraints method [4] and state-space models (Kalman filter) [5].

The success of these approaches is directly related to the prior constraints employed, the amount of noise in the data, and the accuracy of the geometric model obtained in the forward model, which is usually obtained by first segmenting medical images to obtain organ boundaries [6], then using numerical solution techniques such as boundary element method (BEM) [7], finite element method (FEM) [8], or a combination of both [9] to solve the forward ECG problem.

It is well-known that more accurate solutions can be obtained when the amount of noise in the data is lower [5]. Thus, in this work we have aimed to reduce the amount of noise before solving the inverse ECG problem. There are several similar studies in the literature. In [7], the authors provided a theoretical formulation to incorporate the effects of geometric noise in the forward transfer matrix. The authors in [10] compared the performances of CRESO, L-curve, and zero crossing methods in determining the optimum regularization parameter for the zero'th order Tikhonov regularization considering the geometric errors. In a study by our group, effects of geometric errors were reduced by modeling geometric error as an additional noise parameter [5]. All of these studies work with noisy data but try to suppress the effects of noise within the inverse solution framework.

On the other hand, there are studies in the literature whose purpose is to de-noise the ECG signals; wavelet transformation [11], finite impulse response (FIR) filtering [12], infinite impulse response filtering [13], adaptive filtering [14] are among the most widely used ECG filtering techniques. But these techniques usually smooth the data and fine high frequency details are lost as a result of filtering. Anisotropic diffusion filter (ADF) has been widely used in medical imaging and image processing applications for suppressing noise in the medical images [15]. It has the advantage of preserving edges while suppressing noise everywhere else. In a previous study, we proposed using one dimensional ADF to reduce the noise level in BSPM [16]; the noise in each node of the body surface potential measurements were filtered independently using ADF. In this study we have explored the use of two dimensional ADF to reduce the noise effect in the BSPM.

II. METHODS

Traditional low-pass filtering techniques such as Gaussian filters or linear diffusion, blurs the edges in the signal which causes data loss. This blurring effect is reduced in the anisotropic diffusion filter due to the adaptive filtering effect of the nonlinear diffusion filter. In our previous study we used one dimensional AD filter to smooth the BSPM used in inverse ECG problem. In that, we used the nonlinear ADF proposed by Perona and Malik [17]. Figure 1 demonstrates the principle of using one dimensional filter. In this figure it is seen that one dimensional ADF uses different time instances of each node to cancel the noise influence in the current state.

Fig. 1 One dimensional network; the black circle represents the current state of one node, the circle on the left represents the previous state, and the circle on the right represents the future state [15]

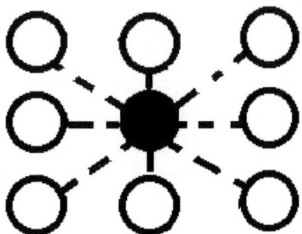

Fig. 2 Two dimensional network; the black circle represents the current state of one node, the circle on the left represents the previous state and the circle on the right represents the future state. The circles above and below represent the current, previous and future states of the neighbor nodes [15]

In this study, we used two dimensional ADF, which is a simple extension of the one dimensional ADF, to reduce the effects of noise. Figure 2 demonstrates the two dimensional network. In this figure, the black circle demonstrates the current state. 2D filter uses eight neighbors (in space and time) in the current window to smooth the signal in corresponding position. In this method the diffusion equation can be written as:

$$\frac{\partial Y}{\partial v} = div(c(t,v)\nabla Y) \quad (1)$$

Where div is the divergence operator, ∇ is the gradient operator, v is the variance, and $c(t,v)$ is the diffusion coefficient defined as:

$$c(t,v) = \frac{1}{1+(\frac{\nabla Y}{K})^2} \quad (2)$$

The diffusion strength is controlled using the diffusion constant $c(t,v)$. When the data gradient is small, the denominator is close to one; therefore smoothing amount is at its maximum. On the other hand, when the data gradient is large (e.g. at the edges), the denominator is large and the diffusion coefficient is small, thus less smoothing is performed. The K constant is chosen by the user and fixed during the application.

III. RESULTS

In this study, we filter the noisy BSPM using ADF before solving the inverse ECG problem to estimate the potentials over the heart surface (*i.e.*, the epicardial potentials). Here, BSPM were simulated from epicardial potentials that were recorded at the University of Utah Nora Eccles Harrison Cardiovascular Research and Training Institute (CVRTI) by R.S. MacLeod and his co-workers [18] from a ventricularly paced canine heart. The forward problem was solved using BEM. The number of epicardial leads used in this study is 490, and the number of simulated BSPM leads is 771. Sampling rate of the signal is 1000 Hz. Torso geometry used for simulating the BSPM included the heart, the lung and the torso surfaces. The BSPM were calculated by multiplying the epicardial potentials with the forward transfer matrix, then by adding Gaussian distributed noise. The forward matrix used in inverse solution was obtained using a homogeneous model. The results were evaluated numerically using the correlation coefficient (CC) and the relative difference measurement star (RDMS) measures.

Tikhonov regularization [1] was used as the inverse solution method. Results were obtained for three cases: 1) no filtering, 2) BSPM were filtered using one dimensional ADF, and 3) BSPM were filtered using two dimensional ADF.

Table 1 The average and standard deviation values of CC over time for different SNR values of the unfiltered data

SNR of noisy BSPM	No filter	1D ADF	2D ADF
5 dB	0.52 ± 0.30	0.62 ± 0.32	0.59 ± 0.25
10 dB	0.58 ± 0.30	0.66 ± 0.28	0.60 ± 0.24
15 dB	0.61 ± 0.27	0.72 ± 0.23	0.65 ± 0.23
20 dB	0.66 ± 0.25	0.73 ± 0.21	0.69 ± 0.23
25 dB	0.70 ± 0.24	0.73 ± 0.21	0.70 ± 0.24

Table 2 The average and standard deviation values of RDMS over time for different SNR values of the unfiltered data

SNR of noisy BSPM	No filter	1D ADF	2D ADF
5 dB	0.88 ± 0.37	0.77 ± 0.34	0.81 ± 0.23
10 dB	0.82 ± 0.35	0.73 ± 0.30	0.80 ± 0.21
15 dB	0.78 ± 0.34	0.67 ± 0.28	0.75 ± 0.22
20 dB	0.72 ± 0.30	0.66 ± 0.28	0.70 ± 0.30
25 dB	0.68 ± 0.29	0.65 ± 0.27	0.68 ± 0.28

In Table 1 and Table 2 the SNR values of the noisy BSPM are varied and the solutions of the three approaches are compared using averages of CC and RDMS values over time. These results show that when the SNR is low, filtering the BSPM using one dimensional or two dimensional ADF before solving the inverse problem enhances the reconstructed epicardial maps compared to using the unfiltered data. However, as the SNR of the BSPMs increases, the solutions with the filtered and unfiltered data produce similar results. Fig 3 demonstrates a graphical representation of the results in Tables 1 and 2. It is clear from these figures that when considering only the filtered solutions, one dimensional ADF produces considerably better results especially for low SNR data while for higher SNR values the performances of both filters are similar.

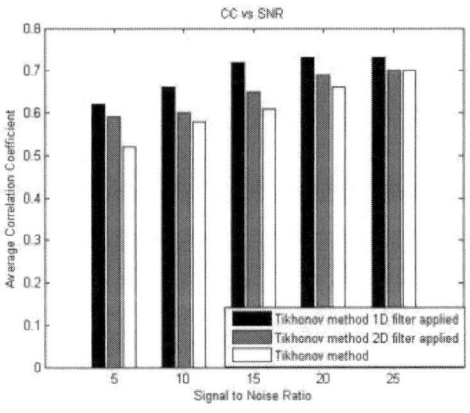

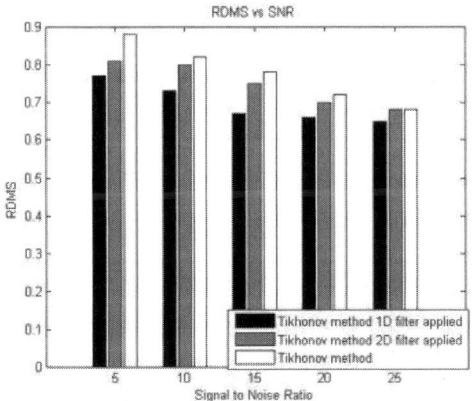

Fig 3 Average CC and RDMS values

To consider the geometric error, we used scaled and shifted heart geometries to calculate the forward matrix used in the inverse solution. Table 3 and Table 4 show the results with scaled and shifted heart geometries, respectively. Measurement noise at 5 dB and 15dB are also added to the BSPM in these simulations. These results show that both the 1D and the 2D filtered data produce better results compared to unfiltered data; 1D ADF produces better results compared to 2D ADF.

Table 3 The average and standard deviation values of CC over time for different amounts of scaling error in the forward model

	Scale amount	No filter	1D ADF	2D ADF
Data at 5 dBs	60%	0.37 ± 0.30	0.43 ± 0.32	0.43 ± 0.23
	90%	0.48 ± 0.30	0.58 ± 0.28	0.54 ± 0.23
	140%	0.54 ± 0.27	0.62 ± 0.23	0.58 ± 0.25
Data at 15 dB	60%	0.48 ± 0.30	0.59 ± 0.31	0.52 ± 0.23
	90%	0.60 ± 0.27	0.68 ± 0.24	0.62 ± 0.21
	140%	0.63 ± 0.22	0.69 ± 0.18	0.65 ± 0.20

Table 4 The average and standard deviation values of CC over time for different amounts of shift error in the +y direction in the forward model

	Shift amount	No filter	1D ADF	2D ADF
Data at 5 dB	+6y	0.49 ± 0.30	0.61 ± 0.20	0.55 ± 0.27
	+10y	0.48 ± 0.30	0.58 ± 0.28	0.53 ± 0.26
	+15y	0.45 ± 0.24	0.54 ± 0.23	0.52 ± 0.24
Data at 15 dB	+6y	0.61 ± 0.33	0.66 ± 0.31	0.62 ± 0.33
	+10y	0.57 ± 0.27	0.62 ± 0.24	0.59 ± 0.19
	+15y	0.54 ± 0.26	0.58 ± 0.24	0.56 ± 0.18

Table 5 computation time for 2.5 GHz system with 4GB RAM

Method	Computation time
No filter	5.824 seconds
1 dimensional ADF	17.603 seconds
2 dimensional ADF	7.803 seconds

The advantage of using 2 dimensional ADF over 1 dimensional ADF is in the computation time. Two dimensional filtering produces the results faster than the one dimensional filtering, as indicated in Table 5.

IV. CONCLUSION

Here we proposed to use two dimensional anisotropic diffusion filter for de-noising BSPM with the goal of improving the reconstructed epicardial potential distributions in the inverse ECG problem. Although two dimensional ADF produces weaker results compared to our previously applied one dimensional ADF, this filter has less computational time compared to the one dimensional case.

ACKNOWLEDGMENT

The Authors would like to thank Dr. Robert S. MacLeod and his co-workers for the data used in this study. A. M. Gavgani thanks The Scientific and Technological Research Council of Turkey (TUBITAK) for supporting his thesis studies.

REFERENCES

[1] Gulrajani RM (1998) The forward and inverse problems of electrocardiography. IEEE Eng Med Bio 17:84–101
[2] Serinagaoglu Y, Brooks DH, MacLeod RS (2006) Improved performance of Bayesian solutions for inverse electrocardiography using multiple information sources. IEEE Trans Biomed Eng 53(10):2024–2034
[3] He B, Wu D (2001) Imaging and visualization of 3-D cardiac electrical activity. IEEE Trans Inf Technol Biomed 5(3):181–186
[4] Brooks DH, Ahmad GF, Macleod RS, Maratos GM (1999) Inverse electrocardiography by simultaneous imposition of mul-tiple constraints. IEEE Trans Biomed Eng 46(1):3–18
[5] Aydin U, Serinagaoglu Dogrusoz Y (2011) A Kalman filter based approach to reduce the effects of geometric errors and the measurement noise in the inverse ECG problem. Medical & Biological Engineering & Computing 49(9): 1003-1013
[6] Ghanem RN, Ramanathan C, Jia P, Rudy Y (2003) Heart-surface reconstruction and ECG electrodes localization using fluoroscopy, epipolar geometry and stereovision: application to noninvasive imaging of cardiac electrical activity. IEEE Trans Med Imaging 22(10):1307–1318
[7] Stanley PC, Pilkington TC, Morrow MN (1986) The effects of thoracic inhomogeneities on the relationship between epicardial and torso potentials. IEEE Trans Biomed Eng 33(3):273–284
[8] Johnson CR, MacLeod RS (1994) Nonuniform spatial mesh adaptation using a posteriori error estimates: applications to forward and inverse problems. Appl Numer Math 14:311–326
[9] Pullan AJ, Bradley CP (1996) A coupled cubic hermite finite element/boundary element procedure for electrocardiographic Problems Comput Mech 18(5):356–368
[10] Johnston P. R, Gulrajani R. M (1997) A new method for regularization parameter determination in the inverse problem of electrocardiography. IEEE Trans Biomed Eng 44:19-39
[11] Erçelebi E (2004) Electrocardiogram signals de-noising using lifting-based discrete wavelet transform. Computers in Biology and Medicine 34(6):479–493
[12] Rossi R (1992) Fast FIR filters for a stress test system. IEEE Proc. Computers in Cardiology, Venice, Italy, 1992, pp 129-132
[13] Pottala E. W, Gradwohl J. R (1988) Comparison of two methods for removing baseline wander in the ECG. IEEE Proc. Computers in Cardiology, Washington, DC, USA, 1988, pp 493-496
[14] Laguna P, Jane R (1992) Adaptive filtering of ECG baseline wander. Engineering in Medicine and Biology Society, Proc. 14th Annual International Conference, Engineering in Medicine and Biology Society, Barcelona, Spain, 1992, pp 508-509
[15] Gerig G, Kubler O, Kikinis R, Jolesz F.A (1992) Nonlinear anisotropic deffusion filtering of MRI data. IEEE Trans Biomed Eng 11: 221-232
[16] Mazloumi Gavgani A, Serinagaoglu Dogrusoz Y (2012) Noise Reduction using Anisotropic Diffusion Filter in Inverse Electrocardiology, IEEE Proc. EMBS 33th Ann Int Conf, San diego, USA, 2012, pp 5919-22
[17] Perona P, Malik J (1987) Scale space and edge detection using anisotropic diffusion. IEEE Transactions on Pattern Analysis and Machine Intelligence 12(7): 629-639
[18] Macleod R. S, Lux R. L, Taccardi B (1997) A possible mechanism for electrocardiographically silent changes in cardiac repolarization. J. Electrocardio. 30:114–121

Author: Alireza Mazloumi Gavgani
Institute: Middle East Technical University
Street:
City: Ankara
Country: Turkey
Email: alireza.gavgani@metu.edu.tr

Detection and Identification of S1 and S2 Heart Sounds Using Wavelet Decomposition and Reconstruction

K. Hassani[1], K. Bajelani[2], M. Navidbakhsh[2], and John Doyle[3]

[1] Department of Biomechanics, Science and Research Branch, Islamic Azad University, Tehran, Iran
[2] Mechanical Engineering Department, Iran University of Science and Technology, Tehran, Iran
[3] Department of General Anesthesiology, Cleveland Clinic Lerner College of Medicine of Case Western Reserve University, Ohio, USA

Abstract—A heart sound segmentation algorithm has been developed which separates the heart sound signal into four parts: the first heart sound, the systolic period, the second heart sound, and the diastolic period. The algorithm uses discrete wavelet decomposition and reconstruction to produce phonocardiographic intensity envelopes. The performance of the algorithm has been evaluated using 14000 cardiac periods from 100 digital phonocardiographic recordings, including normal and abnormal heart sounds. In tests, the algorithm was over 93 percent correct in detecting the first and second heart sounds.

Keywords—phonocardiography, auscultation, first heart sound, second heart sound, systolic period, diastolic period, murmurs, wavelet decomposition, wavelet reconstruction, segmentation, normalized average Shannon energy.

I. INTRODUCTION

Many cardiovascular pathological conditions cause murmurs and other heart sounds aberrations before they are reflected as other findings, such as changes in the electrocardiogram (ECG) [1]. Although ECG and echocardiographic examinations are widely used in cardiac diagnosis, the old-age art of subjective heart sound analysis by auscultation is very often first performed by physicians to evaluate the functional state of the heart. However, cardiac diagnosis is seldom based on the auscultation alone due to the fact that the auscultation is subjective and depends highly on the skills of the interpreter. Still, a simple initial examination done by noncardiologists is often done before a patient suspected of cardiac disease is sent to a cardiologist for further evaluation.

In auscultation the observer tries to listen and analyze the heart sounds components separately, and then mentally synthesize the heart features. The important components of a cardiac cycle which should be identified are: the first heart sound (S1), the systolic period, the second heart sound (S2), and the diastolic period in this sequence. Important features include the rhythm, the timing and relative intensity of the heart sounds, the degree of splitting of S2, and the characteristics of any murmurs or other extra sounds. However, because it is difficult by auscultation alone to quantify these features it has been argued that more objective methods based on the quantified features of the heart sound components might make for a more reliable diagnosis.

Heart sound segmentation algorithms found in literature may be broadly classified into two types: those that require an ECG reference to synchronize the segmentation and those that do not. The latter may be further classified into supervised and unsupervised methods. In an ECG reference based approach, QRS complexes and T-waves are sometimes detected in order to locate the S1 and S2 segments, respectively [2]. With low-quality ECG signals, T-waves are not always clearly visible; in such cases, S2 sounds may be classified by an unsupervised classifier [3].

To avoid extra hardware and data acquisition needs, some researchers have tried to identify S1 and S2 sounds by without the using ECG as a reference. Here, several supervised techniques have been suggested, such as artificial neural networks [4] and decision trees [5]. Another broad approach involves unsupervised techniques such as the envelogram [6], spectrogram quantization methods [7], and self-organizing maps using Mel frequency cepstrum coefficients [8]. In practice, these methods do not always perform well for all types of heart sounds (e.g. patients with severe murmurs). Also, in the case of heart sounds produced by prosthetic valves, it is well known that these sounds are dependent on several factors such as location of implantation and type of prosthetic valve, with the consequence that these methodologies do not always provide the necessary flexibility to be applicable for all kinds of prosthetic heart valves.

One particularly reliable method applied in current state-of-the-art heart sound segmentation algorithms implicitly or explicitly involves the S1-S2 interval regularity. In this paper, a novel method aimed at the detection of S1 and S2 heart sounds is proposed which does not rely on an ECG reference or any other prior clinical information, but instead relies in part on this regularity. The purpose of this study is to develop an automatic heart sound segmentation algorithm which would be less sensitive to ambient noise and suboptimal recording conditions and uses the heart sound signal as the only data source. The paper is structured as follows: In section 2 the proposed method for S1 and S2 heart sound detection is introduced. In section 3 and 4 some results are discussed. Finally, in section 5 some conclusions are drawn.

II. METHODS

The proposed method has four key stages. In the first stage, heart sounds of 100 cases are collected and prepared for the next stage. In the second stage, heart sounds are decomposed into a series of coefficients using the Discrete Wavelet Transform. In the next stage the Shannon energy of the coefficients of the wavelet decomposed heart sounds are calculated. In addition, the identified sound lobes are physiologically validated since many artifacts and ambient noise may be captured during data acquisition. In the final stage, the heart cycles are detected by using the Shannon energy of the coefficients while the S1 and S2 heart sounds are identified based on the estimated systolic interval and the instantaneous heart rate. These steps are illustrated in the block diagram of Fig.1.

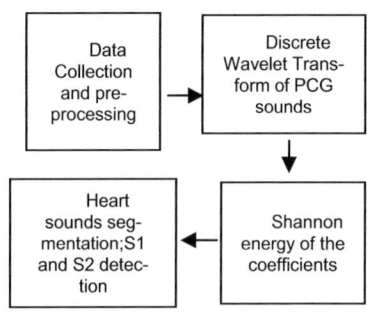

Fig. 1 The block diagram of the method

A. Data Collection and Preprocessing

We collected our heart sounds at the Pediatric Clinic of Modares Hospital in Tehran between 2010 to 2011. Data was collected from a total of 100 cases, including 8 normal cases, with an age range of 1 to 26 years. Written informed consent was obtained from the patients or their parents for publication of the case reports and accompanying images. A copy of the written consent is available for review by the Editor-in-Chief of this journal. The patients (except for the normal controls) had a history of heart murmur, which included 30 with a ventricular septal defect (VSD), 12 with an atrial septal defect (ASD), 6 with both ASD and VSD, 12 with Tetralogy of Fallot (TOF), 6 with pulmonary stenosis (PS), 34 with other kinds of diseases. In all cases, the diagnosis was confirmed by echocardiography. The summary of medical history and diagnostic findings for some of the cases is presented in Table 1.

Table 1 Summary of medical history and diagnostic findings for some of the cases

Number	Patient Age	Patient Weight (kg)	Cardiac disease
Case 1	8	21	Atrial Septal Defect
Case 2	11	25	Pulmonic Stenosis
Case 3	20	65	PDA
Case 4	6	35	Aortic Stenosis
Case 5	24	85	Healthy

Heart sound signals recorded by electronic stethoscopes are often encompassed with high frequency noise, hence preprocessing is essential [9–11]. The signals were filtered to eliminate the noise, and followed by normalization and segmentation. These steps were illustrated in the following:

Filtering: Since heart sound signals are mainly less than 800 Hz, a 2nd order Butterworth low-pass filter with a -3 dB pass band gain as the minimum and cut-off frequency of 800 Hz, was designed by digital finite impulse response (FIR) methods [11]. The Butterworth filter is a type of signal processing filter designed to have as flat a frequency response as possible in the passband. It is also referred to as a maximally flat magnitude filter.

Normalization: The amplitude of different heart sound signals were all normalized and limited to the scale of [-1 1]. The equation of normalization is in the following:

$$x_{norm}[n] = \frac{x_{PCG}[n]}{max|x_{PCG}[n]|} \quad (1)$$

An example of filtering and normalization of a normal heart sound record is shown in Fig 2.

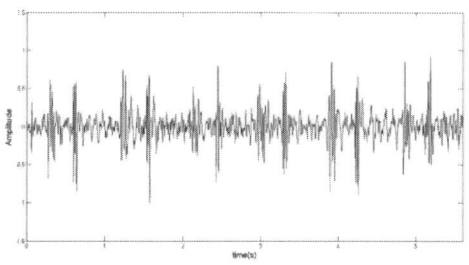

Fig. 2 A normal heart sound after filtering and normalization.

Down sampling: the original sampling rate by the Gold Wave software was set to 44100 Hz. In order to decrease the processing time and in accordance with the Nyquist theorem, the PCG signal was down-sampled using by retaining every 10th sample.

Discrete Wavelet Transform of PCG sounds

In order to identify S1s and S2s correctly, the frequency band in which the majority of power for S1 and S2 located should be used. Earlier studies using FFT to analyze the frequency contents of the first and second heart sounds have indicated that the frequency spectrum of S1 contains peaks in a low frequency range (10 to 50Hz) and a medium-frequency range (50 to 140Hz). The frequency spectrum of S2 was found to contain peaks in a low-frequency range (10 to 80Hz), a medium-frequency range (80 to 200Hz) and a high-frequency range (220 to 400Hz) [1]. However, some samples have most of the energy of S1 and/or S2 in the low-frequency range while other samples have more energy in the medium- or high-frequency range. In addition, ambient noise from different sources have quite different frequency contents and intensity characteristics. It is not possible to select a fixed frequency band filter which would be suitable to eliminate noises in all samples. All these factors made us to use several frequency band signals rather than only one in order to get a good segmentation result.

Before decomposition, the original PCG signal was down sampled by a factor of 10. Since some murmurs having higher frequency than the normal sounds (up to 600Hz) [1] are still below half of the new sampling frequency 4410Hz, any useful events of the heart sounds are not missed. After down sampling, an 11th-level discrete wavelet decomposition of the original signal was done to obtain the coefficients of all the components of the decomposition. Using these coefficients, the details and approximations at desired level were obtained by reconstruction. The details and approximations vary depending on the wavelet families and orders used in the decomposition and reconstruction. Order 2.8 Bi-orthogonal filters, which have 11 taps, were used in our work.

After the reconstruction, the original PCG signal **S** can be expressed as:

$$S = d_1 + \ldots + d_8 + d_9 + a_9 \qquad (2)$$

According to the characteristics of the frequency spectrum of S1, S2 and the possible noises, detail d9 was selected as the source for segmentation. Fig.3 shows an example of the original heart sound recording, its approximation a9 and details d8 and d9.

The selected signal d9 was segmented using a segmentation algorithm based on the envelope calculated from the normalized average Shannon energy. The average

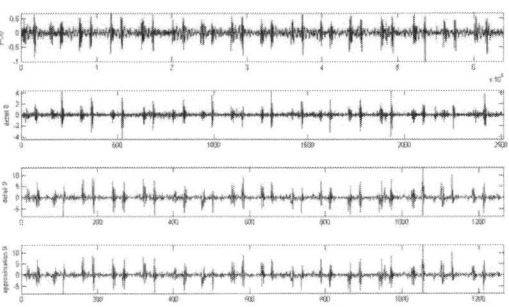

Fig. 3 Phonocardiogram and its details and approximation at level 8 and 9

Shannon energy attenuates the effects of low value noise and makes the low intensity sounds easier to be found. The average Shannon energy was calculated using the following formula:

$$Shannon\ energy\ (x) = -x^2 \cdot \log x^2 \qquad (3)$$

Heart sound Segmentation

Actual heart sound recordings are very complicated vary significantly from recording to recording. This fact preventing one from using a simple threshold to pick up all the S1s and S2s. Additionally, extra 'peaks' due to the second part of the split S2 or other events, like clicks and snaps caused by dysfunction of the heart may be present. Also, some peaks, usually the first heart sounds, can be too weak compared with other peaks to be identified. Moreover, artifacts resembling the real peaks in duration and amplitude might be recorded and can be selected as S1s or S2s.

In order to solve these problems, modifications in the threshold setting and detection rules of picking S1s and S2s must be made. First, a simple threshold was used to mark all the peak locations of continuous segments exceeding the threshold limit. Then time intervals between two adjacent marks were calculated. According to the mean value and standard variation of the intervals, both lower and higher time interval limits were calculated. These limits were used to remove extra peaks and find lost weaker peaks. Fig.4 shows an example of peak detection.

After the suspected S1s and S2s have been marked, it is needed to identify which one is S1 and which is S2. Here, the identification has based on the following facts: (a) the longest time interval between two adjacent peaks in the recording (within 20 seconds) is the diastolic period (from the end of S2 to the beginning of S1); (b) the duration of the systolic period (from the end of S1 to the beginning of S2) is relatively constant compared to the diastolic one. After the longest time interval was found, the start and the end marks of that interval. were set as S2 and S1 respectively. Then the intervals forward and backward from the longest

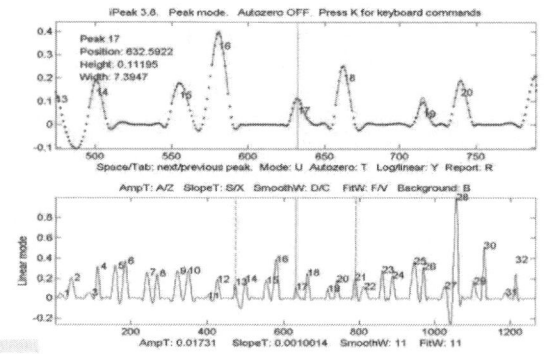

Fig. 4 Peak detection.

interval on were checked. Those marks which destroyed constancy limitations of systolic and diastolic period were discarded and the rest S1s and S2s were identified. The artifacts were discarded in this identifying procedure.

The segmentation results of d8, d9 and a9 were compared to choose a best candidate. In some cases all the three band signals gave correct results, while in others one or two gave better results than others. The chosen criterion was that which identified more S1s and S2s and less discarded peaks. The priority order of these three band signals is d9, d8 and a9, because d9 can give relatively accurate boundaries of S1s and S2s. Fig 5. shows an example of segmentation result.

III. RESULTS

Clinical Case 1

Fig. 5 shows a PCG sample from a healthy 24 year old man with an innocent heart murmur. The corresponding ECG, in order to evaluate the accuracy of proposed method, and the PCG are shown. Note that S1 and S2 sounds are detected and systolic and diastolic periods are shown by red and green markers respectively.

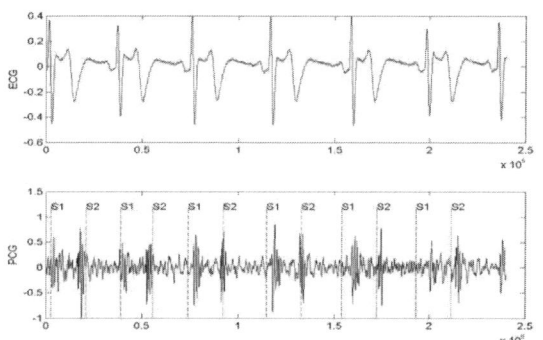

Fig. 5 Electrocardiogram and phonocardiogram without murmur in a healthy individual (a) ECG (b) PCG with detected S1, systolic period, S2 and diastolic period.

Clinical Case 2

Fig. 6 shows PCG samples for a case of ASD (atrial septal defect) in an 8 year old girl. The corresponding ECG, in order to validate the accuracy of proposed method, and the PCG are shown. Note that ASD murmurs exist in the mid-systolic period. However, the S1 and S2 sounds are detected, with the systolic and diastolic period shown by red and green markers respectively.

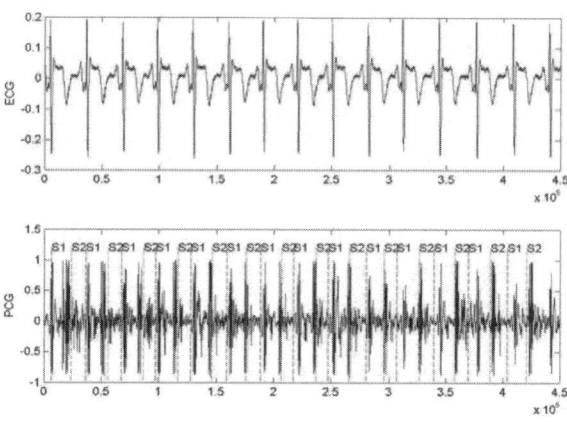

Fig. 6 Sample phonocardiogram from an ASD case (a) ECG (b) PCG with detected S1, systolic period, S2 and diastolic period.

Clinical Case 3

Fig. 7 presents PCG samples for a case of AS (aortic stenosis) in a 6 year old girl. The corresponding ECG, in order to evaluate the accuracy of proposed method, and the PCG are shown. Note that even with midsystolic murmurs, the S1 and S2 sounds are detected, with the systolic and diastolic periods shown by red and green markers respectively.

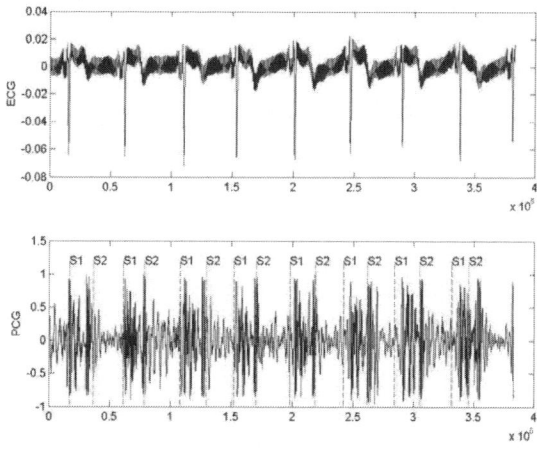

Fig. 7 Sample phonocardiogram from an AS case (a) ECG (b) PCG with detected S1, systolic period, S2 and diastolic period.

Clinical Case 4

Fig. 8 presents PCG samples for a case of PS (pulmonary stenosis) in a 11 year old boy. The corresponding ECG and the PCG are shown. Note that even with midsystolic murmurs, the S1 and S2 sounds are detected. The systolic and diastolic periods are shown by red and green markers respectively.

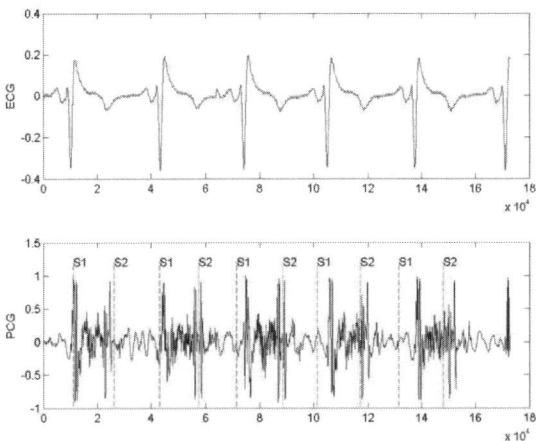

Fig. 8 Sample phonocardiogram from a PS case (a) ECG (b) PCG with detected S1, systolic period, S2 and diastolic period.

Clinical Case 5

Fig. 9 shows PCG samples for a case of PDA (patent ductus arteriosus) in a 20 year old girl. The corresponding ECG and the PCG are shown. Note that even with systolic and diastolic murmurs, the S1 and S2 sounds are detected. Systolic and diastolic period are shown by red and green markers respectively.

For these recording samples, the total cycles for recognition and the results of detection were demonstrated in Table 2. Using the proposed method with iterative recognition, 84.04% of heart beat cycles with S1 were identified, 9.76% were missed and 6.18% were false positive. For S2, 84.04% were detected, 9.76% were missed and 6.18% were false positives (FPs).

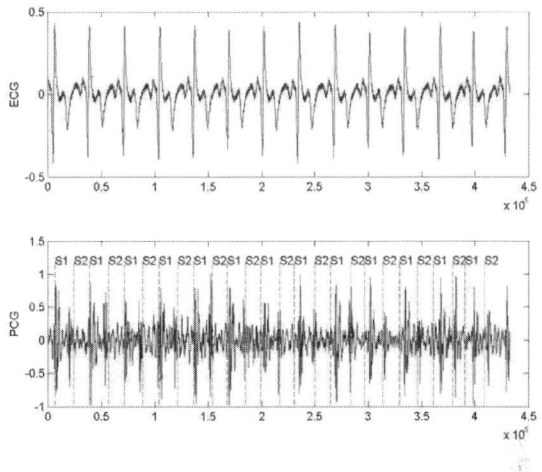

Fig. 9 Sample PCG from a PDA case phonocardiogram (a) ECG (b) PCG with detected S1, systolic period, S2 and diastolic period.

The overall performance of the algorithm was evaluated by sensitivity and precision. The detected and missed component were denoted as true positive (TP) and false negative (FN). The sensitivity and precision (also known as positive predictive value) of the method was calculated as follows:

$$sensitivity = \frac{TP}{TP+FN} \qquad (4)$$

$$precision = \frac{TP}{TP+FP} \qquad (5)$$

The sensitivity of the detection method applied to 100 patients in the database was 89.58% for both S1 and S2, and the precision for both S1 and S2 were 93.14% respectively.

disease	S1			S2			Total Cycles
	Recognized	Missed	FALSE	Recognized	Missed	FALSE	
VSD	3120	360	280	3120	360	280	3760
ASD	1920	220	140	1920	220	140	2280
TOF	1220	360	140	1220	360	140	1720
PS	1980	80	80	1980	80	80	2140
PDA	1720	200	180	1720	200	180	2100
AS	1060	180	60	1060	180	60	1300
Healthy	1200	20	20	1200	20	20	1240

IV. DISCUSSION

The detection of S1 & S2 in a cardiac cycle is first step towards characterizing murmurs and related heart diseases. The work presents a robust generalized segmentation algorithm that needs heart rate as simultaneous analysis auxiliary input and not ECG signal. It is found to work well with over 93% for 100 pathological cases, the 7% error mostly coming for a few diseases where there is an overlap of murmur frequency with normal heart sound signal. The relative position of the murmur with respect to S1 & S2 with its length and shape can be used for further classification of diseases under investigation.

One reason for the incorrect identification of S1s and S2s is contamination by high level interfering signals like speech, crying, or other ambient noises, which may overlapped with heart sounds randomly in location, intensity and frequency band and cannot be deleted in these three band signals. However, the algorithm presented in this paper can eliminate the effect of murmurs. This is a benefit of the algorithm compared to some previous ones used in our work. The sudden release of the stethoscope from the patients for a short time during the recording of data had also resulted in incorrect detection. This can be avoided by improving the recording techniques. Large intensity murmurs overlapping with S1 or S2 will make the correct segmentation and automatic analysis and identification in these three frequency bands difficult. However, because these murmurs are so intense, they cannot be neglected by auscultation or by watching the PCG signal. In our study, we roughly classify all kinds of murmurs that can be met into two classes: innocent ones and pathological ones. But in each main class, there are several subclasses that have their own quite special features.

V. CONCLUSION

The presented automatic segmentation algorithm using wavelet decomposition and reconstruction to select suitable frequency band for envelope calculations has been found to be effective to segment phonocardiogram signals into four parts. The algorithm has shown over 93 percent success in 100 recordings with different kinds of murmurs, which include 14000 cycles of heart sounds.

REFERENCES

[1] Rangayyan RM, Lehner RJ(1988)Phonocardiogram Signal Analysis: A Review. CRC Critical Reviews in Biomedical Engineering, 15: 3, 211-236.
[2] Segaier M E, Lilja O, Lukkarinen S, Srnmo L, Sepponen R, Pesonen E (2005) Computer-Based Detection and Analysis of Heart Sound Murmur. Ann Biomed Eng. 33 937-42.
[3] Carvalho P, Gil P, Henriques J, Antunes M, Eug´enio L (2005) Low Complexity Algorithm for Heart Sound Segmentation using the Variance Fractal Dimension .IEEE Int. Workshop on Intelligent Signal Processing, 593-95.
[4] Hebden J E and Torry I N (1996) Neural Network and Conventional Classifiers to Distinguish between First and Second Heart Sound. IEE Colloquium on Artificial Intelligence Methods for Biomedical Data Processing.,1-6.
[5] Stasis A Ch, Loukis E N, Pavlopoulos S A, Koutsouris D (1996) Using decision tree algorithm as a basis for a heart sound diagnosis decision support system. 4th Int. IEEE EMBS Special Topic Conf. on Information Technology Applications in Biomedicine. ,1-6.
[6] Liang H, Lukkarinen S, Hartimo I (1997) Heart Sound Segmentation Algorithm Based On Heart Sound Envelogram .Computers in Cardiology. ,105-8.
[7] Liang H, Lukkarinen S, Hartimo I (1998) A boundary modification method for heart sound segmentation algorithm. Computers in Cardiology, 593-5.
[8] Kumar D, Carvalho P, Gil P, Henriques J, Antunes M ,Eug´enio L (2006) A new algorithm for detection of S1 and S2 heart sounds. Proc. IEEE International Conference on Acoustics, Speech and Signal Processing., 1180-3.
[9] Omran S, Tayel M (2004) A heart sound segmentation and feature extraction algorithm using wavelets. 1st Int. Symp. on Control, Communications and Signal Processing, 235–8
[10] Gupta CN, Palaniappan R, Swaminathan S. Krishnan S M (2007) Neural network classification of homomorphic segmented heart sounds. J. Applied soft computing., 7 :289-97.
[11] Tseng Y L, Ko PY. Jaw F S (2012) Detection of the Third and Fourth Heart Sounds Using Hilbert-Huang Transform. Biomed Eng Online. 11-8.

Estimation of the Hemoglobin Glycation Rate Constant Based on the Mean Glycemia in Patients with Diabetes

P. Ładyżyński[1], P. Foltyński[1], J.M. Wójcicki[1], M.I. Bąk[2], S. Sabalińska[1],
J. Krzymień[2], J. Kawiak[1], and W. Karnafel[2]

[1] Nałęcz Institute of Biocybernetics and Biomedical Engineering, Polish Academy of Science, Warsaw, Poland
[2] Department and Clinic of Gastroenterology and Metabolic Diseases, Medical University of Warsaw, Warsaw, Poland

Abstract—Glycated hemoglobin A1c (HbA1c) has been the gold standard index of the metabolic control in the diabetes treatment for 30 years. The goal of this study was to estimate the overall glycation rate constant (*k*) in a simple mathematical model of HbA1c formation for patients with type 1 and type 2 diabetes based on the mean glycemia calculated using the continuous glucose monitoring data. The study group consisted of 10 participants including 5 patients with type 1 diabetes (1 women and 4 men aged 50 ± 19 years) and 5 patients with type 2 diabetes (3 women and 2 men aged 57 ± 10 years) with stable glycemic control. The mean *k* in the whole study group was equal to $1.37 \pm 0.25 \times 10^{-9}$ L/(mmol s). The mean *k* values calculated separately for patients with type 1 and type 2 diabetes were equal to $1.40 \pm 0.30 \times 10^{-9}$ L/(mmol s) and $1.34 \pm 0.21 \times 10^{-9}$ L/(mmol s), respectively (p = 0.60). The obtained results show that the hemoglobin glycation rate constant in patients with type 1 and type 2 diabetes with stable metabolic control is close to the value of this parameter in non-diabetic volunteers. However, a higher inter-subject variability of *k* estimated in patients with diabetes suggests that some additional factors might influence a pace of HbA1c formation that are not present in the simplistic model that was studied.

Keywords—Hemoglobin A1c, Mathematical modeling, Continuous glucose monitoring, Diabetes mellitus.

I. INTRODUCTION

Diabetes is one of the most challenging problems of the global health care system. In 2011 it was 366 million patients with diabetes aged from 20 to 79 years around the world. This number will increase up to 552 million in 2030 according to predictions of the International Diabetes Federation [1].

Glycated hemoglobin A1c (HbA1c) has been the gold standard index of the metabolic control in the diabetes treatment for 30 years. HbA1c is the major form of glycated hemoglobin constituting 60-80% of all glycated hemoglobin. HbA1c is a product of the non-enzymatic reaction that bounds a glucose particle to N-terminal valine residue of the β-chain of hemoglobin A0 [2]. HbA1c concentration depends on three processes, i.e. kinetics of the glycation reaction, formation of the red blood cells (RBCs) in bone marrow and elimination of RBCs from circulation.

Mathematical modeling is one of the methodologies that have been used to reproduce these processes and to assess HbA1c formation. The ultimate goal of such a modeling is to better understand a relationship bounding HbA1c and the preceding glycemia and to simulate different clinical scenarios, e.g. a reaction of HbA1c on the sudden deterioration of the glycemic control, an influence of the anemia on HbA1c, etc.

A few HbA1c models have been studied so far [3-6]. The model proposed by B.W. Beach [3] was demonstrated to be able to predict changes of HbA1c with a high accuracy [5]. This is a non-linear chemical model assuming first-order reaction in respect to glucose and hemoglobin, which is characterized by a single overall glycation rate constant (*k*). Since HbA1c is formed during the whole life span of RBCs (i.e. approx. 4 months), it is necessary to approximate glucose concentration over such a long period to estimate *k* for a particular individual.

The goal of this study was to estimate the overall glycation rate constant for patients with type 1 and type 2 diabetes based on the mean glycemia calculated using the continuous glucose monitoring data.

II. MATERIALS AND METHODS

A. Model of HbA1c Formation

According to the model that was used, HbA1c formation in each single RBC can be expressed by a simple differential equation [3]:

$$\frac{d[HbA(t)]}{dt} = -k[HbA(t)][G(t)] \qquad (1)$$

where square brackets denote a measure of concentration and *G* stands for glucose, *HbA* for non-glycated hemoglobin, *k* for the overall glycation rate constant and *t* for time. The following assumptions, that had been validated earlier [5, 6], were used to estimate *k*:

- RBCs' life span is constant and equal to 120 days.
- RBCs' count *in vivo* is constant.
- HbA1c concentration in the newly generated reticulocytes is equal to zero.

- Concentration of all other types of hemoglobin except HbA and HbA1c is negligible.
- Influence of hemoglobin loss from RBCs on HbA1c is negligible.
- Short-term glucose variability does not influence HbA1c formation in individuals with stable long-term metabolic control (i.e. with constant mean glycemia).

Under these assumptions, the total amount of non-glycated hemoglobin (*HbA*) at the time of *HbA1c* testing (t_0) is given by the following formula [6]:

$$[HbA(t_0)] = \int_{t_0-T}^{t_0} \left([HbA_0] e^{-k[G]\int_t^{t_0} dt} \right) d\tau = \frac{[HbA_0]}{k[G]} \left(1 - e^{-kT[G]} \right) \quad (2)$$

where T is a life span of RBCs, $[G]$ is considered to be constant and equal to the mean glycemia over T and HbA_0 is the amount of hemoglobin in RBCs released to the blood stream in each time unit.

Taking into consideration the hemoglobin mass balance and the fact that *HbA1c* is measured and reported as a percentage of the total hemoglobin, a formula expressing *HbA1c* concentration in percents of the total hemoglobin can be obtained [6]:

$$HbA1c(t_0) = 100 \times \left(1 - \frac{1 - e^{-kT[G]}}{kT[G]} \right) \quad (3)$$

Eq. 1 is used to estimate k based on the mean G over RBCs' life span T and *HbA1c* measured at t_0.

B. Continuous Glucose Monitoring

The average plasma glucose concentration over the assumed RBCs' life span of 120 days was estimated based on the continuous glucose monitoring in the interstitial fluid using a Guardian RT system (Medtronic Diabetes, Northridge, CA, USA). Each subject had three glucose sensors applied. Each monitoring period lasted for 5 to 6 days. The assumed time span between the first two monitoring periods was equal to 1 month and the third monitoring period was initiated 2–3 weeks after the second one. Glucose measurements were stored every 5 minutes. The sensors were calibrated at least 4 times a day using the capillary blood glucose concentration as a reference measured with a portable glucometer Accu-Chek Go (Roche Diagnostics, Basel, Switzerland). According to the current recommendation of the International Federation of Clinical Chemistry and Laboratory Medicine, glucose concentrations reported by Guardian RT were rescaled to reflect the corresponding plasma glucose concentrations as if the measurements were done with the gold standard glucose analyzer YSI 2300 Stat Plus (Yellow Springs Instruments Inc., Yellow Springs, Ohio, USA). The exact procedure is described elsewhere [5, 6]. Thus, all glucose concentrations reported in this work reflects plasma glucose equivalents.

C. HbA1c Measurements

The HbA1c concentration was measured repeatedly (4-5 times) after finishing the third glucose monitoring period, using the cation-exchange HPLC method with a D-10 analyzer (Bio-Rad Laboratories, Hercules, CA, USA). All the HbA1c values reported in this work are unaltered raw data, which are in agreement with the HbA1c scale most commonly used in medical practice. However, an identification of k was performed using the HbA1c values corrected according to the master linear equation bounding the results reported by D-10 analyzer with the values calibrated according to the most specific reference methods to obtain unbiased HbA1c concentrations. It was demonstrated earlier that this procedure allowed to avoid overestimation of the glycation rate constant [5].

D. Study Group

The study group consisted of 10 patients with diabetes including 5 patients with type 1 diabetes (1 women and 4 men aged 50 ± 19 years) and 5 patients with type 2 diabetes (3 women and 2 men aged 57 ± 10 years) with stable glycemic control. The stability of glycemic control was confirmed by small differences between the mean glucose concentrations registered during three monitoring periods, which were equal to 0.34 ± 0.22 mmol/L and 0.11 ± 0.10 mmol/L in patients with type 1 and type 2 diabetes, respectively.

The study adheres to the Declaration of Helsinki, the subjects provided the informed written consent for taking part in the study and the Local Ethical Committee approved the study protocol.

III. RESULTS

A. Descriptive Rules

The mean glucose concentrations for 10 study participants are presented in Fig 1. In average glucose concentration was equal to 7.5 ± 1.3 mmol/L in the whole study group. It was higher and more diversified in patients with type 1 diabetes (8.2 ± 1.5 mmol/L) than in patients with type 2 diabetes (6.8 ± 0.7 mmol/L). However, the differences were not significant according to the Mann-Whitney test ($p = 0.12$).

The HbA1c concentrations measured in each patient after finishing the third monitoring period are shown in Fig. 2.

The mean HbA1c was equal to $6.8\% \pm 1.1\%$ in the whole study group. Patients with type 1 diabetes had worse metabolic control as described by HbA1c ($7.4\% \pm 1.4\%$) than

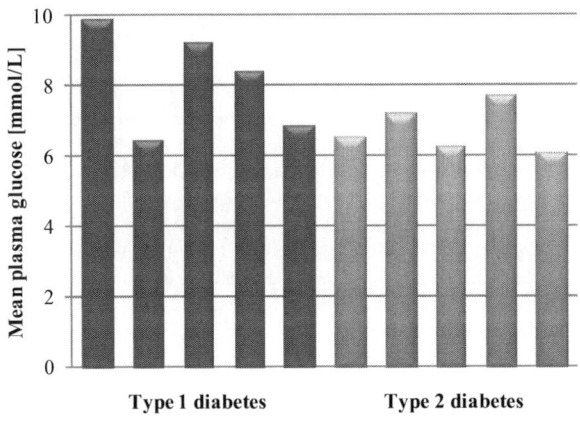

Fig. 1 Mean plasma glucose for 10 study participants

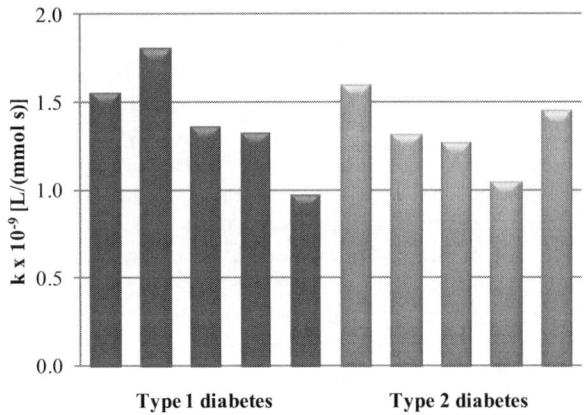

Fig. 3 Hemoglobin glycation rate constant (k) calculated for 10 study participants

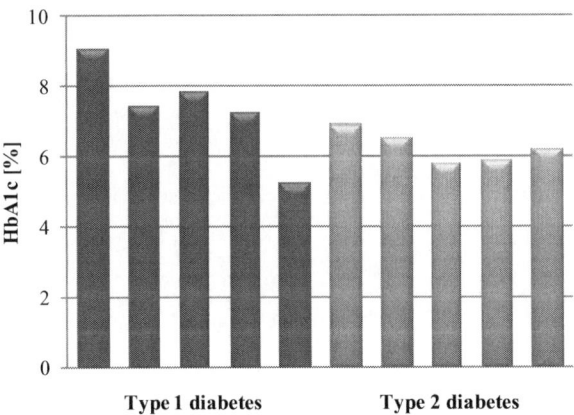

Fig. 2 HbA1c at the end of the third glucose monitoring period for 10 study participants

patients with type 2 diabetes (6.3% ± 0.5%). The mean difference was not significant (p = 0.12) due to a high inter-patient variability.

Based on the mean glucose concentrations estimated using data recorded in the three monitoring periods, the measured HbA1c values and eq. 3 we calculated k, that made it possible to equalize the measured HbA1c with the calculated one for each individual participant of the study (Fig. 3).

The mean glycation rate constant in the whole study group was equal to $1.37 ± 0.25 \times 10^{-9}$ L/(mmol s). The mean values of k calculated separately for patients with type 1 and type 2 diabetes were equal to $1.40 ± 0.30 \times 10^{-9}$ L/(mmol s) and $1.34 ± 0.21 \times 10^{-9}$ L/(mmol s), respectively (p = 0.60).

The mean value of k for the whole study group as well as mean values of k calculated for patients with type 1 and type 2 diabetes were in a good agreement with the value of this parameter ($1.27 ± 0.12 \times 10^{-9}$ L/(mmol s)) in a group of non-diabetic volunteers (p = 0.45 in the Kruskal-Wallis non-parametric ANOVA) that was estimated using the same methodology and reported earlier [6]. However, a trend towards an increase of the mean k for the higher mean glycemia was noted. The inter-subject variability expressed by the variance of k was higher in diabetic than in non-diabetic subject according to the F-test (p = 0.017). The lowest coefficient of variation was reported earlier in the healthy subjects (9.2%) [6] and the highest was registered in this study for patients with type 1 diabetes (21.6%).

IV. DISCUSSION

In this study the overall hemoglobin glyction rate constant was estimated for the simple mathematical model of HbA1c formation in a group of patients with type 1 and type 2 diabetes and with a stable long-term glycemic control. To simplify estimation of this parameter it was assumed that glycemia was constant during the life span of RBCs and equal to the mean plasma glucose approximated based on the continuous glucose monitoring in the interstitial fluid. It was demonstrated earlier that in case of non-diabetic subjects this methodology made it possible to obtain values of k only marginally different from those estimated using methods taking into consideration the continuous course of the plasma glucose [6].

The mean value of k estimated in this study is in good agreement with values of this parameter reported earlier based on *in vivo* and *in vitro* studies, when methodological differences are accounted for [6].

A tendency, which was noted, showing that the higher mean glycemia the higher mean value and the inter-subject variability of *k*, might be related to the limited number of subjects in the study group and a few sources of errors, that were not accounted for, such as the analytical random errors, imprecision of calculations and errors related to validity of the simplifying assumptions. However, this tendency might also show that the glycation model used was to simple and that more accurate model should take into consideration some additional factors influencing the glycation rate such as pH, oxidative stress, enzymatic deglycation or inhibitors of the glycation reaction [7].

V. CONCLUSIONS

The obtained results show that the hemoglobin glycation rate constant in patients with type 1 and type 2 diabetes with stable metabolic control is close to the value of this parameter in non-diabetic volunteers. However, a higher inter-subject variability of the glycation rate constant estimated in patients with diabetes suggests that some additional factors might influence a pace of HbA1c formation that are not present in the simplistic model that was studied.

ACKNOWLEDGMENT

This work was supported by a research grant from the National Centre for Science (Grant no. N N518 289340).

REFERENCES

1. IDF at http://www.idf.org/
2. Hoelzel W, Weykamp C, Jeppsson JO et al (2004) IFCC reference system for measurement of hemoglobin A1c in human blood and the national standardization schemes in the United States, Japan, and Sweden: a method-comparison study. Clin Chem 50:166–174
3. Beach KW (1979) A theoretical model to predict the behavior of glycosylated hemoglobin levels. J Theor Biol 81:547–561
4. Mortensen H B, Christophersen C (1983) Glucosylation of human haemoglobin A in red blood cells studied in vitro. Kinetics of the formation and dissociation of haemoglobin A1c. Clin Chim Acta 134:317–326
5. Ładyżyński P, Wójcicki JM, Bak M et al (2008) Validation of hemoglobin glycation models using glycemia monitoring in vivo and culturing of erythrocytes in vitro. Ann Biomed Eng 36:1188–1202
6. Ładyżyński P, Wójcicki JM, Bak M et al (2011) Hemoglobin glycation rate constant in non-diabetic individuals. Ann Biomed Eng 39:2721–2734
7. Cohen RM, Franco RS, Khera PK et al (2008) Red cell life span heterogeneity in hematologically normal people is sufficient to alter HbA1c. Blood 112:4284–4291

Author: Piotr Ładyżyński
Institute: Nałęcz Institute of Biocybernetics and Biomedical Engineering, Polish Academy of Sciences
Street: 4 Trojdena
City: 02-109 Warsaw
Country: Poland
Email: Piotr.Ladyzynski@ibib.waw.pl

Spatial Dynamics of the Topographic Representation of Electroencephalogram Spectral Features during General Anesthesia

B. Direito[1], C. Teixeira[1], B. Ribeiro[1], A. Dourado[1], M.P. Santos[2], and M.C. Loureiro[2]

[1] Centre for Informatics and Systems, University of Coimbra, Coimbra, Portugal
[2] Anesthesiology Service, Centro Hospitalar e Universitário de Coimbra, Coimbra, Portugal

Abstract—The goal of this study is to identify spatial distributions of the EEG recordings related to the anesthesia depth. Serious adverse effects may result from the inaccurate estimation of anesthesia depth during surgery.

We analyzed the spatio-temporal distribution of spectral electroencephalogram (EEG) collected from patients subjected to general anesthesia. The approach is based on the topographic mapping of spectral features of the EEG recordings, and the analysis of the positions of the maxima and minima over the anesthesia cycle.

We found different patterns depending on the analyzed feature and on the anesthesia phase. The maxima of the spectral edge frequency (SEF) feature are closer to the nasion (frontal and temporal areas) before anesthesia, shift to farther regions during the anesthetic period, and return to regions close to the nasion during the recovery period. The minima of relative power of Gamma-band present the opposite behavior, i.e., closer to the nasion during the anesthetic period, and far apart before anesthesia and during recovery).

Keywords—Anesthesia, EEG, Topographic maps.

I. INTRODUCTION

Traditionally, the monitoring of the anesthetic procedure, i.e., the discrimination of the different depths of anesthesia, has been performed by subjective methods, including the analysis of the autonomic nervous system (temperature, ventilation, pupil diameter, arterial blood pressure, and other hemodynamic parameters), patient response to surgery, and to external stimuli.

In the past decades depth anesthesia monitors began receiving a considerable attention. An appropriate estimation of anesthesia depth would allow the implementation of tailored drug administration solution avoiding awareness and side effects [1]. These anesthesia monitors are based on indexes that relate to the patient consciousness state during surgery. The indexes are computed using features extracted from raw electroencephalogram (EEG) recordings, and usually aimed at quantifying the shift from high to low frequencies and *vice-versa*.

Different monitors are commercially available based on different EEG measures, such as Bispectral Index (BIS), Narcotrend, M-Entropy, Patient State Index, and aepEX [2]. BIS (Aspect Medical Systems) is the most widely used and was approved by the Food and Drugs Administration (FDA) in 1996. The algorithm behind BIS is complex and partially unknown. The known information reports that it is based on a combination of time and frequency domain features of the EEG signal collected in the frontal scalp region. Clinical evidence and literature review show that the actual intra-operative monitoring with 'top' commercial equipment does not guaranty the absence of significant errors, including the BIS monitor [2].

This fact fosters the search for other features from the EEG signal, aiming to improve the reliability of the available EEG monitors.

The objective of our study is to analyze the spatial distribution of spectral features from the EEG signal during the period beginning prior to the anesthesia induction, until the recovery period (after surgery). Based on the topographic mapping of spectral features, we determine the trajectory of specific points (centers of maxima and minima regions) over the anesthesia cycle. The trajectory is quantified over time as the distance between each point and a reference defined as the nasion (depressed area between the eyes) [3]. We hypothesize that both increase and drop in the brain activity may reflect spatial variations, so both the minima and the maxima need to be evaluated.

II. DATA

Patients were subjected to surgery and general anesthesia at the operating rooms of the Centro Hospitalar e Universitário de Coimbra (CHUC), Portugal. The CHUC ethical commission previously approved data collection, and the patient's consent was signed allowing us to analyze their data. Patients were monitored simultaneously by the standard operating room instrumentation and by additional EEG electrodes placed over the scalp, according to the 10-20 system [3]. The standard monitoring included respiratory parameters, hemodynamic parameters, and BIS module parameters (including the BIS index). The EEG electrodes are connected to an amplifier (SD-LTM32, Micromed, Italy) that sampled the analogue signals at 256 Hz.

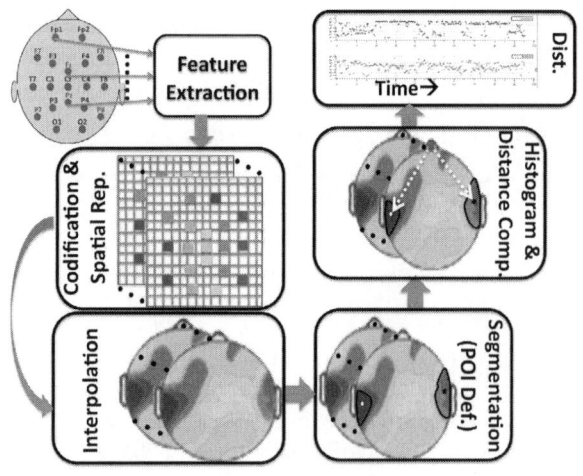

Fig. 1 Procedure to compute the spatio-temporal distribution of spectral electroencephalogram (EEG) features, collected from patients subjected to general anesthesia

III. METHODS

The methodology employed is summarized in Fig. 1. More details on the topographic maps development can be obtained form [4].

The first step is the topographical representation of the data at each instant. Topographic mapping is a neuroimaging technique for representing the spatial brain activity codified as colored contours [5]. They express the amplitude of the EEG or a related feature, and may help in clinical diagnosis by joining the temporal dimension of the EEG with the spatial dimension [6,7]. The topographic representation of the data was based on a subset of 19 electrodes: Fp1, Fp2, F7, F8, F3, F4, Fz, FT9, FT10, T7, T8, C3, C4, Cz, P7, P8, P3, P4, Pz, O1, and O2.

Next, three spectral features were computed for each electrode, aimed at the characterization of the low frequencies, high frequencies and shift between them. The features considered were: relative powers in the Delta (0.5-4Hz) and Gamma (30-128Hz) frequency sub-bands, and the spectral edge frequency (SEF). The spectral edge frequency represents a quantification of the spectral power distribution. Most of the spectral power is comprised in the 0Hz – 40Hz band. We define spectral edge frequency as the frequency below which 50% of the total the total power of the signal is located. The relative power in the Delta and Gamma bands quantify localized changes in the very low and very high frequencies, respectively. SEF measures broader spectral changes, including changes on the intermediate frequencies.

To compute the features over time we considered non-overlapping windows of five seconds each. To obtain the topographic representation we consider a matrix with 67x67 positions, where a color is associated to discrete positions related to the electrode positions. Red corresponds to the highest feature value, while dark blue to the lowest feature value. The color of the rest of the matrix positions was obtained by cubic spline interpolation. As time elapses (i.e. each five seconds), the topographic maps changes, capturing spatial characteristics in the brain activity. The next step is the segmentation of the map to extract the maximum and minimum of the map. To perform the segmentation task, we used the n-cut algorithm (the method treats the map segmentation as a graph partition problem) [8].

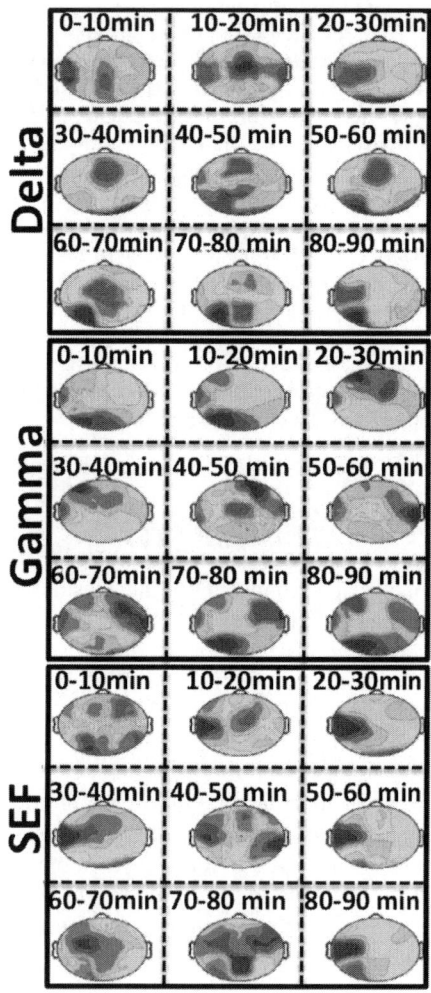

Fig. 2 2D histograms obtained for an exemplary patient.

To analyze the spatial dynamics of the spectral features we considered two approaches. On the one hand, for each topographic map (i.e. every 5 seconds), we compute the Euclidean distance between the maximum (and the minimum) and the matrix cell closer to the nasion. This allowed us to track the maxima and minima over time.

On the other hand, in order to improve visualization, we used intervals of 10 minutes to create 2D histograms based on the positions of the maxima and minima over that period. Each point of the 67 by 67 grid corresponds to a bin in a 2-D histogram. Consequently, each 2-D histogram represents the maxima and minima regions of 240 topographic maps.

IV. RESULTS

Examples of the obtained 2D histograms are depicted in Fig. 2 for an illustrative patient. We can observe several distinctive characteristics before anesthesia administration (0-10min). Considering the Delta-band, higher activity was recorded mostly in the parietal lobe. The maxima steadily shifted from the parietal areas to temporal and frontal areas during the surgery (intervals 10-80 minutes). In the final minutes of the recording, the maxima returned to the temporal and parietal areas. The minima shifted from left temporal regions to the occipital area.

The Gamma-band presents a different pattern. The maxima are predominant in the left temporal region throughout the entire EEG recording. However, the minima shift from occipital regions to temporal and frontal regions while returning to occipital regions by the end of the intervention.

The SEF seems to suggest an occipital accentuation (higher values of SEF in occipital electrodes) during the anesthesia period.

Another interesting approach to understand the spatio-temporal dynamics is to track the distance between a reference point and the maxima and the minima (for each segment, i.e. every five seconds). The idea is to determine if there are differences in the position of maxima and minima prior to drugs administration and during surgery. Since we are looking for variations from frontal to occipital areas instead of laterality, we defined the reference point as the nasion. Fig. 3 illustrates the distances computed for the different features as function of time, for the same patient as presented in Fig. 2. The BIS index, computed by the operating room instrumentation, describing the anesthesia cycle is presented in Fig. 3g. Considering the relative power in the Delta-band, no specific pattern seems to emerge. The minimum (Fig. 3b) presents some fluctuation before anesthesia administration. More significant variations are present in the positions of the SEF and the relative power of Gamma-band. The maxima of SEF (Fig. 3e) are closer to the nasion (frontal and temporal areas) before anesthesia, shift to further regions during the anesthetic period, and return to regions close to the nasion during the recovery period. The minima of relative power of Gamma-band (Fig. 3d) present the opposite behavior (closer to the nasion during the anesthetic period, and far apart before anesthesia and during recovery). To mention that the dispersion on the distance measures during surgery are due to the EEG contamination by artifacts generated by the electronic scalpel.

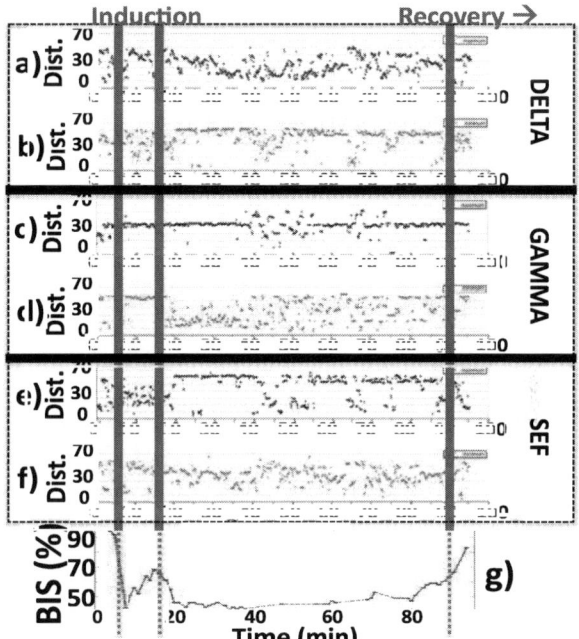

Fig. 3 Distances between the spatial-temporal maximum and minimum of the different features and the nasion, as compared with the BIS curve index (g). a), c) and e) represent the distance of the maximum. b), d) and f) represent the distance of the minimum. The vertical red lines indicate the induction and the starting of the recovering period.

V. DISCUSSION AND CONCLUSIONS

The measure of the depth of anesthesia is still a challenging problem. Our hypothesis is that the analysis of the spatio-temporal dynamics would be an important source of information to determine the consciousness of patient during anesthesia. BIS index (one of the state-of-the-art monitors of anesthesia depth) uses the information of a couple of electrodes connected to the forehead of the patient, possibly ignoring valuable spatio-temporal information available in other regions.

In this study, we showed coherent shifts of the maxima and minima with the anesthesia cycle. These changes occur in space (frontal to occipital regions), time (before and after drug administration) and frequency (the Gamma-band, Delta-band and SEF show particular behaviors).

Further work is necessary to evaluate the statistical significance of these variations. Correlation with other variables should also be faced (e.g., drugs used, age, gender).

ACKNOWLEDGEMENTS

Project iCIS (CENTRO-07-0224-FEDER-002003). CT was hired under the Ciência 2007 program of FCT. BD is supported by a grant from FCTUC.

REFERENCES

1. Bruhn J, Myles P, Sneyd R, Struys M Depth of anaesthesia monitoring: what's available, what's validated and what's next? Br J of Anaesth. 2006; 97(1): 85-94.
2. Voss L, Sleigh J. Monitoring consciousness: the current status of EEG-based depth of anaesthesia mon- itors Best Pract Res Clin Anaesthesiol. 2007;21:313 - 325.
3. Niedermeyer E, Lopes Da Silva FH. Electroencephalography: basic principles, clinical applications, and related fields. Lippincott Williams & Wilkins 2005.
4. Direito B, Teixeira C, Ribeiro B, Castelo-Branco M, Sales F, Dourado A. Modeling epileptic brain states using EEG spectral analysis and topographic mapping J Neurosci Methods. 2012;210:220 - 229.
5. Sandini G, Romano P, Scotto A, Traverso G. Topography of brain electrical activity: a bioengineering approach. Med Prog Technol. 1983;10:5.
6. Caat M, Maurits NM, Roerdink JBTM. Data-Driven Visualization and Group Analysis of Multichannel EEG Coherence with Functional Units Visualization and Computer Graphics, IEEE Transactions on. 2008;14:756-771.
7. Alba A., Marroquín J. L., Arce-Santana Edgar, Harmony T.. Classification and interactive segmentation of {EEG} synchrony patterns Pattern Recognition. 2010;43:530 - 544.
8. Shi J., Malik J.. Normalized cuts and image segmentation Pattern Analysis and Machine Intelligence, IEEE Transactions on. 2000;22:888-905.

Author: César A. D. Teixeira
Institute: Centre for Informatics and Systems (CISUC)
Street: Polo II, Pinhal de Marrocos
City: Coimbra
Country: Portugal

The Properties of the Missing Fundamental of Complex Tones

T. Matsuoka[1,2] and Y. Iitomi[3]

[1] MIT, Cambridge, U.S.A.
[2] Faculty of Engineering, Tohoku University, Sendai, Japan
[3] Faculty of Engineering, Utsunomiya University, Utsunomiya, Japan

Abstract—The existence of the missing fundamental phenomenon is known, but its mechanism is unknown. We showed how the information of the missing fundamental f_0 explicitly appeared on the aggregated autocorrelogram of the output pulse train for input signal f_1 to one cochlear model and the output pulse train for input signal f_2 to another cochlear model (where $f_1=nf_0$ and $f_2=(n+k)f_0$). In practice, we listen to a complex tone of f_1 and f_2 with each ear. In this paper, we try to investigate the influence of the pitch and the number of harmonic components of complex tones on perceiving the missing fundamental.

Keywords—the missing fundamental, complex tone, perception, cochlear models experiment.

I. INTRODUCTION

The frequency band of a telephone line channel is from 300 Hz to 3400 Hz. Although pitch frequency of speech is not in the frequency band, we can perceive the pitch over the telephone. The mechanism is unknown. It is considered that the missing fundamental is produced in the auditory system when we listen to a complex tone of $f_1=nf_0$ and $f_2=(n+k)f_0$ [1]. f_0 is known as the missing fundamental.

We confirmed that subjects were able to perceive the missing fundamental by the significant difference test (level 5 %) of the psycho-acoustic experimental results. In the psycho-acoustic experiments, subjects simultaneously listened to a pure tone of f_1 with his/her one ear and a pure tone of f_2 with his/her another ear [2]. In practice, we listen to a complex tone of f_1 and f_2 with each ear. .Subjects perceived the missing fundamental f_0 more easily in the psycho-acoustic experiment for two complex tones than for two pure tones. We made clear the mechanism and presented it at BIOSIGNAL2010 [3]. We try to investigate the influence of the pitch and the number of harmonic components on perceiving the missing fundamental.

II. METHODS

We made cochlear models shown in Fig. 1 (The figure is the example in the case of two cochlear models. In this paper, we use two cochlear models, three cochlear models, etc. [4], [5]). We showed how the information of the missing

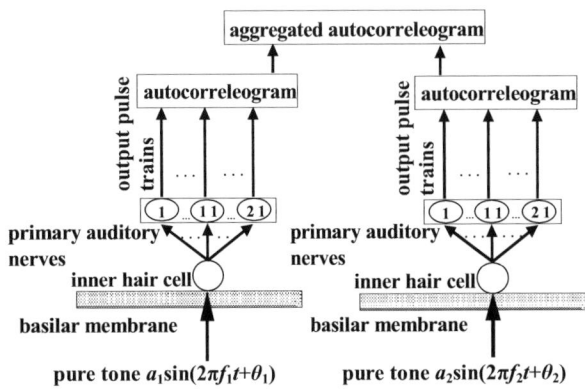

Fig. 1 Cochlea models, autocorrelogram and aggregated autocorrelogram

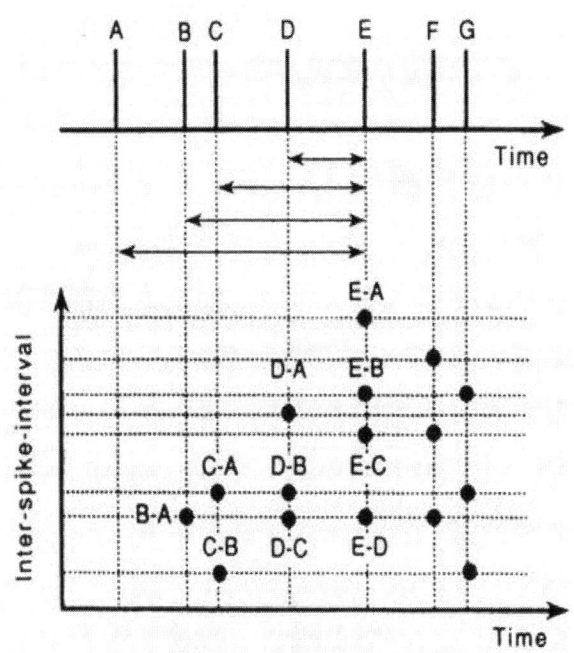

Fig. 2 A pulse train (the upper part) and it's autocorrelogram (the lower part)

fundamental f_0 explicitly appeared on the aggregated autocorrelogram of the output pulse train for input signal f_1 to one cochlear model and the output pulse train for input signal f_2 to another cochlear model [2]. An example of autocorrelogram is shown in Fig. 2. In this paper, we have investigated the properties of the missing fundamental of complex tones having more than two components on cochlear models. We can't carry out the psycho-acoustic experiments for more than two tones, because we have only two ears.

We use the maximum peak except for the missing fundamental peaks (a.b.MPexceptforMFP) in Fig. 3 to see the perceptivity of the missing fundamental.

III. RESULTS AND DISCUSSION

A. The Experiment of the Influence of the Pitch on the Extraction of the Missing Fundamental (Experiment 1)

Table 1 Component frequencies sets for the experiment of the influence of the pitch (for the experiment 1)

	Fundamental Frequency[Hz]	Component Frequencies [Hz]			
①	$f_0 = 150$	450	600	750	900
②	$f_0 = 225$	450	675		900
③	$f_0 = 500$	1500	2000	2500	3000
④	$f_0 = 750$	1500	2250		3000
⑤	$f_0 = 700$	2100	2800	3500	4200
⑥	$f_0 = 1050$	2100	3150		4200
⑦	$f_0 = 1000$	3000	4000	5000	6000
⑧	$f_0 = 1500$	3000	4500		6000

Table 2 The value of the maximum peak except for the missing fundamental peaks in the experiment 1

	Fundamental Frequency [Hz]	Method(a)	Method(b)
①	$f_0 = 150$	0.74	0.72
②	$f_0 = 225$	0.77	0.77
③	$f_0 = 500$	0.99	0.964
④	$f_0 = 750$	0.995	0.989
⑤	$f_0 = 700$	0.998	0.984
⑥	$f_0 = 1050$	0.995	0.993
⑦	$f_0 = 1000$	×	0.993
⑧	$f_0 = 1500$	×	×

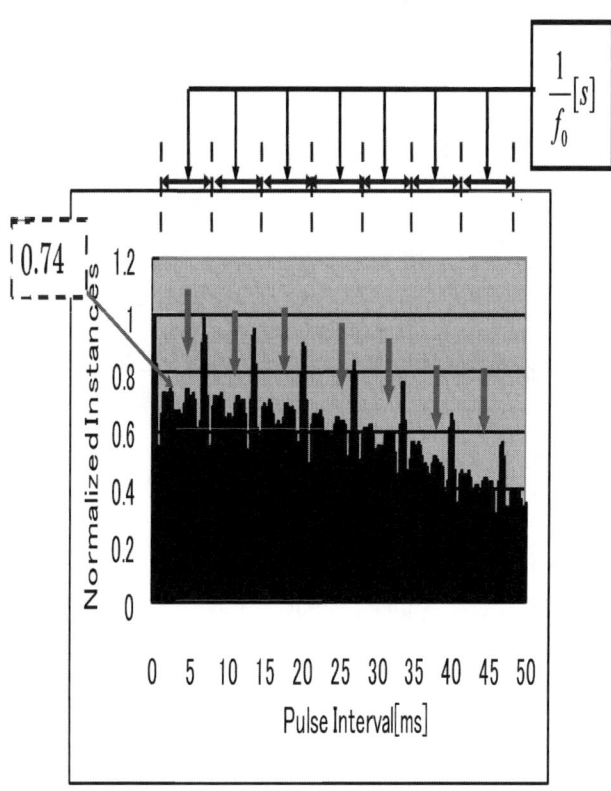

Fig. 3 The maximum peak except for the missing fundamental peaks (In this example, it is the peak of the value 0.74)

We have carried out the following two experiments.
Experiment 1 is for investigating the influence of the pitch on the extraction of the missing fundamental and

experiment 2 is for investigating the influence of the number of harmonic components on the extraction of the missing fundamental.

We have compared the experimental results in (a) with in (b) about each experiment.

(a) The case that each cochlear model is fed a pure tone.
(b) The case that each cochlear model is fed a complex tone.

For experiment 1, the component frequencies sets are shown in Table 1.

Here, the set of ① has the same lowest frequency and the same highest frequency to those of the set of ②. The ratio of the highest frequency to the lowest frequency is 1.5. ③ and ④, ⑤ and ⑥, ⑦ and ⑧ are in the same manner.

The experimental results are shown in Table 2. x means that the extraction of the missing fundamental is impossible. For every set, the value of MPexceptforMFP {the maximum peak except for the missing fundamental peaks} is smaller in (b), on having no concern with the height of the pitch, than in (a). It means that the extraction of the missing fundamental is more easy in (b)(complex tone input) than in (a)(pure tone input).

B. *The Experiment of the Influence of the Number of Harmonic Components on the Extraction of the Missing Fundamental (Experiment 2)*

The component frequencies sets for experiment 2 are shown in Table 3. The experimental results are shown in Table 4 for (a) and Table 5 for (b).

In the case that f_0 is in the range of low frequency (less than 4 components in ① and ②), the perception of the missing fundamental becomes easy not only in (a) but also in (b) according as increasing the number of harmonic components. In the case that f_0 is in the range of high

Table 3 Component frequencies sets for the experiments of the influence of the number of harmonic components (for the experiment 2)

No.	f_0 [Hz]	2 components	3 components	4 components	5 components	6 components
①	150	450	600	750	900	1050	1200
②	225	450	675	900	1125	1350	1575
③	500	1500	2000	2500	3000	3500	4000
④	750	1500	2250	3000	3750	4500	5250
⑤	700	2100	2800	3500	4200	4900	5600
⑥	1050	2100	3150	4200	5250	6300	7350
⑦	1000	3000	4000	5000	6000	7000	8000
⑧	1500	3000	4500	6000	7500	9000	10500

* harmonic components [Hz]

frequency (less than 4 components in ③,④,⑤, and ⑥), the perception of the missing fundamental becomes easy only in (b) according as increasing the number of harmonic components.

Table 4 The value of the maximum peak except for the missing fundamental peaks in the experiment 2 (a)

No	f_0 [Hz]	Number of Harmonic Components				
		2	3	4	5	6
①	150	0.85	0.76	0.72	0.72	0.73
②	225	0.77	0.77	0.73	0.75	0.75
③	500	0.986	0.971	0.964	0.958	0.96
④	750	0.986	0.989	0.979	0.981	0.983
⑤	700	0.991	0.987	0.984	0.988	0.987
⑥	1050	0.997	0.993	0.991	0.991	0.991
⑦	1000	0.997	0.995	0.993	0.993	0.993
⑧	1500	0.997	×	×	×	×

Table 5 The value of the maximum peak except for the missing fundamental peaks in the experiment 2 (b)

No	f_0 [Hz]	Number of Harmonic components				
		2	3	4	5	6
①	150	0.84	0.77	0.74	0.75	0.76
②	225	0.74	0.77	0.76	0.79	0.81
③	500	0.986	0.986	0.99	0.991	0.992
④	750	0.992	0.995	0.996	0.996	0.996
⑤	700	0.997	0.998	0.998	0.998	0.998
⑥	1050	0.994	0.995	0.996	0.997	0.997
⑦	1000	×	×	×	×	×
⑧	1500	×	×	×	×	×

The reason has been made clear by the consideration using cochlear models. i.e. complex tones give the auditory system the information of the missing fundamental and the information of combination tones $\{(f_2 - f_1), (2f_1 - f_2), \text{etc.}\}$, where $(f_2 - f_1)$ equals f_0, and $(2f_1 - f_2)$ equals f_0, etc. The extraction of the missing fundamental f_0 becomes more easily for complex tones, by the information from combination tones, than for pure tones (the figure of the information is abbreviated).

The investigation of the influence of multiples of f_0 (for example, comparing one complex tone (in $f_1=2f_0$, $f_2=3f_0$, and $f_3=5f_0$.) with another complex tone (in $f_1=2f_0$, $f_2=4f_0$, and $f_3=5f_0$.)) is future works.

CONCLUSIONS

The missing fundamental f_0 has been extracted more easily for complex tones in cochlear model experiments than for pure tones. The reason has been made clear by the consideration using cochlear models i.e. the extraction of the missing fundamental f_0 becomes more easily for complex tones by the information from combination tones than for pure tones.

The research results of the mechanism generating the missing fundamental can be applied to the followings. They are the research of forming neural networks for perception in the auditory system, improving cochlear implant, realizing electronic watermark, etc.

ACKNOWLEDGMENT

I am pleased to acknowledge the considerable assistance of Ph.D. Joseph S. Perkell (RLE at MIT).

REFERENCES

1. Greenberg S, Rhode W S (1987) Auditory processing of complex sound. In:Yost W A, Watson C S, editors. Lawrence Erblaum Associates pp225-236.
2. Matsuoka T, Ono Y (1998) Phase-locking by integral pulse frequency modulation and information of missing fundamental in pulse trains. 20th Annual Int Conf IEEE in MBS Proc.,Vol. 20, No. 6, Hong Kong, 1998, pp 3184-3187
3. Matsuoka T, Iitomi Y (2010) The perception of the missing fundamental of complex tones, BIOSIGNAL2010 Proc., Brno, Czech Republic, 2010, p15
4. Iitomi Y (2008 March) Master Thesis (at Utsunomiya Univ.)
5. Patterson R.D.(2000) Auditory images, How complex sounds are represented in the auditory system, JASJ(E)21(4),pp.183-190, (2000).

Use macro [author address] to enter the address of the corresponding author:T.Matsuoka
Institute: Faculty of Engineering,Tohoku University
Street: 6-6-4 Aramaki aza Aoba
City: Sendai
Country: Japan
Email: matsuokat@live.jp

Synchrony Analysis of Unipolar Cardiac Mapping during Ventricular Fibrillation

J. Caravaca[1], E. Soria-Olivas[1], A.J. Serrano-López[1], M. Bataller[1], A. Rosado[1],
L. Such-Belenguer[2], and J.F. Guerrero[1]

[1] Digital Signal Processing Group, ETSE, University of Valencia, Valencia, Spain
[2] Physiology Department, School of Medicine, University of Valencia, Valencia, Spain

Abstract—Ventricular Fibrillation (VF) is one of the main causes of death in developed countries. Recent studies have shown that fibrillation have a complex organization scheme. This work uses three measures of synchrony to characterize three groups of rabbit hearts. These groups consist of rabbits trained with physical exercise (N=7), untrained rabbits treated with a drug (N=13) and a control group of untrained rabbits (N=15). Cardiac mapping records were acquired using a 240-electrode array placed on left ventricle of isolated rabbit hearts, and VF was induced pacing at increasing rates. Two acquisitions were performed: maintained perfusion, and ischemic damage produced by an artery ligation. The used measures are Spatial Correlation (SC), Coupling Index (CI) and Synchrony Index (SI), previously proposed to study fibrillation arrhythmias. These measures were used to quantify the synchrony level between a reference electrode and its neighbors at 4 increasing radiuses, increasing its electrode spacing. The significance of the effects of radius, group and its interaction was tested using the Generalized Estimating Equations methodology. The obtained results have shown that these synchrony measures provide a feasible way to characterize the three groups, where the effects of the interaction between group and radius (p<0.001), and the individual effect of the radius were statistically significant (p<0.01), both for maintained perfusion and with ischemic damage. All the groups decrease its synchrony as electrode spacing increases, as it was observed in the three used measurements. Within this perspective, the drugged group has shown the lowest synchrony reduction. The trained group have shown similarities with the control group.

Keywords—Ventricular Fibrillation, Unipolar Cardiac Mapping, Synchrony analysis.

I. INTRODUCTION

Ventricular fibrillation (VF) is the cause in about 75% and 84% cases of sudden cardiac death in the United States [1]. This arrhythmia consists in an uncoordinated ventricular contraction caused by an irregular electrical activation. VF was classically described as a completely desynchronized arrhythmia but, the evidence of the studies performed during last decade have demonstrated that there is a certain degree of temporal and spatial organization in cardiac fibrillation [2]. VF is mainly analyzed using cardiac mapping techniques, studying the electrical activity simultaneously recorded in hundreds of adjacent locations.

Most of the techniques used to analyze fibrillation rhythms are based on the idea of studying each recording electrode individually, providing maps of a feature extracted from the recorded signal. These features can be extracted from time or frequency domains. The main drawback of these techniques is that they do not provide information regarding the dependence between couples of recording electrodes, e.g. in what measure the activations are produced at the same time on both electrodes. Several features were proposed in the last years to quantify the synchrony of fibrillation rhythms (as atrial and ventricular fibrillations) [3][4][5][6]. This work uses three different synchronization measurements to analyze the effects of physical exercise in VF.

The cardiac adaptations to physical exercise have both extrinsic and intrinsic components. In the extrinsic component, physical exercise increases the parasympathetic activity, producing a decrease in cardiac frequency and other antiarrhythmic effects [7]. On the other hand, some previous works of the authors of this paper have reported intrinsic effects, as the reduction of dominant frequency, or the increases of temporal and frequential regularities [8]. Nevertheless, these intrinsic effects were not analyzed in terms of synchrony.

This paper entails a double purpose; on one hand it will study the capabilities of three different parameters to differentiate among cardiac mapping recordings acquired from three different groups of rabbits (control, trained and drugged) during VF. On the other hand it will compare the synchrony level of these three groups, and how this synchrony is affected by electrode spacing. In that way, the effects of physical exercise in the spatial synchronization will be analyzed, and these effects will be compared to the ones produced by the treatment with a drug.

The remainder of the paper is organized as follows. In Section II a description of the methods including data acquisition, synchrony analysis and statistical methods is provided. Section III presents and discusses the obtained results distinguishing between maintained perfusion conditions and the presence of ischemic damage. Finally, a summary of the main conclusions is given on Section IV.

II. METHODS

A. Data Acquisition

The experiments were performed in the cardiac electrophysiology laboratory at University of Valencia. Rabbit hearts preparations were used, isolated and perfused according to a Langendorff technique [9]. Three groups of rabbits were used: control, trained and drugged. The control group (N=15) consisted in sedentary rabbits. Rabbits in trained group (N=7) were submitted to a running program. Finally, drugged group (N=13) consisted in sedentary rabbits, which were infused with Glibenclamide (10µM) during the experiments.

The recordings of the epicardial activity were acquired using the cardiac mapping system MAPTECH® using a multielectrode array consisting of 240 electrodes (interelectrode distance of 1mm) placed between the anterior and posterior walls of the left ventricle. VF was induced pacing at increasing rates outside of the capturing area.

Two recordings of five minutes were acquired with each heart. The first recording was performed in maintained perfusion conditions, in order to study VF regardless of the metabolic deterioration induced by the absence of the myocardial perfusion caused by VF. After five minutes of acquisition it was stopped, and a ligature was performed in a coronary artery in order to produce ischemic damage. When the ligation was produced, the second recording was acquired.

In all the recordings, a preprocessing step was performed; the recordings were divided in temporal segments of 4 seconds. The signal quality of each segment was analyzed, discarding those electrodes that were noisy or have low amplitudes. The synchrony analysis explained in next subsection was applied to each preprocessed segment.

B. Synchrony Analysis

This work studies the relationship between the spacing between a couple of electrodes and its synchrony. In that way, a reference electrode was chosen on the center on the electrode array, and its synchrony with the neighbor electrodes at four different radiuses was quantified with three different approaches. Figure 1 shows the location of the reference electrode (black) and the electrodes at the four radiuses (red tones).

Three measures were used to quantify synchronization. For every measure, its mean value in the eight electrodes of each radius was given as the synchronization at this radius. The used synchrony measures are described above.

Spatial Correlation (SC): it is based on the idea that two recordings that are synchronized must be correlated [3]. It computes the normalized cross correlation between a

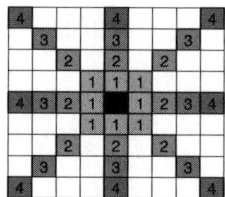

Fig. 1 Electrodes used in the synchrony analysis as were located in the multielectrode array. The black square is the reference electrode (center of the array), squares numbered from 1 to 4 were used as the neighbors of the reference electrode with increasing radiuses.

normalized signal (mean 0, standard deviation 1) and a normalized reference electrode [3]. The maximum of the cross correlation between both electrodes is normalized by the length of the signals, given a measure ranged from 0 to 1.

Coupling Index (CI): it is based on the concept of morphological similarity between pairs of signals [4]. CI provides an estimation of the probability of finding similar activation waves simultaneously in the two electrodes. A detailed formulation of the used algorithm can be found in [4].

Synchrony Index (SI): it quantifies the level of synchronization between two electrodes using the Shannon entropy of the delays between its activations times, a detailed description can be found in [5].

C. Statistical Analysis

The measured synchronies at the four neighborhood radiuses of the three groups are compared with the Generalized Estimating Equations (GEE) methodology for repeated measures [10], where each radius was considered as a repeated measure.

GEE models the mean response (mean measured synchronization) as a linear function of its covariates (group, distance to reference electrode and its interaction). To account for variation in the correlation between repeated measures, GEE allows the specification of its correlation structure [10]. In this case, an autoregressive structure was assumed, in which correlation among repeated measures on a subject depends on their separation in order of measurement. Unlike repeated measures ANOVA, GEE does not require a particular distribution in the outcome variable [10]. A detailed description of GEE and its application to repeated measures analysis can be found on [11].

In this work, GEE was computed using GEEQBOX for Matlab [12]. The significance of the effect of the covariates (group, distance to the reference electrode and its interaction) on the measured synchrony was tested. A level of significance of 5% was considered statistically significant.

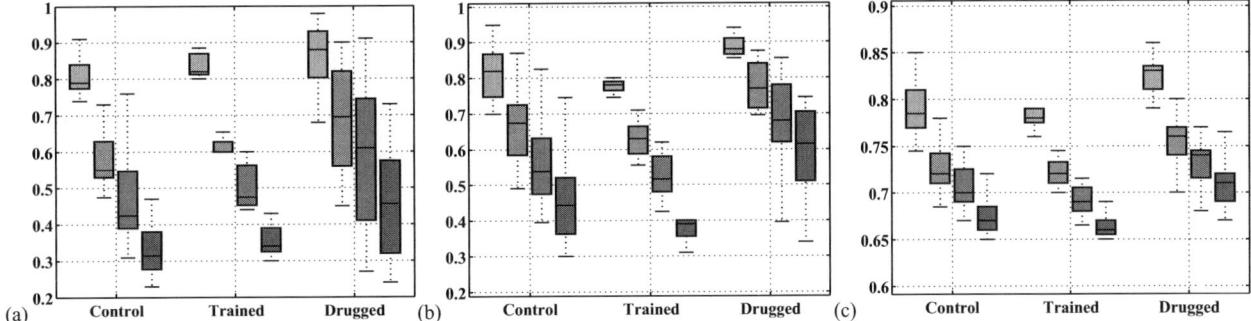

Fig. 2 Grouped boxplots of synchrony under ischemic damage conditions. Each group of boxplots shows the obtained values at increasing radiuses to reference electrode, using the same color legend than Figure 1. (a) Spatial Correlation (b) Coupling Index (c) Synchrony Index

III. RESULTS

A. Maintained Perfusion

The results obtained with the synchrony measures in maintained perfusion are showed in Table 1. Regarding CC, the effects of group, distance and interaction between group and distance are statistical significant ($p=0.03$, $p<0.001$ and $p<0.001$, respectively). Note that drugged group shows the lowest reduction of CC as radius increases (0.86 to 0.46), control group shows the highest (0.82 to 0.34) meanwhile the trained group remains between them (0.82 to 0.36), as can be also seen on Figure 2(a).

Distance and interaction between distance and group are statistical significant for CI ($p<0.001$ in both cases), meanwhile there is not enough evidence to ensure the significance of the individual effect of group with a confidence of 95% ($p=0.08$). In the same way that was observed with CC, CI shows the lowest reduction with the increase of electrode spacing in the case of drugged group (0.82 to 0.57). Figure 2(b) shows the distribution of CI for the four radiuses and the three groups.

Finally, in the case of SI, the discussion about effects of the covariates is the same that the exposed for CI. Distance and interaction are statistical significant ($p<0.001$ in both cases), while the individual effect of group is not significant ($p=0.98$). Using this synchrony measure, the differences between trained and control groups are even less marked. Nevertheless, the drugged group shows the lowest reduction of SI (0.82 to 0.7), as can be noted on Figure 2(c).

B. Ischemic Damage

The results with ischemic damage, shown on Table 2, are similar to those obtained in maintained perfusion. In the case of CC, higher values are obtained for ischemic deterioration than with perfusion, while CI and SI are higher in the case of perfusion.

Regarding the effects of the covariates, group is only significant for CC ($p<0.01$), while it is not significant for CI and SI ($p=0.15$ and $p=0.41$ respectively). On the other hand, the effect of distance, and the joint effect of group and distance are significant to all the synchrony measures ($p<0.001$ in all cases). Figure 3 shows the obtained distribution of the synchrony measures with ischemic deterioration. Note the bigger dispersion of the measures with ischemic damage, compared to the dispersion of the measures in perfusion.

Table 1 Results of synchrony analysis under normal perfusion, expressed as mean±standard deviation. R: radius to reference electrode, SC: Spatial Correlation, CI: Coupling Index, SI: Synchrony Index

		R=1	R=2	R=3	R=4
SC	Control	0.82±0.07	0.59±0.11	0.47±0.12	0.34±0.1
	Trained	0.84±0.04	0.63±0.07	0.51±0.07	0.36±0.06
	Drugged	0.86±0.1	0.69±0.16	0.59±0.2	0.46±0.16
CI	Control	0.82±0.08	0.67±0.11	0.58±0.14	0.48±0.15
	Trained	0.78±0.05	0.63±0.06	0.53±0.08	0.42±0.1
	Drugged	0.82±0.17	0.72±0.19	0.66±0.21	0.57±0.19
SI	Control	0.8±0.04	0.73±0.04	0.71±0.03	0.68±0.03
	Trained	0.8±0.03	0.73±0.03	0.7±0.03	0.68±0.03
	Drugged	0.82±0.06	0.76±0.05	0.73±0.06	0.7±0.05

Table 2 Results of synchrony analysis under ischemic damage. R: radius to the reference electrode (electrode spacing), SC: Spatial Correlation, CI: Coupling Index, SI: Synchrony Index

		R=1	R=2	R=3	R=4
SC	Control	0.85±0.05	0.66±0.09	0.54±0.13	0.4±0.12
	Trained	0.81±0.08	0.6±0.1	0.49±0.1	0.36±0.07
	Drugged	0.88±0.07	0.72±0.12	0.62±0.16	0.49±0.14
CI	Control	0.6±0.15	0.41±0.14	0.32±0.14	0.23±0.1
	Trained	0.6±0.21	0.4±0.22	0.31±0.19	0.23±0.16
	Drugged	0.6±0.17	0.43±0.2	0.36±0.2	0.27±0.15
SI	Control	0.75±0.04	0.68±0.04	0.66±0.03	0.64±0.03
	Trained	0.74±0.05	0.68±0.05	0.65±0.05	0.63±0.04
	Drugged	0.76±0.05	0.69±0.05	0.67±0.05	0.64±0.04

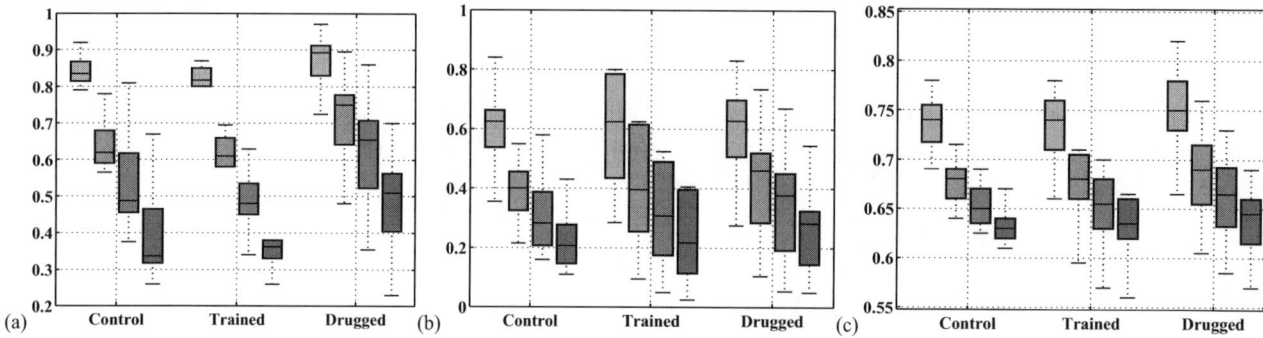

Fig. 3 Grouped boxplots of synchrony under ischemic damage conditions. Each group of boxplots shows the obtained values at increasing radiuses to the reference electrode, using the same color legend than Figure 1. (a) Spatial Correlation (b) Coupling Index (c) Synchrony Index

IV. CONCLUSIONS

Despite fibrillation has been traditionally considered as an unorganized arrhythmia, several measurements of synchrony were proposed in the last years. These measurements quantify the grade of simultaneity of the activations produced in a couple of electrodes.

Three different synchrony measures (SC: Spatial Correlation, CI: Coupling Index, and SI: Synchrony Index) were used to characterize three groups of ventricular fibrillation recordings. These groups consist of rabbits that have performed a training program with physical exercise, untrained rabbits treated with a drug (Glibenclamide) and a control group of untrained rabbits.

The obtained results have shown that all the used synchrony measures decrease as electrode spacing increases. The effect of the electrode spacing is statistically significant with a 99% level of confidence in the three measurements.

The significance of the group has not been observed with all synchrony measurements. Group is individually significant with a confidence of 5% for SC, and with a confidence of 10% for CI. Nevertheless, it is not significant for SI.

Finally, interaction between group and spacing is significant for all the synchrony measures, showing that not all the groups decrease its synchrony with electrode spacing in the same proportion. Within this perspective, the drugged group has shown the lowest reduction of synchrony, while control and trained groups have shown higher reductions.

ACKNOWLEDGMENTS

This work was supported in part by the Spanish Ministry of Science and Innovation: Plan Nacional de I+D+I, Acción Estratégica: "Deporte y Actividad Física", project DEP2010-22318-C02-02.

REFERENCES

1. Chugh S S, Reinier K, Teodorescu C et al (2008) Epidemiology of sudden cardiac death: clinical and research implications. Progress in cardiovascular diseases, 51(3):213-228
2. Jalife J, (2000) Ventricular fibrillation: mechanisms of initiation and maintenance. Annual review of physiology, 62(1):25-50
3. Salama G, Choi B R (2007) Imaging ventricular fibrillation. Journal of electrocardiology, 40(6):S56-S61
4. Faes L, Ravelli F (2007). A Morphology-Based Approach to the Evaluation of atrial fibrillation organization. IEEE Engineering in Medicine and Biology Magazine, 26(4):59-67
5. Masè M, Faes L, Antolini R et al (2005) Quantification of synchronization during atrial fibrillation by Shannon entropy: validation in patients and computer model of atrial arrhythmias. Physiological measurement, 26(6):911-923
6. Calcagnini G, Censi F, Michelucci A et al (2006) Descriptors of wavefront propagation. IEEE Engineering in Medicine and Biology Magazine, 25(6): 71-78
7. Blomqvist C G, Saltin B (1983) Cardiovascular adaptations to physical training. Annual Review of Physiology:45(1), 169-189
8. Guerrero J, Rosado-Muñoz A, Serrano A J et al (2009) Modifications on regularity and spectrum of ventricular fibrillation signal induced by physical training. IEEE Proc. Computers in Cardiology, Utah, 2009, pp. 321-324
9. Skrzypiec-Spring M, Grotthus B, Szeląg A et al (2007) Isolated heart perfusion according to Langendorff—still viable in the new millennium. Journal of pharmacological and toxicological methods 55(2):113-126
10. Davis C S (2002) Statistical methods for the analysis of repeated measurements. Springer
11. Ballinger G A (2004) Using generalized estimating equations for longitudinal data analysis. Organizational research methods 7(2):127-150
12. Ratcliffe S J, Shults J (2008) GEEQBOX: A MATLAB Toolbox for Generalized Estimating Equations and Quasi-Least Squares. Journal of Statistical Software 25(14):1-14

Author: Juan Caravaca Moreno
Institute: GPDS University of Valencia
Street: Avenida Universidad sn, Burjassot
City: Valencia
Country: Spain
Email: juan.caravaca@uv.es

Symbolic Analysis of Heart Period and QT Interval Variabilities in LQT1 Patients

V. Bari[1,2], T. Bassani[3], A. Marchi[4], G. Girardengo[5], L. Calvillo[6], S. Cerutti[1], P.A. Brink[7], L. Crotti[5,8], P.J. Schwartz[5], and A. Porta[9]

[1] Department of Electronics Information and Bioengineering, Politecnico di Milano, Milan, Italy
[2] Gruppo Ospedaliero San Donato Foundation, Milan, Italy
[3] Internal Medicine - Clinical and Research Center, IRCCS Humanitas Clinical Institute, Rozzano, Milan, Italy
[4] Department of Anesthesia and Intensive Care Unit, IRCCS Humanitas Clinical Institute, Rozzano, Milan, Italy
[5] Department of Molecular Medicine, University of Pavia and Fondazione IRCCS Policlinico S. Matteo, Pavia, Italy
[6] Laboratory of Cardiovascular Genetics, IRCCS Istituto Auxologico Italiano, Milan, Italy
[7] Department of Internal Medicine, University of Stellenbosch, Stellenbosch, South Africa
[8] Institute of Human Genetics, Helmholtz Zentrum München, Neuhergerg, Germany
[9] Department of Biomedical Sciences for Health, Galeazzi Orthopedic Institute, University of Milan, Milan, Italy

Abstract—Heart period and QT interval variabilities carry important information about the state of the autonomic nervous system. Autonomic function is impaired in the long QT syndrome type 1 (LQT1) and this impairment plays a central role in triggering fatal arrhythmias. Twenty-four hour Holter recordings from 26 mutation carrier (MC) subjects and 11 non mutation carrier (NMC) coming from the same family with founder effects were analyzed. After the extraction of heart period, approximated as the time distance between two consecutive R peak on the ECG (RR), and QT interval, approximated as the temporal distance between R apex and the apex or the end of T wave (RTa and RTe respectively), we performed symbolic analysis over the obtained RR, RTa and RTe beat-to-beat series. Results showed that, while the two groups could not be discriminated by symbolic analysis of RR series, the same analysis carried out on RTa and RTe series evidenced significant differences reflecting the impairment of the autonomic control directed to ventricles and remarking the different information achieved from RR and QT variabilities.

Keywords—Symbolic analysis, heart rate variability, QT variability, autonomic nervous system, long QT syndrome.

I. INTRODUCTION

The analysis of the spontaneous variations of heart period allows the inference of the state of the autonomic nervous system. More specifically, respiratory sinus arrhythmia assessed as the power of the heart period variability in the high frequency band (HF, from 0.15 to 0.5 Hz) is known to be related to vagal modulation directed to the sinus node [1].

Recently, it has been demonstrated that also the beat-to-beat variations of QT interval, representing on the ECG trace the duration of ventricular depolarization and repolarization, reflects autonomic modulation [2]. More specifically, the QT power in the low frequency band (LF, from 0.04 to 0.15 Hz) has been associated to sympathetic modulation [2,3]. It was supposed that, while spectral analysis of heart period series can provide information on the autonomic influences directed to the sinus node, QT variability can give information on those directed to the ventricles [2].

Long QT (LQT) syndrome is an inherited pathology characterized by a lengthening in ventricular repolarization duration because of a prolonged repolarization of cardiac myocites, leading to arrhythmias, such as torsades de pointes, and to sudden death [4]. Twelve different genetic mutations leading to LQT syndrome have been identified so far. The most common is the mutation of KCNQ1 gene, encoding the potassium channel. This mutation identifies the LQT syndrome type 1 (LQT1) variant, whose patients are characterized by a major risk for arrhythmias during physical exercise and, in general, in situations in which the sympathetic activation is prevalent [4,5]. In this study, 24h Holter recordings from 11 non mutation carrier (NMC) and 26 mutation carrier (MC) subjects belonging to the same South African family have been analyzed. These subjects were characterized by a founder effect and came from the same descendent carrying the KCNQ1 mutation, originally settled in South Africa in approximately 1700 [5].

Symbolic analysis was found helpful to describe heart period variability due to its inherent capability of describing patterns and quantifying their degree of repetition [6]. Its main advantage with respect to tools in the frequency domain relies on the relaxation of linear hypothesis underlying mechanisms generating the dynamics, thus possibly accounting for nonlinearities known to be present in QT variability due to its non linear relation with heart period [7].

The aim of this study is to evaluate cardiovascular control in LQT1 patients through the symbolic analysis of heart period and QT beat-to-beat series. We will exploit a symbolic analysis previously introduced in [8,9]. The contrast between MC and NMC groups will allow a better understanding of the impact of the KCNQ1 gene mutation on autonomic modulation directed to the sinus node and ventricles. Symbolic analysis will be carried out over beat-to-beat sequences extracted from 24h Holter recordings [10].

II. SYMBOLIC ANALYSIS

Symbolic analysis is a non linear method based on the transformation of the time series into symbols. We utilized the approach proposed in [8]. Briefly, given a time series $y=\{y(i), 1 \leq i \leq N\}$, where i is the progressive sample counter and N is the series length, the range of the time series y was divided in ε bins. Each bin has a size of (max-min)/ε where max and min are the maximum and minimum values of y. Each original value of the time series is substituted by an integer coding the bin in which the sample is found, thus becoming a series $y_\varepsilon=\{y_\varepsilon(i), 1 \leq i \leq N\}$ whose values range from 0 to ε-1. Based on the technique of the delayed coordinates [11] the series of patterns $y_{\varepsilon,L}=\{y_{\varepsilon,L}(i), L \leq i \leq N\}$ is constructed where $y_{\varepsilon,L}(i)=(y_\varepsilon(i), y_\varepsilon(i-1), \ldots, y_\varepsilon(i-L+1))$. According to [8] $\varepsilon=6$ and $L=3$ were set. All patterns were grouped in families as follows: i) patterns with no variations among the symbols (0V, all the symbols are equal); ii) patterns with one variation (1V, two consecutive symbols are identical and the remaining one is different); iii) patterns with two like variations (2LV, the symbols are characterized by an ascending or descending trend); iv) patterns with two unlike variations (2UV, the symbols form a peak or a valley). The percentage of appearance of each pattern was evaluated and referred to as 0V%, 1V%, 2LV%, 2UV% in the following. Analysis was carried out iteratively over windows of 250 beats, with an overlap of 200 beats. The distribution of 0V%, 1V%, 2LV%, 2UV% was assessed over the entire series and the median value was taken as representative of the whole series [10].

III. EXPERIMENTAL PROTOCOL AND DATA ANALYSIS

A. Study Population and Experimental Protocol

Twenty-four hour Holter 12 lead ECG (Mortara Instrument Inc., Milwaukee, WI, USA) were recorded from 26 MC LQT1 patients and 11 NMC control subjects with founder effect [5]. The sampling rate was 180 Hz. Analysis was carried out on the lead with the best signal-to-noise ratio. All subjects did not take any drug, including β-blockers, at the moment of recording. All probands and family members provided written informed consent for clinical and genetic evaluations. Protocols were approved by the Ethical Review Boards of the Tygerberg Hospital of Stellenbosh University, Vanderbilt University and University of Pavia. Approved consent forms were provided in English or Afrikaans as appropriate. The study is in keeping with the principles of the Declaration of Helsinki for medical research involving human subjects.

B. Beat-to-Beat Variability Series Extraction

The QRS complex was detected on the ECG and the R-apex was located using parabolic interpolation. Heart period was approximated as the temporal distance between two consecutive R-parabolic apexes (RR interval). QT interval was approximated as the temporal distance between R and T wave apexes (RTa) and as the temporal distance between R peak and T wave end (RTe). T wave apex was fixed with minimum jitters using parabolic interpolation as in the case of QRS complex [12]. T wave end was searched in a time window whose length depended on the previous RR interval duration. The T wave end was detected when the first derivative assessed over the T wave downslope fell below a user-defined threshold (usually the 30% of the absolute maximal first derivative value assessed on the T-wave downslope) [12]. The R peak delimiting the measure of RTa(i) and RTe(i) was the one defining the end of the RR(i) interval. RR, RTa and RTe intervals were automatically computed from 24h Holter recordings, with a continuous update of all the parameters during the analysis. The detections were manually checked to avoid erroneous identifications or missed beats and corrected by means of cubic spline interpolation only in case of ectopic beats or evident arrhythmias. The number of corrections was always lower than the 3% of the values in the series. Series with a length of 5000 beats were extracted during the daytime (from 2 to 6 PM). After the extraction of RR, RTa and RTe series, mean (μ_{RR}, μ_{RTa} and μ_{RTe}) and variance (σ^2_{RR}, σ^2_{RTa} and σ^2_{RTe}) were calculated. Then, symbolic analysis was carried out over the whole series of 5000 beats.

C. Statistical Analysis

We performed unpaired t-test (or Mann-Whitney rank sum test when appropriate) to check the significance of the difference between NMC and MC groups. A $p<0.05$ was always considered as significant.

IV. RESULTS

Table 1 shows mean and variance of RR, RTa and RTe in the two groups. It can be noticed that RR mean and variance was not significantly different b`etween NMC and MC, while RTa and RTe mean resulted significantly higher in MC population, thus confirming the main phenotype of the

Table 1 Mean and variance of the RR, RTa and RTe intervals in NMC and MC individuals.

	NMC	MC
μ_{RR} [ms]	698±110	750±91
σ^2_{RR} [ms^2]	1041±586	1446±1383
μ_{RTa} [ms]	235±33*	324±40
σ^2_{RTa} [ms^2]	70±58*	41±20
μ_{RTe} [ms]	311±30*	400±46
σ^2_{RTe} [ms^2]	130±151	132±87

Values are expressed as mean ± standard deviation. μ = mean; σ^2 = variance; NMC = non mutation carrier; MC = mutation carrier. The symbol * indicates a between-group difference with $p<0.05$.

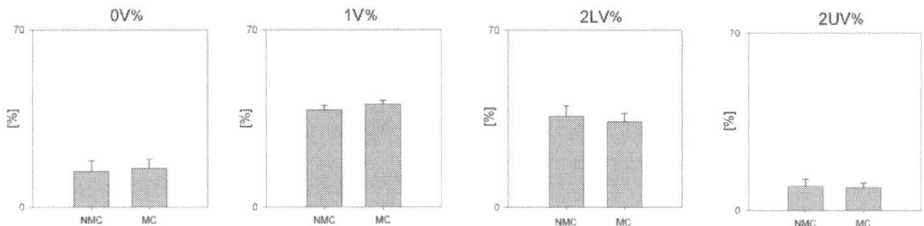

Fig.1 Panels represent 0V%, 1V%, 2UV%, 2LV% evaluated on RR series (from left to right panel, mean±standard deviation). The symbol * indicates p<0.05 between the two populations.

pathology, that leads to a longer ventricular repolarization duration. The RTa variance was significantly larger in NMC group, while RTe variance was not affected by mutation.

Figure 1 shows the results of symbolic analysis on RR series in terms of 0V%, 1V%, 2LV%, 2UV% (mean ± standard deviation). None of the considered parameters distinguished the two groups of subjects.

Figure 2 and Figure 3 show the results of symbolic analysis derived from RTa and RTe series respectively. It is worth noting that, while symbolic analysis of RR series was not able to differentiate the two groups, 0V% and 1V% of RTa series were significantly lower in NMC (p<0.05) and 2LV% of RTa series was significantly higher in NMC with respect to MC (p<0.01). In addition, 1V% of RTe series was significantly lower in NMC with respect to MC (p<0.001).

It is also worth noting that the different symbols had a different percentage of appearance over the series: the pattern 2UV was the most infrequent in all the three series (i.e. RR, RTa and RTe), while the most present was 1V in RR and RTa series of both NMC and MC, followed by 2LV. In RTe series 1V was the most present in MC group, followed by the pattern 0V, while in NMC group 0V was the pattern most frequently found, followed by 1V.

V. DISCUSSION

Symbolic analysis is a helpful tool to investigate cardiovascular control from spontaneous variability [8,9]. In this work this method was applied on RR and QT variability series in subjects belonging to the same founder family with LQT1 positivity, divided in NMC, used as control group, and MC. The analysis was carried out over 24h Holter ECG recordings after extracting automatically RR and QT interval series. Comparisons between NMC and MC were made according to time domain indexes (i.e. mean and variance) and indexes evaluating the probability of finding specific patterns in the RR, RTa and RTe series as classified by symbolic analysis.

Time domain results showed that the two groups resulted indistinguishable based on RR mean, while RTa and RTe mean was increased in MC group, thus confirming the main phenotype of the pathology.

Symbolic analysis of RR series did not show any difference between the two populations, while the two groups could be distinguished according to symbolic analysis of RTa and RTe series. Particularly, 0V% and 1V% of RTa series and 1V% of RTe series resulted significantly increased in MC group, while 2LV% of RTa series significantly decreased. This result confirms the importance of the assessment of the QT variability to better separate the two groups. In addition, this finding suggests that autonomic control, underpinning the RR and QT variabilities, is different [2]. Indeed, neural activity acting on the sinus node is different from that impinging the ventricles, thus producing a fraction of QT changes that cannot be considered a simple mirroring effect of the RR variations [2]. The increase of 0V% can be interpreted as an increase of sympathetic modulation, thus suggesting an increased sympathetic drive directed to the ventricles [8]. Also the increase of 1V% in RTa and RTe series supports this conclusion. The presence of 1V pattern can be interpreted as a result of an incomplete vagal activation due to the inability of vagal system to take control and impose faster changes. Sympathetic drive might be responsible for this incomplete vagal modulation. At this regard it is worth recalling that 1V% was found to be decreased in presence of a reduced sympathetic modulation in humans and animals suffering for chronic heart failure [13,14].

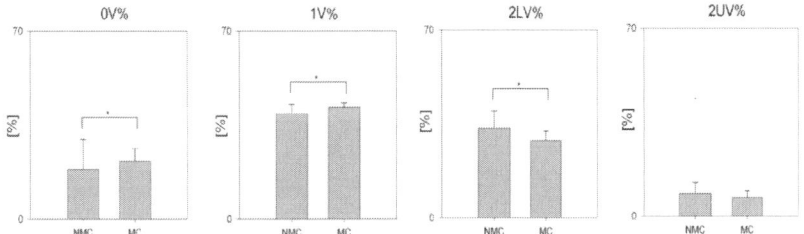

Fig.2 Panels represent 0V%, 1V%, 2UV%, 2LV% evaluated on RTa series (from left to right panel, mean±standard deviation). The symbol * indicates p<0.05 between the two populations.

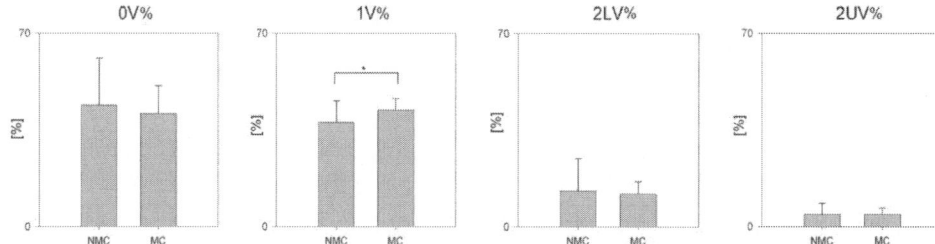

Fig.3 Panels represent 0V%, 1V%, 2UV%, 2LV% evaluated on RTe series (from left to right panel, mean±standard deviation). The symbol * indicates p<0.05 between the two populations.

In healthy subjects the reduction of 2LV% on the RR series is usually interpreted as an index of the decrease of vagal modulation directed to the sinus node [9]. Since the RTa measure can be largely influenced by respiratory-related artifacts due to cardiac axis movements synchronous with respiration [12], it is difficult to interpret the reduction of 2UV% in the RTa series of MC group as a result of the decrease of the parasympathetic modulation. However, this result is worth being investigated further because it could not be easily dismissed as the consequence of an artifact since there is no reason that the respiratory-related cardiac axis movements could affect more importantly NMC group than MC one. Furthermore, additional investigations could be addressed to compare symbolic analysis to another non linear technique, as the evaluation of QT adaptation to RR changes [15] in the same LQT1 population.

VI. CONCLUSION

This work showed that analysis of QT variability allowed the separation between NMC and MC groups in LQT1 syndrome, while the analysis of RR variability series failed. In addition, symbolic analysis of QT variability indicated a larger sympathetic modulation in MC group. The more important sympathetic drive might be responsible for the susceptibility of MC group to fatal arrhythmias. Future analyses distinguishing symptomatic patients from asymptomatic ones could be helpful to clarify further the role of sympathetic control. It is worth noting that this conclusion was reached despite the low temporal resolution of the 24h Holter recordings and the uncontrolled experimental conditions (i.e. recordings were made during daily activities). The analysis suggests that QT variability should always be monitored in addition to the more traditional RR variability to better typify patients affected by LQT syndrome.

ACKNOWLEDGMENTS

This work was supported by the Telethon GGP07016 and GGP09247 grants.

REFERENCES

1. Task Force of the European Society of Cardiology and the North American Society of Pacing and Electrophysiology (1996) Standard of measurement, physiological interpretation and clinical use. Circulation 93:1043-1065
2. Porta A, Tobaldini E, Gnecchi-Ruscone T et al. (2010) RT variability unrelated to heart period and respiration progressively increases during graded head-up tilt, Am J Physiol 298:H1406-H1414
3. Porta A, Bari V, Badilini F et al. (2011) Frequency domain assessment of the coupling strength between ventricular repolarization duration and heart period during graded head-up tilt, J Electrocardiol 44:662-668
4. Schwartz PJ, Priori SG, Spazzolini C et al. (2001) Genotype-phenotype correlation in the long-QT syndrome: gene-specific triggers for life-threatening arrhythmias. Circulation 103:89-95
5. Brink PA, Crotti L, Corfield V et al. (2005) Phenotypic variability and unusual clinical severity of congenital long-QT syndrome in a founder population. Circulation 112:2602-2610
6. Voss A, Kurths J, Kleiner HJ et al.(1995) Improved analysis of heart rate variability by methods of nonlinear dynamics. J Electrocardiol 28:81-88
7. Bazett HC. (1929) An analysis of the time-relations of electrocardiograms. Heart 7:353-370
8. Porta A, Guzzetti S, Montano N et al. (2001) Entropy, Entropy rate, and pattern classification as tools to typify complexity in short heart period variability series. IEEE Trans Biomed Eng 48(11):1282-1291
9. Porta A, Tobaldini E, Guzzetti S et al. (2007) Assessment of cardiac autonomic modulation during graded head-up tilt by symbolic analysis of heart rate variability. Am J Physiol 293:H702-H708
10. Porta A, Faes L, Masé M et al. (2007) An integrated approach based on uniform quantization for the evaluation of complexity of short-term heart period variability: Application to 24 h Holter recordings in healthy and heart failure humans. Chaos 17(1):015117
11. Takens F (1981) Detecting strange attractors in fluid turbulence *in* Lecture notes in mathematics, Rand and Young, Springer, Berlin 366-381
12. Porta A, Baselli G, Lombardi F et al. (1998) Performance assessment of standard algorithms for dynamic R-T interval measurement: comparison between R-Tapex and R-Tend approach. Med Biol Eng Comput 36:35-42.
13. Maestri R, Pinna GD, Accardo A et al. (2007) Nonlinear indices of heart rate variability in chronic heart failure patients: redundancy and comparative clinical value. J Cardiovasc Electrophysiol 18(4):425-433
14. Tobaldini E, Montano N, Wei S-G et al. (2009) Autonomic Cardiovascular Modulation. Symbolic analysis of the effects of central mineralcorticoid receptor antagonist in heart failure rats. IEEE Eng Med Biol Mag 28(6):79-85
15. Pueyo E, Smetana P, Caminal P et al.(2004) Characterization of QT interval adaptation to RR interval changes and its use as a risk-stratifier of arrhythmic mortality in amiodarone-treated survivors of acute myocardial infarction. IEEE Trans Biomed Eng 51(9):1511-1520

Author: Vlasta Bari
Institute: IRCCS Istituto Ortopedico Galeazzi
Street: via R. Galeazzi 4
City: Milan
Country: Italy
Email: vlasta.bari@mail.polimi.it

Mechanical Stimulation and Cardiovascular Control in Parkinson Disease

T. Bassani[1], V. Bari[2], A. Marchi[3], S. Tassin[4], L. Dalla Vecchia[5],
M. Canesi[6], F. Barbic[1,7], R. Furlan[1,7], and A. Porta[8]

[1] Internal Medicine - Clinical and Research Center, IRCCS Humanitas Clinical Institute, Rozzano, Milan, Italy
[2] Gruppo Ospedaliero San Donato Foundation, Milan, Italy and Department of Electronics,
Information, and Bioengineering, Politecnico di Milano, Milan, Italy
[3] Department of Anesthesia and Intensive Care Unit, IRCCS Humanitas Clinical Institute, Rozzano, Milan, Italy
[4] Ecker Technologies Sagl, Switzerland
[5] IRCCS Maugeri Foundation, Milan, Italy
[6] Parkinson Institute, Istituti Clinici di Perfezionamento, Milan, Italy
[7] BIOMETRA Department, University of Milan, Neuroscience Research Association, Milan, Italy
[8] Department of Biomedical Sciences for Health, Galeazzi Orthopedic Institute, University of Milan, Milan, Italy

Abstract—**In addition to motor symptoms such as postural instability, rigidity, and bradykinesia, Parkinson disease (PD) subjects are characterized by autonomic dysfunction. When assessing the degree of the cardiovascular autonomic dysfunction, relevant information can be obtained from entropy-based complexity analysis of short term beat-to-beat fluctuations of heart period (HP) and systolic arterial pressure (SAP). A novel non-invasive procedure has recently been developed to improve conditions of PD patients through the mechanical stimulation of the fore feet. The present study performed an entropy-based complexity analysis of HP and SAP series in 10 subjects affected by PD in supine position and during 75 degrees head-up tilt test before and after the mechanical stimulation procedure. The results suggest an improvement of vascular control as resulted from a reduction of complexity of SAP series. Future studies are necessary to confirm this promising result over a larger database.**

Keywords—**Parkinson's disease, conditional entropy, heart rate variability, complexity, autonomic nervous system.**

I. INTRODUCTION

The neural degeneration of basal ganglia is considered the critical mechanism inducing movement disorders in Parkinson's disease (PD). Nevertheless, the peripheral alterations of the sensory-motor system play a crucial role in promoting motor disability in PD [1]. At this regard the sensory innervation was found to be severely impaired in PD subjects, with an unequal deficit distribution at different sites [1]. Not surprisingly, plantar mechanical or electrical stimulations [2] such as the rhythmic vibratory stimulation of trunk muscles or of the soles ameliorated movement indexes in PD subjects [3,4]. As far as plantar mechanical stimulation is concerned, a novel non-invasive procedure of mechanical stimulation acting on the hallux and on the first metatarsal joint has been recently devised (Ecker Technologies Sagl, Switzerland).

Subjects affected by PD are characterized primarily by motor symptoms, but alterations in cardiovascular autonomic control are detected in 10-40% of patients [5,6]. Autonomic impairment can be detected in PD patients by assessing the short term beat-to-beat fluctuations of heart period (HP) and systolic arterial pressure (SAP) [7]. Usually head-up tilt maneuver is utilized to evoke a sympathetic activation and concomitant vagal withdrawal and to test the autonomic response to a gravitational challenge [8].

The time course of cardiovascular variables is characterized by patterns with different dominant frequencies and shapes. This variety of features is usually referred to as dynamical complexity. At this regard relevant information about cardiovascular control can be obtained from entropy-based complexity analysis of HP and SAP series [9-11]. The relevance of entropy-based complexity indexes has been substantiated by their ability to identify pathological conditions [12,13]. In PD patients the complexity of HP and SAP series was found larger than that of age-matched healthy individuals, thus becoming a hallmark of autonomic impairment [14].

The aim of the present study is to perform an entropy-based complexity analysis of HP and SAP series in subjects affected by PD. Patients will be evaluated at rest in supine position (REST) and during head-up tilt (TILT) before and after the mechanical stimulation of specific points of both fore feet. Two different entropy-based complexity analysis approaches [15,16] were utilized to describe autonomic impairment and quantify the potential improvement induced by mechanical stimulation.

II. CONDITIONAL ENTROPY EVALUATION

A. Corrected Conditional Entropy

The corrected conditional entropy (CCE) was proposed as an entropy-based method for the evaluation of complexity of a time series $x=\{x(i), i=1,...,N\}$, where i is the sample counter and N is the series length [11,15]. It was originally devised to avoid the a priori assignment of the pattern length L: a setting usually necessary to assess approximate entropy [9] and sample entropy [10]. Full details about CCE definition can be found in [11,15]. Briefly, defined as $x_L(i)=(x(i),x(i-1),...,x(i-L+1))$ the ordered sequence of L consecutive samples formed by the most recent value $x(i)$ and by the previous L-1 samples, $x_{L-1}(i-1)=(x(i-1),...,x(i-L+1))$, the approach is based on the evaluation of the conditional entropy (CE) measuring the average amount of information carried by $x(i)$ when $x_{L-1}(i-1)$ is given. When assessed over a limited amount of data, the reliability of the conditional distribution of $x(i)$ given $x_{L-1}(i-1)$ decreases with L. In order to prevent the artificial decline of the CE a corrective term was added to the CE estimate, thus leading to the definition of the CCE. Normalized CCE (NCCE) was obtained by dividing CCE by the Shannon entropy of the series. The NCCE minimum was taken as an index of complexity. It ranged from 0 to 1, where 0 indicates null complexity and 1 indicates the opposite situation.

B. K-Nearest-Neighbor Conditional Entropy

As in case of CCE, the k-nearest-neighbor CE (KNNCE) was proposed [16] to allow entropy-based complexity analysis without a priori fixing the pattern length. At difference with CCE, KNNCE does not exploit a corrective term to compensate the bias of CE estimate. This approach takes advantage of a completely different strategy compared to CCE. Full details about KNNCE definition can be found in [16]. Briefly, the conditional distribution of x given $x_{L-1}(i-1)$ was assessed over the k patterns nearest to $x_{L-1}(i-1)$ regardless their actual distance from it (i.e. the k nearest neighbors of $x_{L-1}(i-1)$ as identified by the Euclidean norm), thus preventing the loss of reliability of the conditional distributions with the pattern length L. Normalized KNNCE (NKNNCE) was obtained by dividing KNNCE by the Shannon entropy of the series. The NKNNCE minimum was taken as an index of complexity bounded between 0 and 1.

III. EXPERIMENTAL PROTOCOL AND DATA ANALYSIS

A. Population

Ten subjects (6 males and 4 females, mean age: 66±6 years; range: 57-78 years), affected by idiopathic PD characterized by a moderate/important motor impairment (Hoehn-Yhar scale 2-3), were enrolled for the study at the Parkinson's Disease Center of Istituti Clinici di Perfezionamento in Milan. The mean disease duration was 13±4 years, the UPDRS III was 25±14. Subjects had similar disorder duration and were free from other diseases on the basis of their reported history, symptoms, physical examination and routine tests. Therapy for Parkinson's disease was maintained unchanged for the 30 days preceding and during the study procedures.

B. Experimental Protocol

All subjects gave their written informed consent. The ethical approval was provided by the Bolognini Hospital Institutional Review Board (# R-156). The protocol adhered to the principles of the Declaration of Helsinki. Recordings were carried out at REST and during TILT with a tilt table inclination of 75 degrees. Each experimental session lasted 20 minutes. During the entire protocol the subjects breathed spontaneously, but they were not allowed to talk. The subjects were recorded before (PRE) and 24 hours after (POST) the mechanical stimulation of the fore feet. The mechanical stimulation refers to the pressure applied on two specific skin points over the plantar surface of both fore feet. The sites of the stimulation corresponded to the tip of the hallux and the lower big toe first metatarsal joint plantar surface. Mechanical stimulation consisted in the application, over the selected site, of a controlled pressure of 0.58±0.04 kg/mm^2 for 5 seconds, by means of a smooth 2 mm diameter tip of a steel stick connected to a dynamometer for instantaneous pressure assessment. In a first stimulating cycle, each of the 2 sites of both feet was stimulated. The procedure was repeated for 3 times in every subject so that the overall stimulation time was approximately 1 minute.

C. Beat-to-Beat Variability Series Extraction

ECG (lead II) and non-invasive arterial blood pressure (Finometer MIDI, Finapres Medical Systems, The Netherlands) were sampled by an analogical-to-digital converter (National Instruments, AT-MIO 16E2). Sampling frequency was 300 Hz. Arterial pressure was measured from the middle finger of the left hand, which was maintained at the level of the heart by fixing the subject's arm to the thorax during the upright position. The arterial pressure signal was cross-calibrated in each session with the use of a measure provided by a sphygmomanometer at the onset of REST. The QRS apex was located over the ECG using parabolic interpolation. HP was approximated as the time distance between two consecutive QRS peaks. Systolic arterial pressure (SAP) was assessed as the maximum arterial blood pressure inside HP. All signals were manually edited to avoid artifacts. HP={HP(i), i=1,..., N} and SAP={SAP(i), i=1,..., N} were extracted on a beat-to-beat basis, where i is the progressive cardiac beat number and N is the series length. Sequences of 250-300 consecutive HP and SAP values (i.e. recordings of few minutes) were analyzed. The mean and variance of HP and SAP were indicated as μ_{HP}, μ_{SAP}, σ^2_{HP}, and σ^2_{SAP} and expressed in ms, mmHg, ms^2 and $mmHg^2$ respectively.

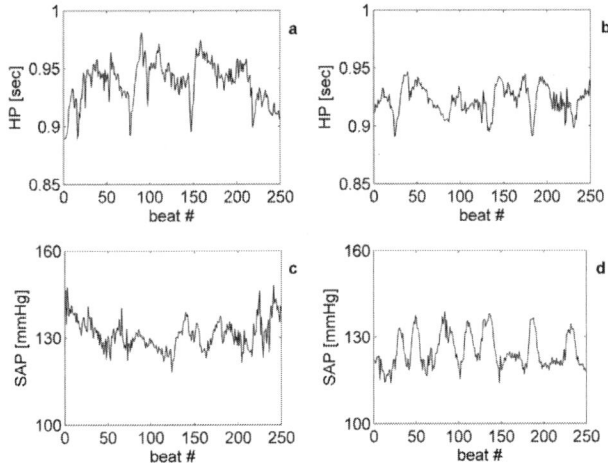

Fig.1 HP and SAP beat-to-beat series relevant to a PD patient in PRE (a,c) and POST (b,d) conditions at REST.

D. Statistical Analysis

Two way repeated measures analysis of variance (one factor repetition, Holm-Sidak test for multiple comparisons) was performed to assess the significance of changes of time domain parameters. Complexity indexes at REST and during TILT were pooled together to check the significance of the differences between PRE and POST. Paired t test was utilized. A p<0.05 was considered as significant.

IV. RESULTS

Table 1 reports the time domain parameters, expressed as mean±standard deviation during PRE and POST conditions at REST and during TILT. The HP mean, μ_{HP}, the HP variance, σ^2_{HP}, and the SAP mean, μ_{SAP}, were smaller during TILT compared to REST in both PRE and POST conditions. The SAP variance, σ^2_{SAP}, was found larger in POST compared to PRE both during REST and TILT.

Figure 2 depicts the minimum of NCCE and NKNNCE (mean+standard deviation) calculated over HP (i.e. $NCCE_{HP}$ and $NKNNCE_{HP}$) and SAP (i.e. $NCCE_{SAP}$ and $NKNNCE_{SAP}$). When the values obtained at REST and during TILT were pooled together, the minimum of $NCCE_{SAP}$ and $NKNNCE_{SAP}$ was significantly lower during POST compared to PRE.

V. DISCUSSION

A recent study found an increased complexity of SAP series in PD patients and suggested that the augmented complexity is the result of the impairment of cardiovascular autonomic regulation [14]. In agreement with this interpretation we hypothesize that, if mechanical stimulation was able to reduce complexity of HP and/or SAP, the stimulation procedure would ameliorate cardiovascular control. Figure 1 shows an example in which more regular SAP oscillations are clearly observable after mechanical stimulation. Indeed, while HP series exhibit similar fluctuations in both PRE and POST (Fig.1a,b), SAP series in POST (Fig.1d) is dominated by the presence of more regular fluctuations compared to PRE (Fig.1c).

In order to assess the differences between PRE and POST, complexity indexes derived at REST and during TILT were pooled together. Lower complexity SAP values were found in POST by both complexity indexes (Fig.2, $NCCE_{SAP}$ and $NKNNCE_{SAP}$). This result suggests a potential improvement of vascular regulation. The reduction of SAP complexity might be related to a more effective sympathetic modulation directed to the vessels. The higher SAP variance (Table 1, σ^2_{SAP}) found in POST condition supports this conclusion.

When evaluating the modifications of the autonomic state induced by the orthostatic stimulus, differences between REST and TILT were found in our PD patients. During TILT slightly lower values of μ_{HP}, μ_{SAP} and HP complexity indexes, $NCCE_{HP}$ and $NKNNCE_{HP}$, were found (Tab.I and Fig.2). These findings are in accordance with the expected sympathetic activation and vagal withdrawal induced by TILT. This tendency appears in both PRE and POST and, thus, it does not seem to be related to the mechanical stimulation.

Table 1 HP and SAP mean and variance before (PRE) and after (POST) mechanical stimulation at REST and during TILT

	PRE		POST	
	REST	TILT	REST	TILT
μ_{HP} [ms]	886.9 ± 89.8	746.1 ± 95.9*	866.7 ± 127.7	735.5 ± 105.4*
σ^2_{HP} [ms^2]	1051.6 ± 1852.8	526.7 ± 978.4	771.2 ± 963.8	308.5 ± 262.1*
μ_{SAP} [mmHg]	124.8 ± 13.3	119.8 ± 20.6	123.9 ± 12.2	113.4 ± 23.6
σ^2_{SAP} [mmHg^2]	18.7 ± 7.7	28.4 ± 15.6	24.7 ± 16.4	40.9 ± 32.9

Values are expressed as mean±standard deviation. The symbol * indicates a significance difference when comparing REST vs TILT with p<0.05.

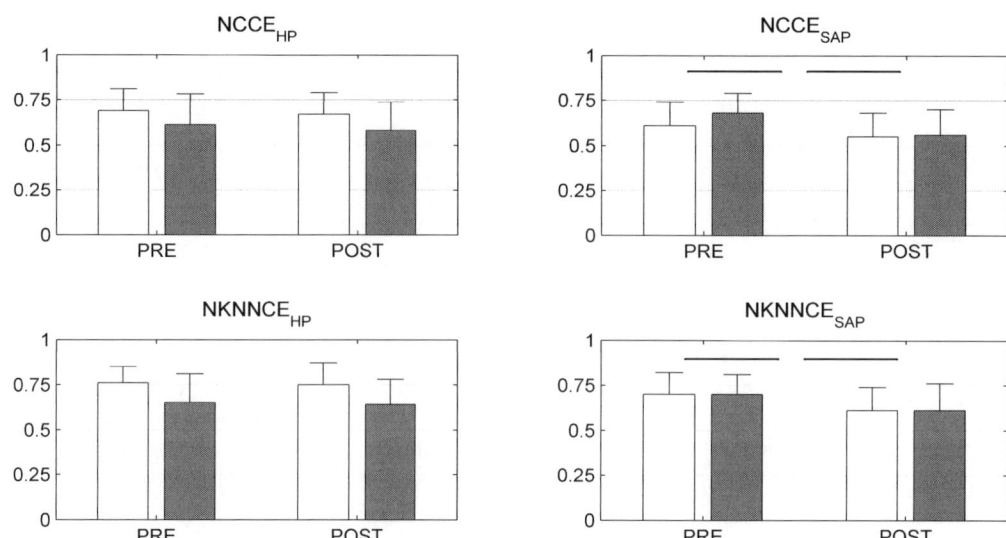

Fig.2 The minimum of NCCE and NKNNCE calculated over HP and SAP at REST (white bars) and during TILT (grey bars) are shown in PRE and POST conditions. The symbol * indicates a significant difference between PRE and POST with $p<0.05$ (data relevant to REST and TILT are pooled

VI. CONCLUSION

The study suggests that plantar stimulation ameliorates vascular control in PD subjects by promoting an increased modulation of sympathetic activity directed to vessels. Future studies are needed to confirm this finding on a larger population with different degrees of motor impairment and to test the recently developed device GONDOLA® (Ecker Technologies Sagl, Switzerland) that makes automatic the mechanical stimulation procedure utilized in this study.

ACKNOWLEDGMENTS

The study was fully financed by Ecker Technologies Sagl, Switzerland.

REFERENCES

1. Pratorius B, Kimmeskamp S, Milani TL (2003) The sensitivity of the sole of the foot in patients with Morbus Parkinson. Neurosci Lett 346(3):173-176
2. Jenkins ME, Almeida QJ, Spaulding SJ et al. (2009) Plantar cutaneous sensory stimulation improves single-limb support time, and EMG activation patterns among individuals with Parkinson's disease. Parkinsonism Relat Disord 15(9):697-702
3. De Nunzio AM, Grasso M, Nardone A et al. (2012) Alternate rhythmic vibratory stimulation of trunk muscles affects walking cadence and velocity in Parkinson's disease. Clin Neurophysiol 121(2):240-247
4. Novak P, Novak V (2006) Effect of step-synchronized vibration stimulation of soles on gait in Parkinson's disease: a pilot study. J Neuroeng Rehabil 3:9
5. Linden D, Diehl RR, Berlit P (1997) Sympathetic cardiovascular dysfunction in long-standing idiopathic Parkinson's disease. Clin Auton Res 7:311-314
6. Ziemssen T, Reichmann H (2010) Cardiovascular autonomic dysfunction in Parkinson's disease. J Neurol Sci 289:74-80
7. Barbic F, Perego F, Canesi M et al. (2007) Early abnormalities of vascular and cardiac autonomic control in Parkinson's disease without orthostatic hypotension. Hypertension 49:120-126
8. Porta A, Gnecchi-Ruscone T, Tobaldini E et al. (2007) Progressive decrease of heart period variability entropy-based complexity during graded head-up tilt. J Appl Physiol 103(4):1143-1149
9. Pincus SM, Cummins TR, Haddad GG (1993) Heart rate control in normal and aborted-SIDS infants. Am J Physiol 33:R638-R646
10. Richman JS, Moorman JR (2000) Physiological time-series analysis using approximate entropy and sample entropy. Am J Physiol 278:H2039-H249
11. Porta A, Guzzetti S, Montano N et al. (2000) Information domain analysis of cardiovascular variability signals: evaluation of regularity, synchronization and coordination. Med Biol Eng Comput 38:180-188
12. Costa M, Goldberger AL, Peng CK (2002) Multiscale entropy analysis of complex physiologic time series. Phys Rev Lett 89:068102
13. Schulz S, Koschke M, Bar KJ et al. (2010) The altered complexity of cardiovascular regulation in depressed patients. Physiol Meas 31:303-321
14. Porta A, Castiglioni P, Di Rienzo M et al. (2012) Short-term complexity indexes of heart period and systolic arterial pressure variabilities provide complementary information. J App Physiol 113:1810-1820
15. Porta A, Baselli G, Liberati D et al. (1998) Measuring regularity by means of a corrected conditional entropy in sympathetic outflow. Biol Cybern 78:71-78
16. Porta A, Castiglioni P, Bari V et al. (2013) K-nearest-neighbor conditional entropy approach for the assessment of the short-term complexity of cardiovascular control. Phys Meas 34:17-33

Author: Tito Bassani
Institute: Clinica Medica, Istituto Clinico Humanitas IRCCS
Street: via Manzoni, 56
City: Rozzano, Milan
Country: Italy
Email: tito.bassani@libero.it

Variability of EEG Theta Power Modulation in Type 1 Diabetics Increases during Hypo-glycaemia

A. Goljahani[1], A.S. Sejling[2], G. Sparacino[1], C. Fabris[1],
J. Duun-Henriksen[3], L.S. Remvig[3], C. Cobelli[1], and C.B. Juhl[3,4]

[1] Department of Information Engineering, University of Padova, Padova, Italy
[2] Department of Cardiology, Nephrology and Endocrinology, Hillerød Hospital, Hillerød, Denmark
[3] Hyposafe, Lyngby, Denmark
[4] Department of Endocrinology, Sydvestjysk Sygehus, Denmark

Abstract—EEG spectral content has been widely investigated to illuminate cognitive processes and assess clinical conditions of patients. In particular, increase of powers in EEG low frequency bands were proved to reflect low levels of glucose concentration in the blood, i.e., hypo-glycaemia states. In the present work we investigate if and how levels of glucose concentrations affect the time course of EEG power modulations in the conventional theta, alpha and beta bands. To this aim, the reactivity index ρ, recently introduced for characterizing individual modulations of alpha rhythms, was utilized to quantify, for each band, EEG power modulations at the P3-C3 channel during induced hypo-glycaemia experiments performed with 10 type-1 diabetic volunteers. Results show that, in any glycemic state, i.e., hyper/eu/hypo-glycaemia, ρ continuously vary in any band, alternating increases and decreases of powers with respect to preceding intervals. In particular, in the theta band, the variability of EEG power modulations during hypo-glycaemia (measured by the ρ sample standard deviation) is significantly higher than in hyper- and eu- glycaemia. This suggests that the variability of ρ in the theta band can be a useful indicator to quantitatively investigate glucose-related EEG changes.

Keywords—EEG, Hypoglycaemia, Reactivity index, Theta band, Diabetes.

I. INTRODUCTION

Many rhythmic activities embedded in human EEG are modulated by changes in mental states and clinical conditions [1]. This is why they have been widely investigated both by cognitive scientists and clinicians. Although visual inspection and counting of zero crossings have often been utilized for characterizing these rhythmic activities, most interesting results have been obtained by relying on information mathematically extracted from the signal, e.g., powers based on power-spectral densities (PSDs), or power variations based on event-related de-synchronizations (ERDs) [2]. Examples of applications are a lot, see e.g. [3,4,5,6,7] for recent specific clinical investigations.

Physiological conditions of particular interest in EEG investigations are those experienced by people affected by diabetes, a pathology which regards about 350 million people around the world, with well known short- and long-term complications caused by excursions of blood glucose concentration (BG) outside its normal range (70–180 mg/dL) [8]. After the pioneering investigations concerning hypoglycaemia-related EEG changes carried out in the 50's [9], several studies, e.g. [10,11], proved that hypo-glycaemia is associated with a power increase in the low frequency bands of the EEG. This suggests that it may be possible to use the brain as a biosensor to detect hypo-glycaemia in real-time by suitable processing of EEG signals [12,13]. Such systems could become a useful addition to minimally-invasive or non-invasive sensors proposed in the last 15 years to continuously monitor glucose levels in real time [14,15] and promptly detecting (or even predicting) hypo-glycemic events, which can be very dangerous in the short term since they may rapidly progress into coma without subject's awareness.

In the present work we investigate if quantifying power modulations in three canonical EEG bands (i.e., theta (4,8) Hz, alpha (8,13) Hz and beta (13,20) Hz) by the reactivity index recently introduced in [16] can allow to enrich information about power-related changes in EEG during hypo-glycaemia. To this aim, EEG data and BG samples collected in parallel in 10 type-1 diabetic volunteers during induced hypo-glycaemia experiments were utilized.

II. MATERIALS AND METHODS

A. Data Base

Data collected in 10 type-1 diabetic subjects studied at Hillerød Hospital, Denmark are considered. In brief, subjects were exposed to insulin-induced hypoglycaemia by intraveneous administration of Actrapid (Novo Nordisk, Bagsværd, Denmark) targeting a one hour period of eu-glycaemia (5-5.5mM) and a subsequent one and a half hour of hypoglycaemia (2mM) during which repeated cognitive testing was performed. These results will be published elsewhere. During the execution of the experiment, blood glucose concentration (BG) and EEG were simultaneously monitored. In particular, BG was frequently determined by a laboratory analyzer YSI (Yellow Springs Instrument Company, Ohio, USA), while 19 EEG channels were recorded by standard cap electrodes placed on the scalp according to the 10/20 international system. EEG was recorded by a digital EEG recorder (Cadwell Easy II, Kennewick, Washington, USA) EEG signals were analogically low-pass filtered in order to avoid aliasing, then digitally acquired and finally down-sampled at 64 Hz. The dynamic range of the EEG was ±4620µV with an amplitude resolution of 0.14 µV. The internal noise level in the analog data acquisition system was estimated to be 1.3 µV RMS. The protocol was approved by the local ethical committee.

The analysis presented herein is developed considering the EEG signal obtained as the subtraction of the recordings from P3 and C3, the rationale being to consider a scalp position close to that investigated in [17] for potential use in hypo-alerts generation.

Fig. 1 shows the P3-C3 EEG signal (upper panel) and the BG time-series (central panel, open bullets) collected in a representative subject (ETW13).

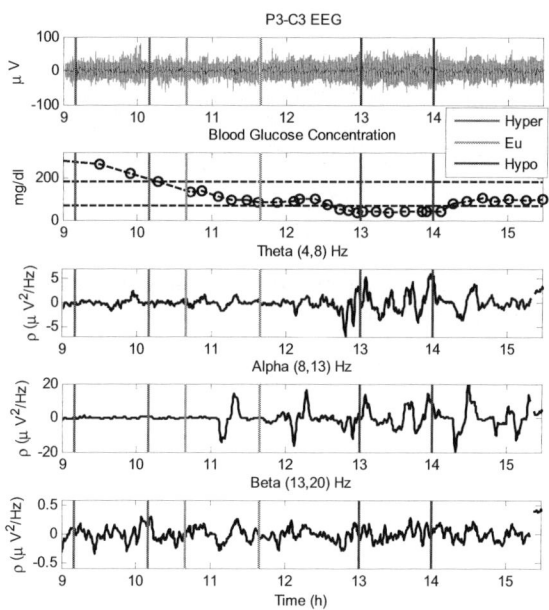

Fig. 1 P3-C3 EEG recording (upper panel) and simultaneous BG measurements (central panel, open bullets) in a representative subject during induced hypoglycaemia experiment (BG samples are smoothed by a spline, dotted line). Horizontal lines in the central panel identify hyperglycaemia and hypoglycaemia thresholds. In the three lower panels, the time courses of the reactivity index ρ relatively to theta, alpha and beta bands are depicted. In all panels, vertical red, green and blue lines refer to hyper-, eu- and hypo- glycaemia intervals, respectively

B. Intervals for the Analysis

For each subject, three 1-hour intervals, corresponding to hyper-, eu- and hypo-glycaemia conditions respectively, were determined by visual inspection of BG time-series. For instance, vertical lines in Fig. 1 mark the selected intervals in the representative subject (hyperglycaemia in red, euglycaemia in green, hypoglycaemia in blue). To facilitate the determination of the three intervals and reduce subjectivity in the analysis, hyper- and hypo-glycaemia thresholds (180 and 70 mg/dl, respectively, denoted by dotted horizontal black lines in the middle panel of Fig. 1) were compared with a smoothing spline previously fitted against the raw BG samples by using the method of [18] (dotted line in the middle panel of Fig. 1).

C. EEG Analysis Based on Reactivity Indexes

The reactivity index, denoted as ρ, was introduced in [16] for measuring the intensity of EEG power decrease between a reference and a test time interval in a generic frequency interval (f_1, f_2). Specifically, given any frequency interval (f_1, f_2) and the PSDs during the reference, $R(f)$, and test, $T(f)$, intervals, as, e.g., in Fig. 2 for two illustrative PSDs, ρ is computed by dividing the average power decrease in (f_1, f_2) (area of the region shaded gray in the figure) by the length of the interval.

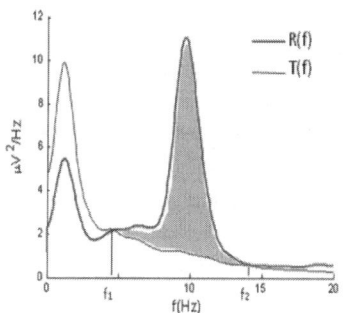

Fig. 2 Illustrative reference, R(f), and test, T(f), PSDs, with boundaries of a generic frequency interval denoted by f_1 and f_2

Formally,

$$\rho = \frac{\int_{f_1}^{f_2} (R(f) - T(f)) df}{(f_2 - f_1)}. \quad (1)$$

Practically, ρ (µV²/Hz) measures the average power modulation per unit of frequency in (f_1, f_2) and a positive value corresponds to a decrease of power, whereas a negative value to an increase.

For each subject and for (f_1,f_2) equal to (4,8) Hz, (8,13) Hz, and (13,20) Hz, series of ρ values were obtained as follows. First, a series of couples of reference (I_{ref}) and test (I_{test}) time intervals was obtained as in Fig. 3. Namely, for each couple, I_{ref} and I_{test} lasted 2 minutes and the distance between their beginnings was set to 10 minutes, which is equal to the integration period of the monitoring system utilized in [17]. The first couple was obtained by aligning the beginning of the corresponding I_{ref} with the first BG. Then, the successive couples were obtained by shifting I_{ref} and I_{test} by 0.5 minutes. The last couple is the one with the beginning of I_{test} aligned with the last BG sample.

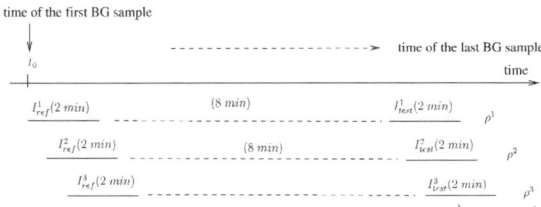

Fig. 3 Sequence of reference (I_{ref}) and test (I_{test}) time intervals for computing ρ series

Couples of $R(f)$ and $T(f)$ were, then, computed by applying the Welch method with Hanning tapered windows lasting 2 seconds without overlap [19]. Finally, by applying eq. (1) to the couples of PSDs, a series of ρ values was obtained for each frequency interval. Time courses of ρ values correspond to evolutions of power modulations between intervals 10 minutes apart. Notably, EEG data portions containing occasional artifacts (amplitude greater than 100 μV) either in the reference or test interval were excluded from the analysis.

III. RESULTS

For each subject, three ρ series were computed relatively to theta, alpha and beta bands as explained in Section II-C. In Fig.1, time courses of ρ values in the three bands are shown for the representative subject ETW13. As apparent, for all the three bands, ρ values oscillate between positive and negative values, meaning that phases of power decrease alternate to phases of power increase, respectively. However, by focusing on each band, interesting differences may be detected. As far as theta is concerned (third panel from the bottom), ρ values vary quite homogeneously during hyper- and eu- glycaemia intervals, whereas, as BG goes under the hypo- glycaemia threshold, variations of ρ values suddenly become more accentuated, as shown, in particular by the area delimited by the blue vertical lines in the panel. For alpha band (second panel from the bottom), the range of variations of ρ values is higher during eu-glycaemia than during hyper-glycaemia. However, intensity of power modulations in eu- glycaemia episodically reaches those in hypo-glycaemia, as, e.g., slightly after 11:00. Finally, in beta band variations of ρ values remain homogenous throughout all the three glycaemia states, meaning that intensities of power modulations lay roughly in the same range. Results obtained in all the other 9 subjects showed a similar overall behavior.

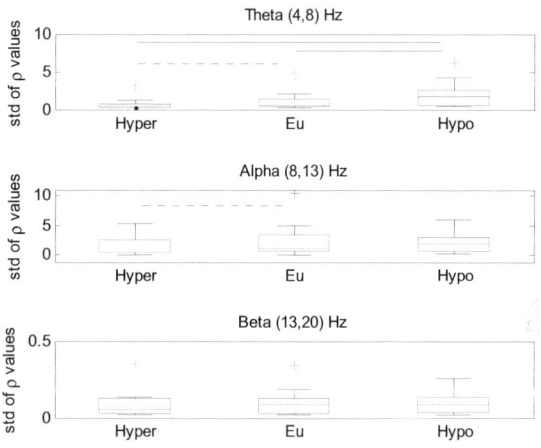

Fig. 4 Boxplots of standard deviations of ρ values in the three glycaemic states from all subjects for theta (top), alpha (middle) and beta (bttom) panels. On each box, the central red line is the median value, the edges of the box are the 75th and 25th percentiles, the whiskers extend to the most extreme data points that are not outliers and the outliers are plotted individually (red cross). Solid and dotted horizontal lines indicate statistically significant differences with p<0.01 and p<0.05, respectively

In order to quantify the variability of power modulations, the standard deviations of ρ values during hyper-, eu- and hypo- glycaemia intervals, defined as in Section II-B, were computed for each band. Then, t-tests were carried out for each band between all couple of states, i.e., hyper- vs eu- glycaemia, eu- vs hypo- glycaemia, hyper- vs hypo-glycaemia, in order to assess if statistically significant differences could be detected. Standard deviations of ρ's obtained from all of the 10 subjects in the three glycemic states are graphically represented in the boxplots of Fig.3 for each of the three considered bands. Confirming the expectations following visual inspection of Fig.1, it was found that, for the theta band (top panel), standard deviations of ρ values during hypo-glycaemia are significantly (p<0.01) higher than those during eu- and hyper- glycaemia (this is highlighted in the top panel of Fig.3 by drawing two horizontal solid lines). Also, standard deviations during eu-glycaemia are slightly higher than those in hyper-glycaemia at a statistical significance of 5% (dotted horizontal lines).

As far as the alpha band is concerned (middle panel of Fig.3), although standard deviations in eu-glycaemia were found to be significantly ($p<0.05$) lower than those in hyper-glycaemia, no statistical differences were found between eu- and hypo-glycaemia. Finally, no statistically significant differences were found in the beta band between the variability of ρ in the three glycaemic states.

IV. CONCLUSIONS

In this paper, time courses of power modulations measured by the reactivity index ρ recently introduced in [16] were analyzed in three conventional EEG bands, i.e., theta, alpha, and beta, in order to assess their possible utility in investigating hypo-glycaemia related changes.

In 10 type-1 diabetic subjects involved in hypo-glycaemia induced experiments, power modulations in the theta band resulted to present a statistically significant increase of variability during hypo-glycaemia with respect to eu-glycaemia. Instead, modulations in alpha band showed statistically higher variability only in eu-glycaemia with respect to hyper-glycaemia. Analysis of the beta band did not lead to any relevant evidence.

In conclusion, the preliminary results presented in this paper suggest that intensities of power modulations in theta band, quantified by reactivity indexes, change with BG levels. Further development of the present research in the short-term will start with assessing results on an extended database. Besides, optimizing parameters such as the duration of reference and test intervals and the distance between them may allow to even more finely and robustly capture hypo-glycaemia related EEG changes. In the mid-term, a finely tuned algorithm based on the reactivity index ρ may be exploited in EEG based devices aimed at detecting hypo-glycaemia, e.g. [17], or at supporting strategies based on co-registrations with CGM systems [18,19].

ACKNOWLEDGMENTS

The authors are grateful to C. Zecchin and A. Facchinetti for their contribution in preliminary data analysis.

REFERENCES

1. Nunez PL, Srinivasan R (2006) Electric fields of the brain, the neurophysics of EEG, 2006, Oxford University Press, Inc., New York
2. Pfurtscheller G, Lopes da Silva FH (1999) Event-related EEG/MEG synchronization and desynchronization: basic principles. Clin Neurophysiol 1842-1857
3. Keogh MJ, Drury PP, Bennet L et al. (2011) Limited predictive value of early changes in EEG spectral power for neural injury after asphyxia in preterm fetal sheep. Pediatr Res 71:345-53
4. Kerr WT, Anderson A, Lau EP et al. (2012) Automated diagnosis of epilepsy using EEG power spectrum. Epilepsia 53(11):189-92
5. Sun L, Grützner C, Bölte S et al. (2012) Impaired gamma-band activity during perceptual organization in adults with autism spectrum disorders: evidence for dysfunctional network activity in frontal-posterior cortices. J Neurosci 32(28):9563-73
6. Israel B, Buysse DJ, Krafty RT et al. (2012) Short-term stability of sleep and heart rate variability in good sleepers and patients with insomnia: for some measures, one night is enough. Sleep 35(9):1285-91
7. Marchetti P, D'Avanzo C, Orsato R et al. (2011) Electroencephalorapy in patients with cirrhosis. Gastroenterology 141(5):1680-9.
8. http://www.idf.org
9. Ross IS, Loeser LH (1951) Electroencephalographic findings in essential hypoglycemia. Electroencephalogr Clin Neurophysiol 3: 141-148
10. Pramming S, Thorsteinsson B, Bendtson I et al. (1990) The relationship between symptomatic and biomechanical hypoglycaemia in insulin-dependent diabetic patients. J Intern Med 228:641-646
11. Hyllienmark L, Maltez J, Dendenell A et al. (2005) EEG abnormalities with and without relation to severe hypoglycaemia in adolescents with type 1 diabetes. Diabetologia 48:412-419
12. Juhl CB, Hojlund K, Elsborg R et al. (2010) Automated detection of hypoglycemia-induced EEG changes recorded by subcutaneous electrodes in subjects with type 1 diabetes-The brain as a biosensor. Diabetes Res Clin Pract 88:22-28
13. Snogdal LS, Folkestad L, Elsborg R, Remvig LS, Beck-Nielsen H, Thorsteinsson B, Jennum P, Gjerstad M, Juhl CB (2012) Detection of hypoglycemia associated EEG changes during sleep in type 1 diabetes mellitus. Diabetes Res Clin Pract, 98: 91-97
14. Sparacino G, Facchinetti A, Cobelli C (2010) "Smart" continuous glucose monitoring sensors: on-line signal processing issues. Sensors 10:6751-6772
15. Sparacino G, Zanon M, Facchinetti A et al. (2012) Italian contributions to the development of continuous glucose monitoring sensors for diabetes management. Sensors 12:13753-13780
16. Goljahani A, D'Avanzo C, Schiff S, Amodio P, Bisiacchi P, Sparacino G 2012 A novel method for the determination of the EEG individual alpha frequency. NeuroImage, 60: 774-786
17. Elsborg R, Remvig LS, Beck-Nielsen H et al. (2011) Biosensors for health, environment and biosecurity, ch. 12: Using the brain as a biosensor to detect hypoglycamia. InTech
18. Facchinetti A, Sparacino G, Cobelli C (2011) Online denoising method to handle intra-individual variability of signal-to-noise ratio in continuous glucose monitoring. IEEE Trans Biomed Eng 58: 2664-2671
19. Welch PD (1967) The use of Fast Fourier Transform for the estimation of power spectra: a method based on time averaging over short, modified periodograms. IEEE Trans Audio Electroacoust, vol. AU-15, 2: 70-73
20. Guerra S, Sparacino G, Facchinetti A et al. (2011) A dynamic risk measure from continuous glucose monitoring data. Diabetes Technol Ther 13:843-85
21. Facchinetti A, Sparacino G, Guerra S et al. (2013) Real-time improvement of continuous glucose monitoring accuracy: the smart sensor concept. Diabetes care 36:793-800

Analysis of Systematic Inter-channel Time Offsets in FECG from Maternal Abdominal Sensing

P. Melillo[1,2], D. Santoro[3], and M. Vadursi[3]

[1] Dipartimento multidisciplinare di specialità medico-chirurgiche – Second University of Naples, Naples, Italy
[2] Dipartimento di Ingegneria dell'Energia Elettrica e dell'Informazione "Guglielmo Marconi", University of Bologna, Bologna, Italy
[3] Dept. of Engineering, University of Naples "Parthenope", Naples, Italy

Abstract—The recently presented spatio-temporal filtering (STF) approach aiming to enhance fetal electrocardiogram (FECG) signals from abdominal sensing is based on an optimized weighted sum of QRS complexes. As the objective function to be maximized is created on the basis of the quality of the synchronization among abdominal channels, the performance of STF can be degraded in presence of uncompensated deterministic time offsets. The paper investigates possible systematic inter-channel time offsets in publicly available FECG signals, as the first step for the introduction of a pre-processing block in the STF implementation aimed at evaluating and compensating such offset.

Keywords—FECG, fetal heart rate monitoring, spatio-temporal filtering, systematic effects compensation.

I. INTRODUCTION

Fetal electrocardiogram (FECG) is a useful tool to evaluate the status of fetus's health. Direct FECG signal is usually acquired by applying probes on fetal scalp, which apart from being potentially dangerous to the fetus's health, is an invasive, yet accurate, technique.

A less invasive technique is represented by the acquisition of electrical activity from maternal abdomen, through a number of electrodes. The acquired signal is in such a case called AECG (abdominal electrocardiogram), which is the superposition of FECG, noise and interference sources, one of which is clearly the maternal electrocardiogram (MECG).

AECG has to be suitably processed in order to measure fetal heart rate (FHR) and possibly the fetal heart rate variability. Due to the mixed signal nature of AECG, and to the negative signal-to-interference ratio, automatic processing of AECG can be a challenging task, which is often started with the extraction of the FECG signal.

Processing techniques proposed in literature to extract the FECG signal from the AECG signal can be classified in four main groups, depending on the different approach they are based on. These are: independent component analysis (ICA), template subtraction, single value decomposition (SVD) and time-frequency filtering.

The main problem that has to be faced is represented by the low value of the fetal cardiac signal, which needs to be properly enhanced before being processed for automatic evaluation of FHR or other parameters. The most common solutions rely on the the repetitiveness of the QRS pattern in the cardiac signal. In presence of multi-channel AECG, the idea of taking advantage of all the available information is another, not necessarily alternative, possibility. Such approach dates its roots back to the so-called spatial filtering, which permits to enhance QRS complexes by a weighted sum of the samples of the same (i.e. synchronous) QRS complexes coming from different abdominal channels[1]. This way, even poor quality QRS complexes, i.e. complexes affected by stronger noise and/or interference, from which it can be hard to detect accurately the peaks, can be exploited to extract the FECG. Similar techniques, such as those based on ICA and SVD, are especially effective when a large number of channels are available. On the contrary, in presence of a reduced number of electrodes, experimental studies have shown that no particular advantage is given by a multi-channel approach over a single channel approach [2].

An improved version of spatial filtering, which includes samples of time-adjacent QRS complexes as terms of the weighted sum has recently been proposed in [3]. Such technique is named spatio-temporal filtering. Weights in the sum are determined by maximizing a objective function, which is basically the ratio of the energy of high-quality QRS complexes to the energy of poor quality ones. The quality of QRS complexes is evaluated according to the quality of synchronization among different abdominal channels. In details, a QRS complex is considered of high quality if the time instants at which peaks are detected (namely, *fiducial points*) on the different channels are either the same or very close to each other.

It is therefore evident that the performance of such an approach strictly depends on the quality of synchronization among channels. In particular, any deterministic temporal offset among channels can influence the quality of the QRS complexes as defined before, and affect the determination of the weights of the spatio-temporal filter, with potential negative consequences on the overall performance of the STF. Despite being potentially detrimental to the application of some spatial filtering methods, and STF in particular, the presence of possible time offset (i.e., relative lag) among signals related to different abdominal channels has never been investigated, to the best of the authors' knowledge.

Stemming from such considerations, we have thoroughly analyzed a set of publicly available AECG signals, with the aim of stating whether deterministic offsets among signals acquired from different abdominal channels can be found, or are even common. Once the deterministic component of relative lags among channels is estimated, such systematic effect can be compensated, prior to the application of STF, and can possibly improve STF performance.

The paper is organized as follows: section II briefly recalls the basic concepts and key tasks of spatio-temporal filtering, whereas section III presents the results of the analysis of public AECG data, aimed to verify whether deterministic time offsets do occur among different channel, and if this is the case, to estimate them. Section III also sets the basis for the definition of a pre-processing algorithm that automatically detects and compensates inter-channel time offsets, which could be inserted as a specific block in the implementation chain of STF. Finally, section IV gives the conclusions.

II. SPATIO-TEMPORAL FILTERING FOR FECG ENHANCEMENT

It is well known that the AECG signal contains the overlap of the following signals: MECG, FECG and other interferences such as power line interference, muscle contractions (electromyogram, EMG), motion artifacts and baseline wander. In addition, although the registration of the fetal heartbeat is performed far from maternal chest, the magnitude of fetal heartbeat is less than the amplitude of maternal heartbeat. FECG signal in AECG signal has thus a negative SNR (maternal heartbeat is considered as noise). Even after MECG suppression, the intensity of FECG is too low and a proper enhancement is needed to measure parameters such as FHR.

The aim of spatio-temporal filtering (STF) is to process FECG signals recorded from different maternal abdominal channels in order to obtain one FECG signal in which QRS complexes have a high SNR. STF for FECG enhancement was originally proposed in [3]: assuming a total of N QRS complexes have been acquired for each of the M abdominal channels, the enhanced k-th QRS complex is obtained as the weighted sum of all the k-th QRS complexes extracted from the M signals, and of the $M \times 2J$ QRS complexes that occur immediately before or after the k-th. For example, if $J = 1$, for each channel, the $(k-1)$-th and the $(k+1)$-th QRS complex are also included in the sum.

The scheme of the overall monitoring system proposed in [3] is depicted in Figure 1. The first block is a filter bank that includes a notch filter and a band-pass filter for interference suppression, followed by an MECG suppression block [4].

The optimal weights of the ST filter are determined by maximizing an objective function with the aim of assigning a greater weight to higher quality QRS complexes. In details, QRS complexes are detected in each channel and marked (possibly through the use of QRS complex detection algorithms [5]), in order to identify the fiducial points $m_{i,j}$, which are the time instants in correspondence of the local peaks (j denotes the channel and i the QRS complex). Then, the mean and the standard deviation of fiducial points $m_{i,j}$ are calculated for each i and the standard deviation is used as a metric of the QRS complex (poor) quality: the noisier the signal is, the more difficult it is to locate QRS complex. Subsequently, the QRS complexes are divided into two groups of the same cardinality, depending on their quality as now defined, and an objective function is created as a parametric signal-to-noise ratio, where the signal to be maximized is the linear combination of the "good" QRS complexes, whereas the noise is represented by the poor quality QRS complexes, and the parameters are the weighting factors of the filter. The filter coefficients that maximize the objective function are finally determined analytically.

It is clear that any time offset among channels, due to the different frequency response of channels, results in an erroneous evaluation of the quality of the QRS complexes, as defined before. From a metrological point of view, the deterministic component of the temporal offsets among the QRS complexes of any two channels can be seen as a systematic error in the difference between the corresponding fiducial points. Indeed, this can affect the performance of spatial-temporal filtering, decreasing the efficacy of the technique.

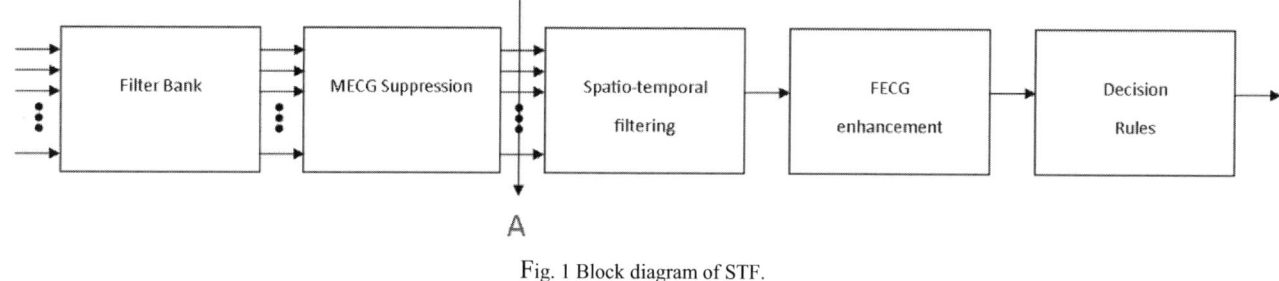

Fig. 1 Block diagram of STF.

It can therefore be worth to investigate whether a systematic effect can be noticed and quantified on publicly available FECG signals, which is the first step to introduce a pre-processing block before the STF (i.e., at section A in Fig. 1) to estimate and compensate such effect.

III. INTERCHANNEL TIME OFFSET ANALYSIS

The following analysis is carried out on two registrations (in the following, signal A and signal B), each of which consisting of four abdominal signals and a direct FECG signal. The duration of each signal is about 60 seconds, and the sampling frequency is equal to 1 kHz. Signals are taken from the publicly available database http://www.physionet.org/physiobank/database/adfecgdb/.

To evaluate the deterministic component of time offsets among different signals, we have first manually located fetal R peaks in all abdominal signals. Each identified R peak is labeled with a value that represents its own fiducial point $r_{i,j}$. The direct FECG signal is exploited to facilitate the identification of the peaks in abdominal signals for those QRS complexes that are more affected by noise. Please note that peaks could also be located automatically, e.g. through specific QRS detection algorithms [5], but the manual solution has been preferred to maximize the accuracy. Fiducial points $r_{i,j}$ related to the same channel (i.e., the same j) are then grouped in four ordered vectors: $\bm{rt_2}$, $\bm{rt_3}$, $\bm{rt_4}$, $\bm{rt_5}$. Then, time offsets between QRS complexes of each pair of channels x and y are calculated as vectorial difference $\bm{drt_{xy}} = \bm{rt_x} - \bm{rt_y}$. Please note that abdominal channels are numbered from 2 to 5, index 1 being reserved to the direct FECG.

Any systematic effects can be evaluated qualitatively by making histograms of the values of each of the six vectors $\bm{drt_{xy}}$: a unimodal distribution with mean not equal to zero and a low variance can be assumed to be the systematic effect of a deterministic offset between channel x and y. This can be estimated as the mean of $\bm{drt_{xy}}$. Fig. 2 and Fig. 3 show the results of the analysis for the two sets of signals. In the figure, the x-axis is expressed in seconds, whereas the y-axis is the absolute frequency of occurrence.

A quick inspection of the histograms related to signal A (Fig. 2) suggests that the channel $ch2$ is in advance with respect to channel $ch3$, which is advance on channel $ch4$. Moreover, channels $ch4$, $ch5$ and $ch2$ can be assumed to be synchronized. It is important to observe that the histograms are concentrated around their own mean values. More detailed results are given in Table 1.

As regards signal B, histograms shown in Fig. 3 show that the channel $ch4$ is advance on channel $ch2$, and channel $ch4$ is advance on channel $ch3$. All the results are reported in Table 2.

Results depicted in Fig. 2 and Fig. 3 confirm that there are some time offsets among FECG signals registered from abdominal signals, whose deterministic components can be easily identified.

On the basis of this information, it is possible to develop an algorithm to detect and compensate possible time offsets among abdominal signals in order to enhance spatio-temporal filtering performance. This could be added as a pre-processing block at section A of the block diagram shown in Fig. 1. The pre-processing would consist in an automatic detection of fiducial points in correspondence of R peaks on FECG signals on the M abdominal channels

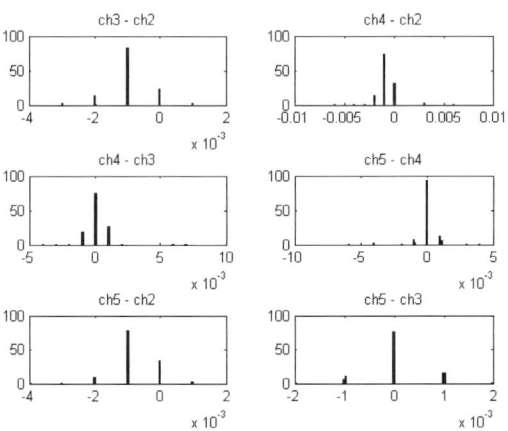

Fig. 2 Histograms of sequences drt_{xy} (signal A).

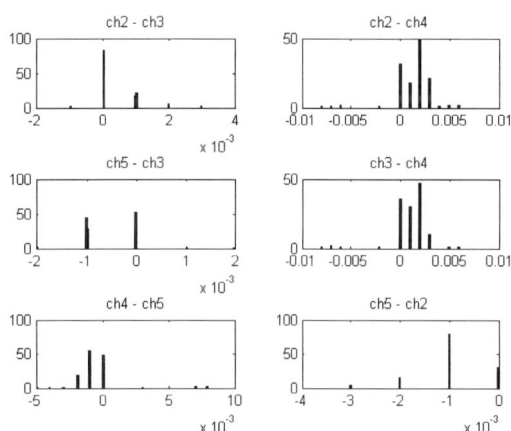

Fig. 3 Histograms of sequences drt_{xy} (signal B).

(after MECG suppression), and the subsequent calculation of vectors drt_{xy} (for $x < y \leq M$). Afterwards, a normality test would be made to determine whether the distribution can be considered a Gaussian and, if this is the case, the sample mean of each vector can be calculated. This would represent the deterministic time offset between channels x and y. Then, an optimal rule should be set to compensate for the offsets. The compensated signals would finally feed the STF.

Table 1 Average value and standard deviation of sequences drt_{xy} (signal A)

drt_{xy}	Average value [ms]	Std. deviation [ms]
drt_{32}	-0.92	0.65
drt_{42}	-0.8	1.1
drt_{43}	0.1	1.1
drt_{52}	-0.80	0.69
drt_{53}	0.11	0.65
drt_{54}	0.0	1.0

Table 2 Average value and standard deviation of sequences drt_{xy} (signal B)

drt_{xy}	Average value [ms]	Std. deviation [ms]
drt_{23}	0.40	0.66
drt_{24}	1.4	1.9
drt_{34}	1.0	1.8
drt_{45}	-0.5	1.7
drt_{52}	-0.95	0.69
drt_{53}	-0.56	0.58

IV. CONCLUSIONS

The paper has investigated the possible presence of inter-channel deterministic time offset in AECG signals, which can degrade the performance of STF techniques for FECG signal enhancement.

The results of the analysis confirm the existence of deterministic time offsets in some publicly available FECG signals, with regard to some channels.

The achieved results open the way to the definition of an algorithm for the automatic detection and compensation of inter-channel time offsets, which is the main focus of the ongoing research activities on this topic.

REFERENCES

1. van Oosterom, A. (1986). "Spatial filtering of the fetal electrocardiogram." Journal of Perinatal Medicine-Official Journal of the WAPM 14(6): 411-419.
2. Kotas M, Jezewski J, Kupka T, Horoba K (2008) Detection of low amplitude fetal QRS complexes, Proc. of 30th IEEE/EMBS Int. Conf., 2008, pp. 4764–4767.
3. Kotas M, Jezewski J, Horoba K, Matonia A (2010) Application of spatio-temporal filtering to fetal electrocardiogram enhancement, Comput Meth Progr Bio 104:1-9 DOI 10.1016/j.cmpb.2010.07.004
4. Ungureanu G M, Bergmans J W M, Oei S G et al (2009) Comparison and evaluation of existing methods for the extraction of low amplitude electrocardiographic signals: a possible approach to transabdominal fetal ECG, Biomed. Tech. 54:66–75.
5. Pan J, Tompkins W J (1985) A real-time QRS detection algorithm, IEEE Trans. Biomed. Eng. 32 (1985) 230–236.
6. Matonia A, Jezewski J, Horoba K et al (2006) The Maternal ECG Suppression Algorithm for Efficient Extraction of the Fetal ECG from Abdominal Signal, Proc. of the 28th IEEE EMBS Annual International Conference, New York City, USA, 2006.
7. Gupta A, Srivastava M C, Khandelwal V et al (2007) A novel approach to Fetal ECG Extraction and Enhancement Using Blind Source Separation (BSS-ICA) and Adaptive Fetal ECG Enhancer (AFE), Proc. of 6th Intern. Conf. on Information, Communications & Signal Processing, 2007.
8. Ming M A, Wang N; Lei S Y, Extraction of FECG Based on Time Frequency Blind Source Separation and Wavelet De-noising, 3rd Intern. Conf. on Bioinformatics and Biomedical Engineering, 2009.
9. Vullings R, Maternal ECG removal from non-invasive fetal ECG recordings (2006) Proc. of 28th Annual Intern. Conf. of the IEEE Engineering in Medicine and Biology Society, 2006.

Author: Paolo Melillo
Institute: Dipartimento multidisciplinare di specialità medico-chirurgiche – Second University of Naples
Street: Via Del Grecchio
City: Naples
Country: Italy
Email: paolo.melillo@unina2.it / paolo.melillo2@unibo.it

Spatiotemporal Brain Activities on Recalling Names of Body Parts

T. Yamanoi[1], Y. Tanaka[1], M. Otsuki[2], H. Toyoshima[3], and T. Yamazaki[4]

[1] Graduate School of Engineering, Hokkai-Gakuen University, Sapporo, Japan
[2] Faculty of Health Sciences, Hokkaido University, Sapporo, Japan
[3] Japan Technical Software Co. Ltd., Sapporo, Japan
[4] Department of Bioscience and Bioinformatics, Kyushu Institute of Technology, Iizuka, Japan

Abstract—The authors measured electroencephalograms (EEGs) from subjects looking at line drawings of body parts and recalling their names silently. The equivalent current dipole source localization (ECDL) method is applied to the event related potentials (ERPs): summed EEGs. ECDs are localized to the primary visual area V1, to the ventral pathway (ITG: Inferior Temporal Gyrus), to the parahippocampal gyrus (ParaHip), the right angular gyrus (AnG), the right supramarginal gyrus (SMG) and the Wernike's area. Then ECDs are localized to the Broca's area, the postcentral gyrus (PstCG) and the fusiform gyrus (FuG), and again the Broca's area. These areas are related to the integrated process of visual recognition of pictures and the recalling of words. Some of these areas are also related to image recognition and word generation. And process of search and preservation in the memory is done from the result of some ECDs to the paraHip.

Keywords—Electroencephalograms, brain activity, event related potentials, equivalent current dipole source localization, word recognition.

I. INTRODUCTION

According to research on the human brain, the primary process of visual stimulus is first processed on V1 in the occipital lobe. In the early stage, a stimulus from the right visual field is processed in the left hemisphere and a stimulus from the left visual field is processed in the right hemisphere. Then the process goes to the parietal associative area [1].

Higher order processes of the brain thereafter have their laterality. For instance, 99% of right-handed people and 70% of left-handed people have their language area in the left hemisphere, in the Wernicke's area and Broca's area [2], [3], [4].

By presenting words written in *kanji* (Chinese characters) and others written in *hiragana* (Japanese alphabet) to the subjects, researchers measured electroencephalograms (EEGs) when those stimuli and the data were summed and averaged according to the type of stimuli and the subjects. As a result event related potentials (ERPs) were obtained. ERPs peaks were detected and analyzed by equivalent current dipole source localization (ECDL) [5] at that latency using the three-dipole model. In both the recognition of the *kanji* and *hiragana* are localized equivalent current dipole (ECD) nodes from early components of ERPs to the V1, V2 and the inferior temporal gyrus (ITG). After that ECDs were localized to the Wernicke's area and the Broca's area. These results agree with the results on MEG, PET, or fMRI [5], [6].

On the other hand, clinical lesion studies have shown that lesions causing disabilities of naming and comprehension of objects are dissociated depending on the target categories, e.g., artificial or biological things. These symptoms are called category-specific disorders [7].

Using the same methodology as that in the above-mentioned research [6], [8], [9], [10], [11], [12], some of the present authors cleared human brain activities during language or image recognition.

In the present study, we measured electroencephalograms (EEGs) in order to investigate the brain activity during watching the line drawings of the body part and recalling a name of body part. And both data were summed and averaged according to the type of the stimuli in order to get event related potentials (ERPs). Each peak of ERPs were detected and analyzed by the equivalent current dipole source localization (ECDL) method [5]. The paper is a continuation of the previous research [13].

II. EEG MEASUREMENT EXPERIMENT

The one subject was a 22-year-old female (MN) and that had normal visual acuity. She was left-handed, and, from the preceding experiment, her dominant language area was considered to be located in the right hemisphere. The other subject was a 22-year-old male (HT) and that had normal visual acuity. He was right-handed. The subjects put on 19 active electrodes and watched a 21-inch CRT 30cm in front of them. The head was fixed on a chin rest on the table.

Each image was displayed on the CRT. Stimuli were simple monochrome images (line drawings) of parts of the human body. Images were of a foot, mouth, finger, ear, and hand (Fig. 1). First, a fixation point was presented, and then a stimulus was presented. Both of these occurred within 3000 msec. EEGs were measured on the multi-purpose portable bio-amplifier recording device (Polymate AP1524;

TEAC) by means of the electrodes; the frequency band was between 1.0 and 2000 Hz. Output was trans-mitted to a recording PC.

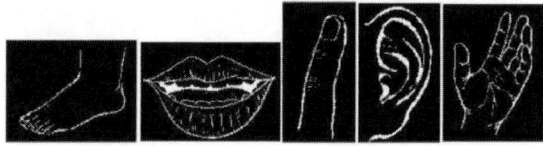

Fig. 1 Presented images of human body parts

We measured the subject's EEGs on each visual stimulus. So as to effectively execute the ECDL method, each visual stimulus data were summed and averaged according to the type of human body part to get event related potentials (ERPs). We analyzed especially the image of a mouth. To each subject, we tried the experiment twice. So as to distinguish these experiments, we labeled MN1, MN2, HT1 and HT2 to each ERP. According to each ERP, the following four characteristics were found: (1) A periodic wave of small amplitude existed until the latency of 350 msec; (2) A negative peak appeared around 450 msec; (3) A positive peak appeared around 500 msec, (4) Amplitude attenuated gradually after 500 msec, and converged around 700 msec (Fig. 3). From the above, we considered that the language process is done after 350msec.

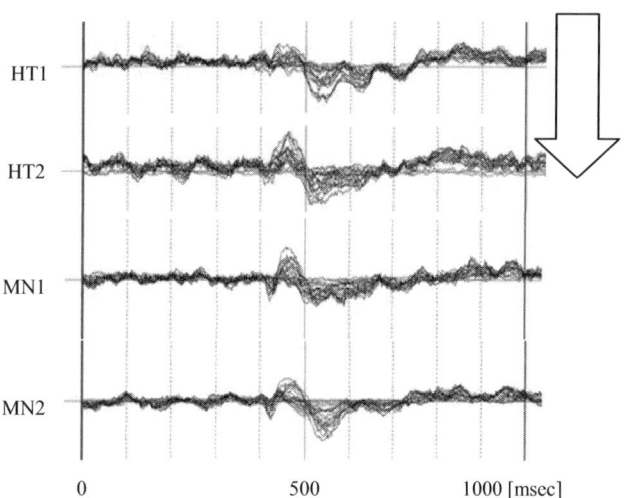

Fig. 3 Event Related Potentials (ERPs) by the present experiment

Then the ECDL method was applied to each ERP. Because the number of recording electrodes were nineteen, three ECDs at most were estimated by use of the PC-based ECDL analysis software "SynaCenterPro [5]" from NEC Corporation. The goodness of fit (GOF) of ECDL was more than 99%.

Some examples of localized ECDs are depicted in Fig. 4 and Fig. 5. In Fig. 4 and Fig. 5, the left pictures are sagittal views, the middle axial views, and the right coronal views. From these three views, one can understand the location of the ECDs in the three-dimensional space. ECDs localized by the ECDL method are indicated by white dots in these figures.

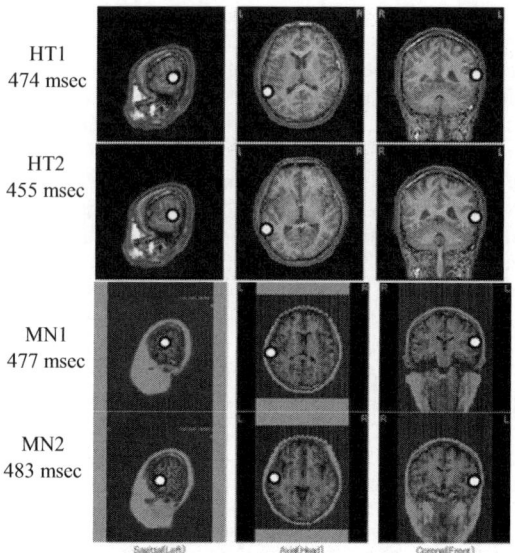

Fig. 4 ECDs localized to the Wernicke's area

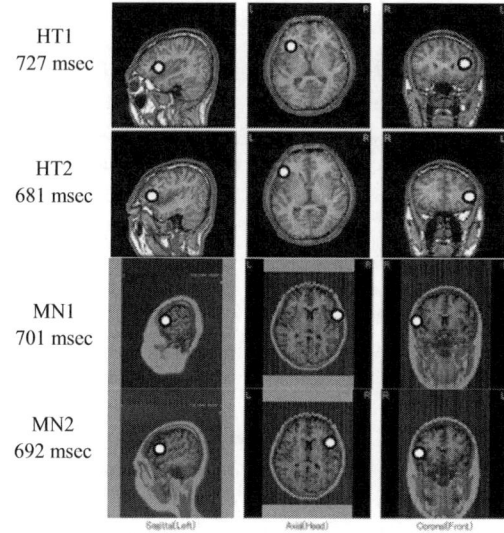

Fig. 5 ECDs localized to the Broca's area and the Broca's area homologue

These processes are done in series or in parallel. The relationship is summarized in Table 1 and Table 2.

Table 1 Relationship between localized source and its latency (HT)

subject	V1	ITG	R ParaHip
HT1	119	326	366
HT2	131	323	393
	R AnG	Wernicke	R Broca
	373	474	524
	427	455	485
	R ParaHip	R PstCG	L FuG
	530	566	537
	506	524	556
	R ParaHip	Broca	
	610	727	
	590	681	[msec]

Table 2 Relationship between localized source and its latency (MN)

subject	V1	ITG	R ParaHip
MN1	127	292	380
MN2	106	334	354
	R SMG	Broca	Wernicke
	443	457	477
	430	455	483
	R ParaHip	R PstCG	R FuG
	481	503	546
	494	547	548
	R ParaHip	R Broca	
	575	701	
	592	692	[msec]

III. RESULTS OF ESTIMATED EQUIVALENT CURRENT DIPOLES

According to the subjects MN and HT, the input pathway is observed as V1 → ITG → the right Parahippocampus (ParaHip) or the right supramarginal gyrus (SMG) → the Wernicke's area (Fig. 6-1, Fig. 6-2). And the out pathway is observed as the Broca's area or the right Broca's area (Broca's homologue) → the right ParaHip → the right Post central gyrus (PstCG) → the right fusiform gyrus (FuG) or the left FuG → the right ParaHip → the Broca's homologue or Brocas'area (Fig. 7-1, Fig 7-2). This pathway includes the PstCG which is supposed as somatosensory area.

Almost the same pathways are found to the subjects MN and HT in case of recalling a name of "mouth." However, the estimated area of the Broca and FuG are opposite to HT and MN each other. It is said that the dominant language area is opposite in some left-handed person. Hence, the dominant language area is supposed to be different in these two subjects.

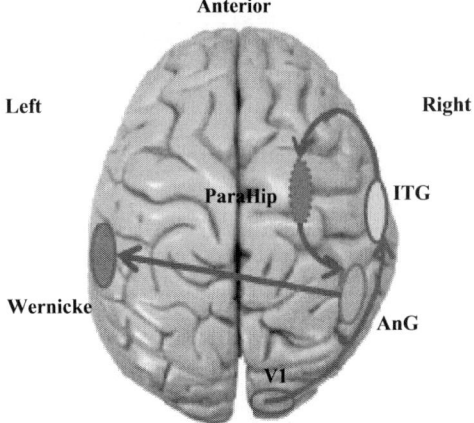

Fig. 6-1 The input pathway (HT)

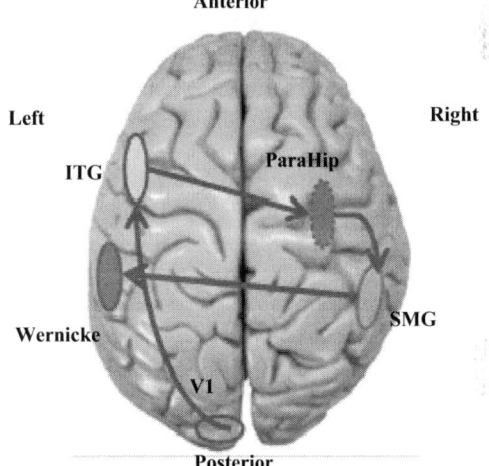

Fig. 6-2 The input pathway (MN)

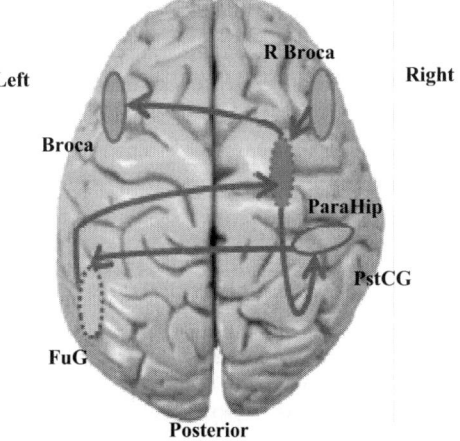

Fig. 7-1 The output pathway (HT)

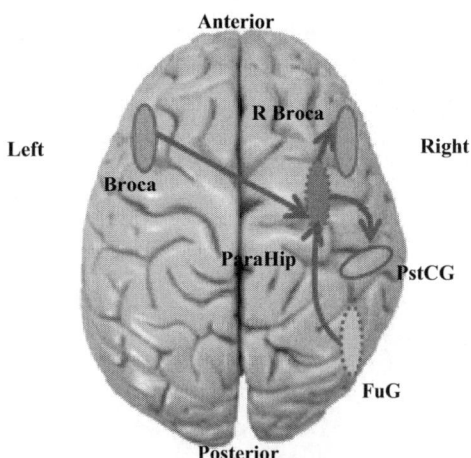

Fig. 7-2 The output pathway (MN)

IV. CONCLUSION

In this study, we estimated human brain activities while a human subject looked at line drawings of parts of the human body and recalled their names silently. ECDs were localized to the word generation area and the image recognition area.

We detected a pathway associated with the recalling of the names of body parts. In case of recalling a name of "mouth," the activities on the AnG, the Broca's area and Wernicke's area, these are so called as the language areas. The ECDs, after localized to the AnG or the SMG, localized to the Wernicke's area, so integration of the input information was done during this period.

In case of the subject HT, estimated activities concentrate to the left hemisphere, e. g. the Broca's area and the Wernicke's area, so his language area is supposed to be the left hemisphere.

In case of the subject MN, the input pathway is the same as HT, however, the out pathway is different from HT. It should be noted that there might be the difference of the dominant hemisphere between input and output of the language, or she might use both hemispheres in language process.

Some activities on the right ParaHip is observed, the process of search and preservation in the memory is done here. Further, we observe activities on the PstCG, which is a part of the somatosensory area, so the subjects made some somatosensory process during recalling the name of "mouth".

ACKNOWLEDGMENT

This research was supported by a project of the High-Tech Research Center of Hokkai-Gakuen University, with the Japanese Ministry of Education, Culture, Sports, Sci- ence, and Technology (MEXT), supported Program for the Strategic Research Foundation at Private Universities, ended at the end of March in 2013.

REFERENCES

1. R. A. McCarthy and E. K. Warrington: Cognitive neuropsychology: a clinical introduction, Academic Press, San Diego, 1990.
2. Geschwind and A. M. Galaburda: Cerebral Lateralization, The Genetical Theory of Natural Selection. Clarendon Press, Oxford, 1987.
3. K. Parmer, P. C. Hansen, M. L. Kringelbach, I. Holliday, G. Barnes, A. Hillebrand, K. H. Singh and P. L. Cornelissen: Visual word recognition: the first half second, NuroImage, Vol. 22-4, pp. 1819-1825, 2004.
4. M. Iwata, M. Kawamura, M. Otsuki et al.: Mechanisms of writing, Neurogrammatology (in Japanese), IGAKU-SHOIN Ltd, pp. 179-220, 2007.
5. T. Yamazaki et al.: "PC-based multiple equivalent current dipole source localization system and its applications", Res. Adv. in Biomedical Eng., 2, pp. 97-109, 2001.
6. T. Yamanoi et al. "Dominance of recognition of words presented on right or left eye -Comparison of Kanji and Hiragana-", Modern Information Processing, From Theory to Applications, Elsevier Science B.V., Oxford, pp. 407-416, 2006.
7. A. Martin, L. G. Ungerleider, J. V. Haxby, "Category specificity and the brain: the sensory/motor model of semantic representations of objects. In Higher Cognitive Functions", in the New Cognitive Neurosciences, ed. MS Gazzaniga, pp. 1023–36. Cambridge, MA: MIT Press, 2000.
8. M. Sugeno, T. Yamanoi, "Spatiotemporal analysis of brain activity during understanding honorific expressions", Journal of Advanced Computational Intelligence and Intelligent Informatics, Vol. 15, No. 9, pp. 1211-1220, 2011.
9. I. Hayashi, H. Toyoshima, T, Yamanoi: "A Measure of Localization of Brain Activity for the Motion Aperture Problem Using Electroencephalogram", Developing and Applying Biologically-Inspired Vision System: Interdisciplinary Concept, Chapter9, pp.208-223, 2012.
10. T. Yamanoi, H. Toyoshima, T. Yamazaki, S. Ohnishi, M. Sugeno and E. Sanchez: Micro Robot Controle by use of Electroencephalograms from Right Frontal Area, Journal of Advanced Computational Intelligence and Intelligent Informatics, Vol. 13, No. 2, pp. 68-75, 2009.
11. H. Toyoshima, T. Yamanoi, T. Yamazaki and S. Ohnishi: "Spatiotemporal Brain Activity During Hiragana Word Recognition Task", Journal of Advanced Computational Intelligence and Intelligent Informatics, Vol. 15, No. 3, pp. 357-361, 2011.
12. T. Yamanoi, H. Toyoshima, S. Ohnishi, M. Sugeno and E. Sanchez: "Localization of the Brain Activity During Stereovision by Use of Dipole Source Localization Method", The Forth International Symposium on Computational Intelligence and Industrial Application, pp. 108-112, 2010.
13. T. Yamanoi, Y. Tanaka, M. Otsuki, S. Ohnishi, T. Yamazaki, M. Sugeno: "Spatiotemporal Human Brain Activities on Recalling Names of Bady Prats", Journal of Advanced Computational Intelligence and Intelligent Informatics, vol. 17, No. 3, 2013.

Author: Takahiro YAMANOI
Institute: Graduate School of Engineering, Hokkai-Gakuen Univ.
Street: West 11-1-1, South 26, Central ward,
City: Sapporo, 064-0926
Country: Japan
Email: yamanoi@lst.hokkai-s-u.ac.jp

Elimination of ECG Artefacts in Foetal EEG Using Ensemble Average Subtraction and Wavelet Denoising Methods: A Simulation

F. Abtahi[1], F. Seoane[1,3], K. Lindecrantz[1,2], and N. Löfgren[4]

[1] School of Technology and Health, KTH Royal Institute of Technology, Stockholm, Sweden
[2] Department of Clinical Science, Intervention and Technology, Karolinska Institute, Stockholm, Sweden
[3] School of Engineering, University of Borås, Borås, Sweden
[4] Neoventa Medical, Mölndal, Sweden

Abstract—Biological signals recorded from surface electrodes contain interference from other signals which are not desired and should be considered as noise. Heart activity is especially present in EEG and EMG recordings as a noise. In this work, two ECG elimination methods are implemented; ensemble average subtraction (EAS) and wavelet denoising methods. Comparison of these methods has been done by use of simulated signals achieved by adding ECG to neonates EEG. The result shows successful elimination of ECG artifacts by using both methods. In general EAS method which remove estimate of all ECG components from signal is more trustable but it is also harder for implementation due to sensitivity to noise. It is also concluded that EAS behaves like a high-pass filter while wavelet denoising method acts as low-pass filter and hence the choice of one method depends on application.

I. INTRODUCTION

Biomedical signals collected from body surface are usually contaminated by noise from different sources such as not desired biological signals. The electrocardiogram (ECG) which shows heart activity is one of strongest electrical signals, and it can be present in a recording of almost any other signals like electroencephalogram (EEG) and electromyogram (EMG). These signals overlap in frequency range and therefore conventional filtering in frequency domain is not applicable. The elimination of ECG as noise is reported by several single or multiple channels methods like Independent Component Analysis (ICA) [1], Adaptive filtering [2], Wavelet denoising, digital filters and ensemble average subtraction [3].

The ST-Analysis in combination of standard cardiotocography (CTG) helps finding general ischemia due to specific fetal positioning that chokes the umbilical cord. The fetal electrocardiogram signals obtain by placing electrode on fetal scalp during labor (after water breaks) referred to an electrode on the mother' thigh. During a feasibility study on fetal EEG recordings from this scalp electrode [4], the need for choosing a proper ECG single channel elimination method has arisen. Neonate EEG with potential similarity to fetal EEG has been selected to compare the performance of implemented methods. In this paper, ensemble average subtraction and Wavelet denoising methods are implemented to reject ECG signal from a simulated neonate EEG signal tainted with ECG.

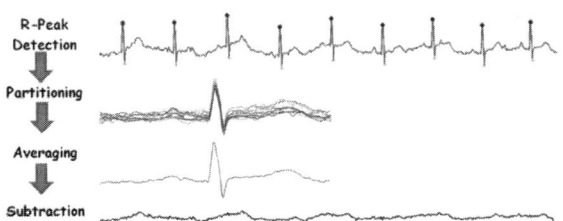

Fig. 1 Procedure of ensemble averaging method, R-Peak detection, synchronous partitioning, averaging and subtraction

II. METHODS

A. Ensemble Average Subtraction Method

The Ensemble Averaging Method (EAS) for elimination of maternal ECG from fetal ECG recorded on the maternal abdomen was introduced by Lindecrantz in [9]. The method was also used by Nakamura in [5] for the removal of ECG artifacts from EEG signals. The elimination method consists of performing R-Peak detection, partitioning of signal to PQRST segments, averaging the segment and subtract the average from each segment. The elimination procedure is depicted in Figure 1 and it is described here.

1) Synchronized Partitioning

The recorded signal is assumed to be the sum of the EEG $x(t)$ and ECG artifacts $z(t)$ according to equation 1.

$$y(t) = x(t) + z(t). \quad (1)$$

The R-Peaks of ECG are used as trigger points. The beginning of PQRST segments in the original method is set to 200msec before the trigger point. Then, raw signal partitions into segments of y_j^* as equation 2.

$$y_j^*(\tau + t_j) = y(\tau + t_j)\{\mu(\tau) - \mu(\tau + t_j - t_{j+1})\}. \quad (2)$$

2) Averaging

The achieved signal segments are averaged as below:

$$\bar{z}^*(\tau) = \frac{1}{L_\tau}\sum_{j=1}^{L} y_j^*(\tau + t_j). \quad (3)$$

The synchronous components $z(t)$ like ECG artifacts become evident in proportion of averaged segments. On the other hand, asynchronous components $x(t)$ will decrease gradually.

3) Subtraction

The average of segments is used as estimate of the artifacts $\bar{z}$ and finally the processed EEG signal $\bar{x}(t)$ is obtained by subtracting the estimated artifact from the raw signal $y(t)$, by using equation 4.

$$\bar{x} = y(t) - \bar{z}(t). \quad (4)$$

The R-Peak detection is a critical part of the EAS method and it has a huge impact on performance. In this work an old fashion derivate-based algorithm, introduced by Jiapu Pan [6] is used. The selection of number of segments that contribute to the average will influence on the estimation of the artifact. The usage of long averaging fragments will destroy some interference details, while very short signal cannot separate the artifact from the underlying signal. The duration of fragmented signal has chosen to ten second empirically. Modifications were made to the original EAS method. First, by using weighted average and giving double weight to the R-Peak that should be eliminated in the signal. The second modification is use of median instead of average which removes outliers from averaging procedure.

B. Wavelets Noise Rejection Method

Wavelet transform is a mathematical tool that can provide a multi-resolution view of the signals. The signal will be first analyzed with the finest resolution consistent of the data and then with coarser and coarser resolution levels. The wavelet probes the structure of signal, using the contribution of different scales. [7]. The general linear time frequency transform can be define as equation 5:

$$s(t) \rightarrow S(a,b) = \int_{-\infty}^{\infty} \psi_{ab}^*(t) dt, \quad (5)$$

Where

$s(t)$ is the signal,

$S(a,b)$ is the transformed signal at scale a and position b,

and $\psi_{ab}^*(t)$ is the kernel function.

The wavelet transform kernel function is defined as equation 6 where a-dependence is an expansion (a > 1) or contraction (a < 1) and the b-dependence is a translation, and hence the wavelets $\psi_{ab}^*(t)$ are self-similar to the $\psi(t)$.

$$\psi_{ab} = \frac{1}{\sqrt{a}} \left(\frac{t-b}{a}\right) \quad (6)$$

The wavelet transform in practice is done by decomposing the signal into an approximation and detail. The approximation is achieved by low-pass filtering and detail is achieved by high-pass filtering. The information content of detail and approximation is same as the information in the original signal. The idea behind Wavelet Noise Rejection (WNR) method is simple; suppose a signal contaminated by white noise, ideally the non-noisy part of signal get

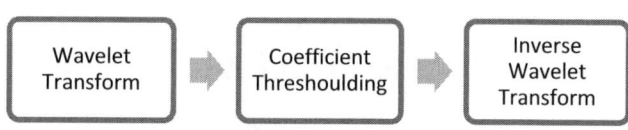

Fig. 2 Wavelet noise rejection steps

concentrated in a few large coefficients while the noise is mapped in a lot of small coefficients. A threshold can be chosen to set all small coefficients to zero and then reconstruct the signal. The noise rejection steps using wavelet are depicted in Figure 2. The proper threshold is critical for good performance of denoising and usually select in an empirical way. The elimination of noise from ECG and QRS detection using wavelets has been reported by Joe-Air Jiang in [3] and P. Sasikala in [8], respectively.

The first step is to select a proper wavelet which has properties and temporal morphology similar to ECG signal. The Coiflet-1, Daub-2 and Symlet-8 are reported in the literature. The Coiflet-1, was reported by [3] is selected in this work because it has spiky and sharp patterns similar to ECG. The implementation was done in MATLAB with the Wavelet Toolbox 2011. Figure 3 shows an eight-level decomposition of ten second neonate EEG contaminated with ECG by using Coiflet-1 wavelet. The repetitive local maxima in wavelet coefficients which belong to R-peaks are obvious in first five details ($d_1, d_2 ... d_5$). The coefficients larger than the selected threshold are set to zero and follows by reconstruction to eliminate ECG artifacts. ECG rejection using wavelets is fundamentally difference from conventional Wavelet Denoising (WD) methods. WD is based on fact that after wavelet decomposition of the noisy signal, the underlying regular part of the signal gets mostly concentrated into a few large wavelet coefficients while noise is mostly spread into many small wavelet coefficients. Therefore, by selecting a proper threshold all of the small coefficients can be set to zero and by reconstruction get the signal almost without noise.

In the present case, the large coefficients belongs to the QRS complexes are considered as noise and hence direct use of the Wavelet Toolbox denoising module is not practical. In the direct use of the WD module, the threshold should be set as high as possible for details ($d_1, d_2 ... d_5$), which will remove all of the details and high frequency components. A modification of thresholding has been made to only set coefficients higher than certain levels to zero.

C. Comparison of Methods

The comparison of ECG noise reduction with WD and EAS methods is done by applying them to the noisy EEG signals. The noisy EEG is synthetized by adding ECG as noise to the known EEG from neonate. Since the actual

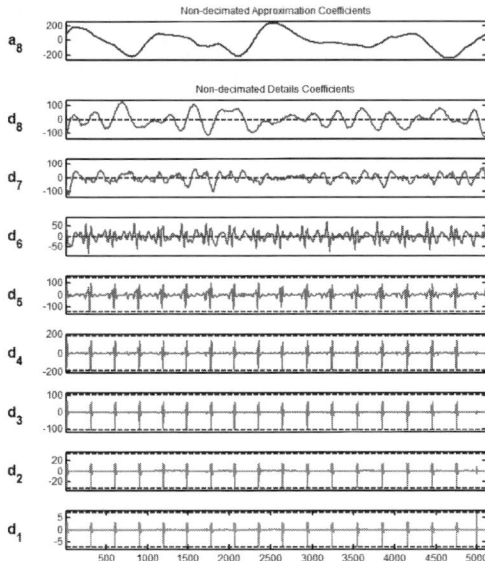

Fig. 3 Decomposition of 10 second recorded signal from fetus head by using Coiflet-1.

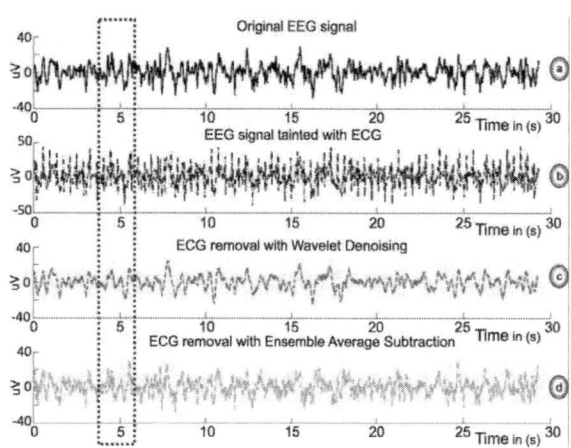

Fig. 4 ECG eliminated signal by using EAS and WD methods. The EEG signal from neonate (a), EEG tainted with ECG (b), ECG eliminated with WD in (c) and EAS in (d) methods are illustrated.

signal is identified, comparison of methods is possible. Neonatal EEG recordings from 10 healthy neonates during quiet wakefulness are used. Comparison of methods is visually done with help of a neurophysiologist at Sahlgrenska university hospital, Goteborg, Sweden.

III. RESULTS

The result of ECG elimination of simulated signals is depicted in Figure 4. It contains an example of EEG signal, ECG signal, simulated signal contaminated by ECG noise, and result of denoising with EAS and WD methods from top to bottom, respectively. In Figure 4.a it is possible to see the thirty-second EEG signal from a healthy neonate, during quiet wakefulness. The contaminated EEG signal with ECG noise is shown in Figure 4.b. The implemented EAS and WD methods are applied to the noisy signal and the results are illustrated in Figure 4.c and 4.d, respectively. It is clearly seen that both signals almost successfully eliminated the added noise. Looking at the signal components of the original and the denoised signals around second five (indicated by dot box), it can be seen that the WD method has removed some details, which are still recognizable in signal resulting from EAS.

The time and frequency domain presentation of the original EEG signal and the signals denoised with the implemented methods are depicted in Figure 5. In this figure, it can be seen that the signal obtained after denoising using the WD method preserves better the morphology of original signal. In the frequency domain it is obvious that signal achieved by the EAS method is attenuated at frequencies below 0.5 Hz. On the other hand, WD also attenuates in frequencies higher than 3 Hz. It is clear that attenuation of frequency components is stronger for WD when compare to the EAS method.

IV. DISCUSSION

The comparison of the results of both noise rejection methods show that the effect of WD method and EAS method as low-pass and high-pass filters, respectively. The wavelet WD has been done by decomposition of noisy signal to approximation and eight levels of detail. The elimination of QRS is achieved by comparing coefficients $(d_1, d_2 ... d_5)$ in detail with certain threshold and set the large coefficients to zero. Since coefficients in detail are set to zero, a smoother signal with lower high frequency components is expected and hence its operation can be considered as low-pass filter. The signal averaging can be consider as smoothing in time domain, which corresponds to low-pass filtering in frequency domain. The subtraction of this low-passed signal from the noisy signal will produce a high-passed signal and hence the EAS method can be considered as high-pass filter. The Cut-off frequency of this filter depends on the number of ECG segments contributes on averaging. The WD method produces a smoother signal and almost none of the added QRS complexes can be found in the result. On the other hand, the signal resulting from applying the EAS method shows some attenuated QRS that are still visible in the signal. Ideally, the average ECG should be equal to all ECG segments but the amplitude of ECG components is variable due to present of the different kinds of noise. Hence, in reality the average ECG is not the same as all PQRST noise. The variety of QRS amplitude

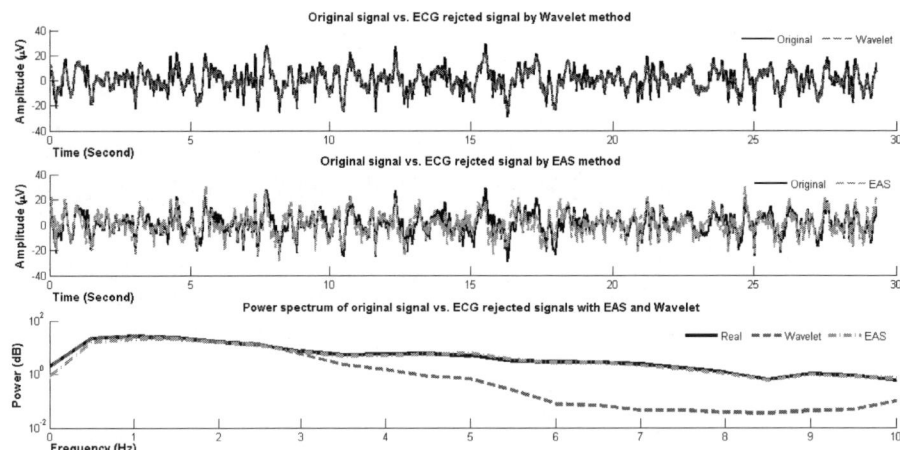

Fig. 5 Comparison of EAS and WD methods with simulated noisy signal.

remains as visible QRS with low amplitude after subtraction of average ECG from signal. The EAS method is also more sensitive to noise because of noise contribution in the average QRS. We have tried to reduce this sensitivity by using median operation instead of averaging because outliers will not contribute in median. The EAS method is more accurate in the sense of estimating different components of ECG like P wave, QRS complex and T wave and subtracts them from noisy signal. Therefore, it is more likely to preserve EEG components which have similar shape and frequency contents of ECG components.

The WD method on the other hand works blind and will remove details, which have shape and frequency contents like ECG components. However, by visual comparison of denoised vs. original EEG signal we can see that the signal denoised by wavelet average performs a better estimation of the original signal. The neonate EEG signals are used in simulation which contains more low frequency activities and hence this conclusion is not general. We have discussed ideas behind comparing wavelet average and ensemble average subtraction methods with low-pass and high-pass filter. Therefore, it is not surprising that by using low-pass filter (Wavelet method) and high-pass filter (EAS) in a signal with high amount of low frequency components, low-pass filter preserve more components of the original signal. The important drawback of both methods is their robustness in real-time applications which is limited by selecting the number of averaged ECG complexes and of level of threshold in WD and EAS methods, respectively.

V. CONCLUSION

The wavelet denoising and the ensemble averaging subtraction methods both successfully eliminate simulated EEG noisy signals. It is also concluded that wavelet denoising method works like low-pass filter while ensemble average subtraction works as high-pass filter. Therefore, eventually the selection among these methods depends on the application but in general the EAS method removes an estimation of all components of the ECG can be considered more trustable.

REFERENCES

[1] Y. Hu, J. Mak, H. Liu, K. Luk, Ecg cancellation for surface electromyography measurement using independent component analysis, in: Circuits and Systems, 2007. ISCAS 2007. IEEE International Symposium on, IEEE, 2007, pp. 3235–3238.

[2] C. Marque, C. Bisch, R. Dantas, S. Elayoubi, V. Brosse, C. Perot, et al., Adaptive filtering for ecg rejection from surface emg recordings., Journal of electromyography and kinesiology: official journal of the International Society of Electrophysiological Kinesiology 15 (3) (2005) 310.

[3] J. Jiang, C. Chao, M. Chiu, R. Lee, C. Tseng, R. Lin, An automatic analysis method for detecting and eliminating ecg artifacts in eeg, Computers in biology and medicine 37 (11) (2007) 1660–1671.

[4] Abtahi, S. F. (2012). Feasibility of fetal EEG recording. Department of Signals and Systems. Göteborg, Chalmers University of Technology. Master Thesis No EX101/2011.

[5] M. Nakamura, H. Shibasaki, Elimination of ekg artifacts from eeg records: a new method of non-cephalic referential eeg recording, Electroencephalography and clinical neurophysiology 66 (1) (1987) 89–92.

[6] J. Pan, W. Tompkins, A real-time qrs detection algorithm, Biomedical Engineering, IEEE Transactions on (3) (1985) 230–236.

[7] A. Romeo, C. Horellou, J. Bergh, N-body simulations with two orders of- magnitude higher performance using wavelets, Monthly Notices of the Royal Astronomical Society 342 (2) (2003) 337–344.

[8] P. Sasikala, R. Wahidabanu, Robust r peak and qrs detection in electrocardiogram using wavelet transform, International Journal of Advanced

[9] K.Lindecrantz. Processing of the fetal ECG; an implementation of a Dedicated Real Time Microprocessor System. Chalmers university of Technology. Technical Report no:135, 1983

Author: Farhad Abtahi
Institute: School of Technology and Health,
KTH Royal Institute of Technology
Street: Alfred Nobels Allé 10
City: Stockholm
Country: Sweden
Email: farhad.abtahi@sth.kth.se

External Uterine Contractions Signal Analysis in Relation to Labor Progression and Dystocia

H. Gonçalves[1], P. Pinto[2,3], D. Ayres-de-Campos[1,3,4,5], and J. Bernardes[1,3,4,5]

[1] Center for Research in Health Technologies and Information Systems (CINTESIS), Faculty of Medicine, University of Porto, Porto, Portugal
[2] Hospital Dr Nélio Mendonça, EPE, Funchal, Portugal
[3] Department of Obstetrics and Gynecology, Faculty of Medicine, University of Porto, Porto, Portugal
[4] Department of Obstetrics and Gynecology, São João Hospital, Porto, Portugal
[5] INEB - Institute of Biomedical Engineering, Porto, Portugal

Abstract—Labor dystocia is a major cause of operative delivery, which is associated with increased risks for both mother and fetus. We assessed linear and non-linear dynamics of external uterine contraction signals, in relation to labor progression and dystocia. Linear time domain, spectral and entropy methods were used to analyze external uterine contraction recordings obtained during the last two hours of labor, in 28 cases with normal and 27 cases with operative deliveries (forceps, vacuum or caesarean). Progression of labor was associated with a statistically significant increase in most linear time domain and spectral indices, both in normal and operative deliveries, whereas most entropy indices increased in normal deliveries, but did not change in operative deliveries. On the other hand, when compared with normal births, operative deliveries were associated with significantly increased entropy indices before the last hour of labor and significantly decreased (a probably associated) sympatho-vagal balance in the last hour of labor. Linear and non-linear analysis of external uterine contraction recordings may provide useful physiopathological and clinical information on the progression of labor and the diagnosis of dystocia.

Keywords—Dystocia, External uterine contractions, Cardiotocography, Spectral analysis, entropy.

I. INTRODUCTION

There have been several attempts in the last decades to improve the evaluation of labor progression and to predict and manage dystocia, a major cause of maternal and fetal trauma [1]. Among these attempts was the development of algorithms for the analysis of the uterine activity using intrauterine manometry, namely with Caldeyro Barcia and Alexandria units [2]. Spectral and entropy analysis of uterine electromyography (EMG) signals have also been developed using prototype research equipment [3]. However, none of these methods has yet proved effective for routine clinical use.

Uterine contractions (UC) signals are routinely acquired in clinical practice by internal or external tocography [4]. They provide important information for the early detection of preterm labor and abnormal labor progress or dystocia. However, research on UC analysis is encompassed by a lack of international guidelines on the interpretation of basic signal features, such as those existent for fetal heart rate signals [5-7].

Internal tocography is considered a very accurate method for UC monitoring [8], but it is an invasive technique that increases the risk of infection and requires previous rupture of the fetal membranes [9]. Moreover, internal tocodynamometry during induced or augmented labor, when compared with external monitoring, has not been shown to reduce the rate of operative deliveries or adverse neonatal outcomes [4]. Therefore, the development of efficient methods for the analysis of the external UC signal is an important research challenge, and one that has a great potential for applicability in routine clinical practice. Some interesting progresses with experimental methods for EMG recordings have also been reported, but this technology has still not entered routine clinical practice.

In this paper, we assess linear and non-linear dynamics of UC recordings, in relation with labor progression and dystocia, using external tocodynamometry (ETOCO), obtained with conventional fetal monitors used in routine clinical practice. We explore signal characteristics not only related to UC but also to other characteristics of uterine activity recordings using ETOCO, such as oscillations related to maternal and fetal movements [2].

II. MATERIALS AND METHODS

A set of 55 UC recordings were acquired, obtained in the same number of women, during the last hours of labor, consecutively selected among UC recordings, and with a mean duration of 328±179 min. The dataset was divided in two groups according to the type of delivery: 28 cases of normal delivery; and 27 cases of operative delivery (forceps, vacuum or caesarean). The main perinatal characteristics of the cases included in the study are presented in Table 1, corresponding to term pregnancies, during labor, with epidural analgesia and favorable outcomes.

Table 1 Main maternal and fetal characteristics of the normal and operative delivery groups included in the study.

Item	Normal (n=28)	Operative (n=27)	P
Maternal data, mean (SD)			
Age (years)	28.4 (4.8)	27.6 (5.7)	0.710
Parity	0.5 (0.6)	0.1 (0.3)	**0.008**
Gestational age (weeks)	39.6 (1.0)	39.9 (1.1)	0.203
Delivery, n (%)			
Normal	28 (100.0%)	0 (0.0%)	
Operative (forceps)	0 (0.0%)	9 (33.3%)	
Operative (vacuum)	0 (0.0%)	12 (44.4%)	
Operative (caesarean)	0 (0.0%)	6 (22.2%)	
Epidural analgesia, n (%)	28 (100%)	27 (100%)	-
Newborn data, mean (SD)			
Birthweight (grams)	3233 (309)	3266 (351)	0.730
1 minute Apgar score	9.3 (0.4)	9.2 (0.7)	0.774
5 minute Apgar score	9.9 (0.3)	9.8 (0.4)	0.210
Umbilical artery blood pH	7.25 (0.07)	7.23 (0.07)	0.304
Gender, n (%)			0.130
Males	13 (46.4%)	18 (66.7%)	
Females	15 (53.6%)	9 (33.3%)	

UC signal was acquired through a conventional external tochodynamometer, applied on the maternal abdomen, linked to a conventional STAN®31 fetal monitor (Neoventa, Gothemburg, Sweden). The UC signal was subsequently exported from the STAN monitor at a sampling rate of 4Hz, via its RS232 port, into the Omniview-SisPorto® 3.5 program (Speculum, Lisbon, Portugal) [10]. The hour before the last (H_1) and the last (H_2) hours of each recording were considered for further processing, where each hour was divided in segments with 5-min length. Each segment was analysed for each case, using time and frequency domain linear, and nonlinear indices, briefly described in the following.

For time domain linear analysis, the following indices were calculated: mean UC (mUC), standard deviation of UC (sdUC), long-term irregularity index (LTI), Delta UC (Δ), short-term variation (STV) and interval index (II). All but II reflect gross changes in UC average and variability, whereas II assesses short-term UC variability taking into account long-term variability [11]. The frequency domain analysis was based on the nonparametric estimation of the spectrum, according to a procedure previously described in detail [11]. Despite the lack of guidelines for UC spectral analysis, a visual interpretation of the spectral analysis suggested the use of the same spectral bands as in heart rate (HR) analysis. The following frequency bands were considered: very low frequency (VLF) at 0–0.04 Hz; low frequency (LF) at 0.04–0.15 Hz; and high frequency (HF) at 0.15–0.40 Hz. Under the scope of HR variability analysis, LF and HF are mainly associated with activity of the sympathetic and parasympathetic systems, respectively [12]. LF_{norm} and HF_{norm} were also computed by normalizing each absolute value by (TP-VLF). The LF/HF index, which reflects the balance between the autonomic nervous system branches, as well as total power (TP, i.e., all the spectrum area), were also considered. For non-linear analysis, approximate entropy (ApEn) [13], and sample entropy (SampEn) [14] were calculated, considering values 0.1 SD, 0.15 SD and 0.2 SD for r and value 2 for m [15], while N was 1200 points (5-min UC segments). The criterion for the selection of the threshold parameter r proposed in Lu et al [16] was also considered. Entropy indices have been associated to complex cortical nervous system activity [15].

For each UC index, the difference between the last two hours of labor (H_1 and H_2) and between the Normal and Operative groups was evaluated using the Spearman correlation coefficient, 95% bootstrap (B=1000) percentile confidence intervals (95% CI) [17] for the median and nonparametric Mann–Whitney and Wilcoxon statistical tests [18], setting significance at p<0.05.

III. Results

Progression of labor was associated with a significant increase of most UC linear (time- and frequency-domain) indices, both in normal and operative delivery groups (Table 2). The exception was the HF_{norm} index, which exhibited a significant decrease from H_1 to H_2 in both groups (Table 2), denoting a significant decrease of the respiratory frequency range. On the other hand, a significant increase of most entropy indices was observed in the normal group when the parameter r was equal to 0.1, 0.15 or 0.2, but the opposite was verified for the operative group considering r_{Lu} (Table 2). The temporal evolution of II, LF/HF and SampEn(2,r_{Lu}) is illustrated in Figure 1.

With respect to the comparison between the normal and operative delivery groups, II and most entropy indices were significantly lower in the normal group in H_1, as well as for HF, HF_{norm} and SD1 in H_2 (Table 2 and Figure 1). The spectral indices LF_{norm} and LF/HF were significantly higher in the normal delivery group during H_2, probably associated with an increase of the sympatho-vagal more detectable in this group (Table 2 and Figure 1).

Table 2 95% confidence intervals and p-values for the UC indices, with respect to the normal (N) and operative (O) groups, in the hour before the last (H_1) and in the last hour (H_2) of labor. For further details please refer to the text.

UC indices	95% confidence interval				p-value			
	H_1		H_2		H_1 vs H_2		N vs O	
	Normal	Operative	Normal	Operative	N	O	H_1	H_2
mUC	17.24-19.89	17.07-20.50	19.90-24.36	21.17-24.91	**0.000**	**0.000**	0.632	0.317
sdUC	13.36-16.68	13.16-15.36	14.84-17.12	15.14-18.39	**0.013**	**0.000**	0.113	0.182
LTI	20.55-25.46	21.21-26.90	24.81-29.70	25.14-33.20	**0.000**	**0.000**	0.831	0.510
Delta	29.50-35.80	28.30-32.00	35.20-42.60	35.40-42.20	**0.000**	**0.000**	0.139	0.415
STV	2.57-2.95	2.60-2.96	3.34-3.78	3.33-4.06	**0.000**	**0.000**	0.704	0.176
II	0.18-0.20	0.20-0.23	0.22-0.25	0.24-0.28	**0.000**	**0.000**	**0.007**	0.259
TP	45.01-58.66	34.32-50.37	66.11-96.30	60.35-92.98	**0.000**	**0.000**	0.077	0.470
VLF	32.10-42.72	26.96-35.89	40.87-60.71	37.81-60.83	**0.000**	**0.000**	0.052	0.864
LF	4.54-6.39	4.39-6.83	9.51-14.41	11.65-20.76	**0.000**	**0.000**	0.790	0.110
LF_{norm}	65.82-70.22	66.37-69.12	73.64-77.12	70.81-74.03	**0.000**	**0.000**	0.315	**0.012**
HF	1.57-2.45	2.15-2.82	3.04-4.54	3.86-5.83	**0.000**	**0.000**	0.233	**0.008**
HF_{norm}	24.78-27.85	25.14-27.96	19.27-22.22	21.25-24.05	**0.000**	**0.000**	0.509	**0.049**
LF/HF	2.41-2.82	2.35-2.71	3.25-4.00	2.98-3.46	**0.000**	**0.000**	0.458	**0.031**
ApEn(2,0.1)	0.34-0.41	0.38-0.44	0.40-0.46	0.37-0.43	**0.001**	0.876	**0.010**	0.846
ApEn(2,0.15)	0.22-0.27	0.24-0.30	0.26-0.31	0.25-0.30	**0.005**	0.737	**0.010**	0.568
ApEn(2,0.2)	0.16-0.20	0.18-0.22	0.20-0.23	0.20-0.24	**0.003**	0.080	**0.014**	0.210
SampEn(2,0.1)	0.17-0.21	0.20-0.26	0.19-0.24	0.18-0.21	**0.021**	0.160	**0.001**	0.615
SampEn(2,0.15)	0.12-0.14	0.13-0.17	0.12-0.16	0.12-0.15	**0.029**	0.225	**0.006**	0.811
SampEn(2,0.2)	0.09-0.11	0.10-0.12	0.10-0.12	0.10-0.12	0.065	0.850	**0.006**	0.718
r_{Lu}	0.05-0.05	0.05-0.06	0.05-0.06	0.06-0.06	**0.000**	**0.000**	0.064	**0.006**
ApEn(2,r_{Lu})	0.62-0.65	0.65-0.68	0.61-0.66	0.61-0.66	0.232	**0.001**	**0.001**	0.201
SampEn(2,r_{LU})	0.40-0.46	0.45-0.55	0.35-0.42	0.30-0.41	0.781	**0.000**	**0.000**	0.146
SD1	0.73-0.82	0.78-0.88	0.92-1.09	1.07-1.32	**0.000**	**0.000**	0.382	**0.003**
SD2	18.60-23.53	18.59-21.67	20.80-24.10	21.14-25.96	**0.013**	**0.000**	0.113	0.182
SD1/SD2	0.04-0.05	0.04-0.05	0.05-0.05	0.05-0.06	**0.000**	**0.000**	0.062	**0.007**

IV. DISCUSSION AND CONCLUSIONS

In this study we observed differences in linear and non-linear analysis of ETOCO recordings obtained in normal and operative deliveries. This shows that these methods are suitable for ETOCO analysis, although further studies are still necessary to assess their clinical usefulness.

To our knowledge, there are no other studies on linear and non-linear analysis of ETOCO recordings. However, other authors have shown that entropy analysis of uterine EMG signals can also help in the identification of cases associated with increased uterine activity, namely in relation to premature delivery [3].

We observed that progression of labor was associated with a statistically significant increase in most linear indices, both in normal and operative deliveries, whereas most entropy indices increased in normal deliveries, but did not change in operative deliveries. On the other hand, operative deliveries were associated with significantly increased entropy indices before the last hour of labor and significantly decreased (a probably associated) sympatho-vagal balance in the last hour of labor. This is consistent with the increase in uterine activity, as well as in maternal movements and heart rate with progression of labor, associated with maternal pushing, more evident in operative deliveries, as well as with an increase in complexity of the uterine activity and of maternal movements, with the progression of normal labors not observable in operative ones, probably in relation to failure of progress or labor obstruction occurring in the latter.

Spectral analysis of external UC recordings usually relies on the evaluation of the signal quality or detection of contractions [19], or on the computation of the spectrum median or maximum [3]. Other authors have analyzed the frequency of the main peak, the median skewness and kurtosis coefficients of the uterine EMG spectrum [20]. However, there is a lack of knowledge in which concerns the spectral bands of interest in external UC recordings, in opposition to the spectral analysis of the uterine EMG [21]. Nevertheless, the spectral bands adopted in this work, similar to those used in HR variability analysis, may be suitable in the spectral analysis of the UC signals, particularly the HF band associated with the respiratory frequency. Further research is warranted in the spectral analysis of external UC signals.

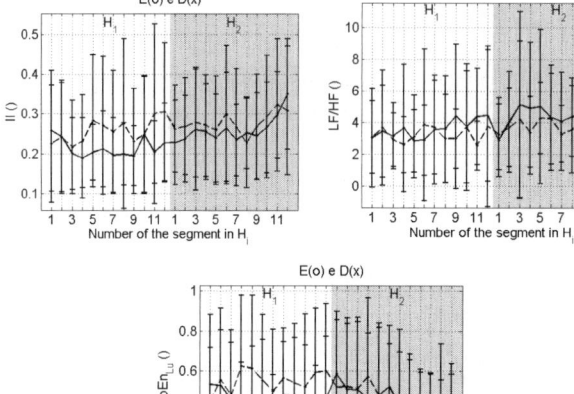

Fig. 1 Temporal evolution of II (upper left plot), LF/HF (upper right plot) and SampEn($2,r_{Lu}$) (lower plot) indices in the form of average±SD, across 5-min segments in the last two hours of labor (H_1 and H_2) in the normal (solid lines) and operative (dashed lines) delivery groups.

In conclusion, linear and non-linear analysis of UC recordings using ETOCO may provide useful physiopathological and clinical information in relation with labor progression and operative delivery. However, further studies are still necessary not only to find out which is the best parameterization for linear and non-linear indices, but also to explore in more depth the clinical usefulness of those indices.

Acknowledgments

Hernâni Gonçalves is financed by a post-doctoral grant (SFRH/BPD/69671/2010) from the Fundação para a Ciência e a Tecnologia (FCT), Portugal.

References

1. Linder N, Linder I, Fridman E et al. (2013) Birth trauma – risk factors and short-term neonatal outcome. J Matern Fetal Neonatal Med DOI 10.3109/14767058.2013.789850
2. Bakker PC, Van Rijswijk S, van Geijn HP (2007) Uterine activity monitoring during labor. J Perinat Med 35:468-477
3. Fele-Zorz G, Kavsek G, Novak-Antolic Z et al. (2008) A comparison of various linear and non-linear signal processing techniques to separate uterine EMG records of term and pre-term delivery groups. Med Biol Eng Comput 46:911-922
4. Bakker JJ, Verhoeven CJ, Janssen PF, et al (2010) Outcomes after internal versus external tocodynamometry for monitoring labor. N Engl J Med 362:306–313
5. ACOG (2005) ACOG Practice bulletin.Intrapartum fetal heart rate monitoring. Obstet Gynecol 106:1453-1461
6. FIGO (1995) FIGO news. Intrapartum surveillance: recommendations on current practice and overview of new developments. Int J Gynecol Obstet 49:213-221
7. RCOG (2001) The use of electronic fetal monitoring. The use of cardiotocography in intrapartum fetal surveillance. Evidence-based Clinical Guideline Number 8. Clinical Effectiveness Support Unit. RCOG Press, London
8. Garfield RE, Maul H, Shi L et al. (2001) Methods and devices for the management of term and preterm labor. Ann NY Acad Sci 943:203-224
9. Soper DE, Mayhall G, Dalton PH (1989) Risk for intraamniotic infection: a prospective epidemiologic study. Am J Obstet Gynecol 161:562-568
10. Ayres-de-Campos D, Sousa P, Costa A et al. (2008) Omniview-Sisporto 3.5 – a central fetal monitoring station with online alerts based on computerized cardiotocogram+ST event analysis. J Perinat Med 36:260-264
11. Gonçalves H, Rocha AP, Ayres-de-Campos D et al. (2006) Internal versus external intrapartum foetal heart rate monitoring: effect on linear and nonlinear parameters. Physiol Meas 27:307-19
12. Task force of the European Society of Cardiology and the North American Society of Pacing and Electrophysiology, Heart rate variability – standards of measurement, physiological interpretation and clinical use. (1996) Circulation 17:354–381
13. Pincus S (1991) Approximate entropy as a measure of system complexity. Proc Natl Acad Sci USA 88:2297-2301
14. Richman JS, Moorman JR (2000) Physiological time-series analysis using approximate entropy and sample entropy. Am J Physiol Heart Circ Physiol 278: H2039-H2049
15. Pincus S, Viscarello R (1992) Approximate Entropy: A regularity measure for fetal heart rate analysis. Obstet Gynecol 79:249-255
16. Lu S, Chen X, Kanters JK et al. (2008) Automatic selection of the threshold value r for approximate entropy. IEEE T Bio-Med Eng 55:1966-1972
17. Martinez WL, Martinez AR (2002) Computational statistics handbook with MATLAB. CRC Press
18. Dudewicz E, Mishra S (1998) Modern mathematical statistics. John Wiley and Sons Inc.
19. Cazares S, Moulden M, Redman CWG et al. (2001) Tracking poles with an autoregressive model: a confidence index for the analysis of the intrapartum cardiotocogram. Med Eng Phys 23:603-614
20. Leman H, Marque C, Gondry J (1999) Use of the electrohysterogram signal for characterization of contractions during pregnancy. IEEE Trans Biomed Eng 46:1222-1229
21. Vinken MPGC, Rabotti C, Mischi M et al. (2009) Accuracy of frequency-related parameters of the electrohysterogram for predicting preterm delivery. A review of the literature. Obstet Gynecol Surv 64:529-541

Quality Based Adaptation of Signal Analysis Software in Pregnancy Home Care System

J. Wrobel[1], K. Horoba[1], J. Jezewski[1], T. Kupka[1], M. Jezewski[2], and T. Przybyla[2]

[1] Institute of Medical Technology and Equipment/Biomedical Signal Processing Department, Zabrze, Poland
[2] Silesian University of Technology/Institute of Electronics, Gliwice, Poland

Abstract—Remote fetal monitoring system ensures a continuous healthcare for high-risk pregnancy patients, whose hospitalization results in in psychological discomfort and a high cost of long hospital stay. In the developed system the bioelectrical fetal signals, recorded indirectly from maternal abdomen, are the source of diagnostic information. Monitoring using mobile instrumentation MI (signal recorder and tablet PC) takes place at patient's home from where the data are transmitted in online mode to the surveillance center (SC).

The software running in MI is responsible for obtaining the fetal electrocardiogram (FECG) of good quality, which is a basis for determination of the fetal heart rate and analysis of QRS morphology. When analysing the abdominal FECG the most crucial is efficient suppression of dominating maternal electrocardiogram. Depending on a noise level a less or more advanced signal analysis software is required to obtain a good quality FECG. Therefore, dedicated signal quality measures control an adaptive modification of the analysis range in the MI. The surveillance center carries out the analysis of variability of the fetal heart rate signal delivered by MI's, and the analysis results provide the final interpretation of the fetal state basing on the so called non-stress test. When the fetal wellbeing is not confirmed, the signal classification process is supported by additional information on QRS morphology changes in the FECG. The amount and content of data transmitted through remote channels to the SC can be managed to ensure the most reliable assessment of the fetal state, while minimizing the MI power consumption.

Keywords—fetal monitoring, fetal electrocardiogram, home care.

I. INTRODUCTION

Fetal monitoring is aimed at evaluation of the fetal condition during pregnancy and in labour. It is accomplished by analysis of characteristic fetal heart rate (FHR) patterns in relation to the uterine contractions and fetal movements. In traditional approach the signals recorded and processed by the bedside monitor are presented as printed waveforms. The use of computerized systems for fetal monitoring assures the required objectivity and reproducibility of the signal interpretation. A need for simultaneous monitoring of many patients leads to a wide use of centralized fetal surveillance systems [1][2]. Signals from fetal monitors along with analysis results are simultaneously presented in the central monitoring station in a form of graphical and numerical data. The currently used systems offer no possibility of monitoring outside the hospital, the patients have to be hospitalized even if there is no direct risk for health. The optimal solution seems to be a remote monitoring of patients with high-risk pregnancy at home.

The highest accuracy and reliability of the FHR signal can be obtained by detection of QRS complexes in the primary bioelectric signal – fetal electrocardiogram (FECG). Additionally a cost of instrumentation is significantly lower in comparison to commonly used Doppler ultrasound monitor. Authors developed mobile measurement instrumentation for acquisition and analysis of the fetal electrocardiogram and uterine contractile activity on a basis of bioelectrical signals recorded from maternal abdominal wall [3]. The signal analysis software running in Mobile Instrumentation (MI) is being adapted to changing measurement conditions. This feature is based on the estimated quality of the bioelectrical signal recorded. Further adaptive modification of the algorithms in the MI is under control of the surveillance center. It is based on a detailed analysis of the FHR variability leading to interpretation of the fetal state.

II. SYSTEM STRUCTURE

The fetal surveillance system comprises a set of bedside devices which transmit the recorded signals to the surveillance center (SC). In case of telemonitoring the remote channel is assumed to work on-line, so the monitoring session will be carried out in the real-time (Fig. 1). Wireless communication is based on the cellular network data transmission service which is used for data transfer between GSM network and the SC located in hospital. The TCP/IP interface enables communication with Mobile Instrumentation via Internet. The SC system software has to be extended to assure continuous error-free data transmission and to enable communication with the patient.

The Surveillance Center has a capability of simultaneous monitoring of up to 24 patients, both remotely and at a hospital. The main tasks are: the analysis of incoming data, dynamic presentation of signals along with analysis results, as well as storing and printing the data. The quantitative parameters describing the acquired signals are used to detect alerting situations. Using the bioelectrical recorder an additional window can be displayed, containing last three averaged fetal P-QRS-T complexes and corresponding T/QRS values. In addition, any time-amplitude relationships within the averaged fetal complexes can be measured and stored with appropriate clinical description.

An instant access to patient's data and acquired signals is offered by optional workstation, connected to the local network. Workstation can be used to set up the system, to create paper documentation as well as to upload and process the signals recorded in the off-line mode (e.g. when communication link is lost). Due to the personal data protection, the access to the given patient's via Internet is permitted only for attending doctor.

The Mobile Instrumentation plays a role of a remote bedside monitor. It comprises a bioelectrical signal recorder and tablet PC with built-in GSM module assuring the wireless connection through the Internet. The tablet software enables acquisition and signals processing, dynamic presentation and on-line evaluation of the signals quality.

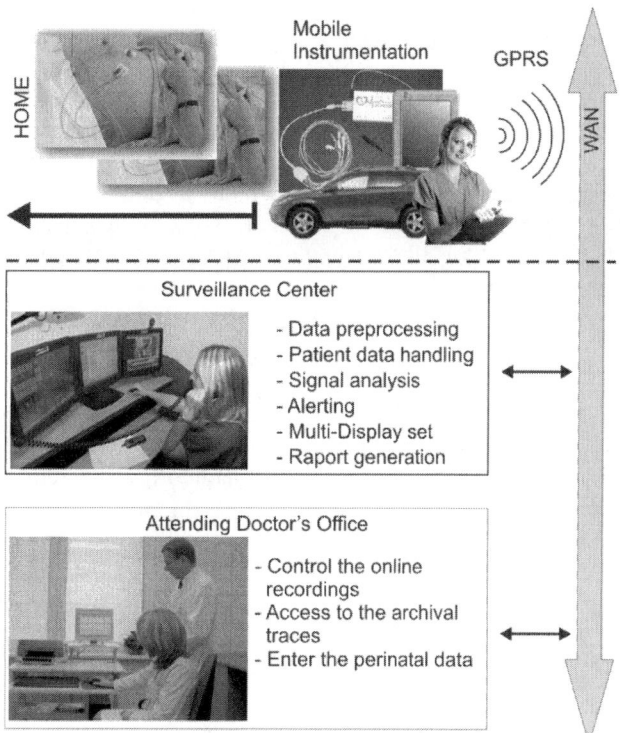

Fig. 1 The structure of telemedical system for home fetal monitoring with online adaptive analysis of bioelectrical abdominal signals

The recorder is equipped with four differential channels for measurement of the abdominal signals [4]. Typical configuration comprises four electrodes placed around the navel and the reference electrode placed above the pubic symphysis. Additionally, the common mode reference electrode is placed on the left leg. Considering abdominal FECG, the basic merit of the recorder unit is a very low level of its own noise which does not exceed 1 μV (peak-to-peak) measured with reference to the inputs, and a high value of CMRR coefficient (120 dB).

Signal recorded from maternal abdomen includes the maternal and the fetal electrocardiograms, the uterine contractile activity from electrical activity of uterine muscle - electrohysterogram (EHG) as well as many unwanted muscle and low frequency components. Suppression of the dominating component – maternal electrocardiogram (MECG) – is the first, and at the same time, the decisive step in abdominal fetal electrocardiography [5]. At first, the spatial filtering based on the generalized singular value decomposition (GSVD) is applied to extract pure dominating MECG from abdominal signal [6]. Then, information on maternal QRS complexes localization is used to suppress the MECG in the abdominal signal, which makes possible further detection of the fetal QRS complexes. In addition, the maternal heart rate signal is determined at this stage. The basic approach to MECG suppression is blanking, where suitably long segment of the abdominal signal comprising the maternal complex is simply replaced by isoline values. In case of coincidence of maternal and fetal complexes the latter one is rejected causing in FHR the signal loss episodes [7]. This disadvantage does not affect clinician's interpretation since for visual analysis the FHR signal is averaged over periods of 2.5 s. Additionally, low computational complexity of the blanking approach leads to significant reduction of power consumption in tablet PC.

In every acquisition channel the blanking is applied initially to the abdominal signal, with simultaneous controlling the quality of the final FECG (Fig.2). The signal quality index (SQI) takes values from 0 to 3, where 0 means very weak signal whereas 3 the best quality. When the best quality signal is obtained in particular channel then the fetal QRS complexes detection and consecutively determination of FHR are carried out using this signal only. The detection function relies on matching filtering and application of a set of decision rules. If in the best channel the SQI=2 (only satisfying signal quality), then additional noise suppression based on projective filtering is applied before the QRS detection starts [8]. If none of the channels provides SQI=2 or 3 this means that the one-channel detection with blanking is not able to ensure good results, which in turn causes significant FHR loss. In that case more precise and advanced MECG suppression method has to be used, i.e. the method

Quality Based Adaptation of Signal Analysis Software in Pregnancy Home Care System

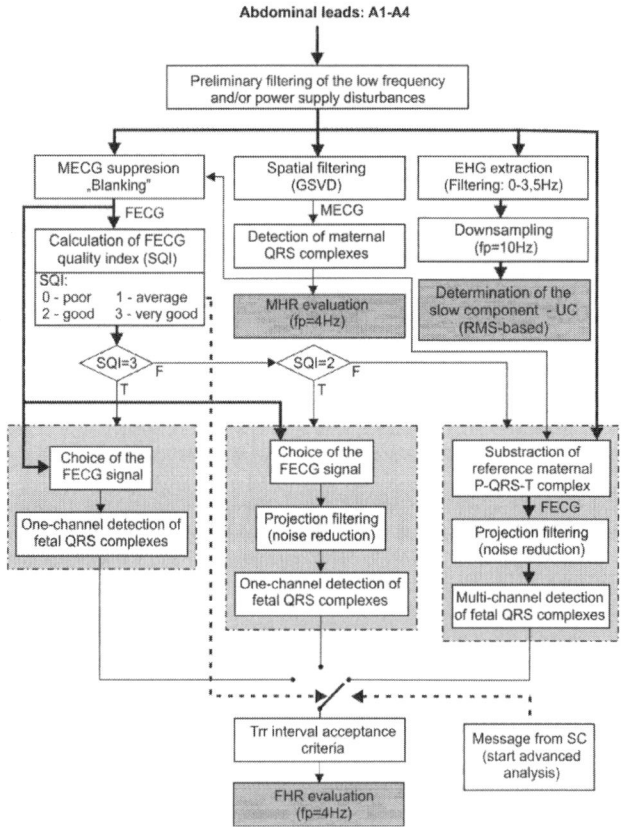

Fig. 2 Structural rearrangement of the signal processing MI software controlled by the FECG quality index (SQI)

based on subtraction of appropriately rescaled and adaptively modified the reference maternal P-QRS-T complexes [3]. Suppression takes place in every abdominal channel, and thus the fetal QRS detection is multichannel. In this approach the additional noise removal procedure is applied to improve the FECG quality before the detection process starts. Any channel with SQI equal to 0 is excluded from the fetal QRS detection.

The analysis of recorded signals performed in the Surveillance Center is consistent with clinical guidelines [9]. The first stage of the FHR signal analysis involves artifacts removal, identification of signal loss episodes and averaging with interpolation of the lost values are carried out. The next stage is the estimation of the FHR baseline, which is a basis for recognition of the acceleration and deceleration patterns as well as for detection of tachycardia and bradycardia episodes. In addition, a set of the indices is determined to describe the instantaneous FHR variability [10]. These indices are calculated for each one-minute segment basing on a heartbeat events obtained from FECG. Analysis of the EHG signal is aimed at recognition of the contraction patterns in relation to the so called basal tone, and then determination of parameters describing the contractions [11][12].

The fetal monitoring session runs according to the procedure presented in Fig. 3. It leads to the classification of the FHR records as normal, abnormal, or suspicious based on the selected analysis parameters defining established in the perinatology so called non-stress test [9].

Usually, the test is carried out for at least 30 minutes. If the record is a classified as normal the monitoring is finished, because the fetal wellbeing is confirmed. Otherwise, the test can be prolonged by next 30 minutes, however it is a decision made by doctor on call. This is justified because the fetal activity can be significantly lower during fetal sleep phase. When the test continues the advanced suppression algorithm basing on subtraction of reference maternal P-QRS-T complex is forced by SC to be carried out in MI (Fig.2). This enables the additional analysis of the fetal P-QRS-T complex morphology. Consecutive fetal complexes obtained in such way undergo the weighted averaging and the relation of the amplitude of T wave to the amplitude of QRS complex is calculated (T/QRS ratio). These values are sent to the Surveillance Center, where they provide the clinician with additional information on the fetal state. The aim of this additional information is to help a clinician in making appropriate decision concerning further patient's treatment and necessity of her hospitalization.

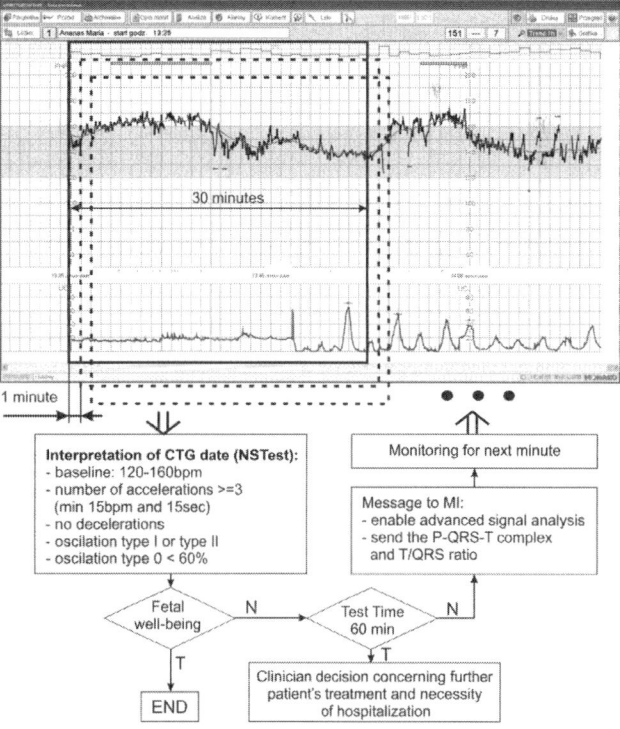

Fig. 3 The CTG data interpretation according to the non-stress test rules

III. CONCLUSIONS

The distributed system for fetal monitoring proposed in this paper will certainly improve a patient's comfort and reduce the cost of medical care, keeping a full functionality of the hospital surveillance system. Additionally, the instant access to the database will make the communication between the patient and her attending doctor much easier. Automated and adaptive adjusting of signal analysis software in Mobile Instrumentation as well as adaptive controlling the fetal monitoring session enable considerable reduction of amount of data that have to be sent to the Surveillance Center from particular MI's. This task sharing has been established taking into account that SC has to manage simultaneously many patients. In most cases, the simplest algorithms and 30-minute monitoring session are sufficient. It is important taking into account the battery operating time in the Mobile Instrumentation. What is more important, reduction of the data transmitted keeps the cost of using GSM/Internet on a level acceptable for hospital and patient.

The measurement instrumentation based on the bioelectrical signals recorder enables additional information on the fetal state to be obtained. It is provided by analysis of the averaged fetal P-QRS-T complex, and particularly with evaluation of the T/QRS value changes. This information enables verification of suspicious recordings.

ACKNOWLEDGMENT

This work was financed in part by the Polish National Science Centre.

REFERENCES

1. Jezewski J, Wrobel J, Horoba K et al. (1996) Computerized perinatal database for retrospective qualitative assessment of cardiotocographic traces in Current Perspectives in Healthcare Computing, Ed. B. Richards B, BJHC Ltd G. Britain:187–196
2. Jezewski J, Wrobel J, Horoba K et al. (2002) Fetal heart rate variability: clinical experts versus computerized system interpretation. Proc. 24th IEEE/EMBS, Huston, 2002, pp 1617-1618
3. Jezewski J, Horoba K, Matonia A, et al. (2003) A new approach to cardiotocographic fetal monitoring based on analysis of bioelectrical signals. Proc. 25th IEEE/EMBS, Cancum, 2003, pp 3145-3149
4. Jezewski J, Matonia A, Kupka T et al. (2012) Determination of the fetal heart rate from abdominal signals: evaluation of beat-to-beat accuracy in relation to the direct fetal electrocardiogram. Biomed Tech 57(5):383-394
5. Kotas M, Jezewski J, Matonia A et al. (2010) Towards noise immune detection of fetal QRS complexes. Comput Meth Prog Bio 97(3): 241-256
6. Callaerts D, De Moor B, Vandewalle J et al. (1990) Comparison of SVD method to extract the fetal electrocardiogram from cutaneous electrode signals. Med Biol Eng Comput 28: 217-224
7. Matonia A, Jeżewski J, Kupka T et al. (2006) The influence of coincidence of fetal and maternal QRS complexes on fetal heart rate reliability. Med Biol Eng Comput 44(5):393-403
8. Kotas M, Jezewski J, Horoba K, et al. (2011) Application of spatio-temporal filtering to fetal electrocardiogram enhancement. Comput Meth Prog Bio 104(1): 1-9
9. NICH (1997) Electronic fetal heart rate monitoring: Research guidelines for interpretations. Am J Obstet Gynaecol 177(12): 1385–1390
10. Czabanski R, Jezewski J, Matonia A et al. (2012) Computerized Analysis of Fetal Heart Rate Signals as the Predictor of Neonatal Acidemia, Expert Syst Appl 39:11846–11860
11. Jeżewski J, Horoba K, Matonia A et al. (2005) Quantitative analysis of contraction patterns in electrical activity signal of pregnant uterus as an alternative to mechanical approach. Physiol Meas 26:753-767
12. Zietek J, Sikora J, Horoba K, et al. (2008) Mechanical and electrical uterine activity. Part I. Contractions monitoring /in Polish/. Ginekol Pol 79: 791-797

Author: Janusz Jezewski
Institute: Institute of Medical Technology and Equipment
Street: Roosevelt 118
City: Zabrze
Country: Poland
Email: jezewski@itam.zabrze.pl

A Recovery of FHR Signal in the Embedded Space

T. Przybyła[1], T. Pander[1], J. Wróbel[2], R. Czabański[1], D. Roj[2], and A. Matonia[2]

[1] Silesian University of Technology/Biomedical Electronics Department, Gliwice, Poland
[2] Institute of Medical Technology and Equipment, Biomedical Signal Processing Department, Zabrze, Poland

Abstract—In the paper we present a proposition of a recovery method for fetal heart rate signals (FHR). Recorded FHR signal often contains interruptions. The interruptions of the signal are caused among others by fetal movements. In such cases, the instantaneous heart rate determination is difficult because parts of the signal are missing. For the reconstruction of missing parts of the FHR signal, an approximation in an embedded space is proposed. The embedded space is created by applying the Taken's theory. Next, the vicinity is determined for the missing samples (missing features in the embedded space). For the vicinity computation, two strategies are chosen: the nearest surrounding of the approximated sample or K nearest neighborhoods. The missing value is computed as a mean or a median. Numerical experiments performed for test (simulated) signals and the real FHR signal show advantages of proposed approach.

Keywords—FHR signal, Embedded space, Takens theory, K-nearest neighbours.

I. INTRODUCTION

Cardiotocography (CTG) is a widely used technique of a fetal monitoring that allows evaluation of the fetal condition during a pregnancy and in a labour [1]. It relies on an analysis of a fetal heart rate (FHR) signal. Nowadays, computerized systems for pregnancy monitoring can recognize the fetal distress basing on automated analysis of some characteristic patterns detected in recorded FHR signals. Key patterns, from clinical point of view, are accelerations and decelerations of fetal heart rate as well as measures of instantaneous variability of heart rate changes. The starting point for automated patterns detection is the estimation of so called FHR baseline. Typical filtering-based algorithms used to determine the baseline require continuous signals with no signal loss segments, which is very difficult to ensure during fetal monitoring sessions [2]. The most popular method for the FHR monitoring, based on monitoring of mechanical activity of the fetal heart using the Doppler ultrasound technique, is particularly susceptible to motion artifacts coming from fetal movements or uterine contractions. These disturbances together with transducer displacement are the common causes of signal loss episodes. Most often the signal loss episodes are being replaced with the mean value of the fetal heart rate estimated from the neighbouring samples or using a linear interpolation of them [3, 4]. In this paper we present method for a reconstruction of the missing parts of the FHR signal. The proposed approach is based on the reconstruction of a dynamic system (that produces FHR signal) described in the phase-space. The phase-space is reconstructed using the Taken's theory [5]. Afterwards, for the corrupted part of the signal similar parts of the signal are searched in the phase-space. The search of similar parts of the corrupted signal is an important problem. Two approaches are proposed to solve this problem. Firstly, we seek for these parts of the signal which lie in the hyper sphere of a given radius. The estimation of the radius is not a trivial task. On the one hand, the radius value should as small as possible i.e. only the most similar (the closest) parts should be chosen. On the other hand, for none of the samples should be select for too small values of the vicinity radius. In the second proposition of the vicinity computation, the well known K-nearest neighbors method is applied.

This paper is organized as follows: the second section contains the description of the FHR signal as well as the problem of the FHR signal interruption and the phase-space reconstruction problems. In the next section we describe the approximation algorithm. Results from the conducted numerical experiments are presented in Section 4. Conclusions complete this paper.

II. FHR SIGNAL AND EMBEDDED SPACE

A. Fetal Heart Rate Signal

The mechanical activity of the fetal heart is characterized by high speed of petal heart valve movements. On the fetal heart the incident ultrasound beam reflects and change its frequency according to Doppler phenomenon [6]. Although the reflected ultrasound beam provides information about all moving structures on its path, the Doppler signal corresponding to the valve petals movements can be obtained by applying a suitable filtering. The instantaneous FHR value is usually determined by applying the autocorrelation function. However, the presence of fetal or maternal movements can deteriorate the quality of the acquired signal, making it impossible to determine the instantaneous fetal heart rate [7]. Another common reasons of signal loss episodes are the accidental change of location of the transducer or the changing position of the fetus. In both situations the health care professional is required to take an action and relocate the transducer, thus the duration of loss episodes can be particularly long. Usually the averaging of known samples is applied to replace the signal loss period [8].

B. Phase–Space Representation

In the proposed method it is assumed that variables describing a system depend on each other in an undetermined way. Observing one variable for a long time makes it possible to reproduce values of the other descripting variables of the system. At the t moment, the value of only variable is known, but other variables affect the observed variable value at the $t+\tau$ moment. The embedding theorem was formulated by Takens [5, 9]. It states, that for a large enough embedding dimension m, the vectors

$$\mathbf{v}_t = [x_t, x_{t+\tau}, x_{t+2\tau}, \ldots, x_{t+(m-1)\tau}], \quad (1)$$

yield a phase space with exactly the same invariant quantities as the original system. The variables $x_t, x_{t+\tau}, x_{t+2\tau}, \ldots, x_{t+(m-1)\tau}$ denote the values of a signal X at times $t, t+\tau, \ldots, t+(m-1)\tau$. The τ is the so–called embedding delay. In many applications, the embedding delay $\tau = 1$ is advantageous [10, 11]. So, $\tau = 1$ will henceforth be used in this study.

Figure 1 depicts an example of the recorded FHR signal. The missing values are marked by $*$. Usually, the missing values are replaced by a mean value determined from a proper parts of the signal. But, for this case, the missing values are determined only once [12]. The analysis in the phase-space allows to compute missing values several times. The signal samples are represented by m-column vectors in the phase-space, where m is the embedded dimension. Hence, the signal which is represented by a vector with m rows in the time domain, in the phase-space domain is represented by the $\mathbf{V}$ matrix with m columns and $n-m$ rows, where n is the length of the signal. An example of such signal representation is presented below.

$$\mathbf{V} = \begin{bmatrix} x & x & \cdots & x & * \\ x & x & \cdots & * & x \\ \vdots & \vdots & \cdots & \vdots & \vdots \\ x & * & \cdots & x & x \\ * & x & \cdots & x & x \end{bmatrix},$$

The x denotes amplitude of the FHR signal. The vectors can be regarded as feature vectors according to the pattern recognition theory. Therefore, a signal sample corresponds to the feature vector in the phase-space and the sample corresponds

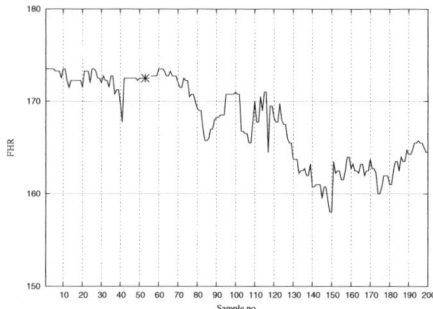

Fig. 1 Example of a FHR signal. The missing values (loss of signal) are marked by $*$.

to the amplitude (s) of the signal at a certain time. The point and the feature vector will be used interchangeably. Similarly, the missing value (the gap) is represented by $*$. It is important, that the symbol $*$ appears in the m subsequent rows in the $\mathbf{V}$ matrix (if only one sample of the signal is missing). A determination of the missing value relies on finding the similar points (points that lie in hypersphere of a given radius), and then computing the missing value as the mean or the median of these points.

C. Vicinity Determination

Let $\mathbf{v}_{t_0}^{(l)}$ be a point of the embedded space that represents a missing value at the l position (feature number). Let $\mathscr{X}$ denotes a set of points that lie in the hypersphere of a given radius ε and the origin at $\mathbf{v}_{t_0}^{(l)}$, i.e

$$\mathscr{X} = \left\{ \mathbf{v}_t \Big|\ \underset{t,\, t \neq t_0}{\forall}\ \|\mathbf{w}_l \mathbf{v}_t - \mathbf{v}_{t_0}^{(l)}\| \leq \varepsilon \right\}, \quad (2)$$

where $\mathbf{w}_l$ is a weight vector and is defined as follows

$$\mathbf{w}_l(k) = \begin{cases} 1 & k \neq l, \\ 0 & k = l \end{cases}.$$

Hence, the missing sample (feature) can be computed as

$$\mathbf{v}_t^{(l)} = \mathrm{median}(\mathscr{X})\,\tilde{\mathbf{w}}_l, \quad (3)$$

where $\tilde{\mathbf{w}}_l$ is the complement to the $\mathbf{w}_l$ and is defined as follows

$$\tilde{\mathbf{w}}_l(k) = \begin{cases} 0 & k \neq l \\ 1 & k = l \end{cases}.$$

In many cases is hard to choose the correct value of the vicinity radius ε. For small values of ε, the set $\mathscr{X}$ might be an empty set. In such a case, the recovery of the missing sample is not possible. Thus, a value of ε should be increased according to an assumed method (e.g. $\varepsilon \leftarrow \varepsilon \cdot \beta$, where $\beta > 1$) and

A Recovery of FHR Signal in the Embedded Space

the vicinity has to be estimated again. This strategy increases the computation time. For large values of ε, the $\mathscr{X}$ might include distant points that can deteriorate the estimation result. Therefore the k–nearest neighbours (kNN) method is proposed as the second way for the vicinity determination. The K-nearest neighbors method is a method for classifying objects based on the closest training examples [13]. Unlike the previous method, the kNN method does not require a radius specification. Instead, the number of vicinity members has to be specified. So, the cardinal number of $\mathscr{X}$ is known and it is equal to the number of specified neighbours.

D. Recovery Algorithm

The proposed recovery algorithm can be described as follows:

1. For a given signal X, fix the embedded dimension m,
2. Compute the phase–space representation of the signal by applying the Takens theory,
3. For missing samples find the vicinity of the point either by searching points in the hypersphere of a given radius ε or by selecting k nearest neighbours,
4. Compute the missing sample as the median of the points from the vicinity,
5. Finally, if the missing sample appears several times in the phase–space representation, the final value of missing sample is a sample mean obtained in the previous step.

III. NUMERICAL EXPERIMENT

In this section we present results obtained from two numerical experiments. In the first experiment a fully deterministic signal which is a sinusoidal signal with amplitude $A = 1$ and frequency $f = 2[Hz]$ is used. This signal is corrupted in the time interval $t \in (0.49, 0.75)[s]$. The corruption lasts $t = 260[ms]$ and for this time period an aplitude of the signal is undefined and unknown. The recovery algorithm is applied for such prepared signal. During the recovery process, the following parameters are fixed: the embedded dimension $m = 21$, hypersphere radius $\varepsilon = 1.0$ and the neighbours number $k = 5$. Figure 2 depicts the corrupted as well as obtained signals. For this experiment, a recovery error is introduced. The recovery error is defined by

$$\xi = \frac{1}{T} \sum_{t \in T} |X_i(t) - X_o(t)|,$$

where $X_i(t)$ is the original (uncorrupted) signal, X_o is the recovered signal and T is the signal duration. For the $260[ms]$ corruption, the proposed algorithm yields negligibly recovery error. The error value is so small due to analytic form

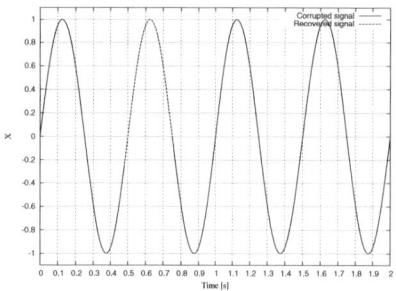

Fig. 2 Example of a recovery of a deterministic signal. The corrupted signal is plotted with solid line while the dashed line represents the recovered signal.

of the signal, i.e. it does not include any noise. In the second experiment, the real FHR signals are used. For the FHR signals, the value of the embedding dimension was estimated by means of the false neighbours method. This method results in $m = 11$. Intuitively, the large number of neighbours results in lower recovery error. However, the vicinity can contain distant points (outliers) that increase the recovery error. Therefore, the number of neighbours is assumed as $k = 15$, for which the smallest value of ξ is obtained. The first FHR signal contains nearly flat heart rate recording. This signal is corrupted and the signal gap lasts for $12.5[s]$. A linear interpolation method is used as the reference method. Figure 3 depicts the original signal (solid line), the recovered signal (dashed line) and the interpolated signal (dotted line). The proposed recovery method yields $\xi = 2.31$ recovery error while the reference method has $\xi = 2.44$. The second used FHR signal contains a deceleration event. As in the previous experiment, this signal is corrupted and the currupion is placed on the falling part of the deceleration. Similarly, the signal gap lasts for $12.5[s]$. The obtained part of the missing signal is presented in Figure 4. As previously, the linear interpolation method is used as the reference method.

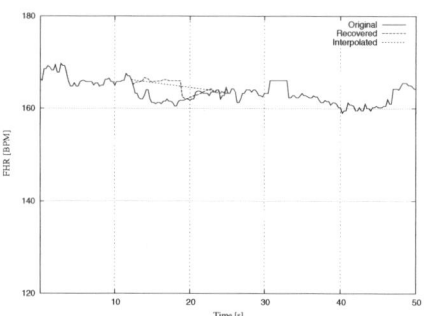

Fig. 3 Recovered signals for the flat FHR. The solid line represents the original signal, dashed line – the recovered signal and the dotted line – the signal obtained from the reference method.

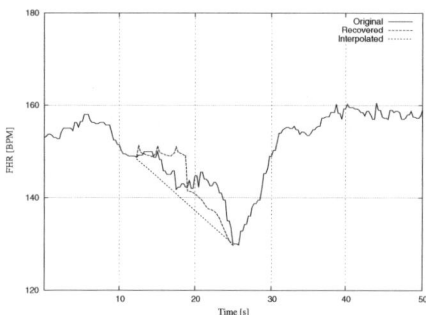

Fig. 4 Recovered signals for the deceleration event. The solid line represents the original signal, dashed line – the recovered signal and the dotted line – the signal obtained from the reference method.

The proposed approach results in $\xi = 3.8$ recovery error, while for the reference method $\xi = 4.7$.

The third used FHR signal includes an acceleration event. The signal is corrupted, and the corruption takes place on rising part of the acceleration. The signal gap lasts for $12.5[s]$. Figure 5 depicts the original signal (solid line), the recovered signal (dashed line) and the interpolated signal (dotted line). The proposed method results in $\xi = 4.01$, while for the reference method $\xi = 5.81$.

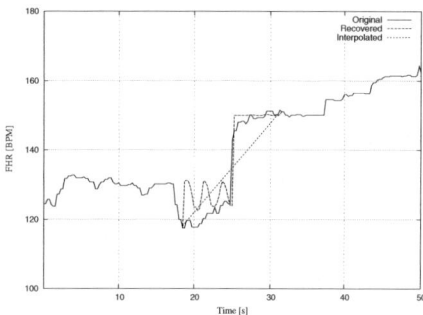

Fig. 5 Recovered signals for the acceleration event. The solid line represents the original signal, dashed line – the recovered signal and the dotted line – the signal obtained from the reference method.

IV. CONCLUSIONS

In this paper we present a fetal heart rate (FHR) signal recovery in an embedded space (phase–space). In many cases, the FHR signal is lost for several seconds. In such cases, an evaluation of the temporary heart rate value is impossible. Therefore, a recovery method of the corrupted FHR signal is desirable. In the proposed approach, the recovery process takes place in an embedded space. The phase–space is determined using the Takens theory. In the proposed approach a very important role plays an estimation of the vicinity of a missing sample. Hence, two approaches are proposed. The first is based on the search of points that lie in a hypersphere of a given radius and origin. The second method of the vicinity detemination uses well–known k nearest neighbours (kNN) method. Results obtained from conducted experiments confirm correctness of the proposed methodology.

REFERENCES

1. Jezewski J., Wrobel J., Horoba K., et al . Fetal heart rate variability: clinical experts versus computerized system interpretation in *24th IEEE/EMBS*(Huston, USA):1617–1618 2002.
2. Jezewski J., Wrobel J., Horoba K., et al . Computerized perinatal database for retrospective qualitative assessment of cardiotocographic traces in *Current Perspectives in Healthcare Computing*(Great Britain):187-196 2003.
3. Jezewski J., Wrobel J., Labaj P., et al . Some Practical Remarks on Neural Networks Approach to Fetal Cardiotocograms Classification in *29th IEEE/EMBS*(Lyon, France):5170-5173 2007.
4. Czabanski R., Jezewski M., Wrobel J., L et al . Predicting the risk of low-fetal birth weight from cardiotocographic signals using ANBLIR system with deterministic annealing and e-insensitive learning *IEEE Trans. Inf. Tech. in Biomed.*. 2010;14:1062–1074.
5. Takens E.. Detecting strange attractors in turbulence *Lecture Notes in Math.* 1981;898:366–381.
6. Kupka T., Jezewski J., Matonia A, et al . Timing events in Doppler ultrasound signal of fetal heart activity in *26th IEEE/EMBS*(San Francisco, USA):337-340 2004.
7. Jezewski J., Horoba K., Gacek A., et al . Analysis of nonstationarities in fetal heart rate signal: Inconsistency measures of baselines using acceleration/deceleration patterns in *7th IEEE/ISSPA*(Paris, France):34-38 2003.
8. Czabanski R., Jezewski J., Matonia A., L et al . Computerized Analysis of Fetal Heart Rate Signals as the Predictor of Neonatal Acidemia *Expert Systems with Applications.* 2012;39:11846–11860.
9. Kantz Holger, Schreiber Thomas. *Nonlinear Time Series Analysis.* Cambridge: Cambridge Univ. Press 2004.
10. Kotas M.. Projective filtering of time–aligned ECG beats *IEEE Trans Biomed. Eng..* 2004;51:1129–1139.
11. Schreiber T., Kaplan D.. Nonlinear noise reduction for electrocardiograms *Chaos.* 1996;6:87–92.
12. Ghannad-Rezaie M., Soltanian-Zadeh H., Ying H., et al . Selection-fusion approach for classification of datasets with missing values *Patt. Rec..* 2010;43:2340–2350.
13. Duda R.O., Hart P.E., Stork D.G.. *Pattern Classification.* New Jersey: Wiley-Interscience 2000.

Author: Janusz Wrobel
Institute: Institute of Medical Technology and Equipment
Street: 118 Roosevelt St
City: Zabrze
Country: Poland
Email: januszw@itam.zabrze.pl

Online Drawings for Dementia Diagnose: In-Air and Pressure Information Analysis

Marcos Faundez-Zanuy[1], Enric Sesa-Nogueras[1], Josep Roure-Alcobé[1], Josep Garré-Olmo[2], Karmele Lopez-de-Ipiña[3], and Jordi Solé-Casals[4]

[1] Escola Universitària Politècnica de Mataró (UPC), Tecnocampus 08302 Mataró, Spain
faundez@eupmt.es
[2] Institut d'assistència sanitària, Salt (Girona, Spain)
[3] UPV/EHU (Basque Country University) Donostia
[4] Digital Technologies Group, University of Vic

Abstract—In this paper we present experimental results comparing on-line drawings for control population (left and right hand) as well as Alzheimer disease patients. The drawings have been acquired by means of a digitizing tablet, which acquires time information angles and pressures. Experimental measures based on pressure and in-air movements appear to be significantly different for both groups, even when control population performs the tasks with the non-dominant hand.

Keywords—**Nonlinear Speech Processing, Alzheimer disease diagnosis, Spontaneous Speech, Fractal Dimensions.**

I. INTRODUCTION

It is well established the use of handwritten tasks for dementia diagnose. In the past, the analysis of handwriting had to be performed in an offline manner. Only the writing itself (strokes on a paper) were available for analysis. Nowadays, modern capturing devices, such as digitizing tablets and pens (with or without ink) can gather data without losing its temporal dimension. When spatiotemporal information is available, its analysis is referred as online. Modern digitizing tablets not only gather the x-y coordinates that describe the movement of the writing device as it changes its position, but it can also collect other data, mainly the pressure exerted by the writing device on the writing surface and also the azimuth, the angle of the pen in the horizontal plane, and the altitude, the angle of the pen with respect the vertical axis (fig. 1 and 2). A very interesting aspect of the modern online analysis of handwriting is that it can take into account information collected when the writing device was not exerting pressure on the writing surface. Thus, the movements performed by the hand while writing a text can be split into two classes:

(a) On-surface trajectories (pen-downs), corresponding to the movements executed while the writing device is touching the writing surface. Each of these trajectories produces a visible stroke.

(b) In-air trajectories (pen-ups), corresponding to the movements performed by the hand while transitioning from one stroke to the next. During these movements the writing device exerts no pressure on the surface.

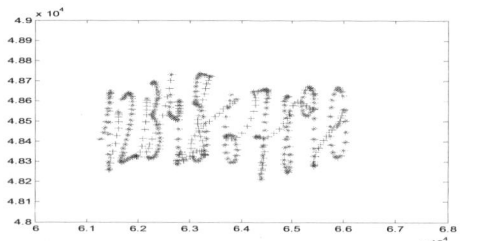

Fig 1 Example of handwritten numerical digits input onto a digitizing tablet. Asterisks (*) represent pen-down information and cross (+) the pen-up.

Fig. 1 shows the acquisition of the ten digits from 1 to 0 using an Intuos Wacom 4 digitizing tablet. The tablet acquired 100 samples per second including the spatial coordinates (x, y), the pressure, and a couple of angles (see Fig. 2). The pen-up in-formation is represented in Fig 1 using "+" while the pen-down is marked with "*". Our experiments on the biometric recognition of people reveal that these two kinds of information are complementary [1] and in fact, contain a similar discriminative capability, even when using a database of 370 users [2-3].

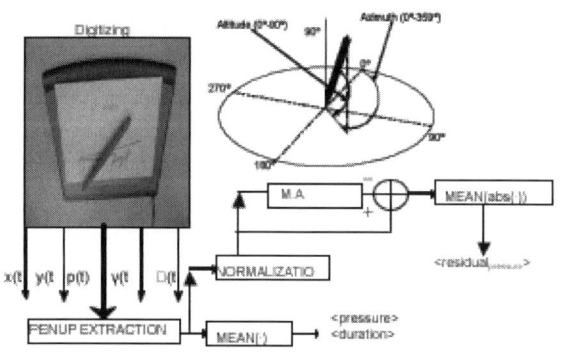

Fig. 2 Information extracted from the digitizing tablet and pressure extracted measures.

Fig. 3 House drawing performed by four individuals with Alzheimer's disease (one per row).
Each column corresponds to pen-down, pen-up and both simultaneously

II. ON-LINE DRAWING APPLIED TO HEALTH ANALYSIS

In the medical field, the study of handwriting has proven to be an aid to diagnose and track some diseases of the nervous system. For instance, handwriting skill degradation and Alzheimer's disease (AD) appear to be significantly correlated [4] and some handwriting aspects can be good indicators for its diagnosis [5] or help differentiate between mild Alzheimer's disease and mild cognitive impairment [6]. Also, the analysis of handwriting has proven useful to assess the effects of substances such alcohol [7] [8], marijuana [9] or caffeine [10]. Aided by modern acquisition devices, the field of psychology has also benefitted from the analysis of handwriting. For instance in [11], Rosenblum et al. link the proficiency of the writers to the length of the in-air trajectories of their handwritings. In the Figure 3 we present one complex drawing with three dimensions performed by individuals with AD of different clinical severity. The visual inspection of the pen down image suggest a progressive degree of impairment, where drawing becomes more disorganized and the three dimensions effect is only achieved in the mild case. The visual information provided by the pen up drawing between AD individuals also indicates a progressive impairment and disorganization when the individuals try to plan the drawing. It is also important to note that the comparison of the pen-up drawing between the mild case of AD and the control (Figures 4) also shows important differences. Besides the increased time on air, there is an increased number of hand movements before decide to put the pen in the surface to drawn. We consider that these graphomotor measures applied to the analysis of drawing and writing functions may be a useful alternative to study the precise nature and progression of the drawing and writing disorders associated with several neurodegenerative diseases. In addition, the pressure is also different. AD people produce softer and simpler strokes. Figure 5 corresponds to the pressure profiles of AD people, figure 6 to control group using the dominant hand and figure 8 to the non-dominant hand of the control group. In the next section we propose several pressure related measures.

Pressure Derived Measures

One important advantage of online drawing is the possibility to evaluate the handwriting pressure from a quantitative point of view. In this paper, we will use the averaged pressure as well as a new proposed measure which consists of the mean absolute value of the residual signal obtained by the difference between the original pressure signal and a smoothed version obtained by a simple moving averaging filter.

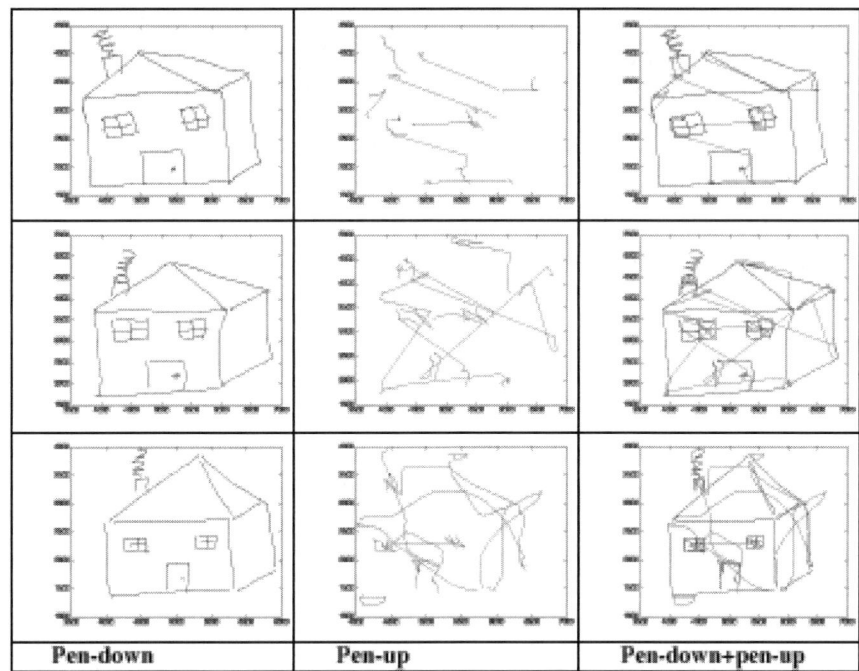

Fig. 4 House drawing performed by three control people (one per row). Each column corresponds to pen-down, pen-up and both simultaneously, performed with the dominant hand

In our experiments, the moving averaging filter consists of averaging 5 consecutive equally weighted samples. Figure 2 represents the diagram of blocks of this algorithm. Figure 9 shows on the top a portion of a sample pressure signal, the output of the averaging filter and the residual signal between both signals on the bottom. The simplest the pressure movements, the smallest the residual signals are (the easiest to predict the signal). Looking at the experimental results of table 1 it is evident the higher time for in-air movements for the AD group, which are around 7 times longer. On the contrary, the time on surface is just around 3 times longer. Thus, there are more differences between control and AD groups when looking at in-air movements. When comparing the non-dominant movements performed by the control group we obtained a 5.6 ratio and 1.5. Again, the in-air times are significantly higher for the AD group than the control group. Looking at the pressures, we observe that the mean pressure is 2.4 times higher for control group and dominant hand, and 2.1 for the non-dominant hand. About the mean absolute value of the residual error, the ratios are respectively 2.2 and 1.3. thus, we can conclude that control group with non-dominant hand are in a middle situation between control group dominant hand and AD group.

Table 1 Statistical analysis/descriptives from the drawings shown in figures 4, 5 and 6.

Measurement	control		pathological
	Dominant hand	Non-dominant hand	Dominant hand
Time in-air	8334	10927	61008
Time on-surface	9680	22177	31521
Total time	18014	33104	92259
<Pressure>	1307.8	1144.4	545.01
<abs(Residual error)>	22.49	12.89	10.27

III. EXPERIMENTAL RESULTS

In this paper we have evaluated the following measures:

a) Time related measures: time in air, time on-surface, and total time (in-air plus on-surface times). The time has been measured as the number of samples.

b) Pressure measures: mean pressure on-surface and mean absolute value of the residual signal between the on-surface pressure and the moving averaged pressure.

Table 1 summarizes some experimental measures of the drawings shown in figures 4, 5 and 6.

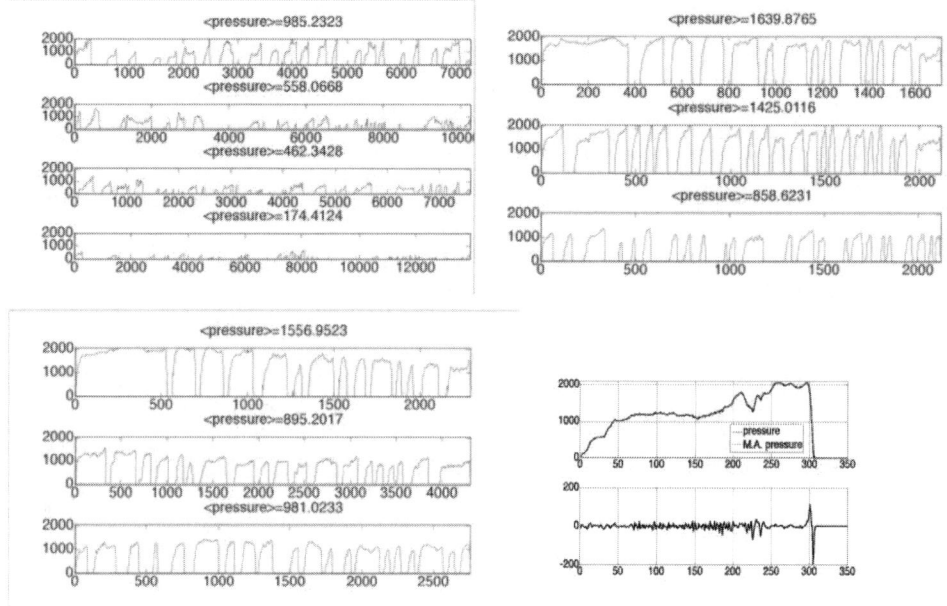

Fig. 5 (a) Pressure profiles for three AD patients, as well as the average pressure (<pressure>), (b) pressure profiles samples for control group, dominant hand, as well as the average pressure, (c) pressure profiles samples for control group, non-dominant hand, as well as the average pressure and (d) a sample of pressure information and the moving average prediction on the top. The difference between both signals on the bottom.

IV. CONCLUSIONS

Although some pathological drawings may look "normal" according to pen-down information, the pen-up information and the pressure look quite entangled and should permit easier diagnose. This observation points out the convenience of online handwriting analysis, which can outperform the classic offline mode, mainly due to the larger amount of available information, as well as the more accurate measurements made by the computer. The differences between control and pathological group do not seem to be related to some physical problem, because the control group, even when using the non-dominant hand performs less entangled pen-up movements. Future work will include some other measurements, with a larger database [12].

ACKNOWLEDGMENT

This work has been supported by FEDER and MEC, TEC2009-14123-C04-04, COST-2102 and SAIOTEK.

REFERENCES

1. E. Sesa-Nogueras, M. Faundez-Zanuy and Jiri Mekyska "An information analysis of in-air and on-surface trajectories in online handwriting" | In press. Cognitive Computation. DOI 10.1007/s12559-011-9119-y
2. E. Sesa-Nogueras and M. Faundez-Zanuy "Biometric recognition using online uppercase handwritten text" Pattern Recognition Vol. 45 (2012) pp 128–144. January 2012
3. E. Sesa-Nogueras and M. Faundez-Zanuy "Writer recognition enhancement by means of synthetically generated handwritten text". Engineering applications of artificial intelligence. In press. Doi: 10.1016/j.engappai.2012.03.010
4. KE Forbes, MF Shanks, A Venneri. The evolution of dysgraphia in Alzheimer's disease, Brain Res.Bull. 63 (2004) 19-24.
5. J Neils-Strunjas, K Groves-Wright, P Mashima, S Harnish. Dysgraphia in Alzheimer's disease: a review for clinical and research purposes, J.Speech Lang.Hear.Res. 49 (2006) 1313-1330.
6. P Werner, S Rosenblum, G Bar-On, J Heinik, A Korczyn. Handwriting process variables discriminating mild Alzheimer's disease and mild cognitive impairment, J.Gerontol.B Psychol.Sci.Soc.Sci. 61 (2006) p228-36.
7. F Asıcıoglu, N Turan. Handwriting changes under the effect of alcohol, Forensic Sci.Int. 132 (2003) 201-210.
8. JG Phillips, RP Ogeil, F Müller. Alcohol consumption and handwriting: A kinematic analysis, Human Movement Science. 28 (2009) 619-632.
9. RG Foley, A Lamar Miller. The effects of marijuana and alcohol usage on handwriting, Forensic Sci.Int. 14 (1979) 159-164.
10. O Tucha, S Walitza, L Mecklinger, D Stasik, T Sontag, KW Lange. The effect of caffeine on handwriting movements in skilled writers, Human Movement Science. 25 (2006) 523-535.
11. S Rosenblum, S Parush, P Weiss. The in Air Phenomenon: Temporal and Spatial Correlates of the Handwriting Process, Perceptual & Motor Skills. 96 (2003) 933.
12. Faundez-Zanuy, M.; Hussain, A.; Mekyska, J.; Sesa-Nogueras, E.; Monte-Moreno, E.; Esposito, A.; Chetouani, M.; Garre-Olmo, J.; Abel, A.; Smekal, Z.; Lopez-de-Ipiña, K. Biometric Applications Related to Human Beings: There Is Life beyond Security. Cognitive Computation 2012, 5(1): 136-151. doi: 10.1007/s12559-012-9169-9.

Spontaneous Speech and Emotional Response Modeling Based on One-Class Classifier Oriented to Alzheimer Disease Diagnosis

K. Lopez-de-Ipiña[1], J.B. Alonso[2], N. Barroso[1], J. Solé-Casals[3], M. Ecay-Torres[4],
P. Martinez-Lage[4], F. Zelarain[5], H. Egiraun[1,6], and C.M. Travieso[2]

[1] Department of System Engineering and Automation, University of the Basque Country, Spain
karmele.ipina@ehu.es
[2] Universidad de Las Palmas de Gran Canaria, IDeTIC
[3] Digital Technologies Group, University of Vic
[4] Neurology Department CITA-Alzheimer Foundation
[5] GuABIAN Association
[6] Plentziako Itsas Estazioa/Estación Marina de Plentzia

Abstract—The purpose of our project is to contribute to earlier diagnosis of AD and better estimates of its severity by using automatic analysis performed through new biomarkers extracted from non-invasive intelligent methods. The methods selected in this case are speech biomarkers oriented to Spontaneous Speech and Emotional Response Analysis. Thus the main goal of the present work is feature search in Spontaneous Speech oriented to pre-clinical evaluation for the definition of test for AD diagnosis by One-class classifier. One-class classification problem differs from multi-class classifier in one essential aspect. In one-class classification it is assumed that only information of one of the classes, the target class, is available. In this work we explore the problem of imbalanced datasets that is particularly crucial in applications where the goal is to maximize recognition of the minority class as in medical diagnosis. The use of information about outlier and Fractal Dimension features improves the system performance.

Keywords—One-class classifier, Nonlinear Speech Processing, Alzheimer disease diagnosis, Spontaneous Speech, Fractal Dimensions.

I. INTRODUCTION

Alzheimer's Disease (AD) is the most common type of dementia among the elderly. It is characterized by progressive and irreversible cognitive deterioration with memory loss and impairments in judgment and language, together with other cognitive deficits and behavioral symptoms. The cognitive deficits and behavioral symptoms are severe enough to limit the ability of an individual to perform everyday professional, social or family activities. As the disease progresses, patients develop severe disability and full dependence. An early and accurate diagnosis of AD helps patients and their families to plan for the future and offers the best opportunity to treat the symptoms of the disease. According to current criteria, the diagnosis is expressed with different degrees of certainty as possible or probable AD when dementia is present and other possible causes have been ruled out. The diagnosis of definite AD requires the demonstration of the typical AD pathological changes at autopsy [1,2,3]. In addition to the loss of memory, one of the major problems caused by AD is the loss of language skills. We can meet different communication deficits in the area of language, including aphasia (difficulty in speaking and understanding) and anomia (difficulty in recognizing and naming things). The specific communication problems the patient encounters depend on the stage of the disease [3,4,5]. The main goal of the present work is feature search in Spontaneous Speech Analysis (ASSA) an Emotional Response Analysis (ERA) oriented to pre-clinical evaluation for the definition of test for AD diagnosis. These features will define control group (CR) and AD disease. Non-invasive Intelligent Techniques of diagnosis may become valuable tools for early detection of dementia. Moreover, these techniques are very low-cost and do not require extensive infrastructure or the availability of medical equipment. They are thus capable of yielding information easily, quickly, and inexpensively [6,7]. This study is focuses on early AD detection and its objective is the identification of AD in the pre-clinical (before first symptoms) and prodromic (some very early symptoms but no dementia) stages. The research presented here is a complementary preliminary experiment to define thresholds for a number of biomarkers related to spontaneous speech. Feature search in this work is oriented to pre-clinical evaluation for the definition of test for AD diagnosis Obtained data will complement the biomarkers of each person [8].

The rest of this paper is organized this way: In Section II, the materials are presented. Section III explains the methodology of the experiments, Section IV shows the experimental results, and finally, conclusions are presented in Section V.

II. MATERIALS

Trying to develop a new methodology applicable to a wide range of individuals of different sex, age, language and cultural and social background, we have built up a multicultural and multilingual (English, French, Spanish, Catalan, Basque, Chinese, Arabian and Portuguese) database with video recordings of 50 healthy and 20 AD patients (with a prior diagnosis of Alzheimer) recorded for 12 hours and 8 hours respectively. The age span of the individuals in the database was 20-98 years and there were 20 males and 20 females. This database is called AZTIAHO. All the work was performed in strict accordance with the ethical guidelines of the organizations involved in the project. The recordings consisted of videos of Spontaneous Speech – people telling pleasant stories or recounting pleasant feelings as well as interacting with each other in friendly conversation. The recording atmosphere was relaxed and non-invasive. The shorter recording times for the AD group are due to the fact that AD patients find speech more of an effort than healthy individuals: they speak more slowly, with longer pauses, and with more time spent on efforts to find the correct word and uttering speech disfluencies or break messages. In the advanced stage of the disease, they find this effort tiring and often want to stop the recording. We complied with their requests. The video was processed and the audio extracted in wav format (16 bits and 16 Khz). The first step was removing non-analyzable events: laughter, coughing, short hard noises and segments where speakers overlapped. Next, background noise was removed using denoiser adaptive filtering. After the pre-processing, about 80% of the material from the control group and 50% of the material from the AD group remained suitable for further analysis. The complete speech database consists of about 60 minutes of material for the AD group and about 9 hours for the control. The speech was next divided into consecutive segments of 60 seconds in order to obtain appropriate segments for all speakers, resulting finally in a database of about 600 segments of Spontaneous Speech. Finally for experimentation from the original database, a subset of 20 AD patients was selected (68-96 years of age, 12 women, 8 men, with a distribution in the three stages of AD as follows: First Stage [ES=4], Secondary Stage [IS=10] and Tertiary stage [AS=6]). The control group (CR) was made up of 20 individuals (10 male and 10 female, aged 20-98 years) representing a wide range of speech responses. This subset of the database is called AZTIAHORE[7].

III. METHODS

In previous work [7] the goal of the experimentation was to examine the potential of the selected features to help in the automatic measurement of the degradation of Spontaneous Speech, Emotional Response and their manifestation in people with AD as compared to the control group. The experiments have analyzed the automatic measurement and definition of appropriate features in Spontaneous Speech, Emotional Response Analysis and Integral Response in the speech of people with AD. Four groups/classes (CR, ES, IS, AS) and four different feature sets were evaluated (section 3.2.1.4). The automatic classification of speech is based on several classifiers. We used five different classifiers: (1) a Support Vector Machine (SVM), (2) a Multi Layer Perceptron (MLP) (3) a KNN Algorithm, (4) Decision Trees (DT) and (5) a Naive Bayes net. The approach's performance was very satisfactory and promising results for early diagnosis and classification of AD patient groups but medical doctors propose new experimentation oriented to detect mainly early stage. The goal of this new experimentation is to detect changes with regard to CR group and outliers which point to presence of AD's symptoms. One class classification will be use for this propose.

A. Feature Extraction

A.1 Automatic Spontaneous Speech Analysis (ASSA)

Spoken language is one of the most important elements defining an individual's intellect, social life, and personality; it allows us to communicate with each other, share knowledge, and express our cultural and personal identity. Spoken language is the most spontaneous, natural, intuitive, and efficient method of communication among people. Therefore, the analysis by automated methods of Spontaneous Speech (SS – free and natural spoken communication), possibly combined with other methodologies, could be a useful non-invasive method for early AD diagnosis. The analysis of Spontaneous Speech fluency is based on three families of features (SSF set), obtained by the Praat software package [9] and software that we ourselves developed in MATLAB. For that purpose, an automatic Voice Activity Detector (VAD) [10] has extracted voiced/unvoiced segments as parts of an acoustic signal. These four families of features (Spontaneous Speech Features, SSF) include: duration, time and frequency domain analysis (pauses, short time energy, centroid).

A.2 Emotional Response Analysis

In this study, we aim to accomplish the automatic selection of emotional speech by analyzing three families of features in speech (Emotional Features, EF): acoustic, voice quality: duration an Emotional Temperature (ET) [7] (pitch, energy, intensity, shimmer, jitter).

A.3 Fractal Dimension

The Fractal Dimension (FD) is one of the most significant features which describe the complexity of a system and could help in the detection of subtle changes for early diagnosis. Moreover this feature has in the ability to capture the dynamics of the system and thus relevant variations in speech utterances. More precisely, an implementation of Higuchi's algorithm [11] in order to add this new feature to the set that feeds the training process of the model. Most of the fractal systems have a characteristic called self-similarity. An object is self-similar if a close-up examination of the object reveals that it is composed of smaller versions of itself.

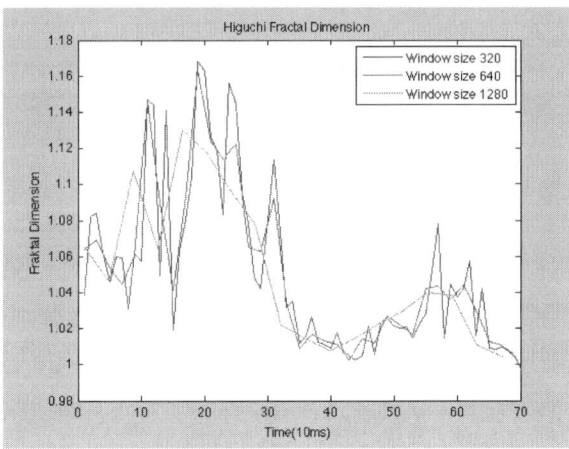

Fig. 1 Higuchi Fractal Dimension (HFD) for an AD case for different window sizes.

Self-similarity can be quantified as a relative measure of the number of basic building blocks that form a pattern, and this measure is defined as the Fractal Dimension. Higuchi was choice because it has been reported to be more accurate in previous works with under-resourced conditions [12]. In this feature set Higuchi Fractal Dimension (HFD), and its max, min, mean and standard deviation have been used in HFD set.

A.4 Automatic Classification by One-Class Classifier

One-class classification problem [13] differs from multi-class classifier in one essential aspect. In one-class classification it is assumed that only information of one of the classes, the target class, is available. This means that just example objects of the target class can be used and that no information about the other class of outlier objects is present. The different terms such as fault detection, anomaly detection, novelty detection and outlier detection originate from the different applications to which one class classification can be applied. The boundary between the two classes has to be estimated from data of only control class. The task is to define a boundary around the target class, such that it accepts as much of the target/control objects as possible, while it minimizes the chance of accepting outlier objects. In the literature a large number of different terms have been used for this problem. The term one-class classification originates from [14]. The application is as follows: The first application for one class classification (also called data description as it forms the boundary around the whole available data) is outlier detection, to detect uncharacteristic objects from a dataset. Secondly, data description can be used for a classification problem where one of the classes is sampled very well, while the other class is severely under sampled. The measurements on the under sampled class might be very expensive or difficult to obtain. Finally, the last possible use of the outlier detection is the comparison of two data sets. Assume that a classifier has been trained (in a long and difficult optimization process) on some (possibly expensive) data [15]. As explained above, the second application of outlier detection is in the classification problem where one of the classes is sampled very well but it is very hard and expensive, if not impossible, to obtain the data of the second class. One of the major difficulties inherent in the data (as in many medical diagnostic applications) is this highly skewed class distribution. The problem of imbalanced datasets is particularly crucial in applications where the goal is to maximize recognition of the minority class. WEKA [16] software was used in carrying out the experiments. The base classifiers to be used in the experimentation were Bagging and MLP, with k-fold cross-validation (k=10).

IV. EXPERIMENTAL RESULTS

The task was Automatic Classification, with the classification targets being: healthy speakers without neurological pathologies and speakers diagnosed with AD. The experimentation was carried out with AZTIAHORE. The results have been analyzed with regard to the feature set described in III. Experimentation has been divided to test One-class classifier only with speech samples from CR group and with information about outliers (patients with AD). The results are shown in Figure 1.

A. One-Class Classifier Only with Information about CR Group:

The results are satisfactory for this study in the case of MLP classifier. Bagging paradigm presents lower computational cost but very poor results. The new fractal features improve the system but they can improve Bagging performance. Window size: For HFD algorithm three different window size have been used: 1280, 640 and 320 samples. The best results have been obtained for a window size of 320 samples. The best results are obtained for integral feature set, which mixes features relative to Spontaneous Speeech and Emotional Response (SSF+HFD+EF+ET).

B. One-class Classifier with Information about Outliers:

The results are satisfactory for this study because they obtained very good results not only for MLP classifier but also for Bagging, which presents lower computational cost. The new fractal features improve the system for both paradigms. The best results have been obtained for a window size of 320 samples in HFD. In this case also the best results are obtained for SSF+HFD+EF+ET set.

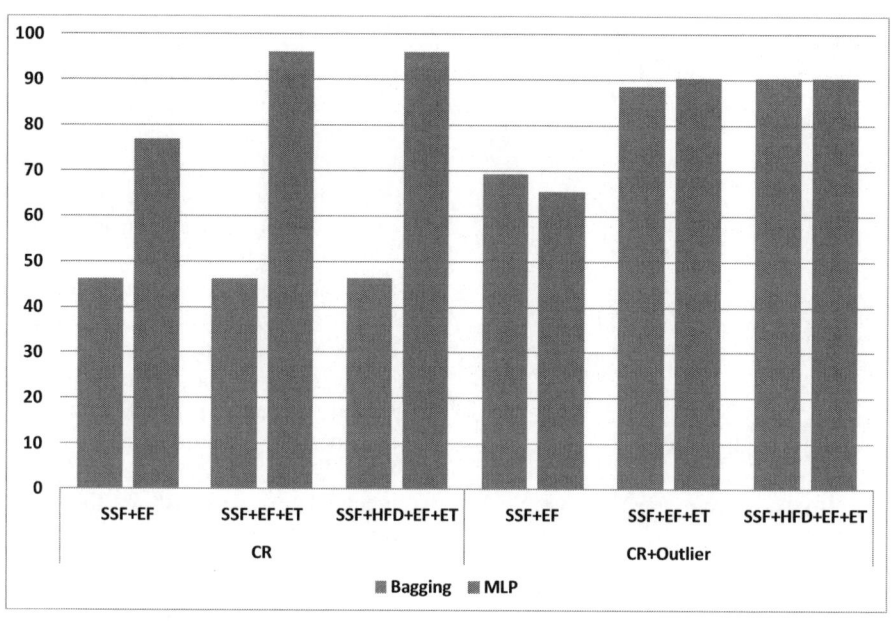

Fig. 2 Higuchi Fractal Dimension (HFD) for an AD signal for different window sizes.

V. CONCLUSIONS

The main goal of the present project is feature search in through new biomarkers extracted from non-invasive intelligent methods as Spontaneous Speech and Emotional Response Analysis oriented to pre-clinical evaluation for the definition of test for AD diagnosis. These features are of great relevance for health specialists to define health people and AD disease symptoms. One-class classification has been used in the experimentation. In one-class classification it is assumed that only information of one of the classes, the target class, is available. In this work we explore the problem of imbalanced datasets by using information about outlier and Fractal Dimension features improves the system performance. The approach of this work complements the previous multi-class modelling and is robust in terms of capturing the dynamics of the whole waveform, and it offers many advantages in terms of computability, and it also makes easier to compare the power of the new features against the previous ones. In future works we will introduce new features relatives to speech modelling oriented to standard medical tests for AD diagnosis and to emotion response analysis.

REFERENCES

1. Mc Kahn G, et al.Clinical diagnosis of Alzheimer's disease: report of the NINCDS-ADRDA Workgroup on AD. 1984; 24:939-944.
2. McKhann GM et al. The diagnosis of dementia due to Alzheimer's disease: Recommendations from the NIAA's Association workgroups on diagnostic guidelines for AD. Alzheimers Dement. 2011 May;7(3):263-9.
3. Van de Pole, L.A., et al., The effects of age and Alzheimer's disease on hippocampal vol-umes, a MRI study. Alzheimer's and Dementia, 2005. 1(1, Supplement 1): p. 51.
4. Morris JC, The Clinical Dementia Rating (CDR): current version and scoring rules. Neu-rology, 1993. 43: p. 2412b-2414b.
5. American Psychiatric Association, 2000. Diagnostic and Statistical Manual of Mental dis-orders, 4th Edition Text Revision. Washington DC.
6. M. Faundez-Zanuy et al. Biometric Applications Related to Human Beings: There Is Life beyond Security, Cognitive Computation, 2012, DOI 10.1007/s12559-012-9169-9
7. K. López de Ipiña, J. B. Alonso, J. Solé-Casals, N. Barroso, M. Faundez, M. Ecay, C. Travieso, A. Ezeiza and A. Estanga Alzheimer Disease Diagnosis based on Automatic Spontaneous Speech Analysis, Proceedings of NCTA 2012. Barcelona, 2012
8. Alzheimer's Association.: http://www.alz.org/
9. Praat: doing Phonetics by Computer. Available online: www.fon.hum.uva.nl/praat
10. Solé J, Zaiats V., A Non-Linear VAD for Noisy Environment. Cognitive Computation. 2010; 2(3):191-198.
11. Higuchi T. Approach to an irregular time series on the basis of the fractal theory. Physica D 1988. 31277:283.
12. K. Lopez-de-Ipiña et al. Feature extraction approach based on Fractal Dimension for Spontaneous Speech modelling oriented to Alzheimer Disease diagnosis. LNAI, Proceeedings of NOLISP, Mons Belgium, 2013.
13. David Martinus Johannes One-class classification, PhD Thesis,Tax. Technische Universiteit Delft, (2001)
14. Moya, M., Koch, M., and Hostetler, L. (1993). One-class classifier networks for target recognition applications. In Proceedings world congress on neural networks, pages 797–801, Portland, OR. International Neural Network Society, INNS.
15. Shehroz S. Khan, Michael G. Madden, A Survey of Recent Trends in One Class Classificatio Lecture Notes in Computer Science Volume 6206, 2010, pp 188-197
16. WEKA. Available online: http://www.cs.waikato.ac.nz/ml/weka/

Complexity of Epileptiform Activity in a Neuronal Network and Pharmacological Intervention

D. Abásolo[1], L. González[1], and Y. Chen[2]

[1] Centre for Biomedical Engineering, Department of Mechanical Engineering Sciences,
Faculty of Engineering and Physical Sciences, University of Surrey, Guildford, UK
[2] Department of Biochemistry and Physiology,
Faculty of Health and Medical Sciences, University of Surrey, Guildford, UK

Abstract—**Neuronal outputs are complex signals of dynamically integrated excitatory and inhibitory components. Decreased synaptic inhibition in a neuronal network increases excitability and multiple spiking in neurons. Synchronized multiple spiking among a neuronal population further generates rhythmic field potentials and this epileptiform activity can propagate in the brain and cause seizures. Pharmacological interventions that reduce rhythmicity of epileptiform activity may have antiepileptic potentials. We evaluated the Lempel-Ziv (LZ) complexity for identifying rhythmicity in population spikes recorded in granule cells of the murine dentate gyrus *in vitro*. Blocking synaptic inhibition by the $GABA_A$ receptor antagonist, bicuculline, caused epileptiform population spikes, and we found that the LZ complexity of the signal was significantly reduced. Moreover, the $GABA_B$ receptor agonist, baclofen, reduced the amplitude of the epileptiform population spike and we found that it increased LZ complexity. The results show that LZ complexity is sensitive to pharmacological interventions that apparently alter rhythmicity of neuronal outputs by desynchronizing neuronal population firing. This novel approach in neuronal signal processing may be used to identify new antiepileptic targets.**

Keywords—**Lempel-Ziv complexity, epileptiform activity, dentate gyrus, excitation, synaptic inhibition.**

I. INTRODUCTION

Epilepsy is one of the most common neurological disorders affecting about 1% of the population. Paroxysmal neuronal population firing of high amplitude and high rhythmicity is generated in the brain and the epileptiform activity propagates in synchrony causing various forms of seizures. This debilitating condition often requires lifelong treatment. However, causes of seizures are often undiagnosed in individual patient. As treatments are limited to a small number of targets, seizures in as many as 30% patients are refractory to medication [1]. A full understanding of cellular events and molecular mechanisms that control the occurrence of epileptiform activity is therefore essential for developing more targeted therapeutic approaches.

Increased excitability of neuronal circuits is thought to underlie generation of epileptiform activity. Decreased expression of the inhibitory neurotransmitter GABA (gamma-aminobutyric acid) and reduced GABAergic synaptic inhibition are some of the major deficits found in patients and in preclinical animal models [2]. In brain tissue isolated *in vitro*, blocking $GABA_A$ receptors significantly reduces synaptic inhibition and neuronal firing pattern is changed from a single spike to repetitive firing following a brief stimulation of excitatory inputs [3]. The repetitive firing is often synchronized within a neuronal population and the resultant rhythmic signals have been predicted to enable widespread synchrony in the brain. Subsequently, pharmacological treatments that reduce rhythmicity of activities may have antiepileptic potentials [4].

In this pilot work the rhythmicity of epileptiform data was studied with Lempel-Ziv (LZ) complexity, a non-linear method of symbolic sequence analysis. We examined population spike (PS) waveforms recorded from granule cells in the dentate gyrus *in vitro*. Our previous results show that PSs with multiple spikes were induced using $GABA_A$ receptor antagonist, bicuculline, and this epileptiform activity was reduced in amplitude by the $GABA_B$ receptor agonist, baclofen [3]. The hypothesis of this study was that LZ complexity would highlight pharmacologically induced changes in rhythms in stimulus evoked PSs.

II. MATERIAL AND METHODS

A. Data Collection

The PSs were recorded from the soma of granule cells in the dentate gyrus [3]. Briefly, transverse sections of the dentate gyrus and the hippocampus were cut at 300 μm thickness using a vibratome. The use of wild-type mice was in accordance with the Animals in Scientific Procedures Act (1986) UK. The brain slices were kept in oxygenated (5% CO_2 / 95% O_2) artificial cerebrospinal fluid, which contained (in mM) NaCl (123), Na_2CO_3 (25), glucose (10), KCl (3.7), $CaCl_2$ (2.5), NaH_2PO_4 (1.4) and $MgSO_4$ (1.2), and maintained at 31°C. Recordings of excitatory synaptic potential and PSs were from a multi-electrode probe (MED-P210A; MED64, Alpha MED Sciences, Osaka, Japan), which consists of 64 indium tin oxide and platinum black electrodes arranged in an 8 by 8 grid with an inter-electrode distance of 100 μm. Electrical pulses (0.2 ms in duration) were delivered every 60 s to stimulate the excitatory inputs, the perforant pathway via one of the electrodes in the array. Stimulation, recording and analysis of extracellular potentials were performed using the Mobius software (version 0.3.7; Alpha MED Sciences, Osaka, Japan).

PS waveforms displayed positively deflected field excitatory synaptic potential superimposed by a large negatively-deflected spike when recorded in the baseline condition. An example of this is shown in Figure 1 (trace identified as B). Following 30 minutes of stable recording of baseline signals, $GABA_A$ receptor antagonist, (-)-bicuculline (Bic, 10μM, Sigma-Aldrich, UK), $GABA_B$ receptor agonist, (±)-baclofen (Bac, 10μM, Sigma-Aldrich, UK), and antagonist, CGP55845 (CGP, 1μM, Tocris Bioscience, UK) were examined by bath application for at least 15 minutes to achieve equilibrium. Example PS recordings under each treatment are shown in Figure 1. The PS after drugs washing out is also shown in Figure 1 (W).

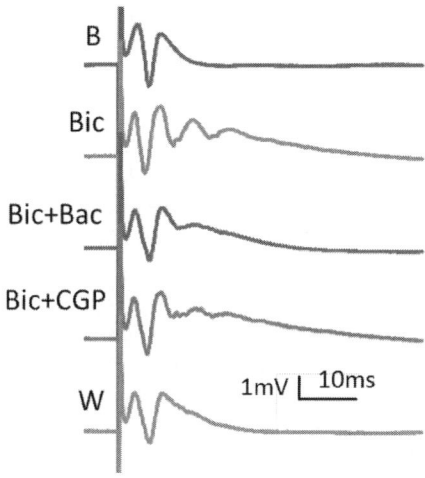

Fig. 1 Example set of PS recordings at baseline (B), and following treatments with bicuculline (Bic), bicuculline and baclofen (Bic+Bac), bicuculline and CGP55845 (Bic+CGP), and after washing out drugs (W). The stimulus artifact is shown at the first 5ms of recording.

B. Lempel-Ziv Complexity

LZ complexity is a non-linear method of symbolic sequence analysis. Originally suggested by Lempel and Ziv [5] for data compression, LZ complexity soon found its way for the analysis of complexity in short time series. There are several reasons for this. First of all, LZ complexity is a non-parametric method, i.e. no input parameters have to be defined for its computation. Complexity in the LZ algorithm is related to the number of distinct substrings and the rate of their recurrence along the given sequence [6]. Furthermore, it does not require long data segments to compute [7]. In addition, LZ complexity is model-independent and only the relevant changes between activity patterns that make a difference to the underlying system itself are considered, regardless of the nature – deterministic chaos, stochastic process, etc. – of the system itself [8]. When applying LZ complexity analysis, a particular model of system dynamics is not tested; data differences are examined on the basis of algorithmic or Kolmogorov's complexity (length of the shortest description of a string in some fixed symbolic language) [9].

These useful properties paved the way for the application of LZ complexity to biomedical signal processing. Biomedical signals are often of short length, contaminated by noise and non-stationary. These characteristics make the blind application of classic non-linear methods derived from chaos theory problematic. LZ complexity offers a fast and accurate estimation of complexity in Kolmogorov's sense for biomedical signals.

The LZ complexity algorithm involves converting the original time series into a discrete sequence of a finite number of symbols. In this study the median was used as the threshold T_d in the sequence conversion, given that the median is robust to outliers [10]. A sequence $P = s(1), s(2),…, s(n)$ is created by comparing the samples from the original sampled signal $x(i)$ with the threshold, with $s(i)$ given by:

$$s(i) = \begin{cases} 0 & \text{if } x(i) < T_d \\ 1 & \text{if } x(i) \geq T_d \end{cases} \quad (1)$$

Once this coarse-grained sequence has been created from the original signal, P is scanned from left to right and the complexity counter $c(n)$ is increased by one unit every time a new subsequence of consecutive characters is encountered. A detailed description of the complexity algorithm can be found in [8].

The complexity algorithm is dependent on the sequence length. For this reason, $c(n)$ should be normalized. For a sequence of length n and an alphabet of α symbols (in the case of a binary conversion, $\alpha = 2$), the upper bound of $c(n)$ is given by [5]:

$$c(n) < \frac{n}{(1-\varepsilon_n)\log_\alpha(n)} \quad (2)$$

where ε_n is a small quantity and $\varepsilon_n \to 0$ ($n \to \infty$). In general,

$$\lim_{n \to \infty} c(n) = b(n) \equiv \frac{n}{\log_\alpha(n)} \quad (3)$$

Therefore, $c(n)$ can be normalized via $b(n)$:

$$C(n) = \frac{c(n)}{b(n)} \qquad (4)$$

$C(n)$ is then the normalized LZ complexity. Greater $C(n)$ values correspond to more complexity in the data.

C. Statistical Analysis

Normality of the LZ complexity results was evaluated using Lilliefors test. Treatment effects were compared using repeated-measures one-way ANOVA followed by Tukey's multiple comparisons. Statistical significance was taken as $p<0.05$. All statistical analyses were performed using Prism 5 (GraphPad Software Inc. La Jolla, CA 92037 USA).

III. RESULTS

LZ complexity was estimated from six datasets of PS waveforms recorded with a sampling frequency of 20 kHz under the 5 different conditions: baseline, bicuculline, bicuculline and baclofen, bicuculline and CGP55845, and after drugs washing out. Thirty minutes of stable recording was usually carried out before any drug treatment during the baseline period (B, Figure 1). To induce epileptiform activity, bicuculline (Bic, 10μM) was applied to the brain slice for 15 minutes (Bic, Figure 1). Multiple spiking PSs were recorded due to repetitive firing in individual neurons, reflecting increased excitability. Then, baclofen (10μM) was added to the bicuculline medium (Bic+Bac, Figure 1) to see if $GABA_B$ receptor activation can affect epileptiform activity. Significant reduction in the amplitude of late PSs was found [3], but whether the rhythm persisted or not was unclear. To determine whether the anti-epileptiform effect of baclofen is due to receptor activation, we also examined CGP55845 in the presence of bicuculline (Bic+CGP, Figure 1), and an increase in late PSs was observed. Finally, washout of both bicuculline and CGP55845 (W, Figure 1) restored the PS to the single spike waveform at baseline. The PS waveforms at these conditions show distinct characteristics in rhythm, so that LZ complexity analysis was performed and results are shown in Table 1. The stimulus artifact at 5ms of each recording was not included in the analysis.

LZ complexity was significantly reduced in bicuculline-treated PS waveforms in all data sets from the baseline ($p<0.01$, B vs. Bic, Table 1), potentially reflecting increased rhythmicity in epileptiform activity. Addition of baclofen to bicuculline increased LZ complexity values significantly ($p<0.05$, Bic vs. Bic+Bac, Table 1) to a level that is not different from the baseline ($p>0.05$, Bic+Bac vs. B, Table 1). $GABA_B$ receptor activation therefore reduced rhythmicity in the epileptiform activity. Blocking $GABA_B$ receptors with CGP55845, moreover, reduced the complexity ($p<0.01$, Bic+Bac vs. Bic+CGP, Table 1) back down again to a level similar to bicuculline alone ($p>0.05$, Bic+CGP vs. Bic, Table 1). It is also noted that the co-treatment of bicuculline and CGP55845 produced more consistently lower LZ complexity that is also significantly different from the baseline ($p<0.001$, Bic+CGP vs. B, Table 1) and washout ($p<0.001$, Bic+CGP vs. W, Table 1). Finally, returning to drug-free medium effectively increased LZ complexity ($p<0.001$, Bic+CGP vs. W, and $p<0.01$, Bic vs. W, Table 1) to the level of the baseline ($p>0.05$, B vs. W, Table 1).

Table 1 LZ complexity results for six datasets of PS waveforms under different pharmacological treatment conditions

Set	B	Bic	Bic+Bac	Bic+CGP	W
1	0.1750	0.0255	0.1640	0.0292	0.2078
2	0.1130	0.0693	0.0765	0.0838	0.1349
3	0.1203	0.0365	0.0729	0.0328	0.0620
4	0.2151	0.0401	0.2406	0.0547	0.2624
5	0.1932	0.1567	0.1859	0.0911	0.1859
6	0.2880	0.1349	0.2114	0.0875	0.2187
Mean ± SD	0.1841 ± 0.0649	0.0772 ± 0.0555	0.1586 ± 0.0698	0.0632 ± 0.0281	0.1786 ± 0.0708

B: baseline period before any treatment; Bic: treatment with bicuculline (10μM); Bic+Bac: co-application of bicuculline (10μM) and baclofen (10μM); Bic+CGP: co-application of bicuculline (10μM) and CGP55845 (1μM); W: washout of drugs.

IV. DISCUSSION AND CONCLUSIONS

The results show that LZ complexity of PS waveforms changes significantly following pharmacological treatments. Firstly, the LZ complexity algorithm has identified significantly reduced complexity following bicuculline treatment compared to the baseline control waveform. The $GABA_A$ receptor antagonist, bicuculline, blocks synaptic inhibition that suppresses recurrent excitation and reduces excitability of neurons [2]. As a result, multiple spiking in burst occurs in neurons. Synchronization among a group of neurons gives rise to multiple spikes at regular intervals in the population field potential and rhythmicity is increased. Conversely, the baseline PS waveforms with single spike show higher levels of complexity, as the signal is the result of complex integration of excitatory and inhibitory actions. Therefore, reduced LZ complexity is found in epileptiform activity associated with increased rhythmicity due to synchronous multiple spiking in neurons.

$GABA_B$ receptors are G-protein coupled receptors for the inhibitory neurotransmitter GABA. They mediate inhibitory actions in the brain by presynaptic inhibition of neurotransmitters and postsynaptic inhibition of membrane excitability [11]. The receptor agonist, baclofen, has been reported to suppress epileptiform activity and seizures in patients and in animal models of seizures, showing an antiepileptic action [12]. In the dentate gyrus, $GABA_B$ receptors are predominantly expressed on inhibitory pathways to modulate GABA release presynaptically, and on granule cells to regulate their excitability postsynaptically. When $GABA_A$ receptors are blocked by bicuculline, the effect of baclofen on granule cells is more likely to be postsynaptic, and this reduced amplitude of the epileptiform activity. Now we show that LZ complexity is significantly increased by baclofen, so that rhythmicity in the epileptiform activity is reduced, indicating desynchronization between neurons. $GABA_B$ receptor activation in the dentate gyrus is therefore potentially antiepileptic by reducing rhythmicity in epileptiform activity.

Accordingly, $GABA_B$ receptor antagonist, CGP55845, significantly reduced LZ complexity to the level of bicuculline, opposite to the effect of baclofen. Reduction in rhythmicity is therefore specific to activation of $GABA_B$ receptors. Moreover, the co-treatment of bicuculline and CGP55845 consistently produced lower LZ complexity, indicating additional rhythmic effect of $GABA_B$ receptor blockade.

Finally, washing out of both bicuculline and CGP55845 restored both the baseline waveform of PSs and LZ complexity. The LZ complexity analysis can therefore distinguish the different pharmacological treatments, indicating its potential for measuring rhythmicity in epileptiform activity.

To the best of our knowledge, this is the first time LZ complexity has been successfully applied to the detection of significant changes induced pharmacologically in short, stimulus-evoked traces with limited number of repetitive cycles. It is, however, noted that the LZ complexity values obtained were scattered with considerable overlap between treatments. For example, some epileptiform traces have higher complexity values than baseline traces in other sets. However, within each dataset the trend is consistent. Further investigation is required to look into the potentially confounding characteristics in these traces. In addition, signal averaging may be applied to further eliminate contamination by rhythmic noises. Furthermore, although results are promising, the sample size was small. As a result, our findings are preliminary and require replication in a larger database of PS waveforms before any conclusion can be made of its potential impact. However, our findings show that the non-linear analysis of stimulus evoked-traces identifies pharmacologically induced changes.

In summary, we found that LZ complexity was significantly reduced in bicuculline-induced epileptiform activity and baclofen restored the reduced LZ complexity to the control level. LZ complexity may therefore be used as a novel analysis to measure rhythmicity in epileptiform activity. Caution should be applied due to the small sample size.

ACKNOWLEDGMENT

This work was supported by the Engineering and Physical Sciences Research Council (EPSRC) [grant number EP/I000992/1] to Ying Chen and Daniel Abásolo, and BBSRC [grant number BB/E010296/1] to Ying Chen. We thank Joshua Foster for data collection.

REFERENCES

1. Romanelli P, Striano P, Barbarisi M et al. (2012) Non-resective surgery and radiosurgery for treatment of drug-resistant epilepsy. Epilepsy Res 99:193–201 DOI 10.1016/j.eplepsyres.2011.12.016
2. Scharfman HE (2007) The Neurobiology of epilepsy. Curr Neurol Neurosci Rep 7:348–354 10.1007/s11910-007-0053-z
3. Foster JD, Kitchen I, Bettler B et al. (2013) $GABA_B$ receptor subtypes differentially modulate synaptic inhibition in the dentate gyrus to enhance granule cell output. Br J Pharmacol 168:1808–1819 DOI 10.1111/bph.12073
4. Löscher W (2011) Critical review of current animal models of seizures and epilepsy used in the discovery and development of new antiepileptic drugs. Seizure 20:359–368 DOI 10.1016/j.seizure.2011.01.003
5. Lempel A, Ziv J (1976) On the complexity of finite sequences. IEEE Trans Inform Theory 22:75–81 DOI 10.1109/TIT.1976.1055501
6. Radhakrishnan N, Gangadhar BN (1998) Estimating regularity in epileptic seizure time-series data. A complexity-measure approach. IEEE Eng Med Biol 17:89–94 DOI 10.1109/51.677174
7. Zhang XS, Zhu YS, Thakor NV et al. (1999) Detecting ventricular tachycardia and fibrillation by complexity measure. IEEE Trans Biomed Eng 46:548–555 DOI 10.1109/10.759055
8. Zhang XS, Roy RJ, Jensen EW (2001) EEG complexity as a measure of depth of anesthesia for patients. IEEE Trans Biomed Eng 48:1424–1433 DOI 10.1109/10.966601
9. Kolmogorov AN (1965) Three approaches to the quantitative definition of information. Infor Trans 1:3–11
10. Nagarajan R (2002) Quantifying physiological data with Lempel-Ziv complexity – Certain issues. IEEE Trans Biomed Eng 49:1371–1373 DOI 10.1109/TBME.2002.804582
11. Bettler B, Kaupmann K, Mosbacher J et al. (2004) Molecular structure and physiological functions of GABA(B) receptors. Physiol Rev 84:835–867 DOI 10.1152/physrev.00036.2003
12. Chandler KE, Princivalle AP, Fabian-Fine R et al. (2003). Plasticity of GABA(B) receptor-mediated heterosynaptic interactions at mossy fibers after status epilepticus. J Neurosci 23: 11382–11391

Author: Daniel Abásolo
Institute: University of Surrey
City: Guildford
Country: United Kingdom
Email: d.abasolo@surrey.ac.uk

Evaluation of Different Handwriting Teaching Methods by Kinematic Analysis

A. Accardo[1], M. Genna[1], I. Perrone[2], P. Ceschia[3], and C. Mandarino[3]

[1] Department of Engineering and Architecture, University of Trieste, Trieste (TS), Italy
[2] Department of Development Age, ULSS 7, Pieve di Soligo (TV), Italy
[3] Primary School, Don Milani, Cernusco sul Naviglio (MI), Italy

Abstract—Development of fine motor skills, especially drawing and handwriting, plays a crucial role in school performance and, more generally, in autonomy of everyday life.

In recent years, the analysis of writing movements that allows to characterize the handwriting process itself, has been directly performed through digital tablets, by measuring parameters extracted from the basic elements of writing, such as components and strokes.

In order to evaluate the handwriting performance in two groups of twenty children each, in which two different teaching methods were used, we examined drawing and handwriting responses by a digital tablet. The dynamic aspects of written traces were studied in five different drawing and writing tasks: a doodle, three graphomotor sequences and a cursive sentence.

Results show differences both in each class across the development and between the two methods.

Keywords—Handwriting, learning, teaching methods, digital tablets.

I. INTRODUCTION

Children spend from 31 to 60% of their school day performing fine motor tasks, in particular handwriting. The combination of digital tablet and of appropriate algorithms permits today to examine both static and dynamic characteristics of writing [1], providing information for the study of fine motor movements also useful for the detection and quantification of both dysgraphia [2] and neurological movement disorders [3-4].

In this work we studied the handwriting process for the evaluation of scholastic teaching methods, using parameters that measure precise kinematic features extracted from digitally recorded writing by means of a commercial tablet [5]. In particular the factors concerning the basic elements of writing, such as components and strokes [6] have shown to be very promising for hand motor performance quantification. A component represents the writing segment between two successive pen lifts, while a stroke is the basic element of writing movements, delimited by the points of minimal curvilinear velocity. The typical kinematic parameters of hand movements, such as duration, length, mean velocity of strokes (and sometime of the components), have been frequently used for handwriting characterization, providing information on the level of automation and fluency achieved by a student.

This paper represents a preliminary study to evaluate in two samples of Primary school Italian students, how different teaching methods could influence these parameters.

II. MATERIALS AND METHODS

A. Study Population and Teaching Methods

Each of the two considered samples consisted of about 20 students, Italian mother-tongue, right-handed and without handwriting problems and organic pathologies. In a sample the experimental Terzi's method [7] was used (S1) while in the other one the traditional teaching method was applied (S2). In the latter teaching approach the child has to solve, without a lot of support, the main instrumental problems of handwriting (space and direction of movements). With no intervention or external help, the risk is that the student automates more and more possibly painful postural habits and ineffective pen handgrips, as well as bad graphic gestures that may foreshadow cases of dysgraphia.

The Terzi's method is a cognitive-motor technique, developed in the first half of the '900 by an Italian teacher, Ida Terzi, to allow blind students to move freely in the great outdoors, thanks to a series of exercises. These exercises act on the organization of the brain, using body movement to arrive at a correct time-space representation in the motor cortex of what the body has made. Over the years, this method has also found several other fields of application in rehabilitation or in schools for the study of mathematics and writing. The exercises improve the perception of space and the time between one movement and another. This method represents a valid method of rehabilitation in cases of children with handwriting and/or learning difficulties.

Acquisitions were performed in two specific phases of learning: the end of first grade of Primary (T1), when the children had learned the grapheme and the end of the second grade of Primary (T2), when they learned the cursive.

B. Tasks

In order to study possible parameter dependence on different teaching methods as well as on motor development, all children undertook a series of five exercises: the first four tests, acquired both at the first and second grade, were totally independent from linguistic aspects differently to the last one that was acquired only at the end of the second grade, after the cursive learning.

The first test (D test) required pupils to draw in fast way, for 7s, a continuous doodle in the oblique direction.

In the second test (M task), the student had to continue as quickly as possible a zigzag sequence (Fig. 1) without pen lifts to the end of the line.

Fig. 1 Representation of the M, Q and S tasks.

The third (Q task) and the fourth (S task) tests required the student to continue as quickly as possible the illustrated symbol sequence (Fig. 1) to the end of the line, at first in a white line (Q task) and then (S task) in a squared line (requiring more accuracy).

In the last test (V task), in which adequate linguistic competences were required, pupils were asked to copy in cursive, as fast as possible, the Italian sentence: *In pochi giorni il bruco diventò una bellissima farfalla che svolazzava sui prati in cerca di margherite e qualche quadrifoglio* (meaning "In few days the caterpillar became a beautiful butterfly that fluttered on the grass in search of daisies and some cloverleaf"). This sentence was constructed in order to contain all the letters of the Italian alphabet.

No indication was given to the students about the posture and the prehension to keep. Data was acquired by means of a commercial digitizing tablet (Wacom, Inc., Vancouver, WA, Model Intuos3), using an ink pen. Pen displacement across the tablet was horizontally and vertically sampled at 200Hz with a spatial resolution of 5 μm.

C. Analysis

Analysis was carried out with a proprietary program written in MATLAB [8]. At first, for each test, the components were identified as the written tracts between two consecutive pen lifts. Then, the horizontal and vertical pen positions were filtered by means of a second order low-pass Butterworth filter (10Hz cut-off frequency) and the curvilinear motion characteristics, i.e. position and velocity curves, were derived. To identify the strokes, an automatic segmentation procedure detected points of minimal curvilinear velocity, hypothesizing that each velocity minimum corresponds to a different motor stroke, as claimed by the bell-shaped velocity profile theory [9]. A series of kinematic and static parameters were calculated and analyzed for each task: the total length and duration of the task; the mean length and duration of each component and stroke; the mean curvilinear (V_c), horizontal (V_x) and vertical (V_y) velocities (the last two in absolute value) of each stroke. In order to study possible changes of these characteristic parameters with schooling as well as between the teaching methods, at first the mean value of each parameter was calculated in each subject and averaged across students of the same sample (S1 or S2) and acquisition time (T1 or T2).

For each task and parameter, the significance of the difference between the two acquisitions (T1 and T2) in each sample (S1 and S2) was evaluated by means of the *Wilcoxon* paired-sample test. Instead, in order to evaluate the two teaching methods, we compare the two samples, in each acquisition, by means of the *Wilcoxon* rank sum test.

III. RESULTS

A. Trends with Age in the Two Samples

Tab.1 shows that in the D task, the only fixed time test, child trace a longer length across the age in both samples due to a higher velocity. As regard the motor planning, in both samples the mean duration of stroke decreases with schooling and also the fragmentation degree (number of stroke normalized on the total numbers of written tracts) is reduced. In the M test, the whole duration decreases only in the S2 although the velocities increase in both samples. This trend can be explained by a greater increment of written path in the S1 rather than in the S2. Despite the child makes a written path longer in the second acquisition for both samples, the velocity growth is so strong that even the total duration is reduced. Also in this task the fragmentation decreases with schooling but the mean stroke duration increases probably because of an increment of their length. In the Q and S tests we can observe similar behavior that in the M test as regard the velocities, the fragmentation, the total duration and the stroke duration. The reduction of the total duration is due to the decrease of both mean pen lift duration and mean component (pen down) duration. Unlike of the Q test, in the S test children present lower speeds due to the increased accuracy required performing the task.

Table 1 Mean±1SD of parameters calculated in all test (except V task, because it was acquired only in T2) in the two acquisitions (T1 and T2) for each sample (S1 and S2). The p-values indicate significance of the difference between the T1 and T2 in each sample (Wilcoxon paired-sample test).

Tests	S1-Terzi's Method			S2-Traditional Method		
	T1	T2		T1	T2	
Test: D	mean±1SD	mean±1SD	p-value	mean±1SD	mean±1SD	p-value
Whole length (mm)	279±127	487±148	0.0003	304±140	701±142	<0.0001
Mean Stroke Duration (ms)	210±40.8	184±55.6	0.03	192±43.3	149±29.2	0.007
Mean Curvilinear Velocity (mm/s)	38.8±17.5	69.6±23.9	0.0003	43.5±20.2	98.2±20.1	<0.0001
Mean Horizontal Velocity (mm/s)	24.8±12.2	44.9±18	0.0005	28.6±14.2	65.2±14.2	<0.0001
Mean Vertical Velocity (mm/s)	29.6±14.3	53.5±17.2	0.0006	32.6±15.2	73.4±15.1	<0.0001
#Strokes/#Tracts	1.36±0.941	1.07±0.199	0.02	1.38±0.733	1.02±0.075	0.01
Test: M	mean±1SD	mean±1SD	p-value	mean±1SD	mean±1SD	p-value
Whole duration (s)	19.20±2.20	18.63±7.88	n.s.	19.99±0.02	14.26±3.93	0.0006
Whole length (mm)	199±54.5	352±44.5	<0.0001	188±39.2	299±43.4	0.0003
Mean Stroke Duration (ms)	183±25.3	233±28	0.0005	170±10.6	219±29.3	<0.0001
Mean Curvilinear Velocity (mm/s)	9.7±2.76	17.1±4.34	0.0001	9.12±1.74	18.8±3.01	<0.0001
Mean Horizontal Velocity (mm/s)	4.39±1.49	6.95±1.87	0.0009	3.83±0.911	9.11±2.46	<0.0001
Mean Vertical Velocity (mm/s)	7.85±2.41	15±4.03	0.0001	7.55±1.57	15.6±2.29	<0.0001
#Strokes/#Tracts	4.28±1.68	1.62±0.583	<0.0001	3.73±1.94	1.8±0.348	0.0004
Test: Q	mean±1SD	mean±1SD	p-value	mean±1SD	mean±1SD	p-value
Whole duration (s)	18.91±2.36	13.63±5.58	0.008	19.11±1.51	11.10±4.69	0.0001
Whole length (mm)	110±28	166±22.2	0.0003	141±20.4	165±31.4	0.008
Mean pen lift duration (s)	1.51±0.64	0.85±0.23	0.0003	1.23±0.25	0.66±0.21	0.0001
Mean Component Duration (s)	2.01±0.60	1.12±0.40	0.0002	1.80±0.35	0.94±0.32	<0.0001
Mean Stroke Duration (ms)	154±10.2	170±20.1	0.004	154±8.5	160±16.4	n.s.
Mean Curvilinear Velocity (mm/s)	8.86±2.36	20.7±9.14	<0.0001	10.8±1.83	24.6±8.57	<0.0001
Mean Horizontal Velocity (mm/s)	4.82±1.61	11.9±5.82	<0.0001	5.85±0.974	15.8±5.95	<0.0001
Mean Vertical Velocity (mm/s)	5.85±1.5	14.1±5.9	0.0001	7.3±1.75	15.8±5.52	<0.0001
#Strokes/#Tracts	3.03±1.16	1.65±0.568	0.0002	2.83±0.68	1.58±0.348	<0.0001
Test: S	mean±1SD	mean±1SD	p-value	mean±1SD	mean±1SD	p-value
Whole duration (s)	18.90±1.85	14.51±5.12	0.004	19.27±1.34	10.76±4.14	<0.0001
Whole length (mm)	99.8±25.9	142±26.7	0.0006	120±21.6	133±14.5	0.03
Mean pen lift duration (s)	1.54±0.48	1.02±0.23	0.001	1.47±0.60	0.82±0.26	0.001
Mean Component Duration (s)	3.44±1.76	1.62±0.62	0.0001	2.30±1.12	1.39±0.69	0.02
Mean Stroke Duration (ms)	158±10.4	162±12	n.s.	156±10.2	163±11.1	n.s.
Mean Curvilinear Velocity (mm/s)	7.06±2.48	15.3±5.6	0.0001	9.1±2.17	19.2±6.45	0.0003
Mean Horizontal Velocity (mm/s)	4.1±1.47	9.12±3.02	0.0001	5.17±1.6	12.5±4.18	<0.0001
Mean Vertical Velocity (mm/s)	4.24±1.86	9.89±4.66	0.0001	5.72±1.66	11.4±4.23	0.001
#Strokes/#Tracts	3.26±1.01	2.32±1.04	0.003	2.35±0.84	1.57±0.55	0.05

B. Comparison between the Two Samples

In the first acquisition (T1) the parameters analyzed in the two samples do not show significant differences in D and M tests (Table 2). In T2, in D test, the S1 has a path length shorter than the S2 probably because of a lower velocity. In the M test lower velocity and longer written path in S1 produce higher duration of execution. For these two tests, the fragmentation is similar in both S1 and S2 while the stroke duration is higher in the S1. In the Q and S tests some differences are found already in the first learning phase, T1: S2 reaches higher speeds than S1 thus it is able to draw a greater number of graphemes (greater total path). In T2, the children of the S2 group are faster than the others and, since at this time both groups write about the same lengths, they also gain in execution time (total duration, mean pen lift duration and mean component duration). For the S1, the mean stroke duration and the fragmentation degree (Nr Stroke/Nr tracts) are generally higher than S2.

The V test, acquired only at the end of second grade, shows that the level of fragmentation does not differ significantly between the two samples. The students of the sample 1 write larger letters (higher values of whole length) but using more time than children of the sample 2.

Table 2 p-values of the difference between the two samples S1 and S2 in T1 and T2 acquisitions (Wilcoxon rank sum test) in the five Tests.

	Test D		Test M		Test Q		Test S		Test V	
	T1	T2	T1	T2	T1	T2	T1	T2	T1	T2
Whole duration (s)	n.s.	0.01	n.s.	n.s.	n.s.	0.02	n.s.			
Whole length (mm)	n.s.	0.0003	n.s.	0.004	<0.0001	n.s.	0.008	n.s.	0.004	
Mean Stroke Duration (ms)	n.s.	0.02	n.s.	n.s.	n.s.	n.s.	n.s.	n.s.	0.0001	
Mean Curvilinear Velocity (mm/s)	n.s.	0.001	n.s.		n.s.	0.004	n.s.		0.005	n.s. n.s.
Mean Horizontal Velocity (mm/s)	n.s.	0.001	n.s.	0.009	0.007	0.02	0.008	0.008	n.s.	
Mean Vertical Velocity (mm/s)	n.s.	0.002	n.s.	n.s.	0.02	n.s.	0.003	n.s.	n.s.	
#Strokes/#Tracts	n.s.	n.s.	n.s.	n.s.	n.s.	n.s.	0.007	0.01	n.s.	
Mean pen lift duration (ms)	n.s.	0.03	n.s.	0.02	<0.0001					
Mean Component Duration (ms)							n.s.	n.s.	0.005	n.s. <0.0001

IV. DISCUSSION

From the comparison between the two acquisitions in each student group it is possible to state that both samples made progresses between the first and the second learning phase in terms of duration, speed and reduction of the degree of fragmentation. Furthermore, when greater accuracy is required, kinematic performances are more contained.

From the second statistical analysis it appears that the sample 2, in which the traditional method was used, is almost always faster than the other one. However, performance in terms of time and length vary greatly depending on the test, even if it is often observed that students in S1, in which the Terzi's method was adopted, reproduce a longer written path than the others belonging to S2 even if both groups had to write the same number of letters.

Since early years of school, Terzi's method makes a more readable and accurate writing and this result could support prevention from dysgraphia, but at the expense of movement fluency. However we are confident that once they reached the last grade, S1 students can still maintain their excellent handwriting quality recovering the observed gap in speed.

It can be also concluded that the tools deployed for the kinematic analysis of handwriting are a way to quantitatively evaluate graphomotor performance and can be also used in the evaluation of teaching methods.

ACKNOWLEDGMENT

Work partially supported by University of Trieste, Master in Clinical Engineering.

REFERENCES

1. A.Accardo, M. Genna and M. Borean, "Development, maturation and learning influence on handwriting kinematics", Human Movement Science vol.32, pp.136-146, 2013.
2. S. Rosenblum, D. Chevion, and P.L. Weiss, "Using data visualization and signal processing to characterize the handwriting process", Pediatric Rehabil., Taylor and Francis Ltd, vol. 4, pp. 404-17, 2006.
3. S.M. Rueckriegel, F. Blankenburg, R. Burghardt et al, "Influence of age and movement complexity on kinematic hand movement parameters in childhood and adolescence", Int J Devl Neuroscience, Elsevier, vol. 26, pp. 655-663, 2008.
4. B.C. Smits-Engelsman, A.S. Niemeijer, and G.P. Van Galen, "Fine motor deficiencies in children diagnosed as DCD based on poor graphomotor ability", Hum Mov Science, North-Holland Pub. Co., vol. 20(1-2), pp. 161-182, 2001.
5. L.P. Erasmus, S. Sarno, H. Albrecht et al, "Measurement of ataxic symptoms with a graphic tablet: standard values in controls and validity in Multiple Sclerosis patients", J.Neursci. Methods, Elsevier, vol. 108, pp. 25-37, 2001.
6. G.P. Van Galen, and J.F. Weber, "On-line size control in handwriting demonstrates the continuous nature of motor programs", Acta Psychol., Elsevier, vol. 100, pp. 195-216, 1998.
7. I. Perrone, A. Accardo, A. Antoniazzi, A. Mina, S. Moro," Rehabilitation of graphomotor disturbances by means of the spatio-temporal Terzi's method", VIII Developmental Coordination Disorder Int. Conf. Proc., Baltimore, MD, USA, p.77, 23-26 june, 2009.
8. M. Genna, A. Accardo, M. Borean, "Kinematic Analysis of Handwriting in Pupils of Primary and Secondary School", 15th IGS 2011 Conference Proc., Cancun, Mexico, Edit. Elena Grassi and Jose L. Contreras-Vidal, pp 193-196, 12-15 june 2011.
9. M. Djioua, R. Plamondon, "A new algorithm and system for the characterization of handwriting strokes with delta-lognormal parameters", IEEE Trans Pattern Anal Mach Intell. Vol. 31(11), pp. 2060-2072, 2009.

Author: Agostino Accardo
Institute: Dept. of Engineering and Architecture, University of Trieste
Street: Via Valerio, 10
City: Trieste
Country: Italy
Email: accardo@units.it

Analysis of MEG Activity across the Life Span Using Statistical Complexity

J. Poza[1], C. Gómez[1], M. García[1], A. Bachiller[1], A. Fernández[2], and R. Hornero[1]

[1] Biomedical Engineering Group, Department TSCIT, ETSI de Telecomunicación,
University of Valladolid, Valladolid, Spain
[2] Department of Psychiatry and Medical Psychology, Faculty of Medicine, Complutense
University of Madrid, Madrid, Spain

Abstract—The aim of this research was to analyze the changes in magnetoencephalographic (MEG) activity across the life span. For this task, 220 healthy subjects with ages ranging from 7 to 84 years were enrolled in the study. A statistical complexity measure based on the Shannon entropy and the Euclidean distance were used to assess changes in MEG oscillations during brain development. An increasing quadratic relationship between entropy and age was found, whereas the opposite behavior was observed between statistical complexity and age. Entropy and statistical complexity significantly increased and decreased, respectively, as a function of age from childhood to adolescence ($p < 0.0071$; Mann-Whitney U-test, Bonferroni corrected). Our results suggest that brain development is accompanied by several changes in the irregularity and statistical complexity patterns. These findings provide new insights into the delimitation of 'normal behavior' of neural dynamics during maturation and ageing.

Keywords—**Brain maturation, magnetoencephalogram, entropy, disequilibrium, statistical complexity.**

I. INTRODUCTION

The brain is likely the most complex organ of the human body. Development across the life span is accompanied by important changes in neural activity [1]. In the last decades, considerable effort has been devoted to explore electroencephalographic changes associated with brain development [2]–[4]. However, only a few works have analyzed magnetoencephalographic (MEG) recordings to study the modifications in brain activity across the life span [5], [6]. Human brain can be affected by several neurological disorders from childhood to senescence [5]. Therefore, the analysis of age-related changes in MEG activity becomes a relevant issue to accurately characterize pathological states and differentiate them from changes associated with normal aging.

During the last decades, several definitions of complexity have been proposed [7], [8]. The present study is focused on the statistical complexity, based on the Shannon entropy and the Euclidean distance. This measure, derived from information theory, provides an alternative description of neural dynamics to that provided by other complexity families, such as algorithmic, dimensional or those directly linked to irregularity [7], [9]. The statistical complexity assumes that states with maximum or minimum entropy (i.e. "randomness" or "perfect order", respectively) do not contain significant information. Therefore, they are considered "trivial" [10]. This notion of complexity introduces a useful framework to characterize complex systems where a delicate interplay between functional segregation and integration can be found, such as the brain [11].

In the present research, we want to analyze the changes in entropy, disequilibrium and statistical complexity of MEG activity during brain development. Our results will provide original insights into the definition of the 'normal behavior' of neural dynamics across the life span.

II. MATERIAL

A. Subjects

Two-hundred and twenty normal controls were enrolled in the study. They were cognitively healthy controls with no history of neurological or psychiatric disorders. Subjects' age ranged from 7 to 84 years. Statistical analyses were carried out by grouping the sample into eight age stages, which correspond to eight decades of life. Table 1 summarizes the socio-demographic data for each age stage. None of the participants were taking any drug that could affect MEG activity at the recording time. All volunteers gave their informed consent to participate in the study, which was approved by the local Ethics Committee.

B. MEG Recordings

Five minutes of spontaneous MEG activity were recorded from the 220 participants using a 148-channel whole-head magnetometer (MAGNES 2500 WH, 4D Neuroimaging, San Diego, USA) placed in a magnetically shielded room at MEG Center Dr. Pérez-Modrego (Spain). Subjects were asked to relax, stay awake and with their eyes closed to avoid blinking and moving. Subjects' behavior and consciousness level were monitored during the recording. The sampling frequency was 678.17 Hz and a hardware band-pass filter of 0.1-200 Hz was applied. MEG recordings were decimated by a factor of 4 (169.549 Hz). Visual inspection and independent component analysis were performed to minimize the presence of artifacts. A mean of 17.8 ± 10.1 artifact-free epochs of 5 s per subject were selected. Finally, epochs were filtered between 1 and 65 Hz.

III. METHODS

A. Spectral Analysis

Measures from information theory are usually computed from a probability distribution function (PDF) that describes the physical processes of the system under study [8]. Previous studies have found that the dynamics associated to brain activity can be characterized using a PDF based on its spectral content [9], [12]. In the present work, the dynamics of MEG activity will be analyzed using the normalized power spectral density function, $PSD_n = \{p_j, j=1,\ldots,N\}$, with N the number of frequency bins between 1 and 70 Hz. PSD_n was calculated for each 5 s artifact-free epoch (848 samples).

B. Entropy

Entropy is a thermodynamic parameter useful to characterize the disorder of a system [8]. Although several entropy definitions can be found, this study is focused on the canonical formulation of statistical mechanics (i.e. the Shannon's entropy, H_S). In the context of signal processing, H_S has been adapted to quantify the irregularity of a signal, based on its power spectrum [12]. Hence, H_S is an information measure that characterizes the uncertainty associated to the physical processes described by PSD_n.

$$H_S[PSD_n] = -\frac{1}{\ln(N)} \cdot \sum_{j=1}^{N} p_j \cdot \ln(p_j), \qquad 0 \leq H_S \leq 1. \quad (1)$$

Eq. 1 includes a normalization by the factor $\ln(N)$. This is the maximum value of H_S, which is obtained when our "ignorance" of the underlying process described by PSD_n is maximized (i.e. all the spectral components are equally probable, $P_e = \{p_j^e = 1/N, j=1,\ldots,N\}$) [10].

C. Disequilibrium

A statistical complexity measure is based not only in the concept of "disorder" or "information", but also in the distance from the given PDF to the PDF that represents the equilibrium (i.e. the uniform distribution, P_e). Thereby, the disequilibrium would detect if there exist "privileged" states among the accessible ones [8]. In the context of the current research, the disequilibrium would reflect if there are "more likely" spectral components in the PSD_n. Certainly, the disequilibrium can be computed using several distances. Nevertheless, the Euclidean norm was selected on the basis of the results obtained in previous MEG studies [9].

$$Q_E[PSD_n] = \sqrt{\frac{N}{N-1}} \cdot \left[\sum_{j=1}^{N} (p_j - p_j^e)^2 \right]^{1/2}, \quad 0 \leq Q_E \leq 1. \quad (2)$$

Table 1 Summary of the socio-demographic data of the sample

Age stage	N	Age (years)	Gender
		Mean ± SD	Males:Females
I (<9 years)	9	8.3 ± 0.9	5:4
II (10-19 years)	20	15.0 ± 3.6	7:13
III (20-29 years)	43	24.4 ± 3.0	20:23
IV (30-39 years)	36	33.0 ± 2.6	18:18
V (40-49 years)	19	44.3 ± 3.1	12:7
VI (50-59 years)	21	55.9 ± 3.0	11:10
VII (60-69 years)	40	64.9 ± 2.7	11:29
VIII (>70 years)	32	75.1 ± 4.1	15:17

It should be noted that Eq. 2 is normalized by the factor $\sqrt{N-1/N}$. This is the maximum possible value of the Euclidean distance, which is reached between P_e and a PDF with only one component different from zero [10].

D. Statistical Complexity

Statistical complexity measures reflect the interplay between the amount of information stored in a system and its disequilibrium [8]. Thereby, two extreme states can be considered as non-informative: (i) maximum foreknowledge (i.e. "perfect order") and maximum ignorance (i.e. "maximum randomness") [10]. In the present study, statistical complexity is defined in terms of H_S and Q_E,

$$C_S^E[PSD_n] = H_S[PSD_n] \cdot Q_E[PSD_n]. \quad (3)$$

C_S^E is not a trivial function of entropy, since it depends on two different PDFs: the one associated to the MEG activity (PSD_n) and the uniform distribution (P_e) [8].

E. Statistical Analysis

An exploratory data analysis was initially performed. Data did not meet parametric test assumptions. Hence, statistical differences between consecutive pairs of age groups (i.e. I vs. II, II vs. III, and so on) were assessed by means of Mann-Whitney U-tests adjusted for multiple comparisons by a Bonferroni correction ($\alpha = 0.05/7 = 0.0071$).

All computations and statistical analyses were performed using Matlab (version 7.14; Mathworks, Natick, MA).

IV. RESULTS

Initially, H_S and Q_E were computed for each 5 s artifact-free MEG epoch. Results were averaged over all channels to obtain a quantitative measure per subject. The evolution across the life span of H_S and Q_E is displayed in Fig. 1. In both cases curve fitting with quadratic models have been included. As depicted in Fig. 1, H_S seems to increase until a certain maximum, decreasing after that value. The opposite behavior can be observed for Q_E.

C_S^E was computed from H_S and Q_E. Fig. 2 illustrates the boxplots corresponding to mean C_S^E values for each age stage. It can be observed that C_S^E shows a decreasing trend until the seventh decade of life and slightly increases during the eight decade. Statistically significant differences were only achieved between the first and second decade of life (i.e. between age stages I and II) for H_S ($Z = -2.8991$, p-value $= 0.0037$) and C_S^E ($Z = 2.7106$, p-value $= 0.0067$). No significant differences were found for the remaining pairwise comparisons (p-values > 0.0071).

V. DISCUSSION

Our findings revealed that H_S tended to increase as a function of age, whereas the opposite behavior was found with Q_E. Likewise, the trend is reversed during the last decades of life (i.e. age stage VIII). These results suggest that the irregularity of the spectral content of MEG activity progressively increases as a function of age, until it reaches a maximum around the seventh decade of life. Several maturational studies also found a quadratic relationship between age and several non-linear parameters [4], [5]. Their findings support the notion that brain activity is associated with an increasing complexity with age [3], until a certain maximum [4], [5]. On the contrary, our findings suggest that C_S^E tended to decrease as a function of age during the first seven decades of life, whereas the opposite trend was observed in later life.

Certainly, the statistical complexity should not be confused with algorithmic complexity, dimensional complexity or the definitions of complexity directly linked to irregularity [7], [9]. As previously mentioned, C_S^E considers the states of maximal order or disorder trivial (i.e. they are non-informative). This fact is illustrated in Fig. 3, where C_S^E is plotted as a function of H_S. In Fig 3.a, the bounds of C_S^E are convex curves, vanishing on global extremes. Fig. 3.b expands the region of Fig. 3.a where the C_S^E values for the 220 controls are located. Likewise, the color scale illustrates the evolution of complexity with age. Fig. 3.b shows that decreasing C_S^E values are obtained with increasing age. This fact can also be observed analyzing the grand-average values for each age group (marked as hexagrams in Fig 3.b). Nevertheless, this trend changes in the age stage VIII. Though a high C_S^E value does not necessarily reflect an optimal information processing, it represents some kind of

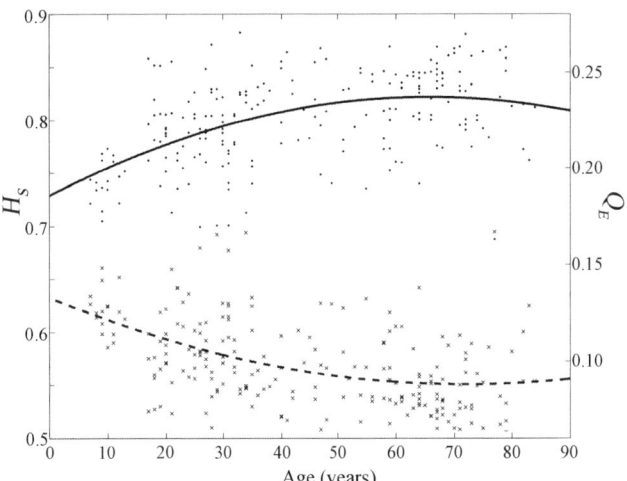

Fig. 1 Scatterplots showing age effects for H_S and Q_E. Each subject is plotted as a filled circle (HS) or a cross (Q_E), whereas fitted models represented as quadratic functions of age are plotted as solid (H_S) and dashed lines (Q_E).

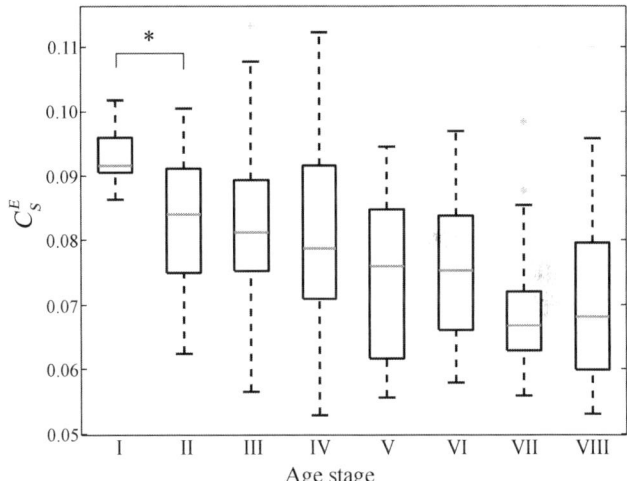

Fig. 2 Boxplots showing the distribution of C_S^E for each age stage. Statistically significant p-values are marked with asterisks (*, $p < 0.0071$).

highly informative processing. Hence, increasing statistical complexity values suggest that MEG oscillations were generated by a system governed by a balanced interplay between functional integration and segregation [11].

The results obtained with H_S, Q_E and C_S^E indicate that the more prominent changes in brain development can be seen between the age stages I and II. Indeed, the first and the second decades of life correspond with the maturation period. Our findings agree with previous studies that found age-related changes of neural activity, specially during

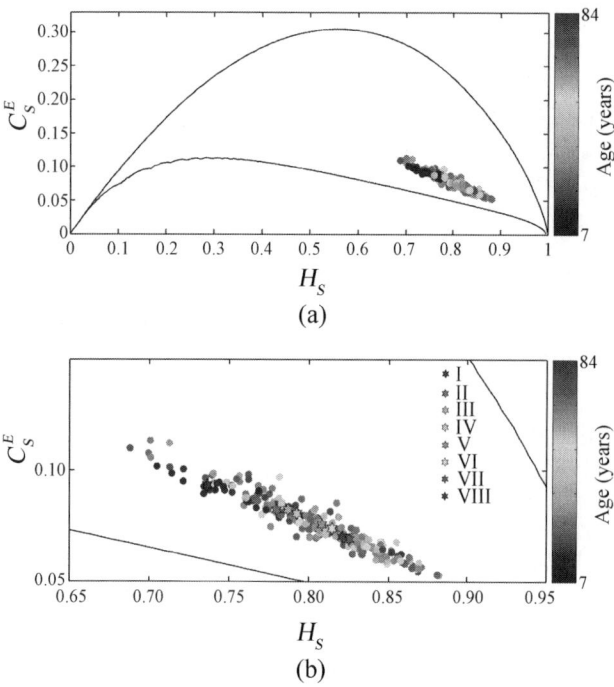

Fig. 3 Complexity *versus* entropy diagram for C^E_S and H_S, as a function of age. (a) $H_S \in [0\ 1]$. (b) $H_S \in [0.65\ 0.95]$ (grand-average values for each age stage are marked as hexagrams). Maximum and minimum possible values of C^E_S are also displayed (continuous curves).

maturation [2]–[5]. Thereby, they seem to reflect the synaptic and axonal modifications that have been reported during the transition from childhood to adolescence [13].

Several issues merit further consideration. Future work should address the assessment of the regional patterns of H_S, Q_E and C^E_S. In addition, further studies should analyze whether different definitions of entropy and disequilibrium could provide alternative statistical complexity measures. Finally, the present research was performed during a resting-state eyes-closed condition. Hence, it would be interesting to assess the neural dynamics elicited during an eyes-open resting condition or during the performance of visual or memory tasks, by means of the statistical complexity.

VI. CONCLUSIONS

Our findings support the notion that neural dynamics of MEG activity are modified across the life span. Brain development is accompanied with a progressive increase of entropy and decrease of statistical complexity until the seventh decade of life, in which a trend change can be found. Furthermore, significant changes in entropy and statistical complexity can be found during brain maturation.

ACKNOWLEDGMENT

This research was supported by: *Ministerio de Economía y Competitividad* and FEDER under project TEC2011-22987; and 'Proyecto Cero 2011 on Ageing' from *Fundación General CSIC, Obra Social La Caixa* and CSIC.

REFERENCES

1. Creutzfeld OD (1983) Cortex cerebri. Berlin: Springer
2. Matthis P, Scheffner D, Benninger C et al. (1980) Changes in the background activity of the electroencephalogram according to age. Electroencephalogr Clin Neurophysiol 49:626–635
3. Anokhin AP, Birbaumer N, Lutzenberger W et al. (1996) Age increases brain complexity. Electroencephalogr Clin Neurophysiol 99:63–68
4. Smit DJ, Boersma M, Schnack HG et al. (2012) The brain matures with stronger functional connectivity and decreased randomness of its network. PLoS One 7:e36896
5. Fernández A, Zuluaga P, Abásolo D et al. (2012) Brain oscillatory complexity across the life span. Clin Neurophysiol 123:2154–2162
6. Schlee W, Leirer V, Kolassa S et al. (2012) Development of large-scale functional networks over the lifespan. Neurobiol Aging 33:2411–2421
7. Gómez C, Hornero R, Abásolo D et al. (2006) Complexity analysis of the magnetoencephalogram background activity in Alzheimer's disease patients. Med Eng Phys 28(9):851–859
8. Kowalski AM, Martín MT, Plastino A et al. (2011) Distances in probability space and the statistical complexity setup. Entropy 13:1055–1075
9. Bruña R, Poza J, Gómez C et al. (2012) Analysis of spontaneous MEG activity in mild cognitive impairment and Alzheimer's disease using spectral entropies and statistical complexity measures. J Neural Eng 9:036007
10. Martin MT, Plastino A, Rosso OA (2006) Generalized statistical complexity measures: Geometrical and analytical properties. Physica A 369:439–462
11. Tononi G, O Sporns, GM Edelman (1994) A measure for brain complexity: relating functional segregation and integration in the nervous system. Proc Natl Acad Sci USA 91:5033–5037
12. Poza J, Hornero R, Escudero J et al. (2008) Regional analysis of spontaneous MEG rhythms in patients with Alzheimer's disease using spectral entropies. Ann Biomed Eng 36:141–152
13. De Bellis MD, Keshavan MS, Beers SR et al. (2001) Sex differences in brain maturation during childhood and adolescence. Cereb Cortex 11:552–557

Author: Jesús Poza
Institute: Biomedical Engineering Group, University of Valladolid
Street: Paseo de Belén, 15
City: Valladolid
Country: Spain
Email: jesus.poza@tel.uva.es

Impact of Device Settings and Spontaneous Breathing during IPV in CF Patients

E. Fornasa[1], A. Accardo[1], M. Ajcevic[1], R. Sartori[2], and F. Poli[2]

[1] Dept. of Engineering and Architecture, University of Trieste, Trieste, Italy
[2] Centro di Riferimento Regionale Fibrosi Cistica FVG, IRCCS Burlo Garofolo, Trieste, Italy

Abstract—**Intrapulmonary Percussive Ventilation (IPV) is a ventilation technique that has been introduced for promoting airway clearance and for recruiting areas of lung in patients with restrictive and obstructive pulmonary diseases.**

Recently, a new transportable device providing noninvasive IPV has been proposed as an alternative to traditional chest physiotherapy for patients with cystic fibrosis (CF). Nevertheless, its clinical effectiveness has not been completely demonstrated yet and further investigations have to be carried out.

The aim of this work was to investigate the functioning of the new proposed IPV system, evaluating the tolerance to this ventilation and assessing flow and pressure modulation when spontaneous breathing was superimposed.

Keywords—**Intrapulmonary Percussive Ventilation, High frequency, Airway clearance, Chest physiotherapy, Cystic fibrosis.**

I. INTRODUCTION

Intrapulmonary Percussive Ventilation (IPV) is a high frequency ventilation technique that delivers short inspiratory bursts of air at high frequency (60–600 cycles per minute) [1]. These air pulses are transmitted into the lung generating intrapulmonary percussions, loosening the mucus and favoring its movement from the distal to the central airways [2]. opThe use of IPV has been proposed for respiratory physiotherapy in patients with restrictive and obstructive pulmonary diseases and, in particular, it has been considered for improving pulmonary function in cystic fibrosis (CF) lung disease [3].

Even if IPV would intuitively appear beneficial, clinical evidence of its efficacy is lacking and further investigations have to be carried out in order to permit its optimal application [3] [4].

This preliminary study has a twofold aim: to assess the feasibility of IPV therapy in CF patients and to quantify changes in flow and pressure delivered to patients, in order to optimize the treatment.

IPV was delivered via a mouthpiece by a transportable device recently introduced for home therapy (Impulsator®, Percussionaire Corporation, Sandpoint, Idaho, USA) and it was proposed to 16 CF patients with different lung conditions. In each subject, three different percussive settings were applied.

The analysis of flow and pressure continuously recorded during the treatment allowed assessing changes in these mechanical variables, according to the device setting and the spontaneous breathing phases.

II. MATERIALS AND METHODS

A. Equipment

Impulsator® is a simplified and transportable version of IPV systems designed for hospital and intensive care (IPV®-1 and VDR-4®, Percussionaire Corporation). A knob in the front panel allows smooth control of the frequency of percussion and determines simultaneous changes in both flow and pressure amplitudes: passing from "easy" ("E") to "hard" ("H") through "average" ("A") positions, frequency decreases and, conversely, flow and pressure amplitudes increase.

The flow and pressure values are mechanically adjusted by a specific mouthpiece called Phasitron® Duo™ (Percussionaire Corporation), which represents the interface between the device and the patient. Thanks to this device, which theoretically operates according to the Venturi effect, the flow delivered to the subject is inversely proportional to the pressure reached at the level of the airways, thus ensuring the adjustment of the percussions depending on the mechanical properties of the subject's respiratory system [5]. Moreover, a vent hole present in the Phasitron® Duo™, which can be manually closed or open with a thumb, largely influences the flow and pressure output: opening the vent hole dramatically drops both flow and pressure. It is important to underline that the circuit can be used in both pediatric and adult patients [6].

Flow and pressure were recorded during IPV using a suitable acquisition and elaboration system [7]. Flow was measured by a mass airflow sensor (AWM730B5, Honeywell, Freeport, Illinois, USA) placed in front of the mouthpiece, while airway pressure was measured with a differential pressure transducer (SCX01DN, Honeywell, Freeport, Illinois, USA) from a side port between the mouthpiece and the connector held in mouth by the patient.

Figure 1 shows a simplified diagram of the respiratory circuit and the acquisition system.

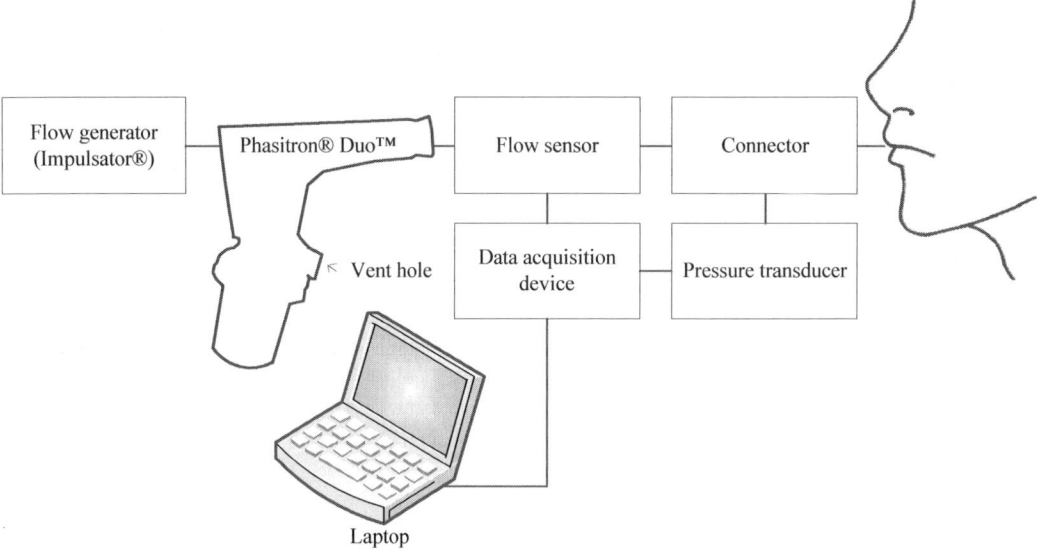

Fig. 1 Box diagram of the respiratory circuit and the acquisition system

B. Patients and Design of the Study

16 patients with CF, who attended the local CF Center and who were using standard chest physical therapy at home, underwent IPV treatment. Inclusion criteria consisted of a diagnosis of CF and an age greater than 10 years. Informed consent was obtained from the patients or their legal guardians. Patients were asked to breathe in and out through the mouthpiece normally, to seal the mouth around the mouthpiece and not to let the cheeks being inflated by the percussed air.

The proposed IPV session examined three percussive settings (in this order: "E", "A" and "H") for 5min each. Progressively the percussive frequency was decreased (from about 700cycles/min in "E", to about 300cycles/min in "A" and 100cycles/min in "H") and, at the same time, flow and pressure amplitudes were increased.

15min is a typical value for the duration of IPV treatment [8] and the change of frequency during the same session permitted to take advantage of both percussive and gas exchange properties of IPV. In fact, high frequencies (more than 300cycles/min) promote the mobilization of secretions and low frequencies (80–200cycles/min) support alveolar ventilation and encourage clearance of secretions [9].

The vent hole in the mouthpiece was always kept closed in order to maximize the percussive characteristics of IPV.

C. Data Analysis

Because the aim of this work was to assess IPV during regular breathing, we excluded from the analysis the parts of recording in which the patient was coughing as well as the first 5 respiratory cycles of every acquisition, necessary to allow the patient to induce a regular breathing pattern.

In order to reduce high frequency noise, flow and pressure signals (sampled at 2000Hz) were digitally low-pass filtered with a second order Butterworth filter with a cutoff frequency of 60Hz. Inspiration and expiration phases of the spontaneous breathing were identified from raw data using a low-pass third order Butterworth filter with a cutoff frequency of 1.5Hz.

In this study, we focused on the inspiratory phase, during which the air was inhaled into lung and the air leak potentially introduced by the nose was negligible. Moreover, we verified that the mouthpiece was properly held by the subjects for the whole duration of the IPV session.

For every inspiratory phase, peak values of pressure and flow were estimated on each cycle and averaged along the whole phase.

III. RESULTS

All the patients tolerated the experimental procedure, except one subject who could not stand the total duration of the treatment. Four patients reported discomfort during the treatment and were not able to breathe through the mouthpiece. In their acquisitions it was not possible to correctly estimate the spontaneous breathing, thus they were discarded.

Figure 2 represents an example of a typical pressure and flow recording when the treatment was properly completed. Pressure and flow peak values, calculated during the inspiratory phase, were generally higher with "H" setting and lower with "E" setting. Table 1 summarizes the averaged values (±1SD) of the peaks of flow and pressure for each subject, calculated over ten consecutive inspiratory phases. The great variability (high SD) of the assessed variables within each subject, not correlated with the respiratory

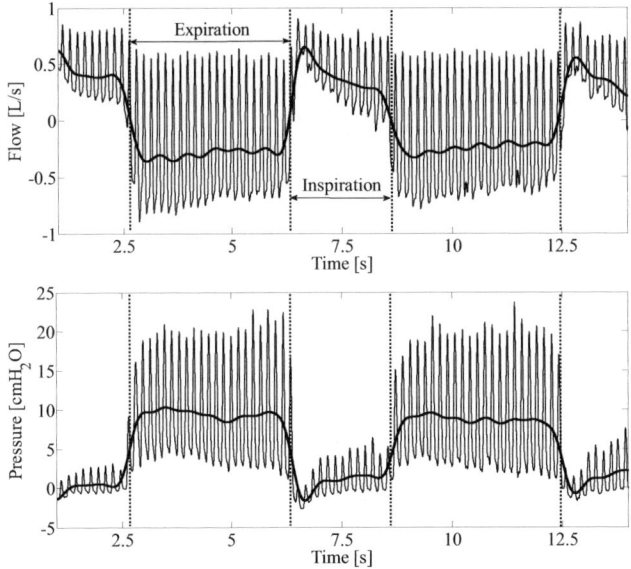

Fig. 2 Example of typical flow and pressure waveforms recorded at the mouthpiece when IPV ("A" setting) was applied and spontaneous breathing superimposed. Bold lines show the signals low-pass filtered with a cutoff frequency of 1.5Hz. Vertical dashed lines mark expiratory and inspiratory phases.

cycles, is only partially due to the subject: some fluctuations in pressure and flow values supplied by IPV were present even if a lung simulator was used. In an "in vitro" study that we conducted (unpublished data), the use of a lung simulator with resistance R=5cmH$_2$O/(L/s) and compliance C=50mL/cmH$_2$O revealed a SD of 1.58cmH$_2$O for a peak pressure of 7.40cmH$_2$O.

The highest difference between "E" and "H" setting was of 7cmH$_2$O for peak pressure and 0.5-0.8L/s for peak flow.

These values are comparable to those that occurred in the "in vitro" analysis.

In any case, peak pressures were limited to at most 10cmH$_2$O during the inspiratory phase. During the expiratory phase, when peak pressures were definitely higher (see Figure 2), the values were still less than 25cmH$_2$O. Moreover, peak flow values were always greater than 0.5L/s with larger values (up to about 1.3L/s) in the "H" position, generally in all patients.

IV. DISCUSSION

The "in vivo" study of IPV permitted to assess how respiratory mechanics variables (flow and pressure) were modulated when spontaneous breathing was superimposed.

In fact, since IPV allows the subjects to breathe regularly in the respiratory circuit, "in vitro" investigations are limited because lung simulators, being totally passive, mimic human lungs imperfectly. For example, on equal device settings ("E", "A" or "H"), flow and pressure amplitudes were amplified during the expiratory phase while during the inspiratory one the percussive effect is virtually absent. This suggests that during expiration the intensity of percussions is greater and, potentially, the mobility of secretions is increased [4].

The great variability of results among subject may be due to the different lung condition but also to how the patient carries out the treatment: in particular, the capability of the subject to seal properly the mouth around the mouthpiece is a crucial factor. Also patients' discomfort emerged during the IPV session seems not to be related to the specific lung condition, but should be considered as a subjective preference instead.

Table 1 Peak values of flow and pressure averaged among ten inspiratory phases

Subj No.	"E" setting		"A" setting		"H" setting	
	Pressure [cmH$_2$O]	Flow [L/s]	Pressure [cmH$_2$O]	Flow [L/s]	Pressure [cmH$_2$O]	Flow [L/s]
1	-0.84±1.04	0.58±0.06	-0.79±0.44	0.56±0.03	-0.14±1.01	0.57±0.03
2	1.15±1.88	0.59±0.05	6.48±2.16	0.69±0.03	6.37±3.83	0.85±0.05
3	2.75±0.89	0.83±0.10	2.01±0.66	1.09±0.06	2.81±0.81	1.23±0.11
4	5.10±1.63	0.87±0.06	7.55±2.01	0.87±0.22	9.82±4.99	1.00±0.09
5	3.22±1.47	0.42±0.14	4.78±2.89	0.60±0.05	5.36±1.93	0.76±0.05
6	1.30±0.68	0.54±0.05	3.74±2.18	1.05±0.08	2.06±4.55	1.31±0.06
7	1.90±1.16	0.53±0.07	3.54±2.09	0.80±0.06	4.02±2.29	0.97±0.08
8	-0.29±2.38	0.61±0.12	-1.45±1.26	0.67±0.08	4.16±2.82	0.72±0.05
9	3.11±0.87	0.58±0.03	1.66±1.93	0.65±0.05	1.23±1.39	0.71±0.08
10	-2.42±2.09	0.54±0.10	0.50±1.58	0.69±0.06	4.59±3.18	0.78±0.20
11	1.85±0.55	0.90±0.09	3.76±0.83	1.19±0.06	3.79±0.54	1.33±0.04

It worth noting that, despite the potential benefits related to the use of IPV, the setting of parameters has a great impact on the mechanical effects produced by IPV and it changes flow and pressure values delivered during the treatment [10]. The device set up should be chosen on the basis of the desired effects on the lung, taking into account the percussive frequency, the impulsive inspiratory to expiratory ratio and the intensity of percussions, quantified (as in the paper) evaluating the peak values of flow and pressure. For example, due to the specific features of the Impulsator® and on the basis of the literature concerning IPV [1], "E" setting may be useful to mobilize secretions, whereas "A" and "H" settings may improve alveolar ventilation.

The presence of only one knob on the Impulsator® does not allow personalized combinations in terms of ratio of expiratory to inspiratory flow, positive expiratory pressure (PEP) and frequency of percussion, unlike other IPV systems (IPV®-1 and VDR-4®).

Further studies are needed to evaluate which device setting is more suitable for patients, according to their specific lung conditions. However, as a first step toward the optimization of the treatment, this study quantified flow and pressure modulation during spontaneous breathing when three different device settings were applied.

The assessment of the role of IPV in secretion removal and in gas exchange improvement was beyond the aim of this work, thus additional studies are clearly necessary to prove IPV usefulness in these fields. The estimation of the effective wash-out during the inspiratory phase represents the next step in order to quantify the actual gas exchange this device allows.

V. CONCLUSIONS

The present results offered a view of the effects of spontaneous breathing and device settings on flow and pressure delivered by Impulsator®. A critical aspect emerged related to the shortage of settings allowed by the device.

Currently, an "in vitro" study on a test lung is being conducted in order to assess IPV variations in relations to different mechanical loads, simulating different lung conditions. Larger studies will be necessary to further define IPV usefulness both in secretion removal and in improvement of gas exchange.

ACKNOWLEDGMENT

Work partially supported by University of Trieste, Master in Clinical Engineering.

REFERENCES

1. Riffard G, Toussaint M. (2012) Indications for intrapulmonary percussive ventilation (IPV): a review of the literature. Rev Mal Respir 29(2):178–190 DOI 10.1016/j.rmr.2011.12.005
2. Chatburn RL. (2007) High-frequency assisted airway clearance. Respir Care 52(9):1224–1235
3. Flume PA, Robinson KA, O'Sullivan BP et al. (2009) Cystic fibrosis pulmonary guidelines: airway clearance therapies. Respir Care 54(4):522–537.
4. Toussaint M, Guillet MC, Paternotte S et al (2012) Intrapulmonary effects of setting parameters in portable intrapulmonary percussive ventilation devices. Respir Care 57(5):735–742 DOI 10.4187/respcare.01441
5. Lucangelo U, Antonaglia V, Zin WA et al (2004) Effects of mechanical load on flow, volume and pressure delivered by high-frequency percussive ventilation. Respir Physiol Neurobiol 142(1):81–91
6. Lucangelo U, Fontanesi L, Antonaglia V et al (2003) High frequency percussive ventilation (HFPV). Principles and technique. Minerva Anestesiol. 69(11):841–848.
7. Riscica F, Lucangelo U, Ferluga M et al (2011) In vitro measurements of respiratory mechanics during HFPV using a mechanical lung model. Physiol Meas 32(6):637–648 DOI 10.1088/0967-3334/32/6/002
8. Birnkrant DJ, Pope JF, Lewarski J et al (1996) Persistent pulmonary consolidation treated with intrapulmonary percussive ventilation: a preliminary report. Pediatr Pulmonol 21(4):246–249
9. Riffard G, Toussaint M (2012) Intrapulmonary percussion ventilation: operation and settings. Rev Mal Respir 29(2):347–354 DOI 10.1016/j.rmr.2011.12.003
10. Nava S, Barbarito N, Piaggi G et al (2006) Physiological response to intrapulmonary percussive ventilation in stable COPD patients. Respir Med 100(9):1526–1533

Author: Elisa Fornasa
Institute: Dept. of Engineering and Architecture, University of Trieste
Street: Via A. Valerio, 10
City: Trieste, I-34127
Country: Italy
Email: elisa.fornasa@phd.units.it

Estimation of Respiratory Mechanics Parameters during HFPV

M. Ajcevic[1], A. Accardo[1], E. Fornasa[1], and U. Lucangelo[1,2]

[1] Department of Engineering and Architecture, University of Trieste, Trieste, Italy
[2] Department of Perioperative Medicine, Intensive Care and Emergency, Cattinara Hospital, University of Trieste, Trieste, Italy

Abstract—The respiratory mechanics parameters (resistance and elastance) are important for adequate mechanical ventilation settings and for a correct clinical diagnosis. High-frequency percussive ventilation (HFPV) is a non-conventional ventilatory strategy characterized by inspiratory high frequency pulsatile flow, while the expiratory phase is passive. It associates the beneficial aspects of conventional mechanical ventilation with those of high-frequency ventilation.

This bench study, as preliminary to patients study, aimed to estimate respiratory mechanics parameters through a non-invasive method, during HFPV at different working pressures, frequencies and mechanical (resistive and elastic) loads. For such purpose, a multiple linear regression method to estimate parameters of the Dorkin high frequency pulmonary model was applied, starting from in vitro measured pressure and flow signals, considering inspiratory phase exclusively.

The results encourage further in vivo studies of this methodology, which has to be clinically assessed. The low rmse% values confirm validity of Dorkin three element model during HFPV.

Keywords—HFPV, respiratory mechanics, parametric identification, Dorkin's pulmonary model.

I. INTRODUCTION

High-frequency percussive ventilation (HFPV) is a non-conventional ventilatory strategy which associates the beneficial aspects of conventional mechanical ventilation (CMV) with those of high-frequency ventilation HFV [1]. HFPV acts as a rhythmic cyclic ventilation with physically servoed flow regulation, which produces a controlled staking tidal volume by pulsatile flow [1-2].

Over the years, HFPV has proven highly useful in the treatment of several different pathological conditions: closed-head injury [3], patients with acute respiratory distress syndrome (ARDS) caused by burns and smoke inhalation [4-5], newborns with hyaline membrane disease and/or ARDS [6], patients with severe gas exchange impairment [7].

The evaluation of visco-elastic properties of the respiratory system during mechanical ventilation is of great importance to detect lung dysfunctions and to evaluate the effect of treatment [8-9].

The evaluation of the resistance and elastance with the end inflation occlusion method is a relatively common practice [8-10] in patients undergoing conventional mechanical ventilation with constant inspiratory flow. This method allows separate estimation of resistive and elastic components of the respiratory system impedance. On the other hand, the occlusion method requires a constant flow, resulting inapplicable during HFPV, characterized by high frequency pulsatile flow.

An alternative method is the parametric identification based on non invasive measurements of flow and pressure signals with a three elements model, proposed by Dorkin, which approximates the respiratory mechanics function at high frequencies [11]:

$$P_{aw}(t) = R \cdot \dot{V}(t) + E \cdot V(t) + I \cdot \ddot{V}(t) + P_0 \quad (1)$$

where, at every time t, $P_{aw}(t)$ represents the pressure applied to the respiratory system, $V(t)$ is the pulmonary volume, $\dot{V}(t)$ is the airflow and $\ddot{V}(t)$ represents the volume acceleration; P_0 represents the pressure offset. The resistance R, elastance E and inertance I parameters describe respectively the viscous, elastic and inertial mechanical properties of respiratory system. This model is suitable for everyday clinical practice because of its simplicity, immediate physiological interpretation of its parameters and its sensitivity to changes in lung mechanics.

In a previous study of our group [12], respiratory parameters during HFPV were estimated in vitro, at different percussive frequencies only at a working pressure (P_{work}) of 30cmH$_2$O.

In this bench study we aimed to extend the previous research, estimating respiratory parameters during HFPV by parametric identification, also to different P_{work} that are commonly applied in clinical practice. Using the aforementioned linear model in 81 different experimental set-up combinations of working pressures, percussive frequencies, resistive and elastic lung loads, we verified the goodness of the model and of the identification method. The parameters were estimated by multiple linear regression (MLR) method starting from measured pressure and flow signals, considering inspiratory phase exclusively in order to simulate the effective clinical condition.

II. MATERIALS AND METHODS

The experimental setup used in this study is shown in Fig.1.

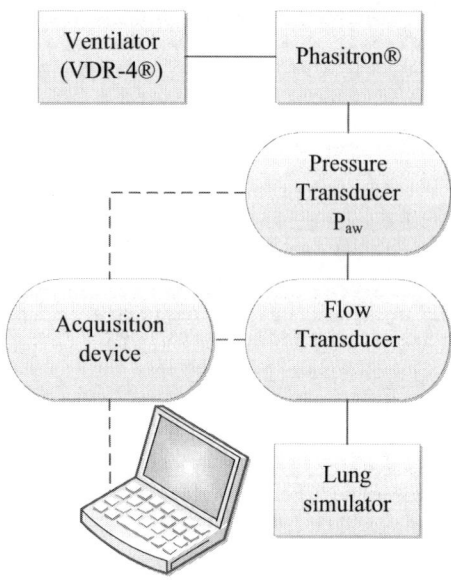

Fig. 1 Box diagram of the experimental set-up

A physical model of respiratory system was provided by a single-compartment lung simulator (ACCU LUNG, Fluke Biomedical, USA). In our measurements the lung simulator was set according to the combinations of resistive loads R of 5 and 20 cmH$_2$O/(L/s) and elastic loads E of 20, 50 and 100 cmH$_2$O/L. High Frequency Percussive Ventilation of the lung simulator was provided by a volumetric diffusive respirator (VDR-4®; Percussionaire Corporation, USA) that delivers minibursts of respiratory gas mixtures in the proximal airways through the Phasitron® that is the heart of this kind of ventilation [1][13]. A ventilator circuit was connected via a dedicated connector to the pressure and flow sensors; the distal part of the latter was linked to the described physical model of respiratory system.

The measurement of flow signal V(t) was performed using Fleisch pneumotachograph (Type 2, Switzerland) connected to a differential pressure transducer (0.25 INCH-D-4V, All Sensors, USA). The pressure signal P$_{aw}$(t) was measured using a differential pressure transducer (ASCX01DN, Honeywell, USA).

The VDR-4® ventilator was set to pulse inspiratory/expiratory (i/e) duration ratio of 1:1, and the overall inspiratory and expiratory duration (I/E) ratio of 1:1 [1][13]. The tele-inspiratory work pressure P$_{work}$ was set to 20, 30 and 40 cmH$_2$O, while the percussive frequency f_P was set to 300, 500 and 700 cycles/min.

Measurements of respiratory signals were performed for the 81 possible combinations of resistive and elastic loads, percussive frequencies, and work pressures during three successive respiratory cycles. Data were acquired at a sampling frequency of 2000Hz with 12 bit resolution (PCI-6023E, National Instruments, USA).

Volume V(t) and volume acceleration $\ddot{V}(t)$ were calculated respectively by numerical integration and differentiation of flow.

The MLR method (based on least squares criterion) was employed to estimate the model coefficients [14] during inspiratory phase of each respiratory cycle.

The adequacy of the model used to describe the measured system was evaluated by means of the normalized residual root mean square error:

$$\mathrm{rmse}\% = \sqrt{\frac{\sum_{k=0}^{N-1}[P_{awm}(kT) - P_{awest}(kT)]^2}{N}} \frac{100}{P_{awpeak}} \quad (2)$$

where P$_{awm}$ represents the measured airway pressure, P$_{awest}$ the estimated airway pressure, T the sampling period, N the number of samples and P$_{awpeak}$ the measured peak pressure.

III. RESULTS

Fig. 2 shows the estimated resistance values (R$_{est}$) versus P$_{work}$, obtained for different set-up combinations of lung elastances and percussive frequencies.

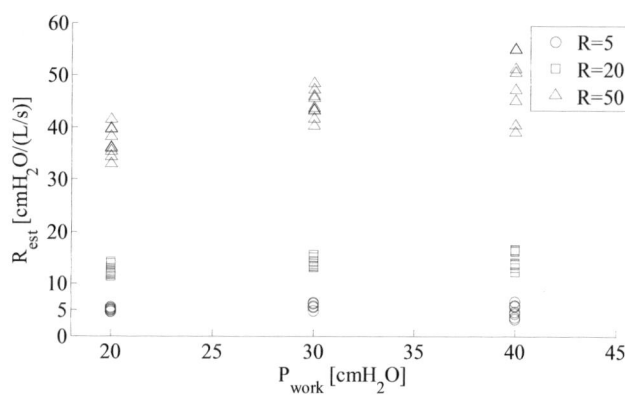

Fig. 2 Estimated resistance values Rest of R values [cmH2O/(L/s)] set on the lung simulator in function of Pwork. Each point represents the result of the parametric estimation in one of the 81 different experimental setting combinations (Pwork, fP, R, E)

Underestimated values were found for a lung simulator R of 20 cmH$_2$O/(L/s) while a good approximation was present for R = 5 cmH$_2$O/(L/s). In both cases R$_{est}$ did not show dependency on P$_{work}$. For R=50 cmH$_2$O/(L/s), R$_{est}$ increased as P$_{work}$ increased, presenting a large variability due to the different elastances and percussive frequencies examined.

Estimated elastance values E$_{est}$ in function of P$_{work}$ are shown in Fig. 3. The estimated values resulted slightly higher than lung simulator nominal ones for E=20 cmH$_2$O/L and E=50cmH$_2$O/L, while for E=100 cmH$_2$O/L overestimated values were present. No dependency on P$_{work}$ was detected. The E$_{est}$ varied for different lung resistances and percussive frequencies, even if without a clear relationship, especially in case of E=100 cmH$_2$O/L.

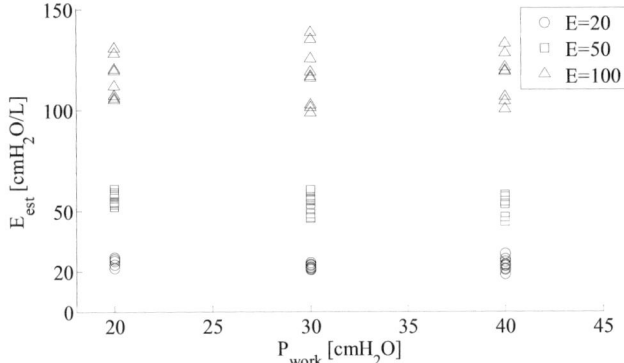

Fig. 3 Estimated elastance values Eest of E values [cmH2O/L] set on the lung simulator in function of Pwork. Each point represents the result of parametric estimation in one of the 81 different experimental setup combinations (Pwork, fP, R, E).

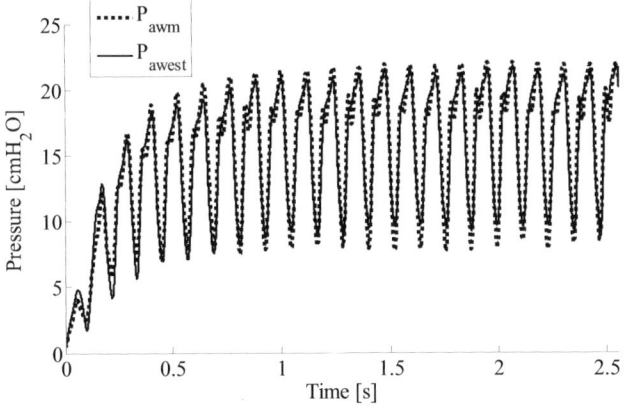

Fig. 4 Example of the measured pressure Pawm curve and the relative comparison with the estimated Pawest curve for R=5cmH2O/(L/s), E=100cmH2O/L, fp =500cycles/min and Pwork=20cmH2O

The estimated inertance values were very low and were considered negligible.

Fig.4 shows an example of the measured P$_{awm}$ and the relative comparison with the estimated P$_{awest}$. Comparing estimated and measured pressures, low rmse% values (4.95±1.51)% were found among the 81 considered combinations.

IV. DISCUSSION

In this in vitro study we aimed to estimate respiratory mechanics parameters during HFPV under different working pressures, percussive frequencies and imposed resistive and elastic lung loads. For such purpose we estimated parameters of the Dorkin linear model by MLR method, from inspiratory pressure and flow signals.

The estimated R and E parameters generally showed low dependency on P$_{work}$. On the other hand, these estimated parameters showed an increasing dependence on f_P and E for higher imposed R values and on f_P and R for higher imposed E values. The values obtained with the parameter estimation procedure were generally near to the nominal ones assessed on the lung simulator. However, the estimated resistance presented underestimation for higher values of simulated R (i.e. R = 20 and 50 cmH$_2$O/(L/s)). This could be due to the fact that the resistance of lung simulator partially depends on the flow, as reported in the manufacturer's specifications.

On the contrary, elastance parameters obtained by parametric identification were slightly higher than the value set on lung simulator, especially for E=100 cmH$_2$O/L. In the latter case the differences respect to the set values did not present a clear relationship with f_P and R, but only a large spread.

Even though some estimated values of R and E did not match the setup ones, it is important to underline the fact that the low rmse% values confirm the adequacy of the model to describe over time the measured system. This fact leads to the assumption that the differences may be mainly due to an imperfect match between nominal and effective values of the lung simulator undergoing high frequency pulsatile flow.

The estimated inertia coefficients were negligible, probably because of the absence of endotracheal tube in the ventilatory circuit. In fact, the presence of endotracheal tube the inertance effect would be more significant [15].

V. CONCLUSIONS

This study aimed to determine the respiratory parameters during HFPV, under different ventilatory conditions, through a non-invasive method. The results encourage further studies of this methodology that has to be clinically assessed. The low rmse% values confirm the validity of the applied linear model during HFPV.

ACKNOWLEDGMENT

Work partially supported by the University of Trieste, Master in Clinical Engineering.

REFERENCES

1. Lucangelo U, Antonaglia V, Zin WA et al. (2004) Effects of mechanical load on flow, volume and pressure delvered by high-frequency percussive ventilation. Respir Physiol Neurobiol 142:81-91
2. Lucangelo U, Accardo A, Bernardi A et al. (2010) Gas distribution in a two-compartment model ventilated in high-frequency percussive and pressure-controlled modes. Intensive Care Med 36:2125-2131 DOI 10.1007/s00134-010-1993-3
3. Hurst JM, Branson RD, De Haven CB (1987) The role of high-frequency ventilation in posttraumatic respiratory insufficiency. J Trauma 27:236–242
4. Lentz CW, Peterson HD (1997) Smoke inhalation is a multilevel insult to the pulmonary system. Curr Opin Pulm Med 3:221–226
5. Reper P, Dankaert R, van Hille F et al. (1998) The usefulness of combined high-frequency percussive ventilation during acute respiratory failure after smoke inhalation. Burns 24:34–38
6. Velmahos GC, Chan LS, Tatevossian R et al. (1999) High-frequency percussive ventilation improves oxygenation in patients with ARDS. Chest 116:440–446
7. Paulsen SM, Killyon GW, Barillo DJ et al. (2002) High frequency percussive ventilation as a salvage modality in adult respiratory distress syndrome: a preliminary study. Am Surg 68:852–856
8. Rossi A, Gottfried SB, Higgs BD et al. (1985) Respiratory mechanics in mechanically ventilated patients with respiratory failure. J Appl Physiol 58:1849-1858
9. Bernasconi M, Ploysongsang Y, Gottfried SB et al. (1988) Respiratory compliance and resistance in mechanically ventilated patients with acute respiratory failure. Intensive Care Med 14:547-553
10. Conti G, De Blasi R A, Lappa A et al. (1994) Evaluation of respiratory system resistance in mechanically ventilated patients: the role of the endotracheal tube. Intensive Care Med 20:421-424
11. Dorkin HL, Lutchen KR, Jackson AC (1988) Human respiratory input impedance from 4 to 200 Hz: physiological and modeling considerations. J Appl Physiol 64:823-831
12. Riscica F, Lucangelo U, Ferluga M et al. (2011) In vitro measurements of respiratory mechanics during HFPV using a mechanical lung model. Physiol Meas 32:637-648 DOI 10.1088/0967-3334/32/6/002
13. Lucangelo U, Antonaglia V, Zin W A et al. (2006) Mechanical loads modulate tidal volume and lung washout during high-frequency percussive ventilation. Respir Physiol Neurobiol 150:44–51
14. Kaczka DW et al. (1995) Assessment of time-domain analyses for estimation of low-frequency respiratory mechanical properties and impedance spectra. Ann Biomed Eng 23:135–151
15. Accardo A, Ajcevic M, Lucangelo U (2013) Flow Resistance Estimation of Endotracheal Tube During High Frequency Percussive Ventilation: Preliminary Results, IFMBE Proc. vol. 39, World Congress on Med. Phys. & Biomed. Eng., Beijing, China, 2012, pp 20–23

Author: Milos Ajcevic
Institute: Dept. of Engineering and Architecture, University of Trieste
Street: Via A. Valerio, 10
City: Trieste, I-34127
Country: Italy
Email: milos.ajcevic@phd.units.it

Signal Source Estimation Inside Brain Using Switching Voltage Divider

Yusuke Sakaue[1], Shima Okada[2], and Masaaki Makikawa[3]

[1] Graduate School of Science and Engineering, Ritsumeikan University, Kusatsu, Japan
[2] Faculty of Science and Engineering, Kinki University, Higashiosaka, Japan
[3] College of Science and Engineering, Ritsumeikan University, Kusatsu, Japan

Abstract—Visualization of neural activities inside the brain is useful for clarifying human brain function. Electroencephalogram (EEG) using a surface electrode is the most popular visualization method in laboratories, and estimation of the signal source using EEG has been extensively studied. Owing to the low spatial resolution of EEG, many surface electrodes are currently required for highly accurate estimates. In an attempt to improve the resolution, we developed a new EEG method in which the position information and potential of the signal source inside a human brain are obtained simultaneously from one signal electrode. A prototype device for proposed method has been developed, and experiments focused on EEG when the eyes were open and closed are conducted to validate our proposed method and device. Results show that the signal source position changes for the two conditions, and our method is plausible.

Keywords—electroencephalograph, voltage divider, human brain, estimation, signal source

I. INTRODUCTION

Many studies have been conducted to clarify human brain function by visualizing neural activities inside the brain. Many visualization methods have been studied, and some of them have been put into practical use [1-5]. Functional magnetic resonance imaging (f-MRI), positron emission tomography (PET), single photon emission computed tomography (SPECT), and near infrared spectroscopy (NIRS) are typical visualization methods that use cerebral blood flow changes to indicate neural activities inside the brain. These methods have relatively high spatial resolution but inferior temporal resolution; therefore, there is an occurrence time delay from real brain activation. Further, large and expensive equipment is required for these methods, making them unpopular in laboratories. In contrast, magnetoencephalography (MEG) and electroencephalogram (EEG) are typical visualization methods using neuroelectronic activities inside the brain. These methods have relatively low spatial resolution and high temporal resolution. The equipment required by MEG is similar to f-MRI; thus, MEG is also difficult to use in laboratories. EEG is easier to measure in comparison and is the most popular visualization method used in laboratories.

Estimation of the signal source position inside the brain using EEG with surface electrodes has been actively studied [6-7]. In previous studies, many surface electrodes were required for highly accurate estimates owing to the low spatial resolution of the EEG. We developed a new EEG method to overcome this disadvantage [8]. In this paper, a prototype device using this method has been developed, and EEG experiments were conducted when the eyes of a human subject were open and closed. The purpose is to clarify the validity of our method and device, i.e., the difference in the signal source position due to brain activity.

II. THEORY

Measurement of the potential inside a human body using surface electrodes is general method in biomedical engineering. This potential is measured as the difference between a signal electrode and a ground electrode. Our proposed method focuses on the position information obtained by connecting any resistance signal and the ground electrodes. Fig. 1 shows a model of our measurement method.

Eqs. (1) and (2) express the output potential of the amplifier when no resistance and additional resistance is connected, respectively. In our method, the resistance voltage divider is configured by the internal resistance inside the human body and this additional resistance, resulting in an

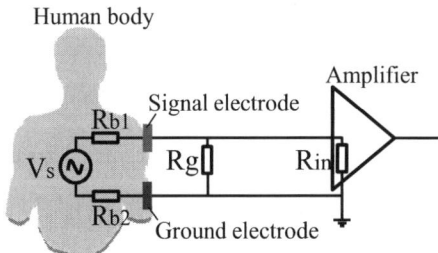

Vs: potential inside the body
Rb1, Rb2: internal resistance inside the body
Rg: additional resistance
Rin: input resistance of the amplifier

Fig. 1 Model of our measurement method

attenuated signal source potential. In Eq. (2), the additional resistance is constant; thus, the output potential of the amplifier depends on internal resistance that originates from physiological tissue and increases as the distance between the signal source position inside the body and the signal electrode increases. Therefore, the attenuation ratio AR calculated from Eq. (3) has the position information of the signal source.

$$V_{out} = V_s \qquad (1)$$

$$V_{out}' = \frac{R_g}{R_{b1} + R_g} V_s \qquad (2)$$

$$AR = \frac{V_{out}'}{V_{out}} \qquad (3)$$

This model is applied to visualize neural activities inside the brain. One signal source is assumed, and the human brain is modeled as an electrical circuit. The medium inside the brain is assumed to be uniform for simplicity. The additional resistance is switched between connected and not connected. Measurement of the same signal source is possible by high-speed switching of the connection condition.

A prototype device for this method was developed in our previous study. This device consisted of two additional resistances, two electrical switches to change the connection condition, two amplifier circuits to amplify the potential obtained from the signal electrode, large external memory to store digitized data, and a microprocessor board (Arduino Mega 2560) to control two electrical switches and digitize the output potential of the amplifier through analog/digital (AD) converter. A microSD card was selected as the external memory. The total gain of the amplifier is 80 dB, and the sampling frequency is 160 Hz. The switching frequency of the additional resistance is also 160 Hz. Fig. 2 shows the system configuration of our developed prototype device driven by batteries.

The internal resistance and capacitance of the electrical components, except for amplifier, in fig. 1 are ignored for simplicity because their values are assumed to be very small compared with a human brain. The polarization potential generated between the skin and surface electrodes is removed at the signal processing step after measurement.

III. METHODS

A. Subjects

Seven healthy adult men (22.7 ± 1.3 year old) with no pathology of any neurological, brain, or eyes disorders were

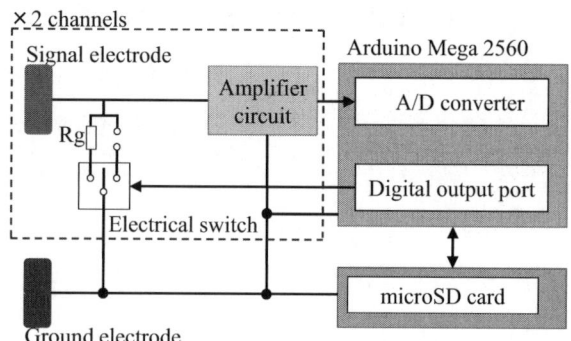

Fig. 2 System configuration for our prototype device

asked to sit on a chair in a shielded room to eliminate electrical artifacts. They were then told to maintain a rest state and to avoid moving their body during EEG measurements. For the eyes-open condition, subjects were instructed to keep their eyes open and look at a point in front of them for 20 s. For the eyes-closed condition, they were instructed to keep their eyes closed and avoid eye movement for as long as possible for 20 s.

B. Data Recording

EEGs were recorded by using two signal electrodes attached to the subject's scalp at the Cz and Pz positions according to the International 10-20 system. One ground electrode was placed on the right mastoid process. Conductive paste was used to attach all electrodes after the removal of dirt and fat on scalp.

C. Data Analysis

Measurement data were converted into EEG potentials. Next, data from 5 to 18 s were extracted owing to artifacts related to eye movement that tended to contaminate the data at the onset and end of each trial. Then, high-frequency artifacts were eliminated by using a low-pass filter (cutoff frequency: 30 Hz). Waves in the gamma band related to higher brain activity were filtered by this step. The task given to subjects was very simple, and higher brain activity was not required. Therefore, gamma waves were assumed to equally affect both conditions and were excluded from data analysis. Finally, the absolute value of the waveform was calculated, and its value for each trial was averaged. AR was calculated from these averaged values.

IV. RESULTS

A. EEG Measurement Results

Fig. 3 shows the measurements for both conditions. Alpha waves are expected to exist for the eyes-closed condition but not in the eyes-open condition because of typical alpha-wave suppression in EEG measurements. Table 1 lists the average amplitude in the two graphs. The results indicate that the average amplitude is attenuated when the additional resistance is connected for both conditions. Similar results are obtained for all subjects.

B. Attenuation Ratio for All Subjects

Fig. 4 shows *AR* for each subject for both conditions. The results indicate that *AR* is different for each condition. Additionally, *AR* is affected by the measurement position.

V. DISCUSSION

A. EEG Measurement Results

Alpha-wave suppression occurs for the eyes-open condition in Fig. 3. This alpha-wave suppression is typical for EEG measurements, and this result is in agreement with measurements using commercial EEG equipment (Polymate AP 1132, DIGITEX LAB. Co., LTD., Japan) conducted prior to each subject's experiment for the same conditions. Therefore, EEG can be measured by using our prototype device. In Fig. 3 and Table 1, the average amplitude is attenuated when an additional resistance is connected. The results indicate that the potential of the signal source is attenuated by connection of additional resistance, and this is consistent with our previous report. Therefore, the measurement results are plausible.

Table 1 Average EEG amplitude when the eyes are opened and closed

Condition	With additional resistance (μV) (mean ± S.D.)	Without additional resistance (μV) (mean ± S.D.)
Eyes open	2.80 ± 2.18	4.64 ± 3.60
Eyes closed	5.15 ± 3.90	10.15 ± 7.68

B. Attenuation Ratio for All Subjects

In Fig. 4, the average amplitude is attenuated for both conditions in all subjects. The results indicate that our proposed method is not affected by individual differences.

There is difference in *AR* for both conditions. For the eyes-closed condition, visual stimulation disappears, and occipital lobe activity is reduced. For the eyes-open condition, there is visual stimulation, and occipital lobe activity is activated. Basically, brain activity changes according to visual stimulation. Therefore, these results suggest that the signal source position changes according to the brain activity state.

Five subjects, except for subjects 3 and 4, exhibit a similar trend at both the Cz and Pz measurement positions. *AR* for the eyes-closed condition is smaller than the eyes-open condition, indicating that the signal source position for the eyes-closed condition is further away than for the eyes-open condition. However, the difference in *AR* is small for each measurement position. These results suggest that the signal source position inside the human brain for the eyes-open and eyes-closed conditions is in a similar position from the two signal electrodes.

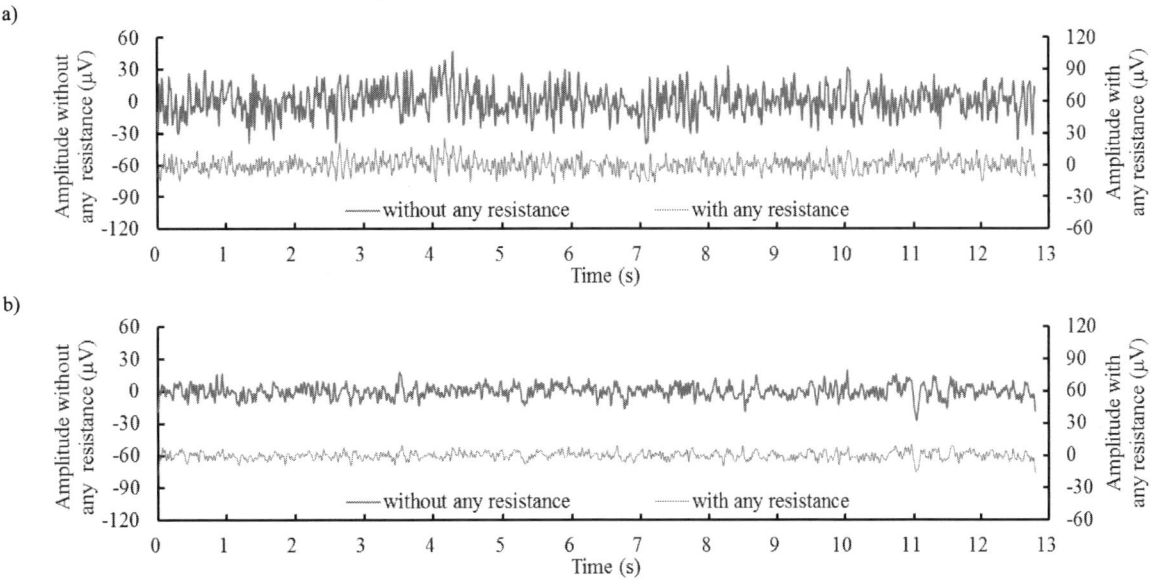

Fig. 3 Measurement example at Cz for the: a) eyes-closed condition and b) eyes-open condition

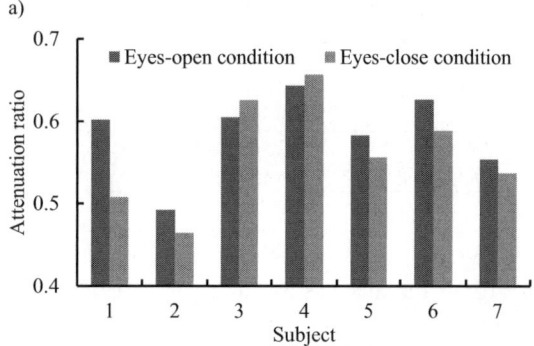

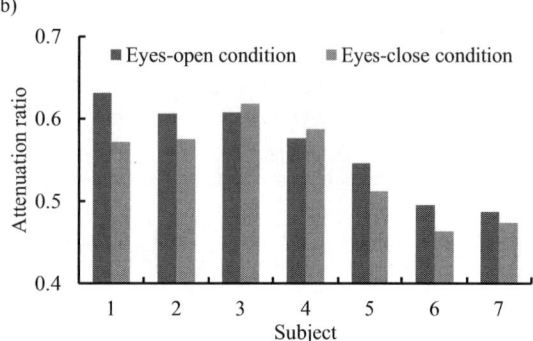

Fig. 4 Attenuation ratio of the EEG when the eyes are open and closed at: a) Cz and b) Pz

VI. CONCLUSIONS

In our previous study, we proposed a new EEG method that can obtain the potential and position information of a signal source inside the human brain and developed a prototype device. In the present study, EEG is measured using our prototype device when the eyes are open and closed. Our results indicate that our proposed method obtains the difference in the signal source position according to the brain activity state.

In our future work, the distance between the signal source position and the signal electrode will be calculated from measurements using our proposed method. This distance could improve the low spatial resolution of EEG, which would be useful for estimating the signal source position using surface electrodes. Our proposed method would contribute significantly to the clarification of human brain functions.

However, our proposed method and prototype device have some drawbacks. One is the number of signal sources, as only one signal source is assumed, but several signal sources would contribute to human brain activities. Therefore, the number of signal sources should be increased. Another drawback is the switching frequency of the additional resistance. It is difficult to apply this method to other biosignals inside the human body: electrocardiogram, electromyogram, and so on. We will consider a new measurement method and develop a new measurement device to alleviate these drawbacks.

ACKNOWLEDMENT

This work was supported by JSPS KAKENHI Grant Number 25282137.

REFERENCES

1. Nemani A K, Atkinson I C, Thulborn K R (2009) Investigating the consistency of brain activation using individual trial analysis high-resolution fMRI in the human primary visual cortex. NeuroImage 47:1417–1424
2. Emri M, Glaub T, Berecz R et al. (2006) Brain blood flow changes measured by positron emission tomography during an auditory cognitive task in healthy volunteers and in schizophrenic patients. Prog Neuropsychopharmacol Biol Psychiatry 30:516–520
3. Fukuyama H, Ouchi Y, Matsuzaki S et al. (1997) Brain functional activity during gait in normal subjects: a SPECT study. Neurosci Lett 228:183–186
4. Ishizu T, Noguchi A, Ito Y et al. (2009) Motor activity and imagery modulate the body-selective region in the occipital–temporal area: a near-infrared spectroscopy study. Neurosci Lett 465:85–89
5. Wheless J W, Venkataraman V, Kim H et al. (2002) Assessing normal brain function with magnetoencephalography. Int Congr Ser 1232:519–534
6. Blenkmann A, Seifer G, Princich J P et al. (2012) Association between equivalent current dipole source localization and focal cortical dysplasia in epilepsy patients. Epilepsy Res 98:223–231
7. Babiloni C, Stella G, Buffo P et al. (2012) Cortical sources of resting state EEG rhythms are abnormal in dyslexic children. Clin Neurophysiol 123:2384–2391
8. Sakaue Y, Okada S, Makikawa M (2012) Development of brain electrical activity estimation device using resistance voltage divider, Proc. Int. Conf. uHealthcare 2012, Seoul National University Hospital. Kyeognju, Korea, 2012, pp 172–174

Author: Yusuke Sakaue
Institute: Graduate school of Science and Engineering, Ritsumeikan University.
Street: 1-1-1, Noji-Higashi
City: Kusatsu
Country: Japan
Email: rr003035@ed.ritsumei.ac.jp

A Phase-Space Based Algorithm for Detecting Different Types of Artefacts

A. Brignol[2] and T. Al-ani[1,2]

[1] LISV-Université de Versailles Saint-Quentin-en-Yvelines, 10-12 Av. de l'Europe, 78140 Vélizy, France
[2] Département Informatique, ESIEE-Paris, Cité Descartes-BP 99, 93162 Noisy-Le-Grand, France

Abstract—The detection and removal of artefact from the recorded biosignals is challenging. The detection and localisation of different types of artefact that may be present in a single-channel biosignal is an important step before their removing. In this work, a new algorithm for semiautomatic detection and localisation of artefacts is proposed. This algorithm is based on *phase space* approach. The originality of our algorithm lies in the fact that it is not limited to one type of artefact but to any artefact characterized by sudden changes with regard to the biosignal of interest. The algorithm was successfully validated on simulated and real-world biosignals contaminated by various types of artefacts.

Keywords— artefact detection, artefact removing, signal processing, phase space.

I. INTRODUCTION

An artefact is a side effect caused by the experimental conditions during data acquisition. There are two possible sources of artefact: external or non-biological environment (e.g., acquisition hardware) and internal or physiological environment (e.g., heart beats). Hence, the recorded electrical activity are obscured by these artefacts, which can influence further signal analysis and decrease the success rate of the diagnosis systems. Regarding biological signals, the difficulty lies in our ability to clearly define what is or is not an artefact.

In many biomedical signals, it is reasonable to say that an artefact is a sudden change (with regard to the biosignal of interest) in amplitude and/or frequency occurring on a part of the signal. What can enable us to characterise an artefact is the presence of extreme values in signal. Indeed, if an artefact is present in the signal, it necessarily implies that there is a remarkable number of extreme values. But all this is very relative; we must find a way to determine what is a sudden change and what is a remarkable number of extreme values.

As an example, in the study of EEG the most commonly encountered internal artefacts during acquisitions are related to muscle activities measured by the electromyogram (EMG), ocular activities (blinking and eye movements) measured by the electrooculogram (EOG), cardiac activities measured by the electrocardiogram (ECG) and ballistic forces on the heart measured by the ballistocardiogram (BCG), event related potential (ERP). The common encountered external artefact is the electrical shift artefacts due to the shift of the electrode on the skin.

Manual identification of artefacts, by an expert, can be difficult, time consuming and tedious and even infeasible for large datasets. Although various automatic or semi-automatic techniques for detection of artefact have been reported [1–17], the need for reliable detection of various artefacts in a single as well as in multiple biosignals is widely acknowledged.

Among many of these works, specific artefacts are sought but in several biomedical signals it is often possible that there are several types of artefacts that their characteristic are not known a priori. It is therefore difficult in this case to use several algorithms to process each artefact separately. The detection of artefact is an important step before its removing. Indeed, there are many algorithms for the removing of the artefact which operate in two generally artefact morphology-dependent steps: detection and removing, see for example. [4, 8, 13, 14]. In this paper, we present only our approach to the artefact detection phase. We leave free the artefact removing phase for a user-specific desired algorithm. Our approach detects various types of artefacts that may be present in a single-channel biosignal, provided that they present a sudden change with regard to the biosignal of interest.

II. METHOD

A. Phase Space

The *phase space* allows to study variations in the signal only with respect to itself. This can be a means to analyse its behavior at different stages of its temporal evolution.

Takens showed [18] that from the evolution of dynamic variables of the process, $x(t_i), i = 1, ..., N$, we can construct a space which is, under certain conditions, topologically equivalent to the original *phase space* by introducing a time delay called *lag*.

To construct a *phase space*, one of the most interesting representation, called *Poincaré plot* or *return map* is used where $x(t_i + lag)$ is plotted against $x(t_i)$, allowing a signal to be analysed in its behaviour in different phases along its evolution in time. *Poincaré plot* is a geometrical representation of a time

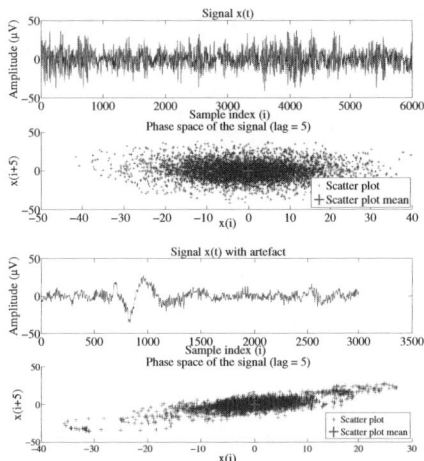

Fig. 1 From top to down, first and second raws: EEG signal without artefact and its phase space ($lag = 5$), third and forth raws: signal with artefact and its phase space ($lag = 5$).

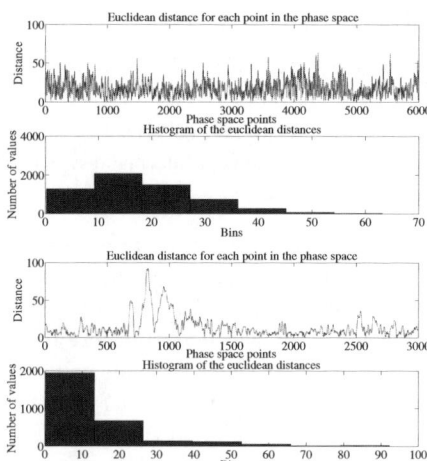

Fig. 2 The Euclidean distance of each point in the phase space and its associated histogram for the EEG signals given in Figure 1. From top to down, first and second raws: case of EEG signal without artefact, third and forth raws: case of EEG signal with artefact.

series in a *phase space*. The parameter *lag* is the time lag choice between x and y coordinates of the *phase space*. The cloud of points obtained (Figure 1) allows clearly to visualise the average and extreme values of the signal. Points near the barycenter of the cloud are related to the average value while those who clearly stand out from the cloud are related to the extreme values. This is even more visible when an artefact is present. To accurately quantify the average and extreme-related points, we calculate the Euclidean distance from each point of the *phase space* to the barycenter. Then, we calculate the histogram of the *Euclidean* distances. Based on this histogram, we can distinguish objectively between the close points and the distant points. Hence, we obtain a statistical means to detect the presence of artefacts in the signal. Figure 2 shows a significant difference between a clean signal and the same signal contaminated by an ocular artefact. In the presence of artefact, the histogram shows a distribution much less uniform than in the case without artefact. This results in a peak in the histogram. By calculating the *kurtosis*, *kurt*, of the histogram values, we have a way to quantify the sharpness (*kurtosis* increases) or the flatness (*kurtosis* decreases) extent of the peak (maximum value of the histogram) compared to the other bands. Hence, depending on the value taken by the *kurtosis* of the histogram, we can determine if an artefact is present or not. It remains to define a threshold, $kurt_{th}$ so that:

- if $kurt(x) \leq kurt_{th}$ then there is no artefact,
- if $kurt(x) > kurt_{th}$ then there is an artefact.

Through extensive trials on various types of artefact in different biosignals, we notice that the more difference there is between a clean and a corrupted part, the more $kurt_{th}$ should be low. As a consequence, it seems that $kurt_{th} = 1$ is satisfactory for simulated data (difference is obviously big) whereas real world data requires a higher threshold such as $kurt_{th}=3$. However, it is up to the user to determine a threshold adapted to their needs. This threshold is mainly used to avoid dealing with signals with no artefact.

B. Location of the Artefacts

In the example shown in Figure 2 for the case of EEG signal with artefact, we note that the histogram shows a peak to the left and decreases quickly as the values increase. The first bar, which is the highest, corresponds to values frequently taken by the signal. These are the average values. In contrast, the last bars are very low relative to the first. These correspond to values infrequently taken by the signal. These are the extreme values. It is necessary to determine a threshold $hist_{th}$ from which we consider that a bar is sufficiently low compared to the highest one, i.e. a bar whose the associated values are extreme. Given the bar height $h(bar)$ and the height of the highest bar h_{max}, then:

- if $h(bar) \geq hist_{th} * h_{max}$ the bar corresponds to the average values,
- if $h(bar) < hist_{th} * h_{max}$ the bar corresponds to the extreme values.

Again, through extensive trials on various types of artefacts in different types of biosignals, we choose from the experience $hist_{th} = 0.1$. This means that, if a bar has a height less than $\frac{1}{10}^{th}$ of h_{max}, then it corresponds to the extreme values

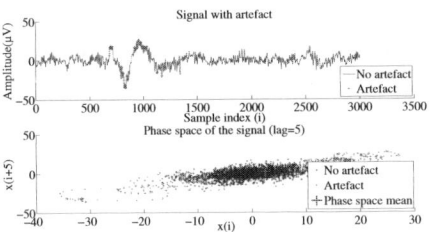

Fig. 3 Correspondence points related to the extreme values of the signal and the phase space for the EEG signal with artefact given in Figure 1.

related to an artefact. We show in Figure 3 the correspondence between the extreme values of the signal and on the phase space. We can see that our method gives consistent results.

C. Optimisation of Our Artefact Detection Approach

The accuracy of the algorithm depends on the quality of the signal and the selection of the segment on the signal to be processed. It may be interesting to apply some preprocessing for better results. It may be appropriate also to center the signal amplitude at zero. The performance of the algorithm is clearly influenced by the centering of the signal or not at the x-axis. Centering the average may be a solution but not necessarily. If the signal is not symmetrical, it does not improve quality of the correction. In the case of a non centered signal, we propose to decompose the signal in high and low frequencies. Experiments have shown that to optimise the detection of artefacts, it is best to decompose the signal into two parts: the low frequencies and high frequencies. One way to process low and high frequencies without losing information contained in the complete signal is to use the average signal envelope $env_{av}(x)$. In order to amplify the low or high frequencies, we add to (for the low frequencies) or subtract from (for the high frequencies) the signal x the average signal envelope $env_{av}(x)$. Finally, we obtain a centered signal easier to process on which we can apply our algorithm. The phase space and the histogram are then extracted from the low and the high frequency parts for localisiton as explained above. Figure 4 summarise our algorithm.

III. RESULTS

We applied our detection algorithm to simulated data and data from the real-world. These test data contained various artefacts-contaminated signals. Because of lack of space in this manuscript, we give only two examples showing our algorithm performance for artefact detection and localisation.

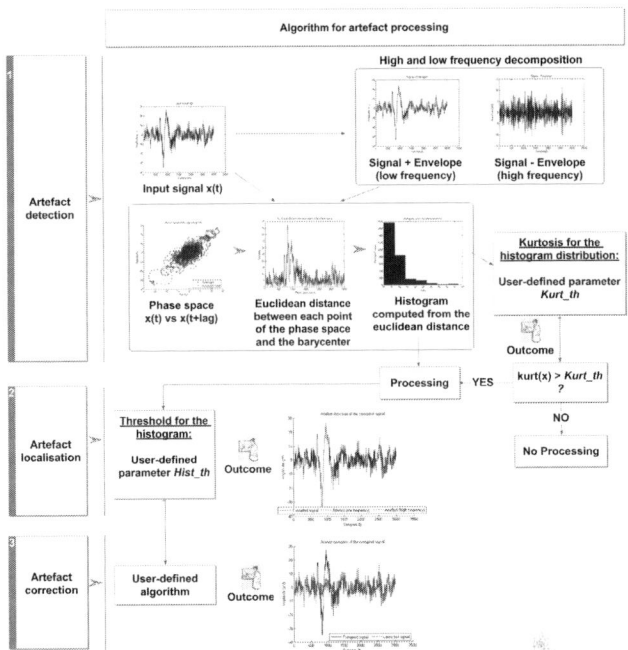

Fig. 4 Diagram summarising the detection algorithm.

A. Simulated Data

The data was generated based on the *classical theory* of event related potentials (ERP) [19] whose computer implementation is freely available at http://www.cs.bris.ac.uk/~rafal/phasereset/. Figure 5 shows a segment of simulated clean EEG signal and the same signal contaminated by two simulated artefacts: ocular artefact and electrical shift artefact.

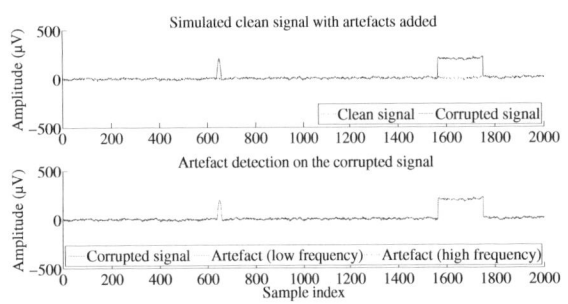

Fig. 5 A segment of simulated EEG signal. Top: the simulated clean signal with two added artefacts: ocular artefact and electrical shift artefact. Down: artefact detection and localisation with their low and high frequencies ($lag=1$, $hist_{th} = 0.1$, $kurt_{th}=1$).

B. Real World Data

The real EEG data was recorded during brain-computer interface session (BCI) on position C3 according to 10/20 international system. These data contain an ocular artefact (Figure 6).

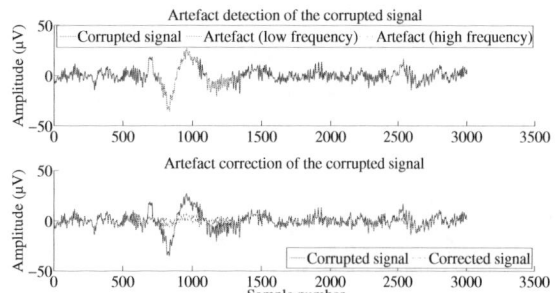

Fig. 6 A segment of real EEG signal with the location of the low and high frequencies of an ocular artefact (top) and the final correction (bottom).

The artefact removing is performed using our own artefact removing algorithm (publication in preparation). As shown in these two examples, the artefacts were detected and localised with success. The evaluation results on all test data showed that all artefacts were detected with very small false detection rate occurred and the algorithm allows to obtain good quality signals.

IV. CONCLUSION

In this paper, a phase spaces-based semi-automatic algorithm to detect and localise various types of artefacts in single-channel biosignal is proposed. The algorithm is applied in the case when artefacts are characterized by sudden changes with regard to the biosignal of interest. The proposed algorithm is based on phase space approach. The algorithm minimise the complexity of the identification and localisation of the artefact zones in order to remove later these artefacts. We concluded that the proposed method is an efficient technique for improving the quality of biosignals in biomedical analysis. Currently, the authors extend this work to optimise more the quality of the results and make the algorithm more automatic by reducing the number of user-defined parameters and seeking automatically the optimal parameter values.

REFERENCES

1. Ktonas P Y, Osorio P L, Everett R L. Automated detection of EEG artifacts during sleep: preprocessing for all-night spectral analysis *Electroencephalogr Clin Neurophysiol.* 1979;46:382-388.
2. Lima P, Leitao J, Paiva T. Artifact Detection in Sleep Eeg Recording (Lisbon, Portugal):273 - 277 1989.
3. Hatskevich Charyton W, Itkis Michael L, Maloletnev Victor I. Off-line methods for detection and correction of EEG artefacts of various origin *International Journal of Psychophysiology.* 1992;12:179-185.
4. Pierre J M, Jan W Jansen, Jan E W Beneken. Artifact detection and removal during auditory evoked potential monitoring *Journal of Clinical Monitoring.* 1993;9:112-120.
5. Klass D W. The continuing challenge of artifacts in the EEG *American Journal of Eeg Technology.* 1995;35:239-269.
6. Velde M V, Erp Gerard V, Cluitmans Pierre J M. Detection of muscle artefact in the normal human awake EEG *Electroencephalography and clinical Neurophysiology.* 1998;107:149-158.
7. Anderer P, Roberts S, Schlögl A et al. Artifact processing in computerized analysis of sleep EEG - a review *Neuropsychobiology.* 1999;40:150-157.
8. Delorme A, Sejnowski T, Makeigb S. Enhanced detection of artifacts in EEG data using higher-order statistics and independent component analysis *NeroImage.* 2007;34:1443-1449.
9. Ottenbacher J, Kirst M, Jatobá L, Huflejt M, Großmann U, Stork W. Reliable motion artifact detection for ECG monitoring systems with dry electrodes in *30th Annual International IEEE EMBS Conference*(Vancouver, British Columbia, Canada):789–792 2003.
10. Campos Viola F, Thorne J, Edmonds B, Schneider T, Eichele T, Debener S. Semiautomatic identification of independent components representing EEG artifact *Clin Neurophysiol.* 2009;120:868-77.
11. Redmond Stephen J, Basilakis J, Celler Branko G, Lovell Nigel H. An Investigation of the Impact of Artifact Detection on Heart Rate Determination from Unsupervised Electrocardiogram Recordings in *ISSC*(Dublin):1-6 2009.
12. Fairley J, Georgoulas G, Stylios C D, Rye David. A Hybrid Approach for Artifact Detection in EEG Data in *Artificial Neural Networks - ICANN 2010 - 20th International Conference, Thessaloniki, Greece, September 15-18, 2010, Proceedings, Part I* (Diamantaras Konstantinos I., Duch Wlodek, Iliadis Lazaros S.. , eds.);6352 of *Lecture Notes in Computer Science*:436-441Springer 2010.
13. Al-ani T, Cazettes F, Palfi S, Lefaucheure J-P. Automatic removal of high-amplitude stimulus artefact from neuronal signal recorded in the subthalamic nucleus *J Neurosci Methods.* 2011;198:135-146.
14. Hoffmann U, Cho W, Ramos-Murguialday A, Keller T. Detection and removal of stimulation artifacts in electroencephalogram recordings in *IEMBS Proc IEEE Eng Med Biol Soc*(Boston, MA):7159-7162 2011.
15. Yang P, Dumont G, Zhang Y-T, Ansermino J M. Artifact Detection and Data Reconciliation in Multivariate VentilatoryVariables Measured During Anesthesia: a Case Study in *Proc of the 2011 International Conference on Advanced Mechatronic Systems*(Zhengzhou, China):11-13 2011.
16. Lee J, McManus David D, Merchant S, Chon Ki H. Automatic motion and noise artifact detection in holter ECG data using empirical mode decomposition and statistical approaches *IEEE Transaction on biomedical engineering.* 2012;59:1499-1506.
17. Nizami S, Green James R., McGregor C. Implementation of artifact detection in critical care: A methodological review *IEEE Reviews in biomedical engineering.* 2013;6:127-142.
18. Takens F.. Detecting strange attractors in fluid turbulence in *Dynamical Systems and Turbulence* (Rand D., Young L.. , eds.):366-381 1981.
19. Akhtar M T, Mitsuhashi W, , James C J. Employing spatially constrained ica and wavelet denoising, for automatic removal of artifacts from multichannel eeg data *SignalProcessing.* 2012;92:401-416.

Corresponding Author: Tarik Al-ani
Institute: ESIEE-PARIS, Département IT
Street: 2 BD Blaise Pascale
City: Noisy-le-Grand
Country: France
Email: tarik.alani@esiee.fr

Automatic Classification of Respiratory Sounds Phenotypes in COPD Exacerbations

Daniel S. Morillo[1], M.A. Fernández Granero[1], A. León[2], and L.F. Crespo[1]

[1] Biomedical Engineering and Telemedicine Lab, University of Cádiz, Cádiz, Spain
[2] Pulmonology and Allergy Unit, Puerta del Mar University Hospital, Cádiz, Spain

Abstract—During the last few years, different COPD phenotypes are being defined. Despite being one of the basic characteristics of exacerbations, studies on changes and peculiarities of respiratory sounds during an exacerbation episode have been barely studied. A computerized analysis of respiratory sounds recorded in patients hospitalized because of acute respiratory symptoms was performed. It was aimed to be applied in the classification of COPD exacerbations only using the auscultation data registered after the admission. The analyzed exacerbations were classified into two categories according to the initial conditions and the evolution of the exacerbation in terms of its acoustic characteristics. Multi-parametric analysis using features extracted in the time-frequency domain was applied and a RBF network was trained and validated for classifying. Based on the cross-validation results, sensitivity of 78.4% and specificity of 81.3% were achieved. The proposed method could contribute to extend the knowledge of respiratory sounds during COPD exacerbations and to provide additional information of the disease as a basis for improving the impact on the patient.

Keywords—COPD, respiratory sounds analysis, exacerbations, neural networks, phenotypes.

I. INTRODUCTION

Chronic Obstructive Pulmonary Disease (COPD) is a major cause of morbidity and mortality worldwide [1]. The prevalence of COPD in subjects over 40 years has been estimated around 10% and this prevalence is increasing progressively with age. The range of mortality associated with this disease has doubled in the last 30 years [2]. COPD is characterized by chronic airflow limitation, usually progressive and associated with an abnormal inflammatory response of the lungs to noxious particles or gases. COPD leads to a marked reduction in physical exercise capacity and eventually to disability.

Exacerbations, or periods of increased symptoms, are a frequent event in the natural history of COPD. Exacerbations have a major negative impact on lung function. The main symptom of the exacerbation is increased dyspnea, often accompanied by increased cough and sputum, change in color of sputum, fever and wheezing, among other signs.

Respiratory sounds have invaluable information about the physiology and pathology of the lung, and therefore also on the obstruction of the airways [3]. The structure changes due to pulmonary lung diseases affect the sound transmission to the surface of the chest wall [4]. In fact, changes in respiratory sounds are one of the clinical signs mentioned in exacerbation episodes. It is generally accepted that the normal vesicular sounds are usually substituted by wheezing sounds.

For almost two centuries, non-invasive listening and interpretation of lung sounds by stethoscope has been an important component of screening and diagnosis of respiratory diseases. However, this practice has always been vulnerable since auscultation is subjective, depends largely on the clinician's experience and has a low reproducibility.

Computer analysis of respiratory sounds could aid in the objective diagnostic and monitoring of lung diseases [5]. In the last few years, computerized lung sounds analysis have been proposed for clinical applications. Automatic detection of wheezing sounds has been widely reported [6-9]. The quantification of wheezing sounds has been proposed, in the study of airflow obstruction, as a possibility for continuous non-invasive monitoring of patients with wheezing [10]. Moreover, respiratory sound intensity, detected by auscultation, has been found strongly correlated with indices quantifying airflow obstruction [11]. Other authors have studied recognition methods with which to discriminate normal lung sounds from pathological sounds and even to classify between different types of adventitious respiratory sounds [12-15].

Non-invasive assessment of differences in the distribution of vibration energy calculated through the computer analysis of acoustic signals has been recently proposed as potentially useful in the distinction of acute dyspnea in patients presenting to the emergency unit with dyspnea due to COPD exacerbations, asthma or heart failure [16].

There is interest in tracheal sounds as possible indicators of airway obstruction [17] and as a source for qualitative and quantitative assessment of airflow. The close relationship between the respiratory airflow and tracheal sounds intensity is well known [18]. Computerized analysis of tracheal sounds has been recently proposed as a simple cost-effective method to aid decision-making in the early diagnosis of pneumonia in COPD patients in resource-constrained settings [19].

In addition, absence of wheezing can be detected in many patients with significant obstruction of the airway [20,21]. But, in these patients, other changes in lung sounds, like the decrease in intensity of normal respiratory sounds, have been also appreciated. In the particular case of COPD, we have recently reported and described two different patterns of presentation and evolution of respiratory sounds during exacerbations.

Patients were followed during their hospital stay and computerized respiratory sound analysis and unsupervised clustering methods were used to characterize their respiratory sounds evolution [22].

The present study was aimed at classifying the COPD exacerbations into the two previously identified subgroups according to the initial conditions and the evolution of the exacerbation in terms of its acoustic characteristics. As a novelty, this classification was performed by analyzing the lung sounds of a single recording, made at presentation to the emergency unit because of acute respiratory symptoms.

II. MATERIALS AND METHODS

A. Subjects

A sample of 53 patients hospitalized for COPD exacerbation in the Pulmonology Unit of the University Hospital Puerta del Mar of Cádiz (Spain) was followed during the hospital stay. Inclusion criteria were physician-diagnosed COPD, a FEV_1/FVC ratio less than 0.7 in a stable phase of disease, admission for hospitalization by acute exacerbation of COPD and signed informed consent before the enrollment. Demographic and clinical characteristics of the sample of patients who participated in the study are presented in Table I. The Hospital's research ethics committee approved the study.

B. Respiratory Sounds Recording

The recording of respiratory sounds was performed with an electronic stethoscope (Thinklabs DS32A) and an electret microphone with a coupling conic chamber and flat response between 50 and 18 kHz. Sounds were simultaneously recorded both in ten different locations on the anterior, side and posterior chest (stethoscope) and over the trachea, on the suprasternal notch (microphone), that has been reported as an effective location to record tracheal sounds [23]. Electronics was embedded in a special housing that was handled by the patients themselves. Sampling rate was 8000 Hz. All recordings were made with the patient in bed placed, with the head of the bed at 45 degrees angle. The first recording, used in this study, was performed within the first 24-hours of admission.

Table 1 Demographic and clinical characteristics of participants

Feature	Value
Age (years)	74.30 (7.72)
Weight (kg)	78.1 (15.8)
Height (cm)	163.9 (7.9)
BMI (kg/m^2)	29.1 (5.8)
Smoking exposure (packs/year)	64.7 (33.1)
Gender (male / female) (%)	90.6 / 9.4
Smoking habits:	
Active smoker (%)	24.5
Former smoker (%)	73.6
COPD Severity (GOLD criteria):	
Stage I (%)	1.9
Stage II (%)	18.9
Stage III (%)	26.4
Stage IV (%)	52.8

Values are given as mean (SD). BMI: body mass index; Smoking habit data was not available for one patient.

C. Signal Processing and Features Selection

Two senior pulmonologists classified each case into two categories according to the initial conditions and the evolution perceived in respiratory sounds from the admission to the discharge. As detailed in [22], patients in the first group (RSE-G_1) were characterized by absence of adventitious sounds and diminished vesicular sounds upon admission, and increased vesicular sounds accompanied by occasional adventitious sounds at the time of discharge. In the second group (RSE-G_2), patients had noteworthy adventitious sounds on admission, mainly wheezes and rhonchi, which decreased or even disappeared at the time of discharge.

A computerized analysis of the respiratory sounds recorded after presentation to the emergency unit was applied. Heart, noise and muscle sounds were filtered by using an equirriple band-pass Finite Impulse Response filter (FIR) and a recursive least squares (RLS) adaptive filter. Respiratory sounds are non-stationary signals. To characterize their spectral time varying properties, STFT was used. The characteristics of tracheal sounds were quantified through a set of 26 features that we have exhaustively described recently [19]. The selection of a small subset but with high discriminatory power generally enhances the classifier accuracy. The search for the best set of features was accomplished by employing the SFS method along with a RBF classifier. Geometric mean of sensitivity and specificity (G_{mean}), derived from the receiver-operator characteristic (ROC) curve was used [24]. SFS strategy comprised two steps. First, a RBF was designed and G_{mean} calculated through leave-one-out cross validation was computed for each of 26 variables. The feature with the highest G_{mean} was selected. Secondly, combinations of two-dimensional vectors that included the preselected feature were explored. A new RBF network was trained and validated. The procedure was repeated and, in each iteration, G_{mean} was calculated and the best subset was selected. The algorithm finished when a new vector did not improve the G_{mean} of the previously trained and validated RFB.

For signal processing and graphical representation, the MathWorks MATLAB® software was used.

D. Classifier, Validation and Performance Metrics

A RBF network was trained and validated to explore automatic classification of patients into the two established groups. RBF neural networks are based on supervised learning and have demonstrated to be highly efficient at modeling nonlinear data and robust to additive noise corruption. The RBF network has a feed forward structure consisting of three layers. The input layer has one neuron for each predictor variable. The optimal number of neurons in the hidden layer is determined by the training process [25,26]. Each neuron consists of a radial basis function. Values coming out of the hidden layer are multiplied by weights associated with the neuron and passed to the summation layer which adds up the weighted values and presents this sum as the output of the network.

LOOCV was used to ensure stability of the results and estimate the performance of the classifier. ROC analysis was performed to select the optimal probability thresholds and to evaluate the model. Sensitivity, specificity, precision and the area under the ROC curve (AUC) were used for an overall estimation of the accuracy of the classifier.

III. RESULTS

The initial feature dataset was comprised of 26 features for each of the 53 subjects. Sequential feature selection (SFS) was used to select the feature subset that best separated exacerbations with a RSE-G_1 profile from RSE-G_2 cases. ROC analysis was carried out for every LOOCV RBF network for every single feature in the initial dataset. As a result, uFB_1 was selected as the feature that maximized G_{mean}. Then, new LOOCV RBF classifiers were trained and tested with added input features. SFS was terminated after seven iterations, when the G_{mean} computed with a LOOCV RBF classifier was lower than the G_{mean} value estimated in the 6_{th} iteration. To sum up, during the SFS algorithm, 161 subsets were evaluated with up to seven features per subset. The best subset found included frequency ratios (uFB_1 and dFB_4), standard frequency features (uSCF, uMNF and uMDF) and entropy measures (uTE and uRE). Table 2 summarizes the classification statistics at each step of the SFS procedure.

G_{mean} increased from 65% in the first iteration to 79.8% in the last one. The overall achieved precision was 88.2% and the AUC 0.83. The corresponding ROC curve of the optimal classifier (validation) is shown in Fig. 1.

Table 2 Performance assessment for the RBF classifier after each step into sequential forward selection process

Features Subset	Se	Sp	G_{mean}	AUC
1:{uFB_1}	67.6	62.5	65.0	0.64
2:{uFB_1, dFB_4}	63.2	68.7	65.9	0.73
3:{uFB_1, dFB_4, uSCF}	70.3	75.0	72.6	0.88
4:{uFB_1, dFB_4, uSCF, uMNF}	67.6	87.5	76.9	0.85
5:uFB_1, dFB_4, uSCF, uMNF, uTE}	75.7	81.2	78.4	0.76
6:{uFB_1, dFB_4, uSCF, uMNF, uTE, uRE}	81.1	75.0	78.0	0.79
7:{uFB_1, dFB_4, uSCF, uMNF, uTE, uRE, uMDF}	78.4	81.3	79.8	0.83

Se: sensitivity (%); Sp: specificity (%)

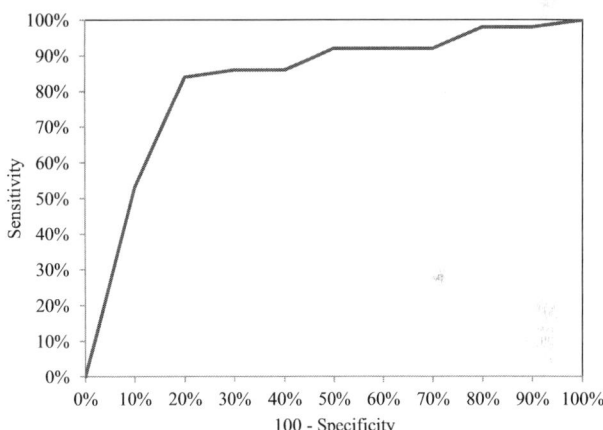

Fig. 1 ROC curve for the validated RBF network

IV. CONCLUSIONS

A novel method based on the computerized analysis of respiratory sounds of patients hospitalized because of acute respiratory symptoms has been described. It is aimed to classify COPD exacerbations. A multi-parametric analysis using features extracted in the time-frequency domain as applied. A RBF classifier able to identify the exacerbation profile only using the respiratory sounds registered after the presentation in the emergency unit was designed. Based on the cross-validation results, sensitivity of 78.4% and specificity of 81.3% were achieved.

The analyzed exacerbations were previously classified into two categories according to their evolution in terms of the acoustic characteristics [22]. Lately, different COPD phenotypes are being defined [27]. The respiratory sounds evolution groups targeted in this study could be related to certain phenotypes. These possible relationships deserve further research.

Despite being one of the basic characteristics of exacerbations, studies on changes of respiratory sounds during an exacerbation have been barely studied. This study aims to extend the knowledge of respiratory sounds during COPD exacerbations and to provide additional information of the disease as a basis for enhancing diagnosis, improving the impact on the patient and reducing costs to health systems.

ACKNOWLEDGMENT

This work was supported in part by the Ambient Assisted Living (AAL) E.U. Joint Programme, by grants from Ministerio de Educación y Ciencia of Spain and Instituto de Salud Carlos III under Projects PI08/90946 and PI08/90947.

REFERENCES

1. Global strategy for the diagnosis, management, and prevention of chronic obstructive pulmonary disease (update February 2013) at http://www.goldcopd.org/
2. Putman-Casdorph H, McCrone S. (2009) Chronic obstructive pulmonary disease, anxiety, and depression: state of the science. Heart Lung 38:34-47.
3. Reichert S, Gass R, Brandt C, Andrès E. (2008) Pulmonary auscultation in the era of evidencebased medicine. Rev Mal Respir 25:674-82.
4. Pasterkamp H, Kraman SS, Wodicka GR. (1997) Respiratory sounds. Advances beyond the stethoscope. Am J Respir Crit Care Med 156:974-87.
5. Gurung A, Scrafford CG, Tielsch JM et al. (2011) Computerized lung sound analysis as diagnostic aid for the detection of abnormal lung sounds: A systematic review and meta-analysis. Resp Med 105:1396-1403.
6. Jain A, Vepa J. (2008) Lung sound analysis for wheeze episode detection. Conf Proc IEEE Eng Med Biol Soc 2008, pp 2582-5.
7. Bentur L, Beck R, Berkowitz D, et al. (2004) Adenosine bronchial provocation with computerized wheeze detection in young infants with prolonged cough: correlation with long-term follow-up. Chest 126:1060-5.
8. Taplidou SA, Hadjileontiadis LJ. (2007) Wheeze detection based on time-frequency analysis of breath sounds. Comput Biol Med 37:1073-83.
9. Guntupalli KK, Alapat PM, Bandi VD, Kushnir I. (2008) Validation of automatic wheeze detection in patients with obstructed airways and in healthy subjects. J Asthma 45:903-7.
10. Baughman RP, Loudon RG. (1984) Quantitation of wheezing in acute asthma. Chest 86:718-22.
11. Bohadana AB, Peslin R, Uffholtz H. (1978) Breath sounds in the clinical assessment of airflow obstruction. Thorax 33:345-51.
12. Alsmadi SS and Kahya YP. (2002) Online classification of lung sounds using DSP, 24th Annual Conference and the Annual Fall Meeting of the Biomedical Engineering Society EMBS/BMES Conference, vol. 2, 2002. pp 1771-1772.
13. Bahoura M (2009). Pattern recognition methods applied to respiratory sounds classification into normal and wheeze classes. Comput Biol Med 39:824-43.
14. Abbas A, Fahim A. (2010) An automated computerized auscultation and diagnostic system for pulmonary diseases. J Med Syst 34:1149-55
15. Feng J, Krishnan S, Sattar F. (2011) Adventitious Sounds Identification and Extraction Using Temporal-Spectral Dominance-Based Features. IEEE Trans Biomed Eng 58(11):3078-3087.
16. Wang Z, Xiong YX. (2012) Lung sound patterns help to distinguish congestive heart failure, chronic obstructive pulmonary disease, and asthma exacerbations. Acad Emerg Med 19:79-84.
17. Noviski N, Cohen L, Springer C, et al. (1991) Bronchial provocation determined by breath sounds compared with lung function. Arch Dis Child 66:952-5.
18. Forgacs P. (1978) The functional basis of pulmonary sounds. Chest 73:399-405.
19. Morillo DS, Jiménez AL and Moreno, SA. (2013) Computer-aided diagnosis of pneumonia in patients with chronic obstructive pulmonary disease. J Am Med Inform Assoc. Published Online First: 8 February 2013 DOI 10.1136/amiajnl-2012-001171
20. Baumann UA, Haerdi E, Keller R. (1986) Relations between clinical signs and lung function in bronchial asthma: how is acute bronchial obstruction reflected in dyspnoea and wheezing?. Respiration 50:294-300.
21. Bohadana AB, Peslin R, Uffholtz H, Pauli G. (1995) Potential for lung sound monitoring during bronchial provocation testing. Thorax 50:955-61.
22. Morillo DS, Astorga S, Fernández Granero MA and León A. (2013) Computerized analysis of respiratory sounds during COPD exacerbations. Comput Biol Med. Published Online First: April 2013. DOI 10.1016/j.compbiomed.2013.03.011
23. Sánchez I and Vizcaya C. (2003) Tracheal and lung sounds repeatability in normal adults, Respir Med. 97:1257-60.
24. Zweig MH and Campbell G. (1993) Receiver-operating characteristic (ROC) plots: a fundamental evaluation tool in clinical medicine. Clin Chem. 39(4):561-577.
25. Chen S, Hong X and Harris CJ. (2005) Orthogonal forward selection for constructing the radial basis function network with tunable nodes. In, IEEE International Conference on Intelligent Computing 2005. Springer-Verlag, pp 777-86.
26. Orr Mark JL. (1995) Regularization in the Selection of Radial Basis Function Centers. Neural Computation 7:606-623.
27. Han MK, Agustí A, Calverley PM, et al. (2010) Chronic obstructive pulmonary disease phenotypes: the future of COPD Am. J. Resp. Crit. Care Med. 182:598–604.
28. Izquierdo-Alonso JL, Rodriguez-González Moro JM, de Lucas-Ramos P et al. (2013). Prevalence and characteristics of three clinical phenotypes of chronic obstructive pulmonary disease (COPD). Respir Med 107(5):724-731.

Corresponding author:
Author: Daniel Sánchez Morillo
Institute: University of Cädiz. School of Engineering
Street: C/Chile,1
City: Cádiz
Country: Spain
Email: daniel.morillo@uca.es

Response Detection in Narrow-Band EEG Using Signal-Driven Non-Periodic Stimulation

M. Cagy and A.F.C. Infantosi

Biomedical Engineering Program / Coppe, Federal University of Rio de Janeiro, Rio de Janeiro, Brazil

Abstract—Statistical Objective Response Detection (ORD) Techniques applied to Evoked Potentials usually depend on the effective Number of Degrees of Freedom (NDOF) of spontaneous EEG. Previous works proposed a method for estimating the NDOF of the *Evoked Potential Detector* (*EPD*) probability distribution under the null hypothesis of no response, based on the EEG autocorrelation function (ACF) when EEG is modeled as a wide-or narrow-band colored noise. Considering narrow-band EEG, the use of Beta distribution based on NDOF estimates via ACF is able to reflect EPD distribution provided that the stimulation rate is non-uniform. The present work assesses a simple real-time criterion based on FFT for randomizing the inter-trial interval using Monte Carlo simulation for several combinations of EEG bandwidths, central frequencies, numbers of epochs and number of samples per epoch. The results suggest such approach as a valid algorithm to be implemented in the development of specific devices to produce random inter-stimuli intervals.

Keywords—Objective Response Detection, Number of Degrees of Freedom, Evoked Potentials.

I. INTRODUCTION

Statistical methods aiming at Objective Response Detection (ORD) have been used in several applications for Evoked Potentials (EP) both in frequency [1-5] and time domains [6-8]. These techniques usually rely on establishing the probability distribution (PDF) for an estimated parameter under the null hypothesis (H$_0$) of no response, assuming spontaneous EEG as normally distributed (Gaussian). Once the estimated value exceeds the critical value for a previously chosen significance level (α), one can reject H$_0$ with a probability of Type-I error (false positive) lower than α.

However, the parameters that define these PDFs usually depend on the number of degrees of freedom (NDOF) of the recorded EEG. The simplest approach to determine the NDOF [1-7] assumes the EEG to be white Gaussian noise (uncorrelated time-samples and epochs). Nevertheless, real EEG suggests an accentuated colored nature hence such assumption is questionable. In this context, Elberling and Don [8] proposed adjusting the NDOF based on fitting the theoretical cumulative distribution function (CDF) for H$_0$ to the cumulative distribution of estimated values from a set of available EEG epochs without sensory stimulation, further adapted by Stürzebecher *et al.* [9] aiming at improving estimation of background noise variance. Alternatively, Lv *et al.* [10] estimated the CDF for H$_0$ related to several ORD techniques using bootstrap.

In previous work [6], CDF-fitted NDOF showed better results when compared to white-band assumption concerning the Evoked Potential Detector (*EPD* – a time-domain ORD technique) for detecting mid-latency auditory EP (MLAEP). However, the methodology based on distribution fitting shows two considerable limitations: (*i*) it demands the availability of a sufficiently large set of EEG epochs without stimulation, and (*ii*) it assumes that the unique initial estimated NDOF can be used for the whole experimental time-series. Aiming at overcoming these limitations, two previous works from our group [11,12] proposed and assessed a method to estimate the NDOF for the probability distribution of *EPD* based on the EEG autocorrelation function (ACF), taken from the same set of EEG epochs used in *EPD*, both by means of simulation and of real data.

Nevertheless, such approach does not strictly apply in cases where the background EEG has narrow band, which commonly occurs with alpha-activity in closed-eyes protocols. Hence, a previous study based on simulated data proposed the use of non-periodic stimulation aiming at overcoming such limitation [13]. Inter-trial intervals with a random component dependent on the EEG central frequency were employed based on *a priori* simulated data characteristics. However, aiming at a real signal application, random intervals should be adjusted during EEG acquisition, which demands real-time adaptation. Thus, the present work investigates, using simulated data, the application of a simple FFT-based central frequency estimator in order to establish the randomization criterion for inter-trial stimulation interval.

II. THEORETICAL BACKGROUND

The Evoked Potential Detector (*EPD*) is defined as:

$$EPD = \sum_{n=n_i}^{n_i+N-1}\left(\sum_{j=1}^{M} x_j[n]\right)^2 \bigg/ M \sum_{n=n_i}^{n_i+N-1}\sum_{j=1}^{M} \left(x_j[n]\right)^2, \quad (1)$$

where $x_j[n]$ represents the jth signal epoch (out of a total of M), within an N-sample time interval of interest ($n = n_i, n_i+1, ..., n_i+N-1$). Some insight about its meaning can be derived from the fact that its expected value is related to the Signal-to-Noise Ratio (SNR) according to $\mathrm{E}(EPD) = SNR/(SNR+1)$.

Thus the *EPD* can be interpreted as an estimate of the degree of response consistency within a set of noisy signal epochs, lying between zero (no response) and unity (no noise corrupting the response). For H_0, and assuming all $x_j[n]$ to be pure Gaussian white noises, the statistical distribution of *EPD* can be related to a Beta distribution (analog to *Magnitude Squared Coherence* except for the shape parameters [1,3]) with shape parameters v_1 and v_2 given by:

$$v_1 = N/2$$
$$v_2 = N(M-1)/2 \qquad (2)$$

Thus, H_0 can be rejected for a significance level α if the estimated *EPD* is higher than a critical value:

$$EPD_{crit} = Beta_{crit\, v_1, v_2, \alpha}, \qquad (3)$$

i.e. the critical value of the Beta distribution for a cumulative probability of $1-\alpha$.

On the other hand, for colored noise (more appropriate for the EEG), such shape parameters do not correspond to the data's NDOF and the resulting critical value (lower than it should be) would imply an actual Type-1 Error (false detection) risk higher than α. Hence, some kind of data-based NDOF correction must be conducted preceding the critical value calculation.

Let $x[n]$ be a stationary Gaussian signal with zero mean and variance σ^2, then for a finite number of samples N, an estimate of its ith moment can be obtained from [14]:

$$\hat{z}_i[N] = \frac{1}{N} \sum_{n=0}^{N-1} x^i[n]. \qquad (4)$$

Then the effective number of independent or uncorrelated random variables $x^i[n]$ within the interval N is:

$$k_i[N] = \frac{\mathrm{var}(x^i[n])}{\mathrm{var}(\hat{z}_i[N])}, \qquad (5)$$

which represents the ith-order NDOF. Based on the ACF ($r_{xx}[m]$), the second-order NDOF (since *EPD* is a quadratic technique) can be estimated from [14]:

$$k_2[N] = N \cdot r_{xx}^2[0] \left\{ \sum_{m=-N+1}^{N-1} \left(\frac{N-|m|}{N} \right) r_{xx}^2[m] \right\}^{-1} \qquad (6)$$

Assuming the correlation is negligible across epochs, values for v_1 and v_2 can be obtained by using $k_2[N]$ instead of N in the expression (2). Such assumption is quite valid when EEG is modeled as a wide-band colored noise, whence such approach yields an estimated Beta-CDF very close to the EPD distribution obtained by Monte Carlo simulation [12]. Nevertheless, such statistics is supposed to fail for non-negligibly correlated successive epochs.

III. MATERIALS AND METHODS

Several sets of simulated EEG signal were generated under different combinations of noise spectra (centered at $F_0 = 6, 8, 10, 12$ or 14 Hz, with varying bandwidths from $F_0/4$ down to $F_0/64$), epoch lengths (N: 100, 80, 60, 40 and 20) and numbers of epochs (M: 2, 4, 8, 16 and 32). Each set consisted of 1,000 trials with M epochs each one of Gaussian colored noise (4th-order band-pass Butterworth filters) "sampled" at $F_s = 100$ Hz. The NDOF were then estimated using unbiased ACF estimator (since it yielded better results than biased estimator [12]), according to the following expression:

$$\hat{r}_{xx}[k] = \hat{r}_{xx}[-k] = \frac{1}{M \cdot (N-k)} \sum_{j=1}^{M} \sum_{n=n_i}^{n_f-k} x_j[n] \cdot x_j[n+k], \qquad (7)$$

for $k = 0, 1, 2, \ldots, N-1$, applied to each trial. The expression (6) was first applied to each estimated ACF, so that statistical descriptors (mean, standard deviation, median and quartiles) of NDOF (denoted as k_2 for simplicity) could be estimated for each set of trials. Further, one global NDOF estimate (k_{2G}) was taken based on the grand-averaged ACF of the whole set of trials.

Based on [13], inter-trial intervals were determined as a fixed epoch length (during N samples) plus a randomized period aiming at decorrelating successive EEG epochs. Randomization followed a uniform distribution between 0 and F_s/F_0 samples, where F_0 represents the central frequency of the narrow-band EEG signal. In this study, F_0 was estimated directly from the peak frequency-bin of the magnitude-squared Fast Fourier Transform (FFT) of each epoch zero-padded up to complete 1,024 samples (in our case, resolution of circa 0.098 Hz). Hence, this estimation is used for determining the interval between the last applied "stimulus" and the next one to be applied, which demands a real-time algorithm embedded in a stimulation device to be developed with this aim.

IV. RESULTS

Initially, for sake of comparison with the proposed inter-trial interval algorithm, one used a blind value of F_0 as the central alpha band (10 Hz – since alpha-activity usually prevails in EEG using closed-eyes protocols). Considering $M = 4$ epochs of $N = 20$ samples, bandwidth was successively narrowed for each of the five values of F_0. Figure 1 shows that the estimated Beta-CDFs based both on the mean value of the trial-by-trial estimated NDOF ($\overline{k_2}$ – dashed lines) and on the grand-averaged ACF (k_{2G} – continuous) virtually superpose. On the other hand, as expected, if EEG has narrow-band spectrum, these CDFs only fit the EPD distribution for $F_0 = 10$ Hz, showing stronger discrepancy for other central frequencies when bandwidth (BW) gets narrower.

Figure 2 shows the results using the proposed inter-trial interval algorithm based on estimating F_0, so that no considerable discrepancy is found between estimated CDFs and EPD distribution for any of the combinations of F_0 and BW.

Fixing simulated EEG with central frequency of 8 Hz and BW = F_0/64, Figure 3 shows the results for several combinations of numbers of epochs (M) and epoch lengths (N). Except for M = 2, all simulated CDF curves could be approximated by the estimated NDOF. Curiously, for M = 2, EPD shows bimodal (N = 100, 80 or 60) or multimodal distributions (remaining values of N). This behavior results from the discrete phase shifts attainable by adding an integer random number of inter-epoch samples, an effect that becomes more noticeable for small number of epochs and of samples per epoch.

Table 1 compares the NDOF values obtained by Beta-curve fitting ($k_{2\,Fit}$) and by the grand-averaged ACF (k_{2G}) to those obtained trial-by-trial (k_2 – showing the mean value, standard deviation, median and quartiles) for M = 32 and 4. One can notice a close resemblance among these values for all combinations of M and N, with higher differences occurring for small values of M. Furthermore, based on the standard variation and the quartiles, the estimation variability has no consistent dependence on N but decreases for higher

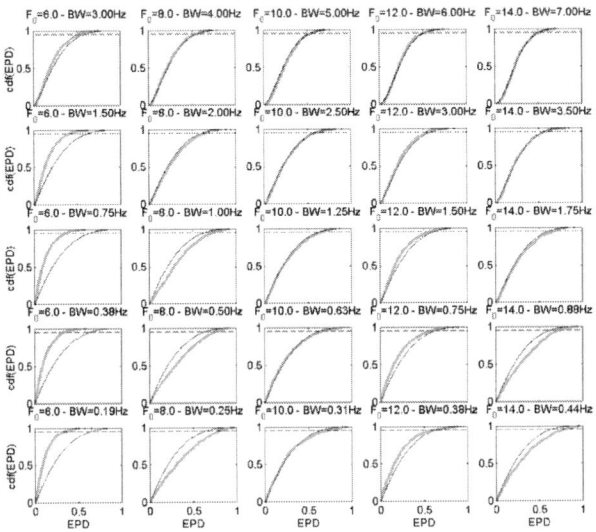

Fig. 1 EPD distribution (thick) and Beta-CDF based on $\overline{k_2}$ (dashed) and on k_{2G} (continuous), varying BW with N = 20 samples, without estimating F_0. Horizontal dashed line refers to 95% cumulative probability (α = 5%).

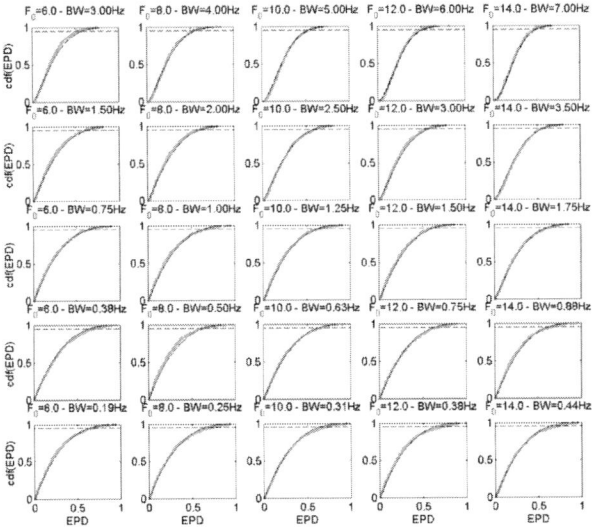

Fig. 2 EPD distribution (thick) and Beta-CDF based on $\overline{k_2}$ (dashed) and on k_{2G} (continuous), varying BW with N = 20 samples, using the proposed inter-trial interval algorithm based on F_0 estimation.

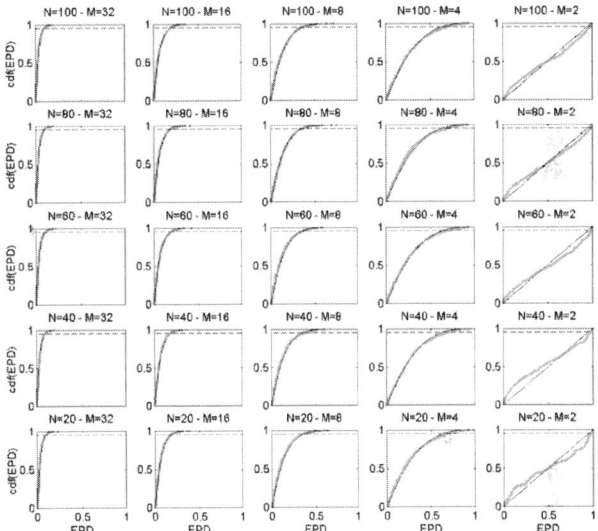

Fig. 3 EPD distribution (thick) and Beta-CDF based on $\overline{k_2}$ (dashed) and on k_{2G} (continuous) with EEG central frequency of 8 Hz and BW = F_0/64 for several combinations of numbers of epochs (M) and epoch lengths (N).

Table 1 Fitted ($k_{2\,Fit}$) vs. estimated NDOF from grand-averaged ACF (k_{2G}) and trial-by-trial ACF (k_2 – showing mean, standard deviation, median and quartiles), for M = 32 and 4.

M	N	$k_{2\,Fit}$	k_{2G}	$\overline{k_2}$	s.d.(k_2)	$P_{25}(k_2)$	Med(k_2)	$P_{75}(k_2)$
	100	2.1123	2.0470	2.0489	0.0164	2.0373	2.0465	2.0583
	80	1.9132	2.0295	2.0311	0.0157	2.0199	2.0300	2.0412
32	60	2.0869	2.0154	2.0169	0.0227	2.0017	2.0161	2.0321
	40	1.8686	2.0024	2.0035	0.0294	1.9844	2.0035	2.0229
	20	1.9882	1.9964	1.9954	0.0376	1.9705	1.9955	2.0239
	100	1.9145	2.0458	2.0655	0.0847	2.0167	2.0368	2.0814
	80	1.8776	2.0275	2.0467	0.0781	2.0052	2.0262	2.0608
4	60	1.7923	2.0123	2.0261	0.0653	1.9819	2.0211	2.0591
	40	1.7496	2.0041	2.0151	0.0799	1.9642	2.0114	2.0651
	20	1.848	2.0010	1.9922	0.0892	1.9292	2.0060	2.0591

M, but keeps relatively small even for $M = 4$ (s.d. between circa 3% and 5% of the mean). Also, the estimated NDOF shows a slight increasing trend as N increases.

V. DISCUSSION AND CONCLUSION

The adjustment of the Beta distribution based on NDOF estimates via ACF was able to reflect EPD distribution provided that the correlation among successive epochs is somehow broken. Such decorrelation can be achieved if stimulation rate is non-uniform, simply adding a criteriously randomized inter-stimuli interval variability over a minimum basal period. This criterion relies on the real-time estimation of the central frequency of background EEG based on Fast Fourier Transform.

The cumulative EPD distribution fitted the estimated Beta-CDF for numbers of epochs (M) as small as 4. Considering that the number of epochs collected in waveform-based EP tests (particularly the mid-latency auditory EP [MLAEP]) is usually higher than 100, the present investigation opens a promising perspective for fast EP detection (small values of M) even when background EEG shows narrow-band spectral distribution, a scenario commonly seen in closed-eyes protocols. Indeed, the NDOF for establishing the statistical distribution of EPD for the null hypothesis of no response, as well as the criterion for the inter-trial interval randomization, could be estimated directly from the collected EEG. Nevertheless, further investigation concerning the Detection Probability (Power of Test) of EPD must be performed considering the alternative hypothesis (presence of response).

Random inter-stimuli intervals have already been applied to auditory EP aiming at reducing the evaluation time by allowing overlap among successive responses (and consequently shorter inter-stimuli intervals) based on a scheme of Maximum-Length Sequences [15,16]. Also, it has been used to minimize the negative effect resulting from the overlap between late potentials and the interesting responses to subsequent stimuli by applying a stimulation interval distribution based on *a priori* and spectral information about the desired EP [17]. Such stimulation schemes should also be investigated within the context of estimating the EEG's NDOF via ACF as possible approaches aiming at breaking inter-epoch correlation.

Although EPD is a time-domain ORD technique, such reasoning is supposed to be valid also for frequency-domain methods. This result supports the development of specific devices to produce random inter-stimuli intervals, since most of the commercially available devices do not implement it. Indeed, it explains why the present study has not assessed real signal, which is its main limitation.

ACKNOWLEDGMENT

This work received financial support from FAPERJ and the Brazilian Research Council (CNPq).

REFERENCES

1. Dobie R A, Wilson M J (1993) Objective Response Detection in the Frequency Domain. Electroencephalography and Clinical Neurophysiology, v.88, p.516-524.
2. Cagy M, Infantosi A F C, Gemal A E (2000) Monitoring Depth of Anaesthesia by Frequency-Domain Statistical Techniques. Brazilian Journal of Biomedical Engineering, v.3, n.2, p.95-107.
3. Cagy M, Infantosi A F C (2007) Objective Response Detection Technique in Frequency-Domain for Reflecting Changes in MLAEP. Medical Engineering & Physics, v.29, n.8, p.910-917.
4. Victor J D, Mast J (1991) A New Statistic for Steady-State Evoked Potentials. Electroencephalography and Clinical Neurophysiology, v.78, p.378-388.
5. Valdés J L, Pérez-Abalo M C, Martín V, Savio G, Sierra C et al. (1997) Comparison of Statistical Indicators for the Automatic Detection of 80 Hz Auditory Steady State Response. Ear and Hearing, v.18, p.420-429.
6. Cagy M, Infantosi A F C, Zaeyen E J B (2007) Detecting the Mid-latency Auditory Evoked Potential during Stimulation at Several Different Sound Pressure Levels. In: IFMBE Proceedings, v.18, p.42-45.
7. Cagy M, Infantosi A F C, Zaeyen E J B (2007) Assessing FSP Index Performance as an Objective MLAEP Detector during Stimulation at Several Sound Pressure Levels. IFMBE Proceedings, v.16, p.492-496.
8. Elberling C, Don M (1984) Quality estimation of averaged auditory brainstem responses. Scandinavian Audiology, v.13 p.187-197.
9. Stürzebecher E, Cebulla M, Wernecke K D (2001) Objective Detection of Transiently Evoked Otoacoustic Emissions. Scandinavian Audiology, v.30, p.78–88.
10. Lv J, Simpson D M, Bell S L (2007) Objective Detection of Evoked Potentials Using a Bootstrap Technique. Medical Engineering & Physics, v.29, p.191-198.
11. Cagy M, Infantosi A F C, Zaeyen E J B (2008) Detection of Evoked Potential Embedded in EEG with Time-Varying Autocorrelation Function. In: Anais do 21o. Congresso Brasileiro de Engenharia Biomédica, p.1631-1634.
12. Cagy M, Infantosi A F C, Miranda de Sá A M F L, Simpson D M (2009) Statistical Evoked Potential Detection with Number of Degrees of Freedom Estimated from EEG Autocorrelation Function. In: IFMBE Proceedings, v.25, n.4, p.2193–2196.
13. Cagy M, Infantosi A F C (2012) Non-Periodic Stimulation for Detection of Evoked Potentials Embedded in Narrow-Band EEG", In: Anais do XXIII Congresso Brasileiro de Engenharia Biomédica, p.1918-1922.
14. Kikkawa S, Ishida M (1988) Number of Degrees of Freedom, Correlation Times, and Equivalent Bandwidths of a Random Process. IEEE Transactions on Information Theory, v.34, n.1, p.151-155.
15. Eysholdt U, Schreiner C (1982) Maximum Length Sequences – A Fast Method for Measuring Brain-Stem-Evoked Responses. Audiology, v.21, p.242-250.
16. Bell S L, Smith D C, Allen R, Lutman M E (2006) The Auditory Middle Latency Response, Evoked Using Maximum Length Sequences and Chirps, as an Indicator of Adequacy of Anesthesia. Anesthesia & Analgesia, v.102, p.495-498.
17. Nait-Ali A M, Adam O, Motsch J-F (1997) On Optimal Aperiodic Stimulation for Brainstem Auditory Evoked Potentials Estimation. In: Proceedings of the 19th International Conference IEEE/EMBS, p.1492-1495.

Author: Antonio Fernando Catelli Infantosi
Institute: Biomedical Engineering Program (COPPE/UFRJ)
Street: P. O. Box 68.510
City: Rio de Janeiro
Country: Brazil
Email: afci@peb.ufrj.br

EMG-Based Analysis of Treadmill and Ground Walking in Distal Leg Muscles

F. Di Nardo and S. Fioretti

Department of Information Engineering, Università Politecnica delle Marche, 60131 Ancona, Italy

Abstract—Differences between treadmill and ground walking have been suspected by some authors. To check this hypothesis, the present study compared treadmill with ground walking in terms of surface electromyographic (sEMG) signal of Tibialis Anterioris (TA) and Gastrocnemius Lateralis (GL), from a large number (hundreds) of strides per subject. The analysis on seven healthy young adults showed no substantial variation in TA activity during treadmill walking, with respect to ground walking. An earlier and increased activity of GL during the transition between Flat foot contact and Push-off phases was detected during treadmill walking with respect to ground walking in terms of both the frequency of GL recruitment and area under the curve of GL profiles; this increase is likely to be related with the reported increase of magnitude of vertical forces during mid-stance in the treadmill compared with ground walking. Incorporating this findings into gait-analysis strategies could lead to a more accurate evaluation of results achieved by the use of treadmill.

Keywords—gait analysis, treadmill, gastrocnemius lateralis, tibialis anterior, surface electromyography.

I. INTRODUCTION

The use of treadmill in gait analysis offers a number of potential advantages: it permits locomotion within a small area, facilitates use of various types of monitoring equipment that require attachments to the subject and, above all, allows to impose well controlled speed constraints for a more meaningful comparison of parameters between sessions. Although several studies have been performed [1]-[3], poor information is available on the differences between treadmill and ground walking in terms of the surface electromyographic (sEMG) signal. The aim of the present study was to compare treadmill walking with ground walking in terms of EMG signal of antagonist muscles of distal lower limb, i.e. tibialis anterior and gastrocnemius lateralis. Statistical analysis of sEMG signal from a very high number (hundreds) of consecutive strides per each subject was performed to this goal. The study is based on the recent availability of robust techniques for the detection of muscle activation intervals, and specific tools for statistical gait analysis [4]-[6].

II. MATERIAL AND METHODS

A. Subjects

Seven healthy adult volunteers were recruited; mean (±SE) values were 23.0±0.5 years for age; 168±3 cm for height; 56.1±3.8 kg for weight and 19.7±0.6 for body mass index (BMI). The EMG activity during gait was recorded in both right and left lower limb of all subjects at self-selected speed over ground and treadmill. Exclusion criteria included history of neurological pathology, orthopedic surgery within the previous year, acute or chronic knee pain or pathology, BMI ≥25, or abnormal gait. Before the beginning of the test, all participants signed informed consent.

B. Recording System: Signal Acquisition and Processing

Signals were acquired by means of a multichannel recording system for statistical gait analysis (Step32, DemItalia, Italy). Each subject was instrumented with foot-switches, goniometers and sEMG probes. Three foot-switches were attached beneath the heel, the first and the fifth metatarsal heads of each foot.

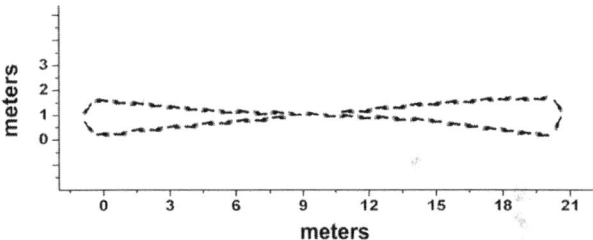

Fig. 1 Schematic representation of the path walked by the recruited subjects during the experiment; subjects walked barefoot over the floor for 5 minutes at their natural pace.

A goniometer (accuracy: 0.5 deg) was attached to the lateral side of each lower limb for measuring the knee joint angle in the sagittal plane. Single differential sEMG probes with fixed geometry were attached over the Tibialis Anterior (TA) and Gastrocnemius Lateralis (GL) of each lower limb, following the SENIAM recommendations [7]. After positioning the sensors, subjects were instructed to walk barefoot over ground for around 4 minutes at their natural pace, following the path schematized in Fig. 1. Natural pace was chosen because walking at a self-selected speed improves the repeatability of EMG data, while variability increases when subjects are required to walk abnormally [8]. Walking speed was measured and reported on the treadmill where subjects were asked to walk barefoot for a training period and then for the trial of around 4 minutes. Foot-switch signals were converted to four levels corresponding to Heel contact (H), Flat foot contact (F), Push off (P), Swing (S) and processed to segment and classify the different gait cycles. During acceleration, deceleration, and changes in direction the strides are different from those of steady state walking. Therefore, goniometric signals (low-pass filtered with cut-off frequency of 15 Hz) along with gait phase durations, were used by a multivariate statistical filter embedded in the Step32 system, to detect and discard outlier cycles, i.e., cycles with the proper sequence of gait phases (H-F-P-S) but with abnormal timing, like those relative to deceleration, reversing, and acceleration [9]-[10]. EMG signals were high-pass filtered (cut-off frequency of 20 Hz) and then processed by a double-threshold statistical detector, embedded in the Step32 system, that provides the onset and offset time instants of muscle activity in a completely user-independent way [4]. To quantify the levels of myoelectric activity during gait, areas under the normalized (i.e. divided for peak value) EMG profiles were calculated over the gait cycle for each subject (in both lower limbs) and for each muscle.

In the present study only gait cycles consisting of the sequence of H–F–P–S were considered.

III. RESULTS

For each subject a mean (±SE) of 497±49 strides during ground walking and 535±27 strides during treadmill walking has been considered. The H-phase lasts 5.9±0.5% of the gait cycle, the F-phase 24.6±1.8%, the P-phase 24.0±1.5% and the S-phase 45.5±0.7%.

For TA (Fig 2. upper panel), a strong correlation was detected between areas under the curve (AUC) of EMG signals collected during ground and treadmill walking (TA-AUC, r = 0.96, p < 0.001); the slope of the regression line was very close to 1 (0.97) and the y-intercept was very close to 0 (-0.77).

Moreover, the TA activation intervals during ground and treadmill walking were substantially superimposed and showed a similar frequency of occurrence.

Also for GL (Fig 2. lower panel), a good correlation was detected between areas under the curve of mean EMG signals collected in each subject during ground and treadmill walking (GL-AUC, r = 0.84, p < 0.001) and the slope of the regression line was close to 1 (0.93); however, the y-intercept was greater than zero (11.8). On average, GL activation interval ranged from 4 to 53 % of gait cycle during ground walking, and from 1 to 64 % of gait cycle during treadmill walking (Fig. 3). In stance phase, the mean muscle activations in function of gait cycle showed an increase of GL-AUC of EMG signals collected during treadmill (dashed line) and ground (continuous line) walking; this increase occurred between 1 and 38 % of gait cycle and

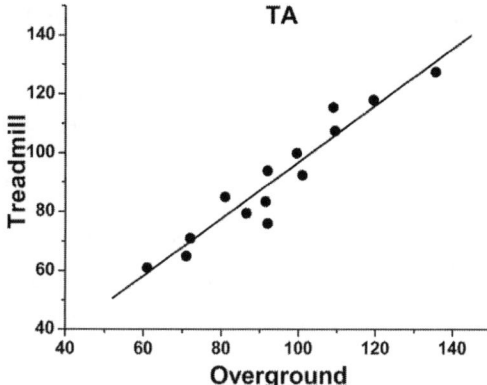

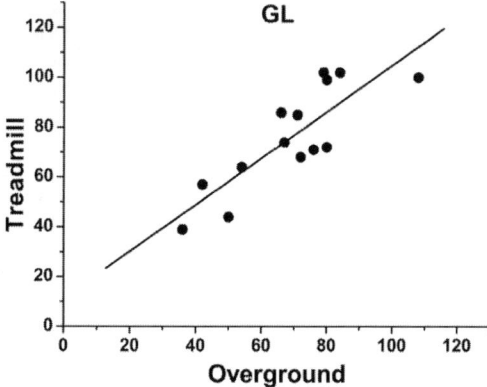

Fig. 2. Linear regression between areas under the curve (AUC) of all sEMG signals collected during ground and treadmill walking for TA (upper panel, r = 0.96, p < 0.001) and GL (lower panel, r = 0.84, p < 0.001).

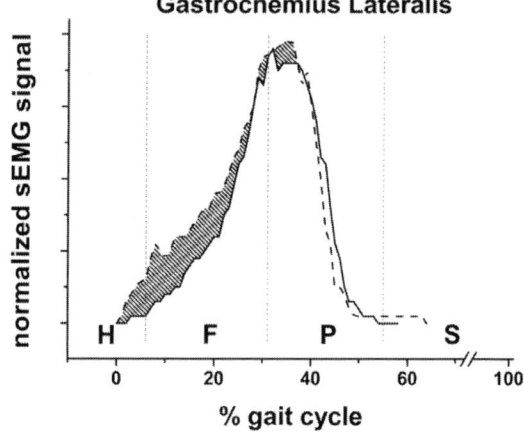

Fig. 3. Average patterns of GL activation vs. percentage of gait cycle during treadmill (dashed line) and ground (continuous line) walking. Grey area represents the difference between dashed and continuous line between 1 and 38 % of gait cycle. H=heel contact; F=flat foot contact; P=push off; S=swing. All average data refer to all recorded signals.

is represented in Fig. 3 as the difference (grey area) between the dashed and the continuous line.

No significant differences in mean (±SE) onset and offset instants of GL activation were detected during stance phase between ground (on: 12.6±2.4; off: 43.2±1.7) and treadmill walking (on: 9.6±2.4; off: 43.5±0.8). Between 3 and 20% of the gait cycle, a GL activity is observed in the 44.9±5.5% of total population for ground and in the 70.2±4.0% for treadmill walking; between 20 and 45% of gait cycle, a GL activity is observed in the 84.0±1.9% of total population for ground and in the 85.1±3.3% treadmill walking. Mean values are computed over the 14 signals recorded in our population, considering data from both right and left lower limb.

IV. DISCUSSION

Some authors suspected that locomotion on a treadmill could be different from ground walking [1]-[3]. To check this hypothesis, the present study compared treadmill walking with ground walking in terms of surface EMG signal of antagonist muscles of distal lower limb, Tibialis Anterioris (TA) and Gastrocnemius Lateralis (GL). A statistical analysis of a large number of strides (approximately 500) for each subject was performed to this aim. The muscle activation intervals, detected in each subject at self-selected walking speed, followed roughly the typical pattern reported for each muscle during gait [11].

Overlapping of TA activation intervals (both in terms of onset and offset instants and frequency of occurrence), strong correlation between ground and treadmill TA-AUC, slope of the regression line very close to 1 and y-intercept of the regression line very close to 0 indicated no substantial variation in TA activity during treadmill walking, with respect to ground walking.

On the other hand, the mean increase of GL-AUC in treadmill data, quantified by a y-intercept of regression line clearly greater than zero, suggests a greater involvement of the GL during treadmill walking with respect to ground walking. The grey area in Fig. 3 shows as this increased EMG activity of GL occurred, above all, during the Flat foot contact phase (F) and in the transition between this phase and the following Push-off phase (P); this increase is likely to be related with the reported 5-9% increase of the magnitude of vertical forces during mid-stance in the treadmill, compared with ground walking [12]. Considering the mean values (±SE) of onset and offset instants of activations, no significant differences were detected between treadmill and ground walking for both TA and GL, as reported in literature [1]. However, the present study compared treadmill and ground walking also in terms of the frequency of recruitment, i.e. the number of times, over total subjects, where the muscle is active in the selected interval of gait cycle (Fig. 4). Results showed that no differences were detected in terms of frequency of TA recruitment between treadmill and ground walking. Although no differences were detected in the 20-45% of gait cycle, the recruitment of GL during 3-20% of gait cycle is 25% more frequent during treadmill than ground walking (Fig. 4). This percentage increases to 35%, when the analysis is limited to the 5-10% interval. This suggests that the modality of walking over a treadmill requires an earlier involvement of GL in the beginning of Flat foot contact phase (F), with respect to ground walking.

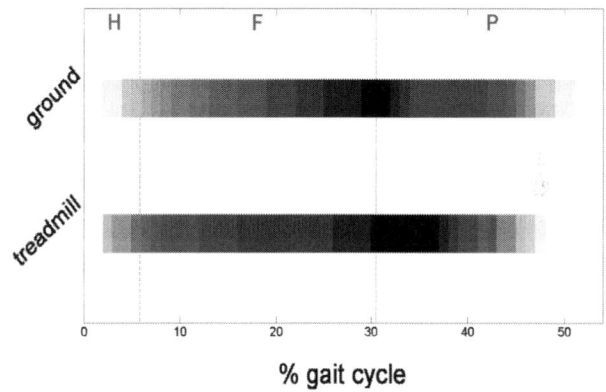

Fig. 4 Patterns of activation intervals for GL during stance phase, in both ground and treadmill walking, as percentage of gait cycle. Horizontal bars are grey-level coded, at each percent of the gait cycle, according to the number of subjects where a certain condition is observed; black: condition observed for all subjects, white: condition never met. The phases of Heel contact (H), Flat foot contact (F) and Push off (P), are delimited by dashed vertical lines.

V. CONCLUSIONS

The study was designed to describe the effect of treadmill walking on the surface electromyographic signal of tibialis anterior and gastrocnemius lateralis. The large number of strides considered for each participant allowed to stress that no substantial variation is observed in TA activity during treadmill walking, with respect to ground walking. However, the present study suggests that the modality of walking over a treadmill requires an earlier and increased involvement of GL in the beginning of Flat foot contact phase (F), with respect to ground walking.

Though these results need to be interpreted with caution, incorporating them into gait-analysis strategies could lead to a more accurate evaluation of results achieved by the use of treadmill.

REFERENCES

[1] Arsenault AB, Winter DA, Marteniuk RG (1986) Treadmill versus walkway locomotion in humans: an EMG study. Ergonomics 29:665-76

[2] Murray MP, Spurr GB, Sepic SB et al. (1985) Treadmill vs. floor walking: kinematics, electromyogram, and heart rate. J Appl Physiol 59:87-91

[3] Stoquart G, Detrembleur C, Lejeune T (2008) Effect of speed on kinematic, kinetic, electromyographic and energetic reference values during treadmill walking. Arch Phys Med Rehabil 89:56-61

[4] Bonato P, D'Alessio T, Knaflitz M (1998) A statistical method for the measurement of muscle activation intervals from surface myoelectric signal during gait. IEEE Trans Biomed Eng 45:287-299

[5] Staude G, Flachenecker C, Daumer M, Wolf W (2001) Onset detection in surface electromyographic signals: a systematic comparison of methods. EURASIP J Appl Signal Process 2:67-81

[6] Di Nardo F, Fioretti S (2013) Statistical analysis of surface electromyographic signal for the assessment of rectus femoris modalities of activation during gait. J Electromyogr Kinesiol 23:56-61

[7] Hermens HJ, Freriks B, Merletti R et al. (1999) European recommendations for surface electromyography. Roessingh Research and Development, Enschede, the Netherlands

[8] Kadaba MP, Ramakrishnan HK, Wootten ME et al. (1989) Repeatability of kinematic, kinetic, and electromyographic data in normal adult gait. J Orthop Res 6:849-860.

[9] Agostini V, Knaflitz M (2012) Statistical Gait Analysis in Distributed Diagnosis and Home Healthcare. Edited by Rajendra Acharya U, Filippo M, Toshiyo T, Subbaram Naidu D, Jasjit SS. Stevenson Ranch, California, (USA):American Scientific Publishers 99–121. Volume 2.

[10] Di Nardo F, Ghetti G, Fioretti S. Assessment of the activation modalities of gastrocnemius lateralis and tibialis anterior during gait: a statistical analysis. J Electromyogr Kinesiol, ahead of print

[11] Perry J (2012) Gait Analysis - Normal and Pathological Function. USA: Slack Inc

[12] White SC, Yack HJ, Tucker CA, Lin HY (1998) Comparison of vertical ground reaction forces during overground and treadmill walking. Med Sci Sports Exerc 30:1537-42

Temporal Variation of Local Fluorescence Sources in the Photodynamic Process

I. Salas-García, F. Fanjul-Vélez, N. Ortega-Quijano, and J.L. Arce-Diego

Applied Optical Techniques Group, TEISA Department, University of Cantabria, Santander, Spain

Abstract—Photodynamic Therapy is an optical treatment modality with promising results to overcome different types of non melanoma skin cancers. Predictive models have become a valuable tool to mimic a real treatment situation and therefore offer the possibility to anticipate the therapeutic response under different clinical scenarios. Regarding photodynamic therapy this is especially important due to actual clinical protocols do not allow to adjust the treatment parameters to provide a custom therapy for non melanoma skin cancer. Despite the fact that inter patients responses vary even under the same fixed dosimetric parameters, nowadays the clinical protocol is applied in the same manner regardless of the patient variability or the type of lesion. This paper presents a predictive model for photodynamic therapy that takes into account the main photophysical and photochemical processes involved in this treatment as well as the generation of the photosensitizer fluorescence. Afterwards the model is applied to different skin tumors and the fluorescence power emission is obtained. The results show the spatial and temporal variation of fluorescence emission sources as the treatment progresses. Future works will deal with the distribution of the fluorescence signal in an heterogeneous media taking into account the sources of fluorescence here obtained.

Keywords—Photodynamic therapy, photosensitizer fluorescence, predictive model, non melanoma skin cancer.

I. Introduction

Photodynamic Therapy (PDT) is an optical treatment modality with promising results to overcome different types of non melanoma skin cancers. Its therapeutic effect lies in the photochemical reactions that take place in the tumor tissue when the light of an optical source interacts with a photosensitized tissue [1]. Despite of its advantages (e.g. non invasive technique, tumor selectivity, use of non ionizing radiation) respect to conventional chemotherapy, radiotherapy or surgery, the biological media complexity makes nowadays difficult to provide a customized treatment. Specific clinical protocols for each photosensitizer are applied under fixed dosimetric parameters that do not take into account the variability among patients or pathologies. However clinical experience has demonstrated that the treatment response varies depending on the type of disease even on the patient. The predictive models have become a valuable tool to mimic a real treatment situation and therefore offer the possibility to anticipate the therapeutic response under different clinical scenarios. Thus it is possible to estimate a priori the distribution of the laser light in the biological tissue [2], the evolution of the molecular components involved in the photochemical reactions [3], the distribution of the photosensitizer in the tissue [4] or the necrotic area produced [5]. On the other hand the photosensitizer ability to emit fluorescence provides a way to track the photosensitizer accumulation in the target tissue and its degradation during PDT [6]. The knowledge of this information is a key factor to ensure that the amount of photosensitizer will be enough for a proper evolution of the photochemical process involved in PDT due to a strong limitation in the concentration of photosensitizer would stop the treatment effects, specifically the generation of singlet oxygen in charge of destroying the tumor tissue. Thus modeling the fluorescence signal emitted by the photosensitizer is especially interesting to predict the evolution of the photosensitizer according to the clinical parameters employed as well as to interpret the photosensitizer fluorescence based monitoring. Both of them are essential to achieve a future personalized treatment which incorporates a proper planning and in situ monitoring of the photodynamic process for each patient.

The present work is devoted to a predictive model that takes into account the main photophysical and photochemical processes involved in PDT as well as the generation of fluorescence during the treatment application. Section II describes the main parts of the model. Then the model proposed is applied to different skin diseases and the results obtained are discussed in Section III. Finally in Section IV we summarize the conclusions of this work.

II. Model

The predictive model employed consists of four main parts. The distribution of the excitation optical radiation was obtained by means of a Monte Carlo approach as it is presented in subsection *A*. Then subsection *B* shows the calculation of the photosensitizer distribution in the target tissue before being irradiated. The photochemical reaction evolution is presented in subsection *C* and finally subsection *D* provides the basis to obtain the photosensitizer fluorescence.

A. Distribution of the Excitation Optical Radiation

In order to obtain the optical radiation distribution in a three-dimensional tissue we used the Radiation Transport Theory (RTT). According to RTT, radiation from a particle attenuates due to absorption and scattering and also gains power because another particle can scatter light in the direction of the particle of interest. In a steady-state situation and without sources inside the tissue, this can be written as in equation (1) where the radiation is expected to be at the point $\vec{r}$, and to follow the direction $\hat{s}$ [1].

$$\hat{s} \cdot \vec{\nabla} I(r,\hat{s}) = -(\mu_a + \mu_s)I(r,\hat{s}) + \frac{\mu_s}{4\pi}\int_{4\pi} p(\hat{s} \cdot \hat{s}')I(r,\hat{s}')d\Omega' \quad (1)$$

$I(r,\hat{s})$ is the specific intensity defined as the light power per unit area per unit solid angle, $p(\hat{s},\hat{s}')$ is the scattering phase function that contains the probabilities of light to be scattered in the different directions μ_a is the absorption coefficient and μ_s the scattering coefficient. The numerical analysis of the RTT was performed by means of the implementation of the Monte Carlo method proposed by Wang and Jacques [2, 7].

B. Endogenous Generation of the Photosensitizer for a Topical Photosensitizer Precursor

Some photosensitizers are endogenously generated in tumor cells once the photosensitizer precursor applied in the skin surface accumulates in the target tissue. This is the case of the photosensitizer Metvix® employed in this work, which contains the prodrug MAL (Methyl aminolevulinate) and is synthesized to PpIX (Protoporphyrin IX) in tumor cells. Both the precursor distribution and the generation of the photoactive element were obtained as Svaasand et al previously proposed in 1996 for a topical photosensitizer [4]. Thus the prodrug distribution can be modeled by means of the Fick's law and the photosensitizer generation taking into account mainly the prodrug concentration and the yield of the conversion process. Equation (2) provides the temporal evolution of the precursor, M, distribution and (3) the photosensitizer concentration, S_0, when the time of removal of the active compound is fast compared to the diffusion time of the precursor. That is, the concentration of PpIX, is proportional to the instantaneous concentration of MAL. In these expressions D is the diffusion coefficient, M_o is the prodrug concentration in the skin surface at t=0, z is the distance from the stratum corneum located at z=0 and characterized by means of its permeability K, τ is the elimination time of the prodrug, τ_p is the time of removal of the active compound, $\tau_{a \to p}$ is the time spent in the generation of the photosensitizer and ε_p is the yield of the conversion process

$$M(t) = M_o \int_0^t \left(\frac{K}{\sqrt{D\pi t'}} e^{-\frac{z^2}{4Dt'}} - \frac{K^2}{D} e^{\frac{K}{D}z} e^{\frac{K^2}{D}t'} \cdot erfc\left(\frac{K}{\sqrt{D}}\sqrt{t'} + \frac{z}{2\sqrt{Dt'}}\right) \right) e^{-\frac{t'}{\tau}} dt' \quad (2)$$

$$S_0(t) = \varepsilon_p \frac{\tau_p}{\tau_{a \to p}} M(t) \quad (3)$$

C. Photochemical Interaction

The photochemical interaction among the photosensitizer, the light delivered by the optical source and the oxygen within the tissue generates the cytotoxic agent or singlet oxygen, in charge of the tumor cells destruction. Such interaction is modeled by means of a stiff differential equations system (4)-(9) that provides the temporal evolution of the molecular components involved such as the photosensitizer in ground state, singlet excited state and triplet excited state, S_0, S_1 and T respectively, the oxygen in ground state $^3O_2$, the singlet oxygen $^1O_2$ and the singlet oxygen receptors R [3, 8]. A detailed description of each parameter can be found elsewhere [5, 8]. The solutions of the stiff differential equations system employed are obtained by means of a differential equation solver within the Matlab® platform.

$$\frac{d[S_0]}{dt} = -\nu\rho\sigma_{psa}[S_0] - kpb[^1O_2][S_0] + \frac{\eta_{10}}{\tau 1}[S_1] + \frac{\eta_{30}}{\tau 3}[T] + \frac{\alpha s}{\tau 3}[T][^3O_2] \quad (4)$$

$$\frac{d[S_1]}{dt} = -\frac{1}{\tau 1}[S_1] + \nu\rho\sigma_{psa}[S_0] \quad (5)$$

$$\frac{d[T]}{dt} = -\frac{\eta_{30}}{\tau 3}[T] - \frac{\alpha s}{\tau 3}[T][^3O_2] + \frac{\eta_{13}}{\tau 1}[S_1] \quad (6)$$

$$\frac{d[^3O_2]}{dt} = -\frac{\alpha s}{\tau 3}[T][^3O_2] + \frac{\eta_0}{\tau 0}[^1O_2] + P \quad (7)$$

$$\frac{d[^1O_2]}{dt} = -kpb[S_0][^1O_2] - kcx[R][^1O_2] - ksc[C]d[^1O_2] - \frac{\eta_0}{\tau 0}[^1O_2] + \frac{\alpha s}{\tau 3}[T][^3O_2] \quad (8)$$

$$\frac{d[R]}{dt} = -kcx[^1O_2][R] + U \quad (9)$$

D. Photosensitizer Fluorescence

The photosensitizer molecules also act as fluorescence sources. This ability is of great interest from the point of view of the treatment planning and monitoring due to the fluorescence emission is directly related to the amount of photosensitizer accumulated in the tissue [8]. The fluorescent power source associated with each of the incremental volumes previously employed to calculate distribution of the excitation optical radiation is obtained as in equation (10). Where v is the light speed in the tissue, ρ is the density of photons present at a point, σ_{psa} is the absorption cross-section of the photosensitizer molecules, η_{10} is the fluorescence quantum yield, $E_{photon_{\lambda em}}$ is the photon energy at the emission wavelength and V is the volume associated with each element of the computation grid employed in a cylindrical coordinate system.

$$P_f(r,z) = v \cdot \rho(r,z) \cdot \sigma_{psa} \cdot [S_0](r,z) \cdot \eta_{10} \cdot E_{photon_{\lambda em}} \cdot V(r,z) \quad (10)$$

III. RESULTS AND DISCUSSION

The model previously presented was used to obtain the fluorescence power emission as the photodynamic treatment progresses and then it was applied to three different skin tumors: basal cell carcinoma (BCC), infiltrative basal cell carcinoma (IBCC) and squamous cell carcinoma (SCC).

The tissue structure considered is composed of an upper layer of tumor tissue of 3 mm lying on normal tissue. The optical properties of the tumor and normal tissues for an excitation wavelength of 633 nm are summarized in Table 1. The photosensitizer accumulation in the target tissue was obtained taking into account the MAL-PpIX diffusion properties and a radius of 1 cm for the superficial lesion extension. The photosensitizer accumulation was assumed to be the same for all the pathologies. The photochemical interaction was also modeled taking into account the photochemical properties of PpIX and assuming an initial homogeneous oxygen distribution in the whole tissue sample.

Table 1 Optical properties of the tissues for λ=633 nm (n: index of refraction, g: anisotropy of scattering)

Type of tissue	n	μ_a [cm^{-1}]	μ_s [cm^{-1}]	g	Layer Depth [mm]
BCC	1.5	1.5	104.76	0.79	3
IBCC	1.5	1.5	142.85	0.79	3
SCC	1.5	2	95.238	0.79	3
Normal	1.37				

The fluorescent power sources associated with each element of the computation grid were obtained for different temporal instants during a treatment time of 10 minutes, a fluorescence quantum yield of 0.2 and considering a cylindrical optical beam with a radius of 0.3 cm perpendicular to the tissue sample to deliver an irradiance of 100 mW/cm². Figure 1 shows the variation of the fluorescence power sources with the depth in the tissue at different temporal instants during the PDT progression in a squamous cell carcinoma. As it can be observed the photosensitizer fluorescence power is highly variable not only with the depth in the tumor tissue but also with the time of treatment.

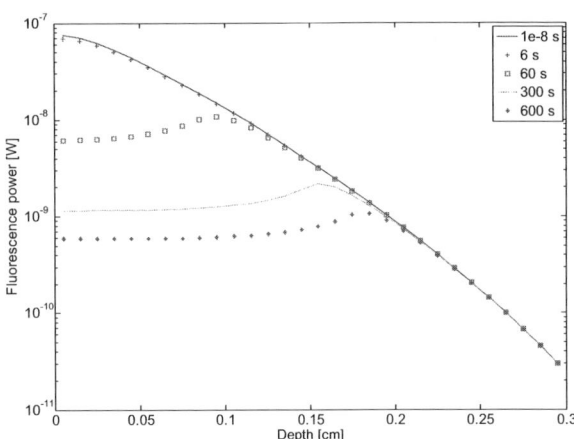

Fig. 1 Photosensitizer fluorescence power [W] emited at different instants of time during PDT vs. depth in the tumor tissue at r=0.0125 cm.

The results clearly show the reduction of the fluorescence power emission in the tumor tissue as PDT progresses and the photosensitizer degradation becomes stronger during the photochemical reaction. However, as the radiation time increases, this fluorescence comes mostly from the deeper tumor layers due to a stronger optical absorption in the superficial layers that accelerates the photosensitizer degradation making it faster than in the deeper tissue. Thus at the beginning of the radiation period the photosensitizer fluorescence power is maximum in the most superficial layers whereas at the end of treatment, t=600 s, it is higher at 0.185 mm from the tumor surface. The relationship between the power of fluorescence sources and the concentration of the photosensitizer provides a way to track its spatial evolution and inherently the potential locations for the production of the cytotoxic agent (i.e. the singlet oxygen). Therefore the singlet oxygen effects will begin in the outer tissues extending to the innermost zones that will be the last to be destroyed whenever a quick degradation of the photosensitizer does not halt the proper supply to trigger the photochemistry involved.

No significant differences were found in the fluorescence power emitted for the three skin diseases previously indicated as it is demonstrated in Figure 2 for the three types of pathology at the beginning a) and the end b) of the treatment. Therefore the results suggest a similar treatment response for the assessed situations as it could be expected due to the optical properties of the three types of tissue that determine the optical energy distribution are almost equal and the photosensitizer distribution and oxygen level were assumed to be the same. However these preliminary results should be interpreted carefully due to the complexity of the biological media and the assumptions taken. Facing this situation, the heterogeneity of different skin tumors could significantly modify a great amount of parameters such as the photosensitizer diffusion, the oxygen level or the light distribution and that is why the application of an accurate predictive modeling as well as a proper election of the implicated parameters gains relevance to anticipate their implications.

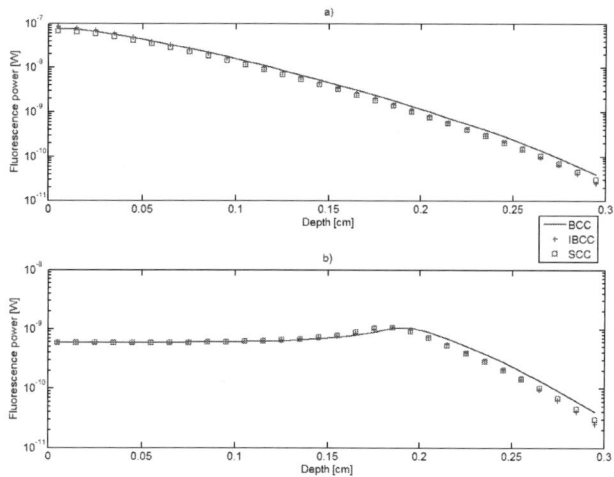

Fig. 2 Photosensitizer fluorescence power [W] emited in three skin tumors (BCC, IBCC and SCC) a) 6 s and b) 600 s after the beginning of the treatment vs. depth [cm] in the tumor.

IV. CONCLUSIONS

Despite of the advantages of PDT for the treatment of non melanoma skin cancer, clinical experience has demonstrated that the treatment response is not always the same even though the clinical protocol is fixed. Predictive models could provide support to treatment planning so as to maximize the therapy. In this work a predictive model for PDT able to depict the photosensitizer fluorescence emission as the treatment evolves was presented. The results clearly show the reduction of the fluorescence power emission in the tumor tissue as PDT progresses and the photosensitizer degradation becomes stronger. Such information is fundamental to ensure that the amount of photosensitizer will be enough for a proper evolution of the photochemical process involved in PDT due to a strong limitation in the concentration of photosensitizer would stop the treatment effects. The model application was shown for three types of skin tumor assuming the same initial photosensitizer accumulation and oxygen level. No significant differences were found in the fluorescence power emitted among them due to their similar optical properties and the assumptions previously indicated. Therefore in these cases similar treatment outcomes are expected for the clinical scenarios and parameters set in this work. Future works will deal with the distribution of the fluorescence signal in an heterogeneous media taking into account the sources of fluorescence here obtained.

ACKNOWLEDGMENT

This work has been partially supported by the project MAT2012-38664-C02-01 of the Spanish Ministery of Economy and Competitiveness, and by San Cándido Foundation.

REFERENCES

1. Vo-Dinh T (2003) Biomedical Photonics Handbook. CRC Press, Boca Raton.
2. Wang L, Jacques S L, Zheng L (1995) MCML – Monte Carlo modeling of light transport in multi-layered tissues. Computer methods and programs in biomedicine 47:131-146.
3. Foster T H, Murant R S, Bryant R G, Knox R S, Gibson S L, Hilf R. (1991) Oxygen Consumption and Diffusion Effects in PDT. Radiation Research 126: 296-303.
4. Svaasand L O, Wyss P, Wyss M T, Tadir Y, Tromberg B J, Berns M W (1996) Dosimetry model for Photodynamic Therapy with topically administered photosensitizers. Lasers in Surgery and Medicine 18:139-149.
5. Salas-García I, Fanjul-Vélez F, Arce-Diego J L (2012) Photosensitizer absorption coefficient modeling and necrosis prediction during photodynamic therapy. Journal of Photochemistry and Photobiology B: Biology 114: 79–86
6. Liu B, Farrell T J, Patterson M S (2010) A dynamic model for ALA-PDT of skin: simulation of temporal and spatial distributions of ground-state oxygen, photosensitizer and singlet oxygen. Phys. Med. Biol. 55:5913-5932.
7. Wang L, Jacques S L, Zheng L (1997) CONV – Convolution for responses to a finite diameter photon beam incident on multi-layered tissues. Computer methods and programs in biomedicine 54:141-150.
8. Salas-García I, Fanjul-Vélez F, Arce-Diego J L (2012) Spatial photosensitizer fluorescence emission predictive analysis for photodynamic therapy monitoring applied to a skin disease. Optics Communications 285: 1581–1588.

Author: J. L. Arce-Diego
Institute: Applied Optical Techniques Group, Electronics Technology, Systems and Automation Engineering Department, University of Cantabria
Street: Avenida de los Castros S/N, 39005, Cantabria, Spain
City: Santander (Cantabria)
Country: Spain
Email: arcedj@unican.es

Statistical Analysis of EMG Signal Acquired from Tibialis Anterior during Gait

F. Di Nardo[1], A. Mengarelli[1], G. Ghetti[2], and S. Fioretti[1]

[1] Department of Information Engineering, Università Politecnica delle Marche, 60131 Ancona, Italy
[2] Posture and Movement Analysis Laboratory, Italian National Institute of Health and Science on Aging (INRCA), 60131 Ancona, Italy

Abstract—Aim of the present study was to identify the different modalities of activation of tibialis anterior (TA) during gait at self-selected speed, by a statistical analysis of surface electromyographic signal from a large number (hundreds) of strides per subject. The analysis on ten healthy adults showed that TA is characterized by different activation modalities within different strides of the same walk. The most recurrent modality consists of three activations observed in 37.4±1.9% of total strides: at the beginning of gait cycle, around stance-to swing-transition and in the terminal swing. Further two modalities differ from the most recurrent one because of 1) the continuous activation during swing; 2) a further activity in the late mid-stance. The study of these different modalities of activation suggested that TA acts as pure ankle dorsi-flexor only in a small percentage (~20%) of total strides, where TA activity occurs in the simpler modality. The increase in the complexity of the recruitment of the muscle introduces an uncommon activity during mid-stance, which does not occur for the flexion of the ankle but is related to the activity of the TA as a foot invertor muscle.

Keywords—statistical gait analysis, ankle flexor muscles, surface EMG.

I. INTRODUCTION

Tibialis anterior plays an important role in everyday activities, especially walking. The main task of tibialis anterior during gait, together with all the ankle dorsi-flexor muscles, is to prevent slapping of the foot on the ground in initial stance, to permit the forefoot to clear the ground in initial swing, and to hold the ankle in position for initial contact [1]. Surface electromyography (sEMG) has been largely used for the assessment of the activation patterns of the ankle flexor muscles during normal and pathological gait [2]-[4]. The general consensus is that tibialis anterior is active mostly during the swing phase, with a continuation of its activity till the next loading response [1,5]. However, activities outside the typical activation intervals for tibialis anterior and, more generally, a large stride-to-stride variability in the EMG profiles have been reported [4,6]. It is, therefore, important to study the natural variability associated with muscle activity during free walking, in order to improve the interpretation of EMG signals in both physiological and pathological conditions. Thus, the aim of the present study was to analyze the variability in the modalities of sEMG activity of tibialis anterior (TA) in healthy adults during gait at self-selected speed. To evaluate the role of TA as antagonist muscle of the gastrocnemius lateralis (GL) in ankle flexion, sEMG activity of GL was also studied. The study is based on the recent availability of robust techniques for the detection of muscle activation intervals, and specific tools for statistical gait analysis, which allow to analyze a large number (hundreds) of strides per subject [7]-[9].

II. MATERIAL AND METHODS

A. Subjects

Ten healthy adult volunteers were recruited; mean (±SE) values were 24.0±0.8 years for age; 174±3 cm for height; 62.5±4.3 kg for weight and 20.5±0.7 for body mass index (BMI); male/female ratio was 5/5. The EMG activity during gait was recorded in both right and left lower limb of all subjects at self-selected speed over ground and treadmill. Exclusion criteria included history of neurological pathology, orthopedic surgery within the previous year, acute or chronic knee pain or pathology, BMI ≥25, or abnormal gait. Before the beginning of the test, all participants signed informed consent.

B. Recording System: Signal Acquisition and Processing

Signals were acquired by means of a multichannel recording system for statistical gait analysis (Step32, DemItalia, Italy). Each subject was instrumented with foot-switches, goniometers and sEMG probes. Three foot-switches were attached beneath the heel, the first and the fifth metatarsal heads of each foot. A goniometer (accuracy: 0.5 deg) was attached to the lateral side of each lower limb for measuring the knee joint angles in the sagittal plane. Single differential sEMG probes with fixed geometry (inte-relectrode distance: 12 mm) were attached over the tibialis anterior and gastrocnemius lateralis of each lower limb following the SENIAM recommendations [10].

After positioning the sensors, subjects were instructed to walk barefoot over ground for around 4 minutes at their natural pace, following the path schematized in Fig. 1. Natural pace was chosen because walking at a self-selected speed improves the repeatability of EMG data, while variability increases when subjects are required to walk

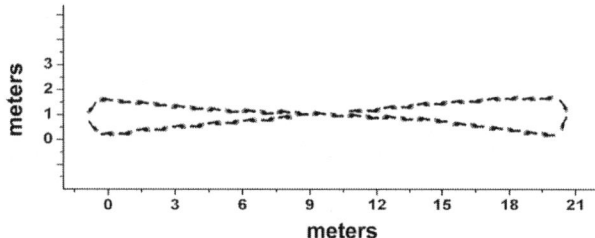

Fig. 1 Schematic representation of the path walked by the recruited subjects during the experiment; subjects walked barefoot over the floor for 5 minutes at their natural pace.

abnormally [11]. Foot-switch signals were converted to four levels corresponding to Heel contact (H), Flat foot contact (F), Push off (P), Swing (S) and processed to segment and classify the different gait cycles. During acceleration, deceleration, and changes in direction the strides are different from those of steady state walking. Therefore, goniometric signals (low-pass filtered with cut-off frequency of 15 Hz) along with gait phase durations, were used by a multivariate statistical filter embedded in the Step32 system, to detect and discard outlier cycles, i.e., cycles with the proper sequence of gait phases (H-F-P-S) but with abnormal timing, like those relative to deceleration, reversing, and acceleration [12]. EMG signals were high-pass filtered (cut-off frequency of 20 Hz) and then processed by a double-threshold statistical detector, embedded in the Step32 system, that provides the onset and offset time instants of muscle activity in a completely user-independent way [7].

During gait, a muscle activates a number of times which is usually variable from cycle to cycle [12]. Thus, muscle on/off instants should be averaged considering each single modality of activation by itself. With modality of activation we mean the number of times when muscle activates during a single gait cycle, i.e. n-activation modality consists of n activation intervals for the considered muscle, during a single gait cycle. In the present study, mean activation intervals (normalized with respect to the gait cycle) for each modality of activation were achieved by means of the Step32 system, according to the following steps. First, muscle activations relative to each gait cycle were identified. Then, for all the gait cycles corresponding to straight line walking, muscle activations were grouped according to the number of activations detected, i.e. relatively to the modalities of activations detected. Finally the on/off time instants were averaged, for each specific modality of activation observed, and relative standard deviation and standard error were computed. In the present study only gait cycles consisting of the sequence of H–F–P–S were considered.

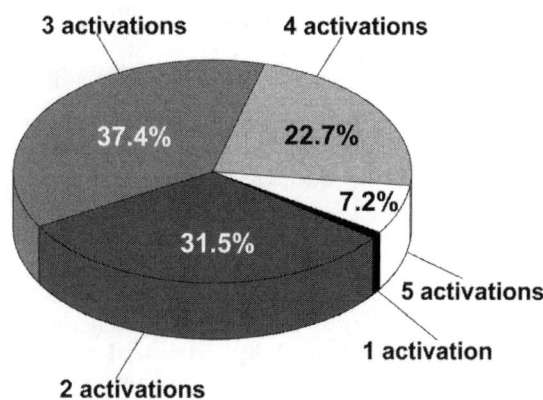

Fig. 2 Mean percentage frequency of each of the five different modalities of activation (from 1 up to 5) detected for TA during walking at self-selected speed.

III. RESULTS

For each subject, a mean (±SE) of 497±22 strides has been considered. The H-phase lasts 5.7±0.3% of the gait cycle, the F-phase 26.8±1.4%, the P-phase 23.3±1.0% and the S-phase 44.2±0.7%.

Statistical analysis of the myoelectric signal in each single subject put in evidence that TA shows a different number of activation intervals in different strides of the same walk; five modalities of activation were observed. Fig. 2 shows the mean percentage frequency for each of the five different modalities of activations detected during walk.

In a 31.5±5.0% of total strides, two activations were observed for TA in the gait cycle: from the beginning up to 8.7±1.1% of the gait cycle, and during the whole swing (2-activation modality) (Fig. 3). In these strides, the most recurrent modality of activation for GL consists of one single activation; in this modality (1-activation modality) for GL, no overlapping between TA and GL activation intervals are detected. In the 2- and 3-activation modalities for GL, superimposition between TA and GL activity are detected during swing.

The most recurrent modality of TA activation during gait cycle consists of three activations (3-activation modality), observed in 37.4±1.9% of total strides: the first occurs at the beginning of gait cycle, the second around stance-to-swing-transition and the third in the terminal swing (Fig. 4).

In these strides, the most recurrent modality of activation for GL consists of two activations (2-activation modality); in this modality for GL, an uncommon activity occurs in swing phase, which overlaps completely the TA activity. Same thing for the 3-activation modality of GL. No

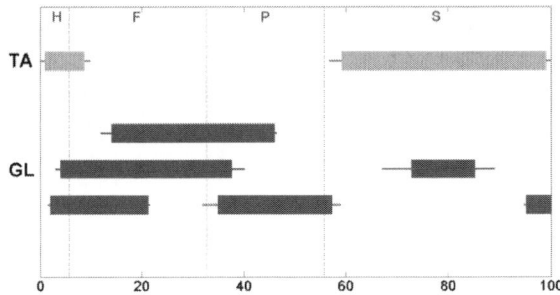

Fig. 3. Mean values (+SE) of GL activation intervals (dark grey bars) vs. percentage of gait cycle, detected in the strides where TA (light grey bars) shows the 2-activation modality. GL activation intervals are reported separately for the modalities with 2, 3 and 4 activations. The heel contact (H), flat foot contact (F), push off (P) and swing phases (S) are delimited by dashed vertical lines.

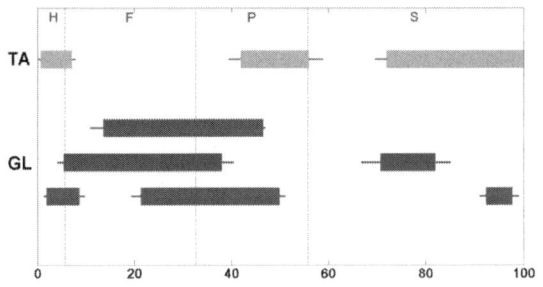

Fig. 4. Mean values (+SE) of GL activation intervals (dark grey bars) vs. percentage of gait cycle, detected in the strides where TA (light grey bars) shows the 3-activation modality. GL activation intervals are reported separately for the modalities with 2, 3 and 4 activations. The heel contact (H), flat foot contact (F), push off (P) and swing phases (S) are delimited by dashed vertical lines.

overlapping between TA and GL activation intervals are detected only in the *1*-activation modality for GL.

The *4*-activation modality for TA (22.7±2.7%) is characterized by a double activity from pre-swing to the end of gait cycle; during stance, it presents the typical activation at the beginning of gait cycle and introduce a further activity in the late mid-stance (Fig. 5). Superimpositions between TA and GL activation intervals are detected both in stance phase (all the GL modalities of activation) and in swing phase (*2*- and *3*-activation modalities for GL).

The remaining 7.2±1.6% of total strides was characterized by five activations, three during the stance and two during the swing (*5*-activation modality).

The 1-activation modality for TA was observed only in 1% of the total stride and showed a large variability ; thus, it is not considered in the present analysis.

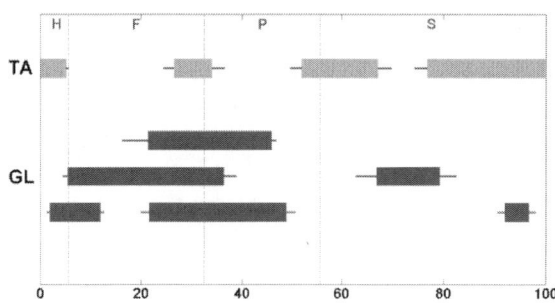

Fig. 5. Mean values (+SE) of GL activation intervals (dark grey bars) vs. percentage of gait cycle, detected in the strides where TA (light grey bars) shows the 4-activation modality. GL activation intervals are reported separately for the modalities with 2, 3 and 4 activations. The heel contact (H), flat foot contact (F), push off (P) and swing phases (S) are delimited by dashed vertical lines.

IV. DISCUSSION

In the present study, sEMG signals were recorded in ten healthy adult subjects, in order to determine and describe the picture of the activation modalities of TA during normal gait. Since the main task of tibialis anterior during gait is to contribute to ankle flexion in antagonism with gastrocnemius [1,5], the GL activity relative to each specific modality of TA activation was also analyzed. A statistical analysis of a large number (approximately 500) of strides for each subject was performed to this aim.

The TA activation intervals, detected in each subject at self-selected walking speed, followed roughly the typical pattern reported for each muscle during gait [1,5]. However, in agreement with [4], the present statistical analysis confirmed that TA showed different modalities in the number of activations and in the timing of signal onset and offset (Fig. 3-5).

During free walking, the most recurrent modality of activation for TA consists of three activations at the beginning of gait cycle, around stance-to-swing transition, and in the terminal swing. The *2*-activation modality differs from the most common modality for the continuous activation during swing. The *4*- and *5*-activation modalities are characterized by a double or a triple activity from pre-swing to the end of gait cycle; during stance, they present the typical activation at the beginning of gait cycle and introduce a further activity in the late mid-stance.

The only strides (22.1±1.9% of the total strides) where TA activity shows no superimposition with GL activation intervals, are those where simultaneously GL presented a single activity and TA showed the 2- or 3-activation modality (Fig. 3 and Fig. 4). Only in these strides TA should be considered performing as pure GL antagonist for ankle flexion. With the increase to 4 (and then to 5) of the number of TA activations during the gait cycle (Fig. 5), a TA activity during mid-stance was detected. This short activity, usually not reported in healthy adults, but observed in children [6,13], overlaps the typical activity of GL as ankle plantar flexor in the 29.9±3.1% of total strides. These findings suggest that the recruitment of TA in this phase does not occur for the flexion of the ankle, as antagonist of GL, but is related to the activity of the TA as a foot invertor muscle for controlling balance during single support and contralateral limb swing.

V. CONCLUSIONS

The analysis of the different modalities of activation suggests that TA acts as pure GL antagonist for ankle flexion only in a small percentage (~20%) of total strides, where TA activity occurs in the simpler modality (*2-* and *3-*activation modality). The increase in the complexity of the recruitment of TA (*4-*activation modality) introduces an uncommon activity during mid-stance, which does not occur for the flexion of the ankle but is related to the activity of the TA as a foot invertor muscle.

REFERENCES

[1] Perry J (1992) Gait Analysis - Normal and Pathological Function. USA: Slack Inc
[2] Petersen TH, Farmer SF, Kliim-Due M, Nielsen JB (2013) Failure of normal development of central drive to ankle dorsiflexors relates to gait deficits in children with cerebral palsy. J Neurophysiol 109:625-639
[3] Stewart C, Postans N, Schwartz MH et al. (2007) An exploration of the function of the triceps surae during normal gait using functional electrical stimulation. Gait Posture 26:482-488
[4] Di Nardo F, Ghetti G, Fioretti S. Assessment of the activation modalities of gastrocnemius lateralis and tibialis anterior during gait: a statistical analysis. J Electromyogr Kinesiol, ahead of print
[5] Sutherland DH (2001) The evolution of clinical gait analysis part l: kinesiological EMG. Gait Posture 14:61-70
[6] Agostini V, Nascimbeni A, Gaffuri A et al. (2010) Normative EMG activation patterns of school-age children during gait. Gait Posture 32:285-289
[7] Bonato P, D'Alessio T, Knaflitz M (1998) A statistical method for the measurement of muscle activation intervals from surface myoelectric signal during gait. IEEE Trans Biomed Eng 45:287-299
[8] Di Nardo F, Fioretti S (2013) Statistical analysis of surface electromyographic signal for the assessment of rectus femoris modalities of activation during gait. J Electromyogr Kinesiol 23:56-61
[9] Staude G, Flachenecker C, Daumer M, Wolf W (2001) Onset detection in surface electromyographic signals: a systematic comparison of methods. EURASIP J Appl Signal Process 2:67-81
[10] Hermens HJ, Freriks B, Merletti R et al. (1999) European recommendations for surface electromyography. Roessingh Research and Development, Enschede, the Netherlands
[11] Kadaba MP, Ramakrishnan HK, Wootten ME et al. (1989) Repeatability of kinematic, kinetic, and electromyographic data in normal adult gait. J Orthop Res 6:849-60
[12] Agostini V, Knaflitz M (2012) Statistical Gait Analysis in Distributed Diagnosis and Home Healthcare. Edited by Rajendra Acharya U, Filippo M, Toshiyo T, Subbaram Naidu D, Jasjit SS. Stevenson Ranch, California, (USA):American Scientific Publishers; 99–121. Volume 2.
[13] Sutherland DH, Cooper L, Daniel D (1980) The role of the ankle plantar flexors in normal walking. J Bone Joint Surg Am 62:354-363

Auto-Mutual Information Function for Predicting Pain Responses in EEG Signals during Sedation

U. Melia[1], M. Vallverdú[1], M. Jospin[2], E.W. Jensen[1], J.F. Valencia[3],
F. Clariá[4], P.L. Gambus[3], and P. Caminal[1]

[1] Dept. ESAII, Centre for Biomedical Engineering Research, Universitat Politècnica de Catalunya, Barcelona, Spain
[2] R&D Department, Quantum Medical SL, Mataró, Barcelona, Spain
[3] Department of Anesthesiology, Hospital Clínic, Universidad de Barcelona, Barcelona, Spain
[4] Dept. IIE, Lleida University, Lleida, Spain

Abstract—The level of sedation in patients undergoing medical procedures evolves continuously, such as the effect of the anesthetic and analgesic agents is counteracted by pain stimuli. The monitors of depth of anesthesia, based on the analysis of the electroencephalogram (EEG), have been progressively introduced into the daily practice to provide additional information about the state of the patient. However, the quantification of analgesia still remains an open problem. The purpose of this work was to analyze the capability of prediction of nociceptive responses based on the auto-mutual information function (*AMIF*). *AMIF* measures were calculated on EEG signal in order to predict the presence or absence of the nociceptive responses to endoscopy tube insertion during sedation in endoscopy procedure. Values of prediction probability of *Pk* above 0.80 and percentages of sensitivity and specificity above 70% and 70% respectively were achieved combining *AMIF* with power spectral density and concentrations of remifentanil.

Keywords—Biomedical signal processing, electroencephalogram, information theory, complexity.

I. INTRODUCTION

The aggression that occurs on patient undergoing surgery triggers a series of responses in the body and in the tissue that may have implications on the outcome of the surgical process. To mitigate the intensity of these responses, a certain level of protection or "anesthetic state" must be achieved. The anesthetic state may be defined as the combination of pharmacological effects that minimize the impact of surgical aggression in the patient.

For several years, various methods have been developed for the noninvasive assessment of the level of consciousness during general anesthesia [1-3]. Since the main action of anesthetic agents occurs in the brain, a reasonable choice is to monitor the electroencephalographic signal (EEG). Changes on the EEG signal are directly related to biochemical variations of a drug induced in the brain and the effects on individual behavior. According to various methods, different EEG monitors have been developed. The three most important monitors consider bispectrum (BIS, A-2000 monitor, Aspect Medical, USA) [4], [5], entropy (SE and RE - State and Response Entropy, S/5 Entropy Module, GE Healthcare, Finland) [6] and auditory evoked potentials (AAI, AEP Monitor/2, Danmeter, Denmark) [7], [8] whereas the most recent is the qCON (Quantum Medical, Spain) [9].

However, it has not been possible to develop a system capable of quantifying analgesia. The classic methods include hemodynamic response, analysis of electrocardiographic waveforms variability, degree of respiratory sinus arrhythmia, plethysmographic response [10], pulse wave, skin conductance [11], the Surgical Stress Index (SSI ®) [12] and the ANI (Metrodoloris, France) [13], based on the heart rate variability. None of them has proven to be clinically useful methods because they are influenced by the response of the autonomic nervous system (ANS) and they are sensitive to other disturbances, such as changes in blood pressure or heart rate due to patient's baseline condition (hypertension, arrhythmias of diverse etiology), sympathomimetic drug delivery or unpredictable situations such as perioperative bleeding.

In this work, indexes based on auto-mutual information function (AMIF) of the EEG signal were proposed in order to assess the prediction of the response to tube insertion during endoscopy procedure. EEG windows of 60 seconds were taken between 30 s and 90 s before the tube insertion in order to avoid the effect of the tube insertion response on the signal (evoked potential, movement artifact, EMG, etc.).

II. MATERIALS AND METHODOLOGY

A. EEG Database and Preprocessing

The database belongs to the Department of Anesthesiology, Hospital Clínic de Barcelona (Spain). This database contains data recorded from more than 300 patients who underwent surgical procedures under general anesthesia with laryngoscopy, intubation or laryngeal mask insertion. For each patient, the following information is available: predicted concentrations of propofol (Ce_{Prop}) and remifentanil (Ce_{Remi}); bispectral Index (*BIS*) and electroencephalogram (*EEG*) signal, presence or absence of gag reflex during endoscopy tube insertion (*GAG*).

All patients belong to 1-3 ASA classification. Patients with altered central nervous system, medicated with analgesics or drugs with central effects on the perception of pain, from moderate to severe cardiomyopathy, neuropathy or hepatopathy that needed control during the anesthetic process were not included in the database.

The EEG was recorded with a sampling frequency of 900 Hz, with a resolution of 16 bits and a recording time of about 60 min. All information Ce_{Prop}, Ce_{Remi}, BIS and GAG were annotated with a resolution of 1 second. After the application of a FIR band pass filter of 100th order, with cut-off frequencies of 0.1-45Hz, the EEG signals were resampled at 128 Hz. Then, the EEG signals were segmented in windows of length of 1 minute between 30 s and 90 s before the annotation of GAG.

The annotated GAG was assigned to the previous 1 minute length window if the differences ΔCe_{Remi} and ΔCe_{Prop} between the first and the last second of the window were ΔCe_{Remi}<0.1 µg/ml and ΔCe_{Prop}<0.1 µg/ml. Otherwise, the window was cut at the sample where the conditions were satisfied. Windows of EEG containing high amplitude peak noise were processed with a filter based on the analytic signal envelope (ASEF) [14]. If the difference between adjacent samples were higher than 10% of the mean of the differences of the previous samples, the remaining samples of the window were eliminated. In this way, the smallest window resulted to be of 50 s. The windows of interests were filtered into the characteristic frequency bands of the EEG signal: δ, 0.1-4 Hz; θ, 4-8 Hz; α, 8-12 Hz; β, 12-30 Hz.

B. Auto-Mutual Information Function

Mutual information (MI) can measure the nonlinear as well as linear dependence of two variables. It is a metric derived from Shannon's information theory to estimate the information gained from observations of one random event on another, and measuring both linear and nonlinear dependences between two time series. It can be regarded as a nonlinear equivalent of the correlation function. Auto-mutual information function (*AMIF*) [15] is defined as

$$AMIF(\tau) = \sum_{x_i \in X} \sum_{x_{i+\tau} \in X} P_{xx}(x_i, x_{i+\tau}) \log_2 \left(\frac{P_{xx}(x_i, x_{i+\tau})}{P_x(x_i) P_x(x_{i+\tau})} \right) \quad (1)$$

The probabilities P_{xx} and P_x were constructed on the series x_i and their delayed series $x_{i+\tau}$, for $\tau = \{1,2,...,128\}$ samples. This function describes how the information of a signal (*AMIF* value at τ =0) decreases over a prediction time interval (*AMIF* values τ>0). In the case of a completely regular and deterministic signal, the *AMIF* would remain at the maximum value of τ=0 for all τ. In the case of an uncorrelated random signal, the *AMIF* would become zero for all τ apart τ=0. Increasing information loss is related to decreasing predictability, and increasing complexity of the signal.

AMIF can be also defined from Rényi information theory as

$$AMIFRe_q(\tau) = \frac{1}{q-1} \log_2 \sum_{x_i \in X} \sum_{x_{i+\tau} \in X} \frac{P_{xx}^q(x_i, x_{i+\tau})}{P_x^{q-1}(x_i) P_x^{q-1}(x_{i+\tau})} \quad (2)$$

where q is the control parameter that defines Rényi information, and was selected as q = {0.1, 0.2, 0.5, 2, 3, 5, 10, 30, 50, 100}. *AMIF* was normalized by the maximum value *AMIF*(0).

Several variables were defined on the AMIF along the delay τ: mean (*m*), first relative maximum (*max*), first relative minimum (*min1*), absolute minimum (*min2*), and decay for τ=1 (*FD*). These variables were calculated from the EEG signal filtered in each one of the characteristic frequency bands. Furthermore, the power spectral density (PSD) for each EEG window was calculated using the Welch method. The power in each band (P_δ, P_θ, P_α and P_β) was calculated as the area under the curve normalized by the total power.

C. Definition of Variables and Statistical Analysis

A non-parametric test, U of Mann-Whitney test, was applied and a significance level *p-value* <0.05 was taken into account. Variables that satisfy this condition were considered for building a linear discriminant function, in order to predict the pain responses. The leaving-one-out method was performed as validation method taking into account the presence or the absence of the gag reflex, GAG=1 and GAG=0, respectively. Sensitivity (*Sen*) and specificity (Spe) were calculated for testing the performance of AMIF variables. *Sen* measures the proportion of responsive state (GAG 1) correctly classified and *Spe* measures the proportion of unresponsive state (GAG 0) correctly classified.

The ability of the variables to describe pain responses was evaluated using prediction probability (P_k), which compares the performance of indicators [16]. The P_k coefficient is a statistic commonly used to measure how well an index predicts the state of the patient. A P_k of 1 represents a perfect prediction and 0.5 is not better than tossing a fair coin. The P_k avoids the shortcomings of other measures being independent of scale units and it does not require knowledge of underlying distributions.

III. RESULTS

Table 1 presents the best variables able to classify both groups of GAG with $p\text{-value}<0.0005$, $Sen>60\%$, $Spe>60\%$ and $Pk > 0.6$. It should be noticed that $AMIF$ could characterized GAG groups in α, β, and δ bands. As it is observed, the best results in α band were obtained for $q>1$, in β band for $q<3$ and δ band for $q<1$. As an example, figure 1 presents the evolution of the averaged $AMIF$ in δ band for all windows of the two groups of GAG, for all q. Also BIS and P_α presented similar classification results (Table 1).

Table 1 Presence and absence of GAG reflex: Single variables

Variables	Pk	$Sen(\%)$ $N=122$	$Spe(\%)$ $N=390$
BIS	0.734	72.1	62.3
P_α	0.744	70.5	65.4
$max(Re_{q=2})_\alpha$	0.716	73.0	61.8
$max(Re_{q=100})_\alpha$	0.704	71.3	65.4
P_β	0.630	--	69.5
$FD(Sh)_\beta$	0.727	68.0	67.4
$FD(Re_{q=2})_\beta$	0.726	66.4	67.9
$FD(Re_{q=0.2})_\beta$	0.723	71.3	65.4
$FD(Re_{q=0.1})_\beta$	0.721	72.1	64.4
P_δ	0.605	--	70.8
$min2(Re_{q=0.1})_\delta$	0.718	--	--
$FD(Re_{q=0.1})_\delta$	0.659	60.7	60.5
$FD(Re_{q=0.2})_\delta$	0.648	--	--

N: number of analyzed windows; P_k: prediction probability; Sen (%), sensitivity; Spe (%), specificity; $p\text{-value}<0.05$

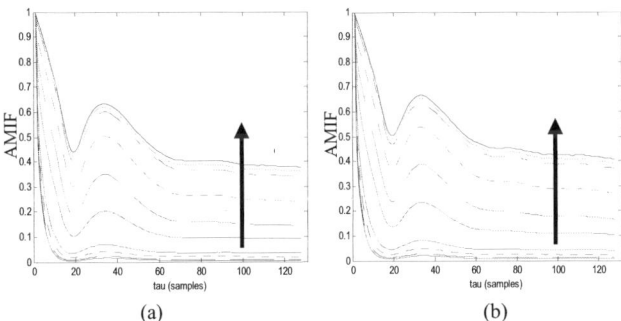

Fig. 1 Averaged $AMIF$ in δ band of all windows of: (a) GAG 0, unresponsive state; (b) GAG 1, responsive state.

Table 2 contains the correlation between the new proposed variables and the BIS, Ce_{Remi} and Ce_{Prop}. As it can be seen, $AMIF$ variables are uncorrelated with Ce_{Remi} and Ce_{Prop} but present correlations with BIS in β band.

In order to increase the percentages of sensitivity and specificity, new variables were built as a function of $AMIF$, P_δ, P_θ, P_α, P_β, Ce_{Remi} and Ce_{Prop}. Each $AMIF$ variable of a band was combined with the same variable of the remaining bands. These new variables were linear combinations of a maximum of three uncorrelated variables. Table 3 shows these variables that give the best classification percentages. As it can be seen in Table 3, combining P_α with Ce_{Remi} and $FD(Re_{q=0.1})_\delta$ it was possible to obtain $Pk >0.8$ and $Sen>70\%$ and $Spe>70\%$. Considering only $AMIF$ variables, the best performance was achieved by $FD(Re_{q=0.1})_\delta$, $FD(Re_{q=0.1})_\beta$.

Table 2 Values of the Pearson's correlation coefficient

Variables	BIS	Ce_{Remi}	Ce_{Prop}
BIS	1	0.196	-0.328
Ce_{Remi}	0.196	1	-0.294
Ce_{Prop}	-0.328	-0.294	1
P_δ	0.312	0.153	-0.228
P_α	-0.643	-0.107	0.267
P_β	0.449	-0.047	-0.053
$min2(Re_{q=0.1})_\delta$	0.043	-0.091	0.002
$FD(Re_{q=0.1})_\delta$	0.236	-0.008	-0.082
$FD(Re_{q=0.2})_\delta$	0.188	-0.012	-0.066
$max(Re_{q=2})_\alpha$	-0.414	-0.016	0.120
$max(Re_{q=100})_\alpha$	-0.345	-0.012	0.096
$FD(Re_{q=0.1})_\beta$	0.695	0.172	-0.278
$FD(Re_{q=0.2})_\beta$	0.713	0.180	-0.284
$FD(Sh)_\beta$	0.753	0.205	-0.295
$FD(Re_{q=2})_\beta$	0.758	0.208	-0.284

Table 3 Presence and absence of GAG reflex: Multi-variables

Variables	Pk	$Sen(\%)$ $N=122$	$Spe(\%)$ $N=390$
P_α, $FD(Re_{q=0.1})_\delta$	0.791	70.4	71.8
P_α, $FD(Re_{q=0.2})_\delta$	0.790	72.4	70.2
$FD(Re_{q=0.1})_\delta$, $FD(Re_{q=0.1})_\beta$	0.747	69.7	68.5
$FD(Re_{q=0.2})_\delta$, $FD(Re_{q=0.2})_\beta$	0.746	69.7	68.2
Ce_{Remi}, P_α, $FD(Re_{q=0.1})_\delta$	0.805	73.5	71.0
Ce_{Remi}, P_α, $FD(Re_{q=0.2})_\delta$	0.803	75.5	70.2

N: number of analyzed windows; P_k: prediction probability; Sen (%), sensitivity; Spe (%), specificity; $p\text{-value}<0.05$

Figure 2 shows the boxplot of the distribution of $FD(Re_{q=0.1})_\delta$ and $FD(Re_{q=0.1})_\beta$. As it can be noted, $FD(Re_{0.1})_\beta$ and $FD(Re_{0.1})_\delta$ present a lower median value for unresponsive state (GAG 0) than responsive state (GAG 1) ($p\text{-value}<0.0005$), what indicates higher complexity behavior in GAG 1 stating a lower sedation level. Comparing the distribution of these variables, it can be denoted strong EEG complexity in β band compared to δ band.

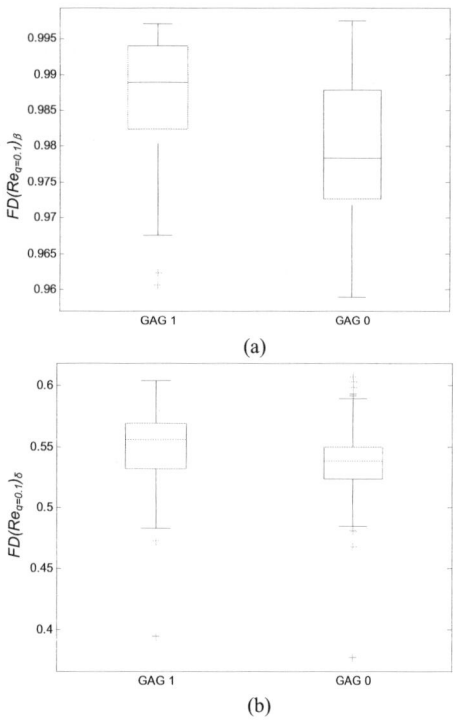

Fig. 2 Distribution of (a) $PDI(Re0.1)_\beta$, (b) $PDI(Re0.1)_\delta$. On each box, the central mark is the median, the edges of the box are the 25th and 75th percentiles. The whiskers are lines extending from each end of the boxes to show the extent of the rest of the data. Values beyond the end of the whiskers are considered outliers and marked with a +.

IV. CONCLUSIONS

Auto-mutual information function (*AMIF*) was applied to one-minute windows of EEG signals recorded during endoscopy procedure in order to predict the response to the tube insertion. Several variables were defined from *AMIF*. The statistical analysis of single variables permitted to obtain values of $Pk > 0.70$.

Values of prediction probability of $Pk > 0.80$ and percentages of sensitivity and specificity of $Sen > 70\%$, $Spe > 70\%$ could be achieved combining *AMIF* with concentration of remifentanil Ce_{Remi} and spectral power in α band P_α.

ACKNOWLEDGMENT

This work was supported within the framework of the CICYT grant TEC2010-20886 and the Research Fellowship Grant FPU AP2009-0858 from the Spanish Government.

REFERENCES

1. Bouillon T W (2004) Pharmacodynamic interaction between propofol and remifentanil regarding hypnosis, tolerance of laryngoscopy, bispectral index, and electroencephalographic approximate entropy. Anesthesiology 100:1353-1372
2. Ferenets R, Lipping T, Anier A et al. (2006) Comparison of entropy and complexity measures for the assessment of depth of sedation. IEEE Trans. On Biomed. Eng 53(6):1067- 1077
3. Gifania P, Rabieea H R, Hashemib M H et al. (2007) Optimal fractal-scaling analysis of human EEG dynamic for depth of anesthesia quantification. J Franklin Institute 344:212–229
4. Rampil I J (1998) A primer for EEG signal processing in anesthesia. Anesthesiology 89:980-1002
5. Sigl J C, Chamoun N G (1994) An introduction to bispectral analysis for the EEG. J Clin Monit 10:392-404
6. Viertiö-Oja H, Maja V, Särkelä M et al. (2004) Description of the Entropy algorithm as applied in the Datex-Ohmeda S/5 Entropy Module. Acta Anaesthesiol. Scand 48(2):154–161
7. Jensen E W, Lindholm P, Henneberg S (1996) Auto Regressive Modeling with Exogenous Input of auditory evoked potentials to produce an on-line depth of anaesthesia index. Methods of Information in Medicine 35:256–260
8. Litvan H, Jensen E W, Galan J et al. (2002) Comparison of conventional averaged and rapid averaged, autoregressive-based extracted auditory evoked potentials for monitoring the hypnotic level during propofol induction. Anesthesiology 97(2):351–358
9. Valencia JF, Borrat X, Gambus P L (2012) Validation of a new index, qcon, for assessment of the level of consciousness during sedation. ASA Annual Meeting Proc., Washington, DC, US, October 13-17 2012, Abstract A640
10. Luginbuhl M (2006) Stimulation induced variability of pulse plethysmography does not discriminate responsiveness to intubation. Br J Anaesth 96(3):323-329
11. Rantanen M (2006) Novel multiparameter approach for measurement of nociception at skin incision during general anaesthesia. Br J Anaesth 96(3):367-376
12. Storm H (2008) Changes in skin conductance as a tool to monitor nociceptive stimulation and pain. Current Opinion in Anaesthesiology 21:796–804
13. Jeanne M, Logier R, De Jonckheere J et al. (2009) Validation of a graphic measurement of heart rate variability to assess analgesia/nociception balance during general anesthesia. IEEE-EMBS Proc., Annual International Conference of the IEEE Eng. in Med. and Biology Society 1840–1843
14. Melia U S P, Claria F, Vallverdu M et al. (2010) Removal of peak and spike noise in EEG signals based on the analytic signal magnitude. IEEE-EMBS Proc., Annual International Conference of the IEEE Eng. in Med. and Biology Society 3523-3526
15. Escudero J, Hornero R, Abasolo D. (2009) Interpretation of the automutual information rate of decrease in the context of biomedical signal analysis application to electroencephalogram recordings. Physiological Measurement 30(2):187–199
16. Smith W D, Dutton R, Smith N T (1996) Measuring the performance of anesthetic depth indicators. Anesthesiology 84:38-51

Author:	Pere Caminal
Institute:	Centre for Biomedical Engineering Research
Street:	Pau Gargallo 5
City:	08028, Barcelona
Country:	Spain
Email:	pere.caminal@upc.edu

Prony's Method for the Analysis of mfVEP Signals

A.J. Fernández-Rodríguez[1], L. de Santiago[1], R. Blanco[2], C. Amo[1], R. Barea[1], J.M. Miguel-Jiménez[1], J.M. Rodríguez-Ascariz[1], E.M. Sánchez-Morla[3], M. Ortiz[1], and L. Boquete[1]

[1] Alcalá University/Department of Electronics, Biomedical Engineering Group Alcalá de Henares, Spain
[2] Alcalá University/Department of Surgey, Alcalá de Henares, Spain
[3] Department of Psychiatry, University General Hospital of Guadalajara, Guadalajara, Spain

Abstract—This research focuses on early detection of multiple sclerosis by analyzing multifocal visual evoked potentials. In this paper, the Prony's method is used for decomposing the evoked potentials in a sum of exponentials. The parameters defining these exponential functions have a significance difference between three groups of patients: healthy, multiple sclerosis with optic neuritis and multiple sclerosis without optic neuritis.

Keywords—Multiple sclerosis, Optic neuritis, mfVEP, Prony method.

I. INTRODUCTION

Multiple sclerosis (MS) is a chronic demyelinating disease of the central nervous system. Globally, the prevalence is 30 per 100.000 (range 5-80) [1], being the more common cause of neuronal disability in young adults. The average age of onset is 29,2 years and is double common in women than in men. This illness reduced quality of life of people with MS, families and society.

The earlier clinical symptom in the most of the cases, is the optic neuritis (ON), it consist in episodes of inflammation and demyelination of the optic nerve [2]. ON typically presents with unilateral periorbital pain exacerbated by movement, decreased visual acuity, and reduced light and color sensitivity.

A method for asses the function of visual pathwaysis the multifocal visual evoked potentials (mfVEP) that can be recorded simultaneously from many regions (usually 60) of the visual field. The development of mfVEP give an opportunities to evaluate the neurodegenerative component involved in the visual pathways.

The typical method for study the mfVEP signals is based on amplitude and latency analysis of the waveform of the patient compared between the both eyes (interocular) and between one eye and data base of control subjects. This kind of analysis of mfVEP signals provided a method to differentiate eyes with different types of disease (Optic neuritis, diabetes, glaucoma).

The aim of this paper is to apply the Prony's method to characterize mfVEP signals into three groups of eyes: multiple sclerosis with optic neuritis (MS-ON), multiple sclerosis without optic neuritis (MS-non-ON) and healthy. This study seeks to determine the usefulness of mfVEP in patients with MS and ON in order to determine the functional changes of the visual pathways.

II. METHOD

A. The Prony's Method

Prony's method allows the decomposition of a set of sampled data as a linear combination of exponential functions. Prony's makes use of least squares analysis to approximately fit the function. Consider the N complex data samples y[1],...y[N]. Prony desired to estimate y[n], by fitting the data with a M-term complex exponential model given by the equation (1)

$$y[n] \cong \sum_{k=1}^{M} A_k e^{(\alpha_k + j\omega_k)(n-1)T_S + j\varphi_k} \quad (1)$$

Where n=1,2,...N, T_S is the sampling period, A_k amplitude, α_k damping factor (s^{-1}), ω_k is angular velocity (radians), and φ_k is the initial phase (in radians) of the kth exponential. These parameters describe with sufficient accuracy the characteristics of the signal y[n], and therefore are a useful feature for signal discrimination and identification.

Figure 1 shows an example of the decomposition of a signal using 13 terms (M = 13).

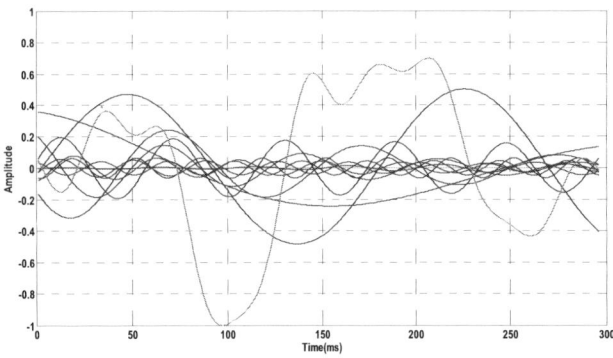

Fig.1 Original signal (orange) and the first 13 terms (blue).

In the example on Fig.1 the mean squared error (MSE) is calculated according (eq.2) between the n values of the original signal (S_o) and the reconstructed signal (S_R) in function of the number of used terms shown in Table 1

$$MSE = \frac{1}{n}\sum_{h=1}^{n}(S_{Oh} - S_{Rh})^2 \quad (2)$$

Table 1 MSE with M terms.

M terms	2	5	9	13
MSE	0.2355	0.0393	0.0054	$1.046 \cdot 10^{-14}$

The reconstructed signals with the numbers of terms of Table 1 are presented in the Figure 2.

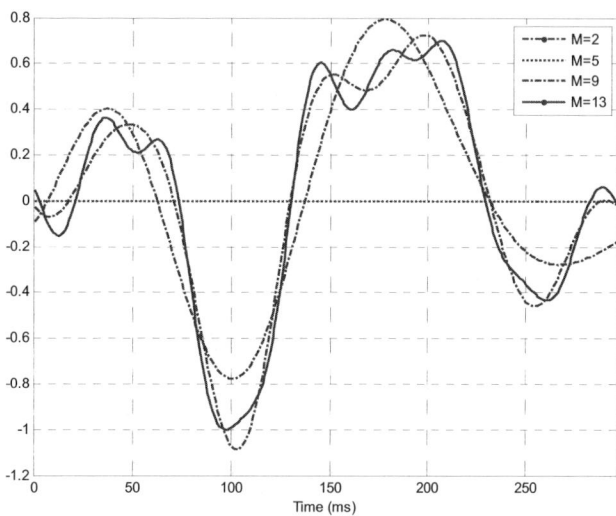

Fig. 2 Reconstructed signal with M terms.

The classic Prony'smethod solves the equation 1 by a series of linear equations. Prony analysis has been used in the detection of interferences in electrical circuits, speech processing and biomedical engineering [3], [4].

B. Patients and Signal Adquisition

The study protocol was approved by the Institutional Review Boards of Alcalá University-affiliated hospitals and adhered to the tenets of the Declaration of Helsinki. All participants provided informed consent.

This study assessed 23 consecutive patients (8 men and 15 women; median age: 30.3 years) with clinical definite MS based on McDonald criteria [5]. All patients had suffered one ON episode at least 6 months before being recruited. ON diagnosis was based on clinical findings such as unilateral visual loss, pain with eye movement, relative afferent pupillary defect and color vision deficiency. There were two study groups: the MS-ON group included 27 eyes (58,7%) that had a history of ON and the MS-non-ON group included 19 eyes (41,3%) with no history of ON.

In addition, 20 age-matched healthy eyes (from 3 men and 16 women; median age: 27.7 years) with a normal neurological and ophthalmologic examination were included as a control group.

Monocular mfVEP recordings were obtained using VERIS software 5.9 (Electro-Diagnostic Imaging, San Mateo, USA). The stimulus was a scaled dartboard with a diameter of 44.5°, containing 60 sectors, each with 16 alternating checks, 8 white (200 cd/m^2), and 8 black (<3 cd/m^2). The sectors were cortically scaled with eccentricity to stimulate approximately equal areas of the visual cortex. The dartboard pattern reversed according to a pseudorandom m-sequence at a frame rate of 75 Hz [6].

Three channels of continuous VEP recordings were obtained using gold cup electrodes. For the midline channel, the electrodes were placed 4 cm above the inion (active), at the inion (reference), and on the forehead (ground). For the other two channels, the same ground and reference electrodes were used, but the active electrodes were placed 1 cm above and 4 cm lateral to the inion on either side. By taking the difference between pairs of channels, three additional "derived" channels were obtained, resulting in effectively six channels of recording.

The records were amplified with the high- and low-frequency cutoffs set at 3 and 100 Hz, respectively (preamplifier P511J; Grass Instruments, Rockland, MA), and sampled at 1.200 Hz. The impedance was <5 KΩ for all electrodes. In a single session, two 7-min recordings were obtained from monocular stimulation of each eye and were averaged for analysis. This averaging, as well as all other analyses, was computed with custom-made programs written in commercial software (Matlab; Mathworks Inc., Natick, MA). The mfVEP responses were filtered offline with the high and low frequency cutoffs set at 3 to 35 Hz using a fast Fourier transformation. The length of each record is 500 ms, although the interval of interest is in the range 45-150 ms.

C. mfVEP Signal Processing

The data analyses were based on 'best channel' responses using software written in MATLAB [7]. The signal-to-noise ratio (SNR) was calculated as the root mean square (RMS) of the sector's waveform in the signal window (45–150ms) divided by the mean RMS from the noise windows (325–430ms) of all 60 sectors. The 60 "best" responses (from the channel with the largest SNR) were used. Records with SNRs<0.23 (log unit) were excluded from the analysis. Due to this, the number of analyzable sectors for each eye (S) is not always 60 (S≤60). The mfVEP signals have been normalized in amplitude.

Each normalized signal corresponding to the sector i ($y_{iN}[n]$) is decomposed according to Prony's Method (eq. 1). As the length of $y_{iN}[n]$ is 296 samples (time interval 5-250 ms), the decomposition allows to obtain M=148 components. This way the following values are obtained: A_{ik}, α_{ik}, ω_{ik} and φ_{ik}, (i=1,…N; k=1,…148).

Due to the principal information components of mfVEP signal are in the low frequencies, only the lowest 13 frequencies are chosen for the study, as it is represented in equation 3.

$$y_i[n] \cong \sum_{p=1}^{13} A_{ip} e^{(\alpha_{ip}+j\omega_{ip})(n-1)T_S+j\varphi_{ip}}; \; w_{i1} < w_{i2} < \cdots w_{i13} \quad (3)$$

For example, in Table 2 are shown the values of the parameter for MS ON and Healthy Subject (right eye, sector 16):

Table 2 Example of Prony's Method parameters.

	MS-ON				HEALTHY			
P	ω	A	α [s^{-1}]	φ	ω	A	α [s^{-1}]	φ
1	0	0,00	3439	3,13	0,00	0,00	5,98	0,00
2	22,3	0,18	-3,06	0,15	37,9	0,06	0,17	0,38
3	42,3	0,23	0,45	-1,62	69,1	0,45	-8,46	-2,57
4	73,4	0,17	-6,28	2,06	86,4	0,55	-13,35	0,85
5	104	0,11	-7,67	0,49	121	0,50	-9,95	-2,31
6	128	0,10	-1,09	-1,26	122	0,15	-31,42	0,47
7	152	0,00	10,6	-1,81	146	0,16	-12,75	0,50
8	173	0,04	-5,51	-2,69	167	0,03	-3,16	-1,15
9	199	0,03	-3,49	-1,97	194	0,01	-0,71	2,86
10	214	0,02	0,49	0,51	213	0,01	0,10	-2,12
11	215	0,00	19,6	1,05	226	0,02	0,00	-2,84
12	226	0,03	-0,01	2,96	492	0,00	-44,26	-2,88
13	376	0,00	-32,2	0,89	500	0,00	-86,80	-0,27

It must by noted that frequency ($\omega/2\cdot\pi$) corresponds with Prony's frecuency, but it is not exactly Fourier frecuency. For example, the frequency calculated with the Prony method can have arbitrary values and the frequency range is different for each signal. Due to this, frequency bands have been defined (Table 3), and Prony' spectrum components have been calculated in each bands.

Table 3 Frequency bands.

BAND	ω_B start	ω_B stop	UNITS
B1	0	$10\cdot 2\cdot\pi$	
B2	$10\cdot 2\cdot\pi$	$20\cdot 2\cdot\pi$	rad/s
B3	$20\cdot 2\cdot\pi$	$35\cdot 2\cdot\pi$	

Each component is classified in one of the three bands (BX). Then the value of amplitude and the damping factor is multiplied and finally all the products on each band (NcBX) are summed:

$$E_{BX} = \sum_{j=1}^{NcBX} A_{jp} \cdot \alpha_{jp} \quad / \omega_{jp} \in BX \quad X = 1,2,3 \quad (4)$$

For each eye, all the EBX parameters associated to each sector are averaged:

$$E_{BX\_eye} = \frac{1}{S} \sum_{se=1}^{S} E_{BXse} \quad (5)$$

The last step is to group the results of each eye depending on the type of disease to which belong (MS-ON, MS-non-ON and HEALTHY).

$$E_{BX\_group} = \frac{1}{e} \sum_{eye=1}^{e} E_{BX\_eye} \begin{cases} e = 27 \, MS \, ON \\ e = 19 \, MS \, NO \, ON \\ e = 20 \, HEALTHY \end{cases} \quad (6)$$

III. RESULTS

The results of applying the method are shown in table4. In all cases, amplitude A is positive, while the damping factor is negative in 92% of analyzed data. This explain why $A\cdot\alpha$ product is negative in all the cases.

Table 4 Mean±SD values

	MS ON	MS NO ON	HEALTHY
E_{B1_group}	-1,83±1,03	-2,55±1,22	-3,77±1,47
E_{B2_group}	-4,60±2,52	-6,27±3,04	-8,43±3,25
E_{B3_group}	-1,22±0,57	-1,61±0,88	-1,88±0,68

The mean of the E_{BX} differed between groups. The values are bigger in the ON patients and smaller on healthy patients.

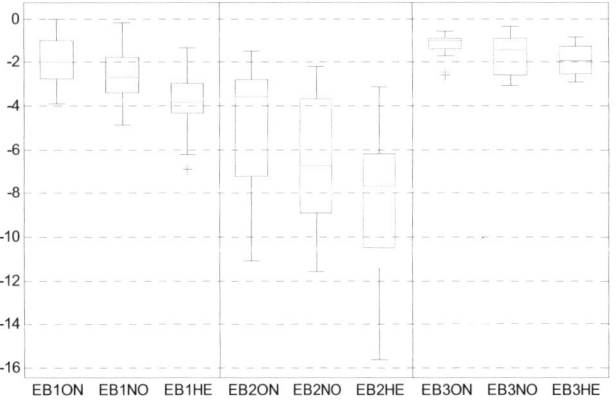

Fig. 3 Boxplot for data.
(ON =Neuritis Optics, NO= No Neuritis Optics, HE=Healthy) Red line corresponds with 50% of values.

The boxplot in Figure 3 represents the degree of dispersion and the interquartile range of the data.

A statistical test of comparison between groups is needed to test for significant differences. The p value has been calculated by the ANOVA function of Microsoft Excel. If the p values are smaller than 5% it has been considered that there is a significant difference between the sets of data.

Table 5 p values

	ALL	ON-NO ON	NO ON-HEALTHY	HEALTHY - ON
E_{B1_group}	0,00%	3,79%	0,75%	0,00%
E_{B2_group}	0,02%	4,84%	3,96%	0,00%
E_{B3_group}	0,80%	7,92%	27,51%	0,08%

In the case of the study for the three group (column ALL) the p values are less than 1%, so there is significance difference between the groups. The statistical analysis performed demonstrated that in the B1 and B2 there is significance difference between the three groups of patients specially in the comparison between Healthy and ON eyes, where the p values are less than 1%. According to these results the best band to classify the eyes is B1 because the p values are smaller than in the other bands.

IV. DISCUSSION AND CONCLUSIONS

In this work is presented the classification of the mfVEP signals using parameters of Prony's Method. Previous work [8] used the Prony's method to evaluate evoked potential analysis, however in the best of our knowledge, is the first time this methodology is used for mfVEP signals. In other Works, the parameters used are amplitudes and latency. For example, [9] identify abnormality (89%) in MS-ON eyes using amplitude/latency. Compared with these techniques, that only analyze amplitudes and latencies, Prony's method allows a complete analysis of the recordings. For example, one advantage is that the Prony's Method is independient of the capture channels used, because is easy to detect the polarity of the signal in the value of the parameters.

Prony's method is a powerful tool applied to the analysis of mfVEP signal. In contrast with other interpolation method (Newton, Lagrange, sinc), Prony's method is more interesting because the decomposition in decreasing exponential signal is closer to the physiological signals recorded in the mfVEP registers.

This study has shown that if the Prony's Method is applied to mfVEP signal, it will be possible to distinguish the state of the eye. Is clear the association between the ON and the alterations in the mfVEP signals, so the Prony's method could provide a marker for measure and study the progress of the disease. Thereby allowing help to provide a more accurate diagnosis in clinical practice and more appropriate treatment of MS patients.

The results presented in this article are obtained by multiplying the parameters of Prony's method, which are less intuitive that the time analysis or other descomposition in frequencies, because these analysis presents parameters (time, amplitude, frequency) which have physic meaning, in contrast with the result of this article, whose parameter (product of Prony's method parameters doesn't have physic meaning). For these reason and to confirm these results, the next step of this project is to design a classification algorithm (ie neuronal networks, fuzzy logic classifier) who allows to: classify the register and to monitor the changes in the progress of the disease.

ACKNOWLEDGMENT

This research has been partially supported by the Ministerio de Ciencia e Innovación (Spain), project: "Advanced analysis of Multifocal ERG and Visual Evoked Potentials applied to the diagnosis of optic neuropathies", with reference: TEC2011-26066.

REFERENCES

1. Atlas multiple sclerosis resources in the world 2008. World Health Organization. 2008
2. Osborne BJ, Volpe NJ. Optic neuritis and risk of MS: Differential diagnosis and management. Cleve Clin J Med. Smith J, Jones M Jr, Houghton L et al (1999) Future of health insurance. N Engl J Med 965:325–329 DOI 10.10007/s002149800025
3. Hansson M, Gansler T, Salomonsson G (1996)Estimation of single event-related potentials utilizing the Prony method. IEEE Trans Biomed Eng. 1996 Oct;43(10):973-81
4. Chen SW. Two-stage discrimination of cardiac arrhythmias using a total least squares-based Prony modeling algorithm. IEEE Trans Biomed Eng, 47:1317-1326
5. McDonald WI, Compston A, Edan G, Goodkin D, Hartung HP, Lublin FD, et al. Recommended diagnostic criteria for multiple sclerosis: guidelines from the International Panel on the diagnosis of multiple sclerosis. Ann Neurol 2001;50:121-7
6. Sutter Erich E, Imaging visual function with the multifocal m-sequence technique. Vision Research Volume 41, Issues 10–11, May 2001, Pages 1241–1255
7. Hood D C, Ohri N, Yang E B, et al (2004) Determining abnormal latencies of multifocal visual evoked potentials: a monocular analysis. Doc Ophthalmol 109: 189–199
8. V Garoosi, Ben H. Development and Evaluation of the piecewise Prony Method for evoked potential analysis. IEEE Trans Biomed Eng. 2000 Dec;47(12):1549-54
9. Laron M,et al (2010) Comparison of multifocal visual evoked potential, SAP and OCT in assessing visual pathway in ms patients. Mult Scler. 2010 Apr;16(4):412-26

EEG Denoising Based on Empirical Mode Decomposition and Mutual Information

A. Mert[1] and A. Akan[2]

[1] Dept.of Marine Eng., Piri Reis University, 34940, Istanbul, Turkey
[2] Dept. of Electrical and Electronics Eng., Istanbul University, 34320, Istanbul, Turkey

Abstract—Empirical mode decomposition (EMD) is a recently introduced decomposition method for non-stationary time series. EMD has an information preserving property so the sum of the decomposed intrinsic mode functions (IMF) can be used to reconstruct the original signal. However, when the signal is corrupted by white Gaussian noise, some of the IMFs may contain most of the noise components. Thus, determining which IMFs have informative oscillations or information free noisy components becomes the main challenge in denoising. In this study, mutual information (MI) is used as a metric to find information free noisy IMFs. Without using an extra reference signal, MI of each IMF and MI of the autocorrelation function (ACF) of each IMF is computed. An adaptive thresholding scheme based on MI scores is applied to decide which IMFs contain most of the white noise. Proposed method is tested on epileptic and normal EEG recordings corrupted by additive white noise to show its denoising capability on stochastic time series.

Keywords—Information theory, Mutual information, White noise reduction, EEG denoising.

I. INTRODUCTION

Processing of signals in a noisy environment is a major problem in biomedical signal processing. Any irrelevant component can be considered as noise which interferes with the informative signal. Especially, white Gaussian noise interference is the foremost problem to remove from the signal.

The simplest method for denoising is to apply some linear filtering methods, in case of the noise and informative signal spectrum are known and non-overlapping. However, the noise and signal should be stationary [1]. Wiener filters are generally used linear techniques for smoothing additive high frequency noise under the assumption of wide sense stationarity [2]. Wavelet denoising approach is considered more appropriate to process non-stationary signals. Thus, a threshold is applied to Wavelet coefficients which are considered having low amplitude levels to remove noisy components [3, 4]. Although it has satisfactory time-frequency resolution, the effect of wavelet function makes it ineffective especially for removal of white noise or noise with sharp edges and impulses of short duration [5]. Besides, empirical mode decomposition (EMD) has been recently introduced by Huang et al. [6] to analyze non-stationary and non-linear processes. The EMD has an advantage of basis function flexibility which is derived from signal itself. Thus, it makes signal analysis adaptive when compared to other methods.

EMD extracts the signal into a few intrinsic mode functions (IMF) that are derived from the oscillations included in the signal. Thus, some of them can be informative or information free noisy components of the original signal. Hence, it is considered as the first step for EMD based denoising methods to investigate informative level of each IMF. Wu and Huang [7] have suggested a hypothesis test-based approach to measure the relevant information level of the IMFs. However, it is reported that this approach becomes inadequate for low frequency IMFs. Hence, several methods using correlation coefficient and relative entropy are investigated to find irrelevant IMFs which are considered to be noise components and extracted from the signal [8, 9].

Electroencephalography (EEG) signals are prone to be corrupted by noise or signals existed by ocular and neural activities [10]. Moreover, the informative EEG signal and additive white noise are both random processes that makes linear filtering to be ineffective, and determining irrelevant IMFs difficult for EMD based denoising. In this study, white noise contaminated EEG signal denoising is investgated using mutual information (MI). It is used to measure the information relationship between corrupted EEG signal and each extracted IMF without using any reference. Moreover, MI is computed between the distribution of amplitude and autocorrelation levels for each IMF and noisy EEG signal. Thus, an effective and stable approach is proposed to determine MI score thresholding to remove irrelevant IMFs.

II. EMPIRICAL MODE DECOMPOSITION

The EMD has been recently introduced by Huang et al. [6] as a tool of data driven and adaptive decomposition method into intrinsic AM-FM components called IMFs so that the sum of IMFs is equal to original signal. Each IMF is required to satisfy two criteria: First, the number of the extrama and the number of zero crossings must be equal or must differ by one at most. Second, the mean of the envelopes determined by the local maxima namely, upper envelope and local minima called lower envelope [11]. The most important algorithm of EMD is to find IMFs called *Sifting*, which is composed of the following steps [12]:

(a) Find local maxima, M_i, $i=1,2,...$, and minima, m_k, $k = 1,2,...$, in $x(n)$.
(b) Compute the corresponding interpolating signals $M(n):=f_M(M_i,n)$, and $m(t):=f_m(m_k,n)$. These are the upper and lower envelopes of the signal.
(c) Let $e(n):=(M(n)+m(n))/2$.
(d) Substract $e(n)$ from the signal: $x(n):=x(n)-e(n)$.
(e) Return to step (a) and stop when $x(n)$ remains nearly unchanged.
(f) After obtaining an IMF, $\varphi_i(n)$, remove IMF from the signal $x(n):=x(n)- \varphi_i(n)$ and return to (a) if $x(n)$ is not constant or trend, $r(n)$.

Thus, the signal can be reconstructed by the sum of IMFs described as follows;

$$x(n) = \sum_{i=1}^{N}\varphi_i(n)+r(n) \quad (1)$$

where N is the total number of extracted IMFs. In this study, we will aim of finding the IMFs containing mostly noise components, and then eliminate them to denoise the EEG signal.

III. MUTUAL INFORMATION

Information theoretic based approaches for information and reduction of uncertainty about a nonlinear and stochastic signal are advantageous over linear measures [13]. MI measures how much the realization of random variable Y defines about the realization of random variable X, denoted by $I(X;Y)$ can be computed by the entropy $H(X)$ and conditional entropy $H(X|Y)$ formulated for a discrete random variable as follows;

$$H(X) = -\sum_{x}P(x)\log P(x) \quad (2)$$

$$H(X|Y) = -\sum_{y}P(y)\sum_{x}P(x|y)\log P(x|y) \quad (3)$$

$$I(X;Y) = H(X)-H(X|Y) = \sum_{x}P(x)\sum_{y}P(x|y)\log\frac{P(x|y)}{P(x)} \quad (4)$$

Thus, $I(X;Y)$ can be considered as an information quantification measure between the signal and its decomposed IMFs. In case the signal is non-linear and stochastic such as EEG signals, it is expected to be a more reliable metric to measure the information level of IMFs' containing white noise.

IV. PROPOSED INFORMATION THEORETIC APPROACH

Decomposition of any noisy signal into IMFs, the extracted IMFs can be relevant oscillations or irrelevant information free oscillations namely noise. Based on observation, the first few IMFs contain mostly white noise components compared to others due to high frequency oscillation behavior of the noise. As such, noise free IMFs can be used to reconstruct the denoised version of the signal.

Similar to Wavelet denoising, a threshold score is required to determine which IMFs refer to noisy oscillations. MI is investigated whether it is reliable metric to distinguish information free IMFs. Further, a threshold is defined based on this MI score to determine the noisy IMFs.

The MI score of $\varphi_i(n)$ can be calculated with respect to either the given signal $x(n)$, or any other reference noise. However, in real applications, the interference or noise is usually unknown. Therefore, we choose to compute MI between $\varphi_i(n)$ and noisy signal, $x(n)$ used as the reference. Now, there are two problems need to be addressed: i) the signal to noise ratio (SNR) of the reference signal, $x(n)$. ii) the type of the noise.

The MI scores between $\varphi_i(n)$ and $x(n)$ yield unreliable detection of noisy IMFs. Thus, we propose a method using autocorrelation function (ACF) of the IMFs before computing the MI scores. ACF based approach gives more distinguished MI scores for the noisy IMFs. The steps of the algorithm are described below:

(a) Let $x(n) = s(n) + w(n)$, be the observed noisy signal where $s(n)$ is the desired signal and $w(n)$ is the additive white noise.
(b) Decompose $x(n)$ into $\varphi_i(n)$, $i=1,2,...,N$, where N is the total number of IMFs.
(c) Compute $R_i(k)$ and $R_x(k)$, which denote ACFs of $\varphi_i(n)$ and $x(n)$ respectively.
(d) Compute $I_i = I(R_i(k); R_x(k))$, where I_i is the MI score between $R_i(k)$ and $R_x(k)$.
(e) Normalize, $I_i = I_i / \max_j(I_j)$, $i=1, 2, ..., N$.
(f) Check if the signal is contaminated by additive noise, i.e., If $I_{max} - I_{min} < 0.8$, then the signal contains noise.
(g) Determine a threshold, $\theta = 0.5(I_3 - I_1) + I_1$.
(i) Then reconstruct a denoised signal, $\hat{s}(n)$ as;

$$\hat{s}(n) = \sum_j \varphi_j(n), \quad j = \{i | I_i > \theta\}.$$

The step (f) is a check point for unknown observed signal to determine if the signal is noisy or not. In case that the signal is decided to be noisy, denoising steps *(g-i)* are applied. During the testing stage of the proposed method, it was observed for signals corrupted by white noise that the max-min range of MI scores is lower than 0.80.

The method is first tested on synthetic signal corrupted by different levels of additive white noise. After verifying the results, we applied the algorithm on EEG signal to check its denoising performance.

V. RESULTS

One channel epileptic and normal EEG recordings [14] with 200 Hz sampling frequency are used to test the method. EEG recordings are decomposed into IMFs using the EMD algorithm developed by Rato et al. [12], and shown in Fig. 1.

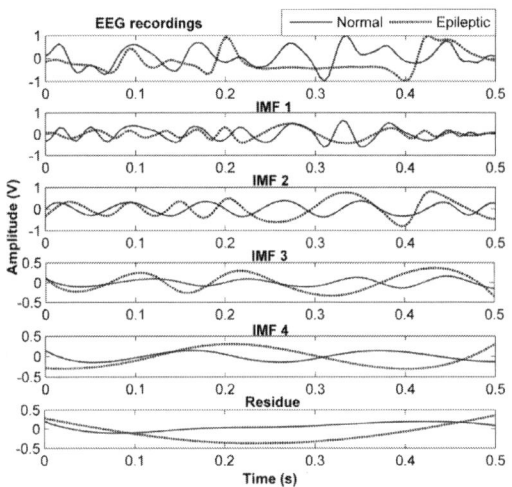

Fig. 1 Original EEG recordings and IMFs

It can be seen that original signals are noise free. The step (f) is proven as all MI scores $I_i = I(R_{signal}; R_{IMFi})$ shown in Fig. 2 are above 0.8.

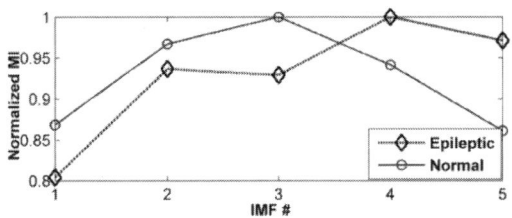

Fig. 2 MI scores of noise free EEG signals

Noise is added to the signals shown in Fig. 1 to test the proposed method at several SNR levels. The result for 20 dB SNR is given in Fig. 3. Notice that, the IMF 1 appears to be mostly white noise dominant oscillation due to its lower MI score than threshold. Noisy normal and epileptic EEG signals and their IMFs are shown in Fig. 4.

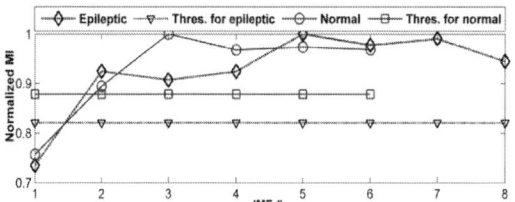

Fig. 3 MI scores of 20 dB white noise added EEG signals

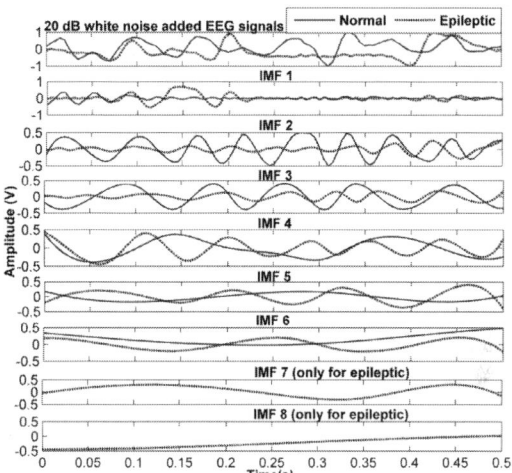

Fig. 4 Noise added EEG signals and their IMFs.

While total number of IMFs is five for noiseless EEG signals, it was six and eight for white noise added normal and epileptic EEG signals, and the first IMF is the extracted oscillation of white noise, and denoised signal is shown in Fig. 5. In order to evaluate the performance with increased noise level, our method is tested for 0 dB SNR, and the MI score graph is given in Fig 6.

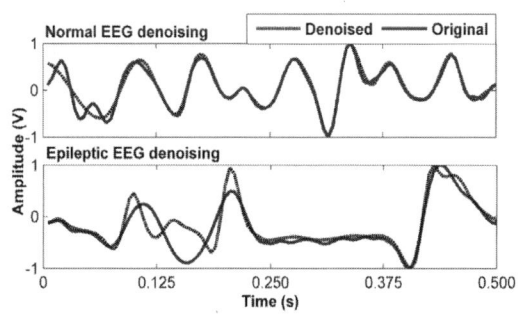

Fig. 5 Denoising of EEG signals corrupted by 20 dB white noise.

As shown in Fig. 6, the first two IMFs of normal and epileptic EEG are due to white noise oscillations that are excluded for denoising. Thus, denoised EEG signals are shown in Fig. 7.

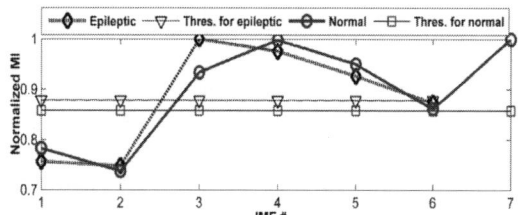

Fig. 6 MI scores of 0 dB white noise added EEG signals.

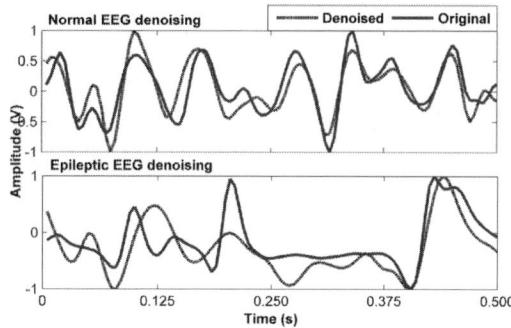

Fig. 7 Denoising of EEC recordings corrupted by 0 dB white noise.

Although the white noise dominant two IMFs are extracted, it seems that noise components exist in the informative EEG related IMFs due to nature of EMD. Finally, the MSE values of the denoised signals contaminated by 20 dB and 0 dB noise are 0.0104 and 0.0531 for normal, 0.0069 and 0.1195 for epileptic EEG respectively.

The resulted MSE values show that the proposed information theoretic approach to EEG denoising based on empirical mode decomposition is successful to determine the IMFs containing mostly white noise and it may be used for reconstruction of a denoised version of the signal.

VI. CONCLUSIONS

In this study, white noise removal on EEG signals using the Empirical Mode Decomposition (EMD) and information theoretic approach based denoising method is proposed. Mutual information (MI) is used as a metric to define the informative level of extracted intrinsic mode functions (IMFs) referencing to the noisy EEG signals. Since white noise is information free, the MI scores of IMFs containing mostly white noise are considered to be lower than the other ones. However, as both signal and IMF contain white noise, it becomes an unstable metric. Hence, the method is improved by computing MI between the autocorrelation function (ACF) of the noisy signal and the ACF of each IMF to make this white noise indicator more obvious. In our simulations, this approach is applied to normal and epileptic EEG signal denoising at several SNR levels to test its performance on non-stationary, stochastic signals. Mean square error (MSE) results indicate that the proposed method can be successfully applied to determine which IMFs contain mostly white noise or which IMFs should be used for reconstruction. Finally, the proposed method has the advantage of not requiring any reference signal as we use the noisy signal itself as a reference to calculate the MI scores.

REFERENCES

1. Chkeir A, Marque C, Terrien J, Karlsson B (2010) Denoising Electrohysterogram via Empirical Mode Decomposition. ISSNIP Biosig. And Biorobot. Conf., pp. 32-35, January.
2. Wiener N (1949) Extrapolation, Interpolation, and Smoothing of Stationary Time Series. Wiley, New York.
3. Donoho D L, Johnstone I M (1994) Ideal spatial adaptation via wavelet shrinkage. Biometrica 81:425-455.
4. Donoho D L (1995) De-noising by soft-thresholding. IEEE Trans. Inform. Theory 41:613-627.
5. Boudraa A O, Cexus J C (2006) Denoising via Empirical Mode Decomposition. Proc. ISCCSP pp1-4 .
6. Huang N E, Shen Z, Long S R et al (1998) The empirical mode decomposition and the Hilbert spectrum for nonlinear and non-stationary Time Series Analysis. Proceeding of Royal Society A 454: 903-995..
7. Wu Z, Huang N E (2005) Statistical Significance Test of Intrinsic Mode Functions. World Scientific Inc, Singapore.
8. Ayenu-Prah A, Attoh-Okine A (2010) A Criterion for Selecting Relevant Intrinsic Mode Functions In Empirical Mode Decomposition. Adv in Adapt Data Analy 2:1-24.
9. Tseng C Y, Lee H C (2010) Entropic Interpretation of Empirical Mode Decomposition and Its Applications in Signal Processing. Adv in Adapt Data Anal 2:429-449.
10. Fei-long Y, Zhi-zeng L (2012) The EEG De-noising Research Based on Wavelet and Hilbert Transform Method. Proc. ICCSEE, pp 361-365.
11. Rilling G, Flandrin P (2008) One or Two Frequencies? The Empirical Mode Decomposition Answers. IEEE Trans Signal Proces 56:85-95.
12. Rato R T, Ortigueira M D, Batista A G (2008) On the HHT, its problems, and some solutions. Mech Syst Signal Pr 22:1374-13894.
13. Mehboob Z, Yin H (2011) Information Quantification of Empirical Mode Decomposition and Its Applications to Field Potentials. Int J Neural Sys 21:49-63.
14. http://eeganalysis.web.auth.gr/dataen.htm#EpilepticEEG, accessed March 2013.

Use macro [author address] to enter the address of the corresponding author:

Author: Ahmet Mert
Institute: Piri Reis University
Street: Tuzla
City: Istanbul
Country: Turkey
Email: amert@pirireis.edu.tr

Decomposition Analysis of Digital Volume Pulse Signal Using Multi-Model Fitting

Sheng-Cheng Huang[1], Hao-Yu Jan[1], Geng-Hong Lin[1], Wen-Chen Lin[1,2], and Kang-Ping Lin[1,2]

[1] Department of Electrical Engineering, Chung-Yuan University, Taoyuan, Taiwan
[2] Holistic Medical Device Development Center, Chung-Yuan University, Taoyuan, Taiwan

Abstract—**This study proposed a mathematical model of combining with Gamma function and Gaussian function, based on the characteristic of DVP waveform, that starts with a steep half and end with flat. This model is distinct from the previous mathematical models, which only use Gaussian function. We performed decomposition of four types of DVP waveform by using the proposed model and Multi-Gaussian model. The results showed that proposed method was suitable for deconstruction of the DVP waveform, and the mean square error(MSE) was 0.2058 ± 0.0979 (10^{-3}). In this study, we succeed in proving that the proposed model is suitable applying on PPG signals.**

Keywords—Pulse Decomposition Analysis, Digital Volume Pulse s, PPG waveform.

I. INTRODUCTION

Cardiovascular Disease(CVD) is one of the major causes of sudden death and critical illness. Effective monitoring cardiovascular status has become a research goal. At present, Doppler Ultrasound is the most clear in noninvasive methods which can directly explore blood vessels including directly revealing the blockage of vessels and blood flow velocity. However, this method has limitations that it can be operated by professionals and with expensive costs.

Additionally, noninvasive pressure and optical methods are more economic which uses tonometry and optical sensor respectively. Among those Photoplethysmography(PPG) is the focus of recent research because of its simple measuring method. It is commonly applied on finger Digital Volume Pulse(DVP). The two noninvasive methods mentioned above use different Physical principles to display the status of blood inside vessels. The waves show traits that indirectly present the characteristics of blood vessels especially applied on Arteriosclerosis study. Previous studies have developed different ways of DVP analysis such as wavelet transform for features extraction and Fourier Transform and Spectral Analysis[7-8]. Nevertheless, Pulse Decomposition Analysis(PDA) is an effective way for features extraction and analysis. Rubins[1] *et al.* and Couceiro[2] *et al.* have proposed using Gaussian Function as mathematical model for PPG signal Decomposition Analysis.

This research proposed a more suitable PPG signal mathematical model according its wave characteristics. It is a Gamma and Gaussian complex model. According to the PPG waveform steep first then flat latter, using the four types of real PPG Dawber *et al*[6]. introduced as samples to perform Decomposition Analysis proposed using Gamma function to fit the steep half and using Gaussian Function to fit the flat half.

II. MATERIALS AND METHODS

A. DVP data

In 1966, Dawber et al. [6] categorized four different types DVP waveforms collected form 1779 individuals. Four types of DVP waveforms were used for decomposition in this study.

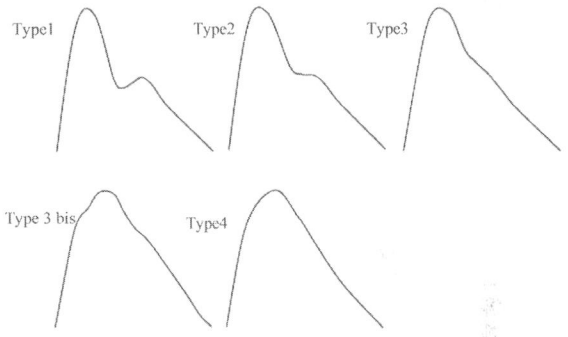

Fig 1. Classification of the digital volume pulse (DVP) waveform according to Dawber et al. [6].

B. Gamma-Gaussian complex Model

This study based on the characteristic of DVP waveform, that start with a steep half and end with flat half, proposed a mathematical model of combining Gamma function and Gaussian function. This following formula was introduced：

$$f(x) = \begin{aligned} & a_1 \cdot \frac{1}{\beta^\alpha \Gamma(\alpha)} x^{\alpha-1} e^{\frac{-x}{\beta}} \\ & + a_2 \cdot \frac{1}{\sqrt{2\pi\sigma_2^2}} e^{-\frac{(x-\mu_2)^2}{2\sigma_2^2}} \\ & + a_3 \cdot \frac{1}{\sqrt{2\pi\sigma_3^2}} e^{-\frac{(x-\mu_3)^2}{2\sigma_3^2}} \end{aligned} \quad (1)$$

$$\Gamma(\alpha) = \int_0^\infty x^{\alpha-1} e^{-x} dx \quad, \alpha > 0 \quad, \beta > 0$$

$$\mu_1 = \alpha \cdot \beta \quad, \sigma_1 = \sqrt{\alpha \cdot \beta^2}$$

Where a_1、a_2、a_3 denote the amplitude that of the Gamma function and the two Gaussian function, respectively, μ_1、μ_2、μ_3 denote the mean value that of the Gamma function and the two Gaussian function, respectively, σ_1、σ_2、σ_3 denote the standard deviation that of the Gamma function and the two Gaussian function, respectively. Fig. 2 shows the estimation results of feature points and decomposed digital volume pulses for a single-cycle DVP.

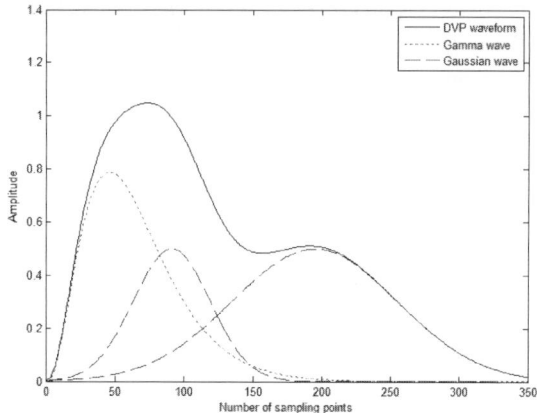

Fig 2. Decomposed digital volume pulses.

C. Parameter Estimation Method

Non-linear Least Squares (NLS) method has been maturely used on parameter estimation for a long time. To find the minimum, Powell algorithm, an iterative numerical algorithm is applied to NLS. Powell was the best method to obtain the optimal solution. It only required evaluating cost function itself without calculating gradient. Although it was strong, the performance depended on the initial condition and the global optimal solution was non-guaranteed.

The drawback could be improved by using a hybrid method combined with simplex method and genetic scheme to solve the sensitive initial condition or by using a multi-resolution algorithm hidden among many local minima to solve the desired global minimum.

It was the fastest method to solve nonlinear minimization problems which was attributed to Quadratic convergence.

D. Study Experiments:

This study used two different models, the proposed model and the Three-Gaussian model, to perform decomposition analysis four types of DVP waveform.

III. RESULTS

Fig 3 showed that the results of decomposition analysis of four types of DVP waveform. In a, b, c, d, e the above figures displayed DVP waveform fitted by our model in. The below showed the residuals of use two different models.

The above figures of (a) to (e) presented decomposition of DVP from one gamma function and two Gaussian functions and four feature points in the DVP waveform signal were obtained. From the bottom figures of (a) to (e), we acquired the residuals of the proposed model in this study which was less than the residuals of three Gaussian model. Table 1 showed the results of different model decomposition analysis. Therefore, the average errors and the standard deviations of using our model and using three Gaussian models were 0.2058 ± 0.0979 (10^{-3}) and 0.5099 ± 0.3803 (10^{-3}) respectively.

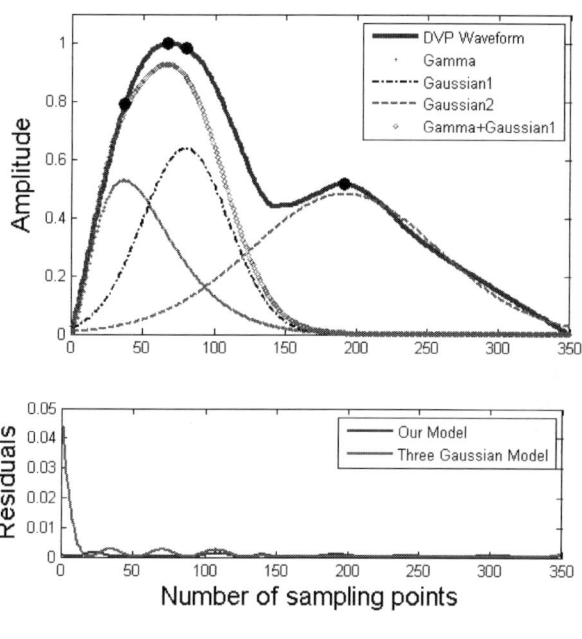

(a)

Fig 3 show that the results of four types DVP waveform decomposition analysis. Top figure shows DVP waveform fitted by Our model in a,b, c, d, e. *Blue* " — " represents the DVP signal. *Green* "×" represents the curve of Gamma function. *Black* " — — " represents the curve of the 1th Gaussian function and *Red* " — " represents the curve of the 2th Gaussian. *Purple* " ◊ " represents the sum of Gamma function and 1th Gaussian function。*Black* " ● " represents the feature points。The bottom figure shows the residuals of use two different model. *Blue* " — " represents the residuals of the proposed model，*Green* " — " represents the residuals of the three Gaussian model.

Decomposition Analysis of Digital Volume Pulse Signal Using Multi-Model Fitting

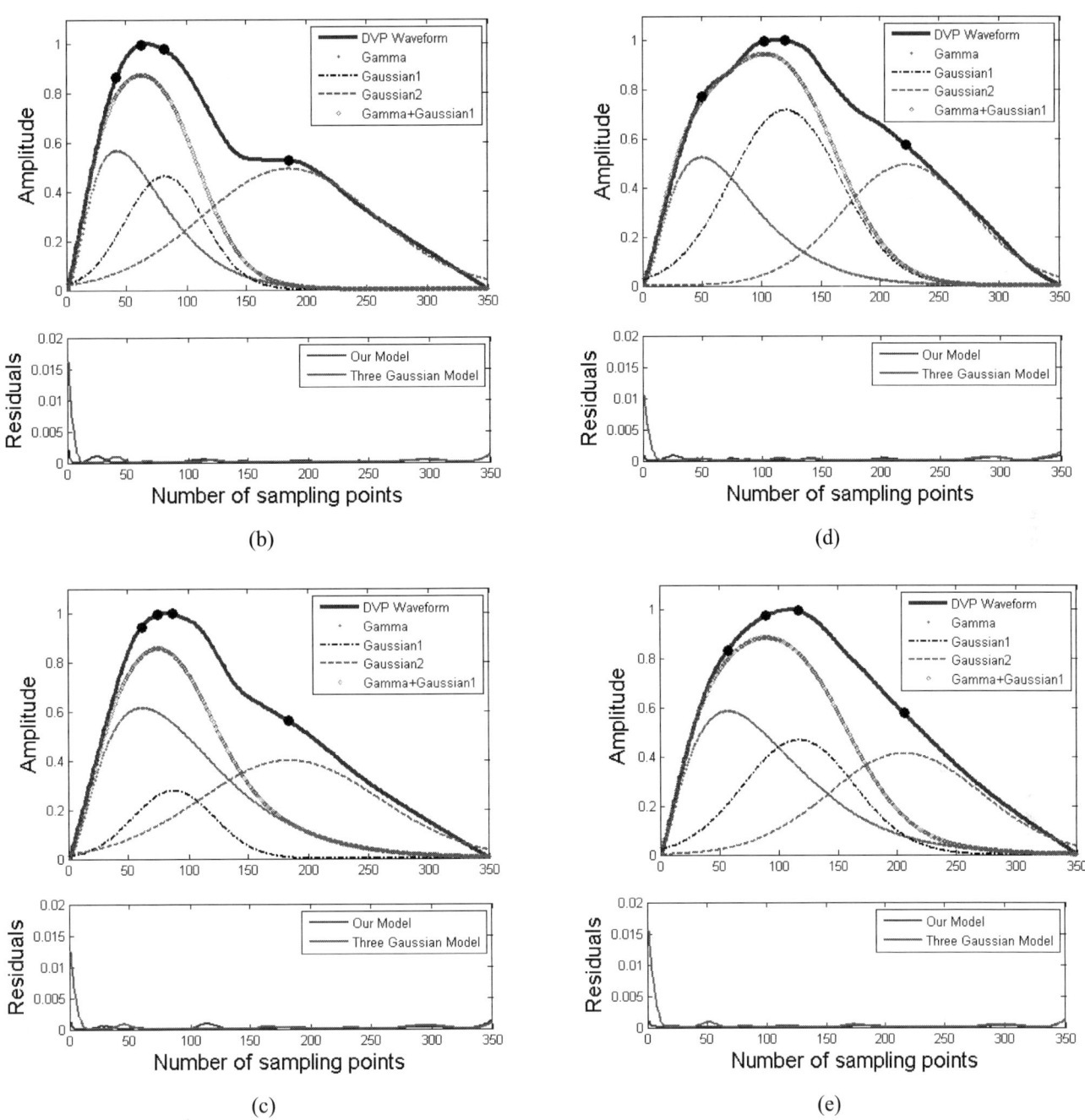

Fig. 3 (*continued*)

Table 1 The results of different model decomposition analysis

	Type 1	Type 2	Type 3	Type 3 bis	Type 4	Mean ±SD
Our Model	0.3733	0.1945	0.1733	0.1708	0.1172	0.2058±0.0979
Three Gaussian Model	1.1885	0.3311	0.3292	0.3143	0.3866	0.5099±0.3803

*Units of (10^{-3})

IV. DISSCUSSIONS AND CONCLUSIONS

Physiologically, the former ascending half area represented the force that the blood vessels' wall took during heart contraction. Also, it revealed important physiological information. Therefore, the discussion is necessary.

Traditionally, used only Gaussian function usually cannot fit very well for its smooth feature. This study proposed a gamma and Gaussian function complex model using the abruptly rising of C that provides a better fitting at the former ascending half of pulse waveform. As showed in Table 1 the comparison of the error of those two methods, the proposed model has smaller error than using gaussian function. Additionally, Fig 3 displayed the same results with different waveforms. As a result, the proposed model is more feasible than traditional Gaussian model applying in PPG signals.

This research has provided proofs that gamma and Gaussian function complex mathematical model is more suitable in the application on PPG signals, as well as various kinds of real PPG, which has accuracy in PPG analysis and decomposition.

ACKNOWLEDGMENT

This study was granted and supported by National Research Program for the Department of Industrial Technology(DoIT) of the Ministry of Economic Affairs (MOEA), R.O.C (101-EC-17-A-19-S1-163).

REFERENCES

1. RubinsU (2008) Finger and ear photoplethysmogram waveform analysis by fitting with Gaussians. Med BiolEngComput. Dec;46(12):1271-6. doi: 10.1007/s11517-008-0406-z. Epub 2008 Oct 15.
2. Couceiro R, Carvalho P, Paiva RP, Henriques J, Antunes M, Quintal I, Muehlsteff J(2012) Multi-Gaussian fitting for the assessment of left ventricular ejectiontime from the Photoplethysmogram. ConfProc IEEE Eng Med Biol Soc. 2012:3951-4. doi: 10.1109/EMBC.2012.6346831.
3. Xu L, Feng S,Zhong Y,Feng C,Meng M., Yan H (2011) Multi-Gaussian Fitting for Digital Volume Pulse Using Weighted Least Squares Method. Information and Automation (ICIA), 2011 IEEE International Conference on, Shenzhen, China, 544 – 549
4. Baruch MC, Warburton DE, Bredin SS, CoteA, Gerdt DW, Adkins CM (2011) Pulse Decomposition Analysis of the digital arterial pulse during hemorrhage simulation. Baruch et al. Nonlinear Biomedical Physics 2011, 5:1
5. Millasseau SC, Ritter JM, Takazawa K, Chowienczyk PJ.(2006) Contour analysis of the photoplethysmographic pulse measured at the finger. J Hypertens. 2006 Aug;24(8):1449-56.
6. Dawber TR, Thomas HE Jr, McNamara PM. Characteristics of the dicroticnotch of the arterial pulse wave in coronary heart disease. Angiology 1973;24:244–255.
7. Cvetkovic D,Übeyli ED, Cosic I.(2008) Wavelet transform feature extraction from human PPG, ECG, and EEG signal responses to ELF PEMF exposures: A pilot study.Digital Signal Processing Academic Press, Inc. Orlando, FL, USA. 2008 Sep:18(5): 861-874
8. Akara SA, Kara S,Latifo˘glu F, Bilgic V. (2013) Spectral analysis of photoplethysmographicsignals:The importance of preprocessing.Biomedical Signal Processing and Control 8 (2013) 16–22

Author: Kang-Ping, Lin

Institute: Department of Electrical Engineering, Chung-Yuan University
Street: 200, Chung Pei Rd, Chung Li,
City: Taoyuan
Country: Taiwan
Email: KpLin@cycu.edu.tw

A Neural Minimum Input Model to Reconstruct the Electrical Cortical Activity

S. Conforto[1,2], I. Bernabucci[1], N. Accornero[3], M. Bertollo[2], C. Robazza[2], S. Comani[2],
M. Schmid[1,2], and T. D'Alessio[1]

[1] Department of Engineering, University Roma TRE, Rome, Italy
[2] BIND - Behavioral Imaging and Neural Dynamics Center, University "G. d'Annunzio", Chieti-Pescara, Italy
[3] Department of Neurology and Psychiatry, Sapienza University of Rome, Italy

Abstract—In recent years, technology has allowed the progressive increase in the number of channels for EEG recording. The scientific rationale is the demand for an increase of the spatial resolution of the recording to better locate the sources of the underlying cortical activity. Despite some papers confirm the improvement of the spatial resolution by using 256 channels we wonder if in fact this density of electrodes on the scalp does not constitute an useless spatial oversampling. Thus we set out to determine whether the amount information derived from a standard 19 channel EEG recording was obtainable with a smaller number of electrodes, in particular with a mounting to 8 channels.

Were used and compared the performance of a Perceptron, a Feed-Forward and a Recurrent neural networks, after supervised training by the back-propagation algorithm. The target was to reconstruct the signals of all the 19 channels starting from only 8 input channels. The data-set was built by using multi-subjects 19 channels recordings containing examples of normal, generalized and focal abnormal EEG activity.

All the types of network have been able to reconstruct the missing channels with an error lower than 1%. From this pilot study seems to conclude that the information content of this 8-channel EEG is equivalent to that obtainable with a number of channels more than double. Further developments will check the optimal ratio between the number of recorded and reconstructed channels and the applicability of the approach in real-life contexts.

Keywords—Artificial Neural Network, EEG, Spectral Analysis, Time Analysis, Amplitude Maps.

I. INTRODUCTION

Recent trends in neurophysiology aim at increasing the number of channels in EEG acquisition systems, looking for an improvement of the measurement's spatial resolution, deemed necessary to better localized the sources of cortical activity.

Some papers claimed for an improvement of the spatial resolution provided by systems using 256 channels [1,2], but the real increase of information has not been demonstrated yet, and it is controversial whether such a high number of channels only give raise to a useless spatial oversampling.

Measurement sessions using a high number of EEG channels are affected by several problems. Among these:

- High cost of the acquisition system and time-consuming mounting of the electrodes and uncomfortable conditions for the subjects undergoing the recordings. These issues are amplified when dealing with patients or with a pediatric population.
- Cumbersome management of the acquired data, high processing time and risk of overcrowding.
- Significant risk of replicated measurements for corruption of the signal quality in some channels (i.e. sweat during physical activity, time-varying noise during the acquisition, etc.).

A few-channel system able to provide the same amount and quality of information obtained by a high-density one could be the solution for all the previous issues. The ideal system should use a limited amount of hardware (i.e., a few EEG channels) and some computational intelligence to derive the information generally acquired by the neglected channels. This approach can be pursued if the available recordings contain information on all (or at least most of) the independent sources of the cortical activity from which the electrical distribution over the scalp derives. After detecting the minimum acquisition set-up, in terms of both number and location of the electrodes, a computational model for the reconstruction of the cortical activity has to be designed and implemented.

In this pilot study, we developed a computational model where a neural approach has been adopted to extend the information provided by a minimum set of measurements to the entire scalp. This is achieved by automatically extracting the signals' features and by generalizing them in the space domain.

In the literature, Artificial Neural Networks (ANNs) have been extensively used for EEG analysis and classification. In particular, ANNs have been tested to automatically recognize normal and pathologic features [3,4], for the assessment of the anesthesia level [5], and also to solve the inverse electromagnetic problem [6].

In this study, different ANN models were analyzed to see if they could be used to reconstruct a whole set of EEG channels on the basis of a subset of them. In particular, three ANN architectures were designed, implemented and compared. The minimal subset of EEG channels was determined by using an informative approach. The principle of the minimal complexity, in terms of both a-priori information (i.e. minimal input subset and training data set) and architecture of the model (i.e. topology, learning rule, training algorithm) is the rationale followed in this study. The Occam's razor ('The simpler of two models, when both are consistent with the observed data, is to be preferred') has been used for a final result adoptable in a real-life context.

This result is a neural model, driven by a reduced number of real measurements, which generates the correct cortical activity over the entire scalp (that is, it replicates the EEG traces recorded in disregarded locations over the scalp). The developed simplified system could be valuable in different fields, from the assessment of sport performance to the development of controllers for human-computer interfaces.

II. MATERIALS AND METHODS

The neural minimum input model has been designed and implemented following this logical flow-chart: *A*. Data acquisition; *B*. Neural model design; *C*. Data set for training and testing; *D*. Model validation.

A. Data Acquisition

The EEG activity was recorded through surface electrodes placed over the scalp according to the International 10-20 System. Nineteen recording channels were acquired (0.05-50 Hz band-pass filtering followed by a 256 samples/s sampling) by a digital system (Micromed, Italy).

B. Neural Model Design

Minimum Input - The 19-channel recordings were processed in order to extract the minimum set of significant channels to be used to drive the neural model. All recordings underwent a Principal Component Analysis (PCA). The first principal component (PC) explained a variance ranging from 55% to 70%. A further part of the signal variance (15-25%) was explained by the second PC, while the other PCs were highly correlated to the recording noise.

8 channels were chosen as the most representative of the first two PCs by using a correlation measure together with a selection of the channels implementing the most uniform spatial sampling of the scalp. The channels respecting both criteria are: Fp1, Fp2, C3, C4, O1, O2, T3, T4.

ANN architecture – The architecture of the network was assessed after implementing and comparing three different topologies characterized by 8 input neurons (i.e. the 8 channels of the minimum input) and 11 output neurons (i.e. the 11 channels recorded in the data set but not included in the minimum input). The analyzed topologies are: 1) perceptron network (P_ANN) with no hidden layer, implementing a mapping between 8 input samples and 11 output samples; 2) feed-forward network (F_ANN) with a 30-neurons hidden layer; 3) recurrent network (R_ANN) with a 30-neurons hidden layer and two 10-lags time delay lines (TDL) connecting the input layer with the hidden one and the hidden layer with the output one respectively.

Training algorithm – All networks have been trained on the same training set by using a supervised criterion implemented by the back-propagation algorithm.

C. Data Set for Training and Testing

The performance of the networks was assessed using different EEG data sets from 5 normal subjects, 4 patients with generalized EEG abnormalities, and 3 patients with focal EEG abnormalities. Training and testing sets were obtained from these data sets as follows.

1. DS1: data recorded from a normal subject: 20 seconds for the training set and 20 seconds for the testing set;

2. DS2: data recorded from a patient. Generalized EEG abnormalities have been segmented to separate anomalous epochs from the normal ones. The training set included the normal epochs (20 seconds) and the testing set included the focal EEG abnormalities (20 seconds), to test the generalization properties of the network on a single patient basis;

3. DS3: data recorded from 5 normal subjects. These were used together to build a unique data set then subdivided into a training and a testing set, to test the generalization properties of the network with respect to different subjects. The training set includes 4 seconds of data from 4 subjects, and the testing set 9 seconds of data from the fifth subject.

4. DS4: data recorded from 2 normal subjects, 2 patients with generalized EEG abnormalities and 2 patients with focal EEG abnormalities have been used to build a training set. The corresponding testing set was obtained using the data from other participants to the experiment (i.e., 2 normal subjects, 2 patients with generalized EEG abnormalities and 1 patient with focal EEG abnormalities).

D. Model Validation

The models have been validated by means of the Mean Square Error (MSE) related to the reconstruction of the 11 EEG channels excluded from the minimum input set. In particular, the training curve has been studied in terms of MSE with respect to the training epochs. The MSE was calculated for the reconstruction of signals belonging to the testing set in both the time and the frequency domains.

The performance of the reconstruction was assessed by comparing the amplitude maps of the original data with those obtained from the data set including both the 8 original EEG data sets and the 11 EEG data sets reconstructed with the ANN. The maps undertook a blind evaluation by a pool of expert neurologists.

III. RESULTS

After training with DS1, all the designed ANNs reconstruct the 11 neglected channels with a global MSE lower then $1*10^{-2}$. The reconstruction was performed sample-by-sample, and the obtained results show the independence of the signal traces from their time course. This feature has been demonstrated also on a surrogate version of the data sets obtained by time shuffling the data.

With regard to DS1, its surrogate and the ANNs:

- P_ANN reconstructed the channel F3 with the minimum error (MSE= $2.4*10^{-3}$), and the channel F8 with the worst performance (MSE= $1*10^{-2}$).
- F_ANN reconstructed the channel F3 with the minimum error (MSE= $2.4*10^{-3}$), and the channel F8 with the worst performance (MSE= $9.5*10^{-3}$).
- R_ANN showed the minimum error for the channel F3 (MSE= $5*10^{-3}$) and the worst performance for channel F4 (MSE= $1*10^{-2}$).

On the basis of these results, we choose the model implemented by P_ANN, because of its simplicity and its performance level, comparable to that of the other models. The reconstruction results in the time and frequency domains are compared with the original data in Figure 1.

From now on, the results refer to the performance of the perceptron P_ANN after training with DS2, DS3, DS4.

DS2 - The training reached convergence in 400 epochs, with MSE=0.002. The reconstruction MSE is minimum for F4 (MSE=$3.9*10^{-3}$) and maximum for F7 (MSE=$2.1*10^{-2}$).

DS3 - The training reached convergence in 100 epochs, with MSE=0.003. The reconstruction MSE is minimum for P4 (MSE=$7.8*10^{-3}$) and maximum for T6 (MSE=$1.8*10^{-2}$).

DS4 - The training exceeds 400 epochs with MSE=0.001. The reconstruction MSE was evaluated using three different testing sets: normal data, generalized EEG abnormalities, focal EEG abnormalities.

For the normal testing data, the reconstruction MSE is minimum for P4 (MSE=$3.3*10^{-3}$) and maximum for C3 (MSE=$1*10^{-2}$).

For the generalized EEG abnormalities, the reconstruction MSE is minimum for P4 (MSE=$1.6*10^{-3}$) and maximum for P3 (MSE=$1.1*10^{-2}$).

For the focal EEG abnormalities, the reconstruction MSE is minimum for F3 (MSE=$7.3*10^{-3}$) and maximum for T6 (MSE=$1.7*10^{-2}$). The quality of the reconstructed T6 channel in the time and frequency domains with respect to the original data can be appreciated from Figure 2.

The panel of expert neurologists who analyzed the reconstructed traces and relative spectra positively evaluated the quality of the reconstruction.

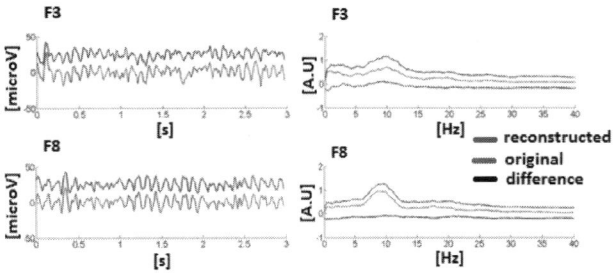

Fig. 1 Results obtained by P_ANN trained by DS1: F3 is the best and F8 is the worst reconstructed channel, respectively. The reconstructed channels are compared with the original data in both time and frequency.

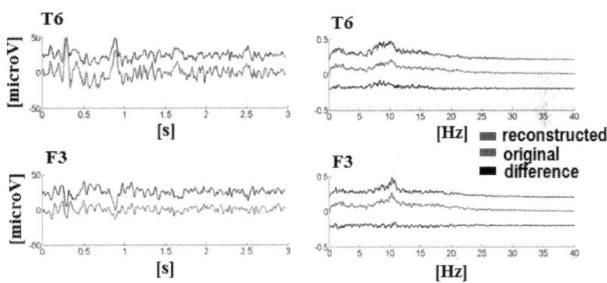

Fig. 2 Results obtained by P_ANN trained by DS4: F3 is the best and T6 is the worst reconstructed channel, respectively. The reconstructed channels are compared with the original data in both time and frequency.

Amplitude Maps - The P_NN trained and tested with DS4, chosen as the most complex and general set of data, was used to reconstruct the amplitude maps of the 11 neglected channels, and to compare them with the same maps calculated with the original data. The panel of expert neurologists defined the reconstructions acceptable and promising. An example of these maps is shown in Figure 3. Amplitude maps are calculated from the original data (upper panel), and from the 8 original channels and 11 reconstructed ones (lower panel, reconstructed traces in red).

IV. CONCLUSIONS

The designed neural models (P_ANN, F_ANN, R_ANN) have the same performance level when tested on DS1 data set. For these reasons P_ANN was chosen as the model to be preferred because of its simple topology.

Then, P_ANN was used to explore the behavior of the model in data set with increased generality, such as DS2,

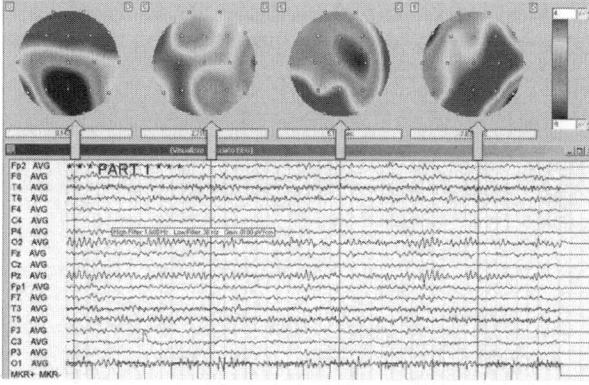

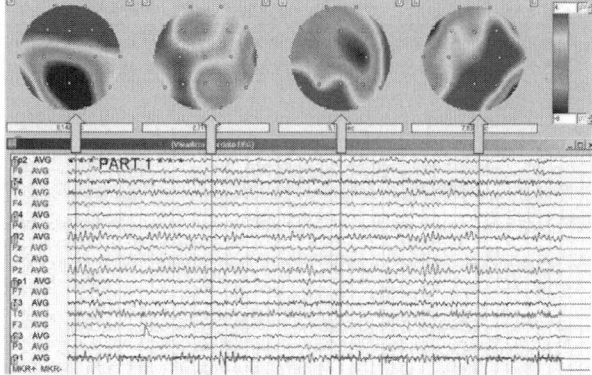

Fig. 3 Amplitude maps extracted at 4 time instants by a 19-channel recording extracted from the data set. Maps are calculated from the original data (upper panel), and from the 8 original and the 11 reconstructed channels (lower panel, reconstructed traces in red).

DS3, DS4. In particular, DS2 aims at testing the generalization capability of the model to recognize normal and abnormal EEG activities on a single patient basis. The P_ANN is trained through 400 epochs and the performance on the testing set is acceptable.

The training process becomes faster for DS3 (100 epochs), which includes data obtained from different normal subjects. The P_ANN seems to specialize in recognizing normal activity (training fast and correct) but has poorer performance than the P_ANN trained by DS2 on the testing set. This could be due to some form of overtraining.

The most important result is the one obtained with the DS4, which is the most general data set. In this case, P_ANN performs well on both training and testing data. Since testing data are composed by different EEG activities and the performance is equivalent for all the testing set, the model generalizes well over different subjects and different activities, giving rise to a promising result.

These preliminary results show that, in contrast with the trend to develop EEG systems with an increasingly higher number of channels, a significant sub-set of them (in this study, the 8 obtained by the PCA analysis) allows a functional analysis that is equivalent to obtained with a larger set of channels (i.e. the all 19 channels). This result could be explained by an information redundancy contained in the acquired signals, which increases with the number of electrodes positioned over the scalp.

The redundancy is very high for DS1. In this case the reconstruction MSE is very low, but increases (even if still acceptable) when the data sets include data from several patients/subjects characterized by different EEG activities. The quality of the reconstructed signals is good enough for the routine clinical analysis.

From the preliminary results obtained in this pilot study, the use a minimum set of recordings seems sufficient to reconstruct the EEG activity over the scalp. The proposed approach thus seems promising for a number of applications, and opens new scenarios for the implementation of smart neural computer-interfaces. Further investigations are needed to deeply test the models and to optimize the choice of the sub-set of channels constituting the minimum input set for the network.

REFERENCES

1. Pflieger ME, Sands SF (1996) 256-channel ERP information growth. Neuroimage 3.3: S10.
2. Suarez E, Viegas MD, Adjouadi M, Barreto A (2000) Relating induced changes in EEG signals to orientation of visual stimuli using the ESI-256 machine. Biomedical Sciences Instrumentation 36:33-38.
3. Liu HS, Tong Z, Fu SY (2002) A multistage, multimethod approach for automatic detection and classification of epileptiform EEG. IEEE Trans on BME 49(12):1557-1566.
4. Peters BO, Pfurtscheller G, Flyvbjerg H (2001) Automatic differentiation of multichannel EEG signals. IEEE Trans on BME 48 (1):111-6.
5. Zhang XS, Rob JR, Erik WJ (2001) EEG complexity as a measure of depth of anesthesia for patients. IEEE Trans on BME 48(12):1424-33.
6. Sun M and Sclabassi RJ (2000) The forward EEG solutions can be computed using artificial neural networks. IEEE Trans on BME 47(8):1044-50.

Author: Silvia Conforto
Institute: Department of Engineering, University Roma TRE
Street: Via Vito Volterra 62
City: Rome
Country: Italy
Email: silvia.conforto@uniroma3.it

Improved Splines Fitting of Intervertebral Motion by Local Smoothing Variation

P. Bifulco[1], M. Cesarelli[1], G. D'Addio[2], and M. Romano[1]

[1] Dept. of Electrical Engineering and Information Technology, University Federico II of Naples, Italy
[2] Fondazione Salvatore Maugeri, Department of Bioengineering, Telese Terme (BN), Italy

Abstract—Knowledge of intervertebral motion provides important information about the level of stabilization of a single spine segment. By using X-ray fluoroscopy it is possible to screen spinal motion during spontaneous patient's movement. Discrete-time experimental measurements can be interpolated by smoothing splines, which provide both noise reduction and continuous-time representation of joint motion. A patient, albeit trained, is unable to perform a spontaneous movement with such regularity to keep constant certain parameters such as the speed. Therefore, spontaneous motion can include different tracts performed at different speed. The use of a single smoothing parameter to fit the entire motion for spline fitting requires a compromise. Alternatively, smaller differences with the experimental data and more stable motion parameters can be obtained by appropriately varying the smoothing parameter: less smoothing is applied in higher velocity tracts. This concept has been applied to the analysis of cervical intervertebral motion as obtained by processing sequences of fluoroscopic images. Preliminary results showed more close representation of the experimental data without missing the regular progression of the joint movement.

Keywords—intervertebral kinematics, spline interpolation, smoothing parameter, joint motion fitting.

I. INTRODUCTION

Common pathologies of the cervical spine such as chronic whiplash dysfunction [1], arthritis [2], segmental degeneration [3], etc., may alter the kinematics of vertebral segments. Hyperkinesias of a spinal segment evidence a laxity of the joint and can be possibly corrected via surgery (spine fixation, disk prosthesis etc.). Detection of spinal instability (degenerant or traumatic) is based on accurate measurement of the intervertebral kinematic [4]. In particular, forward displacement of the vertebrae greater than 3.5 mm and angle between adjacent endplates greater than 11 degree is regarded as a sign of instability [5] and indication for surgery. Also to assess cervical arthroplasty, evaluate performance of disc prosthesis and to measure post-surgery patient's condition it is important to measure the intervertebral kinematic in-vivo [6] [7].

In spite of the importance to measure intervertebral motion in-vivo, only few, non invasive techniques are viable for clinical application. Intervertebral kinematics measurements are currently based on functional flexion-extension radiography [8] [9]. However, this method involves the use of few, end-of-range radiographic projections disregarding the evolution of the motion between these extremes.

X-ray fluoroscopy allows continuous-time screening of cervical vertebrae during spontaneous flexion extension of patient's head. Fluoroscopy is a radiological technique based on the use of image intensifiers to strongly reduce the X-ray dose and allow prolonged recording, but it produces more noisy images than those by conventional radiography. Fluoroscopic device can provide a frame rate enough high to appropriately sample the whole patient's motion.

By means of appropriate image processing, position and orientation of each vertebra can be estimated for each frame of the fluoroscopy sequence. Discrete-time trajectories of two adjacent vertebrae can be combined to obtain an estimation of the intervertebral kinematics.

Experimental measurements of the intervertebral kinematics (e.g. angles and displacements time-series) can be considered as a superposition of the true kinematic signal (i.e., intervertebral motion) and noise (i.e., measurement error). By taking into account the viscoelastic properties of the disk and other soft tissues, which provide a damping effect [10] to the joint, the actual intervertebral kinematics can be only gradual and smooth and therefore, the true kinematic signal is band-limited [11]. On the contrary, measurement errors depends on various factors (e.g., imperfections of algorithms, computation approximation, etc.) and can be considered as additive, white noise (i.e., uncorrelated, band-unlimited) [12]. Therefore, the lower frequency part of the estimated signals is mainly associated with motion, while the remaining (high-frequency content) with noise and can be significantly reduced by low-pass filtering.

Smoothing spline functions have been largely used to represent kinematics (e.g. [13]). They offer a continuous-time representation of kinematics and, at the same time, provide a low-pass filtering of the experimental measurements suppressing the noise. High frequency noise suppression is fundamental when computing the derivatives of the signals (i.e. velocity and acceleration), because derivation intrinsically emphasize high frequency. Spline functions can ensure strong continuity not only to the fitted kinematic data but also to their velocity and acceleration.

A patient, albeit trained, is unable to perform a spontaneous movement with such regularity to keep constant certain parameters such as the speed. Therefore, spontaneous motion can include tracts with a mild velocity and tract performed at more speed. In general, this will imply that kinematic signals will have a smaller bandwidth where the variation are slow and a wider bandwidth where the signal shows fast variations.

This study presents an improved spline-fitting of intervertebral motion by means of smoothing splines with a variable smoothing parameter driven by the velocity of patient's motion.

II. MATERIALS AND METHODS

A 9 inch digital fluoroscopy device (Stenoscop, GE Medical Systems) was used for in-vivo measurement. The X-ray tube parameters were adjusted for each subject, on average they were set to 1 mAs and 50 kVp; the acquisition frame rate was set to 4 frame/sec; the focus-plane length was about 1 m; image pixel resolution was 0.45 mm. Healthy subjects were instructed to spontaneously perform the maximum flexion- extension movement of their neck. A calibration phantom was used to test for geometrical distortions and to measure image noise at different gray-levels.

Being fluoroscopy images obtained by a much reduced number of photons, the image noise is Poisson's distributed and cannot be considered a mere white additive Gaussian noise. Therefore, the concise parameters of the Poisson distribution were estimated from the actual fluoroscopic images [14], to improve noise suppression. Then, the fluoroscopy sequences were preprocessed by using an edge-preserving, adaptive average filter that incorporate the information of noise variance vs. grey intensity [15].

Vertebra tracking was achieved by matching the current vertebra projection with a preselected template of that vertebra, opportunely displaced and rotated (vertebrae were assumed to be rigid and the analysis was limited to the sagittal plane [16] [17]). To automate the process of vertebra registration, the recognition of vertebra displacement and rotation was based on maximization of an image similarity measure [18] [19] [20]: the gradient correlation. The template of the cervical vertebra was chosen by selecting the part of the vertebra projection that does not superimpose with adjacent vertebra along with the whole the patient's motion. In particular, the posterior process was included in the template (in contrast with lumbar spine tracking [11], were only the vertebral body was considered).

The sequences of the values of displacement and rotation of a given vertebra obtained for each radiological frame constitute a description of the trajectory of that vertebra (see Fig. 1). From these data the intervertebral description of motion was obtained (i.e. the trajectory of the upper vertebra with respect to the lower).

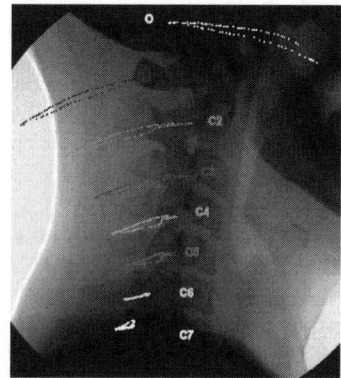

Fig. 1 Estimated trajetories of cervical vertebrae

Discrete-time intervertebral data were firstly interpolated by quintic smoothing spline [21]. These spline do not pass via all the experimental data but provide a certain degree of low-pass filtering depending on their smoothing parameter. This parameter was chosen considering an acceptable experimental data fit within the tracts of mild velocity. The steadier are the experimental data, the more low-pass filtering is recommended. If a larger bandwidth is considered, the splines tend to oscillate between the experimental data (fitting also the noise) and this phenomena will result in extremely variable estimations of velocity.

Spline polynomials also provide a continuous description of motion, which will be used to compute the instantaneous centre of rotation (ICR). ICR were also computed only for absolute angular velocities greater than a preset threshold.

By analyzing the residuals (of the fitting operation) and of the velocity some tracts of the movement cannot be well represented and need more bandwidth. Only within these tracts the residuals were interpolated with a smaller smoothing parameter (see Fig. 4) and the coefficient of the spline polynomials were updated, allowing a closer representation of the experimental data. The standard deviation (SD) of the residuals was computed to quantify the fit.

III. RESULTS

As an example, the intervertebral kinematics of the segment C5-C6 is presented in Fig. 2. Experimental data are depicted as dots, while the first spline interpolation (smoothing parameter=0.97) is shown as a continuous line, while the further interpolation as a dashed. The extension phase develops in the time interval 3-9 s, while the flexion mainly at 21-24 s.

Improved Splines Fitting of Intervertebral Motion by Local Smoothing Variation

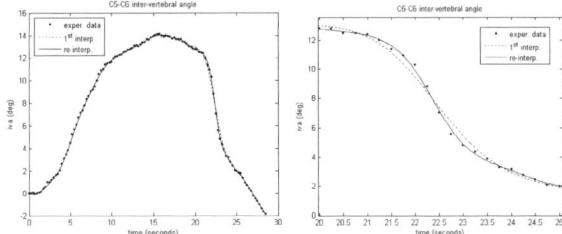

Fig. 2 (a) Intervertebral angle for the segment C5-C6; (b) Enrlagement

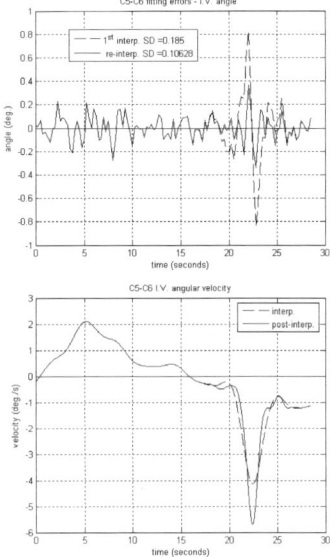

Fig. 3 (a) Residuals (angle); (b) Angular velocity. Dahed line: first spline interpolation; continuous line: re-interpolation

Figure 3 represents the residuals (i.e. spline fitting errors) for the intervertebral angle and the angular velocity, which was analytically computed from the coefficients of the interpolating polynomial. It clearly appears that, during flexion, the patient performed the motion with more speed and the fitting-errors resulted larger and more correlated.

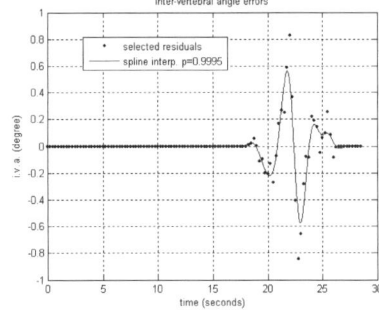

Fig. 4 Selected residuals within the tract of higher velocity and their additional spline interpolant

Figure 4 represents the re-interpolation of the only part of the residuals belonging to flexion with a lesser low-pass filtering (smoothing parameter p=0.9995). The standard deviation of kinematic data was reduced of 23% on average between a minimum of 5% and a maximum of 43%.

As example, trajectories of the ICR of a healthy subject relative to the C2-C3, C3-C4, C4-C5 and C5-C6 segments are represented in Figure 5. ICR trajectory are depicted in red for flexion and in green for extension and are superimposed on a single frame of the fluoroscopic sequence. In addition, the approximate locations of finite centers of rotation (i.e. centre computed between two extremes of motion) obtained by a previous study [16] by averaging results obtained from ten healthy subjects, were also depicted as light blue dashed circles to provide a qualitative comparison.

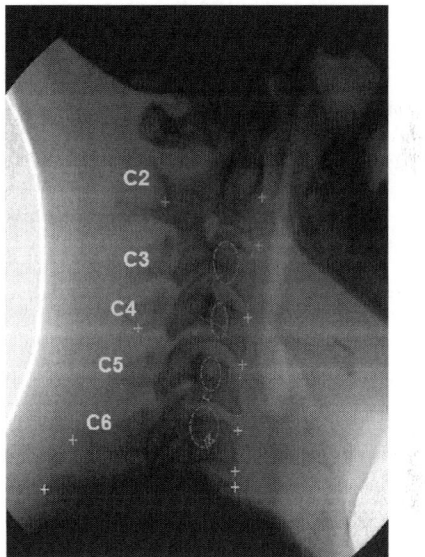

Fig. 5 Trajectories of the instantaneous centers of rotation

IV. COMMENTS AND CONCLUSIONS

By means of fluoroscopy it is possible to describe the whole progress of intervertebral motion in the plane of view. Spline interpolation provides both noise reduction and continuous representation of motion. Joint analysis of errors and velocity can lead to selection of tract of motion that need less smoothing. A finer interpolation within those tracts reduces the residuals only at the place of need, while maintaining a less variant description of the motion elsewhere. In many cases the resulting residuals were white only after re-interpolation.

At moment this technique seems inadequate (because of the frame rate) to measure intervertebral kinematics (e.g. disk deformations) during vibrations [23] [24] and rapid mechanical stress or shock (as in case of car accidents, which are common cause of cervical whiplash).

Previous studies [25] [26] pointed out that the intervertebral centre of rotation is much more sensible to mild degeneration of disk and ligament. Most of the literature presents the finite centre of rotation (easier to compute) but that represents only an approximation of the ICR. ICR trajectories can provide better understanding of the segmental motion in-vivo.

ACKNOWLEDGMENT

This work was partially supported by the DRIVE IN^2 project, founded by the Italian P.O.N. 2007/13 and by the QUAM project, founded by Italian Ministry of Economic Development.

REFERENCES

1. Kristjansson E, Leivseth G, Brinckmann P, Frobin W. Increased sagittal plane segmental motion in the lower cervical spine in women with chronic whiplash-associated disorders, grades I-II: a case-control study using a new measurement protocol. Spine. 2003;28:2215–2221. DOI: 10.1097/01.BRS.0000089525.59684.49
2. Imagama S, Oishi Y, Miura Y, Kanayama Y, Ito Z, Wakao N, Ando K, Hirano K, Tauchi R, Muramoto A, Matsuyama Y, Ishiguro N. Predictors of aggravation of cervical spine instability in rheumatoid arthritis patients: the large joint index. J Orthop Sci. 2010 Jul;15(4):540-6. DOI: 10.1007/s00776-010-1475-z
3. Boselie TF, Willems PC, van Mameren H, de Bie R, Benzel EC, van Santbrink H. Arthroplasty versus fusion in single-level cervical degenerative disc disease. Cochrane Database Syst Rev. 2012 Sep 12;9:CD009173.
4. Panjabi MM, Lydon C, Vasavada A, et al. On the understanding of clinical instability. Spine. 1994;19:2642–2650.
5. White AA, Johnson RM, Panjabi MM, Southwick WO. Biomechanical analysis of clinical stability in the cervical spine. Clin Orthop 1975; 109: 85-95
6. Nabhan A, Steudel WI, Nabhan A, Pape D, Ishak B. Segmental kinematics and adjacent level degeneration following disc replacement versus fusion: RCT with three years of follow-up. J Long Term Eff Med Implants. 2007;17(3):229-36.
7. Nabhan A, Ishak B, Steudel WI, Ramadhan S, Steimer O. Assessment of adjacent-segment mobility after cervical disc replacement versus fusion: RCT with 1 year's results. Eur Spine J. 2011 Jun;20(6):934-41. DOI: 10.1007/s00586-010-1588-2
8. Dimnet J, Pasquet A, Krag MH, Panjabi MM. Cervical spine motion in the sagittal plane: Kinematics and geometric parameters. Journal of Biomechanics. 1982; 15, 959-969.
9. Leone A, Guglielmi G, Cassar-Pullicino VN, Bonomo L. Lumbar Intervertebral Instability: A Review. Radiology. 2007; 245(1), 62-77.
10. Niosi CA, Oxland TR. Degenerative mechanics of the lumbar spine. The Spine Journal 4, 2004.202S–208S
11. Cerciello T, Romano M, Bifulco P, Cesarelli M, Allen R. Advanced template matching method for estimation of intervertebral kinematics of lumbar spine. Journal of Medical Engineering and Physics. 2011; 33(10):1293-1302. DOI: 10.1016/j.medengphy.2011.06.009
12. Challis JH. An examination of procedures for determining body segment attitude and position from noisy biomechanical data. Medical Engineering & Physics. 1995. 17 (2), 83–90
13. Xu X, Chang CC, Faber GS, Kingma I, Dennerlein JT. Interpolation of segment Euler angles can provide a robust estimation of segment angular trajectories during asymmetric lifting tasks. Journal of Biomechanics 2010. 43, 2043–2048. DOI: 10.1016/j.jbiomech.2010.03.010
14. Cesarelli M, Bifulco P, Cerciello T, Romano M, Paura L. X-ray fluoroscopy noise modeling for filter design. International Journal of Computer Assisted Radiology and Surgery. March 2013; 8(2): 269-278 . DOI: 10.1007/s11548-012-0772-8
15. Cerciello T, Bifulco P, Cesarelli M, Fratini A. A comparison of denoising methods for X-ray fluoroscopic images. Biomedical Signal Processing and Control 2012; 7(6): 550-559 . DOI: 10.1016/j.bspc.2012.06.004
16. Bifulco P, Sansone M, Cesarelli M, Allen R, Bracale M. Estimation of out-of-plane vertebra rotations on radiographic projections using CT data: a Simulation Study. Journal of Medical Engineering and Physics. 2002; 24 (4):295-300.DOI: 10.1016/S1350-4533(02)00021-8
17. Bifulco P, Cesarelli M, Allen R, Romano M, Fratini A, Pasquariello G. 2D-3D Registration of CT Vertebra Volume to Fluoroscopy Projection: A Calibration Model Assessment. EURASIP Journal on Advances in Signal Processing Vol. 2010, Article ID 806094. DOI: 10.1155/2010/806094
18. Penney GP, Weese J, Little JA, Desmedt P, Hill DLG, Hawkes DJ. A comparison of similarity measures for use in 2-D-3-D medical image registration. IEEE Trans. Med. Imaging 1998; 17(4): 586–595
19. Wu J, Kim M, Peters J, Chung H, Samant SS. Evaluation of similarity measures for use in the intensity-based rigid 2D-3D registration for patient positioning in radiotherapy. Med Phys. 2009 Dec; 36(12):5391-403. DOI: 10.1118/1.3250843
20. Bifulco P, Cesarelli M, Allen R, Sansone M, Bracale M. Automatic Recognition of Vertebral Landmarks in Fluoroscopic Sequences for Analysis of Intervertebral Kinematics. Journal of Medical and Biological Engineering and Computing. 2001; 39(1):65-75.
21. Bifulco P, Cesarelli M, Cerciello T, Romano M. A continuous description of intervertebral motion by means of spline interpolation of kinematic data extracted by videofluoroscopy. Journal of Biomechanics: 2012 Feb; 45(4): 634-641. DOI: 10.1016/j.jbiomech.2011.12.022
22. Van Mameren H, Sanches H, Beurgens J, Drukker J. Cervical spine motion in the sagittal plane (II) position of segmental averaged instantaneous centers of rotation: a cineradiographic study. Spine. 1992. 17, 467-474.
23. Fratini A, La Gatta A, Bifulco P, Romano M, Cesarelli M. Muscle motion and EMG activity in vibration treatment. Med Eng Phys. 2009 Nov;31(9):1166-72. DOI: 10.1016/j.medengphy.2009.07.014
24. Cesarelli, M., Fratini, A., Bifulco, P., La Gatta, A., Romano, M., Pasquariello, G. Analysis and modelling of muscles motion during whole body vibration. Eurasip Journal on Advances in Signal Processing 2010 , art. no. 972353 . DOI: 10.1155/2010/972353
25. Brown T, Reitman CA, Nguyen L, Hipp JA Intervertebral motion after incremental damage to the posterior structures of the cervical spine. Spine (Phila Pa 1976). 2005 Sep 1;30(17):E503-8.
26. Hwang H, Hipp JA, Ben-Galim P, Reitman CA, Threshold cervical range-of-motion necessary to detect abnormal intervertebral motion in cervical spine radiographs. Spine 2008. 33(8), 261-267. DOI: 10.1097/BRS.0b013e31816b88a4

Author: Paolo Bifulco
Institute: Dept. of Electrical Engineering and Information Technology, University Federico II of Naples
Street: Via Claudio, 21
City: Napoli
Country: Italy
Email: pabifulc@unina.it

Effects of Wavelets Analysis on Power Spectral Distributions in Laser Doppler Flowmetry Time Series

G. D'Addio[1], M. Cesarelli[2], P. Bifulco[2], L. Iuppariello[2], G. Faiella[2], D. Lapi[3], and A. Colantuoni[3]

[1] S. Maugeri Foundation, Rehabilitation Institute of Telese Telese Terme (BN), Italy
[2] Dept. of Biomedical, Electronics and TLC Engineering University of Naples, "Federico II", Naples, Italy
[3] Federico II University of Naples /Department of Clinical Medicine and Surgery, Naples, Italy

Abstract—The evaluation of endothelium function impairments are of great clinical interest in many vascular diseases and laser Doppler flowmetry (LDF) is the gold standard technique for its evaluation. LDF signals show low-frequency oscillations related to heartbeat, respiratory, myogenic and endothelial activities. Although wavelets analysis (WLT) of LDF has been shown as a better technique than Fourier (FFT) and Short Time Fourier Transform (STFT) in the resolution of these oscillatory components, the overall spectral power modifications have not yet been described. Aim of the paper is to study the effects of WLT analysis on power spectral distributions in LDF. We studied 20 min LDF recordings of 20 obese subjects by PeriFlux LDF system. Signals were detrended by moving average algorithms and transformed by FFT, STFT and WLT for spectral analysis. The spectral power has been calculated in bands I (0.6-2 Hz), II (0.145-06), III (0.052-0.145), IV (0.021-0.052), V (0.0095-0.021) and VI (0.005-0.0095) in percent values of the total spectral power. Results of the ANOVA tests between spectral powers in the six bands by the three different spectral analysis' methods showed that WLT exhibited the highest F-value, reflecting significant difference ($p<0.001$) between II vs III and IV, III vs V and VI, IV vs V and VI bands. Moreover, while WLT values in I band are significantly lower of those of FFT and STFT, WLT values, both in IV and III band, are significantly higher than those of FFT and STFT ($p<0.001$). Therefore, the overall effects of WLT analysis on power spectral distributions in LDF time series in the studied population seem to limit the spectral power in the I band moving the power to the III and IV bands with the interesting effect to minimize heart rate variability spectral power enhancing myogenic and neurogenic sympathetic activity spectral power in LDF signals.

Keywords—skin blood flow, laser Doppler flowmetry, wavelets analysis, Fourier Transform, Short Time Fourier Transform.

I. INTRODUCTION

The regulation of tissue perfusion occurs in microcirculation. Peripheral arterioles contract and relax and vary their diameter and the vascular tone through several factors including the autonomic nervous system [1]. The interaction between endothelial and smooth muscle cells plays a key role.

Endothelium function impairments have been shown in several pathological conditions; therefore, an accurate evaluation of the endothelium dysfunctions is of great clinical interest to the diagnosis and therapy of many vascular diseases.

The study of the skin microcirculation is very complex because arterioles are highly reactive to mechanical and thermal stimulus; therefore, the techniques implemented for its evaluation can be highly affected by artifacts [2,3].

Various techniques have been used to evaluate endothelial function, including brachial arterial imaging, plethysmography and laser Doppler flowmetry (LDF) but the latter is the only one allowing us to study the oscillatory components in the blood flow reflecting the endothelial and smooth muscle cell reactivity.

The LDF is a non-invasive method to evaluate microvascular blood flow measuring the velocity of the particles in solution through the interpretation of the light beam undergoing the Doppler effect [4]. Typical low-frequency oscillations of the LDF signals in human skin and their related physiological interpretations have already been described in literature [5].

Although wavelets analysis (WLT) in LDF time series, alternatively to conventional Fourier (FFT) and Short Time Fourier Transform (STFT), facilitates a better resolution of several oscillatory components on a wide frequency interval [6], the overall modifications of the spectral power values in the low-frequency bands of interest by WLT have not been identified. Therefore, the aim of the present paper was to study the effects of Wavelets analysis on power spectral distributions in LDF time series in obese peoples.

II. MATERIALS AND METHODS

A. Blood Flow Microcirculation

The main role of blood flow is the oxygenation of tissue cells. Physiological parameters that permit an adequate oxygen supply to tissues are: Blood Perfusion, Concentration of blood cells moving (CMBC) and mean velocity of the cells. These parameters are correlated by following relation:

$$\text{Perfusion Blood} = \text{CMBC} \times \text{velocity}$$

The blood perfusion represent the most important parameter to study the microvascular alterations: it is a relative value (semi-quantitative), corresponding to the product of the relative number of blood cells in motion, and causes the Doppler effect (relative speed of these cells in the measured volume). The values were expressed in Perfusion Units (PU).

The dynamics of blood flow in the microcirculation were analyzed using spectral techniques. The detailed analysis of each frequency component by the power spectrum provide information on the dynamics of mechanisms in vascular regulation. Table 1 shows the six frequency ranges correlated to specific physiological processes.

Table 1 Spectral bands of interest and their physiological interpretation

Band (Hz)	Physiological interpretation
I (0.6-2)	Heartbeat
II (0.145-0.6)	Respiratory Activity
III (0.052-0.145)	Myogenic activity
IV (0.021-0.052)	Neurogenic sympathetic activity
V (0.0095-0.021)	No-dependent endothelial activity
VI (0.005-0.0095)	Endothelial – related metabolic activity

B. Population and Study Protocol

We studied 20 LDF recordings, lasting 20 min at least in twenty obese subjects recruited for the present study.

Each subject (60 – 75 years old) under investigation was properly prepared in order to avoid erroneous results.

The flow measurement was performed in standardized environmental conditions (constant room temperature at 22-25 °C) with the patient adapted for about 30 min to the environmental conditions, because the skin blood flow is markedly influenced by the external temperature.

These conditions permit the stabilization of the heart rate. The hematocrit value must always be taken into account as its variation determines alterations in reading.

The study was conducted by PeriFlux System 5000, constituted by a main unit that could be equipped with one, two, three or four functional units.

The main unit PeriFlux (PF) 5001 was equipped with a laser diode that emits constantly divergent radiation.

This laser light was transmitted to the unit LDPM PF 5010 that permits the blood flow perfusion measurement in normal tissues.

The device, ultimately, only detects a signal correlated to red blood cells moving in a given volume multiplied by their average speed.

C. Signal Preprocessing

Signals were recorded with a 32 Hz sampling rate and expressed in perfusion unit (PU). The preprocessing includes a linear detraining of the signal which eliminates the average component and a first filtering moving average of a time window of about 200 seconds to remove trends below 0.0025 Hz and a second moving average filtering with a time window of about 0.25 seconds to remove the trends above 2 Hz.

D. Spectral Analysis

The oscillations in skin blood flow are detected by LDF signals through spectral analysis with the main advantage to separately detect and study the presence of each oscillatory component peaks within the six bands of table 1.

Spectral analysis was first simply performed by means of the Fourier Transform (FFT), obtaining the total spectrum of the signal.

The low-frequency oscillations analysis has then been improved through the time-frequency methods using a logarithmic frequency resolution. Short Time Fourier Transform (STFT) has been performed using a time window length of 200 sec with sequences of 6400 samples, calculating spectrogram of short constant segments to evaluate the time-dependent changes. The fixed spectral resolution of STFT analysis however represent an important limitation of this technique.

To achieve a more detailed analysis a variable-resolution technique like the continuous wavelet transform (WLT) has been used [6], with an adaptable window length allowing to contemporary detect both low frequencies by longer windows and high frequencies by short windows. Before calculating the wavelet transform, an array of equally spaced frequencies of 0.0001 Hz has been created and the following relationship has been defined to obtain a vector of scale starting from an array of equally spaced frequencies.

$$a = \frac{F}{f * t}$$

Where a is the scale, f is the frequency to transform in scale, t is the sampling period and F is the center frequency of the mother wavelet used; in our case a Morlet mother whose center frequency is 0.8125 Hz. The result is a vector of scale of 961 samples. It is reported in abscissa the number of samples and not the time of the observations; on the ordinate the scale for each coefficient.

The following Fig. 2 shows the comparison between the different types of spectral analysis on the same LDF time series.

Effects of Wavelets Analysis on Power Spectral Distributions in Laser Doppler Flowmetry Time Series 649

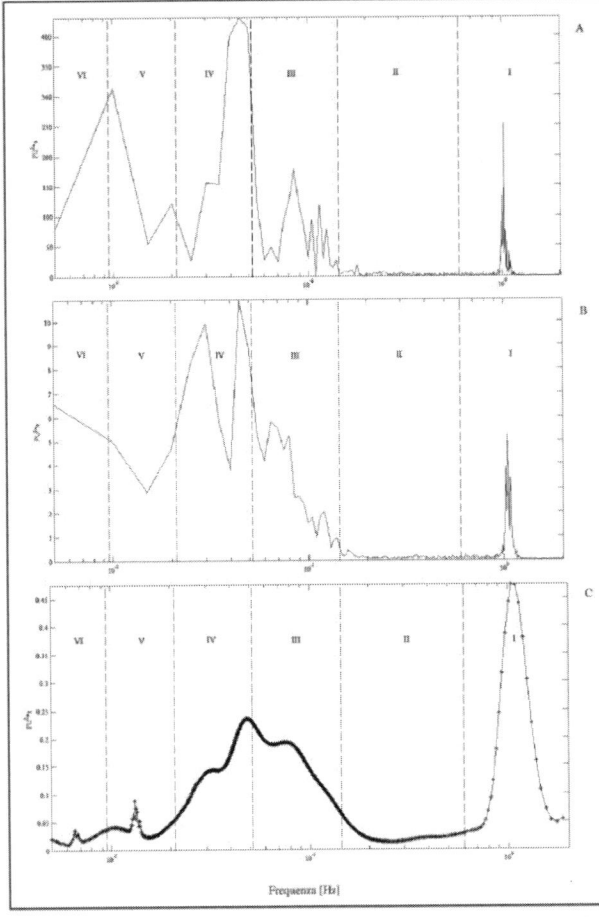

Fig.1 Twenty-minute LDF signal analyzed through FFT (A), STFT (B) and by WLT transform (C). Frequency in logarithm scale on x axis.

Table 2 mean and standard deviation values for spectral analysis.

	VI	V	IV	III	II	I
FFT	7±9	15±1	14±8	11±9	12±9	41±21
STFT	5±5	14±1	16±8	13±9	12±9	40±21
WLT	3±3	10±5	37±8	29±1	11±7	10±8

Table 3 F values of the ANOVA test (upper part of table) and values of the post Tukey's multiple comparison Test (lowe part of table).

	FFT	STFT	WLT
F value	16,27***	14,88***	45,25***
Tukey's Multiple Comparison Test			
I vs II	***	***	ns
I vs III	***	***	***
I vs IV	***	***	***
I vs V	***	***	ns
I vs VI	***	***	ns
II vs III	ns	ns	***
II vs IV	ns	ns	***
II vs V	ns	ns	ns
II vs VI	ns	ns	ns
III vs IV	ns	ns	ns
III vs V	ns	ns	***
III vs VI	ns	ns	***
IV vs V	ns	ns	***
IV vs VI	ns	ns	***
V vs VI	ns	ns	ns

Table 4 p-values of the ANOVA test within each of the three power spectral values.

	VI	V	IV	III	II	I
P value	Ns	Ns	***	***	Ns	***

III. RESULTS

Statistical results of the analyzed signals are shown in the following tables. Table 2 reports mean and standard deviation values respectively for FFT, STFT and WLT spectral analysis for each of the six studied bands, expressed as percent values of the total spectral power. The upper part of Table 3 shows F values of the ANOVA test between spectral powers (values on rows of table 2) for each of the three different spectral analysis' methods, while the lower part of the same table describes significant values of the post Tukey's multiple comparison Test (*** p<0.001) between all couples of bands for each of the three different spectral analysis methods. Finally, the table 4 reports the p-values of the ANOVA test within each of the three power spectral values calculated by the three different spectral analysis methods (values on columns of table 2) for each band.

IV. CONCLUSIONS

Results allow us to point out the following two novel findings.

First, although all the three spectral methods significantly discriminate between the six spectral bands of interest, the WLT analysis is the one showing the highest F-value. Moreover, while the significant F-values of FFT and STFT analysis are mostly due just to the very high Band I values but not to the difference between the spectral powers in the other bands, on the other hand the high F-value of the WLT analysis reflects significant difference also between II vs III and IV, III vs V and VI, IV vs V and VI bands, as clearly shown by Tukey's post test results. This is a quantitative statistical confirmation of the higher spectral resolution features of the WLT analysis in the different six bands of interest of LDF oscillations, compared to the flatter spectral profiles obtained by FFT and STFT techniques.

Second, comparing the resulting spectral power values for each band obtained by the three different spectral analysis' methods, the IV, III and I band were the only bands with significant different values ($p<0.001$). Particularly while WLT values in I band are lower of those of FFT and STFT, on the other hand WLT values, both in IV and III band, are higher than those of FFT and STFT.

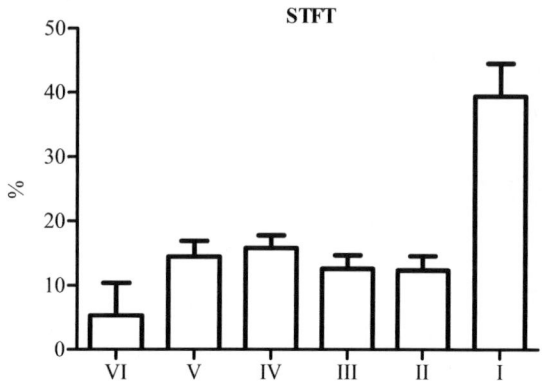

Fig. 2 Barplot of the mean and standard deviation values of the power in the six different studied bands by STFT spectral analysis (y values as percent values of the total spectral power)

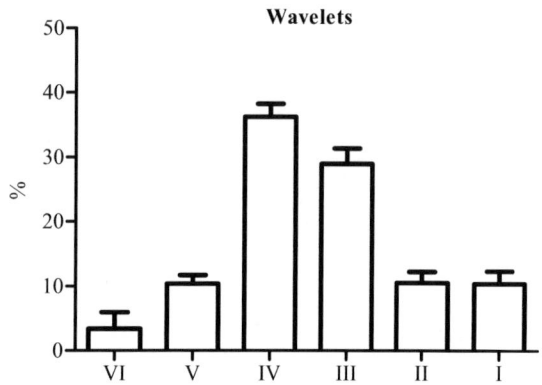

Fig. 3 Barplot of the mean and standard deviation values of the power in the six different studied bands by Wavelets spectral analysis (y values as percent values of the total spectral power)

Therefore, we suggest that the overall effects of WLT analysis on power spectral distributions in LDF time series in the studied population seem to limit the spectral power in the I band moving the power to the III and IV bands. This is a very interesting effect because WLT is able to minimize the portion of the power spectrum related to the heart rate variability enhancing the contribution of myogenic and neurogenic sympathetic activity in LDF signal.

Acknowledgment

This work was partially supported by the DRIVE IN^2 project, founded by the Italian P.O.N. 2007/13 and by the QUAM project, founded by Italian Ministry of Economic Development.

References

1. H. Z.Cummins, N. Knable and Y. Yeh. "Respiratory and cardiac effects on venous return" *Phys. Rev. Lett.,* 12:150-153, 1964.
2. M.D.Stern. "In vivo observation of microcirculation by coherent light scattering, *Nature*, 254:56-58, 1975.
3. D. Watkins and G.A. Holloway. "An instrument to measure cutaneous blood flow using the Doppler shift of laser light", *IEEE Trans Biomed Eng BME*, 25:28-33, 1978.
4. Kvandal, P., Landsverk, S.A., Bernjak, A., Stefanovska, A., Kvernmo, H.D., Kirkebøen, K.A., 2006. Low frequency oscillations of the laser Doppler perfusion signal in human skin. Microvasc. Res. 72, 120–127.
5. Landsverk, S.A., Kvandal, P., Kjelstrup, T., Benko, U., Bernjak, A., Stefanovska, A.,Kvernmo, H., Kirkebøen, K.A. "Human skin microcirculation after brachial plexus block evaluated by wavelet transform of the laser Doppler flowmetry signal", Anesthesiology 105 (3), 478–484, 2006..
6. Bernjak A, Stefanovska A. Importance of wavelet analysis in laser Doppler flowmetry time series. Conf Proc IEEE Eng Med Biol Soc. 2007;2007:4064-7.

Author: Giuliana FAIELLA
Institute: D.I.E.T.I., University "Federico II"
Street: via Claudio, 21
City: Naples
Country: Italy
Email: giuliana.faiella@gmail.com

Outliers Detection and Processing in CTG Monitoring

M. Romano[1], G. Faiella[1], P. Bifulco[1], G. D'Addio[2], F. Clemente[3], and M. Cesarelli[1]

[1] D.I.E.T.I. University of Naples "Federico II", Naples, Italy
[2] S. Maugeri Foundation, Rehabilitation Institute of Telese, Telese Terme, Italy
[3] Istituto di Ingegneria Biomedica - Consiglio Nazionale delle Ricerche (IsIB-CNR), Roma, Italy

Abstract—Cardiotocography is a technique used to assess the foetal wellbeing using the recording of foetal heart rate and uterine contractions. Generally these signals are observed to get information about the foetal development and wellbeing and it represents the only medical report that has a legal value to testify the foetal health. Unfortunately, really accurate prediction of foetal wellbeing is still a goal quite difficult to reach. Some problems are related to the visual interpretation of cardiotocographic traces, others to the acquisition system that, in some conditions, can cause the degradation or the loss of the signal. These anomalies, cardiac arrhythmia and artifacts in foetal heart rate signals do not represent physiological variations, so that they can be defined outliers. In cardiotocography literature the problem of outliers processing is underestimated even if they affect both time and frequency analysis. When faced, the outlier problem is solved with a pre-processing phase that uses algorithms to detect and remove spikes.

In this work, we firstly present in detail the updated version of an algorithm for detection and correction of outliers. Then, we evaluate the impact of outliers and of different correction strategies on foetal heart rate analysis. Obtained results demonstrate that the proper outliers detection is fundamental because they heavily influence the estimation of foetal heart rate variability, calculated by the short term variability, a parameter well known for its diagnostic value.

Keywords—Computerised cardiotocography, outliers, FHR pre-processing, Short Term Variability (STV).

I. INTRODUCTION

Foetal monitoring is a branch of medicine as important as delicate, since, in general, it is not possible to employ direct measurements. Cardiotocography (CTG) is an indirect technique to assess foetal wellbeing, by means of the recording of foetal heart rate (FHR) and uterine contractions (UC), typically employed in clinical environment at the end of pregnancy. Let us recall briefly that cardiotocography was introduced mainly to early detect hypoxic states of the foetus. Nevertheless, despite to initial hopes, really reliable results have not yet been obtained. In fact, while the negative predictive value of CTG is very good, it is known that foetuses compromised not always manifest cardiotocographic changes. This is because the correct CTG acquisition and its accurate reading are two goals quite difficult to reach. In spite of these considerations, it is important to underline that the CTG is currently the most widely used procedure for assessment of the foetal state and, in some countries, is a medical report with a legal value. For this reason, many efforts have been made to attempt to make CTG interpretation more reliable and still make very active this field of scientific research [1, 2].

Computerised CTG was initially introduced to this aim, providing more objective and reproducible evaluations of some parameters of diagnostic interest. Among those, FHR variability (FHRV), being related to the functioning and development of the autonomous nervous system (ANS), is doubtless one of the most important and of the most difficult to estimate by naked eye [3, 4, 5]. Nevertheless, also in computerised CTG, some aspects have not yet been faced with sufficient deepening.

About the acquisition systems, it is known that the FHR signal is recorded with a US Doppler probe and that often relative movements between it and foetus cause the degradation or even the loss of the signal. The erroneous recording of FHR samples may lead to substantial differences in near heart intervals and in turn generate spikes in the CTG trace (signal artifacts). Other anomalous FHR samples can have a physiological nature, being due to foetal cardiac arrhythmias. However, these arrhythmias do not represent an expression of the normal behaviour of the ANS, so, together with artifacts can be considered outliers and can affect CTG processing [6].

Although outliers treatment is an important task of signals processing, in cardiotocography literature it is an extremely underestimated problem. It is very difficult to find indications about this specific topic and, in the case, the only operation described is the deletion of isolated spikes [7].

In our opinion, instead, can be very important to correctly manage outliers, so, in this work, we updated and optimised a previously developed software for outliers detection and correction [6] and tested how outliers presence affects the computation of a fundamental parameter related to FHRV, the short term variability (STV).

II. MATERIALS AND METHODS

A. CTG and Simulated Signal

The software was tested on simulated signals, resembling real FHR, built in accordance with a previously described and employed methodology [6, 8] and on real CTG recordings. Currently, we have a database of CTG recorded in clinical environment (here called real-DB) populated from about 600 signals.

Based on the real-DB, we observed that, on average, about 0.4% of outliers and 3% of signal loss (FHR values much lower than the mean) can be observed. Therefore, the simulated FHR signals, lasting more than 30 minutes, were designed with different combinations of their characteristics (FHR mean, presence of accelerations and decelerations, outliers number) and involving maximum 8 artifacts resembling Premature Ventricular Depolarizations (PVD), 6 isolated outliers and three losses of signal of 20, 25 and 30 samples respectively (figure 1). The signal loss was simulated as abrupt or gradual.

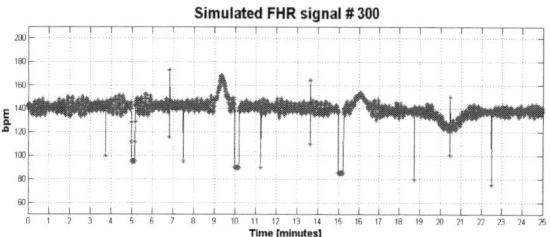

Fig. 1 Example of simulated FHR signal. In red FHR samples which highlight simulated outliers.

B. First Software Proposal

As previously indicated in depth [6], in order to detect outliers in the FHR signal it is not sufficient to know that an outlier is a sample which appears to be inconsistent with the other data but it is necessary to establish for them a quantitative definition.

Very shortly, let us remember that our software consisted into two main steps: outliers detection and correction.

The *detection* step, in turn, can be divided in other two steps: detection of local outliers (cardiac arrhythmias and short time artifacts) and detection of global outliers (signal losses).

For the detection of local outliers, the algorithm does a double scan of the FHR (forward and backward) and it uses two different thresholds according to the quality of the signal[1]. An outlier is a sample that exceeds the threshold with respect to a reference value computed as median on five samples, both previous (backward scan) and next (forward scan).

The global outliers are detected after the local ones because the analysis of many CTG traces showed that the global outlier are often preceded and followed by intervals with opposite monotony (local outliers already called in this case "guard groups"). All the samples between the guard groups are global outliers if they satisfy also a quantitative definition substantially based on the distance from the other samples (please refer to [6] for details).

The *correction* step employs two different approaches, one for local outliers and one for the global ones.

The local outliers are substituted using a fifth order median filter centered on the sample to substitute.

The global outliers are substituted with a linear interpolation that starts from the last valid sample preceding the first guard group and ends at the first valid sample following the second guard group.

C. Software Update

Increasing the number of tests on real CTG signals, we noted that too often our software failed in outliers detection. This happened mainly in presence of signal losses without guard groups (see section Results), a situation that we had not previously considered, so we have updated the software.

The new algorithm is included, in place of the previous, in a software for CTG pre-processing, which segments each recorded signal in a number of reliable continuous tracts (i.e. where the signal quality level is medium or good); and, for each tract, recovers the uneven FHR series [6, 8] and manages outliers. Finally, bad quality tracts, lasting not more than 3 s are interpolated, in order to avoid the excessive FHR fragmentation.

The basic structure of algorithm phase dedicated to outliers treatment is almost the same, so we still distinguish local from global outliers (recognized with new rules) in detection step but then process them with the same strategy.

In detection step, *global outliers* are now defined as all the samples for which the absolute difference respect to the FHR mean is out of a range fixed by a threshold, set as percentage of the FHR mean; in particular, the threshold is 30% for samples of good quality and 20% for samples of medium quality.

The detection of *local outliers* is carried out analyzing only the samples that are not global outliers. A candidate local outlier is a sample for which the difference between its amplitude and the median value calculated on five samples positioned before and after the sample in analysis exceeds a threshold set according to the sample quality. The threshold is increased of 30 % of the set value to analyze the samples positioned at the extreme limits of each FHR signal tract. This choice is due to the impossibility to scan the tract in both directions, procedure that increase reliability of the method.

To avoid that samples belonging to slow baseline variations are detected as outliers, another check has been added. The candidate outlier is elected local outlier only if it likely does not belongs to the line estimated as regression line of the neighbour samples.

The correction step uses for the substitution the median value calculated on 8 samples. The median is computed

---

[1] All commercial cardiotocographs provide in output a signal quality expressed on three levels which represents the reliability associated to each FHR sample, considered good, medium or bad.

centring the vector on the outlier and excluding other eventual outliers already detected.

In the final version of the software, it is possible to set number of consecutive outliers to be substituted with the median procedure (NOC) and length of bad quality segment to be interpolated (LI); however, in default setting, the *correction* procedure substitutes only isolated outliers. Moreover, when consecutive outliers are not substituted their quality is set equal to zero so that in a following step of preprocessing can be interpolated.

D. STV Computation

To verify the importance of outliers processing in FHR signals, we estimated STV by means of the standard deviation (SD) after floatingline subtraction, as previously proposed [9, 10] and processed outliers with the update version of the software.

SD was computed on FHR time-windows of 30 s with an overlap of n-1 samples obtaining a vector of SD values (one for each time-window) aligned with the FHR signal. Since it was not possible to compute SD of the first and last 15 s of each good segment of FHR signal, these samples were set to zero. Mean value of the obtained vector, without null samples, is STV index.

In order to assess how outliers and different correction procedures affect results, different tests, listed below, were carried out simply changing NOC and LI.
I. The outliers were not detected (outliers processing was not used) so that the whole FHR signal, including their values, is considered in STV computation.
II. The outliers are detected and they are all corrected with the median value.
III. The outliers are detected and they are all corrected with a linear interpolation.
IV. The outliers are detected but they are not substituted (in this case their quality is set at zero and FHR signal, and in turn STV computation, results very segmented).
V. The outliers are detected and just the isolated ones are corrected with the median value, while the interpolation is done only for bad quality tracts lasting maximum 7 samples (corresponding to 3 s of CTG trace).

III. RESULTS

First version and update version of the software were tested on different simulated and real signals. As mentioned in the section Materials and Methods, the phase of outliers processing showed comparable performances in detecting outliers except in cases of sudden signal losses where the first version failed totally whereas the update version detected all outliers.

As example, in figure 2 it is possible to observe that local outliers (PVD and isolated outliers) and gradual signal loss (red ovals in figure 2) were detected from both versions of the software but the abrupt signal losses (blue ovals) were recognised only by the update version (marked as green stars in figure 2).

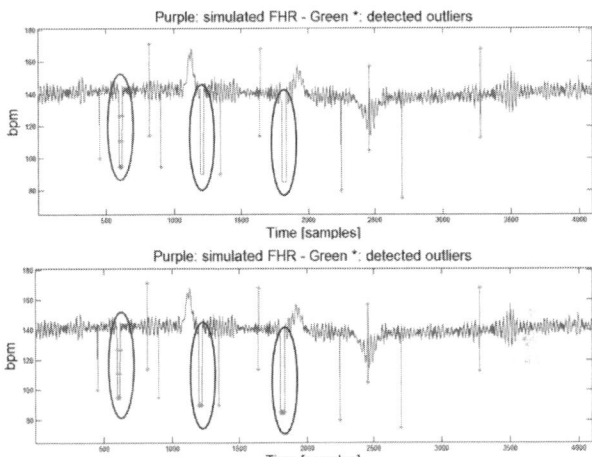

Fig. 2 Green stars: detected outliers. On the top - first version of the software, it is noticeable that only the first gradual signal loss is detected. On the bottom - update version of the software, all the outliers are detected.

About STV assessment, en example of STV computed for a test IV is shown in figure 3. In this figure, it is possible to observe the interruptions in STV time trend, which represent about the 25% of FHR length.

For each FHR signal, the relative percentage error (*erp*) was computed comparing the STV index with respect to the value set in FHR simulation.

Each test was repeated for 30 simulated FHR and the results are shown in table 1 as mean and standard deviation of *erp*.

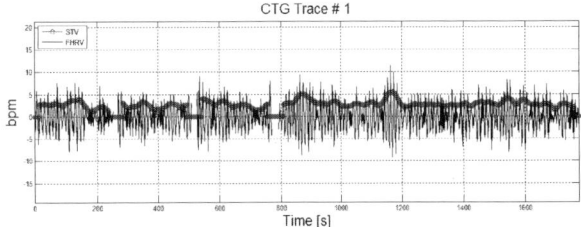

Fig. 3 In red STV time trend superimposed on FHRV in back

Table 1 Average and Standard deviation of *erp* in STV computation

Test	I	II	III	IV	V
Mean	26,60	9,82	9,67	5,98	7,71
Stand. Dev.	1,28	0,79	0,89	1,93	1,27

IV. DISCUSSION

The paper dealt with the problem of outliers processing in CTG monitoring. In general, in signal processing, algorithm treating outliers implement two steps, detection and correction.

Detection is the step dedicated to outliers recognition, after their specific definition. The literature about HRV proposes to define a temporal trend in which are located the inter-beat intervals in presence of a normal cardiac rhythm. The samples which do not belong to the temporal trend are considered outliers [11, 12]. Other approaches involve the use of thresholds set on the basis of a mean value computed with different rules [11, 13].

The correction step can implement two different approaches: elimination of the outliers or their substitution with a sample that fits better to the series trend [11, 12]. The most widely used solutions are median filters and linear and non linear (cubic spline) interpolations.

Following these literature indications, even if not specific for FHR processing, and the results of tests on a large number of real CTG, we developed and then updated an algorithm for outliers detection and substitution that provided very satisfying performances on 450 simulated and 100 real FHR signals.

After that, in order to validate the importance of such an algorithm, we computed STV using different strategies of outliers processing. We chose this index because of its known importance in FHRV evaluation [1, 2, 9] and because, being correlated with the beat-to-beat variability, was likely altered by the presence of outliers.

Obtained results prove that outliers detection is fundamental to perform accurate analysis on FHR signals. The computed *erp*, in fact, is maximum (test I) in presence of non detected outliers.

The errors computed in the other tests, instead, are similar, so demonstrating that *erp* is less sensitive to the correction step.

The choice of the best algorithm, therefore, can be done according to other aspects. In the final version of pre-processing, we adopted the solution proposed as test V, in order to achieve a trade-off between not excessive alterations of local FHR trends (substituting too much samples all with the same value equal to the median) and its high segmentation (to process too short segments could prevent other kinds of analysis, such as the time-frequency).

REFERENCES

1. M. Romano, M. Cesarelli, P. Bifulco, M. Ruffo, A. Fratini, G. Pasquariello. (2009), Time-frequency analysis of CTG signals. Current Development in Theory and Applications of Wavelets, Volume 3, Issue 2, pages 169-192.
2. H.P. Van Geijn. (1996), Developments in CTG analysis. Baillieres Clin Obstet Gynaecol, 10 (2): 185-209.
3. G. Improta, M. Romano, F. Amato, M. Sansone, M. Cesarelli. (2012), Development of a software for automatic analysis of CTG recordings. Paper n. 310, GNB2012, Rome, Italy.
4. J PardeyJ, M Moulden, CWG Redman. (2002)A computer system for the numerical analysis of non-stress tests. Am J Obstet Gynecol, May, vol. 186, n. 5: 1095-1103.
5. M. Cesarelli, M. Romano, P. Bifulco, G. Improta, G. D'Addio. (2013), Prognostic Decision Support using Symbolic Dynamics in CTG Monitoring. Proceedings of EFMI STC "Data and Knowledge for Medical Decision Support".
6. M. Cesarelli, M. Romano, P. Bifulco, and A. Fratini. (2007) Cardiac arrhythmias and artifacts in fetal heart rate signals: detection and correction. Proceedings of Medicon. pp. 789–792.
7. D. Ayres-De-Campos, J. Bernardes, A. Garrido, J. Marques-De-Sa, L. Pereira-Leite. (2000), SisPorto 2.0: a program for automated analysis of cardiotocograms. The Journal of Maternal-Fetal Medicine 9:311 D318.
8. M. Cesarelli, M. Romano, P. Bifulco, F. Fedele, M. Bracale (2007) An algorithm for the recovery of fetal heart rate series from CTG data. Computers in Biology and Medicine, 37 (5): 663-669.
9. M. Cesarelli, M. Romano, P. Bifulco (2009) Comparison of short term variability indexes in cardiotocographic foetal monitoring. Computers in Biology and Medicine, Volume 39, Issue 2, Pages 106-118.
10. M. Cesarelli, M. Romano, G. D'Addio, M. Ruffo, P. Bifulco, G. Pasquariello, A. Fratini. (2011), Floatingline estimation in FHR signal analysis. 5th European IFMBE Conference, Budapest, Hungary. IFMBE Proceedings 37, pp. 179–182.
11. P. A. U. Li and T. I. K. Kanen (1999) Characterization And Application Of Analysis Methods For Ecg And Time Interval Variability Data Application Of Analysis, Oulu University Library.
12. D. Lebrun, (2003) Analysis of neonatal heart rate variability and cardiac orienting responses", University of Florida.
13. J. E. Mietus, C. Peng, P. Ivanov, and A. L. Goldberger (2006) Detection of Obstructive Sleep Apnea from Cardiac Interbeat Interval Time Series Abstract, pp. 1–6.

Author: Giuliana FAIELLA
Institute: D.I.E.T.I., University "Federico II"
Street: via Claudio, 21
City: Naples
Country: Italy
Email: giuliana.faiella@gmail.com

Changes in Heart Rate Variability Associated with Moderate Alcohol Consumption

A. Fratini[1], P. Bifulco[1], F. Clemente[2], M. Sansone[1], and M. Cesarelli[1]

[1] Dept. of Electrical Engineering and Information Technologies, University of Naples Federico II, Naples, Italy
[2] IsIB CNR Institute of Biomedical Engineering, Rome, Italy

Abstract—Technological advances have driven some attempt of vital parameters monitoring in adverse environments; these improvements will make possible to monitor cardiac activity also in automotive environments. In this scenario, heart rate changes associated with alcohol consumption, become of great importance to assess the drivers state during time.

This paper presents the results of a first set of experiments aimed to discover heart rate variability modification induced by moderate assumption of alcoholic drink (i.e. single draft beer) as that typically occurs in weekend among some people. In the study, twenty subjects were enrolled and for each of them two electrocardiographic recordings were carried out: the first before alcohol ingestion and the second after 25-30 minutes. Each participant remained fasting until the second ECG acquisition was completed. ECG signal were analyzed by typical time-domain, frequency and non linear analysis. Results showed a small increase in LF/HF ratio which reflects a dominance of the sympathetic system over the parasympathetic system, and an increase in signal complexity as proven by non linear analysis.

However, the study highlighted the need to monitor HRV starting from alcohol ingestion until its complete metabolization to allow a more precise description of its variation.

Keywords—Electrocardiography, heart rate variability, alcohol consumption.

I. INTRODUCTION

Heart functions are moderated by the intervention of the autonomic nervous systems ANS. [1, 2] Different conditions act on the regulation of heart rate (HR) and heart rate variability (HRV) such as circadian rhythm, level of activity, emotion, medications. [metti biblio] Monitoring of vital parameters is becoming ubiquitous and easier to achieve; new electronic front-end as well as special electrodes [3, 4] are well suited for contactless applications. These improvements will make possible to monitor ECG and more particularly HR even in an automotive environment.

HR and its variability have been strongly investigated in the past, either for disease prognosis or assessment of specific patient conditions also in fetus stress assessment. [1, 5, 6] HRV has also been used in the study of the sleep and its disorders and recently for the assessment of modification in HRV that presents before falling asleep events. [7]

Discover those event could be crucial in automotive environment, in which the physiological condition of the driver affects its way to respond to the external stimuli and in turn, its level of safety and that of its passengers as well as other drivers.

In this scenario, the modification in HR and in its variation due to alcohol consumption, become of great importance to identify specific features to assess the drivers state during time. This study reports the result of a first set of experiments aimed to highlights HRV modification induced by moderate assumption of alcoholic drink (i.e. single draft beer) as that typically occurs in weekend among young people.

II. METHODS

In order to estimate the effects of moderate alcohol consumption on physiological variables a first experimental study was defined. Participants were voluntarily enrolled at a public place during the period of the Oktober Fest, the known Bavarian folk festival, having a large audience attracted by the event.

A. Subjects

Twenty subjects, including 11 men and 9 women, from 24 to 40 years (mean age 30±6.5) participated in the study. The volunteers were duly informed about the scope of the research and the measurement protocol. Those who have chosen to participate in the study gave their written consent to the processing of their data. The subjects were moderate consumer of alcohol; more than a half of the participants was recreationally sport trained and an equal percentage smokers.

B. Measurement Protocol

For each participant two electrocardiographic recordings were taken; the first recording was performed before any alcohol intake while the second 25-30 minutes after the ingestion of a 0.5 cl draft beer. Each participant remained fasting until the second ECG acquisition was completed. Five-minutes recordings for short-term HRV analysis [2] were made using a cardiac monitor (PropaQ Encore 202

EL - Welch Allyn) with subject sit and relaxed. A previously developed software suit was used to transfer data from the monitor to an external desktop computer. The monitor sampling frequency for ECG signal is fixed at 180 Hz. Each subject was interviewed during registration in order to know personal details and main habits.

C. R Wave Extraction and HRV Analysis

R to R intervals extraction was performed using Kubios HRV software suite [2]; two researchers were asked to verify independently the estimated R peaks positions and make corrections whenever appropriate obtaining the normal to normal (NN) beat intervals. NN/RR ratio was then computed to assure data reliability; recordings belonging from one subject were excluded since the NN/RR ratio resulted less than 90%. [8]

HRV was analyzed through time-domain, frequency-domain and a non-linear parameter analysis.

Time-domain analysis: Different time domain measures were used for this study: mean NN-interval in milliseconds, standard deviation of all NN intervals (SDNN) in milliseconds, Mean HR and standard deviation of instantaneous heart rate values (STD HR) in [1/min], square root of the mean of the sum of the squares of differences between adjacent NN interval (RMSSD) in milliseconds (ms), number of adjacent NN intervals differing more than 50 ms (NN50 count), percentage of difference between adjacent NN intervals differing more than 50 ms. (pNN50), the integral of sample density distribution of RR intervals divided by the maximum of the density distribution (RR triangular index).

Frequency-domain analysis: For spectrum analysis we used the typical frequency bands suggested in case of short-term analysis: Very Low Frequency (≤ 0.04 Hz); Low Frequency (0.04-0.15 Hz); High Frequency (≥ 0.15 H).

A 16th order autoregressive model has been used to analyze NN intervals.

Non linear analysis: The nonlinear properties of HRV were analyzed using Poincaré Plot, approximate and Sample Entropy, detrended fluctuations, correlation dimension and recurrence plots. [8]

III. RESULTS

As reported in table 1, from the time-domain analysis it has been possible to note that no significant variations occurred in Mean NN interval nor in Mean HR after the alcoholic drink.

The comparison of standard deviation of instantaneous heart rate values (STD HR) and the square root of the mean of the sum of the squares of differences between adjacent NN interval (RMSSD) showed a small reduction after alcohol ingestion ($p<0.05$). Number of adjacent NN intervals differing more than 50 ms (NN50 count), the percentage of difference between adjacent NN intervals differing more than 50 ms (pNN50) as well as RR triangular index were lower after the beer, however only RR triangular index was significant ($p<0.05$).

Table 1 Comparison of Mean NN and Mean HR values before alcohol ingestion and after alcohol ingestion

	Before	After
Mean NNI [ms]	724.89±98.5	725.26±89.78
Mean HR [1/min]	84.72±10.53	84.54±10.26

Table 2 Comparison of STD HR and RMSSD values before and after alcohol ingestion

	Before	After
STD HR [1/min]	8.02±1.9	6.79±2.11
RMSSD [1/min]	37.71±15.66	32.51±17.68

Table 3 shows the results of frequency-domain analysis: the mean power found in the HF band showed a small decrease with a single beer, while the power contribution of VLF showed an increase; no differences were found for LF. LF/HF ratio showed a significant increase ($p<0.05$).

Table 3 Power contents of classical HRV bands VLF, LF,HF before and after alcohol ingestion

	Before	After
VLF [%]	30.23±11.21	33.64±13.24
LF [%]	57.95±7.98	57.25±11.95
HF [%]	11.81±5.24	9.11±4.79
LF/HF	5.76±2.50	7.73±3.43

Figure 1 reports an example of Pontcaré plot of one subject before and after alcoholic drink. as it is possible to note some variation occur in terms of SD1 and SD2.

Table 4 shows the results of non-linear analysis: small variation can be recognized for SD1 and SD2.

However, the significant results were only found for Recurrence Plot (Lmean), Shannon Entropy and Correlation Dimension (D2) ($p<0.05$).

IV. CONCLUSION

This work has been focussed on the analysis of the heart rate variability in case of ingestion of alcoholic beverages.

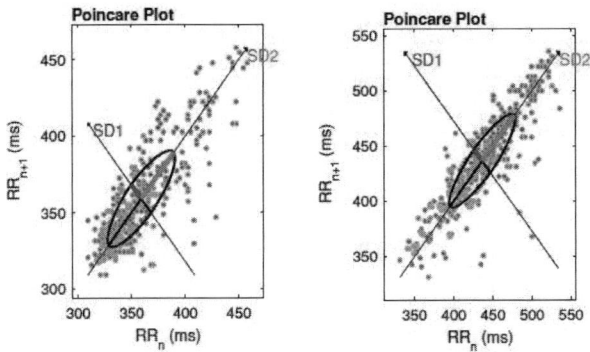

Fig. 1 Example of Pointcaré plot results: the plots reports the state of a subject before alcohol ingestion (right) and after alcohol ingestion (left)

Table 4 Non-linear analysis results

	Before	After
SD1 [ms]	26.71±11.10	23.01±12.53
SD2 [ms]	92.76±22.05	83.88±32.61
Lmean [beat]	10.96±1.86	13.32±3.47
Lmax [beat]	297.16±140.57	302.37±133.63
REC [%]	34.20±4.87	37.08±5.75
DET [%]	98.99±0.55	99.18±0.41
ShanEn	3.19±0.19	3.33±0.98
ApEn	0.99±0.10	0.98±0.10
SampEn	1.12±0.20	1.07±0.20
DFA:[$\alpha 1$]	1.52±0.13	1.55±0.14
DFA:[$\alpha 2$]	0.72±0.17	0.71±0.18
D2	3.07±0.72	2.25±1.00

Results highlighted an overall readjustment of the frequency content, resulting in an increase in the LF/HF ratio, which reflects an imbalance between the sympathetic and parasympathetic systems favoring the first. The condition described is similar to that which occurs in the stages of falling asleep. Non linear analysis showed a small increase in complexity of the signal as reported by the analysis of entropy or correlation dimension. In conclusion, the ingestion of small quantity of alcohol induces variations in heart rate variability and changes of some parameters can be recognized. However, this study highlighted the need to monitor HRV during all the stages of alcohol intoxication (starting from its ingestion until its complete metabolization) to allow a better description of the changes in HRV.

ACKNOWLEDGEMENTS

This work was supported by "DRIVEr monitoring: technologies, methodologies, and IN-vehicle INnovative systems for a safe and ecocompatible driving *DRIVE IN^2* project - founded by the Italian National Program Piano Operativo Nazionale Ricerca e Competitivitá 2007/13.

REFERENCES

1. Coumel P, Maison-Blanche P, Catuli D. Heart rate and heart rate variability in normal young adults. *Journal of Cardiovascular Electrophysiology.* 1994;5:899 - 911.
2. Heart rate variability: standards of measurement, physiological interpretation and clinical use. Task Force of the European Society of Cardiology and the North American Society of Pacing and Electrophysiology. *Circulation.* 1996;93:1043 - 1065.
3. Gargiulo G.D., Bifulco P., Cesarelli M., Jin C., McEwan A., Schaik A.. Wearable dry sensors with Bluetooth connection for use in remote patient monitoring systems *Stud. Health Tech. Info.* 2010;161:57–65.
4. Gargiulo G.D., McEwan A.L., Bifulco P., et al. Towards true unipolar bio-potential recording: a preliminary result for ECG *Physiological measurement.* 2013;34:N1.
5. Cesarelli M., Bifulco P., Bracale M.. Evaluating time-varying heart-rate variability power spectral density *Engineering in Medicine and Biology Magazine, IEEE.* 1997;16:76 -79.
6. Ruffo M., Cesarelli M., Romano M., Bifulco P., Fratini A.. An algorithm for FHR estimation from foetal phonocardiographic signals *Biomedical Signal Processing and Control.* 2010;5:131 - 141.
7. Furman G.D., Baharav A., Cahan C., Akselrod S.. Early detection of falling asleep at the wheel: A heart rate variability approach. in *Computers in Cardiology*;35((1)Tel Aviv University):1109-1112 2008.
8. Goldberger A.L., Amaral L.A.N., Glass L., et al. PhysioBank, PhysioToolkit, and PhysioNet: Components of a new research resource for complex physiologic signals *Circulation.* 2000;101:e215–e220.

Author: Antonio Fratini
Institute: DIETI - University of Naples *Federico II*
Street: Via Claudio, 21
City: Napoli
Country: Italy
Email: a.fratini@unina.it

EEG Rhythm Analysis Using Stochastic Relevance

L. Duque-Muñoz[1], C.A. Aguirre-Echeverry[2], and G. Castellanos-Domínguez[2]

[1] Instituto Tecnolgico Metropolitano, Grupo de Automtica de Electrónica, Medellín, Colombia, leonardoduque@itm.edu.co
[2] Universidad Nacional de Colombia sede Manizales/Signal Processing and Recognition group, Manizales, Colombia

Abstract—The rhythm analysis in advanced signal processing methods has long of interest in application areas such as diagnosis of brain disorders, epilepsy, sleep or anesthesia analysis, and more recently in brain computer interfaces. This work discusses a methodology to evaluate quantitatively each rhythm contribution to neuronal activity. The methodology assesses a set of relevant weights measuring time–variant values of the rhythm waveform variance. Relevance evaluation is based on a time – variant rhythm estimation with Time variant autoregressive models and exponentially damped sinusoidal models, which are compared in terms of better discrimination of dynamic properties of considered neuronal activities. The methodology is evaluated using an electroencephalogram dataset. Qualitative and quantitative results of accomplished relevance weights as well as classifier performance (using an SVM) are presented. As a result, the discussed methodology for relevance evaluation of EEG rhythms gives support to Epileptic seizure identification, with the added benefit of simplicity and interpretability of the assessed set of relevance weights.

Keywords—Exponentially damped sinusoids, Relevance rhythm diagram, Time variant autoregressive model.

I. INTRODUCTION

Since the early days of automatic EEG processing, it had been noted that the EEG spectrum contains some characteristic waveforms with clinical and physiological interests that fall primarily within the following four frequency bands: *Delta* rhythm, (noted δ, with frequencies $f < 4\ Hz$), is a slow brain activity preponderant only in deep sleep stages of normal adults. Otherwise, it may suggest pathologies. *Theta* band, ($\theta, f \in [4,8]\ Hz$), is present in normal infants and children as well as drowsiness and sleep in adults. Only a small amount of θ rhythm appears in the normal awakening adult; on the contrary, the presence of high theta activity suggests abnormal or pathological conditions. *Alpha* rhythm, ($\alpha, f \in [8,13]\ Hz$), is present in normal adults during relaxed and mentally inactive awakeness. The α rhythms are blocked by opening the eyes (visual attention) and other mental efforts such as thinking. *Beta* band, ($\beta, f \in [14,30]\ Hz$), is mostly marked in fronto–central region with less amplitude than α rhythms. It is enhanced by expectancy states and tension. However, there are no strict frequency ranges for determining these different bands. Furthermore, certain inter–individual variability should be taken into account by selecting the frequency band for each subject. So, frequency band ranges used by several references are given in [1].

Nonetheless, according to the rhythm activity (mostly characterized by their amplitudes) different interpretations on whether the patient is healthy can be made. Particularly, the most important human frequency band is assumed to be the α rhythm, or brain natural frequency, appearing when the eyes are closed and one begins to rest. In epilepsy cases, however, δ and θ rhythms exhibiting lower frequencies and higher magnitudes with respect to α waves may take place. The brief and episodic neuronal synchronous discharges occur with dramatically increased amplitude. Moreover, two different types of seizures can have a alike behavior in terms of the raw EEG data, but they may have different underlying time–evolving structures not readily discernable from the recording by visual inspection, as quoted in [2]. The main goal of this work was to study how the rhythm analysis identifies the various events recorded in the EEG. The aim is also to study detection of seizures to determine whether this technique is effective as an auxiliary system diagnosis with epilepsy.

II. MATERIALS AND METHODS

Time–varying AR Analysis: Considered stochastic rhythm–based feature set is estimated based on TVAR modeling of EEG data. Namely, let $\mathbf{y} = \{y(t) : \forall t \in T\}$, be an underlaying non–stationary signal, for which every instantaneous value $y(t) \subset \mathbb{R}$ is modeled by a time–varying autoregressive model of order n_φ, described as follows:

$$y(t) = \boldsymbol{\varphi}^\mathsf{T}(t)\mathbf{h}(t) + \varepsilon(t) \qquad (1)$$

where $\boldsymbol{\varphi}(t) = [\varphi_1(t), \varphi_2(t), \ldots, \varphi_{n_\varphi}(t)], \boldsymbol{\varphi} \subset \mathbb{R}^{n_\varphi}$ is the parameter vector holding the TVAR model coefficients; vector $\mathbf{h}(t) = [y(t-1), y(t-2), \ldots y(t-n_\varphi)], \mathbf{h} \subset \mathbb{R}^{n_\varphi}$, is the observation vector formed by delayed versions of $y(t)$, and $\varepsilon(t) \sim \blacktriangle(0, \sigma_\varepsilon^2(t))$, is a sequence of non observed and uncorrelated innovations with zero-mean and time–varying variances $\sigma_\varepsilon^2(t)$ generated the process $y(t)$. In Eq. (1), the parameter vector, which is made time–dependent to take into account changing with time system dynamics, is usually unknown and must be estimated from the data. Once the current estimate of $\hat{\varphi}_1(t)$ is available, time–frequency information is obtained by computing an spectrogram. In Eq. (1), the parameter vector, which is made time dependent to take into account changing with time system dynamics, is usually unknown and must be estimated from the data. Once the current estimate of $\hat{\varphi}_1(t)$ is available, time–frequency information is obtained by computing a time–varying spectrum, that is, $\boldsymbol{h}(t) = \boldsymbol{G}(t)\boldsymbol{h}(t-1) + \varepsilon(t)$, where $\boldsymbol{G}(t) = \begin{bmatrix} \boldsymbol{\varphi} \vec{i}_1 \ldots \vec{i}_{n_\varphi} \end{bmatrix}^\top$, being vector $\vec{i}_j \in \mathbb{R}^{1 \times n_\varphi}$, with $\vec{i}_j = \{\delta_n : n = 1, \ldots, n_\varphi\}$, where $\delta_n = 1$, if $n = j$, otherwise, $\delta_n = 0$. Square matrix $\boldsymbol{G}(t) \in \mathbb{R}^{n_\varphi \times n_\varphi}$ holds n_φ distinct eigenvalues (including n_c pairs of complex eigenvalues) being reciprocal roots of the autoregressive characteristic equation at moment t. For the class of TVAR, the basic decomposition models state the following sum of $n_c + n_\varphi$ real processes [3]:

$$y(t) = \sum_{j=1}^{n_c} z_j(t) + \sum_{j=2n_c+1}^{n_\varphi} \widetilde{z}_j(t)$$

in which the $z_j(t)$ processes are defined through the complex eigenvalues and $\widetilde{z}_j(t)$ through the real ones. Based on a TVAR(n_φ) model, a set of latent subseries, $\{z_i\}$, is computed from which the corresponding rhythms are extracted. Particularly, each time subseries is associated with a respective rhythm as follows: $z_i \rightarrow \boldsymbol{x}_i : \forall i = i$.

Exponentially Damped Sinusoidals: In case of parametric modeling, to describe the common dynamics in a given set of signal, it is represented as a finite sum of p discrete-time exponentially damped complex sinusoids:

$$y(t) = \sum_{i=1}^{p} a_i \exp(j\phi_i) \exp((-d_i + j2\pi f_k)t\Delta n) + \varepsilon(t) = \sum_{i=1}^{p} z_i(t) \quad (2)$$

where $\Delta n t$ is the time lapse between the origin and the sample $y(t)$, Δn is the sampling interval and $\varepsilon(t)$ is Gaussian white noise. The parameters of this model are the amplitudes a_i, the phases ϕ_i, the damping factors d_i, and the frequencies f_i. Since exponentially damped sinusoids are a set of basis functions, they can be used to model any arbitrary signal sufficiently closely, assuming that the model order is high enough. The EDS model is proven to be useful for modeling non–stationary signals containing transients, where the adding damping helps to lower the model order when modeling transients (compared to undamped sinusoidal modeling). Present study addresses the problem of estimating the model parameters when the EEG signal is embedded in noise by using the subspace–based exponential data fitting proposed in [4], where an EEG signal is stacked in the Hankel data matrix. Then, the signal subspace is estimated by means of the singular value decomposition, from which the model parameters needed in Eq. (2) are estimated. Lastly, each one of the damped sinusoids $x_i(t)$ associated with the EDS model is associated with a frequency range of rhythms: $z_i \rightarrow \boldsymbol{x}_i : \forall i = i$.

Relevance Analysis: Relevance analysis distinguishes variables that represents effectively the subjacent physiological phenomena according to some evaluation measure. Such representative variables are named *relevant features*, whereas the evaluation measure is known as *relevance measure*. Variable selection tries to reject those variables whose contribution to representation target is none or negligible (*irrelevant features*), as well as those that have repeated information (*redundant features*). Thus, the first objective concerning the variable selection stage is to define the concept of relevance [5]. The time-varying relevance measure is evaluated as:

$$\rho_v(\boldsymbol{S}_y; \tau) = [\chi(1) \cdots \chi(\tau) \cdots \chi(pT)]^\top, \quad (3)$$

where $\chi(\tau) = \mathcal{E}\{|\lambda_j^2 v_j(\tau)|\}$, $\{\lambda_j : j = 1, \ldots, q\}$ is the set of most relevant eigenvalues of the matrix $\boldsymbol{S}_y$, and the scalar $v_j(\tau)$ is the respective element in the instant τ, and $\tau = 1, \ldots, pT$ indexes each of the relevance values calculated for the entire set of time-varying data. To determine the relevance related to each of the stochastic variables, the Eq.(3) can be arranged to the relevance matrix $[\rho_{v1}(\boldsymbol{S}_y; t) \cdots \rho_{vf}(\boldsymbol{S}_y; t) \cdots \rho_{vF}(\boldsymbol{S}_y; t)]^\top$, where each row $\rho_{vf}(\boldsymbol{S}_y; t) = [\chi((f-1)T+1) \ldots \chi(t) \ldots \chi(fT)] \in \mathbb{R}^{T \times 1}$ shows the contribution of the $\boldsymbol{s}_{ri}$ stochastic feature along fixed time moments.

III. EXPERIMENTAL SET-UP

A. Electroencephalographic Recording Database

Data Set: This collection is publicly available, described in [6]. The complete data set consists of five subsets (denoted A, B, C, D, and E), each one composed of 100 single–channel EEG segments of 23.6-s duration. Sets A and B have been taken from surface. EEG recordings of five healthy volunteers with eye open and closed, respectively. Signals from sets D and C have been measured in seizure–free intervals from five subjects in the epileptic zone and from the hippocampal formation of the opposite hemisphere of the brain. Set E comprises of epileptic signals recorded during a seizure (ictal) from all recording sites. All EEG signals were digitized at 173.61 Hz with 12 bit resolution. To keep the main features of the EEG of interest, the data were filtered by a low–pass filter of cutoff frequency 40 Hz.

B. Estimation of Relevance Weights for EEG Rhythms

Regarding the TVAR modeling, the number of time subseries calculated in Eq. (1) depends directly on the model order. The frequency of oscillation modes can be associated to the EEG rhythms by computing the angular frequencies ω_i of the autoregressive polynomial roots. In the case of EEG signals, an adequate model order is empirically fixed to be $7 < n_\varphi < 12$. So, each time subseries is associated with a respective rhythm as follows: $[\mathbf{x}_1 + \mathbf{x}_2] \to \delta$, $[\mathbf{x}_3 + \mathbf{x}_4 + \mathbf{x}_5 + \mathbf{x}_6] \to \theta$, $[\mathbf{x}_7 + \mathbf{x}_8 + \mathbf{x}_9 + \mathbf{x}_{10}] \to \alpha$, $[\mathbf{x}_{11} + \mathbf{x}_{12}] \to \beta$.

As a result, for each decomposition technique, the input training space has a dimension $T \times n_r$, where $T = 4096$ is the number of samples per considered EEG segment, and $n_r = 4$ is the number of rhythms at hand. Table 1 shows the rhythm weights for Problem I, computed by the measuring stochastic variability for the considered database. Weights are normalized by the maximum value obtained for each decomposition method. It can be seen, that weight estimates converge, practically, to a same value for each rhythm waveform. However, some important variations can be noted in case of estimated β rhythm weights, and therefore, a lack of their precision should be inferred.

Taking into account that the higher the weight - the more relevant the respective rhythm waveform, then, for normal labeled observations (subset A), the rhythm exhibiting the largest weight is α, having the strongest influence. This waveform has been highly studied because its topographic distribution (maximum amplitude over occipital regions) and their high reactivity (have a strong attenuation when passing to alert state). For the studied dataset, δ and θ waveforms keep a strong enough activity, while β rhythm holds a lowest contribution. In the seizure state, the rhythm exhibiting the largest weight is the one having a lower frequency and higher amplitude (i.e. δ). Although, a low–frequency rhythm may be present in some brain states such as deep sleep or high concentration, its increased activity may be an indication of focal epilepsy on the temporal region of the brain [7]. Besides, the second highest weight becomes the θ waveform, which is also strongly associated with epilepsy (very common in patients with frontal lobe epilepsy, 98.1% of the cases [7]); but this rhythmic activity is not associated with mental activation states or sleep. Since the β relevance weights decrease, it remains holding the lowest contribution.

Assuming the normal brain state as the baseline activity, Table 1 also shows the variations of waveform relevance weights after the brain state changes to Epileptic seizure activity (row termed "Difference"). As seen, δ and θ rhythm waveforms increase their contribution (noted as "+"), whereas α and β reduce their influence on brain activity (noted as "−"), that is, their pattern is strongly affected by patient clinical condition.

Table 1 Estimated rhythm weights from considered time–variant decompositions for both underlaying data bases.

Method	δ	θ	α	β
Seizure				
TVAR	$1 - 0.02$	0.92 ± 0.03	0.53 ± 0.05	0.23 ± 0.06
EDS	$1 - 0.09$	0.86 ± 0.08	0.45 ± 0.11	0.12 ± 0.10
Normal				
TVAR	0.73 ± 0.07	0.89 ± 0.05	$1 - 0.03$	0.35 ± 0.09
EDS	0.72 ± 0.10	0.82 ± 0.08	$1 - 0.05$	0.28 ± 0.13
Difference				
TVAR	+	+	−	−
EDS	+	+	−	±

Nonetheless, due to high measured dispersion of relevance weights, the EDS method fails on determining the correct variation of β waveform (This uncertainty is noted as "±"). It is worth mentioning that the total contribution of all waveform covariance–based weights increases after the brain state changes to Epileptic seizure, specifically, the total power of EEG data becomes greater in about 10%.

Variation of each rhythm energy distribution from one brain state to another can be illustrated by introducing the relevance rhythm diagram (RRD), which is an orthorhombic–shape strip of uniform width, computed in such a way that relevance weights of a given reference brain state must be confined within the narrow region (actually, within a 3σ width of estimated rhythm values), as shown in Figure 1. To get more interpretability of the assessed set of waveform relevance weights for both brain states under comparison, location of rhythms are intentionally changed to increase the perception of difference among obtained relevance weights. So, taking advantage of the known labels of observations, estimated for Problem I (A*–E) RRD is shown in Figure 1. Notation * indicates the reference brain state. In both cases, normal brain activity related events, mostly, are circumscribed within 3σ width for all considered rhythm waveforms. In case of Epileptic events, shape of seizure–related

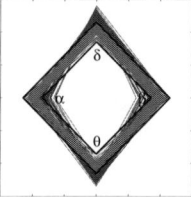

Fig. 1 Obtained A*–E RRD diagram. Reference normal events (A) are depicted with red lines while seizure related events (E) with blue lines.

Table 2 Results for the first classification problem, in terms of considered classifier performance measures (a_{ac}, a_{se}, a_{sp}.)

Method	SVM Classifier Performance		
	Accuracy(%)	Sensitivity(%)	Specificity(%)
TVAR	99.06 ± 1.11	100 ± 0	98.33 ± 1.84
EDS	96.16 ± 3.16	97.18 ± 3.25	93.23 ± 2.28

diamond becomes strained, i.e., vertical axis becomes wider while the horizontal axis tends to be narrower. Therefore, low rhythm bands (δ and θ) increase their power contribution, while high rhythm bands (α and β) diminish their activity. With the aim of validating the proposed training methodology for selecting stochastic features, classification is estimated by employing the support-vector machines-based classifier using the RBF kernel. Both kernel dispersion and classifier tradeoff parameter, c, are tuned for testing with a particle swarm meta-heuristic optimization that is bio-inspired method used for SVM parameter determination [8].

The performed accuracy of the SVM classifier is estimated, and is evaluated by using the conventional crossvalidation procedure, which consists in dividing the database into 10 folds; each one having an equal number of signals per class. The computed average classification measures of performance along with the respective standard deviations are presented in Table 2, for both considered decomposition techniques. The best classifier performance is ($a_{cc} = 99.06\%$, $a_{se} = 100\%$, $a_{sp} = 98.33\%$), which is achieved by the TVAR decomposition technique showing low values of standard deviation.

IV. DISCUSSION AND CONCLUSIONS

This work proposes a methodology to evaluate quantitatively each waveform contribution to neuronal activity related to either normal or Epileptic seizure events. Discussed methodology lies in the hypothesis that providing an appropriate relevance evaluation of time-variant rhythm waveforms one can distinguish neuronal activity related to normal and Epileptic seizure events. While obtained preliminary results are promising, the following aspects are to be taken into consideration: As seen from Table 1, estimated rhythm weights do not change significantly with considered timevariant decompositions. However, the lower dispersion of relevance weight estimates are related to TVAR enhanced representation, while highest variance correspond to EDS. However, that EDS decomposition fails in determining the correct variation of β waveform. From the assessed stochastic relevance analysis, the introduced relevance rhythm diagrams allow to qualify the contribution of waveform dynamics for each patient's condition, enabling better clinical interpretability of the obtained results, as illustrated in Figure 1. Based on estimated relevance weights of timevariant rhythms, classifier performance is carried out. As seen from Table 2 for the dataset, the best obtained classifier performance in the classification Problem ($a_{ac} = 99.06\%$, $a_{se} = 100\%$ $a_{sp} = 98.33\%$) is achieved by the TVAR decomposition technique, indicating a high classification ability in the epileptic seizure detection.

Future work includes the application of the discussed methodology to analyze other brain activity and to determine the feasibility of prediction of seizures.

ACKNOWLEDGEMENTS

This research is carried out under the grant *Instituto Tecnolgico Metropolitano project INAE51, and Programa Jóvenes Investigadores e Innovadores convocatoria 525 de 2011"*, *Convenio Interadministrativo Especial de Cooperación No. 0043 entre COLCIENCIAS y la Universidad Nacional de Colombia Sede Manizales.*

REFERENCES

1. Iscan Zafer, Dokur Zmray, Demiralp Tamer. Classification of electroencephalogram signals with combined time and frequency features *Expert Systems with Applications.* 2011;38:10499–10505.
2. Adeli Hojjat, Zhou Ziqin, Dadmehr Nahid. Analysis of EEG records in an epileptic patient using wavelet transform. *Journal of Neuroscience Methods.* 2003;123:69–87.
3. West Mike, Prado Raquel, Krystal Andrew D.. Evaluation and Comparison of EEG Traces: Latent Structure in Nonstationary Time Series *Journal of the American Statistical Association.* 1999;94:375-387.
4. De Clercq Wim, Vanrumste Bart, Papy Jean-Michel, Van Paesschen Wim, Van Huffel Sabine. Modeling common dynamics in multichannel signals with applications to artifact and background removal in EEG recordings. *IEEE Transactions on Biomedical Engineering.* 2005;52:2006–2015.
5. Sepulveda-Cano L., Acosta-Medina C., Castellanos-Dominguez G.. Relevance Analysis of Stochastic Biosignals for Identification of Pathologies *EURASIP Journal on Advances in Signal Processing.* 2011;2011:10.
6. Andrzejak Ralph G, Mormann Florian, Widman Guido, Kreuz Thomas, Elger Christian E, Lehnertz Klaus. Improved spatial characterization of the epileptic brain by focusing on nonlinearity. *Epilepsy Research.* 2006;69:30–44.
7. Lantz G., Michel C. M., Seeck M., Blanke O., Landis T., Ros?n I.. Frequency domain EEG source localization of ictal epileptiform activity in patients with partial complex epilepsy of temporal lobe origin *Byosistems engineering.* 2009;110:176-184.
8. Lin S, Ying K, Chen S, Lee Z. Particle swarm optimization for parameter determination and feature selection of support vector machines. *Expert Systems with Applications.* 2008;35:1817–1824.

Optimize ncRNA Targeting: A Signal Analysis Based Approach

N. Maggi[1], P. Arrigo[2], and C. Ruggiero[1]

[1] University of Genoa, DIBRIS, Genoa, Italy
[2] CNR, ISMAC, Genoa, Italy

Abstract—Non-coding RNAs (ncRNAs) are ribonucleic acids capable to control different genetic and metabolic functions. Digital signal processing (DSP) has been widely applied for analyze DNA sequences. Herein we present a DSP application to filter out the most probable binding site of ncRNAs applied to a miRNA binding site recognition.

Keywords—Digital Signal Processing, ncRNA, gene target prediction.

I. INTRODUCTION

The discovery of post-transcriptional gene silencing Experiments have disclosed the perspective to study the role of non-coding RNAs on gene expression control. These molecules are continuously acquiring significance in many different cellular processes such as signal transduction or cell-cell communication. The exploitation of interest on biological functionality of ncRNAs has stimulated the development of analytical tools in order to predict binding site on putative target messengers RNA (mRNAs).

The rapid advancement of molecular biology has allowed to demonstrate the existence of many different non-coding RNAs (ncRNAs). These molecules have been classified into different families:

1. tRNAs: they are involved in translational processes
2. Small Interferring RNA (siRNAs): they are involved in RNA silencing
3. microRNA (miRNAs): they are involved in RNA silencing
4. Small nuclear (snRNAs): they are located in the nucleus and are involved in RNA splicing
5. Small nucleolar (snoRNAs): they are in nucleolus and are involved in RNA modification.
6. Small Cajal-body specific (scaRNAs): involved in RNA modification
7. small guide (gRNAs): involved in RNA editing
8. piwi interacting RNA (piRNAs): involved in gene silencing.

MicroRNAs (miRNAs) are, among previously listed classes, the more extensively studied ncRNAs. miRNAs are small RNA that lead to the degradation of messenger RNA or disturb the translation of mRNA in sequence-dependent manner. A miRNA recognize his preferential target by the complementarity between a special nucleotide motif, embedded in mature microRNA, and some specific binding sites displaced along the mRNA. Less information is available about the functionality and recognition site for the other classes of ncRNAs. For several years the interaction of miRNAs with their targets was analyzed taking into account only the 3' untranslated regions (UTR) of mRNAs. This approach was based on the assumption that the main recipients of miRNA activity are these regions, however recent findings also show the presence of many miRNA targets in their amino acid coding sequence (CDS).

We herein propose a signal processing analysis to optimize the search of 'true' miRNA binding sites on mRNAs. The generality of DSP method is a valuable feature also for the analysis of other ncRNAs.

Target identification is one of the major challenges in miRNA analysis, since a miRNA may control many different mRNAs. At present the biophysical processes relating to the recognition and interaction between a miRNA and its potential targets are mostly unknown. As a consequence screening is mainly carried out by bioinformatics analysis. The currently available methods for target recognition originate a redundant set of potential binding sites because the tools are mainly based on pattern matching systems. Several algorithms for miRNA gene target prediction have so far been set up. Most of them are based on sequence complementarity searches between a miRNA and the 3'UTR of the gene target. The use of this algorithms has produced a very high number of potential gene targets, therefore it is not possible to carry out experimental validations for them all. Attempts have been made to identify miRNA gene targets by cDNA microarray [1] but it has recently been shown that a miRNA regulates gene expression at the protein level only, and miRNA are not affected [2]. Moreover, recent findings by [3] have shown that one miRNA can repress the production of hundreds of proteins. This shows the need of proteomic data analysis for the accurate detection of miRNA gene targets. Recent work focusing on the integration of miRNA microarrays, proteomic analysis, miRNA gene target prediction and gene network contruction has shown the role of miRNA in cartilage destruction and related inflammatory and metabolic gene networks. [4].

The result of a previous analysis [5] have underlined the role of compositional features of miRNA in the recognition process. The critical recognition motif, called 'seed', is supported by other stabilizing elements in the functional miRNA. The available method for binding sites identification does not allow to deep investigate the effect of the seed context. Digital Signal Processing (DSP) methods have been applied for a long time in bio sequence analysis [6] and can help the researcher to better study the role of the seed context.

II. METHODS

In order to use DSP tools to identify possible regions of interest in a DNA or RNA sequence, it is necessary to map the sequence from a symbolic form in term of a character string to a numerical form. In order to achieve this a numeral is assigned to to each nucleotide in the sequence. Different techniques have been proposed, each of which aims to enhance information which is hidden in the sequence for further analysis.

In many cases nucleotides have been represented by lexicographic order (A=1,C=2,G=3,T/U=4) or by different binary codes that can represent not only the nucleotide but also its properties such as number of H-bonds or weak (A, T/U) or strong (G, C). A very promising method to code nucleotide is based on electro-ion interaction potential (EIIP) [7], which is defined as the average energy of the localized electrons of the nucleotide. Resonant Recognition Model (RRM) [8, 9] is a physical and mathematical model which interprets biological sequence linear information using signal analysis methods in order to treat the primary sequence as a discrete signal. In this case we have applied RRM to a RNA sequence. From this model it is possible to generate the EIIP values that are assigned to the nucleotides, and that originates a numerical sequence which represents the distributions of the free electrons' energies within in the nucleotide sequence. This methods have been successfully used for several problems, such as the identification of proteins' hot spots, the identification of coding regions and peptide design.

EIIP it is biologically more meaningful than a lexicographic method as it represents a physical property. In that methods the corresponding values only represent the presence or absence of a nucleotide.

EIIP is an estimate of average energy states of all valence electrons in a particular nucleotide. Table 1 shows the values of EIIP for each nucleotide. The nucleotide sequence can be converted into a numerical sequence by replacing each nucleotide with the corresponding EIIP value. For example, if $x[n]$= TGTGCTTGCA, then using the values from Table 1, the corresponding EIIP numerical sequence will be $x[n]$=[0.1335 0.0806 0.1335 0.0806 0.1340 0.1335 0.1335 0.0806 0.1340 0.1260]. The next steps use digital signal analysis methods applied to the obtained numerical series in order to investigate the potential.

Table 1 Electro Ion Interaction pseudo potentials of nucleotides

Nucleotide	EIIP values
A	0.1260
G	0.0806
C	0.1340
T	0.1335

In our study we used a frequency domain method to filter out the possible regions of a sequence that can target miRNAs. The objective of this work is to obtain a reduced set of miRNAs that can be verified experimentally similarly to what is often done in the process of drug discovery [10-13]. Specifically, we have applied a Short Time Fourier Transform (STFT) based method to the mRNA sequence to evidence possible different function related to different position in the sequences. The STFT of the EIIP indicator sequence is employed to predict the miRNA binding site. The STFT introduce time-localization by dividing a signal into a number of short overlapping sections using a sliding window. Fourier transform is not apply to the whole signal, but only to the section selected by the sliding window.

Thus, we have a series of frequency spectra with each spectrum corresponding to a short interval of time. The STFT of the corresponding EIIP numerical sequence and is given by is defined by

$$U[k] = \sum_{m=0}^{L-1} x[m+rR]w[m]e^{\frac{-j\pi km}{N}} \quad (1)$$

where L is the window length, N is the DFT length, R is the shift interval, and r and k are integers such that

$$-\infty < r < \infty \quad \text{and} \quad 0 \leq k \leq N-1$$

Then the spectral content at k^{th} instant is:

$$S[k] = |U[k]|^2 \quad (2)$$

The time and frequency resolution depends by the length of the window. An increase in the window length enhances the frequency resolution at the expense of the time resolution while a decrease in the window length improves the time resolution at the expense of the frequency resolution.

Generally DSP methods have been applied to DNA sequence for the identification of regions that code the protein. It has been observed that there is a prominent power spectrum peak at frequency $f = 1/3$ in the coding regions of a DNA sequence. This characteristic is related to the nucleotide triplet arrangement (or codon) that encodes protein amino acids. In our case we take into account the fact that the beginning of the recognition between a miRNA and its target on a mRNA sequence is based on motifs with length of 6 nucleotides.

The binding regions can be identified by evaluating S[L/6] over a window of L samples, then sliding the window by one samples and recalculating S[L/6]. This process is carried out over the entire mRNA sequence. The peaks in the spectra obtained by the sliding window should correspond to the binding site of miRNAs. The window has been splitted into segmentation element dividing the sampling window by 6. This value has been selected taking the average e dimension of many biological functional motifs such as polyadenylation signals (AATAAA) that is one of more extensively studied regulatory element of mRNA.

Many types of windowing have been set up in the past. It has been shown that for coding region identification the window length affects the results and its optimal value depends on the length and the number of exons in sequence [14-16].

At the moment we have no information about the optimal type and the length of the window. Hence, in order to tune the parameters for the DSP analysis we have selected the window amplitude on the basis of maximal length of a functional miRNA. Their length is comprised between 21-23 bp and we have added a single base for bias. We have decided to use a Blackman window of length $L=24$ that is shown in figure 1.

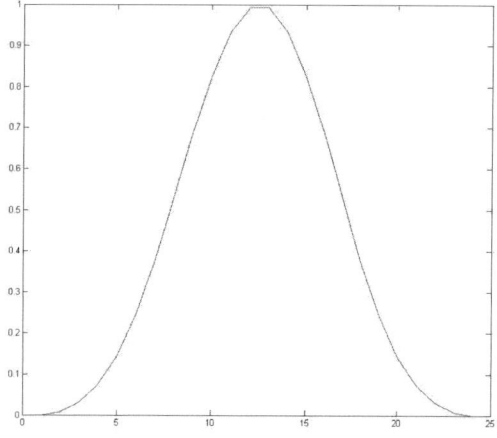

Fig. 1 Blackman window

We have applied this method to ras-related C3 botulinum toxin substrate 3 (RAC3) protein. The protein encoded by this gene is a GTPase which belongs to the RAS superfamily of small GTP-binding proteins. Members of this superfamily appear to regulate a diverse array of cellular events, including the control of cell growth, cytoskeletal reorganization, and the activation of protein kinases (). RAC3 protein is one of a targets for human premirna with highest content of CG dinucleotides [17]. This compositional property seems potentially correlated with mutation propensity (CpG) [18].

For our analysis we have used 3'UTR of RAC3 that it has been extracted form from Ensembl database (www.ensembl.org). [19] under the code ENST00000306897. The result of the applied methods is shown in fig. 2

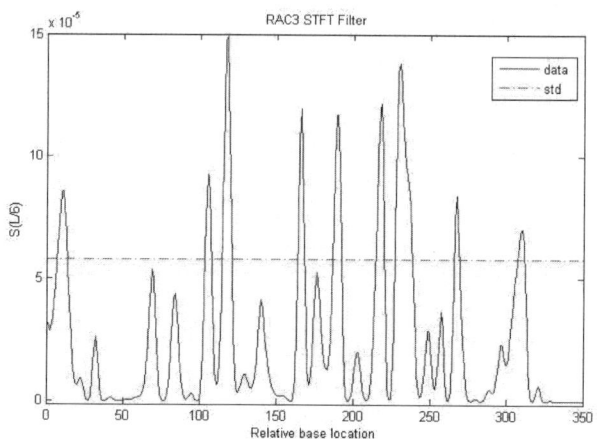

Fig. 2 RAC3 filtering using STFT

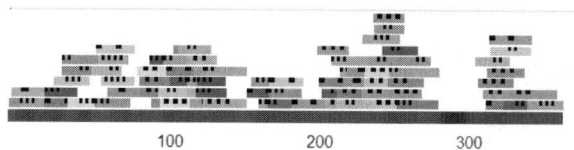

Fig. 3 Microcosm result for RAC3 query

III. RESULTS AND DISCUSSION

The DSP method allows to filter out the binding sites with highest STFT peaks that are correlated with high intermolecular interaction propensity. It should be noticed that there is a similarity with those can be obtained from using a pattern matching methods such as Microcosm (http://www.ebi.ac.uk/enright-srv/microcosm/htdocs/targets/v5/) [20]. Using DSP filtering we can identify 9 regions in RAC3 3'UTR than can be further investigated. It should e possible to query miRNA databases to extract possible miRNA that can bind the target and then proceed to experimental validation.

The method seems to be promising due to the fact that it could be easily applied to other RNA sequences that are not studied at the moment, but that could be potential target for miRNA.

It is important to underline that ncRNAs have acquired great importance in Molecular Medicine, as a consequence knowledge about the involvement of a ncRNA, particular of a siRNA or a miRNA, in pathological process is helpful to develop diagnostic and therapy systems. It is also noticeable that small ncRNAs constitute a possible way to investigate the function of a gene in its particular functional pathway

REFERENCES

1. Huang, J.C., et al. (2007) Using expression profiling data to identify human microRNA targets. Nat Meth. 4(12): p. 1045-1049
2. Ma, L., J. Teruya-Feldstein, and R.A. Weinberg (2007) Tumour invasion and metastasis initiated by microRNA-10b in breast cancer. Nature. 449(7163): p. 682-8
3. Selbach, M., et al. (2008) Widespread changes in protein synthesis induced by microRNAs. Nature. 455(7209): p. 58-63
4. Iliopoulos, D., et al. (2008) Integrative microRNA and proteomic approaches identify novel osteoarthritis genes and their collaborative metabolic and inflammatory networks. PLoS One. 3(11): p. e3740
5. Wang, B. (2013) Base Composition Characteristics of Mammalian miRNAs. Journal of Nucleic Acids. 2013: p. 6
6. Zoete, V., A. Grosdidier, and O. Michielin (2009) Docking, virtual high throughput screening and in silico fragment-based drug design. J Cell Mol Med. 13(2): p. 238-48
7. Nair, A.S. and S.P. Sreenadhan (2006) A coding measure scheme employing electron-ion interaction pseudopotential (EIIP). Bioinformation. 1(6): p. 197-202
8. Pirogova, E., G.P. Simon, and I. Cosic (2003) Investigation of the applicability of dielectric relaxation properties of amino acid solutions within the resonant recognition model. IEEE Trans Nanobioscience. 2(2): p. 63-9
9. Cosic, I. (1994) Macromolecular bioactivity: is it resonant interaction between macromolecules?--Theory and applications. IEEE Trans Biomed Eng. 41(12): p. 1101-14
10. Maggi, N., P. Arrigo, and C. Ruggiero (2012) Comparative analysis of Rac1 binding efficiency with different classes of ligands: morpholines, flavonoids and imidazoles. IEEE Trans Nanobioscience. 11(2): p. 181-7
11. Lavecchia, A. and C. Di Giovanni (2013) Virtual Screening Strategies in Drug Discovery: A Critical Review. Curr Med Chem
12. Lill, M. (2013) Virtual screening in drug design. Methods Mol Biol. 993: p. 1-12
13. Tanrikulu, Y., B. Kruger, and E. Proschak (2013) The holistic integration of virtual screening in drug discovery. Drug Discov Today. 18(7-8): p. 358-64
14. Akhtar, M., J. Epps, and E. Ambikairajah (2008) Signal Processing in Sequence Analysis: Advances in Eukaryotic Gene Prediction. Selected Topics in Signal Processing, IEEE Journal of. 2(3): p. 310-321
15. Akhtar, M., E. Ambikairajah, and J. Epps, *Digital Signal Processing Techniques for Gene Finding in Eukaryotes*, in *Image and Signal Processing*, A. Elmoataz, et al., Editors. 2008, Springer Berlin Heidelberg. p. 144-152.
16. Akhtar, M., E. Ambikairajah, and J. Epps. *Optimizing period-3 methods for eukaryotic gene prediction*. in *Acoustics, Speech and Signal Processing, 2008. ICASSP 2008. IEEE International Conference on*. 2008.
17. Arrigo, P., C. Mitra, and A. Izzotti. *Influence of pre-mirna compositional properties on RISC complex recruitment and target selection*. in *Health Informatics and Bioinformatics (HIBIT), 2012 7th International Symposium on*. 2012.
18. Giuliano, F., et al. (1993) Potentially functional regions of nucleic acids recognized by a Kohonen's self-organizing map. Comput Appl Biosci. 9(6): p. 687-93
19. Hubbard, T.J., et al. (2009) Ensembl 2009. Nucleic Acids Res. 37(Database issue): p. D690-7
20. Griffiths-Jones, S., et al. (2008) miRBase: tools for microRNA genomics. Nucleic Acids Res. 36(Database issue): p. D154-8

Author:	Carmelina Ruggiero
Institute:	DIBRIS – University of Genoa
Street:	Via all'Opera Pia, 14
City:	Genova
Country:	Italy
Email:	carmel@dibris.unige.it

Mathematical Modelling of Melanoma Collective Cell Migration

J.V. Gallinaro[1], C.M.G. Marques[2], F.M. Azevedo[1], and D.O.H. Suzuki[1]

[1] Biomedical Engineering Institute, Federal University of Santa Catarina, Florianopolis, Brazil
[2] Sports and Health Center, State University of Santa Catarina, Florianopolis, Brazil

Abstract—A collective cell migration model is used to simulate the migration of HBL and C8161 melanoma cell lines during a scratch wound assay. The model has four parameters: force resulted from lamellipod creation in the scratch wound assay, elasticity modulus of the cell layer, adhesion constant between cell and extracellular matrix (ECM) and proliferation rate. Parameter values were obtained with a least square nonlinear regression routine and data from an assay performed with both cell lines. Coefficient of determination for cell lines HBL and C8161 was, respectively, $R^2 = 0.79$ and $R^2 = 0.89$. Parameter values were compared to values from other work on the literature and were mostly in accordance with them. C8161 cell line presented smaller values of force from lamellipod creation and adhesion constant, which suggests that more invasive cell lines, in this case represented by C8161 cell line, have less dependence on adhesion molecules, as proposed by other authors.

Keywords—Mathematical model, cell migration, melanoma.

I. INTRODUCTION

Mathematical models can aid tumour growth and cell migration research since they allow simulations that are not easily obtained in in vitro or in vivo experiments. Mathematical models of tumour growth and cell migration have been developed in order to help treatment and diagnosis of several types of cancer [1, 2, 3, 4, 5]. Some of these models have a system of non-dimensionalised equations [1, 2, 5], which complicates quantitative comparison with experimental work. Other models, such as the one developed by Eikenberry *et al.* [4], have dimensionalised equations, however they do not quantitatively compare their results with experimental work. These facts lead to models that may indicate the way cell migration and tumour invasion occur, but not necessarily are capable of describing the experimentally obtained situation.

Most part of current cancer cell migration models are based on a diffusion process [1, 2, 4, 5]. These studies present, generally, a principal variable: cancer cell concentration, which varies in time and space according to a diffusion equation based on Fick's Law, with diffusion coefficient either constant or variable.

It is known that cell migration depends on the cell-to-ECM interaction, through creation of adhesion complexes [6, 7].

Besides, it is also known that melanoma is a type of cancer that presents mostly collective migration [7], which implies in adherent junctions that promote cell-to-cell adhesion [8]. Diffusion processes by themselves do not account for these connections. Therefore, some models propose different manners of including parameters in the diffusion process that can represent cell-to-cell and cell-to-ECM connections, as in the model proposed by Chaplain *et al.* [9].

A study shown by Mi *et al.* [10] models a healing process using an elastic continuum model. This model was based on collective migration principles: during migration cells do not separate from the edges and no wholes are created on the cell layer. Mi *et al.* [10] proposed a mathematical model of enterocyte migration, during a healing process, in which cells undergo movement, deformation and proliferation.

The aim of this work is to propose the use of a collective cell migration mathematical model that could be quantitatively compared to experimental data from two different melanoma cell lines.

II. METHODS

A. Mathematical Model

The model proposed by Mi *et al.* [10] was a one-dimensional model that described migration of a cell layer after the opening of a wound on the layer, as in a *scratch wound* assay. The model followed some considerations: (1) it was based on a monolayer of cells, (2) there was connection between cells in the layer, (3) after the scratch (from the *scratch wound* assay), an external force was created as a result of lamellipod formation, (4) the cells in the interior of the layer did not form lamellipodia and hence were not directly actuating the motion, (5) the cell layer came under deformation, movement and material growth.

According to the considerations presented earlier, Mi *et al.* [10] demonstrated that cell layer migration would occur according to the equation:

$$\frac{\partial x}{\partial t} = \frac{k}{b} \frac{\partial^2 x}{\partial s^2} \left(\frac{\partial x}{\partial s}\right)^{-2} \quad (1)$$

where *s* represents the position of a cell on the original cell layer, $x(t,s)$ represents the position of the cell located on *s* on the original layer at time instant t, k is the elasticity modulus of the cell layer, and b is the adhesion constant of the interaction cell-ECM.

Deformation gradient (ε) is defined as:

$$\varepsilon = \frac{\partial x}{\partial \hat{s}} - 1 \quad (2)$$

where $\hat{s}(t,s)$ is the hypothetical position of cell located at s on the original layer at time instant t, if all deformation from the layer was instantaneously removed.

From Equation 2, one can obtain:

$$\varepsilon = \frac{\partial x}{\partial t}\frac{\partial t}{\partial \hat{s}} - 1 \quad (3)$$

Solving Equation 3 for $\partial \hat{s}/\partial t$, one can obtain the hypothetical position of cell frontier with Equation 4

$$\frac{\partial \hat{s}}{\partial t} = (\varepsilon + 1)^{-1}\frac{\partial x}{\partial t} \quad (4)$$

Initial and boundary conditions used were:

$$x(0,s) = 0, \quad for \ 0 \le s \le L, \quad (5)$$

$$x(t,0) = 0, \quad for \ t \ge 0, \quad (6)$$

$$\frac{\partial x(t,1)}{\partial s} = e^{(F/k)+\rho t}, \quad (7)$$

where F is the force developed as result of lamellipod formation after scratch.

B. Calibration

A routine nonlinear minimization of the least square error was used to estimate the parameters. The constants k, b and F appear on the problem only as the ratios $\kappa = k/b$ and $\varphi = F/k$. Therefore, regression was performed for the constants κ, φ and ρ.

Experimental data used on calibration were based on the work by Marques [11]. Data were obtained from a study using HBL and C8161 human cutaneous melanoma. The human HBL melanoma line was established from a lymph node metastasis of a nodular melanoma. The human C8161 melanoma line was established from an abdominal wall metastasis, which indicated that it was highly invasive.

Cells were seeded in culture plates in culture medium at a concentration of $4 \cdot 10^4 cells/ml$ per well and were incubated for one day under standard culture conditions. On the third day, a *scratch wound* for migration assay was made in each well using a plastic pipette tip, creating a cell free zone in each well. The reduction of distance between the scratch edges at different time points (0, 2, 4, and 8 hours) represented the migration of melanoma cells.

C. Simulation

A script was written in MATLAB to implement the model. Discretization both in time and space was obtained with the finite differences method [10]. Time (dt) and space (ds) steps used were, respectively, 1/120 h and 0.0125 m.

III. RESULTS

The curves obtained with the mathematical model for the HBL and C8161 cell migration after nonlinear regression, Fig. 1, had determination coefficient of, respectively, $R^2 = 0.79$ and $R^2 = 0.89$. The values of the constants obtained are shown on Table 1.

Table 1 Constants obtained with the nonlinear regression for HBL and C8161 cell lines.

Constants	HBL	C8161
$\kappa = k/b [\mu m^2 h^{-1}]$	5.40	131.68
$\varphi = F/k [dimensionless]$	0.70	0.58
$\rho [h^{-1}]$	0.87	0.52

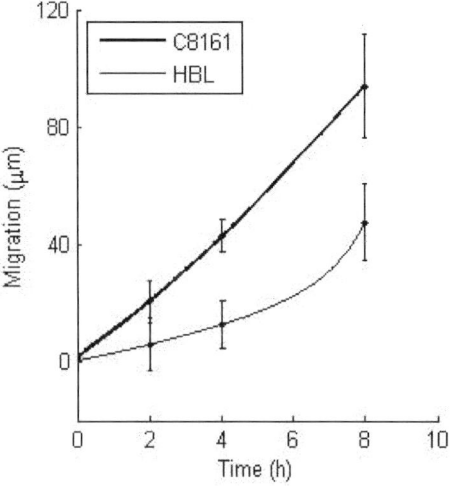

Fig. 1 Curve obtained with the simulation of the mathematical model after nonlinear regression using experimental data from C8161, with $R^2 = 0.89$, and HBL cell line, with $R^2 = 0.79$. Experimental data are represented as mean standard deviation, n=7.

A study was performed to determine the importance of deformation and proliferation on cell migration for both situations, Fig. 2. In order to determine migration due to deformation, for each cell line, a simulation was run with values κ

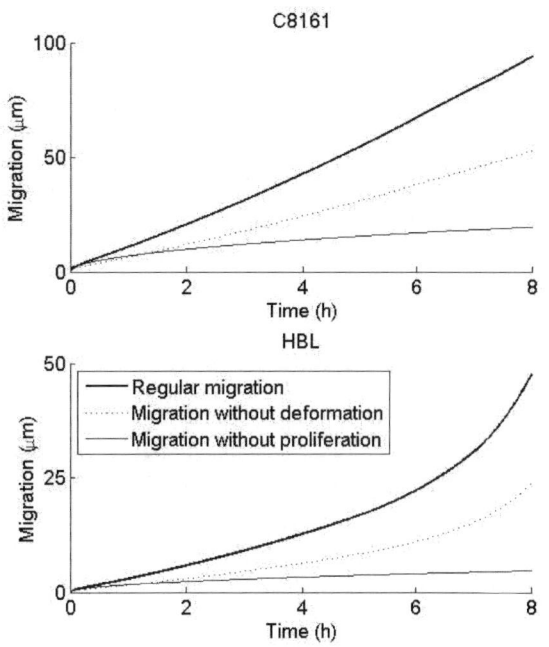

Fig. 2 Cell migration with values from Table 1, cell migration in the hypothetical situations where no deformation and no proliferation occurred, for both cell lines C8161 and HBL.

and φ from Table 1 and $\rho = 0 \ h^{-1}$. Migration due to proliferation was determined with Equation 4.

IV. DISCUSSION

A collective cell migration model [10] was used to simulate melanoma cell migration during a *scratch wound* assay. The model was calibrated with experimental data [11] that was obtained from HBL and C8161 melanoma cell lines.

Cell migration models based on diffusion have been used to model cancer cell migration [1, 2, 4, 5]. However, the coefficient values demonstrated by some of these models, such as by Eikenberry *et al.* [4] and Anderson *et al.* [1], are small ($0.0009 \ mm^2 \ dia^{-1}$ $0.07 \ mm^2 \ dia^{-1}$) when compared to diffusion coefficients experimentally obtained, for instance, by Lyng *et al.* [12] ($50 \ 80 \ mm^2 \ dia^{-1}$). One possible explanation for this difference is that these models do not take the collective migration into account, as it is the case for melanoma cells. The elastic continuum model, proposed by Mi *et al.* [10], take some collective migration aspects into consideration and it is, therefore, a valid option for modeling *scratch wound*assays.

The values obtained during calibration are shown in Table 1. Since the parameters of interest k, b and F are presented on the model only as the ratios $\kappa = k/b$ and $\varphi = F/k$, non-linear regression was performed for κ and φ. However, if a reference value is set for one of the three parameters of interest, it is possible to estimate the value of the other two. Mi *et al.* [10] modeled enterocyte migration during wound healing and showed a value of elasticity modulus k of approximately $44 \ nN$. Considering melanoma cells migrating during the *scratch wound* assay have the same value of elasticity modulus, the HBL cell layer would have force resulted from lamellipod formation F of approximately $30.8 \ nN$ and adhesion constant b of approximately $8.2 \ hnN \ \mu m^{-2}$. C8161 cell layer would have force resulted from lamellipod formation F of approximately $25.5 \ nN$ and adhesion constant b of approximately $0.33 \ hnN \ \mu m^{-2}$.

Less adhesiveness from C8161 cell line, represented by its smaller adhesion constant value, allows its higher migration, when compared to the HBL cell line, although the HBL cell line has greater lamellipod force and proliferation rate. Less adhesiveness from the C8161 cell line is responsible for its greater migration observed on Fig. 2, when proliferation rate was set to zero.

Brunner *et al.* [13] measured cell body motility forces and found values with the same order of magnitude than those found in this work. Considering the elasticity modulus equal for both HBL and C8161 cell lines, HBL cell line, which is the least invasive one, has greater values of both force and adhesion cell-ECM. DiMilla *et al.* [14] showed a bell shaped curve to describe the dependence of cell migration speed on adhesion, *i.e*, an increase on adhesion would lead to an increase on migration speed until a certain value, after which speed would stabilize and afterwards begin to decrease with an increase on adhesion. In this case, the higher adhesion of HBL cell line, as compared to the C8161 line, leads to decreased migration.

The force created by lamellipod formation depends on adhesion molecules that create the adhesion cell-ECM [6] and Mi *et al.*[10] proposed that the lamellipod force should be proportional to the square root of integrin concentration on the ECM. Thus, the higher value of lamellipod force from HBL cell line presented in this work, as compared to C8161 cell line, could indicate a greater dependence on adhesion molecules. Friedl *et al.* [7] showed that collective migration depend more on integrins and proteases when compared to individual migration, such as amoeboid, an individual migration observed in some highly metastatic cancer types. Therefore, the fact that the C8161 showed to be more invasive than the HBL cell line and its lower values of lamellipod force and adhesion constant was in fair agreement with the work on collective cell migration demonstrated by Friedl *et al.* [8].

V. CONCLUSION

The model used in this work may be an option when modeling melanoma cell migration, since it accounts for collective migration characteristics, which is not the case for all cell migration models. The parameter values found were in accordance with values from other works in the literature. Results suggest that cell invasiveness may be related to the cell dependence on adhesion molecules.

ACKNOWLEDGEMENTS

The authors thank CNPq for the financial support.

REFERENCES

1. Anderson ARA, Chaplain MAJ, Newman EL, Steele RJC, Thompson AM. Mathematical modelling of tumour invasion and metastasis *Computational and Mathematical Methods in Medicine.* 2000;2:129–154.
2. Anderson A.R.A.. A hybrid mathematical model of solid tumour invasion: the importance of cell adhesion *Mathematical Medicine and Biology.* 2005;22:163–186.
3. Ciarletta P., Foret L., Amar M.B.. The radial growth phase of malignant melanoma: multi-phase modelling, numerical simulations and linear stability analysis *Journal of The Royal Society Interface.* 2011;8:345–368.
4. Eikenberry S., Thalhauser C., Kuang Y.. Tumor-immune interaction, surgical treatment, and cancer recurrence in a mathematical model of melanoma *PLoS computational biology.* 2009;5:e1000362.
5. Tohya S., Mochizuki A., Imayama S., Iwasa Y.. On rugged shape of skin tumor (basal cell carcinoma) *Journal of theoretical biology.* 1998;194:65–78.
6. Lauffenburger D.A., Horwitz A.F.. Cell Migration: Review A Physically Integrated Molecular Process *Cell.* 1996;84:359–369.
7. Friedl P., Wolf K., others . Tumour-cell invasion and migration: diversity and escape mechanisms *Nature Reviews Cancer.* 2003;3:362–374.
8. Friedl P., Gilmour D.. Collective cell migration in morphogenesis, regeneration and cancer *Nature Reviews Molecular Cell Biology.* 2009;10:445–457.
9. Chaplain M.A.J.. Multiscale mathematical modelling in biology and medicine *IMA Journal of Applied Mathematics.* 2011;76:371–388.
10. Mi Q., Swigon D., Rivière B., Cetin S., Vodovotz Y., Hackam D.J.. One-dimensional elastic continuum model of enterocyte layer migration *Biophysical journal.* 2007;93:3745–3752.
11. Marques C.M.G.. *Tissue engineered human skin models to study the effect of inflammation on melanoma invasion.* PhD thesisThe University of Sheffield 2010.
12. Lyng H., Haraldseth O., Rofstad E.K.. Measurement of cell density and necrotic fraction in human melanoma xenografts by diffusion weighted magnetic resonance imaging *Magnetic resonance in medicine.* 2000;43:828–836.
13. Brunner C.A., Ehrlicher A., Kohlstrunk B., Knebel D., Kas J.A., Goegler M.. Cell migration through small gaps *European Biophysics Journal.* 2006;35:713–719.
14. DiMilla PA, Barbee K., Lauffenburger DA. Mathematical model for the effects of adhesion and mechanics on cell migration speed *Biophysical journal.* 1991;60:15–37.

Author: Julia Vianna Gallinaro
Institute: Biomedical Engineering Institute (IEB-UFSC)
Street: UFSC - CTC Departamento de Engenharia Eletrica
City: Florianopolis, SC
Country: Brazil
Email: juliavg@gmail.com

Synthetic Atrial Electrogram Generator

M.W. Rivolta[1], L.T. Mainardi[2], R. Sassi[1], and V.D.A. Corino[2]

[1] Dipartimento di Informatica, Università degli Studi di Milano, Crema, Italy
[2] Dipartimento di Elettronica, Informazione e Bioingegneria, Politecnico di Milano, Milan, Italy

Abstract—Atrial electrogram (AEGs) recorded invasively inside the atrium can be analyzed to assess the organization of the atrial electrical activity. These organization measures are commonly based either on the repeatability/regularity of the atrial activations or on the correlation/synchronicity among electrograms recorded in different sites. For many applications, it could be useful to have a synthetic AEG generator to test new methods on. So far, models capable to reproduce realistic AEGs do not exist yet. Aim of this study is to propose a unified approach to generate synthetic AEGs during sinus rhythm (SR) and atrial tachyarrhythmias, namely atrial fibrillation (AF) and atrial flutter (AFL). In particular, three different Wells organizations classes and different atrio-ventricular conductions will be considered during AF and AFL, respectively. A database of simulated signals during AF was created, containing AEGs with different degrees of Wells' organization and different dominant frequencies. These AEGs were tested using spectral analysis assessing spectral concentration (SC), and wave-morphology similarity (WMS). Both indexes were in agreement with those presented in the relevant literature.

Keywords—Synthetic atrial electrogram, atrial fibrillation, atrial flutter, sinus rhythm.

I. INTRODUCTION

During the last decades, the innovation of the technology led to an increase in the complexity of the computer simulations, that are now a reliable tool for validating models and methods in many fields. In the electrophysiological field, realistic models for simulations are not always easy to obtain; this is due to the high intra/inter variability of each phenomena in each subject.

A few models for simulating the human electrocardiogram (ECG) have been proposed. Some years ago, a dynamical model which is capable of replicating many of the important features of the ECG was introduced [1]. It can generate different morphologies for the PQRST-complex and it allows the operator to prescribe specific characteristics of the heart rate dynamics such as the mean and standard deviation of the heart rate and spectral properties. This model can be used to generate ECG in sinus rhythm as well as ECG with abnormal morphological waves or during atrial fibrillation [2]. More recently, a Gaussian wave-based approach was introduced to model the temporal dynamics of ECG signals. This model may be used for generating synthetic ECGs as well as separate characteristic waves such as the atrial and ventricular complexes [3].

However, no models exist providing realistic simulated atrial electrograms (AEGs). Aim of this study is to propose a model to generate synthetic AEGs during sinus rhythm (SR) and atrial tachyarrhythmias, namely atrial fibrillation (AF) and atrial flutter (AFL). This is, in part done, mixing some validated models.

II. METHODS

AEGs can be seen as the sum of atrial and ventricular activity

$$\mathrm{AEG}(t) = \mathrm{AA}(t) + \mathrm{VA}(t) \quad (1)$$

where $\mathrm{AA}(t)$ is the electrical activity generated only by the atrial cells and $\mathrm{VA}(t)$ is the ventricular activity that interferes on the measurements.

The atrial activity AA can be rewritten in two subcomponents

$$\mathrm{AA}(t) = \mathrm{AA}_{\mathrm{far}}(t) + \mathrm{AA}_{\mathrm{near}}(t) \quad (2)$$

where $\mathrm{AA}_{\mathrm{far}}(t)$ and $\mathrm{AA}_{\mathrm{near}}(t)$ represent, respectively, the far field and near field effects. Although a clear distinction between these fields effect does not exist, this has been previously used as a reasonable approximation [4].

The ventricular activity VA is normally overlapped to the atrial one; it is mainly due to the far field effects of both ventricles during their depolarization phase.

The simulation of atrial and ventricular activity requires the generation of morphological waves and their positioning in time.

A. Atrial Activity Model

A1 Far Field Effect and Measuring Noise

The necessity of modeling the far field effect and the measuring noise led us to think about their properties during SR and atrial tachyarrhythmias. In SR and AFL, it is basically present only broadband measuring noise instead, during AF, the far field effects become relevant. The high number of far sources produces a noisy effect added to the near field activity. According to the reasons above, the autoregressive model is suite for representing these electrical sources.

The AA_{far} is modeled by means of an autoregressive model as

$$AA_{far}(t) = \sum_{k=1}^{p} c_k AA_{far}(t-k) + w(t) \quad (3)$$

The model parameters (c_k and p) can be derived by fitting a set of real AEG signals. When derived from real AEG signals, segments of AEG between two atrial depolarizations, i.e., the AA_{far}, are considered. Each segment is fitted by autoregressive models of order p, with p ranging between 1 and 40. To pool together a common model, once determined the optimal global order by the Akaike information criterion, the corresponding c_k coefficients among all the models are averaged.

A2 Morphology of close field activity

The close field effects AA_{near} is modeled using the general formula of the electrical dipole moving along a line (not necessarily straight) [5, 4]. The potential generated by this dipole in a uniform infinite medium is

$$\phi(t) = \frac{\mathbf{p} \cdot \mathbf{u_d}}{4\pi\sigma ||\mathbf{r}(t)||^2} \quad (4)$$

where $\mathbf{p}$ is the dipole moment, $\mathbf{u_d}$ is the unit vector directed from the source point to the field point, σ is the electrical conductiviy and $\mathbf{r}(t)$ is the unit vector directed from the source point to the field point vector pointing the evaluation point.

The local field effect can be approximated by a summation of N dipoles moving nearby the measurement point. The close field effects AA_{near} becomes:

$$AA_{near}(t) = \sum_{k=1}^{N} \phi_k(t) \quad (5)$$

where $\phi_k(t)$ is the eletrical potential generated by the $k-$th dipole that travels along a specific direction and N is the total number of dipoles.

A3 Position of close field activity

During SR, the atrial depolarizations precede the ventricular ones, therefore atrial and ventricular rate are the same (see Section B2).

Due to the lack of relevant models, during AF, the position of the close field activity was considered temporal uncorrelated. The position of AA_{near} was so selected assuming that atrial depolarizations' front arrives to the electrodes according to a gamma distribution $AA_{near} \sim \Gamma(k, \theta)$, whose probability density function is close to an exponential (for small values of k) or to a normal (for high values of k) distribution. Depending on the rhythm to simulate, different values should be used, keeping in mind that $k\theta$ is the mean, thus its inverse is the atrial rate.

B. Ventricular Activity Model

B1 Morphology of Ventricular Activity

The ventricular morphology is analogously obtained as the potential generated by a current dipole. Local variations in VA amplitude and width are inserted on a beat-to-beat basis. Since the ventricular activity is a far but strong field effect, the electrical potential can be modeled by a single dipole using Eq.(4).

B2 Position of Ventricular Activity

To generate a realistic sequence of RR intervals different models exist depending on the rhythm to be simulated.

During SR, an autoregressive model was used to model the RR series, as it has been shown that this model is adequate for RR series approximation [6].

During AF, a realistic sequence of RR intervals can be generated using an atrioventricular node model as described in Lian et al. [7] or Corino et al. [8, 9]. The latter will be used in this work and is defined by parameters that characterize the arrival rate of atrial impulses, the probability of an impulse choosing either one of the two atrioventricular nodal pathways, the refractory periods of these pathways, and the prolongation of the refractory periods.

III. SIMULATIONS

In this section, different rhythms are simulated: i) SR; ii) AFL; iii) AF. Figure 1 shows the AEGs for the different simulated rhythms. It is worth noting that in all examples the amplitude of the ventricular activity is much smaller than that of the atrial one. However, this ratio can be changed and the ventricular activity can be bigger, for example when simulating signals recorded closer to the ventricles.

A. Sinus Rhythm

During SR, the internal surface of the atrium chambers is crossed by a wave that can be approximated by a plane wave. We simulated this situations using a set of parallel dipoles (equispaced) that, fixed the direction, travel along a straight line. The simulation could be optimized taking into account the symmetric properties of the dipole potential, in fact, the electrical component transversal to the direction becomes null. Figure 1a shows the simulated AEG during SR.

B. Atrial Flutter

After model identification of the position of the ventricular activity, we defined the flutter rate as the mean heart rate multiplied by, respectively, 2, 3 and 4. Figure 1b shows the simulated AEG during AFL.

C. Atrial Fibrillation

To mimic in the synthetic signals different degrees of atrial organization according to the three Wells' classes, the standard deviation of the white gaussian noise in input to AA_{far} was varied in the range 0.05-0.4 a.u. being a higher variance associated to higher Wells' classes. The position of AA_{near} was selected assuming that atrial depolarizations' front arrives to the electrodes according to a gamma distribution $AA_{near} \sim \Gamma(k,\theta)$, whose probability density function is close to an exponential (for small values of k) or to a normal (for high values of k) distribution. Thus, Wells' type-I was simulated using $k = 255$ and $\theta = 0.56$, Wells' type-II $k = 100$ and $\theta = 1.43$, and Wells' type-III $k = 20$ and $\theta = 7.14$, corresponding to an AF rate of 7 Hz. To have a certain variability in the shape of AA_{near} waves, their morphology was randomly chosen among a set of possible shapes for atrial potentials, obtained using different directions of the dipole (see Section IV).

IV. SIMULATION EVALUATION

To qualitatively assess the proposed generator, a small database of AEGs during AF was built.

In order to build AA_{near} and VA, a bipolar electrode was placed in the origin of the cartesian axes (2-mm pole distance) and two points on circle around it were considered as, respectively, the starting and the ending point of the travelling dipole. We discretized the circle in 30 equispaced points, for a total of 870 possible shapes. In function of the desidered Wells' class, we chose the shapes in a subset of the possible ones.

A spectral and morphology analysis was performed, aiming at evaluating both the dominant frequency and the regularity of the signal. We computed the spectral concentration [10] and the wave-morphology similarity [11]. In particular,

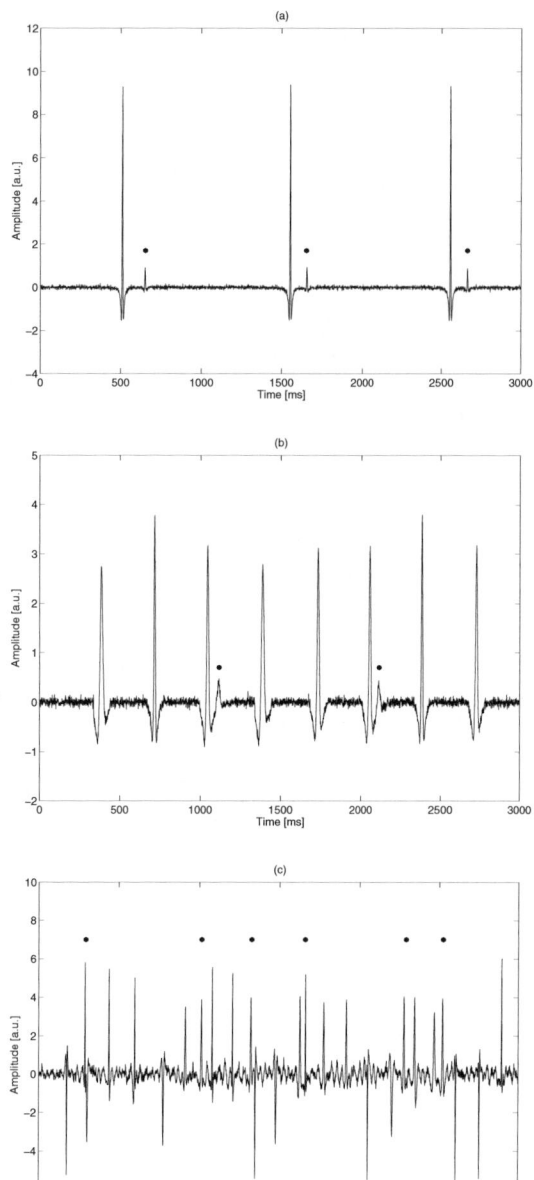

Fig. 1 From the top to bottom, examples of AEG during SR, 3:1 AFL and type-II AF. The circles represent the ventricular activity position. It is worth noting that in AF the ventricular activity cannot be visually identified, as it can during SR and AFL.

Table 1 Mean ± standard deviation of the spectral concentration (SC) and wave-morphology similarity (WMS) for the three AF Wells' classes with 7 Hz dominant frequency. It is also shown the values of real WMSs (third column)

	SC	WMS
type-I	0.67 ± 0.02	0.78 ± 0.02
type-II	0.48 ± 0.02	0.34 ± 0.02
type-III	0.17 ± 0.01	0.11 ± 0.01

50 AEGs for each of the Wells' class and dominant frequency were simulated. One third of the recordings had a dominant frequency of 5 Hz, one third of 7 Hz and the remaining of 9 Hz. The results of each Wells' class and 7 Hz dominant frequency are shown in Table 1.

V. Discussion and Conclusion

A simulator (o generator) of synthetic AEGs has been proposed for the first time to the best of our knowledges. AEGs during SR, AF, and AFL can be obtained.

Briefly, we used autoregressive models for simulating VA locations in SR and AFL, a specific model for the atrioventricular conduction during AF, a gamma distribution for AA_{near} locations and the dipole equation for generating the wave forms. Tuning the model parameters allowed the generation of synthetic AEGs with properties similar to the real ones.

All the synthetic AEGs were created using a single equivalent dipole with the only exclusion of SR in which a planar wave was simulated placing a set of parallel dipoles along a straight line and employing Eq. (5). This is reasonable for SR, AFL and type-I AF. For higher disorganized class of AF, it has been shown [4] that localized activity is given by the summation of both near and mid-field effect. In fact, in real data, simple wave forms are not present for Wells' type-II and type-III. Unfortunately, a realistic model of the inter-times among consecutive close-field effects during AF is not available yet. We proposed to use a gamma distribution for simulating them.

The proposed unified approach for simulating AEGs obtained good results. The SC and WMS increased in function of the Wells' classification, in particular, the values of the latter were in the range of those showed in [11].

References

1. McSharry PE, Clifford GD, Tarassenko L, Smith L. A dynamical model for generating synthetic electrocardiogram signals *IEEE Transactions on Biomedical Engineering.* 2003;50:289-294.
2. Healeyl J, Clifford GD, Kontothanassisl L, McSharry PE. An Open-Source Method for Simulating Atrial Fibrillation Using ECGSYN *Proceedings of Computer in Cardiology.* 2004;31:425-427.
3. Sayadi O, Shamsollahi MB, Clifford GD. Synthetic ECG Generation and Bayesian Filtering Using a Gaussian Wave-Based Dynamical Model *Physiol Meas.* 2010;31:1309–1329.
4. Spach MS, Barr RC, Johnson EA, Kootsey JM. Cardiac Extracellular Potentials : Analysis of Complex Wave Forms about the Purkinje Networks in Dogs *Circ Res.* 1973;33:465-473.
5. Malmivuo J, Plonsey R. *Bioelectromagnetism: Principles and Applications of Bioelectric and Biomagnetic Fields.* Oxford University Press 1995.
6. Soderstrom TS, Stoica PG. *System Identification.* Prentice Hall 1989.
7. Lian J, Müssig J, Lang V. Computer Modeling of Ventricular Rhythm During Atrial Fibrillation and Ventricular Pacing *IEEE Trans. Biomed. Eng..* 2006;53:1512-1520.
8. Corino VDA., Sandberg F, Mainardi LT, Sornmo L. An Atrioventricular Node Model for Analysis of the Ventricular Response During Atrial Fibrillation *IEEE Trans Biomed Eng.* 2011;58:3386-3395.
9. Corino VDA., Sandberg F, Lombardi F, Mainardi LT, Sornmo L. Atrioventricular nodal function during atrial fibrillation: Model building and robust estimation *Biomedical Signal Processing and Control.* 2012;in press.
10. Everett TH, Kok L, Vaughn RH, Moorman JR, Haines DE. Frequency Domain Algorithm for Quantifying Atrial Fibrillation Organization to Increase Defibrillation Efficacy *IEEE Trans Biomed Eng.* 2001;48:969-978.
11. Faes L, Nollo G, Antolini R, Gaita F, Ravelli F. A Method for Quantifying Atrial Fibrillation Organization Based on Wave-Morphology Similarity *IEEE Trans Biomed Eng.* 2002;49:1504-1513.

Author: Massimo Walter Rivolta
Institute: Università degli Studi di Milano
Street: Via Bramante 65
City: Crema
Country: Italy
Email: massimo.rivolta@unimi.it

EMG Topographic Image Enhancement Using Multi Scale Filtering

K. Ullah, B. Afsharipour, and R. Merletti

Laboratory for Engineering of Neuromuscular System (LISiN), Politecnico di Torino, Torino, Italy

Abstract—Digital image enhancement is the process of adjusting image contrast to facilitate further analysis and extraction of the required features. This study presents a multi-scale, Hessian matrix based filtering, for enhancing the motor unit action potential (MUAP) propagation pattern in the multi-channel sEMG images. The proposed filter utilizes the eigenvalues of the Hessian matrix to enhance the MUAP propagation pattern, which appear as line or tubular structure in the sEMG images. The filters are tested using both simulated and experimental sEMG images. The MUAP propagation is enhanced in the images as the response of the filter is high in the direction of MUAP propagation and zero otherwise, thus suppresses the background. The resultant multi-scale filtered sEMG images provide significantly improved visualization of the MUAP propagation. This study is helpful in improving estimation of physiological and anatomical parameters from sEMG images.

Keywords—Hessian Matrix, Eigenvalues, sEMG image, Taylor series.

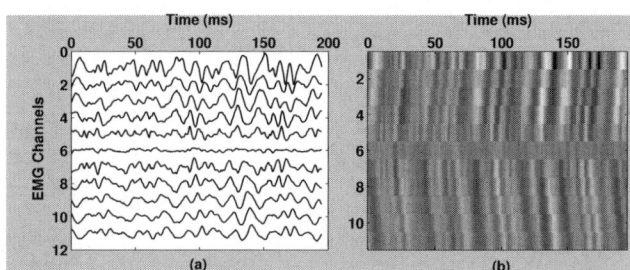

Fig. 1 (a) An eleven channel single differential sEMG acquired with an array of 12 electrodes from Right Upper Trapezius muscle (b) representation of sEMG in (a) as a gray scale topographic image

I. INTRODUCTION

One advantage of the multi-channel surface EMG (sEMG) signal is that it can be treated as a topographic image (Fig. 1b). In these sEMG images, the x dimension corresponds to time and the y dimension corresponds to space i.e. channels and the gray level corresponds to the sEMG amplitude, the MUAP propagation appear as line-like or tubular patterns forming a single line or a V-shape, depending on the location of the innervation zone (IZ) along the motor unit (MU) length. Accurate visualization and quantification of the line or V-shapes, reflected in the sEMG images is an important prerequisite for extraction of physiological and anatomical parameters like locating IZ, estimation of fiber conduction velocity, fiber direction, length of the fiber etc.

EMG signals are the summation of the MUAPs occurring within the detection volume of the electrodes. MUAPs are biphasic or triphasic and are not synchronized; constructive and destructive superimpositions occur. [1] . Due to this intrinsic property of EMG, the MUAP propagation pattern has variations in contrast and often can not be distinguished from the background. The amplitude of the sEMG signal also varies due to the non-uniform thickness of the mucosa or pinnation of the fibers. Thus the MUAP propagation is often asymmetric about the IZ and amplitude of the MUAP above and below the IZ is not the same. Due to these variations in the sEMG signal, MUAP propagation appears with non-uniform contrast often very low, in the sEMG image (Fig. 1b). Thus sEMG image needs to be enhanced to extract information concerning muscle's anatomy or physiology.

There are also several intrinsic and extrinsic sources of low-frequency noise that may contaminate the sEMG signal. One of the intrinsic noise sources is electro-chemical noise, which is generated at the skin-electrode interface [2]. This noise has an ultimate effect on the appearance of MUAP propagation pattern in the sEMG image.

Due to the above mentioned variations in sEMG amplitude and noises, the MUAP propagation pattern which appears as a line or tubular structure, sometimes has low contrast and is hardly visible. This is a main limitation of the sEMG image for further processing, especially segmentation. In contrast, the appearance of the MUAP propagation pattern as line or tubular structure, is significantly improved in the enhanced images.

Literature review reveals significant research activity in the enhancement of line-like and tubular structure in digital images [3] but no applications to sEMG images.

Eigenvalues λ_1 and λ_2 of the Hessian matrix of sEMG images, can be used to determine the local likelihood of MUAP propagation pattern. A similar approach is used by Frangi et al. to enhance linear features in digital images [4]. Some of the methods use fixed scale and (nonlinear) combinations of finite difference operators applied in a set of orientations [5]. All these methods have shown problems to detect line-like structures over a large range of scale, since they perform a fixed scale analysis.

In this paper a multi-scale approach is used to enhance the MUAP propagation in sEMG images, inspired by the work done by Frangi et al. [4] and Lorentz et al [6]. In the proposed method first we convolve the input sEMG image with derivative of Gaussian at multiple scales and then Hessian matrix and its eigenvalues are calculated at each pixel to determine whether a pixel belongs to MUAP propagation structure or not. It is found that a pixel belonging to a MUAP propagation region in sEMG image, has λ_1 smaller (nearly zero) and λ_2 of larger magnitude with negative sign. The norm of the Hessian matrix is then calculated using the eigenvalues, which is smaller in the background where no MUAP propagation is present as the eigenvalues are smaller. In region with higher contrast than background where MUAP propagation exists, the norm value is larger as at least one of the eigenvalues will be larger.

II. Materials and Method

As indicated in Fig. 1b, multichannel single differential sEMG signals can be represented as a topographic image $I(x,y)$. After Taylor expansion to second order of the sEMG image, when the gradient is weak, the eigenvalues of the Hessian matrix express the local variation of the intensity in the direction of the associated eigenvectors. The sEMG image $I(x,y)$ can be represented in the neighborhood of a point p_0 as follows.

$$I(p) \equiv I(p_0) + \delta p^T \nabla_I(p_0) + 0.5 \delta p^T H_I(p_0) \delta p + error \quad (1)$$

where, $\delta p = p - p_0$, $\nabla_I(p_0)$ and $H_I(p_0)$ are respectively the gradient vector and Hessian matrix of the sEMG image computed at p_0. A significant second derivative across a MUAP direction of propagation is expected. Therefore, for white structures on dark background, a linear structure has one negative and high eigenvalue and one with low absolute value. This leads to define a response function which depends on the eigenvalues of the Hessian matrix. To track MUAP propagation with various signal amplitudes and widths, the gradient vector and Hessian matrix are computed at multiple scales (σ) using scale space theory [7]. The gradient is denoted by ∇I, and defined as the vector,

$$\nabla I \equiv \begin{pmatrix} g_x \\ g_y \end{pmatrix} = \begin{pmatrix} \frac{\partial I}{\partial x} \\ \frac{\partial I}{\partial y} \end{pmatrix}. \quad (2)$$

In the framework of scale space theory, the differentiation in (2) is obtained as a convolution with derivatives of Gaussian, which implies: linearity, spatial shift invariance and invariance under rescaling [9]. The gradient vector is now described as,

$$\nabla I \equiv \begin{pmatrix} \frac{\partial I}{\partial x} \\ \frac{\partial I}{\partial y} \end{pmatrix} \equiv \begin{pmatrix} \sigma^\gamma I(x,y) * \frac{\partial}{\partial x} G(x,y,\sigma) \\ \sigma^\gamma I(x,y) * \frac{\partial}{\partial y} G(y,x,\sigma) \end{pmatrix}. \quad (3)$$

where the 2D Gaussian is defined as:

$$G(x,y) = \frac{1}{\sqrt{2\pi\sigma^2}} e^{-\frac{x^2+y^2}{2\sigma^2}}, \quad (4)$$

Here σ is the scaling of the Gaussian kernel in the direction perpendicular to the MUAP propagation and the parameter γ is the normalization parameter introduced by Lindeberg [8], that defines a family of normalized derivatives. This normalization is necessary for comparison of the results of differentiation at various scales, because the derivative is a decreasing function of the scale. In MUAP propagation enhancement applications, where no scale is preferred, γ is usually set to one. As stated earlier, the second order derivative (Hessian) has an intuitive justification in the context of MUAP propagation pattern enhancement. The second order derivative of a Gaussian kernel at scale σ generates a probe kernel that measures the contrast between the regions inside and outside the range $(-\sigma, \sigma)$ in the direction of the derivative. The second order partial derivatives of the sEMG image in the form of Hessian matrix can be expressed as,

$$H(x,y) = \begin{bmatrix} I_{xx} & I_{xy} \\ I_{yx} & I_{yy} \end{bmatrix}, \quad (5)$$

where I_{xx} and I_{yy} represent the second-order partial derivatives of the sEMG image $I(x,y)$ with respect to x and y-axis respectively, and I_{xy} is the mixed partial derivative of the image and $I_{xy} = I_{yx}$. This four-element matrix is approximated by convolving the sEMG image, with the respective derivatives of the Gaussian kernel at the selected scale σ for each pixel of the sEMG image, thus it can be described as,

$$H(\sigma_k) = \begin{bmatrix} \sigma_k^\gamma I(x,y) * \frac{\partial^2}{\partial x^2} G(x,y) & \sigma_k^\gamma I(x,y) * \frac{\partial^2}{\partial x \partial y} G(x,y) \\ \sigma_k^\gamma I(x,y) * \frac{\partial^2}{\partial y \partial x} G(x,y) & \sigma_k^\gamma I(x,y) * \frac{\partial^2}{\partial y^2} G(x,y) \end{bmatrix}, \quad (6)$$

where σ_k has a fixed range depending on the possible widths of the tubular structures.

The MUAP propagation can be modeled as a tubular structure (Fig. 2) with Gaussian profile about its axes (Y-axis) as,

$$I_0(X,Y) = \frac{C}{\sqrt{2\pi\sigma_0^2}} e^{-\frac{Y^2}{2\sigma_0^2}}, \quad (7)$$

where $I_0(X,Y)$ is the intensity of the pixel (X,Y) inside the tubular region, X, Y are the axes of the tubular structure

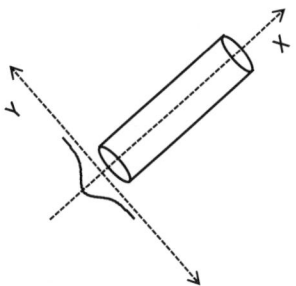

Fig. 2 A Gaussian tubular structure with standard deviation σ_0: Model of the MUAP propagation in sEMG image

and σ_0 is the the standard deviation of the gray level inside the tubular structure as shown in Fig. 2. The Hessian matrix for each of the pixel inside this tubular structure can be calculated as,

$$H_0(X,Y) = \begin{bmatrix} \frac{\partial^2}{\partial X^2}I_0 & \frac{\partial^2}{\partial X \partial Y}I_0 \\ \frac{\partial^2}{\partial Y \partial X}I_0 & \frac{\partial^2}{\partial Y^2}I_0 \end{bmatrix}, \quad (8)$$

The above second order derivatives of I_0 are obtained by differentiating (7) as, $\frac{\partial}{\partial X}I_0 = 0$, $\frac{\partial}{\partial Y}I_0 = -\frac{Y}{\sigma_0^2}I_0$, $\frac{\partial^2}{\partial X^2}I_0 = 0$, $\frac{\partial^2}{\partial X \partial Y}I_0 = 0$, $\frac{\partial^2}{\partial X^2}I_0 = \frac{Y^2 - \sigma_0^2}{\sigma_0^4}I_0$. Substituting these in eq.(8), we get,

$$H_0(X,Y) = \begin{bmatrix} 0 & 0 \\ 0 & \frac{Y^2 - \sigma_0^2}{\sigma_0^4}I_0(X,Y) \end{bmatrix} \quad (9)$$

The MUAP propagation (Line-like) structures can be identified based on the analysis of principal curvatures, which are normally obtained as the eigenvalues of the Hessian matrix [4]. Thus the Hessian matrix is decomposed using eigen values analysis. Assuming the two eigenvalues of the Hessian matrix of the 2D sEMG image are λ_1 and λ_2, such that $|\lambda_1| \leq |\lambda_2|$ with associated eigenvectors $\vec{v}_1$, depicting direction along the MUAP propagation, and $\vec{v}_2$, corresponding to the normal direction to the MUAP propagation at each pixel. The eigenvalues of the Hessian matrix H_0 of the tubular structure are calculated as,

$$\lambda_1 = 0, \lambda_2 = -\frac{\sigma_0^2 - Y^2}{\sigma_0^4}I_0(X,Y) \quad (10)$$

The corresponding eigenvectors are $\vec{v}_1(1,0)$ and $\vec{v}_2(0,1)$. Using the scale space theory for estimating the Hessian matrix, the same values are obtained except that σ_0 is replaced by $\sqrt{\sigma_0^2 + \sigma^2}$, where σ is the scale of the Gaussian kernel used for estimating Hessian matrix. From eq.(10) it is clear that, λ_2 is negative when $Y^2 < \sigma_0^2$, while in scale space theory $Y^2 < \sigma_0^2 + \sigma^2$: which means that the pixel (X,Y) is inside the tubular structure. Thus each pixel can be classified as background pixel or related to MUAP propagation. To design a filter that enhances the pixels related to the MUAP propagation region and suppresses the background, the norm of the Hessian matrix (S) is calculated to quantify the separation between background pixels and MUAP propagation pixels. The norm of the Hessian matrix is calculated as follow.

$$S(x,y) = \sqrt{\lambda_1^2 + \lambda_2^2}, \quad (11)$$

where λ_1 and λ_2 are the eigenvalues of $H(x,y)$ at pixel (x,y) The output of the filter, for enhancing the MUAP propagation region and suppressing the background at a scale σ_k is then defined as [4],

$$V_0(S(x,y),\sigma_k) = \begin{cases} (e^{-\frac{R_B^2}{2\alpha^2}})(1 - e^{-\frac{S^2(x,y)}{2c^2}}) & if \lambda_2 < 0 \\ 0 & otherwise \end{cases} \quad (12)$$

Here $R_B = \lambda_1/\lambda_2$ which is minimum for a tubular structure. $S(x,y)$ is the norm of the eigenvalues of the Hessian matrix for each pixel, α and c are the thresholds to control the sensitivity of the filter. The Hessian matrix and the response of the filter are calculated using scale-space theory, for various scales σ_k of the Gaussian kernel, in a fixed range. When the scale σ_k of the Gaussian kernel, matches the standard deviation σ_0 of the MUAP propagation i.e. tubular structure in the sEMG image, the filter response will be maximum. Therefore, the output of the filter is the max value w.r.t to each scale and is given as:

$$V_0(x,y) = max(V_0(S(x,y),\sigma_k)). \quad (13)$$

III. RESULTS AND DISCUSSIONS

To evaluate the performance of the proposed filter, various simulated and experimental sEMG images are used. Fig. 3 shows a synthetic sEMG image simulated using the EMG model developed at LISiN by Dario Farina et al [10] and its enhanced version. It is obvious from the results that the proposed filter preserves and enhances the tubular structure. Fig. 4 shows a 12 channel sEMG signal from Bicep Brachii muscle recorded under 50% MVC. Although the MUAP propagation has a very low contrast in the original image, it is enhanced and the visualization is improved in the output image. The range of scaling parameter σ of the Gaussian kernel, for both the synthetic and experimental sEMG images of Figs. 3 and 4, is taken from 0.25 to 2 IED.

The motivation for this study was to provide an enhanced image for the algorithms estimating physiological and anatomical parameters of muscle. A MUAP propagation pattern enhancement filter is presented based on local structure

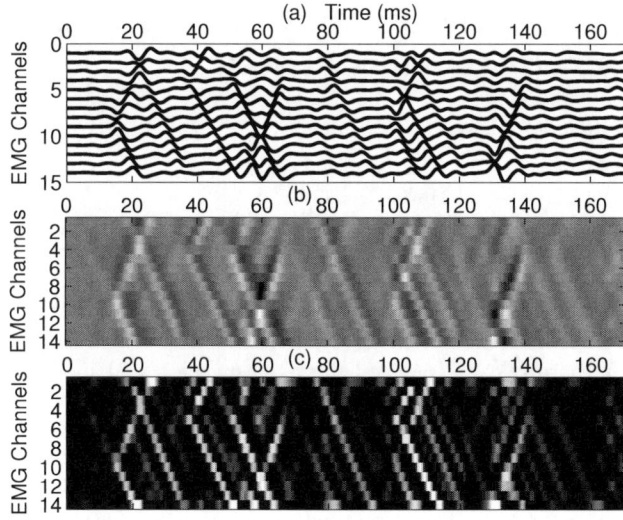

Fig. 3 (a) A 14-channel synthetic sEMG signal with IED of 5mm (b) The topographic sEMG image. (c) The enhanced sEMG image.

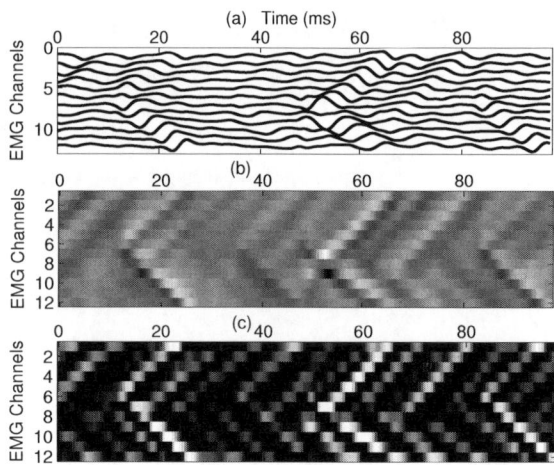

Fig. 4 (a) An 12-channel recorded from Bicep Brachii muscle with an array of 12 electrode with IED=8mm, during 50% MVC. (b) The tomographic sEMG image. (c) Enhanced sEMG image.

of the MUAP propagation pattern. On the basis of eigenvalues of the Hessian matrix, each pixel is classified as related to MUAP region or background. A filter is designed that enhances the MUAP region and suppresses the background. This method does not segment the sEMG image but can estimate the width of the tubular structure i.e the MUAP propagation pattern in terms of the scaling factor σ_k.

IV. CONCLUSION

A MUAP propagation pattern enhancement filter is proposed to boost the performance of EMG feature extraction algorithms by providing them an enhanced sEMG image. The proposed approach first models the MUAP propagation as a tubular structure and then enhances it using the information of its curvature. The proposed filter only enhances the bright part of the MUAP propagation but it can be modified to enhance both the darker and the bright parts of MUAP propagation pattern. The results on the synthetic and experimental signals show, that the proposed filter can be successfully used to enhance the linear and tubular features in sEMG and other images.

REFERENCES

1. Staudenmann, D., et al. Improving EMG-based muscle force estimation by using a high-density EMG grid and principal component analysis *Biomedical Engineering, IEEE Transactions.*2006. 53(4): p. 712-719.
2. Huigen, E., Peper, A., Grimbergen, C.A. Investigation into the origin of the noise of surface electrodes. *Medical and Biological Engineering and Computing.*2002,40, 332338.
3. Yoshinobu Sato, et al. Three-dimensional multi-scale line filter for segmentation and visualization of curvilinear structures in medical images. *Medical Image Analysis.* 1998 volume 2, number 2, pp 143168.
4. Alejandro F. Frangi et al. Muliscale Vessel Enhancement Filtering in *Proc. of the First International Conference on Medical Image Computing & Computer-Assisted Intervention*, p.130-137, October 11-13, 1998
5. Y. P. Du and D. L. Parker. Vessel enhancement filtering in three-dimensional MR angiograms using long range signal correlation. *JMRI* 7(2):447450, 1997.
6. C. Lorenz, et al. Multi-scale line segmentation with automatic estimation of width, contrast andtangential direction in 2D and 3D medical images. In J. Troccaz, E. Grimson, and R. Mosges,beds., *Proc. CVRMed-MRCAS97, LNCS*, pages 233242, 1997
7. Lindeberg, T., *Scale space theory in computer vision.* Netherlands: Kluwer Academic 1994.
8. T. Lindeberg. Edge detection and ridge detection with automatic scale selection. *In Proc. Conf. on Comp. Vis. and Pat. Recog.*, pages 465470, San Francisco, CA, June 1996.
9. Karl , et al. Model Based Detection of Tubular Structures in 3D Images. *Computer Vision and Image Understanding*, 2000, pages 130-171.
10. Farina D, Mesin L, Martina S, Merletti R. A surface EMG generation model with multilayer cylindrical description of the volume conductor. *Biomedical Engineering, IEEE Transactions.*2004. 51(3): p. 415-426.

Corresponding Author: Khalil Ullah
Institute:LISiN, Politecnico di Torino
Street:Via Cavalli 22/H-10138
City:Torino
Country: Italy
Email: khalil.ullah@polito.it

Model Based Estimates of Gain between Systolic Blood Pressure and Heart-Rate Obtained from Only Inspiratory or Expiratory Periods

D.S. Fonseca[1], A. Beda[2], A.M.F.L. Miranda de Sá[1], and D.M. Simpson[3]

[1] Biomedical Engineering Programme, Federal University of Rio de Janeiro, Brazil
[2] Department of Electronic Engineering, Federal University of Minas Gerais, Brazil
[3] Institute of Sound and Vibration Research, University of Southampton, UK

Abstract—Many physiological mechanisms influence in the heart rate variability (HRV), such as the baroreflex response due the compensatory increase in blood pressure, as well as the breathing cycle (with an increase of heart-rate during inspiration). Model-based methods have thus been proposed to unravel this intricate interaction between heart-rate and blood pressure modulated by the baroreflex and breathing cycle. The complex interaction between systolic blood pressure (SAP), heart-rate variability (HRV) and breathing can be studied through closed loop linear models. In previous work it has been suggested that the baroreflex (the change in heart-rate in response to a change in SAP) differs between the inspiratory and expiratory periods. We investigated this on 25 healthy volunteers breathing spontaneously, through a causal closed loop model with distinct estimates for the inspiratory and expiratory periods, using an estimator for gapped data. The results showed no significant difference in baroreceptor gain between the inspiratory and expiratory phases, but there was a significant difference in high frequency (0.15-0.5 Hz) in the gain HRV-> SAP. The proposed method provides a convenient means for system identification with a switched-type non-linearity, with wide applicability in biomedical signals.

Keywords—baroreflex, causal estimates, heart rate variability, systolic blood pressure.

I. INTRODUCTION

The baroreflex response refers to the increase in heart-rate provoked by a decrease in arterial blood pressure [1]. The increase in heart-rate then provokes a compensatory increase in blood pressure, as cardiac output is also increased [1]. Heart-rate also varies according to the breathing cycle (with an increase during inspiration) – leading to the well-known Respiratory Sinus Arrhythmia (RSA), as well as synchronized fluctuations in blood pressure. The origin of RSA has been linked to both the respiratory centre in the brain (central drive) and peripheral mechanisms, including the baroreflex [2, 3], though the former is probably the main contributor[2-4]. There is thus a complex interaction between heart-rate variability (HRV), arterial blood pressure (in particular systolic arterial pressure – SAP) and respiration, and model-based methods have been proposed in an attempt to disentangle the causal pathways [5, 6]. However, the interaction between blood pressure and heart-rate is complicated further by the well-established observation that the baroreflex is reduced during inspiration and increased during expiration [7, 8]. This lead to the proposal that estimators of baroreflex gain should only use the expiratory periods, though this restriction does not appear to have been maintained in more recent work with spontaneous variations in heart-rate[8]. In the original studies the sequence method was employed to estimate baroreflex gain: for this, linear regression is applied to short sequences of increasing or decreasing blood pressure and RR intervals to estimate the average gradient. Since this method only uses repeated short epochs of data for analysis, the restriction to only use expiratory periods can easily be applied. For model-based analysis, using standard system identification methods, this restriction cannot be directly implemented, as these methods require continuous data and cannot handle gaps.

In previous work we investigated closed loop models of the interaction between respiration and heart-rate, using estimates obtained only during inspiration or expiration [9, 10]. We now extend this to provide model-based estimates of gain (SAP → HRV and HRV → SAP) obtained from either all of the recording, or only during the inspiratory or expiratory periods. The aim now is to investigate if estimates of gain differ according to the data selected.

II. CAUSAL GAIN ESTIMATES

In order to obtain the causal gain estimates, a multivariate AR model is implemented which represents a closed loop model [5] relating the signals x_1 (here the HRV signal) and x_2 (SAP in the present work):

$$x_1[t] = \sum_{k=1}^{n} a_{1,1}(k) x_1[t-k] + \sum_{k=1}^{n} a_{1,2}(k) x_2[t-k] + e_1[t] \quad (2)$$

$$x_2[t] = \sum_{k=0}^{n} a_{2,1}(k) x_1[t-k] + \sum_{k=1}^{n} a_{2,2}(k) x_2[t-k] + e_2[t] \quad (3)$$

where $a_{i,j}$ represents the filter coefficents, and e_1 and e_2 are prediction errors. From these models the magnitude of the frequency response can be estimated according to

$$G_{i \to j}(f) = \left| \frac{A_{j,i}(f)}{1 - A_{j,j}(f)} \right| \quad (4)$$

where $A_{i,j} = \sum_{k=0}^{n} a_{i,j}(k) z^{-k}$ and $z = e^{j \cdot 2 \cdot \pi \cdot f / f_s}$, with f_s being the sampling rate. This model is particularly useful for estimating causal interactions (causal coherence), as used in previous work [5, 9]– though this will not be explored in the current paper.

The filter parameters ($a_{i,j}(k)$) can be estimated according to the usual least-squares method. In order to estimate the models for only the inspiratory (expiratory) phase, all the samples during the expiratory (inspiratory) phase are marked as 'missing' by replacing them with Not-a-Number (NaN) – see Fig. 1. The coefficients of the AR model can then be estimated by the least squares method applied over the remaining samples [11, 12]. This may be illustrated for the example for a univariate AR model shown in (11)

$$x[n] = \sum_{i=1}^{M} a_i x[n-i] + e[n] \quad (11)$$

and order M=3, with sample x[4] missing. As shown in (12), the error e[n] can only be calculated for samples e[3], e[8], e[9] and e[10]. In all other lines, either the left side of the equation, or the matrix product are NaN, and hence the residual e[i] cannot be calculated (NaN). In a similar manner, samples prior to x[0] are unknown and therefore set to NaN. The parameters a_1, a_2, a_3 are then estimated by minimizing the mean-square error for the remaining four 'good' samples only (e[3],e[8],e[9],e[10]):

$$\begin{bmatrix} x[1] \\ x[2] \\ x[3] \\ NaN \\ x[5] \\ x[6] \\ x[7] \\ x[8] \\ x[9] \\ x[10] \end{bmatrix} = \begin{bmatrix} x[0] & NaN & NaN \\ x[1] & x[0] & NaN \\ x[2] & x[1] & x[0] \\ x[3] & x[2] & x[1] \\ NaN & x[3] & x[2] \\ x[5] & NaN & x[3] \\ x[6] & x[5] & NaN \\ x[7] & x[6] & x[5] \\ x[8] & x[7] & x[6] \\ x[9] & x[8] & x[7] \end{bmatrix} \begin{bmatrix} a_1 \\ a_2 \\ a_3 \end{bmatrix} + \begin{bmatrix} NaN \\ NaN \\ e[3] \\ NaN \\ NaN \\ NaN \\ NaN \\ e[8] \\ e[9] \\ e[10] \end{bmatrix}$$

The use of gaps in the data, represented by NaN thus permits the analysis to focus on inspiration (or expiration) only, across the whole recording, by setting expiratory (or inspiratory) periods to NaN. It should be emphasized that in this analysis the data from before and after the gaps must not be concatenated, as this would lead to artificial discontinuities in the data.

III. MATERIAL AND METHODS

The signals used in this work came from a database collected in the Biomedical Engineering Program COPPE/Federal University of Rio de Janeiro, Brazil, with local ethical committee approval. Respiratory air flow, ECG and noninvasive blood pressure signals (from a Finapres device) were recorded from a total of 25 volunteers (13 men and 12 women), healthy, non smokers, at rest, breathing spontaneously. HRV was derived from the RR interval of the ECG and SAP from blood pressure. Details on data acquisition protocol and pre processing are provided in[13]. HRV and SAP were then uniformly resampled (4 Hz) by cubic spline interpolation from their beat-to-beat values and high-pass filtered (0.05Hz, 2nd order Butterworth, applied in the forward and reverse direction), in order to reduce the dominance of very low frequency activity. All signals were visually inspected. Signals were then broken into inspiratory and expiratory phases, by replacing samples with NaNs during negative and positive air flow, respectively (see example in Fig. 1). Inspiratory and expiratory intervals were identified from the air flow signals. Gains (equation (4)) were calculated using in-house implementations in Matlab® of the bivariate AR algorithm outlined in section II, for the whole recording and for the gapped data. Mean values of gain (alpha gain for SAP → HRV – see [14]) were obtained for the low-frequency (LF: 0.05-0.15 Hz), and high frequency (HF: 0.15-0.5 Hz) bands for the 25 volunteers, using the whole signal and only inspiration/expiration signals, respectively.

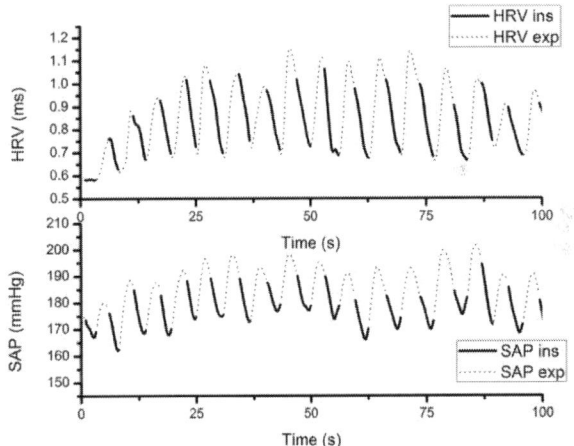

Fig. 1 One example of HRV and SAP signals with gaps for inspiration (ins) and expiration (exp) respectively.

IV. RESULTS

The Akaike's criteria were used for selecting the ARX model order [15], and a sixth order was then used throughout the analysis. The mean respiratory frequency among all subjects was 12.4 (±3.6 std. dev.) breaths per minute (bpm), but breathing was quite irregular in some subjects.

Median causal gains are shown in figures 2 and 3. For the gain HRV→SAP (Fig. 2), the values for inspiration are consistently higher during inspiration than expiration, but for SAP→HRV (alpha gain, Fig. 3) such differences are less consistent. Median values in the LF and HF bands (together with the interquartilic range) are shown in table 1. Friedman tests were applied to the LF and HF bands and showed significant differences (p<0.05) for HF HRV→SAP (post-hoc Tukey tests showed differences between expiration and and both all data and inspiration), marginal significance (p=0.06) for the LF band and no significant difference for HRV→SAP.

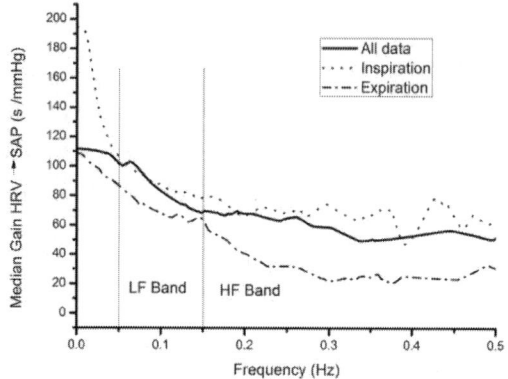

Fig. 2 Mean Causal gain estimates (HRV→SAP) for all data and for inspiration/expiration periods.

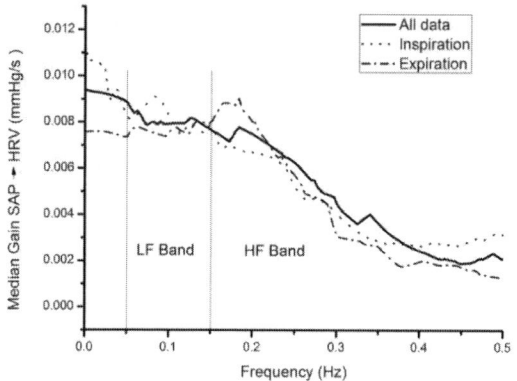

Fig. 3 Median causal gain estimates (SAP→HRV) for all data and for inspiration/expiration periods.

V. DISCUSSION

In previous studies of the interactions between SAP and HRV most commonly a linear, time-invariant model is assumed, that is applied across the whole recordings. The proposed method using gapped data to assess gain from SAP to HRV and vice-versa is less restrictive in terms of the assumption of linearity in the system, i.e. inspiration and expiration may now display different input-output relationships – thus allowing for a switched-type nonlinearity, in accordance with the differing baroreflex gains reported previously for inspiration and expiration [7]. Previously we had used closely related methods in the study of the relationship between EEG and cerebral blood flow [11, 12]. In the latter case the gaps were used primarily to avoid signal epochs contaminated by high levels of artifact or where the EEG pattern of interest was temporarily absent. Clearly this approach has huge potential in a much wider range of biomedical applications, where systems are often time-variant and data editing with the removal of short epochs of poor quality data is very often a requirement.

Table 1 Median and Interquartile range for causal gain estimates

	Band	Median	Interquartile range
$G_{HRV \to SAP}$ All	LF	83.39 s/mmHg	23.56
	HF	56.05 s/mmHg	12.55
$G_{HRV \to SAP}$ Insp	LF	87.02 s/mmHg	13.91
	HF	68.96 s/mmHg	6.76
$G_{HRV \to SAP}$ Exp	LF	68.07 s/mmHg	9.29
	HF	26.09 s/mmHg	8.15
$G_{SAP \to HRV}$ All	LF	$8 \cdot 10^{-3}$ mmHg/s	$2.31 \cdot 10^{-4}$
	HF	$3.9 \cdot 10^{-3}$ mmHg/s	$4.2 \cdot 10^{-3}$
$G_{SAP \to HRV}$ Insp	LF	$8.1 \cdot 10^{-3}$ mmHg/s	$8.98 \cdot 10^{-4}$
	HF	$3.2 \cdot 10^{-3}$ mmHg/s	$3.3 \cdot 10^{-3}$
$G_{SAP \to HRV}$ Exp	LF	$7.7 \cdot 10^{-3}$ mmHg/s	$2.36 \cdot 10^{-4}$
	HF	$2.9 \cdot 10^{-3}$ mmHg/s	$4.4 \cdot 10^{-3}$

We found no significant difference in baroreflex (alpha gain) gain between inspiration and expiration in either the HF or the LF band. In previous work [7], higher baroreflex gain was observed during expiration at slow breathing (12 bpm), but this difference was lost at higher frequencies (24 bpm). The methods of their study and the current one differ considerably. Their calculations were made using the sequence method, and neck suction applied at different stages of the breathing cycle. In the current work we used only spontaneous variations. Furthermore, they found that the stimulus applied during late inspiration and early expiration produced maximal responses, with little sinus node inhibition for stimuli in early and mid inspiration. In the current work we have split the data strictly according to the inspiratory and expiratory phases, which may not be optimal to show any respiratory effects.

What exactly is measured with the spontaneous alpha gain (compared other methods) has been questioned [16], given that respiration provokes changes in HRV though central mechanisms, as well as baroreflex responses, and alpha gain may thus be affected by the former mechanism. Our previous work [9] indicated higher coherence between respiratory air flow and HRV during inspiration than expiration, such that the confounding effect may be stronger in the former.

There was a very striking difference between the gain HRV→SAP for both the HF band (and a notable effect at LF), with consistently much higher values during inspiration. This suggests that a very simplistic linear (mechanical) model of how HRV affects SAP is not appropriate. However, the effect of respiratory centres in the brain on both respiration and HRV may be confounding results.

VI. CONCLUSION

The proposed method for system identification and (causal) gain estimation by least squares fitting of the model over only the available samples is promising for the analysis of the interaction between SAP and HRV with regard to differences between respiratory phases [9, 10]. The initial results showed no significant differences for baroreflex gain between inspiration and expiration (but significant differneces for HRV→SAP), indicating that previously expressed concerns may not be justified. A more detailed analysis will be carried out in a follow-on paper.

ACKNOWLEDGMENT

Thanks for the financial support from the Brazilian agencies, CNPq, CAPES and FAPERJ and the Royal Society (UK), as for the help of Dr. Frederico Jandre, Prof. Antonio Giannella-Neto and Nadja S. Carvalho with data collection, and Aluizio Netto in data processing.

REFERENCES

1. D. Eckberg, "The human respiratory gate," *Journal of Physiology-London,* vol. 548, pp. 339-352, APR 15 2003 2003.
2. D. L. Eckberg, "Point: Counterpoint: Respiratory sinus arrhythmia is due to a central mechanism vs. respiratory sinus arrhythmia is due to the baroreflex mechanism," *Journal of Applied Physiology,* vol. 106, pp. 1740-1742, May 2009.
3. J. M. Karemaker, "Last Word on Point: Counterpoint: Respiratory sinus arrhythmia is due to a central mechanism vs. respiratory sinus arrhythmia is due to the baroreflex mechanism," *Journal of Applied Physiology,* vol. 106, pp. 1750-1750, May 2009.
4. A. Beda, A. Guldner, D. Simpson, N. Carvalho, S. Franke, C. Uhlig, et al., "Effects of assisted and variable mechanical ventilation on cardiorespiratory interactions in anesthetized pigs," *Physiological Measurement,* vol. 33, pp. 503-519, MAR 2012 2012.
5. A. Porta, R. Furlan, O. Rimoldi, M. Pagani, A. Malliani, and P. van de Borne, "Quantifying the strength of the linear causal coupling in closed loop interacting cardiovascular variability signals," *Biological Cybernetics,* vol. 86, pp. 241-251, Mar 2002.
6. A. Porta, "Comments on Point: Counterpoint: Respiratory sinus arrhythmia is due to a central mechanism vs. respiratory sinus arrhythmia is due to the baroreflex mechanism RSA: Origin of the respiratory sinus arrhythmia: insights from causal analysis," *Journal of Applied Physiology,* vol. 106, p. 1, May, 2009 2009.
7. D. Eckberg, Y. Kifle, and V. Roberts, "Phase relationship between normal human respiration and baroreflex responsiveness," *Journal of Physiology-London,* vol. 304, pp. 489-502, 1980 1980.
8. M. Hollow, T. Clutton-Brock, and M. Parkes, "Can baroreflex measurements with spontaneous sequence analysis be improved by also measuring breathing and by standardization of filtering strategies?," *Physiological Measurement,* vol. 32, pp. 1193-1212, AUG 2011 2011.
9. D. S. Fonseca, A. Beda, A. M. F. L. Miranda de Sá, and D. M. Simpson, "Gain and coherence estimates between respiration and heart-rate: Differences between inspirationand expiration," *Autonomic Neuroscience,* vol. in press, 2013.
10. D. S. Fonseca, A. Beda, S. D. M, and A. M. F. L. Miranda de Sá, "Gain and causal coherence estimates between respiration andheart-rate during spontaneous breathing: differences betweeninspiration and expiration," in *7th Conference of the European Study Group on Cardiovascular Oscillations*, Kazimierz Dolny - Poland, April 22-25, 2012.
11. D. M. Simpson, A. F. C. Infantosi, and D. A. B. Rosas, "Estimation and significance testing of cross-correlation between cerebral blood flow velocity and background electro-encephalograph activity in signals with missing samples," *Medical & Biological Engineering & Computing,* vol. 39, pp. 428-433, Jul 2001.
12. D. M. Simpson, D. A. B. Rosas, and A. F. C. Infantosi, "Estimation of coherence between blood flow and spontaneous EEG activity in neonates," *Ieee Transactions on Biomedical Engineering,* vol. 52, pp. 852-858, May 2005.
13. A. Beda, F. C. Jandre, D. I. W. Phillips, A. Giannella-Neto, and D. M. Simpson, "Heart-rate and blood-pressure variability during psychophysiological tasks involving speech: Influence of respiration," *Psychophysiology,* vol. 44, pp. 767-778, Sep 2007.
14. D. Laude, J. Elghozi, A. Girard, E. Bellard, M. Bouhaddi, P. Castiglioni, et al., "Comparison of various techniques used to estimate spontaneous baroreflex sensitivity (the EuroBaVar study)," *American Journal of Physiology-Regulatory Integrative and Comparative Physiology,* vol. 286, pp. R226-R231, JAN 1 2004 2004.
15. H. Akaike, "New look at statistical-model identification," *Ieee Transactions on Automatic Control,* vol. AC19, pp. 716-723, 1974.
16. Y. Tzeng, P. Sin, S. Lucas, and P. Ainslie, "Respiratory modulation of cardiovagal baroreflex sensitivity," *Journal of Applied Physiology,* vol. 107, pp. 718-724, SEP 2009 2009.

Author: Diogo Simões Fonseca
Institute: Federal University of Rio de Janeiro
Street: P. O. Box 68.510
City: Rio de Janeiro
Country: Brazil
Email: diogo.simoes@peb.ufrj.br

Dielectric Properties of Dentin between 100 Hz and 1 MHz Compared to Electrically Similar Body Tissues

T. Marjanović and I. Lacković

University of Zagreb / Faculty of electrical engineering and computing, Zagreb, Croatia

Abstract—Electrical properties of many human tissues have already been measured. However, there is insufficient information about dentin dielectric behavior that can be used for modeling the tooth impedance in the frequency range used by most electrical apex locator devices. Instead, properties of a human cortical bone [1] or own measurements at one or two frequencies were used when needed [2].

The goal of this study is to measure dielectric properties of a human tooth dentin in frequency range of 100 Hz to 1 MHz and to compare them with available data. The measurement was performed on a dentin sample cut out of an extracted human tooth. Dentin sample was electrically interfaced to the measuring equipment using two cotton swabs moisten in physiological saline solution. Conductivity and relative permittivity were calculated at each measured frequency, and the results compared to the dielectric properties of a few similar tissues according to [3] and [2]. The best match was observed with the dielectric properties of a cortical bone according to [3]. However, about 4 times of error was observed in conductivity. Additionally, an excellent agreement of conductivity was observed for dielectric properties used in [2], nevertheless relative permittivity differed about 5 times. Using these data could produce a significant error in calculation of tooth impedance.

Keywords—Dentin impedance, dielectric properties of dentin, root canal length measurement.

I. INTRODUCTION

Dentin is mineralized tissue similar to skeletal bone, but it contains no osteoblasts, osteoclasts, blood vessels or nerves. About 70 % of dentin consists of the mineral hydroxylapatite, 20 % is organic material and 10 % is water. As the result of dentinogenesis, it is rich in Haversian canals, filled with water, 2–5 μm in diameter and extending radially along the root canal. Peritubular dentin surrounding the canals is more mineralized than the intertubular dentin between canals. Therefore, it is reasonable to assume anisotropic values of dentin dielectric properties [4].

The compilation of dielectric properties of body tissues that covers the frequency range 10 Hz to 100 GHz (containing both own measurements and literature data and model with 4 Cole-Cole dispersions) has been published in [1,3,5].

Some recent publications contain additional information on specific tissues at lower frequencies [6]. Unfortunately, some data are not available. For example, ratio of electrical conductivity in the axial and radial directions has been reported as 3.2 : 1 for the hard bone tissue, while the soft bone tissue was reported to be less anisotropic and with longitudinally higher values of conductance. However, no specific information about dentin was given.

Therefore, for the purpose of modeling the tooth impedance in the entire frequency range of interest in root canal length measurement (i.e. frequency range used by most electrical apex locator devices), it is suitable to measure dielectric properties of dentin and to compare them with available and already used data in similar manners.

II. MATERIALS AND METHODS

A dielectric property of dentin was measured on extracted human tooth sample. Tooth was immersed in physiological saline solution (0.9% NaCl) 24 hours prior the experiment. Rectangular prism shaped dentin stick measuring $2.5 \times 2.5 \times 9.0$ mm was cut out of a buccal side of a single rooted incisor tooth with the height extending along the tooth. The sample was cut out from the central part of a dentin between root canal and enamel, leaving at least 1 mm of edge tissue.

The impedance amplitude and phase was measured at 24 frequencies in the range of 100 Hz to 1 MHz using the 100μA current stimuli by the Hewlett Packard HP4284A precise LCR meter. Measurement was performed at the temperature of 23°C with relative humidity about 50%. Two moisten cotton swabs were used as electrodes for interfacing the base of the dentin sample as shown in Fig. 1. Cotton swabs were connected in the measurement circuit with the crocodile clips. The outer surface of dentin sample was dried prior the mounting and the cotton was moisten with the physiological saline until the whole base of a dentin sample was interfaced without grouting at its lower surface. The purpose of saline solution is to maximize the contact area and to simulate the actual environment inside the tooth. As the result, the whole base area of prism-shaped dentin sample was interfaced to the electrode.

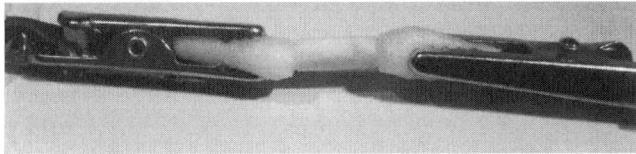

Fig. 1 Interfacing the rectangular prism-shaped dentin stick with the moisten cotton swabs and the crocodile clips as the active and neutral electrodes

The impedance of the cotton swabs connected to the crocodile clips was compensated by measuring them separately and subtracting it from the measured impedances of the dentin sample. The impedance of cotton to dentin interface was neglected in this experiment due to its low value comparing to the impedance of the dentin sample.

Electrical conductivity $\sigma(\omega)$ and relative permittivity $\varepsilon_r(\omega)$ were calculated out of measured impedances $Z(\omega)$ at each frequency:

$$\sigma(\omega) = \frac{d}{S} \cdot \mathrm{Re}\left\{\frac{1}{Z(\omega)}\right\}, \quad (1)$$

$$\varepsilon_r(\omega) = \frac{d}{S \cdot \varepsilon_0} \cdot \frac{1}{\omega} \cdot \mathrm{Im}\left\{\frac{1}{Z(\omega)}\right\}, \quad (2)$$

where d is the height of a dentin prism, S is the base surface area. The complex relative permittivity $\hat{\varepsilon}_r(\omega)$ is given by:

$$\hat{\varepsilon}_r(\omega) = \varepsilon_r(\omega) + \frac{\sigma(\omega)}{j\omega \cdot \varepsilon_0} = \varepsilon_r(\omega) - j\frac{\sigma(\omega)}{\omega \cdot \varepsilon_0}, \quad (3)$$

$$\hat{\varepsilon}_r(\omega) = \frac{d}{S \cdot \varepsilon_0} \cdot \frac{1}{j\omega \cdot Z(\omega)}. \quad (4)$$

Calculated dielectric properties of dentin were then compared with the electrically similar tissues according to the literature data of dielectric tissue properties [5], which were parametrized as 4 Cole-Cole models:

$$\hat{\varepsilon}_r(\omega) = \varepsilon_\infty + \frac{\sigma_0}{j\omega\varepsilon_0} + \sum_{m=1}^{4} \frac{\Delta\varepsilon_m}{1 + (j\omega\tau_m)^{(1-\alpha_m)}}, \quad (5)$$

where ε_∞, σ_0, $\Delta\varepsilon_m$, τ_m and α_m for $m \in \{1,2,3,4\}$ are model parameters covering a wide frequency spectrum (10 Hz – 20 GHz). For the purpose of comparison of measured dentin properties in this study, electrical conductivity and relative permittivity of relevant biological tissues were calculated only at frequencies from 100 Hz to 1 MHz.

III. RESULTS

Figure 2 shows measured impedances of dentin sample. Unfortunately measurement showed relatively poor repeatability after reconnection of dentin sample (up to 40% of initial value) and when measurement setup was left standing for prolonged period of time (up to 75% of initial value after several hours, impedance is rising over the time). Therefore, Fig. 2 represents mean value of three successive measurements with freshly set dentin sample between the cotton swabs, immediately after swabs were moisten

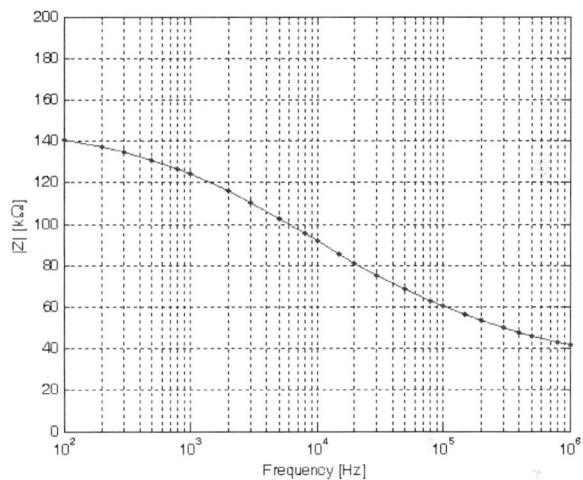

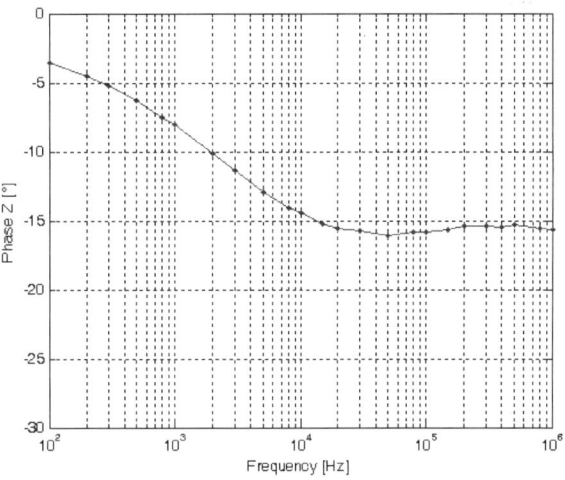

Fig. 2 Measured impedance of the dentin sample 2.5×2.5×9 mm, corrected for the impedance of cotton swab electrodes

and the calibration with shorted cotton swabs was performed (within 10–15 minutes).

After the measurement, 9 mm long dentin sample was cut crosswise into two pieces of 3.5 mm in length using 2 mm tick cutting tool. Impedance measured on one segment appeared to be significantly larger (about 4 times, roughly) than on the other segment. However, the ratio of impedance magnitudes was inconvenient to measure exactly due to increased variability of 3.5 mm long dentin samples comparing to 9 mm sample.

Electrical conductivity and relative permittivity calculated for 9 mm long dentin sample are showed in Fig. 3.

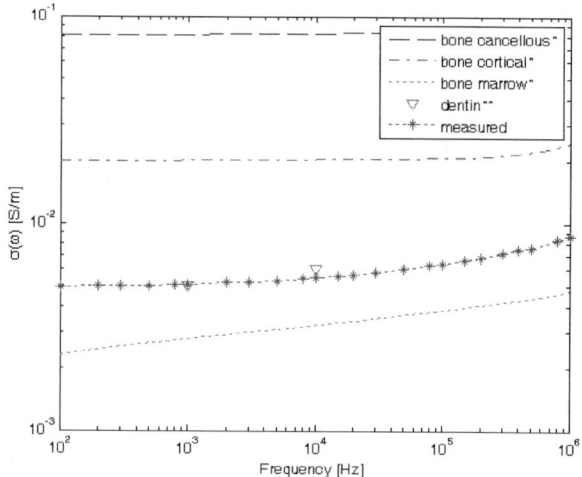

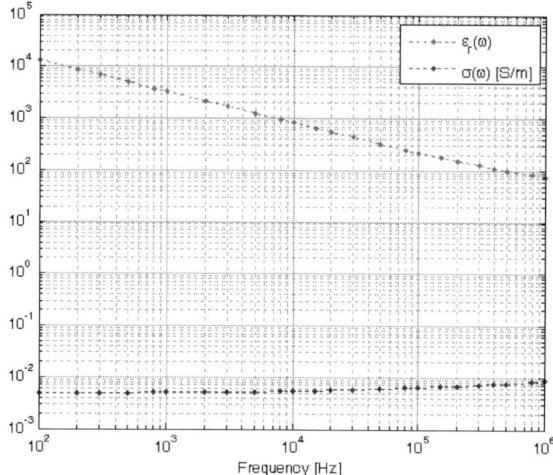

Fig. 3 Measured dielectric properties of dentin sample.

IV. DISCUSSION

Considering the rough surface of hard dentin tissue, sample has been interfaced using the wet cotton swabs rather than using metal electrodes with flat surface for the purpose of maximizing contact area between the electrode and dentin sample. In the case of dry electrodes dentin would be interfaced at several jointing points only, leading to height contact impedance, strongly dependent of applied pressure. On the other hand, using moist cotton swabs has demonstrated problems with grouting and drying, therefore the physiological saline should be dosed carefully according to applied pressure on the dentin sample. Together with drying of cotton swabs, the dentin in the vicinity dries as well causing increase in impedance. This is the reason for lower stability of measured impedance of shorter dentin samples (e.g. 3.5 mm in length instead of 9 mm).

Dielectric properties of measured dentin sample are shown in Fig. 4, together with the data for similar tissues from other authors. Comparison of our data to data in

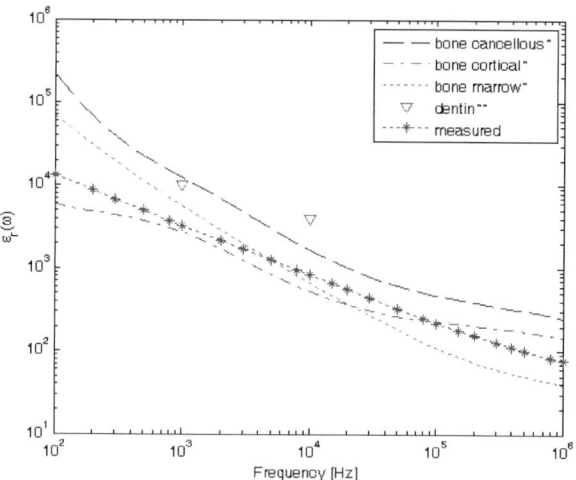

Fig. 4 Measured dielectric properties compared with the literature data for cancellous and cortical bone, bone marrow, and dentin at two frequencies (data from: *[3]., **[2].).

database [3] reveals that measured values of conductivity of dentin behave more like a bone marrow than cortical or cancellous bone tissue. However, obtained values of relative permittivity of dentin behave more like values for the cortical bone.

The values of conductivity and relative permittivity for cancellous and cortical bone and the bone marrow shown in Fig. 4 are calculated according to equation (5). It is important to notice that these models, although covers a wide frequency range, are estimated on the basis of measurements on higher frequencies - mainly in the radiofrequency and microwave band. At frequencies below 1 MHz the accuracy of model parameter estimates are questionable.

On the other hand, in [2] authors have used conductivity and relative permittivity of dentin for numerical FEM simulation of tooth impedances, unfortunately at two discrete frequencies only: 1 and 10 kHz. Conductivity measured in this paper showed agreement with those data, however the measured relative permittivity appeared to be 3 to 5 times lower then in [2]. Using higher values of relative permittivity could increase capacitive behavior of impedance simulated in [2], especially at 10 kHz, and therefore produce additional slope in impedance spectrum and increase observed impedance ratio between two frequencies.

It is also important to notice that dentin impedance in this paper was measured in the direction of root canal propagation, which is between about 45° to 90° to the direction of Haversian canals. Dentin sample also includes dentin of various age and mineralization levels (the sample extends from the less mineralized root apex toward the more mineralized crown). Therefore it represents only typical dentin behavior of the whole root canal. Dentin inhomogeneity and anisotropy should be considered as well for more precise dentin dielectric behavior description.

V. CONCLUSIONS

Complex impedance of human dentin sample was measured in the frequency range of 100 Hz to 1 MHz in this paper. Dielectric properties (in the manner of conductivity and relative permittivity) were calculated and the results compared with the available data in the literature. Overall the best agreement of measured values with various tissue properties according to [3] was achieved for cortical bone. Using the literature model parameters for cortical bone [5] to calculate the dielectric properties of dentin will produce an error of about 4 times in conductivity in the frequency range 100 Hz to 1 MHz according to performed measurement. Measured values of conductivity agree with the values used in [2] at 1 kHz and 10 kHz, however this measurement showed up to 5 times lower relative permittivity, indicating less capacitive behavior of considered root canal impedance FEM model.

Further work should concentrate on measuring the directional dependency of dielectric properties, classified according to the dentin age and mineralization levels.

REFERENCES

1. C. Gabriel, S. Gabriel, E.Corthout, The dielectric properties of biological tissues: I. Literature survey, Phys. Med. Biol., 41, str. 2231-2249, 1996.
2. D. Križaj, J. Jan, V. Valenčič, "Modeling AC Current ConductionThrough a HumanTooth", *Bioelectromagnetics*, 25, str. 185-195, 2004.
3. S. Gabriel, R. W. Lau, C. Gabriel, The dielectric properties of biological tissues: II. Measurements in the frequency range 10 Hz to 20 GHz, Phys. Med. Biol., 41, str. 2251-2269, 1996.
4. E. Barsoukov, J.R. Macdonald, *Impedance spectroscopy: theory, experiment, and applications.* Hoboken, New Jersey: John Wiley and Sons Inc., 2005. ISBN 0471647497, 9780471647492
5. S. Gabriel, R. W. Lau, C. Gabriel, The dielectric properties of biological tissues: III. Parametric models for the dielectric spectrum of tissues, Phys. Med. Biol., 41, 2271-2293, 1996.
6. C. Gabriel, A. Peyman i E.H. Grant, Electrical conductivity of tissue at frequencies below 1MHz, Phys. Med. Biol., 54, 4863-4878, 2009. ISBN doi:10.1088/0031-9155/54/16/002

Author: Tihomir Marjanović
Institute: Faculty of electrical engineering and computing
Street: Unska 3
City: Zagreb
Country: Croatia
Email: tihomir.marjanovic@fer.hr

Hypoglycaemia-Related EEG Changes Assessed by Approximate Entropy

C. Fabris[1], A.S. Sejling[2], G. Sparacino[1], A. Goljahani[1], J. Duun-Henriksen[3],
L.S. Remvig[3], C. Cobelli[1], and C.B. Juhl[3,4]

[1] Department of Information Engineering, University of Padova, Padova, Italy
[2] Department of Cardiology, Nephrology and Endocrinology, Hillerød Hospital, Hillerød, Denmark
[3] Hyposafe, Lyngby, Denmark
[4] Department of Endocrinology, Sydvestjysk Sygehus, Denmark

Abstract—Several studies performed in human beings demonstrated that glucose concentration in blood can affect EEG rhythms, typically evaluated by standard spectral analysis techniques. In the present work, we investigate if EEG complexity assessed by a nonlinear algorithm, Approximate Entropy (ApEn), reflects changes of glucose concentration levels during an induced hypoglycaemia experiment. In particular, in 10 type-1 diabetic volunteers, ApEn was computed from the P3-C3 EEG channel at different temporal scales and then correlated to the three classes of glycaemic states, i.e. hyper/eu/hypo-glycaemia. Results show that, for all considered temporal scales, EEG complexity in hypoglycaemia is lower, with statistical significance, than in eu- and in hyperglycaemia. No statistically significant difference can be evidenced between ApEn values in hyper- and in eu-glycaemic states. In conclusion, in addition to power indexes in the four traditional EEG bands, other indicators, and ApEn in particular, can be used to quantitatively investigate glucose-related EEG changes.

Keywords— EEG, Hypoglycaemia, Multiscale Entropy, Approximate Entropy, Diabetes.

I. INTRODUCTION

Type 1 diabetes is caused by absolute insulin deficiency. Despite the increased use of insulin analogues and insulin pumps for continuous insulin infusion, it is not possible to replace the insulin need exactly. As a result, blood glucose concentration often exceeds the normality range towards hypo- (<70 mg/dl) or hyper- (>180 mg/dl) glycaemic conditions [1]. Although frequent hyperglycaemic events can be dangerous in the long term (complications include nephropathy, retinopathy and cardiovascular diseases), the condition of hypoglycaemia represents an even more severe threat since it may rapidly progress into coma without subject awareness, especially at night. The introduction of minimally-invasive, if not non-invasive, sensors proposed in the 2000's to continuously monitor glycaemia in real time can be of significant help in the timely detection, and even prevention, of dangerous hyper-/hypo-glycaemic events, see [2,3] for recent methodological and technical reviews.

The first studies concerning hypoglycaemia-related EEG changes date back to the 1950's [4]. Since then, several studies, e.g. [5,6], proved that hypoglycaemia is associated with an increased power in the low frequency bands. The hypoglycaemia associated EEG changes are distinguishable from baseline EEG both during daytime and during sleep [7,8], and accordingly the possibility of using the brain as a biosensor to detect hypoglycaemia in real-time from suitable processing of EEG signals has been proposed in [7].

In the present work, we investigate if ApEn, a widely-used indicator determined by the Approximate Entropy nonlinear algorithm [9], can detect and characterize changes of EEG complexity during the transition to hypoglycaemia in 10 type-1 diabetic volunteers.

II. MATERIALS AND METHODS

A. Data Base

Data collected in 10 type-1 diabetic subjects studied at Hillerød Hospital, Denmark, are considered. In brief, subjects were exposed to insulin-induced hypoglycaemia by intravenous administration of Actrapid (Novo Nordisk, Bagsværd, Denmark) targeting a one hour period of eu-glycaemia (90-99 mg/dl) and a subsequent one and a half hour of hypoglycaemia (36 mg/dl) during which repeated cognitive testing was performed. These results will be published elsewhere. During the execution of the experiment, blood glucose concentration (BG) and EEG were simultaneously monitored. In particular, BG was frequently determined by a laboratory analyzer YSI (Yellow Springs Instrument Company, Ohio, USA), while 19 EEG channels were recorded by standard cap electrodes placed on the scalp according to the 10/20 international system. EEG was recorded by a digital EEG recorder (Cadwell Easy II, Kennewick, Washington, USA). EEG signals were analogically low-pass filtered in order to avoid aliasing, then digitally acquired and finally down-sampled at 64 Hz. The dynamic range of the EEG was ±4620 µV with an amplitude resolution of 0.14 µV. The internal noise level in the analog data acquisition system was estimated to be 1.3 µV RMS. The protocol was approved by the local ethical committee.

The analysis presented herein is developed considering the EEG signal obtained as the subtraction of the recordings

P3-C3, the rationale being to consider a scalp position close to that investigated in [10] for potential use in hypo-alerts generation.

Fig. 1 shows the P3-C3 EEG signal (upper panel) and the BG time-series (central panel, open bullets) collected in a representative subject (ETW13).

B. Intervals for the Analysis

For each subject, three 1-hour intervals, corresponding to hyper-, eu- and hypo-glycaemia conditions respectively, were determined by visual inspection of BG time-series. For instance, vertical lines in Fig. 1 mark the selected intervals in the representative subject (hyperglycaemia in red, euglycaemia in green, hypoglycaemia in blue). To facilitate the determination of the three intervals and reduce subjectivity in the analysis, hyper- and hypo-glycaemia thresholds (180 and 70 mg/dl, respectively, denoted by dotted horizontal black lines in the middle panel of Fig. 1) were compared with a smoothing spline previously fitted against the raw BG samples by using the method of [11] (dotted line in the middle panel of Fig. 1).

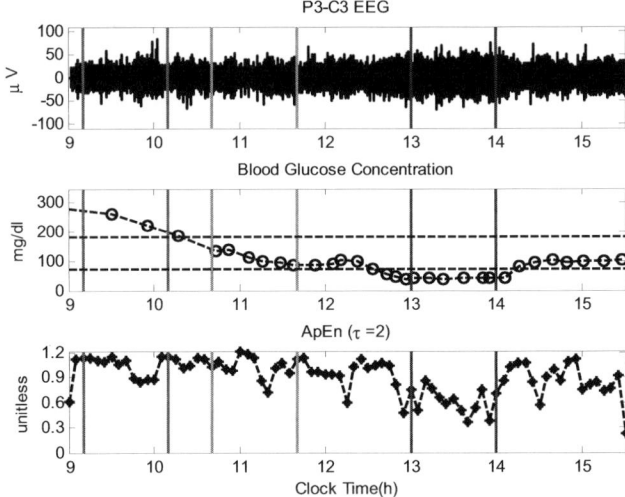

Fig. 1 P3-C3 EEG recording (upper panel) and simultaneous BG measurements (central panel, open bullets) in a representative subject during induced hypoglycaemia experiment (BG samples are smoothed by a spline, dotted line). Horizontal lines in the central panel identify hyperglycaemia and hypoglycaemia thresholds. ApEn values ($m=4$, $r=0.2$, $\tau=2$) obtained from EEG data every 5 min are reported in the lower panel. Vertical red, green and blue lines refer to hyper-, eu- and hypo-interval, respectively.

C. Multiscale Approximate Entropy Analysis

ApEn is an entropy measure introduced by Pincus in the 90's to quantify the complexity/irregularity of time-series [9]. ApEn is applicable to relatively short and noisy data sets and, for this reason, in the last two decades it has been widely used to study biological signals [12], e.g. endocrine-metabolic [13], cardiovascular [14] and EEG [15]. In particular, ApEn quantifies the logarithmic likelihood that two patterns in a sequence that are close (within a certain tolerance) remain close at the next incremental comparison. Thus, for two time-series under comparison, a lower ApEn value indicates more self-similarity and regularity.

More in detail, to calculate ApEn for an evenly sampled time-series $\{u(k)\}, k=1,...,N$, the positive integer m (template length) is first chosen. Then, $N-m+1$ m-dimensional vectors, $x_m(1), x_m(2), ..., x_m(N-m+1)$, are defined as:

$$x_m(i) = [u(i) \quad ... \quad u(i+m-1)]^T \quad (1)$$

with $1 \leq i \leq N-m+1$. Given the positive real r (tolerance), for all $i \leq N-m+1$, the quantity $C_i^m(r)$ is defined as:

$$C_i^m(r) = \frac{\#\text{of vectors } x_m(j): d[x_m(i), x_m(j)] \leq r}{N-m+1} \quad (2)$$

with $1 \leq j \leq N-m+1$.

The distance $d[x_m(i), x_m(j)]$ that appears in eq. (2) is defined for $k=1,...,m$ as:

$$d[x_m(i), x_m(j)] = \max(|u(i+k-1) - u(j+k-1)|) \quad (3)$$

Similarly, for all $i \leq N-m$, the quantity $C_i^{m+1}(r)$ is defined for vectors $x_{m+1}(1), x_{m+1}(2), ..., x_{m+1}(N-m)$ of $m+1$ elements as:

$$C_i^{m+1}(r) = \frac{\#\text{of vectors } x_{m+1}(j): d[x_{m+1}(i), x_{m+1}(j)] \leq r}{N-m} \quad (4)$$

with $1 \leq j \leq N-m$.

Mean values of the logarithms of $C_i^m(r)$ and $C_i^{m+1}(r)$ are then calculated as:

$$\Phi_m = \frac{1}{N-m+1} \sum_{i=1}^{N-m+1} \log(C_i^m(r)) \quad (5a)$$

$$\Phi_{m+1} = \frac{1}{N-m} \sum_{i=1}^{N-m} \log(C_i^{m+1}(r)) \quad (5b)$$

ApEn is finally defined as the difference between Φ_m and Φ_{m+1}:

$$\text{ApEn}(N, m, r) = \Phi_m - \Phi_{m+1}. \quad (6)$$

Notably, in calculating ApEn, the use of "multiscale" approaches can be also considered in order to investigate signal dynamics at different temporal scales, as suggested by Costa et al. in [16]. In such a kind of analysis, from the original time-series $\{x_1, x_2, ..., x_N\}$, a new time-series $\{y_j^{(\tau)}\}$ is defined as:

$$y_j^{(\tau)} = \frac{1}{\tau} \sum_{i=(j-1)\tau+1}^{j\tau} x_i, \qquad 1 \leq j \leq N/\tau, \qquad (7)$$

where the integer parameter τ is the chosen time scale. Then, ApEn is computed for several values of τ. In practice, the multiscale approach reduces the length of the time-series by averaging τ consecutive samples belonging to non-overlapping blocks.

D. Choice of ApEn Parameters

To determine ApEn parameters N, m and r, a preliminary study has been developed. The optimum parameterization allowing to maximize the difference between ApEn mean values in eu and hypo conditions resulted in 5 minute EEG epochs (i.e. $N=19200$ samples), with m set to 4 and r set to 0.2 times the sample standard deviation (SD) of the whole EEG recording. These values are consistent with the empirical suggestions made by Pincus in [9,12], i.e. choose m such that 10^m is of the order of N and r between 0.1 and 0.25 times the signal SD.

As far as the time scale τ is concerned, values ranging from 1 to 8 were investigated (keeping m and r parameters constant).

III. RESULTS

The results reported below are referred to ApEn values obtained with the parameterization which was seen, *a posteriori*, to maximize the difference between ApEn mean values in eu and hypo states, i.e. ApEn index computed on 5 minute EEG segments, with parameters set as follows: $m = 4$ and $r = 0.2 \cdot SD$. For sake of space, only the case $\tau = 2$ is documented, but similar results were obtained for all investigated values of τ.

With these parameters, the sequence of ApEn values obtained for the representative subject is shown in Fig. 1 (lower panel). As one can immediately appreciate by visual inspection, ApEn seems to decrease passing from hyper- and eu-glycaemia (values laying within the red and green lines, respectively) to hypoglycaemia (blue lines), even if the difference between hyper- and eu-glycaemia seems not significant. Since ApEn was calculated from 5-min EEG segments, mean values in the 1-h hyper-, eu- and hypo-glycaemia intervals could be calculated by averaging 12

point values. For the representative subject ETW13 of Fig. 1, these are 1.0300, 1.0176 and 0.6028, respectively, and confirm that, on average, ApEn during hypoglycaemia is lower than in eu-/hyper-glycaemic states. Similar results were obtained for the other 9 subjects.

To evaluate if the difference between ApEn mean values in eu- and hypo-glycaemia conditions has a statistical significance, a paired t-test was used to analyze the values obtained in the entire dataset of 10 subjects. Results, that are graphically summarized in the boxplot of Fig. 2, indicate that ApEn exhibits a significant decrease from eu- to hypo-glycaemia ($p<0.003$). A statistical significant difference was obtained also from the comparison between hyper and hypo ApEn values, but not from that between hyper and eu ($p>0.05$).

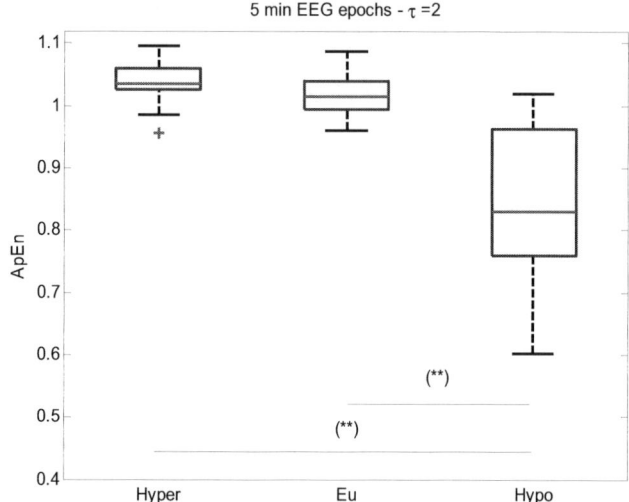

Fig. 2 Boxplot of ApEn (mean) values in the three glycaemic states from all subjects. On each box, the central red line is the median value, the edges of the box are the 75th and 25th percentiles, the whiskers extend to the most extreme data points that are not outliers and the outliers are plotted individually (red cross). Double stars (**) indicate statistically significant differences.

IV. CONCLUSIONS

In this paper, the widely-used complexity indicator ApEn was considered as candidate tool for the investigation of hypoglycaemia related EEG changes.

The study suggest that insulin-induced hypoglycaemia not only results in an increase in the power of low frequency EEG waves, as already shown in [4,5,6], but also in a decrease of EEG complexity, quantifiable with ApEn index. A precise neurophysiological basis of this has not been established, but hypoxemia, as being another condition of reduced brain metabolism, displaces similar reduction in complexity as measured by ApEn [17].

Possible developments of the present research in the brief-term include the extension of the data base, the evaluation of other EEG channels, the refinement of ApEn parameters, the assessment of other nonlinear indicators [18,19] and the comparison with the proposed ApEn, the persistence of EEG changes in complexity after restoration to euglycaemia (see right part traces in Fig. 1), the assessment of the potential influence of factors such as gender, age, hypoglycaemia awareness status, …. In addition, by considering parallel EEG and CGM recordings, it will be useful to investigate the correlation between EEG parameters, e.g. the ApEn presented here but also spectral measures of [7] and [20], and the dynamic risk concept of [21], where information of glucose level and trends is merged in a unique indicator. In the mid-term, future developments of this research can lead to the incorporation of new methods into the device recently proposed in [10] to detect hypoglycaemic events in real-time by processing EEG signals measured by subcutaneous electrodes. In the future, such a device could represent an interesting complement to commercial CGM sensors in order to render them "smart" by suitable online numerical algorithms [22].

ACKNOWLEDGMENTS

The authors are grateful to C. Zecchin and A. Facchinetti for their contribution in preliminary data analysis.

REFERENCES

1. http://www.idf.org
2. Sparacino G, Facchinetti A, Cobelli C (2010) "Smart" continuous glucose monitoring sensors: on-line signal processing issues. Sensors 10:6751-6772
3. Sparacino G, Zanon M, Facchinetti A et al. (2012) Italian contributions to the development of continuous glucose monitoring sensors for diabetes management. Sensors 12:13753-13780
4. Ross IS, Loeser LH (1951) Electroencephalographic findings in essential hypoglycemia. Electroencephalogr Clin Neurophysiol 3:141-148
5. Pramming S, Thorsteinsson B, Bendtson I et al. (1990) The relationship between symptomatic and biomechanical hypoglycaemia in insulin-dependent diabetic patients. J Intern Med 228:641-646
6. Hyllienmark L, Maltez J, Dendenell A et al. (2005) EEG abnormalities with and without relation to severe hypoglycaemia in adolescents with type 1 diabetes. Diabetologia 48:412-419
7. Juhl CB, Hojlund K, Elsborg R et al. (2010) Automated detection of hypoglycemia-induced EEG changes recorded by subcutaneous electrodes in subjects with type 1 diabetes-The brain as a biosensor. Diabetes Res Clin Pract 88:22-28
8. Snogdal LS, Folkestad L, Elsborg R et al. (2012) Detection of hypoglycaemia associated EEG changes during sleep in type 1 diabetes mellitus. Diabetes Res Clin Pract 98:91-97
9. Pincus SM (1991) Approximate entropy as a measure of system complexity. Proc Natl Acad Sci USA 88:2297-2301
10. Elsborg R, Remvig LS, Beck-Nielsen H et al. (2011) Biosensors for health, environment and biosecurity, ch. 12: Using the brain as a biosensor to detect hypoglycemia. InTech
11. Facchinetti A, Sparacino G, Cobelli C (2011) Online denoising method to handle intra-individual variability of signal-to-noise ratio in continuous glucose monitoring. IEEE Trans Biomed Eng 58:2664-2671
12. Pincus SM (2001) Assessing serial irregularity and its application for health. Ann N Y Acad Sci 954:245-267
13. Roelfsema F, Pereira AM, Adriaanse R et a. (2010) Thyrotropin secretion in mild and severe primary hypothyroidism is distinguished by amplified burst mass and basal secretion with increased spikiness and approximate entropy. J Clin Endocrinol Metab 95:928-934
14. Guerra S, Boscari F, Avogaro A et al. (2011) Hemodynamics assessed via approximate entropy analysis of impedance cardiography time series: effect of metabolic syndrome. Am J Physiol 301:H592-H598
15. Anier A, Lipping T, Ferenets R et al. (2012) Relationship between approximate entropy and visual inspection of irregularity in the EEG signal, a comparison with spectral entropy. Br J Anaesth 109:928-934
16. Costa M, Goldberger AL, Peng C-K (2002) Multiscale entropy analysis of complex physiologic time series. Phys Rev Lett 89:068102(1-4)
17. Papadelis C, Kourtidou-Papadeli C, Bamidis PD et al. (2007) The effect of hypobaric hypoxia on multichannel EEG signal complexity. Clin Neurophysiol 118:31-52
18. Cerutti S, Carrault G, Cluitmans PJ et al. (1996) Non-linear algorithms for processing biological signals. Comput Methods Programs Biomed 51:51-73
19. Voss A, Schulz S, Schroeder R et al. (2009) Methods derived from nonlinear dynamics for analyzing heart rate variability. Philos Transact A Math Phys Eng Sci 367:277-296
20. Goljahani A, D'Avanzo C, Schiff S et al. (2012) A novel method for the determination of the EEG individual alpha frequency. Neuroimage 60:774-786
21. Guerra S, Sparacino G, Facchinetti A et al. (2011) A dynamic risk measure from continuous glucose monitoring data. Diabetes Technol Ther 13:843-852
22. Facchinetti A, Sparacino G, Guerra S et al. (2013) Real-time improvement of continuous glucose monitoring accuracy: the smart sensor concept. Diabetes care 36:793-800

Non-linear Indices of Heart Rate Variability in Heart Failure Patients during Sleep

R. Cabiddu[1], S. Mariani[1], J. Henriques[2], S. Cerutti[1], and A.M. Bianchi[1]

[1] DEIB, Dipartimento di Elettronica, Informazione e Bioingegneria, Politecnico di Milano, Milano, Italy
[2] CISUC, Centro de Informática e Sistemas da Universidade de Coimbra, Coimbra, Portugal

Abstract—In recent times researchers have manifested an interest towards non-linear analysis of the HRV signal, which might provide more significant diagnostic and prognostic information than the traditionally used approaches in pathological conditions characterized by reduced variability (such as, among others, Heart Failure, HF). The aim of the present study was to investigate if non-linear HRV derived parameters have clinical relevance in HF, specifically during the night.

The study was conducted on ten normal subjects and ten HF patients. For each subject HRV signal portions corresponding to daytime and nighttime were selected and a set of non-linear parameters were calculated.

Changes were observed between the parameters extracted from the signals acquired during the day and during the night and between the two populations. Significant differences were found between the two populations for some of the parameters evaluated from night recordings. Parameters which were able to discriminate between the two groups during sleep included Detrended Fluctuation Analysis index DFA1 (p-value = 0.0037), Fractal Slope (p-value = 0.0040), Sample Entropy (p-value = 0.0445) and Poincaré indices P1 (p-value = 0.0083) and P2 (p-value = 0.0061). Interestingly, the same parameters were not able to discriminate between the two groups during wakefulness. Our results confirm the possibility of using non-linear parameters derived from HRV signals recorded during sleep to discriminate between normal and pathological subjects.

Keywords—HRV, Heart Failure, Sleep, Non-linear analysis.

I. INTRODUCTION

Cardiovascular disease represents the leading cause of death in developed countries [1]. This has motivated the development of quantitative markers able to assist in the diagnosis and prognosis of cardiac pathologies and in the cardiovascular risk stratification.

Heart failure (HF) was found to be characterized by an abnormal autonomic control [2]. Cardiac autonomic regulation influences cardiovascular rhythmicity; thus, by applying spectral analysis on the heart rate variability (HRV) signal, it is possible to investigate autonomic cardiovascular control and to target function impairment. Spectral methodologies are widely used to quantify the sympatho-vagal balance in physiological and pathological conditions. HRV time and spectral parameters have been consistently shown to provide significant prognostic information in a number of pathological conditions, including HF [4; 5].

Besides other methodologies, an interest towards non-linear analysis of the HRV signal has been manifested: parameters that can be calculated include the Detrended Fluctuation Analysis (DFA) indices, the Sample Entropy (SampEn), the slope of the linear relationship between the log-power and the log-frequency (1/f slope) and the Lempel-Ziv Complexity (LZC) [4]. The development and introduction of new non-linear approaches seem to provide a new perspective in investigating the autonomic regulation of cardiac activity [5]. Specifically, it has been hypothesized that non-linear methodologies might provide more significant and comprehensive information than the spectral traditional approaches in conditions of reduced variability (which characterizes, among others, the cardiac activity of HF patients) and when transient changes or coactivation of the sympathetic and parasympathetic systems occur. In these situations non-linear assessment of autonomic changes may furnish interesting diagnostic, therapeutic and prognostic information.

Differences in HRV non-linear parameters were reported in HF patients with respect to normal subjects. Specifically, the 1/f slope, a measure of the HRV fractal complexity, was found to be higher and the approximate entropy, a measure derived from the analyses of short-term variability, was found to be lower in HF patients [6]. Also, Poincaré plots, representing the heart rate beat to beat variations, were shown to have prognostic value in HF patients [2]; altered Poincaré plots were proved to identify an increased risk for all-cause mortality and sudden cardiac death in these patients.

Although during the last years some studies have been conducted, HRV anomalies in HF patients have not been extensively investigated yet [7]. More extensive studies should be exploited aimed to seek and understand better the correlation between the HRV and the clinical status in these patients. Non-linear dynamic appears as a powerful and reliable tool to analyze HRV non-linear characteristics, but procedures need to be standardized before this technique can be used as a standard tool to characterize the health status of a subject. Non-linear indices might also be combined with traditional time or frequency parameters to allow a more precise characterization of patients.

The aim of this work was to investigate if and how a set of non-linear HRV parameters change between healthy subjects and HF patients, during night and day, and to evaluate if such parameters can be used to discriminate between HF patients and normal subjects, during daytime and nighttime.

II. METHODS

A. Subjects

10 normal subjects and 10 HF patients participated in the study. RR series were extracted from Holter recordings performed within the NOLTISALIS (NOnLinear TIme Series AnaLysIS) research program and were included in a database created with the aim to investigate the nonlinear properties of HRV [8].

B. Elaboration

RR series were visually analyzed to identify and select six hour long signal portions, one within the subject's wakefulness period and one within the subject's sleeping period. A set of non-linear parameters were than calculated for each signal portion.

C. Parameters

a) Detrended Fluctuation Analysis (DFA):

The DFA is a modified root mean square analysis of a random walk, which can be applied to biological data to quantify long-range correlation properties [9].

The RR series B is integrated as follows [5]:

$$y(k) = \sum_{i=1}^{k}[B(i) - B_{ave}] \quad (1)$$

The integrated series $y(k)$ is divided into boxes of equal size n and data are detrended within each box. The root-mean-square $F(n)$ of the detrended time series $y_d(k)$ is calculated as follows:

$$F(n) = \sqrt{\frac{1}{N}\sum_{k=1}^{N} y_d(k)^2} \quad (2)$$

By repeating this calculation for different box sizes it is possible to characterize the relationship between the mean fluctuation $F(n)$ and the box size n. The fluctuation is therefore characterized by a scaling exponent, which is the slope of the line relating $\log F(n)$ to $\log n$. In this work two scaling exponents were estimated: a short-term exponent for $4 \leq n \leq 16$ (DFA1) and a long-term exponent for $16 \leq n \leq 64$ (DFA2).

b) 1/f slope:

The 1/f slope is defined as the power-low regression line fitted to the HRV power spectrum, for $f < 0.01$ Hz [10].

c) Sample Entropy (SampEn):

In order to calculate the SampEn index the time series is compared to a pattern of length m and the outcoming index provides an indication about its regularity, with a tolerance r [5]. In this study $m = 5$ and $r = 0.2$ were adopted.

d) Lempel-Ziv Complexity (LZC):

The LZC characterizes the randomness of finite time series and gives a measure of the algorithmic complexity, defined as the minimum quantity of information needed to define a binary string [5]. The LZC algorithm is described in [11].

e) Poincaré Plots:

Poincaré plots represent beat to beat variations in heart rate and have been suggested to be a useful tool to provide better insight into the abnormal autonomic control of the heart in HF [12]. These bidimensional plots are constructed by plotting each RR interval against its subsequent RR interval. Two descriptors widely used to define a Poincaré plot are P1, which measures the dispersion of points perpendicular to the line of identity (the 45° diagonal line on the Poincaré plot), and P2, which measures the dispersion along the line of identity [12]. P1 and P2 are directly related to basic statistical measures and were computed as follows:

$$P1 = \frac{1}{2}SDSD^2 \quad (3)$$

Where SDSD is the standard deviation of the successive difference of the RR intervals;

$$P2 = \sqrt{2*SDRR^2 - P1} \quad (4)$$

Where SDRR is the standard deviation of the RR intervals.

III. RESULTS

Table 1 shows the average values, along with standard error (SE) values, of all parameters for healthy and pathological subjects, during wakefulness and sleep.

The parameters mean values are plotted in Figure 1, for normal subjects and HF patients, during day and night.

Table 1 Values (Average ± SE) of HRV derived parameters calculated for the normal subjects (NR) and for the HF patients (HF) during day and night.

	NR day	NR night	HF day	HF night
DFA1	1.013±0.022	0.776±0.035	1.019±0.047	0.982±0.051
DFA2	0.960±0.022	1.062±0.028	0.918±0.093	0.951±0.082
1/f	-1.061±0.070	-0.672±0.066	-1.201±0.074	-1.072±0.102
SampEn	0.848±0.079	1.127±0.057	0.740±0.163	0.885±0.096
LZC	0.991±0.004	0.989±0.005	0.984±0.012	0.969±0.022
P1	0.004±0.001	0.005±0.001	0.003±0.001	0.002±0.000
P2	0.103±0.008	0.115±0.011	0.086±0.016	0.077±0.006

DFA1 is lower during the night, for both groups. DFA2 and Sample Entropy are higher for the normal group and appear to be higher during sleep for each group.
1/f slope is higher for HF patients and for both groups it is lower during sleep.
LZC is slightly higher in the normal group and is slightly lower during the night for both populations.
P1 and P2 are overall lower in the pathological subjects and, during sleep, increase in the normal group, while decrease in the HF group.
An ANOVA statistics was applied to investigate the capabilities of the different parameters in discriminating the two groups of subjects, both during day and night.
The ANOVA results are reported in table 2. P-values are never significant (< 0.05) during the day, while they are significant during the night for DFA1, 1/f slope, SampEn, P1 and P2.

IV. DISCUSSION

This work aimed at investigating how a set of non-linear HRV parameters change between normal subjects and HF patients, during wakefulness and sleep and at finding among them those able to discriminate between normal and pathological subjects, during wakefulness and sleep.

For both groups, DFA1 is slightly lower during sleep, indicating a decrease of short-term correlation, while DFA2 is slightly higher during sleep, indicating an increase in long-term correlation. This result confirms the existence of a multiscale regulation of HRV, due to the interaction between cardiac regulatory mechanisms across multiple temporal scales [13].

The 1/f slope appeared to be higher for the HF patients, in line with previous results [5; 6], indicating reduced HRV fractal complexity [6].

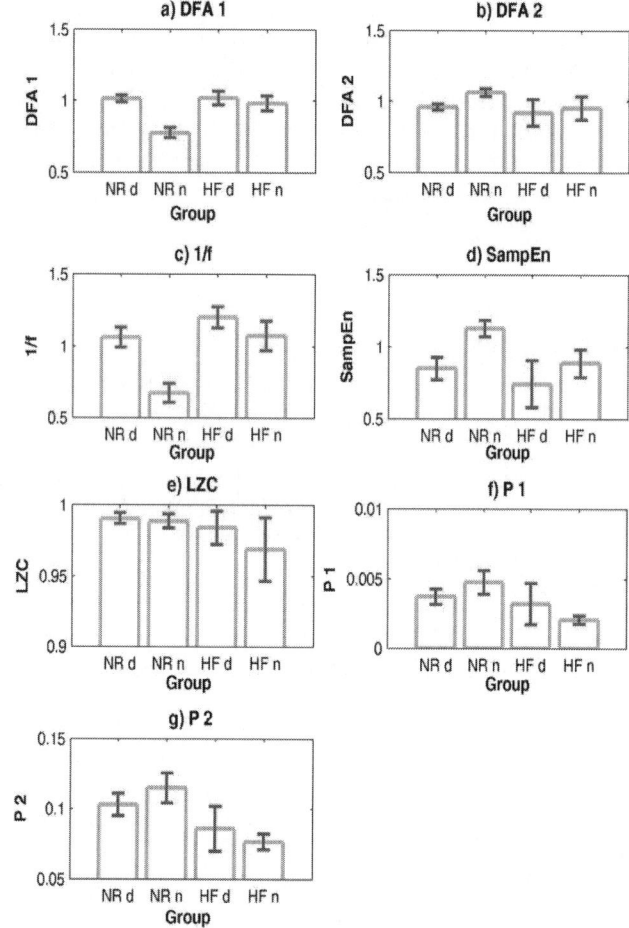

Fig. 1 Average values of DFA 1 (a), DFA 2 (b), 1/f (c), Sample Entropy (d), LZC (e), P1 (f) and P2 (g), calculated for the normal subjects (NR) and for the HF patients (HF) during day and night. Standard error values are indicated in red.

Table 2 P-values resulting from the ANOVA analysis conducted on the parameters evaluated for the two populations, both during day and night (* denotes statistically significant results).

	NR day – HF day	NR night – HF night
DFA1	0.9135	0.0037*
DFA2	0.6697	0.2124
1/f	0.1883	0.0040*
SampEn	0.5564	0.0445*
LZC	0.6041	0.3974
P1	0.7479	0.0083*
P2	0.3466	0.0061*

For both groups it decreased during sleep. The 1/f significance is not completely understood, but there's evidence that lower HRV complexity may be associated with poor prognosis in HF patients [6].

The HF group presented reduced SampEn values, in line with previous evidence showing that lower SampEn characterizes pathological conditions, including HF and obstructive sleep apnea [3; 5]. SampEn increased for both groups during the night, suggesting reduced regularity during sleep. LZC resulted lower for the HF patients, in accordance with previous results [5]. This indicates a lower HRV complexity in pathological conditions. A lower LZC complexity was also found during sleep, for both populations. P1 and P2 were found to be lower in the HF group, indicating a lower dispersion of the RR interval lengths in HF patients. This is in line with previous evidence [2]. During sleep, dispersion, in both directions, increases in the normal population, while decreased in the HF patients.

The ANOVA analysis showed that some of the proposed parameters were able to discriminate between the two populations. Specifically, DFA1, 1/f, SampEn, P1 and P2 were significantly different between the two groups during the night, but not during the day. It was interesting to find that values of these parameters are comparable between the two groups during wakefulness, while they assume significantly different values during sleep.

Our results support the hypothesis of a central role played by the cardiovascular autonomic nervous system in the pathophysiology of HF, especially during the night.

The relationship between HRV nonlinear and linear indices should be further explored; healthy subjects and HF patients can be distinguished using conventional measures based on time and frequency domain analysis; nonetheless, considering that cardiac regulation mechanisms are characterized by a non-linear nature, the use of non-linear methods is believed to provide a more complete description of the dynamical changes of HRV. Moreover, nonlinear approaches provide a quantification of the complexity of HRV, which is accepted as a marker of pathological conditions [14]. The differences between HRV nonlinear parameters evaluated during the night for healthy subjects and HF patients should also be investigated in better detail; the use of nonlinear HRV indices able to distinguish between HF patients and healthy subjects, together with the application of modern devices for the acquisition and evaluation of HRV during sleep, might represent an innovative approach to HRV monitoring and cardiovascular risk stratification.

ACKNOWLEDGMENT

This work was partially supported by the EU project HeartCycle (ICT FP7 216695).

REFERENCES

1. Madeiro JP, Cortez PC, Oliveira FI, Siqueira RS (2007) A new approach to QRS segmentation based on wavelet bases and adaptive threshold technique, Med Eng Phys 29(1):26-37
2. Brouwer J, van Veldhuisen DJ, Man in 't Veld AJ, Haaksma J, Dijk WA, Visser KR, Boomsma F, Dunselman PH (1996) Prognostic value of heart rate variability during long-term follow-up in patients with mild to moderate heart failure. The Dutch Ibopamine Multicenter Trial Study Group, J Am Coll Cardiol 28(5):1183-9
3. Guzzetti S, Mezzetti S, Magatelli R, Porta A, De Angelis G, Rovelli G, Malliani A (2000) Linear and non-linear 24 h heart rate variability in chronic heart failure. Auton Neurosci 86(1-2):114-9
4. Nolan J, Batin PD, Andrews R, Lindsay SJ, Brooksby P, Mullen M, Baig W, Flapan AD, Cowley A, Prescott RJ, Neilson JM, Fox KA (1998) Prospective study of heart rate variability and mortality in chronic heart failure: results of the United Kingdom heart failure evaluation and assessment of risk trial (UK-heart), Circulation 98(15):1510-6
5. Bianchi AM, Mendez MO, Ferrario M, Ferini-Strambi L, Cerutti S (2010) Long-term correlations and complexity analysis of the heart rate variability signal during sleep. Methods Inf Med 49(5):479-83
6. Butler GC, Ando S, Floras JS (1997) Fractal component of variability of heart rate and systolic blood pressure in congestive heart failure, Clin Sci 92(6):543-50
7. Chattipakorn N, Incharoen T, Kanlop N, Chattipakorn S (2007) Heart rate variability in myocardial infarction and heart failure, Int J Cardiol 120(3):289-96
8. Sassi R (2000) Analysis of HRV complexity through fractal and multivariate approaches, PhD thesis, Politecnico di Milano, Italy
9. Perakakis P, Taylor M, Martinez-Nieto E, Revithi I, Vila J (2009) Breathing frequency bias in fractal analysis of heart rate variability. Biol Psychol 82(1):82-8
10. Bigger JT Jr, Steinman RC, Rolnitzky LM, Fleiss JL, Albrecht P, Cohen RJ (1996) Power law behavior of RR-interval variability in healthy middle-aged persons, patients with recent acute myocardial infarction, and patients with heart transplants. Circulation 93(12):2142-51
11. Ferrario M, Signorini MG, Magenes G (2007) Comparison between fetal heart rate standard parameters and complexity indexes for the identification of severe intrauterine growth restriction, Methods Inf Med 46:186-190
12. Karmakar CK, Khandoker AH, Gubbi J, Palaniswami M (2009) Complex Correlation Measure: a novel descriptor for Poincaré plot, BioMedical Engineering OnLine 8:17
13. Ho YL, Lin C, Lin YH, Lo MT (2011) The prognostic value of non-linear analysis of heart rate variability in patients with congestive heart failure--a pilot study of multiscale entropy, PLoS One 6(4):e18699
14. Moraru L, Tong S, Malhotra A, Geocadin R, Thakor N, Bezerianos A (2005) Investigation of the effects of ischemic preconditioning on the HRV response to transient global ischemia using linear and nonlinear methods, Med Eng Phys 27(6):465-73

Corresponding author:
Author: Ramona Cabiddu
Institute: DEIB, Dipartimento di Elettronica, Informazione e Bioingegneria, Politecnico di Milano
Street: via Golgi, 39 - 20133
City: Milano
Country: Italy
Email: ramona.cabiddu@gmail.com

Detrended Fluctuation Analysis of EEG in Depression

M. Bachmann[1], A. Suhhova[1], J. Lass[1], K. Aadamsoo[2], Ü. Võhma[2], and H. Hinrikus[1]

[1] Department of Biomedical Engineering, Technomedicum, Tallinn University of Technology, Tallinn, Estonia
[2] Clinic of Psychiatry, North Estonia Regional Hospital, Tallinn, Estonia

Abstract—Diagnosis of depression is still based mainly on evaluation of the intensity of subjective and clinical symptoms by psychiatrists. This study is aimed to give additional objective information about depression analyzing the electroencephalographic (EEG) signal using the method of detrended fluctuation analysis (DFA). DFA is applied to evaluate the presence and persistence of long range correlations in time in EEG signals. EEG recordings were carried out on the groups of depressive and healthy subjects of 18 female volunteers each. The DFA was calculated on EEG signals from P3-Pz channel at a length of 5 minutes. The DFA method revealed statistically significant difference between healthy and depressive subjects. Resting EEG of healthy subjects exhibited persistent long-range correlation in time. In depression the long-range correlation was less persistent and for about half of the depressed subjects (44%) the EEG revealed long-range anticorrelation in time.

Keywords—EEG, DFA, depression, mental disorder.

I. INTRODUCTION

According to World Health Organization depression is expected to rank second by 2020 and current predictions indicate that by 2030 depression will be the leading cause of disease burden globally calculated for all ages, both sexes [1]. However, the methods of diagnosis of depression and also other mental disorders have been traditionally based mainly on evaluation of the intensity of subjective and clinical symptoms by psychiatrists. Currently no objective indicator in clinical practice exists.

As electroencephalography (EEG) is easily available, cost effective method and reflects the ongoing bioelectric activity. EEG is a valuable method for getting objective information about changes in brain physiology specific in depression.

We have previously successfully differentiated the healthy controls and depressive subjects using spectral asymmetry index (SASI) [2, 3]. However, the coherence analysis performed on the data did not reveal statistically significant difference between groups.

Since EEG exhibits complex behavior [4, 5], nonlinear measures can be a good alternative to more frequently applied linear methods. For instance, a nonlinear method, detrended fluctuation analysis (DFA), a modified root mean square analysis of a random walk allows to detect long-range correlations in a seemingly nonstationary time series [6]. DFA has been previously successfully applied on various physiological data [6, 7, 8, 9, 10]. There exist also EEG studies of depression using DFA [11, 12]. However, the EEG signal amplitude is rarely used as an input to DFA, with the exception of the study by Hosseinifard et al [12]. The main emphasis of the study is classification accuracy of healthy and depressive subjects, which reaches more than 75%. However, there is no information about the values of different parameters needed to calculate DFA nor the outcome of DFA calculations. Therefore, no conclusion can be drawn about the differences in EEG characteristics of healthy and depressed subjects.

The aim of this work is to gain additional objective information about depression by clarifying whether resting EEG exhibits long-range correlation. For this purpose, the character of the EEG long range correlations in depression and health will be evaluated.

II. METHODS AND EQUIPMENT

A. Subjects

The experiments were carried out on a group of 18 female patients with major depressive disorder, mean age 35 years, standard deviation 11 years.

Subjects with depressive disorder without antidepressant treatment were selected from a hospital inpatient unit. Subjects with nonpsychotic depressive disorder as defined by ICD-10 criteria and determined by 17-item Hamilton Depression Rating Scale (HAM-D) score higher than 14 were eligible. The average score for the group was 22.8 (standard deviation 3.3).

The study was conducted in accordance with the Declaration of Helsinki and was formally approved by the local Medical Research Ethics Committee.

B. Experimental Procedure and EEG Recording Equipment

The experimental procedure for a subject included continuous EEG recording during 30 minutes between time interval 9 a.m. to noon.

The experimenter and a subject were in the same laboratory room during the experiments. The room was dark but no other special conditions were provided. The subjects were lying in a relaxed position, eyes closed and ears blocked during the experiments.

Cadwell Easy II EEG measurement equipment was used for the EEG recordings. The EEG was recorded using 19 electrodes, which were placed on the subject's head according to the international 10 20-electrode position classification system. Bipolar parietal channel P3-Pz was chosen for analysis. Raw EEG signals were recorded using the Cadwell Easy EEG data acquisition system within a frequency band of 0.3-70 Hz. The impedance of recording electrodes was monitored for each subject prior to data acquisition and it was always below 5 kΩ. The sampling frequency was 400 Hz. Due to computational load, only the first 5 minutes of each recording was used for further analysis.

EEG Analysis

DFA is calculated directly in the time domain [6, 7]. First the 5 minutes EEG signal was divided into 20 s segments, giving 15 segments for further processing. Next, the DFA was calculated for each segment. After that the median value of the DFA algorithm over all segments was calculated for each subject and the statistical analysis was performed.

The DFA was calculated for all the segments as follows [6]. First, the EEG signal segment x(i), where i is the length of the segment ranging from 1 to N (N=8000), is integrated to generate a new time series y(k).

$$y(k) = \sum_{i=1}^{k}[x(i) - \bar{x}] \quad k = 1,...,N \qquad (1)$$

where $\bar{x}$ is the average of the EEG signal x(i). After that the new time series y(k) is divided into n equal windows. Window length started from 4 samples up to 812 samples varying equidistantly on logarithmic scale (0.01 s up to 2.03 s). The maximum window length was selected as about 1/10 of the signal segment length [13].

In each window n, the least squares line, $y_n(k)$, is fit to the data y(k). The fitting range was chosen from 0.1 s, excluding the prominent alpha frequency [14], to 1.1 s, as brain often suppresses large fluctuations on longer timescales [15]. Next, the local trend $y_n(k)$ is subtracted from the data y(k). The root mean square fluctuation of the demeaned, integrated and detrended signal segment is calculated as:

$$F(n) = \sqrt{\frac{1}{N}\sum_{k=1}^{N}[y(k) - y_n(k)]^2} \qquad (2)$$

Those final steps are repeated for all window sizes giving the average fluctuations as a function of window length. Those fluctuations are expected to increase with the window length. The scaling is present in case on a log-log graph of F(n) vs. n appears a linear correlation. The slope of the line, that is the scaling exponent α, relating logF(n) to logn describes the type of scaling. For white noise, no correlation between consecutive values, the integrated value (y(k)), corresponds to a random walk and α = 0.5 [16]. A scaling exponent between 0.5 and 1 indicates correlation. Large fluctuations are likely to be followed by large fluctuations and small fluctuations are likely to be followed by small. In case a scaling exponent is between 0 and 0.5, the data is anti-correlated. Therefore, large fluctuations are likely to be followed by small fluctuations and vice versa. While scaling exponent 0.5 corresponded to white noise, the scaling exponent 1 corresponds to 1/f noise and scaling exponent 1.5 represent Brownian noise.

Student's t-test was performed to evaluate the differences between the scaling exponents of depressive and control group. The difference was considered statistically significant for p values lower than 0.05.

III. RESULTS

Figure 1 illustrates the DFA results for a control subject. The root mean square fluctuations F(n) are plotted against window length n in a log-log plot marked with 'x'. The solid line represents the linear least squares fit in the predefined region from 0.1-1 seconds. The scaling exponent α is calculated as the slope of the linear fit. For this subject, the scaling exponent is 0.819.

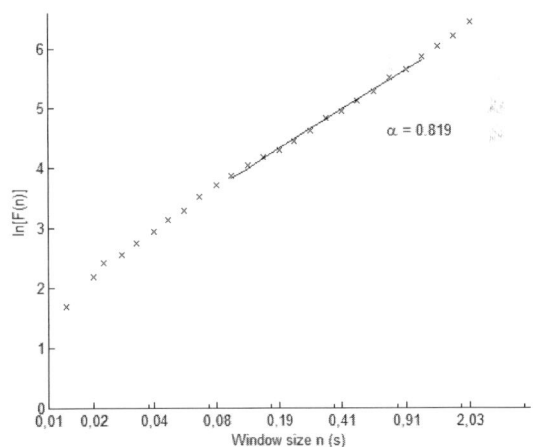

Fig. 1 F(n) versus n (in seconds) in a log-log plot for one control subject. The scaling exponent α was obtained over the fitting range 0.1s-1.1s.

Figure 2 illustrates the DFA algorithm calculation results for a depressive subject. In this case the scaling exponent has a lower value, α = 0.617. In addition, the crossover emerges on larger window length.

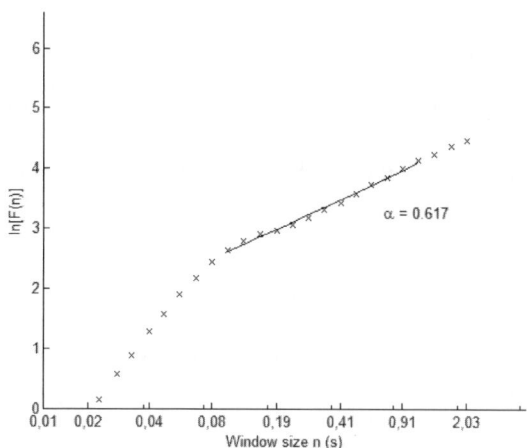

Fig. 2 F(n) versus n (in seconds) in a log-log plot for one depressive subject. The scaling exponent α was obtained over the fitting range 0.1s-1.1s.

Scaling exponents for control group and for depressive group averaged across all subjects and the corresponding standard deviations are presented in Figure 3. While control group has an average scaling exponent with a value of 0.735, the scaling exponent of the depressive group is much lower 0.571. According to the student's t-test, the difference in average scaling exponent between control group and depressive group is statistically significant ($p < 0.05$).

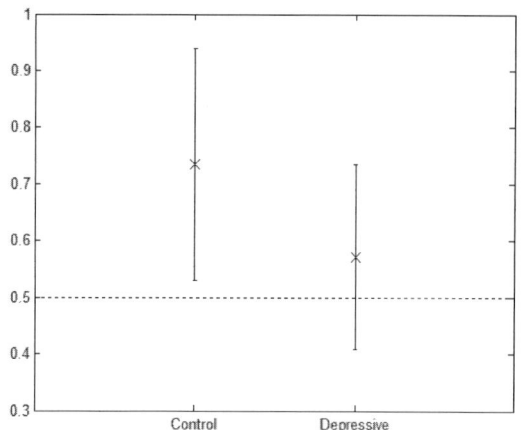

Fig. 3 Scaling exponents for control group and for depressive group averaged across all subjects. Vertical lines represent standard deviation.

The horizontal dotted line in Figure 3 represents the scaling exponent level for uncorrelated white noise (α=0.5). Considering the vertical lines representing the standard deviation of average scaling exponent, we can see that a part of depressive subjects are characterized also by a scaling exponent smaller than 0.5 indicating anti-correlation. To be more precise, about half of the depressive subjects (44.4%) show anti-correlation while only one subject (5.6%) in control group indicate anti-correlation phenomena.

IV. DISCUSSION

The average scaling exponent (α = 0.735) of the control group indicate that for majority of healthy subjects the EEG signal exhibits long-range correlation in time. Consequently, fluctuations are larger for larger time-scales. On the other hand for 44.4 % of depressive subjects the signals are temporally anti-correlated. Thus, for larger time-scales the fluctuations are smaller than in case of white noise. The rest of the depressive group exhibits also long-range correlations in time, but those are not as persistent as for controls.

The linear least squares fit was performed on a log-log plot at window lengths 0.1-1.1 seconds. However, Figure 1 and 2 indicate that at least for control subjects even longer window lengths can be used. Therefore, in future it would be interesting to analyze whether the length of the linear scaling region can characterize the EEG signal difference for control and depressive group. Will the brain in depression suppress the large fluctuations on shorter time-scales compared to controls?

The results of this study cannot be directly compared to other studies due to different window lengths starting from 5 seconds up to 50 seconds [17, 18]. In addition, different studies apply DFA to different EEG parameters - signal amplitude [12], amplitude envelope of different EEG rhythms [18], synchronization likelihood [11] etc. In a recent study, Hosseinifard et al [12] differentiated depression patients from controls by calculating various linear and nonlinear features – DFA of the EEG signal amplitude as one of them – and applying different machine learning techniques for classification. However, only the classification results are explicitly presented leaving out the values of different parameters in use while calculating the DFA. From all the parameters used as an input for classifiers, only the results of a linear method, the highest accuracy power band, were presented. The mean alpha power band was significantly higher in left hemisphere of depressed subjects compared to the left hemisphere of controls. The bipolar channel P3-Pz, used in this study, belongs to the same region. However, for better localization of significant brain areas characterizing depression, all EEG channels need to be analyzed.

Hosseinifard et al [12] used linear and non-linear methods for classification of depression. The classification accuracy was up to 90 % combining only nonlinear measures, as opposed to about 77% with linear power features in use. Our previous study using linear method, the coherence analysis, to differentiate between depressive and control subjects did not reveal statistically significant difference between groups. However, in this study, the nonlinear DFA indicated that healthy EEG exhibits more persistent long-range correlation than EEG in depression. Even more, for about half of the depressive subjects' long-range anti-correlation was revealed.

In future, the length of scaling region for depressive and control subjects will be analyzed as during the current analysis, the scaling region for control group seemed to be even larger than was used in this study and also larger than the scaling region of the depressive group. Also different channels will be of interest.

V. CONCLUSIONS

The results indicate that resting EEG of healthy subjects exhibits persistent long-range correlation in time. In depression the long-range correlation was less persistent than for healthy subjects. In contrast, for about half of the depressed subjects (44%) the EEG revealed long-range anti-correlation.

More specific analysis is needed to consider the extended lengths of scaling region and different brain areas.

ACKNOWLEDGMENT

This study was supported by the European Union through the European Regional Development Fund.

REFERENCES

1. World Health Organization (2012) Global burden of mental disorders and the need for a comprehensive, coordinated response from health and social sectors at the country level. Sixty-fifth World Health Assembly, provisional agenda item 13.2
2. Hinrikus H, Suhhova A, Bachmann M et al. (2009) Electroencephalographic spectral asymmetry index for detection of depression. Med Biol Eng Comp 47:1291–1299
3. Hinrikus H, Bachmann M, Lass J et al. (2009) "Method and device for determining depressive disorders by measuring bioelectromagnetic signals of the brain", US2009/0054801
4. Korn H, Faure P (2003) Is there chaos in the brain? II. Experimental evidence and related models. C R Biol 326(9):787-840
5. Stam CJ (2005) Nonlinear dynamical analysis of EEG and MEG: review of an emerging field. Clin Neurophysiol 116(10):2266-301
6. Peng CK, Buldyrev SV, Havlin S et al. (1994) Mosaic organization of DNA nucleotides. Phys Rev E 49:1685-1689
7. Peng CK, Havlin S, Stanley HE et al (1995) Quantification of scaling exponents and crossover phenomena in nonstationary heartbeat time series. Chaos 5:82-7
8. Montez T, Poil SS, Jones BF et al. (2009) Altered temporal correlations in parietal alpha and prefrontal theta oscillations in early-stage Alzheimer disease. PNAS 106 (5): 1614–1619
9. Kim JW, Shin HB, Robinson PA (2009) Quantitative study of the sleep onset period via detrended fluctuation analysis: Normal vs. narcoleptic subjects. Clin Neurophysiol 120(7):1245-51
10. Wijnants ML, Cox RF, Hasselman F et al. (2012) A trade-off study revealing nested timescales of constraint. Front Physiol 3:116
11. Lee JS, Yang BH, Lee JH et al. (2007) Detrended fluctuation analysis of resting EEG in depressed outpatients and healthy controls. Clin Neurophysiol. 118(11):2489-96
12. Hosseinifard B, Moradi MH, Rostami R (2012) Classifying depression patients and normal subjects using machine learning techniques and nonlinear features from EEG signal. Comput Methods Programs Biomed. 109(3):339-45
13. Hu K, Ivanov PC, Chen Z, Carpena P, Stanley HE (2001) Effect of trends on detrended fluctuation analysis. Phys Rev E 64:111–114
14. Robinson PA (2003) Interpretation of scaling properties of electroencephalographic fluctuations via spectral analysis and underlying physiology. Phys Rev E 67:032902:1-2:4
15. Kim JW, Shin HB (2008) Nonlinear properties of electroencephalograms during nocturnal sleep of narcoleptic patients. Sleep Med 8:S42
16. Montroll EW, Shlesinger MF (1984) Nonequilibrium phenomena II: from stochastics to hydrodynamics. North-Holland, Amsterdam
17. Abásolo D, Hornero R, Escudero J et al. (2008) A study on the possible usefulness of detrended fluctuation analysis of the electroencephalogram background activity in Alzheimer's disease. IEEE Trans Biomed Eng. 55(9):2171-9
18. Nikulin VV, Brismar T (2005) Long-range temporal correlations in electroencephalographic oscillations: Relation to topography, frequency band, age and gender. Neuroscience 130(2):549-58

Author: Maie Bachmann
Institute: Tallinn University of Technology
Street: Ehitajate Rd 5
City: Tallinn
Country: Estonia
Email: maie@cb.ttu.ee

Can Distance Measures Based on Lempel-Ziv Complexity Help in the Detection of Alzheimer's Disease from Electroencephalograms?

S. Simons and D. Abásolo

Centre for Biomedical Engineering, Department of Mechanical Engineering Sciences, Faculty of Engineering and Physical Sciences, University of Surrey, Guildford, UK

Abstract—This pilot study applied three distance-based bivariate Lempel-Ziv complexity (LZC) measures to investigate the changes in electroencephalogram (EEG) signals between 11 patients with Alzheimer's disease (AD) and 11 age matched controls. These methods measure richness of complexity between pairs of signals. Complexity of control subjects' EEGs was richer, i.e. signals were made from a greater number and greater range of subsequences, than those from AD patients in almost all cases in two non-normalized distance-based methods. Only some pairs including electrode T4 (2.1% of the total) occasionally showed the reverse result. Statistically significant differences were found with these two methods in 21 and 18 of 120 tested electrode pairs, respectively (Student's t test, $p<0.01$). Receiver operating curves were used to calculate the sensitivity (number of correctly classified AD patients) and specificity (number of correctly classified controls). Accuracy is the combined correct classification of controls and AD patients. The maximum sensitivity found was 100%, specificity 90.9% and accuracy 86.4% at various electrode pairs with both non-normalized methods. The normalized method showed many electrode pairs with increased richness of complexity for AD patients than controls (67.5% of the total). It was found that this was due to the normalization procedure modifying the distribution of the original complexities from the electrode pairs. These findings suggest non-normalized distance-based bivariate LZC measures can be reliably applied to complex physiological signals such as human EEGs to further understand the effect of AD on the complexity of brain signals of patients. However, care must be taken when normalization procedures are applied.

Keywords—Alzheimer's Disease, Electroencephalogram, Non-linear analysis, Lempel-Ziv Complexity, Distance Bivariate Analysis.

I. INTRODUCTION

Alzheimer's disease (AD) is a neurological condition of complex etiology with progressive symptoms of memory and function loss caused by modification of amyloid β and hyperphosphorated tau in neurons, changing information transmission in the brain [1]. The 'preclinical' phase of the disease, where the AD patient is undiagnosed, can be as long as 20 years [2] as symptoms increase. With the development of more effective treatments for AD and the increase of patients, the need for early diagnosis is imperative to ensure that treatments can be utilized effectively.

There is evidence that the progress of the disease can be detected through changes of brain signals measured with an electroencephalogram (EEG) [3]. The disease must be highly progressed for visual identification with an EEG, signal processing techniques may improve the ease at which changes due to AD can be seen in the earlier stages of the disease. Non-linear signal processing has been shown to reliably identify changes in EEGs caused by AD [3]. Lempel-Ziv Complexity (LZC) is a non-linear method that calculates a signal's complexity given the rate of appearance of distinct substrings within a signal [4]. This method can be extended to selectively study different complexities by investigating combinations of signals. For instance, if applied to two separate recordings from the same system, complexity relates to the interdependence of their dynamics.

In this study we investigated the validity of three distance-based bivariate LZC measures [5] and their ability to distinguish between EEG recordings from AD patients and controls. It is hypothesized that this will show significant differences between the complexity of the EEG in the two groups, with an increased complexity of signals being seen in controls in comparison to AD patients.

This paper is arranged as follows. In section 2 we introduce the database of signals, the methods and the statistical evaluations. Section 3 contains the results. Finally, discussions, comparisons and conclusions are held in section 4.

II. METHODS

A. EEG Signal Database

This database has been described in a number of different studies e.g. [6]. However the pertinent aspects will be reproduced here for completeness. The sample group consisted of 11 probable AD patients (5 men and 6 women, 72.5 ± 8.3 years mean ± standard deviation (SD)) and 11 age matched controls (7 men and 4 women 72.8 ± 6.1 years mean ± SD). The Mini-Mental State Examination scores were 13.1±5.9 and 30±0 (mean ± SD) respectively. Ethical approval was received for the collection of data.

Signals were recorded in an awake but resting state with closed eyes using the international 10-20 electrode placement system (electrodes Fp1, Fp2, F3, F4, C3, C4, P3, P4, O1, O2, F7, F8, T3, T4, T5, T6, Fz, Cz and Pz). Over 5 minutes of data were collected for each subject at 256 Hz with a 12-bit analogue to digital converter. These were then reviewed by a clinician and 5 s epochs (1280 points), free from movement and electrooculographic artifacts and with minimal electromyographic activity were selected and copied for off-line analysis. For each electrode for each subject 30.0 ± 12.5 (mean ± SD) epochs were selected. These were then further filtered with a Hamming window finite impulse response band-pass filter between 0.5 and 40 Hz.

B. Lempel-Ziv Complexity

LZC is a measure of complexity of a one-dimensional, non-linear signal which can be reliably calculated given short data sets [6]. It has been widely applied to medical studies to investigate the changes in complexity different medical pathologies cause to biological signals, including AD [6], Parkinson's disease [7], and coma [8].

Parsing converts the signal to a binary sequence given a threshold T_d. The median was chosen as the threshold due to its robustness in the presence of outliers [9]. Given a signal $X = x(1), x(2), ..., x(n)$, the converted sequence $H = h(1), h(2), ..., h(n)$ with elements $h(i)$ defined by [10]:

$$h(i) = \begin{cases} 0 & \text{if } x(i) < T_d \\ 1 & \text{if } x(i) \geq T_d \end{cases} \quad (1)$$

This new sequence H is then scanned from left to right to identify the different subsequences held within the sequence. The LZC, $c(n)$, is the number of these unique subsequences, identified as follows [4]:

1. S and Q are two subsequences of H and SQ the concatenation. The last character is removed is denoted $SQ\pi$ and $v(SQ\pi)$ the vocabulary of all subsequences of $SQ\pi$. Initially $c(n) = 1$, $S = s(1)$ and $Q = s(2)$, so $SQ\pi = s(1)$
2. Generally $S = s(1), s(2), ..., s(r)$ and $Q = s(r+1)$, so $SQ\pi = s(1), s(2), ..., s(r)$. If Q belongs to $v(SQ\pi)$ then it is not a new sequence but a subsequence of $SQ\pi$
3. With $Q = s(r+1), s(r+2)$ again judge if Q does or does not belong to $v(SQ\pi)$
4. Repeat until Q is a new sequence, then $c(n) = c(n)+1$. In this case $Q = s(r+1), s(r+2), ..., s(r+i)$ is not a subsequence of $SQ\pi = s(1), s(2), ..., s(r+i-1)$
5. Thereafter, $S = s(1), s(2), ..., s(r+i)$ and $Q = s(r+i+1)$. This is repeated until Q is the last character of H

For example, $H = 110101101010$ parses to give 1.10.1011.01010 which has a LZC = 4. A highly complex signal, therefore, with have a higher LZC than a less complex signal of the same length.

C. Distance-Based Bivariate LZC

Distance-based bivariate measures identify differences in two signals which are taken simultaneously from two different points. The distance measure must satisfy three criteria [10]:

1. Identity ($D(R,T) \geq 0$; equality is satisfied if $R = T$)
2. Symmetry ($D(R,T) = D(T,R)$)
3. Triangle inequality ($D(R,U) \leq D(R,T) + D(T,U)$)

A number of distance-based bivariate LZC measures were defined by Otu and Sayood [5], with mathematical proof of their agreement to the three conditions mentioned above. These are [5]:

$$D_1(R,T) = \max[c(RT) - c(R), c(TR) - c(T)] \quad (2)$$

$$D_2(R,T) = \frac{\max[c(RT) - c(R), c(TR) - c(T)]}{\max[c(R), c(T)]} \quad (3)$$

$$D_3(R,T) = c(RT) - c(R) + c(TR) - c(T) \quad (4)$$

Where $c(X)$ and $c(XY)$ are the complexities of sequence X and the concatenation of the two sequences, X then Y, respectively, calculated with LZC. The distance-based bivariate LZC of signal pairs with few subsequences in common would be higher than in signal pairs with a large percentage of subsequences in common.

All possible electrode pairs (120) were tested with each method, each pair being separately investigated for its ability to distinguish between the 11 AD patients and the 11 control subjects.

D. Statistical Analysis

As the results followed a normal distribution, Student's t test ($p < 0.01$) was used to identify statistically significant differences between AD patients and controls at each electrode pair. Two-tailed ANOVA was also calculated with Matlab to evaluate interactions between electrode pairs and the diagnostic groups.

Statistically significant pairs were investigated using Receiver Operating Characteristic (ROC) curves [11] again calculated with Matlab. These plot the sensitivity (proportion of correctly identified AD patients) against 1-specificity (specificity is the proportion of correctly identified controls). ROC accuracy identifies the total number of correctly identified subjects. The area under these curves (AROC) shows the distribution of the data. Two distributions with similar means and an overlapping data spread will have a lower AROC than distributions with significantly different means and distinct data spreads.

III. RESULTS

All three methods were found to satisfy the criteria for distance measures. For D_1 and D_3, controls showed a greater LZC bivariate distance than AD patients in 96.7% and 99.2% of all the electrode pairs, respectively. Those pairs not conforming to this trend all relate to electrode T4 (F4-T4, F7-T4, C3-T4, C4-T4 for D_1 and F4-T4 for D_3) but this is only marginal, with the largest trend reversal accounting for only 2.8% of the control complexity at that electrode pair. On the other hand, only 32.5% electrode pairs with method D_2 showed increased distances for controls.

Differences between electrode pairs obtained with D_1 were statistically significant in 21 pairs with Fp2-P3 and O1-T5 the most significant ($p = 0.0028$). Differences between electrode pairs obtained with D_3 were only statistically significant in 18 cases, however the p-value was slightly lower than that found with D_1 (Fp1-P3 $p = 0.0016$). Again D_2 proved significantly less differentiating than the other methods with only one statistically significant pair, Fp1-C4 ($p = 0.0029$). Two tailed ANOVA further showed that there were significant differences between results given the starting electrode for the electrode pair, given the diagnosis of the subject, and that the two factors did not interact.

The maximum sensitivity and specificity seen were 100% (Fp1-T5 D_3) and 90.9% (Fp1-T5, F3-O1, F3-O2, F8-P3, O1-P3 D_1 and Fp1-P3, Fp1-T6, F3-P3, O1-T6 D_3), respectively. Maximum accuracy was 86.4% for D_1 (Fp2-O2, F3-O2, F3-P3) and 81.8% for D_2 (Fp1-C4) and D_3 (Fp1-O2, Fp1-P3, Fp1-T5, Fp1-T6, F3-O1, F3-P3, O1-P3, O1-P4, O1-T5, O1-T6). AROC values peak at 0.876 for Fp1-P3 with D_3, then 0.860 for Fp2-O1 and Fp2-P3 with D_1 and 0.785 for Fp1-C4 with D_2. Note that the most statistically significant electrodes, shown with lowest p values, do not exactly correspond to the most accurate results defined by the ROC curves.

IV. DISCUSSION

In this pilot study, we have used an extension of LZC for bivariate data to analyze the EEG signals of 22 subjects, 11 AD patients and 11 age-matched controls. This distance-based measure of complexity has shown statistically significant differences between AD patients and controls.

In the case of an AD patient, the number and range of the different subsequences in the two compared signals are small and, as such, when the signals from the different electrodes are concatenated, the complexity of the resulting sequence is similar to that of either original signal. In the case of a control subject, however, the richness of the separate signals is significantly higher, with a larger number of subsequences and a greater range of patterns present in both of the original data sets. When control signals are concatenated, the complexity of the concatenated signal may, therefore, be significantly higher than that of its two constituent signals if a low number of these subsequences are in common. This shows an increased richness in the subsequences of control subjects over the subsequences that define the signals of AD patients. It is this difference in the complexity of the pair of signals that these methods measure.

This correlates with findings with the univariate LZC on this database, where the highest sensitivity and specificity were both 90.9% (O1 and P3 respectively) with an accuracy of 81.8% at both electrodes [6] and also with other studies where other complexity methods have been applied to AD EEG databases (a recent review can be found in [3]). Note univariate LZC applied to this dataset shows reduced detection reliability than the bivariate study. Univariate LZC also showed a decreased signal complexity in AD patients' magnetoencephalogram (MEG) signals, with higher AROC (0.900) but lower sensitivities and specificities (81.0% and 85.7%) [12] than in this study. However these results cannot be directly compared to the bivariate study as the methodology applied measure complexity differently in the two cases.

An MEG database also including subjects with mild cognitive impairment (MCI), which presents similar but less severe impairments and increases the probability of AD in later life [13], was also evaluated using univariate LZC. Significant differences between AD and MCI and control and MCI were only found in the posterior and anterior areas of the brain [14], similar regions to those identified with highly significant complexity changes in this bivariate study. Furthermore, this is in agreement with the findings of other complexity and regularity based measures for both EEG and MEG signals in AD [3, 14].

While the authors know of no other bivariate LZC studies on AD, other bivariate techniques have been applied to characterize changes in brain signals caused by this pathology. One bivariate technique already applied to AD databases is synchrony. Dauwels et al [15] presented a number of these techniques all applied to the same database to allow for easy comparison between methods. Only Granger causality and stochastic event synchrony showed differences between the groups of subjects, with a classification rate of 83%, less successful than the bivariate method tested in this study.

The electrode pairs showing increased complexity in AD patients in D_1 and D_3 in relation to electrode T4 reflect other findings from this database using embedding entropy methods e.g. [6, 16] and detrended moving average [17].

In the case of D_2, there is an increased percentage of electrode pairs where the distance was greater in AD patients than controls. While Otu and Sayood introduced D_2 as a normalized distance measure based on D_1, our results find the distribution is not consistent between the two methods. This is because the denominator in D_2 affects the distance result, with a high complexity signal artificially reducing the complexity of the interaction of the signals.

As a result of the small sample size, the findings of this study should be seen as preliminary and the methodology must be applied to a larger population of subjects to further investigate the possible diagnostic value of this technique. Furthermore, univariate LZC has previously shown similar changes to those seen with AD in Parkinson's disease [7], and indeed MCI [14]. The distance-based bivariate LZC, therefore, needs to be applied to a wider range of pathologies to check whether these changes are unique to AD.

V. CONCLUSIONS

This work has shown an increased richness of complexity in the EEG signals of control subjects in comparison to those signals of AD patients, corroborating the evidence already presented in similar studies and the understanding that AD changes the dynamics of physiological processes within the brain [18]. However, caution must be taken due to the small size of the study group.

REFERENCES

1. Pievani M, de Haan W, Wu T et al. (2011) Functional network disruption in the degenerative dementias. Lancet Neurol 10:829-843 DOI 10.1016/S1474-4422(11)70158-2
2. Reiman E, Quiroz Y, Fleisher A et al. (2012) Brain imaging and fluid biomarker analysis in young adults at genetic risk for autosomal dominant Alzheimer's disease in the presenilin E280A kindred: a case-control study. Lancet neurol 11:1048-1056 DOI 10.1016/S1474-4422(12)70228-4
3. Dauwels J, Vialatte F, Cichocki A (2010) Diagnosis of Alzheimer's disease from EEG signals: Where are we standing? Curr Alzheimer Res 7:487-505 DOI 10.2174/156720510792231720
4. Lempel A, Ziv J (1976) On the complexity of finite sequences. IEEE T Inform Theory 22:75-81 DOI 10.1109/TIT.1976.1055501
5. Otu H, Sayood K (2003) A new sequence distance measure for phylogenetic tree construction. Bioinformatics 19:2122-2130 DOI 10.1093/bioinformatics/btg295
6. Abasolo D, Hornero R, Gomez C (2006) Analysis of EEG background activity in Alzheimer's disease patients with Lempel-Ziv complexity and central tendency measure. Med Eng Phys 28:315-322 DOI 10.1016/j.medengphy.2005.07.004
7. Gomez C, Olde Dubbelink K, Stam C et al (2011) Complexity analysis of resting-state MEG activity in early-stage Parkinson's disease patients. Ann Biomed Eng 39:2935-2944 DOI 10.1007/s10439-011-0416-0
8. Wu D, Cai G, Yuan Y et al. (2011) Application of nonlinear dynamics analysis in assessing unconsciousness: A preliminary study. Clin Neurophysiol 122:490-498 DOI 10.1016/j.clinph.2010.05.036
9. Nagarajan R (2002) Quantifying physiological data with Lempel-Ziv complexity-certain issues. IEEE Trans Biomed Eng 49:1371-1373 DOI 10.1109/TMBE.2002.804582
10. Zhang X, Roy R, Jensen E (2001) EEG complexity as a measure of depth of anesthesia for patients. IEEE Trans Biomed Eng 48:1424-1433 DOI 10.1109/10.966601
11. Fawcett T (2006) An introduction to ROC analysis. Pattern Recogn Lett 27:861-874
12. Gomez C, Hornero R, Abasolo D (2006) Complexity analysis of the magnetoencephalogram background activity in Alzheimer's disease patients. Med Eng Phys 28:851-859 DOI 10.1016/j.medengphy.2006.01.003
13. Petersen R, Knopman D, Boeve B et al. (2009) Mild cognitive impairment: Ten years later. Arch Neurol 66:1447-1455 DOI 10.1001/archneurol.2009.266
14. Fernandez A, Hornero R, Gomez C et al. (2010) Complexity analysis of spontaneous brain activity in Alzheimer disease and mild cognitive impairment: An MEG study. Alzheimer Dis Assoc Disord 24:182-189 DOI 10.1097/WAD.0b013e3181c727f7
15. Dauwels J, Vialatte F, Musha T et al. (2010) A comparative study of synchrony measures for the early diagnosis of alzheimer's disease based on EEG. NeuroImage 49:668-693 DOI 10.1016/j.neuroimage.2009.06.056
16. Abasolo D, Hornero R, Espino P et al (2006) Entropy analysis of the EEG background activity in Alzheimer's disease patients. Physiol Meas 27:241-253 DOI 10.1088/0967-3334/27/3/003
17. Abasolo D, Hornero R, Gomez C et al (2009) Electroencephalogram background activity characterization with detrended moving average in Alzheimer's disease patients. Proc. of 6th IEEE International Symposium on Intelligent Signal Processing, Budapest, Hungary, 2009
18. Jeong J (2004) EEG dynamics in patients with Alzheimer's disease. Clin Neurophysiol 115:1490-1505 DOI 10.1016/j.clinph.2004.01.001

Author: Samantha Simons
Institute: University of Surrey
City: Guildford
Country: United Kingdom
Email: s.simons@surrey.ac.uk

Predictive Value on Neurological Outcome of Early EEG Signal Analysis in Brain Injured Patients

A. Accardo[1], M. Cusenza[1], L. Prisco[2], F. Monti[3], A. Draisci[3], and W. Calligaris[3]

[1] Dept. of Engineering and Architecture, University of Trieste, Trieste, IT
[2] Critical Care Dept, University College Hospital, London, UK
[3] Clinical Neurophysiology Unit, University Hospital of Cattinara, Trieste, IT

Abstract—Electrophysiological examinations constitute objective and accurate measures of cerebral function. They can be recorded at bedside, which is of major value in intensive care units. The objective of the present study considers the linear and non-linear analysis of resting electroencephalography (EEG) signals as predictors for poor outcome and differences between brain regions. We studied 24 brain injured patients (trauma, cerebral anoxia, intracranial haemorrhage, cerebral infections) and 12 healthy controls and compared their EEG power spectra, the zero crossing and fractal dimension splitting the brain in four regions: left hemisphere, right hemisphere, anterior cortex, posterior cortex. In this study early linear and non-linear parameters showed the ability to predict recovery of communicative skills in brain injured patients at 6 months.

Keywords—Coma, EEG, Power spectrum, Zero crossing, Fractal dimension.

I. INTRODUCTION

Consciousness has been defined as a multifaceted concept which can be divided into two main components, arousal and awareness. In clinical practice, assessment of the state of consciousness of patients suffering from severe brain injury becomes an arduous task and misdiagnoses are very common [1]. High-order cerebral areas encompassing lateral and midline frontoparietal networks show no activation in either volunteers under deep anesthesia or in patients in a *vegetative state*. At the opposite, patients in minimally conscious state (MCS) can show activation of the high-order areas sometimes similar to healthy control subjects [2].

A common finding of studies on both pathological and pharmacological coma is an impairment of the activity of a widespread cortical network, encompassing bilateral fronto-parietal associative cortices [3,4].

Patients in coma are unconscious because they cannot be awakened (i.e. they never open the eyes). Following coma, some patients may "awaken" (meaning they open the eyes) but remain unaware. This condition is called the "vegetative state" [5] recently renamed "unresponsive wakefulness syndrome" (UWS) [6]. MCS refers to patients who are unable to reliably communicate but show reproducible albeit fluctuating behavioural evidence of awareness (i.e. non-reflex movements or command following [7]).

A large frontoparietal network encompassing bilateral frontal and temporo-parietal associative cortices has its activity commonly impaired during altered states of consciousness [2,3,4]. This network can be divided into several parts with distinct functions [8]. In particular, a distinction can be made between a network involved in the awareness of self (the "internal" midline default mode network) encompassing precuneus/posterior cingulated, mesasfrontal/anterior cingulated, and temporo-parietal cortices [9,10] and an "external" more lateral and dorsal frontoparietal network involved in the awareness of environment [8]. Paralleling clinical experience, a non-linear correlation was found between this default/"internal" network connectivity and the level of consciousness ranging from healthy volunteers to MCS, UWS and comatose patients [11].

The aim of this study is to evaluate early linear and non-linear EEG parameters as predictors of long-term neurological outcome in brain injured patients and to correlate recovery of brain functions to different brain regions.

II. PATIENTS AND METHODS

A. Patients and Controls

We performed a prospective study in 24 patients (median age 63 years) who experienced acute brain injury (8 head trauma, 3 cerebral anoxia, 9 cerebral hemorrhage, 1 ischemic stroke, 3 cerebral infections) and admitted to Intensive Care Unit of University Hospital of Cattinara, Trieste (Italy) during 2011. Patients met the following selection criteria: a significant degree of impairment of consciousness for more than 24 hours, at least one brain CT scan or MRI on admission has been taken, stabile hemodynamic and ventilation, a motor response according to Glasgow Coma Scale (GCS) lower than M6 or equal to M6 2 times over 4, a core body temperature less than 38°C, and age > 14 years.

There were excluded patients with prior history of neurological or psychiatric problems known to produce EEG abnormalities, patients receiving hypnotic agent or muscle-blocker agent at the time of recording and those diagnosed as brain dead (suppressed EEG).

Clinical and electrophysiological examinations were performed upon the approval by local ethics committee and written informed consent was obtained from each subject.

The patient group was matched for age with 12 normal control subjects: 8 males and 4 females (median age: 56 years). The control participants, free of alcohol and drug dependency, showed no sensory impairments nor any neurological and neuropsychiatric diseases. None of them had ever suffered a brain injury with loss of consciousness.

B. Clinical Evaluation

Clinical assessment included GCS scoring with eyes opening (1-4), speech response (1-5), and motor response (1-6), scoring 15 in total. GCS was evaluated on admission, 30 minutes after EEG recording, at 1 month after injury and at 6 months. Consciousness at 1 and 6 moths was evaluated using the Coma Recovery Scale-Revised [12] and according to the scale patients were defined as UWS, MCS, communicative patients (emerging MCS, EMCS).

C. EEG Acquisition and Analysis

Electrophysiological examinations were recorded at bedside by using disposable needle electrodes as soon as patients were free of sedatives. 19 electrodes were placed according to the International 10-20 system (Galileo System, EBNeuro, Florence, Italy) with A1 and A2 as references. Impedance was kept less than 5 kΩ. The filters were set from 0.4 Hz to 70 Hz and the sampling frequency was of 512Hz. Each recording lasted at least 15 minutes.

Preprocessing: EEG data were digitally offline filtered with a band pass filter (0.5-60 Hz) to remove noises and muscular artefacts. The signal has been divided into 50% overlapping sections having a 10s duration for the whole length of each recording; each segment was detrended and windowed (Hamming). All parameters were calculated in these epochs and averaged among all the derivations of the region of interest, obtaining one value per parameter and side every 5 seconds. Electrodes were grouped in four regions of interest: left hemisphere (PS: T7,P7,P3,O1,C3), right hemisphere (PD: T8,P8,P4,O2,C4), fronto-central regions (ANT: Fp1,Fp2,Fz,F7,F3,F4,F8,Cz,C3,C4) and parieto-temporo-occipital regions (POS: T7,P7,P3,Pz,P4,P8, T8,O1,O2).

Linear analysis: Linear analysis considered some spectral parameters calculated from the power spectral density (PSD) estimated by periodogram method. From PSD, the power in the traditional EEG-ranges (delta 0.5-4Hz, theta 4–8Hz, alpha 8-15Hz, beta 15-30Hz, gamma 30-60Hz) and the power in a low frequency band (LF 0.5-5Hz) and in a high frequency band (HF 8-15Hz) were computed.

Non-linear analysis: Beside these linear parameters, some other non linear factors were evaluated. In particular the zero crossing (ZC) count and the fractal dimension (FD) of each EEG interval were estimated. The ZC count is a non linear parameter used in the analysis of random signals [13]. It can be computed by counting the number of baseline crossings in a fixed time interval. Fruitful connections exist between zero-crossing counts and dominant frequency so that it can be used to identify possible changes in dominant spectral components.

The fractal dimension (FD) represents a measure of the complexity of the EEG signal able to distinguish specific behavioural states (sleep stages, epileptic seizure, etc.). Among the various algorithms available for FD estimation, in this study FD was computed by Higuchi's algorithm which is based on the measure of the mean length of the curve $L(k)$ by using a segment of k samples as a unit of measure [14]. Before FD estimation, the EEG signal coming from each derivation was resampled at 128Hz.

The calculation of ZC and FD parameters is faster than the one of the spectral analysis and both can be correctly estimated on short epochs (down to two seconds).

Statistical analysis: Using MedCalc 11.1.1.0 (MedCalc© Statystical Software, Belgium) stepwise multiple regression has been performed in order to find predictive factors among linear and non-linear parameters in each brain region (alpha to enter/remove: 0.15).

Table 1 Medians power spectra in the high and low frequency bands for all clinical groups

	CONTR (12)	DEC (12)	EMCS (5)	MCS (3)	UWS (4)
PS HF	60555	19323	10938	17626	6434
PS LF	100013	529764	138856	1732935	375624
PD HF	59510	22601	12287	22683	6833
PD LF	103888	534204	115141	3319379	135583
ANT HF	17661	10265	14943	10870	4225
ANT LF	87316	437170	209101	1589814	141128
POS HF	61124	22609	14295	24014	8958
POS LF	112957	617597	78207	4120437	408237

PS=Left Hemisphere, PD=Right Hemisphere, ANT=Frontal regions, POS=Parieto-Temporo-Occipital regions.

III. RESULTS

At 6 months mortality was 50% (due to cerebral causes and other medical diseases which occurred during the recovery phase). The median LF and HF band power spectra averaged per brain regions for all clinical groups (neurological clinical outcome at 6 months) are reported in Table 1.

Significant differences (p<0.02) were found between controls and the other groups as expected. However unexpectedly high values have been found in the low frequency bands in all brain regions (especially in the delta band and in the posterior brain regions) in the MCS group. Non-linear parameters as medians of each brain regions are reported in Table 2. The stepwise multiple regression included the Zero Crossing parameter of the left hemisphere as independent variable (constant 0.7286, T-value 3.76, p<0.01) but with partial explanation of the model (Rsq 29.98) (Fig.1).

Table 2 Medians of non-linear parameters in the different brain regions for all clinical groups

	CONTR	DEC	EMCS	MCS	UWS
PS FD	1,61	1,21	1,32	1,24	1,27
PS ZC	271	79	150	74	82
PD FD	1,54	1,24	1,30	1,26	1,28
PD ZC	215	83	146	77	96
ANT FD	1,63	1,21	1,35	1,19	1,31
ANT ZC	281	84	160	80	86
POS FD	1,49	1,21	1,29	1,24	1,28
POS ZC	228	79	143	74	83

PS=Left Hemisphere, PD=Right Hemisphere, ANT=Frontal regions, POS=Parieto-Temporo-Occipital regions.

IV. DISCUSSION

The quantitative analysis performed, in comparison to controls and EMCS patients but also to deceased and UWS patients, showed that the impact of delta band in MCS patients is considerably high. Among all regions considered, the delta band in the MCS group was significantly higher (p<0.02) in the posterior one (parieto-temporo-occipital region). Other authors have demonstrated an increase in the delta band in MCS in comparison to communicative patients in different brain regions, but they didn't analyse UWS patients [1]. Kotchoubey et al. [15], comparing power spectra of patients in UWS and MCS, found that the latter had less slow activity than the former. This study suggested the existence of a continuum illustrating that the lower the awareness level in a patient, the greater the presence of slow wave activity. Our results demonstrated the highest presence of slow wave activity in MCS patients, but not in the poorest outcome groups (UWS and deceased patients). We hypothesized a possible bias due to the limited number of patients per group and to the intrinsic characteristics of these patients, whose neurological status is often fluctuating during the day. However both controls and EMCS patients

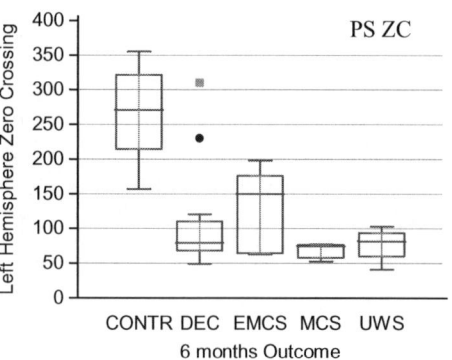
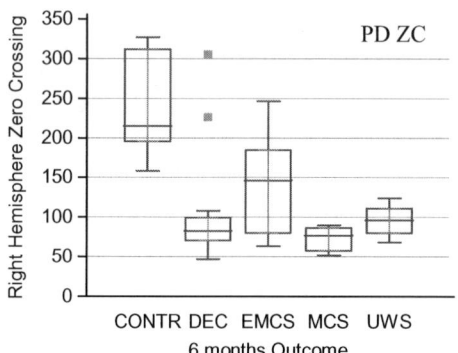
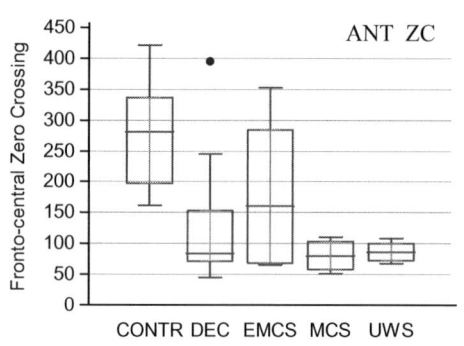
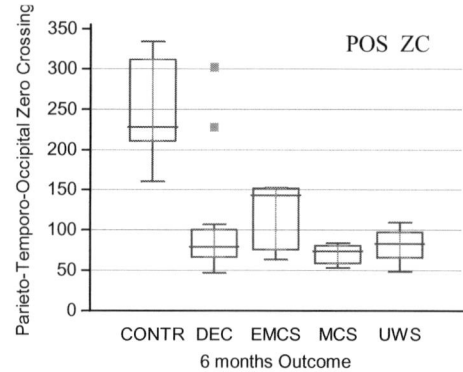

Fig. 1 Medians (25-75° percentiles) of ZC in the different groups.

showed the lowest LF activity in all regions, especially in the posterior associative areas and almost the highest HF activity in the anterior regions. The prefrontal cortex (PFC) has been demonstrated essential when it comes to organizing voluntary activity and it is viewed as the highest order cerebral area in guiding complex behavior [16].

Among non-linear parameters ZC in all brain regions showed a good predictive value of communication recovery after brain injury. Remarkably, left hemispheric ZC shown significant differences between control and all the patients. In fact, this hemisphere is particularly involved in the symbolic communication process, while the right hemisphere plays a special role in spontaneous communication. Damage to the left hemisphere in right handed people is associated with deficits in language expression and comprehension.

The degree of preservation of the prefrontal-parietal network may account for the differences in awareness level between the lowest state of consciousness [17]. Therefore, a possible interpretation of our results is that deceased patients as well as those in MCS and UWS show a higher dysfunction in this network in comparison to communicative patients (EMCS and controls). According to Schiff and Plum [18], damage in this network is the primary basis for the lack of manifestation of awareness in a given patient. The relative significance of each area within this cortico-subcortical network, as well as corresponding connections, remains unclear.

V. CONCLUSION

Despite some limitations and biases of the present study, we can conclude that the predictive and discriminative value of early linear and non-linear EEG analysis in brain injury patients could integrate the admission clinical evaluation in order to identify those patients who will recover communicative skills.

ACKNOWLEDGMENT

Work partially supported by University of Trieste, Master in Clinical Engineering.

REFERENCES

1. Leon-Carrion J, Martin-Rodriguez JF, Damas-Lopez J, Barroso y Martin JM, Dominguez-Morales MR. Brain function in the minimally conscious state: a quantitative neurophysiological study. Clin Neurophysiol. 2008 Jul;119(7):1506-14.
2. Laureys S. The neural correlate of (un)awareness: lessons from the vegetative state. Trends Cogn Sci. 2005 Dec;9(12):556-9.
3. Ragazzoni A, Pirulli C, Veniero D, Feurra M, Cincotta M, Giovannelli F, Chiaramonti R, Lino M, Rossi S, Miniussi C., Vegetative versus minimally conscious states: a study using TMS-EEG, sensory and event-related potentials. PLoS One. 2013;8(2):e57069..
4. Alkire MT, Miller J. General anesthesia and the neural correlates of consciousness. Prog Brain Res. 2005;150:229-44.
5. Jennett B, Plum F. Persistent vegetative state after brain damage. A syndrome in search of a name. Lancet. 1972 Apr 1;1(7753):734-7.
6. Laureys S, Celesia GG, Cohadon F, Lavrijsen J, León-Carrión J, Sannita WG, Sazbon L, Schmutzhard E, von Wild KR, Zeman A, Dolce G; European Task Force on Disorders of Consciousness. Unresponsive wakefulness syndrome: a new name for the vegetative state or apallic syndrome. BMC Med. 2010 Nov 1;8:68.
7. Giacino JT, Ashwal S, Childs N, Cranford R, Jennett B, Katz DI, Kelly JP, Rosenberg JH, Whyte J, Zafonte RD, Zasler ND. The minimally conscious state: definition and diagnostic criteria. Neurology. 2002 Feb 12;58(3):349-53.
8. Boly M, Phillips C, Balteau E, Schnakers C, Degueldre C, Moonen G, Luxen A, Peigneux P, Faymonville ME, Maquet P, Laureys S. Consciousness and cerebral baseline activity fluctuations. Hum Brain Mapp. 2008 Jul;29(7):868-74.
9. Gusnard DA, Raichle ME, Raichle ME. Searching for a baseline: functional imaging and the resting human brain. Nat Rev Neurosci. 2001 Oct;2(10):685-94.
10. Mason MF, Norton MI, Van Horn JD, Wegner DM, Grafton ST, Macrae CN. Wandering minds: the default network and stimulus-independent thought. Science. 2007 Jan 19;315(5810):393-5.
11. Vanhaudenhuyse A, Noirhomme Q, Tshibanda LJ, Bruno MA, Boveroux P, Schnakers C, Soddu A, Perlbarg V, Ledoux D, Brichant JF, Moonen G, Maquet P, Greicius MD, Laureys S, Boly M. Default network connectivity reflects the level of consciousness in non-communicative brain-damaged patients. Brain. 2010 Jan;133(Pt1):161-71.
12. Kalmar K, Giacino JT. The JFK Coma Recovery Scale--Revised. Neuropsychol Rehabil. 2005 Jul-Sep;15(3-4):454-60.
13. Kedem B. Spectral analysis and discrimination by Zero-Crossings. Proceed. of IEEE 1986;74:1477-93.
14. Higuchi T. Approach to an irregular time series on the basis of the fractal theory. Physica D 1988;31:277-83.
15. Kotchoubey B, Lang S, Mezger G, Schmalohr D, Schneck M, Semmler A, Bostanov V, Birbaumer N. Information processing in severe disorders of consciousness: vegetative state and minimally conscious state. Clin Neurophysiol. 2005 Oct;116(10):2441-53.
16. Groenewegen HJ, Uylings HB. The prefrontal cortex and the integration of sensory, limbic and autonomic information. Prog Brain Res. 2000;126:3-28.
17. Boly M, Faymonville ME, Peigneux P, Lambermont B, Damas P, Del Fiore G, Degueldre C, Franck G, Luxen A, Lamy M, Moonen G, Maquet P, Laureys S. Auditory processing in severely brain injured patients: differences between the minimally conscious state and the persistent vegetative state. Arch Neurol. 2004 Feb;61(2):233-8.
18. Schiff ND, Plum F. Cortical function in the persistent vegetative state. Trends Cogn Sci. 1999 Feb;3(2):43-44.

Author: Agostino Accardo
Institute: Dept. of Engineering and Architecture, Univ. of Trieste
Street: Via Valerio, 10
City: Trieste
Country: Italy
Email: accardo@units.it

Lempel-Ziv Complexity Analysis of Local Field Potentials in Different Vigilance States with Different Coarse-Graining Techniques

D. Abásolo[1], R. Morgado da Silva[1], S. Simons[1], G. Tononi[2], C. Cirelli[2], and V.V. Vyazovskiy[3]

[1] Centre for Biomedical Engineering, Department of Mechanical Engineering Sciences,
Faculty of Engineering and Physical Sciences, University of Surrey, Guildford, UK
[2] Department of Psychiatry, University of Wisconsin-Madison, Madison, Wisconsin, United States of America
[3] Department of Biochemistry and Physiology, Faculty of Health and Medical Sciences, University of Surrey, UK

Abstract—Analysis of electrophysiological signals recorded from the brain with Lempel-Ziv (LZ) complexity, a measure based on coarse-graining of the signal, can provide valuable insights into understanding brain activity. LZ complexity of local field potential signals recorded from the neocortex of 11 adult male Wistar-Kyoto rats in different vigilance states – waking, non-rapid-eye movement (NREM) and REM sleep – was estimated with different coarse-graining techniques (median, LZCm, and k-means, LZCkm). Furthermore, surrogate data were used to test the hypothesis that LZ complexity results reveal effects accounted for by temporal structure of the signal, rather than merely its frequency content. LZ complexity values were significantly lower in NREM sleep as compared to waking and REM sleep, for both real and surrogate signals. LZCkm and LZCm values were similar, although in NREM sleep the values deviated in some epochs, where signals also differed significantly in terms of temporal structure and spectral content. Thus, the interpretation of LZ complexity results should take into account the specific algorithm used to coarse-grain the signal. Moreover, the occurrence of high amplitude slow waves during NREM sleep determines LZ complexity to a large extent, but characteristics such as the temporal sequence of slow waves or cross-frequency interactions might also play a role.

Keywords—Lempel-Ziv complexity, k-means, local field potential, surrogate data, sleep.

I. INTRODUCTION

Different brain states are associated with changes in neuronal network function and characterized by specific spatiotemporal patterns of cortical activity [1]. Three main vigilance states are usually identified in mammals including humans: waking, non-rapid-eye movement (NREM) and REM sleep. They are distinguished not only based on animal's behavior but also based on the total amplitude and spectral power of cortical electroencephalogram (EEG) and local field potential (LFP) [2]. The functional significance of the difference in brain activity between vigilance states is not yet understood.

While much has been learned about different brain states using conventional spectral analysis, non-linear analysis metrics are still rarely used. This becomes especially relevant when signals differ drastically in their frequency content and amplitude.

Non-linear analysis of EEG or LFP signals with Lempel-Ziv (LZ) complexity, a measure of complexity based on coarse-graining of the signal, can provide valuable insights into understanding brain activity [3], [4]. LZ complexity is a method of symbolic sequence analysis and, as such, is based on a coarse-graining of the time series being investigated. The signal must be transformed into a finite symbol sequence before estimating its LZ complexity. While the choice of coarse-graining approaches may appear essential for interpretability of LZ complexity values obtained in different vigilance states, no attempts to perform a comparison between coarse-graining techniques have been made. In this study, we applied LZ complexity analysis to cortical LFP recordings collected in freely-moving rats during spontaneous waking and sleep states with different coarse-graining techniques. Furthermore, surrogate data were used to test the hypothesis that LZ complexity results reveal effects purely accounted for by temporal structure, rather than frequency content, and may provide additional insights into the characteristics of the signal that contribute to the information content.

This paper is organized as follows. In section 2 we introduce the LFP signals, the different coarse-graining methods applied, the surrogate data used, and the statistical evaluations carried out. Results are presented in section 3 and our findings are discussed in section 4, where the conclusions of the study are also presented.

II. MATERIAL AND METHODS

A. Materials

LFP recordings and single- and multi-neuron activity were collected with microwire arrays implanted in the frontal cortex of 11 adult male Wistar-Kyoto rats during undisturbed waking and sleep [2]. Three stages – waking, NREM and REM sleep – were identified offline based on the EEG/LFP signals and electromyogram (EMG), as well as frame-by-frame analysis of the video recording. For this analysis one period of stable continuous awake state has been chosen, along with one individual episode of NREM and REM sleep in each rat.

The total amount of vigilance states that contributed to these analyses was 8.5 ± 0.6, 12.2 ± 2.3 and 3.0 ± 0.2 minutes for waking (W), NREM sleep (N) and REM sleep (R), respectively. Examples of the three different vigilance states (8-s epochs) are shown in Figure 1.

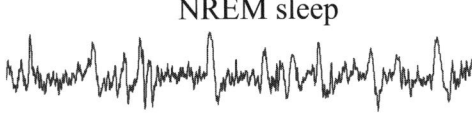

Fig. 1 Representative 8-s LFP traces recorded from the frontal cortex in a freely-behaving rat in spontaneous waking, NREM and REM sleep.

B. Lempel-Ziv Complexity

The LZ algorithm [5] is a method of symbolic sequence analysis that can be used to measure the complexity of finite length sequences. Several aspects of this method make it attractive for biomedical signal processing: it is non-parametric, it can be applied to estimate the complexity of relatively short sequences, and it does not assume a particular model (e.g. chaotic or stochastic) for the system generating the time series [6]. In addition, the computational cost of estimating LZ complexity is significantly lower than that of embedding entropy methods in widespread use, such as approximate or sample entropy. This is a considerable advantage for the analysis of chronic sleep recordings.

LZ complexity is based on the coarse-graining or symbolization of the original time series. This process involves converting the original biomedical signal into a sequence with a finite number of symbols. In most cases, a binary conversion is used. The coarse-graining process determines how much information can be retained from the original signal. In this paper, two different solutions – using the median as the threshold due to its robustness to outliers [7] or k-means in the symbolization of the original signal [8] – are presented and compared.

a) *Median.* The signal is converted into a binary sequence $P = s(1), s(2),..., s(n)$ by comparing each sample $x(i)$ with a threshold, in this case the median of the time series T_d, with $s(i)$ then given by [6]:

$$s(i) = \begin{cases} 0 & \text{if } x(i) < T_d \\ 1 & \text{if } x(i) \geq T_d \end{cases} \quad (1)$$

b) *k-means.* This approach is based on the grouping of data around centroids corresponding to points around which most of the data is agglomerated [8]. The number of centroids k is defined by the user and is equal to 2 for binary sequences.

In the initial iteration of the method, one must set the two initial centroids as follows:

$$z_1(1) = x_m + \varepsilon \cdot x_m \quad (2)$$

$$z_2(1) = x_m - \varepsilon \cdot x_m \quad (3)$$

where we assume $\varepsilon = 0.005$ and x_m is the mean of the data points from the original signal, $x(i)$ [8]. Distances of each data point to centroids are then calculated as:

$$\begin{aligned} D_1^i &= \|x(i) - z_1(1)\|^2 \\ D_2^i &= \|x(i) - z_2(1)\|^2 \end{aligned} \quad (4)$$

The signal can then be converted into a binary sequence $P = s(1), s(2),..., s(n)$ with $s(i)$ given by the following equation [8]:

$$s(i) = \begin{cases} 1 & \text{if } D_1^i < D_2^i \\ 0 & \text{if } D_1^i \geq D_2^i \end{cases} \quad (5)$$

Thus, each data point is set based on a minimum distance criterion. All points assigned with symbol 1 will belong to group 1, and points assigned with symbol 0 will belong to group 2. In a new iteration, two new centroids have to be defined. For group 1, $z_1(2)$ is the average coordinate among all the members in the group. For group 2, $z_2(2)$ is the average coordinate among all members of the group. Equations (4) and (5) are then re-applied in order to find the new distance values and the new symbolic sequence P. The procedure has to be repeated until $z_1(j+1) = z_2(j)$ for all j.

Once the symbolic sequence P has been created, its complexity needs to be estimated. This has been done using the parsing process suggested by Lempel and Ziv [5], which scans P from left to right and adds one unit to a complexity counter $c(n)$ every time a new subsequence of consecutive characters is found. In this way, LZ complexity is related to the number of distinct substrings and the rate of their recurrence along the given sequence [9]. As a result, LZ complexity is very much in the spirit of Kolmogorov's algorithmic complexity [10], [11]. A detailed description of the complexity algorithm can be found in [6].

Last, but not least, the complexity counter has to be normalized to obtain a complexity measure independent of the sequence length. In general, the upper bound of the complexity is given by [5]:

$$b(n) \equiv \frac{n}{\log_\alpha(n)} \quad (6)$$

α is the number of symbols in the alphabet (hence $\alpha = 2$ for a binary conversion). Therefore, the normalized LZ complexity can be defined as:

$$C(n) = \frac{c(n)}{b(n)} \quad (7)$$

C. Surrogate Data

To investigate whether the complexity of signals derived from brain activity is not merely an artifact of the spectral frequency content, surrogate signals were generated from the LFP signals. Firstly, the Fourier transform of the signals was computed and the phases of the Fourier coefficients randomized while at the same time keeping unchanged their magnitude [12]. Then, the inverse Fourier transform into the time domain was performed. As a result, surrogates indistinguishable from naturalistic LFP signals with respect to spectral characteristics were obtained.

D. Statistical Analysis

The LZ complexity values are reported as means ± standard error of the mean (SEM). Two-tailed paired t-tests were used to evaluate significant differences between LZ complexity values obtained in different vigilance states and between different coarse-graining techniques within a state.

III. RESULTS

LZ complexity was computed for the LFP signals with the two coarse-graining methods aforementioned, k-means (LZCkm) and median (LZCm), over consecutive 4-s epochs. We found that average LZ complexity values, computed over consecutive 4-s epochs, were invariably substantially lower in NREM sleep as compared to waking and REM sleep (Figure 2).

We next used two different methods of coarse-graining the signals to compare if vigilance-state specific differences are affected by the algorithm. This appeared not to be the case, as the values were virtually identical between k-means (LZCkm) and median (LZCm) coarse-graining approaches (results are summarized in Table 1). The differences between results obtained with LZCkm and LZCm were not statistically significant.

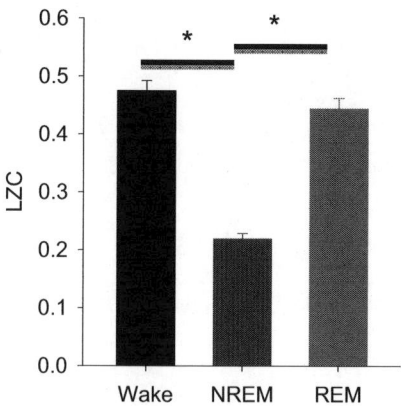

Fig. 2. Average LZ complexity (LZC) values in the three behavioral states (n = 11 rats). The values are obtained with k-means approach. Asterisks above the bars depict significant differences ($p<0.05$)

Table 1 LZ complexity results for waking (W), NREM sleep (N) and REM sleep (R) obtained with k-means (LZCkm) and median (LZCm) coarse-graining approaches

Method	W	N	R
LZCkm	0.46 ± 0.046	0.25 ± 0.031	0.47 ± 0.027
LZCm	0.46 ± 0.046	0.25 ± 0.030	0.47 ± 0.026

However, while on average LZCkm and LZCm values were similar, during a subset of 4-s epochs the values deviated, and that was especially apparent in NREM sleep. On average, the proportion of epochs where absolute difference (either positive or negative) between LZCkm and LZCm values exceeded 10% was 2.7 ± 2.4%, 16.8 ± 4.3% and 1.3 ± 1.2% of all 4-s epochs in waking, NREM and REM sleep respectively. Interestingly, in NREM sleep, where the proportion of epochs where LZCkm and LZCm values differed by more than 10% was substantial, signal variance and LFP power in slow-wave range showed a systematic difference as well. Specifically, the difference in both signal variance and slow wave activity (SWA) showed higher values during those epochs where LZCkm and LZCm were different (variance: 81.8 ± 16.9 vs. 58.3 ± 13.9 μV, $p = 0.0351$, paired t-test; relative SWA: 121.5 ± 2.6 vs. 95.1 ± 1.5 % of mean SWA over all epochs, $p = 7.3030e-004$, paired t-test).

LZ complexity values were significantly lower in NREM sleep as compared to waking and REM sleep for both real and surrogate signals. Interestingly, average LZ complexity values for surrogate signals differed only marginally for waking (increase by 1.3%) and REM sleep (decrease by 1.4%), but were significantly lower in NREM sleep (decrease by 8.4 ± 1.4%). Computing the distribution of LZ complexity values revealed that this difference is accounted by fewer epochs with higher LZ complexity values for the surrogate data than for the real data.

IV. DISCUSSION AND CONCLUSIONS

In this study we performed a detailed analysis of LZ complexity derived from cortical LFP signals in freely behaving rats. Since the patterns of spontaneous cortical activity are different between different vigilance states [2] we hypothesized that the information content of the signals is state-dependent. As expected, we found that the values of LZ complexity were substantially lower in NREM sleep, which is considered a state functionally disconnected from the environmental input, and is characterized by absent or reduced consciousness [13].

We also found that different approaches of coarse-graining have only marginal overall influence on the resulting values of LZ complexity in waking and REM sleep, while the differences in NREM sleep were more pronounced. Interestingly, in those epochs where there was an effect of coarse-graining technique, signals also differed significantly in terms of temporal structure and spectral content. Thus, interpreting the information measures obtained with LZ complexity should take into account the specific algorithm used to coarse-grain the signal. Our analyses also showed vigilance state-specific differences and similarities between LZ complexity values computed from real and surrogate data. The data suggest that the occurrence of high amplitude slow waves during NREM sleep determines LZ complexity to a large extent, but other characteristics, such as the temporal sequence of slow waves or cross-frequency interactions might also play a role.

An important implication of our results is that LZ complexity may not only provide information above and beyond of what can be obtained with conventional power spectral analysis, but, in fact, be the only option for studies in which signals to be compared are inherently different, such as due to a difference in age, gender, ethnic origin, pharmacological treatment, etc. Therefore, introducing LZ complexity as a new tool to investigate brain signals in waking and sleep is important. On one hand, using non-linear analysis can provide sensitive measures, relatively independent from spectral power. On the other hand, LZ complexity values can provide unique insights into the network mechanisms of waking and sleep, and pave the way towards gaining better understanding of their physiological and functional relevance.

Although results are promising, the sample size was small. As a result, our findings are preliminary and require replication in a larger database before any conclusion can be made of its potential impact.

ACKNOWLEDGMENT

This work was supported by the Engineering and Physical Sciences Research Council [grant number EP/I000992/1].

REFERENCES

1. Buzsáki, G (2006) Rhythms of the brain. Oxford University Press, Oxford ; New York, 2006.
2. Vyazovskiy VV, Olcese U, Lazimy YM et al. (2009) Cortical firing and sleep homeostasis. Neuron 63:865–878 DOI 10.1016/j.neuron.2009.08.024
3. Abásolo D, Homero R, Gómez C et al. (2006) Analysis of EEG background activity in Alzheimer's disease patients with Lempel-Ziv complexity and central tendency measure, Med Eng Phys 28:315-322 DOI 10.1016/j.medengphy.2005.07.004
4. Arnold MM, Szczepanski J, Montejo N et al. (2012) Information content in cortical spike trains during brain state transitions. J Sleep Res: 22:13–21 DOI 10.1111/j.1365-2869.2012.01031.x
5. Lempel A, Ziv J (1976) On the complexity of finite sequences. IEEE Trans Inform Theory 22:75–81 DOI 10.1109/TIT.1976.1055501
6. Zhang XS, Roy RJ, Jensen EW (2001) EEG complexity as a measure of depth of anesthesia for patients. IEEE Trans Biomed Eng 48: 1424–1433 DOI 10.1109/10.966601
7. Nagarajan R (2002) Quantifying physiological data with Lempel-Ziv complexity – Certain issues. IEEE Trans Biomed Eng 49:1371–1373 DOI 10.1109/TBME.2002.804582
8. Zhou S, Zhang Z, Gu J (2011) Interpretation of coarse-graining of Lempel-Ziv complexity measure in ECG signal analysis, Proc. 33rd Annual Conference of the IEEE EMBS, Boston, USA, pp. 2716–2719
9. Radhakrishnan N, Gangadhar BN (1998) Estimating regularity in epileptic seizure time-series data. A complexity-measure approach. IEEE Eng Med Biol 17:89–94 DOI 10.1109/51.677174
10. Kolmogorov AN (1965) Three approaches to the quantitative definition of information. Infor Trans 1:3–11.
11. Hu J, Gao J, Principe JC (2006) Analysis of biomedical signals by the Lempel-Ziv complexity: the effect of finite data size. IEEE Trans Biomed Eng 53:2606–2609 DOI 10.1109/TBME.2006.883825
12. Palus M, Hoyer D (1998) Detecting nonlinearity and phase synchronization with surrogate data. IEEE Eng Med Biol Mag 17:40–45 DOI 10.1109/51.731319
13. Tononi, G (2012) Integrated information theory of consciousness: an updated account. Arch Ital Biol 150:56–90 DOI 10.4449/aib.v149i5.1388

Author: Daniel Abásolo
Institute: University of Surrey
City: Guildford
Country: United Kingdom
Email: d.abasolo@surrey.ac.uk

Heart Rate Variability in Pregnant Women before Programmed Cesarean Intervention

Juan Bolea[1], Raquel Bailón[1], Eva Rovira[2], Jose María Remartínez[2], Pablo Laguna[1], and Augusto Navarro[2]

[1] GTC, I3A, IIS Aragón, University of Zaragoza, Zaragoza, Spain, CIBER de Bioingeniería, Biomateriales y Nanomedicina, Spain
[2] Anaesthesia Service, Hospital Miguel Servet, Zaragoza, Spain, Faculty of Medicine, University of Zaragoza, Zaragoza, Spain

Abstract—Background: Heart rate variability (HRV) indices have shown ability for hypotension prediction during spinal anaesthesia in pregnant women programmed for cesarean only the same day of the surgery but not the previous day. Objective: To study changes in linear and nonlinear HRV indices of pregnant women programmed for cesarean between the previous day and the surgery day. Methods: Previous day recordings (PDR) and surgery day recordings (SDR) of 71 pregnant women programmed for cesarean have been studied during the following conditions: lateral decubitus (LD), supine decubitus (SD) and Valsalva maneuver recovery (VR). Linear HRV indices include classical temporal and spectral indices. Nonlinear HRV indices consist of sample and approximate entropy and correlation dimension (D_2). Results: Some linear HRV indices show very significant increases ($p < 0.01$) in SDR with respect to PDR: HRM in VR, SDNN in LD and SD, power in the very low frequency band in LD, power in the low frequency band in LD and SD. On the contrary, nonlinear HRV indices show a decrease in almost every condition and index in SDR with respect to PDR, being only statistically significant for D_2 in SD ($p < 0.01$). Conclusions: The increase in the former linear indices and the decrease in nonlinear indices the surgery day with respect to the previous day can be attributable to the stress induced by the imminent surgery.

Keywords—heart rate variability, pregnancy, spinal anaesthesia, prospective studies, human

I. Introduction

Heart rate variability (HRV) allows noninvasive assessment of autonomic nervous system (ANS) [1], although the exact contribution of its two branches (sympathetic and parasympathetic) is a matter of debate and research [2]. Several HRV indices allow the assessment of the impact that some pathologies have on ANS (diabetes, obesity), medium and long term risk stratification (mortality after AMI), and short term response to drugs (hypotension and hemodynamical instability with general or spinal anaesthesia [3, 4]). The former applications, as well as its potential clinical use, make analysis of HRV specially interesting.

Spinal anaesthesia is the elective technique in cesarean intervention due to its lower maternal risk and lower fetal exposition to depressant drugs than general anaesthesia, despite its high rate of hemodynamic instability and clinical hypotension (>60%) in habitual practice without pharmacological prevention [5]. Repercussion of these hypotension events on the mother and the fetus makes the prediction of this clinical situation specially relevant.

HRV has been studied in pregnant women for hypotension risk prediction after spinal anaesthesia for cesarean intervention. Predictive value has been reported for linear (ratio between power in low frequency (P_{LF}) and high frequency (P_{HF}) bands, LF/HF [6]) and nonlinear (peak correlation dimension [7] and approximate entropy [8]) parameters. In Hans et al work [6] the LF/HF index predictive value is limited to the analysis of ECGs recorded the surgery day, and not the previous day. This fact motivates the study of the differences in HRV between surgery day and the previous day.

Our hypothesis is that these hypotension events may be caused by some disorders of ANS, which can be induced by stress of the imminent surgery.

This work is designed to describe changes in different linear and nonlinear HRV indices, obtained from ECGs recorded the surgery day (SDR) and the previous day (PDR) in pregnant women programmed for elective cesarean. Recordings are made in lateral decubitus rest, in supine decubitus or hemodynamic stress test, and in the recovery period after Valsalva maneuver in sitting position.

II. Database

The database consists of the ECGs of 71 pregnant women programmed for cesarean intervention with inclusion criteria, recorded in the University Hospital Miguel Servet, Zaragoza, Spain, after giving their informed consent.

Three subjects were removed from the study, one due to sinus rhythm disruption, and the other two due to technical problems during the recording.

Study population characteristics (mean±standard deviation) of the 68 valid subjects are: gestational age 38.15±0.95 weeks; age 33.65±4.78 years; body mass index 27.95±4.37 kg/m^2; and height 162.92±6.20 cm. Indications for cesarean were: 27 for iterative cesarean; 20 breech presentation; 6 feet presentation; 2 transversal presentation; 8 placenta praevia; 1 previous uterotomy; and 4 for other causes.

Two-lead ECGs were acquired for each subject the evening before the surgery (PDR), after admission (between 19-20 pm), and just before the surgery (SDR), first thing in the morning (during 8 to 10 am) and patient fasting, in the Surgery Area, with a sampling frequency of 1000 Hz (Biopac Data Acquisition MP System). A protocol was designed which submits the pregnant women to physiological stress in order to enhance ANS alterations. Firstly, the subject was at lateral decubitus (relaxed position, minimum stress) for 7 minutes (LD); then at supine decubitus (hemodynamic stress caused by aorto-cava compression) for other 7 minutes (SD); finally, a Valsalva maneuver of 15 s is performed, and the ECG is recorded during at least one minute of the recovery (RV).

III. METHODS

Linear (temporal and spectral) and nonlinear HRV indices have been studied. First, beat occurrence time series are detected from the ECG using a wavelet-based detector [9]. Ectopic beats and misdetections are corrected applying the integral pulse frequency modulation (IPFM) model [10]. Then, two representations of HRV have been considered: the unevenly sampled RR interval series [1] and the HRV signal obtained from the IPFM model [10], sampled at 4 Hz.

Linear indices Temporal indices, namely, HRM, SDNN, RMSSD, SDSD and pNN50 are measured over RR interval series. Spectral indices are obtained from the power spectral density of the HRV signal obtained from IPFM model. Absolute and normalized powers are computed in the classical bands [1]: very low frequency (P_{VLF}, 0.015-0.04), low frequency (P_{LF}, 0.04-0.15 Hz) and high frequency (P_{HF}, 0.15-0.4 Hz) and also the ratio LF/HF. Normalized P_{LF} is also considered: $P_{LF}^n = \frac{P_{LF}}{P_{LF}+P_{HF}}$.

Nonlinear indices. Nonlinear indices include sample and approximate entropy [11, 12], SampEn and ApEn, respectively, which quantify the irregularity of a time series. Correlation dimension (D_2), which is an estimator of fractal dimension, related to the minimum number of variables needed to model the dynamics of the time series in the phase space [13] is also considered. Larger D_2 values are associated with more complex systems generating the time series. The former nonlinear indices are estimated on RR series. Their computation are based on correlation integrals, which depend on the number of points N and the threshold value/s r used [14]. In this work, $N = 300$ and , $r = 0.1$ for SampEn and ApEn and $r \in [0.01\ 1.2]$ in steps of 0.01 for D_2 are used applied on normalized to unity amplitude signal.

IV. RESULTS

Mean and standard deviation ($\mu \pm \sigma$) of HRV indices from PDR and SDR in LD, SD and VR conditions are shown in Table 1. Paired T test statistical analysis over the results obtained are made with SPSS© software (PDR vs SDR), considering a p value lower than 0.05 as significant.

In the comparison of temporal indices in SDR with respect to PDR, HRM shows no statistical differences in LD and SD, and significant increase in VR ($p < 0.01$); SDNN presents significant increase in LD ($p < 0.001$) and SD ($p < 0.01$); SDSD, RMSSD and pNN50 show significant increase ($p < 0.05$) in LD and SD.

All spectral indices display increases the surgery day with respect to the previous day in all conditions. Parameter P_{LF} shows increases highly significant ($p < 0.001$) in LD and SD, and very significant ($p < 0.01$) in VR; P_{VLF} show highly significant ($p < 0.001$) increase in LD and very significant ($p < 0.01$) in SD, while P_{HF} only shows highly significant differences ($p < 0.001$) in SD. The ratio LF/HF increases in all the conditions but with very significant differences ($p < 0.01$) only in LD and VR.

Entropy indices do not show meaningful differences between SDR and PDR in neither condition. Among nonlinear indices only D_2 shows significant differences in SD ($p < 0.01$), being lower in SDR than in PDR in all conditions.

Figure 1 displays indices P_{VLF}, P_{LF} and the ratio LF/HF and D_2 during the three studied conditions in SDR and PDR.

V. DISCUSSION

Previous day ECG recordings (PDR), made the admission evening, can be considered as basal recordings of term pregnant women, both in LD, SD and VR. Surgery day recordings (SDR), made just before the cesarean, with the patient in the surgical area, and very close in time to PDR, may reflect changes in ANS regulation induced by the stress of the imminent surgery, biological conditions due to the fasting and/or circadian fluctuations in autonomic function.

Comparing SDR with PDR, linear indices (temporal and spectral) show significant and highly significant increases in nearly all positions, reflecting a global increase in variability. Among temporal indices only HRM increases significantly ($p < 0.01$) with respect to the previous day in VR, with a notable increment with respect to LD the same day ($p < 0.001$). The other parameters (SDNN, SDSD, RMSSD, and pNN50) present significantly higher values in LD and SD with respect to the previous day, while lower values in VR (without statistical significance), attributable to the heart rate increase in this

Table 1 HRV indices represented as mean (μ), standard deviation (σ), and p-value (PDR vs SDR), those lower than 0.05 (significant) are marked in bold.

		Lateral Decubitus			Supine Decubitus			Valsalva Recovery		
		μ	σ	p_{value}	μ	σ	p_{value}	μ	σ	p_{value}
HRM [bpm]	PDR	80.20	1.10	0.574	79.93	1.20	0.359	82.48	1.21	**0.007**
	SDR	79.99	1.36		81.61	1.55		85.42	1.36	
SDNN [s]	PDR	39.55	2.46	**1e-04**	44.18	2.04	**1e-04**	49.10	2.17	0.549
	SDR	47.41	2.62		52.59	2.95		49.03	2.03	
SDSD [s]	PDR	29.66	2.50	**0.018**	27.40	1.99	**0.043**	35.02	2.44	0.222
	SDR	34.42	3.13		32.25	2.86		32.01	1.98	
RMSSD [s]	PDR	29.63	2.50	**0.018**	27.37	1.99	**0.043**	34.96	2.43	0.219
	SDR	34.38	3.13		32.22	2.86		31.96	1.97	
pNN50 [%]	PDR	9.89	1.66	**0.012**	8.04	1.26	**0.019**	14.93	1.82	0.200
	SDR	13.62	2.07		11.85	1.97		12.41	1.60	
P_{VLF} [ms^{-2}]	PDR	225.92	25.45	**1e-04**	361.09	43.19	**0.001**	405.57	51.38	0.340
	SDR	342.99	34.28		500.24	51.26		443.09	59.80	
P_{LF} [ms^{-2}]	PDR	141.70	22.39	**1e-04**	160.54	14.96	**1e-04**	242.83	22.36	**0.002**
	SDR	207.08	34.61		235.24	21.19		322.55	29.12	
P_{HF} [ms^{-2}]	PDR	147.80	25.06	0.130	110.11	13.80	**1e-04**	205.19	21.80	0.956
	SDR	179.88	36.02		141.99	17.83		218.93	24.95	
P_{LF}^n (%)	PDR	52.89	2.13	**0.023**	61.52	1.88	0.275	54.68	2.19	**0.008**
	SDR	57.72	2.06		63.43	1.92		60.20	2.02	
LF/HF	PDR	1.49	0.13	**0.009**	2.12	0.18	0.256	1.67	0.17	**0.003**
	SDR	1.94	0.19		2.46	0.23		2.08	0.19	
SampEn	PDR	1.06	0.03	0.969	0.94	0.03	0.369	0.88	0.04	0.347
	SDR	1.07	0.04		0.92	0.04		0.85	0.04	
ApEn	PDR	1.00	0.02	0.547	0.91	0.02	0.291	0.86	0.03	0.273
	SDR	0.99	0.02		0.89	0.03		0.83	0.03	
D_2	PDR	5.30	0.10	0.231	4.99	0.13	**0.006**	4.53	0.14	0.122
	SDR	5.18	0.14		4.48	0.14		4.38	0.14	

position. The increase in frequency domain indices appears in all bands (VLF, LF and HF), and shows a predominant sympathetic activity in autonomic balance, with significant increase in LF/HF ratio both in LD and VR.

Regarding nonlinear parameters, both entropies and D_2 show in SDR systematically lower values than in PDR in all conditions, only reaching statistical significance the decrease in SD ($p < 0.01$).

In situations of stress stimulus in pregnant women, an increase in LF/HF ratio has been reported [15] due to HF power decrease. However, our patients present, the day of the surgery, an increase in both bands at SDR, even though the higher increase in LF determines sympathetic predominance, as revealed by the LF/HF ratio. This difference may be explained by the fact that in Klinkenberg [15] pregnant women are submitted to a psychosocial stress, which can be considered as an isolated stress, while anxiety or fear for the imminent surgery in our work is an intangible stimulus which may last in time [16].

It has also been reported that patients with two or more anxiety symptoms present a higher basal LF/HF ratio, with less reactivity to tilt stimulus [17], while our patients present higher LF/HF ratio the day of the surgery and keep their reactivity with similar increments in SD and VR regarding LD.

Prolongued fasting (48 h.) decreases linear HRV indices during resting and during a tilt test a meaninful decrease of P_{HF}, SDNN and RMSSD have been reported by Mazurak and co-workers [18]. Besides HRV indices from circadian fluctuacion studies, as reported by Nakagawa et. al. [19], show a sinusoidal daily rhythm with acrophase or maximum values, in the middle of the night or of the day. In addition, HRM and P_{LF} fluctuations have inverse phases respect to P_{HF}. Nevertheless, the results obtained show a significant increase in basal conditions (LD) as well as during hemodynamic stress(SD), and the fact of the timing recordings (19-21 pm at PDR and 8-10 am at SDR) suggest that possible fasting and circadian fluctuations effects are masked by physicosocial stress.

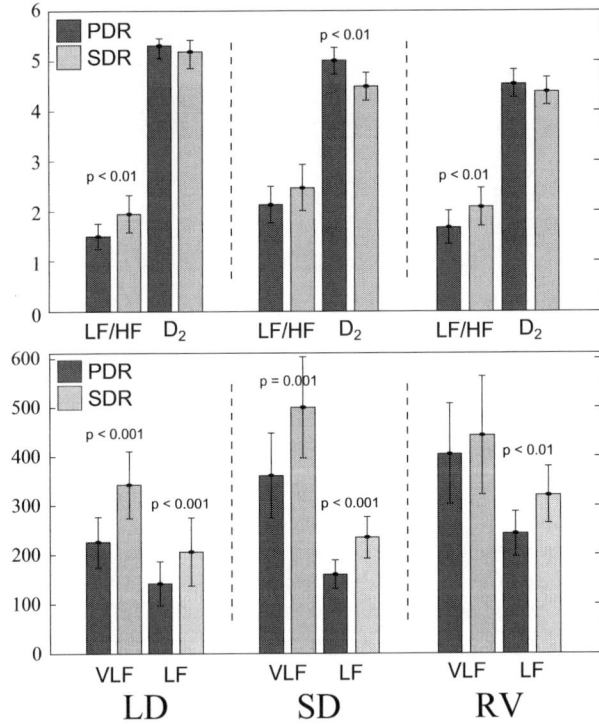

Fig. 1 Significant differences of the computed parameters at three positions (**LD**: Lateral decubitus; **SD**: Supine decubitus; **RV**: Valsalva maneuver recovery). Top panel shows LF/HF ratio and D_2 [adim]; Bottom panel shows VLF and LF [ms^{-2}].

VI. CONCLUSION

In this study HRV variability was evaluated before programmed cesarean section by non-linear and linear methods. The hypotension events that can suffer the women during surgery under spinal anesthesia was hypothesized due to disturbances in the ANS that could be detected previous hours to the surgery. In terms of pregnant women programmed for cesarean delivery there is a significant increase in linear HRV indices the day of the surgery with respect to the previous day, both in temporal indices and spectral indices with sympathetic predominance (HRM, LF/HF), as well as a decrease in nonlinear HRV indices (sample and approximate entropies and D_2), which is only statistically significant for D_2 in SD. Reported changes in the ANS that can cause hypotension episodes may be attributable to the stress of the imminent surgery.

REFERENCES

1. Task Force of the European Society of Cardiology and the North American Society of Pacing and Electrophysiology . Heart Rate Variability: standards of measurement, physiological interpretation, and clinical use *Circulation*. 1996;93:1004–1065.
2. Billman GE. The LF/HF ratio does not accurately measure cardiac sympatho-vagal balance *Front Physiol.* 2013:4–26.
3. Fujiwara Y, Asakura Y, Sato Y, Nishiwaki K, Komatsu T. Preoperative ultra short-term entropy predicts arterial blood pressure fluctuation during the induction of anesthesia *Anesth Analg.* 2007;104:853–856.
4. Hanss R, Bein B, Weseloh H, et al. Heart Rate Variability predicts severe hypotension after spinal anesthesia *Anesthesiology.* 2006;104:537–545.
5. Cyna AM, Andrew M, Emmett RS, Middleton P, Simmons SW. Techniques for preventing hypotension during spinal anaesthesia for caesarean section *Cochrane Database Syst Rev.* 2006;18:CD002251.
6. Hanss R, Bein B, Ledowski T, et al. Heart rate variability predicts severe hypotension after spinal anaesthesia for elective caeserean delivery *Anesthesiology.* 2005;102:1086–1093.
7. Chamchad D, Arkoosh VA, Horrow JC, et al. Using heart rate variability to stratify risk of obstetric patients undergoing spinal anesthesia *Anesth Analg.* 2004;99:1818–1821.
8. Ghabach MB, El-Khatib MF, Zreik TG, et al. Effect of weight gain during pregnancy on heart rate variability and hypotension during caesarean section under spinal anaesthesia *Anaesthesia.* 2012;66:1106–1111.
9. Martínez J. P., Almeida R., Olmos S., Rocha A. P., Laguna P.. Wavelet-based ECG delineator: evaluation on standard databases *Trans on Biomed Engin.* 2004;51:570-581.
10. Mateo J., Laguna P.. Improved Heart Rate Variability Signal Analysis from the Beat Occurrence Times According to the IPFM Model *Trans on Biomed Engin.* 2000;47:985-996.
11. Pincus S.M., Gladstone I.M., Ehrenkranz R.A.. A regularity statistic for medical data analysis *J Clin Monitor.* 1991;7:335-345.
12. Richman J.S., Moorman J.R.. Physiological time-series analysis using approximate entropy and sample entropy *AM J Physiol-Heart C.* 2000;278:H2039-H2049.
13. Grassberger P., Procaccia I.. Characterization of strange attractors *Phys Rev Lett.* 1983;50:346-349.
14. Canga L., Navarro A., Bolea J., Remartínez J.M., Laguna P., Bailón R.. Non-Linear Analysis of Heart Rate Variability and its Application to Predict Hypotension during Spinal Anesthesia for Cesarean Delivery in *Computing in Cardiology 2012, VOL 39*;39 of *Comp. in Cardiol*:413-416 2012.
15. Klinkenberg AV, Nater UM, Nierop A, Bratsikas A, Zimmermann R, Ehlert U. Heart rate variability changes in pregnant and non-pregnant women during standardized psychosocial stress *Acta Obstet Gynecol Scand.* 2009;88:77–82.
16. Kreibig SD. Autonomic nervous system activity in emotion: A review *Biological Psychology.* 2010;84:394–421.
17. Piccirillo G, Elvira S, Bucca C, Viola E, Cacciafesta M, Marigliano V. Abnormal passive head-up tilt test in subjects with symptoms of anxiety power spectral analysis study of heart rate and blood pressure *Int J Cardiol.* 1997;60:121-131.
18. Mazurak N., Günther A., Grau F.S., et al. Effects of a 48-h fast on heart rate variability and cortisol levels in healthy female subjects *Eur. J. Clin. Nutr..* 2013;67:401-406.
19. Nakagawa M., Iwao T., Ishida S., et al. Circadian rhythm of the signal averaged electrocardiogram and its relation to heart rate variability in healthy subjects *Heart.* 1998;79:493-496.

Author: Juan Bolea
Institute: University of Zaragoza; Street: Mariano Esquillor S/N
Email: jbolea@unizar.es

Fractal Changes in the Long-Range Correlation and Loss of Signal Complexity in Infant's Heart Rate Variability with Clinical Sepsis

E. Godoy[1], J. López[2], L. Bermúdez[2], A. Ferrer[1], N. García[2], C. García Vicent[2], E.F. Lurbe[2,3], and J. Saiz[1]

[1] I3BH, Universidad Politècnica de Valencia, Valencia, Spain
[2] Consortium-Hospital General Universitario de Valencia, Universidad de Valencia, Valencia, Spain
[3] CIBERobn. Instituto de Salud Carlos III

Abstract—Sepsis, a critical bacterial infection of the bloodstream, is a frequent cause of illness and death in premature infants in the neonatal intensive care units. A prospective analysis was conducted of the inter–beat-intervals time series (IBI) derived from 1 hour 30 minutes electrocardiogram (ECG) recordings for 14 episodes of clinical sepsis in 90 infants 12-18 hour old in the maternity ward of the Consorcio-Hospital General Universitario de Valencia. The aim was to test the hypothesis that normal beat-to-beat fluctuations in heart rate show fractal long-range correlations and that pathological condition reduce the adaptive capacity of the individual degrading the information carried by their signals and modifying the fractal scaling characteristics. Fractal properties of these series were evaluated by applying the method of Detrended Fluctuation Analysis (DFA) for the quantification of the correlation property in the highly non-stationary IBI time series. The main finding is that heart rate time series from infants with sepsis show a breakdown of the long-range correlation behaviour and consequently complexity-loss. The long-range scaling exponents for the sepsis cases showed a statistically significant deviation (p<0.001) from the long-range scaling exponents of the healthy cases. This method may be of use in distinguishing healthy from pathologic time series of infants with sepsis based on differences in the fractal scaling property, and could be introduced into the clinical practice for the assessment of heart rate patterns of new born infants.

Keywords—sepsis, heart rate variability, detrended fluctuation analysis, scaling exponents, fractal.

I. INTRODUCTION

Sepsis, a critical bacterial infection of the bloodstream, is a frequent cause of illness and death in premature infants in the neonatal intensive care units. The early signs of sepsis are neither specific for the disease, nor very sensitive, and infants are often not diagnosed until they are very ill [1].

Diagnostic testing by blood cultures is usually reserved until signs of illness appear, and even so they have a high false-negative rate, particularly when based on small amounts of blood that can be securely removed and the length of time before results are accessible [2][3]. Early diagnosis, before clinical signs of illness is an important objective but difficult to achieve [4].

During the course of a sepsis, generally before the clinical diagnosis is attainable there exist changes in the cardiac rhythm characterized by reduced variability and transient decelerations. Some authors have interpreted those changes as an alteration of the mechanisms underlying the control of the cardiac rhythm [5][6][7].

The nonlinear interaction between the two branches of the autonomic nervous system, parasympathetic and sympathetic, is the assumed mechanism for the type of unpredictable heart rate variability verified in healthy subjects.

Other studies show that under healthy conditions, interbeat-interval (IBI) time series exhibit long-range power-law correlations classically showed by dynamical systems far from equilibrium [8][9].

On the other hand, certain disease states may be going together with alterations in this fractal (scale invariant) correlation property.

II. METHODS

To test for the statistical significance in using the DFA method in discriminating between healthy neonates and sepsis diagnosed neonates we analyzed the IBI derived from 1 hour and 30 minutes ECG recordings collected by using the General Electric-Dash 3000/4000® bedside monitors obtained from 14 sepsis diagnosed infants and 14 healthy infants at the Consortium-Hospital General Universitario de Valencia.

ECG recordings were processed both manually and automatically using computerized beat recognition algorithms and the IBI time series generated were preprocessed for ectopic interval detection and correction (Fig. 1). The average length of the IBI time series was $N = 12,062$ beats ($\pm 1,922$ beats).

The fractal properties of the data were evaluated by applying the DFA. As detailed elsewhere in a number of contributions [10][11], the IBI time series were integrated as follows:

$$Y(k) = \sum_{i=1}^{k}[B(i) - B_{ave}] \qquad (1)$$

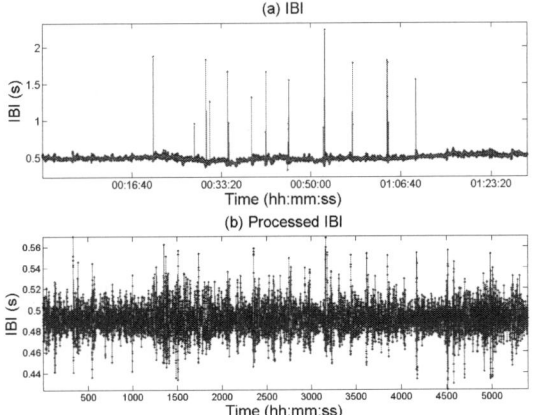

Fig. 1 IBI time series. Ectopic interval detection (a). IBI after ectopic interval correction (b)

where $Y(k)$ is the kth value of the resulting integration ($k = 1, 2, \ldots, N$), $B(i)$ is the ith sample, and B_{ave} is the mean value of the series of length N. Subsequently, the integrated series $Y(k)$ were divided into subsets of independent segments, or boxes, having equal number of n samples. The trends were locally subtracted from $Y(k)$, so reducing the non-stationary artifacts. For each series, the average root-mean-square fluctuations $F(n)$ were then calculated for different values of n (i.e. different time scales):

$$F(n) = \sqrt{\frac{1}{N}\sum_{k=1}^{N}[Y(k) - Y_n(k)]^2} \quad (2)$$

Finally, the relationship on a double-log graph between fluctuations $F(n)$ and time scales n were approximately evaluated by a linear model indicating that $F(n)$ is proportional to n^α, where α is the single exponent describing the correlation properties of the entire range of time scales.

In some cases it was found that the DFA plot shows a not rigorously linear form but rather two different linear regions with dissimilar slopes separated at a break point, (Fig. 2). This observation suggests there is a short range scaling exponent, α_1, and a long-range scaling exponent, α_2, over longer periods.

A complete uncorrelated process, were the value at one inter-beat-interval is completely uncorrelated from any previous values, (e.g. white noise) corresponds to an $\alpha = 0.5$. An α greater than 0.5 and less than or equal to 1.0 indicates persistent long-range power-law correlations such that a large inter-beat-interval is more likely to be followed by a large inter-beat-interval and a short one being followed by a short interval. In contrast, $0 < \alpha < 0.5$ indicates a different type of power-law correlation such that large and small values of the time series are more likely to alternate. The special case of $\alpha = 1$ corresponds to $1/f$ noise. For $\alpha \geq 1$, correlations still exist, but they are not of a power-law form anymore; $\alpha = 1.5$ indicates Brown noise, the integration of white noise. $1/f$ noise can be interpreted as a tradeoff between the complete randomness of white noise (very irregular behavior) and the very smooth behavior of Brownian noise [10].

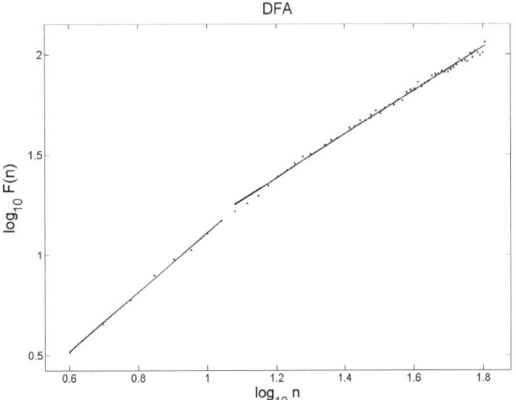

Fig. 2 Detrended Fluctuation Analysis (DFA) of a healthy infant ($N = 12,000$ beats). Double-log graph between F(n) vs. n and the linear model evaluation. The plot shows the break point and the short-range scaling exponent (α_1) and the long-range scaling exponent (α_2)

The Fig. 3 compares the DFA analysis of two 24 hour long inter-beat-interval of a healthy infant and a sepsis diagnosed infant. The healthy subject inter-beat-interval shows almost perfect power-law scaling exponent with $\alpha = 1.0$ (i.e., $1/f$ noise) and for the pathological case $\alpha = 1.3$ (closer to Brownian noise); note the different scaling behavior of the pathological case showing a reverse crossover. This result is consistent with previous work in comparing DFA analysis for healthy and pathological subjects [10].

Statistical comparison between groups was conducted using Student's t test. A $p < 0.05$ was considered statistically significant.

III. RESULTS

The two exponents, short range scaling exponent α_1 and long-range scaling exponent α_2, were calculated for each data set of length $N = 12,000$ beats (~1 hour and 30 minutes). Longer data sets were segmented into multiple subsets (each with 12,000 beats).

For healthy infants we found exponents for the inter-beat-intervals time series: $\alpha_1 = 1.334 \pm 0.1202$ and $\alpha_2 = 1.180 \pm 0.0857$ (mean value ± standard deviation). For the group with the diagnostic of sepsis we found $\alpha_1 = 0.999 \pm 0.082$.

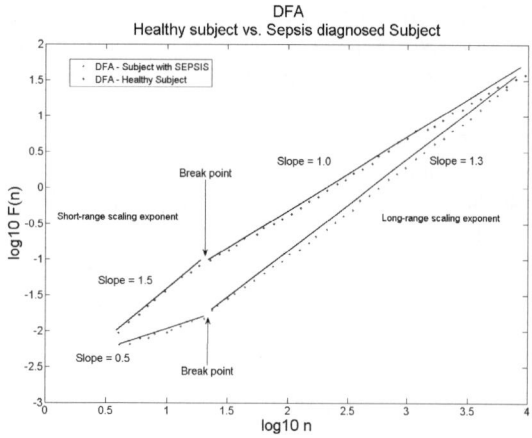

Fig. 3 Detrended Fluctuation Analysis of two 24 hour cases of a healthy infant and a sepsis diagnosed infant. Long-range scaling exponent ($\alpha_{2=1.0}$) for the healthy case and long-range scaling exponent ($\alpha_{2=1.3}$) for the pathological case.

and $\alpha_2 = 1.281 \pm 0.084$ (mean value ± standard deviation), both α_1 and α_2 significantly different from normal with a Student's t test result of $p<0.001$, see Table 1.

Fig. 4 shows that fairly good discrimination can be achieved between these two groups using these scaling exponents. All 14 out of 14 healthy cases show crossover and 14 out of 14 sepsis cases showed reverse crossover.

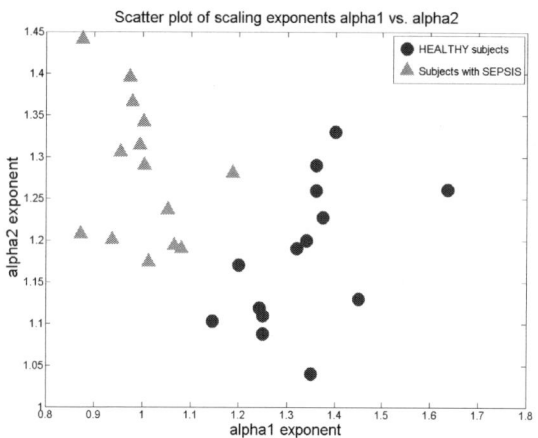

Fig. 4 Scatter plot of scaling exponent's α_1 and α_2 for the healthy subjects and subjects with diagnostic of sepsis. The scaling exponents α were calculated from inter-beat-intervals time series of length 12,0000 beats. Note good separation between healthy subjects and sepsis diagnosed subjects with points clustered in two distinct groups

Table 1: Scaling exponents for healthy and sepsis diagnosed subjects measurements.

Healthy subjects		Subjects with diagnostic of sepsis	
Scaling Exponents			
α_1	α_2	α_1	α_2
1,250	1,088	1,065	1,194
1,635	1,261	1,080	1,190
1,374	1,228	1,187	1,281
1,145	1,103	1,012	1,174
1,243	1,119	0,876	1,441
1,200	1,170	0,979	1,366
1,250	1,110	0,955	1,306
1,360	1,260	1,002	1,342
1,320	1,190	0,974	1,396
1,340	1,200	1,052	1,237
1,400	1,330	1,003	1,290
1,450	1,130	0,994	1,314
1,350	1,040	0,938	1,201
1,360	1,290	0,872	1,208
Mean values and Standard Deviations			
1,334	1,180	0,999	1,281
0,1202	0,0857	0,082	0,084

IV. CONCLUSIONS

We showed that the Detrended Fluctuation Analysis is capable of identifying crossover with different patterns for healthy infants and sepsis diagnosed infants. We note also that data from the normal inter-beat-intervals time series of healthy infants are slightly clustered around the long-range scaling exponent $\alpha_2 = 1.0$ (power-law scaling) and differently the inter-beat-intervals time series from subjects with diagnostic of sepsis have long-range scaling exponent near $\alpha_2 = 1.3$ (closer to Brownian noise), and showing more variation, probably associated to different clinical situations and varying severity of the pathological state.

This apparent loss of fractal organization of the heart dynamics in the cases of sepsis may reflect the degradation of the physiological regulatory system (i.e. neuroautonomic system).

This method may be of use in discriminating healthy from pathologic time series of infants with sepsis based on differences in this scaling property and complexity-loss, and introduced into the clinical practice for the assessment of heart rate patterns of new born infants.

Furthermore, this finding may be of interest from the practical point of view since it could be the method of fluctuation analysis for the development of new devices to be used in bedside and ambulatory monitoring of new born infants.

ACKNOWLEDGMENT

This work was partially supported by the "VI Plan Nacional de Investigación Científica, Desarrollo e Innovación Tecnológica" from the Ministerio de Economía y Competitividad of Spain (TIN2012-37546-C03-01).

We wish to thank the technical support provided by Nuubo (Smart Solutions Technologies, S.L.).

REFERENCES

1. Griffin, M.P., et al., Heart rate characteristics and clinical signs in neonatal sepsis. Pediatr Res, 2007. **61**(2): p. 222-7.
2. Moorman, J.R., et al., Cardiovascular oscillations at the bedside: early diagnosis of neonatal sepsis using heart rate characteristics monitoring. Physiological Measurement, 2011. **32**(11): p. 1821.
3. Kellogg, J.A., et al., Frequency of low level bacteremia in infants from birth to two months of age. Pediatr Infect Dis J, 1997. **16**(4): p. 381-5.
4. Fanaroff, A.A., et al., Incidence, presenting features, risk factors and significance of late onset septicemia in very low birth weight infants. The National Institute of Child Health and Human Development Neonatal Research Network. Pediatr Infect Dis J, 1998. **17**(7): p.593-8.
5. Richman, J.S. and J.R. Moorman, *Physiological time-series analysis using approximate entropy and sample entropy.* American Journal of Physiology - Heart and Circulatory Physiology, 2000. **278**(6): p. H2039-H2049.
6. Griffin, M.P. and J.R. Moorman, Toward the Early Diagnosis of Neonatal Sepsis and Sepsis-Like Illness Using Novel Heart Rate Analysis. Pediatrics, 2001. **107**(1): p. 97-104.
7. Lake, D.E., et al., *Sample entropy analysis of neonatal heart rate variability.* Am J Physiol Regul Integr Comp Physiol, 2002. **283**(3): p. R789-97.
8. Bak, P., C. Tang, and K. Wiesenfeld, *Self-organized criticality: An explanation of the 1/f noise.* Phys Rev Lett, 1987. **59**(4): p. 381-384.
9. Stanley, H.E., *Introduction to phase transitions and critical phenomena.* Introduction to Phase Transitions and Critical Phenomena, by H Eugene Stanley, pp. 336. Foreword by H Eugene Stanley. Oxford University Press, Jul 1987. ISBN-10: 0195053168. ISBN-13: 9780195053166, 1987. **1**.
10. Peng, C.K., et al., Quantification of scaling exponents and crossover phenomena in nonstationary heartbeat time series. Chaos, 1995. **5**(1): p. 82-7.
11. Iyengar, N., et al., Age-related alterations in the fractal scaling of cardiac interbeat interval dynamics. Am J Physiol, 1996. **271**(4 Pt 2): p. R1078-84.

Author: Eduardo J. Godoy
Institute: I3BH, UPV
Street: Camino de Vera, s/n. (46022)
City: Valencia
Country: Spain
Email: egodoy@gbio.i3bh.es

Estimation of Coupling and Directionality between Signals Applied to Physiological Uterine EMG Model and Real EHG Signals

A. Diab[1,3], M. Hassan[2], J. Laforêt[1], B. Karlsson[3], and C. Marque[1]

[1] Université de Technologie de Compiègne, UMR CNRS 7338, Biomécanique et Bioingénierie, Compiègne, France
[2] Université de Rennes1, Campus de Beaulieu, INSERM, Laboratoire Traitement du Signal et de L'Image, Rennes, France
[3] School of Science and Engineering, Reykjavik University, Reykjavik, Iceland

Abstract—Several measures have been proposed to detect the strength and direction of relationships between biosignals. In this paper we study two nonlinear methods that are widely used in functional and effective connectivity analysis: nonlinear correlation coefficient (h^2) and general synchronization (H). The performance of both methods is tested on two dimensional coupled synthetic nonlinear Rössler model. The best method for these synthetic Rössler signals is then applied for the first time to signals generated by a physiological uterine EMG model. It was then applied to real uterine EMG recorded during pregnancy and labor in order to compare their synchronization and propagation direction. The best performing method was tested on the physiological model data, and the results obtained on real signals are encouraging. It may possibly be of use for solving the open questions of the relationships between uterine EMG at different places on the uterus, and may provide a way to localize their sources.

Keywords—functional and effective connectivity, uterine EMG.

I. INTRODUCTION

In most coupled data and systems, such as neuronal areas in human body, the intrinsic and internal variants and the interdependencies among their subsystems are not accessible. Therefore, in order to quantify this interdependencies and its directions, attempts have been made through measuring the relationships between system outputs, represented mostly by time series [1]. As data are practically always nonlinear [2] the application of traditional techniques such as cross-spectrum and cross-correlation analysis [3] has its limitations.

The hypothesis that drives this study is that labor is associated with a great coupling between uterine cells, inducing thus a synchronized uterus that should generate efficient contractions (the whole uterus contracts) in order to push the baby out. This coupling should be evidenced from uterine electromyogram (EMG) by using non linear signal processing tools [4].

The objectives of this paper are to provide evidence for the usefulness and appropriateness of nonlinear directional and coupling analysis for synthetic, simulated and real uterine EMG signals. Two nonlinear methods, (h^2) [5] and (H) [6] were applied to pure synthetic two dimensional nonlinear coupled Rössler model. The method that yields the best result on Rössler data is then tested for the first time by using a uterine EMG physiological model. This method is then applied to a dataset of real uterine EMG measured in pregnancy and labor to see what happens to synchronization and direction when going from pregnancy to labor.

The results on synthetic signals show a superiority of h^2 over H method. h^2 was then tested on physiological uterine EMG model data by using it to estimate the matrix of correlation and to map the direction between all channels. Finally h^2 is used to detect the correlation and the direction between the whole set of channels of a real uterine EMG signals. The results show that when going from 3 weeks before labor (WBL) to labor, synchronization appears more regular, and direction seems more concentrated toward cervix.

II. MATERIAL AND METHODS

A. Data

1. Synthetic Signals

To evaluate our coupling and direction measures, we applied them to a two dimensional coupled Rössler nonlinear stationary model defined as:

$$\begin{aligned}\frac{dx_1}{dt} &= -w_x x_2 - x_3 \\ \frac{dx_2}{dt} &= -w_x x_1 + a x_2 \\ \frac{dx_3}{dt} &= b + x_3(x_1 - c) \\ \frac{dy_1}{dt} &= -w_y y_2 - y_3 + C(x_1 - y_1) \\ \frac{dy_2}{dt} &= w_y y_1 + a y_2 \\ \frac{dy_3}{dt} &= b + y_3(y_1 - c)\end{aligned} \quad (1)$$

where a=0.15, b=0.2 and c=10. The factor C permits us to control the coupling strength between the two oscillators. The two signals that we use to evaluate the relation are x_1 and y_1. We used w_x = 0.95 and w_y = 1.05 which are the values used in [7].

2. Physiological Model

We apply our methods to the electrophysiological model developed by our team which aims at describing the multiscale evolution of the uterine electrical activity, from its genesis at the cell level, then to its propagation at the myometrium level, and up to its projection through the volume conductor tissues to the abdominal surface [8].

3. Real Signals

The methods used here are "bivariate" in that we used all the combinations between the 16 monopolar channels of a 4x4 recording matrix located on the women's abdomen (see [9] for details). The signals were recorded during pregnancy and in labor at the Landspitali University Hospital in Iceland, following a protocol approved by the relevant ethical committee (VSN 02-0006-V2) with sampling frequency of 200 Hz. The uterine EMG signals were segmented manually to extract uterine activity bursts, then filtered using a CCA-EMD method developed by our team [10]. After segmentation and filtering, we got 84 labor bursts and 22 bursts at 3 WBL (Weeks Before Labor). The analysis below was applied to these segmented uterine bursts.

B. Methods

1. Nonlinear Correlation Coefficient (h^2)

The underlying idea is that if the value of X is considered as a function of the value of Y, the value of Y given X can be predicted according to a nonlinear regression curve. The nonlinear correlation coefficient between demeaned signals X and Y is then calculated as follows (for more details see [11]):

$$h_{Y|X}^2 = \frac{\sum_{k=1}^{N} Y(k)^2 - \sum_{k=1}^{N}(Y(k)-f(X_i))^2}{\sum_{k=1}^{N} Y(k)^2} \quad (2)$$

where $f(X_i)$ is the linear piecewise approximation of the nonlinear regression curve. The estimator $h^2_{Y|X}$ ranges from 0 (Y is independent of X) to 1 (Y is fully determined by X). For a nonlinear relationship, $h^2_{x|y} \neq h^2_{y|x}$ and the difference $\Delta h^2 = h^2_{x|y} - h^2_{y|x}$ indicates the degree of asymmetry of the nonlinear coupling. The delay at which the maximum value for h^2 is obtained is used as an estimate of the time delay between the signals, so we can also obtain the difference $\Delta \tau = \tau_{x|y} - \tau_{y|x}$. On combining the information of asymmetry and of time delay in coupling, the following direction index has been recently proposed by Wendling et al. [12] to provide a robust measure of the direction of coupling:

$D_{x|y} = \frac{1}{2}[\text{sgn}(\Delta h^2) + \text{sgn}(\Delta \tau)]$.

If $D_{x|y} = +1$ (or -1) so $Y \rightarrow X$ (or $Y \rightarrow X$). $D_{x|y} = 0$ indicates a bidirectional ($Y \leftrightarrow X$) coupling between signals.

2. General Synchronization (H)

From time series measured in two systems X and Y, let us reconstruct delay vectors $x_n=(x_n, ..., x_{n-(m-1)\tau})$ and $y_n=(y_n, ..., y_{n-(m-1)\tau})$, where $n=1,...,N$; m is the embedding dimension and τ denotes the delay time. Let $r_{n,j}$ and $s_{n,j}$, $j=1,..., k$, denote the time indices of the k nearest neighbors of x_n and y_n,

respectively. For each x_n, the squared mean Euclidean distance to its k neighbors is defined as:

$$R_n^{(k)}(X) = \frac{1}{k}\sum_{j=1}^{k}(x_n - x_{r_{n,j}}) \quad (3)$$

The Y-conditioned squared mean Euclidean distance is:

$$R_n^{(k)}(X|Y) = \frac{1}{k}\sum_{j=1}^{k}(x_n - x_{s_{n,j}})^2 \quad (4)$$

Thus, a nonlinear interdependence measure can be defined accordingly [11]:

$$H^{(k)}(X|Y) = \frac{1}{N}\sum_{n=1}^{N}\log\frac{R_n(X)}{R_n^{(k)}(X|Y)} \quad (5)$$

where $R_n(X)$ is the average distance of a vector x_n to all the other vectors. This measure is close to zero if X and Y are independent, while it is positive if nearness in Y implies also nearness in X for equal time partners. The nonlinear interdependence is an asymmetric measure, in the sense that $H(X|Y) \neq H(Y|X)$ so $H(X|Y) > H(Y|X)$, if $X \rightarrow Y$.

III. RESULTS

A. Results on Synthetic Signals

We try here to test the sensitivity of these methods to direction changes. So, we apply them to synthetic signals generated by the bivariate Rössler model cited above with 0.5 as coupling degree. The given results are the mean of 30 monte-carlo trials.

The length of both generated signals is 6000 points. Initially, the direction is from signal x_1 to y_1 for the whole signal duration. To change the direction, we cut the signals in three equal parts of 2000 points. We keep the two extreme parts of both signals and we exchange their middle parts. So the direction is from x_1 to y_1 for the first and the third part, and from y_1 to x_1 for the middle part, as indicated on Figure 1 by the arrows and by the reference curve. We compute the synchronization and the direction from x_1 to y_1 along the obtained signals. Results are plotted in Figure 1. We see in Figure 1 the superiority of h^2 over H because the synchronization measure given by h^2 in the middle part (2000-4000) is smallest than the ones obtained for the extreme parts which coincide with the reference curve. This is normal since the coupling in the middle part is from y_1 to x_1 (due to the exchange) and the measure of synchronization is computed from x_1 to y_1. But H gives an opposite results which is erroneous. This erroneous result is clearer in the plot of direction, since 1 indicates a direction from x_1 to y_1 and 0 the opposite direction. It is clear that h^2 gives the right direction for all parts while H gives the opposite of the true directions.

B. Results on the Electrophysiological Model

Figure 2 presents the test of the best method h^2 on synthetic signals by applying it to the physiological model data.

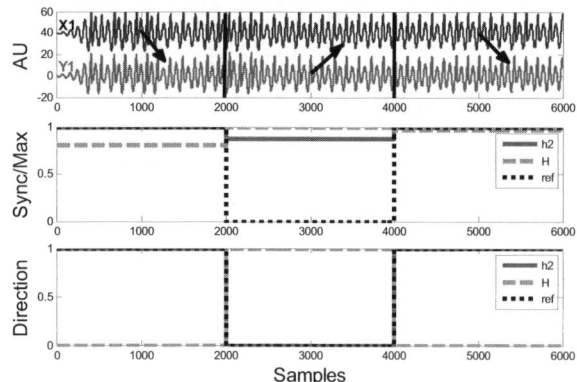

Fig. 1 Sensitivity of methods to the change of direction between signals. (top) signal generated by Rossler model after the exchange of direction; (middle) synchronization estimation and (bottom) direction estimation. Black curve is the reference curve

We apply h^2 once on signal with a planar and circular propagation. After verification of their Gaussian distribution, the results of matrix of correlation are thresholded by $m+2\sigma$, where m is the mean of the correlation matrix and σ its standard deviation, in order to remove the least significant couplings between channels and to increase figure readability. As we see for the map of direction in case of planar propagation, all the arrows reveal the longitudinal propagation of action potential (left to right) illustrated in the top of Figure 2, but with a bidirectionality between most of the electrodes. The correlation between the channels is distributed around the diagonal in the h^2 matrix.

In the case of circular propagation, the direction of arrows is toward the center with some shifting to the left and to the top. Here the correlation between channels is distributed in the whole matrix.

C. Results on Real Uterine EMG Signals

After validation of h^2 on simulated data, h^2 is applied then to real uterine EMG bursts recorded at 3 WBL (22 contractions) and labor (84 contractions). The data presented in Figure 3 are the mean over all available contractions. The results are also thresholded as indicated above.

In Figure 3 we compare the matrix of correlation and the map of direction of two gestation stages, 3 WBL and labor. It is clear that the uterine cells are less synchronized at 3 WBL than during labor. Indeed the labor matrix of synchronization is more regular then the 3 WBL one. The row and the column represented by channel 9 divide the matrix of synchronization during labor in four more or less active regions where we see a high interaction and synchronization between the channels in each region.

In terms of direction, we see in Figure 3 that at 3 WBL there are arrows in all the directions (propagation in the entire matrix) but without any dominant direction. During

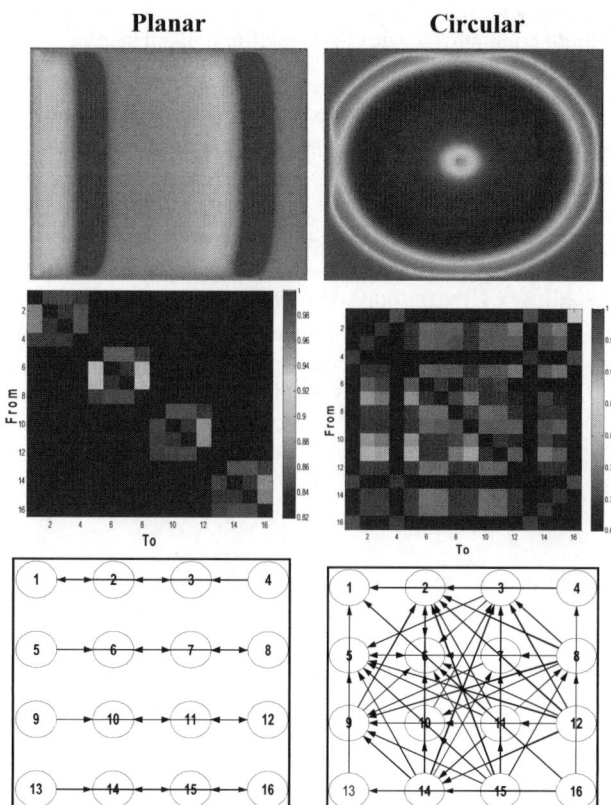

Fig. 2 Propagation wave of action potential (top), quantity (middle) and direction (bottom) of information flow between the 16 monopolar channels of uterine EMGs simulated by the physiological model: in case of planar propagation (left) and circular propagation (right) by using h^2 method.

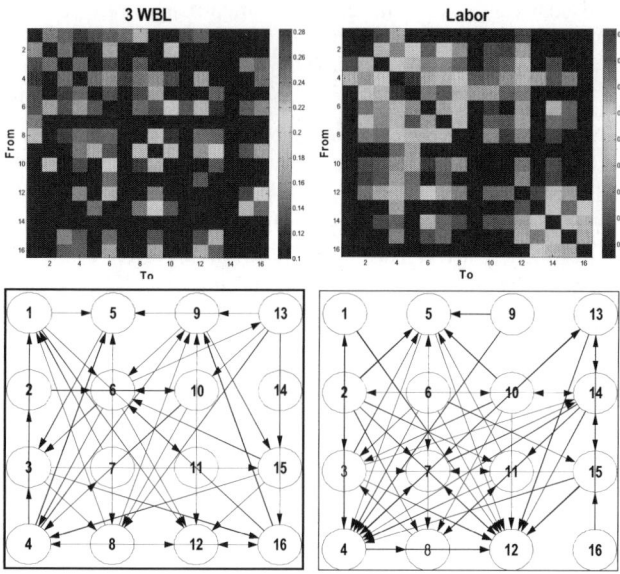

Fig. 3 Quantity (top) and direction (bottom) of information flow between the 16 monopolar channels of uterine EMG recorded at 3 WBL (left) and during labor (right) by using the h^2 method.

labor, the arrows are also in all the directions but with a dominant direction towards the bottom of the matrix.

IV. DISCUSSION

Two methods of relationships and directionality estimation (h^2 and H) are tested on bivariate Rössler synthetic signals in order to test their sensitivity to direction change. h^2 method is able to detect the correct correlation in function of direction change, as well as the right direction, whereas H method fails to detect the correct connectivity and gives the opposite of the true direction.

Since h^2 gives the best result on synthetic signal we validate it on simulated data generated by a physiological uterine EMG model, which is used for the first time in this paper. This model simulates signals closer to real signals than the Rössler model. In planar case h^2 evidences the correct longitudinal direction related to the signal propagation, but with bidirectionality between the electrodes, due to the simultaneous presence of two waves under the electrode matrix. In the circular case, the direction of arrows is towards the center with some shifting to the left and the top. The concentration to the center is due to the fact that the previous action potential should explain the next action potential generated at the center, as seen in the top of Figure 2. The shifting comes from the edge effects. Indeed, the action potential encounters the left and the top edges before the right and the bottom ones.

The synchronization between channels is restricted to the diagonal in planar case because in this case the propagation is only longitudinal. So the connection is only in one direction (between columns). In the case of circular propagation the synchronization is propagated trough the whole matrix. In this case, the electrodes are recruited by the wave in all the directions at the same time. So the synchronization simultaneously spreads to the whole matrix.

Finally we compare the matrix of synchronization and the map of direction for signals recorded at 3WBL and during labor by using only h^2. Results show an increase and a more regular synchronization (4 highly correlated regions) and a more concentration of direction toward the cervix for labor than for 3WBL, which is coherent with our previous studies.

Our ongoing work is to improve the performance of h^2 regarding the nonstationarity of the signals (such as computing h^2 in predefined segments). We are also working on extracting parameters from the connectivity graphs by using a graph theory based approach (considering the electrodes as node and the connectivity lines as edges of the graph). Different parameters such as density of graphs, ingoing and outgoing edges for each node can be extracted which may help to get new discriminant parameters between pregnancy and labor contractions.

V. CONCLUSION

We performed a comparison of two synchronization measures on synthetic signal in order to test their sensitivity to direction change. The best method (h^2) was then tested on physiological uterine model data. This is the first time that a multilevel physiological model is used in this manner. The methods were then applied to real uterine electrical bursts recorded at two pregnancy terms. The results indicate that the synchronization was spread in the whole matrix and in all directions and is more regular and more concentrated toward the cervix when approaching to labor. We will next attempt to validate these results on more data.

ACKNOWLEDGMENT

The authors wish to thank the French Ministry of Research, French ministry of foreign affair and the EraSysBio+ program for funding this research.

REFERENCES

1. M. Ahmadlou, H. Adeli, "Visibility graph similarity: A new measure of generalized synchronization in coupled dynamic systems". *Physica D: Nonlinear Phenomena*, 2011. **241**(4): p. 326–332.
2. M. Hassan, J. Terrien, B. Karlsson, C. Marque, "Comparison between approximate entropy, correntropy and time reversibility: Application to uterine electromyogram signals". *Medical engineering & physics*, 2011. **33**(8): p. 980-986.
3. P.F. Panter, "Modulation, noise, and spectral analysis: applied to information transmission". 1965: *McGraw-Hill New York*.
4. M. Hassan, J. Terrien, B. Karlsson, C. Marque, "Spatial analysis of uterine EMG signals: evidence of increased in synchronization with term". in *Conf Proc IEEE Eng Med Biol Soc*. 2009, p. 6296-6299.
5. J. Pijn, and F. Lopes Da Silva, "Propagation of electrical activity: nonlinear associations and time delays between EEG signals". *Basic Mechanisms of the EEG*, 1993: p. 41-61.
6. J. Arnhold, P. Grassberger, K. Lehnertz, CE. Elger, "A robust method for detecting interdependences: application to intracranially recorded EEG". *Physica D: Nonlinear Phenomena*, 1999. **134**(4): p. 419-430.
7. M.G. Rosenblum, A.S. Pikovsky, and J. Kurths, "Phase synchronization of chaotic oscillators". *Phys Rev Lett*, 1996. **76**(11): p. 1804-1807.
8. J. Laforet, C. Rabotti, J. Terrien, M. Mischi, C. Marque, *Toward a Multiscale Model of the Uterine Electrical Activity*. Biomedical Engineering, IEEE Transactions on, 2011. **58**(12): p. 3487-3490.
9. B. Karlsson, J. Terrien, V. Gudmundsson, T. Steingrimsdottir, and C. Marque, "Abdominal EHG on a 4 by 4 grid: mapping and presenting the propagation of uterine contractions". in *11th Mediterranean Conference on Medical and Biological Engineering and Computing*. 2007. Ljubljana, Slovenia. p. 139-143.
10. M. Hassan, S. Boudaoud, J. Terrien, B. Karlsson, and C. Marque, "Combination of Canonical Correlation Analysis and Empirical Mode Decomposition applied to denoise the labor electrohysterogram". *IEEE Trans Biomed Eng*, 2011 .**85**(9): p. 2441-2447.
11. E. Pereda, R.Q. Quiroga, and J. Bhattacharya, "Nonlinear multivariate analysis of neurophysiological signals". *Progress in Neurobiology*, 2005. **77**(1-2): p. 1-37.
12. F. Wendling, F. Bartolomei, J.J. Bellanger, and P. Chauvel, "Interpretation of interdependencies in epileptic signals using a macroscopic physiological model of the EEG". *Clin Neurophysiol*, 2001. **112**(7): p. 1201-18.

Author: Ahmad Diab
Institute: University of technology of Compiègne.
Street: Rue Dr. Schweitzer
City: Compiègne
Country: France
Email: ahmad.diab@utc.fr

Dynalets: A New Tool for Biological Signal Processing

J. Demongeot, O. Hansen, and A. Hamie

AGIM CNRS FRE 3405, Faculty of Medicine, University J. Fourier of Grenoble, 38B700 La Tronche, France

Abstract—The biological information coming from electro-physiologic signal sensors needs compression for an efficient medical use or for retaining only the pertinent explanatory information about the mechanisms at the origin of the recorded signal. When the signal is periodic in time and/or space, classical compression procedures like Fourier and wavelets transforms give good results concerning the compression rate, but provide in general no additional information about the interactions between the elements of the living system producing the studied signal. Here, we define a new transform called Dynalets based on Liénard differential equations susceptible to model the mechanism at the source of the signal and we propose to apply this new technique to real signals like ECG.

Keywords—Fourier transform, Wavelets, Dynalets, Signal Processing, ECG.

I. Introduction

There are different manners to represent a biological signal aiming to both i) explain the mechanisms having produced it and ii) facilitate its use in medical applications. The biological signals come from electro-physiologic signal sensors like ECG and have to be compressed for an efficient medical use by clinicians or to retain only the pertinent explanatory information about the mechanisms at the origin of the recorded signal for the researchers in life sciences. When the signal is periodic in time and/or space, the classical compression processes like Fourier and wavelets transforms give good results concerning the compression rate, but bring in general no supplementary information about the interactions between elements of the living system producing the studied signal. Here, we define a new transform called Dynalets based on Liénard differential equations, susceptible to model the mechanism that is the source of the signal and we propose to apply this new technique to real signals like ECG.

II. Fourier and Haley Wavelet Transforms

1) The Fourier transform comes from the aim by J. Fourier to represent in a simple way the functions used in physics, notably in the heat propagation (in 1807, cf. [1]). He used a base of functions made of the solutions of the simple pendulum differential equation (cf. a trajectory in Fig. 1): $dx/dt = y$, $dy/dt = -\omega^2 x$, its general solution being:

$$x(t) = k \cos\omega t, \quad y(t) = -k \omega \sin\omega t.$$

By using the polar coordinates θ and ρ defined from the variables x and $z = -y/\omega$, we get the new differential system:

$$d\theta/dt = \omega, \quad d\rho/dt = 0,$$

with $\theta = \text{Arctg}(z/x)$ and $\rho^2 = x^2 + z^2$. The polar system is conservative, its Hamiltonian function being defined by $H(\theta,\rho) = \omega\rho$. The solutions $x(t) = k \cos\omega t$, $z(t) = k \sin\omega t$ have 2 degrees of freedom, k and ω, respectively the amplitude and the frequency of the signal, and they constitute an orthogonal base, when we choose for ω the multiples (called harmonics) of a fundamental frequency ω_0.

After the seminal theoretical works by Y. Meyer [2,3], I. Daubechies [4] and S. Mallat [5], J. Haley defined a simple wavelet transform for representing signals in astrophysics (in 1997, cf. [6]). He used a base of functions made of the solutions of the damped pendulum differential equation (cf. a trajectory in Fig. 1):

$$dx/dt = y, \quad dy/dt = -(\omega^2 + \tau^2) x - \tau y,$$

its general solution being:

$$x(t) = k e^{-\tau t} \cos\omega t, \quad y(t) = -k e^{-\tau t} (\omega \sin\omega t + \tau \cos\omega t)$$

By using the polar coordinates θ and ρ defined from the variables x and $z = -y/\omega - x/\tau$, we get the differential system:

$$d\theta/dt = \omega, \quad d\rho/dt = -\tau\rho,$$

The polar system is dissipative (or gradient), its potential function being defined by $P(\theta,\rho) = -\omega\theta + \tau\rho^2/2$. The solutions $x(t) = k e^{-\tau t} \cos\omega t$, $z(t) = k e^{-\tau t} \sin\omega t$ have 3 degrees of freedom, k, ω and τ, the last parameter being the exponential time constant responsible for pendulum damping.

III. The Van Der Pol System

For Dynalets transform, we propose to use a base of functions made of the solutions of the relaxation pendulum differential equation (cf. a trajectory in Fig. 1 Top), which is a particular example of the most general Liénard differential equation:

$$dx/dt = y, \quad dy/dt = -R(x)x + Q(x)y,$$

which is specified in van der Pol case by choosing $R(x) = \omega^2$ et $Q(x) = \mu(1-x^2/b^2)$. Its general solution being not algebraic, but approximated by a family of polynomials [6-9].

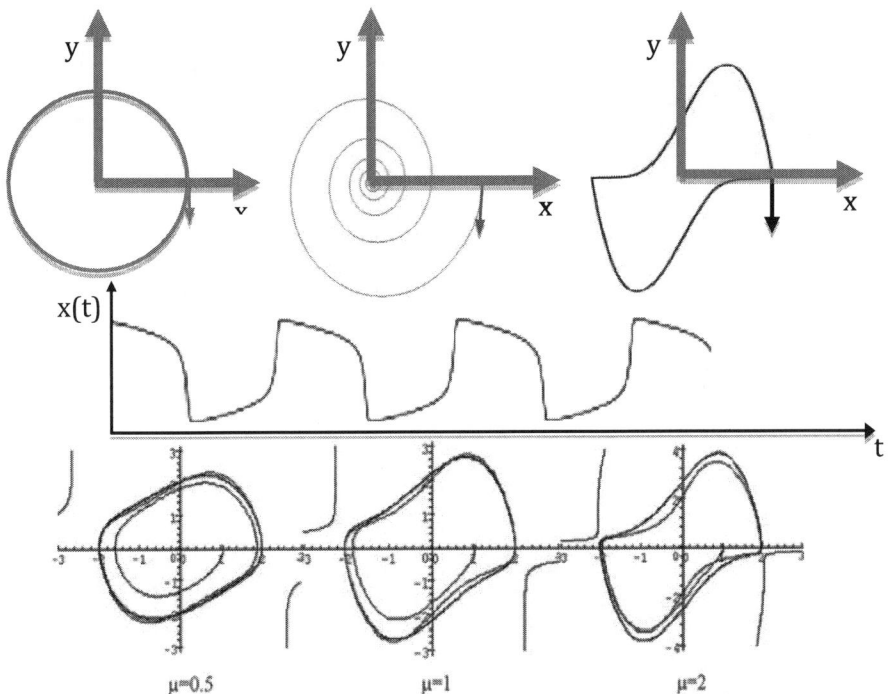

Fig. 1 Top left: a simple pendulum trajectory. Top middle: a damped pendulum trajectory. Top right: van der Pol limit-cycle ($\mu=10$). Middle: relaxation oscillation of van der Pol oscillator without external forcing ($\mu=5$). Bottom: representation of the harmonic contour lines $H(x,y) = 2.024$ for different values of μ.

The van der Pol system is dissipative (or gradient), its potential-Hamiltonian system, with P and H functions (Fig. 2 Top left), H being for example approximated at order 4, when $\omega = b = 1$, by [8,9]:

$H(x,y) = (x^2+y^2)/2 - \mu xy/2 + \mu yx^3/8 - \mu xy^3/8$, which allows to obtain the equation of its limit-cycle (cf. Fig. 1 Bottom): $H(x,y) \approx 2.024$. The van der Pol system has 3 degrees of freedom, b, ω and μ, the last an-harmonic parameter being responsible of the pendulum damping. These parameters receive different interpretations:

- μ appears as an anharmonic reaction term: when $\mu=0$, the equation is that of the simple pendulum, *i.e.*, a sine wave oscillator, whose amplitude depends on initial conditions and relaxation oscillations are observed even with small initial conditions (Fig. 1 & 2 Middle), whose period T near the bifurcation value $\mu=0$ equals $2\pi/\text{Im}\beta$; β is the eigenvalue of the Jacobian matrix J of the van der Pol equation:

$$J = \begin{pmatrix} 0 & 1 \\ -1 & \mu \end{pmatrix},$$

whose characteristic polynomial is equal to: $\beta^2 - \mu\beta + 1 = 0$, hence:

$$\beta = (\mu \pm (\mu^2-4)^{1/2})/2$$

and

$$T \approx 2\pi + \pi\mu^2/8.$$

- b looks as a term of control: when $x > b$, the derivative of y is negative, acting as a moderator on the velocity. The maximum of the oscillations amplitude is about 2b whatever initial conditions and values of the other parameters. More precisely, the amplitude $a_x(\mu)$ of x is estimated by $2b < a_x(\mu) < 2.024b$, for every $\mu > 0$, and when μ is small, $a_x(\mu)$ is estimated by: $a_x(\mu) \approx (2+\mu^2/6)b/(1+7\mu^2/96)$ [6,7]. The amplitude $a_y(\mu)$ is obtained for $dy/dt=0$, that is approximately for $x=b$, then $a_y(\mu)$ is the dominant root of the following algebraic equation: $H(b,a_y(\mu))=2.024$.
- ω is a frequency parameter, when μ is small. When $\mu \gg 1$, the period T of the limit cycle is determined mainly by the time during which the system stays around the cubic function where both x and y are $O(1/\mu)$ and the oscillations period T is roughly estimated to be $T \approx 2\pi\mu/\omega$, and the system can be

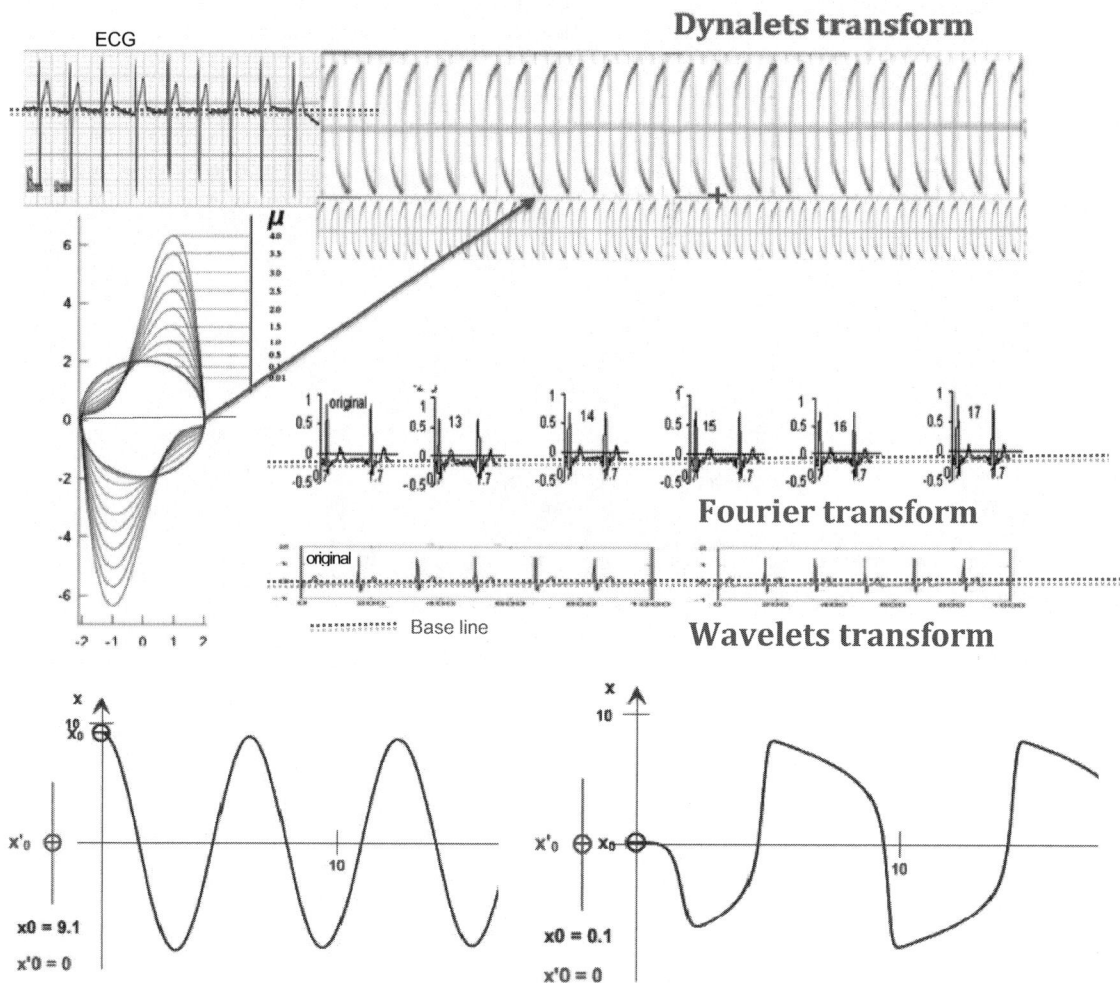

Fig. 2 Top left: original ECG signal (V1 derivation). Top right: decomposition into two temporal profiles respectively of period T and T/2, whose corresponding functions are orthogonal for the integral on [0,T[vector product. Middle left: representation of different van der Pol limit-cycle, for different values of µ (from µ=0,01 in red, to µ=4). Fourier. Middle right: comparison between Fourier and Haley wavelets decompositions of the ECG signal (V5 derivation), showing a more rapid fit with wavelets (until the 4th harmonics) than Fourier (until the 17th harmonics). Bottom: from [10], representation of different waves from van der Pol oscillator simulations (from [10]), from the symmetric type (left, for µ=0.4, ω=1, b=4) to the relaxation type (right, for µ=4, ω=1, b=4).

rewritten as: $d\chi/dt=\zeta$, $d\zeta/dt=-\omega^2\chi+\mu(1-\chi^2/\mu^2)\zeta \approx -\omega^2\chi+\mu\zeta$, with change of variables: $\chi=\mu x/b$, $\zeta=\mu y/b$.

IV. THE DYNALETS TRANSFORM

The Dynalets transform consists in identifying a Liénard system based interactions mechanisms between its variables (well expressed by its Jacobian matrix) analogue to those of the experimentally studied system, whose limit cycle is the nearest (in the sense of the Δ set distance, based on the Hausdorff point to set distance) to the signal in the phase plane (xOy), where y=dx/dt. For example, the Jacobian interaction graph of the van der Pol system contains a couple of positive and negative tangent circuits. Practically, for performing the Dynalets transform it is necessary to choose: i) the parameter µ such as the period of the van der Pol signal equals the mean period (either for the van der Pol or for the signal referential, chosen as the ECG signal in Fig. 2) a translation of the origin of axes, then rotate these axes and do a homothetic change of variables to match the first van der Pol. The whole approximation procedure can be done for the ECG signal (Fig. 2 and [10-16]) involves steps 1-3:

1) to perform a symmetrizing of x axis in the case of derivation V1 in order to get a signal similar the ECG V5

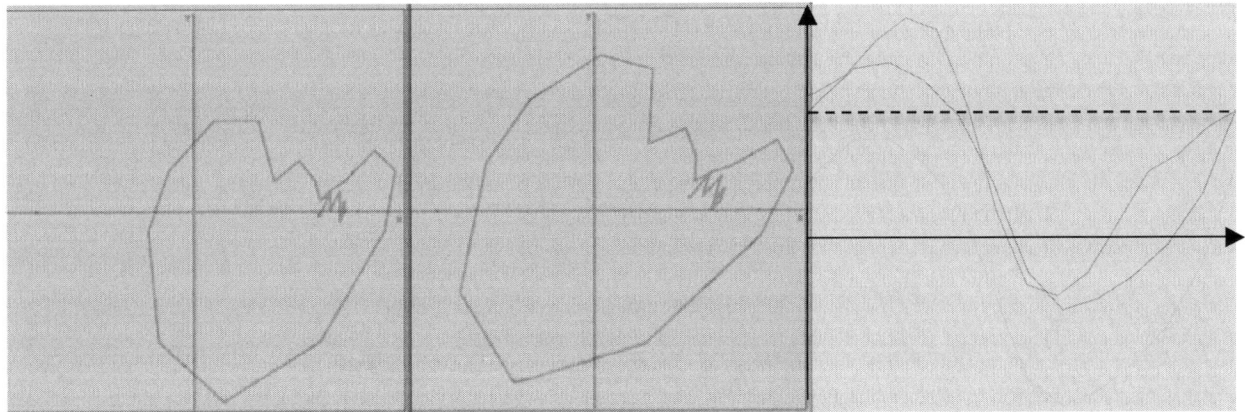

Fig. 3 Calculation of the first harmonics of the ECG signal of Fig. 2. Left: original van der Pol signal (in green) and ECG signal (V5 derivation, in red) in the phase plane (x0y). Middle: geometric matching of the two signals in the phase plane. Right: temporal profiles of the first harmonic of ECG (dark green).

(Fig. 2 and Fig. 3) and a transformation rotating / scaling the x and y axes of the ECG signal, so as to adjust them to maximum and minimum of x and y of the van der Pol,

2) to perform a translation of the origin of axes of the ECG signal by adjusting the base line to a selected phase of a vdP limit cycle of same period T (called pitch period) as the ECG period,

3) to finish the approximation matching the ECG points set to the vdP, by minimizing the difference set distance Δ between the interiors of the ECG and vdP cycles (denoted respectively **ECG** and **VDP**, with interiors **ECG$_o$** and **VDP$_o$**) in the phase plane: $\Delta(\mathbf{ECG_o}, \mathbf{VDP_o})$=Area $[(\mathbf{ECG_o} \backslash \mathbf{VDP_o}) \cup (\mathbf{VDP_o} \backslash \mathbf{ECG_o})]$, by using a Monte-Carlo method for estimating the area of the sets interior to the linear approximation of the ECG and vdP cycles, calculated from the point samples $\{X_i\}_{i=1,100}$ and $\{Y_i\}_{i=1,100}$.

V. CONCLUSIONS

Generalizing compression tools like Fourier or wavelets transforms is possible, for non symmetrical biological signals are often produced by mechanisms based on interactions of regulon type (i.e., possessing at least one couple of positive and negative tangent circuits inside their Jacobian interaction graph). In this case, we can replace the differential systems giving birth to biological signals by a Liénard type equation, like the van der Pol system classically used to model relaxation waves. The corresponding new transform, called Dynalets transform, has been built in the same spirit as the wavelets transform.

ACKNOWLEDGMENT

This work is supported by the ANR Project REGENER.

REFERENCES

1. Fourier J (1808) Propagation de la Chaleur dans les corps solides. Nouveau Bull. Sciences Soc. Philomathique de Paris 6:112-116
2. Lemarié P G, Meyer Y (1986) Ondelettes et bases hilbertiennes. Revista matemática iberoamericana 2:1-18
3. Meyer Y (1989) Wavelets and Operators, Analysis at Urbana, London Math. Soc., Lecture Notes Series 137, London, pp. 256-365
4. Daubechies I (1988) Orthonormal bases of compactly supported wavelets. Comm Pure Appl Math 41:909-996
5. Mallat S (1989) A theory of multiresolution signal decomposition: the wavelet representation. IEEE Transactions on Pattern Analysis and Machine Intelligence 11:674-693
6. Fisher E (1954) The period and amplitude of the van der Pol limit cycle. J. Applied Physics 25:273-274
7. Lopez J L, Abbasbandy S, Lopez-Ruiz R (2009) Formulas for the amplitude of the vd Pol limit cycle. Schol Res Exch 2009:854060
8. Demongeot J., Glade N., Forest L (2007) Liénard systems and potential-Hamiltonian decomposition. I Methodology. Comptes Rendus Mathématique 344:121-126
9. Demongeot J., Glade N, Forest L (2007) Liénard systems and potential-Hamiltonian decomposition. II Algorithm. Comptes Rendus Mathématique 344:191-194
10. Bub G, Glass L, Shrier A (2003) Coupling dependent wavefront stability in heterogeneous cardiac cell cultures. Biophys J 84:408
11. Lind R, Brenner, M, Haley S M (1997) Estimation of Modal Parameters Using a Wavelet-Based Approach. NASA Report TM-97-206300
12. http://www.sciences.univnantes.fr/sites/genevieve_tulloue/Meca/Oscillateurs/vdp_phase.html
13. http://wikimedia.org/wikipedia/commons/7/70/ECG_12derivations
14. Noble D (1962) A modification of the Hodgkin-Huxley equations applicable to Purkinje fibre action and pacemaker potential. J Physiol 160:317-352
15. McAllister R E, Noble D, Tsien R W (1975) Reconstruction of the electrical activity of cardiac Purkinje fibres. J Physiol 251:1-59
16. Fenton F H, Cherry E M (2008) Models of cardiac cell. Scholarpedia 3:1868

Cyclostationarity-Based Estimation of the Foetus Subspace Dimension from ECG Recordings

M. Haritopoulos[1], J. Roussel[1], C. Capdessus[1], and A.K. Nandi[2]

[1] PRISME Laboratory, 21 rue de Loigny La Bataille, Chartres, France
[2] Department of Electronic and Computer Engineering, Brunel University, Uxbridge, Middlesex, UK

Abstract—In this work, a novel method based on the cyclostationary properties of electrocardiogram (ECG) signals is introduced in order to classify independent subspaces into components reflecting the electrical activity of the foetal heart and those corresponding to mother's heartbeats, while the remaining ones are mainly due to noise. This research is inspired from multidimensional independent component analysis (MICA), a method that aims at grouping together into independent multidimensional components blind source separated signals from a set of observations. Given an input set of observations, independent component analysis (ICA) algorithms estimate the latent source signals which are mixed together. In the case of ECG recordings from the maternal thoracic and abdominal areas, the foetal ECGs (FECGs) are contaminated with maternal ECGs (MECG), electronic noise, and various artifacts (respiration, for example). When ICA-based methods are applied to these measurements, many of the output estimated sources have the same physiological origin: the mother's or the foetus' heartbeats. Thereby, we show that a procedure for automatic classification in independent subspaces of the extracted FECG and MECG components is feasible when using a criterion based on the cyclic coherence (CC) of the signal of interest.

Keywords—foetal electrocardiogram, cyclostationarity, cyclic coherence, multidimensional independent component analysis.

I. INTRODUCTION

Electrocardiogram (ECG) is one of the most popular diagnostic tools for heart monitoring. Non-invasive electrical activity measurement devices, like electrodes, offer clinicians a versatile tool, without heavy constraints, for preventing complications or diseases due to heart defects. When these measurements come from the thoracic or abdominal regions of a pregnant woman, detecting and separating the foetal electrocardiogram (FECG) from the maternal one (MECG) is not such an easy task, since the FECG has a very low amplitude voltage signal compared to the MECG, and, moreover, the activity measured by the cutaneous electrodes is disturbed by other noise signals, such as random instrumentation noise, power line signal, breathing or baseline wandering.

In these last decades, various signal processing techniques have emerged and many of them have been customized for FECG extraction purposes ([1], [2], [3]). Among them are blind source separation (BSS) techniques [4] and the underlying mathematical tool which is independent component analysis (ICA). It aims at separating unknown source signals of interest from a mixture of them, observed with the aid of sensors. Numerous methods can be found in the dedicated literature, but only a few make use of the concept of multidimensional independent component analysis (MICA), introduced in the late 90's. Interested readers can find a recent review on FECG signal processing in [5].

In the rest of the paper, we first present the MICA concept. An introduction to the cyclostationary nature of ECGs and a description of a cyclostationarity measure, called cyclic coherence, follows. Next, a novel FECG classification procedure is proposed and validated on a real world FECG dataset; it is based on the cyclic coherence computation of the ICA estimated components from a set of sensor signals. Finally, conclusions are drawn from this work.

II. ICA AND THE MULTIDIMENSIONAL ICA (MICA) CONCEPT

A. Blind Source Separation Problem Formulation and ICA

Jutten *et al* first proposed a simple 2×2 neural network-based self-adaptive algorithm ([6], [7]) as a solution to the separation of independent source signals, and introduced the concept of independent component analysis, which has been theoretically investigated few years later by Comon [8]. The problem in blind source separation (BSS) states as follows: given a set of M observed signals $x_i(t), i \in [1,M]$, find the N unknown sources $s_i(t), i \in [1,N]$, with $N \leq M$, that are hidden in the observations. In matrix form, the problem writes:

$$\mathbf{x}(t) = A\mathbf{s}(t). \quad (1)$$

where $\mathbf{x}(t) = [x_1(t), x_2(t), ..., x_M(t)]^\dagger$ and $\mathbf{s}(t) = [s_1(t), s_2(t), ..., s_N(t)]^\dagger$ denote the mixture vector and the unknown source vector, respectively, with $\dagger$ the transpose operator and with A an unknown $M \times N$ full rank mixing matrix. The mixture model of eq.1 is the *linear instantaneous noiseless* one, which is widely used in the literature and which only assumes statistically independent source components s_i and at most one Gaussian entry.

B. Multidimensional ICA

A source subspace separation method for FECG extraction has been presented in [9]. In this work, authors use second-order statistics based on the singular value decomposition (SVD) of the observed ECG data matrix to estimate the underlying source subspaces, assuming that the latent bioelectric phenomena generating the foetal and the mother heartbeats are statistically independent. Then, they enhance the accuracy of the proposed algorithm by introducing a higher-order singular value decomposition (HOSVD) technique.

A similar technique is proposed by Cardoso in [10] named multidimensional independent component analysis (MICA). It can be seen as a generalization of the ICA model to multidimensional components. The author illustrates his method with real FECG data. First, he estimates the unknown sources by applying the well known ICA algorithm JADE [11] to a 3-channel real ECG data of an expectant mother. After extraction of the independent components (ICs), he groups two of them into a bi-dimensional component corresponding to the MECG; the choice of these two components is done only after visual inspection and, in this case, that works because mother's and foetus' contributions are clearly distinguishable. After computing the orthogonal projection matrices for each one of the two estimated multidimensional components, a back-projection to the original input space permits to reconstruct the mother and foetal signal contributions to the mixtures.

We demonstrate below how to establish an automated procedure based on the cyclostationary properties of the ECG signals, in order to classify ICA-based extracted ICs into multidimensional components.

III. CYCLOSTATIONARITY MEASURES

A. The Cyclostationarity Property

The main assumption for this procedure to succeed, is that the foetus' as well as the mother's heartbeats, if they do not exhibit a strict period, they are at least repetitive, i.e. cyclostationary at the frequency α_0 of the heartbeats. A cyclostationary signal is not strictly periodic, but some of its statistical properties are periodic. This same assumption of FECG cyclostationarity used in previous work [12], together with the *a priori* estimation of the foetus' fundamental cyclic frequency, provided very promising results with regard to the foetal PQRST extraction.

In this context, we use a statistical measure of cyclostationarity as a criterion to decide whether an ICA estimated IC corresponds to the FECG rather than to the MECG subspace. This measure is the cyclic coherence.

B. The Cyclic Coherence Measure

In this work, we use a spectral version of the simple coherence [13] which is a measure of the correlation degree between two signals, $x(t)$ and $y(t)$, at each frequency value f. It is called the cyclic coherence (CC) and it has been introduced in [14]:

$$C_x^\alpha(f) = \frac{E[X(f)X^*(f-\alpha)]}{\left(E\left[|X(f)|^2\right]E\left[|X(f-\alpha)|^2\right]\right)^{\frac{1}{2}}}, \quad (2)$$

where $X(f)$ and $X(f-\alpha)$ are the Fourier Transforms of $x(t)$ and alpha value frequency-shifted $x(t)$, respectively.

The CC measure is normalised. A CC value near to one for a frequency f, indicates a strong coherence between components of signal $x(t)$ at frequencies f and $(f-\alpha)$.

IV. THE PROPOSED METHOD AND ILLUSTRATION WITH REAL WORLD DATA

To illustrate the proposed method, we worked on the same dataset as in the original MICA work [10], i.e. the well-known *DaISy* database [15], but we considered the whole recordings set. This is an eight-channel cutaneous recordings set from an expectant mother. The first five potentials come from electrodes placed at different locations in the abdominal region of the mother, while the last three recordings come from her thoracic region. All eight signals are recorded simultaneously at a sampling rate of $500Hz$ and each one of them lasts $5sec$ (Figure 1).

A. The Classification Procedure

We first applied the JADE algorithm to the whole observations vector and then we computed the CCs of the estimated components. A visual inspection of the obtained CCs leads to a preliminary classification of extracted components $\{3,5\}$ into the FECG subspace, components $\{1,4\}$ into the noise subspace and the remaining components $\{2,6,7,8\}$ into the MECG subspace (Figures 2, 3 and 4, respectively). Indeed, in Figure 2 one clearly notices the high values over all the spectral frequencies range, for cyclic frequency around $4.5Hz$ that corresponds to the fundamental cyclic frequency of the foetus' heart [12], while the second high-valued spectral line corresponds to its second harmonics. On the contrary, no (f, α)

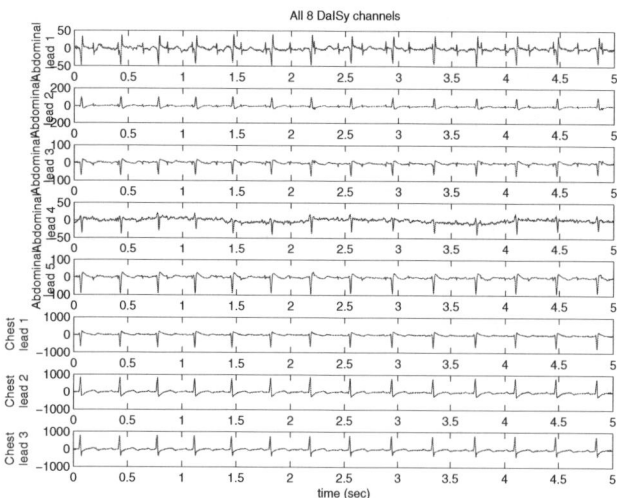

Fig. 1 Eight surface electrodes recordings

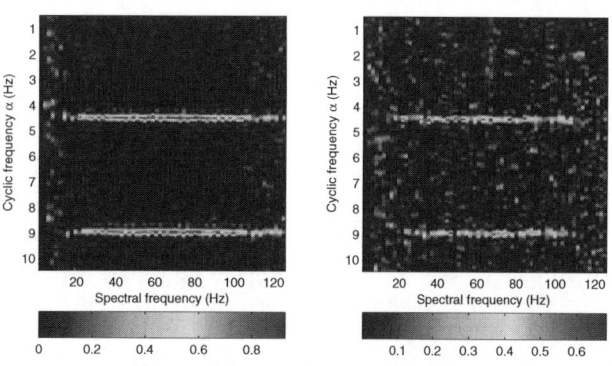

Fig. 2 The 2-dimensional FECG subspace

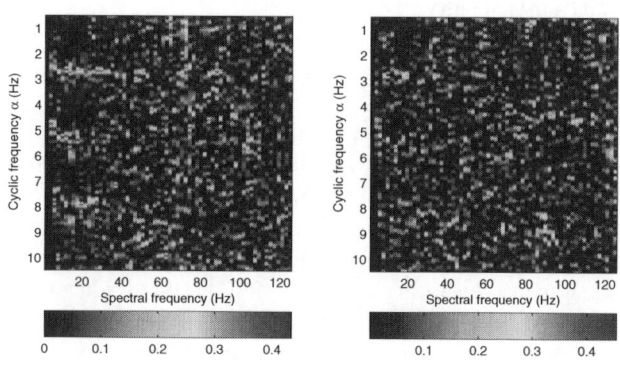

Fig. 3 The 2-dimensional noise subspace

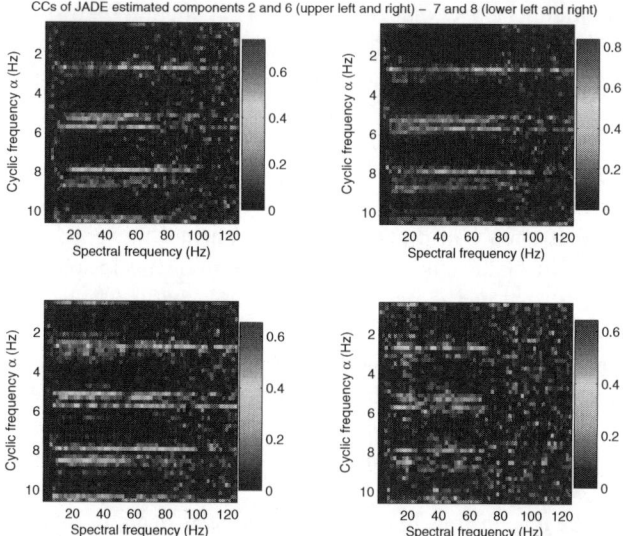

Fig. 4 The 4-dimensional MECG subspace

couple has a significantly high value in the CCs of JADE extracted components 1 and 4, as one can observe in Figure 3. This is in agreement with the kurtosis values of these two CCs: 4.94 and 0.54, for components 1 and 4, respectively. Lower kurtosis values may correspond to noisy background.

Things get a little bit different for the MECG subspace. CCs of JADE estimates 2, 6 and 8 exhibit distinct higher values for cyclic frequency around $2.7Hz$ into the spectral frequency range $[20,80]Hz$ (Figure 4); all three span the subspace of the MECG components. On the other hand, the CC of the JADE estimated component number 7, shows also high values for cyclic frequencies different from the mother's fundamental one and its harmonics. That needs more investigations to interpret, but it is not decisive for the classification procedure of foetal components that this work suggests.

Next, we integrate the cyclic frequencies for each CC of the ICA estimated components over the spectral frequencies to obtain the integrated CC (iCC). At this stage to automate the extraction of the noise subspace components, one can compute the standard deviation of each iCC; standard deviation values for this example are $\{0.68; 3.59; 5.29; 0.69; 2.41; 4.20; 4.44; 1.92\}$ for the 1st through to the 8th JADE extracted IC. Looking for standard deviation values below a low-valued threshold, is an efficient way for isolating ICs whose iCCs do not contain significant information.

Then, from each one of the previously computed iCCs, one subtracts its maximum value as shown in Figure 5. This is implemented only for extracted components that they do not span the noise subspace, i.e., JADE extracted components

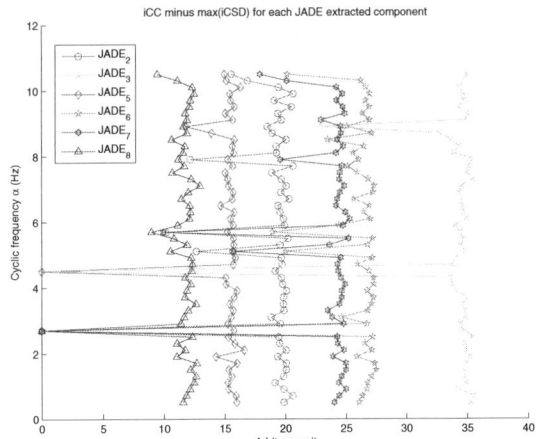

Fig. 5 The two null values of the remaining 6 transformed iCCs do correspond to the FECG's and the MECG's fundamental cyclic frequencies (4.5Hz and 2.7Hz, respectively)

$\{2,3,5,6,7,8\}$. One remarks that the yellow ($JADE_3$) and red ($JADE_5$) data plots of Figure 5 exhibit both a null value for the same cyclic frequency equal to 4.5Hz (i.e., the foetus' heart cyclic frequency) while the four remaining components have a modified iCC of zero value at 2.7Hz (i.e., the mother's heart cyclic frequency). Finally, one can group blind-separated ICs into 3 independent subspaces: the FECG one spanned by ICs $\{3,5\}$, a noise one spanned by ICs $\{1,4\}$ and the MECG subspace spanned by ICs $\{2,6,7,8\}$.

V. CONCLUSION

This work proposes an automated classification procedure of blindly separated independent foetal components from maternal ECG recordings, that consists of four steps:

1. Apply a BSS/ICA algorithm to extract ICs from raw ECG data.
2. Compute the CC and then the iCC for each extracted independent component at the previous step.
3. Classify ICs whose standard deviation value of their iCC is below a threshold into a subspace corresponding to the noise.
4. Compute the difference between each iCC and its maximum value, and group into the same independent source subspaces the corresponding ICs exhibiting null values for the same cyclic frequencies.

The method is tested on the real world DaISy dataset and gave very promising results. Future research work concerns the validation of this method to simulated foetal ECGs, as well as further investigations about the computed CCs of the MECG source subspace components.

ACKNOWLEDGEMENTS

A. K. Nandi would like to thank the Université d'Orléans for the Visiting Professorship, which contributed to this study.

REFERENCES

1. Zarzoso V, Nandi A K. Noninvasive Foetal Electrocardiogram Extraction: Blind Separation Versus Adaptive Noise Cancellation *IEEE Transactions on Biomedical Engineering.* 2001;48(1):12-18.
2. Sameni R, Jutten C, Shamsollahi M B. Multichannel Electrocardiogram Decomposition Using Periodic Component Analysis *IEEE Transactions on Biomedical Engineering.* 2008;55(8):1935-1940.
3. Martín-Clemente R, Camargo-Olivares J L, Ornillo-Mellado S, Helena M, Román I. Fast Technique for Noninvasive Fetal ECG Extraction *IEEE Transactions on Biomedical Engineering.* 2001;58(2):227-230.
4. Herault J, Jutten C, Ans B. Détection de grandeurs primitives dans un message composite par une architecture neuromimétique en apprentissage non supervisé in *Actes 10ème Colloque GRETSI*(Nice, France):1017-1022 1985.
5. Sameni R, Clifford G D. A Review of Fetal ECG Signal Processing Issues and Promising DIrections *The Open Pacing, Electrophysiology & Therapy journal.* 2010;3:4-20.
6. Jutten C, Herault J. Blind separation of sources, part I: An adaptive algorithm based on neuromimetic architecture *Signal Processing.* 1991;24(1):1-10.
7. Jutten C, Herault J. Blind separation of sources, part II: Problems statement *Signal Processing.* 1991;24(1):11-20.
8. Comon P. Independent Component Analysis, a New Concept *Signal Processing, Special Issue on Higher Order Statistics.* 1994;36(3):287-314.
9. Lathauwer L De, Moor B De, Vanderwalle J. Fetal Electrocardiogram Extraction by Source Subspace Separation in *Proceedings IEEE Signal Processing / Athos Workshop on Higher-Order Statistics*:134-138 1995.
10. Cardoso J F. Multidimensional independent component analysis in *Proceedings of the 1998 IEEE International Conference on Acoustics, Speech and Signal Processing*;4:1941-1944 1998.
11. Cardoso J F, Souloumiac A. Blind beamforming for non Gaussian signals in *Radar and Signal Processing, IEE Proceedings F*;140(6):362-370 1993.
12. Haritopoulos M, Capdessus C, Nandi A K. Foetal PQRST Extraction from ECG Recordings using Cyclostationary-Based Source Separation Method in *32 Annual International Conference of the IEEE EMBS*:1910-1913 2010.
13. Max J, Lacoume J L. *Méthodes et techniques de traitement du signal et application aux mesures physiques. Tome 1 : Principes généraux et méthodes classiques.* fifth ed. 1996.
14. Hurd H L. *An investigation of periodically correlated stochastic processes.* PhD thesisDuke University 1970.
15. ESAT/SISTA K. U. Leuven. DaISy; Database for the Identification of Systems at http://homes.esat.kuleuven.be/ smc/daisy/ 1999.

Feature Extraction Based on Discriminative Alternating Regression

C.O. Sakar[1], O. Kursun[2], and F. Gurgen[3]

[1] Department of Computer Engineering, Bahcesehir University, Istanbul, Turkey
[2] Department of Computer Engineering, Istanbul University, Istanbul, Turkey
[3] Department of Computer Engineering, Bogazici University, Istanbul, Turkey

Abstract—Canonical Correlation Analysis (CCA) aims at measuring linear relationships between two sets of variables (views). Recently, CCA has been used for feature extraction in classification problems with multi-view data by means of view fusion. However, the extracted correlated features with CCA may not be discriminative since CCA does not utilize the class labels in its traditional formulation. Besides, the CCA features are computed based on within-set and between-set sample covariance matrices of the views which can be very sensitive to representation-specific details and noisy samples of the two views. In this paper, we propose a method, D-AR (Discriminative Alternating Regression), in which the two above-mentioned problems encountered in the application of CCA for feature extraction are addressed: (1) the class labels are incorporated into the proposed feature fusion framework to explore correlated and also discriminative features, and (2) the use of sensitive sample covariates matrices is avoided while fusing the two views. D-AR is a supervised feature fusion approach based on Multi-layer Perceptron (MLP) implementation of alternating regression. From the neurobiological perspective, the architecture of D-AR is similar to the model of a single neuron in the cerebral cortex which has a function of discovering and representing one of the hidden factors in its sensory environment. The MLP trained on each view aims to predict the class labels and also the hidden factors which are responsible for the correlation. We show that the features found by D-AR on training sets accomplishes significantly higher classification accuracies on test set of an experimental dataset.

Keywords—discriminative CCA, neural implementation of CCA, overfitting and overlearning in covariates, feature extraction.

I. INTRODUCTION

Dimensionality reduction is a crucial step in machine learning problems especially when the number of samples is not sufficient compared to data dimensionality to learn generalizable predictive models. When the data is represented with multiple sets of features (multi-view data), feature fusion methods such as Canonical Correlation Analysis (CCA) [1] are used for reducing the dimensions of the views. The popularity of such datasets is increasing in many fields such bioinformatics and biomedical engineering [2] [3].

Due to its recent popularity, several studies are proposed to improve CCA from different aspects. These studies address two main issues: the use of sensitive sample covariance matrix problem and improvement of its discriminative ability. The studies that deal with the sensitive sample covariance matrix problem can be categorized into three groups: (1) iterative alternating regression approaches which avoid the use of sample covariance matrices, (2) robust CCA approaches which utilize robust estimation of the sample covariance matrices which are used for the extraction of covariates [4] [5] [6], and (3) reducing the dimension of each view independently using Principal Component Analysis (PCA) before feature fusion as a preprocessing step. The second issue addressed by improved CCA approaches is the discriminative ability of CCA. As known, CCA is an unsupervised feature extraction method because it does not incorporate the class information into its framework while fusing the views. To increase the classification performance of CCA features, Sun et al. [7] proposed Discriminative Canonical Correlation Analysis (DCCA) method which employs the class information by maximizing the correlation between feature vectors in the same class and minimizing the correlation between feature vectors belonging to different classes, and showed that DCCA features have higher discriminative power than of CCA. However, it has been recently shown that projective directions explored by DCCA are equal to the LDA projective directions with respect to an orthogonal transformation [8]. Besides, DCCA has still the problems of using sensitive sample covariance matrices.

In this paper, we propose a method called Discriminative Alternating Regression (D-AR), which addresses both of the above aforementioned robust CCA and discriminative CCA issues. The proposed D-AR method is a Multilayer Perceptron (MLP) based alternating regression architecture which avoids the use of sensitive sample covariance matrices and also incorporates the class information into the feature fusion framework. Besides, LDA and obviously DCCA have the limitation of less than number of classes orthogonal projective directions due to the rank deficiency of the between-class scatter matrix whereas D-AR does not suffer from this problem. We compare the proposed method D-AR with CCA, PCA+CCA, alternating regression, and LDA in the aspect of the classification accuracies on an emotion recognition dataset.

II. PROPOSED METHOD: DISCRIMINATIVE ALTERNATING REGRESSION APPROACH

In this paper, we propose a method, D-AR (Discriminative Alternating Regression), which aims at exploring discriminative and robust features by incorporating the class labels into the view fusion framework to increase the discriminative power and using a multilayer perceptron-based alternating regression to avoid the use of sensitive sample covariance matrices. The overall architecture of the method is shown in Figure 1 which is based on the alternating regression approach [9] implemented by a multi-layer perceptron neural network with a "linear" hidden layer. The backpropagation algorithm [10] is used as the learning rule for the two perceptrons. The input layer of the proposed MLP network consists of the view features. The information in the high-dimensional input layer are transformed into a low-dimensional subspace (hidden layer) so that the respective output units in the output layer can pair up maximally with the respective output units of the other view and also the class labels can be predicted. The aim is obtaining a subspace of the input features (p number of hidden neurons where p<n and p<m) in the hidden layer which preserves both the common information with the other view and the class information. From the neurobiological perspective, the MLPs (dendrites) constructed in both side of the D-AR network is similar to the model of a single neuron in the cerebral cortex which has a function of discovering and representing one of the hidden factors in its sensory environment [11] [12]. Forcing the discriminative information to be contained by both of the views increases the reliability and generalization of the extracted features especially on datasets with small sample size and high-dimensional input space. Inhibition is used to decorrelate the output units.

The training signals used for the outputs of the two perceptrons are evolved through an alternating regression procedure. In the alternating regression procedure, we minimize the differences in output covariates computed by the two perceptrons in response to their coincident input vectors from view 1 and 2. We use cascading anti-Hebbian inhibition algorithm for decorrelating the multiple outputs of a perceptron.

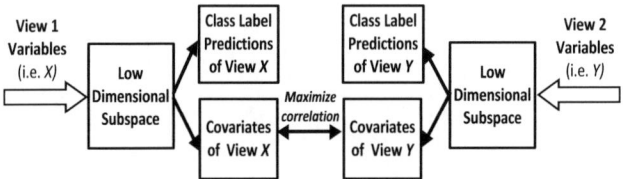

Fig. 1 Architecture of the proposed D-AR method. Dimensions m and n of X and Y are reduced to p, where $m > p$ and $n > p$. Correlated outputs (covariates) are alternated between two views to maximize their correlations

III. EXPERIMENTAL RESULTS

The discriminative power of the covariates extracted with the proposed D-AR method are evaluated and compared against of the traditional CCA, PCA+CCA, AR, and LDA methods on an emotion recognition dataset with various number of training samples. The obtained covariates are fed to linear kernel Support Vector Machines (SVM) and k-Nearest Neighbor (k-NN) classifiers.

The Cohn-Kanade Facial Expression Database (CK+) [13] consists of 327 video clips each along with an emotion label recorded from 118 subjects. The video clips in CK+ dataset belong to 7 different emotions which are (1) anger, (2) contempt, (3) disgust, (4) fear, (5) happiness, (6) sadness, and (7) surprise. Each video clip begin with a neutral expression and proceeds to a peak expression where the emotion is most significant.

Two types of feature representations extracted from CK+ dataset are used as views in our experimental studies. The first view consists of appearance-based features. Each sample in the appearance based-view has 4096 (64x64) features (pixels). The second view consists of the geometric features [14]. Each sample is represented with 134 features in the geometric-based view.

For statistical significance, the dataset is shuffled 10 times, and different training and test sets are generated in each run. In Figure 2, the obtained average SVM and k-NN (k=3) classifier test set accuracies are shown with respect to increasing number of training samples per class. As seen, both view 1 and view 2 covariates of D-AR gave significantly higher accuracies with SVM as well as k-NN (paired t-test, $p < 0.05$). While LDA as a supervised dimensionality reduction has the closest performance to D-AR, the CCA, PCA+CCA, and AR methods showed no significant superiority over each other. The results show that both SVM and k-NN classifiers accomplished 85% classification accuracy with view 2 covariates of D-AR extracted with only 10 samples per class.

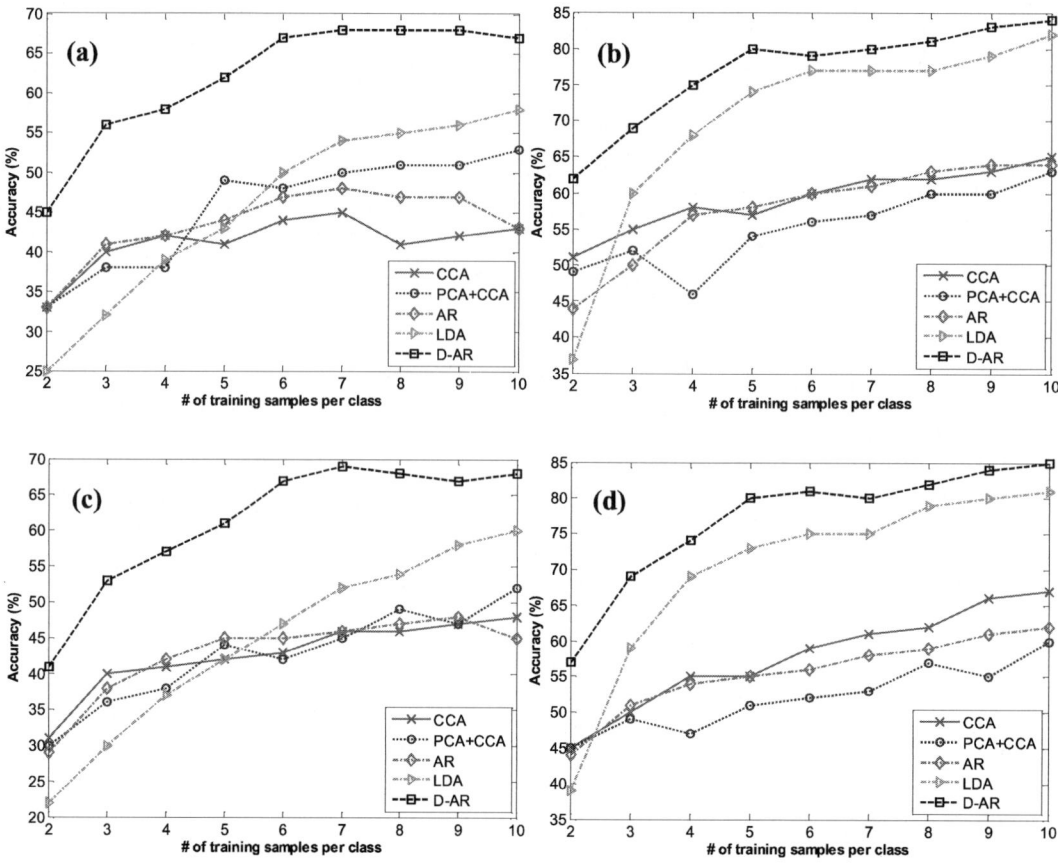

Fig. 2 Number of training samples per class versus accuracies of covariates of (a) view 1 (SVM) (b) view 2 (SVM) (c) view 1 (3-NN) (d) view 2 (3-NN)

IV. CONCLUSIONS

Canonical Correlation Analysis (CCA) is an unsupervised dimensionality reduction technique that aims to extract maximally correlated projections from the two given views. In multi-view machine learning problems in which at least two different representations of the same underlying phenomenon are available, CCA is used as a feature fusion method and the fused features are fed to classification algorithms. However, the use of CCA in classification problems suffers from two problems. Firstly, CCA covariates are computed based on the sample covariance matrices which are known to be sensitive to outlier and noisy samples. Such samples may arise from the temporary corruption of a view or other noisy conditions. Secondly, CCA does not utilize the class labels in its traditional analytical solution. Therefore, CCA tends to preserve highly correlated features instead of less correlated but more discriminative features in the reduced subspace.

We propose a method called D-AR, which deals with both of the above-mentioned problems. D-AR avoids the use of sample covariance matrices and also integrates the class labels into its feature fusion framework. It is based on the alternating regression approach implemented by a multi-layer perceptron neural network with a "linear" hidden layer that transforms each view into low-dimensional subspaces that preserve correlated and also discriminative information of the views. The proposed method also does not have the limitation of less than number of classes orthogonal projective directions which is a limitation of LDA due to the rank deficiency of the between-class scatter matrix. We have applied various methods for fusing the views of CK+ emotion recognition dataset on which the covariates extracted by D-AR have higher classification accuracies than those of CCA, PCA+CCA, alternating regression, and LDA.

ACKNOWLEDGMENT

The work of C. O. Sakar is supported by the Ph.D. scholarship (2211) from Turkish Scientific Technical Research Council (TÜBİTAK). He is a Ph.D. student in the Computer Engineering Department at Bogazici University, Istanbul, Turkey.

REFERENCES

1. Hotelling H (1936) Relations between two sets of variates. Biometrika, 28: 312-377
2. Sakar CO, Kursun O, Seker H, Gurgen F (2013) Combining Multiple Clusterings for Protein Structure Prediction. International Journal of Data Mining and Bioinformatics, accepted.
3. Zhang YN, Yu DJ, Li SS, Fan YX, Huang Y, Shen HB. (2012). Predicting protein-ATP binding sites from primary sequence through fusing bi-profile sampling of multi-view features. BMC Bioinformatics. 31;13:118 doi: 10.1186/1471-2105-13-118
4. Karnel G (1991) Robust canonical correlation and correspondence analysis. In: The Frontiers of Statistical Scientific and Industrial Applications, (Volume II of the proceedings of ICOSCO-I, The First International Conference on Statistical Computing), American Sciences Press, Strassbourg, 335-354
5. Croux C, Dehon C (2002) Analyse canonique basee sur des estimateurs robustes de la matrice de covariance. La Revue de Statistique Appliquge, 2:5-26
6. Taskinen S, Croux C, Kankainen A, Ollila E, Oja H (2006) Canonical Analysis based on Scatter Matrices. Journal of Multivariate Analysis, 97(2): 359-384
7. Sun, T., Chen, S., Yang, J. and Shi, P. (2008). A supervised combined feature extraction method for recognition, Proceedings of the IEEE International Conference on Data Mining, Pisa, Italy, 1043–1048.
8. Shin YJ, Park CH (2011) Analysis Of Correlation Based Dimension Reduction Methods. Int. J. Appl. Math. Comput. Sci., 21(3):549-558.
9. Wold H (1966) Nonlinear estimation by iterative least squares procedures. In: F.N. David (ed.), A Festschrift for J. Neyman, Wiley, New York, 411-444
10. Rumelhart, D.E., Hinton, G.E., & Williams, R.J. (1986). Learning internal representations by error propagation. In: Parallel distributed processing: Explorations in the microstructure of cognition, Rumelhart, D.E., McClelland, J.L., PDP Research Group (eds) MIT Press, Cambridge, Mass, 1:318-362.
11. Kursun O, Favorov OV (2004) SINBAD automation of scientific discovery: From factor analysis to theory synthesis. Natural Computing 3:2 207-233
12. Favorov OV, Ryder D, Hester JT, Kelly DG, Tommerdahl M (2003). The cortical pyramidal cell as a set of interacting error backpropagating networks: a mechanism for discovering nature's order. In: Theories of the Cerebral Cortex, R. Hecht-Nielsen and T. McKenna (eds.), Springer, London, 25-64.
13. Lucey P, Cohn JF, Kanade T, Saragih J, Ambadar Z, Matthews I (2010). The Extended Cohn-Kande Dataset (CK+): A complete facial expression dataset for action unit and emotion-specified expression. Third IEEE Workshop on CVPR for Human Communicative Behavior Analysis (CVPR4HB 2010).
14. Sakar CO, Kursun O, Karaali A, Erdem CE (2012). Feature Extraction For Facial Expression Recognition By Canonical Correlation Analysis, IEEE 20th Signal Processing and Applications Conference (SIU), Mugla, Turkey

Author: C. Okan Sakar
Institute: Bahcesehir University
Street: Ciragan Cad. Besiktas
City: Istanbul
Country: Turkey
Email: okan.sakar@bahcesehir.edu.tr

Influence of Signal Preprocessing on ICA-Based EEG Decomposition

Z. Zakeri[1], S. Assecondi[2], A.P. Bagshaw[2], and T.N. Arvanitis[1]

[1] School of Electronic, Electrical and Computer Engineering, University of Birmingham, Birmingham, United Kingdom
[2] School of Psychology, University of Birmingham, Birmingham, United Kingdom

Abstract—**Independent Component Analysis (ICA) has been widely used for analysis of EEG data and separating brain and non-brain sources from the EEG mixture. In this study, we compared decomposition results of the most commonly applied ICA algorithms: AMICA, Extended-Infomax, Infomax and FastICA. We examined 12 conditions of EEG data pre-processing, and assessed the independence and physiological plausibility of the recovered components. The results demonstrate that, in general, there were no significant differences in the decomposition results, while data pre-processing choices had a much more pronounced effect. In conclusion the efficiency of the ICA decompositions is highly dependent on the pre-processing steps applied to the EEG data submitted to ICA, rather than type of ICA applied.**

Keywords—**EEG, ICA, Pre-processing.**

I. Introduction

Electroencephalography (EEG) provides a tool for clinicians to monitor and analyse brain functions for the diagnosis and treatment of mental and neurological disorders [1]. The EEG data recorded on the scalp is a mixture of brain and non-brain signals (such as body or eye movement artifacts) [2]. Independent Component Analysis (ICA) is a statistical method that has been widely adopted to separate artifacts from EEG sources [3]. ICA is a Blind Source Separation (BSS) technique that employs higher order statistics to estimate underlying sources from multidimensional EEG data [3]. ICA algorithms estimate independent components by maximizing the non-Gaussianity of sources [4]. The non-Gaussianity of sources can be quantified using different metrics, resulting in components that may differ across algorithms. Several studies have compared ICA algorithms based either on the actual independence of the components [5] or on their ability to isolate specific types of artifacts [6]. The reliability of ICA decomposition, which is affected by removing linear trends from each epoch of data, has also been assessed [7]. Recently, performance of ICA algorithms has been evaluated on the basis of the independence and biological plausibility of the estimated sources [8]. The effect of the initialization step of different ICA algorithms, whitening and sphering, and different length of data to be submitted to ICA has also been studied [9]. However, the effect of different pre-processing steps on the output of different ICA algorithms has not received a great deal of attention in comparison studies. The motivation of the current study is to assess the effect of several data preparation and pre-processing steps on the performance of ICA algorithms. We used some of the most commonly applied ICA algorithms, AMICA [10], Extended-Infomax [4], Infomax [11] and FastICA [12], on different types of pre-processed EEG data. We are mainly interested in separating artifactual components, e.g., eye artifacts, therefore we chose those algorithms better suited for such a purpose. In addition to compare ICA algorithms based on the components' independence and biological plausibility [8], several types of pre-processed EEG data are used as a factor to assess the quality of ICA algorithms.

II. Materials and Methods

A. Data Collection and Pre-processing

Ten healthy subjects underwent a visual stimulation task. The task consisted of a presentation of a hemifield checkerboard (with either high or low contrast) for 500 ms, with phase reversal for another 500 ms, followed by a fixation cross. The fixation cross was randomly changed to × to maintain attention. Participants were asked to respond (press button) when they recognize this change. The data was recorded using 64 channels (62 scalp electrodes, following the standard 10-20 system, one ECG and one EOG electrodes, connected to a BrainAmp MR-Plus Amplifier, BrainProducts, Gilching, Germany) at a sampling rate of 5000 Hz. Raw EEG data were band-pass filtered at 0.016-100 Hz and down-sampled to 256 Hz. Before entering ICA decomposition, the pre-processed data underwent further processing steps, which were a combination of the following:

- ExG: the non-EEG channels (ECG and EOG) were included (wExG) or excluded (wnExG) from further analysis;
- FILTER: the EEG data were further band-pass filtered at 0.16-40 Hz (FLT) or no additional filter was applied (noFLT);
- SEGMENTATION: the EEG data were segmented (SEG) into epochs (100 ms before and 900 ms after stimulus

Influence of Signal Preprocessing on ICA-Based EEG Decomposition

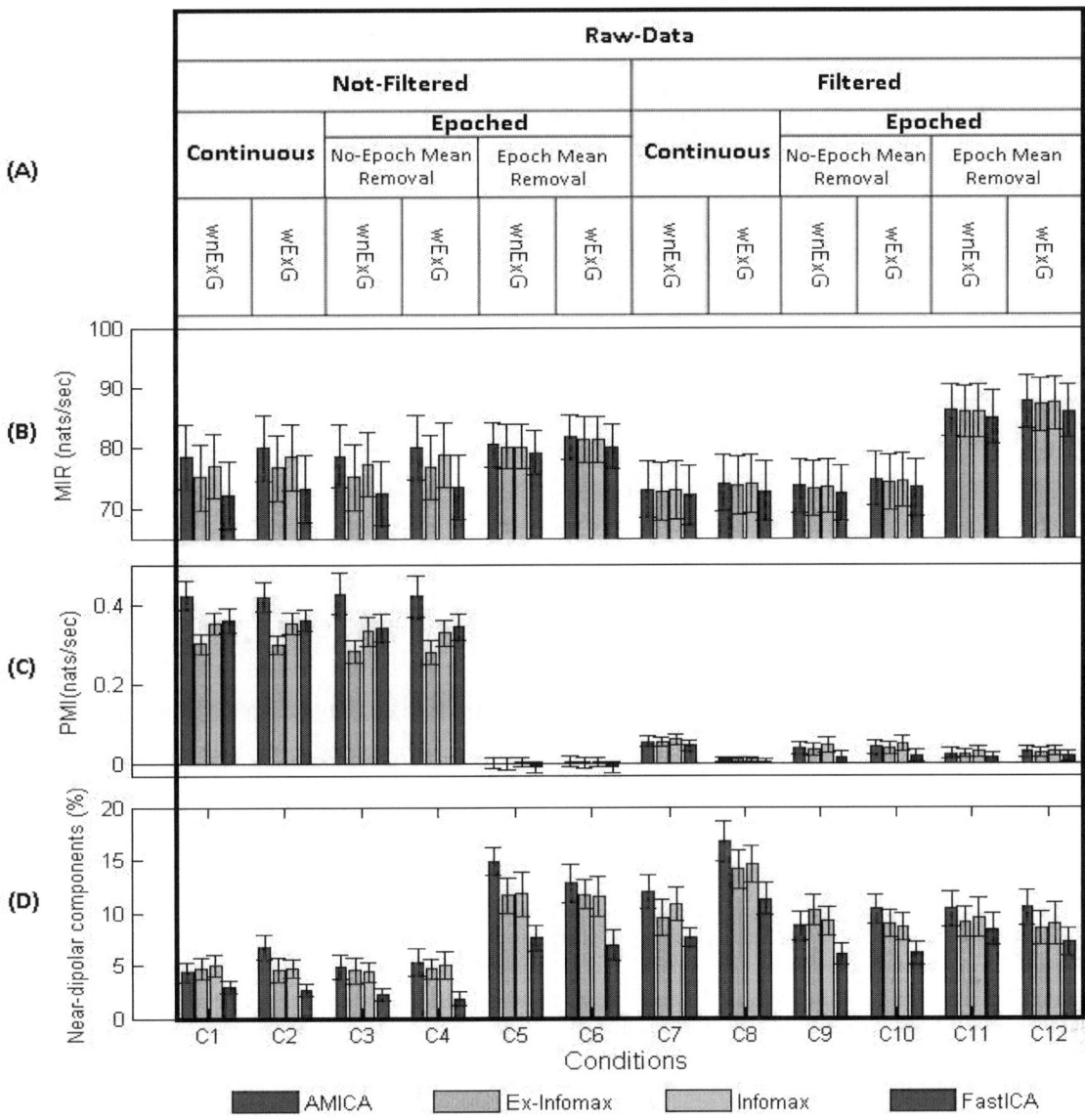

Fig. 1 The pre-processing hierarchy of the raw EEG data and the performance of ICA algorithms based on MIR, PMI and Dipolarity. 1.A. Different types of EEG data pre-processing before running ICA algorithms. The corresponding pre-processing category of each mini bar chart is defined column-wise towards a branch of the hierarchy at the top of the figure (i.e. C5: Not-Filtered-Epoched-mean-removal-wnExG). The mean and standard errors of MIR, PMI and Dipolar sources for each ICA algorithm that was applied on all types of the defined pre-processed data are shown in panels 1.B, 1.C and 1.D, respectively.

onset) or continuous data were used (CNT) for ICA decomposition;
- EPOCH MEAN REMOVAL: in case of epoched data, the mean values across the epoch was removed (RmvMean) or not removed (noRmvMean) before ICA decomposition. Removing the epoch mean is recommended rather than removing the pre-stimulus baseline [7].

We obtained 12 pre-processing categories (conditions) illustrated in Fig 1 (upper part).

B. Performance Evaluation

Independence: ICA algorithms decompose EEG data into independent sources. In order to assess the independence of the decomposition, we considered two measures based on mutual information: Mutual Information Reduction (MIR) and Pairwise Mutual Information (PMI) [8]. MIR is the total reduction of mutual information when moving from the sensor space (EEG channels) to the source space (independent components). A larger MIR indicates a greater reduction of statistical dependencies between EEG channels, therefore higher independence of the components. PMI measures the mutual information between pairs of component time courses. A lower PMI is expected for components which are more independent. The mutual information for both MIR and PMI estimates is bias corrected [13].

Physiological plausibility: The sources estimated by ICA are supposed to be biologically plausible, in addition to being independent [8]. EEG sources are supposed to be spatially localized and appear dipolar in the far-field at the scalp electrodes distance. The components that can be fit with a dipole are considered more physiologically plausible. The Dipolarity is the number of components fitted by a dipole with less than e 5% residual variance. The number of non-dipolar sources can be minimized by the selected 5% threshold while obtaining the components that resemble to the dipole sources.

A repeated measure analysis of variance (ANOVA) with a within-subject factor algorithm was used to analyse the effect of different types of pre-processing on the performance of ICA algorithms. The continuous and epoched data were analyzed separately. A Bonferroni correction was performed for the within and between groups. Also, one-way ANOVA was used to analyse the statistical difference between ICA algorithms within each type of pre-processing, with MIR, PMI and Dipolarity as dependent measures and the algorithms as factor. The level selected to indicate statistical significance was at $P<0.05$.

III. RESULTS

Fig 1 (panels B, C and D) shows the MIR, PMI and Dipolarity for each condition and algorithm. We found that AMICA usually outperforms the other algorithms in terms of MIR and Dipolarity, while the best performance in terms of PMI is obtained by FastICA. However, the differences between these four algorithms fail to reach significance within each condition, except for not-filtered continuous data when EOG-ECG channels are included (condition 2); the difference between AMICA and Extended-Infomax, in terms of PMI, AMICA and FastICA, with respect to the number of dipolar sources, are statistically significant ($P<0.05$). Also, the statistical significant difference between AMICA and FastICa, in terms of Dipolarity, was obtained when the mean removed epoched data was not filtered and EOG-ECG channels were excluded (condition 5, $P<0.05$). When considering the pre-processing factors (ExG, FLT, SEG and RmvMean) we found that including the ExG channels when decomposing continuous filtered data (conditions 7 and 8), reduces PMI and increases Dipolarity without affecting MIR. We found that epoch mean removal has a strong impact on the decomposition: it improves the MIR in both filtered and non-filtered data (conditions 3, 4, 5, 6 and 9, 10, 11, 12) while reducing the PMI. It also improves Dipolarity in non-filtered data (conditions 3, 4 and 5, 6). The PMI values and Dipolarity are improved by filtering for both continuous (conditions 1, 2 and 7, 8) and epoched data (conditions 3, 4, 5, 6 and 9, 10, 11, 12), while MIR has not been affected. The variation between MIR and PMI values of ICA algorithms are larger for not-filtered continuous and not-filtered no-epoch-mean-removed data than other types of pre-processing conditions, in which ICA algorithms are more consistent. Fig 2 shows the relationship between MIR, PMI and Dipolarity (for the best three conditions 12, 5 and 8 in each panel, respectively). As the MIR values increase for all ICA algorithms through the three pre-processing conditions, the number of dipolar sources decrease (Fig 2.A). However, FastICA was more consistent in terms of number of dipolar sources across two pre-prosessing conditions: where EEG data is filtered, epoched, epoch-mean-removed and ExG channels are included (condition 12), and where EEG data is not filtered, epoched and epoch-mean-removed excluding ExG channels (condition 5). It is also shown that the relationship between Dipolarity and MIR, PMI and MIR, Fig 2.B and Fig 2.C respectively, is not linear for ICA algorithms across pre-processing conditions.

IV. DISCUSSION AND CONCLUSION

We studied how four most commonly applied ICA algorithms are affected by the pre-processing of EEG data. We found that including EOG-ECG channels in the filtered continuous data improved PMI and Dipolarity. The reason is that they can provide information for characterizing eye movement artifacts in the EEG data thus improving EEG source separation. The performance of ICA algorithms was affected by the existence of the DC component in the epoched data. When the presence of this component is eliminated by removing the mean of each epoch, the performance of ICA algorithms can improve to extract latent independent components from EEG data. This may increase the accuracy of components' activities and scalp topographies [7]. Therefore, the independence criterion and physiological plausibility of the estimated sources improved when the mean of each epoch was removed in the epoched data. We also found that filtering improved the decomposition of the continuous and epoched data with respect to PMI and Dipolarity: Filtering the data (0.16-40 Hz) can remove slow drifts and high frequency noise in the data. The presence of slow drifts and high frequency noise in the EEG data can occupy degrees of freedom in ICA space and so introducing common information between recovered components. Therefore, the PMI

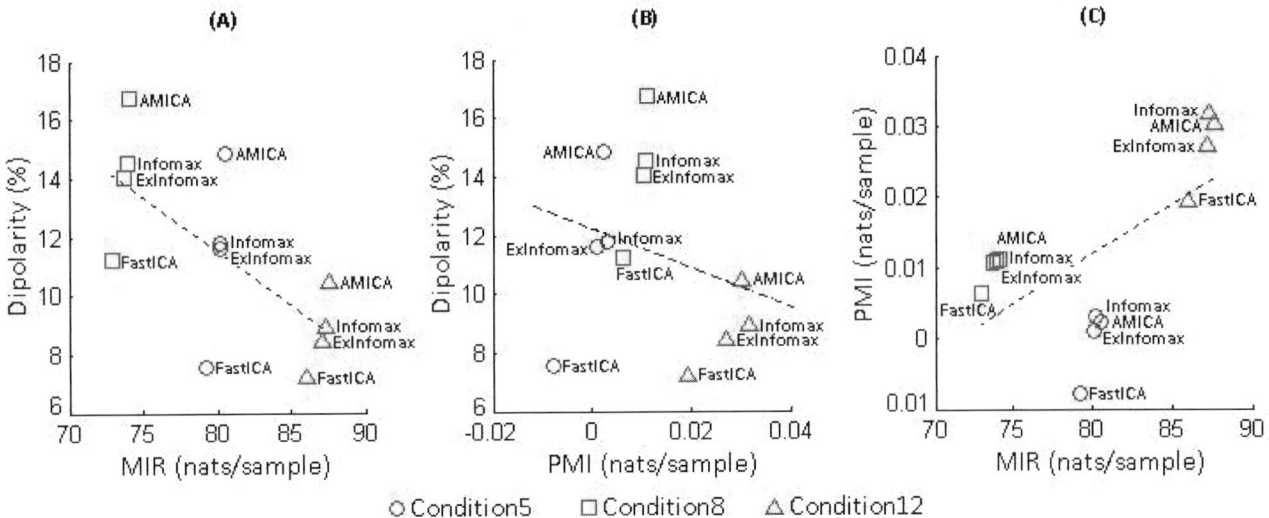

Fig. 2 The Correlation between MIR, PMI and Dipolarity for conditions 12, 5 and 8. The black dashed line represents linear regression. A. The near linear correlation between MIR and dipolar sources through conditions ($R^2 = 0.46$). B. Non-linear relationship between the number of dipolar sources and PMI values for three conditions ($R^2 = 0.07$). C. Non-linear relationship between MIR and PMI values through conditions ($R^2 = 0.40$).

values between estimated sources becomes higher for the continuous and epoched data when they are not filtered. Consequently, the number of dipolar sources decreases as independence of sources is decreased. The existence of DC component, slow drifts and high frequency noise caused more variation between MIR and PMI values of ICA algorithms, when continuous and epoched (no-epoch-mean-removed) data is not filtered. We obtained the near linear relationship between MIR and Dipolarity across the best three pre-processing conditions. However, improving the Dipolarity and MIR across the three pre-processing conditions does not convey the improvement of PMI values necessarily.

Overall, the reliability of estimated sources by ICA is considerably dependent on the type of pre-processing of the given data to ICA algorithms rather than a type of ICA algorithm itself. Although selecting the type of ICA algorithms is of concern in EEG analysis, algorithm type does not make a great difference for artifact removal. The pre-processing type of non-filtered epoch mean removed (with and without EOG-ECG channels) gives overall better results in terms of independence of components and Dipolarity (conditions 5 and 6).

REFERENCES

1. Koubeissi Mohamad Z. Niedermeyers Electroencephalography, Basic Principles, Clinical Applications, and Related Fields *Archives of Neurology*. 2011;68:1481–1481.
2. Sanei Saeid, Chambers Jonathon A. *EEG signal processing*. Wiley-Interscience.
3. Aapo Hyvarinen, Juha Karhunen, Erkki Oja. Independent component analysis 2001.
4. Lee Te-Won, Girolami Mark, Bell Anthony J, Sejnowski Terrence J. A unifying information-theoretic framework for independent component analysis *Computers & Mathematics with Applications*. 2000;39:1–21.
5. Krishnaveni V, Jayaraman S, Kumar PM Manoj, Shivakumar K, Ramadoss K. Comparison of independent component analysis algorithms for removal of ocular artifacts from electroencephalogram *Meas. Sci. Rev. J*. 2005;5:67–78.
6. Delorme Arnaud, Sejnowski Terrence, Makeig Scott. Enhanced detection of artifacts in EEG data using higher-order statistics and independent component analysis *Neuroimage*. 2007;34:1443.
7. Groppe David M, Makeig Scott, Kutas Marta. Identifying reliable independent components via split-half comparisons *NeuroImage*. 2009;45:1199.
8. Delorme Arnaud, Palmer Jason, Onton Julie, Oostenveld Robert, Makeig Scott. Independent EEG sources are dipolar *PloS one*. 2012;7:e30135.
9. Korats Gundars, Le Cam Steven, Ranta Radu et al. Impact of window length and decorrelation step on ICA algorithms for EEG Blind Source Separation in *International Conference on Bio-inspired Systems and Signal Processing-Biosignals 2012* 2012.
10. Palmer Jason A, Kreutz-Delgado Ken, Rao Bhaskar D, Makeig Scott. Modeling and estimation of dependent subspaces with non-radially symmetric and skewed densities in *Independent Component Analysis and Signal Separation*:97–104,Springer 2007.
11. Bell Anthony J, Sejnowski Terrence J. An information-maximization approach to blind separation and blind deconvolution *Neural computation*. 1995;7:1129–1159.
12. Hyvärinen Aapo, Oja Erkki. Independent component analysis: algorithms and applications *Neural networks*. 2000;13:411–430.
13. Miller George A. Note on the bias of information estimates *Information theory in psychology: Problems and methods*. 1955;2:95–100.

Muscle Synergies Underlying Voluntary Anteroposterior Sway Movements

S. Piazza[1], D. Torricelli[1], I.M. Alguacil Diego[2], R. Cano De La Cuerda[2], F. Molina Rueda[2],
F.M. Rivas Montero[2], F. Barroso[1], and J.L. Pons[1]

[1] Spanish National Research Council (CSIC), Madrid, Spain
[2] Movement Analysis, Biomechanics, Ergonomic and Motor Control Laboratory (LAMBECOM),
Faculty of Health Sciences at the Rey Juan Carlos University, Madrid, Spain

Abstract—Stable postural control is a fundamental ability for humans, which requires complex interactions of several neural mechanisms. The deep analysis of this ability presents either diagnostic and rehabilitation potentials.

In this study, seven healthy participants performing voluntary postural sway movements along the anteroposterior direction are analyzed. Muscle activity is decomposed with muscle synergies analysis in order to determine if it presents evidence of an internal modular organization.

Results indicates that two modules are sufficient to account for over the 90% of muscle variability, suggesting a strong evidence for modular control.

Keywords— motor control theories, postural control, muscle synergies.

I. INTRODUCTION

Human ability to maintain stable balance represented always an important topic of scientific interest. On one side, the multiple nervous circuits recruited for the control of such an apparently simple task make postural control analysis a preferred diagnostic tool for the evaluation of neuralogical impairments. On the other side, the fundamental role of balance in all principal lower limb tasks, makes postural rehabilitation of primary importance. When postural control is degraded, all other bipedal functions are affected. Effective techniques of diagnosis and rehabilitation of postural control are of primary importance for the recovery of lower limb functions and the improvement of a subject's quality of life.

From studies on human and animals, novel theories of motor control have been elaborated that the nervous system operates a modular strategy to activate multiple body muscles. These modules, known as muscle synergies, may allow to reduce the complexity on the control task while maintaining the flexibility in movements. Recently, muscle synergies have already been proposed as a potential diagnostic tool [1]. In the realm of postural control, experimental findings [2, 3] support the hypothesis that balance can be described by a delayed feedback control process, in which muscle synergies are temporally recruited according to task-level variables such as CoM or CoP kinematics. This model has been confirmed by experiments in animals [4] and healthy subjects [5, 6, 3] during reactive conditions, e.g. responses to platform translations. Also for voluntary movements, studies demonstrate that mediolateral (ML) rhythmic sway can be described by two muscle synergies modules [7, 8]. In the anteroposterior (AP) plane muscle modular control has been observed in terms of M-Modes [9, 10, 11] but not muscle synergies. The goal of this study is to verify if a reduced number of muscle synergies can be observed during the execution of voluntary postural sway movements in the AP direction.

II. METHODS

A. Participants

Seven healthy subjects (five males and two females) participated in the experiment. Subjects' mean age was 33.2 years (± 10 SD). The research was carried out at the Movement Analysis, Biomechanics, Ergonomic and Motor Control Laboratory (LAMBECOM), Faculty of Health Sciences at the Rey Juan Carlos University (Madrid, Spain). The investigation took place between February and April, 2012. The study has been approved by local Human Ethics Committee and written consent was obtained from all participants.

B. Procedures

Participants were asked to step onto the a platform base of a Computerized Dynamic Platform Posturography system (CDPP), each foot positioned on one force plate according to the manufacturer's instructions. Participants were instructed to stand upright, leaving arms at their side throughout the entire test and with eyes open. All participants wore a safety harness to prevent falling. The tests were conducted in a peaceful atmosphere and with the removal of alcohol, smoke and psychotropic drugs in the previous 24 hours.

We defined a cycle in the AP plane as the movement associated to the displacement of the CoP from the anterior position to the posterior and back. Subjects were asked to perform Rhythmic Weight Shift (RWS) movements in the ML direction, followed by RWS movements in the AP direction. Each exercise was repeated at the sway frequencies of 0.167

Hz (10 bit per minute, bpm), 0.250 Hz (15 bpm), and 0.5 Hz (30 bpm. For each frequency, subjects were allowed to familiarize with the exercise and synchronize their movement with the auditory cues for a minute. Afterwards, sEMG and CoP displacement were recorded during three complete cycles. Trials were repeated four times, resulting in a total of 12 sway cycles for each frequency. Auditory information was provided to the subject by means of a digital metronome to allow for synchronization.

Superficial EMG signals have been recorded with a ZeroWire (Cometa ©, Italy) 16 channel wireless electromyograph equipped with 50 mm diameter surface Ag/AgCl circular electrodes (DORMOR©, Telic S.A., Spain) and synchronized with force data from CDDP platform. sEMG electrodes were placed according to the SENIAM recommendations [12] on the following muscles: gluteus maximus GMAX), gluteus medius (GMED), tensor fascia latae (TFL), rectus femoris (RF), biceps femoris (BF), gastrocnemius lateral (GAS), soleus (SOL), and tibialis anterior (TA).

C. EMG Processing

EMG signals have been sampled at 2000 Hz, passband filtered (20-400Hz, Butterworth 4th order), rectified, and zero-lag low-pass filtered to estimate muscle activity. To obtain the same pattern of smoothing across conditions, the threshold of the LP filter was bind to the sway rate [13], using 0.85 Hz, 1.25 Hz and 2.5 Hz for the 0.17 Hz, 0.25 Hz and 0.5 Hz sway respectively. EMG envelopes of each individual muscle has been then normalized to the peak value of the same muscle during all the trials, considering AP and ML trials, and resampled at 1% of the sway cycle.

Muscle synergies of each subject were extracted from a frequency-specific pool of data Ω and a frequency-generic pool of data Π. Ω is composed by the envelopes from AP sway at each specific frequency. The second pool Π is the combination of the envelopes from the three sway conditions.

For each pool, nonnegative matrix factorization (NNMF) [14] was used to decompose the envelopes. Using different matrix factorization algorithms may provide results with similar consistency [15]. However, NNMF was chosen because his components are constituted only by positive values, closely resembling the mechanical characteristics of human muscles.

The algorithm was initialized with random values and run iteratively for 20 times to avoid local minima [16]. The solution providing better reconstruction (lower squared error between the original and the reconstructed EMG envelopes) was chosen.

$\hat{W}$ and $\hat{H}$ are the muscle synergies vectors and activations extracted from Π, while W_i and H_i are extracted from the i-th frequency from the data pool Ω.

We iterated the analysis of each subject by varying the number of synergies between 1 and 5. At each iteration step, we estimated the quality of the reconstruction by calculating the mean total variability accounted for (VAF) and the variability accounted for per each muscle (VAF_{muscle}) [5].

The optimal number of muscle synergies was chosen by considering a double threshold value for VAF. We chose the lowest number of synergies to which either VAF was greater than 90% [17] or the increment with respect to VAF of previous iteration was <5% [18].

D. Muscle Synergies Sorting

To compare synergies we performed a functional sorting based on the synergy recruitment profile: H was averaged around the middle part of the sway cycle and the following algorithm was used to rate the synergy:

$$\lambda_i = \sum_{t=40\%}^{60\%} (\bar{H}_i)^2(t) \quad (1)$$

Where λ_i is the comparison value for the module W_i, $\bar{H}_i$ is the mean activation and the sum considers the central interval, between 40% and 60% of the sway cycle. The modules were then ordered in ascending order based on the value of λ.

E. Statistical Analysis

To verify that every subject was able to perform sway at the predefined frequencies, we measured the sway frequency from the movements of the CoP, and extracted the mean and SD for each condition.

Similarities between the synergy vectors (W_i) extracted from the frequency-specific data pool Ω and reference synergy vectors $\hat{W}$ extracted from the frequency-generic pool Π have been analyzed by means of Pearson correlation coefficient (r) across subjects and conditions. The same algorithm has been applied to activation coefficients H_i and reference activations $\hat{H}$.

The analysis was done in Matlab (Mathworks ®). A testing for the hypothesis of no correlation has been also performed through the calculation of the P value, meaning the probability of getting a correlation as large as the observed values by random chance. Small P value (< 0.05) means statistically significant correlation.

III. RESULTS

Movements were performed at mean frequencies of 0.18 ± 0.03, 0.25 ± 0.04, 0.46 ± 0.12. Data show high synchronization with reference frequency. Shapes and peak-to-peak amplitudes were also very similar across subjects and conditions: after the process of time scaling from 0 to 100% of each sway cycle, profiles showed high similarities (Figure 1) across subjects, with a correlation coefficient $r = 0.95 \pm 0.07 (P < 0.001)$ across all conditions.

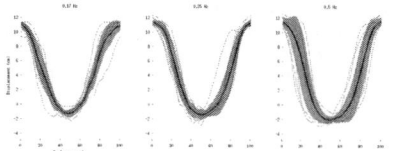

Fig. 1 CoP scaled between 0 and 100% of the sway cycle for a representative subject. Each cycle is represented by a magenta line. Black lines represent median values, The grayed area surrounding represents the standard deviation.

EMG of each individual do not present high variability (fig. 2). Subject 6 present a different pattern of activation in GAS, RF Gmax and GMed indicating a possible outlier from electrodes misplacement.

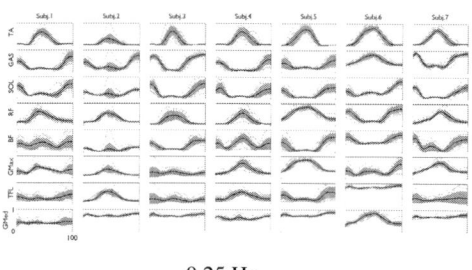

0.25 Hz

Fig. 2 EMG envelops from the 8 leg muscles over the (up to) 12 cycles, performed during sway movements AP direction at a representative condition (0.25 Hz). Bold line depicts the mean muscle activation across trials and the gray shade his standard deviation.

The analysis of dimensionality reveals that two modules are sufficient to reconstruct most of the original EMG envelopes when considering the data pool Ω. Mean VAF for all subjects is $94.98 \pm 2.05\%$, $93.73 \pm 2.24\%$ andsented the reference synergy vectors $\hat{W}$ and activations $\hat{H}$ from the data pool Π. On the second $95.23 \pm 2.33\%$. As depicted in Figure 3, 2 modules account for the variability of each muscle in all conditions, resulting in a mean value of $VAF_{muscle} > 70\%$ in all muscles and conditions.

When considering the frequency-generic data pool Π, two modules still account for the $93.7 \pm 2\%$, satisfying the threshold of 90% of the VAF (fig. 4).

To test the hypothesis that participants use the same synergies across different sway frequencies, the correlations between W_i and $\hat{W}$, and between H_i and $\hat{H}$ were calculated. Results indicate high level of correlation for all exercises (table 1)

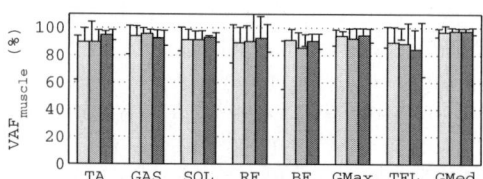

Fig. 3 Mean value and standard deviation of VAF_{muscle} between all patients when using 2 muscle synergies for the factorization of the frequency-specific data-set Ω. Each group of bar refers to a different muscle. Each bar represents a different exercise.

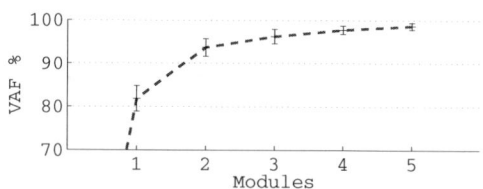

Fig. 4 Variance accounted for (VAF) as a function of number of modules. Modules are extracted from the data pool Π.

Reference synergies and activations from all subjects are represented side-by-side in Figure 5.

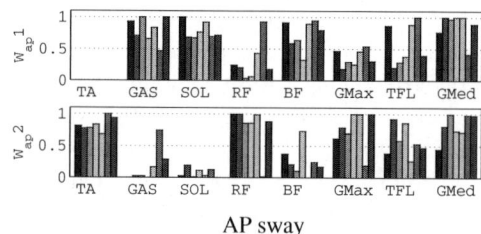

AP sway

Fig. 5 Muscle synergies vectors of each subject extracted from frequency-generic data pool Π. Every color corresponds to one subject. Please note that Subject 6 may present GAS, RF, GMax and GMed inverted.

IV. DISCUSSION

In the experiment we analyzed sway movements at three different frequencies along the AP direction. CoP profiles of each exercise were compared to evaluate kinematic performance. Solutions obtained from NNMF decomposition were tested while varying the number of modules. We found that 2 modules are sufficient to describe more than the 90% of the variance in the original sEMG signal across all the conditions.

We hypothesized that the same modules are used in different sway frequencies. Our observation confirmed that both the modules and their activations are persistent between different sway frequencies.

From fig. 5 we can observe common trends in the values of the synergy vectors $\hat{W}$ of different subjects. GAS, SOL and BF are controlled by $\hat{W}_1$, TA RF and GMax by $\hat{W}_2$.

Table 1 Correlation between frequency-specific vectors W_i, H_i and reference vectors $\hat{W}$, $\hat{H}$.

	W_1			W_2		
Hz	0.17	0.25	0.50	0.17	0.25	0.50
Mean	0.97	0.98	0.96	0.97	0.96	0.96
SD	±0.03	±0.02	±0.02	±0.01	±0.05	±0.03
	H_1			H_2		
Hz	0.17	0.25	0.50	0.17	0.25	0.50
Mean	0.99	0.99	0.98	0.99	0.99	0.97
SD	±0.00	±0.01	±0.01	±0.01	±0.01	±0.02

The contribute to TFL and GMed are equally high in the two modules, which may suggest that the modulation of these two muscles is not important for the task considered, as can confirmed by the flat activation pattern observable in fig. 2 . Our results suggests that different subjects may adopt a common modular organization during postural sway movements.

V. CONCLUSIONS

This study provides experimental support to the idea that anteroposterior balance in healthy subjects is controlled by spatially fixed muscle synergies, modulated by time-varying activations, and that these modules are maintained across different sway frequencies. This modular control strategy may serve to reduce motor control complexity and present similar features between different subjects .

The comparison between different subjects also suggests similarities in the modules compostion. However, a dedicated analysis should be performed to confirm this hypothesis.

ACKNOWLEDGEMENTS

This project was funded by the Spanish Ministry of Science and Innovation CONSOLIDER-INGENIO, project HYPER (Hybrid NeuroProsthetic and NeuroRobotic Devices for Functional Compensation and Rehabilitation of Motor Disorders, CSD2009-00067

REFERENCES

1. Safavynia Seyed A, Torres-Oviedo Gelsy, Ting Lena H. Muscle Synergies: Implications for Clinical Evaluation and Rehabilitation of Movement. *Topics in spinal cord injury rehabilitation.* 2011;17:16–24.
2. Torres-Oviedo Gelsy, Ting Lena H. Subject-specific muscle synergies in human balance control are consistent across different biomechanical contexts. *Journal of neurophysiology.* 2010;103:3084–3098.
3. Safavynia Seyed a, Ting Lena H. Task-level feedback can explain temporal recruitment of spatially fixed muscle synergies throughout postural perturbations *Journal of neurophysiology.* 2012;107:159–77.
4. Torres-Oviedo G, Macpherson J M, Ting L H. Muscle synergy organization is robust across a variety of postural perturbations *Journal of Neurophysiology.* 2006;96:1530–1546.
5. Torres-oviedo Gelsy, Ting Lena H, Kautz Steven A, Bowden Mark G, Clark David J, Neptune Richard R. Muscle synergies characterizing human postural responses. *Journal of Neurophysiology.* 2007;98:2144–2156.
6. Chvatal Stacie A, Torres-Oviedo , G. Safavynia S A, Ting Lena H, Torres-Oviedo Gelsy, Safavynia Seyed a. Common muscle synergies for control of center of mass and force in nonstepping and stepping postural behaviors. *Journal of neurophysiology.* 2011;106:999–1015.
7. Torricelli D, Aleixandre M, Alguacil Diego I M, et al. Modular control of mediolateral postural sway in *34th Annual International Conference of the IEEE Engineering in Medicine and Biology Society (EMBC)*;2012:3632–5 2012.
8. Piazza S, Mansouri M, Torricelli D, Reinbolt J A, Pons J L. A biomechanical model for the validation of modular control in balance in *Converging Clinical and Engineering Research on Neurorehabilitation*:815–819Springer Berlin Heidelberg 2013.
9. Wang Yun, Asaka Tadayoshi, Zatsiorsky Vladimir M, Latash Mark L. Muscle synergies during voluntary body sway: combining across-trials and within-a-trial analyses. *Experimental Brain Research.* 2006;174:679–693.
10. Danna-Dos-Santos Alessander, Slomka Kajetan, Zatsiorsky Vladimir M, Latash Mark L. Muscle modes and synergies during voluntary body sway. *Experimental brain research. Experimentelle Hirnforschung. Expérimentation cérébrale.* 2007;179:533–50.
11. Krishnamoorthy Vijaya, Latash Mark L, Scholz John P, Zatsiorsky Vladimir M. Muscle synergies during shifts of the center of pressure by standing persons. *Experimental brain research. Experimentelle Hirnforschung. Expérimentation cérébrale.* 2003;152:281–92.
12. Hermens H J, Freriks B, Stegeman D, et al. European Recommendations for Surface ElectroMyoGraphy Tech. Rep. 5 2000.
13. Hug François. Can muscle coordination be precisely studied by surface electromyography? *Journal of electromyography and kinesiology : official journal of the International Society of Electrophysiological Kinesiology.* 2011;21:1–12.
14. Lee DD, Seung HS. Algorithms for non-negative matrix factorization *Advances in neural information processing . . . .* 2001.
15. Tresch Matthew C, Cheung Vincent C K, D'Avella Andrea. Matrix factorization algorithms for the identification of muscle synergies: evaluation on simulated and experimental data sets. *Journal of neurophysiology.* 2006;95:2199–212.
16. Cheung Vincent C K, D'Avella Andrea, Tresch Matthew C, Bizzi Emilio. Central and sensory contributions to the activation and organization of muscle synergies during natural motor behaviors. *The Journal of neuroscience : the official journal of the Society for Neuroscience.* 2005;25:6419–34.
17. Hug F., Turpin N.A., Couturier Antoine, Dorel Sylvain. Consistency of muscle synergies during pedaling across different mechanical constraints *Journal of Neurophysiology.* 2011;106:91.
18. Clark David J, Ting Lena H, Zajac Felix E, Neptune Richard R, Kautz Steven A. Merging of healthy motor modules predicts reduced locomotor performance and muscle coordination complexity post-stroke. *Journal of neurophysiology.* 2010;103:844–57.

Recognition of Brain Structures from MER-Signals Using Dynamic MFCC Analysis and a HMC Classifier

Mauricio Holguin[1], German A. Holguin[1], Hernán Darío Vargas Cardona[1], Genaro Daza[2], Enrique Guijarro[3], and Alvaro Orozco[1]

[1] Department of Electrical Engineering, Universidad Tecnológica de Pereira, Pereira, Colombia
[2] Instituto de Epilepsia y Parkinson del Eje Cafetero, Pereira, Colombia
[3] I3BH, Universitat Politécnica de Valencia, Valencia, Spain

Abstract—A novel methodology for the characterization of Microelectrode Recording signals (MER-signals) in Parkinson's patients in order to recognize basal ganglia in the brain is presented in this work. The most common approach of MER signals analysis consists of time-frequency analysis through Short Time Fourier Transform, Wavelet Transform, or Filters Banks. We present an approach based on MEL-Frequency Cepstral Coefficients (MFCC) and K-means clustering to obtain dynamic features from MER-signals. A Hidden Markov Chain (HMC) with 1, 2, 3, and 4 states was used for the classification of four classes of basal ganglia: Thalamus (Tal), Zone Incerta (ZI), Subthalamic Nucleus (STN) and Substantia Nigra reticulata (SNr), achieving a positive identification over 82%. A performance analysis for each HHM model is presented using ROC curves.

Keywords—**Parkinson's Disease, MER signals, MEL-Frequency Cepstral Coefficients (MFCC), Hidden Markov Chain (HMC), Dynamic Features.**

I. INTRODUCTION

Parkinson's Disease (PD) is one of the most studied movement disorders. Deep Brain Stimulation (DBS) of the Subthalamic Nucleus (STN) has showed an improvement on some Parkinson's symptoms. Depending on the microelectrode stimulator location, different symptoms can be disabled [1]. The use of imaging techniques like Magnetic Resonance Imaging (MRI) and Computed Tomography (CT) is necessary for the surgical planning. However, the use of MRI and CT is not enough for validating the location of target points and the navigation trajectory, because they do not provide physiological information of the brain. For this reason, the interpretation of physiological signals known as Microelectrode Recording signals (MER-signals), is crucial [2]. MER signals have a non-stationary behavior due to the contribution of various factors including their own variation of the discharges, which are not exactly the same, or exactly regular in rhythm. Identification of basal ganglia, from processing and classification of MER signals during DBS in PD patients, serves as medical support for the correct location of a target brain area and the respective implantation of microelectrodes. A previous work [3], has addressed processing approaches of MER signals based in temporal analysis of spikes, however the performance of this analysis is significantly reduced in areas with low spike activity, i.e. Zone Incerta (ZI). Another common approach is time-frequency analysis, where the MER signals are transformed to different spaces, for example, the Short-Time Fourier Transform space(STFT) [4], the wavelet space (WT), and also with empirical methods like the Hilbert-Huang Transform (HHT) [5]. The processing of non-stationary and highly oscillatory signals, such as MER, requires the use of special methods. Analysis by adaptive filter banks is one of the most powerful methods for characterization of the wavelet [6]. Adaptive filters are a wavelet decomposition assembly adapted to signals, constructed by Lifting Schemes (LS). This characterization methodology combined with simple classifiers such as a linear or quadratic Bayesian has shown to be a very efficient procedure to extract discriminant information from MER signals [7]. Similarly, more sophisticated classifiers like Support Vector Machines with Polynomial Kernel and Hidden Markov Models Classifier(HMM), achieved excellent results in automatic identification of basal ganglia [8]. In this work, we propose a signal processing methodology based on MEL-Frequency Cepstral coefficients (MFCC) and K-means clustering to obtain dynamic features from MER signals. The cepstral analysis method is viewed as a unified approach to the linear prediction method, in which the model spectrum varies continuously from all-pole to cepstral according to the value of a parameter γ. The efficiency of MEL cepstral analysis in MER-signals recognition is demonstrated using a HMC classifier with 1, 2, 3, and 4 states.

II. MATERIALS AND METHODS

A. MER Signals Database

The database of the Technological University of Pereira (DB-UTP) includes recordings of surgical procedures in patients with Parkinson's disease, whose ages are between

55 ± 6 (3 men, 1 woman). All the patients signed an informed consent form. Microelectrode recordings were obtained using the ISIS MER system (Inomed Medical GmbH). MER signals were labeled by neurophysiology and neurosurgery specialists from the Institute of Parkinson and Epilepsy of the Colombian coffee growing region, located in the city of Pereira. In total, there are 190 neural recordings divided in four classes: 26 signals from Thalamus (TAL), 71 signals from STN, 80 signals from SNr, and 13 from ZI. Each recording lasted 1s with a sampling frequency of 25 kHz and 16-bit resolution.

B. MEL-Frequency Cepstral Coefficients (MFCC)

MEL-Frequency Cepstral Coefficients (MFCC) are parametric features based on the perceptual bandwidth of the human hearing system. MFCC is composed of two kinds of filters linearly spaced for frequencies below 1kHz and logarithmically spaced for frequencies above 1kHz. [9]. This spacing is known as the MEL scale. MFCC has both time-domain and frequency domain information allowing for both linear and non-linear dynamic feature extraction. These dynamic features are a common ground for both voice signals and signal from MER [10]. The general MFCC method consists of seven steps that are shown in Fig. 1 [9].

- **Segmentation**: Is the process of dividing the signal into small frames with N samples. Usually, each frame has some overlap with its adjacent frames.
- **Windowing:** Usually, a Hamming window is applied in order to smooth the sides of the frame allowing for the integration of all lines that are close in frequency.
- **Pre-emphasis:** This is the application of a filter that emphasizes high frequencies. Energy of the signal in high frequencies is increased.
- **FFT:** Fast Fourier Transform using

$$Y_m = \frac{1}{F} \sum_{n=0}^{F-1} x(n)w(n)e^{-j\frac{2\pi}{F}nm} \quad (1)$$

where, F is the number of frames, w the Hamming window, and x the input signal.

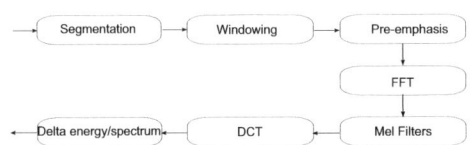

Fig. 1 MFCC extraction and preprocessing.

- **MEL Filter:** A bank of uniformly spaced filters is applied in the Mel scale. This bank of filters has a frequency response with band-pass triangular shape. The magnitude of the response of each filter is equal to 1 right in the center and it decreases linearly to zero in the center of the adjacent filter.
- **DCT:** The Discrete Cosine Transformation is applied, converting the Mel spectrum back into the time domain. The result of these conversion is known as MFCC.
- **Delta Energy/Spectrum:** Given the quasi-stationary property of the processed signals, the frame characteristics change as they overlap while evolving. To represent this phenomenon, 13 deltas in velocity and 39 double deltas in acceleration are added.

C. Clustering BY K-MEANS

As a result of the MFCC feature extraction procedure, a discretization method that can represent the coefficients as discrete observations is required. To achieve this goal, the K-means clustering technique is used [11]. The original procedure defines k centroids according to the existence of k clusters. The centroids are placed such that they are as far apart as possible from each other. Each subsequent observation is associated with the nearest centroid. Then, all centroids are iteratively relocated in order to minimize the distance cost function between observations and centroids.

D. Hidden Markov Chain (HMC)

A Markov chain is a system that can be considered, for all time instants, to be in one of a discrete set of N states $\{S_1, S_2, ..., S_N\}$. At uniformly spaced discrete time intervals, the system performs transitions between states according to a transition probability set, for a given time t and the current state q_t. The Markov chain is said to be of the first order if the transition probability depends only on the current state,

$$\begin{aligned} P\{q_t = S_j | q_{t-1} = S_i, q_{t-2} = S_k, ...\} \\ = P\{q_t = S_j | q_{t-1} = S_i\} = a_{i,j} \end{aligned} \quad (2)$$

with $a_{i,j} > 0$ and $\sum a_{i,j} = 1$.

The model described up to this point is known as Observable Markov Chain since the process output is the same as the state set for each time instant. A Hidden Markov Chain (HMC), is the extension of the observable model for the case where the output is a probabilistic function of the state, thus the model is a double embedded stochastic process not directly observable, but indirectly observable via the set of output sequences, as discussed in [12].

A HMC is characterized by:

- N: The number of states.
- M: The number of different symbols that are observable in the states. The set of symbols is given by $V = \{v_1, v_2, \ldots, v_M\}$.
- $A = [a_{i,j}]$: state transition probability distribution.
- $B = [b_j(k)]$: observation symbol probability distribution in state j, where $b_j(k) = P\{v_k \text{ at } t | q_t = S_j\}$.
- $\pi = [\pi_i]$: initial state distribution, where $\pi_i = P\{q_1 = S_i\}$.
- O: observation sequence, where $O = \{O_1, O_2, \ldots O_T\}$, with O_t one of the symbols from V.
- λ: parameters set for HMC model, where $\lambda = \{A, B, \pi\}$.

HMC are used to solve three classical problems:

- The evaluation problem: Given the observation sequence and the model, how to efficiently compute $P\{O|\lambda\}$, the probability distribution of the observation sequence given the model.
- The decodification problem: Given the observation sequence and the model, how to adequately choose a sequence of states that gives the best explanation of the observations.
- The training problem: Given the observation, how to fit the model parameters to maximize $P\{O|\lambda\}$.

E. Receiver Operating Characteristic (ROC)

ROC analysis is a method to compare the diagnostic accuracy of various tests objectively, and was originally developed in the early 50's in order to analyze the detection of radar signals [13]. For the ideal case of a classification test, the probability distributions of the test do not overlap and have a qualifying threshold in the middle of the two distributions. Under this ideal scenario, the sensitivity and specificity were 100%. However, for the vast majority of test classification probability distributions overlap and the threshold of discernment can lead to misclassification. Changes in the threshold value always involve an inverse relationship between sensitivity and specificity. A ROC curve represents the relationship between sensitivity and specificity, when evaluated for all possible threshold values.

III. EXPERIMENTAL RESULTS

A. Model Selection

To select the best model under a set of HMC with discrete observations, it is necessary to evaluate the diagnostic accuracy of different models. To perform this evaluation, ROC analysis is performed on HMC with 1, 2, 3 and 4 states respectively.

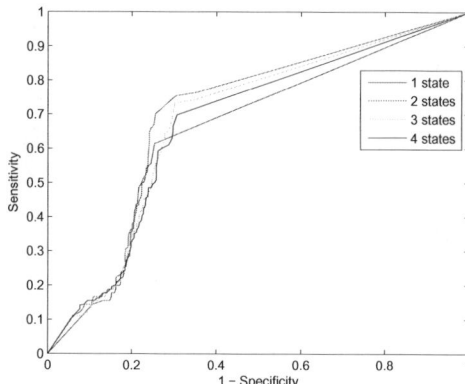

Fig. 2 ROC Curve for HMC with 1(blue curve), 2(red curve), 3(yellow curve) y 4(black curve) states.

Table 1 Area under ROC curves

States	1	2	3	4
Area	0.7018	0.7329	0.7141	0.7091

The initial step, prior to the training of a model is feature extraction by applying MFCC to frames of 200ms, with overlap between frames of 1/3 and applying a Hamming window type.

The database was separated into 60% for training and 40% for validation and discretizes the features observed by K-means clustering. Finally, we used an Expectatio-Maximization training algorithm for HMC according to the number specific states in order to obtain representative ROC curve for each. Fig. 2 shows the curves obtained.

For the evaluation of the different models can use the area under the ROC curve [13]. Table 1 shows these results, giving that the 2-state model has the largest area under the curve with a value of 0.7329.

B. Training and Validation

Once the appropriate number of states for the HMC is known, we proceed with training in order to discern the signals from different brain areas. For the selected model, we get a total 24 MFCCs with a bank of 24 filters and a discretization with 32 centroids. For the trained model and the validation data set, Table 2 shows the results of the recognition accuracy after running the algorithm 10 times.

Table 2 shows the results in percentages, where a row represents an algorithm running and the column Total shows the average for the four brain zones. The bottom row called Total represents the average and the standard devation for the 10 executions of the algorithm for a particular zone. In the lower right corner, we see that the average of the 10 runs in the four brain zones is 82.59%.

Table 2 Recognition accuracy from MER-signals using our approach

Iteration	SNr	STN	Tal	ZI	Total
1	0.9643	0.6250	1	1	0.8973
2	0.9643	0.5000	1	0.4000	0.7161
3	0.9643	0.5000	1	1	0.8661
4	0.9643	0.5938	1	1	0.8895
5	0.9643	0.4375	1	0.4000	0.7004
6	1	0.4375	1	1	0.8594
7	1	0.4375	1	1	0.8594
8	1	0.5000	1	0.8000	0.8250
9	0.8571	0.5938	1	0.6000	0.7627
10	1	0.5313	1	1	0.8828
Total	0.9679±0.0428	0.5156±0.0695	1±0.0	0.82±0.2573	0.8259

IV. CONCLUSIONS AND FUTURE WORK

Our results showed that the MEL-Frequency Ceptral Coefficients, with its bank of filters that allow a continuous frequency scale, provide a dynamic set of features that distinguish between different brain structures. These factors, together with the discretization of features using K-means to optimize the number of centroids required, adequately represent a discrete number of observations for training HMC.

From the ROC curve analysis, Fig. 2 and Table 1, it can be clearly concluded that a HMC model with 2 states presents the best discriminant performance in recognizing brain areas. It also highlights the importance of selecting a 2-state model over models with 3, 4 or more states since training on those models require more computational resources.

From Table 2 it can be seen that areas of the Thalamus and Substantia Nigra Reticulata present the best accuracy in the recognition process with 100% and 96.79% respectively. In turn, the Subthalamic Nucleus area has the lowest accuracy with a 51.56% and is responsible for a greater proportion in the overall accuracy of the process.

As future work, there is a need to delve into the dynamics of the signals related to areas critical to the overall process of recognition such as the STN and ZI. From the results obtained, it can be inferred that improving recognition and uncertainty in these two areas can significantly improve the process to achieve global recognition.

ACKNOWLEDGMENTS

This research is developed under the projects: *"Desarrollo de un sistema automático de mapeo cerebral y monitoreo intraoperatorio cortical y profundo: aplicación neurocirugía"* and *"Desarrollo de un sistema efectivo y apropiado de estimación de volumen de tejido activo para el mejoramiento de los resultados terapéuticos en pacientes con enfermedad de parkinson intervenidos quirúrgicamente"*, both financed by Colciencias with codes 111045426008 and 111056934461 respectively.

REFERENCES

1. O. K. Chibirova, *Unsupervised spike sorting of extracellular electrophysiological recording in subthalamic nucleus of parkinsonian patients.* BioSystems 79(1-3): 159-171, 2005.
2. C. Hamani, E.O. Richter, Y. Andrade, W. Hutchison, J.A. Saint-Cyr and A.M. Lozano, *Correspondence of microelectrode mapping with magnetic resonance imaging for subthalamic nucleus procedures.* Surg Neurol 63: 249ï¿½253, 2005.
3. H. Chan, T. Wu, S. Lee, M. Lin, S. He, P. Chao, and Y. Tsai, *Unsupervised wavelet-based spike sorting with dynamic codebook searching and replenishment.* Neurocomputing, vol. 73, no.7-9, pp. 1513-1527,2010.
4. P. Novak, S. Daniluk, S. Elias, and J. Nazzaro, *Detection of the subthalamic nucleus in microelectrographic recordings in parkinson disease using the high frequency (500 hz) neuronal background.* Neurosurgey, no. 106, pp. 175-179, 2007.
5. R. Pinzon, M. Garces, A. Orozco, and J. Nazzaro, *Automatic identification of various nuclei in the basal ganglia for parkinson's disease neurosurgery,.* in EMBC 2009, Proceedings of the Annual International Conference of the IEEE Engineering in Medicine and Biology Society, Minneapolis, USA, pp. 99-104, 2009.
6. E. Giraldo, G. Castellanos and A.A. Orozco, *Feature Extraction for MER Signals Using Adaptive Filter Banks.* in Electronics, Robotics and Automotive Mechanics Conference, pp. 582-585, 2008.
7. R. Pinzon, A. Orozco, G. Castellanos and H. Carmona, *Feature Selection using an Ensemble of Optimal Wavelet Packet and Learning Machine: Application to MER Signals.* in Communication Systems Networks and Digital Signal Processing (CSNDSP), 7th International Symposium on. pp 25-30, 2010.
8. A. Tahgva, *Hidden semi-markov models in the computerized decoding of microelectrode recording data for deep brain stimulator placement.* in EMBC 2011, Proceedings of the Annual International Conference of the IEEE Engineering in Medicine and Biology Society, pp 758-764, 2011.
9. L. Muda, M. Begam, I. Elamvazuthi, *Voice Recognition Algorithms using Mel Frequency Cepstral Coefficient (MFCC) and Dynamic Time Warping (DTW) Techniques.* Journal of computing, vol 2, issue 3, ISSN 2151-9617. http://sites.google.com/site/journalofcomputing/, 2010.
10. F. V. Nelwamondo, T. Marwala. *Faults detection using Gaussian mixture models, Mel-Frecuency Cepstral Coefficientes and Kurtosis.* IEEE International Conference on Systems, Man, and Cybernetics. Taipei, Taiwan. pp. 290-295, 2006.
11. J. B. MacQueen. *Some Methods for classification and Analysis of Multivariate Observations.* Proceedings of 5-th Berkeley Symposium on Mathematical Statistics and Probability, Berkeley, University of California Press, 1: pp. 281-297, 1967.
12. L. R. Rabiner. *A tutorial on hidden Markov models and selected applications in speech recognition.* Proc. IEEE, vol. 77, pp. 257-286, 1989.
13. A.R. van Erkel, P.M.T. Pattynama, *Receiver operating characteristic (ROC) analysis: Basic principles and applications in radiology.* Elsevier Science Ireland Ltd. European Journal of Radiology, 27, pp. 88-94, 1998.

Feed-Forward Neural Network Architectures Based on Extreme Learning Machine for Parkinson's Disease Diagnosis

P.J. García-Laencina, G. Rodríguez-Bermúdez, and J. Roca-Dorda

Centro Universitario de la Defensa de San Javier (University Centre of Defence at the Spanish Air Force Academy),
MDE-UPCT, Santiago de la Ribera, Murcia, Spain

Abstract—Feed-Forward Neural Networks have been successfully applied for solving many biomedical problems. However, its design stage is far slower than required in practice. Recently, Extreme Learning Machine (ELM) has been proposed to solve this drawback. This paper presents several ELM architectures and its application for a real problem of recognizing Parkinson's disease. Experimental results show the usefulness of the ELM-based neural networks.

Keywords—**Neural Networks, Extreme Learning Machine, Medical Diagnosis, Parkinson's Disease.**

I. INTRODUCTION

Computational Intelligence (CI) methods and, in particular, Feed-Forward Neural Networks (FFNNs), provide proficient and powerful Computer Aided Diagnosis (CAD) systems [1, 2]. A CAD system learns hypotheses from a large amount of diagnosed patient records, i.e., the data collected from a number of necessary medical examinations along with the corresponding diagnoses made by medical experts, in order to assist the practitioners in making future diagnosis. Although FFNNs have shown their usefulness in biomedical applications [1, 2], these learning machines present an important design drawback due to the fact that their traditional training methods do not provide an efficient implementation [3, 4]: many parameters have to be properly tuned by slow (often gradient-based) algorithms. In order to solve this difficulty, G. B. Huang has proposed the Extreme Learning Machine (ELM) algorithm [5]. This method provides extremely fast training speed, easy implementation and good generalization performance. Nowadays, ELM has attracted many attentions and it has been used in different biomedical research areas with great success. However, ELM requires intensive search for sizing the architectures. Several approaches for ELM networks have been recently implemented in order to overcome this drawback [6, 7].

Following our previous research works [7], the main goals of this paper are to present several novel ELM-based neural architectures and, then, to extend these efficient methods for Parkinson's disease diagnosis [8] in order to shown its competitive advantage over the state-of-the-art procedures.

II. NOTATION FOR CLASSIFICATION IN CAD SYSTEMS

Classification problems involve the labeling of unclassified data with a specific output class [3]. For reaching this goal, the classification system has to be designed using a learning set of N samples $(\mathbf{x}_n, t_n)$ in order to $f(\mathbf{x}) \approx t$, where $f()$ denotes the classification system. Each input pattern, $\mathbf{x}_n = [x_{n1}, x_{n2}, ..., x_{nd}]^T \in \mathbb{R}^d$, is labeled as belonging to one of the C different classes, where the target t_n indicates the class of the n-th sample. Note that in CAD systems [1], each pair $(\mathbf{x}_n, t_n)$ is associated to the n-th patient, which is described by the d observed components of $\mathbf{x}_n$ related with the disease under study (p.e. gender, blood pressure, presence/absence of certain symptoms, …) and its corresponding diagnosis, denoted by t_n. As we have already mentioned, in this work, the dataset under study is a biomedical problem about diagnosis Parkinson's disease [8] and the classifier, $f()$, is modeled using FFNNs [4].

III. PARKINSON'S DISEASE DIAGNOSIS: DATASET

Parkinson's Disease (PD) is one of the most common neurodegenerative diseases. This disease is associated with a resting tremor, stiffness and rigidity of the muscles and slowness of movement.

Voice analysis has been used in many researches to detect various illnesses, specifically to the diagnosis of PD [8]. Voice signal recording is easy and non-invasive so it is a notable parameter for detecting and tracking the symptoms and progression.

In particular, the dataset under study has been created by Max Little of the University of Oxford, in collaboration with the National Centre for Voice and Speech, Denver, Colorado, who recorded the speech signals, and provided by UCI Machine Learning Repository on the internet [9]. This dataset is composed of a range of biomedical voice measurements from 195 voice recording of 31 people, 23 with PD. The main aim of the data is to discriminate healthy people from those with PD. From [8], the target is to predict the health status (1, Parkinson's; 0, Healthy) and the following input features are available to reach this goal.

Table 1 Input features of the Parkinson's disease dataset

Feature	Description	
x_1	Average vocal fundamental frequency (Hz)	
x_2	Maximum vocal fundamental frequency (Hz)	
x_3	Minimum vocal fundamental frequency (Hz)	
x_4	Jitter (%)	
x_5	Jitter(Abs)	Several measures of variation in fundamental frequency
x_6	RAP	
x_7	PPQ	
x_8	DDP	
x_9	Shimmer	
x_{10}	Shimmer(dB)	Several measures of variation in amplitude
x_{11}	APQ3	
x_{12}	APQ5	
x_{13}	APQ	
x_{14}	DDA	
$x_{15} - x_{16}$	NHR, HNR (Two measures of ratio of noise to tonal components in the voice status)	
$x_{17} - x_{18}$	RPDE, D2 (Two nonlinear dynamical complexity measures)	
x_{19}	DFA (Signal fractal scaling exponent)	
$x_{20} - x_{22}$	SPREAD1, SPREAD2, PPE (Three nonlinear measures of fundamental frequency variation)	

IV. FEED-FORWARD NEURAL NETWORKS

In general, a FFNN is an adaptive system which consists of an interconnected group of M *artificial neurons* by using *synaptic weights* [4]. Figure 1 depicts a scheme of a FFNN with M hidden neurons.

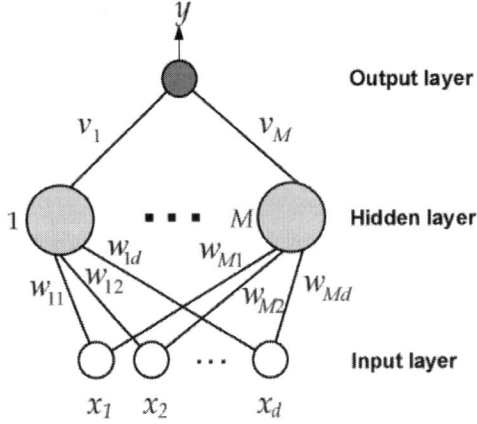

Fig. 1 Standard architecture for a FFNN with a hidden layer of M neurons.

For each input pattern $\mathbf{x}_n$, each artificial neuron receives d inputs and computes a single output by forming a linear combination according to its *input weights* (w_{ji}). The weighted sum, plus an additional bias term (b_j), is passed through a non-linear function $g(\cdot)$, which is known as *activation function*, in order to produce the *hidden neuron output* (h_{jn}). Some appropriate activation functions are the sigmoid and the hyperbolic tangent [3, 4]. After that, these neuron outputs are linear combined by *output weights* (v_j) for computing the *network output* (y_n):

$$y_n = \sum_{j=1}^{M} v_j g\left(\sum_{i=1}^{d} w_{ji} x_{in} + b_j\right) = \sum_{j=1}^{M} v_j h_{jn} \approx t_n \quad (1)$$

with $n = 1, \ldots, N$. Note that, according to (1), the learning procedure for FFNNs has to find appropriate weight parameter values in order to provide accurate estimates for the target variable. It is done by minimizing the measured error between the output network and the desired outputs in a given training dataset. The most popular training procedure is *Back-Propagation* (BP) algorithm [3, 4], which is based on gradient-descent optimization, such as the *Scaled-Conjugate Gradient* (SCG) approach [3]. It should be noted that, according to the literature [3, 4], we use the widely-used term *Multi-Layer Perceptron* (MLP) to refer FFNN with sigmoid functions trained by gradient-based procedures. More details about FFFNs can be found in [3, 4], which are very good references.

Although gradient-descent based methods have been traditionally used for FFNNs, it is well known that they are generally very slow due to improper learning steps or may easily converge to local minima. Moreover, many training repetitions may be required in order to obtain an accurate predictive model and, also, to choose the hidden layer size by means of Cross-Validation (CV) techniques.

V. EXTREME LEARNING MACHINE

In the last years, the *Extreme Learning Machine* (ELM) algorithm has been proposed to solve these limitations [5]. Before explaining ELM, let us to consider a set of N input patterns, and from (1), we can write the following matrix-based equation: $\mathbf{Hv} = \mathbf{t}$, where $\mathbf{H} = \{h_{jn}\}$ is the hidden output matrix with d rows and M columns. The ELM algorithm is based on the concept that, if w_{ji} and b_j are assigned to random values, a FFNN with a single hidden layer can be considered as a linear system and, then v_j can be analytically obtained through simple generalized inverse operation of the hidden layer output matrices:

$$\mathbf{v} = \mathbf{H}^\dagger \mathbf{t}, \quad (2)$$

where $\mathbf{H}^\dagger$ denotes the Moore-Penrose generalized inverse of $\mathbf{H}$. The ELM algorithm tends to give good performance at extremely fast learning speed [5]. Besides, ELM looks much simpler than most learning approaches for FFNNs. It is should be noted that the standard ELM algorithm requires a CV technique, such as 10-fold CV or Leave-One-Out (LOO) CV, to choose an appropriate value for M.

A. OP-ELM: Optimally Pruned Extreme Learning Machine

As an extension of the ELM algorithm [5], Miche et al. [6] recently introduce the OP-ELM (Optimally Pruned ELM) algorithm. It is composed of three main stages:

1. *Random initialization of a large FFNN*. This first step is performed using the standard ELM algorithm for a large enough number of hidden neurons, M.
2. *Ranking of hidden neurons*. In the second stage, *MultiResponse Sparse Regression* (MRSR) is applied for sorting the hidden neurons according to their accuracy. It must be remarked that the obtained ranking by MRSR is exact for linear problems. Since the output is linear with respect to the randomly initialized hidden neurons, MRSR gives an exact ranking of neurons.
3. *Selection of hidden neurons*. Once the ranking has been obtained, the columns of **H** are sorted according to this rank. Then, the best neurons are determined by estimating the LOO error for different numbers of neurons and selecting those M^* neurons (with $M^* \leq M$) such that minimize the LOO error:

$$M^* = arg\ min_{k \in \{1,\ldots,M\}} E_{LOO,k}, \quad (3)$$

where $E_{LOO,k}$ denotes the LOO error computed using the *PREdiction Sum of Squares* (PRESS) statistic for a FFNN composed of $k \leq M$ hidden neurons. In general, the computation of the LOO error is very expensive in computational terms and it can be usually impractical. This drawback is solved using the PRESS statistic, which provides an exact estimation of the LOO error for linear models using a closed matrix-based formula.

On the contrary that the original ELM algorithm, OP-ELM does not need to split up the learning set into training and validation subsets because it directly determines the optimal hidden layer size by computing the LOO error in a fast way using PRESS [6]. Therefore, OP-ELM is faster than ELM and it uses more training data.

B. Sequential Feature Selection in FFNNs Trained with the OP-ELM Algorithm

Once the OP-ELM algorithm automatically determines the hidden layer size in a fast way, there is an open key issue: *the design of the input layer*, i.e., selecting the most discriminative input features for modeling the classification task and discarding irrelevant variables that can distort the classifier design. This stage is known as *Feature Selection* (FS). In order to enhance the generalization performance of the FFNN trained by OP-ELM, this work performs *Sequential Feature Selection* (SFS) [3, 7]. In particular, features are sequentially added to an empty candidate set until the addition of further features does not decrease the LOO error. Several repetitions (L) of the OP-ELM algorithm have to be done for averaging the LOO error in each possible feature subset. It should be noted the SFS is impractical for designing FFNN using traditional methods, but the speed of OP-ELM makes feasible the intensive search of best features.

VI. EXPERIMENTAL RESULTS

From the original dataset of Parkinson's disease [8, 9], 70% of input patterns are randomly selected for designing purposes (training and validation) and the remaining one for testing. Firstly, the original ELM method is compared with the standard BP algorithm for training MLPs by considering from 1 to 100 hidden neurons with sigmoid activation function. For each hidden layer size, 50 different weight initializations of the standard MLP training (in particular, BP with SCG optimization, BP-SCG) and the ELM algorithm are performed. In the case of MLP training, the maximum number of training iterations is established to 1000 and, in order to avoid overfitting, the MLP learning is early stopped using the 10-fold CV procedure. Experiments have been carried out in MATLAB 7.11(R2010b) running in the same machine (4 GB of memory and 2.67 GHz).

Figure 2 shows the evolution of the training time (in seconds) with respect to the hidden layer size with both training procedures: MLP (BP-SCG) and ELM. As it can be observed in this figure, and considering the total training time, ELM runs around 17.5 times faster than MLP. It is straight-forward to see that the training time increases as the hidden layer size is larger.

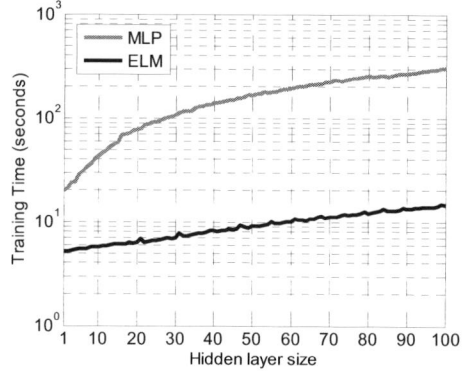

Fig. 2 Evolution of the total training time (in seconds) with respect of the hidden layer size of FFNNs trained using the standard BP with SCG for MLPs and the ELM algorithm based on pseudoinverse computations.

Figure 3 shows the classification accuracy (in %) with different number of neurons in the gradient-based MLP training and the ELM method. Results are average of 50 simulations under 10-fold CV procedure.

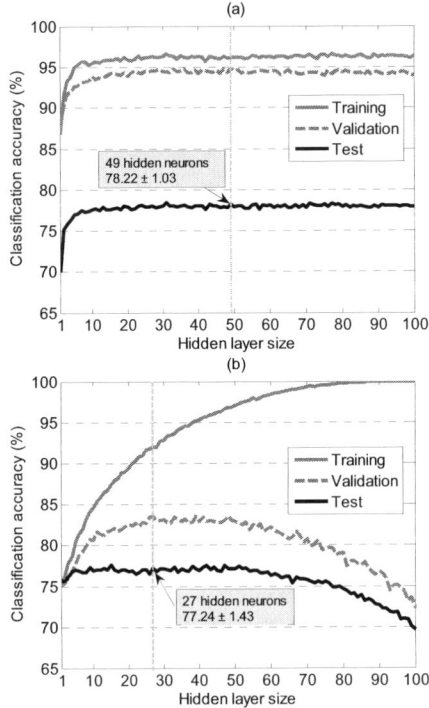

Fig. 3 Evolution of the classification accuracy (in %) with respect of the hidden layer size of FFNNs trained using (a) the BP algorithm with the SCG optimization and (b) the standard ELM algorithm.

From Figures 2 and 3, we can see that MLP provides better performance results than ELM but it requires excessive computational requirements for achieving a small enhancement in the classification accuracy. Besides, it can be observed that ELM tends to overfit the training dataset as the number of neurons increases. It is due to the fact that the original ELM algorithm doesn't incorporate any procedure for avoiding the overfitting phenomenon. Another inconvenient of both approaches, MLP and ELM, is that the optimal hidden layer size has to be determined by incremental search, which entails a high computational cost. For all these reasons, the OP-ELM method has been considered in this paper because it provides an automatic and fast design of FFNNs.

Table 2 gives the obtained experimental results using the different procedures evaluated in this paper for designing FFNNs. According to these results, OP-ELM outperforms MLP and ELM with several orders of magnitude faster than both standard algorithms. With respect to the SFS approach, it significantly increases the classification accuracy up to 86.00%. Note that the training time of OP-ELM with SFS still remains much lower than both MLP and ELM. We have found that the most discriminative features for the Parkinson's disease dataset are the following four variables: x_{22}, x_1, x_{17} and x_{15}; which are ordered as their relevance for solving this medical diagnosis task. The remaining variables (82.61% of the data) are discarded.

Table 2 Obtained results

Algorithm	Classification Accuracy (mean±std, %)	Training Time (seconds)	Hidden layer size
MLP	78.22 ± 1.03	1.648e+4	49
ELM	77.24 ± 1.43	9.420e+2	27
OP-ELM	79.10 ± 5.39	1.179e+1	36.50 ± 11.40
OP-ELM with SFS	86.00 ± 2.95	5.415e+2	39.26 ± 10.94

VII. CONCLUSIONS

This paper analyzes and compares traditional and emergent methods for training FFNNs and its application for diagnosis of Parkinson's disease. The experimental results show that the ELM-based procedures outperform the common gradient-based optimization methods in both computational design time and classification accuracy. As future works, ensemble methods will be analyzed for exploiting the diversity of the different FFNNs and, also, the extension to other studies of Parkinson's disease.

REFERENCES

1. Begg R, Lai DTH, Palaniswami M (2012) Computational Intelligence in Biomedical Engineering, CRC Press, Florida, USA.
2. Schizas CN, Pattichis CS, Bonsett CA (1994) Medical Diagnostic Systems: A Case for Neural Networks, Technology and Health Care, 2: 1-18.
3. Bishop CM (2006) Pattern Recognition and Machine Learning. Springer.
4. Haykin (1998) Neural Networks: A Comprehensive Foundation (2nd Edition), Prentice Hall.
5. Huang GB, Wang DH, Lan W (2011) Extreme learning machines: a survey, International Journal of Machine Leaning and Cybernetics, 2(2): 107-122. DOI 10.1007/s13042-011-0019-y
6. Miche Y, Sorjamaa A, Bas P, Simula O, Jutten C, Lendasse A (2009) OP-ELM: optimally pruned extreme learning machine. IEEE Transactions on Neural Networks, 21(1): 158–162. DOI 10.1109/TNN.2009.2036259
7. García-Laencina PJ (2013), Improving predictions using linear combination of multiple extreme learning machines, Information Technology & Control, 42(1): 86-93. DOI 10.5755/j01.itc.42.1.1667
8. Little MA, McSharry PE, Roberts SJ, Costello DAE, Moroz IM (2007) Exploiting nonlinear recurrence and fractal scaling properties for voice disorder detection, BioMedical Engineering OnLine, 6:23. DOI 10.1186/1475-925X-6-23
9. Bache K, Lichman M, (2013). UCI Machine Learning Repository at http://archive.ics.uci.edu/ml

Breast Tissue Microarray Classification Based on Texture and Frequential Features

M.M. Fernández-Carrobles[1], G. Bueno[1], O. Déniz[1], and M. García-Rojo[2]

[1] E.T.S.I.Industriales, Universidad de Castilla-La Mancha, Ciudad Real, Spain
[2] Dept. Anatomía Patológica, Hospital General Universitario, Ciudad Real, Spain

Abstract—This paper shows a complete and novel study for breast TMA classification based on texture, frequential features and color models information. Thus a relevant set of features for automatic breast TMA classification is found. These features are obtained from 1st and 2nd order Haralick statistical descriptors as well as by filtering with the Fourier and Daubechies Wavelet transforms. Moreover, the discriminant value of the features has been analyzed under different color models, that is, RGB, Lab*, HSV, Hb* and b* channel. Thus, a total of 3133 features with 5 color models were analyzed. The TMA images were divided into four classes: i) benign stromal tissue with cellularity, ii) adipose tissue, iii) benign structures but anomalous and iv) ductal and lobular carcinomas. A statistical study of the features was carried out to provide information about their influence in the tissue classes. Finally, the classification was performed with 10 classifiers using 10-fold and leave-one-out cross validation for all color models selected. Furthermore, the classification results were also analyzed by means of a principal component analysis (PCA). A 95% accuracy and 90% precision with PCA and a Fisher classifier on Hb* image channels is obtained with the selected features.

Keywords—TMA (Tissue Microarray), texture features, statistical descriptors, frequential descriptors, color models.

I. INTRODUCTION

The tissue Microarray (TMA) is an ordered array of up to several hundreds small cylinders of single tissues (core sections) in a paraffin block from which sections can be cut and processed like any other histological section to detect gene expressions or chromosomal alterations.

The first step in the assessment by pathologists of breast TMA core sections is the classification of each spot into different classes, mainly: i) benign stromal tissue with cellularity, ii) adipose tissue, iii) benign and anomalous structures and iv) different kinds of malignity, that is, ductal and lobular carcinomas. Each spot subjected to nuclear staining is then assigned a Quickscore [1] that reflects its immunopositivity. Applying this procedure to breast TMA sections from large numbers of individuals is time consuming and as above mentioned suffers from subjectivity (inter- and intra-observer variability) and misinterpretations [2, 3, 4, 5]. The automatic analysis of TMA data is still a challenge. This paper addresses this need.

Most of the classification problems depend on three factors: 1) the size of the database, 2) the feature descriptors and 3) the classifiers. The different TMA cores may be seen as different textures from the viewpoint of computer vision [6]. The feature descriptors give information about smoothness, rugosity and regularity of the surfaces. There are different theoretical representations of texture analysis models, each extracting different kind of texture information. Among the most common texture analysis models are the statistical models, the signal processing models with spatial domain or frequency space filtering, multi-resolution models and structural models. Studies about automatic diagnosis of prostate tissue are described in the literature, DiFranco [7] and Doyle [8].

Amaral *et al.* [2, 3] proposed an algorithm based on texton histograms with and without color together with differential invariants for breast TMA classification into four classes. Classification performance was assessed using 344 cores. A two-layer neural network (NN) was compared to a nearest-neighbour classifier and a single-layer network. The system based on texton histograms without color achieved a correct-classification rate of 75% and up to 84% using RGB color model.

Wang *et al.* [9] use textural features to separate different tissue regions on a TMA. The system was evaluated with 9 hematoxilina and eosina (H&E) tissue core images of lung carcinoma and 9 bright field immunohistochemistry (IHC) tissue core images of lung carcinoma with a biomarker named BAX. The system was able to identify cancerous cells and achieved 89% accuracy.

Xing *et al.* [4] proposed a sparse reconstruction based classification algorithm for imaged breast TMA. The classification is based on multiple scale texton histogram, integral histogram and AdaBoost learned with SVM. Using standard RGB images, and tested on a dataset with 547 images, they achieved an overall classification accuracy of 88%.

Le [5] uses filtering for texture classification. Quadrature mirror filters (QMF) for Wavelet transform are used for the feature extraction stage together with Support vector machines (SVM) in classification stage. The scoring performance was assessed using 256 TMA cores. The average

Breast Tissue Microarray Classification Based on Texture and Frequential Features

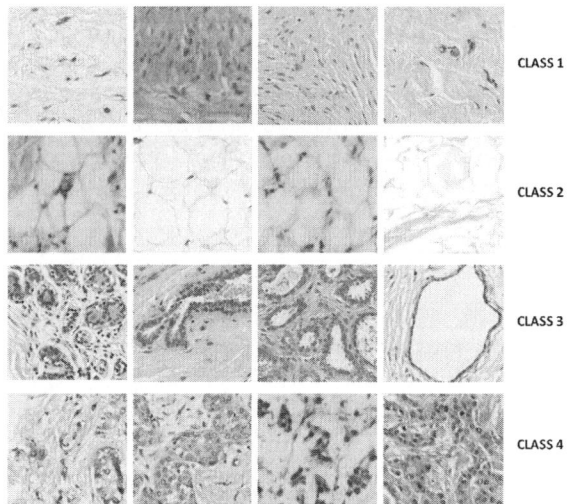

Fig. 1 Breast TMA classes. TMA images stained with H&E. Class 1) benign stromal tissue with cellularity, Class 2) adipose tissue, Class 3) benign structures and anomalous and Class 4) different kinds of malignity.

accuracy of four classes over 50 leave-one-out experiments could be up to 80.42%.

Though the methods reported in the literature are promising further studies are needed. As explained before, one of the principal classification problems is to choose the descriptors and classifier to be used. There are lots of possibilities to combine descriptors and classifiers to improve the results. However, very often the color model selected for the classification is ignored in the studies and it also takes an important part in the final results. At the moment, and as far as the authors know, there are not studies of breast TMA classification based on a combination of texture and frequential features in different color models and color channels. Next section describes the method used for the feature and the classification stages, as well as the materials. Then, Then, the results are reported and finally the main conclusions are drawn.

II. METHODS AND MATERIALS

The TMA images have been acquired by the motorized microscope ALIAS II (LifeSpan Biosciences Inc.) and by Aperio ScanScope T2 at 10x. Once the breast TMA cores were digitalized, 628 representative regions of the 4 tissue classes were selected. The size of these regions was 200 x 200 each one and the TMA tissue classes were: i) benign stromal tissue with low and medium cellularity, ii) adipose tissue, iii) benign structures and anomalous and iv) different kinds of malignity, that is, ductal and lobular carcinomas. The first class (i) is characterized by the pink hue-blue stromal cells prior staining due to tissue with H&E. The second class (ii) is represented in the images as bubbles on the tissue stroma. The third class (iii) shows lobules and other anomalous structures. The types of anomalous benignity represented in class 3 are: sclerosing and adenosis lesions, fibroadenomas, tubular adenomas, phyllodes tumors, columnar cell lesions and duct ectasia. Finally, the fourth class (iv) is characterized by the different kind of malignity. The images of this class show ductal and lobular carcinomas in-situ and invasive. Some examples of the tissue classes are shown in Fig. 1.

A. Feature Extraction and Statistical Analysis

After selecting the images, different types of processing are performed on them. The first is color conversion. The RGB color model is not the only color model used. Also, the intensity images of classes in different color models are extracted: HSV, La*b*, Hb* and b*. The emphasis on the b* channel (belongs to La*b* color model) is due to the facilities of this channel to make a good segmentation on the breast TMA structures. This is essential to improve the results in the classification. Besides this, the intensity image obtained by the RGB model has been filtered through the Fourier bands and Wavelets. This process obtains eight new images, four extracted from four Fourier bands and four by the Wavelets. Thus, we analyze 3133 features, including:

- Haralick coeficients: 1st and 2nd order texture statistics, drawn from the image histogram and the co-occurrence matrix based features.
- Filtering texture features based on the 1st and 2nd order texture statistics calculated on the Fourier image transform.
- Filtering texture features based on the 1st and 2nd order texture statistics calculated on Daubechies wavelet transforms.

Once all the texture information is extracted a statistical study of the features is performed to include only features influenced by the tissue type. The statistical analysis carried out is based on a set of tests namely: Kolmogorov-Smirnov test to asses normality of the variables or extracted features, Levene test for homoscedasticity assesment, Analysis of Variance and non-parametric Kruskall-Wallis test.

B. Statistical Descriptors

In 1973, Haralick presented a general procedure for extracting 2nd order texture features on an image [10]. 1st and 2nd order statistical descriptors are a quantification of the spatial variation in the spatial image shade. The 1st order statistical descriptors are based on the image histogram.

The histograms can extract statistical values of the gray level image distribution like the mean, the variance or the standard deviation. The 2nd order statistical descriptors consider the relationship of the image pixels. They are based on the Grey Level Co-occurrence Matrix (GLCM) of the image. GLCM represents the spatial dependence of the image pixels. Further, these relationships can be defined indicating the distance and the angle between the reference pixel and its neighbour. Distances were taken at 1, 3 and 5 pixel-wide neighbourhoods and at a direction parameter equal to 00°, 45°, 90° and 135° to cover different directions.

C. Fourier and Wavelets Transform

The spectrum of the Fourier transform describes the direction of the repeating structures of the image as high concentrations of energy in the spectrum. This allows the extraction of certain parts of the image with much detail. The Fourier transform also has translational, periodicity, rotation and scaling properties.

The Wavelet transform applied to an image can be described as a low-resolution image at scale n, plus a set of details on it ranging from low to high resolution. The two-dimensional Wavelet transform is implemented by applying low-pass and band pass filters to the rows and columns of the image. In this way, the original image is divided into four sub-images: low resolution, vertical, horizontal and diagonal orientation.

D. Classification

Once textural features are computed, the next issue is how to assign each query case to a pre-established class. The classification performed by the features is divided into two stages. First, the classification is performed with 10 classifiers and the confusion matrix is calculated. A second step, a classification with automatic feature reduction using PCA is performed. Each classification test was performed with an intensity image (RGB, HSV, La*b*, b* or Hb*) and the Fourier and Wavelet images extracted from RGB intensity images.

Thus, the classification tests were carried out with the classifiers: Fisher, feedforward neuronal network by Levenberg-Marquardt rule, Bayesian classifier, support vector machine (SVM), normal densities based quadratic classifier, Normal densities based linear classifier, k-nearest neighbour classifier, binary decision tree, nearest mean classifier and Parzen density based classifier [11]. For each classifier, the total error is stored, the class error and the result obtained by image at the test set. The classification was tested by 10-fold cross validation (10fcv) and leave-one-out (loo) for all color models selected and the Fourier and Wavelets images. Then, the confusion matrix is calculated for each classifier.

Classification with automatic reduction of the features by PCA is applied. PCA generates a new set of features from linear combinations of the original features. The new features are used to perform classification tests by 10-fold cross validation (10fcv) and leave-one-out (loo). For each PCA test, the number of features are increased five by five.

III. RESULTS

Firstly, a total of 628 TMA images in RGB was available. These images were divided into four classes: i) 170 (class 1), ii) 103 (class 2), iii) 163 (class 3) and iv) 192 (class 4). Moreover, the discriminant value of the features have been analyzed for different color models, that is, RGB, Lab*, HSV, Hb* and b* channel as well as for the Fourier and Wavelets transforms. Therefore the dataset contains 8164 images (3140 from color models, 2512 from Fourier and 2512 from Wavelets). The classification test was initially performed with a total of 3133 features (241 features per color model, Fourier bands and Wavelets). After the statistical study, 462 features without tissue information were removed. Finally, a total of 2671 features were analyzed. Then, each classification test was performed with an average of 1748 features. This paper shows the best results obtained.

- The best average results for each of the four classes when using all features (2671) for both 10fcv and loo validation schemes were obtained with Fisher and a neuronal network classifier trained by the Levenberg-Marquardt rule (NN-LM) in Hb* and RGB color space respectively:
  (1) 91.8% agreement with NN-LM in Hb* for 10fcv.
  90.6% agreement with NN-LM in RGB for loo.
  (2) 94.2% agreement with Fisher classifier in Hb* for 10fcv.
  96.2% agreement with Fisher classifier in RGB and Hb* for loo.
  (3) 90.2% agreement with NN-LM in RGB for 10fcv.
  90.2% agreement with Fisher classifier in RGB for loo.
  (4) 90.1% agreement with NN-LM in Hb* for 10fcv.
  89.1% agreement with NN-LM in Hb* for loo.
- The best average results for the global classification when using all features were obtained with a neuronal network classifier trained by the Levenberg-Marquardt rule (NN-LM) for 10fcv in Hb* color space. The confusion matrix of these results are showed in Table 1 with an average of 89.15% agreement.
- This study only has analyzed the results of the classification using PCA for the best two color models of the

Table 1 Best classification with 2671 texture features. Database: 170 (class 1), 103 (class 2), 163 (class 3) and 192 (class 4)

\multicolumn{8}{c}{NN-LM Classifier in Hb* for 10fcv}
Lab	C1	C2	C3	C4	TPR (%)	FPR (%)	Acc (%)	Prec (%)
1	156	2	5	7	91.7	3.0	95.5	91.7
2	4	91	7	1	88.3	2.2	97.1	88.3
3	3	2	141	17	86.5	4.7	92.9	86.5
4	7	2	10	173	90.1	4.4	92.9	90.1

Table 2 Best classification with PCA using 160 texture features. Database: 170 (class 1), 103 (class 2), 163 (class 3) and 192 (class 4)

Fisher Classifier in Hb* for loo								
Lab	C1	C2	C3	C4	TPR (%)	FPR (%)	Acc (%)	Prec (%)
1	158	2	7	3	92.9	2.6	95.8	92.9
2	0	101	1	1	98	0.3	96.8	98
3	2	7	143	11	87.7	4.2	94.4	87.7
4	12	9	7	164	85.4	6.2	93.1	85.4

previous classification. The results obtained when reducing the number of features with PCA improve the average results of the normal classification. The best average result was obtained with Fisher classifier and 160 features for loo validation scheme in Hb*. That is, an average of 90.2% agreement. The confusion matrix of these results are shown in Table 2.

A quantitative evaluation based on True Positive (TPR) and False Positive rates (FPR) together with the accuracy (Acc) and precision (Prec) is also provided on the tables. These results are promising, reaching near 95% accuracy and 90% precision. This improves on previous results reported in the literature.

IV. CONCLUSION

The classification tests showed encouraging results. The fact that the classification uses the features obtained of three types of descriptors and different classifiers allows to compare results and obtain the best classification for breast TMA. Classification results also improves using color models which provides more information about texture. Most of the studies using texture are based in the RGB model. Hence the importance of this study which combines not only the most relevant descriptors for different classifiers but also integrates the color model information. A suitable combination of color channels was found with the H and b* channels. These channels are part of two different color models, that is HSV and La*b*. The use of H and b* channels improve the classification results. We are currently working to combine those classifiers giving better results per each class and analyzing further color models.

ACKNOWLEDGEMENTS

The authors acknowledge partial financial support from the Spanish Research Ministry Project TIN2011-24367.

REFERENCES

1. Detre S, Jotti G Saclani, Dowsett M. A "quickscore" Method for Immunohistochemical Semiquantitation: Validation for Oestrogen Receptor in Breast Carcinomas. *Journal of Clinical Pathology.* 1995;48: 876 - 878.
2. Amaral T, McKenna S, Robertson K, Thompson A. Classification of breast tissue microarray spots using texton histograms in *Medical Image Understanding and Analysis*:144-148 2008.
3. Amaral T, Mckenna S, Robertson K, Thompson A. Classification of Breast- Tissue Microarray Spots Using Colour and Local Invariants. in *IEEE International Symposium on Biomedical Imaging: From Nano to Macro*:999 - 1002 2008.
4. Xing F, Liu B, Qi X, Foran DJ, Yang L. Digital tissue microarray classification using sparse reconstruction *Medical Imaging 2012: Image Processing.* 2012;8314:1-8.
5. Le Trang Kim. Automated method for scoring breast tissue microarray spots using Quadrature mirror filters and Support vector machines in *IEEE International Conference on Information Fusion (FUSION)*:1868–1875 2012.
6. Qi X, Kim H, Xing F, Parashar M, Foran DJ, Yang L. The Analysis of Image Feature Robustness Using Cometcloud *Journal of Pathology Informatics.* 2012:3-33.
7. DiFranco MD, O'Hurleyb G, al EW Kayc. Ensemble based system for whole-slide prostate cancer probability mapping using color texture features *Computerized Medical Imaging and Graphics.* 2011;35:629 645.
8. Doyle S, Feldman M, Tomaszewski J, Madabhushi A. Automated grading of prostate cancer using architectural and textural image features 2007:1284-1287.
9. Ching-Wei Wang. Robust Automated Tumour Segmentation on Histological and Immunohistochemical Tissue Images *PLoS ONE.* 2011;6:e15818.
10. Haralick RM, Shanmugam K, Dinstein I. Textural Features for Image Classification *IEEE Trans. on PAMI.* 1973;9:532-550.
11. Russell SJ, Norvig P. *Artificial Intelligence: A Modern Approach.* Prentice Hall 3rd ed. 2009.

Author: M. del Milagro Fernández Carrobles
Institute: E.T.S.I.Industriales, Universidad de Castilla-La Mancha
Street: Avenida Camilo José Cela N2
City: Ciudad Real
Country: Spain
Email: MMilagro.fernandez@uclm.es

Fuzzy System for Retrospective Evaluation of the Fetal State

R. Czabanski[1], J. Wrobel[2], K. Horoba[2], J. Jezewski[2], and A. Matonia[2]

[1] Division of Biomedical Electronics, Silesian University of Technology,
16 Akademicka Str., 44–100 Gliwice, Poland
[2] Institute of Medical Technology and Equipment, Biomedical Signal Processing Department,
118 Roosevelt Str., 41-800 Zabrze, Poland

Abstract—Cardiotocography is the most common method of biophysical assessment of fetal condition based on the analysis of fetal heart rate (FHR) signal. Due to difficulties with automatic interpretation of recordings, artificial intelligence methods are frequently used for FHR signal classification. However, the problem is the evaluation of the true actual fetal state, that could serve as reference in learning algorithms. The prognostic value of the recorded signal can be objectively verified on the basis of retrospective assessment of the neonatal outcome, which is determined with a help of newborn attributes. In practical applications, only one selected attribute is usually used as the reference. Consequently, the information of the true actual neonatal outcome represented by the remaining attributes is lost. The paper presents a fuzzy method of the neonatal outcome evaluation as a function of all available newborn attributes. The consistency of inference results with the assessment based on single newborn attributes shows the higher effectiveness of the fuzzy system and indicates the possibility of practical application to the objective validation of the learning algorithms.

Keywords—**fuzzy system, newborn outcome, fetal monitoring.**

I. INTRODUCTION

Cardiotocography is based on analysis of a fetal heart rate (FHR) signal recorded using Doppler ultrasound technique. Appropriate FHR variability is an indicator of adequate blood oxygenation, and hence it allows for an indirect diagnosis of fetal distress. In classical approach the printed FHR waveforms are interpreted by a clinician, whose task is to identify the signal characteristics and classify the recordings. High complexity of the signal patterns makes a visual analysis of recordings difficult and results in high level of intra- and interobserver variability [1]. In order to improve the objectivity and repeatability of the assessment the automated signal analysis was introduced.

The computer-aided fetal monitoring systems provide an automated quantitative description of FHR signals. However this information is usually not converted into the knowledge useful for the final medical decision. At the same time it is difficult to implement the heuristic principles of diagnostic procedures used by experienced clinicians in a form of simple algorithm. Hence, computational intelligence is frequently used in the process of the FHR signals classification. Methods that are based on fuzzy logic [2,3], artificial neural networks [4,5,6], neuro-fuzzy systems [7,8,9] and algorithms using the statistical learning theory [10,11,12] can be distinguished among the procedures of qualitative evaluation of the FHR recordings. However, the problem is evaluation of the true actual fetal state to be necessary as a reference in supervised learning algorithms. Several solutions [2,11] used the interpretation of an expert clinician as reference, but the prognostic value of the signal classification can be objectively verified only by retrospective evaluation based on neonatal outcome. The neonatal outcome is defined by several newborn attributes, but in validation usually a single one is used [4,6,9,10,12], or the reference is defined as their Boolean function [6,7]. The lack of simultaneous analysis of all attributes influences the reasoning quality. The paper presents a fuzzy inference system for retrospective evaluation of the fetal state as a function of all newborn outcome attributes. The resulting consistency with the results of the assessment based on single attributes indicate the usability of the proposed method in the learning algorithms for retrospective evaluation of fetal state.

II. NEWBORN OUTCOME ATTRIBUTES

During the fetal monitoring we evaluate the actual fetal state, but the diagnosis can not be verified at the time of the recording session. Such information is to be revealed after the delivery only, and neonatal outcome is retrospectively assigned to the fetal state at the time of monitoring. Such verification is possible, because in perinatology it is assumed that the fetal state can not change rapidly during the course of pregnancy. Hence, the assessment of the fetal state reflects the prediction of neonatal outcome.

Neonatal outcome is defined on the basis of several newborn attributes: the birth weight expressed in percentiles related to the centile chart of a given population of newborns, the Apgar score being the method of visual evaluation of selected newborn characteristics and the negative logarithm of the hydrogen ion activity (gasometry pH) in the umbilical cord blood. The values of each attributes are divided into ranges [13] assigned to three classes of the neonatal outcome: normal, suspicious and pathological (Table 1). Typically, in the retrospective evaluation of the fetal state a single attribute is used. The other approach involves using a Boolean function (usually a logical sum or a logical product) of selected attributes. The proposed method consists in application of fuzzy inference, defining the fetal state as the degree of membership to fuzzy set that represents a class of the neonatal outcome.

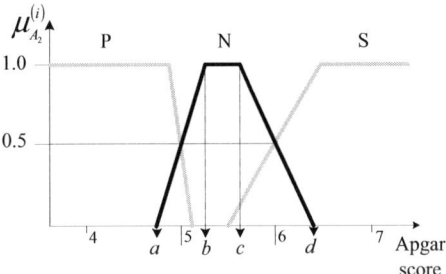

Fig. 1 An example of trapezoid input fuzzy sets defined for the Apgar score (N - denotes normal, S – suspicious, and P – pathological fetal state)

Table 1 The classification of newborn outcome attributes

Neonatal outcome	Birth weight	Apgar score	Gasometry pH
Normal	≥ 10	≥ 7	≥ 7.2
Suspicious	[5, 10)	[5, 6]	[7.1, 7.2)
Pathological	< 5	< 5	< 7.1

Fuzzy assessment is determined in the process of approximate reasoning using a set of conditional fuzzy rules in the form:

$$\forall_{1 \leq i \leq I} R^{(i)} : \text{ if } \left(X_1 \text{ is } A_1^{(i)}\right) \text{ and } \left(X_2 \text{ is } A_2^{(i)}\right) \text{ and } \left(X_3 \text{ is } A_3^{(i)}\right)$$
$$\text{then } Y \text{ is } B^{(i)} \quad (1)$$

where X_1 is the input linguistic variable defining the percentile of the birth weight, X_2 is the linguistic variable related to Apgar score, X_3 is the linguistic variable defining pH measurement, $A_1^{(i)}$, $A_2^{(i)}$ and $A_3^{(i)}$ are the linguistic values (terms) represented by fuzzy sets that are characterized by trapezoid membership functions $\mu_{A_j}^{(i)}(x)$. The input fuzzy sets define a class of neonatal outcome, being the result of the assessment of a single newborn attribute as normal (N), suspicious (S) or pathological (P). Symbol Y indicates the output linguistic variable related to the evaluation of neonatal outcome using the selected newborn attributes. Values of the output linguistic variable are also statements of natural language describing the fetal state, which are represented by fuzzy sets $B^{(i)}$, characterized by triangular membership function $\mu_{B^{(i)}}(y)$.

Shape of trapezoid membership functions in antecedents of fuzzy rules is defined by four parameters denoted as a, b, c and d (Fig. 1). Their values are calculated on the basis of the statistical analysis of the available set of newborn attributes. Parameters b and c are defined as lower and upper quartile of newborn attribute from a given range, while a and d are calculated under the assumption that membership of values defining the boundaries between classes of neonatal outcome are the same for both classes and equal to 0.5.

The membership function of output fuzzy sets are defined as isosceles triangles with a base width $w^{(i)} = 2$. Figure 2 shows an example of membership function of output sets.

Each fuzzy rule has three inputs characterized by three fuzzy sets hence, the complete rule base of the system consists of $I = 3^3 = 27$ rules. The inference result depends on relations between single newborn attributes, which define the assessment of the fetal state. Each of these relations is represented by a single fuzzy rule. We assumed that the neonatal outcome is:

- pathological, if any of the attributes indicates a fetal pathology,
- normal, if two or more attributes indicate a fetal wellbeing,
- suspicious, all the remaining cases.

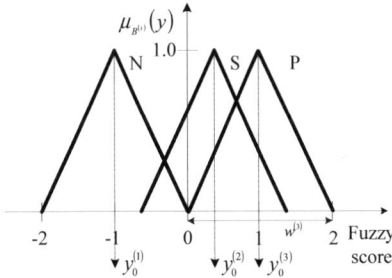

Fig. 2 An example of membership function of output fuzzy sets

The operator **and** in antecedents of fuzzy rules (1) represents the product of input fuzzy sets that is defined as algebraic product (t-norm) of membership functions $\mu_{A_j}^{(i)}$. Hence, the firing strength of the rules is formulated as:

$$\forall_{1 \leq i \leq I} F^{(i)}(\mathbf{x}_0) = \prod_{j=1}^{3} \mu_{A_j}^{(i)}(x_{0j}), \quad (2)$$

where $\mathbf{x}_0 = [x_{01}, x_{02}, x_{03}]^T$ is the vector of values of newborn attributes. The inference is based on a conjunctive interpretation of fuzzy rules using Larsen relation:

$$\mu_{B^{(i)'}}(y) = \mu_{B^{(i)}}(y) \cdot F^{(i)}(\mathbf{x}_0), \quad (3)$$

where $\mu_{B'^{(i)}}$ is the consequent fuzzy set being the result of reasoning with a single fuzzy rule only. Arithmetic mean is applied as an aggregator operator. Consequently, the membership function of the final conclusion B' is defined as:

$$\mu_{B'}(y) = \frac{1}{I}\sum_{i=1}^{I}\mu_{B'^{(i)}}(y) \qquad (4)$$

The crisp output of the system is the result of defuzzification process. In the proposed solution, the representative element of the final fuzzy set is calculated with a center of gravity location operator:

$$y_0 = \frac{\int y\,\mu_{B'}(y)\,dy}{\int \mu_{B'}(y)\,dy} \qquad (5)$$

For output fuzzy sets (with triangular membership functions) we can determine the value of the integral:

$$\int \mu_{B'^{(i)}}(y)\,dy = F^{(i)}(\mathbf{x}_0), \qquad (6)$$

which leads to a simple equation defining the system output:

$$y_0 = \frac{\sum_{i=1}^{I} F^{(i)}(\mathbf{x}_0)\,y_0^{(i)}}{\sum_{i=1}^{I} F^{(i)}(\mathbf{x}_0)}, \qquad (7)$$

where $y_0^{(i)}$ denotes the center of gravity location of the triangle membership function of the i-th rule. Thus, the final, crisp output of the fuzzy system is a weighted mean of single rules outputs, with weights related to firing strengths of the rules.

The center of gravity locations are defined under the assumption that the normal neonatal outcome is related to negative $y_0^{(i)} = -1$, and pathological to positive $y_0^{(i)} = +1$ output of a single fuzzy rule. For the rules related to suspicious neonatal outcome we proposed variable location of the output fuzzy set $y_0^{(i)} = p^{(i)}, |p^{(i)}| < 1$. Hence, values of $p^{(i)} > 0$, increase, and $p^{(i)} < 0$ decrease the possibility of the final assessment of the fetal state as pathological. For $p^{(i)} = 0$ fuzzy rules defining state "suspicious" do not affect the final conclusion. As the result of fuzzy inference we assumed the assignment of neonatal outcome into two classes only: normal or pathological. Consequently, the fetal state is defined as pathological if the crisp output value exceeds predefined threshold Δ. A binary classification does not limit our considerations, since it is possible to determine the class "suspicious" as corresponding to values $|y_0| < \Delta$.

III. RESULTS AND DISCUSSION

The research material used in the numerical experiments is a collection of the newborn attributes from neonatal forms of an archive of computerized fetal surveillance system [14]. Raw data set was analyzed in order to eliminate missing or invalid attribute values (outliers). Finally we obtained a set of attributes from 825 newborns. Table 2 shows the distribution of attributes related to different neonatal outcome classes.

Table 2 The distribution of newborn attributes related to different neonatal outcome classes

Newborn attribute	Normal	Suspicious	Pathological
Birth weight	749	76	0
Apgar score	804	18	3
Gasometry pH	767	50	8

To validate the proposed fuzzy system we examined the consistency between the inference results and the assessment based on the single newborn attributes. Firstly, we assigned the attributes relating to the suspicious class of neonatal outcome as representing a normal fetal state (SaN - optimistic interpretation), and secondly as a fetal pathology (SaP – pessimistic interpretation). Additionally, we applied a complex criterion of the neonatal outcome evaluation based on the Boolean function of newborn attributes. As there were no newborns for which all considered attributes simultaneously indicated pathology we considered only the logical sum (OR). The classification performance was measured with a help of the commonly used sensitivity (SE) and specificity (SP) indices. We defined the classification accuracy (CC) – the number of correctly classified cases expressed as the percentage of the dataset. Additionally we introduced the integrated classification quality index (QI) defined as $QI = \sqrt{SE \cdot SP}$. The evaluation of the neonatal outcome was determined for variable $p^{(i)}$ and Δ. Values of $p^{(i)}$ were changed in the range of [-0.75, +0.75], while Δ in the range of [-0.50, +0.50], both with step 0.25. All calculation were carried out in Matlab® environment using Jacket™ package, allowing the GPU computing power to be applied.

The best results of fuzzy inference using the optimistic interpretation are shown in Table 3 – SaN. The analysis of the birth weight was not possible because there were no pathological cases in the considered dataset. The highest quality of the fuzzy inference, in relation to all considered newborn attributes, was obtained for a high value of the threshold ($\Delta = 0.50$) and the interpretation of the single rules related to suspicious fetal outcome as indicating fetal wellbeing $\left(p^{(i)} = -0.25\right)$. Such interpretation is consistent with the assumed assessment of the reference fetal state.

Table 3 The best results of classification using different interpretations of single newborn attributes

Newborn attribute	QI	SE	SP	CCL
SaN				
Apgar score	99.57	100.0	99.15	99.15
Gasometry pH	99.88	100.0	99.76	99.76
OR function	100.0	100.0	100.0	100.0
SaP				
Birth weight	93.53	98.86	88.65	89.58
Apgar score	90.95	100.0	82.71	83.15
Gasometry pH	93.11	100.0	86.70	87.64
OR function	98.25	99.30	97.22	97.58

The best results for SaP interpretation were obtained for: $\{p^{(i)} = 0.75, \Delta = 0.00\}$ and $\{p^{(i)} = -0.25, \Delta = -0.50\}$ as well as $\{p^{(i)} = 0.25, \Delta = -0.25\}$ (Table 3). It may be noted that for the pessimistic approach we also got satisfactory results of the fuzzy evaluation. The highest quality was obtained for low values of Δ and interpretation of fuzzy rules defining the class "suspicious", which is consistent with the assumed reference outcome.

Fuzzy inference may represent an optimistic as well as pessimistic approach. It is due to variable number of "points" $p^{(i)}$ that are associated with fuzzy rules related to the suspicious neonatal outcome. The evaluation can be also performed without the Δ threshold, taking the values from the unit range as the final result, which can be then interpreted as a degree of certainty of the retrospective fetal state assessment on the basis of newborn attributes.

IV. CONCLUSIONS

In the presented work we investigated the possibility of retrospective evaluation of the fetal state on the basis of the newborn outcome attributes. The neonatal outcome was evaluated using the proposed fuzzy inference system based on the conjunctive interpretation of fuzzy rules defining the relationship between single newborn attributes and the final assessment. To verify the quality of the results we used a real data set consisted of the attributes obtained from newborn forms. The appropriate selection of the inference parameters provided high consistency with the assessment of single newborn attributes. The proposed method allows for combining different newborn attributes in order to determine the proper evaluation of the fetal state, that is necessary to objectively validate the algorithms for retrospective evaluation of fetal state.

ACKNOWLEDGMENT

This work was supported in part by the Polish National Science Center, Cracow, Poland.

REFERENCES

1. Jezewski J, Wrobel J, Horoba K et al. (2002) Fetal heart rate variability: clinical experts versus computerized system interpretation. Proc. 24th IEEE/EMBS, Huston, 2002, pp 1617-1618
2. Skinner J, Garibaldi J, Ifeachor E (1999) A fuzzy system for fetal heart rate assessment. Lect Notes Comput Sc 1625:20–29
3. Huang YP, Huang YH, Sandnes FE (2006) A fuzzy inference method-based fetal distress monitoring system. Proc. IEEE ISIE'06, Canada, 2006, pp 55–60
4. Frize M, Ibrahim D, Seker H et al. (2004) Predicting clinical outcomes for newborns using two artificial intelligence approaches. Proc. 26th IEEE/EMBS, San Francisco, 2004, pp 3202–3205
5. Jezewski M, Wrobel J, Labaj P et al. (2007) Some Practical Remarks on Neural Networks Approach to Fetal Cardiotocograms Classification" Proc. 29th IEEE/EMBS, Lyon, 2007, pp 5170-5173
6. Jezewski M, Czabanski R, Wrobel J et al. (2010) Analysis of extracted cardiotocographic signal features to improve automated prediction of fetal outcome. Biocybern Biomed Eng 30:39–47
7. Leski JM (2003) Neuro-fuzzy system with learning tolerant to imprecision. Fuzzy Set Syst 138:427–439
8. Czabanski R, Jezewski M, Wrobel J et al. (2008) A neuro-fuzzy approach to the classification of fetal cardiotocograms. Proc. vol. 20, IFMBE 14th Int Conf NBC, Latvia 2008, pp 446–449
9. Czabanski R, Jezewski M, Wrobel J et al. (2010) Predicting the risk of low-fetal birth weight from cardiotocographic signals using ANBLIR system with deterministic annealing and ε-insensitive learning. IEEE T Inf Technol B 14:1062–1074
10. Georgoulas G, Stylios C, Groumpos P (2006) Predicting the risk of metabolic acidosis for newborns based on fetal heart rate signal classification using support vector machines. T Bio-Med Eng 53: 875–884
11. Krupa N, Zahedi E, Ahmed S et al. (2011) Antepartum fetal heart rate feature extraction and classification using empirical mode decomposition and support vector machine. Biomed Eng OnLine 10:1–15
12. Czabanski R, Jezewski J, Matonia A et al. (2012) Computerized Analysis of Fetal Heart Rate Signals as the Predictor of Neonatal Acidemia, Expert Syst Appl 39:11846–11860.
13. Sikora J (2001) Digital analysis of cardiotocographic traces for clinical fetal outcome prediction. Perinat Ginekol 21:57–88 (in Polish)
14. Jezewski J, Wrobel J, Horoba K et al. (1996) Computerized perinatal database for retrospective qualitative assessment of cardiotocographic traces in Current Perspectives in Healthcare Computing, Ed. B. Richards B, BJHC Ltd G. Britain:187–196

Author: Janusz Jezewski
Institute: Institute of Medical Technology and Equipment
Street: Roosevelt 118
City: Zabrze
Country: Poland
Email: jezewski@itam.zabrze.pl

Advanced Processing of sEMG Signals for User Independent Gesture Recognition

A. Doswald[1], F. Carrino[2], and F. Ringeval[1]

[1] Document Image and Voice Analysis group, Department of Informatics, University of Fribourg, Fribourg, Switzerland
[2] University of Applied Sciences of Western Switzerland and University of Fribourg, Fribourg, Switzerland

Abstract— While the classification of gestures recorded with sEMG can reach very high recognition rates when the user has trained on the system, performance obtained on unknown users remains low. In this work we attempt to use advanced signal processing and pattern classification methods for improving classification performance of gestures on unknown users. Our approach is to take an existing feature set, add promising features, and use feature selection to prune poor features. For classification we use a support vector machine with a Pearson VII kernel, for which a particle swarm optimization was used to search through its parameter space. Results are presented on the NinaPro database, and show excellent results when the user is known to the system as well as a significant improvement on existing work when the user is unknown.

Keywords—sEMG classification, feature sets, PUK kernel.

I. INTRODUCTION

There are many applications related to the analysis and classification of sEMG signals, such as the manipulation of prostheses or even external devices such as robotic wheelchairs [1]. The usual approach is to extract features from the signal, and then present these features to a classifier. For feature selection, a popular and successful approach is to take sets of simple features. The most commonly used is the Hudgins feature set [2], which uses the mean absolute value (MAV), waveform length (WL), zero crossing and slope sign change of the signal. While this already allows for good classification - 91.8% for four classes - this has been improved on in other studies. In [3] they compare 37 features used in EMG signal classification from the time domain, frequency domain and features using predictive models to craft a better feature set. Their optimal selection for a set considering performance and robustness to noise is: MAV, WL, Wilson amplitude, autoregressive coefficients (AR) of the 4th order, MAV slope, mean frequency and power spectrum ratio, with which they obtain a classification score of over 96% for 6 classes; this set is called the Phinyomark feature set.

From the existing state-of-the-art of classifiers used in conjunction with sEMG signals, the most successful are neural networks (NN), support vector machines (SVM) and linear discriminant analysis (LDA) [4, 5]. And while it is difficult to establish a comparison between studies, in [6] they test those 3 classifiers on a few feature sets and with different settings. They found that SVMs reach the best performance, 95.5% success on their set, followed by the LDA at 94.5%, and the NN coming in last at 93%.

The testing methodology for these studies assumes however that the classifier is trained solely on a single user's data. As such, none of these results are readily applicable to the case where a system is presented with a completely new user. In addition, the signal processing models used to generate the features are mostly linear, even though the EMG signal itself is non-linear and non-stationary due to the random contractions of groups of muscle fibers. However, as the Phinyomark feature set and the SVM classifier seem to provide best performance in the automatic recognition of gestures from sEMG sensors, we use them as the baseline for our own work. The novelty of our work can be divided in three folds with the use of: (i) a user independent evaluation scheme to estimate performance of the system, (ii) advanced signal processing able to take into account the non-linear and non-stationary nature of the sEMG signal and (iii) advanced pattern recognition methods to classify the gestures. Our aim is to optimize the signal processing and classification steps to a point that the system can be directly used on an unknown user, without having to perform a full training session. We present results on the sEMG NinaPro database [7], in order to provide a common point of comparison for other researchers.

II. METHODS

A. Feature Extraction

Our approach to feature selection is to take the very solid Phinyomark feature set and to attempt to improve the results by extending it with some new promising features. We add the following features that were not considered in [3]: statistics on the AR residue, statistics on the Hilbert-Huang transform (HHT) of the signal, and finally higher order statistics (HOS) cumulants. Both HHT and HOS have been successfully used in sEMG analysis [8, 9], and while AR is a good feature for sEMG classification, its residue should hold valuable information about the signal.

Finding the AR-residue is simply a matter of taking the time series generated by the auto-regressive approximation of a signal, and then subtracting this time series from the original data. Since there is no prior work on which information from the residue is meaningful for EMG signal analysis, the tendency is to use a brute-force approach by taking a large range of common statistics. We therefore computed a series of 29 measures for which the reader would find all details in Table II in [10].

Features taken from the frequency domain usually perform poorly when compared to those taken from the time domain [3]. One of the reasons for this may be that these features depend on the Fourier transform, which makes assumptions that do not hold for the EMG signal. The HHT [11] however is able to handle nonlinear and non-stationary signals, and since its computation is data-driven, it is expected to provide closer estimates of the true mean frequency than the one obtained from the Fourier transform. The HHT is performed in two parts: first an empirical mode decomposition is applied to the signal - which will result in intrinsic mode functions (IMFs) that the Hilbert transform can be safely applied to, i.e. the improper integral of the transform is well defined for the limited band of frequencies of the IMFs - followed by the Hilbert transform. The algorithm is fully described in [8].

The mathematical definition of the Hilbert transform $Y(t)$ of a time series $X(t)$ is the following [11, 12]:

$$Y(t) = \frac{1}{\pi} \int_{-\infty}^{\infty} \frac{X(t')}{t-t'} dt' \quad (1)$$

which leads to the analytical signal Z(t):

$$Z(t) = X(t) + iY(t) = a(t)e^{i\theta(t)} \quad (2)$$

where

$$a(t) = \sqrt{X^2(t) + Y^2(t)} \text{ and } \theta(t) = arctan(\frac{Y(t)}{X(t)}) \quad (3)$$

We can now define the pulsatance and frequency as :

$$\omega(t) = \frac{d\theta(t)}{dt} \text{ and } f(t) = \frac{\omega(t)}{2\pi} \quad (4)$$

This transform is done for every IMF and the last process thus consists in extracting the mean frequency from all IMFs (MNFHHT). We obtain it by first taking the mean instantaneous frequency of each IMF [8, 12]:

$$MIF(j) = \frac{\sum_{t=1}^{m} f_j(t) a_j^2(t)}{\sum_{t=1}^{m} a_j^2(t)} \quad (5)$$

where m is the number of IMFs. This gives us the mean frequency:

$$MNFHHT = \frac{\sum_{j=1}^{m} \|a_j\| MIF(j)}{\sum_{j=1}^{m} \|a_j\|} \quad (6)$$

The time series from (3) and (4) may also impart important information about the signal, but as it is not clear which measures would be meaningful, we take the same set of 29 statistics on the sum of $a_j(t)$ and on the sum of $f_j(t)$ as we did for the AR-residue. This leads to a total of 59 features taken for the HHT.

The main interest of HOS for EMG signal analysis is that, being invariant to Gaussian noise, it is able to capture the non-Gaussian component of the signal. Higher order cumulants can be seen as the extension of the auto-correlation function, which is in fact the second order cumulant:

$$R_{xx}(\tau) = C_{2x}(\tau) = E\{x(t)x(t+\tau)\} \quad (7)$$

where $E\{\}$ is the expectation operator, $x(t)$ is a time series and τ is a time-shift. From this equation we can obtain higher order cumulants. The third and fourth order cumulants are defined as:

$$C_{3x}(\tau_1, \tau_2) = E\{x(t)x(t+\tau_1)x(t+\tau_2)\} \quad (8)$$

$$\begin{aligned}&C_{4x}(\tau_1, \tau_2, \tau_3) = \\ &E\{x(t)x(t+\tau_1)x(t+\tau_2)x(t+\tau_3)\} \\ &-C_{2x}(\tau_1)C_{2x}(\tau_2-\tau_3) - C_{2x}(\tau_2)C_{2x}(\tau_3-\tau_1) \\ &-C_{2x}(\tau_3)C_{2x}(\tau_1-\tau_2)\end{aligned} \quad (9)$$

These cumulants are defined under the hypothesis that the mean of the time series is zero, and in practice it is the estimates of these values that we must use. The cumulant-based features we chose are the same as those proposed in [9]: the second-, third-, and fourth-order cumulants for time-shifts of 0 to 3 samples. Since cumulants are symmetric in their arguments, this leads to a list of 19 cumulants.

B. Classification

The purpose of feature selection is to keep the best and least redundant subset of features for the current task, which will often lead to better classification scores. In our case, this step is necessary due to the brute force nature of some of the features. We use correlation-based feature selection (CFS). The details can be found in [13], but the basic hypothesis is: "Good feature subsets contain features highly correlated with (predictive of) the class yet uncorrelated with (not predictive of) each other".

We chose a SVM classifier, as it seems to be the best solution from available comparisons in the state-of-the-art. To account for the complexity of our features we use a generic kernel called the Pearson VII function kernel. It is defined as:

$$K(x,y) = \frac{1}{1 + ((2\sqrt{\|x-y\|^2}\sqrt{2^{1/\omega}-1})/\sigma)^{\omega}} \quad (10)$$

This kernel's mapping power is equal to or stronger than the mapping that can be achieved with the polynomial or RBF kernels. It is able to simulate either of those functions, as well as map to spaces that neither of those kernels can map to [14]. However it has the disadvantage of having an extra parameter when compared to those kernels, and examples show that the optimal values for σ, ω and for the permissiveness of the margins C vary a great deal between cases. Though using a grid search would be prohibitively long, a strong metaheuristic can explore the solution space efficiently. We use particle swarm optimization (PSO) as it has been shown that it performs well on unconstrained linear continuous problems, outperforming genetic algorithms for example [15]. We will be comparing our SVM configuration against a SVM using a polynomial kernel, as studies have found that the linear SVM (polynomial of degree 1) has a performance equivalent to more complex standard kernels [6].

III. RESULTS

We will first present the corpus used, followed by our evaluation methodology before presenting our results.

The NinaPro database's purpose is to provide the biorobotics community with a public comparison and benchmarking tool for hand prosthetics [7]. The sEMG data in NinaPro was recorded with ten electrodes on the forearm. The setup is the following:

- The sensors used are the OttoBock MyoBock 13E200. They have a frequency bandwidth of 90 - 450 Hz. Amplification was set to 14000 times.
- Eight of the sensors are evenly distributed within an armband set just under the elbow, and the remaining two are set just below, one on a targeted wrist extensor and the other on a flexor.
- The data is sampled at 100Hz and directly righted.

The database contains the recording of gestures of the wrist and of the fingers, with each gesture held for 5 sec. followed by 3 sec. of rest. A set of 52 different gestures are recorded in this way, which include the gestures we base our study on. The set of gestures is recorded 10 times by each of 27 subjects, leading to a total of 14k gestures in the database. As the initial application of our study was to use sEMG to guide a mobile robot, we chose only the following wrist gestures for classification: flexion, extension, abduction, adduction, clenching of the fist, and a rest state.

In this work, we use a "leave one subject out" (LOSO) cross-validation scheme, which is commonly used in the classification of speech data. The process is similar to k-fold cross-validation, except that instead of taking a random sampling of the data in order to create the partitions, the data is separated by subjects: for N subjects, N-1 subjects are used for training, while the left-out subject is used for validation. The purpose of this partitioning is to attempt to obtain features and classification parameters that are robust when a new subject wishes to use the EMG system. This is different from the norm where the results are taken from classifying values from each single subject, and then averaged. The LOSO classification is more difficult by comparison, as the characteristics of the EMG signal varies a great deal between subjects.

In our tests we present four levels of complexity as to the features treated and the classifier used. We have:

- C1: The SVM using a polynomial kernel, with the order varied from 1 to 20. The permissiveness of the margins C is kept at a default value of 1. The Phinyomark feature set is used, as well as on the C2 and C3 cases.
- C2: The SVM using the PUK kernel, where the PSO algorithm searches parameter values between 0 and 10000 with 50 particles for 50 iterations.
- C3: The SVM using the PUK kernel + PSO, using CFS feature selection.
- C4: The SVM using PUK + PSO + CFS, using our extended feature set.

We also train two classifiers: the first allows the system to decide if an action is intended by classifying between the rest state or an action, and if necessary the second then detects which action is performed. This system adds a barrier preventing a controlled device starting to take an action when none was intended, and simplifies the following classification by bringing it from 6 classes to 5. In all cases we use the best result provided by the parameter search for comparison.

In table 1 we can see that increasing the sophistication of the method used provides better results, from 73.64% of gestures correctly classified for the simple SVM, to 77.71% for the full classification methodology using our extended feature set. For the detection of gestures, the reverse is true: simpler methods generally achieving better results; the procedure without CFS obtains the best results, with a polykernel of degree 6 getting the best performance.

In [7] they also attempt a LOSO approach on the NinaPro database for 3, 11 and 52 different gestures, obtaining

Table 1 LOSO classification scores of 5 gestures from the NinaPro database for complexity levels C1 to C4

Classification type	C1	C2	C3	C4
Gesture detection	**99.26**	99.19	97.49	98.02
Gesture classification	73.64	75.43	77.36	**77.71**

Table 2 Standard classification scores of 5 gestures from the NinaPro database for complexity levels C1 to C4

Classification type	C1	C2	C3	C4
Gesture detection	99.78	**99.79**	**99.79**	99.44
Gesture classification	96.64	**97.29**	**97.29**	95.86

a success rate of about 55%, 22% and 8% respectively, using only the average of each channel and a least-squares SVM with a RBF kernel. While we did not exactly reproduce their tests, our method provided a much better score for five gestures when compared to the result they obtain using only three.

As a comparison for other work, in table 2 we present our results using the more standard method of evaluating data: by performing a classification on the k-fold distributed data of each individual user, and then averaging the results. The difference in results is remarkable, with the classification of gestures reaching a success rate of 97.29% when using the PUK kernel with or without CFS. As for the detection of gesture, the success rate climbs to 99.79%.

IV. CONCLUSION

In this work we attempt to improve the classification rate of gestures in three manners: new features, feature selection, and an improved classifier. As we expected, the more complex PUK kernel provides better results on our larger feature set than the linear SVM recommended in [6], though for the simpler gesture detection a polykernal can provide better results. We can also see that while our extra features provide a benefit for user independent classification, they diminish performance otherwise: the complexity introduced by the extra features is not beneficial for the simpler case where the user is known. The CFS and the PUK SVM kernel provide the most noticeable improvements, while the contribution of the extra features - HHT, HOS and AR-residue - is smaller. We find that the performance of our system is very good: excellent in the standard testing method and significantly improving on existing work in the case of LOSO testing.

ACKNOWLEDGEMENTS

We would like to thank the NinaPro team and Manfredo Atzori for allowing us access to their corpus.

REFERENCES

1. Moon I., Lee M., Chu J., Mun M.. Wearable EMG-based HCI for Electric-Powered Wheelchair Users with Motor Disabilities in *ICRA*(Barcelona, Spain):2649-2654 2005.
2. Hudgins B., Parker P., Scott R.N.. A New Strategy for Multifunction Myoelectric Control *IEEE Trans. on Biomedical Engineering.* 1993;40:82-94.
3. Phinyomark A., Phukpattaranont P., Limsakul C.. Feature Reduction and Selection for EMG Signal Classification *Expert Syst. with Appl..* 2012;39:7420-7431.
4. Asghari Oskoei M., Hu H.. Myoelectric Control systems - A Survey *Biomedical Signal Processing and Control.* 2007;2:275-294.
5. Ahsan Md R, Ibrahimy Muhammad I, Khalifa Othman O. EMG Signal Classification for Human Computer Interaction: A Review *European Journal of Scientific Research.* 2009;33:480-501.
6. Oskoei M.A., Hu H.. Support Vector Machine-Based Classification Scheme for Myoelectric Control Applied to Upper Limb *IEEE Trans. on Biomedical Engineering.* 2008;55:1956-1965.
7. Atzori M., Gijsberts A., Heynen S., et al. Building the NINAPRO Database: A Resource for the Biorobotics Community in *Proceedings of the IEEE International Conference on Biomedical Robotics and Biomechatronics*:51 2012.
8. Xie H., Wang Z.. Mean Frequency Derived via Hilbert-Huang Transform with Application to Fatigue EMG Signal Analysis *Computer Methods and Programs in Biomedicine.* 2006;82:114-120.
9. Nazarpour K., Sharafat A.R., Firoozabadi S.M.P.. Application of Higher Order Statistics to Surface Electromyogram Signal Classification *IEEE Trans. on Biomedical Engineering.* 2007;54:1762-1769.
10. Ringeval F., Sonderegger A., Noris B., Billard A., Sauer J., Lalanne D.. On the Influence of Emotional Feedback on Emotion Recognition and Gaze Behavior in *Affective Computing and Intelligent Interaction*(Geneva, Switzerland) 2013.
11. Huang N.E., Shen Z., Long S.R., et al. The Empirical Mode Decomposition and the Hilbert Spectrum for Nonlinear and Non-Stationary Time Series Analysis *Proceedings of the Royal Society of London. Series A: Mathematical, Physical and Engineering Sciences.* 1998;454:903-995.
12. Zong C., Chetouani M.. Hilbert-Huang Transform Based Physiological Signals Analysis for Emotion Recognition in *2009 IEEE International Symposium on Signal Processing and Information Technology (ISSPIT)*:334-339 2009.
13. Hall M.A., Smith L.A.. Feature Selection for Machine Learning: Comparing a Correlation-Based Filter Approach to the Wrapper in *Proceedings of the Twelfth International Florida Artificial Intelligence Research Society Conference*:235-239AAAI Press 1999.
14. Üstün B., Melssen W.J., Buydens L.M.C.. Facilitating the Application of Support Vector Regression by Using a Universal Pearson VII Function Based Kernel *Chemometrics and Intelligent Laboratory Systems.* 2006;81:29-40.
15. Hassan R., Cohanim B., De Weck O., Venter G.. A Comparison of Particle Swarm Optimization and the Genetic Algorithm in *46th AIAA/ASME/ASCE/AHS/ASC Structures, Structural Dynamics, and Materials Conference*:1-13 2005.

Adaptive Classification Framework for Multiclass Motor Imagery-Based BCI

L.F. Nicolas-Alonso, R. Corralejo, D. Álvarez, and R. Hornero

Grupo de Ingeniería Biomédica, ETSI Telecomunicación, Universidad de Valladolid, Valladolid, España

Abstract—The non-stationary nature of electroencephalogram (EEG) is a major issue to robust operation of brain–computer interfaces (BCIs). The objective of this paper is to propose an adaptive classification framework whereby a processing stage is introduced before classification to address non-stationarity in EEG classification. Features extracted from EEG signals are adaptively processed before classification to reduce the small fluctuations between calibration and evaluation data. In this way, static classifiers can be employed in non-stationary environments without additional changes. The session-to-session performance of the proposed adaptive approach is evaluated on a multiclass problem posed in the BCI Competition IV dataset 2a. A probabilistic generative model was used as a classification algorithm. The results yields a significantly higher mean kappa of 0.62 compared to 0.58 from the baseline probabilistic generative model without adaptive processing. Also, the proposed approach outperforms the winner of the BCI Competition IV dataset 2a. These results suggest a promising approach separating adaptation-related tasks and classification.

Keywords—**Brain Computer Interface, motor imagery, multiclass, classification, adaptation, non-stationary, EEG.**

I. INTRODUCTION

A brain computer interface (BCI) based on electroencephalogram (EEG) is a system that enables humans to interact with their surroundings, without the involvement of peripheral nerves and muscles, by using control signals generated from EEG activity [1]. BCIs create an alternative non-muscular pathway for relaying a person's intentions to external devices such as computers, speech synthesizers, assistive appliances, and neural prostheses amongst many others.

A major challenge for BCI research is the non-stationarity of brain activity. Diverse behavioral and mental states continuously change the statistical properties of brain signals [2]. BCI systems are usually calibrated by users through supervised learning using a labeled dataset. However, patterns observed in the experimental samples during calibration sessions may be different from those recorded during online sessions. Therefore, adaptive algorithms are a very important issue for improving BCI accuracy.

Several machine learning techniques have been attempted to address the non-stationarity in BCI. In this regard, semi-supervised learning has been suggested to update the classifier in online sessions on a continuous basis [3]. In semi-supervised learning, the classifier is initially trained using a small labeled data set. Next, the classifier is updated with unlabeled data. However, in a realistic BCI scenario, the signal associated with the subject's intentions is not usually known and the labels are not available. Consequently, unsupervised learning was proposed to find hidden structures in unlabeled data. Some unsupervised methods rely on techniques for sequential adaptation of static classifiers [4-9] or Kalman filtering [10, 11].

The design of adaptive classifiers on the basis of the static counterparts follows the strategy of adjusting sequentially one or several parameters in the static classifier [4-9]. However, in the case of complex classifiers, there is no simple algorithm for online sequential learning when class labels are not given. For these reason, a general framework to devise adaptive algorithms but using static classifiers could be more convenient. In this way, static classifiers can be employed without additional changes.

The aim of this study is to propose an adaptive classification framework whereby a processing stage is introduced before classification. The processing stage performs adaptation-related tasks while the classification stage remains unchanged. Features extracted are processed before classification in order to reduce the small fluctuations between training and evaluation data. Afterwards, any classifier can be employed. In this work, a probabilistic generative model is used for multiclass classification. The performances of both adaptive and static approaches are investigated on the BCI Competition IV dataset 2a [12]. The results are compared with the winner of the competition [13].

II. BCI COMPETITION IV DATASET 2A DESCRIPTION

The BCI Competition IV dataset 2a challenges the session-to-session transfer. This dataset comprises two sessions, one for training and the other for evaluation, of EEG data from 9 subjects performing 4 classes of motor imagery, namely, left hand, right hand, feet, and tongue. Each session included 288 trials of data recorded with 22 EEG channels and 3 monopolar electrooculogram (EOG) channels (with left mastoid serving as reference). For more details refer to Tangermann *et al* [12].

III. PROPOSED METHOD

The architecture of the proposed algorithm is illustrated in Figure 1. It is based on Filter Bank Common Spatial Patterns (FBCSP) [13] and comprises five consecutive stages: multiple bandpass filtering using Finite Impulse Response (FIR) filters, spatial filtering using the Common Spatial Patterns (CSP) algorithm, feature selection, adaptive processing and classification of the selected CSP features.

A. Band-Pass Filtering

The first stage employs a filter bank that decomposes the EEG signals into 9 frequency pass bands, namely, 4-8 Hz, 8-12 Hz,..., 36-40 Hz [13]. Every filter has a finite impulse response designed by means of Kaiser Window. The transition bandwidth is set at 1 Hz. Other configurations are as effective, but this transition bandwidth yield a reasonable order filter and discriminative capacity between frequency bands.

B. Spatial Filtering

The second stage of feature extraction performs spatial filtering using CSP algorithm for each band-pass signal. CSP is a successful algorithm for the design of motor imagery-based BCIs [14]. It has been devised for the analysis of multichannel data belonging to 2-class problems. Consequently, it is necessary to set up CSP filters based on the trials for each class versus the trials for all other classes [15].

CSP calculates the spatial filters by solving an eigenvalue decomposition problem that involves the mean spatial covariances for each of the two classes. For more details refer to Ang et al [13]. The spatial filtered signal Z is then obtained from the EEG trial E as

$$Z = W^T E, \quad (1)$$

where W is a matrix containing the spatial filters computed by CSP. Each column of W represents a spatial filter. There are as many spatial filters as EEG channels. For each frequency band, CSP feature vectors are given by

$$x = \frac{\log\left[\text{diag}(\widetilde{W}^T E E^T \widetilde{W})\right]}{\text{trace}(\widetilde{W}^T E E^T \widetilde{W})}, \quad (2)$$

where $\widetilde{W}$ represents a matrix having some spatial filters of W. All spatial filters of W are not relevant for subsequent classification [14]. In accordance with FBCSP [13], the first 2 and the last 2 columns of W are selected. Finally, the 16 features of the 9 frequency bands for a single-trial are concatenated to form a single feature vector of 144 features.

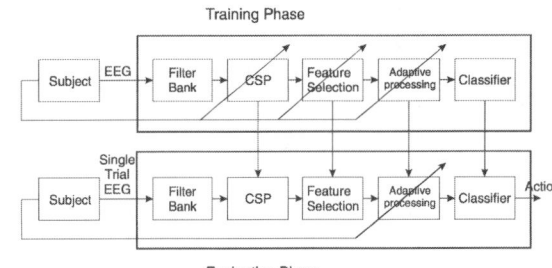

Fig. 1 Architecture of the algorithm for the training and evaluation phases. The architecture is based on FBCSP.

C. Feature Selection

After spatial filtering, a feature selection algorithm is employed to select the most discriminative features. Mutual Information-based Best Individual Feature (MIBIF) algorithm is used [13]. MIBIF involves the computation of the mutual information of each feature and class labels. Then, features with higher mutual information are selected. In this work, selected features include the ones whose mutual information is higher than the half of the highest mutual information.

D. Classification

The classification stage decides the class to which the feature vectors belong. A probabilistic generative model is used [16]. As a novelty, a stage performing adaptation related-processing is introduced before classification.

Adaptive Processing

The adaptive processing stage centers every incoming data by subtracting the global mean. Firstly, the global mean is estimated from the whole training data. Across the evaluation session, the global mean μ_C is updated in a casual manner with the following exponential update rule [8]

$$\mu_C(n, t) = (1 - \eta)\mu_C(n - 1, t) + \eta x(n, t), \quad (3)$$

where $x(n, t)$ is the current input feature vector of the nth evaluation trial at the time t and η is the update coefficient. The update coefficient is fixed to $\eta = 0.05$ for all subjects as was suggested by Vidaurre et al. [8].

Probabilistic Generative Model

We adopt a generative approach to classify each feature vector x into a specific class. Posterior probabilities $P(y|x)$ are computed through the Bayes' rule [16]

$$P(y|x) = \frac{P(y)P(x|y)}{\sum_y P(x|y)P(y)}. \quad (4)$$

It is assumed that the class-conditional densities $P(x|y)$ are Gaussian and all classes have the same probability of occurrence, $P(y = 1,2,3,4) = 0.25$. Then, the density for each class is given by

$$P(x|y) = \frac{1}{(2\pi)^{\frac{N}{2}}|\Sigma|^{\frac{1}{2}}} exp\left[-\frac{1}{2}(x-\mu_y)^T \Sigma^{-1}(x-\mu_y)\right], \quad (5)$$

where μ_i is the estimated mean of class i and Σ is the estimated common covariance matrix. Both of them are computed from the training samples using maximum likelihood [16]. Finally, the feature vector x is assigned to class y with the following maximum a posteriori (MAP) rule [16]

$$y = arg \max_{y=1,2,3,4} p(y|x). \quad (6)$$

IV. RESULTS

The session-to-session transfers of the proposed algorithm with adaptive processing is evaluated and compared with the non-adaptive counterpart as well as the winner of the competition [13]. In other words, the performances with update coefficient $\eta = 0.05$ (adaptive) or $\eta = 0$ (non-adaptive) are measured. As the organizers of BCI Competition IV [12], Cohen's kappa coefficient is used to quantify the performance. The results are presented in Table 1. It can be observed that the adaptive algorithm yields a higher mean kappa value of 0.62 compared to 0.58 for the non-adaptive algorithm and 0.57 for the FBCSP algorithm.

Since the results in Table 1 show that the kappa value of subject 9 is significantly improved from 0.62 for non-adaptive to 0.72 for adaptive classification, a further analysis is performed to investigate the effect of the adaptive processing stage. Figure 2 illustrates the impact of the adaptation on the feature distribution in a 3 dimensional sub-space: the three most discriminative features are depicted. The mean and variances of the feature distributions are represented for the four classes with darker-gray ellipsoids (training session) and lighter-gray ellipsoids (evaluation session). On the left,

we can clearly see an inter-session mean shift and rotation of the distributions. On the right, the same distributions are shown, but the mean shifts decrease on average.

V. DISCUSSION AND CONCLUSIONS

This study addresses the problem of non-stationarity in imaginary motor task classification. An adaptive classification framework with a processing stage before classification is proposed in order to deal with non-stationarity. This new stage reduces adaptively the global mean shift between training and evaluation data. Approaches with or without this stage were assessed on a multiclass problem posed in the BCI Competition IV dataset 2a and compared against the winner of the competition.

Our results suggest that the reduction of global mean shift by unsupervised learning increases the performance of the subsequent classification. This fluctuation is unrelated to tasks or classes and, accordingly, can be addressed in an unsupervised manner. Adaptive processing before classification enables the use of the same classifier for training and evaluation sessions reducing the loss of performance as a result of non-stationarity.

Several adaptive classifiers designed on the basis of static classifiers can be found in the literature [4-9]. However, the separation between adaptation-related tasks and classification has the advantage that any static classifier can be employed to devise adaptive algorithms without the need to make *ad-hoc* changes to them. Moreover, the computational cost needed to adjust complex adaptive classifiers is avoided.

Some limitations of the exponential adaptation have to be considered. First, both classes are required to be equally likely. Vidaurre *et al* [8] showed that the performance decreases as the trials of one class were removed from a balanced evaluation dataset. Nevertheless, it was also showed that this scheme of unsupervised adaptation was rather robust to this issue. Second, the exponential rule presents the difficulty of the proper choice of the update coefficient η [8]. Future work should employ more sophisticated adaptive procedures in addition to the exponential rule. On the other hand, additional classification algorithms could be considered to test the usefulness of our adaptive framework prospectively.

In summary, an adaptive approach that separates adaptive-related tasks from classification has been tested on BCI Competition IV dataset 2a and compared against the winner of the competition. Beyond the limitations, this study suggests a promising direction to the separation between adaptive tasks and classification.

Table 1 Performance in terms of Cohen's kappa coefficient of FBCSP, static classification and adaptive classification

Method	Subjects									Mean
	A1	A2	A3	A4	A5	A6	A7	A8	A9	
FBCSP (winner)	0.68	0.42	0.75	0.48	0.40	0.27	0.77	0.75	0.61	0.57
Static $\eta = 0$	0.76	0.36	0.84	0.55	0.36	0.30	0.71	0.75	0.62	0.58
Adaptive $\eta = 0.05$	0.77	0.39	0.87	0.55	0.47	0.32	0.74	0.79	0.72	0.62

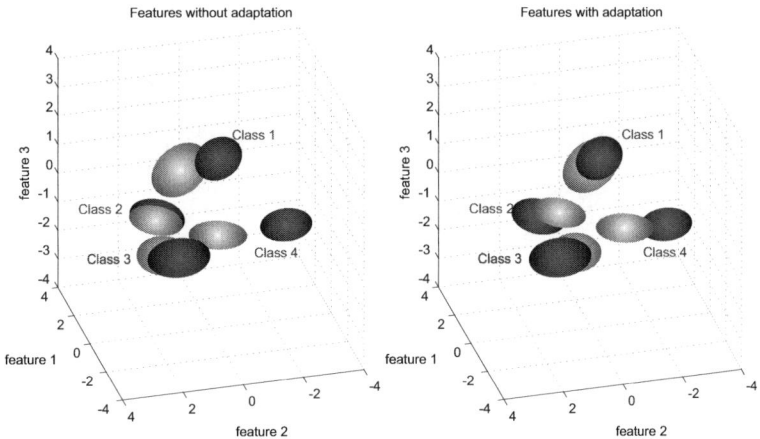

Fig. 2 Representations of the feature distributions for the three most discriminative features. The mean and variances of the feature distributions are represented for the four classes with darker-gray ellipsoids (training session) and lighter-gray ellipsoids (evaluation session). Data from subject 9 from the BCI Competition IV, dataset 2a.

Acknowledgment

This research was supported in part by the Project Cero 2011 on Ageing from Fundación General CSIC, Obra Social La Caixa and CSIC and a grant by the Ministerio de Economía y Competitividad and FEDER under project TEC2011-22987. L. F. Nicolas-Alonso was in receipt of a PIF-UVa grant from University of Valladolid. R. Corralejo was in receipt of a PIRTU grant from the Consejería de Educación de la Junta de Castilla y León and the European Social Fund (ESF).

References

1. Nicolas-Alonso L F and Gomez-Gil J (2012) Brain computer interfaces, a review. Sensors 12:1211-1279
2. Shenoy P, Krauledat M, Blankertz B, et al. (2006) Towards adaptive classification for BCI. J Neural Eng 3:R13
3. Li Y, Guan C, Li H, et al. (2008) A self-training semi-supervised SVM algorithm and its application in an EEG-based brain computer interface speller system. Pattern Recognition Letters 29:1285-1294
4. Ikeda K and Yamasaki T (2007) Incremental support vector machines and their geometrical analyses. Neurocomputing 70:2528-2533
5. Liu G, Huang G, Meng J, et al. (2010) Improved GMM with parameter initialization for unsupervised adaptation of Brain–Computer interface. Int J Numer Meth Biomed Eng 26:681-691
6. Vidaurre C, Sannelli C, Müller K-R, et al. (2010) Machine-learning-based coadaptive calibration for brain-computer interfaces. Neural Comput 23:791-816
7. Yan L, Kambara H, Koike Y, et al. (2010) Application of covariate shift adaptation techniques in brain-computer interfaces. IEEE Trans Biomed Eng 57:1318-1324
8. Vidaurre C, Kawanabe M, von Bünau P, et al. (2011) Toward unsupervised adaptation of LDA for brain–computer interfaces. IEEE Trans Biomed Eng 58:587-597
9. Liu G, Zhang D, Meng J, et al. (2012) Unsupervised adaptation of electroencephalogram signal processing based on fuzzy C-means algorithm. Int J Adapt Control 26:482-495
10. Yoon J, Roberts S, Dyson M, et al. (2008) Adaptive Classification by Hybrid EKF with Truncated Filtering: Brain Computer Interfacing Springer Berlin Heidelberg, ebook
11. Tsui C S L, Gan J Q and Roberts S J (2009) A self-paced brain–computer interface for controlling a robot simulator: an online event labelling paradigm and an extended Kalman filter based algorithm for online training. Med Biol Eng Comput 47:257-265
12. Tangermann M, Müller K-R, Aertsen A, et al. (2012) Review of the BCI Competition IV. Front Neurosci 6:55
13. Ang K K, Chin Z Y, Wang C, et al. (2012) Filter bank common spatial pattern algorithm on BCI competition IV datasets 2a and 2b. Front Neurosci 6:39
14. Blankertz B, Tomioka R, Lemm S, et al. (2008) Optimizing spatial filters for robust EEG single-trial analysis. IEEE Signal Processing Magazine 25:41-56
15. Townsend G, Graimann B and Pfurtscheller G (2006) A comparison of common spatial patterns with complex band power features in a four-class BCI experiment. IEEE Trans Biomed Eng 53:642-651
16. Bishop C M (2006) Pattern recognition and machine learning springer, New York, NY

Author: L. F. Nicolas-Alonso
Institute: ETSI Telecomunicación, Universidad de Valladolid
Street: Paseo de Belén 15, 47011
City: Valladolid
Country: España
Email: lnicalo@ribera.tel.uva.es

Computer Program for Automatic Identification of Artifacts in Impedance Cardiography Signals Recorded during Ambulatory Hemodynamic Monitoring

P. Piskulak[1], G. Cybulski[1,2], W. Niewiadomski[1,2], and T. Pałko[1]

[1] Institute of Metrology and Biomedical Engineering, Department of Mechatronics,
Warsaw University of Technology, Warsaw, Poland
[2] First Department of Applied Physiology Mossakowski Medical Research Centre, Warsaw, Poland

Abstract—The aim of this work was to design and verify a computer program which could detect artifacts in impedance cardiography (ICG) signals. It contains a procedure for creating a pattern of one typical cycle of the ICG trace and enabling the classification of each heart evolution as either a valid one or an artifact. It proposed an index characterizing the shape of the ICG signal pattern. The user can significantly influence the classification process by modifying the initial parameters of the patterns. The program automatically detects the characteristic points in the electrocardiographic (ECG) and ICG signals (dZ/dt) necessary for calculation of hemodynamic parameters and allows the user to browse previously recorded data. The program calculates basic cardiac hemodynamic parameters: heart rate (HR), pre-ejection period (PEP), left ventricular ejection time (ET), stroke volume (SV), cardiac output (CO) and other derivative ones, basing on previously classified data. We described the program structure and the operation of its main functions. The application is written in C++ using the Qt graphic library. It is flexible and can be modified or extended to suit the user's needs. The analysis' findings (based on on example data) are also presented.

Keywords—Impedance cardiography, motion artifacts, Ambulatory monitoring, noise, signal quality.

I. INTRODUCTION

Ambulatory impedance cardiography (AICG) monitoring is a promising technique allowing the estimation of cardiac hemodynamics during transient events. The analysis of hemodynamic parameters during the patient's normal activity would be difficult or even impossible to perform using other, well-established, "classical" methods [1-4]. The users of stationary impedance cardiography systems (ICG) are familiar with the problem of motion artifacts during ICG signal recordings. Those problems are even more pronounced in holter-type ambulatory impedance cardiography systems due to the motion of the patient [2, 4]. Since the amplitude of the first derivative of the ICG signal is roughly two orders smaller than that of ECG, any disturbance in electrode–skin contact results in a significant decrease in signal quality. Usually, the careful placement, positioning, and firm fixing of electrodes can significantly reduce the rate of artifacts [5,6]. Although the problem seems to be known, it has not so far been intensively explored using quantitative methods.

Willemsen et al., as a result of the verification of their system (the VU-AMS ambulatory monitor for impedance cardiography) concluded that it was a valid device for measuring systolic time intervals in real-life situations, but that its applicability for absolute stroke volume and cardiac output determination remained to be established [3]. Their reservations, however, only involved stroke volume and cardiac output values obtained during exercise.

Since some AICG recordings are too noisy to analyze, it is important to know the rate of artifacts during standard impedance holter recordings. This problem was considered in another paper [7], where it was found that the average rate of useful signals was (mean ± SD) 0.63±0.26 with a range of 0.21-0.98. Those rates were obtained as the worst case for five-minute strips taken from the whole period of each day and night recording.

Figure 1 presents examples of noisy and relatively clean recordings obtained in a related study when the subject walked on stairs and remained in a supine position, respectively [8].

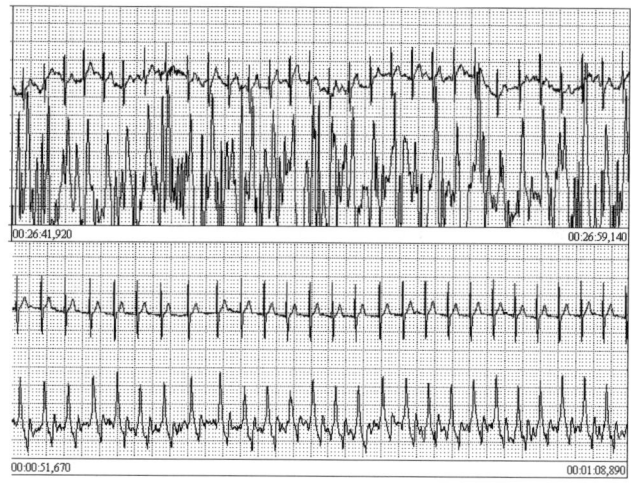

Fig. 1 The noisy (top strip) and artifact-free recordings (bottom strip) obtained in the previous study [8].

The aim of this study was to develop a method for automatic analysis of impedance cardiography signals, allowing distinction between the normal shape of the dZ/dt signal in ICG and any artifacts. In contrast to the previous studies, in which the cardiac evolution was classified as normal basing on the wide range of acceptable values of cardiac parameters [9-12], the idea was to create a pattern of one typical cycle of the dZ/dt signal and compare each heart beat's signal with it. We also intended to provide an index which could characterize the discrepancies between the basic pattern and the analyzed cycle.

II. METHODS

Impedance Cardiography Signal Descriptors

Figure 2 presents the typical impedance cardiography trace - changes in the first derivative of the ΔZ signal denoted as dZ/dt (2nd channel), recorded simultaneously with one ECG lead (1st channel). It also presents a way of describing characteristic points in the dZ/dt trace, which are used to determine the variables necessary to calculate cardiac hemodynamic indices. Characteristic points in the ECG and ICG signals (dZ/dt) can be described as follows:

Q – the beginning of ventricular depolarization (beginning of QRS complex in ECG),

R – the peak of the electrical stimulation of the cardiac chambers in QRS complex;

B – the beginning of ejection from the left ventricle, determined from dZ/dt signal;

C – the moment of maximum flow through the aortic valve, which allows one to determine dZ/dt_{max} (the maximum value of the first derivative of ICG signal)

X – the moment of closure of the aortic valve.

The detection of the characteristic points described above is essential for estimating cardiac parameters such as stroke volume (SV), cardiac output (CO), and various systolic time intervals (STI's). Systolic time intervals, which include pre-ejection period (PEP), left ventricular ejection time (LVET or ET), and electromechanical systole (EMS, the sum of PEP and ET), might be considered as measures of cardiac muscle contractility.

Computer Program for ICG Signal Analysis

The computer program for automatic analysis of ICG signals was partly based on earlier algorithms [9-10]. It consists of the following procedures: detection of the QRS complex, detection of characteristic points on the dZ/dt ICG curve, generation of the ICG evolution pattern, classification of the artifacts, and report generation.

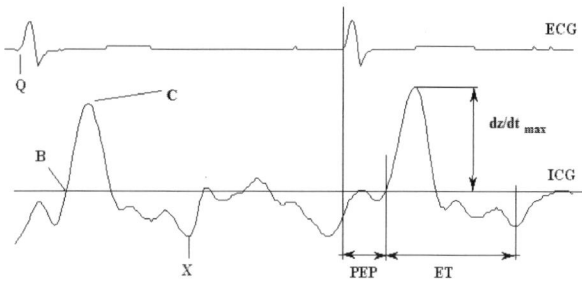

Fig. 2 The typical impedance cardiography trace - two cycles of the dZ/dt (first derivative of ΔZ) signal, denoted as dZ/dt (2nd channel), recorded simultaneously with one ECG lead (1st channel). Please note the way of describing characteristic points in the dZ/dt trace and the way of determining PEP, ET (LVET), and dZ/dt_{max} (adapted from [4]).

The first procedures find the characteristic points Q and R in the QRS complex. The R points in ECG are localized using the detection function with a heuristically determined detection threshold. The detection function exposes the characteristic features of the QRS complex and ignores features of the other components of the ECG signal. It was done through the application of the following operations: derivation, squaring, and average filtration. When the R point was detected for one cardiac cycle, the Q point was found 40ms before the R.

In the ICG signal, the following points were located: the point where the ICG signal crossed the baseline (B), the maximum of the dZ/dt signal (C), and the closing of the aortic valve (X). The importance of proper B point detection was discussed in [13]. The ways of determining the characteristic points and the fluctuation of the baseline (due to respiratory modulations) were described in an earlier paper [9,10] and summarized in a monograph [4].

On the basis of these data, the following parameters are calculated for each cycle: length of the cycle (T), heart rate (HR), stroke volume (SV), cardiac output (CO), distance between Q points (QQ), ejection time (ET), pre-ejection period (PEP), maximal amplitude of ICG signal (dZ/dt_{max}=AMP) and basic chest impedance (Z_0). The program also has a graphical user interface which enables presentation of the data, the pattern, the outcome of the artifact classification, and the numerical results.

The ICG Signal Pattern

The pattern of the first derivative ICG signal (*dZ/dt*) for one heart cycle was created by averaging 100 consecutive ICG cycles, synchronized by ECG R waves. Within the averaging section, the characteristic points (corresponding to Q in ECG, B, C, and X) are detected (and later used to calculate the pattern shape indices). The pattern of the ICG signal, with marked characteristic points, is presented on Fig. 3.

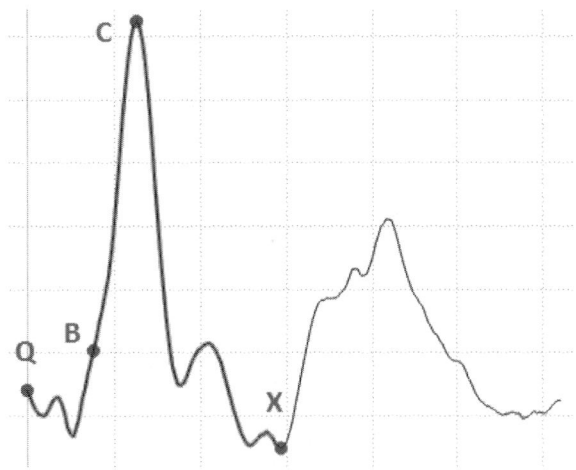

Fig.3 The ICG pattern of a QQ cycle with marked characteristic points.
B – the point where the curve cross the the base-line;
C – the maximum of the segment; X – the minimum of the segment.

The Pattern Shape Descriptors and Their Comparison

The pattern was characterized by the shape descriptors (form factors) used in automatic analysis of ECG ambulatory monitoring. Form factors are simple numerical measures of specific features of the analyzed shape – in our case, the QX period. They allow for objective and easy comparison of one ICG cycle with the pattern. We used a descriptor which relates the length of a curve and the sum of the area between the curve and the baseline during the period QX. The definition of that factor, sometimes called Malinowska's factor (FF_1), is given in equation (1):

$$FF_1 = [0.5L(\pi S)^{-0.5}] - 1 \qquad (1)$$

where:
L = circumference,
S = sum of the surface area above and under the baseline over QX period.
This factor has a value of 0 for a circle.

Classification of a signal as normal or artifact is based on the result of the comparison between the FF_1 values of the pattern and the evaluated cycle.

It is crucial to identify the appropriate level of discrimination for a particular form factor. To estimate this level, a 20-minute segment of the ICG signal was analyzed and each evolution was classified as either normal or artifact. Simultaneously, the absolute values of the differences between FF_1 values for each detected cycle and the pattern were calculated.

The value of the factors' differences and the corresponding classification variable were entered into a receiver operating characteristic (ROC) plotting program (published by Giuseppe Cardillo [14]).

The ROC illustrates the performance of a binary classifier system as its discrimination threshold varies. The area under the curve (AUC) is equal to the probability that a classifier will rank a randomly chosen positive instance higher than a randomly chosen negative one (assuming "positive" ranks higher than "negative") [14].

The cost-effective cutoff is the point on an ROC curve whose distance from the upper left corner of the graph is minimal. It corresponds to the optimal cutoff value for the form factor.

III. RESULTS

Fig. 4 presents the ROC curve obtained for one of the subjects analyzed in this study. The area under the curve (AUC) was 0.768, whereas the cost effective cut for FF_1 was 0.335.

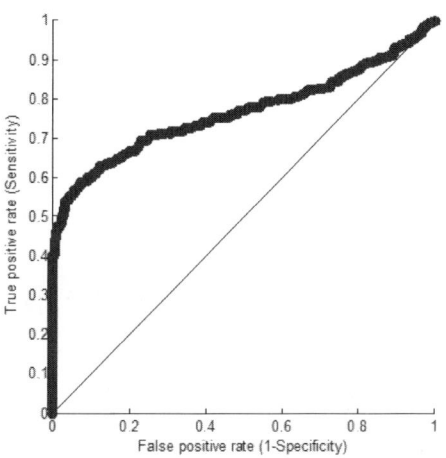

Fig.4 The ROC curve for the FF_1 factor describing the relationship between the length of the signal line and area delimited by the signal line.

Table 1 contains the results of the classification performed by the program and the operator for the analyzed period of the recordings: true positive (TP), true negative (TN), false positive (FP) and false negative (FN). The sensitivity – pTP/(TP+FN) – and specificity – TN/(TN+FP) – were 0.695 and 0.771, respectively. The positive predictive value – TP/(TP+FP) – and negative predictive value – TN/(TN+FN) – were 0.888 and 0.492, respectively.

Table 1 Confusion matrix for the automatic classification

		Manual	
		True	False
Automatic	Positive	608	77
	Negative	259	267

IV. CONCLUSIONS

The study has some limitations. The localization of the B point and the baseline level are crucial for proper calculation of PEP, ET, and the amplitude of (dZ/dt)max [13]. The main source of errors in the ICG-derived parameters is the usage of the simple model of blood flow in the thorax. A second source of errors is associated with uncertainty in the determination of the aforementioned ICG signal descriptors [4]. Also, the assumption of Q wave occurrence 40ms before R could be a source of errors in some pathological conditions (when the QRS complex is longer).

We wrote and checked a program that automatically detects the characteristic points in the ECG and ICG (dZ/dt) signals necessary for calculation of hemodynamic parameters and allows the user to browse previously recorded data. It creates the typical pattern of the ICG signal for one cardiac cycle based on 100 cycles. An index characterizing the shape of the signal and allowing for classification of each cycle as either "normal" or "artifact" was proposed. The "first approach" optimal value of the index was calculated. The proposed index was a better solution than that used in our previous papers [1,3,12].

This study contains only preliminary results; the proposal of the index shows the discrepancies between the shape of the pattern and the analyzed cycle. Although the AUC value was relatively high, we will try to introduce other indices and test them on a larger data set, which we hope will be more efficient. We also intend to validate the classifier more intensively.

ACKNOWLEDGMENT

This study was supported by the research programs of institutions the authors are affiliated with.

REFERENCES

1. Cybulski G, Książkiewicz A, Łukasik W, Niewiadomski W, Pałko T (1995) Ambulatory monitoring device for central hemodynamic and ECG signal recording on PCMCI flash memory cards, Computers in Cardiology, IEEE, pp. 505-507.
2. Nakagawara M and Yamakoshi K (2000). A portable instrument for non-invasive monitoring of beat-by-beat cardiovascular haemodynamic parameters based on the volume-compensation and electrical-admittance method. Medical & Biological Engineering & Computing, vol.38, no.1, Jan., pp.17-25.
3. Willemsen GH. De Geus EJ. Klaver CH. Van Doornen LJ. Carroll D. (1996) Ambulatory monitoring of the impedance cardiogram. Psychophysiology. 33(2): 184-93.
4. Cybulski G (2011). Ambulatory Impedance Cardiography. The Systems and their Applications. Series: Lecture Notes in Electrical Engineering, Vol. 76, 1st Edition, ISBN: 978-3-642-11986-6, 150 pp., Springer-Verlag Berlin and Heidelberg GmbH & Co. KG, DOI: 10.1007/978-3-642-11987-3
5. Nakonezny PA. Kowalewski RB. Ernst JM. Hawkley LC. Lozano DL. Litvack DA. Berntson GG. Sollers JJ 3rd. Kizakevich P. Cacioppo JT. Lovallo WR (2001) New ambulatory impedance cardiograph validated against the Minnesota Impedance Cardiograph. Psychophysiology, 38(3): 465-473.
6. Bogaard HJ, Woltjer HH, Postmus PE, de Vries PM (1997) Assessment of the haemodynamic response to exercise by means of electrical impedance cardiography: method, validation and clinical applications, Physiological Measurement, 18(2): 95-105.
7. Cybulski G, Niewiadomski W, Gąsiorowska A, Kwiatkowska D (2007). Signal quality evaluation in Ambulatory Impedance Cardiography., IFMBE Proceedings vol. 17 (Ed. Scharfetter, Merwa), 13th International Conference on Electrical Bioimpedance, Graz, Austria, Aug.29th-Sept.2nd, Springer, 590-592.
8. Cybulski G, Niewiadomski W, Aksler M, Strasz A, Gąsiorowska A, Laskowska D and Pałko T. Identification of the Sources of Artefacts in the Holter-Type Impedance Cardiography Recordings. IFMBE Proceedings, 5th European Conference of the International Federation for Medical and Biological Engineering, Ed. Akos Jobbagy, Springer, ISBN 978-3-642-23507-8, 2011, vol. 37, pp: 1272-1274,
9. Cybulski G (1988) Computer method for automatic determination of stroke volume using impedance cardiography signals. Acta Physiologica Polonica, 39 (5-6): 494-503.
10. Cybulski G, Miśkiewicz Z, Szulc J, Torbicki A, Pasierski T (1993) A comparison between impedance cardiography and two dimensional echocardiography methods for measurements of stroke volume (SV) and systolic time intervals (STI), Journal of Physiology and Pharmacology, 44(3): 251-258,
11. Gasiorowska A, Nazar K, Mikulski T, Cybulski G,. Niewiadomski W, Smorawinski J, Krzeminski K, Porta S, Kaciuba-Uscilko H (2005) Hemodynamic and neuroendocrine predictors of lower body negative pressure (LBNP) intolerance in healthy young men, J Physiol. Pharmacol. 56, 2, 179-19
12. Cybulski G, E Michalak, E Koźluk, A Piątkowska, W Niewiadomski (2004) Stroke volume and systolic time intervals: beat-to-beat comparison between echocardiography and ambulatory impedance cardiography in supine and tilted positions, Medical and Biological Engineering and Computing, 42: 707-711.
13. Lozano DL, Norman G, Knox D, Wood BL, Miller BD, Emery CF and Berntson GG (2007) Where to B in dZ/dt. Psychophysiology, 44: 113–119. doi: 10.1111/j.1469-8986.2006.00468.x
14. http://www.mathworks.com/matlabcentral/fileexchange/19950-roc-curve/content/roc.m

Author: Gerard Cybulski
Institute: Institute of Metrology and Biomedical Engineering,
Department of Mechatronics, Warsaw Univ. of Technology,
Street: 8 Św. A. Boboli
City: Warsaw
Country: Poland
Email: G.Cybulski@mchtr.pw.edu.pl

Comparison between Artificial Neural Networks and Discriminant Functions for Automatic Detection of Epileptiform Discharges

C.F. Boos, G.R. Scolaro, and F.M. Azevedo

Biomedical Engineering Institute, Federal University of Santa Catarina, Florianópolis, Brazil

Abstract—This study presents the performance analysis between two classifiers when they are used together with mimetic analysis based morphological features to develop a method for automatic detection of epileptiform discharges in EEG signals. We applied mimetic analysis in the form of extracting a set of morphological descriptors, which in this study represent a set of parameters related to morphology features of the EEG waveform. The two tested classifiers are Discriminant Functions (DF) and Artificial Neural Networks (ANN). The DFs are obtained from Discriminant Analysis and are frequently applied in pattern classification problems such as automatic identification epileptiform discharges. On the other hand, the ANNs are an Artificial Intelligence tool commonly used in pattern recognition methods and systems. Simulations showed average efficiency of 84%, sensitivity of 86% and specificity of 82%. While the neural networks presented better sensitivity values, the discriminant functions had better specificity results. Also, it was noticed that the efficiency values for small sized classifiers were equivalent but as the classifier's size increased the neural networks exhibited better results than the discriminant functions.

Keywords—EEG signal, Epileptiform Pattern, Artificial Neural Network, Discriminant Function.

I. INTRODUCTION

The analysis of electroencephalogram (EEG) records can be an important tool for confirmation of clinical diagnosis of epilepsy [1]. This analysis is made by a thorough visual review of the EEG signals and is performed by a specialist with the objective of locating a very specific type of electrographic activity, called epileptiform patterns or discharges. However, since the records for this analysis are acquired from long term monitoring of the subject the reviewing process can be a very long and tiresome task.

Over the years, several studies have proposed methods of automatic EEG review with the intent of facilitating and reducing the time required from the specialists for the EEG record analysis. However, as literature shows, there is still no method with detection performance in accordance with the needs of specialists [2].

Nevertheless, according to literature, the automatic spike detection method proposed by Gotman and Gloor [3] can be considered one of the most used and successful systems so far. Their method is based on mimetic analysis where the EEG signals are decomposed in segments and sequences according to a set of rules based on morphological analysis of the EEG signal.

The methods for automatic epileptiform detection using Artificial Neural Networks (ANN) are very common and their classification performance can be considered relatively good [4]. Generally, the application of neural networks can be performed in conjunction with the parameter extraction or using EEG signals directly. In both cases the data, i.e., the parameters or signals are inserted in the neural networks as input stimuli.

The EEG signals may or not be digitally processed in order to enhance certain characteristics of the waveform and parameters – which in this study are called morphological descriptors – can be extracted from mimetic analysis which is based on waveform morphology analysis.

As an alternative to the use of ANN with parameters as input stimuli, other EEG based studies [5, 6] applied a statistical method called Discriminant Analysis to create Discriminant Functions (DF) that can be used for classification of the analyzed data or be used as a predictor for new data, provided that this new data has the same technical features of the one used to create the functions.

Therefore, the purpose of this paper is to compare the performance of neural networks and discriminant functions when they are used together with mimetic analysis based morphological features to develop a method for automatic detection of epileptiform discharges.

II. MATERIAL AND METHODS

A. Database

The database[1] used in simulations is composed by nine EEG records from seven adult epileptic subjects undergoing long term monitoring for pre-surgical evaluation. From this database we extracted 600 segments of EEG signal containing specific patterns divided in two classes: segments with epileptiform discharges (EEP) and segments with non-epileptiform patterns (NEP). Examples of these two classes of patterns can be observed in Figure 1. For the simulations performed with DF and ANN we divided the 600 patterns in four sets:

- set A – for neural network training;
- set B – for neural network cross-validation;
- set C – for discriminant function modeling;
- set D – for testing both approaches (DF and ANN).

---

[1] The use of the EEG pattern database was approved by the University's Ethics Committee (protocol number 144469).

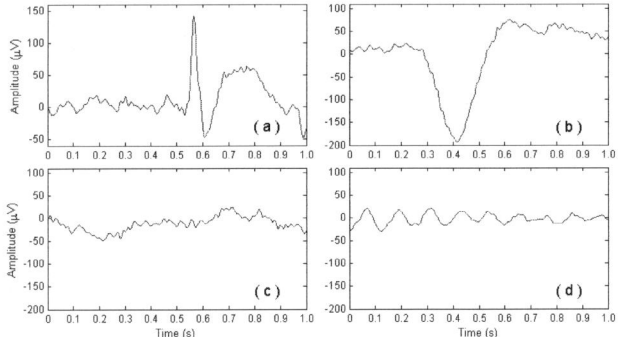

Fig. 1 Examples of the types of patterns in the database: (a) epileptiform, (b) eye blink, (c) normal background activity and (d) alpha waves.

Set A and set B were each formed by manually selecting 60 patterns. Set C, also called "model set", is composed by 120 patterns and is a combination of set A and B. Set D, also called "test set", was formed by random selection of 200 patterns. All sets are balanced, i.e., composed by equal amounts of EEP and NEP.

B. Mimetic Analysis

Several methods of EEG analysis have been proposed in attempt to solve the problem of determining an accurate methodology for automatic epileptiform pattern recognition and one of these methods is called Mimetic Analysis. As the name suggests, this type of analysis attempts to mimic the reasoning used by the specialist at the task of identifying epileptiform patterns.

This work applies mimetic analysis by extracting from the EEG signals a set of morphological descriptors, i.e. a set of parameters related to morphology features of the EEG waveform. The descriptors we used were developed and refined in previous studies [7, 8]. The selected set contains 45 parameters related to statistical (mean, variance, standard deviation, among others) and morphological characteristics (waveform amplitude, segment duration, angle of peaks, among others) that differentiate epileptiform events from the rest of the events that may be present in an EEG signals.

An algorithm was developed for the extraction of the descriptors. The beginning of the algorithm consists of data preparation, in which the signals are interpolated to 1k Hz sampling frequency, filtered to remove eventual power line interference and have their frequency spectrum adjusted to attenuate the frequency components lower than 5 Hz.

After this preparatory step the descriptors are extracted in clusters with similar characteristics. The first cluster is related to amplitude and duration of segments, followed by the extraction of the duration of the pattern under analysis. The second cluster is related to the angle and slope of the peaks of the signal. The third cluster is related to amplitude and duration of segments of the pattern's neighborhood regions, i.e., the portions of the signal adjacent to the pattern.

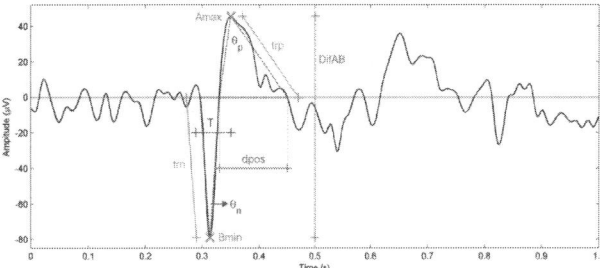

Fig. 2 Examples of the first two clusters of extracted descriptors.

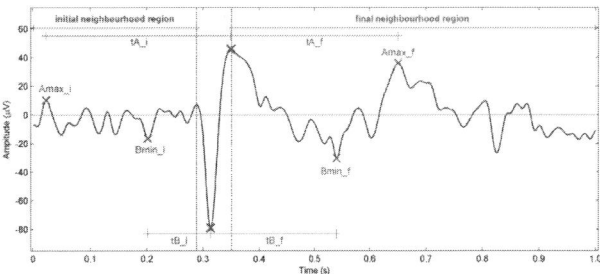

Fig. 3 Examples of the third cluster of extracted descriptors.

The last cluster is related to statistical measurements taken from the segment containing the pattern under analysis. After the extraction is completed, the set of descriptors is used as input stimuli of Artificial Neural Networks and as variables for the design of Discriminant Functions. Some examples of descriptors of the first two clusters are illustrated in Figure 2 and Figure 3 illustrates the descriptors from the third cluster.

C. Discriminant Functions

The Discriminant Functions are frequently used for size reduction, classification and/or feature extraction [9]. They are a linear weighted combination of n independent variables that can be used as a predictor for classification problems such as the automatic EEG signal review for epileptiform pattern identification.

In this study, the functions were obtained by applying Linear Discriminant Analysis on the set of morphological descriptors, i.e., we used the descriptors as the independent variables from which the discriminant functions are modeled.

Since the simulations were made with commercial statistical software, IBM® SPSS Statistics, we were able to perform a stepwise analysis with Mahalanobis Distance method. The application of stepwise method was important because it allowed us to change the reference value of the criteria used for inclusion and exclusion of variables in the discriminant function. Consequently, this variation of these reference values allowed us to find functions that used different groups and quantities of descriptors to classify the EEG patterns.

D. Artificial Neural Networks

Artificial Neural Networks, as mentioned before, is an Artificial Intelligence tool commonly used in methods and/or for automatic detection of epileptiform discharges.

In this study we opted for the use of Feedforward Multilayer Perceptrons with supervised training and Error Backpropagation learning algorithm.

For the proposed analysis and comparison with DF, several ANN were implemented. The networks had the same basic structure: three layers, with decreasing number of neurons in each layer and only one output neuron.

All simulations used the same signals and the descriptors used as input stimuli were selected according to the results of the DA. In other words, only the descriptors that appear in the discriminant functions were used in the ANN simulations, for example: if the discriminant function has only three descriptors, the ANN will be trained with the same three descriptors as input. This process was employed to allow a better comparison between DF and ANN.

The configuration parameters, such as learning rate, initial synaptic weights and activation functions, were varied throughout the simulations in an effort to find the best configuration possible.

III. RESULTS

After all neural networks were trained and discriminant functions were obtained, both classifiers were tested with the same set of patterns. This "test set", as mentioned before, is a balanced set of 200 randomly chosen patterns and the same "test set" was used for ANN and DF testing to allow a better comparison between the two approaches.

The analysis of the results, which will be discussed in the next section, was based on the following metrics: number of True Positives (TP) and Negatives (TN), number of False Positives (FP) and Negatives (FN), and Sensitivity, Specificity and Efficiency values.

The best results achieved after the tests can be observed in Table 1, Table 2 and Figure 4 through Figure 6. Along with the performance results, the tables and figures also present the size of the classifiers. For the ANN, the size refers to the number of neurons in each layer.

Table 1 Results achieved after simulations with Discriminant Functions

Classifier	Size[a]	TP	TN	FP	FN
DF 1	3	83	81	19	17
DF 2	7	85	84	16	15
DF 3	9	90	80	20	10
DF 4	12	90	81	19	10
DF 5	22	89	77	23	11
DF 6	24	90	74	26	10
DF 7	31	85	75	25	15
DF 8	36	86	77	23	14

a Size is the number of variables in the discriminant function.

Table 2 Results achieved after simulations with Artificial Neural Networks

Classifier	Size	TP	TN	FP	FN
ANN 1	03:02:01	85	81	19	15
ANN 2	07:03:01	85	85	15	15
ANN 3	09:04:01	85	86	14	15
ANN 4	12:06:01	85	87	13	15
ANN 5	22:08:01	85	86	14	15
ANN 6	24:08:01	83	84	16	17
ANN 7	31:06:01	87	85	15	13
ANN 8	36:08:01	88	88	12	12

a. Size is the number of neurons (N) in each layer of the network ($N_{input\ layer}:N_{hidden\ layer}:N_{ouput\ layer}$)

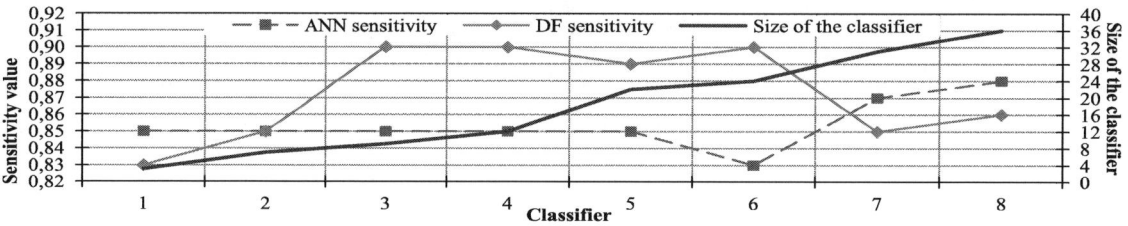

Fig. 4 Sensitivity values of the ANN and DF classifiers and their respective size (number of descriptors used by the classifier).

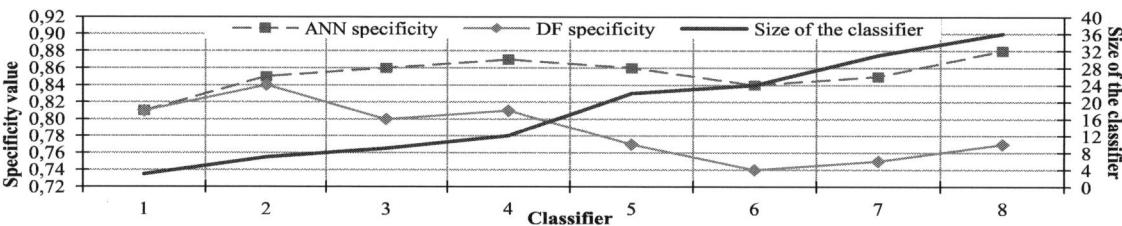

Fig. 5 Specificity values of the ANN and DF classifiers and their respective size (number of descriptors used by the classifier).

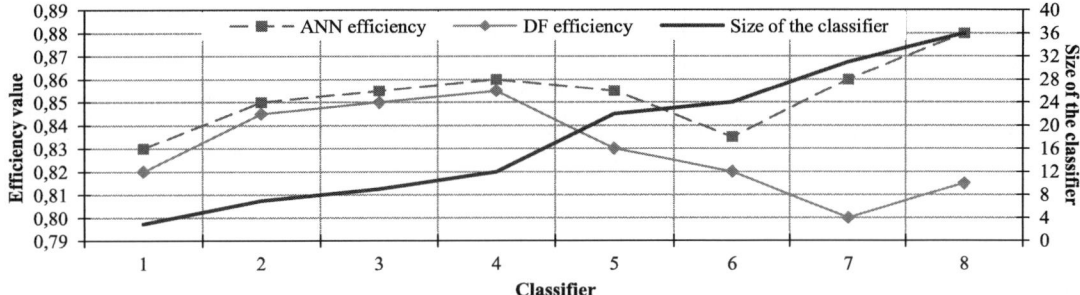

Fig. 6 Efficiency values of the ANN and DF classifiers and their respective size (number of descriptors used by the classifier).

IV. DISCUSSION AND CONCLUSIONS

The number of descriptors in the obtained Discriminant Functions ranged from 3 to 36. Eight different functions were found and all of them exhibited good statistical significance ($p < 0.05$). The functions' sensitivity values varied between 83 and 90% while the specificity was only between 74 and 84%. The efficiency values were all above 80%, with the best function presenting 85,5% efficiency. The function DF6 can be considered the one with the best overall performance, presenting a 5% false negative rate and 19 (9,5%) false positives

The results obtained with Artificial Neural Networks tests exhibited relatively good results with efficiency values ranging from 83 and 88%. The specificity and sensitivity values presented by the networks varied from 81 to 88% and 83 to 88%, respectively. The best network in this study was ANN8, which had an overall performance of 88% and presented only 6% false positive and negative rates.

When comparing the two types of classifier, neural networks and discriminant functions, the ANNs presented better sensitivity values while DFs had better specificity results. An interesting difference could be noticed on the efficiency of the classifiers when taking into account their size, i.e., the number of descriptors involved. When the classifier had relatively small size (≤ 12) the ANNs and DFs had equivalent results but for larger classifiers – with more than 24 descriptors – the neural networks clearly presented better results. This aspect can be an indication that for large sized group of descriptors the use of neural networks is more suitable than the application of discriminant functions. However, further simulations have to be made, especially with a lager EEG database, in order to certainly determine which classifier is better suited for application with Mimetic Analysis.

Finally, the performance results achieved with both classifiers can be considered compatible with the other similar studies and promising results were achieved with of both DA and ANN in combination with Mimetic Analysis.

ACKNOWLEDGMENT

This study was co-sponsored by the National Council for Scientific and Technological Development (CNPq).

REFERENCES

1. Pillai J, Sperling MR (2006) Interictal EEG and the diagnosis of epilepsy. Epilepsia 47 Suppl 1:14–22. doi: 10.1111/j.1528-1167.2006.00654.x
2. Halford JJ, Schalkoff RJ, Zhou J, et al. (2013) Standardized database development for EEG epileptiform transient detection: EEGnet scoring system and machine learning analysis. Journal of Neuroscience Methods 212:308–16. doi: 10.1016/j.jneumeth.2012.11.005
3. Gotman J, Gloor P (1976) Automatic recognition and quantification of interictal epileptic activity in the human scalp EEG. Electroencephalography and Clinical Neurophysiology 41:513–29.
4. Halford JJ (2009) Computerized epileptiform transient detection in the scalp electroencephalogram: obstacles to progress and the example of computerized ECG interpretation. Clinical Neurophysiology 120:1909–15. doi: 10.1016/j.clinph.2009.08.007
5. Chapman RM, Nowlis GH, McCrary JW, et al. (2007) Brain event-related potentials: diagnosing early-stage Alzheimer's disease. Neurobiology of aging 28:194–201. doi: 10.1016/j.neurobiolaging.2005.12.008
6. Romo Vázquez R, Vélez-Pérez H, Ranta R, et al. (2012) Blind source separation, wavelet denoising and discriminant analysis for EEG artefacts and noise cancelling. Biomedical Signal Processing and Control 7:389–400. doi: 10.1016/j.bspc.2011.06.005
7. Pereira M do CV (2003) Avaliação de Técnicas de Pré-Processamento de Sinais de EEG para Detecção de Eventos Epileptogênicos Utilizando Redes Neurais Artificiais. 255.
8. Boos CF, Azevedo FM de, Scolaro GR, Pereira M do C V. (2011) Automatic Detection of Paroxysms in EEG Signals Using Morphological Descriptors and Artificial Neural Networks. In: Laskovski A (ed) Biomedical Engineering, Trends in Electronics, Communications and Software. InTech, pp 387–402
9. Raykov T, Marcoulides GA (2008) An Introduction to Applies Multivariate Analysis. 496.

Author: Christine Fredel Boos
Institute: Biomedical Engineering Institute (Department of Electrical Engineering of the Federal University of Santa Catarina).
Street: Campus Universitário – Trindade – CEP 88040-900.
City: Florianópolis
Country: Brasil
Email: cfboos@gmail.com

Classification of Early Autism Based on HPLC Data

T. Kristensen

Department of Computing, Bergen University College, 5020 Bergen, Norway
CEO of the company Patten solutions ltd, Norway (http://www.patternsolutions.no)

Abstract—In this work a technique for early autism identification has been developed. For classification a Multi-Layered Perceptron (MLP) neural network and a Support Vector Machine (SVM) have been used. Early identification of autism is important because the prognosis for treating autism is then much better. The patterns used to train both systems are extracted from HPLC data of urine. The database consists of two types of samples, from normal and autistic children. The classification rate has been estimated to about 80 % or better for both algorithms. From the experiments we may conclude that the algorithm that gave the best results was SVM. All the software to do the analysis has been developed in Java.

Keywords—Autism, HPLC spectra, MLP, SVM.

I. INTRODUCTION

Early test for identification of children at risk for developing autism may soon be a reality. Children suffering from autism may have a certain chemical signature in their urine. These findings also indicate that there can be certain substances in the urine that may contribute to the onset of autism.

Autism has earlier been linked to metabolic abnormalities, and the hypothesis is that these metabolic changes may be detectable in the children's urine. Using HPLC (High Performance Liquid Chromatography) spectral data [17] we have found that children of a normal group and an autism group seem to have a distinct chemical fingerprint in their urine.

Autism is usually diagnosed by a series of behavioural tests and symptoms [13]. Suffering from autism may be identified early at an age of 5 months. However, a clear diagnosis is usually not possible before the children are 2 or 3 years old. There seems to be a growing evidence that the earlier the behavioural therapies for autism are started, the better the chances are for the children to be able to live relatively normal lives when growing up.

If one is able to identify the risk for developing early autism by a chemical or statistical test rather by observing the manifestation of a full-blown behaviour, the therapy can be started much earlier. There also exist scientists that that are linking autism of children with the production of toxins that may interfere with the brain development [11]. One of the compounds that may be identified in the urine was N-methyl Nicotinamide (NMND) which also has been associated with the Parkinson's disease. Some scientists are also arguing for that autism may be related to metabolic products of certain bacteria which need to be identified [6,7].

However, the current work is about how to use a Multi-Layered Perceptron (MLP) neural network [4,5] and a Support Vector Machine (SVM) [2], to classify between HPLC samples of normal children and children suffering from autism.

The organization of the paper is as follows: In section 2 we describe the data. In section 3 the feature extraction techniques is presented. Section 4 defines a MLP neural network and the algorithm used for training. In section 5 the SVM network and training algorithm for such a network is presented. In section 6 we present the result for both a small scale experiment and a proof of concept experiment. Section 7 gives the conclusion.

II. THE DATA

The HPLC data was recording by the company Tipogen ltd. at Bergen Hightech Centre, Norway. The first delivery of data was based on 30 samples of urine spectra from autistic and normal children. The aim of a first experiment was to verify a so-called *proof of principle*. The intention was to find out if datamining based on machine learning algorithms [3] could be used to verify autism using HPLC spectra from urine of children. The HPLC spectra look like the one given in Fig. 1.

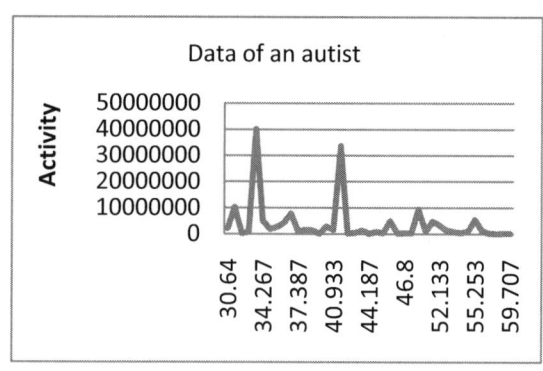

Fig. 1 HLPC spectrum of an autist

The first axis represent what is called '*retention time*' which represents the peak ID. The second axis represents the intensities or the '*peak area*'. The data was delivered in a spread sheet format, one spread sheet for normal children and one for autistic children.

III. FEATURE EXTRACTION

The length of the different samples for both the control data and autistic data varied. This was not so easy to handle in a adequate way in the beginning of the analysis. An example of a pattern that was generated from the data is given in Fig.2. Each sample has a specific number of peaks. The different numbers of peaks belonging to each sample make the analysis more complicated and also represented an essential parameter in the recognition process.

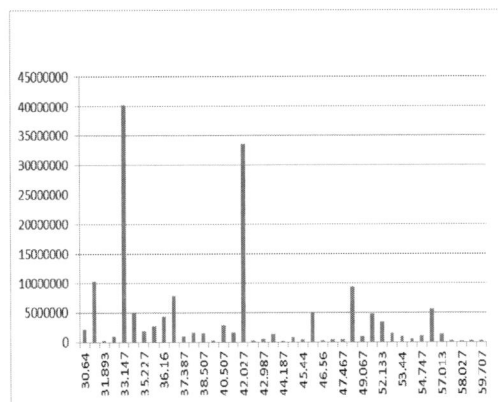

Fig. 2 HLPC peak extraction for generating patterns.

A. Pattern Diagnostics

The ability to discriminate patterns acquired from healthy individuals, from individuals affected from a disease is the most important aspect of such a pattern diagnostic technique [1]. Signatures based on HPLC data in urine [17] or Mass Spectrometry (MS) data [9] in the blood may be used to develop a diagnostic classifier system. Such a pattern diagnostics is based on the analysis of huge amount of data to find patterns to identify the disease.

An alternative method is to use Mass Spectrometry (MS) data in the blood. This has given promising results in detection of early cancer [12], and may also be used to show early autism. Mass Spectrometry data consists of a set of m/z values (m is the atomic mass and z is the charge of the ion) and the corresponding relative intensities of all molecules present with that m/z ratio. The MS data of a chemical sample thus is an indication of the presence or absence of the actual molecules. The data might therefore be used to predict the presence of a disease condition and distinguish it from a sample taken from a healthy individual [10].

IV. MULTI-LAYERED PERCEPTRON

A Multi-Layered Perceptron (MLP) network generally contains three or more layers of processing units [3,5]. Fig 3 shows the topology of a network containing three layers of nodes. The first layer is the input layer, the middle layer or the "hidden" layer consists of "feature detectors"- units that respond to particular features that may appear in the input pattern. Sometimes, there can be more than one hidden layer. The last layer is the output layer. The activities of these units are read as output from the network and may be used to define different categories of patterns.

A. Training of MLP

The MLP network is trained using a supervised learning algorithm. A set of training examples are presented to the network, where a target vector is given for each training example. The output vector of the network is compared to the target vector, and the weights in the network adjusted to make the network perform better according to the training examples and future instances. The training algorithm used in the paper is the backpropagation algorithm.which is a well known learning algorithm to use in classification [6,7].

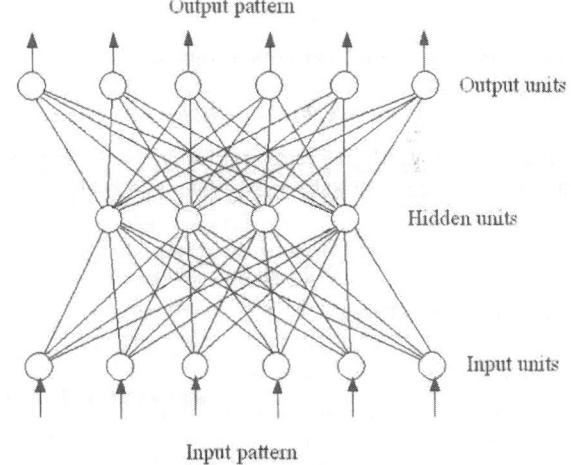

Fig. 3.A MLP Network consisting of three layers of nodes

Each node is activated in accordance with the input to the node and the activation of the node. Then, the difference between the calculated output and the target output is calculated. The weights between the output layer, hidden layers and the input layer are adjusted by using this error function. The output of a node is calculated using the sigmoid function $f(x) = 1/1 + e^{-x}$. The weighted sum $S_j = \sum w_{ji}a_i$ is inserted into the sigmoid function, and the result is the output value from a unit j given by:

$$f(S_j) = 1/1 + e^{-S_j}, \qquad (1)$$

The error value of an output unit j can then be computed by:

$$\delta_j = (t_j - a_j)f'(S_j) \qquad (2)$$

where t_j and a_j are the target and output value for unit j, respectively and f' is the derivative of the function f.

For a hidden node, the error value is calculated as:

$$\delta_j = \sum \delta_k w_{kj} (f'(S_j)) \qquad (3)$$

From the formula we see that the error of a processing unit in the hidden layer is computed by the upper layer. Finally, the weights can be adjusted by:

$$\Delta w_{ji} = \alpha \, \delta_j \, a_i \qquad (4)$$

where α is the learning rate.

Another parameter is very often introduced in the MLP network, called momentum. It has been shown that the use of this additional parameter can be helpful in speeding up the convergence and avoiding local minima [4]. By including this in the equations the next iteration step can be written as:

$$w_{ji}(t+1) = w_{ji}(t) + \alpha \, \delta_j \, a_i + \beta \Delta w_{ji}(t) \qquad (5)$$

where α is the learning rate, β is the momentum and Δw_{ji} the weight change from the previous processing step.

V. SVM THEORY

SVM is a computationally efficient learning technique that is now being widely used in pattern recognition and classification problems in general [2]. This approach has been derived from some of the ideas of statistical learning theory to control the generalization abilities of a learning machine [14,15].

In such an approach the machine learns an optimum hyper-plane that classifies the given pattern. By use of kernel functions, the input feature space, by applications of a non-linear function can be transformed into a higher dimensional space where the optimum hyper-plane can be learnt. This gives great flexibility by using one of many learning models by changing the kernel functions.

A. SVM Classifier

The basic idea of a SVM classifier is illustrated in Fig.4. This figure shows the simplest case in which the data vectors (marked by 'X' s and 'O' s) can be separated by a hyper-plane. In such a case there may exist many separating hyper-planes.

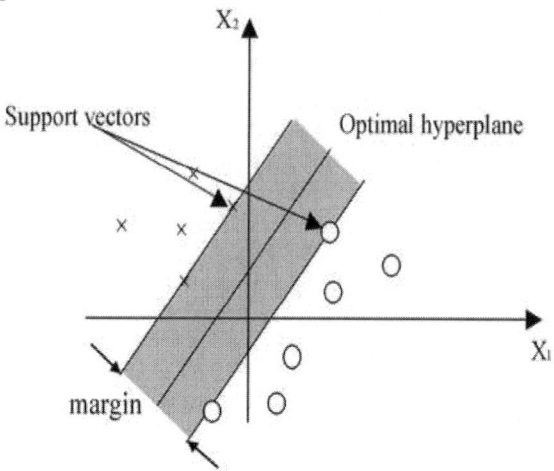

Fig. 4 A SVM classification by maximizing the margins between the classes

Among them, the SVM classifier seeks the separating hyper-plane that produces the largest separation margins. In the more general case, in which the data points are not linearly separable in the input space, a non-linear transformation is used to map the data vectors into a high-dimensional space (called feature space) prior to applying the linear maximum margin classifier.

SVM uses a kernel function in which the non-linear mapping is implicitly embedded. The discriminant function of the SVM classifier can be defined as:

$$f(x) = \sum_{i=1}^{N} \alpha_i y_i K(x_i, x) + b \qquad (6)$$

where $K(\cdot,\cdot)$ is the kernel function, x_i are the support vectors determined from the training data, y_i is the class indicator (e.g. +1 and -1 for a two class problem) associated with each x_i, N is the number of supporting vectors determined during training, α is the Lagrange multiplier for each point in the training set and b is a scalar representing the perpendicular distance of the hyper-plane from origin..

The most commonly used kernel functions are the polynomial kernel given by :

$$K(\mathbf{x_i}, \mathbf{x_j}) = (\mathbf{x_i}^T \mathbf{x_j} + 1)^p \text{, where p > 0 is a constant} \quad (7)$$

or the Gaussian Radial Basis Function (RBF) kernel given by

$$K(\mathbf{x_i}, \mathbf{x_j}) = \exp(-\|\mathbf{x_i} - \mathbf{x_j}\|^2 / 2\sigma^2) \quad (8)$$

where $\sigma > 0$ is a constant that defines the kernel width.

VI. EXPERIMENTS AND RESULTS

A special program was made in Java to generate the patterns from the original HPLC data to be used in the experiments. The generation of optimal selection of patterns were very much dependent on computing the right number of peaks from the HPLC spectra.

Both the training and testing data used were written to file to be read later into a java program. The java program developed for training used the package javANN (java Artificial Neural Network) developed by the company Pattern solutions ltd. [18] in Norway.

A special program in Java was developed to change the format of the data to be used in the LIBSVM toolbox [16] to be able to classify between a normal and an autistic child. Before the training started the regularization parameter C have to be determined. We have to determine the value of C experimentally. Such an approach is not optimal. The performance of the SVM classifier was best for C values from 100 and up to around 200. The SVM algorithm was tested on the same samples of data (unseen) as the MLP algorithm, with a penalty constant C = 100 or 200.

A. MLP and SVM Experiments

a) Small Scale Experiments:

In this experiment 18 samples were used for training and 12 samples were used for testing. In the MLP experiments the number of hidden nodes was selected to 100. The learning rate was set to 0.1 and the momentum to 0.9. The number of iterations were set to 10000.

The MLP algorithm was tested on 12 unseen samples. These data was unknown for the system, but we knew the category of them. It was then easy to calculate the performance of the system. The network was able to correctly classify 11 either as an autist or normal children. The best performance was estimated to 11/12 = 91.7 %.

For the SVM algorithm the best performance was estimated to 83.4 % where 10 of 12 samples were classified correctly. Both algorithms identified one false positive sample on the average where a normal child was classified as an autist. This is a more serious mistake than a false negative classification error where an autist is classified as normal.

b) Large Scale Experiments:

The second delivery of data was based on 62 samples of autistic children and 52 samples of normal children, totally 114 samples. The intention of the second analysis was to see if the proof of principle experiment that was verified in the first delivery of data, could be extended to a *proof of concept* experiment. The training set consists of 71 samples of data and 43 was used for testing.

The best performance for the SVM algorithm used was estimated to 88.4 % with a penalty constant C = 100. This corresponds to 38 of 43 samples were correctly classified. The average performance of using SVM was estimated to about 85 %.

For the MLP network the best performance was estimated to 81.4 % with an average value of 78.3. This corresponds to 35 of 43 samples were classified correctly. The average number of false positive cases in all experiments was found to be two for both algorithms,

VII. CONCLUSIONS

Pattern diagnostics represents a new way to detect early diseases. In this case it has been used to diagnose early autism. The type of analysis requires small amount of urine to generate HPLC spectra. However, Mass Spectrometry data should also be possible to use, we believe. The most promising aspect of such an analysis is a very high throughput sine both HPLC and MS spectra can be determined in short time. One important aspect of such an analysis is that patterns themselves are independent of the identity of proteins as a descriminator. The classification can therefore be done before the identity of proteins is determined.

In the paper we have carried out both a proof of principle and proof of concept experiments. Two quite independent algorithms have been used in the analysis and both algorithms have shown consistently results with repect to early identification of autism from HPLC data.

The experiments are not optimal with respect to the values of all the variables being used. The selection of different parameter values has been carried out by experimenting. Therefore, it remains a lot of tuning of the parameter values to be able to adapt the algorithms to the given data in an optimal way. This belongs to our future work.

REFERENCES

1. Aebersold R, Mann M (2003) Mass spectrometry based on proteomics, Nature 422.
2. Burges CJ (1998) Tutorial on Support Vector Machines for Pattern Recognition.Knowledge Discovery and Data Mining.
3. Haykin S Neural Networks and Learning Machines (2009) Third Edition, Pearson.
4. Mitchell TM (1997) Machine Learning, International Edition, McGraww-Hill Companies.
5. Kristensen T (1997) Neural networks, Fuzzy Logic and Genetic Algorithms, Cappelen Academic Publisher (in Norwegian).
6. Kristensen T, Guillaume F (2013) Classification of DNA Sequences by a MLP and a SVM Network. Proceedings of International Conference on Bioinformatics and Computational Biology, BIOCOMP' 13th July 22-25, Las Vegas, CSCREA Press, USA.
7. Kristensen T, Patel R (2003) Classification of Eukaryotic and Prokaryotic Cells by a Backpropagtion Network, Proceedings of IEEE International Joint Conference on Neural Networks, IJCNN 2003, Portland, Oregon, USA.
8. Kristensen T (2003) Prototypes of ANN Biomedical Pattern Recognition Systems, Proceedings of IASTED International Conference on Simulation and Modeling (ASM 2002), Crete, Greece.
9. Lolita AL, Ferrari M, Petricoin E (2003) Clinical Proteomics written in Blood, Nature 425:905.
10. Lilian RH, Farid H, Donald BR (2003) Probabilistic Disease Classification of Expression-Dependent Proteomic Data from Mass Spectrometry of Human Serum, Journal of Computational Biology Volume 10, Number 6.
11. New Scientist (2010) pp 9, 12th of June.
12. Petricoin E, Liotta AL (2004).Seldi-tof-based serum proteomic pattern diagnostics for early detection of cancer, Current Opinion in Biotechnology, 15: 24-30.
13. Scientific American (2012), pp 11, November 2012.
14. Vapnik VN (1998) Statistical Learning Theory, Wiley, New York.
15. Vapnik, VN (1999) An overview of statistical learning theory, IEEE Transactions on Neural networks, September, 1999.
16. Chih-Chung Chang and Chih-Jen Lin. LIBSVM (2001): a library for Support Vector Machines. available online at http://www.csie.ntu.edu.tw/ cjlin/libsvm,.
17. http://stemedhub.org/resources/714/download/HPLCdata.pdf
18. Kristensen T, javANN (2007): Java Artificial Neural Networks. Pattern Solutions AS, (http://www.patternsolutions.no).

Author: Terje Kristensen
Institute: Computing
Street: Nygårdsgaten 112
City: Bergen
Country: Norway
Email: tkr@hib.no

Feature Selection Techniques in Uterine Electrohysterography Signal

D. Alamedine[1,2], M. Khalil[2], and C. Marque[1]

[1] Université de Technologie de Compiègne, CNRS UMR 7338,
Biomechanics and Bio-Engineering, Compiègne, France
[2] Lebanese University, Azm platform for Research in Biotechnology and its Applications,
LASTRE laboratory, Tripoli, Lebanon

Abstract—To classify between labor and pregnancy contractions, feature extraction from uterine electrohysterography (EHG) signal has been used by many researchers. The number of features (linear and nonlinear) becomes huge when using them for classification techniques. The aim of this paper is to reduce the number of features and use only the pertinent ones to achieve the classification problem. In this paper, we compare the results of three selection methods. The first one is based on measuring the distance between the feature histograms of the 2 studied classes (pregnancy and labor). The second one is the Sequential Forward Selection method (SFS) and the third method is based on the Binary Particle Swarm Optimization techniques (BPSO). All methods give common pertinent features whatever the type of classifiers. These features will be used in future works to classify between labor contractions and pregnancy contractions.

Keywords—Uterine EMG, preterm labor, feature selection, particle swarm optimization, feature extraction.

I. INTRODUCTION

Preterm birth is a major cause of neonatal morbidity and mortality. The medical, physiological and socioeconomic consequences of these prematurity are important [1]. One of the goals of pregnancy monitoring is to differentiate normal contractions of pregnancy, which are ineffective, of those, effective, that could cause dilation of the cervix and induce premature birth. The current methods used in obstetric are not precise enough for the early detection of preterm birth threats. A promising method to monitor the effectiveness of uterine contractions during pregnancy started in the 80s; it is based on the study of uterine electromyographic signal (EHG) [2]. This signal represents the electrical activity triggering the mechanical contraction of the myometrium. Many researchers extracted several features from the EHGs, in order to find specific information to detect preterm birth.

The objective of this paper is to select the most significant feature subset extracted from multiple studies (16 features known as linear, 3 features known as nonlinear), in order to identify, on a given population, the most pertinent ones and to discriminate between pregnancy and labor contractions. In this paper, we focus on three methods of feature subset selection. The first one is based on the measurement of the distance between the feature histograms of the two classes (Pregnancy and Labor) by using Jeffrey divergence (JD) distance. The second one is based on the Sequential Forward Selection (SFS) and the last method is based on the Binary Particle Swarm Optimization (BPSO) technique.

This paper is organized as follow: in the first part we will present the experimental protocol. Then the calculated features are explained and all methods of feature selection are summarized. At the end we will present the results of selection from real EHG signals.

II. MATERIAL AND METHODS

A. Experimental Protocol

The EHGs were recorded by placing an array of 16 electrodes on the woman's abdomen and two reference electrodes on each hip. To increase the signal to noise ratio, we calculated the vertical bipolar signals. In our study we use only one bipolar channels located on the vertical median axis of the uterus. The measurements were performed in two hospitals in France (Center of Obstetrics and Gynecology, Amiens) and Iceland (Landspitali university hospital). We recorded EHG on 48 women: 32 during pregnancy (33-39 weeks of gestation) and 16 during labor (39-42 weeks of gestation) and digitized them with a sample frequency of 100 Hz. After a manual segmentation of the EHG, based on tocodynamometer data (mechanical effect of the contractions), we obtained 133 pregnancy and 133 labor contractions.

B. Feature Extraction

In our study two categories of features (features known as linear and as nonlinear) are extracted from our EHGs.

a) Linear Features

Several frequency features are extracted from the power spectral density (PSD). In our study, we used the Welch Periodogram to calculate the PSD of each contraction. We extracted four frequency features from this PSD: mean frequency *MPF* [3], Peak Frequency *PF* [4], deciles $D1...D9$ [5, 6] which contain the median frequency $D5$. Deciles correspond to the frequencies $D1...D9$ that divide the power spectral density into equal parts containing 10% of the total energy.

Some authors have also used time-frequency methods such as wavelet decomposition to characterize EHG. In our work, we use the wavelet symlet 5. After the decomposition of each contraction into detail coefficients, we calculate the variances on the following details levels 3,4,5,6 and 7 [7].

b) Nonlinear Features

The Lyapunov Exponent: LE studies the stability and the sensibility on the initial conditions of the system. We compute the distance between two trajectories in the phase space of the signal. If we have a small distance so we can say that the signal is stable, if not, we say that the signal is not stable. In our study we have used the equation of LE described in [8] and represented by:

$$\lambda = \lim_{t \to \infty} \lim_{\|\Delta_{y_0}\| \to 0} \left(\frac{1}{t}\right) \log \left(\| \Delta_{y_t} \| / \| \Delta_{y_0} \|\right) \quad (1)$$

Where $\| \Delta_{y_0} \|$ presents the Euclidean distance between two states of the system to an arbitrary time t_0, and $\| \Delta_{y_t} \|$ corresponds to the Euclidean distance between the two states of the system at a later time t.

Sample Entropy: We use SE to identify the regularity of EHG signals. In our work, we use the sample entropy described in [9]:

$$SE_{m,r}(x) = \begin{cases} -\log\left(\frac{C_m}{C_{(m-1)}}\right) : C_m \neq 0 \land C_{m-1} \neq 0 \\ -\log\left(\frac{N-m}{N-m-1}\right) : C_m = 0 \lor C_{m-1} = 0 \end{cases} \quad (2)$$

where the four parameters N, m, r and C_m represent respectively the length of the time series, the length of sequences to be compared, the tolerance for accepting, and the number of pattern matches (within a margin for r) that is constructed for each m.

Variance Entropy: Recent studies [10] used the variance entropy on biological signals but not on the EHG. We are interested in our work to use variance entropy, because this method combines the variance with sample entropy via inverse-variance weighting. For a time series x, variance entropy is defined as:

$$VarEn(x,m,r) = \frac{\sum_{i=1}^{p} SE_{m,r}(x_i) \times w_i}{\sum_{i=1}^{p} w_i} \quad (3)$$

where x_i is the *ith* segment of x, w_i is the inverse variance of x_i, and p is the number of sliding windows.

As a summary, we tested 19 selected features: mean frequency *MPF*, Peak Frequency *PF*, deciles *D1...D9*, which contain the median frequency *D5*, Variance on the details levels 3, 4, 5, 6,and 7 (*W1,W2,W3,W4* and *W5*) after wavelet decomposition, Lyapunov exponent (*LE*), Sample Entropy (*SE*), Variance entropy (*VarEn*).

C. Feature Selection Techniques

a) Feature Selection Based on Jeffrey Divergence Distance

This method consists in calculating, for each feature, the distance between the two histograms (Pregnancy and Labor ones) by using the Jeffrey divergence (JD) represented by:

$$D_{Je}(H, G) = \sum_y \left(h_y \log \frac{h_y}{g_y} + g_y \log \frac{g_y}{h_y}\right) \quad (4)$$

H and G are the two histograms where N bins are defined as $H=\{h_y\}$ and $G=\{g_y\}$, with the bin index $y \in \{1,2,...,N\}$.

This distance between the two histograms permits to measure the similarity/dissimilarity of their corresponding statistical properties: a small distance means a large similarity and a large distance means a small similarity. Therefore this distance will be then used to select the discriminant features. Indeed, the greater the distance between the feature histograms of the pregnancy and labor classes is, the more discriminating the feature is (see [11] for details).

b) Sequential Forward Selection (SFS)

Sequential forward Selection (SFS) is a sequential search algorithm for feature selection. SFS begins with an empty subset. By using a classifier, the value of the criterion objective function (J) is calculated for each feature. The feature with the best performance (best J criterion) is selected (Y_k) and then added to the subset. The next step consists in adding sequentially the feature x^+ that has the highest objective function J ($Y_K + x^+$) when combined with the Y_K features that have already been selected. This cycle is repeated until all features are selected (see [12] for detail).

c) Binary Particle Swarm Optimization (BPSO)

PSO was developed by Eberhart and Kennedy in 1995. PSO is a population based stochastic optimization technique, inspired by social behavior of bird flocking or fish schooling [13]. PSO uses a number of particles (swarm) moving around in the space in order to achieve the best solution. We assume that our search space is n-dimensional and each particle is a point in this space. The position of the ith particle of the swarm is represented as $X_i = (x_{i1},...x_{id},...x_{in})$. Each particle has a best previous position $pbest_i=(p_{i1},...,p_{id},...p_{i,n})$, which corresponds to the best fitness value. The fitness is the objective function to be reached (i.e. correct classification) The global best particle subset among all the particles in the population is represented by $gbest = (p_{g1},...,p_{gd},...p_{gn})$. The velocity of the ith particle is denoted by $V_i = (v_{i1},...v_{id},...,v_{in})$. The particles velocity and position are manipulated according to the following two equations:

$$v_{id}^{k+1} = w\, v_{id}^k + c_1 r_1^k \left(p_{id}^k - x_{id}^k\right) + c_2 r_2^k \left(p_{gd}^k - x_{id}^k\right) \quad (5)$$

$$x_{id}^{k+1} = x_{id}^k + v_{id}^{k+1} \quad (6)$$

Where w is the inertia weight; c_1 and c_2 are positive constant; r_1 and r_2 are two random values in the range [0, 1].

Kennedy and Eberhart also proposed a binary particle swarm optimization (BPSO) in order to solve optimization problems for discrete valued features [14]. In BPSO, the position of each particle is represented as binary string. By comparing PSO and BPSO we found that they have a common velocity equation, and a different particle position equation computed as follows:

$$S(v_{id}^{k+1}) = \frac{1}{1 + e^{v_{id}^{k+1}}} \quad (7)$$

$$x_{id}^{k+1} = \begin{cases} 1 \; if \; r_3 < S(v_{id}^{k+1}) \\ 0 \quad Otherwize \end{cases} \quad (8)$$

where r_3 is a random number in the range [0, 1].

Recently BPSO has been widely used in the literature for feature subset selection. In this case the length of a binary string from each particle is equal to the length of the total number of features and each particle presents a candidate for subset selection. If the bit included in the binary strings has a "1" value, thus the feature is selected, otherwise the feature is not selected. The following steps present the BPSO algorithm:

1. Initialize all particles positions and velocity randomly. Set the number of iterations K and BPSO parameters.
2. Calculate the fitness value $F(X_i)$ of each particle. Fitness presents the percentages of correct classification.
3. Compare the fitness of each particle to its best fitness so far ($pbest_i^k$ of last iteration k):
if $F(X_i^{k+1}) > F(pbest_i^k)$ then $F(pbest_i^{k+1}) = F(X_i^{k+1})$ and $pbest_i^{k+1} = X_i^{k+1}$
Else $F(pbest_i^{k+1}) = F(pbest_i^k)$ and $pbest_i^{k+1} = pbest_i^k$
4. Determine the global best position $gbest^{k+1}$ from all $pbest_i^{k+1}$. Then compare $gbest^{k+1}$ with $gbest^k$:
if $F(gbest^{k+1}) > F(gbest^k)$ then $gbest = gbest^{k+1}$
Else $gbest = gbest^k$
5. Update the position and the velocity of each particle according to equations (5) and (8).
6. Go to step 2, and repeat until the number of iterations is reached.

When the number of iterations is reached, we obtain an optimal solution (best subset of feature selection).

The parameters we use for the BPSO are: 30 particles, the length of each particle is equal to 19 (maximum number of features), K=100 iterations. The acceleration constant c_1 and c_2 were set to 2. We also used a linear descending inertia weight passing from 0.6 to 0.1. These values are classical ones used in the literature.

III. RESULTS

A. Results of Feature Selection by Using JD Distance

In this method, we firstly calculated the 19 features for each existing contraction in the selected database (133 pregnancy (Class1) and 133 labor contractions (Class2)). We then computed the value for each feature in each class and present them as a histogram of values. After calculating the distances between every two corresponding feature histograms in these 1x19 histograms by using JD, we obtain a distance matrix of dimension 1x19. Finally we apply a threshold on the matrix of distance in order to select the feature histograms that present the largest distance between pregnancy and labor contractions (see [11] for more details). This threshold is equal to the mean+1*standard deviation of the distance distribution. Our results indicate that 3 features are selected: Lyapunov exponent (*LE*), Variance entropy (*VarEn*) and feature of wavelet decomposition which is the variances on the following details level 1(*W1*).

B. Results of Feature Selection by Using SFS and BPSO

In this study, three different classifiers (Linear discriminant analysis (LDA) [15], Quadratic discriminant analysis (QDA) [16], and K-nearest neighbor (KNN) with k=11 (the choice of K is based on the number of training set) [15] have been used with the two feature selection algorithms (BPSO and SFS) for fitness evaluation. BPSO algorithm successfully finds the best feature subsets that have the maximum fitness after 100 iterations (1 run). Then, to evaluate the variance of BPSO, many different runs were performed. SFS algorithm also finds the best feature subset but in a sequential manner, then we chose the combination of features that gives the minimal error. The real database consists of two classes (pregnancy and labor). Each class contains 133 contractions and from each contraction we extract the 19 features presented in the previous section. About 70% of the data set is used for training the classifier and the rest of this data (30%) is used for testing.

Table 1 Comparison between BPSO and SFS

Classifier	BPSO (gbest that have best fitness among 200 runs)	SFS (Combination of features with minimal error)
QDA	LE, VarEn, W1, W2, W3, W4, W5 D2, D5, D8, D9 MPF	LE, SE, VarEn, W1, W5, D1, D9, MPF
LDA	LE, SE, VarEn, W2, W3, W4, D2, D3, D5, D6, D7, D8, PF	LE, SE, VarEn, W1, W2, W3, W4, W5, D2, D4, D5, D6, D7, D8, MPF
KNN	LE, VarEn, W1, W2, W5, D1, D6, D7, MPF, PF	VarEn, W1, W2, W3, W4, W5, D4, D5, D7

Table 1 presents the results of BPSO and SFS by using different type of classifiers. It indicates the selected feature subset obtained from BPSO (corresponding to the best *gbest*) for the 3 classifiers. The fitness obtained by these three feature subsets are: 92.48% with QDA, 93.73% with LDA and 91.22% with KNN. For SFS, the combinations of selected features presented in table 1 for each type of classifier give the minimum error.

Whatever the type of classifier (see table1), the common pertinent features with BPSO are: *LE, VarEn,* and *W2*. We follow the same procedure for the second method and the observed common features are: *VarEn, W1,* and *W5*.

IV. DISCUSSIONS – CONCLUSIONS

As observed in our results on real uterine EMG signals, we notice that BPSO and SFS, with the three types of classifiers, select different subsets of features. In our study it would very important to retain the common features selected with different methods. We would indeed suppose that these features are very pertinent. As we have seen in our results, there are three pertinent features selected by the first method (*LE, VarEn,* and *W1*), three selected by BPSO (*LE, VarEn,* and *W2*), whatever the type of classifier, and three other pertinent features selected by SFS (*VarEn, W1,* and *W5*). As a result, we can say that the features *VarEn, LE, W1, W5,* and *W2* are the most pertinent as being common to many selected subsets. Especially, the *VarEn* seems to be very important as being selected by all methods.

As perspectives, we must apply classification methods based on neural network, SVM or KNN by using only the selected features with the aim of classifying labor and pregnancy contractions. Comparison between results with selection and without selection of features must be made. Additionally, we will have to adapt the two algorithms BPSO and SFS to the use of other classifiers such as support vector machine (SVM) and/or feed forward neural network. We will also include in this selection process features related to uterine synchronization and activity propagation.

ACKNOWLEDGMENT

We are indebted to French Regional Council of Picardy and FEDER funds for funding this work.

REFERENCES

1. Marque, J. Gondry, J. Rossi, N. Baaklini, and J. Duchêne, "Surveillance des grossesses à risque par électromyographie utérine". *RBM-News*, 1995. vol. 17, no. 1, pp. 25–31.
2. H. Alvarez and R. Caldeyro, "Contractility of the human uterus recorded by new methods". *Surgery, gynecology & obstetrics*, 1950. vol. 91, no. 1, pp. 1.
3. J. Terrien, T. Steingrimsdottir, C. Marque and B. Karlsson, "Synchronization between EMG at different uterine locations investigated using time-frequency ridge reconstruction: comparison of pregnancy and labor contractions". *EURASIP Journal on Advances in Signal Processing*, 2010.vol. 2010.
4. W. L. Maner, R. E. Garfield, H. Maul, G. Olson, and G. Saade, "Predicting term and preterm delivery with transabdominal uterine electromyography ". *Obstetrics & Gynecology*, 2003.vol. 101, no. 6, pp. 1254–1260.
5. Marque, H. Leman, M. L. Voisine, J. Gondry, and P. Naepels, "Traitement de l'électromyogramme utérin pour la caractérisation des contractions pendant la grossesse". *RBM-News*, Dec. 1999.vol. 21, no. 9, pp. 200–211.
6. J. Sikora, A. Matonia, R. Czabanski, K. Horoba, J. Jezewski, and T. Kupka, "Recognition of premature threatening labour symptoms from bioelectrical uterine activity signals". *Archives of perinatal medicine*, vol.17, no.2, pp.97-103, 2011.
7. M. O. Diab, C. Marque, and M. A. Khalil, "Classification for uterine EMG signals: Comparison between AR model and statistical classification method".*International Journal of computational cognition*, 2007.vol. 5, no. 1, pp.8-14.
8. A. Diab, M. Hassan, C. Marque, and B. Karlsson, "Quantitative performance analysis of four methods of evaluating signal nonlinearity: Application to uterine EMG signals".*34th Annual International IEEE EMBS conference*, Sep. 2012.San Diego, USA.
9. G. Fele-Žorž, G. Kavšek, Ž. Novak-Antolič, and F. Jager, "A comparison of various linear and non-linear signal processing techniques to separate uterine EMG records of term and pre-term delivery groups". *Medical and Biological Engineering and Computing*, 2008. vol. 46, no. 9, pp. 911–922.
10. M. Hu and H. Liang, "Variance entropy: a method for characterizing perceptual awareness of visual stimulus". *Applied Computational Intelligence and Soft Computing*, 2012.vol. 2012, pp.1-6.
11. Alamedine, M. Khalil, C. Marque. "Parameters Extraction and Monitoring in Uterine EMG Signals". Detection of Preterm Deliveries". *Recherche en Imagerie et Technologies pour la Santé (RITS)* ,8-11 avril 2013. Bordeaux, France.
12. L. Ladha and T. Deepa, "Feature selection methods and algorithms". *Int. J. Comput. Sci. Eng.*, 2011.vol. 3, no. 5, pp. 1787–1797.
13. R. Eberhart and J. Kennedy, "A new optimizer using particle swarm Theory". *in Micro Machine and Human Science. MHS'95, Proceedings of the sixth International Symposium on*, 1995.pp.39-43.
14. J. Kennedy and R. C. Eberhart, "A discrete binary version of the particle swarm algorithm". IEEE International Conference *on Systems, Man, and Cybernetics, 1997. Computational Cybernetics and Simulation*, 1997, vol. 5, pp. 4104–4108.
15. S. Dudoit, J. Fridlyand, T.P. Speed, "Comparaison of discrimination Methods for the classification of tumors Gene Expression Data". *Journal of the American Statistical Association*, March 2002.vol. 97, no. 457, pp.77-87.
16. N. Georgiou-Karistianis, M. A. Gray, D. J. F. Domínguez, A. R. Dymowski, I. Bohanna, L. A. Johnston, A. Churchyard, P. Chua, J. C. Stout, and G. F. Egan, "Automated differentiation of pre-diagnosis Huntington's disease from healthy control individuals based on quadratic discriminant analysis of the basal ganglia: The IMAGE-HD study". *Neurobiology of disease* , 2012.

Author: Dima Alamedine
Institute: University of technology of Compiègne
Street: Rue Dr .Schweitzer
City: Compiègne
Country: France
Email: dima.alamedine@utc.fr

Studying Functional Brain Networks to Understand Mathematical Thinking: A Graph-Theoretical Approach

Georgios Bamparopoulos[1], Manousos A. Klados[1], Nikolaos Papathanasiou[2], Ioannis Antoniou[2], Sifis Micheloyannis[3], and Panagiotis D. Bamidis[1]

[1] Lab of Medical Informatics, School of Medicine, Aristotle University of Thessaloniki, Greece
[2] Department of Mathematics, Aristotle University of Thessaloniki, Greece
[3] School of Medicine, University of Crete, Greece

Abstract—Brain function during mathematical thinking is a common concern of scientists from different research fields. The study of functional brain networks extracted from electroencephalographic (EEG) signals using graph theory seems to meet the challenge of neuroscience to understand brain functioning in terms of dynamic flow of information among brain regions. Some studies have found differences between the basic arithmetic operations among brain regions; however, they have not explained the brain function during difficult mathematical tasks and complex mathematical processing. This study investigates the changes of the functional networks' organization among different mathematical tasks. To this end, EEG data from 10 subjects were recorded during three different tasks, Number Looking which was served as the control situation, Simple Addition which refers to the addition of single-digit numbers and Difficult Multiplication with trials of two-digit multiplications. We analyzed weighted graphs so as to provide a more realistic representation of functional brain networks. Mutual information was employed to form the weights among the different channels, while various global and local graph indices were further examined. The results suggest that there are some statistically significant differences between graph theoretical indices among the different tasks and their range of values depend on the particular task.

Keywords—**Mathematical Cognition, Graph Theory, Mathematics, Mutual Information, Functional Connectivity.**

I. INTRODUCTION

Numerical estimation, commonly known as subitizing, refers to the rapid and accurate judgments of number without actual counting. Enumeration of elements results from various cognitive processes and involves different brain regions. The most important components of this system are the cortical structures located in the posterior parietal and parieto-occipital region [1-4].

Left angular gyrus (lAG) seems to incorporate a crucial component of the system which accounts for learning and retrieving math facts [5-7]. According to a recent study [8] the lAG is involved in the retrieval of verbally stored arithmetic facts (such as multiplication facts or simple additions) by operating within the language system through its connections to the left temporo-parietal cortex [9]. Another study found that that lAG is engaged in arithmetic fact retrieval together with frontal structures [10]. Moreover, this region appears to be more active in exact calculation. During exact calculation the lAG shows higher activation for arithmetical operations related to rote verbal memory arithmetic facts. Thus, lAG shows more activation for multiplication tables in relation to subtraction or number comparison [8, 11].

Some studies have found differences between the basic arithmetic operations among brain regions apart from those who focused predominately on the frontal and parietal lobes. However, they have not explained the brain function during difficult mathematical tasks and complex mathematical processing. In fact, what are the series of events in which the brain does advanced mathematical processing and what is the exact mechanism that supports mathematics like fractions, complex numbers and calculus are difficult questions to answer. Nonetheless, Intraparietal sulcus (IPS) is a key region for mathematics which is activated when numbers are presented in various notations or spoken number words. In addition to IPS, parietal frontal networks, some basal ganglia and other networks associated with long-term and working memory, are engaged in difficult problem solving.

A variety of problems that concern brain function can be dealt with network comparison and characterization. The problem of comparing these networks is the task of measuring their topological difference or similarity. In order to investigate the changes of the functional networks' organization among different tasks (number looking, simple additions and two digit multiplications), we analyzed weighted graphs in which the weights were extracted using a nonlinear association measure. To this end, functional networks were extracted from EEG signals, and then we computed global and local graph theoretical indices for the networks so as to investigate if their values depend on different tasks.

II. MATERIALS AND METHODS

A. Participants

Ten right handed medical students of the University of Crete were recruited for this study. Participants with a history of psychiatric or neurological illness or under medication were excluded from the study. All participants had normal or corrected to normal vision. Recordings were performed during the morning, while all participants signed an informed consent form, while the experimental protocol was approved by the ethics committee of the Department of Medicine of the University of Crete.

B. Experimental Design

EEG data were acquired during three different tasks, Number Looking (NL) which was served as the control situation and contained 20 trials, Simple Addition (SA) which refers to the addition of single-digit numbers (e.g. 3+8) with 20 trials as well, and Difficult Multiplication (DM) with 8 trials of two-digit multiplications (e.g. 32x75). All the responses for the calculation tasks were acquired orally. For the purpose of the current study, we chose pieces of 10s duration, without visible artifacts and with correct answers and all the signals were further cleared by ocular artifacts using the REGICA [12] plugin for the EEGLAB toolbox [13]. No response was required during number looking and fixation, while for the SA and DM task students were not restricted to a prescribed time period to provide their answer.

C. Functional Connectivity

Mutual Information index is given by the next formula:

$$I(X;Y) = \sum_{y \in Y} \sum_{x \in X} p(x,y) \log \frac{p(x,y)}{p(x)p(y)}$$

where $p(x,y)$ is the joint probability distribution and $p(x)$, $p(y)$ are the marginal probability distribution functions of X and Y respectively.

The functional connectivity was computed using a normalized variant of mutual information:

$$\frac{I(X;Y)}{\sqrt{I(X) \cdot I(Y)}}$$

where $I(X)$, $I(Y)$ are the entropies [14] of X and Y respectively.

D. Graph Analysis

A graph is a pair of two sets, $G=(V,E)$ where V is a set of objects called nodes and E is a set of a collection of pairs of these objects which are connected with links. In a weighted graph, each link is associated with a number called the weight. In our study, the nodes are the channels and the weights are defined by the value of mutual information of each pair. We constructed 30 (3 tasks x 10 subjects) weighted brain networks from the correlation matrices.

In this study, we used six networks indices which are described next.

1. Clustering Coefficient:

Clustering coefficient is a measure of degree to which vertices in a graph tend to cluster together. Onnela et al [15], introduce subgraph "intensity" as the geometric mean of its link weights and "coherence" as the ratio of the geometric to the corresponding arithmetic mean. Using these measures, motif scores and clustering coefficient can be generalized to weighted networks. Clustering coefficient is given by the following type:

$$c^W = \frac{1}{n} \sum_{i \in N} \frac{2t_i^W}{k_i(k_i - 1)}$$

where $t_i^W = \frac{1}{2} \sum_{j,h \in N} \left(w_{ij} w_{ih} w_{jh} \right)^{\frac{1}{3}}$ is the (weighted) geometric mean of triangles around.

2. Global Efficiency

Efficiency of the path between two nodes is the inverse of their shortest distance. In case where there is not any path to connect two nodes, their distance is infinite so their efficiency is 0. Global efficiency (GE) [16,17,19] is given by the following formula:

$$E^W = \frac{1}{n} \sum_{i \in N} \frac{\sum_{j \in N, j \neq i} \left(d_{ij}^W \right)^{-1}}{n-1}$$

where N is the number of vertices in graph G and $d_{ij}^W = \sum_{a_{uv} \in g_{i \leftrightarrow j}^w} f(w_{uv})$, where f is a map from weight to length and $g_{i \leftrightarrow j}^W$ is the shortest weighted path between i and j.

3. Modularity

The modularity [18] is a statistic that quantifies the degree to which the network may be subdivided into clearly delineated groups (modules) and is given by the next formula:

$$Q^W = \frac{1}{l^W} \sum_{i,j \in N} \left[w_{ij} - \frac{k_i^W k_j^W}{l^W} \right] \delta_{m_i, m_j}$$

where k_i^W is the strength of node i, l^W is the sum of all weights in the network, m_i is the module containing the node i and $\delta_{m_i,m_j} = 1$ if $m_i = m_j$ or 0 otherwise.

4. Strength

Strength of a node is the sum of weights attached to its links. Nodes with more strength tend to have more power:

$$k_i^W = \sum_{j \in \Gamma(i)} w_{ij}$$

where $\Gamma(i)$ is the set of neighbors of a node i.

5. Betweenness Centrality

The betweenness centrality [19] of a node i is the proportion of all shortest paths between pairs of the other nodes that pass through this node. If we imagine information between nodes in the network and always taking the shortest possible path, then betweenness centrality measures the fraction of that information that will flow through a node. Betweenness centrality of node i is defined as:

$$b_i = \frac{1}{(n-1)(n-2)} \sum_{\substack{h,j \in N \\ h \neq j, h \neq i, j \neq i}} \frac{\rho_{hj}(i)}{\rho_{hj}}$$

where ρ_{hj} is the number of shortest paths between h and j, and $\rho_{hj}(i)$ is the number of shortest paths between h and j that pass through i.

6. Local Efficiency

The local efficiency (LE) is the global efficiency computed on node neighborhoods, and is related to the clustering coefficient. It is defined as:

$$E_{loc}^W = \frac{1}{n} \sum_{i \in N} \frac{\sum_{j,h \in N, j \neq i} \left(w_{ij} w_{ih} \left[d_{jh}^W(N_i) \right]^{-1} \right)^{1/3}}{k_i(k_i - 1)}$$

where $d_{jh}^W(N_i)$ is the length of the shortest path between nodes j and h that contains only neighbors of i [20].

E. Statistical Assessments

In order to test if the set of values of a graph index for each task follows the normal distribution, we used Kolmogorov–Smirnov test (K–S test), but the test rejected the null hypothesis at the 5% significance level.

So, Mann–Whitney U test was employed to identify the statistically significant differences between the graph indices of brain networks at different tasks. The difference was considered significant at $p < 0.05$.

III. RESULTS AND DISCUSSION

Taking into consideration the Fig.1, it seems that functional brain networks constructed from the two-digit multiplications tend to have high community structure and more communities (modules) than networks constructed from other tasks. Another important question on brain networks is how tightly clustered they are. We observe that clustering coefficient is higher during number looking and this indicates that there is higher clustering in these networks.

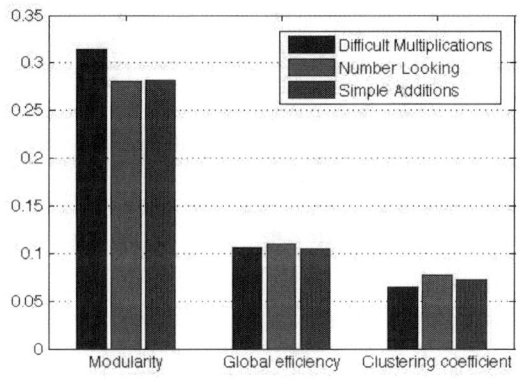

Fig. 1 **Global graph indices**. This figure depicts the values of global graph indices. Each bar represents the mean value of a graph index among the subjects at a specific task.

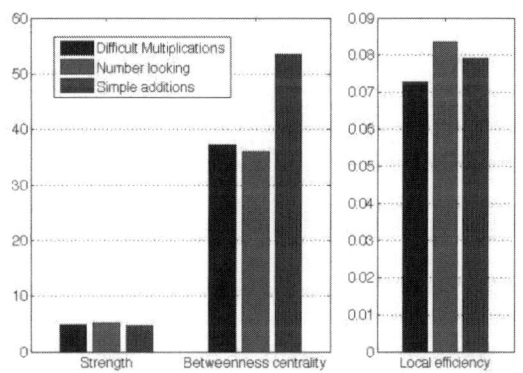

Fig. 2 **Local graph indices**. This figure depicts the values of local graph indices. We computed the average of each local index among the 57 nodes and then the average of this value among the subjects at each task.

Regarding the efficiency of the networks, subjects seem to have more efficient graphs during the task of number looking owing to the fact that both GE and LE are higher (Fig.1, Fig.2). Moreover, higher values of local efficiency suggest a larger level of internal organization and fault tolerance. So it seems that brain networks during number looking is more faults tolerant in contrast to those of other tasks.

As far as weights are concerned, it seems that on the average, nodes in network of NL tend to have more power than those of networks in other tasks. It is also clearly observable that betweenness centrality, a measure of a node's centrality in a network, is on the average higher in networks of SA.

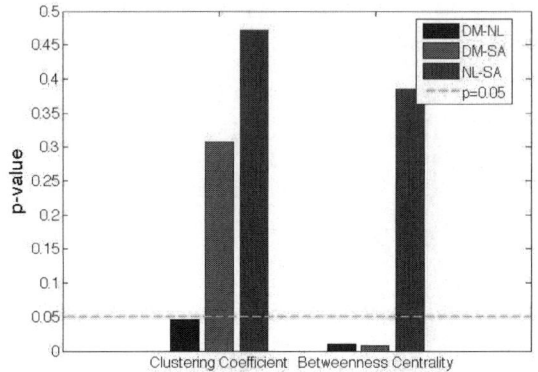

Fig. 3 In this figure we see a bar plot for p-values of Mann–Whitney U test. Each bar depicts the p-value from comparison of two sets, each containing 10 values of a specific graph index, one per subject. We observe that p-value from comparison of Difficult multiplications (DM) and Number Looking (NL) tasks is below the limits of 5%, for both Clustering Coefficient and Betweenness Centrality. In addition, p-value from comparison of (DM) and (SA) is also below 0.05 for Betweenness Centrality.

Finally, we found that there are statistically significant differences between DM and NL tasks for clustering coefficient and betweenness centrality. Furthermore, the networks of SA task had significant different betweenness centrality than networks of NL task.

REFERENCES

1. Piazza M (2010): Neurocognitive start-up tools for symbolic number representations. Trends in Cogn. Science 14: No 12.
2. Piazza M, Mechelli A, ButterworthB, Price CJ (2002a): Are subitizing and counting implemented as separate or functionally overlapping processes? Neuroimage 15: 435-446.
3. Piazza M, Giacomini E, Le Bihan D, Dahaene S (2002b): Single-trial classification of parallel pre-attentive and serial attentive processes using functional magnetic resonance imaging. Proc. R. Soc. Lond, B 270: 1237-1245.
4. Piazza M, Izard V (2009): How humans count: Numerosity and the parietal cortex. The Neuroscientist 15: 261-273
5. Butterwarth B, Walsh V (2011): Neural basis of mathematical cognition. Current biology 21: No 16.
6. Feigenson L, Dahaene S, Spelke E (2004): Core systems of number. Trends in Cogn. Sci. 8: No7.
7. Zamarian L, Ischebeck A, Delazer M (2009): Neuroscience of learning arithmetic evidence from brain imaging studies. Neurosci. Biobeh. Rev. 33: 909-925.
8. Dahaene S, Piazza M, Piel P, Cohen L (2003): Three parietal circuits for number processing. Cogn. Neuropsychol. 20: 487-506.
9. Ansari D (2008): Effects of development and enculturation on number representation in the brain. Nat. Rev. Neurosci. 9: 278-291.
10. Jost K, Khader PH, Burke M, Bien S, Rösler F. (2010), Frontal and parietal contributions to arithmetic fact retrieval: A parametric analysis of the problemsize effect. Human Brain Mapping Doi: 10.1002/hbm.21002.
11. Grabner RH, Ansari D, KoschutningK, Ebner F, Neuper C (2009): To retrieve or to calculate? Left angular gyrus mediates the retrieval of arithmetic facts during problem solving. Neuropsychologia 47: 604-608.
12. Klados MA, Papadelis C, Braun C, Bamidis PD. (2011) REG-ICA: A hybrid methodology combining Blind Source Separation and regression techniques for the rejection of ocular artifacts. Biomedical Signal Processing and Control, 6:3: 291-300
13. Delorme A & Makeig S (2004) EEGLAB: an open source toolbox for analysis of single-trial EEG dynamics (pdf, 0.7 MB) Journal of Neuroscience Methods 134:9-21
14. C. E. Shannon (1948). A mathematical theory of communication, Bell System Technical Journal, Vol. 27, pp. 379-423; pp. 623-656.
15. Jukka-Pekka Onnela, Jari Saramäki, János Kertész, and Kimmo Kaski (2005). Intensity and coherence of motifs in weighted complex networks. PHYSICAL REVIEW E 71, 065103
16. Vito Latora and Massimo Marchiori (2001). Efficient Behavior of Small-World Networks. Phys. Rev. Lett. 87, 198701.
17. Mikail Rubinov, Olaf Sporns (2010). Complex network measures of brain connectivity: Uses and interpretations. NeuroImage 52: 1059–1069.
18. M. E. J. Newman and M. Girvan (2004). Finding and evaluating community structure in networks. PHYSICAL REVIEW E 69, 026113.
19. Freeman, L.C. (1978). Centrality in social networks: conceptual clarification. Soc. Netw. 1,215–239.
20. C Lithari, MA Klados, C Papadelis, C Pappas, M Albani, PD Bamidis. How does the metric choice affect brain functional connectivity networks? Biomedical Signal Processing and Control, 2012 7(3),pp. 228-236.

A Short Review on Emotional Recognition Based on Biosignal Pattern Analysis

Manousos A. Klados, Charalampos Styliadis, and Panagiotis D. Bamidis

Laboratory of Medical Informatics, Medical School, Aristotle University of Thessaloniki, Thessaloniki, Greece

Abstract—Emotional intelligence has been argued to be more important than verbal or mathematical intelligence. So the last decade researchers try to embody the emotional with the artificial intelligence developing machines capable of understanding the human's emotions. This is the core of a modern and rapidly growing research field called Affective Computing. The main goal of the Affective Computing science is to understand the basis of emotions as well as the way they are expressed, and make the machines able to recognize the human's emotions. Recognizing emotional information requires the extraction of meaningful patterns from the gathered biological data. This is done by adopting machine learning techniques using one modality, or the fusion of different modalities. This study comes to introduce the physiological basis of emotional recognition in order to give us further evidence about the use of each modality, while it makes more clear the ways that someone can combine the data from different modalities. Moreover, the most prominent studies for emotional recognition, based only on biological signals, are reported, and despite the fact that their results are encouraging, there are some serious unsolved problems which are addressed and further discussed herein.

Keywords—Emotional Recognition, Biosignal Pattern Analysis, Emotions, Affective Computing.

I. INTRODUCTION

Based on the psychology theories which support the embodiment of emotion, several Affective Computing (AC) applications are focused on detecting patterns of physiological activity related to the detection of different emotions [1]. Physiological psychology studies how physiological variables such as brain stimulation affect other (dependent) variables such as learning or emotional reaction. The main goal of physiological psychology or psychophysiology is the study of behavior which has strong implication for AC research and especially in emotional recogntion.

Behavior extends beyond emotional processing, including cognitive processes like perception, attention, memory and problem solving. Today most of the methods used to monitor the physiological states are noninvasive and they are based on the recordings of electrical signals produced by brain (electroencephalography-EEG), by heart (electrocardiography-ECG), muscles (electromyography-EMG), skin (electrodermal activity-EDA) and eyes (electrooculography-EOG).

Emotional recognition begins with passive sensors which capture data about the user's physical state or behavior without interpreting the input. The data gathered is analogous to the cues humans use to perceive emotions in others. For example, EEG can capture alternation in brain rhythms, ECG can give us the heart rate as well as its variability, from EMG we can extract face postures and movements, while EDA can be used for the detection of the user's engagement to the current emotion [2].

Recognizing emotional information requires the extraction of meaningful patterns from the gathered data. This is done by adopting machine learning techniques using one modality, (e.g. EEG), or the fusion of different modalities (e.g. EEG and EDA), and produce either labels (i.e. 'confused') or coordinates in a valence-arousal space.

II. PHYSIOLOGICAL BASE OF EMOTIONAL RECOGNITION

A. Law of Initial Values (LIVs)

Law of initial values (LIVs) states that the physiological response to a stimulus depends on the pre-stimulus physiological level. For example, for a higher initial level, we expect a lower response if the stimulus produce an increase and a larger response if it produce a decrease respectively. However LIV doesn't seem to be generalized to all measures (e.g. skin conductance), while it can be influenced by other variables.

B. Arousal

Arousal measures indicate if the subject is calm or excited. From the early steps of classical research it was known that performance is a function of arousal [3], [4] (Fig. 1).

Performance is optimal at a critical level of arousal and the arousal-performance relationship is influenced by task constraints [5].

C. Stimulus and Individual Responses

Stimulus-Response (SR) theory supports that for a specific stimulus; subjects will produce specific physiological response patterns. Ekman and his colleagues [6] found that autonomic nervous system specificity allows certain emotions to be recognized.

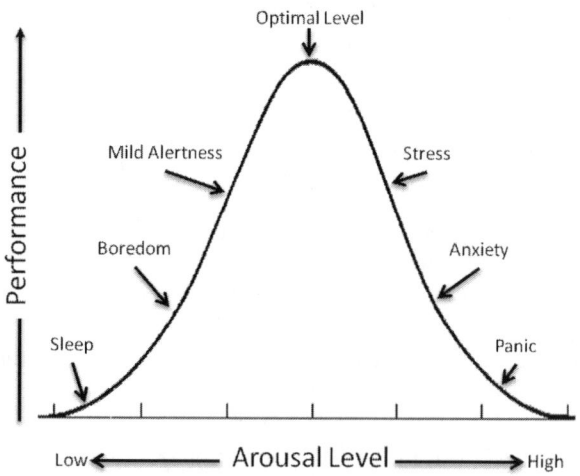

Fig. 1 The relationship between performance and arousal.

Individual-Response (IR) specificity integrates SR specificity but with one crucial difference. This difference lies with the fact that SR specificity claims that the pattern of a certain response is similar for most people, while IR pertains to how consistent the individual's responses are to different stimulations.

D. Cardio-Somatic Features

Cardio-Somatic features refer to the changes of heart responses as the preparation of the body for a certain behavioral response (e.g. fight or fly). This effect might be the result of physiological changes in situation like the dissonance effect described by [7]. In this study, subjects were stimulated with an argument that it is in agreement with their existing attitudes and with an argument in disagreement with them. The results revealed that the first stimulus showed lower arousal in form of electrodermal activity in contrast to the second one.

E. Habituation and Rebound

Habituation and Rebound are two common effects which appear in sequences of stimuli with the same psychological background. When a certain stimulus is presented repeatedly, the physiological responses are decreased (habituation). On the other hand, the rebound effect refers to the return in the prestimulus levels after stimulus's presentation [8].

III. MULTIPLE MODALITIES

It is known that multiple physiological and behavioral responses are occurring during a single emotional episode. For example anger is expected to trigger various responses, such as particular facial, and body expressions, as well as changes in human's physiology like increased heart rate or increased EDA. Responses from multiple channels which are bound in space and time during an emotional experience are essential for the emotion theory [9]. Although that, multi-modal emotion recognition systems are widely advocated, they are rarely implemented [10], because of technical difficulties which are increased in multisensory environments. Nevertheless, the advantages of multi-modal human-computer interaction systems have been recognized as the next-generation approaches [11].

There are three methodologies to fuse signals from different sources/sensors, and each one is depending on when the information from the different channels is combined [11].

A. Data Fusion

Data Fusion is performed on the raw signals exported from the sensors and it can only be applied when these signals have the same temporal resolution. It is not commonly used because of it's sensitivity to noise produced by the malfunction or misalignment of the different sensors.

B. Feature Fusion

Feature Fusion is performed on feature's sets extracted from each channel. This is the approach that is used more multi-modal HCI and it has been used in AC as well (for example the Ausburg Biosignal Toolbox [12]). As features of each signals (EEG, ECG, SCR, etc…) we can consider the mean, median, standard deviation, maximum and minimum alongside with some unique features from each sensor (for example P300 amplitude for EEG, HRV for ECG etc.). These are individually computed for each sensor and then they are combined across the sensors.

C. Decision Fusion

Decision Fusion is performed by combining the outputs of the classifiers for each signal. Thus firstly, the emotional states will be classified from each sensor and then they will be integrated in order to obtain a global view across the various sensors [11].

IV. EMOTIONAL RECOGNITION

Despite the evidence for SR specificity, the accurate recognition of emotions requires models that are personalized, so researchers are now focused in this direction. In this study [13], different algorithms for mapping the physiological signals to certain emotions were compared. More recently [14] proposed a PsychoPhysiological Emotional Map (PPEM) that created personalized mappings between biological signals, like heart rate and skin conductance, and the two dimensions of emotions (valence and arousal).

Nowadays there is an increasing number of algorithms used for the emotional recognition through the EEG signals. These algorithms are composed by two parts; the feature extraction and the classification. For example, in this study [15], a short-time Fourier transform has been used in order to extract some feature from the EEG signals and then the Support Vector Machine (SVM) was adopted for the classification purposes. They tried to discriminate four different emotions (happiness, sadness, angriness and pleasure) with a classification accuracy of 82.37%.

In the same sense, another study [16] has proposed a framework which exploits the role of delta brainwaves in the emotional processing, and found some statistically significant differences among four pictures' groups (IAPS [17]) with different emotional content. Based on this study, Frantzidis and his colleagues [18] have extracted several neuroscientific features and using C4.5 classification algorithm achieved to discriminate the aforementioned categories with accuracy among 57%-92% exploiting the gender differences on emotional perception [19].

Despite the good performance of EEG in the detection of humans' emotions, it has some technical drawbacks for its use in real life scenarios. For this reason, many scientists have studied some other biosignals, like ECG, EDA and EMG, which can be gathered more easily and accurately than the EEG, and they can be used outside the laboratory.

As it was mentioned before, there are very few systems that have implemented multi-modal emotion recognition. In the late 90s, Vyzas and Picard [20] have used EMG, EDA and respiration rate as well, in order to discriminate eight emotions (Neutral, Anger, Hate, Grief, Platonic Love, Love, Joy, and Reverence). They have used three methods for the recognition:Sequential floating forward search (SFFS) feature selection with K-nearest neighbors classification, Fisher projection on structured subsets features with MAP classification, and a hybrid SFFS-Fisher projection method. Their classification accuracy varied from 40-46% in their try to discriminate all the eight emotions, while it reached to 88.33% for Anger, Joy and Reverence. Three years later, Picard and her colleagues [21] employed ECG, EDA and EMG on the aforementioned dataset and they have reached 81.25% accuracy exploiting 40 features with their hybrid SFFS-Fisher projection method.

Wagner et al [22] have also used ECG, EDA and EMG in order to discriminate 4 different emotion induced by self-selected songs. They have totally extracted 120 features, while they have used PCA in order to find the most suitable combination of features describing the participant's emotional state. Three well-known classifiers, linear discriminant function, k-nearest neighbour and multilayer perceptron, are then used to perform supervised classification, giving them classification accuracy among 80%-90%.

More recently, Calvo and his colleagues [23] have also used ECG, EDA and EMG for discriminating 8 different emotional categories. They have extracted 120 features from the three aforementioned modalities, while they have used eight classification algorithms (ZeroR, OneR, Function Trees (FT), Naïve Bayes, Bayesian Network, Multilayer Perceptron, Linear Logistic Regression and SVMs). This is one of the few studies that have tried to make a personalized emotional recognition system achieving 68%-80% classification accuracy for each individual subject, but in their try to make a system for all the subjects their accuracy dropped to 42%.

Another interesting study [24] tried to develop a user-independent system, based on physiological signal databases obtained from multiple subjects. The signals used were ECG, EDA and skin temperature variation. The features used for the classification purposes were devised so that emotional characteristics could be extracted from short-segment signals. A support vector machine was adopted as a pattern classifier. Correct-classification ratios were 78.4% and 61.8%, for the recognition of three and four categories, respectively.

V. CONCLUSIONS AND CONSIDERATIONS

The main goal of the aforementioned approaches is the improvement of the classification accuracy as well as their ability to detect as much emotions as possible. The field of emotional recognition is still an unexplored area, while it is somehow restricted by some unsolved problems.

A serious drawback for an emotional recognition system is the time needed for the feature extraction and the classification process. The number of the modalities as well as the number of channels for each modality makes the problem more complex and more time consuming, resulting to systems for offline use only.

Another common problem regarding the number of electrodes used for gathering the various biosignals is that their setup needs a lot of time and a lot of preparation in order to record accurate and artifact-free signals, while all these sensors make the participant feel uncomfortable. So it is an urgent need to reduce the number, despite that until now we need a lot of sensors in order to detect accurately the human's emotions.

Another drawback is the accuracy, especially when we are called to recognize many emotions. Despite the fact that humans experience innumerable emotions, until now only few of them can be recognized. It is obvious that we have a strong negative relationship among the number of emotions and the classification accuracy.

Considering all the above, we understand that we need a system which will be able to detect as many emotions as it is possible with the use of the least possible sensors. To the best of our knowledge the better classification accuracy that have ever reported is 83.33% using only three channels for the detection of six different emotions [25].

Taking into account that the emotional perception, as well as the reaction to emotional stimuli are not constant and vary among humans, it is an urgent need for researchers to develop and share emotional databases. These databases may include data from several modalities, like EEG, ECG, EDA, EMG etc, during experiments with different ways of emotion elicitation (visual, audio, olfaction etc.).

ACKNOWLEDGMENT

This study was funded by the Operational Program "Education and Lifelong Learning" of the Greek Ministry of Education and Religious Affairs, Culture and Sports (ref. number 2012ΣE24580284)

REFERENCES

1. P. Pinel, Biopsychology. Allyn and Bacon, 2004
2. Garay, Nestor; Idoia Cearreta, Juan Miguel López, Inmaculada Fajardo (April 2006). "Assistive Technology and Affective Mediation" (PDF). Human Technology: an Interdisciplinary Journal on Humans in ICT Environments 2 (1): 55–83. Retrieved 2008-05-12.
3. R. Yerkes and J. Dodson, "The Relation of Strength of Stimulus to Rapidity of Habit-Formation," J. Comparative Neurology and Psychology, vol. 18, pp. 459-482, 1908.
4. R. Malmo, "Activation," Experimental Foundations of Clinical Psychology, pp. 386-422, Basic Books, 1962.
5. E. Duffy, "Activation," Handbook of Psychophysiology, pp. 577-622, Holt, Rinehart and Winston, 1972.
6. P. Ekman, R. Levenson, and W. Friesen, "Autonomic Nervous System Activity Distinguishes among Emotions," Science, vol. 221, pp. 1208-1210, 1983
7. R. Croyle and J. Cooper, "Dissonance Arousal: Physiological Evidence," J. Personality and Social Psychology, vol. 45, pp. 782-791, 1983.
8. L. Andreassi, Human Behaviour and Physiological Response. Taylor & Francis, 2007.
9. P. Ekman, "An Argument for Basic Emotions," Cognition and Emotion, vol. 6, pp. 169-200, 1992.
10. A. Jaimes and N. Sebe, "Multimodal Human-Computer Interaction: A Survey," Computer Vision and Image Understanding, vol. 108, pp. 116-134, 2007.
11. R. Sharma, V.I. Pavlovic, and T.S. Huang, "Toward Multimodal Human-Computer Interface," Proc. IEEE, vol. 86, no. 5, pp. 853- 869, May. 1998.
12. J. Wagner, N.J. Kim, and E. Andre, "From Physiological Signals to Emotions: Implementing and Comparing Selected Methods for Feature Extraction and Classification," Proc. IEEE Int'l Conf. Multimedia and Expo, pp. 940-943, 2005.
13. F. Nasoz, K. Alvarez, C.L. Lisetti, and N. Finkelstein, "Emotion Recognition from Physiological Signals Using Wireless Sensors for Presence Technologies," Cognition, Technology and Work, vol. 6, pp. 4-14, 2004.
14. O. Villon and C. Lisetti, "A User-Modeling Approach to Build User's Psycho-Physiological Maps of Emotions Using BioSensors," Proc. IEEE RO-MAN 2006, 15th IEEE Int'l Symp. Robot and Human Interactive Comm., Session Emotional Cues in HumanRobot Interaction, pp. 269-276, 2006.
15. Lin, Y.P., Wang, C.H., Wu, T.L., Jeng, S.K., Chen, J.H.: EEG-based emotion recognition in music listening: A comparison of schemes for multiclass support vector machine. In: ICASSP, IEEE International Conference on Acoustics, Speech and Signal Processing - Proceedings. pp. 489–492. Taipei (2009)
16. Manousos A. Klados, Christos Frantzidis, Ana B. Vivas, et al., "A Framework Combining Delta Event-Related Oscillations (EROs) and Synchronisation Effects (ERD/ERS) to Study Emotional Processing," Computational Intelligence and Neuroscience, vol. 2009, Article ID 549419, 16 pages, 2009.
17. Lang, P.J., Bradley, M.M., & Cuthbert, B.N. (2008). International affective picture system (IAPS): Affective ratings of pictures and instruction manual. Technical Report A-8. University of Florida, Gainesville, FL
18. Frantzidis, C.A.; Bratsas, C.; Klados, M.A.; Konstantinidis, E.; Lithari, C.D.; Vivas, A.B.; Papadelis, C.L.; Kaldoudi, E.; Pappas, C.; Bamidis, P.D., "On the Classification of Emotional Biosignals Evoked While Viewing Affective Pictures: An Integrated Data-Mining-Based Approach for Healthcare Applications," Information Technology in Biomedicine, IEEE Transactions on , vol.14, no.2, pp.309,318, March 2010
19. C. Lithari, C. A. Frantzidis, C. Papadelis, Ana B. Vivas, M. A. Klados, C. Kourtidou-Papadeli, C. Pappas, A. A. Ioannides, P. D. Bamidis, Are Females More Responsive to Emotional Stimuli? A Neurophysiological Study Across Arousal and Valence Dimensions, Brain Topography, March 2010, Volume 23, Issue 1, pp 27-40,
20. E. Vyzas and R.W. Picard, "Affective Pattern Classification," Proc. AAAI Fall Symp. Series: Emotional and Intelligent: The Tangled Knot of Cognition, pp. 176-182, 1998.
21. Picard, R.W.; Vyzas, E.; Healey, J., "Toward machine emotional intelligence: analysis of affective physiological state," Pattern Analysis and Machine Intelligence, IEEE Transactions on , vol.23, no.10, pp.1175,1191, Oct 2001
22. J. Wagner, N.J. Kim, and E. Andre, "From Physiological Signals to Emotions: Implementing and Comparing Selected Methods for Feature Extraction and Classification," Proc. IEEE Int'l Conf. Multimedia and Expo, pp. 940-943, 2005.
23. R.A. Calvo, I. Brown, and S. Scheding, "Effect of Experimental Factors on the Recognition of Affective Mental States through Physiological Measures," Proc. 22nd Australasian Joint Conf. Artificial Intelligence, 2009.
24. K. Kim, S. Bang, and S. Kim, "Emotion Recognition System Using Short-Term Monitoring of Physiological Signals," Medical and Biological Eng. and Computing, vol. 42, pp. 419-427, May 2004
25. Petrantonakis, P.C., Hadjileontiadis, L.J.: Emotion recognition from EEG using higher order crossings. IEEE Transactions on Information Technology in Biomedicine 14(2), 186–197 (2010)

Author: Manousos A. Klados
Institute: Laboratory of Medical Informatics, Medical School, Aristotle University of Thessaloniki, Greece
Street: P.O. Box 323 54124
City: Thessaloniki
Country: Greece
Email: mklados@med.auth.gr

Matching Pursuit with Asymmetric Functions for Signal Decomposition and Parameterization

K.J. Blinowska[1], W.W. Jedrzejczak[2], and K. Kwaskiewicz[1]

[1] Department of Biomedical Physics, Warsaw University, 00 681 Warszawa, Poland
[2] Institute of Physiology and Pathology of Hearing, 01 943 Warszawa, Poland

Abstract—The high resolution time-frequency method - adaptive approximations by matching pursuit relies on the decomposition of signals into basic waveforms taken from a very large dictionary. Usually Gabor functions are used, however they are often not optimal in describing waveforms present in biological and medical signals. This especially concerns the signals of steep rise and slower decay. For the decomposition of this kind of signals we introduced a dictionary of functions of various degrees of asymmetry. The application of this dictionary to otoacoustic emissions and speech demonstrated the advantages of the proposed method. The approach allows for correct determination of the latencies of the components, removes the "pre-echo" effect generated by symmetric waveforms that do not match the structures of the analyzed signal sufficiently and provides more sparse representation. Additionally, we introduced a time-frequency-amplitude map, which is more adequate for representation of asymmetric waveforms than the conventional time-frequency-energy distribution.

Keywords—matching pursuit, time-frequency distributions, amplitude map, adaptive approximations, asymmetric functions dictionaries.

I. INTRODUCTION

For the analysis of non-stationary signals many methods are used: windowed Fourier Transform, Wigner transform, wavelets and adaptive approximations by matching pursuit (MP). The comparison of the MP with the other time-frequency (t-f) methods may be found e.g. in: [1], [2]. In the MP algorithm introduced by Mallat and Zhang [3], Gabor functions are used, because they provide the t-f resolution close to the one allowed by the indefinity principle. However, some signals contain highly asymmetric components, which are poorly described by Gabor functions. This problem is especially important when the latency of the response is of interest. We have encountered this problem in the analysis of otoacoustic emissions (OAEs) [4].

An OAE is a weak signal generated by the inner ear after the acoustic stimulus and also sometimes spontaneously. The response of the cochlea observed after the broadband stimulus consists of components of short duration located in the t-f plane approximately according to the power law and a class of responses called synchronized spontaneous otoacoustic emissions (SSOAEs), which last much longer than typical evoked components. These components are poorly described by the symmetric Gabor functions [4]. This observation has encouraged us to introduce a dictionary of asymmetric functions for OAE decomposition [5], but the class of functions that we propose may have a much larger field of applications.

The aim of this work is to demonstrate the usefulness of the MP method with the enriched dictionary containing functions of varying asymmetry, to the analysis of time series. Additionally we propose representation of the signals in t-f space by means of the amplitude maps. This kind of representation eliminates the cross-terms and is more intuitive than conventional energy distribution.

II. METHOD

A. Matching Pursuit with Asymmetric Dictionary

The method of adaptive approximation by MP introduced by Mallat and Zhang [3] relies on the decomposition of the signal into an assembly of functions (atoms) from a very large and redundant dictionary of Gabor functions:

$$g(t;\mu,\sigma,\omega) = N \cdot e^{\frac{-(t-\mu)^2}{2\sigma^2}} \cdot e^{i\omega t} \quad (1)$$

where: N – normalization constant, σ - scale, μ - position in time, ω - circular frequency.

Finding an optimal approximation of the signal by selecting functions from such a large family is an NP-hard problem, therefore a suboptimal iterative procedure is applied. In the first step of the procedure the vector is chosen which gives the largest inner product with the signal and then the procedure is repeated on the subsequent residues until the desired percentage of signal's energy is accounted for.

For parameterization of asymmetric functions we have proposed two-sided functions; the ascending part is based on a Gabor function and the descending part on the exponentially decaying sinusoid:

$$\Lambda(\mu,\sigma,\omega,T_m) = N \cdot \begin{pmatrix} \exp\left(-\frac{(t-\mu)^2}{2\sigma^2}\right) & ; & t \leq T_m \\ \exp(-\alpha \cdot (t-\tau)) & ; & t > T_m \end{pmatrix} \cdot \exp(i\omega t) \quad (2)$$

where $\alpha = \frac{T_m - \mu}{\sigma^2}$ and $\tau = \frac{T_m + \mu}{2}$

Additional parameter $T_m > \mu$ controls the asymmetry of the atom. T_m describes the point where the Gaussian envelope changes into exponent. N is normalization constant. The function obtained in this way is continuous up to first order derivative.

B. Amplitude-Time-Frequency Distribution.

Usually the energy density of the signal is visualized in the t-f plane in terms of Wigner-Ville (WV) distribution, which is a quadratic representation. When the signal contains more than one time-frequency component the cross terms occur. In case of MP with Gabor functions the cross-terms are expressed explicitly and can be easily removed [1]. However for asymmetric functions cross terms appear in the energy distribution. In order to counteract this phenomenon we propose to introduce instead of energy distribution the amplitude distribution in time-frequency.

The idea is based on calculation of dot product between scaled to 1 modulus of Fourier transform of an atom and a vector describing its sampled envelope $P(t)$. In this way, we get the amplitude representation of an atom in t-f space $A(\omega,t)$, given by the expression:

$$A(\omega,t) = Z^T(\omega) \cdot P(t) \quad (3)$$

$$Z^T(\omega) = \frac{FT(\Lambda)}{\max(FT(\Lambda))} \quad (4)$$

$$P(t) = \langle R^n f, \Lambda_{I_n} \rangle \cdot \Omega(\mu,\sigma,T_m) \quad (5)$$

where Λ_{I_n} is the atom (described in time-frequency) and $\langle R^n f, \Lambda_{I_n} \rangle$ is the atom amplitude. $Z^T(\omega)$ is scaled to 1 modulus of Fourier transform of an atom, and $P(t)$ the atom's envelope. Time-Amplitude profile of amplitude map of a single atom gives an atom envelope. The t-f representation of the decomposed signal is a sum of distributions given by Eq. 3.

III. RESULTS

In Fig. 1 is shown the decomposition of the simulated asymmetric signal given by the formula:

$$f(t) = \frac{t}{10^{-6} + t^2} \cos(\omega \cdot t) \quad (6)$$

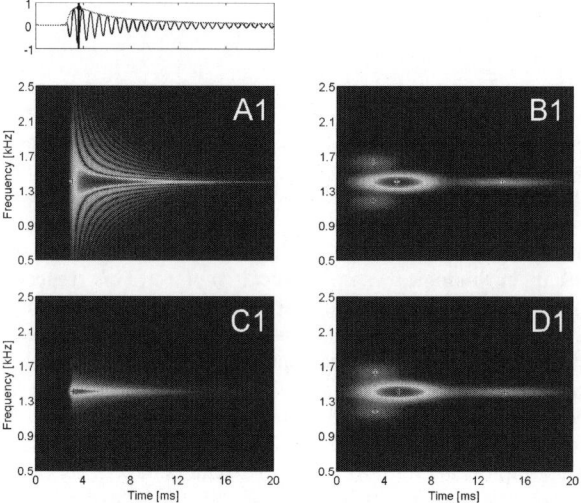

Fig. 1 The decomposition of the function shown in the upper panel. At the left the representation by means of the enriched dictionary, at the right by means of the Gabor dictionary. Upper panels: A1,B1 energy-t-f distribution, panels at the bottom: amplitude-t-f representation for enriched dictionary - C1 and Gabor dictionary - D1.

The simulated signal did not belong to the family of functions contained in the applied dictionary. We can observe that one atom of the enriched dictionary accounted for 97.7% percent of the signal energy. In case of the Gabor dictionary four atoms were required to account for the same percentage of energy. For Gabor representation the energy appears before the start of the signal, this phenomenon is called "dark energy" or "pre-echo". It is absent in case of application of asymmetric functions.

In Fig.1, A1 where VW distribution is shown we may observe, that the squaring procedure leads to the appearance of "ghost structures" in the time- frequency- energy space and the maximum of energy in the t-f map is shifted in relation to the maximum of amplitude. Distribution which alleviates the effect of the shift was introduced by Rihaczek [6], however this approach does not eliminate the cross terms, which obviously are absent in the amplitude-t-f representation.

The time series in which the application of the enriched dictionary appears to be indispensable are otoacoustic emissions (OAE). In Fig. 2 the t-f representations of exemplary transient evoked otoacoustic emission (TEOAE) are shown

By application of MP to the TEOAE signals, it was first possible to identify their basic components and connect them with the resonance modes of the cochlea [7]. Among these components short- and long-lasting waveforms were identified. The long-lasting components are not well described by the Gabor function, since their shape is asymmetric. One can see that the structure at 1.4 kHz, consisting of two atoms, short- and long-lasting was, in case of the Gabor dictionary described by four atoms, and additionally in case of Gabor functions "pre-echo" is observed for some components.

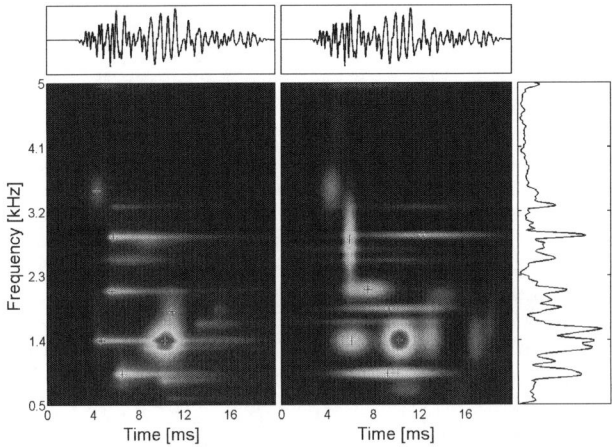

Fig. 2 Amplitude time-frequency representation of the TEOAE signal shown above. At the left – obtained by means of enriched dictionary, at the right obtained by means of Gabor functions. At the very right - power spectrum.

An important parameter for TEOAE analysis is the latency of each component. There were attempts to identify them by means of wavelet transform [8], however, a relevant estimation of latencies became possible only after the application of MP to the TEOAE analysis [4], [9], since in this approach latency is given directly as a parameter of the fitted function. However in case of long-lasting components the latency can be correctly retrieved only by application of enriched dictionary.

The differences between the latencies determined by means of Gabor and enriched dictionary are visible in Fig.2 and also in Fig.3, where an example of MP application to speech signal is presented. It may be observed in Fig.3 that the signal is better represented by the dictionary encompassing asymmetric functions, namely the latencies are correctly determined and the maximum of the amplitude of the signal corresponds with the maximum of t-f distribution, which is not the case for Gabor functions.

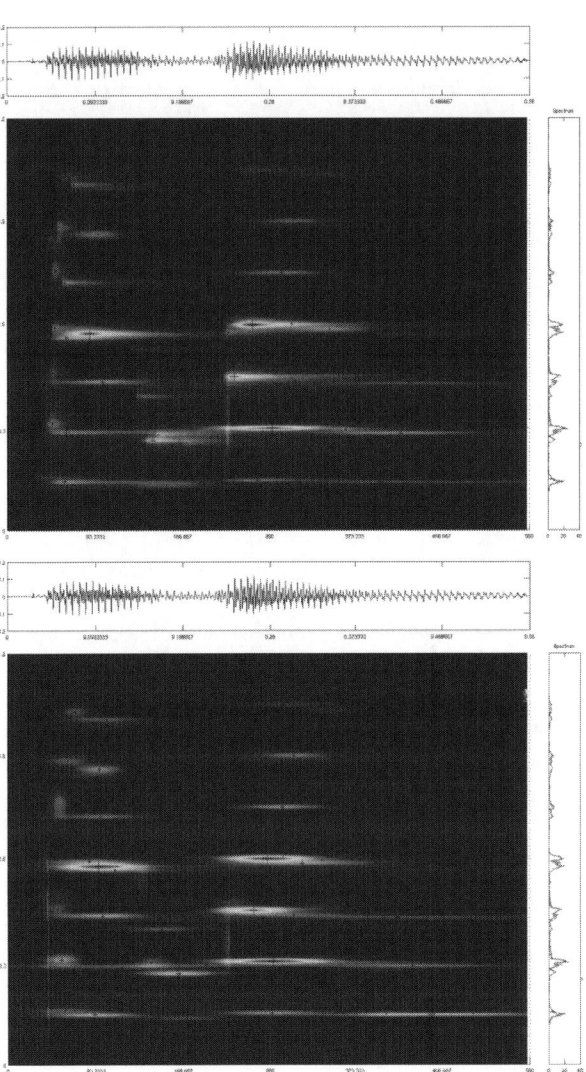

Fig. 3 The amplitude-time-frequency maps of the speech signal (word Adam) shown for convenience above each map. Upper panel - decomposition by means of enriched dictionary, at the bottom by means of Gabor dictionary.

IV. DISCUSSION

Herein we have proposed for representation of time series asymmetric functions of the flexible form. Different kinds of waveshapes may be designed changing one parameter only. The representation by means of asymmetric functions is more sparse than for Gabor functions. As an example may serve the long-lasting OAE components for which usually at least two Gabor atoms were needed. The proposed method based on the application of asymmetric functions in the dictionary of MP functions allowed for the elimination of the "pre-echo" effect – the artifact connected with the occurrence of energy before the start of the actual signal. This effect was caused by the fact that symmetric functions do not sufficiently match the signal structures.

An important aspect of the proposed method is the possibility of accurate determination of the components latency. In case of OAE signals it is a parameter important for clinical diagnosis [4], [8] and also for understanding the mechanisms of OAE generation, namely for comparison of the models with the experimentally estimated latency-frequency dependence.

We have proposed another kind of t-f representation of signals. A broadly applied representation of t-f distribution in terms of energy comes from electrical engineering, where power is of basic interest. Also, more traditional methods of analysis do not allow for easy determination of the amplitude values of signal structures. In case of biomedical signals amplitude is of primary interest, since it is more understandable by medical doctors and biologists, its link with the signal itself is more intuitive.

We have presented herein the advantages of the asymmetric waveforms for analysis of two kinds of signals, however the proposed approach may find the application for a much broader class of biomedical signals e.g.: phonocardiograms or evoked potentials. The proposed method apart from high-resolution of t-f distribution offers also parametric description of the components in terms of their frequency, latency, time span, amplitude and degree of asymmetry. These kinds of parameters have a clear meaning and could be helpful in revealing mechanisms of biomedical signals generation.

Acknowledgment

This work was supported by the grant of Ministry of Science and Higher Education to the Faculty of Physics of Warsaw University.

References

1. Blinowska K. J, Durka P. J. (2001) Unbiased high resolution method of EEG analysis in time-frequency space, Acta Neurobiologiae Experimentalis 61: 157-174.
2. Blinowska, K. J & Zygierewicz, J., 2011. Practical Biomedical Signal Analysis Using Matlab, CRC Press, Boca Raton, London, New York.
3. Mallat S.G, Zhang Z, (1993) Matching pursuit with time-frequency dictionaries, IEEE Trans. Sign. Process. 41: 3397-3415.
4. Jedrzejczak W. W, Hatzopoulos S, Martini A, Blinowska K.J. (2007) Otoacoustic emissions latency difference between full-term and preterm neonates, Hear. Res. 231: 54-62.
5. Jedrzejczak W. W, Kwaskiewicz K, Blinowska K.J, Kochanek K, Skarzynski H, (2009) Use of the matching pursuit algorithm with a dictionary of asymmetric waveforms in the analysis of transient evoked otoacoustic emissions. J. Acoust. Soc. Am. 126: 3137-3146.
6. Rihaczek W. (1968) Signal energy distribution in time and frequency, IEEE Trans. Informat. Theory, IT-14: 369-374.
7. Jedrzejczak W. W, Blinowska K.J, Konopka W, Grzanka A, Durka P.J, (2009) Identification of otoacoustic emission components by means of adaptive approximations, J. Acoust. Soc. Am. 115: 2148-2158.
8. Sisto R, Moleti A, (2002) On the frequency dependence of the otoacoustic emission latency in hypoacoustic and normal ears. J. Acoust. Soc. Am. 111: 297-308.
9. Jedrzejczak W.W, Blinowska, K. J, Konopka W, (2005) Time-frequency analysis of transiently evoked otoacoustic emissions of subjects exposed to noise, Hear. Res. 205: 249-255.

Author: K.J. Blinowska
Institute: Department of Biomedical Physics, Warsaw University
Street: Hoza 69
City: Warszawa 00-681
Country: Poland
Email: kjbli@fuw.edu.pl

Analysis of Intracranial Pressure Signals Using the Spectral Turbulence

M. García[1], J. Poza[1], D. Santamarta[2], D. Abásolo[3], and R. Hornero[1]

[1] Biomedical Engineering Group, TSCIT Department, University of Valladolid, Valladolid, Spain
[2] Department of Neurosurgery, University Hospital of León, León, Spain
[3] Centre for Biomedical Engineering, Department of Mechanical Engineering Sciences,
Faculty of Engineering and Physical Sciences, University of Surrey, Guildford, United Kingdom

Abstract—Hydrocephalus includes a range of conditions characterized by clinical symptoms, enlarged ventricles and disorders in cerebrospinal fluid (CSF) circulation. Infusion tests can be used to analyze CSF dynamics in patients with hydrocephalus. In infusion tests, intracranial pressure (ICP) is artificially raised, while the resulting ICP is measured in order to detect CSF circulation alterations. In this study, we analyzed 77 ICP signals recorded during infusion tests using the spectral turbulence (*ST*). Each signal was divided into four artifact-free epochs. The mean *ST*, <*ST*>, and the standard deviation of *ST*, SD[*ST*], were calculated for each epoch. Statistically significant differences were found between the basal phase of the infusion test and the remaining phases using <*ST*> and SD[*ST*] ($p < 1.7 \cdot 10^{-3}$, Bonferroni-corrected Wilcoxon tests). Furthermore, we found significantly higher <*ST*> and significantly lower SD[*ST*] values in the plateau phase than in the basal phase. These findings suggest that the increase in ICP induced by infusion studies is associated with a significant loss of irregularity and variability of the spectral content of ICP signals. In conclusion, spectral analysis of ICP signals could be useful for understanding CSF dynamics in hydrocephalus.

Keywords—Hydrocephalus, intracranial pressure, spectral turbulence.

I. INTRODUCTION

The term hydrocephalus encompasses a range of disorders characterized by clinical symptoms, abnormal brain imaging and derangement of cerebrospinal fluid (CSF) dynamics [1]. The study of intracranial pressure (ICP) and CSF dynamics can help in the decision about performing shunt placement surgery and can also provide valuable information for shunted patient management [2]. Infusion tests are routinely performed to study CSF circulation disorders in patients showing features of hydrocephalus. In infusion tests, CSF pressure is increased by addition of external volume. The resulting ICP is recorded, usually in the lumbar subarachnoidal space [3].

Some therapies in hydrocephalus are based on the ICP time-averaged mean [3], [4]. However, this parameter does not account for all the information included in the ICP waveform and does not provide a deep insight into cerebral autoregulation mechanisms [3], [4]. Several methods have been developed in order to better understand CSF circulation disorders. Some of them rely on the non-linear [3], [5] or spectral [4], [6], [7] analysis of ICP signals.

In this study we applied the spectral turbulence (*ST*) to characterize ICP signals recorded during infusion tests in patients with hydrocephalus. This is a spectral parameter that quantifies the irregularity of ICP recordings in terms of the degree of similarity between the power spectral density (PSD) of adjacent segments of the ICP signals. *ST* can be regarded as an alternative to other standard irregularity measures, such as entropies or disequilibrium measures [8]. We believe that *ST* could be useful to characterize irregularity patterns during episodes of intracranial hypertension induced by infusion studies. Therefore, the aims of this study were: (i) to analyze the irregularity patterns in ICP signals based on *ST*; (ii) to test whether *ST* could reveal significant differences among phases of the infusion study; and (iii) to introduce an alternative framework to understand brain dynamics in hydrocephalus.

II. SUBJECTS AND ICP DATA RECORDING

ICP signals of 77 patients showing features of hydrocephalus (41 male and 36 female, age 69 ± 14 years, mean ± standard deviation, SD) were recorded during infusion tests at the Department of Neurosurgery of the University Hospital of León (Spain). Infusion tests were performed as a supplementary hydrodynamic study and as a decision aid on the surgical management of patients [3]. Patients or a close relative gave their informed consent to participate in the study, which was approved by the local ethics committee.

Infusion studies were performed using a variant of the Katzman and Hussey method [9]. Under local anesthesia, patients were positioned in the lateral recumbent position and two needles (caudal needle and rostral cannula) were inserted in their lower lumbar region. The caudal needle was connected to an infusion pump (Lifecare® 5000, Abbott Laboratories). The rostral cannula was connected to a pressure microtransducer (Codman® MicroSensorTM ICP transducer, Codman & Shurtleff). The pressure signal from the analog output of the microtransducer monitor was amplified (ML110 Bridge amplifier) and digitized (PowerLab 2/25 Data Recording System ML825, ADI Instruments).

The analog to digital converter was connected to a computer in order to visualize and record the ICP signals. After 5 minutes of baseline recording, a Ringer solution at a constant infusion rate of 1.5 ml/min was infused until a plateau was reached. After the infusion stopped, CSF pressure was recorded until it decreased towards baseline levels [3]. For each recording, a neurosurgeon selected four artifact-free epochs, representative of four phases of the infusion test:

- Epoch 0 (E0) corresponds to the basal phase.
- Epoch 1 (E1) is representative of the early infusion phase and usually describes an ascending slope.
- Epoch 2 (E2) corresponds to the plateau phase.
- Epoch 3 (E3) represents the recovery phase, once the infusion has stopped and the ICP signal returns slowly to basal state levels.

ICP recordings were acquired with a sampling frequency of 100 Hz. The recordings were also processed using a finite impulse response (FIR) bandpass filter with cut-off frequencies 0.1 Hz and 10 Hz. This frequency range was chosen in order to minimize the influence of the DC component and preserve the relevant spectral content [7].

III. METHODS

A. Calculation of Spectral Turbulence

A Fourier analysis was carried out in order to characterize the spectral content of ICP signals. Biomedical signals are intrinsically non-stationary [10]. Thus, we analyzed the ICP signals using the short-time Fourier transform (STFT), since it takes into account their time-varying properties. Each ICP signal was divided into N_T segments of 5 s (length $L=500$ samples) with an overlap of 4 s [3], [7]. The power spectral density (PSD) was calculated as the Fourier transform of the autocorrelation function in each temporal window [8], [11]. The PSD was normalized to a scale from 0 to 1 leading to the normalized PSD, $PSD_n^{(i)}(f)$, in each temporal interval i ($i = 1, ..., N_T$).

From PSD_n different time-frequency parameters can be calculated. ST is a measure that quantifies the spectral changes of the signal over time [12]. It is based on calculating the PSD_n map and comparing adjacent spectra by means of the correlation coefficient [12], [13]. It can be defined as:

$$ST^{(i)} = \rho\left[PSD_n^{(i)}(f), PSD_n^{(i+1)}(f)\right], \quad i = 1, ..., N_T - 1, \quad (1)$$

where $\rho[\cdot]$ denotes the Pearson correlation coefficient between $PSD_n^{(i)}(f)$ and $PSD_n^{(i+1)}(f)$.

The mean (<ST>) and the standard deviation (SD[ST]) of ST were subsequently calculated from the time series formed by the temporal evolution of ST. <ST> summarizes the average degree of similarity between the spectral content of adjacent time slices, while SD[ST] describes the lack of homogeneity in correlation around the mean value [14]. An average value of <ST> and SD[ST] in the four artifact-free epochs was obtained for each subject. We will denote by <$ST0$> to the average value of ST in E0, by <$ST1$> to the mean ST value in E1, by <$ST2$> to the mean ST value in E2 and by <$ST3$> to the mean ST value in E3. Similarly, SD[$ST0$], SD[$ST1$], SD[$ST2$] and SD[$ST3$] represent the value of SD[ST] in E0, E1, E2 and E3, respectively.

B. Statistical Analysis

An initial exploratory analysis was performed to study the data distribution. The Kolmogorov–Smirnov test with Lilliefors significance correction and the Shapiro–Wilk test were used to assess the normality of <ST> and SD[ST] in the four artifact-free epochs. Variables did not meet parametric test assumptions. Thus, the non-parametric Friedman test was used to determine whether statistically significant interactions ($\alpha=0.01$) could be found between epochs of the infusion test [15]. *Post hoc* analyses were carried out using the Wilcoxon signed rank test with Bonferroni correction to account for multiple comparisons ($\alpha=0.01/6=1.7 \cdot 10^{-3}$) [15].

IV. RESULTS

We calculated the PSD for the 5-s segments in which each ICP recording was divided. The evolution of ST over time was then computed from PSD_n. Finally, the mean values (<$ST0$>, <$ST1$>, <$ST2$> and <$ST3$>) and the standard deviations (SD[$ST0$], SD[$ST1$], SD[$ST2$] and SD[$ST3$]) of ST in the four artifact-free epochs were obtained.

Table 1 summarizes the mean values of the epoch length, CSF pressure, <ST> and SD[ST]. These values were averaged over the 77 subjects in our database. These results show that the lowest <ST> value was obtained in the basal phase, then increased during infusion to reach the highest values in the plateau phase and, finally, decreased in the recovery phase. In the case of SD[ST], the highest SD[ST] values were obtained for the basal phase, then decreased during infusion towards the lowest SD[ST] values, obtained in the plateau phase. Finally, SD[ST] increased again during the recovery phase. These evolutions are depicted in Fig. 1.

The Friedman test revealed significant interactions among <ST> values ($\chi^2(3)=55.89$, $p=4.27 \cdot 10^{-12}<0.01$) and SD[$ST$] values ($\chi^2(3)=23.14$, $p=3.78 \cdot 10^{-5}<0.01$). In order to analyze these interactions, *post hoc* analyses were carried out using the Wilcoxon signed rank test with Bonferroni correction. Statistically significant differences were found between the following pairs: <$ST0$> vs. <$ST1$>, <$ST0$> vs. <$ST2$>, <$ST0$> vs. <$ST3$>, SD[$ST0$] vs. SD[$ST1$], SD[$ST0$] vs. SD[$ST2$] and SD[$ST0$] vs. SD[$ST3$]. These results are

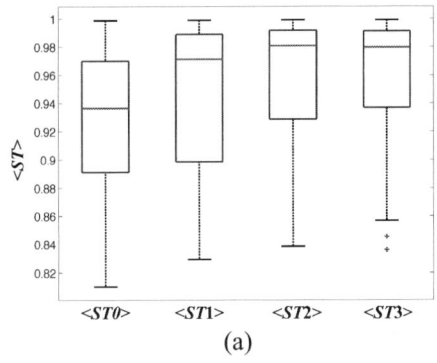
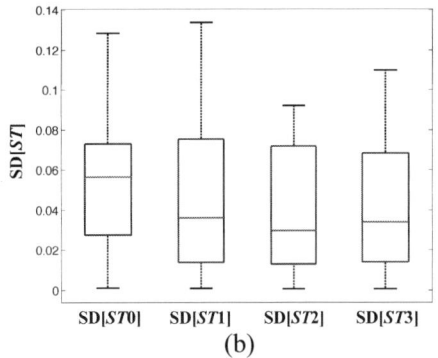

Fig. 1 Boxplots displaying the distribution of $<ST>$ and SD[ST] in the four artifact-free epochs. (a) $<ST>$. (b) SD[ST].

summarized in Table 2, where the statistically significant values have been highlighted.

V. DISCUSSION

In this study, we analyzed the changes produced in ST during infusion studies in patients with hydrocephalus. $<ST>$ and SD[ST] were calculated for the four artifact-free epochs in which the infusion study was divided.

Our results showed that significant differences were found between the basal phase and the remaining phases of the infusion tests using $<ST>$. It is noteworthy that a significant increase in $<ST>$ was found in the plateau phase with respect to the basal phase. This suggests that intracranial hypertension due to volume loading results in a significantly higher degree of similarity in the spectral content. Besides, $<ST>$ can be considered as an indirect measure of the irregularity of the signal [14]. Thus, the increase in $<ST2>$ with respect to $<ST0>$ is associated with an irregularity loss in E2 with respect to E0. This result correlates with the decrease in Lempel-Ziv (LZ) complexity between the plateau and basal phases of infusion studies found in [3]. Decreased complexity during episodes of intracranial hypertension has also been found in pediatric patients with traumatic brain injury [5]. Certainly, irregularity and complexity are linked, focusing on different signal characteristics. They can be regarded as complementary measures to quantify the degree of disorder in infusion tests.

Results in Table 2 also show that significant differences were found between E0 and the remaining phases of the infusion test using SD[ST]. Besides, our results showed a significant decrease in SD[$ST2$] with respect to SD[$ST0$]. These findings suggest that there is a significantly lower variability in the spectral content when CSF pressure reaches the range of intracranial hypertension. This variability reduction could be linked to the decrement in data dispersion found in previous studies on infusion tests [3], where lower SD values in LZ complexity were found in the plateau phase than in the basal phase.

The decreased irregularity and variability during episodes of intracranial hypertension found in this study might be explained from an early Cushing response mediated by a moderate rise in ICP during infusion studies [3], [16]. This early Cushing response elicits a moderate rise in arterial blood pressure, a mild decrease in cerebral perfusion pressure, a reduction of cerebral blood flow and an increase in the heart rate variability without a modification of its mean value [16]. These processes could affect the shape of the

Table 1 Mean values of the epoch length, CSF pressure, $<ST>$ and SD[ST] for the four artifact-free epochs

Parameter	E0	E1	E2	E3
Length (s)	175 ± 54	300 ± 49	497 ± 139	199 ± 50
CSF pressure (mm Hg)	8.247 ± 3.747	16.241 ± 5.446	25.381 ± 8.975	16.381 ± 5.978
$<ST>$	0.930 ± 0.048	0.950 ± 0.049	0.957 ± 0.047	0.956 ± 0.045
SD[ST]	0.055 ± 0.030	0.042 ± 0.033	0.040 ± 0.029	0.039 ± 0.029

Table 2 Z statistics and p-values associated with the Wilcoxon tests for $<ST>$ and SD[ST]

Parameter	E0 vs. E1		E0 vs. E2		E0 vs. E3		E1 vs. E2		E1 vs. E3		E2 vs. E3	
	Z	p	Z	p	Z	p	Z	p	Z	p	Z	p
$<ST>$	-5.69	**1.28·10⁻⁸**	-6.00	**1.93·10⁻⁹**	-6.27	**3.54·10⁻¹⁰**	-2.93	**3.40·10⁻³**	-2.08	**3.74·10⁻²**	-1.37	0.17
SD[ST]	-4.71	**2.49·10⁻⁶**	-4.62	**4.18·10⁻⁶**	-5.00	**5.62·10⁻⁷**	-0.47	0.64	-1.13	0.26	-0.57	0.57

ICP waveform in the plateau phase and influence the irregularity and variability of ICP signals. It has also been suggested that decreased irregularity is associated with greater system isolation [17]. It has been stated that increased hypertension and severity of brain injury produce physiological uncoupling between the cardiovascular and the autonomic system [18] and, thereby, decreased irregularity. These mechanisms have been suggested as a possible explanation for the decreased complexity in E2 reported in [3], which is related to the loss of variability found in our study.

Finally, some limitations merit further consideration. In this study, spectral analysis was performed on a heterogeneous group of patients showing features of hydrocephalus. Since this study is focused on assessing whether ST could derive alternative measures from ICP signals, heterogeneity in patient population should not be regarded as a serious drawback. It would also be desirable to test the proposed method on a larger set of ICP signals.

VI. CONCLUSION

In summary, $<ST>$ and $SD[ST]$ reveal changes in the ICP signal spectrum during stress episodes. Our findings suggest that the evolution of CSF pressure in infusion studies is characterized by a significant loss of irregularity and variability. We conclude that spectral measures, such as ST, may help to gain further insight into brain dynamics in hydrocephalus. Future research should explore other spectral parameters in order to obtain complementary information to better understand CSF dynamics in hydrocephalus.

ACKNOWLEDGMENT

This research was partially supported by: the *Ministerio de Economía y Competitividad* and FEDER under project TEC2011-22987; the 'Proyecto Cero 2011 on Ageing' from *Fundación General CSIC, Obra Social La Caixa* and CSIC; and project VA111A11-2 from *Consejería de Educación (Junta de Castilla y León)*.

REFERENCES

1. Weerakkody RA, Czosnyka M, Schuhmann MU (2010) Clinical assessment of cerebrospinal fluid dynamics in hydrocephalus. Guide to interpretation based on observational study. Acta Neurol Scand 124:85–98
2. Czosnyka M, Czosnyka Z, Momjian S et al (2004) Cerebrospinal fluid dynamics. Physiol Meas 25:R51–R76
3. Santamarta D, Hornero R, Abásolo D et al (2010) Complexity analysis of the cerebrospinal fluid pulse waveform during infusion studies. Childs Nerv Syst 26: 1683–1689
4. Czosnyka M, Pickard JD (2004) Monitoring and interpretation of intracranial pressure. J Neurol Neurosurg Psychiatry 75: 813–821
5. Hornero R, Aboy M, Abásolo D et al (2007) Analysis of intracranial pressure during acute intracranial hypertension using Lempel-Ziv complexity: further evidence. Med Biol Eng Comput 45: 617–620
6. Momjian S, Czosnyka Z, Czosnyka M et al (2004) Link between vasogenic waves of intracranial pressure and cerebrospinal fluid outflow resistance in normal pressure hydrocephalus. Br J Neurosurg 18: 56–61
7. García M, Poza J, Santamarta D et al (2013) Spectral analysis of intracranial pressure signals recorded during infusion studies in patients with hydrocephalus. Med Eng Phys DOI 10.1016/j.medengphy.2013.04.002
8. Poza J, Hornero R, Escudero J et al (2008) Regional analysis of spontaneous MEG rythms in patients with Alzheimer's disease using spectral entropies. Ann Biomed Eng 36: 141–152
9. Katzman R, Hussey F (1970) A simple constant-infusion manometric test for measurement of CSF absorption I: rationale and method. Neurology 20: 534–544
10. Heissler HE, König K, Krauss JK et al (2012) Stationarity in neuromonitoring data. Acta Neurochir Suppl 114: 93–95
11. Akay M (1998) Time, frequency and wavelets in biomedical signal processing. IEEE Press, New Jersey
12. Kelen GJ, Henkin R, Starr AM et al (1991) Spectral turbulence analysis of the signal-averaged electrocardiogram and its predictive accuracy for inducible sustained monomorphic ventricular tachycardia. Am J Cardiol 67: 965–975
13. Barbosa PRB, de Souza A, Barbosa EC et al (2006) Spectral turbulence analysis of the signal-averaged electrocardiogram of the atrial activation as predictor of recurrence of idiopathic and persistent atrial fibrillation. Int J Cardiol 107: 307–316
14. Poza J, Hornero R, Escudero J et al (2008) Analysis of spontaneous MEG activity in Alzheimer's disease using time-frequency parameters. Proc IEEE EMBS Conference, Vancouver, Canada, 2008, pp 5712–5714
15. Jobson JD (1991) Applied Multivariate Data Analysis. Volume II: Categorical and Multivariate Methods. Springer, New York
16. Schmidt EA, Czosnyka Z, Momjian S et al (2005) Intracranial baroreflex yielding an early Cushing response in humans. Acta Neurochir Supp 95: 253–256
17. Pincus SM (1994) Greater signal regularity may indicate increased system isolation. Math Biosci 122:161–181
18. Goldstein B, Toweill D, Lai S et al (1998) Uncoupling of the autonomic and cardiovascular systems in acute brain injury. Am J Physiol 275:1287–1292

Author: María García Gadañón
Institute: Biomedical Engineering Group, TSCIT Department, University of Valladolid
Street: Paseo de Belén 15
City: Valladolid
Country: Spain
Email: maria.garcia@tel.uva.es

Graph-Theoretical Analysis in Schizophrenia Performing an Auditory Oddball Task

A. Bachiller[1], J. Poza[1], C. Gómez[1], V. Molina[2], V. Suazo[3], A. Díez[3], and R. Hornero[1]

[1] Biomedical Engineering Group (GIB), Dpt. TSCIT,
University of Valladolid, Paseo de Belén 15, 47011-Valladolid, Spain
[2] Psychiatry Department, Faculty of Medicine,
University of Valladolid, Av. Ramón y Cajal 7, 47005-Valladolid, Spain
[3] Institute of Biomedical Research (IBSAL), Paseo de San Vicente 58, 37007-Salamanca, Spain

Abstract—The aim of this research was to study the organization of the brain functional network during an auditory oddball task in schizophrenia (SZ). Electroencephalographic (EEG) activity was recorded from 31 schizophrenic patients and 38 healthy controls. In a first step, coherence was used to estimate the similarity between the spectral content of each pair of electrodes. In a second step, a graph was generated from the similarity matrix and two network parameters were computed: the clustering coefficient and the path length. Our results indicate that SZ patients obtained lower clustering coefficient and longer path length variations between the baseline and the P300 response than controls. These findings suggest an abnormal organization of the brain functional network in SZ.

Keywords—Schizophrenia, coherence, graph network, clustering coefficient, path length.

I. INTRODUCTION

Schizophrenia (SZ) is a psychiatric disorder of cognition, characterized by a cognitive processing dysfunction [1]. It is accompanied by hallucinations, delusions, loss of initiative and cognitive impairments. Recent formulations have defined SZ as a disconnection syndrome, associated to a reduced capacity to integrate information between different brain regions [2]. Some studies have addressed the interpretation of functional connectivity in SZ. Magnetic resonance imaging (MRI) studies have shown morphological abnormalities of regional gray matter structures [3].

Neural oscillations are the main mechanism for enabling coordinated activity during normal brain functioning [4]. Oscillations in high frequency ranges (beta and gamma) establish synchronization in local cortical networks, whereas lower frequency ranges (delta, theta and alpha) modulate long-range synchronization [4]. Coherence has been widely applied to electroencephalographic (EEG) signals to analyze the functional connectivity between brain regions [5]. A further approach to study the complex organization of the human brain network is the application of the "network theory" principles. A graph is a mathematical representation of a network, which is essentially reduced to nodes and connections between them. The use of a graph-theoretical approach has been considered potentially relevant and useful, as demonstrated on several sets of brain functional networks [6]. Graph network analysis offers information about integration, segregation, connectivity and overall organization of brain networks. For this regard, its application to study SZ revealed several neural network changes [3, 5–9]. Most of these studies have focused on resting state [8] or two-back working memory task [5–7]; however, the auditory oddball paradigm has been used in few researches [9].

In the present study, the coherence was used to generate connectivity/similarity patterns between the spectral content of EEG activity from different electrodes. The aim of this research was to characterize some neuropathological alterations associated with SZ, during an auditory oddball task, by means of graph network theory.

II. MATERIALS

A. Subjects

Sixty-nine subjects were enrolled in the study. Thirty-one were SZ patients, including 20 chronic stably treated patients (CP) (12 men and 8 women, age = 40.4 ± 10.4 years, mean ± standard deviation, SD) and 11 minimally treated patients (MTP) (7 men and 4 women, age = 33.5 ± 9.9 years, mean ± SD). MTP had not received any previous treatment prior to their inclusion (first episode patients) or they had dropped their medications for longer than 1 month. The diagnosis was made according to Diagnostic and statistical manual of mental disorders (DSM-IV) criteria and the patients' clinical status was scored using the Positive and Negative Syndrome (PANSS). The control group was formed by 38 healthy volunteers (23 men and 15 women, age = 33.7 ± 13.1 years, mean ± SD).

B. EEG Recording

EEG recordings were acquired while subjects were relaxed and with eyes closed. An oddball 3-stimulus paradigm was employed with a 500 Hz-tone target, a 1000 Hz-tone distracter and a 2000 Hz-tone standard stimulus.

The experiment was composed by a random series of 600 tones with probabilities of 0.20, 0.20 and 0.60, respectively.

The EEG was recorded using a BrainVision® equipment (Brain Products GmbH; Munich; Germany) formed by 17 tin sensors mounted in an electrode cap according to the 10/20 International System. Recordings were re-referenced to the average activity of all active sensors and the sampling rate was 250 Hz. EEG signals were filtered between 1 and 70 Hz and a 50 Hz notch filter was applied to remove power line noise. Artifact rejection was conducted following a two-steps approach. Firstly, Independent Component Analysis was carried out to decompose the signal in 17 components. Components related to eyeblinks were discarded. Secondly, artifacts were automatically rejected using an adaptive thresholding method. To complete the preprocessing, each EEG recording was divided into 800 ms–length epochs from -250 ms to 550 ms with respect to the stimulus onset (200 samples per epoch).

III. METHODS

A. Coherence Measure

In order to quantify the differences in the spectral content between EEG sensors, the coherence was applied. Coherence is a measure to assess functional interplays between pairs of cortical regions [10]. It describes the strength of the correlation between two time series as a function of frequency [11]. The mean square coherence (MSC) between two signals corresponds to their cross–spectral density function normalized by their individual auto–spectral density functions [10].

$$MSC_{XY}(t,f) = \frac{|S_{XY}(t,f)|^2}{S_{XX}(t,f) \cdot S_{YY}(t,f)}. \quad (1)$$

Rappelsberger et al. [12] proposed a coherence estimation to retain temporal information, named event-related coherence (ERC). ERC is based on the assumption that the same patterns of physiological activity are repeated at the same latency trial to trial [12]. To obtain a time course of coherence, each EEG epoch (200 samples) was divided into temporal segments of 41 samples with a 90% overlapping. Then, 32 time intervals were obtained and coherence was calculated as described above. Finally, ERC values were averaged in the six conventional frequency bands: δ (1-4 Hz), θ (4-8 Hz), α (8-13 Hz), β_1 (13-19 Hz), β_2 (19-30 Hz) and γ (30-70 Hz).

B. Graph Theory

The brain can be assimilated to a complex anatomical and functional network. Hence, it can be represented by means of a graph. A graph is defined as a number of nodes or vertices and the corresponding edges between them [13]. The value of each edge depends on the importance of the relationship between the nodes [12, 13].

Coherence is limited from 0 to 1. The higher the coherence values, the higher the correlation between the spectral content. Hence, coherence values can be applied directly to the edges of a graph analysis. Then, a network with $N=17$ vertices (corresponding to the 17 electrodes) can be defined and ERC values between two electrodes be used to establish the edges weights (denoted as w^b_{ij}, where b denotes the frequency band). A graph can be characterized using several parameters [11]. In this research, the clustering coefficient and the average path length were used. These parameters measure the nature of structural building blocks and sub-networks, and the sensitivity to the level of integration in a network, respectively [11].

The clustering coefficient of a vertex i, C_i, reflects the presence of triangles (complete subgraphs of three vertices) in networks [11, 14]. It should be noted that symmetry is required ($w^b_{ij} = w^b_{ji}$) and $0 \le w^b_{ij} \le 1$ [14]. These constrains are fulfilled by ERC. Therefore, the clustering coefficient for the vertex i at each frequency band is defined as,

$$C^b_i = \frac{\sum_{\substack{k \ne i \\ l \ne k}} \sum_{l \ne i} w^b_{ik} \cdot w^b_{il} \cdot w^b_{kl}}{\sum_{\substack{k \ne i \\ l \ne k}} \sum_{l \ne i} w^b_{ik} \cdot w^b_{il}}, \quad b = \{\delta, \theta, \alpha, \beta_1, \beta_2, \gamma\}. \quad (2)$$

The average clustering coefficient, C_W, for the whole graph at each frequency band is defined as the average of the clustering coefficient in the 17 nodes.

The path length is defined as the average number of edges of the shortest path between pairs of edges. The length between two vertices i and j is defined as the inverse of the weight between them: $L^b_{ij} = 1/w^b_{ij}$ if $w^b_{ij} \ne 0$, and $L^b_{ij} = +\infty$ if $w^b_{ij} = 0$ [14]. The path length between two vertices is then defined as the sum of the lengths of the edges of this path. The shortest path L^b_{ij} between two vertices i and j is the path between i and j with the shortest length [14]. Equation (3) shows the average path length of the entire graph at each frequency band. It is calculated using the harmonic mean. Hence, it takes into account infinite path lengths between isolated nodes (i.e. $1/\infty \to 0$) [14].

$$L^b_W = \frac{1}{\frac{1}{N \cdot (N-1)} \cdot \sum_{i=1}^{N} \sum_{j \ne i}^{N} \frac{1}{L^b_{ij}}}, \quad b = \{\delta, \theta, \alpha, \beta_1, \beta_2, \gamma\} \quad (3)$$

C. Parameter Baseline Correction

In order to achieve a stimulus-independent characterization, a baseline correction process has been carried out. A pre-stimulus value was obtained by averaging the values in the interval (-250 0) ms. Likewise, a response value was obtained considering the (150 450) ms post–stimulus interval. The baseline correction was then carried out using the "percent change from baseline" method [15]. For that purpose, pre-stimulus mean value was subtracted from response value and the result was divided by the pre–stimulus value for each subject.

D. Statistical Analysis

Initially, the exploratory analysis revealed that data did not meet parametric test assumptions. Afterwards, a two level statistical analysis was made. Firstly, a Wilcoxon signed rank test ($\alpha=0.05$) was used to analyze the evolution in each group. In a second step, statistical significance between groups was assessed by means of Mann-Whitney U–tests ($\alpha=0.05$).

IV. RESULTS AND DISCUSSION

Mean values of clustering coefficient and path length were calculated from each graph, obtaining the temporal evolution of these parameters. In a first step, a Wilcoxon signed rank test was applied to analyze the differences between the baseline and the P300 response. As Table 1 shows, controls reached a statistically significant increase in C_W^δ ($p<0.0001$) and C_W^θ ($p=0.0021$), as well as a decrease in $C_W^{\beta_2}$ ($p=0.0020$), between the P300 response and the baseline. On the other hand, Table 2 shows the path length obtained at each frequency band. A significant decrease of L_W^δ ($p=0.0032$) between the P300 response and the baseline was obtained in CP group. In the control group, the P300 response showed higher values of L_W^δ ($p<0.0001$) and L_W^θ

Table 1 Clustering coefficient (mean ± SD) at baseline and P300 response for each group. Only the frequency bands with statistically significant results are displayed. Significant *p*-values are marked with asterisks (*, $p<0.05$; **, $p<0.01$; ***, $p<0.0001$).

Group	Segment	Frequency band		
		δ	θ	β_2
CP	Baseline	0.32 ± 0.07	0.35 ± 0.08	0.30 ± 0.10
	P300	0.34 ± 0.08 *	0.35 ± 0.07	0.30 ± 0.09
MTP	Baseline	0.32 ± 0.09	0.37 ± 0.10	0.30 ± 0.08
	P300	0.34 ± 0.09	0.38 ± 0.08	0.28 ± 0.07 *
C	Baseline	0.33 ± 0.10	0.34 ± 0.10	0.31 ± 0.12
	P300	0.37 ± 0.09 ***	0.38 ± 0.08 **	0.30 ± 0.11 **

Table 2 Path length (mean ± SD) at baseline and P300 response for each group. Only the frequency bands with statistically significant results are displayed. Significant *p*-values are marked with asterisks (*, $p<0.05$; **, $p<0.01$; ***, $p<0.0001$).

Group	Segment	Frequency band		
		δ	θ	β_2
CP	Baseline	3.45 ± 0.74	3.12 ± 0.59	3.78 ± 0.98
	P300	3.25 ± 0.73 **	3.12 ± 0.47	3.78 ± 0.94
MTP	Baseline	3.60 ± 0.96	3.12 ± 0.65	3.81 ± 0.77
	P300	03.35 ± 0.91	2.95 ± 0.51	3.92 ± 0.68
C	Baseline	3.54 ± 0.94	3.31 ± 0.69	3.71 ± 0.98
	P300	3.15 ± 0.75 ***	3.01 ± 0.51 ***	3.91 ± 1.06 ***

($p=0.0002$), as well as lower values of $L_W^{\beta_2}$ ($p=0.0009$), than the baseline segment. These results suggest that the brain network features are altered during the auditory oddball task [9].

In a second step, baseline correction was calculated for each subject and results were statistically analyzed using Mann-Whitney U–tests. Figs. 1 and 2 summarize the mean clustering coefficient and path length at each frequency band for each group, respectively. They represented the relationship between the P300 response and baseline, expressed in percent of change. A positive value indicates that the P300 value is higher than the baseline value. On the contrary, a negative value indicates a lower P300 value than the baseline value. As Fig. 1 shows, a statistically significant higher ΔC_W^θ ($p=0.0283$) was obtained in controls in comparison with CP. On the other hand, Fig. 2 depicts that controls obtained statistically significant lower ΔL_W^θ ($p=0.0049$) than CP.

Previous studies have reported that statically significant lower C_W values in SZ imply relatively sparse local connectedness of the brain functional networks [5]. Interactions between interconnected brain regions are believed to be a basis of human cognitive processes [8]. Short absolute path lengths have been demonstrated to promote interactions between and across different cortical regions [8]. The higher ΔL_W^θ obtained by controls in comparison with CP, may indicate that information interactions between interconnected brain regions are slower and less efficient in SZ [8]. Thus, the lower clustering coefficients and the longer path length obtained by CP in comparison with controls support the SZ disconnection hypothesis and may indicate an abnormal organization of the brain functional network [5, 9].

Finally, some aspects of the present research merit further consideration. Additional work is required to compute other connectivity measures and to extract other network parameters, like centrality, efficiency or modularity. In addition, it could be interesting to examine the assessment of regional patterns.

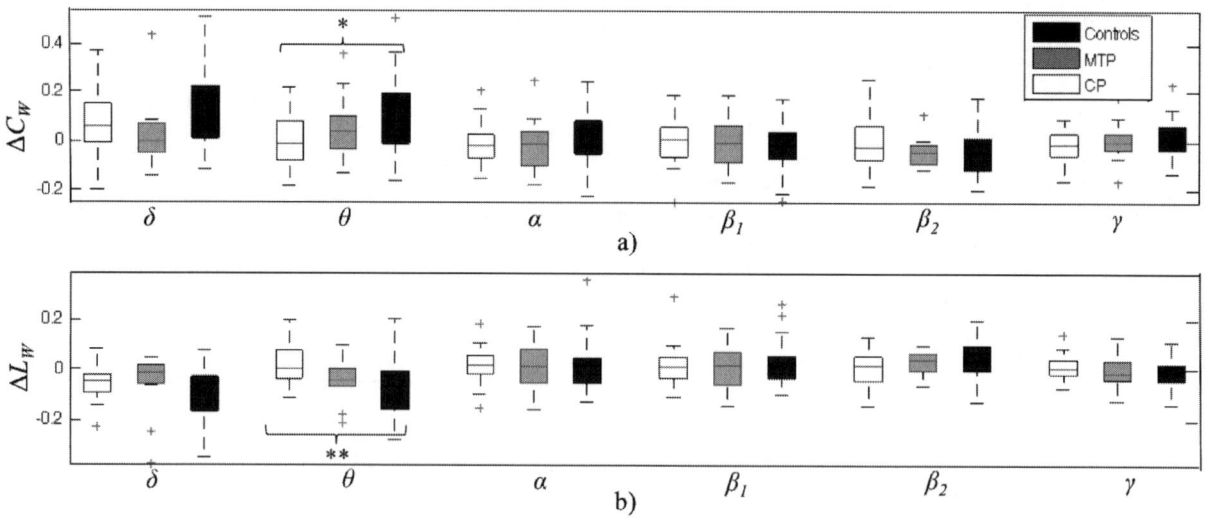

Fig. 1 Boxplot displaying the percent of change for the network parameters in CP, MTP and controls groups. a) Clustering coefficients (ΔC_W) for each frequency band. b) Path length (ΔL_W) for each frequency band. Statistically significant p-values are marked with asterisks (*, $p<0.05$; **, $p<0.01$)

V. Conclusions

Our research analyzes the application of ERC to generate a graph, useful for characterizing the organization of brain networks. Our findings support the notion that SZ is a disconnection syndrome, showing frequency-dependent neural network alterations.

Acknowledgment

This research was supported in part by: *Ministerio de Economía y Competitividad* and FEDER under project TEC2011-22987 and by the 'Project Cero 2011 on Ageing' from *Fundación General CSIC, Obra Social La Caixa* and CSIC. A. Bachiller was in receipt of a PIF-UVa grant from University of Valladolid.

References

1. Weinberger DR, Egan MF, Bertolino A et al. (2001) Prefrontal neurons and the genetics of schizophrenia. Biol Psychiatry 50:825–844
2. Friston KJ (1998) The disconnection hypothesis. Schizophr Res 30:115–125
3. van den Heuvel MP, Mandl RC, Stam CJ et al. (2010) Aberrant frontal and temporal complex network structure in schizophrenia: a graph theoretical analysis. J Neurosci 30: 15915–15926
4. Uhlhaas PJ, Singer W (2010) Abnormal neural oscillations and synchrony in schizophrenia. Nat Neurosci 11:100–113
5. Fallani FDV, Maglione A, Babiloni F et al. (2010) Cortical network analysis in patients affected by schizophrenia. Brain Topogr 23: 214–220
6. Micheloyannis S, Pachou E, Stam CJ et al. (2006) Small-world networks and disturbed functional connectivity in schizophrenia. Schizophr Res 87:60–66
7. Bassett DS, Bullmore ET, Meyer-Lindenberg A et al. (2009) Cognitive fitness of cost-efficient brain functional networks. Proc Natl Acad Sci 106:11747–11752
8. Liu Y, Liang M, Zhou Y et al. (2008) Disrupted small-world networks in schizophrenia. Brain 131:945–961
9. Yu Q, Sui J, Rachakonda S et al. (2011) Altered small-world brain networks in temporal lobe in patients with schizophrenia performing an auditory oddball task. Front Syst Neurosci 5:1–13
10. Nunez PL, Srinivasan R, Westdorp et al. (1997) EEG coherency: I: statistics, reference electrode, volume conduction, Laplacians, cortical imaging, and interpretation at multiple scales. Electroenceph Clin Neurophysiol 103:499–515
11. Stam CJ, van Straaten ECW (2012) The organization of physiological brain networks. Clin Neurophysiol 123:1067–1087
12. Rappelsberger P, Pfurtscheller G, Filz O (1994) Calculation of event-related coherence—a new method to study short-lasting coupling between brain areas. Brain Topogr 7:121–127
13. Boccaletti S, Latora V, Moreno Y et al. (2006) Complex networks: structure and dynamics. Phys Rep 424:175–308
14. Stam CJ, De Haan W, Daffertshofer et al. (2009) Graph theoretical analysis of magnetoencephalographic functional connectivity in Alzheimer's disease. Brain 132:213–224
15. Roach BJ, Mathalon DH (2008) Event-related EEG time-frequency analysis: an overview of measures and an analysis of early gamma band phase locking in schizophrenia. Schizophr Bull 34:907–926

Author: Alejandro Bachiller
Institute: Biomedical Engineering Group.
Street: Paseo de Belén, 15, 47011
City: Valladolid
Country: Spain
Email: alejandro.bachiller@uva.es

Spectral Parameters from Pressure Bed Sensor Respiratory Signal to Discriminate Sleep Epochs with Respiratory Events

Giulia Tacchino[1], Guillermina Guerrero[2], Juha M. Kortelainen[3], and Anna M. Bianchi[1]

[1] DEIB – Dipartimento di Elettronica, Informazione e Bioingegneria, Politecnico di Milano, Milan, Italy
[2] Engineering Faculty, Autonomous University of San Luis Potosí, San Luis Potosí, México
[3] VTT Technical Research Centre of Finland, Tampere, Finland

Abstract—The Pressure Bed Sensor (PBS), which is presented as a contactless sensor for physiological signals recording, allows the acquisition of respiration movements signal. The aim of the present study is to identify spectral parameters from the PBS respiratory signal that allow the discrimination between normal and abnormal breathing epochs.

The nasal airflow and the PBS respiratory signal acquired on 19 subjects were pre-processed in order to obtain their positive envelope signals. Both of them were analyzed by means of an optimized Time-Variant Autoregressive Model (TVAM). Total sleep time was divided into consecutive epochs of 60 s classified as normal and abnormal (at least one apnea or hypopnea). The mean Power Spectral Density (PSD) for each sleep epoch was estimated from the averaged TVAM coefficients. Spectral features were extracted from both the nasal airflow and the PBS respiratory signal. A statistically significant difference (p-value<0.01) between normal and abnormal breathing epochs has been found in all the considered spectral features for both the nasal airflow and the PBS respiratory signal. These results suggest that the discrimination between normal and abnormal breathing epochs is thus possible by using parameters obtained from an easy-to-use, comfortable and non-obtrusive system for sleep monitoring, such as the Pressure Bed Sensor.

Keywords—Sleep-Disordered Breathing (SDB), Pressure Bed Sensor (PBS), non-obtrusive system, Time-Variant Autoregressive Model (TVAM), spectral features.

I. INTRODUCTION

Sleep-Disordered Breathing (SDB) includes a broad array of disorders characterized by abnormalities during sleep of respiratory pattern and ventilation, such as apneas and hypopneas [1]. The assessment of SDB usually requires the recording of the nasal airflow and other physiological signals (polysomnographic examination, PSG) during an overnight investigation in specialized sleep centers [2].

In previous works [3, 4] the nasal airflow signal positive envelope has been investigated to characterize periodic breathing patterns. Normal breathing frequency ranges from 12 to 20 breaths per minute (i.e., 0.20 Hz ... 0.33 Hz), whereas the periodic breathing pattern is characterized by cycle lengths between 25 and 100 s (i.e., 0.01 – 0.04 Hz). Starting from this assumption, spectral features from the Power Spectral Density (PSD) of the positive envelope signal have been computed. A discrimination between healthy subjects and patients suffering from SDB was achieved by accurately combining spectral parameters computed in the frequency band centered around 0.025 Hz. Despite the promising result, a nasal cannula, that is surely intrusive and uncomfortable for the patient, is required for the nasal airflow signal recording.

An interesting alternative for the acquisition of a signal related to respiration is a Pressure Bed Sensor (PBS), which is presented as a contactless sensor for physiological signals recording, with a higher comfort for the patient [5]. Indeed, pressure sensors integrated into a bed mattress allow the non-obtrusive recording of the multichannel ballistocardiographic (BCG) signals, which enable the analysis of heart beat intervals, respiration and body movements [6]. In particular the respiratory signal can be extracted from the BCG signals by applying an adaptive principal component analysis (PCA) model as described in [7, 8].

The aim of the present study is to demonstrate that the above-mentioned spectral parameters derived from the PBS respiratory signal assume statistically different values in sleep epochs with and without respiratory events, thus the PBS can be a non-obtrusive tool for the SDB screening.

II. MATERIALS AND METHODS

A. Recording Protocol

The signals analyzed in the present study were acquired at the Sleep Centre of Tampere University Hospital on 23 subjects, of age between 49 and 68 years, with suspected SDB. For a whole night, all the patients slept on the Pressure Bed Sensor (PBS) and heart beat intervals, respiratory signal and movement activity were recorded [5, 6]. Simultaneously, all participants underwent a full overnight PSG with the acquisition of several signals including nasal airflow. The PSG recordings were scored through the Somnologica software (Medcare, Reykjavik, Iceland) for sleep stages and respiratory events (apneas and hypopneas) identification, providing the reference for the present study. The recordings of 4 subjects were discarded because of the absence of the nasal airflow signal among the acquired ones.

B. Signals Pre-processing

For each one of the remaining 19 subjects, only nasal airflow from PSG and respiratory signal derived from PBS were analyzed. The former signal, sampled at 200 Hz, was anti-aliasing low-pass filtered and then down sampled at 5 Hz, to the same sampling frequency with the PBS respiratory signal. Then, both signals were high-pass filtered in order to remove their baseline: the cut off frequency was set to 0.08 Hz.

As previously done in [3], we used the respiratory signal positive envelope obtained through the identification of the absolute maximum in each inspiratory cycle. Because of the absence of a regular sampling in the positive envelope, this time series was linearly interpolated in order to obtain a sampling frequency of 1 Hz. Lastly, the signals were normalized by subtracting the minimum value and dividing by the maximum-minimum difference. This kind of operation is necessary to directly compare all the signals: in fact, after normalization, they all vary between 0 and 1.

C. Features Extraction

For each subject, we have divided the total sleep time into consecutive 60 seconds-long epochs labeled according to the number of respiratory events (apneas, hypopneas) falling within them. In general an epoch was labeled with the number 1 only if at least the 60% of an event fell within the epoch, otherwise it was labeled with the number 0.

Positive envelope signals extracted from nasal airflow and from PBS respiratory signal were analyzed by means of a Time-Variant Autoregressive Model (TVAM). It was optimized by using a variable forgetting factor (Fortescue) and a method of robustness for AR recursive identification [9]. According to the Akaike Information Criterion (AIC) and to the Anderson's whiteness test, the model order was set to 5. The forgetting factor was allowed to vary in time, following the signal dynamics, from a minimum of 0.965 to a maximum of 0.995. A set of 5 model coefficients was then computed for each sample of the time series; the coefficients that fell within each sleep epoch were then averaged.

The mean PSD was estimated for each 60 seconds-long sleep epoch from the averaged coefficients and then decomposed into its single spectral components, according to the method described in [10]. The frequency and power values associated to rhythmic components in the frequency band ranging from 0.01 Hz to 0.04 Hz were calculated.

For each sleep epoch, 5 spectral features were computed:

- *Tp*: total power in absolute units
- *p*: power in the frequency band ranging from 0.01 Hz to 0.04 Hz in absolute units
- *p%*: power in the frequency band ranging from 0.01 Hz to 0.04 Hz divided by the total power and expressed as percentage
- *pole module*: absolute value of the spectrum pole in the frequency band ranging from 0.01 Hz to 0.04 Hz
- *pole frequency*: frequency associated to the spectrum pole in the frequency band ranging from 0.01 Hz to 0.04 Hz

Fig. 1 represents an example of mean PSD estimated for one sleep epoch of 60 s (panel a), and the corresponding TVAM pole diagram (panel b). Four of the above-mentioned spectral features are also shown in Fig. 1.

D. Statistical Analysis

Spectral features, belonging to all subjects, connected to normal breathing epochs were compared to the ones connected to abnormal breathing epochs. After verifying that the two groups were not normally distributed, by means of a Kolmogorov-Smirnov test, a Kruskal-Wallis one-way analysis of variance was applied in order to highlight which features could significantly discriminate between normal and abnormal breathing epochs (epochs with apneas and hypopneas).

The statistical analysis was done separately for the nasal airflow signal positive envelope and for the PBS respiratory signal positive envelope.

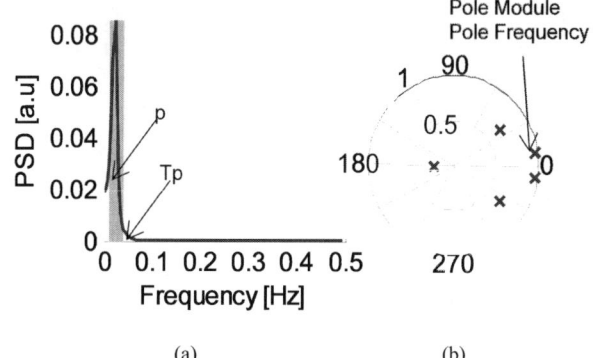

Fig. 1 (panel a) an example of mean PSD (blue line) for a sleep epoch of 60 s. The orange stripe indicates the frequency band ranging from 0.01 Hz to 0.04 Hz. Total Power (*Tp*) is the area under the blue line, while *p* is the area under the blue line that falls within the orange stripe. (panel b) the corresponding TVAM poles diagram. The complex and conjugate poles falling within the frequency band of interest are used to compute the spectral features *pole module* and *pole frequency*. 180 in the unitary circle corresponds to the half of the sampling frequency, that is 0.5 Hz.

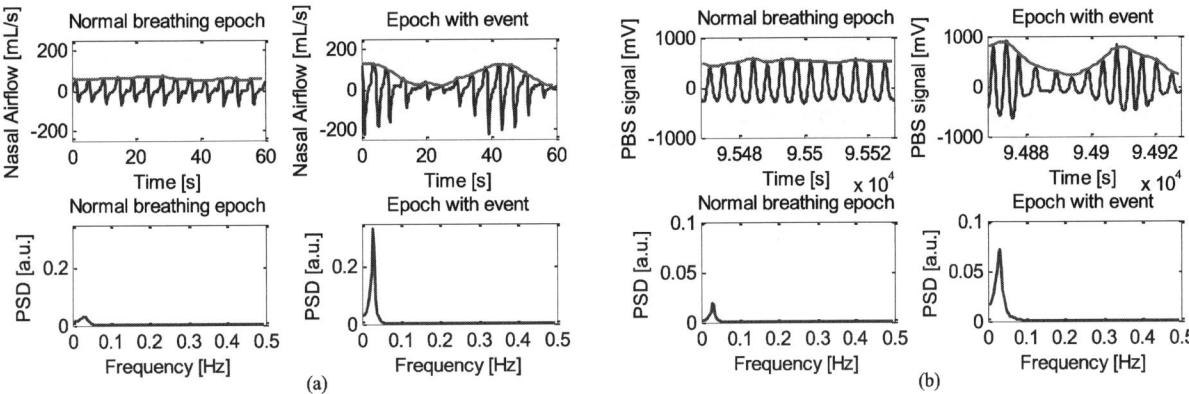

Fig. 2 A segment of nasal airflow signal (blue line) with its positive envelope (red line) from a normal breathing epoch of 60 s (left upper row) and from an epoch of 60 s with an apnea (right upper row). The PSD of each positive envelope segment is shown beneath the corresponding signal (panel a). A segment of PBS respiratory signal (blue line) with its positive envelope (red line) from a normal breathing epoch of 60 s (left upper row) and from an epoch of 60 s with an apnea (right upper row). The PSD of each positive envelope segment is shown beneath the corresponding signal (panel b).

III. RESULTS

Fig. 2 (panel a) upper row shows the nasal airflow signal from PSG and its derived positive envelope in a normal breathing epoch and in an epoch with an apnea, respectively. In the second row the PSDs of the positive envelope signals are shown. Fig. 2 (panel b) upper row represents the PBS respiratory signal and its derived positive envelope in a normal breathing epoch and in an epoch with an apnea, respectively. In the second row the PSDs of the positive envelope signals are shown.

Referring to nasal airflow signal, the median value and the interquartile range (iqr = 75th percentile - 25th percentile) of each spectral feature computed on all abnormal breathing epochs, compared with normal breathing epochs, are reported in Table 1. Similarly, referring to PBS respiratory signal, the median value and the iqr of each spectral feature computed on all abnormal breathing epochs, compared with normal breathing epochs, are reported in Table 2. Power is measured in absolute units because of the positive envelope signal normalization, while frequency associated to the spectrum pole is expressed in Hz. All the features showing a p-value < 0.01 are marked with an asterisk.

IV. DISCUSSION

Normal breathing cycle ranges from 12 to 20 breaths per minute, corresponding to a mean frequency of 0.25 Hz, whereas the breathing amplitude pattern in the presence of a respiratory event (apnea or hypopnea) reveals cycle lengths between 25 s and 100 s, corresponding to a mean frequency value of 0.025 Hz [3]. As shown in Fig. 2 (panel a), the 60 seconds-long segment of the nasal airflow positive envelope signal is clearly modulated by the respiratory event, resulting in a PSD with a power centered around 0.025 Hz much higher than in the normal breathing epoch. Similarly, as shown in Fig. 2 (panel b), also the 60 seconds-long segment of the PBS respiratory signal positive envelope is modulated by the respiratory event, resulting in a PSD with a power centered around 0.025 Hz higher than in the normal breathing epoch. A trivial visual examination of both panels in Fig. 2 reveals that the frequency and power values associated to rhythmic components in the frequency band ranging from 0.01 Hz to 0.04 Hz can be helpful in the discrimination between normal and abnormal breathing epochs. This is also shown in the statistical analysis reported in Tables 1 and 2. A statistically significant difference (p-value < 0.01) between normal and abnormal breathing epochs has been found in all of the considered spectral features for both the nasal airflow and the PBS respiratory signal.

Referring to the both Tables, power in the frequency band ranging from 0.01 Hz to 0.04 Hz (p) in abnormal breathing epochs is much higher than in normal breathing epochs because of the modulation of the positive envelope in the presence of respiratory events. Also the total power (Tp) is significantly higher in epochs with respiratory events, most probably due to the higher variability of the signal when apneas or hypopneas occur. Similar considerations can be done for the normalized power ($p\%$). The absolute value of the spectrum pole in the frequency band ranging from 0.01 Hz to 0.04 Hz (*pole module*) assumes higher values in abnormal breathing epochs rather than in normal breathing epochs, in that the higher the contribution of rhythmic components in the above-defined frequency band, the closer the pole to the unitary circle is. The frequency associated to the spectrum pole in the frequency band ranging from 0.01 Hz to 0.04 Hz (*pole frequency*) is also significant in discriminating epochs with respiratory events. As regards Table 2, the spectral features $p\%$ and *pole module* derived from the PBS respiratory signal are statistically significant in discriminating epochs with respiratory events, but the difference between the median values that they assume in normal and abnormal breathing epochs is not as high as in the corresponding features derived from the nasal airflow signal. This could be explained considering that the nasal airflow positive envelope corresponds with the inhalation phase of the respiratory cycle, while the pressure sensor signal recorded through the PBS does not directly show the inhalation or exhalation. However, the applied PCA model based respiration signal extraction method gives the correct polarity for the respiration phase in more than 90% of the epochs [8]. For this reason, all the spectral features derived from the PBS respiratory signal allow the discrimination between normal and abnormal breathing epochs as well as the spectral features derived from the nasal airflow signal.

Table 1 Nasal airflow spectral features median and (iqr) for normal and abnormal breathing epochs. p-values < 0.01 are marked with *.

	Normal epochs	Abnormal epochs
T_p [a.u.]	0.0033 (0.0057)	0.0096 (0.0114)*
p [a.u.]	0.0012 (0.0037)	0.0067 (0.0112)*
$p\%$	60.1 (87.9)	97.1 (56.9)*
pole module	0.86 (0.11)	0.90 (0.07)*
pole frequency [Hz]	0.034 (0.034)	0.032 (0.016)*

Table 2 PBS respiratory signal spectral features median and (iqr) for normal and abnormal breathing epochs. p-values < 0.01 are marked with *.

	Normal epochs	Abnormal epochs
T_p [a.u.]	0.0027 (0.0140)	0.0068 (0.0146)*
p [a.u.]	0.0019 (0.0117)	0.0064 (0.0140)*
$p\%$	97.7 (5.3)	98.4 (3.1)*
pole module	0.94 (0.04)	0.96 (0.03)*
pole frequency [Hz]	0.024 (0.009)	0.025 (0.007)*

V. CONCLUSION

A statistically significant difference between normal and abnormal breathing epochs has been found in all the considered spectral features for both the nasal airflow and the PBS respiratory signal. As a preliminary study, these results suggest that the discrimination between epochs with respiratory events and normal breathing epochs is thus possible by using parameters obtained from an easy-to-use, comfortable and non-obtrusive system for sleep monitoring, such as the Pressure Bed Sensor.

ACKNOWLEDGMENTS

This work was partially supported by the HeartCycle Project ICT FP7 216695 of the European Community and by the project 2010-1351. Development of a Technology Platform between the South and the North of Europe: Exchange Research Program between Politecnico di Milano and VTT—Technical Research Centre of Finland (S2N)).

REFERENCES

[1] C. Morgenstern, M. Schwaibold, W. J. Randerath, A. Bolz and R. Jané, "An Invasive and a Noninvasive Approach for the Automatic Differentiation of Obstructive and Central Hypopneas," *Biomedical Engineering, IEEE Transactions on*, vol. 57, pp. 1927-1936, 2010.

[2] K. Kesper, S. Canisius, T. Penzel, T. Ploch and W. Cassel, "ECG signal analysis for the assessment of sleep-disordered breathing and sleep pattern," *Med. Biol. Eng. Comput.*, vol. 50, pp. 135-144, 2012.

[3] A. Garde, B. Giraldo, R. Jane, I. Diaz, S. Herrera, S. Benito, M. Domingo and A. Bayes-Genis, "Analysis of respiratory flow signals in chronic heart failure patients with periodic breathing," in *Engineering in Medicine and Biology Society, 2007. EMBS 2007. 29th Annual International Conference of the IEEE*, 2007, pp. 307-310.

[4] A. Garde, B. F. Giraldo, R. Jane and L. Sornmo, "Time-varying respiratory pattern characterization in chronic heart failure patients and healthy subjects," in *Engineering in Medicine and Biology Society, 2009. EMBC 2009. Annual International Conference of the IEEE*, 2009, pp. 4007-4010.

[5] J. M. Kortelainen, M. O. Mendez, A. M. Bianchi, M. Matteucci and S. Cerutti, "Sleep Staging Based on Signals Acquired Through Bed Sensor," *Information Technology in Biomedicine, IEEE Transactions on*, vol. 14, pp. 776-785, 2010.

[6] M. Migliorini, A. M. Bianchi, D. Nisticò, J. Kortelainen, E. Arce-Santana, S. Cerutti and M. O. Mendez, "Automatic sleep staging based on ballistocardiographic signals recorded through bed sensors," in *Engineering in Medicine and Biology Society (EMBC), 2010 Annual International Conference of the IEEE*, 2010, pp. 3273-3276.

[7] G. Guerrero-Mora, P. Elvia, A. M. Bianchi, J. Kortelainen, M. Tenhunen, S. L. Himanen, M. O. Mendez, E. Arce-Santana and O. Gutierrez-Navarro, "Sleep-wake detection based on respiratory signal acquired through a Pressure Bed Sensor," *Engineering in Medicine and Biology Society (EMBC), 2012 Annual International Conference of the IEEE*, pp. 3452-3455, 2012.

[8] J. M. Kortelainen, M. van Gils and J. Parkka, "Multichannel bed pressure sensor for sleep monitoring," in *Computing in Cardiology (CinC), 2012*, 2012, pp. 313-316.

[9] G. Tacchino, S. Mariani, M. Migliorini and A. M. Bianchi, "Optimization of time-variant autoregressive models for tracking REM - non REM transitions during sleep," in *Engineering in Medicine and Biology Society (EMBC), 2012 Annual International Conference of the IEEE*, 2012, pp. 2236-2239.

[10] G. Baselli, A. Porta, O. Rimoldi, M. Pagani and S. Cerutti, "Spectral decomposition in multichannel recordings based on multivariate parametric identification," *Biomedical Engineering, IEEE Transactions on*, vol. 44, pp. 1092-1101, 1997.

Author: Giulia Tacchino
Institute: Politecnico di Milano
Street: Piazza Leonardo da Vinci, 32
City: Milano
Country: Italy
Email: giulia.tacchino@mail.polimi.it

Wavelet Energy and Wavelet Entropy as a New Analysis Approach in Spontaneous Fluctuations of Pupil Size Study – Preliminary Research

W. Nowak, E. Szul-Pietrzak, and A. Hachol

Institute of Biomedical Engineering and Instrumentation, Wroclaw University of Technology, Wrocław, Poland

Abstract—The wavelet transform technique was used to analyze the spontaneous fluctuations in pupil size behavior measured by dynamic pupillometry. The fluctuations show a wide frequency scale and their periods varied in time. Within the frequency range studied, 0–~2 Hz, the characteristic oscillations were revealed, arising from both local and central regulatory mechanism. The study illustrates the potential of dynamic pupillometry combined with dynamical system analysis for studies of pupil dynamic mechanism.

Keywords—dynamic pupillometry, spontaneous fluctuations of pupil size, time-frequency analysis, wavelet transform.

I. INTRODUCTION

Spontaneous fluctuations of pupil size (SFPS) – i.e. random movements and changes of pupil size, even in constant stimuli conditions, reflect the autonomous nervous system activity, the cognitive function and emotional state [1,2,3].

The need to develop a low-cost pupillometry system and a new SFPS analysis method is an effect of promising results of research concerning the use of pupil dynamic study for noninvasive monitoring the physiological systems controlled by ANS [4].

So far, the quantitative characterization of the pupil dynamic in concerned mainly as sleepiness indicator [5,6] and the Fourier transform (FT) is the most widely used method for analyzing their dynamics [7]. However, the power spectrum does not capture the complexities of the fluctuations since it requires stationarity of the signal and it does not give the time evolution of the frequency pattern. SFPS series are nonstationary because their statistical moments (means and variances) are not constant over time. The presence of non-stationary dynamics in pupil behavior suggest the necessity to study this system using another methods and a new quantitative parameters.

The short-time Fourier transform (STFT) approach designed by our group was presented in [8]. With this approach the Fourier transform is applied to time-evolving windows and then the evolution of the frequencies can be followed and the stationary requirements are partially satisfied. However, important limitation of this method is the uncertainty principle.

To overcome this limitation a time evolving parameter can be defined from a time-frequency representation of the signal as provided by the wavelet-transform (WT) [9]. The WT makes no assumptions about record stationarity and the only input needed is the time series. In this case, the time evolution of frequency patterns can be followed with an optimal time-frequency resolution. Therefore, wavelet entropy based on the WT reflects the degree of order/disorder of the signal and is treated as a physical measure to estimate system complexity. A low value of WE indicates predictability, regularity or a less complex state, whereas high WE indicates unpredictability, irregularity and greater complexity. WT has been used mainly in the analysis of EEG signals [10].

The most important aim of this work was to demonstrate the applicability of WT to characterize in a quantitative way functional dynamics of order/disorder states in SFPS signals.

II. MATERIAL AND METHODS

A. Subjects and Experimental Set-up

Three volunteers (females, aged between 30 and 37 years) took part in the experiment. All participants were in good health, without any eye disease and they were non-smokers. SFPS was recorded using laboratory version of the pupillometr equipment at a sampling rate of 60 Hz. The pupillometer is a monocular device enabling recording the size and geometrical centre of the pupil, in response to light and accommodation stimulations as well as spontaneous pupil size fluctuations (pupil behavior in constant light-accommodation conditions). The pupillometer is composed of a PC and measurement head which is consisted of stimuli module (SM), recording module (RM) and control-analysis module (CM).

The measurement head is supported on an ophthalmologic slit lamp base equipped with a joystick control (up/down, forward/backward and left/right). A standard adjustable ophthalmic forehead rest and a chin rest are also fixed to the slit lamp base and provide adequate stabilization of the subject's head during the measurement. The scheme of the system is shown in Fig.1.

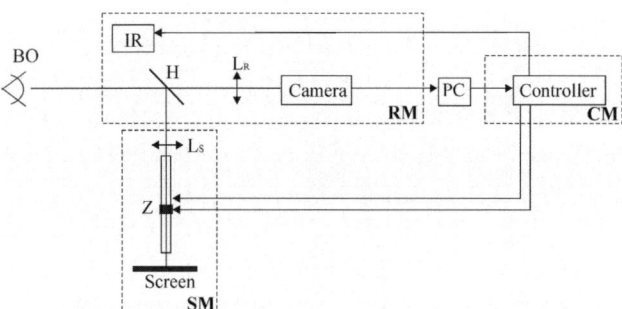

Fig.1 The scheme and the photo of the pupillometr

It is worth to emphasize that luminance level and accommodation level during spontaneous pupil size fluctuation should also be strictly controlled.

In project of the stimuli module the following points had to be considered: (1) Near/Light needs to be stimulated in the eye being imaged, (2) Near/light stimulus system must not interfere with the primary optical path and (3) The level of near/light must be accurately controlled.

To light stimuli, the system is equipped with a white LED diode or alternatively with RGB LED diode which is used as a source of chromatic light. The stimuli light passes through the lens L_S, falls on a dichroic mirror H, and then is projected onto the tested eye. It is so called the Maxwellian-view, i.e. a beam of light stimuli falls on the center of the pupil, and the diameter of the light beam is smaller than the minimum diameter of the pupil. The stimuli parameters are adjusted by operator utilizing the designed software and they can be regulated as follow: wavelength, luminance level and waveform.

To near stimuli, the system incorporate a moveable white LED diode (Z) which is used as a source of fixation point moveable (by means of a simple slider) between near (N) and far (F) positions. The diode is chosen to be small and dimly lit compared to the background illumination from screen so that changes in its position will not elicit light reflex in subject. The focal length of accommodation lens is greater than both N and F, the distances between L_S and the near and far fixation point positions. Subject will at all times view virtual images of Z which are upright and magnified. The virtual position of Z as it is moved between the near and far positions will depend on the focal length of L_S. In the pupillometer the focal length for L_S is 100 mm. Then if N and F are respectively 20 mm and 99 mm, the apparent distance of the fixation point (ie. The virtual image distance in respect of L_S) can be varied within the range 25 mm to 9.9 m.

Before initiating a measurement series, all subjects participated in a training session concerning the pupillometric measuring procedure. At the beginning measurements started with adapting the subject to darkness (5 min) and calibrating the system to individual characteristics of the subject. A subject was asked to keep her eyes open, look at the fixation point (set to 4m), and avoid blinking and head movements during the recording procedure. Subjects were also asked to avoid drinking coffee and alcohol for 24 hours before the measuring session. Measuring sessions were conducted between 6:00 am and 8:00 pm hour. Each measurement lasted for 42.6 seconds, and the signal length used for the analysis contained 2560 samples.

B. Analysis Method

Wavelet decomposition is the basis for the calculation of wavelet energy and wavelet entropy. In order to calculate those parameters of the signal $\{x(t)\}=x(1), x(2), ..., x(N)$, one should follow these steps:
Perform the signal decomposition using wavelet transform

$$x(t) = \sum_{k=1} A_j(k)\, \varphi_{j,k}(t) + \sum_j \sum_k C_j(k)\psi_{j,k}(t) \qquad (1)$$

Where $C_j(k)$ are the discrete wavelet transform coefficients at resolution level j and time k.
Calculate the energy at each resolution level $j=-1, ..., -N$

$$E_j = \sum_k |C_j(k)|^2 \qquad (2)$$

Calculate the total energy

$$E_{tot} = \sum_j E_j \qquad (3)$$

Calculate the relative wavelet energy, which can be considered as a probability distribution of energy across different decomposition levels

$$p_j = \frac{E_j}{E_{tot}} \qquad (4)$$

The wavelet entropy is defined as

$$WE\,(p) = -\sum_j p_j \cdot \ln[p_j] \qquad (5)$$

In order to study temporal evolution, the analyzed signal is divided into nonoverlapping time windows of length L=512 data points = 8.53 seconds. WT was applied to each single sweep using 'db2' function as mother wavelet. Before applying the wavelet transform, the linear trend was removed from the signal.

In this work, we chose nine frequency bands associated with resolution levels appropriate to the wavelet analysis in the scheme of multiresolution. The level j is associated with a frequency band, ΔF, obtained in the following way:

$$2^{j-1}F_s \le \Delta F \le 2^j F_s \qquad (6)$$

where Fs is the sampling frequency.

We denoted these band-resolution levels by β_j (/j/=1,..,9) and their frequency limits are given in Table 1.

Table 1 The frequency bands associated with the different resolution wavelet levels according with sample frequency 60 Hz.

Notation	ω_{min}	ω_{max}	Resolution level
β_1	15	30	-1
β_2	7.5	15	-2
β_3	3.8	7.5	-3
β_4	1.9	3.75	-4
β_5	0.95	1.87	-5
β_6	0.47	0.94	-6
β_7	0.23	0.47	-7
β_8	0.12	0.23	-8
β_9	0.06	0.12	-9

The frequency band between 0.06 and 1.87 Hz was studied. The 1.87Hz was chosen as an upper frequency limit, since in the spectral analysis of SFPS the higher harmonic components were only sometimes found above 2 Hz.

III. RESULTS

Figures 2 (a)-(c) show exemplary three series of SPSF signals and their histograms recorded for the same subject but in different psycho-physical conditions (rest or tired subject). Panel A shows a normal signal with many small both accelerations and decelerations. Panel B shows a reduced signal variability. Panel C shows an abnormal signal with many large decelerations.

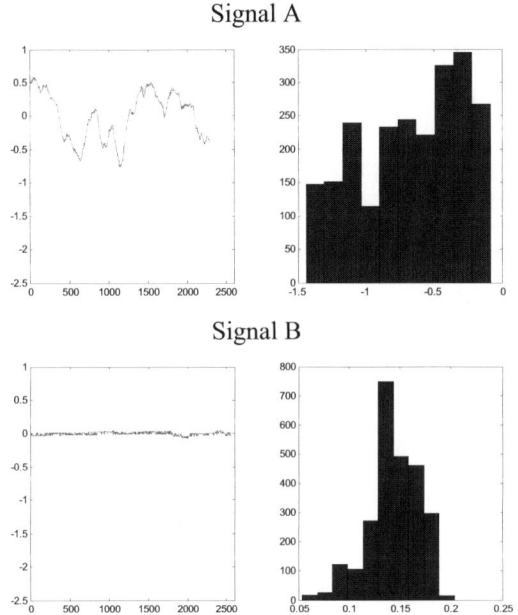

Fig.2. The exemplary three series of SPSF signals and their histograms.

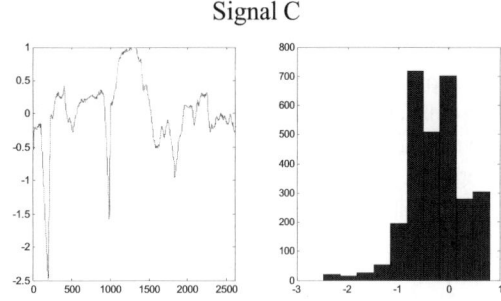

Fig.2 (*continued*)

From Eq. (4), we calculate the energy distribution as a function of the frequency level, j, for all three signals. Table 2 give a comparison of the percentage of energy distribution for the different frequency bands for the three series. Table 3 presents a comparison between the percentage of entropy distribution calculated from Eq. (5).

Table 2. Percentile distribution of the energy for SPSF as a function of the frequency bands.

j	Signal A	Signal B	Signal C
-1	0,0000	0,0008	0,0000
-2	0,0000	0,0052	0,0000
-3	0,0001	0,0158	0,0001
-4	0,0003	0,0197	0,0006
-5	0,0016	0,0322	0,0074
-6	0,0177	0,0355	0,0416
-7	0,1536	0,0762	0,2050
-8	0,3477	0,3143	0,2365
-9	0,4789	0,5004	0,5133

Table 3. Percentile distribution of the entropy for SPSF as a function of the frequency bands.

j	Signal A	Signal 2	Signal 3
-1	0,0001	0,0055	0,0000
-2	0,0002	0,0268	0,0001
-3	0,0007	0,0636	0,0006
-4	0,0022	0,0746	0,0045
-5	0,0103	0,1059	0,0330
-6	0,0684	0,1154	0,0939
-7	0,2280	0,1863	0,2466
-8	0,2516	0,2196	0,3299
-9	0,2674	0,1658	0,3169

The analysis of the obtained results exhibits the following features:

- the major part of the energy is carried by two principal frequency bands corresponding to j=-8 and j=-9 for all analyzed signals,
- the medium and high frequencies (more than or equal to j=-5) are negligible,
- the energy distribution spectrum of the SFPS is similar for the regular (rest) and non-regular (tired) conditions, but the first one show more "wide band" energy spectrum,
- the greater pupil size oscillation shows greater energy and entropy values.

IV. CONCLUSIONS

In this study we applied a wavelet energy and wavelet entropy to characterize the pupil dynamics during spontaneous pupil size fluctuations. We have analyzed three example series recorded for rest and tired healthy subject. Time series which were analyzed represent complex papillary system dynamics, because it shows the fluctuations in the activity of the symphathetic and parasymphathetic innervations of the iris muscles. And that is regulated via neural feedback mechanisms including various types of receptors.

As has been shown in the tables and figures, the WT provides a very powerful technique to detect signal changes by observing changes in the energy spectra and entropy spectra of the series. Results from using WE can be summarized as follows: in rest state the system has lower WE values - the normal state is more ordered condition.

Our results confirm that the possibilities offered by the wavelet transform provides enable the detection and quantification of differences in the spontaneous pupil size behavior. Further studies will aim at determining the possibility of using this method of analysis as a diagnostic technique for distinguishing different types of neuro-ophthalmological diseases.

ACKNOWLEDGMENT

This work was supported by the Polish Ministry of Science and Higher Education (Grant No. 405338)

REFERENCES

1. Calcagnini G, Censi S, Lino S et al. (2000) Spontaneous fluctuations of human pupil reflect central autonomic rhythms. Methods Inf Med, 39:142-145
2. Partala T, Surakka V.(2003) Pupil size variation as an indication of affective processing. Int J Hum-Comput Stud, 59: 185-198
3. Dioniso D P, Granholm E, Hillix W A, Perrine W F (2001) Differentiation of deception using papillary responses as an index of cognitive processing. Psychophysiology, 38: 205-211
4. Ferrari G L, Marques , J L B, Gandhi RA et al. (2010) Using dynamic pupillometry as a simple screening tool to detect autonomic neuropathy in patients with diabetes: a pilot study. Biomed Eng, On-Line, 9:26
5. McLaren J W, Hauri P J, Lin S C, Harris C D. (2002) Pupillometry in clinically sleepy patients, Sleep Med, 3: 347-352
6. Wilhelm H, Lüdtke H, Wilhelm B (1998) Pupillographic sleepiness testing in hypersomniacs and normals. Graefes Arch Clin Exp Ophthalmol, 236:725-729
7. Lüdtke H, Wilhelm B, Adler M et al. (1998) Mathematical procedures in data recording and processing of papillary fatigue waves. Vision Res, 38: 2889-2896
8. Nowak W, Hachol A, Kasprzak H. (2008) Time-frequency analysis of spontaneous fluctuation of the pupil size of the human eye. Opt Applicata, 382
9. Rosso O, Blanco S, Yordanowa J et al (2001) Wavelet entropy: a new tool for analysis of short duration brain electrical signals. J Neurosci Methods, 105: 65-75
10. Quian Quiroga R, Rosso O.A., Basar E et al (2001) Wavelet entropy in event-related potentials: a new method shows ordering of EEG oscillations. Biol Cybern, 84: 291-299

Author: Wioletta Nowak
Institute: Institute of Biomedical Engineering and Instrumentation
Street: Wybrzeze Wyspainskiego 27
City: Wroclaw
Country: Poland
Email: wioletta.nowak@pwr.wroc.pl

Rhythm Extraction Using Spectral–Splitting for Epileptic Seizure Detection

L.M. Sepúlveda[1], J.D. Martínez[1], L. Duque[2], C.D. Acosta[1], and G. Castellanos[1]

[1] Universidad Nacional de Colombia/Signal Processing and Recognition Group, Manizales, Colombia
[2] Instituto Tecnológico Metropolitano, Medellin, Colombia

Abstract—EEG recordings contain dynamic information inherent to its nature, therefore, the accurate estimation of patologies such as epilepsy, is highly dependent on the inclusion of such information into an automatic diagnosis system. In the present work, several approaches based on *t–f* spectral splitting by means of stochastic variability technique, which are aimed to extract dynamic information of EEG recordings for epileptic seizures detection are proposed and compared. Results show accuracy rates over **94 99%** for a two-classes problem, and over **92%** over a five classes problem.

I. INTRODUCTION

The electroencephalogram (EEG) is the most specific method to define epileptogenic cortex [1]. The EEG is an electrical potential recording in the form of synchronous discharges generated by the cerebral cortex nerve cells. The scalp EEG detects the neuronal activity by electrodes attached to the scalp. The electroencephalographic (EEG) signals represent the clinical signs of the synchronous activity of the neurons in the brain, but in case of epileptic seizures, there is a sudden and recurrent mal–function of the brain that exhibits considerable short–term non–stationarities that can be detected analyzing these recordings [2]. Thus, for example, the possibility to automatically detect epileptic seizures from EEG signals is limited by the wide variety of frequencies, amplitudes, spikes, and waves that use to appear along time with no precise localization [3]. Thereby, the performance of automatic decision support systems depends of features parameterizing accurately the non–stationary behaviors. Thus, a current challenging problem is to detect a variety of non–stationary biosignal activities with a low computational complexity, to provide tools for efficient biosignal database management and annotation.[1]

Due to the activity of brain cells, brain–waves (termed *rhythms*) are generated having a frequency of around $3-30$ Hz [4]. In order to obtain clinically interpretable results for medicine, δ, θ, α and β frequency band activities of EEG signals have to been investigated [5]. The meaning of each

one of these waveforms is explained in [3] as follows: $i)\alpha$: is the most important waveform of the EEG and comprises frequencies between $8-12$ Hz. This wave appears when the eyes are closed and one begins to rest, $ii)$ δ, θ: these waves should be seen when the epileptic seizures occurs and have higher amplitudes and lower frequencies respect to α waves ($0-4$ Hz and $4-8$ Hz, respectively), and $iV)\beta$: in addition, brain produces desynchronize waves, which have higher frequency ($13-30$ Hz.) and lower magnitude, called β waves.

The time–frequency (t–f) representations has been proposed, planned to determine the energy distribution along the frequency axis at each time instant, with aiming to investigate the time–variant properties of the spectral parameters during either transient physiological or pathological episodes [6]. In this line of analysis, the use of relevant and non–redundant dynamic features is presented in [7] based on the STFT; nevertheless, there are some normal *t–f* maps whose waveform resembles like pathological ones, and vice versa; so, the spatial distribution of the energy in each sub–band is not clear. In [8] a spectral splitting is proposed for stochastic features extraction based on the Short Time Fourier Transform (STFT); different relevance measures are proved for boundaries selection. However, the study of the behavior of the signal using different decomposition techniques with variant argument in the time and the analysis of the incidence of each decomposition in the classification task is still an open problem.

Present study proposes a methodology aiming to find the more suitable technique for epileptic seizures detection. Particularly, two approaches are proposed: *a*) Epileptic seizure detection using dynamic features extracted directly from the EEG decomposition, *b*) epileptic seizure detection based on dynamic relevance over the EEG TFR maps, *c*) epileptic seizure detection using the knowledge about the EEG–rhythms, process them individually with the best scheme.

II. METHODS

A. Enhanced t-f Representation

Particularly, the short time Fourier transform is employed introducing a time localization concept by means of a tapering window function of short duration, ϕ, that is going along

---

[1] This work was carried out under grants: *"Programa Nacional de Formacion de Investigadores "GENERACION DEL BICENTENARIO", 2011*, research project 111045426008 funded by COLCIENCIAS, and s PhD. scholarship founded by Universidad Nacional de Colombia.

the underlying biosignal, $y(t)$, as follows:

$$S_y(t,f) = \left| \int_T y(\tau)\phi(\tau-t)e^{-j2\pi f\tau}d\tau \right|^2, \quad S_y(t,f) \in \mathbb{R}^+ \quad (1)$$

with $t, \tau \in T, f \in F$.

Based on introduced *spectrogram* of Eq. (1), the corresponding t–f representation matrix, $S_y \in \mathbb{R}^{T \times F}$, can be described by the row vectors, $S_y = [s_1 \ldots s_f \ldots s_F]^\top$, with $s_f \in \mathbb{R}^{1 \times T}$, where vector $s_f = [s(f,1) \ldots s(f,t) \ldots (f,T)]$, with $s(f,t) \in \mathbb{R}$, is each one of the time–variant spectral decomposition components at frequency f, and equally sampled through the time axis.

B. Measure of Stochastic Variability

The amount of stochastic variability of the spectral component set is computed following the approach given in [8], that is based on time-variant decomposition estimated by adapting in time any of commonly used latent variable techniques, upon which a piecewise stationary restriction is imposed [7]. Namely, the time-evolving principal component analysis is extended to the dynamic feature modeling by stacking the input observation matrix in the following manner:

$$\Xi_y = \begin{bmatrix} s_1^1 & s_2^1 & \cdots & s_F^1 \\ s_1^2 & s_2^2 & \cdots & s_F^2 \\ \vdots & \vdots & \vdots & \vdots \\ s_1^M & s_2^M & \cdots & s_F^M \end{bmatrix}, \quad \Xi_y \in \mathbb{R}^{M \times FT} \quad (2)$$

where vector s_f^i corresponds to f-th short–term spectral component estimated from the i-th spectrogram matrix S_y^i which is related to the i-th object, with $i \in M$, where M stands for the total number of observations. Consequently, the truncated singular values decomposition takes place:

$$\Xi_y = \widetilde{U}\widetilde{\Lambda}\widetilde{V}^\top = \sum_{i=1}^{q} \lambda_i u_i v_i^\top, \quad (3)$$

where matrix $\Lambda = \mathrm{diag}(\lambda) \in \mathbb{R}^{q \times q}$ is the singular values matrix, $U = [u_1, \ldots, u_q] \in \mathbb{R}^{no \times q}$ are the eigenvectors of $XX^\top$, and $V = [v_1, \ldots, v_q] \in \mathbb{R}^{p \times q}$ corresponds to the eigenvectors of $X^\top X$. Accordingly, the amount of stochastic variability of the spectral component set is computed by the singular value decomposition calculation over observation matrix in Eq. (2). So, the following time–variant relevance measure is done [7]:

$$g(\Xi_y; \tau) = [\chi(1) \cdots \chi(\tau) \cdots \chi(FT)]^\top, \in \mathbb{R}^{FT \times 1} \quad (4)$$

being $\chi(\tau) = \sum_{i=1}^{q} \lambda_i^2 v_i^\tau$, where v_i is a vector computed by the square of each one of the elements of the τ–row of $\widetilde{V}$.

To determine distinctly the relevance related to each one of the time-variant spectral components, measure vector given in Eq. (4) can be arranged into a matrix, termed *relevance matrix*, as follows:

$$\Gamma(S_y) = [g(s_1) \ldots g(s_f) \ldots g(s_F)], \quad \in \mathbb{R}^{T \times F}, \quad (5)$$

holding stochastic variability, measured for the whole spectral component set, $\{s_f\}$. So, stationarity and non-stationarity components of a biosignal recording $i \in M$, can be derived from the amount of variability of its spectral components.

III. EXPERIMENTAL SETUP

A. Databases

Present study considers two different EEG datasets for epileptic seizure detection. The former, is the PhysioNet available collection consisting of five subsets (denoted Z, O, N, F, and S), each one composed of 100 single–channel EEG segments of 23.6-s duration [3]. Sets Z and O have been taken from surface (extracranial) EEG recordings of five healthy volunteers with eye open and closed, respectively. Signals from sets N and F have been measured in seizure–free intervals from five subjects in the epileptic zone and from the hippocampal formation of the opposite hemisphere of the brain. Set S comprises of epileptic signals recorded during seizure (ictal) from all recording sites. Sets N–S have been recorded intracranially. All EEG signals were digitized at 173.61 Hz with 12 *bit* resolution. In this work, the two-classes problem (Z, O and N, F, S) and five classes problem are considered.

The latter set is obtained from the database of the *Instituto de Epilepsia y Parkinson del Eje Cafetero*, holding the two subsets of Problem I. Each data set contains 160 recorded scalp EEG signals from 20 channels corresponding to the electrodes placed on the head according to the International 10 – 20 System of Electrode Placement Standard. Set Z holds 80 recordings labeled as normal (seizure–free), whereas set S holds 80 recordings labeled as having epilepsy (A neurologist reviewed the EEG to mark all epileptic events). Recordings had been sampled at a frequency of 256 Hz, with 12-*bit* resolution and a 2-*min* duration. All patients underwent clinical examination by a neurologist.

A1 Spectral Enhancement by Spectral Splitting

The algorithm 1 describes the provided spectral splitting along a given f axis by means of the introduced relevance measure. Figure 1 shows the behavior of the relevance for each point in the t–f maps. It has to be noted that the t–f analysis has been carried out by the spectrogram, within a range

Table 1 Frequency band selection

Frequency Bands	# Features
DB1	
0 − 10, 10 − 30, 30 − 40, 40 − 50, 50 − 83 [Hz]	5
DB2	
0 − 13, 13 − 25, 25 − 43.01, 43.01 − 56.2, 56.2 − 60.1, 60.1 − 73.9, 73.9 − 83 [Hz]	7

of 0 to 83 Hz, by using a gaussian window with lengths of 2.9 s and 0.05, for both databses, respectively. Besides, Table 1 shows the frequency bands that are chosen. Each band includes 1 time series of vector cepstral coefficients that is computed by 1 triangular response filter.

Input: $x(t)$
Output: Frequency bands
foreach *Class k* **do**
  foreach *Observation i* **do**
    1. Calculate S_y for t–f decomposition;
  end
end

1. Calculate Ξ_y and $g(\Xi_y, \tau)$;
2. Reshape the relevance vector into a relevance matrix $\Gamma(S_y) \in \mathbb{R}^{\Lambda \times T}$;
3. Calculate a column vector containing the mean value of each row of the relevance matrix $\in \mathbb{R}^{f \times 1}$;
4. Select the frequency bands where the relevance presents significant changes on its behavior i.e. the local minimums of the curve with the lowest variance;

Algorithm 1: Algorithm for the frequency bands selection by relevance analysis

According to the sub–band selection by spectral splitting, both representations of the EEG are decomposed using cepstral coefficients. In this process, for each time a triangular response filter is applied, obtaining for DB1, 5 stochastic features for spectrogram and 3 stochastic features for scalogram; and for DB2, 7 stochastic features for spectrogram and 3 stochastic features for scalogram.

B. Results of Time–Frequency Decomposition

Tuning of the different schemes is carried out by using the average classification accuracy for the automatic saeizure detection. Because of high computational cost of stochastic feature-based training, dimension reduction of the input space is carried out by means of a time–evolving version of the standard PCA, as in [9].

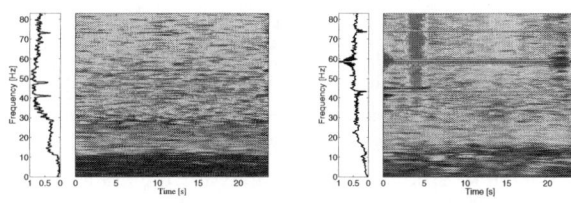

(a) Relevance map for spectrogram DB1 (b) Relevance map for spectrogram DB2

Fig. 1 Selection of the frequency bands using relevance analysis for EEG recordings.

Table 2 Classification accuracy

Approach	# Features	k–nn Acc [%]	SVM Acc [%]
DB1 Five–Class Problem			
Baseline	512	78.93 ± 2.72	86.53 ± 4.20
Spectral splitting	5	78.66 ± 2.35	90.44 ± 2.67
DB1 Two–Class Problem			
Baseline	512	96.33 ± 2.45	98.50 ± 1.22
Spectral splitting	5	99.00 ± 1.00	99.16 ± 1.60
DB2			
Baseline	512	82.00 ± 3.32	83.66 ± 2.17
Spectral splitting	7	94.04 ± 1.02	93.33 ± 1.76

Table 2 summarizes the performed classification accuracy for the different approaches and its respective set of stochastic features. As seen, there is no statistical difference in terms of classification performed by the decomposition technique (t–f and its respective reduced set in the case of five–class problem for DB1. Nevertheless, in the two–class problem for DB1 and for DB2, a significant improvement is achievement by spectral splitting.

C. EEG–Rhythms Analysis

For EEG–rhythms analysis, the parameters of the t–f are the same than for the EEG signal, but treating each rhythm as an individual signal. The dynamic features obtained are mixed in a k–nn–classifier array, aiming to preserve the different dynamics. Besides, Table 3 shows the results for different rhythms combination.

IV. DISCUSSION

This work proposes a comparison between different EEG signal decompositions and its respective enhancements through spectral–splitting for epilepsy events detection. Additionally, a methodology to evaluate the rhythms or brain–waves contribution to neuronal activity related to either normal or epileptic seizure events is proved. The methodology lies in the hypothesis that providing an appropriate spectral sub–band division, it is possible to detect ictal events when the rhythms information is not available.

Present study discusses the introduction of spectral splitting over the EEG decomposition to optimize the epileptic seizures detection, in terms of improving the classification accuracy with the lowest possible complexity but not increasing the computational effort) [8]. Both considered approaches of EEG decomposition achieve an small feature number (7 cepstral coefficients as maximum). In addition, if the rhythms are unknown, the discussed training methodology provides additional benefit of easier clinical interpretation of EEG–derived parameters through by the relevance map, principally in the two–class problem.

Table 3 Classification accuracy for rhythms analysis

Rhythms	k-nn Acc [%]	SVM Acc [%]	Rhythms	k-nn Acc [%]	SVM Acc [%]
DB1 Five–Class Problem					
δ	70.66 ± 4.28	79.60 ± 2.24	θ	75.40 ± 3.61	74.80 ± 2.07
α	68.33 ± 3.06	76.20 ± 2.24	β	74.40 ± 2.08	82.80 ± 2.10
δ, θ	84.33 ± 3.19	85.00 ± 2.95	θ, β	86.93 ± 2.49	87.66 ± 1.15
δ, θ, β	88.46 ± 2.91	90.83 ± 0.63	$\delta, \theta, \beta, \alpha$	89.33 ± 2.77	90.00 ± 3.31
DB1 Two–Class Problem					
δ	87.83 ± 3.77	97.33 ± 3.02	θ	89.33 ± 2.62	87.33 ± 5.34
α	84.33 ± 5.22	89.66 ± 6.12	β	93.33 ± 1.92	99.33 – 0.91
δ, θ	94.00 ± 3.35	98.75 ± 0.83	θ, β	96.33 ± 1.72	99.5 – 0.83
δ, θ, β	98.88 ± 1.10	97.91 ± 1.59	$\delta, \theta, \beta, \alpha$	99.00 ± 0.88	98.33 ± 2.35
DB2					
δ	72.75 ± 2.03	87.66 ± 3.36	θ	70.04 ± 3.12	81.33 ± 3.31
α	81.70 ± 2.47	90.33 ± 3.36	β	84.54 ± 2.60	91.16 ± 3.66
δ, θ	77.70 ± 2.92	82.58 ± 1.70	θ, β	86.83 ± 3.24	87.50 ± 2.33
δ, θ, β	87.87 ± 1.80	89.25 ± 1.22	$\delta, \theta, \beta, \alpha$	90.00 ± 3.88	92.22 ± 1.27

According to Table 3 in almost all cases, the rhythm β, i.e the rhythm with the higher frequency, has a significant impact in the classification accuracy. This behavior is due to the fact that this brainwave represents the non–synchronous waves that occur when an epileptic seizure has place [10]. Additionally, the β rhythm is associated with the preparation and inhibitory control in the motor system.

Concerning to the comparison between the best approach within and without the knowledge of the EEG rhythms, the classification accuracy is improved by the spectral splitting in the case of the two–class problem (99 ± 0.8 % vs 99 ± 1% for DB1 and 94 ± 1 % vs 90 ± 4% for DB2) is achieved.

V. CONCLUSION

The training methodology is explored, which is based on t–f spectral splitting by means of stochastic variability technique for epileptic seizure detection. Regarding to the dimensionality reduction by spectral splitting, the number of dynamic features needed for classification task is around 500 less than the baseline; in this way, the advantage of the method proposed in this paper to get a better estimation of the frequency bands of interest is evident, reaching higher accuracy rates (over 90%). As future work, further efforts finding an alternative for epileptic seizure diagnosing, having the added benefit of low cost and simplicity, should be focused on extended studies over another t–f approaches, as it is the case of Wavelet Transform.

REFERENCES

1. Noachtar Soheyl, Rémi Jan. The role of {EEG} in epilepsy: A critical review *Epilepsy Behavior.* 2009;15:22 - 33.
2. Tarvainen M.P., Georgiadis S., Lipponen J.A., Hakkarainen M., Karjalainen P.A.. Time-varying spectrum estimation of heart rate variability signals with Kalman smoother algorithm :1 -4 2009.
3. Tzallas A., Tsipouras M., Fotiadis D.. Epileptic Seizure Detection in Electroencephalograms using Time-Frequency Analysis *IEEE Transactions on Information Technology in Biomedicine.* 2009;13:To be appeared.
4. Sawant Hemant K., Jalali Zahra. Detection and classification of EEG waves *Oriental Journal of Computer Science and Technology.* 2010;3:207-213.
5. Kiymik M. Kemal, Güler Inan, Dizibüyük Alper, Akin Mehmet. Comparison of STFT and wavelet transform methods in determining epileptic seizure activity in EEG signals for real-time application. *Comp. in Bio. and Med..* 2005;35:603-616.
6. Sun M., Scheuer M., Sclabassi R.. Decomposition of biomedical signals for enhancement of their time–frequency distributions *Journal of the Franklin Institute.* 2000;337:453–467.
7. Sepulveda-Cano Lina, Acosta-Medina Carlos, Castellanos-Dominguez Germán. Relevance Analysis of Stochastic Biosignals for Identification of Pathologies *EURASIP Journal on Advances in Signal Processing.* 2011;2010.
8. Martínez-Vargas J.D., Sepulveda-Cano L.M., Travieso-Gonzalez C., Castellanos-Dominguez G.. Detection of obstructive sleep apnoea using dynamic filter-banked features *Expert Systems with Applications.* 2012;39:9118 - 9128.
9. Sepulveda-Cano L.M., Gil E., Laguna P., Castellanos-Dominguez G.. Selection of Nonstationary Dynamic Features for Obstructive Sleep Apnoea Detection in Children *EURASIP Journal on Advances in Signal Processing.* 2011;2011:10.
10. Akin M., Arserim M.A., Kiymik M.K., Turkoglu I.. A new approach for diagnosing epilepsy by using wavelet transform and neural networks in *Engineering in Medicine and Biology Society, 2001. Proceedings of the 23rd Annual International Conference of the IEEE*;2:1596 - 1599 vol.2 2001.

EEG Metrics Evaluation in Simultaneous EEG-fMRI Olfactory Experiment

Eva Manzanedo[1,2], Ana Beatriz Solana[2], Elena Molina[1,2], Ricardo Bruña[2], Susana Borromeo[1,2], Juan Antonio Hernández-Tamames[1,2,3], and Francisco del Pozo[2]

[1] Universidad Rey Juan Carlos, Móstoles, Madrid, Spain
[2] Center for Biomedical Technology, Universidad Politécnica de Madrid, Pozuelo de Alarcón, Madrid, Spain
[3] Fundación CIEN-Fundación Reina Sofía, Madrid, Spain

Abstract—This work analyses the statistical significant changes in the EEG signal after the supply of olfactory stimuli during simultaneous EEG/fMRI olfactory experiments. Different spectral parameters are evaluated including the spectral power in different EEG bands of interest and spectral morphology measures, such as median frequency or statistical complexity. Beta 2 spectral power increase in frontal, midline and parietal areas has been found to be very significant ($p < 0.00000840$) when an odorant of any type is supplied to a subject in this experiment. Also, this increase has been found to be more significant for trigeminal than for natural odorants.

Keywords—**olfaction, EEG, power spectral density, median frequency, Euclidean distance, Shannon entropy, statistical complexity.**

I. INTRODUCTION

Olfactory deficits can be produced due to several causes such as traumatic events, viral diseases or as an early symptom in neurodegenerative diseases. The reliable evaluation of the olfactory response is a useful way to diagnose or monitor the evolution of these diseases [1].

The EEG olfactory response has already been evaluated showing statistical significant power increases and decreases in different frequency bands and areas of the brain [2]. This study presents significant results in different spectral metrics between pre and post stimulus 3.5 s segments. The evaluated metrics include the power spectral density (PSD) in theta, alpha and beta bands, and other spectral parameters (spectra morphology): median frequency (MF), Euclidean distance (ED), Shannon entropy (SE) and López Ruiz-Mancini-Calbet statistical complexity (LMC).

The MF is the frequency to which half of the PSD area is accumulated. The redefinition of the SE done by Powell & Percival quantifies the degree of uniformity of the PSD of a signal [3]. The ED between the uniform distribution and the PSD of a signal has also been proposed as a measure of the equilibrium of a signal [4]. The LMC [4] is the product of these last two metrics, and combines the information stored in a signal with its disequilibrium. All these metrics has been used in previous studies to characterize different diseases (epilepsy, Alzheimer…) or task reactions (visual, auditory…) [5] [6] [7].

The main aim of this work is to identify EEG features that can be combined with fMRI to improve the objective study of the olfactory function using simultaneous EEG/fMRI.

II. MATERIALS AND METHODS

A. Subjects and Experiment

Five healthy subjects (four women and one man, ages from 30 to 45 years), who had not been diagnosed with any olfactory disorder, were studied.

Olfactory stimulus was administered using a custom-built olfactometer [8]. The odor is delivered to the subject's nose, where the change from clean air to odorized air with rapid temporal features. Three different types of odorants (butanol (middle trigeminal properties), mint (trigeminal) and vanilla or coffee (natural)) were randomly administered during the experiments synchronizing manually the odor supply with the subject inspiration monitored with respiratory gating using a pneumatic belt. The stimulation paradigm was a block-design consisted of 9s activation periods with no less than 25s interstimulus intervals. At least 36 stimuli were administered to each subject (during three fMRI series).

Subjects were instructed to close their eyes during the acquisition in order to avoid eye blinking artifact.

The EEG was registered simultaneously with fMRI, since the results obtained for the EEG will be applied to fMRI analysis in a future work. The olfactometer is fully compatible with MR scanner and EEG system [8].

B. Data Acquisition

EEG data were recorded using a Brain Products MR-compatible EEG system with 32-EEG channels.

fMRI data were collected using a General Electric Signa 3.0 T MR scanner using the body coil for excitation and an 8-channel quadrature brain coil for reception. An optimal synchronization scheme between EEG and MR systems was used to improve the EEG artifact removal.

C. Data Preprocessing

Brain Vision Analyzer 2.0 was used to remove MR scanner EEG artifacts in the EEG.

Gradient artifact was removed using the average artifact subtraction method with an adaptive noise cancelation algorithm (AAS+ANC) [9] using a sliding window of 22 volumes to account for possible movements of the subject in the calculation of the template. Subsequently, a downsampling from 5000 Hz to 500 Hz and a bandpass filter 1-70 Hz was applied. The pulse related artifact was removed using an ICA approach [10]. First, the R peaks were semi-automatically detected from the ECG electrode placed in the back of the subject. Then we applied an Infomax ICA algorithm to the 31 EEG channels (excluding ECG) to obtain 31 independent components (IC). The ICs related to the pulse-related artifact were selected computing the Global Field Power (GFP) distribution on the ICs detecting the contribution of each IC to the artifact. After that, the selected ICs were band-pass filtered between 1-20 Hz to remove possible residual real EEG signal from the ICs (pulse-related artifact affects mainly 1-10 Hz frequency band) and so to avoid the subtraction of this residual signal from the EEG. Finally, each filtered ICs was multiplied by the weight indicated in the inverse of the mixing matrix for each channel and subtracted from it. This procedure was performed for all the EEG channels.

Pre and post stimulus segments that exceeded the mean value of each channel plus/minus four times the standard deviation were rejected for subsequent analysis.

The vanilla results of one of the subjects were discarded due to a lack of enough concentration in the correspondent valve during the experiment.

D. Data Analysis

The sum of the PSD within five bands of interest – theta (4-8 Hz), alpha (8-12 Hz), beta1 (13-16 Hz), beta2 (17-23 Hz) and beta3 (24- 30 Hz)- and four spectral parameters (MF, ED, SE, LMC) were computed for each stimulus (one value per 3.5s pre stimulus segment and one value per 3.5s post-stimulus segment, averaging by seven brain electrode areas: frontal, central, midline, parietal, temporal, occipital and temporo-parietal) using Matlab 2008. Differences in pre and post stimulus for each metric value were analyzed using a Wilcoxon signed-rank test to evaluate it between paired dependent segments. The statistical analyses were considered significant for a p<0.05. The Hodges-Lehmann nonparametric estimator (HLE) was used to estimate the median of the population for each estimated parameter.

III. RESULTS

Table 1 shows the most significant differences between power in each frequency band (theta, alpha, beta1, beta2, beta3) after and before the stimulus onset, distributed in several electrode areas (frontal, central, parietal, temporal, temporal-parietal, and midline electrodes) regarding all odorants stimuli. The most significant changes are in beta 2/ beta 3 bands in frontal, midline, parietal, occipital, central and temporal areas.

Table 1 Most significant EEG spectral power changes (all odorants)

EEG Parameter	p value	HLE (μV^2)
Frontal area – beta 2 band	0.000000828	3.2554
Midline area – beta 2 band	0.00000286	1.5767
Parietal area – beta 2 band	0.00000840	1.8877
Occipital area – beta 2 band	0.0000394	2.5971
Central area – beta 2 band	0.0000646	2.5522
Frontal area – beta 3 band	0.0000826	0.5084
Midline area – beta 3 band	0.00015924	0.28491
Central area – beta 3 band	0.00028405	0.52525
Temporal area – beta 2 band	0.00036663	4.2316

Table 2 shows the most significant power differences per frequency band and electrode area, for each type of odorant analyzed independently. A significant increase in frontal/parietal/midline beta power in trigeminal odorants (butanol and mint) can be seen. Natural odorant results show a significant decrease in central and occipital alpha power. Additionally mint has a significant increase in theta band in temporal and parietal areas.

Table 2 Most significant EEG spectral power changes for each odorant

EEG Parameter	p value	HLE (μV^2)
Butanol – midline area – beta 2 band	0.00010074	17.488
Butanol – temporal area – beta 2 band	0.00018635	5.249
Butanol – parietal area – beta 2 band	0.00030331	21.542
Butanol – frontal area – beta 2 band	0.00038068	32.635
Mint – frontal area – beta 2 band	0.00039182	31.676
Mint – frontal area – beta 3 band	0.0018722	0.5538
Mint – parietal area – theta band	0.0058317	9.749
Mint – midline area – beta 2 band	0.0066346	13.443
Mint – parietal area – beta 2 band	0.0090472	15.601
Mint – temporal area – theta band	0.0097478	30.951
Coffee/vanilla – frontal area – alpha band	0.031725	-8.3422
Coffee/vanilla – occipital area – alpha	0.038394	-14.676

Figure 1 shows the pre and post stimulus paired beta 2 power values in frontal areas for mint. It can be seen that an increase does happen in most of the events.

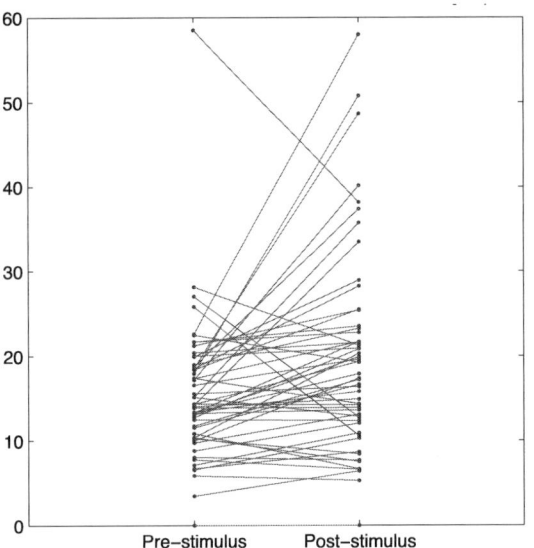

Fig. 1 Wilcoxon's plot of spectral power changes in beta 2 band frontal area (Mint) (μV^2)

Table 3 shows the significant differences between the four spectral parameters (MF, normalized ED, normalized SE, normalized LMC) after and before the stimulus onset distributed in several electrode areas (frontal, central, parietal, temporal, temporal-parietal, and midline electrodes) considering all odorants stimuli. If we analyze each odorant independently (Table 4), only the natural odorant produced a significant decrease in ED, increase in SE and decrease in LMC in the occipital area.

Table 3 Most significant EEG spectral parameter changes (all odorants)

EEG Parameter	P value	HLE
ED – occipital area	0.000094354	-0.0021645
SE – occipital area	0.00036299	0.00896
LMC – occipital area	0.0068396	-0.001374
SE – midline area	0.041717	0.004845

Table 4 Most significant EEG spectral parameter changes for each odorant

EEG Parameter	P value	HLE
Coffee/vanilla ED – occipital area	0.0000263	-0.0047767
Coffee/vanilla SE – occipital area	0.001022	0.01703
Coffee/vanilla LMC – occipital area	0.0026306	-0.0031565

IV. DISCUSSION

In this work, we have shown that several EEG spectral parameters change after olfactory stimuli. In a preliminary study we have used one of this parameters (frontal beta 2 power increase, whose values are shown in figure 1 only for mint odorant) to modulate the BOLD response in a single-trial EEG-informed fMRI analysis. However, there are other ways in which these parameters can be used (i.e. evaluate the inclusion or not of each olfactory event in the fMRI analysis or to extract features in a joint or parallel ICA approach [11]). These ideas will guide our future works.

There are differences in the results considering each odorant independently, especially between natural and trigeminal or middle trigeminal odorants. These differences could be due to the different brain areas activated in each one of these types of odorants, since trigeminal odorants are irritative and activate frontal and parietal areas, which natural odorants don't activate or activate less strongly, and show a greater overall activation [12] [13]. This could explain a much more significant frontal/parietal beta power increase with mint or butanol than with coffee or vanilla.

We have seen in the preliminary EEG-informed fMRI analyses using frontal beta 2 power differences that the statistical power of brain activity in frontal areas and other olfactory related brain areas, such as the insula, increases in mint and butanol, but not in natural odorants (these results are out of the scope of this work).

PSD-based parameters resulted more significant in this study than the parameters based on other spectral features. Natural odorants are coherent between occipital alpha decrease and occipital ED decrease/ LMC decrease/ SE increase.

In a previous work [1], statistically significant differences have been found in alpha band, which we haven't replicated. Both the scarce results in spectra morphological measures and the lack of significant changes in alpha band could be due to the experiment was done with closed eyes. So another study should be carried out with open eyes to check this point.

Furthermore, these results should be replicated in a larger population of healthy subjects and also differences should be observed between them and pathological subjects. Moreover, the results should be replicated inside and outside the scanner for each subject in order to assure that results are independent of the experimental environment and the preprocessing of the signal. Moreover, several ongoing improvements in the olfactory design will lead to more stable results: automatic synchronization between odor supply and respiration, electronic device to measure instant and concentration of the odor delivered directly to subjects' nose, and finally implementing a mixed block-event design with odor supply of cents of milliseconds interleaved with air clean inside a block of events in order to improve the EEG signal and maintain the statistical power of the fMRI.

V. CONCLUSION

We have demonstrated that there is significant EEG information related to olfactory response that could complement fMRI data analysis, even without having evoked potentials.

Beta 2 spectral power increase in frontal, midline and parietal areas has been found to be very significant ($p < 0.00000840$) when an odorant of any type is supplied to a subject in this experiment. These parameters were also more significant for trigeminal than for natural odors.

The use of these parameters in the combined analysis of EEG/fMRI is promising for the objective evaluation of the olfactory function both in healthy and pathological subjects.

REFERENCES

1. Alberts M, Tabert M, Devanand DP. Olfactory dysfunction as a predictor of neurodegenerative disease (2006). Current Neurology and Neuroscience Reports, 6(5): 379-386.
2. E. Bonanni et al. (2006) Quantitative EEG analysis in post-traumatic anosmia. Brain Research Bulletin, 71:69-75
3. Powell GE & Percival IC. (1979) A spectral entropy method for distinguishing regular and irregular motion of Hamiltonian systems, Journal of Physics A: Mathematical and General, 12(11): 2053-2071
4. López-Ruiz R, Mancini HL & Calbet X. (1995) A statistical measure of complexity, Physics Letters A, 209:321-326.
5. Kannathal N, Choob ML, Acharyab UR & Sadasivana PK. (2005) Entropies for detection of epilepsy in EEG. Computer Methods and Programs in Biomedicine, 80(3): 187-194.
6. Abásolo D, Hornero R, Espino P, Álvarez D & Poza J (2006). Entropy analysis of the EEG background activity in Alzheimer's disease patients. Physiological Measurement, 27(3): 241-253.
7. Rosso OA, Martin MT, Figliola A, Keller K & Plastino A. (2006) EEG analysis using wavelet-based information tools. Journal of Neuroscience Methods, 153(2): 163-182.
8. Borromeo et al. (2010) Objective Assessment of Olfactory Function Using Functional Magnetic Resonance Imaging (fMRI). IEEE Transactions, 59-2602-2608.
9. Alen, P. (2000). A method for removing imaging artifact from continuous EEG recorded during functional MRI. Neuroimage, 12(2): 230-239.
10. Srivastava, G, Crottaz-Herbette, S, Lau, KM, Glover, GH, & Menon, V. (2005). ICA-based procedures for removing ballistocardiogram artifacts from EEG data acquired in the MRI scanner. Neuroimage, 24(1): 50-60.
11. Eichele T, Calhoun VD, Debener S. (2009) Mining EEG-fMRI using Independent Component Analysis. Int J Psychophysiol 73(1): 53-61.
12. Hummel T, Doty RL, Yousem DM. (2005) Functional MRI of intranasal chemosensory trigeminal activation. Chemical senses, 30(suppl 1): i205-i206.
13. Boyle JA, Heinke M, Gerber J, Frasnelli J, Hummel T. (2007) Cerebral activation to intranasal chemosensory trigeminal stimulation.Chemical senses, 32(4): 343-353.

Author: Eva Manzanedo Sáenz
Institute: Universidad Rey Juan Carlos
Street: Tulipán s/n
City: Móstoles
Country: Spain
Email: eva.manzanedo@ctb.upm.es

An Emboli Detection System Based on Dual Tree Complex Wavelet Transform

G. Serbes[1], B.E. Sakar[2], N. Aydin[3], and H.O. Gulcur[1]

[1] Biomedical Engineering Institute, Bogazici University, Istanbul, Turkey
[2] Department of Computer Programming, Bahcesehir University, Istanbul, Turkey
[3] Department of Computer Engineering, Yildiz Technical University, Istanbul, Turkey

Abstract—Automated decision systems for emboli detection is a crucial need since it is being done by visual determination of experts which causes excess time consumption and subjectivity. This work presents an emboli detection system using various dimensionality reduction algorithms on Doppler ultrasound signals recorded from both forward and reverse flow of blood transformed via Fast Fourier Transform (FFT), Discrete Wavelet Transform (DWT), and Dual Tree Complex Wavelet Transform (DTCWT). The combined forward and reverse DTCWT based features produced the highest performance when fed to SVMs classifier. As to compare dimensionality reduction algorithms, although PCA and LDA gave comparable accuracies, LDA has accomplished these accuracies only with two components due to its less than the number of classes' orthogonal projective directions limitation. SVMs yielded higher classification accuracies than *k*-NN with all considered dimensionality reduction methods since SVMs classifier is more robust to noise and irrelevant features. With the ability to localize well both in time and frequency, wavelet transform based extracted features gave higher overall classification accuracies than FFT with the more stable classifier SVMs. Additionally, DTCWT accuracies are higher with SVMs than those of DWT since it also has the ability of being shift-invariant.

Keywords—Discrete Complex Wavelet; Embolic Signals; Dimensionality Reduction; Support Vector Machines.

I. INTRODUCTION

The transcranial Doppler ultrasound is a commonly used method to detect asymptomatic embolic signals (ES) in the cerebral circulation [1]. In certain conditions, such as carotid artery stenosis, cardiac valvular disease and atrial fibrillation, asymptomatic ESs are used for the identification of active embolic sources in stroke-prone individuals and the selection of high-risk patients for appropriate treatment.

Traditionally, for detecting ESs, individual spectral recordings are analyzed visually by experts. This type of detection is time consuming and subject to observer's experience. As a consequence of these drawbacks, an automated system is required for a reliable and clinically useful emboli detection technique.

A Doppler ultrasound signal detected by the transcranial Doppler ultrasound system can contain two types of high intensity signals other than the ESs. These signals can be named as the Doppler speckles (signals caused by red blood cell aggregates), and the artifacts (signals caused by tissue movement, probe tapping, speaking, and any other environmental effects).

ESs are resulted because of the reflection of transmitted Doppler ultrasound signals from embolic particles which are bigger than red blood cells. Therefore ESs have some distinctive characteristics when compared to Doppler speckles (DS) and artifacts. ESs appear as increasing and then decreasing in intensity for a short duration, usually less than 300 ms and their bandwidth is usually much less than that of DSs (Therefore, ESs can be considered as narrow-band signals relative to DSs).

The output of a Doppler ultrasound system has two components which are called as in-phase and quadrature-phase components. The information concerning blood flow direction is encoded in the phase relationship between these two components and by using various methods forward and reverse blood flow signals are obtained. Unlike the artifacts, which are bidirectional, ESs and DSs are approximately unidirectional (there can be small leakages in the opposite direction).

Generally the aim is to distinguish ESs from artifacts and DSs using Doppler ultrasound. Subsequently an automated system is tried to be built up with feature extraction from these signals using various methods followed with classification. After a selected feature extraction technique, PCA can be used to reduce the dimensionality of the extracted features that were used in an automated emboli detection system [2].

If we go deep into the feature extraction, the Doppler ultrasound signal can be considered as a narrow-band signal when an embolus appears; therefore frequency analysis based methods are frequently used as feature extraction steps in ES detection systems [3]. In [4] a spectrogram analysis based detection method is proposed. Along with these techniques, fast Fourier transform (FFT) is also commonly used in feature extraction. However, continuous wavelet transform (CWT) based methods perform better than fast Fourier transform (FFT) in describing ESs [5]. In the discrete wavelet transform (DWT) case which is a fast implementation of CWT, an automated system which uses DWT to derive several parameters for detecting ESs was proposed in [6]. In this study Doppler ultrasound signals were decomposed into an optimum number of frequency bands and then these bands were reconstructed. From these reconstructed bands several parameters were obtained and used in detection algorithm.

Dual tree complex discrete wavelet transform (DTCWT), which is an improved version of ordinary DWT with limited redundancy, can also be used in the analysis of ESs. The DTCWT was developed to overcome the lack of shift invariance property of ordinary DWT [7]. This property of DTCWT can be very important when the wavelet coefficients are used as features in machine learning algorithms to detect emboli, because the emboli information is encoded in the phase relationship of the in-phase and quadrature-phase components and any phase-distortion during the analysis steps can reduce the discriminative power of wavelet features. The success of DTCWT in the analysis of non-stationary signals such as ESs was proved before in [8].

In this study, a Doppler ultrasound dataset consisting of 100 samples from all embolic, DS and artifact 1024-point signal pairs – forward and reverse direction – is used. FFT, DWT and DTCWT are applied to these 300 signal pairs and the dimensionality (1024 in this case) of resulted coefficients is reduced with various dimensionality reduction methods for removing signal components which do not carry useful information.

As a result of these processes, the dimensionality reduced features are obtained and they are fed into two classifiers k-NN and SVMs. The obtained results are presented and compared for each transform in detail.

II. MATERIALS AND METHODS

A. Emboli Dataset Description

For this study a Doppler ultrasound dataset consisting of 100 embolic, 100 DS, and 100 artifact signal pairs was created. The Doppler ultrasound signals were recorded using a transcranial Doppler system (EME Pioneer TC4040 which is manufactured by Nicolet Biomedical, Madison, USA). The sampling frequency was 7150 Hz and the data length was 1024 points. The recordings were made from the ipsilateral middle cerebral artery of patients with symptomatic carotid stenosis. The artifacts were created artificially during patient recordings by tapping the probe, speech or coughing and obtained from natural artifacts occurring during patient movement, speech, or coughing during routine patient recordings [6].

B. Overall Emboli Detection System

The collected Doppler ultrasound signals are transformed with FFT, DWT, and DTCWT methods to obtain transform coefficients. The obtained datasets consists of 1024 features (transform coefficients) and 300 samples. In order to overcome the curse of dimensionality problem, before feeding to classifiers, the dimension of the datasets are reduced by applying linear dimensionality reduction techniques. Finally, the obtained reduced dimensionality feature sets are fed to SVMs with linear kernel and k-NN classifiers.

i. Feature Extraction

In FFT feature extraction part, Fourier transform coefficients are found and the absolute values of the coefficients are used as features. For DWT and DTCWT feature extraction parts, both forward and reverse Doppler ultrasound signals are decomposed to 5 scales. In DWT, the filter coefficients given in [9] and for DTCWT the filter coefficients given in [7] were used. After these processes the absolute value of DWT and DTCWT coefficients are used as features for dimensionality reduction steps. For an ES, the extracted FFT (only one side), DWT, and DTCWT features for both forward and reverse channels can be seen in Figure 1; as expected only in forward direction emboli occurs. As it can be seen, in FFT features emboli shows itself as a narrow-band signal pattern. In DWT and DTCWT features emboli patterns can be seen in second and third scales.

ii. Dimensionality Reduction

In this paper, we use PCA as an unsupervised technique and LDA as a supervised technique to reduce the dimensionality of the emboli dataset. For the application of PCA, the dimensionality of each dataset is reduced by preserving the 90% of the data variance. As known, LDA has the limitation of less than number of classes' orthogonal projective directions due to the rank deficiency of the between-class scatter matrix [10]. Therefore, as the emboli dataset includes three classes, we reduced the dimension of the datasets to two with LDA.

Besides, as emboli blood flow is observed in two directions (forward and reverse); our dataset can be treated as a multi-view (two-view) dataset. The term "multi-view" is used to refer multiple sets of features about the same underlying phenomenon. Hence, other than the single view dimension reduction methods (PCA and LDA), we also perform Canonical Correlation Analysis (CCA) [11] method to reduce the dimensions of the forward and reverse views. CCA is a feature fusion method which aims to explore the linear relationships between two different but related views (multidimensional variables).

It is known that the sample covariance matrices used in the formulation of CCA are sensitive to outliers and noisy samples [11]. To solve the sensitivity and also the singular matrix problems, we regularize CCA by applying PCA as a preprocessing step.

III. EXPERIMENTAL RESULTS

The overall SVMs and *k*-NN accuracies and detection rates of emboli, artifact, and speckle classes are presented in Table 1, respectively. The training sets are generated by randomly choosing half of each class samples (50 from each), and use the other half as the test set. The train-test splits are repeated 20 times for statistical significance and average accuracies along with class detection rates are reported. As seen in Table 1, the highest overall accuracies are obtained with DTCWT features for all dimensionality reduction techniques. Overall accuracy is obtained by feeding the components of forward and reverse views as input to the classifier together. The stand-alone accuracies of forward and reverse signals are also higher with DTCWT features.

Besides, it must be noted that both DWT and DTCWT performed higher accuracies than FFT. Another point to be emphasized is the success of DTCWT in the detection of emboli signals. The SVMs classifier accomplished higher accuracies with the components of PCA and LDA single view dimension reduction methods than those of CCA.

As seen in Table 1, although *k*-NN (*k* = 3) is a non-linear local classifier, it performed worse than linear kernel SVMs which shows the superiority of SVMs on *k*-NN in detection of emboli signals. Although FFT features gave higher accuracies than DWT and DTCWT with *k*-NN classifier for PCA and PCA+CCA dimensionality reduction methods, it is also observed that DWT and DTCWT are more successful at detecting the emboli samples. Furthermore, DTCWT is again superior on FFT with LDA features.

The projections of the FFT and DTCWT feature sets on the LDA components of the forward view are shown in Figure 2. It is seen that for both feature sets the artifact samples are well discriminated from the other two classes.

It is also observed that while with FFT features some emboli samples are intermixed with artifact samples, with DTCWT features the emboli and artifact samples are better discriminated. The other remarkable point is that DTCWT features have better discriminated the emboli and speckle samples than those of FFT.

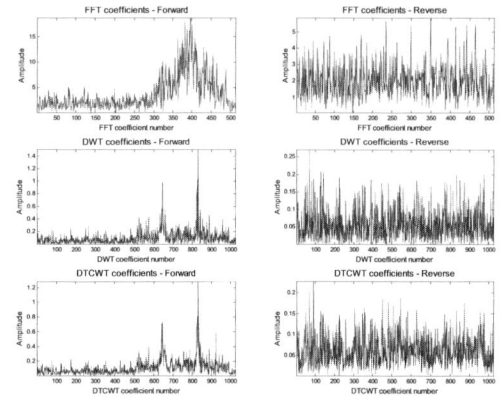

Fig. 1 Extracted features from an embolic signal with FFT, DWT and DTCWT

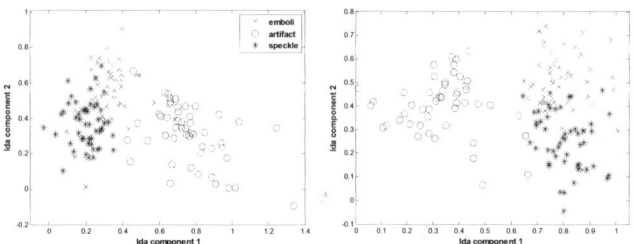

Fig. 2 Projections on the LDA components extracted from (left) FFT and (right) DTCWT data of forward view

Table 1 Overall accuracies (%) and detection rates (%) of each class with SVMs and *k*-NN

		PCA			PCA+CCA			LDA		
		FFT	DWT	DTCWT	FFT	DWT	DTCWT	FFT	DWT	DTCWT
SVM	**Overall Accuracy**	0.83	0.91	**0.93**	0.77	0.79	**0.84**	0.82	0.87	**0.91**
	Forward Accuracy	0.80	0.90	**0.92**	0.77	0.80	**0.83**	0.81	0.87	**0.90**
	Reverse Accuracy	0.73	0.77	**0.78**	0.72	0.69	**0.73**	**0.72**	0.68	0.69
	Emboli Detection Rate	0.75	0.83	**0.87**	0.64	0.67	**0.73**	0.73	0.77	**0.87**
	Artifact Detection Rate	0.82	0.98	0.99	0.76	0.90	0.93	0.80	0.95	0.96
	Speckle Detection Rate	0.95	0.91	0.92	0.91	0.81	0.85	0.95	0.89	0.89
		PCA			PCA+CCA			LDA		
		FFT	DWT	DTCWT	FFT	DWT	DTCWT	FFT	DWT	DTCWT
k-NN	**Overall Accuracy**	**0.71**	0.66	0.70	**0.70**	0.64	0.65	0.77	0.81	**0.83**
	Forward Accuracy	0.71	0.69	**0.77**	**0.67**	0.62	**0.67**	0.47	0.44	**0.51**
	Reverse Accuracy	**0.69**	0.62	0.66	**0.65**	0.62	**0.65**	0.53	0.53	0.53
	Emboli Detection Rate	0.92	**0.98**	**0.98**	0.85	0.93	**0.96**	0.66	0.73	**0.78**
	Artifact Detection Rate	0.94	0.92	0.95	0.92	0.86	0.88	0.94	0.91	0.93
	Speckle Detection Rate	0.28	0.07	0.16	0.33	0.12	0.12	0.71	0.79	0.78

IV. CONCLUSIONS

In this paper, we propose an emboli detection system, in which firstly the Doppler ultrasound signals that belong to both forward and reverse blood flow directions are transformed with DTCWT, then the dimensionality of the obtained feature sets are reduced with PCA and LDA as single view dimensionality reduction methods and CCA as a feature fusion method. The obtained reduced feature sets of both forward and reverse directional signals are fed to SVMs and *k*-NN classifiers individually and also as combined. We compare the success of DTCWT features with those of FFT and DWT by applying the same dimensionality reduction and classification procedures.

First of all, we must note that SVMs based detection methods are superior on *k*-NN based methods for all the dimensionality reduction methods due to the known generalization and sensitivity to noisy samples and irrelevant feature problems of *k*-NN classifier especially on high-dimensional datasets. Accordingly, due to less than the number of classes' orthogonal projective directions limitation of LDA, the dimensionality of the reduced feature space of LDA is comparably lower than those of PCA and CCA. However, since the class labels are incorporated into the dimensionality reduction scheme of LDA, when only the first two components are used, the highest classification accuracies and also emboli detection rates are obtained with LDA.

Secondly, comparing the results of feature extraction methods we see that the highest classification accuracy and emboli detection rate are obtained when the combined forward and reverse DTCWT based features are fed to SVMs as input. Therefore, we can conclude that in the emboli detection the reverse blood flow direction has also significant discriminative information. Besides, DWT has performed better than FFT with SVMs classifier. The projections of the FFT and DTCWT feature sets on the LDA components confirmed the better discriminative ability of DTCWT features. As a conclusion, wavelet transform based extracted features give higher overall classification and emboli detection accuracies than FFT based features in SVMs classifier due to their well localization property in both time and frequency. As known, the wavelet transform provides a time-scale representation of signals which have good frequency resolution at low frequencies, but also have good time resolution at high frequencies. Additionally, due to its shift-invariance property, DTCWT surpasses DWT in overall accuracy and embolic detection rate with SVMs.

REFERENCES

1. Markus HS, Monitoring embolism in real time, Circulation, vol. 102, no. 8, pp. 826-828, 2000.
2. Xu D, Wang Y, An automated feature extraction and emboli detection system based on the PCA and fuzzy sets, Computers in Biology and Medicine, vol. 37, pp. 861-871, 2007.
3. Roy E, Abraham P, Montresor S, Baudry M, and Saumet JL, The narrow band hypothesis: an interesting approach for high-intensity transient signals (HITS) detection, Ultrasound in medicine & biology, vol. 24, no. 3, pp. 375-382, 1998.
4. Roy E, Montrésor S, Abraham S, Saumet JL, Spectrogram analysis of arterial Doppler signals for off-line automated hits detection, Ultrasound in medicine & biology, vol. 25, no. 3, pp. 349-359, 1999.
5. Aydin N, Padayachee S, Markus HS, The use of the wavelet transform to describe embolic signals, Ultrasound in Medicine and Biology, vol. 25, pp. 953-958, 1999.
6. Aydin N, Marvasti F, Markus HS, Embolic Doppler Ultrasound Signal Detection Using Discrete Wavelet Transform, IEEE Transactions on Information Technology in Biomedicine, vol. 8, no. 2, pp. 182-190, 2004.
7. Selesnick IW, Baraniuk RG, Kingsbury NG, The dual-tree complex wavelet transform, IEEE Signal Processing Magazine, vol. 22, no. 6, pp. 123-151, 2005.
8. Serbes G and Aydin N, Denoising Embolic Doppler Ultrasound Signals using Dual Tree Complex Discrete Wavelet Transform, Engineering in Medicine and Biology Society (EMBC), 2010 Annual International Conference of the IEEE, 2010, pp. 1840-1843.
9. Kingsbury NG, Image processing with complex wavelets, Philosophical Transactions of the Royal Society, vol. 357, pp. 2543-2560, 1999.
10. Alpaydin E, Introduction to Machine Learning, 2nd ed.: The MIT Press, 2010.
11. Hardoon DR, Szedmak S, Shawe-Taylor J, Canonical correlation analysis; An overview with application to learning methods, Neural Computation, vol. 16, no. 12, pp. 2639-2664, 2004.

Author: Betul Erdogdu Sakar
Institute: Bahcesehir University
Street: Abide-i Hurriyet, Sisli
City: Istanbul
Country: Turkey
Email: betul.erdogdu@bahcesehir.edu.tr

New Indices Extracted from Fetal Heart Rate Signal for the Assessment of Fetal Well-Being

A. Fanelli[1], G. Magenes[2], and M.G. Signorini[1]

[1] Dipartimento di Elettronica, Informazione e Bioingegneria (DEIB), Politecnico di Milano, Milano, Italy
[2] Dipartimento di Ingegneria Industriale e dell'Informazione, University of Pavia, Italy

Abstract—In this paper, new parameters based on the Phase Rectified Signal Average (PRSA) of the Fetal Heart Rate (FHR) signal are introduced to assess fetal wellbeing. They are defined as the positive and negative slopes of the PRSA curve, namely Acceleration Phase Rectified Slope (APRS) and Deceleration Phase Rectified Slope (DPRS), depending on the sign of the slope. The parameters were computed on FHR time series recorded from 59 healthy and 61 Intra Uterine Growth Restricted (IUGR) fetuses, during CTG non-stress tests. The performance of APRS and DPRS was compared with the performance of other parameters extracted from the PRSA curve already existing in literature, and with other clinical indices that are provided by computerized cardiotocographic systems. The APRS and DPRS performed better than any other parameter considered in this study to distinguish between healthy and IUGR fetuses. We believe that these new indices might provide useful improvements of the quality of fetus early diagnosis.

Keywords—Cardiotocography, Fetal Monitoring, Heart Rate Variability, Phase Rectified Signal Average.

I. INTRODUCTION

Recording Fetal Heart Rate (FHR) and measuring its variability represent a non-invasive way to collect information about the fetal state and the proper development of his/her autonomic nervous system [1]. Moreover, FHR can be directly connected to the level of fetal oxygenation that is essential for fetal well-being. For these reasons FHR monitoring is crucial to identify risky conditions in the fetus. It contributes to a reduction in complications and fetal death as well as it restrains the need for invasive interventions. FHR monitoring is also very important to minimize risks of fetal morbidity and mortality and to evaluate the optimal timing for delivery.

Fetal pathological conditions such as Intra Uterine Growth Restriction (IUGR), diabetes, infections and placental abruption can be diagnosed thanks to Heart Rate Variability (HRV) analyses [2]. In particular IUGR is one of the most severe causes of perinatal morbidity and mortality: it consists of a pathological inhibition of fetal growth, with a consequent failure of the fetus to attain its growth potential. It is strictly connected to fetal hypoxia and asphyxia. The incidence of IUGR is approximately 5% of all pregnancies [3]. Thus, fetal monitoring is extremely important to detect IUGR or other risky conditions in the fetus, helping the decision process of clinicians.

The most used technique to monitor fetal condition is cardiotocography (CTG), which combines the measure of FHR through a Doppler ultrasound probe with uterine contractions using a pressure sensor. Doppler Ultrasound allows the detection of heart beat events by sensing fetal heart movements. Using this approach, a trace reporting the heart rate changes during the screening process is generated [4].

The main problem in CTG analysis resides in the fact that most diagnostic conclusions rely on qualitative eye inspection of CTG traces and, thus, on the clinician experience. This causes a diffused intra-observer and inter-observer variability; thus it is important to improve the reliability of FHR analysis by introducing new indicators related to the pathophysiological condition of the fetal heart.

For these reasons we have been working for many years trying to identify parameters able to quantify fetal well-being in an objective way and to overcome inter and intra observer variability [5] [6]. The purpose of this paper is to improve the quality of early diagnosis of fetal harms by introducing and applying new indices obtained from the Phase Rectified Signal Average (PRSA) curve [7]. Results we present were obtained after applying a set of parameters to a selected population of normal and IUGR fetuses. The performance of the new Acceleration (or Deceleration) Phase Rectified Slope (APRS) parameter is compared with time domain and nonlinear CTG indices already used by our group for FHR analysis, and with other parameters obtained from the PRSA curve as reported in the literature [8][9]. Analyses were done on healthy and uterine-restricted fetuses, in order to identify which parameters most efficiently classify the two populations.

II. MATERIALS AND METHODS

A. Data Collection

Our system consists of a HP-1350 CTG fetal monitor connected to a computer using a RS-232 serial port. However, our computerized system is compatible with any other CTG monitor using the HP data protocol (Agilent, Corometrics 170, Philips 50A, etc.).

CTG recordings were collected at the Azienda Ospedaliera Universitaria Federico II, Napoli, Italy. Ethical committee approved the protocol. Population was composed of 120 subjects (59 healthy and 61 IUGR). The growth-restricted group was defined "a posteriori" after birth by estimated weight below the 10th percentile for its gestational age and estimated abdominal circumference below the 10th percentile. All recordings were acquired in a controlled clinical environment, with the pregnant woman lying on a bed after collecting their informed consent. The gestational age at the date of recording was 32.27±2.79 weeks for the IUGR group and 34.78 ± 0.53 weeks for the normal fetuses, respectively. The average length of the recordings was 3418 ± 1033 sec for IUGR fetuses and 2450 ± 724 sec for healthy subjects.

B. Signal Pre-processing

The HP-1350 quantifies the quality of the acquired FHR signal with a color index: good (green), acceptable (yellow) and bad (red). This quality evaluation is based on the output of the autocorrelation procedure computed during data collection. Each FHR recording is then divided into windows of 360 data points (3 min). The red quality points are replaced by the average of the nearest five FHR points. If the trace contains more than five consecutive red-quality points or a subinterval includes more than 5% red quality values, that subinterval is left off the analysis. This approach allows correcting noisy segments or discarding them when the signal-to-noise ratio is too poor (insufficient) for further analysis. Acquisition and preprocessing procedure is fully described in [9].

C. Phase Rectified Signal Average (PRSA)

Phase Rectified Signal Average (PRSA) is a technique introduced by Bauer et al. in 2006 [7]. It allows detection and quantification of quasi-periodic oscillations in non-stationary signals affected by noise and artifacts, by synchronizing the phase of all periodic components. This method demonstrated its usefulness in FHR signal analysis, when episodes of increasing and/or decreasing FHR appear [8]. In fact, occurrence or absence of such periods can be related to the healthy status of the fetus. In the Ob-Gyn literature increases and decreases of FHR are commonly referred as "accelerations" and "decelerations". These are not the terms we decided to adopt in this study as our indices do not necessary correspond to the definitions adopted by clinicians. Anyway, these modulation events could be studied by evaluating signal oscillations. For this reason, we introduced the PRSA method to quantify fetal well-being states.

Computing the PRSA curve requires a time series $i=1,\ldots,N$, characterized by periodicities and correlations, as well as containing non-stationary and noise events. The first step is the computation of the so-called Anchor Points (AP).

AP are fiducial points, selected according to the average value of the signal before and after a certain instant k, within a selected time window.

In our analysis we define as APs those x_i, belonging to the FHR time series, such as the following inequality stands, within a time window of length $2T$:

$$\frac{1}{T}\sum_{j=0}^{T-1} x_{i+j} > \frac{1}{T}\sum_{j=1}^{T} x_{i-j} \quad (1)$$

Inequality (1) identifies anchor points that mark a signal increase. A similar inequality can be used to identify decreases by replacing the > symbol with the < symbol. According to this definition, around half of all points in the time series are identified as be Anchor Points (Fig 1, upper diagram). The T parameter can be used to control the upper frequency of the periodicities that are detected by PRSA. In our analyses, T values from $T=5$ to $T=100$, step 5, were tested.

Anchor Points can be used to phase-rectify the signal [10], removing noise and preserving only periodic oscillations in the time series. After detecting the AP, windows of $2L$ samples are built around each anchor point. Since many of the anchor points are adjacent, many of the defined windows will overlap. The parameter L should be larger

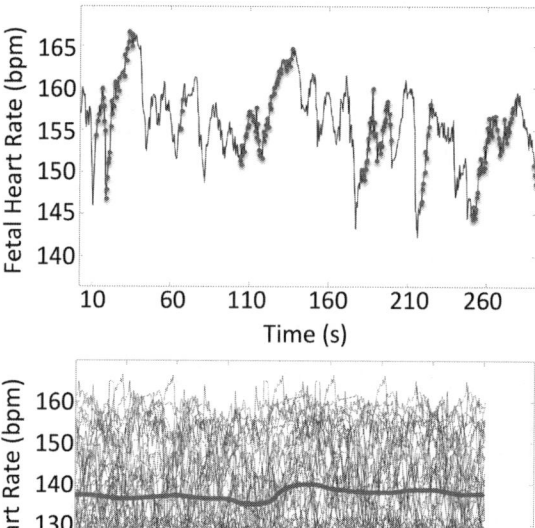

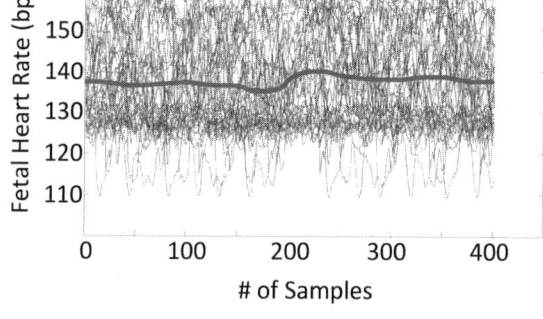

Fig. 1 Fetal heart rate series after the preprocessing steps (up): the Anchor Points are detected using inequality (1) and are highlighted as red points in the graph. Computation of PRSA curve (down): 400 samples windows around each anchor point are synchronized

than the period of slowest oscillation that one wants to detect. We tested several values of L, ranging from 50 to 1000, step 20.

Fig. 1 (lower diagram) shows all $2L$ windows, obtained from the previous step, synchronized in their anchor points and averaged, in order to obtain a single PRSA curve per patient (red curve in the diagram). The averaging process filters out all non-periodic components that are not synchronized, preserving events with a fixed phase relationship with the anchor points only. For a more detailed description of the algorithm, please refer to [7].

After obtaining the PRSA curve, it is useful to summarize the information within a single parameter, which describes the dynamic characteristics of the curve. Bauer et al. [9] employed the Acceleration (or Deceleration) Capacity parameter. They applied $AC(DC)$ index to identify a mortality predictor after myocardial infarction:

$$AC(DC) = [X(0) + X(1) - X(-1) - X(-2)]/4 \quad (2)$$

where $X(0)$ is the sample corresponding to the anchor point. This equation is a quantification of X by Haar wavelet analysis, using the scale of 2.

Huhn et al. [8] were the first applying PRSA to FHR series. They employed a parameter very similar to the AC to identify and classify IUGR fetuses. The Average Acceleration (or Deceleration) Capacity (AAC) they used, corresponds to the integral measure of all periodic acceleration-related oscillations. The AAC was defined as follows:

$$AAC(S) = \frac{1}{2L}\left[\sum_{i=0}^{L-1} X(i) - \sum_{i=-L}^{-1} X(i)\right] \quad (3)$$

In our analysis, we introduced a new parameter computed on the PRSA curve. Indeed, since in FHR signals the diagnostic information is contained in the number and the temporal characteristics of increases and decreases in heart rate, we define the Acceleration (Deceleration) Phase Rectified Slope (APRS or DPRS), as the slope of the PRSA curve computed in the anchor point:

$$APRS = \left|\frac{\partial X(i)}{\partial i}\right|_{X(0')} \quad (4)$$

This parameter is a descriptor of both the average increase (decrease) in FHR amplitude (absolute change of heart frequency) and the time length of the increase (decrease) event. According to the Ob-Gyn literature, these two measures are the most used in clinical practice to quantify fetal well-being.

D. Time Domain and Non-Linear Parameters

In order to provide a precise and complete picture of fetal well being, we also computed a set of time domain parameters that are commonly and traditionally used to quantify fetal well-being such as Short Time Variability (STV) [10] Delta, Long Term Irregularity (LTI) [10] and Interval Index [11], as well as non linear parameters, such as Approximate Entropy [12].

III. RESULTS

All parameters were computed on the cardiotocographic recordings of 59 healthy and 61 IUGR fetuses. For each parameter, we obtained a single value per subject. Table I summarizes values for the two populations. We reported only results obtained using $T=40$ samples (20sec) for AP identification, and $L=200$ (200 sec window) for PRSA construction. Those values showed the best performance in classifying IUGR and healthy subjects.

Before direct comparison, we verified that the two populations had Gaussian distributions for all parameters using Kolmogorov–Smirnov test.

Table 1 Results from the analysis of HRV signals in fetus populations

Parameter	Healthy (mean ± std)	IUGR (mean ± std)
Subject Number	59	61
Time Parameters		
Delta (ms)	42.9 ± 11.97	29.67 ± 9.25
STV (ms)	6.7 ± 2.24	4.29 ± 1.62
Interval Index	0.87 ± 0.07	0.86 ± 0.06
Long Term Irregularity (ms)	21,46 ± 6.53	17.17 ± 5.37
Non Linear Parameter		
Approximate Entropy	1.33 ± 0.2	1.27 ± 0.19
PRSA parameters		
Acceleration Capacity (bpm)	-0.045 ± 0.56	0.012 ± 0.126
Average Acceleration Capacity (bpm)	1.49 ± 1.89	1.873 ± 1.27
Acceleration Phase Rectified Slope (bpm)	0.17 ± 0.04	0.119 ± 0.043
Deceleration Capacity (bpm)	-0.16 ± 0.52	-0.0146 ± 0.129
Average Deceleration Capacity (bpm)	-1.36 ± 1.25	-1.802 ± 1.365
Deceleration Phase Rectified Slope (bpm)	-0.17 ± 0.04	-0.117 ± 0.042

We tested the performance of the parameters in discriminating healthy from IUGR patients using the *t*-test. Table II summarizes the results. Among time parameters, Delta and STV show the best performance in the discrimination task (Delta: p-value of 1.45e-9; STV: p-value of 1.22e-9). LTI is also efficient in the discrimination (p-value: 2.08e-4). II is the only time parameter which fails to reject the null hypothesis of the t-test. The non-linear parameter we employed in the analysis, ApEn, does not allow the rejection of the null hypothesis at 5% significance level.

The parameter we introduced in this paper shows overall the best performance. Indeed the APRS allows the rejection of the null hypothesis with a *p*-value of 1.12e-9. The DPRS behaves even better, with a *p*-value of 9.57e-12. DPRS parameter is the one, in the analyses, which exhibits the smallest *p*-value in the discrimination between healthy and IUGR patients. On the contrary, AC and AAC are not efficient in the discrimination of our data.

Table 2 Results of the T-Test Comparison between Healthy and Iugr Patients

Parameter	t-test	p-value
Time Parameters		
Delta	1	1.45e-9*
Short Time Variability	1	1,22e-9*
Interval Index	0	0,37
Long Term Irregularity	1	2.08e-4*
Non Linear Parameter		
Approximate Entropy	0	0.06
PRSA parameters		
Acceleration Capacity	0	0.44
Average Acceleration Capacity	0	0.20
Acceleration Phase Rectified Slope †	1	1.12e-9*†
Deceleration Capacity	0	0.07
Average Deceleration Capacity	0	0.06
Deceleration Phase Rectified Slope	1	9.57e-12*†

* Statistical significance < 0.05; †Parameter introduced in this study

IV. DISCUSSION AND CONCLUSIONS

We introduced new parameters able to reliably discriminate IUGR fetuses from healthy ones. APRS and DPRS are quantitatively linked to the increase and decrease events (accelerations and decelerations) of the HRV signal. The strength of these indices was verified by comparing their performance with three different sets of parameters that already demonstrated a noticeable level of usefulness in fetal Heart Rate Analysis. The results on a population of 120 subjects (59 healthy and 61 IUGRs, carefully selected) show that the parameters we are proposing perform better than any other considered in the comparison. This is confirmed by the t-tests on the two populations.

According to [7], the PRSA curve describes the main increase and decrease episodes (or patterns) in the FHR time series under analysis. Although these events do not correspond exactly to the definitions of "acceleration" and "deceleration" of FHR signal, usually employed in the clinical routine, they provide almost the same information about the FHR time course.

As a matter of fact accelerations and decelerations are the FHR increase and decrease patterns mostly used by clinicians to quantify fetal well-being. For this reason, we worked to define global indices descriptive of the entity of FHR increases and decreases. The slope of the PRSA curve depends both on amplitude and duration of such events and this is the reason why it has been proposed to distinguish healthy and IUGR patients.

Results we obtained are coherent with the existing literature on fetal monitoring. Indeed, FHR decelerations are more significant than accelerations to quantify fetal risky conditions as their time courses are strictly related to the physiological recovery from possible hypoxic states. The results obtained applying DPRS and APRS confirm this observation. DPRS performed better than APRS in the discrimination task.

In conclusion, APRS and DPRS turned to be simple and reliable indicators of the correct behavior of the autonomic nervous system in the fetus. They are easily computed in "one shot" analysis of the whole tracing, without any need to segment the recording in one minute chunks as it happens for STV and Delta indices, and to identify, separate and analyze individually each acceleration and deceleration.

Obviously they do not represent the "panacea" for reliably assessing fetal wellbeing. These indices can simplify the automatic analysis of FHR recordings and can be included in a set of quantitative parameters helping the clinicians in the hard task of the prenatal diagnosis.

REFERENCES

1. M. V. Kamath, E. L. Fallen, "Power spectral analysis of heart rate variability: a noninvasive signature of cardiac autonomic function," Crit. Rev. Biomed. Eng., vol. 21, no. 3, pp. 245–311, 1993.
2. M. A. Hasan et al., "Detection and Processing Techniques of FECG Signal for Fetal Monitoring", Biological Procedures Online, vol. 11, no. 1, pp. 263-295, 2009.
3. D. Peleg, C. M. Kennedy, S. K. Hunter, "Intrauterine Growth Restriction: Identification and Management", Am Fam Physician, vol. 58, num. 2, pp. 453-460, 1998.
4. M.J. Stout, A.G. Cahill, "Electronic fetal monitoring: past, present and future.", Clin. Perinatol., vol. 38, nn. 1, pp. 127-142, 2011.
5. M. G. Signorini, G. Magenes, S. Cerutti, D. Arduini, "Linear and nonlinear parameters for the analysis of fetal heart rate signal from cardiotocographic recordings", IEEE Trans Biomed Eng, nn. 50, pp. 365–374, 2003.
6. M. Ferrario, M.G. Signorini, G. Magenes, "Complexity analysis of the fetal heart rate variability: early identification of severe intrauterine growth-restricted fetuses", Med Biol Eng Comput, vol. 47, nn. 9, pp. 911-919, 2009.
7. A. Bauer, A. Bunde, P. Barthel, R. Schneider, M. Malik, G. Schmidt, "Phase rectified signal averaging detects quasi-periodicities in non-stationary data", J Phys A, vol. 364, pp. 423–434, 2006.
8. E. A. Huhn, S. Lobmaier, T. Fischer, R. Schneider, A. Bauer, K. T. Schneider and G. Schmidt, "New computerized fetal heart rate analysis for surveillance of intrauterine growth restriction", Prenat Diagn, nn. 31, vol. 5, pp. 509-514, 2011.
9. Bauer, P. Barthel, R. Schneider, "Deceleration capacity of heart rate as a predictor of mortality after myocardial infarction: cohort study", Lancet, nn. 367, vol. 1674–1681, 2006.
10. J. D. Haan, J. V. Bemmel, B. Versteeg, A. Veth, L. Stolte, J. Janssens, J. Eskes, "Quantitative evaluation of fetal heart rate patterns: I. processing methods", Eur J Obstet Gynecol, nn. 3, pp. 95-102, 1971
11. D. Arduini, Rizzo, G. Piana, A. Bonalumi, P. Brambilla, C. Romanini, "Computerized analysis of fetal heart rate: I. description of the system (2ctg)", J Matern Fetal Invest, vol. 3, pp.159-164, 1993.
12. S. M. Pincus, "Approximate entropy (ApEn) as a complexity measure", Chaos, nn. 5, pp. 110-117, 1995.

E-mail: mariagabriella.signorini@polimi.it

A Novel Method of Measuring Autonomic Nervous System Pupil Dynamics Using Wavelet Analysis

G. Leal[1], C. Neves[2], and P. Vieira[1]

[1] Faculty of Sciences and Technology/Department of Physics, New University of Lisbon, Lisbon, Portugal
[2] Santa Maria Hospital/Department of Ophthalmology, Medicine University of Lisbon, Lisbon, Portugal

Abstract—**This paper presents a Pupillometer that has been developed in order to detect the pupil's variation in time in a non-invasive way. The output signal of the equipment, a Pupillogram, was analyzed using Time-Frequency Analysis with the aim of better understanding the brain signals' activity expressed in the human eye.**

The Autonomic Nervous System (ANS) is divided into the Sympathetic and Parasympathetic components and theoretically they are always competing with each other as a two oscillator model.

The goal of this project is to demonstrate that these components can be identified and their frequency bands become more evident.

It is known that the pupil contracts when the Sympathetic component of the ANS is inhibited and the pupil dilates due to the inhibition of the parasympathetic component of the ANS. Because the pupil constrictor muscle is supplied by parasympathetic fibers and the dilator by sympathetic fibers, autonomic disorders may be detected with the appliance of a light stimulus (unresponsiveness to light) or other alarm stimuli.

With the present results we show that the signal from the pupil's variation has three major frequency bands, the first being 0.04-0.08 Hz band (LF - Low Frequency), the second being 0.10-0.70 (MF – Medium Frequency) and the third being 0.80 – 1.2 Hz band (HF - High Frequency). The MF and HF bands showed interesting results, from which we can infer a possible representation of the ANS activity.

Keywords—**Pupillometry, Neuro-Physiology, Autonomic Nervous System, Wavelet Analysis.**

I. INTRODUCTION

The Pupil is an aperture that occupies roughly the central area of the cornea in the eye.

This aperture has several functions, such as optimizing retinal illumination for visual perception, maximizing the number of photons at the surface of the eye's photoreceptors, reducing optical aberrations and reducing refractive error in cases of accommodation.

Pupil size, shape and reactivity to light have been used as indicators of neurological function in brain-injured patients, particularly in comatose patients. Thus, the problem of pupil morphology detection is important for making non-invasive diagnosis of many different diseases, because the pupil of human eye fluctuates either to adapt the amount of light to the retina or when a subject is gazing at a fixed object in absence of light stimulation and/or visual accommodation. [1]

Pupil abnormalities have been widely reported in association with generalized autonomic failure but, except in diabetes, rarely investigated in detail. [2]

Therefore the pupil has a great clinical importance because it serves as an indicator of:

1. Afferent Input;
2. Level of wakefulness (pupil diameter);
3. Autonomic nerve output to each iris;
4. Optical properties of the eye;
5. Monitoring pharmacologic effects.

The pupillary radius response to an external light stimulus might provide an indirect means of assessing the integrity of neuronal pathways controlling pupil size [3]. This is extremely useful in the clinical setting because the amount of transient pupil contraction to a light stimulus, or the steady state diameter of the pupil under constant illumination, can reflect the health of the retina and of the optic nerve and may be used to detect disease. [4]

Under normal conditions, pupil size continuously undergoes small changes due to the pupil's motility. The variance in diameter reflects the balance between tension in the two iris muscles, the iris sphincter (sympathetic inhibition) and the iris dilator (parasympathetic inhibition). At any time, pupil diameter is determined by the relative activity in the nerve fibers that excite these muscles. The parameters that describe pupillary behavior are pupillary unrest ("hippus", PU) and anisocoria (unequal size of the pupil). Measuring dynamic changes in pupil diameter (pupil motility) is a key problem in many studies of the eye and of the nervous system.

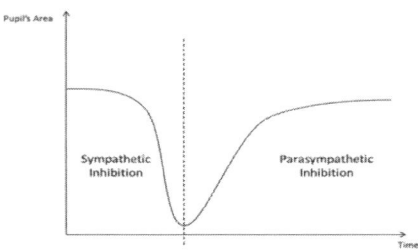

Fig. 1 Pupil's diameter decreases after a light stimulus and then increases because of the ANS compensation mechanisms.

Pupil's activity is mainly controlled by the Autonomic Nervous System (ANS). The ANS is composed by the Sympathetic Nervous System and the Parasympathetic Nervous System (see Fig.1).

The ANS two components have specific working bands of frequencies and if we could isolate them into separated signals we should be able to see the activity of each element and their response to several stimulli (see Fig.2).

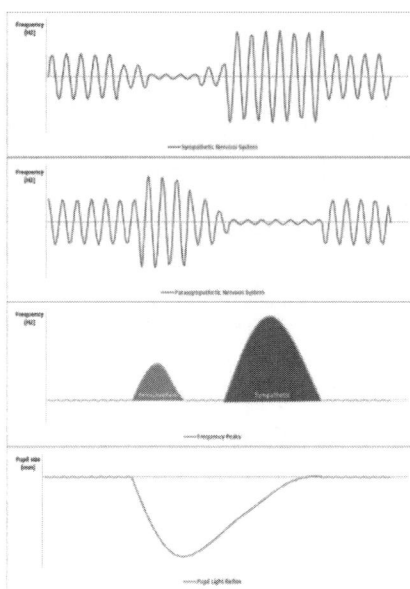

Fig. 2 Sympathetic Nervous System (first plot); Parasympathetic Nervous System (second plot); the third plot shows the frequency peak analysis of the two signals; the last plot shows the pupil's diameter behaving after a light stimulus (Pupil Light Reflex).

The ANS in the eye can also be compared to a two oscillator model (see Fig.3).

We know that a similar study of the heart and lungs has already been made, by decomposing the given signals (ECG and Pneumogram) into a series of sine and cosine functions of different frequencies and amplitudes, allowing the definition of a power spectrum in which three major ranges of frequencies for human subjects can be recognized (see Table 1): very low frequencies (VLF), low frequencies (LF) and high frequencies (HF). [5]

Table 1 Autonomic Nervous System's related frequencies according to the Task Force of the European Society of Cardiology and the North American Society of Pacing and Electrophysiology, 1996.

Frequency type	Frequency Interval
Very Low Frequencies	0.00 – 0.04
Low Frequencies	0.04 – 0.15
High Frequencies	0.15 – 0.40

The ANS behavior in the eye system can be compared to the one in the heart if taken the due precautions: the pupil's mass is much lower than the heart's; the pupil's oscillation frequency is much higher than the heart's and the pupil's reaction time is also much higher than the heart's (see Fig. 3 and Equations (1), (2) and (3)).

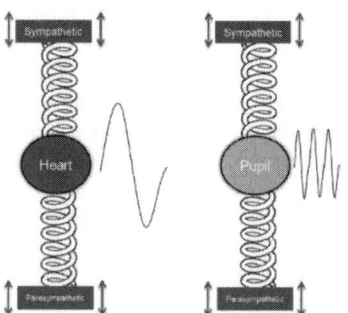

Fig. 3 Autonomic Nervous System model in the heart and the eye.

$$m_{Heart} \gg m_{Pupil} \qquad (1)$$
$$\Delta t_{Heart} \gg \Delta t_{Pupil} \qquad (2)$$
$$f_{Pupil} \gg f_{Heart} \qquad (3)$$

II. EXPERIMENTAL METHODS

A. The Pupillometer

Several commercial pupillometers already exist. They range from specially designed rulers or gauges to highly sophisticated infrared video-based systems. These last ones are used in toxicology, behavioral science and for drug, alcohol screening and assessing brain damage.

With the advent of refractive surgery of the cornea, the role of pupil diameter in controlling aberration in the optical system has become more clinically apparent. [4]

These devices can also be used for making diagnosis and monitoring psychiatric disorders such as depression and schizophrenia. [4]

This project used a Pupillometer described in [3] and [6].

The data presented in this paper was acquired according to four different protocols: a standard protocol, a random stimuli protocol, a random alert protocol and a random acquisition protocol.

The standard protocol includes a Pupillogram acquisition. During the signal acquisition the subject is stimulated with a light flash and before each flash a sound is emitted as a warning. These stimuli occur in a fixed time interval.

In the second protocol sometimes the light flash is not applied and in the third protocol the sound warning is random (in both random stimuli and random alert protocols the Pupillometer user also does not know if the stimulus and the sound are going to be applied).

The random acquisition protocol is similar to the first one, the only difference being that the time interval between a sound and a light flash is not a fixed value (also unknown to the user).

Below, in Fig. 4, the protocol diagram is shown:

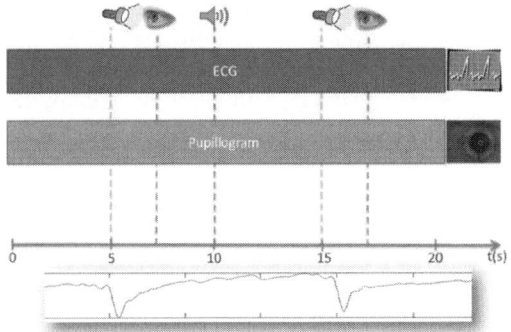

Fig. 4 The protocol diagram. A pupillogram is acquired and a sound warning is emitted before each stimulus. An ECG can be acquired parallel to the pupillogram for better correlation conclusions.

B. Image Processing

After each acquisition the PCD (Pupil Contour Detector) Algorithm was used.

PCD is an algorithm based on the concept of Intensity Threshold/Region Growing that measures the subject's pupil features along time [4].

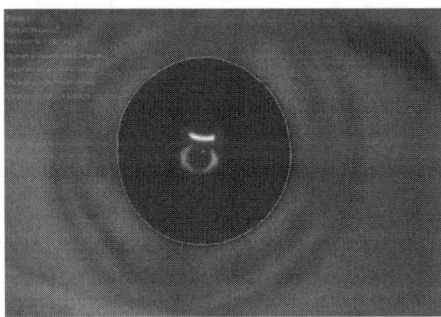

Fig. 5 Example of a PCD output.

Briefly, the PCD measures the spatial attributes of the pupil and its oscillations (pupil area, perimeter, horizontal diameter, vertical diameter, minimum diameter, maximum diameter, best fitted ellipse and centroid). A video file decomposition is made and each frame is processed using threshold and region growing methods (described in [6]).

C. Signal Analysis

After the image Processing, Wavelet Analysis was used to determine the main frequency components of each Pupillogram in order to allow correlation of the acquired signals.

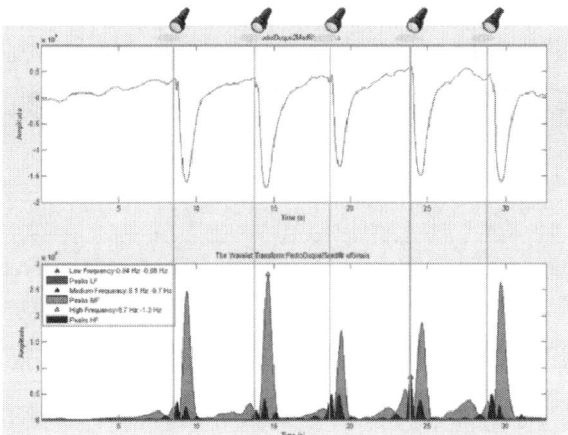

Fig. 6 The first plot shows an example of a pupillogram of a subject that was stimulated five times using the pupil light reflex test. The second plot shows the wavelet analysis of the Pupillogram with the most relevant frequencies of the signal.

Table 2 Amplitude values of each local minimum and their MF and HF local maximums.

Stimulus Peak	1	2	3	4	5
Minimum Amplitude	-1,60E+04	-1,72E+04	-1,32E+04	-1,49E+04	-1,61E+04
Maximum MF Peak	2,47E+08	2,79E+08	1,71E+08	1,86E+08	2,64E+08
Maximum HL Peak (first peak before each MF Peak)	3,50E+07	1,82E+07	5,00E+07	8,33E+07	4,99E+07

III. CONCLUSIONS

Thirty acquisitions were made to 5 subjects, each subject following each one of the protocols. After the image processing every Pupillogram was analyzed using Daubechies Wavelet family of order 5.

Three frequency intervals were defined as starting points of the analysis (see Table 3). According to the thesis presented in our model (Fig. 3), and using the known physical relations (Equations (1), (2) and (3)), higher frequency intervals were used than the ones considered in the Heart Model.

Table 3 Autonomic Nervous System's related frequencies defined as the most likely values for correlating to the ones known in the heart model.

Frequency type	Frequency Interval
Low Frequencies (LF)	0.04 – 0.08
Medium Frequencies (MF)	0.10 – 0.70
High Frequencies (HF)	0.80 – 1.20

The LF chosen value was mostly not visible. Therefore we did not use it t'o infer conclusions.

When the stimulus occurs, the subject's pupil contracts and then dilates, giving the signal a high change in amplitude. During contraction the HF band enhances its amplitude and during dilation the MF band is much more evident. As the specialized literature tells us [5, 7], MF has a greater amplitude than that of the HF.

A detailed analysis of the results shows that we can relate the pupil's fluctuations to the sympathetic and parasympathetic nervous flows in the subject's brain but we cannot yet conclude that the used frequencies are the ones that better describe the studied systems (they are a possible range of frequencies).

Through table 2 we can conclude that along each pupil contraction there is a relation between the amplitude of the local minimum and the MF and HF maximum peaks. When the amplitude is higher, the MF component tends to be lower in the local minimums, which have lower amplitudes. In these cases the HF component tends to be higher (see Table 4).

Table 4 When comparing 2 stimulus their MF and HF maximum peaks have different behavior in amplitude terms.

Comparison of peaks	
Amplitude	>>
MF Peak	<<
HL Peak	>>

We can see two distinct frequency components, one faster than the other, corresponding to the Sympathetic and Parasympathetic Autonomic Nervous Systems, respectively. And we will also be able to correlate the pupil's signal to the ECG from each subject.

As was described in the introduction section, this feature can be used to compare the pupil's fluctuations of a normal subject to the ones of the pupil of a subject that suffers from a neurological disorder (Alzheimer or Narcolepsy, for example).

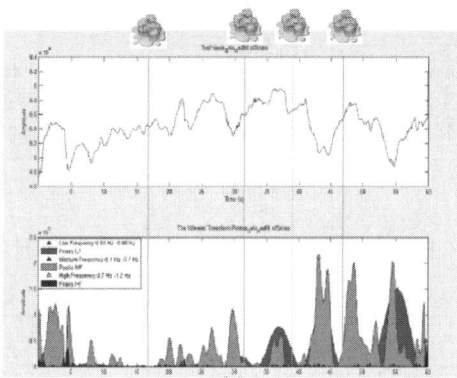

Fig. 7 The first plot shows an example of a pupillogram of a subject that was stimulated four times using the cold pressure test. The second plot shows the wavelet analysis of the Pupillogram with the most relevant frequencies of the signal.

Since the pupil light reflex (applied stimulus) affects mostly the Parasympathetic Nervous System we are now acquiring more data through the application of other type of stimuli, this time to stimulate the Sympathetic component. The cold pressure test was chosen (the subject plunges his/her hand in a bucket of cold/ice water).

In Fig. 7 we can see an example of the cold pressure test. The subject was stimulated unknowingly of the instant of each stimulus.

Simultaneously, we will work on the portability factor by developing the PCD in several platforms such as smartphones, since the PCD can be used not only for medical but also for more recreational purposes (for example as a lie detector).

Acknowledgment

This project has been funded by the Department of Physiology of the Institute of Molecular Medicine and by the Faculty of Sciences and Technology of the New University of Lisbon. It has been approved by the Santa Maria Hospital (Lisbon, Portugal) Ethics Committee (based on the Laser Products – Conformance with IEC 60825-1 and IEC 60601-2-22; Guidance for Industry and FDA Staff).

References

1. Iacoviello, D., Matteo, L. "Parametric characterization of the form of the human pupil from blurred noisy images." *Computer Methods and Programs in Biomedicine* 77 (2005): 39–48.
2. Bremner, F.D., Smith, S.E.: Pupil Abnormalities in Selected Autonomic Neuropathies. Journal of Neuro-Ophthalmology 26(3) (2006): 209-19.
3. Leal, G., P. Vieira and C. Neves. "Pupillometry: Development of Equipment for Studies of Autonomic Nervous System". *DoCEIS 2012*, IFIP AICT 372, 2012. 553 – 562.
4. KARDON, Randy. "Pupil". *Adler's Physiology of the Eye*. Edited by Paul L. Kaufman and Albert Alm. 10th Edition. St Louis: Mosby, 2003. 713-743.
5. Ducla-Soares J. L., M. Santos-Bento, S. Laranjo, A. Andrade, E. Ducla-Soares, J. P. Boto, L. Silva-Carvalho and I. Rocha. "Autonomic Neuroscience: Wavelet analysis of autonomic outflow of normal subjects on head-up tilt, cold pressor test, Valsalva manoeuvre and deep breathing." Experimental Physiology 92, 2007:677-686
6. Leal, Gonçalo, Vieira, Pedro, Neves, Carlos. "Development of an instrument to measure pupil dynamics in subjects suffering from neuro-ophtalmological disorders". O. Dössel and W.C. Schlegel, eds. WC 2009, IFMBE Proceedings 25/XI, 2009. 149–152.
7. Malliani, A. "The sympathovagal balance explored in the frequency domain." *Principles of Cardiovascular Neural Regulation in Health and Disease*. Boston, Mass./Dordrecht, the Netherlands/London, UK: Kluwer Academic Publishers, 2000. 65–108.

Author: Gonçalo Macedo Leal
Institute: Faculty of Sciences and Technology
Street: N377-1
City: Almada
Country: Portugal
Email: gnl@campus.fct.unl.pt

Part VI
Bio-micro and Bio-nano Technologies

Analysis Sensitivity by Novel Needle-Type GMR Sensor Used in Biomedical Investigation

H. Shirzadfar[1], J. Claudel[1], M. Nadi[1], D. Kourtiche[1], and S. Yamada[2]

[1] Institut Jean Lamour, CNRS, UMR 7198/ Département Nanomatériaux Electronique Et Vivant (406), Université de Lorraine, Nancy, France
[2] Institute of Nature and Environmental Technology / Division of Biological Measurement and Application, Kanazawa University, Kanazawa, Japan

Abstract—We have attempted to characterize the new needle-type probe giant magnetoresistance (GMR) as a magnetic micro biosensor. The sensitivity of this sensor has been determined. The high sensitivity of GMR sensor has a good ability to estimate magnetic fluid particles (in low concentrations) and to detect the micro or nano magnetic markers in bio-medical application.

Keywords—GMR Sensor, Sensitivity, Helmholtz coil, Magnetic flux density.

I. INTRODUCTION

Biosensor represents a device in which a sensitive layer containing biological material such as enzymes, DNA, bacteria and etc. [1], directly responding to the presence of defined component. This device generates a signal linked to the concentration of the component. Today, progressive development of a biosensor, as powerful analytical tool, relies mainly on its high sensitivity [2].

It should be mentioned that the magnetic properties of biological matter are still investigated in the frame of research related to nanoparticles applications [3]. Variations of susceptibility and permeability frequency support the development of new biomedical applications.

Fabrication the new biosensor in basic of Giant Magnetoresistance effect is a new way for detecting biological molecules being magnetic or being labeled with magnetic beads or nanoparticles.

In this paper we present the needle-type GMR sensor designed and fabricated in Kanazawa University, which has a high capability for detection and measurement the physical properties of low concentration magnetic fluid in small field range [4, 5].

II. NEEDLE TYPE GMR SENSOR

The two main parts of the GMR sensor are a thin needle of sensor and a Wheatstone bridge circuit. The needle type of this GMR sensor consists of two sensing elements with sensing area 75μm×75μm, the first sensing element is at the tip of the needle and the second sensing element is at the end of the needle, the two sensing elements are connected in a Wheatstone bridge configuration. The needle length of GMR sensor is 30mm by cross section 300μm×300μm. This size of needle is very adequate for injection in small container and distinct this model from other magnetic sensors.

The needle is fabricated from a hard material such as Aluminum Titanium Carbide (AlTiC), and compound materials of Aluminum Oxide and Titanium Carbide (Al2O3/TiC). The scheme of GMR sensor is presented in Fig. 1.

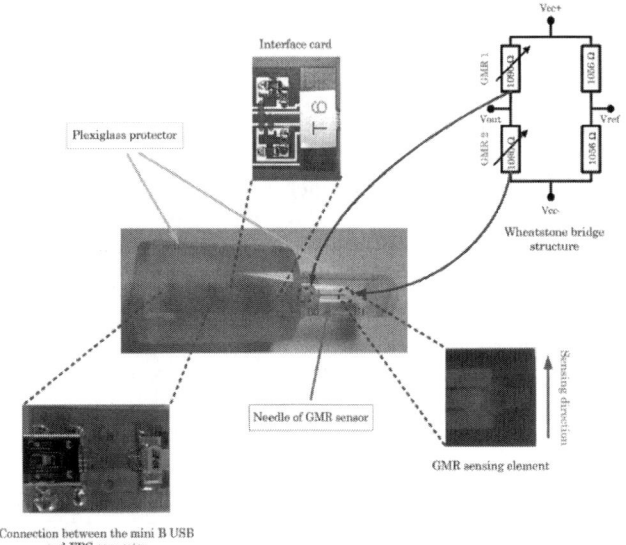

Fig. 1 Real image of needle type GMR sensor

The Wheatstone bridge configuration of sensor has two GMR sensing elements and two fixed resistances. Two GMR sensing elements are sensible to present the magnetic flux density (B), the first GMR estimates the magnetic flux density (B_1) in the container of magnetic fluids and the second determinates the external magnetic flux density (B_0). The detection direction of the GMR sensor is perpendicular to the needle, and this direction is very important for positioning the coil. Moreover, for this type of GMR sensor the maximum current is 9.7mA.

III. CHARACTERIZATION OF GMR SENSOR

A. Determining Sensitivity of GMR Sensor

Determination of sensitivity value of each sensor is main experience part before the final test. For this reason, we tried to estimate accurately the sensitivity value of GMR sensor.

As mentioned in previous part, GMR sensor has two sensing elements at the tip and end of sensor's needle. To characterize correctly the sensor, we should place only one sensing element in uniform magnetic flux density. For this purpose we fabricated Helmholtz coil. The Helmholtz coil is ideal for producing a large area of uniform magnetic field. It should be noted that the measurement magnetic field range by GMR sensor is within tens of nT to few mT.

The small Helmholtz coil was designed and fabricated by authors. It consists of two coils with diameter 18 mm. The coil radius is equal to the distance between two coils, each coil has 9 wire turns.

The measurement setup for characterization of GMR sensor that has been prepared for this research is presented in Fig. 2.

To facilitate the work and to provide protection of needle against any shock (see Fig. 2), we have used the displacement system type RCP2-SA6C produced by Rosier company. It is controlled by controller PCON-CG-42PI that allows backup 512 preset positions with the software. The displacement range of system is 100 mm with positioning repeatability approximately ±0.02mm [6]. The measurement setup is controlled by two laptops, one to conduct the displacement system and second to control the HF2IS impedance spectroscope and also to store measurement data, obtained by the GMR sensor.

The GMR sensor is connected to DC power supply (±6V) by circuit interface where the resistance placed to prevent the current. Differential signal Vout−Vref was supplied to preamplifier AD 620 where output signal is strengthened approximately 1000 times (60dB) and after submitted to HF2IS Impedance Spectroscope. The data (Vrms, phases [deg]) of output of GMR signal was recorded with interface program of HF2IS Impedance Spectroscope. Furthermore, the HF2IS Impedance Spectroscope operates on the basis of a lock-in amplifier and can be used as impedance or lock-in amplifier.

The Helmholtz coil is supported by an amplifier (variable gain). By virtue of two output signal generators of HF2IS Impedance Spectroscope, we can directly generate the supply system of Helmholtz coil without presenting waveform generator that leads to more system integration. Between power amplifier and small Helmholtz coil we have put a multimeter for checking the passing current in coil and for protection the wire coil. The experiment frequency of exciting field of the Helmholtz coil is 60Hz. By change the value of input voltage (mV), the magnetic flux density (B) has been switched approximately from 5μT to 410μT. These results have been obtained by 3D gaussmeter 460 three-channel Hall effect (AC RMS accuracy is ±2% and AC frequency range is 10Hz to 400Hz) [7].

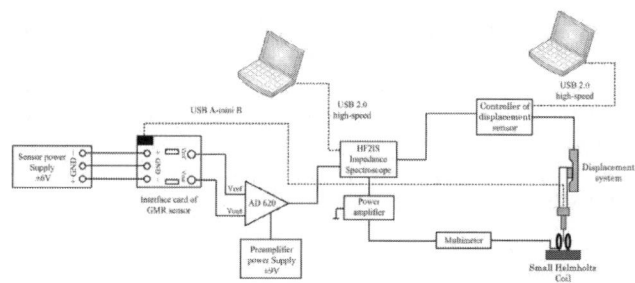

Fig. 2 Measurement setup for characterization the GMR sensor

Two PCB terminal blocks embedded into interface card of the GMR facilitate changing the value of circuit resistance and allow to find the appropriate amount of resistance, needed to safe the GMR sensor and to have the most possible sensitivity. The different values of resistances in interface card of GMR sensor have been tried. Figure 3 presents the experimental results to found the best sensitivity values of GMR needle probe sensor by different resistances.

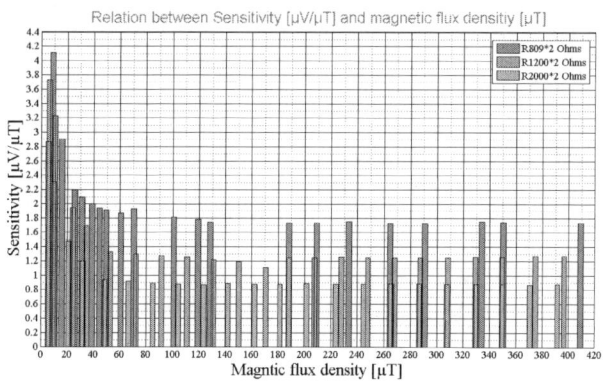

Fig. 3 The relative between sensitivity (μV/μT) and B (μT)

The top choice to obtain high sensitivity is to use two additional resistors in the interface card circuit of the GMR sensor by value 809Ω with magnetic flux density equal approximately to 9μT.

The GMR sensor sinusoidal output voltage (V_{out}–V_{ref}) result that amplified with AD 620 and used two resistors 809Ω in interface supply circuit, shown in Fig. 4.

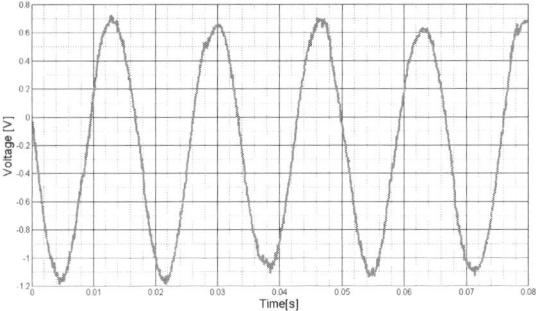

Fig. 4 Sensor output voltage (V) with R=2*809 Ω in interface card

The output voltage (μV) and phase (deg) results obtained by the GMR sensor (only for sensing element at top of sensor's needle) with used two resistors (809Ω) as a function B (μT) is presented in Fig. 5. As shown in results the Vout–Vref (μV) is liner and proportional by change the value of magnetic flux density (μT).

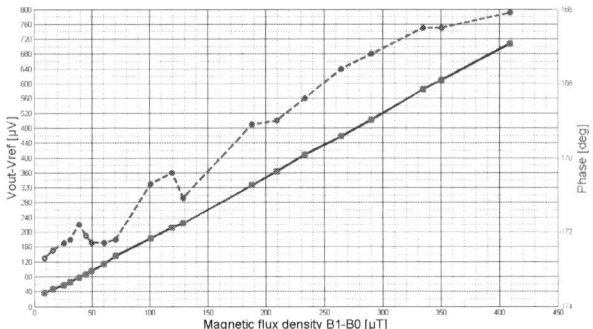

Fig. 5 Output voltage result by probe GMR sensor (μV) and phase calculation (deg)

B. Calculation the Variation R of GMR Sensing Element in Wheatstone Bridge Connection of Sensor

As mentioned the GMR needle type probe consists of two sensing elements and both sensing elements are connecting to Wheatstone bridge of the GMR sensor. The sensing elements at the tip and at the end of needle are equivalent to a potentiometer (ΔR+R), and they used for detecting the exciting magnetic field (potentiometer is variable according to changing magnetic field). Figure 6 presents the resistance variation of GMR element at top of needle as function the magnetic field (the resistance variation for another sensing element is zero, due to the absence of a magnetic field).

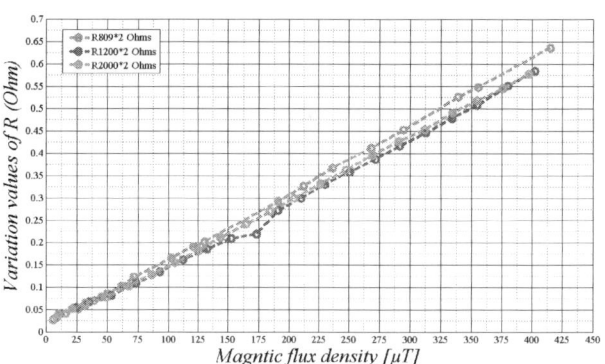

Fig. 6 The values of ΔR according magnetic flux density (B) with different resistances in interface supply card of GMR sensor

We can observe from the Fig. 6, that all results are linear and proportional to values of magnetic flux density (B). In additional all results are similar for different resistances obtained from interface supply card of the GMR sensor. Because in theory, the variation of R for sensing element in Wheatstone bridge of the GMR sensor is changed according to change the magnetic flux density. For each value of resistance (in interface card) the results of variation resistance of sensor are approximately the same. More, according the results achieved from GMR sensor, we conclude that the value of the GMR element sensitivity as function of magnetic flux density (B) is approximately 0.34% mT^{-1}.

C. Distance Calculation of Displacement System

The displacement system adapts for changing the position of the Helmholtz coil according to the sensing direction of each sensor. This system provides us possibility to control carefully the needle of the GMR sensor for two positions, injecting and cleaning the needle.

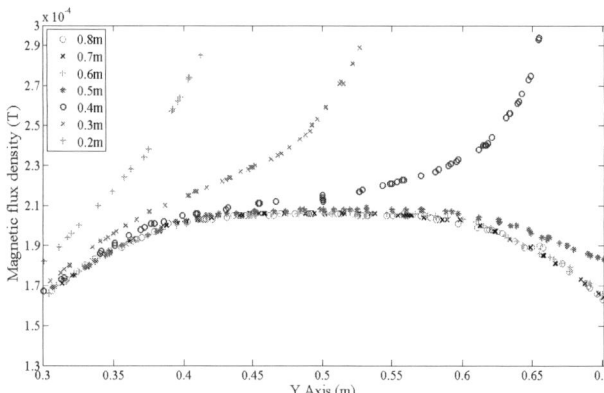

Fig. 7 Changing the magnetic flux density (B) in midpoint of Helmholtz coil considering to displacing the displacement system along the Y axis

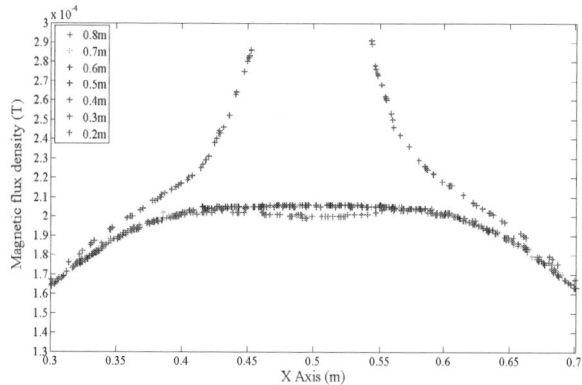

Fig. 8 Changing the magnetic flux density (B) in center of Helmholtz coil considering to changing the position of displacement system along the X axis

Given the moving the displacement system of the GMR sensor, the magnetic flux density disrupts and changes. For this reason, the distance between the displacement system and the Helmholtz coil for both axes (X and Y) is important. Further, the Helmholtz coil with diameter 0.52m is fabricated to use for next experiment with ferrofluids.

We utilized the simulation of COMSOL 4.0 to estimate the appropriate height. We created a displacement system with dimension of 0.4H×0.06W×0.06D and selected the several heights from 0.8m to 0.2m (near the midpoint of the Helmholtz coil). The simulation results obtained for vertical and horizontal positions are presented in Fig. 7 and Fig. 8.

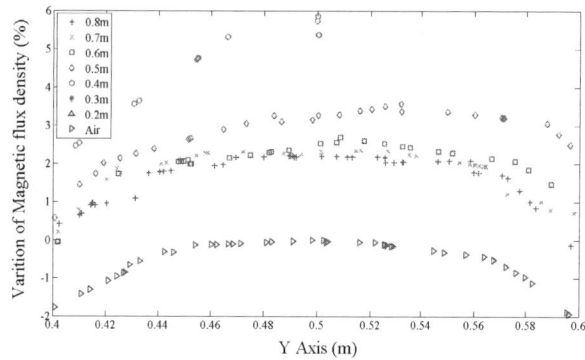

Fig. 9 Calculation of the percentage error (%) of magnetic flux density variation as function of the system position along the Y axis

Also, the variation of magnetic flux density (%) as function of position the displacement system have been calculated and is presented in Fig. 9. The position of displacement system of the GMR sensor in the locations 0.2m and 0.3m are destroyed completely the uniformity of magnetic field and for this reason, they are not observed in Fig. 9. In order to solve the disordering of the magnetic flux density in the center of the Helmholtz coil, the displacement system of sensor is placed in a distance 40cm (approximately the point 0.74m in a simulation) from the upper surface of the Helmholtz coil. The GMR sensor is connected to displacement system with a plexiglass rod.

IV. CONCLUSIONS

In this paper we presented a magnetic sensor appropriate to use in biomedical applications. We completed the characterization of the sensing element, as well as optimizing the total sensitivity of the GMR sensor as a function of the load resistance. Eventually, a value 4.12 µV/µT for maximum sensitivity of sensor has been obtained. Furthermore, we determined the height of displacement system to not disorder the magnetic flux density in the center of the Helmholtz coil. The proposed sensor is ready for next researches to perform measurements on ferrofluids and detection the bacteria with bio-markers.

ACKNOWLEDGMENT

This research is supported by Institute Jean Lamour (IJL) in University of Lorraine in corroboration with Institute of Nature and Environmental Technology of Kanazawa University. The authors would like to thank P. Roth from Institut Jean Lamour (IJL) for his technical support.

REFERENCES

1. Rodriguez-Mozaz S, Marco M P, Lopez de Alda D et al. (2004) Biosensors for environmental applications: Future development trends. Pure Appl Chem 76:723–752
2. Ibrahim M, Claudel J et al. (2013) Geometric parameters optimization of planar interdigitated electrodes for bioimpedance spectroscopy. J Elec Bioim 4:13–22
3. Gupta A K, Gupta M. (2005) Synthesis and surface engineering of iron oxide nanoparticles for biomedical application. Biomaterials 26:3995–4021
4. Gooneratne C P, Kakikawa M, Ueno T et al. (2010) Measurement of minute changes in magnetic flux density by a novel GMR needle probe for application in Hyperthermia Therapy. J Magn Soc Jpn 34:119–122
5. Shirzadfar H, Haraszczuk R, Nadi M et al (2012) Detecting and Estimating Magnetic Fluid Properties by a Needle-Type GMR Sensor. Alushta, Ukraine
6. Rosier at http://www.rosier.fr
7. Lakeshore at http://www.lakeshore.com

Author: Hamidreza Shirzadfar
Institute: Institut Jean Lamour, CNRS, UMR 7198, Département Nanomatériaux Electronique Et Vivant (406)
Street: Rue du jardin botanique
City: Vandœuvre-lès-Nancy
Country: France
Email: hamidreza.shirzadfar@univ-lorraine.fr

Focusing of Electromagnetic Waves by Non-Spherical, Au-Si Nano-particles

A.P. Moneda and D.P. Chrissoulidis

Aristotle University of Thessaloniki/Department of Electrical and Computer Engineering, Thessaloniki, Greece

Abstract—A dyadic Green's function for a sphere with an eccentric spherical inclusion is used to evaluate the electric-field intensity throughout a non-spherical nano-particle, excited by a distant, electric, Herz dipole. The particle comprises gold (Au) and glass (Si) eccentric spherical layers. The theory is analytical and exact. The numerical results suggest that proper combination of the two materials can enhance the electric-field intensity on the surface of the metal by as much as 200 times.

Keywords—dyadic Green's function, non-spherical, nano-particle.

I. INTRODUCTION

A dyadic Green's function (dGf) [1] for a sphere with an eccentric spherical inclusion [2] is used to investigate the focusing properties of nano-particles composed of gold and glass parts. Two cases are considered: (a) a glass sphere with an eccentric, spherical, gold coating (Au/Si particle) and (b) a solid gold sphere with an eccentric, spherical, glass coating (Si/Au particle). Excitation is provided by a distant point source. Our solution to this radiation problem is general and exact because (a) no simplifying assumptions are made in the analytical steps leading to the dGf and (b) there is no theoretical constraint with regard to the physical size, the electrical properties, and the eccentricity of the composite, non-spherical particle.

The numerical application of the aforesaid theory pertains to tissue hyperthermia and it is aimed at finding the geometry of the Au-Si particle which favors the formation of an intensive hot spot on the metallic surface. The electrical properties of noble metals vary with the wavelength in the optical band of the electromagnetic spectrum [3]; the dielectric constant, being complex with negative real part, is known to be responsible for the enhancement of the electric-field intensity on the surface of such metallic bodies [4].

This phenomenon can be optimized by proper combination of the metal with dielectric materials and/or by consideration of roughness or curvature on the metallic surface [5]–[12]. The role of several parameters in composite Au-Si particles, such as the size of a glass core underneath a gold layer, the thickness of a glass coating around a gold core, and the eccentricity between core and coating, is investigated with regard to the intensity of hot spots on the surface.

II. THEORY

The geometry of the problem is shown in Fig. 1. A small dipole, actually a point source, radiates in the presence of a dielectric sphere containing an eccentric, spherical inclusion. The source can be inside or outside the inhomogeneous sphere; the host sphere, the spherical inclusion, or both may be lossy.

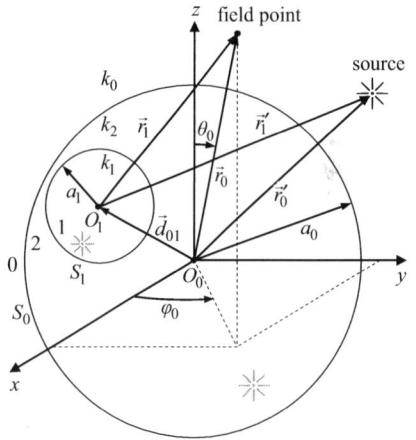

Fig. 1 Geometry

The space outside the composite non-spherical particle is unbounded, free space (region 0). The host sphere (region 2, radius a_0) accommodates the spherical inclusion (region 1, radius a_1) which is centered at the tip of the vector $\vec{d}_{01}$. The inclusion can be anywhere inside the host sphere, but it cannot protrude from the spherical surface S_0; hence, the condition $d_{01} + a_1 \leq a_0$ must be satisfied. The wavenumber of each region is $k_i = \omega\sqrt{\varepsilon_i \mu_0}$, where $i = 0,1,2$. Any field point in space can be defined by a position vector $\vec{r}_0$ or $\vec{r}_1$, the subscript indicating the origin, which is O_0 (host sphere) or O_1 (inclusion); the primed position vectors $\vec{r}_0'$ or $\vec{r}_1'$ point to the source. Harmonic time dependence $e^{-j\omega t}$ is implied and suppressed.

The electric dGf of the geometry described above is determined by superposition of the dGf of the unbounded free-space and the dGf corresponding to the scattered part of the EM field; it has been formulated in [2] as follows:

$$\bar{\bar{G}}_e^{(is)}(\vec{r},\vec{r}') = -\delta_{si}\frac{1}{k_s^2}\hat{r}_i\hat{r}_i\delta(\vec{r}-\vec{r}_i') + j\frac{k_s}{4\pi}\sum_{v,m,n} c_{mn}$$
$$\times \left[\begin{array}{l}\delta_{i0}\vec{F}_{v,mn}^{(3)}(k_i\vec{r})\vec{A}_{v,mn}^{(i)}(\vec{r}') + \delta_{i1}\vec{F}_{v,mn}^{(1)}(k_i\vec{r})\vec{C}_{v,mn}^{(i)}(\vec{r}') \\ + \delta_{si}\vec{F}_{v,mn}^{(1)}(k_s\vec{r})\vec{F}_{v,-mn}^{(3)}(k_s\vec{r}_i')\end{array}\right], \quad (1)$$

$$\bar{\bar{G}}_e^{(2s)}(\vec{r},\vec{r}') = -\delta_{s2}\frac{1}{k_s^2}\hat{r}_i\hat{r}_i\delta(\vec{r}-\vec{r}_i') + j\frac{k_s}{4\pi}\sum_{v,m,n} c_{mn}$$
$$\times \left[\begin{array}{l}\vec{F}_{v,mn}^{(3)}(k_2\vec{r}_1)\vec{A}_{v,mn}^{(2)}(\vec{r}') + \vec{F}_{v,mn}^{(1)}(k_2\vec{r}_0)\vec{C}_{v,mn}^{(2)}(\vec{r}') \\ + \delta_{s2}\vec{F}_{v,mn}^{(1)}(k_s\vec{r})\vec{F}_{v,-mn}^{(3)}(k_s\vec{r}_i')\end{array}\right], \quad (2)$$

wherein $\delta(\cdot)$ is the Dirac delta function and $\delta_{nn'}$ is the Kronecker delta; $v=M,N$; $n=1,2,\ldots$; $m=-n,\ldots,n$; $i=0,1$; $c_{mn}=(-1)^m(2n+1)/n(n+1)=c_{-mn}$; $\vec{F}_{M,mn}^{(t)}(k\vec{r})$, $\vec{F}_{N,mn}^{(t)}(k\vec{r})$ are vector spherical harmonics [2],[13] and $t=1,3$. Eqs. (1) and (2) hold for $r_i < r_i'$; the superscripts 1 and 3, if marked by a tilde, must be interchanged if $r_i > r_i'$. The unknown, vector, wave amplitudes $\vec{A}_{v,mn}^{(i)}(\vec{r}')$ and $\vec{C}_{v,mn}^{(i)}(\vec{r}')$ – $i=0,1,2$ – can be determined [2] by enforcement of the boundary conditions on S_0, S_1. Ultimately, the electric-field intensity in region $f=0,1,2$ – field region – due to a point source in region $s=0,1,2$ – source region – is obtained by use of the following integral:

$$\vec{E}(\vec{r}) = j\omega\mu_0 \iiint_{V'} \bar{\bar{G}}_e^{(fs)}(\vec{r},\vec{r}')\cdot\vec{J}(\vec{r}')dV'. \quad (3)$$

III. APPLICATION

The configurations studied are shown in Fig. 2; a gold sphere of radius $a_0 = 65\,\text{nm}$ accommodates a smaller sphere of glass (Au/Si case) or it is surrounded by a spherical glass coating (Si/Au case). The source is a small, distant, electric Herz dipole on the z axis, oriented along the x axis. The wavelength of the incident radiation spans the range $190-1900\,\text{nm}$ and the refractive index of gold varies accordance to published data [3]. The refractive index of silica glass is 1.5. The electric dGf of Section ii, adapted to each of the aforesaid configurations, yields the electric-field intensity in every part of space. The investigation is aimed at finding the maximum value of the amplitude of the electric-field intensity, i.e. $|\vec{E}|_{\max}$, regardless of where that may be, inside or outside the Au-Si particle. A map of $|\vec{E}|$ on the xO_0z plane is shown in Fig. 7(a) for the marginal case of a naked, solid, gold sphere of radius $a_0 = 65\,\text{nm}$ at $\lambda = 548\,\text{nm}$; $|\vec{E}|$ is given in dB relative to $1\,\text{V/m}$. A black or white circle in Fig. 7 denotes the (outer) surface of glass or gold, respectively.

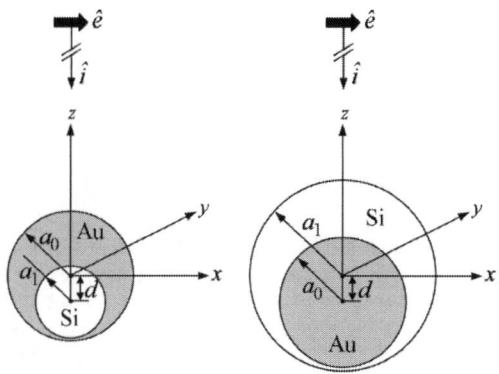

Fig. 2 Gold sphere with glass core (Au/Si particle) or glass coating (Si/Au particle) and distant electric Herz dipole

The effect of glass, core or coating, on the gold sphere is investigated first by consideration of the Au/Si case shown in the left-hand side of Fig. 2. This being the first step, we set $d=0$, which means that the glass core is at the center of the gold sphere. The effect of core size on $|\vec{E}|_{\max}$ is studied by allowing the normalized core radius, i.e. a_1/a_0, to vary in the range $[0,1)$ and by sweeping the wavelength band $190-1900\,\text{nm}$. Results for $|\vec{E}|_{\max}$, in dB relative to $1\,\text{V/m}$, and for the wavelength which gives rise to $|\vec{E}|_{\max}$, i.e. the resonant wavelength, are shown against the normalized radius of the glass core a_1/a_0 in Fig. 3; we use red for $|\vec{E}|_{\max}$ and blue for the resonant wavelength; the same is done in Figs. 4-6.

These calculations suggest that a glass core within a gold sphere may enhance the electric-field intensity, but only for core size $0.6a_0 \leq a_1 < 0.95a_0$; as the glass core grows, the resonant wavelength increases, which ought to be expected according to [6], [8], [9]. The curves depicting our calculations of the resonant wavelength are in very good agreement with the black dashed curve which corresponds to a theoretical estimation, based on a solution of the Laplace equation [8]. The smooth change in the resonant wavelength, as the core grows, makes this configuration attractive for various applications [14], [15]. No information about the location of $|\vec{E}|_{\max}$ is provided by Fig. 3, but the field maximum is always found just outside the gold sphere. Furthermore, $|\vec{E}|_{\max}$ can be enhanced because of the glass core by as much as $10\,\text{dB}$; such enhancement is achieved with $a_1 = 0.93a_0$ at $\lambda = 891\,\text{nm}$.

The effect of eccentricity on $|\vec{E}|_{\max}$ is studied by consideration of the aforesaid Au/Si particle with a glass core of size $a_1 = 0.85a_0$. The results of Fig. 4 manifest that $|\vec{E}|_{\max}$ increases with the eccentricity, but the resonant wavelength is $704\,\text{nm}$ for any value of d. If the core is

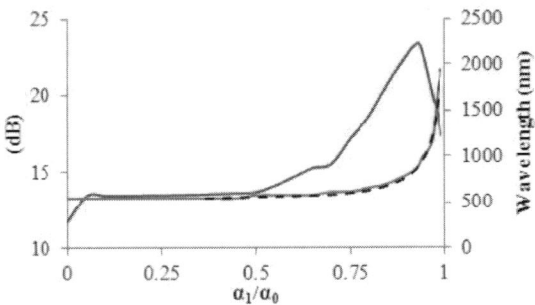

Fig. 3 Concentric Au/Si particle: $|\vec{E}|_{max}$ (red line) and resonant wavelength (blue line)

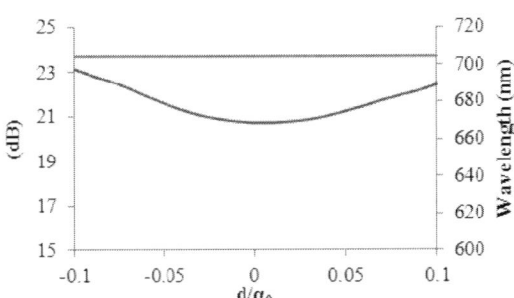

Fig. 4 Eccentric Au/Si particle: $|\vec{E}|_{max}$ and resonant wavelength

placed close to either pole of the host sphere ($d \approx \pm 0.1 a_0$), $|\vec{E}|_{max}$ is found at the surface of the metal and on the side of that pole; the enhancement due to eccentricity can be almost 3 dB.

Maps of $|\vec{E}|$ on the xO_0z plane of the aforesaid Au/Si particle ($a_0 = 65$ nm, $a_1 = 0.85 a_0$) are shown in Figs. 7(b) and 7(c) for $d = 0$ and $d = -0.1 a_0$, respectively, both at $\lambda = 704$ nm. It is interesting that $|\vec{E}|_{max}$ is found on the waist of the particle if the core is concentric, but it shifts to the pole, north or south, which is approached by the core when $d \approx \pm 0.1 a_0$.

Attention is next focused on the Si/Au particle shown in the right-hand side of Fig. 2: a sphere of glass (radius a_1) accommodates an eccentric, solid gold sphere of radius $a_0 = 65$ nm. As in the case treated before, the investigation begins with the concentric configuration. The thickness of the glass coating varies so that $a_0/a_1 \in [0.3, 1]$ and the results are shown in Fig. 5.

A black, dashed line is shown beside the plot of the resonant wavelength in the range $a_0/a_1 \in [0.5, 1]$; that curve, practically a straight line, is an outcome of calculations based on a theoretical solution of the Laplace equation [8],

but it underestimates the resonant wavelength, probably because the assumption of a static electric field, upon which that theory is based, is poor in this case. As far as $|\vec{E}|_{max}$ is concerned, the glass coating results in enhancement of the electric-field intensity on the surface of the metal. As long as $a_0 < 0.8 a_1$, the thicker the glass the higher is the level of $|\vec{E}|_{max}$ and that trend is accompanied by monotonous redshift of the resonant wavelength. If $a_0 \approx 0.3 a_1$, the aforesaid enhancement of $|\vec{E}|_{max}$ is, according to Fig. 7(d), almost 4 dB at $\lambda = 704$ nm, relative to a naked, solid gold sphere of the same size.

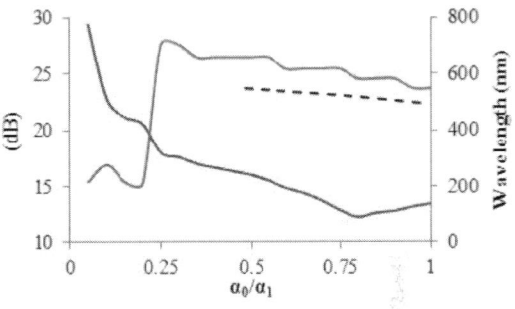

Fig. 5 Concentric Si/Au particle: $|\vec{E}|_{max}$ and resonant wavelength

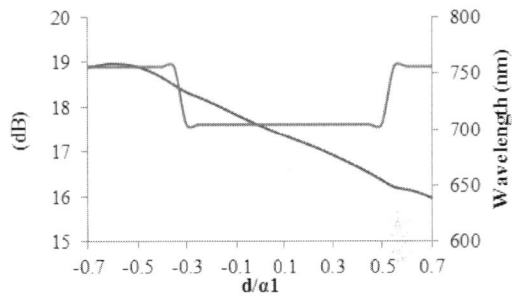

Fig. 6 Eccentric Si/Au particle: $|\vec{E}|_{max}$ and resonant wavelength

The effect of an eccentric glass coating is seen (Fig. 6) by allowing the gold core to move along the z axis, from the south to the north pole of the Si/Au particle; the external radius of the composite particle is $a_1 = a_0/0.3$. The maximum of $|\vec{E}|$ decreases substantially as the core moves towards the north pole (side of incidence.) The change in $|\vec{E}|_{max}$ from pole to pole is almost 3 dB; the corresponding change in the resonant wavelength is negligible. Compared to the concentric particle, the enhancement of $|\vec{E}|_{max}$ due to the eccentricity of the core is up to about 1.5 dB; this value is achieved with $d = -0.65 a_1$ at $\lambda = 755$ nm. A map of $|\vec{E}|$ on the xO_0z plane of the eccentric Si/Au particle is shown by Fig. 7(e).

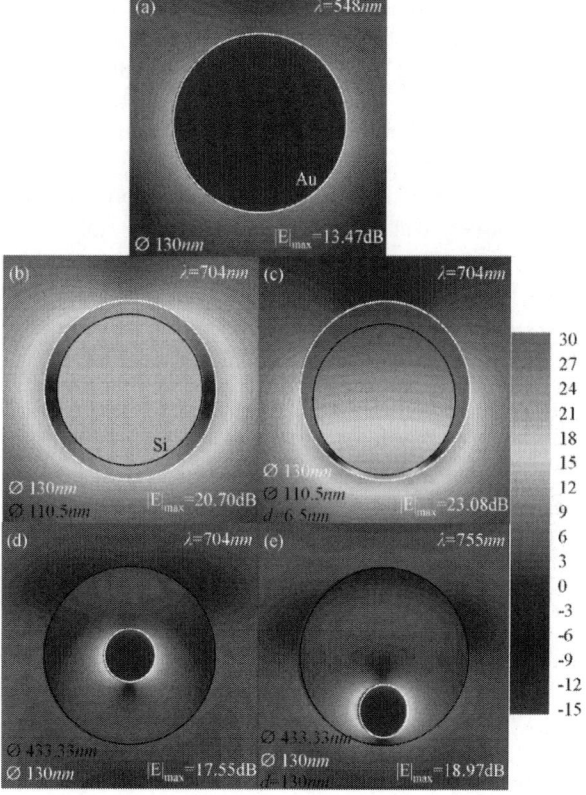

Fig. 7 Maps of the electric-field intensity on xO_0z plane

IV. CONCLUSION

Focusing properties of Au-Si particles have been investigated in the wavelength range 190–1900 nm by use of the dGf of a sphere with an eccentric spherical inclusion. The numerical results suggest that an eccentric glass core within a sphere of gold can enhance the amplitude of the electric-field intensity on the surface of the metal by up to 10 times, relative to a solid gold sphere of the same size. The electric-field intensity within the hot spot on the metal surface of an Au/Si particle can exceed that of the incident wave by almost 200 times. Si/Au particles are less efficient than Au/Si particles, as far as focusing is concerned.

ACKNOWLEDGMENT

This research has been co-financed by the European Union (European Social Fund–ESF) and Greek national funds through the Operational Program "Education and Lifelong Learning" of the National Strategic Reference Framework (NSRF) –Research Funding Program: **THALES**. Investing in knowledge society through the European Social Fund.

REFERENCES

1. Tai C (1993) Dyadic Green Functions in Electromagnetic Theory, 2nd ed. IEEE Press, New York
2. Moneda A, Chrissoulidis D (2007) Dyadic Green's function of a sphere with an eccentric spherical inclusion. *J Opt Soc Am A* 24:1695–1703
3. Johnson P, Christie R (1972) Optical constants of the noble metals. *Phys Rev B*, 6:4370–4379
4. Moskovits M (1985) Surface–enhanced spectroscopy. *Rev Mod Phys*, 57:783–826
5. Campion A, Kambhampati P (1998) Surface–enhanced Raman scattering. *Chem Soc Rev*, 27:241–250
6. Harris N, Ford M, Cortie M (2006) Optimization of plasmonic heating by gold nanospheres and nanoshells. *J Phys Chem B*, 110:10701–10707
7. Averitt R, Westcott S, Halas N (1999) Linear optical properties of gold nanoshells. *J Opt Soc Am B* 16:1824–1832
8. Neeves A, Birnboim M (1989) Composite structures for the enhancement of nonlinear–optical susceptibility. *J Opt Soc Am B* 6:787–796
9. Kerker M, Blatchford C (1982) Elastic scattering, absorption, and surface–enhanced Raman scattering by concentric spheres comprised of a metallic and a dielectric region. *Phys Rev B*, 26:4052–4063
10. Barnickel P, Wokaun A (1989) Silver coated latex spheres: optical absorption spectra and surface enhanced Raman scattering. *Mol Phys*, 67:1355–1372
11. Mühlschleger P, Eisler H, Martin O et al. (2005) Resonant optical antennas. *Science*, 308:1607–1609
12. Hao E, Schatz G (2004) Electromagnetic fields around silver nanoparticles and dimers. *J Chem Phys*, 20:357–366
13. Morse P, Feshbach H, (1953) *Methods of Theoretical Physics*. McGraw-Hill, New York
14. Loo C, Lowery A, Halas N et al (2005) Immunotargeted nanoshells for integrated cancer imaging and therapy. *Nano Lett*, 5:709–711
15. Pissuwan D, Valenzuela S, Cortie M (2006) Therapeutic possibilities of plasmonically heated gold nanoparticles. *Trends Biotechnol*, 24:62–67

Author: D.P. Chrissoulidis
Institute: Aristotle University of Thessaloniki
Street: Faculty of Engineering
City: GR-54006 Thessaloniki
Country: Greece
Email: dpchriss@auth.gr

Hybrid Microfluidic Biosensor for Single Cell Flow Impedance Spectroscopy: Theoretical Approach and First Validations

J. Claudel[1,2], M. Ibrahim[1,2], H. Shirzadfar[1,2], M. Nadi[1,2], O. Elmazria[1,2], and D. Kourtiche[1,2]

[1] Université de Lorraine; Institut Jean Lamour (UMR 7198), F-54506, Vandœuvre les Nancy
[2] CNRS; Institut Jean Lamour (UMR 7198), F-54506, Vandœuvre les Nancy

Abstract—This paper presents a micro-biosensor based on Electrical Bio-Impedance Spectroscopy (EBIS), applied to blood cells diagnosis. Its hybrid conception uses the EBIS system coupled to a microfluidic device for cell by cell dynamic measurements. The microdevice is composed of 20x20µm² section micro channel, with platinum microelectrodes tested for different geometries. A Surface Acoustic Wave (SAW) device allows an accurate control of the fluidic displacement inside the channel. Simulation using athree dimensional Finite Elements Method (3D-FEM) have permitted to determinate the sensitivity of electrodes and to optimize their design, It also demonstrates the ability of our sensor to perform single cell measurements up to several hundred cells per second. First measurements confirms the simulation predictions and proved the ability to measure small impedance variations, less than 0.2%, during passage of particles.

Keywords—Microfluidic, impedance spectroscopy, biosensor, biological cells.

I. INTRODUCTION

Biosensors are based on micro and nanotechnology, using methods such as impedimetric microelectrodes [1] or Giant Magnetoresistance [2]. They allow many applications for biomedical applications as diagnosis and tissues characterization. Impedance spectroscopy permits to determine the physiological status of bio-samples by measurement of their electrical responses, without using any markers. Some pathology induces change in physical structure and chemical composition, thus changing electrical impedance which can be measured and interpreted. These types of measurements are performed since a long time on a large scale of tissues and bio-samples at macroscopic scale, and many models have been developed [3]. The current interest is focused on micro-sample characterization (lab on chip system), now possible thanks to advances in MEMS technologies. Macroscopic measurements provide mean values on a large volume of heterogeneous tissues and cells. As example, blood is a complex biofluid composed of red cells, white cells, plasma, etc... Their electrical properties varies with frequency as well as with the hemaocrit or temperature (references jaspard). White blood cells (WBC) represents less than 1% of blood particles, their impact is usually neglected in whole blood and requires the measurement of a large amount of different cells, one by one, with a good differentiation [4].

This paper, based on previous works [5], presents a new microfluidic sensor design for cell by cell dynamic meaurement. It combines bio-impedance measurement and microfluidic approach based on Surface Acoustic Wave (SAW) [6].

The second section describes the sensor structure with the different parts of sensing area, composed of microelectrodes of different geometries. We describe the possibility to use SAW devices as micro-actuator to create a fluidic movement to leading to move the cells and control with accuracy their displacement. The theories for electric and dielectric properties of cells suspensions will be exposed and an electric equivalent model essential for measurment's interpretations is given.

In the third section, we define a physical model, with itsgeometrical and electrical parameters, in order to simulate it by Finite Element Method (FEM). This allows to consider complex simulations in order to predict the sensor response, to optimize and study the system performances.

Fourth section shows and discuss the firstexperimental results and section 5 concludes about the validity of single cell measurement and the respect of stated assumptions.

II. THEORETICAL

A. Sensor Structure

The microfluidic sensor is composed of an inlet and an outlet tank for liquid deposition, connected by a 20µm² x

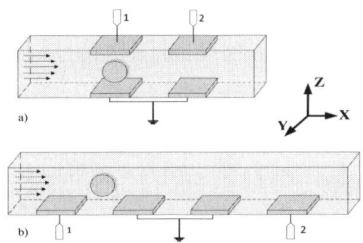

Fig. 1 Design of the measurement area for parallel microelectrodes a) and coplanar microelectrodes

20μm² cross-section microchannel. It contains a measurement area with coplanar and parallel microelectrodes of different geometries, as show in Fig. 1.

There are two identical pairs of electrodes for each kind of geometry, which allows the possibility to perform differential measurements and monitoring of the particles speed.

The innovation of this sensor concerns the integration of a SAW device in delay line configuration, i.e: 2 interdigital transducers (IDT), as shown in Fig. 2 patterned on piezoelectric substrate. This permit, to generate surface acoustic waves when RF signal with appropriate frequency is applied to the IDT. The acoustic wave leads to move liquid sample into the microchannel. The frequency of SAW excitation can be determined using the ratio of digits period λ and typical wave propagation speed v of the substrate. In our case, the piezoelectric substrate considered is in Lithium Niobate ($L_iN_bO_3$) 128° Y cut, which shows a high coupling coefficient (5.6%) and a phase velocity of v=3900 m/s. The spatial period of IDTs was fixed to 24μm IDT leading to a SAW operating frequency around163MHz, which is more higher than measurement frequencies.

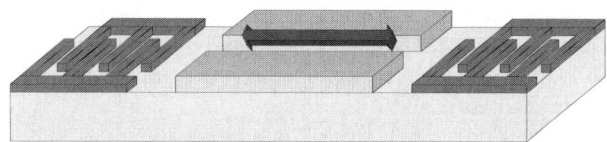

Fig. 2 Design of the measurement area for parallel microelectrodes a) and coplanar microelectodes.

B. Equivalent Model

As mentioned in the introduction, measurements and modeling of tissue were performed at macroscopic scale in the past. Numerous proposed models show difficulties to equate the relation between physiological and electrical properties of tissues and cell suspensions. This modeling part focuses on two of the most used models, based on Maxwell's equations and Fricke's electrical model [7], presented in Fig. 3.

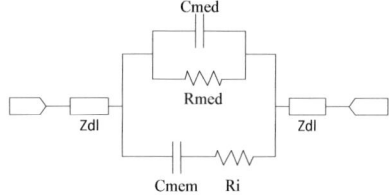

Fig. 3 Electric schematic of Fricke model for cells suspension.

The electric and dielectric properties of medium and particle are respectively represented by R_{med}-C_{med} and R_i-C_{mem}. Hywell Morgan [8] uses simplifications of MMT equations measurement to determine discrete equations using passive equivalent components and optimize them for a single cell. Calculation of each part is possible using (1) to (4), with electric, dielectric and geometrical properties of particles and medium. Each property can be determined at different frequencies, where they are predominant.

$$R_{med} = \frac{1}{\sigma_{med}(1 - 3\phi/2)k} \quad (1)$$

$$C_{mem} = \frac{9\phi r C_{mem,s}}{4}k \quad (2)$$

$$R_i = \frac{4\left(\frac{1}{2\sigma_{med}} + \frac{1}{\sigma_i}\right)}{9\phi k} \quad (3)$$

$$C_{med} = \varepsilon_0 \varepsilon_{med}(1 - 3\phi/2)k \quad (4)$$

III. PHYSICAL SIMULATIONS

In this section, the model of the sensor loaded by a medium and a cell is described. The global system can be implemented by defining electric and dielectric properties of each part composing it, and their geometries. They are discretized into simple geometries and simulated using Finite Element Method (FEM) on COMSOL Multiphysics software. In this part, our interest is focused on R_{med} variation due to cell geometrical properties and position.

A. Modeling Parameters

To simplify the simulations, only the measurement area was defined in Comsol multiphysics. Geometrical parameters of different parallel and coplanar models showed on Fig. 1, are defined in Table 1, where L is the length of electrodes (along x axis), W the width of channel and electrodes (along y axis) and G the gap between electrodes (along z axis for parallel and x axis for coplanar geometries).

Table 1 Font sizes and styles

Name	Type	L (um)	W (um)	G (um)
P-1	Parallel	20	20	20
P-2	Parallel	20	20	40
P-3	Parallel	20	40	20
C-1	Coplanar	20	20	20
C-2	Coplanar	40	40	20
C-3	Coplanar	20	40	20
C-4	Coplanar	20	20	40

Simulations were performed using usual values found in the literature for majority of cells in the order of several µm: 78 for ε_{med}, 1.4 S/m and 0.7 S/m respectively for σ_{med} and σ_i, and 1µF/cm² for $C_{mem,s}$.

B. Results and Discussions

A first step was to study the influence of the cell's sizes and their placement along z axis on impedance. Figure 4 and 5 represent these variations in percent for different geometries.

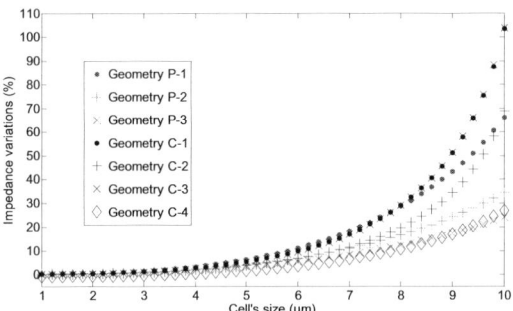

Fig. 4 Impedance variation in percent as function of cell's radius for different electrodes geometries.

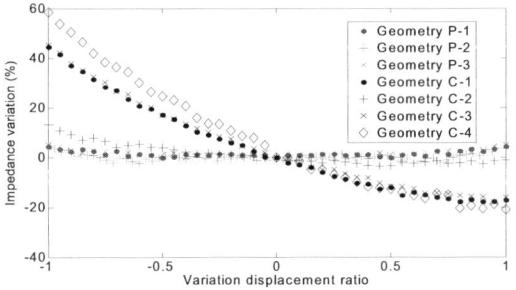

Fig. 5 Impedance variation error in percent, for an 8µm cell as function of its relative position along z axis for different electrodes geometries.

It appears that coplanar electrodes present better impedance variations compared to parallel electrodes, up to two times more. Impedance variations decrease with the increasing of electrodes sizes, and stay very low in all cases for small particles with radius less than 5µm. Parallel electrodes present a really good immunity as function of the cell placement, with a negligible variation in the center of displacement range. *A contrario* Coplanar electrodes present no negligible variations for the second geometry and very high variations for the others, up to 60% for the extreme cases. In all cases, these errors could be reduced by a correct centering of the cells, which can be obtained by a sufficient flow speed.

Dynamic measurements required a large amount of discrete impedance measurement. Only maximum and minimum values, corresponding to presence or absence of a cell are usable. It is necessary to do at least a correct measure in the stable part of measurement area. A second set of simulations was achieved by modifying cell position along x axis. To improve understanding of results, the ratio x_{cell}/x_{tot} was used to define the relative displacement in area measurement, and relative impedance variation (impedance normalized to its maximum). As shown in Fig. 6, results obtained for parallel and coplanar electrodes for same electrodes size are quite similar with no significant differences. Only the increase of electrodes length permit to increase the stable area, but reduce the sensitivity.

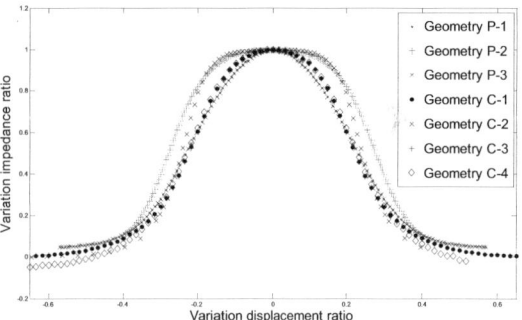

Fig. 6 Variation of impedance ratio, for 8µm cell as a function of it relative position along x axis for different electrodes geometries

Measurement time depends on the frequency of the signal and the number of averaging periods used to increase parasitic immunity, and tolerance. In our case, measurements are performed in the range of 100 kHz to 10MHz. Using 100 kHz as minimum frequency and the precedent relation; the maximum theoretical cell's rate can be deteminated as a function of selected tolerance for stable area and number of measured periods. Results are summarized in Table 2, and confirm our assumptions regarding the ability to measures up to several hundred cells per second.

Table 2 Maximal theoretical range in cell per second.

Tolerance (%)	1 period	4 periods	16 periods	50 periods
1	10667	2667	667	213
3	18667	4667	1167	373
5	23333	5833	1458	467

IV. PRELIMINARY EXPERIMENTS

Experiments were realized with platinum coplanar electrode C-1, using polystyrene micro-beads. The microchannel was realized in SU_8 photoresist closed with Polydimethylsiloxane (PDMS) cover. Impedance measurements were performed using a digital lock-in amplifier and trans-impedance current pre-amplifier in two points configuration, as shown in Fig. 7.

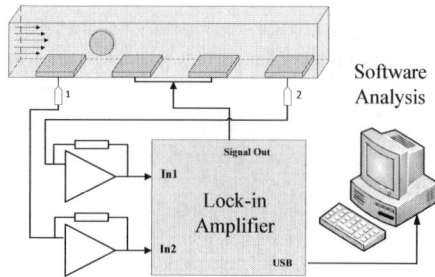

Fig. 7 Measurement setup of the microfluidic device.

Figure 8 shows measured impedance during passage of 2 microparticles at 500 kHz. This results evidences the ability of our system to measure very few impedance variations, less than 0.2% during the passage of beads. The row signal can be improved using digital filtering after acquisition as shown in fig. 8 (red line).

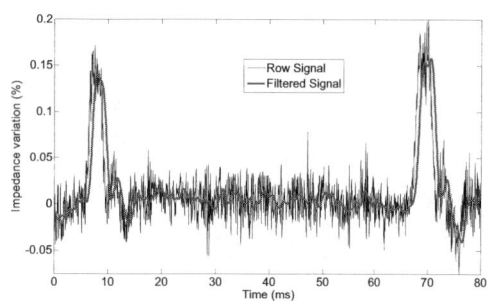

Fig. 8 Impedance variations during passage of two micro-particles. Row and processed signal.

V. CONCLUSION

An hybrid sensor based on impedance spectroscopy and microfluidic device, able to perform dynamic cells measurements was described and characterized. An original fluidic system, based on SAW technology allows to move cells without need for a conventional micropump.

Simulations permited to validate the ability to characterize single cells, with a sufficient impedance variation, during its passage : up to several ten percent. Seven electrodes with different geometries were simulated. Results show that, coplanar electrodes allows the better sensitivity while he parallel electrode allows the better immunity face to cell position. Using impedance variations as a function of cell displacement, we have determined the maximal theoretical measurement, and prove the ability to measure up to several hundred cells per seconds according to our hypothesis.

These experiments had evidenced the ability to realize this kind of measurement, with the capability to detect a very small impedance variations.

ACKNOWLEDGMENT

The authors gratefully thank the MINALOR skill center of Institut Jean Lamour (IJL) in University of Lorraine, for their technical support.

REFERENCES

1. Ibrahim M, Claudel J, Kourtiche D et al (2011) Physical and Electrical Modeling of Interdigitated Electrode Arrays for Bioimpedance Spectroscopy. New Developments and Applications in Sensing Technology 83:169-189
2. Shirzadfar H, Haraszczuk R, Nadi M et al (2012) Detecting and Estimating Magnetic Fluid Properties by a Needle-Type GMR Sensor. Alushta, Ukraine
3. Gabriel S, Lau R, Gabriel C (1996) The dielectric properties of biological tissues: III. Parametric models for the dielectric spectrum of tissues. Phys. Med. Biol. 41:2271
4. Cheung K.C, Gawad S, Renaud P (2005) Impedance spectroscopy flow cytometry: On-chip label-free cell differentiation. Cytometry A 65A, 124–132.
5. Ibrahim M, Claudel J, Kourtiche et al (2011). Optimization of planar interdigitated electrode array for bioimpedance spectroscopy restriction of the number of electrodes. Fifth International Conference on, 2011, pp 612-616.
6. Beyssen D, Le Brizoual L, Elmazria O et al (2006) Microfluidic device based on surface acoustic wave. Sensors Actuators B Chem. 118:380-385.
7. Fricke H, Morse S (1925) The electric resistance and capacity of blood for frequencies between 800 and 4½ million cycles. J Gen Physiol. 9:153-167.
8. Morgan H, Sun T, Holmes D et al (2007) Single cell dielectric spectroscopy. J. Phys. D: Appl. Phys. 40:61-70.

Author: Julien Claudel
Institute: Institut Jean Lamour-Université de Lorraine
Street: rue du jardin botanique
City: Vandoeuvre les Nancy, 54506 Cedex
Country: France
Email: julien.claudel@univ-lorraine.fr

Integrating an Electronic Health Record Graphical User Interface into Nanoelectronic-Based Biosensor Technology

Ana María Quintero[1], Carlos Cavero Barca[1], Carlos Marcos Lagunar[1], César Mediavilla[1], José Luis Conesa[2], Miguel Roncalés[2], Yaiza Belacortu[2], Alejandro Juez[2], Luis Fernandez[3,4], Alexandra Martin[5], and Renzo Dal Molin[5]

[1] ATOS, ATOS Research and Innovation (ARI), Madrid, Spain
[2] AlphaSIP, Zaragoza, Spain
[3] Group of Structural Mechanics and Materials Modelling (GEMM), CIBER-BBN, Zaragoza, Spain
[4] Aragon Institute of Engineering Research (I3A), University of Zaragoza, Zaragoza, Spain
[5] SORIN CRM, Clamart, France

Abstract—Lack of good and standardised Electronic Health Record (EHR) Graphical User Interfaces (GUI) can hinder the acceptance of ground-breaking medical devices devoted to healthcare professionals and / or patients. Usually, the GUIs are designed too late in the process, rather than early and iteratively throughout the development of the device. In addition, end-users should be involved at all stages, from framing the GUI to be developed, to testing and validating it in 'real situations'. Standards permit the interoperability between systems using a common reference model. In this article we present the work done in the development of an EHR GUI for an innovative nanotelectronic-based biosensor technology for cardiovascular diseases diagnosis integrating standardised EHR information into the patient health profile. The challenge was ensuring that the Information Technology developments contributed to the effectiveness and acceptance of the device among medical community. We have achieved this goal based on: (i) a close work with the biosensors developers, in order to gain a better understanding of technology needs; (ii) development of information systems that consider the connection and communication with already widespread hospital information systems using the European standard EN13606 to guarantee the usefulness of the gathered patient data; (iii) a User-Centred Design approach, involving doctors from the very start of the development, producing a more intuitive and acceptable tool for healthcare professionals.

Keywords—**Electronic Health Record, Archetypes, Nanoelectronic, Biosensor, Graphical User Interface.**

I. INTRODUCTION

The challenge presented in the current article was to design an Electronic Health Record (EHR) Graphical User Interfaces (GUI) for the innovative Arcatech biosensor technology [1], an *in-vitro* diagnostic device, built in a robust, user-friendly and cost-effective way. Compared to other biosensor technologies, the electrochemical-based biosensors can be integrated with other low-cost electronics, being really economical at high volumes. Basically, the biosensor consists of an electrode integrated on a microfluidic cartridge with a chemical interface which makes the connection to the clinical sample to be analysed. Subsequently, capturing of bio-targets can be detected by measuring impedance changes in the electrical signal. With on-chip detection electronics, small electrical changes can be detected within milliseconds, enabling massively parallel real-time monitoring of bio-molecule binding events.

In spite of the listed advantages, there is a possibility that Arcatech platform (the biosensor plus the designed GUI) could not be embraced by the physicians, overlooking its relevance to their practices. Some of the reason behind this attitude [2] could be the needed investment of time and energy to learn how to use them; and the awkward to integrate them in the day-by-day established routine. In order to mitigate both facts, we have designed and implemented a GUI through a rigorous development process. The hypothesis is that devoting more attention to the GUI (for personal computers and off-the-desktop applications), we can improve the ways in which information technology contributes to the acceptance of ground-breaking health-care devices.

Thereby, Arcatech EHR-GUI was designed to: (i) make the most of the innovative biosensor; (ii) integrating the patient health information included in the EHR stored in the Hospital Information Systems (HIS) to provide the doctors with an overview of the subject of care health profile; (iii) adding new capabilities beyond the biosensor needs to enhance the platform's appeal. In order to ensure that end-users will certainly accept the developed platform in their clinical practice, we engaged with physicians from the early stages of the design. This approach allows to:

- Reduce risks during the development phase.
- Improve the design based on end-user observations and feedback (User-Centred Design).
- Identify potential use errors or problems through intermediate usability assessments before deploying the final version, saving time and money.
- Create an initial group of interest in the platform.

Basically, the steps that should be integrated with the design development procedure are:

- Researching published literature and available technologies to compare our design.
- Determining and understanding the relative importance of user needs through interviews and focus groups.
- Performing intermediate and final usability evaluations to examine how users interact with a system and use the users' feedback to optimize design.
- Verifying and validating the prototype to ensure that the delivered product meets the original design intent and specifications.

Figure 1 shows the data flow of the overall system. The features of the final software developed for the Arcatech biosensor are reported in the ensuing sections.

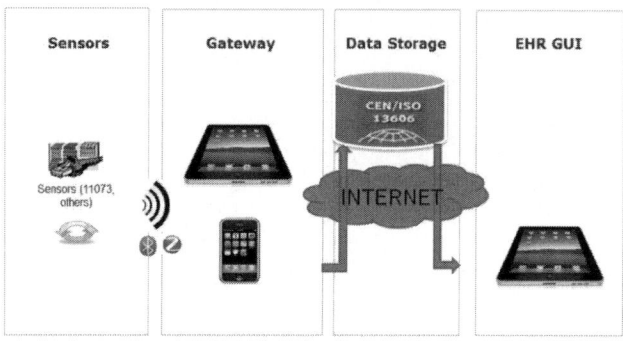

Fig. 1 Data flow of the overall system

II. SENSORS

The complete system is built around a 4 channel sensor capable of simultaneous detection of several biomarkers. To start a test, the blood sample is located on a disposable fluidic chip that is where the chemical reactions take place. This chip incorporates a filter so that only plasma is actually entering the chip reaction chambers.

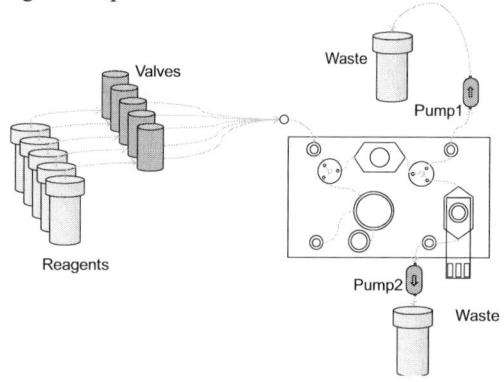

Fig. 2 Microfluidic chip schema

Inside the chip the plasma sample reacts with the marked antibodies, and magnetic particles.

The on board electronics regulates the necessary steps for the reactions and measurements to take place, acting on the valves for the reagent selection and the micro-pumps for mixing and moving the sample trough the fluidic chip.

Once the sample has reacted with the antibodies, the sample is moved to a three terminal sensor located inside the chip, where an amperometric measurement is done using an integrated four channel potentiostat.

The test results are then shown on the on system screen, in case the systems is used stand alone, and also available to be sent wirelessly through a Bluetooth interface to the GUI.

III. GATEWAY

For a proper gathering of sensor data, it is crucial to design a simple although robust intermediate system able to operate in low energy-consumption scenarios, as for example a mobile device.

The Ubiquitous Tele-monitoring Kit [10], developed in Atos, aims to produce exactly that kind of product. For the integration with Arcatech biosensors, a simplified version has been developed. The Gateway acts as an effective intermediate for the retrieval of data from the biosensors maintaining Bluetooth device discovering and Security-enhanced connection. Furthermore it allows uploading the data in a normalised way into the EHR using openEHR archetypes as it was mentioned in the above section. A reduced version of the protocol enables a quick and secure communication between both devices.

For demo purposes, the Gateway has been tested on two different Android devices, tablet and smartphone, both of them with Android version 4.0.3. Anyway, the APIs used to develop this reduced version of the Gateway have been deliberately chosen to be the simplest to ensure the backwards compatibility with old devices.

IV. EHR STORAGE

The standardization of Electronic Health Record (EHR) is crucial to achieve the semantic interoperability in healthcare; it defines the architecture for exchanging properly the health information between EHR coming from different systems [3].

The "dual model approach", also called the two-level modeling [4] defines a clear separation between the volatile medical concepts (archetypes) and the static structure of storage (reference model). The clinical knowledge is built by means of archetypes based on a stable information model [5]; if new information is needed the system could be extended through archetypes keeping the technical solution unaltered. openEHR / ISO/EN EN13606 designs a standard architecture to connect EHR-related data among different Health Information Systems (HIS) [6][7] using the two-level modeling approach.

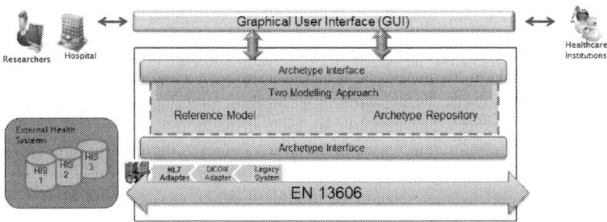

Fig. 3 openEHR / ISO/EN 13606 approach

The main data structure in openEHR / ISO/EN 13606 is entry which contains all the health data. It is organised by means of compositions as the small pieces of EHR exchangeable. In the openEHR four types of entries can be distinguished: observations, evaluations, instructions, and actions. Observations and actions represent information about past events. Evaluations cover current assessment of the health data by the medical staff, including diagnosis, prognosis but also information coming from Decision Support System. Instructions represent future events as for instance medications order [8].

For the proposed research multiple archetypes have been created or reused to cover the clinical information needed. The observation archetypes have been used to retrieve the information coming from the biosensors (NT-proBNP, cTnT and CRP). Observation (measurements, but also tobacco use or allergies), evaluation (diagnoses and diseases) and instruction (medications) archetypes downloaded from the Clinical Knowledge Manager (a repository of validated archetypes) [9] or created ad-hoc have been used to show the measurements, diseases, medications and so on.

V. EHR GRAPHICAL USER INTERFACE

A. Expert System

The interpretation of the data resulting from a point of care measurement of the plasma concentration of one or more cardiac biomarker(s) is difficult in the setup of heart failure. The quantitative result have a more valuable clinical meaning when associated with other clinical data such as structural or functional examination results, signs, symptoms, patient history and habits. In order to take into account this reality, an algorithm designed for demonstration purposes only, has been implemented into the GUI. Although fictive and simplified, this "expert system" is based on literature, especially on clinical studies investigating the impact of some cardiac biomarker level determination on the cardiovascular disease risk assessment [11-17]. The calculation takes into account the plasma concentration values of four cardiac biomarkers (cTnT, H-FABP, CRP and NT-proBNP) as well as the following clinical data: age (under or above 50 years old), echography result, presence or absence of cardiac rhythm abnormalities, systolic blood pressure, body mass index, smoker or not and diabetic or not). Depending on the above mentioned data, five ranges for risk of heart failure events are discriminated aiming at guiding the doctor in the decision process. The expert system should be versatile, configurable by the doctors and adaptable to a larger number of biomarkers.

B. Graphical User Interface

The Graphical User Interface (GUI) is a user-friendly and easy to use tool connected to the EHR that allows the physicians to obtain as much information as possible from the outset.

Once the professional is identified into the system (user/password) the software shows the overall view of the health status of the last loaded patients (with a graph on the right side with meaningful colours). When some patient is selected the specific data of the subject of care is displayed (patient's identification data - full name - and his/her status). Patient's status is the risk factor (from CHF unlikely to very probable CHF) calculated depending on rules defined in "Expert system Configuration" (see section A).

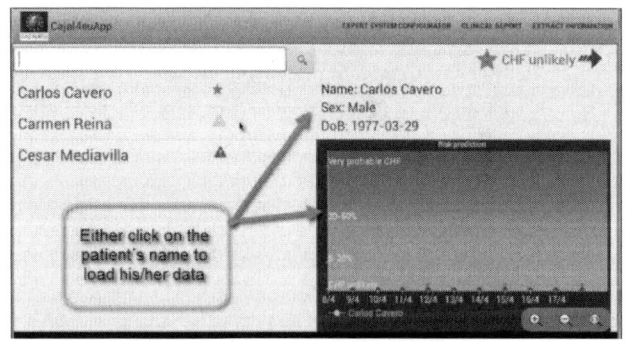

Fig. 4 Patients list (left) and risk factor details (right).

The health status is represented using meaningful colours: low risk in green and high-risk in red (increasing the colour along the intermediate status). Interface options:

1. Expert System Configuration: Allows the professional to configure rules to all the data included in the system (medications, diseases, biomarker measurements and so on) depending on the risk he/she wishes to estimate, by default the system has pre-defined rules that estimate 5 years CHF risk prediction. Each rule has certain score,

for every score values and obviously the total one, the higher the score, the higher the risk.
2. Clinical reports: The doctor can generate PDF clinical reports at any time. This report will contain all patient information, Clinical data, diseases, PoC values, medications, etc., including snapshots of the biomarkers.

The next step is to know the reasons behind the risk estimation, thus the relevant patient data coming from the EHR is shown, divided into three sections (i) Clinical data: diseases, last Point of Care (PoC) and also medications (ii) Demographic data and (iii) Habit Patterns: alcohol, smoker and diet. The EHR data is retrieved using openEHR archetypes in a dynamic way, so include new information does not imply changes in the interface.

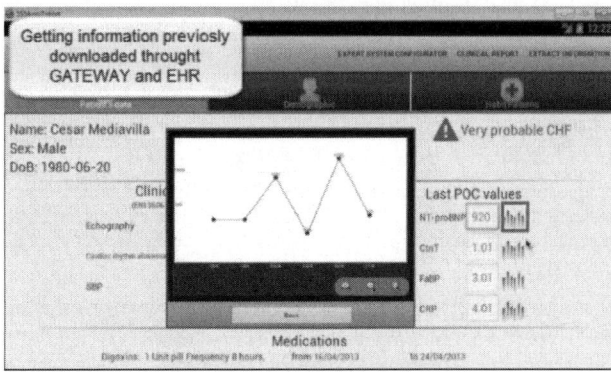

Fig. 5 Patient's data and biomarker view

All the measurements coming from the biosensors (PoC) could be displayed as graphics, showing date and concentration (See Fig. 5).

VI. CONCLUSIONS

Current tools in the market do not integrate EHR information with nano-biosensor technology. The research presented in this article is based on user-centric approach. We have integrated the EHR European standards with the data coming from the PoC, applying analytics tools in order to provide doctors with isolate information useful to their practice.

We have closed the loop from the PoC to the doctor's application to analyse the health status of the patient (EHR GUI), and the Hospital Information System to store the data. The software also offers a risk prediction based on scores provided manually by the medical experts; graphical windows to visualize the data; and tools to generate reports.

To sum up, this GUI allows professionals a fast diagnostic thanks to the knowledge of patient's status and the causes that have generated this result, on a friendly, intuitive, fast and easy to use GUI EHR connected.

ACKNOWLEDGMENT

The research described in this paper was carried out in the CAJAL4EU project (Nanobiosensors in health context). The project is co-funded by the ENIAC Joint Undertaking (JU) (grant agreement #120215) and by national authorities.

REFERENCES

1. Abad L., del Campo F.J., et al. (2012) Design and fabrication of a COP-based microfluidic chip: Chronoamperometric detection of Troponin T. Electrophoresis Volume 33, Issue 21, 3187–3194.
2. Yarbrough (A.K.), Smith T.B. (2007) Technology Acceptance among Physicians: A New Take on TAM. Med Care Res Rev 64:650-672.
3. Muñoz P., Trigo J. D. et al. (2011) The ISO/EN 13606 Standard for the Interoperable Exchange of Electronic Health Records. Journal of Healthcare Engineering. 2(1): 1-24.
4. Beale T. (2002) Archetypes constraint-based domain models for future-proof in-formation systems. s.l. : OOPSLA-2002 Workshop on Behavioural Semantics.
5. Goosen W., Goossen-Baremans A. et al. (2010) Detailed Clinical Models: A Review." Healthcare informatics research. 16(4): 201.
6. CEN/TC251-ISO/TC215 (2010) Electronic Healthcare Record (EHR) Communication. Parts 1: Reference Model, Part 2: Archetype Model, Part 3: Reference Archetypes and Term lists, Part 4:Security and Part 5: Interface Specification.
7. Beale T., Heard S. (2008) Architecture overview. [Online] openEHR Foundation. [Cited: 16 April 2013] www.openehr.org/releases/1.0.2/architecture/overview.pdf.
8. Beale T., Heard S., Kalra D., Lloyd D. (2007) The openEHR Reference Model - EHR Information Model. [Online] openEHR foundation. [Cited: 16 April 2013] www.openehr.org/releases/1.0.1/architecture/rm/ehr_im.pdf.
9. Clinical Knowledge Manager. (2010) [Online] OpenEHR Foundation. [Cited: 16 April 2013] http://www.openehr.org/ckm/.
10. Marcos Lagunar C. et al. (2012) Ubiquitous Tele-Monitoring Kit (UTK): Measuring Physiological Signals Anywhere at Anytime. Lecture Notes in Computer Science 7657, pp. 183-191, 2012.
11. Blankenberg et al (2010) New Biomarkers and CVD Risk, Circulation
12. Y. Sato et al (2012) Cardiac troponin and heart failure in the era of high-sensitivity assays , Journal of Cardiology 60 160–167
13. Ishii et al (2003) cTnT and BNP in Heart Failure, Clinical Chemistry 49, No. 12
14. Januzzi et al (2005) Heart failure/NTproBNP testing in the emergency department, The American Journal of Cardiology Vol. 95
15. O'DONOGHU et al (2005) The Effects of Ejection Fraction on N-Terminal ProBNP and BNP Levels in Patients With Acute CHF: Analysis From the ProBNP Investigation of Dyspnea in the Emergency Department (PRIDE) Study, Journal of Cardiac Failure Vol. 11 No. 5 Suppl.
16. Januzzi et al (2006) International NT-proBNP analysis, European Heart Journal 27, 330–337
17. Harrison and al (2002) Predicting future cardiac events in patients with dyspnea, Annals of emmergency medicine 39:2

Synthesis of Gadolinium-doped Fluorescent Au/Ag Nanoclusters as Bimodal MRI Contrast Agents

Walter H. Chang[1,2], Cheng-An J. Lin[1,2], Ching-Yi Chang[1], and Wen-Fu. Lai[3]

[1] Department of Biomedical Engineering / Center for Nano Bioengineering,
Chung Yuan Christian University, Chung-Li, Taiwan, Republic of China
[2] Institute of Biomedical Technology, Chung Yuan Christian University, Chung-Li, Taiwan, Republic of China
[3] Center for Nano-tissue Engineering and Imaging, Taipei Medical University, Taipei, Taiwan, Republic of China

Abstract—In this presentation synthesized the T_1 contrast agent, combinations of MRI and optical imaging modalities will be introduced. Using bovine serum albumin conjugated with gadolinium and fluorescence gold or silver nanoclusters (Fluorescence gold /silver protein). An average of 34 Gd-DTPA chelates were covalently conjugated to bovin serum albumin. The T_1 relaxivity of BSA-Gd-DTPA was $163.27 mM^{-1}sec^{-1}$. The experimental results shows a new dual-modal imaging agent of fluorescent gold/silver protein with a paramagnetic in vitro and in vivo, which can be hold many potential applications in medical diagnostics and therapy.

Keywords—contrast agent, gadolinium, gold, silver, fluorescent protein.

I. INTRODUCTION

MRI is one of the most powerful and non-invasive diagnostic techniques for living organisms based on the interaction of protons with the surrounding molecules of tissues. MR imaging can offer high spatial resolution and the capacity to simultaneously obtain physiological and anatomical information, whereas optical imaging allows for rapid screening. The current MRI contrast agents are in the form of paramagnetic complexes which are usually gadolinium (Gd^{3+}) chelates , accelerate the longitudinal (T_1) relaxation of water protons and therefore exhibit bright contrast where they localize. Gadolinium diethylene-triaminepentaacetate (Gd-DTPA) has been the most widely used for T_1 contrast agent. This study synthesized the T_1 contrast agent, combinations of MRI and optical imaging modalities. Using bovine serum albumin conjugated with gadolinium and fluorescent gold or silver nanoclusters (Fluorescence gold /silver protein).

II. MATERIALS AND METHODS

In this research, we used bovine serum albumin(BSA) as a template to synthesize the dual-functional gadolinium containing gold/silver fluorescent proteins. At first, we synthesized many kinds of fluorescent proteins , investigated the mechanisms of BSA adsorption and reduction of gold ions, and confirmed the abilities of gold conjugation with BSA will not influence the conjugation of gadolinium. Secondly, we synthesized the MRI contrast agent with fluorescence, tested the fluorescent properties with UV-Vis absorption and photoluminescence spectra to confirmed all the elements were conjugated on the BSA, and calculated their concentrations. Finally, we did experiments in vivo and in vitro to establish firmly this probe can enhance the contrast ability by MRI imaging.

The experiments included the following items:

A. Synthesis dual-function developer gadolinium containing gold/silver fluorescent proteins,
B. Gadolinium-containing albumin, BSA-Gd-DTPA(Diethylenetriaminepentaacetate),
C. Containing the gadolinium gold fluorescent protein synthesis, Au@BSA-Gd,
D. Gadolinium and silver-containing fluorescent protein synthesis, Ag@BSA-Gd.

III. RESULTS AND DISCUSSIONS

Fluorescent proteins are the complex of gold nanoclusters with the protein's molecules. In our experiments we found the human serum albumin can adsorb the gold ions directly. Gold ions can be reductive in the template of protein and formed to a very stable pink fluorescent protein. However we found that not all kinds of proteins can be formed to a pink fluorescence. We have to find a suitable protein template to conjugate the gadolinium. The bovine serum albumin was a good protein as a template due to not only its property was most similar to human serum albumin but its fluorescent intensity was very high.

Bovine serum albumin has blue fluorescence at the first time, but it becomes golden yellow when synthesized with Au@BSA. Using UV to do excitation it will become red fluorescence as shown in Fig 1. The gadolinium-containing bovine serum albumin (BSA-Gd) is transparent in the white light and their absorption and fluorescence spectra is no difference compared with BSA. It looks like pale yellow after synthesizing with gold fluorescent protein. When it excited by UV it still has red fluorescent. When we mixed bovine serum albumin with silver it will become dark brown at one minute and then have the red-orange fluorescent. It will become silver nanoparticles if the reaction time is 24 hours. The fluorescence intensity will lower than gold fluorescent protein. The fluorescence of Ag@BSA-Gd is lower than Ag@BSA as shown in Figure 1.

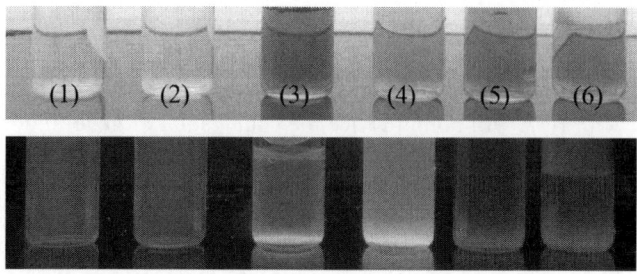

Fig. 1 The upper picture is with white light and the bottom picture is excited by UV of (1)BSA, (2)BSA-Gd, (3)Au@BSA, (4)Au@BSA-Gd, (5)Ag@BSA, and (6)Ag@BSA-Gd.

The UV-Vis absorption spectrum of gadolinium-contain gold and silver with fluorescent protein was shown in Figure 2. Gold fluorescent protein has a peak at 280nm due to the absorption of protein with fluorescent gold nanocluster. Comparing absorption between Au@BSA with Ag@BSA at 280nm the peak of Ag@BSA is higher than that of Au@BSA.

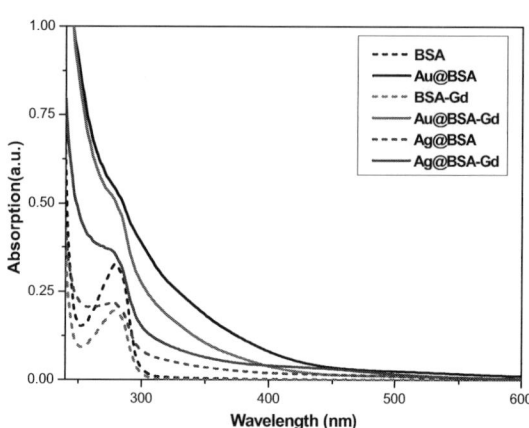

Fig. 2 UV-Vis absorption spectrum of gadolinium-containing gold and silver fluorescent proteins.

Fluorescent spectrum of gadolinium-containing gold and silver fluorescent proteins excited by 350nm was shown in Figure 3. All of the BSA, BSA-Gd, Ag@BSA, and Ag@BSA-Gd has a peak around 400nm. It is the Raman signal due to the dilution of materials. The BSA and BSA-Gd groups both have blue fluorescent. However the four groups of Au@BSA, Ag@BSA, Au@BSA-Gd, and Ag@BSA-Gd have two peaks. It demonstrated that the non-protein fluorescene can success to synthesize gold and silver proteins. The Au@BSA and Au@BSA-Gd emit a red light at 640nm. Ag@BSA emits a red light at 636nm. However Ag@BSA-Gd emits a peak at 610nm. Its color likes organge.

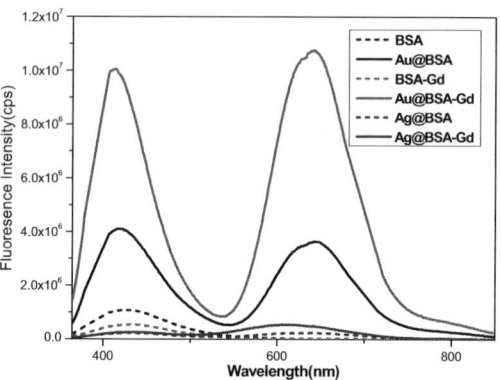

Fig. 3 Emission spectrum of gadolinium-containing gold and silver fluorescent proteins excited by 350nm

The concentration of HepG$_2$ cell is 2×10^4 cell/well and seeded in Chamber silde for 24 hours. The concentration of Au@BSA-Gd and Ag@BSA-Gd is 10 mg/ml were cultured to the cells for 4 hours. The culture was done in the nonserum's environment. Then they will be washed with two liters dH$_2$O for five times. Then they were saved in freeze-drying. In Figure 4, (a1) - (a3) is under the nonserum media to culture, (b1) - (b3) is 10mg/ml concentration of Au@BSA-Gd, (c1) - (c3) is 10mg/ml concentration of Ag@BSA-Gd, (a1), (b1), (c1) is the cell observed in the white light, (a2), (b2), (c2) is excitation by UV, and (a3), (b3), (c3) is the merge of (a) and (b). It was found out that Au@BSA-Gd and Ag@BSA-Gd both will go inside the cell. It means that non-specific binding of the material can be used as biomarkers for hepatoma cells. However it was obviously that the toxicity of Ag@BSA-Gd is high. The cells become apoptosis. Comparing between Au@BSA-Gd with Ag@BSA-Gd, we found out that the cause of apoptosis is not the Gd but the silver ions. When the silver ions get into the cells it is easy to bind with mitochondria and will influence the signal's pathway and the cells become the apoptosis[3,4].

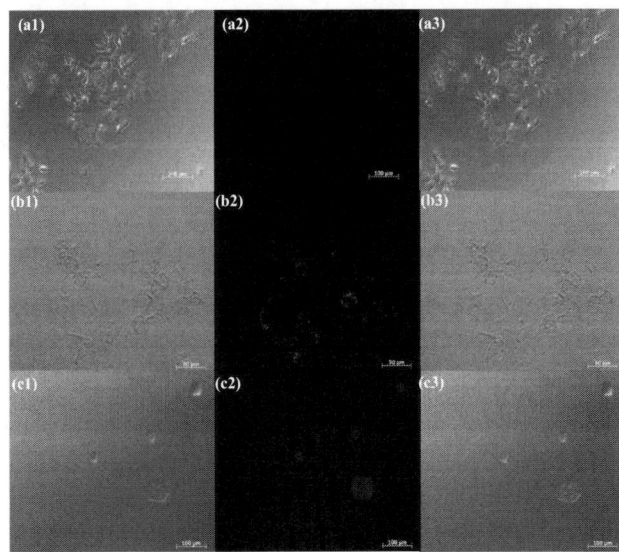

Fig. 4 HepG$_2$ intakes the Au@BSA-Gd and Ag@BSA-Gd. It is cultured with nonserum media for 4 hours. (a1)-(a3) is cultured in the nonserum media. (b1)-(b3) are cultured 10mg/ml concentration of Au@BSA-Gd. (c1)-(c3) are cultured 10mg/ml concentration of Ag@BSA-Gd. (a1), (b1), and (c1) are observed in the white light. (a2), (b2), and (c2) are excited by UV. (a3), (b3), and (c3) are the merge of (a) and (b).

Figure 5 is the image of nude mice after the tail vein injection of 200ul Au@BSA-Gd. The magnetic field is T_1. Comparing the imaging before injection(a) with after 30 minutes injection(b) the higher contrast ratio can be seen. After 4 hours the developing agent will arrive the bladder. It was demonstrated that the developed probe can enhance the contrast in vivo.

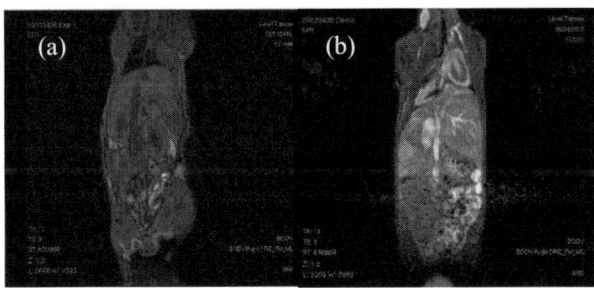

Fig. 5 The MRI image(T_1) of nude mice after the tail vein injection of 200ul Au@BSA-Gd. (a) before injection and (b) after injection of Au@BSA-Gd 30 min.

IV. CONCLUSIONS

In this study, we have developed a new probe of Au@BSA-Gd for MIR and optical imaging modalities. Using bovine serum albumin conjugated with gadolinium and fluorescent gold or silver nanoclusters(fluorescent gold/silver protein). An average of 34 Gd-DTPA chelates were covalently conjugated to bovin serum albumin. The T_1 relaxivity of BSA-Gd-DTPA was 163.27 mM^{-1}sec^{-1}. The experimental results shows a new dual-modal imaging agent of fluorescent gold/silver protein with a paramagnetic in vitro and in vivo, which can hold many potential applications in medical diagnostic and therapy.

ACKNOWLEDGMENT

We gratefully acknowledge the supports of research grants from The National Science Council of Taiwan (NSC 101-2627-M-033-001, 101-2120-M-038-001, 101-2628-E-033-001-MY3, 101-3011-P-033-003-), and partial financial support from Chung Yuan Christian University (107044-12).

REFERENCES

1. Hu, D. H.; Sheng, Z. H.; Zhang, P. F.; Yang, D. Z.; Liu, S. H.; Gong, P.; Gao, D. Y.; Fang, S. T.; Ma, Y. F.; Cai, L. T., Hybrid gold-gadolinium nanoclusters for tumor-targeted NIRF/CT/MRI triple-modal imaging in vivo. Nanoscale 2013, 5 (4), 1624-1628.
2. Mathew, A.; Sajanlal, P. R.; Pradeep, T., A fifteen atom silver cluster confined in bovine serum albumin. J Mater Chem 2011, 21 (30), 11205-11212.
3. Lapresta-Fernandez, A.; Fernandez, A.; Blasco, J., Nanoecotox-icity effects of engineered silver and gold nanoparticles in aquatic organisms. Trac-Trend Anal Chem 2012, 32, 40-59.
4. Mahmoudi, M.; Serpooshan, V., Silver-Coated Engineered Magnetic Nanoparticles Are Promising for the Success in the Fight against Antibacterial Resistance Threat. Acs Nano 2012, 6 (3), 2656-2664.

Author:	Walter H. Chang, Ph.D.
Institute:	Chung Yuan Christian University
Street:	200, Chung Pei Rd.
City:	Chung-Li
Country:	Taiwan, R.O.C.
Email:	whchang@cycu.edu.tw

Microchannel Modification to Enhance the Sensitivity in Biosensors

M. Gomez-Aranzadi[1,2], M. Mujika[1,2], S. Arana[1,2], and D. Hansford[3]

[1] CEIT-IK4 and Tecnun (University of Navarra), Manuel de Lardizabal 15, 20018 San Sebastian, Spain
[2] CIC microGUNE, Goiru Kalea 9, 20500 Arrasate-Mondragon, Spain
[3] Department of Biomedical Engineering, The Ohio State University, Columbus, Ohio 43210, USA

Abstract—A new method to enhance surface-sample interaction in biosensors through the integration of specific 3D microfluidic structures is reported. The method was tested in a device with a microfluidic network made from PDMS. The fabrication process included a specific double layer mold fabrication. As an intermediate step before the tests, a self assembled monolayer (SAM), that acted as a ligand for 252 nm amine-coated fluorescent polystyrene particles, was formed on top of a gold layer. The results showed that the microstructures produced an increase of coverage of the sensitive areas up to 15% which, combined with the property of the SAM of binding with different biological molecules, provide a potential method to enhance the sensitivity of a wide range of multipurpose microfluidic biosensors.

Keywords—microfluidics, PDMS, biosensor, microstructures, sensitivity.

I. INTRODUCTION

In the last few decades there has been a great interest in the development of lab-on-a-chip devices due to the advantages they provide compared to traditional diagnostic methods. The ability to detect biological agents in an economic, easy to use, portable device is one of the main research lines in the health care industry [1]. BioMEMS (biological Micro-Electro-Mechanical-Systems) are part of this line as they provide the possibility to interact with different types of biological components. These devices work in two phases: detection and transduction. First a ligand is immobilized onto a surface and then a sample fluid is put in contact with it. Usually it is necessary to include a microfluidic network in the device to run all the fluid through the designated sensing area. The problem with this approach is that within small (<100 μm) fluidic channels the advective inertial forces in the fluid are small compared to viscous forces, as indicated by the low Reynolds number (Re<<1). The result is a purely laminar flow, where the fluid flows in parallel layers. This is a major limitation for microfluidic systems in which a single surface is the active one. Only the particles contained in the lowest layer interact with the capture molecules, while most of the analyte contained in the fluid is not detected by the sensor. Lateral diffusion could put more molecules in contact with the sensing surface, but in the length scales of standard microfluidic flows it is negligible.

There have been different approaches to avert the effects of the laminar flow in microfluidics [2,3], but one of the most promising was the design proposed by Stroock, et al.[4] and followed by others [5-9].

Given the need to increase the interaction of the bulk of the biofluid with the sensing surface for many detection applications, in this paper the design and testing of embedded 3D microfluidic structures in microchannels are presented. The increased interaction within the system was tested with a standard SAM and amine-coated nanospheres for validation of the concept. Furthermore, a novel process for SAM formation after device fabrication is presented. This versatile solution offers the possibility of using the same design for other applications where another type of coating is needed, whether it is another type of SAM or other specific molecules.

II. MATERIALS AND METHODS

A. Materials and Equipment

Silicon wafers were purchased from Telecom-STC Co. Ltd., Russia. Microposit S1818 positive photoresist was purchased from Rohm and Haas Electronic Materials LLC and Silastic T-4 PDMS silicone elastomer, including base and curing agent was purchased from Dow Corning Corp., USA. SU-8 100 and SU-8 2015 negative photoresists were obtained from MicroChem Corp., USA. The film photo masks were purchased from Micro Lithography Services, Ltd., UK. An Ismatec ISM 937C peristaltic pump was purchased from IDEX Health & Science GmbH, Germany. 3-mercaptopropionic acid (MPA) was purchased from Fluka Sigma-Aldrich Co. LLC, USA. 1-Ethyl-3-(3-dimethylaminopropyl) (EDC) carbodiimide HCl and N-hydroxysuccinimide (NHS) were obtained from Thermo Scientific S.L.U., Spain. 252 nm amine-coated fluorescent polystyrene spheres were obtained from Merck Chimie S.A.S., France. A P-6 profilometer from Tencor-KLA Corporation, U.S.A was used to characterize the mold. Two microscopes were used as imaging platforms, an M205FA from Leica and an Eclipse Ti from Nikon. NIS-Elements (Nikon) software was used to analyze the images.

B. Design and Microfabrication

The general design of the device consisted of a microfluidic network of four channels (Ch1 to Ch4), 13 mm long and 400 μm wide (Fig. 1(b)), placed on a silicon chip containing 12 patterned sensing areas (Fig. 1(a)).

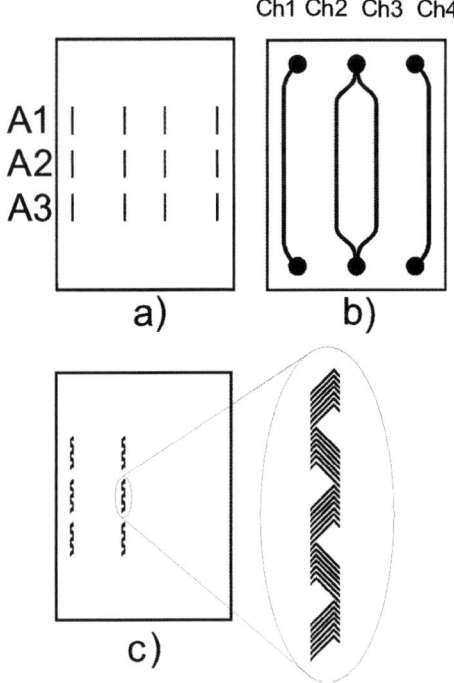

Fig. 1 Design of the masks for the sensitive areas (a), channels (b) and microstructures (c). The herringbone shape of the microstructures is shown in the detail.

The central channels had a common inlet and outlet, providing channels with half the flow rate of the outer channels from the same pump. The two channels on the left included embedded microstructures as seen in Fig. 1(c), while the ones on the right featured standard rectangular cross-sections in order to test the results in the same device. Each channel floor contained three gold patterned areas, A1, A2 and A3. The common inlet and outlet for the two central channels also provided uniform flow rate between the microchannels with (Ch2) and without (Ch3) the 3D flow microfeatures.

The sensing areas were based on the functionalization by means of an amine-sensitive molecular layer that could sense amine-coated microspheres. This chemistry was used as a test case to demonstrate the increased binding to the surfaces under the 3D microfeatures. A 150 nm thick gold layer was used as the substrate for the SAM. Gold was chosen for the substrate surface due to its high affinity to thiol (-SH) groups, the head groups of the self assembled monolayers, and its ability to be patterned to provide highly preferential binding. The gold layer was then patterned and deposited through a lift-off process with standard sputtering and photolithography methods. The sensing areas had a width of 200 μm, a length of 2.9 mm and a separation of 1.7 mm between them. There were two main reasons to include three different areas: to test the behavior along the length of the channels (i.e. effects of depletion) and to check if the effect of the microstructures was localized or not.

Two different masks were used to fabricate the microfluidic network, one for the channels and another for the microstructures, as shown in Fig. 1(b) and (c). The design chosen for the microstructures was a herringbone shaped design, which separates the flow in two different parallel streams rotating in clockwise and counter-clockwise directions at the same time [4,5]. This geometry is effective inducing the mixing and therefore was expected to increase the contact of the nanospheres with the sensitive surface. The asymmetry helped further to induce the mixing as the streams changed in size after each cycle [6]. In this case the asymmetry was set to 20%-80% (index of 0.2), and was displaced each cycle from right to left as seen in Fig.1 (c). The branches of the herringbone formed a 90° angle between them, with grooves 25 μm wide and the same separation between grooves.

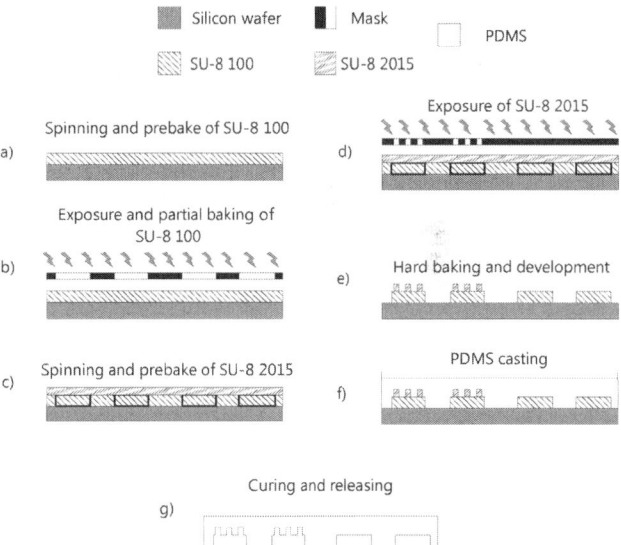

Fig. 2 Fabrication diagram of the microfluidic structure.

The mold fabrication process was based on the patterning of SU-8 negative photoresist onto a substrate of silicon [10-13] as seen in Fig. 2. In this design it was decided to build channels 100 μm high and microstructures 35 μm high. The objective was to obtain a ratio of over 30%, as this parameter has been described [7] as one of the most important factors in the design of micromixers.

C. SAM Formation and Testing

The SAM was formed inside the finished devices just before each test. The composition and formation were prepared following specific works on SAM formation and characterization [14,15].

III. RESULTS AND DISCUSSION

A. Silicon Mold and PDMS Channels

A picture of the mold obtained in the profilometer is shown in Fig. 3, which confirms that the microstructures were defined with precision. The measurements of the height along the extension of the wafer were an average 100±2 μm for the channels and an average of 35±1 μm for the embedded microstructures.

The reproduction of the microstructures in the PDMS microfluidic network mirrored perfectly the mold as the SEM picture presented in Fig. 3 (inset) shows.

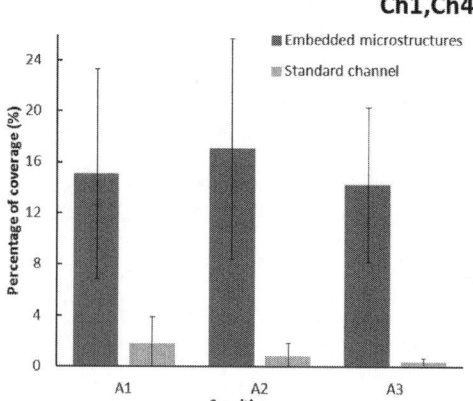

Fig. 5 Percentage of sensitive area covered by fluorescence in lateral channels (Ch1 and Ch4) with embedded microstructures versus standard rectangular cross-section channels. The three pairs of bars correspond to the three sensitive areas (A1, A2 and A3) present in the channels. The values were obtained processing images of six chips taken with an exposure time of 2 seconds and a magnification of 10x.

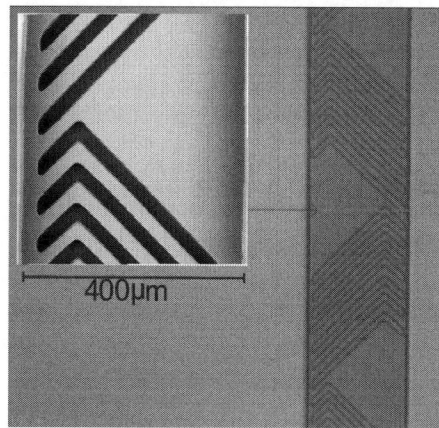

Fig. 3 Mold of a channel with microstructures (right). SEM picture of the microstructures embedded in the roof of the channel (inset).

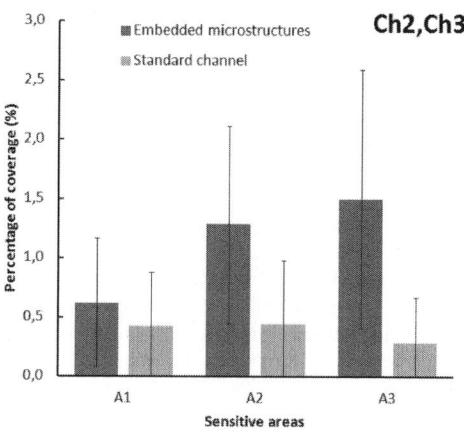

Fig. 6 Percentage of sensitive area covered by fluorescence in central channels (Ch2 and Ch3) with embedded microstructures versus standard rectangular cross-section channels. The three pairs of bars correspond to the three sensitive areas (A1, A2 and A3) present in the channels. The values were obtained processing images of six chips taken with an exposure time of 2 seconds and a magnification of 10x.

B. Surface-Sample Interaction Results

As Ch2 and Ch3 shared the same inlet and outlet, the flow rate in them was half of the one set up in the pump, and therefore their results were studied separately from the lateral channels (Ch1 and Ch4).

Results from lateral channels are displayed on Fig. 5. They were obtained processing the images taken in the Eclipse Ti microscope with the NIS-Elements software, and further confirm that the microstructures considerably enhanced the binding of the nanospheres.

The average coverage of the sensitive areas in channels with embedded microstructures was 15%, whereas in standard cross-section channels an average of 1% was obtained. Although there was a certain variation in the results as seen in the standard deviations, it was moderate enough to differentiate between both types of channel.

In the case of central channels (Ch2 and Ch3), where as stated before the flow rate was 50 μl/min, the results differed as seen in Fig. 6.

In channels with embedded microstructures an average coverage of approximately 1% was obtained, a much lower value than in lateral channels. In standard cross-section channels the value also decreased to about 0.4%, but the most important thing in this case is that the difference between both types of channels was considerably smaller.

While in lateral channels the difference was of 1 to 15, in central channels it was around 1 to 3. In addition to this, the variation in the results was relatively high making it much harder to discern clearly between them. All of this confirmed that the flow rate was a critical parameter.

Tests without the self assembled monolayer were also run as control measures. In lateral channels (Ch1 and Ch4), the fluorescence observed with microstructures was 8.2 times lower than in the same type of channels with the SAM, while for standard cross-section channels it was only approximately 2.3 times lower. These results prove that the microstructures had a potentiating effect on the SAM, and at the same time confirm that the SAM was correctly formed during the principal tests.

IV. CONCLUSIONS

We designed, fabricated and tested embedded microstructures to enhance surface-sample interaction in bioMEMS. We obtained an average coverage of 15% of amine-coated nanospheres on SAM modified sensitive areas, whereas for channels without microstructures the result was 1%. The experimental results also indicated a potentiating effect on the SAM. The correct formation of the SAM after the microfluidic device was fabricated and during the tests confirmed the possibility of using the same design for other applications where other type of coating is required. The focalized effect of the microstructures suggested a possibility of increasing the sensitivity of sensors fabricated with this type of microfluidic network.

Our ongoing research is focused on applying this design to a specific biosensor to test its capability of enhancing the sensitivity on working bioMEMS.

ACKNOWLEDGMENT

The author wishes to thank the Secretary of State of Investigation, Development and Innovation of the Ministry of Economy and Competitivity of Spain for funding this research within the framework of the SIMcell Project DPI 2012-38090-C03-D3.

REFERENCES

1. Chin C D, Linder V, Sia S K (2006) Lab-on-a-chip devices for global health: Past studies and future opportunities. Lab Chip 7:41-57 DOI 10.1039/B611455E
2. Hofmann O, Voirin G, Niedermann P et al. (2002) Three-Dimensional microfluidic confinement for efficient sample delivery to biosensor surfaces. Application to immunoassays on planar optical waveguides. Anal. Chem. 74:5243-5250 DOI 10.1021/ac025777k
3. Vijayendran R A, Motsegood K M, Beebe D J et al. (2003) Evaluation of a three-dimensional micromixer in a surface-based biosensor. Langmuir 19:1824-1828 DOI 10.1021/la0262250
4. Stroock A D, Dertinger S K W, Adjari A et al. (2002) Chaotic mixer for microchannels. Science 295:647-651 DOI 10.1126/science.1066238
5. Aubin J, Fletcher D F, Bertrand J et al. (2003) Characterization of the mixing quality in micromixers. Chem. Eng. Technol. 26:1262-1270 DOI 10.1002/ceat.200301848
6. Aubin, Fletcher D F, Xuereb C (2005) Design of micromixers using CFD modelling. Chem. Eng. Sci. 60:2503-2516 DOI 10.1016/j.ces.2004.11.043
7. Yang J T, Huang K J, Chin Y C (2005) Geometric effects on fluid mixing in passive grooved micromixers. Lab Chip 5:1140-1147 DOI 10.1039/B500972C
8. Stott S L, Hsu C H, Tsukrov D I et al. (2010) Isolation of circulating tumor cells using a microvortex-generating herringbone-chip. Proc. Natl. Acad. Sci. 107:18392-18397 DOI 10.1073/pnas.1012539107
9. Golden J P, Floyd-Smith T M, Mott D R et al. (2007) Target delivery in a microfluidic immunosensor. Biosens. Bioelectron. 22:2763-2767 DOI 10.1016/j.bios.2006.12.017
10. Zhang J, Tang K L and Gong H Q (2001) Characterization of the polymerization of SU-8 photoresist and its applications in micro-electro-mechanical systems (MEMS). Polym. Test 20:693-701 DOI 10.1016/S0142-9418(01)00005-8
11. Jackman R J, Floyd T M, Ghodssi R, Schmidt M A et al. (2001) Microfluidic systems with on-line UV detection fabricated in photo-definable epoxy. J. Micromech. Microeng. 11:263 DOI 10.1088/0960-1317/11/3/316
12. Lin C H, Lee G B, Chang B W et al. (2002) A new fabrication process for ultra-thick microfluidic microstructures utilizing SU-8 photoresist. J. Micromech. Microeng. 12:590 DOI 10.1088/0960-1317/12/5/312
13. Friend J, Yeo L (2010) Fabrication of microfluidic devices using polydimethylsiloxane. Biomicrofluidics 4:026502 DOI 10.1063/1.3259624
14. Ding S J, Chang B W, Wu C C, et al. (2005) Impedance spectral studies of self-assembly of alkanethiols with different chain lengths using different immobilization strategies on Au electrodes. Anal. Chim. Acta 554:43-51 DOI 10.1016/j.aca.2005.08.046
15. Ansorena P, Zuzuarregui A, Perez-Lorenzo E et al. (2011) Comparative analysis of QCM and SPR techniques for the optimization of immobilization sequences. Sens. Actuator B-Chem. 155:667-672 DOI 10.1016/j.snb.2011.01.027

Author: Mikel Gomez-Aranzadi
Institute: CEIT
Street: Manuel de Lardizabal 15
City: San Sebastian 20018
Country: Spain
Email: mgomez@ceit.es

Towards Point-of-Use Dielectrophoretic Methods: A New Portable Multiphase Generator for Bacteria Concentration

B. del Moral Zamora[1], J.M. Álvarez Azpeitia[2], J. Colomer Farrarons[1],
P.Ll. Miribel Català[1], A. Homs Corbera[1,2], A. Juárez[3,4], and J. Samitier[1,2]

[1] Department of Electronics, Faculty of Physics, University of Barcelona,
Martí i Franquès 1, 08028 Barcelona, Spain
[2] Nanobioengineering group, Institute for Bioengineering of Catalonia (IBEC),
Baldiri Reixac 10-12, 08028 Barcelona, Spain
[3] Microbial biotechnology and host-pathogen interaction group,
Institute for Bioengineering of Catalonia (IBEC), Baldiri Reixac 10-12, 08028 Barcelona, Spain
[4] Department of Microbiology, Faculty of Biology, University of Barcelona, Diagonal 643, 08028 Barcelona, Spain

Abstract—This manuscript presents portable and low cost electronic system for specific point-of-use dielectrophoresis applications. The system is composed of two main modules: a) a multiphase generator based on a Class E amplifier, which provides 4 sinusoidal signals (0º, 90º, 180º, 270º) at 1 MHz with variable output voltage up to 10 Vpp (Vm) and an output driving current of 1 A; and b) a dielectrophoresis-based microfluidic chip containing two interdigitated electrodes. The system has been validated by concentrating *Escherichia Coli* at 1 MHz while applying a continuous flow of 5 μL/min. Device functionalities were verified under different conditions achieving a 83% trapping efficiency in the best case.

Keywords—Dielectrophoresis, Cell Concentrator, electronics, lab-on-a-chip (LOC), portable device, low cost, Class E amplifier.

I. INTRODUCTION

Since Pohl [1] discovered the dielectrophoretic effect (DEP) there has been an increase of interest in using this particle manipulation technique, especially in the last decade [2]. The evidence is the large number of scientific publications that appear each year regarding the different applications of DEP: concentrate [3], sort [4], rotate [5] and move [6-7] particles or biological material. In fact, all these works are the consequence of DEP versatility since it can be used in several applications, such as microfluidics [8], medical diagnostics [9] or biosensors [3, 10-11].

Nowadays, these DEP applications are requiring new DEP system functionalities in different aspects. On one hand, many advances have been done in microfluidic systems, usually by new electrodes shapes or materials [12]. On the other, new electronic devices for DEP experiences have been developed [13-16, 6]. However, it is still difficult to find tailor-made electronic devices for DEP applications, combining microfluidics and electronics for a portable DEP system. Usually, regular multiple-use commercial devices have a high cost and large dimensions, which is a disadvantage for many applications.

In this manuscript a portable and low cost system for DEP applications is presented. The system is composed of a microfluidic chip, comprising two interdigitated electrodes, and an electronic multiphase generator specifically designed for DEP applications. The generator allows DEP electric field control by combination of four driving available phases φ1=0º, φ2=90º, φ3=180º and φ4=270º) or by varying the amplitude of each one of these driving signals (Vm).

The system has been tested using bacteria concentration experiments. This application is needed for many biomedical, food-control or environmental analysis situations, where bacteria or cells are recuperated from large volumes of sample and need to be pre-concentrated [3, 10-11]. In our case, *E. Coli* has been concentrated by DEP, using the designed portable system, at its optimal trapping frequency (1 MHz) as it will be explained later on. Experiments have been done at different applied voltages and different phase combinations, maintaining the same DEP effect between the different analyzed cases.

II. THEORY

The dielectrophoresis [1] is the movement of an electrically neutral particle when a non-uniform electric field is applied. This effect is defined in terms of DEP force. If it is considered a homogeneous isotropic particle which is polarized linearly, the force is defined by (1) [17-18].

$$F_{DEP} = \frac{1}{2} V \cdot Re[\alpha^*(\omega)] \nabla |E|^2 \quad (1)$$

Where E is the electric field, V is the volume of the particle and α is the effective polarizability, which involves the permittivities of the medium and the particle. According to the expression (1), α sign and the electric field gradient describe the force direction. When a positive α factor is obtained, the particle is attracted to an electric field maximum or to a region with high electric field intensity (which is called positive DEP or p-DEP). Otherwise, the particle is attracted to an electric field minimum (negative DEP or n-DEP). Hence, DEP allows to control the movement of a particle by modifying the electrical properties of the medium. Moreover, the DEP effect can also be modified by varying the applied signal in terms of phase (φ), frequency (ω) and voltage level (Vm). Furthermore, changing the electrode shape or placing dielectric structures strategically also creates different DEP forces by affecting the electric field uniformity. In this study, DEP force is modified using different phase combinations and by varying the applied voltage.

On the other hand, as previously mentioned, the concentrated particles were *E. Coli* bacteria. Hence, its shape must be considered to determine its behavior in terms of dielectrophoretic forces. For a bacteria model of an ellipsoid shape with two dielectric layers [19], as reported [20], the theoretical range to manipulate *E. Coli* by means of p-DEP is 500 kHz -10 MHz, with the sample conductivity used in our work (section III.B). Also, the maximum *E. Coli* p-DEP trapping force is obtained around 1 MHz, which value is set as the designed generator working frequency.

III. MATERIALS AND METHODS

A. DEP Multiphase Generator

The device is mainly formed of three modules (Fig. 1): a) A low power signal generator which creates four shifted and

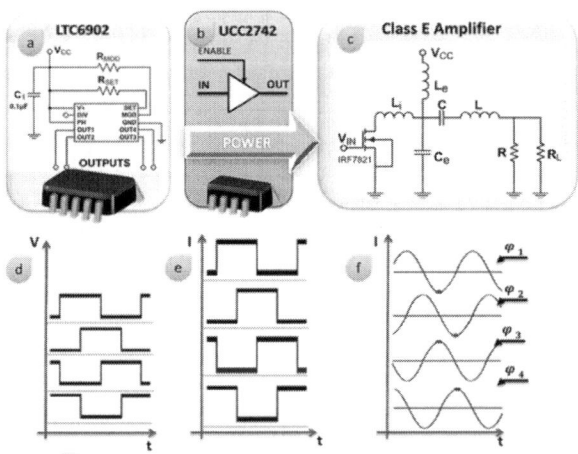

Fig. 1 DEP generator bloc diagram and its related signals.

frequency stable signals to control the output module. b) A driver which adapts the signal to the desired level of current for the next stage. c) A class E amplifier which creates the DEP signals. The first module (a), the low power signal generator, is based on the LTC6902 (Linear Technology), which generates four squared signals (Fig. 1. d) synchronized between them and whose outputs are shifted φ1, φ2, φ3 and φ4 respectively. This chip is prepared for low voltage designs since it only supplies 400 µA. Following our purpose, a Driver UCC27424 (Texas Instruments) is used (module b), which increases the current levels of the signals (up to 4 A) as it is depicted in Fig. 1. e. Finally, the driving signals control the final module (c) which is composed of a sinusoidal generator circuit which is based on a Class E amplifier. This structure (Fig. 1.c circuit) is capable of generating high frequency signals with a stable output voltage [21-22]. The amplifier creates a sinusoidal signal (Fig. 1.f) which follows the frequency of the driving signal created in the previous modules and introduced by a power transistor NMOS. The Class E parameters (Le, Ce, C, L, R) could be tuned using the following expressions (2-5). Note that RL is the equivalent resistance of the microfluidic chip and the inductive element Li is introduced to reduce the instant power demand of the system, due to the commutation from the square signal input.

$$L_e = \frac{0{,}4001R}{\omega_s} \quad (2)$$

$$C_e = \frac{2{,}165}{R\omega_s} \quad (3)$$

$$L = \frac{QR}{\omega_s} \quad (4)$$

$$\omega_s L = \frac{1}{C\omega_s} = 0{,}3533 \quad (5)$$

Finally, four independent channels perfectly synchronized at φ1, φ2, φ3 and φ4 are obtained. The generator outputs are selectable and allow varying output voltage up to 10 Vpp.

B. Bacterial Strains and Culture Media

E. Coli 5K cells were grown overnight in 10 mL Luria–Bertani broth at 37 °C. *E. Coli* experimental samples for DEP were obtained by first pelleting an overnight culture using centrifugation at 5000 rpm for 5 minutes and re-suspending the cells in 10 mL deionized water. Then they were diluted to achieve a final cell concentration around $2.4 \cdot 10^3$ bacteria/µL. The conductivity of the experimental *E. Coli* samples was 11.38 µS/cm. After that, samples at this final concentration were separated and frozen in different 1 mL collecting tubes for further experimental use.

C. Microfluidic Device Fabrication Process

The microfluidic chip fabrication can be divided into three main steps: microchannel molding (a), electrode fabrication (b) and microfluidic chip bonding (c). Each process will be detailed as follows.

a) Microchannel Molding: The SU8 50 microchannels molds were fabricated over glass slides. First, the slides were cleaned and activated by a Piranha attack during 15 minutes. Then, a 50 µm high SU-8 50 was spun over glass slides. Once developed, the microchannel mold was obtained. Soon after, the microchannel was replicated by mixing, degasing and pouring a 10:1 ratio PDMS pre-polymeric solution into the mold. Finally, after a 70°C at 1h curing process, the casted PDMS was peeled off from the master.

b) Electrode Fabrication: In order to fabricate the microelectrodes a lift-off soft lithography process was applied. First, an initial Piranha chemical attack was used. The AZ 1512 photoresist was chosen to act as a sacrificial layer. After exposure and first development of the AZ 1512, two metal layers, formed by 20 nm of Ti and 80 nm of gold, were vapor-deposited onto the surface. Then, the electrodes were obtained by removing the AZ photoresist.

c) Microfluidic chip bonding: In order to seal the microfluidic chip, the electrode surface was cleaned by using an oxygen plasma process. Then, the PDMS microchannels were aligned and attached by contact to the electrodes slide. Later, one cable was welded to each pad by conductive silver paint. Finally, an epoxy glue mix was also applied to the weld and cured at room temperature for 60 minutes. Eventually, two Nanoport Assemblies were attached to the inlet and outlet connections of the chip.

D. Experimental Setup

The experimental setup defined for each experiment of this study is shown in Fig. 2. The complete microfluidic module is composed of a 6 port manual valve (Valco), connected to a 5 mL syringe mounted in an infusion

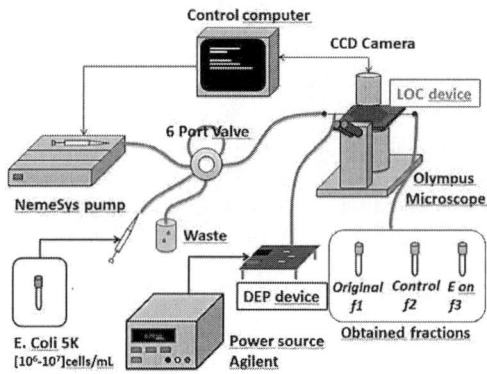

Fig.2 Experimental Setup

micropump (Cetoni NEMESYS). This syringe is connected to the microfluidic chip. This was placed over an inverted microscope stage (Olympus IX71) with a digital camera (Hamamatsu Orca R2).The microfluidic chip electrodes were activated by the DEP multiphase generator defined in section A. A power source (Agilent E3631A) was used to power it.

E. Experimental Protocol

In order to analyze the device trapping efficiency, three fractions were obtained from each experimental process (Fig. 2): the original fraction (f1), the control fraction (f2),

and the field action escaped bacteria's fraction (f3). The fractions f2 and f3 are obtained after introducing 150 μL of *E. Coli* sample (50 μL of original sample diluted in 100 μL of deionized water during the experimental process). Hence, in this particular case, a maximum concentration factor of 3 is possible, since the sample is concentrated in a volume of 50 μL.

The next protocol was followed in order to obtain every fraction: a 1 mL tube was defrosted to collect the first 50 μL sample (f1). The 1 mL tube's remaining content was introduced into the microfluidic module by 50 μL loads through the valve. After the first load injection without electric field activation, the second fraction (f2) was collected. Then, after a second load injection and keeping the electric field activated, a third fraction (f3) was collected. Every fraction was collected at a constant 5 μL/min flow rate by continuous deionized water pumping. Moreover, each 150 μL fraction was diluted again until reaching 200 μL, owing to the cytometer specifications, and immediately frozen to -20 °C. Once the fractions were defrosted, a flow cytometry (Beckman Coulter FC 500) bacteria counting process was applied. In order to improve the analysis accuracy, 1 μL of Green Fluorescent Nucleic Acid Stain (Invitrogen SYTO 13) was added to each 200 μL fraction.

IV. Results

In order to verify the system functionality, five series of concentration experiments were planned. These are defined in Table 1. The functionality of all DEP multiphase generator channels was tested with four different experiments.

Table 1 Experimental Cases

Experimental Case	Applied signals	Resultant potential V_{RMS}	Applied voltage $V_{pp} = 2V_m$	Case 1 Equivalent voltage
Case 1- Single Phase (reference)	E1: $\varphi_1 = 0°$ E 2: GND	$V_m/\sqrt{2}$	10Vpp	--
Case 2 90°diphase	E1: $\varphi_1 = 0°$ E2: $\varphi_2 = 90$	V_m	7Vpp	10Vpp
Case 3 270°diphase	E1: $\varphi_1 = 0°$ E2: $\varphi_4 = 270°$	V_m	7Vpp	10Vpp
Case 4 180°diphase	E1: $\varphi_1 = 0°$ E2: $\varphi_3 = 180°$	$\sqrt{2}V_m$	5Vpp	10Vpp
Case 5 180°diphase	E1: $\varphi_1 = 0°$ E2: $\varphi_3 = 180°$	$\sqrt{2}V_m$	10Vpp	20Vpp

Vm=maximum applied voltage. E1, E2= electrode 1 and 2.

Two signals from the device channels were applied to the pair of interdigitated electrodes, creating different phase combinations which generate the same DEP effect. Each applied voltage was determined by equalizing the resultant potential of each pair of applied signals (V_{RMS}) to Case 1 (see Table 1). Hence, cases 1, 2, 3 and 4, it was expected to obtain similar concentration efficiencies since all of them had equivalent applied fields. Case 5 was defined to verify if counter-phase signals could be the best option for concentration purposes, since it was expected to obtain better results with the same Vm per channel (max mum voltage) applied as in Case 1 and with the double voltage as compared with Case 4. Then, six repetitions of each Case were done to obtain statistics of concentration efficiency. Experimental results are depicted on Fig. 3. The median and the quartiles of each sample group (Cases) are presented in order to give a visual representation of the sample group distribution. To report a statistical analysis, the obtained bacteria counts were introduced in SPSS Statistics Software (IBM). Since the obtained counts are independent samples, a non-parametrical test was done in order to analyze the obtained results. Thus a U-Mann Whitney test was carried out, where equivalent effective-voltage Cases (Case 1, 2, 3 and 4) were analyzed in pairs. Then, no significant differences were obtained for these Cases. Moreover, these Cases were compared with the double voltage Case (Case 5). In this test significant differences between samples were detected ($p<0.01$), since Case 5 presents a higher trapping efficiency due to its equivalent applied voltage of 20 Vpp.

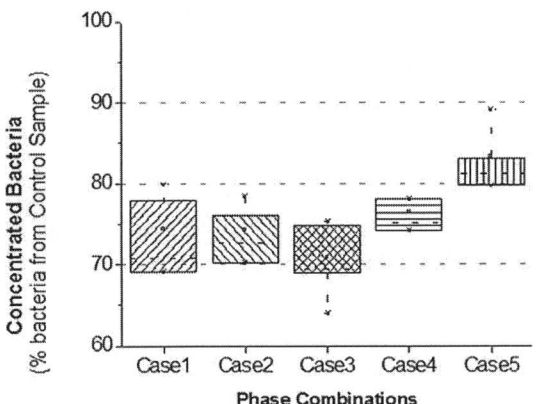

Fig. 3 Counting of trapped bacteria ((f2-f3)/f2) relative to control sample.

V. CONCLUSIONS

A multiphase generator for concentration analysis was designed, which is actually working at the optimal E. Coli p-DEP trapping frequency of 1 MHz. The device was tested and validated in a series of E. Coli concentration experiments with exhaustive cellular counts. The experiments were done at different phase combinations (equivalent DEP effects) with 150 µL of diluted *E. Coli* samples and an average concentration of $2.4 \cdot 10^3$ bacteria/ µL, at 5 µL/min continuous flow rate. From the analysis of these experiments, a concentration efficiency of 75% was obtained. Nevertheless, for counter-phased signals, the trapping efficiency increased to 83% when the applied voltage was doubled.

REFERENCES

1. Pohl, H. A., Hawk, I. Science 1966, 152, 647.
2. Pethig R., Biomicrofluidics 4, 022811, 2010
3. Hamada, R., Takayama, H., Shonishi, Y., Hisajima, T., Mao, L., Nakano M., Suehiro J., J Phys: Conf Ser, 2011, 307, 012031.
4. van den Driesche, S., Rao, V., Puchberger-Enengl, D., Witarski, W., Vellekoop, M. J. Sens Actuat B- Chem. 2011.
5. Ino, K., Ishida, A., Inoue, K. Y., Suzuki, M., Koide, M., Yasukawa, T., Shiku, H., Matsue, T.,Sens Actuat B-Chem. 2011, 153, 468-473.
6. Issadore, D., Franke, T., Brown, K. A., Westervelt, R. M. Lab Chip 2010, 10, 2937-2943.
7. Cheng, I. F., Chung, C. C., Chang, H. C. Microfluid Nanofluid 2011, 10, 649-660.
8. Wang, Y.; Ye, Z.; Ying, Y. Sensors 2012, 12, 3449-3471.
9. Shafiee, H., Sano, M. B., Henslee, E. A., Caldwell, J. L., Davalos, R. V. Lab Chip 2010, 10, 438
10. Swami, N., Chou, C. F., Ramamurthy, V., Chaurey, V. Lab Chip 2009, 9, 3212-3220.
11. Puttaswamy, S. V., Su, Y. J., Sivashankar, S., Yang, S. M. Liu, C.H., Micro Electro Mechanical Systems (MEMS), 2012 IEEE 25th International Conference on,2012, 1305-1308.
12. Baylon-Cardiel, J. L., Jesús-Pérez, N. M., Chávez-Santoscoy, A. V., Lapizco-Encinas, B. H. Lab Chip 2010, 10, 3235-3242.
13. Hunt, T. P., Issadore, D., Westervelt, R. M. Lab Chip 2008, 8, 81-87.
14. Miled, M. A., Sawan, M. Circuits and Systems (ISCAS), 2011 IEEE International Symposium on, 2011, 2349-2352.
15. Qiao, W., Cho, G., Lo, Y. H. Lab Chip 2011, 11, 1074-1080.
16. Jen, C. P., Huang, C. T., Chang, H. H. Microelectron Eng 2011, 88, 1764-1767.
17. Morales, F. H. F., Duarte, J. E., Martí, J. S. Rev Acad Colomb Cienc Exactas, Fís y Nat 1936, 32, 361.
18. Jones, T. B. Electromechanics of particles, Cambridge Univ Pr 2005.
19. Morgan, H. and Green, N.G. Research Studies Press LTD, 51-63 (2002). Morganti, D., Morgan, H. Colloids Surf. Physicochem. Eng. Aspects 2011, 376, 67-71.
20. Castellarnau, M., Errachid, A., Madrid, C., Juárez, A., Samitier, J. Biophys. J. 2006, 91, 3937-3945.
21. Lenaerts, B., Puers. R. Omnidirectional inductive powering for biomedical implants, Springer Netherlands 2008.
22. Rashid, M. H., González, M. H. R. V., Fernández, P. A. S. *Electrónica de potencia: Circuitos, dispositivos y aplicaciones*, Pearson Educación, México 2004.

Author: Beatriz del Moral Zamora
Institute: Universitat de Barcelona. Facultat de Física
Street: Martí I Franquès 1, 2a planta
City: Barcelona
Country: Spain
Email: bdelmoral@el.ub.edu

Flagella – Templates for the Synthesis of Metallic Nanowires

L. Deutscher[1], L.D. Renner[1], and G. Cuniberti[1,2]

[1] Institute for Materials Science and Max Bergmann Center of Biomaterials, TU Dresden, 01062 Dresden, Germany
[2] Division of IT Convergence Engineering, POSTECH, Pohang, Korea

Abstract—The classical approach to synthesize nanowires is based on deposition of conductive metal ions. In this report we present the modification of bacterial appendages in order to form nanowires from renewable biological structures, bacterial flagella. Flagella have many advantageous characteristics of proteinaceous constructs: they possess the ability to self-assemble, to extend over several micrometers in length and possess diameters of 10-20 nm making them ideal candidates for nanowires. We show the isolation of micrometer-sized flagella from three different bacterial strains: *Pseudomonas fluorescens*, *Lysinibacillus sphaericus* and *Shewanella oneidensis*. The dimensions and characteristics of the isolated flagella were investigated by atomic force microscopy and scanning electron microscopy. The bacterial appendages had an average length of approximately 3.7 μm and an average diameter of 10.6 nm. In order to achieve conductive nanowires we applied a methodology to metallize purified flagella. Our results show the applicability of bacterial flagella as scaffolds for nanowire synthesis. We envision that our approach will enable the application of bacterial nanowires to a wide range of biomolecular sensors.

Keywords—**nanowire, biotemplating, flagella, *Lysinibacillus sphaericus*, *Pseudomonas fluorescens*, *Shewanella oneidensis*.**

I. INTRODUCTION

Nanostructures (i.e. nanowires, nanorods and nanotubes) represent a growing field in nanotechnology and are of particular scientific interest in nanosensorics. For example, semiconducting nanowires own characteristics that can be directly applied in sensor devices. Nanowire based field effect transistor sensors are supposed to represent a promising tool for the quick detection of biomolecules and pathogens in medical applications [1]. Much effort in this research area has been invested in the utilization of metal and semiconducting nanowires. However, the attention has rapidly expanded to investigate the practicability of innovative materials for the synthesis of nanowires. Biomolecules represent suitable building blocks for such approaches. Many biomolecules naturally occur in the nanometer size range and can result in broad structural varieties. Additionally, the biosynthesis of these molecules is highly controlled in terms of their sequence, conformation and composition [2]. Furthermore, the self-assembling characteristics of many biomolecules circumvent the need to synthesize them into wires which could enhance the mass production of nanowires based on biomolecules. Hence, the utilization of biological structures that act as nanowire templates is a promising approach. In the past, various biomolecules were used as templates: for example, phospholipid tubules [3], DNA [4], actin filaments [5], viruses [6] or amyloid fibers [2]. The resulting nanowires exhibited electrical properties ranging from insulator through semiconductors to conductors with metallic conductivities covering a wide range of applications [7].

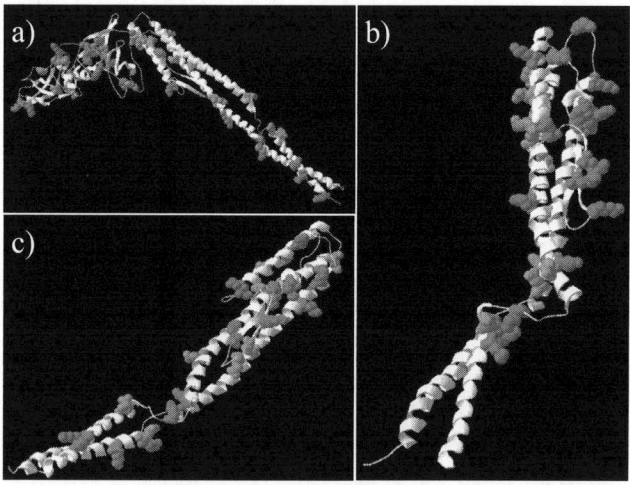

Fig. 1 Predicted 3D models of a) *Pseudomonas fluorescens*, b) *Lysinibacillus sphaericus* and c) *Shewanella oneidensis* flagellin proteins. Potential metal-binding sites are color-coded, green: lysine, orange: glutamic acid.

In addition to the above mentioned biomolecules, another group of biological structures that can be applied as nanowire templates are bacterial flagella. Flagella are extracellular appendages used to facilitate the motility of the bacteria. A flagellum consists of three parts: a basal body, a hook and a helical filament that is 10 – 20 nm wide and several micrometers long. The helical filament is build up from protein monomers, called flagellins. The physical size of these filaments turns them into excellent candidates for nanowire templates. Furthermore, the amino acid structure of the flagellins contains several binding sites for metal ions (Figure 1). Here, we exploit the existence of metal binding sites of bacterial flagella to illustrate the opportunity to generate conductive nanowires.

II. MATERIALS AND METHODS

A. Bacterial Strains

Pseudomonas fluorescens (DSM 50124) and *Lysinibacillus sphaericus* (NCTC 9602) cells were grown for 12 hours under aerobic conditions at 28 °C in 25 ml medium 1 (5 g/l peptone and 3 g/l beef extract). *Shewanella oneidensis* MR-1 cultures were incubated in Luria-Bertani medium (10 g/l peptone, 5 g/l yeast extract and 10 g/l NaCl) at 30 °C for 12 hours.

B. Protein BLAST and 3D Modeling

The amino acid sequences of the flagellins for all three bacterial strains were taken from the NCBI Reference Sequence Database (Table 1) and were compared using the NCBI BLAST tool [8,9].

Table 1 Reference Sequences of used flagellin proteins

Bacterial strain	RefSeq	No. of aa
Pseudomonas fluorescens	AGC13085.1	552
Lysinibacillus sphaericus	YP_001696896.1	270
Shewanella oneidensis MR-1	NP_718793.1	273

To predict the secondary and tertiary structures of the proteins the FASTA formatted sequences have been entered to the online protein homology/analogy recognition engine, (Phyre, version 2.0) [10]. The obtained protein data based formatted structures have further been processed with the Swiss-PDBViewer [11].

C. Purification/Isolation of Bacterial Flagella

To separate cell appendages the bacteria were collected from the nutrient solution by centrifugation, washed in water and resuspended in 0.1 M TRIS HCl (pH 7.5). Then, we used a needle and a syringe and pipetted the cells several times through the needle to mechanically shear off the appendages. Subsequently, we separated the flagella from the cells by centrifugation (980 g, 10 min). Finally we concentrated the flagella by ultra-centrifugation (105,000 g, 30 min) and subsequently resuspended the flagella in 1 ml Tris HCl. We added PMSF (at final concentration of 0.1 mM) to inhibit protease activity. For conductivity measurements the flagella were resuspended in deionized water.

We investigated the results of flagella separation by fixing them on silicon wafers and analyzing the purity of the samples as well as the integrity and the dimensions (length and height) of the appendages with atomic force microscopy (AFM). The obtained images were processed with the software WSxM [12].

D. Metallization

Solutions of isolated flagella (50 µl) were incubated with 10 µl 0.1 M Pd(OAc)$_2$ solution at 25 °C for 2 hours, subsequently reduced with NaBH$_4$ (0.1 M) and incubated for one hour at 25 °C.

III. RESULTS

A. Structural Comparison of Flagellin Proteins

We compared the amino acid sequences of the flagellins from the three bacterial strains in order to identify potential binding sites for metal ions and to interpret potential variations in the conductance behavior. The flagellin from *P. fluorescens* is double the size of the flagellins from *L. sphaericus* and *S. oneidensis* (Table 1). A comparison of the three proteins by using the *P. fluorescens* flagellin as a query sequence (using the NCBI protein BLAST) revealed a sequence identity of 47 % with the *L. sphaericus* flagellin (expect value $5e^{-46}$) and 52 % with the *S. oneidensis* flagellin (expect value $9e^{-54}$). The expect values indicate a significant biological relation between these three proteins.

Furthermore we used the sequence data to generate 3D models of the proteins (Figure 1). The model of *L. sphaericus* flagellin contains no, *S. oneidensis* two and *P. fluorescens* 58 *ab initio* modeled residues. These values illustrate that more than 99 % of the structure of *L. sphaericus* and *S. oneidensis* flagellins are highly reliable. For *P. fluorescens* flagellin 89 % of the protein structure is predicted with confidence.

After obtaining the 3D models we searched for potential metal-binding sites (lysine and glutamic acid) and highlighted them within the model (lysine: green; glutamic acid: orange). We identified a high number of potential metal-binding sites within the flagellin proteins supporting our idea to be able to successfully metalize the structures.

B. Characterization of Isolated Flagella

We investigated the physical characteristics (length, width) of isolated flagella with AFM. *P. fluorescens* flagella range range from 2 to 5.5 µm with an average value of 3.8 µm. We extracted the flagellar diameters from the height profiles. For *P. fluorescens* the flagella measured 11 nm in diameter.

C. Metallization of Isolated Flagella

Based on our promising isolation and purification of flagella we investigated the metallization of flagella from different bacterial strains with palladium. Comparing AFM images of isolated flagella prior to and after the metallization revealed a significant change in the morphological features (Figure 3). The flagella lost their homogeneously characteristics and seemed to be covered and sometimes spotted with bigger particles, as indicated by the brighter color. To verify the metallization, we repeated the measurements of the flagella diameter. We expected that the coverage of the flagella would lead to an increased diameter. Figure 3 indicates the height profiles of two neighbored flagella. The diameter of the coated metallized flagellum is around 45 nm and the uncovered flagellum has an average height of about 10 nm. Thus, as expected the coverage with palladium led to an increase in diameter of about 30 nm.

IV. DISCUSSION

One of the major challenges in nanotechnology is the fabrication of various thin nanowires using an easy, fast and inexpensive method that avoids extensive production processes. Therefore, the application of biomolecules and biomolecular scaffolds represents a suitable alternative to assemble nanowires. In this report we show that unmodified flagella can easily be purified and subsequently metalized in order to fabricate conductive nanowires.

We demonstrated a method to easily isolate flagella from bacterial cells in high concentrations and with high purity. This method appears to be sufficiently gentle to isolate micrometer-sized flagella with average lengths bigger than 3 μm which is several times longer, than it has been shown for γ-prefoldin protein filaments [3]. Our approach facilitates a much better integration of the isolated flagella into electrical devices fabricated by conventional laser lithography. Nevertheless, the precise control of the flagella length within this method is a favorable feature in order to expand the applicability of flagella-template nanowires. Therefore, in further investigations we will study the *in vitro* self-assembling characteristics of flagella. Defining the required parameters for the elongation of the flagella would allow for a directed regulation of the filaments length. Our procedure produces flagella that support the attachment of palladium particles which leads to a complete coverage of the filaments. The coverage of the flagella is indicated by a 2 to 3 times increase in diameter of the treated structures compared to the untreated controls. However, the coverage's composition needs to be further investigated to confirm the presence of palladium attached to the flagella (e.g. by energy dispersive X-ray spectroscopy).

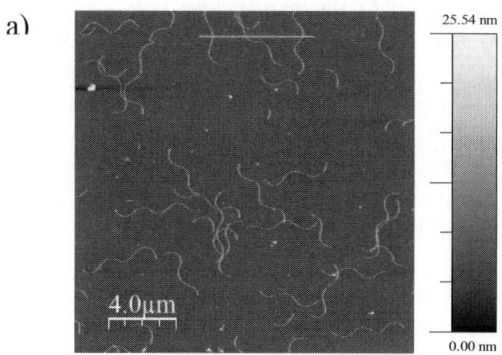
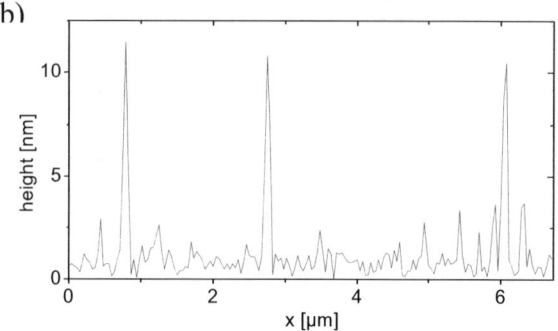
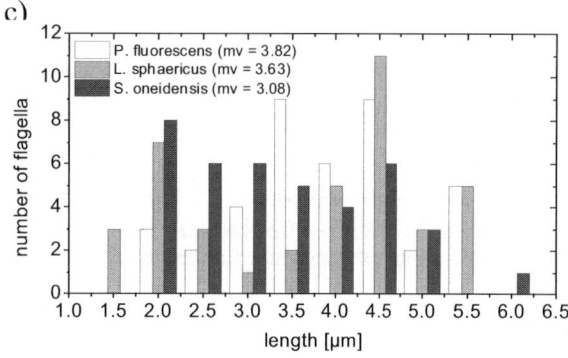

Fig. 2 a) AFM image of isolated flagella from *L. sphaericus* with the blue line representing the route of the measured height profile depicted in b). c) Length distribution of isolated flagella from *P. fluorescens*, *L. sphaericus* and *S. oneidensis*. The length of 40 flagella was measured, respectively.

For *L. sphaericus* the lengths were between 1.5 and 5.5 μm (mean of 3.6 μm). The mean value of the measured flagella diameter was 9.5 nm, making them slightly thinner and shorter than the ones from *P. fluorescens*. Isolated flagella from *S. oneidensis* show an average length of 3.1 μm and measure around 11.4 nm in diameter. Figure 2a depicts a representative AFM image and a representative height profile of the flagella from *L. sphaericus* (Figure 2b).

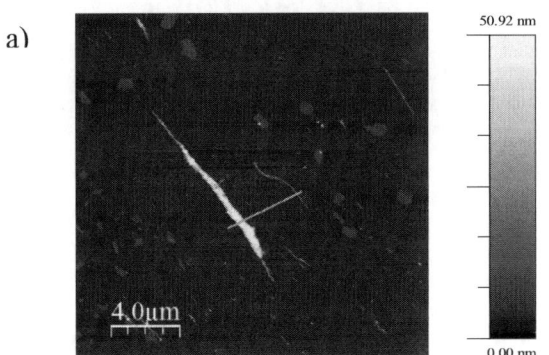

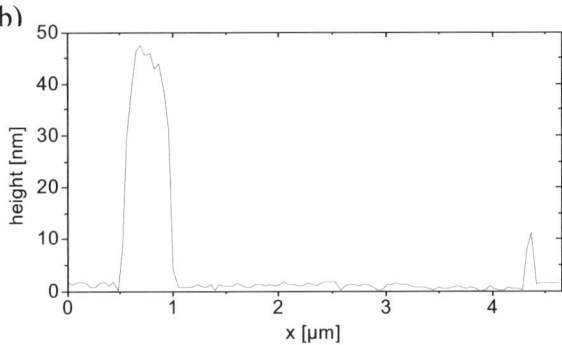

Fig. 3 a) AFM image of isolated and metallized (Pd) flagella from *L. sphaericus* with the blue line representing the route of the measured height profile depicted in b). The diameter of the metallized flagellum is about 45 nm, more than 30 nm higher than the uncoated flagellum.

Bacterial flagella comprise distinct characteristics to be applied as nanowire templates. Except for their convenient physical size the filaments are naturally located outside the bacterial cell. This feature enables the simple, mechanical separation of the filaments from cells without any need to enzymatically digest the bacteria. This step obviously enhances the convenience of the applied method and the entire approach to use flagella as nanowire templates.

V. CONCLUSIONS

In summary, we have successfully implemented a methodology to produce conductive nanowires from unmodified flagellar biotemplates. Our results show the applicability of our method to isolate intact flagella. Furthermore, we present results of flagellar metallization from three different bacterial strains. Our results emphasize the practicability of the applied method. Further investigations will reveal the conductive characteristics of these generated bacterial nanowires and their potentials for the integration into electrical devices.

ACKNOWLEDGMENT

This work is funded by the European Union (ERDF) and the Free State of Saxony via the ESF project 100098212 InnoMedTec. We gratefully acknowledge support from the German Excellence Initiative via the Cluster of Excellence EXC 1056 "Center for Advancing Electronics Dresden" (cfAED). This research was supported by World Class University program funded by the Ministry of Education, Science and Technology through the National Research Foundation of Korea (R31-10100).

REFERENCES

1. Patolsky F, Zeng G, Hayden O et al. (2004) Electrical detection of single viruses. PNAS 101:14017-14022
2. Scheibel T, Parthasarathy R, Sawicki G et al. (2003) Conducting nanowires built by controlled self-assembly of amyloid fibers and selective metal deposition. PNAS 100:4527-4532
3. Schnur J M, Proce R et al. (1987) Lipid based tubule microstructures. Thin Solid Films 152:181–206
4. Deng Z, Mao C (2003) DNA-templated fabrication of 1D parallel and 2D crossed metallic nanowire arrays. Nano Letters 3:1545-1548
5. Patolsky F, Weizmann Y, Willner I (2004) Actin-based metallic nanowires as bio-nanotransporters. Nat Mater 3:692-695
6. Mao C, Solis D J, Reiss B D et al. (2004) Virus-based toolkit for the directed synthesis of magnetic and semiconducting nanowires. Science 303:213-217
7. Glover D J, Giger L, Kim J R, Clark D S (2013) Engineering protein filaments with enhanced thermostability for nanomaterials. Biotechnol J 8:228-236
8. Altschul S F, Madden T L, Schäffer A A et al. (1997) Gapped BLAST and PSI-BLAST: a new generation of protein database search programs. Nucleic Acids Res. 25:3389-3402
9. Altschul S F, Wootton J C, Gertz E M, et al. (2005) Protein database searches using compositionally adjusted substitution matrices. FEBS J. 272:5101-5109
10. Kelley L A, Sternberg M J E (2009) Protein structure prediction on the web: a case study using Phyre server. Nature Protocols 4:363-371
11. Guex N, Peitsch M C (1997) Swiss-model and the Swiss-PdbViewer: An environment for comparative protein modeling. Electrophoresis 18:2714-2723
12. Horcas I, Fernandez R, Gomez-Rodriguez J M et al. (2007) WSXM: A software for scanning probe microscopy and a tool for nanotechnology. Rev. Sci Instrum 78:013705-8

Author: Linda Deutscher
Institute: Institute for Materials Science and Max Bergmann Center of Biomaterials
Street: Budapester Str. 27
City: 01062 Dresden
Country: Germany
Email: g.cuniberti@nano.tu-dresden.de

A Versatile Microfabricated Platform for Single-Cell Studies

A. Benavente-Babace[1,2], D. Gallego-Pérez[3], D.J. Hansford[3],
S. Arana[1,2], E. Pérez-Lorenzo[1,2], and M. Mujika[1,2]

[1] CEIT-IK4 and Tecnun (University of Navarra), Donostia-San Sebastián, Spain
abenavente@ceit.es
[2] CIC microGUNE, Arrasate-Mondragon, Spain
[3] Department of Biomedical Engineering, The Ohio State University, Columbus-OH, USA

Abstract—Here we present the development of a tunable microfluidic device for single cell studies that combines a number of technologies for applications in cell therapy and cancer research. The fabricated device consists of a main trapping module and two clip-on units. In this work optimum fabrication, cell trapping and seeding parameters are described. This versatile microfabricated platform will enable the investigation of fundamental biological processes such as cell migration or intercellular communications without the necessity of any surface functionalization and in a more in-vivo like manner.

Keywords—Single-cell, microfluidics, guided migration, co-culture.

I. INTRODUCTION

'Average cell' has represented always a hitch for a better understanding of the different biological and behavioral processes of cells [1,2]. Traditional methods are based on average responses from a population that often mask the difference among individuals. There is evidence about the heterogeneity of single cells within a genetically identical population that is critical in their function and viability. Therefore single cell analysis is a key factor for a better understanding towards cancer research [3] or tissue regeneration, and opens a lot of opportunities for drug discovery and for more personalized diagnostics. However, multiple individual cells are required to obtain statistically meaningful data, and for that reason high throughput analysis is critical.

Microfluidic platforms have become a key toolbox for studying cell biology. Their inherent capacity for manipulating fluids and molecules at small length scales with high control, high throughput and with minimized use of samples and reagents have made them the preferred choice for researchers in the last two decades. Therefore, a great collection of microsystems for cell-based assays has been developed. Hui and Bathia developed a micromechanically-controlled device for studying cell-cell interactions with application on cell therapy [4]. The Kamn group developed a platform to mimic 3D microenvironments for cell migration studies [5]. When focusing at single cell level based research, different trapping strategies have been reported [6-11] with distinctive applications. In particular, single cell isolation based on hydrodynamic resistance seems to be an attractive way for avoiding cell damage [12,13]. However, while all these are good examples of single-cell level handling, further efforts are required for 'cleaner' approaches, avoiding pre-treatments that could alter cell behavior, more versatile platforms and at the same time devices with simpler fabrication procedures.

In this work we present a multipurpose PDMS microfluidic platform for single cell studies. The device has the ability of trapping cells based on hydrodynamic resistance concept by means of sieve-like structures [11,13]. In addition, it can be easily interfaced with a number of micro-nanofabricated modules that can be used in a host of applications. Herein we explored the development of this platform for potential application in single cell studies related to structurally guided tumor cell migration [14] and cell-cell interactions [15].

II. MATERIALS AND METHODS

A. Microdevice Design and Fabrication

The microdevice is composed of three different entities: the main platform, containing the trapping structures, and other two clip-on modules (see Fig. 1). One of these contains 1D line patterned surface in order to carry out cell migration related studies [14]; the other one has microwells on it which will offer the microenvironment for intercellular communication studies [15]. We will refer to these parts as traps, lines and microwells respectively. In order to manufacture them, three different molds were fabricated using standard UV photolithographic techniques and single-layer soft-lithography (see Fig. 2). Briefly, three 4'' silicon wafers were cleaned in piranha solution (H_2SO_4/ 1:1) at 200°C for 20 minutes and rinsed with deionized water. The mold containing the traps used a 27 μm thick layer of negative photoresist (SU-8 2025, MicroChem Co.) spin coated at 2600 rpm. The wafer was then softbaked, exposed to the specific photomask, postbaked and developed according to manufacturer's instructions. Finally, the wafer was hard baked for 2 minutes at 150°C. For the mold containing different size microwells, an 8 μm thick layer of negative photoresist (SU-8 2005, MicroChem Co.) was spin coated at 1000 rpm onto the wafer. As described before, the wafer was then subsequently processed (prebaked, aligned and exposed to the specific mask, post-exposure baked and developed) according to manufacturer's instructions. This wafer was also hard baked at 150 °C for 2 minutes. For the final pattern of lines (1,5μm tall × 2μm wide ridges spaced by 2 μm grooves), S1813 positive photoresist was spin coated at 3000 rpm for 1 minute. After softbaking the wafer was aligned to the photomask and exposed to UV light. Then it was developed in MF-319 (Shipley Co.) and rinsed with isopropanol.

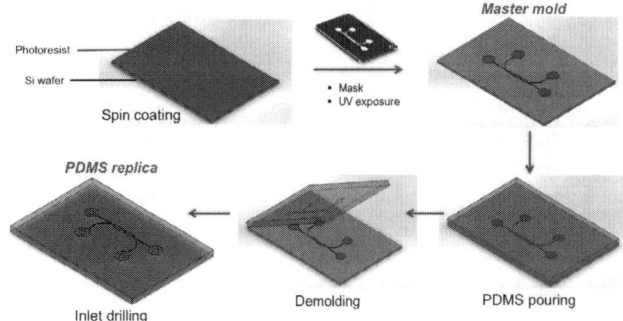

Fig. 2 Photolithography process for the main module: *'traps'* geometry. The *'microwells'* and the *'lines'* modules are fabricated in an analogue way but without drilling the inlets and outlets.

B. Cell Culture

293-GPF expressing cell line was obtained from the American Type Culture Collections (ATCC). Cells were cultured in Dulbecco's modified Eagle's medium (DMEM) containing 10% fetal bovine serum (FBS) and 1% of antibiotics and incubated in a humid atmosphere at 37° C and 5% CO_2.

C. Cell Trapping and Device Operation

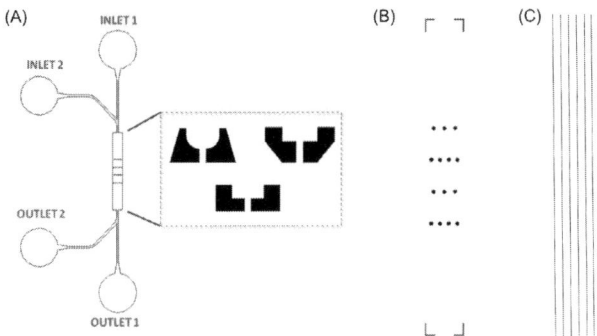

Fig. 1 Designs of the modules of the microfluidic platform. A) Main module, containing the microchannels and the chamber with the traps array. B) Attachable module, containing microwells designs. C) Second attachable module, 1D lines geometry.

Following fabrication of the SU-8 masters, polydymetylsiloxane (Sylgard 184 PDMS, Dow Corning) replicas were prepared according to the manufacturer's instructions – weight ratio of 10 parts of base to one part of curing agent –, degassed in a vacuum desiccator for 20 minutes and then poured onto the patterned silicon wafers before curing at room temperature for 48 hours. Finally, molds were disassembled. The inlets and outlets in the traps containing PDMS replica were punched by a needle for tube connections. Then, depending on the required study, the trapping module was bonded to the lines or the microwells module by a brief – 1 min, 100W – oxygen plasma treatment (Femto®, Diener Electronic).

A preliminary characterization of the microstructures was conducted using optical microscopy and scanning electron microscopy (SEM).

Before use, the devices were sterilized with 70% ethanol. Then, the chip was filled with media and incubated at 37 °C for ≥3 hours before cell loading to allow protein adsorption on the surface and facilitate cell attachment. Afterwards, media is perfused through a syringe pump for one of the two inlets (see Fig. 1a) of the device refreshing old media and eliminating any bubbles inside the microfluidic device. Cell suspensions are then loaded in a 1 mL syringe and flowed into the device through the remaining inlet. The flux rate ratio between both inlets could be adjusted in order to control the specific line of traps to be filled [16]. Once all the traps are filled with cells, the device is put in the incubator allowing cells to settle at the bottom of the microwells or on the micropatterned lines.

D. Imaging

Different SEM (Hitachi S3000-H) images were acquired in order to characterize the molds and the PDMS replicas. Prior to imaging the devices were coated with a thin Au/Pd layer to prevent charging. Cells were imaged during the different experiments stages using a contrast phase microscope (Nikon Eclipse TS100-F).

III. RESULTS AND DISCUSSION

A. Microdevice Fabrication and Assembly

SEM micrographs (Fig. 3) confirmed that traps, wells and lines modules are accurate replicas of the master molds that are 27 µm, 8 µm and 1,5 µm in height respectively. The trapping arrays within the main chamber of the device consist of rows containing pocket-like structures with different geometries that have a small aperture of 5 µm (Fig. 3a). This aperture is designed to regulate the hydrodynamic resistance and to achieve cell trapping. Several diameters (25 µm, 50 µm and 75 µm) of wells were designed for the microwells module in order to control single-cell or cell-cell interaction studies. The center-to-center distance matches that of the traps. Fig. 4 shows the final device assembly after bonding the main module with the lines interface (Fig. 4a) and with the corresponding microwells interface (Fig. 4b). As it can be seen in the micrographs both the lines and the microwells were successfully aligned to the superimposing cell trapping microstructures. All these dimensions were tailored for an average cell diameter of ≈ 20 µm.

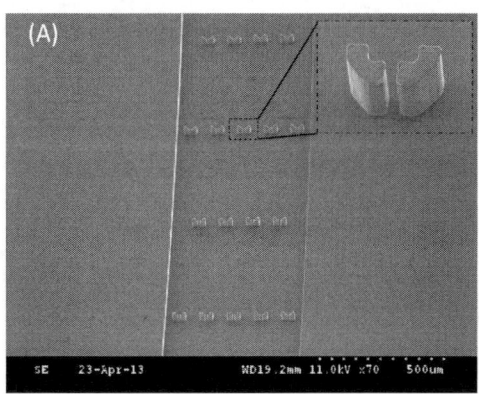

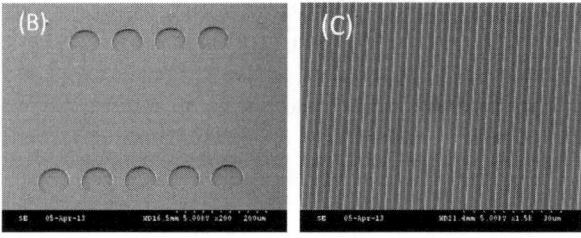

Fig. 3 SEM micrographs of the PDMS modules. A) Main module, traps, micrographs. B) Two rows of 75 um diameter wells. C) Lines module, 2um width by 2um groove separation.

Due to silicon SU-8 masters presented some stress cracks, they were hard baked. In addition, we found that the optimal developing time for the traps containing mold was 3-4 minutes longer than the standard.

The thickness of the PDMS clip-on modules was optimized according to the focal distance of the microscope objective in order to facilitate imaging of the cells.

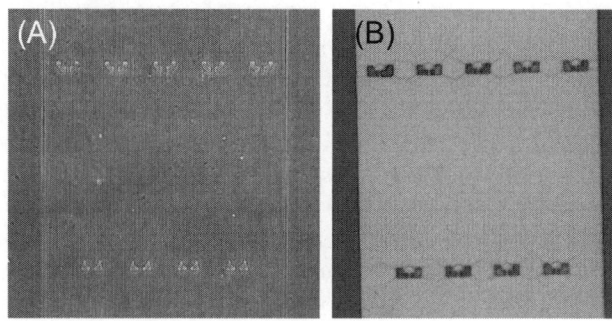

Fig. 4 Traps inside the chamber of assembled devices. A) Traps interfacing lines module. B) Traps interfacing the microwells module.

B. Cell Trapping Procedure Optimization

In order to trap the cells without damage and in an efficient way, different parameters have to be taken into account, such as loading time, flow velocity and cell density. In order to improve the cell viability the cells should be loaded within 5-10 minutes after starting the perfusion of the cell suspension. This helps to minimize the amount of shear stress the cells are exposed to during the trapping stage.

When loading cells at high speed, traps are filled faster and the loading time is minimized. However, cells could be subjected to excessive shear stress under these conditions, which could compromise cell behavior and the integrity of the membrane, resulting in cell death (Fig. 5a). In contrast, if the loading velocity is too low the underlying mechanism of trapping, based on the hydrodynamic resistance of the traps, would not work properly because the cells cannot be directed towards the traps. We found that a flow speed of 1µL/min maximized both viability and trap efficiency.

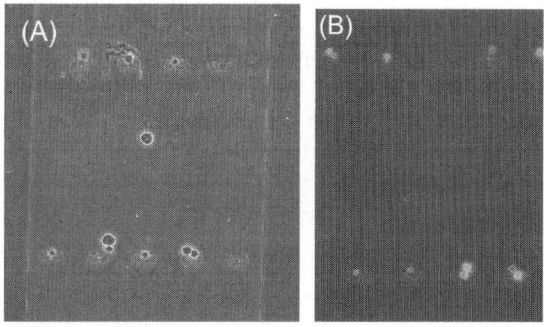

Fig. 5 Common problems during trapping. A) Cell debris, stressed and death cells after being subjected to high shear. B) Multiple trapping.

In addition, cell density also played a crucial role in the trapping dynamics. Increasing cell density leads to shorter loading times. We found that the optimal density was 3.0×10^5 cells/mL. Higher densities lead to clogging of the microchannels and multiple cells lodging per trap (Fig. 5b). These established parameters are equally valid for both device modules: lines and microwells.

C. Cell Seeding Optimization

Once the cells are trapped (Fig. 6), they need to be in static culture for a proper attachment and spreading. If the culture chamber experiences pressure variations or reversing flow, cells are released from the traps and a re-trapping procedure should be implemented, which affects cell viability. We found that in order to let the cells anchor to the surface, they should be kept under static culture for at least 2 hours, followed by the respective cell assay depending on the nature of the module.

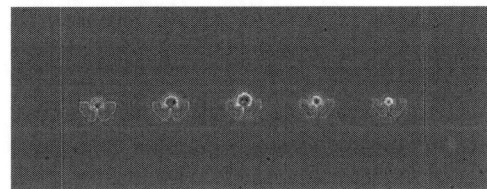

Fig. 6 Trapped cells

IV. CONCLUSIONS

Most of the performed cell biology studies are based on whole cell populations assays. This may result in misleading information based on average data and masking the heterogeneity among individual cells. Microfabricated devices have led to great advances in the biomedical field; however further efforts are required for long-term high throughput studies avoiding the alteration of cell microenvironment. This work presents a single-cell isolation microfluidic platform that can be easily attached to two different modules. These make the device suitable for a number of biomedical applications, including guided tumor cell migration and controlled cell-cell interaction. Optimum fabrication and cell trapping and seeding parameters were described for both applications. Ongoing work in both, cancer cell migration studies and tunable heterotypic and homotypic cell-cell interaction, which will be addressed in future publications. This single-cell level multipurpose platform will facilitate a better understanding in key biological processes by providing well-controlled and versatile microenvironments.

ACKNOWLEDGMENT

The authors acknowledge the Bask Government for the grant promoting doctoral theses for young pre-doctoral researchers and thank the Secretary of State of Investigation, Development and Innovation of the Ministry of Economy and Competitivity of Spain for funding this research within the framework of the SIMcell Project DPI 2012-38090-C03-D3.

REFERENCES

1. Dawson P. S. S. (1988). The 'average cell'—a handicap in bioscience and biotechnology. *Trends in Biotechnology*, 6(5), 87-90.
2. Levsky J. M., & Singer R. H. (2003). Gene expression and the myth of the average cell. *Trends in cell biology*, 13(1), 4-6.
3. Nagrath S., Sequist L. V., Maheswaran S, et al. (2007). Isolation of rare circulating tumour cells in cancer patients by microchip technology. *Nature*, 450(7173), 1235-1239.
4. Hui E. E., & Bhatia S. N. (2007). Micromechanical control of cell–cell interactions. *Proceedings of the National Academy of Sciences*, 104(14), 5722-5726.
5. Chung S., Sudo R., Mack PJ, et al. (2009). Cell migration into scaffolds under co-culture conditions in a microfluidic platform. *Lab on a Chip*, 9(2), 269-275.
6. Neuman K. C., & Nagy A. (2008). Single-molecule force spectroscopy: optical tweezers, magnetic tweezers and atomic force microscopy. *Nature methods*, 5(6), 491-505.
7. Wheeler A. R., Throndset W. R., Whelan RJ, et al. (2003). Microfluidic device for single-cell analysis. *Analytical chemistry*, 75(14)
8. Irimia D., Tompkins R. G., & Toner M. (2004). Single-cell chemical lysis in picoliter-scale closed volumes using a microfabricated device. *Analytical chemistry*, 76(20), 6137-6143.
9. Voldman J., Gray M. L., Toner M., & Schmidt M. A. (2002). A microfabrication-based dynamic array cytometer. *Analytical Chemistry*, 74(16), 3984-3990.
10. Tan, W. H., & Takeuchi, S. (2007). Monodisperse alginate hydrogel microbeads for cell encapsulation. *Advanced Materials*, 19(18)
11. Di Carlo D., Wu L. Y., & Lee L. P. (2006). Dynamic single cell culture array. *Lab on a Chip*, 6(11), 1445-1449.
12. Hong S., Pan Q., & Lee L. P. (2012). Single-cell level co-culture platform for intercellular communication. *Integr. Biol.*, 4(4), 374-380.
13. Skelley A. M., Kirak O., Suh H., et al. (2009). Microfluidic control of cell pairing and fusion. *Nature methods*, 6(2), 147-152.
14. Gallego-Perez D., Higuita-Castro N., Denning L, et al. (2012). Microfabricated mimics of in vivo structural cues for the study of guided tumor cell migration. *Lab on a Chip*, 12(21), 4424-4432.
15. Gallego-Perez D., Higuita-Castro N., Sharma S., et al. (2010). High throughput assembly of spatially controlled 3D cell clusters on a micro/nanoplatform. Lab on a chip, 10(6), 775-782.
16. Benavente-Babace A, Ubarrechena-Bengoechea A, Pérez-Lorenzo E, Mujika-Garmendia M (2012) Caracterización de flujos en microchips destinados a atrapar células individuales, Libro de Actas XXX CASEIB 2012 - 19 a 21 de noviembre - San Sebastián, Spain, 2012

Author: Ainara Benavente-Babace
Institute: CEIT – Centro de Investigaciones Técnicas de Gipuzkoa
Street: Pº Manuel Lardizabal 15, 20018
City: San Sebastián
Country: Spain
Email: abenavente@ceit.es

Photoacoustics of Gold Nanorods under Low Frequency Laser Pulses in Optical Hyperthermia

C. Sánchez[1,2], J.A. Ramos[1], T. Fernández[1,3], F. del Pozo[1,2], and J.J. Serrano[1,2]

[1] Centre for Biomedical Technology (CTB), Universidad Politécnica de Madrid (UPM), Madrid, Spain
[2] Biomedical Research Networking Center in Bioengineering Biomaterials and Nanomedicine (CIBER-BBN), Zaragoza, Spain
[3] Centre for Molecular Biology "Severo Ochoa" (CBMSO) Universidad Autónoma de Madrid (UAM), Madrid, Spain

Abstract—When aqueous suspensions of gold nanorods are irradiated with a pulsing laser (808 nm), pressure waves appear even at low frequencies (pulse repetition rate of 25 kHz). We found that the pressure wave amplitude depends on the dynamics of the phenomenon. For fixed concentration and average laser current intensity, the amplitude of the pressure waves shows a trend of increasing with the pulse slope and the pulse maximum amplitude. We postulate that the detected ultrasonic pressure waves are a sort of shock waves that would be generated at the beginning of each pulse, because the pressure wave amplitude would be the result of the positive interference of all the individual shock waves.

Keywords—**Gold nanorods, laser modulation, pressure waves, thermal expansion, ultrasonic sensors.**

I. INTRODUCTION

Gold nanorods (GNRs) have shown a huge potential for different biomedical applications because of their large light absorption and scattering cross-sections in the near-infrared (NIR) region [1]. Thanks to this optical behavior, gold nanorods are suitable as sensors [2, 3, 4, 5, 6]. Moreover, GNRs are able to transform the absorbed energy into localized heat. This optical effect is used to develop cancer therapies [7, 8, 9, 10, 11] as photothermal tumor destruction either by direct enough increase of temperature or indirectly by co-adjuvant drugs.

Our research group has recently developed an optical hyperthermia device based on irradiation of GNRs with a continuous wave laser in order to induce in vitro death of human brain astrocytoma cells (1321N1). The effectiveness of the method was determined by measuring changes in cell viability after laser irradiation of cells in the presence of GNRs. In accordance to other results in comparable experiments [12, 13, 14], ours indicated that continuous laser irradiation in the presence of the particles induced a significant decrease in cell viability, while no decrease in cell viability was observed with laser irradiation or incubation with GNRs alone [15].

Since the GNRs originate a thermoelastic expansion process when absorbing the light, pressure waves could be generated [16, 17]. In this first work, we study the formation of photoacoustic waves when irradiating an aqueous suspension of GNRs with a NIR laser tuned for their surface plasmon resonance, at low pulsing frequencies and low average power densities, as compared to other works [18, 19, 20]. We want to measure their dependencies on the experiment parameters to learn how to control them to use their possible effects in damaging cells subsequently.

II. MATERIAL AND METHODS

The continuous wave laser (5W MDL H808, PSU-H-LED power source; Changchung New Industries) works at 808 nm. The modulation is made using a pulse generator (50MHz Pulse Generator TG5011 LXI; TTi) connected to the TTL input of the laser. The TTL signal allows a pulsed signal (0 V – ON, 5 V – OFF) of different frequencies (0 – 30 KHz) and duty cycles (0 – 100 %). The Table 1 shows the maximum values of the average current feeding the laser that the device is able to reach for different duty cycles.

Table 1 Maximum average current feeding the laser for different duty cycles

Duty cycle (%)	80	70	60	50	40	30
$I_{MAX-LASER}$ (A)	3.87	3.29	2.64	2.17	1.61	1.10

The plasmon resonance peak of the GNRs (30-10-808 Nanorodz; Nanopartz) is tuned to the laser source. The GNRs are dispersed in deionized water (36 μg/ml) and have an axial diameter of 10 nm and a length of 41 nm.

The laser is connected to the system via a multimode optical fiber (600 μm, 1.5 m MM fiber; Changchung New Industries). The laser light from the fiber irradiates the samples through a collimating lens (78382; Newport).

The ultrasonic device is based on four piezoelectric receivers (25 KHz RCVR; Kobitone Speakers & Transducers, Mouser Electronics), around the irradiated samples. The ultrasound receivers bring their signals to a lock-in amplifier (SR810 Lock-in Amplifier; Stanford Research Systems), whose reference input is the TTL signal. The experimental enclosure is a squared insulated plastic vessel with the four ultrasonic sensors placed on each side of the square and a polyurethane foam cover in order to avoid bounces (Fig. 1).

A precision quartz cell with a light path of 10x10mm (QS-111 SUPRASIL; Hellma GmbH &Co. KG) is placed in the middle of the vessel to hold the samples (3ml).

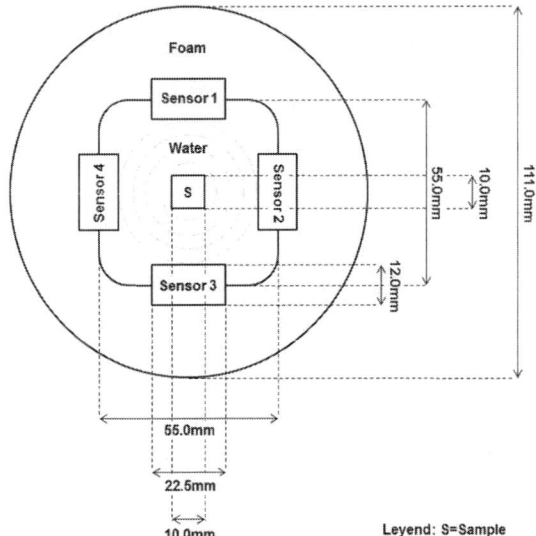

Fig. 1 Experimental enclosure diagram (view from above)

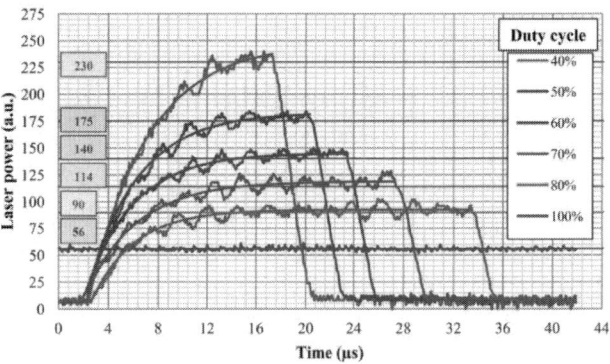

Fig. 2 Shape and peak values of laser power and fitting curves for different duty cycles ($I_{LASER} = 1.5A$) measured with a photoreceiver (Large-area visible photoreceiver 2031; New Focus)

During the experiments, the laser (25KHz) irradiates the samples placed in the quartz cell. The sound waves are transmitted through the water to the ultrasonic sensors. The voltage signal output by the sensors is amplified (factor 100) and fed to the lock-in which displays its value (Sensitivity $= -63dB$ ($0dB = 1$ V/μbar). We can range the average current intensity that feeds the laser, I_{LASER}, up to 5.0 A (the average laser power is linearly proportional to I_{LASER}) and the modulation duty cycle up to 80 %. The GNRs concentration is chosen as 0, 36, 72 and 144 μg/ml.

III. RESULTS AND DISCUSSION

The Fig. 2 shows several laser pulse shapes for a fixed average laser current intensity ($I_{LASER} = 1.5$ A) and the fitting curves for their rising edges. We used an exponential approach: $y = A\left(1 - exp\left(\frac{-t}{\tau}\right) + B\right)$, where A is the maximum amplitude of the exponential curve, B is the offset of the pulsed signals, τ is the pulse rising time constant, t is the time, and y is the laser power. To keep constant the average laser current intensity, the signal peaks grow as duty cycles are shortened.

In the Table 2 we summarized the main parameters from different pulsing regimes and the corresponding for their fitting curves. The slope in $t = 0$s, m, and the maximum amplitude, A (absolute peak value less offset), are mathematically related as: $m = y'[t = 0s] = \frac{A}{\tau}$.

Table 2 Connection between the parameters of the laser pulses and the fitted exponential curves ($I_{LASER} = 1.50$ A, 36 μg/ml)

Laser pulses		Fitted curves		
Duty cycle (%)	Peak power (a.u)	A Max. (a.u)	τ Time Constant (s)	m Slope in t = 0s
40.00	230.00	239.35	4.62	51.81
50.00	175.00	171.29	3.99	42.93
60.00	140.00	132.39	3.82	34.66
70.00	114.00	100.35	4.08	24.60
80.00	90.00	83.94	3.43	17.14

In the Table 3 we show the voltages output by the lock-in for different GNRs concentrations, at fixed duty cycle of 80 % and average laser current intensity ($I_{LASER} = 3.75$ A).

Table 3 Lock-in amplifier output for different values of gold nanorods concentration (duty cycle = 80%, $I_{LASER} = 3.75$ A). S1, S2, S3, S4 = Sensors

	Lock-in output (nV)			
	Water	36 μg/ml	72 μg/ml	144 μg/ml
S1	94	192	321	562
S2	75	148	248	512
S3	98	232	315	546
S4	104	253	345	561

Looking at the Table 3 we can observe that the S2 output is significantly lower than the others. This effect may be due to slight inaccuracies in the manual assembly of the device. Since this behavior is repeated whatever the measuring conditions, its outputs can be ignored

The Fig. 3 shows that for a fixed value of the duty cycle (80%) and for each average laser current intensity (I_{LASER}), the higher is the concentration of the GNRs, the greater is the output voltage, so the higher is the amplitude of the pressure signal. We can also see that there is a linear proportionality between the voltage and the GNRs concentration. This result is consistent with the fact that the greater is the GNRs concentration the higher is their efficiency to absorb the light [21]. Since the absorption phenomenon is independent for each particle [22], the resulting thermoelastic expansion should be also independent.

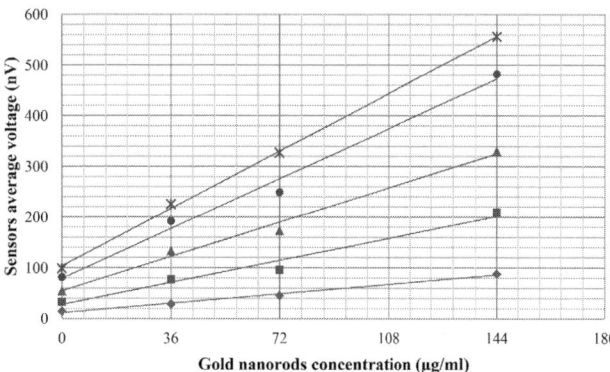

Fig. 3 Average voltage from sensors 1, 3 and 4, and fitting lines for different concentrations. Duty cycle = 80% and I_{LASER} = 0.75 A (orange), 1.25 A (blue), 1.75 A (green), 2.25 A (red) and 3.00 A (dark blue) (Sensors average voltage (nV) = V_{AVG}, GNRs concentration (µg/ml) = C, mean squared error = R^2)

The Fig. 4 shows the sensor 3 output as a function of the duty cycle and of different average laser intensities, when using a fixed GNRs concentration of 36 µg/ml.

We can observe that the voltage levels for each fixed laser current intensity and different duty cycles can be represented using a fitted line growing for decreasing values of the duty cycle. Moreover, the higher is the fixed average laser current intensity the higher pressure signal can be detected for a given duty cycle. Therefore in producing pressure waves at low frequencies, the main cause is not the amount of energy but the dynamics of the phenomenon.

As we can observe in the Table 2, each value of duty cycle corresponds to a maximum amplitude value. Moreover, this maximum amplitude is linearly proportional to the slope in $t = 0s$ (m) and the proportionality factor is given by the time constant τ as we can conclude from the definition of m. The parameter τ depends on the behavior of the instrument in the generation of laser pulses with different requirements of energy, duty cycle or peak laser current intensity. In this case, the values of τ are very similar for all the generated laser pulses, so it can be taken as a constant.

Then, the dependence of the maximum amplitude value on the detected pressure waves is similar to its dependence on the slope in $t = 0s$ taking into account the proportionality factor (τ).

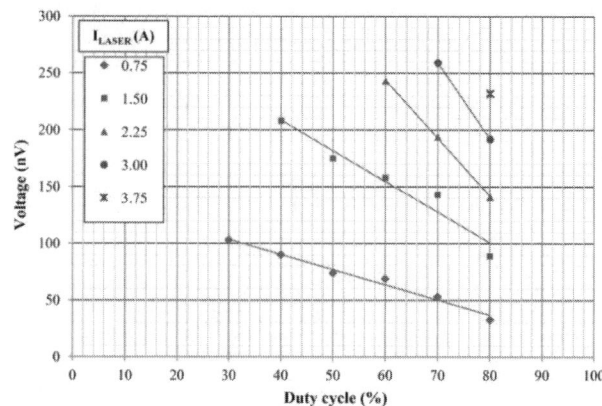

Fig. 4 Sensor 3 output and fitting lines for different laser intensities in a duty cycle sweep (GNRs concentration of 36 µg/ml)

In the Fig. 5 we present the dependence of the detected voltage levels with the maximum amplitude A. The pulses are not ideal, and the slopes change depending on the duty cycle. A (and also m) exhibits a growing trend respect to the detected voltage level, and hence on the waves pressure. As expected, they are similar for different average laser current intensities.

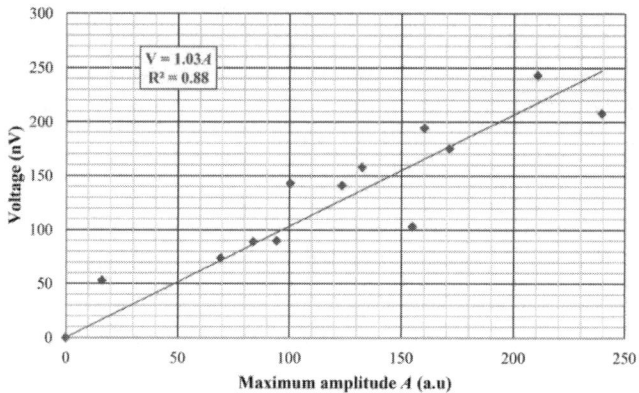

Fig. 5 Dependence of the voltage levels with the maximum amplitude value, A, of the rising edges of the pulses (Voltage (nV) = V, Maximum amplitude value (a.u.) = A, mean squared error = R^2) for different average laser current intensities (I_{LASER} = 0.75, 1.50 and 2.25 A) in a duty cycle sweep (36 µg/ml, sensor 3)

The shown result is consistent with previous theoretical studies of photoacoustic propagation [23, 24] in which is stated that in the limit of almost instantaneous optical absorption, the acoustic pulse shape is approximated by the time derivative of the optical pulse. Nevertheless, according to the results, we cannot determine if the actual linear dependence of the pressure waves is associated to the derivative of the laser pulse or to the maximum amplitude.

IV. CONCLUSION

We have developed a device to measure and characterize the photoacoustic effect that appears when an aqueous suspension of GNRs is irradiated with a low frequency and low power NIR pulsing laser. We have found that the magnitude of the wave pressure depends not only on the total amount of provided energy but on the dynamics of the phenomenon. The pressure magnitude shows a growing trend, probably linear, respect to the slope in $t = 0$s of the pulses and hence with respect to the maximum amplitude value. In our system, the slope and the maximum amplitude are linearly proportional. However, we postulate that the measured phenomenon is the result of the coexistence of all the individual processes from each nanorod and therefore, the addition of all the individual pressure waves generated positively interfere to form the observed transient phenomenon.

In short, we can say that the irradiation of gold nanorods suspended in water with a low frequency exciting laser light generates thermoelastic expansion processes that result in acoustic waves that are quantifiable as proposed. The magnitude of these signals can be controlled through the modification of the laser and modulation parameters. The determination of the real influence of these effects in biomedical applications, and the correct theoretical explanation are important tasks for the next future.

REFERENCES

1. Ni W, Kou X, Yang Z et al. (2007) Tailoring longitudinal surface plasmon wavelengths, scattering and absorption cross sections of gold nanorods. Acs Nano 2:677–686
2. Yu C, Irudayaraj J (2007) Multiplex biosensor using gold nanorods. Anal Chem 79:572–579
3. Huang H, Liu X, Zeng Y et al. (2009) Optical and biological sensing capabilities of $Au_2S/AuAgS$ coated gold nanorods. Biomaterials 30:5622–5630
4. Darbha GK, Rai US, Singh AK et al. (2008) Gold-nanorod-based sensing of sequence specific HIV-1 virus DNA by using hyper-Rayleigh scattering spectroscopy. Chem-Eur J 14:3896–3903
5. Tong L, Wei Q, Wei A et al. (2009) Gold nanorods as contrast agents for biological imaging: Optical properties, surface conjugation and photothermal effects. Photochem Photobiol 85:21–32
6. Ha S, Carson A, Agrawal A et al. (2011) Detection and monitoring of the multiple inflammatory responses by photoacoustic molecular imaging using selectively targeted gold nanorods. Opt Express 2:645–657
7. Von Maltzahn G, Park JH, Agrawal A et al.(2009) Computationally guided photothermal tumor therapy using long circulating gold nanorods antennas. Cancer Res 69:3892–3900
8. Peng CA, Wang CH (2011) Anti-neuroblastoma activity of gold nanorods bound with gd2 monoclonal antibody under near-infrared laser irradiation.Cancers 3:227–240
9. Terentyuk GS, Ivanov AV, Polyanskaya NI et al. (2012) Photothermal effects induced by laser heating of gold nanorods in suspensions and inoculated tumours during in vivo experiments. Quantum Electron+ 42:380–389
10. Alkilany AM, Thompson BL, Boulos SP et al. (2012) Gold nanorods: their potential for photothermal therapeutics and drug delivery, tempered by the complexity of their biological interactions. Adv Drug Delivery Rev 64:190–199
11. Jaskula-Sztul R, Xiao Y, Javadi A et al. (2012) Multifunctional gold nanorods for targeted drug delivery to carcinoids. J Surg Res 172:235–235
12. Tong L, Zhao Y, Huff TB et al. (2007) Gold nanorods mediate tumor cell death by compromising membrane integrity. Adv Mat 19:3136–3141
13. Dreaden EC, Mackey MA, Huang X et al. (2011) Beating cancer in multiple ways using nanogold. Chem Soc Rev 40:3391–3404
14. Choi WI, Sahu A, Kim YH et al. (2012) Photothermal cancer therapy and imaging based on gold nanorods. Ann Biomed Eng 40:1–13
15. Fernández T, Sánchez C, Martínez A et al. (2012) Induction of cell death in a glioblastoma line by hyperthermic therapy based on gold nanorods. Int J Nanomed 7:1511–1523
16. Pustovalov V, Zharov V (2008) Threshold parameters of the mechanisms of selective nanophotothermolysis with gold nanoparticles. In: Optical Interactions with Tissue and Cells XIX: SPIE Proc., the International Society for Optical Engineering, San Jose, California, USA, edited by S. L. Jaques, W. P. Roah, R. J. Thomas (Society of Photo-Optical Instrumentation Engineers) 685412–1
17. Hu M, Wang X, Hartland GV et al. (2003) Vibrational response of nanorods to ultrafast laser induced heating: Theoretical and experimental analysis. J Am Chem Soc 125:14925–14933
18. Letfullin RR, Joenathan C, George TF et al. (2006) Laser-induced explosion of gold nanoparticles: potential role for nanophotothermolysis of cancer. Nanomedicine 1:473–480
19. Letfullin RR, George TF, Duree GC et al. (2008) Ultrashort laser pulse heating of nanoparticles: Comparison of theoretical approaches. Advances in Optical Technologies (Hindawi Publishing Corporation)
20. Shah J, Park S, Aglyamov S et al. (2008) Photoacoustic imaging and temperature measurement for photothermal cancer therapy. J Biomed Opt 13:034024
21. Cole JR, Mirin NA, Knight MW et al. (2009) Photothermal efficiencies of nanoshells and nanorods for clinical therapeutic applications. J Phys Chem B 113:12090–12094
22. Novotny L, Hecht B (2006) Principles of nano-optics. Cambridge University Press, Cambridge, United Kingdom
23. Emelianov SY, Li PC, O'Donnell M (2009) Photoacoustics for molecular imaging and therapy. Phys Today 62:34–39
24. Diebold GJ, Sun T (1994) Properties of photoacoustic waves in one, two, and three dimensions. Acta Acust United Ac 80339–351

Author: Cristina Sánchez
Institute: Center for Biomedical Technology (CTB-UPM)
Street: Campus Montegancedo
City: Pozuelo de Alarcón (28223, Madrid)
Country: Spain
Email: cristina.sanchez@ctb.upm.es

Demonstration of an On-Chip Real-Time PCR for the Detection of *Trypanosoma Cruzi*

R.C.P. Rampazzo[1,2], M. Cereda[3], A. Cocci[3], M. De Fazio[3], M.A. Bianchessi[3], A.C. Graziani[1,2,4], M.A. Krieger[1,2,4], and A.D.T. Costa[1,2]

[1] Instituto de Biologia Molecular do Paraná (IBMP), Curitiba, Brazil
[2] Instituto Carlos Chagas (ICC), Fundação Oswaldo Cruz, Curitiba, Brazil
[3] STMicroelectronics Srl (ST), Agrate Brianza, Italy
[4] Department of Bioprocess Engineering and Biotechnology, Federal University of Paraná (UFPR), Curitiba, Brazil

Abstract—Discovered in 1909, Chagas disease, caused by the parasite *Trypanosoma cruzi*, continues to be an important tropical disease in Latin America and Caribbean, affecting more than 7 million people. Recently, Chagas disease has also been reported in US and Europe mainly because of unscreened blood donations by immigrants from endemic areas. Current Chagas disease diagnostic relies on search for parasites on blood smears using optical microscopes, screening for serological response, xenodiagnosis, hemoculture and, more recently, polymerase chain reaction (PCR). Each of these tests has its own problems, mainly because of the changes in the immunological profile of the patients and number of circulating parasites in the blood throughout the evolution of the disease. In endemic areas, logistic issues are an additional difficulty for delivering the results to the patients, so a rapid and sensitive point of care test is desirable. In this work, we present results of an on-chip test able to detect equivalent amounts of parasites' DNA such as those typically present in both acute and chronic phase of the disease. We developed a silicon chip-based PCR using a portable thermocycler with fluorescent detectors that has many advantages over the conventional PCR. When compared to the conventional PCR, our on-chip reaction shows similar sensitivity (between 1 and 0.1 genome equivalents), but a shorter reaction time (35 min versus 90 min). The results presented herein are the first step towards the development of a portable diagnostic test using a fast, sensitive and specific reaction for the detection of *Trypanosoma cruzi*.

Keywords—Real time PCR, point of care, *Trypanosoma cruzi*, diagnostic, silicon chip.

I. INTRODUCTION

Despite its description after more than 100 years, Chagas disease remains a significant health problem in Latin America and the Caribbean, with more than 7 million people infected. During the last decade, it has been observed an increase in the number of cases in North America, Europe and the Pacific countries, as a result of immigration and the non obligation for Chagas disease screening in blood donations. Chagas disease is caused by the protozoan parasite *Trypanosoma cruzi*, being transmitted mainly by insect members of the Triatominae subfamily, by unscreened blood and organ donations, or from mother to fetus. Ingestion of contaminated food has also been reported [1, 2].

Chagas disease is characterized by three distinct phases: acute, chronic asymptomatic and chronic symptomatic [3]. During the acute phase, there is high parasitemia and low immunological response, and the diagnostic usually relies on identification of the parasite on blood smears using an optical microscope in conjunction with the clinical symptoms [1, 2]. At some point, the immunological response sets in and controls the disease, decreasing the parasitemia to levels that preclude the diagnosis by identification of the parasite in blood samples. From this moment on, antibodies against *T. cruzi* can be used to diagnose the disease. The decrease in parasitemia is the beginning of the chronic asymptomatic phase, which can remain asymptomatic for many decades. Later on, clinical symptoms such as hypertrophy of heart, esophagus or colon may appear on 10-30% of patients, which characterize the chronic symptomatic phase. The chronic phase exhibits a low but detectable level of parasites in the patient's blood [3, 4].

Recent studies have shown that it is possible to use real-time PCR (qPCR) to detect the parasite in blood samples, as part of an ongoing effort to establish a consensus protocol for qPCR detection of *T. cruzi* [5-7]. qPCR is a sensitive and accurate method, with the interesting possibility of target multiplexing, but it is laborious and time-consuming. Furthermore, it requires skilled personnel, a rather well established laboratory setting and an expensive instrument to perform the test, so the cost for the initial setup is high.

Nowadays, point of care diagnostic tests are evolving to develop solutions such as easiness of use, miniaturization, and target multiplexing, bringing great potential to improve patient treatment through fast and accurate diagnostics, either at the doctor's office or at the bedside [8]. An extra bonus is the overall healthcare cost decrease due to faster and more accurate diagnosis, thus improving treatment via the decrease in the social and economic burden inherent to treatment delays or inaccurate diagnosis.

This work shows the first step towards a molecular-based, point of care diagnostic test for Chagas disease. Using a silicon die and a portable, light-weight thermocycler, we developed an on-chip qPCR with sensitivity and accuracy comparable to qPCR performed in conventional thermocyclers. However, due to its intrinsic characteristics, our on-chip qPCR is at least 3 times faster and uses roughly 3-5 times less reagents, thus decreasing the sample-to-result timeframe as well as the costs for molecular tests.

II. MATERIALS AND METHODS

<u>DNA Extraction and Dilution Curves</u>: 10^7 *Trypanosoma cruzi* epimastigote cells (strain Dm28c) were diluted in phosphate saline buffer (PBS). Total DNA was extracted using the High Pure PCR Template Preparation Kit (Roche Applied Science). Extracted DNA from 10^7 cells was regarded as representative of 10^7 genome equivalents (GE), and was diluted in PBS into aliquots ranging from 10^4 to 10^{-1}.

<u>Conditions to conventional qPCR</u>: Reactions contained 5 µL of extracted DNA in a final volume of 20 µL, 0.3 µM of each primer (cruzi1 and cruzi2) and 0.2 µM of probe (cruzi3, FAM-BHQ1). Conventional qPCR was performed on the ABI7500 (Life Technologies), with cycling condition as follows: 50°C/2 min, 95°C/10 min, and 40 cycles of 95°C/15 sec and 60°C/1 min ([6] with minor modifications).

<u>Conditions to on-chip qPCR</u>: On-chip qPCR was performed on a prototypal equipment, developed by STMicroelectronics under the name "Q3" (www.biotobit.com). Q3 is a compact platform designed to be easily used by unskilled personnel as a point-of-care system (Figure 1). Compared with standard lab instrumentation, its affordable price eliminates the need to perform a batch of tests in order to be cost effective. The Q3 platform integrates PCR technology, electrical temperature control, illumination and optical detection for real-time PCR-based molecular testing on a lab-on-chip. The chip is a disposable device composed of a silicon die manufactured using semiconductor technology, sealed with a transparent frame forming 6 reaction chambers. The Q3 instrument drives the lab-on-chip to control the reaction temperature utilizing on-chip integrated heaters and temperature sensor (Figure 2).

On-chip reactions contained 1 µL of extracted DNA in a final volume of 5 µL, 0.9 µM of each primer (cruzi1 and cruzi2) and 0.8 µM of probe (cruzi3, FAM-BHQ1). The cycling conditions were as follows: 95°C/40 sec, and 45 cycles of 95°C/10 sec and 64°C/30 sec. Optimized optical parameters on the Q3 equipment were set to led power at 7, camera gain at 5.31 and picture frame rate at 0.5 f/s.

Fig. 1 Q3 platform for on-chip qPCR.

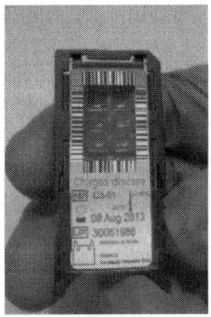

Fig. 2 Q3 disposable silicon-based lab-on-chip.

III. RESULTS

Recent studies [5-7] showed that it is possible to detect less than one GE of *T. cruzi* DNA by using a multicopy target gene. Since qPCR is significantly influenced by the reaction mastermix, we first tried to reproduce the published results using our reaction conditions in a standard qPCR equipment.

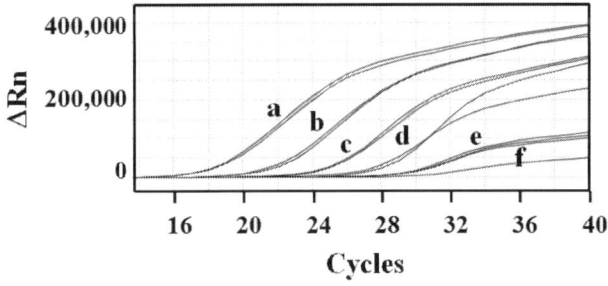

Fig. 3 ABI7500 qPCR curves for the detection of a *T. cruzi* DNA target in a ten-fold dilution series of the parasite's GE, starting at 10,000 GE (line a) down to 0.1 GE (line f). Shown are representative curves in duplicates for each DNA concentration (n = 3 independent experiments).

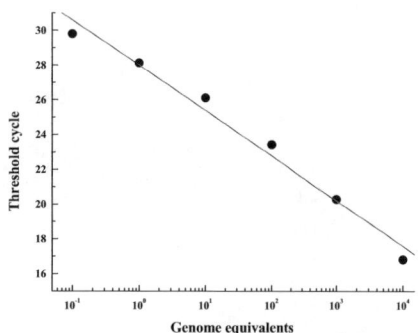

Fig. 4 Linear regression for the mean of the threshold cycles of all ABI7500 experiments data points. Slope = -2.60; $R^2 = 0.982$.

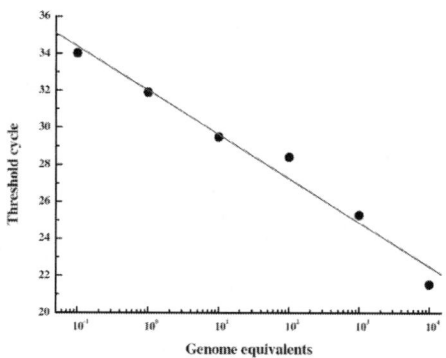

Fig. 6 Linear regression for the mean of the threshold cycles for each data points of all Q3 experiments. Slope = -2.38; $R^2 = 0.974$.

Figure 3 shows representative curves for qPCR detection of a satellite DNA target in a serial dilution of *T. cruzi* GEs, using the ABI7500 equipment. The qPCR reaction was performed using 10,000 GE (line a), down to 0.1 GE (line f) in ten-fold dilutions.

Linear regression of the mean of the threshold cycle of each data point triplicates of Figure 3 experiment shows that the amplification efficiency is maintained over 6 orders of magnitude (slope = -2.60) (Figure 4).

Similar experiments using the same samples were performed using the portable qPCR thermocycler Q3. Figure 5 shows representative curves for an experiment similar to the one shown in Figure 3.

Figure 6 shows the linear regression of the mean of the threshold cycle of each data point triplicates of Figure 5 experiment. As obtained with the standard equipment ABI7500, the amplification efficiency on the Q3 thermocycler is maintained over 6 orders of magnitude (slope = -2.38).

Threshold cycles for the dilution curves obtained in both equipments are summarized and compared in Table 1. Threshold cycles on the Q3 thermocycler are constantly higher than those obtained on ABI7500, with a constant shift in Ct of 3-5 cycles. Interestingly, the limit of detection for both instruments is similar (between 0.1 and 1 GE).

Table 1 Comparison of threshold cycles for the detection of dilution curves of GE on the ABI7500 and the Q3. Results shown are averages ± SD of at least three independent experiments for each GE.

Genome Equivalents	ABI7500	Q3
10,000	16.8 ± 0.4	21.5 ± 0.3
1,000	20.3 ± 0.6	25.3 ± 0.6
100	23.4 ± 0.7	28.4 ± 0.7
10	26.1 ± 0.1	29.5 ± 1.9
1	28.1 ± 2.2	31.9 ± 3.6
0.1	29.8 ± 1.8	34.0 ± 1.9

IV. DISCUSSION

Neglected tropical diseases (NTD) are among the most prominent cause of death and life year losses in developing countries. Chagas disease, as well as helminthic diseases, filariosis, trachoma, schistosomiasis and dengue fever, are the most relevant NTD that, when combined, may be more prevalent than AIDS, tuberculosis and malaria [1, 2].

The results shown allow the comparison between a qPCR assay performed on the widely used ABI7500, and the novel, portable qPCR thermocycler Q3, using as target a satellite DNA sequence from the etiological agent of Chagas disease *Trypanosoma cruzi* [6].

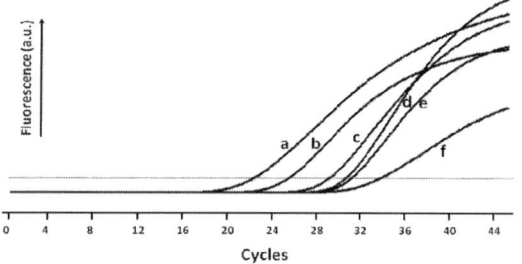

Fig. 5 Q3 qPCR curves for the detection of a *T. cruzi* DNA target in a ten-fold dilution series of the parasite's GE, starting at 10,000 GE (line a) down to 0.1 GE (line f). Shown are representative curves for a minimum of three independent experiments for each DNA concentration.

Both equipments presented similar precision ($R^2 = 0.982$ for the ABI7500 and 0.974 for the Q3) and sensitivity (limit of detection between 0.1 and 1 parasite's GE). Our data are in perfect agreement with recent published results [5-7]. The importance to detect very low concentrations of the parasite is related with the disease phases. The acute phase shows large number of circulating parasites in the patient's blood, whereas patients at chronic phases present low amount of parasites. Furthermore, tests based on the immune response are very prone to cross-reaction with possible co-infections, and molecular-based tests are more specific since the target is a DNA sequence.

The on-chip qPCR system has some advantages over the conventional ABI7500. Besides the user-friendly interface, the Q3 uses a silicon chip to modulate the temperature of the reaction, allowing for fast temperature ramping, and thus shorter total reaction time. The reaction we used to detect *T. cruzi* took 90 min on the ABI7500 and took only 35 min on the Q3. Interestingly, we noted that the cycle thresholds on the ABI7500 were, on average, 4.3 cycles lower than on the Q3 for the same amount of parasite GE. However, despite this apparently significant difference, both tests showed similar sensitivity (same limit of detection).

The on-chip tests shown in this work allow for fast and accurate clinical management decisions, at the point of need, with minimal manipulation of samples and equipment. These features were highlighted previously for other diagnostic needs, such as coronary syndrome, heart failure, human immunodeficiency virus, and tuberculosis [9].

Finally yet importantly, the reaction volume can be 5 to 8 times smaller on the Q3 than on the ABI7500, resulting in a significant economy of reagents and samples. This feature is interesting when we consider that neglected diseases are an important health issue in developing countries. In addition, a smaller reaction volume is interesting when using rare or low volume samples.

V. CONCLUSIONS

The on-chip, fast, sensitive and accurate qPCR reaction described in this work is the first step towards a true molecular-based point of care diagnostic test and device. Results were comparable to those obtained with the standard equipment in the field (ABI7500) and to those published by others [5-7]. Our next steps include the integration of the on-chip qPCR shown in this work in a membrane-based sample preparation microfluidic chip [10], thus getting closer to a true qPCR point-of-care test. Besides the reduction in cost and sample-to-answer time, our reaction, combined with the portable, light-weighted, low cost equipment, opens the possibility of taking sensitive diagnostic tools to areas of difficult access as well as to the patient bedside.

ACKNOWLEDGEMENT

We are thankful to Dr. Otacilio C. Moreira and Dr. Constança Britto for help in the initial experiments. This work was funded by grants from CNPq (590032/2011-9 and 404242/2012-0 to M.A.K.) and FP7 GA-N° 287770 (to M.A.B.).

REFERENCES

1. WHO, T.D.P. (2010) First WHO report on neglected tropical diseases 2010: Working to overcome the global impact of neglected tropical diseases. WHO Press, Geneva
2. Hotez P, Bottazzi M, Franco-Paredes C et al. (2008) The neglected tropical diseases of Latin America and the Caribbean: a review of disease burden and distribution and a roadmap for control and elimination. PLoS Negl Trop Dis 2: e300
3. Tanowitz H, Kirchhoff L, Simon D et al. (1992) Chagas' disease. Clin Microbiol Rev 5: 400-419
4. Tarleton R, Curran J (2012) Is Chagas disease really the "new HIV/AIDS of the Americas"? PLoS Negl Trop Dis 6: e1861
5. Moreira O, Ramirez J, Velazquez E, et al. (2012) Towards the establishment of a consensus real-time qPCR to monitor *Trypanosoma cruzi* parasitemia in patients with chronic Chagas disease cardiomyopathy: a substudy from the BENEFIT trial. Acta Trop 125: 23-31
6. Piron M, Fisa R, Casamitjana N, et al. (2007) Development of a real-time PCR assay for *Trypanosoma cruzi* detection in blood samples. Acta Trop 103: 195-200
7. Schijman A, Bisio M, Orellana L, et al. (2011) International study to evaluate PCR methods for detection of *Trypanosoma cruzi* DNA in blood samples from Chagas disease patients. PLoS Negl Trop Dis 5: e931
8. Kulinsky L, Noroozi Z, Madou M (2013) Present technology and future trends in point-of-care microfluidic diagnostics. Methods Mol Biol 949: 3-23
9. Chan C, et al. (2013) Evidence-Based Point-of-Care Diagnostics: Current Status and Emerging Technologies. Annu Rev Anal Chem (Palo Alto Calif) DOI 10.1146/annurev-anchem-062012-092641
10. Morschhauser A, Stiehl C, Grosse A. et al. (2012) Membrane based sample preparation chip. Biomed Tech (Berl) 57: 923-925

Author: Rita R. C. P. Rampazzo
Institute: Instituto de Biologia Molecular do Parana (IBMP)
Street: Rua Prof Algacyr Munhoz Mader, 3775
City: Curitiba
Country: Brazil
Email: rirampazzo@tecpar.br

A Stand-Alone Platform for Prolonged Parallel Recordings of Neuronal Activity

G. Regalia[1], E. Biffi[1], A. Menegon[2], G. Ferrigno[1], and A. Pedrocchi[1]

[1] Politecnico di Milano, Department of Electronics, Information and Bioengineering,
Neuroengineering and Medical Robotics Laboratory, Milan, IT
[2] Advanced Light and Electron Microscopy Bio-Imaging Centre, Experimental Imaging Centre,
San Raffaele Scientific Institute, Milan, IT

Abstract—In this work we report the development of a compact environmental chamber designed to perform prolonged recordings of the bioelectrical activity exhibited by neuronal networks grown on MicroElectrode Arrays (MEAs). The chamber was specifically designed to house several neuronal networks, in order to improve the experimental throughput. To reproduce an environment suitable for cells growth (in terms of temperature, pH and humidity) the chamber was coupled with a temperature control system and a gas warming and humidifying module. Validation tests demonstrated that the environment inside the portable chamber is comparable to standard cell incubators environment. To collect MEA extracellular signals, custom multichannel pre-processing boards have been developed and placed across the chamber top plate. Simulations and experimental tests demonstrated that the defined circuitry is well suitable to process neuronal spikes with overlapping thermal and biological noise. With this equipment, we were able to recorded spontaneous neuronal electrical activity from hippocampal cultures grown inside the chamber for several hours, which is not possible with the standard MEA recording setup due to environmental fluctuations. This system can collect multichannel data from neuronal cultures over long periods, providing an effective solution for long-term studies of neural activity.

Keywords—Microelectrode arrays, neuronal cultures, long-term recordings.

I. INTRODUCTION

Neuronal cultures grown on MicroElectrode Arrays (MEAs) represent a powerful tool in a large variety of neurophysiological and neuropharmacological studies [1, 2]. *In vitro* neuronal ensembles are extremely sensitive to changes in the surrounding environment (temperature, pH, humidity) [3]. Thus, the establishment of an experimental setup able to maintain stable conditions is an absolute requirement in order to design truly significant experiments and collect reliable data with MEAs.
Nowadays, standard MEA-based experimental platforms are well-established setups for several neurobiological applications where short recordings (i.e., from 10 minutes to a couple of hours) are adequate to gather the information of interest. Nevertheless, the possibility to perform longer investigations of neuronal activity is required to throw light on mechanisms evolving over longer time windows (e.g., precisely localize the onset/offset of a neural mechanism).

Some effort has been already directed at improving standard setups in order to reach this aim [4]. Among the proposed solutions, an effective solution is represented by compact commercial top stage incubators for microscopy analysis (e.g., Ibidi GmbH, Okolab s.r.l.) which provide an effective environment control but they are hardly modifiable to be coupled to MEA technology. On the other hand, custom setups built up for MEA-based experiments, do not include the control of all environmental parameters and they usually provide temporary solutions [5].

Accordingly, our group devised a novel system which merges an effective environmental control and multisite recordings capability. We fabricated indeed an environmental chamber for neuronal cultures grown on MEAs and we coupled it to external commercial electronics [3]. Signals recorded from MEAs in this first prototype suffered from a low quality. Moreover, the prototype was sized to house only one MEA chip, which results in time-consuming experiments.

In order to fulfill the requirements of environmental stability, good quality recordings and multi-MEA format, we have been developing a new version of the environmental chamber. In particular, we present here: (i) the design and a preliminary quantitative environmental characterization of the environmental chamber (ii) the design and validation of a custom multichannel front-end and (iii) preliminary results regarding prolonged recordings with this platform.

II. MATERIALS AND METHODS

A. Multi-MEA Environmental Chamber Realization

The environmental chamber has been realized by assembling PMMA plates in order to create two boxes: an external one (220x220x45 mm) and an internal one (180x180x30 mm), the latter being surrounded by a water jacket. Both boxes are sealed with an airtight top plate, through which openings for air inlet and outlet, the insertion of a pt100 temperature probe and medium exchange from outside have been drilled. In the internal box, a reference well in the center and four 50x50 mm housings for 60 channel MEA chips in the corners have been located. The recording electronics (see paragraph C) contacts the 60 pads of each MEA chip by means of vertical gold spring probes. Signals are carried outside by means of a 68-pin connector inserted and sealed through the top plate.

The chamber heating is obtained by means of a circulating bath (E306, Ecoline, Lauda GmbH) and a feedback control, as previously described [3]. To maintain the pH of the medium in a physiological range (7.2-7.4), an air flow enriched with CO_2 is injected in the chamber. To slow medium evaporation, the flow is warmed and humidified by a bubbling bottle inserted into the bath (Okolab s.r.l.). Figure 1 (top) reports a schematic representation of the chamber connected to the heating bath.

B. Environmental Parameters Characterization

During experiments, the temperature of culturing medium contained in Petri dishes located in the four MEA housings was monitored by thermo-couples to verify its maintenance in a physiological range. Regarding the chamber gaseous atmosphere, experimental tests have been conducted to link the air flow rate to the CO_2 content inside the chamber. 20% O_2, 70% N_2, and 10% CO_2 air flow rates spanning from 40 to 100 ml/min were obtained by means of a flow meter (NG series, Platon SaS) and maintained for 30 minutes. For each flow rate, the gaseous CO_2 percentage was measured by a CO_2 tester (Heraeus, Thermo Scientific). Concerning the water vapor content, measurements of relative humidity (RH%) after the bubbling column and inside the environmental chamber have been performed (HMP233 thermoigrometer, Vaisala Inc.).

C. Recording System Design

To perform electrical recordings from the 60-channel MEA chips, custom pre-processing boards have been designed and coupled to a multichannel commercial data acquisition system (USB-ME64, 50 kHz, 16 bit, Multi Channel Systems, MCS, GmbH), as shown in Figure 1 (bottom). The pre-amplification stage was designed to be implemented on two 30-channel boards (65x65 mm) inside the culture chamber, in order to avoid the degradation of the signal-to-noise ratio (SNR) of recordings, and a following filter stage on two 30-channel external boards (100x100 mm).

The custom front-end was designed to fulfill the main following requirements: (1) high gain (nearly 60 dB), suitable for the amplification of neuronal signals *in vitro* (20-400 μV peak to-peak) (2) 3rd order high pass filter (300 Hz), (3) 3rd order low pass filter (5 kHz), (4) low noise (< 20 μV peak-to-peak, i.e., typical thermal and background biological noise at the microelectrodes), (5) linear phase lag in the filter bandwidth.

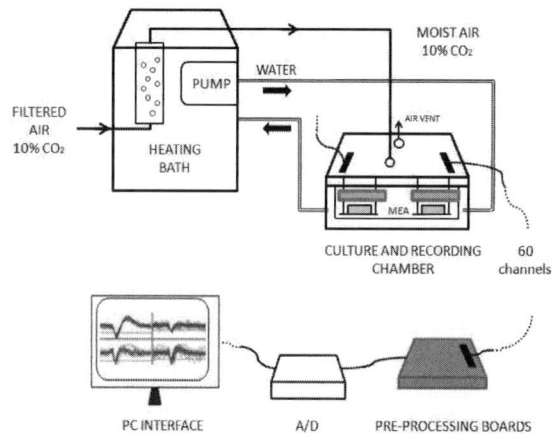

Fig. 1 Schematic rapresentation of the culture chamber connected to the heating bath (top) and the recordin equipment (bottom).

The band-pass filter defined consists of a pre-amplification stage, a Sallen-key high-pass filter and a Sallen-key low pass filter followed by a RC low pass. The performance of the circuit was simulated with Spice software (LTSpice, Linear Technology). Once realized, the assessment of the real frequency response gain of the whole front-end was obtained by providing sine waves (100 μV peak-to-peak amplitude) by means of a sinusoidal wave generator (GFG-8210, Instek), with frequency varying between 1 Hz and 10 kHz, for each channel (analysis performed with Matlab). The input-referred noise (300 Hz - 5kHz) of the custom setup was determined measuring the channel output and dividing it by the overall nominal gain, with all inputs connected to the reference electrode by filling the MEA with a saline buffer. The channel crosstalk was measured by sending a 100 μV @1 kHz controlled sine wave to one channel, and recording from directly adjacent ones.

D. Electrophysiological Recordings of Neuronal Cells

Neuronal cultures on MEAs were obtained with a standard protocol of our lab, described in [5,6]. Short recordings (< 30 min) from hippocampal neurons (CD1 mice, E17.5, 200.000 cells/MEA) grown on standard MEA chips (MCS GmbH) were performed during the 2nd and 3rd week of maturation. Besides, activity of the same networks was recorded by means of the standard equipment (MCS GmbH), in order to compare the two systems in terms of signal quality and spike morphology. Moreover, to evaluate the feasibility of prolonged experiments with the system, we performed recordings lasting several hours (> 3 h). Spikes were detected by a comparison with a threshold based on noise level (as described in [7]). Then time stamps were analyzed by means of standard signal processing implemented with custom algorithms (Matlab®) [5,8]. The SNR for a single channel has been defined averaging the spikes amplitudes over the recording window and dividing by the noise over the first 500 ms.

III. RESULTS

A. Environmental Parameters Validation

To maintain the desired temperature in the MEA housings the set-point has been set to 36.8 °C. After the initial heating phase, temperature measurements compare well among the 4 housings and show small oscillations around 37°C (i.e., <0.5°C peak-to-peak), (Figure 2 A). As a result of the air flow rate characterization, flow rates around 45 ml/min allow to balance CO_2 losses occurring along the tubing system and in the chamber and to reach 5% CO_2, i.e., the value needed to maintain cell culture pH at 7.4 [3] (Figure 2 B). Moreover, with 100 ml/min flow rate RH% in the bubbling column and in the chamber reaches almost 95% and 80%, respectively.

B. Recording Boards Performance Evaluation

The defined front-end circuit is suitable for the analog processing of *in vitro* neuronal signals. The measured gain of pre-amplifiers and filter boards is in agreement with calculations and Spice simulation (absolute error equal to ~1.6 dB in the bandwidth) and phase lag is almost linear over the spikes frequencies (R^2=0.924). Also, noise performances in terms of input-noise and cross-talk are comparable to noise as measured from the commercial equipment (Table 1) and other custom setups [9,10].

C. Electrophysiological Recordings of Neuronal Cells

Neuronal recordings performed with the custom boards are comparable to recordings of the same cultures with the standard setup in terms of SNR (Table 1) and spike morphology (Figure 3).

Moreover, preliminary results show that neuronal cells inside the chamber do not undergo the activity decline, that typically occurs with the standard setup, as observed throughout time windows longer than 2 hours. Figure 4 reports the spike rate of a neuronal culture recorded in the standard setup and immediately after in the environmental and recording chamber. Apart from the effect of the repositioning, neuronal activity inside the chamber is characterized by fluctuations and mean value comparable to the activity exhibited in the standard setup.

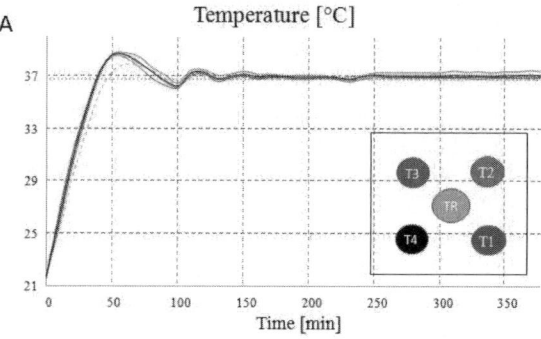

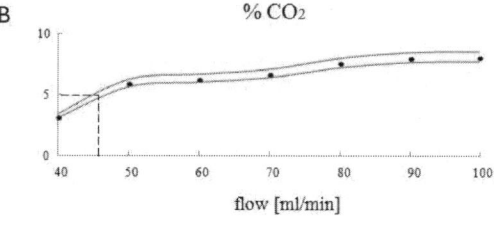

Fig. 2 A: Temperature time-course in the reference well (TR) and in the 4 housings (T1-T4). The inlet reports the positions of the temperature probes. B: CO2 percentage at different air flow rates. Mean values over three measurements (points) and sensor accuracy (lines) are displayed.

Table 1 Noise and SNR comparison (mean values over 5 channels)

	Custom	Commercial	Unit
Input noise	0.56	1.11	µV RMS
Cross-talk	- 36	-42	dB
SNR	5.4	5.46	dB

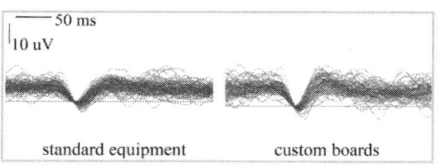

Fig. 3 Comparison of spike waveforms as recorded by the same microelectrode and the same culture with the commercial and custom boards.

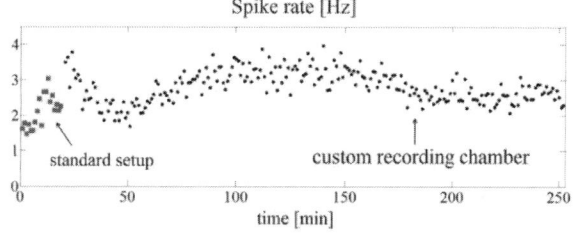

Fig. 4 Spike rate of a neuronal culture recorded in the standard setup and in the environmental recording chamber. Mean spike rates over 1 minute bins are reported.

IV. Discussion

To tackle the issue of environmental stability during *in vitro* electrophysiological experiments, we designed and validated a stand-alone platform aimed at maintaining a controlled environment while growing and recording from neuronal networks on MEAs.

To realize a controlled environment, we connected the chamber to a temperature controller and a system to inject air enriched with CO_2 and water vapor. We demonstrated that the chamber maintains a stable physiological temperature in each MEA housing. Moreover this preliminary measurements suggest that a flow rate of at least 100 ml/min is necessary to obtain (i) a CO_2 percentage almost equal to the quantity contained in the air delivered by the gas source (ii) a high level of RH in the chamber, i.e. 80% RH, which is comparable to other commercial top stage incubators (Ibidi GmbH) but not as high as standard cells incubators. These results lead our work towards a further characterization with a gas mixture containing less CO_2 (e.g., 5-7%), in order to obtain 5% CO_2 while keeping higher flow rates in the chamber.

Regarding the custom front-end, it is suitable to be coupled with the chamber, both in terms of sizes, environmental compatibility and recording performances. Moreover, the realized boards are cheaper and more easily replicable than commercial recording front-end devices or custom CMOS-based systems [10]. Preliminary results assessed the feasibility of performing experiments with MEAs longer than standard ones.

Regarding the throughput of the system, the actual prototype houses four 60-channel MEA chips, which means up to 24 cultures if 6-well MEAs (9 electrodes per well) are used (MCS GmbH). Even if we validated the recording performance of one full front-end, the integration with a multiple MEA data acquisition system (MCS GmbH) is being carried on.

V. Conclusions

In this work we presented the design and preliminary validation of a stand-alone platform for parallel prolonged experiments from neuronal cells grown on MEAs. Our final aim is to provide a compact technological tool for an electrophysiological laboratory, independent from both bulky incubators and expensive front-end equipments. This system provides new perspectives for *in vitro* long-term, high-throughput electrophysiological studies on neuronal cultures on MEAs.

Acknowledgment

Authors would like to thank people from the Alembic facility for their support and Dr. De Ceglia for the dissection of hippocampi. They are also grateful to Dr. Colombo, Dr. Molinaroli and Dr. Lucchini for their support with measurement equipment and useful discussions.

References

1. Johnstone AF, Gross GW, Weiss DG et al. (2010) Microelectrode arrays: a physiologically based neurotoxicity testing platform for the 21st century. Neurotoxicology 31:331-350
2. Rossi S, Muzio L, De Chiara et al. (2011) Impaired striatal GABA transmission in experimental autoimmune encephalomyelitis. Brain Behav Immun 25:947-56
3. Biffi E, Regalia G, Ghezzi D et al. (2012) A novel environmental chamber for neuronal network multisite recordings. Biotechnol Bioeng 109:2553-2566
4. Potter SM, DeMarse TB (2001) A new approach to neural cell culture for long-term studies. J Neurosci Methods 110:17-24
5. Novellino A, Scelfo B, Palosaari T et al. (2011) Development of micro-electrode array based tests for neurotoxicity: assessment of interlaboratory reproducibility with neuroactive chemicals. Front Neuroeng DOI 10.3389/fneng.2011.00004
6. Ghezzi D, Menegon A, Pedrocchi A et al. (2008) A Micro-Electrode Array device coupled to a laser-based system for the local stimulation of neurons by optical release of glutamate. J Neurosci Methods 175:70-78
7. Biffi E, Piraino F, Pedrocchi A et al. (2012) A microfluidic platform for controlled biochemical stimulation of twin neuronal networks. Biomicroflu 6:24106-2410610
8. Biffi E, Ghezzi D, Pedrocchi A et al. (2010) Development and validation of a spike detection and classification algorithm aimed to be implemented on hardware devices. Comput Intell and Neurosci 659050 DOI 10.1155/2010/659050
9. Biffi E, Menegon A, Regalia G et al. (2011) A new cross-correlation algorithm for the analysis of "in vitro" neuronal network activity aimed at pharmacological studies. J Neurosci Methods 199:321-327
10. Bottino E, Massobrio P, Martinoia S et al. (2009) Low-noise low-power CMOS preamplifier for multi-site extracellular neuronal recordings. Microelec J 40:1779-1787
11. Rolston JD, Gross RE, Potter SM (2009). A low-cost multielectrode system for data acquisition enabling real-time closed-loop processing with rapid recovery from stimulation artefact. Front Neuroeng DOI 10.3389/neuro.16.012.2009

Author: Giulia Regalia
Institute: Politecnico di Milano, Department of Electronics, Information and Bioengineering, Neuroengineering and Medical Robotics Laboratory
Street: via G. Colombo 40
City: Milano
Country: Italy
Email: giulia.regalia@mail.polimi.it

Interdigitated Biosensor for Multiparametric Monitoring of Bacterial Biofilm Development

S. Becerro[1,2], J. Paredes[1,2], and S. Arana[1,2]

[1] CIC microGUNE, Goiru Kalea 9, 20500 Arrasate-Mondragon, Spain
[2] CEIT-IK4 and Tecnun (University of Navarra),
Paseo de Manuel Lardizábal 15, 20018 Donostia-San Sebastián, Spain

Abstract—Adhesion and growth of bacterial biofilms causes numerous problems in a wide variety of sectors, and more particularly, in those related to the medical environment or the industry. Treatment and disposal of this kind of infections is often hampered by the antibiotic resistance of biofilms as well as by a total lack of symptoms in the early stages. Therefore, it is necessary to find new methods for the early detection of biofilm development so as to improve the efficiency of treatments and to reduce the health complications suffered by patients. For this purpose, this paper focuses on the design and development of interdigitated microelectrode based biosensors that allow the detection of bacterial adhesion since the first steps of biofilm generation through impedance spectroscopy and electrochemistry. Both techniques have been proved as suitable tools for biofilm sensing and the results of the monitorization of bacterial biofilms of *S. epidermidis* in culture medium are presented. While variations of 40 % within a few hours of incubation have been achieved with impedimetric monitoring, electrochemistry increases both selectivity and sensitivity of the recorded measurements (variations of 60% have been obtained). Moreover, a multiparametric design for electrochemical and temperature measurements that will allow to obtain additional information of bacterial activity is presented.

Keywords—interdigitated microelectrode, impedance microbiology, biofilm, electrochemical detection, temperature sensor.

I. INTRODUCTION

Biofilms are formed when bacteria adhere to surfaces and excrete a sticky and viscous substance. Any surface with a combination of nutrients and humidity is suitable for bacterial adhesion. Once immobilized, bacteria grow, reproduce and produce extracellular polymers that form a matrix which increases the antimicrobial resistance of the adhered microorganisms by blocking the access of antibiotics through it [1].

Biofilm adhesion entails a serious problem in many types of applications and more particularly, in those related with the medical environment or industry. More than 80% of the infections that are diagnosed are associated with the presence of bacterial biofilms [2]. Besides, the main problem lies in the fact that biofilms are only sensitive to the action of antibiotics in the first phases of their development. However, patients do not present symptoms in these early stages so it is very improbable to detect infections, hence to treat them. Therefore, there is a huge necessity of finding new ways of early detection.

Electrical impedance spectroscopy (IS) has been widely used during the last four decades as a tool for microorganism detection. By culturing adherent cells on electrode surface, IS can detect their presence by means of the induced impedance changes [3]. For this purpose, interdigitated microelectrode array based sensors are very commonly used as they offer notable features such as a large sensitive area in a limited space as well as an accurate response.

At the same time, electrochemical impedance spectroscopy (EIS) has emerged in the last decades as another powerful technique for microorganism detection as well as for its use in other types of studies like corrosion reactions or characterization of complex interfaces and electrode processes [4]. Electrochemical studies are focused on the study of the faradaic currents generated by redox reactions. A configuration made of charged electrodes in contact with a liquid or a gas will allow a charge transfer that can be measured in the form of a current. This signal can be used to determine parameters such as the concentration of analytes or to identify the molecules present in the solution. Moreover, so as to enhance the results, redox cycling can be conducted because it allows signal amplification by using two working electrodes: collector and generator. In this way, each molecule contributes with several electrons to the measured current making it possible to detect even single molecules in the solution. The measured current will depend to a large extent on the number of reactions per unit of time that in turn depends on the time expensed by the molecules to cover the distance between the first electrode where they react and the second one. By minimizing the distance between electrodes, the measured current will be increased. This is the reason why interdigitated microelectrode based sensors are very useful for this technique.

Considering the facts described above, this work focuses on the evaluation of electrical impedance spectroscopy and electrochemistry as tools for biofilm detection using interdigitated array microelectrode based biosensors. Moreover, the developments of an interdigitated microelectrode based biosensor for impedance monitoring and a multiparametric chip composed of both an electrochemical sensing part and a temperature resistor are described.

II. MATERIALS AND METHODS

A. Chemicals and Reagents

The culture media used in this work was Tryptic Soy Broth, TSB (BBL™, ref: 211768) enriched with 5% glucose (Dextrose from Difco™, ref: 215530). Saline 4,5 % solution from Panreac (PA-ACS-ISO ref: 131659) and Brain Heart Infusion, BHI (BactoTM, ref: 237500) were also used. All solutions and media were prepared with deionized water (Merck Millipore) and media were also sterilized at 121°C for 1 hour in the autoclave.

Ethanol at 99,5% (Panreac ref: 131659) and 10% Hellmanex II solution (Hellma Analytics ref: 9-307-010-507) were used for cleaning biosensors.

For electrochemical measurements, a 50 mM solution of sulfuric acid (H_2SO_4 at 96% from Sigma-Aldrich ref: 320501) and a 25 mM of potassium ferricyanide ($K_3Fe(CN)_6$ from Sigma-Aldrich ref: 60299) were prepared with distilled water (Merck Millipore).

B. Microorganisms and Culture Protocol

The bacterial strain of *Staphylococcus epidermidis* (S.E) used in this study was *ATCC 35984*. It belongs to the American Type Culture Collection (ATCC) and was purchased from the Spanish Type Culture Collection (CECT). The sample preparation followed the different steps described in a previous manuscript drawn up by the authors of this paper [5].

C. Experimental Setups and Biosensors

Several experimental setups for electrical impedance spectroscopy were developed *ad hoc*: a 96-well plate based system, a CDC-Reactor, a Petri dish based system and a Labtester. Biosensors were fabricated by means of silicon microtechnologies on 3 inches wafers. The 20×30 μm interdigitated microelectrodes consist of a 100 nm thick layer of gold sputtered onto a 15 nm chromium adhesion layer. Each array has an active area of 16 mm² formed by 40 pairs of microfingers.

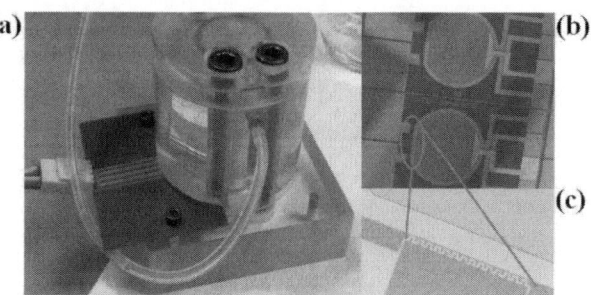

Fig. 1 (a) Experimental setup for electrochemical measurements, (b) biosensor general view and (c) detail of the temperature sensor.

For electrochemical measurements, an experimental setup was also developed *ad hoc* (Fig. 1 (a)). Biosensors were fabricated on thermally oxidized 4 inches silicon wafers with a deposition of titanium as adhesion layer. Biosensors are formed by two 15×15 μm interdigitated working electrodes formed by a 100 nm thick layer of gold, a 250 nm thick platinum counter electrode and a 600 nm thick silver reference electrode (Fig. 1 (b)).

Finally, a temperature sensor has been designed (Fig. 1 (c)). Based on the value of the platinum resistivity used in the sputtering process and on the width of the commercial temperature sensors, a 250 nm thick platinum sensor composed of 163 squares of size 25 μm has been included into the final chip.

D. Impedance and Electrochemical Measurements

Impedance spectral measurements were carried out using a *Solartron 1260* (Solartron Analytical) impedance analyzer together with a multiplexer system [6] that allows an automated sampling in a sequential and cyclic way. Measurements were performed in the frequency range from 10 Hz to 100 kHz with a sinusoidal input signal of 100 mV without polarization.

Regarding electrochemical measurements, these were performed using a multi-potentiostat model *1000B* from CH Instruments. When using $K_3Fe(CN)_6$, potential was scanned from -0.6 V to 0.6 V with a scanning rate of 100 mV/s and a sampling interval of 0.001 V. For H_2SO_4 the parameters were: 0 V to 1.2 V; 100 mV/s and 0.001 V.

For biofilm detection, the same protocol has been followed in all the experiments. Each biosensor was measured once every 30 minutes and the measurements were made inside the incubator at 37°C in an atmosphere of 10% CO_2.

Temperature sensor characterization was carried out in a climatic chamber *CCM-40/81* (Dycometal). A power supply *EX42102* from TTI and a *Keithley 2000 Multimeter* were used for resistance measurements.

III. RESULTS AND DISCUSSION

A. Electrical Impedance Spectroscopy

As it has been mentioned, the authors of this paper have developed different setups for impedance spectroscopy. All of them have been proved as excellent tools for biofilm detection and in Fig. 2 the results obtained by using the system based on Petri dishes as a support structure can be observed as an example. Fig. 2 (a) presents the relative variations of the equivalent series resistance while Fig. 2 (b) illustrates the variations of the capacitance. Fluctuations are studied over time and measurements taken at different frequencies are compared.

The measurement started 30 minutes before the infection of the culture medium in order to obtain a *plateau* that served both to verify that the initial culture medium was not undesirably infected and to obtain a reference baseline.

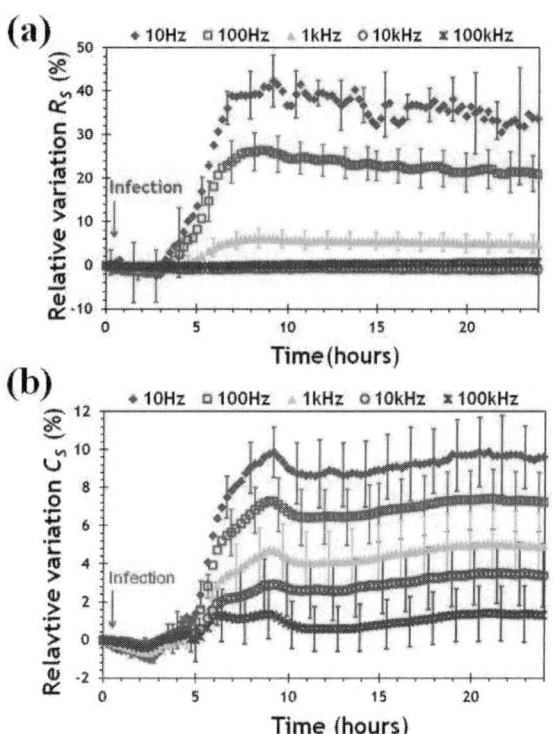

Fig. 2 Impedance measurement during the development of a microbiological culture of S. epidermidis. The initial inoculum had a value of ~1x10^6 CFU/mL. (a) Relative variation of the equivalent series resistance and (b) relative variation of the equivalent series capacitance.

As it was expected, the curves show the characteristic behavior of bacterial growth and development. During the first 4 hours, there was no significant variation in the measured impedance which responds to the first phase of the bacterial growth: the lag phase. During this phase, individual bacteria are maturing and adapting to the media so there is no yet any division neither any aggregation between them. However, once the logarithmic growth phase starts, the impedance suffers an abrupt variation up to 40 % in the case of the resistance. Capacitance variations are almost 4 times smaller than resistance ones. Finally, bacteria reach the stationary phase and no more variations on the impedance measurements occur.

B. Multiparametric Measurements: EIS and Temperature

Once Impedance Spectroscopy was validated as a suitable tool for biofilm detection, a multiparametric chip composed of an electrochemical sensor and a temperature sensor has been developed to enhance both sensitivity and selectivity.

A three-electrode system is used for this kind of experiments: a working electrode in which reactions will be studied, an auxiliary electrode and a reference electrode that controls the potential of the working electrode. On the other hand, redox cycling technique requires at least a supplementary working electrode. This way, the products of the redox reactions generated at one electrode will diffuse to the other electrode and the opposite reaction will take place.

First of all, in order to enhance the electrochemical response of gold electrodes, it is necessary to carry out a surface electrode pretreatment so as to obtain an accurate and reproducible response. There are several pretreatments that can be conducted but electrochemical polishing with H_2SO_4 has been proved to offer good features and to minimize the roughness of polycrystalline gold surface at the same time that it offers reproducible surfaces [7]. The problem of this technique is that H_2SO_4 etches chromium and due to this, the layers of gold, platinum and silver detach from the substrate. It is for this reason that titanium has replaced chromium as adherent layer.

Once pretreatment was carried out, the system has been validated by using ferricyanide as a probe due to its well-known electrochemical behavior. Moreover, by culturing adherent cells into the biosensor, biofilm development and growth has been detected. Fig. 3 shows the relative variations obtained after infecting the culture media with bacteria 15 minutes after the beginning of the experiment. As it can be seen, during the first hours, the electrochemical changes that occur in the culture media after bacteria addition are recorded. Moreover, approximately 3 hours

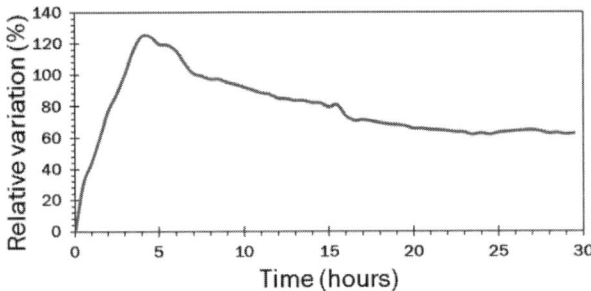

Fig. 3 Current variations recorded during the electrochemical monitoring of the development of a microbiological culture of S. epidermidis. The initial inoculum had a value of ~1x10^6 CFU/mL.

after the infection, current starts to decrease due to bacterial adhesion onto the chip producing relative variations up to 60% in the measured signal. By measuring bacterial adhesion through cyclic voltammetry, results have been enhanced because bacteria have been detected in a shorter time and the obtained variations are higher than those recorded with impedimetric monitoring. Sensitivity has therefore been improved.

Finally, as it has been mentioned, a temperature sensor has been added onto the chip so as to obtain information about the changes of temperature onto the surface where biofilm deposits. In this way, the capacity of detection of the biosensor is improved and information about the metabolic activity of the biofilm can be attained. Moreover, discrimination between different types of bacteria is expected to be achieved. Furthermore, real-time surface temperature measurements can also be used to correct the thermal drifts of the monitoring curves.

A platinum resistor has been designed using thin film techniques as thin film temperature sensors are able to work in an extremely fast manner and with low power consumption [8]. Moreover, platinum has been selected because of its good thermal response and its resistance that exhibits a linear temperature dependency.

The measurements and characterizations carried out have proved the linearity of the temperature sensor that has a TCR value of 1950 ppm/°C. Moreover, preliminary assays have confirmed the accurate response of the sensor as well as its utility in biofilm temperature measurement.

IV. CONCLUSIONS

We have monitored bacterial development and growth by means of impedimetric and electrochemical measurements on interdigitated microelectrode based biosensors. The presence of microorganisms has been detected since the first steps of bacterial adhesion with both techniques. However, electrochemical measurements have been proved to offer better sensitivity in a shorter time. By integrating a biosensor into medical devices, detection of microorganisms in implants and intravascular catheters could be achieved since the early stages when antibiotics are still useful.

Moreover, in order to obtain additional information about biofilm development and growth, a multiparametric device has been designed. By adding a temperature sensor to the final chip, a correlation between the measured temperature during biofilm adhesion and the molecular compounds present on it may be established in order to make a better discernment between bacteria and to study the metabolic activity of the biofilm. Besides, a temperature sensor is essential for the corrections of the thermal drift in the obtained monitoring curves.

Once the optimization of cyclic voltammetry experiments has been completed, new measurements have to be performed in order to improve detection. Temperature measurements in biofilm have to be also carried out. Moreover, a higher specificity will be achieved by adding a third sensor to the final chip: a pH sensor that will allow measuring the difference in pH with and without biofilm as well as the possible difference between different kinds of microorganisms.

REFERENCES

1. Donlan R M (2002) Biofilms: Microbial Life on Surfaces. Emerg Infect Dis 8: 881-890
2. Williams D L, Bloebaum R D (2010) Observing the Biofilm Matrix of Staphylococcus epidermidis ATCC 35984 Grown Using the CDC Biofilm Reactor. Microscopy and Microanalysis 16: 143-152
3. Mamouni J, Yang L (2011) Interdigitated microelectrode-based microchip for electrical impedance spectroscopic study of oral cancer cells. Biomed Microdevices 13: 1075-1088
4. Lasia A (1999) Modern Aspects of Electrochemistry. Plenum Publishers, New York
5. Paredes J et al. (2012) Real time monitoring of the impedance characteristics of Staphylococcal bacterial biofilm cultures with a modified CDC reactor system. Biosensors and Bioelectronics 38: 226-232
6. Becerro S, Benavente A, Paredes J, Arana S (2011) Desarrollo de un sistema de multiplexado para la realización de medidas multiparamétricas en cultivos microbiológicos en laboratorio. XXIX Congreso Anual de la Sociedad Española de Ingeniería Biomédica, Cáceres, Spain, 2011.
7. Carvalhal R F, Sanches Freire R, Kubota L T (2005) Polycrystalline Gold Electrodes: A Comparative Study of Pretreatment Procedures Used for Cleaning and Thiol Self-Assembly Monolayer Formation. Electroanalysis 17: 1251-1259
8. Resnik D et al. (2009) Experimental study of Ti/Pt thin film heater and temperature sensors on Si platform. IEEE Sensors 2009 Conference.

Human Splenon-on-a-chip: Design and Validation of a Microfluidic Model Resembling the Interstitial Slits and the Close/Fast and Open/Slow Microcirculations

L.G. Rigat-Brugarolas[1,2], M. Bernabeu[3], A. Elizalde[3], M. de Niz[3], L. Martin-Jaular[3], C. Fernandez-Becerra[3], A. Homs-Corbera[1,2], H.A. del Portillo[3,5], and J. Samitier[1,2,4]

[1] Nanobioengineering group, Institute for Bioengineering of Catalonia (IBEC), Barcelona, Spain
[2] Centro de Investigación Biomédica en Red de Bioingeniería, Biomateriales y Nanomedicina (CIBER-BBN), Zaragoza, Spain
[3] Barcelona Centre for International Health Research (CRESIB, Hospital Clínic - Universitat de Barcelona), Barcelona, Spain
[4] Department of Electronics, Barcelona University (UB), Martí I Franques, 1, Barcelona, 08028, Spain
[5] Institució Catalana de Recerca I Estudis Avançats (ICREA), Barcelona, Spain

Abstract—Splenomegaly, albeit variably, is a landmark of malaria infection. Due to technical and ethical constraints, however, the role of the spleen in malaria remains vastly unknown. The spleen is a complex three-dimensional branched vasculature exquisitely adapted to perform different functions containing closed/rapid and open/slow microcirculations, compartmentalized parenchyma (red pulp, white pulp and marginal zone), and sinusoidal structure forcing erythrocytes to squeeze through interstitial slits before reaching venous circulation. Taking into account these features, we have designed and developed a newfangled microfluidic device of a human splenon-on-a-chip (the minimal functional unit of the red pulp facilitating blood-filtering and destruction of malarial-infected red blood cells). Our starting point consisted in translating splenon physiology to the most similar microfluidic network, mimicking the hydrodynamic behavior of the organ, to evaluate and simulate its activities, mechanics and physiological responses and, therefore, enable us to study biological hypotheses. Different physiological features have been translated into engineering elements that can be combined to integrate a biomimetic microfluidic spleen model. The device is fabricated in polydimethylsiloxane (PDMS), a biocompatible polymer, irreversibly bonded to glass. Microfluidics analyses have confirmed that 90% of the blood circulates through a fast-flow compartment whereas the remaining 10% circulates through a slow compartment, equivalently to what has been observed in a real spleen. Moreover, erythrocytes and reticulocytes going through the slow-flow compartment squeeze at the end of it through 2µm physical constraints resembling interstitial slits to reach the closed/rapid circulation.

Keywords—organ-on-a-chip, malaria, microfluidics, spleen.

I. INTRODUCTION

Malaria, a mosquito-borne infectious disease caused by a parasite of the genus *Plasmodium*, is a developing world health challenge affecting many lives. Annually, there are between 200 and 300 million documented cases of this pathology, resulting in more than 1 million deaths per year, many of whom are African children under the age of five.

Although *Plasmodium falciparum* is the most prevalent causal parasite species, responsible of nearly the 80% of all infections and mortalities [1], the recent call for the malaria's global eradication has highlighted the importance of *Plasmodium vivax* in the framework of new vaccines development [2]. The principal trait is the fact that its primary host red blood cell is the reticulocyte, which normally accounts for only about 1% of the human red blood cell population. Reticulocytes (immature red blood cells) are not easily obtained routinely, so this hampers any systematic investigation of this malaria parasite [3], which represents the most widespread genre worldwide.

Advances in bioengineering and microfluidics have made possible to achieve the necessary technology to pattern complex microstructures, derived from electronic microfabrication protocols, allowing precise control over the flow of minute volumes of liquid, and thus present new mechanisms for studying the underlying cellular activities, mechanics and, therefore, physiological response to evaluate multiple biological hypotheses. The integration of such mechanisms in a comprehensive cellular analysis system has the potential to enable user-friendly and low-cost experimental platforms on Lab-on-a-chip (LOC) devices, analogously referred to as Micro Total Analysis Systems (µTAS) [4][5].

In this sense, many efforts have been directed toward developing different bio-chip based systems that permit the investigation of both biological and physical features of malaria parasite. In particular, microfluidic devices have been utilized to study reduced deformability of *Plasmodium* parasitized erythrocytes [4][6][7], cytoadhesion events [8][9], interactions of platelets or proteins and erythrocytes with different endothelial cells [10] or, more recently, to evaluate multiple microfluidic culture systems. Also some other devices have been fabricated for the development of new tools for malaria detection using several separation principles, mostly based on malaria-infected red blood cells' intrinsic properties (such as dielectric differences [11], paramagnetic affinity [12] or changes in stiffness [13]).

However, a step forward in this field concerns organ-on-a-chip devices. Development of these microengineering platforms has opened new possibilities to create *in vitro* models that reconstitute more accurately complex 3D organ-like structures allowing to mimic the hydrodynamic behavior of an organ by integrating dynamic mechanical and/or biochemical properties and interactions that can be variable in time and location. This 3D cell-culture models may provide critical insight into the cellular basis for different diseases, aiming for a better approximation and understanding of *in vivo* cellular organization, function, and interactions than simple 2D tissue-culture classical systems. This kind of platforms are still in progress, but advances in cell culturing and manufacturing devices with micro and nanoengineering methodologies mean that they could eventually supplement or supplant animal studies or, at least, help to prioritize them. Further, this novel engineering-type approach should help researchers to discover new biological pathways or validate new hypotheses in those pathologies were no good animal model exists, like for example in malaria, besides helping pharmaceutical companies in the development, testing and evaluation of toxic side effects of new therapeutic substances [14].

Several biomimetic platforms, in the interface between nanobio and tissue engineering have been developed recently, reproducing, for instance, the alveolar-capillary interface [15], kidney proximal tubule system [16], a ventricular myocardium film resembling cardiac tissue [17] or an engineered liver-on-a-chip reconstituting the hepatic microarchitecture [18], to name but a few.

Nevertheless, as far as we are concerned, no one has developed a spleen-like platform for studying the importance of this organ in malaria disease. The Spleen, moreover being the largest lymph organ in our body and, therefore, having an outstanding importance in blood filtering, may play a crucial role in this pathology. The spleen is a complex three-dimensional branched vasculature exquisitely adapted to perform different functions containing closed/rapid and open/slow microcirculations, compartmentalized parenchyma (red pulp, white pulp and marginal zone) [19]. On account of this, the spleen is essential for removing old, abnormal and rigid red blood cells that are unable to pass through the 2μm slits between the endothelial cells of the splenic sinusoids. Likewise, parasitized erythrocytes containing bulky viscous *Plasmodium* parasites either are retained by the red pulp (in the reticular mesh) or lose their parasites by pitting (in the mentioned interendothelial slits, the narrowest constrictions in our body). Notwithstanding, due to technical and ethical constraints the relevance of the spleen in malaria remains vastly unknown.

In an effort to resolve this controversy, yet seeing the great need to implement this kind of platforms, and taking into account the previous features, in this article we present a newfangled multilayered microfluidic device of a human splenon-on-a-chip (the minimal functional unit of the red pulp facilitating blood-filtering and destruction of malarial-infected red blood cells). Our starting point consisted in translating splenon physiology to the most similar microfluidic network, mimicking the hydrodynamic behavior of the organ, to evaluate and simulate the activities, mechanics and physiological responses and, therefore, enable us to study biological hypotheses. Different physiological features have been translated into engineering elements that can be combined to integrate a biomimetic microfluidic spleen model for the study of *Plasmodium vivax* malaria disease.

Thence, a key difference between our bio-mimetic analysis platform and other microfluidic devices, such as, for example, capillary obstruction models, is the capability of providing a continuous whole blood perfusion of the microchannels over several hours allowing the observation of organ-intrinsic properties in a 3D culture device.

II. MATERIALS AND METHODS

A. Device Design

A two-layer microfluidic device was designed to translate a simplified but accurate version of spleen's red pulp physiology using standard photo-lithographic and soft-lithographic techniques customized for fabricating a multilayer platform. The device consists in two main sections that represent both slow-flow and fast-flow channels (open/slow and close/fast microcirculations) as shown in Fig. 1. Therefore, the dimensions of those channels are different. In the first one, the height is 5.5μm, whereas in the rest of the device the height is 35μm. The width is also different (144μm vs. 188μm, Fig. 2). The height of all constrictions is greater than 4μm in order to constrain normal red blood cells in the planar configuration while complicating the passage of parasitized cells. A device height of less than 4μm is found to exclude some erythrocytes from passing through [4]. Those measures were calculated taking into account that, in physiological conditions, the 90% of the blood circulates through the fast-flow section while the remaining 10% circulates through the slow compartment. The adjusted structure was designed in the basis of the classical fluidic resistance (R_H) model by imposing geometric relations on the two LOC branches.

The main required design specifications corresponds to the slow-flow section, where, in order to mimic both the reticular mesh and the interendothelial slits, we have designed a pillar matrix constriction zone and multiple parallel constriction microchannels (2μm width each) connected to the fast microcirculation, simulating the exit towards the sinusoids into the splenic venous system.

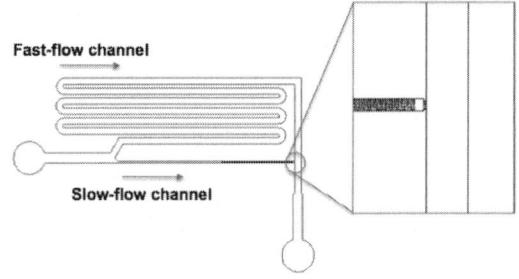

Fig. 1 Layout of the splenon-on-a-chip with a detail of the pitting zone.

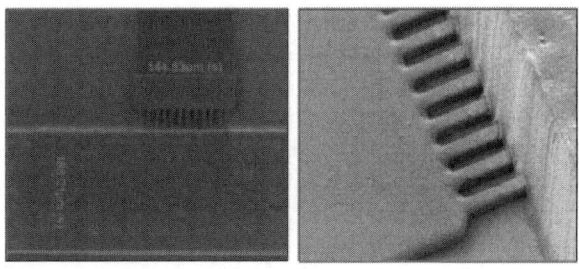

Fig. 2 Measures of both channels (detail)

B. Device Fabrication

The fabrication procedure for the splenon-on-a-chip consists in a multi-step process established to obtain two different height using two different photoresists by means of photo-lithographic techniques. All works were carried out in the clean room facility of the Institute for Bioengineering of Catalonia (IBEC) at University of Barcelona.

All solvents and chemical were obtained from Sigma Aldrich unless otherwise specified. SU-8 photoresists and SU-8 developer were from MicroChem (Newton, MA).

The masters were fabricated over glass slides (Deltalab 100002). A thoroughly cleaning protocol based on chemical baths and surface activation process was used for the slides, starting with an acetone treatment, followed by isopropanol and finally ethanol. After dehydration, the glass surface was subjected to 10W of O_2 plasma power for 10 minutes to improve photoresist adhesion. The first step consists in drawing Ordyl alignment marks for a correct structuring of the bi-layer device in the two axis. Afterwards, the slide must be gyrated in order to start working in the other side, avoiding in that sense mask contact problems and resolution loses, arising from Ordyl features height. Straightaway, SU8 2000.5 was spinned (500nm), after repeating the cleaning protocol, for acting as an interface layer between glass and further photoresist layers. In order to fabricate the first layer of the device (slow-flow channel), a 5.5μm-high SU8-10 negative photoresist was spun over the glass slide applying a three-step spinning protocol to obtain the desired SU8 thickness. The slide was then soft baked on a hot plate before being exposed with a chrome-on-glass photomask to UV Light (4s, 25mW cm^2, 345nm) in a mask aligner (MJB4 aligner, SÜSS Microtec, GE). The exposed slide was given a post-exposure bake in the sequence of 65°C for 1 minute and 95°C for 3 minutes, left to cool for up to 1 hour, and finally developed using the SU8 developer for 30s.

For the second device layer (fast-flow channel), Ordyl SY300 negative photoresist was disposed on top of the last SU8 layer and introduced in a laminator in order to have a smooth attached 35μm-high film surface. Then, the slide was exposed with an acetate photomask to UV Light (5s, 25mW cm^2, 345nm) in the mask aligner, and subsequently placed on a hot plate at 65°C for 3 minutes. At last, the Ordyl film was developed using the Ordyl developer for 3 minutes for finally obtaining the bi-layer master.

To replicate the master's microchannels, a PDMS pre-polymer mixture (curing agent-to-PDMS ratio of 1:10, Sylgard®184, Dow Corning) was placed in a desiccator and vacuum was applied in order to remove bubbles. Then the mixture was poured on top of the SU8-Ordyl master to fabricate a PDMS mould, heat up at 65°C for 3 hours and afterwards kept at room temperature for 24 hours. The casted PDMS was peeled off carefully and inlet and outlet holes were made using a Harris Uni-Core 1mm puncher. After cleaning, both the glass and the PDMS structures were chemically modified in O_2 plasma and immediately pressed together to form a permanent bond.

C. Experimental Setup

Briefly, the experimental Setup consists in a syringe pump (PHP 2000, Harvard Apparatus) used to introduce human whole blood into the microfluidic device at a pre-determined flow rate of 5μL/min. The experiments were carried out at room temperature (25°C) and the sample at physiological temperature of 37°C.

Optical measurements were performed using an inverted optical microscope (Olympus IX71) with an integrated CCD Hamamatsu camera.

Teflon tubing was inserted into the access holes, which were slightly smaller than the outer diameter of the tubing to form a pressure seal between the tubing and the hole.

III. RESULTS

Microfluidics analyses has confirmed that 90% of the blood circulates through a fast-flow compartment whereas the remaining 10% circulates through a slow compartment (Fig. 3). Moreover, erythrocytes and reticulocytes going through the slow-flow compartment squeeze at the end of it through 2μm physical constraints resembling interstitial slits to reach the closed/rapid circulation (Fig. 4).

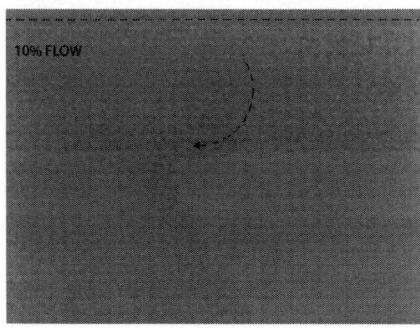

Fig. 3 Blood Flow division (90% Fast channel vs. 10% Slow channel)

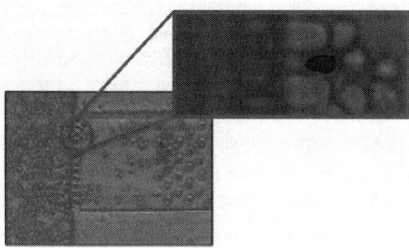

Fig. 4 Erythrocytes squeezing through the 2μm physical constraints

IV. CONCLUSIONS

We have developed a newfangled microfluidic device of a human splenon-on-a-chip that mimics the hydrodynamic behavior of the organ, resembling the interstitial slits and the fast/slow microcirculation. This biomimetic-chip based platform should help in have a deeper understanding of the molecular basis of spleen pathophysiology in human malaria.

ACKNOWLEDGEMENTS

Part of this work was financially supported by the technology transfer program of the Fundación Botín and by the Explora Program of the Ministry of Economy and Competitiveness of the Government of Spain.

REFERENCES

1. Tuteja R. et al. (2007) The future of modern genomics. 274.
2. Alonso PL, Brown G, Arevalo-Herrera M, Binka F, Chitnis C, Collins F, Doumbo OK, Greenwood B, Hall BF, Levine MM et al. (2011) A research agenda to underpin malaria eradication. PLoS medicine 2011, 8(1) DOI: e1000406.
3. Galinski MR, Barnwell JW. (1996) Plasmodium vivax: merozoites, invasion of reticulocytes and considerations for malaria vaccine development. Parasitology today. vol 12. no. 1.
4. Guo Q, Reiling SJ, Rohrbach P, Ma H. (2011) Microfluidic biomechanical assay for red blood cells parasitized by *Plasmodium falciparum*. Lab on a chip. DOI: 10.1039/c2lc20857a.
5. Parra-Cabrera C, Sporer C, Rodriguez-Villareal I, Rodriguez-Trujillo R, Homs-Corbera A, Samitier J. (2012) Selective *in situ* funcionalization of biosensors on LOC devices using laminar co-flow. Lab on a chip. DOI: 10.1039/C2LC40107J
6. Shelby JP, White J, Ganesan K, Rathod PK, Chiu D. (2003) A microfluidic model for single-cell capillary obstruction by *Plasmodium falciparum*-infected erythrocytes. PNAS. DOI: 10.1073/pnas.2433968100.
7. Handayani S, Chiu D, Tjitra E, Kuo J, Lampah D, Kenangalem E, Renia L, Snounou G, Price RN, Anstey NM, Rusell B. (2009) High deformability of *Plasmodium vivax*-infected red blood cells under microfluidic conditions. The Journal of Infectious Disease; 199: 445–50. DOI: 10.1086/596048
8. Herricks T, Seydel KB, Turner G, Molyneux M, Heyderman R, Taylor T, Rathod PK. (2011) A microfluidic system to study cytoadhesion of *Plasmodium falciparum* infected erythrocytes to primary brain microvascularendothelial cells. Lab on a chip. DOI: 10.1039/c11c20131j
9. Antia M, Herricks T, Rathod PK. (2007) Microfluidic modeling of cell–cell interactions in malaria pathogenesis. PLoS Pathog 3(7): e99. DOI:10.1371/journal.ppat. 0030099
10. Usami S, Chen HH, Zhao Y, Chien S, and Skalak R. (1993) Ann. Biomed. Eng., 1993, 21, 77–83.
11. Gascoyne P, Mahidol C, Ruchirawat M, Satayavivad J, Watcharasit P, Becker FF. (2002) Microsample preparation by dielectrophoresis: Isolation of malaria. Lab on a Chip, 2, 70-75. DOI: 10.1039/b110990c.
12. Zimmerman PA, Thomson JM, Fujioka H, Collins WE, Zborowski M. (2006) Diagnosis of malaria by magnetic deposition microscopy. Am. J. Trop. Med. Hyg., 74(4), 2006, pp. 568–572.
13. Bow H, Pivkin IV, Diez-Silva M, Goldfless SJ, Dao M, Niles JC, Suresh S, Han J. (2011). A microfabricated deformability-based flow cytometer with application to malaria. Lab on a chip. 11, 1065-1073.
14. Huh D, Hamilton GA, Ingber DE. (2011) From 3D cell culture to organs-on-chips. rends in Cell Biology, Vol. 21, No. 12. DOI:10.1016/j.tcb.2011.09.005.
15. Huh D. et al. (2010) Reconstituting Organ-Level Lung Functions on a Chip. Science 328, 1662. DOI: 10.1126/science.1188302.
16. Jang KJ, Suh KY. (2010) A multi-layer microfluidic device for efficient culture and analysis of renal tubular cells. Lab on a Chip 10.
17. Grosberg A. Alford PW, McCain ML, Parler KK. (2011) Ensembles of engineered cardiac tissues for physiological and pharmacological study: Heart on a chip. Lab on a chip. DOI: 10.1039/c11c20557a.
18. Nakao, Y. et al. (2011) Bile canaliculi formation by aligning rat primary hepatocytes in a microfluidic device. Biomicrofluidics 5.
19. Del Portillo HA, Ferrer M, Brugat T, Martin-Jaular L, Lacerda MVG. (2012) The role of the spleen in malaria. Cellular Microbiology. 14 (3), 343–355. DOI: 10.1111/j.1462-5822.2011.01741.
20. Oh KW, Lee K, Ahn B, Furlani EP. (2012) Design of pressure-driven microfluidic networks using electric circuit analogy. Lab on a chip. DOI: 10.1039/c2lc20799k.

Microbioreactor Integrated with a Sensor for Monitoring Intracellular Green Fluorescence Protein (GFP)

Godfrey Pasirayi, Meez Islam, Simon M. Scott, Liam O'Hare, and Zulfiqur Ali[*]

School of Science and Engineering, Teesside University, Borough Road, Middlesbrough, TS1 3BA, United Kingdom
Teesside University/School of Science and Engineering, Middlesbrough, United Kingdom
Z.Ali@tees.ac.uk

Abstract—A prototype Polydimethylsiloxane (PDMS) based microbioreactor integrated with a photodiode detector for monitoring intracellular green fluorescent protein ultraviolet (GFPUV) expression is described. The developed system is compact, simple and inexpensive thus making it ideal for an economical on chip detection of intracellular GFP and other biological applications. The detection limit for cell free GFPUV was found to be 4.8×10^{-8} M while for Fluorescein isothiocyanate (FITC) was found to be 1.2×10^{-6} M. The performance of the photodiode detector was benchmarked with a CCD spectrophotometer and results showed favorable comparison. This study demonstrates the quest to develop integrated microsystems that can be used to monitor cellular dynamics in real time.

Keywords—light emitting diode, fluorescence, microbioreactor, GFPUV.

I. INTRODUCTION

Green fluorescent protein (GFP) is a protein which occurs naturally in the jellyfish *Aequorea victoria* and other coelenterates. Since its discovery, GFP and its variants are widely used as fluorescent tags of single proteins, whole organisms, cells and complex protein structures [1]. Moreover, the discovery of GFPs has enabled biologists to examine processes in living cells and study their dynamics aiding developments in neurobiology, and cell biology [2]. Recently, GFP's have made considerable headway into the world of art and commerce where, various transgenic fluorescent animals expressing GFPs have been reported including mice, fish, chickens, mice, dogs, pigs, cats and marmosets as well as decorative transgenic plants [3,4].

GFP monitoring in biological processes is affected by a number of factors including: the requirements of molecular oxygen for fluorescence to occur, the amount of GFP produced and inner filter effect from intracellular light absorption and scattering by cell particles. GFP requires at least 2hrs from transcription to a functional protein [1] consequently this makes it difficult to monitor early developmental events. For example many biological molecules autofluoresce at the same wavelength as GFP, making background noise a significant concern especially when expression of the gene of interest is very low [5]. It has been estimated that at least 1 µmol of GFP is required per mammalian cell to achieve a fluorescence that is twice to that of the background [1].

To overcome these limitations there has been a drive by researchers to develop detectors that have high sensitivity, fast response times and relatively cheap and easy integration with Lab on a chip (LOC) devices, unlike conventional spectrofluorimeters and fluorescence microscopes which are bulky and most expensive. Pioneering efforts to detect GFP expression in intracellular environments have been made by Eichhorn *et al* [6] who reported a sensor for measuring real time GFP production using genetically modified cells. The group developed an on-line GFP sensor for quantitative monitoring of GFP production in high optical density cultures of genetically modified *E.coli* cells. In a further study, the group developed an improved low cost, highly sensitive solid state of the art sensor for real time monitoring of GFP production in a standard flow through cuvette which was coupled to a conventional bioreactor [7]. The sensor consisted of a LED induced excitation source and a pin photodiode as the detector. The dynamic range of the sensor was $7.4 \times 10^{-9} - 3.7 \times 10^{-5}$ M with limit of detection 7.0×10^{-9} M for GFP. Recently Jóskowiak et al [8] developed a microscale GFP detection system using a p-i-n thin film amorphous silicon light sensitive layer microfabricated on a glass substrate integrated with an amorphous silicon carbon alloy absorption filter and an ultra-thin PDMS sheet to detect intracellular expression of GFP in *E.coli*. GFP was both detected in aqueous solution in nano molar concentration and intracellularly for 1×10^6 cells/mL range.

The aim of the present study is to develop a disposable membrane aerated PDMS microbioreactor integrated with photodiode based detector and demonstrate the detection of intracellular GFP expression in genetically modified cells. *Escherichia.coli* (pWTZ594) that constitutively expresses green fluorescent protein GFPUV was used as the model organism. The choice of the GFPUV fluorophore used in this work is attributed to its large amount of fluorescence which is 18 times brighter than the wild type GFP and consequently makes it easy to detect [9].

II. MATERIALS AND METHODS

A. Fabrication of the Microbioreactor

The LOC device was fabricated with soft lithography technique using PDMS, as illustrated in Figure 1.

---
[*] Corresponding author.

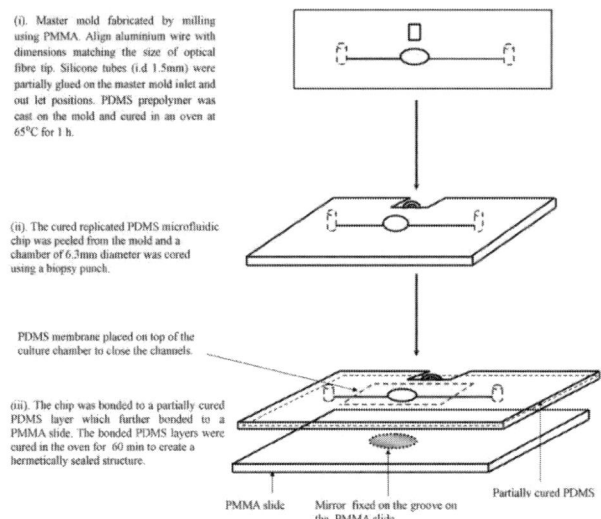

Fig. 1 Step by step fabrication of the PDMS microbioreactor.

B. System Set Up

The schematic of the optical set up is shown in Figure 2. It consists of a UV-LED excitation source with peak wavelength 395 nm (Roithner, Austria), a photodiode (PDA36 EC, Thorlabs UK) as the detector, Lock in amplifier (405 DSP, Scitec Instruments, UK), a PC with LABVIEW 8.6software and a Data Acquisition Board USB 6221, (National Instruments, UK) for acquiring analog signals and converting them to digital signals, LED driver and a single chamber microfluidic bioreactor (Figure 1). The excitation light from the UV-LED was modulated at 1 kHz using a LED driver developed in house, which provided a fixed current source of 600 mA. The excitation light from the UV LED (395 ±10 nm) was not filtered as its spectrum was extremely narrow [10 nm Full-Width Half-Maximum (FWH). The excitation light from the UV–LED is coupled through a custom made optical fibre (core diam, 1200 μm, BFL371000 CUSTOM, Thorlabs UK) and directed at the edge of the culture chamber of the chip at 0° incidences, for exciting the fluorophore. The isotropic fluorescence emission photons from the sample was passed through a Kodak wratten 2 optical filter (Edmund Optics, USA) to block light of wavelengths <500 nm. A mirror, with a higher reflective surface was placed in a grooved position on the PMMA slide, underneath the chip to facilitate the alignment of the sensitive area of the photodiode with the growth chamber of the microbioreactor. The output electrical signals from the photodetector were coupled on to a DSP lock in amplifier (DSPLA) which was under the control of a web programme.

In the DSPLA the fluorescence signals were passed through second order, high pass electrical filter with a cut off frequency of 1 kHz and further amplified and integrated with a time constant of 2 s. The integrated and filtered signals were collected from the DSPLA and channelled to the DAQ-USB 6221 which was linked to a computer via LabVIEW for data acquisition. The developed detection system was evaluated by conducting a series of experiments using standard solutions of FITC and GFPUV. FITC was excited with a blue LED with peak emission wavelength of 460nm while pure GFPUV was excited with a UV-LED with peak emission wavelength of 395 nm.

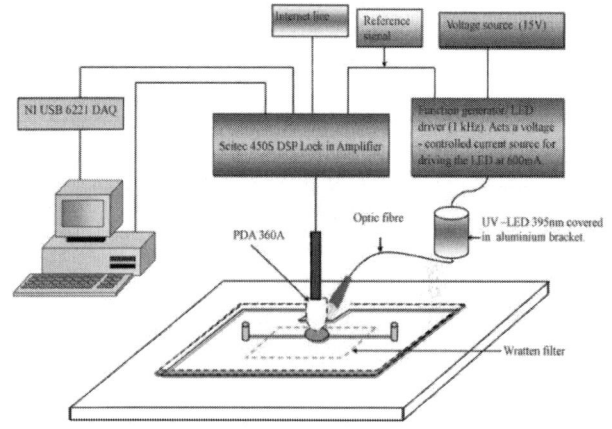

Fig. 2 Schematic showing the experimental set up.

C. Cultivation of Microorganism

The genetically modified *Escherichia coli* MC1060 strain containing the plasmid pWTZ594 obtained from *E.coli* Genetic Centre USA was initially subcultured on Tryptone Soya Agar (TSA) (Oxoid, UK) supplemented with ampicillin and 0.15 mM IPTG, (Sigma Aldrich, UK) for green fluorescent protein expression followed by incubation at 37 °C for 24 h. After incubation, the inoculated TSA surfaces were illuminated on a UV lamp and brightly green fluorescent single colonies were picked and used to inoculate 50mL of sterile modified terrific broth (MTB) supplemented with ampicillin in 250 mL Erlenmeyer flask. The flask was incubated at 37 °C for 2-4 h with shaking (150 rpm) up to an OD_{600} of 0.2. Aliquots of the bacterial suspension were used to inoculate the microbioreactor and shake flask respectively. The fluorescence signals arising from the induced *E.coli* cells were assessed as a function of cell growth and concentration. Samples for shake flask (SF) experiments were diluted appropriately and their fluorescence measured on a Hitachi F2000 Fluorimeter.

III. RESULTS AND DISCUSSION

A. Detection of Standard Solutions

The initial steps of evaluating the performance of the system was carried out by measuring the fluorescence signals generated from known concentration of standard fluorophore solutions of FITC and GFPUV. The sensor output signals (V) of the respective solutions as a measure of fluorescence were plotted as a function of the concentration of the standard solutions (Figure 3 a-b). The results in both cases show excellent linearity with regression correlation coefficients of $R^2 = 0.986$ and 0.989 for GFPUV and FITC respectively. The lowest detectable concentration of GFPUV and FITC on chip was 4.8×10^{-8} M and 1.2×10^{-6} M respectively. The LOD was calculated by using the mathematical expression described in (1).

$$LOD = 3.3\sigma + S_B \qquad (1)$$

Where σ the standard deviation of the response and SB is the background signal.

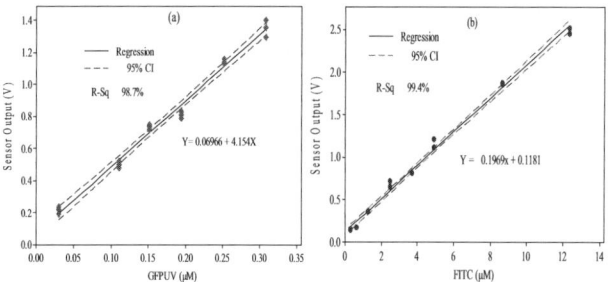

Fig. 3 Detection of standard fluorophore solutions
(a) Linear dependence of the fluorescence signals (V) with increasing GFPUV. (b) FITC standard solution concentration.

The difference in the LOD between FITC and GFPUV can be explained in terms of the large Stokes shift value for GFPUV (218 nm) solutions, in contrast with the smaller Stokes shift value for FITC (28 nm), which has the tendency of reabsorption. In all the measurements, two baselines namely the dark voltage (response of the photodiode to dark conditions i.e. when there is no illumination on the detector) and the background voltage (response of the photodiode with no fluorophore present in the sample volume), were considered. The background voltage could be due to the photodiode dark voltage, autofluorescence from PDMS material and stray light leaking from excitation source. The performance of the PD detector using the standard fluorophore solutions was bench marked with a CCD Ocean Optics spectrometer and results compared favourably.

B. Detection of Intracellular GFPUV Expression

The utility of the integrated microbioreactor chip was further demonstrated by performing a series of experiments using transformed *E. coli* expressing *GFPUV* and detecting the amount of expressed protein as a function of cell growth. IPTG was added to the MTB medium together with *E.coli* cells to ensure uniform protein expression and fluorescence between the cells. Control experiments were performed with non IPTG induced cells to provide a reference which was then subtracted from the overall sensor output. The growth profile of the cells in the microbioreactor was bench marked with SF results and the results show that *E.coli* cells were able to grow under the microfluidic bioreactor environment and efficiently expressing the fluorescent protein (GFPUV) (Figures 4-5). The results of the microbioreactor and shake flask using known concentration of the organism show a linear relationship between the fluorescence intensity of GFPUV and the number of fluorescing bacterial cells. The results in both scales show a steady growth and increase in the amount of GFPUV expression and are consistent with literature reports [7,8].

To ensure oxygen availability to the *E.coli* cells in the growth chamber a permeable PDMS membrane was incorporated on the microfluidic bioreactor device. The reported diffusivity of O_2 in PDMS is 3.4×10^{-5} cm^2/s [10]. Due to the high surface area to volume ratio of the LOC device large interfacial area over which O_2 diffusion can occur is created and this was considered adequate to meet the O_2 requirements of the *E.coli* cells. The LOD for the developed fluorosensor for intracellular GFPUV is several times higher than the previous work [6, 11]. The reduced sensitivity could be explained by the difference in the optical qualities of the materials used. For example, the inherent auto fluorescence arising from the PDMS material used in fabricating the microbioreactor chip could also have contributed to the disparity in contrast to the quartz material used by Randers *et al* [6].

The precision of any detector measurements are influenced by many experimental parameters and interferences. The geometrical arrangement of the set up and positioning of all elements such as the photodetector, microfluidic chip, excitation light, and filters had direct influences on the measurements obtained. The reflecting mirror that was used for aligning the system contributed to the improved collection efficiency of the sensor as fluorescent emission light that would otherwise leave the set-up, was reflected backwards, thus increasing the intensity of light reaching the detector. The photodetector was positioned directly above the microbioreactor growth chamber to maximize the collection of the emitted fluorescence. The amount of excitation light reaching the intracellular GFPUV fluorophore was maximized by placing the optic fibre tip in close proximity (2 mm) of the growth chamber as well as the photodetector (Figure: 2).

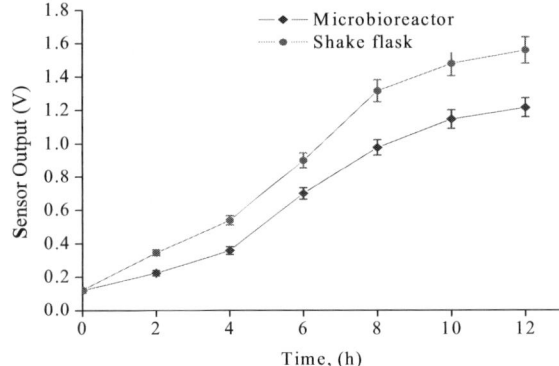

Fig. 4 Photo response of the sensor to growth profile of modified *E.coli* in the microbioreactor and shake Flask (SF).

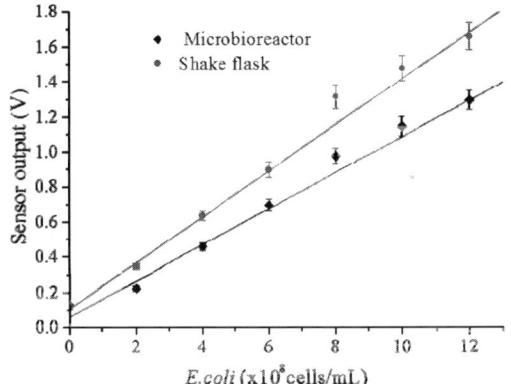

Fig. 5 Linear dependence between cell concentration and sensor output.

IV. CONCLUSIONS

A simple integrated fluorescence detector system for microfluidic applications has been developed in several stages. The developed system was capable of detecting the concentration of pure GFPUV solutions down to 4.8×10^{-8} M while for intracellular GFPUV the lowest detectable cell density that would give a minimum detectable fluorescence signal was 2.7×10^{7} cfu/mL. The advantage of the developed system allows cell growth to be monitored non-invasively and key biological information about cellular developments such as protein synthesis can be obtained in real time. The developed proof of concept fluorescence detection system could be useful in a variety of applications, including environmental, biomedical studies, drug development as well as for on-line fluorescence monitoring of reporter gene expression dynamics and cell culture studies such as in cancer cell progression. Although the system is viewed as operational, additional improvements could be done by incorporating very sensitive wavelength specific photodiodes or avalanche photodiodes but these would be at higher cost.

REFERENCES

[1] Tsien RY: (1998) The green fluorescent protein, Annu.Rev.Biochem.67: 509-544
[2] J. Chuck Harrell, Carol A. Sartorius, Wendy W. Dye,: ZsGreen Labeling of Breast Cancer Cells to Visualize Metastasis (2007) CR742322 US (630662):1-3.
[3] Lai L, Park K, Cheong H, Kühholzer B, et al. (2002) Transgenic pig expressing the enhanced green fluorescent protein produced by nuclear transfer using colchicine-treated fibroblasts as donor cells. Mol Reprod Dev 62:300-306.
[4] Xi JY, Hyo SL, (2008) Generation of cloned transgenic cats expressing red fluorescence protein. Biol Reprod 78:425-431.
[5] March JC, Rao G, Bentley WE (2003) Biotechnological applications of green fluorescent protein. Appl Microbiol Biotechnol 62:303-315.
[6] Randers-Eichhorn L, Albano et al. (1997) On-line green fluorescent protein sensor with LED excitation. Biotechnol Bioeng 55:921-926.
[7] Kostov Y, Albano CR, Rao G (2000) All solid-state GFP sensor. Biotechnol Bioeng 70:473-477.
[8] Jóskowiak A, Santos MS, et al. (2011) Integration of thin film amorphous silicon photodetector with lab-on-chip for monitoring protein fluorescence in solution and in live microbial cells. Sensors and Actuators, B: Chemical (2011)156:662-667.
[9] Chalfie M, Tu Y, Euskirchen G, et al (1994) Green fluorescent protein as a marker for gene expression. Science 263:802-805.
[10] Merkel TC, Bondar VI, et al (2000) Gas sorption, diffusion, and permeation in poly (dimethylsiloxane). J Polym Sci Part B 38:415-434.
[11] Kostov Y, Harms P, Randers-Eichhorn L, Rao G (2001) Low-cost microbioreactor for high-throughput bioprocessing. Biotechnol Bioeng 72:346-352.

Author: Zulfiqur Ali
Institute: Teesside University
Street: Borough Road
City: Middlesbrough
Country: United Kingdom
Email: Z.Ali@tees.ac.uk

Simultaneous Electrochemical Detection of Dopamine, Catechol and Ascorbic Acid at a Poly(acriflavine) Modified Electrode

M. Rashid[1], V. Auger[1], and Z. Ali[2]

[1] Technology Future Institute, University of Teesside. Middleborough. U.K.
[2] Graduate Research School, University of Teesside. Middleborough. U.K.

Abstract—Electro polymerized poly(acriflavine) modified gold (Au) electrode has been prepared and used for the electrochemical detection of dopamine (DA) and catechol (CA) without the interference of ascorbic acid (AA). The experimental results demonstrates that, PAF/Au electrode exhibited favorable electron transfer kinetics and electro catalytic activity towards the oxidation of Dopamine and Ascorbic acid in neutral (pH 7.0) and acidic (pH 5.0) conditions. PAF/Au was found be a good mediator for electrochemical oxidation of Ascorbic Acid, Dopamine and Catechol in pH 7.0 and pH 5.0 PBS buffer solutions. In the Differential Pulse Voltammetry study, PAF/Au electrode offers large peak separation, 120mV between AA and DA at pH 7.0 owing to the introduction of positively charged electrostatic barrier generated by PAF film. Similarly, a large peak separation, 380mV between AA and CA was recorded in this experimental study with a RSD of 6.5%.

Keywords—Modified electrode, Dopamine, Catechol, Ascorbic Acid.

I. INTRODUCTION

Dopamine is one of the crucial naturally occuring catecholamine neurotransmitter molecule and plays an important role in the activities of the central and peripheral nervous systems such as learning, memory formation and message transfer in the central nervous system [1]. In the human body, excessive abnormalities of DA levels are symptoms of several diseases, such as Parkinsonism, schizophrenia and HIV infection [2, 3]. Catechol (CA) (1,2-dihydroxybenzene) is isomer of dihydroxybenzene and often coexists with DA. Recently, the selective and sensitive detection of DA and CA in real samples has attracted much attention for elecroanalysis. Compared to other analytical methods electrochemical detection is more commonly preferred due to reduction in time for analysis, waste generation and reagent consumption as well as offering potential for on-site monitoring. However, the major challenge for the use of electrochemical detection for CA and DA within a biological fluid sample is the interference due to Ascorbic Acid.

Ascorbic acid is a common and vital vitamin in human diet and popular for its antioxidant properties. The anti oxidant property of ascorbic acid helps to improve immune function and has been suggested to protect against a number of malignancies including cancer [4]. Recent clinical studies have also reported that, ascorbate concentration in the biological fluid represents the amount of oxidative stress in human metabolism [5-7].Typically, DA, CA, AA will coexist in a biological sample, so it is very important to develop simple and rapid methods for selective determination of AA, CA and DA in routine analysis.

Recently, several approaches have been used to selectively detect AA, CA and DA using fluorimetry [8], Chemiluminescence [9], ion-exchange chromatography [10], ultraviolet–visible spectroscopy [11] and capillary electrophoresis [12] but these have various drawbacks including higher cost. Several researchers have employed electrochemical methods for the determination of AA, CA and DA in biological samples. However, electrochemical detection of DA with a traditional solid electrode, such as glassy carbon, is a challenge because of the interference of Ascorbic acid (AA) which in biological samples can be 100-1000 times higher concentration than DA. Electrochemical detection of DA for in vivo applications is particularly difficult [13]. AA is oxidized at a potential close to the DA potential region with a consequence of electrode fouling due to the adsorption of oxidation products [14] which result in overlapping oxidation peaks.

Chemical modification of a solid metal working electrode can be used to enhance the selectivity of DA by introducing an electrostatic barrier at the surface of working electrode [15-20]. Dopamine (pKa 8.87) is protonated at physiological pH while Ascorbic Acid (pK_a 4.10) is negatively charged. Consequently, an electrostatic barrier induced betweeen the protonated DA analyte and the negative charge of the film allows selective determination of DA in the presence of AA. The electrostatic barrier generated by the thin film polymer is ineffective for DA and critical for AA selective determination [21]. Several approaches has been reported on the use of an electrostatic barrier at the electrode surface [16, 22]. However, the control of an electrostatic barrier is complex because it depends on a number of factors such as film thickness, porosity, physical and chemical stability. Polymer film modified electrodes can offer a stable electrostatic barrier for analytes in electrochemical detection systems.

Electronically conducting polymer, containing phenyl or amine functional group in structure such as polyaniline, polypyrrole, polythiophene, has been proposed and used in biosensors and electrochemical devices [23, 24]. Compared to poly-aniline, poly-pyrrole polymers, aromatic diamine polymers have a number of novel functions, such as chargeable electro activity, high pre-selectivity to various electro active species, high sensibilities to bio substances at an extremely low concentration, pronounced electro catalytic properties, strong adhesion to metal and high capacitance [25-27]. Poly acriflavine, a typical aromatic diamine polymer containing amine functional group can modify the functionality of the working electrode [25].

S.-M. Chen et al, reported poly (acriflavine) film (PAF) – modified electrode and its electro catalytic properties towards NADH, nitrite and sulfur oxoanions [26, 27]. The polyacriflavine was electro polymerized on a glassy carbon electrode (GCE) for analytical application. The PAF/GCE was reported as a good mediator in the acidic medium. A PAF modified electrode has been used for amperometric detection of Ascorbic and Dopamine in real sample [25]. The electro polymerization process was carried out on glassy carbon electrode (GCE) at pH 1.5. The chemically modified electrode successfully separated AA and DA oxidation peak current in the cyclic voltammetry study with a potential difference of 255mV.

In this experimental work, the PAF-film coated Au modified electrode was prepared by the electrochemical deposition method to separate the oxidation potential of AA, DA and CA in neutral and acidic medium, using both the electro catalytic and the charge properties of PAF. The thickness of the electro polymerized polymer film was controlled with the monomer concentration and number of cycles for electrochemical growth. Cyclic Voltammetry and Differential plus Voltammetry techniques were used to separate and detect AA, DA and CA from a mixed analytes. The interference of DA and CA on the AA at the several pH ranges was investigated.

II. EXPERIMENTALS

A. Chemicals and Instruments

Reagent grade poly-acriflavine, Dopamine hydrochloride, Catechol and Ascorbic acid were all purchased from Sigma(UK). Sodium phosphate dibasic heptahydrate (98%) and sodium phosphate monobasic monohydrate (98%) prepared for the PBS buffer solution (15mM). The supporting elecrolyte, Na_2SO_4 were purchased from Fisher (UK). Ultrapure water was used for the preparation of electrolyte solutions. The gold (Au) working electrode, platinum (Pt) counter electrode were supplied from BASi (UK) whilst Silver/Silver chloride (Ag/AgCl) saturated reference electrode, CHI instrument (USA), were used for the polymerisation and electrochemical measurements. In the experimental work Autolab instrument was used for electrode modification while CHI instruments CH-900 was used for the analytes detection. The electrode polishing was carried out using alumina powder (0.05 μm particle size) and cloth were purchased from CHI instrument (USA).

B. Electrode Preparation

The modified electrode sensitivity and selectivity can be influenced by the working electrode surface contamination and defects. Before electro polymerization, it must be ensured that the working electrode surface is flat and contamination free with low background current. A number of methods have been developed to facilitate the maintenance and cleaning of working electrode. Among the several cleaning methods, mechanical polishing and electrochemical polishing method is more attractive for rapid cleaning process. In the electrochemical polishing methods, two-electrode configurations were employed in the cell with a potential gradient. In this experimental work, we used 0.1μm and 0.05μm alumina powder for the mechanical polishing and 0.05M H_2SO_4 solution for the electrochemical polishing.

C. Fabrication of Polymer Film on the Au Working Electrode

The thin films poly acriflavine were synthesized electrochemically on gold macro electrode substrates using cyclic voltammetry. Electrochemical polymerisation was performed in a three-electrode system. Au working electrode, Pt counter electrode and Ag/AgCl rereference electrode were employed in the electrochemical deposition process. To minimize the electrolytic ohmic drop, the reference electrode was placed in close proximity to the Au working electrode. Controlled polymerisation was performed by applying sequential linear potential scan rate of 50mV/s between 0 and 0.75 V versus Ag/AgCl electrode. A Na_2SO_4 (0.1M) supporting electrolyte was used to rinse the thin film polymer before the analytical investigation. After the electropolymerisation and deposition, the working electrode deposited with the thin film polymer film were first rinsed with a solution of 0.5 M supporting electrolyte followed by DI water. The film deposited working electrode was placed for 1 hour in the oven at 85°C for the water and moisture removal process. The dried film-coated working electrodes were subsequently used for the electrochemical detection of dopamine, catechol, ascorbic acid and mixed solutions.

III. RESULTS AND DISCUSSIONS

A. Synthesis of Poly-Acriflavine in Acidic Medium

The electro-polymerization of poly acriflavine on the Au working electrode has been carried out by oxidative polymerization method by using sodium sulphate as an oxidant in a non-oxidizing protonic acid like HCl. In the electro polymerization process, the pH o was controlled at pH 3.0 with varying acriflavine monomer (1mM, 5mM, 10mM and 15mM) solutions. The reaction temperature was controlled and with cyclic voltammetry, 25 cycles were applied for a reproducible modified electrode.

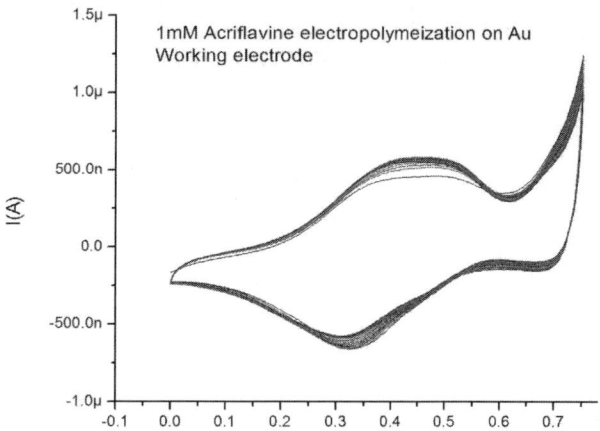

Fig. 1 Cyclic voltammogram (CVs) of electro polymerized gold electrode with 1mM acriflavine. Potential between 0 to 0.75 V (vs. Ag/AgCl) in 0.1 M Na2SO4 at a sweeping rate of 0.05V/s for 25 cycles.

Figure 1 shows the preparation of PAF film electrochemically formed on Au working electrode in a pH 3.0 aqueous phosphate buffer solution in the presence of 0.1M Na_2SO_4 electrolyte and 1mM acriflavine monomer. Before the electro polymerization process, the dissolved oxygen was removed from the monomer solution by passing nitrogen gas. At the applied potential range, dissolved oxygen helps to develop deep yellow coloration of acriflavine and reduce polymeric film formation. From Figure 1, it was observed that, during the first cycle, acriflavine monomer is oxidized at +0.475V and at reverse cycles, it exhibited cathodic peak potential at around +0.300V. The oxidation peak current at +0.475V increased cycle by cycle, which was the indication of poly (acriflavine) formation on Au working electrode.

In this experimental study, three different Au working electrodes with similar active area were prepared to detect ascorbic acid, catechol and dopamine. In the electro polymerization process, for the lower monomer concentration (1mM and 5mM) the measured oxidation peak current of poly acriflavine increases with the increment cycle number. Higher monomer concentration, i.e. 10mM and 15mM acriflavine monomer concentration shows comparatively lower oxidation current at the potential window +0.400 to +0.0475V due to rapid formation of oligimers on the working electrode surface.

B. Simultaneous Determination of Dopamine, Catechol and Ascorbic Acid Using Differential Pulse Voltammetry Method

Electrochemically modified working electrodes offer modified permeation and charge transport characteristics that will allow for selective determination of the DA, CA and AA analytes. In the neutral medium, on the bare Au working electrode, DA, CA and AA at the same concentration (5mM) will oxidize in the same potential region (0.38 to 0.410 V vs. Ag/AgCl). However, electrochemically deposited poly acriflavine polymer film shifts the potential for the oxidation process. Electrochemically deposited poly acriflavine film is positively charged, so in neutral medium, AA which is negatively charge tends to attach the polymer film with better charge transfer. Conversely, DA and CA carry a positive charge in the neutral medium and an electrostatic barrier is generated against the positively charged polymeric film. In summary, oxidation will take places at a more positive potential for DA and CA on the CME electrode surface and at a more negative potential for AA.

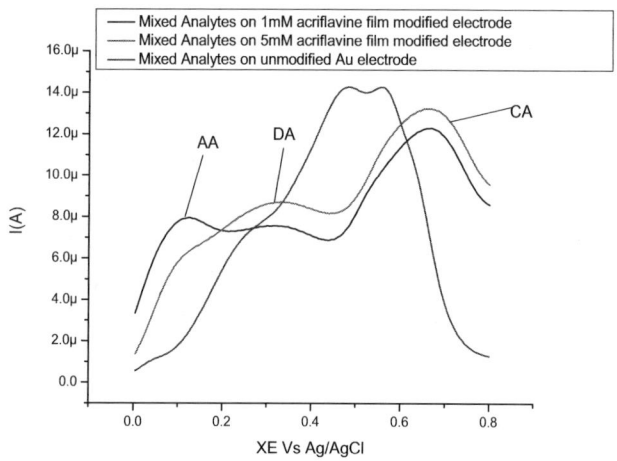

Fig. 2 DPV curve for 5mM Dopamine, Catechol, Ascorbic Acid and mixed analytes at pH 7.0 on polymer film modified and unmodified electrode.

From the Figure 2, the unmodified/bare Au working electrodes at pH 7.0 is unable to separate oxidation peak for AA, DA and CA. Electrochemically deposited thin PAF polymeric film, is able to successfully separate and detect the oxidation peak current of AA, DA and CA over the investigated potential region. The unmodified working electrode shows higher oxidation peak current at the +0.410v to +0.590V range. This is due to overlapping of the mixed analytes peak. In the case of the modified electrode, three different oxidation peaks current appears at +0.110V, +0.320V and +0.700V respectively for AA, DA and CA. The electrostatic attraction and repulsion force of the positively charged poly-acriflavine polymeric films helps to separate and detect the mixed analytes. In neutral medium, the positively charged polymeric matrix allows negatively charged Ascorbic acid to penetrate and oxidize on the modified working electrode. Dopamine and Catechol faces a comparatively repulsive force due to their positive charge and are consequently oxidized at the comparatively higher potential. It is seen that the 5mM monomer concentration offers stable polymeric matrix for the simultaneous detection of Ascorbic acid, Dopamine and Catechol.

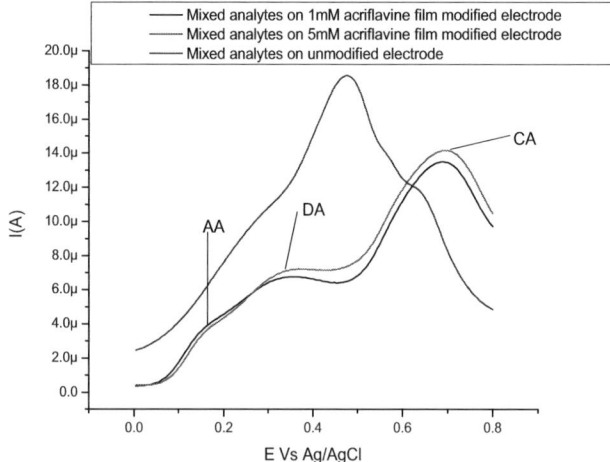

Fig. 3 DPV curve for 5mM Dopamine, Catechol, Ascorbic Acid and mixed analytes at pH 5.0 on polymer film modified electrode.

In a slightly acidic medium (pH 5.0), similar separated peaks are observed for AA, DA and CA mixed analytes. However, due to acidic medium, the active surface area of the modified working electrode reduces with the functionality of the electrostatic barrier. As a result, in the acidic medium lower oxidized current is measured for the separated peaks. 5mM polymeric films offer stable active surface area for the electrochemical detection in the slightly acidic medium like neutral medium.

IV. CONCLUSIONS

In this study, electro polymerized poly acriflavine film coated Au working electrodes were synthesized with various polymer molar concentration in highly acidic medium. Stable thin film conducting polymer was formed and CV was used to study the in situ growth of the film on the working electrode. The PAF/Au modified electrode provided the large peak separation current between AA and DA, of 110mV. Similarly, the same monomer composition (5mM) poly acriflavine offered 590mV peak current separation between AA and CA due to electrostatic attraction and repulsion forces of the selected analytes. The PAF film, prepared in 5mM monomer, had a large peak separation current and potential than those prepared in 1mM, 10mM and 15mM. This is because, at lower monomer concentration the active surface area of the working electrode cannot be covered by PAF at fixed number of cycle (cycle no) and at higher monomer concentration, the synthesized thick film results in higher resistance. The modified electrode offers stable and reproducible detection over a four weeks period with RSD 6.5%. The implemented modification method can be applicable to microelectrode biosensor development for simultaneous detection of AA, DA and CA within a flow cell.

REFERENCES

1. Mark Wightman R. (1988) Analytical Chemistry.60(13):769A-7.
2. Adams RN. (1976) Anal Chem. 48(14):1126A-38.
3. Damier P, Hirsch EC, Agid Y, Graybiel AM. (1999) Brain. 122(8):1437-48.
4. Arrigoni O, De Tullio MC. (2002) Biochim Biophys Acta Gen Subj. 1569(1-3):1-9.
5. Bijur GN, Ariza ME, Hitchcock CL, Williams MV. (1997) Environ Mol Mutagen. 30(3):339-45.
6. Shohami E, Beit-Yannai E, Horowitz M, Kohen R. (1997) Journal of Cerebral Blood Flow and Metabolism. 17(10):1007-19.
7. Koshiishi I, Imanari T. (1997) Anal Chem. 69(2):216-20.
8. Wang HY, Hui QS, Xu LX, Jiang JG, Sun Y. (2003) Anal Chim Acta. 497(1-2):93-9.
9. Hu Y, Li X, Pang Z. (2005) J Chromatogr A. 1091(1-2):194-8.
10. Dai X, Fang X, Zhang C, Xu R, Xu B. (2007) J Chromatogr B Anal Technol Biomed Life Sci. 857(2):287-95.
11. Fraisse L, Bonnet MC, De Farcy JP, Agut C, Dersigny D, Bayol A. (2002) Anal Biochem. 309(2):173-9.
12. Caussé E, Pradelles A, Dirat B, Negre-Salvayre A, Salvayre R, Couderc F. (2007) Electrophoresis. 28(3):381-7.
13. Stamford JA, Kruk ZL, Millar J. (1988) Brain Res. 454(1-2):282-8.
14. Zare HR, Nasirizadeh N, Mazloum Ardakani M. (2005) J Electroanal Chem. 577(1):25-33.
15. Kulagina NV, Shankar L, Michael AC. (1999) Anal Chem. 71(22):5093-100.
16. Suzuki A, Ivandini TA, Yoshimi K, Fujishima A, Oyama G, Nakazato T, et al. (2007) Anal Chem. 79(22):8608-15.
17. Lin L, Chen J, Yao H, Chen Y, Zheng Y, Lin X. (2008)Bioelectrochemistry. 73(1):11-7.
18. Dalmia A, Liu CC, Savinell RF. (1997) J Electroanal Chem. 430(1-2):205-14.
19. Malem F, Mandler D. (1993)Anal Chem. 65(1):37-41.
20. Gonon FG, Fombarlet CM, Buda MJ, Pujol JF. (1981) Anal Chem. 53(9):1386-9.
21. Wang CY, Wang ZX, Zhu AP, Hu XY. (2006) Sensors. 6(11):1523-36.
22. Viry L, Derré A, Poulin P, Kuhn A. (2010) Physical Chemistry Chemical Physics. 12(34):9993-5.
23. Ramanavičius A, Ramanavičiene A, Malinauskas A. (2006) Electrochim Acta. 51(27):6025-37.
24. Sangodkar H, Sukeerthi S, Srinivasa RS, Lal R, Contractor AQ. (1996)Anal Chem. 68(5):779-83.
25. Nien P-, Chen P-, Ho K(2009) Sens Actuators, B Chem. 140(1):58-64.
26. Chen S-, Liu M-, Kumar SA. (2007) Electroanalysis. 19(9):999-1007.
27. Lin K-, Chen S-. (2006) J Electrochem Soc. 153(5):D91-98.

Induced Transmembrane Voltage during Cell Electrofusion Using Nanosecond Electric Pulses

L. Rems, D. Miklavčič, and G. Pucihar

University of Ljubljana, Faculty of Electrical Engineering, Tržaška 25, SI-1000 Ljubljana, Slovenia

Abstract—Electrofusion is a method for facilitating fusion of cells in close contact by using microsecond-duration electric pulses. Electric pulses induce a voltage across cell membranes, which leads to membrane electroporation and brings the membranes into fusogenic state. However, electrofusion efficiency is very low when fusion partner cells considerably differ in size, since the magnitude of the induced transmembrane voltage (TMV) depends proportionally on the cell size when microsecond pulses are applied. Recently, we proposed that the problem of fusing differently sized cells could be overcome by simply reducing the pulse duration to nanoseconds (ns). Namely, during ns pulse exposure the TMV depends less on the cell size and more on the electric properties of the cells and the surrounding medium. To further investigate the possibility of fusing cells with ns pulses, we constructed a finite element model of equally and differently sized cells in contact, mimicking their arrangement during electrofusion. We calculated the time course of TMV for fusion media with two different conductivities (σ_e = 0.01 and 0.1 S/m), which are widely used in existing electrofusion protocols. Our results demonstrate that ns pulses provide the possibility to selectively electroporate the contact areas between cells (i.e. the target areas for electrofusion), regardless of the size of fusion partner cells, and for a relatively wide range of pulse durations. In medium with σ_e = 0.01 S/m, selective contact area electroporation can be achieved with pulses of up to few microsecond duration, whereas in medium with σ_e = 0.1 S/m, shorter pulses with duration below few hundred nanoseconds need to be applied. Electrofusion by means of ns pulses could, therefore, provide a method for improving fusion efficiency in cases where cells of different size need to be fused, such as in hybridoma technology for monoclonal antibody production.

Keywords—electrofusion, electroporation, induced transmembrane voltage, finite element model

I. INTRODUCTION

When a cell is exposed to an electric pulse, charges build on both sides of the cell membrane causing an increase in the transmembrane voltage (TMV). If TMV reaches a sufficiently high value (~1 V), structural rearrangement of the membrane lipid bilayer occurs (a phenomenon termed electroporation), which leads to substantial increase in the membrane permeability [1]. In the highly permeable, electroporated state, cell membranes are also fusogenic; provided that electroporated cells are in close contact, they can be fused [2]. Such method of cell fusion is known as electrofusion and is for example used in hybridoma technology for monoclonal antibody production [3] and in immunotherapy for production of cell vaccines [4]. In contrast to other fusion methods, no viral or chemical additives are required.

Conventionally, electrofusion is performed with 10–100 µs duration pulses, which ensures that cell membranes become fully charged during the exposure. Under such conditions, the magnitude of TMV (and consequently the extent of electroporation) is proportional to the cell size. This presents a challenge for fusing cells, which considerable differ in size, since large cells may not recover from application of electrical pulses required for electroporation of small cells. Ultimately, this leads to extremely low fusion yields. One typical example is hybridoma technology where small B lymphocytes are fused with large myeloma cells to form antibody-producing hybridoma cell lines [5].

Recently, a numerical study by Pucihar and Miklavčič suggested that the problem of fusing differently sized cells could be overcome by using shorter, nanosecond (ns) pulses [6]. Namely, in the nanosecond range, cell membranes are still in the charging phase and their TMV depends less on the cell size. Indeed, experiments on different cell lines demonstrated that cell size and shape play little role in electroporation with ns pulses [7].

In this paper we further investigate the possibility of fusing cells with ns pulses by performing calculations of TMV on a numerical model of two equally and differently sized cells in contact, during exposure to electric pulses. Since TMV in the nanosecond range significantly depends on the extracellular medium conductivity, we performed calculations for media with conductivities of 0.01 S/m and 0.1 S/m, which correspond to conductivities of fusion media widely used in existing electrofusion protocols [2,3,5]. Our results reveal important advantages of fusing cells with ns pulses, specifically in a low conductive medium (e.g. 0.01 S/m).

II. METHODS

Two-dimensional axisymmetric finite element model of cells in contact, mimicking the arrangement of cells during electrofusion, was constructed in Comsol Multiphysics 4.3 (Comsol, Burlington, MA, USA). Two spherical cells with either equal or different radii were placed in a rectangle representing the extracellular medium (Fig. 1). Cell nuclei were also included in the model, since during ns pulse exposure high electric field is also present in the cell interior and could potentially affect larger cell organelles [8].

The cells were exposed to electric field by assigning an electric potential to two opposite sides of the rectangle. The right side was grounded, whereas the left side was excited by a Heaviside function (Comsol functions *flc1hs*) with 1 ns rise time. Other boundary conditions are indicated in Fig. 1.

Electric potential in each subdomain of the model (extracellular medium, cytoplasm, nucleoplasm) was determined by equation:

$$-\nabla \cdot (\sigma_i \nabla V) - \nabla \cdot \frac{\partial (\varepsilon_i \nabla V)}{\partial t} = 0, \quad (1)$$

where σ_i and ε_i denote the conductivity and dielectric permittivity of a given subdomain, respectively. Membranes were included in the model by assigning a current density $\boldsymbol{J}$ through boundaries, representing the membranes [9]:

$$\boldsymbol{n} \cdot \boldsymbol{J} = \frac{\sigma_m}{d_m}(V - V_{ref}) + \frac{\varepsilon_m}{d_m}\left(\frac{\partial V}{\partial t} - \frac{\partial V_{ref}}{\partial t}\right) \quad (2)$$

Here, $\boldsymbol{n}$ is the unit vector normal to the boundary surface, V is the electric potential on the interior side of the boundary, V_{ref} is the potential on the exterior side of the boundary, and σ_m, ε_m, d_m, are the membrane conductivity, membrane dielectric permittivity, and membrane thickness, respectively. The TMV was then determined as the difference between electric potentials on each side of the boundary. The values of the model parameters are given in Table 1.

Segments of cell membranes, that formed the contact areas between cells, were assigned a thickness of two lipid membranes, as they account for part of the membrane of the left cell and part of the membrane of the right cell. We assumed that the TMV distributes equally between both membranes and that twice the TMV is required for electroporation of the contact area. For this reason, we present only half of the TMV calculated over the entire contact area in Fig. 2. The nuclear envelope was considered in the same way, as it consists of two lipid membranes.

III. RESULTS AND DISCUSSION

The model was used to calculate the time course of TMV on cells in contact after the onset of exposure to an external electric field. The TMV is presented for a time window ranging from 1 ns to 100 µs. Since the magnitude of TMV correlates with the extent of membrane electroporation,

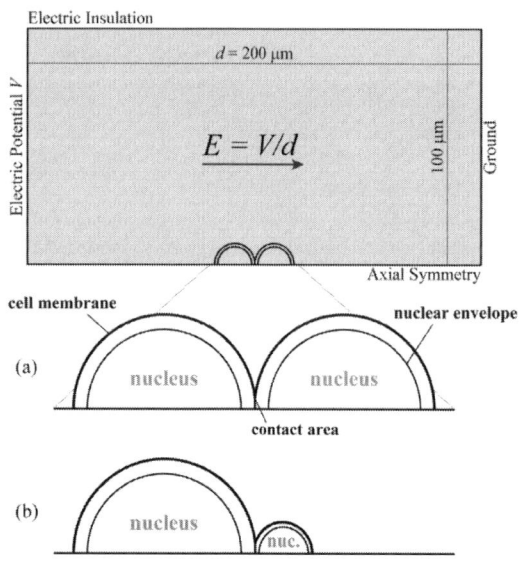

Fig. 1 Model of two cells in contact, exposed to electric field. (a) Two equally sized cells with radii of 9 µm. (b) Two differently sized cells with radii of 9 µm and 3 µm. The magnitude of the electric field was determined as the voltage difference between the left and the right side of the rectangle (the electrodes), divided by the electrode distance. The direction of the electric field is indicated with an arrow.

Table 1 Parameters of the model

Parameter	Symbol	Value
Cell radius	R_c	9 µm, 3 µm [§]
Nuclear radius	R_n	7.6 µm, 2.5 µm [‡]
Extracellular medium conductivity	σ_e	0.01 S/m, 0.1 S/m
Extracellular medium permittivity	ε_e	80 ε_0
Cytoplasmic conductivity	σ_{cp}	0.25 S/m
Cytoplasmic permittivity	ε_{cp}	70 ε_0
Cell membrane thickness	d_{cm}	5 nm
Cell membrane conductivity	σ_{cm}	$5 \cdot 10^{-7}$ S/m
Cell membrane permittivity	ε_{cm}	4.5 ε_0
Nucleoplasmic conductivity	σ_{np}	0.5 S/m
Nucleoplasmic permittivity	ε_{np}	70 ε_0
Nuclear envelope thickness	d_{ne}	10 nm
Nuclear envelope conductivity	σ_{ne}	$1 \cdot 10^{-4}$ S/m
Nuclear envelope permittivity	ε_{ne}	7 ε_0
Vacuum permittivity	ε_0	$8.85 \cdot 10^{-12}$ F/m

[§]Chosen to cover the size range of myeloma cells and B lymphocytes.
[‡]Nucleus occupies 60% of the cytoplasmic volume, as considered typical for lymphocyte cells [10].

this approach can be used to identify "primary targets" for electroporation with respect to the pulse duration. For example, if the highest TMV at time 100 ns is established on

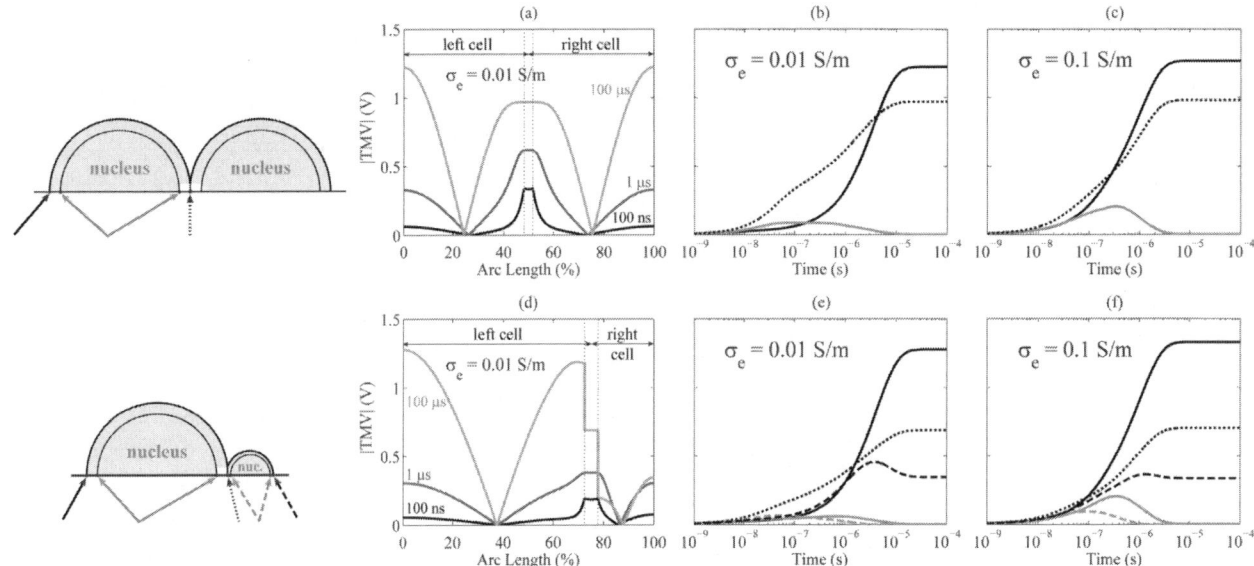

Fig. 2 Calculations of TMV on membranes of two equally sized cells with radii of 9 μm (a–c) and two differently sized cells with radii of 9 μm and 3 μm (d–f) after the onset of exposure to an electric field of 1 kV/cm. (a,d) TMV along cell membranes at times 100 ns, 1 μs, and 100 μs, in medium with conductivity 0.01 S/m. The contact areas are marked with vertical lines. (b,c) Calculated time courses of the absolute value of TMV on the pole of the left cell (black solid line), the point in the middle of the contact area (black dotted line) and the nuclear poles (grey solid line). The TMV on the nucleus was taken as the maximum TMV on either nuclear pole. These points are indicated by arrows under the image of the cell model. Calculations were performed for different extracellular medium conductivities, which are indicated in the top of graphs. (e,f) Calculated time courses of the absolute value of TMV on the pole of the large cell (black solid line), the pole of the small cell (black dashed line), the point in the middle of the contact area (black dotted line), the pole of the larger nucleus (grey solid line), and the pole of the smaller nucleus (grey dashed line). Calculations were performed in the same way as in (b,c).

the contact area between cells, this area can be selectively electroporated by applying a 100 ns pulse.

Results in Figs. 2a–c were obtained for a model of two equally sized cells with radii of 9 μm. Fig. 2a shows an example of the spatial distribution of TMV along the cell membranes, starting from the pole of the left cell and ending at the pole of the right cell, at times 100 ns, 1 μs, and 100 μs. The vertical dotted lines mark the contact area. The TMV on both cells is symmetrical (as cells are equal in size), and reaches the highest value either on the cell poles or on the contact area. The TMV along the nuclear membranes is not shown, but has a similar spatial distribution; the nuclear TMV is always the highest on the nuclear poles (the points where the membrane normal is parallel to the direction of the electric field).

Figs. 2b and c show the time course of the absolute value of TMV on the cell pole (black solid line), the point in the middle of the contact area (black dotted line), and the highest TMV on the nuclear poles (gray solid line). Note that the time is presented on a logarithmic scale as this allows one to study the TMV in the nanosecond and microsecond range simultaneously. Calculations were performed for media with two different conductivities (σ_e); 0.01 and 0.1 S/m. In medium with $\sigma_e = 0.01$ S/m (Fig. 2b) the TMV on the con-

tact area exceeds the TMV on the cell pole during the first 4 μs. In medium with $\sigma_e = 0.1$ S/m (Fig. 2c) the TMV on the contact area still exceeds the TMV on the cell pole, however, to a lesser extent and for a shorter period of time (up to 300 ns). The TMV on the nucleus remains below the TMV on the contact area for the entire time considered.

Figs. 2d–f show similar calculations as Figs. 2a–c, however, they were performed for two differently sized cells with radii of 9 μm and 3 μm. Fig. 2d demonstrates that the TMV, which establishes in the steady state (100 μs), is considerably higher on the large cell than on the small cell and also on the contact area. This indicates the difficulty of fusing differently sized cells with "classical" microseconds pulses; namely, applying pulses that would result in electroporation of the contact area, would at the same time cause extensive electroporation and possibly death of the large cell.

Figs. 2e and f present the time course of TMV on the pole of the large cell (black solid line), pole of the small cell (black dashed line), point in the middle of the contact area (black dotted line), larger nucleus (grey solid line), and smaller nucleus (grey dashed line). In medium with $\sigma_e = 0.01$ S/m the TMV on the contact area again exceeds the TMV on all other membranes for the first 1.5 μs of the exposure. The TMVs on both cell poles are quite similar for times below 1 μs; however, the TMV on the large cell afterwards substantially increases above the TMV on the small cell and also above the TMV on the contact area. In medium with $\sigma_e = 0.1$ S/m, the TMV on the contact area similarly exceeds the TMV on other membranes, but for a shorter time (only up to 100 ns).

The above results suggest that for a certain range of pulse durations selective electroporation of the contact areas between cells (i.e. the target areas for electrofusion) could be achieved. Most importantly, selective contact electroporation could be achieved regardless of the size of fusion partner cells. This presents the possibility for effectively fusing cells of different size without causing any damage to either large or small cells, since it is not expected that other membrane areas (apart from the contact area) would electroporate at all.

The range of pulse durations for which selective electroporation can be observed depends, however, on the extracellular medium conductivity. Our calculations demonstrate that by using medium with $\sigma_e = 0.01$ S/m, selective contact electroporation could be achieved with pulses of up to few μs duration, whereas by using medium with $\sigma_e = 0.1$ S/m, pulses would have to be approximately ten times shorter, i.e. on the order of 100 ns.

The conductivity of the extracellular medium significantly affects membrane charging dynamics, whereas its influence on the TMV becomes less important at steady state (when TMV reaches a constant value). In medium with lower conductivity, the charging of the cell membrane becomes slower. This can be used to explain why the TMV on the contact area exceeds the TMV on other membranes during the membrane charging process. The contact area is surrounded from both sides by relatively highly conductive cytoplasm (0.25 S/m), whereas the rest of the membrane is surrounded from one side with low conductive extracellular medium (0.01 or 0.1 S/m). Higher conductivity of the cytoplasm, therefore, causes charging of the contact area at a faster rate. This effect becomes more pronounced when the difference between the cytoplasmic and the extracellular conductivity is very large, i.e. in the case of medium with $\sigma_e = 0.01$ S/m.

In summary, our study demonstrates significant advantage of fusing cells with ns pulses, especially when fusion partner cells considerably differ in size. In contrast to conventionally used microseconds pulses, which primarily target large cells, nanosecond pulses provide the possibility to electroporate the contact areas between cells only. The conductivity of the fusion medium plays an important role in electrofusion with ns pulses, since it determines the range of pulse durations for which selective electroporation can be observed. In this aspect, using fusion medium with very low conductivity (0.01 S/m) could be potentially advantageous. In such medium, selective contact electroporation could be achieved even with pulses of up to few microseconds duration. Using longer pulses may additionally increase the possibility for successful fusion after electroporation, since both theoretical and experimental evidence suggest that the fusion process may already be initiated during the pulse [2], [11].

ACKNOWLEDGMENT

This work was supported by Slovenian Research Agency (ARRS). Research was conducted within the scope of the European Associated Laboratory for Pulsed Electric Field Applications in Biology and Medicine (LEA EBAM).

REFERENCES

1. Kotnik T, Pucihar G, Miklavčič D (2010) Induced transmembrane voltage and its correlation with electroporation-mediated molecular transport. J Membr Biol 236:3–13
2. Ušaj M, Flisar K, Miklavčič D et al. (2013) Electrofusion of B16-F1 and CHO cells: the comparison of the pulse first and contact first protocols. Bioelectrochemistry 89:34–41
3. Yu X, McGraw PA, House FS et al. (2008) An optimized electrofusion-based protocol for generating virus-specific human monoclonal antibodies. J Immunol Methods 336:142–151
4. Guo W, Guo Y, Tang S et al. (2008) Dendritic cell-Ewing's sarcoma cell hybrids enhance antitumor immunity. Clin Orthop 466: 2176–2183
5. Schmitt JJ, Zimmermann U, Neil GA (1989) Efficient generation of stable antibody forming hybridoma cells by electrofusion. Hybridoma 8:107–115
6. Pucihar G, Miklavčič D (2011) A numerical approach to investigate electrofusion of cells of different sizes, IFBME Proc. vol. 37, Budapest, Hungary, 2011, pp 1326–1329
7. Bowman AM, Nesin OM, Pakhomova ON et al. (2010) Analysis of plasma membrane integrity by fluorescent detection of Tl$^+$ uptake. J Membr Biol 236:15–26
8. Joshi RP, Schoenbach KH (2010) Bioelectric effects of intense ultra-short pulses. Crit Rev Biomed Eng 38:255–304
9. Pucihar G, Kotnik T, Valič B et al. (2006) Numerical determination of transmembrane voltage induced on irregularly shaped cells. Ann Biomed Eng 34:642–652
10. Polevaya Y, Ermolina I, Schlesinger M et al. (1999) Time domain dielectric spectroscopy study of human cells. II. Normal and malignant white blood cells. Biochim Biophys Acta 1419:257–271
11. Sugar IP, Förster W, Neumann E (1987) Model of cell electrofusion. Membrane electroporation, pore coalescence and percolation. Biophys Chem 26:321–335

Author: Lea Rems
Institute: University of Ljubljana, Faculty of Electrical Engineering
Street: Tržaška 25
City: SI-1000 Ljubljana
Country: Slovenia
Email: lea.rems@fe.uni-lj.si

Gene Transfer to Adherent Cells by *in situ* Electroporation with a Spiral Microelectrode Assembly

Tomás García Sánchez[1], Maria Guitart[2], Javier Rosell-Ferrer[1], Anna M. Gomez-Foix[2,3], and Ramon Bragós[1]

[1] Electronic and Biomedical Instrumentation Group, Department of Electronic Engineering, Universitat Politècnica de Catalunya, Barcelona, Spain
[2] Departament de Bioquímica i Biologia Molecular, IBUB, Universitat de Barcelona, Barcelona, Spain
[3] CIBER de Diabetes y Enfermedades Metabólicas (CIBERDEM), Barcelona, Spain

Abstract—In this study, *in situ* electroporation is applied to adherent cell monolayers growing on standard multiwell plates. A new microelectrode assembly based on spiral geometry and fabricated using standard PCB tecnology with slight modifications is used. The system was tested in the electroporation of two different cell lines (CHO and HEK 293). A fluorescent probe was initially used to test the extent of permeabilization and adjust experimental conditions. Subsequently, plasmid DNA encoding green fluorescent protein (GFP) was transfered by electroporation. Together with these experiments, cell viability was studied. We show for permeabilization experiments up to 70 % of fluorescent cells detected in both cells lines. Successful gene electrotransfer was obtained with more than 9 % and 15 % for CHO and HEK 293 respectively. In this work we prove how our device can be used for electroporation of adherent cells under the same standard laboratory conditions as regular biochemical treatments.

Keywords— Electroporation, spiral, microelectrodes, adherent cells, gene electrotransfer.

I. INTRODUCTION

Electroporation is an phenomenon related with a state of increased permeability to different molecules in the plasma membrane of biological structures when high electric field pulses are applied to the sample under treatment [1]. The technique is used for the introduction to the cell cytoplasm of different compounds; small molecules such as certain drugs or macromolecules such as naked DNAs [2]. In the recent past decades gene electrotransfer has become a standard method in the transfection of a wide variety of eukaryotic cell lines [3]. Functional gene therapy by electroporation has been applied not only using naked DNA plasmids but also siRNAs, mRNAs and other active compounds [4][5]. The main distinctive property of this technique is the fact that electroporation is based only on physical phenomena preventing the use of chemical agents.

Most common *in vitro* equipment perform electroporation with cells suspended in cuvettes, these cuvettes include two planar big electrodes positioned on both sides with an electrode gap of various milimiters. Nevertheless, there are some reasons that support the suitability of *in situ* electroporation of adherent cells. The term *in situ* used here is referred to the application of electroporation electric field pulses with adherent cells attached to the surface of standard tissue culture plates. Trypsinization procedure usually performed for detachment of cell monolayers is eliminated, thus preserving the integrity of surface proteins and avoiding alterations in the normal function of cells [6].

Among the previous systems applying electroporation directly to adherent cell monolayers, there are two main approaches; the first one uses big electrodes positioned above the cell monolayer and the second one consisting of microelectrode structures patterned at the bottom of the culture plates where cells are grown on top of these electrodes. The main advantage of the second approach is that the use of microelectrodes requires lower voltages applied to reach high electric field intensities, thus leading to simpler and cheaper pulse generators as well as reduced safety issues. On the other hand, one of the main drawbacks of this systems is the fact that cells need to be cultured as a monolayer on top of these microelectrodes and not on standard culture surfaces leading to biomedical user distrust. With the proposed system, we can apply electroporation using a microelectrode structure with cells growing on the surface of standard culture plates. In this sense, we have previously reported the feasibility of successfully apply electroporation on adherent cell cultures growing on the surface of standard multiwell plates using a set of interdigitated microelectrodes with short interelectrode distances [7]. In this study a new and improved electrode assembly is presented and applied on the electroporation of two different adherent cell lines.

II. MATERIALS AND METHODS

A. Spiral Microelectrode Assembly

The electrode assembly consisted of six equally spaced spiral lines coiled in parallel. Dimensions of the lines were 75 μm width and 150 μm spacing. Printed Circuit Board (PCB) technology was used to pattern the microelectrodes in a circular substrate with dimensions suitable for use within 24 multiwell culture plates. Electrodes were fabricated with copper and then covered with a final Nickel(Ni)/Gold(Au) coating to ensure non cytotoxicity. The complete fabrication process applied to a different electrode geometry is described in detail in [8]. The electrode assembly is positioned automatically above the cell monolayer of each individual well avoiding contact with cells by means of microseparators (10 μm) patterned on the surface of the microelectrodes. This design allows to create a highly uniform electric field on the surface of the culture plates were cells are growing.

This structure is specially conceived to apply the electroporaration pulses to the sample under treatment and also, with the same set of microelectrodes, to perform electrical bioimpedance measurements with a four-electrode configuration on the whole cell monolayer during the electroporation procedure. The information extracted from these measurements show the dynamics of pore formation and resealing and can be used to adjust the electric field parameters for each cell line and each specific molecular compound used [9]. In Fig. 1 a model of the geometry is depicted, both measurement and electroporation connections are shown. As indicated in the figure, the use of this spiral geometry enables the possibility of measuring in a four-electrode configuration but using duplicated electrodes for voltage measurements in two different positions. This explanation is given only in order to help the reader to understand the desing concept behind this spiral geometry, nevertheless, in this study only electroporation results are shown.

B. Cell Culture and Chemicals

HEK 293 human embryonic kidney cell line was cultured as a monolayer in Dulbeccos modified Eagle medium (DMEM) supplemented with 10 % fetal bovine serum (FBS) and supplemented with 1 % penicillin, streptomycin and fungizone (PSF). Chinese hamster ovary (CHO) cell line was cultured in DMEM/Ham's F12 with glutamine medium (PAA: The Cell Culture Company).

Two different low-conductivity electroporation buffers were used in the experiments. Low-conductivity electroporation buffer I (LCEB-I) was used with CHO cell line. LCEB-I consisted of 10 mM Na_2HPO_4 (pH 7.4), 1 mM $MgCl_2$ and 250 mM sucrose [10]. Conductivity was 1.6 mS/cm. The second buffer was low-conductivity electroporation buffer II (LCEB-II) and its composition was: 300 mM sorbitol, 4.2 mM KH_2PO_4, 10.8 mM K_2HPO_4, 1 mM $MgCl_2$, and 2 mM HEPES, pH 7.20 [11]. LCEB-II was used with HEK 293 cell line. Conductivity of 4.35 mS/cm.

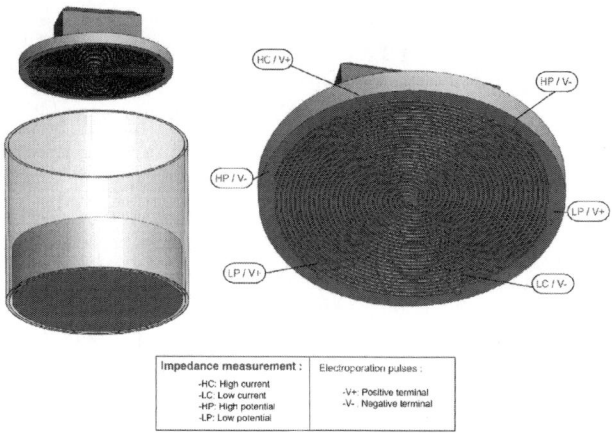

Fig. 1 Schematic representation of the microelectrode assembly and detailed explanation of the different connection configurations both for applying electroporation or for measuring bioimpedance in a four-electrode arrangement

For experiments using fluorescein isothiocyanate-dextran average molecular weight 20,000 Da (FD20S), FD20S (Sigma-Aldrich, Madrid, Spain) was added to the LCEB-I or LCEB-II buffers at a final concentration of 2.5 mg/ml. Plasmid electrotransfection experiments were done adding pGFP (pEGFP-N1) at a final concentration of 10 g/ml to the corresponding buffer.

C. Experimental Procedure

CHO cell line was plated at 7×10^4/well, HEK 293 cell line at 9×10^4/well in 24 multiwell plates at 37 °C in a humidified 5 % CO2 incubator for approximately 24 h, to reach 50–60 % confluence.

Before application of electric pulses, 200 μl of LCEB-I or LCEB-II supplemented with FD20S or pEGFP-N1 were added to each well. Electric field pulse delivery was performed using a self-constructed biphasic stimulator. The stimulator generates bipolar pulses acting as a current fixed current source. Eight biphasic pulses, with time separation between the positive and negative parts of the pulse of 100 μs and frequency repetition of 1 Hz were applied. In our case, and considering that our final objective is the introduction of DNA plasmids, the duration of electric field pulses was fixed at 1 ms. Different current intensities to create electric fields of 0.8, 1, 1.2, 1.6, 1.8 and 2 kV/cm were tested. These electric fields intensities were calculated measuring the voltage drop between adjacent electrodes with an oscilloscope and dividing this value by the distance between them (E=V/d). These intensities were measured in the surface of the electrodes but electric fields affecting the cell monolayer are lower due to the relative distance between electrodes and cells. This effect was studied by finite element simulations not shown in this work. Finally, in control cells no electric pulses were applied, but the electrode was positioned above the cell monolayer for an equivalent period of time. All experiments were held at room temperature ($\approx 20°$ C).

D. Results Assessment

In FD20S experiments, after the electroporation procedure, cells were incubated for an additional 30-min period in the incubator. After this period, cells were rinsed twice with PBS and replaced with fresh culture medium and left for 2 h in the incubator for complete resealing before results assessment. Then, cells were examined under a Leica (Wetzlar, Germany) DMI 4000B inverted microscope for fluorescence, to detect FD20S, which has an excitation wavelength of 485 nm and emission at 510 nm. Images were taken with a digital camera (Leica DFC 300x,10x magnification).

In plasmid electroporation experiments, 1 ml of corresponding fresh culture medium was added to the 200 μl of LCEB 10 min after the electroporation and cells were left for 24 h in the incubator before gene expression was studied using flow cytometry in a Cytomics FC500 MPL flow cytometer (Beckman Coulter, Inc, Fullerton, CA). Results were presented as % of green fluorescent cells with the total number of cells.

Additionally, viability experiments were done applying the same conditions as those of the plasmid electrotransfection experiments but in the absence of plasmid. Subsequent to 24 h electroporation treatment, 200 μl/well of 4 % formaldehyde was added. Following a 10 min incubation period at room temperature, the fixative was removed and cells were washed twice with phosphate-buffered saline (PBS). Afterward, cells were stained with Hoescht 1 μg/ml for 10 min in agitation. Cells were then washed with PBS for 10 min in agitation and finally cells were completely aspirated. Once dried, cells were examined under a Leica (Wetzlar, Germany) DMI 4000B inverted microscope for fluorescence. ImageJ automatic nucleus counter within the particle analysis plugin was used to process the images. Percentage of viable cells was calculated with respect to control were no electric pulses were applied.

III. RESULTS AND DISCUSSION

Fig. 2 shows representative micrographs of FD20s uptake in CHO and HEK 293 cells after eletroporation treatment. Permeabilization rates higher than 70 % were achieved for the best condition (1800 and 1600 V/cm respectively in CHO and HEK 293 cells). The use of this type of fluorescent

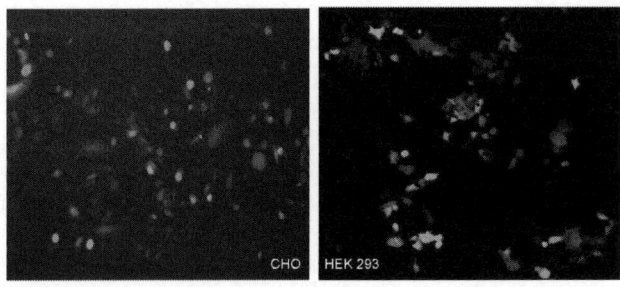

Fig. 2 Micrographs of FITC-dextran 20 kDa uptake after electroporation. Both cell lines were treated with 8 biphasic pulses,1 ms duration at 1 Hz frequency repetition. Electric field intensities were 1800 and 1600 V/cm respectively in CHO and HEK 293 cells

probles is useful as a first step for setting and optimizing the experimental conditions to be used in subsequent electroporation experiments where active molecules are used.

After this first step, gene electrotransfer was assayed in both cell lines. Results in Fig. 3 show, for each cell line, GFP expression levels 24 h after treatment corresponding to four different electric field intensities within the range tested and the corresponding viability for each electric field level.

Flow cytometry results show successful gene electrotransfer in both cell lines. In CHO cells maximum gene expression of 9 % was obtained with an electric field intensity of 1800 V/cm. The resulting viability for this electric field level was around 41 %. For HEK 293 cells up to 15 % of transfection was detected. In this case, viability results were around 80 % of viable cells. Although in the case of HEK 293 higher electric fields could have been applied according to viability results, we observed cell electrofusion when higher electric fields were used. To avoid cell fusion, contact between cells should be minimized.

Analyzing these results we can observe clear differences between permeabilization effect obtained using FITC-dextran, where a high percentage of cells were stained with the fluorescent probe, and results introducing GFP plasmid DNA. These differences have been already studied by other authors [10]. This observation is even more pronounced in adherent cells. The state of adherence is believed to affect in the introduction of plasmid DNA more than other molecules. One of the main reasons for this phenomenon is the different cellular cytosketelon conformation when cells are growing in adherence if compared with cells in suspension. A complex mechanism involving cellular cytoskeleton is responsible for DNA plasmid traslocation into the cell cytoplasm after membrane poration [12]. Some further research in this direction should be done in order to improve and optimize the results obtained in plasmid DNA experiments. Our results show the ability of the new electrode geometry to perform elecroporation to adherent cell monolayers.

Gene Transfer to Adherent Cells by *in situ* Electroporation with a Spiral Microelectrode Assembly

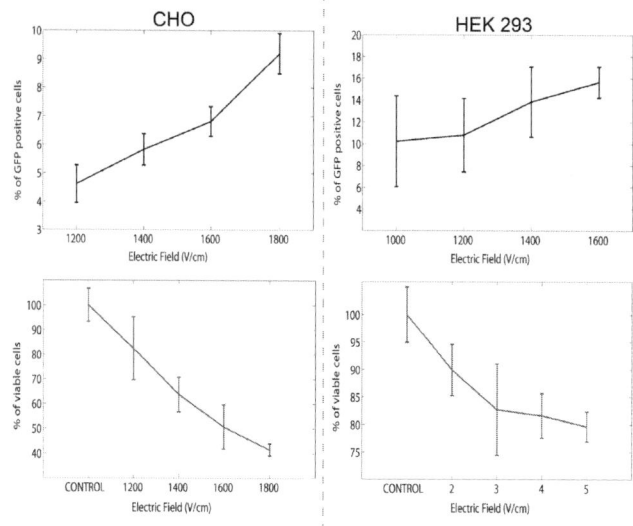

Fig. 3 GFP electrotransfection results 24 h after treatment assesed by flow cytometry are shown in the upper part of the figure. In the lower part, viability results corresponding with the same electroporation conditions are shown. 8 biphasic pulses, 1 ms duration and 1 Hz frequency repetition were applied at different electric field intensities. Results are presented as means±standard desviation of at least three different repetitions.

IV. CONCLUSION

In this study we presented and tested a new microelectrode assembly designed on the basis of two requirements, first using it for *in situ* electroporation of adherent cell monolayers and second, being suitable for performing electrical impedance measurements in a four-electrode configuration.

We can confirm the suitability of our system as a new tool for performing electroporation to adherent cell cultures generating a highly uniform electric field. The use of a microelectrode structure with reduced interelectrode distances implies the advantage of using low voltages and simplified eletric field pulse generation equipment. Also, the concept of *in situ* application enables to culture cells in standard multiwell plates helping in the standarization of the technique.

ACKNOWLEDGEMENTS

We would like to thank Anna Orozco and Alfonso Mendez for their unconditional assistance.

REFERENCES

1. Teissie J., Golzio M., Rols M. P.. Mechanisms of cell membrane electropermeabilization: A minireview of our present (lack of ?) knowledge *Biochimica et Biophysica Acta (BBA) - General Subjects.* 2005;1724:270-280. doi: DOI: 10.1016/j.bbagen.2005.05.006.
2. Teissie Marie-Pierre Rols, Justin . Electropermeabilization of Mammalian Cells to Macromolecules: Control by Pulse Duration *Biophysical Journal.* 1998;75:1415-1423.
3. Cegovnik Urska, Novakovic Srdjan. Setting optimal parameters for in vitro electrotransfection of B16F1, SA1, LPB, SCK, L929 and CHO cells using predefined exponentially decaying electric pulses *Bioelectrochemistry.* 2004;62:73-82. doi: DOI: 10.1016/j.bioelechem.2003.10.009.
4. Huang Huang, Wei Zewen, Huang Yuanyu, et al. An efficient and high-throughput electroporation microchip applicable for siRNA delivery *Lab on a Chip.* ;11:163-172.
5. Van Tendeloo Viggo F. I., Ponsaerts Peter, Lardon Filip, et al. Highly efficient gene delivery by mRNA electroporation in human hematopoietic cells: superiority to lipofection and passive pulsing of mRNA and to electroporation of plasmid cDNA for tumor antigen loading of dendritic cells *Blood.* 2001;98:49-56.
6. Huang Hsiang-Ling, Hsing Hsiang-Wei, Lai Tzu-Chia, et al. Trypsin-induced proteome alteration during cell subculture in mammalian cells *Journal of Biomedical Science.* 2010;17:36.
7. Garcia-Sanchez Tomas, Sanchez-Ortiz Beatriz, Vila Ingrid, et al. Design and Implementation of a Microelectrode Assembly for Use on Noncontact In Situ Electroporation of Adherent Cells *The Journal of Membrane Biology.* 2012;245:617-624. J Membrane Biol.
8. Garcia-Sanchez T., Guitart M., Rosell J., MaGomez-Foix A., Bragos R.. Automatic system for electroporation of adherent cells growing in standard multi-well plates in *Engineering in Medicine and Biology Society (EMBC), 2012 Annual International Conference of the IEEE:*2571-2574.
9. Joachim Wegener Charles R. Keese, Giaever Ivar. Recovery of Adherent Cells after In Situ Electroporation Monitored Electrically *BioTechniques.* 2002;33:348-357.
10. Marjanovi Igor, Haberl Saa, Miklavi Damijan, Kanduer Maa, Pavlin Mojca. Analysis and Comparison of Electrical Pulse Parameters for Gene Electrotransfer of Two Different Cell Lines *The Journal of Membrane Biology.* 2010;236:97-105.
11. De Vuyst Elke, De Bock Marijke, Decrock Elke, et al. In Situ Bipolar Electroporation for Localized Cell Loading with Reporter Dyes and Investigating Gap Junctional Coupling *Biophysical Journal.* 2008;94:469-479. doi: DOI: 10.1529/biophysj.107.109470.
12. Rols Marie-Pierre. *Mechanism by Which Electroporation Mediates DNA Migration and Entry into Cell and Targeted Tissues*:19-34. New Jersey: Humana Press 2008.

Author: Tomás García Sánchez
Institute: Electronic and Biomedical Instrumentation Group, Department
of Electronic Engineering, Universitat Politècnica de Catalunya
Street: Campus Nord, Edifici C4, Jordi Girona 1-3, 08034
City: Barcelona
Country: Spain
Email: tomas.garcia.sanchez@upc.edu

Part VII
Cardiovascular, Respiratory and Endocrine Systems Engineering

Low Contents of Polyunsaturated Fatty Acids in Cultured Rat Cardiomyocytes

D. Sato[1], T. Karimata[1], T. Wakatsuki[1], Z. Feng[2], A. Nishina[3], M. Kusunoki[4], and T. Nakamura[1]

[1] Department of Biomedical Information Engineering, Graduate School of Medical Science,
Yamagata University, Yamagata, Japan
[2] Department of Bio-Systems Engineering, Graduate School of Science and Engineering,
Yamagata University, Yonezawa, Japan
[3] Department of Materials and Applied Chemistry, College of Science and Technology,
Nihon University, Tokyo, Japan
[4] Department of Internal Medicine, Medical Clinic, Aichi Medical University, Nagoya, Japan

Abstract—Reconstructed myocardial tissue still does not have enough contractility. It is well known that fetal and neonatal cardiomyocytes utilize glucose and lipid, respectively, as their energy substrates, and that cultured ones mainly use glucose in spite that their age is comparable to neonate one, probably due to insufficient supply of lipids from culture medium. In the present study, we compared 7 saturated (SFA), 6 monounsaturated (MUFA) and 11 polyunsaturated (PUFA) fatty acid contents in cultured cardiomyocytes (Cul group) with those in fetal (Fet group, approximately 17 d after impregnation) and neonatal (Neo group, 9 d old) rats, where the age of Cul-group cells was set nearly equal to Neo-group one. SFA contents in Cul group were generally lower than those in Fet group, and were close to those in Neo group, except for C12:0 of which content was highest in Neo group. MUFA contents in Cul group were generally lower than those in Fet group but similar to or rather higher than those in Neo group, except for C24:1n-9 of which content was again highest in Neo group. In contrast, most of PUFA contents in Cul group appeared lower than those in both the Fet and Neo groups, and differences in 5 of 10 detected PUFAs were statistically significant between Cul and Neo groups. The results suggest that PUFA contents in cultured cardiomyocytes might be insufficient to exert enough contractile ability. In conclusion, cultured cardiomyocytes may need more lipid, PUFAs in particular.

Keywords—Cardiomyocytes, Fatty acid composition, Energy substrate, Cardiac tissue engineering

I. INTRODUCTION

Some cell sources, such as autologous myoblast and/or bone marrow cells, have already been used clinically to treat ischemic hearts [1]. In those procedures, cell suspensions are often injected into damaged myocardium via coronary arterial or intraventricular approaches. However, the direct injection of dissociated cells has crucial difficulties in migration, resulting in limited efficacy [2].

For cardiac tissue equivalent (CTE), i.e., cardiac tissue reconstructed with cultured cardiomyocytes, is one of the promising alternative methods for salvaging severely damaged myocardium. One well-recognized problem, however, is that the maximal twitch stress CTE can generate is only 1/10 of that in animals and humans. One of the major causes may include difference in metabolism between cultured and natural cardiomyocites [3].

In the present study, we focus on the lipid metabolism in cardiomyocytes. Major energy substrates in the fetal and mature myocardium are known to be glucose and lipids, respectively, whereas cultured cardiomyocytes mainly utilize glucose [3] in spite that their age is close to that in mature, neonatal myocardium.

In general, the culture medium is composed of a variety of water-soluble substances, and fat-soluble ones are rarely taken into account. Lipids provided in culture medium are only a little amount of palmitic, oleic and stearic acids contained in supplemental fetal bovine serum (FBS). Due to the considerable shortage of lipid source in the culture medium, cultured cardiomyocytes might be forced to use glucose as an alternative energy substance. That is, energy source supplied to cultured myocardium may be unnatural.

In addition, polyunsaturated fatty acids (PUFAs) are further important not only as energy substrates but also as ligands in some cytokine functions [4]. Thus, the shortage of PUFAs might result in the impairment of some important cell functions.

In the present study, we compare fatty acid contents in cultured cardiomyocytes with natural myocardium obtained from fetal and neonatal rats, and determine the difference among them.

II. MATERIALS AND METHODS

A. Sample Preparation

All of surgical and experimental procedures described below followed the Guide for the Care and Use of Laboratory Animals of the National Institutes of Health, USA, and were approved by the Yamagata University Animal Research Committee (no. 23067).

Fetal (Fet) group: 17 ± 1 days after impregnation, pregnant rats of approximately 14 weeks of age were anesthetized, and the jugular vein and carotid artery were cut. Fetal rats were taken out at laparotomy, and after lumber fracture, their ventricles were harvested.

Culture (Cul) group: the ventricles of fetal rats were harvested as shown above and washed with PBS supplemented with 0.3 μg/mL insulin and 10 μg/mL heparin sodium. They were cut into pieces and digested at 37 °C with 0.1 % collagenase type I. Cardiomyocytes were purified with a 2-layer (1.082 and 1.059 g/ml) concentration gradient of Percoll (Sigma-Aldrich, St. Louis, MO, USA). After centrifugation at 1600 ×g for 30 min, cardiomyocytes congregating in between the 2 layers was collected and resuspended in culture medium.

Cardiomyocytes were then incubated at 37 °C for 14 days. Culture medium was changed 24 hours later, and every day thereafter. Cardiomyocytes were usually beating spontaneously throughout the culture period. On the 14th day, cardiomyocytes were harvested by scraping them.

Neonatal (Neo) group: neonatal rats of 9 days of age were anesthetized. After the jugular vein and carotid artery were cut, the ventricle was obtained.

B. Fatty Acids Determination

To extract the total lipid, each sample was homogenized in PBS, and the homogenate was suspended in chloroform and methanol mixture (2/1 v/v). After centrifugation, total fatty acid in the lower chloroform layer was collected, and methylated to obtain fatty acid methyl esters (FAMEs). Twenty-four FAMEs were determined with a gas chromatography system.

III. RESULTS

Figs. 1-3 summarize contents of 7 saturated (SFA), 6 monounsaturated (MUFA), and 10 polyunsaturated (PUFA) fatty acids, respectively. Another fatty acid, γ-linolenic acid (C18:3n-6), was not detected in any samples.

Main SFAs obtained were myristic (C14:0), palmitic (C16:0), and stearic (C18:0) acids, and their relationships between groups were very similar to each other (Fig. 1): contents in Fet group were maximal, followed by those in Neo group, and Cul-group SFAs were the lowest. Only C18:0 content in Cul group was significantly lower than that in Neo group ($P<0.05$). The relationship between groups was similar in other small-content SFAs except for lauric acid (C12:0). C12:0 contents in Neo group seemed higher than that in the other groups although any significant difference was not detected.

In MUFAs (Fig. 2), main FAs (palmitoleic (C16:1n-7) and oleic (C18:1n-9) acids) showed similar relationships between groups again, but their order was different from that in SFAs: that is, Fet-group FAs were maximal followed by Cul-group ones, and Neo-group MUFAs were the lowest. Both the C16:1n-7 and C18:1n-9 in Cul group were significantly higher than those in Neo one ($P<0.01$). C16:1n-7 content in Fet group was significantly higher than that in Neo group. Other MUFA contents in Neo group were generally lower than those in Fet and Cul groups, except for nervonic acid (C24:1n-9) where the content in Neo group was the highest ($P<0.01$). Erucic acid (C22:1n-9) content was significantly higher in Fet group than in Neo one ($P<0.01$).

In contrast to the general aspects seen in SFA and MUFA contents, main PUFA contents (linoleic (C18:2n-6), arachidonic (C20:4n-6), docosatetraenoic (C22:4n-6), and docosahexanoic (C22:6n-3) acids) in Cul group were considerably lower than those in Neo group ($P<0.01$) (Fig. 3). In addition, 5-8-11 eicosatrienoic (C20:3n-9) content also was significantly lower than that in Neo group ($P<0.05$). Some PUFAs in Fet group were higher (C22:5n-3, $P<0.05$) or

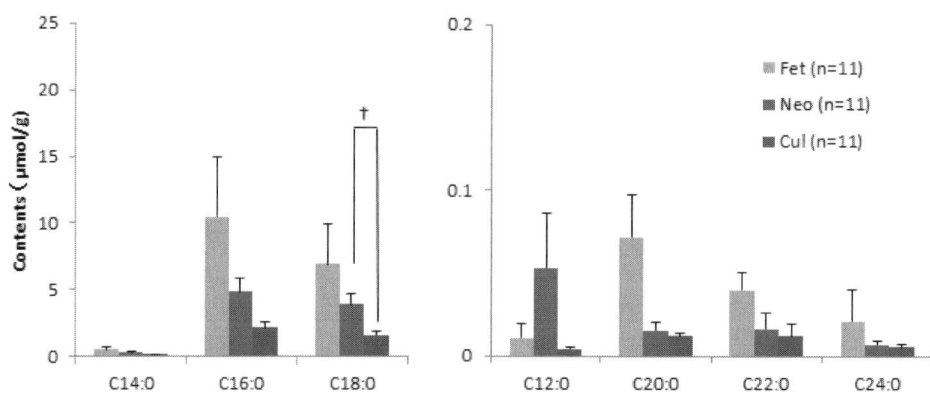

Fig. 1 Saturated fatty acids content. Data are expressed in mean ± SE. †$P<0.05$ (Kruskal-Wallis test).

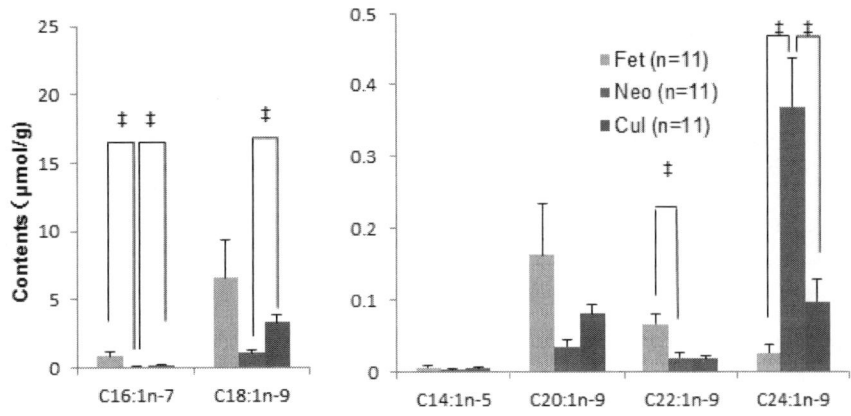

Fig. 2 Monounsaturated fatty acid content. Data are expressed in mean ± SE. ‡$P<0.01$ (Kruskal-Wallis test).

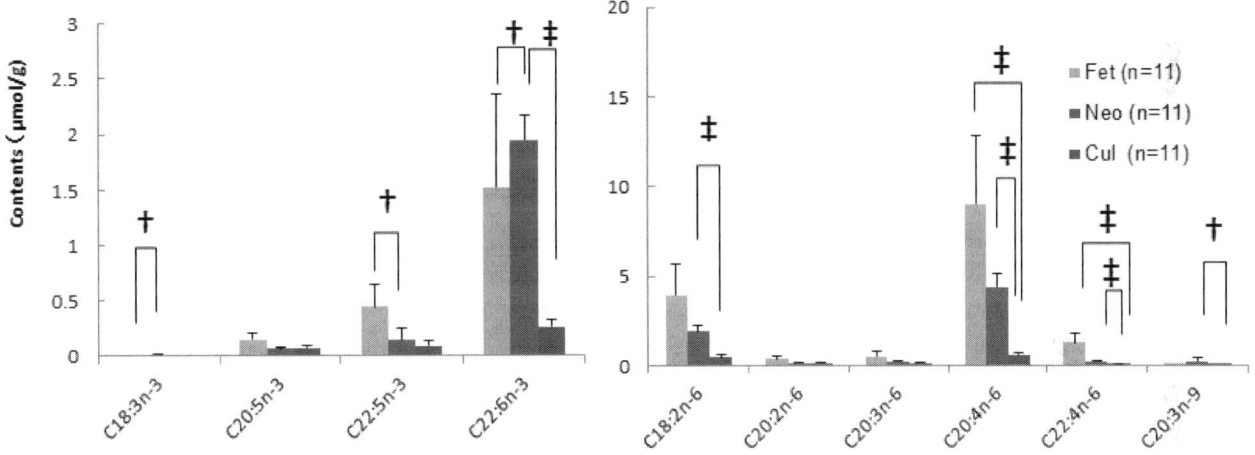

Fig. 3 Polyunsaturated fatty acid content. Data are expressed in mean ± SE. ‡$P<0.01$, †$P<0.05$ (Kruskal-Wallis test).

lower (C18:3n-3, $P<0.05$; C22:6n-3, $P<0.05$) than those in Neo group, and higher than those in Cul group (C20:4n-6 and C22:4n-6, $P<0.01$).

IV. DISCUSSION

In the present study, fatty acid contents in cultured cardiomyocytes were compared with those in fetal and neonatal myocardium. In the cultured cardiomyocytes, spontaneous beating was observed during the culture period, which may indicate that the cardiomyocytes were kept in good condition in culture.

FA contents in Neo and Cul groups were generally lower than those in Fet one. Lipids are the major energy substrate of the myocardium in adults, while glucose in fetus and cultured cardiomyocytes [3]. Therefore, the results obtained between Fet and Neo groups could be attributed to the difference in main energy source of cardiomyocytes. That is, because not lipid but glucose is mainly consumed in fetus, lipid could be left more in fetal cardiomyocytes than in neonatal ones where lipids are consumed much instead.

In case of the difference observed between fetal and cultured cardiomyocytes, the story may be totally different because cultured cardiomyocytes could not use lipids in spite that they are as old as neonatal ones. Obviously, lipid supply to the cardiomyocytes was not enough in Cul group, and it could be the main cause that the cardiomyocytes in culture are forced to use glucose instead of lipid as the main energy source. If the lipid supply were sufficient, much fatty acids might be accumulated in cultured cardiomyocytes, as seen in fetal myocardium, without consumption. Or, the candiomyocytes might change their main energy source from glucose to lipids.

SFA contents in Cul group were generally lower than those in Fet group, and not significantly lower but rather close to those in Neo group (Fig. 1), except for stearic (C18:0) acid. The results may suggest that SFAs in cultured cardiomyocytes could not be deficient much probably because cardiomyocytes can synthesize SFAs from materials in the medium.

Interestingly, MUFA contents in Cul group, in general, were rather higher than those in Neo group (Fig. 2), where the difference in 2 main MUFA (palmitoleic (C16:1n-7) and oleic (C18:1n-9) acids) contents were significant ($p < 0.05$). Although the implication of the results is not clear, they at least indicate that MUFA could sufficiently be equipped in cultured cardiomyocytes, or that MUFAs may not be consumed much. Anyhow, MUFAs again could be synthesized from medium materials, when necessary.

In contrast, PUFA contents in Cul group were lower than those in Neo group (Fig. 3). In n-6 PUFAs, the differences in contents of linoleic (C18:2n-6), arachidonic (C20:4n-6), and docosatetraenoic (C22:4n-6) acids were significant ($p < 0.01$). n-6 PUFA may be very important for cardiomyocytes because they are known to reduce the risks of coronary and cardiovascular diseases. In particular, C18:2n-6 is one of the major fatty acids lowering the low-density lipoprotein cholesterol (LDL-C) metabolism by down-regulating LDL-C production and enhancing its clearance [5].

C20:3n-9 acts as an analogue of C20:4n-6 when n-6 PUFAs are insufficient in human. However, C20:3n-9 content in Cul group was still left lower than that in Neo one in spite that n-6 PUFAs content in Cul group was lower, too. The cause of this discrepancy was not clear in the present study.

n-3 PUFA contents in Cul group were generally lower than those in Neo group, where the difference in C22:6n-3 was significant ($p < 0.01$). n-3 PUFAs have a role in protecting cardiomyocytes [6], and consequently benefit contractile properties of cardiomyocytes [7]. Thus, n-3 PUFA might play some role in changing metabolic mode in cultured cardiomyocytes.

As stated above, low PUFA contents in Cul group were probably due to the shortage of their supply from culture medium. In particular, since most of PUFAs cannot be synthesized in cardiomyocytes, isufficient PUFAs could have considerably affected the cell function. Therefore, more PUFAs should be supplied in medium to be taken up by cardiomyocytes to improve their function.

V. CONCLUSIONS

Insufficient PUFAs in cultured rat cardiomyocytes may result in low CTE function whereas SFAs and MUFAs would be synthesized from water-soluble material in the culture medium. To improve CTE function, lipids, especially PUFAs, should be supplied more to the medium for uptake in cardiomyocytes in culture.

ACKNOWLEDGMENT

The present study was supported financially in part by Grants-in-Aid for Scientific Research (C) (23500539) from the Japan Society for the Promotion of Science.

REFERENCES

1. Fujita T, Sakaguchi T, Miyagawa S et al (2011) Clinical impact of combined transplantation of autologous skeletal myoblasts and bone marrow mononuclear cells in patients with severely deteriorated ischemic cardiomyopathy. Surg Today 41:1029–1036
2. Takehara N, Matsubara H (2011) Cardiac regeneration therapy: connections to cardiac physiology. Am J Physiol 301:H2169–H2180
3. Rajabi M, Kassiotis C, Razeqhi P et al (2007) Return to the fetal gene program protects the stressed heart: a strong hypothesis. Heart Fail Rev 12:331–343
4. Schmitz G, Ecker J (2008) The opposing effects of n-3 and n-6 fatty acids. Prog Lipid Res 47:147–155
5. Wijendran V, Hayes KC (2004) Dietary n-6 and n-3 fatty acid balance and cardiovascular health. Annu Rev Nutr 24:597–615
6. Leroy C, Tricot S, Lacour B et al (2008) Protective effect of eicosapentaenoic acid on palmitate-induced apoptosis in neonatal cardiomyocutes. Biochim Biophys Acta 1781:685–693
7. Szentandrássy N, Pérez-Bido MR, Alonzo E et al (2007) Protein kinase A is activated by the n-3 polyunsaturated fatty acid eicosapentaenoic acid in rat ventricular muscle. J Physiol 582:349–358

Corresponding author:

Author: Daisuke Sato
Institute: Department of Biomedical Information Engineering, Graduate School of Medical Science, Yamagata University
Street: 2-2-2 Iida-nishi
City: Yamagata
Country: Japan
Email: d_sato@yz.yamagata-u.ac.jp

Classification of Chronic Obstructive Pulmonary Disease (COPD) Using Integrated Software Suite

Almir Badnjevic[1], Mario Cifrek[2], and Dragan Koruga[3]

[1] New Technology d.o.o., Sarajevo, Bosnia and Herzegovina
[2] University of Zagreb, Faculty of Electrical Engineering and Computing, Zagreb, Croatia
[3] Army Medical Center Novi Sad, Petrovaradin, Serbia

Abstract—Chronic Obstructive Pulmonary Disease (COPD) is a respiratory disorder characterized by chronic and recurrent airflow obstruction, which increases airway resistance. About 75% of COPD patients do not have established diagnosis, most of them in mild degree, but also 4% in severe and 1% in very severe degree of COPD. The reason for that are slow progression of symptoms as cough and exercise intolerance, as well as development of disease in the elderly. Integrated software suite is developed to assist clinicians in the analysis and interpretation of pulmonary function tests data to better detect, diagnose and treat COPD conditions. A total sum of 385 patient reports with previously diagnosed COPD or normal lung conditions by clinicians was tested with this tool. With diagnosed COPD by clinicians there were 252 patients, and even in 92% the software has performed the classification of COPD in the same way as doctors. The software classification of patients with normal lung function was 90.97%.

Keywords—COPD, diagnosis, integrated software suite, fuzzy logic, neural network.

I. INTRODUCTION

In normal individuals, breathing takes minimal effort from the body but the pulmonary system is sensitive to multiple agents that can trigger reactions or diseases. The most disabling disorders are chronic obstructive pulmonary diseases (COPD), asthma, lung cancer and restrictive lung diseases.

COPD is a respiratory disorder characterized by chronic and recurrent airflow obstruction, which increases airway resistance [1]. The most COPD is caused by long-term smoking [2]. Two main examples of them are obstructive emphysema and chronic bronchitis. About 75% of COPD patients do not have established diagnosis, most of them in mild degree, but also 4% in severe and 1% in very severe degree of COPD [3]. The reasons for that are slow progression of symptoms as cough and exercise intolerance, as well as development of disease in elderly. It is estimated that in Croatia COPD affects up to 10% of the adult population. Approximately 200.000 to 300.000 people die due to COPD in Europe every year [4], more than from lung cancer and breast together [5], [6]. By mortality the COPD is on 4th place in the world, after myocardial infarction, malignant diseases and cerebrovascular insults. There are around 600 million patients with COPD today in the world, double than diabetics. Predicted to 2020, COPD will become the world's third biggest „killer" [7], and thus the main growing public health problem.

Pulmonary function tests the most commonly used to detect COPD are spirometry and Impulse Oscillometry System (IOS). Spirometry as a test of lung function is most commonly used in the diagnosis of chronic obstructive pulmonary disease and asthma. Spirometry is running by measuring forcing maneuver [8].

In contrast to forced spirometry, the forced oscillation technique (FOT) superimposes small air pressure perturbations on the natural breathing of a subject to measure lung mechanical parameters. IOS measures respiratory impedance using short pulses of air pressure [9,10].

The most successful diagnosis is achieved by combination of IOS and spirometry. In this way we obtain a static assessment of the patient. In order to get patient's dynamic assessment of pulmonary function, it is also necessary to take into account the patient's symptoms and allergies, perform auscultation of patient and to do bronchial dilation (BDT) and bronchial provocation tests (BPT). BPT is not necessary for COPD, but for diagnose of other respiratory disease is very helpful.

The actual benefit of the integrated software suite is to help the clinical utilization of impulse oscillometry, and at the same time to assist in the diagnosis of differentiation of Chronic Obstructive Pulmonary Disease (COPD) and asthma. Data which are first entered to software are related to a directory, symptoms and risk factors drawn from two major consensuses for COPD and asthma, Global Initiative for chronic Obstructive Lung Disease (GOLD) and Global Initiative for Asthma (GINA). The parameters of IOS, spirometry, information about the symptoms, allergies and auscultation of the patient, are included in the neuro-fuzzy system, in order to help the software to suggest proper classification of COPD, asthma, or normal lung condition. In the cases where it is not possible to determine the diagnosis the software will indicates to do BDT and/or BPT, after which new tests are required for IOS and spirometry, in order to get a complete diagnosis. This software increases the percentage of correct interpretation of data obtained by measuring tests of impulse oscillometry and spirometry, and friendly guides the user to quickly obtain data to be used in evaluation of lung function.

II. INTEGRATED SOFTWARE SUITE

Integrated software suite described in this chapter is developed as interactive and user-friendly software. The software is using data from IOS and spirometry measurements as inputs. To get the clinical assessment of lung function with this software, user needed to input information's such as symptoms and allergies for COPD and normal subjects. This was achievable by utilizing the object oriented design methodologies, model based feature extraction and an inference system that incorporates the strengths of artificial neural networks and fuzzy decision making. The programming language to develop this software was Microsoft Visual C# Express. The software uses .NET technology to deploy applications, while in the background calls MATLAB functions for the main applications through MATLAB's .NET interface. The software has a feature of automatically reading and parsing standard format for IOS, Spirometry (SP) and Body Plethysmography (BP) recorded in the .pdf files.

Architecture for disease recognition is presented in Figure 1. After the software reads the data from .pdf files each of the reports ceptualise in fuzzy system and performed a preliminary classification of diseases on the basis of individual report.

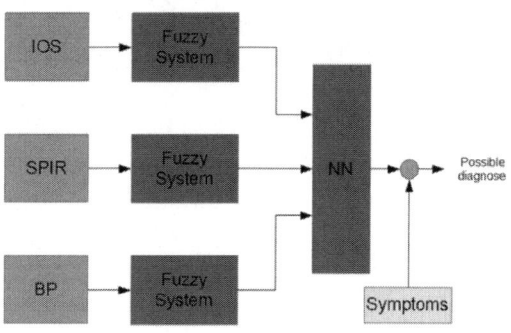

Fig. 1 Architecture for disease recognition

Fuzzy system is designed according to the instructions and experience of experts in the field of medicine and equipment manufacturers for diagnostics of respiratory disease. Example of the implemented fuzzy reasoning for the case of IOS is shown in Figure 2.

The results of the analysis and classification based on the reports of IOS, SP and BP are the input vector of the neural network. Basic test does not take into account the report of BP, because it is not required in the initial test. In the case of BDT or BPT, the result of the report is one of the inputs into the neural network for the next test.

For BP test there is added another input as the result of BDT's (positive or negative) and it is involved in the classification of disease after a period of BDT's.

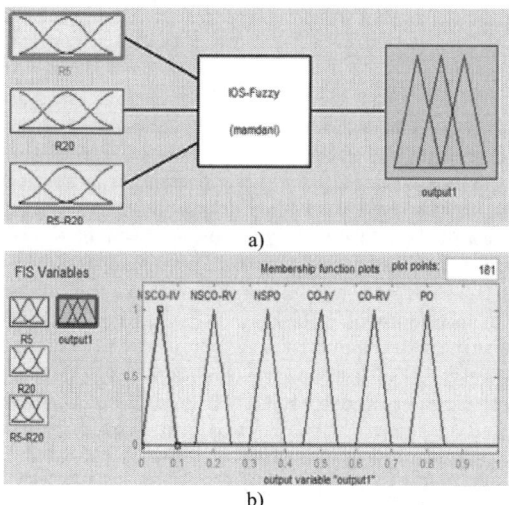

Fig. 2 a) Implemented fuzzy reasoning for IOS, b) Output variables

The basic architecture of the neural network is shown in Figure 3. It is a linear feed forward network that is due to the application expert enough to properly perform the separation, and do not need complex architecture of network due to the presence of additional fuzzy classification system. In our case, due to the small quantity of data and based on the rules of identification systems, it was not necessary for a partition of data on the estimation and validation. All data being used for training, while evaluation is performed by the experts (doctors).

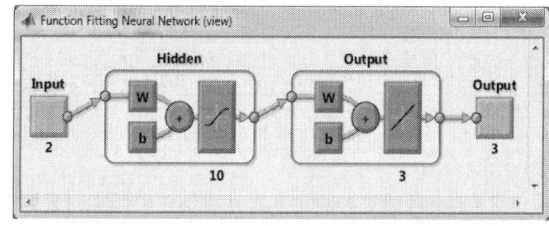

Fig. 3 Basis architecture of implemented neural network

Initial window of integrated software suite is presented in Figure 4. The software has been realized by the actual needs of physicians. Therefore, first the user needs to enter the Patient Info such as symptoms, allergies and perform auscultation of patient. Example of a Patient Info form is shown in Figure 5.

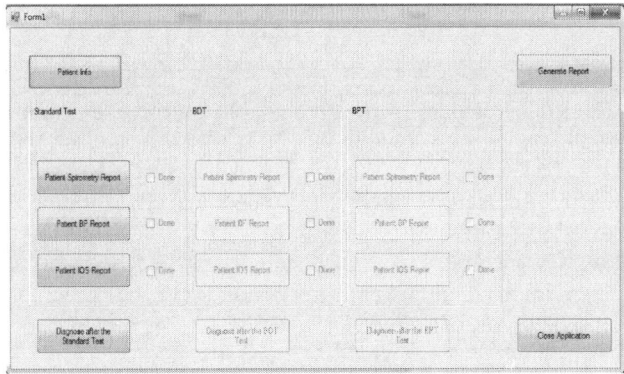

Fig. 4 Initial window of integrated software suite

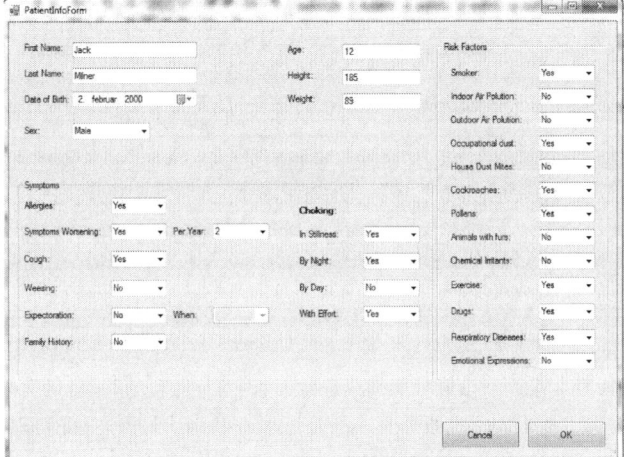

Fig. 5 Example of Patient Info form

Once you have entered information from the Patient Info form, it is possible to access the uploading measurement results of IOS and spirometry. If after the first measuring results software can't classify the disease, it suggests to approach BDT and/or BPT, and then user receive a final classification of diseases.

III. MATERIALS AND METHODS

We enrolled 385 patients who previously visited departments for lung diseases in Clinical Centre University of Sarajevo, Bosnia and Herzegovina and Army Medical Center Novi Sad, Serbia.

In this study, first all patients were asked by physician to respond to questions regarding symptoms, allergies and risk factors of COPD. Second step was measuring lung function with spirometry and IOS. Using Impulse oscillatory system (IOS), total respiratory resistance R5, proximal resistance R20, distal capacitive reactance X5 and resonant frequency Fres, were measured. Using spirometry, forced vital capacity (FVC) and forced expiratory volume in one second FEV1 were measured, while ratio FEV1/FVC was calculated. Patient's demographic variables, history and physical examination were obtained by a trained physician. Subjects were divided into 2 groups of healthy and obstructive lung disease (COPD) based on their medical history, physical examination and GOLD guidelines.

COPD was diagnosed by having a positive history of dyspnea (progressive, exertion or persistent), chronic cough (may be intermittent or non-productive), chronic sputum and history of exposure to tobacco, occupational dusts, chemicals or other smokes after the age of 40. The diagnosis of COPD was confirmed by having a positive history, R5>150%, X5>0.15 and FEV1/FVC<0.7. Patients with these test results were evaluated for bronchial hyper reactivity (BDR) throw BDT, to check in which group of COPD they belong (A, B, C or D) [11]. The bronchial dilation test was performed according to international guidelines using 400 μg of salbutamol. After that, physicians were able to give final diagnosis. Total sum of 385 patients who were enrolled in this study, 252 (65.45%) were grouped as COPD, while 133 (34.55%) were grouped as healthy controls. Integrated software suite was used for both groups of subjects.

IV. RESULTS AND DISCUSSION

Results of this study showed a high efficiency of integrated software suite in the classification of COPD disease (92%) and healthy patients (90.97%). In Table 1 are expressed percentage of hits and misses of the integrated software suite in the classification of COPD and healthy patients who were involved in this study. From table 1 we can see that in the case of patients affected with COPD the software showed a small percentage of false positive results (8%) as compared to physician diagnosis. In these 20 cases, the software classified those patients with other forms of respiratory diseases, not as COPD. In the case of healthy patients the software also showed a small percentage of false positive results (9.03%) as compared to physician's diagnosis. In these 12 cases, the software classified those patients as not healthy.

Table 1 Effectiveness of the integrated software suite in the classification of COPD patients and healthy patients

Patients with COPD		Healthy patients	
252		133	
Hits by software	Misses by software	Hits by software	Misses by software
232 (92%)	20 (8%)	121 (90.97%)	12 (9.03%)

Testing the efficiency of integrated software suite at each single step of classification of COPD patients and healthy controls were obtained from the results of this study. These results are presented in Table 2, while graphical representation is given on Figure 7.

Table 2 Hits and misses of integrated software suite at each single step of classification in COPD and healthy controls

Steps for classification of COPD	Criteria for classification of COPD	Patients with COPD (Hits by software) Σ = 232	Patients with COPD (Misses by software) Σ = 20	Healthy controls (Hits by software) Σ = 121	Healthy controls (Misses by software) Σ = 12
IOS and spirometry test results	R5 > 150% X5 > 0.15% FEV1/FVC < 0.7	210 (90.52%)	12 (60.00%)	117 (96.69%)	9 (75.00%)
Bronchial dilation test results	FEV1/FVC < 0.7	22 (9.48%)	8 (40.00%)	4 (3.31%)	3 (25.00%)

R5 - Total respiratory resistance
X5 - Distal capacitive reactance
FEV1 - Forced expiratory volume in one second

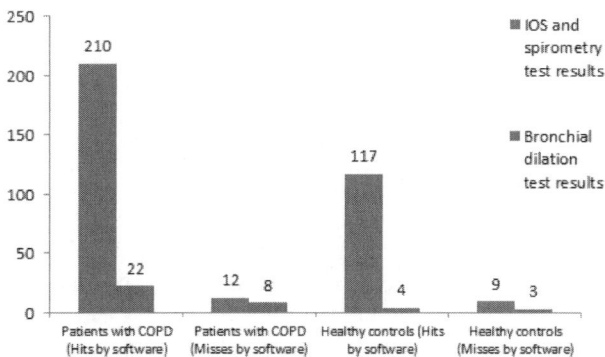

Fig. 9 Graphical representation of integrated software suite in classification of asthmatics and healty patients (Results are expressed in percentages)

Based on the study results shown in Table 2. and Figure 9. it is clear that the software classify the most patients with COPD in the first step, i.e. after IOS and spirometry test results (90.52%). In these cases of patients, R5 was more than 150%, X5 more than 0.15% and FEV1/FVC less than 0.7. In the second step, i.e. after making BDT, the software has classified 9.48% of patients with COPD. For these patients we needed to do BDT to confirm is their FEV1/FVC less than 0.7.

In the case of healthy controls, after the first step software classified 96.69% of patients as healthy controls. These patients had normal IOS and spirometry test results and symptoms also. In the second step, where the test results of the patients contained minor deviations from normal predictive values, the percentage were 3.31%.

V. CONCLUSIONS

In this paper we present results of classification of COPD using integrated software suite. Based on the total number of 385 patients involved in this study, the software correctly classified COPD in 92% of cases, whereas in the case of healthy subjects correctly performed the classification in 90.97% of cases. Also, it is shown that for classification of diseases integrated software suite takes into account the complete dynamic picture of patients. In this paper we also presented that the clinical evaluation of lung function is a more comprehensive with developed software.

REFERENCES

1. Tortora, G J and Grabowski, S R. "Principles of Anatomy and Physiology". Tenth Edition. New York, NY, USA: John Wiley & Sons, 2003.
2. Mayo Clinic. Mayo Clinic COPD Definition. [Online] 2011. http://www.mayoclinic.com/health/copd/DS00916.
3. Anthonisen N. Chronic Obstructive Pulmonary Disease. In: Goldman L, Auseillo D. *Goldman: Cecil Medicine*. Philadelphia, PA: Saunders Elsevier; 2007:chap 88.
4. European Lung Function. COPD Burden in Europe. http://www.european-lung-foundation.org/index.php?id=63. Accessed 28 October 2008.
5. WHO. The World Health Report 2002. Reducing risks, promoting healthy life MDI. WHR. 202. A. Geneva, The World Health Organisation;
6. Ferlay J, et al., GLOBOCAN 2002. Cancer Incidence, Mortality and Prevalence Worldwide. IARC Cancer Base No.5, Version 2.0. IARCPress, Lyon, 2004.
7. Murray CJ, Lopez AD. Alternative projections of mortality and disability ba cause 1990-2020: Global Burden of Disease Study. Lancet 1997;349:1498-504.
8. Nielsen, KG and Bisgaard, H. "The Effect of Inhaled Budesonide on Symptoms, Lung Function, and Cold Air and Methacholine Responsiveness in 2- to 5-year-old Asthmatic Children". Am J Respiratory Critical Care Medicine 2000, Vol. 162, pp. 1500-1506.
9. Song, T W, et al. "Correlation between spirometry and impulse oscillometry in children with asthma". Acta Paediatrica, 2008, Vol. 97, pp. 51-54.
10. Larsen, G L, et al. "Impulse oscillometry versus spirometry in a long-term study of controller therapy for pediatric asthma. J Allergy Clin Immunol. 2009, Vol. 123, 4, pp. 861-867.
11. Global Initiative for Chronic Obstructive Lung Disease, "Pocket guide to COPD Diagnosis, Management and Prevention", 2011.

Use macro [author address] to enter the address of the corresponding author:

Author: Almir Badnjevic
Institute: New Technology Ltd
Street: Paromlinska 53G
City: Sarajevo
Country: Bosnia and Herzegovina
Email: badnjevic.almir@gmail.com

Mathematical Model of Apico Aortic Conduit in Presence of Steno-Insufficiency

G. Fragomeni[1], M. Rossi[2], F. Condemi[1], R. Mazzitelli[1],
G.F. Serraino[2], and A. Renzulli[2]

[1] Magna Graecia University, Bioengineering Unit, Catanzaro, Italy
[2] Magna Graecia University, Cardiac Surgery Unit, Catanzaro, Italy

Abstract—The revival of the apico aortic conduit has attracted new interest towards this alternative treatment for severe aortic stenosis not suitable to conventional valve replacement. The new automated coring and apical connector insertion kit released by Correx (Waltham, MA) made the procedure feasible on beating heart without cardiopulmonary bypass. However, little is known about the changes in the perfusion of the epiaortic vessels after apico aortic conduit implantation, especially when severe aortic stenosis is associated with aortic valve insufficiency.

We constructed a computational model to investigate the perfusion of the epiaortic vessels before and after apico aortic conduit implantation in the particular scenario of a patient with severe aortic valve stenosis associated each time with different grades of aortic insufficiency (mild, medium, moderate and severe).

There were a total of eight combinations analyzed. In all simulations, there was a diminished flow through the epiaortic vessels the more severe the concomitant aortic insufficiency was. After apico aortic conduit implantation, there was an absolute augmentation of the median output in each epiaortic vessels compared with the same combination of mixed aortic valve disease before its implantation. Interestingly, retrograde flow from the descending aorta was trivial and did not contribute to the output improvement of the epiaortic vessels.

The computational analysis suggested a protective effect, rather than a stealing phenomenon, by the apico aortic conduit to the cerebral perfusion.

Keywords—Aorta, Heart valve, Cerebral circulation, Computer applications, Circulatory hemodynamics.

I. INTRODUCTION

The idea of an apico aortic conduit (AAC) to shunt the systemic circulation from the left ventricle to the descending aorta was introduced by Carvel et al in 1910 [1]. Aortic valve bypass (AVB or apico aortic conduit) has been performed since the early 1960s as an alternative surgical approach for symptomatic aortic stenosis (AS) [2,3]. In 2011, Correx (Waltham, MA) released a complete kit for AAC with an installation tool that enables coring and insertion of a left ventricle connector on a beating heart, making the technique of implantation attractive and easily reproducible also for those centers without historical experience about AAC implantation [4]. However, little is known about the systemic hemodynamic changes determined by a second outlet between the heart and the great vessels. Vliek et al [5] and Gammie et al [6] reported a mean value of about 70% of the cardiac output (CO) through the conduit and the remaining 30% through the native aortic valve. Balaras et al [7], showed that after AAC implantation, the perfusion of the epiaortic vessels depended solely on the forward flow via the native aortic valve. However, the perfusion of the epiaortic vessels after AAC implantation for severe AS but with concomitant aortic insufficiency (AI) remains unexplored. The presence of mixed aortic valve disease (i.e. AS and AI) among candidates for AAC is a quite common scenario in our clinical practice, but the effect of AAC on the perfusion of epiaortic vessels has never been studied in this specific subset of patients. We constructed a computational model to study the changes in the epiaortic vessels in cases of mixed aortic valve disease before and after AAC implantation, considering severe AS and four grades of AI: mild, medium, moderate and severe. The numerical discretization used to solve the governing equations of our model was finite elements method (FEM) [8]. Computational fluid dynamic (CFD) analysis resolved the system, providing reliable descriptions of the behavior of real fluids in simulated geometry [9-11]. The aim of this powerful sophisticated computing technique of imaging processing was to support the physician's decisions in difficult or unusual case scenarios, thus avoiding the obstacles of an *in vivo* analysis and allowing realistic and specific patient flow simulation [8,12].

II. MATERIAL AND METHODS

A. The CFD Model

The apical aortic conduit is characterized by three components: a LV connector, a valved conduit with a bioprosthetic valve and a distal conduit for the aortic connection, implanted between the LV apex and the descending aorta. We elaborated a CFD model to describe the characteristic parameters of the blood flowing through a 3D simplified geometry that represented the aorta, the arteries of aortic arc [13], with rigid walls, and the distal conduit [14], positioned in the descending aorta. The ventricle and the LV conduit above its own valve were excluded to simplify the model and to obtain a right compromise between complexity and reliability of the solution. Simulations were performed using the Comsol Multiphysics 4.2a software (COMSOL Inc, Stockholm, Sweden) on a fixed grid of approximately 53,000 tetrahedral elements. The Comsol solved the governing equations (Navier-Stokes equations) that described the conservation of mass and momentum of the fluid. Blood was assumed to be a Newtonian fluid with density $\rho = 1060$ kg/m^3 and dynamic viscosity $v = 0.005$ Pa*s. In order to define realistic boundary conditions [15,16], we used pre- and post- operative echocardiographic data of one of our own female patients with severe aortic stenosis, who underwent AAC implantation at our institution. We assumed that, at the four outlets, there was a resistance R. Then the flow could be written:

$$Q = \frac{P - P_\infty}{R} \quad (1)$$

where P_∞ was the pressure at the beginning and at the end of the cardiac cycle. From the above equation 1, we defined the pressure P above the outlet arteries [17]. After AAC implantation, the average flow profile set in the ascending aorta and in the conduit inlet was 6 l/min: 30% flows through the native valve, while 70% flows through the apical conduit [5-7,18]. The simulation was run through three cardiac cycles, but in the results we reported only the last one [19]. After validation of the CFD model with severe AS and AAC through the literature data [7,19], four grades of IA were introduced: mild (1+), medium (2+), moderate (3+) and severe (4+). We defined as boundaries conditions, a curve of instantaneous flow rate that corresponded to an average flow rate of 1.8 l/min, in the aorta inlet, and of 4.2 l/min at the conduit inlet, with a total average flow of 6 l/min in presence of AAC and a total average flow rate of 3.7 l/min that flowed only through the aortic valve before AAC implantation (CASE V-VIII).

B. Study Design

The aim of the simulation was to observe the flow distribution in the epiaortic vessels (brachiocephalic trunk, left carotid artery and subclavian artery) in the particular scenario of a patient with AAC, severe aortic stenosis and different grades of AI, and to extrapolate indications whether was possible and safe to implant an AAC in this specific subset of patients. Actually, to weigh the role of AAC on the perfusion of the epiaortic vessels, the flow distribution was evaluated in four case-scenarios against control-cases, with the same degrees of aortic steno-insufficiency but without AAC.

C. Definitions

AI was defined as mild (1+), medium (2+), moderate (3+) and severe (4+) according to the regurgitation fraction: 25%, 35%, 45% and 60%, respectively. For each case, we assumed that the flow through the native aortic valve in the diastolic phase had a negative peak, whit a value corresponding to the regurgitant fraction.

The regurgitant fraction (RF) was calculated as follows following, [20,21]:

$$RF = \frac{V_r}{V_f} \times 100 \quad (2)$$

where V_r was the reverse flow volume, V_f was the forward flow volume.

Fixed HR of 75 bpm was used in all simulations. Further we calculated the total average forward flow (Q_T) as:

$$Q_T = (\int_2^{2.4} Q_f(t)dt - \int_{2.5}^3 Q_r(t)dt) \times 75 \quad (3)$$

where $Q_f(t)$ was the function of instantaneous forward flow, while $Q_r(t)$ was that of retrograde flow. The left part of equation allowed to define the effective forward volume from the instantaneous flow rate. So, the effective average flow was derived.

III. RESULTS

The simulation on the control case (AS + AAC) showed, on the epiaortic vessels, a forward flow of 1.12 l/min in the innominate artery; 0.41 l/min in the left carotid artery; 0.41 l/min in the left subclavian artery. As the AI was introduced in the model, there was a progressive reduction in the forward flow through all the epiaortic vessels the greater the AI was. In each scenario the forward flow reduction was similar among the three epiaortic vessels. In AS + AAC + AI (25%) the reduction was 7.14%, 7.32%, 7.32%; in AS + AAC + AI(35%) was 9.82%%, 9.76%, 9.76%; in AS + AAC + AI(45%) was 13.40%, 14.63%,14.63%; in AS + AAC + AI(60%) was 17.86%, 17.07%, 17.07%, respectively (table 1). The medium output through a cardiac cycle in each epiaortic vessel remained positive with the lowest value of 0.92 l/min in the innominate artery; 0.34 l/min in the left carotid artery; 0.34 l/min in the left subclavian artery in the case-scenario with severe AI. There was a retrograde flow from the epiaortic vessels at the diastolic peak with a 2% retrograde flow from the descending aorta towards the native aortic valve that did not contribute to support the cerebral circle. The head-to-head evaluation of the medium forward flow in an entire cardiac cycle of each of the three epiaortic vessels in the four case-scenario described above but before and after AAC implantation (tables 1, 2) showed a neat rise in the forward flow, whenever an AAC was implanted the greater the AI was - maximum rise of 73.9%, 73.53% and 73.53 %, respectively – in the case of severe AI (fig. 1). In mixed aortic valve disease there was a diminished forward flow through the epiaortic vessels compared with isolated severe aortic stenosis.

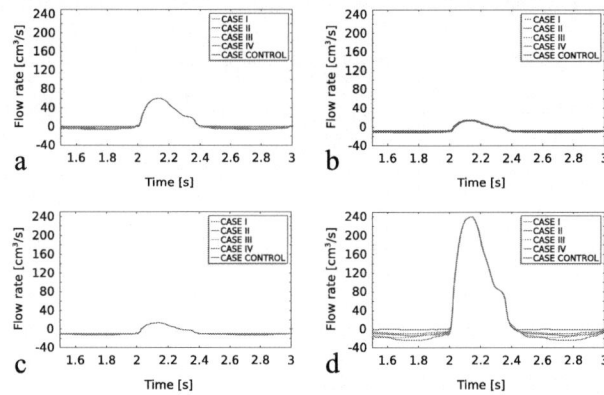

Fig. 1 Variation of flow rate in the diastolic phase from the case control to CASE IV after apico aortic conduit implantation in the three epi-aortic vessels and in the descending aorta: innominate artery (a); left carotid artery (b); subclavian artery (c); descending aorta (d).

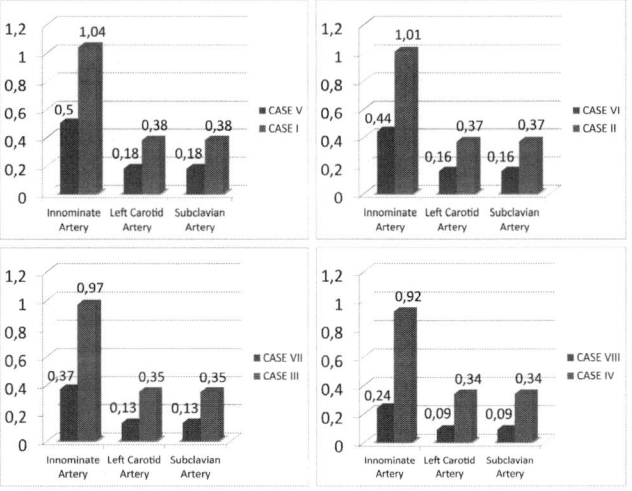

Fig. 2 Head to head evaluation of the average forward flow in the three epiaortic vessel, before and after apico aortic conduit for each combination of mixed aortic valve disease.

The presence of an AAC increased the forward flow through all the epiaortic vessels in every simulation with mixed aortic valve disease. Actually, the flow augmentation was higher the worse the coexisting AI was. Retrograde flow from the descending aorta recorded at the diastolic peak did not contribute to cerebral perfusion.

Table 1 Distribution of flow rate after apico aortic conduit implantation in case control and in different grades of aortic valve insufficiency (Case I-IV).

		CC	CASE I	CASE II	CASE III	CASE IV
Innominate Artery	Q_T (l/min)	1.12	1.04	1.01	0.97	0.92
	RF (%)	-	-7.14	-9.82	-13.40	-17.86
Left Carotid Artery	Q_T (l/min)	0.41	0.38	0.37	0.35	0.34
	RF (%)	-	-7.32	-9.76	-14.63	-17.07
Subclavian Artery	Q_T (l/min)	0.41	0.38	0.37	0.35	0.34
	RF (%)	-	-7.32	-9.76	-14.63	-17.07
Descending aorta	Q_T (l/min)	4.14	3.83	3.71	3.59	3.41
	RF (%)	-	-7.49	-10.38	-13.28	-17.63

AS: Aortic stenosis; AAC: apico aortic conduit; CC: case control (AS with AAC); Q_T: total average flow rate; RF: regurgitant fraction.

Table 2 Distribution of flow rate before apico aortic conduit implantation in case control and in different grades of aortic valve insufficiency.

		CC	CASE I	CASE II	CASE III	CASE IV
Innominate Artery	Q_T (l/min)	0.67	0.50	0.44	0.37	0.24
	RF (%)	-	-25.37	-34.33	-44.78	-64.18
Left Carotid Artery	Q_T (l/min)	0.24	0.18	0.16	0.13	0.09
	RF (%)	-	-25.00	-33.33	-45.83	-62.50
Subclavian Artery	Q_T (l/min)	0.24	0.18	0.16	0.13	0.09
	RF (%)	-	-25.00	-33.33	-45.83	-62.50
Descending aorta	Q_T (l/min)	2.37	1.75	1.57	1.31	0.85
	RF (%)	-	-26.16	-33.76	-44.72	-64.14

AS: Aortic stenosis; AAC: apico aortic conduit; CC: case control (AS without AAC); Q_T: total average flow rate; RF: regurgitant fraction.

IV. CONCLUSION

We analyzed nine models through Comsol Mutiphysics 4.2a. All simulations showed a diminished flow through the epiaortic vessels in patients with AS and concomitant AI. Moreover, the higher AI was the worse the flow in the epiaortic vessels became. Our data confirm that the cerebral perfusion depends solely on the antegrade flow. Furthermore, the comparison of the medium output of each epiaortic vessels before and after AAC implantation showed higher mean value whenever the AAC was present independently from the grade of AI.

We explain this finding with the overall augmentation of the total CO obtained after AAC having bypassed the severe obstruction of the native aortic valve. Remarkably, the retrograde flow from the descending aorta after ACC in every case of mixed aortic valve disease was trivial and did not contribute to the rise of the epiaortic vessels output at all.

The computational analysis suggests that the AAC plays a protective effect towards the cerebral perfusion in cases of mixed aortic valve disease, giving the premises to safely extend the indications for AAC implantation.

ACKNOWLEDGMENT

This study was co-funded by the European Commission, European Social Fund and "Regione Calabria". Authors had full control of the design of the study, methods used, outcome parameters and results, analysis of data and production of the manuscript.

REFERENCES

1. Carrel A. On the Experimental Surgery of the Thoracic Aorta and the Heart. Ann Surg 1910; 52:83-95.
2. Sarnoff S, Donovan T, Case R. The surgical relief of aortic stenosis by means of apical aortic valvular anastomosis. Circulation 1955;11:564–74.
3. Cooley DA, Norman JC, Reul GJ Jr, Kidd JN, Nihill MR. Surgical treatment of left ventricular outflow tract obstruction with apicoaortic valved conduit. Surgery 1976;80:674-80.
4. Adams C, Guo LR et al. Automated Coring and Apical Connector Insertion Device for Aortic Valve Bypass Surgery. Ann Thorac Surg 2012; 93: 290-293.
5. Vliek CJ, Balaras E, Li S, et al. Early and midterm hemodynamics after aortic valve bypass (apicoaortic conduit) surgery. Ann Thorac Surg 2010;90:136-43.
6. Gammie JS, Krowsoski LS, Brown JM, et al. Aortic valve bypass surgery: mid-term clinical outcomes in a high-risk aortic stenosis population. Circulation 2008;118:1460-1466.
7. Balaras E, Cha KS, Griffith BP, Gammie JS. Treatment of aortic stenosis with aortic valve bypass (apicoaortic conduit) surgery: an assessment using computational modeling. J Thorac Cardiovasc Surg 2009;137:680-7.
8. Formaggia L, Quarteroni A, Veneziani A. Cardiovascular Mathematics. Modeling and simulation of the circulatory system. Milan, IT: Springer-Verlag 2009;1: 63-66.
9. Mir-Hossein Moosavi, Nasser Fatouraee , Hamid Katoozian. Finite element analysis of blood flow characteristics in a Ventricular Assist Device (VAD). Simulation Modelling Practice and Theory 2009;17:654-663.
10. Yoshiyuki Tokuda, Min-Ho Song, Yuichi Ueda, et al. Three-dimensional numerical simulation of blood flow in the aortic arch during cardiopulmonary bypass. European Journal of Cardio-thoracic Surgery 2008;33:164-167.
11. Dumont K, Vierendeels J, Kaminsky R, van Nooten G, Verdonck P, Bluestein D. Comparison of the Hemodynamic and Thrombogenic Performance of Two Bileaflet Mechanical Heart Valves Using a CFD/FSI Model. Journal of Biomechanical Engineering 2008;129:558-564.
12. Leuprecht A, Kozerke S, Boesiger P, Perktold K. Blood flow in the human ascending aorta: a combined MRI and CFD study. Journal of Engineering Mathematics 2003;47:387-404.
13. Wright NL. Dissection study and menstruation of the human aortic arch. J Ana.1969;104:377-85.
14. Tsukasa Miyatake, Toshifumi Murashita, Noriko Oyama, Satoshi Yamada, Kaoru Komuro, Keishu Yasuda. Apicoaortic Valved Conduit for a Patient With Porcelain Aorta. Asian Cardiovasc Thorac Ann 2006;14:76-79.
15. Yiemeng Hoi, Yu-Qing Zhou, Xiaoli Zhang, R. Mark Henkelman and David A. Steinman. Correlation Between Local Hemodynamics and Lesion Distribution in a Novel Aortic Regurgitation Murine Model of Atherosclerosis. Annals of Biomedical Engineering 2011; 39: 1414-1422.
16. Kroon W, Huberts W, Bosboom M, van de Vosse F. A Numerical Method of Reduced Complexity for Simulating Vascular Hemodynamics Using Coupled 0D Lumped and 1DWave Propagation Models. Computational and Mathematical Methods in Medicine 2012.
17. Benim AC, Nahavandi A, Assmann A, Schubert D, Feindt P, Suh SH. Simulation of blood flow in human aorta with emphasis on outlet boundary conditions. Applied Mathematical Modelling 2011;35:3175-3188.
18. Vliek CJ, Balaras E, Li S, et al. Early and Midterm Hemodynamics After Aortic Valve Bypass (Apicoaortic Conduit) Surgery. Ann Thorac Surg 2010;90:136-143.
19. Kar B, Delgado RM 3rd, Frazier OH, et al. The effect of LVAD aortic outflow-graft placement on hemodynamics and flow: Implantation technique and computer flow modeling. Tex Heart Inst J 2005;32:294-8.
20. Didier D, Ratib O, Lerch R, Friedli B. Detection and quantification of valvular heart disease with dynamic cardiac MR imaging. Radiographics 2000;20:1279-99.
21. Nichols W W, Pepine C J, Conti C R, Christie L G, Feldman R L. Quantitation of aortic insufficiency using a catheter-tip velocity transducer. Circulation 1981;64:375-380.

Author: Gionata Fragomeni
Institute: Medical and Surgical Science Departement
Street: Campus S. Venuta - Germaneto
City: 88100 - Catanzaro
Country: ITALY
Email: fragomeni@unicz.it

Computational Hemodynamic Model of the Human Cardiovascular System

A. Talaminos[1], L.M. Roa[1,2], A. Álvarez[3], I. Valverde[4], and J. Reina[2,5]

[1] Biomedical Engineering Group (GIB), University of Seville, Seville, Spain
[2] CIBER de Bioingeniería, Biomateriales y Nanomedicina, Zaragoza, Spain
[3] MD, Cardiac Surgeon, Seville, Spain
[4] MD, PhD, Division of Imaging Sciences, King's College London, London, United Kingdom
[5] Department of Signal Theory and Communications, University of Seville, Seville, Spain

Abstract—In this work, the design and development of a computational model of the human cardiovascular system is proposed. The model allows the analysis by simulation of the behavior of some hemodynamic variables not easy to be measured in the clinical practice, in different physio-phatological cardiac situations. The model has been validated comparing the results obtained with hemodynamic values published by other authors. As a result of this work, different simulations of some cardiac pathologies are shown, together with the dynamic behaviors of the different variables considered in the model.

Keywords—Heart, pulmonary circulation, systemic circulation, structural mathematical model, computational modeling.

I. INTRODUCTION

Cardiovascular diseases are the main death cause in the developed world and mean a high health cost [1]. Therefore, it is necessary to invest efforts and research combining theoretical studies and experimentation to promote new cardiovascular diagnosis methods, patients monitorization modalities and more effective therapies.

Nowadays, the clinical management of patients is based on the prior experience of the medical team, statistical population studies about patients samples with pathologies similar to the ones which the sick presents, and diagnostic methods, often invasive. This is why we do not know how reliably the patient will respond to the different pharmacological or surgical interventions. In this regard, computational modeling has been proven to be a useful resource for the analysis and comprehension of the complex biological mechanisms within the vascular system and supplements other experimental studies to understand the cardiovascular physiopathology and to generate new hypothesis to study [2].

Currently, heart models are divided mainly in finite elements (FE) [3]-[4] or of pressure-volume (PV) models [5]-[6]. FE models have a bigger complexity and a greater computational cost, which make their clinical applicability hard, while for PV models it is only required a minimal set of parameters, such as the cavities' elasticity or the arterial resistances; and because of that they are more amenable to development and to understand by the clinical staff.

In this work, it is introduced the design and development of a computational model that simulates the different behaviors associated to some hemodynamic variables of the cardiovascular system from different physio-phatological cardiac conditions. To provide a wider view, it has been necessary to consider some of the mechanisms associated to the cardiovascular system, including the heart as a pump, and the arteries, veins and capillaries that conform the systemic circulation and the pulmonary circulation. The purpose of this model is to serve as a clinical tool to help provide a better understanding of the physio-phatological behaviors of the cardiovascular system.

II. METHODS AND MATERIALS

The methodology used for the construction of the structural model has been the following:

1. Development of the structure of the model.
2. Quantification of the relations established in the previous stage.
3. Codification.
4. Validation.
5. Experiments conducted by computer simulation.

The volumes in the different compartments have been considered as the state variables of the model, in which the constitutive relations observed in the model have been resistance and capacitance. Mitral, tricuspid, aortic and pulmonary valves have been modeled as valves that open or close depending on the blood pressure variations at both sides of the valve. This way, state variables are expressed through ordinary differential equations. The relations between different flows and blood pressures have been established by resistance functions of capacitance.

The model has been implemented using the Matlab tool for system simulation, employing the modified Euler method with an integration period of a one millisecond. Simulations have been done heartbeat by heartbeat, which presents a distinguishing aspect with respect to other works [5]-[6].

A. Model Description

The proposed cardiovascular system computational model has been designed as a lumped-parameter fluid circuit that is divided in four compartments: left heart, right heart, pulmonary circulation and systemic circulation. Furthermore, these four compartments are composed of different cavities accounting for the heart ventricles and the atrium and arteries, veins and blood capillaries. The model takes into account the four heart cavities with equal interest, in spite of the fact that the left heart has historically been more widely studied in the literature [7].

The four heart cavities, right atrium (C_{ar}), left atrium (C_{al}), left ventricle (C_{vl}), and right ventricle (C_{vr}), have been modeled by four variable capacitors. On the other hand, the cavities that constitute the blood vessels and capillaries are modeled as groups of constant capacitors, including the systemic atrial capacitance (C_{as}), the venous systemic capacitance (C_{vs}), the pulmonary atrial capacitance (C_{ap}) and the pulmonary venous capacitance (C_{vp}).

The four heart atrioventricular valves are represented within the model as a valve and a resistor which facilitates or obstructs the blood flow. The resistances of mitral and tricuspid valves are considered as the left resistance (R_l) and the right resistance (R_r) respectively, while the resistances of the aortic and pulmonary valves are given by the systemic left resistance (R_{ls}) and by the right pulmonary resistance (R_{rp}).

The rest of the model considered resistors are part of the pulmonary circulation and the systemic circulation. The R_{as} resistance represents the systemic arterial resistance, while R_{vs} defines the systemic venous resistance within the systemic circulation. In the same way, for the pulmonary circulation it has been considered the arterial pulmonary resistance R_{ap} and the venous pulmonary resistance (R_{vp}). On the other side, capillaries are represented by the systemic capillary resistance (R_{cs}) and the pulmonary capillary resistance (R_{cp}), both considered variable throughout the cardiac cycle.

The blood flowing along the circulatory system is modeled by fluid flows which determine the measured blood flowing between the different cavities. The complete fluid circuit is shown in Figure 1.

The blood volumes in each compartment are considered as state variables in the model and are expressed normally by ordinary diferential equations. This way, the blood volumes of a given compartment are defined as the difference between the magnitude of blood flow in ($\dot{Q}_e$) and out ($\dot{Q}_s$) within the compartment:

$$\frac{dV}{dt} = \dot{Q}_e - \dot{Q}_s \qquad (1)$$

Generally, the flow ($\dot{Q}$) circulating between compartments is determined by the quantity of the difference of blood pressure between the origin compartment (P_1) to the

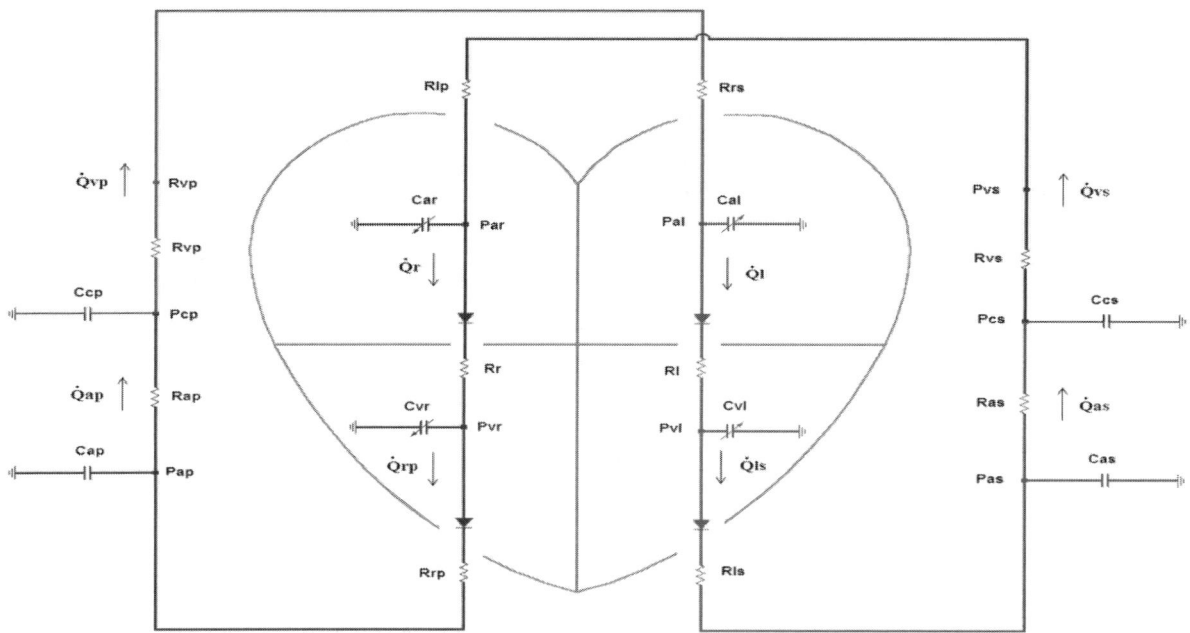

Fig. 1 Proposed model structure of the human cardiovascular system.

destination compartment (P_2), by a resistance (R) associated to the union of both compartments. When the blood pressure in the destination compartment is bigger than the one in the origin compartment, then the blood flow between them is null:

$$\dot{Q} = \begin{cases} (P_1 - P_2)/R, & P_1 > P_2 \\ 0, & P_1 \leq P_2 \end{cases} \quad (2)$$

The blood pressure in each cardiac cavity is defined by the relation between the blood volume in the compartment and its capacitance, which depends on the contraction strength and the cardiac frequency.

B. Validation

To verify the validity of the model, a first simulation has been made to obtain the different hemodynamic variables considered for a healthy adult under rest conditions. This way, the model has been validated comparing the results obtained with the simulation and the data given by other authors' models [8]-[9], considering relations between the blood pressures, flows, volumes, ventricular and atrial capacities, systemic arterial blood pressure and left atrial blood pressure. The time length for this specified simulation has been of 7.5 seconds, running over 9 heartbeats during this period.

The different blood pressures on the left side of the heart are depicted in Figure 2, including the left ventricular blood pressure, systemic blood pressure and left atrial blood pressure. On the other hand, the blood pressure on the right side of the heart is presented in Figure 3, including the right ventricular blood pressure, pulmonary blood pressure and right atrial blood pressure.

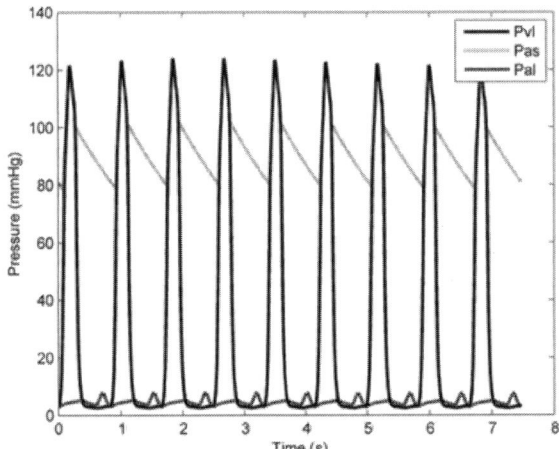

Fig. 2 Left ventricular blood pressure (P_{vl}), systemic arterial blood pressure (P_{as}) and left atrial blood pressure (P_{al}) for a healthy heart under rest conditions.

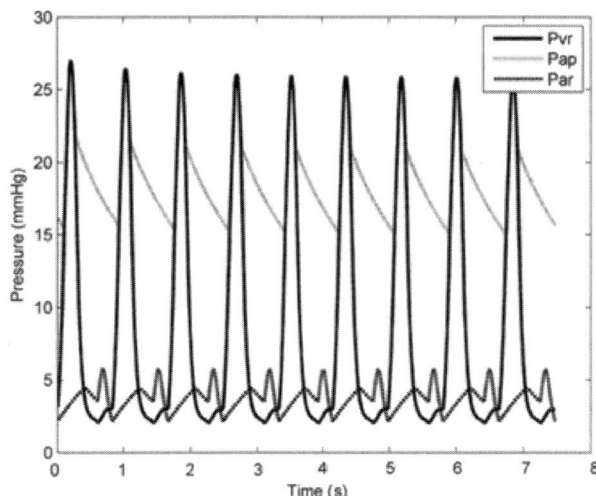

Fig. 3 Right ventricular blood pressure (P_{vr}), pulmonar arterial blood pressure (P_{ap}) and right atrial blood pressure (P_{ar}) for a healthy heart under rest conditions.

In Figure 4, the volume-pressure curve is illustrated for the left side of a healthy heart in normal conditions, checking the different cardiac cycle stages in the representation.

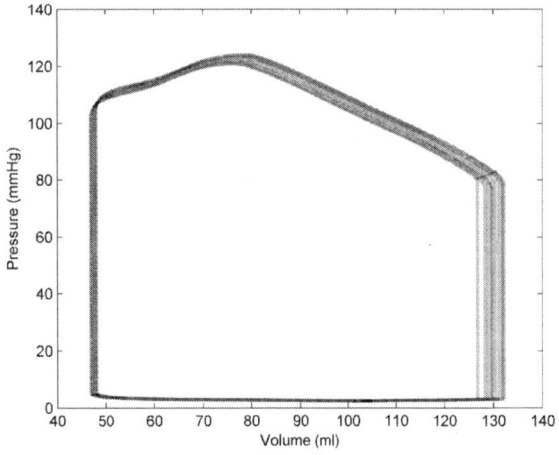

Fig. 4 Volume-pressure curve.

III. RESULTS

Once the model was validated, several experiments were conducted which tried to analyse the possible behaviours for the hemodynamic variables of the cardiovascular system, simulating different pathologies.

One of these experiments is illustrated in Figure 5, where typical conditions of a severe aortic stenosis are simulated increasing the value of the resistance systemic left resistance, by which a narrowing and obstruction of the aortic valve produces a decrease of the blood flow towards the organism.

In normal conditions, the orifice area of the aortic valve fluctuates between 3 and 4 cm^2 [10], while for a severe aortic stenosis, this area is decreased to less than 1 cm^2 [11]. The pressure gradient between the systemic arterial pressure and the left ventricular pressure vary from approximately 40 to 50 mmHg [12], as shown in Figure 5.

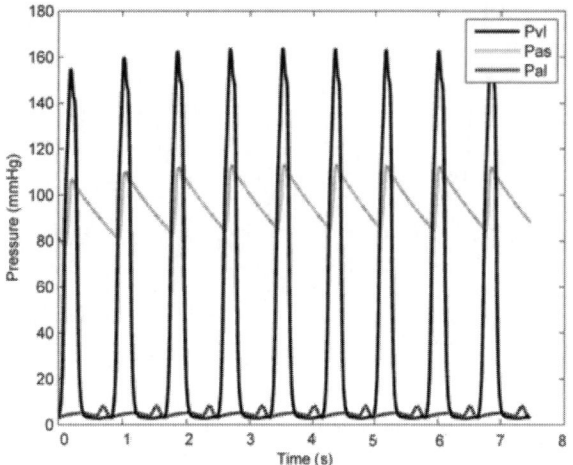

Fig. 5 Left ventricular pressure (P_{vl}), systemic arterial pressure (P_{as}) and left atrial pressure (P_{al}) for an aortic valve stenosis.

As it can be seen, the work load made by the left ventricle rises considerably, reaching a pressure close to 160 mmHg during systole.

IV. CONCLUSIONS

It has been developed a useful tool, easy to use and low-cost which allows the obtaining of some variables associated to the cardiovascular system in humans by computational simulation.

The proposed model allows a better understanding of the dynamical behaviors of some hemodynamic variables and allows the simulation of different physiological or pathological scenarios, serving as a risk predictor in cardiovascular complications.

ACKNOWLEDGMENT

This work has been partially supported by the CIBER-BBN (inside project PERSONA), and a grant from the Fondo de Investigación Sanitaria inside project PI082023. CIBER-BBN is an initiative funded by the VI National R&D&i Plan 2008–2011, Iniciativa Ingenio 2010, Consolider Program, CIBER Actions and financed by the Instituto de Salud Carlos III with assistance from the European Regional Development Fund.

REFERENCES

1. M. Pop, M. Sermesant, T. Mansi, E. Crystal, S. Ghate, J. Relan, C. Pierre, Y. Coudiere, J. Barry, I. Lashevsky, B. Qiang, E. R. McVeigh, N. Ayache, and G. A. Wright (2012) EP Challenge STACOM'11: Forward Approaches to Computational Electrophysiology Using MRI-Based Models and In-Vivo CARTO Mapping in Swine Hearts in Statistical Atlases and Computational Models of the Heart. Imaging and Modelling Challenges, vol. 7085, pp. 1-13
2. J. H. Yang and J. J. Saucerman (2011) Computational Models Reduce Complexity and Accelerate Insight Into Cardiac Signaling Networks, Circ. Res., vol. 108, n.º 1, pp. 85-97
3. S. Sugiura, T. Washio, A. Hatano, J. Okada, H. Watanabe, and T. Hisada (2012) Multi-scale simulations of cardiac electrophysiology and mechanics using the University of Tokyo heart simulator, Prog. Biophys. Mol. Biol, vol.110, pp. 380-389
4. H. Xia, K. Wong, and X. Zhao (2012) A fully coupled model for electromechanics of the heart, Comput. Math. Methods Med, vol. 2012, Article ID 927279
5. X.-Y. Zhang and Y.-T. Zhang (2006) A model-based study of relationship between timing of second heart sound and systolic blood pressure, Proc. Annu. Int. Conf. IEEE Eng. Med. Biol. Soc. Conf., vol. 1, pp. 1387-1390
6. A. Pironet, T. Desaive, S. Kosta, A. Lucas, S. Paeme, A. Collet, C. G. Pretty, P. Kolh, and P. C. Dauby (2013), A multi-scale cardiovascular system model can account for the load-dependence of the end-systolic pressure-volume relationship, Biomed. Eng. Online, vol. 12, p. 8
7. M. K. Rausch, A. Dam, S. Göktepe, O. J. Abilez, and E. Kuhl (2011) Computational modeling of growth: systemic and pulmonary hypertension in the heart,» Biomech Model. Mechanobiol, vol. 10, n. 6, pp. 799-811
8. S. Ravanshadi and M. Jahed (2012) Introducing a distributed model of the heart, in 2012 4th IEEE RAS EMBS International Conference on Biomedical Robotics and Biomechatronics (BioRob), pp. 419-424
9. J. L. Palladino and A. Noordergraaf (2009), Functional requirements of a mathematical model of the heart, Proc. Annu. Int. Conf. IEEE Eng. Med. Biol. Soc. Conf., pp. 4491-4494
10. B. H. Grimard and J. M. Larson (2008) Aortic stenosis: diagnosis and treatment, Am. Fam. Physician, vol. 78, n. 6, pp. 717-724
11. A. Vahanian, O. Alfieri, F. Andreotti, M. J. Antunes, G. Barón-Esquivias, H. Baumgartner, M. A. Borger, T. P. Carrel, M. De Bonis, A. Evangelista, V. Falk, B. Iung, P. Lancellotti, L. Pierard, S. Price, H.-J. Schäfers, G. Schuler, J. Stepinska, K. Swedberg, J. Takkenberg, U. O. Von Oppell, S. Windecker, J. L. Zamorano, and M. Zembala (2012) Guidelines on the management of valvular heart disease (version 2012), Eur. Heart J., vol. 33, n. 19, pp. 2451-2496
12. H. Baumgartner, J. Hung, J. Bermejo, J. B. Chambers, A. Evangelista, B. P. Griffin, B. Iung, C. M. Otto, P. A. Pellikka, y M. Quiñones (2009) Echocardiographic assessment of valve stenosis: EAE/ASE recommendations for clinical practice, Eur. J. Echocardiogr. J. Work. Group Echocardiogr. Eur. Soc. Cardiol., vol. 10, n. 1, pp. 1-25

Author: Alejandro Talaminos Barroso
Institute: Biomedical Engineering Group, University of Seville
Street: Av. Descubrimientos s/n (41092)
City: Seville
Country: Spain
Email: atalaminos@us.es

Object-Oriented Modeling and Simulation of the Arterial Pressure Control System by Using MODELICA

J. Fernandez de Canete[1], J. Luque[2], J. Barbancho[2], and V. Muñoz[1]

[1] Dpt. System Engineering and Automation, Andalucia-Tech University, Malaga, Spain
[2] Dpt. Electronics, Andalucia-Tech University, Sevilla, Spain

Abstract—The modeling of physiological control systems via mathematical equations reflects the calculation procedure more than the structure of the real system modeled, and several simulation environment have been used so far for this task. Nevertheless, object-oriented modeling is spreading in current simulation environments through the use of the individual components of the model and its interconnections to define the underlying dynamic equations. In this paper we describe the use of the MODELICA$_{TM}$ simulation environment in the object-oriented modeling of the cardiovascular control system. The performance of the controlled system has been analyzed by object-oriented simulation in the light of existing hypothesis and physiological data used here for validation purposes.

Keywords—Object-Oriented Modeling, MODELICA$_{TM}$ Simulation Language, Cardiovascular System, Pressure Control.

I. INTRODUCTION

There are a considerable number of specialized and general-purpose modeling software applications available for biomedical studies, which are commonly divided into structure-oriented and equation-oriented. Most of them follow a causal modeling approach and require explicit coding of mathematical model equations or representation of systems in a graphical notation such as block diagrams, which is quite different from common representation of physiological knowledge.

The object-oriented approach can offer many advantages in biomedical research both for the building of complex multidisciplinary models when dynamics are given by a set of differential algebraic equations [1]. The object-oriented approach has been made possible by using the modeling environment MODELICA$_{TM}$ [2] among others, which allows the system, subsystem, or component levels of a whole physiological system to be described in increasing detail. While the MODELICA$_{TM}$ environment has been used for a long time in different fields of engineering [3] there are few results in biomedical system modeling [4-5] particularly in cardiovascular modeling and control.

Cardiovascular modeling and control present a particular challenge and require both a multi-scale and a multi-physics approach [6-7]. Several mathematical models of the closed loop cardiovascular system have been developed [8-9] and software tools based on hierarchical block diagrams have also been applied to the description of cardiovascular systems [10-11].

The regulatory control of the cardiovascular system has been studied intensively, more than any other physiological system. In fact, the high rate of cardiovascular diseases is certainly one of the main reasons for its analysis. The control mechanisms responsible for maintaining arterial blood pressure may be divided into short-term processes, which are effective over a period of seconds to hours, and long-term processes that operate over days to weeks. The former are largely neural based, utilizing receptors in the heart and blood vessels to sense blood pressure and the autonomic nervous system (ANS) to regulate the cardiac function and diameter of resistance vessels. The mid-term control is basically hormonal while the renal system plays the central role at long-term [6].

In this paper we describe the use of the MODELICA$_{TM}$ simulation environment in the object-oriented modeling of the cardiovascular regulatory system considering the short, medium and long-term mechanisms. For this task we have followed an acausal hierarchical structure whose validity has been previously assessed to represent the cardiovascular dynamics as a multi-compartmental system. The results have been obtained under both physiological and pathological conditions, namely, hypertension due to either increased values of peripheral resistance, heart rate or intravascular volume, and hypotension due to either decreased heart rate or severe hemorrhage, assuming the validation tests previously performed with physiological data.

II. MODELING THE CARDIOVASCULAR CONTROL SYSTEM

The cardiovascular model here used can be represented with a set of nonlinear equations describing its dynamics [8]. Nevertheless, the present paper focuses on the multi-compartmental acausal approach so that the whole system will be described by eight compartments, namely, the four cardiac chambers (auricles and ventricles), the pulmonary circulation (arterial and venous) and the systemic circulation (arterial and venous) (Fig. 1).

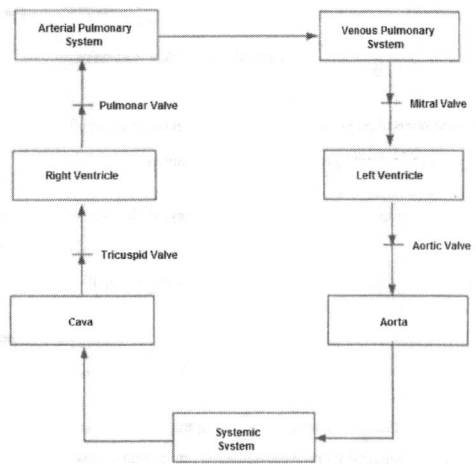

Fig. 1 Compartmental Model of the Circulatory System

Each compartment will be modeled by using a mathematical relationship between pressure $P_i(t)$, volume $V_i(t)$, input flowrate $F_{i_{in}}(t)$ and output flowrate $F_{i_{out}}(t)$ relative to the ith compartment given as

$$\frac{dV_i(t)}{dt} = F_{i_{in}}(t) - F_{i_{out}}(t) \qquad (1)$$

$$P_i(t) = \frac{V_i(t)}{C_i} \qquad (2)$$

while flowrate $F_{ij}(t)$ between compartments i and j is defined by

$$F_{ij}(t) = E_i(P_i(t) - P_j(t)) \qquad (3)$$

$$F_{ij}(t) = E_i \, lim(P_i(t) - P_j(t)) \qquad (4)$$

where C_i and E_i stand respectively for the compliance and admittance of compartment i, while the function $\lim(x) = \max(x, 0)$ is used in case of flowrate through a valve. It is important to highlight the temporal dependence $E_i = E_i(t)$ which exhibits each of the ventricular compartments.

The short-term mechanism to control arterial pressure is represented by the baroreceptors, which as described in [12] regulate the heart rate, the ventricular elastance, the cava elastance and the peripheral system resistance as a function of the mean aortic pressure value through a first-order filter with transfer function $H_j(s)$ and a dead zone (Fig. 2).

The long-term mechanism to control arterial pressure is represented by the renal system through the elimination of urine. In order to model this effect it is assumed as described in [13] that there is a linear relation between the renal excretion flowrate $J_u(t)$ and the plasmatic volume $V_p(t)$ stated as

$$J_u(t) = \begin{cases} K_{ud}\left(\frac{V_p(t)-V_{pn}}{V_{pn}}\right) + J_{un}, & V_p(t) < V_{pn} \\ K_{ue}\left(\frac{V_p(t)-V_{pn}}{V_{pn}}\right) + J_{un}, & V_p(t) \geq V_{pn} \end{cases} \qquad (5)$$

where K_{ud}, K_{ue} and J_{un} represent a measured excretion constant for dehydration and overhydration and the urine flow under normal condition respectively while V_{pn} stands for the plasmatic volume under physiologic conditions. The plasmatic volume $V_p(t)$ is calculated by aggregation of the volume $V_i(t)$ of each of the constitutive compartments applying the hematocrit factor.

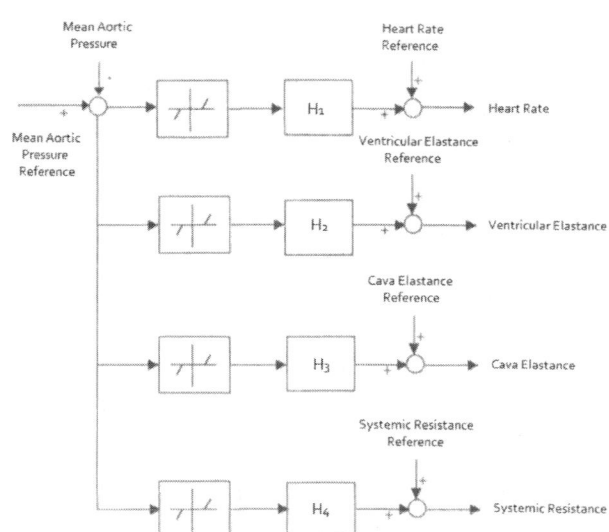

Fig. 2 Diagram of the Baroreceptor Regulatory System

The whole cardiovascular control system is obtained by integration of the circulatory system, the short-term and the long-term arterial pressure control modules described.

III. THE MODELICA$_{TM}$ MODELING APPROACH

MODELICA$_{TM}$ [3] is in general an equation-based object-oriented language for modeling continuous complex physical systems for the purpose of computer simulation, and is primarily based on equations instead of assignment statements. It also has multi-domain modeling capability, meaning that model components corresponding to physical objects from several different domains can be described and connected. MODELICA$_{TM}$ is an object-oriented language whose basic construct is class (model, block, function, type,...), which facilitates reuse of components and evolution of models.

The most important difference with regard to the traditional block-oriented simulation tools is in the different way of connecting components. So, a special-purpose class connector as an interface defines the variables of the model shared with other sub-models, without prejudicing any kind of computational order. In this way the connections can be, besides inheritance concepts, thought of as one of the key features of oriented-object modeling, enabling effective model reuse.

The basic idea of implementation in MODELICA$_{TM}$ is to decompose the described system into components that are as simple as possible and then to start from the bottom up, connecting basic components (classes) into more complicated classes, until the top-level model is achieved. The default integration method is the DASSL code as defined by [14], nevertheless some other methods are available as Runge-Kutta and BDF based.

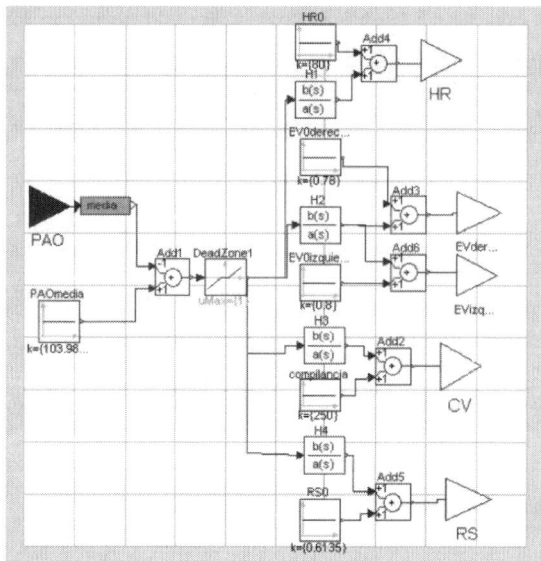

Fig. 4 MODELICA$_{TM}$ Block Diagram of the Baroreceptor Mechanism

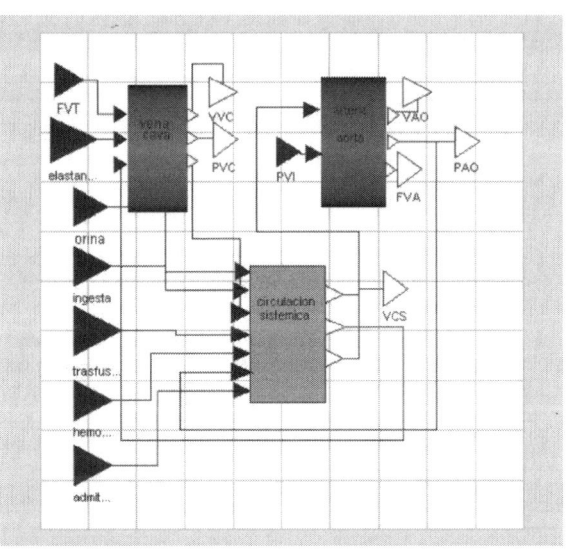

Fig. 3 MODELICA$_{TM}$ Block Diagram of the Systemic Circulation

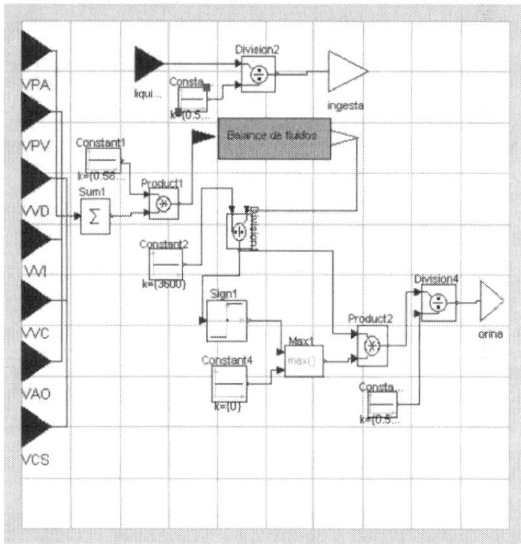

Fig. 5 MODELICA$_{TM}$ Block Diagram of the Renal Control Mechanism

The arterial pressure control system was simulated using MODELICA$_{TM}$ including each of the circulatory compartments above mentioned together with the control mechanisms. The MODELICA$_{TM}$ block diagrams of any of the compartments are illustrated in Fig. 3, while the baroreceptor control system and the renal control system programming are also depicted in Fig. 4 and Fig. 5.

IV. RESULTS

In order to test the performance of the pressure control system several experiences have been accomplished, both under physiologic conditions by varying resistance, elastance or heart rate reference values in the baroreceptor control module and by considering liquid intake, hemorrhage and blood transfusion as external flowrates entering the circulatory system. The results have been obtained by assuming the validation tests previously performed with physiological data by [8].

In Fig. 6 it is depicted the short-term control system response to a system resistance increasing showing the mean aortic pressure regulation to a constant reference level.

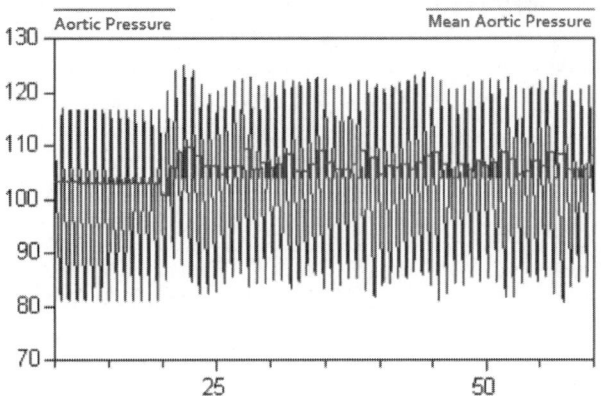

Fig. 6 Baroreceptor response to a sudden 30% change in the system resistance at t = 20 s

In Fig. 7 it is shown the long-term control response to a slight hemorrhage combined with a transfusion applied to the system circulation compartment. It can be observed the systemic pressure value is restored after the transient showing the ability of the renal system to compensate for these abnormal situations.

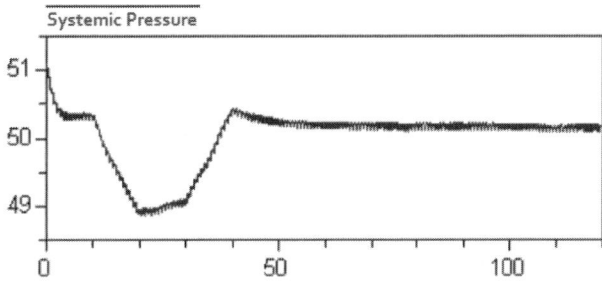

Fig. 7 Renal response to an hemorrhage and subsequent transfusion of 10 ml/min during 10 min at t = 10 s and t = 30 s respectively

V. CONCLUSIONS

An object oriented computer model of the short- and long- term arterial pressure control system under both physiological and pathological condition is presented.

The whole system has been modeled and simulated under the MODELICA$_{TM}$ integrated environment, which uses a hierarchical modeling strategy which aids to a quantitative understanding of the control process.

The described approach represents also a valuable tool in the teaching of physiology for graduate medical students.

Future works consider the application of the proposed approach to the optimized performance of a patient undergoing hemodialysis treatment. Also, the development of a MODELICA$_{TM}$ Physiological Library for the building of cardiovascular and respiratory models is currently under construction.

REFERENCES

1. Hakman M, Groth T. (1999). Object-oriented biomedical system modelling:/The rationale. Comput. Methods Programs Biomed. 59:1-17.
2. Mattsson SE., Elmquist H, Otter M (1998). Physical system modelling with Modelica,. Control Eng. Practice 6:501-510.
3. Tiller M. (2001). Introduction to physical modeling with MODELICA, Kluwer Ac. Press.
4. Cellier FE., Nebot A. Object-oriented modeling in the service of medicine, Proc. 6th Asia Simulation Conference, Beijing, China, 2005, pp 33-40.
5. Fernandez de Canete J, Del Saz Huang P. (2010). First-principles modeling of fluid and solute exchange in the human during normal and hemodialysis conditions. Comput. Biol. Med. 40:740-750.
6. Guyton AC, Hall JE (1996). Textbook of medical physiology, 9th Edition, W.B. Saunders, New York, NY.
7. Berne RM, Levy MN (1997). Cardiovascular physiology, Mosby, St. Louis.
8. Avanzolini G, Barbini P, Cappello A, Cevenini G (1988). CACDS simulation of the closed-loop cardiovascular system. Int. J. Biomed. Comput. 22:39-49.
9. Rothe RF, Gersting JM (2002). Cardiovascular interactions: an interactive tutorial and mathematical model, Am. J. Physiol. Adv. Physiol. Educ. 26 :98–109.
10. Raymond GM, Butterworth E, Bassingthwaighte JB (2003). JSIM: Free software package for teaching physiological modeling and research. Exp. Biol. 280:102-107.
11. Abram SR, Hodnett BL, Summers RL, Coleman TG, Hester RL (2007). Quantitative circulatory physiology. An integrative mathematical model of human mathematical model of human physiology for medical education, Adv. Physiol. Educ. 31:202–210.
12. Kline I (1988). Handbook of biomedical engineering. Academic Press, London.
13. Gyenge C, Bowen BD, Reed RK, Bert JL (2003). Mathematical Model of Renal Elimination of Fluid and Small Ions during Hyper and Hypovolemic conditions. Acta Anaest. Scand. 47:127-137.
14. Brenan KE, Campbell SL, Petzold LR (2011). Numerical solution of initial value problems in differential algebraic equations, SIAM, 2nd edition.

Electrical Remodeling in the Epicardial Border Zone of the Human Infarcted Ventricle

J.V. Visconti, L. Romero, J.M. Ferrero, J. Sáiz, and B. Trenor

Instituto Interuniversitario de Investigación en Bioingeniería y Tecnología Orientada al Ser Humano (I3BH), Universidad Politécnica de Valencia, Valencia, Spain

Abstract—The electrical and structural remodeling of the epicardial border zone (EBZ) after the infarct healing phase is related to an increased incidence of arrhythmias. Indeed, this remodeling affects the action potential duration (APD), refractoriness, vulnerability to conduction block, and conduction safety. Many studies have been undertaken to evaluate the effects of ionic remodeling in the border zone of the ischemic tissue, mainly in dogs and rabbits. In the present study, ionic remodeling in human ventricular cells was simulated applying changes to the model of human cardiac ventricular action potential proposed by O'Hara et al., to analyze the behavior of myocytes in the human EBZ. For this purpose, we introduced the ionic changes proposed by several experimental studies. We analyzed the APD of the remodeled cells, the behavior of the ionic currents, APD rate-adaptation, the action potential amplitude (APA), and finally we performed a sensitivity analysis to investigate the role of the remodeled ionic parameters in the alteration of the main electrophysiological characteristics. Our results show that after remodeling, APD increased by 44%, APD rate-adaptation was also significantly altered, as well as APA. These results are consistent with those obtained in other species. The sensitivity analysis showed that the conductance of the rapid delayed rectifier potassium current (G_{Kr}) has a marked effect on the action potential duration at 90% repolarization (APD_{90}), action potential at 30% repolarization (APD_{30}), and AP triangulation; the changes in APA are mainly determined by the permeability of Ca^{2+} (P_{Ca}) and slightly by Na^+ channels conductance (G_{Na}).

Keywords—epicardical border zone, ion channel remodeling, action potential duration restitution, ischemic zone, infarction.

I. INTRODUCTION

It is well known that myocardial infarction increases the risk of developing ventricular arrhythmias days and weeks after the ischemic event. Reentrant arrhythmias propagating through a thin rim of surviving epicardial tissue surrounding the infarct have been mapped in canine models of transmural infarcted hearts created by ligation of the left anterior descending coronary artery [10, 15]. The thin rim of surviving epicardial tissue surrounding the infarct has been called epicardial border zone (EBZ). Transmembrane ion channels [1, 5, 7, 9 13], Ca^{2+} handling [13], and gap junctions [18] remodeling of the EBZ in the following 3 to 7 days postinfarction has been experimentally reported. Moreover, there is experimental evidence that generation and maintenance of cardiac arrhythmias is favored by structurally and electrophysiologically remodeled substrates [12].

In this paper, we aimed at modeling the EBZ in human epicardial cells by modifying the O'Hara et al. human ventricular action potential (AP) model [11] to reproduce the experimentally observed electrical remodeling. Our remodeling is based on the Decker and Rudy EBZ canine model [4] and on experimental data published in the literature. In addition, a sensitivity analysis of AP biomarkers of arrhythmic risk to the modification of the parameters altered by the remodeling of the EBZ was performed to analyze the impact of the parameters involved in the remodeling of the EBZ.

II. METHODS

A. Human EBZ Model

The maximal conductance (G_{Na}) of the fast sodium current (I_{Na}) was decreased to 86% [9, 13, 15] (Eq. 1) and the steady-state curve of the inactivation gate of I_{Na} (h_∞) was 9 mV rightward shifted [9, 13] ($V_{h,\,shift}$ = -9 mV) (Eq. 2).

$$G_{Na} = G_{Na} \cdot 0.86 \quad (1)$$

$$h_\infty = \frac{1}{\left(\dfrac{((V - V_{h,Shift}) + 82.9)}{6.086}\right)} \cdot 100 \quad (2)$$

The L-type Ca^{2+} current ($I_{Ca,L}$) was also modified. The maximum Ca^{2+} permeability (P_{Ca}) was reduced to 76.3% to reproduce the experimentally observed decreased density [1] (Eq. 3), the I-V relationship was 4 mV rightward shifted ($V_{d,shift}$ = -4 mV) (Eq. 4) and the slow and fast time constants of the voltage dependent inactivation gate (f_τ=0.5 and $f_{1\tau}$=1.7) were reduced (Eq. 5).

$$P_{Ca} = P_{Ca} \cdot 0.763 \quad (3)$$

$$d_\infty = \frac{1}{1 + \exp\left(\dfrac{-((V + V_{d,shift}) + 3.940)}{4.230}\right)} \quad (4)$$

$$\tau_{f,fast} = f_\tau \cdot 7.0 + \cfrac{1}{0.0045 \cdot \exp\left(\cfrac{-(V+20.0)}{10.0}\right) + f_{1\tau} \cdot 0.0045 \cdot \exp\left(\cfrac{V+20.0}{10.0}\right)} \quad (5)$$

$$\tau_{f,slow} = f_\tau \cdot 1000 + \cfrac{1}{0.000035 \cdot \exp\left(\cfrac{-(V+5.0)}{4.0}\right) + f_{1\tau} \cdot 0.000035 \cdot \exp\left(\cfrac{V+5.0}{6.0}\right)}$$

The rapid delayed rectifier potassium current (I_{Kr}) in the EBZ model was modified to reproduce the experimentally observed reduction in density and faster activation. Specifically, the fast and the slow components of the time constant of the I_{Kr} activation gate were reduced [7] (Eq. 6, $f_{\tau,xr}$ =1,667) and the maximal I_{Kr} conductance was decreased to approximately 70% [2, 7, 10] (Eq. 7).

$$\tau_{xr,fast} = 12.98 + \cfrac{1}{f_{\tau,xr} \cdot 0.3652 \cdot \exp\left(\cfrac{V-31.66}{3.869}\right) + 4.123 \cdot 10^{-5} \cdot \exp\left(\cfrac{-(V-47.78)}{20.38}\right)} \quad (6)$$

$$\tau_{xr,slow} = 1.865 + \cfrac{1}{f_{\tau,xr} \cdot 0.06629 \cdot \exp\left(\cfrac{V-34.70}{7.355}\right) + 1.128 \cdot 10^{-5} \cdot \exp\left(\cfrac{-(V-29.74)}{25.94}\right)}$$

$$G_{Kr} = G_{Kr} \cdot 0.70 \quad (7)$$

The approximately 70% reduction in density of I_{K1} recorded in the EBZ [9, 12] was mimicked by reducing the maximal I_{K1} conductance (G_{K1}) (Eq. 8).

The slow delayed rectifier K^+ current conductance (G_{Ks}) was reduced to 30% to match experimental data [7] (Eq. 9).

Finally, the transient outward K^+ current (I_{to}) maximal conductance (G_{to}) was reduced by 90% to reproduce decrease of I_{to} density in the EBZ [9] (Eq. 10).

$$G_{Ks} = G_{Ks} \cdot 0.20 \quad (8)$$

$$G_{K1} = G_{K1} \cdot 0.30 \quad (9)$$

$$G_{to} = G_{to} \cdot 0.1 \quad (10)$$

B. Stimulation Protocol

Steady-state AP duration at 90% and at 30% repolarization (APD_{90} and APD_{30}, respectively), AP triangulation and AP amplitude (APA) were calculated from the simulations following the application of a train of 500 square transmembrane current pulses at a cycle length (CL) of 1000 ms.

APD adaptation curves were constructed by plotting steady-state APD_{90} versus steady-state diastolic interval (DI) for CLs ranging from 380 to 1000 ms.

Finally, a sensitivity analysis [16, 17] of the proposed EBZ model was performed to study the impact of each model parameter involved in EBZ remodeling, namely, G_{Na}, $V_{h,shift}$, P_{Ca}, $V_{d,shift}$, f_τ, $f_{1\tau}$, $f_{\tau,xr}$, G_{Kr}, G_{Ks} and G_{to}. Simulations were carried out by varying one parameter at a time in the proposed EBZ model, by −30% and +30% of the modification introduced in the remodeling, respectively.

C. Data Analysis

The percentage of change ($D_{c,p,a}$) and the sensitivity ($S_{c,p}$) for each electrophysiological characteristic "c" and for each model parameter "p" were calculated as follows:

$$D_{p,c,a} = \frac{c_{c,a} - c_{control}}{c_{control}} \cdot 100 \quad (11)$$

$$S_{c,p} = \frac{\Delta D_{b,c,a}}{\Delta a} = \frac{D_{b,c,+30\%} - D_{b,c,-30\%}}{0.6} \quad (12)$$

with $c_{c,a}$ and $c_{control}$ being the magnitude of the characteristic "c" when the model parameter "p" is increased by "a" and under control conditions, respectively. As 30% reduction of $V_{h,shift}$ compromises I_{CaL} activation, sensitivities of the electrophysiological characteristics to this parameter were calculated as follows:

$$S_{c,p} = \frac{\Delta D_{b,c,a}}{\Delta a} = \frac{-D_{b,c,-30\%}}{0.3} \quad (13)$$

Relative sensitivities ($r_{c,p}$) were also calculated by dividing the sensitivities ($S_{c,p}$) by the maximum absolute value of sensitivity obtained for that specific biomarker.

III. RESULTS AND DISCUSSION

Figure 1A show the APs in myocytes from NZ and EBZ. The changes observed in the ABZ AP are consistent with those obtained in experimental models [4, 9].

Figures 1B to 1D shows the temporal evolution of the main ionic currents and their alteration in the remodeled EBZ. As can be observed in panel B and D, the peak of the Na^+ current and I_{Kr}, respectively, are delayed and significantly reduced. Similar results were observed in [4]. The peak of Ca^{2+} current through L-type channels was increased. This observation differs from the results obtained in [4]. We hypothesize that this effect may be due to the important decrease in I_{Na} that has to be compensated by I_{CaL}. All these changes cause a significant increase in APD_{90} yielding 227.45 ms in the NZ and 329.78 ms in the EBZ, and a slight increase of APA (123.31 mV in the NZ and 125.98 mV in the EBZ).

Our simulations showed also an increase in the effective refractory period (ERP). In the EBZ the shortest diastolic interval (DI) to elicit a second AP was 69.2 ms, while in the NZ the second AP could be elicited with a DI of 7.2 ms. Our results are in agreement with those obtained experimentally [6].

Figure 2 shows the rate dependence of APD in the NZ and the EBZ. Our results were similar with those obtained by Decker et al.

Electrical Remodeling in the Epicardial Border Zone of the Human Infarcted Ventricle

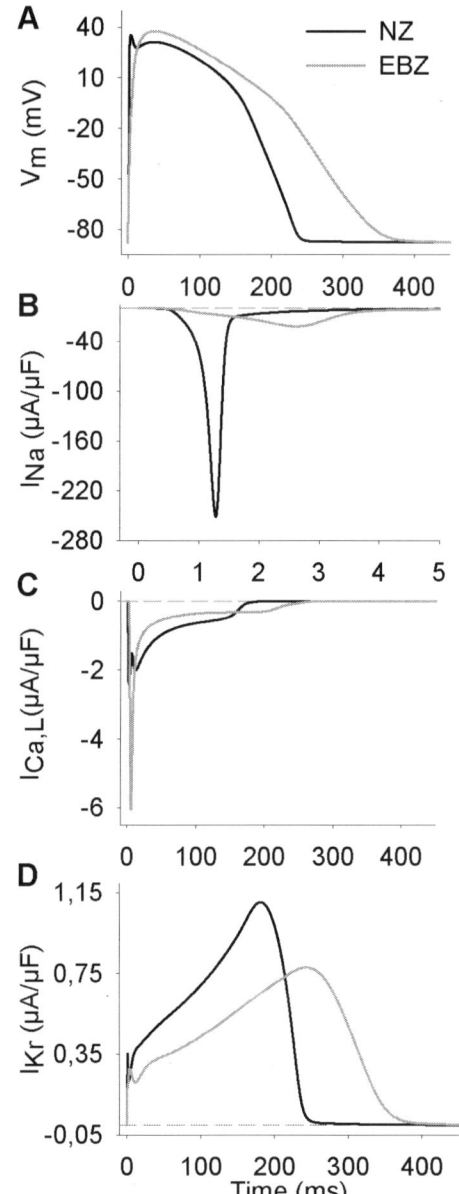

Fig. 1. A: Action potential (AP) in normal cells (NZ) and in the remodeled epicardial border zone (EBZ). B to D: Na$^+$ current (I_{Na}), $I_{Ca,L}$ and I_{Kr} in myocytes from NZ and EBZ.

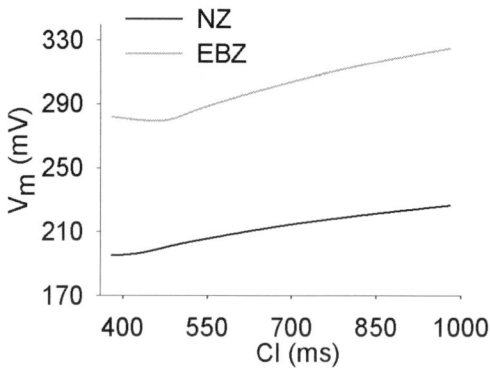

Fig. 2. Action potential duration (APD) rate-adaptation. 500 pulses were applied, with an initial coupling interval (CI) of 380 ms. CI was progressively increased up to 1000 ms. APD was calculated at 90% repolarization for the last AP for every 50 CI.

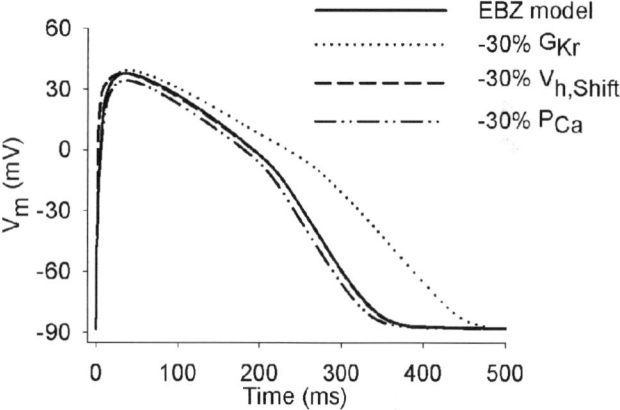

Fig. 3. Action potentials (APs) for in the remodeled EBZ myocytes. Some of the ionic parameters were reduced as indicated in the legend.

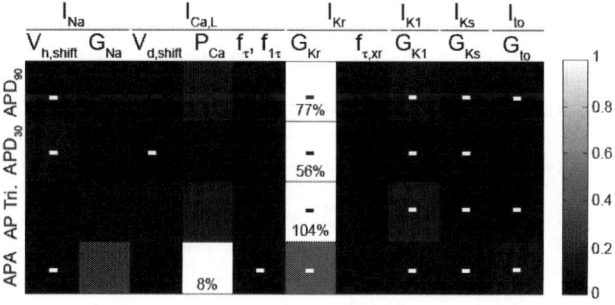

Fig. 4. Relative sensitivities of the arrhythmic risk biomarkers to variations involved in the remodeling of epicardical border zone. Negative sign indicates negative sensitivity.

Figure 3 shows a comparison of AP in EBZ myocytes with the principal parameters involved in the EBZ remodeling reduced in 30%. The reduction of G_{Kr} prolongs APD significantly.

Figure 4 shows a comparison of the relative sensitivities ($r_{c,p}$) of all combinations of electrophysiological characteristics to parameter variations using a gray scale.

Parameters exerting a negligible influence or null on all biomarkers are shown. Negative sensitivities are marked with a negative sign.

APD_{90} and APD_{30} are mainly affected by changes in G_{Kr}, therefore the AP triangulation is affected too, while the APA is affected more by changes in Ca^{2+} permeability and in a lesser extent by changes in h_∞ and G_{Kr}.

IV. CONCLUSIONS

The absence of in vivo studies in human infarct border zone requires the approximation of action potential models to changes according to experimental studies in other species and whose dynamics are similar to the human ventricular cell.

The results found are consistent with those obtained by others authors in others species. The remodeling we have presented serves as a basis to approximate the behavior of the ventricle human myocytes subject to previous infarction and analyze the arrythmogenic consequences.

ACKNOWLEDGMENT

This work was partially supported by the Plan Avanza en el marco de la Acción Estratégica de Telecomunicaciones y Sociedad de la Información del Ministerio de Industria Turismo y Comercio of Spain (TSI-020100-2010-469), "VI Plan Nacional de Investigación Científica, Desarrollo e Innovación Tecnológica" from the Ministerio de Economía y Competitividad of Spain (TIN2012-37546-C03-01), and Dirección General de Política Científica de la Generalitat Valenciana (GV/2013/199).

REFERENCES

1. Aggarwal R, Boyden PA. Diminished Ca^{2+} and Ba^{2+} currents in myocytes surviving in the epicardial border zone of the 5-day infarcted canine heart. Circ Res. 1995 Dec;77(6):1180-91.
2. Cabo C, Boyden PA. Electrical remodeling of the epicardial border zone in the canine infarcted heart: a computational analysis. Am J Physiol Heart Circ Physiol. 2003 Jan;284(1):H372-84.
3. Decker KF, Heijman J, Silva JR, Hund TJ, Rudy Y. Properties and ionic mechanisms of action potential adaptation, restitution, and accommodation in canine epicardium. Am J Physiol Heart Circ Physiol. 2009 Apr;296(4):H1017-26. doi: 10.1152/ajpheart.01216.2008.
4. Decker KF, Rudy Y. Ionic mechanisms of electrophysiological heterogeneity and conduction block in the infarct border zone. Am J Physiol Heart Circ Physiol. 2010 Nov;299(5):H1588-97. doi: 10.1152/ajpheart.00362.2010.
5. Dun W, Baba S, Yagi T, Boyden PA. Dynamic remodeling of K^+ and Ca^{2+} currents in cells that survived in the epicardial border zone of canine healed infarcted heart. Am J Physiol Heart Circ Physiol. 2004 Sep;287(3):H1046-54.
6. Gough WB, Mehra R, Restivo M, Zeiler RH, el-Sherif N. Reentrant ventricular arrhythmias in the late myocardial infarction period in the dog. 13. Correlation of activation and refractory maps. Circ Res. 1985 Sep;57(3):432-42.
7. Jiang M, Cabo C, Yao J, Boyden PA, Tseng G. Delayed rectifier K currents have reduced amplitudes and altered kinetics in myocytes from infarcted canine ventricle. Cardiovasc Res. 2000 Oct;48(1):34-43.
8. Litwin SE, Zhang D. Enhanced sodium-calcium exchange in the infarcted heart: effects on sarcoplasmic reticulum content and cellular contractility. Ann N Y Acad Sci. 2002 Nov;976:446-53.
9. Lue WM, Boyden PA. Abnormal electrical properties of myocytes from chronically infarcted canine heart. Alterations in Vmax and the transient outward current. Circulation. 1992 Mar;85(3):1175-88.
10. McDowell KS, Arevalo HJ, Maleckar MM, Trayanova NA. Susceptibility to arrhythmia in the infarcted heart depends on myofibroblast density. Biophys J. 2011 Sep 21;101(6):1307-15. doi: 10.1016/j.bpj.2011.08.009.
11. O'Hara T, Virág L, Varró A, Rudy Y. Simulation of the undiseased human cardiac ventricular action potential: model formulation and experimental validation. PLoS Comput Biol. 2011 May;7(5):e1002061. doi: 10.1371/journal.pcbi.1002061.
12. Pinto JM, Boyden PA. Electrical remodeling in ischemia and infarction. Cardiovasc Res. 1999 May;42(2):284-97. Review.
13. Pu J, Boyden PA. Alterations of Na^+ currents in myocytes from epicardial border zone of the infarcted heart. A possible ionic mechanism for reduced excitability and postrepolarization refractoriness. Circ Res. 1997 Jul;81(1):110-9.
14. Pu J, Robinson RB, Boyden PA. Abnormalities in Ca(i)handling in myocytes that survive in the infarcted heart are not just due to alterations in repolarization. J Mol Cell Cardiol. 2000 Aug;32(8):1509-23.
15. Rantner LJ, Arevalo HJ, Constantino JL, Efimov IR, Plank G, Trayanova NA. Three-dimensional mechanisms of increased vulnerability to electric shocks in myocardial infarction: altered virtual electrode polarizations and conduction delay in the peri-infarct zone. J Physiol. 2012 Sep 15;590(Pt 18):4537-51. doi: 10.1113/jphysiol.2012.229088.
16. Romero L, Pueyo E, Fink M, Rodríguez B. Impact of ionic current variability on human ventricular cellular electrophysiology. Am J Physiol Heart Circ Physiol. 2009 Oct;297(4):H1436-45. doi: 10.1152/ajpheart.00263.2009.
17. Trenor B, Cardona K, Gomez JF, Rajamani S, Ferrero JM Jr, Belardinelli L, Saiz J. Simulation and mechanistic investigation of the arrhythmogenic role of the late sodium current in human heart failure. PLoS One. 2012;7(3):e32659. doi: 10.1371/journal.pone.0032659.
18. Yao JA, Hussain W, Patel P, Peters NS, Boyden PA, Wit AL. Remodeling of gap junctional channel function in epicardial border zone of healing canine infarcts. Circ Res. 2003 Mar 7;92(4):437-43.

Address for correspondence:

Author: José V. Visconti Gijón
Institute: I3BH. Universidad Politécnica de Valencia
Street: Camino de Vera s/n
City: 46022 Valencia
Country: Spain
Email: jvvisconti@gbio.i3bh.es

Comparing Hodgkin-Huxley and Markovian Formulations for the Rapid Potassium Current in Cardiac Myocytes: A Simulation Study

E. Godoy, L. Romero, and J.M. Ferrero

I3BH, Universitat Politecnica de Valencia, Valencia, Spain

Abstract—Two mathematical approaches have been used to describe ion channel kinetics in cardiac action potential simulations: the classical Hodgkin & Huxley (HH) gating model formalism and state transitions Markov models (MMs). This last approach is more flexible and it is capable of characterizing state transitions dependent on the state of the channel, consequently being able to match experimental observations on gating currents more accurately and mechanistically. However, the high number of differential equations associated to MMs makes multi-cellular whole-organ simulations very costly from a computational point of view. Moreover, the large number of transition rates implicit in the model also adds difficulties in parameter estimation.

The aim of this study is to compare the results obtained with both formalisms when simulating the effects of the delayed rectifier K+ current (I_{Kr}) in the action potential in isolated cardiac cells. For this purpose, a new mathematical model for I_{Kr} based on the HH formalism ($I_{Kr,HH}$) using the Clancy – Rudy ($I_{Kr,CR}$) dynamics for guinea pig ventricular cells was developed. The dynamic characteristics of the CR model were fitted to the HH formalism to obtain the parameters for the new HH current. This new model was incorporated to the Clancy – Rudy action potential model to simulate the steady state action potential and ionic currents under physiological cycle lengths, and the action potential duration (APD) restitution curve.

Our results show similar action potential and ionic current waveforms and APD values with both I_{Kr} models, although some differences are observed on the I_{Kr} time course. We conclude that the HH formalism is appropriate and more efficient than the MM formalism to simulate I_{Kr} current behavior during the AP under the normal physiological conditions considered in this study.

Keywords—Simulation, ionic current, cardiac action potential, Hodgkin-Huxley, Markov.

I. INTRODUCTION

The first computational model of the action potential (AP) was formulated by Hodgkin and Huxley for the squid giant axon [1]. This model mathematically describes how AP in neurons are initiated and propagated. Hodgkin and Huxley models allow the reproduction of macroscopic ionic currents. The conductance of an ion channel is computed as a function of the open probability of each gate of the channel and the maximum conductance of the membrane for that ion. In addition, gate transitions from closed to open and vice versa are independent of the remaining gate states.

However, this formulism is unable to reproduce the later experimental observation of state dependent transitions of ion channels. This fact contributed to the appearance of Markov models in 1999 [2]. In these models transitions between channel states depend on the channel state and not on its past history.

In principle, Markovian models are able to reproduce ion channel behavior more faithfully than Hodgkin and Huxley models. However, Markov models add complexity in parameter estimation and are more computationally demanding. Therefore, both Hodgkin and Huxley and Markov models are currently being used to simulate ion channels [3][4].

The aim of this study is to compare the differences between these two formalisms during steady-state AP development and in the resulting dynamic and static restitution curves. For this purpose, a new model ($I_{Kr,n}$) for I_{Kr} with the same electrophysiological characteristics as the Clancy and Rudy I_{Kr} model ($I_{Kr,CR}$) [5] has been developed using the Hodgkin and Huxley formalism.

II. METHODS

A. New I_{Kr} Model Development

A new mathematical model for I_{Kr} based on the Hodgkin–Huxley formalism (referred to as $I_{Kr,HH}$ hereinafter) was formulated. The model parameters for the $I_{Kr,HH}$ current kinetics were fitted using a Non-linear Least Square – Trust Region algorithm to reproduce the electrophysiological characteristics of the I_{Kr} formulated by Clancy and Rudy using the Markov formulism ($I_{Kr,CR}$) [5], namely, the steady state activation curve, steady state inactivation curve, activation time constant, the deactivation time constant, the inactivation time constant and the recovery from inactivation time constant. The electrophysiological characteristics of the I_{Kr} formulated by Clancy and Rudy were obtained from the simulation of standard patch-clamp protocols [6].

Figure 1 shows the steady state activation curve, steady state inactivation curve, and activation and deactivation time constant of the new model (solid red lines) together

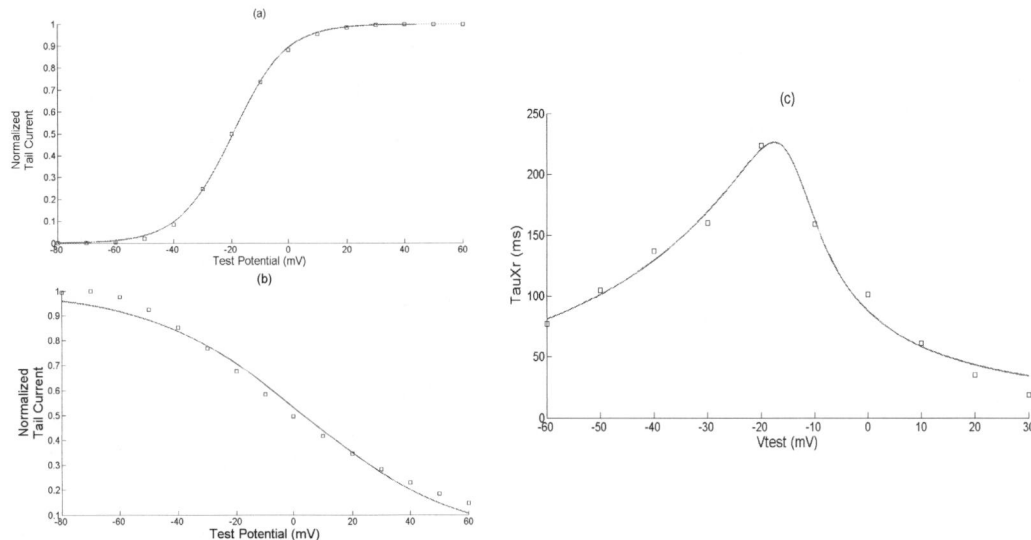

Fig. 1 Fitting of the kinetic parameters of I_{Kr}. (a) Activation curve. (b) Inactivation curve. (c) Time constant for activation.

with the kinetics of $I_{Kr,CR}$ (black circles). In addition, the maximal conductance of $I_{Kr,HH}$ was obtained from scaling the maximal conductance of $I_{Kr,CR}$ to elicit the same steady-state current values as $I_{Kr,CR}$ during the activation protocol.

B. Simulation of Action Potentials and Ionic Currents

In order to simulate the action potential and underlying ionic current, both the CR version of I_{Kr} ($I_{Kr,CR}$) and the HH version ($I_{Kr,HH}$) were respectively incorporated to the Clancy and Rudy model of the guinea-pig action potential [4]. Simulations were conducted at different basic cycle lengths (BCL) as detailed below, and the results using both versions of the current were compared.

C. Stimulation Protocols

Steady-state AP duration at 90% repolarization (APD_{90}), was calculated from the simulations following the application of a train of 300 square transmembrane current pulses at BCL of 1000 ms.

APD adaptation curves were depicted by plotting steady-state APD_{90} versus steady-state diastolic interval (DI) for BCLs ranging from 300 to 3500 ms.

The APD_{90} restitution curves were obtained using the S1S2 restitution protocol. The S1S2 restitution protocol consisted of a train of 300 square current pulses (S1) at a BCL of 1000 ms, followed by an extra stimulus (S2) applied at coupling intervals (CI) ranging from 50 to 5000 ms. The restitution curves were created by plotting the APD_{90} elicited by S2 versus DI for each CI.

III. RESULTS

Figure 2 (top panel) shows a comparison between action potentials simulated using the new $I_{Kr,HH}$ (red lines) and the Markov-formulated $I_{Kr,CR}$ (black lines) at four different BCL values. As seen in the figure, the action potentials elicited using the new HH formulation of I_{Kr} are slightly prolonged when compared to action potentials obtained with the Markovian CR formulation. At a BCL of 1000 ms, for instance, APD_{90} is three milliseconds longer with the HH formulation.

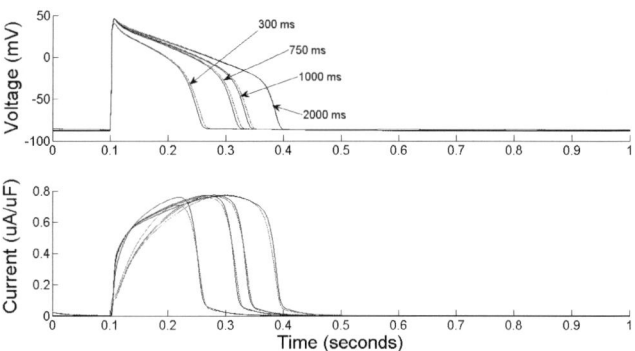

Fig. 2 Top: action potentials at different BCLs for the CR (black) and HH (red) formulation. Numbers indicate BCL value. Bottom: comparison of I_{Kr} time-courses.

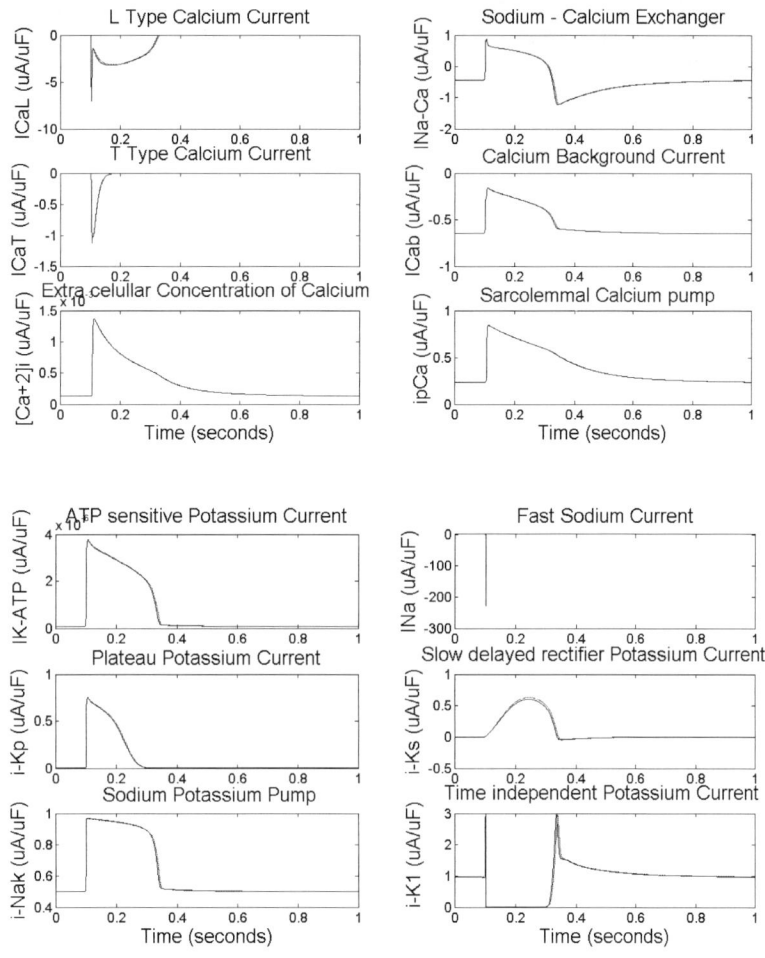

Fig. 3 Time-course of selected ionic currents and concentrations

Figure 2 (bottom panel) shows a comparison between the time-course of I_{Kr} current in both cases analyzed. It is clearly observed that the peak value of the current reaches very similar values in both kinds of simulations (which were expected, as the maximal conductance of $I_{Kr,HH}$ was scaled to elicit the same steady-state current values as $I_{Kr,CR}$ during the activation protocol in the patch-clamp simulations). However, the waveform of $I_{Kr,HH}$ is somewhat different from $I_{Kr,CR}$ in the rising phase of the current. With the Markovian CR formulation, the current activates with a very high rising slope and then decelerates before peaking, while in the HH case the rising phase is significantly slower. After the current reaches its peak, it returns to zero following similar time-courses in both cases.

To further analyze the differences between both formulations, other membrane ionic currents apart from I_{Kr} were plotted and compared. Figure 3 shows selected ionic currents (and also intracellular calcium concentration) obtained with the original CR formulation (black traces) and the HH alternative formulation (red lines) after pacing the cell with a BCL of 750 ms. As seen in the figure, the differences between the time course of currents and calcium concentration are minimal. Only the slow potassium current (I_{Ks}) and the inward rectifier potassium current (I_{K1}) show a small degree of noticeable differences in their peak values (though the difference are less than 5% in both cases).

The dynamic properties of action potentials elicited with both formalisms were also compared. To do so, the APD_{90}

restitution curve was chosen as representative of the dynamic behavior of the cell electrophysiology, as it is an important indicator of arrhythmogenesis [7]. Figure 4 shows the restitution curves obtained with the new HH formulation of I_{Kr} (red trace) and the Markovian CR formulation (black data). As expected (see Figure 2, top panel), APD_{90} values were slightly higher for the HH case. However, the differences were small (less than 7 ms for every BCL tested) and the trend of both curves was very similar. None of them showed a maximum slope higher than unity, which is associated to high risk of alternans and ventricular fibrillation [Qu et. al. 2010].

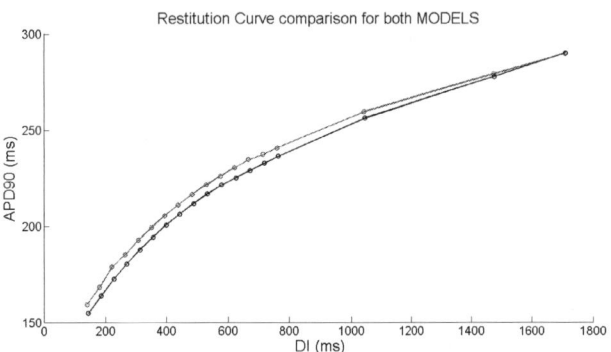

Fig. 4 Comparison between APD restitution curves (APD_{90} vs. Diastolic Interval) using the new HH formulation for I_{Kr} (red) and the Markovian CR one (black).

IV. CONCLUSIONS

We conclude that, under the simulated normal physiological conditions, the HH formulation for I_{Kr} does not yield significant differences in action potential, ionic current and ionic concentrations waveforms when compared to the more realistic (but more computationally costly) Markovian CR formulation. The dynamic properties of the action potential (APD restitution curve) do not show significant differences either.

If the HH formulation of I_{Kr} is realistic enough to model the current under pathological conditions (e.g. mutations that affect ion channel properties) and/or non-physiological conditions (e.g. conditions that cause afterdepolarizations) remains to be established.

ACKNOWLEDGMENT

This work was partially supported by the "VI Plan Nacional de Investigación Científica, Desarrollo e Innovación Tecnológica" from the Ministerio de Economía y Competitividad of Spain (TIN2012-37546-C03-01), by the Programa de Apoyo a la Investigación y Desarrollo (PAID-06-11-2002) de la Universidad Politécnica de Valencia, and by Dirección General de Política Científica de la Generalitat Valenciana (GV/2013/199).

REFERENCES

1. HODGKIN, A. L. & HUXLEY, A. F. 1952. A quantitative description of membrane current and its application to conduction and excitation in nerve. *J Physiol,* 117, 500-44.
2. CLANCY, C. E. & RUDY, Y. 1999. Linking a genetic defect to its cellular phenotype in a cardiac arrhythmia. *Nature,* 400, 566-9.
3. RUDY, Y. & SILVA, J. R. 2006. Computational biology in the study of cardiac ion channels and cell electrophysiology. *Q Rev Biophys,* 39, 57-116.
4. TRENOR, B., ROMERO, L., CARDONA, K., GOMIS, J., SAIZ, J. & FERRERO, J. M. 2011. Multiscale Modeling of Myocardial Electrical Activity: From Cell to Organ.
5. CLANCY, C. E. & RUDY, Y. 2001. Cellular consequences of HERG mutations in the long QT syndrome: precursors to sudden cardiac death. *Cardiovasc Res,* 50, 301-13.
6. SANGUINETTI, M. C. & JURKIEWICZ, N. K. 1990. Two components of cardiac delayed rectifier K+ current. Differential sensitivity to block by class III antiarrhythmic agents. *J Gen Physiol,* 96, 195-215.
7. QU, Z., XIE, Y., GARFINKEL, A. & WEISS, J. N. 2010. T-wave alternans and arrhythmogenesis in cardiac diseases. *Front Physiol,* 1, 154.

Author: Eduardo J. Godoy
Institute: GBio-e, I3BH - UPV
Street: Camino de Vera, s/n. (46022)
City: Valencia
Country: Spain
Email: egodoy@gbio.i3bh.es

Mechanical Properties of Different Airway Stents

A. Ratnovsky[1], N. Mell[1], S. Wald[1], M.R. Kramer[2], and S. Naftali[1]

[1] Afeka, Tel Aviv Academic College of Engineering, Medical Engineering, Tel Aviv, Israel
[2] Rabin Medical Center, Pulmonary Institute, Petach Tikva, Israel

Abstract—Central airway obstruction is frequently increases the morbidity and mortality rates of patients suffering from airway obstruction. Insertion of airway stents improves pulmonary function and the quality of life of these patients. The present study was conducted to examine the effect of the radial forces exerted by four main types of airway stents (silicon stent, balloon dilated metal stent, self- expanding metal stent and covered self expanding metal stent) on the trachea and to compare between their mechanical function. Mechanical measurements were done using experiment system that consists of force gauge and self-made adaptors. Numerical simulations were performed on 8 different stents geometries that were inserted into a 3D trachea model. The results of the simulations clearly showed correlation between the diameter of the stent and the stresses. Stents composed of Cobalt alloy and Stainless steel 316L exerted the highest stress value while stents composed of silicon exerted the lowest value. Covering metal stents with less stiff material as silicon or polyethylene reduce significantly the value of stresses. Stenosis (symmetric or non-symmetric) increases significant the level of stresses in all stents. In conclusion, the low stresses found in the silicon stents may be due to weak contact between the stent and the trachea and can explain its migration. Thus, silicon stents are recommended for use in case of short time therapy. Metal stents covered with silicon or polyethylene reduce the stresses on the trachea while still retain strong contact with it. Therefore, they may reduce formation necrosis of mucosa and fistulas while still prevent stent migration.

Keywords—**Airway, Stent, Airway Obstruction, Stress, Numerical Simulation.**

I. INTRODUCTION

Obstruction of the central airways may be due to a variety of benign processes, malignant stenosis and airway complications following lung transplantation. The obstruction can produce symptoms that affect the patient's quality of life (dyspnea, stridor, etc.) and can bring to significant morbidity and mortality rates. Insertion of airway stents can improve pulmonary function and the quality of life at patients with airway obstruction. The stent are design to restore airway patency and to mimic normal airway anatomy and physiology [1]. In general the airway stents can be classified into four groups: Silicone stent, Balloon dilated metal stent, Self-expanding metal stent and Covered Self expanding metal stent. The silicone stents have the advantage of being easily removed and exchanged when needed. However, their major drawbacks are stent migration and obstruction [2]. Metallic stent is usually made of steel or nitinol and they can be placed via flexible bronchoscopy under local anesthesia. The balloon expandable metallic stent relies on the ballon to diliate it to its correct diameter at the target site. Due to their lack of elastic re-expansion they can be crushed or deformed with coughing and should not be used in the central airways. However, they have often been used in children [3]. A self-expanding stent has a shape memory that enables it to assume its predetermined configuration when released from a constraining delivery catheter [2]. These kinds of stents can be made from steel or nintol and can be covered or uncovered by thin polyurethane membranes [2]. They have less migration effect, less interface with cilia, and they are thinner so their inner diameter is wider. However, expandable metallic stents are difficult to remove, because of the airway epithelium that can grow in and around them, and they also susceptible to metal fatigue and fracture over time. Covered Self expanding metal stent has a decreased risk of tumor ingrowth and granulation tissue formation. The Disadvantages of this type of stents are radial force that may cause necrosis of mucosa and fistula formation and tumor in growth along the length of the stent.

There are several parameters according to which a stent is selected for a particular patient. These include permanent or temporary stent, general or local anesthetic, high or low likelihood to migration, and if it encourages growth of tissue [1]. Many advantages as well as disadvantages are associates with each type of stent. However, no data have shown the benefit of one stent rather than another [4]. The main drawback of airway stents is the growth of tissue around the stent and inside due to radial forces. The effect of the radial forces exerted by the different stents on the airways has not been investigated yet. The objective of this study was to examine the effect of the radial forces exerted by the stents on the trachea and to compare between the mechanical function of the four main types of airway stents.

II. METHODS

The mechanical function of the stents has been inquired using an experimental setup which consists of force gauge and self-made polyester film adaptor (Fig. 1). The adaptor dimensions were fit to each stent circumference. Stress strain curve for each stent under variable radial forces were obtained.

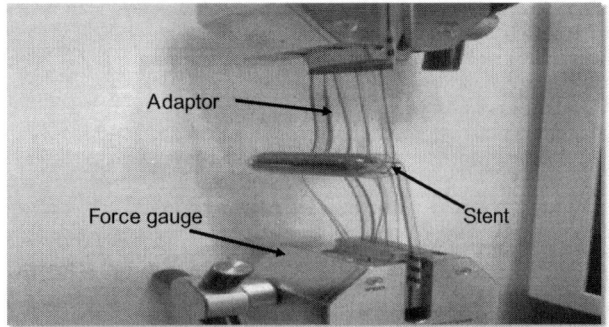

Fig. 1 The experimental setup

Numerical simulations were performed on 8 different stents (smooth silicon and stud silicon stents, 2 types of balloon expandable metallic stents, 2 types of uncover self expandable stents and 2 types of cover self expandable stents) that were inserted into trachea model (Fig.2). The dimensions, geometries and model materials were correlated with existing stents. In order to compare between the different stents a length of 50 mm and diameters of 14, 15 and 16 mm were chosen for all stents. The trachea geometrical model consists of two main layers: smooth muscle and cartilage. The muscle layer was designed as a hollow cylinder. The cartilage layer was composed of twenty C - shaped rings. Three different tracheas model were built: healthy trachea, trachea with 30% symmetric stenosis and trachea with 30% non-symmetric stenosis. The trachea materials were modeled as linear elastic materials. Most of the mechanical properties of the cartilage and the smooth muscle were taken from the literature [5].

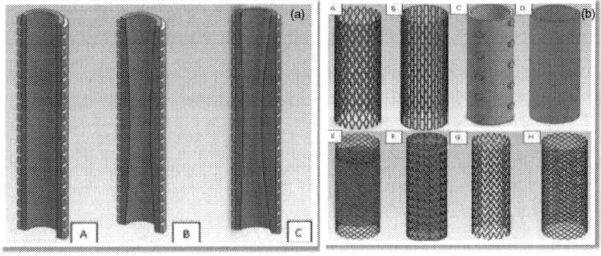

Fig. 2 (a) The three tracheas model: (A) healthy trachea, (B) trachea with 30% symmetric stenosis and (C) trachea with 30% non-symmetric stenosis. (b) The eight types of stent

Some of the values for the smooth muscle were not available and therefore were assumed. The simulations were performed by SolidWorks simulation Ver. 2011 which is commercial software of finite element methods (FEM) using the Shrink Fit feature. The mesh type was defined as Lagrazian, and the elements were defined as 3D, 40,000-180,000 - 4 nodes - tetrahedral elements were used.

The stents materials were assumed homogeneous and isotropic with no temperature dependence. Linear range of stresses and displacements were assumed. The trachea was modeled as an elongated cylinder with a uniform diameter and thickness. Uniform surface load distribution was used.

A steady-state form of dynamic equation was solved (Eq. 1).

$$Ku = F \qquad (1)$$

The boundary conditions that were defined: the force (F) on the external wall of the stent is equal to the force on the inner wall of the trachea (Eq. 2), radial displacements (U) on the external wall of the stent are equals to the inner wall of the trachea (Eq. 3) and zero displacement at the harnessed edges of the trachea (Eqs. 4-5).

$$F_{trachea} = F_{stent} \qquad (2)$$

$$U_{trachea} = U_{stent} \qquad (3)$$

$$u(z,r,\theta)\big|_{harnessed\ area} = 0 \qquad (4)$$

$$u(z,\theta)\big|_{harnessed\ area} = 0 \qquad (5)$$

The influence of different stent diameter, stent geometry and stent material on the radial stresses were investigated using the numerical simulations. In addition, the influence of inserting the stent into model of healthy trachea versus model of trachea with 30% symmetric and non-symmetric stenosis was examined.

III. RESULTS

A. Experimental Measurements

The metal cover stents have both higher linear range and higher stiffness than the metal stents. The silicone stent has relatively narrow linear range but higher stiffness compare to the other stents. At the end of the linear range the radial forces distort the stent shape which explains the sharp reduction in the stress at the end of the linear range (Figs 3-4).

B. Numerical Simulation

As was expected, in all stents, correlation was found between the diameter of the stent and the stress distribution.

The higher differences in the stresses between stents with different diameters were found in the healthy trachea (Fig. 5). Stents composed of stiff material generate higher stresses (Fig. 6), thus, stents composed of cobalt alloy and the smooth silicon stent. Both types exerted lower stress than the metal stents (Fig. 7). Obstruction of the trachea stainless steel 316L obtained the highest value. The silicon stud stent exerts higher stresses on the trachea compare to

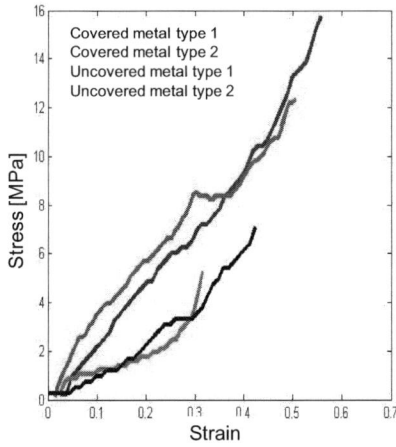

Fig. 3 Stress strain curves of metallic stents

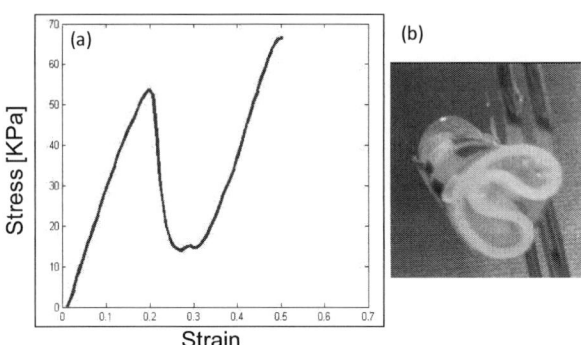

Fig. 4 (a) Stress strain curves of the silicon stents (b) Folding of the silicon stent during experiment

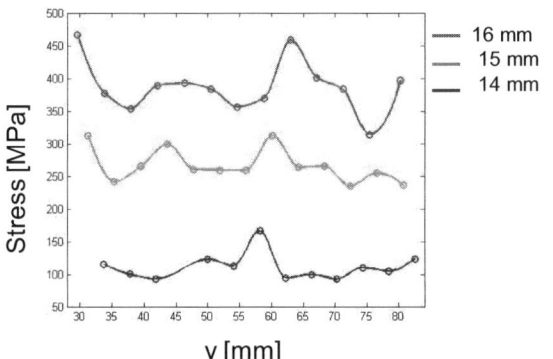

Fig. 5 Stress distribution on healthy trachea along 3 different diameters metal stents axis

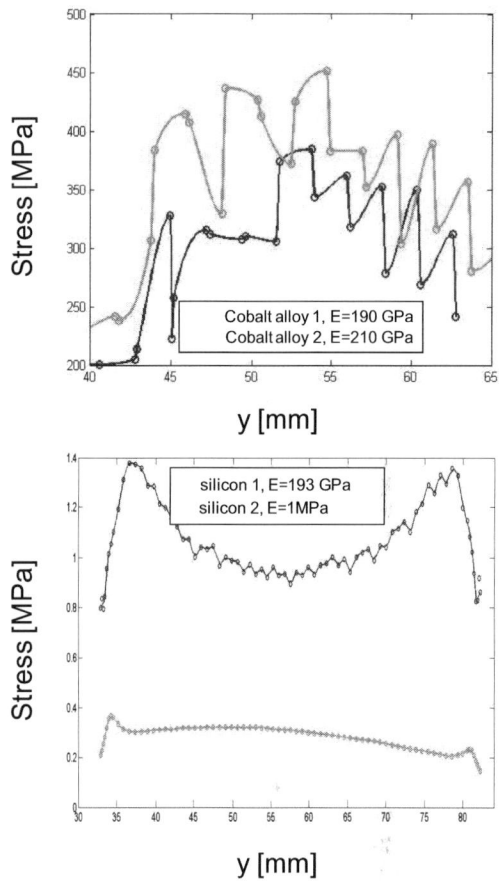

Fig. 6 Stress distribution on trachea with symmetric stenosis along the axis of stents with different material properties

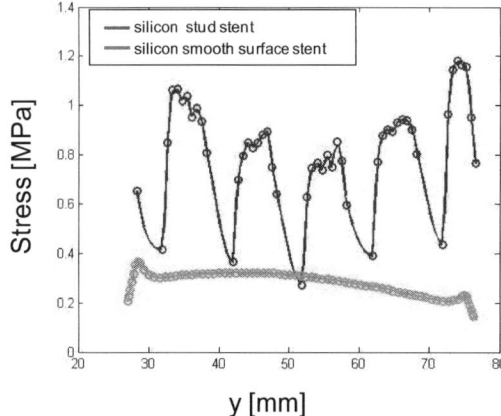

Fig. 7. Stress distribution on trachea with symmetric stenosis along stud and smooth surface silicon stents axis

causes a significant raise to the stress distributions. The covered material absorbed part of the pressure caused due to the interaction between the metal stents and the trachea. Therefore, low stresses were found in the cover stents compare to the uncovered stents. Smooth structure designed stents produced lower stresses. Smooth structure designed stents produced lower stresses.

IV. CONCLUSIONS

The low stresses found in the silicone stents may indicate weak contact with the trachea and can explain their migration. The high stresses found in the metal stent can encourage tissue growth and restenosis. Covered metal stents reduce the stresses on the trachea, while still retain strong contact with it. Therefore, these stents are most recommended for use in the trachea.

REFERENCES

1. Ranu H, Madden B.P (2009) Endobronchial stenting in the management of large airway pathology. Postgrad Med J .85: 682-687
2. Lee P, Kupeli E, Mehta A.C (2010) Airway Stents. Clin Chest Med 31: 141-150
3. Nicolai T (2008) Airway Stents in Children. Pediatr Pulmonol 43: 330–344
4. Chin C.S, Litle V, Yun J, Weiser T, Swanson S.J (2008) Airway Stents.,Ann Thorac Surg 85(2): S792-S796
5. Trablsi O, Perez A, Mena A, Lopex-Villalbos J.L, Ginel A, Doblare M (2011) FE simulation of human trachea swallowing movement before and after the implantation of an endoprothesis. J App Math Model, 35: 4902-4912

Author: Anat Ratnovsky
Institute: Afeka, Tel Aviv Academic College of Engineering
Street: Beny Efraim
City: Tel Aviv
Country: Israel
Email: ratnovskya@afeka.ac.il

Site-Specific Mechanical Properties of Aortic Bifurcation

J. Kronek[1], L. Horny[1], T. Adamek[2], H. Chlup[1], and R. Zitny[1]

[1] Laboratory of Biomechanics, Faculty of Mechanical Engineering, Czech Technical University in Prague,
Technicka 4, 16607, Prague, Czech Republic
[2] University Hospital Na KralovskychVinohradech, Srobarova 50, 10034, Prague, Czech

Abstract—This study describes stiffness of arterial wall near aortic bifurcation. Uni-axial tensile tests with specimens cut out from human abdominal aorta, aortic bifurcation and common iliac arteries were carried out. Material anisotropy was verified as well as site-specific stiffness. The stiffness was described by initial secant modulus of elasticity (E_{INIT}). This parameter was gained from stress/strain data at 12% of deformation. The tissue samples in various regions and directions were concluded as described as follows: region of aorto-iliac bifurcation is significantly stiffer in longitudinal direction than in the circumferential direction in all of three examined parts of arterial tree. The comparison between tissue samples cut and loaded in longitudinal direction indicates that arterial tree stiffens in caudal direction. No significant change in material properties was found in the apex of bifurcation in comparison with adjacent tissue. Stiffening of arterial tree during ageing was verified.

Keywords—aortic bifurcation, mechanical properties, atherosclerosis, anisotropy.

I. INTRODUCTION

The relationship between atherosclerosis and the mechanical state of arterial wall have been studied extensively. For instance [1] reported that regions of arterial wall near ostia of branches, which represent concentrators of the wall stress, are saturated by low-density lipoproteins (LDL) more than straight parts of arteries. Similar results were obtained by Bratzleret in [2]. It suggests that mechano-biological interaction could be a key factor for atherosclerosis development. The question is which one is the main factor and which factors do not play a role. Authors of [3] wonder, why intramyocardial coronary arteries (inside the heart) remain free of atherosclerosis, whereas the epicardial arteries (on the heart surface) do not. The result probably is, that surrounding tissue (myocardium) acts as a support, which decrease the transmural pressure and thus it decrease circumferential stress and strain of these arteries. Previous complex study [4] examined, among others, the distribution of LDL in the vertebral artery. Amount of LDL in the parts of the vertebral artery passing through bone canal is lower than in the parts of the artery between the two bone canals. Surrounding stiff bones also play a role of support protecting the artery from systolic stretch. Both of these studies suggest that the transmural pressure, which causes circular (and partly axial) stress, is a main atherosclerosis generator / accelerator instead of flow conditions. Another evidence of relationship between transmural pressure and atherosclerosis is a well known risk factor of high blood pressure. According to Fuster et al. in [5] the probability of coronary artery disease and stroke increases exponentially with blood pressure.

Stress and strain distribution within arterial wall is next to boundary conditions (pressure, surrounding) given by both geometry (thickness, branch and non-planarity angles) and material properties (nonlinear stiffness, viscoelasticity, anisotropy). For instance the geometry of a FEM model of the artery bifurcation [6] is based on precise medical imaging, material properties are not site-specific. This study and the later publication [7] deal with bifurcations and branches of arteries as stress concentrators.

Physiological principle of homeostasis governs mechano-biological interaction of living tissues with mechanical environment. A response of a tissue to changes in mechanical environment can be mediated by changes in geometry (radius, thickness) but also with a change in internal structure which is then observed as a change in macroscopical mechanical properties. That is why an understanding of site-specific stress-strain relationships is essential for etiology or tissue engineering.

Aorto-iliac bifurcation (AB) is an area dividing high pressure blood flow from descending abdominal aorta into pelvis and lower limbs.

AB seems to be a suitable part of arterial tree for studying phenomena of material non-uniformities because of several reasons: 1) it is large enough to prepare samples for mechanical tests from it, 2) it is geometrically fairly uniform (more than the other bifurcations of large arteries) and 3) obstructed arteries in the area of AB is also clinically serious (decreased blood supply of lower extremities often causes pain and even may lead to amputation).

There are only a few of articles [8-10] dealing with geometrical parameters of AB systematically. Moreover, to the best of our knowledge no study investigating material properties at AB linked to pathological conditions is available in the literature.

II. MATERIAL AND METHODS

Five cadaveric tissue donors were involved in present study (age 22, 50, 51, 58, 64 years, all male). Abdominal aorta (AA), aortic bifurcation (AB) and common iliac arteries (CIA) were resected from each cadaver. The tissue specimens were cut out and prepared for mechanical testing immediately after removing. Rectangular strips were prepared and grouped according to their location and orientation as schematically described in Fig.1.

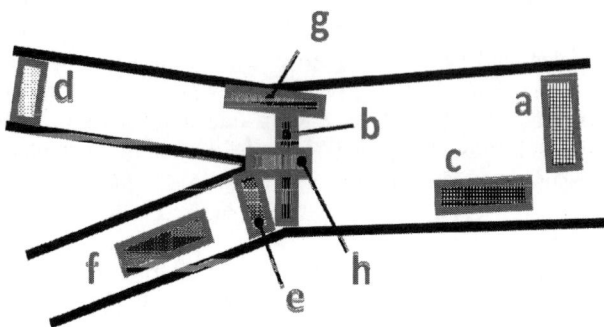

Fig 1 Position and orientation of samples groups: samples cut out from AA in the circumferential direction a) distant from AB (n=6) and b) near the AB (n=3), c) samples cut out from AA in the longitudinal direction (n=6), samples cut out from CIA in the circumferential direction d) distant from the AB (n=3) and e) near AB (n=5), f) samples cut out from CIA in the longitudinal direction (n=4), g) samples cut out from lateral side of AB (n=4) and h) samples cut out from apex (flow divider) of AB (n=4).

A pair of lengthwise and a pair of crosswise lines were painted on intimal surface of the samples to allow us optical recording of deformation during loading. Each sample was loaded in the direction of its longer side in customer-specific tensile testing machine (Zwick/Roell). Four precycles (up to the stretch $\lambda=1.1$) and stress relaxation for 100s were applied prior to final testing (up to $\lambda=1.4$ or a failure; Fig.2). Loading speed was 0.1mm/s (crosshead velocity). It means that stretch rate was between $0.003s^{-1}$ and $0.01s^{-1}$ according to the sample length. Acting force was recorded by force transducers (U9B HBM). The geometry of samples was measured before testing. Nominal engineering stress-strain curves were evaluated from the final loading cycle.

Stress-strain data were fitted by two-parameter Mooney-Rivlin hyperelastic model (not presented in this paper). In order to simplify interpretation of mechanical properties of the tissue, initial secant modulus of elasticity (E_{INIT}) was also evaluated considering it as a slope of the line from the origin to a point [$\varepsilon=0.12$, $\sigma(\varepsilon=0.12)$] or a point of failure initiation. The strain was calculated as a ratio between the elongation (actual length minus reference length) and the reference length. Reference length is the length of a sample in the beginning of a final loading cycle. Also width and thickness of the samples were measured in reference configuration. An example of stress-strain curve is shown in the Fig. 3.

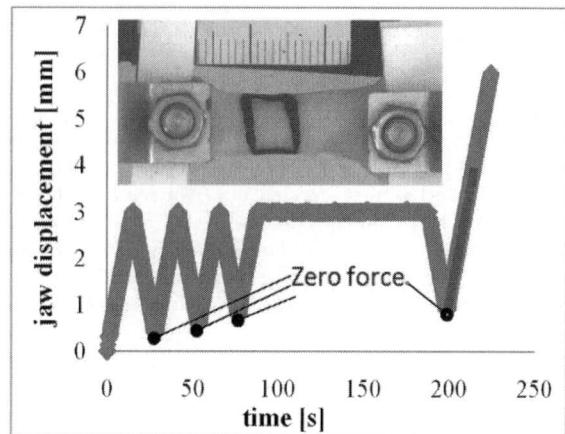

Fig. 2 Loading regime consists in four precycles (up to $\lambda=1.1$), 100s stress relaxation and final loading cycle up to a failure. Red line denotes evaluated zone. The picture of clamped sample is inserted.

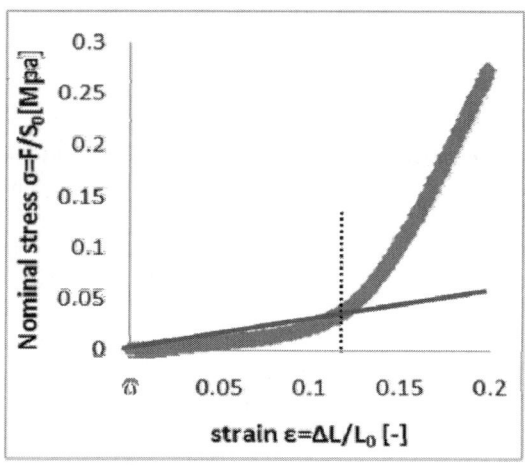

Fig. 3 Example of stress-strain relationship. The slope of red line represents initial secant modulus of elasticity.

Computed E_{INITs} were normalized with respect to mean E_{INITs} calculated for each donor. Subsequently normalized E_{INIT} were grouped with respect to the location and orientation (Fig. 1) and averaged (Fig. 4). To better understand the procedure, there is a sequence of calculation steps: 1) parameters E_{INITs} were assigned to each measured sample, 2) mean E_{INITs} were calculated for each donor separately,

3) normalized E_{INITs} were calculated as E_{INIT} divided by relevant mean E_{INIT}, 4) normalized E_{INIT} from the same sample group were averaged.

III. RESULTS AND DISCUSSION

Considerable differences between parameters E_{INIT} donor by donor were found. Therefore the parameters were normalized. Results (see Fig. 4) suggest that in the region of AB arterial wall is significantly stiffer in longitudinal direction than in the circumferential direction (groups c vs. a, g vs. e or b and f vs. d). Comparison between f, g and c indicates that arterial tree stiffens in caudal direction. This is in accordance with well-known decrease in the number of elastic lamellae in the wall. When a variance of the results is considered, the data does not seem to suggest significant trend or differences in the circumferential direction. It should however be noted that results were obtained only for five donors as they are a part of the feasibility study.

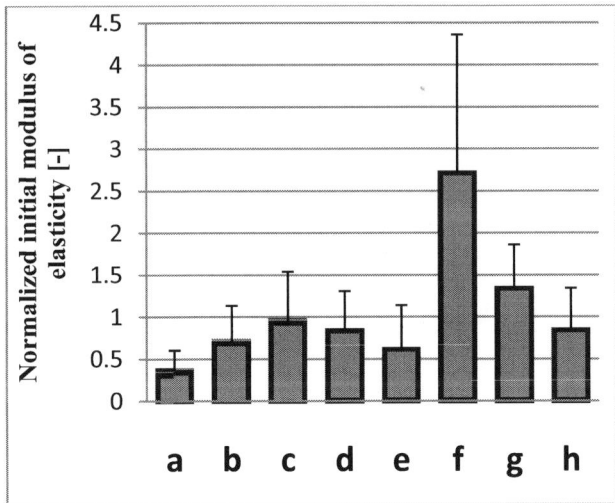

Fig. 4 Normalized initial moduli of elasticity. The figure shows how many times the one group of samples is stiffer than the average. The horizontal tick above each bar represents standard deviation.

Authors are aware of inaccuracies associated with the fact, that naturally anisotropic material was cut and tested in each case only in one direction. This procedure was chosen for two reasons. First geometrical complexity of arterial wall in bifurcations does not allow us to carry out a biaxial tensile test. Second the aim of the study was to illustrate site-specific mechanical properties as clear as possible based on the only one parameter. Normalized E_{INIT} was chosen as an appropriate parameter.

Referring to [7], the stress concentration factors (K_σ) differ only by 7% if an effect of anisotropy is taken into account or is not (evaluated from the FEM model of arterial branch). K_σ is a ratio between max stress (at the branch ostium) and nominal stress (at uniform tube-like part of a main artery). It could justify the performance of uni-axial tensile test with anisotropic material.

The relaxation zone prior to final testing was included in order to evaluate viscoelastic properties of the tissue. Unfortunately these stress-on-time data showed no significant trend and do not provide any additional information.

Finally it should be noted that the basic hypothesis of this approach is that material properties of arterial tree affect (initialize or accelerate) atherosclerosis, but also reverse causality applies. Thus, it is possible, that atherosclerosis consequence was measured rather than its reason. An influence of a plaque occurrence on wall stiffness was discussed in a large number of studies. During first stage of atherosclerosis, LDL enter the intima layer. The intima layer begins to be foam-like and soften. However, at a later stage of atherosclerosis calcification deposits make an arterial wall stiffer.

The positive correlation between the age and probability of arteriosclerosis is known. Our investigation confirmed this correlation. Generally the tissue samples of "older" donors (age 51, 58 and 64) were less ductile and stiffer than the tissue samples from "younger" donors (age 22 and 50). Comparison of E_{INIT} averaged over all samples from each donor can be seen in Tab.1. It is clear at first sight, why it was not possible to compare groups of samples without normalizing parameters E_{INIT} by the averages.

Table 1 Mean E_{INIT} for each donor. It is calculated from all samples for each donor separately. The tab. Illustrate, that arterial wall obtained from youngest donor, was generally weaker than the rest of samples.

age of the donor	mean E_{INIT}
22	0.28
50	0.66
51	2.63
58	2.57
64	2.40

IV. CONCLUSIONS

Mechanical properties of arterial wall were reduced to only one parameter – initial moduli of elasticity in corresponding direction.

This study showed that the region of AB does not differ sharply in mechanical properties from the arteries proximal and distal. It is implicated from above that this area represents a transition region, where mechanical properties change continuously from relatively compliant aorta to relatively stiff arteries of lower extremities. The tissue of arterial wall is stiffer in the longitudinal direction compared to the circumferential direction. No significant trend was found in circumferential direction.

ACKNOWLEDGMENT

The project was supported by the Grant Agency of the Czech Technical University in Prague, grant SGS13/176/OHK2/3T/12 and the Czech Ministry of Health, grant NT 13302.

REFERENCES

1. Thubrikar MJ, Keller AC, Holloway PW, Nolan SP (1992) Distribution of low density lipoprotein in the branch and non-branch regions of the aorta. Atherosclerosis. 97:1-9
2. Bratzler RL, Chisolm GM, Colton CK, Smith KA, Lees RS (1997) The distribution of labelled low-density lipoproteins across the rabbit thoracic aorta in vivo. Arteriosclerosis. 28:289-307
3. Robicsek F, Thubrikar MJ (1994) The freedom from atherosclerosis of intramyocardial coronary arteries: Reduction of mural stress – a key factor. Eur J Cardiothorac Surg. 8:228-235
4. Wolf S, Werthessen NT et al. (1976) Dynamics of Arterial Flow. Advances in Experimental Medicine and Biology, Vol. 115. Plenum, New York, 1976, pp 378
5. Fuster V, Ross R, Topol EJ et al. (1996) Atherosclerosis and Coronary Artery Disease. Volume 1. Lippincott-Raven Publishers, Philadelhia, 1996, pp 210-213.
6. Salzar, RS, Thubrikar MJ, Eppink RT (1995) Pressure-induced mechanical stress in the carotid artery bifurcation: A possible correlation to atherosclerosis, Journal of Biomechanics 28(11), pp 1333 - 1340
7. Thubrikar MJ (2007) Vascular Mechanics and Pathology, Springer Verlag
8. Bargeron C, Hutchins G, Moore G et al. (1986) Distribution of the geometric parameters of human aortic bifurcations, Arteriosclerosis, Thrombosis, and Vascular Biology, 6(1), pp 109-113
9. Shah PM, Scarton HA, Tsapogas MJ et al. (1978) Geometric anatomy of the aortic-common iliac bifurcation, Journal of Anatomy, 126(3), pp 451-458
10. Nanayakkara B, Gunarathne C, Sanjeewa A et al. (2009) Geometric anatomy of the aortic-common iliac bifurcation, Galle Medical Journal, 12(1), pp 8-12

Author: Jakub Kronek
Institute: Laboratory of Biomechanics, Faculty of Mechanical Engineering, Czech Technical University in Prague,
Street: Technicka 4
City: Prague
Country: Czech Republic
Email: jakub.kronek(a)cvut.cz

Cardiac Autonomic Nervous System Activity on Breathing Training

H. Nakamura[1], H. Saito[2], and M. Yoshida[1]

[1] Faculty of Biomedical Engineering, Osaka Electro-Communication Univeristy, Osaka, Japan
[2] Graduate School of Biomedical Engineering, Osaka Electro-Communication University, Osaka, Japan

Abstract—The purpose of this study is to evaluate effects of breathing training on cardiac autonomic nervous system (CANS) activities through Tone-Entropy analysis during a breathing training trial and before and after the training trial. CANS activity is intensely associated with breathing rhythm. In this experiment, the device that makes load to respiratory muscle is used to perform breathing training for twelve male subjects. R-R intervals were severally measured for 5 minutes during trials: the breathing training trial (T), the resting state before T trial (V1), the resting state trials after T trial were V2 and V3 in order. Our results show that parasympathetic activities in all the subjects were excited because Tone-Entropy plots distributed at relatively right-lower positions. HR became higher during T trial than any other trials. After T trials, parasympathetic activity was gradually increased from V2 to V2 trials owing to nervous system recovery. It is indicated that Tone-Entropy analysis show the detailed process of alteration of CANS activity, especially parasympathetic activity during breathing training and its recovery.

Keywords—Tone-Entropy analysis, cardiac autonomic nervous system, breathing training, long-term monitor.

I. INTRODUCTION

The purpose of this study is to evaluate effects of breathing training on cardiac autonomic nervous system (CANS) activities through Tone-Entropy analysis during a breathing training trial and before and after the training trial. CANS activity is intensely associated with breathing rhythm.

In clinical use, breathing training is performed to recover the respiratory function for their patients. Reis et al.[1] reported that cardiac autonomic control of heart rate is associated with inspiratory muscle weakness in chronic obstructive pulmonary diseases like COPD. Nussinovitch et al.[2] also reported that Deep Breath Test (DBT) was examined whether or not DBT can evaluate familial dysautonomia. The fact presents that DBT provides the possibility to diagnose some respiratory diseases. Therefore, Tone-Entropy analysis may help us to evaluate CANS activity during breathing training because Tone-Entropy analysis can absolutely measure CANS activity without breathing control instructions.

For verification of the Tone - Entropy analysis during breathing training, CANS activity during breathing training and its recovery were examined and discussed through Tone-Entropy analysis in this description.

II. METHODS AND MATERIALS

A. Subjects

Twelve young healthy male subjects (21.5+/-1.6 yrs; mean +/-SD) participated in this experiment. Their heights and body weights were 171.9+/-5.5 cm and 70.20+/-9.1 kg. SBP and DBP were also 121.4+/6.1 and 68.0+/-6.9 bpm respectively, which is mostly in normal condition. The subjects were asked not to do intense exercises from the previous day. This experiment was approved by Osaka Elctro-Communication University ethical committee. All the subjects accepted whole experimental protocols on this experiment with informed consent approved the committee.

B. Device or IMT

In this experiment, Threshold-IMT(PHILIPS, USA) were used. This device makes loads to respiratory muscle during inspiratory phase. This device is being used to prevent and cure the postoperative pulmonary complication on clinical scene. The device can adjust loading pressure from $7cmH_2O$ to $41cmH_2O$ during the inspiratory phase. The inspiratory loading pressure is recommended 30% of PImax. Inspiratory load on training trials in this experiment was fixed at $23.1cmH_2O$ because twenties' young people mean distribution indicates that 30% of PImax may be appropriate for respiratory training.

C. Experimental Protocol

The subjects were asked not to drink alcohol and play high-intensity sports since previous experiment day and to prevent caffeine before eight hours to experiment and then not to eat foods and medicines for two hours before.

The subjects seated in a comfortable chair. Before starting records, we were waiting until baseline of heart period becames stable. If the baseline of heart period was judged to be stable, we start to record R-R intervals without a cue because fluctuations of R-R intervals on any other external effects were avoided.

R-R intervals were measured during each trial: voluntary breathing before breathing exercise trial (V1), breathing exercise trial (T) and twice voluntary breathing after breathing exercise trial(V2 and V3). Every recording duration of their trials was fixed in 5 minutes. Non-recording intervals between V1 and T and between T and V2 were taken in

1 minute respectively. At between V2 and V3, there was no time interval and the recordings were being kept.

The subjects controlled breathing with rhythmic sounds with a metronome. Breathing frequency was defined at 10/min in T trials owing to the rhythm close to voluntary breathing rhythm. The rhythmic sounds from sound speaker and visual feedback with an object's movement on PC monitor enabled the subjects easily to breathe with a constant frequency. Before formal recording, the subjects were practically performed breathing training for a few minutes to be familiar to use the device and with breathing rhythm. There was no cue on the onset of recording to prevent heart period fluctuation caused with the cue.

D. The ECG Recording System

The ECG signals of whole data were recorded with CM5 position. The electrodes used in this experiment were disposable (NIHON KOHDEN, R-150). The ECG amplifier was originally manufactured. The effective frequency band is ranging from 10Hz to 100Hz and then its amplification in the effective frequency band mostly becomes 61dB.

The software system for recording ECG signals, which can operate simultaneously to save and display ECG signals on thread based programming technique, is also originally developed. The ECG signals amplified by the ECG amplifier saved to the PC with its sampling rate 1kHz and its resolution 16bit.

E. Tone-Entropy Analysis

Tone-Entropy analysis is a novel method which evaluates statistical property of variability of R-R intervals. For the purpose of examination of heart rate variability, Percentage Index (PI) was defined as the rate of variation between the R-R intervals and next one by Oida et al.[3]. Definitional equation of PI is shown as the equation (1):

$$PI_n = \frac{I_{n-1} - I_n}{I_{n-1}} \times 100 \ (\%) \quad (1)$$

CANS activities are evaluated on the tone and entropy which indicate statistical characteristics of PI. Tone-Entropy analysis needs a PI histogram whose frequencies in each class were accumulated from R-R intervals recorded for more than a couple of minutes for calculation of tone and entropy. Their bin width of PI histogram is determined as 1.0% from PI values. Tone means the expectation of PI distributions. Therefore, the equation of tone is shown as the equation (2).

$$\text{Tone} = \frac{1}{N} \sum_{i}^{N} PI_n \ (\%) \quad (2)$$

Oida et al.[3] showed that the physiological meaning of tone is associated with parasympathetic nervous system activity from the pharmacological blockade experiment.

Event probability p_j of the j-th PI class is defined as the ratio of PI frequency in the j-th class to total number of PI samples. In Tone-Entropy analysis, intensity of variability of R-R intervals is defined as information entropy. In the following sentences, *entropy* means information entropy. Entropy is the expectation of quantity of information in a distribution. Entropy is shown as the equation (3).

$$\text{Entropy} = -\sum_{j=-M}^{M} p_j \log_2 p_j \ (\text{bit}), \quad (3)$$

where M indicates the absolute value of the range of PI histogram. The base of the entropy is set at 2 because we follow a precedent in informatics. Then, entropy presents uniformity of a distribution: the broader a distribution becomes, the higher the entropy becomes. Therefore, entropy physiologically means intensity of whole CANS activities. Much fluctuation of R-R intervals, caused by CANS activities, makes an increase on the broader PI histogram. Therefore, entropy reflects entire CANS activities.

Fig.1 shows Tone-Entropy space which can absolutely evaluate CANS activity on quite different experiment. The small plots present Tone-Entropy plots of twenties' young people [4]. On Fig.1, the small plots form like a curve from left-higher to right-lower direction. If any subjects roughly keep physiological and mentally stationary condition, their plots surely place close to the curvilinear path. Therefore, if any plots are distant from the path, it can be recognized that the data surely have somewhat troubles on recording or analyzing or physiological abnormality like arrhythmia.

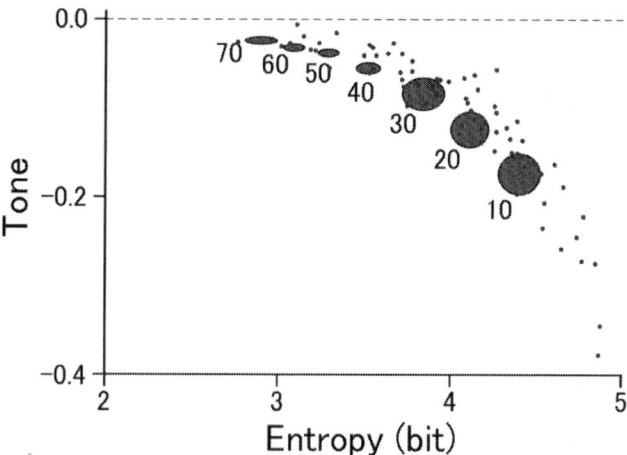

Fig. 1 The small plots show young peoples (twenties) distributes along a curvilinear path [4]. Amano et al.[5] reported that mean distributions of each age group gradually shift from right-lower to left-higher position approximately along the curvilinear path as their age increases in spite of quite different experiments each other.

F. Statistical Analysis

In this experiment, Tone, Entropy and HR were applied with one-way ANOVA to examine statistically significant variances between their trials. The critical ratio for statistical significance was determined at 5%. Also, in each pair of the trials, multiple comparison was also applied to compared whether or not the means of each pair have statistically significant difference in tone, entropy and HR. Here Bonfferoni's method was used at 5% critical ratio.

III. RESULTS

Fig.2 shows examples on alteration of mean HR in four trials severally. At breathing training trial (T), cyclic heart period variability is easily observed. The cyclic period almost corresponds breathing rhythm. Lung stretch reflex causes their cyclic arrhythmia via parasympathetic nervous system activity. Heart period variability in V2 was fluctuating between about 700ms and 1000ms. In contrast, heart period variability in V3 was fluctuating between about 800ms and 1100ms. In brief, HR risen by breathing training was descended from V2 to V3. Therefore, it is indicated that parasympathetic nervous system activities may be activated after breathing training for recovery.

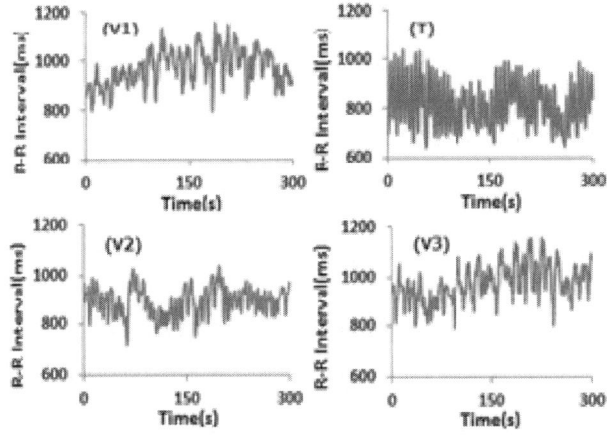

Fig. 2 Examples of R-R interval sequences in each trials.

The means and standard errors of HR at V1, T, V2, V3 are shown Fig.3. In between trials, there are statistically significant variances with one-way ANOVA (p=0.0024). In the next, to find significant differences between their trials, the Bonferroni's method as a multiple comparison was applied. Homogeneity of variance was not rejected. The fact suggests that multiple comparison is effective. Fig.3 shows significant differences between their trials with asterisk(s). T trial has significant differences as compared with any other trials. Therefore, HR in T trial was relatively increased with statistical significance. Any other pairs have no significant difference with critical ratio 5%.

The means of tone in T trial have statistically significant differences as compared with those in any other trials the same as HR. However, entropy between T and V3 trials was slightly accepted with the null hypothesis (p=0.0521) although the means of entropy in the pairs of T vs. V1 and T vs. V2 have significant differences. The significant decrease of tone and the significant increase of entropy indicate that parasympathetic nervous system was relatively activated from the previous facts on Tone-Entropy analysis. Acceptance of the null hypothesis of T trial vs. V3 trial indicates the difference between T and V3 trials was shortened although the p-value was only 0.0521.

Also, Fig.4 shows the individual plots form a curve shown in Fig.1. In this experiment, the mean of participated

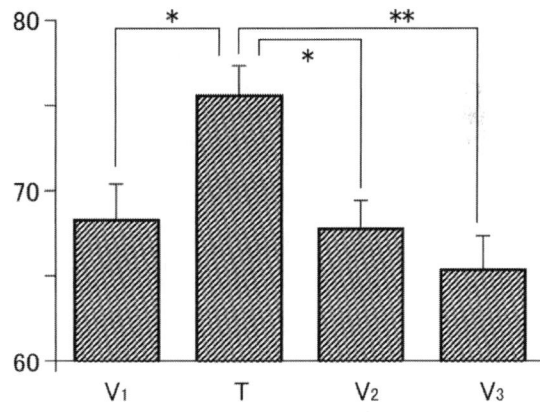

Fig. 3 Heart rates (mean+/-SE) on each trial. In Training trial, the highest heart rate was shown. After that, heart rates gradually decreased from V2 to V3. *p<0.05, **p<0.01 with Bonferroni method.

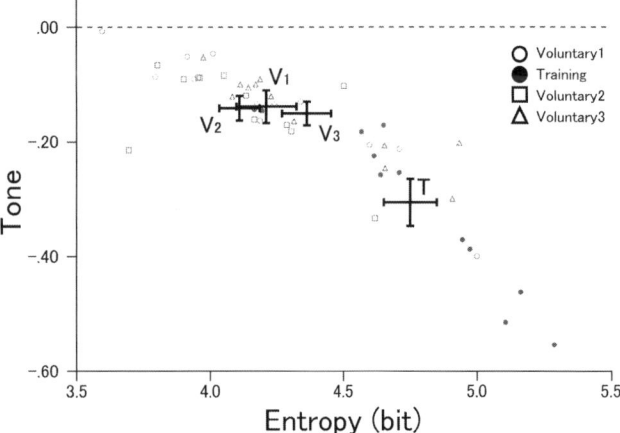

Fig.4 Each distributions of Tone-Entropy plots on each trial. The distribution of T was placed at right-lower position. V1 was the most left-higher position and then V2 shifted to right-lower direction.

subjects' age was 21.5 yrs. The means of the tone and entropy of V1 trial were -0.138 and 4.21bit. Then, those of twenties' plots in Amano et al.[5] reported were -0.124 and 4.11bit. In spite of quite different situations and participants, their mean values are almost the same. The fact indicates that Tone-Entropy analysis has high reproducibility to evaluate CANS activity.

IV. DISCUSSION

Our results indicate that Tone-Entropy analysis may evaluate CANS activities during breathing training and its recovery. There is a respiratory center in the medulla oblongata. Neurons of medulla oblongata were controlled by commands from the respiratory control center which exists pons. During breathing training, Hering-Breuer reflex causes excitation of parasympathetic nervous system activity. Our results support the fact because Tone-Entropy analysis shows the plots during breathing training distributed around right-lower positions on the Tone-Entropy space. From previous publications, it is indicated that, when the tone becomes lower and entropy becomes higher, parasympathetic nervous system is activated. Deep breathing makes a diffusion of serotonin in blood. Serotonin also causes to activate parasympathetic nervous system activity. Therefore, the possibility of chemical receptor action cannot be rejected.

One of significant facts of this experiment is that V3 trial has statistically significant increase of parasympathetic nervous system activity as compare with V1 and V2 trials. Oida et al.[3] stated that the recovery processes after high intensive exercises shift from left-higher to right-lower positions gradually. The alteration of the process is similar to our result in the alteration of the process from V2 trial to V3 trial. Further more detailed studies will be required to determine how rate and how long the recovery of the process affects CANS activity.

Tone-Entropy analysis can evaluate CANS activity without breathing control indication. The merit lets us have the ability to measure long-term CANS activity appropriately as compared with frequency domain analysis. Badra et al.[6] tried to reduce that respiratory variability from heart rate variability with frequency coherence. However, not only the technique is complicated but also coherence technique may have the possibility to decay effective components. Calculation on Tone-Entropy analysis has less computational complexity. Therefore, Tone-Entropy analysis has the affinity to small size instumentaions for long-term monitoring purpose because the algorithm easily leads to microprocessor owing to the low computational complexity.

V. CONCLUSIONS

In this study, twelve young healthy male subjects' cardiac nervous system activity during breathing training and its recovery after the training was evaluated through Tone-Entropy analysis. In this experiment, our results show that HR and parasympathetic nervous system activity are significantly increased during breathing training. Then, slight recovery process after the training may be shown in Tone-Entropy space. Also, our results present that Tone-Entropy has high reproducibility as compared with the previous published results, which performed with quite different situation and participants.

REFERENCES

1. M.S. Reis et al., "Deep breathing heart rate variability is associated with respiratory muscle weakness in patients with chronic obstructive pulmonary disease," Clinics, Vol.65, No.4, 2010.
2. U. Nussinovitch et al., "Deep breath test for evaluation of autonomic nervous system dysfunction in familial dysautonomia," Israel Med Associat J. Vol.11, No.10, pp.615-618, 2009.
3. E. Oida et al., "Tone-Entropy analysis on cardiac recovery after dynamic exercise," J Appl Physiol. Vol. 82, No. 6, pp. 1794–1801, 1997.
4. H. Nakamura et al., "Reproducibility of evaluation on cardiac autonomic nervous system activity through Tone-Entropy Analysis in young subjects," IFMBE proceedings of the International Federation for Medical and Biological Engineering, Vol.22, pp.1406-1409, 2008.
5. M. Amano et al., "A comparative scale of autonomic function with age through the Tone-Entropy analysis on heart period variation," Eur J Appl Physiol 98, 276–283, 2006
6. L.J. Badra et al., "Respiratory modulation of human autonomic rhythms," American J Phyiol Heart & Circ Physiol, Vol.280, No.6, H2674-H2688, 2001.

Author: Hideo Nakamura
Institute: Osaka Electro-Communication Univeristy
Street: 1130-70 Kiyotaki,Shijonawate
City: Osaka
Country: Japan
Email: h-nakamu@isc.osakac.ac.jp

Context Aware Contribution in Ambulatory Electrocardiogram Measurements Using Wearable Devices

F.J. Martinez-Tabares and G. Castellanos-Dominguez

Signal Processing and Recognition Group, Universidad Nacional de Colombia, Campus La Nubia,
Km. 7 Via al Magdalena, Manizales Colombia.
fjmartinezt@unal.edu.co

Abstract—Context aware can improve the measurement accuracy of biosignals that would be uncertain in the case of paying close attention exclusively to the biosignal source. A suitable solution for sensing the effect of context on Electrocardiogram measurements is to make comparisons with acceleration records, since it is strongly related to the patient motion, which is a major cause of disturbance. In this paper the acceleration is used to establish relationships between electrocardiogram and user movements, allowing to reduce interference and improving the quality of the records taken in motion using wearable technologies. Tests were performed in an adult patient during rest and walking periods, demonstrating that dubious signal peaks can be identified using our simple correlation algorithm. By using this technique, the accuracy of the classification process increased from 95% to 98%, proving to be a useful procedure in ambulatory.

Keywords—Context Aware, Acceleration, Biosignal, Electrocardiogram, Electrodoctor.

I. INTRODUCTION

There is a tendency to perform ambulatory, noninvasive and long-term monitoring [1-2], implying to overcome difficulties like disturbance reduction, how to get a comfortable sensor system with low energy consumption, and maintaining low computational cost during the multimodal processing stage.

This trend includes research of small integrated wireless sensor nodes, such as accelerometers and compasses, to capture human movements, which may have many applications in medical care [3]. In [4] multi-parameter measurements of Electrocardiogram (ECG) and acceleration were taken, but these signals were not targeted with the purpose of reducing the disturbance caused by motion.

In pursuit of improve the measurements quality, it is possible to establish correlations between different channels of the same variable [5], allowing to reduce disturbances The multi-sensor measurement, usually called diversity, has been proven in neonates using wearable ECG and acceleration sensors [6], however no results in moving adults were reported.

Respect to the comfort factor, textile electrodes provide a more natural feel than conventional disposable electrodes or metal, but there is a possible reduction in skin contact, especially by the moving patient.

In order to improve the collected data quality, besides knowing the user's state, it is important to have an awareness of the environment. Context information should be associated with measurements of the body; this is called context awareness and allows the behavior modification on the basis of this information [6, 7].

Acceleration can be measured in order to select the channels with higher quality and discarding those with high levels of disturbance [6], since disturbance is not usually present in all channels simultaneously. However, this technique causes that all channel information be discarded due to a short time of disturbance.

In this work we perform simultaneous measurements of ECG and acceleration in three axes, to reduce disturbance during movements that caused release of the textiles sensors, showing that it is possible to use the same measures in conjunction with a simple algorithm, to reduce localized disturbances, preventing high data loss that may occur if the full derivation would have been discarded.

II. MATERIALS AND METHODS

Ambulatory biosignal measurement implies to undergo permanent and sudden body changes in position and speed. This movement is generally reflected in disturbances for the ECG baseline reducing the signal quality.

Disturbance caused by the motion can be reduced based on the acceleration measurement, because acceleration is the physical variable that represents the change of speed in a time interval, and the second derivative of position with respect to time.

$$a = dv/dt \quad (1)$$

Moreover, the force is associated with the change of the amount of movement or what is the same, the product of mass and acceleration.

$$F = \frac{dp}{dt} = \frac{d(mv)}{dt} = m \cdot \frac{dv}{dt} = m \cdot a \qquad (2)$$

In summary, knowing the mass of the patient, and recording the acceleration over a time interval, it is possible to deduce information about the force, trajectory, velocity and position of a given object.

This principle can be exploited to obtain information from the user's interaction with the environment, and establish relationships with ECG because movements in proportion to the speed and strength tend to cause disturbances in ECG, which may cause misdiagnosis.

Additionally, it is possible to know the angle changes and therefore changes in position due to the influence of gravity on the accelerometers.

Being an accelerometer an instrument that detects changes in speed, it is obvious that it also measures the effect of gravitational force toward the earth center. Therefore, the use of a triaxial accelerometers can be very useful for detecting motion angles, since these are proportional to the influence exerted by gravity on the different axes of the accelerometer.

By detecting angle changes is possible to know the position of the person and the type of movements performed.

The proposed methodology aims to conserve the major quantity of ECG signal information as possible. It consist in simultaneously acquire ECG signals and acceleration, and determining the absolute values for their correlation. In case of detecting a high correlation, it means that there is a high possibility that the change in ECG voltage do not be due to heart failure, and that be caused by movement. Therefore, it is suggested to skip two seconds around the point with high correlation to avoid false positives and to keep the remaining of the signal, which can be subsequently improved.

III. EXPERIMENTAL SETUP

Figure 1 shows the implemented experimental configuration. It uses the textile garment with ECG sensors, *ELECTRODOCTOR* (Fig. 2), developed by the Signal Processing and Recognition Group from the Universidad Nacional de Colombia and the CELBIT LTD. Company. This garment transmits the ECG signal to a laptop using Bluetooth.

Acceleration is recorded using a mobile phone Motorola Fire, located at waist level, with a software developed for the Android operating system, that stores in a micro-SD memory, the data provided by the built in accelerometer. The ECG and acceleration data are stored with date-hour-minute and second, for later association with ECG.

A test is performed on an adult with periods of motion and resting with the purpose of obtaining a distinction between a real problem and a disturbance caused by heart movement. The aim is to detect when a false alarm of heart

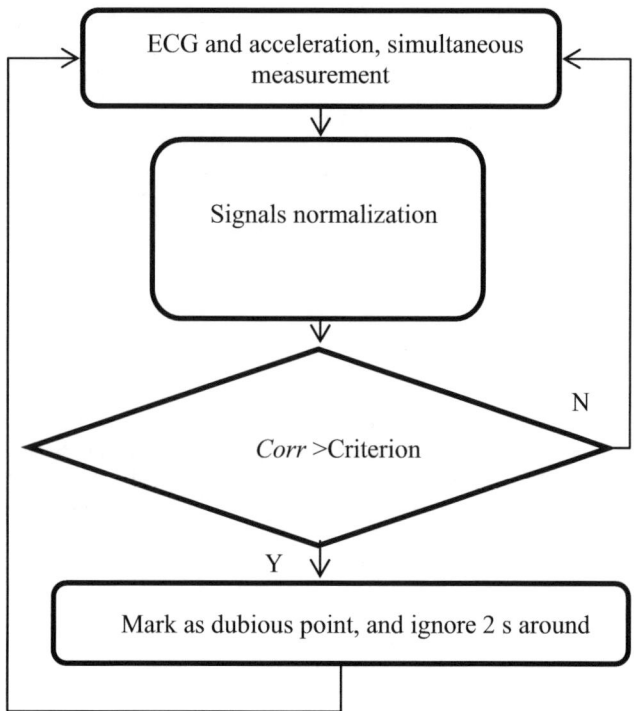

Fig. 1 Experimental diagram based on the proposed methodology

failure can occur due to movement, in order to ignore the signal at that point.

The comparison is made using the Pan-Tompkins beat detection algorithm, marking each beat, and using the correlation algorithm to detect areas of potential false positives, which are also marked. It is expected that the acceleration measurement helps to detect and ignore false positives.

Combined data are normalized, resampled and analyzed with the Matlab software.

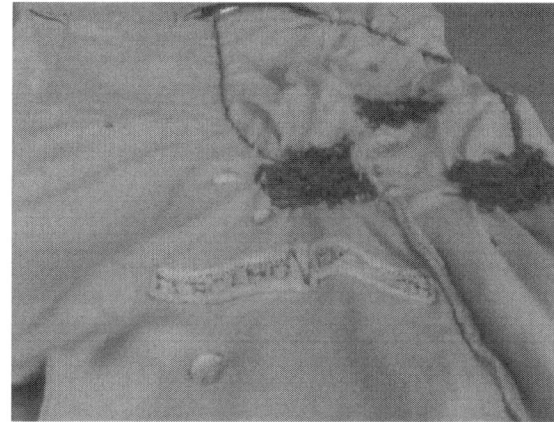

Fig. 2 Inner side of the sensorized garment showing textile electrodes weaved with conductive thread

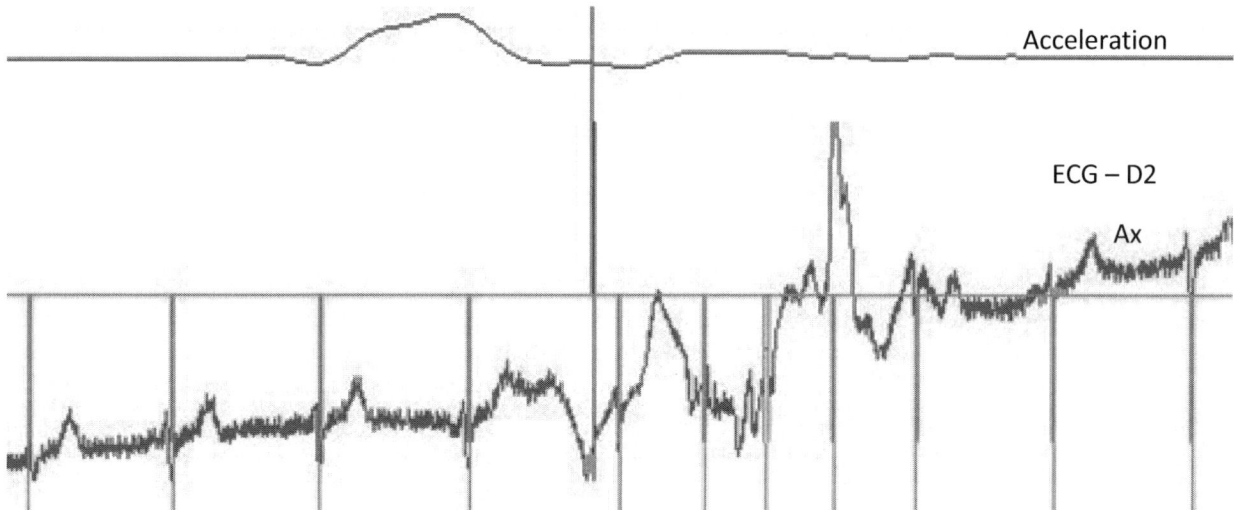

Fig. 3. Simultaneous measurements of acceleration and ECG (derivation D2). There is a change in acceleration above the ECG disturbance.

IV. RESULTS

Fig. 3 shows a measurement during movement (center) with resting periods at its start and end, using the sensorized garment (Fig. 4). It was found that the change in acceleration due to user movement precedes the broader disturbance in the ECG.

After a sudden movement, the electrodes vary slightly their skin contact producing changes in the amount of interference that is introduced to the ECG, the 60 Hz noise can increase or decrease, and the baseline level tends to change by the offset associated with a capacitance.

Blue lines in Fig 3 indicate each beat detected by the Pan-Tompkins algorithm, after an acceleration change can be observed that heartbeat detection tends to fail. ECG section where the disturbance is concentrated is indicated by the proposed correlation algorithm and marked using a green line, around it is the 2 s area to be ignored.

By using only the Pan-Tompkins algorithm, accuracy was 95%, but it increased to 98% implementing the proposed correlation algorithm. There exists an error margin when the patient movement is very strong.

The battery life during continuous monitoring failed to exceed three hours of operation, showing that the power consumption remains high.

V. CONCLUSIONS

A tool to acquire biosignals in moving patients was developed. Acceleration allowed obtaining context awareness, improving ECG measurement. Future work may concentrate on reducing power usage; optimize the computational cost and improving acceleration measures for strong movements, as well as using it like indicator of calories consumption, tremors in patients with neurodegenerative diseases or a fall detector for older adults.

ACKNOWLEDGMENT

The authors acknowledge the support of the Universidad Nacional de Colombia – Convocatoria GTI, as well as COLCIENCIAS with under grant 523 and the doctoral scholarships program – Generación del Bicentenario.

Fig. 4 Sensorized garment used for the signal acquisition with smarth phone

REFERENCES

1. R. S. H. Istepanian, E. Jovanov, and Y. T. Zhang, "Guest editorial introduction to the special section on m-health: Beyond seamless mobility for global wireless healthcare connectivity," *IEEE Trans. Inf. Technol. Biomed.*, vol. 8, no. 4, pp. 405–412, Dec. 2004.
2. R. S. H. Istepanian and Y.-T. Zhang. 4G Health - The Long-Term Evolution of m-health. *IEEE Transactions on Information Technology in Biomedicine*, vol. 16, no. 1, pp. 1-5, 2012.
3. Y. C. Tseng et al., "A Wireless Human Motion Capturing System for Home Rehabilitation," Mobile Data Management: Systems, Services and Middleware, 2009. MDM '09. Tenth International Conference on , vol., no., pp.359,360, 18-20 May 2009.
4. Wan-Young Chung; Young-Dong Lee; Sang-Joong Jung, "A wireless sensor network compatible wearable u-healthcare monitoring system using integrated ECG, accelerometer and SpO2," Engineering in Medicine and Biology Society, 2008. EMBS 2008. 30th Annual International Conference of the IEEE , vol., no., pp.1529,1532, 20-25 Aug. 2008
5. Martinez-Tabares, F.J.; Espinosa-Oviedo, J.; Castellanos-Dominguez, G., "Improvement of ECG signal quality measurement using correlation and diversity-based approaches," *Engineering in Medicine and Biology Society (EMBC), 2012 Annual International Conference of the IEEE* , vol., no., pp.4295,4298, Aug. 28 2012-Sept. 1 2012.
6. Designing for realiable textile neonatal ECG Monitoring using Multi-Sensor Recordings
7. Chiti, F.; Fantacci, R.; Archetti, F.; Messina, E.; Toscani, D., "An integrated communications framework for context aware continuous monitoring with body sensor networks," Selected Areas in Communications, IEEE Journal on , vol.27, no.4, pp.379,386, May 2009.

A Versatile Synchronization System for Biomedical Sensor Development

M. Zakrzewski, A. Joutsen, J. Hännikäinen, and K. Palovuori

Department of Electronics and Communications Engineering, Tampere University of Technology, Tampere, Finland

Abstract—Accurate synchronization between a custom-made sensor and a commercial reference sensor is a common problem while developing novel biomedical sensors for wearable or smart environment applications. This paper presents a widely applicable non-galvanic synchronization system between two or more sensors. The solution is a simple, low-cost infrared link, and it was demonstrated and tested in measurements using a commercial full-polysomnographic recording device and a custom-made radar motion sensor. The system was successfully used in a sleep measurement study for 12 full-night recordings. The synchronization accuracy of the system is 2.2 ms which is adequate for all common biosignals. The hardware and software design are available online. This provides other sensor developers possibility to easily use the system in their research projects. Although the technology used is not novel, the aim of the paper is to ease the practical synchronization problems.

Keywords—Time synchronization, Sensor data fusion.

I. INTRODUCTION

While studying novel biomedical sensors for wearable and smart home applications, we have several times faced the need to synchronize two or more simultaneously measured signals from separate devices. Novel monitoring techniques need to be compared to conventional methods with simultaneous recordings. Typically, the purpose has been to synchronize the data of a custom-made sensor to a reference sensor. However, there are a couple of typical problems. Commercial reference sensors do not have adequate number of free acquisition channels to connect the sensor/sensors under development (SUD) to it. Or the sampling frequency is too low for the SUD. Thus, two data acquisition devices are needed. A synchronization signal could be fed through a synchronization cable forming a galvanic connection. However especially in hospital environment, a galvanic contact between a commercial biomedical reference and a custom-made sensor often requires extra work for assuring the safety procedures. Galvanic contact might not be a good idea in laboratory either especially if the reference sensor contains contact electrodes (such as with ECG (electrocardiography) or EEG (electroencephalography)). Malfunction of the custom-made sensor could potentially cause an electric shock or ground currents to the measurement device. Furthermore, the cables limit the movements of the patient.

A common approach is to use some sort of synchronization action performed by the patient to generate a distortion pattern that can be identified in each monitored signal [1]. Then, the recorded signals are manually synchronized. We, however, wanted a more reliable and automatic synchronization method. One approach is presented by Camacho et al. [2]. They used pressure signal to synchronize a mechanical ventilator and a surface electromyogram (sEMG) data. Pressure signal was chosen based on the only available acquisition channel in the mechanical ventilator. In general case, a pressure signal is not practical for synchronization.

This led us to develop a non-galvanic infrared (IR) -link between the reference sensor and the custom-made SUD. The synchronization link was used in our sleep measurement study. The purpose of the measurement was to study the microwave Doppler radar sensors [3] during sleep. As a reference, a commercial polysomnographic (PSG) recording device was used. The used PSG device is a portable model. Thus, the non-contact link also allowed the user to move freely during the measurement.

The techniques presented in this paper are not novel. However, the synchronization of data measured with separate devices is a common problem in sensor development projects. Although there are several ways to answer the problem, most of them require custom designing. Similar solutions are not, to best of our knowledge, commercially available. Nor are they easily available online. In this paper, a solution to this practical problem is presented. Further, the design (both hardware and software) is provided online and is free for downloading [4]. We hope this will allow researchers to quickly adopt our system in their own studies and thus spend less time with configuring the measurement equipment and more time to concentrate on the real research questions.

In this paper, the design of the synchronization link and signal decoding in Matlab are described. In addition, it is presented how the system was used in our measurements and how reliable the link was.

II. SYSTEM DESIGN

The system consists of at least four parts: a transmitter, a SUD, a reference measurement device, and one receiver for each SUD and reference unit. The transmitter is either an independent device or integrated into one of the SUDs. In our case, the transmitter was integrated into the radar sensor board. One receiver is needed per each measurement device (the SUD or the reference unit). This requires one free channel from the acquisition board. At least 512 Hz sampling frequency is required for accurate synchronization signal decoding. Table 1 sums up the system requirements and specifications.

Table 1 The system specifications and requirements

System specifications	
Synchronization accuracy (maximum of values)	2.2 ms (@ f_s = 512 Hz) 1.5 ms (@ f_s = 800 Hz)
Synchronization message overflow	36 hours
Reference device and SUD requirements	
Number of free acquisition channels	1
Min. sampling frequency	512
Supply voltage for the receiver (alternatively battery powered)	2.7 V to 5.5 V
Supply current for the receiver (alternatively battery powered)	1.5 mA (max.) 1.2 mA (typ.)

A. Transmitter Unit

The transmitter consists of IR leds and a microcontroller (ATtiny13) to control the leds. Four IR leds are used to increase the luminosity, and thus, to double the detection range. The board is powered from the mains supply through a 12 V transformer. The picture of the transmitter unit is shown in Fig. 1a and the schematic diagram in the Fig. 2a.

The transmitter uses modified Philips RC-5 protocol [5] and Manchester coding. This means that a logical zero bit is formed by pulsing the signal during the first half of the bit and the second half being idle. The pulsing frequency is 38 kHz. Similarly, a logical one is formed by the first half bit being idle and the second half being pulsed. Two start bits are used, and the most significant bit (MSB) of the data is sent first. Both the start bits are logical ones. The length of one bit is 10.4 ms, and the length of one message is 16 bits (2 start bits + 14 data bits).

The data sent is an ordinal number starting from zero when the transmitter is turned on. Thus, we can easily differentiate the consecutive messages from each other. The message is sent approximately every 8 seconds. This means that the message body overflows every 36 hour, which is definitely long enough for sleep recordings.

B. Receiver Unit

The size of the receiver board is small, 2.5 cm x 1 cm, so it can be easily attached to patient's forehead or other easily visible place. For proper functioning, it is important to ensure the visibility to the transmitter. The transmitted signal is reflected from walls and other objects, so, the direct line

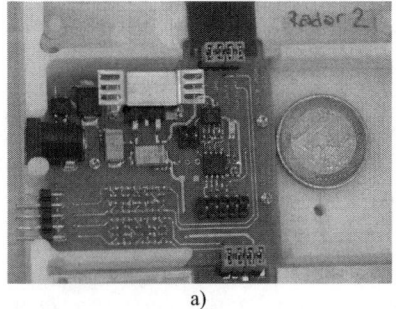

Fig. 1. A picture of a) the transmitter and b) the receiver units. The transmitter board also contains the power supply (in the top left corner) and radar signal preprocessing (in the lower left corner). The transmitter part is located on the right hand side of the board. The receiver board has some extra space on the top side of the board so it can easily be attached with a piece of medical tape.

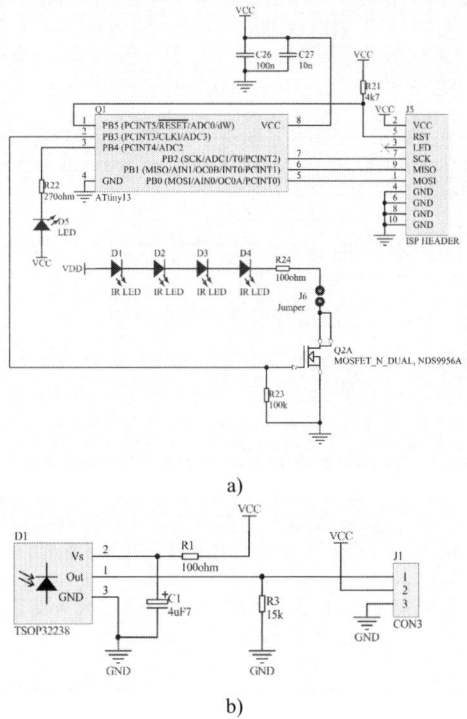

Fig. 2. The schematic diagram of a) the transmitter and b) the receiver units. The transmitter contains 4 IR leds, a microcontroller, and a connector for programming. The receiver design is very simple. Basically, it only contains an IR receiver (TSOP32238 [6]).

of sight is not required. But a thick blanket or an arm covering the full field of view of the receiver will block the signal.

The first version of the receiver was coin battery (CR2025) powered. In the second version, the receiver was powered directly from the measurement device. So, there are two possibilities for powering if the device can drive a small current (1.5 mA) at at least 2.7 V. The picture of the receiver unit is shown in Fig. 1b and the schematic diagram in the Fig. 2b. To make the receiver unit as simple as possible and to minimize the possibility of malfunction, the demodulated signal is sampled directly by the acquisition board of the reference device or the SUD. Thus, one free acquisition channel is needed.

C. Data Synchronization Decoding in Matlab

Matlab is used to decode the digitized synchronization signals measured in the reference device and in the SUD. Firstly, the original ordinal numbers sent from the transmitter are decoded from the sync pulses. Secondly, the sync signals are compared to find the matching ordinal numbers, and the time differences between the starting and ending points of the two (or more) recordings are calculated. The data before the first and after the last matching sync pulse is discarded. If needed in other applications, the timing of this unmatched data could also be extrapolated with a somewhat smaller accuracy. Finally if the signals are sampled with different sampling frequencies, resampling needs to be performed. We resample all data to 512 Hz frequency.

D. Data Synchronization Accuracy

The accuracy of the synchronization system is limited by the sampling frequencies, f_s. Also the delay caused by the IR receiver causes inaccuracy, but that delay is only between 180 µs and 390 µs (with TSOP32238 [6]). The sampling delay is the maximum delay of detecting the start bit which is $1/f_s$. Thus, it is the dominating source of error. With the sampling frequency of 512 Hz, the accuracy is 2.2 ms. On the other hand with the sampling frequency of 800 Hz, the accuracy is 1.5 ms.

This calculation naturally presumes that the transmitted signal is received by the receiver unit. Some obstacles between the transmitter and the receiver, some other source of IR-signals in the measurement environment, or the receiver being too far from the transmitter causes disturbances that might deteriorate the whole synchronization. However, if the sampling frequencies of the data acquisition devices remain constant during the measurement, it is enough if at least a couple of transmitted messages are recorded during the measurement. The rest of the data can be presumed to be evenly distributed.

III. MEASUREMENTS

A. Measurement Set Up

We performed 12 measurements with our system. Two microwave Doppler radar sensors [3] were used as the SUD. The whole-night full PSG recording was used as a reference. The used PSG device was a commercial, portable SOMNOscreen plus, SOMNOmedics GmbH, Germany [7].

The measurements were performed in 12 nights each with a different test patient. The PSG device contained a large set of sensors, for example EEG, ECG, abdominal and thoracic effort, snoring (microphone), naso/oral flow, SpO2 (oxygen saturation), position, and periodic leg movement (PLM) sensors. The sync receiver for the reference was attached to the forehead of the patient with a piece of medical tape.

The radar sensors, the transmitter unit, and the sync receiver for the radars were set up beforehand for the whole measurement period. The radar sensor recording was activated remotely through a VPN (Virtual Private Network) connection. Thus, the beginning and the end of the two recordings could differ from each other by hours. The possibility of the sync message overflow was excluded by asking the researcher on-site to boot the transmitter unit before each measurement.

B. Synchronization Performance

The performance of the synchronization system was analyzed by calculating the success ratio in the Matlab demodulator. This was calculated by the ratio between the amount of correctly decoded messages and the number of messages sent between the first and the last received synchronization message. In some of the measurements, the PSG device recording was started already before the transmitter was turned on, so using the full recording length would give a distorted result. The recording length of the radar sensor and the PSG device are different, thus, the performance is calculated separately for each device. In addition, the synchronization messages recorded in both the devices were matched to see the results for simultaneous recording. The results are presented in a Table 2 for 12 patient recordings.

In most of the measurements the ratio of correctly decoded messages is close to 100 %. However in the test sets 1 and 2, there was a cable breakage in the PSG receiver unit. This disturbed a large part of the synchronization signal. Luckily, some parts of the signal were still recorded correctly, and we were able to synchronize the signal from two devices. For the following measurement sets, the breakage was noticed and corrected. As previously noted, data can synchronized even with very small values of the success ratio.

There are several causes for performance errors: the demodulation algorithm fails to decode a proper synchronization message correctly, a part of a message is distorted or the full message is missing, or there is a bad wire connection caused by cable breakage in the receiver unit. A distortion can be due to several reasons. Some artifacts such as a TV remote controller in the environment cause extra pulses or interferences with the transmitted pulse. Some obstacle (such as a hand or a blanket) is blocking the receiver. Or a patient (and the receiver) is moving out of the field of view of the transmitter due to for example entering to toilet during the measurement. Also, the attachment place of the receiver is important. We attached the receiver to the forehead of the patient, knowing that some sleeping positions may cause the receiver to be between the patients head and the pillow. All of these error sources were faced during the measurements.

Table 2 The performance of the synchronization system

Patient nro.	PSG device		Radar sensor		Both devices combined	
	Recording length / message	Correctly decoded syncs / %	Recording length / messages	Correctly decoded syncs / %	Simultaneous recording length / messages	Correctly decoded and matched syncs / %
1	3679	48	5277	100	3679	48
2	1395	9	5771	74	1395	9
3	3967	96	5233	99	3965	96
4	4191	88	4921	99	4191	87
5	4437	92	5776	97	4220	93
6	3989	97	4887	100	3989	97
7	4060	95	4570	100	3829	94
8	3706	97	5814	100	3706	96
9	4123	97	4725	100	3117	98
10	3901	99	6043	90	3775	99
11	4060	97	4780	100	2537	99
12	4000	96	4838	100	3998	91

IV. DISCUSSION

In this paper, a constant sampling frequency was presumed. This, however, is not always the case. In that case, the system should be modified to send messages more often, and the data resampling should be performed against the device with a constant sampling frequency.

If there were a need for higher synchronization accuracy in the future, it could be obtained with the same hardware using a more thorough analysis of the recorded signal. As the seemingly digital recording is in fact an analog signal, the transition of the IR bits holds embedded information about the phase difference between the transmitter and recorder clocks. Since the accuracy for the current application was very satisfactory, this was left for future implementation.

V. CONCLUSIONS

This paper presents a versatile system that enables the synchronization of data from two or more separate devices. Similar synchronization difficulties are commonly faced while studying novel sensor technologies. Therefore, the system design presented in this paper is available online.

The synchronization system is a low cost IR-link, thus, it allows a non-galvanic connection between the measurement devices. The system was used successfully in 12 whole-night recordings for synchronizing data measured with a commercial PSG device and custom-made radar sensors. The synchronization accuracy is at least 2.2 ms with a sampling frequency of 512 Hz.

ACKNOWLEDGMENT

The authors would like to acknowledge voluntary patients for participating the study, Spa Hotel Peurunka for providing the study environment, and researchers from Department of Health Sciences from University of Jyväskylä, Jyväskylä, Finland, for preparing the PSG recording. This work was supported by the monitoring and treatment of obesity related sleep disorders (MotoSD) – project.

REFERENCES

1. Bannach D, Amft O, Lukowicz P (2009) Automatic Event-Based Synchronization of Multimodal Data Streams from Wearable and Ambient Sensors, Smart Sensing and Context, pp 135–148.
2. Camacho A, Hernandez A.M, Londono, Z, Serna L.Y, Mananas, M.A (2012) A synchronization system for the analysis of biomedical signals recorded with different devices from mechanically ventilated patients, IEEE EMBC, pp. 1944–1947.
3. Zakrzewski M, Raittinen H, and Vanhala J (2012) Comparison of Center Estimation Algorithms for Heart and Respiration Monitoring with Microwave Doppler Radar, IEEE Sensors J., 12:3, pp. 627–634.
4. A synchronization system hardware and software design. Available at: https://wiki.tut.fi/SmartHome/SynchronizationSystem/
5. Philips RC-5 protocol at: http://www.sbprojects.com/knowledge/ir/rc5.php
6. Vishay, IR receiver: www.vishay.com/docs/82489/tsop322.pdf.
7. SOMNOmedics, SOMNOscreen plus: http://www.somnomedics.eu/products/cardio-respiratory-screening-8-15-channels/technical-data-somnoscreentm-plus.html

Author: Mari Zakrzewski
Institute: Tampere University of Technology
Street: Korkeakoulunkatu 3
City: Tampere
Country: Finland
Email: mari.zakrzewski@tut.fi

ECG Acquisition Using Fluid-Repellent, Wearable Electrodes

F.J. Martinez-Tabares[1], G.P. Cardona-Cuervo[2], and G. Castellanos-Dominguez[1]

[1] Signal Processing and Recognition Group, Universidad Nacional de Colombia, Campus La Nubia,
Km. 7 Via al Magdalena, Manizales Colombia
fjmartinezt@unal.edu.co

[2] CELBIT LTDA., Campus La Nubia, Km. 7 Via al Magdalena, Parque Innovación X105, Manizales Colombia
gerencia@celbit.net

Abstract—We developed a conductive textile electrode, coated with Silicon dioxide nanoparticles for easy long-term ECG acquisition, whose behavior is explained by a mathematical and physical model. Results include contact angles exceeding 150°, while the electrical conductivity was as low as 3 Ω cm2.

Keywords—ECG, Holter, Wearable, Fluid-Repellent, Superhydrophobic, Electrodoctor.

I. INTRODUCTION

Cardiovascular diseases (CVD) are the biggest cause of deaths worldwide and often occur during routine daily activities [1]. For this reason, there is an increasing trend toward realizing electrocardiogram (ECG) acquisition using mobile networks [2], [3]. World Health Organization (WHO) encourages the development and use of wearable acquisition systems; however, this call implies an effort to reduce disturbances caused by movement and preserving hygienic conditions.

In this work, we report the development of a garment containing flexible ECG electrodes, manufactured by knitting conductive thread in a random pattern on the textile. Subsequently, Silicon dioxide (SiO2) nanoparticles were applied to the surface, obtaining a conductive, superhydrophobic material, intended for long term ECG measurements in moving patients.

In order to explain this result, we have developed a theoretical model based on the Cassie-Baxter equation [4], which assumes that some conductive fibers can pass between the applied nanoparticles, transporting the patient's biopotentials without getting wet, because droplets roll off easily, due to the support provided by the nanoparticles matrix. The model purposes to enhance the Cassie-Baxter equation, adding a new variable to the equation: the conductive fiber, generating the interface (hydrophobic solid)-fluid-air-(conductive fiber).

To proof our model and method, we observed the surface of the manufactured electrodes using Atomic Force Microscopy (AFM) and Scanning Electron Microscopy (SEM) techniques; chemical composition was verified through Energy-Dispersive Spectroscopy (EDS). A nanoparticles network was clearly observed, we also verified that the electrode conducts electricity and stays dry when water is applied. ECG signal acquisition was performed using the conductive electrode, which transports the patient biopotentials to an Electrodoctor Holter developed by the Signal Processing and Recognition Group from the Universidad Nacional de Colombia and the CELBIT LTDA Company.

EDS spectrum obtained from the conductive fibers showed iron and carbon peaks, while particles spectrum had peaks for SiO2. The garment could acquire high quality ECG during movement; fluids were repealed because contact angles are close to 155°, while the textile feeling is preserved. But there exists a change in color, possibly caused by the nanoparticles distribution leading to light diffraction. Therefore, future works can study how to avoid the color shift in the garment, and determine whether nanoparticles represent health risks for people.

II. MATERIALS AND METHODS

We propose to develop a garment with knitted electrodes using conductive thread. The second step consist in applying a layer of nanoparticles forming a microtextured surface, to generate a superhydrophobic behavior, that guarantee the user hygiene. Our approach is based on the Cassie–Baxter [4] model, but we add a new variable associated to conductive microfibers.

The small fibers of conductive thread come in contact with the patient's skin, and also can pass through the matrix of nanoparticles.

Our model consists in relating the contact angle of a liquid in an air-solid compound interface (θ_c), the fraction of water-solid interface area (f_1) and the water-air interface (f_2). We assume that contact between the fiber surface and the drop (f_3) forms a new air-water-(hydrophobic solid)-(conductive fiber) interface, as follows:

$$\cos \theta_c = f_1 \cos\theta_1 - f_2 \cos\theta_2 - f_3 \cos\theta_3 \quad (1)$$
$$\theta_2 = 180°; \cos\theta_2 = -1 \quad (2)$$
$$f_1 + f_2 + f_3 = 1 \quad (3)$$
$$\cos \theta_c = f_1(\cos \theta_1 + 1) + f_3(\cos \theta_3 + 1) - 1 \quad (4)$$

III. EXPERIMENTAL SETUP

Electrodes are woven using conductive thread over a lycra t-shirt. Silicon Dioxide Nanoparticles prepared using the sol-gel technique, are applied over the garment. Fluid-repellency is verified by observing the angle formed between the liquid/solid interface, while the acquisition quality is measured from the resistance and direct acquisition testing. The location of the nanoparticles on the fabric and microfiber was assessed with an Atomic Force Microspe (AFM).

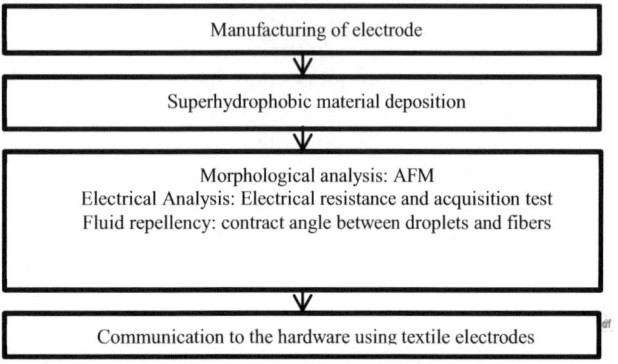

Fig. 1 Block diagram for the experimental procedure

Acquisition was performed by connecting the electrodes to an Electrodoctor Holter developed by the Signal Processing and Recognition Group and CELBIT LTDA Company.

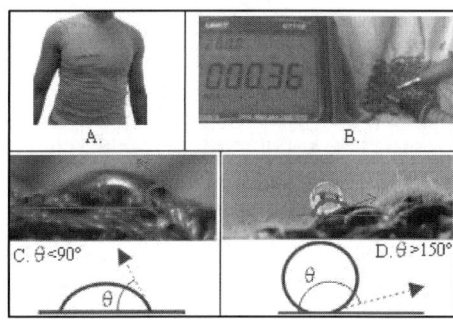

Fig. 2 Results; A. Smart garment; B. Electrical resistance of the coated electrodes; C. Contact angle without surface modification ($\theta < 90°$); D. Contact angle with surface modification ($\theta > 150°$).

Fig. 2-B presents the measured electrical resistance using a digital multimeter UNI-T UT71A/B.

IV. RESULTS

In Fig. 3 can be seen the textile electrodes repealing the water droplets.

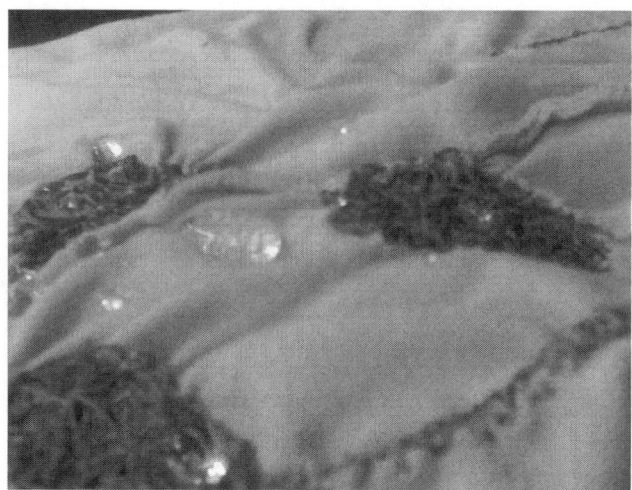

Fig. 3 Textile electoredes covered with nanoparticles showing superhydrophobic properties

Atomic Force Microscopy images of the microfibers are shown in Fig. 4.

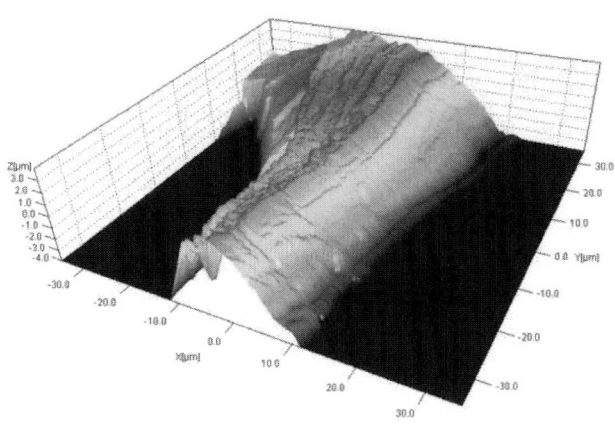

Fig. 4 AFM image of a microfiber of conductive thread.

Images of the nanoparticles mounted over the microfibers were acquired using AFM, Fig. 5 presents an image of a network of nano-SiO2 in a scale of 6 um. the average size of the nanoparticles is 20 microns.

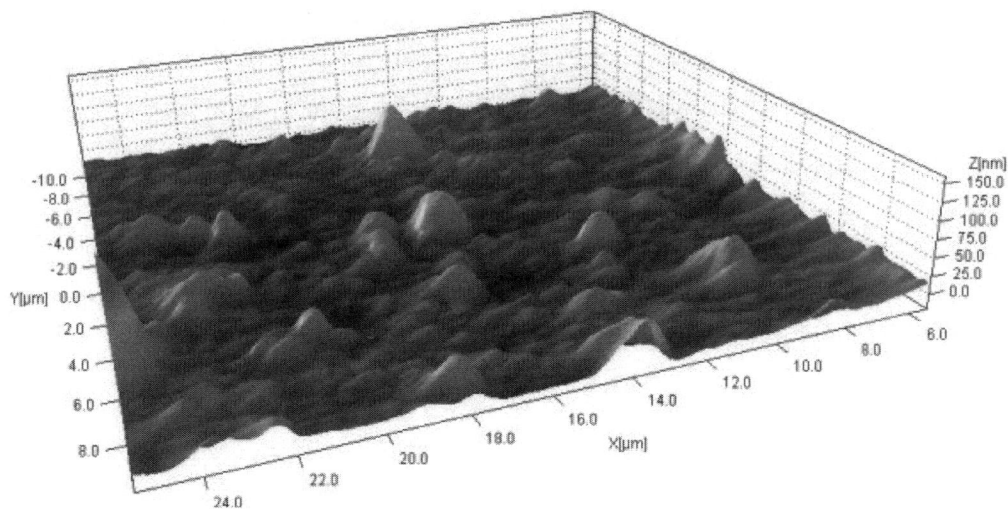

Fig. 5. AFM image of the SiO2 nanoparticles

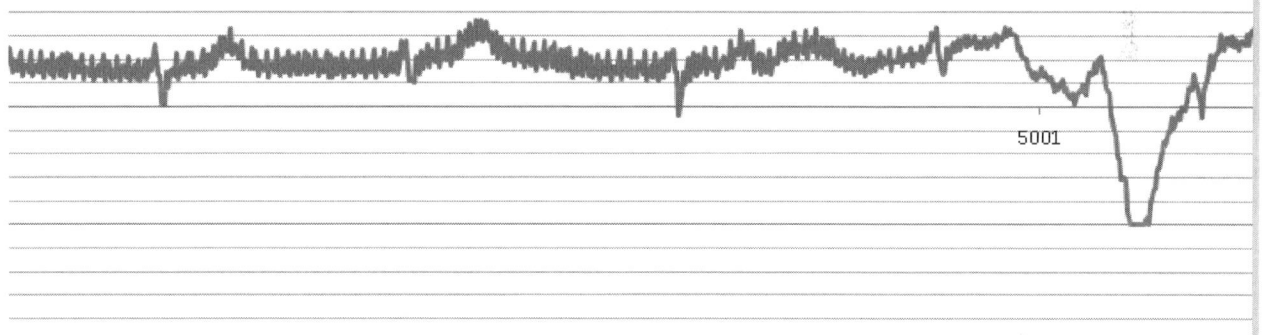

Fig. 6 ECG acquired using the smart garment covered with nano-SiO2

Acquired ECG signals show disturbances due to 60 Hz noise and baseline wandering caused by the movements of the patient (Fig. 6).

V. CONCLUSIONS

A smart clothing for hygienic measurements in moving patients was developed. Signals have considerable 60Hz noise and are susceptible to strong movements causing baseline wandering. The hydrophobicity was tested and the textile feeling is preserved, and a model based on the Cassie-Baxter model was proposed for sensing using conductive fibers.

Future work could focus on improving the signal quality.

ACKNOWLEDGMENT

The authors acknowledge the support of the Universidad Nacional de Colombia – Convocatoria GTI, as well as COLCIENCIAS under grant 523 and the doctoral scholarships program – Generación del Bicentenario. We also would like to thank the Plasma Physics Laboratory for the use of the characterization equipment.

REFERENCES

1. Global Burden of Disease Study 2010 (GBD 2010) Mortality Results 1970-2010. Seattle, United States: Institute for Health Metrics and Evaluation (IHME), 2012. Available at http://www.who.int/healthinfo/global_burden_disease.
2. R. S. H. Istepanian, E. Jovanov, and Y. T. Zhang, "Guest editorial introduction to the special section on m-health: Beyond seamless mobility for global wireless healthcare connectivity," *IEEE Trans. Inf. Technol. Biomed.*, vol. 8, no. 4, pp. 405–412, Dec. 2004.
3. R. S. H. Istepanian and Y.-T. Zhang. 4G Health - The Long-Term Evolution of m-health. *IEEE Transactions on Information Technology in Biomedicine*, vol. 16, no. 1, pp. 1-5, 2012.
4. A. B. D. Cassie and S. Baxter. Wettability of porous surfaces. *Transactions of the Faraday Society,* vol. 40, pp. 546-551, 1944.

Accuracy of the Oscillometric Fixed-Ratio Blood Pressure Measurement Using Different Methods of Characterization of Oscillometric Pulses

J. Talts, R. Raamat, K. Jagomägi, and J. Kivastik

Department of Physiology, University of Tartu, Tartu, Estonia

Abstract—The aim of this study was to assess the accuracy of the oscillometric fixed-ratio blood pressure measurement if different methods for characterization of oscillometric pulses were used. By means of a computer-based simulator, the widely used amplitude-based and also three different integral-based methods of characterization were implemented. We examined to which extent these methods were influenced by variations in pulse pressure, the shape of the arterial pressure pulse and the shape of the artery-cuff pressure/volume relationship. Errors were calculated as differences between the simulated systolic and diastolic blood pressure values and corresponding true values measured from the pressure pulses used as input signals of the model. Simulation demonstrated that the amplitude-based and complementary integral-based characterization methods showed the smallest error ranges for estimating systolic blood pressure. For diastolic blood pressure, the smallest error range was achieved if the full integral-based method was used.

Keywords—Oscillometric blood pressure measurement, characteristic ratio, amplitude-based characterization, integral-based characterization, modelling

I. INTRODUCTION

The oscillometric method is widely used in non-invasive blood pressure (BP) measurement. Owing to the simplicity of this method, oscillometry is employed in the majority of present-day automated BP monitors.

Oscillometric BP is typically determined from the oscillometric envelope (OE), which is based on the use of successive oscillations from the occlusive cuff during its inflation or deflation. Mostly, the recorded cuff pulsations are characterized by their amplitude [1, 2]. However, an integral of each cuff oscillation can also be implemented for characterization, using a full, partial or complementary integral of each oscillometric pulse [2-4], but this approach has been rarely discussed in the literature.

The highest point of the OE is generally regarded as the mean arterial pressure (P_{mean}). If the fixed-ratio estimation is applied, the systolic and diastolic blood pressures (P_{syst} and P_{diast}, respectively) are determined using certain empirical fractions of the maximum oscillation amplitude. These fractions are known as characteristic ratios (k_{syst} and k_{diast}, respectively) [2].

The aim of this model study was to assess the accuracy of the oscillometric fixed-ratio BP measurement if the amplitude-based and integral-based characterization of oscillometric pulses was implemented together with several affecting factors.

II. METHODS

A. Descripton of the Model

We applied the simulation technique to model oscillometric BP measurement similar to that used in our earlier studies [5, 6]. The simulator contained an artery-cuff pressure/volume (P/V) model with arterial pressure pulses as an input signal and cuff volume oscillations as an output signal. There was a possibility to modify the shape of the asymmetric arctangent P/V relationship, and the amplitude and shape of the input pressure pulse.

This model was used with several combinations of pulse and pressure/volume parameters to estimate P_{syst} and P_{diast}. Estimation errors for these combinations were calculated and the range of errors (difference between maximum and minimum error) for every characterization method was found in the final step.

The shape of the artery-cuff P/V relationship was modified by two indices (Fig. 1):

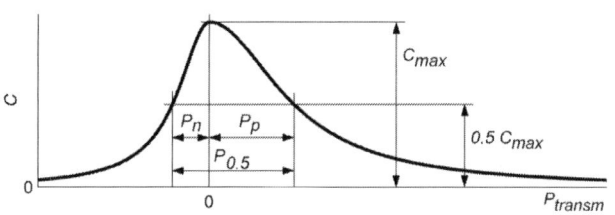

Fig. 1 Schematic diagram of the slope (compliance) of the artery-cuff pressure/volume relationship used to model oscillometric blood pressure measurement

1. steepness index ($P_{0.5}$) – a half-maximum width of the pressure/compliance curve, indicating the rate of decrease of the maximum compliance (C_{max}) to half value,
2. symmetry index (k_{PV}) – can be expressed as following: $k_{PV} = P_n / (P_n + P_p)$, where P_n and P_p are the portions of $P_{0.5}$ for the negative and positive transmural pressure. The transmural pressure (P_{transm}) is defined as $P_{transm} = P_{intr} - P_{cuff}$, where P_{intr} denotes the intra-arterial pressure and P_{cuff} denotes the cuff pressure.

The shape of the arterial pressure pulse was described by the pulse shape index, k_{pulse}, determined as $k_{pulse} = (P_{mean} - P_{diast}) / P_{pulse}$, where P_{pulse} denotes pulse pressure, defined as $P_{pulse} = P_{syst} - P_{diast}$.

B. Set of Model Parameters

It has been reported in earlier studies [6-8] that among the factors considerably influencing the accuracy of oscillometric BP measurement are pulse pressure, the shape of the arterial pressure pulse and the shape of the arterial P/V relationship. We assessed the accuracy of BP measurement by examining errors in P_{syst} and P_{diast} resulting from the variation of four affecting factors:

1. P_{pulse} (varying from 30 to 70 mmHg),
2. k_{pulse} (varying from 0.25 to 0.45),
3. k_{PV} (varying from 0.25 to 0.45),
4. $P_{0.5}$ (varying from 30 to 50 mmHg).

The influence of every individual factor was studied separately, i.e. if one of these factors was varying from its minimum value to maximum value, the rest of the parameters were equal to their basal values.

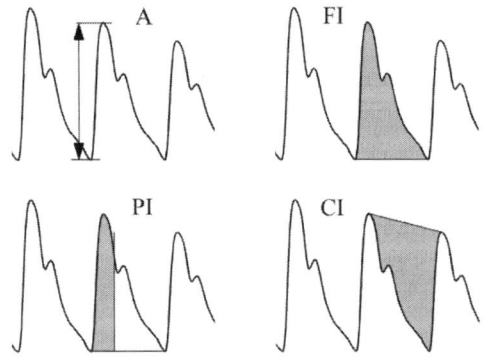

Fig. 2 Amplitude- (A), full integral- (FI), partial integral- (PI) and complementary integral-based (CI) characterization techniques

We regarded the following arterial condition as basal for our model: $P_{pulse} = 50$ mmHg, $k_{pulse} = 0.35$, $k_{PV} = 0.35$ and $P_{0.5} = 40$ mmHg.

In addition to separate influence of factors, the error ranges were calculated for all combinations of minimum, basal and maximum values of four factors ($3^4 = 81$ cases).

C. Characterization of Oscillometric Pulses and Choice of Characteristic Ratios

Oscillometric pulses were extracted by subtracting the baseline from the total pressure curve. This procedure did not alter the shape of the extracted pulses. Once extracted, the pulses were characterized using four following methods (Fig. 2):

1. amplitude-based characterization (A),
2. full integral-based characterization (FI),
3. partial integral-based characterization (PI; integrated during the first 0.25 s time interval),
4. complementary integral-based characterization (CI; area between pulse and the straight line between adjacent peaks)

Whereas the choice of characteristic ratios is empirical, we adjusted values for k_{syst} and k_{diast} for every method in such a manner, that using these values yielded symmetrical errors for P_{syst} and P_{diast} over a used range of alterations in affecting factors.

Errors were calculated as differences between the simulated P_{syst} or P_{diast} values and corresponding references (true values) measured from the pressure pulses as input signals of the model. However, it is worth to pay more attention to the error ranges than to the peak values of positive or negative errors, as the peak values can be reduced on properly centering the error range. To follow this consideration, the characterization methods were finally compared on the basis of error ranges.

III. RESULTS

Table 1 represents error ranges for P_{syst} and Table 2 represents error ranges for P_{diast} if different methods of characterization of oscillometric pulses (A, PI, FI and CI) together with variation of four affecting factors (P_{pulse}, k_{pulse}, k_{PV} and $P_{0.5}$) were modelled. Fig. 3 is a graphical presentation of data in Table 1 and 2.

It is clearly seen that the size of error ranges for P_{syst} as well as for P_{diast} depends on the method of characterization of oscillometric pulses.

Table 1 Error ranges for P_{syst} in mmHg if different methods of characterization of oscillometric pulses together with variation of four affecting factors were modelled

Varying factor	Characterization			
	A	PI	FI	CI
P_{pulse} *	5.6	9.6	16.1	5.2
k_{pulse} *	0.0	1.8	0.8	2.4
k_{PV} *	9.1	10.7	12.8	7.6
$P_{0.5}$ *	3.4	5.5	7.9	3.0
All	18.6	28.8	39.0	18.5

*if one of the factors was varying, the rest of factors were equal to their basal values.

Table 2 Error ranges for P_{diast} in mmHg if different methods of characterization of oscillometric pulses together with variation of four affecting factors were modelled

Varying factor	Characterization			
	A	PI	FI	CI
P_{pulse} *	5.2	4.3	3.1	11.9
k_{pulse} *	0.0	0.4	1.3	6.9
k_{PV} *	7.0	5.8	3.9	8.8
$P_{0.5}$ *	2.9	2.5	2.2	7.3
All	15.2	13.1	10.6	36.2

*if one of the factors was varying, the rest of factors were equal to their basal values.

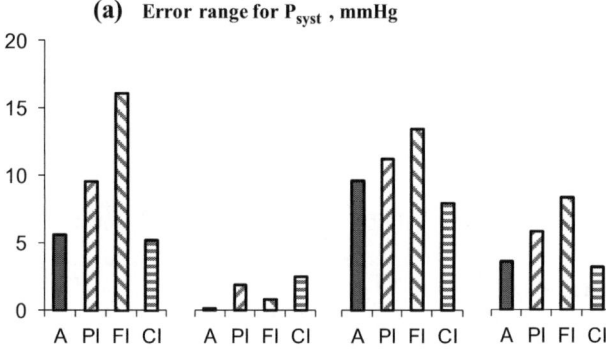

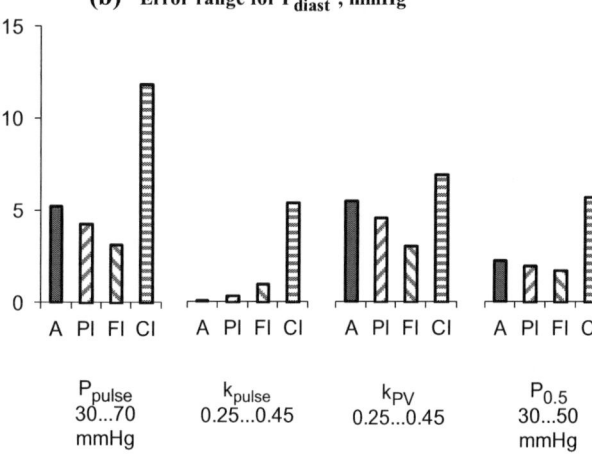

Fig. 3 Error ranges for P_{syst} (a) and P_{diast} (b) resulting from variation of different affecting factors and using different characterization of oscillometric pulses

IV. DISCUSSION

The accuracy of the oscillometric device can be affected by several components, e.g. choosing the cuff, extracting and characterizing the oscillometric pulses, selecting an algorithm. Accuracy can also differ across patient groups: for example, oscillometric waveform shape can change in individuals with stiff arteries and this can influence the measurement.

To describe the nonlinear P/V relationship of the arterial wall, analytical approaches like exponential [7-10], arctangent [5, 6, 11-14] and combined (hyperbolic + logarithmic) [15, 16] functions have been used, but a graphically reconstructed elastogram can also be implemented [17]. From analytical functions the biexponential approximation is most commonly used because of its convenient analytical handling, but the arctangent approximation is also preferred as it yields realistic shapes of the derivative-based as well as delta-based compliance/pressure curves [18]. In this study we applied the asymmetric arctangent model of the artery–cuff P/V relationship.

To separate oscillations, the baseline subtraction was applied in our analysis. Additionally, we tested the 2 Hz high-pass filtering and revealed that it did not change results substantially.

Our simulation study of the accuracy of the oscillometric fixed-ratio BP measurement demonstrated a dependence of error ranges of P_{syst} and P_{diast} on the pulse- and artery-related factors. Generally, the variation of pulse pressure (P_{pulse}) and the shape factor of the arterial P/V relationship (k_{pv}) were mainly responsible for blood pressure estimation errors. The effect of the pulse shape (k_{pulse}) change was the smallest.

The commonly used amplitude-based characterization is relatively well applicable for finding of both systolic and diastolic pressures, while the integral-based method should be chosen separately for systolic and diastolic pressures.

V. CONCLUSIONS

The oscillometric blood pressure measurement was studied from the viewpoint of accuracy, i.e. which characterization method of oscillometric pulses is more robust when the pressure pulse and artery properties are altered. The error range for systolic pressure was smaller if the amplitude-based or complementary integral-based characterization of oscillometric pulses was used. The error range for diastolic pressure was the smallest if the full integral-based characterization was applied.

ACKNOWLEDGMENT

This study was supported by the Estonian Ministry of Education and Research (SF0180125s08).

REFERENCES

1. Amoore JN (2012) Oscillometric sphygmomanometers: a critical appraisal of current technology. Blood Press Monit 17:80–88
2. Ng K, Small C (1994) Survey of automated non-invasive blood pressure monitors. J Clin Eng 19:452–475
3. Ruiter KA (1990) Automatic non-invasive blood pressure reading device. US patent 4,922,918
4. Nelson CH, Dorsett TJ, Davis CL (1989) Method for nninvasive blood-pressure measurement by evaluation of waveform-specific area data. US patent 4,889,133
5. Talts J, Raamat R, Jagomägi K (2006) Asymmetric time-dependent model for the dynamic finger arterial pressure-volume relationship. Med Biol Eng Comput 44:829–834
6. Raamat R, Talts J, Jagomägi K, Kivastik J (2011) Errors of oscillometric blood pressure measurement as predicted by simulation. Blood Press Monit 16:238–245
7. Ursino M, Cristalli C (1996) A mathematical study of some biomechanical factors affecting the oscillometric blood pressure measurement. IEEE Trans Biomed Eng 43:761–778
8. Baker PD, Westenskow DR, Kück K (1997) Theoretical analysis of non-invasive oscillometric maximum amplitude algorithm for estimating mean blood pressure. Med Biol Eng Comput 35:271–278
9. Raamat R, Talts J, Jagomägi K, Länsimies E (1999) Mathematical modelling of non-invasive oscillometric finger mean blood pressure measurement by maximum oscillation criterion. Med Biol Eng Comput 37:784–788
10. Babbs CF (2012) Oscillometric measurement of systolic and diastolic blood pressures validated in a physiologic mathematical model. Biomed Eng Online 11:56
11. Langewouters GJ, Wesseling KH, Goedhard WJ (1984) The static elastic properties of 45 human thoracic and 20 abdominal aortas in vitro and the parameters of a new model. J Biomech 17:425–435
12. Gizdulich P, Wesseling KH (1988) Forearm arterial pressure–volume relationships in man. Clin Phys Physiol Meas 9:123–132
13. Talts J, Raamat R, Jagomägi K, Kivastik J (2011) An Influence of Multiple Affecting Factors on Characteristic Ratios of Oscillometric Blood Pressure Measurement. IFMBE Proceedings 34:73–76
14. Raamat R, Talts J, Jagomägi K, Kivastik J (2013) Accuracy of some algorithms to determine the oscillometric mean arterial pressure: a theoretical study. Blood Press Monit 18:50–56
15. Drzewiecki G, Hood R, Apple H (1994) Theory of the oscillometric maximum and the systolic and diastolic detection ratios. Ann Biomed Eng 22:88–96
16. Liu J, Hahn JO, Mukkamala R (2013) Error Mechanisms of the Oscillometric Fixed-Ratio Blood Pressure Measurement Method. Ann Biomed Eng 41:587–597
17. Antonova ML (2013) Noninvasive determination of arterial elasticity and blood pressure. Part II: elastogram and blood pressure determination. Blood Press Monit 18:41–49
18. Talts J, Raamat R, Jagomägi K (2009) Influence of pulse pressure variation on the results of local arterial compliance measurement: A computer simulation study. Comput Biol Med 39:707–712

Author: Jaak Talts
Institute: Department of Physiology
Street: Ravila 19
City: Tartu
Country: Estonia
Email: jaak.talts@ut.ee

Differences in QRS Locations due to ECG Lead: Relationship with Breathing

M.A. García-González, A. Argelagós, M. Fernández-Chimeno, and J. Ramos-Castro

Group of Biomedical and Electronic Instrumentation, Department of Electronic Engineering,
Universitat Politècnica de Catalunya (UPC), Barcelona, 08034 Spain

Abstract—Results of heart rate variability analysis depend on the accuracy of the RR time series that is measured only in one lead of the ECG. RR time series can subtly change from lead to lead. We have obtained the ECG of 21 healthy subjects in a quiet measurement, sampled at 5 kHz and from the I and II standard leads. For each subject a total of 60 minutes was measured. The QRS complexes in both leads have been detected using a conventional QRS detector plus a further refinement using a matching template. Differences between the locations of the QRS complexes have been quantified and compared with the breathing signal of each subject as well as the derived RR time series. The typical uncertainty in the fiducial points and RR time series is usually below 1 ms and the errors are modulated by breathing.

Keywords—Heart rate variability, ECG lead, Breathing.

I. INTRODUCTION

Heart rate variability (HRV) has become a clinical and research tool to study the effect of the autonomic nervous system on the cardiovascular system [1]. The first step in any analysis of heart rate variability consists on the temporal location of the QRS complexes on the measured electrocardiogram (ECG). These detections assign to each QRS complex a temporal position (or fiducial point) where the heartbeat is considered to occur. After the fiducial points are obtained, the RR time series is defined as the intervals between consecutive fiducial points. But some indexes that characterize the HRV are very sensitive to small errors in the determination of the RR time series especially when the HRV is very small [2],[3].

In a previous work [4] we analyzed the differences in RR time series when measured in different leads of the ECG. Results showed that there were in fact slightly differences in the obtained RR time series and that those differences depended on the lead, QRS detector and noise. The work employed a freely available database sampled at 1 kHz, some recordings were noisier than others and there was only access to the ECG signals. Graphical plotting of the differences of the RR time series suggested a nearly periodic component that was hypothesized to be caused by breathing. Nevertheless, this hypothesis could not be proved at the time because the breathing signals were not available.

The aim of this work is to study the differences in fiducial points of the QRS complex between two standard leads in a best case scenario (relaxed subject, very high sampling frequency), quantify how much of these differences are explained by breathing and how much these differences can differ from subject to subject.

II. MATERIALS AND METHODS

For the study we measured the ECG and breathing of 21 healthy subjects (age: 29 years ± 4 years, sex: 5 females/16 males, body mass index: 25.5 kg/m^2 ± 4.4 kg/m^2). Data was acquired using a Biopac MP36 data acquisition system (Santa Barbara, CA, USA). Channels 1 and 2 of the system were devoted to measure conventional ECG with a bandwidth between 0.05 Hz and 150 Hz while channel 3 was employed to measure the respiratory signal obtained from a thoracic piezoresistive band (SS5LB sensor by Biopac, Santa Barbara, CA, USA) with a bandwidth of 0.05 Hz to 10 Hz. Channel I measured the ECG standard lead I while channel II measured the ECG standard lead II. For the ECG measurement we used monitoring electrodes with foam tape and sticky gel (3M Red Dot 2560). Each channel was sampled at 5 kHz.

During the measurement, the subjects were asked to be very still in supine position on a comfortable conventional single bed and awake. After attachment of sensors, we recorded a total of 60 minutes. From minute 5 to minute 55, the subjects were listening to music. For analysis purposes, the 60 min measurement was segmented in 12 recordings of 5 min. For each of these recordings, the QRS complexes were detected for both leads. A first rough fiducial point was obtained by using the Pan-Tompkins QRS detector [5] but was further refined by maximizing the correlation between any detected QRS complex and the first detected QRS complex using templates of 200 ms duration centered on the rough fiducial point. After this detection and for each recording, two time series are obtained: the R_I corresponds to the fiducial points of the QRS complexes using the standard I lead while R_{II} corresponds to the fiducial points of the QRS complexes using the standard II lead. We define the error in the fiducial point detection as:

$$\Delta R_{I \to II}(n) = R_I(n) - R_{II}(n) \qquad (1)$$

where n is the heartbeat number since the start of the recording. The error has been resampled using as a time reference the R_I time series at 4 Hz. On the other hand and for comparison purposes, the breathing signal has been sampled at the locations of the R_I time series and further resampled at 4 Hz. Figure 1 shows an example of the error in the fiducial point and the derived breathing signal.

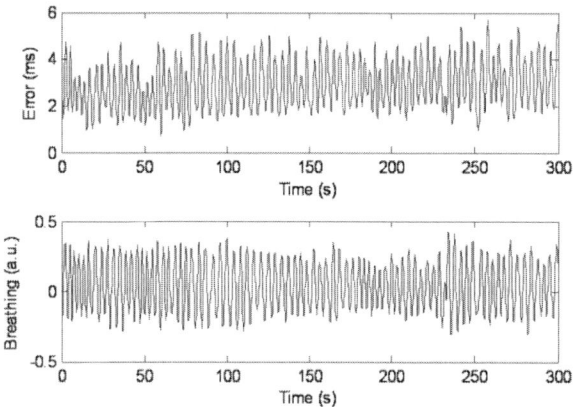

Fig 1 Example of the error in the fiducial point and the simultaneous acquisition of the breathing signal.

The standard deviation of the error in the fiducial point was computed for each recording (*SDER*). Moreover, the spectral coherence between the error in the fiducial point and the breathing signal was estimated by using the Welch's averaged periodogram method and computing the magnitude squared coherence estimate. The estimation is obtained by dividing each recording into eight sections with 50% overlap, windowing with a Hamming window and estimating the average power spectrum of each signal and the average power cross-spectrum. Frequencies with a magnitude squared coherence estimate higher than 0.75 were considered to compute how much error is caused by breathing. This estimation of error due to breathing (*ERB*) is obtained by taking the square root of the integration of the power spectrum of the error in the frequencies that have a high coherence with the breathing signal. Prior to any spectral computation, slow drifts have been removed on the signal by applying a Hodrick-Prescott filter with $\lambda = 10^9$ [6]. Figure 2 shows the coherence and the power spectrum of the error of the time series in figure 1. The ratio *ERB/SDER* was also computed to estimate the relative amount of error due to breathing (*RERB*).

A similar procedure has been followed to evaluate errors in the RR time series. The error in RR time series has been computed by using:

$$\Delta RR_{I \to II}(n) = RR_I(n) - RR_{II}(n) \qquad (2)$$

Then, the standard deviation of the error in the RR time series (*SDERR*), the error due to breathing (*ERRB*) and the ratio of both errors (*RERRB*) has been assessed. The same strategy and threshold for assessing the influence of breathing in the fiducial point has been employed to obtain theses errors for the RR time series.

III. RESULTS

Mean and standard *SDER* computed for the complete set of recordings is 0.45 ms ± 0.25 ms while for *ERB* is 0.24 ms ± 0.24 ms and for *RERB* is 48.0% ± 29.3%. A two-way analysis of variance shows very significant differences in the indexes ($p<0.001$) due to the subject under measurement. Nevertheless, the time of measurement since the monitorization started do not affect the indexes.

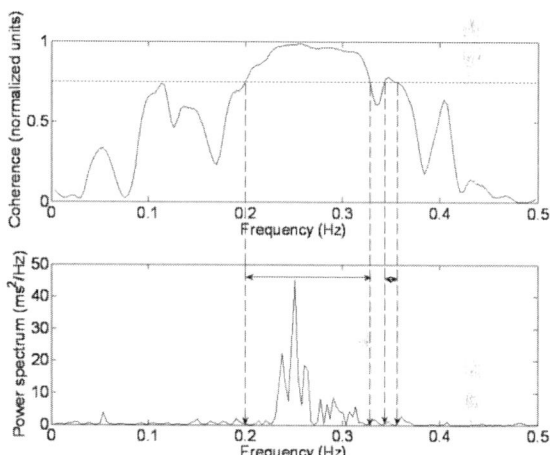

Fig 2. Example of computation of *ERB* by computation of the spectral coherence and power spectrum of the error. The arrows show the considered frequencies for the estimation of *ERB*. In this case, *SDER* is 1.07 ms while *ERB* is 1.02 ms.

There was no significant correlation for *SDER* or *ERB* with age, or body mass index. Nevertheless, there was a significant correlation between *RERB* and body mass index (Pearson's Product Correlation Moment: -0.55 ($p<0.05$)) suggesting that the percentage of error due to breathing is higher in slim people.

As for the errors in the determination of the RR time series, the mean and standard *SDERR* is 0.53 ms ± 0.36 ms while for *ERRB* is 0.31 ms ± 0.35 ms and for *RERRB* is 50.5% ± 26.5%. Once again, a two-way analysis of variance shows very significant differences in the three indexes ($p<0.001$) due to the subject under measurement but no differences associated to the time of measurement since the monitorization started. The only significant correlation found for the errors was for *RERRB* and body mass index (Pearson's Product Correlation Moment: -0.52 (p<0.05)). Table 1 summarizes the results.

Table 1 Summary of errors

	Mean	Standard Deviation
SDER (ms)	0.45	0.25
SDERR (ms)	0.53	0.36
ERB (ms)	0.24	0.24
ERRB (ms)	0.31	0.35
RERB (%)	48.0	29.3
RERRB(%)	50.5	26.5

IV. DISCUSION AND CONCLUSIONS

The errors in the fiducial points and in the RR time series are low but impose a limit in the resolution that can be achieved for the analysis of HRV. The digitizing error due to the sampling frequency (f_s) has a typical uncertainty of [7]:

$$u(RR) = \frac{1}{f_s \cdot \sqrt{6}} \quad (3)$$

The typical uncertainty in the RR time series caused by measuring in the standard lead II instead of in the standard lead I is similar to the typical uncertainty due to sampling the ECG at 770 Hz. And this is in a best case scenario: These errors are expected to rise in a noisier measurement, by lowering the sampling frequency or by changing leads. Moreover, errors are higher in some subjects.

This work has also shown that the errors in the fiducial point and in the RR time series are affected by breathing. The amount of error due to breathing also depends on the subject under measurement and in percentage is higher for slim people.

As summary, typical uncertainty in fiducial points and in the obtained RR time series associated with the chosen lead is generally lower than 1 ms and the associated errors are modulated by breathing.

ACKNOWLEDGMENT

This work was supported by MINECO project PSI2011-29807-C02 and from the Spanish Ministry of Health, Redes de Investigación del Instituto de Salud Carlos III, (REDINSCOR).

REFERENCES

1. Task Force of the European Society of Cardiology and the North American Society of Pacing and Electrophysiology (1999) Heart Rate Variability, Standards of measurement, physiological interpretation, and clinical use. Eur Heart J 17:354–381
2. García-González M A, Fernández-Chimeno M and Ramos-Castro J (2004) Bias and uncertainty in heart rate variability spectral índices due to the finite ECG sampling frequency. Physiol Meas 25:1–16
3. García-González M A, Fernández-Chimeno M and Ramos-Castro J (2009) Errors in the Estimation of Approximate Entropy and Other Recurrence-Plot-Derived Indices Due to the Finite Resolution of RR Time Series. IEEE Trans Biomed Eng 56:345–351
4. García-González M A, Ramos-Castro J and Fernández-Chimeno M (2011) The effect of electrocardiographic lead choice on RR time series, Proc. of the Annual International Conference of the IEEE Engineering in Medicine and Biology Society EMBC, Boston, EEUU, 2011, pp 1933–1936
5. Pan J and Tompkins W J (1985) A real-time QRS detection algorithm. IEEE Trans Biomed Eng 32: 230–236
6. Hodrick R J and Prescott E C (2007) Postwar U.S. business cycles: an empirical investigation. Journal of Money, Credit and Banking 29:1–16
7. Merri M, Farden D C, Mottley J G and Titlebaum E L (1990) Sampling frequency of the electrocardiogram for spectral analysis of the heart rate variability. IEEE Tran. Biomed Eng 37:99–106

Use macro [author address] to enter the address of the corresponding author:

Author: Miguel Ángel García González
Institute: Electronic Engineering Department, Universitat Politècnica de Catalunya (UPC)
Street: Jordi Girona 1-3
City: Barcelona
Country: Spain
Email: miquel.angel.garcia@upc.edu

Synchrosqueezing Index for Detecting Drowsiness Based on the Respiratory Effort Signal

N. Rodríguez-Ibáñez[1], M.A. García-González[1], M. Fernández-Chimeno[1], *Member IEEE*,
H. De Rosario[2], and J. Ramos-Castro[1], *Member IEEE*

[1] Group of Biomedical and Electronic Instrumentation, Electronic Engineering Department,
Universitat Politècnica de Catalunya (UPC), Barcelona, Spain
[2] Instituto de Biomecanica de Valencia, Universidad Politécnica de Valencia (UPV), Valencia, Spain

Abstract—Objective: The aim of this work is to evaluate a new index to assess the alertness state of drivers based on the respiratory dynamics derived from an inductive band using the variability of the respiratory rhythms assessed by Synchrosqueezing (SSQ).

Background: Biomedical variables like abdominal effort, which is related to autonomic nervous system, provides direct information of the driver physiological state, instead of indirect indicia of the participant's behavior.

Method: The respiration data used in this study was recorded by doing 18 simulator tests in a controlled scenario. In order to evaluate the viability of variability of the respiratory rhythms assessed by Synchrosqueezing (SSQ) Mean and the Standard Deviation of the Instantaneous Respiration Frequency (MIRF and SDIRF) obtained in the analysis for awake states and drowsy states individually was calculated.

Results: The results demonstrate the viability of drowsiness detection in simulator using abdominal effort signal analyzed by Synchrosqueezing based on the significant results of the statistical analysis comparing the Mean (MIRF) and de Standard Deviation (SDIRF) of both awake and drowsy intervals instantaneous respiration frequency (MIRF 0.28 Hz ± SDIRF 0.06 Hz (awake) vs. MIRF 0.21 Hz ± SDIRF 0.06 Hz (drowsy), N=18, p<0.001, t=4.88.

Keywords—Alertness state, Synchrosqueezing, Fatigue, Abdominal effort and Inattention.

I. INTRODUCTION

Driver drowsiness is one of the main causes of vehicle accidents. A recent study showed that 20% of crashes and 12% of near-crashes were caused by drowsy drivers. The morbidity and mortality associated with drowsy-driving crashes are high, perhaps because of the higher speeds involved combined with delayed reaction time [1]. One approach for preventing traffic accidents is to develop technological countermeasures for detecting driver drowsiness, so that drivers can be warned before a crash occurs.

Abdominal effort signal provides direct information of the driver physiological state [2], and may be especially useful to collect detailed information of the drowsiness cycle and anticipate risky situations while driving. The aim of the work is to study if a Synchrosqueezing based algorithms can be useful to detect drowsiness while driving. The measurements are intended to monitor drivers in both fully awake and sleep deprived conditions and in a controlled scenario.

II. MATERIALS AND METHODS

A. Subjects

A group of 18 volunteers (9 men and 9 women) with ages between 18 and 60, took part in the study. The experiments were performed in two sessions each in different days. In one session the subjects performed the test supposedly having slept normally the night before. The other session performed it in the early morning deprived of sleep, after their workday and having remained awake during the 24 previous hours.

Fig. 1 Subject performing a test

The subjects were monitored during the driving task through biometric sensors. Bitmed Exea Ultra® (Barcelona, Spain) biomedical monitor was used to record biomedical signals as thoracic, diaphragmatic and abdominal respiration in order to compare the quality concluding that the abdominal signal is the best for the analysis due to its stability. All signals were synchronized by a trigger or by the internal clock of the computers with an infrared camera that recorded the face of the driver.

B. Laboratory Setup

The experiments were conducted in the automotive laboratory of the Institute of Biomechanics of Valencia with controlled light, temperature and background sound. All experimental sessions were performed with the same simulated night environment, with only an artificial dim light, and a stable temperature around 24ºC-26ºC. A monotonous road scenario with sound efects at low volume was played during the simulation, to further induce drowsiness.

The simulator includes a fixed-base, a full car cabin and driving simulator software (Fig. 2). The car has traditional equipment (dashboard, pedals, active steering wheel, automatic transmission and controls) conected to a computer in order to record the driving performance like lane changes, speed and brake use. A beam projector displayed a virtual scenario on a screen in front of the car.

Fig. 2 Virtual Driving Simulator

C. Procedure

Once the subjects were informed and signed the consent forms, they were required to complete a sleep habits questionnaire. After that, they were instrumented and drove during a period between 15 and 20 minutes until they felt familiar with the driving simulator. Then the simulation was restarted, and the subjects had to drive during 1 hour 30 minutes, in a highway with low traffic and smooth curves in a night simulation. During all this period they were constantly monitored. During the driving task, every 20 minutes subjects heard an audio message with the word "Hello" to which the subjects were required to press a button on a tactile screen placed in the dashboard of the car. The reaction time of the subjects was measured by the system software. At the same time, the subjects were required to rate their perceived level of sleepiness on a scale of 5 items shown through the same tactile screen. The scale used to measure the subjective level of sleepiness was a modified version of the 10-point Karolinska Sleepiness Scale (KSS). The used scale consisted of the following alert-sleepiness gradations: 1=extremely alert, 2=rather alert but tired, 3= sleepy but could perform any activity, 4= sleepy, I'm not sure I could perform certain activities and 5=very sleepy, great effort to keep awake, fighting sleep. The tactile screen was located near the driver, within easy view.

After finishing the route, the light and sound were turned off, and the subjects remained seated with closed eyes, in order to have a baseline measurement of the physiological activity in a context similar to drowsiness or sleep without driving.

D. Drowsiness State Classification

In order to classify the state of the driver, a reference or Gold Standard signal (GS) was needed. The GS was obtained by the combination of three different External Observers plus the subjective evaluation of the subjects reporting Karolinska Scale results every 20 minutes.

The External Observer signal was based on the subjective assessment of the behaviour of the driver. Body and face movements were annotated by three external observers that analyzed the video in real time and classified every minute of the test as 'attentive' (value 0) or 'drowsy' (value 2), according to the criteria derived of the investigations reported in reference [3]. The final External Observer signal was also computed with a majority ballot.

E. Synchrosqueezing Algorithm as a Drowsiness Index

The interest of respiration volumetric signals measured by inductive band to detect drowsiness resides in the fact that abdominal effort signal is directly related to sympathovagal system, that controls the awake-sleep cycles and also the sleep onset. The fact that the abdominal effort signal becomes unstable in drowsy states makes it particularly suitable for time-frequency representation based on the analysis with algorithms like Synchrosqueezing (SSQ).

Synchrosqueezing (SSQ) combines wavelet analysis and reassignment. It aims at sharpening the time-scale representation by reallocating the coefficients of the continuous wavelet transform (CWT) to another point depending on the local behavior of the representation [4]. However it differs from classical reassignment methods by allowing the exact reconstruction of the signal from the extracted components.

The time or frequency domain description alone may not provide adequate information for a non-stationary signal but SQ algorithms extract the time–frequency representation which displays frequency components and their amplitudes occurring at any given time points [5].

The signal was previously preprocessed by decimation to sampling frequency of 20Hz of the raw signal, originally sampled at 100 Hz and a Butterworth order 2, band-pass filter with cutoff frequencies of 0.05 and 0.5 to avoid electrode polarization effects, artifacts related to body movements and noise.

The Analysis block comprises a Synchrosqueezing analysis of the preprocessed signal. Synchrosqueezing Toolbox Version 1.1 (posted 2012-06-20) proposed by Brevdo et al. [6] was used. This toolbox implements several time-frequency and time-scale analysis methods including forward and inverse Discrete Wavelet Transform. The SSQ transform is implemented in this Toolbox as:

$$T_f(\omega_l, b) = \int_{\{a:\omega_f(a,b)\epsilon W_l, |W_f(a,b)|>\gamma\}} W_f(a,b) a^{-3/2} da \quad (1)$$

where a is the discrete scale employed for the estimation of the instantaneous frequency. The SSQ transform avoids the spreading of energy over several scales by reallocating the power in discrete frequencies. In a narrowband signal (such as our preprocessed respiratory signal), the frequency that corresponds to the highest energy is an estimate of the instantaneous frequency.

We have estimated the instantaneous frequency of the preprocessed respiratory signal by using a Morlet wavelet with number of voices 64, lambda = $1e^3$ and number of curves (nc) =1.

In order to evaluate the viability of the results obtained with the analysis block two different evaluations where done. First visual evaluation of the signal obtained from the detection block compared with the GS and second, analysis of the Mean and the Standard Deviation of the Instantaneous Respiration Frequency (MIRF and SDIRF) obtained in the analysis for awake states and drowsy states individually.

On the visual analysis, as can be seen in the figure 3 represented by a sleep deprived subject, the abdominal effort signal (upper panel) increases the variability when the GS signal (second panel) indicates a drowsy state.

During the firsts 35 minutes of the test the subject remains awake. The increase in variability after the minute 35 can be seen in the respiration data and also in the Synchrosqueezing Instantaneous frequency (SSQIF) as a decrease in the index accompanied by high variability (lower panel).

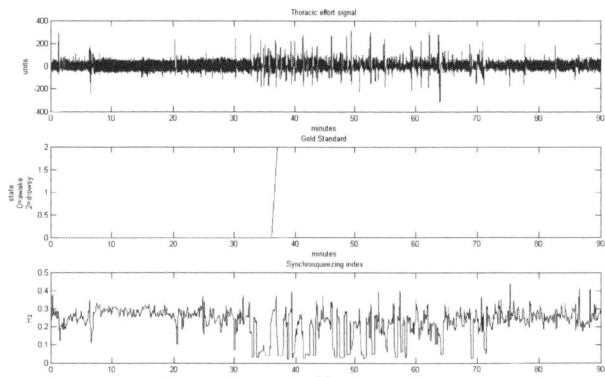

Fig. 3 Visual analysis of a sleep deprived subject

In order to evaluate statistically the Synchrosqueezing algorithm as a drowsiness detector, the Synchrosqueezing instantaneous frequency signal for each of the 18 tests has been divided in two different stages: awake and drowsy, having three intervals of 5 minutes from each phase to extract statistical information as the mean and the standard deviation.

For each test it has been identified three different intervals of awake respiration signal with duration of 5 minutes and three different intervals of drowsy respiration signal with duration of 5 minutes too. Mean and standard deviation of both awake and drowsy Instantaneous respiration frequency intervals (MIRF and SDIRF) of the subjects pool has been calculated.

III. RESULTS

A paired t-test shows that the mean instantaneous respiratory frequency on the analyzed intervals while the subject is fully awake (MIRF awake) is higher than when having a drowsiness episode (MIRF drowsy) (0.28 Hz ± 0.06 Hz vs. 0.21 Hz ± 0.07 Hz, N=54, p<0.001, t=6.68). The standard deviation of the instantaneous respiratory frequency inside the analyzed intervals is lower during the alert state (SDIRF awake) than during the drowsy state (SDIRF drowsy) (39 mHz ± 20 mHz vs. 66 mHz ± 25 mHz, N=54, p<0.001, t=-8.23)

If the indexes are averaged for each experiment (we have selected three realizations for each experiment in each state), the differences are still significant (mean respiratory frequency: 0.28 Hz ± 0.06 Hz (MIRF awake) vs. 0.21 Hz ± 0.06 Hz (MIRF drowsy), N=18, p<0.001, t=4.88; standard deviation of the respiratory frequency: 39 mHz ± 16 mHz (SDIRF awake) vs. 66 mHz ± 22 mHz (SDIRF drowsy), N = 18, p<0.001, t=-7.29).

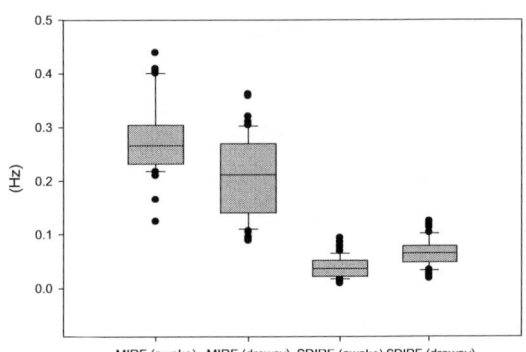

Fig. 4 Drowsiness detection statistical analysis comparing Mean and Standard deviation for awake (MIRF/SDIRF awake) and drowsy (MIRF/SDIRF drowsy) segments of the test

A Two Way Analysis of Variance shows that the mean instantaneous respiratory frequency (MIRF) depends on the state (alert or drowsy, $p<0.001$) and on the subject ($p<0.001$). The same results are obtained for the standard deviation of the instantaneous respiratory frequency (SDIRF).

IV. CONCLUSIONS

The results demonstrate the viability of drowsiness detection in simulator using abdominal effort signal analyzed by Synchrosqueezing based on the significant results of the statistical analysis comparing the Mean and de Standard Deviation of both awake and drowsy intervals instantaneous respiration frequency (MIRF and SDIRF).

The results show MIRF higher in wake intervals that in drowsy intervals and otherwise to SDIRF results. These results are related to the state of the driver but also the subject variability itself.

ACKNOWLEDGMENT

This work has been partially funded by the Spanish MINISTERIO DE CIENCIA E INNOVACIÓN. Proyecto IPT-2011-0833-900000.Healthy Life style and Drowsiness Prevention-HEALING DROP

REFERENCES

1. Faber, J (2004) "Detection of different levels of vigilance by EEG pseudospectra", Neural Network World vol. 14, no. 3-4, pp. 285–290.
2. Rodríguez-Ibáñez N, García-González MA, Fernández-Chimeno M, Ramos-Castro J (2011) . "Drowsiness detection by abdominal effort signal analysis in real driving environments" Conf Proc IEEE Eng Med Biol Soc.2011:6055-8.
3. De Rosario, H; Solaz, J.S.; Rodríguez, N.; and Bergasa, L.M. (2010).Controlled inducement and measurement of drowsiness in a driving simulator IET Intell. Transp. Syst. 4(4): 280-288. DOI:10.1049/ietits. 2009.0110
4. Huang,N-E, Z. Shen, S-R. Long, M-C. Wu, H-H. Shih, Q. Zheng, N-C. Yen, C-C. Tung, and H-H. Liu, "The empirical mode decomposition and the hilbert spectrum for nonlinear and non-stationary time series analysis," Phil. Trans. R. Soc. A, vol. 454, no. 1971, pp. 903−995, 1998.
5. Chuan Li, Ming Liang , "A generalized synchrosqueezing transform for enhancing signal time–frequency representation", Signal Processing 92 (2012) 2264–2274
6. Brevdo, E., Fuckar, N.S., Thakur, G., Wu, H.: The Synchrosqueezing algorithm: a robust analysis tool for signals with time-varying spectrum. ;CoRR(2011)
7. Franco,C, Pierre-Yves Gum´ery1, Nicolas Vuillerme2, Anthony Fleury4 and Julie Fontecave-Jallon1, "Synchrosqueezing to investigate Cardio-Respiratory interactions within simulated volumetric signals", 20th European Signal Processing Conference (EUSIPCO 2012) Bucharest, Romania, August 27 - 31, 2012

Applying Variable Ranking to Oximetric Recordings in Sleep Apnea Diagnosis

D. Álvarez[1], G.C. Gutiérrez-Tobal[1], J. Gómez-Pilar[1], F. del Campo[1,2], M. López[1], and R. Hornero[1]

[1] Grupo de Ingeniería Biomédica, ETSI Telecomunicación, Universidad de Valladolid, Valladolid, España
[2] Servicio de Neumología, Hospital Universitario Río Hortega, Valladolid, España

Abstract—This study is aimed at assessing the usefulness of variable ranking techniques for feature selection in the context of sleep apnea hypopnea syndrome (SAHS) diagnosis from blood oxygen saturation (SpO_2) recordings. Time, frequency, linear, and nonlinear analyses were carried out to compose an initial feature set from oximetry. Principal component analysis (PCA) and fast correlation-based filter (FCBF) were used to derive suitable feature subsets. Support vector machines (SVMs) were applied in the classification stage. A total of 240 subjects suspected of suffering from SAHS composed the population under study. FCBF-based feature subsets significantly outperformed PCA in the test set. A SVM with 5 input features from FCBF achieved the highest performance: 86.5% sensitivity, 83.3% specificity, and 85.4% accuracy. Our results suggest that a suitable analysis of the feature space by means of variable ranking techniques could provide useful information to assist in SAHS diagnosis.

Keywords—sleep apnea hypopnea syndrome, oximetry, variable ranking, support vector machines.

I. INTRODUCTION

The sleep apnea hypopnea syndrome (SAHS) is a sleep-related breathing disorder characterized by frequent breathing cessations (apneas) or partial collapses (hypopneas) during sleep [1]. Undiagnosed SAHS significantly decreases the patients' quality of life. Researchers link SAHS with daytime hypersomnolence, neurocognitive dysfunction, metabolic deregulation, and/or cardiovascular and cerebrovascular diseases [1, 2]. Recent studies suggest that 20% of adults have at least mild SAHS and 7% of adults have moderate-to-severe SAHS [3].

Nocturnal polysomnography (NPSG) in a specialized sleep unit is the gold standard methodology for SAHS diagnosis. However, NPSG is labor-intensive, expensive and time-consuming, which has led to large delays in diagnosis and treatment [4]. Thus, there is a great demand on new techniques aimed at overcoming these limitations. In this way, several studies focus on blood oxygen saturation (SpO_2) from overnight oximetry due to its simplicity, reliability, portability, and lower cost [5, 6].

Different pattern recognition techniques have been applied in the context of SAHS diagnosis from oximetry, such as discriminant analysis [7, 8], logistic regression [9, 10], and neural networks [11, 12]. Multivariate regression techniques have been also successfully applied, such as adaptive regression splines [13] and multivariate linear regression [14, 15]. On the other hand, few studies applied feature selection before classification, which can be advantageous in different ways: reducing measurement, storage and computational requirements, avoiding redundant and noisy information, selecting complementary features, and defying the curse of dimensionality. In the field of SAHS diagnosis from oximetry, stepwise selection has been mainly applied [9, 12, 14]. We hypothesized that an analysis of the feature space by means of different variable selection techniques could provide further knowledge on SpO_2 dynamics.

In the present study, feature extraction, selection, and classification stages were carried out to analyze SpO_2. Time frequency, linear, and nonlinear measures were computed to compose an initial feature set. Next, variable ranking was applied to derive smaller but more suitable feature subsets. Principal component analysis (PCA) and fast correlation-based filter (FCBF) were assessed. PCA is a variable construction approach widely used to select variables in a transformed space [16], whereas FCBF is a ranking procedure that selects relevant and non-redundant variables in the original space [17]. Finally, support vector machines (SVMs), which are high-performance classifiers widely used in different contexts, were applied in the classification stage.

The goal of this study is to assess the usefulness of these algorithms for feature selection in the context of SAHS diagnosis. We hypothesized that a prospective evaluation of different feature subsets could provide further knowledge on SpO_2 dynamics.

II. DATA SET

Two hundred and forty (240) consecutive subjects suspected of suffering from SAHS composed the overall population under study. Patients were recruited from the sleep unit of the Río Hortega Hospital (Valladolid, Spain).

The standard apnea – hypopnea index (AHI) from PSG was used to diagnose SAHS. Apnea was defined as a drop in the airflow signal greater than or equal to 90% from baseline lasting at least 10 seconds. Hypopnea was defined as a drop greater than or equal to 50% during at least 10 seconds, accompanied by a desaturation greater than or equal to 3% and/or an arousal. Subjects with an AHI ≥ 10 events per hour (e/h) were diagnosed as suffering from SAHS. A positive diagnosis of SAHS was confirmed in 160 patients, with an average AHI of 36.6 ± 25.7 e/h (mean ± SD) (average body mass index (BMI) 30.8 ± 4.3 kg/m^2). The remaining 80 subjects composed the SAHS negative group, with an average AHI of 3.9 ± 2.4 e/h (average BMI 27.8 ± 3.7 kg/m^2). This dataset was randomly divided into two independent groups: training set (40%) and test set (60%).

SpO$_2$ recordings from PSG were saved and processed offline to compose the initial oximetric feature set. SpO$_2$ was recorded at a sampling rate of 1 Hz. An automatic pre-processing stage was carried out to remove artifacts.

III. METHODOLOGY

Our methodology was divided into three stages: i) feature extraction, ii) feature selection, and iii) feature classification. Independent training and test populations were used to compose feature subsets and built classifiers and to assess diagnostic performance, respectively.

A. Feature Extraction

Oximetric recordings were parameterized by means of 16 features from different signal processing approaches: time, frequency, linear, and nonlinear. The power spectral density (PSD) was estimated applying the Welch's method (512-sample Hanning window, 50% overlap and 1024-points DFT). All metrics were computed for each whole overnight recording. The following features were obtained [9, 10]:

- Features 1 to 4. First to fourth-order moments ($M1t - M4t$) in the time domain: arithmetic mean ($M1t$), variance ($M2t$), skewness ($M3t$) and kurtosis ($M4t$) were applied to quantify central tendency, amount of dispersion, asymmetry and peakedness, respectively.
- Features 5 to 8. First to fourth-order moments ($M1f - M4f$) in the frequency domain.
- Feature 9. Median frequency (MF), which is defined as the component which comprises 50% of signal power.
- Feature 10. Spectral entropy (SE), which is a disorder quantifier related to the flatness of the spectrum.
- Feature 11. Total spectral power (P_T), which is computed as the total area under the PSD.
- Feature 12. Peak amplitude (PA) in the apnea frequency band, which is the local maximum of the spectral content in the frequency range 0.014 – 0.033 Hz [7].
- Feature 13. Relative power (P_R), which is the ratio of the area enclosed under the PSD in the apnea frequency band to the total signal power.
- Feature 14. Sample entropy ($SampEn$), which quantifies irregularity in time series, with larger values corresponding to more irregular data.
- Feature 15. Central tendency measure (CTM), which provides a variability measure from second order difference plots.
- Feature 16. Lempel – Ziv complexity (LZC), which is a measure of complexity linked with the rate of new subsequences and their repetition along the signal.

B. Feature Selection

A linear re-scaling of each feature was carried out to obtain zero mean and unit variance. The following variable ranking techniques were applied in order to filter features:

1. Principal component analysis (PCA). PCA is aimed at mapping the pattern vector **x** from the original p-dimensional feature space to a new d-dimensional feature space, where $d \leq p$ [16]. New patterns are the projection of the original observations onto the eigenvectors of the original covariance matrix. Each eigenvector accounts for a portion of the total variation of the original data. The variance linked with each eigenvector is represented by its associated eigenvalue λ_d by means of its explained variance (EV):

$$EV = \frac{\lambda_d}{\sum_{k=1}^{p} \lambda_k}. \quad (1)$$

PCA is commonly applied as a filter method to select variables in the transformed space [16]. New variables are ranked according to their EV and the optimum number of components is estimated using some cut-off proportion. In this study, three common thresholds were used: i) *average criterion*, where components exceeding the average variance were selected, ii) 90% threshold, where consecutive components whose cumulative variance exceeds 90% were selected, and iii) 95% threshold, where consecutive components whose cumulative variance exceeds 95% were selected.

2. Fast correlation-based filter (FCBF). FCBF automatically selects relevant and non-redundant features [17]. It relies on symmetric uncertainty (SU), which is a normalized measure of information gain (IG). The method is divided into two steps. Firstly, a relevance analysis is carried out. SU between input features (Xi) and AHI (Y) was computed as follows [17]:

$$SU_i(X_i,Y) = 2\frac{IG_i(X_i,Y)}{H_i(X_i)+H(Y)}, i=1,...,p. \quad (2)$$

where H refers to Shannon's entropy. Features are ranked from more relevant (higher SU_i) to less relevant (lower SU_i). In this study, three different thresholds were applied in order to perform pre-selection of features according to their relevance: i) average criterion, where features exceeding the mean SU were selected, ii) log criterion, where variables exceeding the SU value of the $[N/\log(N)]$-th ranked feature were selected, and iii) all criterion, where the default value $SU = 0$ was used and all features pass to the second step. This step is a redundancy analysis. SU between each pair of pre-selected features ($SU_{i,j}$) is computed starting from the most relevant. When $SU_{i,j} \geq SU_i$, the feature j is discarded due to redundancy.

C. Feature Classification

SVMs are binary classifiers that search for the optimum separating hyperplane in a transformed high dimensional space, resulting in the following mapping function [18]:

$$y(x,w) = w^T z + w_0, \quad (3)$$

where x is the input pattern and $z = \varphi(x)$ performs the transformation to the high dimensional space. w is commonly obtained in terms of the Lagrange multipliers η^n:

$$w = \sum_{n=1}^{N} \eta^n t^n \varphi(x^n), \quad (4)$$

where N is the number of observations and t^n is the target output. The output of a SVM is expressed in terms of the support vectors (those for which $\eta^n \neq 0$) as follows [18]:

$$y = \sum_{n \in S} \eta^n t^n K(x^n, x) + w_0, \quad (5)$$

where S is a subset of the indexes corresponding to the support vectors and $K(\cdot,\cdot)$ represents the inner product kernel function in the transformed space. In the present study, a linear kernel is used. Leave-one-out cross-validation (loo-cv) was applied in the training set to obtain the optimum value of the regularization parameter C.

IV. RESULTS

Feature extraction was carried out for each SpO_2 recording. PCA and FCBF were applied in the training set to compose reduced feature subsets. A SVM was trained for every subset. Table 1 shows input features from each selection procedure and the classification performance of the corresponding SVM classifier in the training set. Similarly, Table 2 summarizes performance assessment of these classifiers in the test set. In addition, a SVM classifier using all features from the initial feature space was trained and tested.

Diagnostic accuracy of SVM classifiers using PCA-derived input variables ranged from 85.4% to 88.5% in the training set. On the other hand, the classification performance significantly decreased in the test set: 78.5% accuracy was reached using 4 variables, 77.8% using 5 variables, and 81.3% using 6 variables.

The number of features derived from FCFB greatly varied using common state-of-the-art filtering thresholds, ranging from 5 to 10 features. Accuracies of SVM classifiers ranged from 83.3% (5 and 6 features) to 86.5% (10 features) in the training set. Performance slightly varied on further assessments in the test set: 83.3% accuracy was reached using 6 features (average criterion) and 85.4% using 4 and 10 features (log and all criteria, respectively). Finally, the classifier using all features obtained 83.3% sensitivity, 87.5% specificity, and 84.7% accuracy in the test set.

V. DISCUSSION

This study assessed the usefulness of variable ranking techniques for feature selection. Two filter approaches independent of the classifier were analyzed. SVMs were used in the classification stage. Input variables from PCA led to high accuracy in the training set. However, classification performance significantly decreased in the test set. On the other hand, input features from FCBF showed higher generalization ability. The highest diagnostic accuracy was obtained using 5 features automatically selected from FCBF, which led to 86.5% sensitivity, 83.3% specificity, and 85.4% accuracy in the test set. The same performance was reached using 10 features automatically selected using a less restrictive criterion for pre-selection of features in the first step of FCBF. Nevertheless, less computational and storage requirements are needed when a smaller number of variables is used. In this way, FCBF slightly outperformed the accuracy obtained by a SVM trained with all features (85.4% vs. 84.7%).

Our results agree with similar studies aimed at improving SAHS diagnosis by means of feature selection. PCA was previously applied to a small set of 3 spectral and 3 nonlinear features [8]. First-to-fifth principal components were selected and 93.0% accuracy (97.0% sensitivity and 79.3% specificity) was reached on the test set. Stepwise selection was also applied to a wide feature set from oximetry, reaching 89.7% accuracy (92.0% sensitivity and 85.4% specificity) using loo-cv [9]. A recent study applied genetic algorithms for exhaustive selection, achieving 87.5% accuracy (90.6% sensitivity and 81.3% specificity) in the test set [10].

Table 1 Feature subsets and its performance in the training set

Feature selection procedure	Se (%)	Sp (%)	Ac (%)
PCA (average): components 1 to 4	81.3	93.8	85.4
PCA (90%): components 1 to 5	81.3	93.8	85.4
PCA (95%): components 1 to 6	85.9	93.8	88.5
FCBF (average): $MF, M2t, M3f, P_R, CTM, SE$	82.8	84.4	83.3
FCBF (log): $MF, M2t, M3f, P_R, CTM$	79.7	90.6	83.3
FCBF (all): $MF, M2t, M3f, P_R, CTM, SE, M4t, LZC, M1t, M3t$	89.1	81.3	86.5
All features	87.5	87.5	87.5

Table 2 Performance assessment of SVM classifiers in the test set

Feature selection procedure	Se (%)	Sp (%)	Ac (%)
PCA (average): components 1 to 4	79.2	77.1	78.5
PCA (90%): components 1 to 5	79.2	75.0	77.8
PCA (95%): components 1 to 6	81.3	81.3	81.3
FCBF (average): $MF, M2t, M3f, P_R, CTM, SE$	87.5	75.0	83.3
FCBF (log): $MF, M2t, M3f, P_R, CTM$	86.5	83.3	85.4
FCBF (all): $MF, M2t, M3f, P_R, CTM, SE, M4t, LZC, M1t, M3t$	83.3	89.6	85.4
All features	83.3	87.5	84.7

Some limitations should be taken into account. Additional variable ranking techniques should be applied in order to exhaustively assess the usefulness of filtering features in the context of SAHS diagnosis. Moreover, additional classifiers should be used to test the generalization ability of our methodology. Finally, further work is needed to test the performance of the proposed methods using portable monitoring.

VI. CONCLUSIONS

In summary, a SVM classifier trained with 5 input variables automatically selected from FCBF achieved a balanced sensitivity – specificity pair and high accuracy in an independent test set. Our results suggest that a suitable feature selection stage performed in the original input space could discern essential information from oximetry.

ACKNOWLEDGMENT

This research was supported in part by project VA111A11-2 from Consejería de Educación (Junta de Castilla y León), the Ministerio de Economía y Competitividad and FEDER under project TEC2011-22987, and the Proyecto Cero 2011 on Ageing from Fundación General CSIC, Obra Social La Caixa, and CSIC.

REFERENCES

1. Young T, Skatrud J, Peppard P E (2004) Risk Factors for obstructive sleep apnea in adults. JAMA 291:2013–2016
2. Patil S P, Schneider H, Schwartz A R et al. (2007) Adult obstructive sleep apnea. Pathophysiology and diagnosis. Chest 132: 325–337
3. Lopez-Jimenez F, Sert F H, Gami A, Somers V K (2008) Obstructive sleep apnea. Implications for cardiac and vascular disease. Chest 133:793–804
4. Whitelaw W A, Brant R F, Flemons W W (2005) Clinical usefulness of home oximetry compared with polysomnography for assessment of sleep apnea, Am J Resp Crit Care, 171:188–193
5. Flemons W W, Littner M R, Rowlet J A et al. (2003) Home diagnosis of sleep apnea: A systematic review of the literature. Chest 124:1543–1579
6. Collop N A, Tracy S L, Kapur V et al. (2011) Obstructive sleep apnea devices for out-of-center (OOC) testing: technology evaluation. J Clin Sleep Med 7:531–548
7. Marcos J V, Hornero R, Álvarez D, Del Campo F, Zamarón C (2009) Assessment of four statistical pattern recognition techniques to assist in obstructive sleep apnoea diagnosis from nocturnal oximetry, Med Eng Phys 31:971–978
8. Marcos J V, Hornero R, Álvarez D, Del Campo F, Aboy M (2010) Automated detection of obstructive sleep apnoea syndrome from oxygen saturation recordings using linear discriminant analysis, Med Biol Eng Comput, 48:895–902
9. Álvarez D, Hornero R, Marcos J V, Del Campo F (2010) Multivariate Analysis of Blood Oxygen Saturation Recordings in Obstructive Sleep Apnea Diagnosis, IEEE Trans Biomed Eng 57 :2816–2824
10. Álvarez D, Hornero R, Marcos J V, del Campo F (2012) Feature selection from nocturnal oximetry using genetic algorithms to assist in obstructive sleep apnoea diagnosis. Med Eng Phys 34:1049–1057
11. Marcos J V, Hornero R, Álvarez D, Del Campo F, López M, Zamarrón C (2008) Radial basis function classifiers to help in the diagnosis of the obstructive sleep apnoea syndrome from nocturnal oximetry. Med Biol Eng Comput 46:323–332
12. Sánchez D, Gross N (2013) Probabilistic neural network approach for the detection of SAHS from overnight pulse oximetry. Med Biol Eng Comput 51:305:315
13. Magalang U J, Dmochowski J, Veeramachaneni S et al. (2003) Prediction of the apnea-hypopnea index from overnight pulse oximetry, Chest 124:1694–1701
14. Olson L G, Ambrogetti A, Gyulay S G (1999) Prediction of sleep-disordered breathing by unattended overnight oximetry. J Sleep Res 8:51–55
15. Marcos J V, Hornero R, Álvarez D, Aboy M, del Campo F (2012) Automated prediction of the apnea-hipopnea index from nocturnal oximetry recordings. IEEE Trans Bio-Med En 59:141–149
16. Jobson J D (1991) Applied multivariate data analysis. Volume II: Categorical and multivariate methods. Springer-Verlag, New York
17. Yu L, Liu H (2004) Efficient feature selection via analysis of relevance and redundancy. J Mach Learn Res 5:1205–1224
18. Vapnik V N (1999) An overview of statistical learning theory, IEEE Trans Neural Netw 10:988–999

Author: Daniel Álvarez
Institute: ETSI Telecomunicación, Universidad de Valladolid
Street: Paseo de Belén 15
City: Valladolid
Country: España
Email: dalvgon@gmail.com

Paroxysmal Atrial Fibrillation Termination Prognosis through the Application of Generalized Hurst Exponents

M. Julián[1], R. Alcaraz[2], and J.J. Rieta[1]

[1] Biomedical Synergy, Electronic Engineering Department, Universidad Politécnica de Valencia, Spain
[2] Innovation in Bioengineering Research Group. University of Castilla–La Mancha, Cuenca, Spain

Abstract— This work evaluates the use of the Generalized Hurst Exponents ($H(q)$) in the prediction of paroxysmal atrial fibrillation (PAF) spontaneous termination from surface electrocardiographic recordings. Although atrial fibrillation (AF) is the most common supraventricular tachyarrhythmia, the mechanisms leading to its spontaneous termination are yet not fully explained. Hurst Exponents, $H(q)$, measure the existence of long-term statistical dependencies in a time series. Therefore, they could provide an estimation of the atrial activity organization during PAF. Since the likelihood of PAF spontaneous termination relates to AF organization, $H(q)$ might predict PAF termination. $H(q)$ has been applied to the prediction of PAF termination using a reference database to allow an easy comparison with previous works. Two established AF organization estimators, the Dominant Atrial Frequency (DAF) and Sample Entropy (SampEn), have been used as a reference. $H(q)$ reached an accuracy of 95.0% in the prediction of PAF termination, thus yielding better results than DAF and SampEn (both obtaining 88.3% accuracy). As a consequence, $H(q)$ could be considered as a promising tool for the prediction of PAF termination.

Keywords— Electrocardiogram, ECG, Atrial Fibrillation, Hurst Exponents.

I. Introduction

Atrial Fibrillation (AF) is characterized by an abnormal excitation of the atria causing an uncoordinated atrial activity [1] and, as a consequence, the atria are unable to pump blood effectively [1]. AF is the most common sustained cardiac arrhythmia, affecting 1–2% of the general population [1]. When the arrhythmia can terminate spontaneously it is denoted as paroxysmal AF (PAF) [1]. To this respect, AF organization is defined as how repetitive the atrial activity (AA) signal pattern is [2], and it correlates with the likelihood of PAF spontaneous termination [3]. The prediction of PAF termination could avoid unnecessary therapy and reduce clinical costs. Therefore, several works have proposed different methods for predicting the spontaneous termination of PAF from noninvasive electrocardiographic (ECG) recordings [4].

On the other hand, the Generalized Hurst Exponents ($H(q)$) [5] are related to the existence of long-term statistical dependencies in a time series. To this respect, when $q = 2$, the exponent relates to the behavior of the autocorrelation function. Hence, the use of $H(2)$ in the analysis of AF signals could provide information about the regularity of the atrial activation patterns and therefore it might provide an estimation of AF organization [6].

The aim of this work is to evaluate the use of $H(2)$ in the prediction of PAF spontaneous termination. To this respect, a reference database [4] has been analyzed in order to allow an easy comparison with previous works. To improve its performance as a predictor of AF termination, $H(2)$ was computed using the optimal computational parameters determined in [7]. The classification results obtained using $H(2)$ were compared with the two most well-known noninvasive AF organization indices: the Dominant Atrial Frequency (DAF) and Sample Entropy (SampEn). The DAF is defined as the frequency corresponding to the highest Power Spectral Density (PSD) amplitude in the 3–9 Hz band [8]. On the other hand, SampEn provides an organization estimation based on the regularity of the signal [9] and it has been applied in different studies as a measurement of AF organization [3].

II. Materials

In order to assess the performance of $H(2)$, the AF Termination Database available at Physionet [10] was analyzed. This database has been described in detail in [4] and it contains 80 two-lead, 1 minute long ECG recordings of PAF, which are classified into three groups: group N (non-terminating) which lasted at least 1 hour in AF after the end of the recording, group T (immediately-terminating), which terminated 1 second after the end of the recording and group S (soon-terminating), whose termination happened 1 minute after the end of the recording.

For the analysis, lead V1 was chosen and the signals were upsampled to 1 kHz using a cubic spline interpolation method to obtain better alignment in QRST subtraction [11]. Then, the signals were preprocessed in order to improve the analysis. First, baseline wander was removed by a bidirectional high-pass filtering with 0.5 Hz cut-off frequency. Then, the signals were filtered using an eight order IIR Chebyshev low-pass filter with 70 Hz cut-off frequency to reduce high frequency noise. Finally, powerline interference was removed through an adaptive notch filter, which preserves the ECG spectral information [12]. Then, the Atrial Activity was extracted using an adaptive QRST cancelation method [13].

III. METHODS

A. Generalized Hurst Exponents

The Generalized Hurst Exponents ($H(q)$) [5] allow to estimate the scaling properties of a time series from the qth order moments of its amplitude increments distribution [5]. $H(q)$ estimates the existence of long-term statistical dependencies in complex and inhomogeneous time series. Thus, $H(q)$ seems an appropriate tool for the analysis of AF signals due to the nonstationary and nonlinear nature of the physiological processes underlying AF [3, 14].

The interpretation of $H(q)$ depends on q. $H(q=2)$ is related to the behavior of the autocorrelation function and, therefore, it is associated with the signal's power spectrum [15]. Since the likelihood of PAF termination relates to AF organization, the use of $H(2)$ as PAF termination predictor has been studied. To this respect, $H(2)$ might provide information about the coordination of the atrial activation patterns during AF and, therefore, it could yield reliable estimations of AF organization.

B. Reference Organization Estimators

Two established noninvasive AF organization metrics were calculated as references. Firstly, SampEn estimates the existence of similar patterns in a time series [3] and larger values of SampEn correspond to more irregular series [3]. This metric is described in detail in [3]. In the present work SampEn was calculated using the computational parameters values $m = 2$ and $r = 0.35$ times the standard deviation of the signal, as is suggested in studies dedicated to this topic [16].

On the other hand, the DAF is related to the atrial cycle length, which depends on the atrial refractoriness [8]. The DAF is defined as the frequency corresponding to the maximum amplitude of the PSD in the 3–9 Hz band [8]. To obtain the DAF, the PSD of each signal was calculated using the Welch Periodogram with a Hamming window of 4096 points in length, 50% overlapping between adjacent sections and 8192-points Fast Fourier Transform.

C. Evaluation of the Methods

The prediction of PAF termination was studied in four different scenarios: I) discrimination between terminating (T) and non-terminating (N) episodes, II) classification into groups T (terminating) and S (soon-terminating), III) discrimination between non-terminating (N) and soon-terminating (S) episodes and IV) identification of all terminating episodes (groups T and S).

Since the the presence of noise and ventricular residua in the extracted AA signal could affect the performance of nonlinear metrics [17], the Main Atrial Wave (MAW) [17] was extracted from the AA through selective band-pass filtering. The MAW was extracted using a band-pass Chebyshev FIR filter with 3 Hz bandwith centered on the DAF [17]. This filter was designed with a linear phase in order to prevent distortion and a relative sidelobe attenuation of 40 dB. $H(2)$ and SampEn were computed on the MAW in order to improve their performance [7, 17].

Regarding the computational data length, $H(2)$ was calculated over the last 15 seconds of each recording, as suggested in [7]. On the other hand, SampEn was computed over all non-overlapping 10-second segments of the signal and then the average value for each recording was calculated [16]. Finally, the DAF was obtained computing the whole recording.

D. Statistical Analysis

First, the signals were tested for stationarity using the runs test [18]. The average value and standard deviation of the signal amplitude were computed on all its non-overlapping 1 second segments. Then, the runs test was applied to each sequence in order to determine whether the process could be considered as stationary.

Then, a study of surrogate data was performed in order to check the nonlinearity of the data because nonlinear indices have been applied to the signals. Since the signals showed a nonstationary behavior, the Truncated Fourier Transform method [19] with cut-off frequency of 3 Hz was applied to generate 40 surrogate data sets from each original recording. This cut-off frequency was selected because the 3–9 Hz band contains most of the atrial activity energy during AF [11]. Then, $H(2)$ was computed over each surrogate data set and compared with the values obtained from their corresponding original signals using the Wilcoxon T test.

Regarding the prediction of PAF termination, the distributions were first tested using the Kolmogorov-Smirnov and Levene tests. Then, the Mann-Whitney U-test was applied to check the differences between groups in each of the analyzed scenarios because the distributions were not normal

Table 1 Classification accuracy of the analyzed metrics in each of the studied scenarios.

	N vs. T	N vs. S	T vs. S	N vs. (T & S)
DAF	88.3%	89.1%	59.3%	86.3%
SampEn-MAW	88.3%	89.1%	61.1%	91.3%
$H(2)$-MAW	95.0%	93.5%	68.5%	93.8%

Table 2 Discrimination results in each of the studied cases through the use of leave-one-out cross-validation.

	N vs. T	N vs. S	T vs. S	N vs. (T & S)
DAF	83.3%	76.1%	59.3%	83.8%
SampEn-MAW	86.7%	84.8%	59.3%	85.0%
$H(2)$-MAW	95.0%	89.1%	64.8%	91.3%

Table 3 Average values and standard deviations of the metrics.

	Group N	Group S	Group T
DAF (Hz)	6.4604 ± 0.6942	5.2796 ± 0.7020	5.0364 ± 0.6866
SampEn-MAW	0.0646 ± 0.0075	0.0495 ± 0.0067	0.0469 ± 0.0091
$H(2)$-MAW	0.9953 ± 0.0008	0.9970 ± 0.0007	0.9975 ± 0.0006

and homoscedastic. A two-tailed value of $p < 0.05$ was considered as statistically significant. Next, the ROC curves were computed in order to determine the optimum classification thresholds and accuracy values in each of the studied scenarios. The optimum classification threshold was defined as the closest point to 100% accuracy. Finally, the consistency of the classification results was improved using leave-one-out cross-validation.

IV. RESULTS

Table 1 contains the accuracy values obtained from the ROC curves, while leave-one-out cross-validation results are presented in Table 2. The best prediction accuracy was achieved by $H(2)$ (see Tables 1 and 2). All the metrics showed statistically significant differences between non-terminating (group N) and terminating PAF (groups S and T). Moreover, a statistically significant difference between groups S and T was found with $H(2)$, while the other metrics did not show statistically significant differences between these groups.

Additionally, the terminating PAF episodes showed higher regularity than the non-terminating ones, which mean that they presented higher $H(2)$ and lower SampEn values. Furthermore, their DAF values were also lower (see Table 3). Figure 1 shows the optimum threshold and classification between groups N and T, while the classification and optimal threshold between groups N and S are showed in Figure 2. Finally, Figure 3 shows the ROC curve obtained in the identification of all terminating episodes.

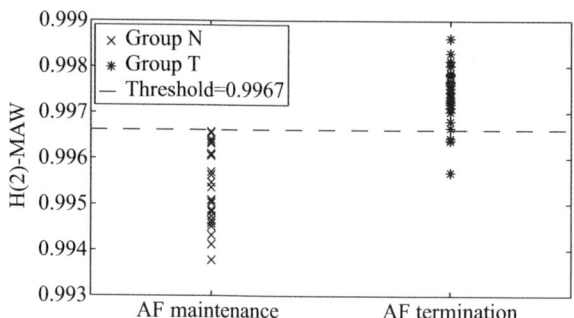

Fig. 1 Classification between non-terminating and immediately-terminating episodes using $H(2)$.

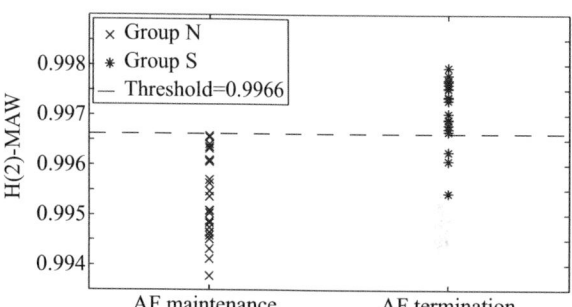

Fig. 2 Discrimination between non-terminating and soon-terminating episodes using $H(2)$.

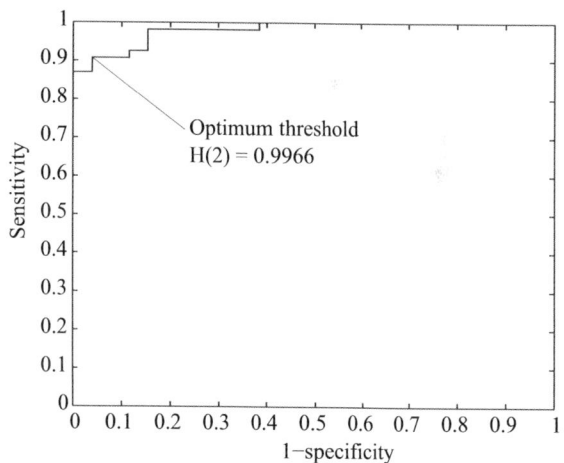

Fig. 3 ROC curve obtained in the identification of all terminating AF episodes (group N vs. groups S and T) using $H(2)$.

Regarding the surrogate data test, Figure 4 shows the values of $H(2)$ obtained for the original signals and their corresponding surrogate data sets. Statistically significant differences were found between original and surrogate data. Therefore, the signals show a nonlinear and nonstationary behavior.

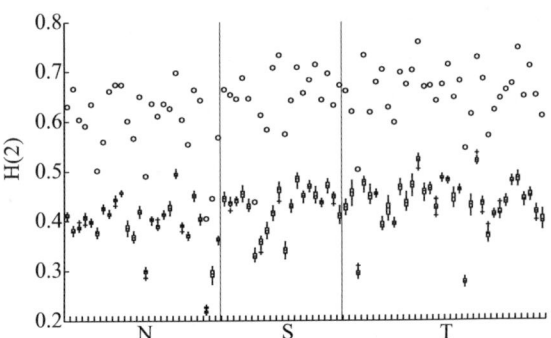

Fig. 4 Values of $H(2)$ obtained for the original (circles) and surrogate (boxplot) signals.

V. Discussion

In this work the performance of $H(2)$ as a predictor of PAF termination has been evaluated. $H(2)$ has been calculated over 15 seconds of the MAW, as recommended in [7]. In this the present work a statistically significant difference between groups S and T has been found, whereas in a previous study [6], where different computational parameters were used, no significant differences between those groups were reported. Moreover, in the present study the accuracy in the discrimination between groups N and T has increased from 93.3% up to 95%. Therefore, the use of 15 seconds segments improved the performance of $H(2)$ in the prediction of PAF termination.

Finally, it would be interesting to focus future works in evaluating the performance of $H(2)$ over segments different from the last 15 seconds of each recording used in the present study.

VI. Conclusions

The Generalized Hurst Exponents have attained a higher accuracy than the well-known AF organization metrics DAF and SampEn in the prediction of PAF spontaneous termination. Therefore, $H(2)$ can be considered as a promising tool for the study of AF organization from noninvasive ECG recordings.

Acknowledgements

Work funded by TEC2010–20633 from the Spanish Ministry of Science and Innovation and PPII11–0194–8121 from Junta de Comunidades de Castilla-La Mancha.

References

1. Camm A. J., Kirchhof P., Lip G. Y. H., et al. Guidelines for the management of atrial fibrillation: the task force for the management of atrial fibrillation of the European Society of Cardiology (ESC). *Europace.* 2010;12:1360-1420.
2. Sih H J, Zipes D. P., Berbari E. J., Olgin J. E.. A high-temporal resolution algorithm for quantifying organization during atrial fibrillation *IEEE Transactions on Biomedical Engineering.* 1999;46 (4):440-450.
3. Alcaraz R, Rieta J J. A review on sample entropy applications for the non-invasive analysis of atrial fibrillation electrocardiograms *Biomedical Signal Processing and Control.* 2010;5 (1):1-14.
4. Moody G B. Spontaneous Termination of Atrial Fibrillation: A Challenge from PhysioNet and Computers in Cardiology 2004 *Computers in Cardiology.* 2004;31:101-104.
5. Barabási A L, Vicsek T. Multifractality of self-affine fractals *Physical Review A.* 1991;44(4):2730-2733.
6. Julián M, Alcaraz R, Rieta J J. Comparative Study of Nonlinear Metrics to Discriminate Atrial Fibrillation Events from the Surface ECG *Computing in Cardiology.* 2012;39:197-200.
7. Julián M, Alcaraz R, Rieta J J. Study on Hurst Exponents Optimized Computational Parameters to Estimate Atrial Fibrillation Organization from the Surface ECG (Submitted to MEDICON 2013).
8. Holm M, Pehrson S, Ingemansson M, et al. Non-invasive assessment of the atrial cycle length during atrial fibrillation in man: introducing, validating and illustrating a new ECG method *Cardiovasc Res.* 1998;38:69-81.
9. Richmann J S, Moorman J. R.. Physiological time-series analysis using approximate entropy and sample entropy *Am. J. Physiol. Heart Circ. Physiol..* 2000;278 (69):H2039-H2049.
10. Goldberger A L, Amaral L A, Glass L, et al. PhysioBank, PhysioToolkit, and PhysioNet: Components of a New Research Resource for Complex Physiologic Signals *Circulation.* 2000;101(23).
11. Bollmann A, Husser D, Mainardi L, et al. Analysis of surface electrocardiograms in atrial fibrillation: techniques, research, and clinical applications *Europace.* 2006;8:911-26.
12. Martens S M M, Mischi M., Oei S. G., Bergmans J. W. M.. An improved adaptive power line interference canceller for electrocardiography *IEEE Transactions on Biomedical Engineering.* 2006;53 (11):2220-2231.
13. Alcaraz R, Rieta J J. Adaptive singular value cancelation of ventricular activity in single-lead atrial fibrillation electrocardiograms *Physiological Measurement.* 2008;29:1351-1369.
14. Bollmann A. Quantification of electrical remodeling in human atrial fibrillation *Cardiovasc Res.* 2000;47:207-9.
15. DiMatteo T. Multi-scaling in finance *Quantitative finance.* 2007;7 (1):21-36.
16. Alcaraz R, Abásolo D, Hornero R, Rieta JJ. Optimal parameters study for sample entropy-based atrial fibrillation organization analysis *Comput Methods Programs Biomed.* 2010;99:124-32.
17. Alcaraz R, Rieta J J. Sample entropy of the main atrial wave predicts spontaneous termination of paroxysmal atrial fibrillation *Medical Engineering & Physics.* 2009;31:917-922.
18. Bendat J S, Piersol A G. *Random Data: Analysis and Measurement Procedures*. Wiley-Interscience, New York2nd ed. 1986.
19. Nakamura T, Small M, Hirata Y. Testing for nonlinearity in irregular fluctuations with long-term trends *Physical Review E.* 2006;74.

Corresponding Author: Matilde Julián Seguí
Institute: Biomedical Synergy
Address: Universidad Politécnica de Valencia,
        Camino de Vera s/n. Building 7F, 5th Floor
City: Valencia
Country: Spain
Email: majuse@upv.es

Evaluation of Laplacian Diaphragm Electromyographic Recordings in a Static Inspiratory Maneuver

L. Estrada[1,2,3], A. Torres[1,2,3], J. Garcia-Casado[4], Y. Ye-Lin[4], and R. Jané[1,2,3]

[1] Institut de Bioenginyeria de Catalunya (IBEC), Biomedical Signal Processing and Interpretation Group, Barcelona, Spain
[2] Universitat Politècnica de Catalunya, Department d'Enginyeria de Sistemes, Automàtica i Informàtica Industrial, Barcelona, Spain
[3] Centro de Investigación Biomédica en Red de Bioingeniería, Biomateriales y Nanomedicina (CIBER-BBN), Spain
[4] Universitat Politècnica de València, Grupo de Bioelectrónica, Valencia, Spain

Abstract—Diaphragm electromyography (EMGdi) provides important information on diaphragm activity, to detect neuromuscular disorders of the most important muscle in the breathing inspiratory phase. EMGdi is habitually recorded using needles or esophageal catheters, with the implication of being invasive for patients. Surface electrodes offer an alternative for the non-invasive assessment of diaphragm activity. Ag/AgCl surface disc electrodes are used in monopolar or bipolar configuration to record EMGdi signals. On the other hand, Laplacian surface potential can be estimated by signal recording through active concentric ring electrodes. This kind of recording could reduce physiological interferences, increase the spatial selectivity and reduce orientation problems in the electrode location. The aim of this work is to compare EMGdi signals recorded simultaneously with disc electrodes in bipolar configuration and a Laplacian ring electrode over chest wall. EMGdi signal was recorded in one healthy subject during a breath hold maneuver and a static inspiratory maneuver based on Mueller's technique. In order to estimate the covered frequency range and the degree of noise contamination in both bipolar and Laplacian EMGdi signals, the cumulative percentage of the power spectrum and the signal to noise ratio in subbands were determined. Furthermore, diaphragm fatigue was evaluated by means of amplitude and frequency parameters. Our findings suggest that Laplacian EMGdi recording covers a broader frequency range although with higher noise contamination compared to bipolar EMGdi recording. Finally, in Laplacian recording fatigue indexes showed a clearer trend for muscle fatigue detection and also a reduced cardiac interference, providing an alternative to bipolar recording for diaphragm fatigue studies.

Keywords—**Laplacian electrode, diaphragm muscle, fatigue, surface electromyography.**

I. INTRODUCTION

Monitoring of muscular activity of respiratory system is an important issue for detecting medical problems, which compromise the normal ventilation process. The breathing consists on an inspiratory phase when air is inhaled and an expiratory phase when air is exhaled. During inspiration, respiratory muscles contribute to expand the thorax, move air into the lungs and develop a negative pressure. In contrast, during expiration, muscles relax, the thorax returns to its starting position and the air is forced out from the lungs. The diaphragm, a large, dome-shaped structure, is the primary muscle involved in the inspiratory phase. Evaluation of inspiratory muscle function is essential to study respiratory diseases as chronic obstructive pulmonary disease, which increases the airway resistance to airflow and compromise the normal function of diaphragm [1]. Respiratory muscles' activity can be assessed by means of intraesophageal electrodes [2] and intramuscular electrodes [3]. However, these techniques have the disadvantage of being invasive and cumbersome for patients [4] with the potential risk of an iatrogenic pneumothorax [5]. Surface electromyography (sEMG), a non-invasive diagnostic tool for measuring muscle electrical activity, is a practical alternative for evaluation of muscle diaphragm function. sEMG is habitually collected by silver/silver chloride (Ag/AgCl) disc electrodes in monopolar or bipolar configuration. On the other hand, Laplacian of surface potential, the second derivative of the surface potential, can be evaluated by means of so-called Laplacian electrodes, a bipolar or tripolar coaxial ring electrodes. It overcomes the reduced spatial resolution and the influence of cardiac activity, drawbacks found in bipolar recordings [6, 7]. The use of Laplacian electrodes has proven promising results in the study of electrocardiogram (ECG) [6], electroencephalogram [8], electroenterogram [7], and sEMG in biceps brachii [9].

This study was undertaken to characterize the diaphragm electromyography (EMGdi) acquired simultaneously by a Laplacian electrode and Ag/AgCl disc electrodes in bipolar configuration over diaphragm muscle. A breath hold and a static inspiratory maneuver were performed.

II. MATERIALS AND METHODS

A. Subject and Experimental Protocol

A nonsmoking subject with no clinical evidence of respiratory disease was included in the investigation. The subject was instructed to sit down in a comfortable chair and to stand straight with the arms beside the body, not to move nor talk, to breathe via a mouthpiece connected to a T-tube, and wore a disposable nose clip to prevent any air exchange through the nostrils. The conducted study consists on the EMGdi signal acquisition during a breath hold maneuver (BHM) and a static inspiratory maneuver (SIM), based on the Mueller's technique [10], to activate the diaphragm in response to a voluntary force inspiration with the T-tube occluded with only a small leak to prevent glottic closure during inspiration maneuver [4]. During the BHM, a disruption of breathing activity was performed for 15 s. Next, during the SIM, the subject was asked to exert three maximal inspiratory pressures (MIP) during 2 s, with 1-min resting intervals. The highest value was taken. Afterward, the subject performed a test exerting a pressure of 80 percent of the chosen MIP for 10 s, and was encouraged to maintain the target pressure using visual feedback.

B. Signal Acquisition

Before placing the electrodes, the skin in the diaphragm area (thoracic region) was mildly abraded with gel (Nuprep, Weaver and Company, USA) and cleansed with alcohol to improve contact impedance. Pre-gelled, disposable circular electrodes of 10 mm diameter (foam electrode 50/PK – EL501, Biopac Systems Inc, Santa Barbara, CA, USA) were placed over diaphragm muscle area between the 7th and 8th intercostal space, lateral to the right midclavicular line, with 50 mm inter-electrode distance. Dry Laplacian electrode (tripolar concentric ring electrode in quasi-bipolar configuration, TCB) of 24 mm outer diameter was plugged into a signal conditioner circuit and placed in the inter-electrode space of two Ag/AgCl electrodes and attached with an adhesive plaster to skin (Fig 1), as it was presented in [9]. A reference electrode was positioned at left ankle. Bipolar electrode configuration and Laplacian electrode were plugged into an amplifier module (EMG 100C, Biopac Systems Inc, Santa Barbara, CA, USA) using a gain of 500. Moreover, Lead-I ECG was collected with an amplifier (ECG 100C, Biopac Systems Inc, Santa Barbara, CA, USA) using a gain of 500. Inspiratory mouth pressure was measured using a mouth-piece attached to a T-tube that was connected to a differential pressure transducer (TSD160, Biopac Systems Inc, Santa Barbara, CA, USA) plugged into a differential amplifier (DAC100C, Biopac Systems Inc, Santa Barbara, CA, USA) using a gain of 50. Amplifiers were connected to an acquisition system (MP150, Biopac Systems Inc, Santa Barbara, CA, USA) and interfaced with a computer to monitoring signals in real time and stored (AcqKnowledge software v.3.2, Santa Barbara, CA, USA). The sampling frequency used was 1000 Hz.

C. Signal Processing and Data Analysis

Both bipolar and Laplacian EMGdi signals were bidirectionally bandpass filtered using a fourth-order, digital Butterworth filter with cut-off frequencies of 10 and 300 Hz.

Estimation of the power spectral density was performed by applying the Welch modified periodogram method (Hamming window, 0.2 seconds length, 4096 - point FFT, and 50 % overlap) in segments of half a second length. For the subsequent signal analysis of EMGdi power spectrum was chosen between $f_1 = 10$ to $f_2 = 300$ Hz.

Signal-to-Noise Ratio (SNR) was evaluated to compare the noise content in bipolar and Laplacian EMGdi spectrums.

SNR was calculated using the energy content in 5-second segment of SIM (E_I) and BHM (E_0), respectively, in bands of 10 Hz.

$$SNR = 10 log_{10}\left(\frac{E_I}{E_0}\right) \quad (1)$$

Cumulative percentage of the power distribution (CmPS), expressed as the running total of power spectrum (PS) over the total power spectrum, was used to compare the spectral distribution of bipolar and Laplacian EMGdi spectrum.

$$CmPS(f) = \frac{\int_{f_1}^{f} PS(f)df}{\int_{f_1}^{f_2} PS(f)df} \cdot 100 \quad (2)$$

To quantify the evolution of diaphragm muscle fatigue, we calculated the following amplitude and frequency parameters:

1) Averaged Rectified Value (ARV), an amplitude parameter, defined as average of the absolute value of the EMGdi signal, which gives an estimation of the total amount of activity over a specific time period T.

$$ARV = \frac{1}{T}\int_0^T |EMGdi(t)|dt \quad (3)$$

2) Median Frequency (MDF), a frequency parameter that divides the EMGdi power spectral density in two regions having the same amount of power.

$$\int_{f_1}^{MDF} PS(f) \cdot df = \int_{MDF}^{f_2} PS(f) \cdot df \quad (4)$$

3) Dimitrov's index, that calculates the ratio of spectral moment of order (-1) and spectral moment of order (5), to highlight lower and higher frequencies respectively, in muscle fatigue.

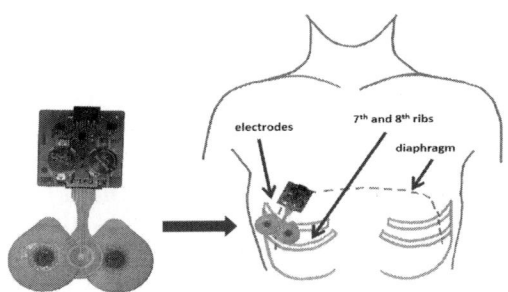

Fig. 1 (a) Two disc electrodes and Laplacian electrode (middle) for recording of EMGdi signal in diaphragm. (b) Electrode placement between 7th and 8th intercostal spaces, lateral to the right midclavicular line

$$FInsm5 = \frac{\int_{f_1}^{f_2} f^{-1} \cdot PS(f) \cdot df}{\int_{f_1}^{f_2} f^{5} \cdot PS(f) \cdot df} \quad (5)$$

Data was analyzed using a moving window of half a second length with steps of 50ms. The signal processing was performed in MATLAB (v. R2011b, Natick, Massachusetts, USA).

III. RESULTS

EMG recording of diaphragm muscle is mostly affected by cardiac activity. Fig 2 shows an excerpt of a 5-sec segment of EMGdi signal, corresponding to a BHM and a SIM, picked up simultaneously by a bipolar disc-electrode configuration and a Laplacian electrode. It is evident that ECG interference greatly affects bipolar EMGdi recording (Fig 2 (a) and (b)) in comparison to Laplacian EMGdi recording (Fig 2 (c) and (d)) during both BHM (diaphragm at rest) and a SIM (high activity of diaphragm). As it is shown in Fig 3, SNR for bipolar recording is higher than Laplacian recording. However, in contrast to that of Laplacian signal, SNR of bipolar signal shows a pronounced decrease in the frequency bandwidth of the ECG components (below 30 Hz). A comparative analysis of the CmPS is shown in Fig 4. Fig 4(a) reveals a rapid increase and an exponential trend in CmPS of bipolar recording during both the BHM and the SIM. The area between the breath hold and the static inspiratory maneuver for CmPS curves of bipolar recordings is greater below 50 Hz, where the cardiac activity is present. In Fig 4(b), at rest (BHM), CmPS of Laplacian signal showed a linear trend; reflecting a spectral distribution similar to white noise. In contrast, from the same figure, it is also clear that the area between the BHM and the SIM for the curves of Laplacian signal is greater above 50 Hz, in comparison to those of bipolar recording. In respect to diaphragm muscle fatigue, ARV, MDF and FInsm5 indexes were calculated. Figs 5 (a) presents the ECG signal corresponding to the lead I. ARV, MDF and FInsm5 fatigue in-

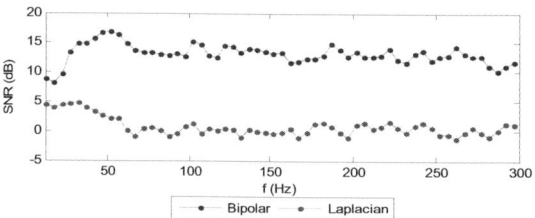

Fig. 3 Spectral distribution of SNR for bipolar and Laplacian EMGdi

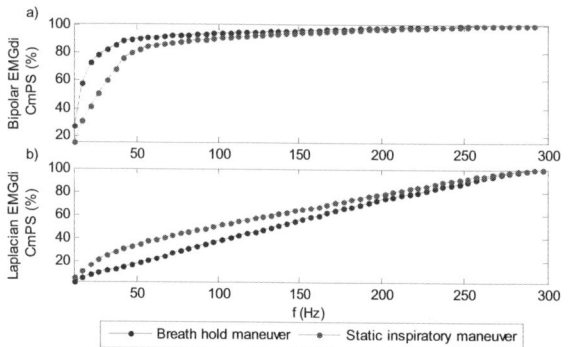

Fig. 4 Cumulative percentage of power for 5-sec EMGdi recording during a breath hold and a static inspiratory maneuver using (a) a bipolar electrode configuration and (b) a Laplacian recordings acquisition

dexes, computed from sEMG signal and recorded over the diaphragm muscle during a SIM, are shown in Fig 5 b, c and d, respectively. According to literature [11], ARV and FInsm5 increases as muscle fatigues, whereas MDF decreases. Our results show that ARV increases during SIM for both bipolar and Laplacian recording. Conversely, the FInsm5 and MDF indexes present clearer trends (rising and decreasing, respectively) in Laplacian recording in contrast to bipolar recording. In addition, the comparison of ECG in Fig 5 (a) with the evolution of fatigue indexes reveals that cardiac activity provokes a non-desired periodic oscillatory component in such evolution. Fatigue indexes in bipolar recording were more influenced by cardiac activity in comparison to Laplacian recording as shown in Fig 5 b, c and d.

IV. DISCUSSION AND CONCLUSIONS

EMGdi signal was recorded simultaneously using a bipolar disc electrode configuration and a Laplacian electrode. Previous works have reported the advantage of using Laplacian electrodes in other anatomical regions of human body [6–9]. In the present study during both BHM and SIM, power spectrum of Laplacian EMGdi signal was more distributed in frequency, contained in a broader frequency range compared to bipolar EMGdi power spectrum. Because of their physical design, Laplacian electrode presents enhanced spatial selectivity, features which are reflected in its spatial transfer function [12]. The use of a Laplacian

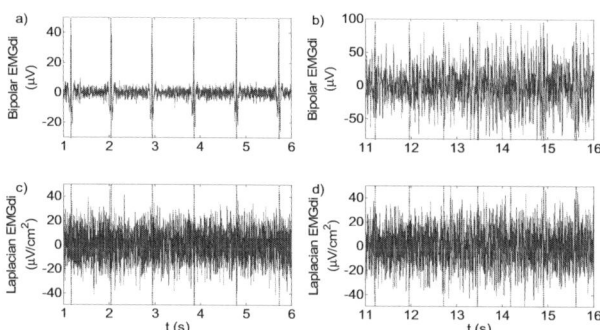

Fig. 2 An excerpt of 5-sec EMGdi recording during a breath hold maneuver and a static inspiratory maneuver: bipolar recording with disc electrodes (a and b) and Laplacian recording with TCB ring electrode (c and d). Vertical red dot lines indicates the R wave of cardiac activity

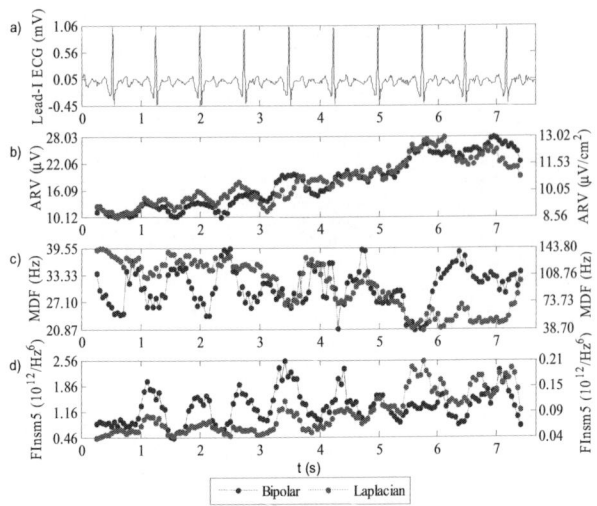

Fig. 5 (a) Lead-I ECG, (b) ARV, (c) MDF and (d) FInsm5 fatigue indexes calculated from EMGdi signal during a static inspiratory maneuver. Left and right scales correspond to parameters from bipolar and Laplacian recordings, respectively

electrode in sEMG recordings can improve the study of motor unit action potential trains [12, 13] and reduce the problem of location and orientation of the fibers related to the use of bipolar disc electrode configuration [12]. On the other hand, signal to noise ratio, was lower in Laplacian recording in comparison to bipolar recording. In this sense, it should be considered that the size of the Laplacian electrode used was smaller than the inter-electrode distance of bipolar recording with disc electrodes. The use of Laplacian electrodes of bigger diameter, or of bipolar configuration instead of TCB could yield signals of higher amplitude and better SNR.

The evolution of fatigue parameters during the SIM was similar to that reported during isometric contraction in the biceps brachii when bipolar and Laplacian electrodes were compared [9], with an increase of ARV and FInsm5 and a decrease of MDF. It was noted that ARV was influenced by cardiac activity in bipolar and Laplacian EMGdi recordings but MDF and FInsm5 were more affected in bipolar EMGdi compared to Laplacian EMGdi. Bipolar recording of EMGdi with conventional disc electrodes are severely affected by cardiac activity which provokes that most energy of this signal is concentrated below 50 Hz.

In conclusion, this work suggests that EMGdi picked up by Laplacian electrodes is less affected by ECG interference, presenting a wider bandwidth of signal. Nevertheless, Laplacian recordings present significantly lower SNR than conventional bipolar recordings. Moreover, during the static inspiratory maneuver, trends of muscle fatigue index from Laplacian signals were much clearer and in agreement with results reported in literature. However, more subjects need to be evaluated to confirm these results and the behavior of Laplacian electrode over diaphragm in subsequent studies.

ACKNOWLEDGMENT

The first author was supported by Instituto para la Formación y Aprovechamiento de Recursos Humanos and Secretaría Nacional de Ciencia, Tecnología e Innovación (IFARHU-SENACYT Program) from the Panama Government under grant 270-2012-273. This work was supported in part by Ministerio de Economía y Competitividad from the Spain Government under grants TEC2010-21703-C03-01 and TEC2010-16945.

REFERENCES

1. Similowski T, Derenne J-PH (1994) Inspiratory muscle testing in stable COPD patients. Eur Respir J 7:1871–1876.
2. Luo YM, Moxham J, Polkey MI (2008) Diaphragm electromyography using an oesophageal catheter: current concepts. Clin Sci (Lond) 115:233–244.
3. Hodges PW, Gandevia SC (2000) Pitfalls of intramuscular electromyographic recordings from the human costal diaphragm. Clin Neurophysiol 111:1420–1424.
4. American Toracic Society/European Respiratoty Society (2002) ATS/ERS Statement on Respiratory Muscle Testing. Am J Respir Crit Care Med 166:518–624.
5. Al-Shekhlee A, Shapiro BE, Preston DC (2003) Iatrogenic complications and risks of nerve conduction studies and needle electromyography. Muscle & Nerve 27:517–526.
6. He B, Cohen RJ (1992) Body surface Laplacian ECG mapping. IEEE Trans Biomed Eng 39:1179–1191.
7. Prats-Boluda G, Garcia-Casado J, Martinez-de-Juan JL, Ye-Lin Y (2011) Active concentric ring electrode for non-invasive detection of intestinal myoelectric signals. Med Eng Phys 33:446–55.
8. Besio WG, Koka K, Aakula R, Dai W (2006) Tri-polar concentric ring electrode development for laplacian electroencephalography. IEEE Trans Biomed Eng 53:926–933.
9. Estrada L, Torres A, Garcia-Casado J, et al. (2013) Characterization of Laplacian surface electromyographic signals during isometric contraction in biceps brachii. Proceedings of the 35th Annual International Conference of the IEEE Engineering in Medicine and Biology Society. Osaka, Japan, 2013, pp 535–538
10. Terris DJ, Hanasono MM, Liu YC (2000) Reliability of the Muller maneuver and its association with sleep-disordered breathing. Laryngoscope 110:1819–1823.
11. González-Izal M, Malanda A, Gorostiaga E, Izquierdo M (2012) Electromyographic models to assess muscle fatigue. J Electromyogr Kinesiol 22:501–512.
12. Farina D, Cescon C (2001) Concentric-ring electrode systems for noninvasive detection of single motor unit activity. IEEE Trans Biomed Eng 48:1326–1334.
13. Farina D, Arendt-Nielsen L, Merletti R, et al. (2003) Selectivity of spatial filters for surface EMG detection from the tibialis anterior muscle. IEEE Trans Biomed Eng 50:354–364.

Author: Luis Estrada
Institute: Institut de Bioenginyeria de Catalunya. Biomedical Signal Processing and Interpretation Group
Street: C/Baldiri Reixac, 4. Torre I, Planta 9
City: Barcelona
Country: Spain
Email: lestrada@ibecbarcelona.eu

Analysis of Normal and Continuous Adventitious Sounds for the Assessment of Asthma

M. Lozano[1,4], J.A. Fiz[2,4,5], and R. Jané[3,4,5]

[1] Innovation Group at Health Sciences Research Institute of the Germans Trias i Pujol Foundation (IGTP), Badalona, Spain
[2] Pneumology Service at the Germans Trias i Pujol University Hospital, Badalona, Spain
[3] Department of Automatic Control (ESAII), Universitat Politècnica de Catalunya (UPC), Barcelona, Spain
[4] Institute for Bioengineering of Catalonia (IBEC), Barcelona, Spain
[5] Biomedical Research Networking Center in Bioengineering, Biomaterials, and Nanomedicine (CIBER-BBN), Spain

Abstract—Assessment of asthma is a difficult procedure which is based on the correlation of multiple factors. A major component in the diagnosis of asthma is the assessment of BD response, which is performed by traditional spirometry. In this context, the analysis of respiratory sounds (RS) provides relevant and complementary information about the function of the respiratory system. In particular, continuous adventitious sounds (CAS), such as wheezes, contribute to assess the severity of patients with obstructive diseases. On the other hand, the intensity of normal RS is dependent on airflow level and, therefore, it changes depending on the level of obstruction. This study proposes a new approach to RS analysis for the assessment of asthmatic patients, by combining the quantification of CAS and the analysis of the changes in the normal sound intensity-airflow relationship. According to results obtained from three patients with different characteristics, the proposed technique seems more sensitive and promising for the assessment of asthma.

Keywords—asthma, bronchodilator response, continuous adventitious sound, respiratory sound intensity, wheezes.

I. INTRODUCTION

Asthma is a complex respiratory disorder that results in a variable, recurring, and often reversible airflow obstruction [1]. Physicians have difficulties in diagnosing asthma, since they have to correlate several aspects: the medical history, a thorough physical examination, and pulmonary function test results. In this context, the bronchodilator (BD) response is a standard pulmonary function test used to control and assess the severity of asthma. It is based on spirometric measurements before and after the administration of a BD. Usually, an improvement in forced expiratory volume in one second (FEV_1) of greater than or equal to 12% is considered to be significant [2, 3]. However, some recent studies have demonstrated that using spirometric criteria alone is inadequate for the diagnosis of asthma [4].

Respiratory sounds (RS) are helpful in understanding the function of the respiratory system. They are classified as normal or adventitious sounds. Due to their clinical interest, many technical studies have tried to detect and characterize continuous adventitious sounds (CAS) [5-8], such as wheezes. CAS are characterized by a pitch of over 100 Hz that lasts more than 100 ms [9], and they are key indicators for assessing the severity of asthma [1]. On the other hand, some other studies have tried to understand the origin of normal RS and their intensity pattern distribution [10-12]. Although results from all previous studies have contributed to the characterization and understanding of RS, there is lack of clinical use and application of these techniques.

In this study, we propose a new approach to the analysis of RS for the assessment of asthmatic patients, by combining the quantification of CAS and the analysis of the normal sound intensity-airflow relationship. A few previous studies have focused on the evaluation of asthma by RS analysis [13-15]. Nevertheless, some were performed on infants and they were based on manual detection of wheezes and their characterization at a fix airflow level or during forced breathing. On the other hand, in [15] they focused on changes in the spatial distribution of breath sound intensity by analyzing dynamic images. Our technique has two major advantages: the automatic differentiation and quantifying of respiratory cycles either with normal sounds or CAS [16], and the analysis of normal RS intensity as a function of airflow level.

II. METHODS

A. Signal Acquisition

RS signals were recorded from asthmatic patients in a sitting position at the Pulmonary Function Testing Laboratory, Germans Trias i Pujol University Hospital, Badalona, Spain. Three piezoelectric contact microphones (TSD108, Biopac, Inc.) were attached to the skin using adhesive rings: two of them on the back at 3 cm below the left/right shoulder blade, and one over the right side of the trachea. Moreover, respiratory airflow signal was recorded using a pneumotachograph (TSD107B, Biopac, Inc.). Each patient was coached to progressively increase the airflow, from shallow breathing to the deepest breaths they could. All signals were sampled at 12500 samples/second using a 16-bit analogue-to-digital converter. After acquisition, RS signals were band-pass filtered using a combination of 8th order Butterworth low-pass and high-pass filters (70 – 2000 Hz). We show a case study with three adult asthmatic patients with different baseline spirometric values and BD response (Table 1). For each patient, we have quantified the percentage of respiratory cycles with CAS at baseline and after BD, for both left and right sides. In addition, we have analyzed the relationship between RS intensity and airflow, before and after BD.

Table 1 Characteristics of asthmatic patients

ID	Total cycles Pre-BD		Total cycles Post-BD		Age	Sex	BMI (Kg/m^2)	FEV_1 (%)	ΔFEV_1 (%)
	Left	Right	Left	Right					
1	46	47	53	53	50	F	24.44	47	26
2	68	66	76	83	60	M	27.08	100	1
3	59	49	53	50	19	M	19.28	59	6

B. Segmentation of RS Signals

After signal acquisition, respiratory phases were obtained using the airflow signal as the reference for automatic sound signal segmentation. Since airflow is positive during inspiration and negative during expiration, respiratory phases were marked off by means of a robust zero crossing detector. In order to avoid detection of false endpoints, only cycles in which the airflow reached at least 0.35 L/s were considered valid cycles. Moreover, two thresholds of 0.2 and 4 seconds were established for minimum and maximum durations of respiratory phases, respectively, according to time duration of normal respiratory cycles. In addition, a threshold of 0.5 seconds was fixed for the maximum time interval between the end of inspiration and the beginning of the corresponding expiration. Two final datasets, pre-BD and post-BD, were obtained for each patient and each side. Each dataset was formed by audio-visual selection of sound signals from the inspiratory cycles, avoiding artifacts such as those from speaking, swallowing, coughing or rubbing. We have focused on inspiratory sounds, which are much louder than expiratory sounds on the back, where we recorded the sound signals.

C. CAS Detection and Inspiratory Cycle Classification

The first step in the analysis of RS is to differentiate respiratory cycles with normal sounds from those with CAS. For that purpose, we made use of an automatic RS classifier, which we had previously developed based on the analysis and feature extraction from instantaneous frequency (IF) of sound signal from each respiratory cycle [16]. Prior to the IF calculation, sound signals were decomposed into narrow-band components by ensemble empirical mode decomposition (EEMD). The core of the proposed classifier is the fact that IF remains almost constant when a CAS signal is within a respiratory cycle. By using this classifier, we quantified the number of CAS cycles from both pre-BD and post-BD datasets, for each patient and from both left and right sides. Then, we evaluated whether the percentage of CAS cycles had significantly increased or decreased with the administration of the BD.

D. RS Intensity – Respiratory Airflow Graphs

In addition to evaluate the changes in the percentage of CAS cycles pre/post-BD, we also analyzed the changes in normal sound intensity. Sound intensity was calculated for inspiratory sound segments corresponding to the top 20% of airflow from each normal sound cycle, from both pre-BD and post-BD datasets. It was defined as the mean power obtained from the power spectral density (PSD) in the frequency band 75-600 Hz. PSD was calculated using Welch's periodogram, with a Hanning window of 80 ms, a 50% overlap between adjacent segments, and 2048 points for the Fast Fourier Transform. Then, each normal sound cycle was determined by its intensity and the maximum airflow reached. Since each patient was characterized by the relationship between normal sound intensity and airflow on both left and right sides, we analyzed the changes in these graphs in order to evaluate the BD effect.

III. RESULTS

In this section we show results from applying the previous techniques to patients shown in Table 1. Firstly, results from respiratory cycle classification and CAS cycle quantification are shown in Fig. 1.

Secondly, we calculated RS intensity from normal sound cycles as a function of the airflow, pre-BD and post-BD, for both left and right sides (Fig. 2).

The patient with severe asthma (ID 1) had 35% (left) and 30% (right) of CAS cycles before BD, as shown in Fig. 1. It is in agreement with her low baseline FEV_1 (47%), which shows that she had a severe bronchial obstruction. After BD, she had not CAS cycles, which is in agreement with her increased FEV_1 (ΔFEV_1 = 26%). Moreover, normal sound intensity significantly increased after BD, in both sides, as can be appreciated from the polynomial regression lines shown in Fig. 2-A.

Contrary to ID1, patient with mild asthma (ID 2) had a normal baseline FEV_1 (100%), and a low number of CAS cycles (<1.5%) before and after BD, as shown in Fig. 1. Moreover, he was a non-responder to BD (ΔFEV_1 = 1%), which agree with a very low increase of normal sound intensity, in both sides (Fig. 2-B). This is reflected in very close polynomial regression lines, pre and post-BD.

Analysis of Normal and Continuous Adventitious Sounds for the Assessment of Asthma

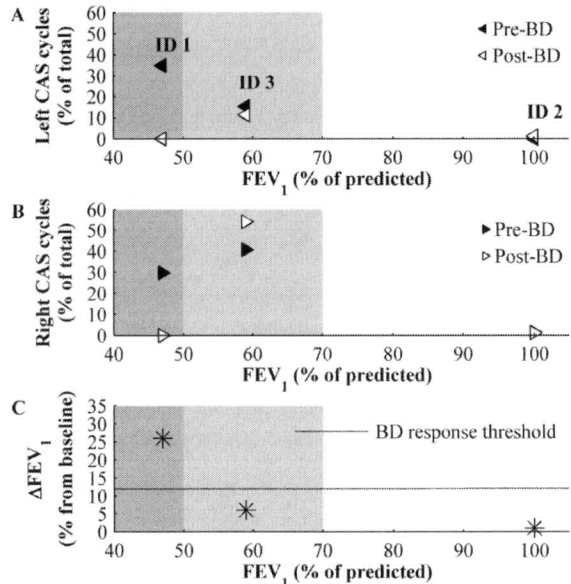

Fig. 1 Percentage of CAS cycles in both left and right sides, at baseline (pre-BD) and after the BD (post-BD) (A, B). Change in FEV_1 after BD, as a per cent from baseline, and threshold for a significant BD response (12%) (C). Three degrees of severity were defined based on baseline FEV_1: mild ($x \geq 70\%$), moderate ($50\% \leq x < 70\%$), and severe ($x < 50\%$).

In addition to the previous extreme cases, we analyzed an intermediate case with moderate asthma (ID 3). He was a slight responder to BD, since he had a ΔFEV_1 of 6%. However, he had a baseline FEV_1 of 59%, which is in the range of moderate-severe asthma. It agrees with his 15% (left) and 41% (right) of CAS cycles at baseline. After BD, the percentages of CAS cycles were 11% (left) and 54% (right), which were maintained high values. Although his BD response was not significant, lines from the polynomial regression in Fig. 2-C show that he had a relevant increase in normal sound intensity, and it was clearer at high airflows.

IV. DISCUSSION

According to the aforementioned results, the baseline FEV_1 is related to the number of CAS cycles. Those patients with a low baseline FEV_1 have high probability of having wheezes, and vice versa.

On the other hand, the changes in the number of CAS and normal sound intensity are more related to the BD response. A positive BD response indicates that the bronchial obstruction has significantly decreased and, therefore, there are less CAS cycles and higher normal sound intensity.

On the contrary, non-responders do not have many changes in their bronchial tree, which implies having few changes in the number of CAS cycles and low increase of their normal sound intensity.

Moreover, we have shown that for a patient with a non-significant BD response (below 12%), the analysis of normal sound intensity is a more sensitive technique, since

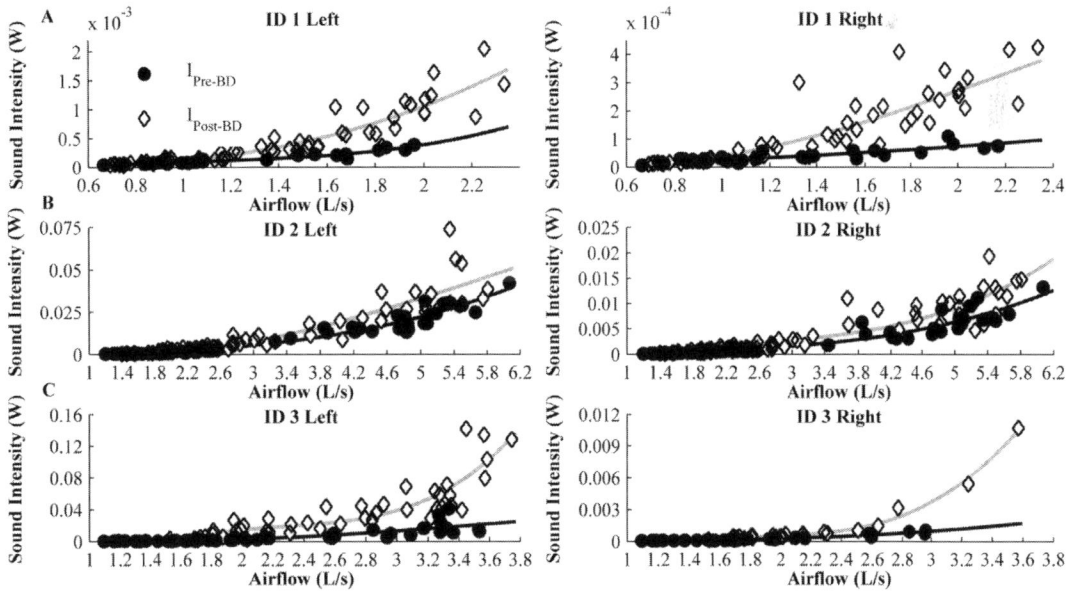

Fig 2 Comparison between normal sound intensity and airflow before and after the administration of a BD. For each patient, results from both left and right sides are shown. Lines black and grey show the 3rd order polynomial regression models that fit pre-BD and post-BD intensities as functions of airflow, respectively.

significant increases can be detected after BD. Furthermore, these increases are more relevant at high airflow levels.

With respect to the proposed technique, it has some advantages: the exploration and evaluation of the pulmonary function at different airflow levels, which are reached during the progressive respiratory maneuver, and the use of two recording channels that allow us to obtain information about the laterality of normal RS and CAS.

V. CONCLUSIONS

In this study we have shown that the combined analysis of normal RS and CAS provides a promising approach to characterize asthmatic patients. It is a simple and non-invasive technique, which seems more sensitive to changes in the pulmonary function. Therefore, it could be a complementary tool in the diagnostic procedure for asthma. However, this work is a case-study with three patients and therefore a higher number of varied cases are required in order to draw final conclusions.

ACKNOWLEDGMENT

This work was made possible thanks to a collaboration agreement between IBEC and IGTP, and it was supported in part by the Spanish Ministry of Economy and Competitiveness under grant TEC2010-21703-C03-01. All authors would like to thank Rosa Gómez, from the pulmonary function testing laboratory at Germans Trias i Pujol University Hospital, for its collaboration in the patient recruitment.

REFERENCES

1. National Heart, Lung, and Blood Institute (2007) National asthma education and prevention program. Expert panel report 3: Guidelines for the diagnosis and management of asthma. http://www.ncbi.nlm.nih.gov/books/NBK7232/
2. Reddel HK, Taylor DR, Bateman ED et al. (2009) An official American Thoracic Society/European Respiratory Society statement: asthma control and exacerbations: standardizing endpoints for clinical asthma trials and clinical practice. Am J Respir Crit Care Med 180:59–99
3. Pellegrino R, Viegi G, Brusasco V et al. (2005) Interpretative strategies for lung function tests. Eur Respir J 26:948–968
4. Gjevre JA, Hurst TS, Taylor-Gjevre RM, Cockcroft DW (2006) The American Thoracic Society's spirometric criteria alone is inadequate in asthma diagnosis. Can Respir J 13:433–437
5. Bahoura M (2009) Pattern recognition methods applied to respiratory sounds classification into normal and wheeze classes. Comput Biol Med 39:824–843
6. Homs-Corbera A, Fiz JA, Morera J, Jané R (2004) Time-frequency detection and analysis of wheezes during forced exhalation. IEEE Trans Biomed Eng 51:182–186
7. Taplidou SA, Hadjileontiadis LJ (2010) Analysis of wheezes using wavelet higher order spectral features. IEEE Trans Biomed Eng 57:1596–1610
8. Xie S, Jin F, Krishnan S, Sattar F (2012) Signal feature extraction by multi-scale PCA and its application to respiratory sound classification. Med Biol Eng Comput 50:759–768
9. Sovijärvi ARA, Dalmasso F, Vanderschoot J et al. (2000) Definitions of terms for applications of respiratory sounds. Eur Respir Rev 10:597–610
10. Murphy R (2007) Computerized multichannel lung sound analysis. Development of acoustic instruments for diagnosis and management of medical conditions. IEEE Eng Med Biol Mag 26:16–19
11. Pasterkamp H, Consunji-Araneta R, Oh Y, Holbrow J (1997) Chest surface mapping of lung sounds during methacholine challenge. Pediatr Pulmonol 23:21–30
12. Torres-Jimenez A, Charleston-Villalobos S, Gonzalez-Camarena R, Chi-Lem G, Aljama-Corrales T (2008) Asymmetry in lung sound intensities detected by respiratory acoustic thoracic imaging (RATHI) and clinical pulmonary auscultation. Conf Proc IEEE Eng Med Biol Soc 2008:4797–4800
13. Sánchez I, Vizcaya C, García D, Campos E (2005) Response to bronchodilator in infants with bronchiolitis can be predicted from wheeze characteristics. Respirology 10:603–608
14. Mazic I, Sovilj S, Magjarevic R (2003) Analysis of respiratory sounds in asthmatic infants. Meas Sci Rev 3:9–12
15. Guntupalli KK, Reddy RM, Loutfi RH et al. (2008) Evaluation of obstructive lung disease with vibration response imaging. J Asthma 45:923–930
16. Lozano M, Fiz JA, Jané R (2013) Estimation of instantaneous frequency from empirical mode decomposition on respiratory sounds analysis. In: Proceedings of the 35th International Conference of the IEEE Engineering in Medicine and Biology Society, Osaka (in press)

Corresponding author:

Author: Manuel Lozano
Institute: Institut for Bioengineering of Catalonia
Street: Baldiri Reixac, 4, Torre I, 9th floor, 08028
City: Barcelona
Country: Spain
Email: mlozano@ibecbarcelona.eu

Automatic Extrasystole Detection Using Photoplethysmographic Signals

A. Solosenko[1] and V. Marozas[1,2]

[1] Biomedical Engineering Institute, Kaunas University of Technology, Kaunas, Lithuania
[2] Signal Processing Department, Kaunas University of Technology, Kaunas, Lithuania

Abstract—**Extrasystoles are a common heart rhythm irregularity. Depending on the frequency of occurrence, extrasystoles might be benign or not benign and may indicate the upcoming heart diseases such as atrial fibrillation, ventricular tachycardia or myocardial infarction. In this paper, we present extrasystoles detection algorithm based on photoplethysmographic (PPG) signals. Seven temporal and amplitude features of the PPG signal are used to describe a single PPG pulse. The heart beat classification is carried out by Naïve Bayes Classifier. Performance of the proposed algorithm was evaluated using the PPG signals from the PhysioNet MIMIC database and signals registered with a portable monitor of physiological processes. The PPG signals were annotated according to the synchronously registered ECG signals. Results show high 96.40 % sensitivity, 99.92 % specificity and 99.89 % accuracy of the algorithm.**

Keywords—**Extrasystoles, photoplethysmography, Naïve Bayes Classifier.**

I. INTRODUCTION

Occurrence of extrasystoles is one of the symptoms that may indicate the cardiac disease. Single extrasystoles usually are not dangerous [1], they occur occasionally in healthy persons. However, frequent extrasystoles, especially premature ventricular contractions (PVCs) [1, 2] may increase the risk of more serious cardiac arrhythmias, e.g. ventricular tachycardia [3]. Recent studies showed that the PVCs are a predictive factor of the cardiac death of individuals without the structural heart disease [4]. Increasing PVCs occurrence especially with the multiform PVCs was associated with the heart failure [5]. Study [6] reported that frequent (more than 7 per minute) PVCs during the recovery after an exercise is a better predictor of an increased risk of death than the ventricular ectopy occurring only during the exercise.

Frequency and morphology of the occurring extrasystoles are evaluated using the Holter monitors. However, electrocardiograph (ECG) electrodes attached to the patient's chest may cause discomfort, limit the freedom of movement or increase the feeling of unhealthiness, especially after wearing the monitor for 24 h. A more convenient in the daily life and cheaper (no electrodes needed) alternative to the Holter monitors might be the method based on photoplethysmography (PPG). The PPG is a non-invasive method measuring blood volume changes in the capillary system by illuminating the tissues with the light. A PPG sensor is more comfortable and can be attached to a finger [7], integrated in the ear phones [8] or a forehead band [9]. Until now, several attempts to detect arrhythmias using PPG signals were accomplished [7, 10, 11]. These methods make use of the temporal and amplitude features extracted from the PPG signal morphology or are based on the continuous wavelet transform.

In this study, we present an extrasystole detection and classification method based on the changes of the temporal and amplitude features of the PPG signal during arrhythmias. The heart beats are classified by using the Naïve Bayes Classifier (NBC) [12]. The NBC classifier is chosen due to the relative simplicity of implementation, efficiency of training and the possibility to implement it in an embedded system retaining high detection sensitivity and specificity. Despite naive assumption of mutual independence among the features, the NBC classifier can compete with the more sophisticated classifiers. The NBC has never previously been applied to investigate extrasystole events measured using PPG technology. The proposed algorithm is able to detect the PVCs and premature atrial contractions (PACs) of various shapes (Fig. 1).

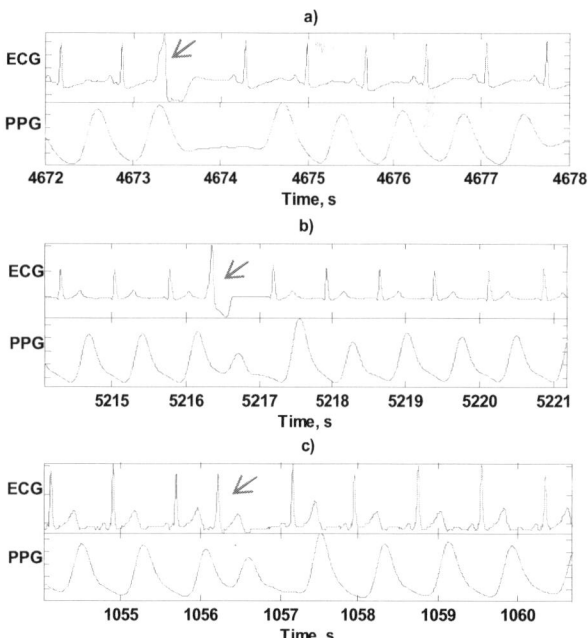

Fig. 1 The most common shapes of extrasystoles in the ECG and PPG signals (recorded on a finger): PVC with the absent extrasystolic peak (a), PVC with the visible extrasystolic peak (b), and PAC (c).

II. METHOD

The block diagram of the proposed algorithm is shown in Fig. 2. In the following sections, the algorithm is explained step by step.

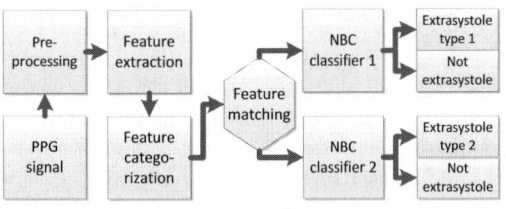

Fig. 2 The block diagram of the proposed algorithm

A. Preprocessing and Feature Extraction

In order to reduce the non-overlapping noises, the PPG signal was preprocessed with 0.5 and 3.5 Hz cutoff frequencies Butterworth 2nd order band-pass filter. The cutoff frequencies were chosen empirically by investigating the performance of the algorithm.

Both negative and positive peaks of the PPG signal were determined with the support of the pulse width modulated (PWM) signal (Fig. 3 b), generated by:

$$PWM_m = \begin{cases} 1, & PPG_m > 0 \\ -1, & PPG_m \leq 0 \end{cases} \quad (1)$$

where m – sample index.

The positive and negative square pulses of the PWM signal defined the intervals in the PPG signal with the approximate locations of the positive and negative peaks. In contrast to the peak detection method based on the signal derivative, the proposed method excludes small insignificant peaks without using the fixed thresholds (see Fig 3. b). The peak detection stage is critical for determining the PPG signal parameters and classification accuracy of the extrasystoles.

The next step is the feature extraction from the PPG signal. The following features were used (see Fig. 4):

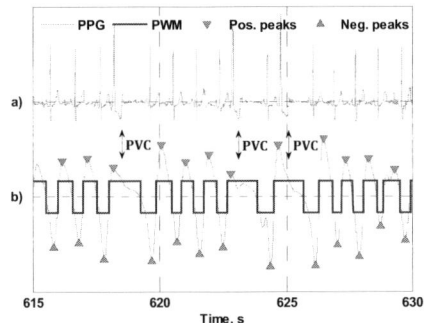

Fig. 3 The ECG signal (a), PPG signal together with the PWM signal, the positive and negative peaks of the PPG signal (b)

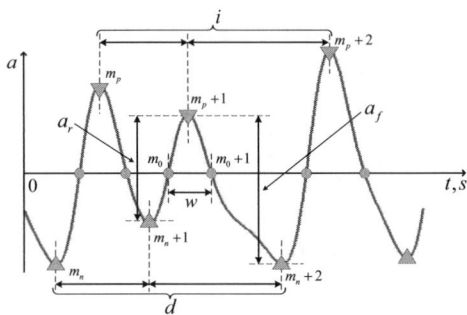

Fig. 4 Feature extraction from PPG signal

amplitudes of the rising and falling fronts, peak-to-peak intervals, pulse durations and pulse widths.

Amplitudes of the rising and falling fronts of the PPG pulse were calculated by (2) and (3) accordingly:

$$a_r(m_p+1) = a(m_p+1) - a(m_n+1) \quad (2)$$
$$a_f(m_p+1) = a(m_p+1) - a(m_n+2) \quad (3)$$

where a_r, a_f – amplitudes of the rising and falling fronts, m_p, m_n – indices of the positive and negative peaks.

Further, the intervals between the adjacent peaks and pulse durations were calculated by (4) and (5), respectively:

$$i(m_p) = (m_p+1) - m_p \quad (4)$$
$$d(m_n) = (m_n+1) - m_n \quad (5)$$

where i –interval between the positive peaks, d –interval between the negative peaks.

Finally, the pulse widths were estimated using (6):

$$w(m_0) = (m_0+1) - m_0 \quad (6)$$

where w – pulse width, m_0 – index of the zero crossing.

The features calculated above comprised the full set of the features used to describe a single beat.

Then the extracted features were processed using normalization and outlier rejection equations. Each type of the features was processed by estimating its ratio to a number of the adjacent features of the same type using (7). Then an array of the ratios $R[n]$ was accumulated:

$$R[n] = F[n]/F[n+m], \quad m = 1...M \quad (7)$$

where F – feature, n and m –indexes of the features, M –analysis window length.

The accumulated array of the ratios was averaged by (8) and a mean value $\overline{R}_1$ of that array was estimated:

$$\overline{R}_1 = \frac{1}{N}\sum_{n=1}^{N} R[n] \quad (8)$$

An array of the ratios $R[n]$ was redefined by excluding the ratios which do not satisfy the conditions in (9). The result acts as an upper threshold.

$$R_1[n] = \begin{cases} R[n], & R[n] < \overline{R}_1 / 0.8 \\ R[n] = \varnothing, & R[n] > \overline{R}_1 / 0.8 \end{cases} \quad (9)$$

Once again, a redefined array $R_1[n]$ was averaged by (10) and a new mean value $\overline{R}_2$ was calculated:

$$\overline{R}_2 = \frac{1}{N}\sum_{n=1}^{N} R_1[n] \quad (10)$$

The feature array $R_1[n]$ was modified again by excluding the values which do not satisfy the conditions in (11) acting as a lower threshold:

$$R_2[n] = \begin{cases} R_1[n], & R_1[n] > \overline{R}_2 \times 0.8 \\ R_1[n] = \varnothing, & R_1[n] < \overline{R}_2 \times 0.8 \end{cases} \quad (11)$$

The upper operations exclude the ratios which would otherwise significantly modify the mean value of the collected feature ratios. At this moment, the standard deviation of the feature ratio array $R_2[n]$ was significantly reduced. The final averaging was carried out by (12) and an array of the ratio averages $\overline{R}[m]$ was accumulated:

$$\overline{R}[m] = \frac{1}{N}\sum_{n=1}^{N} R_2[n] \quad (12)$$

where m – index of the feature ratio average.

Next, the feature categorization was used to create the feature sets for the NBC classifier. In total, 8 categories represented the dynamic ranges of the extracted features (Table 1). The categories were the intervals of the percentage deviations from 1. The features with the ratios close to 1 were considered normal.

Table 1 Feature categories

	Intervals, %	Amplitudes, %	Pulse widths, %
1	30 – 60	15 - 55	10 – 45
2	60 – 80	55 - 70	45 – 75
3	80 – 95	70 - 85	75 – 95
4	95 – 105	85 – 115	95 – 105
5	105 – 120	115 – 135	105 – 135
6	120 – 140	135 – 155	135 – 155
7	140 – 245	155 – 175	160 – 280
8	None of the above intervals		

In case the feature did not fit any category, it was considered undefined. Thus, the beats with at least one undefined feature are excluded from the further investigation.

B. Beat Classification

Two NBC classifiers are used to classify the types of the extrasystoles mentioned above. Classification was made by multiplying the feature probabilities calculated on the basis of the training data and estimating the maximal probability (13):

$$\hat{C} = \arg\max_c P(C)\prod_{i=1}^{N} P(F_i | C) \quad (13)$$

where $\hat{C}$ – class, $P(C)$ – prior probability, $P(F_i|C)$ – likelihood of evidence, i – feature index, N – number of features.

The classifiers were trained using the annotated signals. During classification, the feature set sequences were ascribed to one or another classifier by matching their certain features e.g. the peak-to-peak interval (see Fig. 2) which might belong to a certain type of the extrasystoles.

III. DATA

An algorithm was evaluated by experimenting with 9 real PPG signals from the PhysioNet MIMIC database [13] and one PPG signal, registered on the forehead with a portable monitor of the physiological processes. Synchronously registered ECG signals (Fig. 3 a) served as the reference annotating the PPG signals. In order to reduce the feature estimation errors, both the PPG and ECG signals from the MIMIC database were interpolated from 125 Hz to 500 Hz sampling frequency. The signals registered with a portable monitor were sampled at the 500 Hz sampling rate. The duration of all the signals was 6000 s.

IV. RESULTS

Performance of the algorithm was expressed in sensitivity *Sen*, specificity *Spe* and accuracy *Acc*. The summary of the signals and obtained results is presented in Table 2.

Table 2 The summary of classification results

PhysioNet signal	# beats	# extrasystolic beats	TP	TN	FP	FN
039m	7073	276	264	6785	1	12
221m	7833	11	11	7822	15	0
230m	7269	4	4	7265	0	0
253m	6423	0	0	6423	0	0
439m	8572	15	14	8556	18	1
444m	8830	22	20	8806	1	2
449m	6322	10	10	6312	1	0
482m	7449	132	131	7316	4	1
484m	7912	84	80	7823	4	4
*Forehead	6066	28	28	6037	18	1
Total	73749	576	562	73145	62	21
Sen			96.40 %			
Spe			99.92 %			
Acc			99.89 %			

where TP – true positives, TN – true negatives, FP – false positives, FN – false negatives.

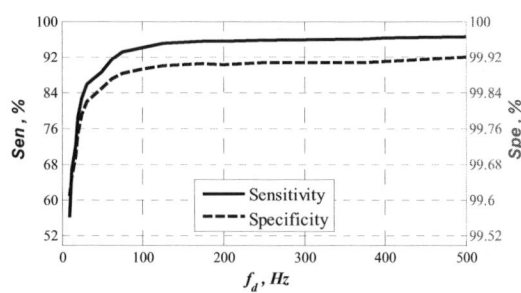

Fig. 5 Sensitivity and specificity as a function of the PPG signal sampling frequency

The PPG signals are low frequency signals thus a lower sampling rate can be used. The lower sampling rate means less use of energy and durable monitoring. Fig. 5 presents the algorithm performance as a function of the PPGs sampling frequency.

V. DISCUSSION AND CONLCUSIONS

Our investigation reveals the possibility to detect the extrasystoles effectively using the PPG signals only. The detection and classification effectiveness mainly depends on the precision of the feature extraction which in turn depends on the quality of the PPG signals because the classifiers rely on the fixed feature categorization thresholds. The algorithm relies on the morphology of the PPG signal thus, detection performance might decrease if the shape of the PPG signal is severely corrupted by noise.

An advantage of our algorithm is its relative implementation and training simplicity, versatility and minimized usage of the computational resources. Both sensitivity and specificity were high at the sampling frequencies equal or higher than 100 Hz (see Fig. 5). Moderate extrasystole detection effectiveness is still possible even at 75 Hz of the sampling frequency. The algorithm is capable to detect non extrasystolic beats accurately even at the high noise levels. These qualities make possible the implementation of the algorithm in the wearable systems.

The main limitation of the investigation is its relatively small database of the signals. There are 121 recordings in the PhysioNet MIMIC database, however, only a limited number of the signals contains PVCs or PACs. In addition, an unequal number of the beats annotated as being non-extrasystolic and extrasystolic influenced the high accuracy values achieved.

Similar specificity (0.990) but lower sensitivity (0.915), was achieved in study [7] though we can not compare these results side by side because the datasets were different.

One of possible future directions is to classify the extrasystoles into the premature ventricular contractions (PVCs) and premature atrial contractions (PACs). At the current stage we classify the beats into two types: PVC beats (Fig. 1 a) and beats associated with both PVCs and PACs (Fig. 1 b and c). We hope that the latter heart beat type might be classified further into two separate classes by taking into account the additional specific features.

ACKNOWLEDGMENT

This work was partially supported by the Lithuanian Agency for Science, Innovation and Technology (Agreement No 31V-24) and by the European Social Fund (Agreement No VP1-3.1-SMM-10-V-02-004).

REFERENCES

1. Strieper M. J., PVCs Benign or not? // Sibley Heart Center Cardiology, 2007; 1 – 3.
2. Rodstein M., Wolloch L., Gubner R. S., Mortality Study of the Significance of Extrasystoles in an Insured Population // Circulation. 1971; 617-625.
3. Kolpan A. B., Stevenson G. W., Ventricular Tachycardia and Sudden Cardiac Death // Symposium on cardiovascular diseases, 2009; 289-297.
4. Hirose H, Ishikawa S, Gotoh T., et.al. Cardiac mortality of premature ventricular complexes in healthy people in Japan // J Cardiol., 2010; 23-6.
5. Ephrem G, Levine M, Friedmann P, et.al. The Prognostic Significance of Frequency and Morphology of Premature Ventricular Complexes during Ambulatory Holter Monitoring // A Noninvasive Electrocardiology. 2013; 118-25.
6. Frolkis, J. P., Pothier, C. E., Blackstone, E. H., et.al. Frequent Ventricular Ectopy after Exercise as a Predictor of Death // N Engl J Med 2003; 348:781 -790.
7. Suzuki T, Kameyama K, Tamura T. Development of the irregular pulse detection method in daily life using wearable photoplethysmographic sensor // Conf Proc IEEE Eng Med Biol Soc., 2009; 6080–6083.
8. Ming-Zher P., Kyunghee K., Goessling, A., et.al. Cardiovascular Monitoring Using Earphones and a Mobile Device // Pervasive Computing, IEEE, 2012; 18-26.
9. Wendelken S., McGrath S., Blike G., et.al. Reflectance Forehead Pulse Oximetry: Effects of Contact Pressure During Walking // EMBS Annual International Conference, 2006; 3529-3532.
10. Gil E., Sornmo L., Laguna P., Detection of Heart Rate Turbulence in Photoplethysmographic Signals // Computing in Cardiology, 2011 665 – 668.
11. Galen. P., Addison P., Watson J., et.al. Systems and Methods for Detecting and Monitoring Arrhythmias Using the PPG // US patent US 20120310100 A1, 2012.
12. Rish I., An empirical study of the naive Bayes classifier // T.J. Watson Research Center; 41- 46.
13. Goldberger AL, Amaral LAN, Glass L., et.al. PhysioBank, PhysioToolkit, and PhysioNet: Components of a New Research Resource for Complex Physiologic Signals // Circulation 101(23) : e215-e220, 2000 (June 13).

Author: Andrius Solosenko
Institute: Kaunas University of Technology
Street: Studentu 65-107
City: Kaunas
Country: Lithuania
Email: andrius.solosenko@ktu.lt

ECG Signal Reconstruction Based on Stochastic Joint-Modeling of the ECG and the PPG Signals

D. Martín-Martínez, P. Casaseca-de-la-Higuera, M. Martín-Fernández, and C. Alberola-López

Laboratory of Image Processing, Universidad de Valladolid, Valladolid, Spain

Abstract—**In this paper, we propose a model–based methodology aimed at reconstructing corrupted or missing intervals of ECG signals acquired together with a PPG signal. To this end, we first estimate a joint–model of the ECG and PPG signals from the largest uncorrupted piece. Then, in a second stage, a set of candidates to replace the corrupted epoch is synthesized by sampling the aforementioned model. Each sample is evaluated respect to a boundary condition in order to select the best candidate. This signal is refined through an iterative method relying on the actual PPG data acquired during the corruption interval. Experiments on real data, show the capability of the proposed methodology to accurately reconstruct ECG pieces, outperforming so far presented solutions not accounting for joint information.**

Keywords—**ECG reconstruction, Joint ECG-PPG modeling, Waveform model, Evolution model, ARMA, PCA.**

I. INTRODUCTION

Nowadays, signal processing of cardiovascular signals such as ECG (electrocardiography) or PPG (photopletysmography) constitutes an invaluable tool for decision support with diagnostic purposes. Most of the signal processing solutions aimed at diagnostic support consist on the development of feature extraction methods so that meaningful indices conveying information on the disease are obtained. A decision on the presence or absence of the disease can be achieved by simply arranging these indices into a feature vector feeding an automatic pattern recognition system.

Statistical modeling of physiological signals can lead to interesting applications beyond simple diagnostic support based on feature extraction. In this case, a parametric model of the acquired signal is obtained by analysis. This model can actually be used for decision support [1] or even for compression purposes [2]. Besides, model sampling can be performed by a synthesis procedure in order to obtain stochastic samples of the model for signal reconstruction. This constitutes a challenging problem to solve, since artifacted or missing epochs are usually ignored and this involves working with (short) pieces of the acquisition and thus wasting some potentially useful information.

In [3] we proposed a reconstruction methodology for cardiovascular signals (suitable for ECG and PPG) based on a shape model of the waveform associated to each beat. Time evolution of the curve parameters is accounted for by means of an AutoRegressive Moving Average (ARMA) model. This model is fitted to each parameter time series after PCA (Principal Component Analysis). Signal reconstruction is performed after parameter estimation from the longest piece of the original signal. The model definition is used together with boundary conditions provided by the available segments to synthesize a new piece to replace the corrupted one. The formal definition of the model has been published in [4].

The proposal in [3] allows for the reconstruction of missing or corrupted epochs by using the model together with information of the available pieces, which are acquired at different time stamps. The ECG and PPG signal are usually acquired together, and they are undoubtedly related. Thus, it seems reasonable to consider the PPG signal to reconstruct the corrupted ECG epoch if the PPG piece is available during this period. A joint modeling approach would be necessary in this case. These kind of approaches have been proved to be successful in reconstructing the respiratory signal from a joint model with the heart rate variability signal [5].

In this paper, a novel methodology for ECG signal reconstruction is proposed. The proposal relies on a model of the ECG and the PPG signal which accounts for information of the joint behavior of both signals. Besides using the uncorrupted ECG intervals for reconstruction, the PPG signal during the interval of interest supports the process. Experimental results show that this approach outperforms the one in [3] which did not consider this joint behavior. The remaining of the paper is organized as follows: Both the modeling methodology and the reconstruction method are presented in section II. Section III discusses the performance of the method by means of an experimental evaluation of the method together with a comparison with the approach in [5]. Finally, section IV closes the paper with the main conclusions obtained from this work.

II. METHODS

A. Formal Approach to the Problem

The problem here addressed mainly consists of the reconstruction of an ECG piece by means of a supporting signal, the PPG, which is acquired simultaneously. We have used the following notation: both the ECG and the PPG signals are referred to as $x_{ECG}(t)$ and $x_{PPG}(t)$ respectively, $I_C \equiv [t_o, t_e]$ denotes the time interval at which the signal is missing or corrupted, and $I_{PRE} = [0, t_o)$ and $I_{POST} = (t_e, T]$ respectively denote the preceding and subsequent intervals to I_C. Thus, a formal description of the problem is: given $x_{ECG}(t)$ and $x_{PPG}(t)$ for $t \in [0, T]$, for which a high intensity artifact deteriorates $x_{ECG}(t)$ during the I_C interval, we want to estimate the ECG signal for this interval —$\hat{x}_{ECG}(t)$— from the the complete PPG signal and the unaltered pieces of the ECG registry.

B. ECG and PPG Joint–Modeling

The proposed model is based on the methodology in [4] (also used for similar purposes in [3]), but with slight modifications; specifically, 1) the parameterization of the ECG beats is far simpler and 2) there is a single temporal model aimed at assessing the temporal evolution of the aggregate parameter vector $\mathbf{w}[n] = [\mathbf{w}_{PPG}^T[n] | \mathbf{w}_{ECG}^T[n]]^T$ instead of one model for each signal. Since the major part of the model remains unchanged from [4], a summary is here presented. For a detailed description the reader is referred to [4].

1. *Preprocessing:* In this stage, both signals are filtered and delineated to divide them in N intervals, one for each beat.
2. *Parametrization:* The shape of each piece from the previous stage is modeled with a parametric curve as follows:

$$B_{ECG}^{(n)}(t) = \Gamma_{ECG}(\mathbf{w}_{ECG}[n], t) + r_{ECG}^{(n)}(t), \quad (1)$$
$$B_{PPG}^{(n)}(t) = \Gamma_{PPG}(\mathbf{w}_{PPG}[n], t) + r_{PPG}^{(n)}(t), \quad (2)$$

where $B^{(n)}(t)$ denotes the n–th beat of the registry, $\Gamma(\cdot, t)$ is the parametric curve, $\mathbf{w}[n]$ is the vector of parameters for the n–th beat, and $r^{(n)}(t)$ is the residue[1] for the same beat. Obviously, the parametrization is not the same for ECG and PPG[2]:

- ECG $\Rightarrow$ $\mathbf{w}$ is composed of both the temporal and the amplitude labels resulting from the delineation process —onset, peak(s) and end of the P, QRS and T complex—; $\Gamma(\mathbf{w}, t)$ is obtained from Hermite's interpolation over the labels forming $\mathbf{w}$.

- PPG $\Rightarrow$ $\mathbf{w}$ is formed by two warping parameters (devoted to ensure that all the pieces have the same length), together with those corresponding to a sum of two gaussian curves and a linear trend; $\Gamma(\mathbf{w}, t)$ is a warped curve resulting from the addition of the sum of gaussians and the linear trend.

Since $\mathbf{w}_{ECG} \in \mathbb{R}^{22}$ and $\mathbf{w}_{PPG} \in \mathbb{R}^{10}$, the joint parameter vector $\mathbf{w}[n] = [\mathbf{w}_{PPG}^T[n] | \mathbf{w}_{ECG}^T[n]]^T \in \mathbb{R}^{32}$.

3. *Temporal Modeling:* This stage is exactly the same as the one in [4], and consists of an ARMA(2,2) model of both the low–pass and the high–pass components of each series yielding the vector $\mathbf{y}[n]$, a PCA transformation of the $\mathbf{w}[n]$ vector aimed at decorrelating its components.

C. Reconstruction Methodology

The reconstruction methodology has been divided into four different stages, namely, model estimation, model sampling (simulation), selection of the optimal sample, and refinement of the optimal solution. A description of each stage is provided below. For the sake of clarity, a schematic diagram of the methodology is provided in Fig. 1.

Analysis: Model Estimation

According to section B, the joint–model ($\mathscr{M}$ hereafter) is estimated from the largest unmodified interval: either I_{PRE} or I_{POST}.

Shyntesis: Simulation of the Model

The objective of this stage is to provide a set of K vectors of parameters ($\{\mathbf{w}_{sim}^{(k)}[n]\}_{k=1}^K$) as plausible candidates to replace the corrupted piece[3]. Despite only the ECG piece is corrupted, the nature of the model is such that no separate synthesis of each signal (ECG or PPG) is viable. Hence the synthetic epochs include parameters of both PPG and ECG. Two issues must be considered at this point:

1. The number of beats to simulate (i.e., the length of the synthetic series; N_B from now on) must be obtained from the I_C interval of the PPG signal. This is an important difference with respect to the methodology proposed in [3, 4].
2. If the corruption is not located at the beginning of the acquisition, the last two beats of the I_{PRE} interval can be used as initial conditions for simulation. The rationale of this number (2) lies on the order of the temporal model, ARMA(2,2).

Selection of the Optimal Solution

The stochastic nature of the model ensures the statistical convergence of the simulations, which does not mean point–

---
[1] That residue might be modeled as zero–mean white noise.
[2] A detailed description is provided in [3, 4].
[3] The model is sampled K times.

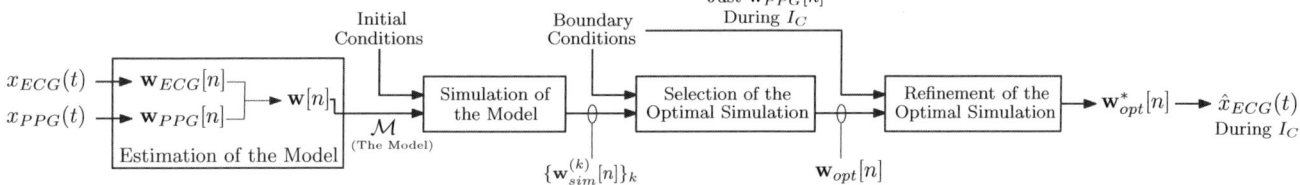

Fig. 1 Pipeline of the proposed methodology for the reconstruction of the ECG signal during the I_C interval. Note that those blocks related to the parametrization are omitted for the sake of simplicity.

to–point convergence. This stage determines which simulation, using a criterium based on this type of convergence, is the best; specifically, this criteria rests upon the point–to–point difference between the simulations and the boundary conditions ($\mathbf{w}_{PPG}[n]$ during I_C and $\mathbf{w}[n]$ during the four first beats of the I_{POST} interval). To address the selection under this criterium, an error term has been defined as $\varepsilon[k] = \varepsilon_C[k] + \varepsilon_{POST}[k]$, $1 \leq k \leq K$, where

- ε_C is the error respect to the boundary condition during I_C. This error is computed as

$$\varepsilon_C[k] = \frac{1}{10} \left\| \sqrt{\frac{1}{N_B} \sum_{n=1}^{N_B} \left(\mathbf{w}_{PPG}[I_c;n] - \mathbf{w}_{Sim,PPG}^{(k)}[n] \right)^2} \right\|_1, \quad (3)$$

where $\|\cdot\|_1$ denotes the ℓ^1 norm of a vector, which has been divided by its size (10 parameters). $\mathbf{w}_{PPG}[I_c;n]$ denotes the n-th sample of the $\mathbf{w}_{PPG}$ series after the beginning of the corruption.

- ε_{POST} is the error respect to the boundary condition during the four beats after the end of the corruption. This error is computed as

$$\varepsilon_{POST}[k] = \frac{1}{32} \left\| \sqrt{\sum_{n=N_B+1}^{N_B+4} \alpha[n] \left(\begin{array}{c} \mathbf{w}[I_C;n] - \\ -\mathbf{w}_{Sim}^{(k)}[n] \end{array} \right)^2} \right\|_1, \quad (4)$$

where 32 is the number of parameters composing $\mathbf{w}$, and $\alpha = [0.5, 0.3, 0.15, 0.05]$ is a weighting parameter aimed at fading out the relevance of the beats forming the boundary condition for the I_{POST} interval.

The optimal simulation is the one that satisfies

$$\mathbf{w}_{opt}[n] = \mathbf{w}_{Sim}^{\mathscr{K}}[n], \quad (5)$$
$$\text{with } \mathscr{K} = \arg\min_k \{\varepsilon[k], 1 \leq k \leq K\}. \quad (6)$$

Refinement of the Optimal Solution

This stage refines the solution obtained from the previous stage. To this end, we have used an iterative method that includes information from the PPG signal during I_C. The algorithm may be described as follows:

1. The series $\mathbf{w}_{opt}[n] = \left[\mathbf{w}_{opt,PPG}^T[n] | \mathbf{w}_{opt,ECG}^T[n] \right]^T$ is modified to include the actual parameter vector of the PPG signal during the I_C interval. Additionally, since these parameters convey all the information related to the heart period, the actual temporal labels of the R peak (t_R) are also plugged into the ECG part of the optimal parameter vector:

$$\mathbf{w}_{opt}[n] \leftarrow \left[\underbrace{\mathbf{w}_{PPG}^T[n]}_{\text{During } I_C} | \mathbf{w}_{opt,ECG}^{(t_R)\,T}[n] \right]^T. \quad (7)$$

2. This novel $\mathbf{w}_{opt}[n]$ is not statistically appropriate, since the correlation between its components is not the same as the one used during the temporal modeling stage (see section B). Thus, we use PCA and Gram–Schmidt orthogonalization to force the right correlation:

$$\mathbf{y}_{opt}[n] = H \cdot \mathbf{w}_{opt}[n] \quad (8)$$
$$\mathbf{y}_{opt}[n] \leftarrow GS\{\mathbf{y}_{opt}[n]\} \quad (9)$$
$$\mathbf{w}_{opt}[n] \leftarrow H^T \cdot \mathbf{y}_{opt}[n], \quad (10)$$

where H is the PCA matrix, a parameter of $\mathscr{M}$, and $GS\{\mathbf{s}\}$ denotes the application of Gram–Schmidt orthogonalization over the vector series $\mathbf{s}$.

3. Once the novel $\mathbf{w}_{opt}[n]$ is obtained, the algorithm returns to step 1).

This procedure is performed until stationary convergence of the algorithm is achieved. Once it has converged, the refined solution $\mathbf{w}^*[n]$ will be used to get the reconstructed ECG piece: $\hat{x}_{ECG}(t)$.

III. EXPERIMENTS AND DISCUSSION

To illustrate the performance of the proposed methodology, we have carried out an experiment with real data obtained from a 25 years old healthy patient under resting conditions. A 5 min registry of the ECG (V5 derivation) and PPG signals was acquired at sampling rates of 200 Hz (ECG) and 66.67 Hz by means of the Omnicrom FT Surveyor device (RGB Medical Devices). The experiment consists in the reconstruction of an ECG epoch that has been manually removed in order to allow for comparisons between the reconstruction and the real (removed) data. In order to provide a

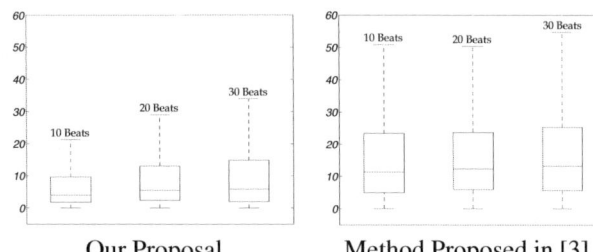

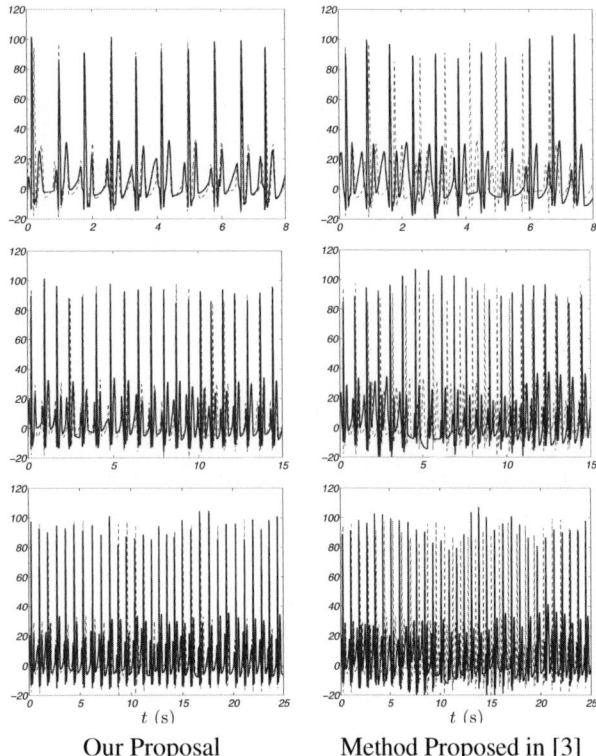

Fig. 3 Boxplots of the absolute error (point–to–point) between the reconstructed and the removed pieces.

Fig. 2 Reconstructed pieces of the ECG signal obtained from our proposal and the method in [3]. The bold line represents the reconstructed piece, while the removed piece is plotted in dashed line. First row: reconstruction for 10 beats; second row: reconstruction for 20 beats; third row: reconstruction for 30 beats.

deeper analysis of the performance, we have selected different lengths (10, 20 and 30 beats) for the removed interval. On the other hand, to emphasize the difference between our method and the one proposed in [3] (point–to–point convergence versus statistical convergence), we have compared the reconstructions obtained with both methods.

We present in Fig. 2 both the real (dashed lined) and the reconstructed (bold line) epochs obtained from our proposal (left column) and the method proposed in [3] (right column). It can be easily noticed that the reconstructions obtained from the proposed method are far more accurate than those yielded by the one in [3]. A numerical comparison is provided in Fig. 3 through the boxplots of the difference (absolute error) between the reconstructions and the real piece.

IV. CONCLUSIONS

We have introduced a methodology for reconstructing ECG registries that have been acquired together with a PPG signal. This methodology, based on a stochastic joint–model of both signals, provides a solution through model sampling and an iterative method aimed at increasing the accuracy of the reconstruction. The main advantage of the proposal lies on the point–to–point convergence of the reconstruction, which makes the solution suitable for visual purposes such as clinical evaluation.

ACKNOWLEDGMENT

This work is partially supported by the Universidad de Valladolid and the Banco Santander under the FPI–UVa fellowship, the Spanish Ministerio de Ciencia e Innovación and the Fondo Europeo de Desarrollo Regional (FEDER) under Research Grant TEC2010-17982, the Spanish Instituto de Salud Carlos III (ISCIII) under Research Grants PI11-01492, PI11-02203 and by the Spanish Centro para el Desarrollo Tecnológico Industrial (CDTI) under the cvREMOD (CEN-20091044) project. Authors would like to thank the support of the company RGB Medical Devices LTD for kindly sharing us the measurement devices.

REFERENCES

1. Clifford G. D., Shoeb A., McSharry P. E., Janz B. A.. Model-based Filtering, Compression and Classification of the ECG *Int. J. of Bioelectromag.*. 2005;7:158–161.
2. Zigel Y., Cohen A., Katz A.. ECG Signal Compression using Analysis by Synthesis Coding *IEEE Trans. Biomed. Eng.*. 2000;47:1308–1316.
3. Martín-Martínez D., Casaseca-de-la-Higuera P., Martín-Fernández M., Alberola-López C.. Cardiovascular Signal Reconstruction based on Parameterization –Shape Modeling– and Non-Stationary Temporal Modeling in *Proceedings of the 20th European Signal Processing Conference*(Bucharest, Romania):1826–1830 2012.
4. Martín-Martínez D., Casaseca-de-la-Higuera P., Martín-Fernández M., Alberola-López C.. Stochastic Modeling of the PPG Signal: A Synthesis–by–Analysis Approach with Applications *IEEE. Trans. Biomed. Eng.*. 2013;In press.
5. Martín-Martínez D., Casaseca-de-la-Higuera P., Martín-Fernández M., Alberola-López C.. Reconstrucción de la Señal Respiratoria Basada en Modelo Conjunto con la Señal de Periodo Cardíaco in *Proceedings of the XXX Congreso de la Sociedad Española de Ingeniería Biomédica*(San Sebastián, Spain) 2012.

Author: D. Martín–Martínez
Institute: E.T.S.I. Telecomunicación, Universidad de Valladolid
Street: Paseo Belén, 15, 47011
City: Valladolid
Country: Spain
Email: dmarmar@lpi.tel.uva.es

Application of Impedance Cardiography for Haemodynamic Monitoring in Patients with Ischaemic Stroke

A. Zielińska[1], H. Dudek[2], and G. Cybulski[1,3]

[1] Institute of Metrology and Biomedical Engineering, Department of Mechatronics,
Warsaw University of Technology, Warsaw, Poland
[2] First Clinic of Neurology, Institute of Psychiatry and Neurology, Warsaw, Poland
[3] Department of Applied Physiology, Mossakowski Medical Research Centre, Warsaw, Poland

Abstract—**Impedance cardiography (ICG) is a noninvasive method of continuous haemodynamic monitoring of cardiac parameters which might be a useful tool in the care of critically ill patients. The aim of this paper was to verify whether impedance cardiography is suitable for monitoring haemodynamic parameters in patients after ischaemic brain stroke. It also investigated whether the results obtained with this method could yield additional data helpful for hypervolemic therapy optimization. haemodynamic data from eight adult patients (5 male and 3 female, age: 33-86 yrs) were included in the analysis. Cardiac parameters were analysed for set times during the time course of the intravenous infusion: at the very beginning and at the end of each quarter. It was concluded that haemodynamic monitoring of cardiac parameters using impedance cardiography might bring benefits during treatment of patients with ischaemic brain stroke.**

Keywords—**Impedance cardiography, Ischaemic stroke, Haemodynamics, Cardiac output, Systolic time intervals.**

I. INTRODUCTION

Brain stroke is one of the world's leading causes of death (9%) and disability. On average, 610,000 patients suffer from ischaemic stroke each year in the United States alone [1]. The problem of the hospital and late mortality rate for stroke patients has not been solved. It seems that monitoring haemodynamic indices may help in decreasing this rate, especially among patients with other accompanying circulatory system disorders. The substantial volume of fluids supplied intravenously during therapy is an additional load for the heart. Cardiac parameter monitoring might be helpful in hypervolemic therapy optimization.

Impedance cardiography (ICG) enables non-invasive, continuous, operator-independent, and automatic measurement of cardiac stroke volume (SV) and systolic time intervals (STI) [2]. The accuracy of ICG was verified by both invasive [3] and non-invasive methods [4]. Because of several controversies, ICG has been used in the past only as an investigational tool, not as a clinical one. This approach has changed since the possibility of reimbursement arose and one of the commercially available systems received FDA approval. This issues was discussed in a monograph [5] and mentioned in a recent review [6].

The aim of this paper was to verify whether the method of impedance cardiography is suitable for monitoring haemodynamic parameters in patients after ischaemic brain stroke.

II. MATERIAL AND METHODS

A. Subjects and Protocol

Eight adult patients (5 male and 3 female, age: 33-86 yrs) admitted to the ICU with a diagnosis of ischaemic stroke based on history and neurological examination (presence of rapidly developing clinical symptoms or signs confirmed by neuroimaging studies) participated in this study. In addition to the routine monitoring methods, ICG and ECG signals were simultaneously recorded using the impedance cardiography ambulatory monitoring system, ReoMonitor.

B. Instrumentation and Analysis

A miniaturised, portable ICG device (ReoMonitor) with a built-in one-channel ECG was used to detect central haemodynamic signals [10]. Heart rate (HR), stroke volume (SV) and pre-ejection period (PEP) were obtained simultaneously by ICG (automatically) in the supine position. Cardiac indices were calculated from the first derivative of the ICG signal (dz/dt) and ECG using the Kubicek formula [7]. The ICG device was verified earlier in patients in both supine and tilted positions[4]. The system had been used successfully in many earlier physiological and clinical investigations [8, 9].

Data were collected continuously but analysed and presented only for five points in time; expressed as percentages of the total time of intravenous treatment, these were 0, 25, 50, 75 and 100. At each point we used data of every parameter, averaged over 10 seconds. Since patients showed high intra-individual variability we decided to present the data as relative changes in the relative time course of intravenous infusion. The relative changes were calculated in reference to resting values obtained before the beginning of the infusion.

III. RESULTS

The cardiac parameters were obtained from different patients and showed high inter-individual variations. The resting HR ranged from 63 to 112 beats·min^{-1}; SV, from 46 to 110ml. Resting CO was estimated from 3.63 to 9.37 l·min^{-1}. The highest CO value was noticed in a patient with a maximal HR and very high resting SV.

The relative changes in the basic cardiac parameters HR, SV, CO (from bottom to top) are presented in Fig. 1, where the horizontal axis values 0, 25, 50, 75 and 100 denote the elapsed percentage of the total infusion time.

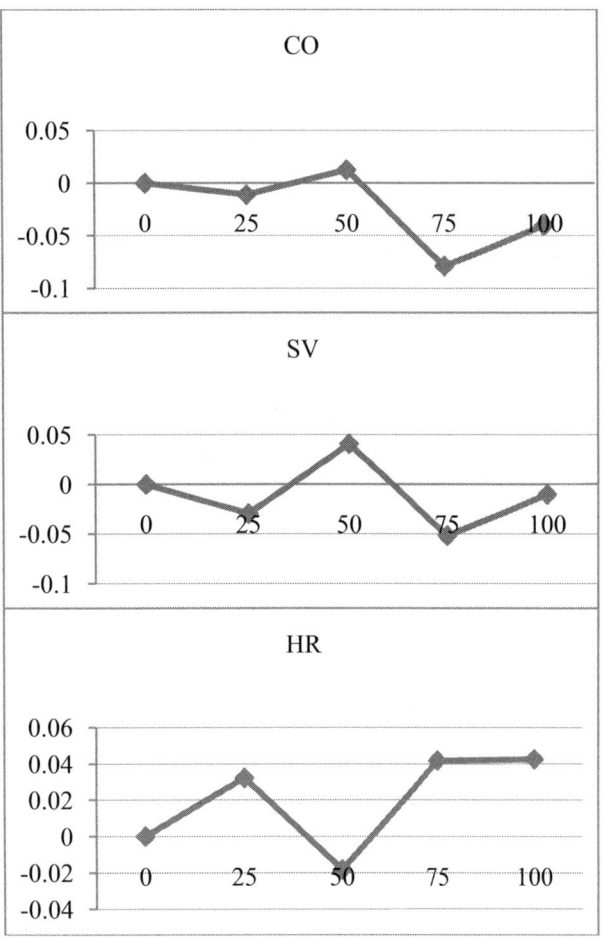

Fig. 1 Relative changes in heart rate (HR), stroke volume (SV) and cardiac output (CO) during the time course of intravenous infusion. The numbers on the horizontal axis denote the elapsed percentage of the total infusion time; 0 denotes the beginning and 100 the end.

Fig. 2 presents basic thoracic impedance (top) and relative changes in pre-ejection period (PEP) and left ventricular ejection time (ET) over two systolic time intervals (middle and bottom respectively), during the course of intravenous infusion.

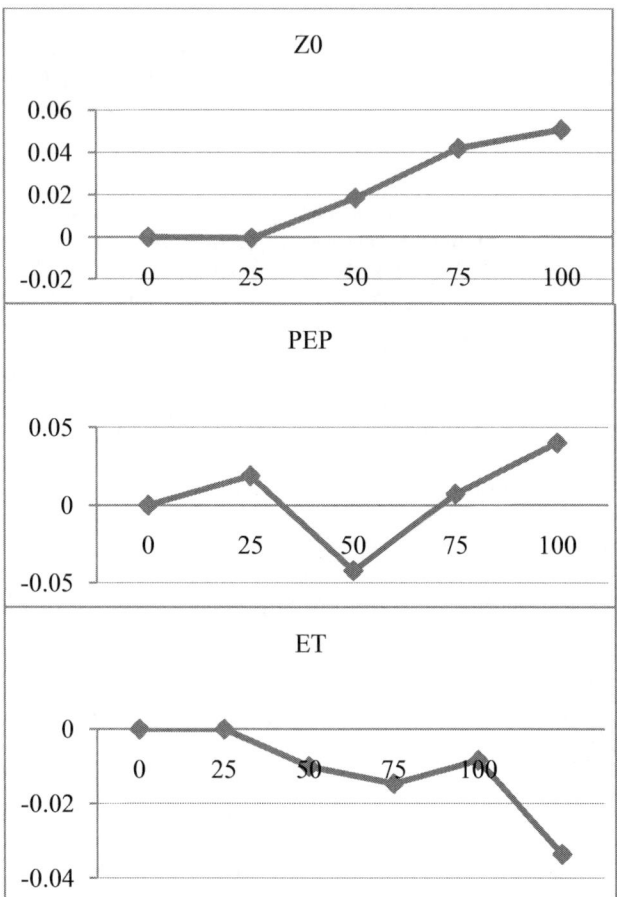

Fig. 2 Relative changes in left ventricular ejection time (ET), pre-ejection period (PEP) and basic thoracic impedance (Z0) during the time course of intravenous infusion. The numbers on the horizontal axis denote the elapsed percentage of the total infusion time; 0 denotes the beginning and 100 the end.

Intra-individual variability was lower in the systolic time intervals. Initial values of ET ranged from 297 to 367 ms whereas PEP ranged from 87 to 104 ms. Initial values of basic thoracic impedance were in the range of 19-33 Ω.

IV. DISCUSSION

A. Haemodynamic Monitoring in Brain Stroke Patients

There is a modest number of papers regarding the applications of ICG method for monitoring patient with brain stroke [10, 11]. The literature is focused on differences between the cardiac indices for ischaemic and haemorrhage stroke and the possibility of gaining a prognostic factor for patients from haemodynamic monitoring. Ramirez et al. [10] found that except for stroke index (SV normalised by dividing by body surface area - BSA) and blood pressure parameters, there was no difference in the noninvasive haemodynamic parameters between the ischaemic stroke group and the haemorrhagic stroke group.

Siebert et al. [11] found a significant association between the systemic vascular resistance, Heather and stroke indices; systolic, diastolic and mean arterial blood pressures; heart rate and mortality in patients with ischaemic stroke. They also used logistic regression analysis to identify thoracic fluid content as the most significant variable correlating with the non-survival among patients with ischaemic stroke and in the whole group (ischaemic and haemorrhagic stroke). They noted that mean arterial pressure and stroke index were also significant parameters in ischaemic stroke, as was heart rate for the whole group. However, they found no significant associations for haemorrhagic stroke.

B. Haemodynamic Effect of Intravenous Infusion

The most surprising result for us was the fact that the base impedance increased by 5% during the intravenous infusion. The direction of the change was contrary to our expectation. We expected that infusion of 500ml would cause an increase in the volume of the central thoracic fluids which should result in a decrease in basic thoracic impedance. However our results might be treated as evidence that that the presence of an extra 500ml of fluids in the circulatory system does not necessarily cause the thoracic segment blood volume to increase.

V. CONCLUSIONS

It was concluded that haemodynamic monitoring of cardiac parameters using impedance cardiography during treatment of patients with ischaemic brain stroke might provide additional useful data.

ACKNOWLEDGMENT

This study was supported by the research programs of institutions the authors are affiliated with.

REFERENCES

1. Donnan GA, Fisher M, Macleod M, Davis SM. Stroke (2008) Lancet; 371: 1612-23 doi: 10.1016/S0140-6736(08)60694-7.
2. Kubicek WG, Karnegis JN, Patterson RP, et al. (1966) Development and evaluation of an impedance cardiac output system, Aviat.Space Environ.Med., 37(10): 1208- 1212.
3. Judy WV, Langley FM, McCowen KD, et al. (1969) Comparative evaluation of the thoracic impedance and isotope dilution methods for measuring cardiac output, Aerospace Medicine, 40 (5): pp. 532-536.
4. Cybulski G, Michalak E, Koźluk E, Piątkowska A, Niewiadomski W (2004) Stroke volume and systolic time intervals: beat-to-beat comparison between echocardiography and ambulatory impedance cardiography in supine and tilted positions, Medical and Biological Engineering and Computing, 42: 707-711, DOI: 10.1007/BF02347554
5. Cybulski G (2011). Ambulatory Impedance Cardiography. The Systems and their Applications. Series: Lecture Notes in Electrical Engineering, Vol. 76, 1st Edition, ISBN: 978-3-642-11986-6, 150 pp., Springer-Verlag Berlin and Heidelberg GmbH & Co. KG, DOI: 10.1007/978-3-642-11987-3
6. Cybulski G, Strasz A, Niewiadomski W, Gąsiorowska A (2012) Impedance cardiography: Recent advancements. Cardiol J 19, 5: 550–556, DOI: 10.5603/CJ.2012.0104
7. Cybulski G, Książkiewicz A., Łukasik W., Niewiadomski W., Pałko T (1996) Central hemodynamics and ECG ambulatory monitoring device with signals recording on PCMCIA memory cards. Medical & Biol. Engin. and Computing, Vol. 34, Suppl.1, Part 1, pp.79-80.
8. Gasiorowska A, Nazar K, Mikulski T, Cybulski G,. Niewiadomski W, Smorawinski J, Krzeminski K, Porta S, Kaciuba-Uscilko H (2005) Hemodynamic and neuroendocrine predictors of lower body negative pressure (LBNP) intolerance in healthy young men, J Physiol. Pharmacol. 56, 2, 179-19
9. Cybulski G, Kozluk E, Michalak E, Niewiadomski W, Piatkowska A (2004) Holter-type impedance cardiography device. A system for continuous and non-invasive monitoring of cardiac haemodynamics. Kardiol Pol. Aug;61(8):138-46.
10. Ramirez MF, Tibayan RT, Marinas CE, Yamamoto ME, Caguioa EV (2005) Prognostic value of hemodynamic findings from impedance cardiography in hypertensive stroke. Am J Hypertens 18: 65S-72S, DOI: 10.1016/j.amjhyper.2004.11.027
11. Siebert J, Gutknecht P, Molisz A, Trzeciak B, Nyka W (2012) Hemodynamic findings in patients with brain stroke, Arch Med Sci. May 9; 8(2):371-4. doi: 10.5114/aoms.2012.28567.

Author: Gerard Cybulski
Institute: Institute of Metrology and Biomedical Engineering, Department of Mechatronics, Warsaw University of Technology,
Street: 8 Św. A. Boboli
City: Warsaw
Country: Poland
Email: G.Cybulski@mchtr.pw.edu.pl

AR versus ARX Modeling of Heart Rate Sequences Recorded during Stress-Tests

J. Holcik[1,2], T. Hodasova[1,3], P. Jahn[4], P. Melkova[4], and J. Hanak[4]

[1] Institute of Biostatistics and Analyses, Masaryk University, Brno, Czech Republic
[2] Institute of Measurement Science, Slovak Academy of Sciences, Bratislava, Slovakia
[3] Department of Mathematics and Statistics, Faculty of Science, Brno, Czech Republic
[4] Equine Clinic, University of Veterinary Sciences and Pharmacy, Brno, Czech Republic

Abstract—This paper presents some ideas about comparison of two fundamental linear approaches (AR and ARX models) for modeling RR interval series and its variability recorded in horses during stress-testing. Their theoretical background is discussed in brief and results of some computational experiments are given and analyzed, as well. In particular, problems of stationarity, determination of the response to changes of load, estimation of model order are examined here.

Keywords—heart rate variability, AR model, ARX model, stress test.

I. INTRODUCTION

Heart rate of an examined subject and in particular its dynamics is determined by a state of both the cardiovascular system and autonomic neural system that controls cardiovascular activity. Quality of a function of the mentioned physiological subsystems is crucial for determining level of fitness. That is why the global aim of our research is to attempt to rate fitness level of horse athletes based on heart rate sequences recorded during stress test. To facilitate the fitness level classification it seems appropriate to describe the heart rate sequences and their variability by means of some mathematical model, parameters of which could be used as features entering the classification algorithms.

There are two significant approaches used for linear mathematical modeling heart rate sequences – Autoregressive (AR) models (e.g. [1] - [5]) and ARX models (Autoregressive, eXtra input) (e.g. [6]).

Both the approaches have their advantages and disadvantages. The greatest disadvantage of the AR model is a requirement for stationarity of data that can be hardly fulfilled under conditions of stress test when the physical load increases. On the other hand, cardiovascular responses in horses usually change with an intensity of load. That is why it seems to be probably difficult to determine adequate optimum parameters of the time-invariant ARX model valid for the whole stress-test examination.

Despite of the mentioned problems both the types of models were used for description of the cardiovascular responses in horses to physical load and the obtained results were compared.

II. DATA

Stress-test in horses consists of several steps. It starts with an approx. 5 minute walk usually followed by lope (5min.) and then by gallop starting with treadmill speed of 7m/s (2 min) that increases after 1 minute by steps of 1m/s up to 10 or 11m/s depending on horse abilities.

ECG signals were recorded from three bipolar chest leads which QRS complexes were detected in. After that the derived RR interval functions were interpolated by piecewise linear function and resampled by a frequency of 10 Hz to obtain equidistant time series.

Altogether 15 data records were taken and processed, 6 of them recorded in a preliminary phase of examination with non-standard experimental arrangement of the load stages.

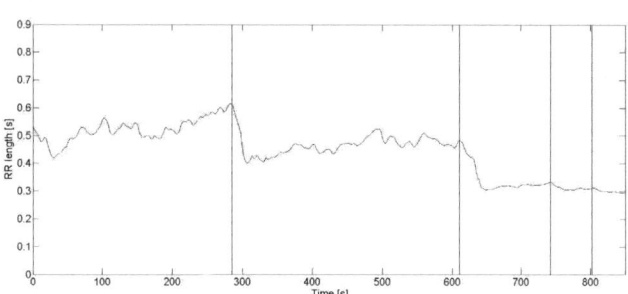

Fig.1 Example of the RR interval sequence recorded during stress-test

III. AR MODELS

As mentioned above the basic disadvantage of the AR model approach is the requirement for data stationarity. Unfortunately, according to an expectation based on practical experiences and also supported by numerous publications (e.g. [1] or [3]) the response to increasing load during stress-test is heavily time variant. Fig.1 depicts sequence of RR intervals determined from ECG signals recorded during the stress-test examination. It can be seen in the figure that the most significant non-stationarity is represented by responses to changes of the load. The moving average low-pass Hamming filter with an impulse response of 600

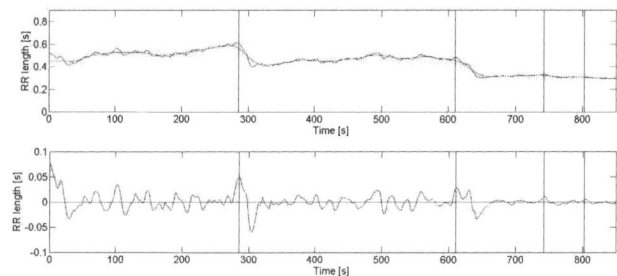

Fig.2 Original RR interval sequence and its drift estimated by the narrow-band low-pass filter (upper part) and the difference sequence between the original sequence and its drift (lower part)

samples was designed and used to remove the non-stationarity. The cut-off frequency was set according to frequency spectrum of the RR interval signal as the frequency separating band of lower frequencies with greater amplitudes from the components with higher frequencies (see Fig.2 - upper part).

It means that it must be valid that

$$E(z) = RR(z) - H_{LP}(z) \cdot RR(z) \quad (1a)$$

or

$$RR(z) = H_{LP}(z) \cdot RR(z) + E(z) \quad (1b)$$

where $RR(z)$ represents Z-transform of the original RR interval sequence, $H_{LP}(z)$ is a transfer function of the used Hamming low-pass filter and $E(z)$ is a Z-transform of a difference sequence $e(k)$ that should represents the stationary behavior of the examined horse during the whole stress-test. The sequence $e(k)$ is modeled by a linear autoregressive system described by a transfer system function $H_{AR}(z) = 1/A(z)$. The function $H_{AR}(z)$ is proportional to power spectral density of sequence $e(n)$ provided that input of the model system is a zero-mean white noise sequence $n(k)$. In such a case it is

$$E(z) = \frac{1}{A(z)} \cdot N(z), \quad (2)$$

where $N(z)$ is the Z-transform of the white noise sequence $n(k)$. Then the eq.(1b) can be rewrite as

$$RR(z) = H_{LP}(z) \cdot RR(z) + \frac{1}{A(z)} \cdot N(z) \quad (3)$$

Weak-sense stationarity (mean only) was verified per partes for partial sequences without drift corresponding to each step of the stress-test load. Because of lack of knowledge about the data statistical distribution non-parametric Kruskal – Wallis test was applied. In this way median of the subsequences proved to be sufficiently time invariant in the vast majority of the analyzed sequences. Small differences from stationarity came to pass in intervals just after the load changes (Fig.2).

Even if the subtracting of the drift roughly ensures the stationarity of analyzed sequence (required for application of the AR model) it unfortunately removes substantial part of information on character of transients tied with the load changes. Although shortening the impulse response and related increasing of the cut-off frequency of the smoothing Hamming filter homogenizes data after subtracting the estimated drift neither the impulse response of 100 samples does not ensure the strict data stationarity.

The principal task for identification of the AR model parameters is to determine its order that provides the best fit of the model to data being processed. Unfortunately, known algorithms for the AR model order estimation are not very reliable. The experimental results ([4], [7]) as well as theoretical studies (e.g. [8]) indicate that the practically used several statistical criteria do not usually yield definitive results, mostly tend to underestimation of the true order of the analyzed AR process. The model order for a partial time-series in every stage of the stress-test examination was searched for in the interval ⟨6, 30⟩. The value of 20 was then determined as the most frequent result computed for the given set of the experimental data sequences by the Akaike information criterion.

Several different approaches are used for the final identification of the AR model parameters, each of them under specific conditions and with specific characteristics. We used two of them – the Yule – Walker method and the unconstrained least square method. The former, based on estimates of autocorrelation function of the analyzed sequence, can be assumed to be one of methods with maximum entropy. Therefore spectral characteristics of the determined model are relatively smooth in comparison with the latter and with relatively poor frequency resolution. On the other hand, parameters determined by the latter method make the resulting linear model much more frequency sensitive, however it does not have to be necessarily stable.

Properties of the determined AR models can be illustrated either by their frequency responses or by distribution of transfer function poles in complex plane (see Fig.3 and Fig.4).

Examples in the figures roughly confirm the above mentioned properties of both methods for identification of the AR model parameters. Frequency responses of the system with parameters obtained by means of the Yule – Walker method are smoother and not very sensitive to particular frequency components (baroreflex, breathing) incorporated in the analyzed signal if compared with results of application of the unconstrained least square method. We can carefully presume the existence of the mentioned harmonic components from the shape of phase characteristic only. However, the smoothness of the frequency responses results in less variability of positions of the transfer function poles.

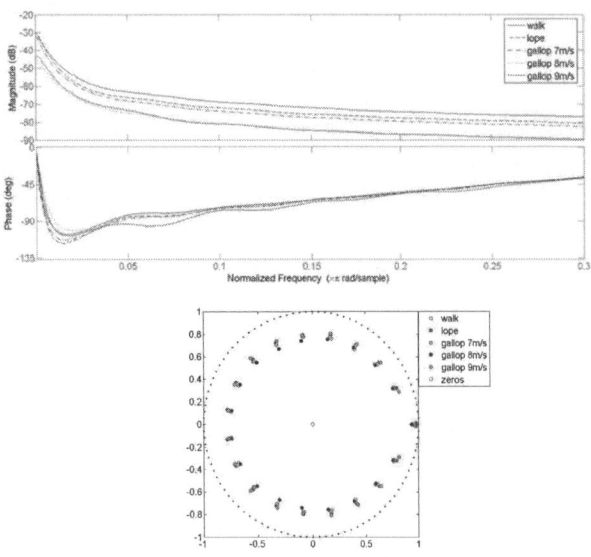

Fig.3 Example of the AR(20) model frequency responses for the stages of the stress-test with parameters identified by means of Yule-Walker method (upper part) and distribution of its poles in complex Z-plane (lower part)

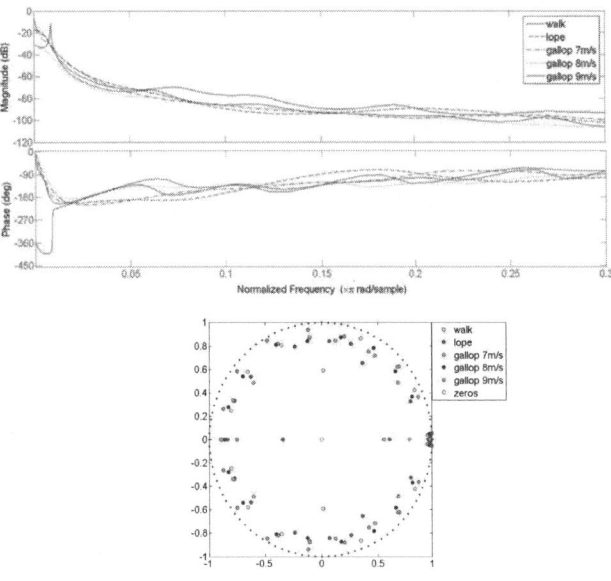

Fig.4 Example of the AR(20) model frequency responses in the stages of the stress-test with parameters identified by means of the unconstrained least square method (upper part) and distribution of its poles in complex Z-plane (lower part)

IV. ARX MODELS

As it was described above the requirements for stationarity can be hardly complied with completely due to the changes of experimental conditions during the stress-tests. However, the load changes can be incorporated into the model structure as it is defined in a class of dynamical systems with external input which do not put any stacionarity demands on the analyzed data. There are three basic structures of these systems that differ in the random part of the definition formula – ARX models (AutoRegressive with eXternal input), ARMAX models (AutoRegressive Moving Average with eXternal input), and OE models (Output Erorr). The ARX systems are defined as

$$RR(z) = \frac{B(z)}{A(z)} \cdot X(z) + \frac{1}{A(z)} \cdot N(z) \quad (4)$$

where $X(z)$ is the Z-transform of a system input sequence that is determined by time-dependency of the load, $A(z)$, and $B(z)$, resp. are polynomials of the system transfer functions. As we can easily compare the formula is very similar to that in eq.(3). As well as in eq.(3) the first member at the right hand side of the formula represents response to the system input, now in more explicit form, and the second member represents response of the system to random interference.

Even if there are quite different conditions under which both the described models could be used, from a theoretical viewpoint they differ in mathematical description of one member of the defining formula only (however, the meaning of both the expressions is essentially the same). Then we can write that

$$\frac{B(z)}{A(z)} \cdot X(z) = H_{LP}(z) \cdot RR(z) \quad (5)$$

The only component in eq.(5) that is not determined on a base of some optimality criterion is the transfer function $H_{LP}(z)$ of the low-pass filter for filtering the experimental RR sequence. However, from the eq.(5) we can simply write

$$H_{LPopt}(z) = \frac{B(z)}{A(z)} \cdot \frac{X(z)}{RR(z)}, \quad (6)$$

that should define both the optimum properties of the low-pass filter used for AR modeling of the RR interval time series and at the same time the model response to the change of the stress-test load that is the most important part of the model behavior. Then both the ways (using AR and ARX system) of the RR sequence analysis recorded during stress-test should be equivalent.

Similarly as in the case of the AR model approach the fundamental task is to determine orders of the both the polynomials used in the eq.(4). Usually a parameter representing time shift between system input and output can be also determined.

The polynomial orders were searched for again by means of the Akaike information criterion in interval of ⟨6, 30⟩ and the values of 20 – the order of the polynomial $B(z)$, 20 – the order of the polynomial $A(z)$, and 1 – the system time shift were chosen as the most frequent values for all the analyzed experimental records.

Coefficients of the polynomials A(z) and B(z) were computed by the unconstrained least square method. Example of frequency responses of the used model transfer function is given in Fig.5. Due to the unconstrained least square optimization models for some experimental data appeared unstable.

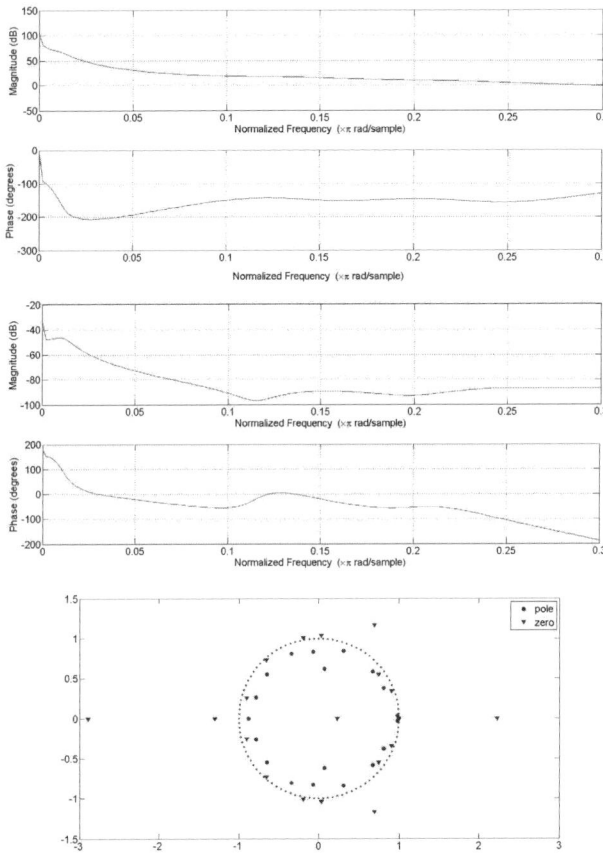

Fig.5 Example of the ARX(20,20,1) model frequency responses – AR subsystem (upper part), ARMA subsystem (in the middle) and the distribution of its nulls and poles in complex Z-plane (lower part)

V. CONCLUSIONS

As it was shown above the mathematical structure of both the model types, the AR as well as the ARX, are basically the same. While the ARX models provide with well established procedure for determination of the model part that models the response to changes of the load, the approach based on the AR systems uses heuristic procedures for this purpose. However, the estimation of this part of the AR model can be done theoretically more precisely if the ARMA system computed for the ARX model is modified according to properties of the external input and the RR interval series (eq.(6)). On contrary, the AR systems look more suitable for description of variations in behavior of the examined subject in the particular stages of the whole stress-test.

To make interpretation of the model description as easy as possible the conversion of the ARMA structure (used in ARX model) or the low-pass MA filter (used in the AR model here) to AR representation can be considered [9].

If relatively simple model is required (necessary for adequate size of a feature space and complexity of the classification algorithms based on the model parameters – model order maximum up to 30) then our experimental results indicate that the spectral description of the RR interval sequence is still too smooth that means the model order is underestimated. A final decision about this fact can be done on the base of classification result only.

ACKNOWLEDGMENT

This research was partially granted by the ESF project No. CZ.1.07/2.2.00/28.0043 "Interdisciplinary Development of the Study Programme in Mathematical Biology" and the project No. APVV-0513-10 „Measuring, Communication and Information Systems for Monitoring the Cardiovascular Risk".

REFERENCES

1. Mainardi L T et al (1995) Pole Tracking Algorithms for Extraction of Time-Variant Heart Rate Variability Spectral Parameters. IEEE Trans. BME, 42:250-259.
2. Aubert A E, Seps B, Beckers F (2003) Heart Rate Variability in Athletes. Sports Med 33:889-919.
3. Orini M et al (2007) Modeling and Estimation of Time-Varying Heart Rate Variability during Stress Test by Parametric and Non Parametric Analysis. Proc. 34 th Conf. **Computers in Cardiology, Durham, U.S.A., 2007, pp.29• 32**
4. Dantas E M et al (2011) Spectral Analysis of Hear Rate Variability with the Autoregressive Method: What Model Order to Choose? Computers in Biology and Medicine 42:164-70. DOI:10.1016/j.compbiomed.2011.11.004
5. Mainardi L T (2009) On the Quantification of Heart Rate Variability Spectral Parameters Using Time-Frequency and Time-Varying Methods. Phil Trans R.Soc. 367:255-275. Doi:10.1098/rsta.2008.0188
6. Perrott M H (1992) An Efficient ARX Model Selection Procedure Applied to Autonomic Heart Rate Variability. MSc Thesis. MIT, 154p.
7. Proakis J G et al (1992) Advanced Digital Signal Processing. Macmillan, New York
8. Boardman A et al (2002) A Study of the Optimum Order of Autoregressive Models for Heart Rate Variability. Physiol. Meas., 23:325-336.
9. Wold, H. (1954) A Study in the Analysis of Stationary Time Series, 2nd revised edition, Almqvist and Wiksell Book Co., Uppsala

Author: Jiri Holcik
Institute: Institute of Biostatistics and Analyses, Masaryk University
Street: Kamenice 126/3
City: Brno
Country: Czech Republic
Email: holcik et iba.muni.cz

Cancellation of Cardiac Interference in Diaphragm EMG Signals Using an Estimate of ECG Reference Signal

A. Torres[1,2,3], J.A. Fiz[1,3,4], and R. Jané[1,2,3]

[1] Institut de Bioenginyeria de Catalunya (IBEC), Biomedical Signal Processing and Interpretation Group, Barcelona, Spain
[2] Universitat Politènica de Catalunya, Department d'Enginyeria de Sistemes, Automàtica i Informàtica Industrial, Barcelona, Spain
[3] Centro de Investigación Biomédica en Red de Bioingeniería, Biomateriales y Nanomedicina (CIBER-BBN), Spain
[4] Servei de Neumologia, Hospital Germans Trias i Pujol, Badalona, Spain

Abstract—The analysis of the electromyographic signal of the diaphragm muscle (EMGdi) can provide important information in order to evaluate the respiratory muscular function. However, EMGdi signals are usually contaminated by the electrocardiographic (ECG) signal. An adaptive noise cancellation (ANC) based on event-synchronous cancellation can be used to reduce the ECG interference in the recorded EMGdi activity. In this paper, it is proposed an ANC scheme for cancelling the ECG interference in EMGdi signals using only the EMGdi signal (without acquiring the ECG signal). In this case the detection of the QRS complex has been performed directly in the EMGdi signal, and the ANC algorithm must be robust to false or missing QRS detections. Furthermore, an automatic criterion to select the adaptive constant of the LMS algorithm has been proposed (μ). The μ constant is selected automatically so that the canceling signal energy equals the energy of the reference signal (which is an estimation of the ECG interference present in the EMGdi signal). This approach optimizes the tradeoff between cancellation of ECG interference and attenuation of EMG component. A number of weights equivalent of a time window that contains several QRS complexes is selected in order to make the algorithm robust to QRS detection errors.

Keywords—**Adaptive Canceller, EMG, Diaphragm muscle.**

I. INTRODUCTION

The electromyographic signal of the diaphragm muscle (EMGdi) can be used to monitor the contractile activity of the diaphragm muscle during respiration. The amplitude of the EMGdi signal is related with the number of active motor units of the muscle during contraction [1]. Also it has been shown that the fatigue of the diaphragm is preceded by shifts from high to low frequency in the EMGdi signal [2].

One of the major concerns in the EMGdi amplitude and frequency analysis is the electrocardiographic (ECG) interference. The ECG interference introduces a distortion in both EMGdi signal power and frequency content. The ECG interference is overlapped with the low frequency range of the EMGdi signal. A traditional method in order to eliminate the ECG interference is simply by means of the detection and removal of the segments of the EMGdi signal with presence of QRS complexes [3]. However, this method segments the signal and excludes portions of the signal that may contain important information. Adaptive filtering techniques have been applied successfully in order to isolate the cardiac component from EMG noise [4], and for heart activity cancellation in EMG signals from different respiratory muscles [5], and from the EMGdi signal [6,7]. The most important part of all these adaptive filtering techniques is the artificial construction of a reference signal that must be highly correlated with the ECG interference but uncorrelated with the EMGdi activity.

In this work, we present an adaptive noise canceler (ANC) based on event-synchronous cancellation scheme with an automatic criterion for selecting the appropriate parameters for the computation of the adaptive filter in order to remove the ECG interference while preserving the essential features of the EMGdi signal, even with the presence of false positives and false negatives in the ECG reference signal.

II. METHODOLOGY

A. Signals and Instrumentation

A mongrel dog was surgically instrumented under general anesthesia via a femoral venous catheter with pentobarbital sodium (25 mg/kg). The EMGdi signal was measured via internal electrodes inserted into the costal diaphragm, as described in [8]. The dog was anesthetized and remained in supine position during the acquisition. The dog breathes against a constant inspiratory resistive load during the study. The EMGdi signal was amplified, analog filtered, digitized with 12 bit A/D system at a sampling rate of 4 kHz, and decimated at 1200 Hz. This protocol and study was approved by the Ethics Committee and performed in accordance with guidelines for Animal Research of the Hôpital Notre-Dame in Montreal (Canada).

Fig 1 shows an example of five respiratory cycles of an EMGdi signal with ECG noise. It can be observed that the cardiac signal overlaps in frequency with the EMG activity.

B. EMGdi Adaptive Noise Canceller

The block diagram of the EMGdi ANC is shown in Fig 2. The utilized scheme is similar to the schemes proposed in [4] and [5]. The primary input to the noise canceller is the original EMGdi signal (corrupted with the ECG interference) filtered in the frequency band of the ECG interference (between 2 and 40 Hz) with a 4th order bidirectional Butterworth filter (d[n]). The reference input (x[n]) must be uncorrelated with the EMGdi activity but correlated with the ECG interference. This signal was artificially generated according to the method described in section C. The output of the adaptive filter is the cancellation signal (y[n]) which is subtracted from the original signal in order to produce the filtered output (EMGdi clean of ECG activity).

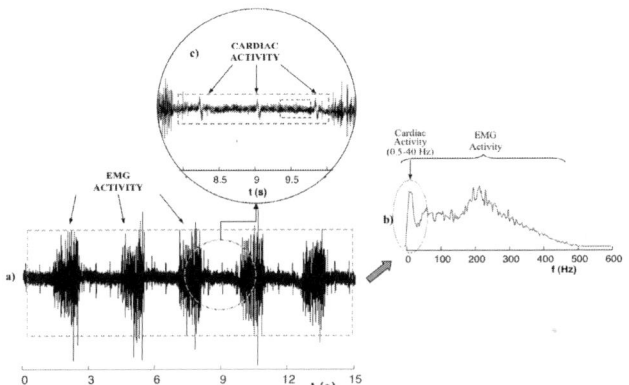

Fig 1 Example of five respiratory cycles of the EMGdi signal
(a) EMGdi signal, (b) EMGdi Power Spectral Density, (c) detail of the cardiac interference present in the EMGdi signal

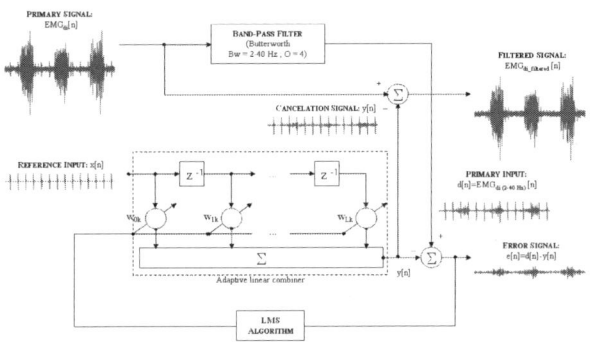

Fig 2 Block diagram of the EMGdi adaptive ECG noise canceller

C. Reference Signal

Fig 3 shows a scheme of the algorithm used to generate the reference signal. As the main ECG frequencies are bounded in the range between 2 and 40 Hz [9], the original EMGdi signal (Fig 4A) was filtered with a 4th order Butterworth band-pass filter with this frequency range. Thus, a great part of the EMGdi activity was eliminated (Fig 4B). Then, the filtered signal is convolved with a manually selected ECG pattern (Fig 4C). The next step is a non-linear transformation based on squaring this convolution (Fig 4D). This transformation makes the signal positive and allows its integration. The local maxima of this signal determine the point of synchronism chosen for the detection of the cardiac beats. The detection of those local maxima is realized by means of two moving average windows on the convolved signal (with lengths of 0.1s and 1s, respectively) (Fig 4E). A train of unitary impulses is generated from those local maxima (Fig 4H). Next, the RMS value is calculated from the EMGdi signal (without filtering), with a window located between 0.3 and 0.1 s before the detection, and another window located between 0.1 and 0.3 s after the detection (Fig 4F). The QRS complexes whose RMS values do not overcome a certain threshold are used to generate an average ECG pattern (Fig 4G). Finally, the unitary train of impulses is convolved with the average ECG pattern in order to generate an estimation of the cardiac activity present in the EMGdi signal (Fig 4I). This signal is considered as the reference input of the adaptive filter.

D. Parameter Selection of the Adaptive Canceller

The scheme of adaptive cancellation (Fig 2) requires the selection of two parameters: the number of weights of the linear combiner (L) and the adjustment or adaptive constant (μ) of the LMS algorithm.

Since it was desired that the adaptive canceller was robust against false positives and false negatives in the detection of the QRS complexes, it has been selected a value of L equivalent to 4 seconds of reference signal. In that way, the reference input canceller will include a sequence of 4 to 10 QRS complexes, even though there were one or two detection errors.

If the adaptive constant μ is too high, the adaptive filtering will remove all the ECG interference in the signal, but it will also perform an undesired attenuation of the EMGdi signal. Furthermore the adaptive algorithm would become unstable. On the other hand, if the μ is too small, the adaptive filtering will not remove completely the ECG interference. Therefore, the choice of the constant μ is a compromise between the complete cancellation of the ECG activity and the preservation of the EMGdi signal. In this work, the constant of adaptation (μ) was chosen so that the energy of the output or cancellation signal matches with the energy of the reference input (that is an initial estimation of the ECG interference). The process of obtaining this value of μ is done through an iterative process. Based on two initial values (points 1 and 2 in Figure 5), we calculate the energy of the cancellation signal using an μ value corresponding to the intersection of the line joining these two points with the energy of the reference signal (point 3 in Figure 6). The following iterations are performed between the new calculated point and the point with nearest μ. The iterative process ends when the cancellation signal energy differs by less than 0.1% of the energy of the reference signal.

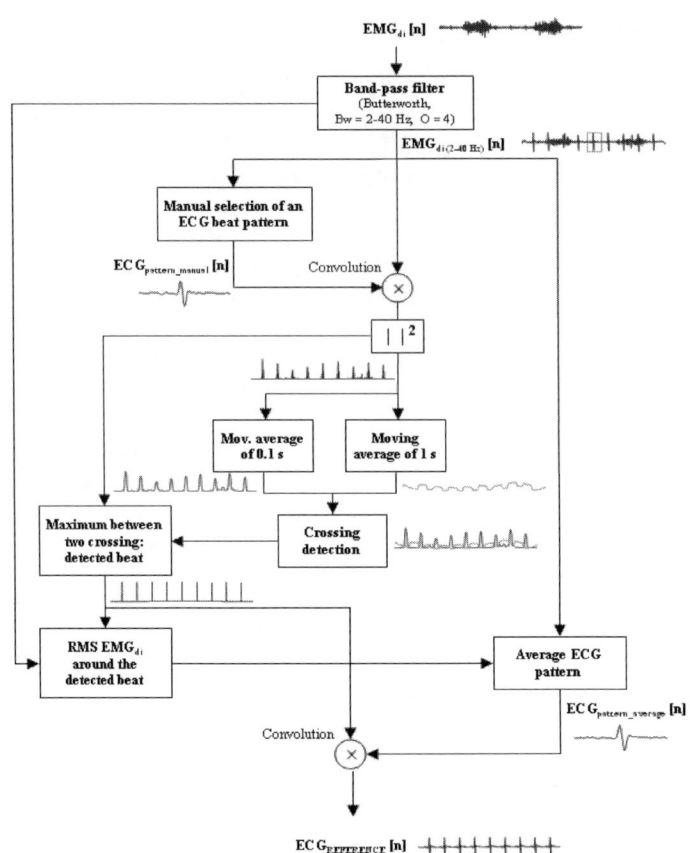

Fig 3 Block diagram of the algorithm for detection and estimation of the ECG activity present in the EMGdi signal (ECG reference)

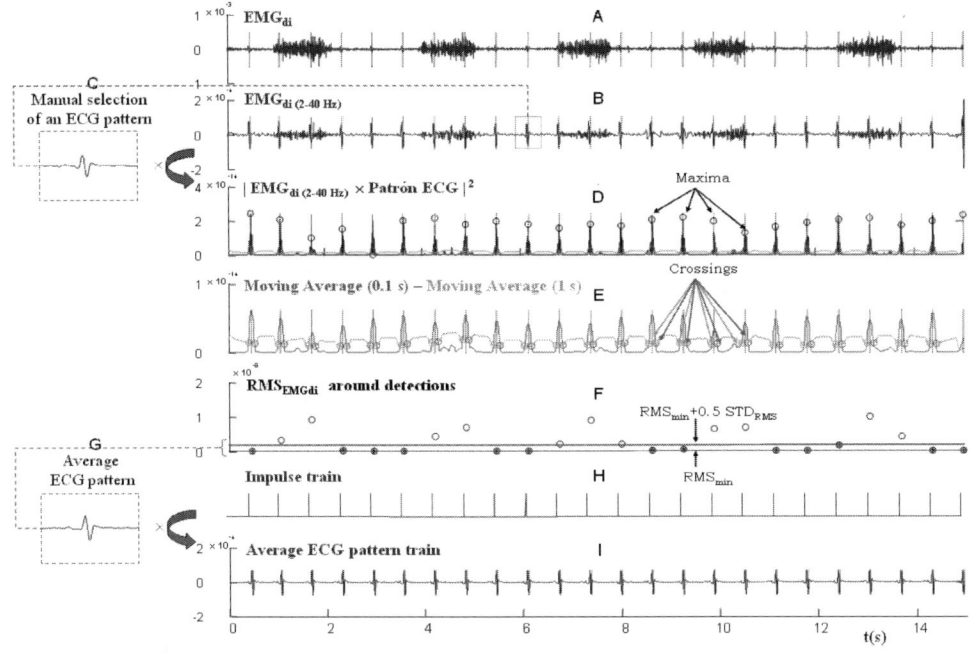

Fig 4 Example of estimation of the cardiac activity (ECG reference) in the EMGdi signal

III. RESULTS

Fig 6 shows an example of application of the proposed adaptive cancellation algorithm over a 4 s segment of an EMGdi signal. Fig 6A shows the 2-40 Hz band pass filtered EMGdi signal. Fig 6B0 shows the generated reference signal. In this example it has been added 3 QRS detection errors: a false negative (FN) and two false positives (FP). Fig 6C0 shows the direct cancellation of the ECG activity by subtraction. The distortion in the EMGdi signal produced by the QRS detection errors can be observed. Fig 6B1-C1, B2-C2 and B3-C3 show the cancellation signal for three different μ values: the optimal μ value (Fig 6B2-C2), a smaller μ value (Fig 6B1-C1), and a greater μ value (Fig 6B3-C3).

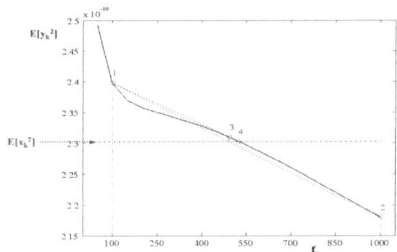

Fig 5 Energy of the reference signal and evolution of the energy of the canceling signal versus the factor μ

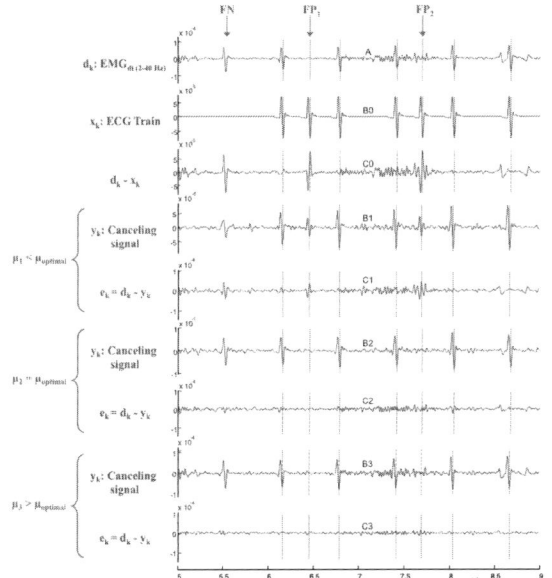

Fig 6 Example of application of the adaptive ECG interference canceller: primary input (A), reference input (B0), cancellation by direct subtraction (C0); and cancellation signals (B1-B2-B3) and error signals (C1-C2-C3), using, respectively a mu value that makes the power of the cancellation signal be less (1), equal (2) and greater (3) than the reference signal

IV. CONCLUSIONS

The reference input used in the proposed ECG interference adaptive cancellation algorithm is an initial estimation of ECG activity present in the EMGdi signal. These detection errors in the QRS complex use to be more frequent when there is presence of high electromyographic activity, and it is precisely in those cases in which it is more desired to cancel the ECG interference. In order to fix possible errors in QRS detector it has been selected a number of weights equivalent to 4 seconds, so that it is combined the information of more than one beat.

Furthermore we have designed an automatic criterion for the selection of the μ constant so that the canceling signal energy equals the energy of the reference signal. Starting from two initial μ values, the proposed automatic criterion reaches the recommended μ value in just 2 or 3 additional iterations. This approach optimizes the tradeoff between cancellation of ECG interference and eliminating EMGdi component.

The adaptive algorithm proposed has not been designed to work in real time. It has been used to eliminate the cardiac interference in the EMGdi of a signal database. It should be noted that, once fixed the number of weights, the algorithm cancels the ECG activity from the EMGdi signal automatically (without any additional input signal or parameter).

The outcomes of this work suggest that the methodology presented in this paper could improve the analysis and interpretation of EMGdi signal, in cases where it is not possible to have a recorded ECG reference signal.

ACKNOWLEDGMENT

This work was supported in part by the Ministerio de Economía y Competitividad from the Spain Government under grants TEC2010-21703-C03-01.

REFERENCES

1. De Luca CJ (1988) Electromyography, Encyclopaedia of Medical Devices and Instrumentation, Webster J.G. ed., John Willey and Sons, New York, pp. 1111-1120
2. De Luca CJ (1984) Myoelectrical Manifestations of localized muscular fatigue in humans, Crit. Rev. Biomed. Eng., 11:251-279
3. Synderby C, Lindström L, Grassino AE (1995) Automatic Asessment of Electromyogram Quality, J. Appl. Physiol., 79(5): 1803-1815
4. Laguna P, Jané R, Meste O, Poon PW, Caminal P, Rix H, Thakor NV (1992) Adaptive Filter for Event-Related Bioelectric Signals using an impulse correlated reference input: comparison with signal averaging, IEEE Trans. Biomed. Eng., vol. 39, no.10, pp. 1032.1044

5. Mañanas MA, Romero S, Topor ZL, Bruce EN, Houtz P, Caminal P (2001) Cardiac Interference in Myographic Signals from Different Respiratory Muscles and Levels of Activitity, Proc. 23rd Annual Int. Conf. IEEE-EMBS, vol. 2, pp. 1115-1118
6. Deng Y, Wolf W, Schnell R, Apple U (2000) New Aspects to Event-Synchronous Cancellation of ECG Interference: an Application of the Method in Diaphragmatic EMG Signals, IEEE Trans. Biomed. Eng., vol. 47, no. 9, pp. 1177-1184
7. E. Aithocine, P.-Y. Guméry, S. Meignen, L. Heyer, Y. Lavault and S. B. Gottfried, "Contribution to Structural Intensity Tool: Application to the Cancellation of ECG Interference in Diaphragmatic EMG", Proc. 28th Annual Int. Conf. IEEE-EMBS, 2006
8. S. Newman, J. Road, F. Bellemere, J.P. Clozel, C.M. Lavigne, and A.E. Grassino, "Respiratory Muscle Length Measured by Sonomicrometry," J. Appl. Physiol.: Resp. Env. Ex. Physiol., vol. 46(3), pp. 753-764, 1984.
9. N. V. Thakor, J. G. Webster and W. J. Tompkins, "Estimation of QRS Complex Power Spectra for Design of a QRS Filter", IEEE Trans. Biomed. Eng., vol. BME-31, pp. 702-705, 1984

Author: Abel Torres
Institute: Institut de Bioenginyeria de Catalunya
Street: C/Baldiri Reixac, 4. Torre I, Planta 9
City: Barcelona
Country: Spain
Email: abel.torres@upc.edu

Relevance of the Atrial Substrate Remodeling during Follow-Up to Predict Preoperatively Atrial Fibrillation Cox-Maze Surgery Outcome

A. Hernández[1], R. Alcaraz[2], F. Hornero[3], and J.J. Rieta[1]

[1] Biomedical Synergy, Electronic Engineering Department, Universidad Politécnica de Valencia, Spain
[2] Innovation in Bioengineering Research Group. University of Castilla-La Mancha, Spain
[3] Cardiac Surgery Department, General University Hospital Consortium, Valencia, Spain

Abstract—Although atrial fibrillation (AF) is the most common cardiac arrhythmia, the knowledge about its causes and mechanisms still is uncompleted. Several studies suggest that structural and electrophysiological remodeling are directly related to its development and perpetuation. To this respect, the surface electrocardiogram (ECG) has been used to assess different aspects of atrial remodeling. However, specific ECG parameters have never been used to provide valuable clinical information in the study of AF aggressive treatments, such as the Cox-Maze surgery. In this work, parameters from the fibrillatory (f) waves, such as their regularity or their amplitude, are studied after a six months follow-up period from the Cox-Maze surgery. This study culminates previous works on Cox-Maze outcome prognosis at discharge and after the blanking period. The results obtained reported a remarkable prediction capability decrease of the preoperative ECG parameters, with respect to similar studies at the time of discharge and after the surgery blanking period. More concretely, the prognosis accuracy of the f waves amplitude, the dominant atrial frequency and the f waves regularity was 58.3%, 62.5% and 66.6%, respectively. Pharmacological and electrical cardioversion treatments administered during the follow-up period could be the main reason to this result. Therefore, atrial substrate remodeling during the follow-up causes a remarkable influence on the preoperative surgery outcome prognosis based on the f waves analysis.

Keywords—ECG, atrial fibrillation, Cox-Maze surgery, organization indices, fibrillatory waves amplitude

I. INTRODUCTION

Atrial fibrillation (AF) is the most common cardiac arrhythmia associated with an adverse prognosis in clinical practice. AF has been shown to increase the risk of systemic embolization and mortality [1]. The causes and mechanisms of atrial fibrillation are still unknown, however, several studies suggest that structural and electrophysiological remodeling play a key role in the development and evolution of AF [2]. To this respect, any persistent change in the atrial structure or function is considered as atrial remodeling [3].

Previous studies on the surface electrocardiogram (ECG) have related the atrial activity (AA) organization both with the number of propagating wavelets in the atria and with the electrical remodeling [4]. In fact, these parameters have been assessed to predict paroxysmal AF spontaneous termination [5] and electrical cardioversion outcome [6]. On the other hand, fibrillatory (f) waves amplitude has been related with the amount of activated atrial tissue, hence it would be related with atrial structural parameters [7]. However, ECG parameters have not been widely used to analyze the atrial substrate remodeling after an aggressive intervention, such as the Cox-Maze surgery. Therefore, the present work analyzes these parameters to predict patient's rhythm after a 6 months follow-up from the Cox-Maze surgery. This study completes previous works predicting Cox-Maze outcome at discharge [8] and after the blanking period [9]. Therefore, it will allow to obtain reliable information about the atrial substrate alteration in patients undergoing concomitant Cox-Maze surgery of AF. This kind of preoperative prognosis could lead to a better understanding of the atrial remodeling mechanisms and, furthermore, could also be used to modulate the patient's therapeutic window depending on the risk of maintaining mid-term AF.

II. MATERIALS

Twenty-four patients (mean age 68.33 ± 6.36 years) composed the database. All patients were in permanent AF for at least 3 months. Cox-Maze surgery was applied concomitantly to another heart surgery. All patients underwent antiarrhythmic and anticoagulant treatment at discharge. At 3 months follow-up, antiarrhythmic treatment was withdrawn in NSR patients whereas electrical cardioversion (ECV) was considered and, in the proper cases, applied to AF patients [10]. After the 6 months follow-up, those patients that were in NSR at 3 months and still maintained the rhythm, anticoagulant drugs administration was stopped. Six months after the Cox-Maze surgery 7 (29%) patients remained in AF while 17 (71%) patients reverted back to NSR.

Twenty seconds segments from preoperative standard 12-lead ECGs were recorded. Lead V_1 was chosen because previous works proved that AA is prevalent in this lead [11]. The signals were processed with an adaptive notch filter at 50 Hz to remove the powerline interference. Then, high frequency noise was reduced with an eight-order forward/backward IIR Chebyshev lowpass filtering with a 70 Hz cut-off frequency. Next baseline wander was removed by high-pass filtering with 0.5 Hz cut-off frequency. Finally, wavelet denoising was applied to reduce muscle noise [12]. After denoising, an adaptive QRST cancelation method was applied to the recordings in order to extract the AA [13].

III. METHODS

A. Atrial Activity Organization

Atrial activity organization was estimated in two ways. Firstly, by means of the dominant atrial frequency (DAF), because its inverse has been related to the atrial cycle length [4]. Secondly, by an nonlinear index such as sample entropy (SampEn), which has been successfully applied previously in different works dealing with AF [6].

The DAF is a frequency domain parameter, therefore, the Power Spectral Density (PSD) must be computed. In this work, the PSD was obtained using the Welch method with a Hamming window of 4096 points in length, 50% overlapping between adjacent windowed sections and a 8192-points fast Fourier transform. Previous studies have demonstrated that the typical AF frequency range is in the 3-9 Hz band [4]. Therefore, the DAF has been defined as the highest amplitude frequency within this range in the AA spectrum.

On the other hand, SampEn measures time series regularity, with larger values corresponding to more irregularity in the data [14]. It is defined as the negative natural logarithm of the conditional probability that two sequences, similar for m points, remain similar at the next point, where self-matches are not included in calculating the probability [14]. This index has been widely used to measure AA organization [6]. Two input parameters must be specified to compute the SampEn, the length of the sequences to be compared, m, and the pattern similarity tolerance, r. As suggested in previous works, optimal values for AF analysis are $m = 2$ and $r = 0.35$ times the standard deviation of the AA [15]. SampEn presents high sensitiveness to noise and ventricular residua [6]. Thus, to reduce this nuisance influence, the SampEn was estimated over the main atrial wave (MAW) [6]. Following previous studies, the MAW was obtained by applying a selective filtering to the AA centered around the DAF [6].

B. Fibrillatory Waves Amplitude

The f waves mean power (fWP) has proven to be a robust estimator of the f waves amplitude [16], therefore, it was used in this work. This index represents the energy carried by the AA signal within the interval under analysis. The AA root mean square value was computed to obtain the fWP [16]. Before extracting the AA signal, each analyzed ECG segment was normalized to its maximum R peak amplitude. This operation avoided all the effects that can modify the ECG amplitude as a function of the different gain factors during recording, electrodes impedance, skin conductivity, etc [16].

C. Nonlinear Analysis Applicability

Given that SampEn is a non-linear metric, data nonlinearity must be tested to confirm its proper applicability. The surrogate data test was selected for testing data nonlinearity. This method consists of obtaining a surrogate dataset from the original data. Then, a nonlinear metric that quantifies some aspect of the series, called discriminating statistic, is computed over the dataset. When the original series discriminating statistic is significantly different than the surrogate data values, non-linearity can be assumed [17]. In the present work 40 surrogate data were generated for each AA signal and the analyzed statistic was SampEn. Wilcoxon T test for paired data was performed to compare the distributions of the SampEn from the original data and surrogate data.

D. Statistical Analysis

Before analyzing prediction results, statistical significant differences between NSR and AF groups were assessed. In order to decide which test should be used, a Shapiro-Wilk test was applied to analyze data distributions and a Levene test was applied to check homoscedasticity of the data. Once the data normality and homoscedasticity were studied, t-Student test was applied to parameters with normal and homoscedastic distribution and the U Mann-Whitney test was applied in the remaining cases. In both tests, a statistical significance $p < 0.05$ was considered as significant.

A threshold was defined for each predictor to classify in NSR and AF to assess the indices prediction capability. The threshold for each index was obtained by computing their receiver operating characteristic (ROC) curve, which is a graphical representation of the tradeoffs between sensitivity and specificity. In this work, sensitivity was the proportion of patients in AF 6 months after surgery correctly identified, whereas specificity was considered as the proportion of patients in NSR correctly classified. The value providing the highest rate of patients properly discerned (i.e., the diagnostic accuracy) was selected as the optimum threshold.

Finally, in order to look for complementary information between f waves amplitude and organization, a classification tree was developed combining the three analyzed parameters.

Table 1 Classification results for DAF, SampEn and fWP together with their statistical significance.

	Sensitivity	Specificity	Accuracy	p
DAF	85.71%	52.94%	62.5%	0.395
SampEn	85.71%	58.82%	66.66%	0.254
fWP	71.43%	52.94%	58.33%	0.975

The used stopping criterion for the tree growth was that each node contained only observations of one class or fewer than 5 observations. Moreover, the impurity-based Gini index was used to look for the best parameter and its threshold for the splitting of each node [18].

IV. RESULTS

Statistically significant differences among original and surrogate data were noticed with the Wilcoxon T test. Therefore, nonlinearity in the AA signal can be assumed and the suitability of the SampEn-based analysis was confirmed.

Table 1 shows sensitivity, specificity and accuracy values for the analyzed indices. As can be seen, all the metrics presented an accuracy below 70%, with specificity values remarkably low. On the other hand, sensitivity values are over 85% for DAF and SampEn but, given the reduced number of patients in AF, its influence in the final diagnostic accuracy was limited. In agreement with this classification result, no statistically significant differences between patient groups were observed for the indices. Moreover, Figures 1-3, representing the computed indices distribution for NSR and AF patients, show a notable overlapping between groups.

Finally, the classification tree was unable to improve the group classification. In a first step SampEn was used to generate two branches, but the remaining parameters were discarded to generate additional nodes.

V. DISCUSSION

The present work has studied f waves parameters to predict the absence of AF after a 6 months follow-up from the surgery. This study culminates previous works analyzing outcome at discharge [8] and after the blanking period [9]. Therefore, reliable information on the atrial substrate postoperative evolution could be obtained. Previous studies showed the fWP as an accurate patient's rhythm predictor. In addition, DAF and SampEn provided accuracy values over 70% and reported statistically significant differences among AF and NSR groups after the blanking period [9]. However, no statistically significant differences have been found in the present

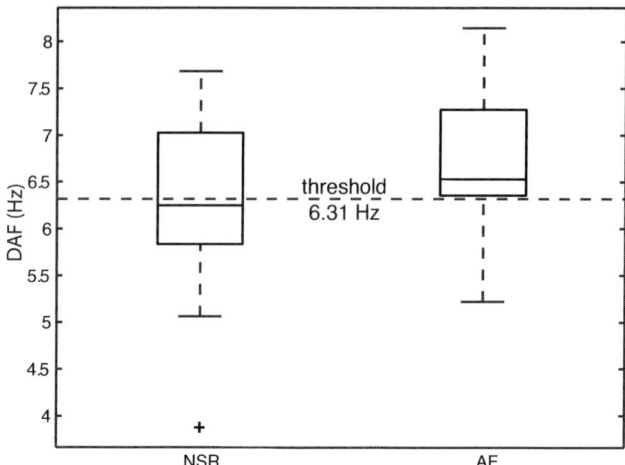

Fig. 1 Box-and-whisker diagram for DAF considering the patient's rhythm at six months after Cox-Maze surgery.

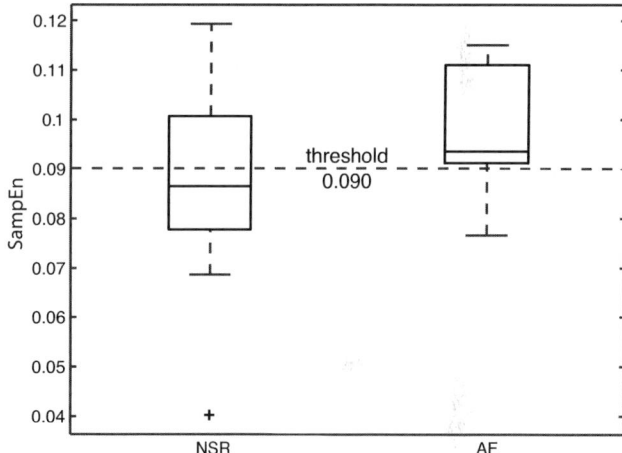

Fig. 2 SampEn box-and-whisker diagram for the patient's rhythm at six months after Cox-Maze surgery.

work. Moreover, the prediction capability has been notably reduced. As a consequence, the atrial substrate remodeling during the follow-up provokes remarkable influences on the surgery outcome prognosis based on the f waves analysis.

From the discharge until six months after the surgery there are several causes that modify the atrial substrate such as heart healing, pharmacologic treatment, etc. However, the most probable major change is produced by ECV applied after the blanking period [10]. In fact, the reduced specificity values observed for the analyzed metrics could be a consequence of the NSR restoration through ECV in a considerable number of patients.

ECV is able to alter atrial electrophysiological substrate to synchronize atrial cells and restore NSR [19]. This alteration affects the number of propagating wavelets in the atria and

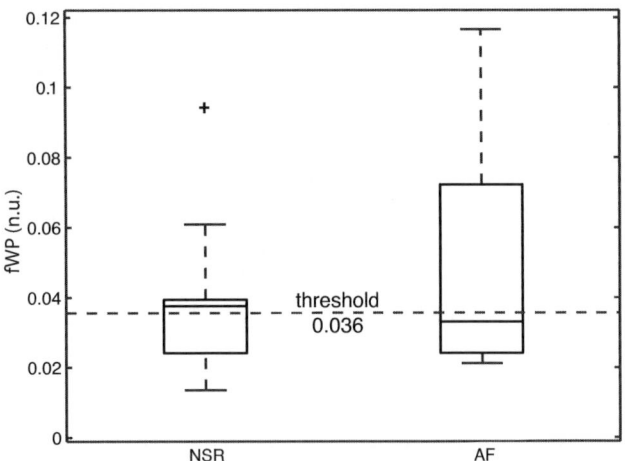

Fig. 3 Box-and-whisker diagram for fWP considering the patient's rhythm at six months after Cox-Maze surgery.

the amount of atrial tissue activated. Since f waves organization and amplitude are respectively related with both parameters [4,7], the discriminant ability of preoperative predictors is hindered by ECV. Thus, the atrial substrate evolution has a response in the f waves parameters prediction capability.

VI. CONCLUSION

This study has proven that the atrial electrophysiological substrate remodeling taking place at mid-term after AF surgery, causes a remarkable decrease on the f waves preoperative prognosis capability. Therefore, further analyses are required in order to establish the influence of antiarrhythmic and cardioversion therapies during the whole follow-up.

ACKNOWLEDGEMENTS

Work funded by TEC2010–20633 from the Spanish Ministry of Science and Innovation and PPII11–0194–8121 from Junta de Comunidades de Castilla-La Mancha.

REFERENCES

1. Benjamin E J, Wolf P A, D'Agostino R B, Silbershatz H, Kannel W B, Levy D. Impact of atrial fibrillation on the risk of death: the Framingham Heart Study. *Circulation.* 1998;98:946–952.
2. Prystowsky E N. The history of atrial fibrillation: the last 100 years. *J Cardiovasc Electrophysiol.* 2008;19:575–582.
3. Nattel S, Burstein B, Dobrev D. Atrial remodeling and atrial fibrillation: mechanisms and implications *Circ Arrhythm Electrophysiol.* 2008;1:62-73.
4. Holm M, Pehrson S, Ingemansson M, et al. Non-invasive assessment of the atrial cycle length during atrial fibrillation in man: introducing, validating and illustrating a new ECG method. *Cardiovasc Res.* 1998;38:69–81.
5. Alcaraz R, Rieta J J. Sample entropy of the main atrial wave predicts spontaneous termination of paroxysmal atrial fibrillation. *Med Eng Phys.* 2009;31:917–922.
6. Alcaraz R, Rieta J J. A review on sample entropy applications for the non-invasive analysis of atrial fibrillation electrocardiograms *Biomedical Signal Processing and Control.* 2010;5:1 - 14.
7. Kamata J, Kawazoe K, Izumoto H, et al. Predictors of sinus rhythm restoration after Cox maze procedure concomitant with other cardiac operations. *Ann Thorac Surg.* 1997;64:394–398.
8. Hernández A, Alcaraz R, Hornero F, Rieta J J. Role of fibrillatory waves amplitude as predictors of immediate arrhythmia termination after Maze surgery of atrial fibrillation in *Computing in Cardiology (CinC), 2012*:661-664 2012.
9. Hernández A, Alcaraz R, Hornero F, Rieta J J. Preoperative Prognosis of Atrial Fibrillation Concomitant Surgery Outcome After the Blanking Period *Submitted to medicon 2013.* 2013.
10. Hornero F, Montero J A, Cánovas S, Bueno M. Biatrial radiofrequency ablation for atrial fibrillation: epicardial and endocardial surgical approach. *Interact Cardiovasc Thorac Surg.* 2002;1:72–77.
11. Petrutiu S, Ng J, Nijm G M, Al-Angari H, Swiryn S, Sahakian A. V.. Atrial fibrillation and waveform characterization. A time domain perspective in the surface ECG. *IEEE Eng Med Biol Mag.* 2006;25:24–30.
12. Sörnmo L, Laguna P. *Bioelectrical Signal Processing in Cardiac and Neurological Applications.* Elsevier Academic Press 2005.
13. Alcaraz R, Rieta J J. Adaptive singular value cancelation of ventricular activity in single-lead atrial fibrillation electrocardiograms. *Physiol Meas.* 2008;29:1351–1369.
14. Richman J S, Moorman J R. Physiological time-series analysis using approximate entropy and sample entropy. *Am J Physiol Heart Circ Physiol.* 2000;278:H2039–H2049.
15. Alcaraz R, Abásolo D, Hornero R, Rieta J J. Optimal parameters study for sample entropy-based atrial fibrillation organization analysis. *Comput Methods Programs Biomed.* 2010;99:124–132.
16. Alcaraz R, Rieta J J. Time and frequency recurrence analysis of persistent atrial fibrillation after electrical cardioversion. *Physiol Meas.* 2009;30:479–489.
17. Palus M, Hoyer D. Detecting nonlinearity and phase synchronization with surrogate data. *IEEE Eng Med Biol Mag.* 1998;17:40–45.
18. Breiman L, Friedman J, Stone C J, Olshen R A. *Classification and Regression Trees.* Chapman & Hall/CRC 1984.
19. Gall N P, Murgatroyd F D. Electrical cardioversion for AF-the state of the art. *Pacing Clin Electrophysiol.* 2007;30:554–567.

Corresponding Author: Antonio Hernández Alonso
Institute: Biomedical Synergy
Address: Universidad Politécnica de Valencia,
        Camino de Vera s/n. Building 7F, 5th Floor
City: Valencia
Country: Spain
Email: anheral@upv.es

A New Method to Estimate Atrial Fibrillation Temporal Organization from the Surface Fibrillatory Waves Repetitiveness

R. Alcaraz[1], F. Hornero[2], and J.J. Rieta[3]

[1] Innovation in Bioengineering Research Group. University of Castilla-La Mancha, Cuenca, Spain
[2] Cardiac Surgery Department, General University Hospital Consortium of Valencia, Spain
[3] Biomedical Synergy, Electronic Engineering Department, Universidad Politécnica de Valencia, Spain

Abstract—Analysis of the electrical activity organization within the atria can reveal interesting clinical information related to atrial fibrillation (AF), which is the most commonly diagnosed arrhythmia in clinical practice. However, a very reduced number of indirect AF organization estimators from the surface electrocardiogram (ECG) have been proposed to date. The present work introduces a new method for direct and short-time AF organization estimation from each surface lead of ECG recordings. The temporal arrhythmia organization was estimated through the computation of morphological variations among single fibrillatory (f) waves. They were delineated and extracted from the atrial activity (AA) signal using an adaptive signed correlation index. The algorithm was tested on real AF recordings to discriminate atrial signals with different organization degrees. Results indicated a notably higher global accuracy (90.3%) in comparison with the two non-invasive AF organization estimates widely used today: the dominant atrial frequency (70.5%) and sample entropy (76.1%). Moreover, due to its ability to assess AA regularity wave-to-wave, the proposed method was also able to pursue AF organization time course more precisely than the aforementioned indices.

Keywords—**Atrial Fibrillation, Electrocardiogram, Organization, Correlation**

I. INTRODUCTION

Atrial fibrillation (AF) is the most common arrhythmia encountered in clinical practice [1]. Despite many years of research and speculation, the physiological mechanisms provoking onset and termination of this arrhythmia still are not fully explained [2]. Nonetheless, previous works have demonstrated that AF is associated with multiple meandering activation waves propagating randomly throughout the atria [3]. The fractionation of the wavefronts, as they propagate, results in self-perpetuating independent wavelets, called reentries. Moreover, a strict correlation between AF organization, defined as how repetitive is the AF signal pattern, and the number of wavefronts wandering the atrial tissue has also been proved previously [4].

However, only two indirect AF organization estimates from the surface ECG recording have been proposed to date. On the one hand, the dominant atrial frequency (DAF), which is defined as the highest amplitude frequency within the 3–9 Hz range of the atrial activity (AA) spectral content [5]. Its inverse has been directly related to atrial refractoriness [6] and, hence, to atrial cycle length [5]. On the other hand, the second way to get a noninvasive estimate of AF organization has been based on a nonlinear regularity index, such as sample entropy (SampEn) [7]. This index has been proposed to estimate the amount of repetitive patterns existing in the fibrillatory (f) waves from the fundamental waveform of the AA signal, which have been named as main atrial wave (MAW) in the literature [8].

The drawback of noninvasive organization estimation is the lack of strict accuracy in the process, given that both SampEn and DAF are only able to assess f waves regularity indirectly. Indeed, both indices require AA signal intervals of several seconds in length, which contain a considerable number of f waves [8]. Thus, it can be considered that these estimators can only yield an average AF organization assessment, thus blurring the possible information that the wave-to-wave analysis over each atrial activation may provide. Overall, this work proposes a new algorithm for direct and short-time evaluation of AF organization from the ECG. The method is able to quantify every single f wave regularity by measuring how repetitive its morphology is along onward atrial activations.

II. METHODS

A. Study Population

The study group consisted of 43 patients undergoing cardiac surgery who developed postoperative AF. For each patient, a standard 12-lead ECG together with two unipolar epicardial recordings were considered for the study. Each recording length varied between 50 and 150 seconds and was digitized at 1 kHz with an amplitude resolution of 0.4 μV. During the surgical procedure two unipolar epicardial electrodes were placed on both left and right atrial appendages.

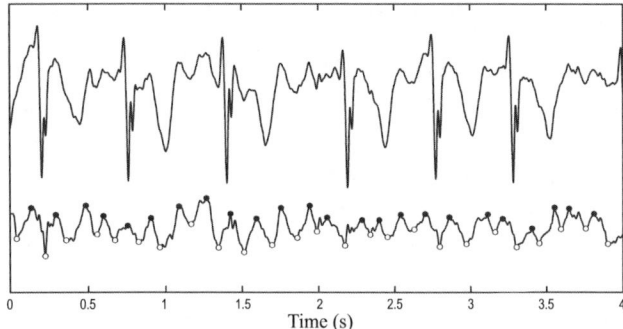

Fig. 1 Individualized f waves detection in a type III AF recording. The ECG (top) and AA (bottom), after QRST cancellation, are displayed. The upper full circles mark the maximum associated to every f wave, whereas lower empty circles indicate their boundaries.

The electrograms provided by these electrodes were exclusively used to yield a manual classification of AF episodes as a function of their organization. Thus, they were inspected separately by four different experienced cardiologists and classified as type I, II, or III following Wells' criteria [9]. Only stable segments lasting more than 15 seconds, classified by all the cardiologists as the same AF type for the two simultaneous epicardial recordings, were included in the analysis. After epicardial classification, the corresponding surface ECG recordings were grouped in the same way to create the database. The final labeled dataset consisted of 129 AF segments of surface ECGs (49 AF-I, 36 AF-II and 44 AF-III) of 15 seconds in length.

B. Signal Preprocessing

Temporal AF organization was estimated only from surface lead V1, because the atrial signal is larger in this lead [10]. To improve later analysis, this lead was preprocessed using different forward/backward filtering approaches to remove baseline wander, high frequency noise and powerline interference [11]. Thereafter, the AA signal was extracted by using an adaptive QRST cancellation method [12] and f waves were delineated making use of a method based on mathematical morphology (MM) [13]. Briefly, MM operators were applied to the AA signal in order to enhance the f waves. Thus, a combination of erosion and dilation operations was applied to the AA with two structuring elements. The first one was adapted to the f waves by an even triangular shape. The second was designed as a rectangular shape to suppress the drift between atrial cycles. Finally, the resulting impulsive signal was used to detect the maximum and boundaries of the atrial activations by peak detection, such as can be appreciated in the example of Fig. 1.

C. Temporal AF Organization Estimation

To quantify f waves repetitiveness over time, thus providing a temporal AF organization estimation, all the possible f waves pairs within the AA segment under analysis were correlated. For this purpose, an adaptive signed correlation index (ASCI) was used. This index is able to detect the amplitude difference between signals and is highly insensitive to impulsive noise [14]. In brief, given two waves, w_i and w_j, they were aligned using their highest amplitude timing. As they were asymmetrically distributed with respect to their maxima, the length of both superimposed waves was determined as $L = L_i \cup L_j$, being L_i and L_j the length of the waves w_i and w_j, respectively. Then, the ASCI, i.e. the morphological similarity, between these two waves was computed as

$$d(w_i, w_j) = \frac{\sum_{k=1}^{L} w_i(k) \otimes w_j(k)}{L}, \quad (1)$$

where $\otimes$ denotes the signed product of two trichotomized scalars, which is defined by

$$w_i(k) \otimes w_j(k) = \begin{cases} 1 & \text{if } |w_i(k) - w_j(k)| \leq T, \\ -1 & \text{if } |w_i(k) - w_j(k)| > T, \\ 0 & \text{if } \nexists\, w_i(k) \text{ and/or } \nexists\, w_j(k), \end{cases} \quad (2)$$

being T the tolerance for accepting that two samples are similar. This tolerance was defined as a percentage of the amplitude range of the wave w_i, i.e., of the difference $\max\{w_i(k)\} - \min\{w_i(k)\}$, for $k = 1, \ldots L_i$. Thus, the ASCI is an index ranging from -1 up to 1. Hence, for signals with similar morphology it approaches 1, due to more concordant sample pairs. In contrast, for signals with different morphology it approaches -1, due to more discordant sample pairs.

Finally, the relative number of similar f wave pairs within the analyzed AA segment and, therefore, the f waves regularity estimation, was computed as

$$\kappa = \frac{2}{N(N-1)} \sum_{i=1}^{N} \sum_{j=i+1}^{N} \mu\left(\frac{d(w_i, w_j) - 1}{r - 1}\right), \quad (3)$$

being N the number of f waves considered for the analysis, r a threshold for evaluating the similarity between pairs of f waves on the basis of their distance, that is, two waves are considered to be similar when their distance is lower than r, and $\mu(x) = e^{-x^n}$. This exponential function is able to establish a continuous boundary between two classes, being n the gradient of the boundary, which acts as the weight of vectors' similarity [15]. In this way, κ ranges from 0 to 1, with higher values indicating more regularity in f waves and, thus, higher temporal AF organization.

Table 1 Mean and standard deviation values of κ, DAF and SampEn computed on the test set of each experiment and averaged for 1000 realizations.

Index	AF-I	AF-II	AF-III
κ	0.722 ± 0.079	0.423 ± 0.108	0.191 ± 0.074
DAF (Hz)	4.921 ± 0.994	5.874 ± 1.221	5.992 ± 1.029
SampEn	0.093 ± 0.012	0.115 ± 0.021	0.121 ± 0.018

Table 2 Mean and standard deviation values of the global accuracy and the percentage of episodes correctly classified for each AF type obtained on the test set of each experiment and averaged for 1000 realizations.

	κ	DAF	SampEn
AF-I	0.846 ± 0.031	0.581 ± 0.039	0.601 ± 0.031
AF-II	0.879 ± 0.034	0.786 ± 0.041	0.883 ± 0.029
AF-III	0.931 ± 0.041	0.702 ± 0.037	0.762 ± 0.033
Global	0.903 ± 0.035	0.705 ± 0.037	0.761 ± 0.032

D. Performance Assessment

The proposed regularity index performance was compared with DAF and SampEn because, as stated before, these are the only existing noninvasive indices for f waves regularity estimation from the standard ECG. Details about the computation of these parameters can be found in [8]. In order to obtain a robust evaluation of the discriminant ability of these three indices, 1000 sets of experiments were carried out. In each experiment, 64 out of 129 labeled patterns were randomly selected as training set, and the remaining 65 patterns defined the test set. In order to assess individually the classification performance of each index for each experiment, two receiver operation characteristics (ROC) curves were computed from the training set to establish the optimal discriminating thresholds between AF-I and AF-II and between AF-II and AF-III. Thereafter, these thresholds were used to determine from the test set the global accuracy, i.e., the percentage ratio between the correctly classified AF segments and the total number of analyzed segments, and the number of patients classified correctly for each AF type. Finally, these values were averaged for the 1000 realizations. Additionally, statistical differences between the three AF types were evaluated by using the Kruskal-Wallis test. Similarly, Mann-Whitney U test was also used to determine the difference between AF-I and AF-II and between AF-II and AF-III.

III. RESULTS

The performance of κ is mainly governed by four parameters. After several tests, optimal values of $T = 0.15$, $r = 0$, $N = 10$ and $n = 2$ were chosen. Thus, mean and standard deviation κ values obtained with these parameters, together with DAF and SampEn values computed for the three AF types on the test set of each experiment and averaged for 1000 realizations, are shown in Table 1. As can be seen, the mean value decreased from AF-I to AF-III for the regularity index and increased for DAF and SampEn. Moreover, clear differences among mean values of κ for AF-I, AF-II and AF-III were noticed. In contrast, for DAF and SampEn, average values of type II and III AF episodes provided to be very close to each other. As a consequence, a notable overlapping between both groups is evidenced.

In Table 2, mean and standard deviation values of the global accuracy and the percentage of episodes correctly classified for each AF type obtained on the test set of each experiment and averaged for 1000 realizations are reported for the proposed regularity index, DAF and SampEn. As can be appreciated, the rate of episodes correctly classified for each AF type and the global accuracy achieved by κ is notably higher than for the DAF and SampEn. Indeed, although statistical tests provided significant differences for all the experiments, the significance was always higher for κ. More precisely, in the worst case, Kruskal-Wallis test provided $p < 0.0001$ for κ, $p < 0.001$ for SampEn and $p < 0.01$ for DAF, whereas Mann-Whitney test reported $p < 0.0001$ for κ, $p < 0.01$ for SampEn and $p < 0.05$ for DAF.

IV. DISCUSSION AND CONCLUSIONS

The proposed regularity index provided a statistically significant decreasing f waves regularity from type I to type III AF. This outcome is in agreement with previous invasive works, which have proved that the difficulty of distinguishing between discrete activations in electrograms is associated to the possible presence of a high number of wavelets wandering simultaneously throughout the atrial tissue and, therefore, to a more complex arrhythmia [4]. Hence, the results reported by κ suggest that ECGs acquired during organized atrial rhythms present f waves with well-defined and repetitive morphology. On the contrary, it could be considered that ECGs recorded during highly disorganized atrial activity with fragmented activations contain surface f waves with very dissimilar morphologies, thus, leading towards zero the probability to consider similar waves.

On the other hand, results also proved that κ was able to reach a notably better performance than SampEn and DAF. This aspect could be assigned to the fact that the proposed regularity index is able to operate directly on the f waves and in short-time, without requiring modification of their shape or averaging the waves during some seconds, such as in the computation of DAF and SampEn. Indeed, its wave-to-wave peformance also allows the algorithm to track fast changes of AF dynamics more accurately than the DAF and SampEn, such as Fig. 2 displays.

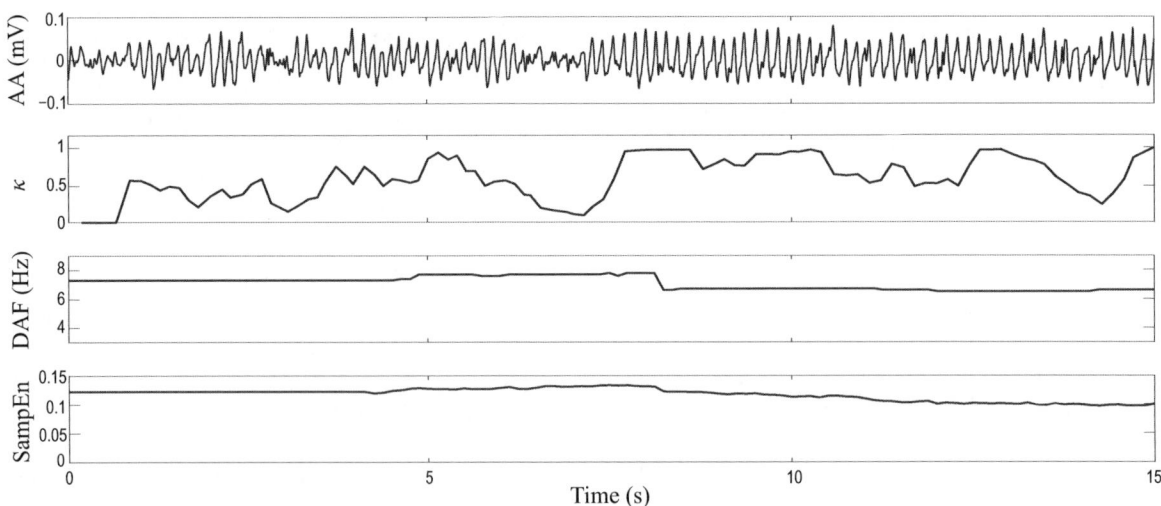

Fig. 2 Dynamic evaluation of the proposed regularity index together with DAF and SampEn during an AF episode with spontaneous changes in regularity. The regularity index was computed for $T = 0.15$, $r = 0$, $N = 5$ and $n = 2$. DAF and SampEn were computed in 4 seconds segments with overlapping of 3.875 seconds, trying to match a wave-to-wave analysis by considering a sliding window of 150 ms, which is the mean duration of a typical f wave.

As a conclusion, the proposed f waves regularity estimator could be considered as the first method presented in the literature able to estimate, in a wave-to-wave fashion, temporal AF organization from single-lead noninvasive recordings.

Acknowledgements

Work funded by TEC2010–20633 from the Spanish Ministry of Science and Innovation and PPII11–0194–8121 from Junta de Comunidades de Castilla-La Mancha.

References

1. Fuster Valentin, Rydén Lars E, Cannom Davis S, et al. 2011 ACCF/AHA/HRS focused updates incorporated into the ACC/AHA/ESC 2006 guidelines for the management of patients with atrial fibrillation: a report of the American College of Cardiology Foundation/American Heart Association Task Force on practice guidelines *Circulation.* 2011;123:e269-367.
2. Sung Ruey J, Lauer Michael R. Atrial fibrillation: can we cure it if we can't explain it? *J Cardiovasc Electrophysiol.* 2005;16:505-7.
3. M A Allessie F I M Bonke, Hollen J. Experimental evauation of Moe's multiple wavelet hypothesis of atrial fibrillation *In: Cardiac electropysiology and arrhythmias*, edited by D. P. Zipes and J. Jalife, New York: Grune and Stratton. 1995:265–276.
4. Konings K. T., Kirchhof C. J., Smeets J. R., Wellens H. J., Penn O. C., Allessie M. A.. High-density mapping of electrically induced atrial fibrillation in humans *Circulation.* 1994;89:1665–80.
5. Holm M., Pehrson S., Ingemansson M., et al. Non-invasive assessment of the atrial cycle length during atrial fibrillation in man: Introducing, validating and illustrating a new ECG method *Cardiovasc Res.* 1998;38:69–81.
6. Capucci A., Biffi M., Boriani G., et al. Dynamic electrophysiological behavior of human atria during paroxysmal atrial fibrillation *Circulation.* 1995;92:1193–202.
7. Richman J. S., Moorman J. R.. Physiological time-series analysis using approximate entropy and sample entropy *Am J Physiol Heart Circ Physiol.* 2000;278:H2039–49.
8. Alcaraz Raúl, Rieta José Joaquín. A review on sample entropy applications for the non-invasive analysis of atrial fibrillation electrocardiograms *Biomed Signal Process Control.* 2010;5:1–14.
9. Wells J L, Karp R B, Kouchoukos N T, MacLean W A, James T N, Waldo A L. Characterization of atrial fibrillation in man: studies following open heart surgery *Pacing Clin Electrophysiol.* 1978;1:426-38.
10. Petrutiu Simona, Ng Jason, Nijm Grace M, Al-Angari Haitham, Swiryn Steven, Sahakian Alan V. Atrial fibrillation and waveform characterization. A time domain perspective in the surface ECG *IEEE Eng Med Biol Mag.* 2006;25:24–30.
11. Sörnmo L., Laguna P.. *Bioelectrical Signal Processing in Cardiac and Neurological Applications*. Elsevier Academic Press 2005.
12. Alcaraz Raúl, Rieta José Joaquín. Adaptive singular value cancellation of ventricular activity in single-lead atrial fibrillation electrocardiograms. *Physiol Meas.* 2008;29:1351–1369.
13. Rieta JJ, Alcaraz R. Fibrillatory Waves Automatic Delineation in Atrial Fibrillation Surface Recordings Based on Mathematical Morphology *Computing in Cardiology (CinC).* 2012;39:805–808.
14. Lian Jie, Garner Garth, Muessig Dirk, Lang Volker. A simple method to quantify the morphological similarity between signals *Signal Processing.* 2010;90:684–688.
15. Chen Weiting, Zhuang Jun, Yu Wangxin, Wang Zhizhong. Measuring complexity using FuzzyEn, ApEn, and SampEn. *Med Eng Phys.* 2009;31:61–68.

Corresponding Author: Raúl Alcaraz Martínez
Institute: Innovation in Bioengineering Research Group
Address: University of Castilla-La Mancha,
         Campus Universitario s/n, 16071
City: Cuenca
Country: Spain
Email: raul.alcaraz@uclm.es

Study on Atrial Arrhythmias Optimal Organization Assessment with Generalized Hurst Exponents

M. Julián[1], R. Alcaraz[2], and J.J. Rieta[1]

[1] Biomedical Synergy, Electronic Engineering Department, Universidad Politécnica de Valencia, Spain
[2] Innovation in Bioengineering Research Group. University of Castilla-La Mancha, Cuenca, Spain

Abstract—The aim of the present work is to determine the optimal use of the Generalized Hurst Exponents ($H(q)$) in the noninvasive organization assessment of atrial arrhythmias. In case of $q = 2$, $H(2)$ relates to the existence of statistical dependencies in the autocorrelation function of a time series. Therefore, it could be used to estimate the pattern repetitivity of atrial arrhythmias. Given that the most common atrial arrhythmia is atrial fibrillation (AF), this work analyzes optimal computational parameters of $H(2)$ with the aim to predict spontaneous AF termination because the probability of AF termination depends on its organization. For this purpose, a reference database containing non-terminating and terminating AF episodes was analyzed. First, $H(2)$ was computed over all non-overlapping segments ranging from 1 up to 30 seconds in order to determine the optimal data length. Then, since the presence of noise and ventricular residua degrades nonlinear metrics performance, the use of a band-pass filtering stage was evaluated. Finally, only the last seconds of each recording were analyzed to determine the applicability of $H(2)$ to short electrocardiographic (ECG) recordings. After optimal computational parameters selection, $H(2)$ yielded a prediction accuracy of 95%, which is a notably better result than previous studies on the same database. Therefore, $H(2)$ can be applied in the study of ambulatory ECG recordings with organization-dependent events.

Keywords—Electrocardiogram, ECG, Atrial Fibrillation, Hurst Exponents.

I. INTRODUCTION

Atrial arrhythmias represent a relevant and common health problem in adults. To this respect, Atrial fibrillation (AF) is the most common atrial arrhythmia and accounts for one third of the hospitalizations caused by cardiac rhythm disturbances [1]. AF is associated with increased long-term risk of stroke, heart failure and mortality. The mechanisms leading to its spontaneous initiation, maintenance or termination have been under intensive investigation in recent years. However, they still remain not fully understood [2]. AF is characterized by a disorganized atrial activity, which causes the atria to be unable to contract in a regular rhythm [2]. When the arrhythmia can terminate spontaneously it is denoted as paroxysmal AF (PAF) [2]. To this respect, the prediction of PAF termination could reduce its associated clinical costs by avoiding unnecessary therapy.

AF organization, which correlates with the likelihood of PAF spontaneous termination [3], can be defined as how repetitive is the atrial activity (AA) signal pattern [4]. Thus, the reliable estimation of AF organization could predict PAF termination. Several ways to predict the spontaneous termination of AF from noninvasive electrocardiographic (ECG) recordings have been proposed in previous works [5]. To this respect, it has been demonstrated that the measurement of the signal regularity through nonlinear metrics like sample entropy can provide an estimation of AF organization [3].

Hurst Exponents (H) [6] measure the existence of long-term statistical dependencies in a time series. Higher values of this metric indicate a slower decrease of the signal's autocorrelation function, and therefore it suggests a higher regularity [6]. This nonlinear index has been proposed as an estimation of Ventricular Fibrillation organization in [7], where the values of H were found to reflect the deterioration of the organization during the first 5 minutes after the onset of this arrhythmia. Besides, the Generalized Hurst Exponents ($H(q)$) [8] are a generalization of the approach proposed by Hurst based on the study of the statistical features of the signal. The use of $H(q)$ as a noninvasive estimator of AF organization has been proposed in [9]. However, no guidelines have been defined for the application of this metric in the study of AF. Therefore, the aim of the present work is to determine the optimal use of $H(q)$ in organization-dependent events of AF, like the prediction of its spontaneous termination.

II. MATERIALS

A. Database

In the present study 60 ECG recordings from the Physionet [10] AF Termination Database were analyzed. This database contains two-lead, 1 minute long Holter ECG recordings of paroxysmal AF and is described in detail in [5]. The studied signals were divided into two groups: the non-terminating group (N) contains AF episodes which lasted at least 1 hour after the end of the recording and the immediately-terminating group (T) contains AF episodes whose termination happened 1 second after the end of the recording. Lead V1 was chosen for the analysis because it presents the highest atrial to ventricular amplitude ratio [11].

B. Signal Conditioning

First, the signals were upsampled to 1 kHz using a cubic spline interpolation method in order improve the alignment in the cancelation of the ventricular activity [12]. Then, the signals were preprocessed to improve the analysis. First, a bidirectional high-pass filter with 0.5 Hz cut-off frequency was applied to remove baseline wander. Then, the signals were filtered using an eight order IRR Chebyshev low-pass filter with 70 Hz cut-off frequency to reduce high frequency noise. Finally, an adaptive notch filter, was applied to remove powerline interference and preserve the ECG spectral information [13]. Then, the ventricular activity was subtracted by applying an adaptive QRST cancelation method [14].

III. METHODS

A. Generalized Hurst Exponents

The Generalized Hurst Exponents ($H(q)$) [8] measure the existence of long-term statistical dependencies in the q-order moments of a time series' amplitude increments [8]. For a stochastic variable $x(t)$ these moments can be represented by the following equation [8]

$$K_q(\tau) = \frac{\langle |x(t+\tau) - x(t)|^q \rangle}{\langle |x(t)|^q \rangle}. \quad (1)$$

Denoting the time between observations as v, $H(q)$ can be defined from $K_q(\tau)$ as [8]

$$K_q(\tau) \sim \left(\frac{\tau}{v}\right)^{qH(q)}. \quad (2)$$

This method was developed for the analysis of complex and inhomogeneous time series having many regions with different properties. In that sense, the application of $H(q)$ as a predictor of AF termination seems appropriate due to the nonstationary and nonlinear nature of the physiological processes underlying AF [15, 16].

The interpretation of $H(q)$ varies with the value of q. In this way, $H(2)$ describes the behavior of the autocorrelation function and therefore it relates to the signal's power spectrum [17]. Regarding the study of AF, $H(2)$ might provide information about its atrial electrical activation patterns and, therefore, it could provide an estimation of AF organization. Thus, $H(2)$ was selected as a predictor of spontaneous AF termination, because this event is associated with changes in the organization of the arrhythmia.

B. Ideal Computational Parameters

Given that no guidelines exist for the application of $H(2)$ to AF signals. First, the optimal data length for $H(2)$ was determined computing this metric over all non-overlapping segments of the signals ranging from 30 seconds to 1 second and then obtaining the average value over the whole recording.

On the other hand, AA signals are often contaminated with noise and ventricular residua, which can degrade the performance of nonlinear metrics, like sample entropy, in the estimation of AF organization [18]. Thus, the use of a previous band-pass filtering step was studied in order to improve the prediction accuracy of the method. Thereby, two different filters were proposed. Firstly, the use of a 3 to 9 Hz band-pass filter was studied because the 3–9 Hz band contains most of the atrial activity power [19]. Secondly, since it has been demonstrated that AF spontaneous termination can be predicted through the estimation of the regularity of the Main Atrial Wave (MAW) [18], the extraction of the MAW by selective filtering was also proposed. To extract the MAW a band-pass FIR filter with 3 Hz bandwith centered on the dominant atrial frequency (DAF) of each signal was applied [18]. Both filters were designed with the Chebyshev approximation, a linear phase in order to prevent distortion and a relative sidelobe attenuation of 40 dB.

The performance of the method in the original and filtered signals was then compared. With this aim $H(2)$ was computed over all non-overlapping optimal length segments of the recordings and then the average value was calculated.

Finally, a gap of the signals' last seconds was analyzed in order to determine the optimal recording length for the prediction of AF termination. To this respect, $H(2)$ was calculated using the optimal previous filtering stage (MAW) and the optimal segment length (15 seconds).

C. Statistical Analysis

The receiver operating characteristics (ROC) curves were calculated to obtain the classification thresholds, specificities and sensitivities in the discrimination between groups N and T. The optimum threshold was determined selecting the closest point to 100% accuracy. Additionally, the area under the ROC curve (AROC) was computed. AROC represents the performance of the metric and varies between 0.5 and 1. Higher values of this parameter represent a better performance. When several parameter values reached the highest prediction accuracy, the one corresponding to the highest value of AROC was selected as optimal.

Table 1 Classification results of $H(2)$ computed on the Main Atrial Wave using different segment lengths.

	Se	Sp	Acc	AROC	Threshold
60 s	94.12%	92.31%	93.33%	0.9751	0.9967
30 s	94.12%	92.31%	93.33%	0.9740	0.9967
20 s	94.12%	96.15%	95.00%	0.9774	0.9967
15 s	91.18%	100.00%	95.00%	0.9796	0.9967
10 s	94.12%	92.31%	93.33%	0.9740	0.9967
5 s	91.18%	96.15%	93.33%	0.9708	0.9967
1 s	88.46%	92.31%	90.00%	0.9649	0.9967

Table 2 Classification results of $H(2)$ computed over the original and filtered signals.

	Se	Sp	Acc	AROC	Threshold
$H(2)$ original	57.69%	82.35%	71.67%	0.6640	0.9542
$H(2)$ 3–9 Hz	97.06%	84.62%	91.67%	0.9559	0.9964
$H(2)$ MAW	91.18%	100.00%	95.00%	0.9796	0.9967

Table 3 Classification results of $H(2)$ computed over the Main Atrial Wave analyzing the last seconds of each recording.

	Se	Sp	Acc	AROC	Threshold
60 s	91.18%	100.00%	95.00%	0.9796	0.9967
45 s	91.18%	100.00%	95.00%	0.9796	0.9967
30 s	91.18%	100.00%	95.00%	0.9796	0.9967
15 s	91.18%	100.00%	95.00%	0.9830	0.9967

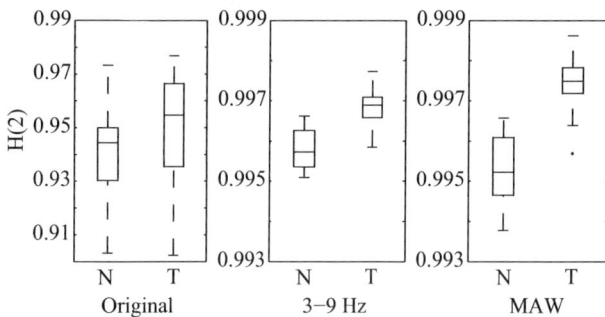

Fig. 1 Values of $H(2)$ computed over the original and filtered signals in the non-terminating (N) and immediately-terminating (T) groups.

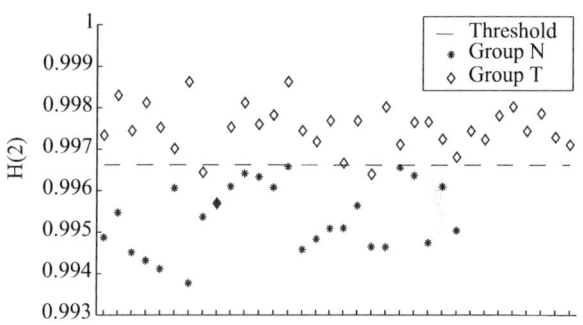

Fig. 2 Values of $H(2)$ computed on the Main Atrial Wave in the non-terminating (N) and immediately-terminating (T) groups.

IV. RESULTS

Since the best classification results were attained using the MAW filtering strategy, Table 1 shows the classification results obtained for different segment lengths computed on the MAW. The use of 20 and 15 seconds segments produced the best classification results. Moreover, the highest AROC value corresponded to the 15 seconds segments, and therefore this computational length was selected as optimal.

Table 2 shows the classification results obtained computing $H(2)$ on the original and filtered signals using the optimal 15 seconds segment length. In addition, the values of $H(2)$ for the original and filtered signals are displayed in Figure 1. The results show that $H(2)$ computed over the filtered signals yielded better prediction accuracy than its direct application to the AA. Moreover, the best prediction accuracy was obtained computing $H(2)$ on the MAW (see Table 2).

Finally, the classification results obtained from the analysis of the last 60, 45, 30 and 15 seconds of each recording are presented in Table 3. An accuracy value of 95% was attained in all cases and the highest value of AROC corresponded to the analysis of the last 15 seconds of each recording. Therefore, this segment of the MAW was selected as optimal for the application of $H(2)$. Figure 2 shows the values of $H(2)$ computed over the last 15 seconds of the MAW in groups N and T and the optimum classification threshold, note that the terminating AF episodes produced higher values of $H(2)$ than the non-terminating ones.

V. DISCUSSION

The use of the Generalized Hurst Exponents to estimate AF organization was proposed in [9]. However, no guidelines for the application of $H(2)$ to AF signals were defined in that previous study. Thus, in the present work the optimal computational parameters for the prediction of AF termination using $H(2)$ have been determined.

Regarding the optimal computational data length, the best results were obtained calculating $H(2)$ on segments of 15 seconds in length. Additionally, a high accuracy was attained analyzing only the last 15 seconds of each recording. Therefore, $H(2)$ can be applied with good accuracy in the study of AF from ambulatory ECG recordings as well as in Holter ECG recordings.

On the other hand, the results presented in this study show that the use of a previous band-pass filtering step improves significantly the performance of $H(2)$ as a predictor of AF spontaneous termination. Moreover, a bandwith of 3Hz centered on the DAF is sufficient to obtain a high accuracy in the prediction of AF termination using $H(2)$, which is consistent with the results presented in previous works [9, 18]. Furthermore, $H(2)$ computed over the MAW achieved better classification results than its application over the signals filtered in the typical AF frequency band, 3–9 Hz [19], which would preserve more information. This may be because most of the essential information from the AA is contained in the DAF and the nearby frequencies.

Regarding the interpretation of the results, the terminating AF episodes showed higher values of $H(2)$ than the non-terminating ones, suggesting a more regular pattern. Since AF organization can be estimated measuring the signal regularity [3], higher values of $H(2)$ correspond to a higher degree of organization.

The main limitation of the use of $H(2)$ on the MAW is that the extraction of the MAW depends on a correct detection of the DAF. Therefore, the performance of the method would be reduced if the amplitude of the atrial activity on the ECG, compared with the ventricular activity and possible interference signals, is too low to allow the detection of the DAF.

VI. CONCLUSIONS

The optimal parameters for the study of atrial arrhythmias using $H(2)$ have been determined in this work. The use of 15 seconds segments allows to obtain the best accuracy in the prediction of AF spontaneous termination. Moreover, the use of a previous band-pass filtering stage improves significantly the performance of the method and the best prediction results were obtained computing $H(2)$ on the MAW. Additionally, the results presented in this work show that $H(2)$ could be applied in the analysis of short ambulatory ECG recordings.

ACKNOWLEDGEMENTS

Work funded by TEC2010–20633 from the Spanish Ministry of Science and Innovation and PPII11–0194–8121 from Junta de Comunidades de Castilla-La Mancha.

REFERENCES

1. Kannel W. B., Wolf P. A., Benjamin E. J., Levy D.. Prevalence, incidence, prognosis, and predisposing conditions for atrial fibrillation: population-based estimates. *Am J Cardiol.* 1998;82:2N–9N.
2. Fuster Valentin, Rydén Lars E, Cannom Davis S, al . 2011 ACCF/AHA/HRS focused updates incorporated into the ACC/AHA/ESC 2006 guidelines for the management of patients with atrial fibrillation: a report of the American College of Cardiology Foundation/American Heart Association Task Force on practice guidelines *Circulation.* 2011;123:e269-367.
3. Alcaraz R, Rieta J J. A review on sample entropy applications for the non-invasive analysis of atrial fibrillation electrocardiograms *Biomedical Signal Processing and Control.* 2010;5 (1):1-14.
4. Sih H J, Zipes D. P., Berbari E. J., Olgin J. E.. A high-temporal resolution algorithm for quantifying organization during atrial fibrillation *IEEE Transactions on Biomedical Engineering.* 1999;46 (4):440-450.
5. Moody G B. Spontaneous Termination of Atrial Fibrillation: A Challenge from PhysioNet and Computers in Cardiology 2004 *Computers in Cardiology.* 2004;31:101-104.
6. Hurst H E. Long-term storage capacity of reservoirs *Trans. Amer. Soc. Civil. Eng..* 1951;116:770-808.
7. Sherman L D, Callaway C W, Menegazzi J J. Ventricular fibrillation exhibits dynamical properties and self-similarity *Resuscitation.* 2000;47:163-173.
8. Barabási A L, Vicsek T. Multifractality of self-affine fractals *Physical Review A.* 1991;44(4):2730-2733.
9. Julián M, Alcaraz R, Rieta J J. Comparative Study of Nonlinear Metrics to Discriminate Atrial Fibrillation Events from the Surface ECG *Computing in Cardiology.* 2012;39:197-200.
10. Goldberger A. L., Amaral L. A., Glass L., et al. PhysioBank, PhysioToolkit, and PhysioNet: components of a new research resource for complex physiologic signals. *Circulation.* 2000;101:E215–E220.
11. Petrutiu Simona, Ng Jason, Nijm Grace M., Al-Angari Haitham, Swiryn Steven, Sahakian Alan V.. Atrial fibrillation and waveform characterization. A time domain perspective in the surface ECG. *IEEE Eng Med Biol Mag.* 2006;25:24–30.
12. Bollmann Andreas, Husser Daniela, Mainardi Luca, et al. Analysis of surface electrocardiograms in atrial fibrillation: techniques, research, and clinical applications. *Europace.* 2006;8:911–926.
13. Martens S M M, Mischi M., Oei S. G., Bergmans J. W. M.. An improved adaptive power line interference canceller for electrocardiography *IEEE Transactions on Biomedical Engineering.* 2006;53 (11):2220-2231.
14. Alcaraz R, Rieta J J. Adaptive singular value cancelation of ventricular activity in single-lead atrial fibrillation electrocardiograms *Physiological Measurement.* 2008;29:1351-1369.
15. Stanley H E, Amaral L A N, Goldberger A L, Havlin S, Ivanov P Ch, Peng C K. Statistical physics and physiology: monofractal and multifractal approaches *Physica A.* 1999;270:309-324.
16. Bers Donald M, Grandi Eleonora. Human atrial fibrillation: insights from computational electrophysiological models *Trends Cardiovasc Med.* 2011;21:145-50.
17. DiMatteo T.. Multi-scaling in finance *Quantitative finance.* 2007;7 (1):21-36.
18. Alcaraz R, Rieta J J. Sample entropy of the main atrial wave predicts spontaneous termination of paroxysmal atrial fibrillation *Medical Engineering & Physics.* 2009;31:917-922.
19. Bollmann Andreas, Husser Daniela, Stridh Martin, et al. Frequency Measures Obtained from the Surface Electrocardiogram in Atrial Fibrillation Research and Clinical Decision-Making *Journal of Cardiovascular Electrophysiology.* 2003;14 (s10):S154-S161.

Corresponding Author: Matilde Julián Seguí
Institute: Biomedical Synergy
Address: Universidad Politécnica de Valencia,
        Camino de Vera s/n. Building 7F, 5th Floor
City: Valencia
Country: Spain
Email: majuse@upv.es

New Sleep Transition Indexes for Describing Altered Sleep in SAHS

O. Urra[1,2] and R. Jané[1,2,3]

[1] Institut de Bioenginyeria de Catalunya (IBEC), Barcelona, Spain
[2] Universitat Politècnica de Catalunya. BarcelonaTech (UPC), Barcelona, Spain
[3] CIBER de Bioingeniería, Biomateriales y Nanomedicina (CIBER-BBN), Spain

Abstract—Traditional Sleep Structure Indexes (TSSIs) are insufficient to identify patterns of altered sleep. TSSIs mainly account for absolute time measures, but different levels of state instability may lead to similar absolute time distribution. Therefore, sleep stability remains beyond the scope of TSSIs. However, recent studies suggest that sleep disorders may be rather influenced by a breakdown in the sleep-stage switching mechanisms. In this study, we propose a set of 11 Sleep Transition Indexes (STIs) that characterize sleep fragmentation and account for the state-stability governed by the ultradian, homeostatic and circadian rhythms. We demonstrate that most of the proposed STIs are potential markers of SAHS severity, while TSSIs are not. In addition, we provide a new framework to analyze sleep disorders from the direct perspective of sleep regulatory mechanisms. In particular, our results indicate that SAHS may be influenced by a dysregulation of homeostatic rhythms but not of ultradian or circadian rhythms.

Keywords—SAHS, Sleep Transitions, Sleep Structure, Polysomnography, Hypnogram.

I. INTRODUCTION

Sleep fragmentation is a general indicator of sleep disorders, and in particular of SAHS [1], but traditional approaches used to quantify sleep disruption have shown to be insufficient to identify altered sleep patterns [2]. Recent findings suggest that the proportion of time spent in a given stage may be less informative than its time distribution across the night [3], [4]. In the case of narcolepsy, patients experience higher state instability, although they have essentially normal amounts of wake and sleep [4].

Transitions between wakefulness/sleep and REM/NREM behave as a flip-flop switch [5],[6]. Switching mechanisms integrate over time slowly accumulating influences including the circadian rhythm determining the sleep onset; the homeostatic balance between wakefulness/sleep; the ultradian interaction between NREM/REM sleep and external influences like noise and convert them into sharp transitions in behavioral state [7]. Intensive research in the field over years evidences that the breakdown of these switching mechanisms may contribute to sleep disorders.

Therefore, we hypothesized that in order to investigate the role of switching mechanisms in the pathogenesis of SAHS, new sleep descriptors that include information about stage transitions would be needed. So, in this study we 1) propose a set of Sleep Transition Indexes (STIs) that account for the effects of the ultradian, circadian and homeostatic rhythms and characterize fragmentation by quantifying shifts to individual stages 2) identify which STIs may detect SAHS severity 3) compare the power of STIs to traditional sleep structure indexes (TSSIs) as potential markers of SAHS severity.

II. METHODS

A. Database

The database used in this study consists of a total of 118 recordings of polysomnography (PSG) (30 females, 88 males) acquired in the Sleep Unity of the Germans Trias i Pujol University Hospital (Table I). Exclusion criteria were low EEG quality; age <18 years or ambiguous diagnosis.

Table 1 Demographic Data (Mean ± Standard Deviation)

	Control	Mild	Moderate	Severe
Age	36.9±15.8	50.3 ± 11.1	52.7 ± 11.1	52.3 ± 9.8
BMI	24.2 ± 3.0	28.4 ± 4.0	28.6 ± 3.2	30.9 ± 4.5
AHI	2.4 ± 1.1	14.3 ± 5.3	31.1 ± 3.2	66.3 ± 19.6
AI	0.5 ± 0.5	4.3 ± 9.9	11.3 ± 15.5	43.3 ± 19.4
ArI	17.8 ±7.7	22.1± 5.3	28.7 ± 15.5	42.5 ± 21.4
N	12	37	24	45

BMI = Body Mass Index in kg/m^2, AHI = Apnea-Hypoapnea Index in h^{-1}, AI = Apnea Index in h^{-1}, ArI = Arousal Index in h^{-1}, N = Number of Subjects

Sleep stages were scored according to the updated recommendations of the American Academy of Sleep Medicine (AASM) [8]: Wake (W), REM (R), N1, N2, N3. But we also kept track of stages S3 and S4 separately in order to study sleep structure in more detail.

B. Diagnosis of SAHS Severity

Patients were evaluated for possible SAHS by an overnight sleep study (PSG). According to established guidelines a preliminary classification of patients was done in basis [8] of AHI: Control, AHI < 5; mild, 5 ≤ AHI < 15; moderate, 15 ≤ AHI < 30; and severe, 30 ≤ AHI. Such diagnosis was adjusted considering the examination carried out by a physician experienced in sleep medicine and the remaining PSG indicators of SAHS such as snoring or sleep structure.

C. Sleep Transition Indexes (STIs)

The set of Stage Transition Indexes (STIs) was designed with the objective of offering a complete set of markers covering the impact of the different sleep regulatory processes underlying those transitions. STIs were directly computed from the hypnogram of the patient. Prior to any computation, the standard 30 second-epochs were clustered into complete sleep bouts (consecutive periods of time spent in a given sleep stage). Then, two types of STIs were defined:

1. Bout Number STIs (bN_X), which counts the number of times a subject enters the stage X. E.g. bN_R states for the times a subject enters REM. Apart from classical stages, we also considered Slow Wave Activity (SWA:N2+N3) [9]. The SWA is specially interesting as it results from the interaction between ultradian and homeostatic rhythms and it is thought to be a key element in the integration of sleep regulation [10], [11]. In the case of the stage S4, we reported the ratio between bN_S4 and the total number of stage bouts (bN_N).

2. Stage Shift STIs (ss_XY), **which count** the number of times a subject shifted between the stages:

- ss_WS: W and any other sleep stage (S).
- ss_RNR: REM (R) and NREM (NR) stages.
- ss_WSWA: W and SWA
- ss_WR: W and R stage.
- ss_N: total number of stage shifts between W, R, S1, S2, S3 and S4.

In all cases, the shifts from stages not considered by a particular STI were ignored. For example, for computing ss_RNR, W bouts were ignored as ss_RNR only considers REM and NREM stages. According to the sleep physiology, ss-type STIs are in direct relationship with the main processes controlling the flip-flop switch in sleep regulation: the circadian process is represented by ss_WS, the ultradian rhythm is represented by ss_RNR and the conjunction of the ultradian and homeostatic processes is represented by ss_WSWA. Note that bN_W is not reported, as it is equivalent to the marker ss_WS.

D. Traditional Sleep Structure Indexes (TSSIs)

The STIs were compared to the traditional sleep structure indexes (TSSIs) provided by the PSG. The TSSIs are widely used in the clinical practice to quantitatively summarize the sleep structure shown in the hypnograms. These measures approach the sleep structure from different perspectives mainly based on absolute timing criteria:

- Total Time of Sleep (TTS) in minutes;
- Sleep Efficiency (E): percent of TTS that the patient stays asleep;
- REM Latency (L): time from the sleep onset needed to reach REM stage for the first time;
- Total amount of minutes (X (min)) and percent of TTS (X (%)) spent in the stage X: W, R, N1, N2, N3. E.g. W (min) would stand for the amount of minutes spent by a patient in W stage.

E. Statistical Analysis

One-way ANOVA tests were carried out on individual STIs and TSSIs against the four severity subpopulations (Control, Mild, Moderate, Severe) to check their association to SAHS severity. Levene's test was used to verify variance homogeneity. Hypothesis tests were carried out at a 5% significance level. All data are represented as means ± 95% Confidence Interval. When ANOVA results were significant ($p<0.05$), the Tukey's *post-hoc* test was implemented in order to determine which populations differ from each other. Tukey's *post-hoc* Test identifies homogeneous subsets of means that do not differ from each other.

III. RESULTS

A. Demographic Data of the Population

Table I provides summary statistics including age, body mass index (BMI), Apnea Hypopnea Index (AHI), Apnea Index (AI), Arousal Index (ArI) and group size (N) of the population of study stratified by SAHS severity subpopulations. Each subpopulation was matched for age and BMI, so little differences are found in these measures among them. Subjects were middle-aged adults (around 50), which is the age-range in which prevalence of SAHS is significantly higher [12]. As expected, SAHS related respiratory indexes (AHI, AI and ArI) increased with severity ($p < 0.001$).

B. STIs and SAHS Severity

Among the 11 proposed STIs, 6 were found to be potential predictors of SAHS severity (Fig 1A). In particular, Tukey's *Post-Hoc* tests determined that the STIs associated to severity showed significant mean differences ($p<0.05$) between severe and mild subpopulations, and severe and control subpopulations. In general, increasing number of sleep bouts undergone by a subject during sleep correlated with increasing severity, regardless the stage for which sleep bouts were computed. Exceptions were found in the case of S3, S4 and REM stages. The number of sleep bouts spent in N1 stage (bN_N1) and in N2 (bN_N2) stage was significantly higher in severe patients compared to moderate, mild and control groups ($p<0.001$ and $p<0.01$ respectively). Similarly, the number of times that a patient entered

Slow Wave Activity (SWA) periods was significantly higher as severity increased. In contrast, the proportion of sleep bouts spent in S4 in respect to the total amount of bouts undergone by patients during sleep (bN_S4/bN_N), was significantly greater in the control group compared to the SAHS affected groups (p<0.01). bN_S4 alone didn't appear to be correlated to SAHS Severity. The number of sleep bouts in S3 and REM (bN_S3, bN_R) did not differ significantly between severity subpopulations.

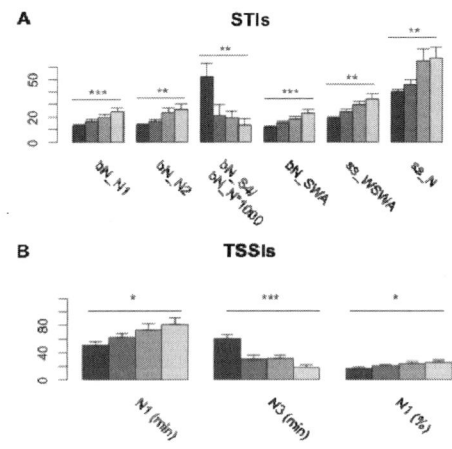

Fig. 1 ANOVA results and mean values of markers significantly influenced by the severity of SAHS (p<0.05). A.- Sleep Transition Indexes (TSIs) B.- Traditional Sleep Structure Indexes (TSSIs). The *p*-value of ANOVA test is indicated for each surface attribute as * p<0.05, ** p<0.01, ***p<0.001. Error bars indicate the 95% confidence interval.

Regarding ss-type STIs, we also found a general tendency for increased amount of shifts (ss_N) if severity was greater (p<0.01). In particular, the number of shifts between wake and SWA states (ss_WSWA) was significantly lower in control groups that in SAHS affected groups (*p<0.01*). However, our results suggest that there is not any significant difference linked to severity in the number of stage shifts between Wake-Sleep states (ss_WS), REM-NREM stages (ss_RNR) and Wake-REM states (ss_WR).

C. TSSIs and SAHS Severity

ANOVA analyses indicated that only 3 TSSIs may be potential independent predictors of SAHS severity: N1 (min), N3 (min) and S1 (%) (*p<0.05*) (Figure 1B). Results indicated that both, the absolute and relative time distribution in stages W, R and N2 was similar among the 4 severity groups (p>0.05). Likewise, no significant differences between groups were found in TTS, Sleep Efficiency and REM latency. The total percent of TTS spent in N3 was found not to be significantly different between groups either (p>0.05). In conclusion, our results suggest that the majority (10/13) of the TSSIs are not associated to SAHS severity.

IV. DISCUSSION

This study evidences that SAHS mainly alters the frequency of sleep stage transitions, while the time distribution across different sleep stages remains almost invariable from controls to patients with different SAHS levels. We demonstrate that the vast majority of markers used in the clinical practice are not able to identify altered patterns of sleep. Therefore we propose 11 alternative markers accounting for sleep stage transitions (STIs) out of which 6 are shown to be potential markers of altered sleep patterns and discriminate among different SAHS severity levels.

Sleep fragmentation is a common characteristic of sleep disorders. However, TSSIs are essentially global time measures that merge alternate stage pieces in a unique index. So patients with highly fragmented sleep may show similar TSSI values than patients with normal sleep if the total time distribution is the same. In contrast, STIs are indirect measures of sleep fragmentation. In particular, ss_N accounts for the total number of shifts in a night. We observed that in general SAHS severity is associated with increased values of shifts, probably due to higher sleep fragmentation. This may explain why most of the STIs are associated to SAHS severity, while most TSSIs are not.

Whether staying in REM or progressing to NREM sleep has a protective of harmful role is still under debate [13]–[15], but in any case differences between severity groups in the distribution of REM and NREM sleep would be expected. Surprisingly, when we compared the frequency entering REM and NREM sleep, the time spent in each state and the shifts between these stages (ss_RNR), we didn't find any significant difference between severity groups. Therefore, our study suggests the fact of staying longer in REM or NREM sleep does not improve or aggravate the pathological condition of SAHS. In fact, a recent study analyzing the temporal relationship between arousal events and upper-airway (UA) opening determined that the responses given to the UA obstruction was similar in REM and NREM stages [16]. Finally, SAHS does not appear to have effects on the ultradian rhythm, since the number of shifts between REM and NREM states remains unaltered.

In contrast, higher frequency of entrances into SWA (ss_WSWA, bN_SWA, bN_N1, bN_N2) was significantly associated to SAHS severity. This is in accordance with the studies reporting that 1) The time needed to recruit enough dilator muscle fibers to counteract UA collapse is similar to the time needed to reach SWA from wakefulness [14]; 2) Reaching deep sleep inhibits the activation of arousals, which are thought to provoke an overshoot in the flow and recurrent ventilator obstruction [16]. Finally it's remarkable that SWA results from the interaction between ultradian and homeostatic processes [10] so, the alteration of these STIs indicates that SAHS may impact these regulatory processes.

The disruption of sleep regulatory systems seems to be involved in sleep disorders [4], [6], [17]. Insomnia and narcolepsy are thought to be caused by the failure of homeostatic and circadian or ultradian rhythms respectively [17], In the case of SAHS, the lack of association of ss_WS and ss_RNR to its severity suggests that the circadian rhythm and ultradian rhythm may not be implicated in the pathophysiology. Therefore, it is likely that the dysregulation of homeostatic processes is a key element behind SAHS.

Clinicians claim that standard markers do not correlate with classic symptoms of SAHS [17], [18]. Consequently, they are obliged to qualitatively consider reported symptoms before making decisions about diagnosis or treatment. Daytime sleepiness and fatigue are common symptoms of SAHS, attributed to the difficulty to reach deep sleep. This study shows that shorter overall stay in deep sleep (N3 (min)) and lower frequency of entrances/exits into/from S4 stage (bn_S4/bN_N) are associated to higher severity. Therefore, the STIs also constitute a quantitative measure of common symptoms that may facilitate clinical decisions.

V. CONCLUSIONS

The proposed set of sleep transition indexes (STIs) are proven to be potential indicators of SAHS severity and overcome the current limitations of traditional sleep structure indexes (TSSIs) for detecting patterns of disturbed sleep. The set of STIs provides a new and necessary framework to analyze sleep disorders from the perspective of the function of the sleep regulatory systems (homeostatic, circadian and ultradian rhythms) since time distribution across diferent sleep stages does not differ from controls. In this context, our results suggest that SAHS is only affected by homeostatic alterations.

ACKNOWLEDGMENT

This work was supported by a grant of the CIBER-BBN and TEC2010-21703-C03-01 of the MINECO, Spain. The authors would like to thank Pulmonology Service of the Hospital Germans Trias i Pujol (Dr. JA Fiz and Dr. J Abad) for providing access to the PSG signals, measurements and hypnograms, acquired in their Sleep Disorders Laboratory.

REFERENCES

[1] Bianchi M, (2010) Obstructive Sleep Apnea Alters Sleep Stage Transition Dynamics. PloS one, 5
[2] Moser D, Anderer P, Gruber G et al. (2009) Sleep classification according to AASM and Rechtschaffen & Kales: effects on sleep scoring parameters. Sleep 32:139–49,
[3] Mariani S, Bianchi AM, Manfredini E et al. (2010) Automatic Detection of A phases of the Cyclic Alternating Pattern during Sleep. Conf Proc IEEE Eng Med Biol Soc, pp. 5085-8
[4] Saper CB, Fuller PM, Pedersen NP et al. (2010) Sleep state switching. Neuron, 68: 1023–42
[5] Zecharia AY, Yu X, Götz T et al. (2012) GABAergic inhibition of histaminergic neurons regulates active waking but not the sleep-wake switch or propofol-induced loss of consciousness. J Neurosci, 32:13062–75
[6] Lu J, Sherman D, Devor M et al. (2006) A putative flip-flop switch for control of REM sleep. Nature, 441:589–94
[7] Achermann P, Borbély A. (1992) Combining different models of sleep regulation. J Sleep Res, 1:144–147
[8] American Academy of Sleep Medicine (ASSM). (2007) The AASM manual for the scoring of sleep and associated events: rules, terminology, and technical specification, 1st ed. Westchester, IL.
[9] Achermann P, Borbély A. (2003) Mathematical models of sleep regulation. Front Biosci, 8:683–93
[10] Terzano MG, Parrino L. (2005) CAP and arousals are involved in the homeostatic and ultradian sleep processes. J Sleep Res, 14:360-8
[11] Vyazovskiy VV, Tobler I. (2012) The Temporal Structure of Behaviour and Sleep Homeostasis. PloS one, 7:1–11
[12] Young T, Palta M, Dempsey J et al. (1993) The occurrence of sleep-disordered breathing among middle-aged adults. N Engl J Med 328:1230–5
[13] Wheatley JR, Tangel DJ, Mezzanotte WS et al. (1993) Influence of sleep on response to negative airway pressure of tensor palatini muscle and retropalatal airway. J Appl Physiol 75:2117-24
[14] White DP. (2005) Pathogenesis of obstructive and central sleep apnea. Am J Resp Crit Care Med, 172:1363–70
[15] Younes M, Loewen AHS, Ostrowski M et al. (2012) Genioglossus activity available via non-arousal mechanisms vs. that required for opening the airway in obstructive apnea patients. J Appl Phys 112:249–58
[16] Younes M. (2004) Role of arousals in the pathogenesis of obstructive sleep apnea. Am J Resp Crit Care Med, 169: 623–33
[17] Chouvarda I, Mendez MO, Rosso V et al. (2012) Cyclic alternating patterns in normal sleep and insomnia: structure and content differences. IEEE Trans Neural Syst Rehabil Eng, 20:642–52
[18] Kulkas A, Tiihonen P, Julkunen P et al. (2013) Novel parameters indicate significant differences in severity of obstructive sleep apnea with patients having similar apnea-hypopnea index. Med Biol Eng Comput

Author: Oiane Urra
Institute: Institut d'Bioenginyeria de Catalunya (IBEC)
Street: Baldiri Reixac, 4, Torre I
City: Barcelona
Country: Spain
Email: ourra@ibecbarcelona.eu

Blood Pressure Variability Analysis in Supine and Sitting Position of Healthy Subjects

B.F. Giraldo[1,2,3], A. Calvo[2], B. Martínez[2], A. Arcentales[1,2], R. Jané[1,2,3], and S. Benito[4]

[1] Department ESAII, Universitat Politècnica de Catalunya (UPC), Barcelona, Spain
[2] Institut de Bioenginyeria de Catalunya (IBEC), Barcelona, Spain
[3] CIBER de Bioingenierìa, Biomateriales y Nanomedicina (CIBER-BBN), Spain
[4] Dept. of Emergency Medicine, Hospital de la Santa Creu i Sant Pau, Dept. of Medicine, Barcelona, Spain

Abstract—Blood pressure carries a great deal of information about people's physical attributes. We analyzed the blood pressure signal in healthy subjects considering two positions, supine and sitting. 44 healthy subjects were studied. Parameters extracted from the blood pressure signal, related to time and frequency domain were used to compare the effect of postural position between supine and sitting. In time domain analysis, the time systolic interval and the time of blood pressure interval were higher in supine than in sitting position ($p = 0.001$ in both case). Parameters related to frequency peak, interquartile range, in frequency domain presented statistically significant difference ($p < 0.0005$ in both case). The blood pressure variability parameters presented smaller values in supine than in sitting position ($p < 0.0005$). In general, the position change of supine to sitting produces an increment in the pressure gradient inside heart, reflected in the blood pressure variability.

Keywords—Blood pressure variability, systolic time intervals, diastolic time intervals.

I. INTRODUCTION

Blood pressure determination continues to be one of the most important measurements in all of clinical medicine and is still one of the most inaccurately performed [1]. It is useful to indicate cardiovascular diseases, such as hypertension and heart attack. Hypertension is a major risk factor for coronary heart disease, stroke, and renal failure, and affects approximately one-third of the American population.

Some factors, such as stature, age, density of blood, posture, resistance of vessel, drugs, diet, disease, etc., could affect the blood pressure [2], [3]. When a person is asleep, blood pressure is lowest and it increases as the person is awake. It can also change with the mood. Blood pressure is so important because it is a valuable indication of body status. Due to the high stress of the modern society, increasing numbers of people have hypertension, especially, the elderly [4].

Pickering *et al.* introduced recommendations for blood pressure measurements in humans, as a statement for professionals from the Subcommittee of the American Heart Association Council on High Blood Pressure Research [1]. An accurate measurement of blood pressure is essential to classify individuals to ascertain blood pressure-related risk, and to guide management.

The gold standard for clinical blood pressure measurement has always been readings taken by a trained health care provider using a mercury sphygmomanometer and the Korotkoff sound technique. But there is increasing evidence that this procedure may lead to the misclassification of some individuals [1]. In elderly subjects, heart rate responses to postural change are attenuated, whereas their vascular responses are augmented [5]. Altered strategy in maintaining blood pressure homeostasis during the upright position may result from age-related cardiovascular autonomic dysfunction.

In the past, several studies have been related to systolic time intervals, obtained from simultaneous recordings as electrocardiograms, carotid pulse tracings, and phonocardiograms, to assess left ventricular function in various clinical settings [6], [7], [8]. Another studies related to ventricular activity and their consequences were exposed by [9], [10], [11]. A review of the physiological and technological background, necessary in understanding the dynamic parameters related to the pulse pressure variation has been presented in Cannesson *et al.* [12].

In this study, we propose the analysis of blood pressure signal in healthy subjects, and compare the change of supine to sitting position. In order to characterize these signals, parameters related to time and frequency domain, and parameters related to blood pressure variability are extracted.

II. MATERIALS AND METHODS

A. Datasets

The study population consisted of 44 healthy subjects (age 21-44 years, 26 male, 18 female). ECG and respiratory flow signal, and synchronized continuously blood pressure were recorded under standardized resting conditions (quiet environment, same place) using BIOPAC Systems Inc. MP150 for the two first signals, and Finometer-Finapres Medical

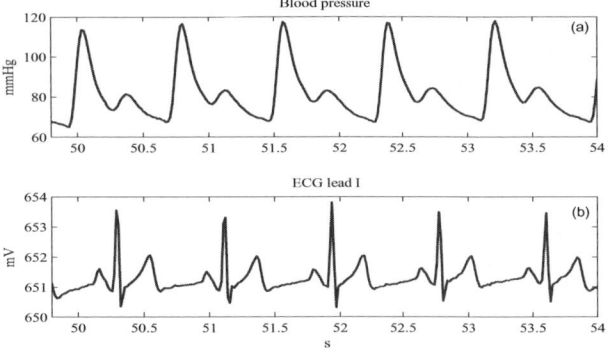

Fig. 1 Excerpts of (a) blood pressure signal and (b) ECG signal of a healthy subject.

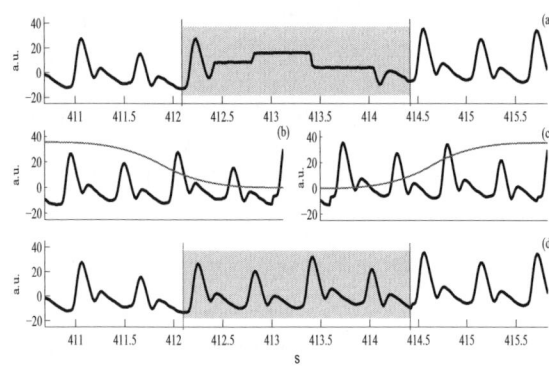

Fig. 2 (a) Original blood pressure signal, (b) and (c) part of signal before and after the calibration segment adjusted in time, and the cross-fade function, and (d) signal reconstructed without the calibration segment.

Systems for blood pressure signals. The subjects were recorded during 30 min in supine position, and after 5 min at rest, 15 min, in sitting position. All signals were recorded at a sampling frequency of 500 Hz. Fig. 1 shows an example of blood pressure and ECG signals of a healthy subject.

B. Signal Preprocessing

The blood pressure signal is recorded using finger cuff. Sometimes the correct detection of the pulse in this measurement is not straightforward. Furthermore, some episodes of calibration system are presents in the record. Periods of constant cuff pressure are used to adjust the correct unloaded diameter of the finger artery. But, the measurement of blood pressure is temporarily interrupted during such a period. When the study consider the dynamic of the signal along time, a possibility is to reconstruct the cycles omitted by the calibration. In our case, we propose the reconstruction of these cycles, considering the neighbor cycles on the left and the right. The original signals at left and right are adjusted in time to fit into the calibration segment, by cubic spline interpolation. The segment is then replaced by a crossfaded of the two extrapolated values, using the following window [13], [14]

$$w(n) = \begin{cases} 1 - \frac{1}{2}(2u(n))^\alpha, & u(n) \leq \frac{1}{2} \\ \frac{1}{2}(2 - 2(2u(n))^\alpha, & u(n) > \frac{1}{2} \end{cases} \quad (1)$$

where $u(n) = (n - n_s)/(n_e - n_s)$, and n_s and n_e the indices of the onset and end of the calibration, respectively. Crossfading is carried out by multiplying the forward extrapolated sequence by $w(n)$ and the backward extrapolated sequence by $1 - w(n)$. A linear down-slope is attained with $\alpha = 1$, whereas a step-like transition results when a $\alpha \to \infty$. The slope of the window is adjusted via the parameter $\alpha = 3$. Fig. 2 illustrates the process applied to this reconstruction.

From the data records the time series of heart rate and of blood pressure were extracted. Ectopic beats as well as other disturbances were removed. To quantify the blood pressure variability several parameters of time and frequency domain were calculated.

C. Parameter Extraction

Dicrotic notches and peaks represent the closure of the aortic valve and subsequent retrograde blood flow. Their locations are used to calculate systolic and diastolic time intervals. The main parameters are extracted through the maximal upward and downward slopes. Several thresholds are defined in order to determine the valid peaks detected. Ectopic values are removed. Table 1 and Fig. 3 show the parameters extracted in time domain.

Table 1 Time Parameter Description

Parameter	Description
P_{pulse} (mmHg)	Difference between systolic and diastolic pressure
tI_{Sys} (s)	Time interval between initial systolic time and the maximum peak of the systole
t_{Sys} (s)	Systolic interval time
t_{Dia} (s)	Diastolic interval time
t_{BP} (s)	Beat-to-beat blood pressure interval
F_c (bpm)	Cardiac frequency

The power spectral density of the signal and the series was estimated using Welch's method and a Hanning window. These were characterized on different spectral bands (see Table 2).

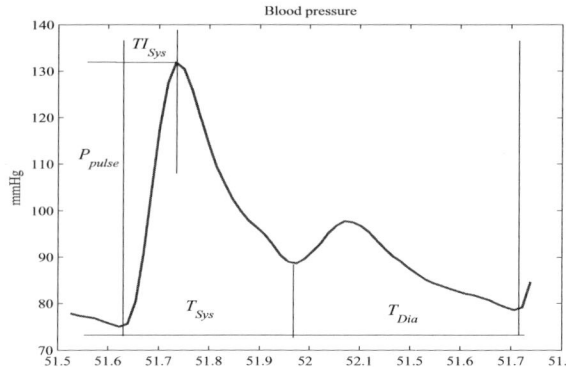

Fig. 3 Characterization of blood pressure signal

Table 2 Description of frequency parameters

Parameter	Description
f_c (Hz)	Central frequency
f_m (Hz)	Mean frequency
f_p (Hz)	Frequency peak
LF/HF	Ratio between low and high frequency bands
f_{Q1} (Hz)	Frequency of the first quartile
f_{Q3} (Hz)	Frequency of the third quartile
f_{max} (Hz)	Frequency over 95% of total power
A_{max} (mV^2/Hz)	Amplitude of f_p
P_w (mV^2/Hz)	Power band

Table 3 Mean and standard deviation of time series parameters extracted of blood pressure signals of healthy subjects in supine and sitting position

Parameter	Supine	Sitting	p-value
T_{Sys} (s)	0.358 ± 0.017	0.312 ± 0.032	0.001
T_{BP} (s)	0.865 ± 0.15	0.827 ± 0.14	0.001
F_c (bpm)	69.5 ± 12	74.2 ± 12	0.001

Mean (M) and statistical dispersion parameters as standard deviation (SD), interquartile range (IQR), kurtosis (K) and skewness (Sk) were calculated for different parameters.

III. RESULTS

In order to study the changes in blood pressure, the tables 3 and 4 show the statistically significant parameters when comparing the healthy subjects in supine and sitting position.

In time domain analysis, frequency cardiac is smaller in supine position than in sitting position. In consequence, the time systolic interval and the time of blood pressure interval are higher in supine than in sitting position. In frequency domain analysis, the most relevant parameters are related to the frequency peak, the interquartile range and the frequency of the third quartile.

Table 5 show parameters related to blood pressure variability that present statistically significance difference. Parameters related to the variation of the mean, kurtosis and skewness are higher in sitting than in supine position.

Table 4 Mean and standard deviation of frequency parameters extracted of blood pressure signals of healthy subjects in supine and sitting position

Parameter	Supine	Sitting	p-value
M_f_m (Hz)	2.25 ± 0.30	2.38 ± 0.27	0.016
M_f_p (Hz)	1.19 ± 0.21	1.23 ± 0.21	0.025
SD_f_p (Hz)	1.62 ± 0.17	1.81 ± 0.18	< 0.0005
M_f_{Q3} (Hz)	2.81 ± 0.31	3.11 ± 0.35	< 0.0005
$IQR\_f(Q3-Q1)$ (Hz)	1.64 ± 0.28	1.92 ± 0.36	< 0.0005
M_f_{max} (Hz)	5.62 ± 0.52	6.08 ± 0.56	< 0.0005

Table 5 Mean and standard deviation of relevant parameters related to blood pressure variability when comparing healthy subjects in supine and sitting position

Parameter	Supine	Sitting	p-value
$SD\_M\_P_{pulse}$	62 ± 11	64 ± 11	< 0.0005
$SD\_Sk\_P_{pulse}$	0.054 ± 0.005	0.074 ± 0.006	< 0.0005
$SD\_K\_P_{pulse}$	0.107 ± 0.009	0.148 ± 0.012	< 0.0005

IV. DISCUSSION

This paper investigates the changes in the blood pressure variability considering two positions of healthy subjects: supine and sitting. Postural changes are a common activity in daily life that represents an important challenge for the cardiovascular control system. Normal response to orthostatism includes decreased baroreflex sensitivity, which is attributed to vagal withdrawal and sympathetic activation towards the heart and the blood vessels [15, 16, 17]. Our results are concordant with those observations, and manifested by a mean decreased of both T_{Sys} and T_{BP} time series, and consequently, an increment in the cardiac frequency.

V. CONCLUSION

In this study, we analyzed the blood pressure signal in healthy subjects, comparing postural position of supine and sitting. The results are according to the expectation in the physiological changes. When a healthy volunteer changes the position of supine to sitting, an increment in the pressure gradient inside heart is generated. This effect is reflected in the decrement of systolic time, and an increment in the cardiac frequency. Furthermore, parameters related to frequency and blood pressure variability increase significantly in sitting position.

ACKNOWLEDGEMENTS

This work was supported in part by the Spanish Government's Ministerio de Economía y Competitividad under grant TEC2010-21703-C03-01. The authors would like to thank all the volunteers for their collaboration in the recording of signals of this database, and the personal of BIOSPIN–IBEC group by their support in the creation of such database.

REFERENCES

1. Pickering TG, Hall JE, Appel LJ., et al. Recommendations for Blood Pressure Measurement in Humans and Experimental Animals : Part 1: Blood Pressure Measurement in Humans: A Statement for Professionals From the Subcommittee of Professional and Public Education of the American Heart Association Council on High Blood Pressure Research Circulation. 2005;111:697–716.
2. Bojanov G. *Handbook of Cardiac Anatomy, Physiology, and Devices*. NJ: Humana Press 2005.
3. Klabunde RE. *Cardiovascular physiology concepts*. Lippincott Williams and Wilkins 2005.
4. WHO. *Global health risks Mortality and burden of disease attributable to selected major risks*. World Health Organization Press 2009.
5. Shannon RP, Maher KA, Santinga JT, Royal HD, Wei JY. Comparison of differences in the hemodynamic response to passive postural stress in healthy subjects greater than 70 years and less than 30 years of age *Am J Cardiol*. 1991;67:1110–1116.
6. Weissler AM, Harris WS, Schoenfeld CD. Systolic time intervals in heart failure in man Circulation. 1968;37:149–159.
7. Eddlemann EE, Swatzell RH, Vancroft WH, Baldone JC, Tucker MS. The use of the systolic time intervals for predicting left ventricular ejection fraction in ischemic heart disease *Am Heart J*. 1977;93:450–454.
8. Gillian RR, Parnes WP, Khan MA, Bouchard RJ, Warbasse JR. The prognostic value of systolic time intervals in angina pectoris patients Circulation. 1979;60:268–275.
9. Klotz S, Hay I, Dickstein ML, et al. Single-beat estimation of end-diastolic pressure-volume relationship: a novel method with potential for noninvasive application *Am J Physiol Heart Circ Physiol*. 2006;291:H403–H412.
10. Vooren H, Gademan MGJ, Swenne CA, TenVoorde BJ, Schalij MJ, Wall EE. Baroreflex sensitivity, blood pressure buffering, and resonance: what are the links? Computer simulation of healthy subjects and heart failure patients *J Appl Physiol*. 2007;102:1348–1356.
11. Reant P, Dijos M, Donal E, et al. Systolic time intervals as simple echocardiographic parameters of left ventricular systolic performance: correlation with ejection fraction and longitudinal two-dimensional strain European *Journal of Echocardiography*. 2010;11:834–844.
12. Cannesson M, Aboy M, Hofer CK, Rehman M. Pulse pressure variation: Where are we today? *Journal of Clinical Monitoring and Computing*. 2010.
13. Esquef PAA, Valimaki V, Roth K, Kauppinen I. Interpolation of long gaps in audio signals using the warped Burgs method in *in 6th Int. Conference on Digital Audio Effects*;2:DAFX 1–5 2003.
14. Garde A., Sornmo L., Jane R., Giraldo B.F.. Breathing Pattern Characterization in Chronic Heart Failure Patients Using the Respiratory Flow Signal *Annals of Biomedical Engineering*. 2010;28:3572–3580.
15. Kardor A., Rudas L., Simon J., Csanady M.. Effect of postural changes on arterial baroreflex sensitivity assessed by the spontaneous sequence method and Valsalva manoeuvre in healthy subjects. *Clin Auton Res*. 1997;7:143–148.
16. Fadel P.J.. Arterial baroreflex control of the peripheral vasculature in humans: rest and exercise *Med Sci Sports Exerc*. 2008;40:2055–2062.
17. Martnez-Garca P., Lerma C., Infante O.. Relation of the baroreflex mechanism with the photoplethysmographic volume in healthy humans during orthostatism *Arch Cardiol Mex*. 2012;82:82–90.

Author: Beatriz F. Giraldo
Institute: Universitat Politècnica de Catalunya
Street: C. Pau Gargallo, 5
City: Barcelona
Country: Spain
Email: Beatriz.Giraldo@upc.edu

Noninvasive Interdependence Estimation between Atrial and Ventricular Activities in Atrial Fibrillation

R. Alcaraz[1] and J.J. Rieta[2]

[1] Innovation in Bioengineering Research Group. University of Castilla-La Mancha, Cuenca, Spain
[2] Biomedical Synergy, Electronic Engineering Department, Universidad Politécnica de Valencia, Spain

Abstract—The present work introduces a novel methodology to quantitatively assess interdependence between real atrial and ventricular activation series during atrial fibrillation (AF), which is the most common arrhythmia in clinical practice. The method is based on a nonlinear index, such as cross-sample entropy (CSE), which estimates the conditional probability to find similar patterns within both activation series. The study has been carried out on patients with paroxysmal and persistent AF in order to be applied over atrial activation series with different properties in their organization. In agreement with previous findings, results showed a statistically significant positive correlation between atrial activity organization and the synchronization between atrial and ventricular activations ($R = 0.53$, $p < 0.01$). Moreover, higher CSE values were also observed for persistent (0.759 ± 0.053) than for paroxysmal AF episodes (0.662 ± 0.091). As a consequence, CSE could be used to reveal clinically useful information in the improvement of current rate control therapies, which are mainly focused on controlling ventricular rate without paying much attention to the atrial fibrillatory process.

Keywords— Atrial Fibrillation, Electrocardiogram, Cross-sample entropy, synchronization.

I. INTRODUCTION

Atrial fibrillation (AF) is the most common arrhythmia in clinical practice, accounting for approximately one-third of hospitalizations [1]. Its mechanisms are still not completely understood today. Nonetheless, theoretical models as well as mapping in humans and animals suggest that this arrhythmia is characterized by the presence of simultaneous multiple wavelets in the atria [2], their number depending on the atrial refractory period, mass, and conduction velocity [1].

This atrial behavior during AF provokes a massive and irregular bombarding of atrial activations into the atrioventricular (AV) node [3, 4]. Since the AV node is unable to conduct such a large number of rapid atrial activations, a remarkable part of them are blocked. This filtering property of the AV node is essential to keep the heart beating at a sustainable rate. However, the mechanisms of atrial impulse blocking remain unclear [5]. To this respect, great efforts have been made to characterize and model both the AV node role and the resulting ventricular response during AF. For this reason, a remarkable number of studies on the AV node electrophysiological properties together with mathematical models of this heart's structure can be found in the literature [4, 6]. However, conflicting results have been reported and, nowadays, still remains a significant debate regarding the AV node behavior as well as the ventricular response during AF [5, 6].

Moreover, recent studies suggest that the rate of atrial activations and their organization during AF could play a key role in the ventricular response [5, 7]. However, the relationship between atrial and ventricular activations has been analyzed to date by considering independently the characterization of both time series [5, 7]. Indeed, these series have been characterized only in a global way by two strategies. On the one hand, the use of statistical features from their probabilistic density function [4, 7]. On the other hand, by indirect estimators computed from ECG recordings of several seconds in length [4, 5]. In contrast to the aforementioned recent works, the present research introduces a method able to quantify interdependence between atrial and ventricular series by assessing the presence and repetitiveness of atrial patterns into the ventricular activation series using a nonlinear index, such as cross-sample entropy (CSE). Thus, this method takes advantage of the instantaneous information provided by each activation, within the atrial and ventricular series, to compute their interdependence.

II. MATERIALS

A total of 91 patients with 24-h Holter ECG recordings suffering paroxysmal or persistent AF were enrolled in the study. AF episodes, defined by irregular ventricular response and absence of P waves [8], were annotated by expert cardiologists. Forty-one patients (mean age 64.3 ± 14.1 years, 16 women) had paroxysmal AF, since episodes of normal sinus rhythm occurred during the 24-h recordings. The remaining 50 patients (mean age 69.8 ± 10.2 years, 23 women) were judged to be in persistent AF, since the arrhythmia was present throughout the 15-day period that immediately preceded the recording, as well as during the entire recording.

III. METHODS

A. Data Preprocessing

The 24-h ECG recordings were acquired with a sampling rate of 125 Hz and 12 bit resolution. For the analysis, a four minute-length segment was extracted from each Holter recording. This segment was randomly chosen within the whole 24-h recording for patients in persistent AF. In contrast, for paroxysmal AF patients, the segment was selected as the central 4 minutes interval of the longest duration AF episode in the Holter. From these segments, only lead V1 was used because the atrial signal is larger in this lead, with respect to the ventricular signal [9]. Furthermore, the atrial fibrillatory activity seems to be mostly reflected in lead V1 [10]. This signal was upsampled to 1 kHz, such as suggested by Bollmann et al. [8], and bidirectionally filtered to remove baseline wander, high frequency noise and powerline interference [11]. Finally, the atrial activity (AA) signal was extracted by an adaptive QRST cancellation method [12].

B. Detection of Atrial and Ventricular Activation Series

R-peaks in the preprocessed lead V1 were considered as the ventricular activations and were detected making use of the Pan and Tompkins' algorithm [13]. The R-peak detection stage was visually supervised by expert cardiologists in order to verify proper detection. In case of false positive or false negative detections, the points were discarded and the corresponding peak was manually identified. Finally, the RR interval series was computed as the difference between successive R-peaks, thus yielding the ventricular activation series.

On the other hand, a direct relationship between the right atrium activations and the f waves in surface lead V1 has been recently proved [10]. Hence, atrial activations were delineated from the surface f waves making use of a method based on mathematical morphology (MM) [14]. Briefly, MM operators were applied to the AA signal in order to enhance the f waves. Thus, a combination of erosion and dilation operations was applied to the AA with two structuring elements. The first one was adapted to the f waves by an even triangular shape. The second was designed as a rectangular shape to suppress the drift between atrial cycles. Finally, the resulting impulsive signal was used to detect the maximum and boundaries of the atrial activations by peak detection.

As for the R-R interval series, differences between successive supervised atrial activations were computed to characterize the AA. In this case, the peak timing of each f wave was used to obtain the atrial activation series, which will be next referred to as FF series.

C. Cross-Sample Entropy

Interdependence between FF and RR series was computed making use of CSE. However, normalization and mean removal of both series were performed prior to CSE computation. Since they have different origin, these operations were required because both series differ in their values and number of occurrences by time interval. Thus, given the FF and RR series of length N and M, respectively, i.e. $FF = [FF(1), FF(2), \ldots, FF(N)]$ and $RR = [RR(1), RR(2), \ldots, RR(M)]$, their normalized samples were obtained as

$$FF^*(i) = \frac{FF(i) - \overline{FF}}{\sigma_{FF}}, \quad (1)$$

and

$$RR^*(i) = \frac{RR(i) - \overline{RR}}{\sigma_{RR}}, \quad (2)$$

$\overline{FF}$ and $\overline{RR}$ being the mean of the atrial and ventricular activation series, respectively, and σ_{FF} and σ_{RR} their corresponding standard deviations.

Thereafter, vectors were composed directly from the normalized m consecutive FF and RR values, such that [15]

$$\mathbf{X}_i^m = \{FF^*(i), FF^*(i+1), \ldots, FF^*(i+m-1)\} \text{ and } \quad (3)$$

$$\mathbf{Y}_j^m = \{RR^*(j), RR^*(j+1), \ldots, RR^*(j+m-1)\}. \quad (4)$$

Hence, these vectors represent m consecutive FF and RR values starting at the ith and jth point, respectively. The distance d_{ij}^m between $\mathbf{X}_i^m$ and $\mathbf{Y}_j^m$ was defined as the maximum absolute difference in their respective scalar components, i.e.

$$d_{ij}^m = d[\mathbf{X}_i^m, \mathbf{Y}_j^m] = \max_{k \in (0, m-1)} |FF^*(i+k) - RR^*(j+k)|, \quad (5)$$

and the probability that RR-patterns were similar to a concrete FF-pattern of window length m, within a tolerance r, was defined as

$$\phi^m(r) = \frac{1}{N-m} \sum_{i=1}^{N-m} \left(\frac{1}{M-m} \sum_{j=1}^{M-m} \Theta(r - d_{ij}^m) \right), \quad (6)$$

where $\Theta(x)$ is the Heaviside function [15]. In a similar way, this probability was computed when patterns of $m+1$ point-length were considered and CSE was estimated as [15]

$$\text{CSE}(m, r, N, M) = -\ln \frac{\phi^{m+1}(r)}{\phi^m(r)}. \quad (7)$$

Hence, this index quantifies the frequency of RR-patterns similar to a given FF-pattern of window length m within a tolerance r, with larger values indicating fewer instances of pattern matches and, thus, greater asynchrony [15].

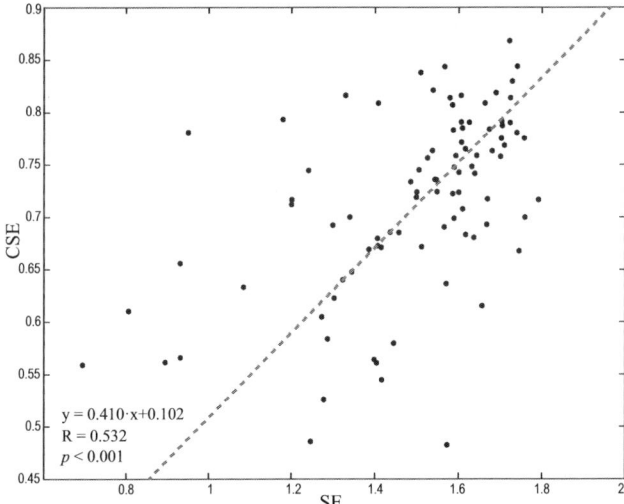

Fig. 1 Relationship between SE values computed from the FF series and CSE values computed from FF and RR series for the AF patients in the database.

IV. RESULTS

To study the interdependence between atrial and ventricular activations, Fig. 1 displays the CSE values, between FF and RR series, as a function of the FF series regularity for all the studied AF episodes. Atrial activation series regularity was estimated with a nonlinear index, such as sample entropy (SE). This metric has provided previously successful ability to estimate AA organization [16]. In a similar way to CSE, SE quantifies the predictability of fluctuations in the values of a time series and assigns a nonnegative number to the sequence, with larger values corresponding to more irregularity in the data [15]. Since its mathematical definition is very similar to the previously described CSE, SE is not detailed here and the reader is referred to the available literature [15, 16]. For both metrics, values of $m = 2$ and $r = 0.25$ were used, given that they are the most widely established parameters for this type of entropies in the literature [17].

Although a notable scatter can be seen in figure, an increasing trend for CSE can be elucidated when the FF series SE also increases. Indeed, the best least means linear fit provided a statistically significant relationship between both entropy estimates ($y = 0.41 \cdot x + 0.102$, $R = 0.53$, $p < 0.01$). Note that the idea of computing this line is to quantify numerically the increasing trend observed in the results and not to adapt them to a linear model. Moreover, persistent AF episodes presented higher average SE and CSE values (1.612 ± 0.093 and 0.759 ± 0.053, respectively) than paroxysmal ones (1.344 ± 0.260 and 0.662 ± 0.091, respectively).

V. DISCUSSION

Although individual analysis of atrial and ventricular activities has proven that they bear prognostic information on long-term AF time course, the interdependence of both activation series has not been adequately explored to date in real-life AF patients [5, 6]. To this respect, only global relationships between parameters, individually computed from atrial and ventricular activities or their probability density functions, have been investigated [4,5,7]. On the contrary, the present work proposes the first alternative able to assess interdependence between atrial and ventricular activities by taking each single-activation information from real ECG recordings in AF. The presented methodology makes use of a nonlinear index such as CSE, which has been originally defined to be applied over time series of the same length [15]. However, given that this index estimates interdependence between two time series by quantifying the probability of finding similar patterns within them, it has been adapted to time series of dissimilar length, such as atrial and ventricular activations, without altering its functionality. In fact, obtained results were in agreement with previous works.

To this respect, the noticed increasing direct relationship between the FF series regularity, estimated via SE, and the CSE values computed by comparing atrial and ventricular activation series suggests that the ventricular response during AF is notably conditioned by the temporal organization of the AA. In this line, by pacing from different sites of the atria in rabbit experiments, Chorro et al. [18] concluded that the ventricular response during AF was not only determined by the AV node properties but also by the rate and irregularity of the atrial fibrillatory process. A similar finding was reported by Kirsh et al. [19], who suggested that the irregular ventricular response during AF is primarily a consequence of the irregularity inherent in the AA, the AV node role being predominantly confined to that of scaling the AA. More recently, Climent et al. [5] have found a positive correlation between the DAF and the predominant RR interval. Given that the DAF has been associated with atrial refractoriness [20] and with the number of active wavelets wandering throughout the atrial tissue [2], it could be considered as an indirect estimator of AA organization. Cosequently, the correlation between DAF and RR intervals also suggests that the ventricular response in AF is influenced by the atrial inputs, with their concrete degree of organization, bombarding the AV node [5].

As expected in light of the aforementioned discussion, paroxysmal AF episodes provided a higher synchronization between atrial and ventricular activities than persistent AF, since CSE presented lower values in the first case and higher in the latter case. As a consequence, the algorithm could be helpful in the improvement of current rate control therapies. In fact, these therapies are mainly focused on controlling ventricular rate, by adjusting the AV node propagation properties, without considering the atrial fibrillatory process [3]. Therefore, the proposed method could be used to assess drug effects both on AA organization and AV node properties, thus gaining relevant information to improve therapy and management of AF patients.

VI. CONCLUSIONS

A pioneer alternative able to assess coupling between atrial and ventricular activities taking each single-activation information from real AF surface recordings has been proposed. According to previous studies, a positive correlation between surface atrial activity organization and synchronization between atrial and ventricular activation series has been quantified through CSE. In this way, the previously suggested atrial activity organization influence on the ventricular response during AF has been non-invasively assessed. Hence, clinically useful information could be yielded to improve current rate control therapies.

ACKNOWLEDGEMENTS

Work funded by TEC2010–20633 from the Spanish Ministry of Science and Innovation and PPII11–0194–8121 from Junta de Comunidades de Castilla-La Mancha.

REFERENCES

1. Fuster Valentin, Rydén Lars E, Cannom Davis S, et al. 2011 ACCF/AHA/HRS focused updates incorporated into the ACC/AHA/ESC 2006 guidelines for the management of patients with atrial fibrillation: a report of the American College of Cardiology Foundation/American Heart Association Task Force on practice guidelines *Circulation.* 2011;123:e269-367.
2. Konings K., Kirchhof C., Smeets J., Wellens H., Penn O., Allessie M. A.. High-density mapping of electrically induced atrial fibrillation in humans *Circulation.* 1994;89:1665–1680.
3. Zhang Youhua, Mazgalev Todor N. Ventricular rate control during atrial fibrillation and AV node modifications: past, present, and future *Pacing Clin Electrophysiol.* 2004;27:382-93.
4. Corino Valentina D A, Sandberg Frida, Mainardi Luca T, Sornmo Leif. An atrioventricular node model for analysis of the ventricular response during atrial fibrillation *IEEE Trans Biomed Eng.* 2011;58:3386-95.
5. Climent Andreu M, Guillem Maria S, Husser Daniela, Castells Francisco, Millet José, Bollmann Andreas. Role of the atrial rate as a factor modulating ventricular response during atrial fibrillation *Pacing Clin Electrophysiol.* 2010;33:1510-7.
6. Platonov Pyotr G, Holmqvist Fredrik. Atrial fibrillatory rate and irregularity of ventricular response as predictors of clinical outcome in patients with atrial fibrillation *J Electrocardiol.* 2011;44:673-7.
7. Climent Andreu M, Atienza Felipe, Millet Jose, Guillem Maria S. Generation of realistic atrial to atrial interval series during atrial fibrillation *Med Biol Eng Comput.* 2011;49:1261-8.
8. Bollmann Andreas, Husser Daniela, Mainardi Luca, et al. Analysis of surface electrocardiograms in atrial fibrillation: techniques, research, and clinical applications *Europace.* 2006;8:911-26.
9. Petrutiu Simona, Ng Jason, Nijm Grace M, Al-Angari Haitham, Swiryn Steven, Sahakian Alan V. Atrial fibrillation and waveform characterization. A time domain perspective in the surface ECG *IEEE Eng Med Biol Mag.* 2006;25:24-30.
10. Alcaraz Raúl, Hornero Fernando, Rieta José J. Assessment of non-invasive time and frequency atrial fibrillation organization markers with unipolar atrial electrograms *Physiol Meas.* 2011;32:99-114.
11. Sörnmo L., Laguna P.. *Bioelectrical Signal Processing in Cardiac and Neurological Applications.* Elsevier Academic Press 2005.
12. Alcaraz Raúl, Rieta José Joaquín. Adaptive singular value cancelation of ventricular activity in single-lead atrial fibrillation electrocardiograms *Physiol Meas.* 2008;29:1351-69.
13. Pan J., Tompkins W. J.. A real-time QRS detection algorithm. *IEEE Trans Biomed Eng.* 1985;32:230–236.
14. Alcaraz Raúl, Hornero Fernando, Martínez Arturo, Rieta José J. Short-time regularity assessment of fibrillatory waves from the surface ECG in atrial fibrillation *Physiol Meas.* 2012;33:969-984.
15. Richman J S, Moorman J R. Physiological time-series analysis using approximate entropy and sample entropy *Am J Physiol Heart Circ Physiol.* 2000;278:H2039-49.
16. Alcaraz Raul, Rieta Jose Joaquin. A review on sample entropy applications for the non-invasive analysis of atrial fibrillation electrocardiograms *Biomed Signal Process Control.* 2010;5:1–14.
17. Pincus S M. Assessing serial irregularity and its implications for health *Ann N Y Acad Sci.* 2001;954:245-67.
18. Chorro F J, Kirchhof C J, Brugada J, Allessie M A. Ventricular response during irregular atrial pacing and atrial fibrillation *Am J Physiol.* 1990;259:H1015-21.
19. Kirsh J A, Sahakian A V, Baerman J M, Swiryn S. Ventricular response to atrial fibrillation: role of atrioventricular conduction pathways *J Am Coll Cardiol.* 1988;12:1265-72.
20. Capucci A., Biffi M., Boriani G., et al. Dynamic electrophysiological behavior of human atria during paroxysmal atrial fibrillation *Circulation.* 1995;92:1193–202.

Corresponding Author: Raúl Alcaraz Martínez
Institute: Innovation in Bioengineering Research Group
Address: University of Castilla-La Mancha,
    Campus Universitario s/n, 16071
City: Cuenca
Country: Spain
Email: raul.alcaraz@uclm.es

Prototype Development of a Computerized System for Interpretation of Heart Sounds Using Wavelet

I.S.G. Brites[1] and N. Oki[2]

[1] State of Sao Paulo University/Department of Electrical Engineering, Graduate Student, Ilha Solteira, Brazil
[2] State of Sao Paulo University / Department of Electrical Engineering, Associate Professor, Ilha Solteira, Brazil

Abstract—Monitoring of the changes of heart sounds provide information about the heart condition. Compared to other methods of monitoring heart condition, heart sound monitoring is cheaper and safer. This paper aims to design a system for automatic analysis of heart sounds using the Discrete Wavelet Transform for extracted its features. Experiments with database compose of heart sound healthy and with pathologies show that the system proposed was able to discern the normal or abnormal sound with the error of 90%.

Keywords—Heart sounds, heart auscultation, classification of signals, discrete wavelet, database.

I. INTRODUCTION

Public policies used by the Brazilian government in health care could be characterized by its access as primary, medium and high complexity. These three levels are used as metrics in the evaluation of public health spending.

The increase of medium and high complexity medical procedures cause increased spending to the Brazilian public health care.

As example of medical procedures, the cardiovascular problems as myocardial infarction, cardiac arrest, arrhythmia, among others, could be detected with invasive and noninvasive exams.

1. Invasive:
   - Transesophageal ultrasound.
   - Scintigraphy.
   - Cardiac catheterization.
2. Non-invasive:
   - Electrocardiogram (ECG).
   - Chest X-ray.
   - Monitoring ECG Holter.
   - Phonocardiogram.
   - Stress test.
   - Tomography of the heart and vessels
   - Magnetic resonance imaging of the heart and vessels (MR)
   - Digital angiography.

All the invasive exams and some noninvasive have need of specialized personnel to carry out and normally have a high cost.

According official definition, the primary health care are based on practical methods, using scientifically proven technology and socially acceptable, made accessible to the general population and with low cost. The primary health care could be consider an investment, because the prevention of diseases discovered in early stage have a better chance of successful treatment and this can even reduce the total cost treatment to the state.. Therefore, the main scope of this type of health care is provides universal access to the community. This first level of contact to the population with health care, must to be close as possible to where people live and work, establishing a process for continued attention to health.

There are currently several researches in this area with the goal of developing automated systems that can help doctors diagnose with low cost. One possibility is explorer the heart sound (phonocardiogram) to detect anomaly and disease.

The objective of this paper is to present a design of a computer system for the classification of heart sound signals as normal or abnormal using Discrete Wavelet Transform. The structure of this paper presents: Section 2 - State of the Art, commenting on the studies in this area, Section 3 - Cardiac Cycle, where is explains how heart sounds are produced, in Section 4 is describing the design of the system proposed and some experimental results are presented and analyzed and finally, in Section 5 the conclusions are presented.

II. STATE OF THE ART

In this section are presented some references who have worked with the interpretation and analysis of sound signals of the human heart in order to direct the development of this research.

The subject of the references can be divided into two parts: one use the signal as the identification process and other use the signal as biomedical pre-diagnosis, but both use the Wavelet Transform decomposition as a method for processing digital cardiac signals.

Using two physiological signals, namely the Electrocardiogram (ECG) and the Phonocardiogram (PCG,) Fatemian, Agrafioti and Hatzinakos [1] presents a feasibility of use these signal to implement a biometric recognition. For signal processing was used the 5th order Daubechies wavelet

Ramos, Carvalho, Paiva, Vale and Henriques [2] have demonstrated the use of software MatLab for create a user-friendly interface to assist physicians in the procedure of cardiac auscultation.

A novel method to determine features of heart sound is present by Kurnaz and Ölmer [3] where Wavelet Transform are used to processing the phonocadiogram signals.

Karmakar [4] presents wavelet transform to extract the features of the phonocardiogram signal (PCG) to a biometric authentication system.

Taplidou and Hadjileontiadis [5] showed that the wavelet transform could reveal and analyze nonlinear characteristics of heart sounds. Through a database of sounds of hearts with cardiovascular pathologies structured the time-frequency domain to quantify the evolution of heart murmurs.

III. Cardiac Cycle

Cardiac events that occur in the beginning of each heartbeat to the beginning of the next are called the cardiac cycle. Each cardiac cycle begins with the spontaneous generation of an action potential in the sinus node. This node is located on the upper side of the right atrium near the opening of the superior vena cava, and the action potential propagates rapidly through the atria and then to the ventricles. Because of this special provision of the conduction system of the atria to the ventricles there is a delay of more than 1/10s, during the passage of the cardiac impulse from the atria to the ventricles. This allows the atria to contract before the ventricles up, pumping blood into the ventricles before the beginning of the strong ventricular contraction. Thus, the atria act as pumps brush to the ventricles, the ventricles and these, in turn, provide most of the force will propel blood through the vascular system. The cardiac cycle consists of a period of relaxation, called the diastole during which the heart is filled with blood, followed by a period of contraction called systole [6].

Listen to the sounds of the body, usually with the aid of the stethoscope is called auscultation. Figure 1 shows the areas of the thoracic wall in which different sounds valve can be better distinguished. Although all the valves sounds can be heard in all of these areas, the cardiologist

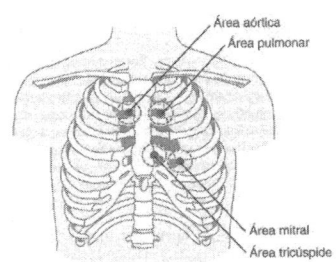

Fig. 1 Human thorax

distinguishes the sounds of the different valves by a process of elimination. That is, when the sounds in different areas and gradually selecting the audible component of each valve [6].

IV. Design of the System

The classification system of heart sounds has been constructed and implemented following the model presented in Figure 2. The details of each block are shown in sequence.

Capture the signal: First of all was used an electret microphone which was connected to computer via sound card for stereo connector P2. To amplify the signal was necessary to use classic stethoscope, however the goal of this prototype is use only the electret microphone with amplifier electronics. However, due the difficult to obtain good results and to facilitate the work we used databases of cardiac sounds available in [1], [7] and [8].

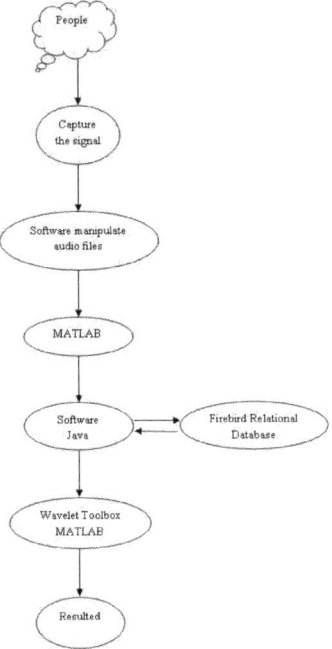

Fig. 2 Prototype System

Software for edit audio files: For manipulation of audio signal the software Audacity was chosen because is free, can amplify the signal, is easy to use and is able to create projects for exploration on file in *.WAV. The WAVE audio file (*. WAV) is one of the most popular media files, and is necessary for the next stage of the prototype. The Figure 3 shows the windows when the software was processing phonocardiographic signal.

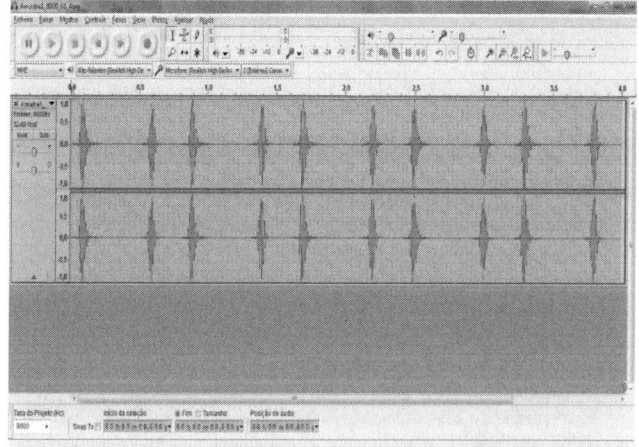

Fig. 3 AudaCity

Software for signal processing: For signal processing the software Matlab was choose. To use MatLab, in reading the audio vectors, it was necessary to convert the files to WAVE format (*. WAV). Necessary feature for the software MatLab, using their media functions, could extract the frequency / time of the audio file. These vectors can be analyzed in MatLab via graphs, spreadsheets and / or data exported to text files (TXT). The results were written as files *.TXT for posterior analysis.

Software Java: The software implemented in the Java language is used for storage the vectors related to audio files generated by Matlab in the database. In next stage of the system proposed is necessary to processing the audio vectors, in order to remove noise and find patterns with witch was possible to classify the signal as normal or abnormal. The Wavelet Transform was chose for decomposing the signal and allow extract some features.

Wavelet Toolbox MATLAB: For performing the decompositions of signals was used the Wavelet Toolbox – Matlab. The Figure 4 shows the decompositions of a audio file.

To understand the MatLab decompositions, following is given a synthetic mathematical representation of Wavelet Transform.

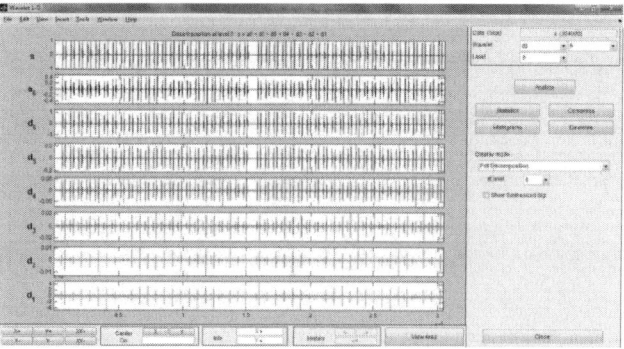

Fig. 4 MatLab

The equation below describes the Discrete Wavelet Transform of a signal x (t) is given by (1).

$$W_f(a,b) = |a|^{-1/2} \int f(t)\psi^*\left(\frac{t-b}{a}\right) dt \quad (1)$$

where f (t) is the function to be analyzed, ψ is the wavelet ψ ((tb) / a) is shifted and scaled version of the wavelet at time b and scale. An alternate form of the equation used in its place and b:

where *f(t)* is the function to be analyzed, ψ is the wavelet, $\psi((t-b)/a)$ is shifted and scaled version of the wavelet at time b and scale. An alternate form of the equation used in its place a and *b*:

$$W_f(s,u) = \int_{-\infty}^{\infty} f(t)\frac{1}{\sqrt{s}}\psi^*\left(\frac{t-u}{s}\right) dt \quad (2)$$

again where ψ is the wavelet, shifted by u scaled by s. One can rewrite the wavelet transform as an inner product..

$$W_f(s,u) = \langle f(t), \frac{1}{\sqrt{s}}\psi\left(\frac{t-u}{s}\right)\rangle \quad (3)$$

The mother-wavelet decomposition chosen for this work was the Daubechies 6 ('db6'), with 6 levels. With the help of software developed in Java, we observed the following items:

- Maximum Absolute Value,
- Maximum Value,
- Minimum,
- Average,
- Intersection point is 0.

Firebird Relational DataBase: The vectors in the heart sound signals have been stored in data base for future project of interpreting heart sounds analysis.

According to the methodology proposed for classification of heart sounds, was used as a model five (5) normal heart sounds, and from them made the comparison for classification. Esteemed achieve over 90% accuracy.Below is a table showing the classification of 15 samples.

Table 1 Results Obtained with 15 samples

Item	Classification
Amostra1_8000_01.wav	Normal
Amostra1_8000_02.wav	Normal
Amostra1_8000_03.wav	Normal
Amostra1_8000_04.wav	normal
Amostra1_8000_05.wav	normal
Amostra1_8000_06.wav	normal
Amostra1_8000_07.wav	normal
Amostra1_8000_08.wav	normal
Amostra1_8000_TrackNo07.wav	abnormal
Amostra1_8000_09.wav	normal
Amostra1_8000_10.wav	normal
Amostra1_8000_TrackNo08.wav	abnormal
Amostra1_8000_11.wav	normal
Amostra1_8000_12.wav	normal
Amostra1_8000_TrackNo09.wav	abnormal

V. CONCLUSIONS

The results were as expected for the development of research. The identification and classification between normal and abnormal heart sounds. This case is very complex and can be investigated in future studies.

ACKNOWLEDGMENT

The authors Fatemian, S. Z., Agrafioti, F. Hatzinakos and for having given their sound database for my analysis in this work.

Thanks to the *Secretaria de Estado de Saúde de Mato Grosso* for understanding and encouraging the development of this research.

REFERENCES

1. Fatemian, S. Z. A Wavelet-based Approach to Electrocardiogram (ECG) and Phonocardiogram (PCG) Subject Recognition. Tese. Canadá: 2009.
2. Goswami, J. C.; Chan, A. K. Fundamentals Of Wavelets. Theory, Algorithms, and Applications. 2 ed. New Jersey: John Wiley & Sons, 2011.
3. Karmakar, A. Biometric Identification and Verification based on time-frequency analysis of Phonocardiogram Signal. Tese. Índia: 2012.
4. Martinez, J. C.; Alajarin, C. V. Contribuciones al Desarrollo de un Sistema Electrónico de Ayuda al Telediagnóstico de Enfermedades Cardiovasculares basado en El Análisis de Fonocardiogramas. Tese. Colômbia: 2006.
5. Guyton, A.C.; Hall, J. E. Tratado de Fisiologia Médica. 10. ed. Rio de Janeiro: Guanabara Koogan, 2006.
6. Tilkian, Ara G.; Conover, Mary Boudreau. Entendendo os Sons e Sopros Cardíacos - (8572414843). Roca Editora: 2004.

Use macro [author address] to enter the address of the corresponding author:

Author: Ivo Sérgio Guimarães Brites
Institute: University State Paulista
Street: Avenida Brasil, 56
City: Ilha Solteira
Country: Brazil
Email: ivobrites@yahoo.com.br

Oxygen Dynamics in Microcirculation of Skeletal Muscle

M. Shibata, S. Hamashima, S. Ichioka, and A. Kamiya

Dept. Bioscience and Engineering, Shibaura Institute of Technology
Tokyo, Japan

Abstract—Capillaries were believed to be the sole source of O_2 supply to tissue. However, the recent studies have demonstrated a longitudinal PO_2 drop in arterioles, suggesting the possibility of O_2 supply from arterioles. Furthermore, analysis of the diffusion process based on the PO_2 drop has shown that the O_2 diffusivity in the arteriolar wall was dramatically greater than in the tissue; thus, such a large O_2 loss from arterioles seemed unlikely to be solely attributable to simple diffusion. To explain this discrepancy we hypothesized the O_2 consumption by arteriolar walls was much higher than previously thought. In this study, we quantified the O_2 consumption rate in arteriolar walls to evaluate its impact on the PO_2 drop in the arterioles. Phosphorescence quenching microscopy was used to determine the intra- and perivascular PO_2 values of rat cremaster arterioles both under normal condition and during vasodilation. Using the measured PO_2 values, we calculated the O_2 consumption rate of the arteriolar wall. Our results showed that 100 times more O_2 is consumed by arteriolar walls, compared with *in vitro* vascular segments; consequently, O_2 consumption by arteriolar walls could be the main cause of the PO_2 drop in arterioles. Furthermore, we found the O_2 consumption rate of the arteriolar walls under normal condition to be higher than during vasodilation and the O_2 consumption to be dependent on the mechanical work of vascular smooth muscle. These findings suggested important roles of arterioles for O_2 transport to tissue. Under resting skeletal muscle, to ensure blood supply to other organs with low systemic blood flow, the arterioles consume a large amount of O_2 to restrict blood flow into skeletal muscle. While During exercise, arteriolar O_2 consumption decreases as a result of vasodilation, thereby efficiently supplying O_2 to the skeletal muscle of high O_2 demand.

I. INTRODUCTION

The systemic dimensions, for example, the heart weight and aortic diameter depend on the body size, while the microcirculatory dimensions, intracapillary distance and capillary diameter, have almost same geometric designs (Fig. 1). These facts suggest that the O_2 is supplied from capillary to tissue by simple diffusion. However, the recent studies [1,2] demonstrated a significant drop of O_2 levels in the arterioles (Fig. 2). These findings raise the question to where these O_2 have gone. We consider that these large O_2 drop in arterioles would be caused by (a) O_2 is diffused from arterioles, and (b) O_2 is consumed by arteriolar walls [3] (Fig. 3). In this study, we have tried to demonstrate how much arteriolar walls consume O_2 under *in vivo* physiological conditions.

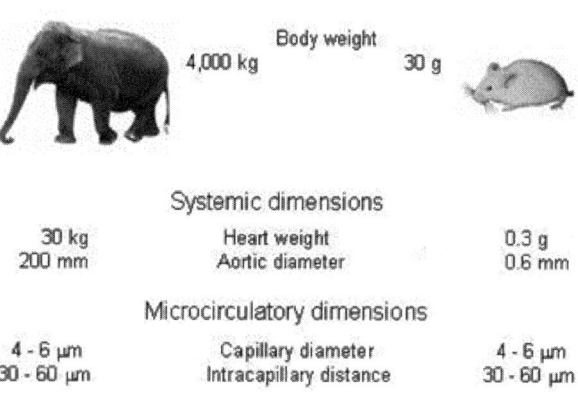

Fig. 1 Comparison of cardiovascular dimensions with elephants and mice.

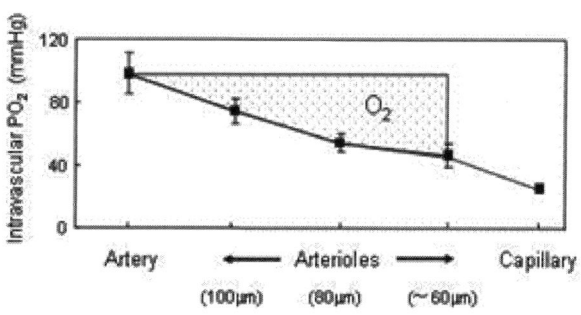

Fig. 2 O_2 levels in the arterioles in rat skeletal muscle.

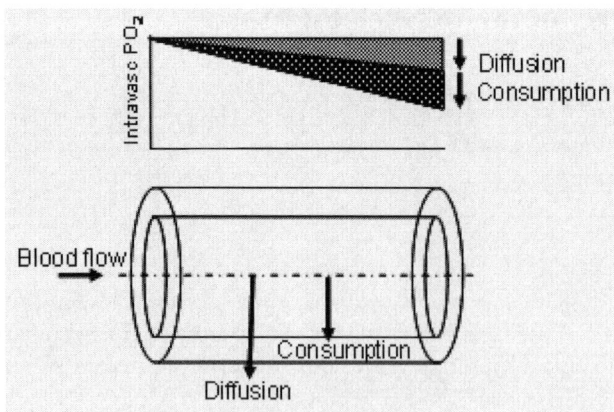

Fig. 3 Is O_2 drop in the arterioles caused by diffusion or consumption.

II. MATERIALS AND METHODS

The study was approved by Shibaura Institute of Technology ethics committee and all of subjects gave written informed consent. Rats were anesthetized with urethane, and the cremaster muscle was spread out in a special bath chamber with an optical port for transillumination, and the surface of the cremaster muscle was suffused with a 37°C Krebs solution. Intra- and perivascular PO_2 in the rat cremaster muscle was measured by phosphorescence quenching technique [4]. Figure 4 shows our developed intravital microscopy to measure microvascular PO_2. Pd-porphyrin was used as a phosphorescent O_2 tracer, and PO_2 values were determined from the phosphorescence decay constant based on Stern-Volmer relationship.

The PO_2 measurements were performed on arterioles of different sizes, classified as shown in Fig 5. Arterioles were classified according to their branching order in the microscopic field. Large arterioles with a diameter of approximately 80 - 120 μm branching from the central cremasteric artery were designated as first order (1A). Branches in first order arterioles were designated as second order (2A, 50 – 80 μm in diameter). Third order arterioles (3A) had a diameter of less than 60 μm and branched from second order arterioles. The intra- and peri-vascular PO_2 measurements were carried out at resting condition and during vasodilation produced by topical application of papaverine (1mM) at the same sites.

The basic model used in this study to estimate the O_2 consumption rate of the arteriolar wall was as follows. The Krogh Cylinder Model of the capillary-tissue system for O_2 delivery in skeletal muscles was modified to suit the arteriolar vascular wall. In the present model, shown in Fig. 6, the O_2 consumption rate per tissue volume per unit time in the arteriolar wall (QO_2) was expressed by the modified Krogh-Erlang equation. PO_2^{peri} and PO_{2i}^n represent the PO_2 values of the outer surface on the arteriolar wall and within the arteriole, respectively. α_t and D represent O_2 solubility and O_2 diffusivity in the arteriolar wall, respectively. R_o and R_i represent the outer and inner radii of the arterioles, respectively. The O_2 consumption rate of the arteriolar wall was determined by utilizing the measured intra and perivascular pO_2 values of the arterioles.

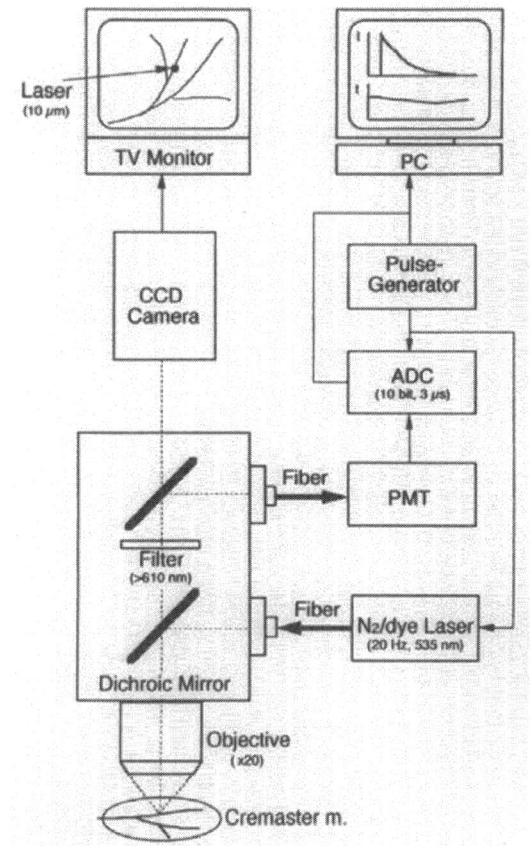

Fig. 4 The entire arrangement of the intravital laser microscope with oxygen-dependent quenching of phosphorescence technique.

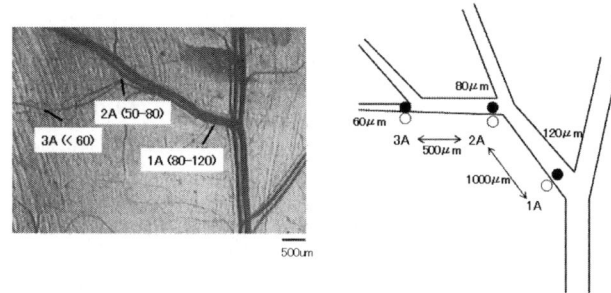

Fig. 5 Classification of arterioles in rat cremaster muscle and measuring points of intra- and perivascular PO_2.

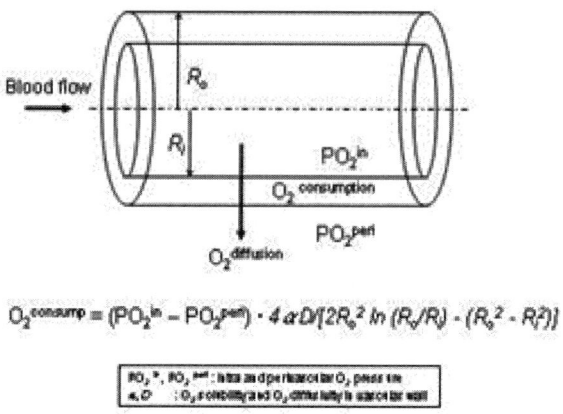

Fig. 6 Theoretical model to estimate vascular wall oxygen consumption rate

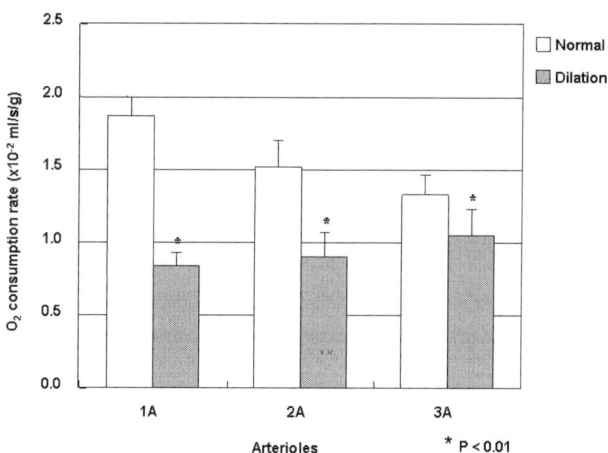

Fig. 8 The oxygen consumption rates of arteriolar walls under normal conditions and during vasodilation

III. RESULTS AND DISCUSSION

The intra- and perivascular PO_2 values of the 1A, 2A, and 3A arterioles at rest and during the papaverine-induced vasodilation are shown in Fig. 7. The intravascular PO_2 values of the 1A arterioles in both conditions were lower than the systemic arterial PO_2 level. The intravascular PO_2 values of the arterioles at rest decreased significantly along the vessel from 1A to 3A, and the perivascular PO_2 values of the different arteriole orders were also significantly lower than the intravascular PO_2 values. During vasodilation, all the PO_2 values of the intra- and periarterioles were significantly higher than the PO_2 values at rest, possibly because of the increased regional blood flow, but the longitudinal and radial rate of decrease was lower than those at rest.

Figure 8 shows the estimated O_2 consumption rates in 1A, 2A, and 3A arteriolar walls at rest and during vasodilation using the intra- and perivascular PO_2 data shown in Fig. 7. The O_2 consumption rate of the arteriolar walls at rest was significantly higher than that during vasodilation, excluding the 3A arterioles. In the resting state, the O_2 consumption rate of the 1A arteriolar wall, located furthest upstream, was the highest, and the O_2 consumption rate sequentially decreased downstream in 2A and 3A arterioles. The workload of the arteriolar smooth muscle of 1A vessels is thought to be greater than that of the smooth muscles of 2A and 3A arterioles, since the intravascular pressure is highest in 1A arterioles. During papaverine-induced vasodilation, the O_2 consumption rates of the arteriolar walls in the 1A, 2A, and 3A vessels all decreased to approximately the same level. The estimated O_2 consumption rates of the arteriolar walls ranged from $1 - 2 \times 10^{-2}$ at rest to $7 - 8 \times 10^{-3}$ ml/s/g during vasodilation. These levels are 100 - 1000 times higher than seen in *in vitro* experiments.

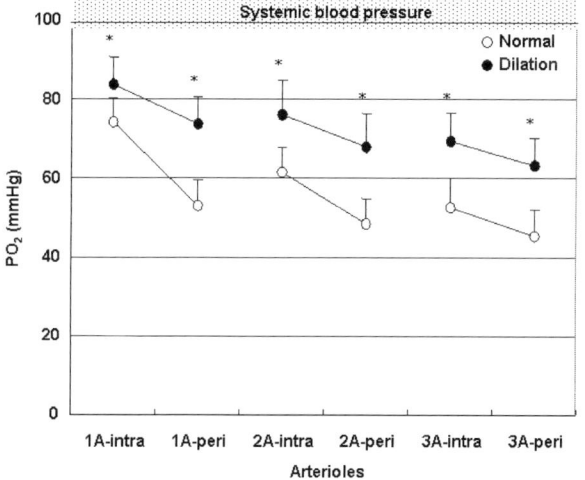

Fig. 7 The intra- and perivascular PO_2 of 1A, 2A and 3A arterioles under normal conditions and during vasodilation

IV. CONCLUSION

1. O_2 consumption rate of functional arterioles was 100 times greater than that seen in *in vitro* experiments.
2. O_2 consumption rate of arteriolar walls in resting skeletal muscle was significantly higher than that during vasodilation.
3. O_2 consumption rate of arteriolar walls depends on the mechanical works of vascular smooth muscle.

REFERENCES

1. Shibata M, Ichioka S, and Kamiya A (2005) Estimating oxygen consumption rates of arteriolar walls under physiological conditions in rat skeletal muscle. Am J Physiol 289: H295-H300
2. Tsai AG, Johnson PC, and Intaglietta M (2002) Oxygen gradients in the microcirculation. Physiol Rev 83: 933-963
3. Shibata M, Ichioka S, Togawa T, and Kamiya A (2006) Arterioles' contribution to oxygen supply to skeletal muscles at rest. Eur J Appl Physiol 97: 327-331
4. Shibata M, Ichioka S, Ando J, and Kamiya A (2001) Microvascular and interstitial pO_2 measurements in rat skeletal muscle by phosphorescence quenching. J Appl Physiol 91: 321-327

Computational Study on Aneurysm Hemodynamics Affected by a Deformed Parent Vessel after Stenting

W. Jeong[1], M.H. Han[2], and K. Rhee[1]

[1] Department of Mechanical Engineering, Myongji University, Yongin, South Korea
[2] Department of Radiology, Seoul National University College of Medicine, Seoul, South Korea

Abstract—To investigate the hemodynamic alterations of a deformed parent vessel after stenting, flow field change after parent vessel stenting was analyzed using computational fluid dynamics. Effects of branch angle change in the vessel bifurcation after stenting on hemodynamic parameters were considered. The results showed that inflow rate, mean velocity, and mean kinetic energy in an aneurysm decreased in the stented vessel comparing to those in the vessel without a stent, which showed flow diversion effects of a stent. Inflow rate, mean velocity, mean kinetic energy in an aneurysm, and maximum wall shear stresses in the parent vessel and in the aneurysm dome increased in the deformed vessel model due to branching angle increase. Parent vessel deformation after stenting should be considered because it could provide unfavorable hemodynamic environment for aneurysm embolization.

Keywords—Aneurysm, Hemodynamics, Stent, Computational fluid dynamics (CFD).

I. INTRODUCTION

Cerebral aneurysms generally occur at arterial curves and bifurcations in the circle of Willis, and rupture of aneurysms causes intracranial hemorrhage, which is associated with high mortality and morbidity [1-3]. Recently, interventional thromboembolization by endovascular insertion of coils has been used to prevent the rupture of aneurysms. Thin platinum coils are filled inside of the aneurysmal sac so that they obliterate the aneurysm sac by thromboembolization. It is difficult to fill the aneurysm sac with coils for wide neck aneurysms because of herniation of coils. Therefore, aneurysm coils are inserted with the flexible stent which prevents aneurysm coils from herniation. Stent can divert the flow into the aneurysm sac as well as prevent herniation of inserted coils.

Currently, the effect of a stent on flow diversion is still controversial. Many studies of numerical simulation and in vitro flow visualization have been conducted using idealized or patient-specific models to study the hemodynamic alterations inside the aneurysm caused by stents [4-12]. Several studies have also shown that the placement of a stent across the aneurysm neck improved the efficiency of flow diversion by reducing the aneurysmal inflow. Meng et al. [9] and Wang et al. [10] investigated the hemodynamic alterations using idealized saccular aneurysm models - side wall and terminal aneurysm models - before and after stent placement. Canton et al. [6] measured changes in flow dynamics using particle image velocimetry (PIV) in bifurcating cerebral aneurysm models after a Neuroform® stent placement and concluded that the magnitude of the velocity of the jet entering the sac was reduced by as much as 11%. Tateshima et al. [11] studied the hemodynamic effect of Neuroform® stent placement across the necks of patient-specific aneurysm models and concluded that the stents significantly altered flow velocity and flow structure in aneurysms. Tang et al. [12] studied the hemodynamic effects of aneurysm neck size after stenting and concluded that the volume flow rate of blood entering into the aneurysm over the entire cardiac cycle could be reduced by more than 50 % after endovascular operation. Most of these studies focused on the hemodynamic characteristics before and after stenting. Although, some studies have been investigated on vascular bifurcation angle change after stenting [13, 14], studies on the hemodynamic alterations by geometrical change of a parent vessel after stenting have not been investigated.

The aim of this study is to investigate the hemodynamic change in an aneurysm after stenting. The change of hemodynamic parameters affected by parent vessel geometry changes by stenting is computationally calculated.

II. METHODS

A. Aneurysm Models

Idealized cerebral aneurysm models were created by 3D CAD software (SolidWorks, Dassault Systems, Concord, U.S.A.). We modeled a terminal aneurysm (Y-typed) located at the tip of the parent vessel with two bifurcating branches to simulate a terminal aneurysm occurred at basilar artery (Fig. 1). Average values of neck width, aneurysm diameter, parent artery diameter were obtained from the measurements reported by Parlea et al. [15]. The diameters of the parent artery (PA), the two daughter branches (DB), the aneurysm sac (AS), and the aneurysm neck width (NW) are 2.7 mm, 2.7 mm, 8.6 mm, 6 mm, respectively. The length of the stent is 10 mm, and the strut width and

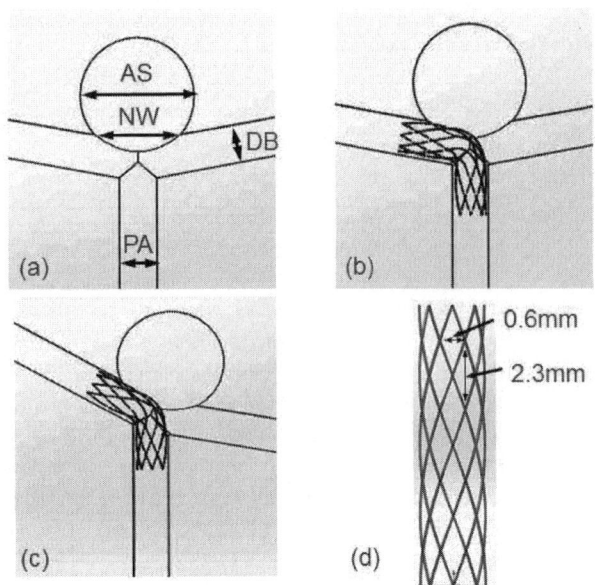

Fig. 1 The geometries of basilar tip aneurysm model. (a) before stenting, (b) after stenting (100°), (c) deformed model after stenting (120°), (d) stent geometry

Table 1 Hemodynamic parameters in the basilar tip aneurysm model before and after stenting.

	No Stent_100°	Stent_100°	Stent_120°
IFR an aneurysm neck	0.0025 kg/s	0.0020 kg/s	0.0022 kg/s
MV inside an aneurysm sac	0.215 m/s	0.137 m/s	0.179 m/s
MKE inside an aneurysm sac	0.029 m²/s²	0.012 m²/s²	0.022 m²/s²
MAXWS at parent vessel	37.2 Pa	40.6 Pa	117.0 Pa
MAXWS at an aneurysm sac	18.3 Pa	10.8 Pa	31.7 Pa

thickness are 0.06 mm and 0.065 mm, respectively. The angle between the parent and daughter artery is 100°. In order to simulate deformation of blood vessel after inserting a stent, the deformed parent vessel geometry after stenting was modeled based on the measurements reported by Gao et al. [13]. The angle between the parent artery and left daughter branch was increased from 100° to 120° by vessel deformation caused by straightening of a stent.

B. Numerical Methods

Numerical analysis was performed using a commercial computational fluid dynamics (CFD) package (Fluent, ANSYS, Inc., Canonsburg, U.S.A.) based on the finite volume method. The fluid domain was meshed into tetrahedral elements using ICEM-CFD (ANSYS, Inc., Canonsburg, U.S.A.) program. The total number of grids was about 2,300,000. Three-dimensional incompressible laminar flow fields were obtained by solving continuity and Navier-Stokes equations computationally. Fully developed velocity profiles were applied at the inlet boundary, and pressure outlet boundary condition was applied at the outlet boundary. We assumed the blood vessel had a rigid wall because cerebral arteries were less elastic than other large arteries [16, 17]. The mean blood flow in the basilar artery was 195 mL/min as reported in the reference [18] for healthy human subjects. The mean velocity in a parent vessel was 0.57 m/s, and Reynolds number was 466. The density of blood was assumed to be 1060 kg/m³. The non-Newtonian viscosity characteristics of blood were modeled by using the Carreau model. The equation for the Carreau model was expressed as [19]:

$$\eta = \eta_\infty + (\eta_0 - \eta_\infty)\left[1 + \lambda^2 \gamma^2\right]^{\left(\frac{q-1}{2}\right)}$$

where rheological parameters of human blood are $\eta_0 = 0.056$ Pa·s, $\eta_\infty = 0.00345$ Pa·s, $\lambda = 3.313$ s, $q = 0.356$. η is viscosity, γ is shear rate.

III. RESULTS

To investigate the hemodynamic changes by vessel deformation after stenting, the flow fields were calculated for the vessel geometries before and after vessel deformation. The quantitative hemodynamic parameters, such as velocity, kinetic energy, and wall shear stress were analyzed since these parameters were suspected to be related to thromboembolization of aneurysms.

Flow velocity in an aneurysm was reduced after stenting comparing to that in the aneurysm model without a stent as shown Fig. 2. The inflow rate (IFR) into an aneurysm neck was decreased after stenting comparing to that before stenting, and the IFRs of the models with a stent were increased as the bifurcation angle of parent vessel increased (Table. 1). The mean velocity (MV) and mean kinetic energy (MKE) inside the aneurysm sac were also decreased after stenting comparing to those of the aneurysm model without a stent. MV and the MKE of the stented vessel model after vessel deformation were also increased as the bifurcation angle of parent vessel increased (Table. 1).

The maximum wall shear stress (MAXWS) of the parent vessel near aneurysm neck was measured. The MAXWS in the parent vessel near aneurysm neck was not noticeably different from that in the stented vessel before vessel deformation. However, the MAXWSs increased as the branching angle increased after vessel deformation (Table. 1).

Fig. 3 showed that aneurysm model without a stent showed high wall shear stress distribution at tip of aneurysm dome. Wall shear stresses in the aneurysm model after stenting decreased comparing to those in the aneurysm model without a stent overall. However, wall shear stress in the parent vessel near the aneurysm neck and the aneurysm sac increased as the angle of parent vessel geometry increased. The MAXWS in an aneurysm sac was calculated to quantify the flow impingement near the aneurysm fundus. The MAXWS in an aneurysm sac was decreased in the stented model compared to that in the aneurysm model without a stent, but the MAXWS in an aneurysm sac in the deformed vessel model was increased (Table 1).

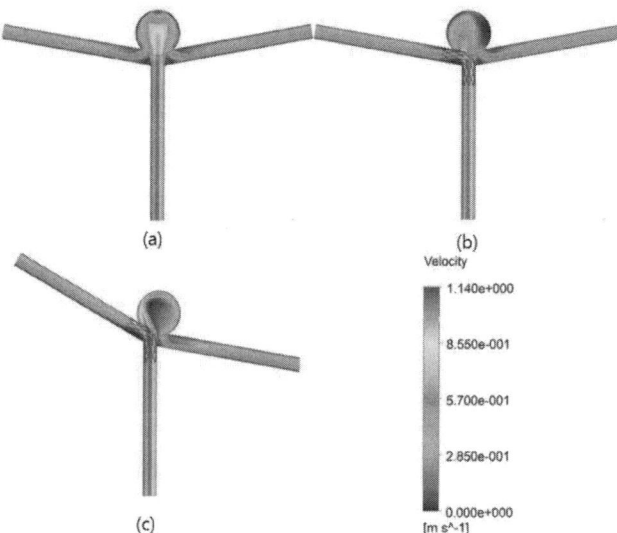

Fig. 2 Velocity contour. (a) before stenting, (b) after stenting (100°), (c) deformed model after stenting (120°)

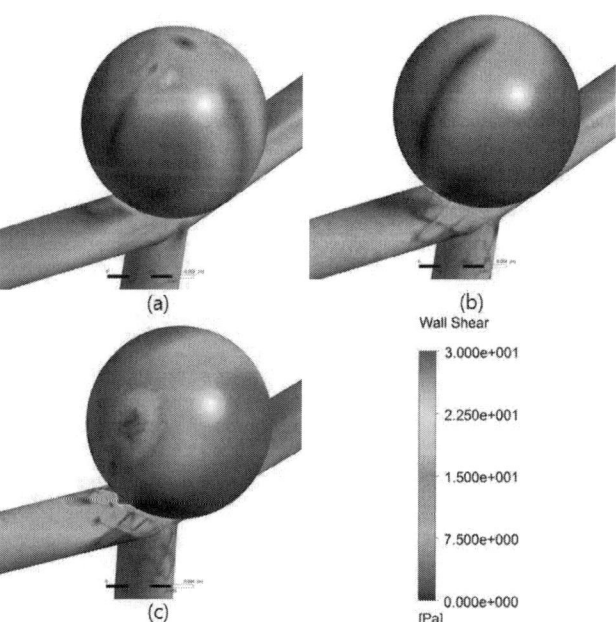

Fig. 3 Wall shear stress contour. (a) before stenting, (b) after stenting (100°), (c) deformed model after stenting (120°)

IV. CONCLUSIONS

Deformation of vessel geometry after stenting could change the hemodynamics in an aneurysm. In order to investigate this hypothesis, computational analysis of flow fields on the deformed vessel geometry after stenting was performed in terminal aneurysm models with a stent. Hemodynamic parameters (inflow rate, mean velocity, mean kinetic energy, maximum wall shear stress) in an aneurysm decreased after stenting comparing to those in the aneurysm model without a stent, which showed flow diversion effect of a stent. The vessel branch angle increase caused by deformation of vessel after stenting increased inflow rate, mean velocity, mean kinetic energy, maximum wall shear stress; therefore, it could provide unfavorable hemodynamic environment for aneurysm embolization.

ACKNOWLEDGMENT

This study was supported by a grant from the Korean Health Technology R&D Project, Ministry of Health and Welfare, Republic of Korea (A111101).

REFERENCES

1. Kaminogo M, Yonekura M, Shibata S, (2003) Incidence and outcome of multiple intracranial aneurysms in a defined population. Stroke 34:16–21
2. Linn F, Rinkel G, Algra A, Van Gijn J et al (1996) Incidence of Subarachnoid Hemorrhage Role of Region, Year, and Rate of Computed Tomography: A Meta-Analysis. Stroke 27:625-629.
3. Winn HR, Jane JA, Taylor J et al (2002) Prevalence of asymptomatic incidental aneurysms: review of 4568 arteriograms. J Neurosurg 96: 43-49
4. Appanaboyina S, Mut F, Lohner R et al (2008) Computational modelling of blood flow in side arterial branches after stenting of cerebral aneurysms. Int J Comput Fluid Dyn 22(10):669-676.
5. Babiker MH, Gonzalez LF, Ryan J et al (2012) Influence of stent configuration on cerebral aneurysm fluid dynamics. J Biomech 45:440-447.
6. Cantón G, Levy DI, Lasheras JC et al (2005) Hemodynamic changes due to stent placement in bifurcating intracranial aneurysms. J Neurosurg 103:146-155.
7. Dorn F, Niedermeyer F, Balasso A et al (2011) The effect of stents on intra-aneurysmal hemodynamics: in vitro evaluation of a pulsatile sidewall aneurysm using laser Doppler anemometry. Neuroradiology 53:267-272.

8. Mulder G, Bogaerds AC, Rongen P et al (2009) On automated analysis of flow patterns in cerebral aneurysms based on vortex identification. J Eng Math 64:391-401
9. Meng H, Wang Z, Kim M et al (2006) Saccular aneurysms on straight and curved vessels are subject to different hemodynamics: implications of intravascular stenting. Am J Neuroradiol 27:1861-1865.
10. Wang S, Ding G, Zhang Y et al (2011) Computational haemodynamics in two idealised cerebral wide-necked aneurysms after stent placement. Comput Methods Biomech Biomed Eng 14:927-937
11. Tateshima S, Tanishita K, Hakata Y et al (2009) Alteration of intraaneurysmal hemodynamics by placement of a self-expandable stent: Laboratory investigation. J Neurosurg 111:22-27.
12. Tang AY-S, Lai S-K, Leung K-M et al (2012) Influence of the aspect ratio on the endovascular treatment of intracranial aneurysms: A computational investigation. J Biomed Sci Eng 5:422-431.
13. Gao B, Baharoglu M, Cohen A et al (2012) Stent-assisted coiling of intracranial bifurcation aneurysms leads to immediate and delayed intracranial vascular angle remodeling. Am J Neuroradiol 33:649-654.
14. Huang Q-H, Wu Y-F, Xu Y et al. (2011) Vascular geometry change because of endovascular stent placement for anterior communicating artery aneurysms. Am J Neuroradiol 32:1721-1725.
15. Parlea L, Fahrig R, Holdsworth DW et al (1999) An analysis of the geometry of saccular intracranial aneurysms. Am J Neuroradiol 20:1079-1089.
16. Hayashi K, Handa H, Nagasawa S et al (1980) Stiffness and elastic behavior of human intracranial and extracranial arteries. Journal of biomechanics 13:175-184.
17. Scott S, Ferguson GG, Roach MR (1972) Comparison of the elastic properties of human intracranial arteries and aneurysms. Canadian journal of physiology and pharmacology 50:328-332.
18. Valencia A and Solis F (2006) Blood flow dynamics and arterial wall interaction in a saccular aneurysm model of the basilar artery. Comput Struct 84:1326-1337.
19. Banerjee RK (1992) A study of pulsatile flows with non-Newtonian viscosity of blood in large arteries. Ph.D. Thesis (Drexel University, Philadelphia)

Author: Kyehan Rhee
Institute: Myongji University
Street: San 38-2, Nam-dong, Cheoin-gu
City: Yongin
Country: South Korea

Email: khanrhee@mju.ac.kr

Inflation Tests of Vena Saphena Mangna for Different Loading Rates

J. Veselý[1], L. Horný[1], H. Chlup[1], and R. Žitný[2]

[1] Department of Mechanics, Biomechanics and Mechatronics, Faculty of Mechanical Engineering Czech Technical University in Prague, Prague, Czech Republic
[2] Department of Process Engineering, Faculty of Mechanical Engineering Czech Technical University in Prague, Prague, Czech Republic

Abstract—The aim of the study was to verify whether constitutive model parameters obtained from quasistatic inflation-extension test could be used for *in-vivo* computational simulation of venous graft used as bypass at arterial conditionns. Inflation tests of human *vena saphena magna* at three loading rates were performed: a quasistatic (slow), moderate, and fast, which was similar to human heart rate. The experimental data were fitted by hyperelastic nonlinear anisotropic constitutive model (Holzapfel-Gasser-Ogden) in order to obtain material parameters. It was found that parameters differ significantly with respect to loading rate. Especially the parameter related to large strain stiffening (k_2) increased with loading rate. The results suggest that constitutive parameters are strongly influenced by experimental conditions which we interpret as viscoelasticity of the vein grafts.

Keywords—inflation test, saphenous vein, constitutive model, CAGB, loading rate.

I. INTRODUCTION

The number of coronary-artery and peripheral-artery-bypass grafts using the saphenous vein as a replacement is increasing [1]. After bypass surgery, the vein is subjected to arterial conditions (pulsations) which are different from venous one. The change of the mechanical environment initiates remodeling process. The most studied effect of graft adaptation is the development of intimal hyperplasia [2]. Large attention was paid to flow conditions and wall shear stress which are suspected to be responsible for hyperplasia which may cause the graft failure (see e.g. [3], [4]). Varicose veins were also studied from the mechanical and histological point of view [5], [6].

However, papers dealing with biomechanics of the graft wall and constitutive modeling are rare in the literature. Probably the most detailed study dealing with mechanics of saphenous veins was performed by Donovan and coworkers [7]. They carried out uniaxial tensile test with 45 saphenous veins and described their mechanical properties in terms of material and structural parameter: vessel diameter, tensile stiffness, failure and ultimate forces, and tensile modulus, failure stress, and strain. Pressure-diameter data from inflation tests with saphenous veins from upper and lower limbs were published in [8], but without identification of constitutive parameters.

The material parameters are necessary for computations (e.g. FEA) to obtain stress and strain field. To the best of our knowledge there is only one paper describing behavior of coronary-artery bypass in terms of constitutive parameters [2]. The aim of this preliminary study was to extend the available set of material parameters.

II. METHODS

The sample of human vena saphena magna was harvested during coronary-artery bypass graft surgery in the General University Hospital in Prague. The usage of the biological material was approved by the Ethics committee of the Hospital. The vein was immediately stored in the physiological solution and tested within four hours after the excision.

A. Inflation Tests

The specimen was mounted in the setup for inflation-extension test (Fig. 1) and inflated by internal pressure as follows. The sample of vena saphena was inflated four times up to approx. 4kPa (vein pressure) and then four times up to approx. 16kPa (systolic pressure).

The sample of the vein underwent pressurization cycles at three loading rates which can be characterized by the frequency of pulsations (Fig. 2). The frequency of loading-unloading cycle (0-16-0 kPa) was:

A. quasistatic slow loading – frequency 0.04 Hz
B. moderate loading – frequency 0.5 Hz
C. fast loading - frequency 1 Hz, which is similar to a heart rate.

The intraluminal pressure was recorded during the test by the pressure transducer (Cressto, Cressto s.r.o. Czech Republic). Sample was recorded by the CCD camera during the loading. The longitudinal (λ_z) and circumferential (λ_t) deformations of the tube were evaluated using the edge detection method and computed via equations (1) and (2) assuming the incompressibility of the venous wall expressed by equation (3).

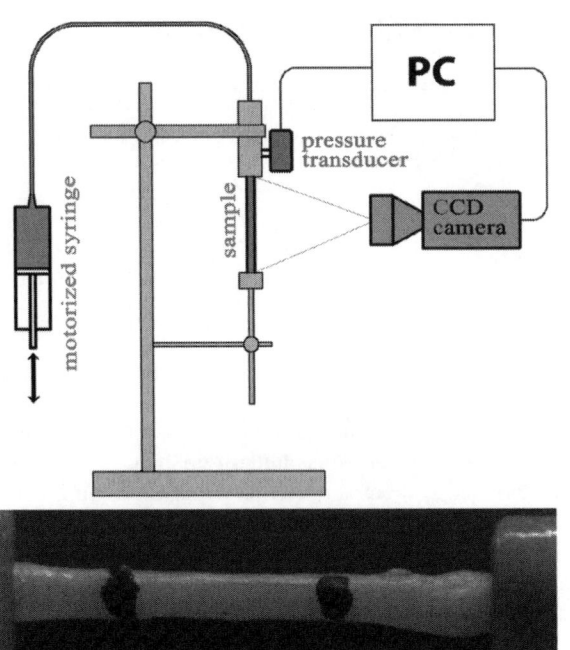

Fig. 1 Up: Inflation test set-up. Down: Photograph of the sample in front of contrasting background from CCD camera. Black markers were used to evaluate the longitudinal deformation of the sample.

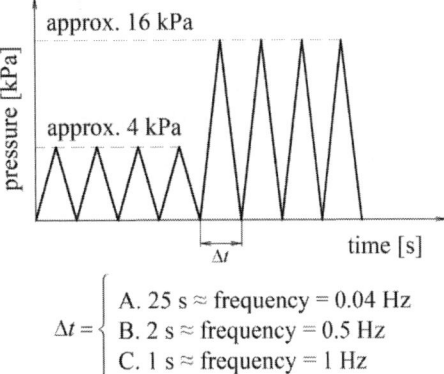

Fig. 2 The loading protocol during inflation tests.

$$\lambda_t = \frac{2\pi r_m}{2\pi R_m} \quad (1)$$

$$\lambda_t = \frac{l}{L} \quad (2)$$

$$\lambda_r = \frac{1}{\lambda_t \lambda_z} \quad (3)$$

Here r_m and R_m are middle radii of the tube in deformed and reference geometry. l and L is longitudinal distance of marks in deformed and reference geometry and λ_r is the radial deformation.

The reference geometry and the distance of longitudinal marks are listed in Table 1.

Table 1 Reference geometry of the sample

Outer radius	R_o	2.77 mm
Thickness	H	0.80 mm
Distance of axial marks	L	11.20 mm

The experimental circumferential (σ_{tt}) and longitudinal (σ_{zz}) stress during the loading was computed adopting presumption of the thin-wall geometry with closed ends which results in Laplace's Law (4), (5).

$$\sigma_{tt}^{EXP} = P\frac{r_m}{h} = P\frac{R_m \lambda_t^2 \lambda_z}{H} \quad (4)$$

$$\sigma_{zz}^{EXP} = \frac{\sigma_{tt}}{2} \quad (5)$$

Here P is intraluminal pressure, H is the reference thickness of the sample.

B. Numerical Model

The data for each loading rate were fitted separately. The hyperelastic Holzapfel-Gasser-Ogden (HGO) nonlinear anisotropic constitutive model [10] was used to fit the experimental data. The strain energy density function is expressed by equation (6).

$$W = \frac{C}{2}(I_1 - 3) + \frac{k_1}{k_2}\left(e^{k_2(I_4 - 1)^2} - 1\right) \quad (6)$$

In (6) C and k_1 are stress-like parameters, k_2 is dimensionless parameter. I_1 is the first invariant of the right Cauchy-Green strain tensor and I_4 is additional invariant arising from material anisotropy and has the meaning of square of the stretch in preferred (fiber) direction. I_1 and I_4 are defined in (7) and (8).

$$I_1 = \lambda_r^2 + \lambda_t^2 + \lambda_z^2 \quad (7)$$

$$I_4 = \lambda_t^2 \cos^2\beta + \lambda_z^2 \sin^2\beta \quad (8)$$

In (8) β defines preferred direction within the material measured from circumferential axis of the tube in cylindrical configuration of the vein. Parameter β was considered to be the same for all loading rates.

According to [9], the stress in the sample is then computed using W using equations (9), (10), (11).

$$\sigma_{rr}^{MOD} = \lambda_r \frac{\partial W}{\partial \lambda_r} - p \qquad (9)$$

$$\sigma_{tt}^{MOD} = \lambda_t \frac{\partial W}{\partial \lambda_t} - p \qquad (10)$$

$$\sigma_{zz}^{MOD} = \lambda_z \frac{\partial W}{\partial \lambda_z} - p \qquad (11)$$

Here p is the Lagrange multiplier which plays a role of the reaction to incompressibility constrain. The equation (9) was combined with $\sigma_{rr} = -P/2$ (radial equilibrium equation) and then used to eliminate p. p is substituted into (10) and (11).

Objective function Q in (12) was minimized and the constitutive parameters of the model were calculated in Maple 16 (Maplesoft, Canada).

$$Q = \sum_{\substack{i=t,z \\ k=1..n}} \left(\sigma_{ii}^{EXP} - \sigma_{ii}^{MOD} \right)_k^2 \qquad (12)$$

III. RESULTS

Pressure-stretch (P-λ) and stress-stretch (σ-λ) curves are plotted to illustrate the effect of the loading rate in Fig. 3. It is shown that increasing loading rate is manifested as the shift of the curves to the left. This can be interpreted in the way that material stiffens with increasing loading rate. Estimated parameters of HGO constitutive model for each inflation rate are listed in Table 2. Considering results in Table 2 parameter k_2 depend strongly on loading rate (the higher the rate, the higher k_2). Parameter C is slightly increasing and k_1 is slightly decreasing with pressurizing rate but it should be noted that W depends linearly on them.

Table 2 Parameters of HGO model for different loading rates

loading frequency	C [kPa]	k_1 [kPa]	k_2 [-]	β [rad]
A. 0.04 Hz	4.93	3.43	19.13	0.75
B. 0.5 Hz	7.66	0.28	46.82	0.75
C. 1 Hz	13.04	0.24	60.30	0.75

IV. DISCUSSION AND CONCLUSIOS

The inflation tests of human vena saphena magna at different loading rates were performed. The experimental data were fitted by anisotropic HGO constitutive model to obtain

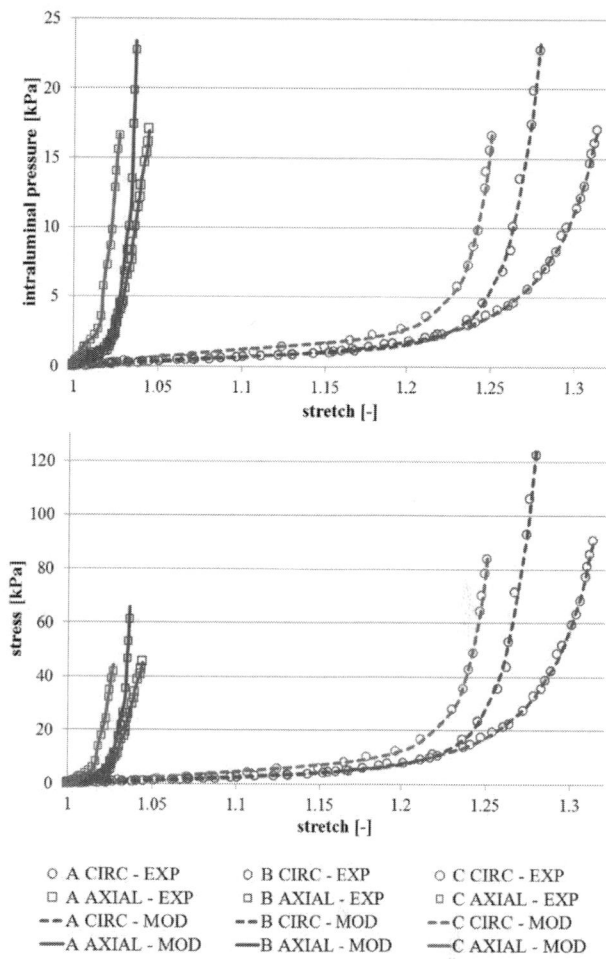

Fig. 3 Experimental data and computed model curves for the sample of vena saphena magna. CIRC and AXIAL denote circumferential and longitudinal direction, EXP and MOD denote experimental and model data.

material parameters for different pressurizing rates. As shown in Table 2 the model parameters differ significantly, especially parameter k_2 which appears in the exponent of the strain energy density function. The results suggest that constitutive parameters are strongly influenced by experimental conditions which we interpret as viscoelasticity of the vein grafts.

The limitation of this preliminary study is related to the number of tested specimens. Only one sample obtained from one donor was tested. More samples from donors of different age and pathology should be studied. We are evaluating other inflation-extension tests in order to extend the available set of material parameters of vena saphena magna.

ACKNOWLEDGMENT

This study has been supported by the Czech Ministry of Health under project NT 13302 and by the Czech Technical University in Prague under project SGS13/176/OHK2/3T/12.

REFERENCES

1. Krasiński Z, Biskupski P, Dzieciuchowicz L, Kaczmarek E et al. (2010) The influence of elastic components of the venous wall on the biomechanical properties of different veins used for arterial reconstruction. Eur J Vasc Endovasc 40:224-229
2. Horny L, Chlup H, Zitny R et al. (2009) Constitutive modelling of coronary artery bypass graft with incorporated torsion. Metalurgia 49:273-277
3. Tran-Son-Tay R, Hwang M, Garbey M et al. (2008) An experiment-based model of vein graft remodeling induced by shear stress. Ann Bimed Eng 36:1083-1091
4. Fernandez MC, Goldman DR, Jiang Z et al. (2004) Impact of shear stress on early vein graft remodeling: A biomechanical analysis. Ann Bimed Eng 32:1484-1493
5. Clarke GH, Vasdekis SN, Hobbs JT, Nicolaides AN (1992) Venous wall function in the pathogenesis of varicose veins. Surgery 111:402-408
6. Wali MA, Dewan M, Eid RA. (2003) Histopathological changes in the wall of varicose veins. Int Angiol 22:188-193
7. Donovan DL, Schmidt SP, Townsend SP et al. (1990) Material and structural characterization of humansaphenous vein. J Vasc Surg 12:531-537
8. Stooker W, Gok M, Sipkema P et al. (2003) Pressure–diameter relationship in the human greater saphenous vein. Ann Thorac Surg 76:1533–1538
9. Holzapfel GA, Gasser TC, Ogden RW (2000) A new constitutive framework for arterial wall mechanics and a comparative study of material models. J Elast 61:1–48

Author: Jan Veselý
Institute: Faculty of Mechanical Engineering CTU in Prague
Street: Technicka 4
City: Prague
Country: Czech Republic
Email: jan.vesely1@fs.cvut.cz

Modeling of Stent Implantation in a Human Stenotic Artery

G.S. Karanasiou[1], A.I. Sakellarios[1], E.E. Tripoliti[1], E.G.M. Petrakis[1], M.E. Zervakis[1], Francesco Migliavacca[2], Gabriele Dubini[2], Elena Dordoni[2], L.K. Michalis[3], and D.I. Fotiadis[4]

[1] Department of Electronic and Computer Engineering, Technical University of Crete, Chania, Crete, Greece
[2] Laboratory of Biological Structure Mechanics (LaBS), Department of Chemistry, Materials and Chemical Engineering "Giulio Natta" Politecnico di Milano, Milano, Italy
[3] Department of Cardiology, Medical School, University of Ioannina, Ioannina, Greece
[4] Unit of Medical Technology and Intelligent Information Systems, Department of Materials Science and Engineering, University of Ioannina, Ioannina, Greece

Abstract—The aim of this work is to introduce a methodology to study the stent expansion and the subsequent deformation of the arterial wall towards the outside direction in order arterial lesion to be rehabilitated and blood flow to be restored. More specifically, a coronary artery and the plaque are reconstructed using intravascular ultrasound and biplane angiography. The finite element method is used for the modeling of the interaction between the stent, balloon, arterial wall and plaque. Appropriate material properties and boundary conditions are applied in order to represent the realistic behavior of each component. We observe that stresses are increased at the region of the first contact between the stent and the wall, which may be considered crucial for plaque rupture. Furthermore, the average calculated stress on the plaque is higher than the average stress on the arterial wall. Thus, stent positioning and deployment depends on a considerable degree on the plaque properties rather than the general arterial geometry. Results indicate that numerical modeling can provide a prediction of the arterial behavior during stent implantation.

Keywords—Stent, Human artery, Atherosclerosis, Finite element method.

I. INTRODUCTION

The cardiovascular disease affects the quality of life of patients and can lead to severe health problems or even death. Atherosclerosis involves arterial wall thickening and blood flow reduction, caused by the buildup of the plaque. Plaque consists of calcium, cholesterol crystals and other types of cells like smooth muscle cells. Plaque grows and causes artery occlusion [1]. Several different treatment techniques and procedures are available for rehabilitation including angioplasty, stent implantation and artery bypass. Angioplasty is an invasive procedure where a balloon catheter is placed into the artery, expands and re-opens the artery [2]. Angioplasty is not the best choice for all plaque types since lesions vary in composition, size and affected area. These difficulties, complications and short comings, which may presented during angioplasty, were faced using an invasive endovascular device, which is called stent.

A stent is a small metallic tube consisted of wires, being initially in a crimped condition, mounted on a balloon catheter device. Once in place, the balloon is inflated, the stent expands and compresses the arterial plaque reserving the inner artery wall open, after balloon removal. Like every mechanical procedure, there is a dependency on the mechanical properties, the geometry and the morphology of the involved components (artery, plaque, stent and balloon) [3].

Stent geometry and design could result in a different arterial behavior. Plaque types vary and are classified according to their stiffness. Thus, different plaque types demonstrate different response when they undergo the same stenting process.

The finite element method (FEM) can be used to study the interaction between the stent, the artery and the plaque components during stent placement. Stent and artery interaction was examined by Berry *et al.* [4]. The plaque component was also considered in more recent studies utilizing idealized arterial geometries such as stenosed straight [5-7] or curved [8] geometries. Arterial wall stresses caused by the stent implantation in stenotic arteries depend on the different plaque types [9]. A finite element model of a real patient artery was developed with the presence of stent, ignoring however the presence of plaque morphology [10].

In this study, the stent implantation procedure is simulated in a reconstructed patient's stenotic artery including the plaque component in the area of stenosis. The effect of the stent implantation device is represented in relation to stress distribution and deformation caused in the arterial wall and plaque.

II. MATERIAL AND METHODS

A. 3D Reconstruction

Data from a 62-year old smoker overweight male (BMI 27.8) with high levels of cholesterol in the blood (hypercho-lesterolemia) and hypertension problems were used. Intra-vascular ultrasound (IVUS) and angiography were employed. The reconstruction of the coronary artery was conducted adopting the approach introduced by Bourantas et al. [11]. This method establishes a particular segmentation methodology which detects the lumen and media adventitia borders and afterwards the detected borders are placed on the 3D catheter path, which is extracted by the processing of bi-plane angiography. The output of this procedure is two point clouds, which represent the arterial wall and lumen geometry. The plaque component is reconstructed after automated plaque characterization using the methodology presented in [12].

B. Computational Simulation

The 3D finite element model was developed in its unexpanded state and it consists of the artery, the plaque, the stent and the balloon. ANSYS 12.1 (ANSYS, Canonsburg, PA) was used for pre- and post-processing. Based on the Open Stent design [13], the geometry of the stent device was appropriately shaped for this specific human artery. The balloon was modeled as a plane cylinder and it was positioned in such a way that the inner surface of the stent was in initial contact with the outer surface of the balloon. The balloon-stent device was not in initial contact with the arterial lumen (Fig. 1). The plaque was lying in the inside area of the arterial wall (Fig. 1). The thickness of the plaque varied from 1.05mm to 1.40mm. The arterial segment was 17.54mm in length and the length of the plaque was 3.27mm. Figure 1 presents the whole model consisted of the artery, the plaque, the stent and the balloon in their initial condition. The mesh density was chosen based upon the elimination of the existing penetration between the contact pairs appearing in this model.

As far as modeling techniques are concerned, there are mainly two different approaches regarding the stent simulation procedure. The first method uses a consistent uniform pressure which increases and is placed directly in the inside surface of the stent or the balloon-stent device [14-16]. In the second method radial displacement is enforced directly on the nodes of the inner cylindrical balloon surface [17]. In this study the simulation of stent deployment was carried out with the method described in [17].

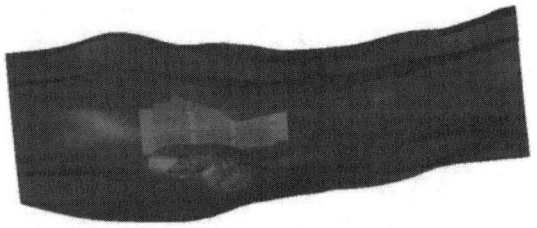

Fig. 1 Geometry of the model: Red, green, blue and yellow colors represent the arterial wall, plaque, stent and balloon geometry, respectively

C. Material Properties

Several components constitute the arterial tissue some of which are collagen and elastin cells. The artery and the plaque can be modeled using several material models. The complexity of the behavior of the arterial tissue can be more accurately described using hyperelastic material models. A Mooney-Rivlin hyperelastic equation was used to define the arterial tissue which depicts the non-linear behavior presented between stress and strain of the arterial tissue and it is defined by a polynomial form [15]. Maurel et al. [18] described the strain energy density function, for an isotropic hyperelastic material, in terms of the strain invariants:

$$W(I_1, I_2, I_3) = \sum_{p,q,r=0}^{n} C_{pqr} (I_1 - 3)^p (I_2 - 3)^q (I_3 - 3)^r, \quad (1)$$

where, W is the strain energy density function of hyperelastic material, C_{pqr} is the hyperelastic constants, $C_{000}=0$, $\lambda_1, \lambda_2, \lambda_3$ are the principal stretches of material and I_1, I_2, I_3 are the strain invariants. The strain invariants are defined in Eqs. (2)-(4):

$$I_1 = \lambda_1^2 + \lambda_2^2 + \lambda_3^2, \quad (2)$$

$$I_2 = \lambda_1^2 \lambda_2^2 + \lambda_1^2 \lambda_3^2 + \lambda_2^2 \lambda_3^2, \quad (3)$$

$$I_3 = \lambda_1^2 \lambda_2^2 \lambda_3^2. \quad (4)$$

In this study a third order, five parameters Mooney Rivlin model (Eq. (5)) was used:

$$W = C_{10}(I_1 - 3) + C_{01}(I_2 - 3) + C_{20}(I_1 - 3)^2 + C_{11}(I_1 - 3)(I_2 - 3) + C_{30}(I_1 - 3)^3, \quad (5)$$

described by Eshghi et al. [19] which is a specific form of Eq. (1). The hyperelastic constants of Eq. (5) are given in Table 1. Stress components were derived by differentiating the strain energy function with respect to strain variables.

Table 1 Artery hyperelastic coefficients.

Coefficients	C_{10}	C_{01}	C_{20}	C_{11}	C_{30}
Artery	0.0189	0.00275	0.08572	0.5904	0

The plaque was assumed to be calcified and was modeled as a linear isotropic elastic material with Young modulus 2.7MPa and Poisson's ratio 0.4913. The stent was made of 304 Stainless Steel and modeled as a bi-linear elasto-plastic material with Elastic Modulus 193GPa, Poisson's ratio 0.27 and Tangent modulus 0.692GPa [20]. Regarding the balloon, it was assumed to be made of polyurethane described as a hyperelastic material. The Mooney–Rivlin model was used with $C_{10}=1.0318$MPa and $C_{01}=3.6927$ MPa [6].

Modeling of Stent Implantation in a Human Stenotic Artery

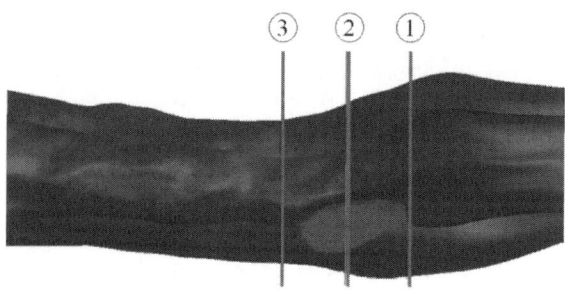

Fig. 2 The model with the selected cross sections

D. Boundary Conditions

In order to perform a steady-state analysis and prevent rigid body motion, certain specific areas must be fixed by applying displacement limitations upon the model's nodes. The artery was fixed at its ends and the stent was tethered in such a way that only radial displacement was allowed. The balloon was in initial contact with the stent and radial displacement (0.8mm) was applied in the inner surface of the balloon. Frictionless contact was assumed for the balloon and the stent contact pair, while artery and plaque were assumed to be bonded. Since the simulation involved large displacements and complex contact analysis, special attention must be taken in parameters like time step, contact algorithm, stiffness and penetration factors.

III. RESULTS AND DISCUSSION

The results are obtained in terms of the stress and the deformation that occur in the model's components. Figure 2 depicts the cross sections where the results are presented. Three different cross sections along longitudinal axis are selected: 1) before the maximum stenotic location, 2) in the maximum stenotic area, and 3) after the stenotic area. The arterial wall deformation and the corresponding von Mises contour maps, of the stent deployment procedure in these cross sections are presented in Figure 3. It is observed that due to the plaque stiffness the artery is deformed more than the plaque, especially in the maximum stenotic area. The stresses caused in the artery wall have a descending ratio when going from the artery's lumen interface to the outer wall surface. The maximum von Mises stresses are observed at the region across the plaque component.

Figure 4 shows the von Mises stress in the arterial wall for a vertical cross section. It is noticed that higher stresses appear in the area behind the stent struts. Figure 5 presents the plaque's von Mises stresses; the maximum plaque stress appears in the contact area between the stent and the artery, in the point where the stent expands and pushes the plaque.

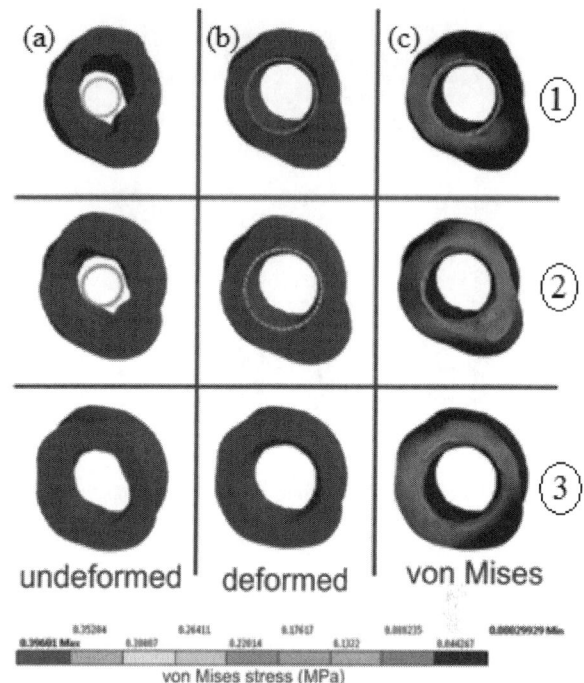

Fig. 3 (a) Undeformed arterial wall and plaque, (b) Deformed arterial wall and plaque, (c) Arterial wall and plaque Von Mises stress, for the three cross sections indicated in Fig.2

Plaque fracture phenomena are most likely to occur in areas associated with high stress [6]. Figure 6 presents the plaque's and artery's volume distribution of von Mises stress. The graph shows that the percentage of plaque volume in the stress range of 0.1-0.15MPa is almost equal to the percentage volume of 0.15MPa-0.2MPa stress range. High von Mises stresses (>0.25MPa) exist in less than 1% of the plaque's stented volume. Regarding the artery it is worth to mention that on average the stresses in the arterial tissue are significantly lower than those in the plaque.

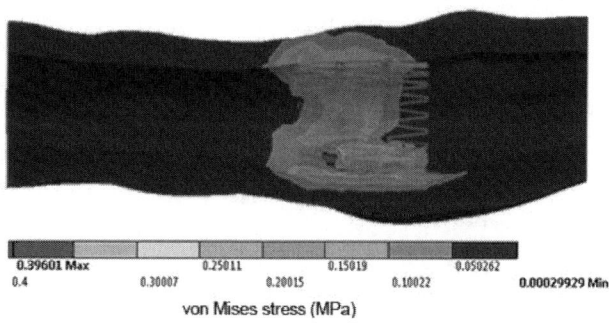

Fig. 4 Arterial wall von Mises stress results

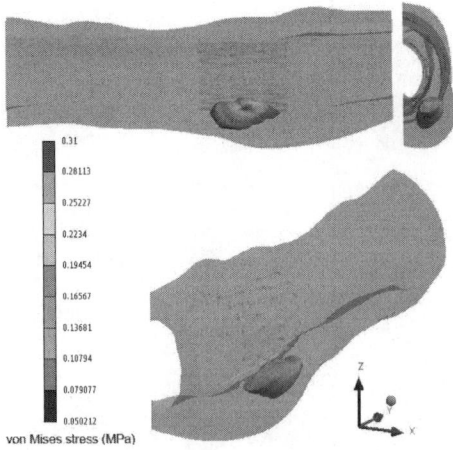

Fig. 5 Von Mises stress results on the plaque component

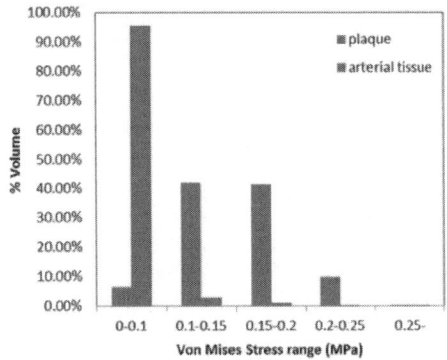

Fig. 6 Von Mises stress percentage volume distribution for the plaque and the arterial wall tissue

IV. CONCLUSION

In this study the finite element method was used to illustrate the stent implantation and revealed that higher stress occurs in the contact area where the stent pushes the underlying plaque. For the first time, realistic geometries of artery and plaque composition are utilized for simulating stent implantation. For future work, the behavior of different plaque types must be examined, as well as the hemodynamic effect of stent deployment on the wall shear stress distribution which plays a significant role to plaque progression and rupture.

ACKNOWLEDGMENT

The authors are grateful to Prof. I.E. Lagaris for his remarkable comments and suggestions. This work is part funded by European Commission (Project "RT3S: Real Time Simulation for Safer vascular Stenting" FP7- 248801).

REFERENCES

1. Barboriak JJ, Rimm AA, Anderson AJ, et al. (1974) Coronary artery occlusion and blood lipids. American heart journal 87:716-21
2. Silingardi R, Tasselli S, Cataldi V, et al. (2013) Bifurcated coronary stents for infrapopliteal angioplasty in critical limb ischemia. Journal of vascular surgery 57:1006-13
3. David Chua SN, MacDonald BJ, Hashmi MSJ (2004) Effects of varying slotted tube (stent) geometry on its expansion behaviour using finite element method. J Mater Process Tech 155–156:1764-71
4. Berry JL, Manoach E, Mekkaoui C, et al. (2002) Hemodynamics and wall mechanics of a compliance matching stent: in vitro and in vivo analysis. Journal of vascular and interventional radiology : JVIR 13:97-105
5. Auricchio F, Di Loreto M, Sacco E (2001) Finite-element Analysis of a Stenotic Artery Revascularization Through a Stent Insertion. Computer Methods in Biomechanics and Biomedical Engineering 4:249-63
6. Chua SND, MacDonald BJ, Hashmi MSJ (2004) Finite element simulation of slotted tube (stent) with the presence of plaque and artery by balloon expansion. J Mater Process Tech 155:1772-9
7. Migliavacca F, Petrini L, Auricchio F, et al. (2003) Deployment of an intravascular stent in coronary stenotic arteries: A computational study Summer Bioengineering Conference, Sonesta Beach Resort in Key Biscayne, Florida, 2003, pp 0169-70
8. Wu W, Wang W-Q, et al. (2007) Stent expansion in curved vessel and their interactions: A finite element analysis. J Biomech 40:2580-5
9. Pericevic I, Lally C, Toner D, et al. (2009) The influence of plaque composition on underlying arterial wall stress during stent expansion: The case for lesion-specific stents. Med Eng Phys 31:428-33
10. Zahedmanesh H, Kelly DJ, Lally C (2010) Simulation of a balloon expandable stent in a realistic coronary artery-Determination of the optimum modelling strategy. J Biomech 43:2126-32
11. Bourantas CV, Kourtis IC, Plissiti ME, et al. (2005) A method for 3D reconstruction of coronary arteries using biplane angiography and intravascular ultrasound images. Comput Med Imag Grap 29:597-606
12. Athanasiou LS, Karvelis PS, Tsakanikas VD, et al. (2012) A Novel Semiautomated Atherosclerotic Plaque Characterization Method Using Grayscale Intravascular Ultrasound Images: Comparison With Virtual Histology. Ieee T Inf Technol B 16:391-400
13. Bonsignore C, Open Stent Design at: http://www.nitinol.com
14. Dumoulin C, Cochelin B (2000) Mechanical behaviour modelling of balloon-expandable stents. J Biomech 33:1461-70
15. Lally C, Reid AJ, Prendergast PJ (2004) Elastic behavior of porcine coronary artery tissue under uniaxial and equibiaxial tension. Ann Biomed Eng 32:1355-64
16. Migliavacca F, Petrini L, Montanari V, et al. (2005) A predictive study of the mechanical behaviour of coronary stents by computer modelling. Med Eng Phys 27:13-8
17. Gervaso F, Capelli C, Petrini L, et al. (2008) On the effects of different strategies in modelling balloon-expandable stenting by means of finite element method. J Biomech 41:1206-12
18. Maurel W (1998) Biomechanical models for soft tissue simulation. Springer-Verlag, Berlin ; New York
19. Eshghi N, Hojjati M.H, Imani M,et al (2011) Finite element analysis of mechanical bechaviors of coronary stent. Procedia Engineering 10 (2011), pp 3056-3061
20. Hibbeler RC (1994) Mechanics of Materials. Macmillan, New York

Author: G.S. Karanasiou
Institute: Technical University of Crete
Street: Akrotiri Campus
City: Chania
Country: Greece, GR 73132
Email: g.karanasiou@gmail.com

In Vitro High Resolution Ultrasonography Measurements of Arterial Bifurcations with and without Stenosis as Inputs for In-Silico CFD Simulations

D. Suárez-Bagnasco [1], G. Balay[1], L. Cymberknop[2], R.L. Armentano[3], and C. Negreira[1]

[1] Laboratorio de Acústica Ultrasonora, Instituto de Física, Facultad de Ciencias,
Universidad de la República, Montevideo, Uruguay
[2] Departamento de Electrónica, Facultad Regional Buenos Aires, Universidad Tecnológica Nacional, Argentina
[3] Departamento de Fisiología, Facultad de Medicina, Universidad de la República, Montevideo, Uruguay

Abstract—Quantification of fluid-structure interactions in arterial walls requires to achieve a complete characterization of flow, shear stress in the interface between blood and endothelium, wall elasticity and wall stresses distribution. Diameter change vs. intravascular pressure is the basis to estimate biomechanical properties (e.g. elasticity, viscosity) of arterial walls in-vivo or in-vitro, being sonomicrometry the gold standard for vessel diameter measurement. High resolution ultrasonography (HRU) is used here in-vitro (as an alternative to sonomicrometry) to get an adequate estimation of the diameter without disturbing the vessel dynamics.

Some experimental in-vitro results are shown for a physical model of arterial bifurcation surgically implemented from a sample of fresh porcine aorta, with and without asymmetric stenosis. Near-physiologic pulsated flow conditions were reproduced in-vitro using a specially designed hemodynamic work bench simulator (HWBS).

In-silico results obtained from a simple CFD model of the abovementioned bifurcation are presented.

Boundary conditions were applied from in-vitro HWBS experimental measurements. Levels of stenosis were reproduced in-silico.

Properly tuned digital simulation based on in-vitro experimental data using our HWBS allows the determination of physical magnitudes (like shear stress) that can't be obtained easily in experimental in-vitro measurements.

Keywords—high resolution ultrasonography, hemodynamic simulator, CFD modeling, arterial bifurcation, stenosis.

I. INTRODUCTION

In-vivo biomechanical properties of the arterial are influenced by the interaction between blood flow and the arterial wall endothelium, arterial wall elasticity and the stress-strain distribution inside the arterial wall and surrounding tissues, amongst other factors (e.g. viscosity).

Quantification of fluid-structure interactions in arterial walls requires to achieve a complete characterization of flow, shear stress in the interface between blood and endothelium, wall elasticity and wall stresses distribution. Characterization can be done in-vivo and in-vitro. In-vivo by means of invasive and non-invasive (tonometry, pulsed Doppler, pulse wave velocity, elastographic) methods. In invasive methods, suitable instrumented animal specimens are used to perform the measurements on a given arterial segment. In-vitro, the characterization can be done by means of a work bench simulator in which arterial segments are intercalated in closed fluid circuits that emulates the main features of systemic arterial circulation [1] [2] [3] [4]. An adequate in-vitro characterization of arterial segments may furnish valuable information to be applied in-vivo [5]. Measurements obtained from cryopreserved [6] and fresh arterial bifurcation samples subjected to different hemodynamic regimes allows the characterization of the biomechanical properties of arterial segments (by parameter estimation). These properties can be employed in a digital simulation model of the interaction between structure and fluid. Diameter change vs. intravascular pressure is the basis to estimate biomechanical properties (e.g. elasticity, viscosity) of arterial walls in-vivo or in-vitro, being sonomicrometry the gold standard for vessel diameter measurement. High resolution ultrasonography can be used in-vitro as an alternative to sonomicrometry to get an adequate estimation of the diameter without disturbing the vessel dynamics [7].

In order to work with arterial bifurcations under near physiological pulsated flow conditions, a specific experimental set-up is needed.

The main purposes of the present paper are:
-To describe the setup specially developed in order to work with arterial bifurcations under near physiological pulsated flow conditions.
-To show some representative experimental results for a physical model of arterial bifurcation surgically implemented from a sample of fresh porcine aorta, with and without constructed asymmetrical stenosis.
-To compare some in-vitro with in-silico results obtained from a simple CFD model of the abovementioned bifurcation.

II. MATERIALS AND METHODS

We used an hemodynamic work bench simulator (HWBS) specially designed to work with arterial bifurcations. Its closed circuit is composed by a pump, the tubing between the pump exit and the entrance to the arterial bifurcation, the arterial bifurcation being studied, tubing between each branch of the bifurcation exit and the entrance to a fluid reservoir, adjustable constrictions on each branch, a fluid reservoir with mean pressure adjustment, and a returning tubing from the reservoir to the pump (Fig. 1).

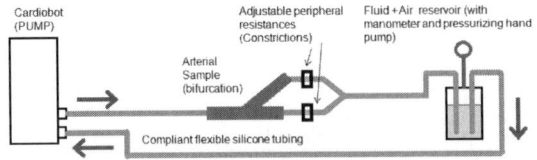

Fig. 1 HWBS fluid circuit schematic diagram. Compliant silicone tubing is used for the circuit.

An artificial heart named Cardiobot is used as the pulsated pump, with frequency and ejection volume adjusted at demand. Cardiobot is an electronically controlled programmable pump, specially developed in our laboratory (see authors affiliations) [2] [8] in order to generate different flow and pressure patterns in the HWBS. For each branch, there is an independent restriction located far away from the corresponding branch output. Each restriction introduces localized mechanical impedance mismatch that produces reflected waves. The combination of upstream and downstream pressure pulses produce global pressure patterns in the arterial sample. Measurements are done after a steady regime of pulsated flow is attained in the circuit. The already established dynamic regime and a time reference signal generated by the Cardiobot (trigger at the beginning of each systole), allows relocation and exchange of sensors during measurements (can be done in the same region alternating probe placement). Samples are mounted on an arterial bifurcation fixing system (couplers and adjustable arms) and submerged in a physiological solution pool. This pool is rounded by an external water pool, where temperature control can be done (Fig. 2).

The circuit can be filled with physiological solution or blood treated with EDTA sodic. Temperature control can be done, with a set-up point of 37°C. The reservoir is a glass flask containing the fluid and air. It's mean pressure can be adjusted at constant values by means of a hand pump and a differential pressure sensor. With no flow and near no transmural pressure in the circuit, high resolution ultrasonic morphometric measurements are done over the bifurcations with a Panametrics V313 transducer probe (15MHz) and with a 64 element array probe (6,99MHz). The probes are fixed on a mechanical graduated positioning system (Fig. 2). A custom algorithm in used to obtain internal and external diameters and wall thickness [1].

After that, several hemodynamic conditions are simulated and measured, adjusting the flow pattern of the Cardiobot, reservoir pressure and the variable restrictions. External diameter measurements are taken at suitable points in each segment of the arterial bifurcation using ultra resolution techniques [5]. Pressure waveforms are obtained using Königsberg sensors located near diameter measurement points. Sensors are sutured so that sensing surface points to the lumen in order to obtain the hydrostatic component of the pressure. Velocity profiles at different locations are obtained using a multi-gate Doppler system (DOP, Signal Processing SA). A physiological solution with micro-spheres powder (scatterers) is used in the. Cross sectional flow is measured using perivascular transit time flowmeters (Triton flowmeters). A Statham pressure sensor (located in the tubing near the common segment) is used for monitoring and initial adjustments procedures. Pressure and flow signals are acquired using data acquisition module (NI USB-6009) operated from a specially developed Matlab application. Pressure and diameter samples are processed in a computer running a specific developed algorithm in order to obtain biomechanical properties of the wall by means of model parameter adjustment [10].

For the reported experiments, a physical model of arterial bifurcation was used, surgically implemented from a sample of fresh porcine aorta (Fig. 3).

Asymmetric stenosis (hemodynamic moderate level stenosis, 20 to 49%) was simulated introducing soft material with ellipsoidal shape in the outer face of the wall of one of the branches. Measurements were done before and after inclusion placement.

 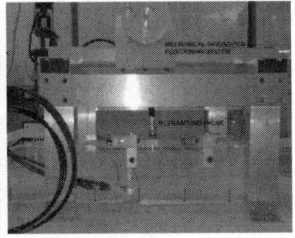 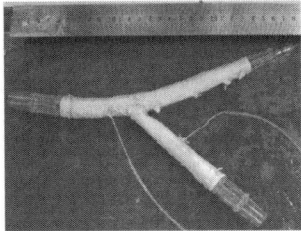

Fig. 2 Left: Arterial bifurcation in the physiological solution pool. Right: Mechanical graduated positioning system (in x, y and z axes) and probe during morphometric measurements

Fig. 3 Surgically constructed physical model of arterial bifurcation. The common segment and the two branches are in the same plane, with a bifurcation angle of 60 degrees between branches. Tubing couplers and two Konigsberg sensors are shown.

III. EXAMPLES OF DATA OBTAINED IN-VITRO

Diameter, pressure, flow and velocity profiles were measured at different places in the common segment and branches of the bifurcations. Fig. 4 shows the simultaneous acquisition of diameter, pressure and cross sectional flow for a location in the common segment.

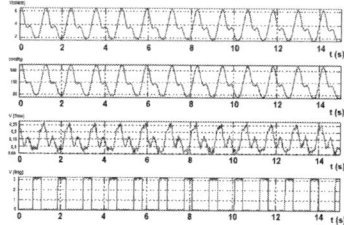

Fig. 4 Diameter, pressure and flow for common segment, with a pulsation frequency of 60 beats per minute. The bottom rectangular signal represents the time reference generated by Cardiobot at each systole.

Velocity profiles were obtained using pulsed Doppler and when the circuit was filled with physiological solution loaded with micro-spheres (scatterers) (Fig.5).

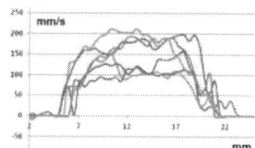

Fig. 5 Raw velocity profiles obtained for the common segment at 40mm from the entrance. Vessel pulsation can be seen through the variation of the profiles width over the x axis (distance in mm). The segment presented a radius of 7,5 mm at zero transmural pressure.

IV. EXAMPLES OF DATA OBTAINED IN-SILICO

In order to compare in-vitro with in-silico results, we constructed a simplified CFD model as a first approximation.

Geometrical measurements with high resolution ultrasound techniques were done to the surgically constructed physical model in order to determine the spatial domain to be used for finite-element digital simulation [11] (Fig. 6).

A mean radius of 7,5mm (obtained form morphometric measurements) was used for the common segment and each branch and a rigid wall condition was imposed (non elastic wall). Lengths were taken from the physical model as well as the bifurcation angle (implemented to be 60 degrees). Time dependent Navier-Stokes equations were solved with a suitable viscosity coefficient, using a commercial software. Fixed and impervious wall conditions were imposed. Pressure boundary conditions were applied to the open ends (common segment entrance and each branch exit) using waveforms obtained during in-vitro experimental measurements [12]. Fig. 7 shows the velocity profiles obtained at the same instant at different cross sections. Flow inversion near the outer walls of the bifurcation branches can be seen as well as detachment of flow past the stenosis.

Fig. 8 (left) shows the relative magnitudes of the velocities for the whole bifurcation in the plane x = 0 at the same

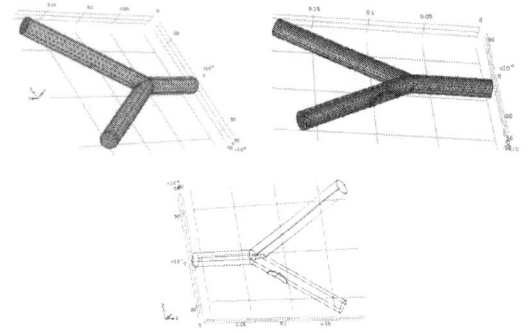

Fig. 6 Top Left: Perspective view of the normal biffurcation grid used for CFD simulations. Top Rigth: Perspective view of the assymetrically stenosed biffurcation grid used for CFD simulations. Stenosis can be seen in the outer face of the shorter branch. Bottom: wireframe rendering showing a moderate level stenosis (of approximately 30%).

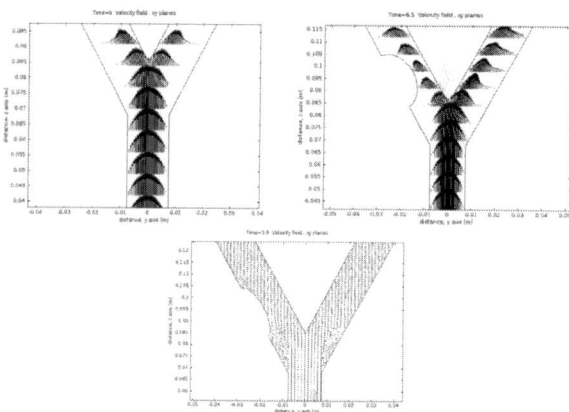

Fig. 7 Top: Velocity profiles near the bifurcation for a given time (left: normal biffurcation, right: stenosed 40%). Bottom: Normalized velocity vectors for 25% stenosis. Flow inversion can be seen at both branches past the output of common segment. Flow inversion and flow detachment can be seen past the stenosis.

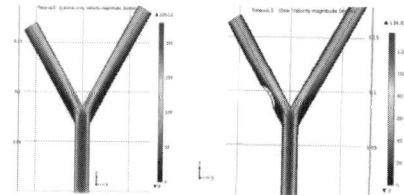

Fig. 8 Relative magnitude of velocity (mm/s) for a given time, in the plane x=0 (left: withouth stenosis, right: with stenosis).

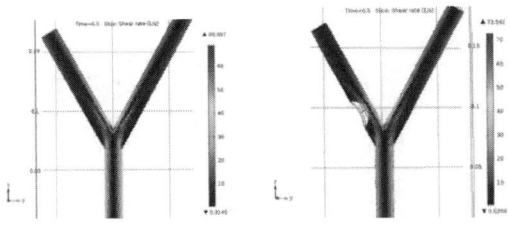

Fig. 9 Shear magnitudes in the plane x=0 (left: withouth stenosis, right: with 40% stenosis).

time instant of Fig.7. Flow inversion can't be seen because this figure represents absolute values.

Fig. 9 shows shear rate magnitude for the same time instant. Lower values can be seen past the stenosis and in flow inversion region in both branches past common segment output.

V. DISCUSSION

Near-physiological flow and pressure patterns can be obtained by means of adjusting Cardiobot pulsation patterns, compliant tube lengths and hydraulic resistances. Realistic flow and pressure patterns are necessary to study fluid-structure interactions in scenarios as near as possible to in-vivo ones. Although physical models (like the one presented here surgically constructed from a fresh artery) are not completely equivalent to a real bifurcation, they enable us to change at will the bifurcation angle, the segment lengths, tapering and level of stenosis. The dimensions of the model facilitated the execution of the in-vitro measurements, which can be later compared with digital simulation results.

In-vitro experimental measurements afforded realistic waveforms of pressure that were used as boundary conditions at the entrance and exits of the bifurcation CFD model.Velocity profiles obtained in-vitro shown a time variation in their widths due to the radius variation of the arterial wall, as consequence of pulsed pressure and elastic compliance. Centerline velocities obtained in-silico (210 to 270 mm/s) for the common segment is greater than the one measured at the same point in-vitro (90 to 211 mm/s). This difference must be assessed taking into account experimental and numerical computation errors, pulsed Doppler measurement limitations in resolution, wall radius variation, energy dissipation in the arterial tissues, and the rigid wall condition of the simplified CFD model. Digital simulation based on in-vitro experimental data allowed the determination of physical magnitudes (like shear stress) that can't be obtained easily in experimental in-vitro measurements. Future work will include in-silico fluid-structure interaction simulations of normal arterial bifurcations and stenosed ones. Biomechanical properties will be obtained from experimental data (adjusting the parameters of a suitable viscoelastic model of the arterial wall to pressure-diameter data) [8] [13] obtained in-vitro. Time reversal and impulsional elastography will be used to get further and complementary mechanical characterization tissues and inclusions (simulating atherosclerotic plaque). Properly fine-tuned in-silico CFD+FSI digital simulation based on in-vitro experimental data obtained with our HWBS will allow the determination of physical magnitudes and hemodynamic indexes (like oscillatory shear index (OSI), temporal wall shear stress gradient (TWSSG), spatial wall shear stress gradient (SWSSG), etc. [8] [14]) that cannot be easily measured experimentally.

ACKNOWLEDGMENT

Authors thanks ANII for grant BE_POS_2010_1_2588 and PEDECIBA for research support.

REFERENCES

[1] J. Brum, D. Bia, N. Benech, G. Balay, R.L. Armentano, C. Negreira, "Setup of a cardiovascular simulator: application to the evaluation of the dynamical behavior of atheroma plaques in human arteries", Physics Procedia, 2010, vol. 3, pp.1095-1101.
[2] G. Balay, J, Brum, D. Bia, R.L. Armentano, C. Negreira 2010, Improvement of artery radii determination with single ultra sound channel hardware & in vitro artificial heart system, Conf. Proc. IEEE Eng. Med. Biol. Soc., pp. 2521-4, 2010.
[3] D. Bia, R.L. Armentano, Y. Zocalo, W. Barmak, E. Migliaro, E.I Cabrera Fischer, "In vitro model to study arterial wall dymamics through pressure-diameter relationship analysis", Latin American Applied Research, 2005, vol. 35, pp. 217-225.
[4] C.D. Bertram, F. Pythoud, N. Stergiopulos, J.J. Meister, "Pulse wave attenuation measurement by linear and nonlinear methods in nonlinearly elastic tubes", Medical Engineering and Physics, 1999, vol 21, pp 155–166.
[5] W. Nichols, M. O'Rourke, Ch. Vlachopoulos, McDonald's blood flow in arteries: theoretical, experimental and clinical principles, Hodder-Arnold, London, UK, 2011.
[6] R.L Armentano, D. Bia, E. Cabrera Fischer et al., "An in-vitro study of cryopreserved and fresh human arteries: a comparison with ePTFE prostheses and human arteries studied non-invasively in vivo", Cryobiology, 2006, 52(1), pp17-26.
[7] J Brum, D Bia, N Bencch, G Balay, RL Armentano, C Negreira, "Arterial diameter measurement using high resolution ultrasonography: In vitro validation", Conf Proc IEEE Eng Med Biol Soc. 2011;2011:203-6.
[8] G. Balay, "Elasticity in arterial tissues, design of an in-vitro artificial heart and a new ultrasonic method of arterial elesticity assessment, M.Sc. thesis (Physics), LAU-FCIEN-UdelaR, July 2012.
[9] K. Chandran, S. Rittgers, A. Yoganathan, "Biofluid mechanics: the human circulation", 2nd ed, CRC Press, Boca Raton, FL, 2012.
[10] L.G. Gamero, R.L. Armentano, J.G. Barra, A. Simon, J. Levenson, "Identification of arterial wall dynamics in conscious dogs", Exp Physiol, 2001, vol 86 pp 519-528.
[11] W. Press, S. Teukolsky, W. Vetterling, B. Flannery, "Numerical recipes: the art of scientific computing", 3rd Ed., Cambridge Univ. Press, N.Y., 2007.
[12] J. Anderson, "Computational fluid dynamics: the basics with applications", McGraw-Hill, N.Y., 1995.
[13] Y.C. Fung, "Biomechanics: mechanical properties of living tissue", Springer Verlag, N.Y., 1981.
[14] T. Hsiai, B. Blackman, H. Jo (Eds), "Hemodynamics and mechanobiology of endotelium", World Scientific, Singapore, 2010

Study on the Dynamic Behavior of Arterial End-to-End Anastomosis

P.C. Roussis[1], A.E. Giannakopoulos[2] and H.P. Charalambous[3]

[1] University of Cyprus/ Department of Civil & Environmental Engineering, Assistant Professor, Nicosia, Cyprus
[2] University of Thessaly/ Department of Civil Engineering, Professor, Volos, Greece
[3] University of Cyprus/ Department of Civil & Environmental Engineering, Graduate Student, Nicosia, Cyprus

Abstract—Vascular disorders, such as atherosclerosis and aneurysms are treated by vascular surgery operations, and involve the stitching of disconnected human arteries with themselves or with artificial grafts. Stitching techniques and related suture materials are of great importance. The aim of this study is to assess the mechanical behavior of the connecting region of two, initially separated, human arteries (end-to-end anastomosis) and to provide useful design parameters for a large number of problems, with different geometrical and mechanical properties.

Keywords—Anastomosis, end-to-end, suture-artery interaction.

I. INTRODUCTION

Vascular surgeries treat vascular diseases, traffic related and other serious injuries that lead to violent artery fractures. Several studies [1–3], have examined the induced arterial-wall stresses in vascular anastomosis models. Most of them limit their research to specific arterial geometries; others ignore the stress concentrations due to suture-artery interaction or the important axial-circumferential deformation coupling of artery response. Additionally, little work has been published on the dynamic analysis of stitched arteries. Related review articles clearly point out the lack of such analyses [4]. In this study a mathematical model based on structural analysis is proposed, aiming to provide a better understanding of the influence of stitching characteristics on the response of an end-to-end artery anastomosis. The present study reaches to important suggestions for safer, cost-efficient angio-surgery operations, in order to improve the healing process and recovery time of the patients.

II. MATERIAL AND METHODS

A. Mathematical Model of the Radial Direction Response

The blood vessel was modeled as an elastic circular cylindrical pipe with a cross section of a circular ring consisting of one layer. The mathematical model developed herein is based on the following assumptions: (a) the cross-sectional dimensions of the arterial ring are small compared to the radius of the centerline of the ring; (b) the centerline of the ring in the undeformed state forms a full circle with radius R; (c) the arterial wall has constant thickness; (d) no boundary constraints are applied on the ring; (e) the effects of rotary inertia and shear deformation are neglected; (f) the arterial tissue consists of a single homogeneous layer and behaves as an orthotropic linear elastic material; (g) viscous effects are ignored.

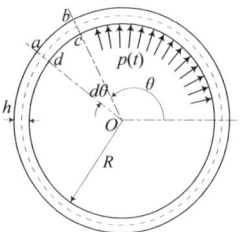

Fig. 1 Arterial model cross section

The equation governing the radial displacement $u(t)$ of the vibrating ring is derived by considering the equilibrium of forces acting on an infinitesimal arterial element (Fig. 1):

$$\rho h \frac{d^2 u(t)}{dt^2} + \frac{E_r h}{R^2} u(t) = p(t) \quad (1)$$

in which $p(t)$ denotes the uniformly distributed wall pressure, ρ is the density of the arterial tissue, h is the arterial wall thickness, and E_r is the radial Young's modulus. The second-order differential equation (1), governing the in-plane vibration of a ring, is identical to the equation of motion governing the forced vibration of an undamped single-degree-of-freedom system. The natural circular frequency of the system is given by $\omega_n = \sqrt{E_r / \rho R^2}$.

B. Dynamic Response of Human Artery Due to Pulse-Type Loading

During a vascular operation the blood flow is interrupted. The first loading cycle, immediately after the flow is restored, is approximated by the loading shown in Fig. 2. In such a case, the internal pressure is abruptly increased from zero to the maximum systolic pressure. The time interval $0 \leq t \leq t_s$ represents the aortic systolic phase, during which the arterial walls inflate due to the maximum overstress pressure. The time interval $t_s < t < t_{cp}$ represents the aortic diastolic phase. The parameters p_d and p_s denote the diastolic pressure and the maximum systolic pressure, respectively.

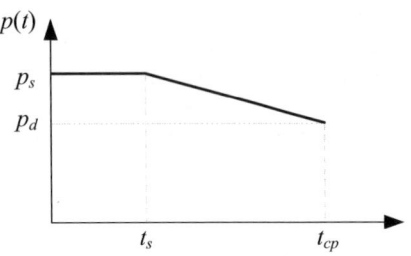

Fig. 2 Arterial pulse time-history approximation.

The radius R is measured at zero blood pressure conditions and in vivo length. During the surgery, longitudinal residual stresses are released, forcing the vessel to decrease its length and increase its diameter. When, subsequently the stitching takes place, the arterial diameter and length return to their prior condition. The residual stress effect is taken into account as a free vibration with initial displacement $u(0) = u_0$, equal to the difference of the increased radius (relieved from axial residual stresses) to radius R, and initial velocity $\dot{u}(0) = 0$. Therefore, the total response of the system is:

$$u(t) = \begin{cases} \dfrac{p_s R^2}{E_r h}\left(1-\cos\omega_n t\right) + u_0 \cos\omega_n t,\ 0 \leq t \leq t_s \\ \dfrac{R^2}{E_r h}\left[\dfrac{p_s - p_d}{t_{cp} - t_s}\left(t_s - t + \dfrac{\sin\omega_n(t-t_s)}{\omega_n}\right)\right] \\ + p_s\left(1-\cos\omega_n t\right) + u_0 \cos\omega_n t,\ t_s \leq t \leq t_{cp} \end{cases} \quad (2)$$

The expression $u_{st} = p_s R^2 / E_r h$ is the static response of the system, representing the displacement caused when the maximum pressure p_s is applied statically. Fig. 3 shows the maximum radial displacement for $p_s = 120$ mmHg, $p_d = 80$ mmHg, $t_s = 0.35$ sec and $t_{cp} = 1$ sec. $T_n = 2\pi/\omega_n$ is the natural period of the system.

B. Suture-Tissue Dynamic Interaction

The interaction of sutures with the arterial tissue may lead to post-surgery complications. The undesirable conditions can be discretized in three failure modes: suture failure, arterial wall tearing, and thrombosis due to blood leaking at the anastomosis interface. Fig. 4a shows the end-to-end anastomosis model. The separate artery parts are connected together with a total of N_s stitches. Each part has length L, radius R, and Young's modulus (in the longitudinal direction) E_L. The stitches have radius r_s, lenght l_s, cross-sectional area $A_s = \pi r_s^2$, and Young's modulus E_s. Different stitching patterns are considered, resulting in different suture loading. The particular loading condition associated with each stitching pattern will be accounted for by means of a participation factor a.

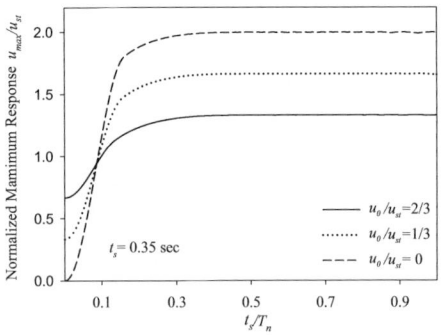

Fig. 3 Normalized maximum deformation for $t_s = 0.35$ sec.

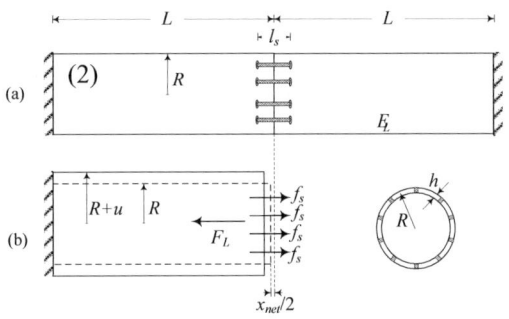

Fig. 4 (a) Model of artery anastomosis (b) Tensile forces of sutures and artery tissue.

On account of the fact that blood is an incompressible fluid, the radial and longitudinal modes of response are coupled. Under the applied blood-pressure, the artery distends, with its radius being increased from R to $R+u(t)$, and, in order for the blood volume to be maintained, its axial length is decreased. With reference to Fig. 4b, the resultant tensile force is determined as $F_L = aN_s f_s$. After several calculations, the net gap between the edges of the two anastomosed artery parts is obtained as

$$x_{net} = \frac{2\pi L E_L l_s h(R+u)\left[(R+u)^2 - R^2\right]}{\pi E_L l_s h(R+u)^3 + aN_s A_s E_s L R^2} \quad (3)$$

The tensile force developed in one suture is

$$f_s = \frac{A_s E_s}{l_s} x_{net} \quad (4)$$

The embedding stress induced on the arterial wall, at the stitching holes region, is approximately:

$$\sigma_s = \frac{af_s}{2r_s h} \quad (5)$$

It is worth noting that, although based on a linear-elastic model, the solution contains as many as sixteen input parameters, related to the geometrical and mechanical properties of the sutures, the geometrical and mechanical properties of the arterial walls, the number of sutures, the loading characteristics and longitudinal residual stresses.

III. RESULTS

The three response quantities, the net gap x_{net}, the suture tensile force f_s, and the embedding stress σ_s are directly connected to the aforementioned failure modes. The normalized response quantities are shown graphically in Fig. 5a. The normalized net gap $x_{net}/2L$ depends on the product of terms aE_s/E_L, L/l_s, A_s/Rh, N_s. Fig. 5a plots the normalized gap $x_{net}/2L$ as a function of the normalized radial displacement u/R, for different values of the normalized parameters. Equation (4) suggests that the normalized tensile suture force (suture strain) depends linearly on the normalized net gap and increases as the ratio $2L/l_s$ is increased (Fig. 5b). It can be seen from Fig. 5c that in order to reduce the embedded stress, we must increase the number of stitches, whereas the parameter aE_s/E_L plays an insignificant role.

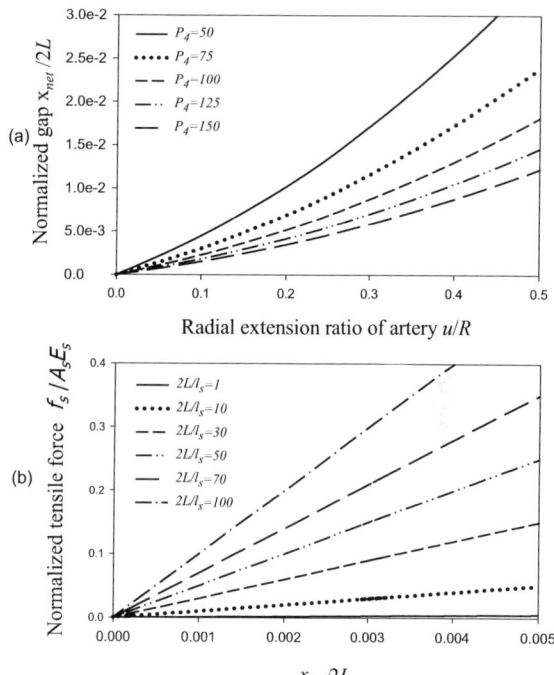

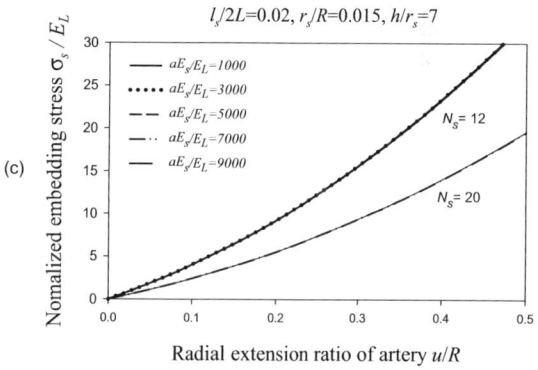

Fig. 5: (a) Normalized gap versus radial extension ratio for different values of the product $P_4 = (aE_s/E_L)N_s(A_s/\pi Rh)(L/l_s)$. (b) Normalized tensile force of each stitch versus the normalized net gap (c) Normalized embedding stress versus radial extension ratio of artery.

Table 1 summarizes the results from a numerical example, in which the response quantities are calculated for given set of artery and suture parameters. The three response quantities of interest are within the accepted range

Table 1 Numerical example

Parameter	Value
R [cm]	0.3
L [cm]	1.5
h [mm]	1
ρ [kg/m3]	1160
u_0 [mm]	0.08
E_r [MPa]	0.4
E_L [MPa]	0.6
E_s [GPa]	1.8
N_s	12
r_s [mm]	0.13
l_s [mm]	2
a	1.7
p_s [mmHg]	120
p_d [mmHg]	80
t_s [sec]	0.35
t_{cp} [sec]	1
Results	
u_{max} [mm]	0.48
x_{net} [μm]	3.86
f_s [N]	0.18
σ_s [Pa]	1.21

of values [5, 6], with the calculated net gap $x_{net} = 3.86$ μm being smaller than the diameter of one red blood cell. Therefore, the anastomosis scheme will not fail.

IV. DISCUSSION

An analytical formulation of the extensional response of an axially symmetric arterial ring was performed in the first part of the study. The analysis results (Fig. 2) suggest that the maximum radial displacement appears to depend primarily on the static displacement of the artery and the pre-stress conditions.

Findings obtained from Section II.C highlight the influence of each parameter on the anastomosed artery response. All response quantities increase monotonically with the radial extension $u(t)$. Fig. 5a underlines the nonlinear relationship between the net gap and the radial extension of the artery. For a given artery configuration, the normalized net gap is decreased when: (a) stronger stitches are used; (b) the total number of sutures is increased; (c) the suture cross-sectional area is increased; and/or (d) the suture length is decreased. Concerning the mechanical response of stitches, it is verified that higher tension is developed when shorter stitches are applied (Fig. 5b). In regard to the artery wall failure (tearing), it has been shown that the normalized embedding stress is decreased as the number of sutures and/or the suture length is increased. Figure 5c indicates insignificant influence of the ratio of suture-to-artery elastic modulus on the embedding stress, as well as the important effect of the total number of stitches applied.

V. CONCLUSIONS

A comprehensive mathematical model was developed from first principles to assess the dynamic behavior of arterial end-to-end anastomosis. The analysis reveals useful interrelations among the problem parameters, thus making the proposed model a valuable tool for the optimal selection of materials and improved functionality of the sutures. The obtained results highlight the dependency of the maximum extensional deformation on the static displacement of the artery and the pre-stress conditions.

ACKNOWLEDGMENT

The authors gratefully acknowledge the constructive suggestions and feedback provided by Dr. Georgios P. Georgiou, American Medical Center.

REFERENCES

1. Ballyk PD, Walsh C et al. (1998) Compliance mismatch may promote graft-artery intimal hyperplasia by altering suture-line stresses. J Biomech 31:229–237
2. Cacho F, Doblaré M et al. (2007) A procedure to simulate coronary artery bypass graft surgery. Med Biol Eng Comput 45:819–827
3. Perktold K, Leuprecht A et al. (2002) Fluid Dynamics, Wall Mechanics, and Oxygen Transfer in Peripheral Bypass Anastomoses. Ann Biomed Eng 30:447–460
4. Migliavacca F, Dubini G. (2005) Computational modeling of vascular anastomoses. Biomech Model Mechanobiol 3:235–250.
5. Brouwers J, Oosting D et al. (1991) Dynamic loading of surgical knots. Surg Gynecol Obstet 173:443–448
6. Mohan D, Melvin JW (1983) Failure properties of passive human aortic tissue. II—Biaxial tension tests. J Biomech 16:31–44

Author: Panayiotis Roussis, Assistant Professor
Institute: University of Cyprus, Department of Civil & Environmental Engineering
Street: 75, Kallipoleos
City: Nicosia
Country: Cyprus
Email: roussis@ucy.ac.cy

Printed by Publishers' Graphics LLC